D1000534

Handbook of Heat and Mass Transfer

Volume 2: Mass Transfer and Reactor Design

Gulf Publishing Company
Book Division
Houston, London, Paris, Tokyo

Handbook of Heat and Mass Transfer

Volume 2: Mass Transfer and Reactor Design

Nicholas P. Cheremisinoff, Editor

in collaboration with:

P. H. Au-Yeung
M. H. I. Baird
S. Carra
H. N. Chang
B. H. Chen
P. N. Cheremisinoff
R. M. Counce
A. L. Cukierman
V. D. Dang
F. J. Desilva
G. Dogu
L. K. Doraiswamy
N. V. K. Dutt
P. Gandhidasan
E. Gonsalves
M. G. Gonzalez
S. Goto
G. Grevillot
P. A. M. Grootscholten
G. Grossman
J. Hanika
W. J. Hatcher, Jr.
R. Hughes
T. Ishikawa
T. Iwata
S. J. Jancic
H. Knapp
R. Krishna
B. D. Kulkarni
J. Levec
M. Morbidelli
H. Mori
Y. Naka
M. Nishikawa
J. Nyvlt
J. K. Park
V. S. Patwardhan
L. Pellegrini
J. J. Perona
A. B. Ponter
E. N. Ponzi
D. H. L. Prasad
E. Ranzi
P. Sauro
H. Seko
V. Stanek
V. V. Tarasov
R. Taylor
J. R. Thome
S. Tone
S. W. Tsai
Y. Y. Wang
G. A. Yagodin
I. Yamada
H. M. Yeh
S. M. Yih

Handbook of Heat and Mass Transfer

Volume 2: Mass Transfer and Reactor Design

ISBN 0-87201-412-6

Library of Congress Cataloging in Publication Data
Main entry under title:

Handbook of heat and mass transfer.

Includes indexes.
Contents: v. 1 Heat transfer operations-v.2. Mass transfer and reactor design.
1. Heat-Transmission-Handbooks, manuals, etc. 2. Mass transfer-Handbooks, manuals, etc.
I. Cheremisinoff, Nicholas P.
TJ260.H36 1986 v.2 621.402′2 84-25338
ISBN 0-87201-411-8 (v. 1)
ISBN 0-87201-412-6 (v. 2)

CONTENTS

Section I: Mass Transfer Principles

Section II: Distillation and Extraction

Section III: Multiphase Reactor Systems

Section IV: Special Applications and Reactor Topics

VOLUME 1: HEAT TRANSFER OPERATIONS

CONTENTS

Section I: Heat Transfer Mechanisms

VOLUME 1: HEAT TRANSFER OPERATIONS

Section II: Industrial Operations and Design

CONTRIBUTORS TO THIS VOLUME

P. H. Au-Yeung, Dept. of Chemistry and Chemical Engineering, Michigan Technological University, Houghton, Michigan, USA.

M. H. I. Baird, Chemical Engineering Dept., McMaster University, Ontario, Canada.

S. Carra, Dipartimento di Chimica Fisica Applicata del Politecnico di Milano, Milano, Italy.

H. N. Chang, Dept. of Chemical Engineering, Korea Advanced Institute of Science & Technology, Seoul, Korea.

B. H. Chen, Dept. of Chemical Engineering, Technical University of Nova Scotia, Halifax, Nova Scotia, Canada.

P. N. Cheremisinoff, Dept. of Civil & Environmental Engineering, New Jersey Institute of Technology, Newark, New Jersey, USA.

R. M. Counce, Dept. of Chemical Engineering, University of Tennessee, Knoxville, Tennessee, USA.

A. L. Cukierman, Programa de Investigacion y Desarrollo de Fuentes Alternativas de Materias Primas y Energia, Ciudad Universitaria, Buenos Aires, Argentina.

V. D. Dang, Dept. of Chemical Engineering, The Catholic University of America, Washington, DC, USA.

F. J. Desilva, Water Process Division, Belco Pollution Control Corp., Parsippany, New Jersey, USA.

G. Dogu, Dept. of Chemical Engineering, Middle East Technical University, Ankara, Turkey.

L. K. Doraiswamy, National Chemical Laboratory, Poona, India.

N. V. K. Dutt, Regional Research Laboratory, Hyderabad, India.

P. Gandhidasan, Dept. of Mechanical Engineering, The University of the West Indies, St. Augustine, Trinidad, West Indies.

E. Gonsalves, Dept. of Civil & Environmental Engineering, New Jersey Institute of Technology, Newark, New Jersey, USA.

M. G. Gonzalez, Cindeca, Centro de Investigacion y Desarrollo en Procesos Cataliticos, La Plata, Argentina.

S. Goto, Dept. of Chemical Engineering, Nagoya University, Nagoya, Japan.

G. Grevillot, Laboratoire des Sciences du Genie Chimique, Nancy, France.

P. A. M. Grootscholten, Unilever Research Laboratory, Vlaardingen, The Netherlands.

G. Grossman, Faculty of Mechanical Engineering, Technion-Israel Institute of Technology, Israel.

J. Hanika, Dept. of Organic Technology, Institute of Chemical Technology, Prague, Czechoslovakia.

W. J. Hatcher, Jr., Dept. of Chemical Engineering, The University of Alabama, University, Alabama, USA.

R. Hughes, Dept. of Chemical Engineering, University of Salford, Salford, United Kingdom.

T. Ishikawa, Tokyo Metropolitan University, Tokyo, Japan.

T. Iwata, Mitsui Engineering and Shipbuilding Co., Ltd., Tokyo, Japan.

S. J. Jancic, Sulzer Brothers, Winterthur, Switzerland.

H. Knapp, Institute of Thermodynamics and Plant Design, Technical University of Berlin, Federal Republic of West Germany.

R. Krishna, Indian Institute of Petroleum, Dehradun, India.

B. D. Kulkarni, National Chemical Laboratory, Poona, India.

J. Levec, Dept. of Chemistry and Chemical Technology, E. Kardelj University, Ljubljana, Yugoslavia.

M. Morbidelli, Dipartimento di Chimica Fisica Applicata del Politecnico di Milano, Milano, Italy.

H. Mori, Nagoya Institute of Technology, Nagoya, Japan.

Y. Naka, Tokyo Institute of Technology, Yokohama, Japan.

M. Nishikawa, Dept. of Nuclear Engineering, Kyushu University, Fukuoka, Japan.

J. Nyvlt, Institute of Inorganic Chemistry, Czechoslovak Academy of Sciences, Prague, Czechoslovakia.

J. K. Park, Dept. of Chemical Engineering, Korea Advanced Institute of Science & Technology, Seoul, Korea.

V. S. Patwardhan, National Chemical Laboratory, Poona, India.

L. Pellegrini, Dept. of Chemical Engineering, Politecnico of Milan, Piazza Leonardo da Vinci, Milano, Italy.

J. J. Perona, Dept. of Chemical Engineering, University of Tennessee, Knoxville, Tennessee, USA.

A. B. Ponter, Office of the Dean of Engineering, Cleveland State University, Cleveland, Ohio, USA.

E. N. Ponzi, Cindeca. Centro de Investigacion y Desarrollo en Procesos Cataliticos, La Plata, Argentina.

D. H. L. Prasad, Regional Research Laboratory, Hyderabad, India.

E. Ranzi, Dept. of Chemical Engineering, Politecnico of Milan, Piazza Leonardo da Vinci, Milano, Italy.

P. Sauro, Dept. of Chemical Engineering, Politecnico of Milan, Piazza Leonardo da Vinci, Milano, Italy.

H. Seko, Research Laboratory of Applied Biochemistry, Tanabe Seiyaku Co., Ltd., Osaka, Japan.

V. Stanek, Institute of Chemical Process Fundamentals, Czechoslovak Academy of Sciences, Prague, Czechoslovakia.

V. V. Tarasov, Mendeleev Institute of Chemical Technology, Moscow, U.S.S.R.

R. Taylor, Dept. of Chemical Engineering, Clarkson University, Potsdam, New York, USA.

J. R. Thome, Dept. of Mechanical Engineering, Michigan State University, East Lansing, Michigan, USA.

S. Tone, Dept. of Chemical Engineering, Osaka University, Toyonake, Japan.

S. W. Tsai, Chemical Engineering Department, National Cheng Kung University, Tainan, Taiwan, Republic of China.

Y. Y. Wang, Dept. of Chemical Engineering, National Tsing Hua University, Hsinchu, Taiwan, Republic of China.

G. A. Yagodin, Mendeleev Institute of Chemical Technology, Moscow, U.S.S.R.

I. Yamada, Nagoya Institute of Technology, Nagoya, Japan.

H. M. Yeh, Chemical Engineering Department, National Cheng Kung University, Tainan, Taiwan, Republic of China.

S. M. Yih, Dept. of Chemical Engineering, Chung Yuan University, Chung Li, Taiwan.

ABOUT THE EDITOR

Nicholas P. Cheremisinoff heads the product development group in the Elastomers Technology Division of Exxon Chemical Company. Previously, he led the Reactor and Fluid Dynamics Modeling Group at Exxon Research and Engineering Company. He received his B.S., M.S., and Ph.D. degrees in chemical engineering from Clarkson College of Technology, and he is also a member of a number of professional societies including AIChE, Tau Beta Pi, and Sigma Xi.

PREFACE

Transport phenomena involve the exchange of momentum, energy, and mass within and across defined system boundaries. Understanding these mechanisms and describing them in the universal language of mathematics enable the forces of nature to be harnessed for mankind's benefit. This is essentially accomplished every day in countless manufacturing operations that produce an endless stream of products and life-sustaining articles throughout the world.

Despite industry's success, our understanding of transport phenomena is limited. Most process industries handle complex flows of mixtures, process streams involving phase changes, and streams with multiple chemical reactions, often requiring energy to be transferred between equipment and process streams. Because the most up-to-date data are fragmented, scale-up is difficult, and all too often results in operations that perform below expectations resulting in costly upsets and shutdowns. This is largely because the literature on heat and mass transfer is so diverse and often contradictory for systems involving phase changes and chemical reactions. Clearly, a more unified approach to designing process operations can only be gained through better understanding of the physical mechanisms of mass and energy transport as well as the chemical kinetics.

This handbook is aimed at unifying heat and mass transport concepts for the practicing engineer. The overall theme of this work is to provide fundamental understanding of the physical processes and detailed guidance in applying principles to designing complex unit operations and chemical reactors involving multiphase flows. Emphasis is placed on more advanced industrial problems, although ample references are given to provide the user with reviews of basic principles.

The work is divided into two volumes.

Volume 1: Heat Transfer Operations, contains two sections. The first section, "Heat Transfer Mechanisms," devotes 20 chapters to discussions of principles and applications of transport phenomena. Momentum transport and fluid dynamics are related to various subjects of heat transfer. In addition, several chapters are devoted to more recent areas of research. The second section, "Industrial Operations and Design," provides 23 chapters on critical unit operations involving heat exchangers, drying operations, combustion, and heat transfer problems in specialized reactors.

Volume 2: Mass Transfer and Reactor Design, contains four sections. Section I, "Mass Transfer Principles," emphasizes multicomponent mass transfer in nine chapters. Discussions also cover mass transfer in reacting flows. Section II, "Distillation and Extraction," is composed of eight chapters, with emphasis given to design-related problems. Again, the references provide further information on the fundamental concepts. Section III, "Multiphase Reactor Systems," devotes 15 chapters to various types of reactors of industrial importance. Many of these chapters relate to reaction kinetics principles. Section IV, "Special Applications and Reactor Topics," provides five chapters on the subjects of industry crystallization, thermal diffusion columns, parametric pumping, and some special topics of interest.

To ensure the highest degree of reliability, the services of more than one hundred specialists were enlisted. This work presents the efforts of these experts. In addition, it incorporates the experience and opinions of scores of engineers and researchers who aided with advice and suggestions in reviewing and refereeing the material presented. Each contributor is to be regarded as responsible for the statements and recommendations in his/her chapter. These individuals are to be congratulated for devoting their time and efforts to producing these volumes. Without their efforts this work could not have been a reality. Special thanks is also expressed to Gulf Publishing Company and, in particular, to Melissa Lewis for her assistance in copyediting and styling.

Nicholas P. Cheremisinoff

SECTION I

MASS TRANSFER PRINCIPLES

CONTENTS

CHAPTER 1

EFFECT OF TURBULENCE PROMOTERS ON MASS TRANSFER

Ho Nam Chang and Joong Kon Park

Department of Chemical Engineering
Korea Advanced Institute of Science & Technology
Seoul, Korea

CONTENTS

INTRODUCTION

Mass transfer operations naturally involve two heterogeneous phases such as gas-liquid, liquid-liquid, gas-solid, and liquid-solid phases. The transfer of substances takes place through their common interfaces whenever the two phases are brought into contact. Initially material near or at the interface moves first, but soon after, a situation arises when the substance at the interface is depleted such that the overall mass transfer is limited. This situation is called "mass-transfer controlled." Also there are occasions when the transport of the substance from the bulk fluid to the interface is not important at all. In this chapter our discussion will be limited to the former mass-transfer controlled transport of the substance in fluids (gases and liquids) to a solid surface.

Consider the mass transfer problem in a straight channel where the fast reaction occurs at the wall. Thus the concentration of the substance at the wall is assumed zero. At the inlet the concentration is one everywhere. As the fluid passes through the channel, the concentration near the wall gradually decreases due to the depletion by the reaction at the wall. In other words, the concentration boundary layer grows. Further downstream from the inlet the mass transfer is not effective at all. This is the famous Graetz-type problem in mass transfer. It is not difficult to imagine the role of convection as well as that of diffusion in mass transfer. Both play important roles in mass transfer, sometimes singly, but in most cases coupled.

Because of the ineffective mass transfer in a laminar flow as seen in the previous example, turbulence conditions are desirable because they enhance mass transfer. However, turbulence accompanies high costs for pumping energy and requires equipment having large capacities. One way of avoiding this is to use a turbulence promoter which enhances mass transfer effectively in a laminar flow condition.

Strictly speaking, various packing materials used in packed beds are a form of turbulence promoters used to improve the efficiencies in gas-liquid contacting. Discussions on this subject can be found in much chemical engineering literature [1–3]. Thus, here the turbulence promoters refer to those used between fluids and solids, more specifically in channels. The shapes of turbulence promoters can be divided into three forms. First is a wrinkled surface [4–7]. The surface-wrinkled turbulence promoters are commonly employed in heat exchangers [4, 5] and in a blood oxygenator or an artificial kidney [6, 7]. The upper part of the wrinkled surface serves as the turbulence promoter by generating a recirculating flow below. This action helps micro-mixing near the solid surface. Bellhouse et al. [7] reported that a vortex-mixing hemodialyzer with a furrowed blood channel could reduce the membrane area required for hemodialysis by a factor of two-thirds. The second form is to attach a rectangular duct or a cylinder in the direction perpendicular to the flow [8, 9]. This type is generally used in heat exchangers by carving a slot in a flat-plate heat exchanger [10] or by attaching fins around the heat exchanger pipe [11]. A mat of hollow fibers used in an artificial kidney is of this type which enhances mass transfer in the dialysate side [12]. The third type promoter is the net type. This promoter is commonly used in membrane processes such as electrodialysis, reverse osmosis, and artificial kidney systems inserted between the membranes [13–16]. Here the promoter has dual functions of membrane support as well as being a turbulence promoter. In electrodialysis turbulence promoters enhance mass transfer by breaking the developing boundary layer and preventing the membrane from contacting. It has the disadvantage of increasing the electrical resistance of the solution and the pressure drop in the channel.

It is very difficult to theoretically analyze the fluid motion and the mass transfer around the promoter due to the complexity of the hydrodynamics. The fluid motions are qualitatively studied through flow visualization and the mass transfer mechanisms must be investigated experimentally. The naphthalene sublimation method is used in measuring mass transfer in gas systems [17], and the limiting current method is used in liquid systems [18–20]. Recently a new approach was initiated by Chang and his colleagues [21, 22], who performed computer simulation on the fluid motion and mass transfer around promoters. This new approach aims at understanding a local mass transfer in one unit of the promoter in terms of fluid motion.

The turbulence promoter increases mass transfer at the expense of the increased pressure drop that results in increased pumping energy. In this connection Isaccson and associates [22–24] studied process optimization on the advantages gained by turbulence promoters in terms of increased limiting currents and their disadvantages in terms of increased pumping energy.

This chapter examines both the fluid motion and mass transfer around turbulence promoters in gas and liquid systems, and the method and the results of numerical simulation will be introduced. Finally, the increase of pumping costs in compensation for the enhanced mass transfer is discussed.

ENHANCEMENT MECHANISMS

While mass transfer in laminar flow is attributable mainly to diffusion near the liquid-solid boundary, mass transfer in turbulent flow relies on vigorous mixing and eddies that penetrate deep into the solid surface. Turbulent flow consists of a turbulent core, a turbulent boundary layer, and a viscous sublayer. According to Levich [26] in turbulent mass transfer, eddy diffusion plays a more important role than pure diffusion even inside the viscous sublayer, due to the small diffusivity of liquids in comparison to the kinematic viscosity. Ruckenstein [27] and Hanratty [28] proposed that the liquid layer in the immediate vicinity of the wall is continuously renewed from the bulk fluid. In turbulence Carino and Brodkey [29] and Nichas et al. [30] confirmed through visualization studies that the main bulk flow penetrated to the wall.

The boundary-layer thickness can be reduced in a laminar flow by installing a turbulence promoter in the channel, which increases the local velocity near the surface without increasing main flow rate. Also downstream from the turbulence promoter the wakes formed excite Tollmein-Schlichting waves and promote early transition to turbulence [31]. Thus the enhancement mechanism can be summarized as first, facilitating the mixing of the fluids near the solid surface with those in the bulk that effectively break down the concentration boundary layer; and secondly, increasing local shear rate to result in a mass transfer rate.

EXPERIMENTAL APPROACHES

Two-Dimensional Gas System

The studies on the mass transfer and fluid motion near the promoter are largely concerned with liquid systems, and examples on gas systems are rare. Thomas [17, 31, 32] investigated enhanced mass transfer in the wake region behind detached cylindrical promoters in a wind tunnel, using the naphthalene sublimation technique (refer to Figure 1). A single detached cylindrical promoter in a duct gives a pronounced maximum in mass transfer rate directly beneath the cylinder as shown in Figure 2. The interaction between the wake and boundary layer causes instability in the boundary layer to result in a small peak in the wake region behind the promoter together with the effect of increasing slowly with distance from the cylinder in contrast to the gradual decrease of mass transfer by the Pohlhausen solution. The maximum increase in local mass transfer through laminar-boundary layers obtained with a single cylinder was 240%. With two cylinders two peaks were observed right beneath the cylinders. Furthermore, there was a marked increase in the rate of forced convection yielding a large and wide peak in the wake region. Two additional promoters further downstream from the first two promoters gave two peaks as expected, and the effect was directly additive upon the effect of the first pair.

The effects of the gap between the promoter and the wall and the fluid velocity on mass transfer are very complex (see Figures 3 and 4). Since the gap between the promoter and the wall is small at a low flow rate, two peaks are formed in the wake region with one promoter. Downstream from the promoter the effect is more pronounced as the gap is smaller and the velocity is larger. When one pair of cylinders are placed near the wall, the average mass transfer including the cylinder and the wake region is given as the ratio

$$\frac{[(\mathrm{St})_{\mathrm{cylinder}}]_{\mathrm{average}}}{(\mathrm{St})_{\mathrm{average}}} = \frac{\int_{l_0}^{L} (\mathrm{St})_{\mathrm{cylinder}}\, dx}{\int_{l_0}^{L} \mathrm{St}\, dx} \tag{1}$$

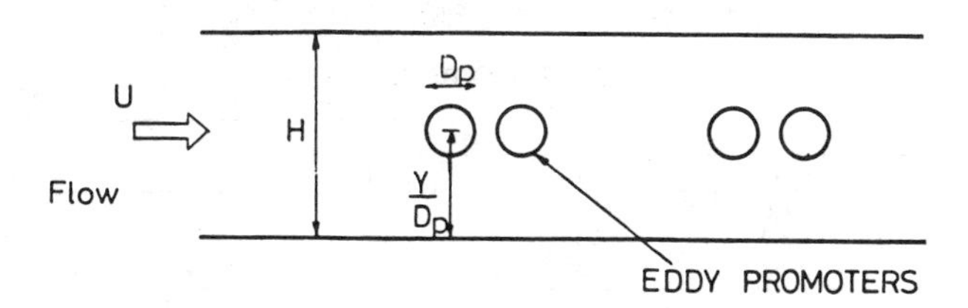

Figure 1. Coordinates and notations of two-dimensional detached cylinder promoter.

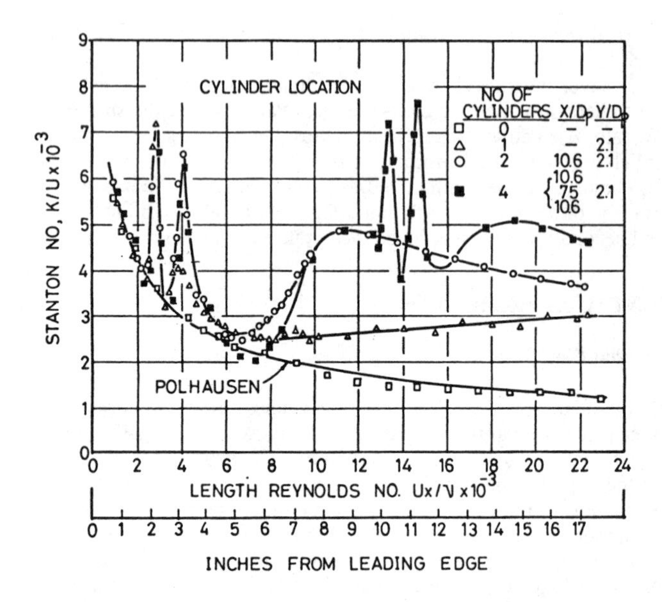

Figure 2. Effect of placing one, two, or four 0.093-in.-diameter cylinders near the edge of the laminar-boundary layer on the value of the local mass transfer Stanton number (Sc = 2.44). (From D.G. Thomas, "Forced Convection Mass Transfer: Part III," *AIChE J.*, Vol. 12, p. 126.)

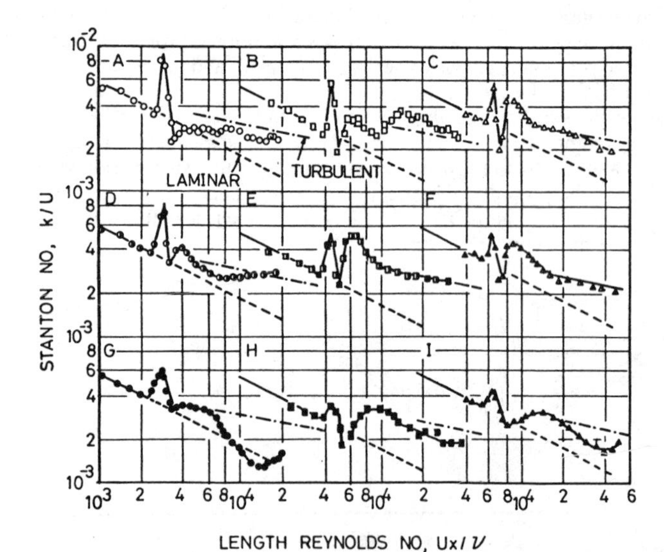

	U ft/sec		
Y/Dp	3.5	4.4	6.2
1.6	A	B	C
2.2	D	E	F
3.2	G	H	I

Figure 3. Effect of a single 0.093-in.-diameter cylinder located near the edge of the laminar-boundary layer on the rate of mass transfer from a flat plate (Sc = 2.44). (From D.G. Thomas, "Forced Convection Mass Transfer: Part III," *AIChE J.*, Vol. 12, p. 127.)

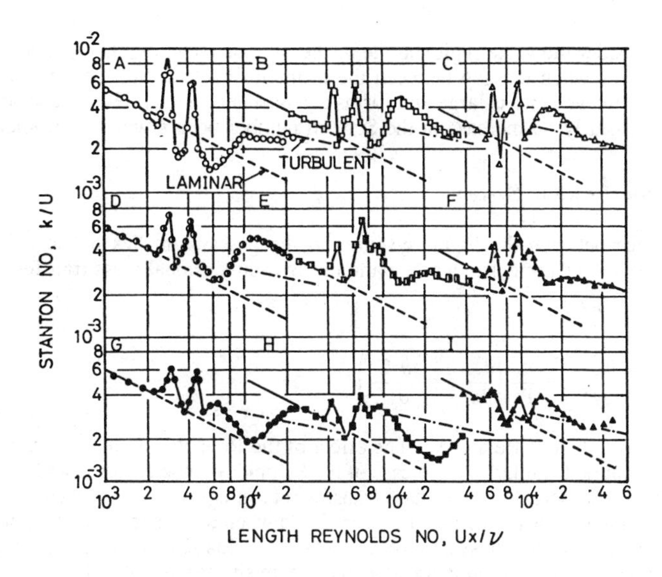

Figure 4. Effect of two 0.093-in.-diameter cylinders located near the edge of the laminar-boundary layer on the rate of mass transfer from a flat plate (Sc = 2.44). (From D.G. Thomas, "Forced Convection Mass Transfer: Part III," *AIChE J.*, Vol. 12, p. 127.)

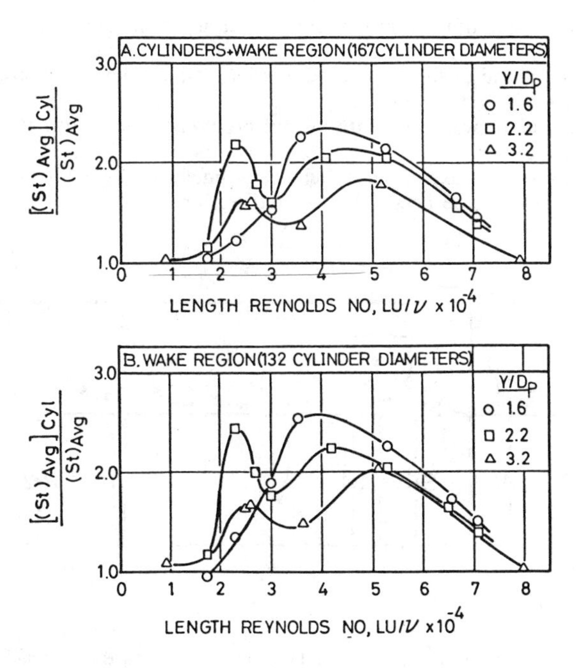

Figure 5. Effect of two 0.093-in.-diameter cylinders spaced 10.6 diameters apart near the edge of the laminar-boundary layer on the average Stanton number (Sc = 2.44). (From D.G. Thomas, "Forced Convection Mass Transfer: Part III," *AIChE J.*, Vol. 12, p. 128.)

Whether we consider the wake region alone or the regions including the cylinders together, two maxima occur when the gap between the promoter and the wall becomes larger (see Figure 5). Since the velocity is large and forms a disturbance even in the absence of a turbulence promoter, nearly no increase in the Stanton number occurs with the promoter.

Two-Dimensional Liquid System

The difference between the liquid system and the gas system lies in the difference of Schmidt numbers. The dimensionless equation in two-dimensional mass transfer systems is as follows.

$$V_x' \frac{\partial C'}{\partial x'} + V_y' \frac{\partial C'}{\partial y'} = \frac{1}{Pe}\left(\frac{\partial^2 C'}{\partial x'^2} + \frac{\partial^2 C'}{\partial y'^2}\right) \tag{2}$$

where Pe is the product of Re and Sc. The effect of the Reynolds-number increase on mass transfer can be applied similarly to the case of the Schmidt number. Watson and Thomas [33] observed a two-fold increase in the mass transfer rate in aqueous systems (Sc = 1,850) as compared to that in gaseous systems (air, Sc = 2.44) when the experiment was performed in the same turbulence promoter system. Miyashita et al. [34–37] studied the mass transfer increase using flow visualization in an aqueous system similar to that in Figure 6. The position of the eddies, separation points, and stagnation points were found to be nearly independent of the Reynolds number, being linear functions of the diameter of the turbulence promoter. The flow characteristics associated with the turbulence promoter are identifiable, namely [34]:

- T − 1: The point ahead of the turbulence promoter at which the boundary layer first appears.
- S − 1: The point ahead of the turbulence promoter which shows the size of the eddies produced.
- S + 1: The point in the wake of the turbulence promoter which shows the size of the eddies produced.
- E + 1: The point in the wake of the turbulence promoter where the backflow, which has now appeared, is perpendicular to the direction of flow.

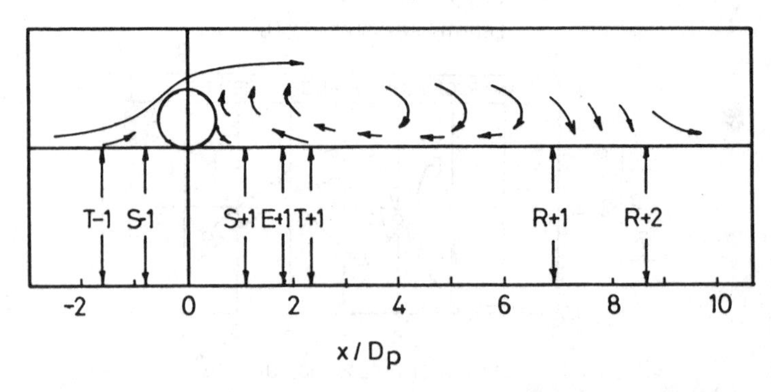

Figure 6. Typical flow pattern with x/Dp. (From H. Miyashita et al., "Flow Behavior and Augmentation of Mass Transfer Rates Using a Turbulence Promoter in Rectangular Duct," *Kagaku Kogaku Ronbunshu,* Vol. 6, p. 153.)

T + 1: The point in the wake of the turbulence promoter where the backflow boundary layer first appears.
R + 1: The leading edge of the reabsorption zone.
R + 2: The trailing edge of the reabsorption zone.

Distribution of shear stress is very important in analyzing mass transfer since a larger shear stress at the surface means a thinner boundary layer for easier mass transfer. At the point T + 1 where separation occurs shear stress becomes zero. In the region between T + 1 and R + 1 where strong back flow occurs the shear stress has the negative value (refer to Figure 7). But further downstream from the promoter the absolute value of the shear stress is larger than that upstream. The rate of shear stress increase with Re in the presence of a turbulence promoter is larger at low Re numbers, since at high Re, turbulence is maintained even without the promoter. The rate of Sh increase with the turbulence promoter is large in the region between R + 1 and R + 2 where shear rate is low, but the convection effect is dominant (see Figure 8). Also the absolute value of the shear stress right behind the promoter was

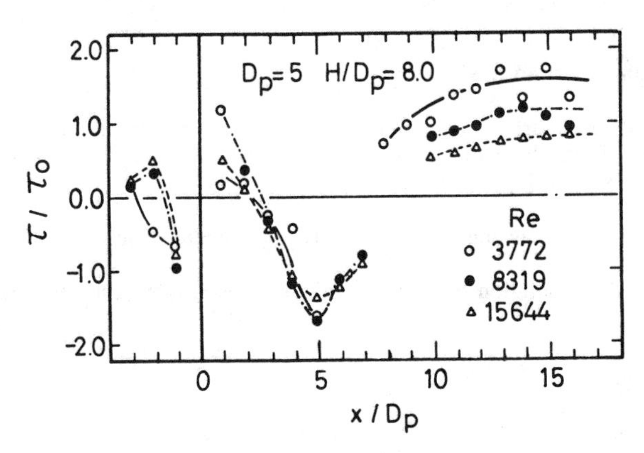

Figure 7. Typical variation of shear stress with x/Dp. (From H. Miyashita et al., "Flow Behavior and Augmentation of Mass Transfer Rates Using a Turbulence Promoter in Rectangular Duct," *Kagaku Kogaku Ronbunshu,* Vol. 6, p. 155.)

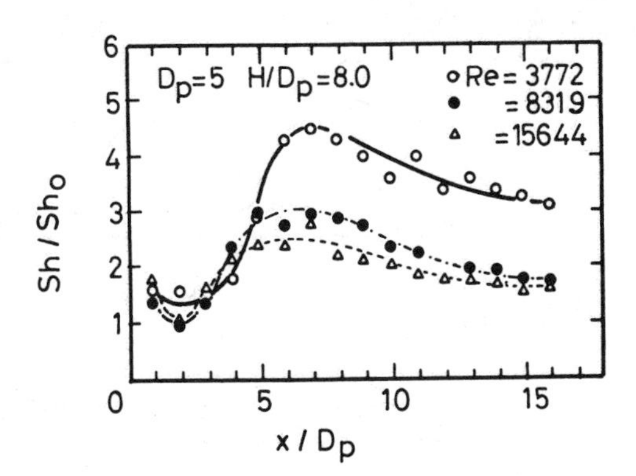

Figure 8. Typical variation of mass transfer with x/Dp. (From H. Miyashita et al., "Flow Behavior and Augmentation of Mass Transfer Rates Using a Turbulence Promoter in Rectangular Duct," *Kagaku Kogaku Ronbunshu,* Vol. 6, p. 155.)

lower than that in the absence of the promoter, but the Sherwood number was larger owing to the mixing and convection by the promoter. It can be concluded that the contribution of the turbulence promoter to the mass transfer is that immediately behind the promoter the effect of turbulent intensity from mixing and convection is predominant; downstream from the promoter the shear stress due to the instability occurring in the boundary layer is more important. The rate of mass-transfer increase with Re is larger at low Reynolds numbers as in the case of the shear stress.

Miyashita et al. [9] studied the system where the turbulence promoter was detached from the wall as shown in Figure 1. When e/Dp is 5 and the promoter is attached to the wall, the fluid forms recirculating flow between the two promoters as illustrated in Figure 9. However when the gap between the promoter and the wall becomes larger, the eddy forms behind the promoter and this eddy forms a Karman vortex downstream. The rate of mass transfer increase due to the promoter depends on the turbulence intensity if no gap exists between the promoter and the wall. If there is a gap, it depends on the shear stress immediately beneath the promoter and turbulence intensity between the promoters (see Figure 10). If the second promoter is placed 7–10 diameters of Dp from the first one, the fluid motion and the mass transfer from the first one is negligible or not influenced at all by the second one [38].

Three-Dimensional Liquid System

The most widely used turbulence promoter in spiral-type artificial kidneys or electrodialysis is the net type. The turbulence promoter should satisfy the following conditions for use in an electrodialysis system [39]:

- Have the capability of causing few stagnant regions where scale formation may be initiated.
- Cause mixing of the solution between the membranes without causing a high hydraulic pressure drop.

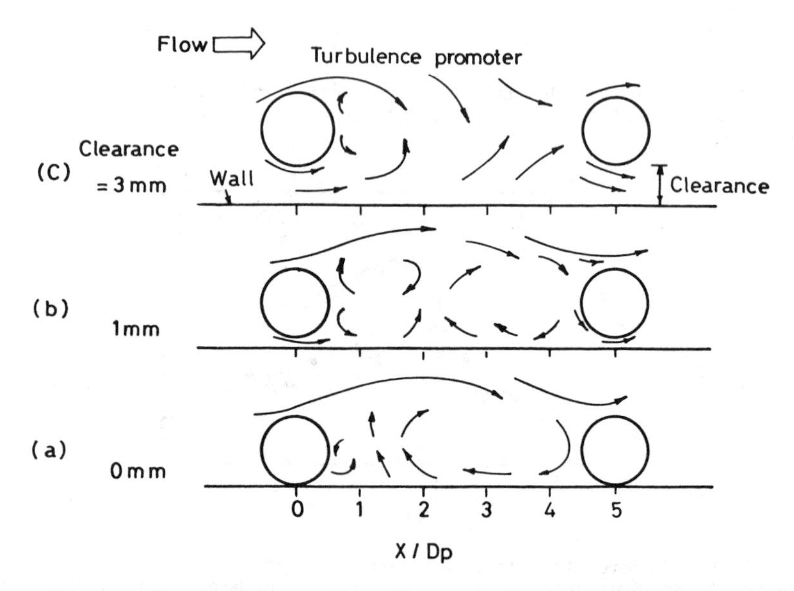

Figure 9. Flow pattern for e/Dp = 5 (Re = 1.55×10^4). (From H. Miyashita et al. [9].)

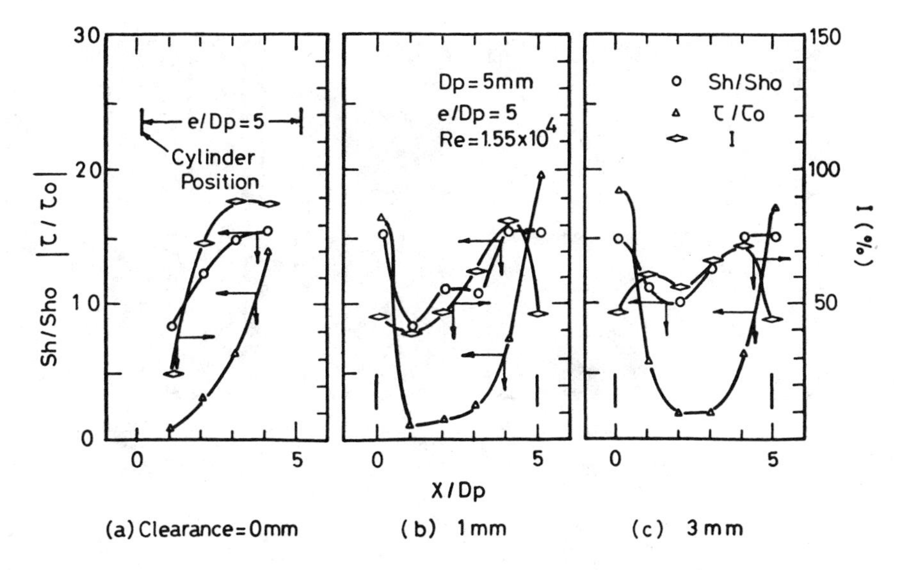

Figure 10. Profiles of transport factors for e/Dp = 5. (From H. Miyashita et al. [9].)

- Have a very low cross-section facing the current to minimize the electrical resistance.
- Be dimensionally stable under the conditions of electrodialysis.
- Be available in a range of thicknesses.
- Be available at a low cost.

Figure 11 shows several net-type promoters which are in commercial use. The study of the fluid motion and mass transfer in one unit experiencing three-dimensional mass transfer is very difficult since the eddies generated by the promoter decay irregularly and behave as in turbulence. Thus it becomes necessary to establish models to explain the mass transfer. Solan et al. [15, 16] proposed the "mesh step" model which is characterized by a mixing factor k and a mesh size. In laminar fluid flow the boundary layer grows in a spacer mesh until it reaches the next mesh and then partial mixing occurs. The thickness of the boundary layer becomes thinner by a factor of $(1 - k')$ and begins to grow further. This growth and reduction step is repeated to the subsequent promoters. The commercial promoter employs a turbulence promoter composed of many units. Thus the boundary layer in each cell will be constant. The boundary-layer thickness and the Sherwood number are given as follows;

$$\frac{\delta c}{H} = 1.33\ Sc^{1/6}\left(\frac{\Delta\bar{x}}{Pe}\right)^{1/2}\left[\frac{1-(1-k')^2}{k'^2}\right]^{1/2} \tag{3}$$

$$Sh = 0.753\ Sc^{-1/6}\left(\frac{Pe}{\Delta\bar{x}}\right)^{1/2}\left[\frac{k'}{2(1-k'/2)}\right]^{1/2} \tag{4}$$

The Sherwood number depends on the 1/3 power of Re in laminar flow and on the 1/2 power of Re in turbulent flow. The mass transfer in the promoter follows the turbulent mode. The "mesh step" model was tested for the various promoters shown in Figure 11 and found to be in good agreement with the experimental observations. Results are shown in Figure 12. In

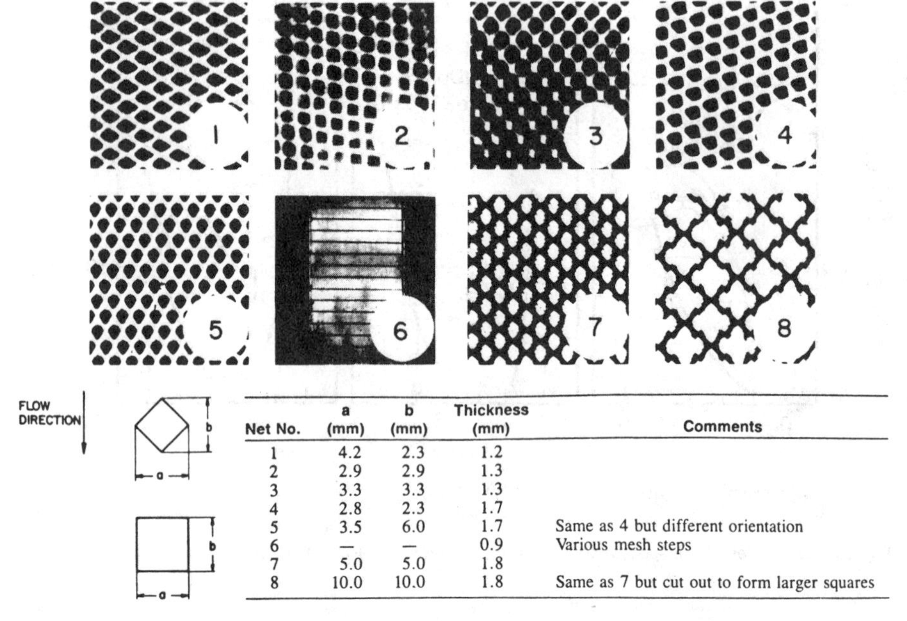

Net No.	a (mm)	b (mm)	Thickness (mm)	Comments
1	4.2	2.3	1.2	
2	2.9	2.9	1.3	
3	3.3	3.3	1.3	
4	2.8	2.3	1.7	
5	3.5	6.0	1.7	Same as 4 but different orientation
6	—	—	0.9	Various mesh steps
7	5.0	5.0	1.8	
8	10.0	10.0	1.8	Same as 7 but cut out to form larger squares

Figure 11. Net-type turbulence promotors. (From Y. Winograd et al., "Mass Transfer in Narrow Channels in the Presence of Turbulence Promoters," *Desalination,* Vol. 13, p. 173.)

mode 6 there is only slight deviation from theory due to the experiment's similarity to the two-dimensional system rather than the three-dimensional system. This model is useful in predicting mass transfer rates in larger systems using laboratory data since only the mixing efficiency "k′," which depends on mesh shape, need be known.

Sherwood numbers obtained experimentally using several types of turbulence promoters are given in Table 1.

NUMERICAL ANALYSIS

It is advantageous to approach the problem of turbulence promoters by analyzing the fluid motion and mass transfer around the promoter quantitatively from a theoretical basis. Chang and co-workers [20, 21] performed a numerical analysis on a two-dimensional model promoter system. The two two-dimensional model promoters, referred to as the zigzag type and cavity type were chosen for the numerical simulation. The systems are illustrated in Figure 13. The governing equations for the study are as follows:

$$\Delta^2\psi = -\omega \tag{5}$$

$$\Delta^2\omega = \mathrm{Re}\left(\frac{\partial\psi}{\partial y'}\frac{\partial\omega}{\partial x'} - \frac{\partial\psi}{\partial x'}\frac{\partial\omega}{\partial y'}\right) \tag{6}$$

$$\Delta^2 C' = \mathrm{Pe}\left(\frac{\partial\psi}{\partial y'}\frac{\partial C'}{\partial x'} - \frac{\partial\psi}{\partial x'}\frac{\partial C'}{\partial y'}\right) \tag{7}$$

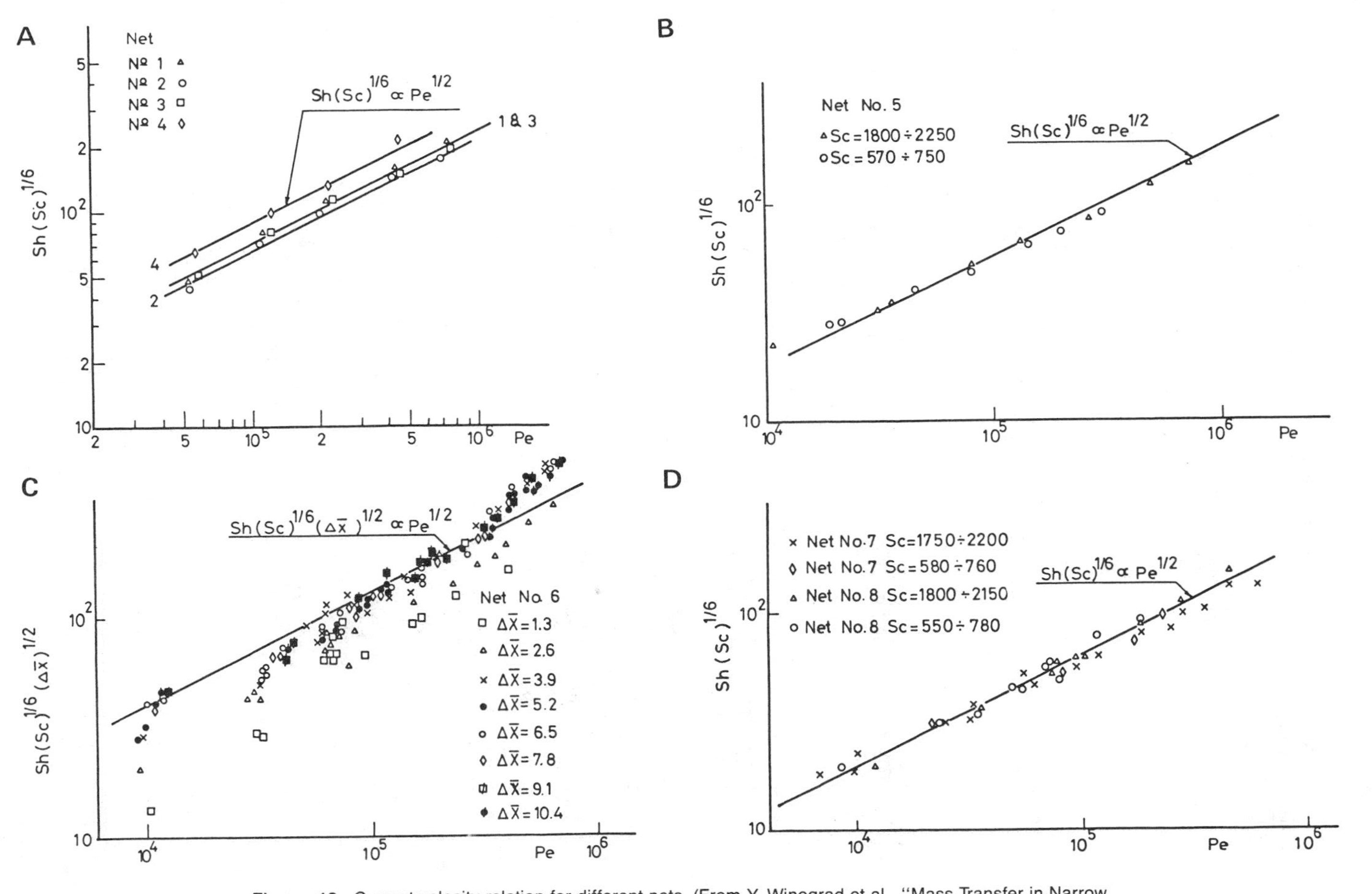

Figure 12. Current-velocity relation for different nets. (From Y. Winograd et al., "Mass Transfer in Narrow Channels in the Presence of Turbulence Promoters," *Desalination*, Vol. 13, p. 177.)

Table 1
Literature Survey of the Most Important Experimental Investigations

Reference	Method	Cells and Turbulence Promoters	Range of Variations	Correlations of the Results
[47]	Electrochemical method	Rectangular channel (cross section: 1.02 × 6.25mm) Rectangular, triangular, and circular promoters $10^{-2} < \frac{D_e}{e} < 0.2$	$250 < Re < 2{,}000$ $Re = \frac{uD_e}{\nu}$ $Sh = \frac{\bar{k}D_e}{D}$ $0.25 < U < 1 m/s$ $Sc = 1{,}000$	(Circular and triangular promoters) $2\lambda = \frac{a}{Re} + 0.175 \frac{D_e}{e} \frac{1-\beta^2}{\beta^2}$ β = contraction coefficient $\beta = 0.8$ $Sh = 4.2\ 10^{-3}\ Sc^{1/3}\ Re'^{0.094} \left(\frac{D_e}{e}\right)^{0.153}$ $\beta = 0.34$ $Sh = 1.33\ Sc^{1/3}\ Re'^{0.404} \left(\frac{D_e}{e}\right)^{0.34}$ $\beta = 0.5$ $Sh = 0.272\ Sc^{1/3}\ Re'^{0.631} \left(\frac{D_e}{e}\right)^{0.357}$
[42]	Electrochemical method—recovery of copper	Swiss roll cell: electrode dimensions 15 × 0.013 × 80 cm	$5 < Re' < 80$ $80 < Re' < 1{,}000$ $Re' = \frac{uD_e}{\nu}$ $Sh' = \frac{\bar{k}D_e}{D}$	$Sh = 1.27\ Re'^{0.34}\ Sc^{1/3}$ $Sh = 0.15\ Re'^{0.82}\ Sc^{1/3}$ Applicable for all cloth separators
[43]	Electrochemical method—reduction of ferricyanide ions	Swiss roll cell: electrode dimensions 186 × 15 × 0.01 cm Polypropylene separator open area: 31% thread diameter: 200 μm mesh count per cm: 22.2	$0 < Q < 30$ ml/s Q = volumetric flow rate (ml/s) $Sc \simeq 10^3$	$k(cm/s) = 1.76\ 10^{-4} Q^{0.84}$
[15]	Electrochemical method	Rectangular channel (cross section: 0.3 × 100 mm) Net-type turbulence promoters	$Pe = Re\ Sc$ $10^4 < Pe < 10^6$ $570 < Sc < 2{,}250$	$Sh = 1.06\ Sc^{1/3} \left(\frac{Re}{e}\right)^{1/2} \left(\frac{K}{2\left(1-\frac{K}{2}\right)}\right)^{1/2}$ K : mixing efficiency K = 1: perfect mixing K = 0: absence of mixing
[24]	Electrochemical method—electrodialysis	Rectangular channel (thickness 1 mm) Cylindrical promoters (diameters 0.5 mm) $16 < \frac{e}{H} < 67$	$Sc = 500$ $300 < Re < 2{,}000$	$Sh = 1.9\ Sc^{1/3} \left(\frac{H}{e}\right)^{0.5} Re^{-0.5}$ $\lambda = \frac{\Delta P}{\frac{L}{H}\rho\frac{u^2}{2}} = 20 \left(\frac{H}{e}\right)^{0.6} Re^{-0.5}$
[25]	Electrodialysis	Rectangular channel 0.38 × 3.7 cm Cylindrical promoters d = 0.167 cm $2 < \frac{e}{H} < 63$	$Sc = 500$ $30 < Re < 2{,}000$ $\lambda = \frac{\Delta P}{\frac{L}{H}\rho\frac{u^2}{2}}$	$\frac{e}{H} = 63$ $Sh = 0.36\ Sc^{1/3}\ Re^{0.45}$ $300 < Re < 800$ $\lambda = \frac{26}{Re} + 0.033$ $\frac{e}{H} = 16$ $Sh = 0.45\ Sc^{1/3}\ Re^{0.51}$ $400 < Re < 1{,}000$ $\lambda = \frac{42}{Re} \pm 0.107$ $\frac{e}{H} = 2$

(Table continued on next page)

Table 1 Continued

Reference	Method	Cells and Turbulence Promoters	Range of Variations	Correlations of the Results
				$Sh = 0.39\ Sc^{1/3}\ Re^{0.62}$ $350 < Re < 1{,}400$ $\lambda = \frac{176}{Re} + 0.57$
[44]	Ultrafiltration	Rectangular channel 0.38 × 7.6 cm Cylindrical promoters $d = 0.46\ H$ $2 < \frac{e}{H} < 63$	$Sc = 15{,}000$ $100 < Re < 1{,}500$	$\frac{e}{H} = 63$ $Sh = 0.15\ Sc^{1/3}\ Re^{0.491}$ $450 < Re < 1{,}500$ $\lambda = 3.92\ Re^{-0.614}$ $\frac{e}{H} = 16$ $Sh = 0.888\ Sc^{1/3}\ Re^{0.649}$ $100 < Re < 1{,}500$ $\lambda = 2.07\ Re^{-0.383}$ $\frac{e}{H} = 2$ $Sh = 0.117\ Sc^{1/3}\ Re^{0.661}$ $200 < Re < 1{,}500$ $\lambda = 6.00\ Re^{-0.3}$ $Sh = 0.189\ Sc^{1/3} \left(\frac{H}{e}\right)^{0.234} Re^{-0.623}$ $\lambda = 14.98 \left(\frac{H}{e}\right)^{0.705} Re^{-0.379}$ } All types
[4]	Heat transfer measurements	Circular tubes $dt = 1$ cm Promoters: Granular-type $2.4\ 10^{-3} < \frac{d_p}{dt} < 0.049$	$1.2 < Pr < 5.94$ $1.4 < 10^4 < Re = \frac{udt}{\nu} < 1.2\ 10^5$	$\frac{\sqrt{\lambda}}{St} = 5.19\ dt^{0.2}\ Pr^{0.44}$
[45]	Electrochemical method and heat transfer measurements	Parallelepipedic channel $H = 0.085$ cm Promoters: corrugated, perforated, P.V.C. (void fraction $\epsilon = 0.870$)	$Sc = 2500$ $3.5 < u < 26$ cm/s	$\frac{\bar{k}}{u}\ Sc^{2/3} = 0.121 \left(\frac{uS}{\nu a \epsilon}\right)^{-0.49}$ S: surface area per unit volume a: viscous energy constant or $Sh = 0.418\ Sc^{1/3}\ Re^{0.51}$ For promoters No. 5, $S/a\epsilon = 6.75\ 10^{-3}$ cm
[46]	Electrochemical method at the macroscopic and microscopic scales	Rectangular channel Cylindrical and mesh-type $2.5 < \frac{e}{H} < 10$ $d_p = 8, 5, 3, 1.5,$ and 0.7 mm	$Sc = 1{,}000$ 1 cm/s $< u <$ 1 m/s	$Sh = 1.190\ Re^{0.61}\ Sc^{1/3} \left(\frac{H}{e}\right)^{0.5} \left(\frac{d_p}{H}\right)^{1.5}$ $450 < Re < 3{,}700$ (circular promoters) $\bar{k} = 8.8\ 10^{-6} \times \left(u\ \frac{\Delta P}{L}\right)^{0.22}$ (m/s) (energetic correlation)

Source: Electrochimica Acta, *Vol. 26, pp. 129–131.*

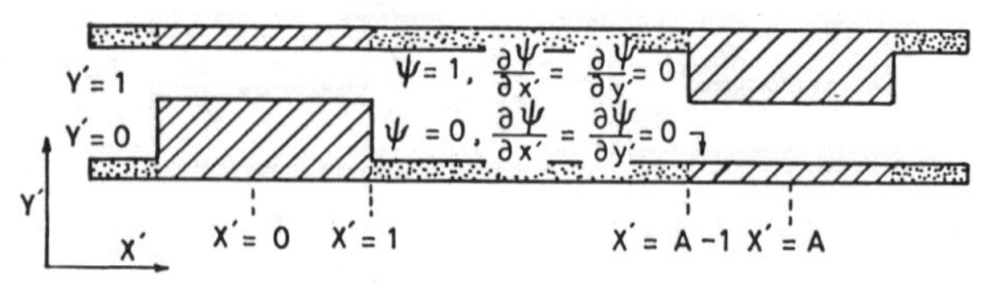

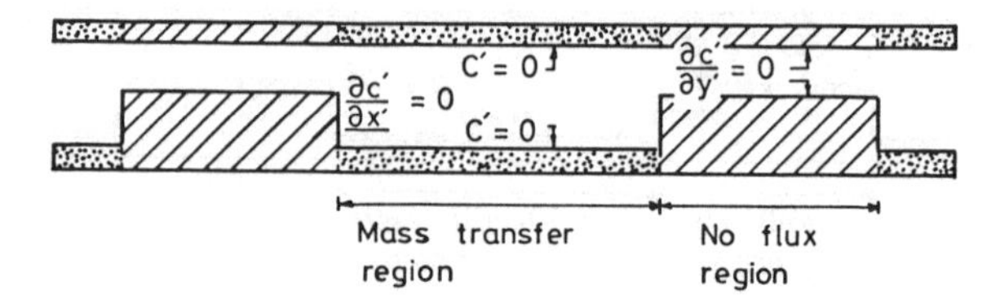

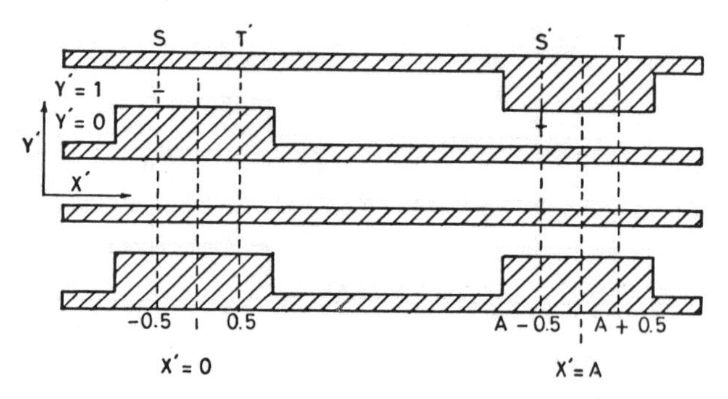

Figure 13. Modeled systems for turbulence promotors. (From I.S. Kang and H.N. Chang, "The Effect of Turbulence Promoters on Mass Transfer—Numerical Analysis and Flow Visualization," *Int. J. Heat Mass Transfer*, Vol. 25, p. 1169.)

where Equation 1 is the dimensionless stream function equation and Equation 6 the vorticity transport equation. Equation 7 refers to the mass transfer equation, which is heavily dependent on the velocity profiles obtained from Equations 5 and 6. The boundary condition assumes that the turbulence promoter is inert and no reaction occurs on the wall of the promoter. The boundary conditions for the wall where high polarization occurs are as follows:

At the upper walls

$$\psi = 1$$

$$\frac{\partial \psi}{\partial x'} = \frac{\partial \psi}{\partial y'} = 0 \tag{8}$$

At the lower walls

$$\psi = 0,$$

$$\frac{\partial \psi}{\partial x'} = \frac{\partial \psi}{\partial y'} = 0 \tag{9}$$

At conducting walls

$$C' = 0 \tag{10}$$

At nonconducting walls

$$\frac{\partial C'}{\partial x'} = 0,$$

or

$$\frac{\partial C'}{\partial y'} = 0 \tag{11}$$

The boundary values for the inlet and the outlet of the repeating units can be calculated using Mitchell's method [40]. In order to prevent numerical instability the upwind-difference method was used [41]. In the case of the zigzag promoter configuration with the aspect ratio AR = 5, two eddies formed: a small eddy is formed at the corner adjacent to the upper wall and the promoter; a relatively large eddy is formed at the bottom wall and the promoter (refer to Figure 14). The eddies formed at the lower wall become larger and move downstream with an increase of Reynolds numbers from 50 to 500. In the cavity-type promoter with AR = 5, two eddies are formed between the two promoters at Re = 50 (see Figure 15). However, as the Reynolds number increases, two eddies coexist under the streamline of $\psi = 0$ (Re = 100). Further increases of Reynolds numbers result in two eddies coalescing and the center of the eddies moving towards the downstream (Re = 200, 500). When the shear stress at the wall becomes large, the boundary-layer thickness becomes thin and the mass transfer rate increases as a result. The dimensionless velocity gradient at the wall increases with Re in a channel with the turbulence promoter while it is a constant independent of Re in a straight channel without a promoter (refer back to Figure 14). In a straight channel the dimensionless shear stress decreases inversely in proportion to the Re. But with the promoter the decrease is less severe than in the straight channel. Thus the overall mass transfer increases with the turbulence promoter. There exists a point in the lower wall of the zigzag-type promoter where shear stress becomes zero and also separation occurs. However, mass transfer is increased by the convective recirculating flow from the bulk flow. Accordingly, local Sh increases with the distance in the case of the zigzag-type promoter as shown in Figure 16. However, on the upper wall the direction of the eddies is clockwise and thus the convection effect near the separation point is very weak. In front of the promoter the mass transfer increases due to the convection. In a cavity-type promoter the mass transfer on the lower plate is influenced by the recirculating flow rather than the shear stress on the wall. Thus the mass transfer is higher downstream from the promoter, and the magnitude is lower than with the zigzag-type promoter. But on the upper plate the local velocity is higher downstream from the promoter to give a higher mass transfer rate by the shear stress than at the zigzag-type promoter. The mean Sherwood number in a single-unit promoter is higher in the zigzag-type promoter than in the cavity-type promoter. The difference becomes more significant as Re and Sc increase (refer to the data in Table 2 for comparisons).

Numerical results were confirmed by Kim et al. [22], using the limiting current method. There was 6% error at Re = 10 and 20% error at Re = 200 between theory and experiments. This may be attributed to the artificial viscosity in improving the stability of the upwind method. The Sh numbers obtained numerically and experimentally are as follows:

Zigzag-type promoter:

$$Sh = 0.519\ Sc^{0.376}\ Re^{0.475} \tag{12}$$

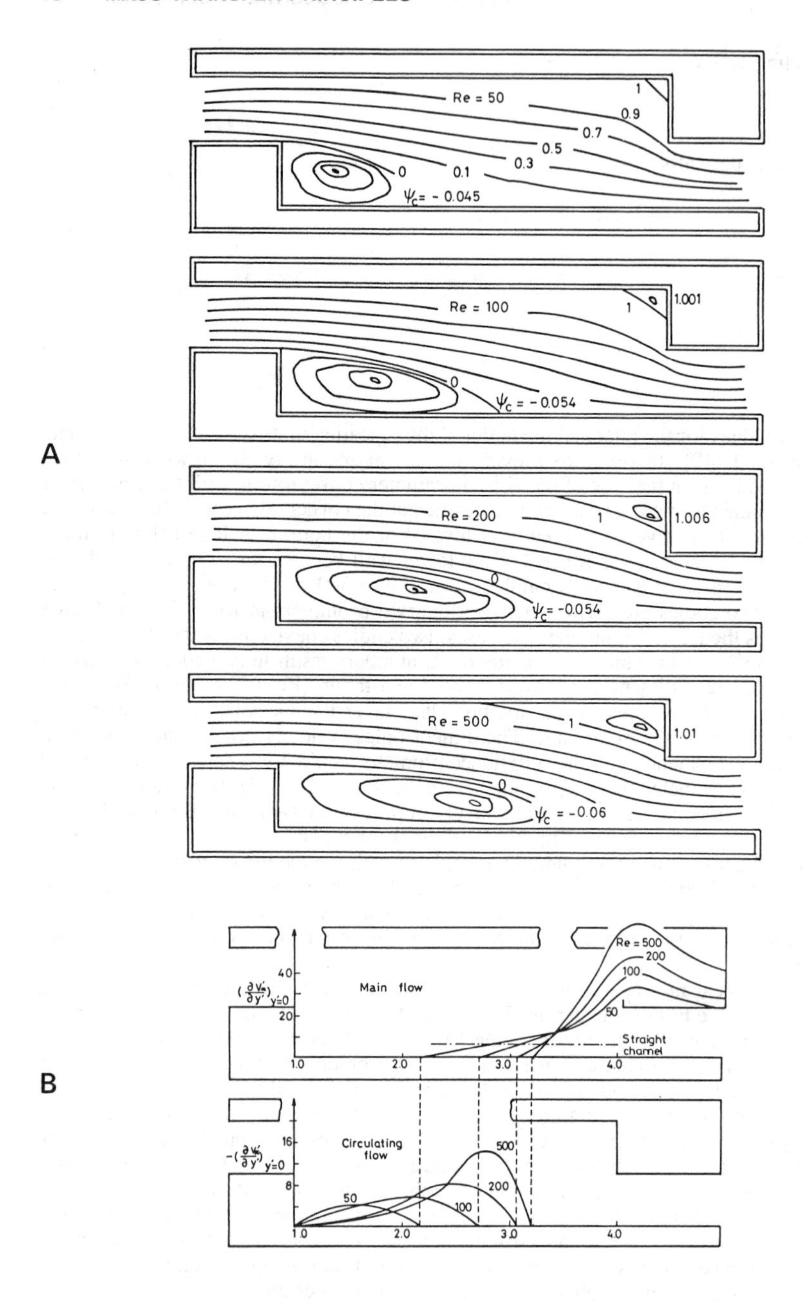

Figure 14. (A) Streamline distributions (AR = 5) for the zigzap-type promoter; (B) dimensionless velocity gradient along the lower wall for the zigzag-type promoter. (From I.S. Kang and H.N. Chang, "The Effect of Turbulence Promoters on Mass Transfer—Numerical Analysis and Flow Visualization," *Int. J. Heat Mass Transfer,* Vol. 25, p. 1171.)

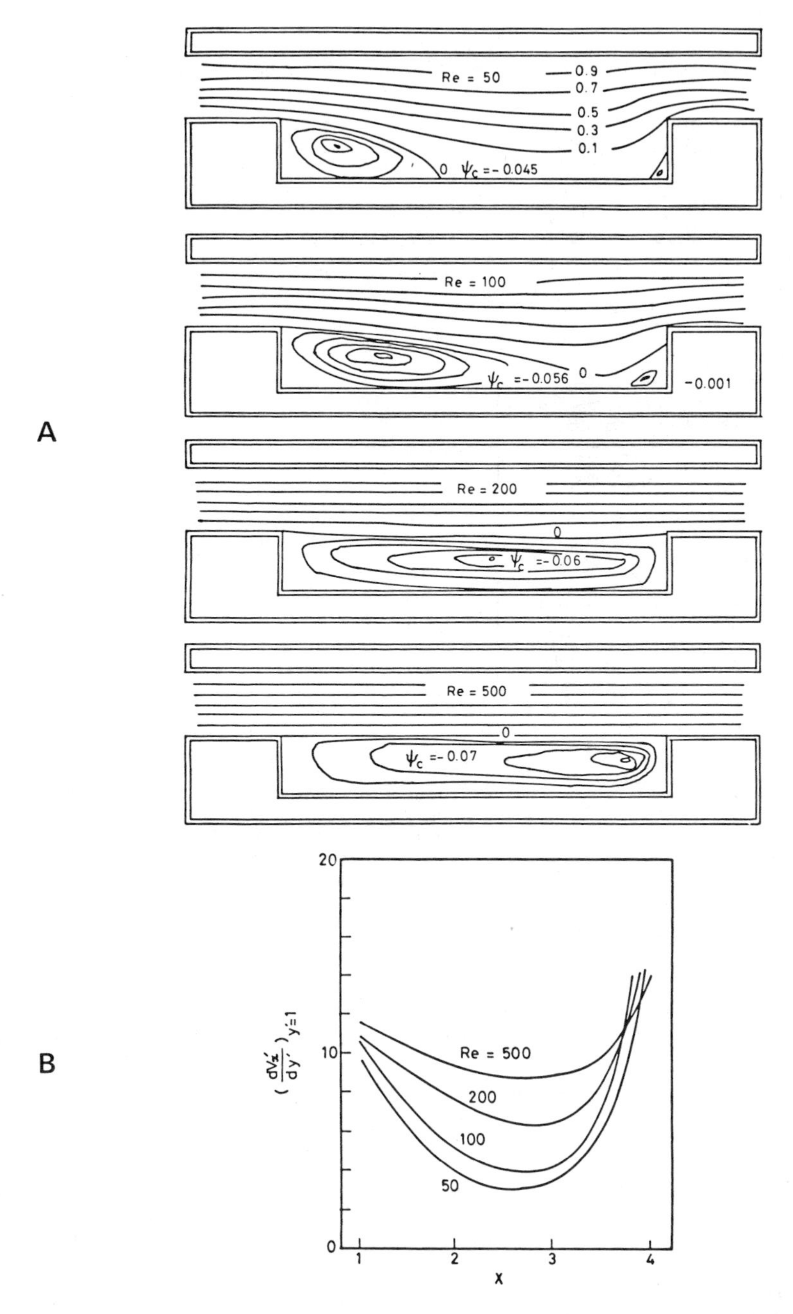

Figure 15. (A) Streamline distributions (AR = 5) for the cavity-type promoter; (B) dimensionless velocity gradient along the upper wall of the cavity-type promoter. (From I.S. Kang and H.N. Chang, "The Effect of Turbulence Promoters on Mass Transfer—Numerical Analysis and Flow Visualization," *Int. J. Heat Mass Transfer,* Vol. 25, p. 1172.)

Table 2
Comparison of the Mean Sherwood Numbers for Each Promoter

	Re							
	50		100		200		500	
Sc	Zigzag	Cavity	Zigzag	Cavity	Zigzag	Cavity	Zigzag	Cavity
20	24.4	22.9	31.9	26.5	42.8	31.0	67.1	37.5
60	35.7	33.3	46.8	38.6	64.3	46.5	104.9	58.5
200	53.9	49.9	72.7	59.2	103.6	73.2	169.3	95.8
600	80.6	73.6	112.7	89.7	160.7	113.7	249.3	151.3
2,000	130.8	117.8	182.1	144.6	248.5	185.9	343.1	239.6

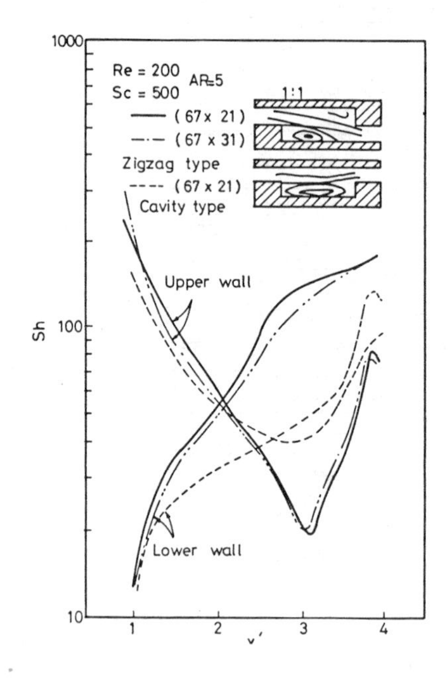

Figure 16. Local Sherwood number distributions (Re = 200, Sc = 600). (From I.S. Kang and H.N. Chang, "The Effect of Turbulence Promoters on Mass Transfer—Numerical Analysis and Flow Visualization," *Int. J. Heat Mass Transfer*, Vol. 25, p. 1173.)

Cavity-type promoter:

$$Sh = 1.069\ Sc^{0.376}\ Re^{0.294} \tag{13}$$

where $10^3 \leq Sc \cdot Re \leq 4 \times 10^5$ and the aspect ratio (AR) = 5

From numerical analysis:

$$Sh_m = 31.6\ Re^{0.420}\ AR^{-0.776} \text{ for the lower wall} \tag{14}$$

$$Sh_m = 24.6\ Re^{0.327}\ AR^{-0.527} \text{ for the upper wall} \tag{15}$$

From experimental analysis (zigzag type):

$$Sh_m = 27.9\ Re^{0.376}\ AR^{-0.656} \text{ for both walls} \tag{16}$$

$Sc = 1{,}690,\ Re \leq 300$

As seen in the correlation, the dependence of the Sh on the Re is larger in the zigzag-type promoter than in the cavity-type promoter. Thus the zigzag-type is more effective as a promoter.

PROCESS OPTIMIZATION

Although a turbulence promoter favorably increases the mass transfer phenomenon, it has the disadvantage of increasing pressure drop. Thus it is important to select the optimum channel geometry and flow conditions for the promoter. Sonin et al. [23–25] studied the optimization of the total cost in a electrodialysis system. The total cost (K) for the system given in Figure 17 is divided into those of electrical energy, mechanical energy, and capital cost:

k_e = cost of electrical energy delivered to the system (\$/joule)
k_p = cost of the mechanical (pumping) energy delivered to the fluid in the system (\$/joule).
k_c = capital cost per unit time, per unit current carrying area A (\$/m²-sec)

$$K = k_e + \frac{k_c}{j}\left(1 + \frac{k_p}{k_c}\frac{\Delta p}{1} VH\right) \text{ (\$/coulomb)} \tag{17}$$

$$K_v = Q(c_{feed} + c_{product})K \text{ (\$/m}^3\text{)} \tag{18}$$

The last term is related to the fluid condition such as the fluid velocity and pressure drop which affects the production cost directly. The second term represents the degree of polarization of current-voltage influenced by the fluid velocity. If the current density is not further improved by the flow condition and follows the Ohm's law, and the pumping cost contributes

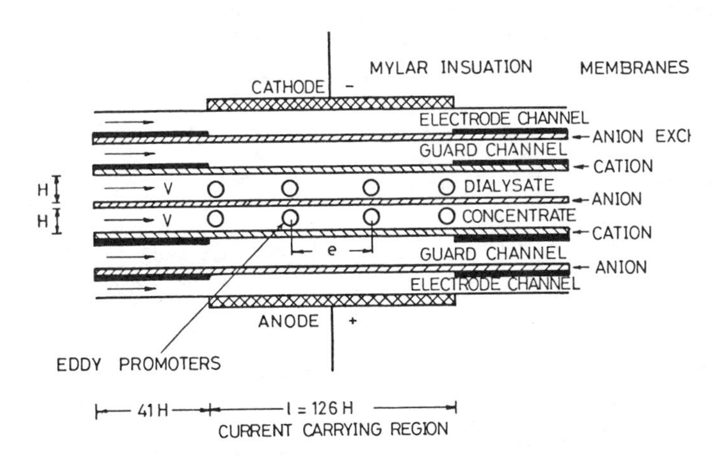

Figure 17. Schematic showing experimental simulation of dialysate-concentrate channel pair in electrodialysis system, with eddy promoters. (From M.S. Isaacson and A.A. Sonin, "Sherwood Number and Friction Factor Correlations for Electrodialysis Systems, with Application to Process Optimization," *I & EC Process Des. and Devel.*, Vol. 15, p. 314.)

negligibly to the production cost, the promoter is said to have attained a hydrodynamically ideal performance.

$$j_{design} \leq \alpha j_{lim}$$

$$\frac{k_p}{k_c}\frac{\Delta p}{1} VH \leq \beta \qquad (19)$$

The optimum product cost in a hydrodynamically ideal condition is given as [25]:

$$(K_v)_{optimum,ideal} = 2(c_{feed} + c_{product}) \left[\frac{k_e k_c \; RTH}{ZD(c_{feed} + c_{product})}\right]^{1/2} (\$/m^3) \qquad (20)$$

Sh is given in terms of Re and Sc. Using a modified friction factor function f, Sh can be expressed as a function of f.

$$f = \lambda^{1/3} Re$$

$$\lambda = \frac{P}{\frac{1}{H}\frac{v^2}{2}} \qquad (21)$$

Figure 18 shows the region of hydrodynamically ideal performance in an electrodialysis system. The systems A,B cannot be operated in a region of hydrodynamically ideal performance, but for the system C it is possible. Thus for this system no further improvements in channel design are necessary to improve the channel design from a hydrodynamic viewpoint. In the cases of systems A and B it is economically more feasible when the points on A,B are closer to the hydrodynamically ideal region. However if the system is operated at a point above point C on curve B, it is less economical since the pumping energy cost exceeds the cost reduction by the current increase. Thus point c is economical. If we express Sh as $Sh = af^n$, we find the following optimum condition is obtained [25]:

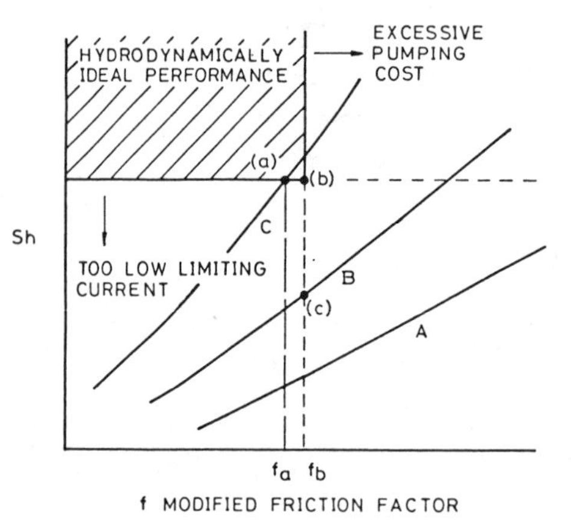

Figure 18. System performance characterized as a plot of Sh versus f. (From A.A. Sonin and M.S. Isaacson, "Optimization of Mass Transfer and Frictional Characteristics of Electrodialysis," 4th Int. Symp. on Fresh Water from the Sea, Vol. 3, p. 243.)

$$\frac{Sh_{optimum}}{Sh^*} = \frac{\theta^2}{\sqrt{6}}\left[\sqrt{1 + \frac{12}{4}} - 1\right]^{1/2} \tag{22}$$

$$\theta = \frac{a\, f^{*n}}{Sh^*}$$

$$\frac{(K_v)_{optimum}}{(K_v)_{optimum,ideal}} = \frac{1}{2}\left[\frac{3 - 2n}{3 - n}\left(\frac{Sh_{optimum}}{Sh^*}\right) + \frac{3}{3 - n}\left(\frac{Sh^*}{Sh_{optimum}}\right)\right]^{1/2} \tag{23}$$

where

$$(K_v)_{optimum,ideal} = 2(c_{feed} + c_{product})\left[\frac{k_e k_c\ RTH}{ZD(c_{feed} + c_{product})}\right]^{1/2} (\$/m^3) \tag{24}$$

$$Sh^* = \frac{2}{\alpha}\left[\frac{Zk_cH}{k_eDRT(c_{feed} + c_{product})}\right]^{1/2} \tag{25}$$

$$f^* = \left(\frac{\rho^2 k_c}{k_p}\right)^{1/3}\frac{H}{\mu} \tag{26}$$

The results obtained from the optimization of the electrodialysis system in Figure 17 are shown in Figure 19 using the following variables:

$Z = 1$

$D = 2 \times 10^{-9} m^2/sec$

$\rho = 10^3 kg/m^3$

$\mu = 10^{-3} kg/sec\text{-}m$

$T = 300°K$

molecular weight $= 60$

and assumed that

$h = 10^{-3} m$

$k_e = 0.01\$/kWh$

$k_c = 50\$/m^2\text{-year}$

$k_p = 0.02\$/kWh$

$\alpha = 0.5$

Sh is the largest in case $e/H = 2$ (refer back to Figure 20). In a brackish feed concentration $e/H = 4$ retains the lowest optimum cost as shown in Figure 19. Thus it is necessary to study the current and pressure drop concurrently for future system development.

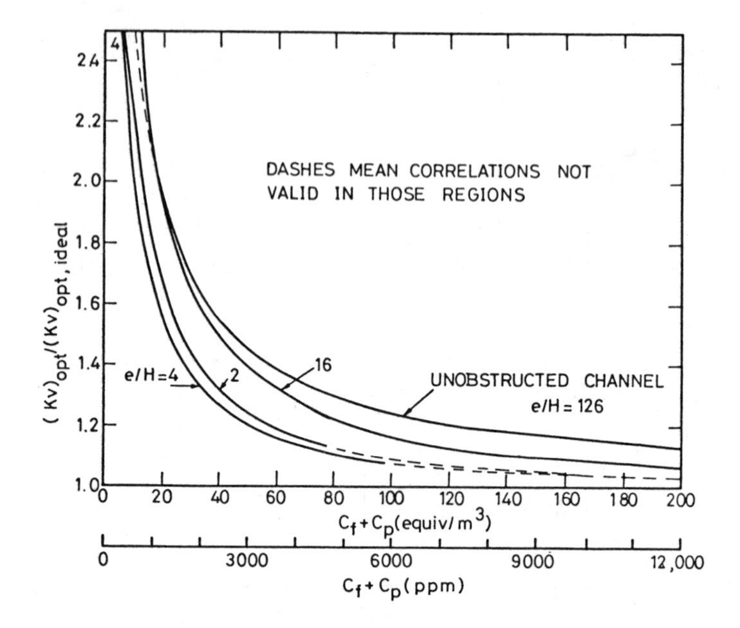

Figure 19. The optimum product cost for the promoter systems and typical electrodialysis process parameters. (From M.S. Isaacson and A.A. Sonin, "Sherwood Number and Friction Factor Correlations for Electrodialysis Systems, with Application to Process Optimization," *I & EC Process Des. and Devel.*, Vol. 15, p. 318.)

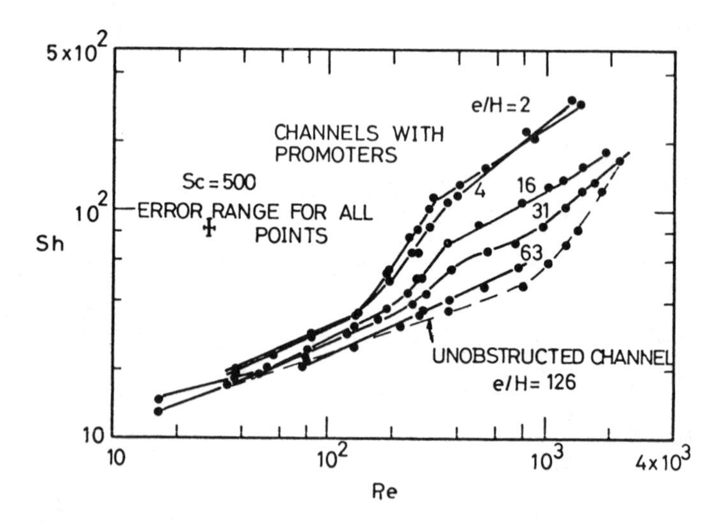

Figure 20. Experimental data for Sh as a function of Re in electrodialysis system with and without eddy promoters (Sc = 500). (From M.S. Isaacson and A.A. Sonin, "Sherwood Number and Friction Factor Correlations for Electrodialysis Systems, with Application to Process Optimization," *I & EC Process Des. and Devel.*, Vol. 15, p. 315.)

NOTATION

- a — dimensionless coefficient in performance characteristic ($Sh = a\ f^n$)
- AR — aspect ratio, e/H
- c' — dimensionless concentration of solute
- c — concentration of solute ($equiv/m^3$)
- c_{feed} — concentration of saline feed in electrodialysis ($equiv/m^3$)
- $c_{product}$ — product concentration in electrodialysis ($equiv/m^3$)
- D — salt diffusion coefficient (m^2/sec)
- D_e — cell equivalent diameter (m)
- D_p — diameter of cylindrical turbulence promoter (m)
- D_t — tube inside diameter (m)
- e — spacing between consecutive promoters (m)
- F — Faraday constant, 9.649×10^4 coulomb/equiv
- f — $\lambda^{1/3}$ Re, modified friction factor defined by Equation 21
- f* — defined by Equation 26
- H — channel thickness (m)
- j — average current density over current carrying area ($coulomb/m^2$)
- j_{lim} — limiting value of j ($coulomb/m^2$)
- j_{design} — design value of j ($coulomb/m^2$)
- K — cost of the product per coulomb of charge passed ($/coulomb)
- k — mass transfer coefficient (mol/cm-sec)
- $\overline{k}$ — average mass transfer coefficient over the electrode plate (mol/cm-sec)
- k' — mixing coefficient
- k_c — capital cost ($\$/m^2/sec$)
- k_e — cost of electrical energy ($/joule)
- k_p — cost of mechanical energy ($/joule)
- K_v — cost per unit volume of product water in electrodialysis system ($\$/m^3$)
- L — length of mass transfer surface (m)
- l_0 — lower limit of integral (Equation 1)
- l — length, in flow direction, of current carrying section (m)
- n — index in performance characteristic ($Sh = a\ f^n$)
- Δp — pressure drop in flow direction (N/m^2)
- Pe — Peclet number
- Q — volumetric flow rate (m^3/sec)
- R — universal gas constant, 8.31 J/K mol
- Re — Reynolds number
- Sh — Sherwood number with turbulence promoter
- Sh_0 — Sherwood number without turbulence promoter
- Sh* — defined in Equation 25
- St — Stanton number
- T — absolute temperature of fluid (°K)
- u — mean liquid (superficial) velocity (m/sec)
- V — velocity of fluid (m/sec)
- v' — dimensionless velocity of fluid
- x — axial distance (m)
- x' — dimensionless axial distance
- $\overline{\Delta x}$ — dimensionless mesh step ($2\Delta x/H$)
- y — distance from flat plate (m)
- y' — dimensionless distance from flat plate
- Z — ion charge number, assumed same for positive and negative ions in electrodialysis

Greek Symbols

- α — a safety factor, defined with regard to Equation 19, taken to be 0.5 in numerical example
- β — a safety factor defined with regard to Equation 19
- δ_c — diffusional boundary layer thickness (m)
- θ — a dimensionless hydrodynamic performance parameter defined in Equation 22
- λ — conventional friction factor
- μ — fluid viscosity coefficient (kg/m-sec)
- ν — kinematic viscosity (m^2/sec)
- ρ — fluid density (kg/m^3)
- τ — shear stress ($kg/m\text{-}sec^2$)
- τ_0 — shear stress without promoter ($kg/m\text{-}sec^2$)
- ψ — stream function
- ω — vorticity

Subscripts

optimum	optimum value, that is, value which minimizes product cost	optimum, ideal	optimum value which would be attained under hydrodynamically ideal conditions

REFERENCES

1. Sherwood, T. K., Pigford, R. L., and Wilke, C. R., *Mass Transfer,* New York: McGraw-Hill, 1979.
2. Treybal, R. E., *Mass Transfer Operations,* 3rd ed., New York: McGraw-Hill, 1980.
3. Perry, R. H., and Chilton, C. H., *Chemical Engineers' Handbook,* 5th ed., New York: McGraw-Hill, 1973.
4. Dipprey, D. F., and Sabersky, R. H., "Heat and Momentum Transfer in Smooth and Rough Tubes at Various Prandtl Numbers," *Int. J. Heat Mass Transfer,* Vol. 6 (1963), pp. 329–353.
5. Aggarwal, J. K., and Talbot, L., "Electrochemical Measurements of Mass Transfer in Semi-cylindrical Hollows," *Int. J. Heat Mass Transfer,* Vol. 22 (1979), pp. 61–75.
6. Zborowski, M., "Theoretical Prediction of the Oxygen Output for a Dialysis Membrane with a Catalyst, in a Flat Plate Dialyzer," *Biorheology,* Vol. 20 (1983), pp. 129–140.
7. Bellhouse, B. J., et al., "A Practical Vortex-mixing Hemodialyzer," *Proceedings of the Third Meeting of ISAO,* Vol. 5 (Suppl) (1981), pp. 686–690.
8. Shiina, Y., et al., "Flow Around Turbulence Promoters in Parallel Channel and Flow Patterns Around Cylinder-Type Turbulence Promoters," *Journal of Nuclear Science and Technology,* Vol. 19 (1982), pp. 720–728.
9. Miyashita, H., et al., "Transport Phenomena Among Turbulence Promoters at/on Wall Surface in Rectangular Duct," *Proceedings of PACHEC '83,* Seoul, Korea. Vol. 1 (1983), pp. 1–6.
10. Focke, W. W., "Plate Heat Exchangers," CSIR Report CENG 445, Pretoria, South Africa, 1983.
11. Thomas, D. G., "Enhancement of Forced Convection Heat Transfer Coefficient Using Detached Turbulence Promoters," *I & EC Process Des. and Devel.,* Vol. 6 (1967), pp. 385–390.
12. ENKA Wuppertal-Barmen Plant, Cuprophan Hollow Fibers Brochure, West Germany.
13. Hicks, R. E., "The Effect of Structure and Orientation of Turbulence Promoters on Pressure Drop, Mass Transfer Correlations and Cost," CSIR Special Report CHEM 218, Pretoria, South Africa, 1970.
14. Schwager, F., et al., "The Use of Eddy Promoters for the Enhancement of Mass Transport in Electrodialysis Cells," *Electrochimica Acta,* Vol. 25 (1980), pp. 1655–1665.
15. Winograd, Y., Solan, A., and Toren, M., "Mass Transfer in Narrow Channels in the Presence of Turbulence Promoters," *Desalination,* Vol. 13 (1973), pp. 171–186.
16. Solan, A., et al., "An Analytical Model for Mass Transfer in an Electrodialysis Cell with Spacer of Finite Mesh," *Desalination,* Vol. 9 (1971), pp. 89–95.
17. Thomas, D. G., "Forced Convection Mass Transfer: Part II. Effect of Wires Located Near the Edge of the Laminar Boundary Layer on the Rate of Forced Convection from a Flat Plate," *AIChE J.,* Vol. 11 (1965), pp. 848–852.
18. Lin, C. S., et al., "Diffusion-Controlled Electrode Reactions," *Ind. Eng. Chem.,* Vol. 43 (1951), pp. 2136–2143.
19. Bradley, P. H., et al., "Effect of Turbulence Promoters on Local Mass Transfer," U.S. Dept. of the Interior, Resource and Development Progress Report No. 597, 1970.

20. Chang, H. N., et al., "Effect of Inert Regions on Local Mass Transfer Rate Measurements using the Limiting Diffusion Current Technique—Case of Poiseuille Flow," *Int. J. Heat Mass Transfer* (in press).
21. Kang, I. S. and Chang, H. N., "The Effect of Turbulence Promoters on Mass Transfer—Numerical Analysis and Flow Visualization," *Int. J. Heat Mass Transfer,* Vol. 25 (1982), pp. 1167–1181.
22. Kim, D. H., et al., "Experimental Study of Mass Transfer Around a Turbulence Promoter by the Limiting Current Method," *Int. J. Heat Mass Transfer,* Vol. 26 (1983), pp. 1007–1015.
23. Sonin, A. A., and Isaacson, M. S., "Optimization of Mass Transfer and Functional Characteristics of Electrodialysis Systems," 4th Int. Symp. on Fresh Water from the Sea, Vol. 3 (1973), pp. 237–251.
24. Sonin, A. A., and Isaacson, M. S., "Optimization of Flow Design in Forced Flow Electrochemical Systems with Special Application to Electrodialysis," I & EC Process Des. and Devel., Vol. 13 (1974), pp. 241–248.
25. Isaacson, M. S., and Sonin, A. A., "Sherwood Number and Friction Factor Correlations for Electrodialysis Systems with Application to Process Optimization," *I & EC Process Des. and Devel.*, Vol. 15 (1976), pp. 313–321.
26. Levich, V. G., *Physicochemical Hydrodynamics,* Englewood Cliffs: Prentice-Hall, 1962, pp. 139–183.
27. Ruckenstein, E., "Some Remarks on Renewal Models," *Chemical Engineering Science,* Vol. 18 (1963), pp. 233–241.
28. Hanratty, T. J., "Turbulent Exchange of Mass and Momentum," *AIChE J.*, Vol. 2 (1956), pp. 359–362.
29. Corino, E. R., and Brodkey, R. S., "A Visual Investigation of the Wall Region in Turbulent Flow," *Journal of Fluid Mechanics,* Vol. 37 (1969), pp. 1–30.
30. Nichas, S. G., et al., "A Visual Study of Turbulent Shear Flow," *Journal of Fluid Mechanics,* Vol. 61, pp. 513–540.
31. Thomas, D. G., "Forced Convection Mass Transfer," *AIChE J.*, Vol. 12 (1966), pp. 124–130.
32. Thomas, D. G., "Forced Convection Mass Transfer," *AIChE J.*, Vol. 11 (1965), pp. 520–525.
33. Watson, J. S., and Thomas, D. G., "Forced Convection Mass Transfer," *AIChE J.*, Vol. 13 (1967), pp. 676–677.
34. Miyashita, H., et al., "Flow Behavior and Augmentation of the Mass Transfer Rate in a Rectangular Duct with a Turbulence Promoter," *Kagaku Kogaku Ronbunshu,* Vol. 6 (1980), pp. 152–156.
35. Miyashita, H., et al., "Transfer Coefficients and Flow Behavior in Back of Turbulence Promoters," Bulletin of Faculty of Engineering, Toyama Univ., Vol. 30 (1979), pp. 51–56.
36. Miyashita, H., et al., "Augmentative Mechanism of Mass Transfer Using a Turbulence Promoter in Rectangular Duct," *Kagaku Kogaku Ronbunshu,* Vol. 7 (1981), pp. 349–354.
37. Miyashita, H., et al. "Experimental Study of the Improvement of Heat Transfer Coefficient Using Turbulent Promoter," *Kagaku Kogaku Ronbunshu,* Vol. 2 (1976), pp. 200–204.
38. Storck, A., and Hutin, D., "Mass Transfer and Pressure Drop Performance of Turbulence Promoters in Electrochemical Cells," *Electrochimica Acta,* Vol. 26 (1983), pp. 127–137.
39. Belfort, G., and Gutter, G. A., "An Experimental Study of Electrodialysis Hydrodynamics," *Desalination,* Vol. 10 (1972), pp. 221–262.
40. Mitchell, N., in *Heat and Mass Transfer in Recirculating Flows,* Gosman, A. D., et al. (Eds.) London: Academic Press, 1969, p. 201.
41. Gosman, A. D., *Heat and Mass Transfer in Recirculating Flows,* London: Academic Press, 1969, pp. 101–103.

42. Robertson, P. M., and Ibl, N., "Electrolytic Recovery of Metals from Waste Waters with the Swiss-roll Cell," *Journal of Applied Electrochemistry,* Vol. 7 (1977), pp. 323–330.
43. Robertson, P. M., et al., "A New Cell for Electrochemical Processes," *Journal of Electroanalytical Chemistry,* Vol. 65 (1975), pp. 883–900.
44. Shen, J. S., and Probstein, R. F., "Turbulence Promotion and Hydrodynamic Optimization in an Ultrafiltration Process," *I & EC, Process Des. and Devel.*, Vol. 18 (1979), pp. 547–554.
45. Hicks, R. E., and Mandersloot, W. G. B., "The Effect of Viscous Forces on Heat and Mass Transfer in Systems with Turbulence Promoters and in Packed Beds," *Chem. Eng. Sci.*, Vol. 23 (1968), pp. 1201–1210.
46. Storck, A., and Hutin, D., "Energetic Aspects of Turbulence Promotion Applied to Electrolysis Processes," *Can. J. Chem. Engng.*, Vol. 58 (1980), pp. 92–102.
47. Leitz, F. B., and Marincic, L., "Enhanced Mass Transfer in Electrochemical Cells Using Turbulence Promoters," *Journal of Applied Electrochemistry,* Vol. 7 (1977), pp. 473–484.

CHAPTER 2

MASS TRANSFER PRINCIPLES WITH HOMOGENEOUS AND HETEROGENEOUS REACTIONS

Vi-Duong Dang

Department of Chemical Engineering
The Catholic University of America
Washington, DC

CONTENTS

INTRODUCTION

Mass transfer principles and practice have been an important area of chemical engineering. Theories and applications up to 1975 can be found from Sherwood et al. [1]. Many industrial operations such as absorption of chlorine by sodium hydroxide solution; and absorption of sulfur dioxide, hydrogen sulfide, and carbon dioxide by water involve mass transfer accompanied by chemical reactions [2]. Recent publications in the literature indicate continuous progress in this area has been made. General mass transfer theory with second-order or higher-order chemical reactions are reported in References 3–19. Researches on mass transfer with heterogeneous reactions are not that extensive as compared to those of mass transfer with homogeneous reactions [3, 20]. Other related research works discuss the effect of mixing on the chemically reactive systems [21–27]. Although the general theories are essential, it is sometimes of particular interest to consider a special geometrical system. Hence, the objectives of this chapter are to analyze (1) the general theory and (2) a tubular reactor system of mass transfer with chemical reactions.

Early development of mass transfer without chemical reactions was often based on the film theory, the penetration theory, or the surface renewal theory [1] all of which are discussed in detail in later chapters. These theories have been extended to mass transfer with chemical reactions, and relatively small deviations in the enhancement factors are found among these three theories [28].

MASS TRANSFER WITH HOMOGENEOUS REACTIONS

Recent works performed on mass transfer are mostly with second-order homogeneous chemical reactions [4–6, 8–14]. Chang and Rochelle [13] developed the enhancement factor in mass transfer by the film and the surface renewal theory for second-order reversible reaction in an effort to model SO_2 absorption into $CaO/CaCO_3$ slurries. They considered particularly the following reaction scheme:

$$A \rightleftarrows B + C$$

This reaction is important in the absorption of SO_2 with dissociation to H^+ and HSO_3^-. The enhancement factor is calculated by the film theory and the surface renewal theory. The surface renewal theory is approximated within 10% by using the film theory solutions with diffusivity ratios replaced by their square roots. DeCoursey [4] derived an approximate analytical expression for the enhancement factor for the absorption with a second-order reversible chemical reaction with corresponding stoichiometry and equal diffusivities of reactants, using the Danckwerts model of mass transfer. When the equilibrium constant $K = 1$ occurs in the bulk-liquid phase, the approximate analytical expression does not agree with the previous numerical results. DeCoursey [4] required additional numerical results for the kinetics, stoichiometry, and diffusivities of the problem he considered. Astarita and Savage [9] analyzed simultaneous absorption of two gases with instantaneous chemical reaction for arbitrary stoichiometry. The reaction scheme they considered was

$$A + B_1 \rightleftarrows B_2$$

$$A' + B_1 \rightleftarrows B_3$$

where A, A′ are the two gases, B_1 is the liquid phase reactant, and B_2, B_3 are the reaction products. The analysis takes into account the shift reaction, which for the simple case above is

$$A + B_3 \rightleftarrows A' + B_2$$

This reaction takes place in the region near the interface. Their analysis shows that the conditions where the physical driving forces for absorption of both gases are large and positive do not imply that the chemical driving forces are both positive. In fact, it is shown that cases arise where one component may desorb even though its physical driving force is positive. Astarita and Savage [9] also applied their analysis to simultaneous absorption of CO_2 and H_2S in monoethanolamine solution in an absorption tower. Aiken [10] is interested in selective removal of hydrogen sulfide from much larger quantities of carbon dioxide using conventional wet scrubbing. Aiken [10] used a film theory model to describe simultaneous transport of two gases into a reactive liquid to examine the conditions that optimized selectivity. One component reacts instantaneously, the other at a finite rate. Selectivity of the component of interest is increased by lowering the concentrations of both gases, but particularly the component of interest. Furthermore, high selectivity for the desired component is shown to be consistent with a high rate of mass transfer of that component, regardless of the interfacial concentration of this component. Zarzycki et al. [5, 6] considered simultaneous absorption and irreversible second-order chemical reaction of two gases in an inert liquid. For particular cases, the concentration profiles can be expressed in terms of the Weierstrass elliptic function and Airy functions. Cornelisse et al. [15] applied a discretization technique to calculate simultaneous absorption or desorption of CO_2 and H_2S in amine solution. Their solutions compared well with the previous analytical exact solutions. Negative enhancement factors could occur for conditions when the absorption of one of the gases forced the other to desorb by induced reversibility in the liquid transfer zone, though the driving forces would predict that both gases would absorb. At simultaneous absorption of two gases accompanied by interfering reactions the sign of the driving forces alone did not ensure the direction of mass transfer.

MASS TRANSFER WITH HETEROGENEOUS REACTIONS

Stowe and Shaeiwitz [3] analyzed mass transfer with both first-order, irreversible, homogeneous and an arbitrary heterogeneous reaction using film theory, subject to the assumption of a linear profile of reactant between the bulk phase and the heterogeneous reaction surface. The enhancement factor depends on the product of two dimensionless groups. One is associated with mass transfer and first-order, irreversible reaction and the other is the Damkohler number representing the transport of reactant to and reacting on the surface. This solution is applied to the problem of solubilization of an insoluble species by an aqueous surfactant solution. Flytzani-Stephanopoulos and Schmidt [20] developed models to incorporate surface reaction and diffusion of volatile products through a boundary layer in order to calculate the effective rates of evaporation and local surface profiles on surfaces having active and inactive regions. The coupling between surface heterogeneities with respect to a particular reaction and external mass transfer may provide a mechanism of the surface rearrangement and metal loss of catalysts encountered in several systems of practical interest.

EFFECT OF MIXING ON MASS TRANSFER WITH CHEMICAL REACTIONS

Ou and Ranz [23, 24] developed a model to analyze simultaneously the complex interactions of mechanical mixing, diffusion, and reaction. The model assumed that reactants diffuse from the adjacent sheets of fluid undergoing a stretching motion, contact one another, and react. For competitive reactions better mixing will generally favor products of reactions with higher reaction-rate constants until reactions reach the reaction control limit when the so-called perfect mixing prevails. Ou et al. [21] proposed a radial mixing scheme by injecting two fluids tangentially in opposite directions into a tubular reactor. They found that the flow-induced swirling in a tubular reactor enhanced conversion over that attainable in an ideal laminar flow reactor, ideal plug flow behavior was approached as the swirling speed increased up to a point beyond which no further enhancement could be achieved. Angst et al. [27] developed a model to describe the influence of diffusive and fine-scale convective mixing on product distribution and compared their results with experimental values. Later, Bourne et al. [22, 26] applied this model to study selectivity of competitive, consecutive reactions. Jensen [25] also applied the same model for mixing with fast chemical reactions.

MASS TRANSFER WITH HOMOGENEOUS AND HETEROGENEOUS REACTIONS IN TUBULAR REACTORS

Recent research in mass transfer with homogeneous and heterogeneous reactions in a tubular reactor has been on simple, irreversible, first-order reactions in the bulk-fluid phase and at the tube wall [29] as well as consecutive irreversible, first-order reactions in the bulk fluid [31] or at the tube wall [30]. A laminar-flow reactor is considered in Reference 29 while a plug-flow reactor is considered in References 30, 31. One particular interest in the system is the effect of axial diffusion on mass transfer.

Consider reactant A entering a laminar-flow reactor from $x = -\infty$ with an inlet concentration C_{A0}. Because of the effect of axial diffusion, the entire reactor is separated into a negative region ($-\infty < X \leq 0$) and a positive region ($0 \leq X < \infty$). Only convective diffusion takes place in the negative region while convective diffusion with consecutive, irreversible, first-order reaction ($A \rightarrow B \rightarrow C$) takes place in the bulk fluid or at the tube wall in the positive region. When the physical properties of the fluid are kept at constant values, the convective diffusion equations and the boundary conditions for the components A and B in the dimensionless forms can be written as:

$$(1 - \eta^2)\frac{\partial \theta_{ij}}{\partial \xi} = \frac{1}{\eta}\frac{\partial}{\partial \eta}\left(\eta \frac{\partial \theta_{ij}}{\partial \eta}\right) + \frac{1}{Pe^2}\frac{\partial^2 \theta_{ij}}{\partial \xi^2}$$

$$+ (-1)^i K_1 \delta(j - 2)\theta_{1j} - K_2 \delta(j - 2)\theta_{2j};\ i = 1,\ 2;\ j = 1,\ 2 \tag{1}$$

$$\theta_{i1}(-\infty, \eta) = \theta_{B_0}\delta(i - 2) \tag{2}$$

$$\frac{\partial \theta_{ij}(\xi,\ 0)}{\partial \eta} = 0 \tag{3}$$

$$\frac{\partial \theta_{i1}(\xi,\ 1)}{\partial \eta} = 0 \tag{4}$$

$$\frac{\partial \theta_{12}(\xi,\ 1)}{\partial \eta} + \alpha \theta_{12}(\xi,\ 1) = 0 \tag{5}$$

$$\frac{\partial \theta_{22}(\xi,\ 1)}{\partial \eta} + \alpha' \theta_{22}(\xi,\ 1) = \alpha \theta_{12}(\xi,\ 1) \tag{6}$$

$$\theta_{i2}(\infty, \eta) = 0 \tag{7}$$

where the first subscript $i = 1$ refers to component A, and $i = 2$ refers to component B; the second subscript $j = 1$ refers to the negative region and $j = 2$ refers to the positive region; $\delta(i - 2)$ is the Dirac delta function which is equal to one when $i = 2$ for component B and equal to zero otherwise. Similarly $\delta(j - 2)$ is the Dirac delta function which is equal to one when $j = 2$ for the positive region and equal to zero otherwise. The continuity equations for the concentration and the axial diffusive fluxes for both components A and B at the connectiong plane $x = 0$ are then:

$$\delta(i - 1) + (-1)^i \theta_{i1}(0, \eta) = \theta_{i2}(0, \eta) \tag{8}$$

$$\frac{\partial \theta_{i1}(0, \eta)}{\partial \xi} = (-1)^i \frac{\partial \theta_{i2}(0,\ \eta)}{\partial \xi};\ i = 1,\ 2 \tag{9}$$

Analytical solutions for Equations 1 through 9 are difficult to obtain. However when the reactor is a plug-flow reactor, solutions of Equations 1 through 7 for component A in the positive and the negative regions and component B in the negative region can be readily obtained by the methods of separation of variables and the orthogonal series expansions as:

$$\theta_{11}(\xi, \eta) = \sum_{n=1}^{\infty} E_n \exp\left[\frac{Pe^2}{2}\left(1 + \sqrt{1 + \frac{4\mu_n^2}{Pe^2}}\right)\xi\right] J_0(\mu_n \eta) \tag{10}$$

$$\theta_{12}(\xi, \eta) = \sum_{n=1}^{\infty} F_n \exp\left\{\frac{Pe^2}{2}\left[1 - \sqrt{1 + \frac{4(K_1 + \gamma_n^2)}{Pe^2}}\right]\xi\right\} J_0(\gamma_n \eta) \tag{11}$$

$$\theta_{21}(\xi, \eta) = \theta_{B0} + \sum_{n=1}^{\infty} G_n \exp\left[\frac{Pe^2}{2}\left(1 + \sqrt{1 + \frac{4\delta_n^2}{Pe^2}}\right)\xi\right] J_0(\delta_n \eta) \tag{12}$$

where the eigenvalues μ_n, γ_n, and δ_n are determined by the following equations, respectively:

$$J_1(\mu_n) = 0 \tag{13}$$

$$\gamma_n J_1(\gamma_n) = \alpha J_0(\gamma_n) \tag{14}$$

$$J_1(\delta_n) = 0 \tag{15}$$

The solution for $\theta_{22}(\xi, \eta)$ can be obtained by superposition as:

$$\theta_{22}(\xi, \eta) = \frac{\alpha}{\alpha'} \sum_{n=1}^{\infty} F_n \exp\left\{\frac{Pe^2}{2}\left[1 - \sqrt{1 + \frac{4(K_1 + \gamma_n^2)}{Pe^2}}\right]\xi\right\} J_0(\gamma_n)$$

$$+ K_n \left\{\exp\left\{\frac{Pe^2}{2}\left[1 - \sqrt{1 + \frac{4(K_2 + \epsilon_n^2)}{Pe^2}}\right]\xi\right\} + \exp\left[\frac{Pe^2}{2}\left(1 - \sqrt{1 + \frac{4\epsilon_n^2}{Pe^2}}\right)\xi\right]\right.$$

$$- \frac{\alpha Pe^2}{J_0(\epsilon_n)(\alpha'^2 + \epsilon_n^2)} \sum_{\substack{m=1 \\ m \neq n}}^{\infty}$$

$$\left[F_m \left[1 - \sqrt{1 + \frac{4(K_1 + \gamma_m^2)}{Pe^2}}\right] \left\{\frac{1}{2}\left[1 - \sqrt{1 + \frac{4(K_1 + \gamma_m^2)}{Pe^2}}\right] - 1\right\}\right.$$

$$J_0(\gamma_n) \left(\exp\left\{\frac{Pe^2}{2}\left[1 - \sqrt{1 + \frac{4(K_1 + \gamma_m^2)}{Pe^2}}\right]\xi\right\}\right.$$

$$\left.\left.- \exp\left[\frac{Pe^2}{2}\left(1 - \sqrt{1 + \frac{4\epsilon_n^2}{Pe^2}}\right)\xi\right]\right) \Big/ \left(\gamma_m^2 - \epsilon_n^2\right)$$

$$- \left\{ \left(2F_n \left[1 - \sqrt{1 + \frac{4(K_1 + \gamma_n^2)}{Pe^2}} \right] \left\{ \frac{1}{2} \left[1 - \sqrt{1 + \frac{4(K_1 + \gamma_n^2)}{Pe^2}} \right] - 1 \right\} \right. \right.$$

$$\left. J_0(\gamma_n) \xi \exp \left[\frac{Pe^2}{2} \left(1 - \sqrt{1 + \frac{4(\epsilon_n^2)}{Pe^2}} \right) \xi \right] \right) \Bigg/ \left[\sqrt{1 + \frac{4(K_1 + \gamma_n^2)}{Pe^2}} \right.$$

$$\left. \left. + \sqrt{1 + \frac{4(\epsilon_n^2)}{Pe^2}} \right] \right\} J_0(\epsilon_n \eta) \quad - \frac{2K_1}{J_0(\epsilon_n) \left(1 + \dfrac{\alpha'^2}{\epsilon_n^2} \right)}$$

$$\sum_{j=1}^{\infty} \left[F_j \left(\exp \left\{ \frac{Pe^2}{2} \left[1 - \sqrt{1 + \frac{4(K_1 + \gamma_j^2)}{Pe^2}} \right] \xi \right\} \right. \right.$$

$$\left. \left. - \exp \left\{ \frac{Pe^2}{2} \left[1 - \sqrt{1 + \frac{4(K_2 + \epsilon_n^2)}{Pe^2}} \right] \xi \right\} \right) \left(\frac{\alpha - \alpha'}{\gamma_j^2 - \epsilon_n^2} \right) J_0(\gamma_j) \right]$$

$$\Bigg/ \left(K_1 + \gamma_j^2 - K_2 - \epsilon_n^2 \right) J_0(\epsilon_n \eta) \qquad (16)$$

where the eigenvalues ϵ_n are determined by the following equation:

$$\alpha' J_0(\epsilon_n) - \epsilon_n J_1(\epsilon_n) = 0 \qquad (17)$$

The coefficients E_n, F_n, G_n and K_n in Equations 10 through 12 and 16 respectively are determined from the continuity equations, Equations 8 and 9.

The dimensionless average concentration and Sherwood number are defined as

$$\theta_{ij_b}(\xi) = \frac{\int_0^a uC_{ij} r dr}{\int_0^a u r dr} = 2 \int_0^1 C_{ij} \eta d\eta \qquad (18)$$

$$Sh = \frac{2k_L a}{2} = \frac{-2a \left. \dfrac{\partial C_i}{\partial r} \right|_{r=a}}{C_{i_b}(\xi) - C_i(\xi, 1)} \qquad (19)$$

Numerical results have been obtained for simplified cases of Equations 1 through 9. For example, when only simple, irreversible, first-order decompositions of component A are considered (i.e. $K_2 = \alpha' = 0$); the equation of continuity of component B need not be solved at all. It is possible to apply the method presented in Reference 29 to solve Equations 1 through 9 for component A. The concentration distribution of component A is now in terms of the eigenfunction series expansions. A typical example of the eigenvalues, the coefficients of the series expansion for the positive and the negative regions, and other related values is tabulated in Table 1 for Pe = 5, K_1 = 1, and α = 100. To obtain good accuracy at a high value of the wall-reaction parameter α, fifty terms of the series expansion are necessary.

Figure 1 shows the dimensionless radial concentration of component A at two different axial positions (x/a = 10^{-4}, 1) in the positive region for Pe = 10, K_1 = 1, α = 1, 10, 100. The concentration of component A decreases progressively as A flows downstream along the reactor due to the homogeneous reaction in the bulk fluid and the catalytic reaction at the tube wall. At a higher value of the wall-reaction parameter, the radial concentration gradient

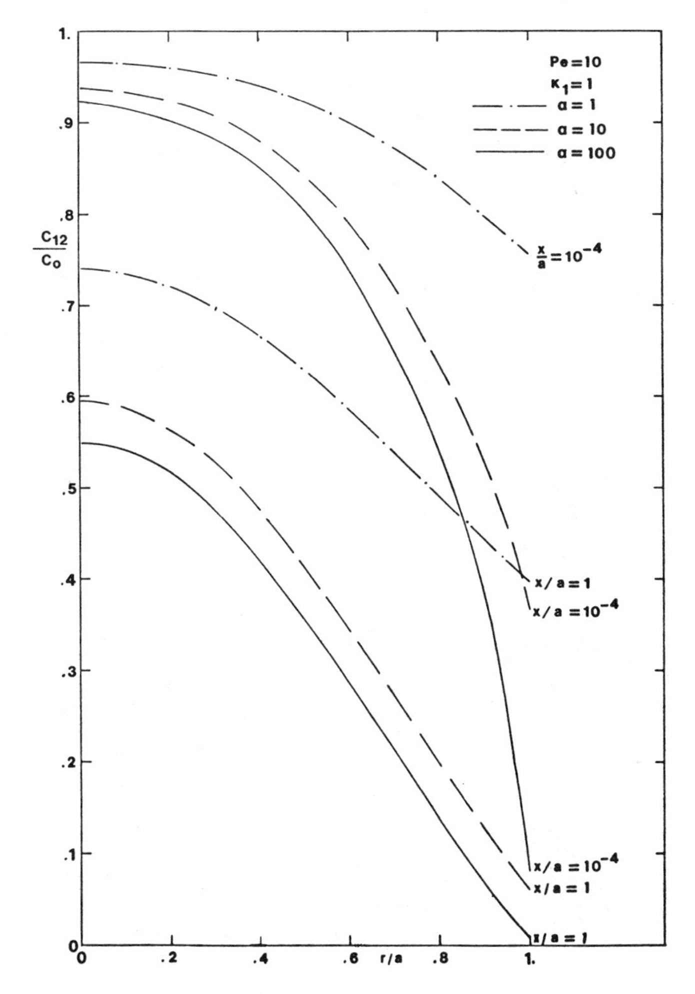

Figure 1. Dimensionless radial concentration distribution in a laminar-flow reactor at two different axial positions ($0 \leq x/a < \infty$) for various values of wall reaction parameters (α = 1, 10, 100) and Pe = 10, K_1 = 1.

Table 1
Eigenvalues and Related Constants for Pe = 5, K = 1, α = 100

n	α_n^2	β_n^2	A_n	B_n	$\int_0^1 \eta Y_n d\eta^*$	$\int_0^1 \eta R_n d\eta^*$	$R_n(1)$
1	11.199085	5.334295	−.353858	.908694	.808583	.202318	.010329
2	30.454712	17.147944	.125811	−.314481	.839997(1)	−.767114(1)	−.016354
3	44.766453	29.710927	−.114351	.189427	−.137446(2)	.396326(1)	.021298
4	60.084926	42.453200	.086073	−.137011	.137157(2)	−.241432(1)	−.025388
5	75.611420	55.276739	−.067761	.108002	−.600395(3)	.163942(1)	.028882
6	91.213322	68.144714	.055344	−.089463	.310314(3)	−.119720(1)	−.031938
7	106.851493	81.039834	−.046441	.076578	−.177893(3)	.919440(2)	.034653
8	122.509982	93.953031	.039811	−.067033	.110047(3)	−.732262(2)	−.037088
9	138.181029	106.879156	−.034667	.059793	−.721252(4)	.599282(2)	.039283
10	153.860404	119.815110	.030602	−.053812	.494834(4)	−.500872(2)	−.041270
11	169.545602	132.758939	−.027281	.049025	−.352005(4)	.425665(2)	.043071
12	185.235047	145.709364	.024555	−.044995	.258281(4)	−.366670(2)	−.044703
13	200.927706	158.665522	−.022245	.041596	−.194103(4)	.319399(2)	.046183
14	216.622875	171.626819	.020300	−.038633	.149353(4)	−.280832(2)	−.047521
15	232.320064	184.592833	−.018605	.036077	−.116689(4)	.248898(2)	.048729
16	248.018927	197.563262	.017154	−.033790	.930897(5)	−.222103(2)	−.049817
17	263.719217	210.537883	−.015860	.031788	−.748046(5)	.199380(2)	.050793
18	279.420761	223.516530	.014738	−.029958	.614645(5)	−.179910(2)	−.051666
19	295.123438	236.499078	−.013718	.028342	−.503895(5)	.163100(2)	.052442
20	310.827170	249.485429	.012827	−.026838	.425020(5)	−.148461(2)	−.053128
21	326.531909	262.475511	−.012002	.025505	−.353016(5)	.135645(2)	.053731
22	342.237637	275.469264	.011279	−.024244	.305430(5)	−.124337(2)	−.054256
23	357.944356	288.466644	−.010598	.023127	−.255289(5)	.114330(2)	.054709
24	373.652087	301.467619	.009999	−.022052	.226934(5)	−.105405(2)	−.055094
25	389.360870	314.472165	−.009425	.021104	−.189498(5)	.974375(3)	.055417
26	405.070757	327.480267	.008921	−.020177	.173760(5)	−.902650(3)	−.055681

(Table continued on next page)

Table 1 Continued

n	α_n^2	β_n^2	A_n	B_n	$\int_0^1 \eta Y_n d\eta$*	$\int_0^1 \eta R_n d\eta$*	$R_n(1)$
27	420.781817	340.491920	−.008430	.019365	−.143758(5)	.838180(3)	.055890
28	436.494130	353.507127	.008000	−.018556	.136851(5)	−.779661(3)	−.056049
29	452.207789	366.525897	−.007572	.917858	−.111087(5)	.726773(3)	.056159
30	467.922899	379.548251	.007200	−.017147	.110775(5)	−.678396(3)	−.056224
31	483.639574	392.574217	−.006822	.016545	−.872128(6)	.634490(3)	.056246
32	499.357943	405.603830	.006495	−.015914	.921510(6)	−.594029(3)	−.056228
33	515.078142	418.637137	−.006156	.015396	−.694307(6)	.557195(3)	.056171
34	530.800322	431.674193	.005865	−.014831	.788059(6)	−.522998(3)	−.056077
35	546.524641	444.715061	−.005554	.014389	−.559762(6)	.491803(3)	.055948
36	562.251270	457.759816	.005290	−.013880	.693040(6)	−.462617(3)	−.055784
37	577.980388	470.808542	−.004998	.013507	−.456671(6)	.435968(3)	.055585
38	593.712187	483.861333	.004755	−.013044	.626731(6)	−.410830(3)	−.055354
39	609.446869	496.918292	−.004472	.012739	−.376922(6)	.387881(3)	.055090
40	625.184645	509.979535	.004239	−.012310	.582392(6)	−.366039(3)	−.054792
41	640.925739	523.045187	−.003951	.012076	−.314816(6)	.346126(3)	.054461
42	656.670383	536.115381	.003718	−.011668	.555251(6)	−.326986(3)	−.054097
43	672.418821	549.190265	−.003403	.011518	−.266255(6)	.309582(3)	.053699
44	688.171308	562.269994	.003147	−.011106	.541839(6)	−.292671(3)	−.053266
45	703.928108	575.354735	−.002757	.011074	−.228213(6)	.277355(3)	.052797
46	719.689496	588.444666	.002421	−.010598	.539572(6)	−.262290(3)	−.052292
47	735.455757	601.539973	−.001802	.010803	−.198381(6)	.248723(3)	.051749
48	751.227187	614.640855	.001178	−.009963	.546448(6)	−.235195(3)	−.051167

* Values of a(b) = a × 10^{-b}

of component A close to the wall is much steeper. Increasing the value of the Peclet number will decrease the concentration of component A because of the effect of axial diffusion.

The dimensionless average concentration of component A along the axial coordinate of the reactor for $K_1 = 1, 100$; Pe $= 1, 5, 10$ and $\alpha = 1, 10, 100$ is shown in Figures 2 and 3. The effect of backmixing can be observed from (1) the penetration depth of component A in the negative region and (2) the deviation of the dimensionless average concentration from unity at the plane $x = 0$. The effect of backmixing is more significant when the Peclet number decreases or the wall reaction parameter α increases. When the homogeneous reaction rate is very fast, component A decomposes very rapidly and the effect of the reaction rate at the tube wall is insignificant as shown in Figure 3.

The Sherwood number of component A along the positive region of the reactor for Pe $=$ 5, 10; $K_1 = 1, 10, 100$ and $\alpha = 1, 10, 100$ is shown in Figures 4 and 5. The Sherwood

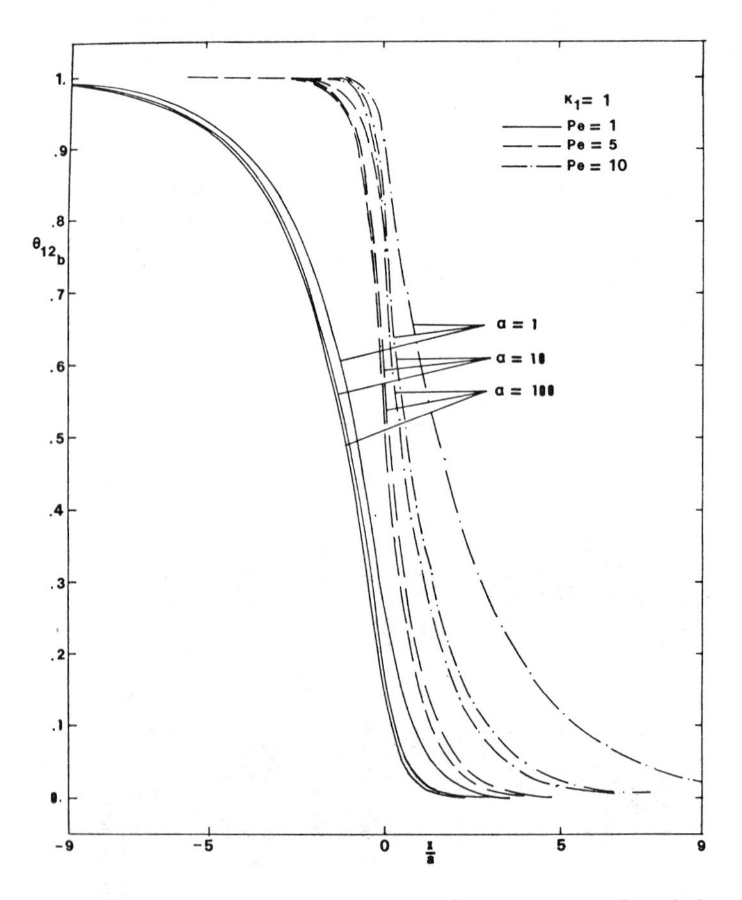

Figure 2. Dimensionless average concentration in a laminar-flow reactor versus dimensionless axial distance ($K_1 = 1$; Pe $= 1, 5, 10$; $\alpha = 1, 10, 100$).

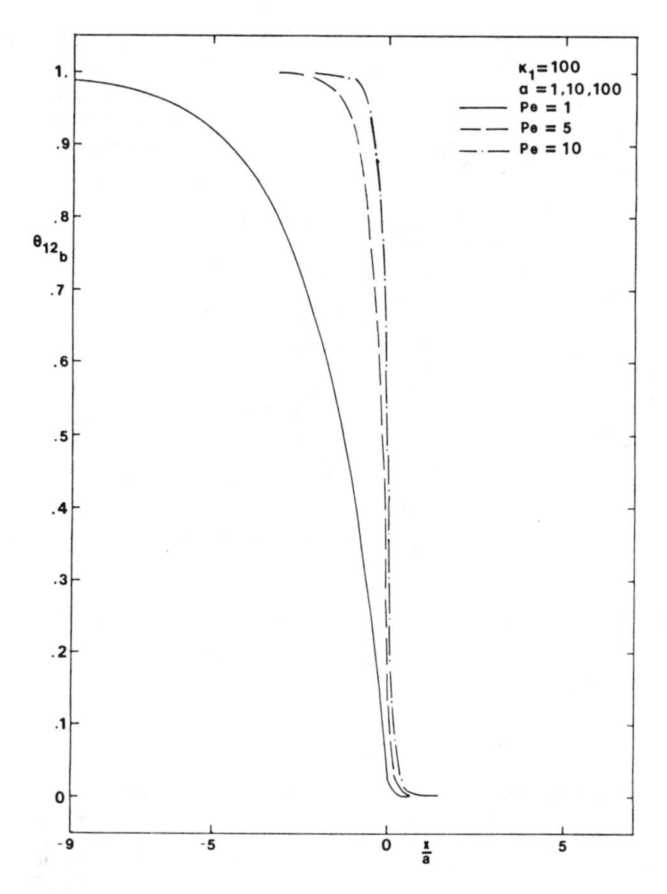

Figure 3. Dimensionless average concentration in a laminar-flow reactor versus dimensionless axial distance ($K_1 = 100$; Pe = 1, 5, 10; α = 1, 10, 100).

number decreases as x/a increases. The cross-over of some of the curves in Figures 4 and 5 indicate the complicated interaction of the homogeneous and heterogeneous reactions and convective diffusion on the Sherwood number or the mass transfer rate.

Another simplified case is the consideration of first-order, irreversible, consecutive reactions only at the tube wall of the plug-flow reactor; i.e. $K_1 = K_2 = 0$ in Equation 1. In this case, simplified solution can be obtained from Equations 10 through 17. Since the intermediate component B is the desired product, the dimensionless average and wall concentrations of component B along the tubular reactor for Pe = 1, $\alpha = 1$, and different values of α' are shown in Figure 6. The maximum yield of component B exists at a certain location of the reactor. Hence it is useful to determine the optimal yield of B and the optimal reactor length for design purposes. Figure 7 shows the optimal yield of B and the optimal reactor length for different values of the wall reaction kinetic parameters α and α'. It is seen that for the range of variables studied in Figure 7; the maximum yield of B is 0.69 while the optimal reactor length x/a is 0.58.

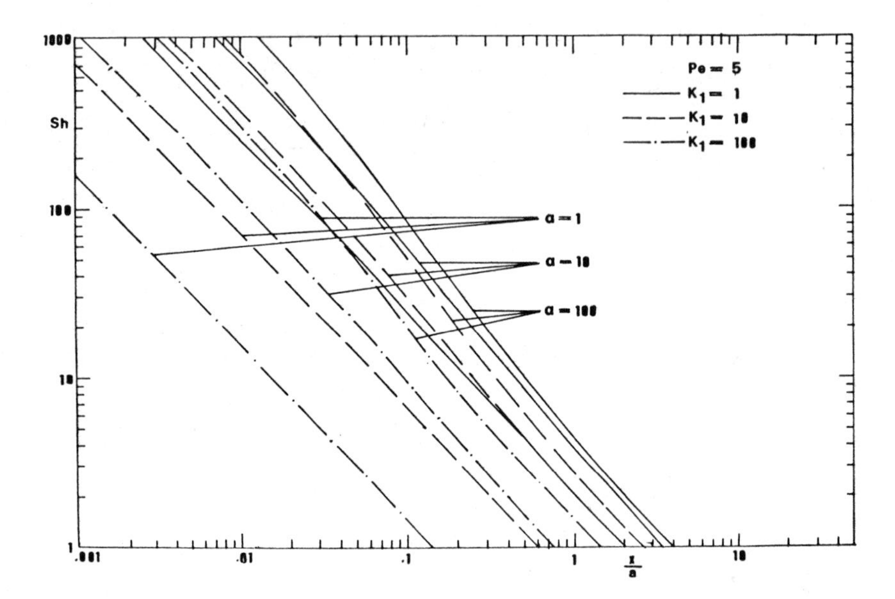

Figure 4. Sherwood number in a laminar-flow reactor versus axial distance for Pe = 5; K_1 = 1, 10, 100 and α = 1, 10, 100.

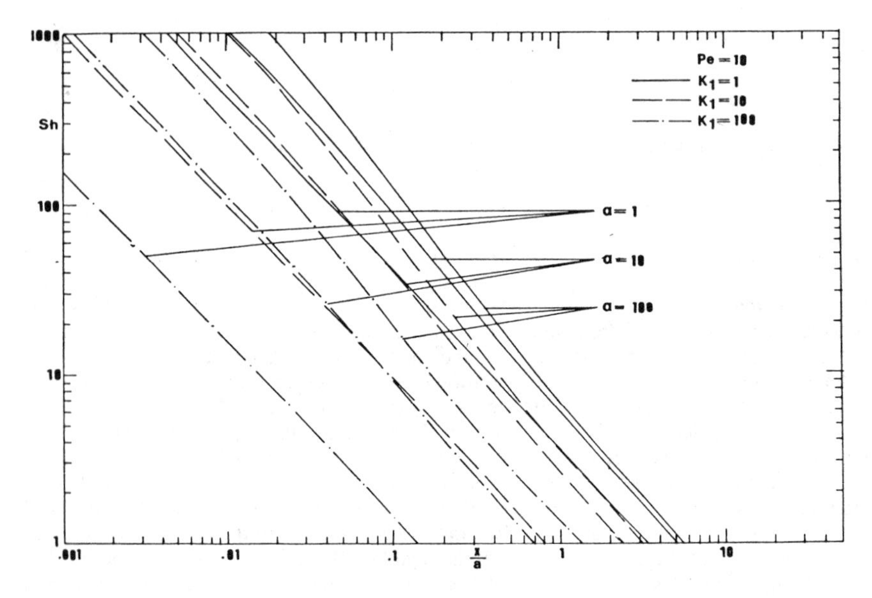

Figure 5. Sherwood number in a laminar-flow reactor versus axial distance for Pe = 10; K_1 = 1, 10, 100 and α = 1, 10, 100.

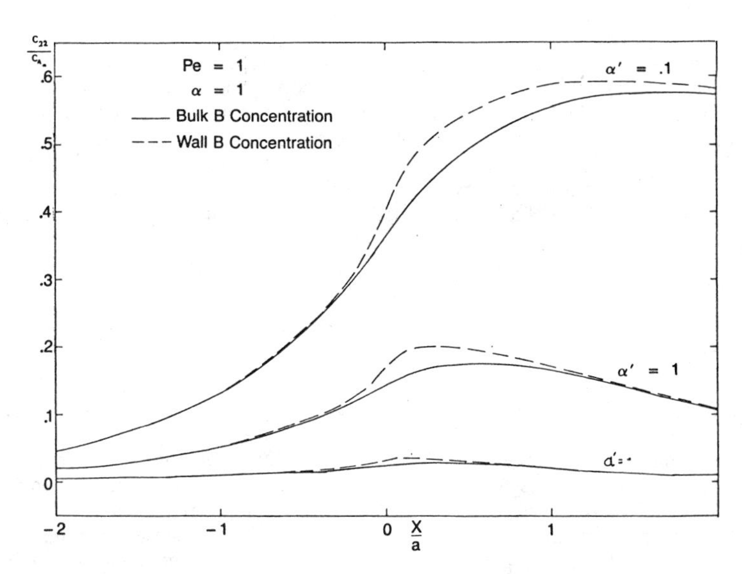

Figure 6. Dimensionless average and wall concentration of component B in a plug-flow reactor versus axial distance for different wall reaction parameter of component B.

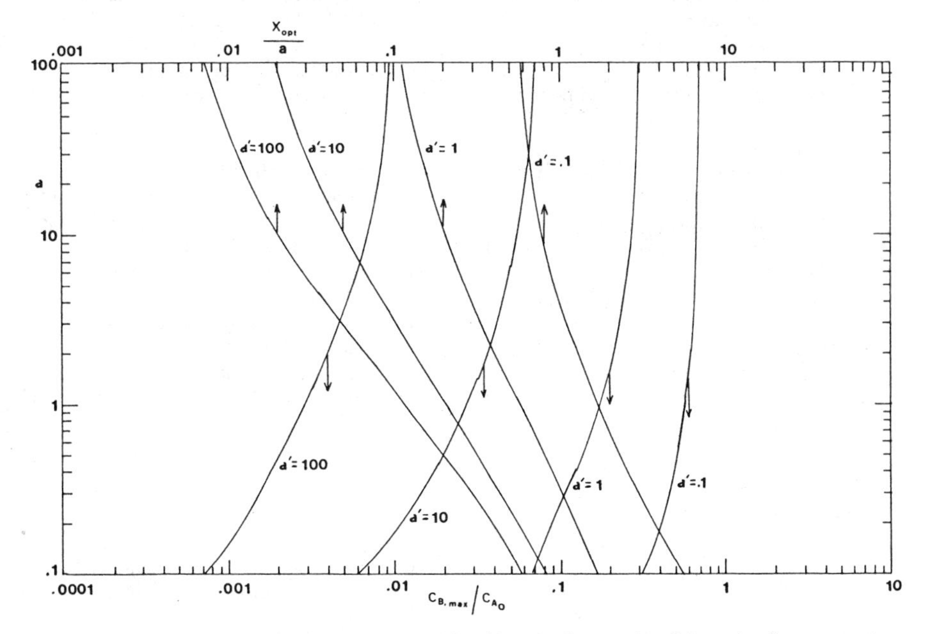

Figure 7. Optimal reactor length and maximum yield of intermediate product B in a plug-flow reactor for different consecutive wall reaction of component A (Pe = 1, α' = 0.1, 1, 10, 100).

SUMMARY

Recent investigations on mass transfer with homogeneous and/or heterogeneous reactions are reviewed. The effect of macromixing, micromixing, and chemical kinetics on mass transfer is investigated. For a laminar-flow reactor with simple, first-order reactions in the fluid and at the tube wall, the interactive effect of convective diffusion and chemical reactions on the concentration field and the Sherwood number is examined. For a plug-flow reactor with consecutive, first-order, irreversible reactions at the tube wall, the optimal yield of the intermediate component B and the optimal reactor length are graphically presented for design purposes.

NOTATION

a	radius of the tube
A	chemical component A
A_n	coefficient of series expansions for component A in the negative region defined in Reference 29
B	intermediate component B
B_n	coefficients of series expansions for component A in the positive region defined in Reference 29
C	final product of consecutive reaction
C_{A0}	inlet concentration of component A
C_{11}	concentration of component A in the negative region
C_{12}	concentration of component A in the positive region
C_{B0}	inlet concentration of component B
C_{21}	concentration of component B in the negative region
C_{22}	concentration of component B in the positive region
D	molecular diffusivity
E_n	coefficient of series expansions in Equation 10
F_n	coefficient of series expansions in Equation 11
G_n	coefficients of series expansions in Equation 12
J_0	Bessel function of the first kind of zero order
J_1	Bessel function of the first kind of first order
k_1	first-order consecutive reaction rate constant for component A in the fluid in the positive region
k_2	first-order consecutive reaction rate constant for component B in the fluid in the positive region
k_{w_1}	first-order reaction rate constant of component A at the tube wall in the positive region
k_{w_2}	first-order consecutive reaction rate constant of component B at the tube wall in the positive region
k_L	mass transfer coefficient
K_n	coefficients of series expansions in Equation 16
Pe	aU/D
r	radial coordinate
R_n	eigenfunctions in series expansions of component A in the negative region defined in Reference 29
Sh	Sherwood number defined in Equation 19
U	average velocity of fluid
x	axial coordinate
Y_n	eigenfunctions in series expansion of component A in the positive region defined in Reference 29

Greek Symbols

α	$k_{w_1}a^2/D$
α'	$k_{w_2}a^2/D$
α_n	eigenvalues of component A in the negative region defined in Reference 29
β_n	eigenvalues of component A in the positive region defined in Reference 29
γ_n	eigenvalues defined by Equation 14
δ	Dirac delta function

δ_n eigenvalues defined by Equation 15
ϵ_n eigenvalues defined by Equation 17
η r/a
θ_{11} $(C_{A_0} - C_{A_1})/C_{A_0}$
θ_{12} C_{A_2}/C_{A_0}
θ_{21} C_{B_1}/C_{A_0}
θ_{22} C_{B_2}/C_{A_0}
θ_{B_0} C_{B_0}/C_{A_0}
K_1 k_1a^2/D
K_2 k_2a^2/D
μ_n eigenvalues defined by Equation 13
ξ x/aPe

Subscripts

i 1 for component A, 2 for component B
j 1 for negative region, 2 for positive region
b average concentration

REFERENCES

1. Sherwood, T. K., Pigford, R. L., and Wilke, C. R., *Mass Transfer,* New York: McGraw-Hill Book Company, 1975.
2. Danckwerts, P. V., *Gas-Liquid Reactions,* New York: McGraw-Hill Book Company, 1970.
3. Stowe, L. R., and Shaeiwitz, J. A., "Mass Transfer with Heterogeneous and First Order Homogeneous Reaction Using the Film Theory," *Chem. Eng. Sci.*, Vol. 38 (1983), p. 635.
4. DeCoursey, W. J., "Enhancement Factors for Gas Absorption with Reversible Reaction," *Chem. Eng. Sci.*, Vol. 37 (1982), p. 1483.
5. Zarzycki, R., Ledakowicz, S., and Starzak, M., "Simultaneous Absorption of Two Gases Reacting Between Themselves in a Liquid—I. Exact Solutions," *Chem. Eng. Sci.*, Vol. 36 (1981), p. 105.
6. Zarzycki, R., Ledakowicz, S., and Starzak, M., "Simultaneous Absorption of Two Gases Reacting between Themselves in a Liquid—II. Approximate Solutions," *Chem. Eng. Sci.*, Vol. 36 (1981), p. 113.
7. Nagy, E., et al., "Mass Transfer Accompanied by First Order Intermediate Reaction Rate in Two Phase Co-current Flow Axial Dispersion," *Chem. Eng. Sci.* Vol. 37 (1982), p. 1817.
8. Boyadjiev, C., "On the Influence of Irreversible Chemical Reactions on Gas Absorption," *Chem. Eng. Sci.*, Vol. 38 (1983), p. 641.
9. Astarita, G., and Savage, D. W., "Simultaneous Absorption with Reversible Instantaneous Chemical Reaction," *Chem. Eng. Sci.* Vol. 37 (1982), p. 677.
10. Aiken, R. C., "Selectivity in Gas Purification: Film Theory, Semi-Infinite Medium," *Chem. Eng. Sci.*, Vol. 37 (1982), p. 1031.
11. Carmichael, G. R., and Chang, S. C., "Mass Transfer Accompanied by Equilibrium and Second-Order Irreversible Reactions," *Chem. Eng. Sci.*, Vol. 35 (1980), p. 2459.
12. Cornelisse, R., et al., "Numerical Calculation of Simultaneous Mass Transfer of Two Gases Accompanied by Complex Reversible Reactions," *Chem. Eng. Sci.*, Vol. 35, (1980), p. 1245.
13. Chang, C. S., and Rochelle, G. T., "Mass Transfer Enhanced by Equilibrium Reactions," *Ind. & Eng. Chem. Fund.*, Vol. 21 (1982), p. 379.
14. Huang, D. T. J., Carberry, J. J., and Varma, A., "Gas Absorption with Consecutive Second-Order Reactions," *AIChE J.*, Vol. 26, (1980), p. 832.
15. Li, C. H., "False Diffusion in Convection and Diffusion with Chemical Reaction," *Int. J. Heat and Mass Transf.*, Vol. 26, (1983), p. 1063.

16. Bauer, R., Friday, D. K., and Kirwan, D. J., "Mass Transfer and Kinetic Effects in an Electrode-Driven Homogeneous Reaction," *Ind. & Eng. Chem. Fund.*, Vol. 20 (1981), p. 141.
17. Ramage, M. P., and Eckert, R. E., "Interactions Between Kinetics and Mass Transfer in the Liquid Phase Chlorination of n-Dodecane," *Ind. & Eng. Chem. Fund.* Vol. 18 (1979), p. 216.
18. Giudice, S. D., and Trotta, A., "Transient Diffusion with n-Order Irreversible Chemical Reaction: A Finite Element Approach," *Chem. Eng. Sci.*, Vol. 33 (1978), p. 697.
19. Torney, D. C., and McConnell, H. M., "Diffusion Limited Reaction Rate Theory for Two-Dimensional Systems," *Proc. Roy. Soc. Series A*, Vol. 387 (1983), p. 147.
20. Flytzani-Stephanopoulos, M., and Schmidt, L. D., "Evaporation Rates and Surface Profiles on Heterogeneous Surfaces with Mass Transfer and Surface Reaction," *Chem. Eng. Sci.*, Vol. 34 (1979), p. 365.
21. Ou, J. J., Lee, C. S., and Chen, S. H., "Mixing of Chemically Reactive Fluids by Swirling in a Tubular Reactor," *Chem. Eng. Sci.*, Vol. 38 (1983), p. 1323.
22. Bolzern, O., and Bourne, J. R., "Mixing and Fast Chemical Reactions—VI," *Chem. Eng. Sci.*, Vol. 38 (1983), p. 999.
23. Ou, J. J., and Ranz, W. E., "Mixing and Chemical Reactions, a Contrast between Fast and Slow Reactions," *Chem. Eng. Sci.*, Vol. 38 (1983), p. 1005.
24. Ou, J. J., and Ranz, W. E., "Mixing and Chemical Reactions, Chemical Selectivities," *Chem. Eng. Sci.*, Vol. 38 (1983), p. 1015.
25. Jenson, V. G., "A Model for Mixing with Fast Chemical Reactions," *Chem. Eng. Sci.*, Vol. 38 (1983), p. 1151.
26. Angst, W., Bourne, J. R., and Sharma, R. N., "Mixing and Fast Chemical Reaction V, Influence of Diffusion within the Reaction Zone on Selectivity," *Chem. Eng. Sci.*, Vol. 37 (1982), p. 1259.
27. Angst, W., Bourne, J. R., and Sharma, R. N., "Mixing and Fast Chemical Reaction—IV. The Dimensions of the Reaction Zone," *Chem. Eng. Sci.*, Vol. 37 (1982), p. 585.
28. Wellek, R. M., Brunson, R. J., and Law, F. H., "Enhancement Factors for Gas-Absorption with Second-Order Irreversible Chemical Reaction," *Canadian J. Chem. Eng.*, Vol. 56 (1978), p. 181.
29. Dang, V. D., "Steady-State Mass Transfer with Homogeneous and Heterogeneous Reactions," AIChE J., Vol. 29 (1983), p. 19.
30. Dang, V. D., "Diffusive Transport with Consecutive Catalytic Wall Reactions," AIChE Summer Meeting, paper No. 59e, Denver, Colorado, August, 1983.
31. Dang, V. D., "Axial Diffusion and Selectivity in a Plug Flow Reactor," AIChE Winter Meeting, Atlanta, Georgia, March, 1984.

CHAPTER 3

CONVECTIVE DIFFUSION WITH REACTIONS IN A TUBE

Vi-Duong Dang

Department of Chemical Engineering
The Catholic University of America
Washington, DC

CONTENTS

INTRODUCTION

Investigations concerning convective diffusion with reactions in a tube are extensively reported in the chemical engineering literature because of the importance to reaction engineering. When there are no chemical reactions occurring in the tube, the problem can be considered as physical transport processes and is itself a complicated one to analyze. Since Taylor [1-3] published his seminal ideas on the dispersion theory in a tube, many other researchers worked on this problem with different approaches [4-28]. When chemical reactions take place in a tube, the analysis of this problem can be treated by a dispersion approximation approach [32-39] or a nondispersion approximation approach [40-54]. The objectives of this chapter are to investigate convective diffusion with chemical reactions in a tube with dispersion and nondispersion approximation methods.

Taylor [1-3] proposed a one-dimensional dispersion model based on an area average concentration to account for the distribution of soluble matter flowing slowly through a tube. Aris [4] applied the method of moments to remove the restrictions imposed on some of the parameters in the analysis of Taylor [1-3]. Gill and Subramanian [6-8] considered different inlet concentration modes for the dispersion processes in a tube by generalizing the model of Taylor to an infinite series. For small-time applications, the exact analytical solution of Lighthill [5] does not include the effects either of longitudinal diffusion or of the existence of the walls. Chatwin [10] relaxed the restriction of Lighthill [3] by considering the longitudinal diffusion. Tseng and Besant [23, 24] proposed a particular solution method. De Gance and Johns [13] used the Hermite polynomials to calculate the first three time-dependent dispersion coefficients. Yu [19-21] examined the same problem by representing the local concentration by a series in terms of the zero-order Bessel function. Smith [15-17] considered not only the dispersion coefficient to be time-dependent, but also a memory of the concentration distribution at earlier times in describing the dispersion process. Wang and Stewart [27] used the orthogonal collocation method to study the dispersion phenomena. Recent application of the dispersion theory to field-flow fractionation was reported in References 18, 25, and 28. Reis et al. [25] characterized hollow-fiber electropolarization chromatography by applying a polarizing field across laminar flows. Smith [18] used a one-dimensional delay-diffusion model to investigate the extent to which the separation of a different contaminant species can be improved by pretreating the sample in a stationary fluid before being eluted into the shear flow. Lightfoot et al. [28] reviewed the process of field flow fractionation by different force fields (electrical, magnetic, thermal, centrifugal, and gravitational) transverse to the flow of the species.

Applying the dispersion model to study convective diffusion with reactions in a tube is approximated by several methods [32–39]. Subramanian et al. used the series expansion method to study dispersion processes with homogeneous chemical reactions [32] and heterogeneous reactions [33] in a tube. DeGance and Johns [34–36] investigated the chemically reactive systems by Hermite polynomials. Smith [38] used the delay-diffusion method to examine dispersion phenomena with an absorption wall. Wan and Ziegler [39] studied the yield of the intermediate component of a consecutive reaction in a tubular reactor. Aris [37] examined both reactive and nonreactive dispersion processes for various systems.

CONVECTIVE DIFFUSION WITH REACTIONS

Recent works on convective diffusion with reactions in a tube, using methods other than the dispersion theory, are not extensive. The conventional approach is to represent the concentration distribution of the chemical species in terms of series expansions [40–54]. When the reactor is a plug-flow reactor, the concentration distribution can be expressed as a sum of the infinite series of Bessel functions [48, 53, 54]. For a laminar-flow reactor neglecting the effect of axial molecular diffusion, the concentration distribution of the chemical species can be written in terms of the sum of the infinite series of orthogonal functions [42, 44, 46]. However, when axial diffusion is considered in the laminar-flow reactor, the functions in the infinite series for the concentration distribution become non-orthogonal and functional analysis is needed to relate the non-orthogonal functions to the orthogonal functions [50, 52]. A simplified one-dimensional dispersion model for the consecutive first-order, irreversible reactions in a tube was developed in terms of the exponential functions and the validity criteria were also established [52–54]. Recent research interests are on the effect of axial molecular diffusion and chemical kinetics on the concentration field and mass transfer. In the following, a summary of the previous findings in this subject is presented.

LAMINAR FLOW CHEMICAL REACTOR

Convective diffusion with first-order irreversible reaction in the fluid and on the tube wall has been studied in Reference 52. Due to the effect of axial diffusion, a negative and a positive region of the reactor are considered. The dimensionless local and average concentrations in the negative and the positive regions and Sherwood number of the reacting species in the positive region are respectively [27]:

$$\theta_1(\xi, \eta) = \sum_{n=1}^{\infty} A_n Y_n(\eta) \exp(\alpha_n^2 \xi) \tag{1}$$

$$\theta_2(\xi, \eta) = \sum_{n=1}^{\infty} B_n R_n(\eta) \exp\left[-\left(K + \frac{1}{Pe}\right)\beta_n^2 \xi\right] \tag{2}$$

$$\theta_{1b}(\xi, \eta) = 1 + 4\sum_{n=1}^{\infty} A_n \exp(\alpha_n^2 \xi) \frac{\alpha_n^2}{Pe^2} \int_0^1 \eta Y_n d\eta \tag{3}$$

$$\theta_{2b}(\xi) = 4 \sum_{n=1}^{\infty} \frac{B_n \exp\left[-\left(K + \frac{1}{Pe}\right)\beta_n^2 \xi\right]}{\beta_n^2 \left(K + \frac{1}{Pe}\right)} \left\{\alpha R_n(1) + \left[K - \frac{\beta_n^4}{Pe^2}\left(K + \frac{1}{Pe}\right)^2\right] \int_0^1 \eta R_n d\eta\right\} \tag{4}$$

and

$$Sh = \frac{2K_L a}{D} = \frac{\frac{2\alpha}{(K + 1/Pe)} \sum_{n=1}^{\infty} B_n R_n(1) \left[\frac{1 - e^{-\left(K+\frac{1}{Pe}\right)\beta_n^2 \xi}}{\beta_n^2}\right]}{\xi[\theta_{2b}(0) - \theta_{2b}(\xi)]} \tag{5}$$

Another simple way to analyze this system is to apply a simple one-dimensional dispersion model which lumps both the homogeneous and the heterogeneous reactions in the dispersion equation. Applying the Danckwerts' inlet boundary condition, the concentration distribution is:

$$\frac{C}{C_0} = \frac{2}{1 + \sqrt{1 - \frac{4K_2}{K_1^2}(K_0 - K)}} \exp\left[\frac{-1 + \sqrt{1 - 4\frac{K_2}{K_1^2}(K_0 - K)}}{2K_2/K_1} x\right] \tag{6}$$

where the effective surface wall-reaction parameter is K_0, $K_1 = -U_{av}$, and $K_2 = D + a^2U_{av}^2/48D$. Comparison between the exact analysis, Equation 4, and the dispersion model, Equation 6, is shown in Tables 1 and 2. In this system, there are three dimensionless groups (xU_{av}/D_L, Kx/U_{av}, and ka^2/D) that can influence the validity conditions of the dispersion model. The term xU_{av}/D_L can be considered as the ratio of the residence time of the chemical species in the reactor to the combined time of the axial convective transport and the radial diffusive transport together. When the chemical reactor is long so that the chemical species have sufficient time to diffuse radially in the reactor, it is expected that the simple one-dimensional dispersion model is valid. Kx/U_{av} can be considered as the ratio of the residence time to the chemical reaction time. When the residence time of the species is longer than the reaction time, the chemical species have sufficient time to diffuse radially in the tube before they are reacted in the fluid. The dispersion model will then be valid for large values of Kx/U_{av}. The term ka^2/D can be considered as the ratio of the radial diffusion time to the reaction time in the fluid. When the radial diffusion time is short compared to the chemical reaction time, the species readily diffuse radially and the dispersion model which is based on the cross-sectional average concentration is then expected to be valid. In summary, when xU_{av}/D_L and Kx/U_{av} are large values and ka^2/D is small, the dispersion model is a good approximation to the reactor behavior. Results in Tables 1 and 2 validate the physical interpretations just discussed.

When both homogeneous and heterogeneous reactions occur in the tube, the one-dimensional dispersion model combines the wall reactions and the homogeneous fluid kinetics in the dispersion equation. In this case, one has to consider the wall-reaction parameter α (=

Table 1
Comparison Between the Present Analysis and the Dispersion Model with Homogeneous Chemical Reaction (K = 10)

$\frac{x}{a}$	Exact Analysis Equation 4	Dispersion Model Equation 6	$\left(\frac{xU_{av}}{D_L} = \frac{x/U_{av}}{\frac{D}{U_{av}^2} + \frac{a^2}{48D}} \right)$	$\frac{Kx}{U_{av}}$
Pe = 1				
0.	0.1486	0.1458	0.	0.
0.2	0.0832	0.0814	0.0995	4.
0.4	0.0465	0.0454	0.1990	8.
0.6	0.0260	0.0253	0.2984	12.
0.8	0.0145	0.0142	0.3979	16.
1.0	0.0081	0.0079	0.4974	20.
1.5	0.0019	0.0018	0.7461	30.
2.0	0.0004	0.0004	0.9948	40.
Pe = 5				
0.	0.5535	0.5169	0.	0.
0.08	0.4516	0.4203	0.2212	0.08
0.24	0.3001	0.2780	0.6636	0.24
0.4	0.1991	0.1838	1.106	0.4
0.8	0.0712	0.0654	2.212	0.8
1.2	0.0254	0.0233	3.318	1.2
1.6	0.0091	0.0083	4.424	1.6
Pe = 10				
0.	0.7745	0.7010	0.	0.
0.25	0.5460	0.4938	0.822	0.5
0.5	0.3847	0.3478	1.644	1.
1.	0.1911	0.1725	3.288	2.
2.	0.0473	0.0425	6.576	4.

Table 2
Comparison Between the Present Analysis and the Dispersion Model with Homogeneous Chemical Reaction (K = 100)

$\frac{x}{a}$	Exact Analysis Equation 4	Dispersion Model Equation 6	$\left(\frac{xU_{av}}{D_L} = \frac{x/U_{av}}{\frac{D}{U_{av}^2} + \frac{a^2}{48D}} \right)$	$\frac{Kx}{U_{av}}$
Pe = 1				
0.	0.0502	0.0486	0.	0.
0.2	0.0072	0.0070	0.0995	40.
0.4	0.0010	0.0010	0.1990	80.
0.6	0.0001	0.0001	0.2984	120.
Pe = 5				
0.	0.2414	0.2091	0.	0.
0.02	0.1033	0.0906	0.2212	0.8
0.06	0.0189	0.0170	0.6636	2.4
0.1	0.0035	0.0032	1.106	4.
0.2	0.0001	0.0001	2.212	8.
Pe = 10				
0.	0.4300	0.3315	0.	0.
0.25	0.0717	0.0632	0.822	0.25
0.5	0.0122	0.0121	1.644	0.5
1.	0.0004	0.0004	3.288	1.

$k_w a/D$ equivalent to K_0 in Equation 6) in addition to the three parameters, xU_{av}/D_L, kx/U_{av} and ka^2/D, previously discussed in order to determine the validity of the dispersion approximation. The term $k_w a/D$ can be considered as the ratio of the radial diffusion time of the species in the tube to the wall-reaction time. When the radial diffusion time is short, the radial concentration profile is then uniform. So the dispersion model is a good approximation to the real reactor behavior when $k_w a/D$ is a small value. Comparison between Equations 4 and 6 for $\alpha = 10, 100$; $K = 1$ is given in Tables 3 and 4, respectively. With the results presented in Tables 3 and 4 and those in Reference 52, one can see that the validity of the dispersion model follows closely to the values of the four dimensionless groups just discussed.

PLUG-FLOW REACTOR

Recent work on convective diffusion in a plug-flow reactor is on consecutive, first-order, irreversible, chemical reactions either in the bulk fluid [54] or at the tube wall [53]. The dimensionless average concentration and Sherwood number of components A and B in the positive region with consecutive reaction in the fluid are, respectively [54]:

$$\theta_{12_b}(\xi) = 2\sum_{n=1}^{\infty} E_n \exp\left[\frac{Pe^2}{2}\left(1 - \sqrt{1 + \frac{4(K_1 + \gamma_n^2)}{Pe^2}}\right)\xi\right]\frac{\alpha_2}{\gamma_n^2} J_0(\gamma_n) \tag{7}$$

$$\theta_{22_b}(\xi) = 2\sum_{n=1}^{\infty}\left\{K_n \exp\left[\frac{Pe^2}{2}\left(1 - \sqrt{1 + \frac{4(K_2 + \epsilon_n^2)}{Pe^2}}\right)\xi\right] - \frac{2K_1}{J_0(\epsilon_n)\left(1 + \frac{\alpha_4^2}{\epsilon_n^2}\right)}\right.$$

$$\left[B_j\left(\exp\left\{\frac{Pe^2}{2}\left[1 - \sqrt{1 + \frac{4(K_1 + \gamma_j^2)}{Pe^2}}\right]\xi\right\} - \exp\left\{\frac{Pe^2}{2}\left[1 - \sqrt{1 + \frac{4(K_2 + \epsilon_n^2)}{Pe^2}}\right]\xi\right\}\right)\right]$$

$$\left.\left[\left(\frac{\alpha_2 - \alpha_4}{\gamma_j^2 - \epsilon_n^2}\right)J_0(\gamma_j)\right] \Big/ (K_1 + \gamma_j^2 - K_2 - \epsilon_n^2)\right\} \quad \frac{\alpha_4 J_0(\epsilon_n)}{\epsilon_n^2} \tag{8}$$

$$Sh_{12} = \frac{2\alpha_2\theta_{12}(\xi, 1)}{\theta_{12_b}(\xi) - \theta_{12}(\xi, 1)} \tag{9}$$

$$Sh_{22} = \frac{2\alpha_4\theta_{22}(\xi, 1)}{\theta_{22_b}(\xi) - \theta_{22}(\xi, 1)} \tag{10}$$

Table 3
Comparison Between the Present Analysis and the Dispersion Model with Homogeneous and Heterogeneous Reaction at the Wall (K = 1, α = 10)

$\frac{x}{a}$	Exact Analysis Equation 4	Dispersion Model Equation 6	$\left(\frac{xU_{av}}{D_L} = \frac{x/U_{av}}{\frac{D}{U_{av}^2} + \frac{a^2}{48D}}\right)$	$\frac{Kx}{U_{av}}$
Pe = 1				
0.	0.1736	0.2635	0.	0.
0.509	0.0613	0.0925	0.2532	0.509
1.009	0.0219	0.0331	0.5019	1.009
2.109	0.0023	0.0035	1.049	2.109
3.109	0.0003	0.0004	1.546	3.109
Pe = 5				
0.	0.6040	0.7493	0.	0.
0.510	0.3259	0.4132	1.126	0.102
1.010	0.1803	0.2303	2.232	0.202
2.11	0.0497	0.0636	4.665	0.422
3.11	0.0154	0.0198	6.877	0.622
4.11	0.0048	0.0061	9.089	0.822
Pe = 10				
0.	0.7945	0.8965	0.	0.
0.51	0.5355	0.6280	1.674	0.051
1.01	0.3725	0.4426	3.318	0.101
2.11	0.1714	0.2051	6.934	0.211
3.11	0.0851	0.1019	10.222	0.311
4.11	0.0422	0.0506	13.510	0.411
5.11	0.0210	0.0252	16.798	0.511
9.11	0.0052	0.0062	23.374	0.911

Table 4
Comparison Between the Present Analysis and the Dispersion Model with Homogeneous and Heterogeneous Reactions at the Wall (K = 1, α = 100)

$\frac{x}{a}$	Exact Analysis Equation 4	Dispersion Model Equation 6	$\left(\frac{xU_{av}}{D_L} = \frac{x/U_{av}}{\frac{D}{U_{av}^2} + \frac{a^2}{48D}}\right)$	$\frac{Kx}{U_{av}}$
Pe = 1				
0.	0.1467	0.2589	0.	0.
0.509	0.0462	0.0836	0.2532	0.509
1.009	0.0152	0.0275	0.5019	1.009
2.109	0.0013	0.0024	1.049	2.109
3.109	0.0001	0.0003	1.546	3.109
Pe = 5				
0.	0.5459	0.7462	0.	0.
0.51	0.2723	0.3890	1.126	0.102
1.01	0.1422	0.2051	2.232	0.202
2.11	0.0347	0.0502	4.665	0.422
3.11	0.0096	0.0140	6.877	0.622
Pe = 10				
0.	0.7450	0.8990	0.	0.
0.51	0.4739	0.6072	1.674	0.051
1.01	0.3174	0.4129	3.318	0.101
2.11	0.1351	0.1768	6.934	0.211
3.11	0.0624	0.0818	10.222	0.311
4.11	0.0289	0.0378	13.510	0.411
5.11	0.0133	0.0175	16.798	0.511

When consecutive chemical reactions occur at the tube wall, the dimensionless average concentration and Sherwood number of components A and B in the positive region of the reactor are:

$$\theta_{12_b}(\xi) = 2\sum_{n=1}^{\infty} A_{n_2} \exp\left[\frac{Pe^2}{2}\left(1-\sqrt{1+\frac{4\beta_n^2}{Pe^2}}\right)\xi\right]\frac{\alpha_2}{\beta_{n_2}^2} J_0(\beta_{n_2}) \tag{11}$$

$$\theta_{22_b}(\xi) = \frac{\alpha_2}{\alpha_4}\sum_{n=1}^{\infty} A_{n_2} \exp\left[\frac{Pe^2}{2}\left(1-\sqrt{1+\frac{4\beta_{n_2}^2}{Pe^2}}\right)\xi\right] J_0(\beta_{n_2})$$

$$+ 2\sum_{n=1}^{\infty} K_n \exp\left[\frac{Pe^2}{2}\left(1-\sqrt{1+\frac{4\epsilon_n^2}{Pe^2}}\right)\xi\right] - \frac{\alpha_2 Pe^2}{J_0(\epsilon_n)(\alpha_4^2+\epsilon_n^2)}\sum_{\substack{m=1\\ m\neq n}}^{\infty}$$

$$\left[\left(A_{m_2}\left(1-\sqrt{1+\frac{4\beta_{m_2}^2}{Pe^2}}\right)\left[\frac{1}{2}\left(1-\sqrt{1+\frac{4\beta_{m_2}^2}{Pe^2}}\right)-1\right]J_0(\beta_{m_2})\right.\right.$$

$$\left.\left.\left\{\exp\left[\frac{Pe^2}{2}\left(1-\sqrt{1+\frac{4\beta_{m_2}^2}{Pe^2}}\right)\xi\right]-\exp\left[\frac{Pe^2}{2}\left(1-\sqrt{1+\frac{4\epsilon_n^2}{Pe^2}}\right)\xi\right]\right\}\right)\Big/(\beta_{m_2}^2-\epsilon_n^2)\right]$$

$$+\frac{A_{m_2}\left(1-\sqrt{1+\frac{4\beta_{n_2}^2}{Pe^2}}\right)\left(1+\sqrt{1+\frac{4\beta_{n_2}^2}{Pe^2}}\right)J_0(\beta_{n_2})\xi\exp\left[\frac{Pe^2}{2}\left(1-\sqrt{1+\frac{4\epsilon_n^2}{Pe^2}}\right)\xi\right]}{\sqrt{1+\frac{4\beta_n^2}{Pe^2}}+\sqrt{1+\frac{4\epsilon_n^2}{Pe^2}}} \tag{12}$$

$$Sh_{12} = \frac{2\alpha_2\theta_{12}(\xi, 1)}{\theta_{12_b}(\xi) - \theta_{12}(\xi, 1)} \tag{13}$$

$$Sh_{22} = \frac{2[\alpha_2\theta_{12}(\xi, 1) - \alpha_4\theta_{22}(\xi, 1)]}{\theta_{22}(\xi, 1) - \theta_{22_b}(\xi)} \tag{14}$$

Some of the results of the Sherwood number of components A and B along the positive region of the reactor are shown in Figures 1 and 2. Figure 1 is the Sh of components A and B in the positive region of the reactor when consecutive chemical reactions occur in the bulk fluid and catalytic simple reactions occur at the tube wall for some typical values of the parameters of the system. When the catalytic wall-reaction rate of component A changes, both the Sh of components A and B will change. However, when the catalytic wall-reaction rate of component B changes as in Figure 1, the Sh of component A remains the same value but the Sh of component B will vary. Steady state values of Sherwood numbers of components A and B are reached downstream of the reactor. When the wall-reaction rates are large, the reactive species react more rapidly so their concentration levels reach lower values downstream of the reactor. Hence the consumption rate of the reactive component at the wall is smaller and the downstream steady-state Sh becomes smaller in value when the wall-reaction parameters of components A or B increase.

Figure 2 shows the Sh of components A and B in the positive region of the reactor when consecutive reactions occur at the tube wall for some typical values of the parameters of the system. Since the catalytic wall reaction of component B, α_4, has no effect on component A, the Sherwood number of component A does not change as the value of α_4 changes. However, the Sh of component B decreases to negative values along the reactor length. It is particularly interesting to find a discontinuity of the Sh of component B to occur when the wall concentration is equal to the average concentration in Equation 14. The physical phenomenon can occur because component B is the intermediate product of the consecutive reactions. It is also found that the numerator and the denominator of Equation 14 are not zero at the same location. In other words the net zero production rate of component B in the numerator of Equation 14 does not occur at the same location where the wall concentration of component B is equal to the average concentration of component B in the denominator. Far downstream of the reactor, a steady-state Sh component B is achieved.

One particular interest in the study of consecutive chemical reactions is to find the optimal yield of component B and the optimal reactor length for the purpose of design. Detailed examinations of plug-flow reactor design for consecutive chemical reactions are given in References 53 and 54. Figure 3 shows the optimal reactor length and the optimal yield of compo-

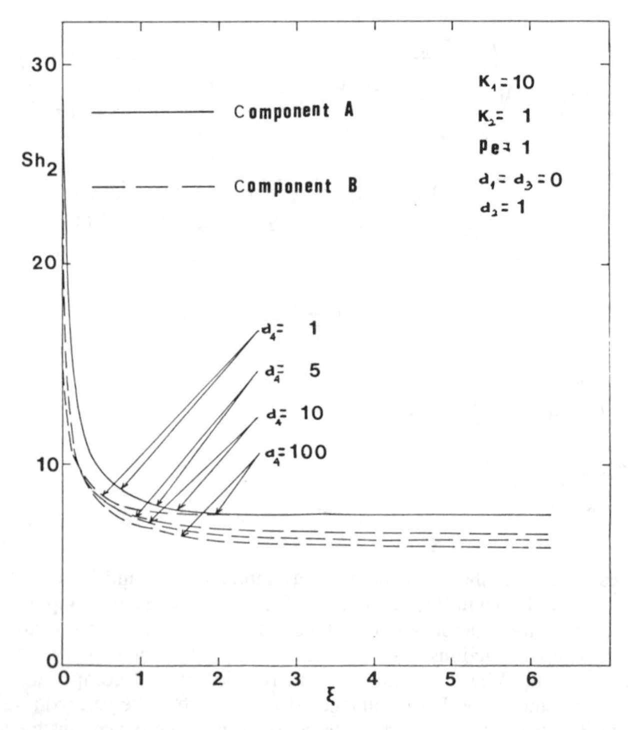

Figure 1. Sherwood number of components A and B versus dimensionless axial coordinate in positive region for different values of α_4, K_1 = 10, K_2 = 1, Pe = 1, α_2 = 1. Plug-flow reactor with consecutive reactions in the bulk-fluid phase.

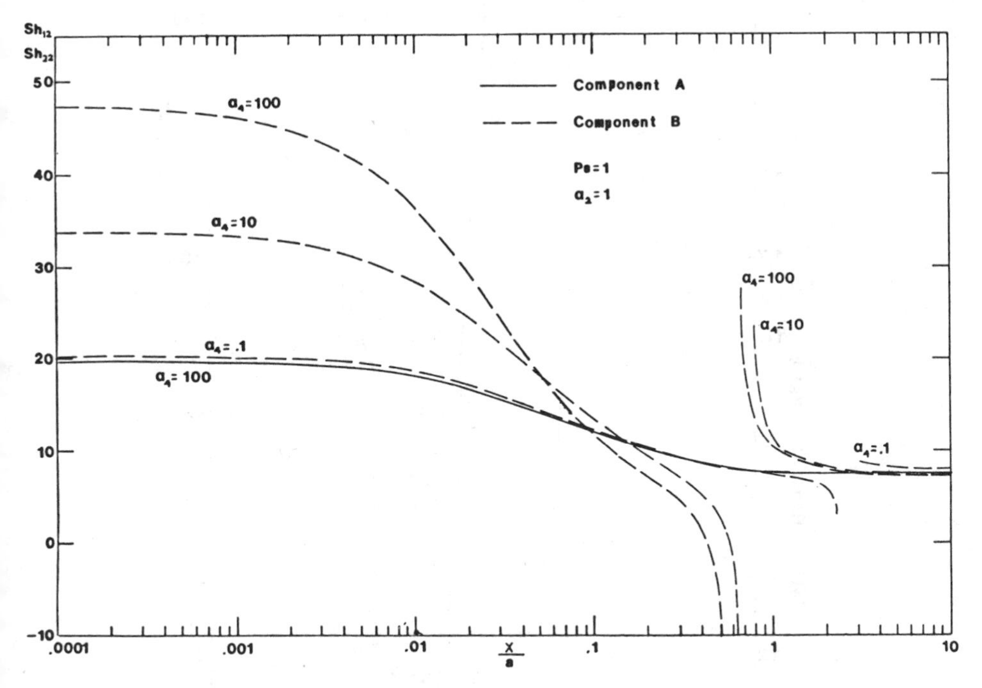

Figure 2. Sherwood number of components A and B versus axial distance for different wall-reaction parameter of component B. Plug-flow reactor with consecutive reactions at the tube wall (Pe = 1, α_2 = 1, α_4 = 0.1, 10, 100.).

nent B for different values of the wall reaction parameters of components A and B when the consecutive reactions occur in the bulk fluid and the catalytic reactions of both components A and B occur at the tube wall. The optimal yield of component B increases when either the decomposition rate of components A or B at the wall decreases. The optimal reactor length increases when the wall-reaction rate of component B is kept constant while the wall-reaction rate of component A increases. On the other hand, the optimal reactor length decreases and then increases when the wall-reaction rate of component A is kept constant while the wall-reaction rate of component B increases.

SUMMARY

Investigation of convective diffusion with reactions in a tube can be treated by a dispersion approximation or nondispersion analysis. For dispersion-model calculation, both complicated theory and simple one-dimensional approximation are used. The latter approximation has been examined for its validity conditions. For nondispersion analysis, the concentration distributions and Sh of reactive species for homogeneous and heterogeneous simple first-order reactions in a laminar-flow reactor and first-order consecutive reactions in a plug-flow reactor are obtained. Some results on the optimal yield of the intermediate component and the optimal reactor length for the consecutive reactions in the bulk fluid of a plug-flow reactor are graphically presented for design purposes.

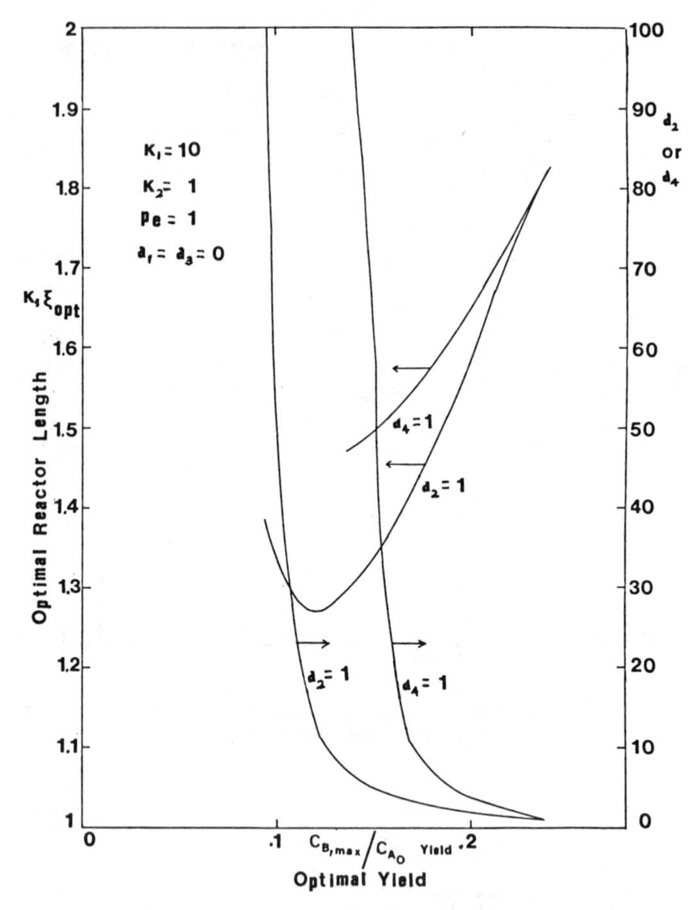

Figure 3. Optimal reactor length and wall reactions of components A and B versus optimal yield. Plug-flow reactor with consecutive reactions in the bulk-fluid phase (K_1 = 10, K_2 = 1, Pe = 1).

NOTATION

- a radius of tube
- A_n coefficients of series expansion in Equations 1 and 3
- A_{n_2} coefficients of series expansion in Equations 11 and 12
- B_n coefficients of series expansion in Equations 2 and 4
- C concentration of reactant
- C_{A_0} inlet concentration of component A at $x = -\infty$
- C_{A_1} concentration of component A in the negative region
- C_{A_2} concentration of component A in the positive region
- C_{B_0} inlet concentration of component B at $x = -\infty$
- C_{B_1} concentration of component B in the negative region
- C_{B_2} concentration of component B in the positive region
- C_0 inlet concentration
- D molecular diffusivity
- E_n coefficients of series expansion in Equation 7

k first-order homogeneous reaction-rate constant
k_{w_2} wall-reaction-rate constant of component A in the positive region
k_{w_4} wall-reaction-rate constant of component B in the positive region
k_w surface wall-reaction-rate constant
k_1 first-order reaction-rate constant of component A in the bulk phase
k_2 first-order reaction-rate constant of component B in the bulk phase
K ka^2/D
K_L mass transfer coefficient
K_0 effective wall-reaction coefficient in the dispersion model, Equation 6
K_1 effective convection coefficient in the dispersion model, Equation 6; also k_1a^2/D
K_2 effective dispersion coefficient in the dispersion model, Equation 6; also k_2a^2/D
K_n coefficients of series expansion in Equation 8
Pe aU_m/D
r radial coordinate
R_n eigenfunctions in Equations 2 and 4
Sh $2K_La/D$, Sherwood number
Sh_{12} Sherwood number of component A in the positive region
Sh_{22} Sherwood number of component B in the positive region
U_m maximum velocity
x axial coordinate
Y_n eigenfunctions in Equations 1 and 3

Greek Symbols

α k_wa/D
α_2 $k_{w_2}a/D$
α_4 $k_{w_4}a/D$
α_n eigenvalues in Equations 1 and 3
β_n eigenvalues in Equations 2 and 4
β_{n_2} eigenvalues in Equations 11 and 12
γ_n eigenvalues in Equation 7
ϵ_n eigenvalues in Equation 8
ξ $\frac{x}{aPe}$
η r/a
θ_{11} $(C_{A_0} - C_{A_1})/C_{A_0}$
θ_{12} C_{A_2}/C_{A_0}
θ_{21} C_{B_1}/C_{A_0}
θ_{22} C_{B_2}/C_{A_0}
θ_{B_0} C_{B_0}/C_{A_0}

Subscripts

b average concentration
i 1 for component A, 2 for component B
j 1 for negative region, 2 for positive region

REFERENCES

1. Taylor, G. I., "Dispersion of Soluble Matter in Solvent Flowing Slowly Through a Tube," *Proc. Roy. Soc.*, A 219:186 (1953).
2. Taylor, G. I., "The Dispersion of Matter in Turbulent Flow Through a Pipe," *Proc. Roy. Soc.*, A 223:466 (1954).
3. Taylor, G. I., "Conditions Under Which Dispersion of a Solute in a Stream of Solvent Can Be Used to Measure Molecular Diffusion," *Proc. Roy. Soc.*, A 225:473 (1954).
4. Aris, R., "On the Dispersion of a Solute in a Fluid Flowing Through a Tube," *Proc. Roy. Soc.*, A 235:67 (1956).
5. Lighthill, M. J., "Initial Development of Diffusion in Poiseuille Flow," *J. Inst. Math. Applic.*, 2:97 (1966).
6. Gill, W. N., "A Note on the Solution of Transient Dispersion Problem," *Proc. Roy. Soc.*, A 298:335 (1967).
7. Gill, W. N., and Subramanian, R., "Exact Analysis of Unsteady Convective Diffusion," *Proc. Roy. Soc.*, A 316:341 (1970).

8. Gill, W. N., and Subramanian, R., "Dispersion from a Prescribed Concentration Distribution in Time Variable Flow," *Proc. Roy. Soc.*, A 329:479 (1972).
9. Chatwin, P. C., "The Approach to Normality of the Concentration Distribution of a Solute in a Solvent Flowing Along a Straight Pipe," *J. Fluid Mech.*, 43:321 (1970).
10. Chatwin, P. C., "The Initial Dispersion of Contaminant in Poiseuille Flow and the Smoothing of the Snout," *J. Fluid Mech.*, 77:593 (1976).
11. Chatwin, P. C., "The Initial Development of Longitudinal Dispersion in Straight Tube," *J. Fluid Mech.*, 80:33 (1977).
12. Fife, P. C., and Nicholes, K. R. K., "Dispersion in Flow Through Small Tubes," *Proc. Roy. Soc.*, A 344:131 (1975).
13. DeGance, A. E., and Johns, L. E., "On the Dispersion Coefficient for Poiseuille Flow in a Circular Cylinder," *Appl. Sci. Res.*, 34:227 (1978).
14. Brenner, H., "A General Theory of Taylor Dispersion Phenomena," *J. Physicochemical Hydrodynamics*, 1:91 (1980).
15. Smith, R., "A Delay-Diffusion Description for Contaminant Dispersion," *J. Fluid Mech.*, 105:469 (1981).
16. Smith, R., "The Early Stages of Contaminant Dispersion in Shear Flows," *J. Fluid Mech.*, 111:107 (1981).
17. Smith, R., "The Contraction of Contaminant Distributions in Reversing Flows," *J. Fluid Mech.*, 129:137 (1983).
18. Smith, R., "Field-Flow Fractionation," *J. Fluid Mech.*, 129:347 (1983).
19. Yu, J. S., "On Laminar Dispersion for Flow Through Round Tubes," *J. Applied Mech.*, 46:750 (1979).
20. Yu, J. S., "An Approximate Analysis of Laminar Dispersion in Circular Tubes," *J. Appl. Mech.*, 43:537 (1976).
21. Yu, J. S., "Dispersion in Laminar Flow Through Tubes by Simultaneous Diffusion and Convection," *J. Appl. Mech.*, 48:217 (1981).
22. Hunt, B., "Diffusion in Laminar Pipe Flow," *Int. J. Heat and Mass Transf.*, 20:393 (1977).
23. Tseng, C. M., and Besant, R. W., "Dispersion of a Solute in a Fully Developed Laminar Tube Flow," *Proc. Roy. Soc.*, A 317:91 (1970).
24. Tseng, C. M., and Besant, R. W., "Transient Heat and Mass Transfer in Fully Developed Laminar Tube Flows," *Int. J. Heat and Mass Transf.*, 15:203 (1972).
25. Reis, J. F. G., Ramkrishna, D., and Lightfoot, E. N., "Convective Mass Transfer in the Presence of Polarizing Fields: Dispersion in Hollow Fiber Electropolarization Chromatography," *AIChE J.*, 24:679 (1978).
26. Purtell, L. P., "Molecular Diffusion in Oscillating Laminar Flow in a Pipe," *Phys. Fluids.*, 24:789 (1981).
27. Wang, J. C., and Stewart, W. E., "New Descriptions of Dispersion in Flow Through Tubes: Convolution and Collocation Methods," *AIChE J.* 29:493 (1983).
28. Lightfoot, E. N., Chang, A. S., and Noble, P. T., "Field Flow Fractionation (Polarization Chromatography)," *Ann. Rev. Fluid Mech.*, 13:351 (1981).
29. Wan, C., and Ziegler, E. N., "On the Axial Dispersion Approximation for Laminar Flow Reactors," *Chem. Eng. Sci.*, 25:723 (1970).
30. Bischoff, K. A., "Accuracy of the Axial Dispersion Model for Chemical Reactor," *AIChE J.* 14:820 (1968).
31. Wissler, E. H., "On the Applicability of the Taylor-Aris Axial Diffusion Model to Tubular Reactor Calculations," *Chem. Eng. Sci.*, 24:527 (1969).
32. Subramanian, R. S., Gill, W. N., and Marra, R. A., "Dispersion Models of Unsteady Tubular Reactors," *Canadian J. Chem. Eng.*, 52:563 (1974).
33. Subramanian, R., and Gill, W. N., "Unsteady Convective Diffusion with Interphase Mass Transfer," *Proc. Roy. Soc.*, A 333:115 (1973) and A 341:407 (1974).
34. Johns, L. E., and Degance, A. E., "Dispersion Approximations to the Multicomponent Convective Diffusion Equation for Chemically Active Systems," *Chem. Eng. Sci.*, 30:1065 (1975).

35. DeGance, A. E., and Johns, L. E., "The Theory of Dispersion of Chemically Active Solutes in a Rectilinear Flow Field," *Appl. Sci. Res.*, 34:189 (1978).
36. DeGance, A. E., and Johns, L. E., "On the Construction of Dispersion Approximations to the Solution of the Convective Diffusion Equation," *AIChE J.*, 26:411 (1980).
37. Aris, R., "Hierarchies of Models in Reactive Systems," *Dynamics and Modelling of Reactive Systems,* W. E. Stewart, W. H. Ray, and C. C. Conley (Ed.), Academic Press (1980), p. 1.
38. Smith, R., "Effect of Boundary Absorption upon Longitudinal Dispersion in Shear Flows," *J. Fluid Mech.*, 134:161 (1983).
39. Wan, C., and Ziegler, E. N., "Effect of Mixing on Yield in Isothermal Tubular Reactors," *Ind. Eng. Chem. Fundam.*, 12:55 (1973).
40. Cleland, F. A., and Wilhelm, R. H., "Diffusion and Reaction in Viscous Flow Tubular Reactors," *AIChE J.*, 2:489 (1956).
41. Katz, S., "Chemical Reactions Catalyzed on a Tube Wall," *Chem. Eng. Sci.*, 10:202 (1959).
42. Walker, R. E., "Chemical Reaction and Diffusion in a Catalytic Tubular Reactor," *Phys. of Fluids.*, 4:1211 (1961).
43. Ulrichson, D. L., and Schmitz, R. A., "Chemical Reaction in the Entrance Length of a Tubular Reactor," *Ind. Eng. Chem. Fundam.*, 4:2 (1965).
44. Lupa, A. J., and Dranoff, J. S., "Chemical Reaction on the Wall of an Annular Reactor," *Chem. Eng. Sci.*, 21:861 (1966).
45. Hudson, J. L., "Diffusion with Consecutive Heterogeneous Reactions," *AIChE J.*, 11:943 (1965).
46. Solomon, R. L., and Hudson, J. L., "Heterogeneous and Homogeneous Reactions in a Tubular Reactor," *AIChE J.* 13:545 (1967).
47. Li, C. H., "False Diffusion in Convection and Diffusion with Chemical Reaction," *Int. J. Heat and Mass Transfer,* 26:1063 (1983).
48. Huang, D. T., and Varma, A., "Yield Optimization in a Tube Wall Reactor," ACS Symposium Series No. 124 (1980), p. 467.
49. Dang, V. D., and Steinberg, M., "Laminar Flow Mass Transfer with Axial Diffusion in a Tube with Chemical Reaction," *Chem. Eng. Sci.*, 32:326 (1977) and 33:1297 (1978).
50. Dang, V. D., "Mass Transfer with Axial Diffusion and Chemical Reaction," *Chem. Eng. Sci.*, 33:1179 (1978).
51. Dang, V. D., and Steinberg, M., "Convective Diffusion with Homogeneous and Heterogeneous Reactions in a Tube," *J. Phys. Chem.*, 84:214 (1980).
52. Dang, V. D., "Steady-State Mass Transfer with Homogeneous and Heterogeneous Reactions," *AIChE J.*, 29:19 (1983).
53. Dang, V. D., "Diffusive Transport with Consecutive Catalytic Wall Reactions," AIChE Summer Meeting, paper No. 59e, August 1983.
54. Dang, V. D., "Axial Diffusion and Selectivity in a Plug-Flow Reactor," AIChE Winter Meeting, March 1984.

CHAPTER 4

TRANSIENT MASS TRANSFER ONTO SMALL PARTICLES AND DROPS

Sergio Carra and Massimo Morbidelli

Dipartimento di Chimica Fisica Applicata
Politecnico di Milano
Milano, Italy

CONTENTS

INTRODUCTION

Small particles are of great importance in many natural and industrial activities. There are various systems among engineering equipment and industrial processes where small particle dispersions are present. The following examples can be given:

- *Solid particles dispersed in fluids* are involved in pneumatic conveyors, dust collectors, fluidized beds, heterogeneous gas-solid, flotation, and sedimentation processes.
- *Liquid drops dispersed in gases* are involved in atomizers, empty absorption columns, dryers, combustors, air pollution, and evaporation processes.
- *Liquid drops dispersed in liquids* are involved in extraction and emulsifying processes.

In the case of liquid drops there are basically three major periods in their lifetime during which mass transfer takes place [1]: formation of the drop in the continuous phase, free rise or fall of the drop through the continuous phase, and coalescence of the drops at the end of the free-motion period. The first period is particularly important in spray columns, while it is usually less important than the others in a multistage perforated-plate column.

In this chapter only particles and liquid drops in free rise or fall through the continuous phase will be considered. The main objective of the chapter is to provide a guide for the estimation of mass transfer rates to or from such particles. To this aim, it is first necessary to review the most important aspects of the fluid mechanics of a single particle in free motion through a continuous fluid phase. This has been accomplished in the section on fluid mechanics of this chapter, with particular emphasis on those aspects directly related to mass transfer, such as particle velocity relative to the continuous phase, external surface area, and internal motions, when present. To this respect, and in general for all the hydrodynamic aspects of these systems, the research of Clift, Grace, and Weber [2] is an important reference point.

In the section entitled "Mass Transfer," the mass transfer processes to and from particles have been examined, while the case involving chemical reactions in the dispersed phase has been examined in the last section.

Throughout the entire chapter the movement of the particle with respect to the continuous phase has been assumed constant with time, while the mass transfer phenomena are intrinsically transient, because the concentration of at least one of the involved components within the particle changes during time.

Transient phenomena can also be involved in the particle fluid mechanics, particularly in the region of drop formation. This aspect has not been considered here, and reference should be made to the review papers by Kintner [3], Heertjes and De Nie [4], and Schügerl [4] where also mass transfer during drop formation has been analyzed in some detail.

It should be pointed out that the present analysis is devoted only to single particles, and no reference is made to the interference between particles which can occur when their concentration is significant, as in some industrial applications [5, 6].

The dimensions of the particle considered fall in the range 0.5 μm − 10 cm. Qualitative correlations between the particle dimension and some physical quantities are shown in Figure 1.

In general, fluid particles can be divided into rigid, noncirculating and circulating particles. Solid particles are considered rigid, while liquid drops can be considered circulating or noncirculating particles, depending on the presence or not of movements of the internal liquid relative to axes fixed to the particle.

It should be mentioned that, although beyond the scope of this chapter, most of the reported correlations apply also to gaseous bubbles in free motion through a continuous liquid. In general, the correlations reported for drops can be extended to bubbles in the limit of very small density and viscosity ratios ($\gamma = \rho_p/\rho$ and $\varkappa = \mu_p/\mu$).

FLUID MECHANICS

Basic Principles

Any hydrodynamic consideration about a fluid or solid particle moving through a continuous fluid phase starts with the Navier-Stokes equations of motion. Using suitable dimensionless variables, this can be written as follows:

$$\frac{D\underset{\sim}{v}^*}{Dt^*} = \frac{\partial \underset{\sim}{v}^*}{\partial t^*} + \underset{\sim}{v}^* \cdot \nabla \underset{\sim}{v}^* = - P^* + \frac{1}{Re} \nabla^2 \underset{\sim}{v}^* \tag{1}$$

coupled to the continuity equation

$$\frac{\partial \rho^*}{\partial t^*} + \nabla \cdot \rho^* \underset{\sim}{v}^* = 0 \tag{2}$$

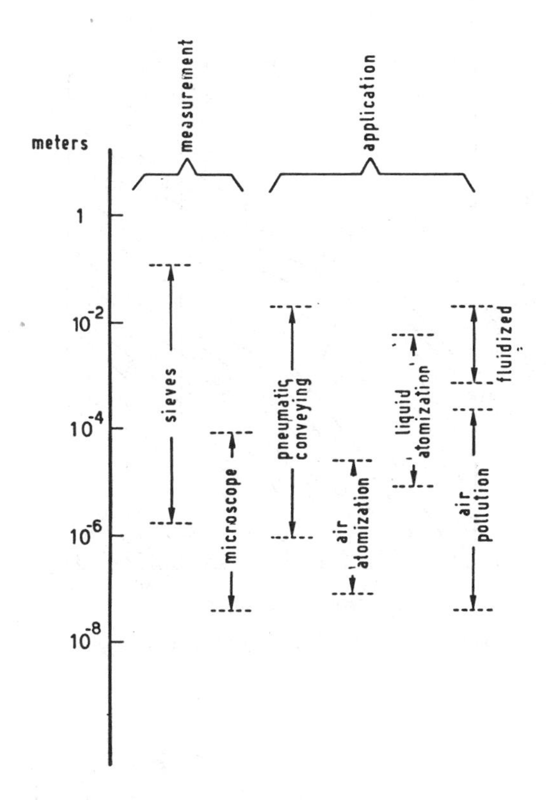

Figure 1. Particle dimensions in relation to measurement methods and applications fields.

which derives from the application of the mass conservation principle.

In the case of incompressible fluids, the continuity equation (Equation 2) reduces to

$$\nabla \cdot \underset{\sim}{v}^* = 0 \tag{3}$$

From the solution of Equations 1 and 3, coupled to suitable boundary conditions, the velocity $\underset{\sim}{v}^*$ and pressure P* field can be determined. Part of the boundary conditions refers to the velocity field remote from the particle and to the existence of some symmetry relationships depending on the geometry of the particular problem under consideration. The remaining boundary conditions refer to the interface between the particle and the external fluid. They state that the normal and tangential velocities are equal in both phases at the interface.

In the case of a rigid particle the system of Equations 1 and 3 apply only to the external fluid phase, and then the previously mentioned boundary conditions are sufficient to fully determine the external velocity and pressure field. In this case the normal and tangential velocities at the interface, relative to axes fixed to the particle, are equal to zero.

On the other hand, for a fluid particle, due to the presence of internal circulation motions, the equations of motion (Equations 1 and 3) must be extended also to the internal phase. Two additional boundary conditions are then required at the interface. These are obtained by stating that the normal and shearing stresses are balanced at the interface. Namely, it is the

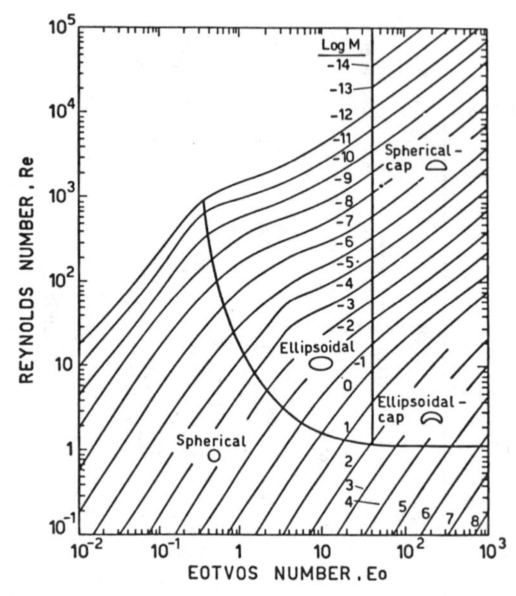

Figure 2. Shape of fluid particles in motion through liquids (M indicates the Morton number).

equivalence of the shearing stresses in the two phases at the interface that causes the occurrence of internal motions of the fluid inside the particle. This constitutes the fundamental difference between solid and liquid particles, which leads to a different behavior both from the hydrodynamic and the mass transfer point of view.

It is noticeable that since the shearing stress balance at the interface involves the surface tension between the two fluids in contact, the internal motions are highly affected by all those variables which affect the interfacial tension. Namely, surface agents, or often even impurities present to such a low extent that the bulk properties of the liquid are not changed in a measurable way, are collected at the particle interface during motion and can lead to substantial changes in internal circulation motions. In particular, surface agents tend to damp internal circulation so that the drop approaches the rigid particle behavior. It is noticeable that the complete elimination of surface agents requires an extremely careful purification procedure of the involved liquids, which is not performed for fluids usually used. Therefore, in the following, special attention will be devoted to the behavior of the so called contaminated systems, which are those actually employed in practice. The study of the behavior of highly pure fluids deserves attention only from a more fundamental point of view, such as the verification of circulation models developed under the assumption of no surface tension alteration.

Shape of Solid and Fluid Particles

Some of the manufactured solid particles are symmetric with the shape of a sphere, a cylinder or a regular polyhedra. However, most particles of practical interest are irregular in shape so that empirical factors need to be devised in order to describe their flow behavior.

Usually, the particle is approximated with an equivalent sphere defined as the sphere with the same volume of the particle. Using this approximation the diameter and the surface area of the equivalent sphere can be computed as a function of the particle volume as follows:

$$d_e = (6V_p/\pi)^{1/3}$$

$$A_e = (36\pi)^{1/3} V_p^{2/3} \tag{4}$$

Further details on empirical shape factors of common use are reported by Clift et al. [2].

In the case of fluid particles in free motion through a continuous phase, the possible shapes are much more limited. Namely, the shape of a moving drop is defined by the balancing of interfacial tension, viscous forces, and inertia forces at the interface. In general the fluid particle shapes can be classified as follows:

- *Spherical:* This shape is attained when inertia forces are negligible with respect to interfacial tension and viscous forces.
- *Ellipsoidal:* Indicates the shape obtained by rotating an ellipse about one of its principal axes. If this is the minor one, the ellipsoid is said to be oblate, otherwise it is prolate. Usually, ellipsoidal drops are characterized by vigorous circulation or oscillations.
- *"Spherical-cap" or "ellipsoidal-cap":* This shape is quite similar to the upper segment cut from a sphere or an ellipsoid, with a flat and sometimes indented base. This shape is peculiar of large drops.

A generical graphic correlation of the possible shapes of a drop in movement in infinite media can be prepared using the Reynolds number, the Eötvos number and the Morton number as shown in Figure 2. This correlation does not apply to liquid drops falling through gases, i.e., for very large values of the density and viscosity ratios γ and $\varkappa$. The effect of the walls of the container on the particle hydrodynamics will be briefly discussed later.

As expected, in the region of low Re and Eo the drop is spherical. At higher Re and intermediate Eo the shape is ellipsoidal, while the spherical or ellipsoidal cap shapes require large values of both Re and Eo. This last regime is generally attainable only by drops moving through liquid continuous phases, because drops falling in gases usually break up before the value Eo = 40 is reached.

A more general procedure for the evaluation of the maximum drop diameter of a free-falling drop has been proposed by Grace et al. [7]. This is based on the assumption that the drop surface becomes unstable when the wave length λ of a disturbance at the interface exceeds a critical value

$$\lambda_{cr} = 2\pi\,(\sigma/g\Delta\rho)^{1/2} \tag{5}$$

Now, as a first approximation, a perturbation of wave length λ has a time t_a available to grow:

$$t_a = \frac{d_e}{4v_p}\left(\frac{2 + 3\varkappa}{2}\right) \ln\,[\cot\,(\lambda/4\,d_e)] \tag{6}$$

where v_p indicates the drop free-falling velocity relative to remote continuous phase and can be evaluated, as a function of the drop diameter, from the relationships reported in the next paragraphs. On the other hand, the time t_g required by the perturbation to grow can be estimated from the following approximation:

$$[\sigma k'^3 + gk'\Delta\rho + \alpha^2(\rho + \rho_p)]\,[k' + m + \varkappa(k' + m_p)] + 4\alpha k'\mu(k' + \varkappa m_p)\,(\varkappa k' + m) = 0 \tag{7}$$

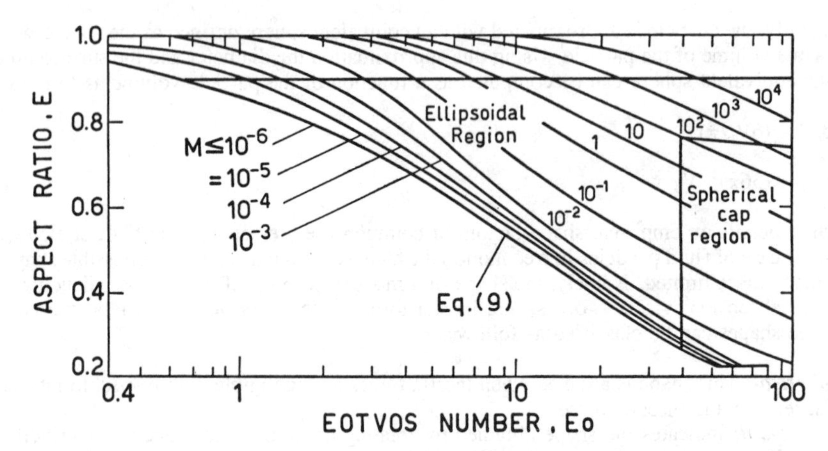

Figure 3. Mean aspect ratio E as a function of Eotvos number, for various values of the Morton number.

Figure 4. System of spherical coordinates.

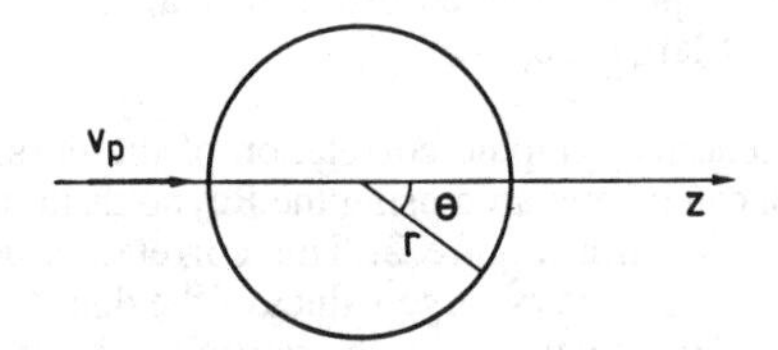

where $\alpha = 1/t_g$
$k' = 2\pi/\lambda$
$m = (k'^2 + \alpha/\nu)^{1/2}$
$m_p = (k'^2 + \alpha/\nu_p)^{1/2}$

Since it is reasonable to assume that λ can not be larger than half the drop circumference, the maximum t_g value is obtained from Equation 6 for $\lambda_{cr} < \lambda < \pi d_e/2$. The maximum diameter value of a stable drop is then obtained by comparison of the time t_g required by the critical perturbation to grow and that t_a actually available for its growth. Comparisons with experimental data have indicated that drop splitting occurs for $t_a > 1.4t_g$.

In the case of $\varkappa > 0.5$, that is for liquid drops falling in gases or in slightly viscous media, the previously reported procedure leads to

$$(d_e)_{max} = 4(\sigma/g\Delta\rho)^{1/2} \tag{8}$$

which can be rewritten in the equivalent form $(Eo)_{max} = 16$.

Thus, the spherical, ellipsoidal-cap-shape regime illustrated in Figure 2 is actually attainable only by liquid drops whose viscosity is significantly lower than the external medium, i.e. $\varkappa < 0.5$.

The data reported in Figure 2 allow a quick evaluation of the shape of a moving drop and a rough estimation of its terminal velocity. More accurate expressions of the latter quantity will be given in the following. On the other side a quantitative definition of the shape of a drop in the ellipsoidal regime can be obtained using Figure 3, where a generalized graphic correlation of the mean aspect ratio E (defined as the ratio between the maximum vertical dimension and the maximum horizontal one) as a function of the Eotvos and the Morton numbers is shown. Note that the reported curves apply to drops of contaminated liquid (i.e., to which no specific purification procedure has been applied) moving in an infinite media.

In the case of high-viscous continuous phase ($\mu \geq 2$ poise), Wellek et al. [8] have proposed a correlation in terms of the Weber number

$$E = 1/(1 + 0.091\ We^{0.95}) \tag{9}$$

From the shape parameter E, the ratio between the external surface area of the ellipsoidal particle and that of the equivalent sphere can be obtained as follows:

$$\frac{A_p}{A_e} = \frac{1}{2E^{2/3}}\left[1 + \frac{E^2}{2e}\ln\left(\frac{1 + e}{1 - e}\right)\right] \tag{10}$$

where $e = (1 - E^2)^{1/2}$.

Approximate Solutions of the Equations of Motion

As mentioned, the velocity and pressure field of a particle in free motion through an infinite medium can be obtained through the solution of the system of Equations 1 and 3. However, due to the nonlinearity of the Navier-Stokes equation (Equation 1), arising from the convective acceleration term $\underset{\sim}{v}\,(\nabla\underset{\sim}{v})$, its solution can only be obtained through suitable numerical techniques. Analytical solutions can only be obtained for some symmetric geometries and under particular approximations which correspond to limiting physical situations of particular importance for the description of hydrodynamic and mass transfer phenomena.

Particularly important is the case of an axisymmetric flow of incompressible fluids around a spherical particle of radius R_p. The geometry of the system suggests employment of the system of spherical coordinates, with the origin placed at the origin of the particle and fixed to the particle, as shown in Figure 4. Moreover, it is convenient to introduce the Stokes stream function ψ, which in this case is related to the velocity components as follows:

$$v_r = \frac{1}{r^2 \sin\theta}\frac{\partial\psi}{\partial\theta} \tag{11a}$$

$$v_\theta = -\frac{1}{r \sin\theta}\frac{\partial\psi}{\partial r} \tag{11b}$$

Creeping Flow

At very low values of the Reynolds number (Re < 1), where the inertia forces are negligible with respect to the viscous ones, the Navier Stokes equation (Equation 1) can be simplified as follows:

$$\nabla P = \mu\nabla^2\underset{\sim}{v} \tag{12}$$

The analytical solution of this equation, coupled to the continuity equation, has been obtained independently by Hadamard [9] and Rybczynski [10] in terms of the stream functions ψ and ψ_p, relative to the continuous and dispersed phase respectively, as follows:

$$\psi = -\frac{v_p r^2 \sin^2\theta}{2}\left[1 - \frac{R_p(2 + 3\varkappa)}{2r(1 + \varkappa)} + \frac{\varkappa R_p^{\,3}}{2r^3(1 + \varkappa)}\right] \tag{13}$$

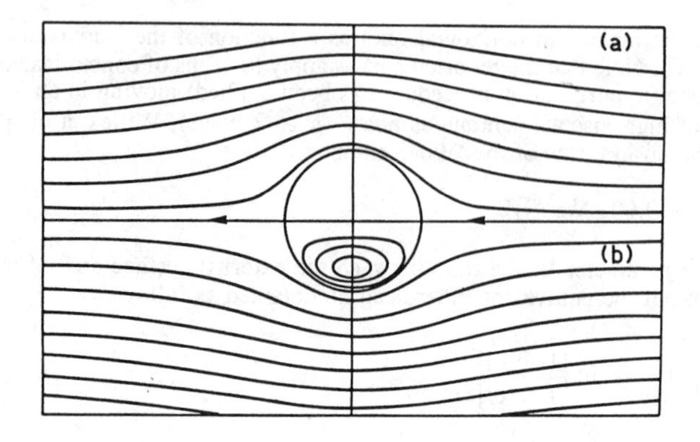

Figure 5. Calculated streamlines relative to the particle at low Re: (a) Stokes solution ($\varkappa = \infty$); (b) Hadamard-Rybczynski solution ($\varkappa = 0$).

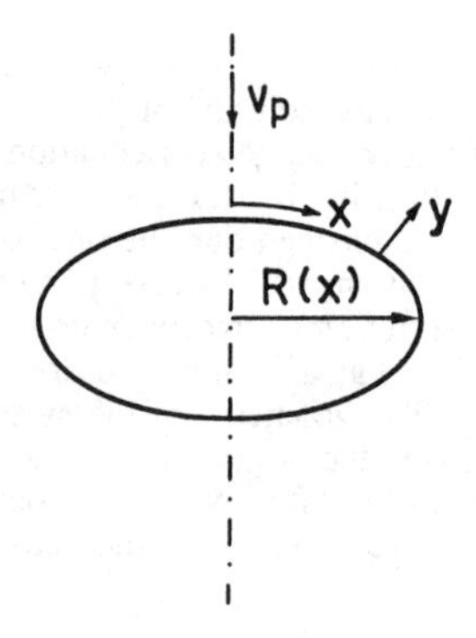

Figure 6. System of boundary-layer coordinates for an axisymmetric particle.

$$\psi_p = \frac{v_p r^2 \sin^2\theta}{4(1 + \varkappa)}\left(1 - \frac{r^2}{R_p^2}\right) \tag{14}$$

The adopted boundary conditions imply uniform flow with velocity v_p in the z-direction remote from the particle, no flow across the interface, continuity of tangential velocity, and tangential stress across the interface.

These equations give rise to stream lines symmetric about the equatorial plane as illustrated in Figure 5. The pressure fields in the continuous and dispersed phases are given by

$$P = P_o + [\mu v_p R_p \cos\theta(2 + 3\varkappa)/2r^2(1 + \varkappa)] \tag{15}$$

$$P_p = P_{op} - [5\, \mu_p v_p r \cos\theta/R_p^2(1 + \varkappa)] \tag{16}$$

From Equations 13 through 16 it can be shown that the force balance at the interface along the normal direction turns out to be exactly satisfied. This means that the sperical-shape approximation is consistent with all the previously mentioned natural boundary conditions of the problem. This allows the conclusion that the actual shape of moving drops in creeping flow is spherical, and significant deformations occur only when the inertia forces become important.

The drag coefficient C_D can be derived from the pressure field, leading to the classic expression

$$C_D = \frac{8}{Re}\left(\frac{2 + 3\varkappa}{1 + \varkappa}\right) \tag{17}$$

For a rigid particle, $\varkappa \rightarrow \infty$ and the flow field can be characterized by solving Equations 3 and 12. In this case the boundary conditions imply a uniform flow with velocity v_0 remote from the particle, and zero normal and tangential velocity at the interface. The solution can be derived directly from Equations 13 to 16 with $\varkappa \rightarrow \infty$. In particular, the following expression for the drag coefficient can be derived from Equation 17:

$$C_{Dr} = 24/Re \tag{18}$$

which constitutes the well known Stokes law. A comparison between the calculated stream lines for a rigid and a circulating particle in viscous flow is shown in Figure 5.

Inviscid flow

This approximation corresponds to the opposite case of the creeping flow, and implies zero viscosity for the continuous phase, i.e. $\varkappa \rightarrow \infty$. Such a fluid is obviously an idealization (corresponding to the classic Eulerian fluid), but it provides some good approximations in the limit of very large Re. In this case the dispersed phase behaves as a rigid body (i.e., a solid particle or a noncirculating drop). The velocity field can be determined by solving Equation 1, where the last term in the r.h.s. has been neglected. In the case of a spherical particle, this leads to

$$v_r = -v_p \cos\theta(1 - R_p^3/r^3) \tag{19}$$

$$v_\theta = v_p \sin\theta(1 + R_p^3/2r^3) \tag{20}$$

while the surface pressure is given by

$$P = P_0 + (\rho v_p^2/2)\left(1 - \frac{9}{4}\sin^2\theta\right) \tag{21}$$

These expressions imply nonzero tangential velocity at the interface and net force acting on the particle equal to zero (d'Alambert's paradox).

Boundary-Layer Approximation

The steady flow at a high Reynolds number of a fluid past an immersed rigid particle or the equivalent opposite movement of the particle through the fluid can be described by means of the boundary-layer approximation. In some way this approximation improves the inviscid flow case, because it assumes that viscous effects can be neglected in large regions of the fluid remote from the particle, while they are taken into account in a thin layer adjacent to the interface. Such a boundary layer is schematically illustrated in Figure 6, where y indicates the coordinate normal to the interface and x the one tangent to the interface, with x = 0 at the front stagnation point. Moreover, R(x) indicates the distance of the interface from the center of the particle, and v_t and v_n the velocity components tangential and normal to the interface, respectively.

Usually, because of the boundary-layer thinness, the derivatives with respect to the streamwise coordinate x can be neglected relative to those in the transverse direction y. Under this approximation, Equations 1 and 3 within the boundary layer reduce to

$$v_x \frac{\partial v_x}{\partial x} + v_y \frac{\partial v_x}{\partial y} = v^\infty \frac{dv^\infty}{dx} + \nu \frac{\partial^2 v_x}{\partial y^2} \tag{22}$$

$$\frac{\partial (v_x R)}{\partial x} + \frac{\partial (v_y R)}{\partial y} = 0 \tag{23}$$

with the boundary conditions

$$v_x = 0 \text{ at } y = 0 \tag{24a}$$

$$v_x = v^\infty, \frac{\partial v_x}{\partial y} = \frac{\partial^2 v_x}{\partial y^2} = 0 \text{ at } y = \delta \tag{24b}$$

where v^∞ represents the fluid velocity at the edge of the velocity boundary layer, and δ is the boundary layer thickness, given by

$$\delta = \left(\frac{90\ \nu}{v^{\infty 8.5} R_p^2} \int_0^x R^2 v^{\infty 7.5} dx \right)^{1/2} \tag{25}$$

By solving this system of equations the velocity field around the particle can be determined. In particular, the following fourth-order approximation for the tangential velocity component can be derived:

$$(v_x/v^\infty) = 4(y/\delta) - 6(y/\delta)^2 + 4(y/\delta)^3 - (y/\delta)^4 \tag{26}$$

Estimation of Terminal Velocity

Knowledge of particle velocity is of fundamental importance for the description of mass transfer processes. For a particle moving through an infinite medium with constant velocity, a global force balance leads to the following definition of the drag coefficient:

$$C_D = \frac{4\ \Delta\rho\ g\ d_e}{3\rho\ v_t^2} \tag{27}$$

Equation 27 is valid for a spherical particle. In the case of an ellipsoidal particle it is convenient to replace the diameter of the sphere with the length of the ellipsoid axis parallel to the line of motion. Its evaluation requires the knowledge of the ellipsoid aspect ratio E as a function of the equivalent sphere diameter, as given in Figure 3.

Recommended relationships for the evaluation of the terminal velocity, or equivalently of the drag ratio, will be given in the following two different situations: low and high Reynolds numbers. Wall effects are neglected here, thus referring to an infinite continuous phase, and will be considered in detail in the next section.

Low Reynolds Number: Re < 1

In the viscous flow regime the drag coefficient for a liquid drop can be evaluated from the Hadamard-Rybczynski model, Equation 17, for any value of the viscosity ratio $\varkappa$. This predicts values of the terminal velocity up to 50% higher than those given by the Stokes law (Equation 18) for a rigid drop ($\varkappa = \infty$).

However, from most experimental measurements it appears that the terminal velocity of both solid and fluid particles, particularly of small diameter, is very close to the value predicted by the Stokes law, regardless to the viscosity ratio $\varkappa$.

This is most probably due to the presence of surface-active substances which tend to accumulate at the interface between the two fluids [11, 12]. Due to the drop movement such substances migrate towards the rear region of the drop, where they accumulate, while the front region remains almost uncontaminated. The resulting concentration gradient leads to a corresponding gradient of surface tension and then to a tangential stress which retards surface motion. Since the purfication required to fully eliminate such contaminates is extremely stringent, it should be assumed that surface active components are present in most fluids of practical interest. This finding makes quite difficult a rigorous estimation of the terminal velocity of fluid particles, because in most cases neither the amount nor the type of the present contaminants can be exactly specified. However, Equations 17 and 18 constitute a close estimation of the upper and lower bounds of the terminal velocity, as they differ at most by 50%.

In order to account for the effect of contaminants a "degree of circulation" Z can be defined [13, 14]:

$$v_t = v_{tr} (1 + Z/2) \tag{28}$$

where v_{tr} indicates the Stokes terminal velocity, and Z can be evaluated as

$$Z = [2/(2 + 3\varkappa)] [2(Y - 1)] \tag{29}$$

The parameter $Y = v_t/v_{tr}$ in Equation 29 accounts for surface effects so that for $Y = 3/2$, Equation 28 reduces to the Hadamard-Rybczynski equation (i.e., pure components), while for $Y = 1$ it reduces to the Stokes law (i.e., no circulation).

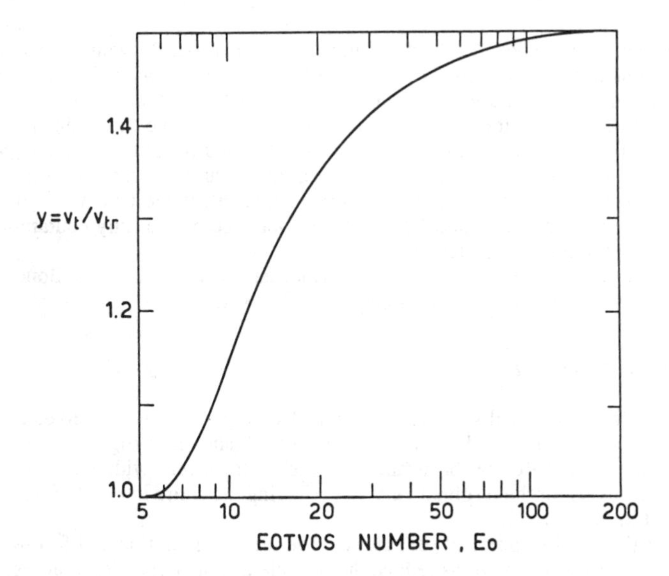

Figure 7. Savic correlation for the evaluation of the terminal velocity as a function of the Eotvos number.

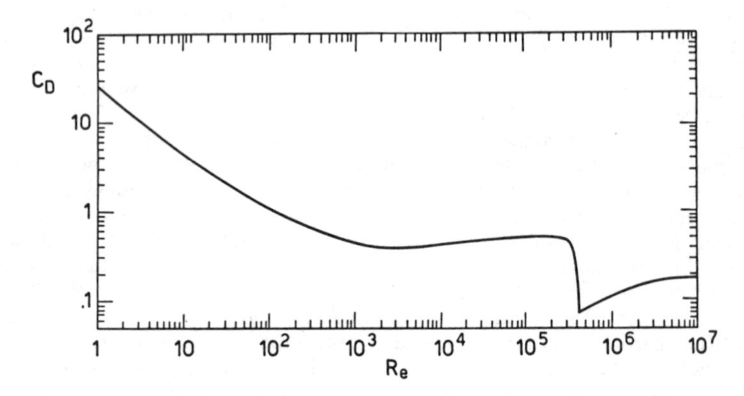

Figure 8. Drag coefficient of a rigid sphere as a function of Reynolds number.

Table 1
Correlation for the Evaluation of Terminal Velocities of Rigid Spheres [2]

Correlation	Range
$\text{Re} = B/24 - 1.7569 \times 10^{-4} B^2 + 6.9252 \times 10^{-7} B^3 - 2.3027 \times 10^{-10} B^4$	$\text{Re} \leq 2.37$
$\text{Log}_{10} \text{Re} = -1.7059 + 1.33438 W - 0.11591 W^2$	$2.37 < \text{Re} \leq 12.2$
$\text{Log}_{10} \text{Re} = -1.81391 + 1.34671 W - 0.12427 W^2 + 0.006344 W^3$	$12.2 < \text{Re} \leq 6.35 \times 10^3$
$\text{Log}_{10} \text{Re} = 5.33283 - 1.21728 W + 0.19007 W^2 - 0.007005 W^3$	$6.35 \times 10^3 < \text{Re} \leq 3 \times 10^5$

Note: $Re = v_t d/\nu$; $B = 4g\Delta\rho\, gd^3/3\mu^2$; $W = Log_{10}B$.

Several attempts have been made in the literature to correlate the value of Y to the type and amount of contaminants [15, 16]. When this information is not available, as in most industrial applications, the correlation of Savic [17], as a function of the Eotvos number, shown in Figure 7, can be adopted. It describes the limiting case where the contaminant concentration is so large that the surface tension goes from its value relative to the pure component at the front of the drop to zero at the rear. Thus, for a given fluid (characterized by a known Eo value) with intermediate but unknown contaminant content the true value of Y is upper bounded by the pure system value 1.5 and lower bounded by the fully contaminated system value given by the Savic correlation.

A simple, although quite approximate criterion, has been proposed by Bond and Newton [18]. It states that internal circulation occurs at $Eo > 4$.

High Reynolds Number: Re > 1

The evaluation of terminal velocity of spherical solid particles has received great attention in the literature. The so-called "standard drag curve" shown in Figure 8, where C_D is reported as a function of Re, has been fully established for a very wide interval of Re values. Explicit expressions for the terminal velocity, covering the entire interval of Re numbers, are summarized in Table 1 [2].

At large values of Re spherical particles exhibit a rocking motion and follow a spiral trajectory. These secondary motions reduce the particle terminal velocity to an extent which is inversely proportional to the density ratio γ. Using the data reported by Sheth [19], Clift et

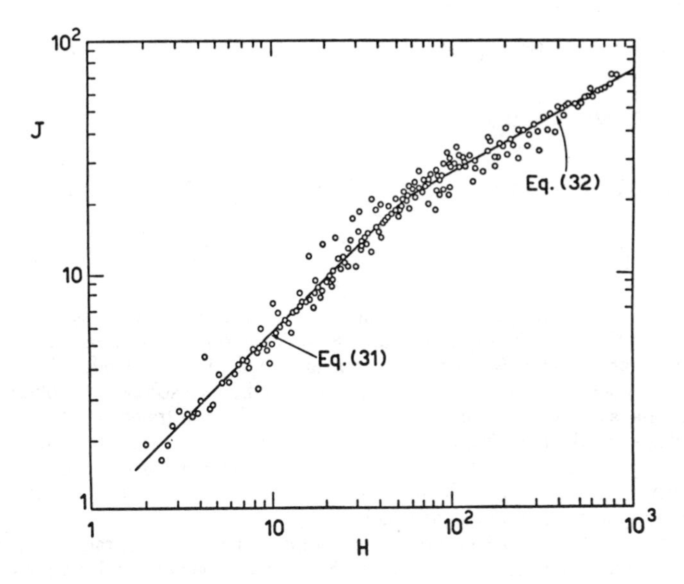

Figure 9. Schematic illustration of the comparison of Equations 31 and 32 with experimental data relative to liquid drops collected by various authors.

al. [2] proposed the following relationship for evaluating the C'_D value, corrected for the effect of secondary motion, as a function of the C_D value reported in Figure 8:

$$C'_D = C_D [1 + 0.13/(2.8\gamma - 1)] \quad (30)$$

which applies for $10^3 < Re < 2 \times 10^5$ and $\gamma \geq 1$.

In the case of fluid drops, contaminants play a major role also at high Re. Several theoretical investigations have been reported to investigate the behavior of fluid drops at high Re, under the assumptions of spherical shape and no surface-agent adsorbed at the interface. These are either based on the numerical solution of the equations of motion or on the application of the boundary-layer theory (see Clift et al. [2] for a detailed analysis). However, due to the adopted assumptions, these equations can only be applied to absolutely pure systems and to strict spherical geometries. Since both the deformation from the spherical shape and the presence of contaminants tend to increase the drag coefficient, it follows that the terminal velocities predicted by the previously mentioned theories constitute an upper limit for the observed values. This situation is similar to that encountered in the low Re regime for the Hadamard-Rybczynski solution.

The most convenient approach for developing terminal velocity correlations valid for contaminated fluids is then to empirically correlate the large body of data collected for various fluids. The most reliable correlation is shown in Figure 9, and can be analytically represented as follows:

$$J = 0.94\, H^{0.757} \text{ for } 2 \leq H \leq 59.3 \quad (31)$$

$$J = 3.42\, H^{0.441} \text{ for } H > 59.3 \quad (32)$$

$$\text{where } H = \frac{4}{3}\,\text{Eo}\,M^{-0.149}\,(\mu/\mu_w)^{0.14} \tag{33a}$$

$$J = \text{Re}\,M^{0.149} + 0.857 \tag{33b}$$

and $\mu_w = 0.0009$ Kg/m-s indicates water viscosity. The validity of this relationship is restricted to

$$M < 10^{-3},\ \text{Eo} < 40,\ \text{Re} > 0.1 \tag{34}$$

and to infinite continuous media. The value of the drop terminal velocity can be explicitly evaluated from Equations 31 or 32, and the r.m.s. deviations between experimental and calculated values are about 15% for the first one and 11% for the second.

It should be noted that ellipsoidal drops usually exhibit two secondary motions. The first one is the same as spherical rigid particles: rocking and spiral trajectory; the second one is given by shape oscillations. The occurrence of such oscillations is most probably related to wake shedding, because onset of oscillations is usually coincident with the onset of vortex shedding from the wake. For contaminated drops oscillations occur at about Re = 200, while in pure systems they are significantly delayed.

The general effect of secondary motions is to improve mass transfer rates to and from fluid particles and to increase the drag coefficient. Namely, the experimental v_t versus Re and C_D versus Re curves exhibit a maximum and a minimum point respectively, which corresponds approximately to the onset of oscillations.

Finally, it is worthwhile mentioning the case of drops characterized by an ellipsoidal-spherical-cap shape, which corresponds to Eo > 40 and Re > 1.2 as shown in Figure 2. This situation often occurs for very large bubbles and drops of the order of 3 cm^3. It has been already noted that this cannot be the case of drops falling in gases, due to their limited critical diameter value, given by Equation 8. On the whole the following relationships can be recommended:

1. Ellipsoidal-cap drops, 1.2 < Re < 40 [20,21]

$$v_t = f\,(g\,b\,\Delta\rho/\rho)^{1/2} \tag{35}$$

   where b is the vertical semiaxes of the ellipsoid and f is given as a function of eccentricity $e = (1 - E^2)^{1/2}$ as follows:

$$f = (1/e^3)\,[\sin^{-1} e - e(1 - e^2)^{1/2}] \text{ for oblate ellipsoid}$$

$$f = [(1/e^2)^{1/2}/e^3\,[e - (1 - e^2)\tanh^{-1} e] \text{ for prolate ellipsoid}$$

2. Spherical-cap drops, 40 < Re < 150 [22]

$$v_t = \frac{2}{3}\,(g\,R_{sc}\,\Delta\rho/\rho)^{1/2} \tag{36}$$

   where R_{sc} is the spherical-cap radius.

3. Spherical-cap drops, Re > 150 [23]

$$v_t = 0.792\,(g V_{sc}^{1/3}\Delta\rho/\rho)^{1/2} \tag{37}$$

   where V_{sc} indicates the spherical-cap volume.

Wall Effects

The movement of solid and fluid particles usually takes place in limited containers, whose walls may affect the field motion. They should be accounted in solving the equations of motion Equations 1 and 3 by replacing the condition of uniform flow remote from the particle, with the condition relative to the presence of an immobile rigid surface at a definite distance from the moving particle.

The effect of the walls can be expressed in terms of the ratio between the actual drag coefficient and that relative to the same system but with infinite continuous phase, $C_{D\infty}$:

$$\zeta = C_D/C_{D\infty} \tag{38}$$

In the following the case of a particle moving along the axis of a cylindrical tube is considered. The drag coefficient ratio is reported separately for solid and fluid particles, limiting the analysis to the case of $\lambda = d_e/d_t < 0.6$. For larger particles the slug flow regime is established. Its study is beyond the scope of this chapter.

Solid Particles

For a spherical rigid particle in the creeping flow regime ($Re < 1$) the empirical relationship proposed by Francis [24] is recommended

$$\zeta = [(1 - 0.475\lambda)/(1 - \lambda)]^4 \tag{39}$$

For higher Re, the values of C_D and of the drag factor ζ are plotted in Figure 10 as a function of Re and the diameter ratio λ. Analytical expressions approximately simulating the reported curves are [25]:

$$\zeta = 1 + (24/ReC_{D\infty})(\zeta_v - 1) \text{ for } 0.1 < Re < 80 \tag{40}$$

where ζ_v refers to the value of ζ in viscous flow, as given for example by Equation 39 [2, 26]:

$$\zeta = 1/(1 - 1.6\lambda^{1.6}) \text{ for } 80 < Re < 10^4 \tag{41}$$

and

$$\zeta = (1 + 1.45\lambda^{4.5})/(1 - \lambda^2)^2 \text{ for } Re > 10^5 \tag{42}$$

Fluid Particles

Fluid particles moving in small cylinders maintain the spherical shape up to Re values larger than those characteristic of fluid particles moving in infinite media (see Figure 2). In fact, drops moving through infinite media tend to be flattened in the vertical direction, while the cylinder walls tend to cause elongation.

Some useful general criteria for wall effects to be negligible (less than 2%) can be given as follows:

$$\begin{aligned} \lambda &\leqslant 0.06 && \text{for } Re < 0.1 \\ \lambda &\leqslant 0.08 + 0.02 \log_{10} Re && \text{for } 0.1 < Re < 100 \\ \lambda &\leqslant 0.12 && \text{for } Re \geq 100 \end{aligned} \tag{43}$$

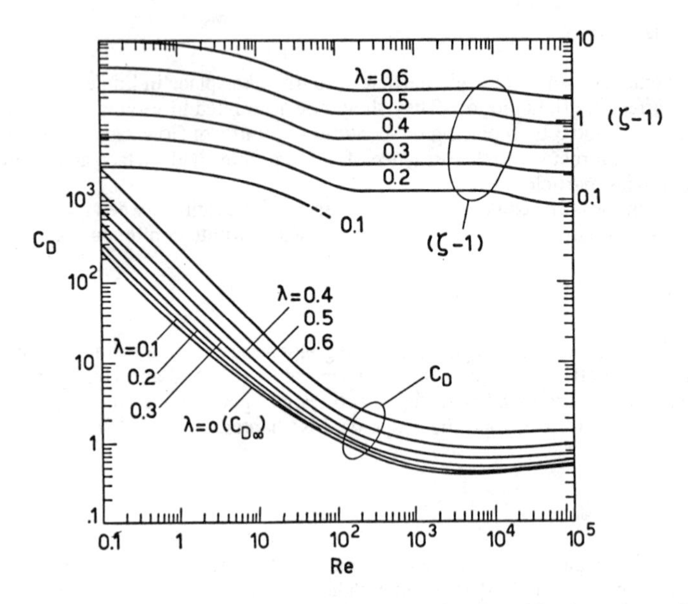

Figure 10. Drag coefficient C_D and drag factor $\zeta = C_D/D_\infty$ as a function of Re, for various values of the diameter ratio $\lambda = d_e/d_t$.

If the wall effects are present, they can be estimated using the same prior correlations reported for rigid particles, because the presence of contaminants components highly reduces internal circulation.

Thus, in the creeping flow regime (Re < 1), Equation 39 can be used; similarly, for intermediate size drops (Eo < 40 and 1 < Re < 200) the values plotted in Figure 10 are recommended. For higher Re (Eo < 40 and Re > 200) the terminal velocity correction factor becomes independent of Re, and can be estimated through the relationship

$$v_t/v_t\infty = (1 - \lambda^2)^{3/2} \tag{44}$$

For the case of large drops (Eo > 40), the relationship proposed by Wallis [27], although originally developed for gaseous bubbles, can be used:

$$v_t/v_t\infty = 1.13 \exp(-\lambda) \quad \text{for } 0.125 \leqslant \lambda < 0.6 \tag{45}$$

MASS TRANSFER

The mass balance of the i-th component in an incompressible fluid and in the absence of chemical reaction, can be written as follows

$$\frac{DC_i}{Dt} = \frac{\partial C_i}{\partial t} + \nabla \cdot (\underset{\sim}{v}\, C_i) = D_i\, \nabla^2 C_i \tag{46}$$

where mass transport is assumed to occur both through convective transport due to fluid flow and through mass diffusion. The latter process is described by means of the diffusion coefficient D_i, which has been assumed constant, i.e. independent of the mixture composition. Moreover, it is assumed that the mass flux does not alter the fluid motion. The assumption allows one to decouple the mass transfer equations from those of fluid motion. In particular, the mass balance Equation 46 is usually solved using the fluid velocity field given by the Navier-Stokes equations as described in the previous section. These approximations are quite accurate in the case of dilute solutions, while for diffusion in multicomponent systems with high mass fluxes specific analyses should be applied [28–30].

In dimensionless form, Equation 46 reduces to

$$\frac{DC_i^*}{Dt^*} = \frac{1}{Pe} \nabla^2 C_i^* \tag{47}$$

$$\text{where } C_i^* = (C_i - C_i^0)/(C_i^i - C_i^0) \tag{48}$$

and the Peclet number is defined as

$$Pe = \frac{L_o v_o}{D_i} = Re\ Sc \tag{49}$$

In general, Equation 47 must be applied to both the continuous and the dispersed phases. Usually, the initial and boundary conditions are given by uniform initial concentration in both phases, uniform concentration in the continuous phase remote from the particle and uniform concentration at the interface. Moreover, the other two conditions are applied at the interface: the two phases are in equilibrium, so that

$$C_{pi}^i = H_i C_i^i \tag{50}$$

and the mass flux across the interface is continuous

$$D_{pi} \frac{\partial C_{pi}}{\partial r} = D_i \frac{\partial C_i}{\partial r} \qquad \text{at } r = 0 \tag{51}$$

where the subscript p refers to the dispersed phase and r is the direction normal to the interface.

The mass transfer process in systems involving a dispersed phase is intrinsically transient due to solute accumulation or depletion in the fluid particles. For this reason the general solution of the problem is not attainable, and suitable approximations must be introduced in order to derive simple solutions valid under specific situations of particular practical interest.

First of all, when the resistance to mass transfer is concentrated in one phase, the solute concentration in the other one can be assumed to be uniform. Rough criteria for the relative importance of the external and internal mass transfer resistances are reported by Clift et al. [2]. Internal resistance is negligible, and thus the concentration within the particle is uniform (i.e., $C_{pi} = H_i C_i^i$) when

$$H(D_p)/D)^{1/2} > > 1 \quad \text{, for short times} \tag{52a}$$

$$H(D_p/D) > > Sh \quad \text{, for long times} \tag{52b}$$

If conditions (52a) and (52b) are reversed, then internal resistance controls and the concentration outside the particle is uniform (i.e., $C_i^i = C_i^\infty$).

In these two cases, the transport equation (Equation 46) needs to be solved only in one phase. If the resistances in the two phases are comparable, they can be both accounted using the additivity rule of individual phase resistances. King [31] has pointed out that this approximation implies that the ratio (Hk_p/k) is constant at all points of the interface. Usually, in the situations encountered in practice the individual resistance additivity rule is sufficiently accurate for engineering purposes.

Another simplification which can be quite useful in practical applications is the assumption of pseudosteady-state conditions for the transport processes in the continuous phase. It consists in neglecting the time derivative term in Equation 46, so highly simplifying its solution. Lochiel and Calderbank [32] suggested that such conditions are established as soon as the mobile-interface particle travels through a distance equal to its own equivalent diameter; for a rigid-interface particle the corresponding distance is about ten diameters.

By comparing the characteristic times of the diffusion processes in the external and internal phases, it can be seen that pseudosteady-state conditions are established in the continuous phase when

$$H\,Sh >> 1 \tag{53}$$

In the following the mass transfer process will be examined separately in the continuous and in the dispersed phase (except for a few particular cases). Moreover, since Equation 53 is likely to be satisfied in most applications, pseudosteady-state conditions will be assumed in the continuous phase.

An approximation of the full transient problem can be obtained only for very short times. Namely, in this case the fluid motion in the two phases does not affect the mass transfer process, which then occurs as if it were between two immobile semi-infinite media. This leads to

$$F = \frac{6(\tau_p/\pi)^{1/2}}{1 + H(D_p/D)^{1/2}} \tag{54}$$

where F is the extraction efficiency (although it applies for transfer both into and out of the particle), defined as

$$F = 1 - \overline{C_i^*} \tag{55}$$

where $$\overline{C_i^*} = \int_{V_p} C_i^* dV_p/V_p \tag{55a}$$

is the particle volume average value of the dimensionless concentration C_i^*, defined by Equation 48.

External Resistance

Similarly as in the case of the momentum boundary layer, as well as for the mass transfer process, the thin-concentration boundary-layer approximation can be applied for large Peclet number values. In this case, the concentration derivative parallel to the particle surface can be neglected with respect to that normal to the surface, because the entire concentration gradient is concentrated in a very thin layer adjacent to the particle surface. With reference to the generic axisymmetric body of revolution shown in Figure 6, and assuming steady-state conditions, Equation 46 can be rewritten as follows

$$v_x \frac{\partial C_i}{\partial x} + v_y \frac{\partial C_i}{\partial y} = D_i \frac{\partial^2 C_i}{\partial_y{}^2} \tag{56}$$

with boundary conditions

$$C_i = C_i^I \quad \text{at} \quad y = 0$$

$$C_i = C_i^{\infty} \quad \text{at} \quad y = \infty$$

$$C_i = C_i^{\infty} \quad \text{at} \quad x = 0$$

As mentioned, the thin concentration boundary-layer approximation can be applied for large Pe, which means large Sc values for any given finite Re. Thus, this approximation can be applied also in the case of moderately low Re, where the momentum boundary-layer approximation is not reliable. Moreover, in practice the thin-concentration boundary layer gives satisfactory results down to Sc equal to about one, even at moderately low Re.

In the case of Sc > > 1, the concentration boundary layer is much thinner than the velocity one, so that a linear profile can be assumed for the tangential velocity v_x along the direction normal to the particle interface

$$v_x = v_x^0 + v_x' y \tag{57}$$

This relationship can be substituted in the continuity Equation 23 to evaluate the normal velocity v_y, which then allows calculation of the concentration profile and mass fluxes around the particle through Equation 56.

The mass flux across the element of area d A is given by

$$d\,N = k_1(x)\,(C^{\infty} - C^i)\,d\,A = D\left(\frac{\partial C}{\partial y}\right)_{y=0} d\,A \tag{58}$$

which leads to the following definition of the local Sherwood number:

$$Sh_1(x) = \frac{k_1(x) d_e}{D} = 2\left(\frac{\partial C^*}{\partial \xi}\right)_{\xi=0} \tag{59}$$

where $\xi = 2y/d_e$.

The average Sherwood number and mass transfer coefficient can be readily derived from the local values through integration over the total surface area of the particle. This leads to

$$Sh = -\frac{A_e}{A}\int_0^{\zeta_m}\left(\frac{\partial C^*}{\partial \zeta}\right)_{\zeta=0} R^*\,d\,\zeta \tag{60}$$

where $\zeta = 2x/d_e$, ζ_m is the maximum value of ζ, $R^* = 2R/d_e$ and A_e represents the external surface area of the equivalent sphere. The evaluation of the average Sherwood number through Equation 60 requires the knowledge of the interfacial concentration gradient, which follows directly from Equations 23, 56 and 58. An analytical solution is available only for two limiting physical situations: rigid and rapidly moving interface [32].

Rigid Interface

From the well known no-slip condition at the interface it follows that $v_x^0 = 0$. With this simplification Equations 23, 57, and 58 can be solved leading to

$$Sh = 0.641\,\frac{A_e}{A}\,Pe^{1/3}\left(\int_0^{\zeta_m}\left(\frac{v_x' d_e R}{2\,v_p}\right)^{1/2} R^*\,d\,\zeta\right)^{2/3} \tag{61}$$

More generally, Equation 61 applies when

$$v_x^0 << v_x' y/2 \qquad (62)$$

This, recalling that the largest possible value of y is the penetration depth of the solute into the solvent $\delta = d_e/Sh$, reduces to

$$Sh << v_x' d_e / 2v_x^0 \qquad (63)$$

Rapidly Moving Interface

This situation occurs when the interface velocity is larger than the particle translation velocity, so that the condition in Equation 62 is reversed, i.e.

$$v_x^0 >> v_x' y/2 \qquad (64)$$

Thus, the tangential velocity component in the entire concentration boundary layer can be approximated as $v_x \simeq v_x^0$, which substituted in Equations 23, 58, and 60 leads to

$$Sh = \left(\frac{2}{\pi}\right)^{1/2} \frac{A_e}{A} Pe^{1/2} \left(\int_0^{\zeta_m} \frac{v_x^0}{v_p} R^{*2} \, d\zeta\right)^{1/2} \qquad (65)$$

On the whole, the evaluation of Sh for the two examined situations requires simply the evaluation of two integrals depending on the particular geometry and flow dynamics under examination. However, it can be noted that the Sh number is a function of the Pe number raised to a power equal to $1/3$ and $1/2$ for rigid and mobile interfaces, respectively. For partly moving interfaces the exponent of Pe should then be expected to lie between $1/3$ and $1/2$.

It is worthwhile pointing out that Equations 61 and 65 cannot cover all possible physical situations because they have been derived under limiting conditions. Thus, in order to develop relationships, for the evaluation of Sh which can be applied in a wider range of operating conditions, other considerations must be introduced, such as matching of asymptotic solutions and semi-empirical correlations of experimental data. The following outlines recommended relationships which cover most situations occuring in practice will be presented and briefly analysed.

Creeping Flow (Re < 1)

For a rigid sphere immersed in a creeping flow, the following equation represents the numerical solution of the mass balance Equation 47, coupled to the Stokes velocity field, with errors which never exceed 2%

$$Sh = 1 + (1 + Pe)^{1/3} \qquad (66)$$

Equation 66 leads to the well known limit for stagnant continuous phase, i.e., Sh = 2 as Pe → 0. On the other hand as Pe → ∞, Equation 66 closely approaches the equation which can be derived from Equation 61 by evaluating the velocity field around the sphere through the Stokes flow:

$$Sh = 0.991 \, Pe^{1/3} \qquad (67)$$

For a fluid sphere, no general solution valid for all Pe and $\varkappa$-values is available in analytic form. However, a close approximation of Sh as a function of Pe is reported in Figure 11 for

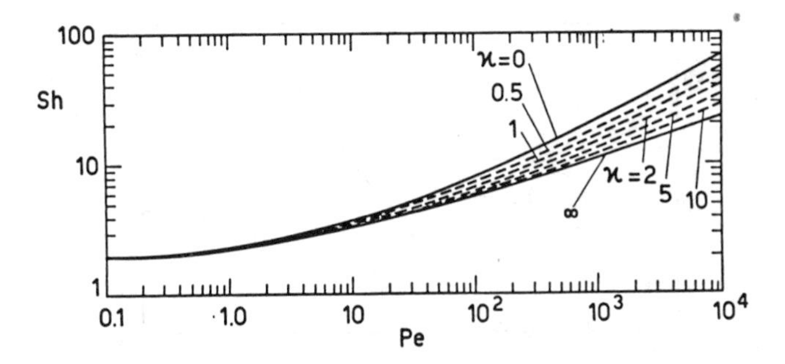

Figure 11. External Sh for fluid spheres in creeping flow regime, as a function of Pe.

various values of the viscosity ratio $\varkappa$. These curves are recommended for predicting Sh. It is noticeable that the curves shown in Figure 11 indicate that as $\varkappa$ increases, the variation of Sh at high Pe changes from $Pe^{1/2}$ to $Pe^{1/3}$. This agrees with Equations 61 and 65 for rigid and circulating particles, respectively.

The asymptotic solution for a circulating particle as $Pe \to \infty$, can be obtained from Equation 65 by evaluating the velocity field around the particle through the Hadamard-Rybczynski model. This procedure leads to

$$Sh = 0.651\ Pe^{1/2}/(1 + \varkappa)^{1/2} \tag{68}$$

which obviously reproduces quite closely the values plotted in Figure 11 in the region of large Pe.

The effect of contaminants on mass transfer can be roughly estimated using Savic's approach previously described for the evaluation of the terminal velocity. The value of Sh is given by the sum of two contributions: the first one, Sh_M, relates to the portion of the spherical particle which is assumed fully circulating, while the second one, Sh_s, refers to the remaining portion of the sphere which is assumed rigid:

$$Sh = Y_M Sh_M + Y_S Sh_S \tag{69}$$

The two contributions Sh_M and Sh_S are evaluated using the data plotted in Figure 11 and Equation 66, respectively. The weighting factors Y_M and Y_S can be estimated as

$$Y_M = [3(Y - 1)/Y]^{1/2} \tag{70}$$

and

$$Y_S = 1 - [(\theta_0 - \sin\theta_0 \cos\theta_0)/\pi]^{1/2} \tag{71}$$

where the velocity ratio Y is calculated as a function of Eo, from the curve shown in Figure 7, while the angle θ_0, which indicates the angle excluded by the rigid portion of the spherical particle, can be evaluated from the curve shown in Figure 12. This equation allows to evaluate an upper bound for Sh in contaminated fluid particles. The lower bound is given by the Sh value for rigid particles, which is attained when the amount of surface active agents is such that all internal motions are fully damped.

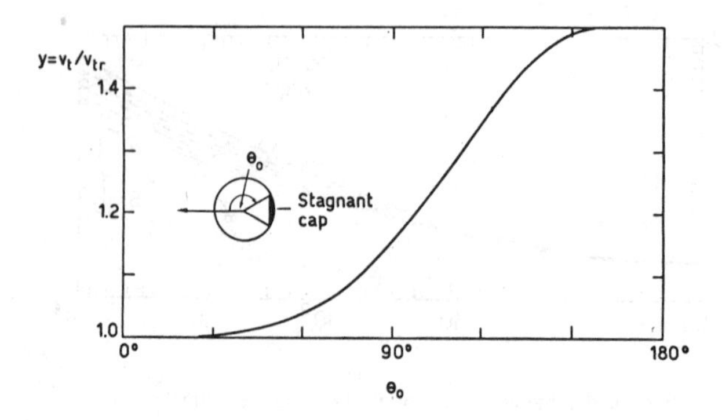

Figure 12. Savic correlation [17] for the effect of the angle excluded by stagnant cap θ_0 on terminal velocity.

High Re (Re > 1)

For rigid spherical particles a large number of experimental data are available. Accurate empirical correlations of such data in a wide range of Re values have been reported by Clift et al. [2]:

$$\mathrm{Sh} = 1 + [1 + (1/\mathrm{Pe})]^{1/3}\,\mathrm{Re}^{0.41}\,\mathrm{Sc}^{1/3} \qquad \text{for } 1 < \mathrm{Re} \leqslant 100 \tag{72a}$$

$$\mathrm{Sh} = 1 + 0.752[1 + 1/\mathrm{Pe}]^{1/3}\,\mathrm{Re}^{0.472}\,\mathrm{Sc}^{1/3} \qquad \text{for } 100 < \mathrm{Re} \leqslant 2 \times 10^3 \tag{72b}$$

$$\mathrm{Sh} = 1 + [1 + 1/\mathrm{Pe}]^{1/3}[0.44\,\mathrm{Re}^{1/2} + 0.034\,\mathrm{Re}^{0.71}]\mathrm{Sc}^{1/3} \qquad \text{for } 2 \times 10^3 < \mathrm{Re} < 10^5 \tag{72c}$$

It is noticeable that Sh tends to depend on Re and Sc raised to a power equal to $1/2$ and $1/3$, respectively.

These results can be confirmed by solving Equation 61, which derives from the thin-concentration boundary-layer approximation, using the velocity field calculated through the momentum boundary-layer approximation, as for example that described by Equation 26. However, accurate quantitative results are difficult to obtain through this procedure because of the impossibility of deriving an analytical solution of the velocity field around a rigid spherical particle moving at high Reynolds numbers. Note that the presence of $\mathrm{Re}^{0.71}$ term in Equation 72c, refers to the contribution of that portion of the sphere where fluid separation occurs, which is significant only at high Re.

For a freely circulating spherical fluid particle, Equation 65 can be solved using the velocity field derived by Abdel-Alim and Hamielec [33] from the numerical solution of the pertinent fluodynamics equations. The obtained expression of the Sh number, which can be applied to a wide range of $\varkappa$ and Re values, is given by:

$$\mathrm{Sh} = \frac{2}{\sqrt{\pi}}\,\mathrm{Pe}^{1/2}\left(1 - \frac{\dfrac{2 + 3\varkappa}{3(1 + \varkappa)}}{\left\{1 + \left[\dfrac{(2 + 3\varkappa)\,\mathrm{Re}^{1/2}}{(1 + \varkappa)(8.67 + 6.54\varkappa^{0.64})}\right]^{n}\right\}^{1/n}}\right)^{1/2} \tag{73}$$

where $n = 4/3 + 3\varkappa$. The values of Sh predicted by Equation 73 do not agree with the experimental data because of the deformation of the fluid particle geometry and the presence of impurities which affect the particle surface tension so reducing internal circulation.

In order to account for particle deformation, the expression derived by Lochiel and Calderbank [32] for an oblate spherical can be used:

$$Sh = \frac{2}{\sqrt{\pi}} Pe^{1/2} \left[\frac{8e^3E^{1/3}}{3(\sin^{-1}e - eE)}\right]^{1/2} \Bigg/ \left[1 + \frac{E^2}{2e} \ln\left(\frac{1+e}{1-e}\right)\right] \qquad (74)$$

where $e = (1 - E^2)^{1/2}$, and the aspect ratio E can be estimated from the data plotted in Figure 2. Moreover, the ratio between the external surface area of the spheroid and that of the volume equivalent sphere is given by Equation 10. It is noticeable that Equation 74 has been derived from Equation 65 using the velocity field around the particle as calculated through the inviscid fluid approximation. Such an approximation is expected to yield accurate results for freely circulating particles moving at high Reynolds numbers.

A closer inspection of Equation 74 reveals that the aspect ratio E does not affect the Sh value to a significant degree. More important is the effect of surface-active agents which significantly reduce internal circulation.

For contaminated systems the value of Sh should be equal to or above the value of rigid spherical particles given by Equation 72. However, it cannot exceed the value relative to pure systems, i.e., spherical particles freely circulating, given by Equation 73. Usually, knowledge of the lower and upper bounds allows even for the partially contaminated systems, one to obtain an adequate estimate of Sh.

The occurrence of secondary motion, such as oscillations, can affect the mass transfer rate to or from fluid particles. Namely, oscillations cause continuous changes of the particle external surface which require an alternate movement of the interior fluid towards the external surface and vice versa. Several models have been proposed to describe mass transfer to and from oscillating fluid particles. The more realistic attempts to describe the surface stretch during particle oscillation [34, 35] (which are examined in detail in the section devoted to internal mass transfer) lead to the same expression for the Sherwood number:

$$Sh_e = 2\nu d_e(f_N/D\pi)^{1/2} \qquad (75)$$

where Sh_e is the external surface area corrected Sherwood number, f_N is the natural drop frequency and ν is a function of the amplitude of the area oscillations ϵ, which depend on the specific model considered. In particular, $\nu = (1 + \epsilon + 3\epsilon^2/8)^{1/2}$ in the Angelo et al. [34] model, and $\nu = (1 + 0.687\epsilon)$ in the Brunson and Wellek [35] model. Clift et al. [2] proposed an empirical value of $\nu = 1.06$, which allows us to represent available experimental data with an average deviation equal to 6% for drops in liquids and 30% for drops in gases.

The natural drop frequency f_N appearing in Equation 75 can be estimated from Lamb's equation [36]:

$$f_N = \left[\frac{48\,\sigma}{\pi^2 d_e^3 \rho(2 + 3\gamma)}\right]^{1/2} \qquad (76)$$

An empirical correlation widely used in the case of oscillating drops has been proposed by Garner and Tayeban [37]:

$$\frac{k\,d_e}{D} = 50 + 0.0085\left(\frac{d_e v_p \rho}{\mu}\right)\left(\frac{\mu}{\rho D}\right)^{0.7} \qquad (77)$$

It is worthwhile pointing out that the effect of oscillations is more significant for the internal mass transfer resistance than for the external one. In particular, when

$$d_e f_N / v_t < 0.15$$

it can be assumed that oscillations do not affect external mass transfer.

Finally, the case of spherical-cap fluid particle deserves mention. As stated before, this particular shape is characteristic of large fluid particles (equivalent diameter larger than 1.8 cm). Assuming potential flow around the particle, Lochiel and Calderbank [32] derived the following relationship, valid for high Re:

$$\mathrm{Sh} = 1.79 \frac{(3E + 4E^3)^{2/3}}{1 + 4E^2} \mathrm{Pe}^{1/2} \tag{78}$$

where the aspect ratio E represents the ratio between the particle height and width.

Internal Resistance

Mass transfer within a fluid or solid sphere can be described by the transient mass balance inside the particle, which assuming constant diffusion coefficient, can be written in dimensionless form as follows:

$$\frac{\partial C_p^*}{\partial \tau_p} = \frac{1}{\phi^2}\left[\frac{\partial}{\partial \phi}\left(\phi^2 \frac{\partial C_p^*}{\partial \phi}\right) + \frac{1}{\sin\theta}\frac{\partial}{\partial \theta}\left(\sin\theta \frac{\partial C_p^*}{\partial \theta}\right)\right] - \frac{1}{2}\mathrm{Pe_p}\left[v_r^* \frac{\partial C_p^*}{\partial \phi} + \frac{v_\theta^*}{\phi}\frac{\partial C_p^*}{\partial \theta}\right] \tag{79}$$

where C_p^* is the dimensionless particle concentration defined by Equation 48 (with uniform initial concentration in the particle), and v_r^* and v_θ^* are the dimensionless components of the internal velocity field.

The usual initial and boundary conditions for Equation 79 are

$$C_p^* = 1 \qquad \text{at } \tau_p = 0,\ \phi \epsilon [0, 1],\ \theta \epsilon [0, \pi] \tag{80a}$$

$$\frac{\partial C_p^*}{\partial \theta} = 0 \qquad \text{at } \theta = 0, \pi,\ \tau_p > 0,\ \phi \epsilon (0, 1) \tag{80b}$$

$$\frac{\partial C_p^*}{\partial \phi} = 0 \qquad \text{at } \phi = 0,\ \tau_p > 0,\ \theta \epsilon (0, \pi) \tag{80c}$$

$$\frac{\partial C_p^*}{\partial \phi} = -\mathrm{Bi}\, C_p^* \ \text{at } \phi = 1,\ \tau_p > 0,\ \theta \epsilon (0, \pi) \tag{80d}$$

In the formulation of boundary condition (80d) it has been assumed that the concentration in the bulk continuous phase remote from the particle and the external mass transfer coefficient do not change during the transfer process. These assumptions are consistent with the pseudosteady-state approximation adopted in the preceding section to describe the external mass transfer process.

In the case of negligible mass transfer resistance in the external phase ($Bi \rightarrow \infty$), the boundary condition (80d) reduces to

$$C_p^* = 0 \tag{80e}$$

indicating that the concentration in the external phase, from the interface to the bulk, is constant and equal to C^∞.

The solution of Equation 79 requires the knowledge of the velocity field inside the particle (i.e., v_r^* and v_θ^*) which can be obtained a priori, due to the independency of the fluid motion equations from the mass transfer process as discussed at the beginning of this section. The solutions of Equations 79 and 80 will be presented in the next paragraph for the various possible fluid dynamic conditions inside the particle.

First of all, it is useful to indicate the significant overall parameters which identify the particle behavior and therefore deserve particular attention. To this aim it is convenient to introduce the following lumped model, which actually constitutes the definition of the overall mass transfer coefficient K_p:

$$V_p \frac{d\overline{C}_p}{dt} = - A_p K_p (\overline{C}_p - C_e) \tag{81}$$

where $\overline{C}_p$ is the particle volume average concentration

$$\overline{C}_p = \frac{1}{V_p} \int_{V_p} C_p \, dV_p \tag{82}$$

and $C_e = HC^\infty$ is the concentration value in equilibrium with the bulk external fluid phase.

Equation 83 may be solved through the introduction of the time average overall mass transfer coefficient

$$\overline{K}_p = \frac{1}{t_1} \int_0^{t_1} K_p \, dt \tag{83}$$

leading to the classical dimensionless expression for the extraction efficiency parameter defined by Equation 55:

$$F = 1 - \overline{C}_p^* = 1 - \exp\left(- \frac{3}{2} \overline{Sh}_p \tau_p\right) \tag{84}$$

where the time average Sherwood number $\overline{Sh}_p$ is based on the time average coefficient $\overline{K}_p$ and for nonspherical particles becomes $\overline{Sh}_{pe} = \overline{K}_p A_p d_e / D_p A_e$, so including the external surface area variation. Thus, it is convenient to report the results of the mass transfer models in terms of either F or $\overline{Sh}_p$.

When available, the solution of the model inclusive of a constant external resistance, i.e. with boundary condition (Equation 80d), will be presented. If this is not possible, then the problem without external resistance is considered, i.e. with boundary condition (Equation 80e), so that the time average internal mass transfer coefficient $\overline{k}_p$ is actually calculated. Such results can be extrapolated to cases involving a significant contribution of the external mass transfer resistance by evaluating the overall mass transfer coefficient through the additivity rule of individual phase resistances

$$\overline{K}_p^{-1} = \overline{k}_p^{-1} + Hk^{-1} \tag{85}$$

where the external mass transfer coefficient k can be estimated through an appropriate relationship among those reported in the previous paragraph under the assumption of steady-state conditions. In the following, two cases will be examined separately, depending on the occurrence of oscillating movements inside the particle which significantly affect the mechanism of mass transport.

Rigid Particles and Nonoscillating Drops

Rigid particles and stagnant drops. In the rigid sphere model the fluid inside the particle is considered stagnant, so that $v_\theta = v_r = 0$ (or $Pe_p = 0$) and Equation 79 can be solved analytically. When the external mass transfer resistance is negligible, the following expression of the extraction efficiency as a function of time has been obtained by Newman [38]:

$$F = 1 - \frac{6}{\pi^2} \sum_{n=1}^{\infty} \frac{1}{n^2} \exp\left(-n^2\pi^2\tau_p\right) \tag{86}$$

Vermeulen [39] has shown that this expression is accurately approximated by the empirical relationship

$$F = \left[1 - \exp\left(-\pi^2\tau_p\right)\right]^{1/2} \tag{87}$$

The extension of this model to the case involving a significant external resistance to mass transfer has been reported by Groeber [40]

$$F = 1 - 6 \sum_{n=1}^{\infty} A_n \exp\left(-\lambda_n^2\tau_p\right) \tag{88}$$

where A_n and λ_n are functions of the external mass transfer coefficient k. In Figure 13 the values of the extraction efficiency F as a function of dimensionless time are shown for various values of the Biot number.

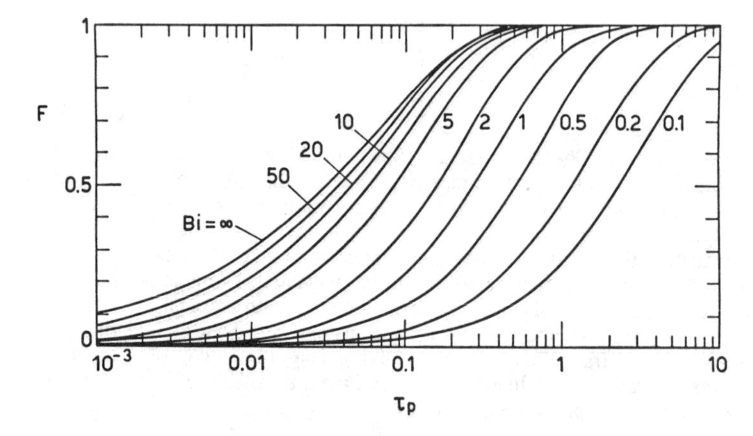

Figure 13. Extraction efficiency for a rigid particle ($Pe_p = 0$) for various external resistance values.

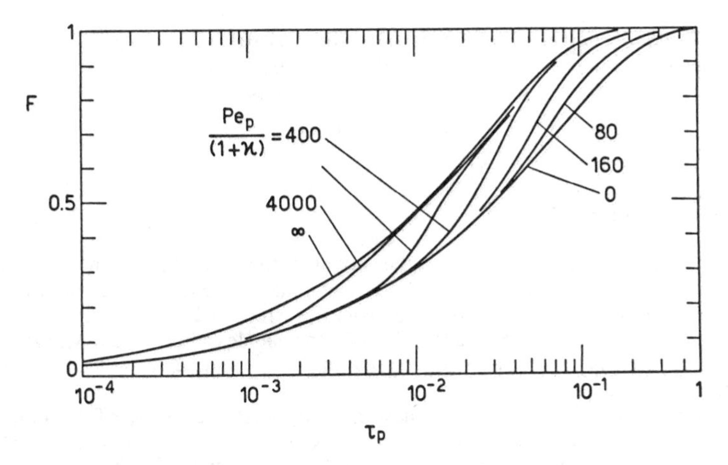

Figure 14. Extraction efficiency for a laminar circulating particle for various Peclet number values.

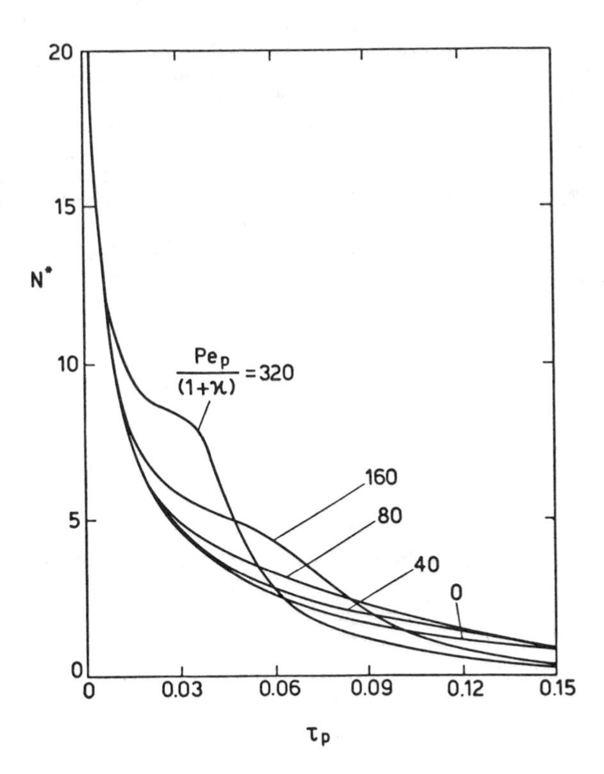

Figure 15. Dimensionless instantaneous mass flux N* as a function of dimensionless time τ_p.

Laminar circulating drops. In the case of a fluid particle moving in viscous regime (Re < 1), the internal motion can be described through the Hadamard and Rybczynski fluid dynamic model. In particular, the axially symmetric dimensionless velocity components are given by

$$v_r^* = -(1 - \phi^2)\cos\theta/2(1 + \varkappa) \tag{89a}$$

$$v_\theta^* = (1 - 2\phi^2)\sin\theta/2(1 + \varkappa) \tag{89b}$$

which substituted in Equation 79 allow for the solving of the general problem of mass transfer in a laminar-circulating spherical drop. The obtained system of equations cannot be solved analytically, thus numerical solutions have been reported in the literature [41]. In Figure 14 the obtained values of F as a function of dimensionless time are shown for various Pe_p values. It appears that as Pe_p increases the internal motions increase, so leading to larger rates of mass transfer.

Internal circulation leads to a peculiar behavior of the instantaneous mass flux at the interface during time. The values shown in Figure 15 exhibit, at sufficiently large Pe_p values, significant oscillations damped in time. These are due to the interaction between the diffusional and the internal convective fluxes. For high Pe_p, the fluid element which is initially located in the surface region quickly traverses the interior region of the drop, where it is enriched in solute (in the case of an extraction process), and reappears once again in the surface region. Such an element can now produce a mass flux of the solute larger than it would have been if there were no diffusion from the interior region of the drop. This leads to a maximum in the instantaneous total mass-flux-versus-time curve which disappears once the circulation of the internal fluid has occurred several times.

Johns and Beckmann [41] have shown that the time frequency between two maxima increases for increasing Pe_p, and that the first inflection point in the curves shown in Figure 15 occurs at a time value which approaches, as $Pe_p \to \infty$, the time required by the internal fluid to traverse the interior region of the drop. These observations are consistent with the qualitative justification of the oscillatory behavior just reported. The dimensionless concentration profiles in a laminar circulating drop calculated by Johns and Beckmann [41] are shown in Figure 16 for various values of the Peclet number.

The just-described model admits two limiting models of great utility. The first one, at $Pe_p \to 0$, is the rigid sphere model. The second one, at $Pe_p \to \infty$, refers to the case where internal circulation motions are fully developed. It has been presented by Kronig and Brink [42], and admits the following analytical solution:

$$F = 1 - \frac{3}{8}\sum_{n=1}^{\infty} B_n^2 \exp(-16\,\lambda_n \tau_p) \tag{90}$$

Table 2
Parameter values for the Kronig and Brink model (Equation 90)

n	B_n	λ_n
1	1.33	1.7
2	0.60	8.5
3	0.36	21.1
4	0.35	38.5
5	0.28	63.0
6	0.22	89.8
7	0.16	123.8

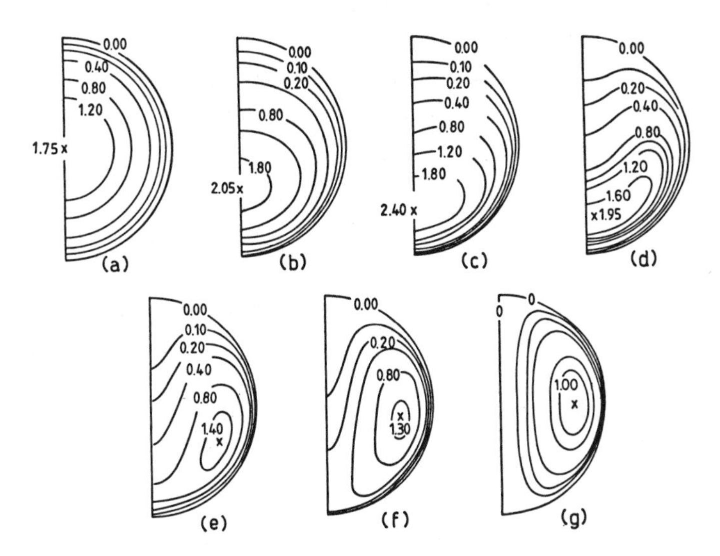

Figure 16. Evolution of the dimensionless concentration profile in a laminar circulating particle for increasing values of the Peclet number. $Pe_p/(1 + \varkappa) = 0$ (a), $= 20$ (b), $= 40$ (c), $= 80$ (d), $= 160$ (e), $= 320$ (f), $= \infty$ (g).

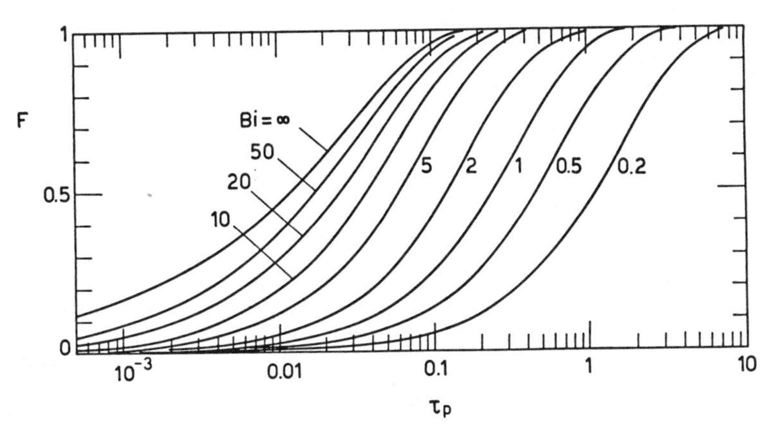

Figure 17. Extraction efficiency for a fully circulating particle ($Pe_p = \infty$) for various external resistance values.

where the values of B_n and λ_n, reported by Heertjes et al. [43], have been summarized in Table 2, for the case of negligible external mass transfer resistances.

Equation 90 can be closely approximated by the following empirical expression developed by Calderbank and Korchinski [44]:

$$F = [1 - \exp(-2.25\pi^2\tau_p)]^{1/2} \tag{91}$$

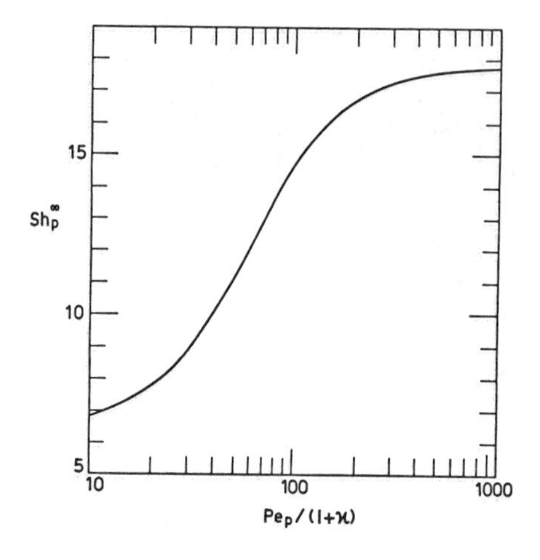

Figure 18. Asymptotic value of the instantaneous Sherwood number, Sh_p, as a function of the Peclet number characteristic of the laminar-circulating particle.

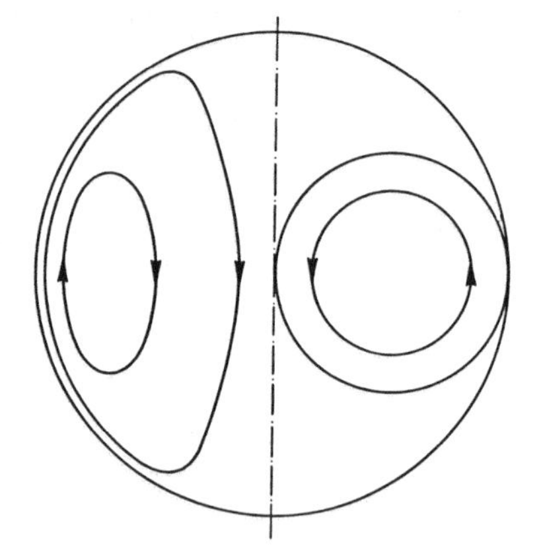

Figure 19. Schematic representation of the circulation pattern in drops (left) and of its approximation through a system of tori adopted in the Handlos and Baron model (right).

which compared with the analogous expression for the rigid sphere model (Equation 87) shows that the internal circulation described by the Hadamard and Rybczinsky model can be represented by a constant effective diffusivity equal to 2.25 times the molecular value.

The extension of Equation 90 to the case of finite external resistance has been developed by Elzinga and Banchero [45] leading to an expression identical to Equation 90. The values of the new parameters B_n and λ_n have been tabulated as a function of the external mass transfer coefficient.

The solution of the Kronig and Brink [42] model, for various values of the Biot number, is shown in Figure 17 in terms of the extraction-efficiency-versus-time curves. The curve indicated by Bi = ∞ identify with the original Kronig and Brink solution (Equation 90).

It is now convenient to briefly indicate when the two limiting models, rigid sphere and Kronig and Brink, can be used with confidence to replace the complete laminar-circulation model. At $Pe_p > 10^3(1 + \varkappa)$, the instantaneous mass flux is within 15% of the Kronig and Brink value. The asymptotic value of the instantaneous Sherwood number is shown in Figure 18 as a function of Pe_p. It appears that for the rigid model ($Pe_p = 0$) $Sh_p^\infty = 6.58$, while for the Kronig and Brink one ($Pe_p = \infty$) $Sh_p^\infty = 17.66$. Moreover, the Sh_p^∞ value for $Pe_p < 10 (1 + \varkappa)$ is within 15% of the rigid model value, and for $Pe_p > 250 (1 + \varkappa)$ is within 5% of the Kronig and Brink value. Note that as $\tau_p > 0.15$, the Sh_p value is very close to Sh_p^∞ for all Pe_p.

Finally, it is worthwhile pointing out that for very low dimensionless time values ($\tau_p < 10^{-3}$), the effect of the Peclet number becomes negligible and the solution obtained through the penetration theory, given by Equation 54, leads to a rather good approximation of the model for all Pe_p.

Turbulent circulating drops. This situation is characteristic of fluid particles in the high Reynolds number region. Handlos and Baron [46] have proposed a particular model of the turbulent internal motion, which takes into account the vibrations of the drop as well as the circulation pattern. The internal streamlines are approximated by the system of tori (as shown in Figure 19), whose average circulation time is estimated through the Kronig and Brink model. Moreover, random radial vibrations are superimposed upon the circulation movement, which cause strong mixing between different streamlines. This constitutes the core of the postulated eddy diffusion mechanism.

The mathematical treatment of these assumptions leads to a radial dependent expression of the effective diffusion coefficient

$$D_e = D_p \, 4.88 \times 10^{-4} \frac{Pe_p}{(1 + \varkappa)} (6 \bar{r}^2 - 8 \bar{r} + 3) \tag{92}$$

where $\bar{r}$ is the dimensionless radius of the considered terms. On the whole the solution of the Handlos and Baron model, in the case of negligible external limitations, may be expressed in the following series form:

$$F = 1 - 2 \sum_{n=1}^{\infty} D_n^2 \exp(-\delta_n \beta \tau_p) \tag{93}$$

$$\text{where } \beta = \frac{Pe_p}{512 (1 + \varkappa)} \tag{93a}$$

For very high time values, Handlos and Baron [46] have approximated Equation 93 by truncating the series at the first term with $D_1^2 = 1/2$ and $\delta_1 = 2.866$.

By direct comparison with the numerical solution of the model, Olander [47] proposed the following empirical relationship:

$$F = 1 - 0.64 \exp(-2.80 \beta \tau_p) \tag{94}$$

which gives accurate results in a larger region of dimensionless time values, approximately given by

$$\tau_p > 12.8/\beta \tag{94a}$$

A comparison of Equation 94 with the original approximation of Handlos and Baron shows that the major difference is in the coefficient of the exponential term (i.e., 0.64 instead of 1).

Wellek and Skelland [48] extended this solution to the case involving finite external mass transfer resistances, and reported the values of the first four δ_n as a function of k.

Patel and Wellek [49] solved numerically the complete Handlos and Baron model; the obtained results are shown in Figure 20 in terms of the extraction efficiency as a function of the time-dependent dimensionless parameter $\beta\tau_p$, for various values of the external mass transfer coefficient k, represented through the dimensionless parameter h defined as

$$h = 2 \, Bi/\beta \tag{95}$$

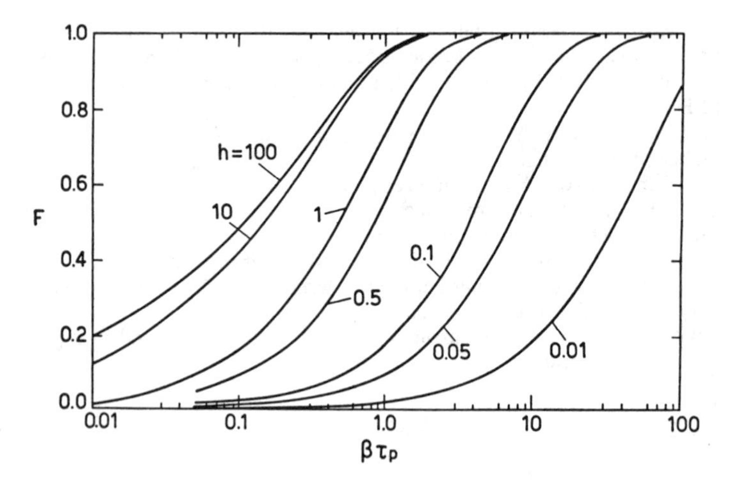

Figure 20. Extraction efficiency for turbulent circulating drops, according to Handlos and Baron model for various external resistance values ($h = 2\ Bi/\beta$).

Alternatively, the introduction of the external mass transfer resistance can be accomplished through the individual resistances addivitity rule (Equation 85), with k_p estimated from Equation 94 and k from one of the expressions reported in the preceding section.

A final consideration deserves mention about the use of the rigid sphere model with an effective diffusion coefficient $D_e = R\ D_p$ to simulate circulating particles. Calderbank and Korchinski [44] have shown that a fully circulating laminar drop can be simulated using a correction factor $R = 2.25$ for the diffusion coefficient (i.e., the internal circulation motion causes an effective increase of the molecular diffusion coefficient of 2.25 times). In the case of a turbulent circulating drop, Johnson and Hamielec [50] have estimated the average correction factor by averaging the effective diffusion coefficient value, given by Equation 92, along the streamline

$$R = \frac{< D_e >}{D_p} = 4.88 \times 10^{-4}\ Pe_p/(1 + \varkappa) \tag{96}$$

This indicates that the effective turbulent diffusion coefficient is directly proportional to the Peclet number.

Regions of applicability of the models. The rigid and laminar circulating models apply to solid particles and to fluid particles moving in the viscous regime, i.e., for $Re < 1$. In particular, the rigid sphere model can be used for fluid particles which do not circulate, as for $Pe_p \rightarrow 0$ or for impure systems where surface-active agents completely damp internal circulations even at quite large Pe_p. On the other hand, the laminar circulation model applies to cases where the drop is freely circulating.

A detailed model of the effect of surface-active agents on mass transfer inside drops has been presented by Huang and Kintner [51]. It is essentially based on the Savic stagnant-cap theory, where the drop is divided in two regions: one free of surfactant where internal circulation is present, and the other one containing so much surfactant that circulation is fully damped. This model gives accurate results when the instantaneous fall or rise velocity of the drop can be accurately measured.

At higher Re numbers (Re > 1), for fully circulating drop ($Pe_p \rightarrow \infty$), the Kronig and Brink model has been proved to give accurate predictions up to Re = 50 by Johnson and Hamielec [50], and even up to Re = 3,000 by Garner and Lane [52]. This finding can be explained noticing that, due to the prevailing effect of convective diffusion with respect to molecular diffusion at high Pe_p, the concentration profiles at high Re approach the streamlines of the Hill spherical vortex [53]. Since these have the same form of the Hadamard and Rybczynski solution in viscous flow, it follows that the rate of mass transfer at high Re is accurately predicted by the Kronig and Brink model.

Thus, in the case of ellipsoidal drops at high Re, the extraction efficiency can be calculated from the Kronig and Brink model, using the sphere equivalent diameter and evaluating the correct external surface area, as a function of the aspect ratio E, through the relationship reported in the fluodynamics section.

The Handlos and Baron model has been often used to describe mass transfer in turbulent and/or oscillating drops. However, Rose and Kintner [54], by means of photographic study of oscillating drops, showed that internal fluid motion approaches a type of random mixing with very little tendency for internal circulation. Thus, the Handlos and Baron model should be considered as applying for drops with vigorous internal circulation, that is in the high Reynolds number region, without oscillations.

Finally, it is worth mentioning the empirical relationship for nonoscillating drops proposed by Skelland and Wellek [55]:

$$\overline{Sh}_p = 31.4\ \tau_p^{-0.338}\ Sc_p^{-0.125}\ We^{0.371} \tag{97}$$

This equation has been derived by direct comparison with experimental data in the range 37 < Re < 546. It leads to a little improvement with respect to the Kronig and Brink model; namely, the average absolute deviation goes from 34% to 46%.

Oscillating drops

For the drop Reynolds number exceeding about 200, the drop may oscillate from an oblate (or spherical) to a more oblate shape. Such oscillations lead to a stretch of the interfacial surface and to a significant internal mixing which greatly enhances the rate of mass transfer.

In order to describe mass transfer inside oscillating drops it is necessary to account for the continuous change of the drop external surface area. Three models are reported which account for this phenomenon in a somehow different way. Next a critical comparison between these and other empirical models will be presented in order to define their accuracy in predicting experimental extraction efficiencies of free falling or rising oscillating drops.

Rose and Kintner model [54]. The fluid particle is assumed to oscillate from a nearly spherical to an oblate ellipsoidal shape and back to a spherical shape during one period of the oscillation. Moreover, the core of the drop is assumed to be well mixed, while all resistance to mass transfer is concentrated in a thin zone δ_x near the interface.

Assuming constant drop volume, the mass balance in the drop reduces to

$$\frac{d\overline{C}_p}{dt} = -\frac{\overline{D}}{\delta_x}\frac{A_p}{V_p}(\overline{C}_p - C_e) \tag{98}$$

which integrated leads to the following expression of the extraction efficiency:

$$F = 1 - \exp\left[-\frac{\overline{D}}{V_p}\int_0^{t_c}\frac{A_p}{\delta_x}\,dt\right] \tag{99}$$

where $\overline{D}$ is an average value between D_p and D weighted on the fraction of resistance in each phase, t_c indicates the residence time of the drop, and A_p is the external surface area which oscillates in time. Assuming an oblate ellipsoidal shape for the drop, the surface area A_p is given by

$$A_p = 2 \pi a^2 + \frac{\pi b^2}{\alpha} \ln \left(\frac{1 + \alpha}{1 - \alpha} \right) \tag{100}$$

where $\alpha^2 = (a^2 - b^2)/a^2$, and a and b are the two ellipsoid semiaxes. The first one is estimated assuming the following oscillation law:

$$a = a_0 + a_p \mid \sin \omega t \mid \tag{101}$$

where a_p is the oscillation amplitude and ω is one half the angular frequency of oscillation ω' as given by the Schroeder and Kintner [56] modification of Lamb's equation (Equation 76):

$$(\omega')^2 = \frac{8\sigma B}{d_e^3} \left[\frac{n(n + 1)(n - 1)(n + 2)}{(n + 1)\rho_p + n\rho} \right] \tag{102}$$

where $n = 2$ and $B = 0.805\ d_e^{0.225}$.

The second semiaxis b is then given by

$$b = 3V/4\pi a^2 \tag{103}$$

which derives from the assumption of constant drop volume.

The thickness of the interfacial resistance zone is estimated assuming that the zone volume remains constant during oscillation of the drop:

$$\delta_x = \frac{[a_0^2 b - (a - \delta_{x0})(b_0 - \delta_{x0})] - 2a\, b\delta_{x0} + b\delta_{x0}^2}{a^2 - 2a\, \delta_{x0} - \delta_{x0}^2} \tag{104}$$

where a_0, b_0, and δ_{x0} indicate the initial values. This last quantity is evaluated through the two-film theory as follows:

$$\delta_{x0} = \overline{D}/K_p \tag{105}$$

K_p is given by the additivity rule (Equation 85).

The local mass transfer coefficient in the continuous phase can be estimated from one of the just-reported relationships (for example, Equation 77), and that in the dispersed phase from the penetration theory

$$k_p = \frac{2}{\pi} (D_p \omega)^{1/2} \tag{106}$$

where the characteristic time has been taken equal to the time for one oscillation cycle.

By substituting Equations 100 through 106 in Equation 99 the extraction efficiency can be evaluated through the numerical calculation of the involved integral.

Angelo et al. model [34]. This model is based on the extension of the penetration theory to surfaces which undergo expansion or contraction during their lifetimes. Under usual approximations, such as negligible diffusional and convective transport parallel to the surface and short exposure times, the behavior close to any surface element may be considered one dimensional. Increasing surface area stretches the surface and causes a velocity normal to the surface which increases the rate of diffusion, and must be accounted in the mass balance of the solute species. It can be shown that such a velocity is given by

$$v_y = -y \frac{\partial \ln A_p}{\partial t} \tag{107}$$

where y is the direction normal to the surface and the derivative represents the local fractional rate of change of interfacial area for a moving surface element.

The penetration theory equation, modified by means of Equation 107, can be integrated leading to the product between the internal mass transfer coefficient and the external surface area averaged over an entire oscillation cycle (i.e., $t = \pi/\omega = 2\pi/\omega'$).

$$\frac{\overline{k_p A_p}}{A_0} = \frac{2}{\pi}\left(\frac{D_p \omega}{\pi}\right)^{1/2}\left[\int_0^{\pi}\left(\frac{A_p(\tau)}{A_0}\right)^{1/2} d\tau\right]^{1/2} \tag{108}$$

where A_p (τ) indicates the interfacial surface area, which changes during time according to the following oscillation law:

$$A_p(\tau) = A_0(1 + \epsilon \sin^2 \tau) \tag{109}$$

where $\tau = \omega t$ and ϵ is the amplitude of the surface stretch, calculated from the maximum and minimum surface area per cycle, $\epsilon = (A_{max} - A_o)/A_o$.

It is worthwhile pointing out that Equation 109 is substantially identical to the surface-time relation used by Rose and Kintner [54]; but, in addition, it allows an analytic solution of the model. Thus, substituting Equation 109 in Equation 108, it follows

$$\frac{\overline{k_p A_p}}{A_0} = \frac{2}{\pi}\left[D_p\, \omega\left(1 + \epsilon + \frac{3}{8}\epsilon^2\right)\right]^{1/2} \tag{110}$$

If the surface variation in time is ignored, i.e., $A(\tau) = A_o$ and $\epsilon = 0$, then Equation 110 reduces to the classical penetration theory result, as given by Equation 106.

In the case of finite external limitations, Angelo et al. [34] suggested the following relation for the average overall mass transfer coefficient:

$$\overline{K_p A_p} = \overline{K_p A_p}/[1 + H(D_p/D)^{1/2}] \tag{111}$$

This equation derives from the assumption of equal surface lifetimes for the dispersed and continuous phases, which is not true in the Rose and Kintner model where the continuous phase resistance is described by means of the expression proposed by Garner and Tayeban [37].

Brunson and Wellek model [35]. In the model developed by Angelo et al. it is assumed that all the surface elements remain in the surface during the entire oscillation cycle. A different oscillation model has been proposed by Beek and Kramers [57]; it is assumed that the interfacial surface is not stretched, but that the increase or decrease of the interface is due to new elements brought from the interior of the drop to the surface or to old elements removed from the surface. Thus, the various surface elements have different lifetimes, and in particular it is assumed that the first part of the time-variable surface to form would be the last to disappear.

Using this oscillation model, combined with Equation 109 and the penetration theory for mass transfer into each surface element, and averaging over the surface and the oscillation time, the following expression for the average of the product between the internal mass transfer coefficient and the external surface area can be derived:

$$\frac{\overline{k_p A_p}}{A_0} = \frac{2}{\pi} (D_p \omega)^{1/2} (1 + 0.687 \epsilon) \tag{112}$$

Experimental verification of the models. The three models just reported attempt to describe the mass transfer process inside oscillating drops through a more or less physically sound description of drop oscillations. Other models of a much more empirical nature can often be of some utility.

Skelland and Wellek [55] have developed the following empirical expression for the average Sherwood number:

$$\overline{Sh}_p = 0.320 \, \tau_p{}^{-0.141} \, Re^{0.683} \, M^{-0.10} \tag{113}$$

where t_c is the time of contact during free fall.

This expression represents data of oscillating drops, for the systems water-ethyl acetate and acetoacetate-water, with average absolute deviation equal to 15% in the range $330 < Sh_p < 3{,}600$ and $400 < Re < 3{,}100$.

The model of Handlos and Baron [46] has often been used to describe oscillating drops. Since oscillations nearly eliminate internal circulatory patterns of the type assumed by Handlos and Baron, this model seems physically unsound in the case of oscillating drops. Moreover, it usually leads to poor predictions of the experimental data. It is worth reiterating that this model seems more appropriate in the case of vigorously circulating drops.

Brunson and Wellek [35] have presented a very detailed comparison of all the previously mentioned models, and few others, with the experimental data for oscillating drops collected by various authors [50, 55, 58, 59].

In summary it can be concluded that the empirical model given by Equation 113 provides the most accurate predictions of the extraction efficiency values, with an average absolute percentage deviation (AAPD) equal to 15.6%. Among the more physically sound models, that proposed by Rose and Kintner is the most accurate leading to AAPD = 22.3%. However, since this model requires numerical calculations to be applied it can be profitably replaced by Equations 110 and 112. Namely, the first one leads to AAPD = 30% and the latter to 26.4%.

It is noticeable that the Angelo et al. model gives the more accurate results, in the presence of finite external mass transfer limitations, when $\underline{k}$ is calculated through Equation 77 and then coupled to $\overline{k}_p$, given by Equation 110, through the individual resistance additivity rule, than when it is used in its original form as given by Equation 111.

The superior accuracy of Equation 113 is even more evident noticing that this is the only relation whose predictions do not show any systematic error. On the other hand, all the other models predict extraction efficiencies systematically smaller than the experimental observations. Moreover, the use of Equation 113 does not require the knowledge of the drop oscillation frequency and amplitude.

MASS TRANSFER WITH CHEMICAL REACTION

The study of mass transfer with chemical reaction in small fluid particles is of interest in two major fields of industrial applications. The first one is related to liquid-liquid chemical reactors where the reactants are in mutually immiscible phases. Examples are given by nitrations, sulfonations, and ester saponification reactions. The second field involves solvent extraction processes. In this case the reaction is merely regarded as a tool to increase the rate of extraction. Usually, unless the flow rate of the extracting phase is much less than that of the raffinate one, it is convenient to disperse the extracting phase [60]. In this case the solute is transferred from the continuous phase to the dispersed phase where it undergoes a chemical reaction. This is the situation which will be examined in the following.

Since the reaction in the dispersed phase significantly delays the particle saturation by maintaining a low solute concentration, in reaction-extraction processes the driving force for mass transfer decreases with time much slower than in unreacting processes. Quantitative estimations of this effect will be given shortly for various fluodynamic conditions, with reference to the case of a single fluid particle rising or falling at constant velocity through a continuous fluid phase.

It is worthwhile pointing out that in order to extend these single-particle models to be able to simulate industrial liquid-liquid reactors or extractors, it is necessary to adequately describe the complex fluodynamics of swarms of drops which continuously undergo coalescence and breakup. However, at the present such approach is not yet feasible and industrial units are usually simulated through mascroscopic heterogenous models where mixing effects are accounted through empirical dispersion terms [61–66].

In the next two paragraphs some of the models previously analyzed in the case of purely physical extraction processes will be extended to the case involving a chemical reaction. In this case the transient mass balance in a spherical particle with internal circulation reduces to

$$\frac{\partial u_p}{\partial \tau_p} = \frac{1}{\phi^2}\left[\frac{\partial}{\partial \phi}\left(\phi^2 \frac{\partial u_p}{\partial \phi}\right) + \frac{1}{\sin\theta}\frac{\partial}{\partial \theta}\left(\sin\theta \frac{\partial u_p}{\partial \theta}\right)\right] - \frac{1}{2}\mathrm{Pe}_p\left[v_r^*\frac{\partial u_p}{\partial \phi} + \frac{v_\theta^*}{\phi}\frac{\partial u_p}{\partial \theta}\right] - \Phi_p^2 r \tag{114}$$

$$\text{where } u_p = C_p/C_{ps},\quad r = R/R_0,\quad \Phi_p^2 = R_0 R_p^2/D_p C_{ps} \tag{115}$$

Note that R indicates the reaction rate of the solute, and R_o its reference value. For example, using as reference condition $C_{po} = C_{ps}$, in the case of an isothermal first-order reaction ($R = k_r C_p$), it follows $r = u_p$ and $\Phi_p^2 = k_r R_p^2/D_p$. This last parameter is usually called reaction factor or Thiele modulus.

The appropriate boundary and initial conditions for Equation 114 are given by

$$\begin{aligned} u_p &= 0 \text{ at } \tau_p = 0,\ \phi \epsilon (0, 1)\ \theta \epsilon (0, \pi) \\ u_p &= 1 \text{ at } \phi = 1,\ \tau_p > 0,\ \theta \epsilon (0, \pi) \\ \frac{\partial u_p}{\partial \phi} &= 0 \text{ at } \phi = 0,\ \tau_p > 0,\ \theta \epsilon (0, \pi) \\ \frac{\partial u_p}{\partial \theta} &= 0 \text{ at } \theta = 0, \pi,\ \tau_p > 0,\ \phi \epsilon (0, \pi) \end{aligned} \tag{116}$$

where external mass transfer resistances have been assumed negligible. Note that Equation 116 is identical to the boundary conditions (Equation 80) adopted in the absence of reaction, except for the definition of the dimensionless concentration.

In order to define the behavior of a single fluid particle in the presence of a chemical reaction the following overall parameters are used:

- Volume average dimensionless concentration of solute.

$$\bar{u}_p = \frac{1}{V_p}\int_{V_p} u_p dV_p \tag{117}$$

- Dimensionless instantaneous surface average mass flux entering the particle

$$N^* = \frac{Nd_e}{C_{ps}D_p} \tag{118}$$

- Ratio between the total mass transferred into the particle (at a given time) and the total mass present in the particle at saturation

$$A^*_{mt} = \frac{3}{2}\int_0^{t_p} N^* d\tau_p \tag{119}$$

- The reaction enhancement factor

$$E^* = \frac{N^*\ (\Phi_p > 0)}{N^*\ (\Phi_p = 0)} \tag{120}$$

The solution to the mass transfer problem requires knowledge of the fluodynamics inside the particle, which appears in Equation 114 through the velocity components v_r^* and v_θ^*. Since these two problems are decoupled (at least under the assumption of dilute solution), it is convenient to first evaluate the velocity field inside the particle and then solve Equations 114 through 116 to get the solute concentration profiles and the mass fluxes values. For the sake of simplicity, the solute concentration in the bulk continuous phase (C^∞) is usually assumed constant in time, as well as the velocity field inside both the dispersed and continuous phases.

Rigid Particles and Non Oscillating Drops

As discussed, in this case the velocity components of the flow field inside the particle are given by the Hadamard-Rybczynski model (see Equation 89). Recall that this model, although it is rigorously valid only in the creeping flow regime (Re < 1), actually allows satisfactory predictions of mass transfer up to Re values well above unity.

First-order Kinetics

In the case of a first-order reaction (i.e., $R = k_r C_p$), the transient mass balance equations have been solved numerically by Watada et al. [67] and Wellek et al. [68], for various values of the involved dimensionless parameter: time τ_p, Peclet number Pe_p, and reaction factor Φ_p.

As in the case of purely physical extraction the model under examination admits two limiting situations: the rigid sphere model as $Pe_p = 0$ and the fully circulating model as $Pe_p = \infty$. The latter corresponds to an extension of the Kronig and Brink model to the case involving chemical reaction. For such limiting situations the model can be solved analytically [68–70]. The obtained results in terms of the unreacted solute concentration $\bar{u}_p$ and the surface average instantaneous mass flux N^* have been summarized in Table 3, for the case of negligible external resistance to mass transfer.

It is worthwhile pointing out that the series S_1 and S_2 appearing in the solution of the rigid sphere model reported in Table 3 are quite slowly convergent, particularly at low values of τ_p. It can be shown [71] that these series can be rewritten in the following form

$$S_1 = \frac{1}{2}\left(\frac{1}{\sqrt{\pi\tau_p}} - 1\right) + \frac{1}{\sqrt{\pi\tau_p}}\sum_{n=1}^{\infty} \exp(-n^2/\tau_p) \tag{121}$$

$$S_2 = \frac{\exp(\Phi_p^2\tau_p)}{2\Phi_p}\left[1 - \frac{2}{1-\exp(2\Phi_p)} - \operatorname{erf}\left(\Phi_p\sqrt{\tau_p}\right) - \frac{\exp(-\Phi_p^2\tau_p)}{\Phi_p} + S_3\right] \tag{122}$$

$$\text{where } S_3 = \sum_{n=1}^{\infty}\left\{[1 - \operatorname{erf}(\Phi_p\sqrt{\tau_p} + n/\sqrt{\tau_p})]\exp(2n\,\Phi_p) - [1 + \operatorname{erf}(\Phi_p\sqrt{\tau_p} - n/\sqrt{\tau_p})]\exp(-2n\,\Phi_p)\right\} \tag{123}$$

The series appearing in these new expressions are much more rapidly convergent. In particular, by using only the first two terms of both of them, the mass flux N^* can be estimated with an error never exceeding 1% for all τ_p values in the range of most practical interest, i.e., $0 < \tau_p < 2.5$.

The solution to the rigid-sphere model in cases involving external resistance to mass transfer has been examined in detail by Morbidelli et al. [71].

The numerical solution of the circulating model has been reported for three values of Pe_p in Figures 21, 22, and 23 in terms of the total mass of solute transferred into the particle (A^*_{mt}) as a function of time (τ_p) for various values of the reaction factor (Φ_p^2) [68]. It should be pointed out that for $Pe_p > 400(1 + \varkappa)$ the circulation model identifies to the fully circulating model ($Pe_p = \infty$) reported in Table 3. Moreover, for reaction factor values $\Phi_p^2 > 640$ the Pe_p value does not affect the behavior of the particle. Namely, at high reaction rate most

Table 3
Solutions of the Rigid and the Fully Circulating Sphere Models

Model	Rigid sphere ($Pe_p = 0$)	Fully circulating sphere* ($Pe_p = \infty$)
$\bar{u}_p =$	$\frac{3}{\Phi_p^2}(\Phi_p \operatorname{Cth}\Phi_p - 1) - 6S_2\exp(-\tau_p\Phi_p^2)$ Where $S_2 = \sum_{n=1}^{\infty}\frac{\exp(-\tau_p n^2\pi^2)}{(\Phi_p^2 + n^2\pi^2)}$	$1 - \frac{3}{8}\sum_{n=1}^{\infty} B_n^2 \frac{\Phi_p^2 + 16\lambda_n\exp[-\tau_p(\Phi_p^2 + 16\lambda_n)]}{\Phi_p^2 + 16\lambda_n}$
$N^* =$	$2(\Phi_p \operatorname{Cth}\Phi_p - 1) + 4(S_1 - \Phi_p^2 S_2)\exp(-\tau_p\Phi_p^2)$ Where $S_1 = \sum_{n=1}^{\infty}\exp(-\tau_p n^2\pi^2)$	$4\sum_{n=1}^{\infty} B_n^2\lambda_n \frac{\Phi_p^2 + 16\lambda_n\exp[-\tau_p(\Phi_p^2 + 16\lambda_n)]}{\Phi_p^2 + 16\lambda_n}$

* *The values of the parameters B_n and λ_n are reported in Table 2.*

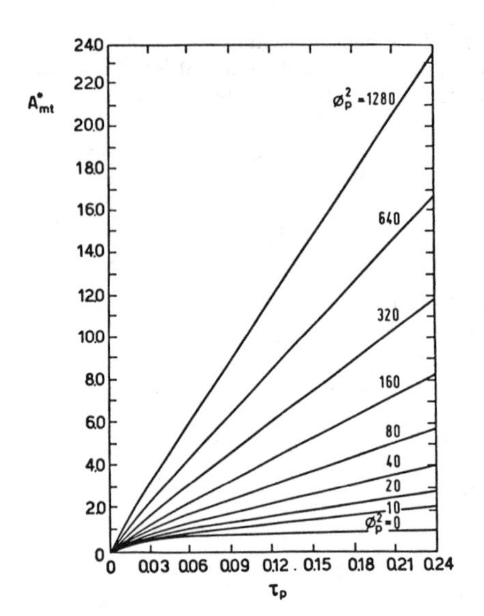

Figure 21. Dimensionless total mass transferred as a function of τ_p (Pe_p = 0) [68].

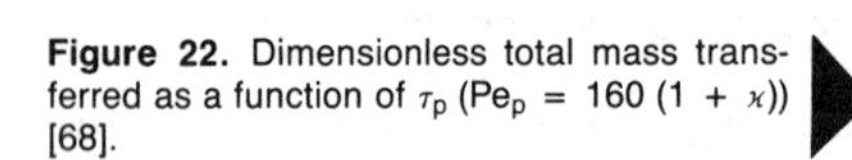

Figure 22. Dimensionless total mass transferred as a function of τ_p (Pe_p = 160 (1 + $\varkappa$)) [68].

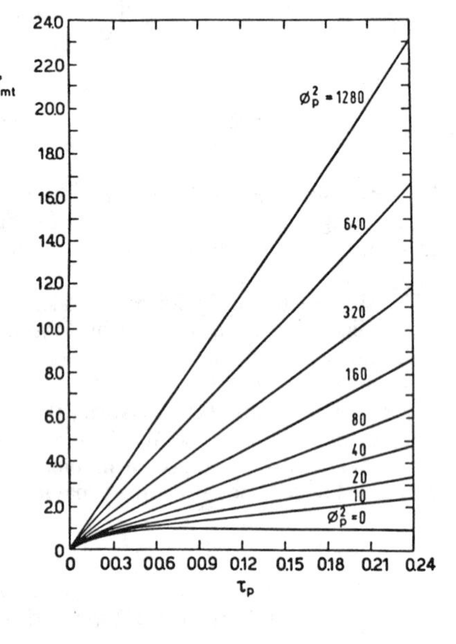

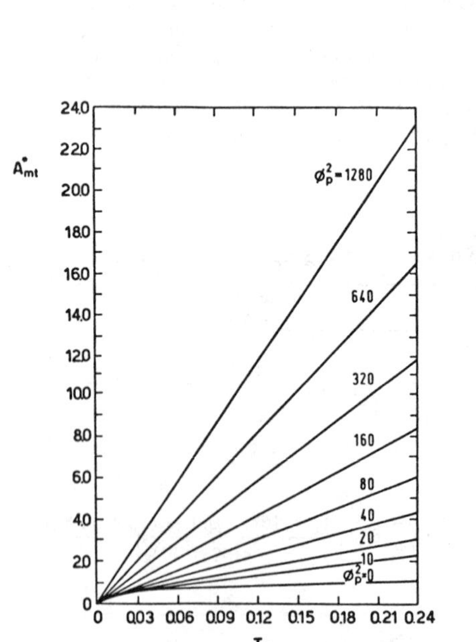

Figure 23. Dimensionless total mass transferred as a function of τ_p (Pe_p = 400 (1 + $\varkappa$)) [68].

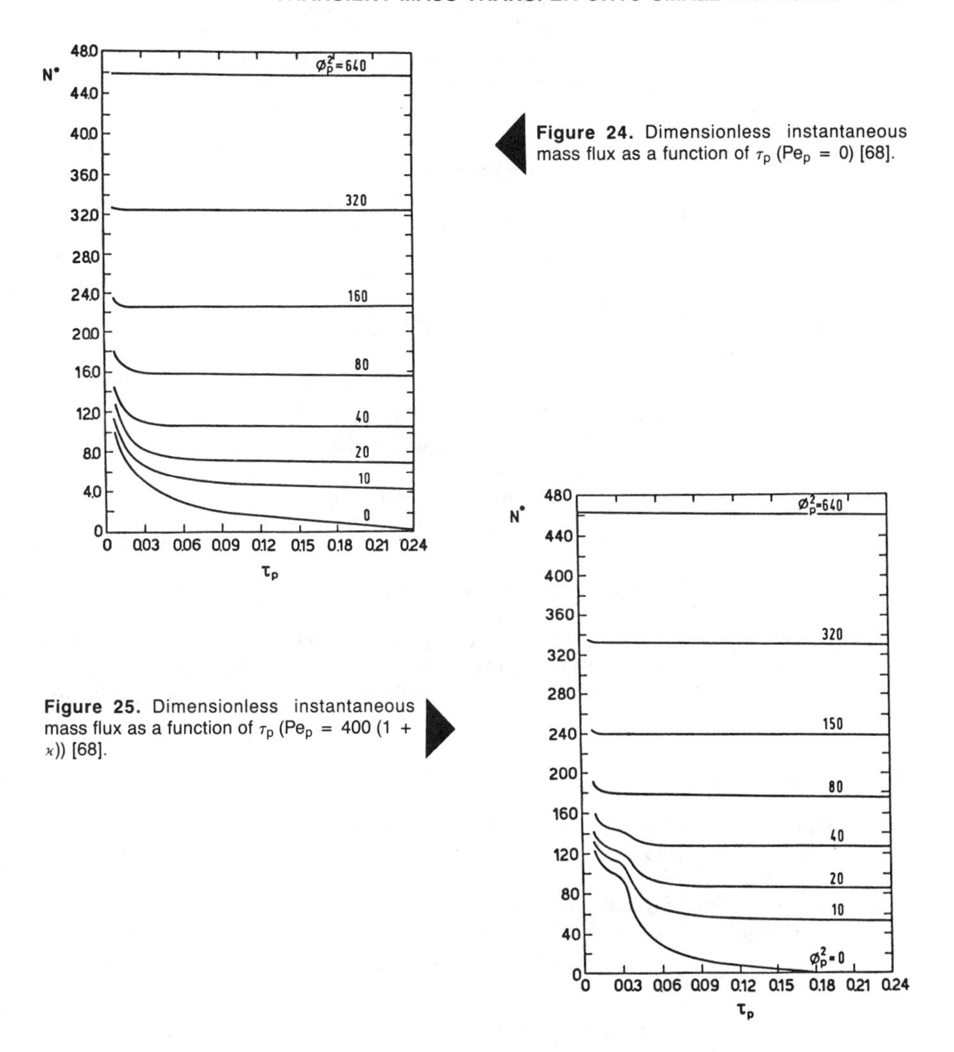

Figure 24. Dimensionless instantaneous mass flux as a function of τ_p (Pe_p = 0) [68].

Figure 25. Dimensionless instantaneous mass flux as a function of τ_p (Pe_p = 400 (1 + $\varkappa$)) [68].

of the solute does not circulate inside the particle before it reacts, so that convective transport does not affect mass transfer. On the other hand, at very short times (i.e., $\tau_p < 10^{-3}$) the circulation inside the particle has not yet occurred. So, at $\tau_p < 10^{-3}$ the value of Pe_p does not affect the solute extraction rate. In these two limiting situations it is convenient to use the rigid sphere model, which, particularly in the form of Equations 121 through 123, is the most simple from the computational point of view.

The behavior of the surface average instantaneous mass flux N* as a function of time is shown in Figures 24 and 25. It can be seen that the N* versus τ_p curves exhibit oscillations for large Pe_p value, similarly to the just discussed case of purely physical extraction. However, in the presence of a reaction it appears that for increasing values of the reaction rate factor these oscillations tend to disappear. This can be explained recalling that for increasing reaction rates, the effect of convective transport inside the particle, which is responsible for the oscillating values of N* in time, decreases rapidly. In Figure 26 the asymptotic values of

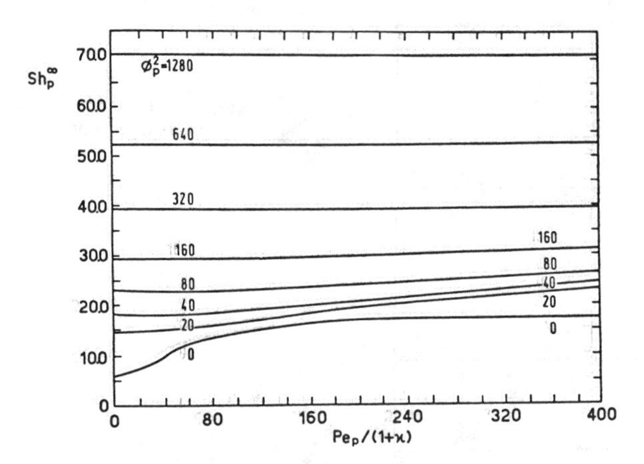

Figure 26. Asymptotic values of the instantaneous Sherwood number [68].

Sh_p as $\tau_p \rightarrow \infty$ are shown. Again it appears that for $\Phi_p^2 > 640$, Pe_p does not affect the value of Sh_p^∞.

Finally, in Figure 27 the behavior of the reaction enhancement factor as a function of time is shown. The reported values clearly emphasize the role of chemical reactions in enhancing the extraction rate in reactive-extraction processes with respect to unreactive-extraction processes.

Second-Order Kinetics

First-order kinetic models can usually be adopted when the second reactant is present in the dispersed phase in large excess. If this is not the case, then a second-order kinetic model should be used.

Brunson and Wellek [72] have examined the case where the following bimolecular irreversible reaction takes place in the dispersed phase.

$$A + B \rightarrow \text{Products} \tag{124}$$

where A is the solute, which migrates from the continuous to the dispersed phase, and B is the second reactant, which is assumed to be insoluble in the continuous phase. The consumption rate of both reactants is given by $R = k_r' C_{pA} C_{pB} (R_0 = k_r' C_{pAs}^2;\ C_{po} = C_{pAs})$, and the reaction factor (Equation 115) reduces to $\Phi_p^2 = k_r' C_{pAs} R_p^2 / D_{pA}$

The solution of the mass transfer problem requires the solution of two transient mass balance equations of the type of Equation 114, one for each reactant. This leads to the introduction of two new dimensionless parameters

$$R_c = C_{pAs}/C_{pB}^o$$

$$R_D = D_{pB}/D_{pA} \tag{125}$$

In Figure 28 is shown the behavior of the total solute mass transferred into the particle as a function of time for various values of the reaction factor. It appears that at low τ_p the instantaneous mass flux is larger for larger Φ_p^2. However, in the particles where the reaction rate is higher, the reactant B is depleted faster than in those particles where the reaction rate is

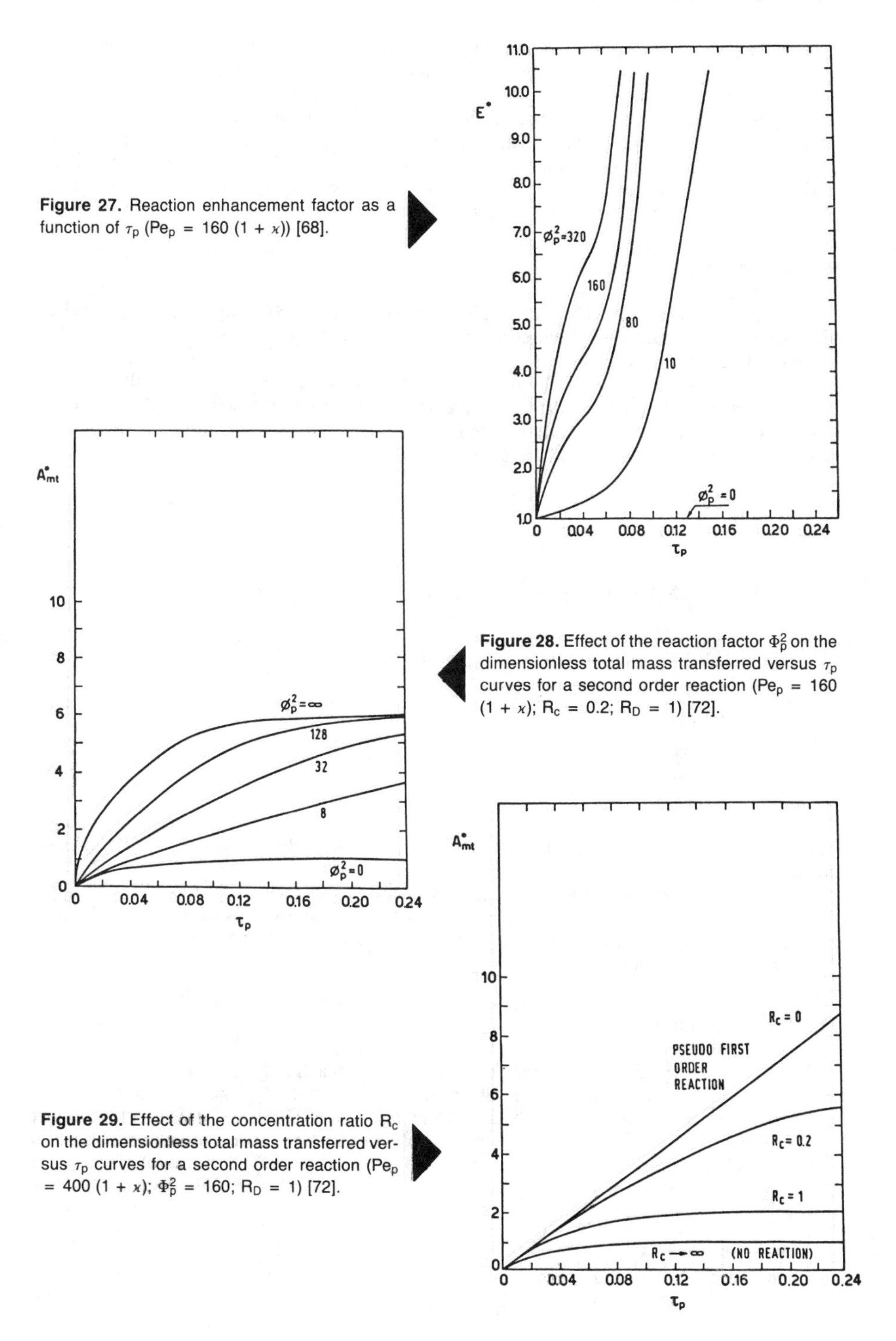

Figure 27. Reaction enhancement factor as a function of τ_p ($Pe_p = 160\ (1 + \varkappa)$) [68].

Figure 28. Effect of the reaction factor Φ_p^2 on the dimensionless total mass transferred versus τ_p curves for a second order reaction ($Pe_p = 160\ (1 + \varkappa)$; $R_c = 0.2$; $R_D = 1$) [72].

Figure 29. Effect of the concentration ratio R_c on the dimensionless total mass transferred versus τ_p curves for a second order reaction ($Pe_p = 400\ (1 + \varkappa)$; $\Phi_p^2 = 160$; $R_D = 1$) [72].

lower. Then, at larger τ_p the instantaneous mass flux might be larger for those particles where the reaction rate is lower. Nevertheless, it is clear that at any given time the total mass transferred is always an increasing function of the reaction factor.

In Figure 29 the effect of the concentration ratio R_c on the total mass transferred is shown. At $R_c = 0$ ($C^0_{pB} >> C_{pAs}$) the kinetics reduces to pseudo first order, while at $R_c = \infty$ ($C^0_{pB} \simeq 0$) the reaction rate vanishes, so that the case of purely physical extraction is attained. Curves at intermediate values of R_c indicate that the total amount of solute transferred is lower than in the corresponding case with first-order kinetics.

Oscillating Drops

The oscillating drop model for purely physical extraction proposed by Brunson and Wellek [35] has been extended by Wellek et al. [68] to the case where a first-order reaction takes place inside the drop. This model is based on the assumption that elements of surface area are continuously formed and destroyed in order to account for the surface-area oscillation with time. For each element of the surface the mass flux during its lifetime is calculated through the penetration theory. Accounting for the presence of a first-order reaction, the instantaneous mass transfer coefficient is then given by

$$k_p(t) = \left(\frac{D_p}{\pi t}\right)^{1/2} (1 + k_r t) \tag{126}$$

This leads to the following expressions for the surface average instantaneous mass flux:

$$N^* = \frac{2}{\tau}\left(\frac{d_e^2 \omega}{D_p}\right)^{1/2} \left[1 + 0.687\,\epsilon + \frac{k_r}{\omega}\left(\frac{\pi}{3} + 0.396\,\epsilon\right)\right] \tag{127}$$

and the reaction enhancement factor

$$E^* = 1 + \frac{k_r}{\omega}\left(\frac{\pi/3 + 0.396\,\epsilon}{1 + 0.687\,\epsilon}\right) \tag{128}$$

As expected, these relationships reduce to the Brunson and Wellek model as $k_r \rightarrow 0$. Namely, as $k_r \rightarrow 0$, $E^* \rightarrow 1$.

It is worthwhile pointing out that for rapidly oscillating droplets, i.e., $w \rightarrow \infty$, the effect of chemical reaction vanishes, i.e., $E^* \rightarrow 1$. In fact, at large oscillations frequency the contact time of each surface element is so low that the reaction has "no time" to significantly increase the extraction rate.

Alternative Models

Two classical theories of mass transfer with chemical reaction have been extensively studied in the literature: the film and the penetration theories [73, 74]. They refer to the simple case of two phases in contact through a planar interface in the absence of convective motions.

Following Brunson and Wellek [72], the film and penetration models for a single bimolecular reaction can be rewritten as follows:

$$\epsilon \frac{\partial C_{pi}}{\partial t} = D_{pi} \frac{\partial^2 C_{pi}}{\partial z^2} - k_r C_{pa} C_{pb}$$

$$i = A, B \tag{129}$$

where z is the distance from the interface and the boundary conditions are given by

$$C_{pA} = C_{pAs}, \frac{\partial C_{pB}}{\partial z} = 0 \text{ at } z = 0$$
$$C_{pA} = 0, \quad C_{pB} = C^{o}_{pB} \text{ at } z = \delta$$
$$C_{pA} = 0, \quad C_{pB} = C^{o}_{pB} \text{ at } t = 0$$

For the film theory $\epsilon = 0$ (i.e, steady-state conditions) and δ indicates the film-thickness, while for the penetration theory $\delta \rightarrow \infty$ and $\epsilon = 1$.

These theories can be used to describe mass transport with chemical reaction in a spherical fluid particle only under specific circumstances. In particular, for low-contact times, i.e., $\tau_p < 10^{-3}$, we have seen that the fluid circulation inside the drop does not affect the solute mass transfer rate. Similarly, for such low-contact times, only the most external region of the drop is involved in the process, so that the geometry of the problem can be approximated as planar. These two models have been numerically compared with the previously mentioned rigid and circulating models described by Wellek et al. [68] in the case of a first-order reaction. It was concluded that in the range $\tau_p < 10^{-3}$, there is no significant difference among any of them.

In the more general case of a bimolecular second-order reaction, the film and penetration theories involve one more assumption: the reactant B is not depleted at a distance δ from the interface. As it appears from the numerical values reported by Brunson and Wellek [72], this approximation leads to errors which never exceed 5% as long as $\tau_p < 10^{-3}$, at least for the reaction-rate values usually encountered in practice.

From these observations, recalling that in most processes where the fluid particle is in the liquid phase the contact times are actually very low, it can be concluded that the film and penetration theories provide a quite useful tool. It is worthwhile pointing out that these two theories lead in most cases (except when the diffusivities of the two reactants are quite different [75]) to very similar results [74]. Therefore the choice among the two is usually only a matter of convenience. In the case of liquid-liquid reactions, the use of the film theory is recommended because it is much more simple from a computational point of view and more accurate empirical correlations are available for its calibration in the absence of chemical reaction. Another important feature of the film model is that a large number of approximate solutions are available for various situations of practical interest. These include: pseudo first-order kinetics [76], bimolecular second-order reactions [77], bimolecular general-order reactions [78], and various complex kinetic schemes [79].

More recently Sharma [80] has reported in detail the extension of the film theory, which has been largely developed for gas-liquid systems, to the case of liquid-liquid extraction with reaction. Depending on the relative rate of diffusion and chemical reaction, the system behavior has been divided into four regimes; for each regime analytical expressions for the extraction rate and experimental methods for the controlling regime identification have been reported.

Wellek and Brunson [81] have compared the predictions of the film model with the experimental data relative to water drops containing sodium sulphate and sodium hydroxide freely falling in a column filled with pure n-penthyl formate. The film model predictions were obtained by combining the mass transfer coefficient value in the absence of reaction, calculated through suitable empirical relationship [35, 55], with the enhancement factor value given by the approximate solution of the film model developed by van Krevelen and Hoftijzer [77]. The comparison with the experimental results, all taken for very low contact times (i.e., $\tau_p < 10^{-3}$), showed that the error in the prediction of the total solute mass transferred in the presence of chemical reaction is comparable to the one intrinsically present in the empirical correlations adopted for the evaluation of the mass transfer coefficient without chemical reaction. Although this conclusion seems promising with respect to the reliability of the film model, it is worthwhile pointing out that the error of the entire procedure is still excessive,

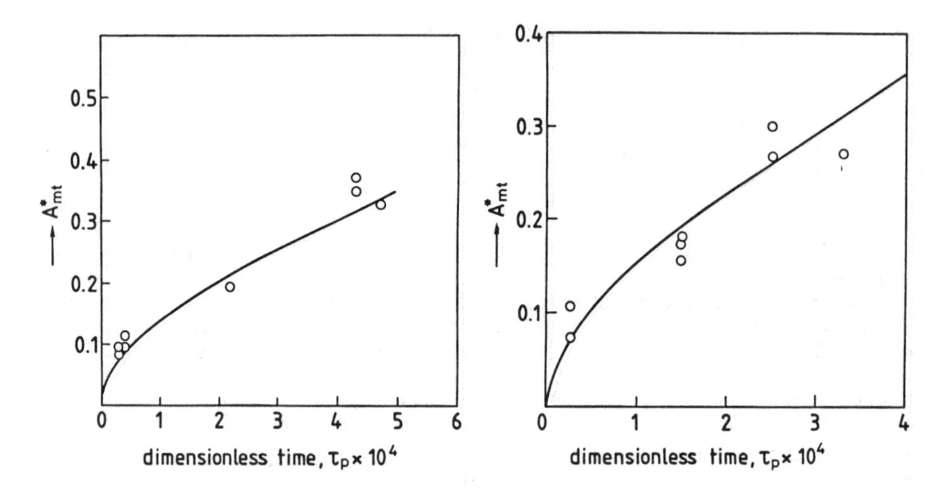

Figure 30. Plot of dimensionless total mass transferred versus dimensionless time.

i.e., of the order of 30% to 50%. These findings indicate that the a priori simulation of diffusion-reaction processes in a fluid particle in free motion has not yet been achieved.

A different approach has been proposed by Morbidelli et al. [82], based on the use of the rigid drop model, where the molecular diffusivity D_p is replaced by an effective diffusivity D_e given by

$$D_e = R\, D_p \tag{130}$$

As mentioned, Calderbank and Korchinski [44] showed that this model with R = 2.25 reproduces the results of the Kronig and Brink model for a fully circulating drop without chemical reaction. In the case of an oscillating drop, the effective diffusion coefficient D_e can be estimated through the Einstein formula, where the characteristic time is estimated as the circulation time along a streamline of the Kronig and Brink model. This leads to the relationship

$$R = a\, Pe_p/(1 + \varkappa) \tag{131}$$

where a is an adjustable parameter which depends on the particular system under consideration. In Figure 30A and B a comparison between the results of this model (with $a = 0.42 \times 10^{-4}$) and the experimental data reported by Wellek and Brunson [81] is shown. In general, a significant improvement over the film and the penetration theories, used without adjustable parameters, has been found (the absolute average percent deviation is reduced by a factor between two and five).

It is noticeable that, following a different approach based on the Handlos and Baron [46] model, Johnson and Hamielec [50] derived for the factor R the same expression (Equation 131) in the case of a nonreacting oscillating drop. However, the value of the derived constant does not agree with their experimental data relative to the water-ethylacetate system, as well as with the previously reported value. Thus, on the whole, it can be concluded that while the form of Equation 131 seems to be confirmed, an a priori exact evaluation of the effective diffusivity is not possible due to the uncertainties involved in the evaluation of the adjustable parameter a, which can be confidently estimated only by comparison with experimental data.

Several works have appeared in the literature on the experimental investigation of the reaction zone in liquid-liquid systems. Cho and Ranz [83, 84] measured the movement of the interface between an aqueous and an organic phase both in the case of instantaneous and slow reaction. The position of the interface was determined by taking advantage of the discontinuity of the concentration gradients at the interface, which leads to a discontinuity of the solution refractive index. The obtained results were interpreted with the model for reaction zone movement in liquid-liquid systems developed by Scriven [85]. Tyroler et al. [86] have reported a similar study in the case of a free-falling drop with fast second-order reaction. For noncirculating drops the observed movement was in excellent agreement with the rigid drop model. More recently Halwachs and Schügerl [87] developed a sophisticated experimental technique based on liquid scintillation for measuring mass transfer coefficients relative to a freely suspended drop. The obtained results were qualitatively explained in terms of a migrating front reaction model.

Finally, it is worthwhile mentioning a particular case of liquid-liquid mass transfer with chemical reaction, which occurs when the solute exists in different chemical form in the two liquid phases. In such case a chemical reaction occurs at the interface, leading to an additional resistance to mass transfer. An example of such situation is given by the extraction of carboxylic acids from polar solvents (where they are mainly present in the monomeric form) to nonpolar solvents (where the dimeric form prevails). In particular, the transfer of benzoic and caprilic acid from toluene to water (with and without NaOH) has been experimentally investigated by Schürgerl and Dimian [88]. Their data were satisfactorily reproduced using some of the models just discussed by introducing an adjustable interfacial resistance, which empirically accounts for the interfacial reaction.

Other phenomena are known to enhance the interphase mass transfer by leading to interfacial turbulence, such as the occurrence of an instantaneous reaction at the interface [83, 88] or the presence of interfacial tension gradients (Marangoni effect). However, quantitative informations about these effects in the case of liquid-liquid reacting systems are quite scarce in the literature (See Pratt [89]).

NOTATION

A_e external surface area of the equivalent particle, defined by Equation 4

A^*_{mt} dimensionless total mass transferred into the particle, defined by Equation 119

A_p external surface area of the particle

Bi Biot number, $k\ R_p/H\ D_p$

C, C_p concentration in the continuous and dispersed phase, respectively

C^*, C^*_p dimensionless concentration, defined by Equation 48

$\overline{C^*}$ volume average dimensionless concentration, defined by Equation 55a

C_D drag coefficient for fluid particle at its terminal velocity, defined by Equation 27

C_e equilibrium concentration, HC^∞

$\overline{C_p}$ volume average concentration in the dispersed phase, defined by Equation 82

C_{ps} concentration at the particle surface

d_e diameter of the volume equivalent sphere

d_t diameter of the containing vessel or tube

D, D_p diffusion coefficient in continuous and dispersed phase, respectively.

D_e effective diffusion coefficient

e eccentricity, $(1 - E^2)^{1/2}$

E aspect ratio, vertical length/horizontal length

E^* reaction enhancement factor, defined by Equation 120

Eo Eotvos number $g\Delta\rho d_e^2/\sigma$

f_N natural frequency of shape oscillation

F extraction efficiency, defined by Equation 55

g gravitational acceleration

h dimensionless parameter, $2\ Bi/\beta$

h_v coordinate directed vertically upwards

H equilibrium ratio between dispersed and continuous phase
k, k_p mass transfer coefficient in the continuous and dispersed phase, respectively
$\underline{k}_l$ local mass transfer coefficient
$\overline{k}_p$ time average mass transfer coefficient, defined by Equation 81
k_r, k_r' first- and second-order reaction rate constants, respectively
K_p overall mass transfer coefficient, defined by Equation 81
$\overline{K}_p$ time average overall mass transfer coefficient, defined by Equation 83
L reference length
M Morton number, $g\ \mu^4\ \Delta\rho/\rho^2\sigma^3$
N surface average instantaneous mass flux,

$$\frac{1}{A_p}\int_{A_p} D_p \left(\frac{\partial C_p}{\partial r}\right)_{r=R_p} d\ A_p$$

N^* surface average dimensionless instantaneous mass flux, $d_eN/C'\ D_p$ (where $C' = C_{ps}$ for reacting particles, $C' = C_p^o - C_{ps}$ for nonreacting particles)
P pressure
P^* dimensionless pressure $(P - P_o - \rho\ g\ h_v)/\rho\ v_o^2$
Pe Peclet number in the continuous phase, $d_e\ v_p/D$
Pe_p Peclet number in the dispersed phase, $d_e v_p/D_p$
r radial coordinate in spherical coordinates; coordinate normal to the interface; dimensionless reaction rate, R/R_o
R correction factor, D_e/D; distance to surface of axisymmetric particle, Figure 6; reaction rate
R^* dimensionless parameter, $2\ R_p/D_e$
R_c concentration ratio, C_{pAs}/C_{pB}^o
R_D diffusivity ratio, D_{pB}/D_{pA}
R_p particle radius
Re Reynolds number, $\rho\ L\ v_o/\mu$ or $\rho\ d_e v_p/\mu$
Sc Schmidt number, ν/D
Sh Sherwood number in the continuous phase, $k\ d_e/D$
Sh_e Sherwood number in the continuous phase based on the volume equivalent sphere, $(k\ A/A_e)d_e/D$
Sh_l local Sherwood number, $k_l d_e/D$
Sh_p Sherwood number in the dispersed phase, $k_p d_e/D_p$
Sh_{pe} Sherwood number in the dispersed phase based on the volume equivalent sphere, $(k_pA_p/A_e)d_eD_p$
$\overline{Sh}_e$ time average Sh_e, $(\overline{kA}/A_e)d_e/D$
$\overline{Sh}_p$ time average Sh_p, $\overline{k}_p\ d_e/D_p$
$\overline{Sh}_{pe}$ time average Sh_{pe}, $(\overline{k_pA_p}/A_e)\ d_e/D_p$
t time
t^* dimensionless time, $t\ v_o/L$
u_p dimensionless concentration in the dispersed phase, C_p/C_{ps}
$\bar{u}_p$ volume average of u_p, defined by Equation 117
$\underset{\sim}{v}$, v local fluid velocity vector and scalar, respectively
$\underset{\sim}{v}^*$ dimensionless local fluid velocity vector, $\underset{\sim}{v}/v_o$
v_θ, v_r θ- and r- component of $\underset{\sim}{v}$ in spherical coordinates.
v_θ^*, v_r^* v_θ/v_p and v_r/v_p, respectively
v_x, v_y velocity in continuous phase parallel and normal to surface, respectively
v_p velocity of particle relative to remote continuous phase
v_t particle terminal velocity
v_{tr} Stokes terminal velocity
V_p particle volume
We Weber number, $d_e v_p^2\ \rho/\sigma$
x cartesian boundary layer coordinate parallel to interface
y cartesian boundary layer coordinate normal to interface
Y dimensionless parameter, v_t/v_{tr}
Z degree of circulation, defined by Equation 28

Greek Symbols

β dimensionless parameter, $Pe_p/512\ (1 + \kappa)$
γ density ratio, ρ_p/ρ
$\Delta\rho$ absolute value of density difference, $[\rho_p - \rho]$
ζ dimensionless parameter defined

by Equation 38; dimensionless coordinate $2x/d_e$

θ angular coordinate in spherical coordinates

κ viscosity ratio, μ_p/μ

λ wavelength disturbance; diameter ratio d_e/d_t

μ, μ_p viscosity of the continuous and dispersed phase, respectively

ν, ν_p kinematic viscosity of the continuous and dispersed phase, respectively

ξ dimensionless coordinate, $2y/d_e$

ρ, ρ_p density of the continuous and dispersed phase, respectively

ρ^* dimensionless density, ρ/ρ_o

σ interfacial or surface tension

τ_p dimensionless time, $4D_pt/d_e^2$

τ dimensionless time, ωt

ϕ dimensionless radial coordinate, r/R_p

Φ_p reaction factor or Thiele modulus, $R_oR_p^2/D_pC_{ps}$

ψ, ψ_p stream function in continuous and dispersed phase, respectively

ω $\omega'/2$

ω' angular frequency of oscillation

Subscripts

$_e$ volume equivalent sphere

$_l$ local

$_p$ dispersed phase or particle

$_0$ reference conditions

^{i} interface

^{0} initial conditions

∞ bulk continuous phase remote from the particle

* dimensionless quantity

time or volume average quantity

REFERENCES

1. Licht, W. and Conway, J., *Ind. Eng. Chem., 42:* 1151 (1960).
2. Clift, R., Grace, J. R., and Weber, M. E., *Bubbles, Drops and Particles,* Academic Press, New York, 1978.
3. Kintner, R. C., *Adv. Chem. Eng., 4:* 51 (1963).
4. Heertjes P. M., and De Nie, L. H., in *Recent Advances in Liquid-Liquid Extraction,* C. Hanson (Ed.), Pergamon Press, Oxford, 1971, p. 367.

4a. Schügerl, K., et al., *Recent Developments in Separation Science,* Part A, N.N.Li (Ed.), Vol. 3, CRC Press, Cleveland, 1977, p. 71.

5. Bapat, P. M., Tavlarides, L. L., and Smith, G. W., *Chem. Eng. Sci, 38:* 2003 (1983).

5a. Hsia M. A., and Tavlarides, L. T., *Chem. Eng. J., 26:* 189 (1983).

6. Sovova, H., *Chem. Eng. Sci., 36:* 1567 (1981).
7. Grace, J. R., Wairegi, T., and Brophy, J., *Can. J. Chem. Eng., 56:* 3 (1978).
8. Wellek, R. M., Agrawal, A. K., and Skelland, A. H. P., *AIChE J., 12:* 854 (1966).
9. Hadamard, J. S., *C. R. Acad. Sci., 152:* 1735 (1911).
10. Rybczynski, W., *Bull. Int. Acad. Pol. Sci. Lett., Cl. Sci. Math. Nat.,* Ser. A. 1911, p. 40
11. Frumkin, A., and Levich, V. G., *Zh. Fiz. Khim., 21:* 1183 (1947).
12. Levich, V. G., *Physicochemical Hydrodynamics,* Prentice-Hall, New York, 1962.
13. Davies, J. T., *Adv. Chem. Eng., 4:* 1 (1963).
14. Davies, J. T., and Mayers, G. R. A., *Chem. Eng. Sci., 16:* 55 (1961).
15. Griffith, R. M., *Chem. Eng. Sci., 17:* 1057 (1962).
16. Davis, R. E., and Acrivos, A., *Chem. Eng. Sci., 21:* 681 (1966).
17. Savic, P., *Natl. Res. Counc. Can. Rep.,* n. MT-22 (1953).
18. Bond, W. N., and Newton, D. A., *Philos. Mag.,* 5: 794 (1928).
19. Sheth, R. B., M. S. Thesis, Brigham Young Univ., Provo, Utah, 1970.
20. Grace, J. R., and Harrison, D., *Chem. Eng. Sci., 22:* 1337 (1967).
21. Wairegi, T., and Grace, J. R., *Int. J. Multiphase Flow, 3:* 67 (1976).

22. Davies, R. M., and Taylor, G. I., *Proc. R. Soc.*, Ser. A 200: 375 (1950).
23. Davenport, W. G., Richardson, F. D., and Bradshaw, A. V., *Chem. Eng. Sci., 22:* 1221 (1967).
24. Francis, A. W., *Physics, 4:* 403 (1933).
25. Fayon, A. M., and Happel, J., *AIChE J., 6:* 55 (1960).
26. Achenbach, E., *J. Fluid. Mech., 65:* 113 (1974).
27. Wallis, G. B., *One-Dimensional Two-Phase Flow,* McGraw-Hill, New York, 1969.
28. Skelland, A. H. P., *Diffusional Mass Transfer,* Wiley, New York, 1974.
29. Sherwood, T. K., Pigford, R. L., and Wilke, C. R., *Mass Transfer,* McGraw-Hill, New York, 1975.
30. Cussler, E. L., *Multicomponent Diffusion,* Elsevier, Amsterdam, 1976.
31. King, C. J., *AIChE J., 10:* 671 (1964).
32. Lochiel, A. C., and Calderbank, P. H., *Chem. Eng. Sci., 19:* 471 (1964).
33. Abdel-Alim, A. H., and Hamielec, A. E., *Ind. Eng. Chem. Fundam., 14:* 308 (1975).
34. Angelo, J. B., Lightfoot, E. N., and Howard, D. W., *AIChE J., 12:* 751 (1966).
35. Brunson, R. J., and Wellek, R. M., *Can. J. Chem. Eng., 48:* 267 (1970).
36. Lamb, H., *Hydrodynamics,* Cambridge Univ. Press, London, 1932.
37. Garner, F. H., and Tayeban, M., *An. Real. Soc. Espan. Fis. Quim., B56:* 479 (1960).
38. Newman, A. B., *Trans. Am. Inst. Chem. Eng., 27:* 203 (1931).
39. Vermeulen, T., *Ind. Eng. Chem., 45:* 1664 (1953).
40. Groeber, H., *Z. Ver. Deutsch. Ing., 69:* 705 (1925).
41. Johns, L. E., and Beckmann, R. B., *AIChE J., 12:* 10 (1966).
42. Kronig, R., and Brink, J. C., *Appl. Sci. Res., A2:* 142 (1950)
43. Heertjes, P. M., Holve, W. A., and Talsma, H., *Chem. Eng. Sci., 3:* 122 (1954).
44. Calderbank, H., and Korchinski, I. J. O., *Chem. Eng. Sci.*, 6: 65 (1956).
45. Elzinga, E. R., and Banchero, J. T., *Chem. Eng. Progr. Symp. Ser. No. 29, 55:* 149 (1959).
46. Handlos, A. E., and Baron, T., *AIChE J., 3:* 127 (1957).
47. Olander, D. R., *AIChE J., 7:* 175 (1961).
48. Wellek, R. M., and Skelland, A. H. P., *AIChE J., 11:* 557 (1965).
49. Patel, J. M., and Wellek, R. M., *AIChE J., 13:* 384 (1967).
50. Johnson, A. I., and Hamielec, A. E., *AIChE J., 6:* 145 (1960).
51. Huang, W. S., and Kintner, R. C., *AIChE J., 15:* 735 (1969).
52. Garner, F. H., and Lane, J. J., *Trans. Inst. Chem. Eng., 37:* 162 (1959).
53. Hill, M. J. M., *Philos. Trans. Real Soc. London Ser. A, 185:* 213 (1894).
54. Rose, P. M., and Kintner, R. C., *AIChE J., 12:* 530 (1966).
55. Skelland, A. H. P., and Wellek, R. M., *AIChE J., 10:* 491 (1964).
56. Schroeder, R. R., and Kintner, R. C., *AIChE J., 11:* 5 (1965).
57. Beek, W. J., and Kramers, H., *Chem. Eng. Sci., 16:* 909 (1962).
58. Garner, F. H., and Skelland, A. H. P., *Ind. Eng. Chem., 46:* 1255 (1954).
59. Kopinsky, S., M.S. Thesis, Mass Inst. of Technology (1949).
60. Johnson, H. F., and Bliss, H., *Trans. Am. Inst. Chem. Eng., 42:* 331 (1946).
61. Trambouze, P. J., Trambouze, M. T., and Piret, E. L., *AIChE J., 7:* 138 (1961).
62. Piret, E. L., Penney, W. H., and Trambouze, P. J., *AIChE J., 6:* 394 (1960).
63. Pavlica, R. T., and Olson, J. H., *Ind. Eng. Chem., 62:* 45 (1970).
64. Hanson, C., in *Recent Advances in Liquid-Liquid Extraction,* C. Hanson (Ed.), Pergamon Press, Oxford, 1971, p. 429.
65. Laddha, G. S., and Degaleesan, T. E., *Transport Phenomena in Liquid-Liquid Extraction,* TATA McGraw Hill, New Delhi (1976).
66. Van Landeghem, H., *Chem. Eng. Sci., 35:* 1912 (1980).
67. Watada, H., Hamielec, A. E., and Johnson, A. I., *Can. J. Chem. Eng., 48:* 255 (1970).
68. Wellek, R. M., Andoe, W. V., and Brunson, R. J., *Can. J. Chem. Eng., 48:* 645 (1970).

69. Danckwerts, P. V., *Trans. Faraday Soc.*, *47:* 1014 (1951).
70. Crank, J., *The Mathematics of Diffusion,* Clarendon Press, Oxford, 1956.
71. Morbidelli, M., et al. *Chem. Eng. Sci.*, *37:* 1645 (1982).
72. Brunson, R. J., and Wellek, R. M., *AIChE J.*, *17:* 1123 (1971).
73. Astarita, G., *Mass Transfer with Chemical Reaction,* Elsevier, Amsterdam (1967).
74. Danckwerts, P. V., *Gas-Liquid Reactions,* McGraw-Hill, New York (1970).
75. Brian, P. L. T., Hurley, J. F., and Hasseltine, E. H., *AIChE J.*, *7:* 226 (1961).
76. Kramers, H., and Westerterp, K. R., *Elements of Chemical Reactor Design and Operation,* Academic Press, New York (1963).
77. van Krevelen, D. W., and Hoftijzer, P. J., *Rec. Trav. Chim.*, *67:* 563 (1948).
78. Hikita, H., and Asai, S., *Int. Chem. Eng.*, *4:* 332 (1964).
79. Morbidelli, M., et al., in *Recent Advances in the Engineering Analysis of Chemically Reacting Systems,* L. K. Doraiswamy (Ed.), Wiley Eastern, New Delhi, p. 336 (1984).
80. Sharma, M. M., in *Handbook of Solvent Extraction,* T.C.L. Lo, M.H.I. Baird, and C. Hanson (Eds.), John Wiley, New York, 1983, p. 37.
81. Wellek, R. M., and Brunson, R. J., *Canad. J. Chem. Eng.*, *53:* 150 (1975).
82. Morbidelli, M., et al., *Chem. Eng. Sci.*, *37:* 1653 (1982).
83. Cho, D. H., and Ranz, W. E., *Chem. Eng. Prog. Symp. Ser.*, *No. 72, 63,* 37 (1968).
84. Cho, D. H., and Ranz, W. E., *Chem. Eng. Prog. Symp. Ser. No. 72, 63:* 37 (1968).
85. Scriven, L. E., *AIChE J.*, *7:* 524 (1961).
86. Tyroler, G., et al., *Can. J. Chem. Eng.*, *49:* 56 (1971).
87. Halwachs, W., and Schügerl, K., *Chem. Eng. Sci.*, *38:* 1073 (1983).
88. Schugerl, K., and Dimian A., *Chem. Eng. Sci.*, *35:* 963 (1980).
89. Pratt, H. R. C. in *Handbook of Solvent Extraction,* T.C.L. Lo, M.H.I. Baird, and C. Hanson (Eds.), John Wiley, New York, 1983, p. 91.

Chapter 5

MODELING HEAT AND MASS TRANSPORT IN FALLING LIQUID FILMS

Siu-Ming Yih

Department of Chemical Engineering
Chung Yuan University
Chung Li, Taiwan

CONTENTS

INTRODUCTION

Liquid film flows are frequently employed in industrial equipment such as wetted-wall distillation columns, gas absorbers, and in various types of coolers, condensers, evaporators and humidifiers. A description of the several important applications are summarized here in Table 1. In this chapter, we will confine mainly to falling liquid films flowing down a solid surface, with only one free surface adjacent to a gas or vapor phase. For large capacity commercial service, tube-bundle columns are used. In these columns, the liquid is introduced to

Table 1
Industrial Applications of Falling Film Shell – and – Tube Heat or Mass Exchangers [1]

Equipment	Applications
• Fluid in pressure flow on shell side, cooling water at atmospheric pressure in film flow in the tubes.	
Condensers	Condensing steam, organic vapor, refrigerants
Liquid coolers	Cooling circulating jacket water or process fluids, e.g. ammoniated brine cooler
Acid coolers	Cooling sulfuric acid, preventing leakage and dilution of acid
Sea water film coolers	Cooling plant circulating water and process fluids for plants located near the ocean, large quantities of cheap water pumped at low power cost, keeps the tubes clean
• Process fluids in film flow on the tube-side.	
Evaporators	Evaporating moisture from urea, concentration of ammonium nitrate; advantages are lack of static head, possible to evaporate without boiling point elevation as in vacuum evaporation, the very short residence time of liquid in contact with the heating surface is especially ideal for heat-sensitive material.
Devolatilizer	Remove residual volatile monomer in the liquid polymer.
Freezers	Continuous production of sized tube ice.
Molders	Selective freezing of isomers, e.g., para-dichlorobenzene from ortho-isomers, product molded and cut to desired shape and size.
Absorber coolers	Absorption of gases by liquid film inside tube, removing heat of absorption or reaction by shell-side coolant, elevated temperature in liquid reduces its absorption capacity and may lead to undesirable product or color change as in sulfonation reactions, e.g., HCl absorption in water, NH_3 absorption in water, SO_3 absorption in dodecyl benzene.
Wetted-wall columns	For distillation, absorption of gases, as a laboratory model for the measurement of molecular diffusivity, gas solubility and reaction rate constant, short columns or spheres also ideal for simulation of flow over packings.
• Other potential uses.	
Biological film reactor (trickling filter)	Liquid in laminar flow over microbiological growing film adhering to mechanical support for anaerobic studies

the top of the vertical tubes and distributed, by special distributing devices, as a thin film on the inner or outer surface of the tubes from where it falls by gravity or is carried downwards by mechanical agitation and wiping. The liquid films are characterized by a large and known interfacial area per unit volume with essentially simple geometrical configuration. The special advantage is the thin film thickness which allows the amount of liquid circulation to be small while keeping the heat or mass transfer rate high. High gas flow rate can be accommodated during cocurrent or countercurrent flow of gas and liquid film with comparatively small pressure drops. However, the difficulty of distributing the liquid uniformly to the perimeter of a large number of tubes in parallel often poses a design problem.

Fluid-Fluid Contacting Patterns

According to the various applications of the film-type equipment shown in Table 1, we can broadly classify the contacting mode into two groups:

1. Indirect contact in which the process fluid is in pressure flow on the shell side in cocurrent or countercurrent flow with the coolant in single-phase film flow inside the tubes (this is mainly for cooling purposes).
2. Direct contact or two-phase flow in which the process fluid is in film flow inside the tubes with the gas or vapor flowing cocurrently or countercurrently, this is mainly for condensation, evaporation, and gas absorption.

In some cases, a double film may form such as film evaporation inside the tube with film condensation outside the tube. The ease of design may sometimes dictate the flow pattern undertaken. For example, in film absorbers, it is much easier to introduce the gas from the bottom of the tube to produce countercurrent flow because of the difficulty in distributing the liquid and simultaneously admitting the gas at the top of the tubes. On the other hand, the phenomenon of flooding sometimes poses a problem and limits the use of countercurrent flow. In cases of two-phase gas/film flow inside the tubes, the contacting pattern can further be subdivided into four types according to flow regime classification:

1. Laminar gas/laminar liquid
2. Laminar gas/turbulent liquid
3. Turbulent gas/laminar liquid
4. Turbulent gas/turbulent liquid

In flow types 1 and 3, the liquid film surface may be smooth or wavy.

Gas Entry Methods

During cocurrent gas/liquid film flow, the gas is usually admitted from the top of the wetted-wall tube through a concentric inner tube which ends at a certain distance from the top of the wetted-wall tube while the liquid is supplied through an annular slot and flows down on

Table 2
Average Film Thickness Correlation (No Gas Flow) in the Form of
δ **(meter)** $= a\,(\nu^2/g)^{1/3}\,(4\Gamma/\mu)^b$

Author	Region	a	b	Theory or Experiment	Equation
Nusselt [18]	Laminar	0.91	1/3	Theory	(12)
Kapitza [23]	Wavy laminar	0.8434	1/3	Theory	(29)
Lukach et al. [54]	Wavy laminar	0.805	0.368	Experiment	(49)
Brotz [55]	Turbulent	0.0682	2/3	Experiment	(50)
Brauer [6]	Turbulent	0.2077	8/15	Experiment	(51)
Feind [2]	Turbulent	0.266	1/2	Experiment	(52)
Zhivaikin [56]	Turbulent	0.141	7/12	Experiment	(53)
Ganchev et al. [57]	Turbulent	0.1373	7/12	Theory	(54)
Kosky [58]	Turbulent	0.1364	7/12	Experiment	(55)
Takahama and Kato [59]	Turbulent	0.2281	0.526	Experiment	(56)
Mostofizadeh [60]	Turbulent	0.1721	0.562	Theory	(57)

the inner surface of the tube. The early chimney-type distributors are ideally suited to this simultaneous feeding of liquid and gas. For countercurrent gas/film flow, the gas can be easily admitted at the bottom of the tubes without special designs. Feind [2] has provided five different forms of the gas entry shapes and their corresponding flooding limit gas Reynolds number-correlating equations during counterflow. These entry shapes are shown in Figure 1.

Liquid Distributors and Distribution Methods

A critical factor in the successful application of the falling film principle is the design of liquid distributors in which the liquid can be distributed evenly to each tube in sufficient amount to completely wet the inner or outer walls and flows with uniform film thickness around the tube periphery. The early distributors are of the chimney type which are usually more suitable to handle two-phase flows. Today many types of new feeding devices and film producing elements are available. Some of them are shown by Sack [1]. These distributors can be classified, according to Malewski [3], into four types of design:

1. The overflow
2. The ring slot
3. The free film
4. The free jet as shown in Figure 1

Another type of liquid distributor commonly used in laboratory single-tube experiments employs a porous, sintered metal tube. However, clogging problems may arise with this type of distributor. The overflow method is the simplest but has a number of disadvantages. A difference in liquid loads often results in varying liquid levels. The rate of flow of the liquid film and the liquid distribution into the tubes are much influenced by fluctuations in the

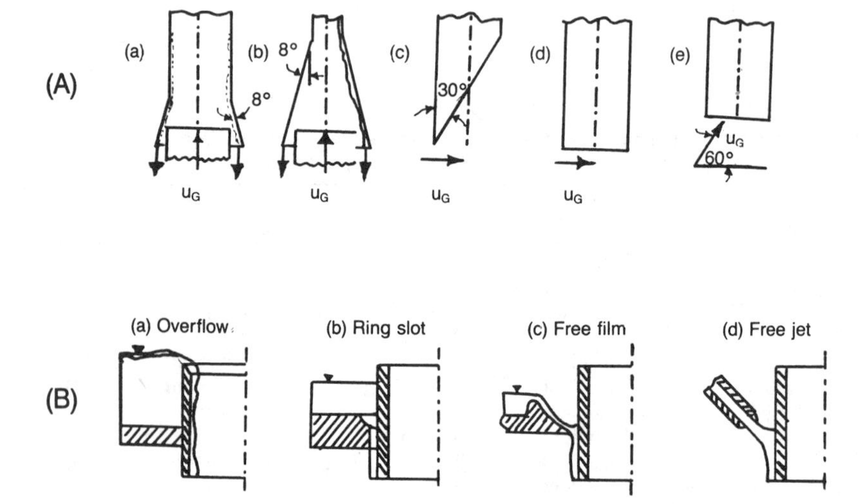

Figure 1. (A) Types of gas entry shapes; (B) types of liquid feeding devices.

liquid level at the front of the tubes and imperfect vertical alignment of the tubes. The ring slots, although more desirable from the viewpoint of uniform distribution, have considerable design and manufacture difficulties. High precision is required because very narrow slots on the order of 1 mm have to be fabricated and installed in order that they are concentric to the wetted-wall tube in perfect alignment. Contaminants or solid particles within the liquid must be avoided to prevent clogging the slot. Even greater design difficulties arise when the outer side of the tubes are to be wetted by using the ring slot. With the free film or the free jet distributor the liquid strikes the tube wall with all surfaces free. Usually only three jets are needed to space over the periphery of the tube in order to form a continuous film downstream. The diameter of the jet nozzle is larger than the annular gap of the slot, thus lessening the clogging problems. In addition, a larger tolerance is permissible in the vertical alignment of the tube. However, at high gas loads, problems of change in angle of impact or atomization of liquid before the jet reaches the tube surface may arise. A new screw-type distributor was designed by Beccari et al. [4] and compared with a conventional cone-type distributor (similar to a ring slot). The screw-type distributor was found to be substantially better in guaranteeing an adequate head loss and a proper distribution of the feed on the tube periphery and among tubes in a bundle. Moreover, the heat transfer coefficients accomplished in film evaporators are appreciably higher with this type of distributor than with a cone-type one.

HYDRODYNAMICS

Types and Regimes of Film Flow

Various types of film flow are observed in practice. The shape of the liquid distributor has a profound influence on the later development of the film down the tube. As the liquid issues from the liquid distributor such as in overflow or through the ring slot, the liquid film surface will accelerate and the film will become thinner down the tube until it reaches a fully developed velocity profile and a constant equilibrium thickness where steady uniform flow occurs. This acceleration zone, in which the flow is steady but nonuniform, is called the hydrodynamic entrance region. On the other hand, beyond a certain liquid Reynolds number when the mean film thickness exceeds the slot width, an accelerated flow becomes a retarded one. Whereas accelerated flows are very stable against disturbances, with surface ripples developing only after the acceleration zone, retarded flows tend to be unstable. Substantial differences in entrance lengths can be obtained for different liquid distributors.

The properties of film flow may depend on the Reynolds, Weber, and Froude numbers, a dimensionless interfacial shear, and various geometrical ratios. Characterization of the flow regime seems to be best based on the Reynolds number which is defined as

$$Re = 4\, u_{av}\delta\rho/\mu = 4\Gamma/\mu = 4Q/\pi d\mu \tag{1}$$

Some authors preferred to use a definition 1/4 times less than Equation 1. In general, as Fulford [5] has pointed out, smooth laminar flow occurs only after the acceleration zone at small Reynolds numbers of the order of ten. The length of this region, however, is rather short, of the order of a few centimeters. At higher Reynolds number, small amplitude waves of the regular sinusoidal form appear on the free surface. The onset of the sinusoidal waves, as determined by Brauer [6], occurs at

$$(\Gamma/\mu)_{w1} = 0.306\, K_F^{1/10} \tag{2}$$

where K_F is the film number defined as

$$K_F = \rho\sigma^3/g\mu^4 \tag{3}$$

At still higher Reynolds number of

$$(\Gamma/\mu)_i = 0.72\ K_F^{1/10} \tag{4}$$

the regular sine waves break down into swell waves with distorted wavefront and gradually pass into the roll wave region. At $(\Gamma/\mu)_{w2}$, the slopes of all the curves for the maximum film thickness, δ_{max}, as a function of Re changes shape where

$$(\Gamma/\mu)_{w2} = 1.35\ K_F^{1/10} \tag{5}$$

In the roll wave region between $(\Gamma/\mu)_i$ and $(\Gamma/\mu)_{w3}$, gravity-type surface disturbance predominates but the base film is still laminar in nature. There is a tendency of the surface waves to emerge and form a large secondary wave form called a surge wave. Above a certain critical Reynolds number of $(\Gamma/\mu)_{cr} \simeq 400$, wavy laminar flow passes through a transition and becomes turbulent, with the surface being roughened by capillary waves. The first capillary wave appears at

$$(\Gamma/\mu)_{w3} = 35\ K_F^{1/10} \tag{6}$$

Finally, above $(\Gamma/\mu) \simeq 600$, swell-like liquid rings appear and descend at a far greater speed than the residual film itself. Thus, according to the type of liquid distributor used, the flow rate and tube length employed, different flow regimes may appear on the tube. The hydrodynamics of these different flow regimes have a different and marked influence on the heat and mass transfer rate. Therefore, it is not surprising to find that experimental results on fluid, heat, and mass transfer often differ among research workers because of the different hydrodynamic conditions attained in their experiments.

Entrance Region Film Flow

Film flow in the entrance region has been studied by a number of authors for different liquid distributors: horizontal slot and overflow [7], cone-type annular slot [8], free overflow [9–11], parallel plate slot [12–14]. In general, the two-dimensional boundary layer equations are set up with the appropriate boundary conditions. For a parallel plate slot, it is assumed that the liquid is in full parabolic flow at the exit of the slot. In this way, Lynn [12] solved the boundary layer equations by neglecting the pressure gradient $\partial P/\partial x$. His theory is in good agreement with his own experimental data performed on an apparatus inclined at 6° to the horizontal with slot width of 2.57 mm and using a mixture of about 50% glycerine in water. Numerical solutions considering variable film depth and the boundary-layer approximation were given by Cerro and Whitaker [13] whose calculations agreed with the data of Lynn [12] and their own. Numerical solutions by Ault and Sandall [14] showed that the term $\partial P/\partial x$ is of negligible importance. Integral methods together with a sine velocity profile proposed by Stucheli and Ozisik [15] gave analytical results which were in close agreement with the numerical solution and data of Cerro and Whitaker [13]. For a free overflow entry method, Bruley [9] assumed that the vertical free stream velocity at the leading edge is uniform and can be approximated by a free-fall equation. His numerical solutions were later extended by Hassan [10] and Haugen [11]. Yilmaz and Brauer [16] used accelerated, wave-free flow down an incline as a model for the description of the fluid dynamic behavior in packed beds. The origin of the flow length x coincides with the transition point between two packing pieces. At this point, a uniform velocity occurs over the film thickness as a result of strong mixing. After the acceleration is complete, smooth laminar flow results from equilibrium of gravity and friction forces. By comparing with Fulford's results [17], they obtained an equation for calculating the dimensionless entry length x_e^* for the formation of smooth laminar flow

$$x_e^* = x_e(g \sin \theta/3\nu^2)^{1/3} = 0.67(\Gamma/\mu)^{4/3} \tag{7}$$

Since waves appear after the complete development of smooth laminar flow, another entry length x_{ew}^* can be defined as the length beyond which waves begin to form and $x_e^* < x_{ew}^*$. This can be calculated from

$$x^*_{ew} = 2{,}100/[(\Gamma/\mu)/(\Gamma/\mu)_{w1} - 1]^{3/4} + 10(\Gamma/\mu) \tag{8}$$

where $(\Gamma/\mu)_{w1}$ is given by Equation 2 modified by adding a term $5/(6 \tan \theta)$ for an incline. In addition, these authors provided a diagram depicting zones for various states of accelerated film flows.

These investigations show that in general:

- For equal Re, the entrance lengths are longer for the more viscous liquids because of the proportionally larger volume flow rate.
- For equal volume flow rate, however, the entrance length is less for the more viscous liquids.
- The entrance length increases with an increase in Re.

Smooth Laminar Flow

For steady uniform laminar flow of a smooth film on a vertical plane, Nusselt [18] obtained the velocity profile u(y), the average velocity u_{av}, the ratio of surface velocity to average velocity u_{max}/u_{av}, and the film thickness δ as follows:

$$u(y) = \frac{\rho g \delta^2}{\mu}\left[\frac{y}{\delta} - \frac{1}{2}\left(\frac{y}{\delta}\right)^2\right] \tag{9}$$

$$u_{av} = \rho g \delta^2/3\mu = 0.08667\ Re^{2/3} \tag{10}$$

$$u_{max}/u_{av} = 1.5 \tag{11}$$

$$\delta = 0.91(\nu^2/g)^{1/3}\ Re^{1/3} \tag{12}$$

The friction factor can be derived as

$$f = 24/Re \tag{13}$$

For film flow on the inside of a vertical tube of radius R, Jackson [19] derived the velocity profile and the ratio u_{max}/u_{av}, respectively as

$$u(r) = \frac{g}{4\nu}(R^2 - r^2) + 2(R - \delta)^2 \ln\left(\frac{r}{R}\right) \tag{14}$$

$$\frac{u_{max}}{u_{av}} = 2\left[\frac{R^2 - (R - \delta)^2 + 2\,(R - \delta)^2 \ln\left(\frac{R - \delta}{R}\right)}{R^2 - 3\,(R - \delta)^2 - \frac{4\,(R - \delta)^4}{R^2 - (R - \delta)^2}\ln\left(\frac{R - \delta}{R}\right)}\right] \tag{15}$$

On the other hand, for film flow on the outside of the tube, the velocity profile is

$$u(r) = \frac{g}{4\,\nu}(R^2 - r^2) + \frac{g}{2\,\nu}(R + \delta)^2 \ln\frac{r}{R} \tag{16}$$

and the flow rate per unit periphery is a complex function involving δ/R as given by Feind [2] and when expanded in powers of δ/R gives

$$\Gamma = u_{av}\delta = \frac{\rho g \delta^3}{3\,\mu}\left[1 + \frac{\delta}{R} + \frac{3}{20}\left(\frac{\delta}{R}\right)^2 - \frac{1}{40}\left(\frac{\delta}{R}\right)^3 + \frac{1}{140}\left(\frac{\delta}{R}\right)^4 \cdots\right] \tag{17}$$

The terms in the brackets represent a correction of the film thickness due to the curvature effect of the tube. For example, for $\delta/R = 0.1$, the film thickness on a tube is about 4% smaller than the one on a plane wall and Equation 12 can be used with negligible error. On the other hand, for $\delta/R = 1$, the film thickness is about 22% smaller. Therefore, when the tube radius is so small such that it is about the same order of magnitude as the film thickness, the use of the implicit equation Equation 17 is abolutely necessary. The importance of this arises recently in a new and innovative design of cylindrical supports for the liquid film to replace the conventional tubes whose diameters have finite minimum dimensions. This new design was described by Lefebvre et al. [20, 21] in which the liquid is fed to the top of a vertically stretched textile cable (for example, polyamide or polyester yarn, 1 to 2 mm diameter) and a cylindrical film is set up around the cable. A bundle of such cables constitutes an outstanding packing employing the wetted-wall principle. They established a film thickness correlation for such cables:

$$\delta/R = 1.33\ Re^{0.35}(gd^3/\nu^2)^{-0.34} \tag{18}$$

Wavy-Laminar Flow

As the flow rate exceeds a certain initial value, waves begin to appear on the free surface. The type of wave and wave structure generally vary with the distance along the flow direction and are strongly influenced by the liquid and gas Reynolds number. Dukler [22] has reviewed the different approaches of modeling the wave structure. The more common approach is either to use a linear stability theory where small perturbation expansions for the stream function are performed to solve the momentum equation, or to search for a periodic solution to the boundary-layer equations employing integral methods. The governing equations for the two-dimensional film flow are the continuity equation

$$\frac{\partial u}{\partial x} + \frac{\partial v}{\partial y} = 0 \tag{19}$$

and the Navier-Stokes equations

$$\frac{\partial u}{\partial t} + u\frac{\partial u}{\partial x} + v\frac{\partial u}{\partial y} = -\frac{1}{\rho}\frac{\partial p}{\partial x} + \nu\left(\frac{\partial^2 u}{\partial x^2} + \frac{\partial^2 u}{\partial y^2}\right) + g\sin\theta_1 \tag{20}$$

$$\frac{\partial v}{\partial t} + u\frac{\partial v}{\partial x} + v\frac{\partial v}{\partial y} = -\frac{1}{\rho}\frac{\partial p}{\partial y} + \nu\left(\frac{\partial^2 v}{\partial x^2} + \frac{\partial^2 v}{\partial y^2}\right) + g\cos\theta_1 \tag{21}$$

with the boundary conditions

$$y = 0\ \text{(wall)} \qquad u = v = 0 \tag{22}$$

$$y = \delta \text{ (interface) } \partial u/\partial y + \partial v/\partial x = 0 \tag{23}$$

$$p = 2\,\mu \frac{\partial v}{\partial y} + \sigma \frac{\partial^2 \delta}{\partial x^2} \tag{24}$$

Kapitza Method of Wavy Film Flow Description

The first comprehensive theoretical and experimental study on the hydrodynamics of wave formation on thin liquid films was made by Kapitza [23]. Certain simplifications were made by neglecting the y-component momentum equation and the terms $v(\partial u/\partial y)$ and $\nu(\partial^2 u/\partial x^2)$ in Equation 20. The boundary conditions also simplify to $\partial u/\partial y = 0$ and $p = \sigma(\partial^2\delta/\partial x^2)$. The velocity distribution is assumed to be still given by the laminar profile

$$u(x, y, t) = 3u_{av}(x, t)(y/\delta - y^2/2\delta^2) \tag{25}$$

The wavelength is assumed to be considerably larger than the film thickness so that $\delta = \delta_0(1 + \phi)$ where δ_0 is the mean film thickness and ϕ is the local deviation. Integrating Equation 20 over the film thickness and assuming that the waves are periodic permit the time dependence to be eliminated through the new variable $(x - ct)$. After various arrangements, a third-order ODE for ϕ results. To the first approximation for an undamped periodic solution to exist, he obtained

$$u_{av} = \delta_0{}^2 g \sin\theta/3\nu \tag{26}$$

$$c/u_{av} = 3 \tag{27}$$

In the second approximation Kapitza determined, by considering that the energy supplied to the film by the gravity force is balanced by the viscous energy dissipation, that

$$(\Gamma/\mu)_{w1} = 0.61(K_F \sin\theta)^{1/11} \tag{28}$$

$$\delta_0 = [(3\nu Q/g)\Phi]^{1/3} = 0.93(3\nu Q/g)^{1/3}, \; \Phi = 0.8 \tag{29}$$

$$c/u_{av} = 2.4 \tag{30}$$

$$\lambda = 7.5(\nu\sigma/\rho g Q)^{1/2} \tag{31}$$

Thus the Kapitza treatment predicts that in a steady periodic flow at the same flow rate, the average film thickness in wavy flow is about 7% less than the laminar film thickness. This has been confirmed in some of the experiments by Portalski [24]. Also Kapitza showed that a critical Reynolds number for wave initiation, $(\Gamma/\mu)_{w1}$, exists. In the same work, Kapitza obtained experimental results for the average film thickness, wave amplitude, wavelength, and wave velocity. Comparison between his theory and data shows that the agreement is only good near the point of wave inception. This is because Kapitza's theory is restricted only to long waves and to low flow rates where regular sinusoidal wave motion exists. Later experimental findings for higher flow rates [25–27] show that Kapitza's theory is inadequate. In particular, data of equilibrium amplitude and ratio of wave celerity to average velocity show variation with the flow rate and the liquid physical properties while Kapitza's theory predicts them to be constants. Kapitza also predicts that the wavelength will vary with the flow rate but experiments show that it is practically constant and independent of flow rate.

A number of refinements on the Kapitza theory have appeared [28–33]. The notable ones are the nonlinear theory of Shkadov [28] which gave much better agreement with Kapitza's data on wavelength and wave velocity, and the work of Penev et al. [32] who solved the complete two-dimensional Navier-Stokes equations without linearization or order-of-magnitude analysis.

Hydrodynamic Stability Theory

Another commonly used approach to predict wave inception and wave characteristics is based on the hydrodynamic stability theory. Infinitesimal two-dimensional waves which are harmonic in space and time and grow or decay temporally are described by a disturbance stream function of the form

$$\phi(x, y, t) = \phi(y) \exp [i\alpha_r(x - ct)] \quad (32)$$

where $\phi(y)$ is the dimensionless amplitude, α_r is the dimensionless real wave number, $c = c_r + ic_i$ is the dimensionless complex wave velocity. The temporal amplification factor is $\alpha_r c_i$. With small perturbations imposed on the velocity and pressure, and substitution for the stream function leads to a temporal form of the Orr-Sommerfeld (O-S) equation:

$$\phi'''' - 2\alpha_m^2\phi'' + \alpha_m^4\phi = i\alpha_m R_l[(\overline{U} - c)(\phi'' - \alpha_m^2\phi) - \overline{U}''\phi] \quad (33)$$

with the boundary conditions

- $\phi(0) = 0$ at wall (34)
- $\phi'(0) = 0$ at wall (35)
- $\phi''(1) + \{\alpha_m^2 + \overline{U}''|_{\eta=1}/[c - \overline{U}(1)]\}\phi(1) = 0$ (36)

  shear stress at interface

- $$-\left\{\alpha_m\left[\frac{R_l \cos\theta_l/Fr^2 + \alpha_m^2 R_l S}{C - \overline{U}(1)} - R_l\overline{U}'(1)\right]\right\}\phi(1) +$$

  $$\alpha_m[R_l(c - \overline{U}(1)) + 3\,i\alpha_m]\phi'(1) - i\phi'''(1) = 0 \quad (37)$$

  normal stress at interface

where
$$\overline{U} = u/u_{av} = 3(\eta - \eta^2/2)$$
$$\eta = y/\delta$$
$$R_l = u_{av}\delta/\nu$$
$$u_{av} = g\delta^2 \sin\theta_l/3\nu$$
$$S = \sigma/\rho u_{av}^2\delta$$
$$Fr = u_{av}/(g\delta)^{1/2} \quad (38)$$

Equation 33 and the boundary conditions Equations 34 through 38 constitute an eigenvalue problem. The equation $c_i(R_l, Fr, \alpha_m) = 0$ defines a relation between R_l and α_m for a given value of θ_l, and the graph describing this relationship is the neutral stability curve. Various approximate analytical and numerical methods have been used to solve the O-S equation [34–41]. Benjamin [34] used a power series expansion and Yih [35] used a successive perturbation method which are restricted to long wavelengths and small Reynolds numbers. Whitaker [36] provided a direct numerical solution which could be used for any Reynolds number and wave number. Since Yih has pointed out that the stability characteristics are governed by the surface waves, this has led Anshus and Goren [37] to replace the semi-parabolic velocity profile by the free surface velocity in the O-S equation while the second derivative of the velocity is kept at its true value. This simplifies the O-S equation to a constant coefficient fourth-order ODE which permitted the eigenvalues to be determined by a simple technique. A comparison with the exact numerical solution of Whitaker shows that this approximation is quite accurate.

Krantz and Owens [42], recognizing that after wave inception, wave amplitude increased spatially and not temporally, questioned the validity of the temporal formulation of the linear stability theory. They proposed a spatial formulation of the O-S equation in which the disturbance stream function is of the form

$$\phi(x, y, t) = \phi(y) \exp [i(\alpha_m x - \omega_r t)] \tag{39}$$

where $\alpha_m = \alpha_r + i\alpha_i$ is the dimensionless complex wave number, α_i is the spatial amplification factor, and ω_r is the dimensionless real angular frequency. The spatial formulation differed from the classical temporal formulation only in that α_m replaced α_r, and ω_r/α_m replaced c. However, the complex wave velocity is the eigenvalue and the real wave number is a parameter in the temporal formulation, whereas the complex wave number is the eigenvalue and the real wave frequency is a parameter in the spatial formulation. The solution to these two eigenvalue problems, according to what Krantz and Owens claimed, will differ. Lin [43], however, maintained that the two formulations and solutions are the same through the use of Gaster's theorem.

To illustrate the usefulness of the prediction methods, Figure 2 shows a comparison between several different predictions with a large amount of wave number and wave velocity data collected by different authors. The numerical solutions of Pierson and Whitaker [41] and Graef [39] seem to represent the data very well even at relatively large Reynolds number. The approximate solution of Anshus and Goren [37] is also very satisfactory in this respect. The predictions using the Kapitza approach, e.g. Gollan and Sideman [30], Penev et al. [32], are less accurate in predicting these wave properties.

Deformation Turbulence Theory for Wavy Liquid Films

Stochastic deformation of the interface produces random motion in the fluids on both sides of the interface. This type of stochastic fluid motion is referred to as deformation turbulence by Brauer [44]. Deformation turbulence in liquid films arises as a result of the fast-changing character of the film interface resulting in a rapid local change in the film thickness. Such changes induce stochastic variations in the magnitude and direction of the local velocity. The stochastic nature of the surface deformation can be confirmed by the measurement of wave frequency using an electrical probe [6]. The deformation turbulence theory as applied to wavy liquid films is based on two assumptions:

1. The wavy film may be substituted by a film with a smooth surface.
2. The effect of deformation turbulence may be expressed by an appropriate parameter such as an effective diffusivity to be introduced into the transport equations.

It should be noted that deformation turbulence is different from Reynolds turbulence which occurs at $Re > 1{,}600$.

Turbulent Flow

Turbulent film flow has attracted a great deal of attention recently because most heat and mass transfer processes in film flow have been conducted in the turbulent flow region. Brauer [6] was the first to show, by means of shadowgraphs, that the film structure consists essentially of two portions: a base film portion and on top of it large turbulent waves. He used an electrical probe to measure the maximum and minimum film thickness. More re-

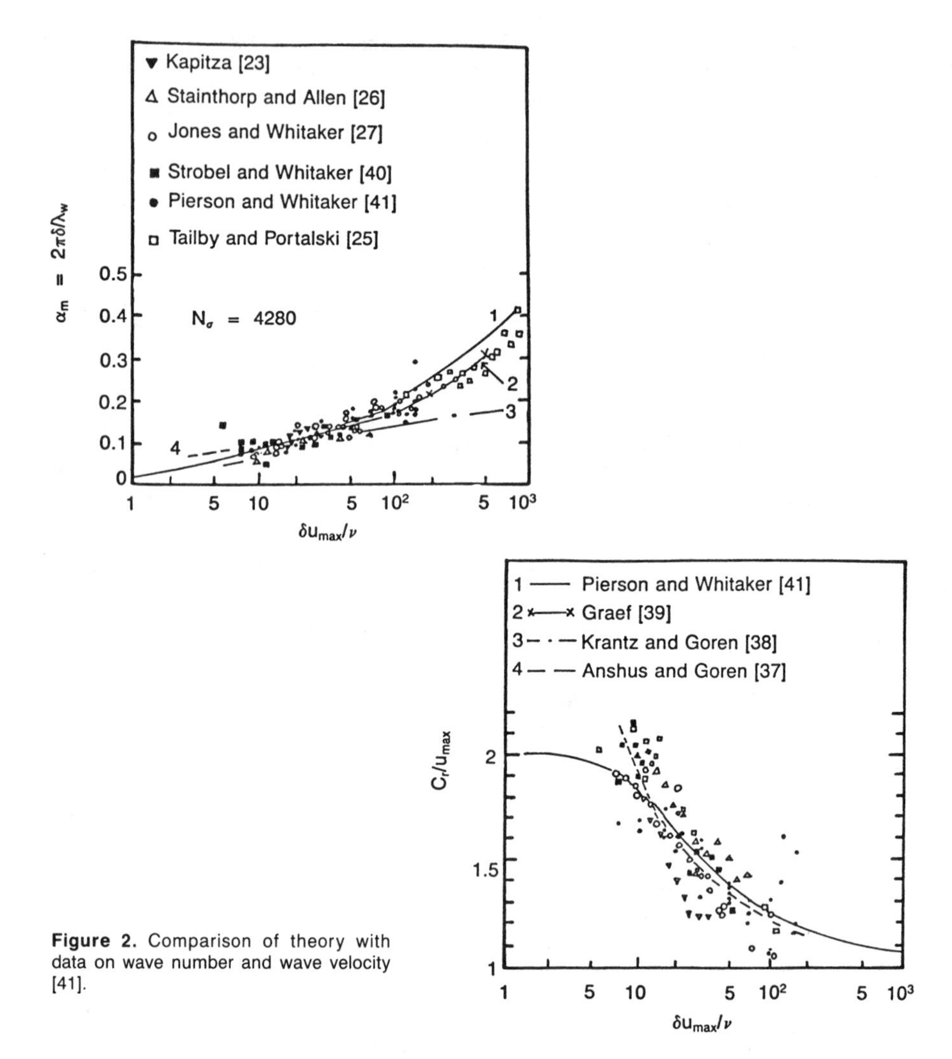

Figure 2. Comparison of theory with data on wave number and wave velocity [41].

cently, Telles and Dukler [45] analyzed the time-varying film thickness obtained from electrical conductivity measurements and showed that the structure of the gas-liquid interface was random in character and behaved as a "two-wave system" as shown in Figure 3. The large turbulent waves carry a significant portion of the total flow and move over the base film with relatively no change in speed or shape. Smaller waves coexisting on this large wave structure lose their identity over small distances. As the wave moves at a much higher velocity passing over the base film, it continuously overtakes the liquid in the base film at the wave front while an equivalent amount of liquid is left behind to make up the laminar base film at the back of the wave. The turbulent waves have an extremely large base-to-height ratio, move independently of each other, and look like solitary large lumps of liquid. Telles and Dukler modeled the large wave shape by a truncated Gram-Charlier series. Methods for calculating statistically meaningful wave celerity, wave separation distance, amplitude, fre-

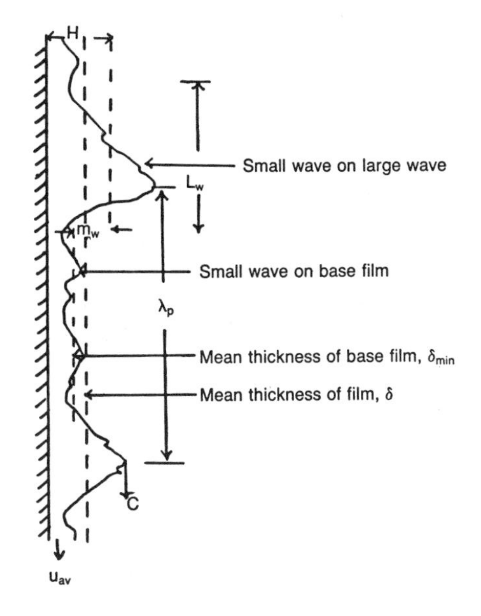

Some Wave Properties

Re = 1600

	Large	Small	
Amp.	0.812	0.051	mm
Length	15.54	1.17	cm
Sep.	32.54	1.17	cm
Veloc.	1.58	0.45	m/s
$\frac{\text{Amp.}}{\delta}$	1.12	0.15	

Figure 3. Identification of wave class [46].

quency, and wave shapes were presented for liquid Reynolds number from 900 to 6,000 and gas Reynolds number from 0 to 60,000. Chu and Dukler [46, 47] subsequently examined the statistics and structure of the thin base film and the large waves respectively. From typical film thickness traces, it was revealed that the base film was not smooth, but covered on its surface by small waves. In addition, there existed another small wave structure which rode on the large waves. The amplitude of all large waves was not constant. The amplitude of the waves on the base film was observed to be insensitive to gas interfacial shear while the large wave structure was strongly influenced by gas flow. They concluded that the large wave structure controlled the hydrodynamics and associated transfer processes in the liquid film while the small waves on the base film controlled the transfer processes in the gas phase.

The statistical approach, sometimes called the variable film thickness model, although providing a more realistic view of the turbulent film structure, is nevertheless still not well established and too complicated for practical engineering applications. Up to now, only meager statistical data for the system water-air at several liquid and gas Reynolds numbers are available. Predictions of heat or mass transfer as presented by Brumfield and Theofanous [48, 49], using the statistical data measured by Dukler and co-workers, have also been limited to those cases where the hydrodynamic data are available. Besides the statistical approach, nearly all remaining major investigations on turbulent film flow are based on the constant film thickness model which assumes that the film can be treated as smooth so that it can be represented by an average film thickness and that some form of a dimensionless velocity profile or eddy diffusivity profile valid for single-phase flow can be modified or directly applied to turbulent film flow.

Film Thickness

The degree of success of the constant film thickness model in predicting heat or mass transfer depends to a large extent on whether the mean film thickness can be predicted or measured accurately. For smooth laminar flow, the film thickness can be predicted from Nusselt's equation, Equation 12. For wavy laminar flow, some investigators found that Kapitza's equation, Equation 29, is more accurate. However, a majority of measurements showed that Nusselt's equation still applied in this region, especially in the higher Reynolds number region.

For turbulent flow, the film thickness can be calculated from the following procedure. The shear stress distribution in the liquid film is given by

$$\tau = \rho(\delta - y)g/g_c + \tau_i \tag{40}$$

where τ_i is the interfacial shear exerted by the gas. The wall shear is

$$\tau_w = \rho\delta g/g_c + \tau_i \tag{41}$$

If we define $\delta^+ = \delta u^*/\nu$, $u^* = (\tau_w g_c/\rho)^{1/2}$, $\tau_i^* = \tau_i g_c/\rho(\nu g)^{2/3}$, $s^3 = (\rho\delta g/g_c)/(\tau_i + \rho\delta g/g_c)$, then Equation 40 becomes

$$s^3 + \tau_i^* s^2/\delta^{+2/3} - 1 = 0 \tag{42}$$

The universal velocity distribution can be calculated from

$$\frac{du^+}{dy^+} = \frac{1 - s^3 y^+/\delta^+}{1 + \epsilon_M/\nu} \tag{43}$$

if the eddy diffusivity profile for momentum, ϵ_m, is specified. The procedure is as follows: a typical value of τ_i^* and δ^+ is chosen and s is calculated from Equation 42 by a root-finding procedure. Then the velocity profile $u^+(y^+)$ can be solved by numerical integration of Equation 43. In Equation 43, $u^+ = u/u^*$, $y^+ = yu^*/\nu$. When there is no interfacial shear, $\tau_i^* = 0$ and $s = 1$, then this equation reduces to the nonsheared film case. The film Reynolds number is calculated from

$$Re = 4 \int_0^{\delta^+} u^+ \, dy^+ \tag{44}$$

Since $\delta^* = s\delta^{+2/3}$ where $\delta^* = \delta/(\nu^2/g)^{1/3}$, a plot of the dimensionless film thickness, δ^*, versus Re for different values of τ_i^* can be obtained. For the nonsheared film case, Dukler and Bergelin [50] were the first to use the three-layer universal velocity profile equations of Nikuradse-von Karmen to calculate the film thickness. For the turbulent flow region, they obtained

$$\Gamma/\mu = \delta^+(3 + 2.5 \ln \delta^+) - 64, \qquad \delta^+ > 30 \tag{45}$$

For $\delta^+ \leq 30$, Portalski [24] obtained, for the laminar sublayer plus buffer layer:

$$\Gamma/\mu = 5\delta^+ \ln \delta^+ - 8.05\delta^+ + 12.5, \qquad 5 \leq \delta^+ \leq 30 \tag{46}$$

and for the laminar flow region

$$\Gamma/\mu = \delta^{+2}/2, \qquad 0 \leq \delta^+ < 5 \tag{47}$$

Portalski noted that the universal velocity profile led to predictions of the average film thickness in fair agreement with experimental data. Since the work of Dukler and Bergelin, there have been many other more refined models proposed which can predict both the average film thickness and the heat or mass transfer coefficients simultaneously. A summary of these turbulence models will be given later.

Film thickness is one of the most important properties of film flow and has been subjected to intensive experimental studies. Fulford [5] has summarized the various experimental techniques used by different authors up to 1964. They are

- Micrometer gauge and pointer
- Improved probe detection by electrical-operated feeler
- Photography
- Weighing the channel and film
- Drainage technique
- Electrical resistance method
- Electrical capacitance method
- Light-beam absorption method
- Radioactive tracer method

Most of these techniques are still currently being used due to their simplicity or good accuracy. The first method is rather obsolete nowadays since it is not accurate in the presence of waves. The fourth and fifth methods are simple but can only give average film thickness. The other methods may be used for local film thickness measurements. Recently a new technique called laser beam refraction has been described by Marschall [51] and Salazar and Marschall [52, 53] which seems to be quite promising.

To illustrate some of the typical measurements, Figure 4 shows the data of the maximum, minimum, and average film thickness of a water film as measured by Brauer [6] using an electrical-operated probe. The minimum film thickness corresponding to the trough of the

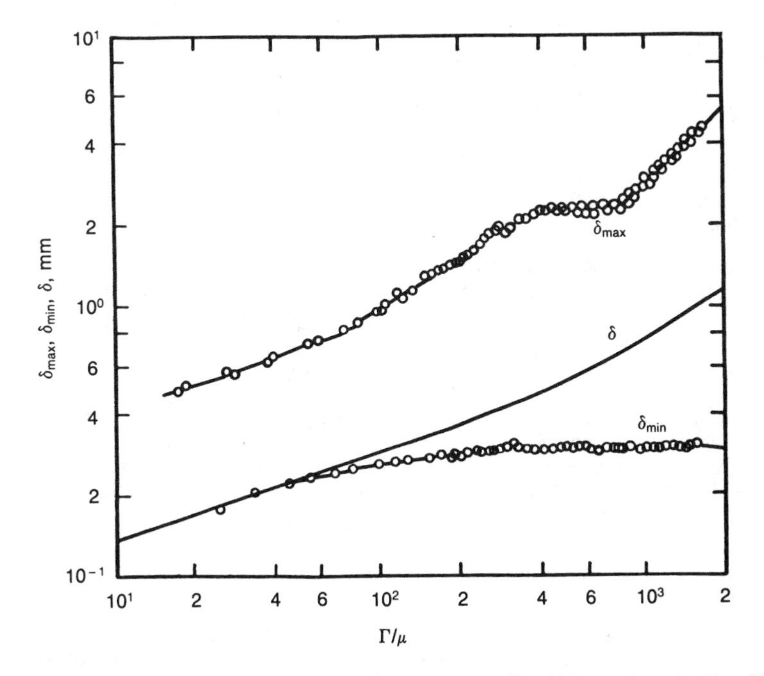

Figure 4. Measurements of maximum, minimum, and average film thickness for water films [6].

wave rises slowly with increase in the Reynolds number and becomes approximately constant after $\Gamma/\mu = 400$. The maximum film thickness corresponding to the crest of the wave, however, increases continuously with an increase in the Reynolds number. Average film thickness correlations have been proposed by a number of investigators in the form of

$$\delta = a(\nu^2/g)^{1/3}(4\Gamma/\mu)^b \tag{48}$$

where the constants a and b are tabulated in Table 2 and δ in meters. In general, the equations of Nusselt, Kapitza, Brauer, or Brotz are the most commonly used for the laminar, wavy-laminar, and turbulent flow regions, respectively.

To illustrate the applicability of the prediction methods, a modified van Driest model used by Yih and Liu [61], as shown later in Table 3, can predict the dimensionless film thickness, δ^+, within 5% when compared with the Nusselt equation at Re < 1,600 and with the Brauer equation at Re > 1,600, Figure 5. In the presence of cocurrent gas interfacial shear, the model also predicts fairly well the film thickness data of Ueda and Tanaka [62], Figure 6. It can be seen that increasing interfacial shear substantially reduces the film thickness.

HEAT TRANSFER

Heating

Laminar and Wavy-Laminar Flow

Under the assumption of fully developed laminar flow with heat transfer from or to the wall and negligible heat transfer between the film interface and the adjoining gas, the governing convection-conduction equation is

$$u(y)\frac{\partial T}{\partial x} = K\frac{\partial^2 T}{\partial y^2} \tag{49}$$

where u is given by the semi-parabolic velocity profile, Equation 9, and the boundary conditions are

$$x = 0 \text{ (entrance)} \qquad T = T_{in} \tag{50}$$
$$y = 0 \text{ (wall)} \qquad T = T_w \text{ or } q_w = \text{constant} \tag{51}$$
$$y = \delta \text{ (interface)} \qquad \partial T/\partial y = 0 \tag{52}$$

At short contact times between the wall and the falling film or when the heating length is short, the liquid temperature changes appreciably only in the immediate vicinity of the wall and by the Leveque assumption, the velocity profile near the wall can be approximated by

$$u(y) = \rho g \delta y/\mu \tag{53}$$

In heating or cooling, the local heat transfer coefficient is based on the difference between the wall and local bulk temperature of the film. For constant wall temperature, the Nusselt number for the Leveque asymptote valid for short contact time is obtained by solving Equation 49 subjected to Equation 53 and the boundary conditions with Equation 52 changed to $T = T_{in}$ at $y \rightarrow \infty$:

$$Nu_x = h_x\delta/k = 0.678x^{*-1/3} \tag{54}$$

where $x^* = x\alpha/\delta^2 u_{max}$ is the dimensionless axial length. For intermediate to long contact times, Equation 49 has to be solved with the complete semi-parabolic velocity profile to give the Nusselt number in the thermal entrance and fully developed regions. A solution employ-

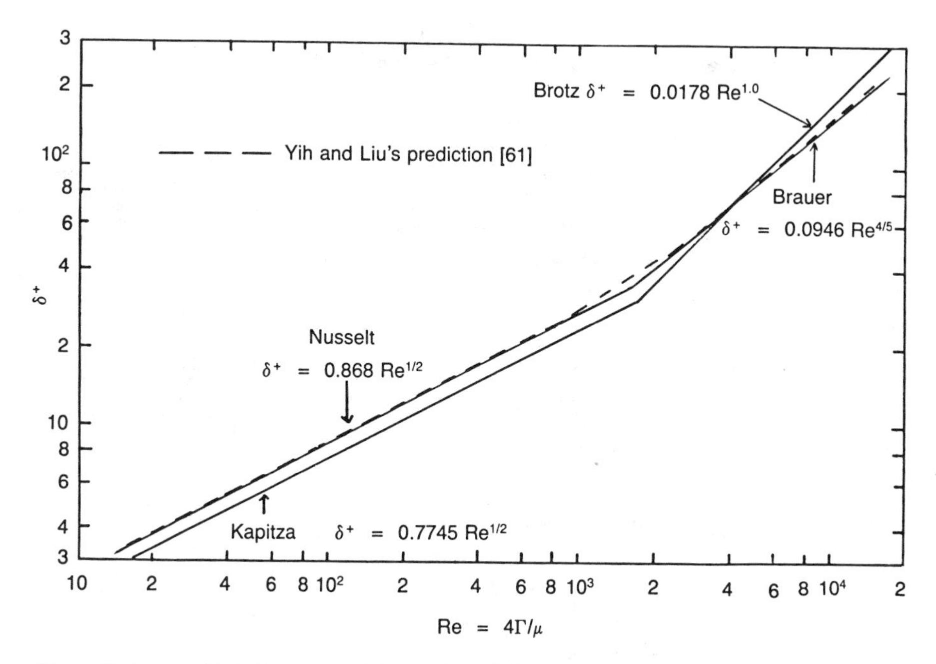

Figure 5. Average film thickness correlations and prediction.

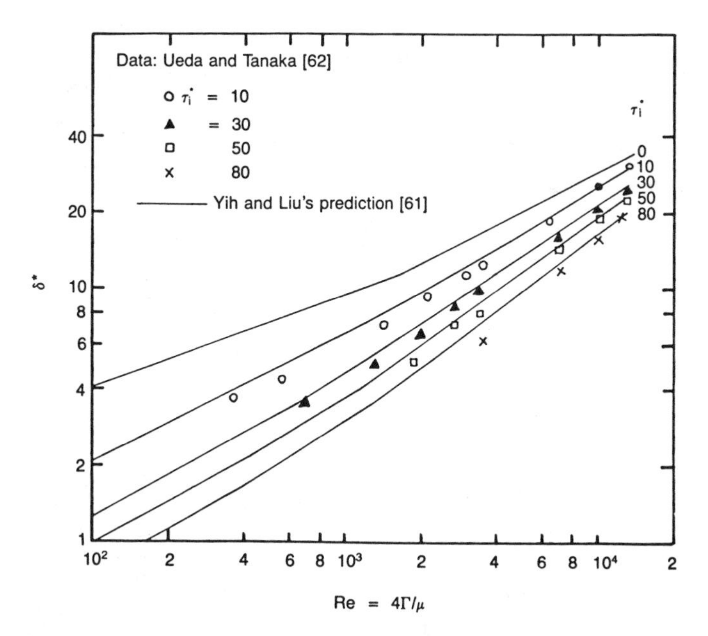

Figure 6. Average film thickness prediction as compared to the data of Ueda and Tanaka [62] for heating with interfacial shear [61].

Table 3
Turbulence Models Used in Film Heating († with Interfacial Shear)

Year	Author	Turbulence Model Used	Comparison with Data
1973	Limberg [67]	Modified van Driest model for $y/\delta \le 0.6$ $\epsilon_M/\nu = -0.5 + 0.5\,(1 + 0.64y^{+2}(\tau/\tau_w)\{1 - \exp[-y^+ (\tau/\tau_w)^{1/2}/25.1]\}^2 f^2)^{1/2}$ $\tau/\tau_w = 1 - y^+/\delta^+$; $f = \exp(-1.66\,\tau/\tau_w)$; $\epsilon_M/\nu = \epsilon_M/\nu \mid_{y/\delta=0.6} =$ constant for $0.6 < y/\delta \le 1$ $Pr_t = 0.89$	δ^+ versus Re conform to Brauer's correlation as shown by Seban and Faghri [70]; Nu_x versus Re slightly higher than Wilke's data [71] at high Re but much higher than Wilke's data at low Re.
1974	Ishigai, Nakanisi, Takehara, and Oyabu [68]	Spalding model for all y^+ $\epsilon_M/\nu = 0.1108\,B\,[\exp(Bu^+) - 1 - Bu^+ - (Bu^+)^2/2 - (Bu^+)^3/6]$ $Pr_t = 1$, $B = 0.4$	δ^* versus Re 10% to 20% higher than Brauer's correlation; Nu_x versus Re slightly higher than Wilke's data [71] at high Re but much higher than Wilke's data at low Re.
1974	Gimbutis [69]	Modified van Driest model $\epsilon_M/\nu = -0.5 + 0.5\,[1 + 0.64y^{+2}(1 - y^+/\delta^+)N^2\alpha_R]^2$ $N = \exp[-(1 - y^+/25)^n/\beta_1]$ for $y^+ \le 25$ $N = 1$ for $y^+ > 25$ n and β_1 are functions of Re, α_R is film curvature factor $Pr_t = 0.9$	δ^+ versus Re in satisfactory agreement; h_x^* versus Re gets better agreement with data when expressed together with a correction factor $(Pr/Pr_w)^{0.25}$.
1976	Faghri [64], Seban and Faghri [70],	Modified van Driest model as proposed by Limberg [67] for $y/\delta \le 0.6$ $\epsilon_M/\nu =$ same as Limberg for $y/\delta \le 0.6$ $\epsilon_M/\nu = \epsilon_M/\nu \mid_{y/\delta=0.6} =$ constant for $0.6 < y/\delta \le y_e/\delta$ Modified Lamourelle and Sandall [72] gas absorption eddy diffusivity as proposed by Mills and Chung [73] for $y_e/\delta < y/\delta \le 1$ $\epsilon_M/\nu =$ same as Mills and Chung for outer region $Pr_t = 0.9$ (Model 3A) $Pr_t = 0.9\,[1 - \exp(-y^+/A^+)]/[1 - \exp(-y^+/B^+)]$ (Model 3) $A^+ = 25.1$, $B^+ = Pr^{-1/2} \sum_{i=1}^{i=5} b_i\,(\log_{10}Pr)^{i-1}$ $b_1 = 34.96$, $b_2 = 28.97$, $b_3 = 33.95$, $b_4 = 6.33$, $b_5 = -1.186$	δ^+ versus Re conform to Brauer's correlation; h_x^* versus Re slightly higher than Wilke's correlation for heating [71] at high Re but much higher than the correlation at low Re for Pr = 5.4.
1982 1983	Yih and Chen [74], Yih and Liu [61]†	Modified van Driest model as proposed by Limberg [67] for $y/\delta \le 0.6$ $\epsilon_M/\nu =$ same as Limberg for $y/\delta \le 0.6$ but $\tau/\tau_W = 1 - s^3y^+/\delta^+$ including the effect of interfacial shear $\epsilon_M/\nu =$ same as Limberg for $0.6 < y/\delta \le 1$ $Pr_t = \{1 - \exp[-y^+(\tau/\tau_W)^{1/2}/A^+]\}/\{1 - \exp[-y^+(\tau/\tau_W)^{1/2}/B^+]\}$	δ^+ versus Re agrees within 5% with Brauer's correlation; better agreement with Wilke's correlation [71] on h_x^* versus Re than Limberg's [67] or Ishigai et al.'s [68] predictions especially for Re > 2,400.

ing the separation of variables method was given by Yih and Huang [63]. Faghri [64] also provided a numerical solution. The fully developed asymptotic Nusselt number and dimensionless heat transfer coefficient are

$$Nu_{\infty} = 1.88 \tag{55}$$

$$h^*_{\infty} = \frac{h_{\infty}}{K}\left(\frac{\nu^2}{g}\right)^{1/3} = 2.066\ Re^{-1/3} \tag{56}$$

Nusselt [18] noted that at small heating lengths of $x < 0.5$ m, the influence of the entrance flow could not be neglected and he gave a theoretical equation valid for the entire domain of Reynolds number:

$$Nu_x = 0.0942\ Re\ Pr\ \delta/L + 1.88 \tag{57}$$

The physical properties of liquid are to be evaluated at the average liquid temperature.

For constant wall heat flux, Yih and Lee [65] and Faghri [64] also provided analytical and numerical solutions respectively. The results for the case of fully developed flow are

$$Nu_{\infty} = 2.059 \tag{58}$$

$$h^*_{\infty} = 2.262\ Re^{-1/3} \tag{59}$$

When waves appear on the free surface, experiments indicated that the waves enhanced the heat transfer rate above that calculated according to laminar flow. Kapitza [23] was the first to present an approximate analysis of heat transfer in wavy liquid films with sinusoidal motion. He showed that probably two properties of wavy flow enhanced the heat transfer rate. One is the decrease in the effective average film thickness below that of the smooth film for the same flow rate, the other is the convective motion generated by the periodic motion. Using the values of wave amplitude and average film thickness from his hydrodynamic model solution, Kapitza has shown that the heat transfer coefficient for a wavy film is 21% greater than that of the equivalent smooth film, a result which agrees with some of the data reported.

Earlier, Bays and McAdams [66] presented data for heating of laminar oil films down the inner surface of vertical tubes ranging in length from 12.2 to 182.9 cm. The Reynolds number ranged from 1.7 to 2,000, which means that wavy flow is also present. The heat transfer coefficients ranged from 227 to 795 watt/m^{20}K. They used steam heating which approximated the constant wall temperature boundary condition. Their data were correlated as

$$\left(\frac{h_x\delta}{K}\right)\left(\frac{\mu_w}{\mu}\right)^{1/4} = \left(\frac{c_p\delta\Gamma}{K\ x}\right)^{1/3} = 0.873\ \chi^{*-1/3},\ x^* < 2/3 \tag{60}$$

where h_x was based on the arithmetic mean of the entrance and exit ΔT between the wall and liquid. Liquid physical properties were evaluated at the arithmetic mean of the entrance and exit liquid temperatures. Equation 60 shows the same trend as the Leveque solution, Equation 54, but is about 30% higher due to the effect of wave motion.

Turbulent Flow

Theoretical treatments of heating in turbulent falling films have been described by a number of investigators [61, 67–70] all using the constant film thickness model and assuming a modified type of eddy diffusivity model originally developed for single-phase flow to be applied to falling films also. The theory assumes that the velocity profile is fully developed and axial heat conduction is negligible for large Peclet number film flow. The energy equation can be represented, in dimensionless form, by

$$u^+ \frac{\partial \theta}{\partial x^*} = \frac{\partial}{\partial \eta}\left[\left(1 + \frac{\epsilon_H}{\nu} Pr\right)\frac{\partial \theta}{\partial \eta}\right] \tag{61}$$

where $x^* = x\alpha/\delta^2 u^*$ and ϵ_H is the eddy diffusivity for heat. The dimensionless temperature, θ, and the respective boundary conditions are

Constant Wall Heat Flux

$$\theta = (T - T_{in})/(q_w \delta/K) \tag{62}$$

$$x^* = 0 \text{ (inlet)} \qquad \theta = 0 \ (T = T_{in}) \tag{63}$$

$$\eta = 0 \text{ (wall)} \qquad \partial\theta/\partial\eta = -1 \tag{64}$$

$$\eta = 1 \text{ (interface)} \qquad \partial\theta/\partial\eta = 0 \tag{65}$$

Constant Wall Temperature

$$\theta = (T_w - T)/(T_w - T_{in}) \tag{66}$$

$$x^* = 0 \text{ (inlet)} \qquad \theta = 1 \tag{67}$$

$$\eta = 0 \text{ (wall)} \qquad \theta = 0 \tag{68}$$

$$\eta = 1 \text{ (interface)} \qquad \partial\theta/\partial\eta = 0 \tag{69}$$

The solution of this problem requires a specification of a turbulence model for ϵ_M and ϵ_H. Various turbulence models have been proposed as shown in Table 3 together with a summary of the comments concerning their comparison with data. To calculate the heat transfer coefficient for both the thermal entrance and fully developed regions, select a Prandtl number and evaluate ϵ_H/ν. Then together with the results for δ^+, Re and $u^+(\eta)$ as computed from Equations 43 and 44, they are substituted into Equation 61 and solved either by a method of separation of variables [74] or by numerical finite difference method [64]. In this way, the Nusselt number, Nu_x, or the dimensionless heat transfer coefficient, $h_x^* = Nu_x/\delta^{+2/3}$, can be calculated as a function of x^*, Re and Pr for different values of τ_i^*. For constant wall heat flux, the local Nusselt number was derived by Yih and Chen [74] as

$$Nu_x = 1/(\theta_w - \theta_m) = \left[\int_0^1 u^+ (\theta_w - \theta) d\eta \Big/ \int_0^1 u^+ d\eta\right]^{-1} \tag{70}$$

$$\text{where } \theta = \sum_{i=1}^{\infty} A_i N_i(\eta) \exp(-\lambda_i^2 x^*) + \frac{4\,\delta^+ x^*}{Re} + \frac{4\,\delta^+}{Re}$$

$$\left[\int_0^{\eta} \frac{\int_0^{\eta} u^+ d\eta - Re/4\,\delta^+}{(1 + \epsilon_H Pr/\nu)} d\eta\right] \tag{71}$$

λ_i, N_i, A_i are respectively the eigenvalue, eigenfunction, and series coefficient. The asymptotic Nusselt number becomes

$$Nu_\infty = \left[\frac{16\delta^{+2}}{Re^2} \int_0^1 u^+ \left(\int_0^{\eta} \frac{Re/4\delta^+ - \int_0^{\eta} u^+ d\eta}{(1 + \epsilon_H Pr/\nu)} d\eta\right) d\eta\right]^{-1} \tag{72}$$

For constant wall temperature, the local and asymptotic Nusselt number were derived respectively as [74]:

$$Nu_x = \left[\sum_{i=1}^{\infty} A_i \exp(-\lambda_i^2 x^*) Ni'(0)\right] \Big/ \left\{\left[\sum_{i=1}^{\infty} A_i \exp(-\lambda_i^2 x^*) \cdot Ni'(0)/\lambda_i^2\right] / (Re/4\delta^+)\right\} \quad (73)$$

$$Nu_\infty = \lambda_i^2 \, Re/4\delta^+ \quad (74)$$

where λ_i is the first eigenvalue of the series solution.

Among the turbulence models used as shown in Table 3, the van Driest model modified with a damping factor seems to be the best one. Figure 7 shows the comparison of the analytical solution of Yih and Chen [74] with the heat transfer coefficient data and numerical solution of Faghri [64] for heating with constant wall heat flux in the thermal entrance region. Evidently good agreement is obtained. Another comparison is made on the asymptotic Nusselt number with the experimental data of Wilke [71]. Although a number of authors have reported experimental investigations on film heating [57, 62, 68, 69, 75–80, 82–84], the most comprehensive and reliable data encompassing a wide range of Prandtl and Reynolds numbers seem to be due to Wilke [71]. He measured local heat transfer rates to a falling film of water and mixtures of water and diethylene glycol solutions on the outside of 4.2 cm diam-

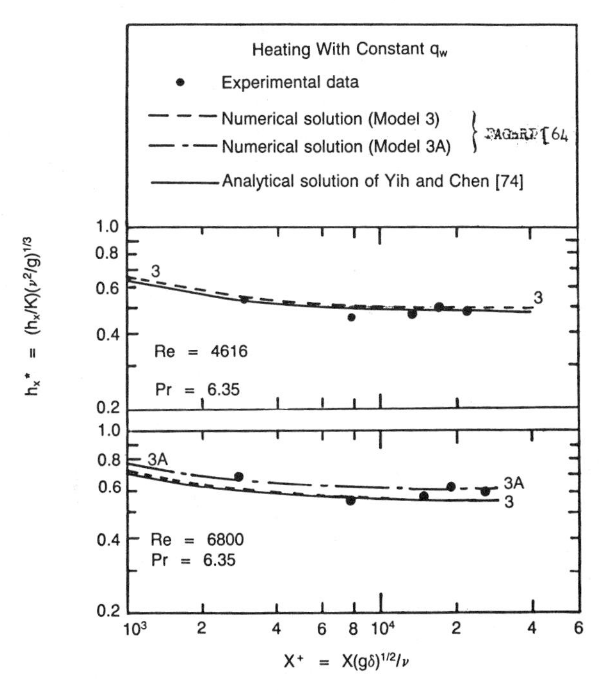

Figure 7. Comparison of theory with data on heating in the thermal entrance region of a turbulent falling film.

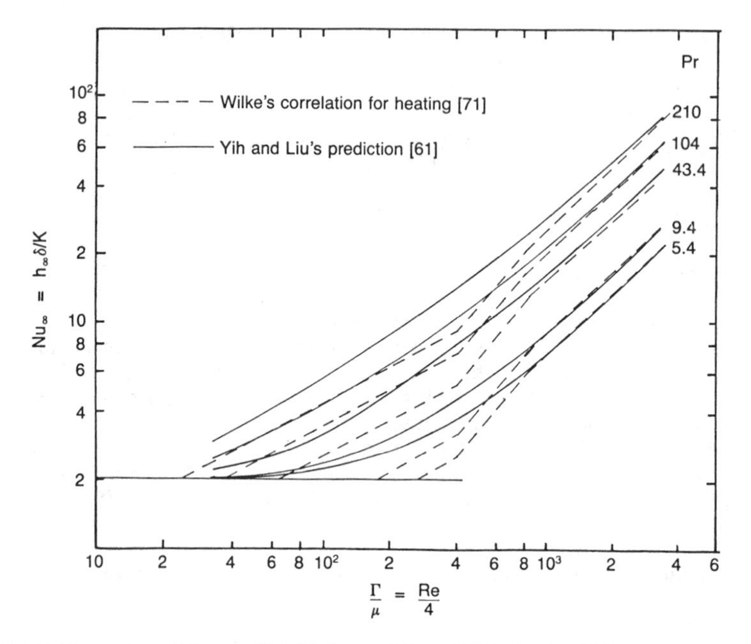

Figure 8. Comparison of theory with Wilke's correlations [71] for heating [74].

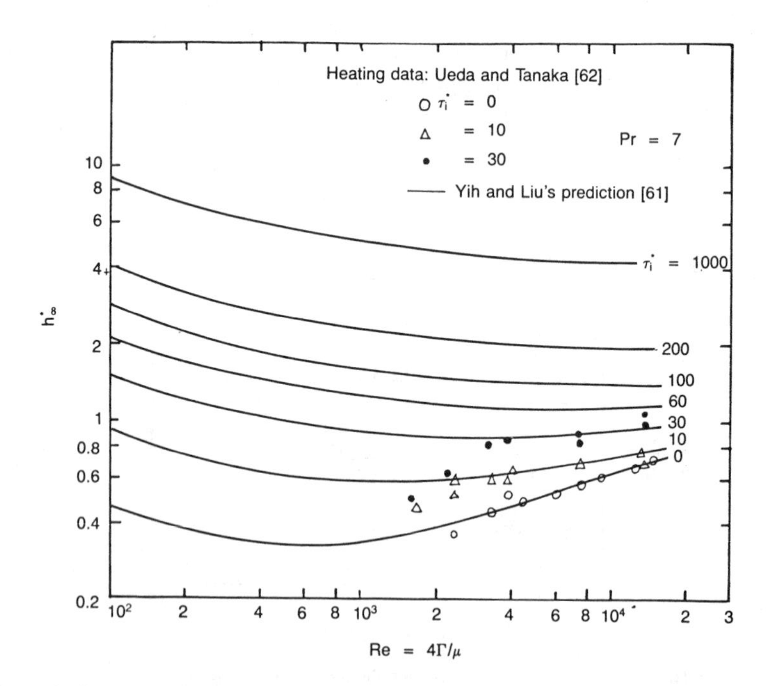

Figure 9. Comparison of theory with data of Ueda and Tanaka [62] for heating with interfacial shear [74].

eter, 2.4 m long tube which was heated internally by hot water. A sampling device was employed which could measure temperature and velocity profiles within the film. Local heat transfer coefficients were measured which reflected the influence of the simultaneously developing temperature and velocity profiles in the thermal entrance region. Most of his measurements were obtained at distances farther down the tube such that asymptotic conditions were reached. He presented the following correlations for the average Nusselt number in the fully developed region:

$$Nu_\infty = a(\Gamma/\mu)^b \, Pr^e \quad (75)$$

	a	b	e	
$\Gamma/\mu < 615\ Pr^{-.646}$	1.88	0	0	(76)
$615\ Pr^{-.646} \leq \Gamma/\mu < 400$	0.0614	8/15	0.344	(77)
$400 \leq \Gamma/\mu < 800$	0.00112	6/5	0.344	(78)
$800 \leq \Gamma/\mu$	0.0066	14/15	0.344	(79)

Figure 8 shows a comparison of the prediction of Yih and Liu's model [74] with Wilke's correlations. Comparisons have also been made with Limberg's [67] and Ishigai et al.'s [68] predictions. The Limberg and Ishigai predictions were both higher than Wilke's data and the best agreement is obtained between Yih and Liu's prediction and Wilke's correlations for $\Gamma/\mu > 600$. For heating with cocurrent gas interfacial shear, the prediction also matches the data of Ueda and Tanaka [62] on dimensionless film thickness and heat transfer coefficients, the latter of which is shown in Figure 9. It can be seen that at large τ_i^*, h_x^* becomes more dependent on τ_i^* than on Re.

Theoretical predictions and experimental results of the asymptotic heat transfer coefficient were presented by Gimbutis [69] with good agreement between the data and his theory which was based on a somewhat different kind of modification of the van Driest model. A correlation was established in the range of $1{,}800 < Re < 45{,}000$ and $4.3 < Pr < 8.4$:

$$h_\infty^* = (0.165\ Re^{0.16} - 0.4)\ Pr^{0.34}\ (Pr/Pr_w)^{0.25} \quad (80)$$

which included the effect of variation of physical properties with heat transfer through the term $(Pr/Pr_w)^{0.25}$. Within its limit of applicability, Equation 80 is found to be in good agreement with Wilke's correlation and also verified some later measurements made by Gimbutis [79].

The measurements taken by Wilke and Gimbutis can be considered as in the low heat flux region. Fujita and Ueda [80] reported that their low heat flux data showed a trend similar to that presented by Wilke, but the measured values were slightly lower than Wilke's. However, in a high heat flux region near the film breakdown heat flux, the heat transfer coefficients were comparatively low and widely scattered. This is caused probably by distortion of the film around the tube. Therefore, the applicability of Wilke's correlations should be limited to liquid films with a uniform thickness along the periphery and with a small temperature difference between the wall and the liquid film.

Schnabel and Schlunder [81] made a comprehensive analysis of the experimental data of various authors on film heating and proposed the following correlations for the dimensionless heat transfer coefficient which seemed to represent the majority of the available data quite well:

For laminar, hydrodynamically and thermally developed flow

$$h_\infty^*/(\mu/\mu_w)^{0.25} = C_\infty(\Gamma/\mu)^{-1/3} \quad (81)$$

For laminar thermal entrance flow

$$h_m^*/(\mu/\mu_w)^{0.25} = C_m[(\Gamma/\mu)^{1/3}\ Pr\ (\nu^2/g)^{1/3}/L]^{1/3} \quad (82)$$

$$h^*_\infty/(\mu/\mu_w)^{0.25} = \begin{cases} 0.0425(\Gamma/\mu)^{0.2}\,Pr^{0.344} & (\Gamma/\mu)_t < \Gamma/\mu < (\Gamma/\mu)_c \text{ (transition region)} \quad (83) \\ 0.0136(\Gamma/\mu)^{0.4}\,Pr^{0.344} & (\Gamma/\mu) > (\Gamma/\mu)_c \text{ (turbulent flow region)} \quad (84) \end{cases}$$

with

$$C_\infty = 1.3 \text{ and } C_m = 0.912 \text{ for } T_w = \text{constant} \quad (85)$$

$$C_\infty = 1.43 \text{ and } C_m = 1.1 \text{ for } q_w = \text{constant} \quad (86)$$

and

$$(\Gamma/\mu)_t = 615\ Pr^{-0.646} \text{ or } (\Gamma/\mu)_t = 2.5\ K_F^{0.176} \quad (87)$$

$$(\Gamma/\mu)_c = 300 \quad (88)$$

Equation 81 came from the theoretical derivation Equations 56 and 59, and Equation 84 is similar to Equation 79 of Wilke. The influence of the two different wall heating boundary conditions does not seem to have great effects on h^*_m. Equations 81 and 82 correlate well the data of various authors both in the thermal entrance and fully developed regions of laminar film flow within $\pm 15\%$ error as shown in Figure 10. Figure 11 shows that in the transition region, Equation 83 also fits the data well. However, Equation 84 can only correlate the data in the turbulent flow region within an error of $\pm 25\%$. The essence of Schnabel and Schlunder's correlation lies in the inclusion of the viscosity ratio correction factor which accounts for the variation of physical properties with temperature variation across the liquid film. This is essentially true in cases of cooling in which the measured results are different from heating. Measurements by Horn [85] showed that the heat transfer coefficients in film heating conformed with the correlations of Wilke. In Wilke's experiments, the difference of viscosity between the core region and the wall region is slight and its variation can be neglected. On the other hand, for cooling experiments, Horn showed that, due to the increased liquid viscosity at the cooled wall, Wilke's correlation has to be modified by including a viscosity ratio correction term such that

$$Nu_\infty = a(\Gamma/\mu)^b\ Pr^e\ (\mu/\mu_w) \quad (89)$$

	a	b	e	
$\Gamma/\mu < 800$	0.00105	6/5	0.344	(90)
$800 \le \Gamma/\mu$	0.00621	14/15	0.344	(91)

The liquid viscosity at the wall, μ_w, is evaluated at the average wall surface temperature while all other physical properties are evaluated at the average liquid temperature. Equation 89 shows a strong influence of the viscosity ratio, (μ/μ_w) to a first power, on Nu_∞. As pointed out by Schnabel and Schlunder [81], this is normally not observed. Instead, their correlation used a 0.25 power which is similar to the correlation factor $(Pr/Pr_w)^{0.25}$ put forward by Gimbutis [69]. A comparison of the two correlations of Horn and Wilke shows that, since $(\mu/\mu_w) < 1$ in cooling, Nu_∞ in heating will be larger than in cooling.

Evaporation

Falling fill evaporators have been extensively used in the process industries in concentrating food (e.g. juice [86], milk [87], sugar [88]) or chemicals (e.g. urea [1]) and in desalina-

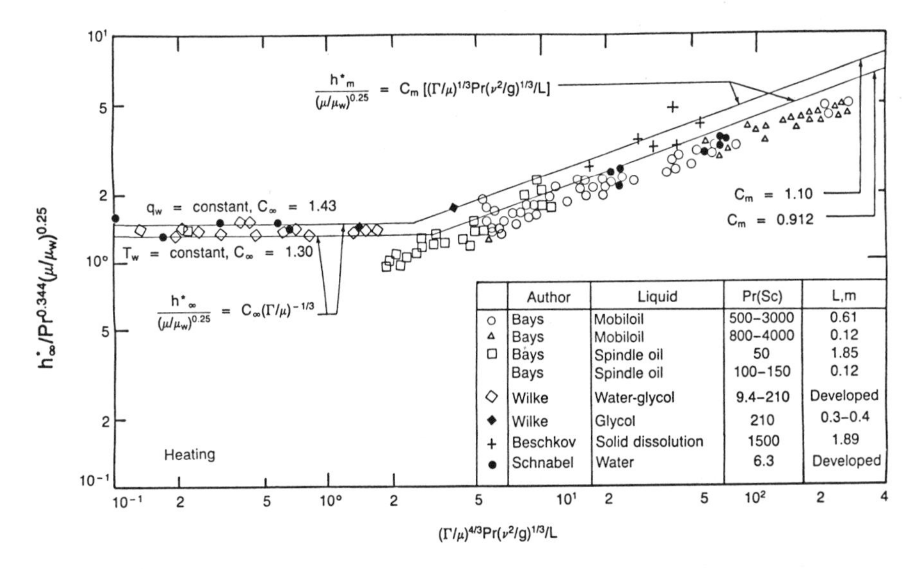

Figure 10. Heat transfer coefficient for film heating in laminar flow in the thermal entrance and fully developed regions [81].

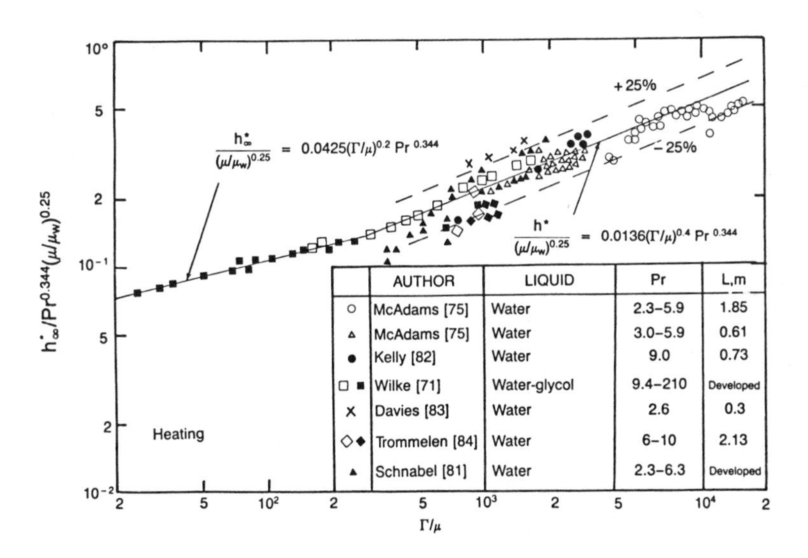

Figure 11. Correlation of asymptotic heat transfer coefficient for film heating in the transition and turbulent flow regions [81].

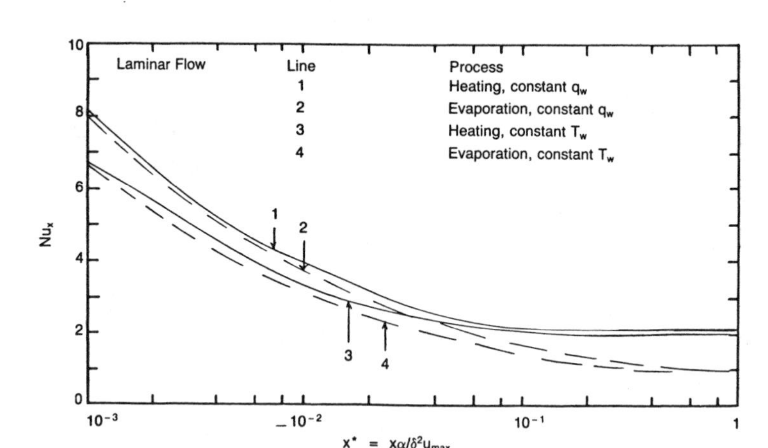

Figure 12. Local Nusselt number as a function of dimensionless axial distance for laminar film heating and evaporation.

tion processes [89, 90]. Generally, the evaporators operate at moderate heat flux densities (less than 50 KW/m^2) and small temperature differences of less than 10°C. Vapor velocities are usually moderate or small. The advantages of falling film evaporators are inherent in the characteristics of film flow which are

- Short contact time between the liquid and the heating surface, a particular necessity for heat-sensitive materials.
- Lack of static head, which makes it possible to evaporate without elevation of boiling point, a case of crucial significance in vacuum evaporation.
- The only mechanical energy needed is the power to pump the feed to the top of the evaporator.

Visual observation and analysis of available information indicates that:

- At low temperature differences of the order of few degrees, there is only surface evaporation with no nucleate boiling.
- At higher temperature differences of the order of about or greater than ten degrees, nucleation and breakoff of vapor bubbles occur.

In this section, we consider only surface evaporation.

Laminar and Wavy-Laminar Flow

For laminar flow, the governing convective-conduction equation under fully developed hydrodynamic conditions is the same as that in heating (Equation 49), but the boundary conditons are changed to

$$x = 0 \quad \text{(entrance)} \quad T = T_{sat} \tag{92}$$

$$y = 0 \quad \text{(wall)} \quad T = T_w \quad \text{or } q_w = \text{constant} \tag{93}$$

$$y = \delta \quad \text{(interface)} \quad T = T_{sat} \tag{94}$$

In evaporation, the local heat transfer coefficient is now based on the difference between the local wall temperature and the saturation temperature of the vapor. Solutions of this problem were provided by Yih and Lee [65] analytically and by Faghri [64] numerically. Figure 12 shows the local Nusselt number versus the dimensionless axial distance x^* for both of the boundary conditions of constant wall temperature and constant wall heat flux. In addition, a comparison is made with heating. As Seban [91] has shown, the evaporation data of Chun [92] in the range of Re from 320–800 and Pr from 5.1–5.9 lie in average 50% higher than the prediction curve. This is attributed most probably to the enhancement effect of the waves. The asymptotic Nusselt number derived from the laminar flow theory is, for both boundary conditions,

$$Nu_{\infty} = 1.0 \tag{95}$$

or

$$h^{*}_{\infty} = 1.095\ Re^{-1/3} \tag{96}$$

As to the theoretical treatment of evaporation in a wavy-laminar falling film, little work has been done. Faghri [64] made an analysis by formulating the following equation:

$$u\frac{\partial T}{\partial x} + v\frac{\partial T}{\partial y} = \alpha\left(\frac{\partial^2 T}{\partial x^2} + \frac{\partial^2 T}{\partial y^2}\right) \tag{97}$$

with the boundary condition of constant wall temperature. The form of the velocities u and v were those predicted by the linear hydrodynamic stability analysis. However, the wave properties came from the work of Kapitza [23] and Rogovan et al. [93]. Numerical solutions showed that the variation of the heat transfer coefficient with the Reynolds number followed the trend of the experimental data of Chun and Seban [94]. Recently, Hirshburg and Florschuetz [33, 95], using the hydrodynamic parameters predicted from a Kapitza-type analysis, formulated a theory which gave quite a close agreement with the data of Struve [96] and Chun and Seban [94].

Turbulent Flow

The prediction of heat transfer in the turbulent flow region is based on a similar method as in heating. A majority of the work uses the "constant film thickness model" by employing a turbulence model derived from single-phase flow and assuming that the film thickness along the tube length remains nearly constant for low rate of evaporation. An inspiring but more complex approach, the so-called "variable film thickness model," used the recent film and wave hydrodynamic data to predict the heat transfer coefficient.

Constant film thickness model. In the "constant film thickness model," the governing energy equation is the same as Equation 61 for heating but the dimensionless temperature, θ, and the respective boundary conditions are now changed to

Constant Wall Heat Flux

$$\theta = (T - T_{sat})/(q_w\delta/K) \tag{98}$$

$$x^{*} = 0\ \text{(inlet)} \qquad \theta = 0\ (T = T_{sat}) \tag{99}$$

$$\eta = 0\ \text{(wall)} \qquad \partial\theta/\partial\eta = -1 \tag{100}$$

$$\eta = 1\ \text{(interface)} \qquad \theta = 0 \tag{101}$$

Constant Wall Temperature

$$\theta = (T_w - T)/(T_w - T_{sat}) \tag{102}$$

$$x^{*} = 0\ \text{(inlet)} \qquad \theta = 1 \tag{103}$$

$$\eta = 0\ \text{(wall)} \qquad \theta = 0 \tag{104}$$

$$\eta = 1\ \text{(interface)} \qquad \theta = 1 \tag{105}$$

Table 4
Turbulence Models Used in Film Evaporation († with Interfacial Shear)

Year	Author	Turbulence Model Used	Comparison with Data
1960	Dukler [116, 117]†	Same as in condensation, Table 5.	Measured total heat transfer rate by Dukler and Elliot [90] lower than theory by 20%–50%.
1967	Domanskii and Sokolov [97]	von-Karman-Nikuradse model $\epsilon_M/\nu = 0$ for $0 \le y^+ \le 5$ $\epsilon_M/\nu = y^+/5$ for $5 \le y^+ \le 30$ $\epsilon_M/\nu = y^+/2.5$ for $30 < y^+$ $Pr_t = 1$	δ^+ versus Re is the same as Dukler [50] and Portalski [24] and compare well with data; h_x^* versus Re agree well with their own data for Pr = 1.76–29.5 and Re = 240–8,000 but substantially higher than other authors' data.
1973	Murthy and Sarma [98]	Sleicher's eddy diffusivity model $\epsilon_H/\nu = (0.091)^2 y^{+2}$ for $y^+ < 50$	Kosky's equation [58], $\delta^+ = 0.0504\ (Re^{7/8})$, is used directly for computing δ^+; h_x^* versus Re lies between the correlating equation of Wilke [71] and Chun and Seban [94] with a 15% deviation.
1973	Mills and Chung [73]	Original van Driest model for inner region $\epsilon_M/\nu = -0.5 + 0.5\{1 + 0.64y^{+2}[1 - \exp(-y^+/26)]^2\}^{1/2}$ Modified Lamourelle and Sandall [72] gas absorption eddy diffusivity for outer region $\epsilon_M/\nu = 6.47 \times 10^{-4}(\rho g^{1/3}\nu^{4/3}/\sigma\delta^{+2/3})(\delta^+ - y^+)^2 Re^{1.678}$ $Pr_t = 0.9$	δ^+ versus Re higher than Brauer's correlation by 15% as shown by Seban and Faghri [70]; h_x^* agrees well with data of Chun and Seban [94].
1975	Beccari, Di Pinto, Santori, Biasi, Prosperetti, Tozzi [99]†	Sleicher's model in inner region $\epsilon_M/\nu = (0.091)^2 y^{+2}$ for $y^+ \le 26$ von Karman's model in outer region $\epsilon_M/\nu = 0.16(\partial u^+/\partial y^+)^3/(\partial^2 u^+/\partial y^{+2})^2$ for $y^+ > 26$	No comparison of δ^+ or h_x^* versus Re presented; evaporation lengths calculated less than pilot plant data by about 20%.
1976	Hubbard, Mills and Chung [100]†	Modified van Driest model for inner region $\epsilon_M/\nu = -0.5 + 0.5\{1 + 0.64y^{+2}(\tau/\tau_w)[1 - \exp(-y^+/26)]^2\}^{1/2}$ $\tau = \rho g(\delta - y) + \tau_i$ Empirical Chung's [101] gas absorption eddy diffusivity for outer region $\epsilon_M/\nu = 8.13 \times 10^{-17}[(\nu g)^{2/3}/Cb\nu^{*2}]Re^{2n}[1 + b(\tau_i/\tau_w^0)]^2 (\delta^+ - y^+)^2$ $Pr_t = 0.9$	Slope of h_x^* versus Re curve lower than slope of data curve of Chun and Seban [94] for evaporation, h_x^* versus Re for condensation higher than data of Ueda et al. [102] for Pr = 2 and $10 \le \tau_i^* \le 200$.
1976	Seban and Faghri [70], Faghri [64]	Same as in heating, Table 3.	h_x^* versus Re agree with evaporation data of Chun and Seban [94] for Pr = 1.77 but lower than the data for Pr = 5.7.
1977	Murthy and Sarma [103]	Sleicher's model in inner region $\epsilon_M/\nu = (0.091)^2 y^{+2}$ for $y^+ \le 30$ Reichardt's model in outer region $\epsilon_M/\nu = 0.667y^+(2 + y^+/\delta^+)[3 + 4(y^+/\delta^+) + 2(y^+/\delta^+)^2]$ for $y^+ > 30$ $\epsilon_H/\nu = 0$ for $0 \le y^+ \le 5$ $\epsilon_H/\nu = (0.091)^2 y^{+2}$ for $5 \le y^+ \le 30$ $\epsilon_H/\nu = 0.36y^+$ for $y^+ \ge 30$	δ^+ versus Re agree well with the data of Belkin et al. [104], Dukler and Bergelin [50], and their own; h_x^* versus Re at Pr = 5 is higher than the data of Chun and Seban [94].

(Table continued on next page)

Table 4 Continued

Year	Author	Turbulence Model Used	Comparison with Data
1980	Wassner [105]	Same as Domanskii and Sokolov's model [97] but with $\tau/\tau_w = 1 - y/\delta$ instead of 1.	h_x^* versus Re is 0.7 times lower than that predicted by Domanskii and Sokolov [97] but agree much better with literature data.
1980 1981	Mostofizadeh [60]† Mostofizadeh and Stephen [106]†	van Driest model modified by Nikuradse mixing length $\epsilon_M/\nu = -0.5 + 0.5\{1 + 4[0.14 - 0.08(1 - y/\delta)^2 - 0.06(1 - y/\delta)^4]^2\delta^2[1 - \exp(-y^+/26)]^2\tau/\tau_w\}^{1/2}$ $\epsilon_H/\nu = 1.875(\bar{\epsilon}_M/\nu)\exp[-0.9/(0.64\Pr\bar{\epsilon}_M/\nu) - \Pr^{-0.1}] \cdot [\exp(1.2564y/\delta - 1)]$	δ vs Re in good agreement with literature data and correlations; h_x^* agree with their own data within $\Pr = 1.14 - 1.56$; h_x^* agree with data of Seban and Faghri [70] for $\Pr = 5.16$ but under-estimate the data for $\Pr = 1.75$.
1983	Yih and Liu [61]†	Same as in heating, Table 3.	h_x^* versus Re in good agreement with data of Chun and Seban [94] for $\Pr = 5.1$ and 5.7 but the prediction is slightly higher than the data for $\Pr = 2.91$ and 1.77.

Various proposed turbulence models are shown in Table 4 together with a summary of the comments concerning their comparison with data. The calculation procedure is analogous to that in heating. For constant wall heat flux, the local Nusselt number was derived by Yih and Chen [74] as

$$Nu_x = \left[\sum_{i=1}^{\infty} A_i N_i(0) \exp(-\lambda_i^2 x^*) + \int_0^1 d\eta/(1 + \epsilon_H \Pr/\nu)\right]^{-1} \tag{106}$$

For constant wall temperature, it is

$$Nu_x = \sum_{i=1}^{\infty} A_i \exp(-\lambda_i^2 x^*) N_i'(0) + \left[\int_0^1 d\eta/(1 + \epsilon_H \Pr/\nu)\right]^{-1} \tag{107}$$

The asymptotic Nusselt number for both boundary conditions is

$$Nu_\infty = \left[\int_0^1 d\eta/(1 + \epsilon_H \Pr/\nu)\right]^{-1} \tag{108}$$

Among the turbulence models proposed as shown in Table 4, the modified van Driest model used by Mills and Chung [73], Hubbard et al. [100], Seban and Faghri [70], and Yih and Liu [61] seems to be the most promising one. However, certain empiricism was called upon in using the first two models. They employed the original van Driest model for the near wall region and a modified eddy diffusivity deduced from gas absorption experiments for the outer region near the gas-liquid interface. The two eddy diffusivity profiles were run until they intersected at a point somewhere from 0.3 δ^+ to 0.85 δ^+ depending on the Reynolds number. Without the interface eddy diffusivity, the predictions using the original van Driest model would have given a too high heat transfer coefficient. Seban and Faghri [70] and Yih

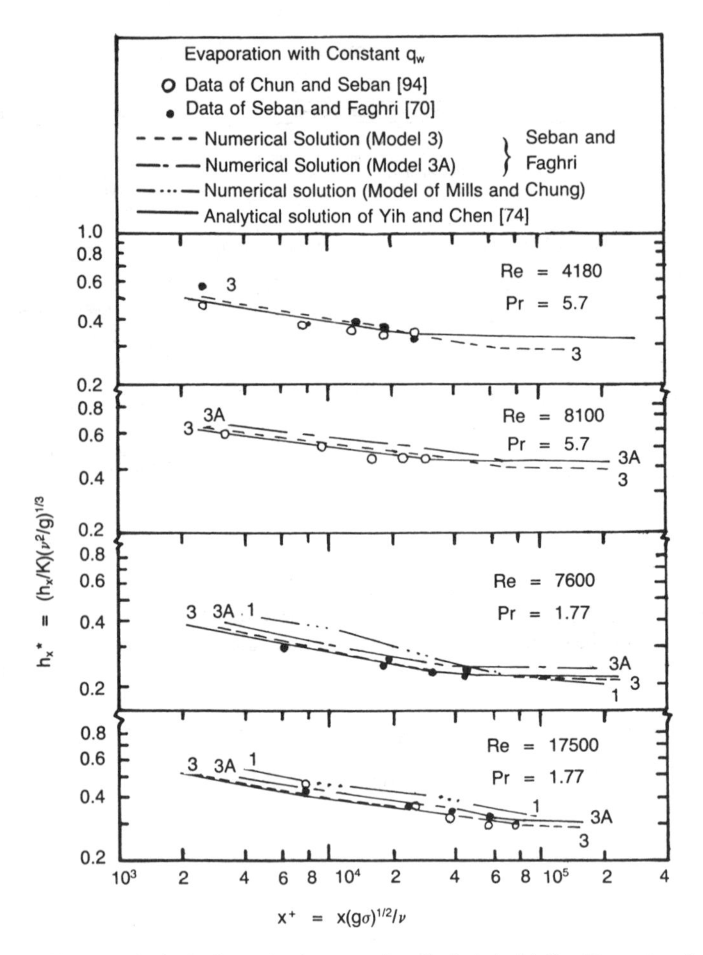

Figure 13. Heat transfer in the thermal entrance region of a turbulent falling film undergoing evaporation with constant wall heat flux [74].

and Liu [61] further incorporated a damping factor, f, in the van Driest model as suggested by Limberg [67], which gave better agreement with the experimental data.

Film evaporation data have been presented by a number of investigators [107–111]. Notably were those by Struve [96] who made comprehensive measurements on an evaporating film of refrigerant R11 (Pr = 4.16). The measured local heat transfer coefficient, obtained by varying the film length, was found to depend on the flow rate, the heat flux density, and the film length. The dimensionless heat transfer coefficient h_x^*, is influenced by the ring slot width at small distances of L = 15–30 cm. In the transition region, h_x^* depends on length L also. In the fully developed region, h_x^* is independent of L and the heat load supplied. In the pseudolaminar region, the data can be correlated as

$$h_\infty^* = 0.71(\Gamma/\mu)^{-0.282} \tag{109}$$

which is about 25%–30% higher than Nusselt's equation, Equation 96, and is a consequence of wave enhancement. For the transition region, the flow is assumed to be fully developed at 90 cm from the distributor and independent of Re:

$$h_x^* = 0.12 + 0.067(L/90) \quad \text{for } 0.2 < L/90 < 0.67 \tag{110}$$

$$h_x^* = 0.045 + 0.18(L/90) \quad \text{for } 0.67 < L/90 < 1 \tag{111}$$

$$h_x^* = 0.22 \quad \text{for } 1 < L/90 \tag{112}$$

For the turbulent flow region:

$$h_\infty^* = 0.014(\Gamma/\mu)^{0.41} \tag{113}$$

Chun [92] and Chun and Seban [94] have presented data on heat transfer coefficient for evaporation of water films from the outside surface of a vertical tube being heated electrically. About one half of the 61-cm long tube was used to permit hydrodynamic development above the heated section. The saturation temperature and hence the Prandtl number was varied by varying the vacuum level. The range of Pr is 1.77–5.7 and the range of Re is 320–21,000. The asymptotic heat transfer coefficient was established by measuring the heat flux, wall temperature, and saturation vapor temperature. Most of the data occurred in the turbulent flow region with a little occurring in the wavy flow region. In the wavy flow region, the data can be correlated well with Zazuli's empirical formula which was originally formulated for condensation [112]:

$$h_\infty^* = 0.606(\Gamma/\mu)^{-0.22} \tag{114}$$

In the turbulent flow region, the data were correlated by:

$$h_\infty^* = 0.0038\, Re^{0.4} Pr^{0.65} \tag{115}$$

The intersection of Equations 114 and 115 denoted a departure from the wavy-laminar region at the Reynolds number:

$$Re_{cr} = 5{,}800\, Pr^{-1.06} = 0.215\, Ka^{-1/3} \tag{116}$$

Where Ka is the Kapitza number $= g\mu^4/\rho\sigma^3 = K_F^{-1}$

The measurements of Fujita and Ueda [113] for evaporation of water at Pr = 1.75 lie about 10% above the Chun and Seban correlation, Equation 115. Schnabel and Schlunder [81] have made a comprehensive analysis of the data of several authors [94, 96, 113–115] and proposed the following correlations:

$$(h_\infty^*)_{lam} = 1.4287\, Re^{-1/3} \quad \text{for Re} < 400 \tag{117}$$

$$(h_\infty^*)_{turb} = 0.00357\, Re^{0.4} Pr^{0.65} \quad \text{for } 3{,}200 < \text{Re} \tag{118}$$

$$(h_\infty^*)_{tran} = [(h_\infty^*)^2_{lam} + (h_\infty^*)^2_{turb}]^{1/2} \quad \text{for } 400 < \text{Re} < 3{,}200 \tag{119}$$

Equation 117 is about 30% higher than the Nusselt theory, Equation 96, indicating wave enhancement while Equation 118 is essentially the same as Equation 115 of Chun and Seban. Equation 119 satisfactorily represents the data in the wavy transition region and so verifies its validity. To illustrate the applicability of the prediction methods, Figure 13 shows the comparison of the predictions of Seban and Faghri [70] and Yih and Chen [74] based on a modified van Driest model with the local heat transfer coefficient data of Seban and co-

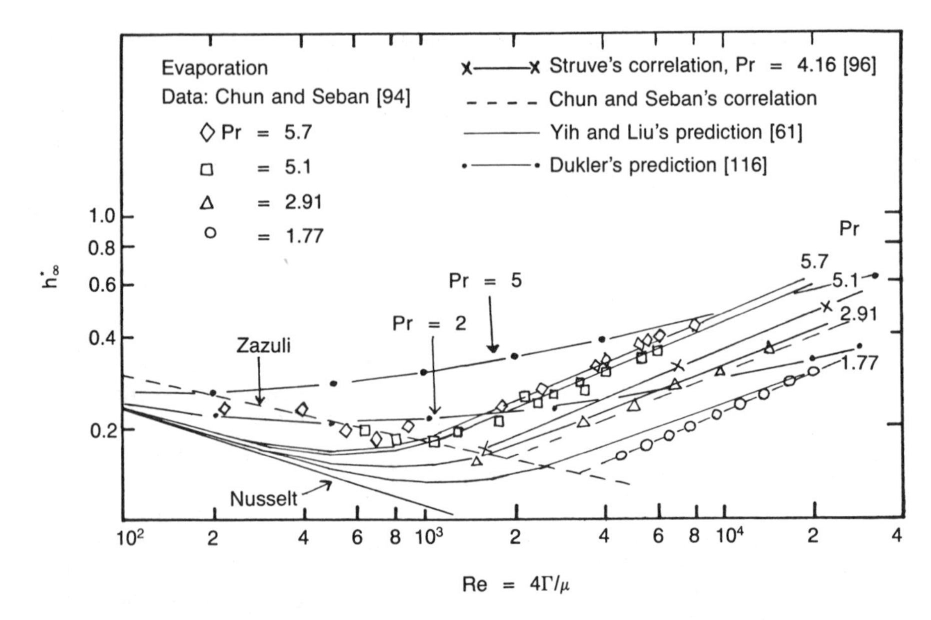

Figure 14. Asymptotic heat transfer coefficient of a turbulent falling film undergoing evaporation with constant wall heat flux.

workers [70, 94] in the thermal entrance region. The models fit the data very well. Comparisons were also made with the asymptotic data of Chun and Seban [94] as shown in Figure 14. Good agreement was obtained for Pr = 5.1 and 5.7 but the prediction was slightly higher than the data for Pr = 2.91 and 1.77 using Yih and Liu's model [61]. Their model gave predictions slightly higher than Seban and Faghri's model because they did not use the gas absorption eddy diffusivity in the interface region. Included in Figure 14 is the correlation of Struve, Equation 113, for Pr = 4.16 which seems to agree well with the trend of Chun and Seban's correlation in the turbulent flow region. At Pr = 4.16, Equation 113 is about 12% lower than Equation 115. Also in Figure 14, the predictions of Dukler [116, 117], using a Deissler and a von Karman eddy diffusivity, respectively, for the inner and outer region, were shown. Dukler's predictions were found to be too high and gave a slope that did not agree with the trend of the data in the turbulent flow region. In this respect, the modified van Driest model with a damping factor f is much more superior.

It is of interest to compare the results on heating and evaporation at the same Reynolds and Prandtl numbers. Figure 15 shows that the entrance length for evaporation is always longer than that for heating. Some sample estimates of entrance length were given in Table 5 based on calculations by Yih and Chen [74]. The film thickness was computed from Brauer's equation, Equation 51. From these computations, it can be concluded that the entrance effect is very important for a turbulent film with small Re and Pr when the heating or evaporation length is short (e.g. shorter than 1 meter). At large Re and Pr, the entrance length becomes rather short and usually can be neglected if the heating or evaporation length is sufficiently long.

Variable film thickness model. The previous models proposed so far essentially assume a constant average film thickness with the effects of the waves and turbulence considered to be lumped into an empirically determined eddy diffusivity. However, recent measurements by

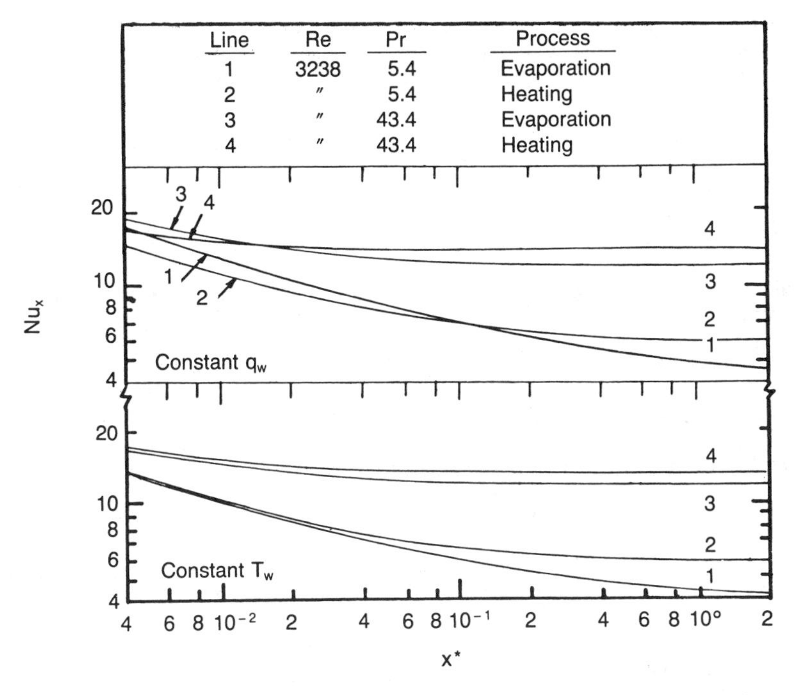

Figure 15. Comparison of local Nusselt number for heating and evaporation in the thermal entrance region [74].

Table 5

Sample Calculations of Thermal Entrance Length (cm) for Turbulent Film Heating and Evaporation [74]

	Re = 1,602 Pr = 5.4 (water) (δ = 0.0423 cm)	Re = 17,500 Pr = 9.4 (water) (δ = 0.1976 cm)	Re = 1,602 Pr = 210 (ethylene glycol) (δ = 0.0828 cm)
Heating, constant q_w	14.29	11.86	8.28
Heating, constant T_w	12.6	9.88	7.45
Evaporation, constant q_w	22.8	20.5	17.4
Evaporation, constant T_w	15.5	13.8	11.5

Dukler and co-workers [45, 46, 47] have revealed that turbulent films consist of a large wave structure and a thin base film, Figure 3. They determined the parameters which characterize this structure for liquid Reynolds number up to 6,000 and gas Reynolds number up to 57,600. Brumfield et al. [48] have made a successful attempt using such a wave model together with standard heat transfer derivations to predict the asymptotic heat transfer coefficient. The prediction is divided into two parts, one for the base film and the other for the turbulent waves. The base film can be either in laminar or turbulent motion with the transition taken at a base film Reynolds number of 2,000. When the base film is laminar,

$$h_b = K/\delta_{min} \tag{120}$$

where δ_{min} is the average thickness of the base film. For the turbulent waves, the van Driest eddy diffusivity was chosen and used for the entire film. Integration of the heat flux equation over the mean wave height $H = \delta_{min} + m_w$, or the base film thickness, whatever applicable, gives

$$h_w = \frac{\rho c_p u^*}{\int_0^{H^+} \frac{dy^+}{1/Pr + (\epsilon_M/\nu)/Pr_t}} = \frac{\rho c_p u^*}{F(H^+)} \tag{121}$$

The turbulent Prandtl number was taken as 0.9. For a given H^+ and τ_i, the shear velocity u^* is found from

$$u^{*3} - \tau_i u^*/\rho - g\nu H^+ = 0 \tag{122}$$

The base film Reynolds number Re_b, can be obtained from integration of the velocity profile as found from either a laminar or turbulent model over the base film thickness δ_{min}. The wave Reynolds number, Re_w, can then be found from the total Reynolds number:

$$Re = (1 - L_w/\lambda_p)Re_b + (L_w/\lambda_p)Re_w \tag{123}$$

where L_w is the base length of an "average wave" and λ_p is the average wave separation distance. The heat transfer coefficient for the entire film is then given by the following area average:

$$h_\infty = (1 - L_w/\lambda_p)h_b + (L_w/\lambda_p)\rho c_p u^*/F(H^+) \tag{124}$$

where h_b for the base film is given by Equation 120 when the base film is laminar and by Equation 121, with $H = \delta_{min}$, when the base film is turbulent. Values of L_w, were obtained from the truncated Gram-Charlier series of Telles and Dukler [45] whereas λ_p, δ_{min}, and wave celerity, c, were obtained from the data of Telles and Dukler also. Very good agreement between the theory and the experimental data of Chun and Seban [94] was obtained. The same kind of predictions were carried out using a more detailed picture of the thin-film structure and experimental large-wave amplitude histograms provided by Chu and Dukler [46, 47] who extracted the statistics of the base film and the large waves from a time-series analysis of the film thickness measurements. Again the agreement is good. Although the "variable film thickness model" is a more realistic method for predicting heat transfer across turbulent films based on an actual structure of the liquid film, the procedure is quite tedious for practical engineering application and predictions can only be made in certain limiting cases where the wave parameter data are available. Presently, only a meager amount of data for water is obtained and extension to other liquids besides water must await a larger and further collection of data. By comparison, the "constant film thickness model" is much easier to use and wave parameter data are not required. This is a substantial advantage which supports its widespread popularity.

Condensation

Film-type condensation on a vertical surface occurs commonly in industrial practice. The condensate first flows laminarly at the top portion of the surface, then changes to wavy and finally to turbulent flow if the length is long enough so that the critical Reynolds number is surpassed. Modern condensers usually employ film rather than drop condensation due to its easy control. The heat transfer coefficients in film condensation of steam, for example, are of the order of 7,000–47,000 watt/m^2°K. In general, several heat transfer resistances can be identified during film condensation. They are the condensate film resistance, the interface resistance at the liquid-vapor interface, and resistances due to the presence of noncondensable gases in the vapor. Minor ones are the resistances at the wall-condensate interface and that due to system pressure. With pure saturated or superheated vapor condensing on a cooled wall, the heat transfer resistance lies mainly in the liquid condensate film whose thickness grows with increasing distance down the solid surface. The effect of gravity, surface tension, surface waves, turbulence, vapor shear, and physical properties of the liquid are important in determining the average heat transfer coefficient. For condensation on the outside of the tubes, the pressure drop or interfacial shear is usually negligible even when the vapor velocity is appreciable. However, for high vapor velocity condensation inside the tubes, the two-phase pressure drop is very significant and interfacial shear becomes a dominant factor in influencing heat transfer. If noncondensable gas such as air is present in the vapor, then resistances to heat and mass transfer will be mainly in the vapor phase.

Although a large number of experimental results have been gathered to date, they are rather inconsistent and incomplete. This is because in practice, the flow is in a mixed mode regime with the top portion of the tube in laminar flow, then further down the tube, wavy flow and finally turbulent flow may appear if the tube is long enough so that the critical Reynolds number for transition is exceeded before the liquid film leaves the tube exit. Since the heat transfer rate for each of the three flow regimes is different, the total heat transfer rate over the tube is an integration of individual heat transfer in each regime. Traditionally, in most cases only overall heat transfer coefficients are measured which can be changed by varying conditions of the fluid on either side of the metal boundary. Then the condensing-side coefficient can be calculated if the coolant-side coefficient is known. However, such calculations are often inaccurate and may lead to erroneous results. This is because the overall temperature difference between the vapor and the coolant is not constant along the tube. For correct and rigorous design purposes, the heat transfer coefficient of the condensing film should be measured directly, and this necessitates the measurement of the saturation vapor temperature and the wall temperature of the condensing surface. To reduce experimental effort, accurate methods for the prediction of condensate film heat transfer coefficient in each of the three regimes are necessary.

Laminar and Wavy-Laminar Flow

As common to other film flow problems, laminar flow only exists at very low condensate Reynolds number beyond which waves will appear. In many circumstances, the thickness of the liquid and vapor boundary layers will be small so that laminar boundary-layer equations can be used to describe the velocity and temperature distribution in the condensate film as well as in the vapor layer (for the vapor layer, the buoyancy term is zero):

$$\frac{\partial u}{\partial x} + \frac{\partial v}{\partial y} = 0 \tag{125}$$

$$\rho\left(u\frac{\partial u}{\partial x} + v\frac{\partial u}{\partial y}\right) = \mu\frac{\partial^2 u}{\partial y^2} - (\rho - \rho_v)g \tag{126}$$

$$\rho c_p\left(u\frac{\partial T}{\partial x} + v\frac{\partial T}{\partial y}\right) = K\frac{\partial^2 T}{\partial y^2} \tag{127}$$

with the boundary conditions

$$y = 0 \text{ (wall)} \qquad u = 0,\ v = 0,\ T = T_w \tag{128}$$

$$y \rightarrow \infty \qquad u \rightarrow 0 \tag{129}$$

$$y = \delta \text{ (interface) } u = u_v,\ T = T_v = T_{sat},\ \mu \frac{\partial u}{\partial y} = \mu_v \frac{\partial u_v}{\partial y},\ K \frac{\partial T}{\partial y} = h_{fg} \frac{d\Gamma}{dx} \tag{130}$$

where the subscript v denotes the vapor phase. A similar set of equations can be written for the vapor phase. Gravity-driven laminar film condensation of pure saturated vapor was first analyzed by Nusselt [18]. He neglected the vapor-layer equations, the inertia force, and the convective heat transport in the liquid layer. A solution was obtained giving a semi-parabolic velocity distribution and a linear temperature profile. The local heat transfer coefficient can be derived as, for the boundary condition of constant wall temperature,

$$h_x = K/\delta = [\rho g K^3 h_{fg}/4\mu(T_{sat} - T_w)x]^{1/4} \tag{131}$$

and the dimensionless local and average heat transfer coefficient can be written as:

$$h_x^* = 1.095\ Re^{-1/3} \tag{132}$$

$$h_m^* = 1.47\ Re^{-1/3} \tag{133}$$

where $h_m = \int_0^L h_x dx/L$, $Re = 4\Gamma/\mu = 4qL/h_{fg}g\mu$. For constant wall heat flux, Kutateladze [112] gave a derivation which is approximately 13% higher than Equation 133:

$$h_m^* = 1.65\ Re^{-1/3} \tag{134}$$

To allow for the fact that the condensate is sometimes cooled from the saturation temperature to the film temperature, Bromley [118] showed that the latent heat, h_{fg}, should be multiplied by $(1 + 0.4c_p(T_{sat} - T_w)/h_{fg})^2$. An improvement on Nusselt's analysis was made by Rohsenow [119] who included energy convection in the energy equation. A modified integral analysis was performed whose results differed from Nusselt's only in the term h_{fg} which was replaced by $[h_{fg} + 0.68c_p(T_{sat} - T_w)]$. A number of theoretical analyses employing the boundary-layer theory was given by Sparrow and co-workers. By means of similarity variable transformation, they were able to solve the boundary-layer equations. Sparrow and Gregg [120] included energy convection and inertia forces and showed that the inertia forces on heat transfer were negligible for Prandtl numbers greater than 1. Sparrow and Eckert [121] further investigated the effects of superheated vapor and noncondensables. For a given temperature difference across the film, superheating increases the heat transfer to the surface, but only to a modest extent. The analysis also showed that the presence of only a few percent of noncondensable gas in the vapor caused a large reduction in the condensation heat transfer coefficient. Koh et al. [122] found that the effect of interfacial shear, in which the downflow condensate induced drag motion within the vapor, on heat transfer was negligible even for a Prandtl number of one. Minkowycz and Sparrow [123] further considered, besides noncondensables, also the presence of interfacial resistance, superheating, variable properties, thermal diffusion and diffusion thermo. Skekriladze and Gomelauri's analysis [124] showed that the effect of momentum transfer caused by the mass of the condensing vapor in the direction normal to the flow is important.

These analyses were based on the premise of a smooth laminar condensate layer which could only occur at very low condensate Reynolds number of the order of $Re < 10$. For the majority of industrial conditions encountered, the condensate layer will be either in a wavy or turbulent flow. Experiments have shown that considerably larger heat transfer coefficients of the order of 25% above that of the Nusselt prediction have been observed in wavy flow and even larger coefficients measured in turbulent flow. The critical Reynolds number for transition from laminar to wavy and wavy to turbulent flow can also be described by the equations described in the section on "Hydrodynamics." However, when the vapor flow is significant, these critical Reynolds numbers are strongly influenced by the vapor velocity and the direction of vapor flow [125].

A number of hydrodynamic stability analyses on condensate flow down a vertical wall [126–128] have shown that the critical Reynolds number above which disturbances will be amplified is so small that in all practical situations, a laminar gravity-driven vertical condensate film is unstable and waves will appear. However, the overall effect of the condensation mass transfer tends to stabilize the film. According to Kapitza's [23] simple analytical treatment of heat transfer in a wavy film based on the assumption of regular sinusoidal wave and considering only heat conduction across the film, the enhanced heat transfer rate is 21% above that for the smooth film. Leonard and Estrin [129] have retained the heat convection terms in their extended analysis of Kapitza's problem. However, the average Nusselt number over a wavelength do not differ too much. Also there is very little effect of wave celerity on the average Nusselt number. Using the wave parameters taken from Massot et al.'s [29] experimental data, this model gives a rather nice prediction of the heat transfer coefficient data at $Pr = 2$ and for Re as large as 500. Another model was proposed by Sofrata [130] who assumed that the waves were sinusoidal and moved with a constant phase velocity. The average heat transfer coefficient derived is

$$h_m = (h_m)_{Nu}/(1 - \beta_1^2)^{3/8} \tag{135}$$

where β_1 is the dimensionless amplitude in $\delta_x = \delta_0\beta_1 \sin n(x - ct)$.
Sofrata extrapolated the theoretical relation between β_1 and the Weber number as derived by Penev et al. [32]. The resulting theoretical correlations were in the form of $Nu = f(Re)$ with $S_1 = (3K_F)^{1/5}$ as a parameter. Comparisons with the data of several authors [131–133] showed that the model underpredicts the data but is better than Nusselt's theory. Subsequently, Sofrata [134] extended this model to take into account the variations in wavelength and amplitude as a function of the Weber number, $We = ((\Gamma/\mu)^5/S_1)^{1/3}$, ie., a function of Γ/μ and S_1, the results of which were taken from Penev et al.'s stability theory. Graphical solutions were presented and compared with literature data at low Reynolds numbers. The new model shows improvement over that obtained from the previous model. Kutateladze et al. [135] have presumed that the main thermal resistance lies in the base film whose thickness is virtually constant within a wide range of Reynolds numbers as have been measured by Brauer [6]. Figure 4 shows that up to some value of $(\Gamma/\mu)^*$, the base film thickness coincides with the average film thickness and can be determined from the equation

$$\delta_{min} = \delta = 1.4422\,(\nu^2/g)^{1/3}(\Gamma/\mu)^{*1/3} \tag{136}$$

After this value of $(\Gamma/\mu)^*$ is surpassed, the base film thickness becomes independent of the Reynolds number. This value of $(\Gamma/\mu)^*$ was taken as

$$(\Gamma/\mu)^* = 2.3Ar^{1/5} \tag{137}$$

based on Brauer's measurements where Ar is the Archimedes number ($Ar = K_F^{-1/2}$). Then the dimensionless average heat transfer coefficient for wavy flow can be determined by substituting Equations 136 and 137 into the relation $h_m = K/\delta_{min}$ to give

$$h_m^* = 0.527Ar^{-1/15} \tag{138}$$

Equation 138 can be used up to a value of $(\Gamma/\mu)_{cr} = 35Ar^{1/5}$ and shows that h_m^* is independent of Re, a result which conforms with the experimental trend within $400 < Re < 4{,}000$.

Experimental data on condensation in the wavy flow region have been reported by a number of investigators [112, 136, 137] in terms of average heat transfer coefficients. Mc-

Adams, in view of the observed enhancement of heat transfer, has recommended that the Nusselt relation be multiplied by a factor of 1.28 for Re < 1,800 to give

$$h_m^* = 1.28(h_m^*)_{Nu} = 1.882\ Re^{-1/3} \tag{139}$$

Kutateladze [112] also recommended multiplying the Nusselt equation by a factor of 1.2. According to the experiments of Zazuli as cited by Kutateladze [112], the Nusselt equation should be modified in the range of 40 < Re < 2,000:

$$h_m^* = 0.8(\Gamma/\mu)^{0.11}(h_m^*)_{Nu} = 1.01Re^{-0.22} \tag{140}$$

The local coefficient, h_x^*, is given by Equation 114. Another modification, as cited by Gogonin et al. [138] and Kutateladze and Gogonin [139] is to incorporate an empirical correction term in the Nusselt equation to account for the effect of wave enhancement on heat transfer in the range of 20 < Re < 400:

$$h_m^* = (\Gamma/\mu)^{0.04}(h_m^*)_{Nu} = 1.39\ Re^{-0.293} \tag{141}$$

Onda et al. [137] have proposed a theoretical model which takes into account the surface ripples, and the derived result is

$$h_m^* = \{(2\delta_{min}/\delta)^{-1} + [\ln(\delta_{max}/\delta_{min})]/[2(\delta_{max} - \delta_{min})/\delta]\}(h_m^*)_{Nu} \tag{142}$$

where δ_{max} and δ_{min} were obtained from Kirkbride's [140] measurements using a pointer and gauge. The theory agreed with their data on condensation of pure saturated vapor of methanol (Pr = 4.6) both for cocurrent and countercurrent flow.

Turbulent Flow

Turbulent flow is usually attained at the lower portion of the condenser tube if the tube is longer than, say, two meters. Kirkbride [140] earlier proposed, on dimensional ground, an equation that can correlate the steam condensation data of Badger et al. [141] quite well:

$$h_m^* = 0.0076\ Re^{0.4} \tag{143}$$

This equation is supposed to represent results for Pr = 5. Inspired by Kirkbride's analysis, Colburn [142] developed a method for calculating the mean heat transfer coefficient when the upper portion of the condensate layer is in laminar flow and the lower portion in turbulent flow. The dimensionless local heat transfer coefficient was calculated from the Nusselt equation and j-factor analogy, respectively, for the two flow regimes:

$$h_x^* = \left(\frac{\mu}{3\Gamma_x}\right)^{1/3} \tag{144}$$

$$h_x^* = 0.056\ Pr^{1/3}\left(\frac{4\Gamma_x}{\mu}\right)^{0.2} \tag{145}$$

The differential rate of condensation, $d\Gamma$, over a differential length, dL, can be expressed as

$$d\Gamma_x = h_x \Delta T dL/h_{fg} \tag{146}$$

Substituting h_x from Equations 144 and 145 into Equation 146 and integrating for the combined laminar and turbulent flow regions, assuming the critical Reynolds number to be 1,600 we obtain

$$h_m^* = Re/[22(Re^{0.8} - 364)/Pr^{1/3} + 12{,}800] \tag{147}$$

or

$$h_m^* = (h_{fg}\mu/L\Delta T)(\nu^2/K^3g)^{1/3}[0.059\ Pr^{1/3}(L\Delta T/h_{fg}\mu)(K^3g/\nu^2)^{1/3} - 190\ Pr^{1/3} + 120]^{1.25} \tag{148}$$

Equation 147 is useful when the number and diameter of tubes are assumed and L is to be calculated. The reverse is true for Equation 148.

With downward flow of a condensing vapor at high velocity inside a tube, the measured film thickness was found to be substantially smaller than the case of negligible vapor velocity while the measured heat transfer coefficient may be as much as ten times higher. Experimental observations have led Carpenter [143] and Carpenter and Colburn [144] to hypothesize that in the presence of large interfacial shear of the vapor, the condensate layer will become turbulent at a much lower Reynolds number of around 960 instead of the usual case of around 1,600. They obtained data for local as well as average heat transfer coefficient for the condensation of five different vapors (steam, methanol, ethanol, toluene, trichloroethylene) inside a 253-cm long tube with a small diameter of 1.17 cm. High vapor velocities up to 152.4 m/s were reached. The Prandtl numbers ranged from 2 to 5. By evaluating the slope of the curve of cooling water temperature versus length, the local value of the heat load was determined as

$$q_x = \frac{w_c c_{pc}}{\pi d}\left(\frac{dT_c}{dL}\right) \tag{149}$$

The local condensing film coefficient was then evaluated by subtracting the cooling water and metal wall resistances from the overall resistance calculated from Equation 149. The length of the condensing section was also estimated.

A theory was developed which assumed that the entire thermal resistance was offered by the laminar sublayer having a linear velocity gradient and the shear force at the outer boundary of the laminar sublayer, τ_s, to be composed of the forces due to:

- Vapor friction, τ_i
- Gravity, τ_g,
- Momentum change of the vapor which was condensed and brought to rest, τ_M:

$$\tau_i = f_v G_v^2/2\rho_v,\ \tau_g = \rho g \delta,\ \tau_M = (G_v/\rho_v)(d\Gamma/dx) \tag{150}$$

$$\tau_s = \tau_i + \tau_g + \tau_M \tag{151}$$

A rigorous design method required the local heat transfer coefficient along the tube be known together with a graphical solution. Three different regions are encountered. At the top, the condensate is laminar with a linear velocity gradient:

$$h_x = K/\delta = K(\rho\tau_s/2\mu\Gamma)^{1/2} \tag{152}$$

At values of Re > 1,000, the condensate is turbulent with a laminar sublayer:

$$h_x = 0.043\ Pr^{1/2}\tau_s^{1/2}K\rho^{1/2}/\mu \tag{153}$$

Finally, near the tube bottom, the vapor velocity becomes negligible, and the Colburn equation for turbulent flow applied.

For an approximate design, the average value of the condensing heat transfer coefficient for the entire tube may be estimated from the following equation based upon the data of Carpenter and Colburn [144]:

$$(h_m/c_pG_m)(c_p\mu/K)^{0.5} = 0.065(f_v\rho/2\rho_v)^{1/2} \tag{154}$$

where G_m is the mean mass velocity of vapor with

$$G_m = [(G_1^2 + G_1G_2 + G_2^2)/3]^{1/2} \tag{155}$$

G_1 and G_2 are the values of G_v at the top and bottom of the tube, respectively. In formulating Equation 154, a simplification was made in that the friction factor of vapor, f_v, was taken as that corresponding to single-phase flow. Actually f_v should be calculated from a two-phase flow correlation since the presence of condensate increases the friction factor considerably over that for an empty tube. Equation 153 can also be converted to a form showing that h_m^* is independent of Re in the presence of large interfacial shear:

$$\text{where } h_m^* = 0.065 \, Pr^{1/2}\tau_i^{*1/2}$$

$$\tau_i^* = \tau_i g_c/g\rho(\nu^2/g)^{1/3} \tag{155a}$$

Although Carpenter and Colburn provided a useful approach for correlating the condensing heat transfer coefficient at high vapor velocity, the equation they used to evaluate the effect of momentum on wall shear stress is incorrect. In addition, their methods of evaluating the friction and gravity terms can be improved upon. This has led Soliman et al. [145] to derive analytical forms of each of these forces and incorporate them into a correlation similar to Equation 153:

$$h_x = 0.036 \, Pr^{0.65}\tau_w^{1/2}K\rho^{1/2}/\mu \tag{155b}$$

where τ_w is approximately equal to τ_s due to the assumption of a linear velocity profile in the laminar sublayer. The change in the exponent of Pr from 0.5 to 0.65 was made to accommodate newer data by other authors. This correlation covers a range of Pr from 1 to 10, vapor velocities from 6 to 305 m/s, and steam qualities from 0.99 to 0.03.

The modeling of turbulent film condensation with or without significant vapor shear is generally based on the same premise as in heating or evaporation that the wall law valid for single-phase flow is equally applicable to the condensate flow. If the vapor velocity is high, such as greater than 60 m/s, and the pressure drop is significant, then the interfacial shear becomes an important parameter that must be considered in the model. The pressure drop in this case must be determined experimentally or evaluated from a two-phase flow correlating equation. The calculation methods resemble those in evaporation. Various turbulent models for the eddy diffusivity of momentum and heat have been proposed for solving the velocity profile, film thickness and the local or average heat transfer coefficient. They are summarized in Table 6. These models can be broadly classified into four main groups:

1. The von Karman-Nikuradse model as used by Seban [146], Rohsenow et al. [147], Shekriladze and Mestvirishvili [148], Ueda et al. [102, 149].
2. Kutateladze's model [112, 139].
3. The Deissler model as used by Dukler [116, 117], Razavi and Damle [150].
4. The modified van Driest model as used by Kunz and Yerazunis [151], Hubbard et al. [100], Blangetti and Schlunder [152], Mostofizadeh and Stephen [106], Blangetti et al. [153], Yih and Liu [61]. Explicit derivations of the heat transfer coefficient in the pres-

ence of vapor shear are possible in the model of Rohsenow et al. [147]. They established equations for calculating the transition from laminar to turbulent flow for a sheared film assuming that the critical Reynolds number is 1,800 for a nonsheared film:

$$(\delta^*)_{cr} = [11.05/(1 - \rho_v/\rho)^{1/3}] - \tau_i^* \tag{156}$$

and

$$(Re)_{cr} = 1{,}800 - 246(1 - \rho_v/\rho)^{1/3}\tau_i^* + 0.667(1 - \rho_v/\rho)(\tau_i^*)^3 \tag{157}$$

for small values of $\tau_i^* < 11.05/(1 - \rho_v/\rho)^{1/3}$

$$(Re)_{cr} = [4(\delta^*)_{cr}^3/3 + 2\tau_i^*(\delta^*)_{cr}^2](1 - (\rho_v/\rho) \tag{158}$$

for large values of τ_i^*

where $(\delta^*)_{cr}^3 + (\delta^*)_{cr}^2\tau_i^* = 36/(1 - \rho_v/\rho)$ (159)

When $(\delta^*)_{cr}$ calculated from Equation 156 is less than its value calculated from Equation 158, the latter is used for calculating $(Re)_{cr}$. Predictions of heat transfer coefficient were compared with the correlation of Carpenter and Colburn [144], Equation 154, and good agreement was obtained in the range of Pr from 2 to 3 and τ_i^* from 5 to 50.

Kutateladze [112] employed the Prandtl two-layer model and derived an equation for the local heat transfer coefficient valid for the region $\delta^+ > 11.6$ (Re > 267) which fits a majority of the condensation data in the wavy and turbulent flow region of Re > 400 if a deliberately decreased $(Re)_{cr} = 400$ is assumed:

$$h_x^* = 0.4\,Pr\delta^{+1/3}(\ln\{[\delta^{+1/2} + (\delta^+ - 11.6)^{1/2}]/[\delta^{+1/2} - (\delta^+ - 11.6)^{1/2}]\} + 4.65\,Pr)^{-1} \tag{160}$$

However, this assumption is incorrect because $(Re)_{cr}$ is usually around 1,600. In Equation 160, the value of δ^+ is calculated as a function of the Reynolds number by integrating the logarithmic velocity profile to give

$$\Gamma/\mu = \delta^+(3 + 2.5\ln\delta^+) - 39 \tag{161}$$

This equation is similar to the relation derived by Dukler and Bergelin [50] and Portalski [24].

Kutateladze et al. [135] further proposed that with a mixed mode of condensate flow, the length of the three flow regimes, x_l, x_w, x_t, should be determined and the integral-average heat transfer coefficient calculated from

$$h_m^* = (x_l/L)(h_m^*)_{Nu} + (x_w/L)(h_m^*)_{wavy} + (x_t/L)(h_m^*)_{turb} \tag{162}$$

Apparently the first important study on the effect of vapor velocity on the transition point from wavy-laminar to turbulent flow for descending water films with cocurrent flow of noncondensing gas (air), as well as of condensing steam, was made by Skekriladze et al. [125]. Their experimental results indicated a marked effect of condensation upon the stability of laminar-wavy flow. For example, while the transition Reynolds number was reduced from 1,600 to 300 for air velocities up to 40 m/s, the condensing steam moving at velocities up to 80 m/s reduced the transition Reynolds number only to 1,000. Ueda et al. [102] also estimated from their theory and experiments in condensation with high interfacial shear that the transition Reynolds number is 970, a result which is in remarkable agreement. These find-

Table 6
Turbulence Models Used in Film Condensation († with Interfacial Shear)

Year	Author	Turbulence Model Used	Comparison with Data
1954	Seban [146]	von Karman-Nikuradse model $\epsilon_M/\nu = 0 \quad 0 \le y^+ \le 5$ $\epsilon_M/\nu = y^+/5 - 1 \quad 5 \le y^+ \le 30$ $\epsilon_M/\nu = y^+/2.5 \quad 30 \le y^+$ $Pr_t = 1$	h_m^* agree with Colburn's analogy results for $Pr = 5$; h_m^* lower than data by 10% to 15% for $2 < Pr < 5$, $Re < 40{,}000$.
1956	Rohsenow, Webber and Ling [147]†	von Karman-Nikuradse model	h_m^* agree fairly with the data of Carpenter and Colburn [144] for $2 < Pr < 3$, $5 < \tau_i^* < 50$.
1960	Dukler [116, 117]†	Deissler model for $y^+ \le 20$ $\epsilon_M/\nu = (0.125)^2u^+y^+\{1 - \exp[-(0.125)^2u^+y^+]\}$ von Karman model for $y^+ > 20$ $\epsilon_M/\nu = (0.4)^2(du^+/dy^+)^3/(d^2u^+/dy^{+2})^2$ $Pr_t = 1$	δ^* versus Re slightly higher than data for $0 \le \tau_i^* \le 200$; h_m^* agree fairly well with Carpenter's data [143].
1963	Kutateladze [112]	Prandtl two-layer model $\epsilon_M/\nu = 0 \quad$ for $y^+ < 11.6$ $\epsilon_M/\nu = 0.16y^{+2}du^+/dy^+ \quad$ for $y^+ > 11.6$	$\delta^†$ versus Re agree fairly well with the Nusselt and Brauer equations; h_x^* agrees with data well for $Re > 400$.
1969	Kunz and Yerazunis [151]†	van Driest model modified by Nikuradze's variation of mixing length and further modified for two-phase flow $\epsilon_M/\nu = \{[0.14 - 0.08(\tau/\tau_w)^2 - 0.06(\tau/\tau_w)^4]^2y^{+2}[1 - \exp(-y^+/A^+)]^2/(1 - \tau/\tau_w)^2\}(du^+/dy^+)$ $Pr_t = \{1.5\exp[-0.9/Pr\epsilon_M/\nu)^{0.64}]\}^{-1}$	δ^+ versus Re slightly lower than Nusselt's equation for laminar flow but agrees well with Dukler's prediction for turbulent film; h_x agrees fairly well with the theories of Colburn [142], Kutateladze [112], Seban [146] and Dukler [116] for condensation; however h_x is about 30% higher than the data of Carpenter [143], Goodykoontz and Dorsch [154] for condensation with interfacial shear.
1973	Shekriladze and Mestvirishvili [148]†	von Karman-Nikuradse model	h_x agrees fairly well with their own data.
1972 1974	Ueda, Kubo, and Inoue [102, 149]†	Rohsenow, Webber, and Ling's model $Pr_t = 1$	δ^* versus Re slightly higher than their own data for $0 \le \tau_i^* \le 200$; h_x^* versus Re higher than their data which are quite scattered.
1976	Hubbard, Mills, and Chung [100]†	Same as in evaporation, Table 4	h_x^* versus Re agree quite well with data of Ueda et al. [102] for $Pr = 2$ and $10 \le \tau_i^* \le 200$.
1978	Blangetti and Schlunder [152]†	van Driest model for inner wall region $\epsilon_M/\nu = -0.5 + 0.5\{1 + 0.64y^{+2}[1 - \exp(-y^+/26)]^2\}^{1/2}$ Modified gas absorption eddy diffusivity for outer region similar to Mills and Chung [73]; $\epsilon_M/\nu = 0.016(\nu^4\rho^3g/\sigma^3)^{1/3}(\Gamma/\mu)^{1.34}[\tau_i^* + \delta^*(1 - y^+/\delta^+)](\delta^+ - y^+)$ $Pr_t = 0.9$	h_x^* versus Re agree with their own data for small τ_i^* of 1.18 and 2.82 and $Re > 4{,}000$.

(Table continued on next page)

Table 6 Continued

Year	Author	Turbulence Model Used	Comparison with Data
1978	Razavi and Damle [150]†	Deissler model for $y^+ \le 26$ as proposed by Dukler [116] ϵ_M/ν = same as Dukler for $y^+ \le 26$ von Karman model for $y^+ > 26$ ϵ_M/ν = same as Dukler for $y^+ > 26$ $Pr_t = 1$ Momentum change of condensing vapor included in evaluating interfacial shear	h_x^* versus Re agrees better with data of Carpenter [143] and Goodykoontz and Dorsch [154, 155] than the predictions of Nusselt, Dukler [116], Kunz and Yerazunis [151], Shekriladze and Mestvirishvili [148].
1980 1981	Mostofizadeh [60]† Mostofizadeh and Stephen [106]†	Same as in evaporation, Table 4.	h_x^* agrees well with their own data on steam condensation for Pr = 1.14 − 1.56.
1982	Blangetti, Krebs, and Schlunder [153]†	Same as Blangetti and Schlunder's model [152]	h_x^* in fair agreement with their data and is better than Rohsenow et al.'s model [147].
1982	Kutateladze [156]	$\epsilon_M/\nu = 0$ for $0 < y^+ < 6.8$ $\epsilon_M/\nu = 0.4(y^+ - 6.8)(1 - y^+/\delta^+)^{1/2}$ for $6.8 < y^+ < 5.44 + 0.2\delta^+$ $\epsilon_M/\nu = 0.08(y^+ - 6.8)(1 - y^+/\delta^+)^{1/2}$ for $5.44 + 0.2\delta^+ < y^+ < \delta^+$	h_x^* lower than that calculated by the author's own Prandtl two-layer model [112] at the same Re and Pr, but this model fits own data better especially at high Re.
1983	Dobran [157]†	von Karman-Nikuradse model for the base film layer near the wall; An eddy diffusivity determined from literature data for the outer wavy layer; $\epsilon_M/\nu = 1 + 1.6 \times 10^{-3}(\delta^+ - \delta_T^+)^{1.8}$ $\delta_r^+/d^+ = 140\{[gd^3\,\rho(\rho - \rho_g)/\mu^2]^{1/2}\}^{0.433}\,(u_G\rho_G d/\mu_G)^{-1.35}$	δ^* versus Re at different τ_i^* agrees well with data of Chien and Ibele [158], Ueda and Tanaka [62]; h_x^* versus Re agrees quite well with data of Ueda and Tanaka [62] for annular-mist flow and Ueda et al. [102] for condensation.
1983	Yih and Liu [61]†	Same as in heating, Table 3.	h_x^* lower than Dukler's predictions for Pr = 0.5 to 10 but is higher for Pr = 0.1; h_x^* agrees well with the data of Kutateladze [139] and the low interfacial shear data of Blangetti and Schlunder [152] for Re > 2,400; also agrees well with the high interfacial shear data of Ueda et al. [102].

ings are contrary to the belief of Rohsenow et al. [147] or Kutaleladze [112] that the transition Reynolds number is reduced to a large extent in the presence of high velocity cocurrent vapor flow.

Dukler [116, 117] used a turbulence model composed of a Deissler eddy diffusivity near the wall and a von Karman eddy diffusivity for the outer region. Numerical calculations and graphical plots of the dimensionless film thickness δ^* and heat transfer coefficient h_x^* (and h_m^*) as a function of Re with $\tau_i^*(= (dp/dL)_{TP} dg^{1/3}/4\rho^{1/3}\mu^{2/3})$ as parameter were presented for a wide range of Pr from 0.1 to 10 and τ_i^* up to 4,000. The dimensionless interfacial shear was evaluated from an empirical two-phase flow correlation for pressure drop, which in turn can be calculated from the usual single-phase friction factor-pressure drop relation by

$$\left[\frac{(dp/dL)_{Tp}}{(dp/dL)_G}\right]^{1/2} = 3.3\left\{\left[\frac{(dp/dL)_L}{(dp/dL)_G}\right]^{1/2}\right\}^{0.22} \tag{163}$$

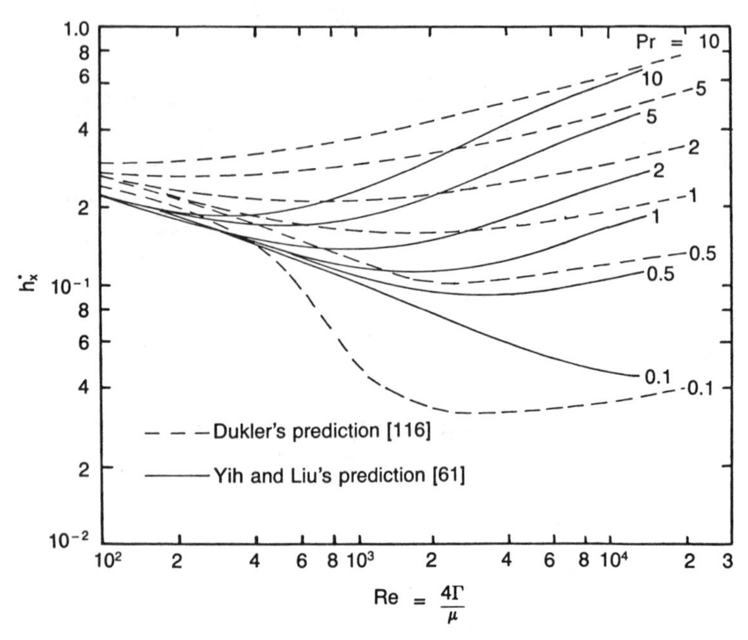

Figure 16. Comparison of the Deissler and von Karman model by Dukler [116] with the modified van Driest model by Yih and Liu for condensation [61].

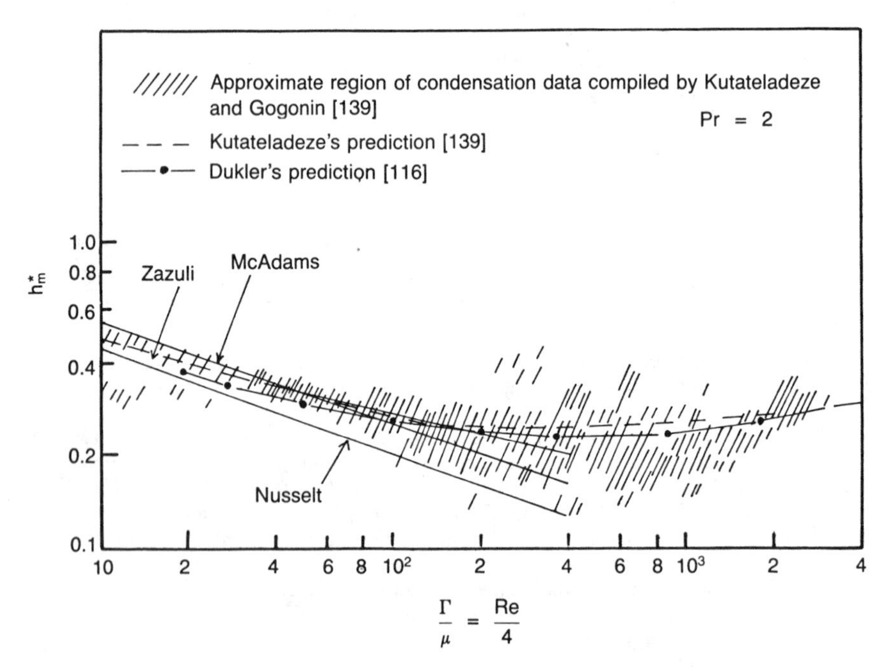

Figure 17. Dimensionless average heat transfer coefficient as a function of Reynolds number.

where $(dp/dL)_G$ and $(dp/dL)_L$ were calculated by assuming each phase flows alone in the tube. Equation 163 is only valid when there is negligible mass flow changes along the tube and the two-phase pressure drop is approximately constant. If there are significant flow changes along the tube, the pressure drop varies and so does τ_i^*. In this case, the variation of τ_i^* along the tube can be calculated from

$$\tau_i^* = A(Re_T - Re_x)^{0.25} Re_x^{0.4} \qquad (164)$$

where Re_T is the Reynolds number at the tube outlet and A is a dimensionless group

$$A = 0.25\mu^{1.173}\mu_G^{0.16}/g^{2/3}d^2\rho^{0.553}\rho_G^{0.78} \qquad (165)$$

The average condensing coefficient can be obtained by integrating the local coefficient over the tube:

$$h_m^* = Re_T/\left(\int_0^{Re_T} dRe_x/h_x^*\right) \qquad (166)$$

Verification of the model was made by comparing it with the data of Carpenter and Colburn [144] in which good agreement was obtained. However, Lee [159] and Blangetti et al. [153] have noted several inconsistencies in Dukler's model. A notable one is that the computed velocity profile is discontinuous at the intersection of the inner and outer regions at $y^+ = 20$ and the velocity profile fails to satisfy the boundary condition at the gas-liquid interface.

In contrast, the modified van Driest model gives a velocity profile which is continuous and an equally good heat transfer prediction as well. An example is shown in Figure 16 in which the prediction of Yih and Liu [61] using a van Driest model modified by a damping factor f and a variable turbulent Prandtl number Pr_t is compared with Dukler's prediction whose Pr_t is set equal to 1. Yih and Liu's predictions are lower but approach Dukler's prediction at increasing Reynolds number. It has been shown in previous sections that the modified van Driest model also predicts the data of heating and evaporation well and gives a slope that is more in accordance with the experimental trend than Dukler's prediction.

To illustrate the applicability of the prediction methods, comparison with pertinent data is necessary. Although a large number of data is available [141, 154, 155, 160–162], these data are often at variance with each other. Kutateladze and Gogonin [139] have compiled extensive data for Pr = 2 and 3 from various authors. In Figure 17 is indicated the approximate region of condensation data compiled by them for Pr = 2. Obviously, Nusselt's prediction in the laminar and wavy-laminar region underestimates the data. McAdams' correlation, Equation 139 and Zazuli's correlation, Equation 140 are both within the envelope of the data in the wavy flow regime and can be regarded as satisfactory in representing the data of mean heat transfer coefficient in the wavy flow regime of $10 < Re < 1{,}600$. Also shown are the predictions of Kutateladze using Equations 141 and 160, and Dukler's prediction from his graphical plot. These two predictions are satisfactory within the envelope of the data in both the wavy-laminar and turbulent flow regions.

Some high interfacial shear condensation data were presented by Ueda et al. [102, 149]. They measured the condensate film thickness and the local heat transfer coefficient during condensation of steam inside a tube. The film Reynolds number ranged from 500–5,000 and vapor velocities up to 220 m/s were achieved which corresponds to a range of τ_i^* from 10 to 300. In Figure 18, these data were compared with the predictions of Rohsenow et al. [147], Hubbard et al. [100], and Yih and Liu [61]. Although some differences exist among these predictions, they can be considered in satisfactory agreement with the data in view of the large scattering of the data. Obviously, more accurate data are needed for further verification.

MASS TRANSFER

Gas Absorption—Liquid-Side Mass Transfer

Wetted-wall apparatus, such as wetted-wall columns, spheres, cones, and strings of spheres, have been commonly used as laboratory models in studying physical or chemical absorption. Short wetted-wall columns or single spheres have the advantage of well-defined laminar flow conditions and known mass transfer interfacial areas. The contact time in this kind of apparatus can be varied, by varying the column length, between approximately 0.03 to 1 second, which is also the contact time practically encountered in industrial packed-column absorbers. The uses of the laboratory short wetted-wall apparatus are to:

1. Simulate the flow over packings and gas absorption in a packed column as a method for simulating the performance of an industrial absorber using a laboratory model.
2. Investigate the liquid-side mass transfer resistance and interpretation of the absorption rate in terms of the film or penetration theory for physical absorption in order to determine the liquid-side mass transfer coefficient.
3. Determine the product of the gas solubility and diffusion coefficient, $C^*D^{1/2}$, or the diffusion coefficient of gases in liquid alone from physical absorption rate data.

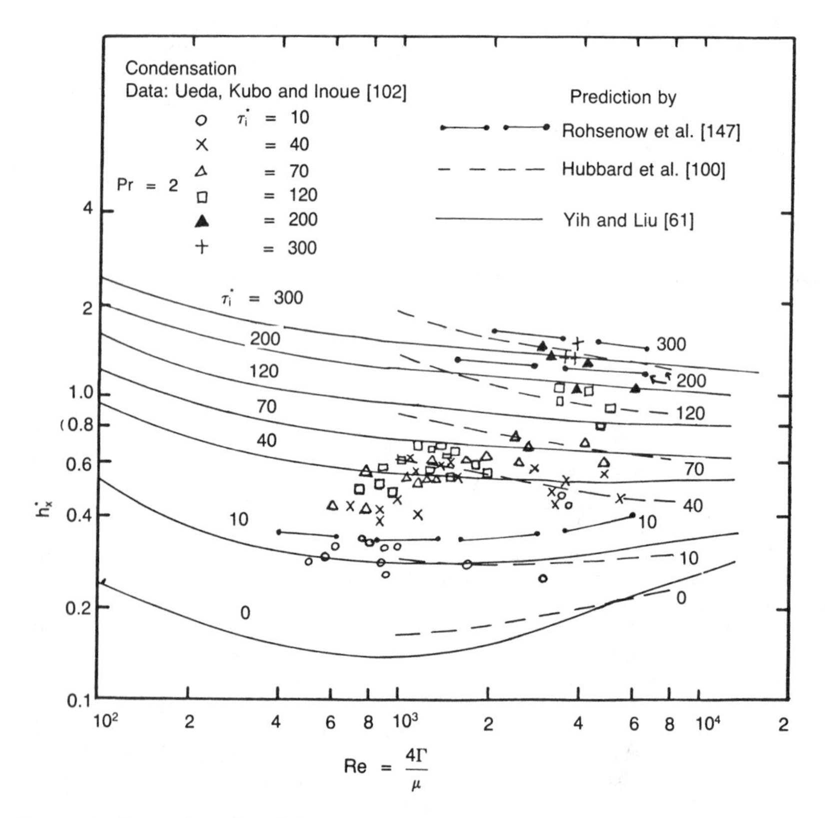

Figure 18. Comparison of predictions with data for condensation with interfacial shear [61].

4. Investigate the gas-side mass transfer resistance and to determine the gas-side mass transfer coefficient from vaporization or absorption experiments.
5. Investigate gas absorption kinetics and/or mechanism and to determine kinetic rate constants and order of reaction from chemical absorption rate data which are interpreted in terms of the film or penetration theory with simultaneous chemical reaction.
6. Determine the kinetics of, for example, the oxidation of a concentrated aqueous sulfite solution with gaseous oxygen in the presence of cobaltous sulfate catalyst as a chemical method for evaluating the interfacial area in a gas-liquid contactor from measured chemical absorption rates.

Laurent and Charpentier [163] have compiled a list of the range of contact times, interfacial area per unit volume, a′/v, mass transfer coefficient for the liquid-side, k_L, and for the gas-side, k_G, for laboratory wetted-wall apparatus (Table 7).

In many cases, long wetted-wall columns, of the order of greater than 1 meter, have been used to:

1. Investigate the liquid-side mass transfer resistance in turbulent film flow and to interpret the results in terms of turbulent mass transfer models such as the film, penetration, surface renewal theory and eddy diffusivity models with the result that the models can be extended to study chemical reaction,
2. Investigate the gas-side mass transfer resistance in turbulent gas flow and to determine the gas-side mass transfer coefficient from vaporization, distillation, or absorption experiments,
3. Be used as an industrial tube-bundle column absorber, particularly attractive for absorption with high heats of reaction and absorption in which the heat may be removed by cooling in the jacket side.

Theoretical mass transfer coefficients in falling liquid films have been determined generally by two approaches:

1. A direct integration of the mass balance equation using a velocity profile predicted from hydrodynamic analysis, and in turbulent flow also using an eddy diffusivity determined either from analogy methods or from experiments
2. Using hypothetical models which involve unknown or adjustable parameters. The first approach is more rigorous because no empirical parameters are needed. However, the velocity profile in a falling film is not always predictable and the second approach may then be useful.

Laminar and Wavy-Laminar Flow

The film theory and the penetration theory have generally been proposed for the studying of liquid-side resistance to gas absorption in packed columns. The film theory predicts that the liquid-side mass transfer coefficient is proportional to the first power of the molecular

Table 7
Range of Contact Times and k_L, k_G, a′/V for Laboratory Wetted-Wall Apparatus [163]

Laboratory Model	Contact Time (s)	Interfacial Area (cm²)	k_L cm (s⁻¹)	$k_G \times 10^6$ (mol cm⁻²) s⁻¹ atm⁻¹	a′/V cm⁻¹	Simulate Industrial Absorber
Cylinder	0.1–2	10–100	0.0036–0.016	1–9	25–60	Packed
Sphere	0.1–1	10–40	0.005–0.016		20–60	"
Cone	0.02–1	80	0.005–0.011		40–70	"
String of spheres	0.1–2	30–360	0.0036–0.016	1–25	20–60	" (Trickle bed)

diffusivity of gas in liquid which is contrary to experimental findings. Higbie's penetration theory of absorption in the absence of gas-phase resistance is probably more reasonable than the film theory in describing gas absorption in a packed column. The liquid is visualized to be flowing over each piece of packing in laminar flow and partially or completely mixed at the junction in flowing from one piece of packing to the next. During the short contact between the liquid and the packing, the dissolved gas diffuses only a short distance into the liquid film by an unsteady diffusion mechanism so that the liquid can be regarded as semi-infinite. The diffusion equation is

$$\frac{\partial C}{\partial t} = D \frac{\partial^2 C}{\partial Y^2} \quad (167)$$

with the boundary conditions:

$$t = 0 \qquad C = C_0 \text{ (initial concentration of gas in liquid)} \quad (168)$$

$$Y = 0 \text{ (interface)} \qquad C = C^* \text{ (saturation concentration)} \quad (169)$$

$$Y \rightarrow \infty \qquad C = C_0 \quad (170)$$

The solution of the penetration theory yields the mass transfer coefficient:

$$k_L = 2(D/\pi t)^{1/2} \quad (171)$$

The contact time can be calculated from the Nusselt theory as

$$t = L/u_{max} = 2L(3/g\nu)^{1/3}(\Gamma/\mu)^{-2/3}/3 \quad (172)$$

Hence

$$k_L = 0.75(D/L)^{1/2}(\nu g)^{1/6}Re^{1/3} \quad (173)$$

The average absorption rate per unit area becomes

$$N_A = N''_A/2\pi(R + \delta)L = 2(C^* - C_0)\,(3D\Gamma/2\pi\delta L)^{1/2} \quad (174)$$

For longer columns, ie., longer contact times in which there is a high degree of liquid saturation at the exit, Higbie's assumption of diffusion into a semi-infinite medium is no longer valid because of the finite thickness of the liquid film and the complete semi-parabolic velocity profile has to be taken into account in solving the convective-diffusion equation:

$$u_{max}\,[1 - (Y/\delta)^2]\,\frac{\partial C}{\partial x} = D \frac{\partial^2 C}{\partial Y^2} \quad (175)$$

with boundary conditions

$$x = 0 \qquad C = C_0 \quad (176)$$

$$Y = 0 \qquad C = C^* \quad (177)$$

$$Y = \delta \qquad \partial C/\partial Y = 0 \quad (178)$$

The solution of this equation has been given by Pigford [164] and Johnstone and Pigford [165]:

$$(C^* - C_1)/(C^* - C_0) = 0.7857 \exp(-5.1213x^*) + 0.1001 \exp(-39.318x^*) + 0.03599 \exp(-105.64x^*) + 0.01811 \exp(-204.75x^*) + \ldots \quad (179)$$

where $x^* = xD/\delta^2 u_{max}$, and C_1, C_0 are the average exit and inlet concentration, respectively. The mass transfer coefficient can then be calculated as

$$k_L = (u_{av}\delta/x) \ln [(C^* - C_1)/(C^* - C_0)] \quad (180)$$

A correlating equation which is easier to use and agrees well with experimental data is given by Brauer [166]:

$$k_L\delta/D = 3.4145 + 0.267(1.5x^*)^{-1.2}/[1 + 0.2(1.5x^*)^{-1.2}] \quad (181)$$

At sufficiently downstream where the concentration profile is fully developed, k_L approaches an asymptotic value given by

$$k_L = 3.4145D/\delta = 3.758Dg^{1/3}\nu^{-2/3}Re^{-1/3} \quad (182)$$

These equations generally represent experimental data quite well provided that entrance acceleration and exit effects are eliminated. Lynn et al. [167] have used short columns (1 to 5 cm) to study SO_2 absorption into water. In their experiments, they found that at a certain height above the surface of the stagnant liquid level in the bottom receiver, an exit effect occurred in which a narrow horizontal band of ripples appeared at the top of a thickened portion of the film which is apparently stagnant. The rate of absorption by this part of the film is relatively low. The height of this stagnant layer does not depend upon the height of the column but is influenced by surface active agents. The height of this layer is evaluated from a plot of the square of the absorption rate, $N_A''^2$, versus the length L using Equation 174 and finding the intercept value on the ordinate as shown by Nijsing and Kramers [168]. The absorption data then become well correlated by the penetration theory when the height of this layer is subtracted from the length of the column. Experiments by Vivian and Peaceman [169] who used several extremely short glass wetted-wall columns 1.9 to 403 cm long to simulate the assumptions of the penetration theory fell below the penetration theory prediction with a deviation of 10%–30%. Since ripples were absent for such a short length and could not be the cause, they attributed this deviation to the entrance acceleration and exit stagnant layer effect which were particularly important in such short columns. Indeed, Scriven and Pigford [7] found that lower absorption rate than that predicted by the penetration resulted when absorption in the entrance acceleration region was taken into account.

Lynn et al. [167], Davidson and Cullen [170] have reported that the use of small wetted-wall spheres can overcome the difficulties of the end effects. Lynn et al. ignored the stretching and contraction effect of the film on the rate of absorption and assumed that the film thickness at any latitude on the sphere is the same as it would be for the same flow rate per unit length on a plane surface making the same angle with the vertical. Using the penetration theory, they derived the rate of absorption as

$$N_A'' = (C^* - C_0)D^{1/2}(18g\pi/\nu)^{1/6}Q^{7/6} \int_0^\pi \left[\sin\theta d\theta \Big/ \left(\int_0^\theta \sin^{1/3}\theta d\theta\right)^{1/2}\right] \quad (183)$$

where the integral term has the value of 2.33. For a string of n spheres with complete mixing at the junction between the spheres

$$(N_A'')_n = nN_A'' \quad (184)$$

while for no mixing it is

$$(N''_A)_n = n^{1/2} N''_A \tag{185}$$

Davidson and Cullen [170] derived the diffusion equations more rigorously and obtained a simplified form of the diffusion equation by using the streamlines as a coordinate framework. For short contact times, the rate of absorption is derived as

$$N''_A = 4.1966(C^* - C_0)D^{1/2}(g\pi/\nu)^{1/6}Q^{1/3}R^{7/6} \tag{186}$$

which is similar to Lynn et al.'s equation whose coefficient becomes 3.772 when converted to the form of Equation 186. For long contact times, Pigford's solution, Equation 179, can be applied as

$$N''_A = Q(C^* - C_0)[1 - 0.7857 \exp(-3.414\alpha^*) - 0.1001 \exp(-26.21\alpha^*) - 0.03599 \exp(-70.43\alpha^*) - 0.01811 \exp(-136.5\alpha^*) - \ldots] \tag{187}$$

where $\alpha^* = 3.36\pi(2\pi g/3\nu)^{1/3}DR^{7/3}/Q^{4/3}$ (188)

This equation should be used in preference to Equation 186 when the outlet liquid concentration exceeds 40% of the saturation value.

Olbrich and Wild [171] generalized Davidson and Cullen's approach and solved the convective-diffusion equation for various flow geometries that have a certain degree of symmetry such as a cone and a revolved cycloid. Laplace transform and power series solution were employed and ten sets of eigenvalues and coefficients of the series solution could be obtained. The additional eigenvalues calculated represent an improvement over Pigford's four-term solution which will no longer be adequate for short to intermediate contact times. Analytical solutions of the convective-diffusion equation have also been given by Davis [172] in terms of the confluent hypergeometric function and Yih and Huang [173] in terms of a series solution. The latter authors showed that the asymptotic mass transfer coefficient can be found easily from the first eigenvalue λ_1 of the series:

$$k_L = 2D\lambda_1^2/3\delta \tag{189}$$

where $\lambda_1 = 2.26311$. This result is in agreement with Equation 182. Other related problems include consideration of thermodynamic nonequilibrium at the gas-liquid interface as a result of addition of surface active agent [174] and low Peclet number flow by including the axial diffusion term [175].

As can be seen from Equations 174 and 183, the product of the gas solubility and diffusivity, $C^*D^{1/2}$, or the diffusivity alone, can be determined from measured absorption rates either in a short wetted-wall column or a sphere. Short wetted-wall columns have been used to measure the diffusion coefficients of carbon dioxide in several organic liquids and hydrocarbon mixtures [176], of carbon dioxide in ethanol-water at 25°C [177], of helium, hydrogen, and carbon dioxide in water at 25°C [178]. Wetted-wall spheres have been used to measure the diffusion coefficients of carbon dioxide, nitrous oxide, chlorine, oxygen, and hydrogen in water within 15°–30°C [170], of nitrogen in cyclohexane at 21°C [171], of carbon dioxide in methanol, ethanol, aqueous propanol and ethylene glycol solutions [179], of nitrogen in liquid cyclohexane within 21°–140°C and 10–15atm [180].

In general, in order to determine accurately the absorption rate, care must be taken to eliminate, besides the entrance and exit effects, surface ripples which occur and grow at even very small flow rates. Very short columns of the kind used by Vivian and Peaceman [169], although sometimes free of ripples, are difficult to construct and suffer from the disadvantage of large entrance and exit effects. Therefore in using a column of, say, longer than 5 cm, surface active agents (SAA) are usually added to eliminate the ripples. In the presence of SAA, the absorption rate usually conformed with the theory predicted by the penetration theory for laminar, wave-free flow. However, some authors [181] found that the measured value lies slightly below the theoretical curve and they attributed this to an interfacial mass transfer resistance caused by the SAA.

The literature values on measured liquid-side mass transfer coefficients in the wavy-laminar flow region show substantial disagreement among various investigators. A fundamental reason for these anomalies is that the mechanism in which the waves affect the transfer processes is not clear. Another reason is that the way the film is generated and the diameter and length of the tubes have a strong influence on mass transfer. Moreover, most of the data on transfer rates in wavy film are reported in terms of overall rate, which integrates transfer rate in the smooth entry region, in the wave developing region and in the fully developed region. This makes it difficult to evaluate which part of the total transfer rate is influenced by the waves and by what types of waves. Usually there is a strong dependence of k_L on the length of the column for short columns less than, for example, 1 meter.

Emmert and Pigford [181] were the first to show the severalfold, usually about 200%, enhancement of mass transfer due to waves during absorption and desorption of O_2 and CO_2 in a water film 1.14 meter long and within the range of Re = 100–1800. By adding a surface active agent, the mass transfer coefficient returned to values predicted by the penetration theory for smooth laminar flow. The mechanism in which the waves increase the mass transfer rate to such a larger extent, much larger than the approximately 20%–30% increase in heat transfer rate, is not clear. Careful experiments by Brauer [6] and Portalski and Clegg [182] have shown that the increase in interfacial area due to waves is less than 10% and cannot explain the more than 200% increase in mass transfer rate. So other mechanisms, such as local mixing, surface renewal, or convective motion, are speculated to strongly influence the mass transfer rate. Calculations have been made by Portalski on examining the velocity distribution derived by Kapitza [183] that regions of reversed flow may exist under the troughs of a periodic traveling wave. The formation of circulating eddies in the flow thus promotes mixing and surface renewal. Massot et al. [29] have solved the two-dimensional linearized integral momentum equations using Kapitza's method and draw flow streamlines indicating the possibility of a reversed flow structure. A number of attempts have been made to calculate the enhanced transport due to waves for very low flow rates where the waves can be regarded as sinusoidal. Levich [184], and later, Ruckenstein and Berbente [185] have used the Kapitza-type velocity distribution to integrate the species mass balance equation:

$$\frac{\partial C}{\partial t} + u\frac{\partial C}{\partial x} + v\frac{\partial C}{\partial y} = D\frac{\partial^2 C}{\partial y^2} \tag{190}$$

The predicted enhancement was found to be only 15%–30%, which was far below the measured 200% increase. They attributed this to the inaccuracy of Kapitza's velocity profile. Subsequently, Berbente and Ruckenstein [31] developed a solution of the boundary-layer momentum equations using a triple series expansion for the velocity which was described by a power series in normal distance up to the sixth degree. They found better agreement between the experimental and theoretical wave properties, and a better mass transfer prediction as well [186]. However, the linearization procedure invoked in solving the momentum equations also restricts its validity to small flow rates. Beschkov and Boyadjiev [187] numerically solved the convective-diffusion equation for two-dimensional waves propagating on the film surface with a constant phase velocity. The dimensionless velocity components and film thickness were represented by a truncated Fourier series where the coefficients were derived from Penev et al.'s work [32]. Comparison of the computed solution with measured data for gas desorption from wavy water films supported the adequacy of the numerical solution. They concluded that the mass transfer enhancement was not caused by surface renewal but by the convective mass flux normal to the interface and the periodic decay and formation of the concentration field due to the wavy motion.

A different approach has been undertaken by Javdani [188] in which the concentration fluctuations, induced by the fluctuating velocity field, are assumed to be a periodic quantity proportional to the mean concentration gradient. A periodic time average of the governing equation, Equation 190, introduces an apparent diffusivity term which varies approximately linearly with the distance from the interface and is a function of flow rate and wave properties such as the wave number, equilibrium amplitude, and wave velocity. A simplified solution based on the penetration theory was presented and compared with a limited amount of data obtained under controlled wave conditions for carbon dioxide absorption into a mineral oil. The theory fails for large values of x^* and tends to give a too high value of the mass transfer coefficient. Faghri [64] later used the complete velocity profile and apparent diffusivity profile deduced by Javdani to solve the convective-diffusion equation. The comparison with data was not much better which means that the discrepancy was not due to the use of the penetration theory.

The above investigations are usually limited to low Reynolds number of the order $Re < 100$. For larger Reynolds number in the wavy and transition regions of $100 < Re < 1{,}600$, an approach based on the assumption of an apparent turbulent flow was suggested. Banerjee et al. [189] proposed an eddy flow model in which Harriott's random eddy modification of the penetration theory was used in conjunction with Levich's estimate of the distance of approach of an eddy towards the interface to evaluate the mass transfer coefficient semi-quantitatively. Brauer [44] recently assumed that, in analogy to molecular diffusion, a mass diffusivity due to deformation turbulence, ϵ_m, can be formulated to describe the increased mass transfer due to the wavy motion. The convection-diffusion equation for either gas absorption or solid dissolution may be described by

$$(1 - \eta^2)\frac{\partial C}{\partial x^*} = \frac{\partial}{\partial \eta}\left[\left(1 + \frac{\epsilon_m}{D}\right)\frac{\partial C}{\partial \eta}\right] \tag{191}$$

but with oppositely-directed boundary conditions. It was assumed that the velocity profile in the low Reynolds number range of a wavy film would not be affected by deformation turbulence and could be still described by the Nusselt velocity profile. Further, ϵ_m/D was assumed to be proportional to $Sc(\Gamma/\mu)^m\eta^n$ for the gas-liquid interface and proportional to $Sc(\Gamma/u)^m(1 - \eta)^n$ for the solid-liquid interface where $\eta = Y/\delta$ is the dimensionless distance measured from the gas-liquid interface. Experimental data for mass transfer from absorption and solid dissolution have been used to obtain empirical equations for the ratio ϵ_m/D by Carrubba [190]:

$$\epsilon_m/D = F_1 - (F_1^{10} + F_2^{10})^{1/10} + F_2 - (F_2^{10} + F_5^{10})^{1/10} + F_5 - (F_5^{10} - F_6^{10})^{1/10} + F_6 \quad \text{for } 12 \le \Gamma/\mu \le 70 \tag{192}$$

$$\epsilon_m/D = F_3 - (F_3^{10} + F_4^{10})^{1/10} + F_4 - (F_4^{10} + F_5^{10})^{1/10} + F_5 - (F_5^{10} + F_6^{10})^{1/10} + F_6 \quad \text{for } 70 \le \Gamma/\mu \le 400 \tag{193}$$

The functions F_1 to F_6 are given by the following equations:

$$F_1 = 1.24 \times 10^{-3}(\Gamma/\mu)^{1.6}Sc\eta^2 \tag{194}$$

$$F_2 = (\{[2.13 \times 10^{-4}(\Gamma/\mu)^{1.6}Sc]^4 + (4.94)^4\}^{1/4} - 4.94)\eta^{1.5} \tag{195}$$

$$F_3 = 1.58 \times 10^{-2}(\Gamma/\mu)Sc\eta^2 \tag{196}$$

$$F_4 = (\{[2.72 \times 10^{-3}(\Gamma/\mu)Sc]^4 + (4.94)^4\}^{1/4} - 4.94)\eta^{1.5} \tag{197}$$

$$F_5 = (\{[2.5 \times 10^{-4}(\Gamma/\mu)^{1.5}Sc]^4 + (3.94)^4\}^{1/4} - 3.94)(1 - \eta)^{2.5} \tag{198}$$

$$F_6 = 7.52 \times 10^{-4}(\Gamma/\mu)^{1.5}Sc(1 - \eta)^3 \tag{199}$$

Very close to the gas-liquid interface, ϵ_m/D is proportional to Y^2 while close to the solid-liquid interface, $\epsilon_m/D\eta$ approaches zero.

A number of authors have reported experimental data on absorption in wavy films. The earliest was Brotz [55], who conducted absorption of CO_2 in water with various film lengths of 32, 40, 46, 144 cm and diameters of 11, 16 and 26 mm. The water temperature ranged from 10° to 80°C. For $(\Gamma/\mu) < 300$, the measured data agreed with the penetration theory while for $(\Gamma/\mu) \geq 300$, Brotz introduced the mixing coefficient $D(\Gamma/\mu)/300$ instead of D in the penetration theory solution. As a result of calculations, Brotz obtained

$$k_L = (6D/\pi L)^{1/2}(g\nu/3)^{1/6}(\Gamma/\mu)^{1/3} \quad \text{for } \Gamma/\mu < 300 \tag{200}$$

$$k_L = 5.77 \times 10^{-2}(6D/\pi L)^{1/2}(g\nu/3)^{1/6}(\Gamma/\mu)^{5/6} \quad \text{for } 300 \leq \Gamma/\mu \leq 590 \tag{201}$$

Kamei and Oishi [191] absorbed CO_2 in water at different temperatures between 8.5°–50°C in a wetted-wall column 2.5 meters long. Their data showed different Reynolds number dependence in the wavy flow regions Re = 50 – 200 and Re = 200–2,000. Brauer [192], in analogy to heat transfer in film cooling and film condensation, assumed that the major mass transfer resistance resides in a wall vicinity layer of thickness δ_1 in which the velocity profile could be taken as linear so that

$$k_L = D/\delta_1 = D\tau_w/\mu u' \tag{202}$$

where u' denotes the velocity at the edge of the diffusion sublayer δ_1. With the aid of expressions for the wall shear stress τ_w from an earlier study [6] inserted, a set of correlations was obtained:

$$k_L = 2.02\ (D/u')(g^2/\nu)^{1/3}(\Gamma/\mu)^{1/3} \quad \text{for } (\Gamma/\mu)w_2 \leq \Gamma/\mu \leq (\Gamma/\mu)_c \tag{203}$$

$$k_L = 3.605\ (D/u')(g^2/\nu)^{1/3}(\Gamma/\mu)^{1/5} \quad \text{for } (\Gamma/\mu)_c \leq \Gamma/\mu \leq (\Gamma/\mu)_{cr} \tag{204}$$

where $(\Gamma/\mu)_c = 0.018K_F^{1/3}$ according to Brauer [6]. By fitting these equations to the comprehensive data of Kamei and Oishi [191], a value of u' giving the best fit is 0.066 m/s.

Hikita et al. [193] made a comprehensive study on the absorption of 6 different gases into a variety of solvents encompassing a wide range of Schmidt numbers from 73 to 2,600, surface tension from 23 to 75 dyne/cm, Re from 30 to 2,000, Re_G from 50 to 10,000, temperatures from 10° to 45°C. Four different lengths, 0.2–1.03 m, and four different column inside diameters, 7–43.5 mm, were used. They also identified that the mass transfer coefficient had different Reynolds number dependence in the two regions of Re = 30–200 and Re = 200–2,000. Their correlation, originally expressed in HTU, is transformed into the dimensionless mass transfer coefficient:

$$k_L^* = (k_L/D)(\nu^2/g)^{1/3} = 0.011\ Re^{0.5}Sc^{0.62}Ga^{-0.04}(\sigma/72)^{-0.15} \quad \text{for } 30 < Re < 200 \tag{205}$$

$$k_L^* = 0.00694\ Sc^{0.5} \quad \text{for } 200 < Re < 2{,}000 \tag{206}$$

Their data indicate that there is a slight influence of surface tension on k_L^* in the region 30 < Re < 200. They observed that the gas rate had no effect on k_L^* when the Re_G is below 7,000 and the difference in k_L noted by Emmert and Pigford [181] between absorption and desorption could not be found. Also their data seem to indicate that, except for d = 7 mm, there is no effect of diameter on k_L in the range of d = 13 – 43.5 mm. However, others, such as Malewski [194, 195], found a strong influence of diameter on k_L for d between 10 to 30 mm in which a decrease of diameter decreases k_L. Malewski simultaneously measured the wave frequency and the gas absorption rate along a column with different column diameters in order to find the relation between mass transfer and wave structure. He interpreted the enhanced mass transfer in the wavy flow region in terms of a degree of mixing, p, and an average residence time, $(t_m)_p$, over the film length L. The series solution of Pigford [164], Equation 179, for laminar flow was assumed to apply to wavy flow also if in Equation 179, $t = x/u_{max}$ is replaced by

$$(t_m)_p = (1 + pZ_L)^{-1}(L/u_{max}) \tag{207}$$

and x is replaced by

$$x' = L/(1 + pZ_L) \tag{208}$$

where Z_L is the wave number determined from

$$Z_L = \int_0^L Z_m(L)dL \tag{209}$$

and

$$Z_m = \left[\frac{1}{(\delta_{max} - \delta_{min})} \int_{\delta_{min}}^{\delta_{max}} f(\delta)d\delta\right] /u_{max} \tag{210}$$

The degree of mixing, however, must be determined from actual mass transfer measurements in wavy flow:

$$p = \{[(k_L)_{wave}/(k_L)_{lam}]^2 - 1\}/Z_L \tag{211}$$

where $(k_L)_{lam}$ is the short contact time solution, Equation 171. The degree of mixing is a function of Re and column diameter. It increases with Re, reaches a maximum value at approximately Re = 320 and then decreases. It also increases with d and approaches p = 1 (complete mixing) when d is large. Conceivably, a decrease in diameter decreases the degree of mixing and hence k_L. This is illustrated in Figure 19, which is drawn from the work of Malewski [195] where the data of Malewski along with three other authors are shown. Also shown are the theoretical predictions for laminar flow given by the short contact time, Pigford's, and long contact time equations, Equations 171, 179, 182, respectively. Data in the wavy flow region are substantially higher than these predictions. As evident, a strong effect of diameter is noticed. The data of Kamei and Oishi [191] and Hikita et al. [193] in which larger diameters were used, can be correlated by the equations proposed by Brauer, Equations 203 and 204. These data also indicate that the mass transfer coefficient is nearly independent of Reynolds number for $100 < \Gamma/\mu < 400$, a result which is similarly found in heat transfer in the wavy flow region. The change of slope at $\Gamma/\mu = 100$ for the k_L versus Γ/μ curve is quite interesting, the reason of which is still unknown. Most probably it is due to a change in the wave structure at this $(\Gamma/\mu)_c$.

Turbulent Flow

When the Reynolds number is larger than approximately 1,600, the wavy flow becomes turbulent and there is a large increase in the mass transfer coefficient which is several times higher than those in the wavy flow regime. This means that a more vigorous mixing or renewal process is present in turbulent flow. At present, there are two possible approaches to the prediction of mass transfer in turbulent flow. The first attempts to relate the mass transfer coefficient to certain hydrodynamic or wave parameters characterizing the turbulence struc-

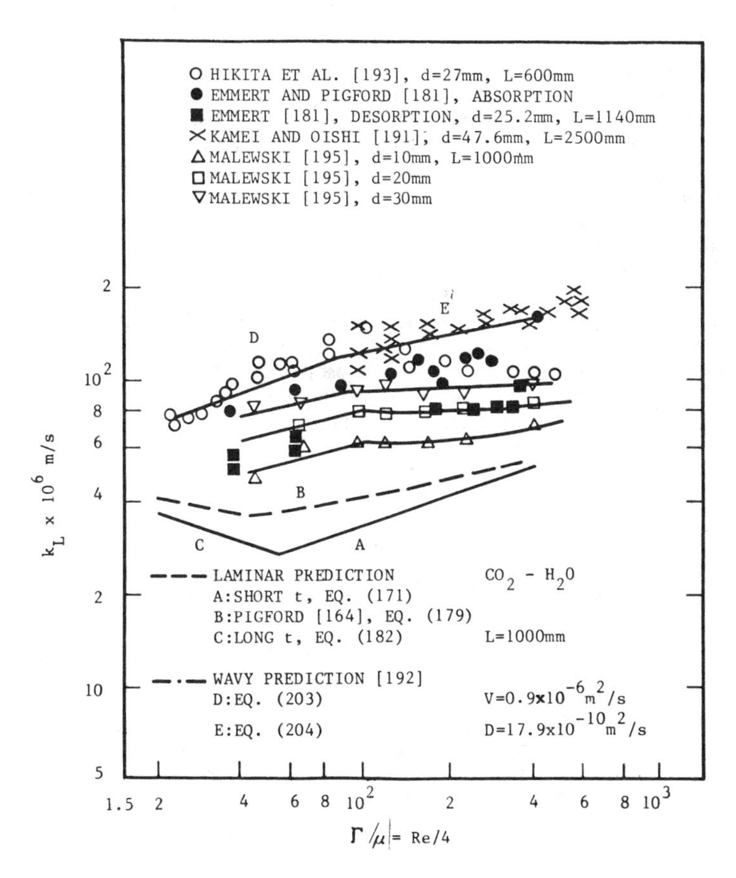

Figure 19. Comparison of theory with data for gas absorption in the wavy flow region [195].

ture near the interface but these parameters must be determined from experiments. The second is to use an empirical or semi-empirical expression for an eddy diffusivity and then to integrate the species mass balance equation to obtain the mean concentration or mass transfer coefficient. Most of the treatments neglect the surface waves and assume the liquid film to be in smooth turbulent motion. They are called the constant film thickness model. As in heat transfer, a variable film thickness model has been developed by Brumfield et al. [49] which gives predictions of mass transfer coefficient quite well.

Film, penetration, surface renewal, boundary layer, and eddy cell models. A number of well-known models considering mass transfer at a gas-liquid interface are available. The earliest two-film theory assumed two stagnant films adjacent to the interface in each phase where stationary diffusion occurs and the mass transfer coefficient is proportional to the molecular diffusion coefficient to a first power, $k_L \sim D^{1.0}$. The penetration theory of Higbie assumed unsteady diffusion into a liquid element residing at the interface for a short exposure time while Danckwerts assumed a statistical distribution of the ages of the liquid element in his surface renewal theory. These penetration theories predict that $k_L \sim (D/t)^{0.5}$. A combination of the film and penetration model by Toor and Marchello [196] give $k_L \sim D^{0.5\text{-}1.0}$. Tadaki and Maeda [197] have compared their data with these theories and found that the film-penetration theory is better. A boundary-layer model was developed by Brauer [192] who assumed that the major mass transfer resistance for gas absorption resides in a layer near the wall analogous to heat transfer. This model cannot be correct because for gas absorption, the major mass transfer resistance is near the gas-liquid interface. Fortescue and Pearson [198] assumed that mass transfer across the surface of a turbulent flowing liquid is determined by the relatively large-scale energy-containing eddies. Lamont and Scott [199], on the other hand, postulated that the mass transfer is due primarily to mixing by small eddies within large eddies that flow to the surface and are deflected along the surface and then return to the bulk liquid. As a result, the mass transfer is dependent upon the energy dissipation, E_v, and liquid viscosity near the interface as well as the molecular diffusivity but independent of surface tension and can be expressed as $k_L \sim (\nu/D)^{-1/2}(E_v\nu)^{1/4}$. Banerjee et al. [200] showed that the mass transfer rate at a free interface is likely to be controlled by small eddies for which viscous dissipation is important and can be estimated as a function of wavelength, amplitude, and velocity from a consideration of the time-averaged vorticity equation. Using various literature data on average values of wavelength, amplitude, and velocity for water films, the mass transfer coefficient was derived as

$$k_L = 2.93 \times 10^{-3} D^{1/2} Re^{0.933} \tag{212}$$

Equation 212 predicts the data of Emmert and Pigford [181] and Kamei and Oishi [191] with a deviation of about $\pm 30\%$ within the range $1{,}500 \leq Re \leq 8{,}000$. The exponent of Re, however, seems to be too high as compared with the experimental trend of 0.6–0.8.

Eddy diffusivity models. A majority of theoretical attempts have used the concept of eddy diffusivity to describe the enhanced transport in turbulent film flow. Recent experimental visualization studies reported by Davies [201] show that there are important eddy movements very close to the gas-liquid interface. In view of the fact that the liquid Schmidt numbers are large, the resistance to mass transfer lies mainly in a region very close to the interface where the concentration gradient is steep and it is necessary to know the transport coefficient accurately only in this region. Accordingly, the eddy diffusivity in this region can be approximated by

$$\epsilon_D = aY^n \tag{213}$$

where Y is the distance measured from the interface. The problem then is to determine the value of the exponent n and the parameter a from turbulence characteristics.

Levich [184], and later Davies [201], postulated that there is a zone of damped turbulence, λ_t, near the interface where the normal fluctuating velocity component, v', decays towards the interface due to surface tension damping while the u' component remains constant. Beyond this zone, the turbulence is assumed normal and v' is equal to the friction velocity v_0. The eddy scale of motion and the velocity fluctuations, v', are assumed to vary linearly with the distance from the interface so that $\epsilon_D \sim Y^2$. The thickness of this zone of damped turbulence is determined by a force balance between the dynamic pressure fluctuations of an eddy which causes deformation of the interface and the surface tension force which opposes the deformation. The mass transfer coefficient was derived by considering steady-state molecular diffusion through a diffusion sublayer:

$$k_L = 0.32 D^{1/2} \delta^{3/4} (g \sin \theta_1)^{3/4} \rho^{1/2} \sigma_{equiv}^{-1/2} \tag{214}$$

where σ_{equiv} = surface tension + gravitational pressure. Equation 214, when compared with the data of Davies [201], was found to underestimate the mass transfer coefficient by about five times. This is mainly because Equation 214 predicts that $k_L \sim Re^{0.4-0.5}$ while experiments show that $k_L \sim Re^{0.6-0.8}$.

King [202] generalized the concept of surface renewal and damped eddy diffusivity profile in the vicinity of the interface into a model which describes the mass transfer behavior of a fluid element at the interface as

$$\frac{\partial C}{\partial t} = \frac{\partial}{\partial Y}\left[(D + aY^n)\frac{\partial C}{\partial Y}\right] \tag{215}$$

with the boundary conditions

$$t = 0 \qquad C = C_0 \tag{216}$$

$$Y = 0 \qquad C = C^* \tag{217}$$

$$Y = \infty \qquad C = C_0 \tag{218}$$

If the surface age is high enough and n is sufficiently large, steady-state mass transfer occurs and the solution of Equation 215 is obtained as

$$k_L = (n/\pi)\sin(\pi/n)a^{1/n}D^{1-(1/n)} \tag{219}$$

King presumed, in analogy with turbulent mass transfer to and from pipe walls, that $n = 4$ for a free surface and surface tension has little effect on the mass transfer. The value of a is evaluated through a modification of a correlation relating the rate of energy dissipation to the mass transfer coefficient at a fixed interface. The analysis shows that, for long columns, k_L should increase with $Re^{1/6}$ and varies with $D^{3/4}$. This, however, does not conform with the observed dependence of k_L which varies as $Re^{0.6-0.8}$ and $D^{1/2}$. Moreover, the data of various authors used by King in developing his theory and correlation all lie in the wavy-flow region. Thus the theory he proposed would not be expected to apply.

In thin films, the size of eddies can be expected to be limited to the order of the film thickness. Therefore, Prasher and Fricke [203] presumed that surface tension and film thickness should have a significant effect in damping the eddies at the interface and at the solid surface. Also the energy of dissipation and viscosity play a role in determining the mass transfer coefficient because the flow and the frequency of flow of the smaller eddies are dominated by viscous effect. In Lamont's model, $k_L = f(E_v,\nu)$ while in the Levich-Davies model, $k_L = f(\rho, v_o, \sigma)$. The Prasher-Fricke model essentially incorporated the hydrodynamic quantities found in both of these models to give $k_L = f(\rho, \sigma, \nu, \delta, E_v, g)$ and arrived at:

$$k_L = (0.108/\pi)(\rho/\sigma)^{1/2}(\delta^3E_v^3/g\nu)^{1/4}D^{1/2} \tag{220}$$

The rate of energy dissipation for a falling film is

$$E_v = u_{av}g = g\nu Re/4\delta \tag{221}$$

This model assumed that viscosity and surface tension are both important in influencing mass transfer at the turbulent gas-liquid interface and gave a dependence $k_L \sim Re^{3/4}D^{1/2}$ which is consistent with experimental trend.

Carrubba [190] has postulated that ϵ_D/D is proportional to $ScRe^m(Y/\delta)^n$ and by a correlation of the data of various authors for long column lengths and $Re \geq 400$, he obtained a relationship for the parameter a, Equation 213, which can be approximated by $a \sim \delta^{0.5}v_0^{2.5}/\nu^{1.5}$.

Henstock and Hanratty [204] recently proposed a viscosity-damped turbulence model (VDTM) which seems to describe the transport through a gas-liquid interface quite well. They postulated that the eddy y-component fluctuating velocity and mixing length are both approximately constant within the bulk liquid. Across a damped turbulence region near the interface, λ_t, they are damped linearly from their bulk value v_0 and l_0 to 0 and ϵ_D is given by

$$\epsilon_D \sim v_y' l_D \sim (v_0 Y/\lambda_t)(l_0 Y/\lambda_t) = aY^2 \tag{222}$$

In the turbulent flow region, it can be assumed that the order of magnitude of l_0 is as large as δ and v_0 is approximately equal to the friction velocity. If the damping of eddies near the gas-liquid interface is due mainly to viscosity, then λ_t must be a function of v_0, l_0, and ν. The VDTM assumes that λ_t is independent of l_0 and, by dimensional analysis, establishes that $\lambda_t v_0/\nu$ = constant so that $a \sim \delta v_0^3/\nu^2$. It can be seen that this expression for a is larger by only a factor of $(\delta v_0/\nu)^{0.5}$ than the results derived by Carrubba. By specifying an empirical turbulent film thickness correlation such as the one by Brauer, Equation 51, Yih [205] was able to deduce that

$$k_L^* = (k_L/D)(\nu^2/g)^{1/3} \sim Re^{2/3} Sc^{1/2} \text{ for } Re > 1{,}600 \tag{223}$$

which agrees well with experimental trend. Similarly if we assume that an eddy diffusivity for wavy flow can be similarly defined, then the order of magnitude of l_0 is not as large as δ but is close to that of λ_t. By specifying the Nusselt equation for laminar film thickness, Equation 12, Yih also deduced that

$$k_L^* \sim Re^{1/6} Sc^{1/2} \qquad \text{for } 300 < Re < 1{,}600 \tag{224}$$

which agrees well in form with the empirical correlation of Carrubba [190] shown in Table 8. Besides gas absorption, Yih has shown that the VDTM can describe the transport through a solid-liquid interface quite well too.

Lamourelle and Sandall [72] absorbed four different gases into water in a long wetted-wall column and interpreted their results in terms of the eddy diffusivity model. From the measurements with water at 25°C, they determined that

$$\epsilon_D = 7.9 \times 10^{-5} Re^{1.678} Y^2 \tag{225}$$

Results obtained by Carrubba [190] are:

$$\epsilon_D = 1.95 \times 10^{-6} \nu (\Gamma/\mu)^{2.5} (Y/\delta)^2 \sim Re^{1.5} Y^2 \tag{226}$$

which is similar in form to Equation 225.

By comparing two types of empirical correlation for the parameter a, one involving the group $(\rho g \nu/\sigma)$ and the other group $(g^2/\nu)^{1/3}$, with about 520 data points, Bin [207] concluded that ignoring the effects of surface tension give better agreement with the data of many authors collected from different gas-liquid systems. The equation of Carrubba, Equation 226, was recommended as the most reliable. However, greater deviations from Bin's correlation still exist for the data of aqueous solutions of ethylene glycol, propanol, and methanol reported by Won [208], Won and Mills [209].

As can be seen from Table 8, theoretical explanation of the mechanism of eddy mass transport at the gas-liquid interface can be classified into three groups:

1. Surface tension-damped turbulence
2. Viscosity-damped turbulence
3. Both surface tension and viscosity-damped turbulence. The first group does not represent the data well, so most probably it is viscosity, and not surface tension, that constitutes the major damping mechanism.

Models of the latter two groups predict that

$$\epsilon_D \sim Y^2 \tag{227}$$

Table 8
Relationships Suggested for a and n in $\epsilon_D = a Y^n$

Year	Author	Relationship for a	n	$k_L \sim$ (for $\delta \sim Re^{0.526}$)
1962 1972	Levich [184] Davies [201]	$a \sim v_0/\lambda_\tau = \rho v_0^3/\sigma_{equiv}$	2	$Re^{0.395}$
1966	King [202]	$a = 0.006E_v\rho/\mu^2$	4	$Re^{0.12}$
1974	Prasher and Fricke [203]	$a = 0.2916(\rho/\sigma)(\delta^3 E_v^3/g\nu)^{1/2}$	2	$Re^{3/4}$
1977	Kishinevskii and Korniyenko [206]	$a = 15\rho v_0^3/\sigma_{equiv.}$	2	$Re^{0.711}$
1976	Carrubba [190]	$a = 6.09 \times 10^{-8} Re^{2.5}\nu/\delta^2$	2	$Re^{0.724}$
1979	Henstock and Hanratty [204]	$a \sim v_0^3\delta/\nu^2$	2	$Re^{0.658}$
1972	Lamourelle and Sandall [72]	$a = 7.9 \times 10^{-5} Re^{1.678} sec^{-1}$ (from own experiments)	2	$Re^{0.839}$
1983	Bin [207]	$a = 1.17 \times 10^{-6}(g^2/\nu)^{1/3} Re^{1.448}$ (By fitting Carruba's equation to about 520 data points of various authors)	2	$Re^{0.724}$

and

$$k_L \sim Re^{0.6\text{-}0.8} \tag{228}$$

which are in line with experimental trend. The validity of these models depends on the choice of selecting the appropriate velocity and length scales that characterize the turbulence.

Extensive measurements on gas absorption in turbulent falling films have been conducted either to determine the dependence of the liquid-side mass transfer coefficient on Reynolds number (e.g., [55, 72, 191, 204, 209–213]), or the dependence of k_L on the diffusion coefficient (e.g.[72, 214, 215]). To measure the mass transfer coefficient, the bulk concentration of the gas in the inlet and outlet liquid together with the saturation concentration must be measured and k_L calculated from

$$k_L = (\Gamma/\rho_L)[(d/2)/(d/2 \pm \delta)] \ln [(C^* - C_0)/(C^* - C_1)] \tag{229}$$

Besides correlating one's own data, some authors (e.g., [190, 213, 216]) have proposed correlating equations which are established by regression analysis of a large number of data collected in the literature. These correlations are usually presented in the following four forms as a function of Re, Sc, and Ga:

1. $H_L(g/\nu^2)^{1/3}$ (230)
2. H_L/L (231)
3. $Sh = k_L\delta/D$ (232)
4. $k_L^* = (k_L/D)(\nu^2/g)^{1/3}$ (233)

The height of a mass transfer unit, H_L, is defined by

$$H_L = \Gamma/\rho k_L \tag{234}$$

Table 9
Constants and Exponents of the Equation $k_L^* = b\,Re^m\,Sc^n\,Ga^e$

Year	Author	Range of Re	b	m	n	e	Range of Others
1954	Brotz [55]	$Re < 1{,}200$	0.725	1/3	1/2	1/2	
		$1{,}200 < Re < 2{,}360$	0.0209	5/6	1/2	1/2	
		$2{,}360 < Re$	0.076	2/3	1/2	1/2	
1955	Kamei and Oishi [191]	$2{,}000 < Re$	0.018	0.7	0.444	0.083	$8.5° < T < 50°C$
1958	Brauer [192]	$5.4K_F^{1/10} \le Re \le 0.072K_F^{1/3}$	0.0346	1/3	0	0	
		$0.072K_F^{1/3} \le Re \le 1{,}600$	0.1293	1/5	0	0	
		$1{,}600 < Re$	0.00235	2/3	0	0	
1959	Hikita et al. [193]	$30 \le Re \le 150$	0.011	0.5	0.62	0.04	$10° < T < 45°C$
		$150 < Re \le 2{,}000$	0.106	0	1/2	0	
1968	Banerjee et al. [200]	$340 < Re < 10{,}900$	0.0001344	0.933	1/2	0	
1971	Johannisbauer [216]	$60 < Re < 300$	0.408	0.4	0.5	−0.05	$60 < Sc < 1{,}100$
		$300 < Re < 1{,}600$	1.28	0.2	0.5	−0.05	$10^8 < Ga < 10^{15}$
		$1{,}600 < Re < 10{,}800$	0.069	0.6	0.5	−0.05	
1972	Lamourelle and Sandall [72]	$1{,}300 < Re < 8{,}300$	0.0002593	0.839	1/2	0	
1976	Carrubba [190]	$48 \le Re \le 280$	0.00812	0.4667	0.5	0	for $Sc \ge 2.13 \times 10^5/Re^{1.6}$
		$48 \le Re \le 280$	3.7522	−1/3	0	0	for $Sc \le 2.13 \times 10^5/Re^{1.6}$
		$280 \le Re \le 1{,}600$	0.04396	0.1667	0.5	0	for $Sc \ge 7.28 \times 10^3/Re$
		$280 \le Re \le 1{,}600$	3.7522	−1/3	0	0	for $Sc \le 7.28 \times 10^3/Re$
		$1{,}600 \le Re$	0.0007575	0.7167	0.5	0	for $Sc \ge 4.59 \times 10^5/Re^{2.5}$
		$1{,}600 \le Re$	3.7522	−1/3	0	0	for $Sc \le 4.59 \times 10^5/Re^{2.5}$
1976	Bakopoulos [217, 218]	Same as Carrubba except that at $1{,}600 \le Re$ and for $Sc \ge 4\,59 \times 10^5/Re^{2.5}$ included a term $\{1 + [11.3(L/d)^{-0.5}]\}^{0.5}$					$14° < T < 40°C$
1980	Koziol et al. [212]	$170 < Re < 335$	1.668	0.39	0 5	−0.1667	$844 \le Sc \le 1{,}085$
		$335 < Re < 1{,}080$	3.882	0.24	0.5	−0.1667	$1.7 \times 10^{13} \le Ga \le 2.7 \times 10^{13}$
		$1{,}080 < Re < 2{,}513$	0.0008923	0.71	0.5	0	
1982	Yih and Chen [213]	$49 \le Re \le 300$	0.01099	0.3955	1/2	0	$8.5° \le T \le 50°C$
		$300 < Re \le 1{,}600$	0.02995	0.2134	1/2	0	$148 \le Sc \le 981$
		$1{,}600 < Re \le 10{,}500$	0.0009777	0.6804	1/2	0	

$k_L^* = (k_L/D)\,(\nu^2/g)^{1/3}$, $Re = 4\Gamma/\mu$, $Sc = \nu/D$, $Ga = gL^3/\nu^2$

and Ga is the Galileo number defined by

$$Ga = gL^3/\nu^2. \qquad (235)$$

Some authors (e.g. [193]) also found an influence of surface tension on k_L in a certain range of Reynolds number in the wavy-flow regime and incorporated a group $(\sigma/\sigma_{\text{water at } 25°C})$ into the correlation. Others, such as Bakopoulos [217, 218], found an influence of L/d on k_L in the turbulent flow region and incorporated a group consisting of L/d in the correlation. The use of k_L is more preferred over Sh because Sh contains the film thickness δ which has different dependencies on Re in different flow regimes. In Table 9, some of the more important correlations are compiled and tabulated in the form of

$$k_L^* = b\,Re^m Sc^n Ga^e \qquad (236)$$

for the wavy- and turbulent-flow regions. To illustrate, the recent correlations of Yih and Chen [213] are shown in Figure 20, which were obtained by fitting their own data and the data of ten other authors with a total of 846 data points and an average deviation of about 15%. These data were presumably taken at fully developed conditions such that the effect of Ga is negligible. Equation 236, without the term Ga^e, is identical in form to the predictions of the VDTM, Equations 223 and 224. The Reynolds number exponent obtained from the fit is 0.6804 in the turbulent-flow region and 0.2134 in the wavy-transition flow region which

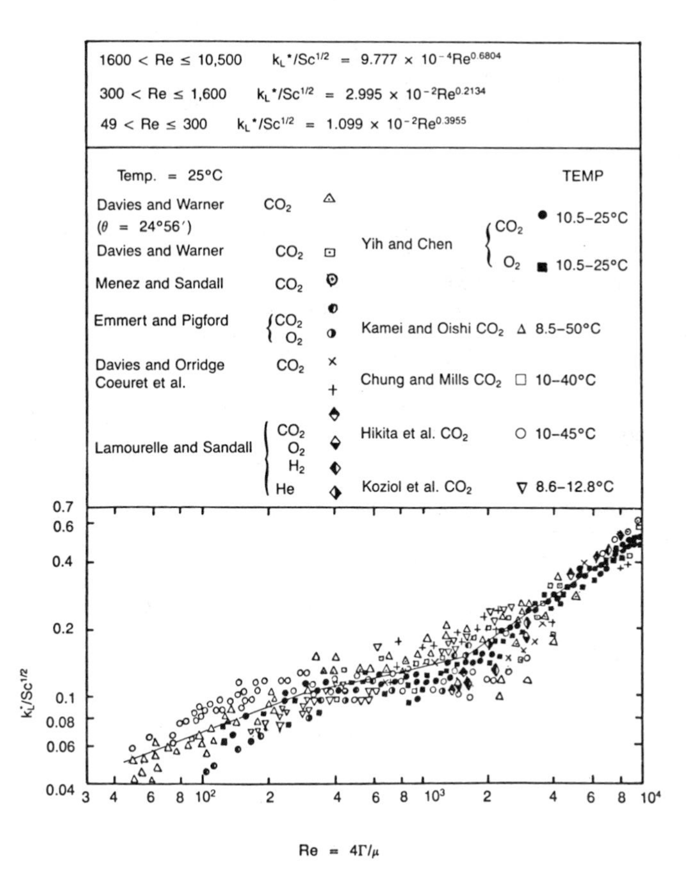

Figure 20. Correlation of mass transfer coefficients in wavy and turbulent falling films [213].

agree quite well with the predictions of 2/3 and 1/6 respectively, thus supporting the hypothesis that viscosity is the major cause of turbulence damping at the gas-liquid interface. In the wavy-flow region of Re ≤ 300, the data are quite scattered and exhibited a different slope from that in the wavy-flow region of 300 < Re < 1,600. The reason for this is still unclear but probably is due to a change in the wave structure. As is evidenced from a comparison in Table 9, there is quite a large discrepancy among correlations proposed by different investigators. This lack of agreement can be attributed to several important causes. The design of the feeding device and its strong effect on the flow and mass transfer in the entrance region is a major cause. Further, geometrical parameters such as tube length and diameter have a great effect. The influence of tube length is only negligible at long film lengths. When the tube diameter is reduced the waviness of the film is damped by the increased curvature of the tube wall, thus decreasing the mass transfer coefficient. Different temperatures used by different authors affect all the dimensionless groups containing physical properties. Other effects, such as surface tension lowering due to impurities and the properties of the tube wall on the wetting characteristics, cannot be expressed qualitatively. Difference in methods of concentration measurement and possible errors in concentration analysis may also be a decisive factor. Inaccurate vertical alignment of the tube axis may disturb the uniform film thickness distribution and cause a decrease in mass transfer. Since most of the data collected are some kind of average coefficient which incorporates transfer rates in the entrance region, in the wavy-flow region and in the turbulent-flow region, it is not surprising to find that measurements often differ among the investigators unless the geometrical parameters and experimental procedures are alike.

Variable film thickness model. All of the models previously proposed have assumed a constant average film thickness, with the effect of the waves considered to appear as an empirically determined quantity such as the eddy diffusivity. However, as shown recently by Dukler and co-workers [45–47], the turbulent gas-liquid interface is random and possesses a large-wave structure which moves over the thin base film with no change in speed or shape, Figure 3. This has led Brumfield et al. [49] to take a more realistic view in modeling the mass transport rate through turbulent falling films. The penetration theory with exposure time $t = (\lambda_p - L_w)/(c - u_{av})$ was used to estimate the absorption rate into the laminar base film where λ_p is the average wave separation distance, L_w is the base length of large waves, c is the wave celerity and u_{av} is the average film velocity. The velocity scale was taken as the shear velocity. The total height of the turbulent wave (base film thickness δ_{min} + mean height of wave above base film m_w) and the wave Reynolds number, $4c(\delta_{min} + m_w)/\nu$, were used to estimate the macroscale from which mass transfer in the turbulent portion of the film can be calculated from a turbulence-centered model. Calculations of average mass transfer coefficient were carried out for four values of Re, covering the range for which statistical wave hydrodynamic data for water were given by Telles and Dukler [45]. The results agreed quite well with the data of three authors [72, 181, 191] within Re = 200–6,000.

Entrance region mass transfer. For liquid flow in packed columns, for example, the film lengths or contact times between complete mixing at the junction of two packings may be very short so that fully developed condition may not be achieved in the liquid film. Therefore, mass transfer in the entrance region must be considered and can be formulated as

$$u(Y) \frac{\partial C}{\partial x} = \frac{\partial}{\partial Y}\left[(D + \epsilon_D(Y) \frac{\partial C}{\partial Y}\right] \tag{237}$$

with boundary conditions

$$x = 0 \quad \text{(inlet)} \qquad C = C_0 \tag{238}$$

$$Y = 0 \quad \text{(interface)} \qquad C = C^* \tag{239}$$

$$Y = \delta \quad \text{(wall)} \qquad \partial C/\partial Y = 0 \tag{240}$$

Sandall [219] numerically solved these equations using a finite difference method by assuming that u can be replaced by u_{max} at short contact times and that the eddy diffusivity determined by Lamourelle and Sandall [72], Equation 225, is valid over the entire film thickness. For very short contact times or short films, the eddy diffusivity does not have any effect on the rate of mass transfer and Higbie's penetration theory is valid:

$$Sh = k_L\delta/D = (\pi x^*)^{-1/2} \tag{241}$$

For long contact times or long films, the concentration profile becomes fully developed and

$$Sh = (2/\pi)\beta^{1/2} \tag{242}$$

where $\beta = a\delta^2/D$ and a is the coefficient in the eddy diffusivity expression. Sandall presented an empirical equation based on his numerical computations for the entrance length x* and within the range of $500 < \beta < 500{,}000$:

$$Sh = 0.564/x^{*1/2} + 0.1957\beta x^{*1/2} \text{ for } \beta x^* \leq 2.5 \tag{243}$$

The entrance length for mass transfer is also estimated to be

$$x_e^* = 2.56/\beta \tag{244}$$

The use of the numerical finite difference method to solve the convective-diffusion equation, especially near the gas-liquid interface, will require a very fine grid size to accommodate the steep concentration gradient there. This has stimulated a number of analytical solution methods by Subramanian [220], Kishinevskii and Kornienko [206], Gottifredi and Quiroga [221], Yih and Seagrave [222] which are more efficient. Kishinevskii and Kornienko's results can be expressed as

$$Sh = (\pi x^*)^{-1/2}(2/\pi^{1/2})(\beta x^*)^{1/2}\{\coth[0.564(\beta x^*)^{1/2}] - 0.282 \cdot (\beta x^*)^{1/2}\sinh^{-2}[0.56(\beta x^*)]^{1/2}\} \tag{245}$$

Gottifredi and Quiroga's solution is

$$Sh = \beta^{1/2}\{\exp(-4\beta x^*/(\pi^2)/(\pi\beta x^*)^{1/2} + (2/\pi)\operatorname{erf}[4/(\pi^2\beta x^*)]^{1/2}\} \tag{246}$$

Yih and Seagrave's solution is in terms of eigenvalues and eigenfunctions:

$$Sh = \sum_{i=1}^{\infty} A_i \left.\frac{\partial N_i}{\partial \eta}\right|_{\eta=1} \exp(-\lambda_i^2 x^*) \tag{247}$$

Concentration profile measurements in gas absorption. In order to understand the mechanism of transport in wavy and turbulent falling films, a disturbance-free measurement method for the concentration profile is absolutely necessary. Interferometric, fluorescence, and spectroscopic methods have been proposed.

Jepsen et al. [223] observed the local concentration profile for CO_2 absorption into water films flowing down an inclined plate of 9° 44′ angle within the range of $732 < Re < 1{,}834$ using an interferometric method. This method is based on the change in index of refraction of the liquid film which is directly proportional to the concentration of CO_2 absorbed. The position of the interference fringes which appears within the film perpendicular to the plate has to be photographed. Their observation showed that the shape of the interference pattern near the interface did not change significantly with the appearance of waves, but the shift of the pattern near the bottom indicated CO_2 penetration to the bottom which could be aided by turbulence in the central region. By assuming a wave model similar to that described by Equation 237 where the velocity profile is still assumed to be parabolic, they calculated the eddy diffusivity due to waves from the concentration profile measured. The total diffusivity calculated (molecular + eddy) showed a marked increase in the central region of the film, with the value tending toward the molecular diffusivity at the two boundaries. This trend is similar to that found by Carrubba [190] who also calculated the eddy diffusivity due to waves from measured mass transfer coefficient data both in gas absorption and solid dissolution.

Hiby [224] and Hiby et al. [225] have criticized that the interferometric method does not give distinct wave pictures near the vicinity of the interface. Instead, they suggested a photometric fluorescence method for measuring the concentration distribution C(x, y) during gas absorption in falling films. Their method was based on the fact that the fluorescence emission of certain pH-indicators changes according to a step function at a particular pH value. By means of this property, they were able to observe, when the film is exposed to ultraviolet radiation, the penetration of an alkaline or acidic gas component into an acidic or an alkaline falling film such as the absorption of NH_3 into HCl solution where the gas undergoes an instantaneous neutralization reaction with the solution containing the indicator. From the measured intensity of fluorescence, the distance between the phase boundary and the neutralization plane can be evaluated. The concentration distribution and hence the variation of the total diffusivity with distance from the interface are then calculated in the range of $1{,}200 < Re < 2{,}800$ using a model similar to Equation 237 which is also used by Jepsen et al. [223]:

$$\epsilon_D(Y) = -\frac{u_{max}}{x_2 - x_1}\frac{\partial Y}{\partial C}\bigg|_{(x_1 + x_2)/2} \int_Y^{\delta}\left(1 - \frac{Y^2}{\delta^2}\right)\left[C(Y)_{x_2} - C(Y)_{x_1}\right]dY \tag{248}$$

The calculated eddy diffusivity increases with Re and the film length and has a maximum near one-third of the distance from the wall, then it decreases and tends toward the molecular diffusivity at both boundaries. From flash photographs, it can be observed that absorption is not uniform across the surface but shows characteristic patterns of spots and strips.

Petermann et al. [226] have criticized the limitations of the fluorescence method in that the fluorescence indicators are only soluble in acid or alkaline solution but not in water. In addition, an instantaneous reaction is an essential condition for the determination of concentration gradient which makes this method unsuitable for systems undergoing a relatively slow chemical reaction. This method is also restricted to absorption processes with relatively high mass transfer rates since the absorption is only noticeable when the concentration of the absorbed gas in a boundary layer of the order of 0.01 mm has reached a value of about 0.01 equiv/l. They proposed instead the use of a spectroscopic recording method employing pH color indicators. The difference between their method and Hiby's lies in the method of measurement which is light extinction by the former and light emission by the latter. The advantage of the spectroscopic method is that there are many more available colored pH indicators than pH fluorescence indicators to be selected from, particularly when water is used as the solvent. In addition, this method does not require an instantaneous reaction between the gas and the solvent.

The total diffusivity calculated from the concentration profile by Hiby [227] showed a maximum value of about $2 \times 10^{-7} m^2/s$ at $Re = 2{,}800$ for a water film. Hiby assumed that if the Reynolds analogy holds for $\epsilon_D = \epsilon_H = \epsilon_M$, then one can write for a water film:

$$D + \epsilon_D = (2 \times 10^{-9} + 2 \times 10^{-7}) m^2/s \tag{249}$$

$$\alpha + \epsilon_H = (1.4 \times 10^{-7} + 2 \times 10^{-7}) m^2/s \tag{250}$$

$$\nu + \epsilon_M = (10^{-6} + 2 \times 10^{-7}) m^2/s \tag{251}$$

Thus, when compared with the molecular diffusivity D, the eddy diffusivity for mass ϵ_D is about 100 times larger and so has a profound influence on mass transfer. The eddy diffusivity for heat ϵ_H is only slightly larger than the thermal diffusivity α and so the enhancement of heat transfer is much smaller while the eddy diffusivity for momentum has apparently very little influence. This may help to explain why the waves enhance the mass transfer rate much more than heat or momentum transfer rate in wavy and turbulent liquid films.

Effect of gas velocity. Relatively little work has been reported on the effect of interfacial shear exerted by the high velocity gas on gas absorption rate. Chung [101], Chung and Mills [211, 228] observed that with cocurrent gas flow up to 14 m/s, interfacial shear resulted in a marked increase in k_L and their results can be correlated as

$$k_L/(k_L)_{no\ shear} = 1 + (0.9 + 1.73 \times 10^{12}\nu^2)[\tau_i/(\tau_w)_{no\ shear}] \tag{252}$$

where ν is in units of m^2/s. With countercurrent gas flow at low velocities, k_L was found to decrease with increasing gas velocity. Henstock and Hanratty [204] have also made measurements for cocurrent gas flow within $Re_G = 3{,}500-34{,}100$.

Gas Absorption with Chemical Reaction

With well-defined hydrodynamics and a range of small contact times between 0.1 to 2 seconds, the short wetted-wall column is ideally suited as a means of studying gas absorption kinetics and/or mechanism. The kinetic rate constants and the order of reaction can be determined from chemical absorption rates which are interpreted in terms of the penetration theory with simultaneous chemical reaction. Other types of laboratory absorbers commonly used are the laminar jet ($t = 10^{-3} - 10^{-1}s$) the quiescent absorber ($t = 2 \times 10^{-2} - 10^3$) and the stirred tank ($t = 6 \times 10^{-2} - 10s$). A detailed treatment on this subject is given in the books of Danckwerts [229] and Astarita [230]. Most of the work on chemical absorption with negligible heat effects in a short wetted-wall column have been concentrated on absorption of:

1. CO_2 in alkaline solutions of carbonate-bicarbonate, hydroxide, sulfite, and ammonia solutions
2. CO_2 and H_2S in carbonate-bicarbonate and hydroxide solutions
3. CO_2 in amines such as mono-, di-, and tri-ethanolamines
4. Cl_2 in water
5. NO_2 in water, acids or alkalis
6. Carbonyl sulfide, COS, in alkaline solutions

Most of these processes are of industrial interest as described in the books of Kohl and Riesenfeld [231] and Astarita et al. [232]. Measurement of absorption rate is usually accomplished by determining the rate of decrease of the gas volume in the absorber by means of a soap-film meter connected to the gas outlet. By shutting off the gas inlet to the absorber and observing the velocity at which a soap bubble moves downward in the calibrated tube as a consequence of gas absorption by the flowing film, the absorption rate can be determined. The exposure time of the liquid to the gas can be varied by varying the liquid flow rate or film length.

Laminar and Wavy-Laminar Flow

Pseudo first-order/irreversible reaction. The governing equation can be formulated as

$$\frac{\partial A}{\partial t} = D_A \frac{\partial^2 A}{\partial Y^2} - k_1 A \tag{253}$$

with boundary conditions

$$t = 0 \qquad A = 0 \tag{254}$$

$$Y = 0 \qquad A = A^* \tag{255}$$

$$Y = \infty \qquad A = 0 \tag{256}$$

where $k_1 = k_2 B_0$ and B_0 is the bulk concentration of liquid reactant B. The solution for the absorption rate per unit area has been given by Danckwerts [229]:

$$N_A = A^*(D_A k_1)^{1/2}[\text{erf}\,(k_1 t)^{1/2} + [\exp\,(-k_1 t)]/(\pi k_1 t)^{1/2}] \tag{257}$$

For short times of exposure,

$$N_A = A^*(D_A/\pi t)^{1/2}(1 + k_1 t) \qquad \text{for } k_1 t << 1 \tag{258}$$

For long times of exposure,

$$N_A = A^*(D_A k_1)^{1/2} \qquad \text{for } k_1 t >> 1 \tag{259}$$

These equations may be used as a basis for the experimental evaluation of $A^*D_A^{1/2}$ and k_1 if $B_0/A^* >> 1$ combined with not too high values of $k_1 t$ to make the depletion of B at the interface small enough to approach the pseudo first-order reaction. Under certain conditions such as a slow reaction or a long contact time, the complete velocity profile and the film thickness may have to be considered in the formulation:

$$u_{max}\left(1 - \frac{Y^2}{\delta^2}\right)\frac{\partial A}{\partial x} = D_A \frac{\partial^2 A}{\partial Y^2} - k_1 A \tag{260}$$

with boundary conditions

$$x = 0 \qquad A = 0 \tag{261}$$

$$Y = 0 \qquad A = A^* \tag{262}$$

$$Y = \delta \qquad \partial A/\partial Y = 0 \tag{263}$$

Stepanek and Achwal [233], Best and Horner [234], Yih and Huang [173], Pedersen and Prenosil [235] have given analytical solutions to these equations. Best and Horner concluded, based on plots of enhancement factor for first-order reaction, that only when the dimensionless reaction rate parameter k_1^* ($= k_1\delta^2/D$) is less than 5 will the film thickness and true velocity profile have any influence on the absorption rate. In a spherical falling film, this problem was solved by Ratcliff and Holdcroft [236] following the method used by Davidson and Cullen [170]. However, Astarita [237] pointed out that there was an error in their derivation and results.

Rates of absorption of CO_2 by solutions of carbonate-bicarbonate buffer solutions were measured in a short wetted-wall column by Nijsing and Kramers [168], Roberts and Danckwerts [238]. The diffusion of CO_2 in the liquid is accompanied by the reactions:

- $CO_3^{2-} + H_2O \rightleftharpoons OH^- + HCO_3^-$ instantaneous reversible   (264)
- $CO_2 + OH^- \rightarrow HCO_3^-$ irreversible second-order   (265)
- $CO_2 + CO_3^{2-} + H_2O \rightarrow 2HCO_3^-$ overall   (266)

where the second step is rate-controlling. According to Nijsing and Kramers, this reaction becomes pseudo first-order under the following conditions: a small partial pressure of CO_2 resulting in a low value of CO_2^*, a small contact time, a low temperature, a high HCO_3^- concentration, and a not-too-high ratio of $[CO_3^{2-}]/[HCO_3^-]$. By fitting Equation 257 to the measured chemical absorption rate, $CO_2^*D_{CO_2}^{1/2}$ and $k_1 = k_2[OH^-]$ may be determined. Since the solubility of a gas in an aqueous salt solution at a given temperature is dependent on the ionic strength, the pseudo first-order rate constant also varies with the ionic strength of the solution. This has also been found by Roberts and Danckwerts [238]. These two investigations have been performed under conditions of high $[HCO_3^-]/[CO_3^{2-}]$ ratios of approximately 0.4 to 2. Hikita et al. [239] studied lower $[HCO_3^-]/[CO_3^{2-}]$ ratios of approximately 0.01 to 0.2. The experimental results can be interpreted by the penetration theory for absorption accompanied by a two-step reaction, $A + B \rightarrow C$ followed by $C + B \rightleftharpoons E$, indicating that the reversibility of the first step reaction is possible. Values of the kinetic rate constant of the rate-determining step, Equation 265, are usually known only up to a temperature of 40°C in the open literature. Savage et al. [240] recently provided values of this rate constant up to 110°C which is the temperature used in industrial practical operation. The data reported were measured using a single-sphere absorber and the absorption rate interpreted by the film theory model.

The absorption of nitrogen dioxide-nitrogen tetroxide gas mixtures (NO_2/N_2O_4) into water or nitric acid solution is an important industrial process in the production of nitric acid or the removal of NO_2 from waste gases. The essential reactions which accompany the absorption in water are, according to Kameoka and Pigford [241]:

- $2NO_2 \rightleftharpoons N_2O_4$ (gas phase equilibrium) (267)
- $2NO_2(N_2O_4) + H_2O \rightarrow HNO_2 + HNO_3$ (reaction in solution) (268)
- $N_2O_4 \rightarrow NO^+ + NO_3^-$ (fast reaction in solution) (269)
- $4HNO_2 \rightarrow 2NO + N_2O_4 + 2H_2O$ (slow reaction in solution) (270)

It is generally agreed that N_2O_4, which is continuously in equilibrium with NO_2, is the active species preferentially absorbed in the liquid and undergoes a rapid pseudo first-order reaction with water. The rate-controlling step is the rate of hydrolysis of N_2O_4 in water. Values of the product $C^*_{N_2O_4}(k_1D)^{1/2}$ were measured in a short wetted-wall column by Wendel and Pigford [242], Dekker et al. [243], Kameoka and Pigford [241], Weisweiler and Deib [244]. Absorption of N_2O_4 into dilute nitric acid was found by Lefers and van den Berg [245] to be a rapid pseudo first-order reaction with water.

Instantaneous reaction of the form $A + zB = qP$. In this case the dissolved gas reacts instantaneously with the reactant at a plane beneath the liquid surface of distance $2\beta^* t^{1/2}$ where the concentration of both become zero, and the rate of reaction is equal to the rate at which A and B can diffuse to the reaction plane. The actual kinetics of the reaction is immaterial. The solutions of the governing equations have been given by Danckwerts [229]:

$$N_A = E_i A^* (D_A/\pi t)^{1/2} \tag{271}$$

where

$$E_i = [\mathrm{erf}\,(\beta^*/D_A^{1/2})]^{-1} \tag{272}$$

is the instantaneous reaction enhancement factor and β^* is calculated from

$$\exp\,(\beta^{*2}/D_B)\,\mathrm{erfc}\,(\beta^*/D_B^{1/2}) = (B_0/zA^*)(D_B/D_A)^{1/2}\exp\,(\beta^{*2}/D_A)\,\mathrm{erf}\,(\beta^*/D_A^{1/2}) \tag{273}$$

When $E_i >> 1$,

$$E_i = (D_A/D_B)^{1/2} + (B_0/zA^*)(D_B/D_A)^{1/2} \tag{274}$$

The instantaneous reaction is approached at small values of B_0/A^* and $kB_0t >> 1$.

When hydrogen sulfide is absorbed in a hydroxide solution, the following ionic reactions may occur:

- $H_2S + OH^- = HS^- + H_2O$ (275)
- $HS^- + OH^- = S^{2-} + H_2O$ (276)

The kinetics of this reaction has been determined by Astarita and Gioia [246] in a wetted-wall column as instantaneous irreversible on the basis of the penetration theory. On the other hand, when hydrogen sulfide is absorbed in a carbonate solution, the reactions may have the form:

- $H_2S + CO_3^{2-} = HS^- + HCO_3^-$ (277)
- $H_2S + HCO_3^- = HS^- + H_2O + CO_2$ (278)
- $CO_3^{2-} + H_2O = HCO_3^- + OH^-$ (279)

The third step is rate-controlling and can be considered as instantaneous irreversible.

Irreversible second-order reaction of the form $A + 2B \xrightarrow{k_2}$ *products.* The governing equations are

$$\frac{\partial A}{\partial t} = D_A \frac{\partial^2 A}{\partial Y^2} - k_2 AB \tag{280}$$

$$\frac{\partial B}{\partial t} = D_B \frac{\partial^2 B}{\partial Y^2} - 2k_2 AB \tag{281}$$

with the boundary conditions

$$t = 0 \qquad B = B_0 \qquad A = 0 \tag{282}$$

$$Y = 0 \qquad \partial B/\partial Y = 0 \qquad A = A^* \tag{283}$$

$$Y = \infty \qquad B = B_0 \qquad A = 0 \tag{284}$$

These equations have been solved numerically by Brian et al. [247] who presented their results by an approximate equation:

$$E = (M^{1/2}\eta^*)/\tanh(M^{1/2}\eta^*) \tag{285}$$

Hikita and Asai [248] also gave an approximate solution:

$$E = [M^{1/2}\eta^* + \pi/(8M^{1/2}\eta^*)] \operatorname{erf}(2M^{1/2}\eta^*/\pi^{1/2}) + \exp(-4M\eta^{*2}/\pi)/2 \tag{286}$$

where

$$M = \pi k_2 B_0 t/4 \tag{287}$$

and

$$\eta^* = (B^*/B_0)^{1/2} = [(E_i - E)/(E_i - 1)]^{1/2} \tag{288}$$

E_i is the enhancement factor for instantaneous reaction as given by Equations 272–274. To illustrate, Equation 285 is presented graphically in Figure 21. The value of E_i can be found from Equations 272–274. Several cases can be identified in Figure 21. When $M^{1/2} << 1$, E is close to 1. This means that the time of contact is very short or the reaction is very slow so that it is virtually physical absorption. When $M^{1/2} << E_i$, the point representing E lies on the envelope of the family of curves which is pseudo first-order reaction with $k_1 = k_2B_0$. Physically, this means that the reactant diffuses from the bulk towards the interface fast

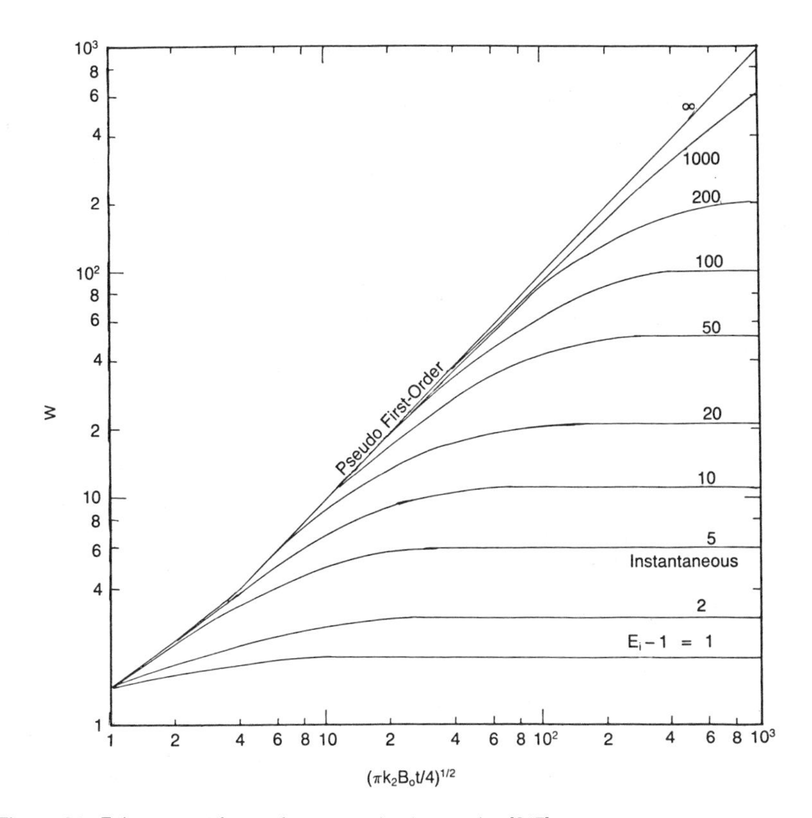

Figure 21. Enhancement factors for a second-order reaction [247].

enough to prevent the reaction from causing important depletion there. When $1 << M^{1/2} << E_i$, the point representing E lies on the straight portion of the envelope which corresponds to a long contact time or a fast reaction, then $E = M^{1/2}$. The absorption rate is independent of contact time in this region. When $M^{1/2} >> E_i$, then $E = E_i$ which is an instantaneous reaction. This occurs for a fast reaction, a long contact time, or when the concentration of reactant is small. The reactant is depleted in the vicinity of the surface such that it is diffusion-controlled. By using a laminar jet, a wetted-wall column and a quiescent absorber which encompass a wide range of contact times such as that used by Sada et al. [249], the data of the laminar jet will fall on the curves of small $M^{1/2}$, the wetted-wall column data in the small to intermediate range of $M^{1/2}$ and the quiescent absorber data in the large $M^{1/2}$ range.

For CO_2 absorption into hydroxide solutions, the reaction mechanism is

- $CO_2 + OH^- \rightarrow HCO_3^-$ irreversible second-order (289)

- $HCO_3^- + OH^- \rightleftharpoons CO_3^{2-} + H_2O$ instantaneous reversible (290)

This reaction has been studied by Nijsing et al. [250], Roberts and Danckwerts [238], Hikita and Asai [251] and Hikita et al. [252] who indicated that the reaction is effectively irreversible second-order. Equation 286, which was derived by Hikita and Asai [248], was found to fit the wetted-wall column data of their own and Nijsing et al. [250] very well.

Hikita et al.'s [239] experiments on CO_2 absorption into NaOH confirmed that the second step of the reaction, Equation 290, is instantaneous reversible and the data can be interpreted by the penetration theory with a two-step chemical reaction of the form $A + B \rightarrow C$ followed by $C + B \rightleftharpoons E$. Hikita et al. [239] derived a solution for the enhancement factor which can be closely approximated by Equation 286 but η^* is given by

$$\eta^{*4} + \left[\frac{E - E_i}{E_i - 1} + \frac{\left(\frac{D_c}{D_A}\right)^{1/2}\frac{C_0}{A^*}}{\left(\frac{D_B}{D_A}\right)^{1/2}\frac{B_0}{A^*}} + \frac{\left(\frac{D_c}{D_A}\right)^{1/2}\frac{C_0}{A^*}}{\left(\frac{D_E}{D_A}\right)^{1/2}\frac{E_0}{A^*}}\right]\eta^{*2}$$

$$+ \frac{E - E_i}{E_i - 1}\left[\frac{\left(\frac{D_c}{D_A}\right)^{1/2}\frac{C_0}{A^*}}{\left(\frac{D_B}{D_A}\right)^{1/2}\frac{B_0}{A^*}} + \frac{\left(\frac{D_c}{D_A}\right)^{1/2}\frac{C_0}{A^*}}{\left(\frac{D_E}{D_A}\right)^{1/2}\frac{E_0}{A^*}}\right] = 0 \tag{291}$$

This equation was found to fit their data well.

Other reactions which fall in the category of second-order are the reaction of carbonyl sulfide with an alkaline solution [253]:

$$COS + OH^- = HCSO_2^- \tag{292}$$

and the reaction of chlorine with ferrous chloride solution in aqueous hydrochloric acid [254].

Kojima et al. [255] took into account the complete semi-parabolic velocity profile and the finite wall boundary condition where the concentration gradient of the absorbed gas and reactant both become zero. This is more appropriate for longer contact times in which the penetration theory introduces deviation. The equations were solved numerically by a finite difference method. Experiments were carried out using a wetted-wall column of 3 different lengths for the systems CO_2 − NaOH, CO_2 − NaOH/Na_2CO_3 and CO_2 − NH_3 solution. Their data were compared with their own numerical solution and with the approximate penetration theory solution of Hikita and Asai, Equation 286. It was concluded that the data agreed with both solutions at short contact times, but those data at long contact times deviated from Hikita and Asai's solution and agreed with their own numerical solution indicating the importance of including the complete velocity profile and finite wall-boundary condition.

Another case of second-order reaction investigated is the reaction between gases A and B when both are absorbed in water such as the absorption of CO_2 and NH_3 in water. The governing equation is the same as Equations 280 and 281 but the boundary conditions are now

$$t = 0 \qquad B = 0 \qquad A = 0 \tag{293}$$

$$Y = 0 \qquad B = B^* \qquad A = A^* \tag{294}$$

$$Y = \infty \qquad B = 0 \qquad A = 0 \tag{295}$$

This problem has been solved by Teramoto et al. [256] using a finite difference method.

Absorption of CO_2 in amine solutions is a process of great industrial interest. The amine solutions usually used are MEA (monoethanolamine), DEA (diethanolamine), and TEA (triethanolamine). For MEA, the overall reactions occurring in the liquid phase are, according to Sada et al. [257]:

- $CO_2 + 2RNH_2 \rightarrow RNHCOORNH_3$ (carbamate)

$$CO_2 + RNH_2 \rightarrow H^+ + RNHCOO^- \quad (296a)$$

$$H^+ + RNH_2 \rightarrow RNH_3^+ \quad (296b)$$

- $CO_2 + RNHCOORNH_3 + 2H_2O \rightarrow 2HCO_3RNH_3$ (297)

where R refers to $HCOH_2CH_2$. With short contact times in a wetted-wall column or a laminar jet, the amount of carbamate can be neglected. Hence the chemical absorption can be analyzed by considering only reaction (1).

Analogously, the overall reactions of CO_2 with diethanolamine are:

- $CO_2 + 2R_2NH \rightarrow R_2NCOOR_2NH_2$ (298)
- $CO_2 + R_2NCOOR_2NH_2 + 2H_2O \rightarrow 2HCO_3R_2NH_2$ (299)

Because of the absence of a hydrogen atom attached to nitrogen, the reaction of CO_2 with triethanolamines are:

- $CO_2 + 2R_3N + H_2O \rightleftharpoons (R_3NH)_2CO_3$ (300)
- $CO_2 + CO_3^{2-} + H_2O \rightarrow 2HCO_3^-$ (301)

For MEA, Emmert and Pigford [258], and Astarita et al. [259] have determined that the reaction mechanism is fast second-order with respect to the first reaction with a stoichiometric coefficient of 2 moles of MEA per mole CO_2 when the carbonation ratio is less than 0.5. Chemical absorption kinetics of CO_2 into MEA, DEA, and TEA over a wide range of contact times were studied by Sada et al. [249, 257]. At short contact times such as in a laminar jet and a wetted-wall column, only the first step of the reaction, Equations 296, 298, and 300, is controlling. However, for long contact times such as in a quiescent absorber, the second step of the reaction, Equations 297, 299, 301, becomes important and is not negligible. Therefore the process was analyzed by the theory of gas absorption accompanied by a consecutive reaction of the form $A + 2B \xrightarrow{k_2} R$ and $A + R \xrightarrow{k_2'}$ products. The governing equations and boundary conditions were solved numerically by a finite difference method and the enhancement factor is given by

$$E = [\pi/(4k_2B_0t)]^{1/2}\{A^{*-1} \int_0^{k_2B_0t} (-\partial A/\partial t \,|_{Y=0})\, dt \quad (302)$$

The second-order rate constant of the first reaction k_2 was estimated from the laminar jet and wetted-wall column absorption data under presumably pseudo first-order reaction regime. By comparing the data with the theoretical calculation curves, the rate constant for the second reaction was found to be quite small. Hikita et al. [260] commented that the rate constants determined by Sada et al. [249] from pseudo first-order reaction regime were higher than that obtained by conventional kinetic method. They found that when CO_2 is absorbed into uncarbonated MEA solution, the reaction is irreversible second-order with a stoichiometric coefficient of 2 if the variation of the physical solubility of CO_2 due to the change in composition of the solute during absorption is taken into account. On the other hand, when CO_2 is absorbed into highly carbonated MEA solutions, the liquid bulk concentration of free

CO_2 is not negligible compared to its interfacial concentration and the reaction becomes reversible second-order having a large value of the equilibrium constant.

Alvarez-Fuster et al. [261] determined the kinetics of the reaction of CO_2 with MEA, DEA, and an organic solution of cyclohexylamine (CHA) in a toluene + 10% isopropanol solution. The reaction was found to be first-order with respect to CO_2, second-order with respect to DEA and CHA and first-order with respect to MEA. Similar study [262] was also performed with two polar organic medium and three viscous organic medium.

Simultaneous absorption of two gases with two reaction planes in the liquid. Simultaneous absorption of H_2S and CO_2 in alkaline solutions is an important industrial process. The kinetics of the reaction of CO_2 and H_2S with a hydroxide solution may be accompanied by one or more of the following instantaneous irreversible reactions as suggested by Astarita and Gioia [263]:

- $CO_2 + 2OH^- \rightarrow CO_3^{2-} + H_2O$ (303)
- $H_2S + OH^- \rightarrow HS^- + H_2O$ (304)
- $H_2S + CO_3^{2-} \rightarrow HS^- + HCO_3^-$ (305)
- $HCO_3^- + OH^- \rightarrow CO_3^{2-} + H_2O$ (306)

They have assumed that there exist two reaction planes in the liquid: the primary reaction plane where H_2S reacts with CO_3^{2-} ion instantaneously by Reaction 3 and the secondary reaction plane where CO_2 reacts with OH^- ion instantaneously by Reaction 1. Onda et al. [264] derived the equations corresponding to this case:

$$\frac{\partial A}{\partial t} = D_A \frac{\partial^2 A}{\partial Y^2} \tag{307}$$

$$\frac{\partial B}{\partial t} = D_B \frac{\partial^2 B}{\partial Y^2} \tag{308}$$

$$\frac{\partial C}{\partial t} = D_C \frac{\partial^2 C}{\partial Y^2} \tag{309}$$

$$\frac{\partial E}{\partial t} = D_E \frac{\partial^2 E}{\partial Y^2} \tag{310}$$

$$\frac{\partial F}{\partial t} = D_F \frac{\partial^2 F}{\partial Y^2} \tag{311}$$

$$\frac{\partial G}{\partial t} = D_G \frac{\partial^2 G}{\partial Y^2} \tag{312}$$

where $A = [CO_2]$, $B = [H_2S]$, $C = [OH^-]$, $E = [CO_3^{2-}]$, $F = [HCO_3^-]$ and $G = [HS^-]$ with the boundary conditions of

$$t = 0 \quad A = B = F = G = 0, \quad E = E_0, \quad C = C_0 \tag{313}$$

$$Y = 0 \quad A = A_i = A^*, \quad B = B_i = B^*, \quad F = F_i, \; G = G_i \tag{314}$$

$$Y = Y_I \quad \text{(first reaction plane)} \qquad B = E = 0 \tag{315}$$

$$-D_B \frac{\partial B}{\partial Y} = D_{E\,II}\left(\frac{\partial E}{\partial Y}\right)_{II} = -D_F \frac{\partial F}{\partial Y} = -D_G \frac{\partial G}{\partial Y} \tag{316}$$

$$Y = Y_{II} \quad \text{(second reaction plane)} \qquad A = C = F = 0 \tag{317}$$

$$D_C \left(\frac{\partial C}{\partial Y}\right) = -2D_A \frac{\partial A}{\partial Y} - D_F \frac{\partial F}{\partial Y} \tag{318}$$

$$D_{E\,II}\left(\frac{\partial E}{\partial Y}\right)_{II} - D_{E\,III}\left(\frac{\partial E}{\partial Y}\right)_{III} = -D_F\left(\frac{\partial F}{\partial Y}\right) - D_A \frac{\partial A}{\partial Y} \tag{319}$$

$$Y \to \infty \quad A = B = F = G = 0, \quad E = E_0, \quad C = C_0 \tag{320}$$

The analytical solution of this problem was obtained by using a method similar to that given by Sherwood and Pigford [265] and the calculated enhancement factor agreed with the data of Onda et al. [264] for the system H_2S and CO_2/aqueous NaOH solution.

If one of the reactions, such as CO_2 with $2OH^-$ ions, is second-order instead of instantaneous, then the equations have to be modified as have been done by Imanara et al. [266] who presented a numerical finite difference solution. In this case, two reaction planes still exist but the CO_2 now do not vanish at the secondary reaction plane as in an instantaneous reaction. Experiments were carried out using the system CO_2 and H_2S/NaOH solution and the absorption rate of each is in good agreement with the numerical solution.

Sulfite-oxidation methods. Measurement of the rate of absorption of oxygen into aqueous sodium sulfite solution in the presence of a catalyst, usually cobaltous sulfate, has been widely used for the determination of the mass transfer properties in gas-liquid dispersions. Two objectives in using this method are: to determine the physical mass transfer coefficient per unit volume, k_La', in order that the device can be characterized for use with an entirely different system and to determine the interfacial area, a', of the bubbles in the dispersion.

The fundamental idea of this approach is to determine the kinetics of this rather rapid reaction from absorption rates measured in a laboratory absorber such as a wetted-wall column. This has been done by a number of authors [267–271]. In the general case where the absorbed O_2 undergoes a rapid reaction with the liquid, the kinetics of which is zero order in sulfite concentration and nth order in oxygen concentration, the total absorption rate may be calculated from

$$N_{O_2} = C_{O_2}^* \{[2/(n+1)] D_{O_2} k_n (C_{O_2}^*)^{n-1}\}^{1/2} \tag{321}$$

$$\text{for} \quad Ha = [2/(n+1)] k_n (C_{O_2}^*)^{n-1} D_{O_2}/k_L > 3 \text{ (fast reaction)} \tag{322}$$

By measuring the absorption rate per unit area N_{O_2} at several values of the partial pressure P_{O_2}, the reaction order n and the rate constant k_n may be determined. Generally, the experimental evidence of the majority of authors reveals that the reaction order with respect to oxygen is 2 instead of 1 as has been assumed by some authors. Once the kinetic data (ie., n and k_n), the solubility and diffusivity are known, one can calculate the interfacial area in any other gas-liquid contactor from Equation 321 by multiplying both sides by a' and from the total absorption rate measured. On the other hand, k_La' and a' can be determined simultaneously based on measurement of the overall specific absorption rate $N_{O_2}a'$ with different values of k_n (by varying the catalyst concentration) in the region where the effects of hydrodynamics and chemical reaction on the mass transfer coefficient are comparable. For a reaction to be of pseudo nth order in O_2, it is necessary to fulfill the condition of

$$Ha << (C_B/zC_{O_2}^*)(D_B/D_{O_2})^{1/2} \text{ and } k_L a' C_{O_2}^* << k_n C_{O_2}^{*n} \tag{323}$$

For a pseudo first-order reaction,

$$N_{O_2}a' = a'C_{O_2}^*(D_{O_2}k_1 + k_L^2)^{1/2} \tag{324}$$

and for a pseudo second-order reaction,

$$N_{O_2}a' = a'C_{O_2}^*(2D_{O_2}k_2C_{O_2}^*/3 + k_L^2)^{1/2} \tag{325}$$

A plot of the total absorption rate, $(N_{O_2}a')^2$ versus varying values of k_1 or k_2 allows k_La' and a' to be determined as suggested by Linek and Vacek [272].

It must be remembered that all these analyses are based on the assumption of smooth laminar flow and most of the experiments are conducted with the addition of surface active agents to suppress ripple formation in order to bring agreement with the theory. Apparently the waves will enhance the chemical absorption rate to about the same extent as they increase the physical absorption rate. Relatively little work has been reported. The only work of importance seems to be that of Kojima and Inazumi [273]. They assumed the existence of a laminar sublayer adjacent to the gas-liquid interface which presents the major mass transfer resistance. The thickness of this layer was estimated by Levich's theory based on Banerjee et al's [189] assumption that the eddy velocity is equal to the surface velocity. The chemical absorption rate was derived by solving the governing nonlinear partial differential equations both numerically and by an approximate method. Experiments were conducted for CO_2 absorption into NaOH and NH_4OH, respectively in a wetted-wall column of 0.5m and 1m long within the range of Re between 100 and 1,200. The experimental data agreed well with the theoretical prediction for a second-order reaction.

Turbulent Flow

The eddy diffusivity model was used by Kayihan and Sandall [274], Menez and Sandall [275], Stepanek and Achwal [276], Yih and Seagrave [222], and Gottifredi and Quiroga [277] to study gas absorption with an irreversible first-order reaction:

$$\frac{\partial C^+}{\partial x^*} = \frac{\partial}{\partial \eta}\left[1 + \beta(1-\eta)^2 \frac{\partial C^+}{\partial \eta}\right] - k_1^* C^+ \tag{326}$$

with the boundary conditions

$$x^* = 0 \text{ (inlet)} \qquad C^+ = 0 \tag{327}$$

$$\eta = 0 \text{ (wall)} \qquad \partial C^+/\partial \eta = 0 \tag{328}$$

$$\eta = 1 \text{ (interface)} \qquad C^+ = 1 \tag{329}$$

where $C^+ = C/C^*$, $\eta = y/\delta$, $k_1^* = k_1\delta^2/D$. Kayihan and Sandall [274] solved Equation 326 numerically by a finite difference method and presented an empirical formula that fits the numerical results within 6.5%:

$$Sh = k_1^{*1/2}\{1 + (\beta/k_1^*) + [\exp(-2k_1^*x^*)]/(\pi k_1^*x^*)\}^{1/2} \tag{330}$$

They also derived some asymptotic solutions which were tabulated by Yih and Seagrave for first-order as well as zero-order reaction:

		First-Order		*Zero-Order*	
• Short t, small k*	Sh	$= (\pi x^*)^{-1/2}$	(331)	$(\pi x^*)^{-1/2}$	(332)
• Long t, small k*		$= (2/\pi)\beta^{1/2}$	(333)	$(2/\pi)\beta^{1/2}$	(334)
• Short t, large k*		$= \dfrac{\exp(-k_1^* x^*)}{(\pi x^*)^{1/2}}$	(335)	$(\pi x^*)^{-1/2} + 2k_0^* x^{*1/2}$	(336)
• Long t, large k*		$= k_1^{*1/2}$	(337)	k_0^*	(338)

Menez and Sandall [275] absorbed CO_2 into an aqueous carbonate-bicarbonate buffer solution in the presence of an arsenite catalyst. The reaction is pseudo first-order with respect to CO_2. Data for the absorption rate were obtained in a wetted-wall column of 183 cm film length and within the range of Re = 1,700–8,500 and for arsenite concentration within 0.05–0.2M. These data agreed well with the solution of Equation 326 when the eddy diffusivity used is that derived from the physical absorption data of Lamourelle and Sandall [72], Equation 231.

An interesting point was pointed out by Gottifredi and Quiroga [277] that Danckwerts' surface renewal model can be exactly described by the time-averaged mass balance differential equations when the convective transport parallel to the interface can be neglected (when $\beta x^* \geq 1.5$) and when the eddy diffusivity is expressed as $\epsilon_D = aY^2$. The surface renewal rate S is related to the eddy diffusivity by

$$S = (2/\pi)^2 \beta \tag{339}$$

Mendez and Sandall [278] again used the eddy diffusivity model to study gas absorption with an instantaneous reaction of the form $mA + nB \rightarrow A_mB_n$. Both A and B diffuse toward a plane of distance Y_r from the interface and their concentration vanish at that plane which can be calculated from the following equation:

$$(\pi/2) - \tan^{-1}[Y_r(a/D_A)^{1/2}](D_A/D_B)^{1/2} = (nB/mA^*)(D_B/D_A)^{1/2} \tan^{-1}[Y_r(a/D_A)^{1/2}] \tag{340}$$

The enhancement factor is derived as

$$E = (\pi/2)\{\tan^{-1}[Y_r(a/D_A)^{1/2}]\}^{-1} \tag{341}$$

CO_2 was absorbed in aqueous NaOH solutions in the range of CO_2 partial pressure of 1–1.5atm and NaOH in the range of 0.02–0.22 N over the range of Re of 1,400–5,400. The data were in good agreement with the theory supporting the validity of the eddy diffusivity model. Stepanek and Achwal [279] have also treated theoretically the same problem of instantaneous reaction.

For a second-order reaction having a nonlinear kinetic expression, the eddy diffusivity approach will yield a nonlinear term $\overline{kA'B'}$ in the time-averaged diffusion equation arising from concentration fluctuations. So Haimour and Sandall [280] abandoned this approach and instead numerically solved the time-dependent diffusion equation. The absorption rate is then calculated according to Danckwerts' surface renewal theory by integrating the numerical results. For the CO_2-NaOH system, the experimental conditions were chosen as 0.3–0.5M NaOH solution with the ratio $[OH^-]/[CO_2]^*$ varied from 9 to 29 such that the reaction can be considered as second-order. The column is 0.69 m long and the Reynolds number range from 1,400 to 5,300. The data were interpreted in terms of the surface renewal rate which was determined from the physical absorption data of Lamourelle and Sandall [72].

Gas Absorption with Heat Effects

Most published works on gas absorption with or without chemical reaction in falling films have neglected heat effects. However, certain processes such as absorption refrigeration or absorption heat pump for heating or cooling involve simultaneous heat and mass transfer. Moreover, a number of industrially important reactions accompanying absorption are strongly exothermic and appreciable liquid temperature rises occur. In most cases, these generated heat loads must be removed for two reasons: one is the deterioration of mass transfer performance due to the decrease of solubility of gases with increase in liquid temperature, the other reason is to avoid possible damaging of reaction product due to rising temperature in the liquid such as in detergent-making by sulfonation of organic alkylates. One of the best solutions to these problems is to conduct gas absorption in a film cooler-absorber in which the gas is absorbed in the tube-side and the heat of reaction and absorption is removed through the coolant either in annular or in film flow in the jacket side. These types of apparatus have been applied commercially to several important industrial processes such as:

1. Hydrogen chloride absorption in water, e.g. Coull et al. [281], Dobratz et al. [282], Gaylord and Miranda [283].
2. Ammonia absorption in water, e.g. Pagani and Zardi [284].
3. Water vapor absorption in lithium bromide solutions as in absorption refrigeration, e.g. Matsuda et al. [285].
4. Sulfur dioxide absorption in alkyl benzene or alpha olefins, e.g. vander Mey [286], Marquis [287].

Due to the rather complex coupling between heat and mass transfer, as the heat effects arising from the heats of absorption and reaction are dependent on mass transfer which in turn is influenced by temperature-dependent physical properties and reaction rate constants, the modeling of the combined transfer problem is not easy. Two approaches have commonly been used: one is to solve the appropriate transport equations and the other is to make use of heat and mass transfer coefficients obtained from existing correlations. Industrial practice usually favors using the latter approach because, as noted by Guerreri [288], these particular absorption processes usually involved high flux mass transfer in which the effect of the mean molar velocity across the direction of the flow is highly important. Differential equations including this term could not be solved analytically. Also turbulent transport is important and correlation for heat and mass transfer coefficients are readily available for both the gas phase and the liquid phase. The disadvantages of this approach, on the other hand, are that the maximum rise of temperature in the liquid and hence the temperature profile along the reactor, which is usually of crucial importance, cannot be predicted by this method whereas this is possible in the first method.

Modeling by Solving Transport Equations

The effect of a linear temperature gradient on the laminar film absorption process with or without first-order or zero-order chemical reaction was analyzed by Yih and Seagrave [289]. Variation of the liquid physical properties, such as viscosity, gas solubility, molecular diffusivity, and kinetic rate constant with temperature are explicitly taken into account. Reaction enhancement factors are calculated and, when compared with an isothermal film, may be greater, less, or even equal to one depending on the relative importance of the parameters representing the effects of heat transfer on the above physical properties and rate constants. Grossman [290] included the axial heat convection term in his analysis but the physical properties are temperature-independent except the gas solubility and no reaction is considered. Inazumi et al. [291] derived the transport equations based on the penetration theory for non-isothermal gas absorption accompanied by a (1, 1)-th order reaction in a laminar falling film. Sensible heat transfer between the gas and the liquid, vaporization of solvent, the heat of absorption and reaction, and the change of ionic strength, I, on the rate constant are considered. The governing equations are:

$$\frac{\partial A}{\partial t} = D_A(T)\frac{\partial^2 A}{\partial Y^2} - k(I, T)AB \tag{342}$$

$$\frac{\partial B}{\partial t} = D_B(T)\frac{\partial^2 B}{\partial Y^2} - 2k(I, T)AB \tag{343}$$

$$\frac{\partial T}{\partial t} = K(T)\frac{\partial^2 T}{\partial Y^2} + \frac{\Delta H_R}{\rho C_p}k(I, T)AB \tag{344}$$

with the boundary conditions:

$$t = 0 \qquad A = 0,\ B = B_0,\ T = T_0 \tag{345}$$

$$Y = 0 \qquad A = A^*(I, T) \tag{346}$$

$$Y = \delta \qquad \partial B/\partial Y = 0,\ \partial T/\partial Y = 0 \tag{347}$$

Numerical solutions obtained were in agreement with the experimental absorption rate for absorption of CO_2 into NaOH solution containing a small amount of surface active agent in a wetted-wall column of length ranging from 5 to 200 cm.

Numerical solution of the two-dimensional continuity, energy, and diffusion equations were presented by Matsuda et al. [285] in their analysis on absorption of water vapor by lithium bromide. Similar work was also performed by Nakoryakov and Grigor'yeva [292].

Heat effects in gas absorption accompanied by a rapid first-order reaction in turbulent liquid films have been analyzed by Sandall [293]. The temperature rise due to the heat of absorption and reaction is separately calculated from the energy equation which incorporated an eddy diffusivity for heat of the form $\epsilon_H = aY^2$ which is similar to that used in gas absorption. The calculations made for the case of carbon dioxide absorption into a carbonate-bicarbonate buffer solution with arsenite as catalyst, which came from a data point of Menez and Sandall [275], showed that the temperature rise is very small in this case.

Modeling by Using Transfer Coefficients

Modeling and design of falling film absorbers have been carried out by Guerreri [288] and Guerreri and King [294] for the absorption of NH_3 into water and by Johnson and Crynes [295] for the absorption of SO_3 into alkylbenzene. Since both of the absorption processes are gas-phase controlled, there is negligible mass transfer resistance in the liquid film but still considerable heat transfer resistance in the liquid film. Guerreri [288] used the Nusselt equation for laminar film condensation and Grigull's equation [296] for turbulent film condensation to calculate the liquid film heat transfer coefficient. Obviously, this is not satisfactory because heating or cooling a falling film is not the same as film condensation whose film thickness increases down the tube. On the other hand, Johnson and Crynes [295] employed an annular flow correlation which was based on steam-water and air-water data. Davis et al. [297] questioned the validity of the approach. They developed a mathematical model to predict the temperature distribution in the laminar falling film with adjoining turbulent gas stream via solution of the velocity and temperature distribution in the film including the effects of the heats of reaction and absorption. The predictions for temperature were in good agreement with the experimental data which were much lower than the predictions from Johnson and Crynes' model.

Gas-Side Mass Transfer

Mass transfer from a falling film to a cocurrent or countercurrent flowing gas stream is important for the understanding of the transport process in the gas phase of a humidification, or distillation, or absorption column. In general there are four possible combinations of flow regimes in the liquid film (laminar with or without rippling, turbulent) and in the gas phase (laminar, turbulent). These four cases have been studied theoretically by a number of authors by solving the basic transport equations. Experimentally, the gas-side mass transfer coefficient can be measured usually by three techniques. The first is to absorb one component of a binary gas mixture under conditions where the liquid-side resistance can be neglected such as the absorption of a highly soluble gas. The second is to measure the gas-side resistance by means of binary distillation. The third is to measure the rate of vaporization of different liquids into a gas. In these techniques, the gas flow is usually turbulent and the liquid film is laminar or wavy.

Laminar Gas-Laminar Liquid Film

This case has been studied by Aihara et al. [298], Chandra and Savery [299] and Hikita and Ishimi [300]. The equation of motion for the gas and the liquid film at fully developed conditions are derived by Hikita and Ishimi [300]:

$$\mu_G\left(\frac{d^2u_G}{dr^2} + \frac{1}{r}\frac{du_G}{dr}\right) = -\left(\frac{\Delta p}{L} \pm \rho_G g\right) \tag{348}$$

$$\mu_L\left(\frac{d^2u_L}{dr^2} + \frac{1}{r}\frac{du_L}{dr}\right) = -\left(\frac{\Delta p}{L} \pm \rho_L g\right) \tag{349}$$

with the boundary conditions

$$r = 0 \text{ (tube axis)} \qquad du_G/dr = 0 \tag{350}$$

$$r = r_i \text{ (interface)} \qquad u_G = u_L,\ \mu_G\frac{du_G}{dr} = \mu_L\frac{du_L}{dr} \tag{351}$$

$$r = r_w \text{ (wall)} \qquad u_L = 0 \tag{352}$$

The $\pm$ sign denotes cocurrent and countercurrent flow, respectively. For cocurrent flow, the gas phase velocity profile was solved as

$$u_G = u_m(2 - U) - 2u_m(1 - U)(r/r_i)^2 \tag{353}$$

where $U = u_i/u_m$, u_m and u_i are the average gas velocity and the interfacial velocity, respectively. The mass transport equation in the gas phase is

$$u_G\frac{\partial C_G}{\partial x} = \frac{1}{r}\frac{\partial}{\partial r}\left(rD_G\frac{\partial C_G}{\partial r}\right) \tag{354}$$

with the boundary conditions

$$x = 0 \text{ (inlet)} \qquad C_G = C_{GO} \tag{355}$$

$$r = r_i \text{ (interface)} \qquad C_G = C_G^* \tag{356}$$

$$r = 0 \text{ (tube axis)} \qquad \partial C_G/\partial r = 0 \tag{357}$$

For cocurrent flow with $U \leq 2$, ie. for the case of no circulation in the gas phase, a solution was obtained by the method of separation of variables which was expressed in terms of the confluent hypergeometric function. The local gas phase Sherwood number is calculated as a function of the Graetz number, Gz,

$$Sh_G = k_G(2r_i)/D_G = -(Gz_G/\pi)\ln\left[\sum_{i=1}^{\infty} A_i \exp(-\pi\lambda_i^2/Gz_G)\right] \tag{358}$$

where

$$k_G = W(C_{G2} - C_{G1})/2\pi r_i \rho_G x(\Delta C_G)_{lm} \tag{359}$$

$$Gz = W/\rho_G D_G x \tag{360}$$

When the value of Gz is very large, ie., short contact time, the penetration theory is valid and can be derived as

$$Sh_G = (4/\pi)(UGz_G)^{1/2} \tag{361}$$

On the other hand, at large contact time for which Gz is small, the asymptotic value is

$$Sh_G = \lambda_1^2 \tag{362}$$

For $U > 2$, circulation will exist in the gas phase. Mass transfer in the annular downflow region was assumed to take place by axial convection and radial diffusion. Mass transfer in the core circulation region, on the other hand, was assumed to occur only by radial diffusion since the mean residence time of the circulating gas in the core region is infinitely long. The mass transport equation is formulated for each of the two regions separately and a solution was obtained in an analogous procedure by Hikita et al. [301]. Sh_G was found to increase with increasing value of U. Similar treatment for countercurrent flow was also given by Hikita and Ishimi [302].

The absorption of ammonia from air into aqueous sulfuric acid solution and methanol vapor from air into water, which were considered to be gas-phase controlled, were studied by Hikita and co-workers [300–302]. For cocurrent flow, the experimental results for absorption of methanol vapor into water at 20°C are shown in Figure 22. The value of the average gas-phase Sherwood number Sh_G increases with increasing Re_L and decreases with increasing Re_G until a minimum value of Sh_G is reached. The solid lines below the dashed line are the theoretical predictions for the case of cocurrent flow without circulation in the gas phase, $U \leq 2$, whereas the solid lines above represent predictions for the case of circulation in the gas phase.

Turbluent Gas-Laminar or Wavy Liquid Film

Hikita et al. [303] derived the same set of transport equations as in Equations 354–357 but now replaced D_G by $(D_G + \epsilon_{DG})$ where ϵ_{DG} is the eddy diffusivity for mass in the gas phase.

A Prandtl mixing length model and a turbulent Schmidt number of 0.8 were used to derive ϵ_{DG} and Sh_G was calculated by the same analytical procedure as in the laminar gas case. Two columns of different diameter and lengths were used to study the absorption of ammonia from air into an aqueous sulfuric acid solution with the addition of SAA to eliminate surface waves. The gas phase Reynolds number was varied from 3,000 to 20,000. It was found that the effects of circulation in the gas and the length-diameter ratio L/d on Sh_G were small and could be neglected. Moreover, for both cocurrent and countercurrent flow, the data on Sh_G

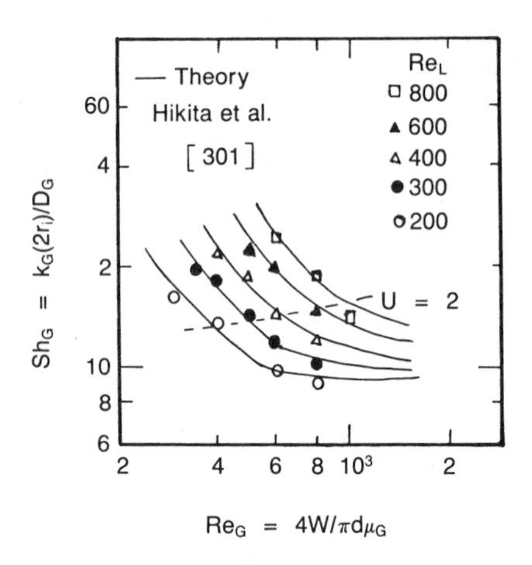

Figure 22. Effects of Re_G and Re_L on Sh_G for absorption of laminar methanol vapor into a laminar water film at 20°C [301].

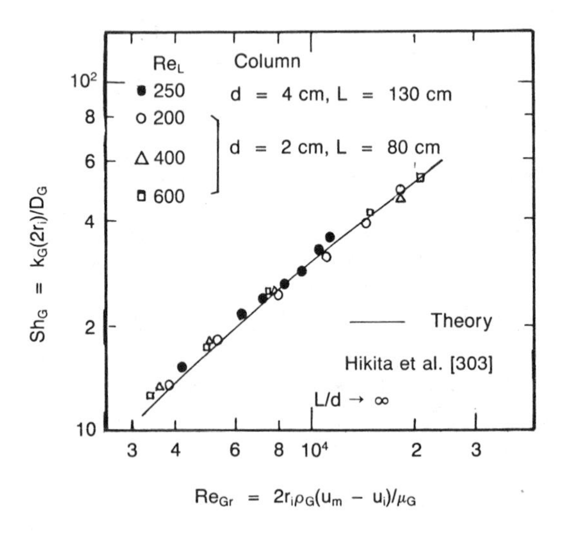

Figure 23. Gas-phase Sherwood number as a function of the gas-phase Reynolds number based on the gas velocity relative to the liquid surface velocity for cocurrent turbulent gas flow in a wetted-wall column [303].

can be correlated as a single function of a gas-phase Reynolds number Re_{Gr} which is based on the gas velocity relative to the liquid surface velocity, Figure 23. This is in agreement with the work of Kafesjian et al. [304] who investigated the vaporization of water from rippling liquid films into a turbulent air stream in countercurrent flow. They also found that their Sh_G data can be correlated as a function of Re_{Gr} alone and so independent of Re_L. Hikita et al. concluded that their Sh_G value increases with increasing Re_G in both cocurrent and countercurrent flow. With increasing Re_L, the Sh_G value decreases in cocurrent flow and increases in countercurrent flow.

Much data and experimental correlations for the gas-phase Sherwood number have appeared. The earliest comprehensive work is by Gilliland and Sherwood [305] who vaporized nine different liquids into a turbulent air stream in either cocurrent or countercurrent flow. Their data were well correlated by the equation

$$Sh_G(p_{BM}/P) = 0.023\ Re_G^{0.83} Sc_G^{0.44} \tag{363}$$

within the range $0.6 < Sc < 2.17$, $150 < Re < 1{,}560$, $1{,}800 < Re_G < 30{,}000$. p_{BM} is defined as the logarithmic mean of the partial pressure of B.

Johnstone and Pigford [165] performed distillation of four binary systems and absorption of ethylene dichloride vapor by benzene. Their results can be correlated in terms of Re_{Gr} as

$$Sh_G(p_{BM}/P) = 0.0328\ Re_{Gr}^{0.77} Sc_G^{0.33} \tag{364}$$

or

$$j_D = 0.0328\ Re_{Gr}^{-0.23} = f/2 \tag{365}$$

within the range $3{,}000 < Re_{Gr} < 40{,}000$ and $0.5 < Sc < 3$.

The extensive data of Jackson and Ceaglske [306] on the vaporization of three liquids into air generally fall below Gilliland and Sherwood's results and j_D is generally less than f/2. The distillation results for 2-propanol-water, however, agreed closely with the distillation data of Johnstone and Pigford and Equation 365, thus confirming the validity of the Chilton-Colburn analogy in this case.

Kafesjian et al. [304] noted that the discrepancies in wetted-wall vaporization data on rippling films of water can be explained by using Re_L to characterize liquid rippling and liquid surface velocity effects when Re_G is used. The surface velocity effect for the nonrippling case can be removed by using Re_{Gr}. Correlation of 220 data points of various investigators on vaporization of water from rippling films gave

$$Sh_G(P_{BM}RT/P) = 0.0065\ Re_G^{0.83} Re_L^{0.15} \tag{366}$$

within the range $25 < Re_L < 1{,}200$.

Data on nonrippling film were represented by

$$Sh_G(P_{BM}RT/P) = 0.013\ Re_{Gr}^{0.83} \text{ for } Re_L < 1{,}000 \tag{367}$$

Sherwood et al. [307] have rearranged this equation to a form similar to Equation 363 by including a Schmidt number dependence:

$$Sh_G(p_{BM}/P) = 0.00814\ Re_G^{0.83} Re^{0.15} Sc_G^{0.44} \tag{368}$$

These investigations were conducted in columns longer than 1 meter such that the effect of column length on the gas-side mass transfer coefficient cannot be observed. Kast [308] and Braun and Hiby [309] observed a strong dependence of Sh_G on the length-to-diameter ratio.

In Braun and Hiby's experiments, ammonia is absorbed from lean ammonia-air mixture into a sulfuric acid film in a wetted-wall column of 4 cm diameter and a L/d ratio between 5 to 35. They measured the local and average Sh_G both in cocurrent and countercurrent flow and obtained, within 10% error, the following correlations which are also shown in Figure 24:

For counterflow

$$Sh_{Gm} = 0.015\, Re_G^{0.75}(\Gamma/\mu)^{0.16}Sc^{0.44}[1 + 5.2(L/d)^{-0.75}] \qquad (369)$$

For cocurrent flow

$$Sh_{Gm} = 0.18\, Re_G^{0.4}(\Gamma/\mu)^{0.16}Sc^{0.44}[1 + 6.4(L/d)^{-0.75}] \qquad (370)$$

within the range $1 \times 10^3 \leq Re_G \leq 1.4 \times 10^4$, $2 \times 10^2 \leq \Gamma/\mu \leq 7 \times 10^2$, and $5 \leq L/d \leq 35$. The exponent for Sc, 0.44, was taken to be the same as the other investigators and p_{BM}/P is approximately equal to one in their experiments.

Simultaneous Heat and Mass Transfer in the Gas Phase

Hikita and Ishimi [311], and Hikita et al. [312] have studied simultaneous heat and mass transfer in laminar and turbulent gas streams respectively in a wetted-wall column by adiabatic vaporization of water into air. The similarity of the form of the basic transport equations for mass and heat makes it possible to apply the results of the mass transfer analyses to

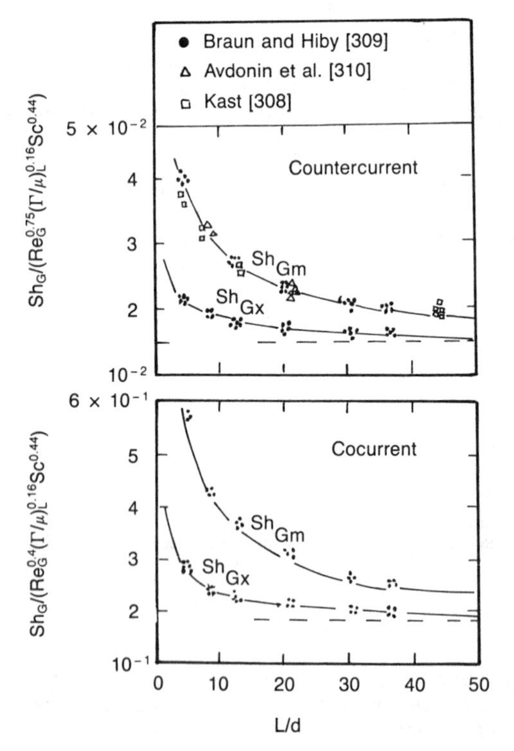

Figure 24. Correlation of Sh_G as a function of the relative film length for turbulent gas flow and wavy film flow [309].

the heat transfer analyses. Therefore the average gas-phase Nusselt number can be calculated from an equation analogous to Equation 358 if Nu_G replaces Sh_G and Gz'_G replaces Gz_G where

$$Nu_G = h_G(2r_i)/K_G \tag{371}$$

$$Gz_G = Wc_{pG}/K_G x \tag{372}$$

$$h_G = Wc_{pG}(T_1 - T_2)/2\pi r_i x(\Delta T)_{lm} \tag{373}$$

The experimental results were in good agreement with this kind of prediction and further, the effect of rippling on the gas-phase mass and heat transfer rates was shown to be negligible. This is contrary to the results of Chandra and Savery [299] whose experiments showed that rippling did enhance the mass transfer rate in the gas phase.

Solid-Liquid Mass Transfer

Although not as important as the corresponding gas absorption process, there are several occasions where solid-liquid mass transfer in falling films may assume special significance, e.g., solid-liquid mass transfer on catalyst particles in a trickle-bed reactor [313] and mass transfer of substrate to a biological film adhering to supporting particles in a trickling filter [314]. The interest in heat transfer between the solid wall and the falling film has also stimulated a number of analogous mass transfer measurements or theoretical predictions so that the result may be applied to heat transfer. Another area of importance is to use solid dissolution in laminar falling films as an experimental method for measuring the molecular diffusivity of a solute in a solvent.

Laminar and Wavy-Laminar Flow

When a section of the vertical or inclined wall coated with a soluble solute is in contact with the flowing solvent film, the solute will dissolve and the process can be described by the convective-diffusion equation:

$$u_{max}[1 - y^2/\delta^2)] \frac{\partial C}{\partial x} = D \frac{\partial^2 C}{\partial y^2} \tag{374}$$

with the boundary conditions

$$x = 0 \text{ (inlet)} \qquad C = C_0 \tag{375}$$

$$y = 0 \text{ (wall)} \qquad C = C^* \tag{376}$$

$$y = \delta \text{ (interface)} \qquad \partial C/\partial y = 0 \tag{377}$$

If the contact time is short, the penetration of the solute into the falling film is very small and the velocity profile in the thin concentration boundary layer near the wall can be regarded as linear with

$$u(y) = \rho g \delta y/\mu \tag{378}$$

The boundary condition, Equation 377, can be changed to

$$y \rightarrow \infty \qquad C = C_0 \tag{379}$$

Then the solution obtained is the Leveque short content time approximation for the local Sherwood number:

$$Sh_x = 0.6787x^{*-1/3} \tag{380}$$

For long contact times, an equation is obtained which is analogous to that corresponding to heating with constant wall temperature, Equation 55,

$$Sh_\infty = 1.88 \tag{381}$$

Analytical solutions were presented for the complete range of contact times by Mashelkar and Chavan [315] for solid dissolution in falling films on a plate, on a sphere, and on a conical surface, by Yih and Huang [316] for a plate, by Hirose et al. [317] for a sphere, and by Beschkov et al. [318] for a plate.

Relatively little experimental work is being done at truly laminar flow conditions for solid dissolution. Hirose et al. [317] measured the dissolution rates of benzoic acid in water and metallic copper in sulfuric acid containing potassium dichromate on single spheres of six different sizes ranging from 1.9 to 7.62 cm in diameter. The liquid flow rate ranged from 0.1 to 20 cm^3/s which encompassed the laminar and wavy-laminar flow region. In the long and short contact time regions, the mass transfer results can be represented by the theoretical results for laminar wave-free flow. However, in the third region of short contact time with a wavy film surface, the measured mass transfer coefficients are higher than the theory which they attributed to the reduction in film thickness due to wave motion. Their results indicated that if measurements were made in the short contact time smooth laminar flow region, the spherical film can be used for the measurement of diffusion coefficient of a dissolved solute because of the relatively small end effects of a sphere.

A number of investigations have been conducted on solid-liquid mass transfer measurements in the wavy flow region with the objective of determining the influence of waves on the solid-liquid mass transfer rate. The method of measurement can be classified mainly into two types:

1. Solid dissolution by weighing the difference in mass of solid
2. Electrochemical method

Solid dissolution methods. Based on the understanding that there is a difference in magnitude of the molecular diffusivities for momentum, heat, and mass, Stirba and Hurt [319] were able to show, with the same reasoning as suggested by Hiby [227] that while the increase in transfer rate due to waves is slight for momentum and heat transfer, this increase is much higher for mass transfer. Dissolution experiments for four solid-liquid systems were conducted within the range of Re from 308 to 2,410. The apparent mass diffusivity due to waves was calculated from the solution of Johnstone and Pigford [165], Equation 179, by replacing the molecular diffusivity by the apparent diffusivity. The apparent diffusivity was found to increase with increasing Re and many times larger than the molecular diffusivity but about the same order of magnitude as the momentum and thermal diffusivities.

Oliver and Atherinos [320] have made solid dissolution runs for β - naphthol and benzoic acid in water films on an inclined plate. The measured mass transfer coefficients for benzoic acid in the range of Re from 240 to 1,200 agreed well with the theoretical solution for smooth laminar flow at short contact time derived by Hikita et al. [321] who assumed that the velocity profile can be approximated by $u(y) = 2u_{max}y/\delta$ near the wall. This coincidence means that the waves have no appreciable effect on enhancing the solid-liquid mass transfer rate. However, in the same wavy flow region, the waves enhance the gas-liquid mass transfer rate by over 200%. Dye tracer experiments made by the authors indicated that the waves exert negligible effect on the flow of liquid adjacent to the solid wall but exert considerable influence near the gas-liquid interface. This may help to explain why there is such a large and different degree of enhancement between solid dissolution and gas absorption.

Measurements by Tanaka et al. [322] for the dissolution of gypsum plate by HCl solutions on an inclined plate of 27.6° angle and within a range of Re from 400 to 5,000 showed that the average mass transfer coefficient is about 40% higher than the Leveque solution. They attributed this enhancement as due to the effect of waves. However, the dissolution data of Beschkov et al. [318] within a range of Re from 100 to 800 showed good agreement with the Leveque solution. This again demonstrates that the enhancement in solid-liquid mass transfer by the wave motion is much less or even negligible when compared with the significant enhancement in gas-liquid mass transfer. Apparently the wave motion propagates into the concentration boundary layer near the wall to a much lower degree than expected.

Molecular diffusivity measurements of 2-naphthol, benzoic acid and salicylic acid in water have been carried out by Nigam et al. [323] by means of the short contact solid dissolution experiment. They found that the measured values are in good agreement with the other methods of measurement for benzoic and salicylic acid but the molecular diffusivity for 2-naphthol in water is about 60% lower than literature values. However, this lower value is possibly more accurate than the others because when this value is substituted into the theoretical expression for solid dissolution in laminar film flow, much closer agreement between the theory and the data of Oliver and Atherinos [320] is obtained.

Electrochemical method. A difficulty for the measurement of heat transfer from a falling film to a wall is that after a short length, thermal equilibrium will be nearly reached. On the other hand, concentration equilibrium is attained only at much longer lengths due to the much larger Schmidt number encountered than the Prandtl number. Therefore an alternative to the heat transfer measurement is by an analogous solid-liquid mass transfer measurement employing an electrochemical technique. This technique has several advantages as have been discussed by Grassmann [324]:

1. An accuracy of about 1%, generally not possible in other heat and mass transfer measurements, may be attained.
2. Local measurements are possible which is specially important if entrance region heat or mass transfer is significant.
3. The possibility of instantaneous measurement allows investigation of small-scale variations in hydrodynamic behavior and the structure of turbulence or the mass transfer of unsteady and fluctuating wavy films.
4. It is difficult to measure accurately the heat transfer coefficient of liquids with high Prandtl numbers at high Reynolds numbers because of the large amount of heat generated by viscous dissipation, the electrochemical method is free of this limitation.
5. Since redox-systems are mostly used as electrolytes, the concentration and the driving force both remain constant in the presence of mass transfer.
6. The local and the instantaneous wall shear stress can be measured by this method.

The electrochemical method has been used by a number of investigators [325–329] to study solid-liquid mass transfer in falling films. The basic interactions between the waviness and the associated transport phenomena were studied by Wragg and Einarsson [325] in the range of Re from 40–1,700. They recorded simultaneously the frequencies of the electrochemical fluctuations and, by means of electrical probe with a thin wire tip, the frequencies of the surface waves. These two frequencies were observed to be closely correlated. The % increase in solid-liquid mass transfer in the wavy flow region, as reflected by the magnitude of the electrochemical fluctuations, is observed to be small (less than 5%). These measurements were further extended to study the effect of cocurrent and countercurrent gas flow by Einarsson and Wragg [326]. Some sample oscilloscope traces of the electrochemical mass transfer fluctuation frequency are reproduced in Figure 25. For cocurrent flow, increasing gas flow is seen to smooth out the fluctuations gradually and finally becomes rather smooth at very high gas Reynolds numbers. This is probably due to the damping of surface waves and eddy motion within the film by the cocurrent gas flow which exerts large interfacial shear. Stainthorp and Batt [330] have also reported a decrease in surface wave amplitude

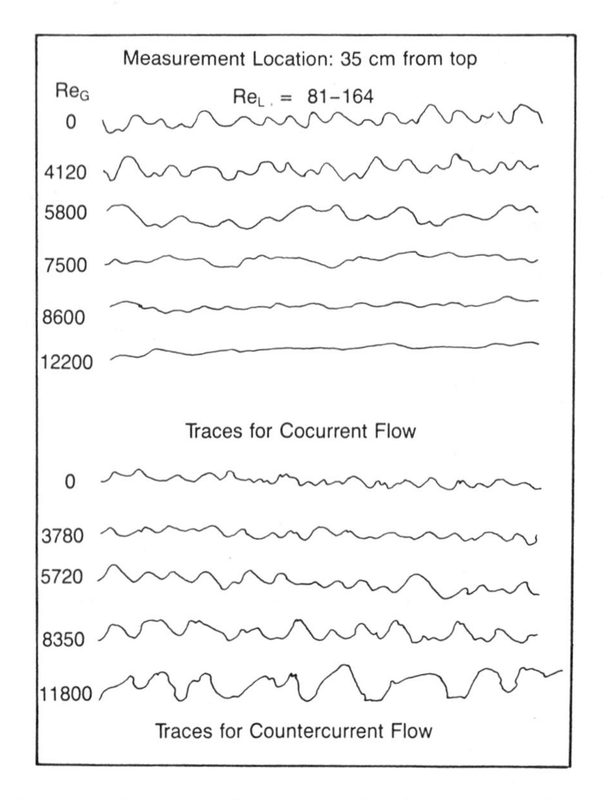

Figure 25. Sample traces of electrochemical mass transfer fluctuation frequency for cocurrent and countercurrent gas flow [326].

with increasing Re_G. For countercurrent flow, however, the opposite trend is observed. The amplitude of the fluctuations increases but their frequency decreases at very large Re_G. Close agreement was obtained between the fluctuation frequency data and the wave frequency data of Stainthorp and Batt [330] indicating a close correlation between the wave and mass transfer motion. The zone of the roll wave tends to move up the column with increasing gas counterflow. The percent mass transfer also increases with increasing Re_G.

Both the instantaneous film thickness and the local liquid-solid mass transfer rate were measured simultaneously by Ito et al. [327] using the electrical resistance and electrochemical technique, respectively. A distinct time lag between the peak mean concentration and the peak of the wave was observed which was similar to the observations of Stainthorp and Wild [331] using an optical system. Recently, Brauner and Moalem Maron [329] also made the same kind of measurements on an inclined plate and found that increasing waviness at high Re only resulted in mild increase in the transfer fluctuation although the disturbance in the film thickness continued to grow with the flow rate (Figure 26). The peak of the transfer rate was observed to be somewhat lagging behind the peak of the wave. However, the major increase in the transfer rate was supposed to occur in the wave front region. They concluded that only the large roll waves will affect the mass transfer characteristics at the solid-liquid boundary.

Turbulent Flow

Predictions of mass transfer from a solid surface into a turbulent falling film have been made by Iribarne et al. [332] based on an adaptation to film flow of a theory given earlier for heat or mass transfer through boundary layers. They solved the basic mass transport equation incorporating an eddy diffusivity profile of van Driest and a turbulent Schmidt number of 0.9. The relation between δ^+ and Re was calculated by using Spalding's eddy diffusivity profile. The theory agreed well with the electrochemical measurements of the mass transfer coefficient within the range of Reynolds number 2,800–4,000 and Schmidt number 1,400–18,400. For small contact times, the Leveque solution represents the data satisfactorily. Similar experiments, but with a different apparatus, were performed by Wragg et al. [333] whose data lay below the data of Iribarne et al. [332]. They observed no effect of turbulence on mass transfer up to Re = 2,200. This means that the ionic mass transfer is controlled mainly by the region of flow close to the wall which is still laminar in nature in spite of the high Reynolds number encountered. Measurements by Kramers and Kreyger [334] on the rate of dissolution of short surfaces (5–80 mm long and 100 mm wide) of benzoic acid in a water film on an inclined plane within Re = 1,500-7,000 have confirmed that the Leveque solution is still applicable even when the film is turbulent. This also means that solid-liquid mass transfer is not appreciably influenced by turbulent motion when the penetration of the solute from the wall into film is small.

Recently, Yih [205] proposed a viscosity-damped turbulence model (VDTM) for the prediction of solid-liquid heat or mass transfer in a falling film. The major mass transfer resistance is assumed to locate near the wall. If it is assumed that the eddy diffusivity for mass can

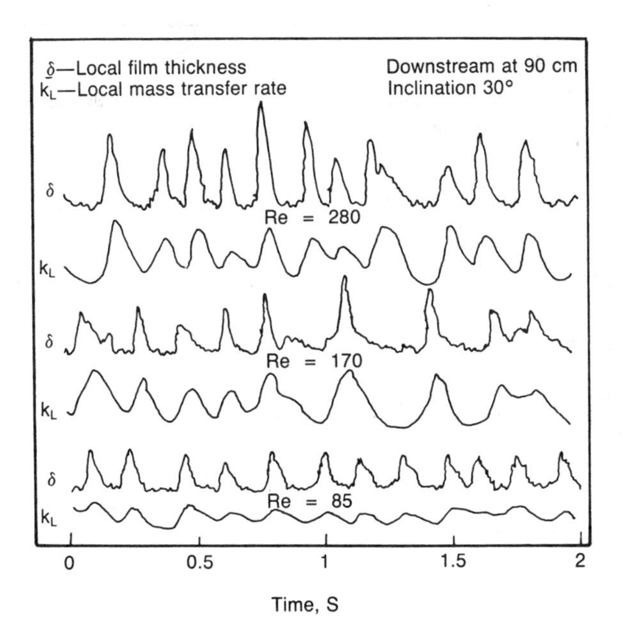

Figure 26. Simultaneous time traces of local instantaneous film thickness and mass transfer rate [329].

be represented by $\epsilon_D = ay^n$ where y is the distance from the wall, then the mass flux equation can be integrated:

$$\int_0^{\infty} \frac{dy}{D + \epsilon_D} = -\frac{1}{N_A}\int_{C^*}^{C_b} dC \tag{382}$$

The mass transfer coefficient derived is the same as that given by Equation 219. Since the wall is more rigid, eddies approaching the wall will be damped more rapidly than approaching the mobile gas-liquid interface. It is assumed that the y-component fluctuating velocity and mixing length are both approximately constant in the bulk liquid. Across a damped turbulence region λ_t near the interface, the fluctuating velocity is damped quadratically from its bulk value v_0 to zero while the mixing length is damped linearly from its bulk value l_0 to zero, i.e.

$$v_y' = v_0 y^2/\lambda_t^2, \quad l_D = l_0 y/\lambda_t \tag{383}$$

ϵ_D as given by the VDTM is $\epsilon_D \sim v_0 l_0 y^3/\lambda_t^3 = ay^3$ and Equation 219 becomes, with $n = 3$:

$$k_L \sim a^{1/3}D^{2/3} = (v_0 l_0/\lambda_t^3)^{1/3}D^{2/3} \tag{384}$$

The fact that ϵ_D varies as y^3 is in line with the deductions of Carrubba [190]. In the turbulent flow region, Equation 51 and the expression $\lambda_t v_0/\nu$ = constant are substituted into Equation 384 to give, for $v_0 \doteq v^* = (\rho g\delta/g_c)^{1/2}$ and $l_0 \doteq \delta$,

$$k_L^* \sim Re^{8/15}Sc^{1/3} \text{ for } Re > 1{,}600 \tag{385}$$

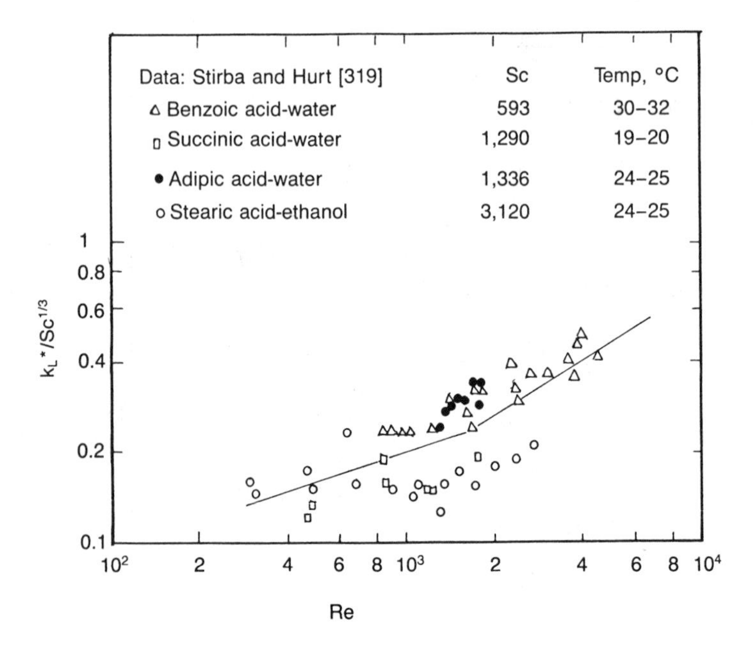

Figure 27. Correlation of solid dissolution data.

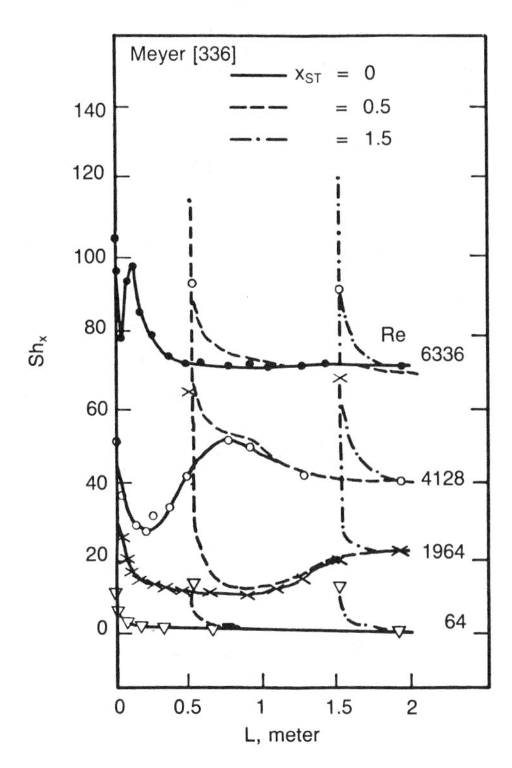

Figure 28. Local Sherwood number as a function of the distance L from top of the tube, x_{ST} denotes the length where mass transfer starts [336].

In the wavy-flow region, Equation 12 and the expression $\lambda_t v_0/\nu$ = constant are substituted into Equation 384 to give, for $v_0 \doteq v^*$ and $l_0 \doteq \lambda_t$,

$$k_L^* \sim Re^{1/6}Sc^{1/3} \tag{386}$$

The predictions that k_L^* varies with $Sc^{1/3}$ are in conformity with the boundary layer treatment and the analogous heat transfer measurements that h^* varies with $Pr^{1/3}$. Due to the scarcity of data, the verification of this model is not easy. A comparison of the model prediction with the data of Stirba and Hurt [319] is shown in Figure 27. In view of possible errors in the data, the prediction seems to be satisfactory. Moreover, support of this model came from its good representation of the heating data of Wilke [71].

Data for the measurement of local mass transfer coefficient along the film length is very scarce. Meyer [328, 336] has provided such a kind of data using the electrochemical method. The local mass transfer coefficient was found to be a function of the measuring length, the starting point for measurement and the liquid height above the ring-slot distributor. The range of Reynolds number investigated is 16–10,000. Typical results are reproduced in Figure 28. For $Re < 160$ and up to $L \leq 0.5$ m, the development of the Sherwood number follows the smooth laminar flow theory.

As the Reynolds number increases, the curves exhibit minimum and maximum points when the data are taken with the starting length at the top of the tube. This is because the hydrodynamics are not fully developed in these regions. When the starting length for measurement is moved downstream, e.g. at x_{ST} = 1.5 m, no such maximum is observed indicating fully developed hydrodynamic conditions.

Carrubba [190] has proposed that the correlations of Wilke [71] for heating may be slightly modified to predict the solid-liquid mass transfer coefficient in the developed flow region:

$$Sh_\infty = 0.0752(\Gamma/\mu)^{1/2}Sc^{1/3} \quad \text{for } 12 \le \Gamma/\mu \le 400 \text{ and } Sc \ge 1.56 \times 10^4/(\Gamma/\mu)^{3/2} \tag{387}$$

$$Sh_\infty = 1.88 \quad \text{for } 12 \le \Gamma/\mu \le 400 \text{ and } Sc \le 1.56 \times 10^4/(\Gamma/\mu)^{3/2} \tag{388}$$

$$Sh_\infty = 0.00114(\Gamma/\mu)^{1.2}Sc^{1/3} \quad \text{for } 400 \le \Gamma/\mu \le 800 \text{ and } Sc \ge [165.12/(\Gamma/\mu)^{1.2}]^3 \tag{389}$$

$$Sh_\infty = 1.88 \quad \text{for } 400 \le \Gamma/\mu \le 800 \text{ and } Sc \le [165.12/(\Gamma/\mu)^{1.2}]^3 \tag{390}$$

$$Sh_\infty = 0.00606(\Gamma/\mu)^{0.95}Sc^{1/3} \quad \text{for } 800 \le \Gamma/\mu \text{ and } Sc \ge [310.23/(\Gamma/\mu)^{0.95}]^3 \tag{391}$$

$$Sh_\infty = 1.88 \quad \text{for } 800 \le \Gamma/\mu \text{ and } Sc \le [310.23/(\Gamma/\mu)^{0.95}]^3 \tag{392}$$

From these correlations, he deduced that near the solid wall,

$$\epsilon_D/D = 7.52 \times 10^{-4}(\Gamma/\mu)^{3/2}Sc(1 - Y/\delta)^3 \quad \text{for } 12 \le \Gamma/\mu \le 400 \tag{393}$$

$$\epsilon_D/D = 2.62 \times 10^{-9}(\Gamma/\mu)^{3.6}Sc(1 - Y/\delta)^3 \quad \text{for } 400 \le \Gamma/\mu \le 800 \tag{394}$$

$$\epsilon_D/D = 3.93 \times 10^{-7}(\Gamma/\mu)^{2.85}Sc(1 - Y/\delta)^3 \quad \text{for } 800 \le \Gamma/\mu \tag{395}$$

The dependence of ϵ_D on y^3 near the solid wall is predicted by the VDTM of Yih [205].

CONCLUSIONS

In this chapter, we have collected and reviewed the various methods of modeling the heat and mass transport process in falling liquid films. For laminar flow, it is evident that the simple penetration theory and the Leveque theory can predict satisfactorily the mass or heat transfer coefficient respectively at the gas-liquid and solid-liquid interface. For wavy flow, increased understanding of the mechanism of enhancement of transport due to waves is possible because of the development of more refined methods for the measurement of concentration or temperature profiles and also methods for measuring instantaneous local fluctuations in film thickness and transfer rate. However, modeling and prediction of transfer rates in wavy films continue to be difficult and not very successful. This is because of the complex and time-varying nature of the waves. In this region, empirical or semi-empirical methods seem to be a better solution. For turbulent flow, the constant film thickness model employing the eddy diffusivity concept seems to be very promising in predicting average film thickness and transfer coefficients. More refined turbulence models will continue to be developed.

Due to space limitations, certain topics in film flow which are of interest e.g. wave and film hydrodynamics, two-phase gas/liquid film pressure drop, flooding, minimum wetting rate in the presence or absence of heat transfer, film breakdown, interfacial turbulence during gas absorption, design and scale-up of tube-bundle columns, enhanced tubes, and non-Newtonian film flow, will not be presented here. These will be available in a forthcoming publication [337]. It is hoped that, by bringing together all the scattered information in the literature, future researchers can be guided towards a thorough and state-of-the-art understanding of film flow so that one can concentrate more on the important aspects of film flow which still remain insufficiently studied. Also, by describing the principles and advantages of falling film exchangers, it is hoped that engineers will become more aware of and appreciate the importance of film flow in film type equipments, thus stimulating further advance in this art.

NOTATION

- a' interfacial area for mass transfer, cm^2
- a constant
- Ar Archimedes number $K_F^{-1/2}$
- A^+ constant, 25.1
- A_i series coefficient
- b constant
- b_i constants
- B_0 bulk concentration of reactant B
- B^+ a van Driest parameter
- c wavy velocity, cm/s
- c_p heat capacity at constant pressure, J/kg-°K
- C_0, C_1 inlet and outlet concentration of gas or solid in liquid, mole/l
- C^* solubility of gas or solid in liquid, mole/l
- C^+ C/C^*
- C_G gas phase concentration, mole/l
- C_m, C_∞ constants
- d diameter of tube, cm
- D molecular diffusivity of gas or solid in liquid, m^2/s
- D_G gas phase molecular diffusivity
- e constant
- E_v energy dissipation
- E, E_i enhancement factor; for instantaneous reaction
- f a damping factor
- f_v friction factor of vapor
- Fr Froude number, $u_{av}/(g\delta)^{1/2}$
- g, g_c gravitational acceleration constant, m/s^2; proportional factor, g-m/N-s^2
- G_v, G_m mass velocity of vapor; mean value, kg/hr-m^2
- Ga Galileo number, gL^3/ν^2
- Gz Graetz number, $W/\rho_G D_G$
- h_{fg} latent heat, kJ/kg
- h_x heat transfer coefficient, W/m^{2}K
- h_x^* $(h_x/K)(\nu^2/g)^{1/3}$
- h_m average heat transfer coefficient
- h_∞ asymptotic heat transfer coefficient
- h_b, h_w heat transfer coefficient of base film; of wave
- H mean wave height
- H_L height of a mass transfer unit, $\Gamma/\rho k_L$
- k_G gas-side mass transfer coefficient, mol/cm^2-s-atm
- k_L liquid-side mass transfer coefficient, m/s
- k_L^* $(k_L/D)(\nu^2/g)^{1/3}$, dimensionless
- k_1, k_2 reaction rate constant for first-order reaction; for second-order reaction
- K thermal conductivity, W/mK
- K_F film number $= \rho\sigma^3/g\mu^4$
- Ka Kapitza number $= g\mu^4/\rho\sigma^3 = K_F^{-1}$
- l_D mixing length of eddy, cm
- l_0 mixing length of eddy in bulk liquid, cm
- L absorption length, cm
- L_w base length of an "average wave"
- M $\pi k_2 B_0 t/4$
- N_i eigenfunction
- Nu_x Nusselt number, $h_x\delta/k$
- N_A, N_A'' absorption rate per unit area, mole/cm^2-s; absorption rate, mole/s
- p degree of mixing
- p_{BM} log mean of partial pressure of B
- P Pressure, kPa
- Pr, Pr_t Prandtl number; turbulent Prandtl number, ϵ_M/ϵ_H
- Pr_w Prandtl number at wall
- q, q_w heat flux; at wall, W/m^2
- Q volumetric flow rate, l/min
- r radial coordinate
- r_i, r_w r at interface; at wall
- R outside radius of tube, cm
- R_1 $u_{av}\delta/\nu$
- Re film Reynolds number, $4\Gamma/\mu = 4Q/\pi d\,\mu = 4qL/h_{fg}g\mu$

Re_b, Re_w Reynolds number of base film; of wave
Re_G gas phase Reynolds number
s $[(\rho\delta g/g_c)/\tau_i + \rho\delta g/g_c)]^{1/3}$
S $\sigma/\rho u_{av}{}^2\delta$
S_1 $(3K_F)^{1/5}$
Sc Schmidt number, ν/D
Sh, Sh_G Sherwood number, $k_L\delta/D$; gas phase Sherwood number, $k_G(2r_i)/D_G$
t contact time of gas and liquid or solid and liquid, s
$(t_m)_p$ average residence time
T, T_{sat} liquid temperature; at saturation
T_w, T_c wall temperature, cooling water temperature
u, u^*, u^+ x-component liquid velocity; friction velocity, $(\tau_w g_c/\rho)^{1/2}$; u/u^*
u_{av}, u_{max} average liquid velocity; maximum value, cm/s
u_G gas velocity, m/s
U u_i/u_m, interfacial velocity/average gas velocity
$\overline{U}$ u/u_{av}
v y-component liquid velocity, cm/s
v_0 eddy velocity in the bulk liquid, approximately equal to v^*, cm/s
v_y' y-component eddy fluctuating velocity, cm/s
v^* friction velocity, $(\tau_w g_c/\rho)^{1/2}$, cm/s
w_c mass flow rate of cooling water, kg/hr
We Weber number, $((\Gamma/\mu)^5/S_1)^{1/3}$
W mass flow rate of gas, kg/hr
x, x^* axial film distance; $x\alpha/\delta^2u_{max}$ for heat transfer, xD/δ^2u_{max} for mass transfer
x_e, x_e^* entrance length; $x_e(g\sin\theta/3\nu^2)^{1/3}$, dimensionless
y, y^+ distance measured from the wall; yu^*/ν
Y distance measured from the gas-liquid interface, cm
Z_L wave number
z stoichiometric coefficient

Greek Symbols

α thermal diffusivity, m^2/s
α_m, α_r dimensionless wave number; real part, $= 2\pi\delta/\lambda_w$
β a δ^2/D
β^* a parameter defined for instantaneous reaction
β_1 dimensionless amplitude
Γ mass flow rate per unit periphery, kg/m-s
δ, δ^*, δ^+ film thickness; $\delta/(\nu^2/g)^{1/3}$; $\delta u^*/\nu$
δ_0 average film thickness in wavy film, cm
δ_{min}, δ_{max} minimum and maximum film thickness, cm
ϵ_M, ϵ_H eddy diffusivity for momentum; for heat, m^2/s
ϵ_D eddy diffusivity for mass, m^2/s
ϵ_m eddy diffusivity due to waves, m^2/s
η y/δ, or y^+/δ^+, or Y/δ
η^* $(B^*/B_0)^{1/2} = [(E_i - E)/(E_i - 1)]^{1/2}$
θ_1 angle of incline plane
θ, θ_w, θ_m dimensionless temperature, $(T - T_{in})/(q_w\delta/K)$ for heating and $(T - T_{sat})/(q_w\delta/K)$ for evaporation with constant q_w, $(T_w - T)/(T_w - T_{in})$ for heating and $(T_w - T)/(T_w - T_{sat})$ for evaporation with constant T_w; at wall; at bulk mean
λ_i eigenvalue
λ_p average wave separation distance, cm
λ_w wave length, cm
λ_t thickness of the zone of damped turbulence, cm
μ, μ_w liquid viscosity; at wall, kg/m-s
ν kinematic viscosity, m^2/s
ρ liquid density, kg/m^3
σ surface tension, N/m
τ, τ_w, τ_i shear stress; at wall; at interface, N/m^2
τ_i^* $\tau_i g_c/\rho(\nu g)^{2/3}$, dimensionless
τ_s shear force at the outer boundary of laminar sublayer
τ_M momentum change of the vapor
ϕ dimensionless amplitude of the

	local deviation from average film thickness
Φ	a constant
ψ	stream function for velocity
ω_r	dimensionless real angular frequency

Subscripts

c	cooling water
cr	critical value
G	gas
in	inlet
L	liquid
m	mean, average
sat	saturated
TP	two-phase
x	local value
w	wall

REFERENCES

1. Sack, M., *Chem. Eng. Progr., 63*:55 (1967).
2. Feind, K., *VDI-Forschungsh., 481* (1960).
3. Malewski, W., *Chemie-Ing.-Techn., 40*:201 (1968).
4. Beccari, M., A. C. di Pinto, and L. Spinosa, *Desalination, 29*:295 (1979).
5. Fulford, G. D., in *Advances in Chemical Engineering,* T. B. Drew, J. W. Hoopes, Jr., and T. V. Vermeulen (Eds.) Vol. 5, Academic Press, N.Y., 1964, p. 151.
6. Brauer, H., *VDI-Forschungsh., 457* (1956).
7. Scriven, L. E., and Pigford, R. L., *AIChE J., 4*:382 (1958).
8. Wilkes, J. O., and Nedderman, R. M., *Chem. Eng. Sci., 17*:177 (1962).
9. Bruley, D. F., *AIChE J., 11*:945 (1965).
10. Hassan, A., *J. Appl. Mech., Trans. ASME, 34*:535 (1967).
11. Haugen, R., *J. Appl. Mech., Trans. ASME, 90*:631 (1968).
12. Lynn, S., *AIChE J., 6*:703 (1960).
13. Cerro, R. L., and Whitaker, S., *Chem. Eng. Sci., 26*:785 (1971).
14. Ault, J. W., and Sandall, O. C., *Can. J. Chem. Eng., 50*:318 (1972).
15. Stucheli, A., and Ozisik, M. N., *Chem. Eng. Sci., 31*:369 (1976).
16. Yilmaz, T., and Brauer, H., *Int. Chem. Eng., 19*:32 (1979).
17. Fulford, G. D., Ph. D. Thesis, Birmingham Univ., England, 1962.
18. Nusselt, W., *VDI-Zeitschrift, 54*:1154 (1910).
19. Jackson, M. L., *AIChE J., 1*:231 (1955).
20. Lefebvre, S., et al., *Chemie-Ing.-Techn., 51*:330 (1979), p. 330.
21. Lefebvre, S., *U.S. Patent No. 3748, 828,* July 31, 1973.
22. Dukler, A. E., in *Progress in Heat and Mass Transfer: Proc. Int. Symp. on Two-Phase Systems,* G. Hetsroni, S. Sideman, and J. P. Hartnett (Eds.), Pergamon Press, England, 1972, p. 207.
23. Kapitza, P. L., in *Collected Papers of P. L. Kapitza,* Macmillan, N.Y., 1964, p. 662.
24. Portalski, S., *Chem. Eng. Sci., 18*:787 (1963).
25. Tailby, S. R., and Portalski, S., *Trans. Instn. Chem. Engrs., 40*:114 (1962).
26. Stainthorp, F. P., Allen, J. M., *Trans Instn. Chem. Engrs., 43*:85 (1965).
27. Jones, L. O., and Whitaker, S., *AIChE J., 12*:525 (1966).
28. Shkadov, V. Ya., *Izv. An SSSR. Mekhanika Zhidkosti i Gaza, 2*:43 (1967).
29. Massot, C., Irani, F., and Lightfoot, E. N., *AIChE J., 12*:445 (1966).
30. Gollan, A., and Sideman, S., *AIChE J., 15*:301 (1969).
31. Berbente, C. P., and Ruckenstein, E., *AIChE J., 14*:772 (1968).
32. Penev, V., et al., *Int. J. Heat Mass Transfer, 15*:1395 (1972).
33. Hirshburg, R. I., and Florschuetz, L. W., *J. Heat Transfer, Trans. ASME, 104*:452 (1982).
34. Benjamin, T. B., *J. Fluid Mech., 2*:554 (1957).
35. Yih, C. S., *Phys. Fluids, 6*:321 (1963).

36. Whitaker, S., *Ind. Eng. Chem. Fundam.*, *3*:132 (1964).
37. Anshus, B. E., and Goren, S. L., *AIChE J.*, *12*:1004 (1966).
38. Krantz, W. B., and Goren, S. L., *Ind. Eng. Chem. Fundam.*, *10*:91 (1971).
39. Graef, M., *Mitteilungen ausdem Max-Planck Inst. fur Stromungsforschung, No. 36*, Gottingen (1966).
40. Strobel, W. J., and Whitaker, S., *AIChE J.*, *15*:527 (1969).
41. Pierson, F. W., and Whitaker, S., *Ind. Eng. Chem. Fundam.*, *16*:401 (1977).
42. Krantz, W. B., and Owens, W. B., *AIChE J.*, *19*:1163 (1973).
43. Lin, S. P., *AIChE J.*, *21*:178 (1975).
44. Brauer, H., *Ger. Chem. Eng.*, *3*:149 (1980).
45. Telles, A. S., and Dukler, A. E., *Ind. Eng. Chem. Fundam.*, *9*:412 (1970).
46. Chu, K. J., and Dukler, A. E., *AIChE J.*, *20*:695 (1974).
47. Chu, K. J., and Dukler, A. E., *AIChE J.*, *21*:583 (1975).
48. Brumfield, L. K., and Theofanous, T. G., *J. Heat Transfer, Trans. ASME, 98*:496 (1976).
49. Brumfield, L. K., Houze, R. N., and Theofanous, T. G., *Int. J. Heat Mass Transfer, 18*:1077 (1975).
50. Dukler, A. E., and Bergelin, O. P., *Chem. Eng. Progr.*, *48*:557 (1952).
51. Marschall, E., *Int. Chem. Eng.*, *17*:38 (1977).
52. Salazar, R. P., and Marschall, E., *Int. J. Multiphase Flow, 4*:405 (1978).
53. Salazar, R. P., and Marschall, E., *Int. J. Multiphase Flow, 4*:487 (1978).
54. Lukach, Yu. Ye., Radchenko, L. B., and Tananayiko, Yu. M., *Int. Chem. Eng.*, *12*:517 (1972).
55. Brotz, W., *Chemie-Ing.-Techn.*, *26*:470 (1954).
56. Zhivaikin, L. Ya, *Int. Chem. Eng.*, *2*:337 (1962).
57. Ganchev, B. G., Kozlov, V., and Lozovetskiy, V., *Heat Transfer-Soviet Research, 4*:102 (1972).
58. Kosky, P. G., *Int. J. Heat Mass Transfer, 14*:1220 (1971).
59. Takahama, H., and Kato, S., *Int. J. Multiphase Flow, 6*:203 (1980).
60. Mostofizadeh, Ch., Ph. D. Thesis, Univ. Stuttgart, F. R. G., 1980.
61. Yih, S. M., and Liu, J. L., *AIChE J.*, *29*:903 (1983).
62. Ueda, T., and Tanaka, T., *Bulletin of the JSME, 17*:603 (1974).
63. Yih, S. M., and Huang, P. G., *J. Chin. Inst. Chem. Engrs.*, *11*:71 (1980).
64. Faghri, A., Ph. D. Thesis, Univ of Calif. at Berkley, 1976.
65. Yih, S. M., and Lee, M. W., *unpublished work*, 1983.
66. Bays, G. S., and McAdams, W. H., *Ind. Eng. Chem.*, *29*:1240 (1937).
67. Limberg, H., *Int. J. Heat Mass Transfer, 16*:1691 (1973).
68. S. Ishigai, et al., *Bulletin of the JSME, 17*:106 (1974).
69. Gimbutis, G., *Proc. 5th Int. Heat Transfer Conf.*, *2*:85 (1974).
70. Seban, R. A., and Faghri, A., *J. Heat Transfer, Trans. ASME, 98*:315 (1976).
71. Wilke, W., *VDI-Forschungsh.*, *490* (1962).
72. Lamourelle, A. P., and Sandall, O. C., *Chem. Eng. Sci.*, *27*:1035 (1972).
73. Mills, A. F., and Chung, D. K., *Int. J. Heat Mass Transfer, 16*:694 (1973).
74. Yih, S. M., and Chen, C. H., *Proc. 7th Int. Heat Transfer Conf.*, *3*:125 (1982).
75. McAdams, W. H., Drew, T. B., and Bays, G. S., *Trans. ASME, 62*:627 (1940).
76. Herbert, L. S., and Sterns, U. J., *Can. J. Chem. Eng.*, *46*:401 (1968).
77. Pike, J. G., Smith , G. A. J., and Thompson, R. G., *Proc. 4th Int. Heat Transfer Conf.*, *2*:FC1.12 (1970).
78. Ganchev, B. G., Bokov, A. Ye., and Musvik, A. B., *Heat Transfer-Soviet Research, 8*:9 (1976).
79. Gimbutis, G. J., *Proc. 6th Int. Heat Transfer Conf.*, *1*:321 (1978).
80. Fujita, T., and Ueda, T., *Int. J. Heat Mass Transfer, 21*:97 (1978).
81. Schnabel, G., and Schlunder, E. U., *Verfahrenstechnik, 14*:79 (1980).
82. Kelly, E., M. S. Thesis, Univ. of Oklahoma, U.S.A., 1951.
83. Davies, J. T., and Shawki, A. M., *Chem. Eng. Sci.*, *29*:1801 (1974).

84. Trommelen, A. M., *Chem. Eng. Sci.*, *22*:1152 (1967).
85. Horn, R. K., Ph. D. Thesis, Univ of Karlsruhe, F. R. G., 1970.
86. Walker, L. H., and Patterson, D. C., *Ind. Eng. Chem.*, *40*:534 (1951).
87. Keville, J. F., *Chem. Eng. Progr*, *54*:83 (1958).
88. Wiegand, J., *J. Appl. Chem. Biotechnol.*, *21*:351 (1971).
89. Sinek, J. R., and Young, E. H., *Chem. Eng. Progr.*, *58*:74 (1962).
90. Dukler, A. E., and Elliott, L. C., *U.S. Dept. of Interior, Office of Saline Water R & D Report No. 287* (1967).
91. Seban, R. A., *Proc. 6th Int. Heat Transfer Conf.*, *6*:417 (1978).
92. Chun, K. R., *Ph. D. Thesis*, Univ. of Calif. at Berkeley, U.S.A., 1969.
93. Rogovan, I. A., Olevskii, V. M., and Runova, N. G., *Theo. Found. Chem. Eng.*, *3*:164 (1969).
94. Chun, K. R., and Seban, R. A., *J. Heat Transfer, Trans. ASME*, *93*:391 (1971).
95. Hirshburg, R. I., and Florschuetz, L. W., *J. Heat Transfer, Trans. ASME*, *104*:459 (1982).
96. Struve, H., *VDI-Forschungsh.*, *534* (1969).
97. Domanskii, I. V., and Sokolov, V. N., *J. Appl. Chem. USSR*, *40*:56 (1967).
98. Murthy, V. N., and Sarma, P. K., *J. Chem. Eng. Japan*, *6*:457 (1973).
99. Beccari, M., et al., *Int. J. Multiphase Flow*, *2*:357 (1975).
100. Hubbard, G. L., Mills, A. F., and Chung, D. K., *J. Heat Transfer, Trans. ASME*, *98*:319 (1976).
101. Chung, D. K., Ph. D. Thesis, Univ. of California at Los Angeles, 1974.
102. Ueda, T., Kubo, T., and Inoue, M., *Proc. 5th Int. Heat Transfer Conf.*, *3*:304 (1974).
103. Murthy, V. N., and Sarma, P. K., *Can. J. Chem. Eng.*, *55*:732 (1977).
104. Belkin, H. H., et al., *AIChE J.*, *5*:245 (1959).
105. Wassner, L., *Warme-und Stoffubertragung*, *14*:23 (1980).
106. Mostofizadeh, Ch., and Stephen, K., *Warme-und Stoffubertragung*, *15*:93 (1981).
107. Unterberg, W., and Edwards, D. K., *AIChE J.*, *11*:1073 (1965).
108. Thoma, R., Ph. D. Thesis, Technische Hochscule Karlsruhe, F. R. G., 1966.
109. Haase, B., *Chem. Techn.*, *22*:283 (1970).
110. Gazit, E., and Hasson, D., *Desalination*, *17*:339 (1975).
111. Meyer, K., Ph. D. Thesis, Technische Univ. Braunschweig, F. R. G., 1975.
112. Kutateladze, S. S., *Fundamentals of Heat Transfer*, Ch. 15, Academic Press, N.Y., 1963.
113. Fujita, T., and Ueda, T., *Int. J. Heat Mass Transfer*, *21*:109 (1978).
114. Haase, B., Ph. D. Thesis, Technische Hochschule Magdeburg, F. R. G., 1966.
115. Elle, C., Ph. D. Thesis, Technische Univ. Dresden, F. R. G., 1970.
116. Dukler, A. E., *AIChE Symp. Ser.*, *56*:1 (1960).
117. Dukler, A. E., *Petro/Chem. Engr.*, *33*:222 (1961).
118. Bromley, L. A., *Ind. Eng. Chem.*, *44*:2966 (1952).
119. Rohsenow, W. M., *Trans. ASME*, *78*:1645 (1956).
120. Sparrow, E. M., and Gregg, J. L., *J. Heat Transfer, Trans. ASME*, *81*:13 (1959).
121. Sparrow, E. M., and Eckert, E. R. G., *AIChE J.*, *7*:473 (1961).
122. Koh, J. C. Y., Sparrow, E. M., and Hartnett, J. P., *Int. J. Heat Mass Transfer*, *2*:69 (1961).
123. Minkowycz, W. J., and Sparrow, E. M., *Int. J. Heat Mass Transfer* *9*:1125 (1966).
124. Shekriladze, I. G., and Gomelauri, V. I., *Int. J. Heat Mass Transfer*, *9*:581 (1966).
125. Shekriladze, I. G., Mestvirishvili, Sh. A., and Mikashavidze, A. N., *Heat Transfer-Soviet Research*, *3*:120 (1971).
126. Marschall, E., and Lee, C. Y., *Int. J. Heat Mass Transfer*, *16*:41 (1973).
127. Marschall, E., and Lee, C. Y., *Warme-und Stoffubertragung*, *1*:32 (1973).
128. Unsal, M., and Thomas, W. C., *J. Heat Transfer: Trans. ASME*, *100*:629 (1978).
129. Leonard, W. K., and Estrin, J., *AIChE J.*, *18*:439 (1972).
130. Sofrata, H., *Warme-und Stoffubertragung*, *14*:201 (1980).
131. Gregorig, R., Kern, J., and Turek, K., *Warme-und Stoffubertragung*, *7*:1 (1974).

132. Ratiani, G. V., and Shekriladze, I. G., *Teploenergetika, 11*:78 (1964).
133. Rasche, H., Ph. D. Thesis, Technical Univ. Berlin, F. R. G., 1977.
134. Sofrata, H., *Warme-und Stoffubertragung, 15*:117 (1981).
135. Kutateladze, S. S., et al., *Thermal Eng. (Teploenergetika), 27*:184 (1980).
136. Shea, F. L. Jr., and Krase, N. W., *Trans. AIChE, 36*:463 (1940).
137. Onda, K., et al., *Kagaku Kogaku, 32*:1215 (1968).
138. Gogonin, I. I., Dorokhov, A. R., and Sosunov, V. I., *Heat Transfer-Soviet Research, 13*:51 (1981).
139. Kutateladze, S. S., and Gogonin, I. I., *Int. J. Heat Mass Transfer, 22*:1593 (1979).
140. Kirkbride, C. G., *Trans. AIChE, 30*:170 (1934).
141. Badger, W. L., Monrad, C. C., and Diamond, H. W., *Ind. Eng. Chem., 22*:700 (1930).
142. Colburn, A. P., *Trans. AIChE, 30*:187 (1934).
143. Carpenter, F. G., Ph. D. Thesis, Univ. of Delaware, 1948.
144. Carpenter, F. G., and Colburn, A. P., *Proc. General Discussion of Heat Transfer, Inst. Mech. Engrs. and ASME,* July 1951, p. 20.
145. Soliman, M., Schuster, J. K., and Berenson, P. J., *J. Heat Transfer, 90*:267 (1968).
146. Seban, R. A., *Trans. ASME, 76*:299 (1954).
147. Rohsenow, W. M., Webber, J. H., and Ling, A. T., *Trans. ASME, 78*:1637 (1956).
148. Shekriladze, I., and Mestvirishvili, Sh., *Int. J. Heat Mass Transfer, 16*:715 (1973).
149. Ueda, T., et al., *Bulletin of the JSME, 15*:1267 (1972).
150. Razavi, M. D., and Damle, A. S., *Trans. Instn. Chem. Engrs., 56*:81 (1978).
151. Kunz, H. R., and Yerazunis, S., *J. Heat Transfer, Trans. ASME, 91*:413 (1969).
152. Blangetti, F., and Schlunder, E. U., *Proc. 6th Int. Heat Transfer Conf., 2*:437 (1978).
153. Blangetti, F., Krebs, R., and Schlunder, E. U., *Chem. Eng. Fundam., 1*:20 (1982).
154. Goodykoontz, J. H., and Dorsch, R. G., *U.S. NASA Report TN D-3953,* 1967.
155. Goodykoontz, J. H., and Dorsch, R. G., *U.S. NASA Report TN D-3326,* 1966.
156. Kutateladze, S. S., *Int. J. Heat Mass Transfer, 25*:653 (1982).
157. Dobran, F., *Int. J. Heat Mass Transfer, 26*:1159 (1983).
158. Chien, S., and Ibele, W., *J. Heat Trasnfer, Trans. ASME, 86*:80 (1964).
159. Lee, J., *AIChE J., 10*:540 (1964).
160. Badger, W. L., *Trans. AIChE, 33*:441 (1937).
161. Goodykoontz, J. H., and Brown, W. F., *U.S. NASA Report TN D-3952,* 1967.
162. Borishanskiy, V. M., et al., *Heat Transfer-Soviet Research, 11*:35 (1979).
163. Laurent, A., and Charpentier, J. C., *Int. Chem. Eng., 23*:265 (1983).
164. Pigford, R. L., Ph. D. Thesis, Univ. Illinois, 1941.
165. Johnstone, H. F., and Pigford, R. L., *Trans. AIChE, 38*:25 (1942).
166. Brauer, H., *Fortschritte der Verfahrenstechnik, VDI-Verlag, 19*:81 (1981).
167. Lynn, S., Straatemeier, J. R., and Kramers, H., *Chem. Eng. Sci.,* 4:49 (1955).
168. Nijsing, R. A. T. O., and Kramers, H., *Chem. Eng. Sci., 8*:81 (1958).
169. Vivian, J. E., and Peaceman, D. W., *AIChE J., 2*:437 (1956).
170. Davidson, J. F., and Cullen, E. J., *Trans. Instn. Chem. Engrs., 35*:51 (1957).
171. Olbrich, W. E., and Wild, J. D., *Chem. Eng. Sci., 24*:25 (1969).
172. Davis, E. J., *Can. J. Chem. Eng., 51*:562 (1973).
173. Yih, S. M., and Huang, P. G., *Chem. Eng. Sci., 36*:387 (1981).
174. Tamir, A., and Taitel, Y., *Chem. Eng. Sci., 26*:799 (1971).
175. Rotem, Z., and Neilson, J. E., *Can J. Chem. Eng., 47*:341 (1969).
176. Davis, G. A., Ponter, A. B., and Crane, K., *Can. J. Chem. Eng., 45*:372 (1967).
177. Simons, J., and Ponter, A. B., *J. Chem. Eng. Japan, 8*:347 (1975).
178. Mazarei, A. F., and Sandall, O. C., *AIChE J., 26*:154 (1980).
179. Won, Y. S., Chung, D. K., and Mills, A. F., *J. Chem. Eng. Data, 26*:140 (1981).
180. Lynch, D. W., and Potter, O. E., *Chem. Eng. J., 15*:197 (1978).
181. R. E., Emmert, and Pigford, R. L., *Chem. Eng. Progr., 50*:87 (1954).
182. Portalski, S., and Clegg, A. J., *Chem. Eng. Sci., 26*:773 (1971).
183. Portalski, S., *Ind. Eng. Chem. Fundam., 3*:49 (1964).

184. Levich, V. G., *Physicochemical Hydrodynamics,* Prentice-Hall, N.J., 1962, p. 692.
185. Ruckenstein, E., and Berbente, C., *Chem. Eng. Sci., 20*:795 (1965).
186. Ruckenstein, E., and Berbente, C., *Int. J. Heat Mass Transfer, 11*:743 (1968).
187. Beschkov, V., and Boyadjiev, C., *Chem. Eng. Commun., 20*:173 (1983).
188. Javdani, K., *Chem. Eng. Sci., 29*:61 (1974).
189. Banerjee, S., Rhodes, E., and Scott, D. S., *Chem. Eng. Sci., 22*:43 (1967).
190. Carrubba, G., Ph. D. Thesis, Tech. Univ. Berlin, F. R. G., 1976.
191. Kamei, S., and Oishi, J., *Mem. Fac. Kyoto Univ., 17*:277 (1955).
192. Brauer, H., *Chemie-Ing.-Techn., 30*:75 (1958).
193. Hikita, H., Nakanishi, K., and Kataoka, T., *Kagaku Kogaku, 23*:459 (1959).
194. Malewski, W., Ph. D. Thesis, Tech. Univ. Berlin, F. R. G., 1963.
195. Malewski, W., *Chemie-Ing.-Techn., 37*:815 (1965).
196. Toor, H. L., and Marchello, J. M., *AIChE J., 4*:97 (1958).
197. Tadaki, T., and Maeda, S., *Kagaku Kogaku, 27*:66 (1963).
198. Fortescue, G. E., and Pearson, J. R. A., *Chem. Eng. Sci., 22*:1163 (1967).
199. Lamont, J. C., and Scott, D. S., *AIChE J., 16*:513 (1970).
200. Banerjee, S., Scott, D. S., and Rhodes, E., *Ind. Eng. Chem. Fundam., 7*:22 (1968).
201. Davies, J. T., *Turbulence Phenomena,* Academic Press, N.Y., 1972.
202. King, C. J., *Ind. Eng. Chem. Fundam., 5*:1 (1966).
203. Prasher, B. D., and Fricke, A. L., *Ind. Eng. Chem. Process Des. Develop., 13*:336 (1974).
204. Henstock, W. H., and Hanratty, T. J., *AIChE J., 25*:122 (1979).
205. Yih, S. M., *Proc. 2nd World Congress Chem. Eng., 5*:168 (1981).
206. Kishinevskii, M. Kh., and Kornienko, T. S., *J. Appl. Chem. USSR, 50*:2520 (1977).
207. Bin, A. K., *Int. J. Heat Mass Transfer, 26*:981 (1983).
208. Won, Y. S., Ph. D. Thesis, Univ. Calif. at Los Angeles, U.S.A., 1977.
209. Won, Y. S., and Mills, A. F., *Int. J. Heat Mass Transfer, 25*:223 (1982).
210. Coeuret, F., Jamet, B., and Ronco, J. J., *Chem. Eng. Sci., 25*:17 (1970).
211. Chung, D. K., and Mills, A. F., *Int. J. Heat Mass Transfer, 19*:51 (1976).
212. Koziol, K., Broniarz, L., and Nowicka, T., *Int. Chem. Eng., 20*:136 (1980).
213. Yih, S. M., and Chen, K. Y., *Chem. Eng. Commun., 17*:123 (1982).
214. Davies, J. T., and Warner, K. V., *Chem. Eng. Sci., 24*:231 (1969).
215. Gildenblat, I. A., Rodinov, A. I., and Demcheno, B. I., *Doklady Academii Nauk USSR, 198*:1149 (1971).
216. Johannisbauer, W., *Diploma-Arbeit,* Institute fur Verfahrenstechnik, Tech. Univ. Aachen, F. R. G., 1971.
217. Bakopoulos, A., Ph. D. Thesis, Tech. Univ. Berlin, F. R. G. 1976.
218. Bakoupoulos, A., *Ger. Chem. Eng., 3*:241 (1980).
219. Sandall, O. C., *Int. J. Heat Mass Transfer, 17*:459 (1974).
220. Subramanian, P. S., *Int. J. Heat Mass Transfer, 18*:334 (1975).
221. Gottifredi, J. C., and Quiroga, O. D., *Chem. Eng. J., 16*:199 (1978).
222. Yih, S. M., and Seagrave, R. C., *Chem. Eng. Sci., 33*:1581 (1978).
223. Jepsen, J. C., Crosser, O. K., and Perry, R. H., *AIChE J., 12*:186 (1966).
224. Hiby, J. W., *Warme-und Stoffubertragung, 1*:105 (1968).
225. Hiby, J. W., Braun, D., and Eickel, K. H., *Chemie-Ing-Techn., 39*:297 (1967).
226. Petermann, J., Broecker, H. C., and Sinn, H., *Ger. Chem. Eng., 1*:312 (1978).
227. Hiby, J. W., *Chemie-Ing-Techn., 45*:1103 (1973).
228. Chung, D. K., and Mills, A. F., *Letters in Heat and Mass Transfer, 1*:43 (1974).
229. Danckwerts, P. V., *Gas-Liquid Reactions,* McGraw Hill, N.Y., 1970.
230. Astarita, G., *Mass Transfer with Chemical Reaction,* Elsevier, Amsterdam, 1967.
231. Kohl, A., and Riesenfeld, F., *Gas Purification,* 3rd ed., Gulf Publishing, Texas 1979.
232. Astarita, G., Savage, D. W., and Bisio, A., *Gas Treating with Chemical Solvents,* John Wiley, N.Y., 1983.
233. Stepanek, J. B., and Achwal, S. K., *Can. J. Chem. Eng., 54*:545 (1976).
234. Best, R. J., and Horner, B., *Chem. Eng. Sci., 34*:759 (1979).

235. Pedersen, H., and Prenosil, J. E., *Int. J. Heat Mass Transfer, 24*:299 (1981).
236. Ratcliff, G. A., and Holdcroft, J. G., *Chem. Eng. Sci., 15*:100 (1961).
237. Astarita, G., *Chem. Eng. Sci., 16*:708 (1962).
238. Roberts, D., and Danckwerts, P. V., *Chem. Eng. Sci., 17*:961 (1962).
239. Hikita, H., Asai, S., and Takatsuka, T., *Chem. Eng. J., 11*:131 (1976).
240. Savage, D., Astarita, G., and Joshi, S., *Chem. Eng. Sci., 35*:1513 (1980).
241. Kameoka, Y., and Pigford, R. L., *Ind. Eng. Chem. Fundam., 16*:163 (1977).
242. Wendel, M. M., and Pigford, R. L., *AIChE J., 4*:249 (1958).
243. Dekker, W. A., Snoeck, E., and Kramers, H., *Chem. Eng. Sci., 11*:61 (1959).
244. Weisweiler, W., and Deib, K. H., *Ger. Chem. Eng., 4*:79 (1981).
245. Lefers, J. B., and van den Berg, P. J., *Chem. Eng. J., 23*:211 (1982).
246. Astarita, G., and Gioia, F., *Ind. Eng. Chem. Fundam., 4*:317 (1965).
247. Brian, P. L. T., Hurley, J. F., and Hasseltine, E. H., *AIChE J., 7*:226 (1961).
248. Hikita, H., and Asai, S., *Int. Chem. Eng., 4*:332 (1964).
249. Sada, E., Kumazawa, H., and Butt, M. A., *AIChE J., 22*:196 (1976).
250. Nijsing, R. A. T. O., Hendriksz, R. H., and Kramers, H., *Chem. Eng. Sci., 10*:88 (1959).
251. Hikita, H., and Asai, S., *Kagaku Kogaku, 28*:1017 (1964).
252. Hikita, H., Asai, S., and Himukashi, Y., *Kagaku Kogaku, 31*:818 (1967).
253. Smirnov, I. N., and Aerov, M. F., *Theo. Found. Chem. Eng., 10*:751 (1976).
254. Gilliland, E. R., Baddour, R. F., and Brian, P. L. T., *AIChE J., 4*:223 (1958).
255. Kojima, H., Inazumi, H., and Imanara, T., *Kagaku Kogaku, 38*:648 (1974).
256. Teramoto, M., Hashimoto, K., and Nagata, S., *Kagaku Kogaku, 34*:1296 (1970).
257. Sada, E., Kumazawa, H., and Butt, M. A., *Can. J. Chem. Eng., 54*:421 (1976).
258. Emmert, R. E., and Pigford, R. L., *AIChE J. 8*:171 (1962).
259. Astarita, G., Marrucci, G., and Gioia, F., *Chem. Eng. Sci., 19*:95 (1964).
260. Hikita, H., et al., *AIChE J., 25*:793 (1979).
261. Alvarez-Fuster, C., et al., *Chem. Eng. Sci., 35*:1717 (1980).
262. Alvarez-Fuster, C., et al., *Chem. Eng. Sci., 36*:1513 (1981).
263. Astarita, G., and Gioia, F., *Chem. Eng. Sci., 19*:963 (1964).
264. Onda, K., et al., *J. Chem. Eng. Japan, 5*:27 (1972).
265. Sherwood, T. K., and Pigford, R. L., *Absorption and Extraction,* 2nd ed., McGraw Hill, N.Y. 1952, p. 332.
266. Imanara, T., Inazumi, H., and Haruyama, Y., *Kagaku Kogaku Ronbunshu, 3*:277 (1977).
267. de Waal, K. J. A., and Okeson, J. C., *Chem. Eng. Sci., 21*:559 (1966).
268. Onda, K., et al., *Kagaku Kogaku, 34*:187 (1970).
269. Wesselingh, J. A., and van't Hoog, A. C., *Trans. Instn. Chem. Engrs., 48*:69 (1970).
270. Reith, T., and Beek, W. J., *Chem. Eng. Sci., 28*:1331 (1973).
271. Onken, U., and Schalk, W., *Ger. Chem. Eng., 1*:191 (1978).
272. Linek, V., and Vacek, V., *Chem. Eng. Sci., 36*:1747 (1981).
273. Kojima, H., and Inazumi, H., *Kagaku Kogaku, 38*:157 (1974).
274. Kayihan, F., and Sandall, O. C., *AIChE J., 20*:402 (1974).
275. Menez, G. D., and Sandall, O. C., *Ind. Eng. Chem. Fundam., 13*:72 (1974).
276. Stepanek, J. B., and Achwal, S. K., *Can. J. Chem. Eng., 53*:517 (1975).
277. Gottifredi, J. C., and Quiroga, O. D., *Int. J. Heat Mass Transfer, 22*:839 (1979).
278. Mendez, F., and Sandall, O. C., *AIChE J., 21*:534 (1975).
279. Stepanek, J. B., and Achwal, S. K., *Chem. Eng. J., 10*:49 (1975).
280. Haimour, N., and Sandall, O. C., *AIChE J., 29*:277 (1983).
281. Coull, J., Bishop, C. A., and Gaylord, W. M., *Chem. Eng. Progr., 45*:525 (1949).
282. Dobratz, C. J., et al., *Chem. Eng. Progr., 49*:611 (1953).
283. Gaylord, W. M., and Miranda, M. A., *Chem. Eng. Progr., 53*:139 (1957).
284. Pagani, G., and Zardi, U., *Hydrocarbon Processing, 51*:106 (1972).
285. Matsuda, A. et al., *Kagaku Kogaku Ronbunshu, 6*:157 (1980).

286. vander Mey, J. E., *U.S. Patent No. 3,328,460,* June 27, 1967.
287. Marquis, D. M., *Hydrocarbon Processing, 47*:109 (1968).
288. Guerreri, G., in *Encyclopedia of Chemical Processing and Design,* J. J. Mcketta (Ed.), Vol. 1, 1976, p. 88.
289. Yih, S. M., and Seagrave, R. C., *Int. J. Heat Mass Transfer, 23*:749 (1980).
290. Grossman, G., *Int. J. Heat Mass Transfer, 26*:357 (1983).
291. Inazumi, et al., *Kagaku Kogaku Ronbunshu, 7*:595 (1981).
292. Nakoryakov, V. Ye., and Grigor'yeva, N. I., *Heat Transfer-Soviet Research, 12*:111 (1980).
293. Sandall, O. C., *Can. J. Chem. Eng., 53*:702 (1975).
294. Guerreri, G., and King, C. J., *Hydrocarbon Processing, 53*:131 (1974).
295. Johnson, G. R., and Crynes, B. L., *Ind. Eng. Chem. Process. Des. Develop., 13*:6 (1974).
296. Grigull, U., *Forsch. Ing. Wes., 13*:49 (1942).
297. Davis, E. J., van Ouwerkerk, M., and Venkatesh, S., *Chem. Eng. Sci., 34*:539 (1979).
298. Aihara, K., et al., *Int. Chem. Eng., 16*:494 (1976).
299. Chandra, V., and Savery, C. W., *AIChE 17th National Heat Transfer Conference,* S. L. C., Utah, Aug. 14–17, 1977, p. 15.
300. Hikita, H., and Ishimi, K., *J. Chem. Eng. Japan, 9*:362 (1976).
301. Hikita, H., Ishimi, K., and Shoda, N., *J. Chem. Eng. Japan, 12*:68 (1979).
302. Hikita, H., and Ishimi, K., *Chem. Eng. Commun, 2*:181 (1978).
303. Hikita, H., et al., *J. Chem. Eng. Japan, 11*:96 (1978).
304. Kafesjian, R., Plank, C. A., and Gerhard, E. R., *AIChE J., 7*:463 (1961).
305. Gilliland, E. R., and Sherwood, T. K., *Ind. Eng. Chem., 26*:516 (1934).
306. Jackson, M. L., and Ceaglske, N. H., *Ind. Eng. Chem. 42*:1188 (1950).
307. Sherwood, T. K., Pigford, R. L., and Wilke, C. R., *Mass Transfer,* McGraw Hill, 1975, p. 213.
308. Kast, W., *Chem. Techn., 18*:152 (1966).
309. Braun, D., and Hiby, J. W., *Chemie-Ing.-Techn., 42*:345 (1970).
310. Avdonin, Yu. A., Olevskii, V. M., and Popov, D. M., *Int. Chem. Eng., 7*:258 (1967).
311. Hikita, H., and Ishimi, K., *Chem. Eng. Commun., 3*:547 (1979).
312. Hikita, H., et al., *Can. J. Chem. Eng., 57*:578 (1979).
313. Satterfield, C. N., Pelossof, A. A., and Sherwood, T. K., *AIChE J., 15*:226 (1969).
314. Atkinson, B., Daoud, I. S., and Williams, D. A., *Trans. Instn. Chem. Engrs., 46*:245 (1968).
315. Mashelkar, R. A., and Chavan, V. V., *J. Chem Eng. Japan, 6*:160 (1973).
316. Yih, S. M., and Huang, P. G., *J. Chin. Inst. Chem. Engrs., 11*:71 (1980).
317. Hirose, T., Mori, Y., and Sato, Y., *J. Chem. Eng. Japan, 7*:19 (1974).
318. Beschkov, V., Boyadjiev, C., and Peev, G., *Chem. Eng. Sci., 33*:65 (1979).
319. Stirba, C., and Hurt, D. M., *AIChE J., 1*:178 (1955).
320. Oliver, D. R., and Atherinos, T. E., *Chem. Eng. Sci., 23*:525 (1968).
321. Hikita, H., Nakanishi, K., and Asai, S., *Kagaku Kogaku, 23*:28 (1959).
322. Tanaka, H., Tago, O., and Sasakura, T., *Kagaku Kogaku, 37*:88 (1973).
323. Nigam, K. K., et al., *Indian J. Techn., 16*:45 (1978).
324. Grassmann, P. P., *Int. J. Heat Mass Transfer, 22*:795 (1979).
325. Wragg, A. A., and Einarsson, A., *Chem. Eng. Sci., 25*:67 (1970).
326. Einarsson, A., and Wragg, A. A., *Chem. Eng. Sci., 26*:1289 (1971).
327. Ito, R., et al., *J. Chem. Eng. Japan, 12*:483 (1979).
328. Meyer, K., Ph. D. Thesis, ETH Zurich, Switzerland, 1977.
329. Brauner, N., and Moalem Maron, D., *Int. J. Heat Mass Transfer, 25*:99 (1982).
330. Stainthorp, F. P., and Batt, R. S. W., *Trans. Instn. Chem. Engrs., 45*:372 (1967).
331. Stainthorp, F. P., and Wild, G. J., *Chem. Eng. Sci., 22*:707 (1967).
332. Iribarne, A., Gosman, A. D., and Spalding, D. B., *Int. J. Heat Mass Transfer, 10*:1661 (1967).

333. Wragg, A. A., Serafimidis, P., and Einarsson, A., *Int. J. Heat Mass Transfer, 11*:1287 (1968).
334. Kramers, H., and Kreyger, P. J., *Chem. Eng. Sci., 6*:42 (1956).
335. Yih, S. M., *Proc. 2nd World Congress Chem. Eng.*, Montreal, Canada, *5*:168 (1981).
336. Meyer, K., *Warme-und-Stoffubertragung, 12*:121 (1979).
337. Yih, S. M., in *Advances in Transport Processes*, (A. S. Mujumdar and R. A. Mashelkar (Eds.), Vol. 5, Wiley Eastern, 1985.

CHAPTER 6

HEAT AND MASS TRANSFER IN FILM ABSORPTION

Gershon Grossman

Faculty of Mechanical Engineering
Technion-Israel Institute of Technology

CONTENTS

INTRODUCTION

Absorption of gases and vapors in liquid films is encountered in numerous applications in the chemical technology. Some broad areas of application include the formation of gas-liquid solutions in various reactors, the separation of gases from liquids in distillation processes, the removal of impurities from streams of gases, and the creation of heating and cooling effects in absorption heat pumps. Of the different types of gas-liquid contactors which may be employed in some of these processes, liquid film absorbers are of particular interest. In contrast with sprays or packed towers, where gas-liquid interaction takes place somewhat in isolation, the film absorber makes it possible to transfer heat into or out of the fluids during the process through the wetted solid surface supporting the film. This feature, along with the more regular, controllable liquid flow and configuration, gives the designer of film absorbers extra flexibility in achieving the desired fluid interaction.

Film absorption normally involves simultaneous heat and mass transfer in the gas-liquid system. The heat of absorption gives rise to temperature gradients leading to the transfer of heat; the temperature influences the vapor pressure-composition equilibrium between the two phases which in turn affects the exchange of mass. In some cases, such as those involving sparingly soluble gases, the heat interaction is small and the process may be considered isothermal. In others, where the absorbate vapor has a large heat of absorption, heat effects associated with the transfer of mass are significant. Further complication arises when the absorption process is accompanied by chemical reaction between the components.

A typical film absorption system is illustrated in Figure 1. A film of liquid solution flows over a solid surface of given geometry. In a practical design, the surface may be that of a horizontal tube, vertical tube, or inclined tray. The solution consists of at least two components—the absorbent and the absorbate. Other substances not affecting the absorption may be contained in the liquid. The film is in contact with a gas containing at least the absorbate component, and often also some of the absorbent, when the latter has a certain volatility under the given conditions. The gas may contain other "inert" components not participating in the absorption process. As an example one may consider a hygroscopic lithium bromide-water solution absorbing moisture from an air stream. Here the nonvolatile absorbent (LiBr) exists only in the liquid phase, the absorbate (H_2O) is present both in the liquid and gas phases and the air is an inert gas not taking part in the process.

The temperature and composition of the liquid and gas phases determine the vapor pressures of the absorbate in them. When the vapor pressure in the liquid is lower than that in the gas, an absorption process takes place at the interface. The nonuniform composition of the absorbate thus created causes its diffusion from the interface into the film and from the bulk of the gas phase toward the interface. The heat generated in the absorption gives rise to temperature gradients which result in heat transfer. Typical temperature and absorbate concentration profiles in the liquid and gas phases are depicted in Figure 1.

Several factors influence the absorption process, and determine not only the rate but also the mechanism for the heat and mass transfer. Physical properties of the fluids, particularly thermal and mass (molecular) diffusivities, control the transfer at the gas-liquid interface. The flow regime—laminar, wavy, or turbulent—affects the transfer rate toward and away from the interface. The thermodynamic equilibrium properties of the participating substances, particularly their pressure-temperature-composition relation and the heat of absorption determine the relative influence of heat and mass transfer. Local instabilities and secondary flows at the interface due to surface tension gradients, surface impurities, buoyant forces, and the like can enhance the transfer considerably. Other enhancement or retardation effects are due to chemical reactions and surface resistance. Finally, the heat transfer at the

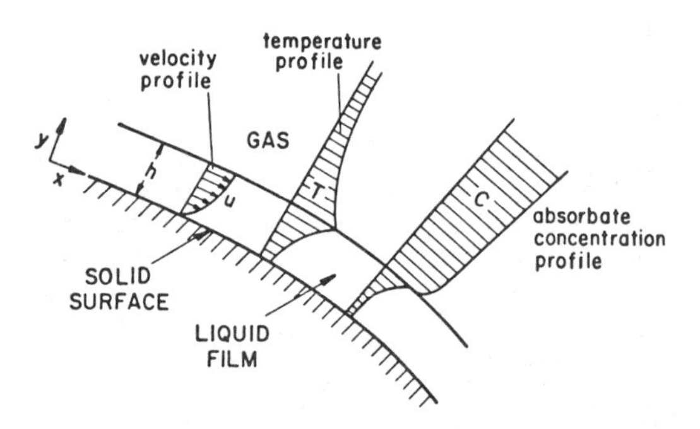

Figure 1. Film absorption system with typical velocity, temperature, and concentration profiles.

solid wall controls the film temperature, which in turn affects the vapor pressure driving force. The overall process is a combined result of these factors and their relative importance varies from one case to another.

The following section discusses the film hydrodynamics. Heat and mass transfer processes have been classified according to their flow regime. Stagnant films are discussed first, followed by films in laminar flow, in wavy flow, and finally in transition and turbulent flow. The last section contains the fundamental transfer equations governing the film absorption process. There, the interested reader can find the theoretical basis for the analytical results given in the rest of the chapter.

FILM HYDRODYNAMICS

The driving force creating the film flow is, in most cases, gravity. This is the reason for the frequent use of the term "falling film" in the film absorption literature. The gravitational force is opposed by the viscous shear force between the film and the solid surface supporting it. When the gas in contact with the film is also in motion, an additional shear force is exerted on the film at the free surface.

If the gas surrounding the film is stagnant and the liquid flow velocity is small, the free surface of the film is observed to be smooth. Increasing the liquid velocity leads to disturbances manifested by waves on the surface, which increase in frequency and amplitude and become more irregular at larger velocities. Further increase in flow makes the film turbulent. Disturbances and instabilities enhancing the turbulence at these high liquid flow rates (with stagnant or slow moving gas) originate mainly at the solid wall, as is the case in turbulent pipe or channel flow. In the opposite extreme of low liquid velocity and high gas velocity, disturbances due to shear originate at the interface. They are in the form of small ripples often appearing in local patches which grow in size with increasing gas velocity until the interface is disturbed completely. Liquid droplets may be detached from the crests of the waves and carried away with the gas flow. If the gas velocity is very large, the liquid film might break up completely.

Chien and Ibele [1] distinguish between two kinds of stabilities governing the liquid film flow. The first relates to the stability of flow within the liquid film proper, and is called "flow stability." The second is related to disturbances on the free surface and is called "interface stability." Flow stability is similar to that involved in pipe flow, while interface stability is peculiar to the film with a free surface. Flow stability is quite well characterized in terms of the liquid flow Reynolds number, while interface stability in the presence of gas flow is influenced by different factors, including geometry, surface properties, gas shear, and the like, and is not as well understood. While the two forms of stability are related to each other, in many cases of practical interest in film absorption the gas flow is small and the flow stability is the dominant factor in determining the film hydraulics.

The film Reynolds number is defined as

$$\mathrm{Re} = \frac{4\Gamma}{\mu} = \frac{4\bar{u}\bar{h}}{\nu} \tag{1}$$

where Γ = mass flow rate per unit breadth
$\bar{u}$ = average flow velocity
$\bar{h}$ = average film thickness
ν = kinematic viscosity
μ = dynamic viscosity.

(The factor 4 in the definition is due to the choice of the hydraulic diameter as the characteristic length; other definitions based on the film thickness and omitting that factor are sometimes found in the literature). The effect of gravity is expressed in terms of the

Froude number, and that of surface tension in terms of the Weber number. The Reynolds number in a falling film is generally related to the former, and in some cases also to the latter.

Experiments by many researchers have aimed at finding the Reynolds number ranges for the different flow regimes. In an excellent review on the flow of liquids in thin films, Fulford [2] gave a comprehensive account of these experiments in chronological order, up to 1964. In examining the results of some of the more extensive ones [3–14], and despite some differences in their specification of the exact limiting Reynolds number values, there seems to exist an agreement on the following hydrodynamic behavior:

1. Laminar flow with a smooth surface exists at $Re < 20$.
2. Flow with surface waves of a partially laminar, partially turbulent nature exists at $20 < Re < 1{,}600$, the transition to turbulence and the surface disturbances becoming more significant with increasing Reynolds number.
3. At $Re > 1{,}600$ the flow is fully turbulent.

Fulford [2] points out that contrary to pipe or channel flow, in thin films a large part of the total film thickness is occupied by the relatively nonturbulent "laminar sublayer" even at very large Reynolds numbers. Hence, the transition from laminar to turbulent flow cannot be expected to be as sharply marked as in the case of pipe flow. The intermediate range of wavy flows may therefore be further subdivided according to the first appearance of a certain type of wave (gravity/capillary), the first onset of turbulence, and similar other criteria.

Further insight into the behavior of falling films may be obtained from analytical solutions of the Navier–Stokes equations governing the flow. Levich [15] has formulated a reduced form of the momentum and continuity equations for thin films. These equations are valid for the laminar and wavy-laminar regimes and are described in the section on fundamental transfer equations later in this chapter. The following is a discussion of some analytical results obtained along with experimental verification for the different flow regimes.

Laminar Flow

For a laminar film flowing at a sufficiently low Reynolds number over a vertical or inclined flat plate, with constant liquid properties and no interfacial shear, the film surface is smooth and its thickness is constant. Nusselt [16] in 1916 treated this flow problem analytically and found the following expression for the film thickness:

$$h = \left(\frac{3\Gamma\mu}{\varrho^2 g \sin\varphi}\right)^{1/3} = 0.91\left(\frac{\nu^2}{g \sin\varphi}\right)^{1/3} Re^{1/3} \tag{2}$$

where φ is the angle of inclination of the plate, measured from the horizontal. The velocity profile in the film is parabolic and the velocity at the surface is 1.5 times the average one. Nusselt's results have been confirmed by numerous experiments [3–5, 7, 8, 10, 11]. It is interesting to note that while the surface velocity changes at $Re > 20$ with the appearance of waves, the thickness Equation 2 has been observed to remain valid up to $Re \cong 1{,}500$, with some researchers setting the limit even higher.

Surface geometries other than a flat plate have been treated analytically for laminar flow. Chien [17] solved the Navier–Stokes equations for the flow of a falling film on the inner and outer surface of a vertical, circular cylinder. When the film thickness is small compared to the radius of the curved surface, the results are well approximated by Nusselt's Equation 2. Stepanek and Colquhoun-Lee [18] solved for the flow of a laminar film on a sphere. Their results show the parabolic velocity profile to represent the flow very closely. Laminar falling films on the outer surface of horizontal tubes were treated by Moalem and Sideman [19].

At the limit of very thin films, the continuous flow may break up into thin streams or separate drops, due to the action of capillary forces. The disruption depends on wetting conditions [15].

Wavy Flow

In the low Reynolds number range of the wavy flow regime ($20 < Re < 200$) the flow is still essentially laminar. The parabolic velocity profile remains in effect with slight variations near the interface. Also, the average film thickness is properly described by Nusselt's Equation 2.

A number of stability models have been published concerning the conditions leading to the first appearance of waves. The general method has been to impose small perturbations on the Navier–Stokes equations and to solve for the conditions under which stability exists. Among those studies one may note the work of Kapitsa [20], Benjamin [21, 22], Binnie [11], and Hanratty and Hershman [23]. The details of these theories are lengthy and cannot be discussed here. The results are generally in agreement with the experimental finding of $Re \cong 20$ for the onset of wavy flow.

Waves in the wavy-laminar range are generally of capillary nature, where surface tension forces play a principal role. These forces appear when the film surface is deformed as a result of a disturbance, and they become important when their magnitude is comparable to that of the other active forces—gravity and viscosity. Levich [15] presents a theoretical analysis of the flow in this regime, following earlier studies by Kapitsa [20]. By assuming the wave amplitude to be much smaller than the film thickness h, the latter is shown to vary like a sinusoidal wave:

$$h = \bar{h}\left[1 + a \sin \frac{2\pi}{\lambda'}(x - ct)\right] \tag{3}$$

where λ' is the wavelength:

$$\lambda' = \frac{2\pi}{\bar{u}}\left(\frac{\sigma\bar{h}}{4.2\varrho}\right)^{1/2} = \frac{3.066}{\Gamma}(\varrho\sigma\bar{h}^3)^{1/2} \tag{4}$$

and c is the wave propagation velocity

$$c = 3\bar{u} \tag{5}$$

These results were verified experimentally by P. L. Kapitsa and S. P. Kapitsa [24] by means of optical techniques applied to films of alcohol and water. They are subject to the limitation of small amplitude ($a \ll 1$) and also require that the wavelength be more than 13.7 times larger than the film thickness.

More elaborate analyses of the flow in wavy regime have followed the work of Kapitsa [20] and Levich [15], which specified the velocity distribution and the wave parameters more accurately in terms of flow conditions. Some of these analyses are described further in the section "Absorption in the Wavy Flow Regime" in relation to the mass transfer problem in wavy flow. A detailed discussion of this work is outside the scope of the present survey.

Other theoretical analyses are available in the literature for larger disturbances and thicker films, which are essentially wavy flows in open channels. Under these conditions, surface tension plays a less important role and waves are mainly of the gravity type. As the uniform wave pattern becomes more irregular, it is no longer possible to obtain simple and reliable results by analytical methods. Most of the available information on the wavy-transition flow regime (Re > 200) is of empirical or semi-empirical nature. The approach taken by investigators has been to ignore the details of surface disturbances and treat the film as smooth, with representative time averages for the thickness and velocity profile.

Transition and Turbulent Flow

Dukler and Bergelin [5] used the universal velocity profile equations of Nikuradse [25] to relate the average film thickness to the Reynolds number. Nikuradse's equations express the dimensionless velocity in terms of the dimensionless distance from the wall for the laminar sublayer, the buffer layer, and the turbulent core. Both the velocity and the distance are normalized with respect to the liquid properties and to the wall shear, which in the case of a falling film is known in terms of the film thickness and gravity. By integrating the dimesionless velocity over the film thickness, Dukler and Bergelin [5] were able to find an expression for the film thickness for the different flow regimes which was in good agreement with Nusselt's Equation 2 for laminar flow. Later, Dukler [26] performed an improved analysis of this type and obtained velocity profiles in the film as well as a flow thickness —Reynolds number curve. Plotting the film thickness itself against the Reynolds number is not very convenient, since it yields a different curve for every liquid. Dukler [26] therefore used a dimensionless film thickness which was termed the Nusselt film thickness parameter N_T:

$$N_T = h\left(\frac{g \sin \varphi}{\nu^2}\right)^{1/3} \tag{6}$$

Nusselt's Equation 2 for the laminar regime yields, in terms of this parameter:

$$N_T = 0.909\ Re^{1/3} \tag{7}$$

whereas Kapitsa's theory [20] for the wavy laminar regime gives:

$$N_T = 0.843\ Re^{1/3} \tag{8}$$

Other researchers have proposed empirical and semi-empirical correlations of the film thickness in terms of the Reynolds number for the wavy-transition and turbulent flow regimes. Some of the important ones are:

The Brötz Equation:

$$N_T = 0.0682\ Re^{2/3} \tag{9}$$

The Zhivaykin and Volgin Equation:

$$N_T = 0.141\ Re^{7/12} \tag{10}$$

The Brauer Equation:

$$N_T = 0.2077\ Re^{8/15} \tag{11}$$

The Feind Equation:

$$N_T = 0.266\ Re^{1/2} \tag{12}$$

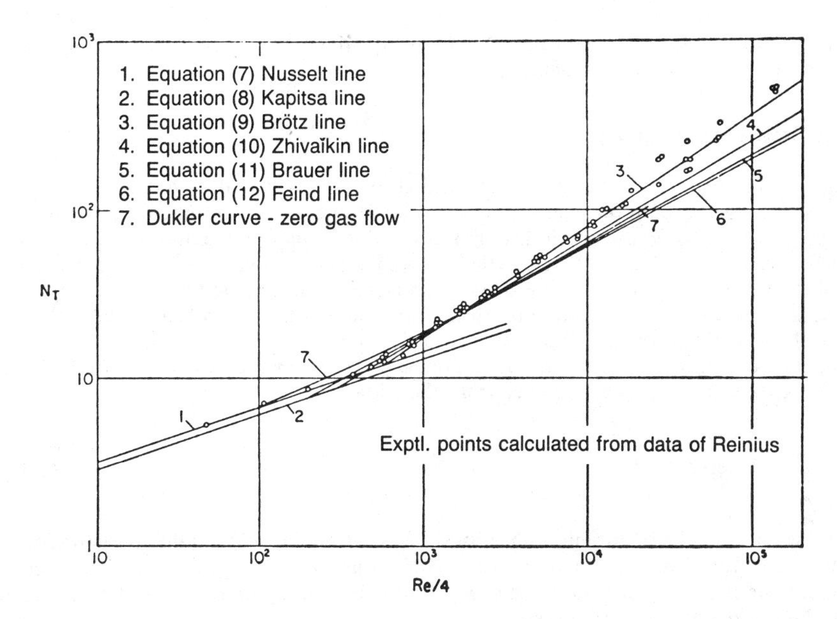

Figure 2. Comparison of various correlations for the film thickness as a function of the Reynolds number (from Fulford [2]).

Figure 2 shows Fulford's [2] plot of Equations 7 through 12, and of the curve given by Dukler's theory [26]. The experimental film thickness data obtained by Reinius [13] is plotted along for comparison and it is clear that the Brötz relation, Equation 9, is in excellent agreement with the experimental values over a wide range of Reynolds numbers. Fulford [2] points out that at the smaller values of Re normally encountered in falling films there is little difference between the predictions of the different relationships; the Dukler curve, which avoids a sharp transition between the laminar and turbulent regimes, lies close to many of the data in the transition and lower turbulent zones.

Another point concerning the transition and turbulent flow regime is the critical Reynolds number at which turbulent flow begins. The critical value has been determined by different researchers from discontinuities and changes in slope which appear in curves of film thickness, surface velocity of the film, heat and mass transfer coefficients, and the like, when plotted against Re. Fulford [2] has tabulated some of the critical values proposed by various investigators and suggested a value of Re between 1,000 to 1,600 for the onset of turbulence. Recently the work of Yih and Chen [14] who compiled over 800 mass transfer data points from the work of ten authors in addition to their own, seems to put the turbulence onset point quite close to Re = 1,600. As pointed out earlier, the transition to turbulence in a thin film is likely to be a gradual process, more so than in pipe or channel flow.

ABSORPTION IN STAGNANT FILMS

Before considering the heat and mass transfer in a moving film, it is of interest to examine the absorption process in a stagnant one. The transfer process is governed by the energy and diffusion equations. As in the discussions of film hydrodynamics, only the results of analytical solution of these equations will be given here, and the interested reader is referred to the section entitled "Fundamental Transfer Equations" for further details on the equations and their boundary conditions.

Process Description and Analysis

Figure 3 illustrates a stagnant film with typical absorbate concentration profiles. Initially, the entire film is at a uniform temperature, T_0, and absorbate concentration, C_0. At time $t = 0$ contact with the gas is established and absorption begins. In the absence of natural convection and other localized flows, the absorbate penetration rate into the film is controlled by molecular diffusion only. This process lends itself easily to theoretical analysis. For the simplest case of isothermal absorption, the solution to the problem is equivalent to that of the well-known heat conduction in a semi-infinite wall. It is referred to in the literature as the Penetration Theory, due to Higbie [29]. The absorbate concentration C in the film varies with time t and distance from the surface y as:

$$\frac{C - C_0}{C_e - C_0} = 1 - \mathrm{erf}\left(\frac{y^2}{4Dt}\right)^{1/2} \tag{13}$$

where D is the diffusion coefficient of the absorbate in the liquid and C_e is the concentration at saturation, determined by vapor pressure equilibrium with the gas at the given temperature. Note that at the interface, the concentration is C_e at all times. The instantaneous liquid film coefficient for mass transfer is:

$$K_L = \frac{-D(\partial C/\partial y)_{y=0}}{(C_e - C_0)} = \left(\frac{D}{\pi t}\right)^{1/2} \tag{14}$$

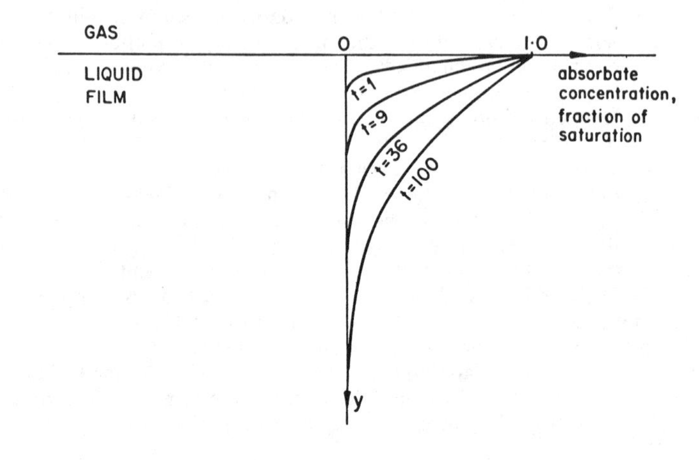

Figure 3. Absorbate concentration profiles at different relative times during absorption in a stagnant film.

and the one averaged over time t is:

$$\bar{K}_L = 2\left(\frac{D}{\pi t}\right)^{1/2} \tag{15}$$

In the case of an absorbate with a significantly large heat of absorption, a combined heat and mass transfer process takes place where the temperature and concentration at the surface are interrelated by thermodynamic equilibrium. The problem was treated analytically by Nakoryakov and Grigoreva [30], under the assumption of a constant heat of absorption and a linear temperature-concentration relation in equilibrium. Their solution was done for laminar slug flow and may be adapted to the case of a stagnant film by replacing $x/\bar{u}$ in their results by t. The temperature T and absorbate concentration C in the film vary with time and distance from the surface as follows:

$$\frac{T-T_0}{T_e-T_0} = \theta = \left(\frac{\lambda}{\lambda+Le^{1/2}}\right)\left[1-\text{erf}\left(\frac{y^2}{4\alpha t}\right)^{1/2}\right] \tag{16}$$

$$\frac{C-C_0}{C_e-C_0} = \gamma = \left(\frac{Le^{1/2}}{\lambda+Le^{1/2}}\right)\left[1-\text{erf}\left(\frac{y^2}{4Dt}\right)^{1/2}\right] \tag{17}$$

where θ and γ are normalized temperature and concentration; α and D are the molecular thermal and mass diffusivities, respectively; Le is the Lewis number, expressing the ratio D/α; and λ is the dimensionless heat of absorption, defined through the physical heat of absorption $\bar{H}a$ as:

$$\lambda = Le\,\frac{\overline{Ha}(C_e-C_0)}{\varrho c_p(T_e-T_0)} \tag{18}$$

T_e is the equilibrium temperature of the liquid at concentration C_0 with the gas, and C_e is the concentration of the solution at temperature T_0 in equilibrium with the gas. T_e and C_e have the following physical significance: T_e is the temperature which could be reached in the film if thermodynamic equilibrium could be achieved without change in concentration; C_e is the concentration that could be reached if thermodynamic equilibrium could be achieved without change in temperature. Both are limiting cases to what actually happens in the simultaneous heat and mass transfer process. The heat and mass transfer film coefficients are obtained from Equations 16 and 17 by calculating the fluxes at the surface and dividing by the temperature and concentration differences, respectively:

$$H_L = \frac{-k(\partial T/\partial y)_{y=0}}{[T(y=0)-T_0]} = \varrho c_p\sqrt{\frac{\alpha}{\pi t}} \tag{19}$$

$$K_L = -\frac{-D(\partial C/\partial y)_{y=0}}{[C(y=0)-C_0]} = \left(\frac{D}{\pi t}\right)^{1/2} \tag{20}$$

Note that the mass transfer coefficient is the same as for isothermal mass transfer; the concentration profiles are of the same shape, with the exception that the concentration at the surface is not C_e but a lower value depending on the relative magnitudes of λ and Le. In the limiting case of negligible heat of absorption $\lambda \rightarrow 0$, Equation 16 yields $T = T_0$ throughout the film and Equation 17 reduces to Equation 13.

In many real cases of absorption in stagnant films, penetration rates have been observed which are significantly different from the analytical prediction of Equations 14 and 20. This is due to surface convection, a phenomenon associated with instabilities near the surface, which generates local liquid motion. A brief survey of these effects is given next.

Surface Convection Phenomena

Even as early as 1855, James Thompson had reported on convection flows at the surface of stagnant pools of liquids, which he described as "certain curious motions observable at the surfaces of wine and other alcoholic liquors." He attributed these flows to local variations of surface tension associated with the evaporation of the liquid. In 1882, Thompson described similar flows being due to buoyancy forces [31]. Both the buoyancy and surface tension mechanisms are known today as being responsible for local convection at the surface of films, a phenomenon sometime referred to in the literature as "interface turbulence."

In 1900, Bénard [32] discovered a cellular pattern which formed at the free surface of thin horizontal layers of liquids heated from below, and a circulating flow associated with this pattern (Figure 4). In his experiments, Bénard observed that the motion occurred only when a critical temperature gradient across the film thickness was exceeded, and that the stable flow pattern was one of hexagonal cells. Bénard's observations motivated Lord Rayleigh [33] to perform a linear hydrodynamic stability analysis, in which he attributed the motion to the buoyancy resulting from thermal expansion of the heated liquid. A minimum film depth was found to be required for the motion to occur. Rayleigh's analysis was extended and improved later by Jeffreys [34], Low [35], and Pellew and Southwell [36]. The instability wave number, indicating the size of the cells was calculated. Both the maximum depth and the minimum temperature gradient requirements have been expressed in terms of a critical value for a single dimensionless number known as the Rayleigh number:

$$Ra = \frac{g\beta\Delta T h^3}{\nu\alpha} \tag{21}$$

where β is the thermal expansion coefficient of the liquid, and ΔT is the temperature difference between the topside and underside of the film. $\beta\Delta T$ in Equation 21 may be replaced by $\Delta\varrho/\varrho$ (where $\Delta\varrho$ is the density span in the film) to provide a more general expression for the Rayleigh number, covering cases in which the density variation inducing buoyancy is not due to temperature gradients only. Concentration gradients, for example, could be responsible for density variations, in which case the thermal diffusivity α in Equation 21 should be replaced by the mass diffusivity D.

Later linear stability studies by Pearson [37] and Nield [38] showed the buoyancy model to be inadequate for explaining Benard's results. Although buoyancy does provide a driving force for the flow, it became clear that the cells in Benard's experiments were induced primarily by surface tension gradients resulting from temperature variations across the free surface. The critical Rayleigh number at the onset of flow due to buoyancy is replaced by a critical value of another dimensionless quantity, the Marangoni number, when the flow is caused by surface tension gradients. The Marangoni number is defined as:

$$Ma = \frac{\Delta\sigma h}{\varrho\nu\alpha} \tag{22}$$

where $\Delta\sigma$ is the surface tension span in the film, the change in σ being caused by temperature variations or, in the more general case, by concentration differences, surfactants, etc. Increasing temperature normally lowers surface tension.

In an attempt to investigate the coupling between surface tension and buoyancy effects on surface convection, Nield [38] took both mechanisms into consideration in his stability

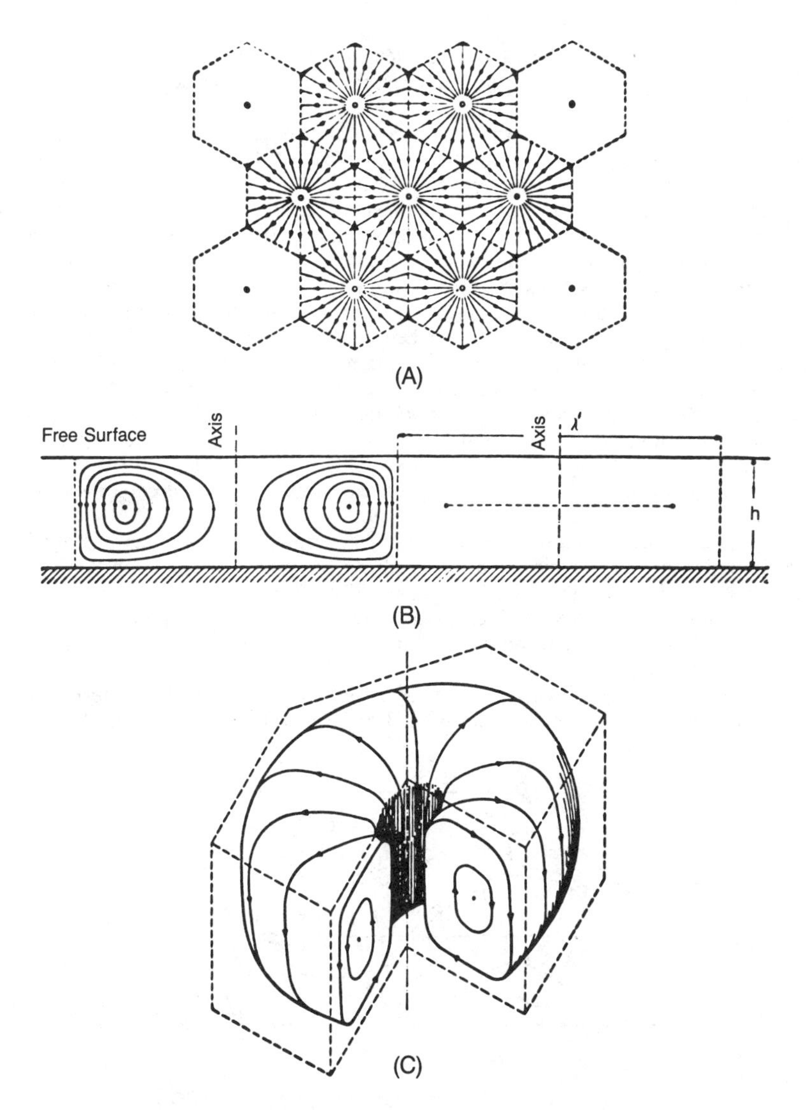

Figure 4. Surface convection of hexagonal cellular pattern as observed by Benard: (A) Top view; (B) side view; (C) perspective view (after Berg, Boudart, and Acrivos [31]).

analysis. He found that the driving force for the motion at the onset of convection was approximately equal to the sum of the surface tension and buoyancy forces in the layer. In thin films such as in Benard's experiments surface tension is dominant; as the depth increases buoyancy plays a more important role. Nield calculated some critical values of the Rayleigh and Marangoni numbers for different heating rates, as well as the size of the cells, and obtained good agreement of his theoretical results with experiments. Further extension of the linear stability analysis was given by Scriven and Sternling [39] who in contrast with the earlier studies allowed for the possibility of shape deformation of the free surface and obtained a criterion for distinguishing visually the dominant force in surface convection: If the motion is driven by surface tension, there is upflow beneath depressions and downflow beneath elevations of the free surface; the converse is true in buoyancy-driven flows.

In order to explore the flow patterns associated with the different convection mechanisms, several researchers have employed optical techniques [31, 40–42]. Noteworthy among those is the extensive study by Berg, Boudart, and Acrivos [31], who defined the nature of the convection flow structures from pictures obtained by Schlieren photography. They investigated evaporating films of several pure substances and binary solutions. It was found that generally, the observed flow structures were largely dependent on the depth of the evaporating layer and to a much lesser extent on the properties of the liquid. The reason for that becomes apparent by examining the expressions for the Rayleigh and Marangoni numbers Equations 21 and 22. In experiments with benzene (a typical liquid with respect to its evaporative convection), thin films of about 0.5 mm produced the typical surface-tension-induced Bénard cells (Figure 4). As the depth of the film was increased to 3–4 mm, the size of the cells increased, as predicted by the stability analyses, and a secondary, unstable rib-like structural flow form became superimposed on the background cell pattern. At depths of about 10 mm the cells disappeared and a pattern of loose lines termed "streamers" (Figure 5A) took form. This flow structure is typical of buoyancy-induced convection. No further change was observed with increasing depths.

Films of water showed an exception to the just-described behavior. This was attributed to contamination by surface-active agents which reduce surface tension substantially and eliminate its variation and thus the Marangoni effect. Surfactants added to the various evaporating liquids in small quantities changed the convective flow pattern completely. Bénard cells did not appear and no convection at all occurred at depths up to 1 mm. As depth was increased to 2 mm, convection occurred in a new form, in two-dimensional worm-like roll cells as illustrated in Figure 5B. Berg et al. [31] showed that surfactants in quantities less than 1/100 of the amount needed to form a close-packed monolayer on the

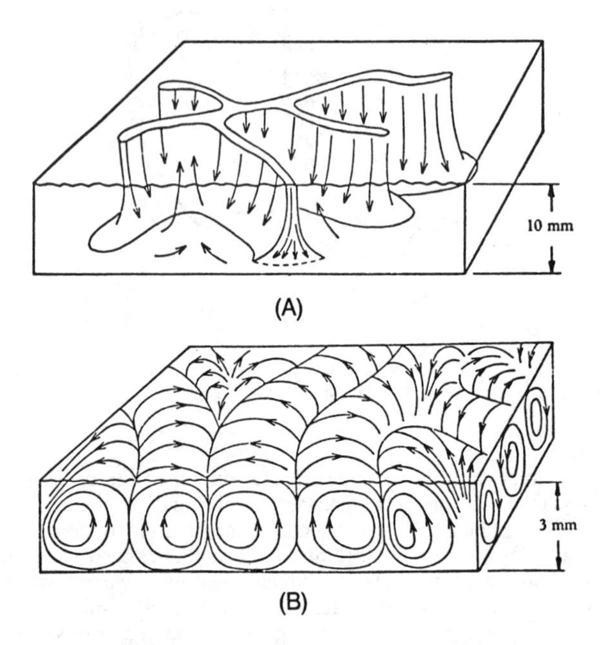

Figure 5. Surface convection patterns: (A) Streamers; (B) roll cells (from Berg, Boudart, and Acrivos [31]).

surface can stabilize a water film up to a thickness of about 10 mm by effectively removing the surface tension mechanism for evaporative convection, thereby leaving the buoyancy mechanism as the only vehicle for setting the fluid in motion.

Most of this discussion of surface convection and the research described has dealt with liquid motion set by evaporation in films heated from below. Historically, this has been the path chosen by investigators of this important phenomenon, who followed the early work of Bénard and attempted to explain his findings. The same type of surface convection originates, however, from other mechanisms which are of greater interest in film absorption. Based on a pioneering, general stability analysis of the interface between two films, Sternling and Scriven [43] list the following effects as promoters of surface convection:

- Solute transfer out of the phase of higher viscosity.
- Solute transfer out of the phase in which its diffusivity is lower.
- Large differences in kinematic viscosity and solute diffusivity between the two phases.
- Steep concentration gradients near the interface.
- Interfacial tension highly sensitive to solute concentration.
- Low viscosities and diffusivities in both phases.
- Absence of surface active agents.
- Interfaces of large extent.

Some of these effects have been observed in the laboratory, which lends credence to the theory.

Most recent research on the fluid mechanical aspects of surface convection concentrated on improvements in the stability analyses and attempts to describe the fluid motion quantitatively. In the first category, Smith [44] extended the analysis of Scriven and Sternling [39] by considering the effect of gravitational forces at the surface in addition to the surface tension. Vidal and Acrivos [45] considered the effect of nonlinear preconvective temperature profiles for surface-tension-driven convection in an extension of Pearson's [37] linear analysis. Scanlon and Segel [46] used a nonlinear theory to predict the emergence and the stability range of the hexagonal flow pattern. Berg and Acrivos [47] accounted for the effect of surface active agents in another extension of Pearson's [37] model and demonstrated theoretically their strong stabilizing effect on surface-tension-induced convection. Brian [48, 49] analyzed the effect of Gibbs adsorption on the Marangoni instability. Imaishi et al. [50] considered the effect of oscillatory instability in a model including both buoyancy and surface tension effects. In the second category, Levich [15] studied the flow due to a linear surface tension gradient. Yih [51] has found several inconsistencies in Levich's analysis and presented a corrected solution to the same problem. Ostrach and Pradhan [52] performed an experiment to find the flow pattern produced by capillary forces created by a temperature variation on the surface in the presence and absence of gravitational forces. Birikh [53] found an exact solution for the flow induced by free convection and surface tension in a planar horizontal layer of liquid with a constant horizontal temperature gradient. Chifu and Stan [54] presented a solution for a laminar film on an inclined plane with a surface tension gradient on the surface in the direction of flow.

The purpose of this discussion has been to describe the phenomenon and explain the mechanisms and pattern of surface convection. The relation between this fluid motion and the transfer process in absorption is definite, although not quantitatively understood as yet. Several studies aiming at better insight into this relation have been conducted with laminar films and will be discussed in the next section.

ABSORPTION IN LAMINAR FILMS

Laminar films ($Re < 20$) in the absence of significant gas flow leading to interfacial shear, are characterized by a smooth interface and a parabolic velocity profile:

$$u = u_s\left[2\left(\frac{y}{h}\right) - \left(\frac{y}{h}\right)^2\right] \tag{23}$$

where u_s is the velocity at the surface, $u_s = (3/2)u$. The film thickness on a flat plate is given by Nusselt's Equation 2. Due to the regular and steady nature of the flow, the heat and mass transfer process lends itself easily to theoretical analysis.

Many cases of interest in laminar film absorption fall into a basic category which may be characterized by the following conditions:

- The liquid solution is Newtonian and its physical properties are constant and independent of temperature and concentration.
- The mass of gas absorbed per unit time is small compared to the mass flow rate of the liquid. Therefore the latter can be assumed constant, as are the film thickness and average flow velocity.
- There is no heat transfer in the vapor phase.
- There are no natural convection or surface convection effects.
- Diffusion thermo-effects are negligible.
- Vapor pressure equilibrium exists between the gas and the liquid at the interface.

Absorption in laminar films conforming to these conditions will be examined first. The effects of chemical reactions, surface resistance, changing properties, etc. will be discussed later in this section.

Transfer Process Under Basic Conditions

Figure 6 describes qualitatively the variations in concentration and temperature in the laminar film resulting from the absorption process. The wall in the case shown is adiabatic and impermeable to the absorbate. The two variables are shown in their dimensionless form, θ and γ respectively, as defined in Equation 16 and 17. It is convenient to introduce also normalized coordinates as follows:

$$\zeta = \frac{1}{Re}\frac{x}{h} \tag{24a}$$

$$\eta = \frac{y}{h} \tag{24b}$$

The liquid at $\zeta = 0$ is at a state of nonequilibrium with the gas. As a result, absorption leading to simultaneous heat and mass transfer begins at the interface and extends its effect gradually into the film. Thermal and concentration boundary layers begin to develop and grow in thickness until they fill the entire depth of the film. The relative growth rate of the boundary layer depends on the Lewis number, $Le = D/\alpha$. For $Le > 1$, the concentration boundary layer develops faster, for $Le < 1$ the thermal one develops faster, and if an absorbent liquid is found with $Le = 1$, the two layers develop at the same rate. Once fully developed, the temperature and concentration profiles continue to vary over the entire film thickness as long as the transfer process at the interface continues. After sufficient distance in the direction of flow, the profiles become uniform across the film as equilibrium is reached.

For films with short exposure time (small ζ) the boundary layers are only partially developed and very thin. Within that small thickness, the velocity may be assumed uniform and the effect of the wall is not felt, which simplifies the problem considerably. The situation is adequately described by Higbie's penetration theory [29], as confirmed experimentally by

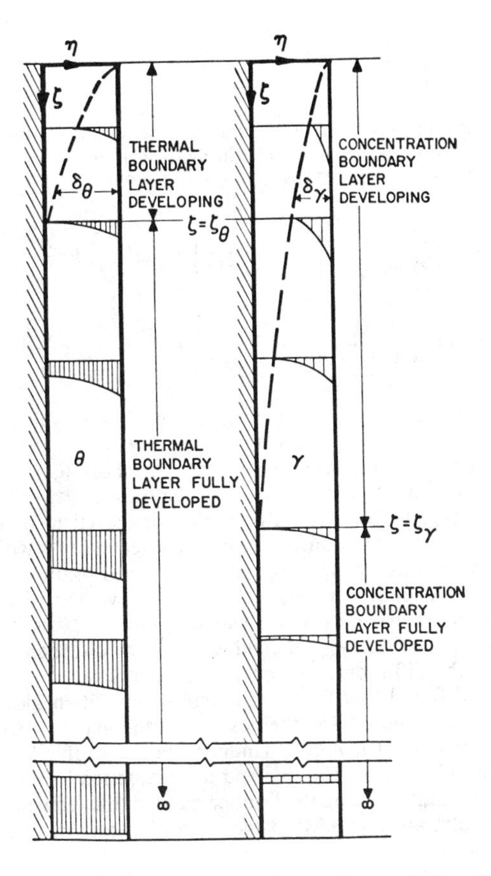

Figure 6. Typical development of the normalized temperature and concentration profiles in laminar absorption with combined heat and mass transfer.

Emmert and Pigford [7]. Equations 13 through 20 may be used with t replaced by x/u_s. Thus, the local mass and heat transfer film coefficients are:

$$K_L = \sqrt{\frac{3}{2\pi} D \frac{\bar{u}}{x}} \tag{25}$$

$$H_L = \varrho c_p \sqrt{\frac{3}{2\pi} \alpha \frac{\bar{u}}{x}} \tag{26}$$

which may be expressed in dimensionless form in terms of the Sherwood and Nusselt numbers as:

$$Sh = \frac{K_L h}{D} = 0.345 \sqrt{Sc/\zeta} \tag{27}$$

$$\mathrm{Nu} = \frac{H_L h}{k} = 0.345\sqrt{\mathrm{Pr}/\zeta} \tag{28}$$

where $\mathrm{Sc} = \nu/D$ and $\mathrm{Pr} = \nu/\alpha$ are the Schmidt and Prandtl numbers, respectively. The dimensionless temperature and concentration distributions in the film, subject to the same conditions as Equations 16 and 17 are given by [30, 55]:

$$\theta = \frac{\lambda}{\lambda + \mathrm{Le}^{1/2}}\left[1 - \mathrm{erf}\left(\frac{3y^2\bar{u}}{8\alpha x}\right)^{1/2}\right] = \frac{\lambda}{\lambda + \mathrm{Le}^{1/2}}\left[1 - \mathrm{erf}\left(\frac{3}{32}\mathrm{Pr}\frac{\eta^2}{\zeta}\right)^{1/2}\right] \tag{29}$$

$$\gamma = \frac{\mathrm{Le}^{1/2}}{\lambda + \mathrm{Le}^{1/2}}\left[1 - \mathrm{erf}\left(\frac{3y^2\bar{u}}{8Dx}\right)^{1/2}\right] = \frac{\mathrm{Le}^{1/2}}{\lambda + \mathrm{Le}^{1/2}}\left[1 - \mathrm{erf}\left(\frac{3}{32}\mathrm{Sc}\frac{\eta^2}{\zeta}\right)^{1/2}\right] \tag{30}$$

The case of isothermal absorption is represented by $\lambda \to 0$ which yields $\theta = 0$ in Equation 29.

When the film contact time with the gas is not short (ζ is not small) the velocity distribution in the liquid must be taken into consideration. Furthermore, if ζ is large enough for the boundary layers to become fully developed, the effect of the wall must also be considered. This requires a solution of the diffusion equation—with the convection term included—for isothermal absorption, and a simultaneous solution of the energy equation with the same term if heat effects are not negligible. Following an early analysis by Vyazovov [56] several investigators have used the classical Graetz–Nusselt formulation of these equations, which neglects diffusion and conduction in the direction of flow [57]. Rotem and Nielson [58] solved the diffusion equation taking into account the diffusion terms in the direction of flow and found that for most practical cases its neglect is indeed justified.

An exact eigenvalue solution for the case of isothermal absorption was obtained by Pigford [59] using the correct parabolic laminar velocity profile and an impermeable wall boundary condition. Grossman [55] obtained a more general solution, allowing for combined heat and mass transfer. Using the Fourier method, the temperature and concentration distributions in the film are expressed in the form of two infinite series of eigenfunctions, $F_n(\eta)$ and $G_n(\eta)$:

$$\theta = \sum_{n=0}^{\infty} A_n F_n(\eta) e^{-\alpha_n^2 \zeta} \tag{31}$$

$$\gamma = \sum_{n=0}^{\infty} B_n G_n(\eta) e^{-\alpha_n^2 \zeta} \tag{32}$$

where the corresponding eigenvalues α_n and coefficients A_n and B_n are determined from the diffusion and energy equations and their boundary conditions. In this analysis, the wall was considered impermeable to mass transfer. Concerning heat transfer, two cases of practical interest were considered: in one the wall was kept at a constant temperature equal to the inlet temperature of the film; in the other, the wall was adiabatic. A boundary condition for a linear absorbent was imposed at the interface, which assumes a linear relation between temperature and concentration in equilibrium and a constant heat of absorption. The eigenvalue solution was extended and complemented by a numerical one for the range of small ζ, for which a large number of terms in Equations 31 and 32 was required.

The solution in normalized form depends on two parameters: the Lewis number, $\mathrm{Le} = D/\alpha$, and the dimensionless heat of absorption λ as defined in Equation 18. Figure 7 describes the general behavior of the temperature and concentration in the film as they vary with the nomarlized length ζ for a typical set of values of Le and λ. Curves are given for θ and γ at the wall (θ_w, γ_w), the liquid bulk ($\bar{\theta}$, $\bar{\gamma}$), and the interface (θ_i, γ_i). The solid lines describe the results for the constant temperature wall and the broken lines for the adiabatic wall. Initially, for small ζ the behavior is the same for both types of wall conditions, with the results identical to those given by the penetration theory Equations 29 and 30. As ζ increases, for the adiabatic wall case the wall, bulk, and interface temperatures increase monotonically toward a final common value and become closer and closer to each other. This steady increase is due to the fact that the heat of absorption is not removed from the

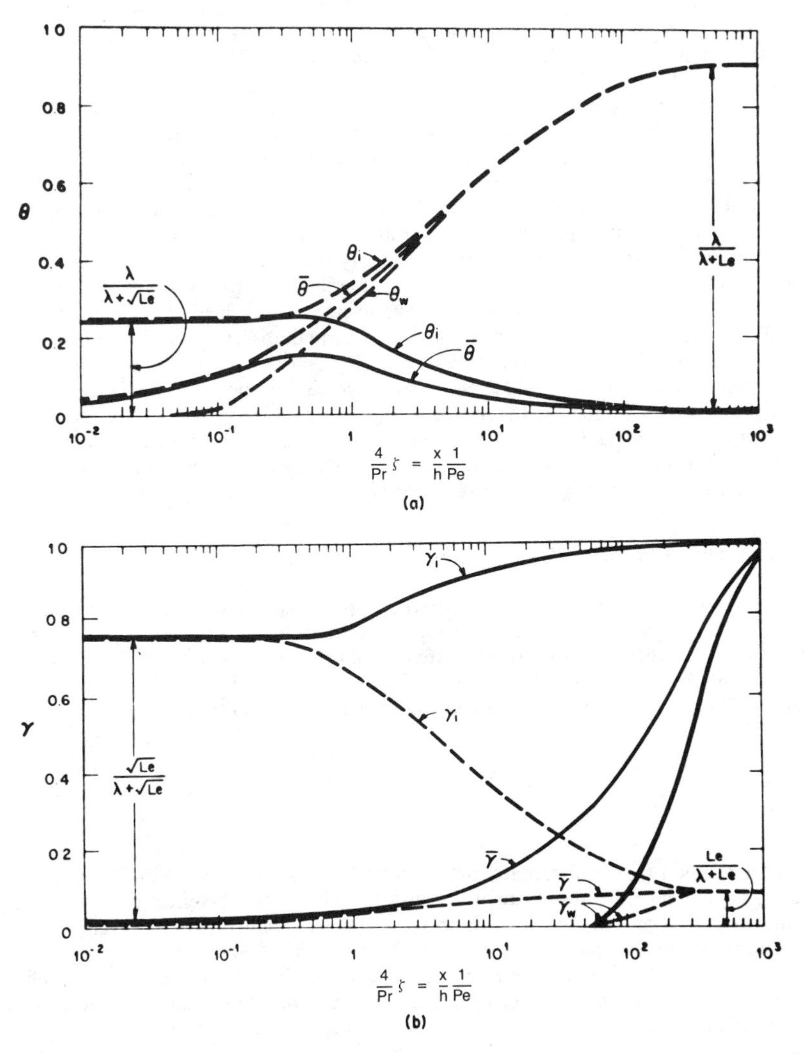

Figure 7. Dimensionless wall, liquid bulk, and interface temperatures and concentrations as functions of the normalized length ζ for Le = 0.001 and λ = 0.01. (from Grossman [55]).

film. With the constant temperature wall, the interface temperature increases slightly following the trend of small ζ, and the bulk temperature attempts to approach it as heat is conducted from the interface into the film. Thus, both temperatures decrease toward zero as heat is removed from the film through the wall. The interfacial concentration in both cases follows a trend opposite to that of the interfacial temperature. The bulk concentration increases in both cases toward a final value equal to γ_i. It is interesting to note that in the adiabatic wall case $\bar{\gamma}$ increases with ζ while γ_i decreases. Figure 7 also shows the asymptotic values of the normalized temperature and concentration: $\theta = 0$ and $\gamma = 1$ with the constant temperature wall, $\theta = \lambda/(Le+\lambda)$ and $\gamma = Le/(Le+\lambda)$ with the adiabatic wall. The isothermal absorption case is obtained as a particular case of Grossman's solution [55] with $\lambda \to 0$.

It should be mentioned that an earlier analysis of combined heat and mass transfer in laminar flow was done by Grigoreva and Nakoryakov [60], who assumed, however, a uniform velocity profile rather than the correct parabolic one. This assumption leads to a deviation of about 20% in the heat and mass transfer coefficients and to underprediction by about 40% of the distance required for boundary-layer development.

Mass transfer coefficients have been calculated from the solutions for the concentration distribution. Grossman [55] calculated the local film coefficient as defined by Equation 20 whereas Pigford [59] derived an average coefficient based on the logarithmic mean concentration difference. In the general case of absorption with combined heat and mass transfer, the Sherwood number Equation 27 variation with ζ is a function of the parameters λ and Le. Figure 8 describes the variation for some typical values of these parameters [55]. The particular case of isothermal absorption is described by the solid line of $\lambda = 0$ for $Le = 10^{-3}$. The asymptotic value of the Sherwood number (defined either way) for very large ζ with a constant temperature wall and/or isothermal absorption is found to be 3.41.

Experimental data are scarce in the truly laminar region ($Re < 20$), and most researchers who attempted measurements in that flow regime have had Reynolds numbers greater than 20 and at least some rippling on the surface. Rippling could be reduced considerably by wetting agents [59], and as long as it was kept to a minimum, experimental results were in reasonable agreement with the laminar theory. Comparison of some isothermal absorption data with theory is presented in Reference 57.

Absorption Under Other Conditions

Chemical reaction, surface convection, variable properties, and surface resistance are some of the mechanisms causing a deviation in laminar film absorption from the basic conditions just described. In order to evaluate the influence of these effects, it is convenient to define an enhancement factor Φ expressing the ratio between the actual absorption mass transfer coefficient (K_L^*) and that obtained under the basic conditions (K_L):

$$\Phi = \frac{K_L^*}{K_L} \tag{33}$$

Chemical reaction of the absorbate in the liquid leads to an increase in the absorption rate due to depletion of the absorbate at the surface as it reacts. The effect of various types of chemical reaction on laminar absorption from the viewpoint of the penetration theory has been reviewed in Reference 57, along with some practical examples.

The order of a chemical reaction in the context of the present problem is defined according to the rate at which the absorbate is consumed per unit volume of liquid. Thus, a reaction of the type

$$A \rightleftarrows P \tag{34}$$

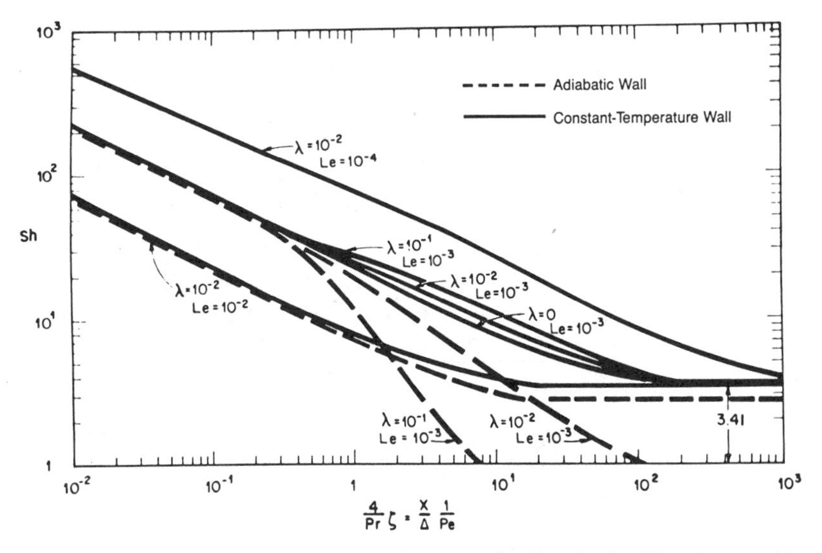

Figure 8. Local Sherwood number as a function of the normalized length ζ for different values of Le and λ (from Grossman [55]).

(where absorbate A is converted to product P) is said to be of order n if the rate r is related to the concentration of A according to the relation:

$$r = k_n C_A^n \tag{35}$$

A more complex case is that of a bimolecular reaction of the type:

$$A + bB \rightleftarrows P \tag{36}$$

where the absorbate A reacts with a solute B already in the liquid. Then, the reaction rate may be dependent on the concentrations of both A and B, in the form:

$$r = k_{nm} C_A^n C_B^m \tag{37}$$

If the dependence of r on C_B is weak, as in the case where C_B is constant throughout the liquid, then $m \to 0$ and the reaction of Equation 36 becomes a pseudoreaction of the type of Equation 34.

Early models of laminar absorption with simultaneous chemical reaction [57, 61] have considered the short exposure time (small ζ) problem under isothermal conditions, for which an analytical solution may be obtained and compared with the results of pure physical absorption from Higbie's [29] penetration theory (described earlier). For the simplest case of a first-order irreversible reaction, of the type of Equations 34 and 35 with n = 1, Danckwerts [62] obtained:

$$\bar{\Phi} = \frac{\bar{K}_L^*}{\bar{K}_L} = \left(\frac{\pi}{2}\right)^{1/2} (k, t)^{1/2} \left\{ \left(1 + \frac{1}{2k, t}\right) \operatorname{erf}(k, t)^{1/2} + \frac{e^{-k, t}}{\sqrt{\pi k, t}} \right\} \tag{38}$$

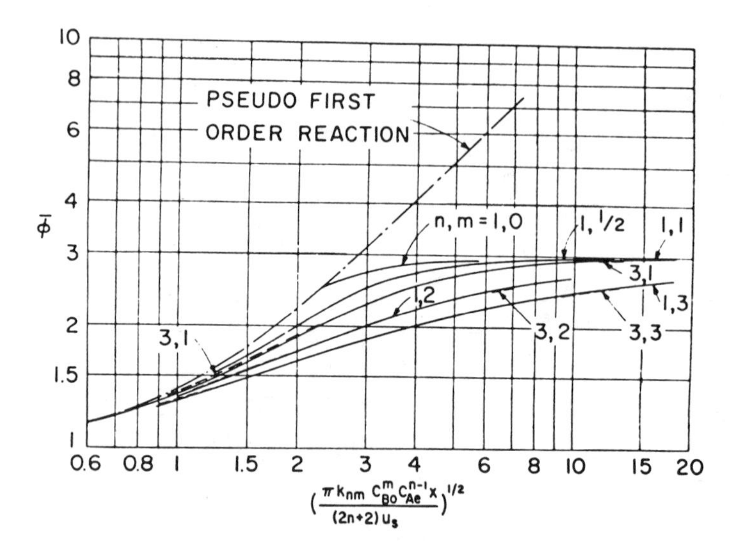

Figure 9. Enhancement factor due to chemical reaction as a function of dimensionless exposure time for different bimolecular reaction orders (n,m). $D_B = D_A$ and $C_{B0}/bC_{Ae} = 2$ (from Brian [66]). Results are valid for short exposure times within the penetration theory approximation.

where t is the exposure time, $t = x/u_s$. Sherwood and Pigford [63] extended these results to the case of a reversible first-order reaction, where the presence of the product P in the liquid affects the reaction rate according to the equilibrium constant K of the two components A and P. Their results were presented graphically [57, 63] in the form:

$$\bar{\Phi} = f(kt, K) \tag{39}$$

The more general case of a second-order reaction according to Equation 36 was treated theoretically by Perry and Pigford [64]. A numerical solution was obtained with $n = m = 1$ in Equation 37. It was shown that in the limit when the reaction is very fast and $D_A \cong D_B$, $\bar{\Phi}$ may be approximated by:

$$\bar{\Phi} = \frac{\sqrt{D_A}}{\sqrt{D_B}}\left(1 + \frac{D_B}{D_A}\,\frac{C_{B0}}{bC_{Ae}}\right) \tag{40}$$

where D_A and D_B are the diffusion coefficients of substances A and B; C_{B0} is the initial concentration of B in the liquid and C_{Ae} is the concentration of the absorbate A in the liquid in equilibrium with the gas. Brian et al. [65] extended the theoretical analysis for a second-order irreversible reaction of the same type to a wider range of parameters, and presented it graptically in the form:

$$\bar{\Phi} = f(k_{11}C_{B0}t, D_B/D_A, C_{B0}/bC_{Ae}) \tag{41}$$

Brian [66] later generalized the solution for an irreversible bimolecular reaction of general order according to Equations 36 and 37, giving curves of the type of Equation 41 for different combinations of n and m. Figure 9 shows an example of this graphical representation for $D_B/D_A = 1$ and $C_{B0}/bC_{Ae} = 2$.

The case of intermediate and long exposure times in which ζ is not small was treated by Jameson, Burchell, and Gottifredi [67]. Here, the effect of the parabolic velocity profile and the wall were taken into consideration. An enhancement factor $\bar{\Phi}$ was calculated, which expressed the ratio of the actual mass transfer film coefficient to that obtained in pure physical absorption at short exposure times Equation 15. The solution was presented graphically in the form:

$$\bar{\Phi} = \frac{\bar{K}_L}{2\left(\frac{Du_s}{\pi x}\right)^{1/2}} = f\left[\frac{\pi k_{nm} C_{B0}^m C_{Ae}^{n-1} x}{(2n+2)u_s},\ h\left(\frac{k_{nm} C_{B0}^m C_{Ae}^{n-1}}{D_A}\right)^{1/2},\ \frac{C_{B0}}{bC_{Ae}},\ \frac{D_B}{D_A}\right] \tag{42}$$

Figure 10 shows an example for $D_B/D_A = 1$ and $C_{B0}/bC_{Ae} = 2$, the same condition as in Figure 9, and for $n = m = 1$. Brian's [66] results for this case are plotted along for comparison. As expected, the results are for short exposure time and a thick film (small ζ). For longer times the enhancement factor decreases sometimes to values smaller than 1.0, due to the influence of the wall.

Sada and co-workers [68, 69] considered the case when the absorbate A, after being absorbed in the liquid, may react not only with the solute B but also with a product of this first reaction. This second reaction (A with P) will then influence the absorption rate only after an exposure time long enough for the product of the first reaction of form. One example of such an absorption process with consecutive chemical reaction is with carbon dioxide absorbed in monoethanolamine, which was studied experimentally by Sada et al. Pedersen and Prenosil [70] analyzed laminar absorption with a first-order reaction considering a resistance to mass transfer in the gas phase.

Another factor causing a deviation from the basic conditions of pure physical absorption is interface resistance, which was studied by Tamir and Taitel [71]. This additional resistance

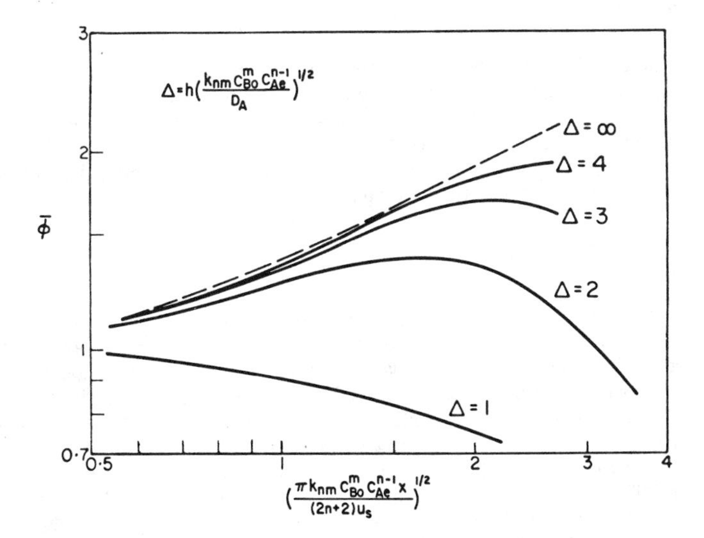

Figure 10. Enhancement factor due to chemical reaction as a function of dimensionless exposure time for different dimensionless film thicknesses. $n = m = 1$, $D_B = D_A$; $C_{BO}/bC_{Ae} = 2$ (from Jameson, Burchell, and Gottifredi [67]). The broken line for infinite thickness represents the penetration theory solution according to Brian [66].

may be explained in terms of lack of thermodynamic equilibrium at the interface. The transfer of absorbate molecules across the interface occurs at a finite rate and of those striking the surface only the ones with sufficient energy to overcome the energy barrier can be absorbed. Therefore, only a fraction of the gas molecules can penetrate into the liquid. This fraction has been termed "accommodation coefficient" [7] and used to calculate a mass transfer coefficient K_i for the interfacial resistance. Thus, the total resistance to mass transfer is the sum of the resistances:

$$\frac{1}{K_{total}} = \frac{1}{K_L} + \frac{1}{K_i} \tag{43}$$

Tamir and Taitel [71] note that although in general interfacial resistance to absorption is quite low, surface contamination in various forms can increase it considerable. They have considered a laminar film with the fully developed parabolic velocity profile and solved the diffusion equation under isothermal conditions with a fixed interfacial resistance. An exact solution was obtained in terms of eigenfunctions for which ten eigenvalues were tabulated for different values of the normalized interface resistance:

$$B = \frac{K_i h}{D} \tag{44}$$

In addition, an approximate solution based on an integral method was obtained and found to be in very good agreement with the exact one. Figure 11 describes the variation of the Sherwood number with the normalized distance ζ for different values of B. From the integral solution, a simple expression was found for the large ζ and asymptotic Sherwood number:

$$Sh = \frac{1}{\frac{11}{40} + \frac{1}{B}} \tag{45}$$

The effect of variable film properties on the absorption process was considered by several investigators. Shah [72] analyzed the simultaneous heat and mass transfer problem with a first-order chemical reaction for short exposure time (penetration theory approach) when the diffusion coefficient, equilibrium concentration at the surface and reaction rate coefficient are dependent upon temperature. Shair [73] assumed an exponential dependence of the viscosity and diffusion coefficients on the temperature in the form:

$$\mu = \mu_0 e^{d/T} \qquad D = D_0 e^{-d/T} \tag{46}$$

Assuming a linear temperature profile across the film, he calculated the resulting velocity profile and showed that it varied significantly from the parabolic profile when interfacial shear was present. Concentration profiles were also calculated. Yih and Seagrave [74] extended this work with the same assumptions of linear temperature variation, μ and D according to Equation 46, to include the effect of first- and zero-order chemical reactions, whose rates varied exponentially with temperature. An increase in the mass transfer coefficient was found to be due to the interfacial shear and increase in molecular diffusivity with temperature. The effect of interfacial shear was also studied by Gottifredi et al. [75]. Chavan and Mashelkar [76] and Mashelkar, Chavan, and Karanth [77] studied the absorption into a film of a non-Newtonian fluid.

Surface convection, also known as interfacial turbulence or Marangoni instability, is one of the important mechanisms of heat and mass transfer enhancement in laminar film absorption. This phenomenon, which has been known for years in connection with film

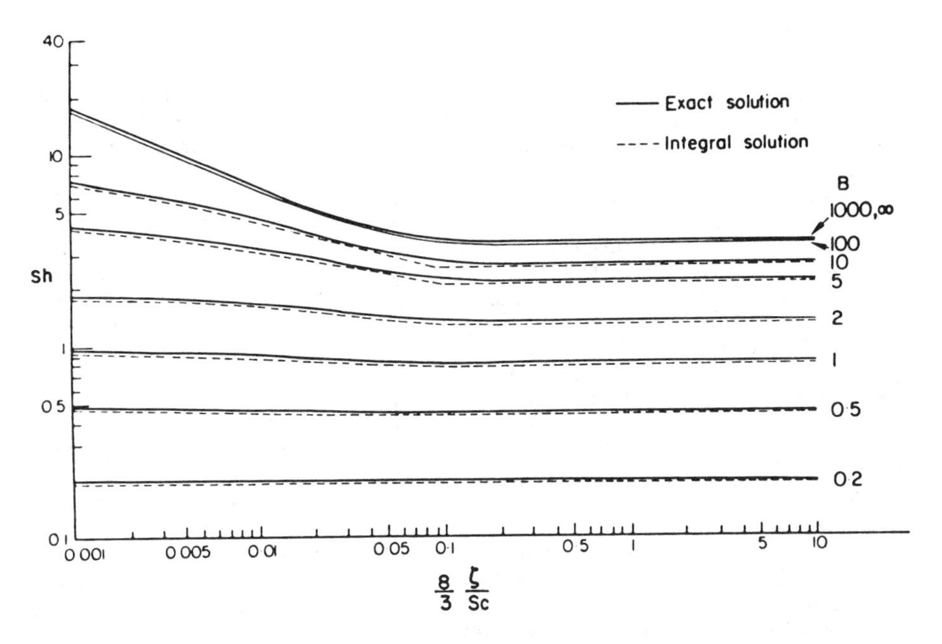

Figure 11. Local Sherwood number as a function of dimensionless length for different values of the normalized surface resistance (from Tamir and Taitel [71]).

evaporation was recognized as having a significant effect on the transfer rate by creating localized motion of the liquid at the surface due primarily to surface tension gradients.

Absorption experiments with short-exposure-time laminar jets and wetted-wall columns have often given good agreement with theoretical predictions based on the penetration theory. In some instances, however, the experimental data were found to deviate sharply from the theoretical predictions. This was noted by Brian et al. [78] who analyzed experimental data from several sources on the absorption of carbon dioxide in aqueous monoethanolamine (MEA) solutions [61, 79–81]. In their own experiments, Brian and co-workers [78, 82] monitored the mass transfer rate by desorbing an inert tracer simultaneously with the absorption of the gas in the liquid in a short wetted-wall column. They found a substantial increase in the mass transfer coefficient over that predicted by the penetration theory and related it to surface convection due to Marangoni instability.

Experiments by other researchers [68, 69, 83–86] have related the surface gradients causing the Marangoni effect to the presence of surface-active reaction products, and to a lesser extent to temperature variations produced by the absorption. Sada et al. [87] noted the correlation between the liquid flow condition and the Marangoni instability. As the liquid flow increases small ripples appear on the surface which interact with the surface tension-induced motion.

Observations of surface convection accompanying chemical absorption were made by Imaishi and Fujinawa [88] using a Schlieren optical system. They studied the absorption of CO_2 in MEA and other liquid solutions in a plane, vertical, wetted-wall column. The interpretation of the Schlieren photographs revealed the convection pattern, although only qualitatively, as a roll-cell-type secondary flow superimposed on the primary flow, with the axis of rotation parallel to the main flow.

Few and limited attempts were made to quantify the Marangoni effect on the mass transfer process and devise a theoretical model for it. Ruckenstein and Berbente [89] and

Ruckenstein [90] proposed a model representing the flow structure generated by the surface convection as roll cells propogating in the main direction of motion with the average velocity of the liquid. An analytical expression was given for these roll cells and on its basis the unsteady convective diffusion equation was solved, to give the enhancement factor with respect to the penetration theory. However, Imaishi and Fujinawa's Schlieren experiments [88] showed this model to be unrealistic, as it assumes the axis of rotation of the roll cells to be perpendicular to the main flow, in contradition with the observations. Imaishi, Hozawa, and Fujinawa [91] proposed a Danckwerts-type surface-renewal model, with a steady-state age distribution function. To confirm the validity of this model was one of the goals of an extensive study by Imaishi et al. [92], where results from carefully controlled mass transfer experiments in a liquid jet and wetted-wall column [93] were analyzed in order to determine the conditions for the initiation of interfacial turbulence and examine its mechanism. The unsteady diffusion and Navier-Stokes equations were solved numerically, yielding the two-dimensional concentration profile and flow pattern in the liquid layer, from which the first appearance of interfacial turbulence and the time dependence of the liquid mass transfer coefficient could be obtained. Reasonable agreement was obtained with the experimental results [93]. The time at which the increase in mass transfer rate could first be detected experimentally was larger than the time for the beginning of instability as predicted by Brian's stability analyses [48, 49]. This was attributed to the time required for the growth of the microturbulence.

In summary, it appears that quantitative understanding of the effect of surface convection on the transfer process in absorption is beginning to emerge, but is still in a premature stage. Further studies will have to define the dimensionless parameters characterizing the interfacial turbulence, and relate the enhancement factor to them.

ABSORPTION IN THE WAVY FLOW REGIME

Although many practical absorption systems operate in the wavy flow regime, the amount of research into the transfer process under this type of flow has to date been relatively limited, in comparison with the studies on laminar or turbulent films. As noted earlier, regular waves on the film surface first appear at $Re \cong 20$ and the laminar wavy pattern prevails aproximately up to $Re \cong 200$, at which the first signs of turbulence appear and the transition to fully turbulent flow begins. The wavy flow regime is considerably more complex than the laminar, and theoretical modeling of the momentum, heat, and mass transfer processes is more difficult. On the other hand, the flow is not as completely irregular as in turbulent films where an empirical or semi-empirical approach must be taken.

As long as the flow Reynolds number is small, the wave pattern is quite regular and the deviation from the laminar behavior is small. This has raised the assumption that the observed increase in mass transfer rate in wavy films compared to the smooth laminar ones is due solely or primarily to the increase in interfacial area. In a one-dimensional sinusoidal wave (Figure 12) with wavelength λ' and amplitude $\bar{h}a'$ (Equation 3) the ratio of the actual interface area, s, to that of a smooth surface, s_0, can be easily calculated:

$$\frac{s}{s_0} = \frac{1}{\lambda'}\int_0^{\lambda'}\left[1+\left(\frac{2\pi\bar{h}a'}{\lambda'}\right)^2\cos^2\left(\frac{2\pi x}{\lambda'}\right)\right]^{1/2} dx \cong 1+\left(\frac{\pi\bar{h}a'}{\lambda'}\right)^2 - \frac{3}{4}\left(\frac{\pi\bar{h}a'}{\lambda'}\right)^4 + \ldots \tag{47}$$

for $\bar{h}a' \ll \lambda$, where a' and λ' may be estimated from the models by Kapitsa [20] and Levich [15]. Portalski [94] derived an expression for the increase in interfacial area for regular two-dimensional waves starting with Kapitsa's [20] equation of the surface profile. The results obtained from this equation, based on the theoretical wavelength, have been generally greater than experimentally determined values. A modified version of Portalski's equation was then developed by Protalski and Clegg [95], which agreed with their experi-

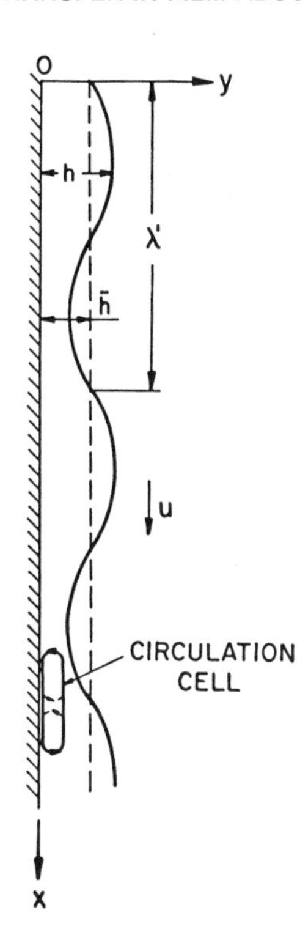

Figure 12. Falling film in the wavy flow regime.

mental results obtained by computer analysis of photographs of dyed liquid films flowing down glass wetted-wall columns. The results were given in the form:

$$\frac{s - s_0}{s_0} = \text{const} \times \text{Re}^{2/3} \tag{48}$$

where the value of the constant varies with the type of liquid.

The increase in area, which is of the order of a few percent cannot, however, account for the increased mass transfer rate with respect to laminar films, observed at somewhat higher Reynolds numbers. Another mechanism must therefore be responsible.

Levich [15] considered the effect of the time variation of the velocity at the surface on the periodic steady-state. He used the velocity distribution obtained from his and Kapitsa's [20] hydrodynamic models to solve the diffusion equation for isothermal mass transfer and thus obtained the concentration distribution from which the mass transfer rate could be calculated. Levich expressed the concentration as a sum of two terms, $C_1 + C_2$. The first, C_1, represented the distribution in the case of steady, laminar film with a smooth interface; the second, C_2, described the correction due to the wave motion. Substituting in the diffusion equation for short exposure (penetration theory type) and averaging with respect to time, Levich obtained an equation for the time average of the correction C_2. The resulting mass transfer coefficient showed an increase by 15% with respect to the laminar, still far short of the experimentally observed values.

Banerjee, Rhodes, and Scott [96] proposed a theory suggesting that the enhanced mass transfer is caused by large circulating eddies induced by the wave motion. The liquid

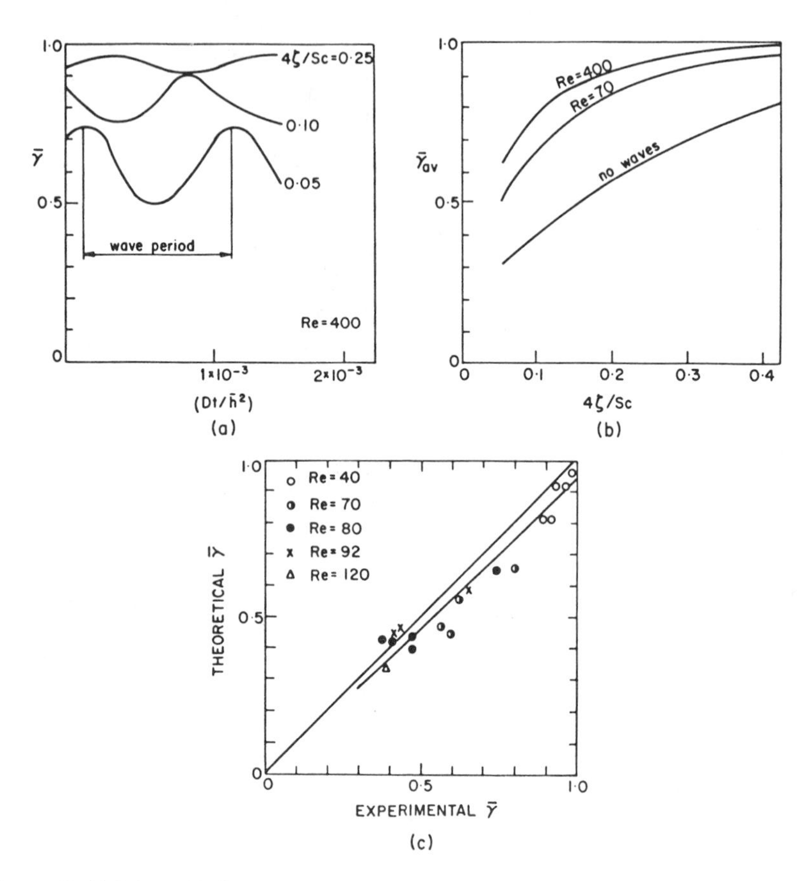

Figure 13. Variations of bulk concentration in film absorption under wavy flow: (A) Time variation at different distances; (B) variation with distance of the time-averaged concentration for different Reynolds numbers; (C) comparison of theoretical and experimental values (after Beschkov and Boyadjiev [105]).

particles move along closed trajectories, as illustrated in Figure 12, and hence a surface renewal occurs. This eddy motion has been demonstrated theoretically by Portalski [97] based on the Kapitsa velocity profile and supported also by Massot et al. [98], who gave an improved and more general analysis of the wave motion and resulting velocity distribution. Banerjee et al. [96], developed a theory based on this phenomenon where the number of eddies generated per unit time and their distance from the surface were estimated and used to calculate an average mass transfer coefficient. The results indicated $\bar{K}_L \sim D^{5/8}$ and were presented graphically in the form:

$$\bar{K}_L/D^{5/8} = f(Re) \tag{49}$$

A comparison with experiments by Emmert and Pigford [7] and Kamei and Oishi [99] showed good agreement for the range $150 < Re < 800$.

The main shortcoming of this type of model is its being based mainly on intuition, rather than evolving from the fundamental transfer laws and equations. Moreover, Penev et al.

[100] have shown by actually solving the complete Navier–Stokes equations, that two-dimensional waves propagating on the surface of a flowing film with a constant-phase velocity make surface renewal impossible. Also, the Levich intuitive surface-tension model used by Banerjee et al. [96] to predict the distance of the eddy from the surface was shown to be inadequate when used in studies of turbulent flows, as will be discusssed further in the next section. It appears therefore that despite its agreement with two sets of experimental results [7, 99] the model of Banerjee et al. [96] cannot be regarded as fundamental and general.

Perhaps the most plausible explanation of the mass transfer enhancement due to wave motion relates the effect of the waves to the transverse convection created by a velocity component in the direction normal to the surface, which is not present in laminar waveless flow. Ruckenstein and Berbente [101] noted that this transverse component was not properly accounted for in Levich's analysis [15]. These authors [101, 102] solved the diffusion equation within the limitations of the penetration theory using the Kapitsa velocity profile and showed that its use in this form led to an increase in the mass transfer rate induced by the wave motion beyond that predicted by Levich [15]. However, this increase was no more than 30% of the laminar, still insufficient to explain experimental results. In a later model, Ruckenstein and Berbente [103] employed a nonlinear treatment of the hydrodynamic problem. Using the velocity distribution thus obtained, they calculated mass transfer rates and showed that waves may amplify the local mass tranfer rate by about 100% compared with laminar flow. The results were expressed in terms of the enhancement factor (see second section) and presented graphically as a function of a flow parameter:

$$\Phi = f\left(\frac{g^{1/6}\Gamma^{11/6}}{\varrho^{4/3}\nu^{7/6}\sigma^{1/2}}\right) \tag{50}$$

Reasonable agreement with the data of Kamei and Oishi [99] was obtained. The model is self-consistent, based on the transfer equations and seems to contain the correct fundamental physics of the problem. Another model based on a similar approach of solving the diffusion equation for short contact time was developed by Javdani [104]. His results describe the enhancement factor as a function of a system parameter including the Reynolds and Peclet numbers, as well as wave parameters.

The penetration theory approach has some limitations when applied to wavy films, more so than in laminar films. As pointed out by Beschkov and Boyadjiev [105], the assumption of short exposure time replaces the developed wavy velocity profile by the components of the surface velocity vector and treats the concentration at a distance from the film as constant. It is therefore clear that this approach overestimates the mass transfer rates, particularly when ζ is not small. These authors [105] have developed a model suitable for both short and long exposure times. The convective diffusion equation was solved numerically for isothermal absorption in a wavy film, with two dimensional waves propagating on the surface with a constant phase velocity. The results were expressed in terms of the dimensionless bulk concentration $\bar{\gamma}$ (defined as in Equation 17 with C replaced by the bulk concentration $\bar{C}$) and deserve some further discussion.

Figure 13A describes the variation of $\bar{\gamma}$ with time within one period of the waves. There are significant oscillations of $\bar{\gamma}$ with respect to its time-averaged value at small ζ (short-exposure time) where the driving force is large. The oscillations decay as the average $\bar{\gamma}$ increases with increasing ζ. The period of the oscillations is close to that of the waves, but their phase is displaced in time. A similar behavior has been observed experimentally by Brauner and Maron [106]. Figure 13B compares the time averaged values of $\bar{\gamma}$ as they vary with ζ for different Reynolds numbers. Two curves are shown for wavy flow and one for waveless flow. For the latter, the increase in $\bar{\gamma}$ at small ζ is less pronounced than with wavy flow. Beschkov and Boyadjiev [105] point out, that the significant increase in mass transfer rate with wavy flow has been obtained from their calculations using a velocity profile not admitting surface renewal [100]. This seems to support the concept that the effect of waves on the transfer process is due to the transverse convection rather than to surface renewal by circulating eddies. Figure 13C compares the theoretical results with experimental ones, obtained by the same authors with carbon dioxide in water films. It is unfortunate that the calculations [105] were not pursued further to give the mass transfer coefficient and its dependence on the operating parameters.

A summary of experimental work on absorption in the wavy flow regime was given by Oliver and Atherinos [107]. This includes the already-mentioned experiments of Kamei and Oishi [99], those of Hikita et al. [108], and the authors' own experiments [107]. In addition, one should note the CO_2 absorption data of Emmert and Pigford [7]. Oliver and Atherinos [107] plotted the measured mass transfer coefficient against the Reynolds number and compared it with theoretical values for smooth laminar films and also with several of the theoretical models for wavy flow described earlier in this section.

Yih and Chen [14] suggested the following correlation, based on 121 data points from the work of ten investigators in addition to their own, for absorption of different gases into water in the Reynolds number range $49 < Re < 300$:

$$\frac{\bar{K}_L}{D}\left(\frac{\nu^2}{g}\right)^{1/3} = 1.099 \times 10^{-2}\, Re^{0.3955}\, Sc^{0.5} \tag{51}$$

The standard deviation in this correlation is 18.21%. The form of the correlation, including the Schmidt number dependence, was chosen by Yih and Chen on the basis of a mass transfer mechanism associated with eddy dissipation at the surface, which is the prevailing mechanism in turbulent flow.

ABSORPTION IN THE TRANSITION AND TURBULENT FLOW REGIME

Turbulent films are characterized by irregular disturbances and instabilities at the interface, which grow in amplitude and frequency as the flow velocity increases. The first signs of turbulence appear at $Re \cong 200$, and the film becomes completely turbulent at $Re \cong 1{,}600$. Due to the highly irregular nature of the flow, present understanding of turbulent films is largely empirical, although some insight has been provided by researchers into the turbulent transfer mechanism near the gas liquid interface which is the controlling factor in absorption.

The transfer of momentum, heat and mass in turbulent flow is known to be affected by the motion of turbulent eddies, the intensity of which is described in terms of the eddy diffusivity ε. The dimensionless parameters ε/ν, ε/α, and ε/D describe the ratio of the eddy to the molecular diffusivity for the three types of transfer, respectively, and are considerably larger than unity in the turbulent flow [109]. The turbulent film may be divided into three main regions across its thickness [110], as illustrated in Figure 14: the wall region, including the laminar and buffer sublayers, the turbulent core in the middle part of the film, and the interface region. The wall and the interface both have a damping effect on the eddies. In the turbulent core, the eddies produce considerable mixing, which provides for low transfer resistance. Figure 14A describes a typical eddy diffusivity profile composed of three semi-empirical formulas [110] for the three regions. The details of these formulas will be discussed further later. Figure 14B describes velocity profiles for several turbulent Reynolds numbers, calculated with the aid of the eddy diffusivity formulas (see Chapter 2 of Volume 1 of this series for further discussions). The corresponding Froude numbers are listed in parentheses. The shape of the velocity profile—almost uniform throughout the film except for a sharp gradient at the wall—reflects the turbulent momentum transfer mechanism: strong mixing in the middle region and damping of the eddies in the wall region, down to laminar flow in the sublayer near the wall. The absence of shear at the film surface allows for the uniform velocity profile of the turbulent core to extend into the interface region.

When interfacial shear is present, an additional sharp velocity gradient is added at the film surface. It is evident from Figure 14B that the velocity profiles become flatter as the Reynolds number increases, and higher Froude numbers are required to produce them.

The turbulent transfer mechanism in the wall region and in the turbulent core is quite well understood, both qualitatively and quantitatively from the work of numerous investigators on turbulent flow in pipes and channels and in boundary layers over solid surfaces. A good summary of this work appears in Kays' book [109], where expressions for the eddy diffusivity may be found. The transfer mechanism near the interface is less well understood and still under investigation. For absorption processes, the behavior at the interface region of the film is of primary importance, and the dominates heat and mass transfer rate for short exposure times (penetration theory approximation).

The following discussion will first describe the turbulent transfer mechanism at the gas-liquid interface. Theoretical models for the overall film behavior and experimental results will be described next.

Turbulent Interface Transfer Mechanism

In the course of numerous studies on heat and mass transfer from and into a turbulent liquid film, several theoretical models have been proposed for the transfer mechanism at the interface. Most of the investigators have adopted the eddy diffusivity approach, in which the molecular diffusion D in the laminar diffusion equation is replaced by the sum $(D+\varepsilon)$. ε is for the most part much larger than D but is damped down to a zero value at the interface

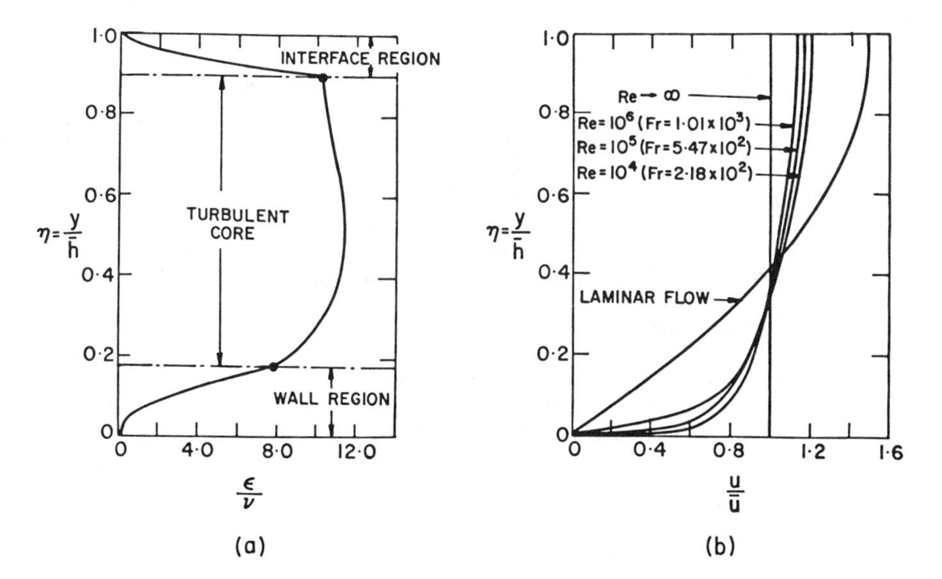

Figure 14. Hydrodynamics of turbulent falling film: (A) Eddy diffusivity profile for Re = 10^4; (B) velocity profiles for different Reynolds numbers and corresponding Froude numbers (from Grossman and Heath [110]).

as ilustrated in Figure 14. The eddy diffusivity close to the surface may be approximated by

$$\varepsilon = ay^n \tag{52}$$

where y is the normal distance from the interface, and a and n are constants. This approximation may be regarded as the first term in a series expansion of ε over a wider range of y.

For long contact times (long films, large ζ), when the concentration profiles become fully developed, it is possible to derive an expression relating the asymptotic mass transfer coefficient to the eddy diffusivity profile given by Equation 52 [111, 112]:

$$\bar{K}_L = \frac{n}{\pi} a^{1/n} \sin\left(\frac{\pi}{n}\right) D^{(n-1)/n} \tag{53}$$

This expression has been very helpful in determining both constants a and n, the former from experimental measurements of the mass transfer coefficient and the latter from experiments as well as from physical considerations on the dependence of $\bar{K}_L$ on D.

One of the earliest models proposed to describe the interface transfer mechanism was an extension of the stagnant-laminar penetration theory due to Higbie [29]. The model assumes eddies to be exposed at the surface for short times while diffusing into the liquid. The exposure time is taken to be uniform for all eddies and the mass transfer coefficient is given by a modified version of Equation 15:

$$\bar{K}_{LM} = 2\sqrt{D/\pi t_c} \tag{54}$$

where t_c is the contact time. Danckwerts [113] proposed a variation of Higbie's model in assuming a distribution of surface ages of the eddies resulting from a constant rate of surface renewal. The diffusion of eddies of varying ages is described by the penetration theory, and the resulting mass transfer coefficient is

$$\bar{K}_L = \sqrt{DS} \tag{55}$$

where S is the rate of surface renewal. Different modifications of the penetration and surface renewal models have been proposed by other investigators. The resulting dependence of $\bar{K}_L$ on the square root of the diffusivity sets the value of n, from Equation 53, at $n = 2$. No clues are given yet as to the value of a.

Levich [15] was followed by other investigators in suggesting that the turbulent transfer process at the interface occurred in much the same way as the momentum transfer process at a solid surface described by Prandtl in his mixing length model. A fluid element with a velocity v travels a distance ℓ in the y direction, carrying with it the characteristic properties of the flow at its starting position. The difference in concentration between the initial and final locations of the element results in mass transfer with an effective diffusivity $\varepsilon \sim v\ell$. v is the characteristic turbulent fluctuations velocity and ℓ is the turbulent mixing length. v and ℓ are assumed to be approximately constant in the bulk of the flow ($v = v_0$ and $\ell = \ell_0$) and to be damped across the same region of thickness δ near the interface from their bulk value down to zero at the surface. Considerations of continuity require that the damping of v near the interface should be linear. Dimensional analysis shows that ℓ should be proportional to the distance to the free surface. Thus

$$\varepsilon \sim \frac{v_0 \ell_0}{\delta^2} y^2 \tag{56}$$

which indicates once more, by virtue of Equation 52, that $n = 2$.

The n = 2 result, indicating a proportionality ratio between the mass transfer coefficient and the square root of the diffusivity Equation 53 has been supported by numerous experiments in which $\bar{K}_L$ was measured. These experiments will be discussed further later in this section. Thus, combining Equations 56 and 53 with n = 2 yields:

$$\bar{K}_L \sim \left(\frac{v_0 \ell_0 D}{\delta^2}\right)^{1/2} \tag{57}$$

In order to determine the value of a in the eddy diffusivity expression, some assumptions must be made regarding the dependence of v_0, ℓ_0, and δ on the flow parameters. This dependence expresses the mechanism of eddy damping near the interface, for which different postulations were made by various researchers. A comparison of the different physical models was given by Henstock and Hanratty [114] and more recently by Bin [115]. Table 1 lists the relations suggested for a by different authors.

Levich [15] assumed the eddy damping at the interface to be due to surface tension. In this respect, the surface acts as a flexible membrane which deforms as it is impinged upon by an eddy and retards its motion. Thus, the thickness of the damping layer is calculated as $\delta \sim \sigma/\varrho v_0^2$. Levich took $\ell_0 \sim \delta$, and v_0 was substituted by the friction velocity

$$v^* = \sqrt{g\bar{h}} \tag{58}$$

Thus, a is found from Equation 56 to be proportional to $\varrho v^{*3}/\sigma$. Levich's model [15] was extended later by Davies [116, 117] to include the effect of surface active substances on the interface. The surface tension σ in Levich's formula for a is then substituted by an equivalent value taking into consideration the stronger damping effect of those substances. It should be noted that the term $\varrho v^{*3}/\sigma$ has the dimensions of 1/t and may be considered as the surface renewal rate S Equation 55 in Danckwerts' [113] model.

Kishinevskiy and Korniyenko [118] attempted to link the Levich–Davies model with Equation 53 for n = 2. They estimated the value of the constant multiplying the term $\varrho v_0^3/\sigma$ in the expression for a and found it to be about 15 by fitting their model equation to experimental data from different authors.

King [111] solved the diffusion equation with the eddy diffusivity given by Equation 52 for various values of n and showed by dimensional analysis that a dimensionless mass transfer coefficient $(K_L/a^{1/n}D^{1-(1/n)})$ is a function of the dimensionless exposure time $(a^{2/n}t/D^{(2/n)-1})$ and n only. For large exposure times he derived Equation 53. He suggested a probable value of n somewhere between 2 and 4 based on experimental data on gas absorption across the free interface in a turbulent agitated vessel, where $\bar{K}_L$ varied as D to a power ranging form 0.5 to 0.75. By assuming a to depend on the energy dissipation rate per unit volume e, the density ϱ, and the kinematic viscosity ν, relations between a and those parameters were derived for specific values of n. For a general n, dimensionless analysis yields $a \sim (e^n/\nu^{3n-4}\varrho^n)^{1/4}$. The energy dissipation rate in a falling film is given by

$$e = \varrho g\bar{u} \tag{59}$$

Note that this value of a, when substituted into Equation 53 yields $\bar{K}_L \sim (e/\varrho)^{1/4}$, independent of n.

Prasher and Fricke [119] proposed an eddy diffusivity model where damping of the eddies is affected by both the surface tension and the viscosity of the liquid. Thus, the features of both the Levich–Davies and King models are incorporated. Form dimensional considerations, they proposed $a \sim (\varrho/\sigma)(\bar{h}^3e^3/\varrho^3 g\nu)^{1/2}$. By correlation with experimental data, they were able to convert this proportionality relation to the equation given in Table 1.

Carrubba [120] proposed a model based on the Prandtl mixing length and related a to the film thickness and the kinematic viscosity of the liquid from dimensional considerations.

By fitting the experimental data of different researchers obtained for long exposure times (long films) with large Schmidt numbers, he obtained the empirical relation given in Table 1.

Henstock and Hanratty [114] proposed the viscosity damped turbulence model (VDTM) in which eddy damping is due completely to viscosity and surface tension does not play a role. They assumed $\ell_0 \sim \bar{h}$, $v_0 \sim v^*$ and from dimensional considerations, $\delta \sim \nu/v_0$. Thus, from Equation 56, they obtained $a \sim v^{*3}\bar{h}/\nu^2$.

Several investigators [121–124] proposed applying a circulation cell model to turbulent flow of the type discussed in the section entitled "Absorption in the Wavy Flow Regime" with regard to wavy flow. As pointed out by Bin [115], these authors had to make numerous assumptions concerning characteristic turbulence scales and wave parameters in order to estimate mass transfer coefficients from their models. In other respects the circulation cell models are equivalent to the eddy diffusivity models. However, the latter approach seems preferrable as it does not involve as many uncertainties, which are introduced into the former by the various assumptions.

Bin [115] proposed comparing the different eddy diffusivity models by relating the mass transfer coefficient from each of them to the flow Reynolds number. The turbulent film thickness may be calculated from a Nusselt film thickness-type formula

$$\bar{h} = \text{const} \times \left(\frac{\nu^2}{g}\right)^{1/3} \text{Re}^m \tag{60}$$

This makes it possible to evaluate the friction velocity and the energy dissipation rate from Equations 58 and 59 respectively. Substituting the resulting quantities in the expressions for those listed in Table 1 yields the dependence of the mass transfer coefficient on the Reynolds number in terms of the free parameter m. Bin [115] used $m = 0.526$, following the work of Takahama and Kato [125] and calculated the powers of the Reynolds numbers explicitly in the $\bar{K}_L$ expression. The results are given in Table 1 for each of the models described. A discussion of these results in the light of correlated experimental data is given next.

Experimental Correlations

Experimental data in the transition and turbulent flow regime are available from about a dozen sources, which enabled researchers to formulate correlations and draw conslusions regarding the validity of the turbulent models discussed. In order to be able to use Equation 53 for the relation between the eddy diffusivity parameter a and the mass transfer coefficient, it is important to choose data collected for fully developed concentration profiles.

Mass transfer data for the Reynolds number range in question has been obtained for the most part in wetted-wall columns under isothermal conditions. Measured quantities included liquid and gas flow rates, inlet and outlet concentrations, column diameter and height, and vapor pressures and temperatures. The latter were used to calculate the values of the physical and transfer properties and the equilibrium concentration at the surface. The film thickness was calculated by most researchers from Brotz's formula Equation 9. Average mass transfer coefficients were computed from this data based on the difference (or logarithmic mean difference) between the surface and bulk concentrations.

Henstock and Hanratty [114] and Bin [115] provided summary listings of the significant experimental studies and their conditions. Most studies were concerned with the absorption of various gases, particularly carbon dioxide, in water. They include the work of Kamei and Oishi [99], Emmert and Pigford [7], Lamourelle and Sandall [112], Menez and Sandall [126], Coeuret et al. [127], Davies and Warner [128], Chung and Mills [129], Henstock and Hanratty [114], Oliver and Atherinos [107], Stirba and Hurt [130], Vyazovov [56], Yih and Chen [14], Hikita et al. [108], and Koziol et al. [131]. Recent studies by Mills and co-workers [129, 132] considered the absorption of CO_2, H_2 and O_2 in organic liquids—various alchohols and aqueous solutions of propanol and ethylene glycol.

The most extensive correlations of experimental data are the recent ones by Bin [115] and Yih and Chen [14]. Bin's correlations are limited to the turbulent flow regime whereas Yih and Chen have also covered the transition and part of the wavy Reynolds number range. Both researchers found it convenient to correlate $K_L D^{-1/2}$ with Re, following the behavior predicted by the turbulence models. The results of Yih and Chen [14] are shown in Figure 15 showing clearly, by the change of slope, the regions of wavy, transition, and fully turbulent flow.

Bin's [115] correlations showed a dependence of the mass transfer coefficient on the Reynolds number to a power of approximately 0.7. A comparison of this experimental value of the Reynolds exponent to the values resulting from the models listed in Table 1 shows that the Levich–Davies and King models yield significantly lower power exponents of Re. Thus, it can be assumed that these models are inadequate for predicting mass transfer coefficients in the turbulent falling film. Of the remaining models, the one suggested by Carrubba [120] was found by Bin to give the best agreement with the 520 experimental data points from the different sources. Thus,

$$\frac{K_L}{D}\left(\frac{v^2}{g}\right)^{1/3} = 0.689 \times 10^{-3}\, Re^{0.724}\, Sc^{1/2} \tag{61}$$

While the data from absorption in water agreed well with the correlation (Equation 61), greater deviations from it were noted with some of the organic solvents. These may be attributed, according to Bin, to possible effects of interfacial turbulence, different wavy structure and inaccuracies in molecular diffusion coefficients. It should be noted that Won and Mills [132] who performed the experiments with the organic solvents, obtained a deviation from the linear relation between K_L and $D^{1/2}$, and suggested a correlation of the mass transfer coefficient with D to a power exponent dependent on surface tension. There is presently no other source of data of similar type with organic solvents to support or contradict these findings.

Another important conclusion from Bin's correlations of experimental data is that surface tension does not seem to affect mass transfer coefficients in turbulent film flow. This conclusion has been reached independently by Yih and Chen [14] who based their correlations on the viscosity damped turbulence model (VDTM) [114]. Based on 846 experimental data points on absorption of different gases in water, from the work of ten investigators

Table 1
Comparison of Eddy Damping Mechanisms at the Gas-Liquid Interface

Author	Damping Mechanisms	a	n	$\bar{K}_L$	$\bar{K}_L$ with m = 0.526
Levich [15] Davies [116, 117]	Surface tension	$\sim\left(\frac{\rho v^{*3}}{\sigma}\right)$	2	$\sim Re^{3m/4}$	$\sim Re^{0.395}$
Kishinevskiy and Korniyenko Davies [118]	Surface tension	$=15\left(\frac{\rho v^{*3}}{\sigma}\right)$	2	$\sim Re^{3(1-m)/2}$	$\sim Re^{0.711}$
King [111]	Viscous dissipation	$\sim\left(\frac{e^n}{\nu^{3n-4}\rho^n}\right)^{1/4}$	Various	$Re^{(1-m)/4}$	$\sim Re^{0.12}$
Prasher and Fricke [119]	Surface tension/ viscosity	$=0.293 \times 10^{-2}\left(\frac{\bar{h}^3 e^3}{\rho g \nu \sigma^2}\right)^{1/2}$	2	$Re^{3/4}$	$\sim Re^{0.75}$
Carrubba [120]	Viscosity	$=6.09 \times 10^{-8}\left(\frac{\nu}{\bar{h}^2}\right) Re^{2.5}$	2	$Re^{1.25-m}$	$\sim Re^{0.724}$
Henstock and Hanratty [114]	Viscosity	$\sim(\bar{h}v^{*3}/\nu^2)$	2	$Re^{5m/4}$	$\sim Re^{0.658}$

After 13 in [115].

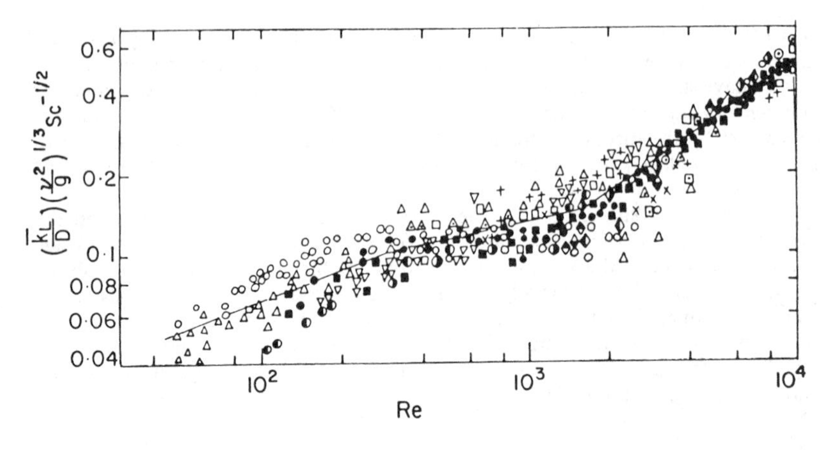

Figure 15. Experimental results for normalized mass transfer coefficient as a function of Reynolds number (from Yih and Chen [14]).

in addition to their own, they proposed the following empirical relations for the mass transfer coefficients:

For 49 < Re < 300

$$\frac{K_L}{D}\left(\frac{v^2}{g}\right)^{1/3} = 1.099 \times 10^{-2}\, Re^{0.3955}\, Sc^{1/2}$$

For 300 < Re < 1,600

$$\frac{K_L}{D}\left(\frac{v^2}{g}\right)^{1/3} = 2.995 \times 10^{-2}\, Re^{0.2134}\, Sc^{1/2}$$

For 1,600 < Re < 10,500

$$\frac{K_L}{D}\left(\frac{v^2}{g}\right)^{1/3} = 9.777 \times 10^{-4}\, Re^{0.6804}\, Sc^{1/2} \tag{62}$$

Theoretical Models

The experimental verification of the eddy diffusivity models previously described by measurements of mass transfer coefficients has invalidated some of them and has shown the others to provide a realistic mathematical expression of the turbulent transfer mechanism at the gas-liquid interface. These eddy diffusivity formulas may now be used in theoretical analyses to predict the heat and mass transfer behavior in turbulent absorption under different conditions. It should be recalled that the experimental correlations Equations 61 and 62 for the mass transfer coefficient, as well as Equation 53 relating K_L to the eddy diffusivity parameters a and n, are only valid for the case of fully developed concentration profiles. Many cases of interest deviate from these conditions. They include turbulent films with short and intermediate exposure times, absorption accompanied by chemical reaction, nonisothermal absorption, and more. The purpose of the theoretical analyses by numerous researchers described in this subsection has been to predict the behavior under those conditions.

The analytical studies are generally based on the solution of the diffusion equation in the two dimensional film coordinates, with the diffusion coefficient expressed as the sum of the molecular and eddy diffusivities. In nonisothermal cases a simultaneous solution of the energy equation is required. Assumptions similar to those listed for laminar films have been made, i.e., constant properties, thermodynamic equilibrium at the interface, and impermeable wall.

Theoretical modeling of pure physical absorption in a turbulent film under isothermal conditions was first discussed by King [111]. He showed that the turbulent diffusion equation could be solved without a priori specification of the values of a or n in the eddy diffusivity expression Equation 52. By dimensional analysis, King showed that a solution may be obtained in terms of two parameters—a normalized mass transfer coefficient ψ and a normalized length X in the form

$$\psi = f(X, n) \tag{63}$$

where

$$\psi = K_L/(a^{1/n}D^{1-(1/n)}) \tag{64}$$

$$X = (a^{2/n}x)/(D^{(2/n)-1}u_s) \tag{65}$$

He obtained two asymptotic solutions, for the cases of very short contact times and very long contact times. In the former case, the eddy diffusivity term is negligibly small compared to the molecular diffusivity, and the classical penetration theory solution Equation 24 results. In the latter case, the effect of X in Equation 63 becomes unimportant and the solution is given by Equation 53.

Experimental data as well as physical considerations regarding the eddy diffusivity behavior near the interface indicate strongly that the exponent n is equal or at least close to 2. This evidence prompted several researchers to obtain the solution in the form of Equation 63 for $n = 2$, not only for the two limiting cases of very short and very long contact times but for the entire range of X. Sandall [133] presented results of numerical calculations in terms of a single formula, good for intermediate contact times, $X \leqq 2.5$. Subramanian [134] employed a method of solution for short contact times, which gave more accurate results than Sandall's approximation. Kishinevskiy and Korniyenko [118] proposed an equation for the mass transfer coefficient valid in the entire range of $X > 0$. Yih and Seagrave [135] presented an eigenvalue solution in the form of a series of eigenfunctions, and Gottifredi and Quiroga [136] obtained an approximate solution covering the entire range of the dimensionless length X. The results obtained by these investigators were tabulated by Bin [115] for values of X from 0.01 to 1,000 and are plotted in Figure 16. The solution by Yih and Seagrave [135] seems to yield less accurate results for small and large values of X, apparently because of the limited number of terms in the series calculation.

Most of the investigators have made two simplifying assumptions in their solution of the turbulent diffusion equations. First, the velocity was taken to be uniform throughout the film and equal to its surface value u_s; second, the eddy diffusivity profile of Equation 52 which is strictly correct for the interface region only was taken to be valid over the complete film thickness. These assumptions do not lead to any significant error in the case of isothermal absorption with an impermeable wall, since the major resistance to mass transfer is concentrated close to the interface, particularly with large Schmidt numbers. These assumptions cannot be made, however, in nonisothermal absorption with heat transfer through the wall. In this case, the correct eddy diffusivity behavior across the film thickness must be taken into account. Mills and Chung [137] and later Hubbard et al. [138] calculated the overall heat transfer coefficient for evaporation from the surface of a falling using an eddy diffusivity profile composed of two different expressions—one which was valid near the wall and the other near the interface. They did not find it necessary to accurately specify the eddy diffusivity in the middle region of the film in view of the low thermal resistance there. Their calculated results were in reasonable agreement with experimental data by Chun and Seban [139]. Later, Seban and Faghri [140] solved the same problem with an eddy diffusivity profile composed of three parts—for the wall, interface, and middle regions of the film—obtained better agreement with the experiments, and demonstrated the importance of modeling correctly not only the regions of low eddy diffusivity near the wall and the surface but also the turbulent core.

Grossman and Heath [110] used these findings in analyzing the nonisothermal absorption process in turbulent films involving combined heat and mass transfer. The turbulent diffusion and energy equations were solved simultaneously with an eddy diffusivity profile over the film thickness composed of three parts, as shown in Figure 14. Expressions for the wall, middle, and interface regions were taken from the work of Van Driest [141], Reichardt [142], and Won and Mills [132], respectively. The temperature and concentration distributions in the film were obtained for two cases of interest. In one, the wall was kept at a constant temperature equal to the inlet temperature of the film. In the other, the wall was adiabatic. Assumptions of constant properties, thermodynamic equilibrium at the surface and impermeable wall were made, as in the isothermal case. An analytical solution was obtained for very short contact times, and the equations were solved numerically for intermediate and long times.

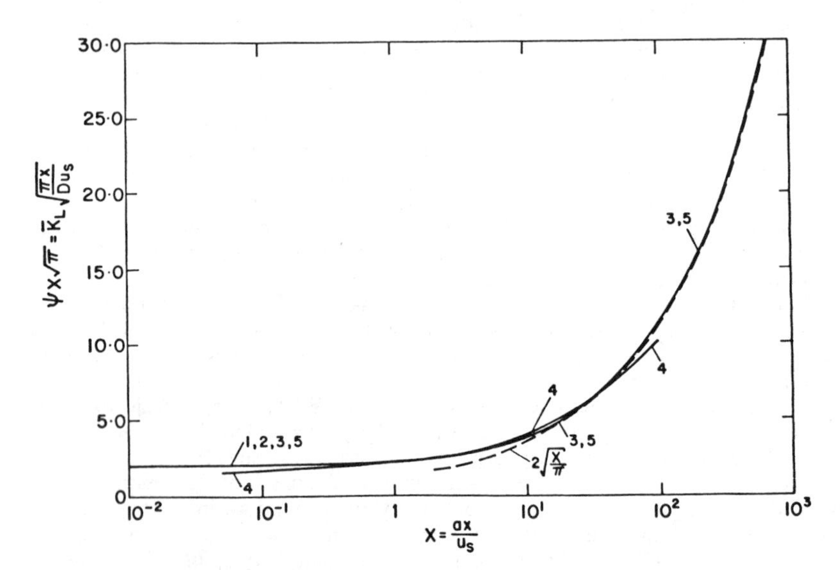

Figure 16. Normalized mass transfer coefficient in turbulent films as a function of dimensionless exposure time from solutions of various authors: (1) Sandall [133], $X \leq 2.5$; (2) Subramanian [134], $X \leq 1.0$; (3) Kishinevskiy and Kornienko [118]; (4) Yih and Seagrave [135]; (5) Gottifredi and Quiroga [136].

The normalized temperature and concentrations θ and γ, defined by Equations 16 and 17 as functions of the dimensionless length (or exposure time) ζ (Equation 24), were found to depend on four characteristic parameters of the problem: the Reynolds, Prandtl, and Schmidt numbers, and the dimensionless heat of absorption λ Equation 18. The behavior is qualitatively the same as in the laminar flow case [55], illustrated in Figure 17. However, the development rate of the profiles is much more rapid, and the resulting transfer coefficients considerably greater. Figure 17 shows the effect of the Reynolds number on the dimensionless bulk concentration for typical values of λ, Pr, and Sc. The laminar flow case is plotted along for comparison. For very short contact times, or very small length, the effect of the turbulence is small and the behavior is of the penetration-theory type, identical to that in laminar flow Equations 16 and 17.

The effect of chemical reaction on the turbulent absorption process was analyzed by several investigators. Menez and Sandall [126] considered isothermal absorption accompanied by a first-order chemical reaction in long films with fully developed concentration profiles. They used the eddy diffusivity expression formulated earlier by Lamourelle and Sandall [112] for the near-surface region, and assumed it to be valid over the entire film thickness. As mentioned earlier in the discussion of pure physical absorption, this assumption does not lead to a significant error for an isothermal process with large Schmidt numbers. These authors compared their results with experimental data of carbon dioxide absorption in aqueous solutions of carbonate-bicarbonate in the presence of an arsenite catalyst and obtained good agreement. Kayihan and Sandall [143] extended their model [126] to the entire range of film length, assuming a uniform velocity profile. They obtained asymptotic analytical solutions for very short and very long films and a numerical solution for the intermediate range. They correlated their results and proposed the following expression for the enhancement factor due to the first-order reaction:

$$\Phi = \frac{K_L^*}{K_L} = \left[\frac{\exp(-2k_1x/\bar{u}) + \pi k_1 x/\bar{u}(1+4a/k_1\pi^2)}{1+4ax/\pi\bar{u}}\right]^{1/2} \tag{66}$$

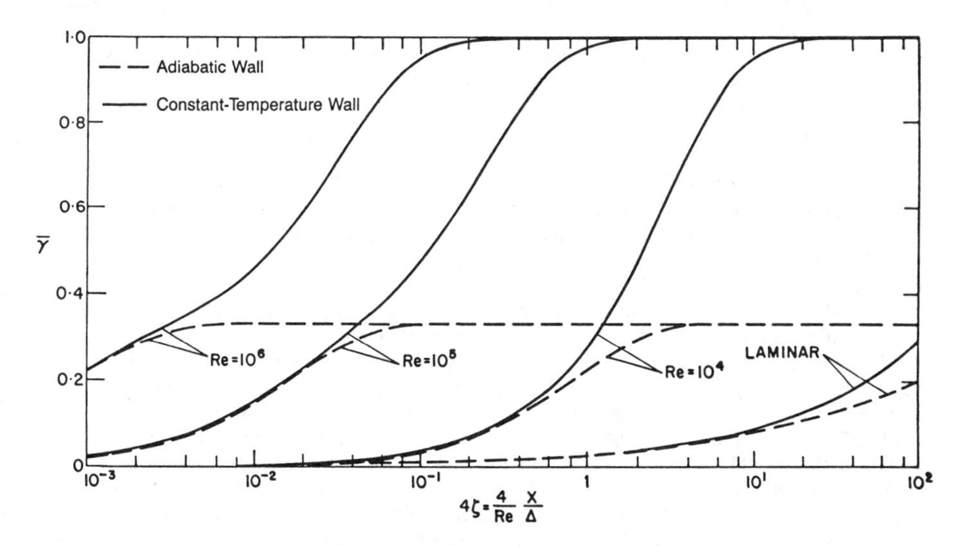

Figure 17. Dimensionless bulk concentration as a function of the normalized length, for λ = 0.01, Pr = 10, Sc = 2,000, and different values of the Reynolds number (from Grossman and Heath [110]).

where a is the eddy diffusivity coefficient in Equation 52, k_1 is the first-order reaction constant Equation 35 and x is the distance along the interface. Yih and Seagrave [135] developed an analytical eigenvalue solution to the same problem with zero-order and first-order chemical reaction, which is in good agreement with the numerical results of [143]. Gottifredi and Quiroga [144] solved the same problem with further extension to include both reversible and irreversible first-order reactions.

Sandall and co-workers next considered bimolecular reactions of the second order Equation 37. Mendez and Sandall [145] solved the turbulent diffusion equations under the same assumptions and with the same eddy diffusivity profile described earlier for absorption with an instantaneous bimolecular chemical reaction, in the limit of a long film (fully developed concentration profiles). Matheron and Sandall [146] analyzed the case of a finite rate bimolecular reaction of the second order in the short contact time limit (penetration and surface renewal models). It was pointed out that the surface renewal theory of Danckwerts is equivalent in many ways to the eddy diffusivity model. The former theory was used by Haimour and Sandall [147] to analyze turbulent absorption accompanied by an irreversible second-order reaction, where the eddy diffusivity approach cannot account for certain additional terms arising in the time-averaged diffusion equation due to the nonlinear kinetics. The authors used experimental data from pure physical absorption of carbon dioxide in water to determine the surface renewal rate S in Danckwerts' formula Equation 55, then applied it to the case with the second-order chemical reaction. The enhancement factor for the mass transfer coefficient can thus be calculated.

Factors Enhancing Turbulent Absorption

Chemical reaction accompanying the turbulent absorption is only one of the factors contributing to the enhancement of heat and mass transfer. As in laminar and stagnant films, surface convection due to the Marangoni effect can increase the transfer coefficients considerably. Other factors, which cannot be discussed in detail under the present scope, will be mentioned briefly.

Geometrical irregularities and protrusions in the surface of the wall cause the formation of flow instabilities which, under the large Reynolds numbers of turbulent flow, can grow and raise the turbulence level in the film. Davies and Warner [148] have obtained experimentally enhancement factors up to 3.5 in the mass transfer coefficients due to wall roughness. Koziol et al. [149] and other researchers have studied this effect quantitatively and obtained empirical correlations for the increased Sherwood number under the effect of wall roughness.

The effect of interfacial shear due to gas flow was considered by Banerjee, Scott, and Rhodes [150], Kasturi and Stepanek [151], Chung and Mills [129] and Henstock and Hanratty [114]. Their experiments have shown a considerable transfer enhancement obtained in the case of cocurrent flow, presumably due to the increased eddying effect near the free surface. The state of knowledge of this effect at this point is purely empirical.

From the previous discussion of eddy diffusivity models for the near-surface region and their comparison with experimental data it appears that the viscosity-damped turbulence model proposed by Henstock and Hanratty [114], with slight modifications by others, is the closest to reality. This model and the experimental results indicate that surface tension has little or no effect in damping turbulence. Yet, the experiments by Mills and co-workers [129, 132] with organic absorbents do indicate a surface tension influence, although not of a nature that could be explained by the Levich–Davies model. Despite the lack of an adequate theoretical model at this point, it seems reasonable to assume on the basis of these experiments that surface tension could have a turbulence damping effect under some conditions, perhaps secondary to the effect of viscosity. The surface tension influence is no doubt one of the important subjects requiring further investigation.

FUNDAMENTAL TRANSFER EQUATIONS

Earlier sections of this chapter have described theoretical models for film absorption under different flow conditions, which were often based on the solution of the momentum, energy, and diffusion equations. Reference to these equations has been made without discussing them in detail. The present section describes these fundamental transfer equations which govern the film absorption process. Here, the interested reader can find the theoretical basis for the analytical results presented in the rest of the chapter.

Flow Equations

The flow of a gravity-driven film of a viscous incompressible liquid of constant physical properties is governed by the Navier–Stokes (momentum) and continuity equations. In the two-dimensional rectangular coordinate system described in Figure 1, these equations, in their general form, may be written as:

$$\frac{\partial u}{\partial t} + u\frac{\partial u}{\partial x} + v\frac{\partial u}{\partial y} = -\frac{1}{\varrho}\frac{\partial p}{\partial x} + g\sin\varphi + \nu\left(\frac{\partial^2 u}{\partial x^2} + \frac{\partial^2 u}{\partial y^2}\right) \tag{67}$$

$$\frac{\partial v}{\partial t} + u\frac{\partial v}{\partial x} + v\frac{\partial v}{\partial y} = -\frac{1}{\varrho}\frac{\partial p}{\partial y} - g\cos\varphi + \nu\left(\frac{\partial^2 v}{\partial x^2} + \frac{\partial^2 v}{\partial y^2}\right) \tag{68}$$

$$\frac{\partial u}{\partial x} + \frac{\partial v}{\partial y} = 0 \tag{69}$$

where u and v are the x and y components of the velocity vector, respectively, and p is the pressure. For the case of thin films, these equations may be simplified substantially by considerations similar to those of boundary-layer flow. First, it is realized that the flow is quasi-one-dimensional, and that the order of magnitude of the y-momentum Equation 68, is much smaller than that of the x-momentum Equation 67. Second, the velocity derivatives in the y direction (across the film) are considerably larger than in the x direction (the direction of flow). Thus, a single equation may be obtained for the x-component of the velocity by combining Equations 67 and 69:

$$\frac{\partial u}{\partial t} + u\frac{\partial u}{\partial x} - \left(\int \frac{\partial u}{\partial x}\,dy\right)\frac{\partial u}{\partial y} = -\frac{1}{\varrho}\frac{\partial p}{\partial x} + g\sin\varphi + \nu\frac{\partial^2 u}{\partial y^2} \tag{70}$$

whereas Equation 68 indicates a constant pressure at each x across the film. The boundary conditions are:

$$\mu\frac{\partial u}{\partial y} = \tau_g \quad \text{at} \quad y = h(x) \tag{71}$$

$$u = v = 0 \quad \text{at} \quad y = 0 \tag{72}$$

where Equation 71 relates the velocity gradient at the gas-liquid interface to the shear which may be imposed by the gas flow.

Since the film is thin, its surface curvature cannot be large. Therefore, the pressure at the surface, which is the same all across the thickness, may be related to the surface tension as [15]:

$$p = -\sigma\frac{d^2h}{dx^2} \tag{73}$$

Substitution into Equation 70 yields:

$$\frac{\partial u}{\partial t} + u\frac{\partial u}{\partial x} - \left(\int \frac{\partial u}{\partial x} dy\right)\frac{\partial u}{\partial y} = \frac{\sigma}{\varrho}\frac{\partial^3 h}{\partial x^3} + g \sin \varphi + \nu\frac{\partial^2 u}{\partial y^2} \tag{74}$$

Equation 74 is the general equation for the liquid flow in a thin film.

In the case of a steady laminar flow with a smooth interface, h = constant, $u = u(y) \neq u(x, t)$, and Equation 74 reduces to:

$$\nu\frac{\partial^2 u}{\partial y^2} + g \sin \varphi = 0 \tag{75}$$

which can be integrated twice with respect to y with the boundary conditions of Equations 71 and 72 to give a parabolic velocity profile. In the absence of interfacial shear ($\tau_g = 0$), the result is Equation 23. Another integration over the film thickness to find the total flow rate yields Nusselt's Equation 2.

For the case of a wavy-laminar flow, the parabolic velocity profile still holds, and Equation 74 may be averaged over y [15], to replace the x-velocity component u(x, y, t) by its local average $\bar{u}(x, t)$:

$$\frac{\partial \bar{u}}{\partial t} + \frac{9}{10}\bar{u}\frac{\partial \bar{u}}{\partial x} = \frac{\sigma}{\varrho}\frac{\partial^3 h}{\partial x^3} - \frac{3\nu\bar{u}}{h^2} + g \sin \varphi \tag{76}$$

For small-amplitude waves of the type given by Equation 3, Equation 76 may be averaged over time and the wave parameters related to the flow characteristics, as discussed earlier in this chapter.

In the case of turbulent flow, an eddy diffusivity model version of Equation 74 may be used. u is replaced by its time-average value (eliminating the turbulent fluctuations) and the kinematic viscosity ν is replaced by the sum of the molecular and eddy viscosities. Thus:

$$g \sin \varphi + \frac{\partial}{\partial y}\left[(\nu + \varepsilon)\frac{\partial u}{\partial y}\right] = 0 \tag{77}$$

and in the absence of interfacial shear,

$$u = g \sin \varphi \int_0^y \frac{(h - y)}{(\nu + \varepsilon)} dy \tag{78}$$

Clearly, the solution of Equation 78 requires the eddy diffusivity distribution with y.

The Diffusion Equation

The transport of the absorbate solute in a liquid absorbent film is governed by the diffusion equation, which expresses a mass balance for a differential liquid element. Written in the two-dimensional rectangular film coordinates (Figure 1) the diffusion equation has the general form:

$$\frac{\partial C}{\partial t} + u\frac{\partial C}{\partial x} + v\frac{\partial C}{\partial y} = D\left(\frac{\partial^2 C}{\partial x^2} + \frac{\partial^2 C}{\partial y^2}\right) + r \tag{79}$$

The first term on the left-hand side describes the increase in absorbate concentration with time inside the liquid element; the second and third terms describe the convective mass flux due to the velocity field. The terms on the right-hand side describe the diffusive mass transfer due to the concentration gradients and the rate, r, at which absorbate is generated or depleted by chemical reaction, when present. Under most conditions of film flow the diffusion in the direction of motion may be neglected with respect to that in the transverse direction since $\partial^2 C/\partial y^2 \gg \partial^2 C/\partial x^2$. The boundary conditions generally describe a uniform concentration $C = C_0$ at $t = 0$ and/or at $x = 0$. In addition, the wall is impermeable to diffusion, hence

$$\partial C/\partial y = 0 \quad \text{at} \quad y = 0 \tag{80}$$

and at the surface the gas and the liquid are in thermodynamic equilibrium

$$C = C_e \quad \text{at} \quad y = h \tag{81}$$

For a stagnant film under pure physical absorption Equation 79 becomes:

$$\frac{\partial C}{\partial t} = D \frac{\partial^2 C}{\partial y^2} \tag{82}$$

Under isothermal conditions, the equilibrium concentration at the surface is independent of temperature and the penetration theory solution, Equation 13 results. Note that in Equation 13 y is measured from the surface, as shown in Figure 3.

For steady-state laminar flow with a smooth interface, the velocity distribution is given by Equation 23 and Equation 79 becomes

$$u_s \left[2\left(\frac{y}{h}\right) - \left(\frac{y}{h}\right)^2 \right] \frac{\partial C}{\partial x} = D \frac{\partial^2 C}{\partial y^2} + r \tag{83}$$

Solutions of this equation have been described earlier for pure physical absorption ($r = 0$) and for various kinds of chemical reaction kinetics.

For the case of wavy laminar flow, the velocity characteristics obtained from the momentum equation may be substituted in Equation 79, which is then averaged over time. Solutions of this type have been described two sections earlier.

For the case of turbulent flow, D in Equation 79 is replaced by $(D+\varepsilon)$ and the time average is taken to eliminate the turbulent fluctuations. Then

$$u \frac{\partial C}{\partial x} = \frac{\partial}{\partial y} \left[(D+\varepsilon) \frac{\partial C}{\partial y} \right] + r \tag{84}$$

The equation may be solved with turbulent profiles for the velocity u and eddy diffusivity ε, of the type shown in Figure 14.

The Energy Equation

The energy equation expresses an energy balance for a differential liquid element. Neglecting diffusion thermo-effects [110] and viscous dissipation and considering the liquid to be incompressible, the equation may be written in the two-dimensional rectangular coordinates in the following general form:

$$\varrho c_p \left(\frac{\partial T}{\partial t} + u \frac{\partial T}{\partial x} + v \frac{\partial T}{\partial y} \right) = k \left(\frac{\partial^2 T}{\partial x^2} + \frac{\partial^2 T}{\partial y^2} \right) + rq \tag{85}$$

where q is the heat of reaction. The equation is similar in form to the diffusion Equation 79 with the terms on the left-hand side describing the change with time in energy content of the element and the convective heat flux, and the terms on the right-hand side describing the conductive heat flux and local heat generation by chemical reactions. Generally, $\partial^2T/\partial y^2 \gg \partial^2T/\partial x^2$ and the heat conduction in the direction of flow may be neglected with respect to that across the thickness.

The boundary conditions describe a uniform temperature $T = T_0$ at $t = 0$ and/or at $x = 0$. Thermodynamic equilibrium at the gas-liquid interface mandates a relation between the temperature, concentration and vapor pressure there, based on the properties of the absorbent-absorbate pair. At the wall, the temperature, its gradient, or a combination of both must be given. The cases of combined heat and mass transfer described in this chapter [55, 110] have dealt with an adiabatic wall ($\partial T/\partial y = 0$) and a constant-temperature wall at a temperature equal to that of the entering liquid ($T = T_0$), but the possibility also exists for a constant heat flux, constant heat transfer coefficient, or a constant temperature different from that of the liquid to be prescribed at the wall.

For a stagnant film under pure physical absorption Equation 85 becomes:

$$\frac{\partial T}{\partial t} = \alpha \frac{\partial^2 T}{\partial y^2} \qquad (86)$$

where α is the thermal diffusivity. Under nonisothermal conditions, Equation 86 coupled with Equation 82 yields the solution given by Equations 16 and 17.

For steady-state laminar flow with a smooth interface, Equation 85 with the velocity distribution Equation 23 becomes:

$$u_s\left[2\left(\frac{y}{h}\right)-\left(\frac{y}{h}\right)^2\right]\frac{\partial T}{\partial x} = \alpha \frac{\partial^2 T}{\partial y^2} + \frac{rq}{\varrho c_p} \qquad (87)$$

A solution of Equation 87 coupled with Equation 83 in the absence of chemical reaction [55] has been described in the section entitled "Absorption in Laminar Films."

For the case of turbulent flow, an eddy diffusivity model may be introduced in the same way as done with the diffusion equation. Equation 85 then becomes:

$$u\frac{\partial T}{\partial x} = \frac{\partial}{\partial y}\left[(\alpha+\varepsilon)\frac{\partial T}{\partial y}\right] + \frac{rq}{\varrho c_p} \qquad (88)$$

A solution of Equation 88 coupled with Equation 84 in the absence of chemical reaction [110] has been described in the previous section.

NOTATION

a	eddy diffusivity coefficient, Equation 52 (m^{-1})
a'	wave amplitude normalized with respect to average film thickness
A_n	coefficients in Equation 31
b	coefficient in Equation 36
B	normalized surface resistance, Equation 44
B_n	coefficients in Equation 32
c	wave propagation velocity (m/sec)
c_p	specific heat of liquid (J/kg-°C)
C	concentration of absorbate in solution ($mole/m^3$)
C_A, C_B	concentration of substances A, B ($mole/m^3$)
C_0	initial concentration of absorbate in solution ($mole/m^3$)

- C_e equilibrium concentration of solution at temperature T_0 (mole/m^3)
- C_{Ae} equilibrium concentration of substance A (mole/m^3)
- C_{B0} initial concentration of substance B (mole/m^3)
- d coefficient in Equation 46 (°K^{-1})
- D, D_A, D_B mass diffusivity in the liquid of the absorbate, component A, component B (m^2/sec)
- e turbulent energy dissipation rate, Equation 59 (J/m^3-sec)
- F_n eigenfunctions in Equation 31
- Fr Froude number $\bar{u}^2/g\bar{h}$
- g gravity (m/sec^2)
- G_n eigenfunctions in Equation 32
- h, $\bar{h}$ local/instantaneous, average film thickness (m)
- $\bar{H}a$ heat of absorption (J/mole)
- H_L, $\bar{H}_L$ local, average heat transfer film coefficient in liquid (W/m^2-°C)
- k thermal conductivity (W/m-°C)
- k_n, k_{mn} reaction rates coefficients, Equations 35 and 37
- K equilibrium constant
- K_i mass transfer coefficient for surface resistance (m/sec)
- K_L, $\bar{K}_L$ local, average mass transfer film coefficient in liquid (m/sec)
- K_L^*, $\bar{K}_L^*$ enhanced local, average mass transfer film coefficient in liquid (m/sec)
- ℓ, ℓ_0 mixing length, Equation 56 (m)
- Le Lewis number, D/α
- m power exponent in Equation 60; also order of chemical reaction
- Ma Marangoni number, Equation 22
- n power exponent in Equation 52; also order of chemical reaction
- N_T Nusselt film thickness parameter, Equation 6
- Nu Nusselt number, $H_L h/k$
- p pressure (N/m^2)
- Pr Prandtl number, ν/α
- q heat of reaction (J/mole)
- r chemical reaction rate (mole/m^3-sec)
- Ra Rayleigh number, Equation 21
- Re Reynolds number, Equation 1
- s, s_0 film surface area in presence, absence of waves (m^2)
- S surface renewal rate, Equation 55 (sec^{-1})
- Sc Schmidt number, ν/D
- Sh Sherwood number, $K_L h/D$
- t time (sec)
- t_c contact time, Equation 54 (sec)
- T temperature of liquid (°C)
- T_0 initial temperature of solution (°C)
- T_e equilibrium temperature of solution at concentration C_0 (°C)
- u, $\bar{u}$ local/instantaneous, average velocity in x-direction (m/sec)
- u_s value of u at the film surface (m/sec)
- v velocity component in y-direction (m/sec)
- v, v_0 characteristic turbulent fluctuations velocity (m/sec)
- v* friction velocity, Equation 58 (m/sec)
- x coordinate in direction of film flow (m)
- X dimensionsless film length, Equation 65
- y coordinate in direction normal to film flow (m)

Greek Symbols

- α thermal diffusivity (m^2/sec)
- α_n eigenvalues in Equations 31 and 32
- β thermal expansion coefficient (°C^{-1})
- γ normalized concentration $(C-C_0)/(C_e-C_0)$
- γ_i, γ_w, $\bar{\gamma}$ normalized concentration at interface, wall, bulk

- Γ liquid mass flow rate per unit breadth (Kg/sec-m)
- δ thickness of eddy damping region near interface (m)
- ε eddy diffusivity (m^2/sec)
- ζ normalized coordinate in direction of flow, Equation 24a
- η normalized coordinate in direction normal to flow, Equation 24b
- θ normalized temperature, $(T-T_0)/(T_e-T_0)$
- θ_i, θ_w, $\bar{\theta}$ normalized temperature at interface, wall, bulk
- λ dimensioness heat of absorption, Equation 18
- λ' wavelength (m)
- μ dynamic liquid viscosity (Kg/m-sec)
- ν kinematic liquid viscosity (m^2/sec)
- ϱ liquid density (Kg/m^3)
- σ surface tension (N/m)
- τ, τ_g shear stress, shear exerted by gas flow (N/m^2)
- φ angle of inclination of wall, measured from the horizontal
- Φ, $\bar{\Phi}$ local, average enhancement factor, Equation 33
- ψ dimensioness mass transfer coefficient, Equation 64

REFERENCES

1. Chien, S. F., and Ibele, W. E., *Int. J. Mech. Sci., 9:* 547 (1967).
2. Fulford, G. D., *Adv. Chem. Engng., 5:* 151 (1964).
3. Friedman, S. J., and Miller, C. O., *Ind. Engng Chem., 33:* 885 (1941)
4. Grimley, S. S., *Trans. Inst. Chem. Engrs, 23:* 228 (1945).
5. Dukler, A. E., and Bergelin, O. P., *Chem. Engng Prog., 48,* 557 (1952).
6. Brötz, W., *Chemie-Ingr-Tech., 26:* 470, (1954).
7. Emmert, R. E., and Pigford, R. L., *Chem. Engng Prog., 50:* 87, (1954).
8. Jackson, M. L., *A.I.Ch.E. J.,* 1, 231 (1955).
9. Brauer, H., *VDI Forsch., 457* (1956).
10. Thomas, W. J., and Portalski, S., *Ind. Engng Chem., 50,* 1081 (1958).
11. Binnie, A. M., *J. Fluid Mech., 5:* 561 (1959).
12. Lilleleht, L. U., and Hanratty, T. J., *A.I.Ch.E. J., 7:* 548 (1961).
13. Reinius, E., *Trans. Roy. Inst. Technol. Stockholm, 179* (1961).
14. Yih, S. M., and Chen, K. Y., *Chem. Eng. Commun., 17:* 123 (1982).
15. Levich, V. G., *Physicochemical Hydrodynamics,* Prentice-Hall, Englewood Cliffs, New Jersey (1962).
16. Nusselt, W., *Z. Ver. dt. Ing., 60:* 541 (1960).
17. Chien, S. F., *Trans ASME, J. Appl. Mech., 33:* 222 (1966).
18. Stepanek, J. P., and Colquohoun-Lee, I., *Chem. Engng J, 8:* 21 (1974).
19. Moalem, D., and Sideman, S., *Int. J. Heat Mass Transfer, 19:* 259 (1976).
20. Kapitsa, P. L., *Zh. Eksperim. i. Teor. Fiz., 18:* 3 (1948).
21. Benjamin, T. B., *J. Fluid Mech., 2:* 554 (1957).
22. Benjamin, T. B., *J. Fluid Mech., 10:* 401 (1961).
23. Hanratty, T. J., and Hershman, A., *A.I.Ch.E. J., 7:* 488 (1961).
24. Kapitsa, P. L., and Kapitsa, S. P., *Zh. Eksperim. i. Teor. Fiz., 19:* 105 (1949).
25. Nikuradse, J., *VDI-Forsch, 356* (1932). In W. M. Kays, *Convective Heat and Mass Transfer,* McGraw-Hill, New York (1966), p. 69.
26. Dukler, A. E., *Chem. Engng Prog., 55:* 62 (1959).
27. Zhivaykin, L. Y., and Volgin, B. P., *Zh. Prikl. Khim., 34:* 1236 (1961).
28. Feind, K., *VDI-Forsch, 481* (1960).
29. Higbie, R., *Trans. A.I.Ch.E., 31:* 365 (1935).
30. Nakoryakov, V. E., and Grigoreva, N. I., *Teor. Khim. Tekhnol. 14:* 483 (1980).
31. Berg, J. C., Boudart, M., and Acrivos, A., *J. Fluid Mech., 24:* 721 (1966)
32. Benard, H., *Ann. Chim. Phys., 23:* 62 (1901).

33. Strutt, J. W. S., (Lord Rayleigh), *Phil. Mag., 32:* 529 (1916) (also *Scientific Papers, 6,* 432).
34. Jeffreys, H., *Proc. Roy. Soc.,* A *118:* 195 (1928).
35. Low, A. R., *Proc. Roy. Soc.,* A *125:* 180 (1929).
36. Pellew, A., and Southwell, R. V., *Proc. Roy. Soc.,* A *176:* 312 (1940).
37. Pearson, J. R. A., *J. Fluid Mech., 4:* 489 (1958).
38. Nield, D. A., *J. Fluid Mech., 19:* 341 (1964).
39. Scriven, L. E., and Sternling, C. V., *J. Fluid Mech., 19:* 321 (1964).
40. Orell, A., Westwater, J. W., *Chem. Eng. Sci., 16:* 127 (1961).
41. Bakker, C. A. P., Van Buytenen, P. M., and Beek, W. J., *Chem. Eng. Sci., 21:* 1039 (1966).
42. Thomas, W. J., and McNicholl, E. K., *Trans. Inst. Chem. Engrs.,* London, *47:* T325 (1969).
43. Sternling, C. V., and Scriven, L. E., *A.I.Ch.E. J., 5:* 514 (1959).
44. Smith, K. A., *J. Fluid Mech., 24:* 401 (1966).
45. Vidal, A., and Acrivos, A., *Ind. Eng. Chem. Fundamentals, 7:* 53 (1968).
46. Scanlon, J. W., and Segel, L. A., *J. Fluid Mech., 30:* 149 (1967).
47. Berg, J. C., and Acrivos, A., *Chem. Eng. Sci., 20:* 737 (1965).
48. Brian, P. L. T., *A.I.Ch.E. J., 17:* 765 (1971).
49. Brian, P. L. T., and Ross, J. R., *A.I.Ch.E. J., 18:* 582 (1972).
50. Imaishi, N., Fujinawa, K., and Tadaki, T., *J. Chem. Eng. Japan, 13:* 360, (1980).
51. Yih, C. S., *Physics of Fluids, 11,* 477 (1968).
52. Ostrach, S., and Pradhan, A., *AIAA J., 16:* 419 (1978).
53. Birikh, R. V., *J. Appl. Mech. and Tech. Physics, 7* (3): 43 (1966).
54. Chifu, E., and Stan, I., *Revue Roumaine de Chimie, 27:* 703 (1982).
55. Grossman, G., *Int. J. Heat Mass Transfer, 26:* 357 (1983).
56. Vyazovov, V. V., *J. Tech. Phys.(U.S.S.R.), 10:* 1519 (1940).
57. Sherwood, T. K., Pigford, R. L., and Wilke, C. R., *Mass Transfer,* McGraw-Hill, New York (1975).
58. Rotem, Z., and Nielson, J. E., *Canad. J. Chem. Engng, 47:* 341 (1969).
59. Pigford, R. L., Ph.D. Thesis, Univ. Illinois, 1941. In R. E. Emmert and R. L. Pigford, *Chem. Engng Prog., 50:* 87 (1954).
60. Grigoreva, N. I., and Nakoryakov, V. E., *Inzh. Fiz. Zh., 33:* 893 (1977).
61. Emmert, R. E., and Pigford, R. L., *A.I.Ch.E. J., 8:* 171 (1962).
62. Danckwerts, P. V., *Trans. Faraday Soc., 46:* 300 (1950).
63. Sherwood, T. K., and Pigford, R. L., *Absorption and Extraction,* 2nd ed., McGraw-Hill, New York (1952).
64. Perry, R. H., Pigford, R. L., *Ind. Eng. Chem., 45:* 1247 (1953); *ibid, 49,* 1400 (1957).
65. Brian, P. L. T., Hurley, J. E., and Hasseltine, E. H., *A.I.Ch.E. J. 7:* 226 (1961).
66. Brian, P. L. T., *A.I.Ch.E. J., 10:* 5 (1964).
67. Jameson, G. J., Burchell, S. R. C., and Gottifredi, J. C., *Int. J. Heat Mass Transfer, 13:* 1629 (1970).
68. Sada, E., Kumazawa, H., and Butt, M. A., *A.I.Ch.E. J. 22:* 196 (1976).
69. Sada, E., Kumazawa, H., and Butt, M. A., *Canad. J. Chem. Engng., 54:* 421 (1976).
70. Pederson, H., and Prenosil, J. E., *Int. J. Heat Mass Transfer, 24:* 299 (1981).
71. Tamir, A., and Taitel, Y., *Chem. Eng. Sci., 26:* 799 (1971).
72. Shah, Y. T., *Chem. Eng. Sci., 27:* 1469 (1972).
73. Shair, F. H., *A.I.Ch.E. J., 17:* 920 (1971).
74. Yih, S. M., and Seagrave, R. C., *Int. J. Heat Mass Transfer, 23:* 749 (1980).
75. Gottifredi, J. C., Yeramian, A. A., and Ronco, J. J., *Chem. Engng. J., 3:* 163 (1972).
76. Chavan, V. V., and Maskelkar, R. A., *Chem. Engng J., 4:* 223 (1972).
77. Mashelkar, R. A., Chavan, V. V., and Karanth, N. G., *Chem. Engng J., 6:* 75 (1973).
78. Brian, P. L. T.; Vivian, J. E., and Matiatos, D. C., *A.I.Ch.E. J., 13:* 28 (1967).
79. Astarita, G., *Chem. Eng. Sci., 16:* 202 (1961).

80. Astarita, G., Marrucci, G., and Gioia, F., *Chem. Eng. Sci., 19:* 95 (1964).
81. Clarke, J. K. A., *Ind. Eng. Chem. Fundamentals, 3:* 239 (1964).
82. Brian, P. L. T., Vivian, J. E., and Mayr, S. T., *Ind. Eng. Chem. Fundamentals, 10:* 75 (1971).
83. Vijayan, S., and Ponter, A. B., *A.I.Ch.E. J., 18:* 647 (1972).
84. Sada, E., et al., *Canad. J. Chem. Engng., 55:* 293 (1977).
85. Sada, E., et al. *J. Chem. Engng. Japan, 10:* 487 (1977).
86. Fujinawa, K., Hozawa, M., and Imaishi, N., *J. Chem. Engng. Japan, 11:* 107 (1978).
87. Sada, E., et al. *A.I.Ch.E. J., 28:* 864 (1982).
88. Imaishi, N., and Fujinawa, K., *Int. Chem. Engng., 20:* 226 (1980).
89. Ruckenstein, E., and Berbente, C., *Chem. Eng. Sci., 25:* 475 (1970).
90. Ruckenstein, E., *A.I.Ch.E. J., 16:* 1098 (1970).
91. Imaishi, N., Hozawa, M., and Fujinawa, K., *J. Chem. Engng. Japan, 9:* 499 (1976).
92. Imaishi, N., et al. *Int. Chem. Engng., 23:* 466 (1983).
93. Imaishi, N., et al. *Int. Chem. Engng., 22:* 660 (1980).
94. Portalski, S., Ph. D. Thesis, Univ. London, 1960. In S. R. Tailby and S. Portalski, *trans. Inst. Chem. Engrs., 38:* 324 (1960).
95. Portalski, S., and Clegg, A. J., *Chem. Eng. Sci., 26:* 773 (1971).
96. Banerjee, S., Rhodes, E., and Scott, D. S., *Chem. Eng. Sci., 22:* 43 (1967).
97. Portalski, S., *Ind. Eng. Chem. Fundamentals, 3:* 49 (1964).
98. Massot, C., Irani, F., and Lightfoot, E. N., *A.I.Ch.E. J., 12,* 445 (1966).
99. Kamei, S., and Oishi, J., *Mem. Fac. Engng. Kyoto Univ., 17:* 277 (1955).
100. Penev, V., et al., *Int. J. Heat Mass Transfer, 15:* 1395 (1972).
101. Ruckenstein, E., and Berbente, C., *Chem. Eng. Sci., 20:* 795 (1965).
102. Ruckenstein, E., and Berbente, C., *A.I.Ch.E. J., 13:* 1205 (1967).
103. Ruckenstein, E., and Berbente, C., *Int. J. Heat Mass Transfer, 11:* 743 (1968).
104. Javdani, K., *Chem. Eng. Sci., 29:* 61 (1974).
105. Beschkov, B., and Boyadjiev, C., *Chem. Eng. Commun., 20:* 173 (1983).
106. Brauner, N., and Maron, D. M., *Int. J. Heat Mass Transfer, 25:* 99 (1982).
107. Oliver, D. R., and Atherinos, T. E., *Chem. Eng. Sci., 23:* 525 (1968).
108. Hikita, H., Nakanishi, K., and Kataoka, J., *Chem. Engng. Tokyo, 23:* 459 (1959).
109. Kays, W. M., *Convective Heat and Mass Transfer,* McGraw-Hill, New York (1966).
110. Grossman, G., and Heath, M. T., *Int. J. Heat Mass Transfer, 27:* 2365 (1984).
111. King, C. J., *Ind. Eng. Chem. Fundamentals, 5:* 1 (1966).
112. Lamourelle, A. P., and Sandall, O. C., *Chem. Eng. Sci., 27:* 1035 (1972).
113. Dankwerts, P. V., *Ind. Eng. Chem., 43:* 1460 (1951).
114. Henstock, W. H., and Hanratty, T. J., *A.I.Ch.E. J., 25:* 122 (1979).
115. Bin, A. K., *Int. J. Heat Mass Transfer, 26,* 981 (1983).
116. Davies, J. T., *Proc. Royal Soc.* (London) *A290:* 515 (1966).
117. Davies, J. T., *Turbulence Phenomena,* Academic Press, New York (1972).
118. Kishinevskiy, M. Kh., and Korniyenko, T. S., *Zh. Prikl. Khim., 50,* 2520 (1977).
119. Prasher, B. D., and Fricke, A. L., *I/EC Process Des. Dev., 13:* 336 (1974).
120. Carrubba, G., Ph. D. Thesis, Technische Universität Berlin. In A. K. Bin, *Int. J. Heat Mass Transfer, 26:* 981 (1983).
121. Fortescue, G. E., and Pearson, J. R. A., *Chem. Eng. Sci., 22:* 1163 (1967).
122. Banerjee, S., Scott, D. S., and Rhodes, E., *Ind. Eng. Chem. Fundamentals, 7:* 22 (1968).
123. Lamont, J. C., and Scott, D. S., *A.I.Ch.E. J., 16:* 513 (1970).
124. Brumfield, L. K., Houze, R. N., and Theofanous, T. G., *Int. J. Heat Mass Transfer, 18:* 1077 (1975).
125. Takahama, H., and Kato, S., *Int. J. Multiphase Flow, 6:* 203 (1980).
126. Menez, G. D., and Sandall, O. C., *Ind. Eng. Chem. Fundamentals, 13:* 72 (1974).
127. Coeuret, F., Jamet, B., and Ronco, J. J., *Chem. Eng. Sci., 25:* 17 (1970).
128. Davies, J. T., and Warner, K. V., *Chem. Eng. Sci., 24:* 231 (1969).
129. Chung, D. K., and Mills, A. F., *Int. J. Heat Mass Transfer, 19:* 51 (1976).

130. Stirba, C., and Hurt, D. M., *A.I.Ch.E. J., 1:* 178 (1955).
131. Koziol, K., Broniarz, L., and Nowicka, T., *Int. Chem. Engng., 20:* 136 (1980).
132. Won, Y. S., and Mills, A. F., *Int. J. Heat Mass Transfer, 25:* 223 (1982).
133. Sandall, O. C., *Int. J. Heat Mass Transfer, 17:* 459 (1974).
134. Subramanian, R. S., *Int. J. Heat Mass Transfer, 18:* 334 (1975).
135. Yih, S. M., and Seagrave, R. C., *Chem. Eng. Sci., 33:* 1581 (1978).
136. Gottifredi, J. C., and Quiroga, O. D., *Chem. Eng. J., 16:* 199 (1978).
137. Mills, A. F., and Chung, D. K., *Int. J. Heat Mass Transfer, 16:* 694 (1973).
138. Hubbard, G. L., Mills, A. F., and Chung, D. K., *Trans. ASME J. Heat Transfer, 98:* 319 (1976).
139. Chun, K. R., and Seban, R. A., *Trans. ASME J. Heat Transfer, 93:* 391 (1971).
140. Seban, R. A., and Faghri, A., *Trans. ASME J. Heat Transfer, 98:* 315 (1976).
141. Van Driest, E. R., *J. Aero. Sci., 23:* 1007 (1956).
142. Reichardt, H., *ZAMM, 31:* 208 (1951). In W. M. Kays, *Convective Heat and Mass Transfer,* McGraw-Hill, New York (1966) p. 71
143. Kayihan, F., and Sandall, O. C., *A.I.Ch.E. J., 20:* 402 (1974).
144. Gottifredi, J. C., and Quiroga, O. D., *Int. J. Heat Mass Transfer, 22:* 839 (1979).
145. Mendez, F., and Sandall, O. C., *A.I.Ch.E. J., 21:* 534 (1975).
146. Matheron, E. R., and Sandall, O. C., *A.I.Ch.E. J., 24:* 552 (1978).
147. Haimour, N., and Sandall, O. C., *A.I.Ch.E. J., 29:* 277 (1983).
148. Davies, J. T., and Warner, K. V., *Chem. Eng. Sci., 24:* 231 (1969).
149. Koziol, K., and Broniarz, L., *Int. Chem. Engng., 20:* 143 (1980).
150. Banerjee, S., Scott, D. S., and Rhodes, E., *Canad. J. Chem. Eng., 48:* 542 (1970).
151. Kasturi, G., and Stepanek, J. B., *Chem. Eng. Sci., 29:* 1849 (1974).

CHAPTER 7

MULTICOMPONENT MASS TRANSFER: THEORY AND APPLICATIONS

R. Krishna

Indian Institute of Petroleum
Dehradun, India

R. Taylor

Department of Chemical Engineering
Clarkson University
Potsdam, New York, USA

CONTENTS

The authors would like to thank R. Krishnamurthy, J. S. Furno, and D. J. Vickery for their help in the preparation of this chapter.

INTRODUCTION

Walter and Sherwood [1] in 1941, on the basis of an extensive experimental study of Murphree vapor and liquid plate efficiencies for absorption, desorption, and rectification operations, concluded "the results indicate that different efficiencies should be used for each component in the design of absorbers for natural gasoline and refinery gases." To the best of our knowledge this is the first published statement concerning the significant, and fundamental differences between the transfer behavior of two-component (i.e., binary) and multicomponent (three or more components) systems. Chemical engineers had to wait until 1957 for further enlightenment when Toor [2] in a pioneer paper described clearly the peculiar characteristics of ternary gas diffusion, i.e., the phenomena of *reverse diffusion* (diffusion against a driving force), *diffusion barrier* (zero diffusion flux even for non-zero driving force) and *osmotic diffusion* (finite diffusion flux despite a vanishing driving force).

Later in 1960 Toor and Burchard [3] worked out a design example to show that the differing component efficiencies, a peculiarity of multicomponent systems, could have a significant effect on column design. A general approach to the solution of multicomponent diffusion problems was made available in 1964 due to the pioneering, independent efforts of Toor [4] and Stewart and Prober [5]. This subject lay dormant until the early seventies, when a concerted program of experimental and theoretical research was initiated at the University of Manchester Institute of Science and Technology. The broad thrust of the efforts at UMIST was to determine the significance to chemical engineering design of multicomponent mass transfer "interaction" or "coupling" effects. Distillation, absorption, extraction, and condensation were the major areas of interest to the UMIST research group lead by the late Professor George Standart. The published work of the UMIST school catalyzed many parallel efforts in other institutions; an early state-of-the-art review [6] traces the development of the theory and applications of multicomponent mass transfer. Though it was shown, quite convincingly, that multicomponent mass transfer effects could, in some cases, have a significant effect in design; computer-efficient design procedures incorporating these advanced theories and models have been developed only within the last five or so years. Time is ripe to summarize the developments to date, but rather than present a historical account of the subject, we aim to provide a step-by-step pedagogical approach stressing the fundamental points and presenting usable design procedures for the industrial researcher or design engineer.

The discussions in this chapter will be limited to fluid-fluid transfer processes in the absence of chemical reactions and electrical or magnetic force fields. Throughout the discussions we have applications in the areas of distillation, absorption, extraction, evaporation, and condensation in mind; though, the formalism presented could with little or no modification be used for the description of transfer processes in heterogeneous reacting systems.

The subject of interphase mass transfer must logically begin with the *description of the interface* itself. The following section will discuss various approaches to the problem of modeling the interface. The four sections after that will consider how to set up the proper relations *describing the diffusion process* (molecular and turbulent) in the region close to the interface. This consists of choosing suitable concentration measures, driving forces, reference velocity frames, constitutive relations, etc. The published literature on this subject is largely restricted to binary (two component) systems; we shall be discussing in some detail the description of diffusion processes in multicomponent systems (i.e., mixtures of three or more components) because these systems are commonly encountered in industrial practice. After that, the next and usually the most important step—to *model the flow (hydrodynamics)* in the region close to the interface—will be discussed.

Once the choice is made with regard to the model for the interface, the constitutive relations describing the diffusion process, and the hydrodynamic flow model; it remains to *solve the diffusion equations* to obtain the mass transfer rates and the mass transfer coefficients. These mass transfer rates and mass transfer coefficients, of course, refer to local values, i.e., to conditions at any position in the separation or reaction equipment. Finally, we consider the use of these models in the simulation and design of practical chemical engineering equipment. Let us begin with the description of interfaces.

PHYSICAL DESCRIPTION OF THE INTERFACE

We shall be considering mainly fluid-fluid interfaces, though a greater part of the discussions will be valid for solid-fluid interfaces as well. Let us define a phase interface to be that region separating two bulk phases, phase "x" and phase "y" (see Figure 1). From a physical point of view, the boundary between two bulk phases is not a mathematical surface. Rather than a strictly two-dimensional surface on either side of which the bulk phases extend, the boundary is actually a three-dimensional, albeit thin, onion-skinlike, region across which the system's properties change rapidly, but not necessarily monotonically, in value from those of the bulk phase on one side to those of the other. This boundary between fluid phases has finite thickness—usually only a few molecular diameters thick but sometimes much thicker—and can have physical properties quite different from either of the contiguous bulk phases. For example, it is possible for the interfacial region to exhibit non-Newtonian behavior, even when the adjoining bulk phases exhibit Newtonian character. Another complication is that the interfacial region may be anisotropic, with grossly different properties in the normal and tangential directions. Even though the interfacial region occupies a minute fraction of the system volume, the behavior of the system is often strongly influenced by the properties of the interface. Stability of foams and falling films, behavior of drops in another fluid, and performance of gas-liquid dispersions are influenced by interfacial effects and properties of the interfacial region.

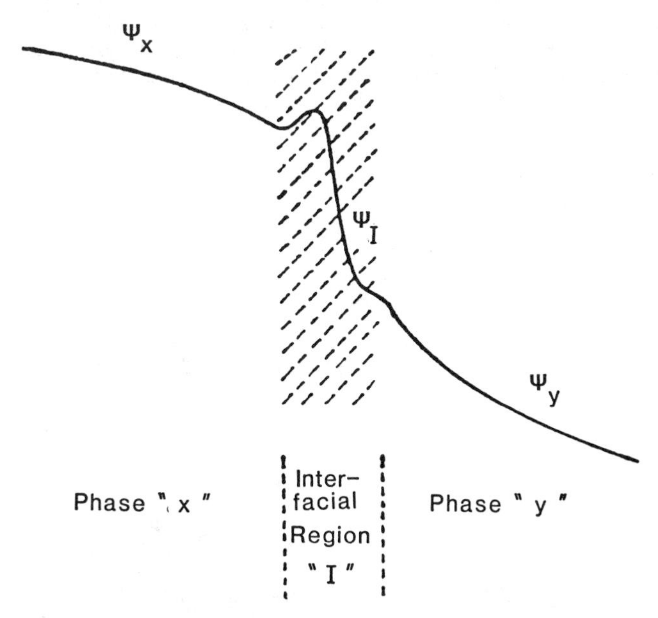

Figure 1. Profiles for property ψ in the bulk fluid phases "x" and "y" and the interfacial region "I". ψ is any specific property (e.g., mass, momentum, energy . . .).

Any description of a boundary region is really a mathematical approximation of the actual physical zone. For a detailed consideration of the physics of interfaces see References 7 through 16. Here we shall be content with a brief discussion of three major methods of describing the interfacial region:

1. Surface of discontinuity.
2. Singular surface.
3. Transition zone or 'stratum', containing or not containing a singular surface.

In the first, most simple, method the boundary is merely a *surface of discontinuity*, possessing no special properties (see Figure 2). In this picture what happens on one side of the boundary is transmitted directly to the other. If, for example, mass enters one side of the boundary, it must come out of the other. There can be no transport of mass along the boundary, neither can mass be stored at the boundary (no surface adsorption). The interface is therefore, literally, an immaterial, mathematical surface with no storage properties for mass, momentum, energy, etc. If ψ represents a specific (meaning per unit mass of mixture) property, then in this model ψ varies continuously in either bulk phase but at the interface itself ψ_I is undefined (Figure 2). This is a highly simplified picture and does not even allow an interfacial tension in the boundary.

The second method is the *singular surface* approach wherein the boundary is again assumed to be a mathematical surface, but with special properties, e.g., surface mass, surface entropy, interfacial tension, etc. This approximation leads to discontinuous changes in properties at the boundary (see Figure 3). This second mathematical description, also called the *surface excess* approach, involves selecting in the transition region a surface that characterizes the orientation and location of the interfacial zone. The bulk phases are then extrapolated up to this reference surface as if uninfluenced by each other and any difference in properties between the actual and hypothetical extrapolated systems assigned to the reference surface. For an equilibrium, or static, system the procedure for selecting the reference surface was put forward by Gibbs. For a nonequilibrium system, as encountered in mass transfer processes, the extrapolation of bulk phase properties is ambiguous. (See References 7 and 11 for a discussion on this subject.)

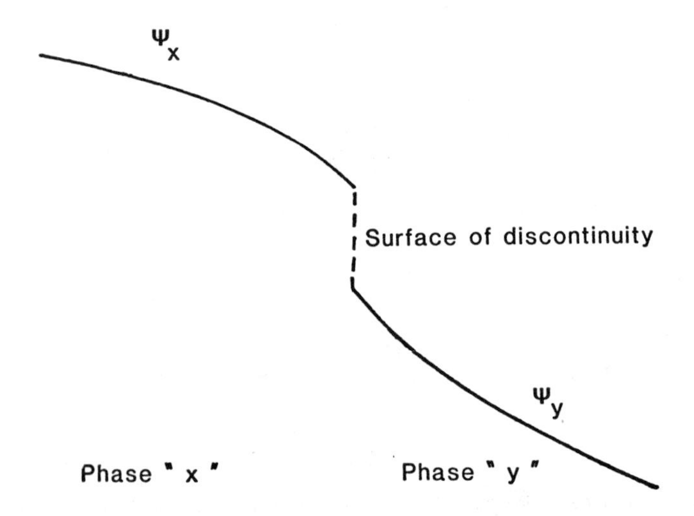

Figure 2. The interface model as a surface of discontinuity. ψ_I is undefined.

The singular surface model is the simplest model possible, able to take account of interfacial effects on mass transfer. Thus, for example, the description of Marangoni effects on mass transfer will require the consideration of surface tension variations along the interface and the resulting flow patterns (e.g., roll cells). The singular surface model is the simplest model which allows the existence of an interfacial tension at the interface.

The third approach is to consider the boundary as a *stratum* or *transition zone* which possesses a finite thickness (see Figure 4). This is the most faithful model of a three dimensional interfacial region. An almost complete analysis of interfacial mechanics using the stratum model is presented by Eliassen [10]. A variant of the stratum model is to allow the existence of a singular surface embedded within the three dimensional zone; this is the model used by Slattery [12] and Deemer and Slattery [17].

Clearly, the model chosen for the description of the interfacial region must depend to a large extent on the problem at hand. If it is desired to study the effect of surfactants on interfacial mass transfer then, clearly, a model using a surface of discontinuity is unlikely to be adequate. Ly, Carbonnell and McCoy [18], for example, discuss the effect of surfactants on interfacial mass transfer using a three-dimensional interfacial region.

In this review, we shall not be dealing specifically with the effects of interfacial phenomena on mass transfer rates. Thus, for the purposes of much of this review it is sufficient to adopt the simplest model for the interface, i.e., a surface of discontinuity. We will therefore develop theories to describe interphase mass transfer using this model, fully appreciative of the fact that if we wish to include surface tension effects in the analysis we will have to choose a more appropriate physical model (e.g., singular surface or stratum model) to describe the interface.

Having chosen the model to describe the interfacial region, the next step is to consider the (differential) equations describing the variations of compositions (concentrations) in either bulk phase and the relations to be applied at the phase boundary.

BALANCE RELATIONS FOR TWO-PHASE SYSTEM INCLUDING A SURFACE OF DISCONTINUITY

Let us consider a two-phase system including a surface of discontinuity (phase interface). Let "x" and "y" represent the two phases. For example, "y" may refer to the gas phase

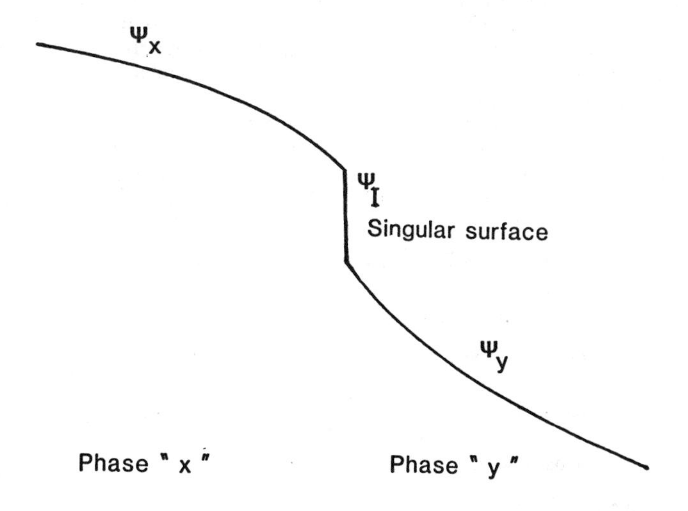

Figure 3. The interface modeled as a singular surface. ψ_I is the value of property ψ at the interface.

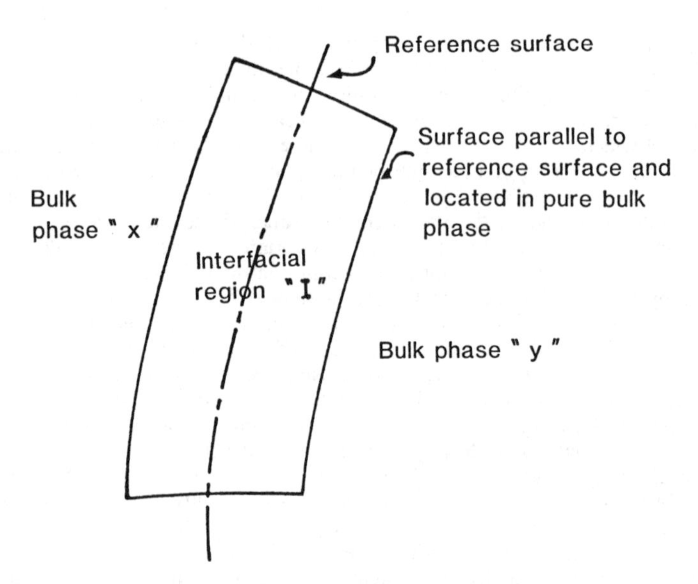

Figure 4. Stratum model for interfacial region, including a reference surface.

and "x" to the liquid phase in a two-phase system. Let the number of components in each phase be n. Let "I" represent the phase interface and ξ the unit normal directed from phase "x" to "y." The system considered is shown pictorially in Figure 5. Our immediate task is to develop the balance relations describing the interphase transport processes taking place in this system.

During interphase mass transfer, concentration gradients will be set up across the interface. The concentration variations in the bulk phases "x" and "y" will be described by *differential equations;* whereas at the interface "I," we will have *jump conditions* or *boundary conditions*. Standart [19] and Slattery [14] give detailed discussions of these relations for the transport of mass, momentum, energy, and entropy. It will not be possible to give here the complete derivations and the reader is therefore referred to these sources; a masterly treatment of this subject is also available in the article by Truesdell and Toupin [20] which is compulsory reading for a serious researcher in transport phenomena.

In the description of the interphase mass transfer process, a variety of measures for constituent concentrations, mixture reference velocities, and diffusion fluxes (with respect to the arbitrarily defined mixture velocity) are used; and are summarized in the following:

Concentration Measures and Other Thermodynamic Mixture Parameters

c_i, kmol/m^3 – molar density of i; $c_i = \varrho_i/M_i$

c_t, kmol/m^3 – mixture molar density; $c_t = \sum_{i=1}^{n} c_i$

x_i, – mole fraction of i; $x_i = c_i/c_t$; $\sum_{i=1}^{n} x_i = 1$

ϱ_i, kg/m^3 – mass density of i; $\varrho_i = c_i M_i$

ϱ_t, kg/m^3 – mixture mass density; $\varrho_t = \sum_{i=1}^{n} \varrho_i$

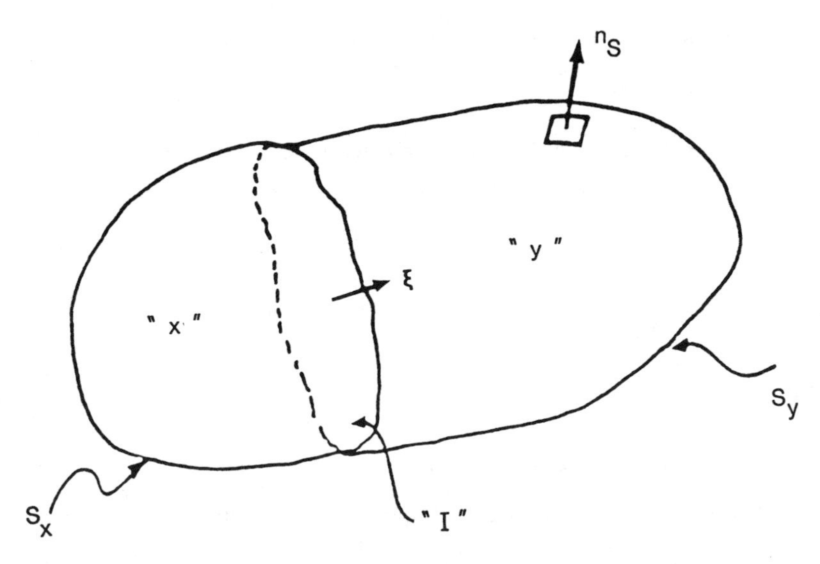

Figure 5. Open finite heterogeneous flow system consisting of phases "x" and "y" with interface "I". S is the external bounding surface with n_S the outward directed normal to S. $\underset{\sim}{\xi}$ is the unit normal directed from "x" to "y" on "I".

ω_i, – mass fraction of i; $\omega_i = \varrho_i/\varrho_t$; $\sum_{i=1}^{n} \omega_i = 1$

M_i, kg/kmol – molecular weight of i

$\bar{V}_i$, m³/kmol – partial molar volume of species i; $\sum_{i=1}^{n} x_i\bar{V}_i = \bar{V}_t$

$\bar{V}_t$, m³/kmol – mixture molar volume; $\bar{V}_t = 1/c_t$

f_i, N/m² – fugacity of i

μ_i, J/kmol – molar chemical potential of species i

a_i, – activity of component i

Some Commonly Used Reference Velocities

$\underset{\sim}{v}$, m/s – mass averaged mixture velocity; $\underset{\sim}{v} = \sum_{i=1}^{n} \omega_i \underset{\sim}{u}_i$

$\underset{\sim}{u}_i$, m/s – velocity of diffusing species i in mixture

$\underset{\sim}{u}$, m/s – molar averaged reference velocity; $\underset{\sim}{u} = \sum_{i=1}^{n} x_i \underset{\sim}{u}_i$

$\underset{\sim}{u}^v$, m/s – volume averaged reference velocity; $\underset{\sim}{u}^v = \sum_{i=1}^{n} c_i\bar{V}_i \underset{\sim}{u}_i$

Some Commonly Used Diffusion Fluxes Relative to Mixture Reference Velocity

$\underset{\sim}{j}_i$, kg/m²/s – mass diffusion flux relative to mass average velocity

$$\underset{\sim}{j}_i = \varrho_i(\underset{\sim}{u}_i - \underset{\sim}{v}); \qquad \sum_{i=1}^{n} \underset{\sim}{j}_i = 0$$

$\underset{\sim}{J}_i$, kmol/m²/s– molar diffusion flux relative to molar average velocity;

$$\underset{\sim}{J}_i = c_i(\underset{\sim}{u}_i - \underset{\sim}{u}); \qquad \sum_{i=1}^{n} \underset{\sim}{J}_i = 0$$

$\underset{\sim}{J}_i^V$, kmol/m²/s– molar diffusion flux relative to volume average velocity;

$$\underset{\sim}{J}_i^V = c_i(\underset{\sim}{u}_i - \underset{\sim}{u}^V); \qquad \sum_{i=1}^{n} \bar{V}_i \underset{\sim}{J}_i^V = 0$$

Equations of Change

In each of the bulk fluid phases "x" and "y", we have the differential equation of change for any conserved property given by:

$$\frac{\partial(\varrho_t \psi)}{\partial t} + \underset{\sim}{\nabla} \cdot (\varrho_t \psi \underset{\sim}{v}) + \underset{\sim}{\nabla} \cdot \underset{\sim}{\Phi} = \zeta \tag{1}$$

where ψ = an arbitrary field quantity per unit mass of mixture
ζ = the rate of production of field per unit volume of bulk phase
$\underset{\sim}{\Phi}$ = a nonconvective flux of the field quantity through external bounding surface S
ϱ_t = the mass density of bulk fluid mixture
$\underset{\sim}{v}$ = the mass average velocity of fluid mixture

All of the foregoing quantities are functions of position and time. The tensorial order of the flux $\underset{\sim}{\Phi}$ is one higher than that of the field quantity ψ.

In addition to Equation 1, which is valid in the bulk fluid phases, we have the following relation to be satisfied at the interface "I":

$$\underset{\sim}{\xi} \cdot \{\underset{\sim}{\Phi}_y + \varrho_{ty}\psi_y(\underset{\sim}{v}_y - \underset{\sim}{u}_I) - \underset{\sim}{\Phi}_x - \varrho_{tx}\psi_x(\underset{\sim}{v}_x - \underset{\sim}{u}_I)\} = \zeta_I \tag{2}$$

where $\underset{\sim}{\xi}$ = the unit normal to "I" directed from the "x" to "y" phase
ζ_I = the rate of production of field quantity per unit area at the interface "I"
$\underset{\sim}{u}_I$ = the velocity of the interface "I"

All the foregoing quantities are functions of position and time. Equation 2 is also called the jump balance condition at the interface.

Equations 1 and 2 are valid for any specific field quantity. We can now write particular balance relations for mass, momentum, and energy by noting the following:

- *Balance of species i* (no chemical reactions in bulk phase or at interface)

$$\psi = \omega_i; \quad \underset{\sim}{\Phi} \equiv \underset{\sim}{j}_i; \quad \zeta \equiv 0; \quad \zeta_I = 0 \tag{3}$$

- *Conservation of total mass of mixture*

$$\psi \equiv 1; \quad \underset{\sim}{\Phi} \equiv 0; \quad \zeta \equiv 0; \quad \zeta_I \equiv 0 \tag{4}$$

- *Conservation of linear momentum*

$$\psi \equiv \underset{\sim}{v}; \quad \underset{\sim}{\Phi} \equiv \underset{\approx}{p}; \quad \zeta \equiv \sum_i \rho_i \underset{\sim}{F}_i; \quad \zeta_I \equiv 0 \tag{5}$$

where $\underset{\approx}{p} = p\underset{\approx}{I} + \underset{\approx}{\tau}$ = the pressure tensor
$\underset{\approx}{\tau}$ = the stress tensor

$\underset{\approx}{I}$ = the unit tensor
ζ = the sum of body forces $\underset{\sim}{F}_i$
p = the thermodynamic pressure

- *Energy balance*

$$\psi \equiv U + \frac{1}{2}\underset{\sim}{v}\cdot\underset{\sim}{v}; \quad \underset{\sim}{\Phi} \equiv \underset{\sim}{q} + \sum_i H_i\underset{\sim}{j}_i + \underset{\approx}{p}\cdot\underset{\sim}{v}; \quad \zeta = \sum \varrho_i \underset{\sim}{F}_i\cdot\underset{\sim}{u}_i; \quad \zeta_I \equiv 0 \tag{6}$$

where U = the internal energy per unit mass of mixture
$\underset{\sim}{q}$ = the conductive heat flux
H_i = the partial specific enthalpy of component i
$\sum H_i\underset{\sim}{j}_i$ = the total enthalpy transferred with the diffusing species i with flux $\underset{\sim}{j}_i$

Our immediate interest is the description of the interfacial mass transfer process and therefore we shall examine further the equations for continuity of species i and the equation for conservation of total mass of mixture.

The differential balance relation for continuity of mass of species i is:

$$\frac{\partial\varrho_i}{\partial t} + \underset{\sim}{v}\cdot\{\varrho_i\underset{\sim}{v}\} = -\underset{\sim}{\nabla}\cdot\underset{\sim}{j}_i, \quad i = 1, 2,\ldots, n \tag{7}$$

For total mixture we have

$$\frac{\partial\varrho_t}{\partial t} + \underset{\sim}{\nabla}\cdot\{\varrho_t\underset{\sim}{v}\} \equiv \frac{\partial\varrho_t}{\partial t} + \varrho_t\underset{\sim}{\nabla}\cdot\underset{\sim}{v} + \underset{\sim}{v}\cdot\underset{\sim}{\nabla}\varrho_t = 0 \tag{8}$$

If we denote the mass average velocity following derivative as:

$$\left.\frac{d}{dt}\right|_{\underset{\sim}{v}} \equiv \frac{\partial}{\partial t} + \underset{\sim}{v}\cdot\underset{\sim}{\nabla} \tag{9}$$

then it is easy to show that Equation 7 can be simplified to the form

$$\varrho_t\left.\frac{d\omega_i}{dt}\right|_{\underset{\sim}{v}} \equiv \varrho_t\left[\frac{\partial\omega_i}{\partial t} + \underset{\sim}{v}\cdot\underset{\sim}{\nabla}\omega_i\right] = -\underset{\sim}{\nabla}\cdot\underset{\sim}{j}_i \tag{10}$$

We have analogously in terms of molar units

$$c_t\left.\frac{dx_i}{dt}\right|_{\underset{\sim}{u}} \equiv c_t\left[\frac{\partial x_i}{\partial t} + \underset{\sim}{u}\cdot\underset{\sim}{\nabla}x_i\right] = -\underset{\sim}{\nabla}\cdot\underset{\sim}{J}_i \tag{11}$$

where we use the mole average velocity following derivative. The conservation of total moles of mixture gives (recall that we do not consider chemical reactions occurring in the bulk phases)

$$\frac{\partial c_t}{\partial t} + \nabla\cdot c_t\underset{\sim}{u} = 0 \tag{12}$$

In this chapter we will consistently use the differential balance relations Equation 11 in molar units and use the molar average reference velocity frame. A completely parallel treatment can be carried out with the mass average velocity $\underset{\sim}{v}$ and Equation 10.

If we choose to represent the diffusion fluxes with respect to the volume average velocity $\underset{\sim}{u}^V$, then the differential balance relations take the form

$$\frac{\partial c_i}{\partial t} + \underset{\sim}{\nabla} \cdot c_i \underset{\sim}{u}^V = -\nabla \cdot \underset{\sim}{J}_i^V \tag{13}$$

It must be emphasized here that Equation 13 cannot be simplified to a form analogous to Equation 10 or 11 because there is no law of conservation of volume.

In general, for applications in chemical engineering, it is preferable to use mole fractions as composition measures, the molar average reference velocity $\underset{\sim}{u}$, and the molar diffusion fluxes $\underset{\sim}{J}_i$ (with respect to $\underset{\sim}{u}$) to describe the diffusion process within a given phase. The volume average reference velocity $\underset{\sim}{u}^V$ is a favorite amongs physical chemists who use this reference in the interpretation of diffusion data in, for exomote, stirred cells. However, there is no conservation of volume, in general, and this choice is not convenient for chemical engineering design purposes; witness the relative simplicity of the differential Equation 11, using $\underset{\sim}{u}$, in comparison to Equation 13 for the corresponding choice of $\underset{\sim}{u}^V$. We take up this subject again for discussion later on when we consider various choices of the driving force for diffusion.

There is one situation, however, when the molar average velocity is not particularly convenient. This arises when we have to solve the equations of continuity of mass in conjunction with the equations of motion; the latter are best expressed in the mass average frame. We shall switch to the choice of the mass average reference velocity $\underset{\sim}{v}$ in our discussion of turbulent mass transfer.

The final working differential equation describing diffusion is (see Equation 11):

$$\frac{\partial x_i}{\partial t} + \underset{\sim}{u} \cdot \underset{\sim}{\nabla} x_i = -\frac{1}{c_t} \underset{\sim}{\nabla} \cdot \underset{\sim}{J}_i, \quad i = 1, 2, \ldots, n \tag{11}$$

Only n-1 of Equation 11 is independent due to the fact that the mole fractions x_i sum to unity and the molar diffusion fluxes $\underset{\sim}{J}_i$ sum to zero (see previous listings of concentration measures, reference velocities, and diffusion fluxes). Exactly analogous relations will hold for the mole fractions y_i in phase "y."

In addition to the differential Equation 11 which apply to the bulk phases the following boundary conditions must be satisfied at the interface "I," provided there are no interface (surface) chemical reactions.

$$\underset{\sim}{\xi} \cdot c_{ix}(\underset{\sim}{u}_{ix} - \underset{\sim}{u}_I) = \underset{\sim}{\xi} \cdot c_{iy}(\underset{\sim}{u}_{iy} - \underset{\sim}{u}_I), \quad i = 1, 2, \ldots, n \tag{14}$$

i.e., the normal component of the flux of component i with respect to the interface must be continuous across the phase boundary. If the interface itself is stationary (i.e., $\underset{\sim}{u}_I = 0$), then Equation 14 can be written as

$$\underset{\sim}{\xi} \cdot \underset{\sim}{N}_{ix} = \underset{\sim}{\xi} \cdot \underset{\sim}{N}_{iy}, \quad i = 1, 2, \ldots, n \tag{15}$$

where $\underset{\sim}{N}_i$ is the molar flux of component i in a stationary, laboratory fixed, coordinate reference frame. Equation 15 merely states that the normal component of the flux $\underset{\sim}{N}_i$ is a phase invariant. These fluxes $\underset{\sim}{N}_i$ are the ones which appear in the design equations and the first objective of this review is, essentially, to consider ways in which they may be calculated from a knowledge of the hydrodynamics and transport properties of the system. The boundary conditions Equation 14, or the simplified version Equation 15, are of course well known to chemical engineers (see for example Reference 21), but it is instructive to follow the general derivations of these relations [14, 19]. The differential Equation 11 applies at any instant of time t. If the conditions prevailing in the phase under consideration are turbulent, then it will be necessary to consider the time-averaged equations. The time-averaging procedure is discussed by, for example, Bird, Stewart and Lightfoot [21]. On time averaging, we obtain

$$\frac{\partial \bar{x}_i}{\partial t} + \bar{\underset{\sim}{u}} \cdot \nabla \bar{x}_i = -\frac{1}{c_t} \nabla \cdot (\bar{\underset{\sim}{J}}_i + \underset{\sim}{J}_{i,turb}), \quad i = 1, 2, \ldots, n \tag{16}$$

where $\underset{\sim}{J}_{i,turb}$ is the turbulent diffusion flux caused by the turbulent eddies present in the system which also contribute to the mass transfer process. The overbars in Equation 16 denote time-averaged quantities; in subsequent discussions we shall omit writing the overbars and take it as understood that time-smoothed variables are considered.

In the core of the bulk fluid phases in turbulent flow, the turbulent diffusion flux, $\underset{\sim}{J}_{i,turb}$, predominates over the molar diffusion flux, $\underset{\sim}{J}_i$, due to molecular diffusion. Close to the interface the turbulence is damped and the molecular diffusion flux, $\underset{\sim}{J}_i$, predominates. Interphase mass transfer modeling consists in seeking suitable simplifications of the actual physical transfer mechanism (by means of molecular and turbulent diffusion) so that the equations may be solved to yield the interfacial transfer fluxes.

We now turn to a discussion of constitutive relations for the molecular and turbulent diffusion fluxes, $\underset{\sim}{J}_i$ and $\underset{\sim}{J}_{i,turb}$.

CONSTITUTIVE EQUATIONS FOR MOLECULAR AND TURBULENT DIFFUSION

Molecular Diffusion in Binary Mixtures

We shall first consider a simple two-component system made up of components 1 and 2. Let $\underset{\sim}{u}_1$ and $\underset{\sim}{u}_2$ represent the velocities of transfer of components 1 and 2 and $\underset{\sim}{u} = x_1\underset{\sim}{u}_1 + x_2\underset{\sim}{u}_2$ represent the molar average velocity of the mixture. If c_1 and c_2 are the molar concentrations of 1 and 2 and c_t is the total mixture molar concentration, then the diffusion flux J_1 is usually related to the mole fraction gradient by the constitutive relation:

$$\underset{\sim}{J}_1 \equiv c_1(\underset{\sim}{u}_1 - \underset{\sim}{u}) \equiv -c_t D_{12} \underset{\sim}{\nabla} x_1 \tag{17}$$

which is Fick's first law of diffusion. D_{12} is the molecular diffusion coefficient. An analogous relation may also be written for component 2:

$$\underset{\sim}{J}_2 \equiv c_2(\underset{\sim}{u}_2 - \underset{\sim}{u}) \equiv -c_t D_{21} \underset{\sim}{\nabla} x_2 \tag{18}$$

It is easy to confirm that since $\underset{\sim}{J}_1 + \underset{\sim}{J}_2 = 0$ and $x_1 + x_2 = 1$ we must have

$$D_{12} = D_{21} = D \text{ (say)} \tag{19}$$

i.e., there is only one diffusion coefficient describing the molecular diffusion process in a binary mixture. There is also only one independent driving force, ∇x_1, and only one independent flux, $\underset{\sim}{J}_1$. Equation 17 defines the diffusion coefficient.

Alternatively, we may choose the diffusion fluxes, $\underset{\sim}{J}_i^V$, in the volume average reference velocity frame, $\underset{\sim}{u}^V$. The Fick's law takes the form

$$\underset{\sim}{J}_1^V \equiv -D \underset{\sim}{\nabla} c_1 \tag{20}$$

where we use the molar concentration gradient driving force: $\underset{\sim}{\nabla} c_1$. Equation 20 represents the most commonly used form of the binary constitutive relationship. However, this form is not the most convenient to use in practical design problems because under nonisothermal conditions, the molar concentration gradients will vary with composition *and* temperature, thus:

$$\underset{\sim}{\nabla} c_1 = \left(c_t + x_1 \frac{\partial c_t}{\partial x_1}\right) \underset{\sim}{\nabla} x_1 + x_1 \frac{\partial c_t}{\partial T} \underset{\sim}{\nabla} T$$

$$= c_t \frac{\bar{V}_2}{\bar{V}_t} \underset{\sim}{\nabla} x_1 + x_1 \frac{\partial c_t}{\partial T} \underset{\sim}{\nabla} T \tag{21}$$

where $\bar{V}_i$ is the partial molar volume of component i and $\bar{V}_t$ is the mixture molar volume. The use of molar concentration gradients as driving forces is not to be recommended because:

- Molar concentrations c_i are not suggested by solution theories as convenient concentration variables (*even* in ideal solutions) to represent the thermodynamically based activity, a_i.
- It is *not* true that $\bar{V}_2/\bar{V}_t \rightarrow 1$ for small concentration gradients, i.e., the simple relation $\underset{\sim}{\nabla} c_1 = c_t \underset{\sim}{\nabla} x_1$ holds *if and only if* c_t = constant and is not even a good approximation *for a dilute* solute.
- The presence of the temperature gradient term, $\underset{\sim}{\nabla} T$, is indeed disturbing and the second term of Equation 21 can be very large for gases—leading to the "hot radiator paradox" mentioned by Sherwood, Pigford, and Wilke [22]. Thus, as pointed out by these authors, use of the molar concentration gradient driving force will predict the existence of a diffusion flux, $\underset{\sim}{J}_1^V$, in a system of uniform composition subject to a temperature gradient.

It is surprising that in spite of the fundamental objections against the use of the molar concentration gradient driving forces and the volume average reference velocity these have been so widely used by so many textbook writers on chemical engineering. The most notable recent example is the book by Carberry [23] which sets up design equations with $\underset{\sim}{\nabla} c_1$ driving forces for the gas phase even for nonisothermal conditions. We believe it is physical chemists who have foisted the volume average reference velocity on us chemical engineers (or is it petroleum technologists who measure production in barrels?).

In the following discussions, we shall consistently prefer the formulation of Equation 17 for binary diffusion. This formulation may be termed practical in the sense that the diffusion coefficient D defined here can be obtained from measured composition profiles in a diffusion apparatus. There are, however, other ways in which the constitutive relations may be formulated using fundamental thermodynamic driving forces. The theory of irreversible thermodynamics provides us with guidelines for setting up the proper fundamental relations. From the theory of irreversible thermodynamics the chemical potential gradients emerge as the proper gradients to describe the diffusion process as we shall see later. This is not very surprising because diffusion arises when there is a departure from equilibrium and the condition of equilibrium is described by an absence of chemical potential gradients. Even if we use chemical potential gradients as driving forces, there are many possibilities of writing the flux-driving force relationship. We prefer the form known as the generalized Maxwell–Stefan (GMS) equation. The driving force is here taken to be $\underset{\sim}{d}_1 = x_1 \underset{\sim}{\nabla}_{T,p} \mu_1/RT$. It is easy to check that for ideal mixtures we have the simplification $\underset{\sim}{d}_1 = \underset{\sim}{\nabla} x_1$, as in the Fick's law formulation. The subscript T,p on the chemical potential gradient is used to emphasize that the gradients are obtained under conditions of constant temperature and pressure. This generalized driving force, $\underset{\sim}{d}_1$, is related to the fluxes by:

$$\underset{\sim}{d}_1 \equiv \frac{x_1}{RT} \underset{\sim}{\nabla} \mu_1 = -\frac{x_1 x_2 (\underset{\sim}{u}_1 - \underset{\sim}{u}_2)}{D'_{12}} \tag{22}$$

where D'_{12} is the Generalized Maxwell–Stefan diffusion coefficient. For ease of presentation, we have omitted writing the constant T and p restriction explicitly; this is taken as understood. The physical interpretation of Equation 22 is as follows. If (and only if) the constituents (1 and 2) are in relative motion to one another and thus moving at different velocities ($\underset{\sim}{u}_1$ and $\underset{\sim}{u}_2$), may we expect chemical potential gradients to be set up in the system as a result of the frictional drag of one set of molecules moving through the other. It does not matter whether this frictional drag arises purely from intermolecular collisions as in the simple (or not so simple) kinetic theory of gases or additionally from intermolecular forces acting between the two sets of molecules, forces which become dominant in diffusion in liquids and solids.

Thus the mutual relative motion of the pair of constituents drags one constituent one way in setting up its potential gradient and, hence, the other constituent will be dragged in the other direction

$$\underset{\sim}{d}_2 \equiv \frac{x_2}{RT}\underset{\sim}{\nabla}\mu_2 = -\frac{x_1x_2(\underset{\sim}{u}_2-\underset{\sim}{u}_1)}{D'_{21}} \tag{23}$$

The Gibbs–Duhem restriction gives $\underset{\sim}{d}_1+\underset{\sim}{d}_2 = \underset{\sim}{0}$, and therefore, the two diffusion coefficients defined by Equations 22 and 23 are identical:

$$D'_{12} = D'_{21} \tag{24}$$

The factor x_1x_2 in Equation 22 multiplying the relative velocity can be easily understood physically as the frictional drag may be expected to be proportional to each of the constituents concerned and hence to their product as well as the relative velocity. Indeed, we may modify this product in any way we please to suit our conception of what constitutes a proper constituent concentration for expressing frictional drag or more commonly, to suit our convenience in defining diffusion transfer fluxes; we have only to multiply and divide by appropriate factors. If desired, the factors in the denominator may be absorbed into the diffusion coefficient. An example already encountered is Fick's law relation Equation 17 which can be equivalently written as

$$\underset{\sim}{\nabla}x_1 = -\frac{x_1x_2(\underset{\sim}{u}_1-\underset{\sim}{u}_2)}{\left[\frac{x_1}{RT}\right]\frac{\partial\mu_1}{\partial x_1}D'_{12}} = -\frac{x_1x_2(\underset{\sim}{u}_1-\underset{\sim}{u}_2)}{D} \tag{25}$$

The Fickian diffusion coefficient D is therefore related to the GMS diffusion coefficient $D'_{12}(= D'_{21})$ by

$$\begin{aligned} D &= D'_{12}\frac{x_1}{RT}\frac{\partial\mu_1}{\partial x_1} \\ &= D'_{12}x_1\partial \ln a_1/\partial x_i \\ &= D'_{12}\left(1+x_1\frac{\partial \ln \gamma_1}{\partial x_1}\right) \end{aligned} \tag{26}$$

where a_i represents the activity of component i, and γ_i is the activity coefficient of i in solution. It is easy to check that for ideal solutions we must have

$$D = D'_{12} \quad \text{(ideal solution)} \tag{27}$$

Thus, in an ideal gas mixture, the Fickian diffusion coefficient is identical to the GMS diffusion coefficient.

Combining Equations 17 and 25 we get

$$\underset{\sim}{d}_1 = -\frac{\underset{\sim}{J}_1}{c_t D'_{12}} = -\frac{x_1(\underset{\sim}{u}_1 - \underset{\sim}{u})}{D'_{12}} \tag{28}$$

which further simplifies to

$$(\underset{\sim}{u}_1 - \underset{\sim}{u}) = -D'_{12}\underset{\sim}{\nabla} \ln a_1 \tag{29}$$

On the left hand side of Equation 29, we have the velocity of motion of component 1 with respect to the mixture. This relative velocity is thus linearly related to the gradient of the logarithmic activity; the constant of proportionality is the GMS diffusion coefficient. In an exactly parallel manner, the Fick's law form Equation 17, may be written as

$$(\underset{\sim}{u}_1 - \underset{\sim}{u}) = -D\underset{\sim}{\nabla} \ln x_1 \tag{30}$$

which shows that the Fick's law coefficient D is a constant of proportionality between the relative velocity with respect to the mixture and the gradient of the logarithmic composition $\underset{\sim}{\nabla} \ln x_1$. The advantage of using the GMS diffusion coefficient is that these coefficients are more easily interpreted in terms of the molecular collision processes which take place within the liquid mixture. Also the composition variation of D'_{12} in a nonideal liquid mixture is more easily describable. For example, as shown by Vignes [24], a plot of log D'_{12} vs. x_1 yields a straight line in many instances. Thus, if we have information on the solution thermodynamics (i.e., activity coefficient variation with composition), it is possible to describe the variation of the Fickian diffusion coefficient D over the entire composition range. The fundamental thermodynamic formulation is, therefore, not only just another formal presentation but can be useful in relating the practical coefficients (Fickian D) to the molecular collision processes and the intermolecular interactions taking place in the mixture.

The setting up of the constitutive relations for a binary system is a relatively easy task because, as pointed out earlier, there is only one independent diffusion flux, only one independent driving force, and therefore, only one independent constant of proportionality (diffusion coefficient). The situation gets quite a bit more complicated when we turn our attention to systems containing 3 or more components. For a ternary mixture there are two independent driving forces ($\underset{\sim}{\nabla} x_1$, $\underset{\sim}{\nabla} x_2$), two independent fluxes ($\underset{\sim}{J}_1$, $\underset{\sim}{J}_2$) and it is not immediately obvious how one might generalize Equation 17, or the fundamental Equations 22 and 28.

Molecular Diffusion in Multicomponent Mixtures

Having set the scene for multicomponent diffusion, let us see how the constitutive relations may be developed for this case. We believe that the best way to introduce these relations is to start with the generalized Maxwell–Stefan formulation, Equation 22. For a three-component (ternary) mixture made up of components 1, 2, and 3, we may consider the potential gradient driving force for component 1, $\underset{\sim}{d}_1$, to be proportional to the relative velocities between 1 and 2 and between 1 and 3, i.e., we may write

$$\underset{\sim}{d}_1 = -\frac{x_1 x_2(\underset{\sim}{u}_1 - \underset{\sim}{u}_2)}{D'_{12}} - \frac{x_1 x_3(\underset{\sim}{u}_1 - \underset{\sim}{u}_3)}{D'_{13}} \tag{31}$$

There is no reason to expect that the GMS diffusion coefficient D'_{12}, defined for the ternary system by Equation 31, will be the same as in the binary system, described by

Equation 22. For nonideal fluid mixtures, they will be different due to the intermolecular forces exerted between each of the components 1 and 2 and component 3. Put another way, the introduction of component 3 into a binary system of 1 and 2 will alter the interactions between 1 and 2 in this ternary system. For ideal gas mixtures, we can use the kinetic gas theory to show that the coefficient D'_{12} in the ternary mixture is identical to D'_{12} in a binary mixture (and each of these in turn is identical to the Fickian diffusion coefficient D_{12}) (see, for example, Reference 25).

We may also write the corresponding equations for components 2 and 3 as:

$$\underset{\sim}{d}_2 = -\frac{x_2x_1(\underset{\sim}{u}_2-\underset{\sim}{u}_1)}{D'_{21}} - \frac{x_2x_3(\underset{\sim}{u}_2-\underset{\sim}{u}_3)}{D'_{23}} \quad (32)$$

and

$$\underset{\sim}{d}_3 = -\frac{x_3x_1(\underset{\sim}{u}_3-\underset{\sim}{u}_1)}{D'_{31}} - \frac{x_3x_2(\underset{\sim}{u}_3-\underset{\sim}{u}_2)}{D'_{32}} \quad (33)$$

Of the three Equations 31–33 only two are independent due to the Gibbs–Duhem restriction $\underset{\sim}{d}_1+\underset{\sim}{d}_2+\underset{\sim}{d}_3 = 0$. It is interesting to note that for the binary system, this restriction is sufficient to *prove* that $D'_{12} = D'_{21}$. For a ternary system we *postulate* that

$$D'_{ij} = D'_{ji}, \quad ij = 12, 13, \text{ and } 23 \quad (34)$$

Equation 34 is essentially a statement of the Onsager Reciprocal Relations discussed in depth in textbooks of irreversible thermodynamics (see for example the elegant presentation in the book by De Groot and Mazur [26]. If you are interested in learning more about these relations we can recommend the book by Truesdell [27]. The Onsager Reciprocal Relations (ORR) have been verified experimentally for a large number of systems and for the purposes of our treatment here we shall assume these to be valid.

Let us return to Equation 31 and consider its physical significance. This equation states that the potential gradient driving force, $\underset{\sim}{d}_1$, of the component 1 arises from the frictional drag of molecules of the first constituent moving past (through) those of the constituent 2 with a relative velocity $(\underset{\sim}{u}_1-\underset{\sim}{u}_2)$, concentration weight factor x_1x_2 and drag coefficient $1/D'_{12}$ *and* of the molecules of the first constituent moving past (through) those of constituent 3 with a relative velocity $(\underset{\sim}{u}_1-\underset{\sim}{u}_3)$, concentration weight factor x_1x_3 and drag coefficient $1/D'_{13}$. As the molecules of all three constituents are, in general, in relative motion with average velocities $\underset{\sim}{u}_i$, it is hard to see how any simpler formulation will suffice. Thus, this relation reduces to the proper binary equation in the limits $x_3 \rightarrow 0$ and $x_2 \rightarrow 0$ for the 1–2 and 1–3 binaries respectively, so that both terms are necessary. It is to be noted that Equation 31 does not include a term $(\underset{\sim}{u}_2-\underset{\sim}{u}_3)$ for the first constituent as it is not reasonable to assume that the relative velocity of these constituents *alone* will produce a potential gradient of the first constituent (Since there would be no direct drag on the molecules of constituent 1). More formally, if a ternary term of the form $x_1x_2x_3(\underset{\sim}{u}_2-\underset{\sim}{u}_3)/E_{123}$ were to be introduced into Equation 31, it could be split up into $x_1x_3(\underset{\sim}{u}_1-\underset{\sim}{u}_3)x_2/E_{123}-x_1x_2(\underset{\sim}{u}_1-$ $-\underset{\sim}{u}_3)x_3/E_{123}$ and these terms absorbed into the existing ones with concentration dependent drag coefficients $1/D'_{13}$ and $1/D'_{12}$.

Thus Equations 31 through 33 are the only consistent generalizations of Equation 22 to a ternary mixture, assuming a linear relation between the potential gradients and the constituents' relative velocities.

There is no reason to expect the pair diffusivities D'_{12} and D'_{13} to be equal, unless the molecules 2 and 3 are very similar in size (molecular weight), shape, polarity, polarizability, hydrogen bonding with 1, etc. The larger the differences in these properties between 2 and 3, the larger will be the difference in the values of D'_{12} and D'_{13}.

If, however, $D'_{12} = D'_{13}$ for similar 2 and 3, we easily find that Equation 31 can be reduced to the binary form, Equation 28, i.e.,

$$\underset{\sim}{J}_1 = -c_t D'_1 \underset{\sim}{d}_1 = \underset{\sim}{N}_1 - x_1 \underset{\sim}{N}_t \quad (D'_1 = D'_{12} = D'_{13}) \tag{35}$$

If (and only if) all the diffusivities D'_{ij} are equal, i.e., for mixtures of similar molecules, may we expect the simple form for a ternary:

$$\underset{\sim}{J}_i = -c_t D' \underset{\sim}{d}_i, \quad i = 1, 2, 3 \quad (D' = D'_{ij = 12,13,23}) \tag{36}$$

Of course, if all the components in the mixture are similar in size and nature then thermodynamic interactions will also be absent and Equation 36 will further simplify to

$$\underset{\sim}{J}_i = -c_t D' \underset{\sim}{\nabla} x_i = -c_t D \underset{\sim}{\nabla} x_i, \quad i = 1, 2, 3 \tag{37}$$

where there will only be one characteristic diffusion coefficient, D', (equal in this case to the Fickian diffusion coefficient D) describing the ternary diffusion process. This is a circumstance to be expected for close boiling hydrocarbons and cannot be a general case. We are interested in developing here a general approach to the modeling of mass transfer and, therefore, will not discuss in great detail various approximate models used to describe diffusion. In many cases, it is obvious what simplifications to use.

Equations 31 through 33 hold for a ternary system, and it is easy to see that they can be generalized to an n-component system in the following manner:

$$\underset{\sim}{d}_i \equiv \frac{x_i}{RT} \nabla_{T,p} \mu_i = -\sum_{\substack{k=1 \\ k \neq i}}^{n} \frac{x_i x_k (\underset{\sim}{u}_i - \underset{\sim}{u}_k)}{D'_{ik}} \tag{38}$$

which equations may also be written in the form:

$$\underset{\sim}{d}_i = -\sum_{\substack{k=1 \\ k \neq i}}^{n} \frac{x_k \underset{\sim}{N}_i - x_i \underset{\sim}{N}_k}{c_t D'_{ik}} = -\sum_{\substack{k=1 \\ k \neq i}}^{n} \frac{x_k \underset{\sim}{J}_i - x_i \underset{\sim}{J}_k}{c_t D'_{ik}} \tag{39}$$

In deriving the generalized Maxwell–Stefan diffusion Equation 38, we asserted, without formal proof, that for diffusion in nonideal fluid mixtures the chemical potential gradients, $\underset{\sim}{\nabla}_{T,p} \mu_i$; or rather

$$\underset{\sim}{d}_i \equiv (x_i/RT) \underset{\sim}{\nabla}_{T,p} \mu_i$$

are the fundamentally correct driving forces to use. Actually, the theory of irreversible thermodynamics (IT) provides the fundamental rationale for this choice. We give below a brief sketch of the IT approach (See De Groot and Mazur [26] and Haase [28] for further details). Another concise treatment of the theory relevant to multicomponent diffusion can be found in the paper by Stewart and Prober [5].

Irreversible Thermodynamics of Diffusion

The purpose of the study of IT is to extend classical thermodynamics to include systems in which irreversible (nonequilibrium) processes are taking place. Such an extension is made possible by assuming that for systems "not too far" from equilibrium the postulate of "local equilibrium" applies:

"Departures from local equilibrium are sufficiently small that all thermodynamic state quantities may be defined locally by the same relations as for systems at equilibrium."

Actually it may be shown that this assumption of local equilibrium follows from the assumption of a linear relation between the fluxes and driving forces [27].

With the help of this postulate, one is able to obtain an explicit expression for σ, the rate of entropy production per unit volume due to various irreversible processes taking place within the system. For isothermal, isobaric processes in the absence of external force fields, the rate of entropy production due to diffusion is given by

$$\sigma = -\frac{1}{T}\sum_{i=1}^{n-1} \underset{\sim}{J}_i \cdot \underset{\sim}{\nabla}_{T,p}(\mu_i - \mu_n) \geqq 0 \tag{40}$$

The second law of thermodynamics requires σ to be positive definite.

σ is seen to be a sum of scalar, or dot, products of two quantities; one of these is the diffusion flux and the other, the chemical potential gradient, may be interpreted as the "driving force" for diffusion. At equilibrium both the fluxes and the forces vanish simultaneously giving

$$\sigma = 0 \quad |\text{ equilibrium }| \tag{41}$$

The chemical potential gradient arises in the theory of IT as the proper driving force for diffusion. This is not surprising because the condition for diffusion equilibria is that the chemical potentials be equal in each phase and departures from equilibrium must be measured as deviations from such an equality.

Linear constitutive relations for the diffusion fluxes may be set up in the form:

$$c_t \underset{\sim}{\nabla}_{T,p}(\mu_i - \mu_n) = -\sum_{k=1}^{n-1} H_{ik} \underset{\sim}{J}_k \tag{42}$$

where the coefficients H_{ik} are symmetric (Onsager reciprocal relations) and [H] is positive definite because of the second law restriction $\sigma \geqq 0$.

The gradients $\underset{\sim}{\nabla}_{T,p}(\mu_i - \mu_n)$ can be expressed in terms of the mole fraction gradients $\underset{\sim}{\nabla}(x)$ by the relation:

$$\underset{\sim}{\nabla}_{T,p}(\mu - \mu_n) = [G]\underset{\sim}{\nabla}(x) \tag{43}$$

where the elements G_{ij} are given by

$$G_{ij} = \frac{\partial^2 G}{\partial x_i \partial x_j} = \frac{\partial(\mu_i - \mu_n)}{\partial x_j} = \frac{\partial(\mu_j - \mu_n)}{\partial x_i} = G_{ij} \tag{44}$$

For thermodynamic stability in the system under consideration, the matrix [G] must be positive definite [29]. The matrix [G] is symmetric.

Substituting Equation 43 in Equation 42 and rearranging we obtain

$$(\underset{\sim}{J}) = -c_t[H]^{-1}[G]\underset{\sim}{\nabla}(x) \tag{45}$$

Let us define a matrix of diffusion coefficients [D], of dimension $n-1 \times n-1$, by

$$(\underset{\sim}{J}) = -c_t[D]\underset{\sim}{\nabla}(x) \tag{46}$$

It is then easy to see from Equations 45 and 46 that

$$[D] \equiv [H]^{-1}[G] \tag{47}$$

which shows that the matrix [D], a generalization of the Fickian coefficient D for multicomponent systems, is equal to a product of two matrices, $[H]^{-1}$ and [G], each of which is symmetric and positive definite.

It follows directly from matrix theory that the eigenvalues of [D] must be real and positive [30]; the eigenvalues of [D] being the roots of the determinantal equation

$$\det |[D]-\hat{D}[I]| = 0 \tag{48}$$

For an n-component system, Equation 48 reduces to an n−1-th order polynomial in $\hat{D}$, giving n−1 eigenvalues: $\hat{D}_1, \hat{D}_2, \ldots, \hat{D}_{n-1}$. For a ternary system, the two roots $\hat{D}_1$ and $\hat{D}_2$ can be found explicitly from

$$\hat{D}_{1,2} = \frac{(D_{11}+D_{22}) \pm (D_{11}-D_{22})\sqrt{1+4D_{12}D_{21}/(D_{11}-D_{22})^2}}{2} \tag{49}$$

The condition for real and positive eigenvalues, $\hat{D}_1$ and $\hat{D}_2$, for a ternary system can therefore be expressed as [30, 31]:

$$D_{11}+D_{22} > 0$$

$$D_{11}D_{22}-D_{12}D_{21} > 0$$

$$(D_{11}-D_{22})^2+4D_{12}D_{21} > 0 \tag{50}$$

It is interesting to note that thermodynamic stability considerations do not require the diagonal elements D_{11} and D_{22} to be individually positive. If recourse is made to the kinetic theory of gases, it can be shown that the main coefficients are individually positive, i.e.

$$D_{11} > 0; \quad D_{22} > 0 \tag{51}$$

All available experimental measurements of the D_{ik} suggest the general validity of the requirement of Equation 51 (see Cussler [31a] and Dunlop et al. [31b] for a comprehensive survey of experimental techniques and results).

The cross coefficients D_{ik} $(i \neq k)$ can be of either sign; indeed it is possible to alter the sign of these cross-coefficients by altering the numbering of the components.

The theory of IT, therefore, provides some very useful information on the structure of the matrix of practical diffusion coefficients, [D].

It is interesting to note that although the Fickian matrix, [D], is a product of two symmetric matrices, it is generally nonsymmetric except for the trivial case where [D] reduces to the form of a scalar, D, times the identity matrix [I].

An alternative way of writing Equation 40 for the rate of entropy production is in terms of the driving force $\underset{\sim}{d}_i \left(\equiv \frac{x_i}{RT} \underset{\sim}{\nabla}_{T,p}\mu_i\right)$

$$\sigma = -c_t R \sum_{i=1}^{n} \underset{\sim}{d}_i \cdot (\underset{\sim}{u}_i - \underset{\sim}{u}) \geq 0 \tag{52}$$

Clearly, at equilibrium the driving forces, $\underset{\sim}{d}_i$, vanish giving $\sigma = 0$.

Substituting the GMS equation for $\underset{\sim}{d}_i$ into the second law restriction, Equation 52, we obtain a very neat and compact expression:

$$\sigma = \frac{c_t R}{2} \sum_{\substack{i,j \\ j \neq i}}^{n} \frac{x_i x_j}{D'_{ij}} (\underset{\sim}{u}_j - \underset{\sim}{u}_i)^2 \geq 0 \tag{53}$$

which is quite remarkable because of the apparent absence of thermodynamic factors. It is even more remarkable that though the Equation 53 was derived much earlier for ideal gas mixtures [32] the generalization to nonideal fluid mixtures, as discussed, was carried out quite recently by Standart, Taylor and Krishna [33], using the Hirschfelder, Curtiss, and Bird treatment as a consistent basis.

The Equation 53 requires that

$$D'_{ij} \geq 0 \tag{54}$$

which is the generalization of the Hirschfelder, Curtiss, and Bird result for ideal gas mixtures.

One of the many advantages of the GMS formulation Equation 38 over the Onsager formulation Equation 42 for multicomponent diffusion is that since the D'_{ij} are defined in terms of $\underset{\sim}{d}_i$ and component velocity differences, $\underset{\sim}{u}_i - \underset{\sim}{u}_j$, both of which are independent of the arbitrary choice of the reference velocity frame, the D'_{ij} are also reference-frame independent. By contrast, the Onsager coefficient matrix [H] and the Fickian matrix [D] are dependent on the choice of the reference velocity of the mixture. Further, we note that although any unsymmetric part of D'_{ij} cancels in Equation 53, we must utilize the Onsager reciprocal relations to conclude that

$$D'_{ij} = D'_{ji} \tag{55}$$

It must be further noted that the coefficients D'_{ii} are undefined.

We see from before that the GMS diffusion equation, derived earlier in a heuristic manner, has a firm foundation in irreversible thermodynamics.

Let us now try to relate the GMS diffusion coefficients D'_{ij} to the Fickian diffusivities D_{ij}, defined by Equation 46.

In view of the relations $\underset{\sim}{N}_i = c_i \underset{\sim}{u}_i = \underset{\sim}{J}_i + c_i \underset{\sim}{u}$, we may also write Equation 39 in the form:

$$\frac{1}{RT} x_i \nabla_{T,p} \mu_i = \sum_{\substack{k=1 \\ k \neq i}}^{n} \frac{x_i \underset{\sim}{N}_k - x_k \underset{\sim}{N}_i}{c_t D'_{ik}} = \sum_{\substack{k=1 \\ k \neq i}}^{n} \frac{x_i \underset{\sim}{J}_k - x_k \underset{\sim}{J}_i}{c_t D'_{ik}} \qquad i = 1, 2, \ldots, n \tag{56}$$

By defining the following matrices [B] and [Γ]

$$B_{ii} = \frac{x_i}{D'_{in}} + \sum_{\substack{k=1 \\ k \neq i}}^{n} \frac{x_k}{D'_{ik}}; \qquad B_{ij} = -x_i(1/D'_{ij} - 1/D'_{in}) \qquad i \neq j = 1, 2, \ldots, n-1 \tag{57}$$

$$\Gamma_{ij} = \delta_{ij} + x_i \frac{\partial \ln \gamma_i}{\partial x_j} \qquad i = 1, 2, \ldots, n-1 \tag{58}$$

we may write Equation 39 in $n-1$ dimensional matrix notation as:

$$(\underset{\sim}{J}) = -c_t [B]^{-1} [\Gamma] \underset{\sim}{\nabla}(x) \tag{59}$$

which, on comparison with the generalized Fick's law formulation, shows that the matrix of practical diffusion coefficients is given by

$$[D] = [B]^{-1}[\Gamma] \tag{60}$$

For ideal gas mixtures, the matrix $[\Gamma]$ reduces to the identity matrix and we obtain

$$[D] = [B]^{-1} \tag{61}$$

For a ternary ideal gas mixture, for example, the elements D_{ij} of the Fickian matrix are given explicitly by the relationships:

$$D_{11} = D'_{13}(x_1D'_{23}+(1-x_1)D'_{12})/(x_1D'_{23}+x_2D'_{13}+x_3D'_{12})$$

$$D_{12} = x_1D'_{23}(D'_{13}-D'_{12})/(x_1D'_{23}+x_2D'_{13}+x_3D'_{12})$$

$$D_{21} = x_2D'_{13}(D'_{23}-D'_{12})/(x_1D'_{23}+x_2D'_{13}+x_3D'_{12})$$

$$D_{22} = D_{23}(x_2D'_{13}+(1-x_2)D'_{12})/(x_1D'_{23}+x_2D'_{13}+x_3D'_{12}) \tag{62}$$

The diffusion coefficients of the binary pairs D'_{ij} can be estimated from the kinetic theory of gases [32, 34, 35] to a reasonable degree of accuracy, particularly for nonpolar molecules. The matrix of diffusion coefficients may therefore be calculated using Equation 62. It may be verified from Equation 62 that if $D'_{12} = D'_{13} = D'_{23} = D'$, then the matrix of diffusion coefficients degenerates to a scalar times the identity matrix, i.e.

$$[D] = D'[I] \quad |\text{ special }| \tag{63}$$

It is interesting now to compare these Equation 63 with Equation 36 and the discussions following it.

For nonideal liquid mixtures, the D'_{ik} are strongly composition dependent. Prediction methods for liquid-phase multicomponent diffusivities are still a subject for research, but by analogy with the corresponding binary relationship, Equation 26, we might expect $[B]^{-1}$ to exhibit a more predictable composition dependence than the Fickian matrix [D]. Indeed, available methods for predicting [D] all rely on the separation of the "kinetic" and "thermodynamic" contributions, $[B]^{-1}$ and $[\Gamma]$; see the works of Bandrowski and Kubaczka [36], Cullinan [37], and Kosanovich [38]. Unfortunately, the prediction methods are still in the very early stages of development. More experimental data with nonideal ternary (or quarternary) systems will be required as an aid to model development; this is an area for future research. It is worth noting here that the UNIFAC group-contribution method allows estimation of $[\Gamma]$ even when no thermodynamic data are available [39]. When vapor-liquid or liquid-liquid equilibrium data are available and correlated using models such as Wilson, NRTL, and UNIQUAC, calculation of Γ_{ik} is possible.

It is clear that the presence of the cross-coefficients D_{ij} $(i \neq j)$ in the Fickian matrix lends to the multicomponent system characteristics quite different from the corresponding binary system. This is best illustrated by considering a ternary mixture; Equation 46 for component 1 may be written explicitly as

$$\underset{\sim}{J}_1 = -c_tD_{11}\underset{\sim}{\nabla}x_1 - c_tD_{12}\underset{\sim}{\nabla}x_2 \tag{64}$$

Since, in general, D_{12} (which should not be confused with the binary D_{12}) is nonzero, we may have a diffusion flux of component 1 even in the absence of a composition gradient for component 1, $\underset{\sim}{\nabla}x_1$. This effect, usually called a coupling effect or a diffusional interaction effect, is large if the cross-coefficient D_{12} is large. Figure 6 shows some calculated values for the ratio D_{12}/D_{11} for an ideal gas mixture for which $D'_{12} = D'_{23}$ (i.e., 1 and 3 are similar). It can be easily seen that when $D'_{12}/D'_{13} = 1$ the ratio $D_{12}/D_{11} = 0$ regardless of the gas composition. This in fact corresponds to the case described by Equation 31. When the ratio D'_{12}/D'_{13} is different from unity, then the cross-coefficient may be significant with respect to D_{11}, especially for high values of the composition x_1. The ratio D_{12}/D_{11} may also assume negative values and the absolute values of this ratio may exceed unity. When the last happens, the coupling effects will be strong indeed. Generally speaking, coupling effects will be strong in multicomponent gas mixtures containing components of widely varying molecular weights; in this case the constituent binary diffusivities D'_{ij} will be widely different and, therefore, will lead to large cross coefficients. For example, in the catalytic dehydrogenation of ethanol to acetaldehyde, the binary gas phase diffusivities of the three binary pairs encountered are (temperature = 548° K; pressure = 1.013 bar); D' (ethanol-hydrogen) = 142.4 mm^2/s; D' (acetaldehyde-hydrogen) = 147.7 mm^2/s; D' (ethanol-acetaldehyde) = 25 mm^2/s. The ratio of the largest coefficient to the smallest one is about 6. Since the cross coefficient, D_{12}; for example; is proportional to the difference between D'_{13} and D'_{12} Equation 62, we must expect the fluxes of ethanol and acetaldehyde to be strongly coupled to each other.

Fickian Diffusion Coefficients and Their Dependence on the Reference Velocity Frame

We have repeatedly expressed our preference for the GMS formulation of the diffusion equations. For solution of many practical diffusion problems, especially involving nonideal fluid mixtures, we must resort to the Fickian formulation. However, the GMS formulations

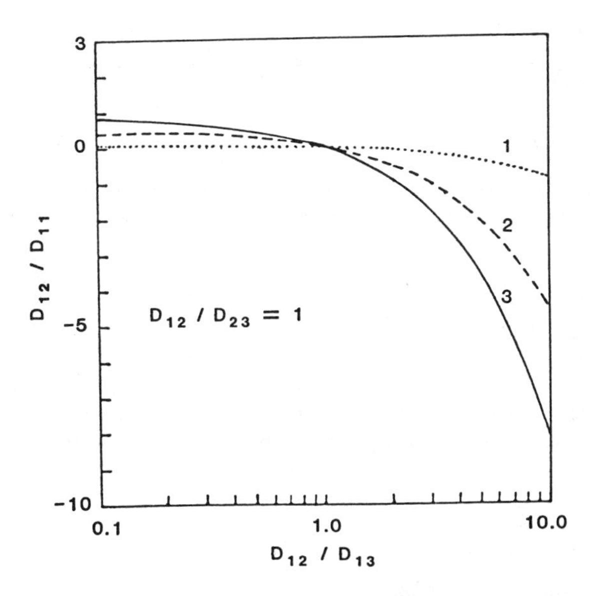

Figure 6. The ratio D_{12}/D_{11} (which are elements of the Fickian matrix [D]) as a function of the binary pair Maxwell-Stefan diffusion coefficients D'_{12}/D'_{13}. Calculations for ideal gas mixture for three different values of x_1 as follows: 1, $x_1 = 0.1$; 2, $x_1 = 0.5$; 3, $x_1 = 0.9$.

still have their utility because they provide a means for separating the Fickian coefficients into their respective "kinetic" and "thermodynamic" contributions.

We now take up in this section the question of the dependence of the Fickian coefficients on the choice of the reference velocity.

For binary systems the situation is rather simple. If the constitutive relations are written in one of the three forms:

$$\underset{\sim}{J}_1 = -c_t D \underset{\sim}{\nabla} x_1 \quad \text{(molar flux with respect to molar average velocity)}$$

$$\underset{\sim}{j}_1 = -\varrho_t D \underset{\sim}{\nabla} \omega_1 \quad \text{(mass flux with respect to mass average velocity)}$$

$$\underset{\sim}{J}_1^v = -D \underset{\sim}{\nabla} c_1 \quad \text{(molar flux with respect to volume average velocity)} \tag{65}$$

then the three diffusion coefficients, D, defined in Equation 65 are all identical [21].

For the general multicomponent case, we may consider the analogous matrix relations

$$(\underset{\sim}{J}) = -c_t[D]\underset{\sim}{\nabla}(x) \quad \text{(molar flux with respect to molar average velocity)}$$

$$(\underset{\sim}{j}) = -\varrho_t[D^m]\underset{\sim}{\nabla}(\omega) \quad \text{(mass flux with respect to mass average velocity)}$$

$$(\underset{\sim}{J}^v) = -[D^v]\underset{\sim}{\nabla}(c) \quad \text{(molar flux with respect to volume average velocity)} \tag{66}$$

In this case, the three matrices, just defined are, in general, different from one another. Cullinan [40] has shown, however, that the eigenvalues of these three matrices are, correspondingly, equal to one another.

We had earlier shown that for thermodynamic stability these eigenvalues must be real and positive. This is a useful and important property which can be exploited in the solution of the diffusion equations.

Equations 60, 61, and 62, derived earlier, hold for the molar average reference velocity frame. To transform [D] to the mass average reference frame $[D^m]$ we have the following similarity transformation

$$[D^m] = [B^m]^{-1}\,[\omega]\,[x]^{-1}\,[D]\,[x]\,[\omega]^{-1}\,[B^m] \tag{67}$$

where [x] is a diagonal matrix whose nonzero elements are the mole fractions x_i; $[\omega]$ is also a diagonal matrix with nonzero elements that are the mass fractions ω_i and $[B^m]$ is a square matrix with elements

$$B_{ik}^m = \delta_{ik} - \omega_i\left(\frac{x_k}{\omega_k} - \frac{x_n}{\omega_n}\right) \qquad i,k = 1, 2, \ldots, n-1 \tag{68}$$

Since [x] and $[\omega]$ are diagonal matrices, their inverses are easy to compute, $[x]^{-1}$, for example, is diagonal with elements that are the reciprocals of the mole fractions: $1/x_i$. The inverse of $[B^m]$ may also be calculated explicitly using the Sherman–Morrison formula [41]. The result is

$$B_{ik}^{m-1} = \delta_{ik} - \omega_i\left(1 - \frac{\omega_n}{x_n}\cdot\frac{x_k}{\omega_k}\right) \tag{69}$$

To convert [D] to the volume average velocity frame we may use another similarity transformation

$$[D^v] = [B^v]^{-1}\,[D]\,[B^v] \tag{70}$$

where the matrix $[B^v]$ has elements

$$B^v_{ik} = \delta_{ik} - c_i(\bar{V}_k - \bar{V}_n) \tag{71}$$

The inverse of $[B^v]$ also may be calculated explicitly

$$B^{v-1}_{ik} = \delta_{ik} - x_i(1 - \bar{V}_k/\bar{V}_n) \tag{72}$$

The proof that the eigenvalues of [D], $[D^m]$ and $[D^v]$ are equal follows immediately from Equations 67 and 70; the eigenvalues of two matrices, [A] and [B] say, that are related by a nonsingular similarity transformation $[A] = [P]^{-1}[B][P]$ are equal [40, 42].

Most experimental data are reported for $[D^v]$. This matrix must be transformed to [D] or $[D^m]$ in order for it to be useful in the applications we consider later in this review. It is clear from Equation 69 that $[D^v]$ is not, in general, equal to [D].

Effective Diffusivity Approaches

What we have seen so far is that a logical extension of the binary diffusion Equation 22 is the generalized Maxwell–Stefan formulation Equation 38, which is consistent with IT theory. If the GMS formulation is shoe-horned into a Fickian-type frame Equation 46, then the resultant matrix of Fickian diffusion coefficients must be nondiagonal in the general case, except for the trivial situation where the mixture is ideal and the components are similar in size and nature [37]. Let us now try a different approach right from the start; consider a naive extension of the binary relation Equation 17 obtained by postulating that each diffusion flux in the mixture is proportional only to its own composition gradient driving force. Thus

$$\underset{\sim}{J}_i = -c_t D_{i,eff} \underset{\sim}{\nabla} x_i. \quad i = 1, 2, \ldots, n \tag{73}$$

where $D_{i,eff}$ is some characteristic diffusion coefficient of species i in the mixture. Summing Equation 71 over the n species; we find, in view of the restriction $\sum_{i=1}^{n} \underset{\sim}{J}_i = 0$, that

$$c_t \sum_{i=1}^{n} D_{i,eff} \underset{\sim}{\nabla} x_i = 0 \tag{74}$$

Eliminating the n-th gradient $\underset{\sim}{\nabla} x_n$ from Equation 74 we obtain

$$\sum_{i=1}^{n-1} (D_{i,eff} - D_{n,eff}) \underset{\sim}{\nabla} x_i = 0 \tag{75}$$

If we now demand of the $D_{i,eff}$ that they be independent of the composition gradients then, since each of the $n-1$ gradients in Equation 75 can be varied independently, the only solution possible to Equation 75 is that all the effective diffusivities are equal to one another

$$D_{i,eff} = D_{n,eff}, \quad i = 1, 2, \ldots, n-1 \tag{76}$$

Equation 76 requires each species in the n-component mixture to have equal facility for transfer irrespective of its molecular size and nature. We may expect such a simple result to be true when all the species making up the mixture are of similar nature and the solution is thermodynamically ideal. It has been found experimentally, for example, that diffusion in the ternary mixture toluene-chlorobenzene-bromobenzene is described adequately by one characteristic diffusion coefficient [43].

In general, however, the $D_{i,eff}$ *do* depend on the composition gradients in the mixture (as shown later) and we must reject the simplification Equation 73.

Turbulent Diffusion

We have thus far considered the constitutive relations for the molecular diffusion flux, J_i. If turbulent conditions prevail, there will be an additional transport contribution by the turbulent eddies. The understanding of turbulent (eddy) momentum transport under these conditions is a prerequisite to the understanding of turbulent mass transport. There is an astronomical amount of literature on turbulence modeling; for a chemical engineer who is interested in learning something about the state of affairs here, we can recommend the book by Launder and Spalding [44]. An excellent review of turbulent heat and mass transfer at interfaces is given by Sideman and Pinczewski [45]. This latter work reviews turbulent heat and mass transfer from the viewpoint of a chemical engineer and is, therefore, particularly useful.

For turbulent mass transport across fluid-fluid interfaces, it is generally only possible to use the simplest description of turbulent diffusion, namely Boussinesq's hypothesis:

$$\underset{\sim}{j}_{i,turb} = -\varrho_t D_{turb} \underset{\sim}{\nabla} \omega_i, \quad i = 1, 2, \ldots, n \tag{77}$$

where D_{turb} is the turbulent eddy diffusivity of mass. We do not have any coupling effects in turbulent diffusion because turbulent eddy mass transport is not species specific or, in other words, all the components are transported by the same mechanism. We use mass units and the mass average reference velocity frame because the use of the eddy diffusivity approach requires consideration of the equations of motion.

Even with this simple constitutive relationship Equation 77 for the turbulent diffusion flux, $\underset{\sim}{J}_{i,turb}$, the problem remains as to the value of the turbulent diffusivity, D_{turb}. For flows in pipes and over flat plates there are sufficient data to predict the value of the eddy momentum diffusivity, ν_{turb}. The most usual procedure for the prediction of D_{turb} is to proceed via the knowledge of ν_{turb}. We therefore define a turbulent Schmidt number:

$$Sc_{turb} = \nu_{turb}/D_{turb} \tag{78}$$

If we accept the analogy between heat and mass transfer, Sc_{turb} must be equal to Pr_{turb}, the turbulent Prandtl number. There are many models for the prediction of Pr_{turb} [45]. It is interesting to note that even for the simple case of pipe flow there is a five-fold spread in the value of Pr_{turb} in the region of the wall [46].

In practice, a correlation for Sc_{turb} (or Pr_{turb}) as a function of the molecular Schmidt number will have to be used; see Reference 45.

By defining Sc_{turb}, we have reduced the problem of estimating D_{turb} to one of estimating ν_{turb}, the turbulent kinematic viscosity. To estimate ν_{turb}, we require information on the velocity profiles between the interface and the bulk fluid phase. Such detailed information is available for some simple flow situations e.g. pipe flow and flow over flat plates. The Prandtl mixing-length theory gives the following expression for ν_{turb}

$$\nu_{turb} = \ell^2 \left| \frac{d\bar{V}_x}{dy} \right| \tag{79}$$

where $\bar{V}_x$ = the time-averaged velocity in the direction of mean-flow
y = the position coordinate, representing distance from the interface

ℓ = the Prandtl mixing length, defined as the distance a fluid eddy (lump of fluid) will travel before losing its identity.

Although it may appear from Equation 79 that the only result of the introduction of ℓ has been to replace one empirical, noncomputable quantity, ν_{turb}, with another, the mixing length is easier to estimate than ν_{turb}. For instance, ℓ cannot be greater than the dimensions of the channel, and it should approach zero near the wall.

So, the steps involved in estimating D_{turb} are:

1. Estimate Sc_{turb} (= Pr_{turb}) from an appropriate correlation [45].
2. Estimate ℓ from a model for the mixing length.
3. Estimate $|d\bar{V}_x/dy|$ from knowledge of velocity profile.

We return to the subject of turbulent mass transfer when we develop some explicit expressions for the calculation of the mass transfer coefficients under turbulent flow conditions, making use of specific empirical models for Sc_{turb}, ℓ, and $|d\bar{V}_x/dy|$.

Simultaneous Transfer of Mass and Energy in Multicomponent Systems

In the foregoing sections, we have developed the general formulation for multicomponent diffusion, both molecular and turbulent, without considering the effects of simultaneous energy transfer. Perfectly isothermal systems are rare in chemical engineering practice, and many processes, such as distillation, condensation, and evaporation, involve simultaneous transfer of mass and energy across fluid-fluid interfaces. The presence of a temperature gradient in a multicomponent system introduces two additional complications:

- Variations in physical, thermodynamic and transport properties due to differences in temperature.
- Large temperature gradients may give rise to material fluxes (thermal diffusion).

The property variations due to temperature differences are taken care of quite simply by introducing temperature-dependent property functions or by use of suitably averaged properties, as is commonly done; the basic mass transfer analysis remains essentially unchanged. The second complication arising out of "coupling" between mass and energy transfer may have more important ramifications.

The interactions between thermal and mass fluxes have long been recognized. A great deal of effort has been devoted to the study of thermal diffusion and a number of reviews are available on the subject [47–49]. Its practical application has been highlighted by the successful application of a thermal diffusion process to the separation of isotopes [50]. It must be said at once that the coupling between thermal and mass fluxes is of no consequence whatever in the classical separation processes such as distillation and absorption. Even in condensation, the temperature gradients (which may sometimes be as high as several tens of degrees per millimeter) are rarely high enough or sustained long enough to make the thermal contributions to the mass fluxes worth considering. Strong coupling effects may, however, be found in processes involving very steep temperature and/or concentration gradients, such as those involved in ablation cooling during rocket reentry, injection of low-molecular-weight gases into boundary-layer flows, chemically reacting flows, and some membrane transport processes (see, for example, Tambour and Gal-Or [51, 52, 53]). The related phenomenon of thermophoresis (migration of small particles in a temperature gradient) seems to be important in the deposition of particles on gas turbine blades (Rosner and coworkers [54–56], for example, review and discuss this subject at some length). A review of thermal diffusion (Soret) effects on interfacial mass transport rates is available from Rosner [57]. The Dufour effect, production of a heat flux due to concentration gradients, has received little attention in the literature, partly due to difficulties in experimentation (see, however, Green [58]).

For transfer in either fluid phase of the two-phase system considered in Figure 5, the differential energy balance relation Equation 6, provides the additional physical law necessary to determine the temperature profiles and energy fluxes. This balance relationship may be rewritten in several alternative, equivalent forms [21]. Two useful forms of the energy balance relation, assuming mechanical equilibrium, are in terms of the partial molar enthalpies:

$$\frac{\partial\left\{\sum_{i=1}^{n} c_i \bar{H}_i\right\}}{\partial t} = -\underset{\sim}{\nabla}\cdot\left\{\underset{\sim}{q} + \sum_{i=1}^{n} \bar{H}_i \underset{\sim}{N}_i\right\} \tag{80}$$

and in terms of the temperature

$$\frac{\partial\{c_t C_p T\}}{\partial t} = -\underset{\sim}{\nabla}\cdot\underset{\sim}{q} - \underset{\sim}{\nabla}\cdot\{c_t C_p T \underset{\sim}{u}\} - \sum_{i=1}^{n} \underset{\sim}{J}_i \cdot \underset{\sim}{\nabla}\bar{H}_i \tag{81}$$

where $\underset{\sim}{q}$ is the conductive heat flux.

In addition to the differential balance relationships Equations 80 and 81, valid in the bulk phases, we have the jump balance condition at the interface "I," Equation 2, which simplifies to

$$\underset{\sim}{\xi}\cdot\underset{\sim}{E}_x = \underset{\sim}{\xi}\cdot\underset{\sim}{E}_y \tag{82}$$

where $\underset{\sim}{E}_x$ and $\underset{\sim}{E}_y$ are the energy fluxes in the adjoining phases at the interface. Equation 82 states that the normal component of the energy fluxes must be continuous across the phase boundary "I" and is entirely analogous to Equation 15 for mass transfer. In either fluid phase the energy flux takes the form

$$\underset{\sim}{E} = \underset{\sim}{q} + \sum_{i=1}^{n} \bar{H}_i \underset{\sim}{N}_i \tag{83}$$

$\underset{\sim}{E}$ plays an analogous role to the molar fluxes $\underset{\sim}{N}_i$ in the interphase energy transfer process. The conductive heat flux, $\underset{\sim}{q}$, plays a role analogous to the molar diffusion fluxes, $\underset{\sim}{J}_i$. We shall be making further use of these analogies later on.

For turbulent flow conditions, on time averaging the differential energy balance relations, we note that the time-smoothed heat flux $\underset{\sim}{q}$ (caused by molecular transport processes) is augmented by the turbulent contribution $\underset{\sim}{q}_{turb}$.

We now seek constitutive relations for $\underset{\sim}{q}$ and $\underset{\sim}{q}_{turb}$.

The most appropriate starting point for setting up the constitutive relation for $\underset{\sim}{q}$ is the theory of irreversible thermodynamics and the expression for the rate of production of entropy due to mass and energy transfer processes [33]

$$\sigma = \underset{\sim}{q}\cdot\underset{\sim}{\nabla}\left(\frac{1}{T}\right) - c_t R \sum_{i=1}^{n} \underset{\sim}{d}_i \cdot (\underset{\sim}{u}_i - \underset{\sim}{u}) \tag{84}$$

In the general case, we must allow for coupling between the transfer processes of heat and mass. With this allowance, the complete expression for the conductive heat flux is

$$\underset{\sim}{q} = -\lambda\underset{\sim}{\nabla}T + \frac{c_t RT}{2}\sum_{\substack{i,j=1\\ j\neq i}}^{n} \frac{x_i x_j}{D'_{ij}}\left(\frac{D_j^T}{\varrho_j} - \frac{D_i^T}{\varrho_i}\right)(\underset{\sim}{u}_j - \underset{\sim}{u}_i) \tag{85}$$

where λ is the thermal conductivity of the mixture and D_i^T is the thermal diffusion coefficient of component i. It is convenient to define multicomponent thermal diffusion factors

$$\alpha_{ij} \equiv \frac{1}{D'_{ij}}\left(\frac{D_i^T}{\varrho_i} - \frac{D_j^T}{\varrho_j}\right), \quad i, j = 1, 2, \ldots, n \tag{86}$$

which has the anti-symmetric property

$$\alpha_{ij} = -\alpha_{ji} \tag{87}$$

α_{ii} is undefined.

With this definition of α_{ij}, we may write

$$\underset{\sim}{q} = -\lambda \underset{\sim}{\nabla} T - \frac{c_t RT}{2} \sum_{\substack{i,j=1 \\ i \neq j}}^{n} x_i x_j \alpha_{ij} (\underset{\sim}{u}_j - \underset{\sim}{u}_i) \tag{88}$$

The term on the right-hand side of the equation gives the contribution of the mass fluxes to the heat flux, $\underset{\sim}{q}$; this contribution is commonly referred to as the Dufour effect.

The inverse of the Dufour effect is the production of mass fluxes due to temperature gradients; this is referred to as thermal diffusion or the Soret effect. To account for this effect, we need to augment the generalized Maxwell–Stefan diffusion equations in the following manner

$$\underset{\sim}{d}_i = \sum_{\substack{j=1 \\ j \neq i}}^{n} \frac{x_i x_j}{D'_{ij}} (\underset{\sim}{u}_j - \underset{\sim}{u}_i) - \sum_{\substack{j=1 \\ j \neq i}}^{n} x_i x_j \alpha_{ij} \frac{\underset{\sim}{\nabla} T}{T} \tag{89}$$

where the generalized driving force $\underset{\sim}{d}_i$ is given, as before, by

$$\underset{\sim}{d}_i \equiv \frac{x_i}{RT} \underset{\sim}{\nabla}_{T,\varrho} \mu_i \tag{90}$$

Equations 88 and 89 are the complete forms of the constitutive relations for simultaneous mass and energy transport. The reader is referred to the paper by Standart, Taylor and Krishna [33] for further background to these derivations. Except where explicitly stated, the Dufour and Soret effects, quantified by the right members of Equations 88 and 89, will be ignored in the design applications to be considered in this review.

For the turbulent heat flux, $\underset{\sim}{q}_{turb}$, we may write, analogous to Equation 77,

$$\underset{\sim}{q}_{turb} = -\lambda_{turb} \underset{\sim}{\nabla} T \tag{91}$$

The estimation procedure for λ_{turb} is analogous to that of D_{turb} and the approach is to proceed via the turbulent Prandtl number

$$Pr_{turb} \equiv \frac{C_p \mu_{turb}}{\lambda_{turb}} \tag{92}$$

For want of better information, we must take $Pr_{turb} = Sc_{turb}$.

DEFINITIONS OF MASS AND HEAT TRANSFER COEFFICIENTS

Thus far we have considered the problem of choosing a physical model for the interfacial region; we have developed the differential equations describing the variation of the compositions in the region close to the interface together with the appropriate boundary conditions to be applied at the interface itself. We have also discussed, in some detail, the constitutive relations for the molecular and turbulent diffusion fluxes of mass and energy.

The next step is to solve the differential equations describing the continuity of mass and energy within each of the bulk phases "x" and "y" for the hydrodynamic conditions prevalent in these phases. It is usually necessary to use simplified hydrodynamic models to describe the flow field in the region of the interface; these models will be considered later. Once the equations have been solved to obtain the composition and temperature profiles, the diffusion fluxes $\underset{\sim}{J}_i$ and $\underset{\sim}{q}$ can be calculated and the interfacial transfer rates $\underset{\sim}{N}_i$ and $\underset{\sim}{E}$ determined. It is customary to determine on the basis of the chosen hydrodynamic model, mass and heat transfer coefficients which reflect the overall transfer facility (molecular and turbulent transport) of the phase under consideration. Before proceeding to discuss the hydrodynamic models, we will provide a formal definition of the mass transfer coefficients.

Let us consider the two-phase—"x" and "y"—system as shown in Figure 7 where the composition profiles have been shown. The cup-mixing compositions in the phases "x" and "y" are denoted by $\langle x_i \rangle$ and $\langle y_i \rangle$ for component i. The interface compositions are x_{iI} on the "x" side of the interface and y_{iI} on the "y" side of the interface. Exactly on the interface itself, the compositions are undefined, keeping with the surface of discontinuity model chosen for the interfacial region. Since the composition variations are usually restricted to a very thin region (not to be confused with the interfacial region) close to the interface, the cup-mixing compositions will, in many practical situations, correspond to the bulk phase mole fractions, denoted by x_{ib} and y_{ib} for the two phases. The starting point for discussions will naturally be Equation 14 describing the continuity of molar fluxes with respect to the interface. The simpler form, Equation 15, will suffice for a majority of the cases in which the interface remains stationary. We proceed with Equation 15 with the understanding that if the interface moves, the fluxes, $\underset{\sim}{N}_i$, must be referred to the interface and not a stationary coordinate reference frame. Further, interphase mass transfer usually takes place in a direction normal to the interface and, therefore, it is sufficient to use the scalar form of Equation 15, i.e.,

$$N_i^x = N_i^y = N_i, \quad i = 1, 2, \ldots, n \tag{93}$$

where N_i^x is the normal component of $\underset{\sim}{N}_i^x$ and is directed from phase "x" to phase "y". N_I^y is directed from the interface "I" to the bulk phase "y."

The Equation 93 may be summed over the n species to yield

$$N_t^x = N_t^y = N_t = \sum_{i=1}^{n} N_i \tag{94}$$

The mass transfer coefficient k in phase "x" for a binary system is best defined in a manner suggested by Bird, Stewart and Lightfoot [21] (p. 639):

$$k_b = \operatorname*{Limit}_{\substack{N_1 \to 0 \\ N_2 \to 0}} \frac{N_1 - x_{1b} N_t}{c_t(x_{1b} - x_{1I})} = \frac{J_{1b}}{c_t \Delta x_1} \tag{95}$$

where the driving force for mass transfer, Δx_1, is taken to be the difference between the bulk phase mole fraction, x_{1b}, and the composition of component 1 at the interface. As already mentioned, the bulk phase composition will in most cases equal the cup-mixed compositions when the interfacial region occupies only a minute fraction of the system volume. The diffusion fluxes used in Equation 95 are the bulk diffusion fluxes, and the mass transfer coefficients obtained in Equation 95 are the bulk phase mass transfer coefficients. By using the interface compositions in the calculations of the convective term $x_1 N_t$ in

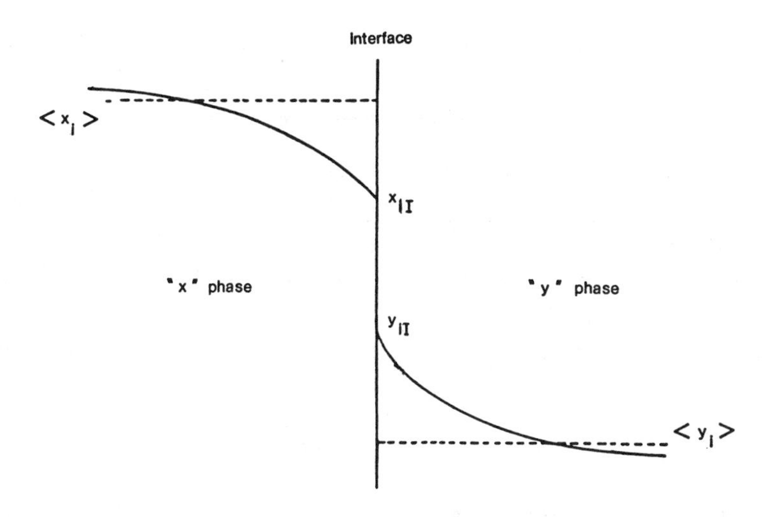

Figure 7. Mole fraction profiles in the region close to the interface during interphase mass transfer. The cup-mixing compositions in either fluid phase are denoted by $<x_i>$ and $<y_i>$.

Equation 95 we shall obtain the interface transfer coefficients. In general the distinction between k_b and k_I (or equivalently between J_{1b} and J_{1I}) is important in the description of mass transfer effects in reactors. We have omitted writing the subscript x (or superscript x) on these last four mentioned quantities because we shall be first considering what happens on the "x" side of the boundary. When we wish to describe the overall mass transfer behavior of the "x" – "y" transfer process we shall distinguish between the quantities on either side of the phase boundary. In proceeding further, it must, therefore, be remembered that analogous relations will hold for the "y" phase using the mole fraction difference $\Delta y_1 \equiv (y_{1I} - y_{1b})$.

A word now about the units of k_b. With the fluxes, N_i, expressed in kmol/s/(m^2 interfacial area), c_t in the units kmol/(m^3), the units of k_b are m/s. But is it really a velocity? To examine this further, we follow the manipulations of Equation 28 to obtain (omitting the limit sign)

$$k_b = \frac{(u_1 - u)}{\Delta x_1 / x_{1b}} \tag{96}$$

The numerator on the right-hand side of Equation 96 is the velocity of transfer of component 1 with respect to the molar average reference velocity of the mixture, u. For a binary system (but *not* always for a multicomponent system), the quotient in Equation 96 is positive and the coefficient k_b is positive. The denominator of Equation 96 can never exceed unity, i.e. $|\Delta x_1 / x_{1b}| \leqq 1$. Therefore we may write

$$k_b \geqq (u_1 - u) \tag{97}$$

which gives a physical significance to the mass transfer coefficient k_b: it is the maximum velocity (relative to the mixture) at which a component can be transferred in the binary mixture. The actual velocity at which a component is transferred is given by

$$(u_1 - u) = k_b \frac{\Delta x_1}{x_{1b}} \tag{98}$$

which may be compared to the Equation 30. k_b appears to be directly relatable to the diffusion coefficient, as will be shown to be the case when we examine various specific hydrodynamic models for mass transfer.

Let us now turn to the more difficult problem of explaining why the limits $N_1 \rightarrow 0$, $N_2 \rightarrow 0$ appear in Equation 95. During the actual mass transfer process itself the composition (and velocity) profiles are distorted by the flow (diffusion) of 1 and 2 across the interface. The mass transfer coefficients defined in Equation 95 correspond to conditions of vanishingly small mass transfer rates, when such distortions are not present. These low-flux or zero-flux coefficients are the ones which are usually available from mass transfer correlations which are usually obtained under conditions where the transfer rates are low. (In some cases we use correlations developed for heat transfer in order to calculate the mass transfer coefficients; in this case the zeroflux coefficients are the ones which are calculated by this procedure). For the actual situation under conditions of finite transfer rates, we may write

$$k_b^{\bullet} = \frac{N_1 - x_{1b}N_t}{c_t \Delta x_1} = \frac{J_{1b}}{c_t \Delta x_1} \tag{99}$$

where the superscript • serves to remind us of the fact that the transfer coefficients $k_b^{\bullet}$ correspond to conditions of finite mass transfer rates. (For further reading on this point we recommend Bird, Stewart and Lightfoot [21]).

Clearly, for calculations of the flux N_1, we need the finite flux coefficients $k_b^{\bullet}$; these are usually related to the zero-flux coefficients by a general relation of the form:

$$k_b^{\bullet} = k_b \Xi \tag{100}$$

where Ξ is a correction factor to account for the effect of finite fluxes on k_b. The exact form of the correction factor depends, of course, on the composition profiles and hence on the chosen hydrodynamic model to describe this process; more about this later.

The analysis for multicomponent mixtures is best carried out by using $n-1$ dimensional matrix notation. We therefore *define* the finite flux mass transfer coefficient matrix $[k_b^{\bullet}]$ by:

$$(J_b) \equiv (N) - (x_b)N_t = c_t[k_b^{\bullet}]\,(x_b - x_I) \tag{101}$$

where the finite flux coefficients are related to the zero-flux coefficients by a relation of the form

$$[k_b^{\bullet}] = [k_b]\,[\Xi] \tag{102}$$

where $[\Xi]$ is the matrix of correction factors. Generally speaking, the calculation of the mass transfer coefficient matrices and the correction matrices for a multicomponent system is very sensitive to the mass transfer model chosen. For example; even for the simplest film model, different approximations lead to different values for the matrices $[k_b]$, $[\Xi]$, and $[k_b^{\bullet}]$. There is a further problem with the multicomponent mass transfer coefficients, that of nonuniqueness the following explains this.

In Equation 101 we have defined $n-1 \times n-1$ elements of the mass transfer coefficients with the help of $n-1$ linear equations. It should be obvious that the elements $k_{ijb}^{\bullet}$ are not unique; i.e., another set of these coefficients can also lead to the same value of the fluxes, N_i. Put another way, making mass transfer measurements in a multicomponent system for the fluxes, N_i, and Δx_i does not uniquely determine the values of the mass transfer coefficients. A large set of measurements of N_i and Δx_i will be necessary to obtain one set of coefficients. In practice, we proceed in a different manner. We try to predict the values of the *multicomponent* mass transfer coefficients from *binary* mass transfer correlations, using as a basis, the generalized Maxwell–Stefan equations. However, various approximations are necessary before tractable solutions are obtained; more about this later.

Let us now turn our attention to the definition of heat transfer coefficients.

By analogy to Equation 95, we may define heat transfer coefficients, in either fluid phase by

$$h = \lim_{N_i \to 0} \left[\frac{E - \sum_{i=1}^{n} \bar{H}_i N_i}{c_t C_p (T_b - T_I)} = \frac{q}{c_t C_p \Delta T} \right] \tag{103}$$

where h is a heat transfer coefficient. We have defied convention in this definition, with one very good reason: the heat transfer coefficient, h, has the units of m/s, just as for the mass transfer. The physical significance of h is simple and straightforward: It represents the maximum speed at which heat can be transferred in the phase under consideration. It is more conventional to define the heat transfer coefficients as

$$h = \left[\frac{q}{\Delta T} \right], \quad \lim_{N_i \to 0}, \quad (i = 1, 2, \ldots, n) \tag{104}$$

but with this definition the parallelism with the mass transfer coefficient definition is lost somewhat. We shall proceed further with the conventional definition, firmly enshrined in all textbooks. The coefficient defined by Equation 104 corresponds to the heat transfer coefficient under conditions of negligible mass fluxes, i.e. zero flux heat transfer coefficients.

Just as finite mass fluxes distort the composition profiles during the interphase mass transfer process, they exert a similar effect on the temperature profiles and the interfacial heat fluxes. Witness the presence of the term $\sum_{i=1}^{n} N_i \bar{H}_i$ in the differential energy balance relation Equation 80.

For the actual conditions of finite mass transfer rates, we have

$$h^{\bullet} = \frac{q}{\Delta T} \tag{105}$$

where the superscript • serves to remind us of the fact that the transfer coefficient corresponds to conditions of finite mass transfer rates. The finite-flux heat transfer coefficients are related to the zero-flux coefficients by a relationship of the form

$$h^{\bullet} = h^{\bullet} \Xi_H \tag{106}$$

where Ξ_H is the correction factor to account for the effect of finite mass fluxes on the heat transfer coefficient h. Both h and Ξ_H depend on the temperature and composition profiles in the region adjacent to the interface, which in turn is affected by the hydrodynamics prevalent in the phase.

INTERPHASE MASS AND ENERGY TRANSFER—SOME GENERAL CONSIDERATIONS

Determination of Interfacial Mass and Energy Fluxes

Having defined the mass and heat transfer coefficients for the interphase transport processes of mass and energy, we now consider how the interfacial fluxes may be calculated.

At the interface between the "x" and "y" phases, we have continuity of mass and energy fluxes:

$$N_i^x = J_{ib}^x + x_{ib}N_t^x = N_i = J_{ib}^y + y_{ib}N_t^y = N_i^y; \quad N_t^x = N_t = N_t^y \tag{107}$$

$$E^x = q^x + \sum_{i=1}^{n} \bar{H}_i^x N_i = E = q^y + \sum_{i=1}^{n} \bar{H}_i^y N_i = E^y \tag{108}$$

Using the definitions of the mass and heat transfer coefficients, we may relate the diffusion and heat conduction fluxes to their respective driving forces:

$$(J_b^x) = c_t^x[k_{xb}^\bullet](x_b - x_I); \quad (J_b^y) = c_t^y[k_{yb}^\bullet](y_I - y_b) \tag{109}$$

$$q^x = h_x^\bullet(T_b^x - T_I); \qquad q^y = h_y^\bullet(T_I - T_b^y) \tag{110}$$

where, for definiteness, we consider transfer from the "x" phase to the "y" phase as leading to a positive flux.

In design applications, the bulk compositions x_{ib}, y_{ib} and the bulk temperature T_b^x and T_b^y are determined by column material and energy balances (discussed later). To complete the picture, we need only say something about the state of interface. In the absence of any better information, it is common to assume that equilibrium prevails at the interface; thus the usual equations of phase equilibrium relate the mole fractions y_{iI} and x_{iI}:

$$y_{iI} = K_i x_{iI}, \quad i = 1, 2, \ldots, n \tag{111}$$

The K_i are the equilibrium ratios or "K-values." A host of methods exist for calculating K-values (see, for example, Henley and Seader [59]). In general, K-values are complicated functions of temperature, pressure, and composition: $K_i = K_i(T_I, P_I, x_{jI}, y_{jI}; j = 1, 2, \ldots, n)$. Finally, we assume, not unreasonably, that the two phases are in mechanical equilibrium; $P_b^x = P_I = P_b^y = P$.

The equations just summarized are nonlinear and must, therefore, be solved by some iterative process. There are two steps in the development of an algorithm for solving the equations:

- The selection of particular numerical methods (see, for example, Ortega and Rheinholt [41]).
- The selection of the order in which the equations are to be solved (see, for further discussion, Myers and Seider [60]).

The following discusses each of these topics, beginning with the latter. Our remarks pertain to any set of nonlinear algebraic equations (particularly those with a physical origin) and should be reread prior to studying the last section.

To our knowledge, most of the algorithms described in the literature employ some kind of tearing strategy to solve the multicomponent mass transfer equations. Equation tearing involves iteratively solving a subset of the complete set of governing equations for a subset (the tear variables) of the complete set of unknowns. Within the loop that determines how the tear variables are to be reestimated, all other equations are satisfied exactly. Thus, if the remaining subset of equations also requires iteration (or, possibly, further tearing), the end result is a series of nested iteration loops. For example, we might solve the mass and energy transfer rate equations (by repeated substitution of the N_i) and the vapor-liquid equilibrium calculations (bubble point calculations [59]) within an outer loop on the interface state (x_{iI}, T_I).

One problem with tearing is the number of times physical properties must be evaluated (several times per outer loop iteration) if temperature- and composition-dependent physical properties are used. It is the physical properties calculations that generally dominate the cost of chemical design problems [61] (the present problems are no exception). A second problem can arise if the inner iteration loops are difficult to converge. We note in the next section that the mass transfer rate Equations 107 and 109 sometimes suffer from convergence problems and the difficulties of converging $y_{iI} = K_i x_{iI}$ (all i) are well known. We wonder what point there is to requiring subsets of the complete set of equations to be satisfied exactly (by iteration) when the estimate of the tear variables does not pertain to the solution. We suggest that a better strategy is to solve *all* of the equations simultaneously by guessing enough unknown variables to permit *all* other quantities appearing in Equations 107 through 111 to be calculated explicitly *and* by not requiring any of the equations to be satisfied until complete convergence has been achieved. This approach is followed in an algorithm to be described later which converges the vapor-liquid equilibrium calculations and the mass and energy transfer rate calculations at the same time. The approach was used earlier by Taylor et al. [62] for solving a particular subset of Equations 107 through 111 and is extended here to the general problem of simultaneous interphase mass and energy transfer.

The equations just presented can be combined to yield a set of $3n-1$ independent equations. Here we rewrite them in the form $(F(X)) = (0)$:

- $n-1$ mass transfer rate equations in the liquid phase

$$(R^L) \equiv (N) - c_t^*[k_{xb}^\bullet](x_b - x_I) - N_t(x_b) = (0) \tag{112}$$

- $n-1$ mass transfer rate equations in the vapor phase

$$(R^V) \equiv (N) - c_t^*[k_{yb}^\bullet](y_I - y_b) - N_t(y_b) = (0) \tag{113}$$

- n interfacial equilibrium equations

$$Q_i^I \equiv K_i x_{iI} - y_{iI} = 0, \quad i = 1, 2, \ldots, n \tag{114}$$

- 1 energy continuity equation ($E^y - E^x = 0$)

$$E^I \equiv q^y - q^x + \sum_{i=1}^{n} N_i(\bar{H}_i^y - \bar{H}_i^x) = 0 \tag{115}$$

The unknown quantities to be determined from these equations are the n molar fluxes, N_i; $2(n-1)$ interfacial mole fractions, x_{iI} and y_{iI}; and the interface temperature T_I. It is assumed that methods of obtaining the heat and mass transfer coefficients are available (see next section).

The independent Equations 112 through 115 are ordered into a vector of functions (F) as follows

$$(F)^T \equiv (R_1^V, R_2^V, \ldots, R_{n-1}^V, E^I, R_1^L, R_2^L, \ldots, R_{n-1}^L, Q_1^I, Q_2^I, \ldots, Q_n^I)$$

The unknown variables corresponding to this set of equations are ordered into a vector (X) as follows

$$(X)^T \equiv (N_1, N_2, \ldots, N_{n-1}, N_n, x_{1I}, x_{2I}, \ldots, x_{n-1,I}, y_{iI}, y_{2I}, \ldots, y_{n-1,I}, T_I)$$

We wish to find the vector of variables, (X^*), that gives $(F) = (0)$.

The first step of the algorithm to find the N_i, x_{iI}, y_{iI} and T_I involves the guessing of the unknowns; the heart of it is the method used to reestimate them if the initial guess was wrong. Newton's method (or the Newton–Raphson method) is one of the oldest yet most reliable algorithms for solving a system of nonlinear equations. The direct prediction Newton correction is given by a solution of the equations linearized about the current estimate, (X_k), of (X)

$$[J_k]\,(X_{k+1}-X_k) = -(F(X_k)) \tag{116}$$

where $[J_k]$ is the Jacobian matrix with elements

$$J_{ij} = \partial F_i/\partial X_j \tag{117}$$

If a problem with Newton's method exists, it is the computation of [J]. For most engineering applications of reasonable complexity, complete derivative information is rarely available in analytical form. Consequently, finite differences must be used to approximate all or part of [J]. In most cases, the unavailable derivatives involve derivatives of physical properties (e.g., $\partial K_i/\partial T$, $\partial K_i/\partial x_i$]. Finite difference approximations of these derivatives can enormously increase the cost of solution because many more physical property calculations are required. Further, neglect of these derivatives generally increases the number of iterations or can even cause failure.

One way to avoid the repeated calculation of [J] is to use a "quasi-Newton" method (see Dennis and More [63] for a recent review of these methods). Broyden's method, for example, builds an approximation $[J'_k]$ to $[J_k]$ from the formula

$$[J'_{k+1}] = [J'_k]+[((Y)-[J'_k]\,(S))\,(S)^T]/(S)^T(S) \tag{118}$$

where $(Y) = (F(X_{k+1})-F(X_k))$ and $(S) = (X_{k+1}-X_k)$. Schubert's method is an extension of the Broyden method that preserves any known sparseness in [J]. The quasi-Newton correction is given by a solution of the linear system Equation 116 with $[J'_k]$ replacing $[J_k]$.

Unfortunately, quasi-Newton methods are not without problems. The first is that they are *not* scale invariant (Newton's method is) and may, as a result, perform poorly or fail on problems that are ill-conditioned. This is an important consideration in chemical engineering applications where material and energy balance equations almost always are part of the model. This particular subset of equations usually exhibit poor scaling. To a lesser extent, the need to supply an initial approximation to [J] can sometimes be a problem. If the initial approximation is poor, it can increase the number of iterations or even cause failure. Quasi-Newton methods appear to be more susceptible than is Newton's method to this problem.

A judicious combination of Newton's method and a quasi-Newton update can, as shown recently by Lucia and co-workers [64–66], be a very efficient means of solving the kinds of equations found in chemical process calculations. In what they call the hybrid method, the Jacobian matrix is divided into two parts; a computed part, $[C_k]$, and an approximated part, $[A_k]$. Thus,

$$[J_k] = [C_k]+[A_k] \tag{119}$$

$[C_k]$ consists of any partial derivatives for which analytical expressions exist. $[A_k]$ is made up of any terms that contain unavailable partial derivative information such as $\partial K_i/\partial x_j$. In the hybrid approach, $[C_k]$ is computed in each iteration while $[A_k]$ only is updated from an initial approximation using the Broyden or Schubert methods. The correction to (X_k) again is provided by Equation 116 with $[J_k]$ given by Equation 119.

We have used the algorithm of Lucia and co-workers to solve the nonlinear algebraic Equations 112 through 115 and found it to be a reliable and highly efficient method. The step-by-step procedure that we use can be found in References 56–58 and is summarized in the following. (Some numerical results and notes on using the hybrid method to solve multicomponent mass transfer problems are given by Taylor et al. [62].)

Newton–Like "Hybrid" Algorithm of Lucia and Machietto

1. Set k, the iteration counter, to zero. Initialize (X_0) and $[A_0]$ ($[A_0]$ may safely be set to the null matrix in most cases). Decide upon a convergence tolerance, t, say.
2. Calculate $(F(X_k))$.
3. Calculate $[C_k]$.
4. Check for convergence. If $\|(F)\| \leq t$, stop. Otherwise continue.
5. Set $[J_k] = [C_k]+[A_k]$ and solve the linear system $[J_k](S_k) = -(F_k)$ for (S_k).
6. Set $(X_{k+1}) = (X_k)+(S_k)$
7. Compute $(F(X_{k+1}))$ and $[C_{k+1}]$ as in Steps 2 and 3 above.
8. Define $(S) = (X_{k+1}-X_k)$ and $(Y) = (F_{k+1}-F_k)-[C_{k+1}](S)$ and update [A] using Broyden's or Schubert's quasi-Newton methods.
9. Increment k and return to Step 4.

The "Bootstrap" Problem and Its Solution

In the sections that follow, it will be necessary to relate the invariant molar fluxes, N_i, to the diffusion fluxes, J_i. The problem here is that all n of the N_i are independent, whereas only $n-1$ of the J_i are independent. Thus one further relation is required in order to make the problem determinate. The additional equation that allows us to relate the N_i directly to the J_i is the interfacial energy balance Equation 108. First, Equation 107 is combined to give for each of the $n-1$ independent species

$$N_t = -\frac{J_i^x - J_i^y}{x_i - y_i}, \quad i = 1, 2, \ldots, n-1 \tag{120}$$

Also, Equation 108 may be rewritten in view of Equation 107

$$\begin{aligned} q^x - q^y &= \sum_{i=1}^{n} (\bar{H}_i^y - \bar{H}_i^x) N_i = \sum_{i=1}^{n} \lambda_i N_i = \\ &= \sum_{i=1}^{n} \lambda_i J_i^y + \sum_{i=1}^{n} \lambda_i y_i N_t = \sum_{i=1}^{n} \lambda_i J_i^x + \sum_{i=1}^{n} \lambda_i x_i N_t \\ &= \sum_{i=1}^{n-1} (\lambda_i - \lambda_n) J_i^y + \bar{\lambda}_y N_t = \sum_{i=1}^{n-1} (\lambda_i - \lambda_n) J_i^x + \bar{\lambda}_x N_t \end{aligned} \tag{121}$$

where we define

$$\bar{\lambda}_y = \sum_{i=1}^{n} \lambda_i y_i; \quad \bar{\lambda}_x = \sum_{i=1}^{n} \lambda_i x_i \tag{122}$$

Equations 120 and 121 may be combined to give $n-1$ independent relations for the total flux, N_t

$$-\frac{J_i^x - J_i^y}{x_i - y_i} = \frac{q^x - q^y}{\bar{\lambda}_y} - \frac{\sum_{k=1}^{n-1} (\lambda_k - \lambda_n) J_k^y}{\bar{\lambda}_y} = N_t, \quad i = 1, 2, \ldots, n-1 \tag{123}$$

In Equations 123 we use the vapor-phase diffusion fluxes. Alternatively we may use the liquid-phase diffusion fluxes to obtain

$$-\frac{J_i^x - J_i^y}{x_i - y_i} = \frac{q^x - q^y}{\bar{\lambda}_x} - \frac{\sum_{k=1}^{n-1} (\lambda_k - \lambda_n) J_k^x}{\bar{\lambda}_x} = N_t, \quad i = 1, 2, \ldots, n-1 \tag{124}$$

We now combine $n-1$ of the Equation 107 with Equation 123 to give

$$N_i = (1 - \Lambda_i y_i) J_i^y - y_i \sum_{\substack{k=1 \\ k \neq i}}^{n-1} \Lambda_k J_k^y + y_i \frac{\Delta q}{\bar{\lambda}_y}, \quad i = 1, 2, \ldots, n-1 \tag{125}$$

where we have defined the parameters

$$\Lambda_k = (\lambda_k - \lambda_n)/\bar{\lambda}_y; \quad \Delta q = q^x - q^y \tag{126}$$

Equation 125 may be rewritten in matrix notation ($n-1$ dimensional) as

$$(N) = [\beta^y](J^y) + (y)\frac{\Delta q}{\bar{\lambda}_y} \tag{127}$$

where the matrix $[\beta^y]$, dubbed the "bootstrap" matrix by Krishna and Standart [6], has the elements:

$$\beta_{ik}^y = \delta_{ik} - y_i \Lambda_k, \quad i, k = 1, 2, \ldots, n-1 \tag{128}$$

An expression analogous to Equation 127 may be written in terms of the liquid-phase diffusion fluxes. Thus

$$(N) = [\beta^x](J^x) + (x)\frac{\Delta q}{\bar{\lambda}_x} \tag{129}$$

Equation 127 or 129 may be used to determine the $n-1$ total fluxes; N_i, $i = 1, 2, \ldots, n-1$; given the $n-1$ diffusion fluxes J_i. The n-th total flux is determined from

$$N_n = N_t - \sum_{i=1}^{n-1} N_i \tag{130}$$

and the total energy flux obtained from Equation 108.

To illustrate the usefulness of this formalism developed, we consider its application to multicomponent distillation.

Multicomponent Distillation

In the treatment of transport processes during distillation, most textbooks (e.g., Sherwood et al. [22]) assume that conditions of equimolar counter transfer, i.e.,

$$N_t = 0 \tag{131}$$

hold. It is clear from Equation 127 that the requirement of Equation 131 will be realized if the following two conditions are satisfied:

1. $\Delta q = q^x - q^y = h_x^\bullet(T_I - T_b^x) - h_y^\bullet(T_b^y - T_I) = 0$ (132)

2. $\lambda_i = \lambda_n, \quad i = 1, 2, \ldots, n-1$ (133)

The first requirement (1) is often met in practice because the heat transfer coefficients in the vapor and liquid phases are such as to wipe out any temperature gradients locally, say on a tray.

Equation 133 requires the molar latent heats of vaporization of the constituent species to be identical. In practice, molar latent heats of many compounds are close to one another but the differences will not be zero. Typically we may expect the term Λ_i, defined in Equation 126 to be of the order of magnitude 0.1.; the sign of Λ_i can be either positive or negative. Let us examine the effect of such small differences in the latent heats on the interfacial rates of transfer.

For a three-component system, Equation 125 may be written explicitly as

$$N_1 = (1 - y_1\Lambda_1)J_1^y - y_1\Lambda_2 J_2^y + y_1\Delta q/\bar{\lambda}_y \tag{134}$$

and

$$N_2 = -y_2\Lambda_1 J_1^y + (1 - y_2\Lambda_2)J_2^y + y_2\Delta q/\bar{\lambda}_y \tag{135}$$

The diffusion fluxes J_1^y and J_2^y need not have the same magnitude for nonideal ternary mixtures. It is quite possible to have a situation in which we have

$$J_2^y = 5\,J_1^y \tag{136}$$

Further, let us suppose that the following values hold

$$y_1\Lambda_1 = +0.05; \quad y_1\Lambda_2 = +0.05 \tag{137}$$

Equation 134 will then give in view of values in Equations 136 and 137

$$N_1 \cong 0.70\,J_1^y \tag{138}$$

The assumption of equimolar counter transfer Equation 131 will of course give $N_1 = J_1^y$, 30% in error. In multicomponent systems, it is quite feasible to have the diffusion fluxes differ by a factor of five as assumed for the purposes of the preceding illustration. Therefore, extremely small cross coefficients in the matrix $[\beta^y]$ can give rise to large deviations from the condition Equation 131. Such effects are not present to such a dramatic extent for binary systems because for this case we must necessarily have

$$J_1^y = -J_2^y$$

and there will be no enhancement of any differences in latent heats. For binary systems Equation 131 represents a good approximation and this explains its enshrinement in so many texts. However, for systems with three or more species large deviations from the equimolar counter transfer condition may be experienced, thus underlining the fundamental differences between the transport characteristics of binary and multicomponent systems.

Other "Bootstrap" Solutions

We have considered the general problem of simultaneous interphase mass and energy transport. In many problems in chemical engineering, mass transfer takes place under nearly isothermal conditions and the "bootstrap" solution derived in the preceding will not be applicable. How then do we calculate the n interfacial fluxes, N_i, from a knowledge of only $n-1$ diffusion fluxes? Various special cases may be identified as shown in the following.

1. *Equimolar diffusion.* Here the conditions of equimolar transport, $N_t = 0$, may be imposed by the physics of the problem. Such is the case, for example, for isobaric diffusion in closed systems (for example in two-bulb diffusion cells). Another situation in which $N_t = 0$ arises in diffusion with heterogeneous chemical reaction in which the reaction stoichiometry is such that there is no net production (or consumption) of moles. The application of $N_t = 0$ for distillation has already been examined. The "bootstrap" solution for equimolar diffusion may be written

$$(N) = [\beta](J) \quad \text{with } \beta_{ik} = \delta_{ik} \tag{139}$$

2. *Stefan diffusion.* In this case, we have diffusion of $n-1$ components in the presence of an "inert," or nontransferring, component species n. The inertness of the n-th component may arise, for example, due to its insolubility in one of the phase. The condition $N_n = 0$ makes the problem determinate and the $n-1$ nonzero fluxes can be calculated from

$$(N) = [\beta](J)$$

$$\text{where } \beta_{ik} = \delta_{ik} + x_i/x_n \tag{140}$$

If the bulk diffusion fluxes, J_{ib}, are utilized in Equation 131, then $\beta_{ik} = \delta_{ik} + x_{ib}/x_{nb}$; on the other hand if interfacial fluxes, J_{iI}, are employed then $\beta_{ik} = \delta_{ik} + x_{iI}/x_{nI}$. The N_i are, of course, phase invariant.

3. *Diffusion with heterogeneous chemical reaction.* Here the flux ratios

$$z_i \equiv N_i/N_t$$

are constrained by the reaction stoichiometry, and known a priori. In this case, a "bootstrap" solution is

$$(N) = [\beta](J)$$

$$\text{with } \beta_{ik} = \frac{\delta_{ik}}{1 - x_i/z_i} \tag{141}$$

4. *Diffusion in a porous medium.* The component fluxes for diffusion in a porous medium under isobaric conditions ($\nabla P = 0$) are related by a generalization of Graham's law [67, 68]

$$\sum_{i=1}^{n} \sqrt{M_i}\, N_i = 0 \tag{142}$$

In this case the bootstrap solution is

$$(N) = [\beta]\,(J) \qquad \text{with } \beta_{ik} = \delta_{ik} - x_i \frac{(\sqrt{M_k} - \sqrt{M_n})}{\sum_{j=1}^{n} x_j \sqrt{M_j}} \tag{143}$$

A More General Result

In all of the particular cases just examined with the exception of the case where $\Delta q \neq 0$ Equation 127, the fluxes are constrained by a linear relation of the form

$$\sum_{i=1}^{n} \nu_i N_i = 0 \tag{144}$$

The ν_i can be regarded as determinacy coefficients. Thus, for equimolar counter transfer ($N_t = 0$), $\nu_i = \nu_n$; for Stefan diffusion ($N_n = 0$), $\nu_i = 0$, $i = 1, 2, \ldots, n-1$ and for Grahams law, $\nu_i = \sqrt{M_i}$. Even for diffusion-controlled heterogeneous chemical reaction, it is possible to find a relation between the fluxes of the form in Equation 144. Finally, for multicomponent distillation with $\Delta q = 0$ we can identify the ν_i with the latent heats, $\nu_i = \lambda_i$. The bootstrap solution is

$$(N) = [\beta]\,(J)$$

$$\text{with } \beta_{ik} = \delta_{ik} - x_i \frac{(\nu_k - \nu_n)}{\sum \nu_j x_j} \tag{145}$$

Equation 145 applies at all points along the diffusion path. Usually, however, we would evaluate the fluxes either at the interface or in the bulk fluid:

$$(N) = [\beta_b]\,(J_b) = [\beta_I]\,(J_I) \tag{146}$$

$$= c_t\,[\beta_b]\,[k_b^\bullet]\,(\Delta x) = c_t\,[\beta_I]\,[k_I^\bullet]\,(\Delta x) \tag{147}$$

One important distinction between the binary and the general multicomponent case is worth recording here. For a binary mixture we must have:

$$\beta_I k_I^\bullet = \beta_b k_b^\bullet = \frac{N_1}{c_t\,\Delta x_1} \tag{148}$$

but we do not have the corresponding equality for the multicomponent system, i.e.

$$[\beta_I]\,[k_I^\bullet] \neq [\beta_b]\,[k_b^\bullet] \tag{149}$$

because in matrix algebra it is *not true* that when $[A]\,(x) = [B]\,(x)$ then $[A] = [B]$. Equation 149 actually spells out explicitly the problem of nonuniqueness of mass transfer coefficients in multicomponents, mentioned earlier. Actually, an exact solution of the Maxwell–Stefan equations will reveal that the two terms on either side of Equation 149 will generally be unequal except for the case of vanishingly small transfer fluxes. It must be realized that the *product* of these two unequal terms with the column matrix of composition driving forces (Δx) will lead to identical values for the transfer fluxes, N_i. The inequality Equation 149 has only recently been pointed out [69].

Addition of Phase Resistances for Nonisothermal Multicomponent Mass Transfer

Addition of mass transfer resistances for *binary* systems is discussed widely in the literature [21, 22]; here we review extensions of conventional treatments to the multicomponent case [4, 70]. The interfacial resistance is ignored in the present analysis.

We consider a vapor-liquid system with n components in either phase, maintained at constant pressure. Equation 147 first is inverted to give, for each phase

$$(x_b - x_I) = \frac{1}{c_t^x}[k_{xb}^{\bullet}]^{-1}[\beta_b^x]^{-1}(N) \tag{150}$$

$$(y_I - y_b) = \frac{1}{c_t^y}[k_{yb}^{\bullet}]^{-1}[\beta_b^y]^{-1}(N) \tag{151}$$

If the vapor-liquid equilibrium relationship Equation 111 is linearized over the range of compositions in passing from the bulk (b) to the interface (I) conditions, we can write at the interface for isobaric conditions

$$(y_I) = [M](x_I) + (b) \tag{152}$$

where [M] is the matrix of equilibrium constants with elements

$$M_{ij} = \partial y_i^* / \partial x_j, \quad i, j = 1, 2, \ldots, n-1 \tag{153}$$

and is diagonal only for thermodynamically ideal mixtures. (b) is a column matrix of "intercepts." For the evaluation of M_{ij} see Krishna [71].

If we wish to combine mass transfer driving forces and resistances of each phase, we must require that at least one phase be saturated; this is necessary for eliminating the partial driving force for that phase. The assumption concerning saturation is implicit in conventional approaches to the binary transport problem and will be used here for the multicomponent case. Here we assume that the liquid phase is saturated; this allows us to calculate the composition of the vapor which would be in equilibrium with the bulk liquid phase as

$$(y_b^*) = [M](x_b) + (b) \tag{154}$$

For binary systems all matrices contain just one element, and we have no difficulty in *deriving* the well-known formula for the addition of resistances

$$\frac{1}{c_t^y K_{oyb}^{\bullet} \beta_b^y} = \frac{1}{c_t^y k_{yb}^{\bullet} \beta_b^y} + \frac{M}{c_t^x k_{xb}^{\bullet} \beta_b^x} \tag{155}$$

where $K_{oyb}^{\bullet}$ is an overall mass transfer coefficient *defined* by

$$N_1 = c_t^y \beta_b^y K_{oyb}^{\bullet}(y_{1b} - y_{1b}^*) \tag{156}$$

The multicomponent generalization of Equation 156 is

$$(N) = c_t^y [\beta_b^y][K_{oyb}^{\bullet}](y_b - y_b^*) \tag{157}$$

but we cannot *derive* an expression for the matrix of overall mass transfer coefficients $[K^{\bullet}_{oyb}]$ from Equations 147 through 154 without first demanding that [A] = [B] follows from the equation [A] (x) = [B] (x). This, as we saw earlier, definitely is not the case. The matrix $[K^{\bullet}_{oyb}]$ may be *defined* by a generalization of Equation 155

$$\frac{1}{c_t^y}[K^{\bullet}_{oyb}]^{-1}[\beta_b^y]^{-1} = \frac{1}{c_t^y}[k^{\bullet}_{yb}]^{-1}[\beta_b^y]^{-1} + \frac{1}{c_t^x}[M]\,[k^{\bullet}_{xb}]^{-1}[\beta_b^x]^{-1} \tag{158}$$

This is, perhaps, the only logical way of defining $[K^{\bullet}_{oyb}]$ but it should be remembered that there are an infinite number of matrices $[K^{\bullet}_{oyb}]$ which statisfy Equations 150, 151 and 157, but only one which statisfies Equation 158. There is no ambiguity at all about the corresponding result for binary systems.

The rate Equation 157, is useful in the development of expressions for calculating multicomponent efficiencies in distillation on a tray or for defining "numbers of transfer units" in continuous contact equipment. Other than in these and similar cases, the concept of the addition of resistances for coupled multicomponent systems is an unnecessary complication in separation process modeling. We shall, indeed, abandon the idea when describing how the models of multicomponent mass transfer we describe next can be used in the simulation and design of multicomponent mass transfer rate-governed processes.

HYDRODYNAMIC MODELS FOR MASS AND HEAT TRANSFER

We have finally reached the stage in which the most important problem in mass and heat transfer modeling has to be tackled, namely the choice of a simplified picture to describe the actual hydrodynamic conditions prevailing in the region of the interface. You may of course wonder why do we need to use a *simplified model*? Well, the difficulty is that though the basic governing differential equations are well understood, in many situations which the chemical engineer has to deal with the fluid flow is so complicated that the governing equations cannot be solved without gross over-simplifications. In essence, the models which are used are grossly simplified transport phenomena equations; in this sense, they are mathematical models. But there is another view to take of the hydrodynamic models and that is to picture them as physical constructs which are more easily analyzed than the actual situation, but which show, in some way, the characteristic behavior of the actual situation. From a pragmatic point, the flow field is simplified to the extent that the equations describing the diffusion process can be solved, preferably analytically.

Film Theory

The film model is the simplest model for mass (and heat) transfer, and the oldest and somewhat maligned. In our opinion, it is probably the most useful model for describing multicomponent mass transfer; we shall devote more space to the multicomponent film model than to any other model. We first consider the problem of mass transfer and take up later the question of heat transfer.

According to the film model, we imagine that in turbulent flow all the resistance to mass transfer is concentrated in a thin film, or layer, adjacent to the phase boundary; that transfer occurs within this film by steady-state molecular diffusion alone and that outside this film, in the bulk fluid, the level of turbulence is so high that all composition gradients are wiped out by turbulent eddy mixing. See Figure 8. Thus, we have a fully turbulent, uniform core of bulk phase flow adjacent to a thin film in laminar flow parallel to the interface. Mass transfer occurs through this film essentially in the direction normal to the interface; that is, any constituent molecular diffusion or convection in the laminar flow parallel to the surface due to composition gradients along the interface are negligible in comparison to

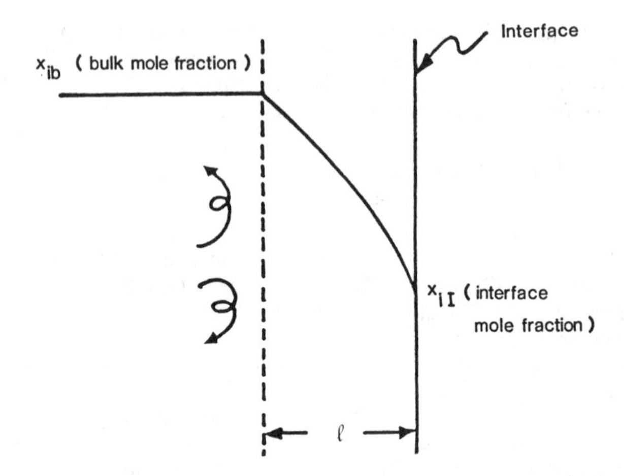

Figure 8. Film model for transfer in phase "x". Turbulent eddies wipe out composition gradients in the bulk fluid phase. Such composition variations are restricted to a layer ("film") of thickness ℓ adjacent to the interface. Due to Lewis and Whitman.

the normal transfers. The laminar film does not, of course, really exist; but if it did, its thickness would be about 0.01–0.1 mm for liquid-phase transport and in the range 0.1–1 mm for gas-phase transport.

Having made the appropriate simplification to the hydrodynamics, it remains to solve the relevant differential equations describing the molecular diffusion process in the diffusion layer. The diffusion process is fully determined by

1. The one-dimensional steady-state form of Equation 11
2. The constitutive relations, Equation 39 or Equation 46
3. The determinacy condition, Equation 144.

We shall develop the solution for the simple case of a two-component mixture before proceeding to the general n-component case.

Film Model for Binary Mass Transfer

Let us consider a planar film between the position coordinates $r = r_0$ and $r = r_\delta$. Mass transfer occurs between the two edges of the film purely by molecular diffusion under steady-state conditions. The thickness of the film is $\ell = r_\delta - r_0$. The equation of continuity of moles of species (we consider a binary system with species 1 and 2) can be written in the form:

$$\frac{dN_1}{dr} = 0; \quad \frac{dN_2}{dr} = 0; \quad \frac{dN_t}{dr} = 0 \tag{159}$$

showing that N_1, N_2 and N_t $(= N_1 + N_2)$ are r-invariant.

The generalized Maxwell–Stefan diffusion Equation 56 reduce to

$$\frac{dx_1}{dr} = \frac{x_1 N_2 - x_2 N_1}{c_t D} \tag{160}$$

where $D = D'\left(1 + \frac{\partial \ln \gamma_1}{\partial \ln x_1}\right)$ is the Fickian diffusion coefficient. We assume in the film model development following that D is constant; this is true for ideal gas mixtures at constant temperature and pressure and a fair approximation for small concentration changes in nonideal fluid mixtures.

We rewrite Equation 160, by substituting $x_2 = 1 - x_1$, as

$$\frac{dx_1}{dr} = \frac{x_1(N_1 + N_2)}{c_t D} - \frac{N_1}{c_t D} \tag{161}$$

It is convenient to define the following parameters:

- A dimensionless distance

$$\eta = \frac{r - r_0}{r_\delta - r_0} = \frac{r - r_0}{\ell} \tag{162}$$

where ℓ is the thickness of the diffusion layer or "film"

- A dimensionless mass transfer rate factor

$$\Phi \equiv \frac{N_1 + N_2}{c_t D/\ell} = \frac{N_t}{c_t D/\ell} \tag{163}$$

- $\zeta \equiv -\frac{N_1}{c_t D/\ell}$ (164)

With these definitions Equation 161 may be written as

$$\frac{dx_1}{d\eta} = \Phi x_1 + \zeta \tag{165}$$

which is to be solved subject to the boundary conditions of a film model:

$$\begin{array}{llll} r = r_0; & \eta = 0 & x_1 = x_{10} & \text{(bulk fluid phase)} \\ r = r_\delta; & \eta = 1 & x_1 = x_{1\delta} & \text{(interface)} \end{array} \tag{166}$$

The linear differential Equation 165 can be solved quite straightforwardly to yield the composition profiles

$$\frac{x_1 - x_{10}}{x_{1\delta} - x_{10}} = \frac{e^{\Phi\eta} - 1}{e^{\Phi} - 1} \tag{167}$$

The diffusion flux at $\eta = 0$, J_{10}, is given by

$$J_{10} = -\frac{c_t D}{\ell} \frac{dx_1}{d\eta}\bigg|_{\eta = 1} = c_t \frac{D}{\ell} \cdot \frac{\Phi}{\exp(\Phi) - 1} \cdot (x_{10} - x_{1\delta}) \tag{168}$$

Similarly the diffusion flux at $\eta = 1$, $J_{1\delta}$ can be obtained:

$$J_{1\delta} = -\frac{c_t D}{\ell}\frac{dx_1}{d\eta}\bigg|_{\eta=1} = c_t\frac{D}{\ell}\cdot\frac{\Phi\exp(\Phi)}{\exp(\Phi)-1}\cdot(x_{10}-x_{1\delta}) \tag{169}$$

which on comparison with the basic definition of the mass transfer coefficient, Equation 99, shows

$$k_b = k_0 = k_\delta = k_I = \frac{D}{\ell} \tag{170}$$

with the correction factors given by

$$\Xi_b \equiv \Xi_0 \equiv \frac{\Phi}{\exp(\Phi)-1}; \qquad \Xi_1 \equiv \Xi_\delta = \frac{\Phi\exp(\Phi)}{\exp(\Phi)-1} \tag{171}$$

If we have mass transfer into a given phase ($\Phi < 0$), then the correction factor, Ξ_0, is greater than unity (and therefore $k_b^\bullet > k_b$). On the other hand, if there is net mass transfer out of the phase under consideration ($\Phi < 0$), then the correction factor Ξ is less than unity (and therefore $k_b^\bullet < k_b$). Bird, Stewart and Lightfoot [21] give graphs for the evaluation of the correction factor, Ξ_0.

The flux N_1 can be calculated from the "bootstrap" solution

$$N_1 = \beta_0 J_{10} = c_t\beta_0 k\frac{\Phi}{\exp(\Phi)-1}\cdot(x_{10}-x_{1\delta}) \tag{172}$$

$$= \beta_\delta J_{1\delta} = c_t\beta_\delta k\cdot\frac{\Phi\exp(\Phi)}{\exp(\Phi)-1}\cdot(x_{10}-x_{1\delta}) \tag{173}$$

Various special cases can now be identified.

1. *Equimolar counter-diffusion.* This situation can arise, for example, in a distillation column when the molar heats of vaporization of the two components are equal to each other or when there is no net molar change during diffusion with heterogeneous chemical reaction. For this special case, we have:

$$N_t = 0; \quad \Phi = 0; \quad \Xi_0 = \Xi_\delta = 1; \quad \beta_0 = 1; \quad \beta_\delta = 1$$

$$N_1 = c_t k(x_{10}-x_{1\delta}); \quad N_2 = -N_1 = c_t k(x_{20}-x_{2\delta}) \tag{174}$$

2. *Stefan diffusion.* In this case we have diffusion of component 1 in the presence of an inert or stagnant gas. This situation arises very often during absorption or condensation operations and when reaction takes place in the presence of inerts or diluents. For this special case, we have:

$$N_2 = 0; \quad \beta_0 = 1+x_{10}/x_{20}; \quad \beta_\delta = 1+x_{1\delta}/x_{2\delta}; \tag{175}$$

$$\Phi = \ell n(x_{2\delta}/x_{20}) = \ell n\frac{1-x_{1\delta}}{1-x_{10}} = \frac{N_t}{c_t k} = \frac{N_1}{c_t k} \tag{176}$$

or

$$N_1 = c_t k\,\ell n\frac{1-x_{1\delta}}{1-x_{10}} \tag{177}$$

3. *Flux ratios fixed by the stoichiometry of a surface reaction.* When the flux ratios $z_1 = N_1/N_t$ and $z_2 = N_2/N_t$ are fixed by the stoichiometry of the surface reaction, the mass transfer rate factor reduces to

$$\Phi = \ell n \frac{1 - x_{1\delta}/z_1}{1 - x_{10}/z_1} = \frac{N_t}{c_t k} = \frac{N_1/z_1}{c_t k} = \frac{N_2/z_2}{c_t k} \tag{178}$$

which expression allows calculation of the fluxes N_1 and N_2. It must be noted that in Equation 178 one of the surface compositions, say x_{10}, will be unknown and given by an expression for the heterogeneous surface reaction at that plane (e.g., $N_{10} = k_s c_t x_{10}$ where k_s is the surface reaction rate constant).

Generalization of the Film Model to Other Geometries

The equations derived for the fluxes across a planar film model can be generalized to cylindrical and spherical films by appropriate definitions of the characteristic length ℓ; see Figure 9.

The expression for the calculation of the flux N_1 for binary diffusion is

$$N_{10} = c_t \beta_0 k \frac{\Phi}{\exp \Phi - 1}(x_{10} - x_{1\delta}) \tag{179}$$

with the definitions

$$\Phi \equiv N_t/c_t k; \quad k \equiv D/\ell \tag{180}$$

Special Cases

- Equimolar diffusion, $N_t = 0$, arising from no molar change during diffusion with heterogeneous reaction:

$$\Phi = 0; \ \beta_0 = 1; \quad N_{10} = c_t k(x_{10} - x_{1\delta}) \tag{181}$$

- Species 1 is infinitely dilute in 2:

$$\Phi \approx 0; \ \beta_0 \approx 1; \quad N_{10} = c_t k(x_{10} - x_{1\delta}) \tag{182}$$

A very important practical application of this film model is to determine the external mass transfer resistance to catalyst particles. For spherical catalysts, for example, it should be clear that the Sherwood number, Sh, defined by $Sh \equiv k2r_0/D$ will be equal to 2 *only* when either of the two special cases just described hold *and* when we have $r_0 \ll r_\delta$. For the general case of nonzero N_t (more common than is usually supposed), we must use Equations 179 and 180 for the calculation of N_1.

Film Model for Multicomponent Mass Transfer in Ideal Gases

Let us turn our attention to n-component mixtures. An exact analytical solution for the film model can be obtained for a mixture of ideal gases, for which the GMS diffusion coefficients D'_{ik} are independent of composition and identical to the diffusivity of the binary gas pair i-k. The multicomponent film model for ideal n-component gas mixtures was first developed by Krishna and Standart [72], and with its appearance a host of solutions for special cases, already available in the literature, have become redundant (see the review of Krishna and Standart [6] for sources). (The relationships between the solution of Krishna and Standart and the many special case solutions are explored by Taylor [73, 74].)

The starting point for the ideal-gas multicomponent film model are the Maxwell–Stefan diffusion equations, written for the "y" phase, here taken to be gaseous

$$\frac{dy_i}{dr} = \sum_{\substack{j=1 \\ j \neq i}}^{n} \frac{y_i N_j - y_j N_i}{c_t D'_{ij}}, \quad i = 1, 2, \ldots, n \tag{183}$$

To solve the set of linear differential Equation 183 we define the following parameters

- Dimensionless distance coordinate (see Figure 9)

$$\eta = \frac{r - r_0}{\ell} \tag{162}$$

- A square matrix, $n-1 \times n-1$ dimensional, of mass transfer rate factors, $[\Phi]$, with elements given by

$$\Phi_{ii} = \frac{N_i}{c_t D'_{in}/\ell} + \sum_{\substack{j=1 \\ j \neq i}}^{n} \frac{N_j}{c_t D'_{ij}/\ell}, \qquad i = 1, 2, \ldots, n-1$$

$$\Phi_{ij} = -N_i\left(\frac{1}{c_t D'_{ij}/\ell} - \frac{1}{c_t D'_{in}/\ell}\right), \qquad i, j = 1, 2, \ldots, n-1 \quad (i \neq j) \tag{184}$$

- An $n-1$ dimensional column vector (φ) with elements

$$\varphi_i \equiv -\frac{N_i}{c_t D'_{in}/\ell}, \quad i = 1, 2, \ldots, n-1 \tag{185}$$

With these definitions, the Maxwell–Stefan diffusion equation can be written in $n-1$ dimensional matrix form as

$$\frac{d(y)}{d\eta} = [\Phi](y) + (\varphi) \tag{186}$$

The boundary conditions are:

$$\begin{aligned} r = r_0; \quad \eta = 0: \quad y_i = y_{i0} \\ r = r_\delta; \quad \eta = 1: \quad y_i = y_{i\delta} \end{aligned} \tag{187}$$

The solution to Equations 186 and 187 is found to be an exact matrix analog of the binary solution, Equation 167; see Krishna and Standart [72] and Taylor [73, 74] for derivations. Thus we obtain the composition profiles

$$(y - y_0) = \{\exp[\Phi]\eta - [I]\}\{\exp[\Phi] - [I]\}^{-1}(y_\delta - y_0) \tag{188}$$

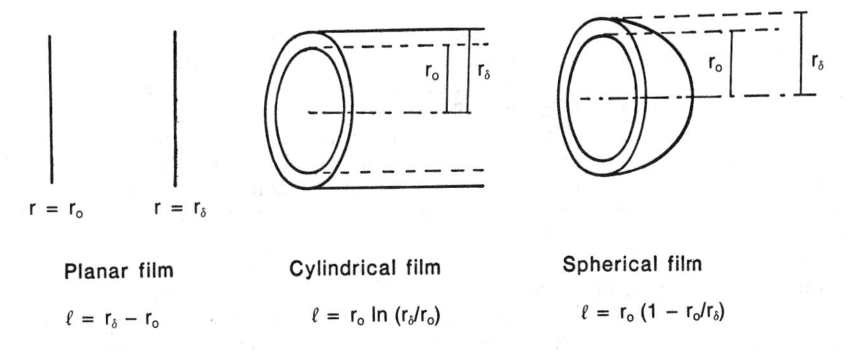

Figure 9. Generalization of film model to other geometries by appropriate definition of the characteristic length ℓ.

The diffusion fluxes J_{i0} can be evaluated from

$$(J_0) = -\frac{c_t[D_0]}{\ell}\frac{d(y)}{d\eta}\bigg|_{\eta=0} = \frac{c_t[D_0]}{\ell}[\Phi]\{\exp[\Phi]-[I]\}^{-1}(y_0 - y_\delta) \tag{189}$$

where the elements of the matrix $[D_0]$ are obtained from Equation 61, with the elements B_{ik} evaluated using the mole fractions y_{i0}.

The matrix of zero-flux mass transfer coefficients $[k_0]$ is thus

$$[k_0] \equiv \frac{[D_0]}{\ell} \tag{190}$$

and the matrix of correction factors, $[\Xi_0]$, is given by

$$[\Xi_0] \equiv [\Phi]\{\exp[\Phi]-[I]\}^{-1} \tag{191}$$

Alternatively, we may proceed via the diffusion fluxes at $z = \delta$:

$$(J_\delta) = -\frac{c_t[D_\delta]}{\ell}\frac{d(y)}{d\eta}\bigg|_{\eta=1} = \frac{c_t[D_\delta]}{\ell}[\Phi]\exp[\Phi]\{\exp[\Phi]-[I]\}^{-1}(y_0 - y_\delta) \tag{192}$$

giving the zero-flux mass tranfer coefficients

$$[k_\delta] \equiv \frac{[D_\delta]}{\ell} \tag{193}$$

and the matrix of correction factors

$$[\Xi_\delta] \equiv [\Phi]\exp[\Phi]\{\exp[\Phi]-[I]\}^{-1} = [\Xi_0]\exp[\Phi]. \tag{194}$$

By invoking the "bootstrap" solution, the fluxes, N_i, can be evaluated by means of the two equivalent expressions:

$$(N) = c_t[\beta_0][k_0][\Xi_0](y_0 - y_\delta) = c_t[\beta_\delta][k_\delta][\Xi_\delta](y_0 - y_\delta) \tag{195}$$

The computation of the fluxes, N_i, from either of the equations in Equation 195 necessarily involves an iterative procedure since the N_i themselves are needed for the evaluation of the matrix of correction factors. A number of numerical methods may be used to solve the set of $n-1$ nonlinear Equation 195. However, Newton's method and its relatives (which are strongly recommended elsewhere in this review) are not recommended here because the derivatives needed for the calculation of the Jacobian matrix must be evaluated numerically, a quite time-consuming operation. In any case, the method of successive substitution, if implemented as described in the following, can be a very effective way of computing the N_i from Equation 195.

Taylor and Webb [75] found that the left-hand member of Equation 195 converges most rapidly if the eigenvalues of $[\Phi]$ are positive in sign (this will be the case if the N_i are positive) whereas the right-hand member converges most rapidly if the eigenvalues of $[\Phi]$ are negative in sign (corresponding to negative fluxes). In fact, the left-hand member of Equation 195 may *never* converge if, in addition to being negative in sign, the eigenvalues, $\hat{\Phi}_i$, are "large" in absolute value. Conversely, the right-hand member may never converge if the $\hat{\Phi}_i$ are "large" and positive. The explanation for this behavior is given by Taylor and Webb [75] and not repeated here. It should be noted that "large" $\hat{\Phi}_i$ are a characteristic of problems involving mass transfer at high rates. Thus the occurrence of large $\hat{\Phi}_i$ in problems of practical importance may be quite rare. Nevertheless, the possibility must be allowed for when drawing up a general purpose algorithm such as that summarized in the following flowchart.

Algorithm for the Calculation of the Molar Fluxes from an Exact Solution of the Maxwell–Stefan Equations

Given: (y_0), (y_δ) c_t, D'_{ij}, ℓ, convergence tolerance ε.
Find: (N), N_t, $[k]$, $[k^\bullet]$.

Step 1. Set iteration counter (k) to zero.
Step 2. Estimate the fluxes $N_i^{(0)}$ (see text).
Step 3. Calculate $[\Phi]$ using current estimate of the fluxes, $N_i^{(k)}$.
Step 4. Calculate $\hat{\Phi}_{av} = \sum_{i=1}^{n-1} \Phi_{ii}/(n-1)$
Step 5. If $\hat{\Phi}_{av} \geqq 0$ follow Steps 6a to 9a.
If $\hat{\Phi}_{av} < 0$ follow steps 6b to 9b.
Step 6a. Calculate $[k_0]$ (first time through only).
Step 6b. Calculate $[k_\delta]$ (first time through only).
Step 7a. Calculate $[\beta_0]$ (first time through only).
Step 7b. Calculate $[\beta_\delta]$ (first time through only).
Step 8a. Calculate $[\Xi_0]$
Step 8b. Calculate $[\Xi_\delta]$
Step 9a. Obtain next estimate of the fluxes from

$$(N^{(k+1)}) = c_t[\beta_0]\,[k_0]\,[\Xi_0]\,(\Delta y)$$

Step 9b. Obtain next estimate of the fluxes from

$$(N^{(k+1)}) = c_t[\beta_\delta]\,[k_\delta]\,[\Xi_\delta]\,(\Delta y)$$

Step 10. Convergence check

If $|(N_i^{(k+1)} - N_i^{(k)})/N_i^{(k)}| > \varepsilon$ (any i)
continue with Step 11. Otherwise, stop.

Step 11. Reestimate the N_i from

$$(N_i^{(k+1)}) = tN_i^{(k+1)} + (1-t)N_i^{(k)}$$

$t\varepsilon(0,1)$ and return to Step 3.

Several aspect of the computational scheme presented here require further clarification; which is provided in the following.

The success or failure of any iterative scheme is very closely related to the goodness of the initial guess of the independent variables (the N_i here) (Step 2). Krishna and Standart [72] suggested taking the N_i as zero in Step 2, in which case $[\Phi]$ reduces to the null matrix and $[\Xi]$ to the identity matrix [I]. Substitution of the N_i calculated in Step 9 back into Step 3 (corresponding to taking $t=1$ in Step 11) may lead to very slow convergence from a null starting guess; it is recommended that, instead, damped substitution be employed with $t = 0.5$ (Taylor and Webb [76]). Convergence will generally be obtained within about ten iterations. (Later in this section we will discuss a method of obtaining much better initial estimates of the fluxes; estimates from which convergence can usually be obtained in two or three iterations even in "difficult" cases.)

The more appropriate member of Equation 195 is determined from the sign of the dominant eigenvalue of $[\Phi]$ (the eigenvalue with the largest absolute value). Computing eigenvalues can be quite time consuming and, in this case, can be avoided altogether if we recognize that the sum of the diagonal elements, Φ_{ii}, is equal to the sum of the eigenvalues. Thus the sign of Φ_{av}, which is easily calculated in Step 4, suffices to determine which member of Equation 195 should be used. In cases where the $\hat{\Phi}_i$ are of mixed sign, they are generally low in value and both members could equally well be used.

For the evaluation of $[\Xi_0]$ and $[\Xi_\delta]$ (Step 9) we may use Sylvester's theorem [42]. For a ternary mixture the elements of $[\Xi]$ are given by [72]:

$$[\Xi] = \frac{(\hat{\Xi}_1 - \hat{\Xi}_2)[\Phi] - (\hat{\Xi}_1\hat{\Phi}_2 - \hat{\Xi}_2\hat{\Phi}_1)[I]}{\hat{\Phi}_1 - \hat{\Phi}_2} \tag{196}$$

where $\hat{\Phi}_1$ and $\hat{\Phi}_2$ are the eigenvalues of $[\Phi]$ and $\hat{\Xi}_1$ and $\hat{\Xi}_2$ are given, respectively for $[\Xi_0]$ and $[\Xi_\delta]$, by

$$\hat{\Xi}_{i0} = \frac{\hat{\Phi}_i}{\exp(\hat{\Phi}_i) - 1}; \quad \hat{\Xi}_{i\delta} = \frac{\hat{\Phi}_i \exp(\hat{\Phi}_i)}{\exp(\hat{\Phi}_i) - 1}, \quad i = 1, 2 \tag{197}$$

Krishna and Standart [72] give further simplified expressions for equimolar diffusion and Stefan diffusion in ternary mixtures.

For systems of more than three components, Sylvester's expansion formula demands excessive amounts of computer time. Alternative approaches to the calculation of $[\Xi]$ have been considered by Taylor and Webb [76] who recommend the use of a method due to Buffham and Kropholler [77] to first calculate $\exp[\Phi]$ from a truncated power series:

$$\exp[M] = [\exp\{[M]/2^q\}]^{2q}$$

$$= \left[[I] + \sum_{k=1}^{\infty} \frac{\{[M]/2^q\}^k}{k!}\right]^{2q} \tag{198}$$

where q is a positive integer calculated from

$$q = \text{Int}\left(\frac{\ln|\hat{M}_d|}{\ln 2}\right)$$

and where

$$|\hat{M}_d| \le \max_i \sum_{k=1}^{n-1} |M_{ik}|$$

is an *estimate* of the dominant eigenvalue of [M] obtained from Gershgorin's theorem [76]. If q is calculated as described, then very few terms of the infinite summation in Equation 198 will be needed; six terms would be typical. Following the calculation of exp [Φ], the correction factor matrix can be computed using only the elementary operations of addition, multiplication, and inversion (Equations 191 and 194).

If the total flux is zero ($N_t = 0$), then [Φ] is singular and [Ξ] cannot be computed as just described. For such cases the inverse of [Ξ] is most efficiently computed from [76]:

$$[\Xi_0]^{-1} = [I] + \sum_{k=1}^{\infty} \frac{[\Phi]^k}{(k+1)!} \tag{199}$$

$$[\Xi_\delta]^{-1} = [I] + \sum_{k=1}^{\infty} (-1)^k \frac{[\Phi]^k}{(k+1)!} \tag{200}$$

In practice, few terms in the power series Equations 199 and 200 will be required. (The power series Equations 199 and 200 may be used if [Φ] is nonsingular; however, the calculation of [Ξ] via exp [Φ] is the faster of the various methods [76].)

In order to illustrate the application of the Krishna–Standart multicomponent film model, we consider a numerical example involving diffusion of acetone (1) and methanol (2) through stagnant air (3) in a Stefan diffusion tube (see Figure 10). The pressure and temperature in the vapor phase are 99,376 Pa and 328.5° K respective. The length of the diffusion path $\ell = 0.238$ m. The Maxwell–Stefan diffusivities are $D'_{12} = 8.48\ \text{mm}^2\ \text{s}^{-1}$; $D'_{13} = 13.72\ \text{mm}^2\ \text{s}^{-1}$; $D'_{23} = 19.91\ \text{mm}^2\ \text{s}^{-1}$, evaluated at 99,376 Pa and 328.5° K. The compositions of acetone and methanol in the vapor phase at the vapor-liquid interface, $\eta = 0$, calculated from equilibrium data, is $y_{10} = 0.319$, $y_{20} = 0.528$. In the bulk air stream, $\eta = 1$, the compositions of acetone and methanol are vanishingly small, i.e. $y_{1\delta} = 0$; $y_{2\delta} = 0$ giving $y_{3\delta} = 1$. We are required to determine the fluxes, N_1 and N_2 and the composition profiles $y_{1\eta}$, $y_{2\eta}$.

Either of the equalities Equation 195, can be used in an iteration procedure for the evaluation of N_1 and N_2; note that for this case $N_3 = 0$ for the "stagnant" air. The results of the computations of the various matrices are summarized in Table 1.

The nonuniqueness property of multicomponent mass transfer coefficients, referred to earlier, is clearly demonstrated by the values given in Table 1. The final values of the fluxes, N_1 and N_2, are the same, whether evaluated using the left- or right-hand equality of Equation 195. The strong diffusional coupling exhibited in this system, acetone-methanol-air, is highlighted by the value of $k^\bullet_{12}\Delta y_2/k^\bullet_{11}\Delta y_1$ and $k^\bullet_{21}\Delta y_1/k^\bullet_{22}\Delta y_2$. The ratio $k^\bullet_{12}\Delta y_2/k^\bullet_{11}\Delta y_1$ represents the contribution to the diffusive flux J_1 from the component 2 driving force. For J_{10} this "cross" or "off-diagonal" contribution is more than 100%. Since both $k^\bullet_{12}$ and Δy_i can take on any sign depending on the physical constraints imposed on the system (and the component numbering), we could encounter one of the three situations sketched below:

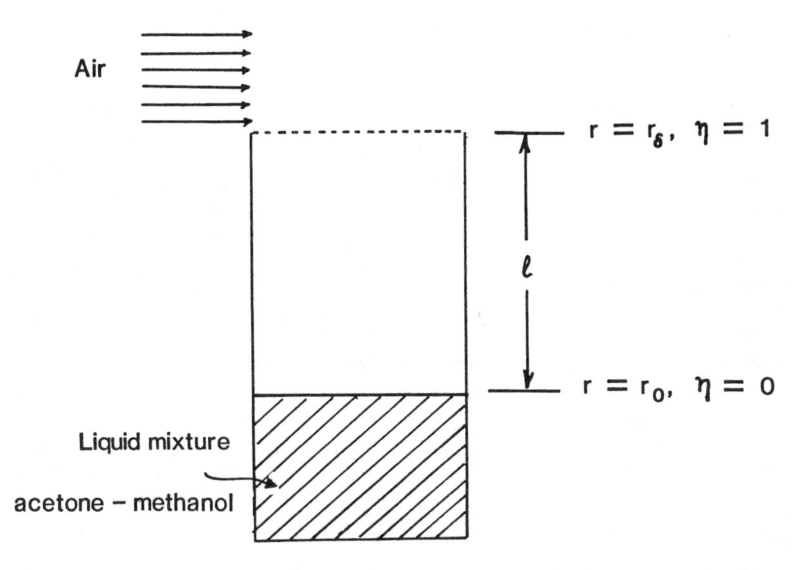

Figure 10. Evaporation of a mixture of acetone and methanol through stagnant air in a Stefan diffusion apparatus.

- Even when its constituent driving force, Δy_1, is zero, we could have a nonvanishing flux J_1,

$$J_1 \neq 0; \quad \Delta y_1 = 0 \tag{201}$$

This situation has been termed *osmotic diffusion* [2].

- Under a certain set of operating conditions and system properties the term $k^{\bullet}_{12}\Delta y_2$ may be of same magnitude and of opposite sign to $k^{\bullet}_{11}\Delta y_1$ leading to

$$J_1 = 0; \quad \Delta y_1 \neq 0 \tag{202}$$

A *diffusion barrier* is considered to exist for component 1 [2].

- It is also conceivable that the term $k^{\bullet}_{12}\Delta y_2$ overshadows $k^{\bullet}_{11}\Delta y_2$ in magnitude and is of opposite sign giving rise to

$$\frac{J_1}{\Delta y_1} < 0 \tag{203}$$

Component 1 experiences *reverse diffusion* in this case [2].
It must be stressed that for a two-component system for which

$$J_1 = c_t k^{\bullet} \Delta y_1 \tag{204}$$

with $k^{\bullet} > 0$, none of the three phenomena, just sketched for a ternary mixture, can take place. These phenomena, called *interaction* phenomena, have been just discussed in the context of ternary gas diffusion but are typical of the general multicomponent case. The practical implications of these interaction phenomena include such interesting possibilities as negative Murphree point efficiencies and unequal component HTU's for multicomponent distillation, as we shall see later. Even though none of the three interaction phenomena are observed for the Stefan diffusion in the system acetone-methanol-air being considered here, is should be clear that the effects of coupling between species transfers, quantified by the terms $k^{\cdot}_{12}\Delta y_2/k^{\cdot}_{11}\Delta y_1$ and $k^{\cdot}_{21}\Delta y_1/k^{\cdot}_{22}\Delta y_2$, are significant and cannot, in general, be ignored. Uncoupled, binary-like, approaches could lead to serious errors.

At first sight it might appear that the second law of thermodynamics is violated for reverse mass transfer to occur. This is not so. One process may depart from equilibrium in such a sense as to consume entropy provided it is coupled to another process which produces entropy even faster. This is of course the basic principle of any pump, whether it moves water uphill or moves heat towards a higher temperature region [78]. For the second law requirement, $\sigma > 0$, to hold it is allowable for σ_1 to be <0, corresponding to reverse diffusion for 1, provided σ_2 and σ_3, due to species 2 and 3 diffusion, be such that the overall entropy production rate is positive ($\sigma_1+\sigma_2+\sigma_3 > 0$).

Duncan and Toor [79] experimentally detected the three interaction phenomena for diffusion in the gaseous system: nitrogen(1)-hydrogen(2)-carbon dioxide(3) in a two-bulb diffusion cell. Their experimental results also confirm the applicability of the Maxwell–Stefan diffusion equations; this conclusion was also reached by Bres and Hatzfeld [80] and Hesse and Hugo [81].

The ratio of driving forces plays an important role in enhancing diffusional interaction effects in multicomponent mass transfer. Thus a small cross-coefficient k_{12} may be linked to a large Δy_2 to result in large interaction effects.

Actually, the Stefan diffusion problem involving acetone-methanol-air correspond to the conditions in an experimental study by Carty and Schrodt [82]. Figure 11 shows a comparison of the experimentally measured composition profiles with the theoretical profiles Equation 188 predicted by the Maxwell–Stefan formulation on the basis of information on the transport properties of the constituent *binary* pairs, D'_{ik}. The excellent correspondence between theory and experiment is to be expected because for ideal-gas mixtures the film model of Krishna and Standart [72] is an exact solution.

A final word about the correction factor matrix $[\Xi]$. As noted, for vanishingly small $N_i \rightarrow 0$, the rate factor matrix $[\Phi]$ tends to the null matrix, $[0]$, and $[\Xi]$ tends to the identity matrix, $[I]$. Departures of $[\Xi]$ from $[I]$ portray the influence of finite fluxes, N_i, on the composition fluxes and transfer coefficients. From Table 1 it can be seen that, in this example, these flux corrections are indeed of importance. Typically, in multicomponent systems, the effect of $[\Xi]$ on the fluxes cannot be ignored as shown by Krishna [83] for a few cases. Interestingly, *even for equimolar transfer*, $N_t = 0$, the correction factor $[\Xi]$ does *not* reduce to $[I]$, as pointed out by Krishna and Standart [72]. This means that the composition profiles will not be linear even though $N_t = 0$. Contrast this with a binary system for which $N_t = 0$ leads to linear composition profiles.

Multicomponent Film Model Based on the Assumption of Constant [D] Matrix—The Linearized Theory of Toor, Stewart, and Prober

In 1964 Toor [4] and, independently, Stewart and Prober [5] put forward a general approach to the solution of multicomponent diffusion problems. Their method relies on the assumption of constancy of the Fickian matrix [D] along the diffusion path. We emphasize this difference between the Toor–Stewart–Prober approach and the exact method previously considered here by using a subscript av; thus $[D_{av}]$. For practical calculations, this means that $[D_{av}]$ has to be evaluated by employing suitable average mole fractions, $y_{i,av}$, in the definition of the B_{ik} Equation 198. The arithmetic average mole fraction ($y_{i,av} = 1/2\,(y_{i0}+y_{i\delta})$) is most commonly used in the calculation of $[B_{av}]$ [5, 84, 85]. Thus, for gas mixtures

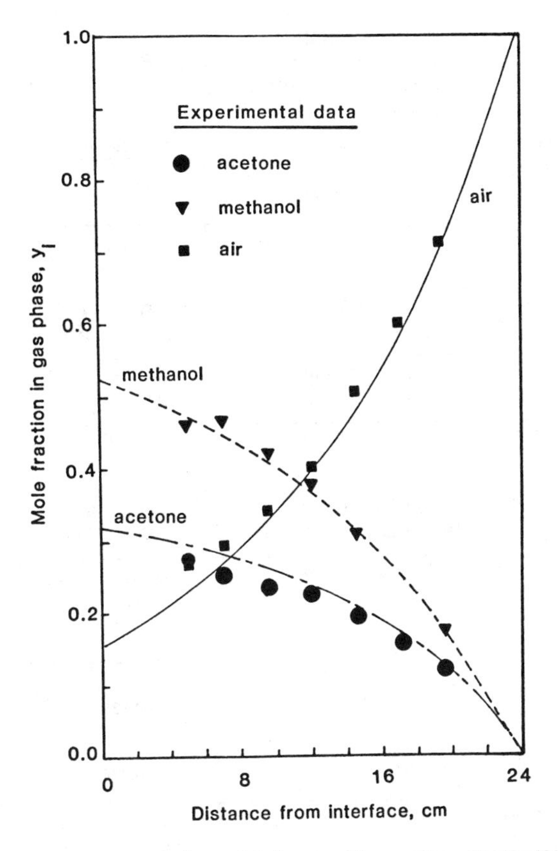

Figure 11. Comparison of theoretically predicted composition profiles, using the Krishna-Standart method, for Stefan diffusion in the system acetone-methanol-air with the experimental data of Carty and Schrodt.

$$[D_{av}] = [B_{av}]^{-1} \tag{205}$$

is assumed constant over the diffusion path.

The so-called "linearized theory" of Toor, Stewart, and Prober is not limited to describing steady, one-dimensional diffusion in ideal gas mixtures (as we shall soon demonstrate); however, for this particular situation

$$(N) = -c_t[D_{av}]\frac{d(y)}{dz} + N_t(y) \tag{206}$$

is constant over the diffusion path.

Let us define a matrix of mass transfer rate factors

$$[\Phi] \equiv \frac{N_t\ell}{c_t}[D_{av}]^{-1} = \frac{N_t\ell}{c_t}[B_{av}] \tag{207}$$

Table 1
Calculation of Acetone and Methanol Fluxes in Stefan Diffusion Tube

Parameters	$\eta = 0$ Interface	$\eta = 1$ Bulk Gas Phase
Compositions (y)	$(y_o) = \begin{pmatrix} 0.319 \\ 0.528 \end{pmatrix}$	$(y_\delta) = \begin{pmatrix} 0 \\ 0 \end{pmatrix}$
Driving force (Δy)	$(\Delta y) = \begin{pmatrix} 0.319 \\ 0.528 \end{pmatrix}$	
Zero flux mass transfer coefficients [k], μm s^{-1}	$\begin{bmatrix} 46.9 & 9.4 \\ 23.4 & 63.2 \end{bmatrix}$	$\begin{bmatrix} 57.6 & 0 \\ 0 & 83.7 \end{bmatrix}$
Bootstrap matrix [β]	$\begin{bmatrix} 3.085 & 2.085 \\ 3.451 & 4.451 \end{bmatrix}$	$\begin{bmatrix} 1 & 0 \\ 0 & 1 \end{bmatrix}$
Rate factor matrix [Φ]	$\begin{bmatrix} 3.262 & -0.5252 \\ -1.385 & 2.402 \end{bmatrix}$	
Correction factor [Ξ]	$\begin{bmatrix} 0.1569 & 0.0691 \\ 0.1822 & 0.2700 \end{bmatrix}$	$\begin{bmatrix} 3.419 & -0.4561 \\ -1.203 & 2.672 \end{bmatrix}$
Finite-flux mass transfer coefficients [k•] μm s^{-1}	$\begin{bmatrix} 9.076 & 5.779 \\ 15.178 & 18.67 \end{bmatrix}$	$\begin{bmatrix} 197.053 & -26.287 \\ -100.646 & 223.545 \end{bmatrix}$
$\dfrac{k^\bullet_{12}\,\Delta y_2}{k^\bullet_{11}\,\Delta y_1}$	1.053	−0.221
$\dfrac{k^\bullet_{21}\,\Delta y_1}{k^\bullet_{22}\,\Delta y_2}$	0.491	−0.272
"Overall" coefficient [W•] ≡ $c_t[\beta][k][\Xi]$ mmol m^{-2} s^{-1}	$\begin{bmatrix} 2.17 & 2.065 \\ 3.597 & 3.749 \end{bmatrix}$	$\begin{bmatrix} 7.171 & -0.957 \\ -3.66 & 8.134 \end{bmatrix}$
Molar fluxes (N) mmol m^{-2} s^{-1}	$\begin{pmatrix} 1.783 \\ 3.127 \end{pmatrix}$	

where ℓ is the length of the diffusion path. Note that we use the same symbol, [Φ], to denote the matrix of rate factors in both the Krishna–Standart and Toor–Stewart–Prober methods.

Further, we define a column matrix [φ] by

$$(\varphi) = -\frac{\ell}{c_t}[B_{av}](N) \tag{208}$$

With these definitions, Equation 206 may be rewritten as follows

$$\frac{d(y)}{d\eta} = [\Phi](y) + (\varphi) \tag{209}$$

which is a matrix differential equation of order $n-1$. With the assumption of constant $[D_{av}]$, the matrices [Φ] and (φ) will also be constant. It can be seen that Equation 209 is identical in form to Equation 186 obtained earlier in presenting the exact solution. Thus expressions for the composition profiles and for the fluxes can be written directly; simply replace $[D_0]$ (or $[D_\delta]$) by $[D_{av}]$, and so on. The composition profiles, for example, are given by [86]:

$$(y - y_0) = \{\exp[\Phi]\eta - [I]\}\{\exp[\Phi] - [I]\}^{-1}(y_\delta - y_0) \tag{210}$$

The diffusion fluxes at $\eta = 0$ are evaluated from

$$(J_0) = -\frac{c_t}{\ell}[D_{av}]\left.\frac{d(y)}{d\eta}\right|_{\eta=0} = \frac{c_t}{\ell}[D_{av}][\Phi]\{\exp[\Phi] - [I]\}^{-1}(y_0 - y_\delta) \tag{211}$$

and the diffusion fluxes at $\eta = 1$ are given by

$$(J_\delta) = \frac{c_t}{\ell}[D_{av}][\Phi]\exp[\Phi]\{\exp[\Phi] - [I]\}^{-1}(y_0 - y_\delta) \tag{212}$$

The invariant molar fluxes follow from a combination of Equations 146 and 211 or 212:

$$(N) = c_t[\beta_0][k_{av}][\Xi_0](y_0 - y_\delta) = c_t[\beta_\delta][k_{av}][\Xi_\delta](y_0 - y_\delta)$$

$$\equiv c_t[\beta][k^{\bullet}_{av}](y_0 - y_\delta) \tag{213}$$

with the matrix of zero flux and finite flux mass transfer coefficients given by

$$[k_{av}] \equiv [D_{av}]/\ell = [B_{av}]^{-1}/\ell; \quad [k^{\bullet}_{av}] = [k_{av}][\Xi] \tag{214}$$

and the correction factor matrices defined by

$$[\Xi_0] = [\Phi]\{\exp[\Phi] - [I]\}^{-1} \tag{215}$$

$$[\Xi_\delta] = [\Phi]\exp[\Phi]\{\exp[\Phi] - [I]\}^{-1} \tag{216}$$

Equations 213 through 216 should be contrasted with Equations 190 through 195, their counterparts in the exact method. It can be seen that $[B_{av}]$, $[k_{av}]$ and $[\Phi]$ (Equation 207) correspond exactly with $[B_0]$ (or $[B_\delta]$), $[k_0]$ (or $[k_\delta]$) and $[\Phi]$ Equation 184. Furthermore, the similarity between the two methods extends right down to the calculation of the individual elements of these matrices as is emphasized in Table 2. This one-to-one correspondence between respective matrices and elements of matrices facilitates a comparison of the two methods and means that a simple modification of the Algorithm for Calculating Molar fluxes from the Exact Solution of the Maxwell–Stefan Equation, can be employed to compute the fluxes from Equations 213, [87] (see, however, the following).

The assumption of constant $[D_{av}]$ in the Toor–Stewart–Prober approach gives rise to two basic, fundamental differences between this approach and the method of Krishna and Standart.

1. The method of Toor–Stewart–Prober is approximate *even for ideal gas mixtures,* for which the Krishna–Standart method represents an *exact* solution.
2. For vanishing total total flux, $N_t = 0$, the correction factor $[\Xi]$ in the Toor–Stewart–Prober approach reduces to the identity matrix because $[\Phi] \rightarrow [0]$ for this case. The composition profiles are, therefore, linear. As stressed earlier, in the exact Krishna–Standart model, the condition $N_t = 0$ does not lead to linear composition profiles. Only when each individual flux tends to vanish, does the exact model lead to $[\Xi] \rightarrow [I]$ and linear composition profiles.

Table 2
Matrices in the Exact Solution and the Approximate Solution of the Maxwell-Stefan Equations

Matrix [M]	Meaning	m_i
Exact Solution Due to Krishna and Standart		
$[B_o] = [D_o]^{-1}$	Inverted matrix of Fickian diffusion coefficients evaluated at the origin.	y_{io}
$[k_o]^{-1} = \ell\,[B_o]$	Inverted matrix of low flux mass transfer coefficients	$y_{io}\,\ell$
$[\Phi]$	Matrix of mass transfer rate factors	$N_i\ell/c_t$
Linearized Method of Toor-Stewart-Prober		
$[B_{av}] = [D_{av}]^{-1}$	Inverted matrix of Fickian diffusion coefficients evaluated at the average composition	$y_{i,av}$
$[k_{av}]^{-1} = \ell\,[D_{av}]^{-1}$	Inverted matrix of low flux mass transfer coefficients	y_{iav}
$[\Phi]$	Matrix of mass transfer rate factors	$y_{iav}N_t\ell/c_t$

A matrix [M] has elements

$$M_{ii} = \frac{m_i}{D_{in}} + \sum_{\substack{k=1\\k\neq i}}^{n} \frac{m_k}{D_{ik}} \quad i = 1, 2, \ldots n - 1$$

$$M_{ij} = -m_i\left(\frac{1}{D_{ij}} - \frac{1}{D_{in}}\right) \quad i \neq j = 1, 2, \ldots n - 1$$

Despite these differences both solutions of the multicomponent diffusion equations will give identical results if:

- All the binary diffusivities are equal; that is

$$D'_{ij} = D', \quad i \neq j = 1, 2, \ldots, n \tag{217}$$

Then $[k_0]$ and $[k_{av}]$ are equal (see Table 2) with

$$k_{0ii} = k_{avii} = D'/\ell; \quad k_{0ij} = k_{avij} = 0, \quad i \neq j = 1, 2, \ldots, n-1 \tag{218}$$

The rate factor matrices, $[\Phi]$, are also equal with elements

$$\Phi_{ii} = N_t\ell/c_tD'; \quad \Phi_{ij} = 0, \quad i \neq j = 1, 2, \ldots, n-1 \tag{219}$$

The correction factor matrices will therefore be equal. In this case, however, the diffusion equations are not coupled, and the i-th diffusion flux is given by

$$J_{0i} = c_tk_{ii}\Xi_{ii}(y_{0i} - y_{\delta i}) = \frac{c_tD'}{\ell}\,\frac{N_t\ell/c_tD'}{\exp(N_t\ell/c_tD') - 1}(y_{0i} - y_{\delta i}), \quad i = 1, 2, \ldots, n-1 \tag{220}$$

and hence the matrix formulations discussed above are not required.

- $n-1$ species are present in vanishingly low concentrations, in an n-th component ($y_{0i} \cong 0, i = 1, 2, \ldots, n-1$; $y_{on} \cong y_{\delta n} \cong 1$). In this case the matrices [k] become diagonal (see Table 2); i.e.

Table 3
Calculation of Acetone and Methanol Fluxes in Stefan Diffusion Tube—Method of Toor-Stewart-Prober

Parameters	$\eta = 0$ Interface		$\eta = 1$ Bulk Gas
Compositions (y)	$\begin{pmatrix}0.319\\0.528\end{pmatrix}$		$\begin{pmatrix}0.0\\0.0\end{pmatrix}$
Driving force (Δy)		$\begin{pmatrix}0.319\\0.528\end{pmatrix}$	
Zero flux mass transfer coefficients $[k_{av}]$, $\mu m\ s^{-1}$		$\begin{bmatrix}50.82 & 5.98\\14.89 & 70.61\end{bmatrix}$	
Bootstrap matrix $[\beta]$	$\begin{bmatrix}3.085 & 2.085\\3.451 & 4.451\end{bmatrix}$		$\begin{bmatrix}1 & 0\\0 & 1\end{bmatrix}$
Rate factor matrix $[\Phi]$		$\begin{bmatrix}2.755 & -0.235\\-0.581 & 1.983\end{bmatrix}$	
Correction factor $[\Xi]$	$\begin{bmatrix}0.193 & 0.039\\0.098 & 0.323\end{bmatrix}$		$\begin{bmatrix}2.948 & -0.194\\-0.483 & 2.306\end{bmatrix}$
Finite-flux mass transfer coefficients $[k^\bullet]$, $\mu m\ s^{-1}$	$\begin{bmatrix}10.415 & 3.935\\9.790 & 23.426\end{bmatrix}$		$\begin{bmatrix}146.93 & 3.936\\9.792 & 159.95\end{bmatrix}$
$\dfrac{k^\bullet_{12}\,\Delta y_2}{k^\bullet_{11}\,\Delta y_1}$	0.625		0.044
$\dfrac{k^\bullet_{21}\,\Delta y_1}{k^\bullet_{22}\,\Delta y_2}$	0.256		0.037
"Overall" coefficients $[W^\bullet] = c_t[\beta][k^\bullet]$ mmol $m^{-2}\ s^{-1}$	$\begin{bmatrix}1.911 & 2.219\\2.893 & 4.288\end{bmatrix}$		$\begin{bmatrix}5.346 & 0.143\\0.356 & 5.820\end{bmatrix}$
Molar fluxes (N) mmol $m^{-2}\ s^{-1}$		$\begin{pmatrix}1.781\\3.187\end{pmatrix}$	

$$k_{0ii} = k_{avii} = D'_{in}/\ell; \quad k_{0ij} = k_{avij} = 0, \quad i \neq j = 1, 2, \ldots, n-1 \tag{221}$$

The respective matrices $[\Phi]$ also become diagonal with elements given by

$$\Phi_{ii} = N_t\ell/c_tD'_{in}; \quad \Phi_{ij} = 0, \quad i \neq j = 1, 2, \ldots, n-1 \tag{222}$$

The validity of approximation Equation 222 for the exact solution depends further on the observation that if $y_{0i} \cong y_{\delta i} \cong 0$ $i = 1, 2, \ldots, n-1$; then the first $n-1$ N_i will be vanishingly low. For these conditions Equations 221 and 222 are valid whatever variation may exist between the D'_{ij}.

We have outlined situations where the two methods may be expected to be in close agreement. It would seem that large differences between the methods are particularly likely in mixtures of constituents with quite different D'_{ij} at high molar concentrations. Nevertheless, even in such cases the errors caused by linearizing the equations are not always large as the following example demonstrates.

In Table 3 we present the results of the computations of the fluxes for the diffusion of acetone and methanol through air in a Stefan tube; the same example used earlier to

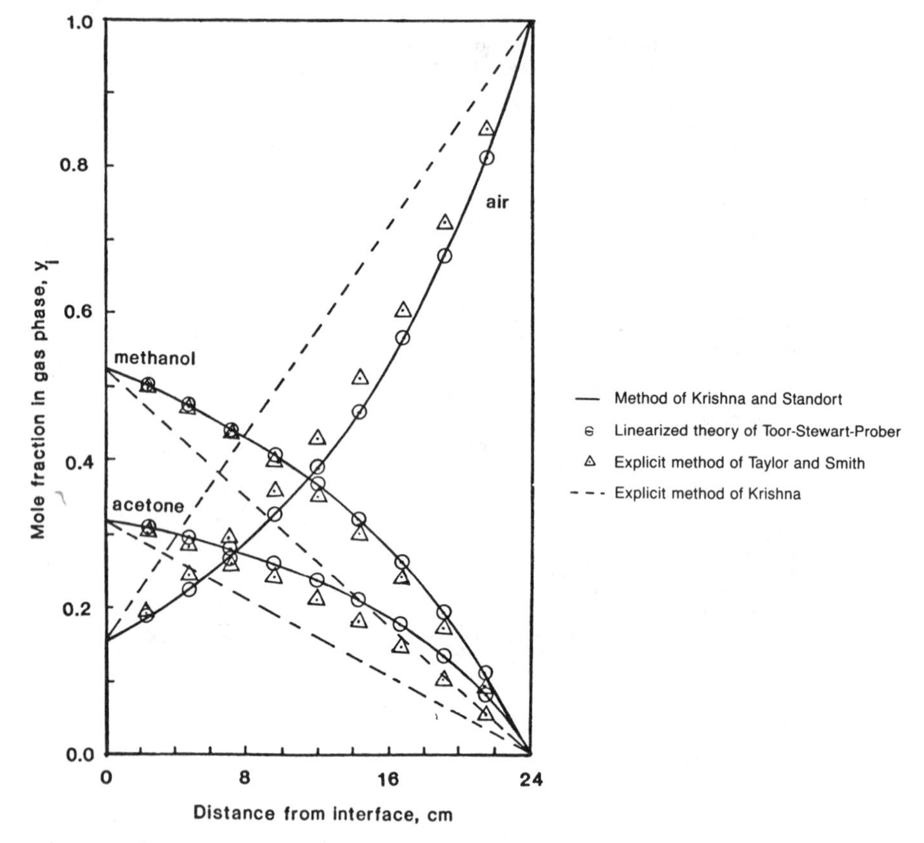

Figure 12. Comparison of theoretically predicted composition profiles for Stefan diffusion in the system acetone-methanol-air. _____ method of Krishna and Standart; ⊙ linearised theory of Toor-Stewart-Prober; △ explicit method of Taylor and Smith; ---- explicit method of Krishna.

illustrate the application of the Krishna–Standart method. The results in Table 3 should be compared to the results in Table 1. Almost every matrix element differs, sometimes quite considerably, from its counterpart in Table 1. Yet the final estimates of the N_i are very close to those obtained from the exact solution. In fact, this behavior is typical of the Toor–Stewart–Prober approach. Figure 12 provides a comparison of the composition profiles computed from the exact solution and those computed from Equation 210; the two profiles are almost identical here.

In the practical applications of these methods (to be considered in the next section) we shall see that it is the fluxes themselves that are of first importance. Thus, it does not matter much if the various coefficient matrices needed to estimate the fluxes differ considerably from their counterparts in another method. It is more important that the estimates of the fluxes are good ones. This certainly is the case in the example considered here. Later, we shall provide evidence that the estimates of the N_i computed from Equation 213 nearly always are in good agreement with the N_i computed from Equation 195.

Equations 214 through 216 can be written in the following alternative form [4, 5, 6, 88], which is more useful from the computational standpoint.

$$[k^{\bullet}_{av}] = [P]\,[\hat{k}^{\bullet}]\,[P]^{-1} \tag{223}$$

where $[\hat{k}^{\cdot}]$ is a diagonal matrix whose nonzero elements are the eigenvalues of $[k^{\cdot}_{av}]$. [P] is a matrix, called the modal matrix, whose columns are the eigenvectors of the matrix [D]. The eigenvalues of $[k^{\bullet}_{av}]$ are related to the eigenvalues of [D] as follows:

$$\hat{k}^{\bullet}_i = \hat{k}_i \hat{\Xi}_i, \quad i = 1, 2, \ldots, n-1 \tag{224}$$

where

$$\hat{k}_i = \hat{D}_i/\ell; \quad \hat{\Phi}_i = N_t\ell/c_t\hat{D}_i = N_t/c_t\hat{k}_i \tag{225}$$

$$\hat{\Xi}_{0i} = \frac{\hat{\Phi}_i}{\exp \hat{\Phi}_i - 1}; \quad \hat{\Xi}_{\delta i} = \frac{\hat{\Phi}_i \exp \hat{\Phi}_i}{\exp \hat{\Phi}_i - 1} \tag{226}$$

are the eigenvalues of $[k_{av}]$, $[\Phi]$, $[\Xi_0]$ and $[\Xi_\delta]$ respectively.

It is important to recognize that this representation of $[k^{\bullet}]$ is possible only because the matrix [D] was assumed constant along the diffusion path. Thus the modal matrix [P] which diagonalizes [D] by the transformation

$$[\hat{D}] = [P]^{-1}[D]\,[P] \tag{227}$$

is also constant along the diffusion path and the transformation which reduces [D] to a diagonal matrix also diagonalizes [k], $[\Phi]$, and $[\Xi]$. The exact solution due to Krishna and Standart [72] does *not* have a similar respresentation; the matrix which diagonalizes [B] is *not,* in general, equal to the matrix which diagonalizes $[\Phi]$ (see Taylor [73]). In fact, the method used by Toor [4] and by Stewart and Prober [5] to solve the multicomponent diffusion equations was first to reduce the matrix differential equations to an equivalent binary-like form using the transformation of Equation 227 (see, also, Krishna and Standart [6] for a summary of the method). It is not always necessary to first uncouple the diffusion equations in order to solve them [4, 86, 89]; it is, however, advantageous from the computational point of view to adopt the original formulation of Toor [4] and Stewart and Prober [5] ([90]).

The relation that makes it possible to compute the fluxes from Equations 212 and 213 is the determinacy condition (Equation 144), which is usefully rewritten as

$$N_t = \sum_{i=1}^{n-1}\left(1 - \frac{v_i}{v_n}\right) N_i = \sum_{i=1}^{n-1}\left(1 - \frac{v_i}{v_n}\right)(J_i + N_t y_i) \tag{228}$$

where the components have been numbered so that $v_n \neq 0$.

Defining a column matrix $(\Delta\hat{y})$ by $(\Delta\hat{y}) = [P]^{-1}(\Delta y)$ and combining Equations 228 with 212 gives

$$N_t = \sum_{i=1}^{n-1}\left(1 - \frac{v_i}{v_n}\right) c_t \sum_{j=1}^{n-1} P_{ij}\hat{k}_j\hat{\Xi}_j\Delta\hat{y}_j + y_i N_t \tag{229}$$

This equation, nonlinear in the single unknown, N_t; may be written in the form [90]:

$$F(N_t) = \sum_{i=1}^{n-1}\left(1 - \frac{v_i}{v_n}\right)\left\{\sum_{j=1}^{n-1} P_{ij}\hat{f}_j\Delta\hat{y}_j + y_i\right\} - 1 = 0 \tag{230}$$

where

$$\hat{f}_j = \hat{\Xi}_j/\hat{\Phi}_j, \quad j = 1, 2, \ldots, n-1 \tag{231}$$

This last step, which involves division by N_t, is possible only if $N_t \neq 0$. For $N_t = 0$, the linearized equations are explicit with $\hat{\Xi}_j = 1$.

The function $F(N_t)$ is very easily solved using Newton's method. An algorithm summarizing the relevant calculations is shown in the following.

Algorithm for the Calculation of the Molar Fluxes From the Film Model of Toor–Stewart–Prober

1. Read in data; D_{ij}, y_{0i}, $y_{\delta i}$, c_t, ℓ, convergence limit ε, set $k = 0$ ($k \equiv$ iteration counter).
2. Calculate [D] at the arithmetic average composition (Equation 61).
3. Calculate $\hat{D}_i$ and [P], invert [P].
4. Obtain an initial estimate of $N_t^{(0)}$ from Equation 235.
5. Calculate $F(N_t^{(k)})$ from Equation 230.
6. Calculate dF/dN_t from Equation 232.
7. Calculate next value of N_t from

   $$N_t^{(k+1)} = N_t^{(k)} - F(N_t^{(k)})/(dF/dN_t^{(k)})$$

8. Check for convergence.

   If $|(N_t^{(k+1)} - N_t^{(k)})/N_t^{(k+1)}| < \varepsilon$, stop.

   Otherwise, increment k and return to Step 5.

   First derivatives of F with respect to N_t (needed in Step 6) are obtained from

$$\frac{dF}{dN_t} = \sum_{i=1}^{n-1}\left(1 - \frac{v_i}{v_n}\right)\sum_{j=1}^{n-1} P_{ij}\frac{d\hat{f}_j}{dN_t}\Delta\hat{y}_j \tag{232}$$

If the fluxes are being evaluated at $\eta = 0$, use y_{j0} in the calculation of $F(N_t)$ and

$$\frac{d\hat{f}_j}{dN_t} = \frac{-1}{c_t\hat{k}_j}\frac{\exp\hat{\Phi}_j}{\exp\hat{\Phi}_j - 1} \tag{233}$$

in the calculation of dF/dN_t.

Once N_t is obtained from the solution of Equation 230, the first $\frac{n-1}{\alpha}$ N_i are calculated directly from

$$N_i = c_t\sum_{j=1}^{n-1}\hat{P}_{ij}\hat{k}_j\hat{\Xi}_j\hat{\Delta}y_j + y_iN_t \tag{234}$$

As shown by Vickery et al. [90], the function $F(N_t)$ possesses many desirable properties which make very rapid convergence a virtual certainty. Initial guesses of N_t are best calculated from

$$N_t = \frac{\sum_{i=1}^{n-1} (\nu_n - \nu_i) c_t \sum_{j=1}^{n-1} P_{ij} \hat{k}_j \Delta \hat{y}_i}{\sum_{i=1}^{n} \frac{\nu_i}{\nu_n} y_{avi}} \quad \text{(initial estimate)} \tag{235}$$

In closing this section on the linearized theory, we note that the N_i can be computed approximately ten times faster from the method just outlined than from a modification of the Krishna–Standart algorithm [87, 90].

Simplified Methods of Krishna, Taylor, and Smith

Both of the methods just discussed require an iterative approach to the calculation of the fluxes. The two methods discussed in this section are explicit (that is, they require *no* iterations; the fluxes can be computed directly). Further, no matrix functions other than the inverse need be computed.

The starting point for developing the explicit methods is the exact matrix differential equation

$$(N) = -\frac{c_t}{\ell}[\beta][B]^{-1}\frac{d(y)}{d\eta} \tag{236}$$

(Equation 236 is nothing more than a combination of Equations 145 and 59). Equation 236 can be written in the form

$$(N) = -\frac{c_t}{\ell}[A]^{-1}\frac{d(y)}{d\eta} \tag{237}$$

where the matrix [A] has elements

$$A_{ii} = \frac{y_i(\nu_i/\nu_n)}{D'_{in}} + \sum_{\substack{k=1 \\ k \neq i}}^{n} \frac{y_k}{D'_{ik}} \tag{238}$$

$$A_{ij} = -y_i\left(\frac{1}{D'_{ij}} - \frac{(\nu_j/\nu_n)}{D'_{in}}\right) \tag{239}$$

It can be shown by inverting $[\beta]$ using the Sherman–Morrison formula [41] that, in fact

$$[A] = [B][\beta]^{-1} \quad \text{or} \quad [A]^{-1} = [\beta][B]^{-1} \tag{240}$$

that is; the coefficient matrices in Equations 236 and 237 are *equal,* and not simply equivalent.

The explicit method of Krishna [91] is based on the assumption that the matrix $[\beta][B]^{-1}$ (or $[A]^{-1}$) can be considered constant over the diffusion path. This, of course, means that both matrices $[\beta]$ and [B] (or [A]) must be evaluated at an average composition. The consequence of this assumption is that the composition profiles are linear (as can easily be proved by integrating Equation 236 over the diffusion path).

The molar fluxes are given by

$$(N) = \frac{c_t}{\ell}[\beta_{av}][B_{av}]^{-1}(y_0 - y_\delta) = \frac{c_t}{\ell}[A_{av}]^{-1}(y_0 - y_\delta) \tag{241}$$

If the v_i coefficients are available it is recommended that the right-hand member of Equation 241 be used in the calculations since it involves less work to evaluate one matrix and its inverse than it does to evaluate the product of a matrix with the inverse of another.

The method of Taylor and Smith [92] is a generalization of an equation derived by Burghardt and Krupiczka [93] for the special case of Stefan diffusion. The generalized method is based on the assumption that the product of $[A]^{-1}$ with a weighted mole fraction $y^* \left(= \sum_{i=1}^{n} v_i y_i\right)$ (i.e. $\{y^*[A]^{-1}\}$) can be considered constant over the diffusion path. The equations to be solved are Equation 237 multiplied by unity in the form of y^*/y^*:

$$(N) = -\frac{1}{y^*}\frac{c_t}{\ell}\{y^*[A]^{-1}\}\frac{d(y)}{d\eta} \tag{242}$$

together with a differential equation describing the variation of y^* through a film [82]:

$$\frac{dy^*}{d\eta} = \Phi y^* \tag{243}$$

Φ, a mass transfer rate factor, is a constant and is given by the solution of Equation 243

$$\Phi = \ln(y_\delta^*/y_0^*) \tag{244}$$

where $y_0^* = \sum v_i y_{i0}$ and $y_\delta^* = \sum v_i y_{i\delta}$. The variation of y^* through the film is given by

$$\frac{y^* - y_0^*}{y_\delta^* - y_0^*} = \frac{e^{\Phi\eta} - 1}{e^{\Phi} - 1} \tag{245}$$

Now, assuming the product $\{y^*[A]^{-1}\}$ constant in Equation 242 (this means evaluating the product at the arithmetic average composition), substituting for y^* from Equation 245 and solving the resulting differential equations gives the composition profiles

$$\frac{y_i - y_{i0}}{y_{i\delta} - y_{i0}} = \frac{y^* - y_0^*}{y_\delta^* - y_0^*} = \frac{e^{\Phi\eta} - 1}{e^{\Phi} - 1} \tag{246}$$

The molar fluxes, N_i, can now be calculated from

$$\begin{aligned}(N) &= -\frac{c_t}{\ell}\{y_{av}^*[A_{av}]^{-1}\}\frac{1}{y_0^*}\frac{d(y)}{d\eta}\bigg|_{\eta=0} \\ &= -\frac{c_t}{\ell}\{y_{av}^*[A_{av}]^{-1}\}\frac{1}{y_\delta^*}\frac{d(y)}{d\eta}\bigg|_{\eta=1}\end{aligned} \tag{247}$$

Evaluating the mole fraction gradients from Equation 246 leads to the final working equation

$$(N) = \frac{c_t}{\ell}[A_{av}]^{-1}\Xi(y_0 - y_\delta) = \frac{c_t}{\ell}[\beta_{av}][B_{av}]^{-1}\Xi(y_0 - y_\delta) \tag{248}$$

where Ξ, a *scalar* correction factor, is given by

$$\Xi = \frac{\Phi}{2}\cdot\frac{e^{\Phi}+1}{e^{\Phi}-1} = \frac{1}{2}\frac{(y_0^* + y_\delta^*)}{(y_0^* - y_\delta^*)}\ln\left(\frac{y_0^*}{y_\delta^*}\right) \tag{249}$$

Table 4
Calculation of Acetone and Methanol Fluxes in Stefan Diffusion Tube: Explicit Methods of Krishna and Taylor and Smith

Parameters	Values
Compositions (y_o)	$\begin{pmatrix} 0.319 \\ 0.528 \end{pmatrix}$
(y_δ)	$\begin{pmatrix} 0.0 \\ 0.0 \end{pmatrix}$
Driving force (Δy)	$\begin{pmatrix} 0.319 \\ 0.528 \end{pmatrix}$
Bootstrap matrix $[\beta_{av}]$	$\begin{bmatrix} 1.277 & 0.277 \\ 0.458 & 1.458 \end{bmatrix}$
Mass transfer coefficient matrix $[B_{av}]^{-1}/\ell$ (μm/s)	$\begin{bmatrix} 50.82 & 5.98 \\ 14.89 & 70.61 \end{bmatrix}$
Overall coefficient matrix $\frac{1}{\ell}[A_{av}]^{-1} = \frac{1}{\ell}[\beta_{av}][B_{av}]^{-1}$ (μm/s)	$\begin{bmatrix} 69.00 & 27.17 \\ 49.97 & 105.68 \end{bmatrix}$
Molar fluxes from method of Krishna (mmol m^{-2} s^{-1})	$\begin{pmatrix} 1.323 \\ 2.552 \end{pmatrix}$
$y_o^* = y_{on}$	0.243
$y_\delta^* = y_{\delta n}$	1.0
$\Phi = \ln (y_\delta^*/y_o^*)$	1.415
Ξ	1.2778
Molar fluxes from method of Taylor and Smith* (mmol m^{-2} s^{-1})	$\begin{pmatrix} 1.690 \\ 3.261 \end{pmatrix}$

* *For this special case only, the general method of Taylor and Smith is equivalent to method of Burghart and Krupiczka.*

The *only* difference between Equation 241 obtained by Krishna [91] and Equation 249 obtained by Taylor and Smith [92] is the inclusion of the scalar correction factor Ξ. For equimolar counter transfer ($N_t = 0$; $v_i = v_n$; $\Phi = 0$; $\Xi = 1$), the two methods are equal and, indeed, equal to the limiting form of Equation 213. In all other instances, the two explicit methods give results that differ only by the scalar factor Ξ. However, this correction factor can result in a clear improvement in the predicted fluxes as the following example demonstrates.

Consider, for the third time, the diffusion of acetone and methanol through air in the Stefan tube. Table 5 summarizes the calculations needed to obtain the N_i from the two explicit methods; notice how few matrices need be computed (one, $[A]^{-1}$, is all that need be computed although we have provided the results for $[\beta_{av}]$ and $[B]_{av}^{-1}$ as well). In this particular example, the explicit method of Krishna severely underpredicts both nonzero fluxes, N_1 and N_2. However, the method of Taylor and Smith yields much better estimates of the fluxes (Tables 1 and 3). The scalar correction factor, Ξ, alone is responsible for the improvement in the results. Why?

A clue to the performance of the two methods is provided by Figure 12, where the composition profiles predicted by the two methods are compared to the profiles obtained from the exact and linearized methods. Krishna's explicit method always yields linear profiles and is unable to account for large deviations from linearity such as are encountered in this example. The method of Taylor and Smith yields identical dimensionless profiles for all three species Equation 246 of the correct (exponential) form (as is evident in Figure 12). One of the reasons for the better performance of the Taylor–Smith method is that Equation 248 represents an exact solution of the Maxwell–Stefan equations if all the D'_{ij} are equal (it is an interesting exercise to prove this). Krishna's explicit method is exact only if all the D'_{ij} are equal *and* the total flux is zero (the v_i are equal). There is, however, more to it. It is apparent from our numerical example that the scalar correction factor Ξ does a pretty good job of correcting what we might consider as the "low-flux" estimates obtained from Equation 241. In fact, the rate factor Φ defined in Equation 244 is an exact eigenvalue of the matrix $[\Phi]$ for three cases:

- All diffusivities equal (regardless of the values of the v_i).
- Stefan diffusion ($N_n = 0$) (regardless of the values of the D'_{ij} and v_i).
- Equimolar counter transfer ($N_t = 0$, $\Phi = 0$) (regardless of the values of the D'_{ij} and v_i).

It is the eigenvalues (literally; "characteristic-values") of $[\Phi]$ that characterize the correction factor matrix $[\Xi]$ [72, 75]. Thus the scalar rate factor Φ and correction factor Ξ when multiplied by identity matrices frequently are quite good models for the behavior of the complete matrices $[\Phi]$ and $[\Xi]$ in the exact solution.

One use for the Taylor–Smith method is to generate initial estimates of the fluxes for use with the Krishna–Standart or Toor–Stewart–Prober methods. Krishnamurthy and Taylor [94] found that very few (1 to 3) iterations are required by the Krishna–Standart method if Equation 248 is used to generate initial estimates of fluxes (Step 2 in the *Algorithm for Computing Molar Fluxes from the Maxwell–Stefan Equations*) even in problems that would otherwise require a great many iterations.

The Effective Diffusivity Methods

Approximate solutions of the Maxwell–Stefan equations are of considerable historical importance in chemical engineering. The oldest, simplest and most widely used methods, pioneered by Hougen and Watson [95] and by Wilke [96], employ the concept of an effective diffusion coefficient. In one of the two categories of effective diffusivity methods, the effective diffusion coefficient, D'_{ieff}, is defined with respect to the diffusion flux:

$$J_i = -c_t D'_{ieff} \frac{dy_i}{dr} = N_i - y_i N_t, \quad i = 1, 2, \ldots, n-1 \tag{250}$$

The effective diffusivity, D'_{ieff}, is then obtained by forcing Equation 183 or 46 into the form of Equation 250. Two general expressions for this coefficient are [21]:

Table 5
Diffusion of Acetone and Methanol in Stefan Tube—Calculation of the Fluxes Using the Effective Diffusivity Method

Parameters	Values			
Compositions (y_o), (y_δ)	$\begin{pmatrix}0.319\\0.528\end{pmatrix}$			$\begin{pmatrix}0.0\\0.0\end{pmatrix}$
Driving force (Δy)		$\begin{pmatrix}0.319\\0.528\end{pmatrix}$		
Method(*)	ED1	ED2	ED3	ED4
k_{1eff} (μms^{-1})	57.65	48.27	49.56	60.73
k_{2eff} (μms^{-1})	83.66	64.74	68.53	79.60
Φ_{1eff}	2.427	2.289	2.349	2.252
Φ_{2eff}	1.672	1.707	1.691	1.718
Ξ_{1eff}	0.235	0.258	0.248	0.265
Ξ_{2eff}	0.387	0.378	0.382	0.376
N_1 (mmol m^{-2} s^{-1})	1.781	1.423	1.494	1.774
N_2 (mmol m^{-2} s^{-1})	3.309	2.594	2.792	3.201

* *Key to methods*
ED1—$D'_{ieff} = D'_{in}$
ED2—D'_{ieff} from Equation 255
ED3—D'_{ieff} from Equation 256
ED4—D'_{ieff} from Equation 252

$$D'_{ieff} = \frac{N_i - y_i N_t}{N_i \sum\limits_{\substack{j=1\\j\neq i}}^{n} \frac{y_j}{D'_{ij}} - \sum\limits_{\substack{j=1\\j\neq i}}^{n} \frac{N_j}{D'_{ij}}}, \quad i = 1, 2, \ldots, n-1 \tag{251}$$

and (Stewart [97])

$$D'_{ieff} = \sum_{k=1}^{n-1} D_{ik} \frac{dy_k}{dy_i}, \quad i = 1, 2, \ldots, n-1 \tag{252}$$

Actually, Stewart presents only the expanded form of Equation 252 for a ternary mixture. Stewart suggests that for practical calculations the relative mole fraction gradients dy_k/dy_i be replaced by the ratio of differences $(y_{k0} - y_{k\delta})/(y_{i0} - y_{i\delta})$. The D_{ik} are calculated from Equation 60 with [B] evaluated at the arithmetic average composition.

Some limiting cases of Equation 251 are of particular importance.

- All binary diffusion coefficients equal

$$D'_{ieff} = D_{ij} = D'_{ij} = D' \tag{253}$$

- $n-1$ species present in such low concentrations that the approximations $y_i \cong 0$, $i = 1, 2, \ldots, n-1$; $y_n \cong 1$ can be made. In this case

$$D'_{ieff} = D'_{in}, \quad i = 1, 2, \ldots, n-1 \tag{254}$$

- When species i diffuses through $n-1$ stagnant gases, $N_j = 0$, $j \neq i$ ([86]):

$$D'_{ieff} = \frac{1-y_i}{\sum_{j=1}^{n} \frac{y_j}{D'_{ij}}} \tag{255}$$

An alternative formula for D'_{ieff} has been used by Burghardt and Krupiczka [93]. Their approach is to set

$$\frac{1}{D'_{ieff}} = \frac{y_i}{D'_{in}} + \sum_{\substack{k=1 \\ k \neq i}}^{n} \frac{y_k}{D'_{ik}} \tag{256}$$

which is equivalent to setting $D'_{ieff} = 1/B_{ii}$ with B_{ii} defined by Equation 57. This really amounts to neglecting the off-diagonal terms in the matrix [B]. A similar approach has been taken by Sato et al. [98], who write

$$D'_{ieff} = D_{ii} \tag{257}$$

which amounts to neglecting the off-diagonal elements of the Fickian matrix [D] (note that this is not the same as neglecting the off-diagonal elements of [B], although the results would probably be similar).

If D'_{ieff} can be assumed constant at some suitably averaged composition then the composition profiles are easily obtained as

$$\frac{y_i - y_{i0}}{y_{i\delta} - y_{i0}} = \frac{\exp(\Phi_{ieff}\eta) - 1}{\exp(\Phi_{ieff}) - 1}, \quad i = 1, 2, \ldots, n-1 \tag{258}$$

where

$$\Phi_{ieff} = N_t \ell / c_t D'_{ieff}, \quad i = 1, 2, \ldots, n-1 \tag{259}$$

The diffusion fluxes at $\eta = 0$ are calculated from

$$J_{i0} = \frac{c_t D'_{ieff}}{\ell} \frac{\Phi_{ieff}}{\exp(\Phi_{ieff}) - 1} (y_{i0} - y_{i\delta}), \quad i = 1, 2, \ldots, n-1 \tag{260}$$

with the N_i then given by Equation 145. Equation 260 can be rewritten as

$$J_{i0} = c_t k^{\bullet}_{ieff}(y_{i0} - y_{i\delta}) = c_t k_{ieff} \Xi_{ieff}(y_{i0} - y_{i\delta}) \tag{261}$$

where we have defined an effective low-flux mass transfer coefficient k_{ieff} by

$$k_{ieff} = D'_{ieff}/\ell, \tag{262}$$

effective correction and rate factors by

$$\Xi_{ieff} = \frac{\Phi_{ieff}}{\exp(\Phi_{ieff}) - 1}; \quad \Phi_{ieff} = \frac{N_t}{c_t k_{ieff}} \tag{263}$$

and an effective high flux mass transfer coefficient by

$$k^{\bullet}_{ieff} = k_{ieff} \Xi_{ieff} \tag{264}$$

In order to compute the fluxes, we note first that Equation 260 requires a knowledge only of N_t. In the limit that $N_t \rightarrow 0$, $\Phi_{ieff} \rightarrow 0$, $\Xi_{ieff} \rightarrow 1$ and the fluxes are obtained directly from

$$N_i = k_{ieff}(y_{i0} - y_{i\delta}) \quad [N_t = 0], \quad i = 1, 2, \ldots, n-1 \tag{265}$$

If the total flux, N_t, is nonzero, an iterative approach is required. Damped repeated substitution will work (see the *Algorithm for Computing the Molar Fluxes from the Exact Solution of the Maxwell–Stefan Equation*), but a more effective procedure is to adapt the method of Vickery et al. [90] for calculating the fluxes from the linearized equations. We proceed as before (paralleling the development of Equation 230), noting that ($\hat{D}_i \equiv D'_{ieff}$; $[P] \equiv [I]$, $\hat{\Xi}_i \equiv \Xi_{ieff}$; $\hat{\Phi}_i \equiv \Phi_{ieff}$) to obtain a single nonlinear equation for the unknown flux, N_t:

$$F(N_t) \equiv \sum_{i=1}^{n-1} \left(1 - \frac{v_i}{v_n}\right)\left(\frac{\Xi_{ieff}}{\Phi_{ieff}}(y_{i0} - y_{i\delta}) + y_{i0}\right) - 1 = 0 \tag{266}$$

This function is easily solved for N_t using Newton's method. The derivative of $F(N_t)$ is needed for this procedure and can be calculated from Equation 232

$$\frac{dF}{dN_t} = -\sum_{i=1}^{n-1} \left(1 - \frac{v_i}{v_n}\right)\left(\frac{1}{c_t k_{ieff}} \frac{\exp(\Phi_{ieff})}{\exp(\Phi_{ieff}) - 1}\right) \tag{267}$$

Convergence is guaranteed from an initial guess obtained from a solution of the low flux limit of Equation 266

$$N_t = \sum_{i=1}^{n-1} \left(1 - \frac{v_i}{v_n}\right) c_t k_{ieff}(y_{i0} - y_{i\delta}) \quad \text{(initial guess)} \tag{268}$$

To illustrate the application of the effective diffusivity method, we consider, for the fourth time, diffusion of acetone and methanol through air in a Stefan tube. Table 5-b summarizes the results of our calculations. In this particular example, methods 1 and 4 (Equations 254 and 252) give excellent estimates of the N_i; as good or better than the explicit methods just discussed. Methods 2 and 3 (Equations 255 and 256) severely underestimate the fluxes. Although we have not drawn the composition profiles obtained from the four methods, it is to be noted that all four methods yield profiles that, on the scale of Figure 11, are almost indistinguishable from the profiles obtained using other approximate methods. We would caution our readers *not* to draw any conclusions regarding the relative merits of different approaches from this solitary example problem. We will return to a comparison of the various methods later in this section.

In the other category of this class of methods, the effective diffusion coefficient is defined with respect to the *molar* flux, N_i. That is

$$N_i = -c_t D'_{ieff} \frac{dy_i}{dr}, \quad i = 1, 2, \ldots, n-1 \tag{269}$$

Force fitting the Maxwell–Stefan Equation 183 into the form Equation 269 gives an expression for the effective diffusivity [99]

$$\frac{1}{D'_{ieff}} = \sum_{\substack{j=1 \\ j \neq i}}^{n} \frac{y_j}{D'_{ij}}\left(1 - \frac{y_i}{y_j}\frac{N_j}{N_i}\right), \quad i = 1, 2, \ldots, n-1 \tag{270}$$

If the flux ratios are known or can be approximated in some way (as is the case when we have diffusion-controlled chemical reactions taking place on catalyst surfaces) then the D'_{ieff} defined in Equation 270 may be calculated.

Now, if D'_{ieff} can be considered constant at some average value, then integration of Equation 269 yields linear composition profiles and a simple expression for calculating the fluxes without iteration:

$$N_i = \frac{c_t D'_{ieff}}{\ell}(y_{i0} - y_{i\delta}), \quad i = 1, 2, \ldots, n-1 \tag{271}$$

Examples illustrating the application of this method are given by Kubota et al. [99] and by Geankoplis [100].

The primary advantage of the effective diffusivity definitions is their simplicity. The primary disadvantage of the use of D'_{ieff} is that these parameters are not, in general, system properties except for the limiting cases noted above. Further, they depend on the magnitude of the fluxes not always known in advance.

Generally in multicomponent systems, effective diffusivities do not have the physical significance of a diffusion coefficient since they may assume values ranging from minus to plus infinity [101]. The effective diffusivity defined by Equation 250 or 252 changes along the diffusion path; is zero at a diffusion barrier, negative in the region of reverse diffusion and infinite at the osmotic diffusion point. Care must, therefore, be taken in drawing analogies between this quantity and a binary diffusion coefficient. Only when the effective diffusivity is positive, bounded, and not a strong function of composition or fluxes is it possible to draw useful analogies. More complicated variations on the effective diffusivity theme; some requiring iteration on D'_{ieff}, others postulating a variation of D'_{ieff} on position or on composition have been discussed by Wilke [96], Toor [2], Shain [102] and by Hsu and Bird [103]. In view of the many better and, indeed, simpler methods that have been developed since then, they have not been included in this review.

Comparison of Methods

All of the methods just described with the exception of the explicit method of Krishna [81] and the effective diffusivity method of Kubota et al. [89] are exact for systems where all the D'_{ij} are equal (this includes binary systems by default). The two exceptions noted are exact if, further, the total flux is zero. Any comparison of the methods must, therefore, focus on systems in which the binary D'_{ij} and or the ν_i differ. A thorough computational comparison of methods has been carried out by Smith and Taylor [85] who solved many thousands (more than ten thousand) of example problems for *each* of twenty-three real ternary gas systems covering the entire range of ratios of diffusion coefficients and determinacy coefficients (the ν_i) likely to be encountered in practice. A summary of their findings is the subject of this section.

For each problem solved, the discrepancy between the molar fluxes predicted from the exact solution and the fluxes calculated from an approximate method was calculated from [86]

$$\varepsilon_i = (N_{i\,Exact} - N_{i\,Approx}) / \sum_{k=1}^{n} \frac{1}{n} |N_{k\,Exact}|, \quad i = 1, 2, \ldots, n \tag{272}$$

Smith and Taylor reported their results in terms of an average discrepancy $\bar{\varepsilon}$, defined for each problem by

$$\bar{\varepsilon} = \frac{1}{n} \sum_{k=1}^{n} |\varepsilon_k| \tag{273}$$

Table 6
Discrepancies Between Approximate and Exact Solutions of the Maxwell-Stefan Equations for a Film Model*

Method	ED1	ED2	ED3	ED4	KR79	TS(BK)	LIN
$\bar{\epsilon}$	0.544	0.260	0.319	0.059	0.140	0.045	0.011
ϵ_t	> 3.50	0.558	> 2.00	0.271	0.372	0.025	0.030
$\lvert \epsilon_{imax} \rvert$	41.5	12.6	44.2	1.50	1.54	0.710	0.310

* *Results for the system acetone-methanol-air. Source: Smith and Taylor [85].*

Key
ED1—Effective diffusivity given by Equation 254
ED2—Effective diffusivity given by Equation 255
ED3—Effective diffusivity given by Equation 286
ED4—Effective diffusivity given by Equation 252
KR79—Explicit method of Krishna (1979)
TS(BK)—Explicit method of Taylor and Smith (1982) and, for the case of Stefan diffusion only, of Burghardt and Krupiczka (1975)
LIN—Linearized theory of Toor-Stewart-Prober (1964)

the discrepancy between the predicted total fluxes, ε_t (this is just the sum of the ε_i) and the maximum individual discrepancy encountered in any problem solved, $\lvert \varepsilon_{i\,max} \rvert$. The first two of these quantities, $\bar{\varepsilon}$ and ε_t were averaged arithmetically over all problems considered. We cannot reproduce all of their results here but a flavor of them can be found in Table 6 where the average discrepancies for the system acetone-methanol-air are reported. The fact that ε_t is sometimes larger than $\bar{\varepsilon}$ is due to the way in which these quantities are defined. The remarks which follow are made in the light of *all* of the calculations made by Smith and Taylor [85], not just those summarized in Table 6.

In general, the solution of the linearized equations can be expected *always* to provide good estimates of the individual fluxes. The largest discrepancies occur when inert species are present, and for this particular case, this solution is by a significant margin the better of the approximate methods. We regard this finding as most important. The implication is that if the assumption of constant [D] is a good one for a film model of steady-state diffusion, then it should also be good for other models of mass transfer for which exact analytical solutions of the Maxwell–Stefan equations have not yet been obtained.

For determinacy conditions such as Graham's law and nonequimolar distillation, the explicit method of Taylor and Smith is actually the better of the approximate methods although the advantage over the linearized theory is negligibly small. This explicit formulation rates second best (among the approximate methods) if stagnant components are in the mixture. The average discrepancy in the predicted total flux is comparable to the discrepancies in this flux obtained from the linearized theory. The great advantage of an explicit method is, of course, rapidity in computation. In fact, we strongly recommend using Equation 248 to generate initial estimates of the fluxes for use with the iterative methods. The reduction in computation time made possible in this way is quite marked, and convergence difficulties are unlikely to be encountered [84].

The explicit method of Krishna [91] is successful only if the ν_i are close together and, therefore (or for other reasons) the total transfer rate is low. At high rates of mass transfer, the assumption of constant $[A]^{-1}$ (or of $[\beta][B]^{-1}$) is a poor one, particularly in cases involving an inert species. The reason for this failure lies in the fact this explicit formulation is *not* exact for multicomponent mixtures with all D'_{ij} equal (except when $N_t = 0$).

The simple effective diffusivity methods Equations 254 through 256 can be used with confidence for systems where the binary D'_{ij} display little or no variation. They must provide good results for all systems if conditions are such that the appropriate limiting forms of

Equations 251 and 252 apply. However, the identification of these conditions requires some a priori knowledge which none of the more rigorous methods require. To knowingly, or unknowingly, use these formulas in situations such that these limiting cases do not apply would stand about as much chance of successfully predicting the fluxes as the throw of a die to predict the magnitude and the toss of a coin to determine the sign. (Strictly speaking, *none* of the limiting cases Equations 253 through 255 applies to the illustrative examples in Table 5. Thus the good results obtained with methods ED1 and ED4, as should be clear from the *averages* quoted in table 6 and Reference 85 are fortuitous.)

The effective diffusivity formula of Stewart [97] is *by far* the best of this class of methods. This should not come as a surprise since this method is capable of correctly identifying the various interaction phenomena possible in multicomponent systems. Indeed, for equimolar counter transfer, this effective diffusivity method becomes equivalent to the linearized theory and to both explicit methods just discussed. In fact, for some systems Stewart's effective diffusivity method is superior to Krishna's explicit method. However, since the explicit methods are simpler to use (requiring the same basic data) than Stewart's effective diffusivity method and, in general, provide such superior results we see no reason for any further use of the effective diffusivity. Another caution is in order here: *all* of the approximate methods are somewhat less successful in predicting, accurately, the rates of osmotic or reverse mass transfer. It is quite simple to find problems in which effective diffusivity methods predict the wrong *direction* of transfer. It is somewhat harder but still possible to find problems in which the explicit methods or the linearized equations predict directions of transfer different from the results given by an exact solution as shown by the calculations of Taylor and Krishnamurthy [104]. However, disagreement on sign is likely only for species whose transfer rate is low relative to the quantity [86]

$$\sum_{k=1}^{n} |N_k|$$

In general, we have no reservations about recommending the use of solutions derived on the basis of constant [D] when a more exact method is not available or too time consuming to use.

Multicomponent Film Model for Mass Transfer in Nonideal Fluid Systems

The starting point for the analysis of mass transfer in nonideal fluid mixtures is the set of generalized Maxwell–Stefan Equation 56, which for one-dimensional transfers, may be written as

$$\sum_{k=1}^{n-1} \Gamma_{ik} \frac{dx_k}{dr} = \sum_{\substack{j=1 \\ j \neq i}}^{n} \frac{x_i N_j - x_j N_i}{c_t D'_{ij}} = \sum_{\substack{j=1 \\ j \neq i}}^{n} \frac{x_i J_j - x_j J_i}{c_t D'_{ij}} \tag{274}$$

where the thermodynamic factors, Γ_{ik}, are given Equation 58. Since the Γ_{ik} and D'_{ik} are, in general, composition dependent, the differential Equation 274 cannot be solved analytically without making further assumptions. In this section we briefly describe how the solutions developed previously for mass transfer in ideal gases can be extended to the more general case.

Krishna [105] developed a solution to Equation 274 by paralleling the ideal gas analysis of Krishna and Standart. Equation 268 is first cast into $n-1$ dimensional matrix form as

$$[\Gamma] \frac{d(x)}{d\eta} = [\Phi](x) + (\varphi) \tag{275}$$

where η, $[\Phi]$ and (φ) are as defined before for the ideal gas case. Krishna [105] made the assumption that the coefficients Γ_{ik} and D'_{ik} could be considered constant along the diffusion path; this means in practice that average values for these coefficients have to be used. With these assumptions, Equation 275 now represents a linear matrix differential equation the solution of which can be written down in a manner exactly analogous to the ideal gas case. Thus the fluxes, N_i, can be calculated from

$$(N) = c_t[\beta_0]\,[k_0]\,[\Xi_0]\,(x_0 - x_\delta)$$

$$= c_t[\beta_\delta]\,[k_\delta]\,[\Xi_\delta]\,(x_0 - x_\delta) \tag{276}$$

where the matrices of mass transfer coefficients are given by:

$$[k_0] = \frac{[D_0]}{\ell} = \frac{[B_0]^{-1}\,[\Gamma_{av}]}{\ell}; \qquad [k_\delta] = \frac{[D_\delta]}{\ell} = \frac{[B_\delta]^{-1}\,[\Gamma_{av}]}{\ell} \tag{277}$$

The subscripts 0 and δ remind us that the appropriate compositions, x_{i0} and $x_{i\delta}$ respectively, have to be used in the defining equations for B_{ik}, Equation 57. The subscript av on $[\Gamma_{av}]$ emphasizes the assumption of constancy of Γ_{ik} along the diffusion path.

The correction factors $[\Xi_0]$ and $[\Xi_\delta]$ are given by

$$[\Xi_0] \equiv [\theta]\,\{\exp[\theta] - [I]\}^{-1}; \qquad [\Xi_\delta] \equiv [\theta]\exp[\theta]\{\exp[\theta] - [I]\}^{-1} \tag{278}$$

where $[\theta]$ is the augmented matrix of rate factors

$$[\theta] = [\Gamma_{av}]^{-1}\,[\Phi]. \tag{279}$$

It is interesting to note the direct influence of the thermodynamic factors Γ_{ik} on the mass transfer correction factors $[\Xi]$, which really reflect their direct influence on the composition profiles which are given by

$$(x - x_0) = \{\exp[\theta]\eta - [I]\}\,\{\exp[\theta] - [I]\}^{-1}\,(x_\delta - x_0) \tag{280}$$

The interaction phenomena discussed earlier for the ideal gas case will also be possible for nonideal fluid mixtures, for which $[\Gamma]$ contribute additionally to the nondiagonality of the matrix $[k^{\cdot}]$ by means of its separate influence on $[k]$, the zero-flux matrix and $[\Xi]$, the correction factor matrix. To illustrate the influence of Γ_{ik} on $[D]$, consider a ternary liquid mixture. The cross-coefficient D_{12} will be given by

$$D_{12} = \frac{\Gamma_{12}D'_{13}(x_1D'_{23} + (1 - x_1)D'_{12}) + \Gamma_{22}x_1D'_{23}(D'_{13} - D'_{12})}{x_1D'_{23} + x_2D'_{13} + x_3D'_{12}} \tag{281}$$

We see, therefore, that the thermodynamic factors Γ_{ij}, given by Equation 58 also contribute to $D_{ij}(i \neq j)$. The large cross coefficients possible in nonideal liquid mixtures are exemplified by the system polystyrene (1)-cyclohexane (2)-toluene (3) for which Cussler and Lightfoot [106] measured $|D_{12}/D_{11}| > 1$ in certain concentration regions.

The linearized theory of Toor [4] and of Stewart and Prober [5] is based on the assumption that the Fickian matrix $[D]$ is constant along the diffusion path. For nonideal mixtures

$$[D_{av}] = [B_{av}]^{-1}\,[\Gamma_{av}] \tag{282}$$

is assumed constant.

By following the development of Equations 210 through 216 we can obtain an expression for the composition profiles Equation 210

$$(x - x_0) = \{\exp[\theta_{av}]\eta - [I]\}\{\exp[\theta_{av}] - [I]\}^{-1}(x_\delta - x_0) \tag{283}$$

and two expressions for the molar fluxes

$$(N) = c_t[\beta_0][k_{av}][\Xi_0](x_0 - x_\delta) \tag{284}$$

$$= c_t[\beta_\delta][k_{av}][\Xi_\delta](x_0 - x_\delta) \tag{285}$$

where we have defined a matrix of low flux mass transfer coefficients

$$[k_{av}] = [D_{av}]/\ell \equiv [B_{av}]^{-1}[\Gamma_{av}]/\ell \tag{286}$$

a matrix of rate factors

$$[\theta_{av}] = \frac{N_t}{c_t}[k_{av}]^{-1} \tag{287}$$

and matrices of correction factors Equation 215, 216

$$[\Xi_0] = [\theta_{av}]\{\exp[\theta_{av}] - [I]\}^{-1} \tag{288}$$

$$[\Xi_\delta] = [\theta_{av}]\exp[\theta_{av}]\{\exp[\theta_{av}] - [I]\}^{-1} \tag{289}$$

The explicit methods can be generalized in similar ways; Krishna [91], for example, assumes constancy of the matrix product $[\beta][B]^{-1}[\Gamma]$. Thus we obtain Equation 241

$$(N) = \frac{c_t}{\ell}[\beta_{av}][B_{av}]^{-1}[\Gamma_{av}](x_0 - x_\delta) \tag{290}$$

For liquid mixtures, effective diffusion coefficients may be defined using the generalized Maxwell–Stefan equations [102].

The computational methods just discussed for use with the simpler forms of the general relations previously presented could also be employed here. Thus, repeated substition to calculate the N_i from Equation 276 (with initial guesses obtained from an explicit method) and Newton's method for calculating N_t from the linearized equations. Obviously, the explicit methods need no iteration.

Generalization of Multicomponent Film Models to Other Geometries

Simply by defining the characteristic diffusion length, ℓ, as shown in Figure 9, the results for the multicomponent film model can be generalized to the cylindrical and spherical geometries. For example, for the Krishna [95] model we have the general expression

$$(N_0) = c_t[\beta_0][k_0][\Xi_0](x_0 - x_\delta) \tag{291}$$

where (N_0) is the column matrix of fluxes at $t = r_0$. The matrix of zero-flux transfer coefficients is given by

$$[k_0] = \equiv \frac{[D_0]}{\ell} \tag{292}$$

where $\ell = r_0 \ln(r_0/r_\delta)$ for the cylindrical geometry and $\ell = r_0(1-r_0/r_\delta)$ for the spherical geometry. Similar modifications are possible for all of the other methods previously discussed.

The Film Model for Heat Transfer with Simultaneous Mass Transfer Neglecting Soret and Dufour Effects

Having tackled the problem of calculating the interfacial fluxes N_i using the film model, let us consider the problem of calculating the interfacial energy fluxes E. We ignore the Soret and Dufour effects in the ensuing analysis.

For steady-state heat transfer within a planar film, the energy balance relation Equation 80 simplifies to

$$E = -\lambda \frac{dT}{dr} + \sum_{i=1}^{n} N_i \bar{H}_i = E_0 = E_\delta = \text{constant} \tag{293}$$

If the reference state for the calculation of the partial molar enthalpies $\bar{H}_i$ is taken to be the pure component at temperature T_{ref}, then we may write

$$\bar{H}_i = C_{pi}(T - T_{ref}) \tag{294}$$

where C_{pi} represents the molar heat capacity of the species i.

In proceeding further, it is convenient to define a heat transfer rate factor

$$\Phi_H = \frac{(\sum N_i C_{pi})\ell}{\lambda} \tag{295}$$

where ℓ is the characteristic diffusion path length, equal to the film thickness for planar geometries.

With definitions in Equations 294 and 295, we may write Equation 293 in the form:

$$\frac{d(T - T_{ref})}{d\eta} = \Phi_H(T - T_{ref}) - \frac{E\ell}{\lambda} \tag{296}$$

which differential equation can be solved with the boundary conditions

$$\begin{aligned} r = r_0,\ \eta = 0&: T = T_0 \\ r = r_\delta,\ \eta = 1&: T = T_\delta \end{aligned} \tag{297}$$

to yield the temperature profile

$$(T - T_0) = \frac{\exp \Phi_H \eta - 1}{\exp \Phi_H - 1}(T_\delta - T_0) \tag{298}$$

The conductive heat flux at $\eta = 0$, q_0 may be evaluated as

$$q_0 = -\frac{\lambda}{\ell} \frac{dT}{d\eta}\bigg|_{\eta=0} = h\Xi_H(T_0 - T_\delta) = h^\bullet(T_0 - T_\delta) \tag{299}$$

where the zero flux heat transfer coefficient is found to be

$$h = \frac{\lambda}{\ell} \tag{300}$$

and the heat transfer correction factor, called the Ackermann correction factor,

$$\Xi_H = \Phi_H/(\exp \Phi_H - 1) \tag{301}$$

which reduces to unity, leading to linear temperature profiles, for vanishingly small mass transfer fluxes, i. e. $N_i \to 0$.

The energy flux E can be calculated from

$$E = q_0 + \sum N_i \bar{H}_i$$
$$= h^{\bullet}(T_0 - T_\delta) + h\Phi_H(T_0 - T_{ref}) \tag{302}$$

which shows that the value of E depends on the arbitrary choice of the reference temperature T_{ref} for the calculation of enthalpies. This is no drawback because in practice we apply the interfacial energy balance in the form

$$E = E^x = E^y \tag{303}$$

and the choice of T_{ref} is of no importance. The film model for estimation of the conductive flux q must actually be seen in the context of our earlier, model independent analysis, of the determination of N_i and E (see beginning of previous section).

Film Model for Coupled Heat and Multicomponent Mass Transfer

Our aim here is to show how the Soret and Dufour effects, neglected in the preceding analysis, can be accounted for. The starting point for our development is the set of generalized constitutive Equation 39 which, for one dimensional transport, can be written as

$$d_i = \frac{x_i}{RT}\frac{d\mu_i}{dr} = \sum_{k=1}^{n} \frac{x_i J_k - x_k J_i}{c D'_{ik}} - x_i \sum_{\substack{j=1 \\ j \neq i}}^{n} x_j \alpha_{ij} \frac{d \ln T}{dr}, \qquad i = 1, 2, \ldots, n \tag{304}$$

$$q = -\left(\lambda + \frac{1}{2} c_t R \sum_{i,j} x_i x_j D'_{ij} \alpha_{ij}^2\right)\frac{dT}{dr} - c_t RT \sum_{i=1}^{n-1} \alpha_{in} D'_{in} d_i \tag{305}$$

The nonlinear coupling in Equations 304 and 305 means that an exact solution must be obtained by numerical methods. The Maxwell–Stefan formulation is particularly useful in this context since the composition and temperature gradients can be related directly to the invariant molar and energy fluxes as

$$\sum_{k=1}^{n-1} \Gamma_{ik} \frac{dx_k}{dr} = \sum_{\substack{k=1 \\ k \neq i}}^{n} \frac{x_i N_k - x_k N_i}{c_t D'_{ik}} - x_i \sum_{\substack{j=1 \\ j \neq i}}^{n} x_j \alpha_{ij} \frac{d \ln T}{dr}, \qquad i = 1, 2, \ldots, n-1 \tag{306}$$

$$q = -\lambda \frac{dT}{dr} - \frac{RT}{2} \sum_{i,j=1}^{n} \alpha_{ij}(x_i N_j - x_j N_i) = E - \sum_{i=1}^{n} N_i \bar{H}_i \tag{307}$$

where the thermodynamic factors Γ_{ik} are defined by Equation 58.

A solution of Equations 306 and 307 is sought subject to the boundary conditions

$$r = 0,\ x_i = x_{i0},\ T = T_0; \qquad r = \ell,\ x_i = x_{i\delta},\ T = T_\delta \tag{308}$$

Expressions for the composition gradients are most readily obtained by first combining the $n-1$ Equation 10 in $n-1$ dimensional matrix form as

$$\frac{d(x)}{d\eta} = [\Gamma]^{-1}[\Phi](x) + [\Gamma]^{-1}(\varphi) - [\Gamma]^{-1}(k_T)\frac{d \ln T}{d\eta} \tag{309}$$

where η is, again, defined by $\eta = r/\ell$. The matrices $[\Gamma]$, $[\Phi]$, and (φ) have their usual meanings. In addition (k_T) is a column matrix of independent thermal diffusion ratios with elements

$$k_{Ti} = x_i \sum_{j=1}^{n} x_j \alpha_{ij}, \quad i = 1, 2, \ldots, n \tag{310}$$

The temperature gradient can now be obtained from Equation 307

$$\frac{dT}{d\eta} = \frac{1}{h}\left(\sum_{k=1}^{n} N_k \bar{H}_k - E\right) - \frac{RT}{2h} \sum_{i,j=1}^{n} \alpha_{ij}(x_i N_j - x_j N_i) \tag{311}$$

where h is a heat transfer coefficient defined by Equation 300. Equation 311 is then inserted into Equation 309 for the composition gradients.

The evaluation of the fluxes consists essentially of numerically integrating Equations 309 and 311 across the film starting from the known boundary conditions at $\eta = 0$. A multidimensional Newton–Raphson procedure can be used to search for the set of fluxes that yields the known compositions and temperature at $\eta = 1$.

Considerable computation time is required to obtain just a single estimate of the fluxes using this procedure. This may be avoided by linearizing Equations 306 and 307 and developing an approximate analytical solution as described by Taylor [117]. It can be seen in Equations 306 and 307 that the heat flux q is proportional to the temperature gradient whereas the diffusion fluxes depend only on the gradient of the logarithm of temperature. Further, these fluxes are of different magnitudes and units (typically $q \cong 10^3\ Jm^{-2}s^{-1}$, $J_i \cong 10^{-3}$ kmol $m^{-2}s^{-1}$). This rules out simply assuming constant coefficients in Equations 306 and 307 and combining them in a form that relates q and the J_i to dT/dz and the dx_i/dz. The difficulty may be overcome by defining "reduced" heat and energy fluxes by

$$\mathcal{J}_0 = q/H_{ref}, \qquad \mathcal{N}_0 = E/H_{ref} \tag{312}$$

(where H_{ref} is some (necessarily nonzero) reference enthalpy) then $\mathcal{J}_0$ and $\mathcal{N}_0$ have the units of a molar flux (hence the choice of symbols). The subscript zero is henceforth associated with thermal fluxes and coefficients. The choice of H_{ref} is arbitrary but it should be invariant over the film. If a reference state is chosen to be fluid at temperature T_{ref}, the constituent molar enthalpies may be found from Equation 294. A convenient choice of H_{ref}, therefore, is

$$H_{ref} = C_{Pm} T_{ref} \tag{313}$$

where C_{Pm} is a mean heat capacity of the mixture defined by

$$N_t C_{Pm} = \sum_{i=1}^{n} N_i C_{Pi} \tag{314}$$

It is important to note that H_{ref} in Equation 313 is a definition of convenience and does not affect the choice of *reference state* in any way.

On observing that $dT/dr = d(T - T_{ref})/dr$, a "constitutive equation" for the flux J_0 is readily obtained by dividing Equation 305 by $C_{Pm}T_{ref}$ to give

$$J_0 = -\frac{\lambda'}{C_{Pm}}\frac{d}{dr}\left(\frac{T-T_{ref}}{T_{ref}}\right) - \frac{c_tRT}{C_{Pm}T_{ref}}\sum_{i,k=1}^{n}\alpha_{kn}D'_{kn}\Gamma_{ki}\frac{dx_i}{dr} \tag{315}$$

where λ' is the coefficient multiplying dT/dr in Equation 305. Equation 315 may be linearized by assuming the coefficients of the gradients are constant over the film. This suggests that a good choice of T_{ref} is the mean film temperature $T_m = 1/2\,(T_0 + T_\delta)$. With this choice of T_{ref}, the temperature dependence of Equation 315 is considerably reduced.

Turning to the diffusion fluxes; Equation 304 may be combined in $n-1$ dimensional matrix form and inverted to give the independent J_i as

$$(J) = -c_t[B]^{-1}[\Gamma]\frac{d(x)}{dr} - c[B]^{-1}(k_T)\frac{d\ln T}{dr} \tag{316}$$

where the matrix [B] has its usual meaning. Equations 315 and 316 are of the same form but include differing dependencies on the temperature gradient. However, with $T_{ref} = T_m$

$$\frac{d\ln T}{dr} = \frac{1}{T}\frac{dT}{dr} = \frac{1}{T_m}\frac{dT}{dr} \tag{317}$$

is a good approximation even for quite high temperature gradients. Any other choice of T_{ref} (T_0 or T_δ say) would weaken the approximation. Thus by defining column matrices of the n independent fluxes $(\mathcal{N})$ and $(\mathcal{J})$ with elements

$$\mathcal{N}_0 = E/H_{ref};\ \mathcal{N}_i = N_i;\ \mathcal{J}_0 = q/H_{ref};\ \mathcal{J}_i = J_i,\quad i = 1, 2, \ldots, n-1 \tag{318}$$

and a column matrix of n independent fluid properties (τ) with elements

$$\tau_0 = (T - T_m)/T_m;\ \tau_i = x_i,\quad i = 1, 2, \ldots, n-1 \tag{319}$$

then Equations 315 through 317 may be combined in n dimensional matrix form as

$$(\mathcal{J}) = -\frac{c_t}{\ell}[D]\frac{d(\tau)}{d\eta} = -c_t[k]\frac{d(\tau)}{d\eta} \tag{320}$$

$$(\mathcal{N}) = (\mathcal{J}) + N_t(\tau) = -c_t[k]\frac{d(\tau)}{d\eta} + N_t(\tau) \tag{321}$$

where the matrices of coefficients are the partitioned matrices

$$[D] = \begin{bmatrix} \dfrac{\lambda'}{c_tC_{Pm}} & \dfrac{R}{C_{Pm}}(D^T)^T[\Gamma] \\ [B]^{-1}(k_T) & [B]^{-1}[\Gamma] \end{bmatrix};\quad [k] = [D]/\ell \tag{322}$$

Equations 320 and 321 may be linearized by assuming $c_t[D]$ remains constant over the film. The inverse of Equation 321 is a first-order matrix differential equation with constant coefficients. The solution simply is an extension of Equation 210

$$(\tau - \tau_0) = \{\exp[\theta]\eta - [I]\}\{\exp[\theta] - [I]\}^{-1}(\tau_\delta - \tau_0) \tag{323}$$

where $[\theta] = N_t[k]^{-1}/c_t = N_t\ell[D]^{-1}/c_t$ (324)

The "diffusion" fluxes at the origin ($\eta = 0$) are obtained from

$$(J_0) = \frac{c_t}{\ell}[D][\theta]\{\exp[\theta]-[I]\}^{-1}(\tau_0-\tau_\delta) = c_t[k^\bullet](\tau_0-\tau_\delta) \quad (325)$$

The film invariant molar and energy fluxes follow from Equations 107 and 108 on specification of an appropriate determinacy condition as described in detail in the previous section.

Equation 325 contains a number of other solutions as special cases. For example; if $\alpha_{12} = 0$ then Equation 325 for two-component mixtures simplifies to give Equations 168 and 299. This result again contains no errors for the heat flux in a mixture with any number of constituents provided all the α_{ij} are zero. The diffusion fluxes remain coupled however and Equation 325 reduces to Equation 211.

Surface Renewal Theory

In the classic penetration model of Higbie, fluid elements (or eddies) are pictured as arriving at the interface from the bulk fluid phase and residing at the interface for a period of time t_e (exposure time). During the time t_e that the fluid element resides at the interface, mass exchange takes place with the adjoining phase by a process of unsteady-state diffusion. The fluid element is quiescent during this exposure period at the interface and the diffusion process is purely molecular. The element may, however, move in plug flow along the interface. A pictorial representation of this model is given in Figure 13 and has been adapted from Scriven [108]. After exposure, and consequent mass transfer, the fluid elements return to the bulk fluid phase and are replaced by fresh eddies and the process is repeated.

The governing differential equation for the unsteady-state diffusion process experienced by the fluid element during its residence at the interface is Equation 11 which may be written for one-dimensional diffusion in a planar coordinate system as

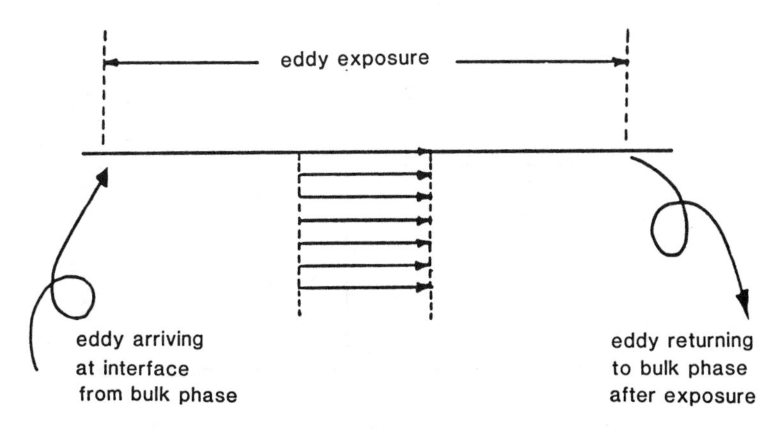

Figure 13. General representation of the surface renewal model. An eddy arrives at the interface and resides there for randomly varying periods of time. During this period, there is plug flow of fluid elements. The bulk fluid is considered to be located at an infinite distance from the interface. Pictorial representation is adapted from Scriven.

$$\frac{\partial x_i}{\partial t} + u_z \frac{\partial x_i}{\partial y} = -\frac{1}{c_t} \frac{\partial J_{iy}}{\partial y}, \quad i = 1, 2, \ldots, n-1 \tag{326}$$

where y represents the direction coordinate for diffusion. The molar diffusion flux J_{iy} is given by Equation 17 for a binary system and either Equation 39 or Equation 46 for a multicomponent system.

The assumptions of the model are incorporated into the initial and boundary conditions. During the diffusion process, the interface has the composition x_{iI} which is usually assumed to be constant (this assumption is not essential to the penetration model), and we have the boundary condition

$$y = 0, \ t > 0: \quad x_i = x_{iI} \tag{327}$$

Before the start of the diffusion process, the compositions are everywhere uniform in the phase under consideration, and equal to the bulk fluid composition, x_{ib}. Thus we have the initial condition

$$y \geqq 0, \ t = 0: \quad x_i = x_{ib} \tag{328}$$

Finally, we have the last boundary condition which is valid for short contact times, i.e.

$$y \to \infty, \ t > 0: \quad x_i = x_{ib} \tag{329}$$

which essentially states that the diffusing component has not penetrated into the bulk fluid phase.

For $u_y = 0$, i.e. for vanishingly small mass transfer rates and no convection perpendicular to the interface, the solution of Equation 326 for a binary system leads to the following expression for the instantaneous value of the mass transfer coefficient

$$k(t) = \left[\frac{D}{\pi t}\right]^{1/2} \tag{330}$$

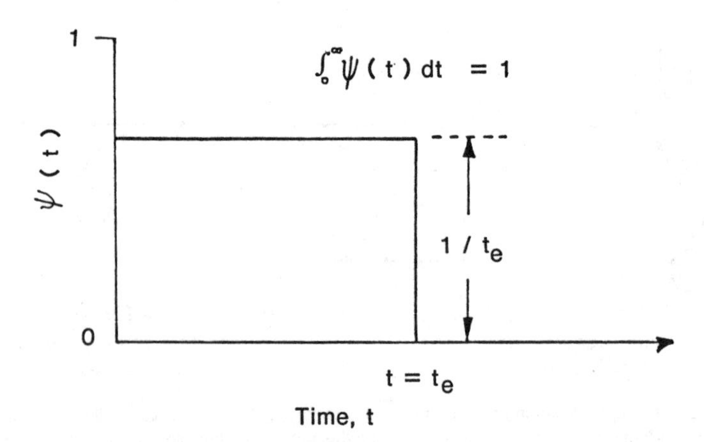

Figure 14. Surface-age distribution function (t) in which each fluid element stays the same period of time at the interface before being replenished from the bulk fluid (Higbie model).

The Higbie penetration model assumes that all the fluid elements have identical residence times at the interface. If one defines a surface age distribution function $\psi(t)$ representing the fraction of elements having ages between t and t+dt at the surface, then the Higbie surface age distribution is $\psi(t) = 1/t_e$ for all $t \leqq t_e$ and $\psi(t) = 0$ for $t > t_e$; see Figure 14. The average mass transfer coefficient over the exposure period t_e is, therefore, given by

$$k = \int_0^\infty k(t)\psi(t)\,dt \tag{331}$$

and can be evaluated with the help of Equation 237 as

$$k = 2\left[\frac{D}{\pi t_e}\right]^{1/2} \tag{332}$$

The Danckwerts surface renewal model supposes that the chance of an element of surface being replaced with fresh liquid from the bulk is independent of the length of time for which it has been exposed. The age distribution function assumed is $\psi(t) = s\exp(-st)$ and is depicted in Figure 15. Here s is the fraction of the area of surface which is replaced with fresh liquid in unit time. The Danckwerts surface age distribution leads to an average value of the mass transfer coefficient given by

$$k = (Ds)^{1/2} \tag{333}$$

For finite mass transfer rates, the zero-flux mass transfer coefficient has to be multiplied by a flux correction factor Ξ, i.e.

$$k^{\bullet} = k\Xi \tag{334}$$

where Ξ is given by [21]

$$\Xi \equiv \frac{\exp(-\Phi^2/\pi)}{1+\mathrm{erf}\left(\dfrac{\Phi}{\sqrt{\pi}}\right)} \tag{335}$$

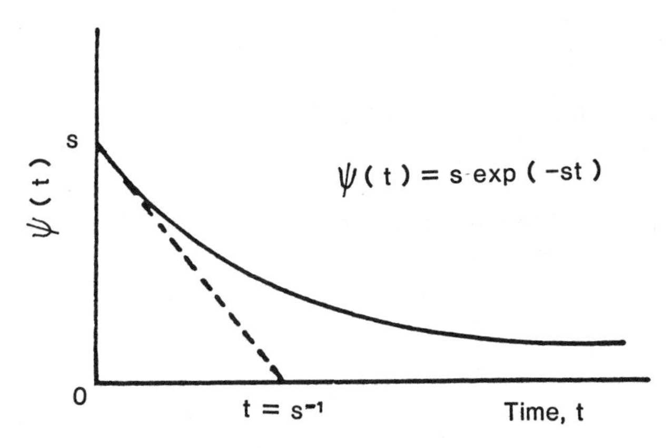

Figure 15. Surface-age distribution according to Danckwerts; the elements undergo random surface renewal at frequency s.

in which the mass transfer rate factor Φ is defined as

$$\Phi \equiv \frac{N_t}{c_t k} \tag{336}$$

Bird, Stewart and Lightfoot [21] give a graphical comparison of the film and penetration model correction factors. From their results, it can be deduced that at a given Φ the film and penetration predictions of Ξ are close to each other and we, therefore, recommend the use of the film model Ξ in design calculations because of the relative simplicity in computations; for the use of the penetration model Ξ, we need to evaluate error functions which are more time consuming.

Let us now turn our attention to multicomponent systems. Here, even for ideal gas mixtures, an exact analytic solution á la Krishna–Standart film model is not available and the practical approach is to use Toor–Stewart–Prober approximation of constant $[D_{av}]$. With the assumption of constant $[D_{av}]$, the general multicomponent solution can be obtained as matrix analogs of the binary solution given above; see Toor [4], Stewart and Prober [5], Krishna and Standart [6] and Taylor [89] for details of the derivations. The final results are summarized below.

The matrix of time-averaged zero-flux mass transfer coefficients are given for the Higbie model as

$$[k] = 2[D_{av}]^{1/2}/(\pi t_e)^{1/2} \tag{337}$$

and the matrix of correction factors $[\Xi]$ has the form

$$[\Xi] \equiv \left\{1 + \operatorname{erf}\frac{[\Phi]}{\sqrt{\pi}}\right\}^{-1}\left\{\exp\left[\frac{-[\Phi]^2}{\pi}\right]\right\} \tag{338}$$

$$\text{where } [\Phi] \equiv \frac{N_t}{c_t}[k]^{-1} \tag{339}$$

For a ternary system, Sylvester's expansion formula offers a convenient means of evaluating $[k]$ and $[\Xi]$. Thus we have, explicitly,

$$[k] = \frac{(\hat{k}_1 - \hat{k}_2)[D] - (\hat{k}_1\hat{D}_2 - \hat{k}_2\hat{D}_1)[I]}{\hat{D}_1 - \hat{D}_2} \tag{340}$$

where $\hat{D}_i$ $(i = 1, 2)$ are the eigenvalues of $[D_{av}]$ and where

$$\hat{k}_i \equiv 2\hat{D}_i^{1/2}/(\pi t_e)^{1/2}, \quad i = 1, 2, \ldots, n \tag{341}$$

are the eigenvalues of $[k]$. The matrix of high flux mass transfer coefficients $[k^\bullet] = [k][\Xi]$ may be calculated from Equation 340 with $[k^\bullet]$ replacing $[k]$ and $\hat{k}_i^\bullet$ replacing $\hat{k}_i$ where

$$\hat{k}_i^\bullet = \hat{k}_i\hat{\Xi}_i \tag{342}$$

are the eigenvalues of $\hat{k}_i^\bullet$ and where

$$\hat{\Xi}_i = \frac{\exp\left(-\dfrac{\hat{\Phi}_i^2}{\pi}\right)}{1 + \operatorname{erf}\dfrac{\hat{\Phi}_i}{\sqrt{\pi}}}, \quad i = 1, 2, \ldots, n \tag{343}$$

$$\hat{\Phi}_i = N_t/c_t\hat{k}_i \tag{344}$$

are the eigenvalues of the correction factor matrix $[\Xi]$ and the rate factor matrix $[\Phi]$ respectively. $[k^\bullet]$ may also be obtained from Equation 223 with $k_i^\bullet$ calculated from Equations 341 through 344.

With these equations for $[k]$ and $[\Xi]$, the diffusion fluxes at the interface ($y = 0$) are obtained from

$$(J_0) = c_t[k]\,[\Xi]\,(x_I - x_b) = c_t[k^\bullet]\,(x_I - x_b) \tag{345}$$

and the molar fluxes at the interface follow from

$$(N_0) = c_t[\beta_I]\,[k^\bullet]\,(x_I - x_b) \tag{346}$$

Notice that it is not useful to derive a similar expression for the bulk fluid ($y \to \infty$) where all of the composition gradients vanish.

Since only N_t is required in order that $[k^\bullet]$ be calculable, the method of Vickery et al. [90], described in the section on film theory, may be employed to compute the fluxes from Equation 345. With the diffusion fluxes given by Equation 345 and $[k^\bullet]$ given as described in the sentence following Equation 344 inserted into Equation 228, we may easily rederive Equations 230 and 232, but now with

$$\hat{f}_j = \hat{\Xi}_j/\hat{\Phi}_j = \frac{1}{\hat{\Phi}_j}\,\frac{\exp(-\hat{\Phi}_j^2/\pi)}{1+\operatorname{erf}(\hat{\Phi}_j/\sqrt{\pi})} \tag{347}$$

The first derivative $d\hat{f}_j/dN_t$ may be derived from Equation 347.

The calculation of $[k]$ using Equation 337 requires a priori estimation of the exposure time t_e or the surface renewal rate s. In some cases, this is possible. For bubbles rising in a liquid the exposure time is the time the bubble takes to rise its own diameter; in other words the jacket of the bubble is renewed every time it moves a diameter. If we consider the flow of a liquid over a packing, when the liquid film is mixed at the junction between the packing elements, then t_e is the time for the liquid to flow over the packing elements. For flow of liquid in laminar jets and in thin films, the exposure time is known but in these cases it may be important to take into account the distribution of velocities along the interface; in the penetration model, this velocity profile is assumed to be flat (i.e. plug flow). For gas-liquid mass transfer in stirred vessels, the renewal frequency in the Danckwerts model, s, may be related to the speed of rotation [22].

One very important restriction in the development of the surface renewal models is the assumption implicit in the boundary condition Equation 329, which assumes that the penetrating, or diffusing, component does not "see" the bulk fluid, which to all intents and purposes is located at an infinite distance from the interface. This assumption is strictly true for short Fourier times

$$Fo \equiv \frac{Dt}{4\delta^2} < 0.05 \tag{348}$$

where δ is the distance from the interface to the "core" of the fluid phase. For example, if we consider diffusion inside rigid droplets, the distance δ corresponds to the radius of the drop.

For long contact times and/or short distances between the interface and the "core," δ, the solution just given for the zero-flux coefficient does not apply. This situation may arise for mass transfer inside very small gas bubbles which stay for a long time in contact with the surrounding liquid phase, as might be the case for gas-liquid dispersions in a distillation or absorption tray column. For such long contact times, the diffusing species will penetrate deep into the heart of the bubble (or drop), and it is important in such cases to define the mass transfer coefficient in terms of the driving forces $\Delta x_i = x_{iI} - \langle x_i \rangle$ where x_{iI} represents the interface composition and $\langle x_i \rangle$ is the cup-mixing composition of the spherical dispersed phase. For a binary system, under conditions of small mass transfer fluxes, the unsteady-state diffusion equations may be solved to give the fractional approach to equilibrium (see Clift, Grace and Weber [109])

$$\frac{x_{1,t=0} - \langle x_1 \rangle}{x_{1,t=0} - x_{1I}} = 1 - \frac{6}{\pi^2} \sum_{n=1}^{\infty} \frac{1}{n^2} \exp(-n^2\pi^2 \, Fo) \tag{349}$$

where $x_{1,t=0}$ = the initial composition within the particle
x_{1I} = the composition at the surface of the particle (held constant for the duration of the diffusion process)
Fo = the Fourier number, Dt/r_0^2
r_0 = the radius of the particle (bubble or droplet)

Clearly when $t \rightarrow \infty$, equilibrium is attained, and the averaged, cup-mixing composition $\langle x_i \rangle$ will equal the surface composition x_{1I}. The Sherwood number at time t defined by taking the driving force to be $x_{1I} - \langle x_1 \rangle$, is [109]:

$$\frac{k \cdot 2r_0}{D} \equiv Sh = \frac{2\pi^2}{3} \frac{\sum_{n=1}^{\infty} \exp(-n^2\pi^2 \, Fo)}{\sum_{n=1}^{\infty} \frac{1}{n^2} \exp(-n^2\pi^2 \, Fo)} \tag{350}$$

Figure 16 shows the variation of Sh with Fo. For large values of Fo, Sh approaches the asymptotic value $\frac{2\pi^2}{3} = 6.58$. For this steady-state limit, the zero-flux mass transfer coefficient is

$$k = \frac{\pi^2}{3} \frac{D}{r_0} \tag{351}$$

showing, as for the film model discussed earlier, a unity-power dependence on the Fickian coefficient D. This is to be contrasted with the square-root dependence for small values of Fo; see the variation in the curvature in Figure 16.

This binary analysis can be extended to multicomponent systems by using the Toor–Stewart–Prober approximation of constant $[D_{av}]$. The fractional approach to equilibrium is given by the $n-1$ dimensional matrix analog of Equation 349:

$$(x_{t=0} - \langle x \rangle) = \left[[I] - \frac{6}{\pi^2} \sum_{n=1}^{\infty} \frac{1}{n^2} \exp[-n^2\pi^2 [D'] \, Fo] \right] (x_{t=0} - x_I) \tag{352}$$

where the Fourier number Fo is defined as $D_{ref}t/r_0^2$; D_{ref} is an arbitrary reference diffusivity; [D′]is the normalized Fickian matrix: $[D'] = (1/D_{ref})[D_{av}]$

$$[Sh] \equiv [k] \cdot 2r_0 [D_{av}]^{-1} =$$

$$\frac{2\pi^2}{3} \left[\sum_{n=1}^{\infty} \exp[-n^2\pi^2 [D'] \, Fo] \right] \left[\sum_{n=1}^{\infty} \frac{1}{n^2} \exp[-n^2\pi^2 [D'] \, Fo] \right]^{-1} \tag{353}$$

In the limit Fo → ∞, [Sh] approaches the asymptotic limit

$$[\mathrm{Sh}] = \frac{2\pi^2}{3}[\mathrm{I}] \tag{354}$$

or

$$[\mathrm{k}] = \frac{\pi^2}{3}\frac{[\mathrm{D_{av}}]}{r_0} \tag{355}$$

A very interesting difference between the short-contact-time value: $[k] = 2[D_{av}]^{1/2}/(\pi t_e)$ and the long-contact-time steady-state value: $[k] = (\pi^2/3r_0)[D_{av}]$ is the variation in the influence of diffusional coupling. The influence of molecular diffusional coupling will be maximum when [k] is proportional to [D] as is the case for the long-time asymptote. This influence is reduced as the contact time is reduced, and it is at its lowest for the short-contact-time square-root dependence. We shall illustrate this by considering mass transfer inside a droplet containing a mixture of acetone(1)-benzene(2)-methanol(3). For this system the elements of [D], the Fickian matrix of diffusion coefficients at 25° C have been measured by Alimadadian and Colver [110]. The experimental data is given in Table 7. Assume the droplet is 5 mm in diameter and initially has the composition $x_1 = 0.52$, $x_2 = 0.28$; $x_3 = 0.2$. Suddenly, at t = 0, the droplet is brought into contact with a surrounding vapor phase which maintains the surface of the droplet at the fixed composition. We shall investigate the influence of the contact time between vapor and liquid (say in the spray regime on a distillation tray) on the ratios k_{12}/k_{11} and k_{21}/k_{11} for the elements of the zero-flux matrix [k].

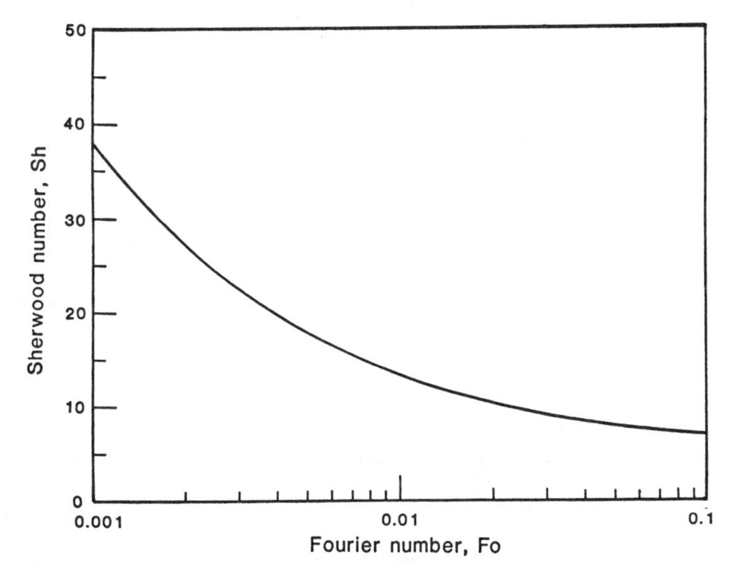

Figure 16. Sherwood number for mass transfer within a rigid spherical particle of radius r_0 as a function of the Fourier number, $Fo = D\dagger/r_0^2$.

Table 7
Molecular Diffusion Coefficients for the system acetone (1)—benzene (2)—methanol (3) at 25°C

Mole Fraction		Elements of the Fickian Diffusion Coefficient matrix [D], $\mu m^2 s^{-1}$			
x_1	x_2	D_{11}	D_{12}	D_{21}	D_{22}
0.350	0.302	3819	420	−561	2133
0.766	0.114	4400	921	−834	2680
0.553	0.193	4472	962	−480	2569
0.400	0.500	4434	1866	−816	1668
0.299	0.150	3192	277	−191	2368
0.206	0.548	3513	665	−602	1948
0.102	0.795	3502	1204	−1130	1124
0.120	0.132	3115	138	−227	2235
0.150	0.298	3050	150	−269	2250

* *From Alimadadian and Colver [110].*

The first step is to estimate $[D_{av}]$ representing some average over the time of the diffusion process. From the experimental data in Table 7 we see that

$$[D_{av}] = \begin{bmatrix} 4{,}280 & 1{,}040 \\ -670 & 2{,}260 \end{bmatrix} \quad \mu m^2 s^{-1}$$

is a reasonably good representation for the composition range under consideration.

For short-contact times, $[k] \approx [D_{av}]^{1/2}$ and using Sylvester's theorem we find $k_{12}/k_{11} = 0.139$; $k_{21}/k_{22} = -0.122$. In the other extreme of long-contact times, the ratio $k_{12}/k_{11} = 0.243$ and $k_{21}k_{22} = -0.295$. The influence of contact time between the liquid droplet and the surrounding gas phase is illustrated in Figure 17 where k_{12}/k_{11} and $-k_{21}/k_{11}$ are plotted against Fo; the reference diffusivity D_{ref} is taken to be D_{11} in the calculations. The consequences of this variation of the ratios of cross to diagonal coefficients could be dramatic in some cases. Let us assume, arbitrarily, that the ratio of driving forces of acetone to the driving force for benzene is

$$\frac{\Delta x_1}{\Delta x_2} = -0.2$$

Under these conditions, we observe that under conditions of low transfer fluxes for the steady-state asymptotic limit

$$\frac{k_{12}\Delta x_2}{k_{11}\Delta x_1} = -1.215$$

or, in other words, acetone will experience reverse mass transfer; while in the other limit of short-contact times

$$\frac{k_{12}\Delta x_2}{k_{11}\Delta x_1} = -0.695$$

we see that though coupling effects are still significant, they are insufficient to drive acetone against its driving force. Thus, the effect of contact time could be to alter the *direction* of mass transfer. This effect was first pointed out by Krishna [111, 112] who showed that the film and penetration times could predict different directions of mass transfer. As we shall see later on in this review, one of the important consequences of coupling effects in multicomponent mass transfer is that component Murphree efficiencies in multicomponent distillation can be unequal to one another and become unbounded (i.e., assume values < 0

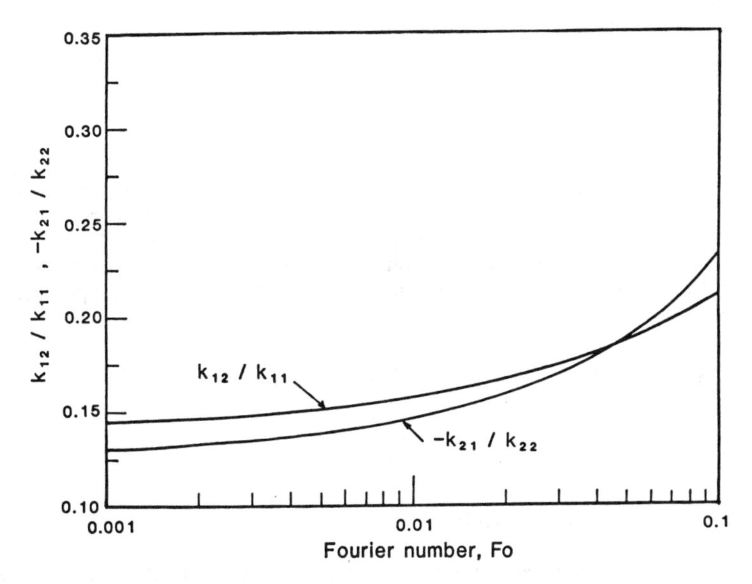

Figure 17. Ratios of k_{12}/k_{11} and $-k_{21}/k_{22}$ for transfer inside a spherical rigid droplet containing a mixture of acetone (1), benzene (2), and methanol (3). Variation with the Fourier number, $Fo = D_{ref}t/r_0^2$. D_{ref} is taken equal to D_{11} in the calculations.

or > 1). From the preceding calculations, it should be clear that this coupling effect will be influenced not only by the structure of the Fickian matrix $[D_{av}]$ but also by the hydrodynamics of two-phase contacting, which factor will influence *inter alia* the contact time between the phases. Apart from the contact time, the distribution of sizes of droplets, or bubbles, will also affect the extent of coupled mass transport. Thus very small bubbles (or drops) may approach the steady-state limit (largest influence of coupling) while larger bubbles will transfer mass in the "short-contact regime" (least influence of coupling). Another influence of phase hydrodynamics is that when the size of the bubble (or droplet) exceeds a certain limit the dispersed phase may begin to circulate or oscillate [109].

The Kronig–Brink [113] model for circulation within the dispersed phase can be generalized to n-component systems to give the following expression for the fractional approach to equilibrium

$$(x_{t=0} - \langle x \rangle) = \left[[I] - \frac{3}{8} \sum_{n=1}^{\infty} A_n^2 \exp\{-16\lambda_n [D'] \,Fo\} \right] (x_{t=0} - x_I) \tag{356}$$

and the matrix of Sherwood numbers

$$[Sh] \equiv [k] \cdot 2r_0 [D_{av}]^{-1} =$$

$$\frac{32}{3} \left[\sum_{n=1}^{\infty} A_n^2 \lambda_n^2 \exp\{-16\lambda_n [D'] \,Fo\} \right] \left[\sum_{n=1}^{\infty} A_n^2 \exp\{-16\lambda_n [D'] \,Fo\} \right]^{-1} \tag{357}$$

The eigenvalues A_n and λ_n have been tabulated by Sideman and Shabtai [114]. For Fo → ∞, the asymptotic value of [Sh] is

$$[Sh] = \frac{32}{3}\lambda_1[I] = 17.66[I] \tag{358}$$

which is 2.7 times the corresponding limit for noncirculating particles, demonstrating the enormous influence of the system hydrodynamics on the mass transfer behavior.

Another situation which is of practical importance is radial, unsteady, diffusion in a cylindrical geometry. This provides a simple physical model for gas issuing from holes on a distillation tray and is representative of the formation zone [115]. The approach to equilibrium is given by

$$(x_{t=0} - \langle x \rangle) = \left[[I] - 4\sum_{n=1}^{\infty} \frac{\exp[-\lambda_n^2 r_0^2 [D'] \, Fo]}{\lambda_n^2 r_0^2}\right](x_{t=0} - x_I) \tag{359}$$

where λ_n are the roots of the zero-order Bessel function $J_0(\lambda_n r_0) = 0$.

The Penetration Model for Heat Transfer

Our discussions on the film model for heat transfer showed an exact parallel with the corresponding mass transfer problem. The same parallel holds for unsteady-state transfer. Thus for $Fo \equiv \lambda t/(\varrho C_p r_0^2) \to 0$, (short—contact—time limit), the time-averaged heat transfer coefficient is given by

$$h = 2\varrho C_p \sqrt{\frac{\lambda}{\varrho C_p \pi t_e}} \tag{360}$$

and the asymptotic, steady-state limit for heat transfer within a rigid spherical body is

$$Nu \equiv \frac{h 2 r_0}{\lambda} = \frac{2\pi^2}{3} \quad \text{or} \quad h = \frac{\pi^2 \lambda}{3 r_0} \tag{361}$$

with instantaneous value of Nu given by Equation 251 in which we use $\lambda/\varrho C_p$ in place of the diffusion coefficient D.

The correction factor for high fluxes is given by Equation 335 in which we use the heat transfer rate factor

$$\Phi_H \equiv \frac{\sum_{i=1}^{n} N_i C_{pi}}{h} \tag{362}$$

in place of the mass transfer rate factor Φ. For chemical engineering design purposes, we recommend the use of the film theory correction factor given by Equation 301.

The Penetration Model for Coupled Heat and Multicomponent Mass Transfer

A penetration-type model has been used by Delancey and Chiang [116–119] to evaluate the significance of the Dufour and Soret effects in multicomponent gas asorption. While circumstances do exist when the Dufour and Soret contributions are a significant part of the interfacial heat and mass fluxes, it is doubtful that these effects are important in the vast majority of situations encountered in practice.

Turbulent Eddy Diffusivity Models

For simple flow situations such as flow over flat plates and inside circular tubes, sufficient information is available concerning velocity profiles and turbulent eddy diffusivities to allow calculation of the mass and heat transfer coefficients and fluxes between the "wall" and the flowing stream. For condensation of mixed vapor mixtures flowing inside a vertical tube, for example, the "wall" can be considered to be the liquid condensate film. We examine first the problem of determining the mass transfer coefficients and mass fluxes and take up later the discussion on the analogous problem of heat transfer.

For definiteness, we consider the transfer processes between a cylindrical wall and a turbulently flowing n-component fluid mixture. We examine the phenomena occurring at any axial position in the tube, assuming that fully developed flow conditions are attained. For steady-state conditions, the equations of continuity of mass of component i (assuming no chemical reactions) takes the form

$$\frac{d}{dr}(rn_{ir}) = 0 \tag{363}$$

where r represents the radial coordinate and n_{ir} is the mass flux of component i; we choose to describe the mass transfer process using mass fluxes and the mass average reference velocity frame because of the need later on to solve the equations of continuity of mass in conjunction with the equations of motion, which are most conveniently expressed in the mass average frame. Equation 363 shows that rn_{ir} is r-invariant. The description of the turbulent transport processes is much more conveniently carried out in terms of a coordinate system measuring the distance from the wall, y:

$$y = R_w - r \tag{364}$$

where R_w represents the radius of the circular tube. We consider n_{iy} to be positive if the flux is directed in the positive y-direction, i.e. from the wall towards the flowing fluid mixture. The mass flux n_{iy} can be written in terms of the diffusive mass flux, j_{iy}, with respect to the mass average mixture velocity, and a bulk-flow contribution, $\omega_i n_{ty}$:

$$n_{iy} = j_{iy} + \omega_i n_{ty}, \quad i = 1, 2, \ldots, n \tag{365}$$

where n_{ty} is the total mixture mass flux. Since the conditions inside the tube are considered to be turbulent, we use time-smoothed fluxes and compositions in Equation 365. The constitutive relation for j_{iy}, taking account of the molecular diffusion and turbulent eddy contributions, is

$$(j_y) = -\varrho_t \{[D] + [D_{turb}]\} \frac{d(\omega)}{dy} \tag{366}$$

where [D] is the matrix of Fickian diffusion coefficients in the mass average frame and $[D_{turb}]$ is the turbulent eddy diffusivity of mass. Since eddy mass transport is not species specific, $[D_{turb}]$ reduces to the form of a scalar times the identity matrix [88, 119].

$$[D_{turb}] = D_{turb}[I] \tag{367}$$

In proceeding further, it is convenient to define the following parameters and variables which incorporate information on the flow:

- Friction velocity u^*

$$u^* = (\tau_0/\rho_t)^{1/2} = (f/2)^{1/2}\bar{u} \tag{368}$$

- A dimensionless distance from the wall

$$y^+ = yu^*\varrho_t/\mu = yu^*/\nu \tag{369}$$

- A dimensionless velocity

$$u^+ = u/u^* = (2/f)^{1/2}u/\bar{u} \tag{370}$$

- A dimensionless tube radius

$$R_W^+ = R_W u^*/\nu = \text{Re}\,(f/8)^{1/2} \tag{371}$$

where Re is the Reynolds number for flow, $2R_W\bar{u}/\nu$.

- A matrix of Schmidt numbers [Sc]

$$[\text{Sc}] = \nu[D]^{-1} \tag{372}$$

- The turbulent Schmidt number

$$\text{Sc}_{turb} = \nu_{turb}/D_{turb} \tag{373}$$

where ν_{turb} is the turbulent eddy kinematic viscosity.

With these definitions, we may write the following expression for the mass flux at the wall, n_{i0}:

$$(n_0) = -\varrho_t u^* \left\{[\text{Sc}]^{-1} + \text{Sc}_{turb}^{-1}\frac{\nu_{turb}}{\nu}[I]\right\}(1 - y^+/R_W^+)\frac{d(\omega)}{dy^+} + (\omega)n_{t0} \tag{374}$$

The ratio y^+/R_W^+ is negligibly small for problems in which the major concern is the estimation of heat and mass transfer rates. This can be seen from the fact that most of the resistance to mass and heat transfer is concentrated in a thin zone of thickness $y^+ = 5$. The term R_W^+ has a value exceeding 300 for Re = 10,000. We shall, therefore, approximate the term $(1 - y^+/R_W^+)$ by unity in the ensuing analysis, though its incorporation into the theoretical development does not pose insurmountable problems.

Let us proceed with the integration of Equation 374 in order to obtain the composition profiles and fluxes. In solving the matrix differential Equation 374 it is convenient to define the matrix integral:

$$[\psi] \equiv \int_0^{y+} \frac{n_{t0}}{\varrho_t u^*}\left\{[\text{Sc}]^{-1} + \text{Sc}_{turb}^{-1}\frac{\nu_{turb}}{\nu}[I]\right\}^{-1} dy^+ \tag{375}$$

which is position dependent, i.e., on the value of the upper limit y^+ in Equation 325. Taking this upper limit to be y_b^+, the reduced distance from the wall at which the bulk compositions (ω_b) are attained, we define

$$[\Phi] \equiv \int_0^{y_b^+} \frac{n_{t0}}{\varrho_t u^*} \left\{[Sc]^{-1} + Sc_{turb}^{-1} \frac{\nu_{turb}}{\nu} [I]\right\}^{-1} dy^+ \tag{376}$$

where τ_0 = the shear stress at the interface or wall
$\bar{u}$ = the average velocity of flow of the multicomponent fluid mixture inside the tube
f = the Fanning friction factor

Comparison with the previous film theory development shows that $[\Phi]$ defined by Equation 376, reduces to $[\Phi]$ defined in Equation 207, when the transfer process is assumed to be purely molecular in nature. The matrix $[\psi]$ assumes the role of $[\Phi]\eta$ used earlier Equation 210.

When the integrand is assumed to be a function of position only and not of composition, an analytical solution to Equation 374 may be obtained. For the elements of [Sc], this means that suitably composition-averaged values must be used (equivalent to the Toor–Stewart–Prober assumption of constant [D]). The turbulent diffusivities of mass and momentum, D_{turb} and ν_{turb}, are strong functions of position. They have their maximum values in the turbulent core of the fluid and reduce to zero at the wall. The discussion of particular models for turbulent transport will be taken up later. Using the method of successive substitution, Equation 374 may be solved for constant [Sc] to yield the composition profile (see Taylor [120] and Krishna [121], for mathematical details)

$$(\omega - \omega_0) = \{\exp[\psi] - [I]\}\{\exp[\Phi] - [I]\}^{-1}(\omega_b - \omega_0) \tag{377}$$

where (ω_0) and (ω_b) are the compositions at the wall $(y^+ = 0)$ and in the bulk fluid $(y^+ = y_b^+)$. The composition profile may be differentiated at $y^+ = 0$ to yield the diffusion fluxes j_{i0}

$$(j_0) = n_{t0}\{\exp[\Phi] - [I]\}^{-1}(\omega_0 - \omega_b) \tag{378}$$

Equation 378 shows that, on comparison with the definition of $[k^\bullet]$,

$$(j_0) = \varrho_t[k^\bullet](\omega_0 - \omega_b), \tag{379}$$

where

$$[k^\bullet]^{-1} = \{\exp[\Phi] - [I]\}\frac{\varrho_t}{n_{t0}} \tag{380}$$

whose limit for vanishing total mixture flux gives the inverted zero flux mass transfer coefficients

$$[k]^{-1} = \frac{\varrho_t}{n_{t0}}[\Phi] \tag{381}$$

which on combination with Equation 380 leads to the following expression for the matrix of correction factors

$$[\Xi] = [\Phi]\{\exp[\Phi] - [I]\}^{-1} \tag{382}$$

which is formally identical to the corresponding film theory result, Equation 215. This result strengthens our recommendation that the "film theory" correction factor be used for all cases involving finite mass transfer rates; we shall later see an analogous result for heat transfer.

By introducing the appropriate "bootstrap" matrix $[\beta_0]$, the final expression for the calculation of the interfacial mass fluxes is

$$(n_0) = \varrho_t[\beta_0]\,[k]\,[\Xi]\,(\omega_0 - \omega_b) \tag{383}$$

Having completed the formal development of an expression for the evaluation of the interfacial fluxes n_{i0}, we turn to the actual evaluation of the elements of the matrix of zero-flux mass transfer coefficients [k] for some specific models for turbulence. It is usual in such developments to define a matrix of (zero-flux) Stanton numbers

$$[St] = [k]/\bar{u} = (f/2)^{1/2}\,[k]/u^* \tag{384}$$

We see from Equations 376, 381 and 384 that the inverted matrix of Stanton numbers is given by the following expression

$$[St]^{-1} = \left(\frac{2}{f}\right)^{1/2} \int_0^{y_b^+} \left\{[Sc]^{-1} + Sc_{turb}^{-1}\frac{\nu_{turb}}{\nu}[I]\right\}^{-1} dy^+ \tag{385}$$

The film theory discussed earlier falls neatly into the framework provided by Equation 385. Thus if we assume that the mass transfer process is governed by molecular transport within an "effective" film of thickness y_b^+ and that the level of turbulence is such as to wash out completely all composition gradients beyond this distance, then we see that

$$[St]^{-1} = \left(\frac{2}{f}\right)^{1/2} [Sc]y_b^+ = \bar{u}[D]^{-1}y_b \tag{386}$$

where we have reintroduced the definition $y_b^+ = y_b(f/2)^{1/2}\bar{u}/\nu$. The matrix of zero-flux mass transfer coefficients $[k] = [D]/y_b$, the classic film theory result.

In general, the integral in Equation 385 is a function of y_b^+, [Sc], Sc_{turb} and the ratio ν_{turb}/ν. We now consider some particular models for turbulence which will aid the estimation of [St].

The first difficulty we encounter is that the position y_b^+, at which the bulk fluid phase compostitions ω_{ib} are reached, is not known precisely. To overcome this shortcoming in our knowledge, we proceed to divide the region $0 - y_b^+$ as follows:

- A region $0 - y_I^+$ in which both molecular and turbulent eddy contributions to mass transfer are important
- $y_I^+ - y_b^+$ in which region, the contribution of turbulent eddy transport is much larger than the molecular contribution and we have the following simplication in this region

$$\frac{\nu_{turb}}{\nu} \gg 1 \tag{387}$$

We thus write Equation 385 as a sum of two integrals

$$[St]^{-1} = \left(\frac{2}{f}\right)^{1/2} \int_0^{y_1^+} \left[[Sc]^{-1} + Sc_{turb}^{-1} \frac{\nu_{turb}}{\nu} [I] \right]^{-1} dy^+$$

$$+ \left(\frac{2}{f}\right)^{1/2} \int_{y_1^+}^{y_b^+} \left[\frac{\nu_{turb}}{\nu} [I] \right]^{-1} dy^+ \qquad (388)$$

We now turn our attention to the estimation of the ratio ν_{turb}/ν. Our starting point here is the equation of motion which for steady-state conditions gives the shear stress profile as (see for examples Bird, Stewart and Lightfoot [21])

$$(\tau_r + \tau_{r,turb}) = \frac{r}{R_W} \qquad (389)$$

Transforming to the y-coordinate system and introducing the constitutive relations for the shear stress due to the molecular contribution ($\tau_y = \mu du/dy$) and due to the turbulent eddy contribution ($\tau_{turb} = -\mu_{turb}\, du/dy$) we can write Equation 389 in terms of reduced parameters as follows

$$\left(1 + \frac{\nu_{turb}}{\nu}\right) \frac{du^+}{dy^+} = (1 - y^+/R_W^+) \qquad (390)$$

For the reasons discussed earlier, we shall assume that the right member of Equation 390 is unity. With this simplification, we obtain the following expression for ν_{turb}/ν:

$$\frac{\nu_{turb}}{\nu} = \frac{1}{du^+/dy^+} - 1 \qquad (391)$$

which can be estimated if the functional form $u^+(y^+)$ is known. For the von Karman turbulent velocity profile, for example, we have

$$\frac{\nu_{turb}}{\nu} = 0, \quad 0 < y^+ \leq 5, \qquad (392)$$

for the viscous sublayer and

$$\frac{\nu_{turb}}{\nu} = \frac{y^+}{5} - 1, \quad 5 < y^+ \leqq 30, \qquad (393)$$

for the buffer zone in which both molecular and turbulent contributions play a role. y_1^+ is taken to be 30 in the von Karman development; beyond this distance from the wall the transport is purely turbulent. Various other models for $u^+(y^+)$ are available; see the review of Sideman and Pinczewski [45].

For another class of models, the mixing length models, an alternative approach is used. Here the turbulent eddy viscosity is assumed to be of the form

$$\nu_{turb} = \ell^2 \left| \frac{du}{dy} \right| \qquad (394)$$

where u is the velocity in the direction of mean flow. ℓ is known as the mixing length and is analogous to the mean free path in the kinetic theory of gases. The physical interpretation of ℓ is that it is the distance over which a turbulent eddy retains its identity. The absolute value of du/dy is required in Equation 394 to ensure that the shear stress changes when the flow field changes direction. The problem of estimating ν_{turb} now rests with a method for calculating the mixing length ℓ. From physical considerations, ℓ must be a function of distance from the wall. The simplest model for ℓ is to take it to be proportional to distance from the wall, as hypothesized by Prandtl:

$$\ell_+ = \kappa y^+ \tag{395}$$

where we have, additionally, defined a reduced mixing length

$$\ell_+ = \ell u^*/\nu \tag{396}$$

The constant κ in Equation 395 is the von Karman constant, equal to 0.4. Though the Prandtl mixing length hypothesis Equation 396 works for conditions in the turbulent core (y^+ greater than about 30), it greatly overestimates the values of ℓ_+ closer to the wall, where the solid surface hinders the mixing mechanisms. An important modification to the Prandtl development was added by van Driest [122], in 1956, who introduced a damping factor:

$$\ell_+ = \kappa y^+(1-\exp(-y^+/A^+)) \tag{397}$$

A^+ is a damping length constant which is interpreted as the distance from the wall beyond which viscous effects are negligible; in the notation used here, A^+ corresponds to y_I^+. Von Driest empirically determined the value of $A^+ = y_I^+ = 26$.

Let us now see how the mixing length models can be used to estimate ν_{turb}/ν. Introducing the definition from Equation 394 into Equation 390 we find a quadratic expression for du^+/dy^+

$$\ell_+^2\left(\frac{du^+}{dy^+}\right)^2 + \frac{du^+}{dy^+} - 1 = 0 \tag{398}$$

and so du^+/dy^+ is obtained explicitly as

$$\frac{du^+}{dy^+} = \frac{-1+\sqrt{1+4\ell_+^2}}{2\ell_+^2} \tag{399}$$

which equation, in combination with Equation 391, allows estimation of ν_{turb}/ν.

Consider now the evaluation of the second integral on the right-hand side of Equation 388. We integrate Equation 390 as follows

$$\int_0^{y_b^+} du^+ = u_b^+ = \int_0^{y_I^+} \frac{dy^+}{1+\dfrac{\nu_{turb}}{\nu}} + \int_{y_I^+}^{y_b^+} \frac{dy^+}{\dfrac{\nu_{turb}}{\nu}} \tag{400}$$

This equation can be used for a second integral on the right-hand side of Equation 388. With this substitution and using $u_b^+ = (2/f)^{1/2}$, the reduced bulk flow velocity, which will closely equal average flow velocity $\bar{u}$, we obtain the final working expression for the estimation of [St]:

$$[St]^{-1} = \left(\frac{2}{f}\right)[I] + \left(\frac{2}{f}\right)^{1/2} \int_0^{1^+} \left[\left\{ [Sc]^{-1} + Sc_{turb}^{-1} \frac{\nu_{turb}}{\nu} [I] \right\}^{-1} - \left\{ [I] + \frac{\nu_{turb}}{\nu} [I] \right\}^{-1} \right] dy^+ \tag{401}$$

The only parameter which awaits discussion is Sc_{turb}, the turbulent Schmidt number. It is usual to assert the equality $Sc_{turb} = Pr_{turb}$, the turbulent Prandtl number:

$$Pr_{turb} = \frac{C_p \mu_{turb}}{\lambda_{turb}} \tag{402}$$

where λ_{turb} is the turbulent eddy thermal conductivity. Experimental values of Pr_{turb} show a marked dependence on the molecular Prandtl number and vary between 0.2 and 2.5 (see the compilation in Blom's thesis [46]). Not surprisingly, there are a variety of models available for Pr_{turb} (see Sideman and Pinczkewski [45]) but a choice of the "best" model cannot be easily made, and for most practical design purposes we are forced to assume $Pr_{turb} = 1$, for want of more reliable information.

Some special cases may be derived for the general expression for [St], Equation 401. When $Sc_{turb} = 1$ and the matrix of Schmidt number is assumed to be equal to the identity matrix, i.e., [Sc] = [I], then we have the Reynolds analogy for multicomponent mass transfer:

$$[St] = \frac{f}{2}[I] \tag{403}$$

The requirement [Sc] = [I] for a multicomponent system is a much more special case than for a corresponding binary system for it requires that all binary pair diffusivities in the multicomponent system be equal to one another and further that $\nu/D = 1$, a situation realizable only for ideal gas mixtures made up of species of similar size and nature.

With the von Karman turbulent velocity profile, Equations 392, 393 and 401 can be combined to yield

$$[St]^{-1} = \frac{2}{f}[I] + 5\left(\frac{2}{f}\right)^{1/2} \left[[Sc] - [I] + \ln\left\{ [I] + \frac{5}{6}[Sc - I] \right\} \right] \tag{404}$$

Inserting a mixing length model (for example the one due to van Driest, Equation 397 for ℓ_+ into Equation 399 and combining with Equations 391 and 401 allows calculation of [St]; in this case, however, numerical integration is required but the calculations are straightforward.

The numerical computation of the fluxes is best carried out by yet another application of the general method of Vickery et al [90], developed earlier for the film model equations of Stewart and Prober [5]. As with that simpler model, the turbulent eddy diffusivity model involves the assumption of a constant matrix of multicomponent diffusion coefficients, [D] (relative to the mass average velocity in this section, relative to the molar average velocity shown earlier). It can be shown that a similarity transformation similar to Equation 227 will uncouple the matrix differential Equation 374. Thus the matrix of mass transfer coefficients $[k^\bullet] = [k][\Xi] = \bar{u}[St][\Xi]$ can be written in a form identical to Equation 223; only the particular expression for the eigenvalues of [k] would be different. By paralleling the development of Equations 228 through 235 (simply replace molar quantities by the corresponding mass quantities), we may easily derive the analogue of Equation 230. The resulting expression, a function only of n_{t0}, is easily solved using Newton's method. The required computations are summarized in the following flow chart. (See Figure 13.)

Algorithm for Calculating the Mass Transfer Rates from Turbulent Eddy Diffusivity Models

Given: (ω_b), (ω_0), $\bar{u}$, $f = f(Re)$, μ, ρ, [D] in the mass frame of reference, (Equation 67).
Find: (n_0), n_{t0}, [k], $[k^\bullet]$.

Step 1. Calculate the eigenvalues, $\hat{D}_i$, of [D]. Compute a modal matrix [P] and its inverse $[P]^{-1}$

Step 2. Calculate $[P]^{-1}(\omega_0 - \omega_b) = (\Delta\hat{\omega})$.

Step 3. Calculate the eigenvalues, $\hat{k}_i$ of [k] from

$$\hat{k}_i = \bar{u}\widehat{St}_i$$

where $\widehat{St}_i$ are the eigenvalues of [St] estimated from, for example:

$$\frac{1}{\widehat{St}_i} = \frac{2}{f} + 5\left(\frac{2}{f}\right)^{1/2}\left\{\widehat{Sc}_i - 1 + \ln\left(1 + \frac{5}{6}(\widehat{Sc}_i - 1)\right)\right\}$$

where $\widehat{Sc}_i = \mu/\rho\hat{D}_i$

Step 4. Set the iteration counter (k) to zero. Estimate $n_{t0}^{(k)}$.

Step 5. Calculate $\hat{\Phi}_i = n_{t0}/\varrho_t\hat{k}_i$; $\hat{\Xi}_i = \hat{\Phi}_i/(e^{\hat{\Phi}_i} - 1)$; $\hat{f}_i = \hat{\Xi}_i/\hat{\Phi}_i$

Step 6. Calculate the function $F(n_{t0})$

$$F(n_{t0}) \equiv \sum_{i=1}^{n-1}\left(1 - \frac{v_i}{v_n}\right)\left\{\sum_{j=1}^{n-1} P_{ij}\hat{f}_j\Delta\hat{\omega}_j + \hat{\omega}_{i0}\right\} - 1$$

Step 7. Calculate the derivative

$$\frac{dF}{dn_{t0}} = -\sum_{i=1}^{n-1}\left(1 - \frac{v_i}{v_n}\right)\left\{\sum_{j=1}^{n-1} P_{ij}\frac{d\hat{f}_i}{dn_{t0}}\Delta\hat{\omega}_j\right\}$$

where $\dfrac{d\hat{f}_i}{dn_{t0}} = -\dfrac{\hat{\Xi}_j e^{\hat{\Phi}_j}}{n_{t0}}$

Step 8. Make a new estimate of n_{t0} from $n_{t0}^{(k+1)} = n_{t0}^{(k)} - F(n_{t0}^{(k)})/(dF/dn_{t0}^{(k)})$

Step 9. Check for convergence. If $|(n_{t0}^{(k+1)} - n_{t0}^{(k)})/n_{t0}^{(k+1)}| > \varepsilon$ increment k and return to Step 5. Otherwise continue.

Step 10. Calculate the mass fluxes from $n_{i0} = \varrho_t \sum_{j=1}^{n-1} P_{ij}\hat{k}_j\hat{\Xi}_j\Delta\hat{\omega}_j + \omega_{j0}n_{t0}$.

Step 11. If and only if [k] and $[k^\bullet]$ are required for some other calculations, they may be calculated from $[k] = [P][\hat{k}][P]^{-1}$ and $[k^\bullet] = [P][\hat{k}^\bullet][P]^{-1}$ where $\hat{k}_i^\bullet = \hat{k}_i\hat{\Xi}_i$. The calculations are now complete.

Before proceeding to an analysis of the problem of simultaneous heat and multicomponent mass transfer, let us pause to consider the application of the turbulent eddy diffusivity models; specifically, of Equation 404. Isopropanol and water are condensing in the presence of nitrogen in a vertical tube. At the vapor inlet, the gas/vapor phase has a composition $y_{1b} = 0.1123$ $y_{2b} = 0.4246$. The composition of the vapor in equilibrium with the condensate is $y_{10} = 0.1457\, y_{20} = 0.1640$. The *binary* diffusion coefficients, D'_{ij}, density and viscosity of the gas/vapor phase are $D_{12} = 15.988$, $D_{13} = 14.439$, $D_{23} = 38.732\ mm^2/s$, $\varrho_t = 0.882\ kg/m^3$, $\mu = 1.606 \times 10^{-5}\ Pa \cdot s$ (estimated at the temperature, pressure and composition of the bulk phase). The friction factor is calculated from $f/2 = 0.023\ Re^{-.17}$ and the vapor phase Reynolds number is 9574. The results of the computations of the *mass* fluxes are summarized in Table 8. We refrain from commenting on the results as this stage (except to point out that the isopropanol is transferring against its intrinsic driving force; $n_{10}/(\omega_{10} - \omega_{1b}) < 0$).

Heat Transfer with Simultaneous Mass Transfer

Having tackled the problem of the calculation of the mass fluxes n_i in the multicomponent mixture, we now take up the problem of estimating the heat transfer coefficients and the energy flux E, across the interface. We ignore Soret and Dufour effects in the ensuing analysis.

Taking $(1 - y^+/R_w^+)$ as unity, the energy balance relation can be derived in a manner analogous to the corresponding Equation 374 for the continuity of mass. The energy flux E_0 at the wall, or interface, is given by

$$E_0 = -\varrho_t C_p u^* \left(Pr^{-1} + Pr_{turb}^{-1} \frac{\nu_{turb}}{\nu} \right) \frac{d(T - T_{ref})}{dy^+} + \sum_{i=1}^{n} n_i C_{pi} (T - T_{ref}) \tag{405}$$

where T_{ref} is the reference temperature for the calculation of the partial specific enthalpy of component i:

$$H_i = C_{pi}(T - T_{ref}) \tag{406}$$

with C_{pi} the specific heat capacity for component i.

We define the integral ψ_H as follows:

$$\psi_H = \int_0^{y^+} \frac{\sum_{i=1}^{n} n_i C_{pi}}{\varrho_t C_p u^* \left(Pr^{-1} + Pr_{turb}^{-1} \frac{\nu_{turb}}{\nu} \right)} dy^+ \tag{407}$$

with

$$\Phi_H = \int_0^{y_{bH}^+} \frac{\sum_{i=1}^{n} n_i C_{pi}}{\rho_t C_p u^* \left(Pr^{-1} + Pr_{turb}^{-1} \frac{\nu_{turb}}{\nu} \right)} dy^+ \tag{408}$$

With these definitions of the heat transfer rate factors, analogous to Equations 375 and 376, we can solve the differential Equation 405 for constant Pr with the boundary conditions $y^+ = 0, T = T_0$; $y^+ = y_{bH}^+$(bulk fluid), $T = T_b$, to obtain the temperature profile

$$\frac{T - T_0}{T_b - T_0} = \frac{\exp \psi_H - 1}{\exp \Phi_H - 1} \tag{409}$$

The integral Φ_H is a heat transfer rate factor which, for purely molecular heat conduction ($Pr_{turb}^{-1} = 0$), reduces to Φ_H defined in Equation 295 (with $\ell \equiv y_{bH}$). The integral ψ_H is equivalent to $\Phi_H\eta$ in Equation 298.

Equation 409 can be differentiated to obtain q_0, the conductive heat flux at the wall:

$$q_0 = h^{\bullet}(T_0 - T_b) = \frac{\sum_{i=1}^{n} n_i C_{pi}}{(\exp \Phi_H - 1)}(T_0 - T_b) \tag{410}$$

or

$$h^{\bullet} = \frac{\sum_{i=1}^{n} n_i C_{pi}}{\exp \Phi_H - 1} \tag{411}$$

Taking the limit of $h^{\bullet}$ as n_i ($i = 1, 2, \ldots, n$) are all vanishingly small, we obtain the zero-flux heat transfer coefficient:

$$h = \frac{\sum_{i=1}^{n} n_i C_{pi}}{\Phi_H} \tag{412}$$

and so the correction factor for high mass fluxes can be derived to be

$$\Xi_H = \frac{\Phi_H}{\exp \Phi_H - 1} = \frac{h^{\bullet}}{h} \tag{413}$$

which is formally identical to the Ackerman film model correction factor derived earlier (see. Equation 301).

Proceeding in an exactly analogous manner to the corresponding mass transfer problem, we may derive the following explicit expression for the Stanton number for heat transfer $St_H(\equiv h/\varrho_t \bar{C}_p \bar{u})$:

$$St_H^{-1} = \frac{2}{f} + \left(\frac{2}{f}\right)^{1/2} \int_0^{y_1^+} \left\{ \frac{1}{Pr^{-1} + Pr_{turb}^{-1} \frac{\nu_{turb}}{\nu}} - \frac{1}{1 + \frac{\nu_{turb}}{\nu}} \right\} dy^+ \tag{414}$$

The previously discussed turbulence models can be used for the evalution of the integral in Equation 412. For example, with the von Karman velocity profile, Equations 392 and 393, we obtain

$$St_H^{-1} = \frac{2}{f} + 5\left(\frac{2}{f}\right)^{1/2} \left(Pr - 1 + \ln\left(1 + \frac{5}{6}(Pr - 1)\right)\right) \tag{415}$$

Equations 410 ~ 412 allow the calculation of the conductive heat flux q_0; the total energy flux E_0 is the sum of the conductive and the bulk flow enthalpy contributions:

$$E_0 = q_0 + \sum_{i=1}^{n} n_i \bar{H}_i \tag{416}$$

Empirical Methods

In order that the mass transfer rates be calculable from the turbulent eddy diffusivity methods described earlier, the friction factor, f, must be available. Thus these models may, as will be demonstrated in the next sectrion, be utilized directly in the simulation of, for example, in-tube condensation of multicomponent vapors or interphase mass transfer in a wetted-wall column. For such geometries and for a few other cases only, the friction factor can be calculated. In a limited number of practical cases, a priori estimates of the film thickness, ℓ, or of the contact time for mass transfer, t_e, can be made and the film and penetration models discussed earlier may be used to estimate the transfer rates. However, in many cases, such a priori estimates are not possible and we must resort to empirical methods of estimating the multicomponent mass transfer coefficients.

Most experimental works in the past have tended to concentrate on two component systems; binary mass transfer data usually are correlated by use of dimensionless groups such as the Sherwood number, the Stanton number and the Chilton–Colburn j-factors [22].

$$Sh = \frac{kd}{D'}; \quad St = \frac{k}{\bar{u}}; \quad j_D = St\, Sc^{2/3}; \quad j_H = St_H Pr^{2/3} \tag{417}$$

where d is some characteristic dimension of the mass transfer equipment.

The Gilliland–Sherwood [22] correlation for gas-phase binary mass transfer in a wetted-wall column is, for example,

$$Sh = 0.023\, Re^{0.83}\, Sc^{0.44} \tag{418}$$

For heat and/or mass transfer from a pipe wall to a turbulently flowing fluid phase, the Chilton–Colburn analogy takes the form

$$j_H = j_D = f/2 \quad \text{or} \quad St = f\, Sc^{-2/3}/2 \tag{419}$$

Equation 419 provides a way of estimating k from an expression for the friction factor f as an alternative to the turbulent eddy diffusivity models *or* from a separate correlation of the heat transfer coefficient, h, if that and not f is available. Provided f is based on shear friction and not on total drag, Equation 419 has been found to hold remarkably well for many types of flow systems and geometries [22].

For multicomponent systems, Toor [4] and Stewart and Prober [5] suggested that correlations of the type given by Equations 418 and 419 be generalized by replacing the binary diffusivity D' by the matrix [D] and the binary mass transfer coefficient k by the matrix [k]. Thus, we have the generalization of Equation 418

$$[Sh] = [k]d[D]^{-1} = 0.023\, Re^{0.83}\, [Sc]^{0.44} \tag{420}$$

and the generalization of Equation 419

$$[St] = [k]/\bar{u} = \frac{f}{2}[Sc]^{-2/3} \tag{421}$$

Note that the matrix of Schmidt numbers $[Sc] = \nu[D]^{-1}$ may be calculated using [D] relative to the molar average velocity Equation 60 or [D] relative to the mass average velocity Equation 67. If the former, we must calculate the molar fluxes from

$$(N) = c_t[\beta_0]\,[k^\bullet]\,(\Delta x) = c_t[\beta_0]\,[k]\,[\Xi]\,(\Delta x) \tag{422}$$

The mass transfer coefficients obtained from correlations in Equations 418 through 421 are the low flux coefficients. These coefficients must be corrected using an appropriate correction factor. We recommend the film theory correction factor matrix:

$$[\Xi_0] = [\Phi]\{\exp[\Phi]-[I]\}^{-1} \tag{423}$$

with

$$[\Phi] = \frac{N_t}{c_t}[k]^{-1} \tag{424}$$

Alternatively, we may prefer to calculate mass fluxes from

$$(n) = \varrho_t[\beta][k][\Xi](\Delta\omega) \tag{425}$$

with $[\Xi]$ given by Equation 423 again but now with $[\Phi] = n_t[k]^{-1}/\varrho_t$ and $[\beta]$ calculated using mass fractions rather than mole fractions.

For the numerical computations we once again suggest the approach of Vickery et al. [90]. (The *Algorithm for Computing Molar Fluxes from the Film Model* of Toor–Stewart–Prober.) If molar fluxes are being calculated, we need only replace the first part of Equation 225 by, for example:

$$\frac{\hat{k}_i d}{\hat{D}_i} = 0.023\,Re^{0.83}\,\hat{Sc}_i^{0.44} \quad \text{or} \quad \hat{St}_i = \frac{f}{2}\hat{Sc}_i^{-2/3} \tag{426}$$

Exactly the same approach is possible if mass fluxes are to be calculated; simply replace the expression for $\hat{k}_i$ in Step 3 of the algorithm chart just mentioned by an expression similar in form to Equation 426.

Another procedure for estimating multicomponent mass transfer coefficients from binary correlations can be used with the Krishna–Standart method, the explicit methods of Krishna and Taylor and Smith and, providing an alternative to the method above, with the Toor–Stewart–Prober method. The procedures are particularly suitable for gas-phase transport. Using the Krishna–Standart method for illustrative purposes, we rewrite Equations 190 and 193 as

$$[k] = [D]/\ell = [B]^{-1}/\ell = [\mathit{B}]^{-1} \tag{427}$$

where the matrix $[B]$ has elements

$$B_{ii} = \frac{y_i}{k_{in}} + \sum_{\substack{k=1 \\ k\neq i}}^{n} \frac{y_k}{k_{ik}}, \qquad i = 1, 2, \ldots, n-1$$

$$B_{ij} = -y_i\left(\frac{1}{k_{ij}} - \frac{1}{k_{in}}\right), \qquad i \neq j = 1, 2, \ldots, n-1 \tag{428}$$

and where the k_{ij} defined by

$$k_{ik} = D'_{ij}/\ell \tag{429}$$

are the mass transfer coefficients that would be calculated if the mixture contains species i and j only. In situations where the film thickness ℓ is not known, Krishna and Standart

[6] suggested that the *binary* k_{ij} be calculated from the appropriate correlation (for example, Equations 418 and 419) and used directly in the calculation of the multicomponent mass transfer coefficients [k] Equations 427 and 428. The finite-flux corrections to [k] in Equation 427 are provided by Equations 191 and 194 with the elements of [Φ] calculated from

$$\Phi_{ij} = \frac{N_i}{c_t k_{in}} + \sum_{\substack{j=1 \\ i \neq j}}^{n} \frac{N_j}{c_t k_{ij}}, \quad i = 1, 2, \ldots, n-1$$

$$\Phi_{ij} = \frac{-N_i}{c_t}\left(\frac{1}{k_{ij}} - \frac{1}{k_{in}}\right), \quad i \neq j = 1, 2, \ldots, n-1 \tag{430}$$

where, once again, we have replaced the quantity D'_{ij}/ℓ by the binary mass transfer coefficients k_{ij}.

It should be clear from Equations 236 through 245 that similar modifications can be made to the explicit methods. It can also be seen from Table 2 that the binary mass transfer coefficients can be used in the same way to estimate [k] in the Toor–Stewart–Prober method. The results of such an approach are not usually very different from matrix generalizations of the binary correlations as shown later.

We now return to the problem involving the condensation of isopropanol and water in the presence of nitrogen; this example was used earlier to illustrate the turbulent eddy diffusivity models. Tables 9 and 10 summarize the computations of the fluxes using the Chilton–Colburn analogy to estimate [k] in the molar frame of reference. The method of Toor–Stewart–Prober is used and [k] is calculated using both of the approaches just discussed. There are some small differences between the respective [k] matrices but the final estimates of the fluxes agree almost exactly. For comparison, we have also included in Table 10 the estimates of the molar fluxes calculated using the Krishna–Standart, Krishna [91], and Taylor–Smith methods—all five methods give virtually identical results here.

It is important to recognize that the results in Tables 9 and 10 cannot be compared directly with our earlier calculations for this system using the turbulent eddy diffusivity model. In order to provide a more meaningful comparison with the results in Table 8, we have solved the problem again in the mass frame of reference; the results are summarized in Table 11. It is noteworthy that the use of the Chilton–Colburn analogy leads to estimates of the mass fluxes that are in excellent agreement with the results of the more fundamental turbulent eddy diffusivity method.

A clue to the success of the Chilton–Colburn analogy can be found in the analysis carried out recently by Fletcher, Maskell, and Patrick [123]. Using a modified von Driest mixing length model, Fletcher et al. showed that for Pr (Sc) > 1, this model predicted a $-2/3$ power dependence of the St on the Prandtl (Schmidt) number, in agreement with the empirical Chilton–Colburn assumption. It is, however, interesting to note that the studies of Fletcher et al. shows that for Pr (Sc) < 0.1, the Stanton number varies inversely as the Prandtl (Schmidt) number for the von Driest model. It may, therefore, be expected that for high molecular diffusivities of a particular binary pair, the predictions of the Chilton–Colburn analogy may well be in error. As multicomponent interaction phenomena are usually important is systems where there is a wide variation in the binary pair diffusivities, the applicability of the Chilton–Colburn analogy to such cases requires careful investigation. We shall return to this subject in the next section.

A further fundamental shortcoming in the Chilton–Colburn approach is that the assumed dependence of [k] on [Sc] takes no account of the variations in the level of turbulence, embodied by ν_{turb}/ν, with variations in the flow conditions. Increase in the system Reynolds number will both affect the friction factor and have a direct influence on the reduced mixing length since at a given distance from the wall, y, the reduced distance $y^+ = (y/R_w)(f/8)^{1/2}$ Re. The increase in the turbulence intensity should be reflected in a relative decrease in the

Table 8
Calculation of Mass Fluxes During the Condensation of Isopropanol and Water in the Presence of Nitrogen Turbulent Eddy-Diffusivity Method

Mass fractions (ω_o), (ω_b)	$(\omega_o) = \begin{pmatrix} 0.2821 \\ 0.0951 \end{pmatrix}$	$(\omega_b) = \begin{pmatrix} 0.2467 \\ 0.2794 \end{pmatrix}$
Matrix of diffusion coefficients [D] from equation 67 ($mm^2\ s^{-1}$)	$\begin{bmatrix} 15.790 & 5.412 \\ 2.678 & 31.585 \end{bmatrix}$	
Eigenvalues: $\hat{D}_1$, $\hat{D}_2$ ($mm^2\ s^{-1}$)	32.454	14.920
$[Sc] = \frac{\mu}{\rho_t}[D]^{-1}$	$\begin{bmatrix} 1.187 & -0.2034 \\ -0.101 & 0.594 \end{bmatrix}$	
Eigenvalues: $\hat{Sc}_1$, $\hat{Sc}_2$	0.561	1.220
$[St] \times 10^3$ (from Equation 404)	$\begin{bmatrix} 4.402 & 0.853 \\ 0.422 & 6.892 \end{bmatrix}$	
Eigenvalues: $\hat{St}_1$, $\hat{St}_2 (\times 10^3)$	7.029	4.265
$[k] = \bar{u}[St]$ ($mm\ s^{-1}$)	$\begin{bmatrix} 33.354 & 6.464 \\ 3.199 & 52.221 \end{bmatrix}$	
Eigenvalues: $\hat{k}_1$, $\hat{k}_2$ ($mm\ s^{-1}$)	53.260	32.315
Eigenvalues: $\hat{\Phi}_1$, $\hat{\Phi}_2$	0.3277	0.5400
Eigenvalues: $\hat{\Xi}_1$, $\hat{\Xi}_2$	0.8451	0.7542
Eigenvalues: $\hat{k}_1^\bullet$, $\hat{k}_2^\bullet$ ($mm\ s^{-1}$)	45.010	24.372
$[k^\bullet] = [k][\Xi]$ ($mm\ s^{-1}$)	$\begin{matrix} 25.395 & 6.370 \\ 3.152 & 43.987 \end{matrix}$	
$(n) = \rho_t[\beta_o][k^\bullet](\Delta\omega)$ ($g\ m^{-2}\ s^{-1}$)	$\begin{pmatrix} 4.041 \\ 11.351 \end{pmatrix}$	
$n_t = n_1 + n_2$ ($n_3 = 0$) ($g\ m^{-2}\ s^{-1}$)	15.392	
$(N) = (n/M)$ ($mmol\ m^{-2}\ s^{-1}$)	$\begin{pmatrix} 67.25 \\ 630.62 \end{pmatrix}$	

Table 9
Calculation of Molar Fluxes During the Condensation of Isopropanol and Water in the Presence of Nitrogen*

Parameters		
Mole fractions (y_o), (y_b)	$\begin{pmatrix} 0.1457 \\ 0.1640 \end{pmatrix}$	$\begin{pmatrix} 0.1123 \\ 0.4246 \end{pmatrix}$
Matrix of diffusion coefficients, [D] Equation 57 ($mm^2\ s^{-1}$)	$\begin{bmatrix} 14.796 & -0.4192 \\ 5.234 & 32.579 \end{bmatrix}$	
Eigenvalues: $\hat{D}_1$, $\hat{D}_2$ ($mm^2\ s^{-1}$)	14.920	32.454
$[Sc] = \frac{\mu}{\rho}[D]^{-1}$	$\begin{bmatrix} 1.2248 & 0.01576 \\ -0.1968 & 0.5563 \end{bmatrix}$	
Eigenvalues: $\hat{Sc}_1$, $\hat{Sc}_2$	1.2201	0.5609
Eigenvalues: $\hat{k}_1$, $\hat{k}^2$ (Equation (7.268); ($mm\ s^{-1}$)	31.173	52.348
Eigenvalues: $\hat{\Phi}_1$, $\hat{\Phi}_2$	0.693	0.413
Eigenvalues: $\hat{\Xi}_1$, $\hat{\Xi}_2$	0.693	0.808
Eigenvalues: $\hat{k}_1$, $\hat{k}_2$ ($mm\ s^{-1}$)	21.605	42.283
[k] (Equation 421) ($mm\ s^{-1}$)	$\begin{bmatrix} 31.023 & -0.506 \\ 6.32 & 54.298 \end{bmatrix}$	
$[k_o^\bullet]$ ($mm\ s^{-1}$)	$\begin{bmatrix} 21.488 & -0.494 \\ 6.172 & 42.427 \end{bmatrix}$	
$(N) = c_t[\beta_o][k_o^\bullet](\Delta y)$ ($mmol\ m^{-2}\ s^{-1}$)	$\begin{pmatrix} 52.50 \\ 665.51 \end{pmatrix}$	

* *Use of Chilton-Colburn analogy in the molar reference frame. Toor-Stewart-Prober empirical method.*

influence of the molecular transport processes. So, for a given multicomponent mixture the increase in the Reynolds number should have the direct effect of reducing the effect of the phenomena of molecular diffusional coupling. Krishna [121] illustrated this shortcoming of the Chilton–Colburn analogy with some calculations for the gas-phase mass transfer between a vapor mixture containing acetone (1)—benzene (2)—helium (3) and a binary liquid mixture containing acetone and benzene. His calculations for the ratio k_{12}/k_{11} as a function of the gas-phase Reynolds number are given in Figure 18. The Chilton–Colburn analogy predicts that this ratio would be independent of Re and equal to 0.189; the von Karman turbulent model, taken as an example to illustrate the application of Equation 401 for the estimation of [k], predicts that the influence of coupling should decrease with increase in Re. The latter trend is in accord with our physical intuition. The influence of turbulence intensity on the extent of diffusional coupling, as portrayed in Figure 18, is exactly analogous to the influence of the Fourier number, Fo, on the extent of diffusional coupling for transfer inside a rigid droplet, considered earlier. The precise description of the mechanism of mass transfer can be important in the estimation of mass transfer rates in multicomponent systems, especially for systems in which the [D] matrix has sizable off-diagonal elements. For the gas-phase mass transfer problem considered in Figure 18, Krishna [2] showed that the Chilton–Colburn and the von Karman turbulent models predict different *directions* of transfer of acetone. The calculation of the flux of any component with a relatively small driving force, is particularly sensitive to the choice of the model for describing the multicomponent mass transfer process.

Table 10
Calculation of Molar Fluxes During the Condensation of Isopropanol and Water in the Presence of Nitrogen*

Parameters		
Mole fractions (y_o), (y_δ)	$\begin{pmatrix}0.1457\\0.1640\end{pmatrix}$	$\begin{pmatrix}0.1123\\0.4246\end{pmatrix}$
Binary diffusion coefficients D_{12}, D_{13}, D_{23} ($mm^2\ s^{-1}$)	15.988, 14.439, 38.732	
Binary Schmidt numbers Sc_{12}, Sc_{13}, Sc_{23}	1.1386, 1.2607, 0.4700	
Binary mass transfer coefficients k_{12}, k_{13}, k_{23} ($mm\ s^{-1}$) (Equation 419)	32.645, 34.102, 58,900	
$[k_{av}] = [B_{av}]^{-1}$ (Equations 427)($mm\ s^{-1}$)	$\begin{bmatrix}31.043 & -0.461\\6.657 & 53.265\end{bmatrix}$	
$[\Phi] = N_t[k_{av}]^{-1}/c_t$	$\begin{bmatrix}0.7063 & -0.0061\\-0.0883 & 0.4116\end{bmatrix}$	
$[\Xi_o] = [\Phi]\{\exp[\Phi] - [I]\}^{-1}$	$\begin{bmatrix}0.6882 & -0.0025\\0.0350 & 0.8082\end{bmatrix}$	
Molar fluxes $(N) = c_t[\beta_o][k_{av}][\Xi_o](\Delta y)$ ($mmol\ m^{-2}\ s^{-1}$)	$\begin{pmatrix}54.43\\675.36\end{pmatrix}$	

Fluxes ($mmol\ m^{-2}\ s^{-1}$)	Table 9	From above	Krishna-Standart	Krishna[91]	Taylor-Smith
N_1	52.60	54.43	54.57	54.54	55.26
N_2	665.51	675.36	675.98	665.82	674.65

* *Use of Chilton-Colburn analogy to calculate binary mass transfer coefficients (molar reference frame).*

Table 11
Calculation of Mass Fluxes During the Condensation of Isopropanol and Water in the Presence of Nitrogen Chilton-Colburn analogy (Mass Reference Frame)

Parameters	
Mass fractions (ω_o), (ω_b)	$(\omega_o) = \begin{pmatrix} 0.2821 \\ 0.0951 \end{pmatrix}$ $(\omega_b) = \begin{pmatrix} 0.2467 \\ 0.2794 \end{pmatrix}$
Matrix of diffusion coefficients [D], Equation (67) $(mm^2\ s^{-1})$	$\begin{bmatrix} 15.790 & 5.412 \\ 2.678 & 31.585 \end{bmatrix}$
Eigenvalues: $\hat{D}_1$, $\hat{D}_2$ $(mm^2\ s^{-1})$	14.920 32.454
$[Sc] = \frac{\mu}{\rho_t}[D]^{-1}$	$\begin{bmatrix} 1.187 & -0.203 \\ -0.101 & 0.594 \end{bmatrix}$
Eigenvalues: $\hat{Sc}_1$, $\hat{Sc}_2$	1.2201 0.561
$[St] \times 10^3$ (Equation 421)	$\begin{bmatrix} 4.382 & 0.889 \\ 0.440 & 6.976 \end{bmatrix}$
Eigenvalues: $\hat{St}_1$, $\hat{St}_2$	4.239 7.119
$[k] = \bar{u}[St]$ $(mm\ s^{-1})$	$\begin{bmatrix} 33.207 & 6.735 \\ 3.332 & 52.864 \end{bmatrix}$
Eigenvalues: $\hat{k}_1$, $\hat{k}_2$ $(mm\ s^{-1})$	32.125 53.946
Eigenvalues: $\hat{\Phi}_1$, $\hat{\Phi}_2$	0.5528 0.3291
Eigenvalues: $\hat{\Xi}_1$, $\hat{\Xi}_2$	0.7489 0.8442
Eigenvalues: $\hat{k}_1$, $\hat{k}_2$ $(mm\ s^{-1})$	24.060 45.553
$[k^\bullet] = [k][\Xi]$ $(mm\ s^{-1})$	$\begin{bmatrix} 25.126 & 6.633 \\ 3.282 & 44.487 \end{bmatrix}$
$(n) = \rho_t[\beta_o][k^\bullet](\Delta\omega)$ $(g\ m^{-2}\ s^{-1})$	$\begin{pmatrix} 4.159 \\ 11.504 \end{pmatrix}$
$n_t = n_1 + n_2$ $(n_3 = 0)$ $(g\ m^{-2}\ s^{-1})$	15.603
$(N) = (n/M)$ $(mmol\ m^{-2}\ s^{-1})$	$\begin{pmatrix} 69.21 \\ 639.12 \end{pmatrix}$

* *Chilton-Colburn analogy (mass reference frame).*

APPLICATIONS OF THE THEORY OF MULTICOMPONENT MASS TRANSFER

Our review of the theory of multicomponent mass transfer would be incomplete if we failed to discuss how the models described in the preceding section can be used in the simulation and design of multicomponent mass transfer rate-governed processes. This is the purpose of this section. Applications to the conventional unit operations of distillation, absorption, condensation, and extraction are considered in some detail. Accompanying the development of the models are comparisons of their predictions with the results of experiments. Several times in this section, we will demonstrate the superiority of the more rigorous methods over the simpler "effective-diffusivity" methods. Reasons of space have kept us from discussing applications to multicomponent mass transfer with heterogeneous or homogeneous chemical reactions and in porous media, membranes, and electrochemical systems, even though many interesting effects are to be found there.

Multicomponent Distillation in Stagewise Contactors

Distillation retains its position of supremacy among chemical engineering unit operations despite the emergence of many new separation techniques (e.g. membranes) in recent years; in fact, when choosing a separation scheme the first question which is usually asked is: "Why

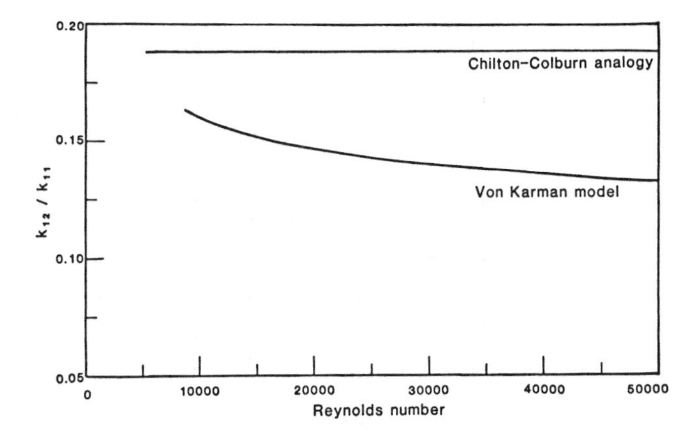

Figure 18. Ratio k_{12}/k_{11} (which are elements of the zero-flux matrix of mass transfer coefficients [k]) as a function of the gas-phase Reynolds number. Mass transfer between a gaseous mixture of acetone (1), benzene (2) and helium (3) and a liquid film containing acetone and benzene. Calculations based on the von Karman turbulent film model and the Chilton-Colburn analogy.

not distillation?" Krishna [124] reviewed the various possibilities of achieving a separation in a multicomponent mixture using the generalized Maxwell–Stefan equations as a theoretical pivot. Despite being well established in industrial practice, there are still many gaps in our understanding of distillation behavior. The most glaring lack of knowledge concerns the prediction of tray efficiencies in multicomponent systems; most of the published data on mass transfer in tray columns are restricted to binary or two component systems. From the discussions in the previous section, it should be clear that for distillation of nonideal fluid mixtures with significant differences in component sizes (molecular weights) and nature, coupling between species transfers (introduced via nondiagonal elements of [β], [D], and [Γ] in *both vapor* and *liquid* phases) can cause interesting effects such as *reverse transfer, mass transfer barrier,* and *osmotic transfer,* as already noted. As far back as 1960, Toor and Burchard [3] performed some theoretical calculations to show that the neglection of interaction effects in design calculations could lead to severe underdesign. Conclusive experimental evidence concerning the significance of diffusional coupling on distillation behavior has only become available during the last five or so years; a comprehensive survey of available experimental data is given by Krishna et al. [125] and Krishna and Standart [6]. We refer the reader to these articles for background information. Here we summarize some of the key aspects relevant to column design.

Let us begin by considering the classic problem of determining the Murphree point efficiencies, in the absence of entrainment or weeping aspects. Figure 19 pictures vapor-liquid contacting on a tray. We concentrate our attention on the mass transfer process in the vertical slice shown in the Figure 19. If v_i represents the molar flow rate of component i in the vapor phase, $V = \sum_{i=1}^{n} v_i$ the total vapor flow rate, a' the interfacial area per unit dispersion (froth) volume; h_f the froth height and A the active bubbling area, then the component material balance may be written

$$\frac{dv_i}{dh_f} = N_i a' A \tag{431}$$

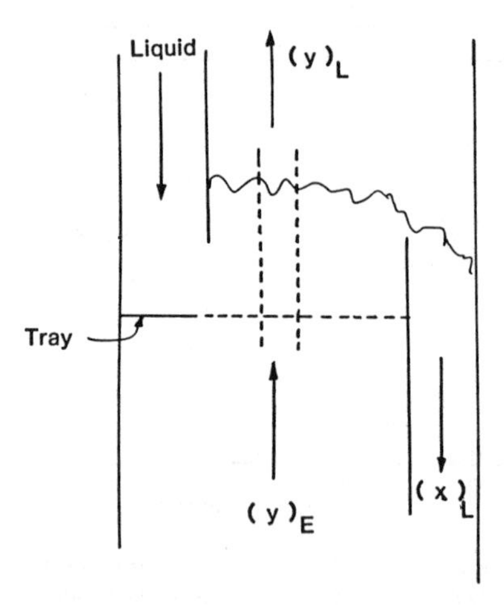

Figure 19. Distillation of a multicomponent mixture in a tray column. $(y)_E$ is the column matrix of entering vapor-phase mole fractions; $(y)_L$ is the column matrix of exiting vapor-phase mole fractions.

where N_i is the molar flux of species i across the gas-liquid interfacial area. For the total mixture we have

$$\frac{dV}{dh_f} = N_t a'A \tag{432}$$

Substituting $N_i = J_i^y + y_i N_t$ in Equation 431 and writing $v_i \equiv y_i V$ we obtain

$$V\frac{dy_i}{dh_f} + y_i\frac{dV}{dh_f} = J_i^y a'A + y_i N_t a'A \tag{433}$$

In view of Equation 432, the right members on both sides of the equality 433 cancel each other and we get

$$V\frac{dy_i}{dh_f} = J_i^y a'A \tag{434}$$

as an *exact* result valid even when $N_t \neq 0$. In $n-1$ dimensional matrix notation, Equation 434, takes the form

$$V\frac{d(y)}{dh_f} = (J^y)a'A \tag{435}$$

The liquid phase in the vertical slice will be assumed to be well-mixed and have a bulk phase composition x_i. Thermal equilibration occurs very rapidly on distillation trays and the assumption of uniform temperatures in both vapor and liquid phases is usually a good one (one exception to this is the case where the gas phase contains nontransferring inerts in which case some temperature gradients may exist in the gas phase). Let y_i^* represent the composition in equilibrium with the bulk liquid phase within the vertical slice. Introducing the rate-relations Equation 157 for (J) we obtain

$$V \frac{d(y)}{dh_f} = c_t^y [K_{0y}] (y^* - y) a' A \tag{436}$$

where $[K_{0y}]$ is the matrix of overall mass transfer coefficients referred to the vapor phase. Equation 436 may be written in the form

$$\frac{d(y)}{d\xi} = [NTU_{0y}] (y^* - y) \tag{437}$$

where we define ξ as the fractional distance along the dispersion height and $[NTU_{0y}]$, the overall vapor phase number of transfer units

$$[NTU_{0y}] \equiv \frac{c_t^y [K_{0y}] a' h_f A}{V} \tag{438}$$

Integrating Equation 437 over the dispersion height, assuming constant $[NTU_{0y}]$, we obtain [125a]:

$$(y^* - y_L) = [\exp[-NTU_{0y}]] (y^* - y_E) \tag{439}$$

where $(y)_E$ and $(y)_L$ denote bulk vapor composition of the vapor stream entering and leaving the vertical slice. The matrix $[Q] \equiv \exp[-NTU_{0y}]$ can be evaluated by use of Sylvester's theorem. Let us call $(y^* - y_L) \equiv (\Delta y)_L$ and $(y^* - y_E) \equiv (\Delta y)_E$. Equation 439 may, therefore, be written as

$$(\Delta y)_L = [Q] (\Delta y)_E \tag{440}$$

For a binary mixture, $[Q]$ reduces to the scalar $Q = \exp(-NTU_{0y})$ and the Murphree point efficiency for component 1, i.e.

$$E_{0y1} = \frac{y_{1L} - y_{1E}}{y_1^* - y_{1E}} = 1 - \frac{(\Delta y_1)_L}{(\Delta y_1)_E} = 1 - Q = 1 - e^{-NTU_{0y}} \tag{441}$$

has a value ranging between 0 and 1 because NTU_{0y} is always a real and positive quantity. Also, for a two component system there is *only one* independent composition (y_1) and *only one* independent point efficiency E_{0yi}, which equals the component efficiency of component 2: $E_{0y1} = E_{0y2}$. The situation is quite different for a multicomponent system. We shall illustrate this by considering the specific example of a ternary mixture. Now, we have two independent compositions and two independent efficiencies. There is *no* theoretical requirement, akin to the binary case, that $E_{0y1} = E_{0y2} = E_{0y3}$. In fact we can show using Equation 440 that the three-component efficiencies are given by

$$E_{0y1} = \frac{y_{1L} - y_{1E}}{y_1^* - y_{1E}} = 1 - \frac{(\Delta y_1)_L}{(\Delta y_1)_E} = 1 - Q_{11} - Q_{12} \frac{(\Delta y_2)_E}{(\Delta y_1)_E} \tag{442}$$

$$E_{0y2} = \frac{y_{2L} - y_{2E}}{y_2^* - y_{2E}} = 1 - \frac{(\Delta y_2)_L}{(\Delta y_2)_E} = 1 - Q_{22} - Q_{21} \frac{(\Delta y_1)_E}{(\Delta y_2)_E} \tag{443}$$

$$E_{0y3} = \frac{y_{3L} - y_{3E}}{y_3^* - y_{3E}} = 1 - \frac{(\Delta y_1)_L + (\Delta y_2)_L}{(\Delta y_1)_E + (\Delta y_2)_E} = \frac{\Delta y_1 E_{0y1} + \Delta y_2 E_{0y2}}{\Delta y_1 + \Delta y_2} \quad (444)$$

Only when [Q] reduces to the form Q[I] will the three-component efficiencies E_{0y1}, E_{0y2} and E_{0y3} be equal to one another. As discussed in earlier sections, we must generally expect $[NTU_{0y}]$ to have significant nondiagonal elements; when this is the case [Q] will also have significantly large cross coefficients and the component efficiencies will be unequal to one another. Examination of Equations 442 through 444 shows that the ratio of driving forces $(\Delta y_1)_E/(\Delta y_2)_E$ plays a key role in determining the relative magnitudes of E_{0yi}. Since in distillation, $\Delta y_1/\Delta y_2$ can in principle take any values in the range $-\infty$ to $+\infty$ (contrast this with a binary system for which $\Delta y_1/\Delta y_2 = -1$), the component E_{0yi} are unbounded and could exhibit values ranging from $-\infty$ to $+\infty$.

Distillation experiments carried out by Krishna et al. [125] with the system ethanol-*tert* butanol-water confirm the expectations voiced previously and the point efficiency of *tert* butanol exhibited values > 1 and < 0 in certain concentration regions. A portion of their experimental data is shown in Table 12.

We shall now try to provide a quantitative explanation of the curious behaviour of the efficiency of *tert* butanol. The first step is to construct a realistic physical model for the tray hydrodynamics. The experimental data of Krishna et al. [125] were obtained under the spray regime for all the runs; we may, therefore, anticipate that the interface mass transfer will be governed by the gas-phase diffusion process. The proper modeling of the tray behavior under spray regime conditions is quite a formidable task for it requires *inter alia* knowledge of the (random) droplet trajectories. The simplest conceivable model for the passage of gas through the dispersion is by assuming that the gas streams through in the form of a cylindrical "jet." This model has in fact been used to model the initial "formation" zone in the froth regime of tray operation [115]; we shall assume here that in the spray regime the "formation" zone extends over the whole dispersion height. For plug flow of gas through the cylindrical tube, with surrounding liquid at composition x, the earlier presented Equation 359 can be applied to the current situation by identifying t as the residence time of the gas phase in the dispersion and r_0 as the radius of the cylindrical tube. Also, $(x_{t=0})$ is to be identified with the inlet compositions $(y)_E$ whereas $\langle x \rangle$ is to be identified with the outlet compositions $(y)_L$. Comparison of Equation 359 with Equation 440, leads then to the following relation for [Q]:

$$[Q] \equiv \exp[-NTU_{0y}] = 4 \sum_{n=1}^{\infty} \frac{\exp[-\lambda_n^2 r_0^2 [D'] \, Fo]}{\lambda_n^2 r_0^2} \quad (445)$$

The estimation of [Q] requires knowledge of $[D_{av}]$, t, r_0 and λ_n. Now, λ_n are the roots of $J_0(\lambda_n r_0) = 0$ and found in standard mathematical tables (e.g., Abramowitz and Stegun [126]).

The gas-phase $[D_{av}]$ can be estimated from the binary pair diffusivities D'_{ik}. For the temperature and pressure conditions prevailing in the distillation experiments these coefficients have been estimated as follows [127]:

$$D'_{12} = 7.99 \text{ mm}^2 \text{ s}^{-1}; \quad D'_{13} = 21.4 \text{ mm}^2 \text{ s}^{-1}; \quad D'_{23} = 16.5 \text{ mm}^2 \text{ s}^{-1}.$$

The tray hydrodynamics determine the value of the gas residence time t and the diameter of the cylindrical jet r_0. Neither of these is known for the experimental conditions; so we shall treat these parameters as "fitting" parameters.

Since we are primarily interested in explaining the curious behavior of *tert* butanol we shall examine the values of the coefficients Q_{21} and Q_{22}. Figure 20 gives the variation of Q_{22} and $-Q_{21}/Q_{22}$ with the Fourier number, Fo (we choose the reference diffusivity value

Table 12
Murphree Point Efficiencies for Distillation of Ethanol-*tert* Butanol-Water in a 76 mm sieve tray column*

Run Number	y_{1E}	y_{2E}	y_1^*	y_2^*	$\frac{(\Delta y_1)_E}{(\Delta y_2)_E}$	E_{oy1}	E_{oy2}	E_{oy3}
S1A	0.1520	0.1405	0.2792	0.3343	0.6563	0.6829	0.6923	0.6886
S2B	0.1466	0.4088	0.1741	0.4628	0.5093	0.6259	0.5273	0.5660
S4A	0.1829	0.1412	0.3193	0.2994	0.8903	0.6019	0.5928	0.5971
S11A	0.1791	0.4203	0.1891	0.4637	0.2304	0.0699	0.6779	0.5638
M24	0.2324	0.3575	0.2606	0.4077	0.5618	0.4793	0.8334	0.7061
M32	0.0824	0.0801	0.2435	0.3363	0.6288	0.6268	0.5445	0.5762
S6C	0.2711	0.3488	0.3050	0.3753	0.2792	0.5013	0.8461	0.6525
S11C	0.1919	0.4627	0.2149	0.4639	19.1667	0.6491	2.0019	0.7162
M3	0.1924	0.4616	0.2143	0.4638	9.9545	0.6169	1.3577	0.6848
M6	0.1932	0.4579	0.2178	0.4601	11.1818	0.6455	1.2058	0.6906
M9	0.2280	0.4359	0.2490	0.4377	11.6667	0.5144	2.9244	0.7092
M34	0.3071	0.3499	0.3389	0.3533	9.0857	0.4463	1.0491	0.5050
M41	0.4733	0.1817	0.5317	0.1835	32.4	0.4433	1.5891	0.4780
M47	0.5242	0.1310	0.5928	0.1318	85.7	0.4605	5.2700	0.5170
M49	0.2198	0.4209	0.2417	0.4403	1.1289	0.5331	1.0735	0.7865
M1	0.1916	0.4646	0.2153	0.4645	−237.0	0.6509	−27.9812	0.7202
M12	0.3334	0.3297	0.3699	0.3291	−60.8	0.4790	−1.2888	0.5095
M40	0.4992	0.1846	0.5250	0.1833	−7.65	0.4811	0.1853	0.5256
M46	0.5558	0.1353	0.6040	0.1335	−26.8	0.3546	−2.9400	0.4800

* *Krishna et al. [125].*

Table 13
Comparison of "Cylindrical Gas Streaming" and "Spherical Dispersed Bubbles" Model Predictions of Murphree Point Efficiencies in the System Ethanol-*tert* Butanol-Water

Run Number	Model Assuming Gas Streams Through Dispersion in Form of Cylindrical "Jet" of Radius r_o. t Is Contact Time.				Model Assuming Gas Is Dispersed in Form of Uniform Spherical Bubbles of Radius r_o. t Is The Gas Residence Time (Assumed Uniform).			
	$\frac{D_{ref}t}{r_o^2}$	E_{oy1}	E_{oy2}	E_{oy3}	$\frac{D_{ref}t}{r_o^2}$	E_{oy1}	E_{oy2}	E_{oy3}
S1A	.070	.550	.476	.505	.035	.555	.483	.511
S2B	.070	.498	.496	.496	.035	.504	.502	.503
S4A	.070	.545	.477	.509	.035	.550	.484	.515
S11A	.070	.592	.473	.495	.035	.596	.480	.502
M24	.070	.534	.483	.501	.035	.539	.490	.508
M32	.070	.547	.478	.504	.035	.551	.485	.510
S6C	.070	.496	.519	.506	.035	.502	.525	.512
S11C	.070	.431	1.912	.505	.035	.439	1.881	.511
M3	.070	.433	1.210	.504	.035	.441	1.198	.511
M6	.070	.434	1.297	.504	.035	.442	1.283	.511
M9	.070	.439	1.289	.506	.035	.447	1.275	.512
M34	.070	.460	.987	.511	.035	.468	.980	.517
M41	.070	.496	1.436	.524	.035	.502	1.416	.529
M47	.070	.506	2.390	.528	.035	.512	2.341	.533
M49	.070	.476	.530	.501	.035	.483	.536	.508
M1	.070	.429	−17.619	.505	.035	.437	−17.137	.511
M12	.070	.458	−2.933	.514	.035	.465	−2.833	.520
M40	.070	.479	.167	.526	.035	.486	.182	.531
M46	.070	.500	−.235	.528	.035	.506	−.208	.533

to be 20 $mm^2\ s^{-1}$) for one particular Run, M46. For values of Fo in the range 0.01–0.2, the influence of coupling, $|Q_{21}/Q_{22}|$, increases with increasing Fo. This is to be expected as the variation of $[k] \propto [D]^n$ changes from the $n = 1/2$ regime to $n = 1$ regime. As $t \to \infty$, the matrix [Q] must reduce to the null matrix [0], and the component efficiencies will all be unity. The effect of increasing coupling with increasing Fo will have a direct influence on the component efficiency of *tert* butanol (see Equation 445). For Run M46, $\Delta y_1/\Delta y_2 = -26.8$ and so the contribution of $Q_{21}\ \Delta y_1/\Delta y_2$ will be to *reduce* E_{0y2}. This effect is shown in Figure 21 where E_{0y2} is plotted as a function of Fo. As $D_{ref}t/r_0^2$ increases from 0.01 to 0.10, E_{0y2} decreases from -0.12 to -0.245, a manifestation of reverse diffusion. Another curious phenomena takes place beyond Fo = 0.10. The E_{0y2} values now *increase*. This can be understood by referring to Figure 20; though $|Q_{21}/Q_{22}|$ increases, we have a simultaneous decrease in Q_{22}. By a detailed examination of the values of Q_{22} and Q_{21}/Q_{22}, it is found that Q_{22} decreases faster beyond Fo = 0.10 than the increase in $|Q_{21}/Q_{22}|$. Therefore E_{0y2} increases in value and eventually, $t \to \infty$, all efficiencies will tend to unity.

The measured experimental value of E_{0y2} for Run M46 is $E_{0y2} = -2.94$. Due to the smallness of the driving force Δy_{2E} for this run, 0.0018, the calculation of E_{0y2} is subject to a large error, even though the experimental composition determinations aimed and achieved high accuracies [25]. A value of Fo = 0.07 was chosen to fit the measured *binary efficiency* for the system ethanol-water in the same column under similar hydrodynamic conditions. The theoretical predictions for all the experimental results of Table 12 taking Fo = 0.07 are given in Table 13. The theory is able to shadow the experimentally observed negative E_{0y2} for Runs M1, M12, and M46, and the values of $E_{0y2} > 1$ for Runs S11C, M6, M9, M34, M41 and M47 (a manifestation of osmotic diffusion). For Runs S1A, S2B, S4A, S11A, M24, and M32 the theory predicts $E_{0y1} \approx E_{0y2} \approx E_{0y3}$, largely in agreement with the experimental findings. It may be concluded that the experimental results can be rationally, and partly quantitatively, explained on the basis of the theories we have developed here. Also, the fitting parameter Fo = 0.07 is not unreasonable if we assume a radius of 4 mm for the gas stream and a gas residence time of 0.15 s. We do not offer these arguments as a *proof* of the validity of the cylindrical gas-stream model. In fact on the right

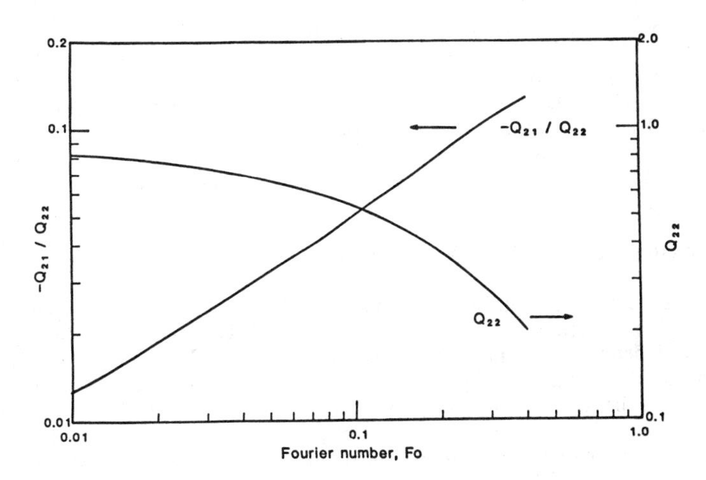

Figure 20. Shows the variation of Q_{12} and $-Q_{21}$ (elements of the Matrix [Q]) as functions of the Fourier number, Fo = $D_{ref}t/r_0^2$, for vapor-phase-controlled mass transfer during distillation of ethanol (1), *tert* butanol (2) and water (3) on a well-mixed tray. Calculations for Run M46 of Krishna et al. [125] with $[D_{av}]$ evaluated at the mean vapor composition between inlet and outlet. Cylindrical gas stream model: D_{ref} = 10 mm^2/s.

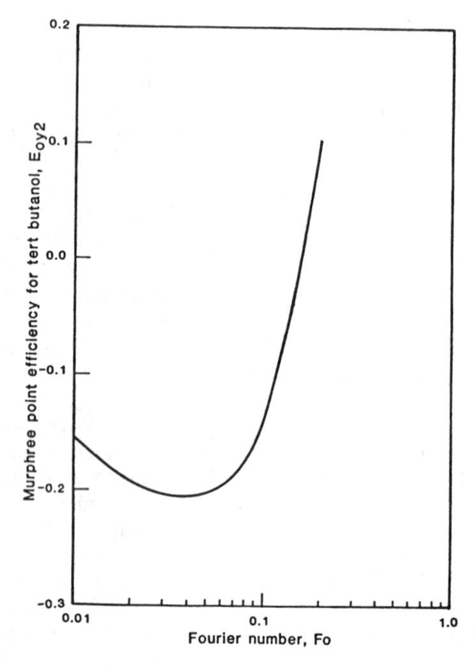

Figure 21. The Murphree point efficiency for *tert* butanol, E_{oy2}, as a function of the Fourier number, $Fo = D_{ref}t/r_0^2$. Calculations for Run M46 of Krishna et al. [125] with $[D_{av}]$ calculated at the mean vapor composition between inlet and outlet: $D_{ref} = 20\ mm^2/s$.

hand side of Table 13, we present the predictions of the uniform rigid spherical bubbles model described by Equation 352, with a fit parameter $Fo = 0.035$, again chosen to "fit" measured binary data. The two sets of predictions are remarkably similar to each other despite their widely different physical basis. Both models do a good job in predicting the trends in the experimentally observed efficiencies. For completeness, we show the E_{0y2} vs. Fo graph for Run M46 using the rigid spherical bubbles model; see Figure 22.

We turn now to the problems of simulating the performance of a *multistage* column. In the course of the discussion, we shall make several references to the results just discussed.

A Nonequilibrium-Stage Model of Multicomponent Separation Processes

Rigorous simulation of multistage processes such as distillation or absorption is, more often than not, based upon the "equilibrium stage" model. This model is well enough known not to need a detailed description here (see, for example, the textbooks by King [128], Henley and Seader [51] and Holland [129, 130]). Briefly, of course, the model includes the assumption that the streams leaving any particular stage are in equilibrium with each other. Component *M*aterial balances, the equations of phase *E*quilibrium, *S*ummation equations and *H*eat balance for each stage (the so-called MESH equations) are solved using one of the very many ingenious algorithms presently available to give product distributions, flow rates, temperatures, and so on.

In actual operation, stages rarely, if ever, operate at equilibrium despite attempts to approach this condition by proper design and choice of operating conditions. The usual way

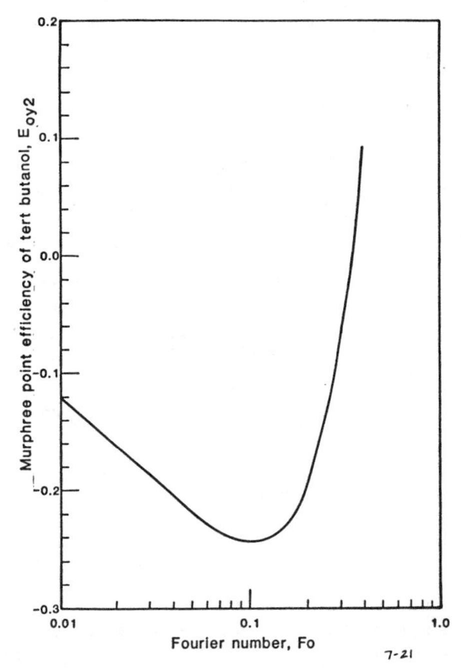

Figure 22. The Murphree point efficiency for *tert* butanol, E_{oy2}, as a function of the Fourier number, Fo = $D_{ref}t/r_0^2$. Calculations for Run M46 of Krishna et al. [125] using a model in which the gas is dispersed in the form of uniform rigid bubbles of radius r_0: D_{ref} = 20 mm^2/s.

of dealing with departures from equilibrium is by incorporating a "stage efficiency" into the equilibrium relations. It is with the introduction of this quantity that the problems begin.

One problem is that there are several different definitions of stage efficiency; Murphree [131], Hausen [132], generalized Hausen (Standart [133]), vaporization (Holland [129]), and others. There is by no means a consensus on which definition is "best." Arguments for and against various possibilities are presented by, among others, Standart [133, 134], Holland [129], Holland and McMahon [135], King [128], and by Medina et al. [136, 137]. Possibly the most soundly based (in a thermodynamic sense), the generalized Hausen efficiencies, are ridiculously complicated to calculate; the least soundly based, the Murphree efficiency, is the one most widely used because it is easily combined with the equilibrium equations. Thermal efficiencies may also be defined but almost always are taken to have a value of unity (except, perhaps, when inert species are involved). Whichever definition of stage efficiency is adopted, it must either be specified in advance or calculated from an equation derived by dividing by some reference separation by the actual separation obtained from a solution of the component material balance equations for each phase. Many different models of stage efficiency (i. e., different solutions of the component phase balances) have been proposed for binary systems (see, again, King [128]); we considered the extension to multicomponent systems of one of the simplest models in the preceding section.

In an n component system there are n − 1 independent component efficiencies which, for lack of anything better to do, have usually been taken to be the same for all components (King [128]).

The experimental data of Krishna et al. [125] just discussed, together with other evidence available in the literature (see the surveys in Krishna and Standart [6] and Krishna et al.

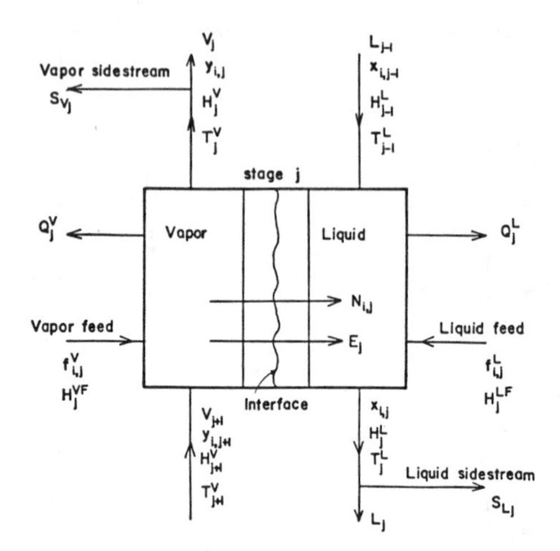

Figure 23. Schematic diagram of a nonequilibrium stage.

[125]), point to the shortcomings of the use of the classical concept of component efficiencies to describe the mass transfer behavior in tray columns; these efficiencies are unbounded and show a complicated dependence on the operating conditions vis à vis composition regions *as well as* on the tray hydrodynamics.

Two major practical conclusions arise from the analysis presented in the previous subsection:

1. Component efficiencies in multicomponent distillation are in general different from one another, as concluded in 1941 by Walter and Sherwood [1]. The design engineer who assumes equal efficiencies may suffer a severe embarassment as a result of neglecting cross (or coupling) effects as this may lead to a severe underdesign as first shown in an interesting theoretical study made by Toor and Burchard [3]. For systems under complete vapor-phase mass transfer control, equations were developed based on a generalized driving force pseudobinary approach (Toor [2]), for the effects of diffusional interactions among the components on their respective plate efficiency. A design calculation was made for the separation of methanol from isopropanol and water. For the hypothetical case in which the binary efficiencies were assumed to be 40%, consideration of interactions gave a column requiring 117 plates compared with 84 plates for the case where diffusional interactions are negligible.
2. The concepts of point and tray efficiencies as developed for binary systems assume a dubious significance when applied to multicomponent systems. In the general case, it appears best to do away with these concepts altogether and attempt to model a distillation column by solving the material and energy balance equations *simultaneously* with the appropriate rate relations for heat and mass transfer. This *nonequilibrium stage model* approach, as developed by Krishnamurthy and Taylor [138–141], is the subject of this section.

A schematic representation of a nonequilibrium stage is shown in Figure 23. Multistage columns consist of a sequence of such stages. Vapor and liquid streams from adjacent stages are brought into contact on the stage and allowed to exchange mass and energy across their common interface represented in the diagram by the vertical wavy line. The stage is assumed to be at mechanical equilibrium; $P_j^V = P_j^L = P_j$. Provision is made for vapor and liquid feed streams, sidestream drawoffs of vapor and liquid and for the addition or removal of heat. Steady-state operation is assumed.

Conservation Relations

As with all models of chemical processes, the analysis of the nonequilibrium stage starts with the construction of material and energy balances. The mass balance for component i on stage j is

$$\mathcal{M}_{ij} \equiv (1+r_j^V)v_{ij} + (1+r_j^L)\ell_{ij} - v_{i,j+1} - \ell_{i,j-1} - f_{ij} = 0 \tag{446}$$

where

$$r_j^V = S_{vj}/V_j; \quad r_j^L = S_{Lj}/L_j \tag{447}$$

The energy balance for the stage is

$$E_j \equiv (1+r_j^V)V_jH_j^V - V_{j+1}H_{j+1}^V + (1+r_j^L)L_jH_j^L - L_{j-1}H_{j-1}^L + Q_j - F_jH_j^F = 0 \tag{448}$$

In equilibrium-stage calculations, Equations 446 through 448 are solved subject to the requirement that the vapor and liquid streams leaving stage j are in complete thermal, mechanical, and chemical equilibrium. In the nonequilibrium stage model of Krishnamurthy and Taylor, the stage balances are split into two parts, one for each phase. For the vapor phase, the component mass balance is

$$\mathcal{M}_{ij}^V \equiv (1+r_j^V)v_{ij} - v_{i,j+1} - f_{ij}^V + \mathcal{N}_{ij}^V = 0 \tag{449}$$

and for the liquid phase

$$\mathcal{M}_{ij}^L \equiv (1+r_j^L)\ell_{ij} - \ell_{i,j-1} - f_{ij}^L - \mathcal{N}_{ij}^L = 0 \tag{450}$$

The last terms in Equations 449 and 450 represent the net loss or gain of species i due to interphase transport. Formally, we may write

$$\mathcal{N}_{ij}^V = \int N_{ij}^V da_j; \quad \mathcal{N}_{ij}^L = \int N_{ij}^L da_j \tag{451}$$

where N_{ij} is the molar *flux* of species i at a particular point in the two phase dispersion and da_j represents the small amount of interfacial area through which that flux passes. We adopt the convention that transfers from the vapor phase to the liquid phase are positive.

It follows directly from Equations 446, 449 and 450 that

$$\mathcal{M}_{ij}^I \equiv \mathcal{N}_{ij}^V - \mathcal{N}_{ij}^L = 0 \tag{452}$$

a result which may also be derived straightforwardly by constructing a material balance around the entire interface (hence the notation in Equation 452). Equation 452 is a statement of the assumption that there is no accumulation of mass at the interface.

The energy balance for the vapor phase is

$$E_j^V \equiv (1+r_j^V)V_jH_j^V - V_{j+1}H_{j+1}^V + Q_j^V - F_j^VH_j^{VF} + \mathcal{E}_j^V = 0 \tag{453}$$

and for the liquid phase the energy balance reads

$$E_j^L \equiv (1+r_j^L)L_jH_j^L - L_{j-1}H_{j-1}^L + Q_j^L - F_j^LH_j^{LF} - \mathcal{E}_j^L = 0 \tag{454}$$

Here, $\mathcal{E}_j$ represents the net loss or gain of energy due to interphase transport. This term may be defined by

$$\mathcal{E}_j^V = \int E_j^V da_j; \quad \mathcal{E}_j^L = \int E_j^L da_j \tag{455}$$

where E_j is the energy flux at some particular point in the dispersion. An energy balance around the interface yields

$$E_j^I \equiv \mathcal{E}_j^V - \mathcal{E}_j^L = 0 \tag{456}$$

Transport Relations

It is worth emphasizing that Equations 446 through 456 hold regardless of the models used to calculate the interphase transport rates $\mathcal{N}_{ij}^V$, $\mathcal{N}_{ij}^L$, $\mathcal{E}_j^V$, $\mathcal{E}_j^L$. With a mechanistic model of sufficient complexity it is possible, at least in principle, to account for mass transfer from bubbles in the froth as well as to entrained droplets and to liquid weeping through the tray floor (in these last two cases the material balance relations would have to be modified). Here, we discuss the determination of the interphase transport rates in a fairly general way; the precise form the rate relations may take is discussed later, when the model is applied to some specific separations.

As shown in earlier sections of this review, a rigorous treatment of mass transfer in multicomponent systems leads to the rate Equation 107

$$(N^V) = c_t^V[k^V][\Xi^V](y^V - y^I) + N_t^V(y^V) \tag{457}$$

$$(N^L) = c_t^L[k^L][\Xi^L](x^I - x^L) + N_t^L(x^L) \tag{458}$$

where the k_{ik}^V, k_{ik}^L are multicomponent low flux mass transfer coefficients. Since the correction factors $[\Xi]$ depend on the mass transfer coefficients and on the mass transfer rates themselves, we represent the flux Equations 457 and 458 by implicit and nonlinear relations of the general functional form

$$N_i^V = N_i^V(k_{ik}^V, y_k^I, y_k^V, N_k^V; k = 1, 2, \ldots n), \quad i = 1, 2, \ldots, n-1 \tag{459}$$

$$N_i^L = N_i^L(k_{ik}^L, x_k^I, x_k^L, N_k^L; k = 1, 2, \ldots n), \quad i = 1, 2, \ldots, n-1 \tag{460}$$

The local energy flux, E, is made up of a conductive heat flux and a convective contribution due to the transport of enthalpy by interphase transport Equation 108

$$E^V = q^V + \sum_{i=1}^{n} N_i^V\bar{H}_i^V; \quad E^L = q^L + \sum_{i=1}^{n} N_i^L\bar{H}_i^L \tag{461}$$

where the $\bar{H}_i$ are the partial molar enthalpies of species i. The conductive heat fluxes, q, are driven by temperature gradients in the fluid

$$q^V = h^V\Xi_H^V(T^V - T^I); \quad q^L = h^L\Xi_H^L(T^I - T^L) \tag{462}$$

where h^V and h^L are the low flux heat transfer coefficients. For the moment, we represent Equations 461 and 462

$$E^V = E^V(h^V, T^V, T^I, N_k^V) \tag{463}$$

$$E^L = E^L(h^L, T^L, T^I, N_k^L) \tag{464}$$

The calculation of the *total* mass and energy transfer rates $\mathcal{N}_{ij}$ and $\mathcal{E}_j$ requires the integration of the point flux relations 457, 458, and 461 over some model flow path. (This, of course, is what's done in the efficiency models but with a different result in mind). In order to simplify the integrations required by Equations 451 and 455, it is assumed that the interface state is the same throughout the dispersion on any stage j. Further, we assume that the mass and heat transfer coefficients can be considered constant on any stage. (Similar assumptions were made in deriving Equations 439 et seq. for the point efficiency.) Then, by imposing a particular shape on the bulk phase composition and temperature profiles we find that the integrated total transport rates can be calculated as the product of the *average* fluxes and the *total* interfacial area, a_j:

$$\mathcal{N}_{ij}^V = \bar{N}_{ij}^V a_j \equiv N_{ij}^V(k_{ik}^V a_j, y_{kj}^I, \bar{y}_{kj}^V, \bar{N}_{kj}^V, k = 1, 2, \ldots, n) \tag{465}$$

$$\mathcal{N}_{ij}^L = \bar{N}_{ij}^L a_j \equiv N_{ij}^L(k_{ik}^L a_j, x_{kj}^I, \bar{x}_{kj}^L, \bar{N}_{kj}^L, k = 1, 2, \ldots, n) \tag{466}$$

$$\mathcal{E}_j^V = \bar{E}_j^V a_j \equiv E_j^V(h_j^V a_j, \bar{T}_j^V, T_j^I, \bar{N}_{kj}^V) \tag{467}$$

$$\mathcal{E}_j^L = \bar{E}_j^L a_j \equiv E_j^L(h_j^L a_j, \bar{T}_j^L, T_j^I, \bar{N}_{kj}^L) \tag{468}$$

where the mole fractions $\bar{y}_{ij}^V$ and $\bar{x}_{ij}^L$ and temperatures, $\bar{T}_j^V$ and $\bar{T}_j^L$ represent the integrated average bulk-phase conditions. If the bulk composition is assumed to be constant throughout the dispersion (corresponding to a well-mixed bulk phase), then the average mole fractions simply are equal to the mole fractions of the streams leaving the stage; $\bar{y}_{ij}^V = v_{ij}/V_j$, $\bar{x}_{ij}^L = \ell_{ij}/L_j$. On the other hand, if the bulk composition varies linearly between the entering and leaving values (corresponding to constant molar fluxes in each phase), then the average mole fractions are the arithmetic averages; $\bar{y}_{ij}^V = (1/2)(v_{ij}/V_j + v_{i,j+1}/V_{j+1})$, $\bar{x}_{ij}^L = (1/2) \times$ $\times (\ell_{ij}/L_j + \ell_{i,j-1}/L_{j-1})$. Finally, if we impose an exponential profile on the bulk mole fractions (as suggested by Equation 439, then the average composition is the logarithmic average of the entering and leaving values. Average temperatures can be calculated in an analogous fashion.

In their simulations of laboratory and pilot-plant scale tray columns, Krishnamurthy and Taylor [139] used the arithmetic average composition for the vapor phase (the more correct logarithmic average gives virtually identical results) and the outlet composition for the liquid phase. This corresponds almost exactly with the conventional model of plug flow of vapor through a well-mixed liquid (see, for example, King [128]). In their simulations of small-scale wetted-wall columns, they use the arithmetic average for both phases. These simple models give good agreement with the experimental data (given later), better even than some more complicated models.

In writing Equations 465 through 468 we have combined the interfacial area term directly with the heat and mass transfer coefficients. This is because many correlations (e.g., the AIChE method for bubble cap trays; see, again, King [128]) give the coefficient-area product. It is, of course, possible to use separate sources for these quantities; Zuiderweg [142] presents separate correlations for k and a for distillation on sieve trays. Notice also that Equations 465 through 468 are implicit in the mass transfer rates but not in the energy transfer rates; there are n − 1 of Equation 465, n − 1 of Equation 466 and one each of Equations 467 and 468.

All this is done so that the rate Equations 465 through 468 need be solved only once for each stage j. The assumptions that were made previously may be relaxed if desired by further dividing *each phase* into a number of regions and writing separate balance and rate equations for each region. The penalty is a large increase in the number of equations to be solved with a corresponding increase in the cost of obtaining a solution.

Interface Model

The interface is assumed to be a surface offering no resistance to transport and where equilibrium prevails. The usual equations of phase equilibrium relate the mole fractions on each side of the interface:

$$Q_{ij}^{I} \equiv K_{ij}x_{ij}^{I} - y_{ij}^{I} = 0 \tag{469}$$

$$S_j^V \equiv \sum_{i=1}^{n} y_{ij}^{I} - 1 = 0; \quad S_j^L \equiv \sum_{i=1}^{n} x_{ij}^{I} - 1 = 0 \tag{470}$$

where $K_{ij} \equiv K_{ij}(x_{ij}^I, y_{ij}^I, T_j^I, P_j)$ are the equilibrium ratios. As noted previously, the interfacial state is assumed to be uniform throughout the dispersion.

Variables and Functions for a Single Nonequilibrium Stage

Given the state of all feed streams and the flow rate of all side streams, heat loads, and pressures on the stage, then there is a total of 6n + 5 unknown quantities for each stage j. These are: the component vapor flow rates (v_{ij}: n in number), the component liquid flow rates (ℓ_{ij}: n), the vapor temperature (T_j^V), the liquid temperature (T_j^L), the interface temperature (T_j^I), the vapor composition at the interface (y_{ij}^I: n), the liquid composition at the interface (x_{ij}^I: n), the mass transfer rates (N_{ij}^V: n) and the energy transfer rates (E_j^V, E_j^L). The 6n + 5 independent equations that permit the calculation of these unknowns are as follows: component material balances for the vapor (M_{ij}^V: n), component material balances for the liquid (M_{ij}^L: n), component material balances around the interface (M_{ij}^I: n), the vapor-phase energy balance (E_j^V), the liquid-phase energy balance (E_j^L), the interface energy balance (E_j^I), the interface equilibrium relations (Q_{ij}^I: n), the summation equation (S_j^V; S_j^L), the vapor-phase mass transfer rate Equation 465 (n − 1), the liquid-phase mass transfer rate Equation 466 (n − 1), and the energy transfer rate Equations 467 and 468.

The nonlinearity in these algebraic equations stems from the presence of the K-values and enthalpies as well as the mass and energy transfer rate terms.

A reduction in this rather large number of equations and unknowns can be obtained by eliminating certain equations which are simple linear combinations of variables. Recognizing that there is really only one set of n strictly independent transfer rates, the N_{ij} ($= N_{ij}^V = N_{ij}^L$) say, we can eliminate the N_{ij}^V from Equation 465 and the N_{ij}^L from Equation 466 using the balance Equation 452:

$$R_{ij}^V \equiv N_{ij} - N_{ij}^V(k_{ikj}^V a_j, \bar{y}_{kj}^I, y_{kj}^V, N_{kj}, k = 1, 2, \ldots, n) = 0 \quad i = 1, 2, \ldots, n-1 \tag{471}$$

$$R_{ij}^L \equiv N_{ij} - N_{ij}^L(k_{ikj}^L a_j, x_{kj}^I, \bar{x}_{kj}^L, N_{kj}, k = 1, 2, \ldots, n) = 0 \quad i = 1, 2, \ldots, n-1 \tag{472}$$

The energy transfer rate Equations 467 and 468 are substituted into the interface energy balance to give

$$E_j^I \equiv \mathscr{E}_j^V(h_j^V a_j, \bar{T}_j^V, T_j^I, y_{kj}^V, N_{kj}) - \mathscr{E}_j^L(h_j^L a_j, \bar{T}_j^L, T_j^I, x_{kj}^L, N_{kj}) = 0 \tag{473}$$

Note that only n mass transfer rates N_{ij} appear in Equations 471 through 473; the transfer rates N_{ij}^L (or N_{ij}^V), $\mathscr{E}_j^V$, and $\mathscr{E}_j^L$ are eliminated from the list of independent variables. The balance equations (M_{ij}^I: n) and the energy transfer rate Equations 467 and 468 are removed from the set of independent functions.

Two more variables, x_{nj}^I and y_{nj}^I can easily be eliminated from the set of variables. They can be computed directly from the summation equations (S_j^V, S_j^I) which, therefore, are dropped from the list of independent equations.

The final set of 5n + 1 independent variables per stage is ordered into a vector (X_j) as follows

$$(X_j)^T \equiv (v_{1j}, v_{2j}, \dots v_{nj}, T_j^V, T_j^L, \ell_{1j}, \ell_{2j}, \dots \ell_{nj}, N_{1j}, N_{2j}, \dots N_{nj}, y_{1j}^I, y_{2j}^I, \dots y_{n-1,j}^I, T_j^I, x_{1j}^I, x_{2j}^I, \dots x_{n-1,j}^I)$$

The 5n + 1 independent equations corresponding to this set of variables are ordered into a vector (F_j) as

$$(F_j)^T \equiv (M_{1j}^V, M_{2j}^V, \dots M_{nj}^V, E_j^V, E_j^L, M_{1j}^L, M_{2j}^L, \dots M_{nj}^L, R_{1j}^V, R_{2j}^V, \dots R_{n-1,j}^V, E_j^I, Q_{1j}^I, Q_{2j}^I, \dots Q_{nj}^I, R_{1j}^L, R_{2j}^L, \dots R_{n-1,j}^L)$$

This concludes the development of the "MERQ equation" (an acronym for *M*aterial balances, *E*nergy balances, *R*ate equations and e*Q*uilibrium relations) for a single stage.

Solving the MERQ Equations

Methods of solving the equilibrium stage MESH equations fall into two groups: tearing methods (where subsets of the complete set of equations are solved in sequence) and "simultaneous correction" (SC) methods (where all of the equations are solved simultaneously).

Krishnamurthy and Taylor [138] tried several different approaches to solving the MERQ equations and found solving all of the equations simultaneously using Newton's method or the hybrid method of Lucia and co-workers [64–66] to be most effective.

Application to a Multistage Process: An Absorber Perhaps

Absorbers frequently are modeled as simple sequences of equilibrium stages [59, 128–130]. If, instead, we choose to model the column by a sequence of s nonequilibrium stages, the vectors of variables and functions corresponding to the entire column are

$$(X)^T \equiv ((X_1)^T, (X_2)^T \dots (X_s)^T)$$

$$(F)^T \equiv ((F_1)^T, (F_2)^T \dots (F_s)^T)$$

where the variables and functions have been grouped by stage as done by Naphtali and Sandholm [143] in their formulation of the equilibrium stage MESH equations. When grouped in this way, the vector of stage functions, (F_j), depends only on the variables for the three adjacent stages, $j-1$, j and $j+1$. Thus the Jacobian matrix has the familiar block tri-diagonal structure shown in Figure 24. The linear system Equation 118 is easily and efficiently solved when [J] is block tri-diagonal using the matrix generalization of the well known Thomas algorithm (Henley and Seader [59], for example, give the steps necessary to implement what is just Gaussian elimination of submatrices). The submatrices $[U_j]$, $[V_j]$, $[W_j]$ are extremely sparse; $[U_j]$ and $[W_j]$ are almost empty. The structure of these submatrices can be derived by straightforward differentiation of the MERQ equations.

It is interesting to compare the total of $5n+1$ variables per stage with the $2n+1$ variables per stage in the formulation of the MESH equations due to Naphtali and Sandholm [143] (see, also, Henley and Seader [59]; Holland [130]). For example, for a ten stage column separating a ternary mixture, the equilibrium-stage model comprises a mere 70 equations compared to 160 equations for the nonequilibrium-stage model. For a fifty-stage, five-component separation process, the equilibrium-stage model boasts 550 equations. This compares to the nonequilibrium stage model with no less than 1,300 equations. Clearly, there is a considerable difference in the size of the problem being solved. However, with a suitable sparse matrix handling computer codes, the large size of the system can easily be accommodated.

Multicomponent Distillation in a Multinonequilibrium-Stage Column

Consider a distillation column equipped with a total condenser and reboiler. A total condenser/reboiler may be described by a set of $2n+1$ equations: n component material

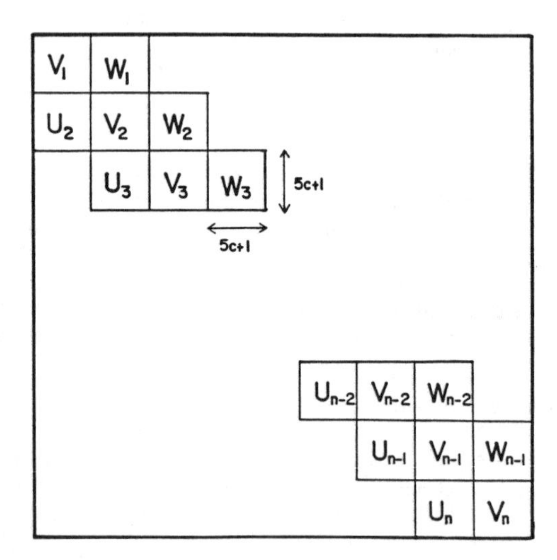

Figure 24. Block tridiagonal structure of the Jacobian matrix for a column modeled by a sequence of non-equilibrium stages; an absorber for example.

balances, a specification function (to replace the energy balance), $n-1$ equations equating the mole fractions in the product and reflux streams and an equation specifying the temperature of the product streams (e.g., $T = T_{bub}$, the bubble point temperature of the product).

For a column modeled in this way, the Jacobian matrix retains its block band structure but the submatrices $[W_1]$, $[U_2]$, $[W_{s-1}]$, and $[U_s]$ are no longer square. Rather, they are rectangular with dimensions $(2n+1)\times(5n+1)$ or $(5n+1)\times(2n+1)$ as indicated in Figure 25. It should be recognized that the steps of the matrix generalization of the Thomas algorithm apply unchanged to the triangularization of the matrix in Figure 25. Provision need only be made for multiplication of rectangular matrices.

Distillation at Total Reflux

The maximum separation possible in a given number of stages is attained at total reflux when the overhead vapor is completely condensed and returned to the column. This situation is also attractive for experimental purposes as no material is lost from the column. Using a mass balance around any top section of the column it is easy to show that

$$v_{i,j+1} = \ell_{ij}$$

This suggests that either the set of component vapor flows, v_{ij}, or the component liquid flows, ℓ_{ij}, can be removed from the set of variables representing the j-th nonequilibrium stage. The component material balance equations (M_{ij}^V or M_{ij}^L) would then be removed from the set of equations for that stage. We will then be left with $4n+1$ variables and functions per stage.

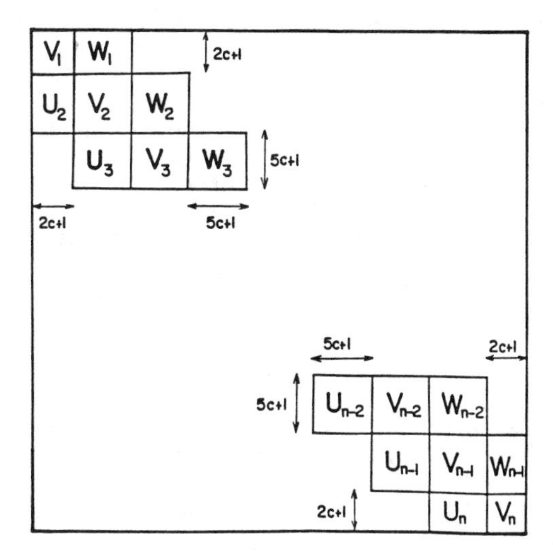

Figure 25. Block tridiagonal structure of the Jacobian matrix for a distillation column modeled as a sequence of nonequilibrium stages together with an equilibrium reboiler and condenser or with a total condenser and total reboiler.

Single-Phase Control

In some separation processes the resistance to mass transfer lies predominantly in one phase. Distillation is a process which, sometimes, is controlled by the vapor-phase resistance [125, 144–147]. If we *assume* that the liquid-phase resistance can be neglected, then the interface liquid composition is that of the bulk liquid and it is no longer necessary to consider the $n-1$ x_{ij}^{I} as independent variables. We are then left with $4n+2$ variables per stage which, of course, means that the number of equations must also be reduced by $n-1$. As there is no resistance to mass transfer in the liquid phase, it is no longer possible to calculate the mass transfer rates in the liquid from the $n-1$ rate equations R_{ij}^{L} which, therefore, are dropped from the set of equations for the j-th stage.

For liquid-phase control, the y_{ij}^{I} and R_{ij}^{V} would be eliminated from the sets of variables and equations.

Vapor-Phase-Controlled Distillation at Total Reflux

For a vapor-phase controlled distillation at total reflux, we have to deal with only $3n+2$ equations and variables per nonequilibrium stage:

$$(X_j)^T \equiv (v_{1j} \ldots v_{nj}, T_j^V, T_j^L, N_{1j} \ldots N_{nj}, y_{1j}^I \ldots y_{n-1,j}^I, T_{jI})$$

$$(F_j)^T \equiv (M_{1j}^V \ldots M_{nj}^V, E_j^V, E_j^L, R_{1j}^V \ldots R_{n-1,j}^V, E_j^I, Q_{1j} \ldots Q_{nj})$$

We note that this situation is encountered in many of the experimental simulations discussed later.

Comparison with Experiment

The development of a new model of a process should be followed by a comparison of its predictions with the results of experiments and with the predictions of rival models. Such a comparison helps to determine if the model is a good one. Krishnamurthy and Taylor [139] have compared the predictions of the nonequilibrium stage model with the experimental data of Vogelpohl [144] (distillation of acetone, methanol and water and of methanol, isopropanol, and water in a bubble-cap-tray column), Free and Hutchison [148] (distillation of acetone, methanol, and ethanol in a bubble-cap-tray column) and Nord [149] (distillation of benzene, toluene, and m-xylene in a bubble-cap-tray column). In all cases, the model did a *very* commendable job of predicting the composition profiles measured for these systems; average absolute differences between predicted and measured mole fractions were seldom greater than 4 mole percent and were often very much less. The simulations of the experiments by Vogelpohl [144] are particularly interesting and are discussed in more detail later.

Vogelpohl [144] has reported some results for the distillation of two ternary systems acetone, methanol, and water and methanol, isopropanol, and water in a 38-tray pilot-plant scale bubble-cap column of 0.3 m diameter and 0.2 m between trays. The experiments were carried out at total reflux. Due to the ease of separating these particular systems, only up to 13 trays were active for the experimental runs for which composition profiles and F-factors are reported. The experiments clearly show that the component Murphree efficiencies are unequal; indeed, in the acetone-methanol-water system, the composition of methanol passes through a maximum in the column and the efficiency for this component becomes unbounded (more on this later).

Krishnamurthy and Taylor [139] tried to predict the measured composition profiles using a model composed of a sequence of $s-2$ nonequilibrium stages ($s-2$ is the number of trays) together with a total condenser (numbered as stage 1) and a reboiler (numbered as stage s). For the special case of vapor-phase-controlled distillation at total reflux, the nonequilibrium stages are modeled by just 3n + 2 equations as shown earlier. For the total condenser, only the component liquid flow rates and the temperature of the reflux are included in the list of variables. The equations corresponding to this set are

$$v_{i2} - \ell_{i1} = 0, \quad i = 1, 2, \ldots, n \tag{474}$$

and

$$T_1^L - T_{bub} = 0 \tag{475}$$

where T_{bub} is bubble point temperature of a liquid with composition ℓ_{i1}/L_1. Equation 475 follows from the fact that the vapor condensed in stage 1 was returned to the column as saturated reflux.

The lack of any feed to the experimental columns meant that it was necessary to tear flow streams at the reboiler end and specify the component vapor flow rates and the temperature of the vapor leaving the reboiler (or a stage above the reboiler). Thus the variables for the last stage are the component vapor flow rates and vapor temperature. The corresponding functions are

$$v_{is} - v_{is,spec} = 0, \quad i = 1, 2, \ldots, n \tag{476}$$

and

$$T_s^V - T_{s,spec}^V = 0 \tag{477}$$

The component vapor flows, $v_{s,spec}$, were calculated from the reported F-factors and the measured composition at the reboiler outlet (not the composition of the batch feed to the reboiler). If the vapor temperature, $T_{s,spec}^V$, was not measured, then they assumed that the vapor was saturated and replaced $T_{s,spec}^V$ in Equation 477 by T_{dew}, the dew point temperature of a vapor with composition v_{is}/V_s.

Mass transfer rates were calculated from equations based on a film model. Several interactive methods were used in the simulations (Krishna–Standart, Toor–Stewart–Prober, Krishna [91], Taylor–Smith); all of them predicted identical composition profiles. In addition, the simulations were repeated using Wilke's effective diffusivity method to calculate the fluxes and an "equal diffusivity" method (in which an average value of all of the binary diffusion coefficients was applied to all of the species). The results of these simulations will be discussed later.

For the Krishna–Standart and Toor–Stewart–Prober methods, the vapor-phase rate equations can be written in n − 1 dimensional matrix form as (see Equations 113)

$$(R_j^V) \equiv (N_j) - c_t^V [k_j^V] a_{kj} [\Xi_j^V] (\bar{y}_j^V - y_j^I) - N_{tj}(\bar{y}_j^V) = (0) \tag{478}$$

Since the film thickness is not known, the matrices $[k_j^V]$ and $[\Phi_j]$ were calculated as described in the previous section Equations 427 through 430. The AIChE correlation of "Numbers of Transfer Units" for gas-phase-controlled mass transfer on trays was rewritten to express the binary mass transfer coefficient-interfacial area product ($k_{ik}a$) directly as a function of the physical and operating characteristics of the system:

$$c_t^V k_{ik} a = \frac{(0.776 + 4.567\,W - 0.2377\,F + 87.319\,L_F)V}{(\mu/\varrho_t^V D'_{ik})^{0.5}} \tag{479}$$

The film model of simultaneous heat and mass transfer in multicomponent systems leads to the following expression for the energy transfer rate through the vapor film (Equation 302 multiplied by the interfacial area a)

$$\mathscr{E}^V = h^V a \frac{\Phi_H^V}{\exp(\Phi_H^V) - 1}(T^V - T^I) + \sum_{i=1}^{n} N_i \bar{H}_i^V \tag{480}$$

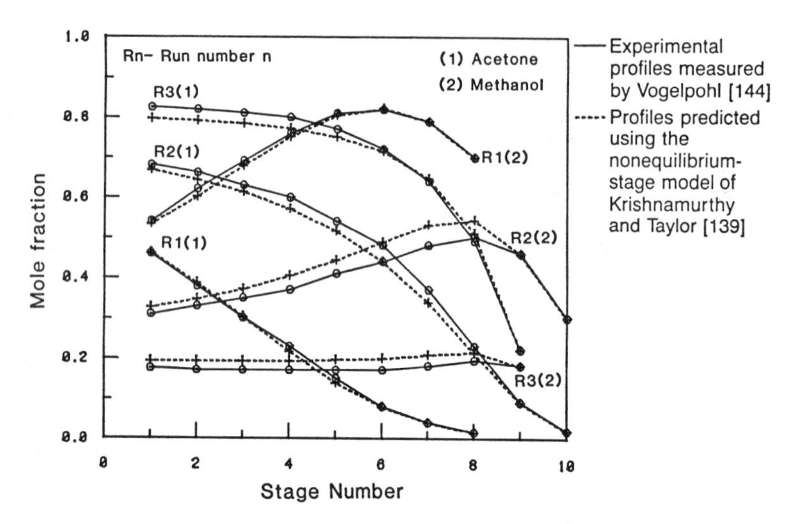

Figure 26. Composition profiles obtained during the distillation of a mixture of acetone (1), methanol (2), and water (3) at total reflux.

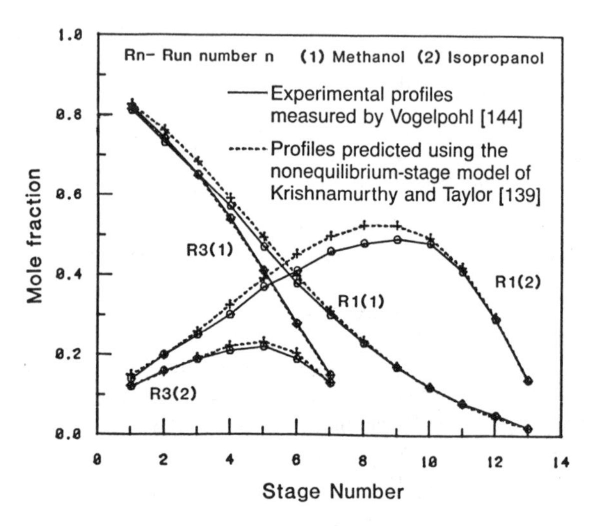

Figure 27. Composition profiles obtained during the distillation of a mixture of methanol (1), isopropanol (2), and water (3) at total reflux.

where Φ_H^V is defined by

$$\Phi_H^V = \sum_{i=1}^{n} N_i C_{pi}^V / h^V a \tag{481}$$

and h^V is the low-flux vapor-phase heat transfer coefficient.

For the liquid phase, a relation similar to Equation 480 can be written

$$\mathscr{E}^L = h^L a(T^I - T^L) + \sum_{i=1}^{n} N_i \bar{H}_i^L \tag{482}$$

where h^L is the liquid-phase heat transfer coefficient. The high flux correction to h^L has been ignored; the resistance to heat transfer in the liquid phase is very much smaller than in the vapor phase and the correction factor $\Phi_N^L/(1-e^{\Phi_N^L})$ usually is close to unity. Equation 482 can be combined with Equations 480 and 473 to give, for each stage j.

$$E_j^I \equiv h_j^V a_j \frac{\Phi_{Hj}^V}{\exp(\Phi_{Hj}^V) - 1} (\bar{T}_j^V - T_j^I) - h_j^L a_j (T_j^I - \bar{T}_j^L) + \sum_{i=1}^{n} N_{ij}(\bar{H}_{ij}^V - \bar{H}_{ij}^L) \tag{483}$$

In the absence of specific methods of evaluating individual phase heat transfer coefficients, the Chilton–Colburn analogy was used to relate k_{ik} to h^V.

$$h^V = k_{av} C_{pm}^V (Le)^{2/3} \tag{484}$$

The liquid-phase heat transfer coefficient was arbitrarily set to 1,000 h^V in order to keep the liquid-phase saturated (a condition observed in all of the experiments). It was found, however, that the composition profiles predicted by the model were virtually independent of the numerical value of the heat transfer coefficients.

Predicted and measured composition profiles for the systems acetone-methanol-water and methanol-isopropanol-water are compared in Figures 26 and 27. The average error for the acetone-methanol-water system is 1.122 mole percent and the maximum error is 2.92 mole percent [139]. These errors are well within acceptable limits. The errors for the system methanol, isopropanol, and water are slightly higher. However, Figure 27 shows the observed profiles are followed quite closely.

It is worth emphasizing once more that the nonequilibrium-stage concept does away entirely with the need to calculate or even to define stage efficiencies. Nonetheless, we may, if we wish, back out values for these quantities from the results of a simulation. In Figures 28 and 29 we show the component efficiencies predicted by the nonequilibrium-stage model for two of the experiments with the system acetone-methanol-water (Vogelpohl has similar figures). It is to be noted that the predicted efficiencies of acetone and water are not very different and remain more or less constant over the height of the column (Figure 28). The stage efficiency of methanol, however, varies considerably and takes values greater than unity and less than zero. In complete contrast, mass trasfer coefficients usually are much less concentration dependent than are stage efficiencies. Krishnamurthy and Taylor [139] observed that the mass transfer coefficients calculated from the AIChE correlation varied very little from tray to tray. A similar observation was made by Krishna et al. [125].

In view of this wide range in the individual component efficiencies, it is pertinent to wonder how the "equilibrium stage" model together with a single value of the stage efficiency (the conventional approach) would fare in these situations. Figure 30 shows how, with a comparison of the interactive and equal diffusivity models for two of Vogelpohl's experiments; there are "large" differences between the two models for the acetone-methanol-water system. To give a feel for the magnitude of the discrepancy, we note that for run number 1 of the acetone-methanol-water system, the measured mole fractions of acetone and methanol on the top stage are 0.46 and 0.54 respectively. The mole fractions predicted using an interactive model of mass transfer are 0.4613 and 0.5334 respectively whereas the equal diffusivity method (equivalent to the conventional equilibrium stage model with a common value for the component efficiencies on any stage j) predicts mole fractions of 0.5215 and 0.4716 respectively. The interactive models are clearly superior in this case.

Implications for Design

A note of caution should be introduced here. We do not believe that a comparison of composition profiles obtained in total reflux experiments is the best way to establish the significance of interaction effects. It is better to compare predicted product distributions in operating problems or, alternatively, the required numbers of stages in design problems—the kind of calculation first made by Toor and Burchard [3] and, using the more rigorous model described in this section, by Krishnamurthy and Taylor [140].

In considering a large number of simulation and design calculations, it must be said that the conventional equilibrium-stage model modified by a stage efficiency that is the same for all components (but may vary from tray to tray) often compares very favorably with the more realistic nonequilibrium-stage model. This, perhaps, goes some way towards explaining the continued use of the equilibrium-stage concept for so many years. However, the discrepancies between the two fundamentally different models can be very large indeed. Figure 31 shows the composition profiles predicted by the nonequilibrium- and equilibrium-stage models for, the distillation of ethanol, tert-butanol and water in a 79-stage column (recall that this is the same ternary system used in the experiments of Krishna et al. [125] discussed at the beginning of this section). The problem specification called for a distillate, 90% of which was ethanol. The equilibrium-stage model together with a stage efficiency that

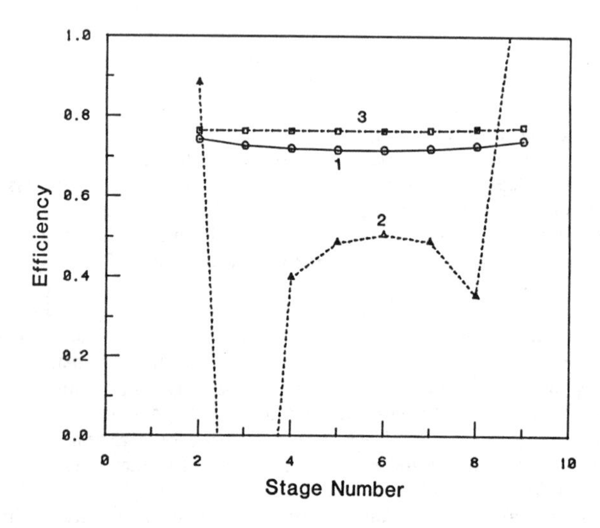

Figure 28. Component efficiency profiles computed from the results of the nonequilibrium-stage model for Run 3 with the system acetone-methanol-water. Experimental composition profile shown in Figure 26.

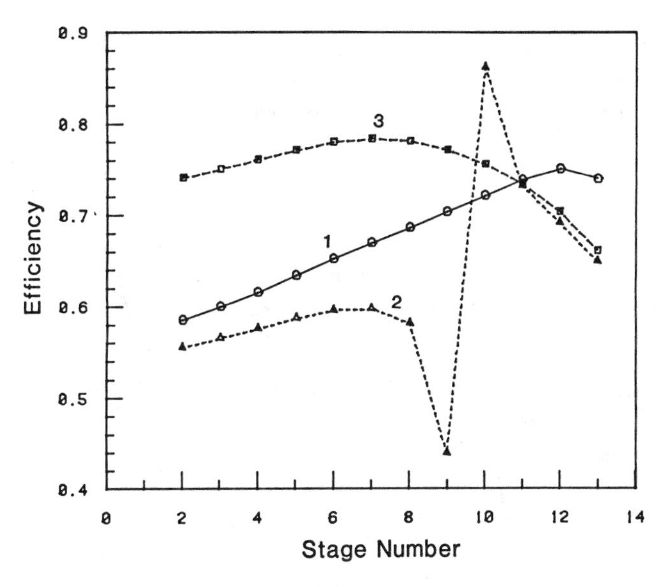

Figure 29. Component efficiency profiles computed from the results of the nonequilibrium-stage model for Run 1 with the system methanol-isopropanol-water. Experimental composition profile shown in Figure 27.

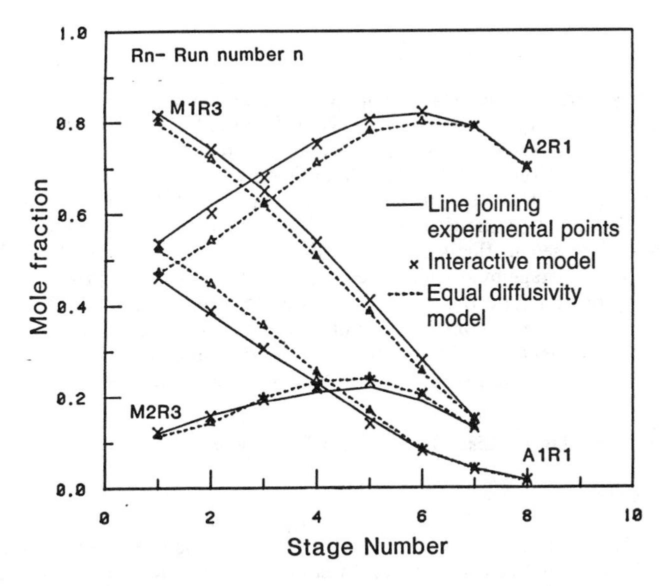

Figure 30. Comparison of experimental composition profiles with the profiles predicted by the interactive and equal diffusivity methods: Am, component m of the acetone-methanol-water system; Mm, component m of the methanol-isopropanol-water system.

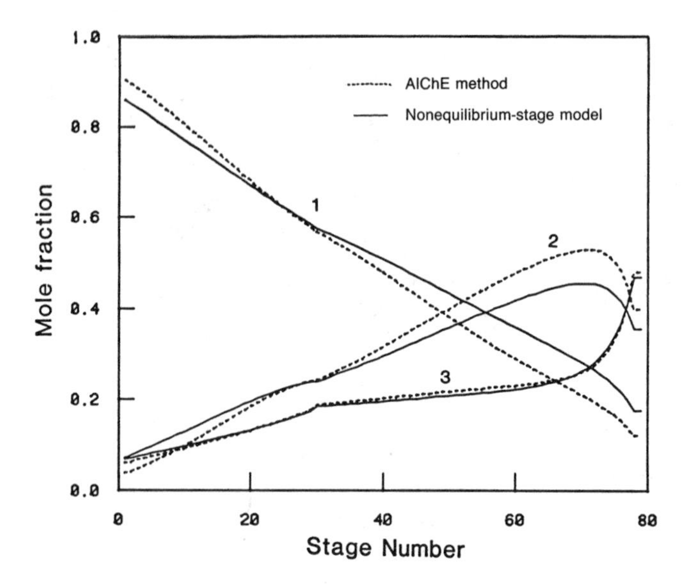

Figure 31. Composition profiles predicted by the equilibrium-stage model (with an efficiency computed from the AIChE method) compared with the composition profiles predicted by the nonequilibrium-stage model for the distillation of ethanol (1), *tert* butanol (2), and water (3) in a 79-stage column (from Krishnamurthy and Taylor [140]).

was allowed to vary from stage to stage (computed from the AIChE method using the key components) led to the total of 79 stages as shown in Figure 31. The nonequilibrium-stage model together with an interactive model of mass transfer predicts that the mole of ethanol in the distillate produced from the 79-stage column is only 0.85. This might not seem like a large difference (only five mole percent); but to achieve a distillate 90% of which is ethanol using the nonequilibrium-stage model requires no less than 121 stages [140].

The few results discussed in this section should serve to undermine the hold that the conventional equilibrium-stage model has on computer-aided separation process design. In the next section we demonstrate the versatility of the nonequilibrium-stage approach by applying it to the simulation of continuous contact equipment. In particular, we provide evidence that the model does a very commendable job of predicting the product distributions that were measured in full-scale packed distillation and absorption columns.

Multicomponent Distillation in Continuous-Contact Equipment

Heights and Numbers of Transfer Units

Component HTU's and NTU's are traditionally used to describe the mass transfer characteristics of multicomponent distillation in equipment in which we have continuous contact between the vapor and liquid phases (e.g., packed columns, wetted-wall columns, etc.). By a treatment parallel to the one on tray columns, the following conclusions can be drawn concerning the component numbers of transfer units and height of transfer units:

1. In general, for multicomponent systems made up of species of dissimilar size and nature:

$$HTU_{0y1} \neq HTU_{0y2} \neq HTU_{0y3} \neq \ldots \neq HTU_{0yn} \quad (485)$$

$$NTU_{0y1} \neq NTU_{0y2} \neq NTU_{0y3} \neq \ldots \neq NTU_{0yn} \quad (486)$$

   and the larger the system nonideality, the larger the differences in the component values.
2. Component HTU's and NTU's depend not only on the system hydrodynamics but also on the operating conditions. More specifically, these component values will show a direct dependence on the ratio of driving forces $\Delta y_i/\Delta y_j$. Depending on the magnitude and sign of the ratio $\Delta y_i/\Delta y_j$, a component NTU or HTU could become unbounded, i.e., assume values anywhere in the range $-\infty$ to $+\infty$.

In order to confirm the above conclusions, Krishna et al. [127] obtained experimental data for distillation of ethanol (1) tert butanol (2) water (3) in a 55.6 mm diameter column with a test section of height 1.013 m. Their experimental data is summarized in part in Table 14.

Krishna et al. [127] present a detailed theoretical analysis of the experimental data in Table 14; here we shall be content with a summary of the major points. According to their calculations, the elements of the matrix of overall vapor-phase transfer units $[NTU_{0y}]$ has the typical structure

$$[NTU_{0y}] = \begin{bmatrix} 0.7 & 0.1 \\ 0.1 & 0.7 \end{bmatrix} \quad (487)$$

Table 14
Experimental Data on Component Heights of Transfer Units of Distillation of the Mixture Ethanol (1)—*Tert* Butanol (2)—Water (3)*

Run Number	Driving Force $\frac{\Delta y_1}{\Delta y_2}$ (Bottom of Column)	Component Heights of Transfer Units HTU_{oy1}	HTU_{oy2}	HTU_{oy3}
1	0.4494	0.96	1.23	1.13
2	0.5152	0.98	1.38	1.23
3	0.4676	1.04	1.25	1.17
4	0.4872	1.32	1.36	1.34
5	0.8774	1.29	1.28	1.29
6	0.8725	1.27	1.28	1.28
7	0.8217	1.22	1.40	1.36
8	0.8143	1.22	1.34	1.28
9	0.7793	1.19	1.37	1.28
10	0.7723	1.21	1.43	1.33
11	0.7918	1.08	1.32	1.20
12	0.3053	1.05	1.25	1.19
13	0.2867	1.18	1.41	1.35
14	0.2752	1.13	1.35	1.29
15	0.2527	0.98	1.37	1.26
16	0.2351	1.07	1.27	1.22
17	0.2251	0.87	1.22	1.13
18	0.2571	0.79	1.10	1.02
19	2.9210	1.36	2.27	1.49
20	1.7088	1.57	1.12	1.39
21	1.5591	1.54	1.27	1.43

* *In 55.6 mm diameter wetted-wall column of height 1.013 m. Data from Krishna et al. [127].*

The variation of the bulk vapor composition along the column height is described by the differential equation [see Krishna et al. [127]):

$$\frac{d(y)}{d\xi} = [NTU_{0y}]\,(y^* - y) = [NTU_{0y}]\,(\Delta y) \tag{488}$$

where we use the notation $\Delta y_i = y_i^* - y_i$.

If we define pseudo-binary height of transfer unit for component i by the integral

$$HTU_{0yi} = \frac{Z}{\displaystyle\int_{Bottom}^{Top} \frac{dy_i}{y_i^* - y_i}} \tag{489}$$

then it is easy to derive the following expressions for the pseudo-binary HTU_{0yi} for small changes in the vapor composition

$$HTU_{0y1} = \frac{Z}{NTU_{0y11} + NTU_{0y12}\dfrac{\Delta y_2}{\Delta y_1}} \tag{490}$$

$$HTU_{0y2} = \frac{Z}{NTU_{0y22} + NTU_{0y21}\dfrac{\Delta y_1}{\Delta y_2}} \tag{491}$$

$$HTU_{0y3} = \frac{\Delta y_1 + \Delta y_2}{\dfrac{1}{NTU_{0y1}}\Delta y_1 + \dfrac{1}{NTU_{0y2}}\Delta y_2} \tag{492}$$

With the value of Z = 1.013 and $[NTU_{0y}]$ given by Equation 487 the following relative trends in the values of HTU_{0y1}, HTU_{0y2}, and HTU_{0y3} can be deduced depending on the value of the ratio of the driving forces $\Delta y_1/\Delta y_2$:

- If $\Delta y_1/\Delta y_2 \approx 1$, then $HTU_{0y1} \approx HTU_{0y2} \approx HTU_{0y3}$

  This is found to be the case for Runs 5 and 6.

- If $\Delta y_1/\Delta y_2 \ll 1$, then $HTU_{0y2} > HTU_{0y3} > HTU_{0y1}$

  This trend is confirmed experimentally by Runs 1, 3, 4, and 7−17.

- If $\Delta y_1/\Delta y_2 \gg 1$, then $HTU_{0y2} < HTU_{0y3} < HTU_{0y1}$

  This trend is confirmed experimentally by Runs 20 and 21.

The only experiment in which the observed trend does not meet with our expectation is Run 19. Krishna et al. [127] suggest that this deviation of theoretical expectation from observation is due to the fact that the driving force Δy_2 is the same order as the accuracy of the experimental composition determination. In other words, the calculations for Run 19 are quite sensitive to the experimental errors.

If the interphase mass transfer formulation is chosen in which the crosscoefficients are ignored, then the observed dependence of the component HTU_{0yi} on the ratio of driving forces $\Delta y_1/\Delta y_2$ cannot be explained. Krishna et al. [127] have presented a detailed simulation of the wetted-wall distillation column in which the matrices $[k_y^\bullet]$ and $[k_x^\bullet]$ in the vapor and liquid phases are estimated on the basis of binary correlations; they were able to predict the outlet vapor composition within the accuracy restraints of the vapor-liquid equilibrium fit and experimental composition determinations.

The broad conclusions to be drawn from the study of Krishna et al. [127] is that pseudo-binary approaches (e.g., use of HTU_{0yi} or NTU_{0yi}) are unsuitable for describing the behavior of multicomponent distillation. The most rational approach to design is to attack the problem head-on by using the appropriate matrices of transfer coefficients in the adjoining fluid phases.

Application of the Nonequilibrium-Stage Model to Continuous Contact Equipment

It is straightforward to apply the nonequilibrium model, described in the previous subsection for simultating multistage equipment, to continuous contact equipment such as wetted-wall and packed columns [139, 140]. To simulate such a device, the column is arbitrarily divided into a number of sections, each of which is modeled just as though it were a nonequilibrium stage. It might be supposed that a large number of sections would be required in order to obtain an accurate numerical solution. In practice, it is found that relatively few sections are needed. For example, only eight or ten sections suffice to model a packed absorber that is thirty feet high [141]. The description of the model in the previous subsection applies almost verbatim to a packed column modeled by a number of nonequilibrium stages (sections). The *only* differences between models for the two classes of separating equipment are the correlations/expressions that must be used to estimate the binary mass transfer coefficients and the interfacial area. For a wetted-wall column, the interfacial area would be known from the geometry and the binary mass transfer coefficients could be estimated from the Gilliland–Sherwood correlation (for example) Equation 418. Correlations for k and a for packed absorbers and distillation columns are discussed by, for example, Bravo and Fair [150].

Comparison with Experiment

This nonequilibrium-stage approach to the modeling of continuous contacting equipment has been verified by Krishnamurthy and Taylor [139, 141]. The experiments of Johnstone and Pigford [151] (distillation of the benzene-toluene, ethanol-water and acetone-chloroform binaries in a wetted-wall column) and Dribicka and Sandall [146] (distillation of the ternary system benzene-ethylbenzene-toluene in a wetted-wall column) were simulated using a model composed of ten nonequilibrium stages. In the total of thirty-two experiments that were simulated, the average absolute discrepancy between predicted and measured mole fractions was less than 1 1/2 mol percent and the maximum discrepancy only 5 mole percent! Figure 32 provides a comparison between predicted and measured composition profiles for three of the experiments carried out by Dribicka and Sandall [151]; note the excellent agreement between model prediction and experimental measurement.

Simulation of Large-Scale Packed Columns

It is one thing to simulate laboratory-scale experiments, quite another to simulate the performance of commercial-scale equipment. Data taken on such equipment are scarce indeed. McDaniel, Bassyoni, and Holland [152–154] conducted field tests on a packed distillation column and a packed absorber operating at the Zoller Gas Plant in Refugio,

Texas. The distillation column, cosisting of two sections 5.2 m in height randomly packed with two-inch metallic Pall rings, was used to separate a mixture of eleven components (straight chain hydrocarbons and their isomers). The absorber was 7.01 m in height and filled with two-inch metallic Pall rings. The gas feed to the absorber contained fourteen species (the components in the distillation column plus nitrogen, carbon dioxide, and methane). McDaniel et al. used their measurements to show that the individual component efficiencies are different. Krishnamurthy and Taylor [141] simulated their field tests using nonequilibrium-stage model previously described. We refer readers to their paper for complete details of the simulations and results. We must content ourselves with the observation that, in view of the quite large uncertainties in the data [153, 154], the results (as can be seen from our summary in Table 15) are very encouraging, lending further support to the concepts underlying the nonequilibrium-stage model.

Condensation of Vapor Mixtures

Condensation of vapor mixtures is an operation of great significance in the chemical process industries. The two words "vapor mixture" cover a wide range of situations. One limit of this range is one in which all components have boiling points above the maximum coolant temperature; in this case, the mixture can be totally condensed. The other limit is a mixture in which at least one component in the initial vapor stream has a boiling point lower than the minimum coolant temperature and, also, is negligibly soluble in the liquid condensate formed from the other components and hence cannot be condensed at all. Examples of such components include nitrogen and helium. An intermediate case of some importance is typified by a mixture of light hydrocarbons in which the lightest members often cannot be condensed as pure components at the temperatures encountered in the condenser but, instead, will dissolve in the heavier components. In each of these cases the vapor mixture may form a partially or totally immiscible condensate.

Existing methods for *designing* heat exchangers to condense multicomponent mixtures are of two basic kinds: equilibrium methods, such as those of Kern [155], Silver [156], and Bell and Ghaly [157], and the differential or nonequilibrium methods that have developed following the original work of Colburn and Drew [158]. In the latter class of methods, a set of one-dimensional differential material and energy balances is integrated numerically along the length of the condenser. Each step of the integration requires the local mass and energy transfer rates to be calculated using, for example, equations based on a film model [158–170], and/or turbulent eddy diffusivity model [171–172]. Still more sophisticated nonequilibrium models based on boundary-layer theory are limited primarily to describing the condensation of binary vapors or of one vapor in the presence of a noncondensing gas. Extensions of the boundary-layer models to multicomponent systems are few in number [173–175] and have not been developed to the point where they could be used in the design of heat exchangers of complex geometry. While the equilibrium methods are widely used (the reasons being their simplicity, rapidity in computation, and because there is no need to compute intermediate vapor compositions or to obtain diffusivity data), the one-dimensional nonequilibrium methods are more soundly based and appear to be attracting increasing interest from condenser designers (see, for example, McNaught [176, 177]; and Butterworth [178]; and Owen and Lee [179]). It is interesting to observe that Butterworth has included a summary of both the equilibrium and the one-dimensional nonequilibrium approaches in his chapter on condensation in Volume 2, which covers the fundamentals of fluid mechanics and heat transfer, of the recently published *Heat Exchanger Design Handbook* (Schlunder [180]) whereas Volume 3, which presents design procedures for heat exchangers of all kinds, includes only a procedure based on the equilibrium methods in the chapter on condensers. It is our belief that it will be only a short time before the more fundamental nonequilibrium models are accorded a section of their own in future editions of Volume 3.

It is our objective here to describe the nonequilibrium models of condensation, to discuss efficient methods of solving the nonlinear equations that constitute the model and to present some experimental support for the models.

A Model of Multicomponent Condensation

A schematic representation of a short section of a single condenser tube is shown in Figure 33. The heat lost by the vapor, thereby causing some of it to condense, is transferred through the condensate, through the tube wall, and into the coolant. The coolant may flow cocurrently with or counter-currently to the vapor and liquid streams which are flowing cocurrently along the tube.

Material and Energy Balance Relations

As always with chemical process calculations, we start from the appropriate material and energy balances which, here, are written around a section of condenser tube of differential area (Figure 32). For the vapor phase, the component material balance reads:

$$\frac{dv_i}{dA} = -N_i^V, \quad i = 1, 2, \ldots, n \tag{493}$$

and for the liquid phase

$$\frac{d\ell_i}{dA} = N_i^L, \quad i = 1, 2, \ldots, n \tag{494}$$

The terms on the right-hand sides of Equations 493 and 494 are the molar fluxes of species i in the vapor and liquid phases respectively; we assume that transfers from the vapor phase to the liquid phase are positive. From a component material balance around the entire differential section, we conclude that

$$N_i^V = N_i^L = N_i, \quad i = 1, 2, \ldots, n \tag{495}$$

Table 15
Summary of Simulations of Packed Distillation and Absorbtion Columns

Parameters	Distillation Column		Absorber	
Height (m)	10.363		7.0104	
Diameter (m)	0.9144/1.2192		0.9144	
†Pressure (Pa)	1.2×10^6		5.5×10^6	
†Reflux ratio	0.90–1.50		-------	
†Feed rate (kmol s^{-1})	0.045–0.056		0.330–0.400 (gas); 0.020–0.029 (liquid)	
Reboiler/bottom temp. (K)	525.0–535.0		264.0–265.0	
Condenser/top temp. (K)	300.0–315.0		268.0–270.0	
	Exptl.	*Predicted*	*Exptl.*	*Predicted*
Errors in product temps.				
Reboiler/bottom (ΔK)	1.120	5.322	1.120	2.596
Condenser/top (ΔK)	2.800	0.480	2.800	1.170
Errors in product compositions (mf)				
Mole fraction > 0.1	0.001	0.001469	0.001	0.024760
Mole fraction 0.01–0.1	--	0.002666	--	0.003638
Mole fraction < 0.01	0.0001	0.001186	0.0001	0.0008891

* *Adapted from Krishna Murthy and Taylor [141].*
† *Specified quantities*

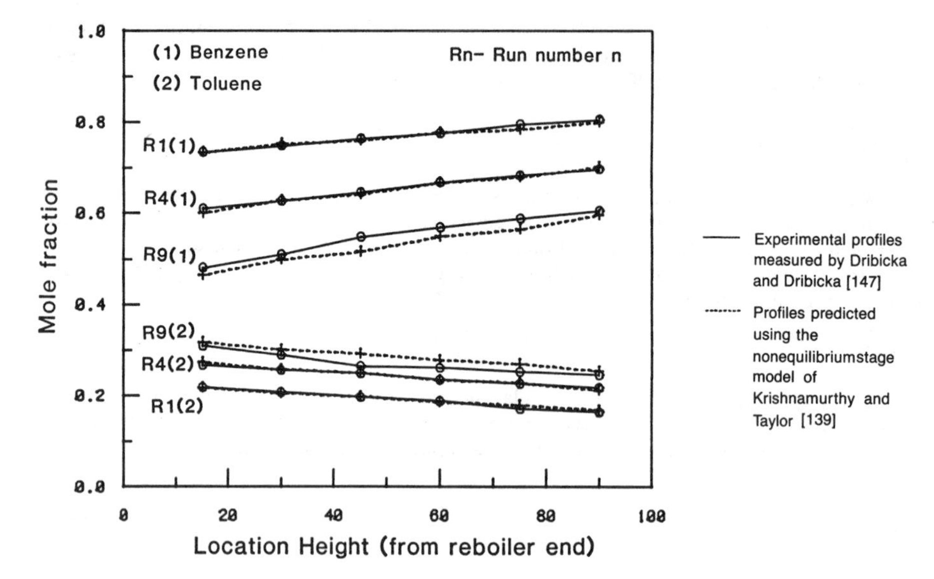

Figure 32. Composition profiles obtained during the distillation of a mixture of benzene (1), toluene (2), and ethylbenzene (3) in a wetted-wall column at total reflux.

Equation 495, which expresses the requirement that the fluxes be continuous across the vapor/liquid interface, will prove useful when the nonlinear equations from which the mass transfer rates are calculated are solved numerically; more on this later.

The differential energy balance for the vapor phase is

$$VC_P^V \frac{dT^V}{dA} = -q^V \tag{496}$$

where q^V is the conductive heat flux out of the bulk vapor. The energy balance for the liquid phase is

$$LC_P^L \frac{dT^L}{dA} = q^L - q^W \tag{497}$$

where q^L is the heat flux into the liquid and q^W is the heat flux across the tube wall into the coolant. The energy balance for the coolant in this section of the condenser is

$$L_c C_P^c \frac{dT^c}{dA} = \pm q^W \quad \begin{array}{l} + \text{ cocurrent} \\ - \text{ countercurrent} \end{array} \tag{498}$$

From an energy balance around the entire differential section we find that

$$E^V = E^I = E^L = E^W = E \tag{499}$$

where E is the *energy* flux given by Equation 108 and repeated here

$$E = q + \sum N_i \bar{H}_i \tag{500}$$

$\bar{H}_i$ is the partial molar enthalpy of component i. Equation 500 can be written for each phase; vapor, condensate, and coolant (for which the convective term drops out). Equations 499 express the requirement that the energy fluxes be continuous across a phase boundary. Differences between the area of the vapor/liquid, liquid/wall, and wall/coolant interfaces are ignored in the present analysis; it would, however, be easy to allow for such variations.

The balance equations just presented are quite independent of the methods used to calculate the mass and energy transfer rates. The equations that permit this calculation were the subject of the previous section; in preparation for the discussion on solving the equations, we briefly pause to remind our readers of the *form* the rate equations take.

Mass Transfer in Multicomponent Gas/Vapor Mixtures

There are a number of methods that could be used to calculate the mass transfer rates in the vapor phase. Most prior work in this area has been carried out using rate equations based on a film model of steady-state one-dimensional transfer (see the list of citations in the introduction to this section). As we have already seen, even for this the simplest of all models of mass transfer, there is rather more than one way of performing the calculations. The methods fall into three categories.

1. Methods of the effective diffusivity type (which neglect interaction effects) with the molar fluxes calculated from

$$N_i^V = c_t^V k_{i,eff}^V \Xi_{i,eff}(y_i^V - y_i^I) + y_i^V N_t^V, \quad i = 1, 2, \ldots, n \tag{501}$$

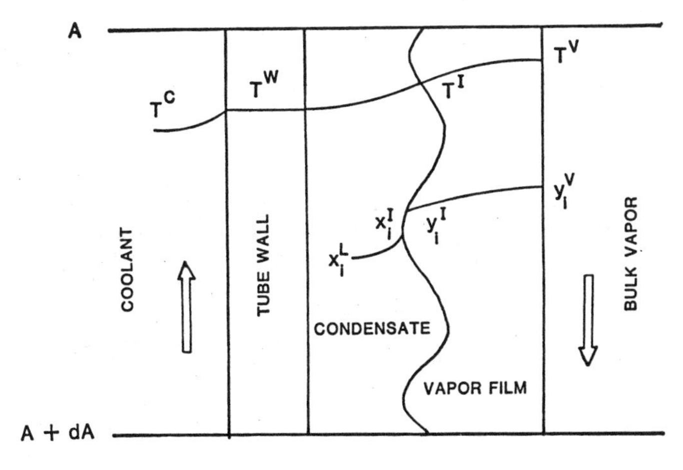

Figure 33. Typical temperature and composition profiles during the condensation of a multicomponent vapor ("film" model).

where k^V_{ieff} is an effective mass transfer coefficient, a function of an effective diffusivity, D'_{ieff}. The effective diffusivity itself may be defined in a number of different ways (see the previous section for a comparison of methods); a common choice, because it represents an exact definition of the effective diffusivity for dilute gases—a situation sometimes approached at the condenser outlet—is to take $D'_{ieff} = D'_{in}$. Webb et al. [181] discuss the conditions which must prevail in order that Equation 501 be an adequate representation of the more rigorous methods of the next category.

2. Methods which take interaction effects into account and which are implicit in the molar fluxes N^V_i. The method of Krishna and Standart [72], based on an exact solution of the Maxwell–Stefan equations, and the solution of the linearized equations due to Toor [4] and to Stewart and Prober [5] are in this category. In these methods, the mass transfer rates are obtained from a matrix equation of the form

$$(N^V) = c^V_t[k^{\bullet}_V](y^V - y^I) + N^V_t(y^V)$$

$$= c^V_t[k^V][\Xi^V](y^V - y^I) + N^V_t(y^V) \quad (502)$$

3. Methods which account for interaction effects but which do not require an a priori knowledge of the mass transfer rates themselves. There are two methods in this category, due to Krishna [91] and to Taylor and Smith [92] (as a generalization and modification of the method of Burghardt and Krupiczka [93]).

In applications to condenser simulation of these three categories of methods, we must estimate the low flux mass transfer coefficients (k^V_{ieff}, $[k^V]$, etc.) using the empirical methods described in the previous section.

The turbulent eddy diffusivity models, developed in the preceding section may also be used to predict the condensation rates. The expression from which the *mass* fluxes, n_i, are calculated is

$$(n^V) = \varrho_t[k^{\bullet}_m](\omega^V - \omega^I) + n^V_t(\omega^V)$$

$$= \varrho_t[k^V_m][\Xi^V](\omega^V - \omega^I) + n^V_t(\omega^V) \quad (503)$$

For the evaluation of $[k^V_m]$ and $[\Xi^V]$ see the preceding section.

Mass Transfer in the Liquid Phase

A description of the mass transfer process in the liquid phase is not often included in condensation calculations because the principle resistance to mass and energy transfer is thought to reside in the vapor phase. For completeness, we provide here a brief summary of approaches to modeling the liquid-phase resistance. We assume that the condensate is completely miscible (see, for example, Sardesai and Webb [182], if it is not).

All of the methods just described for calculating mass transfer rates in gas/vapor mixtures may be extended to deal with the liquid phase. Thus, for example, the effective diffusivity approach would lead to a rate equation of the form

$$N^L_i = c^L_t k^L_{ieff}\Xi^L_i(x^I_i - x^L_i) + x^L_i N^L_t, \quad i = 1, 2, \ldots, n-1 \quad (504)$$

A more rigorous approach to the calculation of the mass transfer in the liquid phase would lead to the rate equation

$$(N^L) = c_t^L[k^L][\Xi^L](x^I - x^L) + N_t^L(x^L) \tag{505}$$

which is a generalization of Equation 502. Methods of calculating the various coefficient matrices involved may be derived as extensions of the methods presented above for multicomponent gas mixtures (see, for example, Equations 276, 284 and 290). The surface renewal (or penetration) models might also be appropriate for estimating $[k^L]$. The most uncertain part of the calculation is obtaining the liquid-phase diffusivity matrix; this is still a subject for research. We note particularly two simple situations here; the first in which the condensate is a binary liquid (for which the calculation of a diffusion coefficient and a mass transfer coefficient pose no real problem) and second, a condensate formed from members of a homologous series of hydrocarbons for which the effective diffusivity approach is quite adequate.

Two limiting cases of condensate behavior may be derived from Equations 504 and 505:

1. The liquid phase is completely mixed with regard to composition (but not to temperature), corresponding to infinitive liquid-phase mass transfer coefficients, and the liquid composition calculated from a material balance along the flow path

$$x_i^I = x_i^L = \ell_i/L \tag{506}$$

2. The liquid phase is completely unmixed, corresponding to zero liquid-phase mass transfer coefficients. In this case the interfacial composition is given by the relative rates of condensation

$$x_i^I = N_i/N_t \tag{507}$$

Most prior work in this area has used one or the other of these two limiting cases (see, for example, Schrodt [159]; Krishna and Panchal [162]). The former is applicable to vertical condensers where the two phases remain in close proximity, the latter to horizontal condensers where the condensate is continuously separated from the vapor (see Schrodt [159]; Webb and McNaught [166]; Butterworth [178] for further discussion of this point). Since there is no condensate at the top of a vertical condenser, initial condensation rates have usually been calculated using the no-mixing limiting case. In fact, it is easy to show that the two limiting cases Equations 506 and 507 are equivalent at the vapor inlet. There is some evidence that the final design is insensitive to whichever extreme is chosen if inert species are present in the vapor mixture [167]; we shall return to this topic later.

Energy Transfer

As noted, the local energy flux is made up of a conductive heat flux and a convective contribution due to the transport of energy by interphase mass transport.

$$E^V = q^V + \sum_{i=1}^{n} N_i\bar{H}_i^V(T^V) = E^I = q^{IL} + \sum_{i=1}^{n} N_i\bar{H}_i^L(T^I)$$

$$= E^L = q^L + \sum_{i=1}^{n} N_i H_i^L(T^L) = E^W = q^W \tag{508}$$

The film model of simultaneous mass and energy transfer leads to the following expression for the heat flux out of the bulk vapour, q^V; Equation 299 again

$$q^V = h^V \frac{\Phi_H^V}{e^{\Phi_H^V} - 1}(T^V - T^I) \tag{509}$$

where Φ_H^V is defined by Equation 295. In practice, the low flux heat transfer coefficient h^V is estimated from an appropriate correlation, the j_H half of the Chilton–Colburn analogy, for example.

The turbulent eddy diffusivity model of simultaneous heat and mass transfer may also be used to calculate q^V Equation 410.

For the liquid phase, a relation similar to Equation 509 can be written

$$q^{IL} = h_0^L(T^I - T^W); \quad q^L = h^L(T^I - T^L) \tag{510}$$

where h_0^L is the heat transfer coefficient that accounts for the resistance to heat transfer in the entire condensate film. Here, we have ignored the high flux correction to h^L; the resistance to heat transfer in the liquid phase is very much smaller than in the vapor phase and the correction factor would be very close to unity. The heat transfer coefficient for the entire condensate h_0^L can be estimated from, for example, Nusselt's equation; other methods are discussed by Webb and McNaught [166] and by Butterworth [178].

There is no mass transfer through the tube wall into the coolant; thus the energy flux E^W is given by

$$E^W = q^W = h^C(T^W - T^C) \tag{511}$$

where h^C is the heat transfer coefficient in the coolant and can be estimated from correlations applicable to the geometry of the condenser.

If we substitute Equations 509, 510 and 511 into the energy flux continuity Equation 508, we find, noting that the choice of reference temperature (T_{ref}) is immaterial:

$$h^V \frac{\Phi_H^V}{e^{\Phi_H^V} - 1}(T^V - T^I) + h^V \Phi_H^V (T^V - T^I) + \sum_{i=1}^{n} N_i \lambda_i$$

$$= h_0^L(T^I - T^W) = h^C(T^W - T^C) = h_0(T^I - T^C) \tag{512}$$

where h_0 is an overall heat transfer coefficient accounting for the resistances to conductive heat transfer in the condensate and in the coolant (the resistance in the tube walls will normally be neglible):

$$\frac{1}{h_0} = \frac{1}{h_0^L} + \frac{1}{h^C} \tag{513}$$

Interface Model

Once again, we adopt the conventional model of a phase interface; a surface offering no resistance to mass transfer and where equilibrium prevails. The usual equations of phase equilibrium relate the *mole* fractions on each side of the interface

$$y_i^I = K_i x_i^I, \quad i = 1, 2, \ldots, n \tag{514}$$

This completes the formal development of the model; next, we discuss how the equations can be solved.

The Computational Problem

The set of differential and algebraic equations just given must be solved numerically in general. The calculations start at the inlet to the condenser where the vapor temperature, pressure, and composition are known and proceed until either a specified area has been reached (a simulation problem) or until a specified amount has been condensed (a design problem). Each time the derivatives are calculated, the nonlinear algebraic equations from which the fluxes are obtained must be solved. If the simple Euler method of integration is used, then the rate equations,need to solved only once per step. However, the solution so obtained pertains to the conditions at the beginning of the step where driving forces and, consequently, condensation rates are highest. Thus, the use of only a "few" Euler step by, for example, Webb and McNaught [166], may result in overprediction of the condensation rates and to an underdesigned condenser. For greater accuracy, a larger number of steps will have to be taken or, alternatively, a higher order integration method will have to be used (for example, fourth order Runge-Kutta method used by Webb and Sardesia [167] and by Schrodt [159]). In either case, the nonlinear rate equations will need to be solved quite a number of times in order to obtain a safer design, thereby considerably increasing the computational cost. One way to obtain a more conservative design using a relatively

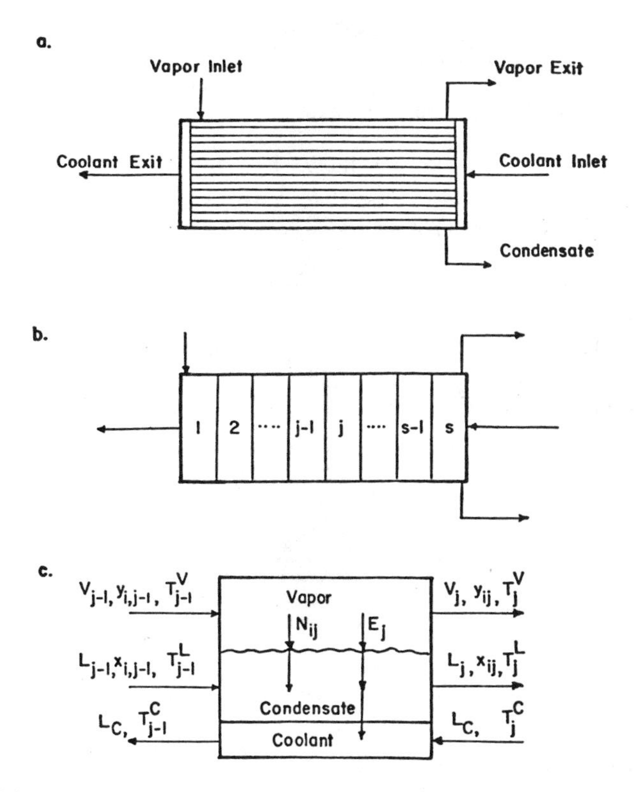

Figure 34. Dividing a condenser into sections: (A) Schematic diagram of a shell-and-tube condenser; (B) the condenser divided into a total of s sections; (C) model of the j-th section.

inexpensive "first-order" method would be to calculate the fluxes at the conditions pertaining to the *end* of the step where driving forces and condensation rates will normally be lowest. Perhaps a compromise solution in which the transfer rates are evaluated at some *average* conditions over the length of the step would be best. Either of these two approaches would lead to a more conservative design with relatively few integration steps. However, neither approach can be effectively implemented if the differential balance equations are solved using available software without repeating the downstream integration over and over until convergence is obtained on the outlet conditions. A solution to this problem, proposed by Taylor et al. [171], is to replace the derivatives in Equations 493 through 498 by finite difference approximations (as is done in the simple Euler method). Thus the balance equations become algebraic equations which can be solved simultaneously with the rate equations to give the conditions at the end of the step. The difference between this approach and a more mathematically correct implementation of the Euler method of integration is that we no longer need a separate routine for advancing the differential balance equations and that the fluxes can be calculated using, as boundary conditions, any combination of the conditions at the beginning and end of the step. In effect, we are dividing the condenser into a number of sections, s say, as shown in Figure 34. We describe this approach in more detail later.

Equations and Variables for a Section of the Condenser

The balance equations for the j-th section are as follows (derived from Equations 493 through 498 as described previously:

1. A material balance for the vapor phase

$$M_{ij}^{V} \equiv v_{i,j-1} - v_{i,j} - N_{ij}\Delta A_j = 0, \qquad i = 1, 2, \ldots, n \tag{515}$$

2. A material balance for the liquid phase

$$M_{ij}^{L} \equiv \ell_{i,j-1} - \ell_{i,j} + N_{ij}\Delta A_j = 0, \qquad i = 1, 2, \ldots, n \tag{516}$$

3. An energy balance for the vapor phase

$$E_j^{V} \equiv \bar{V}_j C_{pj}^{V}(T_{j-1}^{V} - T_j^{V}) + q_j^{V}\Delta A_j = 0 \tag{517}$$

$$\text{with } \bar{V}_j \cong \frac{1}{2}(V_j + V_{j-1})$$

4. An energy balance for the liquid phase

$$E_j^{L} \equiv \bar{L}_j C_{pj}^{L}(T_{j-1}^{L} - T_j^{L}) - (q^{L} - q^{W})\Delta A_j = 0 \tag{518}$$

$$\text{with } \bar{L}_j \cong \frac{1}{2}(L_j + L_{j-1})$$

5. An energy balance for the coolant

$$E_j^{C} \equiv L_c C_{pj}^{C}(T_{j-1}^{C} - T_j^{C}) \mp q_j^{W}\Delta A_j = 0 \tag{519}$$

The subscript j denotes the conditions at the end of the j-th section (second subscript if component properties are involved) unless it is attached to a physical property or to a quantity with an overline (the significance of these quantities is discussed later).

These equations are augmented by:

6. Rate equations for the vapor phase. For example:

$$R_{ij}^{V} \equiv N_{ij} - c_{tj}^{V} \sum_{k=1}^{n-1} k_{ikj}(\bar{y}_{k}^{V} - y_{k}^{I}) - N_{tj}\bar{y}_{ij}^{V} = 0, \quad i = 1, 2, \ldots, n-1 \tag{520}$$

if the Krishna–Standart or Toor–Stewart–Prober film models are used or

$$R_{ij}^{V} \equiv N_{ij} - \{\varrho_{tj}^{V} \sum_{k=1}^{n-1} k_{mikj}(\bar{\omega}_{k}^{V} - \bar{\omega}_{k}^{I}) - n_{tj}\bar{\omega}_{ij}^{V}\}/M_{i} = 0, \quad i = 1, 2, \ldots, n-1 \tag{521}$$

if the turbulent eddy diffusivity models are used.

7. Rate equations for the liquid phase. For example:

$$R_{ij}^{L} \equiv N_{ij} - k_{ieff}^{L}\Xi_{i}^{L}(x_{i}^{I} - \bar{x}_{i}^{L}) - N_{tj}\bar{x}_{i}^{L} = 0, \quad 1, 2, \ldots, n-1 \tag{522}$$

if an effective diffusivity approach is used.

8. An energy balance at the vapor-liquid interface

$$E_{j}^{I} \equiv h_{j}^{V} \frac{\Phi_{Hj}^{V} e^{\Phi_{Hj}^{V}}}{e^{\Phi_{Hj}^{V}} - 1} (\bar{T}_{j}^{V} - T_{j}^{I}) + \sum N_{i}\lambda_{i} - h_{0j}(T_{j}^{I} - \bar{T}_{j}^{C}) \tag{523}$$

9. An energy balance at the wall

$$E_{j}^{W} \equiv h_{j}^{C}(T_{j}^{W} - \bar{T}_{j}^{C}) - h_{j}^{L}(T_{j}^{I} - T_{j}^{C}) = 0 \tag{524}$$

10. Equilibrium equations for the interface

$$Q_{ij}^{I} \equiv K_{ij}x_{ij}^{I} - y_{ij}^{I} = 0, \quad i = 1, 2, \ldots, n \tag{525}$$

In compiling this list of independent equations, we have used the flux continuity Equation 495 to eliminate one set of molar fluxes (N_i^V or N_i^L). The bulk phase conditions denoted by the overlines, $\bar{y}_{ij}^{V}$, $\bar{T}_{j}^{V}$, $\bar{x}_{ij}^{L}$, $\bar{T}_{j}^{L}$, $\bar{T}_{j}^{C}$, appearing in the rate equations may be calculated using the inlet conditions (e.g., $\bar{y}_{ij}^{V} = v_{ij-1}/v_{j-1}$), the outlet conditions (e.g., $\bar{y}_{ij}^{V} = v_{ij}/V_{j}$) or at some average condition (e.g., $\bar{y}_{ij}^{V} = 1/2(v_{ij-1}/V_{j-1} + v_{ij}/V_{j})$). All physical properties are evaluated at the bulk conditions used in the determination of the mass and energy transfer rates and are considered constant in the section. The interface state and wall temperature also are considered to be uniform in the j-th section; thus the mass and energy transfer rate equations as well as the equilibrium equations need be solved only once per section.

Given the state of all streams leaving section $j-1$, then there is a total of $5n+3$ unknown quantities for each section j. These are: the component vapor flow rates (v_{ij}: n in number), the component liquid flow rates (ℓ_{ij}: n in number), the vapor temperature (T_j^V), the liquid temperature (T_j^L), the interface temperature (T_j^I), the wall temperature (T_j^W), the coolant temperature (T_j^C), the vapor composition at the interface (y_{ij}^I: n − 1 in number), the liquid composition at the interface (x_{ij}^I: n − 1 in number) and the mass transfer rates (N_{ij}: n). The 5n + 3 equations that permit the calculation of these unknowns are as follows: component material balances for the vapor phase (M_{ij}^V : n), component material balances for the liquid phase (M_{ij}^L: n), the vapor-phase energy balance (E_j^V), the liquid-phase energy balance (E_j^L), the interface energy balance (E_j^I), the coolant energy balance (E_j^C), the wall energy balance (E_j^W), the interface equilibrium equations (Q_{ij}: n), the vapor-phase mass transfer rate equations (R_{ij}^L: n − 1 in number).

The independent equations are ordered into a vector of functions as follows:

$$(F_j)^T \equiv (M_{1j}^V, M_{2j}^V, \ldots, M_{nj}^V, M_{1j}^L, M_{2j}^L, \ldots, M_{nj}^L, E_j^V, E_j^L, E_j^C, E_j^W,$$

$$R_{1j}^V, R_{2j}^V, \ldots, R_{n-1,j}^V, E_j^I, Q_{1j}^I, Q_{2j}^I, \ldots, Q_{n-1j}, Q_{nj}^I, R_{1j}^L, R_{2j}^L, \ldots, R_{n-1j}^L)$$

The vector of variables corresponding to this set of equations is

$$(X_j)^T \equiv (v_{1j}, v_{2j}, \ldots, v_{nj}, \ell_{1j}, \ell_{2j}, \ldots, \ell_{nj}, T_j^V, T_j^L, T_j^C, T_j^W,$$

$$N_{1j}, N_{2j}, \ldots, N_{n-1,j}, N_{nj}, y_{1j}^I, y_{2j}^I, \ldots, y_{n-1,j}^I, T_j^I, x_{1j}^I, x_{2j}^I, \ldots, x_{n-1,j}^I)$$

Solving the MERQ Equations

We now address the problem of solving the MERQ (an acronym for *M*aterial balance, *E*nergy balance, *R*ate and e*Q*uilibrium) equations for multicomponent condensation represented by the function vectors (F_j). Most of the methods described in the literature employ some kind of tearing strategy to solve the condensation equations. (Equation tearing and its drawbacks were described in an earlier section). For example, Krishna et al. [161] solve the mass and energy transfer rate equations (by repeated substitution of the N_i) and the vapor/liquid equilibrium equations (bubble point calculations) within an outer loop that, in effect, solves the liquid mixing equations using Newton's method. The outer loop tear variables were the n − 1 interfacial compositions x_i^I. Other tearing algorithms, those of Price and Bell [160] and of Webb and co-workers [166, 168, 183] involve up to three levels of iteration loop.

An algorithm for solving the mass and energy transfer rates and equilibrium equations simultaneously was proposed by Taylor, Lucia, and Krishnamurthy [62] and extended to deal with the balance equations by Taylor et al. [171]. As discussed earlier Newton's method or the hybrid method are very effective methods for solving the kinds of equations found in chemical process problems.

Calculation Procedure

We are now in a position to summarize our recommended calculation procedure. The conditions of the entering streams V_0, v_{i0}, T_0^V and the pressure must be specified (L_0 and ℓ_{i0} must also be specified if they are non-zero). The coolant temperature at the vapor entrance end of the condenser is required if the coolant temperature is to be included as a variable (or else specified it if is not). The number of sections, s, must be known in advance as well as the area of each section, ΔA_j, (this can be calculated from the geometry of the condenser and must be fixed prior to performing a simulation calculation).

The MERQ equations for section 1 are solved in order to obtain the conditions at the end of section 1 and at the beginning of section 2 and so on. If Newton's method or the hybrid approach are used this will require the calculation of the function vector (F) at least once per iteration. A possible route is summarized in the following. Note that at any time the discrepancy functions (F) are to be calculated using the current values of the independent variables (X). This means that an initial guess of each quantity in (X) must be supplied for section 1. Thereafter, good initial estimates of the unknowns for section j (j = 2, ..., s) become available from the solution to the equations for step $j-1$.

1. Using the current values of the component flows at the end of the section, v_{ij} and ℓ_{ij}, calculate the compositions of the bulk vapor and liquid streams; any combination of the following may be used as desired

   - Inlet $\bar{y}_{ij}^{V} = v_{ij-1}/V_{j-1}$; $\bar{x}_{ij}^{L} = \ell_{ij-1}/L_{j-1}$ (526)

   - Average $\bar{y}_{ij}^{V} = \frac{1}{2}(v_{ij-1}/V_{j-1} + v_{ij}/V_j)$; $\bar{x}_{ij}^{L} = \frac{1}{2}(\ell_{ij-1}/L_{j-1} + \ell_{ij}/L_j)$ (527)

   - End $\bar{y}_{ij}^{V} = v_{ij}/V_j$; $\bar{x}_{ij}^{L} = \ell_{ij}/L_j$ (528)

2. In a similar way, calculate the bulk temperatures, $\bar{T}_j^V$, $\bar{T}_j^C$, $\bar{T}_j^L$.
3. Calculate all physical and transport properties at the required conditions. In some cases, film average temperatures and compositions may need to be used. These average conditions can easily be calculated with the information presently available.
4. Calculate the mass and heat transfer coefficients using whatever method is preferred (see prior discussion).
5. Calculate K-values and latent heats of vaporization at the current estimate of the interfacial conditions.
6. Complete the calculation of the MERQ Equations 514–525.

Following the calculation of (F), the Jacobian matrix must be calculated and the next estimate of (X) computed from the solution of the linear system Equation 116. The calculations are then repeated until convergence of all functions has been obtained. In practice, this means converging the energy balances since these functions have numerical values far larger than any of the others. Once convergence has been obtained, the solution vector (X^*) is used to initialize the calculations for the next section. In this way we proceed until the end of the condenser is reached.

Some Variations on a Theme

There are many instances in which the equation and variable sets just presented can be simplified somewhat. We now consider several special cases.

1. One approximation that might be worth consideration is to set the condensate temperature to the arithmetic average of the wall and interface temperatures. If this is done, the liquid temperature and liquid-phase energy balance can be removed from the set of independent variables and equations respectively. This leaves us with a total of $5n+2$ independent equations and variables per section:

$$(F_j)^T \equiv (M_{1j}^V, \ldots, M_{nj}^V, M_{1j}^L, \ldots, M_{nj}^L, E_j^V, E_j^W, E_j^C, R_{1j}^V, \ldots, R_{n-1,j}^V,$$
$$E_j^I, Q_{1j}^I, \ldots, Q_{n-1,j}^I, Q_{nj}^I, R_{1j}^L, \ldots, R_{n-1,j}^L)$$

$$(X_j)^T \equiv (v_{1j}, \ldots, v_{nj}, \ell_{1j}, \ldots, \ell_{nj}, T_j^V, T_j^W, T_j^C, N_{1j}, \ldots, N_{n-1,j},$$
$$N_{nj}, y_{1j}^I, \ldots, y_{n-1,j}^I, T_j^I, x_{1j}^I, \ldots, x_{n-1,j}^I)$$

It should be noted that this approximation is made in many of the numerical simulations of multicomponent condensers described in the literature and in all the other following special cases.

2. If we choose to approximate the mass transfer process in the liquid phase by one of the two limiting cases Equations 506 and 507 then the rate equations R_{ij}^L need to be replaced by the mixing equations, X_{ij}^L:

$$X_{ij}^L \equiv x_{ij}^I - \bar{x}_{ij}^L = 0 \qquad \text{(mixed)} \tag{529}$$

$$X_{ij}^L \equiv x_{ij}^I - N_{ij}/N_{tj} = 0 \qquad \text{(unmixed)} \tag{530}$$

In either case, we might choose to delete the x_{ij}^I and ℓ_{ij} from the set of variables and compute them directly from Equations 529 and 530 which, therefore, would be removed from the set of independent equations.

$$(F_j)^T \equiv (M_{1j}^V, \ldots, M_{nj}^V, E_j^V, E_j^W, E_j^C, R_{1j}^V, \ldots, R_{n-1,j}^V, E_j^I, Q_{1j}^I, \ldots, Q_{n-1,j}^I, Q_{nj}^I)$$

$$(X_j)^T \equiv (v_{1j}, \ldots, v_{nj}, T_j^V, T_j^W, T_j^C, N_{1j}, \ldots, N_{n-1,j}, N_{nj}, y_{1j}^I, \ldots, y_{n-1,j}, T_j^I)$$

3. If one component in the vapor is truly noncondensable (or the calculations are to be made assuming that it is noncondensable), then the equations pertaining to that component (let's call it component n) will need to be modified. To start with, component n cannot, by definition, be present in the condensate which, therefore, contains n − 1 species. The set of mass transfer rate equations in the liquid phase must be reduced in number by one and the flux N_n dropped from the set of variables. Also, the equilibrium equation corresponding to component n and the liquid-phase composition x_{ij}^I must be removed from the set of equations and variables. If, in addition, the condensate is considered to be well mixed or completely unmixed, we are left with the following vectors of equations and variables:

$$(F_j)^T \equiv (M_{1j}, \ldots, M_{nj}, E_j^V, E_j^W, E_j^C, R_{1j}^V, \ldots, R_{n-1,j}^V, E_j^V, Q_{1j}^I, Q_{2j}^I, \ldots, Q_{n-1,j}^I)$$

$$(X_j)^T \equiv (v_{1j}, \ldots, v_{nj}, T_j^V, T_j^W, T_j^C, N_{1j}, \ldots, N_{n-1,j}, T_j^I, y_{1j}^I, y_{2j}^I, \ldots, y_{n-1,j}^I)$$

Notice that the interface temperature, T_j^I, now is paired with the interface energy balance function E_j^I, *not* the equilibrium equation Q_{nj}^I, which has been dropped. As in all other cases, the interface mole fraction y_{nj}^I is computed from the requirement that the mole fractions sum to unity.

4. If the wall temperature profile is known, as is sometimes the case, then it is convenient to remove the coolant temperature and coolant energy balance from the sets of variables and equations.

$$(F_j)^T \equiv (M_{1j}^V, \ldots, M_{nj}^V, E_j^V, R_{1j}^V, \ldots, R_{n-1,j}^V, E_j^I, Q_{1j}^I, \ldots, Q_{n-1,j}^I)$$

$$(X_j)^T \equiv (v_{1j}, \ldots, v_{nj}, T_j^V, N_{1j}, \ldots, N_{n-1,j}, T_j^I, y_{1j}^I, \ldots, y_{n-1,j}^I)$$

where we have also incorporated the simplifications that were made in case 3.

5. One further simplification that can be made is to remove the vapor and liquid-phase material balance equations (M_{ij}^V, M_{ij}^L) from the set of independent equations. These equations are linear if the fluxes N_{ij} are included in the set of variables (X) (which, of course, they must be if the rate equations are to solved simultaneously with the other equations). Thus, the balance Equations 515 and 516 can be solved directly for the component flow rates at the end of the j-th section given estimates of the molar fluxes and the bulk conditions calculated as desired Equations 526 through 528. The inclusion of linear equations in the vector of functions (F) does nothing to affect the rate of convergence since they are always satisfied on every iteration after the first. Including the linear equations in (F) increases the computational cost slightly but it does reduce the initial effort of deriving and coding expressions for the partial derivatives of (F) and, in our experience, can sometimes improve the stability of the hybrid method by allowing more complete expressions for the partial derivatives to be included in the computed part [C] and fewer derivatives in the approximated part [A].

The model just presented may also be used, with only two small changes, to describe a *cocurrent* separation process in, for example, an adiabatic wetted-wall column. In this case, the energy balance equations for the coolant Equation 498 and at the wall are dropped from the set of model equations and T^W and T^C are dropped from the set of variables. The energy balance for the liquid phase, Equation 497, must be included here but is simplified by deleting the heat flux q^W from the right hand side. Other ways to simplify the general model may also be derived.

Simulation of Multicomponent Condensation-A Discussion of Research Results

There is a great shortage of experimental data on mass transfer in multicomponent vapor (+ inert gas)-liquid systems. Most of the published works deal with absorption (or condensation or evaporation) or a single species in the presence of an inert nontransferring component. A set of ternary mass transfer experiments were carried out by Toor and Sebulsky [183] and Modine [184] in a wetted-wall column and also in a packed column. These authors measured the simultaneous rates of transfer between a vapor-gas mixture containing acetone (1)–benzene (2)–nitrogen (3) or helium (3) and a binary liquid mixture of acetone and benzene. The vapor and liquid streams were in cocurrent flow for the wetted-wall column and counter-current for the packed column. Their experimental results show that diffusional interaction effects were significant in the vapor phase, especially for the runs with helium as carrier gas. The theoretical model used by Toor and Sebulsky and Modine to explain their results was based on the generalized driving force approach of Toor [3]. More recently, Krishna [185, 186] used the wetted-wall column experimental data of Modine [184] to test the applicability of the Krishna and Standart [72] multicomponent film model and also the linearized theory of Toor [4] and Stewart and Prober [5]. Furno et al. [172] have used the same data to evaluate the turbulent eddy diffusivity models.

Webb and Sardesai [167] report the results of a number of experiments involving the condensation of isopropanol (1) and water (2) in the presence of nitrogen (3) or freon-12 (3) as noncondensing gases. The experiments were carried out in a 1-m-long vertical tube of 0.023 m internal diameter. Webb [168] reviews other experimental data obtained by his co-workers, including the results of Deo [188] obtained in a condenser of annular geometry. Numerical simulations of these experiments discussed by Webb and Sardesai [167] and Webb [168] (who used the Krishna–Standart, Toor–Stewart–Prober and effective diffusivity

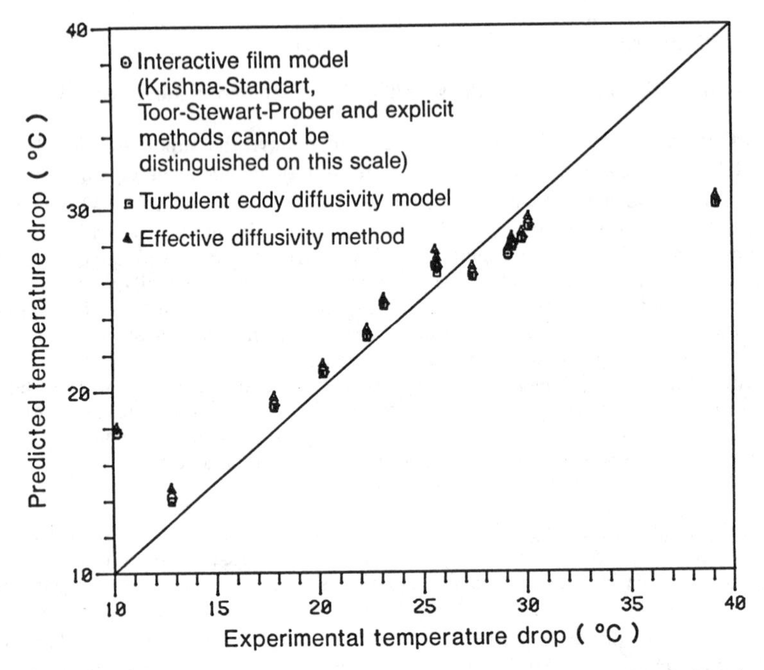

Figure 35. Comparison of predicted and experimental temperature drop for the condensation of isopropanol (1) and water (2) in the presence of nitrogen or freon 12 (3). Experimental data of Webb and Sardesai [167] obtained in a vertical tube; theoretical predictions by Furno et al [172].

methods to calculate the condensation rates), by McNaught (who, in addition to these models, used the equilibrium method of Silver [156]) and by Furno, Taylor, and Krishna [172] (who employed the turbulent eddy diffusivity models (as well as the methods based on film theory)).

Simulation of condensation on tube bundles is discussed by Bandrowski and Kubaczka [165], Shah and Webb [170] and by McNaught [117].

In addition to the few experimental studies, there are many papers cited throughout this section that consider only the theoretical/computational problems of multicomponent condensation. Indeed, from this standpoint, multicomponent condensation is, perhaps, the most studied of all of the applications of multicomponent mass transfer theory. Thus, despite the shortage of suitable data, it is possible to make some fairly definite statements. In the discussion which follows, we focus attention on the following:

1. A comparison of the noninteractive film models (the effective diffusivity methods) with the film models that take multicomponent interaction effects into account (Krishna–Standart, Toor–Stewart–Prober, Krishna and Taylor–Smith).
2. A comparison of the interactive film models that use the Chilton–Colburn analogy to obtain the heat and mass transfer coefficients with the turbulent eddy diffusivity models.
3. The influence of the model used to approximate the mass transfer behavior in the liquid phase (i.e., mixed, unmixed, or rate model).
4. The importance of step size and choice of bulk phase conditions on the conservatism and accuracy of the simulation.

Significance of Interaction Effects

In considering the many condenser simulations that have been reported in the literature, we are unable to find an application where the differences between any of the multicomponent film models that account for interaction effects (Krishna–Standart; Toor–Stewart–Prober; Explicit methods) are really significant. However, effective diffusivity methods may yield profiles that can differ quite markedly from the profiles obtained using an interactive model. Evidence for the significance of interaction effects in condensation can be found in the papers by Krishna et al. [161], Krishna, Rohm, Webb and Sardesai [167] and Taylor and co-workers [171, 172]. Figures 35 through 37 compare the temperature drops, total condensation rate, and condensate composition measured in the experiments of Sardesai and Webb [167] to the corresponding quantities predicted using several mass transfer models. It can be seen that the effective diffusivity method yields reasonably good predictions of the overall temperature drops and total condensation rates. However, the composition of the condensate (which is an accurate measure of the *relative* rates of condensation and, therefore, of the *individual* condensation rates) is quite poorly predicted by the effective diffusivity methods. In complete contrast, the interactive film models do a much better job of predicting the condensate composition (as well as the temperature drops and total condensation rates). Interaction effects are particularly pronounced in some of the experiments of Modine [184]; Krishna [186] has shown that reverse diffusion takes place in some of the runs. Figure 38 [172] provides a comparison between the measured mass

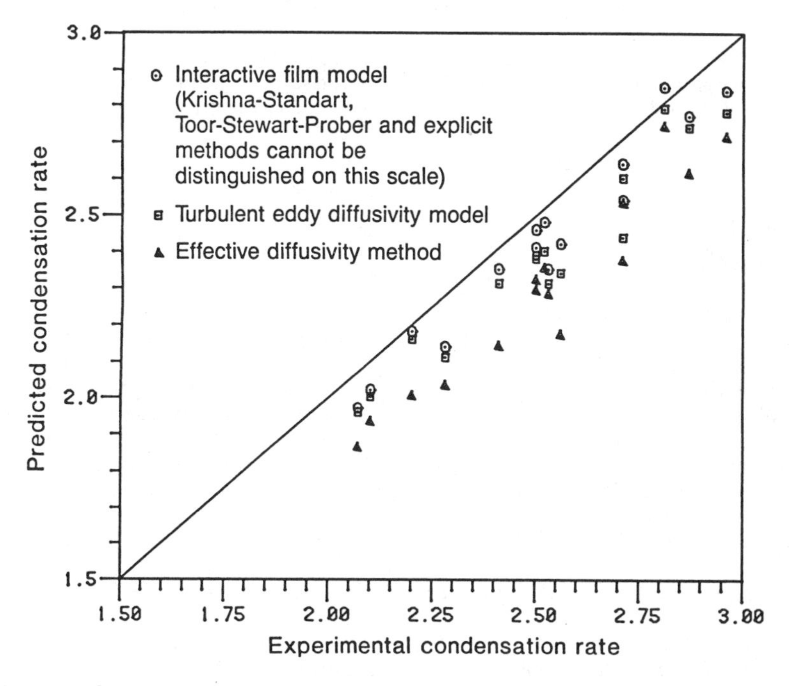

Figure 36. Comparison of predicted and experimental condensation rates for the condensation of isopropanol (1) and water (2) in the presence of nitrogen or freon 12 (3). Experimental data of Webb and Sardesai [167]; theoretical predictions by Furno et al. [172].

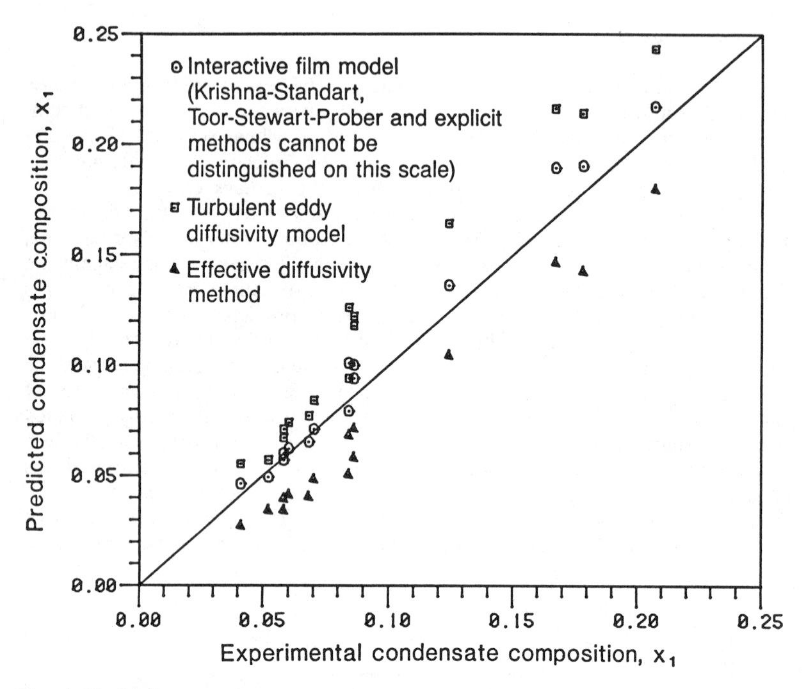

Figure 37. Comparison of predicted and experimental condensate composition (mole fraction of isopropanol) for the condensation of isopropanol (1) and water (2) in the presence of nitrogen or freon 12 (3). Experimental data of Webb and Sardesai [167]; theoretical predictions by Furno et al. [172].

transfer rates and the rates predicted by a number of models of vapor-phase transport. The models that account for interaction effects are clearly superior here.

Figures 39 through 41 show selected composition and temperature profiles predicted by several mass transfer models for a simulation of a fairly typical industrial condensation problem; a mixture of hydrocarbons—five of them—condensing in the presence of hydrogen (which is considered to be sparingly soluble in the condensate). The profiles predicted by the effective diffusivity methods are quite different from those obtained using the interactive models. We should also emphasize that there are *no* computational advantages to using an effective diffusivity approach; if the calculations are done as suggested, then the effective diffusivity and interactive film models take roughly the same amount of time. Indeed, it is quite simple to devise problems in which the effective diffusivity methods require more computer time than the more rigorous methods.

Interactive Film Model or Turbulent Eddy Diffusivity Model

A comparison between the turbulent eddy diffusivity model and the interactive film models that use the Chilton–Colburn analogy is provided by Taylor and co-workers [171, 172]. It must be noted that there are two ways of using the analogy; the first in the molar frame of reference, the second in the mass frame of reference. In most cases, there is very little difference between the turbulent eddy diffusivity model (which is necessarily

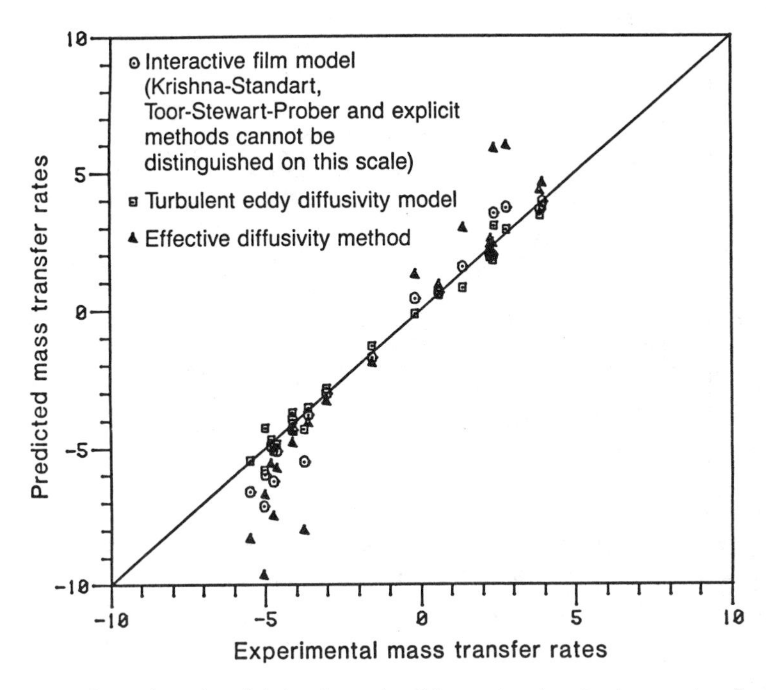

Figure 38. Comparison of predicted and experimental mass transfer rates for mass transfer between a gaseous mixture of acetone (1), benzene (2), and nitrogen or helium (3) and a liquid film containing acetone and benzene. Experimental data obtained by Modine [185] in a wetted-wall column; theoretical predictions by Furno et al. [172]

based in the mass reference frame) and the Chilton–Colburn-based film model in the mass frame. Even at very high vapor flow rates, corresponding to inlet Reynolds numbers in the range of sixty to one hundred thousand, it is difficult to find situations where the turbulent eddy diffusivity models differ from the mass frame Chilton–Colburn methods to any extent. This is an important result for it indicates that the Chilton–Colburn analogy, already widely used in design calculations, is probably unlikely to lead to large discrepancies when compared to more sophisticated turbulent eddy diffusivity models. This is also important from the computational viewpoint; the Krishna–Standart, Toor–Stewart–Prober, and the explicit methods are somewhat less demanding of computer time than are the turbulent eddy diffusivity models. This is due to necessity of computing the eigenvalues of [D] and the matrix function in Equations 401 and 404 using Sylvester's theorem. The computation of the matrix functions required in the Krishna–Standart and Toor–Stewart–Prober methods can be computed very much more efficiently (particularly if there are more than three or four components in the mixture) from a power series [76, 87].

The differences between the mass reference frame and molar reference frame models are a little more marked, at least in Figures 39 through 41. We should remember that the two models are based on quite different assumptions; Equation 520 is based on the assumptions of constant molar density and constant [D] whereas Equation 521 is based on the assumptions of constant mass density and constant $[D^m]$. Quite often, there is little to tell these methods apart [171]. The mass frame models are *marginally* better than the molar frame models at predicting the overall mass transfer rates observed in the experiments of Modine

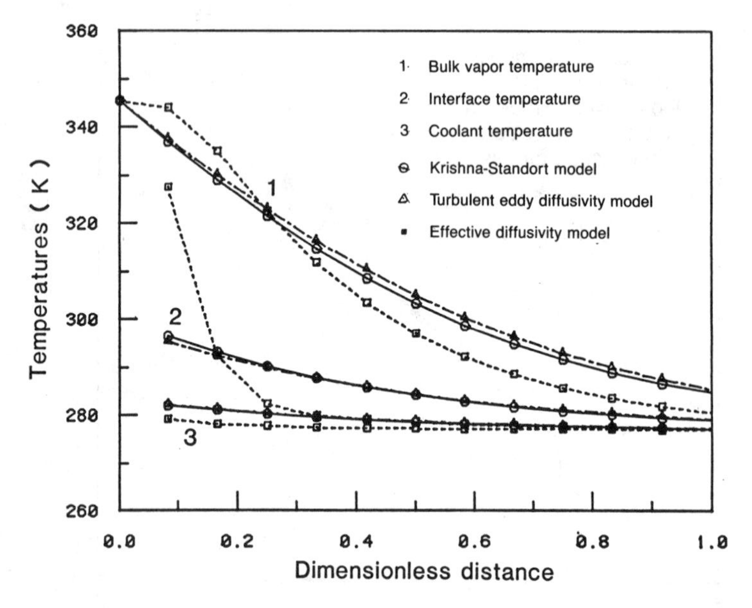

Figure 39. Temperature profiles obtained from a simulation of the condensation of a six component mixture; n-octane (1), n-heptane (2), n-hexane (3), n-butane (4), propane (5), and hydrogen (6). Calculations by Taylor et al. [171].

[184]. However, the mass frame models are noticeably poorer than the molar frame models at predicting the condensate composition measured in the experiments of Webb and Sardesai [167] (Figure 37) (although the temperature drops and total condensation rates are well enough predicted). It is difficult to draw any conclusions regarding the merits of the various *interactive* models from these few results; all that can be said at this point is that the assumptions underlying the turbulent eddy diffusivity models (i.e., constant ϱ_t and $[D^m]$) do not appear to be as good here as the assumption of constant c_t and $[D]$.

One reason for the close agreement between the interactive film models and the turbulent eddy diffusivity models was discussed in the previous section. Another reason is the tendency for different models to predict very similar total fluxes and fluxes of the major transferring species. Even the effective diffusivity models benefit from this. Serious discrepancies between the interactive models generally occur only for species whose fluxes are small relative to the major flux. In such cases, it is very likely that two different models will predict different directions of mass transfer for the minor transferring species; i.e., one model might predict that a species is evaporating while another may predict that the same species is condensing. The calculations made by Krishna [121] are a case in point. There is, however, more to it; the close agreement between composition and temperature profiles is largely due also to the constraints of material and energy balance which we must impose on each section of the device. In order to illustrate this point, let us assume that two different mass transfer models predict total fluxes and fluxes of the major transferring species that are in close agreement, with, say, a five percent relative discrepancy (this is quite typical; see Smith and Taylor [85]). Let us further assume that the two methods predict different directions of transfer for a minor transferring species and that the magnitude of this flux is about ten times smaller than the major contribution to the total flux (in our experience, these are the circumstances in which different interactive models sometimes predict different directions of transfer for a species). It is easy to see with reference to the material balance Equation 515 that the component flow rates (or the mole fractions) of *all* species will still be in reasonably good agreement at the end of the current section, even though one model has predicted a small increase in the flow rate of one component whereas another model predicted a small decrease. Moreover, it is also easy to check (with a few elementary calculations) that this conclusion is not affected by the relative amounts of the various species present in the vapor.

There is yet another reason why the various interactive models are usually in excellent agreement; also related to the requirements of mass and energy conservation. If one mass transfer model predicts a high flux of one species in one section of the condenser, more of that component will condense in that section leading to a lower driving force and, therefore, lower condensation rate in the next section. On the other hand, if another model predicts a lower condensation rate in one section of the condenser, not as much of that component is predicted to condense leading to higher driving forces and condensation rates in the next section. In this way, overprediction in one section is compensated by underprediction in the next (or, of course the opposite). To a very large extent, the effective diffusivity methods benefit from this as well; witness the relatively good predictions of the total amount condensed and of the total heat load in many of the simulations of Webb and Sardesai's experiments.

All these factors taken together result in the smoothing out of the column profiles that can be observed in Figures 39 and 41.

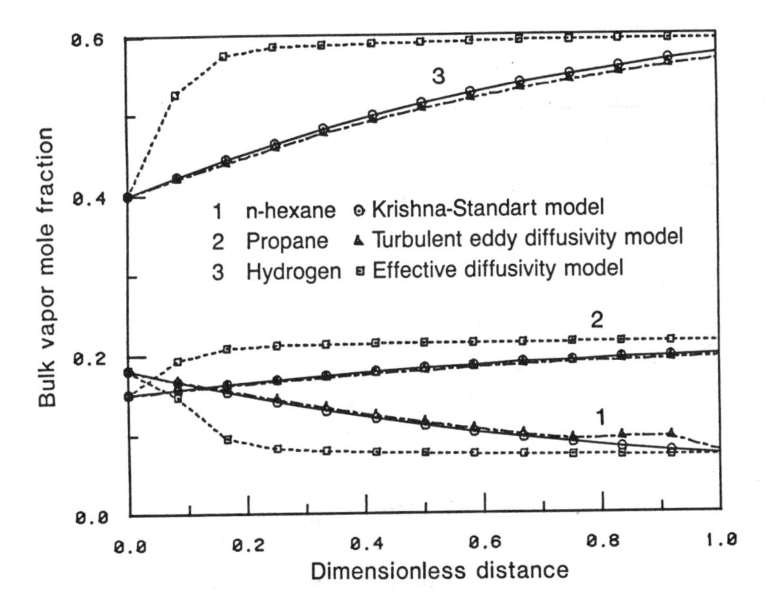

Figure 40. Vapor composition profiles obtained from a simulation of the condensation of a six component mixture; n-octane (1), n-heptane (2), n-hexane (3), n-butane (4), propane (5), and hydrogen (6). Calculations by Taylor et al. [171].

Liquid-Phase Models

There is no significant difference between the two extremes of condensate mixing (i.e., completely mixed or completely unmixed) if there is a noncondensing or sparingly soluble gas present in the vapor stream. This conclusion was reached by Krishna et al. [161], Webb and Sardesai [167], Taylor and co-workers [169, 171]. However, if all of the components are condensable, there can be a very considerable difference between these two extremes. McNaught [176] writes that it is common practice to assume complete mixing of the condensate when applying the Silver [156] (equilibrium) method and "it is well established that this can lead to underdesign if significant separation of the phases occurs" (as in a horizontal condenser); the same thing can be said if complete mixing of the condensate is assumed in the nonequilibrium models described above. The more conservative design is, in general, obtained with the condensate assumed to be completely unmixed. We would emphasize that there is no computational penalty for adopting either one of the two extreme cases. Interestingly enough, it is impossible to predict situations where some components condense while others evaporate using the no-mixing option. It is quite possible to predict this situation with a rate model or with the well-mixed condensate.

The Number of Steps to Take

We now come to one of the most important questions (at least from the point of view of computational efficiency); just exactly how many sections should we divide the condenser into? The answer depends on the choice of bulk phase conditions used in the rate equations.

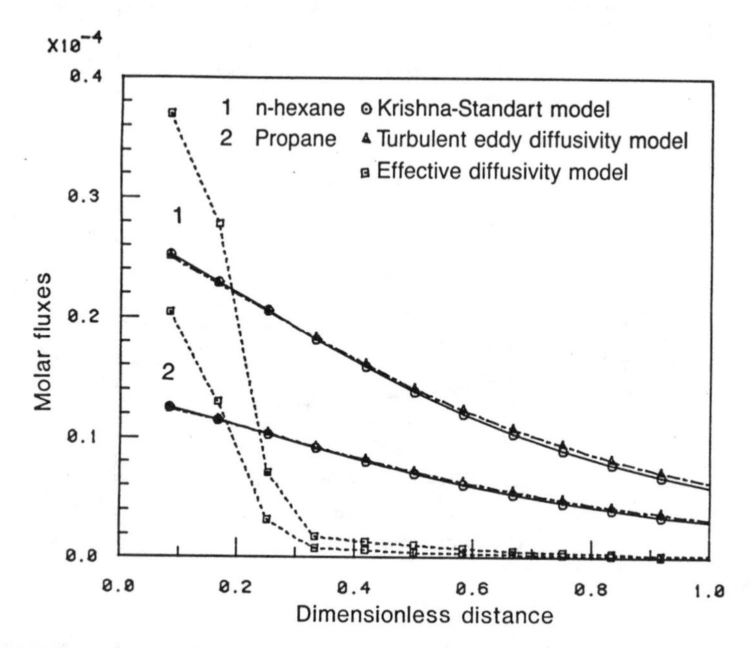

Figure 41. Molar flux profiles obtained from a simulation of the condensation of a six component mixture; n-octane (1), n-heptane (2), n-hexane (3), n-butane (4), propane (5), and hydrogen (6). Calculations by Taylor et al. [171].

As noted previously, there are three possibilities; using the inlet conditions, the outlet conditions, or some average of the two. The discussion here is after Taylor et al. [171].

The greater the number of steps, the smaller the discrepancy between the three averaging procedures. This, of course, is only to be expected since a large number of steps reduces the magnitude of the change in any particular quantity over any one step to the point where it does not matter much which conditions are used in the rate equations. For a given number of steps, all three choices yield quite similar predictions of the total amount condensed and of the heat load. The use of the end-of-section conditions in the rate equations leads to the most conservative design (i.e., the lowest condensate flow or a larger area to condense a given amount). Also, the fewer the number of steps taken, the more conservative the design.

It is in the prediction of the composition of the exit stream that we find the largest discrepancies between the different approaches. Using the *arithmetic* average of the mole fractions and temperatures at the beginning and at the end of the section in the rate equations for section j often give good estimates of the composition and temperature of the exit stream with as few as *four* sections (or, sometimes, only two) if noncondensing or sparingly soluble gases are present. This is found to be the case in all of the example problems considered by Taylor and Noah [169] (see Taylor et al. [171]). If all components condense, it appears to be better to use the *end* of section conditions in the rate equations. In fact, it is sometimes impossible to obtain solutions using either the beginning or average conditions with fewer than about four sections. This is due to the overprediction of the fluxes in the first sections of the condenser with the result that negative amounts of some species may be present in downstream sections. This usually has disastrous consequences in the subroutines used to calculate the physical properties. Moreover, the results obtained with as few as two sections are remarkably close to the results obtained using a very large number of steps (more or less equivalent to an exact numerical solution). The situation is completely different if the beginning conditions are used; with only four steps, the total amount condensed and heat load may be well enough predicted but the mole fractions in the exit vapor may be in error by as much as an order of magnitude.

Summary

In this section, we have described in some detail a nonequilibrium model of multicomponent condensation. Ways of solving the model equations have been considered and a calculation procedure recommended (in which the differential conservation equations are approximated by finite differences and the resulting set of algebraic equations solved *simultaneously* with the nonlinear equations representing the processes of interphase transport and interfacial equilibrium). A number of special cases of the general model were also identified.

With regard to the various models of vapor-phase mass transfer we conclude that:

1. Effective and equal diffusivity models should *not* be used in the determination of the rates of mass transfer in the vapor phase. They are not justified on theoretical grounds nor on experimental grounds, and their use offers no reduction in the cost of obtaining a solution or any increase in the ease by which that solution is obtained.
2. The film models that take interaction effects into account [4, 5, 72, 91, 92] yield temperature and composition profiles that, for all practical purposes, are indistinguishable.
3. There is little to choose between the film models that use the Chilton–Colburn analogy to obtain the heat and mass transfer coefficients and the turbulent eddy diffusivity methods when they are used to predict the performance of multicomponent condensers.

With regard to the liquid phase; there is very little to distinguish the results obtained using the two extremes of condensate mixing if noncondensing or sparingly soluble gases are present in the vapor phase. There can be a very considerable difference if all species condense. More conservative design is obtained with the liquid phase assumed to be completely unmixed.

We conclude from calculations of Taylor et al. [171] that the end-of-section conditions should be used in the determination of the rates of heat and mass transfer if all species condense. If noncondensing or sparingly soluble gases are present, the arithmetic average of the beginning and end conditions should be used. The possibility of being able to complete a design or simulation calculation with fewer than ten sections (possibly with as few as two or four sections), with a consequent large reduction in the computer time requirements (as compared to the numerical integration of the differential equations that we started with), overcomes one of the published objections to the use of the more realistic nonequilibrium models of condensation [176, 178].

Multicomponent Mass Transfer In Liquid-Liquid Systems

Since thermodynamic nonidealities are of the essence for phase separation in liquid-liquid systems, and such nonidealities contribute to multicomponent interaction effects, it may be expected that liquid-liquid extraction would offer an important industrial application of the theories presented in the previous sections. This is indeed found to be the case. We present some experimental evidence to show the significance of interaction effects in liquid-liquid extraction, and on the basis of this evidence we will draw some broad conclusions regarding their importance in design calculations. The evidence we present is largely based on experiments carried out in a modified Lewis batch extraction cell [189–192]. The analysis

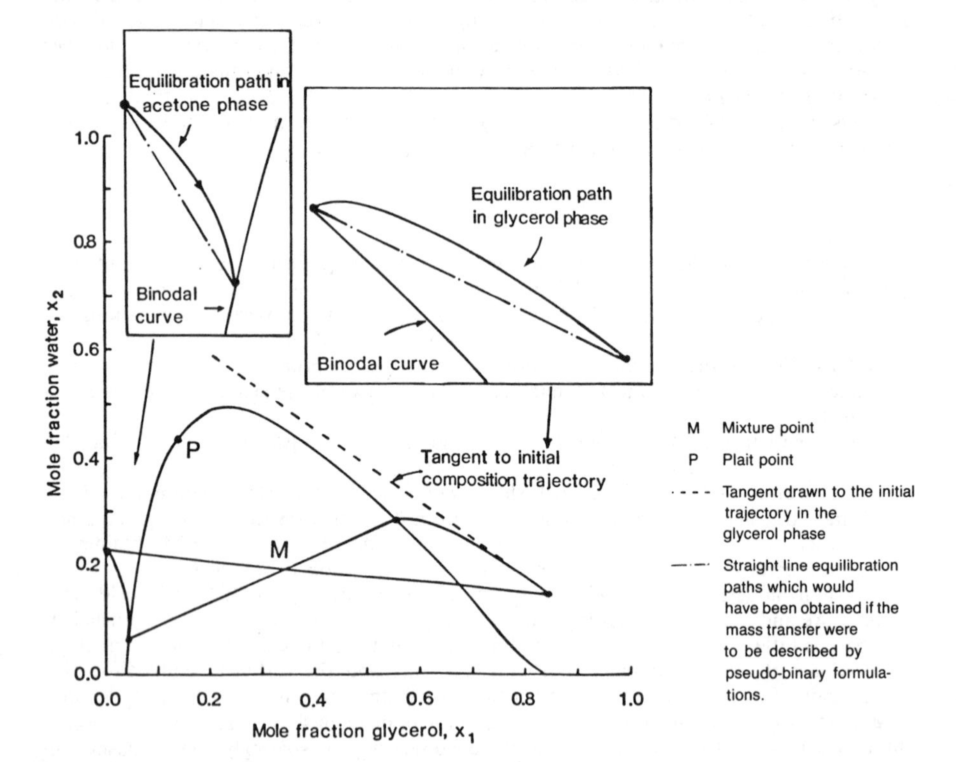

Figure 42. Equilibration paths during mass transfer in the system glycerol (1), water (2), and acetone (3) in a batch extraction cell [192]. Experimental data correspond to Run C.

we present here is due to Krishna et al. [192]. The experimental system which will be used to demonstrate multicomponent interaction effects is glycerol (1)—water (2)—acetone (1); this system is of Type I.

Equilibration Paths

The modified Lewis batch extraction cell used to obtain mass transfer data for the system glycerol—water—acetone is simply a single glass cylinder with mixing, contacting, and sampling facilities, with a net capacity of around 6 litres. The heavier phase (i.e. glycerol or ″ phase) is introduced at the bottom of the cell, and a horizontal ring and disc is placed on the liquid surface. The lighter phase (acetone or ′ phase) is charged on top. Two turbine stirrers and a set of vertical baffles in each compartment provide the necessary agitation for complete mixing of the phases. Representative samples are withdrawn through sample tubes at suitable intervals (typically 30 minutes), and their compositions determined.

From the experimental-phase compositions at different intervals, the "equilibration paths" can be determined on a ternary diagram. These are parametric curves, one for each phase and with points in the two curves in pairwise correspondence. In the general case where both phases are initially unsaturated, the equilibration paths obtained experimentally [192] are shown typically by Run C in Figure 42. Given enough contact time, the phases will approach and attain equilibrium, becoming mutually saturated. The equilibration paths will, therefore, terminate on the corresponding ends of a tie-line, since metastable states are prevented from arising by the constant agitation.

Now, the initial amounts and compositions of each phase are fixed and known from experiment. If the comparatively small amounts of sample withdrawn from either phase are neglected, the extraction cell can be considered to be a closed system in which the total and constituent mass are constant and equal to the initial amounts M_{t0}, M_{i0}. At any given time, therefore

$$M_t' + M_t'' = M_{t0} \tag{531}$$

$$M_i' + M_i'' = M_{i0} \tag{532}$$

or in terms of the "bulk" or average compositions of the phases, defined by

$$x_i' M_t' = M_i'; \quad x_i'' M_t'' = M_i'' \tag{533}$$

the following equations of conservation are obtained

$$M_t' + M_t'' = M_{t0}$$

$$x_i' M_t' + x_i'' M_t'' = M_{i0} = x_{i0} M_{t0}, \quad i = 1, 2 \tag{534}$$

These are the only independent equations of material balance that can be written down for the ternary system. All the terms in the left-hand side of Equation 534 are functions of time. Therefore, Equation 534 is a set of three equations in two unknowns M_t' and M_t'', and this redundancy provides for statistical check of the measurements and allows "best values" to be calculated.

Eliminating M_t' and M_t'' from between the previous equations a form of the lever rule is obtained

$$\frac{x_2'' - x_2'}{x_1'' - x_1'} = \frac{x_2'' - x_{20}}{x_1'' - x_{10}} \tag{535}$$

This shows that the straight line that joins any two light and heavy phase compositions corresponding to a same instant during the experiments must also pass through the point that represents the overall composition of the system (point M in Figure 42). This must, in particular, be true of the terminal compositions of each phase in mutual equilibrium. Hence, we see that the equilibrium tie-line is uniquely fixed by the initial masses charged to the cell, since there is only one tie-line which passes by M (otherwise, more than two phases in equilibrium could coexist, which is not possible in a system of Type I as is the case here).

If A represents the interfacial area between the two liquid phases, we can calculate the interfacial fluxes as follows

$$N_i = \frac{1}{A}\frac{dM_i}{dt}; \quad N_t = \frac{1}{A}\frac{dM_t}{dt} \tag{536}$$

From the material balance relations (Equations 531 and 532) we see that

$$N_i' + N_i'' = 0$$

$$N_t' + N_t'' = 0, \tag{537}$$

i.e., the rates of transfer of one phase must be negative of the corresponding rate for the other. The diffusive fluxes ($J_i = N_i - x_i N_t$) can be calculated for either phase from

$$J_i = \frac{M_t}{A}\frac{dx_i}{dt}, \quad i = 1, 2 \tag{538}$$

where only two of the J_i are independent.

If the driving forces for mass transfer are taken to be the difference in compositions between the interface (x_{iI}) and bulk fluid (x_i), then the constitutive relations for (J) may be written as

$$(J) = c_t[k^{\bullet}](x_I - x) \tag{539}$$

where $[k^{\bullet}]$ represents the matrix of mass transfer coefficients. We may combine Equations 538 and 539 and write

$$\frac{d(x)}{dt} = [K](x_I - x) \tag{540}$$

where we have additionally defined a "volumetric" transfer coefficient matrix

$$[K] = c_t \frac{A}{M_t}[k^{\bullet}] \tag{541}$$

whose elements have the dimensions of $[\text{time}]^{-1}$.

If we assume, for the moment, that $[K]$ is time invariant, the differential Equation 539 can be solved with the initial condition $t = 0$, $(x) = (x_0)$, to obtain the transient composition trajectories

$$(x - x_0) = [\exp[-[K]t]](x_I - x_0) \tag{542}$$

Let [P] represent the modal matrix of $[K]$, i.e., [P] has the property that

$$[P]^{-1}[K][P] = [\hat{K}] \tag{543}$$

where $[\hat{K}]$ represents the diagonal matrix

$$[\hat{K}] = \begin{bmatrix} \hat{K}_1 & 0 \\ 0 & \hat{K}_2 \end{bmatrix} \tag{544}$$

The structure of [P] is such that its columns are the eigenvectors, (e_1) and (e_2), of the matrix $[K]$, i.e.

$$[P] = [(e_1)(e_2)] \tag{545}$$

From the definition of eigenvectors, we note that (e_i) satisfies the matrix equation

$$[[K] - \hat{K}_i[I]](e_i) = (0) \tag{546}$$

and so the eigenvectors are given by

$$(e_1) = \begin{pmatrix} 1 \\ -\dfrac{K_{11} - \hat{K}_1}{K_{12}} \end{pmatrix} \quad (e_2) = \begin{pmatrix} 1 \\ -\dfrac{K_{11} - \hat{K}_2}{K_{12}} \end{pmatrix} \tag{547}$$

Premultiplying Equation 542 by $[P]^{-1}$, we obtain

$$(\hat{x} - \hat{x}_0) = \exp[-[\hat{K}]t](\hat{x}_I - \hat{x}_0) \tag{548}$$

which represents a set of two uncoupled equations

$$\hat{x}_i - \hat{x}_{i0} = \exp[-\hat{K}_i t](\hat{x}_{iI} - \hat{x}_{i0}), \quad i = 1, 2 \tag{549}$$

in pseudo-compositions, defined by

$$(\hat{x}) = [P]^{-1}(x).$$

In the pseudo-composition space $(\hat{x}_1, \hat{x}_2)$, the equilibration paths Equation 549 are straight lines approaching, the equilibrium state.

To recover the composition profiles in the "real" (x_1, x_2) space we premultiply Equation 548 by [P] giving

$$(x - x_0) = e^{-\hat{K}_1 t}\Delta\hat{x}_{10}(e_1) + e^{-\hat{K}_2 t}\Delta\hat{x}_{20}(e_2) \tag{550}$$

where the $\Delta\hat{x}_{i0}$ are (pseudo-) initial driving forces:

$$(\Delta\hat{x}_0) = [P]^{-1}(x_I - x_0). \tag{551}$$

Examination of Equation 550 shows that the initial trajectory will be dictated by the dominant eigenvalue, say $\hat{K}_2$, and the initial path will lie along (e_2). As equilibrium is approached, the equilibration path will lie along the "slow" eigenvector (e_1).

If the matrix $[K]$ is diagonal (i.e., with zero cross-coefficients), then [P] will reduce to the identity matrix and the composition profiles in real (x_1, x_2) space will be straight lines:

$$x_i - x_{i0} = \exp(-K_{ii}t)\Delta x_{i0}, \quad i = 1, 2 \tag{552}$$

In other words, if we were to adopt a pseudo-binary mass transfer formulation: $J_i = c_t k_{i,eff}(x_{iI} - x_i)$, then the consequence of this must be that the equilibration paths are rectilinear. The actual observed equilibration paths are highly curvilinear (see Figure 42) and, therefore, it appears that we must reject at the very outset the pseudo-binary mass transfer formulation for the diffusion fluxes.

The only flaw in the reasoning leading to this conclusion is our initial assumption that $[K]$ is time-invariant. Now M_t varies with time and the matrix $[k^\bullet]$ must be expected to be composition dependent and we must expect, in general, $[K]$ to be time-varying. For short time-intervals and small composition changes, $[K]$ may, however, be assumed to be constant. With this, less restricted, assumption let us examine the slope of the composition trajectory for the equilibration path in the glycerol-rich phase on the right-hand side of Figure 40. We observe that the tangent to the initial trajectory fails to intersect the binodal curves completely and so a diagonal, pseudo-binary, mass transfer formulation must be rejected for the initial stages of the equilibration process, since no real interface composition x_{iI} will satisfy the initial trajectory.

A quantitative analysis of the experimental data indeed shows that the matrix $[K]$ has significant off-diagonal elements and that a pseudo-binary formulation must be rejected for the system glycerol-water-acetone [192].

Mass Transfer in the Vicinity of a Plait Point

We now consider some further peculiarities of liquid-liquid mass transfer which manifest themselves in the vicinity of the plait point (P in Figure 42). It is here that the theory of irreversible thermodynamics is indispensible in our analysis of the diffusion process. Our starting point will be the following relation for the Fickian matrix of diffusion coefficients:

$$[D] = [H]^{-1}[G] \tag{553}$$

For thermodynamic stability in the system under consideration, the matrix [G] must be positive definite as noted earlier in this chapter. At the plait point itself, the determinant of [G] must vanish identically (see Modell and Reid [29]). For a ternary system we must have

$$G_{11}G_{22} - G_{12}G_{21} = 0 \tag{554}$$

where $G_{ij} = \dfrac{\partial(\mu_i - \mu_n)}{\partial x_j} = G_{ji}$. Since from Equation 47 we must also have

$$|D| = |G|/|H| \tag{555}$$

we derive the follow consequence for the Fickian matrix at the critical point P:

$$|D| = 0 \tag{556}$$

i.e., the determinant of [D] vanishes identically at P. It follows from matrix algebra that one of the eigenvalues of the matrix [D], say $\hat{D}_1$, is equal to zero. So we must have at the critical point

$$\hat{D}_1 = 0; \quad \hat{D}_2 = D_{11} + D_{22} \tag{557}$$

since $D_{11}D_{22} - D_{12}D_{21} = 0$.

The eigenvectors corresponding to $\hat{D}_1$ and $\hat{D}_2$ are, respectively:

$$(e_1) = \begin{pmatrix} 1 \\ -\dfrac{D_{11}}{D_{12}} \end{pmatrix} \text{ and } (e_2) = \begin{pmatrix} 1 \\ \dfrac{D_{22}}{D_{12}} \end{pmatrix} \tag{558}$$

We shall now show that the first eigenvector, (e_1), is in a direction parallel to the limiting tie line in the vicinity of P. To do this, we begin with the condition of equilibrium between two phases "and" for which the following relation must be satisfied

$$\mu_i(x_1', x_2') = \mu_i(x_1'', x_2''), \quad i = 1, 2, 3 \tag{559}$$

Consider now two phases, also in mutual equilibrium, the compositions of which differ from the previous corresponding phases by infinitesimal amounts. Considered as variations of these phase, the chemical potentials of these new phases must satisfy

$$d\mu_i' = d\mu_i'', \quad i = 1, 2, 3 \tag{560}$$

At constant temperature and pressure, we have from the Gibbs–Duhem restriction

$$x_1' d\mu_1' + x_2' d\mu_2' + x_3' d\mu_3' = x_1'' d\mu_1'' + x_2'' d\mu_2'' + x_3'' d\mu_3'' \tag{561}$$

Writting $x_3 = 1 - x_1 - x_2$ we may derive the following relationships

$$d(\mu_i' - \mu_3') = d(\mu_i'' - \mu_3''), \quad i = 1, 2 \tag{562}$$

and

$$x_1' d(\mu_1' - \mu_3') + x_2' d(\mu_2' - \mu_3') = x_1'' d(\mu_1'' - \mu_3'') + x_2'' d(\mu_2'' - \mu_3'') \tag{563}$$

which relations can be rearranged in the form

$$\frac{x_2'' - x_2'}{x_1'' - x_1'} = -\frac{d(\mu_1' - \mu_3')}{d(\mu_2' - \mu_3')} = -\frac{d(\mu_1'' - \mu_3'')}{d(\mu_2'' - \mu_3'')} \tag{564}$$

The differentials $d(\mu_i - \mu_3)$ may be written in terms of the composition fluctuations as:

$$d(\mu - \mu_n) = [G]\, d(x) \tag{565}$$

and so the Equations 561 through 564, written in terms of dx_i take the form

$$[G']\, d(x') = [G'']\, d(x'') \tag{566}$$

and

$$\frac{x_2'' - x_2'}{x_1'' - x_1'} = -\frac{G_{11}' dx_1' + G_{12}' dx_2'}{G_{21}' dx_1' + G_{22}' dx_2'} = -\frac{G_{11}'' dx_1'' + G_{12}'' dx_2''}{G_{21}'' dx_1'' + G_{22}'' dx_2''} \tag{567}$$

with $G_{12} = G_{21}$.

As the critical point is approached, the following limits are obtained

$$[G'] = [G''] = [G] \tag{568}$$

and

$$dx_i' = dx_i'' = dx_i. \tag{569}$$

So Equation 567 takes the limiting form

$$\frac{dx_2}{dx_1} = -\frac{G_{11} + G_{12}\dfrac{dx_2}{dx_1}}{G_{12} + G_{22}\dfrac{dx_2}{dx_1}} \tag{570}$$

which represents the slope of the limiting tie-line. Solving the resultant quadratic equation for dx_2/dx_1 we obtain

$$\frac{dx_2}{dx_1} = -\frac{G_{12}}{G_{22}} \pm \sqrt{\frac{G_{12}^2 - G_{11}G_{22}}{G_{22}}} = -\frac{G_{12}}{G_{22}} = -\sqrt{\frac{G_{11}}{G_{22}}} \tag{571}$$

because $G_{11}G_{22} - G_{12}^2 = 0$ at P.

Now, using the relation $[D] = [H]^{-1}[G]$ we can show that

$$\frac{D_{21}}{D_{22}} = -\frac{D_{11}}{D_{12}} = -\sqrt{\frac{G_{11}}{G_{22}}} \tag{572}$$

after invoking the requirement of $|H| = 0$ at equilibrium. Since the slope of the eigenvector (e_1) is $-D_{11}/D_{12}$ (see Equation 558) it follows that (e_1) is parallel to the limiting tie-line.

Let us now consider the equilibration paths in the vicinity of the point P. The matrix $[\underline{K}]$ which determines that the equilibration process will be a function of [D] and so the modal matrix of $[\underline{K}]$ will have the following structure

$$[P] = [(e_1)(e_2)] = \begin{bmatrix} 1 & 1 \\ -D_{11}/D_{12} & D_{22}/D_{12} \end{bmatrix} \tag{573}$$

where (e_1) and (e_2) are the eigenvectors of [D] as shown earlier.

With (e_1) and (e_2) as given, the equilibration paths Equation 550 simplify to give (recall that $\hat{D}_1 = 0$):

$$\hat{x}_1 - \hat{x}_{10} = \Delta\hat{x}_{10} + \exp(-\hat{D}_2 t)\Delta\hat{x}_{20} \tag{574}$$

$$x_2 - x_{20} = -\frac{D_{11}}{D_{12}}\Delta\hat{x}_{10} + \frac{D_{22}}{D_{12}}\exp(-\hat{D}_2 t)\Delta\hat{x}_{20} \tag{575}$$

Eliminating $\Delta\hat{x}_{10}$ from the two prior equations we obtain the relation between the compositions x_1 and x_2:

$$(x_2 - x_{20}) = -\frac{D_{11}}{D_{12}}(x_1 - x_{10}) + \frac{D_{11}+D_{22}}{D_{12}}\exp(-\hat{D}_2 t)\Delta\hat{x}_{20} \tag{576}$$

which represents a straight line in the (x_1, x_2) space with a slope $dx_2/dx_1 = -D_{11}/D_{12}$ equal to that of the limiting tie-line. Figure 43 shows diagramatically the results of actual equilibration runs [192] obtained with the initial phases on the tangent to the binodal curve at the critical point P. The path was indeed linear, and the rates of mass transfer were significantly lower than in other regions of the ternary diagram. The slowness of the mass transfer can be understood from the fact that the equilibration process is dictated by the smaller of the two eigenvalues. As the critical point is approached, the smaller eigenvalue tends to vanish.

Thus far, we have presented experimental evidence on the system glycerol-water-acetone to illustrate the applications in liquid-liquid mass transfer; the analysis presented is of

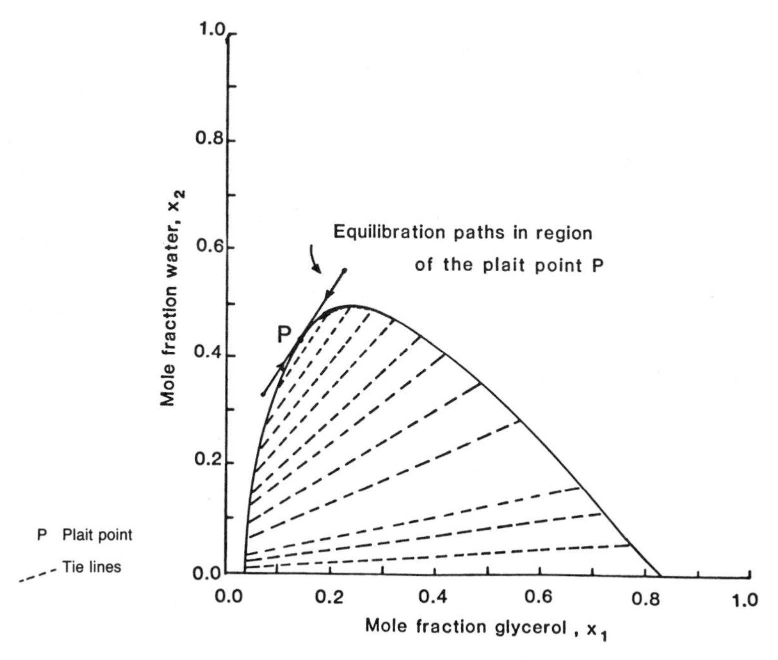

Figure 43. Mass transfer in the region of the critical point in the system glycerol (1), water (2), and acetone (3). The actual equilibration paths measured experimentally are rectilinear and parallel to the limiting tie-line [192].

general applicability. There is a shortage of experimental data on mass transfer in other systems but the available data (e.g., for the system water-chloroform-acetic acid; see Vitagliano et al. [193]) confirm the conclusions that multicomponent interaction effects should be expected to be very significant, especially in the vicinity of the critical point.

Due to the significant magnitude of the nondiagonal elements of $[\underline{K}]$, component "extraction efficiencies" or "heights of transfer units" will show very odd, unbounded, behavior entirely analogous to that previously pointed out for multicomponent distillation in tray and packed columns. The most rational approach to the design of extraction columns is to do away with dubious concepts such as efficiencies and HTU's and work directly in terms of sets of partial mass transfer coefficient matrices in either phase. The nonequilibrium-stage model, previously described for distillation is in principle usable for extraction problems. This is an area for future work.

CLOSURE

We have attempted to provide a rigorous fundamental framework for treating simultaneous heat and mass transfer in multicomponent systems. On the basis of available experimental evidence, we have shown that such systems can display transfer characteristics completely different from a simple binary system, which is exclusively treated in standard text books. Restricting ourselves to fluid-fluid, nonreacting transfer processes, we have shown how the formalisms developed in this review could be applied in a rigorous and computationally efficient manner in the design of distillation, absorption, and condensation equipment.

To aid further development in this industrially important area we feel that additional research efforts in the following subjects are essential:

- Experimental measurements for diffusion coefficients in nonideal ternary and quaternary mixtures; improved prediction methods for these diffusivities.
- Experimental measurements with ternary and quaternary mixtures in industrial contactors for distillation, absorption, extraction, and condensation equipment to verify the design procedures which have so far been developed and tested.
- A rigorous design procedure for multicomponent extraction columns using the theories presented in this review; the non-equilibrium stage model would need to be extended to take account of backmixing characteristics of the dispersed and continuous phases.
- A closer look is warranted for the transfer processes in extractive distillation especially when there is a possibility of obtaining two liquid phases on a particular tray. In view of the interesting phenomena which can occur in the region of the critical point, it remains to be seen what the practical implications are in column design.

NOTATION

a_i activity of component i in solution $[-]$, Table 1

a' interfacial area per unit volume of dispersion $(m^2 m^{-3})$, Equation 431

a interfacial area (m^2), Equation 451

ΔA_j interfacial area in the j-th section of a condenser (m^2), Equation 515

A_n eigenvalue in Kronig–Brink model, Equation 357

A^+ damping constant in van Driest mixing length model $[-]$, Equation 397

A active bubbling area on tray (m^2), Equation 431; also, interfacial area in batch extraction cell (m^2), Equation 536

$[A]$ matrix defined by $[A]^{-1} = [\beta][B]^{-1}$ $(m^{-2} s)$, Equations 238 through 240; also, approximated part of Jacobian matrix (various), Equation 119

[B] inverted matrix of diffusion coefficients (m^{-2} s), Equation 57

[B^m] transformation matrix [−], Equations 67 through 69

[B^V] transformation matrix [−]. Equations 70 through 72

[*B*] matrix of inverted binary mass transfer coefficients (m^{-1} s), Equation 428

c_i molar density of component i (kmol m^{-3}),

c_t mixture molar density (kmol m^{-3}),

C_{pi} specific heat of component i (J kg^{-1}), Equation 406; also, molar heat capacity of component i (J $kmol^{-1}$), Equation 294

$\bar{C}_p$ specific heat of mixture (J kg^{-1}), Equation 92; also, molar heat capacity of mixture (J $kmol^{-1}$), Equation 81

[C] computed part of Jacobian matrix (various), Equation 119

d characteristic length of contacting device (m), Equation 417

d_i generalized driving force for mass diffusion (m^{-1}), Equations 22 and 38

D Fickian diffusivity for binary mixture ($m^2 s^{-1}$), Equation 19

D'_{ij} Maxwell–Stefan diffusivity for pair i−j ($m^2 s^{-1}$), Equation 22

D'_{ieff} effective diffusivity of component i in multicomponent mixture ($m^2 s^{-1}$), Equations 73 and 250

D_{turb} turbulent eddy diffusivity ($m^2 s^{-1}$), Equation 77

D_i^T thermal diffusion coefficient (kg $m^{-1} s^{-1}$), Equation 85

D_{ref} reference value for diffusion coefficient ($m^2 s^{-1}$), Equation 352

[D] matrix of Fickian diffusion coefficients (m^2 s^{-1}), Equation 46

$\hat{D}_i$ i-th eigenvalue of [D] ($m^2 s^{-1}$), Equation 49

[$\bar{D}$] normalized Fickian diffusivity matrix [−], Equation 352

[$\mathbb{D}$] augmented matrix of transport coefficients ($m^2 s^{-1}$), Equation 322

(e_i) i-th eigenvector of (D), [−], Equation 546

E energy flux in stationary coordinate frame of reference (W m^{-2}), Equation 83

$\mathscr{E}$ energy transfer rate (W), Equation 455

E_{0yi} Murphree point efficiency of component i [−], Equations 441 through 444

E energy balance equation (W m^{-2}, W), Equations 115 and 448

f Fanning friction factor [−], Equation 368

f_i fugacity of component i (Pa); also, molar flow of component i in feed stream (kmol s^{-1}), Equation 446

$\hat{f}_i$ function of total molar flux [−], Equations 23 and 347

$F(N_t)$ function of total molar flux [−], Equation 230

F F-factor, product of gas flow rate per unit bubbling area and the square root of the gas density ($kg^{1/2} m^{-1/2} s^{-1}$), Equation 479

$\underset{\sim}{F}_i$ body force acting per unit mass of i (N kg^{-1}), Equation 6

(*F*) vector of independent equations (various)

Fo Fourier number [−], Equation 348

G_{ij} chemical potential—composition derivative (J $kmol^{-1}$), Equation 44

h heat transfer coefficient (W $m^{-2} K^{-1}$), Equation 104

h heat transfer coefficient (m s^{-1}), Equation 103

h_f froth or dispersion height (m), Equation 431

$\bar{H}_i$ partial specific enthalpy (J kg^{-1}), Equation 406; also,

	partial molar enthalpy of component i ($J\,kmol^{-1}$), Equation 80
HTU_{oyi}	pseudo-binary height of a transfer unit for component i (m), Equation 485
I	referring to interphase or interface
[H]	matrix of transport ceofficients, Equation 42
[I]	identity matrix [−], Equation 188
$\underset{\sim}{I}$	unit tensor [−], Equation 5
j_D	Chilton–Colburn j-factor for mass transfer [−], Equation 417
j_H	Chilton–Colburn j-factor for heat transfer [−], Equation 417
j_i	mass diffusion flux relative to the mass average velocity ($kg\,m^{-2}\,s^{-1}$)
J_i	molar diffusion flux relative to the molar average velocity ($kmol\,m^{-2}\,s^{-1}$)
J_i^V	molar diffusion flux relative to the volume average reference velocity ($kmol\,m^{-2}\,s^{-1}$)
$j_{i,\,turb}$	turbulent diffusion flux of component i ($kg\,m^{-2}\,s^{-1}$), Equation 77
$J_0(\,)$	zero order Bessel function [−], Equation 359
(J)	matrix of diffusion and reduced heat conduction fluxes ($kmol\,m^{-2}\,s^{-1}$), Equation 318
[J]	Jacobian matrix (various), Equation 117
k	mass transfer coefficient in a binary mixture ($m\,s^{-1}$), Equation 95
K_0	overall mass transfer coefficient in a binary mixture ($m\,s^{-1}$), Equation 156
[k]	matrix of multicomponent mass transfer coefficients ($m\,s^{-1}$), Equation 101
$\hat{k}_i$	eigenvalue of [k] ($m\,s^{-1}$), Equation 224
k_{ieff}	pseudo-binary (effective) mass transfer coefficient of component i in a mixture ($m\,s^{-1}$), Equation 261
$[K_0]$	matrix of multicomponent overall mass transfer coefficients ($m\,s^{-1}$), Equation 157
$[K]$	matrix of rate coefficients (s^{-1}), Equation 541
$\hat{K}_i$	eigenvalue of (K) (s^{-1}), Equation 544
K	equilibrium ratio (K-value) [−], Equation 111
k_T	thermal diffusion ratio [−], Equation 310
ℓ	generalized characteristic length (m), Figure 10; also, mixing length describing turbulent transport (m), Equation 79
ℓ_i	molar flow rate of component i in liquid phase ($kmol\,s^{-1}$), Equation 446
Le	Lewis number [−], Equation 484
L_F	volumetric liquid flow rate per unit average liquid flow path width ($m^2\,s^{-1}$), Equation 479
M_i	molecular weight of component i ($kg\,kmol^{-1}$); also, moles of i in batch extraction cell (kmol), Equation 532
M_t	total moles of mixture in batch extraction cell (kmol), Equation 531
[M]	symbol for a general matrix (various), Table 2 Equation 198; also, matrix of equilibrium constants [−], Equations 152 and 153
M	material balance equation ($kmol\,s^{-1}$), Equation 446
n	number of components in the mixture [−], Equation 7
n_i	mass flux of component i referred to a stationary coordinate reference frame ($kg\,m^{-2}\,s^{-1}$), Equation 365
n_t	mixture total mass flux referred to a stationary coordinate reference frame ($kg\,m^{-2}\,s^{-1}$), Equation 365
N_i	molar flux of component i referred to a stationary coordinate reference frame ($kmol\,m^{-2}\,s^{-1}$), Equations 15 and 93

N_t mixture molar flux of component i referred to a stationary coordinate reference frame ($kmol\,m^{-2}\,s^{-1}$), Equation 94

Nu Nusselt number [−], Equation 361

[NTU] matrix of numbers of transfer units Equation 438

NTU_{0yi} pseudo-binary number of transfer units of component i [−], Equation 486

N mass transfer rate ($kmol\,s^{-1}$), Equation 449

(N) matrix of molar and reduced energy fluxes ($kmol\,m^{-2}\,s^{-1}$), Equation 318

p thermodynamic pressure (Pa), Equation 5

$\tilde{p}$ pressure tensor (Pa), Equation 5

[P] model matrix of [D], [−], Equations 223 and 227

Pr molecular Prandtl number [−], Equation 405

Pr_{turb} turbulent Prandtl number [−], Equation 92

q conductive heat flux ($W\,m^{-2}$), Equation 6; also, integer parameter [−], Equation 198

q_{turb} turbulent contribution to the conductive heat flux ($W\,m^{-2}$), Equation 91

[Q] matrix describing unaccomplished equilibrium [−], Equation 440

Q equilibrium equation [−], Equations 114 and 469

r coordinate direction (m), Equation 159

r_0 inner edge of film (m), Equations 162 and 166; also, radius of spherical particle (m), Equation 350

r_d outer edge of film (m), Equations 162 and 166

r_j ratio of sidestream flow to interstage flow [−], Equation 447

R gas constant (8414.4 $J\,kmol^{-1}\,K^{-1}$), Equation 22

Re Reynolds number [−], Equation 418

R_w inner radius of tube wall (m), Equation 364

R_W^+ reduced tube radius [−], Equation 371

R mass transfer rate equation ($kmol\,m^{-2}\,s^{-1}$, $kmol\,s^{-1}$), Equations 112 and 471

s surface renewal frequency (s^{-1}), Equation 333

S summation equation [−], Equation 470

Sc Schmidt number [−], Equation 418

[Sc] matrix of Schmidt numbers [−], Equations 372 and 420

Sc_{turb} turbulent Schmidt number [−], Equation 78

Sh Sherwood number [−], Equations 350 and 417

[Sh] matrix of Sherwood numbers [−], Equations 353 and 420

St Stanton number [−], Equation 417

[St] matrix of Stanton numbers [−], Equations 384 and 421

t time (s), Equation 1

t_e exposure time (s), Equation 332

T temperature (K), Equation 22

T_{ref} reference temperature (K), Equation 294

$\tilde{u}_i$ velocity of diffusion of species i ($m\,s^{-1}$)

$\tilde{u}$ molar average reference velocity ($m\,s^{-1}$)

$\tilde{u}^V$ volume average reference velocity ($m\,s^{-1}$)

$\tilde{u}_I$ velocity of the interface ($m\,s^{-1}$), Equation 2

u* friction velocity ($m\,s^{-1}$), Equation 368

$\bar{u}$ average flow velocity ($m\,s^{-1}$), Equation 368

u^+ reduced velocity ($m\,s^{-1}$), Equation 370

v mass average mixture velocity ($m\,s^{-1}$)

V molar flow rate of mixture ($kmol\,s^{-1}$), Equation 432

V_i partial molar volume ($m^3\,kmol^{-1}$)

V_t mixture molar volume ($m^3\ kmol^{-1}$)

v_i molar flow rate of component i ($kmol\ s^{-1}$), Equation 431

W weir height (m), Equation 479

[W] matrix of mass transfer coefficients ($kmol\ m^{-2}\ s^{-1}$), Table 3

x_i mole fraction of component i [−]

"x" referring to phase "x" [−]

[*X*] matrix of independent variables (various)

y_i mole fraction of component i [−]

y^* weighted mole fraction in explicit methods [−], Equation 242

y distance from wall or interface (m), Equation 364

y^+ reduced distance from wall or interface [−], Equation 369

y_l distance from wall beyond which viscous effects are negligible (m), Equation 388

y_l^* reduced distance from wall beyond which viscous effects become negligible [−], Equation 388

y_b distance from wall at which the bulk phase properties attained (m), Equation 386

y_b^+ reduced distance from wall at which the bulk phase properties are attained [−], Equation 385

z_i ratio of flux of i to total flux of mixture [−], Equation 141

Z height of contacting device (m), Equation 489

Greek Symbols

α_{ij} multicomonent thermal diffusion factors [−], Equation 86

[β] bootstrap matrix [−], Equations 127 and 145

γ_i activity coefficient of component i in solution [−], Equations 26 and 58

[Γ] matrix of thermodynamic factors [−], Equation 58

δ distance from interface (m), Equation 348

δ_{ik} Kronecker delta, 1 if i = k, 0 if i = k [−], Equation 128

ε_i discrepancy between mass transfer models [−], Equation 272

ζ rate of production of field quantity in buik fluid mixture (various), Equation 1

ζ_I rate of production of field quantity at the interface (various), Equation 2

ζ dimensionless mass transfer parameter for binary system [−], Equation 164

η dimensionless distance [−], Equation 162

[Θ] augmented matrix of mass transfer rate factors [−], Equation 287

κ von Karman constant = 0.4 [−], Equation 395

λ_i molar heat of vaporization of component i ($J\ kmol^{-1}$), Equation 121

Λ_i dimensionless thermal parameters [−], Equation 126

λ molecular thermal conductivity ($W\ m^{-1}\ K^{-1}$), Equation 85

λ_{turb} turbulent thermal conductivity ($W\ m^{-1}\ K^{-1}$), Equation 91

λ_n eigenvalue in the Kronig–Brink model [−], Equation 357

μ_i molar chemical potential of component i ($J\ kmol^{-1}$)

μ molecular (dynamic) viscosity of mixture (Pa s), Equation 369

μ_{turb} turbulent eddy viscosity (Pa s), Equation 92

ν molecular kinematic viscosity of mixture ($m^2\ s^{-1}$), Equation 369

ν_{turb} turbulent eddy kinematic viscosity ($m^2\ s^{-1}$), Equation 78

ν_i determinacy coefficients for species i (various)

$\underset{\sim}{\xi}$ unit normal directed from phase "x" to phase "y" [−], Equation 2;

also, dimensionless distance along dispersion or column height [−], Equation 437

Ξ — correction factor for high fluxes in binary mass transfer [−], Equation 100; also, correction factor for high fluxes in explicit methods [−], Equation 249

Ξ_{ieff} — correction factor for high fluxes in pseudo-binary (effective diffusivity) methods [−], Equation 263

Ξ_H — correction factor for the effect of high fluxes on the heat transfer coefficient [−], Equation 106

$[\Xi]$ — matrix of high flux correction factors [−], Equation 102

$\hat{\Xi}_i$ — i-th eigenvalue of $[\Xi]$, [−], Equation 224

ϱ_i — mass density of component i $(kg\,m^{-3})$

ϱ_t — mixture mass density $(kg\,m^{-3})$

σ — rate of entropy production, Equation 40

(τ) — matrix of independent fluid properties [−], Equation 319

τ — shear stress (Pa), Equation 368

$\underset{\approx}{\tau}$ — stress tensor (Pa), Equation 5

(φ) — column matrix of dimensionless mass transfer parameters [−], Equations 185 and 186

$\underset{\sim}{\Phi}$ — nonconvective flux of field quantity (various), Equation 1

Φ — mass transfer rate factor for binary mass transfer [−], Equation 163; also, mass transfer rate factor for explicit methods [−], Equation 244

Φ_{ieff} — mass transfer rate factor in pseudo-binary (effective diffusivity) methods [−], Equation 259

Φ_H — heat transfer rate factors [−], Equation 295

$[\Phi]$ — matrix of mass transfer rate factors [−], Equations 184, 207, 339 and 381

$\hat{\Phi}_i$ — i-th eigenvalue of $[\Phi]$ [−], Equation 196

$\psi(t)$ — surface-age distribution (s^{-1}), Equation 331

ψ — referring to any field variable (various), Equation 1

$[\psi]$ — rate factor matrix [−], Equation 375

ψ_H — heat transfer rate factor [−], mass fraction of component i [−]

Subscripts

av — denotes that suitably averaged properties are used in the determination of the indicated parameter

b — bulk phase property

d — dominant eigenvalue

E — quantity entering zone under consideration

eff — pseudo-binary or "effective" parameter

H — parameter relevant to heat transfer

I — referring to the interface

i — component i property or parameter

i, j, k — component indices, stage, or section numbers (j only)

L — quantity leaving zone under consideration

m — mean value; also refers to the mass average velocity

n — n-th component

0 — overall parameter; also denotes reduced energy and heat conduction fluxes

0y — overall parameter referred to the vapor phase

r — radial position

ref — denotes reference quantity

spec — denotes specified quantity

t — referring to total mixture

T, p — constant temperature and pressure

turb — turbulent parameter or property

w — referring to the wall

x — referring to the "x" phase

y — referring to the "y" phase

Superscripts

C	referring to the coolant
F	referring to the feed
I	referring to the interface
(k)	denotes iteration number
L	liquid phase
m	referring to the mass average velocity
V	referring to the vapor/gas phase
v	referring to the volume average velocity
W	referring to the wall
x	referring to the "x" phase
y	referring to the "y" phase
+	reduced quantity
•	denotes parameter under conditions of high transfer rates
*	equilibrium value; also friction velocity; also appears on a weighted mole fraction
′	referring to phase ′
″	referring to phase ″

Miscellaneous

	overbar denotes partial molar property; also averaged parameter
⟨ ⟩	referring to cup-mixing parameter
	referring to pseudo- or transformed variable; eigenvalue of corresponding matrix

Mathematical symbols

d/dt	derivative moving with the average fluid velocity
∂/∂t	derivative at a fixed point in space
$\underset{\sim}{\nabla}\cdot$	divergence
$\underset{\sim}{\nabla}$	gradient
$\tilde{\Delta}$	difference operator
lim	limit

Matrix operations and notation

( )	column matrix
[ ]	square matrix
$[\ ]^{-1}$	inverse of a square matrix
$(\ )^{T}$	row matrix
| |	determinant of a square matrix

REFERENCES

1. Walter, J. F., and Sherwood, T. K., "Gas Absorption in Bubble-Cap Columns," *Ind. Eng. Chem.*, Vol. 33 (1941), pp. 493–501.
2. Toor, H. L., "Diffusion in Three Component Gas Mixtures," *AIChE J.*, Vol. 3 (1957), pp. 198–207.
3. Toor, H. L., and Burchard, J. K., "Plate Efficiencies in Multicomponent Distillation," *AIChE J.*, Vol. 6 (1960), pp. 202–206.
4. Toor, H. L., "Solution of the Linearized Equations of Multicomponent Mass Transfer," *AIChE J.*, Vol. 10 (1964), pp. 448–455, 460–465.
5. Stewart, W. E., and Prober, R., "Matrix Calculation of Multicomponent Mass Transfer in Isothermal Systems," *Ind. Eng. Chem. Fundam.*, Vol. 3 (1964), pp. 224–235.
6. Krishna, R., and Standart, G. L., "Mass and Energy Transfer in Multicomponent Systems," *Chem. Eng. Commun.*, Vol. 3 (1979), pp. 201–275.
7. Bupara, S. S., Ph. D., Dissertation in Chemical Engineering, University of Minnesota, 1965.
8. Clarke, R., Ph. D. Dissertation in Engineering Thermodynamics, Rice University, 1970.

9. Hopke, S. W., Ph. D. Dissertation in Chemical Engineering, Northwestern University, 1970.
10. Eliassen, J. D., Ph. D. Dissertation in Chemical Engineering, University of Minnesota, 1970.
11. Murphy, C. L., Ph. D. Dissertation in Chemical Engineering, University of Minnesota, 1966.
12. Slattery, J. C., "General Balance Equation for a Phase interface," *Ind. Eng. Chem. Fundam.,* Vol. 6 (1967), pp. 108–115.
13. Slattery, J. C., Corrections to article in Reference 12, *Ind. Eng. Chem. Fundam.,* Vol. 7, 1968, p. 672.
14. Slattery, J. C., *Momentum, Energy and Mass Transfer in Continua,* 2nd et. Huntington, New York: Krieger Publishing Company, 1981.
15. Swartz, C. M., Ph. D. Dissertation in Chemical Engineering, University of Minnesota, 1972.
16. Walker, R.M., Ph. D. Dissertation in Engineering Thermodynamics, University of Connecticut, 1974.
17. Deemer, A. R., and Slattery, J. C., "Balance Equations and Stuctural Models for Phase Interfaces," *Int. J. Multiphase Flow,* Vol. 4 (1978), pp. 171–192.
18. Ly, L.–A. N., Carbonell, R. G., and McCoy, B. J., "Diffusion of Gases Through Surfactant Films: Interfacial Resistance to Mass Transfer." *AIChE J.,* Vol. 25 (1979), pp. 1015–1024.
19. Standart, G. L., "The Mass, Momentum and Energy Equations for Heterogeneous Flow Systems," *Chem. Eng. Sci.* Vol. 19 (1964), pp. 227–236.
20. Truesdell, C., and Toupin, R., "The Classical Field Theories," *Handbuch der Physik,* S. Flugge, (Ed.), Vol. III/1, Berlin: Springer-Verlag, 1960.
21. Bird, R. B., Stewart, W. E., and Lightfoot, E. N., *Transport Phenomena,* New York: Wiley, 1960
22. Sherwood, T.K., Pigford, R. L., and Wilke, C. R., *Mass Transfer,* New York: McGraw-Hill, 1975.
23. Carberry, J. J., *Chemical and Catalytic Reaction Engineering,* New York: McGraw-Hill, 1977.
24. Vignes, A., "Diffusion in Binary Solutions," *Ind. Eng. Chem. Fundam.,* Vol. 5 (1966), pp. 189–199.
25. Present, R. D., *Kinetic Theory of Gases,* New York: McGraw-Hill, 1958.
26. De Groot, S. R., and Mazur, P., *Nonequilibrium Thermodynamics,* Amsterdam: North-Holland, 1962.
27. Truesdell, C., *Rational Thermodynamics,* New York: McGraw-Hill, 1969.
28. Hasse, R., *The Thermodynamics of Irreversible Processes,* London: Addison-Wesley, 1969.
29. Modell, M. M., and Reid, R.C., *Thermodynamics and its Applications,* 2nd ed., Englewood Cliffs, N. J.: Prentice-Hall, 1983.
30. Kirkaldy, J. S., "Isothermal Diffusion in Multicomponent Systems," *Advances in Materials Research,* Vol. 4 (1970), pp. 55–100.
31. Yao, Y. L., "Algebraical Analysis of Diffusion Coefficients in Ternary Systems," *J. Phys. Chem.,* Vol. 45 (1966), pp. 110–115.
31a. Cussler, E. L., *Multicomponent Diffusion,* Amsterdam: Elsevier, 1976.
31b. Dunlop, P. J., Steel, B. J., and Lane, J. E., "Experimental Methods for Studying Diffusion in Liquids, Gases and Solids." Chapter in *Techniques of Chemistry,* Vol. I, *Physical Methods of Chemistry,* New York: Wiley, 1972.
32. Hirschfelder, J. O., Curtiss, C. F., and Bird, R. B., *Molecular Theory of Gases and Liquids,* New York: Wiley, 1954.
33. Standart, G. L., Taylor, R., and Krishna, R., "The Maxwell–Stefan Formulation of Irreversible Thermodynamics for Simultaneous Heat and Mass Transfer," *Chem. Eng. Commun.,* Vol. 3 (1979), pp. 277–289.

34. Chapman, S., and Cowling, T. G., *The Mathematical Theory of Non-Uniform Gases*, 3rd ed., prepared in cooperation with D. Burnett, Cambridge: Cambridge University Press, 1970.
35. Reid, R. C., Prausnitz, J. M., and Sherwood, T.K., *The Properties of Gases and Liquids*, 3rd ed., New York: McGraw-Hill, 1977.
36. Bandrowski, J., and Kubaczka, A., "On the Prediction of Diffusivities in Multicomponent Liquid Systems," *Chem. Eng. Sci.*, Vol. 37 (1982), pp. 1309–1313.
37. Cullinan, H. T., and co-workers, "Prediction of Multicomponent Diffusion Coefficients," *Ind. Eng. Chem. Fundam.*, Vol. 5 (1966), pp. 281–283; Vol 6 (1967), pp. 72–77; Vol. 6 (1967), pp. 616; Vol. 7 (1968), pp. 317–319; Vol. 7 (1968), pp. 331–332; Vol. 7 (1968), pp. 519–520; Vol. 9 (1970), pp. 84–93; Vol. 10 (1971), pp. 600–603; *Can. J. Chem. Eng.*, Vol. 45 (1967), pp. 377–381; Vol. 49 (1971), pp. 130–133; Vol. 49 (1971), pp. 632–636; Vol. 49 (1971), pp. 753–757; *AIChE J.*, Vol. 13 (1967), pp. 1171–1174; Vol. 21 (1975), pp. 195–197
38. Kosanovich, G. M., Ph. D. Dissertation in Chemical Engineering, State University of New York at Buffalo, 1975.
39. Prausnitz, J. M., et al. Computer Calculations for Multicomponent Vapor-Liquid and Liquid-Liquid Equilibria, Englewood Cliffs, NJ: Prentice-Hall, 1980.
40. Cullinan, H. T., "Analysis of the Flux Equations of Multicomponent Diffusion." *Ind. Eng. Chem. Fundam.*, Vol. 4 (1965), pp. 133–139.
41. Ortega, J. M., and Rheinbolt, W. C., *Iterative Solution of Nonlinear Equations in Several Variables*, New York: Academic Press, 1970.
42. Amundson, N. R., *Mathematical Methods in Chemical Engineering. I: Matrices and their Applications*, Englewood Cliffs, NJ: Prentice-Hall, 1966.
43. Burchard, J. K., and Toor, H. L., "Diffusion in an Ideal Mixture of Three Completely Miscible Non-Electrolytic Liquids—Toluene, Chlorobenzene, Bromobenzene." *J. Phys. Chem.*, Vol. 66 (1962), pp. 2015–2022.
44. Launder, B. E., and Spalding, D. B., *Mathematical Models of Turbulence*, London: Academic Press, 1972.
45. Sideman, S., and Pinczewski, Z., Chapter in *Topics in Transport Phenomena* (C. Gutfinger (Ed.), New York: Halstead Press, 1975.
46. Blom, J., Ph. D. Dissertation, Eindhoven University of Technology, 1970.
47. Grew, K. E., and Ibbs, T. L., *Thermal Diffusion in Gases*, Cambridge: Cambridge University Press, 1962.
48. Grew, K. E., "Thermal Diffusion" Chapter in Transport Phenomena in Fluids H. J. M. Hanley (Ed.), New York: Marcel Dekker, 1969.
49. Mason, E. A., Munn, R. J., and Smith, F. J. "Thermal Diffusion in Gases." *Advances in Atomic and Molecular Physics*, Vol. 2 (1966), pp. 33–87.
50. Glasstone, S., *Sourcebook on Atomic Energy*, 2nd ed., Princeton, NJ: Van Nostrand, 1958.
51. Tambour, Y., and Gal-Or, B., "Phenomenological Theory of Coupling in Multicomponent Compressible Laminar Boundary Layers," *The Physics of Fluids*, Vol. 19 (1976), pp. 219–226.
52. Tambour, Y., and Gal-Ov, B., "Coupled Heat and Mass Transfer in Surface Reacting Boundary Layers," *Proc. Levich Conference*, Oxford (1977), pp. 349–369.
53. Tambour, Y., "On Thermal-Diffusion Effects in Chemically Frozen Multicomponent Boundary Layer with Surface Catalytic Recombination Behind a Strong Moving Shock," *Int. J. Heat Mass Transfer*, Vol. 23 (1980), pp. 321–327.
54. Srivastava, R., and Rosner, D. E., "A New Approach to the Correlation of Boundary Layer Transfer Rates with Thermal Diffusion and/or Variable Properties," *Int. J. Heat Mass Transfer*, Vol. 22 (1979), pp. 1281–1294.
55. Gokoglu, S., and Rosner, D. E., "Correlation of Thermophoretically Modified Small Particle Deposition Rates is Forced Convection Systems with Variable Properties, Transpiration Cooling and/or Viscous Dissipation," *Int. J. Heat Mass Transfer* (1984), in press.

56. Rosner, D. E., and Fernandez de la Mora, J., "Small Particle Transport Across Turbulent Nonisothermal Boundary Layers," *J. Eng. Power* (ASME), Vol. 104 (1982), pp. 885–894.
57. Rosner, D. E., "Thermal (Soret) Diffusion Effects on Interfacial Mass Transport Rates" *Physicochemical Thermodynamics,* Vol. 1 (1980), pp. 159–185.
58. Green, S. J., Ph. D. Dissertation in Chemical Engineering, University of Pittsburgh, 1968.
59. Henley, E. J., and Seader, J. D., *Equilibrium Stage Separation Operations in Chemical Engineering,* New York: Wiley, 1981.
60. Myers, A. L., and Seider, W. E., *Introduction to Chemical Engineering and Computer Calculations,* Englewood Cliffs, NJ: Prentice-Hall, 1976.
61. Westerberg, A. W., et al. *Process Flowsheeting,* Cambridge: Cambridge University Press, 1979.
62. Taylor, R., Lucia, A., and Krishnamurthy, R., "A Newton-Like Algorithm for the Efficient Estimation of Rates of Multicomponent Condensation by a Film Model" *I. Chem. E. Symposium Series,* No. 75 (London) (1983), pp. 380–397.
63. Dennis, J. E., and More, J. J., "Quasi-Newton Methods—Motivation and Theory," *SIAM Rev.,* Vol. 19 (1977), pp. 46–89.
64. Lucia, A., and Machietto, S., "New Approach to Approximation of Quantities Involving Physical Properties Derivatives in Equation-Oriented Process Design." *AIChE J.,* Vol. 29 (1983), pp. 705–712.
65. Westman, K. R., Lucia, A., and Miller, D., "Flash and Distillation Calculations by a Newton-Like Method," *Comput. Chem. Eng.* (1984), pp. 219–228.
66. Lucia, A., and Westman, K. R., "Low Cost Solutions to Multistage, Multicomponent Separation Problems by a Hybrid Fixed Point Algorithm," Proc. Conference on the Foundations of Computer Aided Chemical Process Design (Snowmass, Colorado), 1984.
67. Mason, E. A., et al. "Grahams Laws of Diffusion and Effusion," *J. Chem. Educ.,* Vol. 44 (1967), pp. 740–744; "Grahams Laws: Simple Demonstrations of Gases in Motion" *J. Chem. Educ.,* Vol. 46 (1969), pp. 358–364; Vol. 46 (1969), pp. 423–427.
68. Jackson, R., *Transport in Porous Catalysts,* Amsterdam: Elsevier, 1977.
69. Krishna, R., "Binary and Multicomponent Mass Transfer at High Transfer Rates" *Chem. Eng. J.,* Vol. 22 (1981), pp. 251–257.
70. Krishna, R., and Standart, G. L., "Addition of Resistances for Non-Isothermal Multicomponent Mass Transfer" *Letts. Heat Mass Transfer,* Vol. 3 (1976), pp. 41–48.
71. Krishna, R., "A Simplified Film Model Description of Multicomponent Interphase Mass Transfer," *Chem. Eng. Commun.,* Vol. 3 (1979), pp. 29–39.
72. Krishna, R., and Standart, G. L., "A Multicomponent Film Model Incorporating an Exact Matrix Method of Solution to the Maxwell–Stefan Equations," *AIChE J.,* Vol. 22 (1976), pp. 383–389.
73. Taylor, R., "On Exact Solutions of the Maxwell–Stefan Equations for the Multicomponent Film Model," *Chem. Eng. Commun.,* Vol. 10 (1981), pp. 61–76.
74. Taylor, R., "More on Exact Solutions of the Maxwell–Stefan Equations for the Multicomponent Film Model," *Chem. Eng. Commun.,* Vol. 14 (1982), pp. 361–362.
75. Taylor, R., and Webb, D. R., "Stability of the Film Model for Multicomponent Mass Transfer," *Chem. Eng. Commun.,* Vol. 6 (1980), pp. 175–189.
76. Taylor, R., and Webb, D. R., "Film Models for Multicomponent Mass Transfer: Computational Methods I—the Exact Solution of the Maxwell–Stefan Equations," *Comput. Chem. Eng.,* Vol. 5 (1981), pp. 61–73.
77. Buffham, B. A., and Kropholler, H. W., "The Evaluation of the Exponential Matrix," Proc. Conf. On Line Computer Methods Relevant to Chemical Engineering, University of Nottingham, 1971, pp. 64–70.
78. Wei, J., "Irreversible Thermodynamics in Engineering," *Ind. Eng. Chem.,* Vol. 58 (October 1966), pp. 55–60.

79. Duncan, J. B., and Toor, H. L., "An Experimental Study of Three Component Gas Diffusion," *AIChE J.*, Vol. 8 (1962), pp. 38–41.
80. Bres, M., and Hatzfield, C., "Three Gas Diffusion—Experimental and Theoretical Studies," *Pflugers Arch.*, Vol. 371 (1977), pp. 227–233.
81. Hesse, D., and Hugo, P., "Untersuchung der Mehrkomponenten—Diffusion in Gemisch Wasserstoff, Athylen und Athan mit einem Stationarem Messrerfahren." *Chemie-Ing.-Techn.*, Vol. 44 (1972), pp. 1312–1318.
82. Carty, R., and Schrodt, T., "Concentration Profiles in Ternary Gaseous Diffusion," *Ind. Eng. Chem. Fundam.*, Vol. 14 (1975), pp. 276–278.
83. Krishna, R., "Effect of High Transfer Rates on the Diffusion Behavior of Multicomponent Systems," *Letts. Heat Mass Transfer,* Vol. 3 (1976), pp. 393–402.
84. Arnold, K. R., and Toor, H. L., "Unsteady Diffusion in Ternary Gas Mixtures." *AIChE J.*, Vol. 13 (1967), pp. 909–914.
85. Smith, L. W., and Taylor, R., "Film Models for Multicomponent Mass Transfer—A Statistical Comparison," *Ind. Eng. Chem. Fundam.*, Vol. 22 (1983), pp. 97–104.
86. Taylor, R., and Webb, D. R., "On the Relationship Between the Exact and Linearised Solutions of the Maxwell–Stefan Equations for the Multicomponent Film Model," *Chem. Eng. Commun.*, Vol. 7 (1980), pp. 287–299.
87. Taylor, R., "Film Models for Multicomponent Mass Transfer: Computational Methods II—The Linerised Theory," *Comput. Chem. Eng.*, Vol. 6 (1982), pp. 69–75.
88. Stewart, W. E., "Multicomponent Mass Transfer in Turbulent Flow." *AIChE J.*, Vol. 19 (1973), pp. 398–400.
89. Taylor, R., "Solution of the Linearised Equations of Multicomponent Mass Transfer," *Ind. Eng. Chem. Fundam.*, Vol. 21 (1982), pp. 407–413.
90. Vickery, D. J., Taylor, R., and Gavalas, G. R., "A Novel Approach to the Computation of Multicomponent Mass Transfer Rates from the Linearized Equations," *Comput. Chem. Eng.*, (1984), pp. 179–184.
91. Krishna, R., "A Simplified Mass Transfer Analysis for Multicomponent Condensation," *Letts. Heat Mass Transfer,* Vol. 6 (1979), pp. 439–448.
92. Taylor, R., and Smith, L. W., "On Some Explicit Approximate Solutions of the Maxwell–Stefan Equations for the Multicomponent Film Model," *Chem. Eng. Commun.*, Vol. 14 (1982), pp. 361–370.
93. Burghardt, A., and Krupiczka, R., "Wnikanie Masy W Ukladach Wielaskladnikowych–Teortyczna Analiza zagadnienia i Okreslenie Wspolczynnikow Wnikania Masy," *Inz. Chem.*, Vol. 5 (1975), pp. 487–510.
94. Krishnamurthy, R., and Taylor, R., "Calculation of Multicomponent Mass Transfer at High Transfer Rates," *Chem. Eng. J.*, Vol. 25 (1982), pp. 47–54.
95. Hougen, O. A., and Watson, K. M., *Chemical Process Principles,* Vol. III, 1st ed., New York: Wiley, 1947.
96. Wilke, C. R., "Diffusional Properties of Multicomponent Gases," *Chem. Eng. Prog.*, Vol. 46 (1950), pp. 95–104.
97. Stewart, W. E., N. A. C. A. Tech. Note 3208, 1954.
98. Kato, S., Inazumi, H., and Suzuki, S., "Mass Transfer ia a Ternary Gaseous Phase," *Int. Chem. Eng.*, Vol. 21 (1981), pp. 443–452.
99. Kubota, H., Yamanaka, Y., and Dalla Lana, I. G., "Effective Diffusivity of Multicomponent Gaseous Reaction System," *J. Chem. Eng. Jap.*, Vol. 2 (1969), pp. 71–75.
100. Geonkoplis, G. J., *Mass Transport Phenomena,* New York: Holt, Rinehart and Winston Inc., 1972.
101. Toor, H. L., and Sebulsky, R. T., "Multicomponent Mass Transfer. I—Theory." *AIChE J.*, Vol. 7 (1961), pp. 558–565.
102. Shain, S. A., "A Note on Multicomponent Diffusion," *AIChE J.*, Vol. 7 (1961), pp. 17–19.
103. Hsu, H. W., and Bird, R. B., "Unsteady Multicomponent Diffusional Evaporation," *AIChE J.*, Vol. 6 (1960), pp. 551–553.

104. Taylor, R., and Krishnamurthy, R., "Film Models for Multicomponent Mass Transfer. —Diffusion in Physiological Gas Mixtures," *Bull. Math. Biol.*, Vol. 44 (1982), pp. 361–376.
105. Krishna, R., "A Generalized Film Model for Mass Transfer in Non-Ideal Fluid Mixtures," *Chem. Eng. Sci.*, Vol. 32 (1977), pp. 659–667.
106. Cussler, E. L., and Lightfoot, E.N., "Multicomponent Diffusion Involving High Polymers—I. Diffusion of Monodisperse Polystyrene in Mixed Solvents," *J. Phys. Chem.*, Vol. 69, pp. 1135–1144.
107. Lightfoot, E. N., and Scattergood, E. M., "Suitability of the Nernst-Planck Equations for Describing Electrokinetic Phenomena," *AIChE J.*, Vol. 11 (1965), pp. 175–192.
107a. Taylor, R., "Coupled Heat and Mass Transfer in Multicomponent Systems: Solution of the Maxwell–Stefan Equations," *Letts. Heat and Mass Transfer*, Vol. 8 (1981), pp. 405–416.
108. Scriven, L. E., "Flow and Transfer at Fluid Interfaces," *Chem. Eng. Educ.*, (Fall 1968), pp. 150–155; (winter 1969), pp. 26–29; (spring 1969), pp. 94–98.
109. Clift, R., Grace, J. R., and Weber, M. E., *Bubbles, Drops and Particles*, London: Academic Press, 1978.
110. Alimadadian, A., and Colver, C. P., "A New Technique for the Measurement of Ternary Diffusion Coefficients in Liquid Systems," *Can. J. Chem. Eng.*, Vol. 54 (1976), pp. 208–213.
111. Krishna, R., "A Note on the Film and Penetration Models for Multicomponent Mass Transfer," *Chem. Eng. Sci.*, Vol. 33 (1978), pp. 765–767.
112. Krishna, R., "Penetration Depths in Multicomponent Mass Transfer." *Chem. Eng. Sci.*, Vol. 33 (1978), pp. 1495–1497.
113. Kronig, R., and Brink, J. C., "The Theory of Extraction From Falling Droplets," *Appl. Sci. Research*, Vol. A2 (1950), pp. 142–154.
114. Sideman, S., and Shabtai, H., "Direct Contact Heat Transfer Between a Single Drop and an Immiscible Liquid Medium," *Can. J. Chem. Eng.*, Vol. 42 (1964), pp. 107–117.
115. Lockett, M. J., Kirkpatrick, R. D., and Uddin, M. S., "Froth Regime Point Efficiency for Gas-Film Controlled Mass Transfer on a Two-dimensional Sieve Tray," *Trans. I. Chem. E.*, Vol. 57 (1979), pp. 25–34.
116. Delancey, G. B., and Chiang, S. H., "Dufour Effect in Liquid Systems," *AIChE J.*, Vol. 14 (1968), pp. 664–665.
117. Delancey, G. B., and Chiang, S. H., "Role of Coupling in Nonisothermal Diffusion." *Ind. Eng. Chem. Fundam.*, Vol. 19, pp. 138–144.
118. Delancey, G. B., and Chiang, S. H., "Analysis of Nonisothermal Multicomponent Diffusion with Chemical Reaction," *Ind. Eng. Chem. Fundam.*, Vol. 19 (1970), pp. 344–349.
119. Delancey, G. B., "The Effect of Thermal Diffusion in Multicomponent Gas Absorption," *Chem. Eng. Sci.*, Vol. 27 (1972), pp. 555–566.
119a. Toor, H. L., "Turbulent Diffusion and the Multicomponent Reynolds Analogy." *AIChE J.*, Vol. 6 (1960), pp. 525–527.
120. Taylor, R., "On Multicomponent Mass Transfer in Turbulent Flow," *Letts. Heat and Mass Transfer*, Vol. 8 (1981), pp. 397–404.
121. Krishna, R., "A Turbulent Film Model for Multicomponent Mass Transfer," *Chem. Eng. J.*, Vol. 24 (1982), pp. 163–172.
122. Van Driest, E. R., "On Turbulent Flow Near a Wall," *J. Aeronaut. Sci.* Vol. 23 (1956), pp. 1007–1011.
123. Fletcher, D. F., Maskell, S. J., and Patrick, M. A., "Theoretical Investigation of the Chilton—Colburn Analogy," *Trans. I. Chem. E.*, Vol. 60 (1982), pp. 122–125.
124. Krishna, R., "A Thermodynamic Approach to the Choice of Alternatives to Distillation," *I. Chem. E. Symposium Series No.* 54 (1978), pp. 185–214.
125a. Toor, H. L., "Prediction of Efficiencies and Mass Transfer on a Stage with Multicomponent Systems," *AIChE J.*, Vol. 10 (1964), pp. 545–547.

125. Krishna, R., et al., "Murphree Point Efficiencies in Multicomponent Systems." *Trans. I. Chem. E.*, Vol. 55 (1977), pp. 178–183.
126. Abramowitz, M., and Stegun, I. A., *Handbook of Mathematical Functions*, Washington: National Bureau of Standards, 1964.
127. Krishna, R., Salomo, R. M., and Rahman, M. A., "Ternary Mass Transfer in A Wetted Wall Column. Significance of Diffusional Interactions Part II. Equimolar Diffusion." *Trans. I. Chem. E.*, Vol. 59 (1981), pp. 44–53.
128. King, C. J., *Separation Processes*, 2nd ed., New York: McGraw-Hill, 1980.
129. Holland, C. D., *Fundamentals and Modeling of Separation Processes*, Englewood Cliffs NJ: Prentice-Hall, 1975.
130. Holland, C. D., *Fundamentals of Multicomponent Distillation*, New York: McGraw-Hill, 1981.
131. Murphree, E. V., "Rectifying Column Calculations," *Ind. Eng. Chem.*, Vol. 17 (1925), pp. 747–750; pp. 960–964.
132. Hausen, H., "The Definition of the Degree of Exchange on Rectifying Plates for Binary and Ternary Mixtures," *Chemie Ingr. Tech.*, Vol. 25 (1953), pp. 595–597.
133. Standart, G. L., "Studies on Distillation—V. Generalized Definition of a Theoretical Plate or Stage of Contacting Equipment," *Chem. Eng. Sci.*, Vol. 20 (1965), pp. 611–622.
134. Standart, G. L., "Comparison of Murphree Efficiencies with Vaporization Efficiencies," *Chem. Eng. Sci.*, Vol. 26 (1971), pp. 985–988.
135. Holland, C. D., and McMahon, K. S., "Comparison of Vaporization Efficiencies with Murphree—Type Efficiencies in Distillation—I," *Chem. Eng. Sci.*, Vol. 25 (1970), pp. 431–436.
136. Medina, A. G., Ashton, N., and McDermott, C., "Murphree and Vaporization Efficiencies in Multicomponent Distillation," *Chem. Eng. Sci.*, Vol. 33 (1978), pp. 331–339.
137. Medina, A. G., Ashton, N., and McDermott, C., "Hausen and Murphree Efficiencies in Binary and Multicomponent Distillation," *Chem. Eng. Sci.*, Vol. 34 (1979), pp. 1105–1112.
138. Krishnamurthy, R., and Taylor, R., "A Nonequilibrium Stage Model of Multicomponent Separation Processes. I—Model Development and Method of Solution," *AIChE J.*, (1984), pp. 449–456.
139. Krishnamurthy, R., and Taylor, R., "A Nonequilibrium Stage Model of Multicomponent Separation Processes. II—Comparison with Experiment," *AIChE J.*, (1984), pp. 456–465.
140. Krishnamurthy, R., and Taylor, R., "A Nonequilibrium Stage Model of Multicomponent Separation Processes. III—Implications for Design." *AIChE J.*, (1984), under review.
141. Krishnamurthy, R., and Taylor, R., "Simulation of Existing Packed Distillation and Absorption Columns," *Ind. Eng. Chem. Process Des. Dev.* (1984), under review.
142. Zuiderweg, F. J., "Sleve Trays—A View of the State of the Art," *Chem. Eng. Sci.*, Vol. 37 (1982), pp. 1441–1464.
143. Naphtali, L. M., and Sandholm, D. P., "Multicomponent Separation Calculations by Linearization," *AIChE J.*, Vol. 17 (1971), pp. 148–153.
144. Vogelpohl, A. "Murphree Efficiencies in Multicomponent Systems," *I. Chem. E. Symposium Series*, No. 56 (1979), pp. 2.1/25–31.
145. Kayihan, F., Sandall, O. C., and Mellichamp, D. A., "Simultaneous Heat and Mass Transfer in Binary Distillation. I—Theory," *Chem. Eng. Sci.*, Vol. 30 (1975), pp. 1333–1339.
146. Kayihan, F., Sandall, O. C., and Mellichamp, D. A., "Simultaneous Heat and Mass Transfer in Binary Distillation. II—Experiment," *Chem. Eng. Sci.*, Vol. 32 (1977), pp. 747–754.
147. Dribicka, M. M., and Sandall, O. C., "Simultaneous Heat and Mass Transfer for Multicomponent Distillation in a Wetted Wall Column," *Chem. Eng. Sci.*, Vol. 34 (1979), pp. 733–739.

148. Free, K. W., and Hutchison, H. P., "Three Component Distillation at Total Reflux." International Symposium on Distillation, *I. Chem. E.* (London) (1960), pp. 231–237.
149. Nord, M., "Plate Efficiencies of Benzene–Toluene–Xylene Systems in Distillation," *Trans. Am. Inst. Chem. Engrs.*, Vol. 42 (1946), pp. 863–881.
150. Bravo, J. L., and Fair, J. R., "Generalized Correlation for Mass Transfer in Packed Distillation Columns," *Ind. Eng. Chem. Process Des. Dev.*, Vol. 21 (1982), pp. 162–170.
151. Johnstone, H. F., and Pigford, R. L., "Distillation in a Wetted Wall Column," *Trans. Am. Inst. Chem. Engrs.*, Vol. 38 (1942), pp. 25–51.
152. McDaniel, R., Bassyoni, A. A., and Holland, C. D., "Use of the Results of Field Tests in the Modeling of Packed Distillation Columns and Packed Absorbers—III," *Chem. Eng. Sci.*, Vol. 25 (1970), pp. 633–651.
153. Bassyoni, A. A., Ph. D. Dissertation in Chemical Engineering, Texas A. M. University, 1969.
154. McDaniel, R., Ph. D. Dissertation in Chemical Engineering, Texas A. M. University, 1969.
155. Kern, D. Q., *Process Heat Transfer*, New York: McGraw-Hill, 1950.
156. Silver, L., "Gas Cooling with Aqueous Condensation," *Trans. I. Chem. E.*, Vol. 25 (1947), pp. 30–42.
157. Bell, K. J., and Ghaly, M. A., "An Approximate Generalized Design Method for Multicomponent/Partial Condensers," *AIChE Symposium Series*, Vol. 69 (1972), pp. 72–79.
158. Colburn, A. P., and Drew, T. B., "The Condensation of Mixed Vapors." *Trans. Am. Inst. Chem. Engrs.*, Vol. 33 (1937), pp. 197–215.
159. Schrodt, J. T., "Simultaneous Heat and Mass Transfer from Multicomponent Condensing Vapor—Gas Systems," *AIChE J.*, Vol. 19 (1973), pp. 753–759.
160. Price, B. C., and Bell, K. J., "Design of Binary Vapor Condensers Using the Colburn Equations," *AIChE Symp. Series*, Vol 70 (1974), pp. 163–171.
161. Krishna, R., et al. "An Ackermann–Colburn–Drew Type Analysis for Condensation of Multicomponent Mixtures," *Letts. Heat and Mass Transfer*, Vol. 3 (1976), pp. 163–172.
162. Krishna, R., and Panchal, C. B., "Condensation of Binary Mixtures in the Presence of an Inert Gas," *Chem. Eng. Sci.*, Vol. 32 (1977), pp. 741–745.
163. Krishna, R., "Effect of Nature and composition of Inert Gas on Binary Vapor Condensation," *Letts. Heat and Mass Transfer*, Vol. 6 (1979), pp. 137–147.
164. Rohm, H. J., "The Simulation of Steady State Behavior of the Dephlegmation of Multicomponent Mixed Vapors," *Int. J. Heat Mass Transfer*, Vol. 23 (1980), pp. 141–146.
165. Bandrowsky, J., and Kubaczka, A., "On the Condensation of Multicomponent Vapors in the Presence of Inert Gases," *Int. J. Heat Mass Transfer*, Vol. 24 (1981), pp. 147–153.
166. Webb, D. R., and McNaught, J. M., "Condensers," Chapter in *Developments in Heat Exchanger Technology*, (D. Chisolm (Ed.), Barking, Essex, England: Applied Science Publishers, 1980.
167. Webb. D. R., and Sardesai. R. G., "Verification of Multicomponent Mass Transfer Models for Condensation Inside a Vertical Tube," *Int. J. Multiphase Flow*, Vol. 7 (1981), pp. 507–520.
168. Webb, D. R., "Heat and Mass Transfer in Condensation of Multicomponent Vapors," Proc. Seventh International Heat Transfer Conference, Munich, Vol. 5 (1982), pp. 167–174.
169. Taylor, R., and Noah, M. K., "Simulation of Binary Vapor Condensation in the Presence of an Inert Gas," *Letts. Heat and Mass Transfer*, Vol. 9 (1982), pp. 463–472.
170. Shah, A. K., and Webb, D. R., "Condensation of Single and Mixed Vapors from a Non-Condensing Gas in Flow Over a Horizontal Tube Bank," *I. Chem. E. Symp. Series No. 75* (1983), pp. 356–371.
171. Taylor, R., et al., "Condensation of Vapor Mixtures. I—A Mathematical Model and a Design Method," *Ind. Eng. Chem. Proc. Des. Dev.* (1984), Under review.

172. Furno, J. S., Taylor, R., and Krishna, R., "Condensation of Vapor Mixtures. II—Simulation of some Test Condensers," *Ind. Eng. Chem. Process Des. Dev.* (1984), under review.
173. Taitel, Y., and Tamir, A., "Film Condensation of Multicomponent Mixtures," *Int. J. Multiphase Flow,* Vol. 1 (1974), pp. 697–714.
174. Tamir, A., and Merchuk, J. C., "Verification of a Theoretical Model for Multicomponent Condensation," *Chem. Eng. J.,* Vol. 17 (1979), pp. 125–139.
175. Sage, F. E., and Estrin, J., "Film Condensation From a Ternary Mixture of Vapors upon a Vertical Surface" *Int. J. Heat Mass Transfer,* Vol. 19 (1976), pp. 323–333.
176. McNaught, J. M., "An Assessment of Design Methods for Condensation of Vapors from a Noncondensing Gas," in *Heat Exchangers—Theory and Practice,* Taborek, J., Hewitt, G. F., and Afgan, N. (Eds.), Washington: Hemisphere Publishing Corporation, 1983.
177. McNaught, J. M., "An Assessment of Design Methods for Multicomponent Condensation Against Data from Experiments on a Horizontal Tube Bundle," *I. Chem. E. Symp. Series No. 75* (1983), pp. 447–458.
178. Butterworth, D., "Condensation," chapter in *Heat Exchanger Design* Handbook (Reference 180), Vol. 2 (1983).
179. Owen, R. G., and Lee, W. C., "A Review of Some Recent Developments in Condensation Theory," *Chem. Eng. Research and Design, Trans. I. Chem. E.,* Vol. 61 (1983), pp. 335–361.
180. Schlunder, E. U., (editor in chief), *Heat Exchanger Design Handbook,* Washington: Hemisphere Publishing Corporation, 1983.
181. Webb, D. R., Panchal, C. B., and Coward, I. "The Significance of Multicomponent Diffusional Interactions in the Process of Condensation in the Presence of a Non Condensable Gas," *Chem. Eng. Sci.,* Vol. 36 (1981), pp. 87–95.
182. Sardesai, R. G., and Webb, D. R., "Condensation of Binary Vapors of Immiscible Liquids," *Chem. Eng. Sci.,* Vol. 37 (1982), pp. 529–537.
183. Webb, D. R., and Taylor, R., "The Estimation of Rates of Multicomponent Condensation by a Film Model," *Chem. Eng. Sci.,* Vol. 37 (1982), pp. 117–119.
184. Toor, H. L., and Sebulsky, R. T., "Multicomponent Mass Transfer. II—Experiment," *AIChE J.,* Vol. 7 (1961), pp. 565–573.
185. Modine, A. D., Ph. D. Dissertation in Chemical Engineering, Carnegie Institute of Technology, 1963.
186. Krishna, R., "Comparison of Models for Ternary Mass Transfer," *Letts. Heat Mass Transfer,* Vol. 6 (1979), pp. 73–76.
187. Krishna, R., "Ternary Mass Transfer in a Wetted–Wall Column—Significance of Diffusional Interactions. I—Stefan Diffusion," *Trans. I. Chem. E.,* Vol. 59 (1981), pp. 35–43.
188. Deo, P. V., Ph. D. Thesis in Chemical Engineering, University of Manchester Institute of Science and Technology, 1979.
189. Standart, G. L., et al., "Ternary Mass Transfer in Liquid-Liquid Extaction," *AIChE J.,* Vol. 21 (1975), pp. 554–559.
190. Sethy, A., and Cullinan, H. T., "Transport of Mass Ternary Liquid–Liquid System," *AIChE J.,* Vol. 21 (1975), pp. 571–582.
191. Cullinan, H. T., and Ram, S. K., "Mass Transfer in a Ternary Liquid-liquid System," *Can. J. Chem. Eng.,* Vol. 54 (1976), pp. 156–159.
192. Krishna, R., et al., "Ternary Mass Transfer in Liquid-Liquid Extraction," *Chem. Eng. Research and Development, Trans. I. Chem. E.,* (1984), under review.
193. Vitagliano, V., et al., "Diffusion in a Ternary System and the Critical Mixing Point," *J. Solution Chemistry,* Vol. 7 (1978), pp. 605–621.

CHAPTER 8

DIFFUSION LIMITATIONS FOR REACTIONS IN POROUS CATALYSTS

Gulsen Dogu

Department of Chemical Engineering
Middle East Technical University
Ankara, Turkey

CONTENTS

INTRODUCTION

The mechanism of mass transport within a porous solid is extremely complex and depends on the nature of the pore structure. Better understanding of mass transport

---

I am grateful to those who have given me the permission to use material from their earlier publications. My special thanks are due to Prof. J. M. Smith. I am most grateful to Prof. Timur Doğu, who was more than helpful in his criticisms and comments.

processes is necessary in the study of reactions catalyzed by porous solids. In such reactions, the reactant molecules must be transported through the catalyst pores to the active sites on the catalyst surface. In some cases, the mass transfer rather than the reaction rate is the controlling step. Complexities in heterogeneous catalysis arise in cases in which transport phenomena significantly influence the over-all surface rate process. Such complexities may also arise in noncatalytic gas-solid reaction systems in which either the solid is the reactant and product gas diffuses in the pores to the bulk or the gaseous reactant diffuses into the solid reactant.

In packed-bed reactors filled with catalyst pellets, the following processes take place:

1. Dispersion in the main fluid phase.
2. Mass transfer of the reactant from the main fluid phase to the external surface of the catalyst pellet.
3. Internal diffusion of the reactant in the pores of the catalyst.
4. Adsorption on active sites on the pore surface.
5. Surface reaction followed by the back diffusion of the reaction products.

For reactions catalyzed by porous solids, the slowest of the rate processes controls the overall reaction rate. If the rate of pore diffusion is smaller than the intrinsic reaction rate, concentration gradients develop within the porous pellet and consequently the observed rate becomes different as compared to the rate evaluated at the surface concentration. The ratio of the observed rate to the rate corresponding to the surface condition is called the effectiveness factor. Under certain conditions the external mass transfer resistance could also affect the observed rate of reaction.

For reacting systems having high heat of reaction, considerable interphase and intrapellet temperature gradients might develop. The overall effectiveness factor is expected to depend upon the thermal gradients as well as the concentration gradients.

Most catalyst pellets are formed by the agglomeration of porous particles. Such catalysts have bidisperse pore structures. Pores within the particles are usually called micropores and pores between the agglomerated particles are called the macropores. For such systems, the diffusion resistance in both macro and micropore regions might have a significant effect on the observed rate.

Prediction of the effects of transport processes on the observed reaction rates requires rigorous mathematical analysis. A comprehensive review of the mathematical theory of diffusion and reaction in porous catalysts is given by Aris [1]. The fine books of Smith [2], Satterfield [3], Petersen [4], Froment and Bischoff [5], and Carberry [6] are some of the important references in this area. In this chapter, a comprehensive and up-to-date coverage of diffusional limitations for reactions in porous catalysts is presented without going into the details of the extensive mathematical derivations. The objective has become to present the results in a clear and readily usable form. The mechanism of diffusion in porous media is reviewed and the methods of prediction and measurement techniques of effective diffusion coefficients are presented. Relative importance of concentration and thermal gradients on the observed rates are covered. Special emphasis is given to transport processes in bidisperse systems and to the criteria to test the importance of diffusional limitations and the relative importance of mass diffusion and thermal effects on the observed rate.

DIFFUSION IN POROUS MEDIA

The pore structure of catalyst pellets is very complex and is not well defined. The geometry of pores are not regular. The pores are interconnected to each other with variable sizes and some pores are dead-ended.

In order to understand the phenomena of transport of species in porous solids, first it is necessary to develop the theory of transport processes in the capillaries. When the system is maintained at a uniform pressure, the flux is diffusive in nature. Diffusion of a gas through a capillary may take place basically by two mechanisms.

1. Molecular diffusion.
2. Knudsen diffusion.

If the equivalent radius of the capillary is large relative to the mean free path of the molecules, then transport takes place by ordinary (molecular) diffusion. On the other hand, if the radius of the pore is small relative to the mean free path, Knudsen diffusion dominates. A region exists in which both of these diffusion processes become important, and it is called the transition region. It is costumary to consider that the transition region prevails for a pore-radius-to-mean-free-path ratio ranging from 0.1 to 10. In addition to the two mechanisms of diffusion just mentioned, the transport by movement of adsorbed molecules over the surface may be important in some cases. This phenomenon is called the surface diffusion.

If the diffusion is in the molecular region, the flux of component A in a binary mixture of A and B is expressed as

$$\bar{N}_A = y_A(\bar{N}_A + \bar{N}_B) - D_{AB}\nabla C_A \tag{1}$$

where D_{AB} is the molecular diffusion coefficient of A in B. When the diffusion is in the Knudsen region, the transport rate of the molecules of A is governed only by the collisions with the capillary wall. In this region, the flux of A is given by

$$\bar{N}_A = -D_{K_A}\nabla C_A \tag{2}$$

where D_{K_A} is the Knudsen diffusion coefficient of species A.

In the transition region, the transport of A is due both to collisions between the molecules and collisions with the capillary wall. By making a momentum balance in the capillary, it has been shown by Scott and Dullien [7] that the flux of A in a capillary in the transition region can be given as

$$\bar{N}_A = -\frac{1}{\dfrac{1-\alpha_A y_A}{D_{AB}} + \dfrac{1}{D_{K_A}}}\nabla C_A \tag{3}$$

The parameter α_A is,

$$\alpha_A = 1 - (M_A/M_B)^{1/2} \tag{4}$$

at constant pressure [8, 9]. Equation 4 is the result of Graham's law, which states that

$$\frac{N_A}{N_B} = -\left(\frac{M_B}{M_A}\right)^{1/2} \tag{5}$$

The composite diffusivity, D_T, in a capillary is a function of pore radius as well as molecular properties of the diffusing species and can be expressed as

$$D_T = \frac{1}{\dfrac{1-\alpha_A y_A}{D_{AB}} + \dfrac{1}{D_{K_A}}} \tag{6}$$

Equation 6 reduces to the well-known Bosanquet formula

$$\frac{1}{D_T} = \frac{1}{D_{AB}} + \frac{1}{D_{K_A}} \tag{7}$$

for equimolar counter-current diffusion and also for dilute systems.

Prediction of Molecular and Knudsen Diffusion Coefficients

Experimental values of binary molecular diffusion coefficients for some gas and liquid pairs are reported by Satterfield [3] and Bird et al. [10]. In the absence of the experimental data, molecular diffusivities of gases can be predicted from the Chapman and Enskog equation [10] at moderate temperatures and pressures.

$$D_{AB} = 0.0018583 \frac{\sqrt{T^3[(1/M_A)+(1/M_B)]}}{p\sigma_{AB}^2 \Omega_{D,AB}} \tag{8}$$

where D_{AB} = the molecular diffusivity (cm^2/s)
p = the pressure (atm)
σ_{AB} = the Lennard–Jones parameter (Å)
$\Omega_{D,AB}$ = the collision integral for the molecular pair AB.

Constants in the Lennard–Jones potential-energy function, σ_A and ε_A/K, for various species are given in Table 1. The values of these constants for an unlike molecular pair can be estimated from

$$\sigma_{AB} = \frac{1}{2}(\sigma_A + \sigma_B) \tag{9}$$

and

$$\varepsilon_{AB} = (\varepsilon_A \varepsilon_B)^{\frac{1}{2}} \tag{10}$$

Table 2 gives $\Omega_{D,AB}$ values as a function of KT/ε_{AB}. Equation 8 gives good estimates of molecular diffusivity, especially for nonpolar gases and for pressures below 0.5 critical pressure. The effect of composition on the molecular diffusion coefficient is small, and consequently these equations can also be used as a first approximation in multicomponent systems. Molecular diffusion in multicomponent systems is given by Stefan–Maxwell equations. For a detailed discussion on this topic the reader should refer to Hirschfelder et al. [11], Bird et al. [10], Chapman and Cowling [12] and Wilke [13]. Prediction methods for diffusion coefficients in liquids is also summarized by Bird et al. [10].

The following equation, which is derived from the gas kinetic theory, can be used for evaluating the Knudsen diffusivity of species A.

$$D_{K_A} = 9{,}700a\left(\frac{T}{M_A}\right)^{1/2} \tag{11}$$

where a = the pore radius (cm)
T = the temperature (°K)
M_A = the molecular weight of the diffusing species.

As it is evident from Equation 11 the Knudsen diffusivity is independent of the total pressure. Youngquist [38] and Henry et al. [39] reported further discussions on this topic.

Effective Diffusivities in Porous Solids

Equation 3 gives the flux of A in the pores of a porous solid based on pore cross-sectional area. In the process of diffusion of a gaseous species through a porous solid it will be advantageous to define an effective flux based on the open pore area plus a solid area perpendicular to the direction of diffusion.

A schematic representation of a porous catalyst is given in Figure 1. The ratio of the open pore area to the total area at any cross-section of the pellet may be taken equal to the void-volume-to-total volume ratio. Therefore, in evaluating effective flux, N_A should be multiplied by the porosity. The pores in the solid are of irregular size and shape, and they are interconnected. The actual diffusion path through the solid depends upon the pore structure. In order to account for these factors tortuosity factor, τ, is introduced. Thus, the effective flux can be written as

$$N_{A_e} = N_A \frac{\varepsilon}{\tau} \tag{12}$$

and the effective diffusivity is related to the composite diffusivity by

$$D_e = D_T \frac{\varepsilon}{\tau} \tag{13}$$

The composite diffusivity, D_T, is a function of pore radius if the Knudsen diffusion is important. For a porous catalyst, Knudsen diffusivity is commonly evaluated using the average pore radius in Equation 11. This procedure gives satisfactory results especially for catalysts with narrow pore size distributions. As it has been pointed out in the recent paper of Wang and Smith [14], this procedure may be improved by summing the Knudsen and bulk contributions over the complete range of pore sizes. Thus, the effective diffusivity based upon total pore plus solid area can be evaluated from

$$D_e = \frac{1}{\tau} \int_0^{\infty} D_T(a)\, f(a)\, da \tag{14}$$

where f(a) da is the void volume in pores of radius a and a + da per unit of total pellet volume. This approach is especially important for pellets with broad pore size distributions.

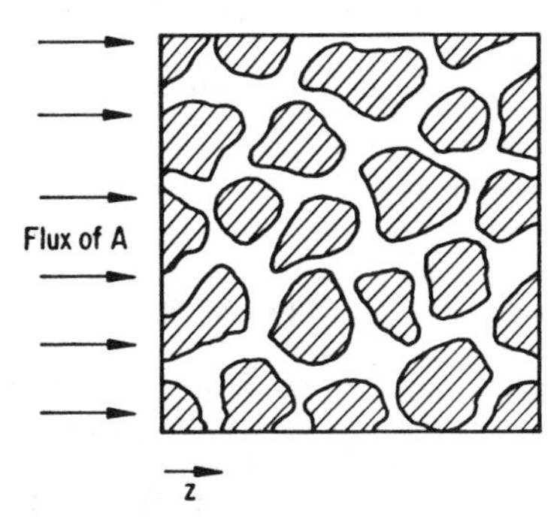

Figure 1. Schematic representation of a porous catalyst.

Table 1
Intermolecular Force Parameters and Critical Properties

Substance	Molecular Weight	Lennard-Jones* Parameters σ (Å)	ϵ_A/K (°K)	Critical Constants(†) T_c(°K)	P_c (atm)	V_c(cm³/g·mole)
Light elements						
H_2	2.016	2.915	38.0	33.3	12.8	65.0
He	4.003	2.576	10.2	5.26	2.26	57.8
Noble gases						
Ne	20.183	2.789	35.7	44.5	26.9	41.7
Ar	39.944	3.418	124.	151.	48.0	75.2
Kr	83.80	3.498	225.	209.4	54.3	92.2
Xe	131.3	4.055	229.	289.8	58.0	118.8
Simple polyatomic substances						
Air	28.97(††)	3.617	97.0	132.(††)	36.4(††)	86.6(††)
N_2	28.02	3.681	91.5	126.2	33.5	90.1
O_2	32.00	3.433	113.	154.4	49.7	74.4
O_3	48.00	-	-	268.	67.	89.4
CO	28.01	3.590	110.	133.	34.5	93.1
CO_2	44.01	3.996	190.	304.2	72.9	94.0
NO	30.01	3.470	119.	180.	64.	57.
N_2O	44.02	3.879	220.	309.7	71.7	96.3
SO_2	64.07	4.290	252.	430.7	77.8	122.
F_2	38.00	3.653	112.	-	-	-
Cl_2	70.91	4.115	357.	417.	76.1	124.
Br_2	159.83	4.268	520.	584.	102.	144.
I_2	253.82	4.982	550.	800.	-	-
Hydrocarbons						
CH_4	16.04	3.822	137.	190.7	45.8	99.3
C_2H_2	26.04	4.221	185.	309.5	61.6	113.
C_2H_4	28.05	4.232	205.	282.4	50.0	124.
C_2H_6	30.07	4.418	230.	305.4	48.2	148.
C_3H_6	42.08	-	-	365.0	45.5	181.
C_3H_8	44.09	5.061	254.	370.0	42.0	200.
n-C_4H_{10}	58.12	-	-	425.2	37.5	255.
i-C_4H_{10}	58.12	5.341	313.	408.1	36.0	263.
n-C_5H_{12}	72.15	5.769	345.	469.8	33.3	311.
n-C_6H_{14}	86.17	5.909	413.	507.9	29.9	368.
n-C_7H_{16}	100.20	-	-	540.2	27.0	426.
n-C_8H_{18}	114.22	7.451	320.	569.4	24.6	485.
n-C_9H_{20}	128.25	-	-	595.0	22.5	543.
Cyclohexane	84.16	6.093	324.	553.0	40.0	308.
C_6H_6	76.11	5.270	440.	562.6	48.6	260.
Other organic compounds						
C_4	16.04	3.822	137.	190.7	45.8	99.3
CH_3Cl	50.49	3.375	855.	416.3	65.9	143.
CH_2Cl_2	84.94	4.759	406.	510.	60.	-
$CHCl_3$	119.39	5.430	327.	536.6	54.	240.
CCl_4	153.84	5.881	327.	556.4	45.0	276.
C_2N_2	52.04	4.38	339.	400.	59.	-
COS	60.08	4.13	335.	378.	61.	-
CS_2	76.14	4.438	488.	552.	78.	170.

* *Values of σ and ϵ/K are from J.O. Hirschfelder, C.F. Curtiss, and R.B. Bird,* Molecular Theory of Gases and Liquids, *John Wiley and Sons, Inc., New York, 1954, pp. 1110–1112; also Addenda and Corrigenda, p. 11. The above values are computed from viscosity data and are applicable for temperatures above 100°K. Values for Kr are due to E.A. Masor,* J. Chem. Phys., *Vol. 32 (1960), pp. 1832–1836.*

† *Values of T_c, P_c, and V_c are from K.A. Kobe and R.E. Lynn, Jr.,* Chem. Rev., *Vol. 52 (1952), pp. 117–236 and* Amer. Pet. Ins. Res. Proj., *Vol. 44, F.D. Rossini (Ed.), Carnegie Inst. of Technology (1952).*

†† *For air, molecular weight and pseudocritical properties T_c, P_c, V_c have been calculated from average, composition of dry air, as given in International Critical Tables, Vol. I (1926), p. 393.*

Source: Permission granted by, R.B. Bird, W.E. Stewart, and E.N. Lightfoot, Transport Phenomena, *John Wiley and Sons., Inc., New York, 1960.*

For catalyst pellets with bidisperse pore structures the tortuosity factors may be evaluated either by using the total range of pore volumes or by considering only the macropores. Wang and Smith [14] reported that the use of a total range of pore volumes in Equation 14 gives reasonable tortuosity values. As discussed later the diffusion in bidisperse systems should better be described by two different diffusivities for the macro and micropore regions.

Some tortuosity factor values reported for different systems are tabulated in the next section (Table 3).

Table 2
Values of the Function $\Omega_{D,AB}$ for Prediction of Diffusivities of Gases at Low Densities

KT/ϵ_{AB}	$\Omega_{D,AB}$	KT/ϵ_{AB}	$\Omega_{D,AB}$
0.30	2.662	1.75	1.128
0.35	2.476	2.0	1.075
0.40	2.318	2.5	1.000
0.45	2.184	3.0	0.949
0.50	2.066	3.5	0.912
0.55	1.966	4.0	0.884
0.60	1.877	5.0	0.842
0.65	1.798	7.0	0.790
0.70	1.729	10.0	0.742
0.75	1.667	20.0	0.664
0.80	1.612	30.0	0.623
0.85	1.562	40.0	0.596
0.90	1.517	50.0	0.576
0.95	1.476	60.0	0.560
1.00	1.439	70.0	0.546
1.10	1.375	80.0	0.535
1.20	1.320	90.0	0.526
1.30	1.273	100.0	0.517
1.40	1.233	200.0	0.464
1.50	1.198	300.0	0.436

Source: Permission granted by, J.O. Hirshfelder, C.F. Curtiss, and R.B. Bird, Molecular Theory of Gases and Liquids, *John Wiley and Sons, Inc., New York, 1954.*

Table 3
Tortuosity Factors for Some Catalysts*

Porous Solid	Total Porosity (ϵ)	Tortuosity Factor (τ)	Reference
Hydrodesulfurization Catalyst, HDS 20-A	0.677	(6.1–9.6)†	[14]
"	0.677	(0.52–1.9)††	[14]
"	0.677	(12–19)§	[14]
Nickel oxide	0.66–0.03	(2.78–107.5)§	[129]
Alumina		∼ 3.0†	[47]
γ-Alumina	0.384	3.7†	[131,3]‖
"	0.384	0.45††	[131,3]‖
Alumina	0.812	0.85††	[132]
Silica-alumina cracking catalyst	0.464	2.1††	[133]
Pt-Al_2O_3	0.74	3.7§	[37]
6.8% WO_3 on SiO_2	0.59	9.9	[130]
14% $CoMoO_4$ on Al_2O_3	0.56	5	[130]

* *Also see Table 11.*

† *Based on total pore size distribution (Equation 14).*

†† *Based on average pore radius calculated from* $\bar{a} = 2V_g/s_g$.

§ *Based on average pore radius calcualted from* $\bar{a}. = \int_0^{V_g} a dV/Vg$.

‖ *Tortuosity values for selected commercial catalysts are reported by Satterfield [3].*

Surface Diffusion

Transport of the adsorbed species on the catalyst surface is called surface diffusion. Contribution of surface diffusion to effective flux in porous solids is small unless appreciable adsorption occurs and the adsorbed molecules are mobile. It can be significant for catalysts with very fine pores for which effective diffusivities are small and surface area is large. The theory and the experimental studies on surface diffusion are reviewed by Barrer [28], Dacey [29], Satterfield [3], and Luss [30]. Reed and Butt [31], and Schneider and Smith [32] have reported experimental surface diffusion values for different systems. The effective surface diffusivities reported by Schneider and Smith [32] for ethane, propane, n-butane, and isobutane on different catalysts such as Pt-alumina, silica-alumina, silica gel, and carbon black are in the range of 10^{-4} to 10^{-6} cm^2/s. Most of the values reported for other systems are also in this range.

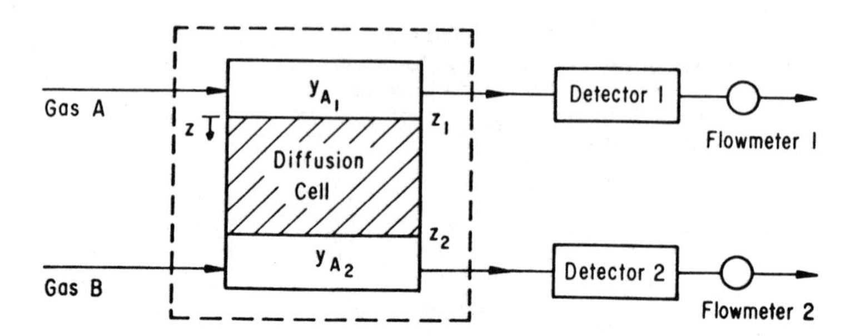

Figure 2. Flow diagram of a Wicke-Kallenbach diffusion apparatus.

Effective Diffusivity Measurement Techniques

The experimental methods employed for the study of diffusion in porous solids can be classified into two groups.

1. Steady-state methods.
2. Unsteady-state methods.

Steady-State Methods

In the steady-state methods, the diffusivities are either determined directly from diffusion data [15, 16] or indirectly from combined reaction and diffusion experiments with large pellets [17, 18].

In the well-known steady-state diffusion measurement method, a Wicke–Kallenbach-type of diffusion cell is used (Figure 2). While gas A flows through one face of the catalyst pellet, gas B flows through the other face. The pressures on both sides are kept the same. Therefore, counter-diffusion of species A and B takes place through the porous solid. Detectors connected to the outlet streams give the compositions of the streams, and the flow rates are determined separately. Knowing the compositions and the flow rates, the fluxes through the pellet are calculated.

For equimolar counter-diffusion, the effective diffusivity can be predicted from Equation 7 and 13. For this constant-pressure diffusion process, the effective flux through the pellet is expressed as

$$N_{A_e} = -D_e \frac{P}{RT} \frac{y_{A2} - y_{A1}}{z_2 - z_1} \tag{15}$$

The effective diffusivity is directly evaluated from Equation 15. Some effective diffusivity values determined by this technique are given in Table 4. For nonequimolar counter-current diffusion Equation 6 should be used instead of Bosanquet formula for composite diffusivity. For this case the effective diffusivity depends on the concentration of the diffusing species. A detailed discussion on this topic is given by Satterfield [3].

Unsteady-State Methods

An earlier attempt to measure diffusivities by an unsteady-state method is reported by Barrer [19]. In this time-lag method, the rate-of-pressure increase with time is measured in

Table 4
Effective Diffusivites in Some Porous Catalysts

Diffusing Species	Porous Solid	Total Porosity(ϵ)	Macro-Porosity (ϵa)	Temperature °C	Pellet Density, ρ_p (g/cm³)	Effective Diffusivity D_e (cm²/s)	Method	Reference
$n\text{-}C_4H_{10}\text{-}He$	Silica gel	0.486		50	1.13	0.00293	Chromatography	[20]
$D_2\text{-}H_2$	60% Co-kieselguhr	0.64	0.39	23.5-(-34)	1.74	0.03	Chromatography	[21]
He-Ar	58% Ni-kieselguhr (Harshaw Ni-0104P)	0.49	0.27		1.83	0.052	Steady-state	[94]
$CO_2\text{-}He$	Nickel oxide	0.438		25		0.016	Steady-state	[129]
$N_2\text{-}He$	Nickel oxide	0.66-0.03		25.5		0.063-0.000022	Steady-state	[129]
$H_2\text{-}N_2$	γ-alumina			25		0.0094-0.124	Steady-state	[47]
$He\text{-}N_2$	alumina	0.667-0.771	0.223-0.303	25.8 (pressure: 0.458-600mmHg)		0.447-0.0485	Steady-state	[39]
$H_2\text{-}N_2$	γ-alumina	0.384		25		0.0293	Steady-state	131,3*
Benzene-H_2	Ni-alumina	0.55-0.74	0.092-0.49		1.53-0.86	0.0043-0.0754	Temperature-difference	[44]
$He\text{-}N_2$	Alumina (Boehmite)	0.771	0.48	24	0.581	0.079	Single-pellet moment	[34]
$He\text{-}N_2$	Alumina (Boehmite)	0.748-0.58	0.427-0.11	45	0.612-1.03	0.050-0.012	Single-pellet moment	35,36
Cyclopropane-N_2	Alumina (Boehmite)	0.748-0.58	0.417-0.11	45	0.612-1.03	0.031-0.0056	Single-pellet moment	35,36
Benzene-H_2	Pt-alumina	0.74	0.31	202	0.65	0.056	Single-pellet moment	[37]
$H_2\text{-}N_2$	Hydrodesulfurization catalyst HDS-20A	0.677-0.569		229-25	0.956	0.090-0.0094	Single-pellet moment	[14]
He-Ar	14% $CoMoO_4$-on Al_2O_3	0.56		100		0.040	Single-pellet moment	[130]
He-Ar	6.8% WO_3 on SiO_2	0.59		100		0.028	Single-pellet moment	[130]

* *Diffusivity values for various selected commercial catalysts are reported by Satterfield and Cadle [131].*

the originally evacuated closed chamber on the outgoing side of the porous pellet. Since the pressures on both sides of the pellet are not the same, this method gives effective diffusivities for cases where Knudsen diffusion dominates in which the effective diffusion flux is independent of total pressure.

Gas chromatography provided another unsteady-state technique to determine the effective diffusion coefficients for gas-solid systems. Schneider and Smith [20] improved the gas chromatography theory, and they have shown that the effective diffusivities can be obtained from the chromatographic experiments with packed-bed systems. The method is based on the fact that the moments of the response peaks at the exit of the system are functions of transport and rate parameters [21]. The n-th moment of the concentration function C(t) is defined as

$$m_n = \int_0^\infty C t^n \, dt \tag{16}$$

The first absolute moment (μ_1) about the origin, which corresponds to the mean residence time, and the second central moment (μ_2), which is commonly called, the variance, are given by the following equations:

$$\mu_1 = \frac{\int_0^\infty C t \, dt}{\int_0^\infty C \, dt} = \frac{m_1}{m_0} \tag{17}$$

$$\mu_2' = \frac{\int_0^\infty C(t-\mu_1)^2 \, dt}{\int_0^\infty C \, dt} = \frac{m_2}{m_0} - \left(\frac{m_1}{m_0}\right)^2 \tag{18}$$

The experimental moment values for the residence-time distribution curves at the bed outlet for a given input pulse, are calculated by numerical integration from the observed chromatographic curves.

For a fixed-bed system, the mathematical expressions for the moments are obtained through the simultaneous solution of Equations 19 and 20. The species conservation equation for the adsorbing species for the bed:

$$\varepsilon_b \frac{\partial C_b}{\partial t} = E_a \frac{\partial^2 C_b}{\partial x^2} - U_0 \frac{\partial C_b}{\partial x} - \frac{3D_e}{R}(1-\varepsilon_b)\left.\frac{\partial C}{\partial r}\right|_{r=R} \tag{19}$$

The pseudo-homogeneous equation for the pellet:

$$\varepsilon \frac{\partial C}{\partial t} = D_e\left(\frac{\partial^2 C}{\partial r^2} + \frac{2}{r}\frac{\partial C}{\partial r}\right) - \varrho_p \frac{\partial C_{ads}}{\partial t} = 0 \tag{20}$$

where

$$\frac{\partial C_{ads}}{\partial t} = k_a\left(C - \frac{C_{ads}}{K_A}\right) \tag{21}$$

In writing Equations 20 and 21, spherical geometry is assumed and adsorption is taken to be linear. The theoretical moment expressions are derived as

$$\mu_1 = (X\varepsilon_b/U_0)\,[1+\delta_0] \tag{22}$$

and

$$\mu_2' = (2X\varepsilon_b/U_0)\,[\delta_1 + (E_a/\varepsilon_b)\,(1+\delta_0)^2\,(\varepsilon_b^2/U_0^2)] \tag{23}$$

where

$$\delta_0 = [(1-\varepsilon_b)\varepsilon/\varepsilon_b]\,[1+(\varrho_p/\varepsilon)K_A] \tag{24}$$

$$\delta_1 = [(1-\varepsilon_b)\varepsilon/\varepsilon_b]\left[\frac{\varrho_p}{\varepsilon}\frac{K_A^2}{k_a} + \frac{r_0^2\varepsilon}{15}\left(1+\frac{\varrho_p}{\varepsilon}K_A\right)^2\left(\frac{1}{D_e}+\frac{5}{k_f r_0}\right)\right] \tag{25}$$

The second moment data gives the effective diffusivity in the porous catalyst as well as the axial dispersion and external mass transfer coefficients. The adsorption equilibrium constant is evaluated from the first moment analysis.

Schneider and Smith [20] used this chromatographic moment technique for the evaluation of intrapellet and interpellet rate parameters for the adsorbing systems of ethane, propane, and n-butane on silica gel. Their diffusivity values are given in Table 4. Pulse-response methods are well developed for measuring diffusivities and adsorption rates together with axial dispersion coefficients in beds packed with catalyst pellets [22–25]. Dougharty [26] studied the effect of surface nonuniformity on the response of packed chromatographic beds. McCoy and co-workers [27] have applied the moment technique to gel permeation chromatography.

Doğu and Smith [34] have developed a single-pellet chromatographic technique for the study of intrapellet rate processes. In this dynamic method, the Wicke–Kallenbach-type of diffusion cell is used. The advantage of the single-pellet technique over bed chromatography is the elimination of interpellet effects and the related parameters [33, 35, 36]. In this way, in principle, intrapellet rate processes can be studied more effectively. The schematic diagram of diffusion cell used in this dynamic single-pellet method is given in Figure 3. Both end faces of the pellet are exposed to the flow of the same carrier gas B. A pulse of sample gas is introduced into the carrier gas at the top of the pellet and the response at the other end of the pellet is measured with a suitable detector. The pressures at both sides of the pellet have been kept equal and a dilute sample is injected to eliminate the concentration dependence of the effective diffusivity and to approximate linear adsorption for an adsorbing tracer. The method has been tested experimentally for nonadsorbing [34], adsorbing [36], and reacting [37] systems. For the diffusion of nonadsorbed gas A through the pellet, the species conservation equation is

$$\varepsilon \frac{\partial C_A}{\partial t} = D_e \frac{\partial^2 C_A}{\partial z^2}, \tag{26}$$

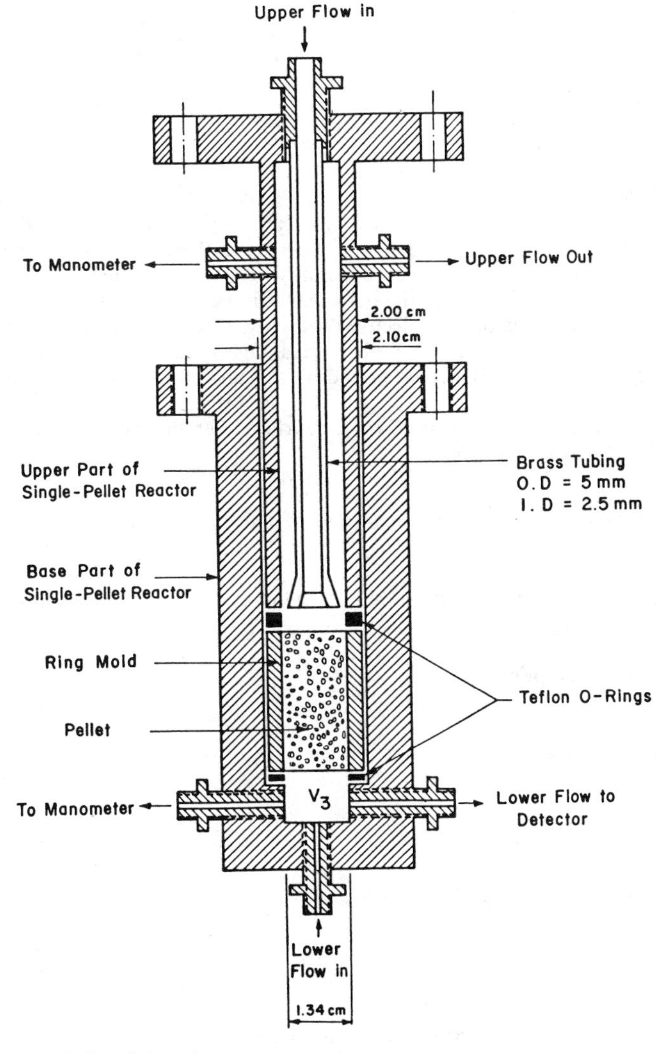

Figure 3. Details of single-pellet reactor.

where z is the direction of diffusion. The initial and boundary conditions for the system shown schematically in Figure 3 are:

$$C_A = 0 \quad \text{at} \quad t = 0 \quad \text{for} \quad 0 < z < L_p \tag{27}$$

$$C_A = M\delta(t) \quad \text{at} \quad z = 0, \tag{28}$$

$$-AD_e\left(\frac{\partial C_A}{\partial z}\right)_{z=L_p} = FC_{AL_p} \tag{29}$$

where δ(t) represents a Dirac delta pulse input and F is the flow rate of the lower stream. From the solution of Equation 26 with Equations 27, 28 and 29 in Laplace domain, the first moment expression is derived as,

$$\mu_1 = \left(\frac{L_p^2\varepsilon}{6D_e}\right)\frac{3\frac{A}{L_p}D_e + F}{\frac{A}{L_p}D_e + F} \tag{30}$$

Experimental values of the first moment are determined from the observed response peaks by numerical integration according to Equation 16. In this single-pellet dynamic technique, first moment is a function of the effective diffusivity. Therefore, first moment data give the effective diffusion coefficient. On the other hand, in bed chromatography second moments are needed for the evaluation of effective diffusivity. Due to the higher accuracy of the first moments, this is considered as another advantage of single-pellet technique over bed chromatography. Thus, this method provides a fast and accurate way of evaluating effective diffusivities.

If the diffusing species is adsorbed on the surface, then the species conservation equation of A within the pellet becomes

$$\varepsilon\frac{\partial C_A}{\partial t} = D_e\frac{\partial^2 C_A}{\partial z^2} - \varrho_p\frac{\partial n_A}{\partial t}, \tag{31}$$

where n_A is the adsorbed concentration of A (g moles/g catalyst). The expressions for $\partial n_A/\partial t$ for different adsorption mechanisms are,

Irreversible adsorption

$$(A + S \xrightarrow{k_a} A - S); \quad \partial n_A/\partial t = k_a C_A$$

Equilibrium adsorption

$$(A + S \xleftrightarrow{K_A} A - S); \quad \partial n_A/\partial t = K_A \partial C_A/\partial t$$

Reversible adsorption

$$(A + S \underset{k_d}{\overset{k_a}{\rightleftharpoons}} A - S); \quad \partial n_A/\partial t = k_a[C_A - (n_A/K_A)]$$

The first absolute and the second central moment expressions for the pellet, for the general case of reversible adsorption mechanism, are given by Doğu and Smith [36].

$$\mu_1 = \left(\frac{L_p^2\varepsilon + L_p^2\varrho_p K_A}{6D_e}\right)\frac{3\dfrac{A}{L_p}D_e + F}{\dfrac{A}{L_p}D_e + F} \tag{32}$$

$$\mu_2' = \left(\frac{L_p^2\varepsilon + L_p^2\varrho_p K_A}{6D_e}\right)\left[\frac{\left(\dfrac{A}{L_p}D_e\right)^2 + \dfrac{2}{5}\dfrac{A}{L_p}D_eF + \dfrac{1}{15}F^2}{6\left(\dfrac{A}{L_p}D_e + F\right)^2}\right] + \frac{L_p^2\varrho_p K_A}{3D_e k_a}\left[\frac{3\left(\dfrac{A}{L_p}D_e\right)^2 + 4\dfrac{A}{L_p}D_eF + F^2}{\left(\dfrac{A}{L_p}D_e + F\right)^2}\right] \tag{33}$$

The analysis of first moment data gives the effective diffusivities and adsorption equilibrium constant, while the second moment data give the adsorption rate constant. The method has been tested experimentally by Doğu [35] for the adsorption of cyclopropane on alumina and it has been shown that diffusivities can be obtained together with adsorption equilibrium and rate constants from the single-pellet dynamic experiments. Some of the effective diffusivity results are tabulated in Table 4.

It has also been shown that the single-pellet pulse response technique provides an effective way of evaluating effective diffusivities together with reaction rate parameters. Simultaneous evaluation of diffusion and rate parameters has been achieved for the reaction of benzene hydrogenation on Pt-Al_2O_3 catalyst [37]. It has also been shown that, by this technique the analysis of the mechanism of the reaction is also possible.

Asaeda et al. [41] and later McGreavy and Asaeda [42] have introduced a nonisobaric model to account for the nonequimolar fluxes to interpret the data obtained by a modified Wicke–Kallenbach-type of apparatus. The isobaric model is considered to be adequate for the evaluation of transport parameters for sufficiently small concentrations and short pulses. Transient diffusivity measurements have been carried out by Chou and Hegedus [43] in a pellet-string reactor. They have applied the pulse-diffusivity measuring technique to catalyst pellets with two zones of differing diffusivities. Jiratová and Horák [44] have introduced a method of effective diffusivity determinations based on the temperature differences within a catalyst pellet.

Some of the effective diffusivity values evaluated by different techniques are listed in Table 4.

Models for the Prediction of Effective Diffusivities

In chemical process industries dealing with gas-solid reaction systems, predictions of diffusive transport in porous media are often required. A completely predictive approach without reference to an experimental data is not possible. Several models are proposed in the literature for the description of diffusive fluxes in porous solids. Better models should need the least amount of experimental data. The random pore model of Wakao and Smith [45] requires only porosity and pore size distribution data for the estimation of effective diffusivity. The model proposed by Johnson and Stewart [47] utilizes the whole pore size distribution function. Another structural model is by Foster and Butt [48], which holds for binary systems at constant pressure. Wheeler [49] approximates the complex pore structure with a model that consists of parallel cylindrical pores of equal radius.

The random pore model of Wakao and Smith [45, 46] is proposed for predicting diffusion rates at constant pressure through a bidisperse porous media. The model predictions can also be reduced to systems with monodisperse pore structures. In the random pore model, the diffusion through a pellet is considered to take place by three parallel paths:

1. Through the macropores with an area of ε_a^2.
2. Through the micropores with an area of $(1-\varepsilon_a)^2$
3. Through macropores and micropores in series.

Based on the three mechanisms, the effective diffusion flux expression for the bidisperse pellet is obtained as follows:

$$N_{A_e} = \left[-\varepsilon_a^2 D_{T_a} - (1-\varepsilon_a)^2 (\varepsilon_i^2/(1-\varepsilon_a)^2 D_{T_i}) - 2\varepsilon_a(1-\varepsilon_a)\frac{2}{(1/D_{T_A}) + (1/D_{T_i}(\varepsilon_i^2/(1-\varepsilon_a)^2)}\right] \left[\left(\frac{P}{RT}\right)\frac{dy_A}{dz}\right] \quad (34)$$

where

$$D_{T_a} = \frac{1}{[(1-\alpha_A y_A)/D_{AB}] + (1/D_{K_a})} \quad (35)$$

$$D_{T_i} = \frac{1}{[(1-\alpha_A y_A)/D_{AB}] + (1/D_{K_i})} \quad (36)$$

Equation 34 is written for a binary gas system of A and B at uniform pressure. D_{K_a} and D_{K_i} are the Knudsen diffusivity of diffusing species A in macropores and micropores and they are evaluated at the average macro and micropore radii respectively. Wakao and Smith have also used the random pore model for the study of diffusion under reaction conditions [46].

Foster and Butt [48] propose a diffusion model which holds for isobaric binary mixtures. In this model, the void volume within the porous solid is considered to be consisting of one centrally converging and one centrally diverging pores. The exact shape of the arrays of these pores is determined from the pore size distribution of the porous solid.

In the method proposed by Johnson and Stewart [47] the diffusion rate through the porous solid is predicted by the integration of diffusion rate in the capillaries over the whole pore-size distribution. For this case the effective diffusivity can be predicted from Equation 14. In their original derivation Johnson and Stewart introduced a parameter K which corresponds to the reciprocal of the tortuosity of the pellet.

Evans et al. [40] propose the dusty gas model by considering the particles making up the porous pellet as dust particles. Some of the other references in this area are Feng and Stewart [50], Pakula and Greenkorn [51], Patel and Butt [52].

EFFECT OF DIFFUSION RESISTANCE ON THE OBSERVED RATE

Effectiveness Factor

Diffusion resistance in porous catalysts causes concentration profiles of reactants and products. Depending upon the relative magnitudes of diffusion and reaction times, concentration profiles are established within the pellet. As a result, the observed rate, as compared to the reaction rate evaluated at the external surface conditions of the pellet, decreases for positive-order reactions. This fact is originally considered by Thiele [55] and an effectiveness factor is defined as,

$$\eta = \frac{\text{observed rate of the reaction}}{\text{rate of reaction at the surface conditions}}$$

$$= \frac{\int_{V_p} R(C, T)\, dV}{V_p R(C_0, T_0)}$$

$$= \frac{\int_{A_e} (D_e \nabla C) \cdot \vec{n}\, dA_e}{V_p R(C_0, T_0)} \tag{37}$$

where $\vec{n}$ is a unit normal vector from the surface of the pellet. Considering a symmetrical catalyst pellet (spherical, cylindrical, or slab) Equation 37 can be simplified as

$$\eta = \frac{1}{\varphi^2}\left(\frac{d\psi}{d\xi}\right)_{\xi_0} \tag{38}$$

where ψ and ξ are the dimensionless concentration (C/C_0) and the dimensionless coordinate in the pellet $(w/(V_p/A_e))$ respectively; and φ is a dimensionless group which is called the Thiele modulus.

$$\varphi = \frac{V_p}{A_e}\left(\frac{R(C_0, T_0)}{D_e C_0}\right)^{1/2} \tag{39}$$

Here V_p and A_e are the volume and external area of the pellet respectively. The value of the Thiele modulus is determined by the ratio of the relative magnitudes of reaction rate to diffusion rate.

The observed reaction rate can then be written as

$$R_{obs} = \eta R(C_0, T_0) \tag{40}$$

The prediction of the effectiveness factor from Equation 38 requires the evaluation of the concentration profiles in the pellet. Assuming that the effective diffusivity is independent of concentration, the one-dimensional species conservation equation for an isotropic catalyst pellet can be written as,

$$\varepsilon \frac{\partial C}{\partial t} = D_e \nabla^2 C - R(C, T) \tag{41}$$

As it has been shown by Whitaker and co-workers [53, 54] this pseudohomogeneous transport equation for the pellet can be obtained by volume averaging of the homogeneous species conservation equation. They have also indicated that for a nonisotropic porous catalyst, an additional term should appear in this equation. Considering an isotropic pellet and an n-th-order reaction, Equation 41 reduces to Equation 42 for the simple slab geometry under steady-state conditions

$$\frac{d^2\psi}{d\xi^2} - \varphi^2\psi^2 = 0 \tag{42}$$

The solution of Equation 42 with the boundary conditions of,

$$\frac{d\psi}{d\xi} = 0 \quad \text{at} \quad \xi = 0 \tag{43}$$

and

$$\psi = 1 \quad \text{at} \quad \xi = 1.0 \tag{44}$$

gives the concentration profiles in the catalyst pellet. For a first-order reaction, the effectiveness factor expression derived from the solution of the pseudohomogeneous species conservation equation for spherical, cylindrical, and slab geometries are given in Table 5. As predicted from these effectiveness-factor expressions, for small values of φ, η approaches unity. On the other hand for large values of the Thiele modulus, diffusion is the controlling mechanism and the limiting form of the effectiveness factor expression becomes

$$\eta = \frac{1}{\varphi} \quad \text{for} \quad \varphi > 3.0 \quad \text{(first-order reaction)} \tag{45}$$

for all three geometries.

The derivation of the effectiveness-factor expressions for different orders [4, 5] gives that the limit of the effectiveness factor expression in the diffusion controlling region becomes

$$\eta = \frac{1}{\hat{\varphi}} \quad \text{for} \quad \hat{\varphi} > 3.0 \tag{46}$$

where $\hat{\varphi}$ is the order-generalized Thiele modulus and defined as,

$$\hat{\varphi} = \left(\frac{n+1}{2}\right)^{1/2} \frac{V_p}{A_e} \left(\frac{kC_0^{n-1}}{D_e}\right)^{1/2} \tag{47}$$

Table 5
Effective Factor Expressions for Different Geometries For a First-Order Reaction

Geometry	ξ	Thiele Modulus φ	ψ	Effectiveness Factor η
Slab	z/L	$L\left(\frac{k}{D_e}\right)^{1/2}$	$\frac{\cosh \varphi\xi}{\cosh \varphi}$	$\eta_f = \frac{\tanh \varphi}{\varphi}$
Infinite cylinder	$r/(r_o/2)$	$\frac{r_o}{2}\left(\frac{k}{D_e}\right)^{1/2}$	$\frac{I_o(\varphi\xi)}{I_o(2\varphi)}$	$\eta_c = \frac{I_1(2\varphi)}{\varphi I_0(2\varphi)}$
Sphere	$r/(r_o/3)$	$\frac{r_o}{3}\left(\frac{k}{D_e}\right)^{1/2}$	$\frac{3 \sinh (\varphi\xi)}{\xi \sinh (3\varphi)}$	$\eta_s = \frac{1}{\varphi}\left(\coth 3\varphi - \frac{1}{3\varphi}\right)$

The variation of the effectiveness factor with the order-generalized Thiele modulus for various reaction orders is given in Figure 4.

The evaluation of the Thiele modulus requires the knowledge of the actual order and the actual rate constant of the reaction. Such data might not be available for many reaction systems. Considering this, the effectiveness factor values are also reported as a function of an observable modulus Φ defined as

$$\Phi = \eta\varphi^2 = \left(\frac{V_p}{A_e}\right)^2 \frac{R_{obs}}{D_e C_0} \tag{48}$$

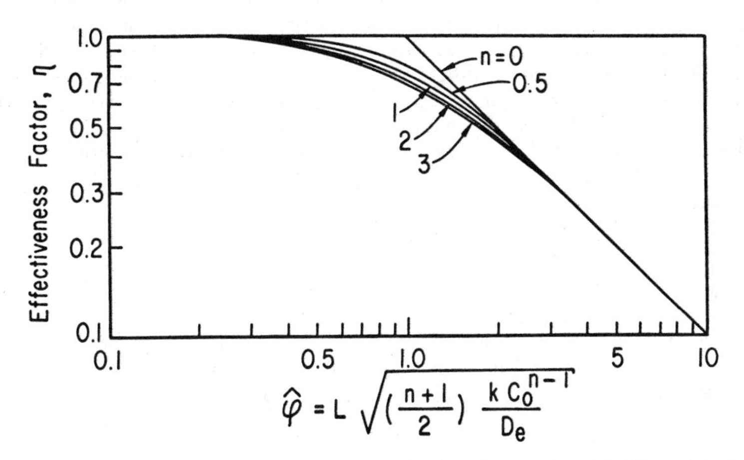

Figure 4. Generalized plot of effectiveness factor for simple-order reactions (slab-like catalyst). Source: Permission granted by Froment, G.F., and Bischoff, K.B., *Chemical Reactor Analysis and Design*, John Wiley and Sons, Inc., New York, 1979.

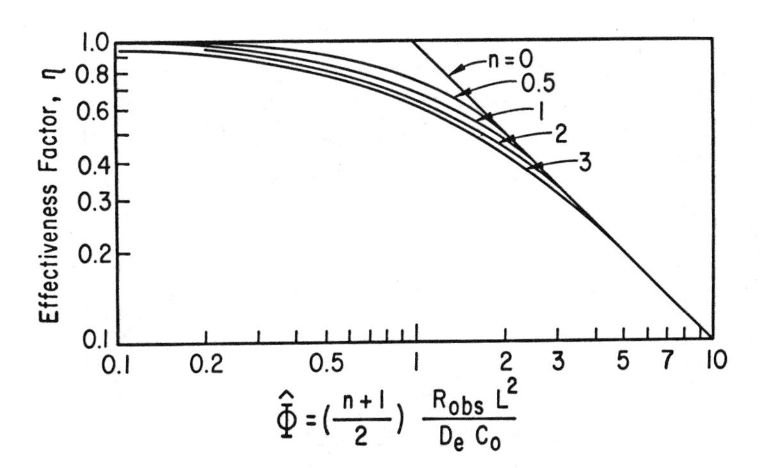

Figure 5. Effectiveness-factor plot in terms of observable modulus (slab-like pellet). Source: Permission granted by Froment, G.F., and Bischoff, K.B., *Chemical Reactor Analysis and Design*, John Wiley and Sons, Inc., New York, 1979.

Variation of the effectiveness factor with respect to the order-generalized observable modulus

$$\hat{\Phi} = \frac{n+1}{2}\Phi \tag{49}$$

for different reaction orders is given in Figure 5. For a first-order reaction, effectiveness-factor values are plotted as a function of the observable modulus for different geometries in Figure 6.

As predicted from Equations 45 and 46 and Figure 6, the product of the effectiveness factor and the Thiele modulus approaches unity in the diffusion controlling region ($\varphi > 3$) for all shapes of catalysts. This is the result of considering V_p/A_e as the characteristic length in the derivation of dimensionless conservation equations and, consequently, in the definition of the Thiele modulus. On the other hand, for intermediate values of φ ($0.3 < \varphi < 3$) the deviation of effectiveness-factor values for different geometries is not negligible. In the recent paper of Miller and Lee [56] a shape normalization for a catalyst pellet is introduced which brings the η-φ curves for all pellet shapes to a single curve corresponding to slab geometry for all φ. As it has been shown by Miller and Lee, the definition of the characteristic dimension is

$$L_G = \frac{V_p}{A_e}\left[1 + \eta_f\left(\frac{L_a}{L}\right)\right] \tag{50}$$

which brings the effectiveness-factor curves for all shapes into a single curve with a maximum deviation of 3%. In Equation 50, η_f is the effectiveness factor for the slab-like catalyst pellet and L_a/L values are given for different geometries as,

$L_a/L = 0$ (slab)
$L_a/L = 1/4$ (infinite cylinder)
$L_a/L = 8/27$ (sphere)

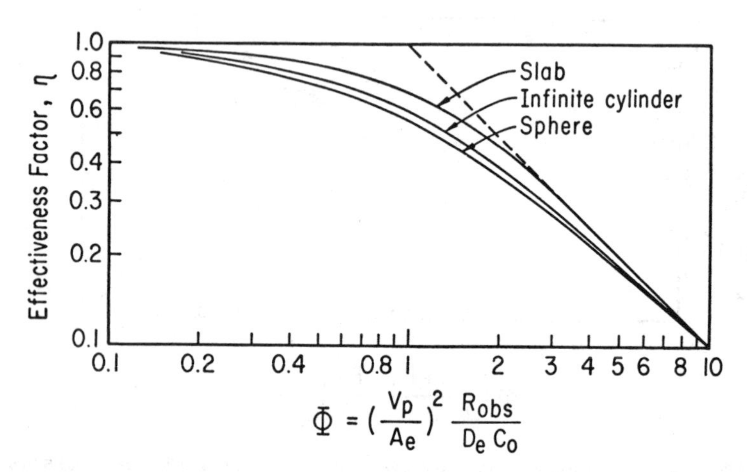

Figure 6. Effectiveness-factor plot in terms of observable modulus for a first-order reaction for different pellet geometries.

They have also shown that a finite cylindrical pellet for which the length is equal to the diameter can be treated as a sphere and L_a/L can be taken as 8/27. Miller and Lee have shown that for $\varphi < 1.5$ the equivalent length can be approximated as

$$L_G = \frac{V_p}{A_e}\left(1 + \frac{L_a}{L}\right) \tag{51}$$

With this method of shape normalization it is possible to predict the effectiveness-factor values for any geometry from the effectiveness factor of slab-like geometry by replacing L by L_G in the Thiele modulus.

As it has been described by Aris [57, 1], Petersen [58], Bischoff [59], and Froment and Bischoff [5], the effectiveness factor for a general rate expression can be obtained from

$$\eta = \frac{\sqrt{2}}{LR(C_0)}\left[\int_{C_c}^{C_0} D_e R(C)\, dC\right]^{1/2} \tag{52}$$

for the slab geometry. Here, C_c is the concentration of the reactant at the centerline of the pellet and can be determined implicitly from

$$L = \int_{C_c}^{C_0} \frac{D_e\, dC}{[2\int_{C_c}^{C} D_e R(C^*)\, dC^*]^{1/2}} \tag{53}$$

Equations 52 and 53 are derived by the integration of the species conservation equation.

Effectiveness factors for the general n-th-order case are treated by Mehta and Aris [60] and Kulkarni and Doraiswamy [61]. For a variety of catalytic reactions, the analysis of the reaction mechanisms lead to the Langmuir–Hinshelwood-type rate laws, which in general can be described as

$$R = \frac{kC^n}{|1 + KC|^m} \tag{54}$$

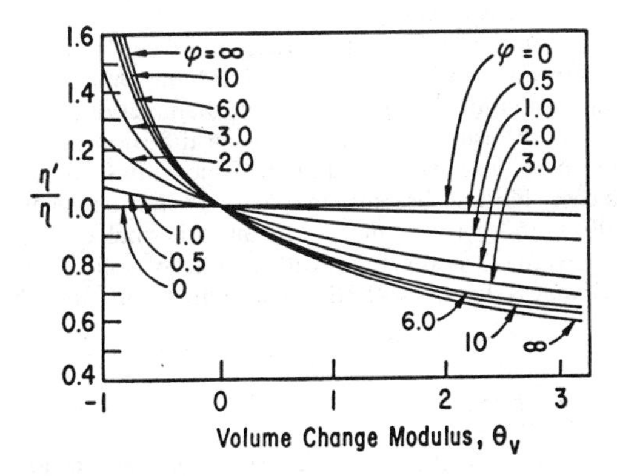

Figure 7. Influence of volume change on gas-phase reactions (first-order reactions—spherical particles). Source: Permission granted by Weekman, V.W., Jr., and Gorring, R.L., "Influence of Volume Change on Gas Phase Reactions in Porous Catalysts," *J. Cat.*, Vol. 4, No. 2, (April, 1965), pp. 260–270.

A comprehensive discussion on the prediction of effectiveness factors for these types of rate laws is given by Aris [1]. For monomolecular reaction mechanisms ($n = m = 1.0$) the effectiveness-factor curves lie between the first-order and zero-order effectiveness-factor curves, depending upon the relative magnitude of KC_0 with respect to unity. On the other hand, for a bimolecular Langmuir–Hinshelwood kinetics with $n = 1$ and $m = 2$, the rate of reaction can be approximated by a negative first-order reaction for large values of KC_0 and by a first-order reaction for small values of KC_0. An interesting behavior of the effectiveness factor has been observed for large values of KC_0. For such a case, effectiveness-factor values of greater than unity are possible. Even, multiple solutions of the effectiveness factor for a given Thiele modulus are determined if the value of KC_0 is very large. Some of the important references related to the Langmuir–Hinshelwood kinetics are References 62–69. Diffusional effects for the complex reaction systems and selectivity is an important area of research and comprehensive discussions on these topics are given by Aris [1], Carberry [6], Smith [2], Petersen [4], Froment and Bischoff [5], and Satterfield [3].

Effect of Volume Change

For catalytic reactions with significant diffusion resistance, effectiveness factors may be significantly effected if the volume of reactants and products differ. For a gas-phase reaction of

$$A \rightarrow nB$$

Weekman and Gorring [70] investigated the effect of volume change for an isothermal pellet considering molecular diffusion in the pores. Their results show that under constant pressure, the effectiveness factor is a function of the Thiele modulus and a volume change modulus defined as

$$\theta_v = (n - 1)y_{A,0} \tag{55}$$

where $y_{A,0}$ is the mole fraction of reactant at the external pellet surface. A typical set of curves that they obtained for a first-order reaction for spherical catalyst pellet is given in Figure 7. As expected, an increase in the number of moles ($\theta_v > 0$) causes a net molar flux out of the catalyst pellet and consequently increases the diffusion resistance. As a result, the effectiveness factor decreases as compared to its value, with no volume change. On the other hand for negative values of θ_v just the opposite is true. If one reactant or an inert is present in the system in great excess, the effect of volume change can be neglected and equimolar diffusion approximation can be made. If the diameter of the pores is smaller than the mean free path of the diffusing gas, then Knudsen diffusion becomes the controlling mechanism. For such a case, a change in number of moles has no effect on the effective diffusivity, but in this case the total pressure may vary within the pellet. Aris [1] discussed the volume change effect under isothermal and nonisothermal conditions. Lin and Lih [71, 72] presented solutions for several systems with significant volume change and discussed the variation of effectiveness factor with the Thiele modulus and the volume change modulus.

External Mass Transfer Resistance

In the previous sections, only the interpellet diffusional effects are considered. In a catalytic reactor, the resistance to mass transfer between the fluid stream and the external pellet surface could also be quite significant. The transport resistances through the boundary layer which has developed over the surface of the catalyst pellet may give rise to appreciable differences between the bulk and surface values of concentrations as well as temperatures. These effects would cause differences in the observed rate of reactions as compared to the rate evaluated at the bulk fluid concentration and temperature. In this section, only the mass transfer limitations are reviewed. Effects of external and internal temperature gradients will be discussed in the next section.

If the external mass transfer resistance is significant, the boundary condition stated in Equation 44 has to be replaced by

$$Bi_m (1-\psi_b)\Big|_{\xi=\xi_0} = \frac{d\psi_b}{d\xi}\Big|_{\xi=\xi_0} \tag{56}$$

where $\psi_b = C/C_b$

$$Bi_m = \frac{k_f(V_p/A_e)}{D_e} \tag{57}$$

The solution of the species conservation equation includes the Biot number (defined by Equation 57) in addition to the Thiele modulus. The magnitude of Biot number gives us the relative importance of the intraparticle diffusion resistance with respect to the external mass transfer resistance. Another dimensionless group which is frequently used to describe the relative importance of the reaction rate with respect to external mass transfer rate is the Damköhler number defined as

$$Da = \frac{\varphi_b^2}{Bi_m} = \frac{(V_p/A_e)R(C_b, T_b)}{k_f C_b}. \tag{58}$$

The solution of the species conservation equation with the boundary condition stated in Equation 56 for slab-like, infinite cylinder and spherical geometries and the corresponding effectiveness-factor expressions are summarized in Table 6. Variation of the effectiveness factor with the Thiele modulus for different values of Damköhler number are illustrated for spherical geometry in Figure 8.

Table 6
External Mass Transfer and Intrapellet Diffusion Effects on the Effectiveness Factor for a First-Order Reaction

Catalyst Geometry	Da	Bi_m	η
Slab-like	$L\left(\frac{k}{k_f}\right)$	$L\left(\frac{k_f}{D_e}\right)$	$\sinh\varphi/[\varphi\cosh\varphi + Da\sinh\varphi]$
Infinite-cylinder	$\frac{R}{2}\left(\frac{k}{k_f}\right)$	$\frac{R}{2}\left(\frac{k_f}{D_e}\right)$	$I_1(2\varphi)/[\varphi I_o(2\varphi + Da I_1(2\varphi)]$
Sphere	$\frac{R}{3}\left(\frac{k}{k_f}\right)$	$\frac{R}{3}\left(\frac{k_f}{D_e}\right)$	$\frac{3\varphi - \tanh 3\varphi}{3\varphi^2\tanh 3\varphi + Da(3\varphi - \tanh 3\varphi)}$

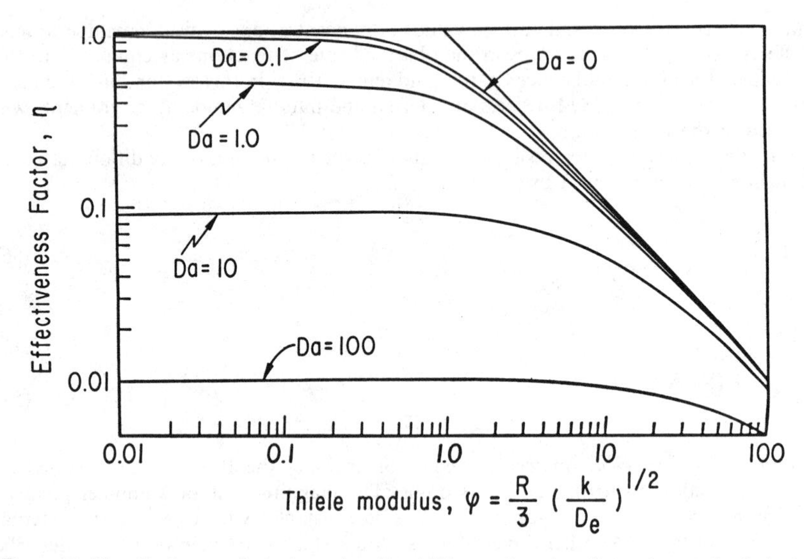

Figure 8. The effectiveness factor for various Damköhler numbers for a first-order reaction (spherical catalyst pellet).

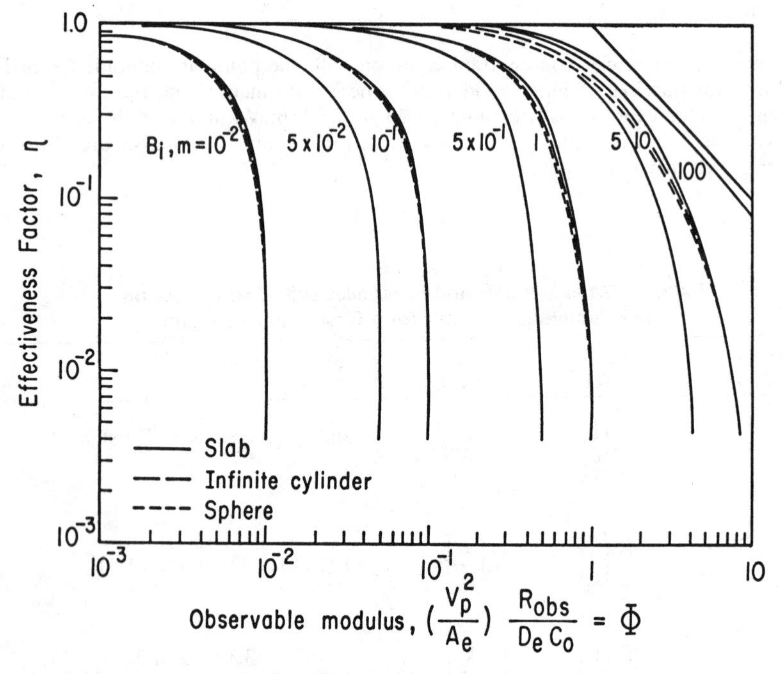

Figure 9. Effectiveness-factor plot in terms of observable modulus for various Biot number for different geometries (first-order reaction).

For large values of the Damköhler number, the limit of the effectiveness factor approaches to

$$\lim_{Da \to \infty} \eta = \frac{1}{Da} \tag{59}$$

instead of unity due to the significant external mass transfer resistance.

Effectiveness-factor curves can also be illustrated by taking the Biot number as the parameter. Figure 9 shows the variation of the effectiveness factor with respect to the observable modulus for different Biot numbers, for slab-like, cylindrical and spherical pellet geometries.

Observed Activation Energy and Observed Order of the Reaction

Considering an Arrhenius type of temperature dependence for the rate constant, the observed activation energy is obtained from

$$E_{obs} = -R_g \frac{d \ln R_{obs}}{d\left(\frac{1}{T}\right)}. \tag{60}$$

As shown by Weisz and Prater [73], if the observed rate is expressed in terms of the effectiveness factor and the rate evaluated at the surface conditions, the ratio of the observed activation energy to the actual activation energy can be obtained from

$$\frac{E_{obs}}{E_a} = 1 + \frac{1}{2} \frac{d \ln \eta}{d \ln \varphi} \tag{61}$$

As illustrated in Figure 10, E_{obs} approaches to one half of the actual activation energy if diffusion is the controlling resistance and external mass transfer limitations are negligible. If the external mass transfer resistance is large, then the observed activation energy approaches to zero.

Considering an n-th-order reaction, the observed rate can be written in the diffusion controlling region as

$$R_{obs} = \eta R(C_0) = \frac{1}{\hat{\varphi}} R(C_0) = \frac{A_e}{V_p}\left(\frac{2kD_e}{n+1}\right)^{1/2} C_0^{(n+1)/2} \tag{62}$$

Thus, the observed order of the reaction becomes $(n+1)/2$ if diffusion is the controlling mechanism.

THERMAL EFFECTS IN POROUS CATALYSTS

Effects of Intrapellet Temperature Gradients

For gas-solid catalytic reactions, in addition to the concentration gradients, significant temperature gradients may develop within the porous catalyst. Especially for systems with high heat of reaction and low thermal conductivity of the pellets, considerable variation of the observed rate from the rate evaluated at the surface temperature is expected. The analysis of the behavior of the nonisothermal reaction requires the simultaneous solution

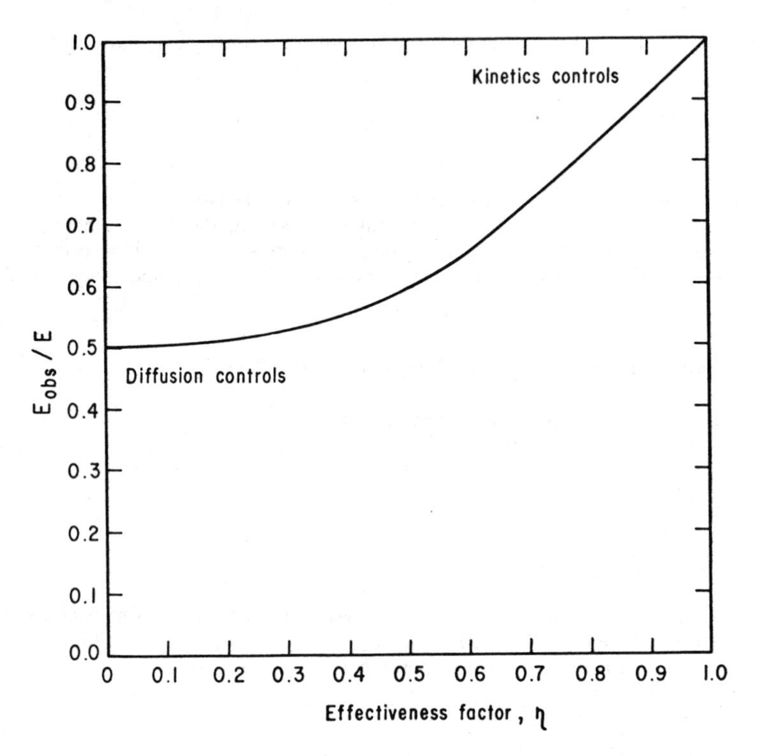

Figure 10. Effect of intraparticle diffusion resistance on the observed activation energy.

of the conservation of mass and energy equations. For a spherical catalyst pellet, steady-state conservation equations are

$$\frac{1}{r^2}\frac{d}{dr}\left(r^2 D_e \frac{dC}{dr}\right) - R(C, T) = 0 \tag{63}$$

$$\frac{1}{r^2}\frac{d}{dr}\left(r^2 \lambda_e \frac{dT}{dr}\right) - R(C, T)\Delta H = 0 \tag{64}$$

These equations have to be solved with the boundary conditions of

$$\text{at } r = 0; \quad \frac{dC}{dr} = \frac{dT}{dr} = 0 \tag{65}$$

$$\text{at } r = r_0; \quad k_f(C_b - C_0) = D_e \left.\frac{dC}{dr}\right|_{r = r_0} \tag{66}$$

$$\text{at } r = r_0; \quad h_f(T_b - T_0) = \lambda_e \left.\frac{dT}{dr}\right|_{r = r_0} \tag{67}$$

Combining Equations 63 and 64, it can be shown that the temperature at any point in the pellet can be related to the concentration [74] as

$$T - T_0 = \frac{D_e(-\Delta H)}{\lambda_e}(C_0 - C) \tag{68}$$

It had originally been shown by Prater [75] that the maximum temperature rise in a pellet can be predicted by taking C = 0 in Equation 68.

$$\frac{\Delta T_{max}}{T_0} = \frac{(-\Delta H)D_e C_0}{\lambda_e T_0} \tag{69}$$

The order of magnitude of the dimensionless group which appear on the righthand side of Equation 69

$$\beta = \frac{(-\Delta H)D_e C_0}{\lambda_e T_0} \tag{70}$$

gives an idea about the relative importance of heat generation within the pellet with respect to the heat removal by conduction through the pellet.

One of the earlier papers on the prediction of effectiveness factors in nonisothermal pellets is of Tinkler and Metzner [74]. Neglecting the external heat and mass transfer resistances and considering small values of $(T - T_0)/T_0$ they have developed effectiveness-factor-versus-

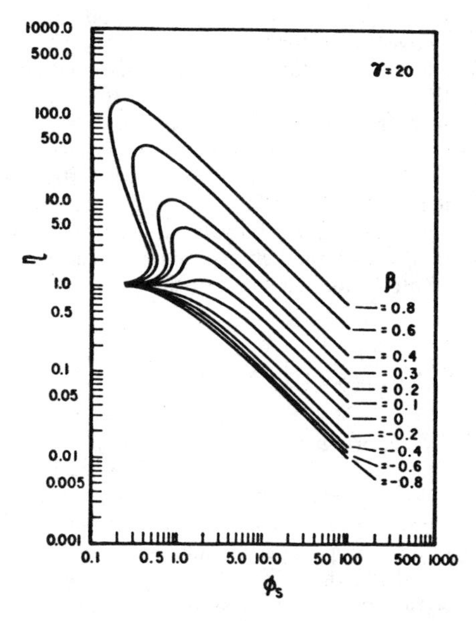

Figure 11. Effectiveness factor for a first-order reaction in a nonisothermal pellet (γ = 20). Source: Permission granted by, Weisz, P.B., and Hicks, J.S., "The Behaviour of Porous Catalyst Particles in View of Internal Mass and Heat Diffusion Effects," *Chem. Eng. Sci.,* Vol. 17, No. 4, (April, 1962), pp. 265–275.

Thiele-modulus curves for different values of a dimensionless parameter which is the product of β (Equation 70) with γ (Equation 71).

$$\gamma = \frac{E_a}{R_g T_0} \tag{71}$$

$$\beta\gamma = \frac{(-\Delta H) D_e C_0 E_a}{\gamma_e T_0^2 R_g} \tag{72}$$

For exothermic reactions ($\beta\gamma > 0$) effectiveness-factor values of greater than unity are possible. Carberry [76] has also presented nonisothermal effectiveness-factor results as a function of the product $\beta\gamma$.

If the energy and mass conservation equations are written in dimensionless form, three dimensionless groups, namely, φ (Thiele modulus), β, and γ appear in these equations for steady-state conditions. Considering a first-order reaction, and assuming D_e and γ_e values are independent of the radial position in the pellet, these equations can be written as

$$\frac{1}{\xi^2}\frac{d}{d\xi}\left(\xi^2 \frac{d\psi}{d\xi}\right) - \varphi^2 \psi \exp\left[\gamma\left(\frac{\theta - 1}{\theta}\right)\right] = 0 \tag{73}$$

$$\frac{1}{\xi^2}\frac{d}{d\xi}\left(\xi^2 \frac{d\theta}{d\xi}\right) + \beta\varphi^2 \psi \exp\left[\gamma\left(\frac{\theta - 1}{\theta}\right)\right] = 0 \tag{74}$$

Prediction of the effectiveness factor in nonisothermal pellets requires the simultaneous solution of Equations 73 and 74. Weisz and Hicks [77] were the first to develop the effectiveness-factor curves for a first-order irreversible reaction in spheres for β values varying between -0.8 to 0.8 and for γ values of -10, 20, 30, and 40. A typical diagram for the variation of the effectiveness factor with respect to the Thiele modulus for $\gamma = 20$ and various values of β is shown in Figure 11. For endothermic reactions both diffusion and heat effects decrease the value of η below unity. On the other hand, for exothermic reactions heat effects might more than offset the diffusional limitation, and η values of greater than one are possible. This point has also been discussed in a later section. It is even possible to have multiple effectiveness-factor values for certain values of φ for highly exothermic reactions. As it is indicated by the shape of the curves, small changes in the operating conditions and consequently in the values of φ and β could cause significant changes in the observed rate if the thermal effects are important. Combining Equations 73 and 68, the following differential equation is obtained:

$$\frac{1}{\xi^2}\frac{d}{d\xi}\left(\xi^2 \frac{d\psi}{d\xi}\right) - \varphi^2 \psi \exp\left|\gamma \frac{\beta(\psi - 1)}{1 + \beta(\psi - 1)}\right| = 0 \tag{75}$$

For many of the industrially important reactions the value of β is smaller than 0.1. Some typical values of β and γ for selected reactions are summarized in Table 7 [41]. For such small values of β, the term $\beta(\psi - 1)$ can be neglected with respect to 1 in the denominator of the exponential term in Equation 75. Under these conditions the effectiveness factor is expected to depend upon φ and the product $\beta\gamma$, and the results practically reduce to the values reported by Tinkler and Metzner [74].

As it was discussed in the previous section from the practical point of view, effectiveness factors can also be plotted as a function of the observable parameter $\Phi = \eta\varphi^2$. Considering the limit for $\beta < 1$, the variation of η with respect to Φ for different values of $\beta\gamma$ is given in Figure 12.

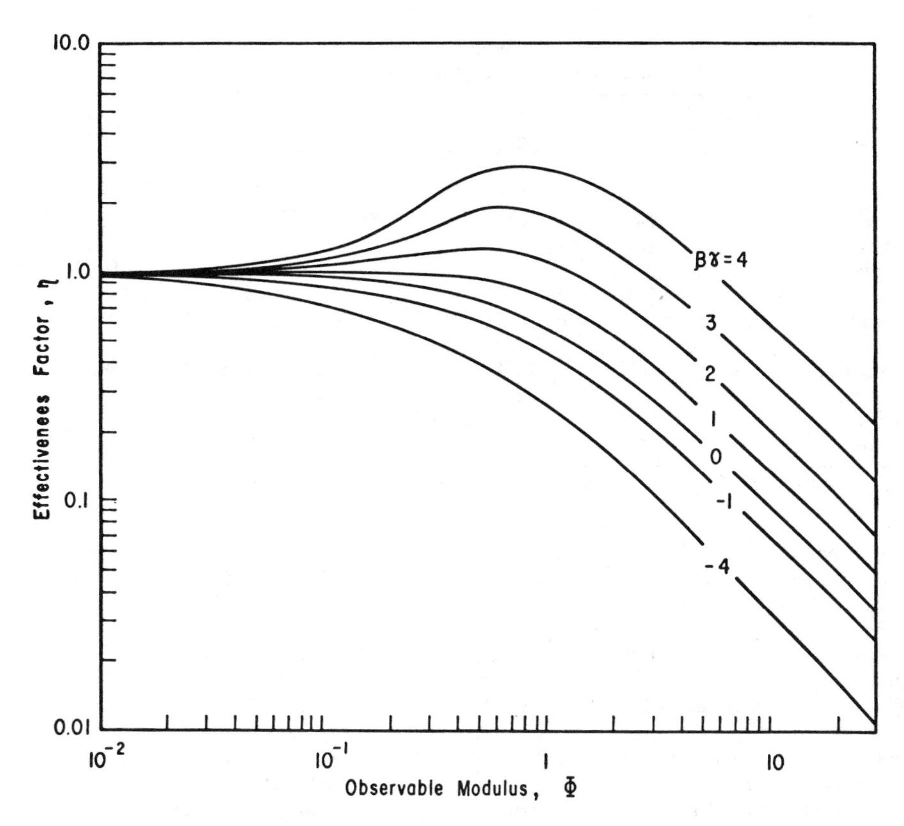

Figure 12. Variation of the effectiveness factor with the observable modulus in a non-isothermal pellet for small β (first-order reaction spherical geometry).

Solutions of mass and energy conservation equations for reactions other than first order are also reported in the literature. Maymo and Smith [79] investigated the heat and mass transfer limitations in the catalytic oxidation of hydrogen using a platinum-alumina catalyst pellet. The rate of this reaction is found to be proportional with the 0.804-th power of the partial pressure of oxygen. By changing the values of β, γ, and φ, Maymo and Smith [79] determined the experimental and theoretical effectiveness-factor values. Their results show very good agreement with the theory. Computed effectiveness factor values for $\gamma = 7.0$ are reported in their work. Tinkler and Metzner [74] have also reported effectiveness-factor values for a second-order reaction, in addition to the first-order reaction. As expected, as the reaction order increases the influence of the diffusion limitations on the observed rate becomes more significant. Some of the other important references in this area are References 80–86. Smith et al. [86] have considered the CO-oxidation reaction over supported -Pt catalyst which confirms a rate expression of the form $R = k(CO)/[1 + K(CO)]^2$. Computed effectiveness factors for different β values are reported.

Effective Thermal Conductivity

Heat conduction through porous catalysts takes place through the solid phase and gas phase in parallel with interchange of heat between the two phases. The model proposed

Table 7
Parameters of Some Exothermic Catalytic Reactions*

Reaction	γ	β	$\beta\gamma$	Lw	φ_s
NH_3 Synthesis [87]	29.4	6.1×10^{-5}	0.0018	2.6×10^{-4}	1.2
Synthesis of higher alcohols from CO and H_2 [87]	28.4	8.5×10^{-4}	0.024	2.0×10^{-4}	-
Oxidation of CH_3OH to CH_2O [87]	16.0	1.1×10^{-2}	0.175	4.5×10^{-3}	1.1
Synthesis of vinylchloride from acetylene and HCl [87]	6.5	0.25	1.65	0.1	0.27
Hydrogenation of ethylene [84]	23–27		2.7–1.0	0.11	0.2–2.8
Oxidation of H_2 [79]	6.75–7.52	0.03–0.34	0.21–2.3	3.6×10^{-2}	0.8–2.0
Oxidation of ethylene to ethyleneoxide [87]	13.4	0.13	1.76	6.5×10^{-2}	0.08
Dissociation of N_2O [88]	22.0		1.0–2.0	-	1–5
Hydrogenation of benzene [89]	14–16		1.7–2.0	6.0×10^{-3}	0.05–1.9
Oxidation of SO_2 [87]	14.8	1.2×10^{-2}	0.175	4.2×10^{-2}	0.9

* *After Hlavacek, V., Kubicek, M., and Marec, M.,* J. Cat., *Vol. 1 (1969) pp. 17–30.*

by Butt [90] for the prediction of effective thermal conductivities is reported to be in good agreement with some of the data reported in the literature. Butt's model is an extension of the random pore model of Wakao and Smith [45] to the thermal conduction in porous pellets.

As it is discussed by Satterfield [3], the thermal conductivity of a catalyst pellet depends upon the geometrical factors and porosity rather than the thermal conductivity of the solid. Most of the effective thermal conductivity values reported in the literature are in the range of 10^{-3} to 10^{-2} J/s cm °C. Some of the effective thermal conductivity values reported in the literature are summarized in Table 8. Effective thermal conductivities of some commercial catalysts are reported by Sehr [91] (Table 9). Most of the thermal conductivity values are measured with catalyst pores filled with air. Since the thermal conductivities of most of the organic vapors are greater than air, somewhat higher effective thermal conductivities are expected under reaction conditions. Mischke and Smith [92] have measured the effective thermal conductivities of alumina catalyst pellets in vacuum and with air and He present in the pores. Thermal conductivities measured in vacuum are 25% to 50% smaller than the values measured with air being present in the pores. Their results showing the effect of macro porosity on the effective thermal conductivity is shown in Figure 13.

The effective thermal conductivities are measured by different techniques in the literature. Sehr [91] reported results obtained by steady state and transient techniques. His analyses show that consistent results are obtained with different techniques. An excellent review of effective heat conduction in porous catalysts is given by Satterfield [3].

Interphase and Intraphase Temperature Gradients

The overall effectiveness factor is expected to depend upon the external heat and mass transfer resistances, in addition to intraphase diffusion and heat transfer limitations. Weisz and Hicks [77] have neglected such effects in their analysis of the nonisothermal effectiveness of catalyst pellets. For many of the catalytic reaction systems external heat transfer resistance may be more significant than the internal resistance [6]. On the other hand, intraphase diffusion limitations are usually more significant than external mass transfer resistance.

For systems with significant external heat and mass transfer limitations boundary conditions (as in Equations 66 and 67) should be used in the simultaneous solution of

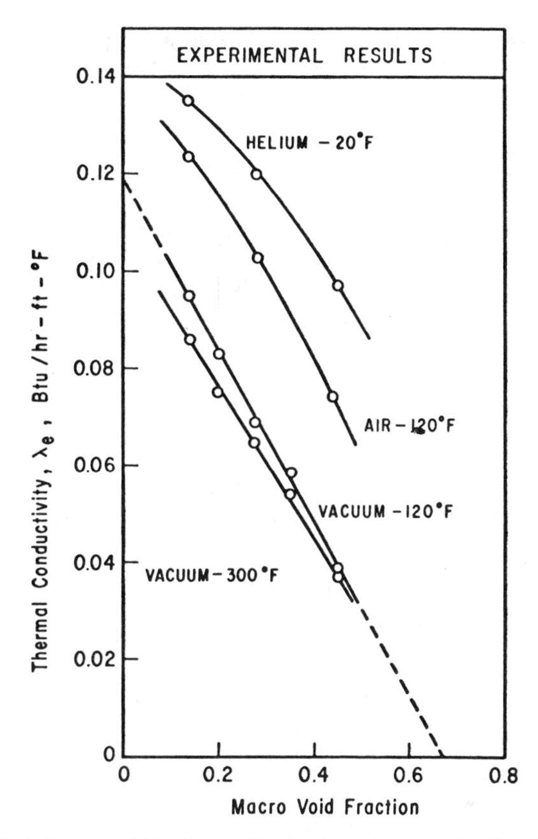

Figure 13. Effect of macro void fraction on effective thermal conductivity. Note: λ_e values should be multiplied by 1.73×10^{-2} in order to convert their units to J/cm s°C. Source: Permission granted by Mischke, R.A., and Smith, J.M., "Thermal Conductivity of Alumina Catalyst Pellets," *Ind. Eng. Chem. Fund.*, Vol. 1, No. 4, (Nov., 1962), pp. 288–292.

Table 8
Effective Thermal Conductivities of Some Porous Catalysts

System	λ_e (J/s cm°C)	Physical Properties ρ_p (g/cm³)	ϵ	T(°C)	Reference
58% Ni on kieselguhr	1.5×10^{-3}	1.88			[94]
25% Ni-25% Graphite					
50% γ-Al_2O_3	1.5×10^{-2}	1.57			[94]
Pt-Al_2O_3	2.6×10^{-3}	0.57	0.78	68.5	[79]
Silica-alumina	3.1×10^{-3} 3.6×10^{-3}	1.25			[91]
Alumina (boehmite)	1.8×10^{-3}	0.896	0.634	49	[92]
10% NiO-alumina	1.2×10^{-3}	0.66		300	[83]
10% NiO-alumina	1.7×10^{-3}	1.14		300	[83]

Equations 63 and 64. As discussed in earlier, the temperature at any point in a catalyst pellet can be related to the concentration by Equation 68. Considering external heat and mass transfer resistances

$$\frac{dT}{dr} = -\left(\frac{D_e(-\Delta H)}{\gamma_e}\right)\frac{dC}{dr} \tag{76}$$

the relation between the temperature and concentration in the pellet can be derived as

$$\frac{T - T_b}{T_b} = \left(\frac{D_e(-\Delta H)C_b}{\lambda_e T_b}\right)\left[\left(\frac{C_b - C}{C_b}\right) - \left(\frac{D_e}{k_f} - \frac{\lambda_e}{h_f}\right)\frac{dC}{dr}\bigg|_{r = r_0}\right] \tag{77}$$

Noting that

$$V_p R_{obs} = A_e D_e \frac{dC_A}{dr}\bigg|_{r = r_0} \tag{78}$$

the following expression is obtained between the temperature and concentration at any point in the pellet.

$$\frac{(T - T_b)}{T_b} = \beta_b\left[\frac{C_b - C}{C_b} + \Phi_b\left(\frac{1}{Bi_h} - \frac{1}{Bi_m}\right)\right] \tag{79}$$

$$\text{where } Bi_h = \frac{h_f(V_p/A_e)}{\lambda_e} \tag{80}$$

$$\beta_b = \frac{(-\Delta H)D_e C_b}{\lambda_e T_b} \tag{81}$$

Table 9
The Thermal Conductivity of Catalyst Particles

Catalyst	λ_e(pellet) (J/s cm°C)	λ_e(powder) (cal/s cm°C)	Density Pellet g/cm³	Density Powder g/cm³
Ni/W	4.7×10^{-3}	3.1×10^{-3}	1.83	1.48
Co/Mo (dehydrogenation catalyst*)	3.5×10^{-3}	2.1×10^{-3}	1.63	1.56
Cr/Al (Reforming catalyst)	2.9×10^{-3}	1.8×10^{-3}	1.4	1.06
Co/Mo (dehydrogenation catalyst†)	2.4×10^{-3}	1.4×10^{-3}	1.54	1.09
Si/Al (cracking catalyst)	3.6×10^{-3}	1.8×10^{-3}	1.25	0.82
Pt/Al_2O_3 (reforming catalyst)	2.2×10^{-3}	1.3×10^{-3}	1.15	0.88
Activated carbon	2.7×10^{-3}	1.7×10^{-3}	0.65	0.52

* *3.6% CoO and 7.1% MoO_3 supported on α-alumina 180 m²/g.*
† *3.4% CoO and 11.3% MoO_3 supported on β-alumina 128 m²/g.*
Source: Permission granted by Sehr, R.A., Chem. Eng. Sci., *Vol. 9, Pergamon Press, (1958), pp. 145–152*

$$\Phi_b = \frac{\left(\frac{V_p}{A_e}\right)^2 R_{obs}}{D_e C_b} \tag{82}$$

The maximum temperature in the pellet corresponds to C = 0.

$$\frac{\Delta T_{max}}{T_b} = \beta_b\left[1 + \Phi_b\left(\frac{1}{Bi_h} - \frac{1}{Bi_m}\right)\right]. \tag{83}$$

Following the procedure of Carberry and Kulkarni [93] it can be shown that the ratio of the external temperature difference to the total temperature difference is

$$\frac{T_b - T_0}{T_b - T} = \frac{\Phi_b/Bi_h}{1 + \frac{\Phi_b}{Bi_m}\left(\frac{Bi_m}{Bi_h} - 1\right)} \tag{84}$$

Kehoe and Butt [94] reported experimental inter and intraphase temperature gradients for Ni-kieselguhr catalyzed benzene hydrogenation reaction. Their results indicate that 0% to 37% of the total temperature difference is due to external resistance. It is shown by Carberry [76,93] that theoretical predictions of Equation 84 are in good agreement with the experimental results of Kehoe and Butt [94].

Carberry and Kulkarni [95] have investigated the effectiveness of an isothermal pellet dictated by external heat transfer limitations. Considering a first-order reaction, the effectiveness factor for such a system can be expressed [6,95] as

$$\eta = \frac{\tanh \varphi_b}{\varphi_b[1 + (\varphi_b \tanh \varphi_b)/Bi_m]} \exp\left(-\gamma_b\left(\frac{T_b}{T_0} - 1\right)\right) \tag{85}$$

Taking $T = T_0$ in Equation 79 and combining with Equation 85 the following expression is obtained.

$$\eta = \frac{\tanh \varphi_b}{\varphi_b[1 + \varphi_b(\tanh \varphi_b)/Bi_m]} \exp\left[\gamma_b\left(\frac{\beta_b \Phi_b Bi_h^{-1}}{1 + \beta_b \Phi_b Bi_h^{-1}}\right)\right] \tag{86}$$

Carberry and Kulkarni [95] reported predicted effectiveness-factor values for different γ_b, Bi_m a nd $\bar{\beta} = \beta_b(Bi_m/Bi_h)$ values. Equation 86 is a useful expression for the prediction of the effectiveness factor of first-order reactions if the external heat transfer resistance is much greater than the internal heat transfer resistance.

External Mass and Heat Transfer Coefficients

As discussed in the previous sections, external mass and heat transfer limitations might be quite significant for certain catalytic reactions. The correlations of the mass and heat transfer coefficients are usually presented in terms of the J_D and J_H factors respectively. These factors are defined as

$$J_D = \frac{k_f}{U_0} \varepsilon_b (Sc)^{2/3} \tag{87}$$

$$J_H = \frac{h_f}{C_p U_0 \varrho} \varepsilon_b Pr^{2/3} \tag{88}$$

A boundary-layer analysis shows that these factors are expected to depend upon the Reynold's number. From the theoretical considerations these J factors can be related to the Reynold's number for a fixed bed as [6]

$$J_D \cong J_H = \frac{1.15}{Re_p^{1/2}} \tag{89}$$

where

$$Re_p = \frac{d_p U_0 \varrho}{\varepsilon_b \mu} \tag{90}$$

Experimental fixed-bed mass transfer data and the prediction of Carberry's boundary-layer model for the J factor is given in Figure 14.

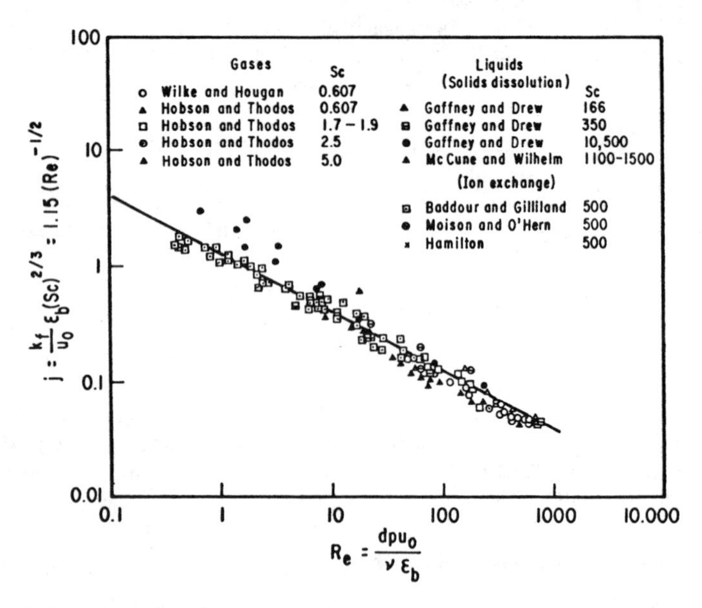

Figure 14. Fixed-bed mass-transfer data contrasted with Carberry's boundary-layer model. Source: Permission granted by Carberry, J.J., *Chemical and Catalytic Reaction Engineering,* McGraw Hill Book Co., New York, 1976. Data from C.R. Wilke and A. Hougen, *Trans. Am. Inst. Chem. Eng.,* 41:445 (1945); M. Hobson and G. Thodos, *Chem. Eng. Prog.,* 47:370 (1951); B.W. Gaffney and T.B. Drew, *Ind. Eng. Chem.* 42:1120 (1950); L. K. McKune and R.H. Wilhelm, *Ind. Eng. Chem.*, 41:1124 (1949), R.F. Baddour and F.R. Gilliland, *Ind. Eng. Chem.,* 45:330 (1953); R.L. Moison and H.A. O'Hern, *Chem. Eng. Prog. Symp. Ser.,* (24)55:71 (1959); P.B. Hamilton, Alfred I. duPont Institute, private communication, 1959; J.J. Carberry, *AIChEJ.,* 6:460 (1960).

DIFFUSIONAL EFFECTS IN CATALYSTS WITH BIDISPERSE PORE STRUCTURE

Effectiveness of Bidisperse Catalysts

Many catalysts have bidisperse pore structures. In the catalyst pellets which are formed by compacting fine porous particles, macropores are formed between the agglomerated particles. On the other hand the fine pores within the particles are called micropores. The macroporosity and the average radius of macropores are significantly affected by the compacting pressure, while the micropore structure is essentially unchanged. In such catalysts the surface area of the micropores is usually much greater than the macropore surface area, and most of the active sites lie within the microporous region. Typical bidisperse pore size distribution curves for catalyst pellets prepared from α-alumina are shown in Figure 15.

One approach to analyze diffusion and reaction processes in such catalysts involves the assumption that macropores are cylindrical tubes, and from the macropores branched cylindrical micropores extend into the pellet. This approach can be considered as an extension of the Thiele's single-pore model which is originally proposed for catalyst pellets composed of only cylindrical macropores. A more realistic approach is to consider the catalyst pellet as an agglomeration of spherical micoporous grains (particles) and to predict the effectiveness factor by the solution of the pseudohomogeneous transport equations in the pellet and in the microporous particles.

Some of the earlier papers on the effectiveness of bidisperse porous catalysts are reported by Mingle and Smith [96], Carberry [97] and Wakao and Smith [46]. Smith and co-workers have defined pellet and particle effectiveness factors seperately.

A schematic diagram of a bidisperse porous catalyst is shown in Figure 16. The observed rate of a reaction catalyzed by a catalyst with a bimodal pore structure depends upon the rate of diffusion of reactants and products both in the macro and micropore regions as well as the reaction rate. The work of Örs and Doğu [98] showed that, in addition to the Thiele modulus, the effectiveness factor depends upon another dimensionless parameter, α, which is defined for a spherical pellet as

$$\alpha = 3(1-\varepsilon_a)\frac{D_i}{D_a}\frac{r_0^2}{r_p^2} \tag{91}$$

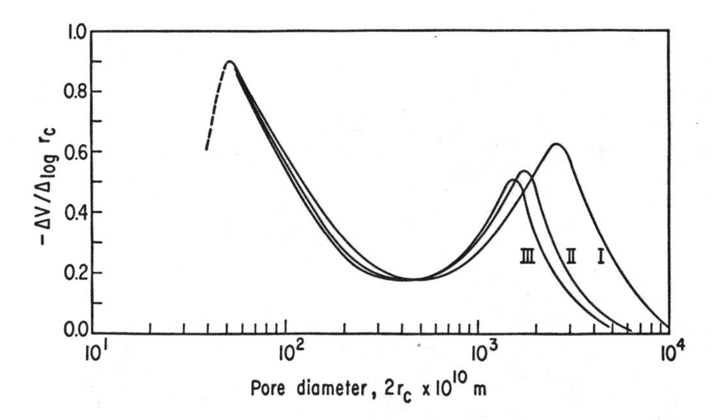

Figure 15. Differential pore size distribution curves of $\alpha-Al_2O_3$ pellets. Pores smaller than 3.5×10^{-8}m considered as micropores. (I) $\epsilon_T = 0.73$, $\epsilon_a = 0.32$; (II) $\epsilon_T = 0.68$, $\epsilon_a = 0.27$; (III) $\epsilon_T = 0.67$, $\epsilon_a = 0.25$.

The magnitude of α is determined by the ratio of diffusion times in the macro and micropore regions. Kulkarni et al. [99] have generalized the Örs and Doğu parameter α, considering different catalyst geometries.

$$\alpha = (1+p)(1-\varepsilon_a)\frac{D_i}{D_a}\frac{r_0^2}{r_p^2} \quad (92)$$

Here, p represents the geometry of the pellet; p = 0, 1, and 2 for slab, cylinder and sphere respectively.

Considering an n-th-order reaction, the pseudohomogeneous conservation equations for the macroporous spherical pellet and for the microporous particle are written respectively as:

Pellet

$$\frac{D_a}{r^2}\frac{d}{dr}\left(r^2\frac{dC_a}{dr}\right) - \frac{3}{r_p}(1-\varepsilon_a)D_i\left(\frac{dC_i}{dr_i}\right)_{ri = r_p} = 0 \quad (93)$$

Particle

$$\frac{D_i}{r_i^2}\frac{d}{dr_i}\left(r_i^2\frac{dC_i}{dr_i}\right) - kC_i^n = 0 \quad (94)$$

It is shown by Örs and Doğu [98] that the effectiveness factor in this type of a catalyst can be predicted from,

$$\eta = \frac{9}{\varphi_i^2\alpha}\left(\frac{d(C_a/C_0)}{d(r/r_0)}\right)_{(r/r_0) = 1.0} \quad (95)$$

where φ_i is the particle Thiele modulus defined as

$$\varphi_i = r_p\left(\frac{k\,C_0^{n-1}}{D_i}\right)^{1/2} \quad (96)$$

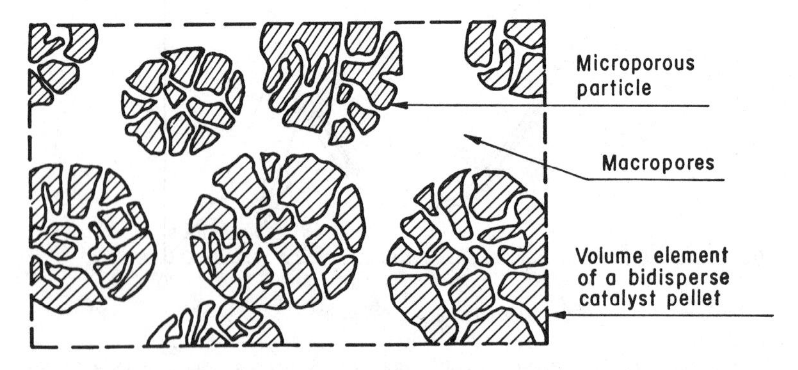

Figure 16. Schematic diagram of a bidisperse porous catalyst.

Equations 93 and 94 are subject to the following boundary conditions.

$$r_i = 0 \qquad \frac{dC_i}{dr_i} = 0 \tag{97}$$

$$r_i = r_p \qquad C_i = C_a \tag{98}$$

$$r = 0 \qquad \frac{dC_a}{dr} = 0 \tag{99}$$

$$r = r_0 \qquad C_a = C_0 \tag{100}$$

For a first-order reaction the following expression is derived for the effectiveness factor [98, 100].

$$\eta = \frac{9}{\varphi_i^2 \alpha}\left(\frac{\left[\alpha\left(\dfrac{\varphi_i}{\tanh \varphi_i} - 1\right)\right]^{1/2}}{\tanh\left[\alpha\left(\dfrac{\varphi_i}{\tanh \varphi_i} - 1\right)\right]^{1/2}} - 1\right) \tag{101}$$

The variation of the effectiveness factor with the particle Thiele modulus, φ_i, for different values of α is shown in Figure 17. Effectiveness-factor values are also plotted as a function of an observable parameter $\eta\varphi_i^2\alpha$ for different values of α (Figure 18). Note that

$$\eta\varphi_i^2\,\alpha = \Phi\alpha = \left(\frac{R_{obs}}{D_a C_0}\, r_0^2\right) 3(1-\varepsilon_a) \tag{102}$$

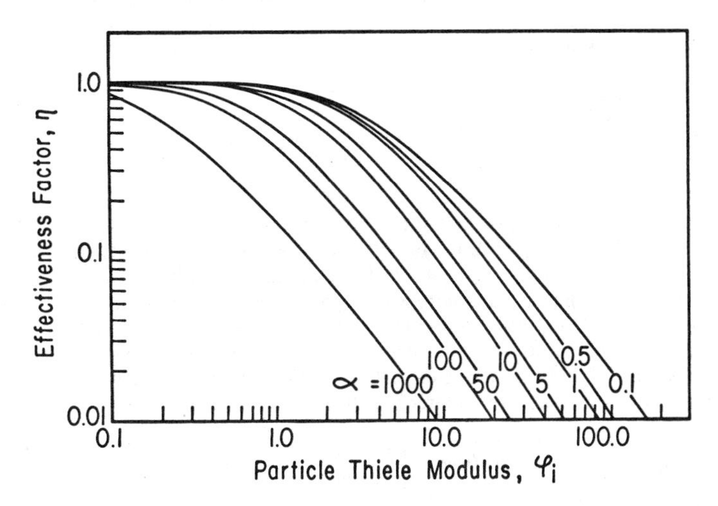

Figure 17. Variation of effectiveness factor with the Thiele modulus and α for a bidisperse catalyst (first-order reaction).

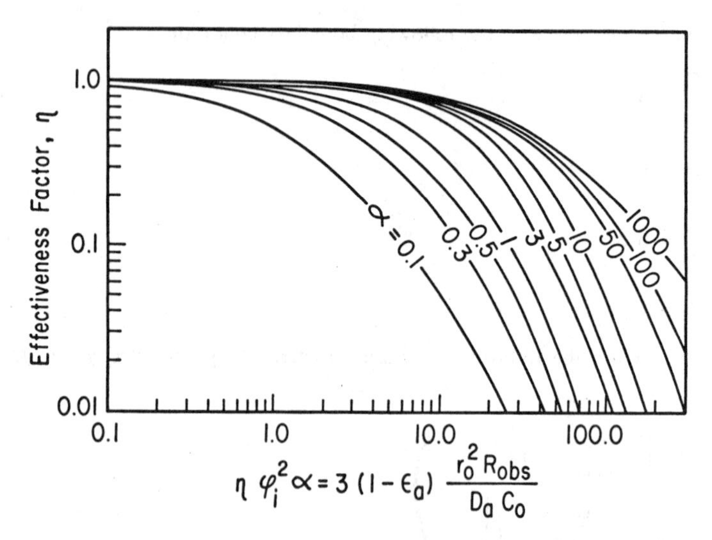

Figure 18. Variation of effectiveness factor with the observable parameter, for a bidisperse catalyst (first-order reaction).

As the value of α increases, the difference between the effectiveness factor predicted for a bidisperse catalyst and the effectiveness factor of a pellet with a monodisperse pore structure becomes smaller and smaller. These results show that the prediction of the effectiveness factor without considering the bimodal pore size distribution would give overestimated values especially for small α.

Külkarni et al. [99] and Jayaraman et al. [101] have extended the approach reported by Örs and Doğu to nonlinear rate forms covering power law. For high values of φ_i, the asymptotic expression for η becomes,

$$\eta = \frac{\left(\frac{2}{n+1}\right)^{1/2} / \left(\frac{n+3}{4}\right)}{\varphi_i^{3/2}\alpha^{1/2}}(1+p)^2 \tag{103}$$

The pseudohomogeneous transport equations for bidisperse porous catalysts are derived by the volume averaging of the conservation equation. Neogi and Ruckenstein [102] have shown that the point sink assumption made in the analysis of bidisperse catalysts is valid if the ratio between the radius of the microporous particle and the pellet is less than 0.05.

Furusawa and Smith [103] have separately defined the effectiveness factors for the particle, macropore diffusion, and for the pellet, and have developed a model to predict the effective diffusivity of bidisperse porous catalysts under reaction conditions. Some of the other references in this area are References 104–106.

Macro and Micropore Diffusion Coefficients

Although many of the porous catalysts have a bidisperse pore structure, in many cases they are treated as monodisperse catalysts and a single effective diffusion coefficient is used to describe the diffusion process in such catalysts. Experimental measurement techniques of effective diffusion coefficients are discussed in the first main section. For some of the bidisperse catalysts, such as molecular sieve pellets which are prepared by use of a binder, diffusion resistance in the micropores could be much greater than in the macropores. On the other hand in many instances macropore diffusion resistance has a substantial effect on the observed rate of the reaction.

Hashimoto and Smith [107, 108] and Furusawa and Smith [109] have shown that both macro and micropore diffusion coefficients can be determined from chromatographic experiments similar to the procedure described in the first section. For bidisperse catalyst pellets, the moments of the response peak depend upon the macro and micropore diffusion coefficients as well as axial dispersion coefficient in the packed-bed, gas to pellet mass transfer coefficient and adsorption equilibrium and rate constants. Macropore and micropore diffusivities of n-butane on alumina powder [108], and nitrogen and n-butane diffusivities in 5A molecular sieves [107] are determined using this technique. Some of the results are given in Table 10. It is shown by Hashimoto and Smith that this technique is well suited in measuring rapid processes. Haynes and Sarma [105] have suggested a similar procedure for the measurement of macro and micropore diffusivities.

As discussed in the first section, the single-pellet pulse-response technique eliminates the interpellet transport processes and the intrapellet parameters can be more accurately determined. Doğu and Uyanık [110], and Doğu and Ercan [111] extended the single-pellet technique to study diffusion and adsorption in bidisperse systems. For the single-pellet system, the mass conservation equations for the adsorbing species in the macroporous pellet and in the microporous particles are

$$\varepsilon_a \frac{\partial C_a}{\partial t} = D_a \frac{\partial^2 C_a}{\partial x^2} - \left[3\frac{(1-\varepsilon_a)}{r_p}\right] D_i \left(\frac{\partial C_i}{\partial r}\right)_{r=r_p} \tag{104}$$

$$\frac{\varepsilon_i}{(1-\varepsilon_a)} \frac{\partial C_i}{\partial t} = \frac{D_i}{r^2}\frac{\partial}{\partial r}\left(r^2 \frac{\partial C_i}{\partial r}\right) - \varrho_p K \frac{\partial C_i}{\partial t} \tag{105}$$

Using the boundary conditions discussed in the first section for the pellet and also using $(\partial C_i/\partial r)_{r=0} = 0$, and $C_i = C_a$ at $r = r_p$ the moment expressions for this system are derived.

$$\mu_1 = \frac{L_p^2}{6D_a}((\varepsilon_a + \varepsilon_i) + (1-\varepsilon_a)\varrho_p K) \frac{\left[3\dfrac{AD_a}{L_p} + F\right]}{\left[\dfrac{AD_a}{L_p} + F\right]} \tag{106}$$

$$\mu_2' = \frac{L_p^4}{6D_a^2}[(\varepsilon_a + \varepsilon_i) + (1-\varepsilon_a)\varrho_p K]^2 \frac{\left[\left(\dfrac{AD_a}{L_p}\right)^2 + \dfrac{2}{5}\left(\dfrac{AD_a}{L_p}\right)F + \dfrac{1}{15}F^2\right]}{\left[\dfrac{AD_a}{L_p} + F\right]^2}$$

$$+ \left[\frac{L_p^2(\varepsilon_i + \varrho_p K)^2 r_p^2}{15 D_i D_a (1-\varepsilon_a)} + \frac{L_p^2 K (1-\varepsilon_a)}{k_a D_a}\right] \frac{\left[\left(\dfrac{AD_a}{L_p}\right)^2 + \dfrac{4}{3}\left(\dfrac{AD_a}{L_p}\right)F + \dfrac{1}{3}F^2\right]}{\left[\dfrac{AD_a}{L_p} + F\right]^2} \tag{107}$$

Table 10
Effective Diffusivities in the Macro and Micropores and Intercrystalline Diffusion Coefficients in Zeolites

Diffusing Species	Porous Solid	Macroporosity ϵ_a	Microporosity ϵ_i	Effective Macro Diffusion Coefficient, D_a (cm^2/s)	Effective Micro Diffusion Coefficient, D_i (cm^2/s)	Intercrystalline Diffusion Coefficient for Zeolites, D_c	Method	Reference
n-butane -He	Alumina (Boehmite)	0.362	0.415	3.2×10^{-2}	4.0×10^{-4}	–	Chromatography (T=30°C)	[108]
n-butane -He	5A molecular sieve pellet	0.321	0.253	3.18×10^{-2}		–	Chromatography (T=200°C)	[107]
N_2-He	5A molecular sieve pellet	0.321	0.253	4.21×10^{-2}	10^{-6}	–	Chromatography (T=0°C)	[107]
C_2H_4-He	α-alumina	0.32	0.41	3.34×10^{-2}	8.9×10^{-6}	–	Single-Pellet moment (T=45°C)	[111]
n-butane -He	Na Y zeolite					2.38×10^{-8}	Modified Bed Chromatography (T=105°C)	[23]
i-C_4H_{10}-He	4A molecular sieve	~ 0.3		2.1×10^{-2}			Chromatography (T=125°C)	[116]
N_2-He	4A molecular sieve	~ 0.3				8.8×10^{-10}	Chromatography (T=90°C)	[116]
NH_3	Amberlyst 21 resin	0.475				10^{-15}–10^{-16}	Gravimetric uptake	[113]
CO_2	Amberlyst 21 resin	0.475				10^{-18}–10^{-19}	Gravimetric uptake	[113]
n-butane -He	Ca × Na	0.5		6.85×10^{-2}–7.78×10^{-2}		7.65×10^{-15}–8.44×10^{-14}	Uptake (T=5°C-60°C)	[106]
i-butane -He	Ca × Na	0.5		7.64×10^{-2}		1.67×10^{-13}	(T=60°C)	[106]
ethane	T zeolite					4.5×10^{-11}	Uptake (T=300°C)	[115]
n C 8	T zeolite					10^{-13}	"	[115]
n C 12	T zeolite					10^{-11}	"	[115]

In this technique, the macropore diffusion coefficient is evaluated from the first moment data and the micropore diffusion coefficient from the second moment. For an inert tracer the high flow rate limits of μ_1 and μ_2' become

$$\mu_{1_\infty} = \frac{L_p^2 \varepsilon}{6D_a} \tag{108}$$

$$\mu'_{2_\infty} = \frac{L_p^4 \varepsilon^2}{90D_a^2} + \frac{L_p^2 \varepsilon_i^2 r_p^2}{45D_i D_a(1-\varepsilon_a)} \tag{109}$$

These expressions can be used for the fast and accurate evaluation of D_a and D_i. Macro and micropore diffusion coefficients and the adsorption equilibrium constant of ethylene on α-Al_2O_3 is measured by Doğu and Ercan [111] with this technique. Some of the results are given in Table 10.

For catalysts with bimodal pore distribution macropore and micropore tortuosity factors might be separately defined. The macropore tortuosity factor (τ_a) can be estimated from

$$D_a = D_{T_a} \frac{\varepsilon_a}{\tau_a} \tag{110}$$

Here D_{T_a} is the composite diffusivity for the macropores. The micropore tortuosity factor can be defined in two different ways; namely based on microporosity of the pellet (ε_i), or porosity of the microporous particle ($\varepsilon_i/(1-\varepsilon_a)$).

$$D_i = D_{T_i} \frac{(\varepsilon_i)/(1-\varepsilon_a)}{\tau_i} \tag{111}$$

Some of the macro and micropore tortuosity factors calculated from the experimental data reported in the literature are given in Table 11.

Diffusion in Zeolites

The use of zeolites as catalysts for the cracking and isomerization of hydrocarbons has become an active area of research in recent years. In his review, Weisz [112] defined a third diffusion regime, namely the configurational regime in addition to the molecular and Knudsen diffusion mechanisms, in the micropores of zeolites. Since the micropore diameters of the zeolites are usually in the same order of magnitude of the molecular diameters of diffusing species, the diffusion is often characterized by a strong molecule-wall interaction. Considering this, most investigators did not consider the adsorbed phase and the gas phase separately but characterized the adsorption and diffusion process with an intercrystalline diffusivity, D_c. The value of the diffusivity defined this way is expected to be orders of magnitude smaller than the micropore diffusion coefficient D_i defined in the previous sections. In setting up the conservation equations in the microporous particles and macroporous region Ruckenstein et al. [113] considered gas-phase diffusion in the pores and adsorption separately. Their results indicate that intercrystalline diffusivity in the micropores, D_c, can be related to the effective micropore diffusion coefficient D_i in the following form

$$D_c = \frac{D_i}{(\varepsilon_i + S_i K)} \tag{112}$$

where S_i = the micropore surface area per unit volume
K = the adsorption equilibrium constant for the linear adsorption

Most of the intercrystalline diffusivities in zeolites are determined from the gravimetric or volumetric experiments which are carried out by measuring the uptake of the zeolite particles when a step increase of the adsorbing gas is introduced.

In these studies adsorption and diffusion processes are characterized by a single diffusivity, D_c, and it is shown in the literature [113, 114] that the uptake can be related to this diffusivity. In such studies external mass transfer resistance from the bulk gas to the surface of the catalyst pellet is usually neglected. Gorring [115] determined diffusion coefficients from sorption rate experiments for n-paraffins and found out a periodic behavior of diffusivity with the length of the molecule. The values of intercrystalline diffusivities obtained by Gorring are in the range of 10^{-11} to 10^{-13} cm^2/s. Both macropore diffusivity, D_a, and the intercrystalline diffusivity, D_c, values of n-butane, iso-butane and 1-butene in synthetic CaX(Na) pellets are determined by Ma and Lee [106] in a constant-volume, well-stirred system. Some of the diffusivity values reported in the literature are summarized in Table 10.

Effective diffusion coefficients in the macropores, D_a, and the intercrystalline diffusivity, D_c, in the micropores of zeolites are also determined by the chromatographic technique which is described in the previous sections. Hsu and Hynes [23] modified the chromatographic method described by Hashimoto and Smith [107] considering an intercrystalline diffusivity in the micropores and a reversible adsorption process on the external surface of the microporous particles. They have also included the zeolite crystallite size distribution into their analysis. Diffusivity values obtained from chromatographic experiments are also summarized in Table 10. Temperature dependence of intercrystalline diffusivities of some hydrocarbons in 5A-sieve are reported by Kumar et al. [116]. Diffusion and reaction in zeolites is also reviewed by Satterfield [3].

The order of magnitude of diffusion coefficients reported in the literature for zeolites differ significantly. These differences are mainly due to the differences of definition of the micropore diffusion coefficient in different studies, and differences of the assumptions of different measuring techniques.

CRITERIA TO TEST THE IMPORTANCE OF DIFFUSIONAL LIMITATIONS IN POROUS CATALYSTS

In the investigation of fluid-solid catalytic reactions and for the interpretation of the experimental data, it is very important to know the effects of transport processes on the observed rate. Since the concentration and temperature dependence of the rate expressions are generally nonlinear and in many cases solid-catalyzed reactions are not simple single-reaction systems, it is difficult to derive analytical expressions for the observed rates. Thus, a number of criteria have been derived to test the importance of pore diffusion and heat transfer for reactions in porous catalysts.

Considering an isothermal pellet, Weisz and Prater [117] presented a criterion to test the diffusional effects in catalyst pellets. For a first-order reaction on a flat slab their criterion for absence of appreciable diffusion effects can be written as

$$R_{obs}\frac{L^2}{D_eC_0} < 1.0 \qquad (113)$$

Some important references in this area are Petersen [118], Schneider and Mitschka [64], Hutchings and Carberry [119], Bischoff [120] and Narshimhan and Guha [121].

Table 11
Macropore and Micropore Tortuosity Factors

Porous Catalyst	ϵ_a	ϵ_i†	τ_a*	τ_i†	Reference
Alumina (Boehmite)	0.362	0.415	1.4–1.7	4.5–6.1	[108]
5A-sieve	0.321	0.253	3.4–4.1	–	[107]
α-alumina	0.32	0.41	2.4	–	[111]
Ca × Na	0.5		4.5	–	[106]
Alumina (Boehmite)	0.48	0.291	3.34		[34]

* *Using Equation 110*
† *Using Equation 111*

Hudgins [122] developed a more general criterion for absence of appreciable diffusion effects, applicable also for reactions having other than power-type rate expressions. For a spherical pellet, Hudgins criterion can be written as

$$\Phi_s = R_{obs}\frac{r_0^2}{D_e C_0} < \frac{1}{C_0}\frac{R(C_0)}{R'(C_0)} \tag{114}$$

Here $R(C_0)$ and $R'(C_0)$ denote the reaction rate and the derivative of reaction rate (dR/dC) evaluated at the external surface concentration of the catalyst pellet.

Both Weisz and Hudgins criteria are formulated considering a single independent reaction. In many of the catalytic reaction systems we are encountered with multiple reactions and the rate expressions of these reactions might depend upon the concentrations of more than one species in the system. A general criterion is derived by Doğu and Doğu [123] which is applicable to any multiple-reaction system and to reactions conforming to any rate law. Doğu and Doğu have considered a system containing m species and n independent reactions. For such a system, the rate of formation of species j, R_j, can be written in terms of the intrinsic rates of independent reactions, $R_{(i)}$, and the stoichiometric coefficients of j-th species in the i-th reaction, v_{ij} (v_{ij} is positive for products and negative for reactants).

$$R_j = \sum_{i=1}^{n} v_{ij}R_{(i)} \ldots \quad j = 1, 2, 3, 4, \ldots, m \tag{115}$$

By expanding the intrinsic reaction rate in a Taylor series about the external surface concentrations and assuming second- and higher-order terms are negligible, $R_{(i)}$ is expressed as

$$R_{(i)} = R_{(i)}|_{r=r_0} + \sum_{j=1}^{m}\left(\frac{\partial R_{(i)}}{\partial C_j}\right)_{r=r_0}(C_j - C_{j,0}) \tag{116}$$

Also assuming parabolic concentration profiles within the catalyst pellet, the effectiveness factor expression for a spherical pellet becomes [123]:

$$\eta = 1 + \frac{r_0^2}{15}\sum_{j=1}^{m}\left(\left(\frac{\partial R_{(i)}}{\partial C_j}\right) / R_{(i)}\right)\Bigg|_{r=r_0}(D_{j,e})^{-1}\left[\sum_{i=1}^{n} R_{(i)obs}v_{i,j}\right] \tag{117}$$

These assumptions are considered to be acceptable if diffusional effects are not significant. By setting up $\eta > 0.95$ for negligible diffusional effects, a general criterion has been obtained for stoichiometrically complex systems. This criterion can be expressed as

$$-\left(\frac{V_p}{A_e}\right)^2 \sum_{j=1}^{m} \frac{(\partial R_{(i)}/\partial C_j)_{r=r_0}}{R_{(i)_{r=r_0}} D_{j,e}} \left[\sum_{i=1}^{n} R_{(i)_{obs}} v_{i,j}\right] < A \tag{118}$$

where $A = 0.083$ for sphere
$A = 0.1$ for infinite cylinder
$A = 0.15$ for slab

The selection of η being greater than 0.95 for negligible diffusion limitation is certainly arbitrary. If this is not appropriate for a given reaction, one can easily set up this limit higher than 0.95. In such a case only the right-hand side of Equation 118 should be multiplied by $(1-\eta_{min})/0.05$, where η_{min} is the minimum value of the effectiveness factor that we allow for negligible diffusion effects. Application of this criterion to some specific reaction systems is reported by Doğu and Doğu [123].

Following a similar approach, Doğu and Doğu [124] have derived a criterion to test diffusional effects on the observed rates for bidisperse porous catalysts. Considering a single reaction and a spherical catalyst pellet which is composed of spherical particles (grains), absence of diffusion effects can be tested using the following criterion.

$$\frac{R'(C_0)}{R(C_0)}\left[R_{obs}\frac{r_0^2}{D_a}\right]\frac{(\alpha+3)}{\alpha} - \frac{1}{15}\frac{R''(C_0)}{R(C_0)}\left[R_{obs}\frac{r_0^2}{D_a}\right]^2\frac{3}{\alpha} < 0.75 \tag{119}$$

where $R''(C_0) = (\partial^2 R/\partial C^2)_{r=r_0}$ and α is the dimensionless parameter which is previously defined by Örs and Doğu [98] Equation 91. The magnitude of α is determined by the diffusion times in the macro and micropore regions. Another approach in this area is reported by Brown [125]. Intraparticle diffusion effects in branched-pore systems are analyzed in this paper.

In all those criteria the catalyst pellet is assumed to be isothermal. On the other hand, thermal effects might have an important role on the observed rates. For endothermic reactions the effectiveness factor is expected to decrease further compared to isothermal case. On the other hand, for exothermic reactions, the increase of the rate due to temperature rise within the pellet can more than offset the decrease in rate due to diffusional effects. Without considering the diffusional effects Anderson [126] suggested the following criterion to test the isothermal behavior of a catalyst pellet.

$$|\Phi_s(\beta\gamma)| < 0.75 \tag{120}$$

where β and γ are defined by Equations 70 and 71.

Recently Doğu and Doğu [127] derived a criterion to test the relative importance of diffusion and thermal effects on the observed rates. Considering a multiple reaction system with n-independent reactions and m species Doğu and Doğu have shown that the following criterion should hold for $1.05 > \eta > 0.95$.

$$\left| -r_0^2\left\{\sum_{j=1}^{m}\frac{\left(\dfrac{\partial R_{(i)}}{\partial C_j}\right)_{r=r_0}}{R_{(i)_{r=r_0}} D_j}\sum_{i=1}^{n} R_{(i)_{obs}} v_{ij} + \frac{\left(\dfrac{\partial R_{(i)}}{\partial (1/T)}\right)_{r=r_0}}{R_{(i)_{r=r_0}}\lambda_e T_0^2}\sum_{i=1}^{n} R_{(i)_{obs}}\Delta H_{(i)}^0\right\}\right| < 0.75 \tag{121}$$

In the derivation of all these criteria, higher than first-order terms in the Taylor expansion of the intrinsic reaction rate are neglected. This assumption can be considered to be acceptable for small deviations of the concentration and temperatures from the surface values. Considering that temperature dependence of the reaction rate is highly nonlinear, this assumption might not be completely appropriate especially when heat effects are considerable. This point has been considered by Doğu [128]. For a single reaction, by taking into account the terms up to third order in the Taylor expansion, the intrinsic reaction rate is written as

$$
\begin{aligned}
R = {} & R(C_0) + \left(\frac{\partial R}{\partial C}\right)_{r=r_0}(C-C_0) + \left(\frac{\partial R}{\partial\left(\frac{1}{T}\right)}\right)_{r=r_0}\left(\frac{1}{T}-\frac{1}{T_0}\right) \\
& + \frac{1}{2!}\left[\left(\frac{\partial^2 R}{\partial C^2}\right)_{r=r_0}(C-C_0)^2 + 2\left(\frac{\partial^2 R}{\partial C\,\partial\left(\frac{1}{T}\right)}\right)_{r=r_0}(C-C_0)\left(\frac{1}{T}-\frac{1}{T_0}\right)\right. \\
& \left. + \left(\frac{\partial^2 R}{\partial\left(\frac{1}{T}\right)^2}\right)_{r=r_0}\left(\frac{1}{T}-\frac{1}{T_0}\right)^2\right] \\
& + \frac{1}{3!}\left[\left(\frac{\partial^3 R}{\partial C^3}\right)_{r=r_0}(C-C_0)^3 + 3\left(\frac{\partial^3 R}{\partial C^2\,\partial\left(\frac{1}{T}\right)}\right)_{r=r_0}(C-C_0)^2\left(\frac{1}{T}-\frac{1}{T_0}\right)\right. \\
& + 3\left(\frac{\partial^3 R}{\partial C\,\partial\left(\frac{1}{T}\right)^2}\right)_{r=r_0}(C-C_0)\left(\frac{1}{T}-\frac{1}{T_0}\right)^2 \\
& \left. + \left(\frac{\partial^3 R}{\partial\left(\frac{1}{T}\right)^3}\right)_{r=r_0}\left(\frac{1}{T}-\frac{1}{T_0}\right)^3\right]
\end{aligned}
\tag{122}
$$

For this type of expansion for a first-order reaction, the criterion for $1.05 > \eta > 0.95$ becomes

$$
\left|\Phi_s(1-\beta\gamma) + \frac{2}{21}\Phi_s^2\left(\beta\gamma - \frac{1}{2}(\beta\gamma)^2\right) + \frac{1}{189}\Phi_s^3\left((\beta\gamma)^2 - \frac{1}{3}(\beta\gamma)^3\right)\right| < 0.75 \tag{123}
$$

First, second and third terms on the left-hand side of this expression correspond to the first-, second- and third-order terms in the Taylor expansion. For endothermic reactions and for small values of $\beta\gamma(\beta\gamma < 1)$ only the first term is sufficient to test whether η is close to one or not. As the value of $\beta\gamma$ becomes larger second- and higher-order terms become significant

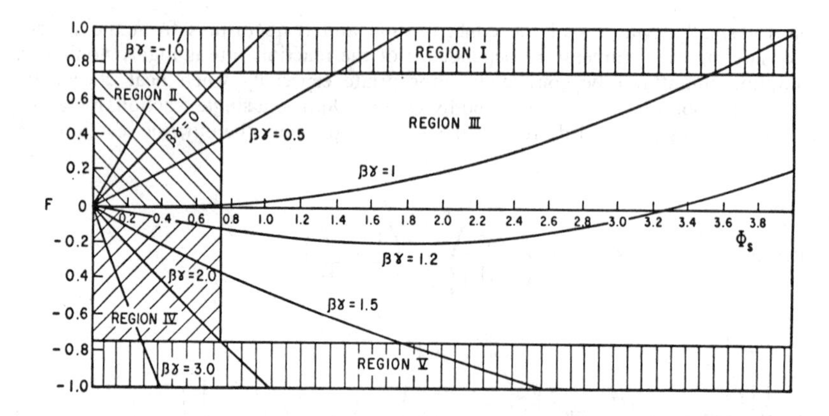

Figure 19. Relative importance of diffusion limitations and thermal effects on the observed rate. In order to have $0.95<\eta<1.05$, the criterion in Equation 123 requires $|F|<0.75$ for a first-order reaction. After Dogu, T., Can. J. Chem. Eng., Vol. 62 (Feb 1985).

$$F = \Phi_s(1-\beta\gamma) + 2/21\,\Phi_s^3\,(\beta\gamma - 1/2\,(\beta\gamma)^2) + 1/189\,\Phi_s^2\,((\beta\gamma)^2 - 1/3\,(\beta\gamma)^3)$$

Region I:	$\eta<0.95$	Significant diffusion limitations
Region II:	$0.95<\eta<1.00$	Negligible diffusion limitations
Region III:	$0.95<\eta<1.05$	Thermal effects compensate diffusion limitations
Region IV:	$1.0<\eta<1.05$	Negligible diffusion limitations
Region V:	$\eta>1.05$	Thermal effects are more important than diffusion effects

especially for $\Phi_s > 1$. A graphical representation of this criterion is shown in Figure 19. As illustrated in this figure, for $\Phi_s < 0.75$ diffusion limitations can be considered to be negligible if there were no thermal effects. For $\beta\gamma < 0$ (endothermic reactions), the value of η drops below 0.95 at much lower Φ_s values. For instance, for $\beta\gamma = -1.0$, in order to have $\eta < 0.95$ it is necessary to have $\Phi_s < 0.4$. In Region I we expect to have $\eta < 0.95$. For exothermic reactions the increase of rate due to temperature rise within the pellet might compensate the diffusional limitations. In Region III we expect to have $0.95 < \eta < 1.05$ due to this effect. For instance for $\beta\gamma = 1$ the value of the effectiveness factor is expected to be greater than 0.95 for Φ_s values up to 3.5. If there had been no heat effect, at such large Φ_s values η would be much smaller than 1.0. Regions II and IV correspond to the zones where we don't expect significant diffusional limitations. For $\beta\gamma$ values of greater than 1.0 thermal effects become even more significant. For $\beta\gamma = 1.5$ effectiveness factor becomes greater than 1.05 for $\Phi_s > 1.8$. Region V corresponds to $\eta > 1.05$.

NOTATION

a	pore radius
$\bar{a}$	mean pore radius
A_e	external area of the catalyst pellet
Bi_m	Biot number for mass transfer defined by Equation 57
Bi_h	Biot number for heat transfer defined by Equation 80
C_A	concentration of species A
C_{ads}	adsorbed concentration of the adsorbing species
C_b	concentration in the bulk fluid in the bed
C_c	concentration at the center-line of the slab

C_0 concentration at the surface of the pellet
C_p specific heat of fluid
$C_{p,s}$ specific heat of pellet
d_p particle diameter
D_{AB} molecular diffusion coefficient of A in B
D_a effective macropore diffusivity
D_c intercrystalline diffusivity for zeolites
D_e effective diffusion coefficient
D_{K_A} Knudsen diffusion coefficient of species A
D_{K_a} Knudsen diffusivity, in macropores evaluated at the average macropore radius
D_{K_i} Knudsen diffusivity in micropores evaluated at average micropore radius
D_i effective micropore diffusivity
Da Damköhler number defined by Equation 58
D_T composite diffusivity defined by Equation 6
D_{T_a} composite diffusivity in macropores, defined by Equation 35
D_{T_i} composite diffusivity in micropores, defined by Equation 36
E_a axial dispersion coefficient
F flow rate (of lower stream in single pellet moment technique)
J_D factor defined by Equation 87
J_H factor defined by Equation 88
k_a adsorption rate constant
k_f film mass transfer coefficient
k reaction rate constant
K_A adsorption equilibrium constant
L half-thickness of slab
L_p pellet length for the single pellet moment technique
L_G characteristic dimension defined by Equation 50
L_w Lewis number, $(= \lambda_e/\rho_p C_{p,s} D_e)$
m_n n-th ordinary moment
M_i molecular weight of species i
n_A adsorbed concentration of A
$\bar{N}_i$ flux of i in the capillaries
$\bar{N}_{i_e}$ effective flux of species i
p pressure
r radial coordinate
r_0 pellet radius
r_p particle radius
R(C,T) reaction rate per pellet volume
R_{obs} observed reaction rate per pellet volume
$R'(C_o)$ the value of $\partial R(C,T)/\partial C$ at $r = r_0$
R_j reaction rate of [reaction i.] species j
$R_{(i)}$ intrinsic rate of reaction i.
$R_{(i)obs}$ observed intrinsic rate of reaction i
R_g gas constant
S_g pore surface area per unit mass of pellet
S_i micropore surface area per volume
T_b bulk temperature
T_0 surface temperature
u_0 superficial velocity
V_p pellet volume
V_g pore volume per unit mass of pellet
w diffusion coordinate in the pellet ($w = z$ for slab, $w = r$ for sphere and infinite cylinder)
x axial direction in the fixed bed
X length of the fixed bed
y_A mole fraction of A
z direction of diffusion in the slab-like pellet

Greek Symbols

α_A defined by Equation 4
α parameter defined by Equation 91
β dimensionless group defined by Equation 70
β_b parameter defined by Equation 81
$\bar{\beta}$ $\beta_b(Bi_m/Bi_h)$
γ dimensionless group defined by Equation 71
γ_b value of γ evaluated at the bulk temperature, T_b
δ_o defined by Equation 24
δ_1 defined by Equation 25
ΔH heat of reaction
ε total porosity

ε_a macroporosity of the pellet
ε_b bed void fraction
ε_i microporosity
ε_A Lennard–Jones parameter
ε_{AB} Lennard–Jones parameter for binary mixture of A and B, defined by Equation 10
η effectiveness factor, defined by Equation 38
(η_f: for slab, η_c: for cylinder, η_s: for sphere)
θ dimensionless temperature, T/T_0
θ_v volume change modulus defined by Equation 55
λ_e effective thermal conductivity
μ_1 first absolute moment
μ_2' second central moment
ξ dimensionless position, $w/(V_p/A_e)$
ξ_0 value of ξ at the surface of the pellet ($\xi_0 = 1$ for slab, $\xi_0 = 2$ for cylinder, $\xi_0 = 3$ for sphere)
ρ fluid density
ρ_p density of pellet
σ_A Lennard-Jones parameter
σ_{AB} Lennard-Jones parameter for mixture A and B defined by Equation 9
Equation 9
τ tortuosity factor
τ_a tortuosity factor based on macropores
τ_i tortuosity factor based on micropores
φ shape generalized Thiele modulus defined by Equation 39
φ_b shape generalized Thiele modulus evaluated at the bulk concentration, C_b, and temperature, T_b
φ_s Thiele modulus for spherical catalyst
φ_i particle Thiele modulus defined by Equation 96
$\hat{\varphi}$ order generalized Thiele modulus
Φ observable shape generalized Thiele modulus defined by Equation 48
$\hat{\Phi}$ observable shape and order generalized Thiele modulus defined by Equation 49
Φ_b observable shape generalized Thiele modulus evaluated at the bulk concentration and temperature (Equation 82)
ψ dimensionless concentration, C/C_0
ψ_b dimensionless concentration, C/C_b

REFERENCES

1. Aris, R., *The Mathematical Theory of Diffusion and Reaction in Permeable Catalysts,* Clarendon Press, Oxford, 1975.
2. Smith, J. M., *Chemical Engineering Kinetics,* 2nd ed., McGraw Hill Book Co., New York, 1970.
3. Satterfield, C. N., *Mass Transfer in Heterogeneous Catalysis,* MIT Press, Cambridge, 1970.
4. Petersen, E. E., *Chemical Reaction Analysis,* Prentice-Hall Inc., New Jersey, 1965.
5. Froment, G. F., and Bischoff, K. B., *Chemical Reactor Analysis and Design,* John Wiley and Sons, New York, 1979.
6. Carberry, J. J., *Chemical and Catalytic Reaction Engineering,* McGraw Hill Book Co., New York, 1976.
7. Scott, D. S., and Dullien, F. A. L., "Diffusion of Ideal Gases in Capillaries and Porous Solids," *AIChE J.,* Vol. 8, No. 1 (March 1962), pp. 113–7.
8. Dullien, F. A. L., and Scott, D. S., "The Flux Ratio for Binary Counter Diffusion of Ideal Gases," *Chem. Eng. Sci.,* Vol. 17, No. 10 (Oct. 1962), pp. 771–5.
9. Mason, E. A., and Evans, R. B., III, "Graham's Laws: Simple Demonstrations of Gases in Motion," *J. Chem. Ed.,* Vol. 46, No. 6 (June 1969), pp. 358–64.
10. Bird, R. B., Stewart, W. E., and Lightfoot, E. N., *Transport Phenomena,* John Wiley and Sons, Inc., New York, 1960.

11. Hirschfelder, J. O., Curtiss, C. F., and Bird, R. B., *Molecular Theory of Gases and Liquids,* John Wiley and Sons, Inc., New York, 1954.
12. Chapman, S., and Cowling, T. G., *Mathematical Theory of Nonuniform Gases,* Cambridge University Press, 1951.
13. Wilke, C. R., "Diffusional Properties of Multicomponent Gases," *Chem. Eng. Prog.,* Vol. 46, No. 2 (1980), pp. 95–104.
14. Wang, C. T., and Smith, J. M., "Tortuosity Factors for Diffusion in Catalyst Pellets," *AIChE J.,* Vol. 29, No. 1 (Jan.1983), pp. 132–136.
15. Brown, L. F., Haynes, H. W., and Manogue, W. H., "The Prediction of Diffusion Rates in Porous Materials at Different Pressures," *J. Catalysis,* Vol. 14, No. 3 (July 1969), pp. 220–225.
16. Rothfeld, L. B., "Gaseous Counter Diffusion in Catalyst Pellets," *AIChE J.,* Vol. 9, No. 1 (Jan. 1963), pp. 19–24.
17. Rao, M. R., and Smith, J. M., "Diffusion Resistences in Alumina and Silica Catalysts," *AIChE J.,* Vol. 9, No. 3 (May 1963), pp. 419–421.
18. Satterfield, C. N., Ma, Y. H., and Sherwood, T. K., "The Effectiveness Factor in a Liquid-Filled Porous Catalyst," *Chem. Eng. Ser.,* No. 28 (Br), (1968), pp. 22.
19. Barrer, R. M., "A New Approach to Gas Flow in Capillary Systems," *J. Phys. Chem.,* Vol. 57, No. 1 (Jan. 1953), pp. 35–40.
20. Schneider, P., and Smith, J. M., "Adsorption Rate Constants from Chromatography," *AIChE J.,* Vol. 14, No. 5 (Sept. 1968), pp. 762–771.
21. Adrian, J. C., and Smith, J. M., "Catalysis with Supported Metals, Part II, Chemisorption Rates by Chromatography-Hydrogen on Cobalt," AIChE 67th National Meeting, Georgia (Feb. 1970).
22. Suzuki, M., Smith, J. M., "Kinetic Studies by Chromatography," *Chem. Eng. Sci.,* Vol. 26, No. 2 (Feb. 1971), pp. 221–35.
23. Hsu, L.K. P., and Haynes, H. W., Jr., "Effective Diffusivity by the Gas Chromatography Technique-Analysis and Application to Measurements of Diffusion of Various Hydrocarbons in Zeolite NaY," *AIChE J.,* Vol. 27, No. 1 (Jan. 1981), pp. 81–91.
24. Haynes, H. W., Jr., "The Determination of Effective Diffusivity by Gas Chromatography. Time Domain Solutions," *Chem. Eng. Sci.,* Vol. 30, No. 8 (1975), pp. 955–961.
25. Shah, D. B., and Ruthven, D. M., "Measurement of Zeolitic Diffusivities and Equilibrium Isotherms by Chromatography," *AIChE J.,*Vol. 23, No. 6 (Nov. 1977), pp. 804–809.
26. Dougharty, N. A., "Gas Solid Chromatography with Nonuniform Adsorbent Surfaces," *AIChE J.,* Vol. 19, No. 2 (March 1973), pp. 379–381.
27. Mehta, R. V., Merson, R. L., and McCoy, B. J., "Moment Analysis of Experiments in Gel Permeation Chromatography," *AIChE J.,* Vol. 19, No. 5 (Sept. 1973), pp. 1068–1070.
28. Barrer, R. M., "Surface and Volume Flow in Porous Media," in *The Solid Gas Interface,* Vol. 2, E. A. Flood (Ed.), Dekker, New York, 1967.
29. Dacey, J. R., "Surface Diffusion of Adsorbed Molecules," *Ind. Eng. Chem.,* Vol. 57, No. 6 (June 1965), pp. 27–33.
30. Luss, D., "Steady State and Dynamic Behavior of a Single Catalyst Pellet," in *Chemical Reactor Theory,* L. Lapidus, and N. R. Amundson (Eds.), Prentice Hall Inc., New Jersey, 1977.
31. Reed, E. M., and Butt, J. B., "Surface Diffusion of Single Sorbates at Low and Intermediate Surface Coverages," *J. Phys. Chem.,* Vol. 75, No. 1 (Jan. 1971), pp. 133–141.
32. Schneider, P., and Smith, J. M., "Chromatographic Study of Surface Diffusion," *AIChE J.,* Vol. 14, No. 6 (Nov. 1968), pp. 886–895.
33. Suzuki, M., and Smith, J. M., "Dynamics of Diffusion and Adsorption in a Single Catalyst Pellet," *AIChE J.,* Vol. 18, No. 2 (March 1972), pp. 326–332.

34. Doğu, G., and Smith, J. M., "A Dynamic Method for Catalyst Diffusivities," *AIChE J.*, Vol. 21, No. 1 (Jan. 1975), pp. 58–61.
35. Doğu, G., "Intrapellet Rate Parameters by Single Pellet Chromatography," Ph. D. Dissertation, University of California, Davis 1974.
36. Doğu, G., and Smith, J. M., "Rate Parameters from Dynamic Experiments with Single Catalyst Pellets," *Chem. Eng. Sci.*, Vol. 31, No. 2 (Feb. 1976), pp. 123–135.
37. Doğu, G., "Dynamic Single Pellet Technique as a Tool to Analyze the Mechanism Benzene Hydrogenation on $Pt-Al_2O_3$," 8th Int. Cong. Catalysis, Berlin (July 1984).
38. Youngquist, G. R., "Diffusion and Flow of Gases in Porous Solids," *Ind. Eng. Chem.*, Vol. 62, No. 8 (Aug. 1970), pp. 52–63.
39. Henry, J. P., Cunningham, R. S., and Geankoplis, C. J., "Diffusion of Gases in Porous Solids Over a Thousand–Fold Pressure Range," *Chem. Eng. Sci.*, Vol. 22, No. 1 (Jan. 1967), pp. 11–20.
40. Evans, R. B., III, Watson, G. M., and Mason, E. A., "Gaseous Diffusion in Porous Media at Uniform Pressure", *J. Chem. Phys.*, Vol. 35, No. 6 (Dec. 1961), pp. 2076–2083.
41. Asaeda, M., Watanabe, J., and Kitamoto M., "Effect of Total Pressure Gradient on Unsteady–State Diffusion Rates of Gases in Porous Media," *J. Chem. Eng. Japan*, Vol. 14, No. 2 (1981), pp. 129–135.
42. McGreavy, C., and Aseda, M., "Characterization of Nonisobaric Diffusion Due to Nonequimolar Fluxes in Catalyst Particles," ACS Symposium Series, Wei, J. and C. Georgakis (Eds.), No. 196 (1981), pp. 473–88.
43. Chou, T., and Hegedus, L. L., "Transient Diffusivity Measurements in Catalyst Pellets with Two Zones of Differing Diffusivities," *AIChE J.*, Vol. 24, No. 2 (March 1978), pp. 225–260.
44. Jiratová, K., and Horák, J., "Determination of Effective Diffusion Coefficients On the Basis of Temperature Differences Within a Catalyst Pellet," *Int. Chem. Eng.*, Vol. 18, No. 2 (April 1978), pp. 297–304.
45. Wakao, N., and Smith, J. M., "Diffusion in Catalyst Pellets," *Chem. Eng. Sci.*, Vol. 17, No. 11 (Nov. 1962), pp. 825–834.
46. Wakao, N., and Smith, J. M., "Diffusion and Reaction in Porous Catalysts," *Ind. Eng. Chem. Fund.*, Vol. 3, No. 2 (May 1964), pp. 123–27.
47. Johnson, M. F. L., and Stewart, W. E., "Pore Structure and Gaseous Diffusion in Solid Catalysts," *J. Cat.*, Vol. 4, No. 2 (Apr. 1965), pp. 248–252.
48. Foster, R. N., and Butt, J. B., "A Computational Model for the Structure of Porous Materials Employed in Catalysis," Vol. 12, No. 1 (Jan. 1966), pp. 180–185.
49. Wheeler, A., "Reaction Rates and Selectivity in Catalyst Pores," *Catalysis 2, Fundamental Principles*, Pt2, Reinhold Publ.Corp., New York, 1955, pp. 105–165.
50. Feng, C., and Stewart, W. E., "Practical Models for Isothermal Diffusion and Flow of Gases in Porous Solids," *Ind. Eng. Chem. Fund.*, Vol. 12, No. 2 (1973), pp. 143–147.
51. Pakula, R. J., and Greenkorn, R. A., "An Experimental Investigation of a Porous Medium Model with Nonuniform Pores," *AIChE J.*, Vol. 17, No. 5 (Sept. 1971), pp. 1265–1268.
52. Patel, P. V., and Butt, J. B., "Multicomponent Diffusion in Porous Catalysts," *Ind. Eng. Chem. Proc. Des. Dev.*, Vol. 14, No. 3 (1975), pp. 298–304.
53. Ryan, D., Carbonell R. G., and Whitaker, S., "A Theory of Diffusion and Reaction in Porous Media," *AIChE*. Symp. Ser. Vol. 77, *Transport with Chemical Reactions*, P. Stroeve, and W. J. Ward (Eds.), (1981), pp. 46–62.
54. Ryan, D., Carbonell, R. G., and Whitaker, S., "Effective Diffusivities for Catalyst Pellets Under Reactive Conditions," *Chem. Eng. Sci.*, Vol. 35, No. 1/2 (Jan. 1980), pp. 10–6.
55. Thiele, E. W., "Relation between Catalytic Activity and Size of Particle," *Ind. Eng. Chem.*, Vol. 31, (1939), pp. 916–920.

56. Miller, D. J., and Lee, H. H., "Shape Normalization of Catalyst Pellet," *Chem. Eng. Sci.*, Vol. 38, No. 3 (March 1983), pp. 363–366.
57. Aris, R., "A Normalization for the Thiele Modulus," *Ind. Eng. Chem. Fund.*, Vol. 4, No. 2 (May 1965), pp. 227–229.
58. Petersen, E. E., "A General Criterion for Diffusion Influenced Chemical Reactions in Porous Catalysts," *Chem. Eng. Sci.*, Vol. 20, No. 6 (June 1965), pp. 587–591.
59. Bischoff, K. B., "Effectiveness Factors for General Reaction Rate Forms," *AIChE J.*, Vol. 11, No. 2 (March 1965), pp. 351–355.
60. Mehta, B. N., and Aris, R., "Communications on the Theory of Diffusion and Reaction–VII. The Isothermal p-th Order Reaction," *Chem. Eng. Sci.*, Vol. 25 (1971), p. 1699.
61. Kulkarni, B. D., and Doraiswamy, L. K., "Effectiveness Factors in Gas–Liquid Reactions: The General n-th-Order Case," *AIChE J.*, Vol. 22, No. 3 (May 1976), pp. 597–600.
62. Morbidelli, M., and Varma, A., "Isothermal Diffusion-Reaction in a Slab Catalyst With Bimolecular Langmuir–Hinshelwood Kinetics," *Chem. Eng. Sci.*, Vol. 38, No. 2 (Feb. 1983), pp. 289–296.
63. Van den Bosh, B., and Luss, D., "Pitfalls in the Prediction of Steady State Multiplicity by Negative Order Kinetics," *Chem. Eng. Sci.*, Vol. 32, No. 5 (May 1977), pp. 560–562.
64. Schneider, P., and Mitschka, P., "Effect of Internal Diffusion on Catalytic Reactions II," *Chem. Eng. Sci.*, Vol. 21, No. 5 (May 1966), pp. 455–463.
65. Schneider, P., and Mitschka, P., "Effect of Internal Diffusion on Catalytic Reactions IV. Reversible Second Order Reaction with Langmuir–Hinshelwood Type of Rate Equation," *Colln. Czech. Chem. Comm. Engl. Edn.* Vol. 31, No. 9 (1966), pp. 3677–3701.
66. Satterfield, C. N., and Roberts, G. W., "Effectiveness Factor for Porous Catalysts. Langmuir–Hinshelwood Kinetic Expressions for Bimolecular Surface Reactions," *Ind. Eng. Chem. Fund.*, Vol. 5, No. 3 (Aug. 1966), pp. 317–325.
67. Moo–Young, M., and Kobayashi, T., "Effectiveness Factors for Immobilized Enzyme Reactions," *Can. J. Chem. Eng.*, Vol. 50 (Apr. 1972), pp. 162–167.
68. Pereira, C. J., and Varma, A., "Uniqueness Criteria of the Steady State in Automotive Catalysis," *Chem. Eng. Sci.*, Vol. 33, No. 12 (Dec. 1978), pp. 1645–1658.
69. Valdman, B., and Hughes, R., "A Simple Method of Calculating Effectiveness Factor for Heterogeneous Catalytic Gas–Solid Reactions", *AIChE J.*, Vol. 22, No. 1 (Jan. 1976), pp. 192–193.
70. Weekman, V. W., Jr., and Gorring, R. L., "Influence of Volume Change on Gas–Phase Reactions in Porous Catalysts", *J. Cat.*, Vol. 4, No. 2 (April 1965), pp. 260–270.
71. Lin, K., and Lih, M. M., "Concentration Distribution, Effectiveness Factor, and Reactant Exhaustion for Catalytic Reaction with Volume Change," *AIChE J.*, Vol. 17, No. 5 (Sept. 1971), pp. 1234–1240.
72. Lih, M. M., and Lih, K., "Simultaneous Heat and Mass Transfer with and without Reactant Exhaustion in Catalyst Particles for Reaction with Volume Change," *AIChE J.*, Vol. 19, No. 4 (July 1973), pp. 832–839.
73. Weisz, P. B., and Prater, C. D., "Interpretation of Measurements in Experimental Catalysis," *Adv. Catalysis,* Vol. 6, Academic Press, New York, 1954.
74. Tinkler, J. D., and Metzner, A. B., "Reaction Rates in Nonisothermal Catalysts," *Ind. Eng. Chem.*, Vol. 53, No. 8 (Aug. 1961), pp. 663–668.
75. Prater, C. D., "The Temperature Produced by Heat of Reaction in the Interior of Porous Particles," *Chem. Eng. Sci.*, Vol. 8 (1958), pp. 284–286.
76. Carberry, J. J., "The Catalytic Effectiveness Factor Under Nonisothermal Conditions," *AIChE J.*, Vol. 7, No. 2 (June 1961), pp. 350–351.
77. Weisz, P. B., and Hicks, J. S., "The Behavior of Porous Catalyst Particles in View of Internal Mass and Heat Diffusion Effects," *Chem. Eng. Sci.*, Vol. 17, No. 4 (April 1962), pp. 265–275.

78. Hlavacek, V., Kubicek, M., and Marec, M., "Analysis of Nonstationary Heat and Mass Transfer in a Porous Catalyst Particle," *J.Cat.*, Vol. 15, No. 1 (Sept. 1969), pp. 17–30.
79. Maymo, J. A., and Smith, J. M., "Catalytic Oxidation of Hydrogen–Intrapellet Heat and Mass Transfer," *AIChE J.*, Vol. 12, No. 5 (Sept. 1966), pp. 845–854.
80. Schilson, R. E., and Amudson, N. R., "Intraparticle Diffusion and Conduction in Porous Catalysts I, Single Reactions," *Chem. Eng. Sci.*, Vol. 13, No. 4 (Apr. 1961), pp. 226–237.
81. Gunn, D. J., "Nonisothermal Reaction in Catalyst Particles," *Chem. Eng. Sci.*, Vol. 21, No. 5 (May 1966), pp. 383–390.
82. Michelsen, M. L., and Villadsen, J., "Diffusion and Reaction on Spherical Catalyst Pellets," *Chem. Eng. Sci.*, Vol. 27, No. 4 (Apr. 1972), pp. 751–762.
83. Otani, S., and Smith, J. M., "Effectiveness of Large Catalyst Pellets," *J.Cat.*, Vol. 5, No. 2 (April 1966), pp. 332–347.
84. Cunningham, R. E., Carberry, J. J., and Smith, J. M., "Effectiveness Factors in a Nonisothermal Reaction System," *AIChE J.*, Vol. 11, No. 4 (July 1965), pp. 636–643.
85. Maymo, J. A., Cunningham, R. E., and Smith, J. M., "Thermal Effectiveness Factor," *Ind. Eng. Chem. Fund.*, Vol. 5, No. 2 (May 1966), pp. 280–281.
86. Smith, T. G., Zahradnik, J., and Carberry, J. J., "Nonisothermal Inter-Intraphase Effectiveness Factors for Negative Order Kinetics-CO Oxidation Over Pt," *Chem. Eng. Sci.*, Vol. 30, No. 7 (July 1975), pp. 763–767.
87. Slinko, M. G., Malinovskaja, O. A., and Beskov, V. S., *Chim. Prom.*, Vol. 42, (1967), pp. 641.
88. Hugo, P., and Müller, R., "Experimentelle Untersuchung der Kornürberhit-zung beim N_2O-Zerfall an Einem Modell Katalysatör," *Chem. Eng. Sci.*, Vol. 22, No. 7 (July 1967), pp. 901–910.
89. Irving, J. B., and Butt, J. B., "An Experimental Study of Effect of Intraparticle Temperature Gradients on Catalytic Activity," *Chem. Eng. Sci.*, Vol. 22, No. 12 (Dec. 1967), pp. 1859–1873.
90. Butt, J. B., "Thermal Conductivity of Porous Catalysts," *AIChE J.*, Vol. 11, No. 1 (Jan. 1965), pp. 106–112.
91. Sehr, R. A., "The Thermal Conductivity of Catalyst Particles," *Chem. Eng. Sci.*, Vol. 9, No. 2 (1958), pp. 145–152.
92. Mischke, R. A., and Smith, J. M., "Thermal Conductivity of Alumina Catalyst Pellets," *Ind. Eng. Chem. Fund.*, Vol. 1, No. 4 (Nov. 1962), pp. 288–292.
93. Carberry, J. J., and Kulkarni, A., "On the Relative Importance of External-Internal Temperature Gradients in Heterogeneous Catalysis," *Ind. Eng. Chem. Fund.*, Vol. 14, No. 2 (1975), pp. 129–131.
94. Kehoe, J. P. G., and Butt, J. B., "Interactions of Inter and Intraphase Gradients in a Diffusion Limited Catalytic Reactions," *AIChE J.*, Vol. 18, No. 2 (March 1972), pp. 347–355.
95. Carberry, J. J., and Kulkarni, A., "The Non-Isothermal Catalytic Effectiveness Factor for Monolith Supported Catalysts," *J. Catalysis*, Vol. 31, No. 1 (Oct. 1973), pp. 41–50.
96. Mingle, J. O., and Smith, J. M., "Efectiveness Factors for Porous Catalysts," *AIChE J.*, Vol. 7, No. 2 (June 1961), pp. 243–249.
97. Carberry, J. J., "The Micro-Macro Effectiveness Factor for the Reversible Catalytic Reaction," *AIChE J.*, Vol. 8, No. 4 (Sept. 1962), pp. 557–558.
98. Örs, N., and Doğu, T., "Effectiveness of Bidisperse Catalysts," *AIChE J.*, Vol. 25, No. 4 (July 1979), pp. 723–725.
99. Kulkarni, B. D., Jayaraman, V. K., and Doraiswamy, L. K., "Effectiveness Factors in Bidispersed Catalysts: The General n-th-Order Case," *Chem. Eng. Sci.*, Vol. 36, No. 5 (May 1981), pp. 943–945.
100. Örs, N., "Effectiveness of Bidisperse Porous Catalysts," M. Sc. Thesis, Middle East Techincal University, Ankara-Turkey (May 1977).

101. Jayaramán, V. K., Kulkarni, B. D., and Doraiswamy, L. K., "A Simple Method For the Solution of a Class of Reaction Diffusion Problems," *AIChE J.*, Vol. 29, No. 3 (May 1983), pp. 521–523.
102. Neogi, P., and Ruckenstein, E., "Transport Phenomena in Solids with Bidispersed Pores," *AIChE J.*, Vol. 26 No. 5 (Sept. 1980), pp. 787–794.
103. Furusawa, T., and Smith, J. M., "Effective Diffusivity of Bidisperse Porous Catalysts Under Reaction Conditions," *J. Chem. Eng. Japan*, Vol. 7, No. 6 (Dec. 1974), pp. 470–472.
104. Jury, H. S., "Diffusion in Tabletted Catalysts," *Can. J. Chem. Eng.*, Vol. 55 (Oct. 1977), pp. 538–543.
105. Haynes, H. W., Sarma, P. N., "A Model for the Application of Gas Chromatography to Measurements of Diffusion in Bidisperse Structured Catalysts," *AIChE J.*, Vol. 19, No. 5 (Sept. 1973), pp. 1043–1046.
106. Ma, Y. H., and Lee, T. Y., "Transient Diffusion in Solids with a Bipore Distribution," *AIChE J.*, Vol. 22, No. 1 (Jan. 1976), pp. 147–152.
107. Hashimoto, N., and Smith, J. M., "Macropore Diffusion in Molecular Sieve Pellets by Chromatography," *Ind. Eng. Chem. Fundam.*, Vol. 12, No. 3 (1973), pp. 353–359.
108. Hashimoto, N., and Smith, J. M., "Diffusion in Bidisperse Porous Catalyst Pellets," *Ind.Eng.Chem. Fundam.*. Vol. 13. No. 2 (May 1974), pp. 115–120.
109. Furusawa, T., and Smith, J. M., "Diffusivities from Dynamic Adsorption Data," *AIChE J.*, Vol. 19, No. 2 (March 1973), pp. 401–403.
110. Uyanık, Ö., and Dogu, G., Analysis of Gaseous Diffusion in Bidisperse Porous Catalysts, by the Moment Technique," TBTAK VI. Science Conf. Proc. (1977), pp. 357–365.
111. Doğu, G., and Ercan, C., "Dynamic Analysis of Adsorption on Bidisperse Porous Catalysts," *Can. J. Chem. Eng.*, Vol. 61, No. 5 (Oct. 1983), pp. 660–664.
112. Weisz, P. B., "Zeolites-New Horizons in Catalysis," *Chem. Technol.*, Vol. 13, No. 8 (Aug. 1973), pp. 498–505.
113. Ruckenstein, E., Vaidyanathan, A. S., and Youngquist, G. R., "Sorption by Solids with Bidisperse Pore Structures," *Chem. Eng. Sci.*, Vol. 26, No. 9 (Sept. 1971), pp. 1305–1318.
114. Ruthven, D. M., Vanvlitis, A., and Loughlin, K., "Diffusion of n-Decane in 5A Zeolite Crystals," *AIChE J.*, Vol. 28, No. 5 (Sept. 1982), pp. 840–841.
115. Gorring, R. L., "Diffusion of Normal Paraffins in Zeolite T," *J. Catal.*, Vol. 31, No. 1 (Oct. 1973), pp. 13–26.
116. Kumar, R., Duncan, R. C., and Ruthven, D. M., "A Chromatographic Study of Diffusion of Single Components and Binary Mixtures of Gases in 4A Zeolites," *Can. J. Chem. Eng.*, Vol. 60 (Aug. 1982), pp. 493–499.
117. Weisz, P. B., and Prater, D. B., "Interpretation of Measurements in Experimental Catalysis," *Adv. Catalysis*, Vol. 6, Acad. Press Inc., New York, N.Y., 1954, pp. 143–196.
118. Petersen, E. E., "A General Criterion for Diffusion Influenced Chemical Reactions in Porous Solids," *Chem. Eng. Sci.*, Vol. 20, No. 6 (June 1965), pp. 587–591.
119. Hutchings, J., and Carberry, J. J., "The Influence of Surface Coverage on Catalytic Effectiveness and Selectivity. The Isothermal and Non-isothermal Cases," *AIChE J.*, Vol. 12, No. 1 (Jan. 1966), pp. 20–24.
120. Bischoff, K. B., "An Extension of the General Criterion for Importance of Pore Diffusion with Chemical Reactions," *Chem. Eng. Sci.*, Vol. 22, No. 4 (April 1967), pp. 525–530.
121. Guha, B. K., and Narsimhan, G., "Control Regimes in Experimentation of Heterogeneous Kinetics," *Chem. Eng. Sci.*, Vol. 27, No. 4 (April 1972), pp. 703–708.
122. Hudgins, R. R., "A General Criterion for Absence of Diffusion Control in an Isothermal Catalyst Pellet," *Chem. Eng. Sci.*, Vol. 23, No. 1 (Jan. 1968), pp. 93–94.

123. Doğu, G., and Doğu, T., "A Note on Diffusion Limitations for Multiple Reaction Systems in Porous Catalysts," *AIChE J.*, Vol. 28, No. 6 (Nov. 1982), pp. 1036–1038.
124. Doğu, G., and Doğu, T., "A General Criterion to Test the Importance of Diffusion Limitations in Bidisperse Porous Catalysts," *AIChE J.*, Vol. 26, No. 2 (March 1980), pp. 287–288.
125. Brown, L. F., "Test for Absence of Intraparticle Diffusion Effects in Branched–Pore Systems," *Chem. Eng. Sci.*, Vol. 27, No. 2 (Feb. 1972), pp. 213–219.
126. Anderson, J. B., "A Criterion for Isothermal Behavior of a Catalyst Pellet," *Chem. Eng. Sci.*, Vol. 18, No. 2 (Feb. 1963), pp. 147–148.
127. Doğu, T., and Doğu, G., "Testing the Relative Importance of Temperature and Concentration Gradients in Catalyst Pellets," *AIChE J.*, Vol. 30, No. 6 (Nov. 1984).
128. Doğu, T., "A General Criterion to Test the Relative Importance of Intraparticle and External Heat and Mass Transfer Resistances in Porous Catalysts," *Can. J. Chem. Eng.*, in press.
129. Kim, K. K., and Smith, J. M., "Diffusion in Nickel-Oxide Pellets, Effects of Sintering and Reduction," *AIChE J.*, Vol. 20, No. 4, (July 1974), pp.
130. Moffatt, A. J., "A Dynamic Method for Measuring Tortuosity and Knudsen Diffusion Contributions to Catalyst Diffusivities," *J. Catal*, Vol. 54, No. 2 (Sept. 1978), pp. 107–115.
131. Satterfield, C. N., and Cadle, P. J., "Diffusion in Commercially Manufactured Pelleted Catalysts," *Ind. Eng. Chem. Proc. Des. Dev.*, Vol. 7, No. 2 (April 1968), pp. 256–260.
132. Henry, J. P., Chennakesavan, B., and Smith, J. M., "Diffusion Rates in Porous Catalysts," *AIChE J.*, Vol. 7, No. 1 (March 1961), pp. 10–13.
133. Villet, R. H., and Wilhelm, R. H., "Knudsen Flow-Diffusion in Porous Pellets," *Ind. Eng.Chem.* Vol. 53, No. 10 (Oct. 1961), pp. 837–840.

CHAPTER 9

KINETICS AND MECHANISMS OF CATALYTIC DEACTIVATION

Esther N. Ponzi and Maria G. Gonzalez

CINDECA. Centro de Investigacion y Desarrollo
en Procesos Cataliticos
Univ. Nac. La Plata. CONICET
La Plata, Argentina

CONTENTS

INTRODUCTION

Heterogeneous catalytic reaction are usually accompanied by a lowering of the catalytic activity. Catalyst deactivation mechanisms have been classified as poisoning, fouling, or sintering.

Transport effects influence the performance and durability of catalyst. It is convenient to describe the results by a pellet effectiveness factor for the main reaction. The effectiveness factor represents the ratio of the actual reaction rate on a partially deactivated pellet and the kinetic rate at bulk conditions on a fresh pellet. This relation takes care of the internal and external mass and heat transfer resistances and the activity decline, so that there is no necessity to solve the differential equations for transport of heat, mass, and activity of each particle in the bed.

Numerical and analytical solutions are given for a single catalyst pellet, which are applicable to the design of reactor.

KINETICS AND MECHANISMS OF CATALYTIC DEACTIVATION

Catalyst Fouling—Parallel and Series Mechanisms

Many reactions of the petroleum and petrochemical industry are accompanied by side reactions. The fouling reaction proceeds simultaneously with the main one in a parallel reaction scheme or consecutively to the main one in a series reaction scheme.

The deactivation of porous catalyst can be explained in terms of active-site coverage or pore blockage.

Catalyst Deactivation by Active-Site Coverage

The reaction schemes employed for the analysis of parallel and/or series mechanisms of fouling can be considered as

$$A(g) \rightarrow B(g) + D(g) \quad \text{main reaction} \tag{1}$$

$$A(g) \rightarrow C(s) \quad \text{parallel fouling} \tag{2}$$

$$B(g) \rightarrow C(s) \quad \text{series fouling} \tag{3}$$

The component C is deposited on the active site, making it inactive.

The effective diffusivities, in general, would be a function of the concentration of component C deposited in the catalyst surface. However in this treatment the effective diffusivities will be assumed constant.

Transient Model—Nonisothermal Pellet

For parallel and/or series fouling the continuity equations for mass and energy balances are described by

$$\varepsilon_p \frac{\partial C_A}{\partial t'} = D_A \nabla^2 C_A - R_A(C_A, T, \theta) - R_d(C_A, T, \theta) \tag{4}$$

$$\varepsilon_p \frac{\partial C_B}{\partial t'} = D_B \nabla^2 C_B + R_A(C_A, T, \theta) - R_f(C_B, T, \theta) \tag{5}$$

$$\varrho C_p \frac{\partial T}{\partial t'} = k \nabla^2 T + (-\Delta H) R_A(C_A, T, \theta) \tag{6}$$

where

$$\nabla^2 = \frac{1}{\xi'^{\alpha}} \frac{\partial}{\partial \xi'} \xi'^{\alpha} \frac{\partial}{\partial \xi'} \tag{7}$$

with $\alpha = 0,1$, and 2 for slab, cylindrical, and spherical particles.

As the coke (C) remains on the active site, the effective diffusivity is equal to zero, and the mass balance for C for the combined parallel and series deactivation may be written

$$\frac{\partial C_C}{\partial t'} = R_d(C_A, T, \theta) + R_f(C_B, T, \theta) \tag{8}$$

For parallel fouling $R_f(C_B, T, \theta)$ is equal to zero in Equation 8 and for series fouling $R_d(C_A, T, \theta)$ is equal to zero in Equations 4 and 8.

When the catalyst decay occurs by site coverage the activity θ is related to the sites occupied by coke

$$\theta = (C_T(\xi) - C_{fC})/C_T(\xi) \tag{9}$$

Therefore, the mass balance of product C, can be expressed in the following form

$$-\frac{\partial \theta}{\partial t'} = \frac{y}{C_T(\xi)}[R_d(C_A, T, \theta) + R_f(C_B, T, \theta)] \tag{10}$$

When an energetically uniform catalytic surface is assumed, the reaction rate (R) may be written in terms of a separable activity factor, Butt et al. [1]:

$$R(Ci, T, \theta) = r(Ci, T)f(\theta) \tag{11}$$

Here, r(Ci, T) represents the reaction rate in absence of catalyst deactivation, and $f(\theta)$ a rate factor associated with the activity level of the catalyst. For r(Ci, T) some power law or Langmuir–Hinshelwood model are used and $f(\theta)$ is given by θ^j.

The classical parameters in the dimensionless continuity equations are obtained if the reaction rates are written as follows:

$$R_A(C_A, T, \theta) = k_A g(C_A, T)\theta^m \tag{12}$$

$$R_d(C_A, T, \theta) = k_d g_d(C_A, T)\theta^d \tag{13}$$

$$R_f(C_B, T, \theta) = k_f g_f(C_B, T)\theta^f \tag{14}$$

In dimensionless form the mass and energy balances Equations 4 through 8 for parallel or series schemes are

$$\frac{\partial A}{\partial \tau} = \nabla^2 A - \varphi^2 \exp[\gamma(1 - 1/T^*)]g(A, T^*)\theta^m - \varphi_d^2 \exp\{\gamma_d(1 - 1/T^*)\}g_d(A, T^*)\theta^d \tag{15}$$

$$\frac{\partial B}{\partial \tau} = \frac{D_B}{D_A}\nabla^2 B + \frac{C_{A0}}{C_{B0}}\varphi^2 \exp[\gamma(1 - 1/T^*)]g(A, T^*)\theta^m - \varphi_f^2 \exp[\gamma_f(1 - 1/T^*)]g_f(B, T^*)\theta^f \tag{16}$$

$$\mathcal{L}\frac{\partial T^*}{\partial \tau} = \nabla^2 T^* + \beta\varphi^2 \exp\{\gamma(1 - 1/T^*)\}g(A, T^*)\theta^m \tag{17}$$

$$-\frac{\partial \theta}{\partial \tau} = \varepsilon_\varrho \frac{C_{A0}y}{C_T(\xi)}\varphi_d^2 \exp\{\gamma_d(1 - 1/T^*)\}g_d(A, T^*)\theta^d \quad \text{(parallel fouling)} \tag{18}$$

$$-\frac{\partial \theta}{\partial \tau} = \varepsilon_\varrho \frac{C_{B0}y}{C_T(\xi)}\varphi_f^2 \exp[\gamma_f(1 - 1/T^*)]g_f(B, T^*)\theta^f \quad \text{(series fouling)} \tag{19}$$

The initial conditions are

$$A = B = 0; \quad \theta = T^* = 1; \quad \tau = 0; \quad \xi_s \geq \xi \geq 0 \qquad \text{(20a, b, c, d)}$$

At the center of the catalyst, the concentration gradients are all equal to zero

$$\frac{\partial A}{\partial \xi} = \frac{\partial B}{\partial \xi} = \frac{\partial T^*}{\partial \xi} = 0; \quad \tau \geq 0; \quad \xi = 0 \tag{21a, b, c}$$

The flux of A and B just equal the rate at which they flow into the pellet, at the surface of the catalyst. For heat transfer the boundary condition is similar

$$Sh(1 - A_s) = \partial A/\partial \xi|_{\xi = 1}; \quad \tau \geq 0; \quad \xi = 1$$

$$\frac{D_A}{D_B} Sh(1 - Bs) = \partial B/\partial \xi|_{\xi = 1}$$

$$Bi(1 - T_s^*) = \partial T^*/\partial \xi|_{\xi = 1} \tag{22a, b, c}$$

When there is no external resistance for mass and energy transport, the boundary conditions at the surface are:

$$A = B = T^* = 1; \quad \tau \geq 0; \quad \xi = 1 \tag{23a, b, c}$$

Transient Model—Isothermal Pellet

When the fouling reaction is slower than the main reaction, the pseudo steady-state analysis is an appropriate model. However, for relatively fast fouling reaction, the complete mass and energy continuity Equations 15 through 17 must be used.

If the temperature profile is uniform within the pellet the continuity Equations 15, 16, 18 and 19 may be written as follows:

$$\frac{\partial A}{\partial \tau} = \nabla^2 A - \varphi^2 g(A, 1)\theta^m - \varphi_d^2 g_d(A, 1)\theta^d \tag{24}$$

$$\frac{\partial B}{\partial \tau} = \frac{D_B}{D_A} \nabla^2 B + \frac{C_{A0}}{C_{B0}} \varphi^2 g(A, 1)\theta^m - \varphi_f^2 g_f(B, 1)\theta^f \tag{25}$$

$$-\frac{\partial \theta}{\partial \tau} = \varphi_d^2 \frac{C_{A0} y}{C_T(\xi)} g_d(A, 1)\theta^d \tag{26}$$

$$-\frac{\partial \theta}{\partial \tau} = \varphi_f^2 \frac{C_{B0} y}{C_T(\xi)} g_f(B, 1)\theta^f \tag{27}$$

Pseudo Steady-State Model—Nonisothermal Pellet

If the parallel and/or consecutive reactions (R_d and/or R_f) are slower than the main reaction (R_A), the time necessary to reach steady state with respect to the accumulation of mass is negligible with respect to the time required for the catalyst activity (θ) to change significantly [2]. Hence the first terms in Equations 15 through 17 can be neglected, just as the last one in Equations 15 and 16.

The mass balances for the pseudo steady-state model are given by the following expressions:

$$\nabla^2 A - \varphi^2 \exp[\gamma(1 - 1/T^*)]g(A, T^*)\theta^m = 0 \tag{28}$$

$$-\frac{\partial\theta}{\partial t_d} = \frac{1}{u(\xi)} \exp\left[\gamma_d(1-1/T^*)\right] g_d(A, T^*)\theta^d \tag{29}$$

$$-\frac{\partial\theta}{\partial t_f} = \frac{1}{u(\xi)} \exp\left[\gamma_f(1-1/T^*)\right] g_f(B, T^*)\theta^f \tag{30}$$

where

$$t_d = \tau\varphi_d^2\varepsilon_p y C_{A0}/C_T^* = t'k_{d0} y C_{A0}/C_T^* \quad \text{and}$$

$$t_f = \tau\varphi_f^2\varepsilon_p y C_{B0}/C_T^* = t'k_{f0} y C_{B0}/C_T^* \tag{31}$$

The relationship between A and B, is similar to that of A and T^*, depending on the boundary conditions. With external film resistance

$$B = B_s + \frac{D_A C_{A0}}{D_B C_{B0}}(A_s - A) \tag{32}$$

$$T^* = 1 + \beta\frac{Sh}{Bi}(1 - A_s) + \beta(A_s - A) \tag{33}$$

Without external transport resistance

$$B = 1 - \frac{D_A C_{A0}}{D_B C_{B0}}(A - 1) \tag{34}$$

$$T^* = 1 - \beta(A - 1) \tag{35}$$

The initial conditions Equation 20 for the pseudo steady-state model are

$$A = A^+; \quad B = B^+; \quad T^* = T^+; \quad \theta = 1; \quad t_d \text{ or } t_f = 0; \quad 1 \geq \xi \geq 0 \tag{36a, b, c, d}$$

where A^+, B^+, and T^+ are concentration and temperature profiles within a fresh catalyst ($\theta = 1$) for steady-state model. The boundary conditions are given by Equations 21 through 23 with t_d or t_f as dimensionless time.

Pseudo Steady-State Model—Isothermal Pellet

Since the temperature profile is uniform within the pellet ($T^* = 1$) the continuity Equations 28 through 30 take the form

$$\nabla^2 A - \varphi^2 g(A, 1)\theta^m = 0 \tag{37}$$

$$-\frac{\partial\theta}{\partial t_d} = \frac{1}{u(\xi)} g_d(A, 1)\theta^d \tag{38}$$

$$-\frac{\partial\theta}{\partial t_f} = \frac{1}{u(\xi)} g_f(B, 1)\theta^f \tag{39}$$

The following summarizes numerical solutions to the pseudo steady-state and transient models.

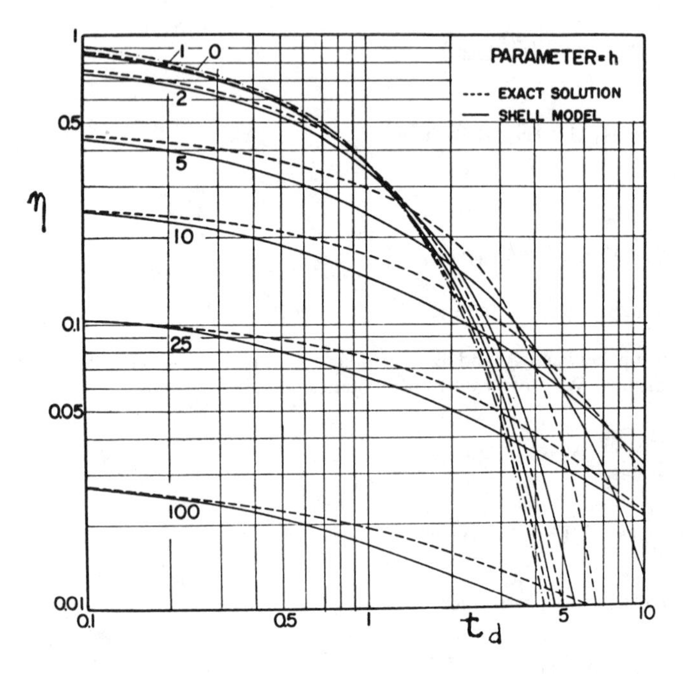

Figure 1. Effectiveness factor for parallel fouling. Isothermal, first-order main reaction, spherical particle. $h = \varphi$; $\theta_A = t_d$;---$\varphi = 0$, equation 116. (From Masamune and Smith [2].)

Pseudo Steady-State Model—Numerical Solutions

Isothermal First-Order Systems (Reaction in a Sphere)

Assumptions:

Main reaction:	$g(A, 1) = A,$	$m = 1$
Parallel fouling:	$g_d(A, 1) = A,$	$d = 1$
Series fouling:	$g_f(B, 1) = B,$	$f = 1$
Spherical particle,	$Sh \rightarrow \infty$,	$u(\xi) = 1$

Mathematical formulation: $\xi = r$

Parallel fouling:
- mass balance (A): Equation 37
- activity balance (θ): Equation 38
- initial and boundary conditions: Equation 21a and 23a
- with $\tau = t_f$ Equation 36a and d

Series fouling:
- mass balance (A): Equation 37
- relationship between A and B: Equation 34
- activity balance (θ): Equation 39
- initial and boundary conditions: Equations 21a and 23a
- with $\tau = t_f$ Equation 36a and d

The equations were solved numerically by Masamune and Smith [2], and the results expressed as effectiveness factor η

$$\eta = \frac{3}{\varphi^2}\frac{\partial A}{\partial r}\bigg|_{r=1} \tag{40}$$

The effectiveness factor for parallel or series mechanism is a function of the following parameters

$$\eta = \eta(t_d, \varphi) \quad \text{parallel fouling} \tag{41}$$

$$\eta = \eta(t_f, \varphi, C_{A0}D_A/C_{B0}D_B) \quad \text{series fouling} \tag{42}$$

Figure 1 shows the effect of the Thiele modulus φ, on the effectiveness factor for parallel fouling. Fresh catalyst with small φ are fouled more rapidly than those with great diffusion resistance.

For series fouling there is interaction between A and B. Figures 2 and 3 give the effectiveness factor as a function of t_f and φ for $D_B/D_A = 1$ and different C_{B0}/C_{A0} values. For series fouling mechanism the extent of deactivation increases with diffusion resistance (φ).

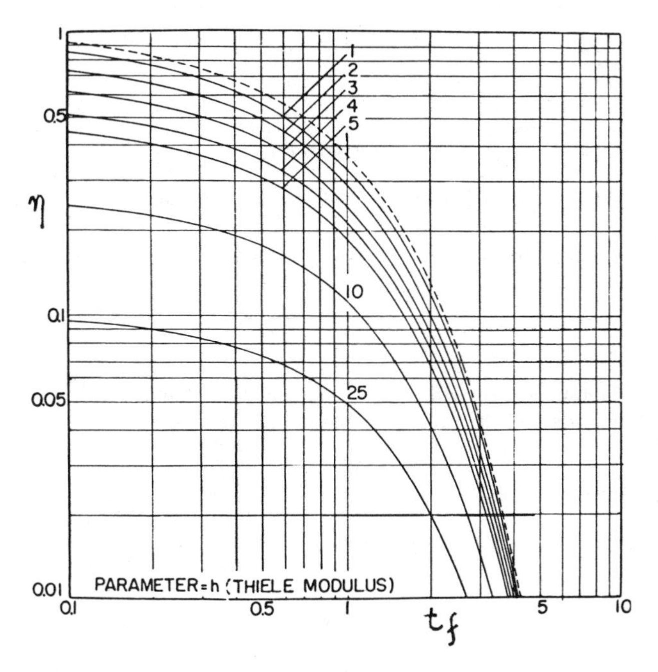

Figure 2. Effectiveness factor for series fouling. Isothermal, first-order main reaction, spherical particle $C_{Bo}/C_{Ao} = 10$, $D_B/D_A = 1$, $h = \varphi$, $\theta_B = t_f$;---$\varphi = 0$, equation 117. (From Masamune and Smith [2].)

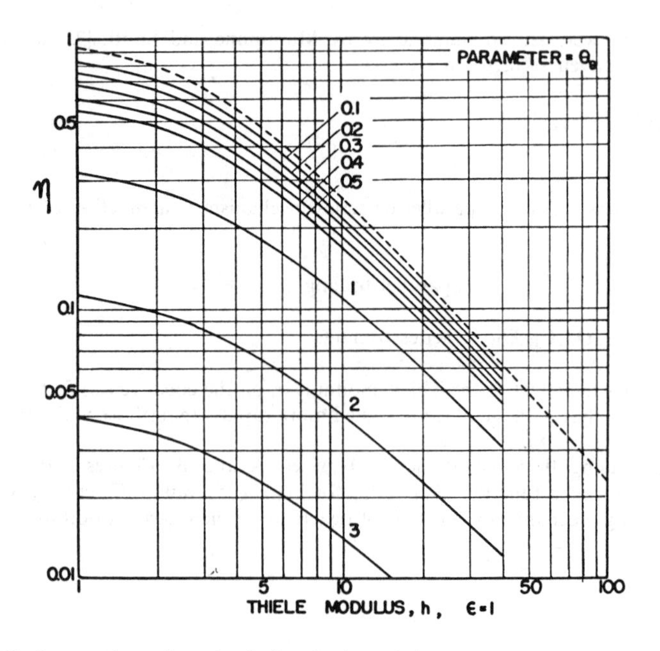

Figure 3. Effectiveness factor for series fouling. Isothermal, first-order main reaction, spherical particle. $C_{Bo}/C_{Ao} = 1$; $D_B/D_A = 1$; $h = \varphi$; $\theta_B = t_f$; $\epsilon = p$; ---$t_f = 0$ (steady state for clean catalyst $\eta = \frac{3}{\varphi^2}(\varphi \coth \varphi - 1)$).(From Masamune and Smith [2].)

Isothermal First-Order Systems (Reaction in a Slab)

The continuity equations and boundary conditions for parallel fouling are the same as with a change only in the operator ∇^2.

For a slab geometry, the effectiveness factor is given by

$$\eta = \frac{1}{\varphi}\frac{\partial A}{\partial x}\bigg|_{x=1} \tag{43}$$

Lamba and Dudukovic [3] reduce the system of partial differential equations by an integral transformation and the results are plotted in Figure 4.

Isothermal Langmuir–Hinshelwood Reactions

Assumptions:

Main reaction: $$g(A, 1) = \frac{K_A'A}{1 + K_A'A + K_B'B}, \; m = 1$$

Parallel fouling: $$g_d(A, 1) = \frac{K_A'A}{1 + K_A'A + K_B'B}, \; d = 1$$

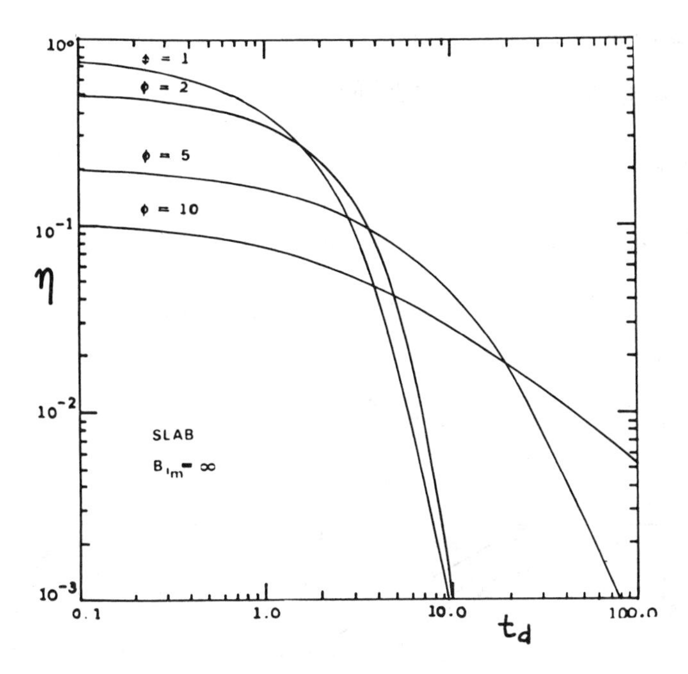

Figure 4. Effectiveness factor for parallel fouling. Isothermal first-order main rection, slab particle. $\theta = t_d$; Bim = Sh. (From Lamba and Dudukovic [3].)

Series fouling: $$g_f(B, 1) = \frac{K_B' B}{1 + K_A' A + K_B' B}, \quad f = 1$$

Spherical geometry, Sh $\rightarrow \infty$, $u(\xi) = 1$

Mathematical formulation: $\xi = r$

Parallel fouling:
- mass balance (A): Equation 37
- activity balance (θ): Equation 38
- initial and boundary conditions: Equations 21a and 23a with $\tau = t_d$ Equation 36a and d

Series fouling:
- mass balance (A): Equation 37
- relationship between A and B: Equation 34
- activity balance (θ): Equation 39
- initial and boundary conditions: Equations 21a and 23a with $\tau = t_f$ Equation 36a and d

The effectiveness factor η is defined as

$$\eta = \frac{3}{\varphi^2} \left.\frac{\partial A}{\partial r}\right|_{r=1} \frac{1 + K_A' + K_B'}{K_A'} \tag{44}$$

and depends on the following parameters

$$\eta = \eta(t_d, \varphi, K'_A, K'_B) \qquad \text{parallel fouling} \tag{45}$$

$$\eta = \eta\left(t_f, \varphi, K'_A, K'_B, \frac{D_A C_{A0}}{D_B C_{B0}}\right) \qquad \text{series fouling} \tag{46}$$

The differential equations were solved by Chu [4] and the results are given in Figure 5.

Isothermal Langmuir–Hinshelwood Model (External Film Resistance)

Assumptions:

$$\text{Main reaction:} \qquad g(A, 1) = \frac{A}{1 + K'_A A + K'_D d_A}, \qquad m = 1$$

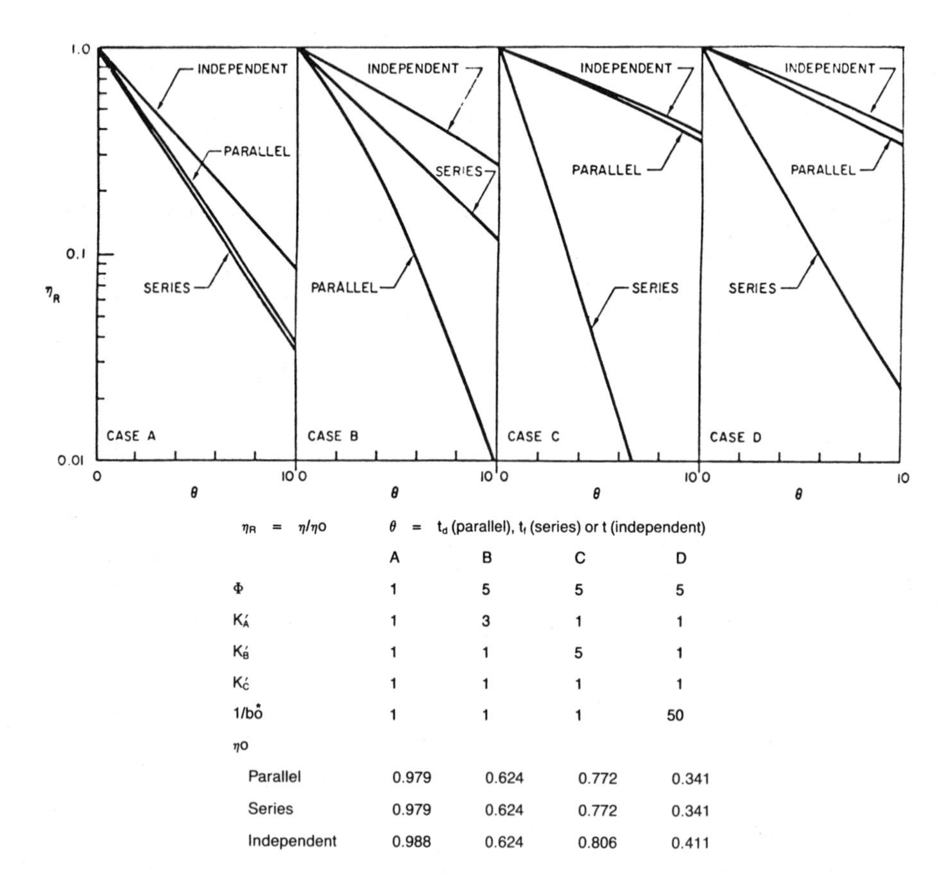

	A	B	C	D
Φ	1	5	5	5
K'_A	1	3	1	1
K'_B	1	1	5	1
K'_C	1	1	1	1
$1/b_0^*$	1	1	1	50
η_0				
Parallel	0.979	0.624	0.772	0.341
Series	0.979	0.624	0.772	0.341
Independent	0.988	0.624	0.806	0.411

Figure 5. Relative effectiveness factor parallel, series, and independent deactivation. Isothermal, Langmuir-Hinshelwood reactions, spherical particle. (From Chu [4].)

Parallel fouling: $g_d(A, 1) = A,$ $d = 1$

Spherical particle, external film resistance, $u(\xi) = 1$

Mathematical formulation: $\xi = r$

Parallel fouling: mass balance (A): Equation 37 $d_A = d_{AS} + (D_A/D_D)(A - A_S)$ activity balance (θ): Equation 38 initial and boundary conditions: Equations 21a and 23a with $\tau = t_d$ Equation 36a and d

The effectiveness factor is given by

$$\eta = \frac{3}{\varphi^2} \left. \frac{\partial A}{\partial r} \right|_{r=1} [1 + K'_A + K'_D d_{A0}] \tag{47}$$

and for parallel fouling is a function of the following parameters

$$\eta = \eta(t_d, \varphi, Sh, K'_A, K'_D, d_{A0}, D_A/D_D) \tag{48}$$

The influence of the external mass transfer resistance (modified Sherwood number, Sh) was examined by Kam et al. [5]. The original paper does not provide values for D_A/D_D and C_{D0}; we suppose that equimolal diffusivities and bulk concentration equal to zero were assumed. The results reported as the ratio of the effectiveness factor at any process time to that at zero time and without fouling are given in Table 1, for $K'_A = K'_D = 10$ and different values of t_d, φ and Sh.

They found no significant difference in the decrease of catalyst activity for the three values of Sherwood number investigated.

However, a decrease in the extent of catalyst fouling with increase in Thiele modulus φ was observed.

Table 1
Parallel Fouling—Ratio of Effectiveness Factor to Any Time to That at Zero Time and Fresh Catalyst*

τ	ϕ	Sh 50	Sh 150	Sh 250
0·5	1	0·9515	0·9516	0·9518
1·0		0·9055	0·9054	0·9072
1·5		0·8616	0·8615	0·8621
2·0		0·8196	0·8167	0·8210
0·5	5	0·9619	0·9615	0·9513
1·0		0·9250	0·9241	0·9239
1·5		0·8893	0·8880	0·8877
2·0		0·8546	0·8530	0·8527
0·5	10	0·9741	0·9729	0·9727
1·0		0·9487	0·9464	0·9459
1·5		0·9238	0·9205	0·9198
2·0		0·8993	0·8951	0·8947

* $\tau = t_d$, $K'_A = K'_C = 10$
From Kam et al[5]

Nonisothermal First-Order Reactions

Assumptions:

Main reaction: $g(A, T^*) = A$, $m = 1$
Parallel fouling: $g_d(A, T^*) = A$, $d = 1$
Series fouling: $g_f(B, T^*) = B$, $f = 1$
Spherical particle, $Sh \to \infty$, $Bi \to \infty$, $u(\xi) = 1$

Mathematical formulation

Parallel fouling:
- mass balance (A): Equation 28
- activity balance (θ): Equation 29
- relationship between T^* and A: Equation 35
- initial and boundary conditions: Equations 21a and 23a
- with $\tau = t_d$ Equation 36a and d

Series fouling:
- mass balance (A): Equation 28
- relationship between B and A: Equation 34
- relationship between T^* and A: Equation 35
- activity balance (θ): Equation 30
- initial and boundary conditions: Equations 21a and 23a
- with $\tau = t_f$ Equation 36a and d

The numerical solution of these equations is complicated because the exponential terms makes convergence difficult. Sagara et al. [6] found the quasilinearization technique to be helpful.

The effectiveness factor Equation 40 depends upon time and the following parameter.

$$\eta = \eta(t_d, \varphi, \gamma, \gamma_d, \beta) \qquad \text{parallel fouling} \tag{49}$$

$$\eta = \eta\left(t_f, \varphi, \gamma, \gamma_f, \beta, \frac{D_A C_{A0}}{D_B C_{B0}}\right) \qquad \text{series fouling} \tag{50}$$

Figure 6 shows the effect of Thiele modulus φ, on the effectiveness factor, for parallel fouling with $\gamma = 20$, $\gamma_d = 30$ and $\beta = 0.1$. The isothermal scheme is represented by dotted lines. At $t_d = 0$ there is no fouling at $\theta = 1$; for nonisothermal pellet at $t_d = 0.01$ this condition is approached. The effectiveness factor here agrees with those presented by Weiz and Hicks [7] for nonisothermal, clean catalyst behavior.

The heat of reaction has a strong influence on the effectiveness factor (Figure 7.)

Effectiveness factors for series fouling are given in Figure 8 for the same values of the parameters, and $D_A C_{A0}/(D_B C_{B0}) = 1$.

Nonisothermal First-Order Reactions (External Film Resistance)

Assumptions:

Main reaction: $g(A, T^*) = A$ $m = 1$
Parallel fouling: $g_d(A, T^*) = A$ $d = 1$
Series fouling: $g_f(b, T^*) = b$ $f = 1$

Spherical particle, external film resistance, exothermic reaction, $C_{B0} = 0$, $u(\xi) = 1$

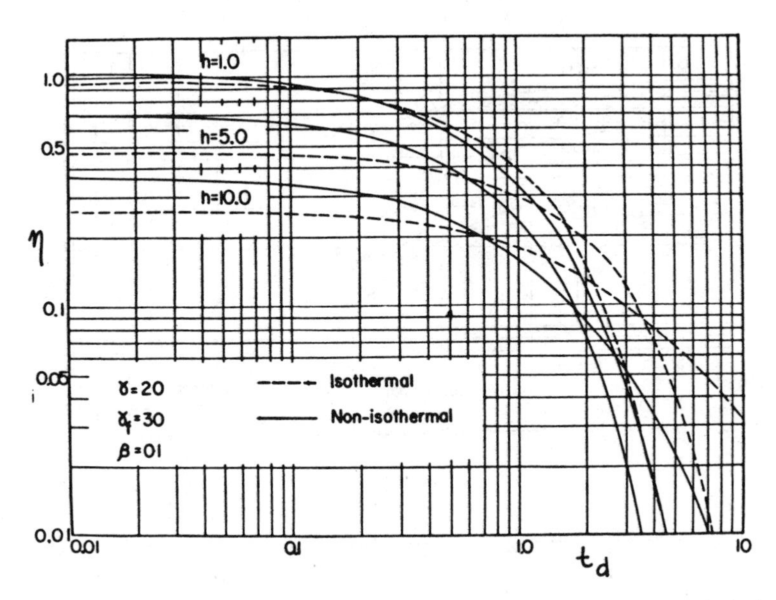

Figure 6. Effectiveness factor parallel fouling. Nonisothermal first-order main reaction, spherical particle. Exothermic main reaction. $\theta_A = t_d$; $\gamma_f = \gamma_d$; $h = \varphi$. (From Sagara et al. [6].)

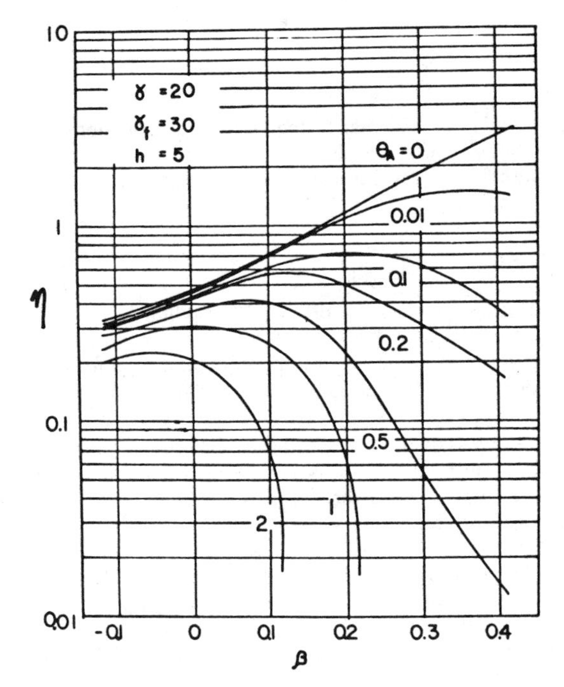

Figure 7. Effectiveness factor parallel fouling. Nonisothermal first-order main reaction, spherical particle. Influence of heat of reaction. $\theta_A = t_d$; $\gamma_f = \gamma_d$; $h = \varphi$. (From Sagara et al. [6].)

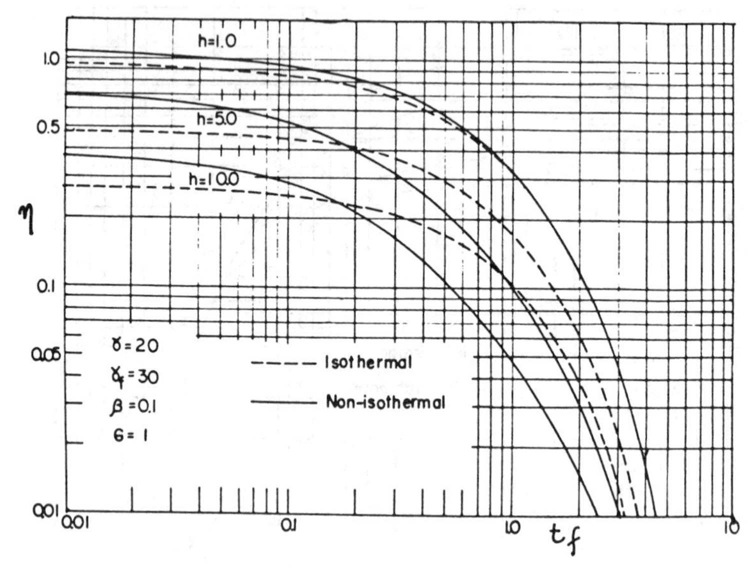

Figure 8. Effectiveness factor for series fouling. Nonisothermal first-order main reaction, spherical particle. ϵ = l/bo*; h = φ. (From Sagara et al. [6].)

The mathematical formulation is similar to the above but Equations 33 and 22a should be used insteadof Equations 35 and 23a.

The dependence of the effectiveness factor Equation 40 is similar, however the parameters Sh and Bi need to be considered.

$$\eta = \eta(t_d, \varphi, \gamma, \gamma_d, \beta, Sh, Bi) \quad \text{parallel fouling} \tag{51}$$

$$\eta = \eta(t_f, \varphi, \gamma, \gamma_f, \beta, D_A C_{A0}/(D_B C_{B0}), Sh, Bi) \quad \text{series fouling} \tag{52}$$

Kam et al. [5] studied the effect of the activation energy parameter for the fouling reaction (γ_d) and the external heat transport resistance (Bi) on catalyst deactivation for parallel fouling.

The results for values of γ_d equal to 5 and 10 are shown in Figures 9 and 10 as ηvsφ for t_d = 0, 0.2 and 0.5, and in Figure 11 as ηvst_d for φ = 1 and 16. The activation energy for fouling reaction γ_d has no effect on the extent of fouling at low values of the Thiele modulus, but it is important at high values of φ.

The influence of the external heat transport resistance is shown in Figure 12 where the effectiveness factor is plotted against φ for Biot number equal to 1, 5 and 15. At low φ, the deactivation is identical for all values of Bi, but in the asymptotic region the effect of fouling is more pronounced at high values of Bi.

For series fouling with C_{B0} = 0, the relationship between A and B Equation 32 is written as

$$b = b_s + (D_A/D_B)(A - A_s) \tag{53}$$

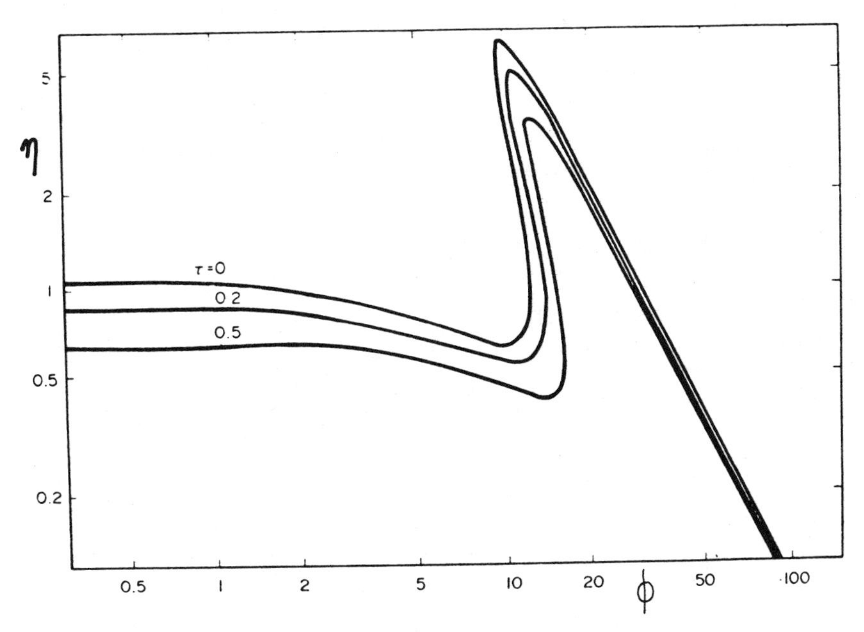

Figure 9. Effectiveness factor for parallel fouling. Nonisothermal first-order main reaction, spherical particle, external film resistance. $\tau = t_d$; Sh = 250; Bi = 5; $\beta = 0.02$; $\gamma = 20$; $\gamma_d = 5$. (From Kam et al. [5].)

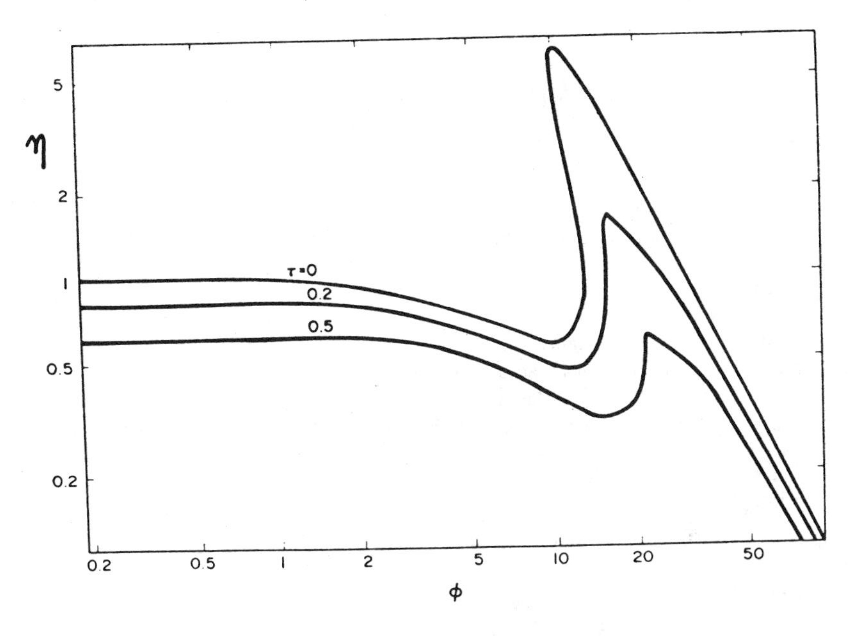

Figure 10. Effectiveness factor for parallel fouling. Nonisothermal first-order main reaction, spherical particle, external film resistance. $\tau = t_d$; Sh = 250; Bi = 5; $\beta = 0.02$; $\gamma = 20$; $\gamma_d = 10$. (From Kam et al. [5].)

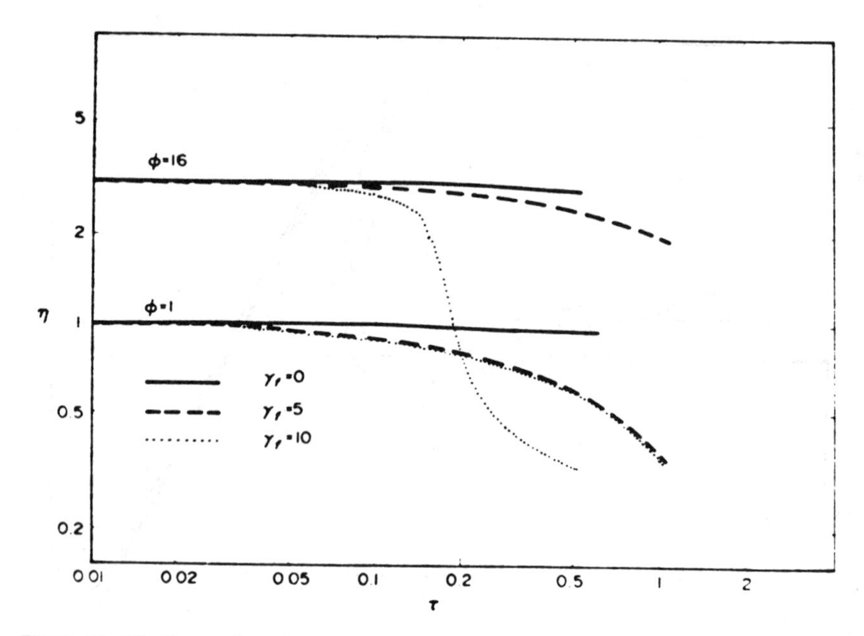

Figure 11. Effectiveness factor for parallel fouling. Nonisothermal first-order main reaction, spherical particle, external film resistance. Effect of dimensionless activation parameter. $\tau = t_d$; Sh = 250; Bi = 5; β = 0.02; γ = 20; $\gamma_f = \gamma_d$. (From Kam et al. [5].)

In this case the balance of active sites Equation 30 takes the following form.

$$-\frac{\partial\theta}{\partial t_{fA}} = \exp\left[\gamma_f(1 - 1/T^*)\right]b\theta \tag{54}$$

For this model the effectiveness factor Equation 40 depends on

$$\eta = \eta(t_{fA}, \varphi, \gamma, \gamma_f, \beta, D_A/D_B) \tag{55}$$

The results are given in Figure 13. For $\varphi = 1$ the variation of η with t_{fA} is negligible, but for $\varphi = 16$ the effectiveness factor decreases as t_{fA} increases and intersects the curve for $\varphi_A = 1$. These results show that for the parameter range investigated, intersection of the curves may occur for series fouling where external resistances are considered. For isothermal and nonisothermal without external resistance systems the intersection of the curves only occur for parallel fouling.

Isothermal First-Order Reactions (Nonuniform Initial Activity)

Assumptions:

Main reaction: $g(A, 1) = u(r)A, \quad m = 1$
Series fouling: $g_f(b^*, 1) = u(r)b^*, \quad f = 1$

$$u(r) \begin{cases} 4r^9 \\ 4r/3 \\ 1 \\ 2.5-2r \end{cases}$$

Spherical particle, $Sh \rightarrow \infty$

Mathematical formulation, $\xi = r$

Series fouling
- mass balance (A): Equation 37 $\nabla^2 b^* + u(r)\theta(\varphi^2 A - \varphi_B^2 b^*) = 0$
- activity balance (θ): Equation 39
- initial and boundary conditions: Equations 21a, b, and 23a
- with $\tau = t_f^*$ Equation 36a, b, d and $b^* = C_{B0}D_B/C_{A0}D_A$; $t_f^* = 0$; $r = 1$

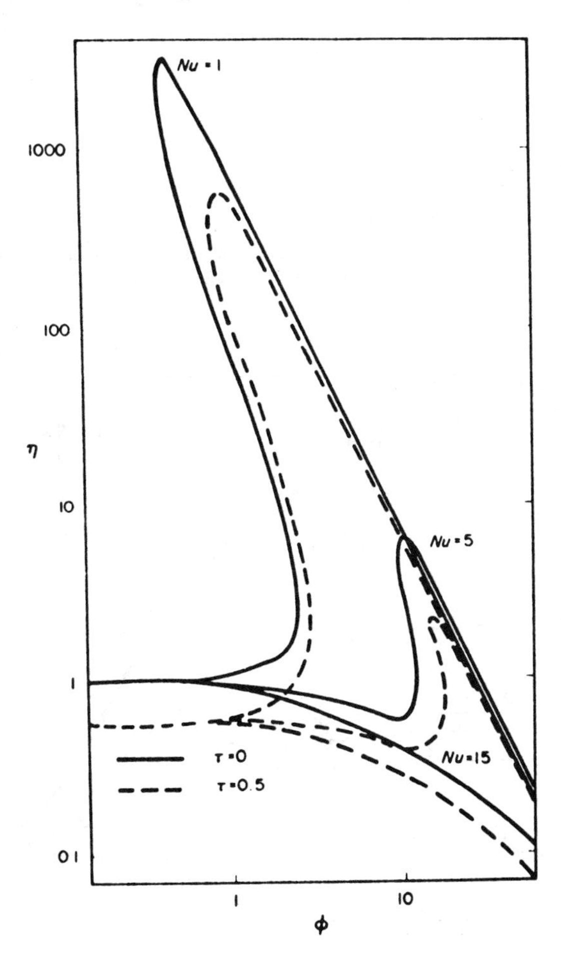

Figure 12. Effectiveness factor for parallel fouling. Nonisothermal first-order main reaction, spherical particle, external film resistance. Effect of Biot number. $\tau = t_d$; Sh = 250; $\beta = 0.02$; $\gamma = 20$; $\gamma_d = 5$ Nu = Bi. (From Kam et al. [5].)

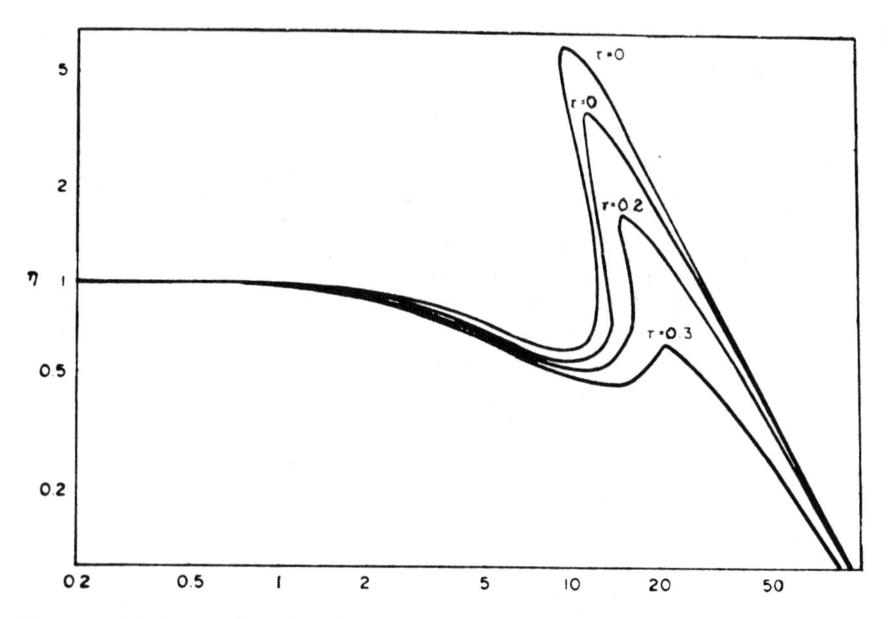

Figure 13. Effectiveness factor for series fouling. Nonisothermal first-order main reaction, spherical particle, external film resistance. $\tau = t_d$; Sh = 250; Bi = 5; $\beta = 0.02$; $\gamma = 20$; $\gamma_f = 5$; $C_{Bo} = 0$; $D_A/D_B = 1$. (From Kam et al. [5].)

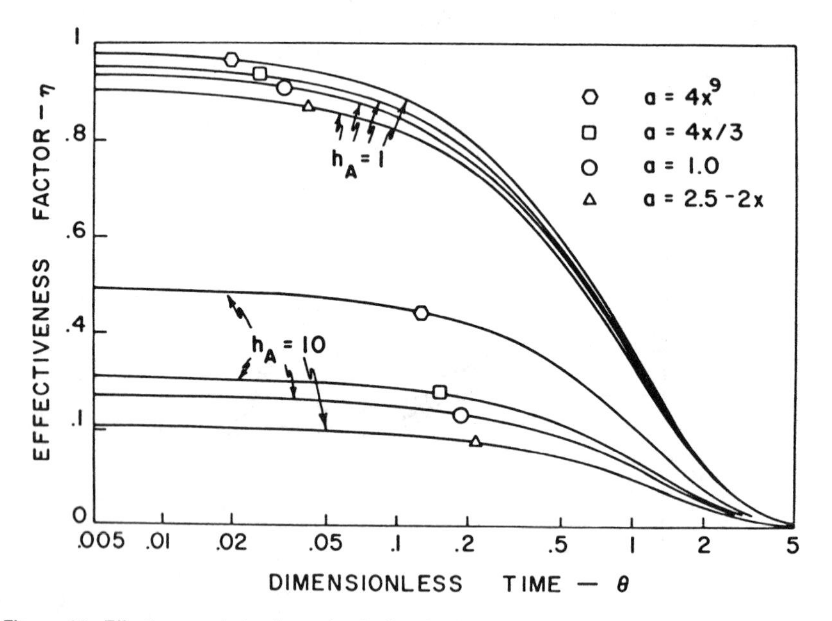

Figure 14. Effectiveness factor for series fouling. Isothermal first-order reactions, initial activity nonuniform, spherical particle. $\theta = t_f^*$; $\phi_B = 0.01\varphi$; $h_A = \varphi$; a = u (r); x = r; $b_0^* = 1$. (From Corbett and Luss [8].)

The effectiveness factor Equation 40 depends on the following parameters

$$\eta = \eta\left(t_f^*, \varphi, \frac{C_{B0}D_B}{C_{A0}D_A}, \varphi_B\right) \tag{56}$$

Corbett and Luss [8] investigated the influence of nonuniform distribution of catalytic activity. Figure 14 gives the deactivation for several initial distribution for two values of Thiele modulus.

Isothermal First-Order Reactions (Nonuniform Activity Distribution. Adsorption on catalyst and support)

This problem was investigated by Shadman–Yazdi and Petersen [9] and Corbett and Luss [8].

They assume that the component B is adsorbed to the same extent on both catalyst and support. For main and fouling reactions the local activity decreases linearly with the local concentration of coke.

For fresh catalyst the intrinsic rate constant for fouling reaction is given as a function of the activity distribution, although the adsorption occurs every where. This mathematical formulation does not agree with physical description.

We rewrite the equations as follows:

Mass balance equations:

$$\nabla^2 A - u(\xi)(1-\omega)\varphi^2 A = 0$$

$$\nabla^2 b^* - u(\xi)(1-\omega)(\varphi^2 A - \varphi_B^2 b^*) = 0$$

$$\frac{\partial \omega}{\partial t_f^*} - u(\xi)(1-\omega)b^* = 0$$

Boundary conditions:

$$A = 1, \quad b^* = C_{B0}D_B/C_{A0}D_A \qquad \xi = 1$$

$$\partial A/\partial \xi = \partial b^*/\partial \xi = 0 \qquad \xi = 0$$

$$\omega = 0 \qquad t_f^* = 0$$

Typical values of η for spherical pellet are given in Figure 15 [8]. For slab geometry and $u(\xi) = x^z$, the results are shown in Figures 16 through 18, Shadman–Yazdi and Petersen [9].

Transient Model—Numerical Solutions:

Isothermal First-Order Reactions

Assumptions:

Main reaction:	$g(A, 1) = A$	$m = 1$
Parallel fouling:	$g_d(A, 1) = A$	$d = 1$
Series fouling:	$g_f(B, 1) = B$	$f = 1$
Spherical particle,	$C_{B0} = 0$	$u(\xi) = 1$

Mathematical formulation: $\xi = r$

Parallel fouling $\begin{cases} \text{mass balance (A): Equation 24} \\ \text{activity balance } (\theta)\text{: Equation 26} \\ \text{initial and boundary conditions: Equations 20a, c, 21a, and 22a} \end{cases}$

When $C_{B0} = 0$, C_{A0} is taken as reference for dimensionless concentration of product B. The mass balance for product B and the series fouling Equations 25 and 27 are modified as:

$$\frac{\partial b}{\partial \tau} = \frac{D_B}{D_A}\nabla^2 b + \varphi^2 g(A, 1)\theta^m - \varphi_f^2 g_f(b, 1)\theta^f \tag{57}$$

$$-\frac{\partial \theta}{\partial \tau} = \varphi_f^2 \varepsilon_p \frac{C_{A0}}{C_T(r)} g_f(b, 1)\theta^f \tag{58}$$

Series fouling $\begin{cases} \text{mass balance (A): Equation 24 with } \varphi_d = 0 \\ \text{mass balance (B): Equation 57} \\ \text{activity balance } (\theta)\text{: Equation 58} \\ \text{initial and boundary conditions: Equations 20a, b, c, 21a, b, and 22a, b} \end{cases}$

The equations were solved numerically by Murakami et al. [10], and the results are given as effectiveness factor for the product B.

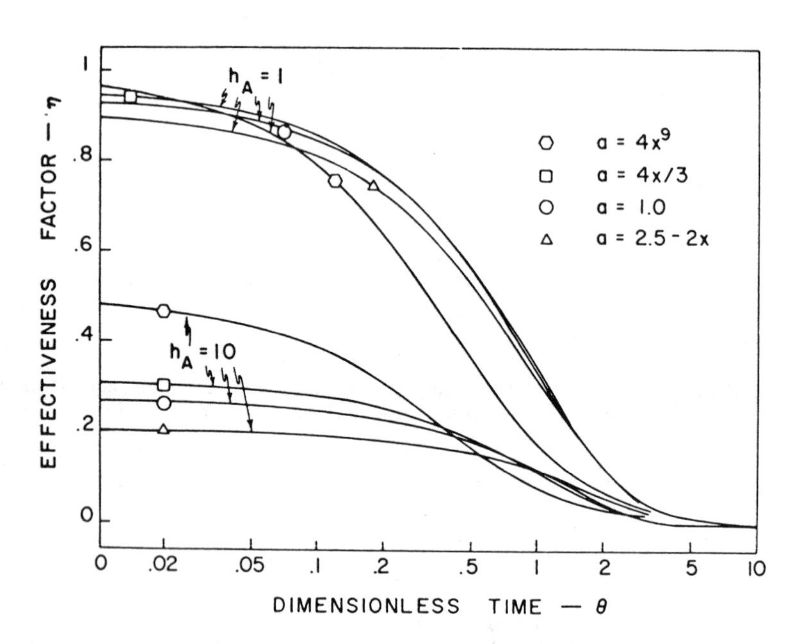

Figure 15. Effectiveness factor for series fouling. Isothermal first-order reactions, initial activity nonuniform, adsorption on catalyst and support. $\theta = t_f^*$; $\varphi_B = 0.01\varphi$; $h_A = \varphi$; $a = u(r)$; $x = r$; $b_0^* = 1$. (From Corbett and Luss [8].)

$$\eta_B = \frac{-3p}{\varphi^2(1+s)} \left.\frac{\partial b}{\partial r}\right|_{r=1} \qquad \text{parallel fouling} \tag{59}$$

$$\eta_B = \frac{-3p}{\varphi^2} \left.\frac{\partial b}{\partial r}\right|_{r=1} \qquad \text{series fouling} \tag{60}$$

and as ratio of effectiveness factor at any time η_B, to the initial one η_{Bi}

$$\varepsilon = \eta_B/\eta_{Bi}$$

Where η_{Bi} takes the following forms for parallel fouling

$$\eta_{Bi} = \frac{3Sh^2}{2\varphi^2 p(\varphi \coth\varphi + Sh - 1)} \left\{ \frac{(Sh/p)\coth\varphi + \varphi}{\varphi\coth\varphi + (Sh/p) - 1} - \coth\varphi \right\} \qquad \text{for } s = p \tag{61}$$

$$\eta_{Bi} = \frac{3Sh^2}{p\varphi^2[(s/p)-1](\varphi\coth\varphi + Sh - 1)} \left\{ 1 - \frac{\varphi\coth\varphi + (Sh/p) - 1}{(s/p)^{0.5}\varphi\coth[\varphi(s/p)^{0.5}] + (Sh/p) - 1} \right\} \qquad \text{for } s = p \tag{62}$$

The effectiveness factor is function of the following parameters

$$\eta_B = \eta_B(\tau, \varphi, \varphi_d, C_{A0}y/C_T^*, Sh) \qquad \text{parallel fouling} \tag{63}$$

$$\eta_B = \eta_B(\tau, \varphi, \varphi_f, D_A/D_B, C_{A0}y/C_T^*, Sh) \qquad \text{series fouling} \tag{64}$$

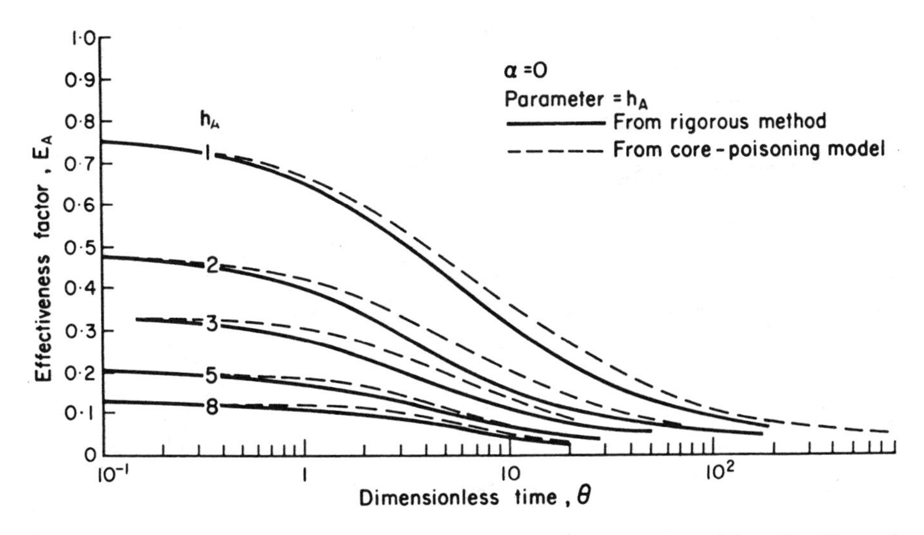

Figure 16. Effectiveness factor for series fouling. Isothermal first-order reactions, slab particle. $E_A = \eta$; $\theta = t_f^*$; $\alpha = Z$; $h_A = \varphi$. (From Shadman-Yazdi and Petersen. [9].)

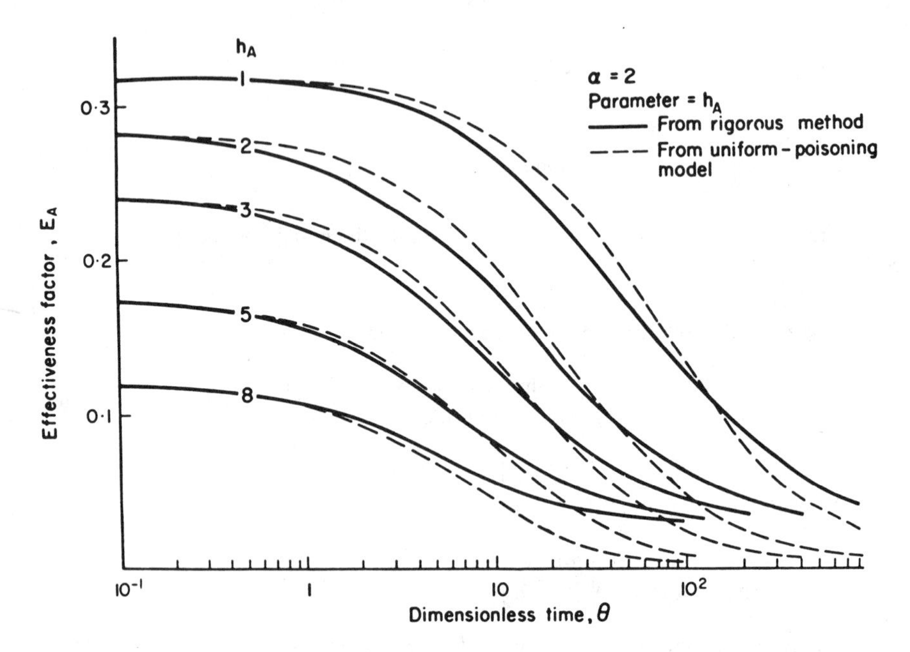

Figure 17. Effectiveness factor for series fouling. Isothermal first-order reactions, initial activity nonuniform adsorption on catalyst and support, slab particle. $E_A = \eta$; $\theta = t_f^*$; $\alpha = Z$; $h_A = \varphi$. (From Shadman-Yazdi and Petersen. [9].)

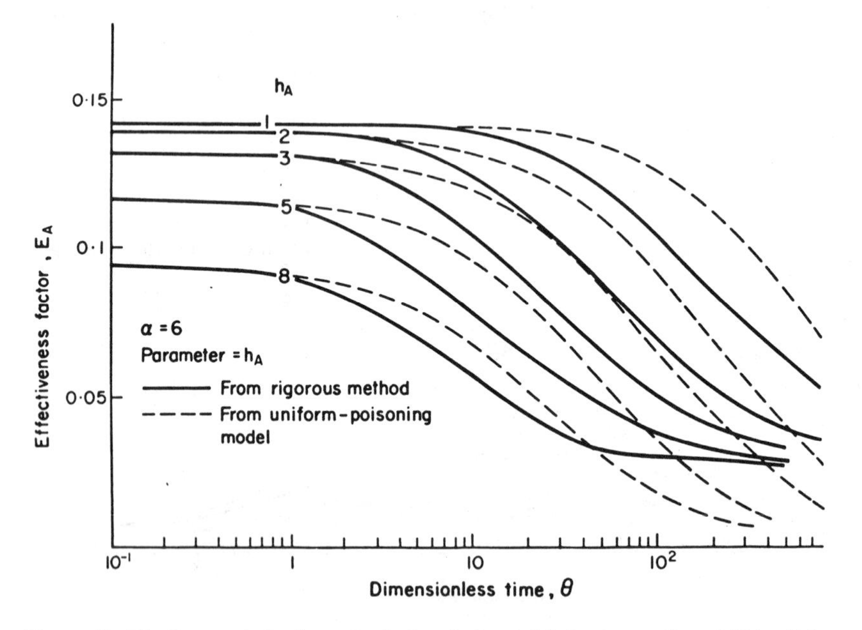

Figure 18. Effectiveness factor for series fouling. Isothermal first-order reactions, initial activity nonuniform, adsorption on catalyst. $E_A = \eta$; $\theta = t_f^*$; $\alpha = Z$; $h_A = \varphi$. (From Shadman-Yazdi and Petersen. [9].)

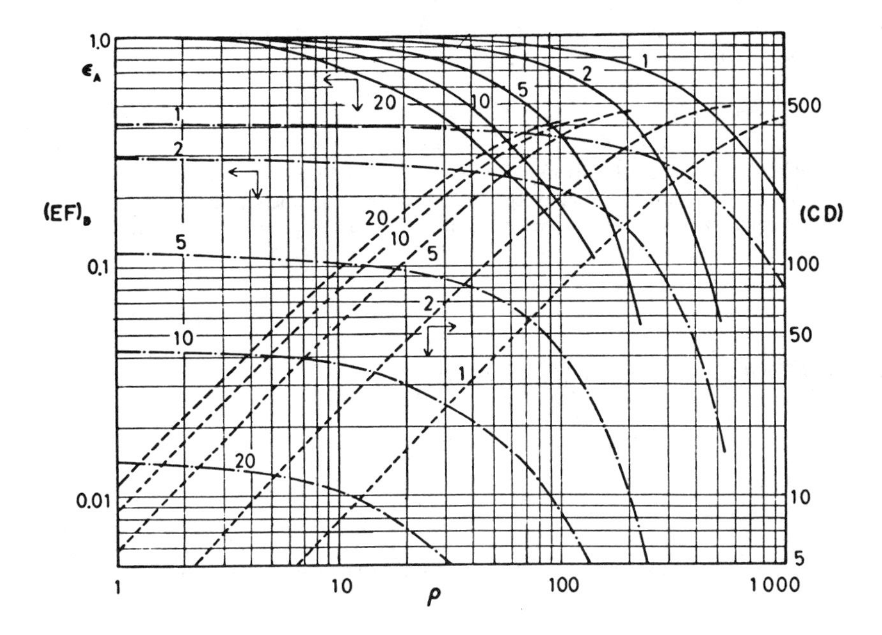

Figure 19. Effectiveness factor for fast parallel fouling. Isothermal first-order reactions, spherical particle, external film resistance. $(EF)_B = \eta_B$; $\epsilon_A = \eta_B/\eta_{Bi}$; $\rho = \tau$; $D_B/D_A = 1$; $\varphi_d = \varphi$; Sh = 10; $\epsilon_p C_{Ao} y/C_T^* = 1/500$. (From Murakami et al. [10].)

The results showing the effect of Thiele modulus are given in Figures 19 and 20 for parallel and series fouling respectively. Fouling proceeds faster in the parallel reaction scheme than in the series reaction scheme under the same conditions. Figure 21 shows the effect of selectivity for series fouling. Figure 22 shows the effect of the Sherwood number in a parallel scheme. A large Sh value causes faster fouling, while the effect of Sh on the fouling rate in a series reaction scheme is not so large. The effect of D_B/D_A is small in both scheme of reactions, Murakami et al. [10].

Nonisothermal First-Order Reactions.

Assumptions:

Main reaction: $g(A, T^*) = A, \quad m = 1$
Parallel fouling: $g_d(A, T^*) = A, \quad d = 1$
Spherical particle, external film resistance, $u(r) = 1$

Mathematical formulation

Parallel fouling:
- mass balance (A): Equation 15
- activity balance (θ): Equation 18
- heat balance (T^*): Equation 17
- initial and boundary conditions: Equations 20a, c, d, 21a, c, and 22a, c

The differential equations were solved by Kam et al. [11] and the results given as effectiveness factor

$$\eta = 3\int_0^1 r^2 g(A, T^*)\theta \, dr \tag{65}$$

The effect of reciprocal Lewis number on parallel fouling is shown in Figure 23(a), (b), and (c) for $\theta = 1$, 5 and 9 respectively. The system reaches the steady state values sooner at lower values of the reciprocal Lewis number ($\mathscr{L}$). The parameter σ is given by

$$\sigma = \frac{C_T^*}{\varepsilon_p C_{A0} \varphi_d y} \tag{66}$$

Nonisothermal Langmuir–Hinshelwood Model

Assumptions:

Main reaction: $m = 1$

$$g(A, T^*) = \frac{A \exp[-\delta_A(1 - 1/T^*)]}{1 + K_A' \exp[-\delta_A(1 - 1/T^*)]A + K_D' \exp[-\delta_A(1 - 1/T^*)]d_A}$$

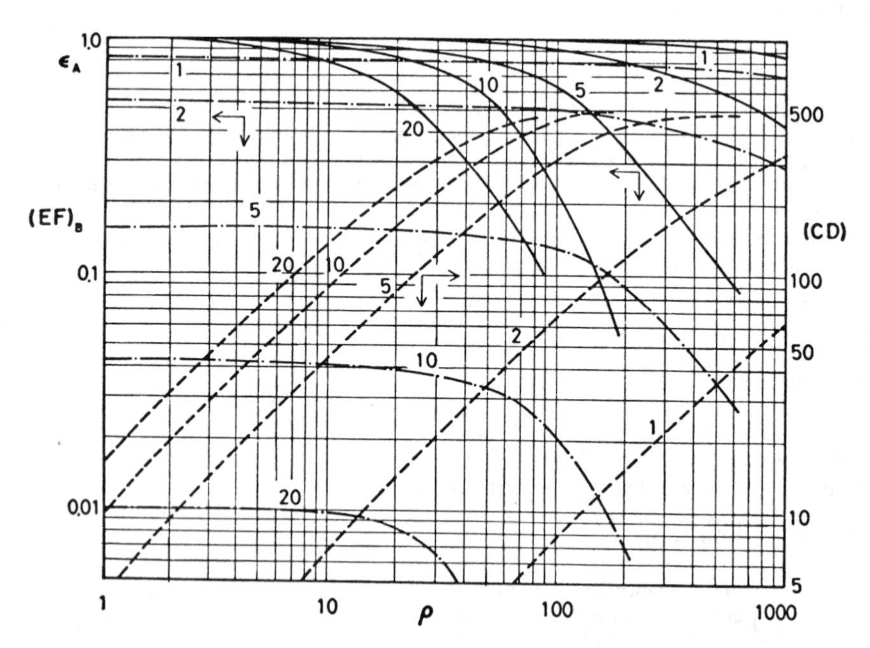

Figure 20. Effectiveness factor for fast series fouling. Isothermal first-order reactions, spherical particle, external film resistance. $(EF)_B = \eta_B$; $\epsilon_A = \eta_B/\eta_{Bi}$; $\rho = \tau$; $D_B/D_A = 1$; $\varphi_d = \varphi$; $Sh = 10$; $\epsilon_p C_{Ao} y/C_T^* = 1/500$. (From Murakami et al. [10].)

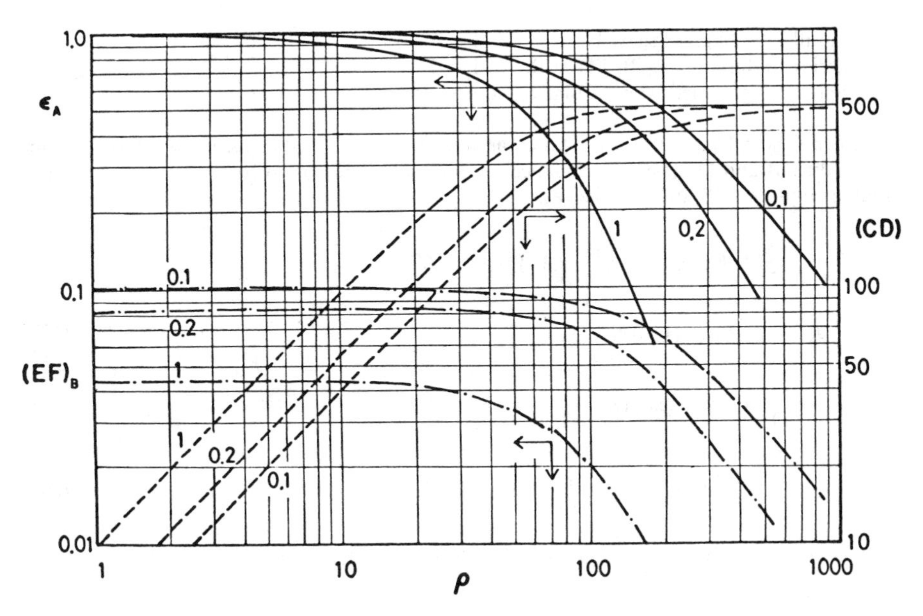

Figure 21. Effectiveness factor for fast series fouling. Isothermal first-order reaction, spherical particle, external film resistance. Influence of selectivity. $(EF)_B = \eta_B$; $\epsilon_A = \eta_B/\eta_{Bi}$; $\rho = \tau$; $D_B/D_A = 1$; $\varphi = 10$; $\varphi_f = S^{0.5}\varphi$; $\epsilon_p C_{Ao} y/C_T^* = 1/500$; Sh = 10. (From Murakami et al. [10].)

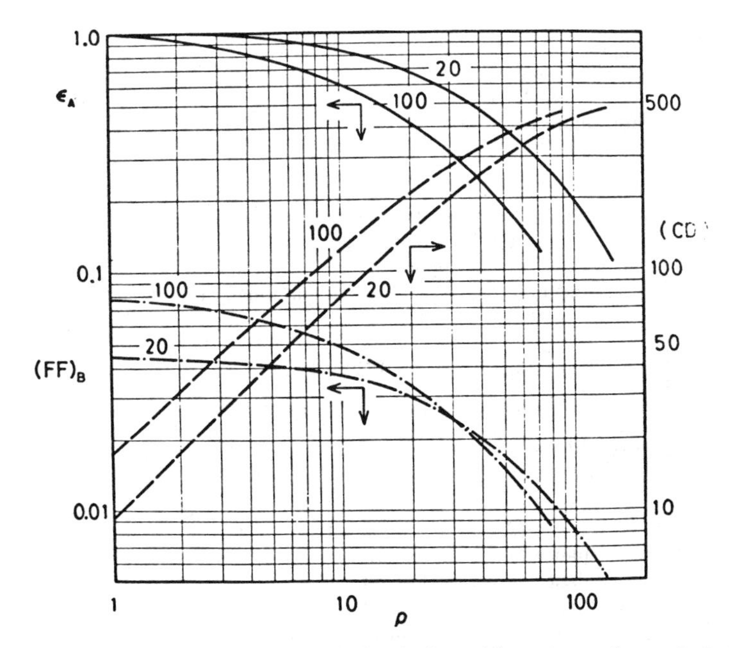

Figure 22. Effectiveness factor for fast parallel fouling. Isothermal first-order reactions, spherical particle, external film resistance. Influence of Sherwood number. $(EF)_B = \eta_B$; $\epsilon_A = \eta_B/\eta_{Bi}$; $\rho = \tau$; $D_B/D_A = 1$; $\varphi_d = \varphi$; $\epsilon_p C_{Ao} y/C_T^* = 1/500$; Sh = 10. (From Murakami et al. [10].)

Parallel fouling: $g_d(A, T^*) = A$, $d = 1$
Series fouling: $g_f(B, T^*) = B$, $f = 1$
Spherical particle, external film resistance, $u(r) = 1$

When $C_{B0} = C_{D0} = 0$, C_{A0} is taken as reference to make dimensionless concentration of product B and D. The mass balance for products B Equation 16 and D and the series fouling Equation 18 are modified as:

$$\frac{\partial b}{\partial \tau} = \frac{D_B}{D_A}\nabla^2 b + \varphi^2 \exp[\gamma(1-1/T^*)]g(A, T^*)\theta^m - \varphi_f \exp[\gamma_f(1-1/T^*)]g_f(b, T^*)\theta^f \tag{67}$$

$$\frac{\partial d}{\partial \tau} = \frac{D_D}{D_A}\nabla^2 d - \varphi^2 \exp[\gamma(1-1/T^*)]g(A, T^*)\theta^m \tag{68}$$

$$-\frac{\partial \theta}{\partial \tau} = \varepsilon_p \frac{C_{A0}y}{C_T(r)}\varphi_f^2 \exp[\gamma_f(1-1/T^*)]g_f(b, T^*)\theta^f \tag{69}$$

The initial and boundary conditions Equations 20b, 21b, 22b are modified as

$$\begin{aligned} b &= 0 && \tau = 0 \quad 1 \geqq r \geqq 0 \\ (\partial b/\partial r) &= 0 && \tau = 0 \quad r = 0 \\ -\frac{D_A}{D_B}\mathrm{Sh}\, b_s &= \left.\frac{\partial b}{\partial r}\right|_{r=1} && r = 1 \end{aligned} \tag{70a, b, c}$$

Mathematical formulation: $\xi = r$

Parallel fouling:
- mass balance (A): Equation 15
- mass balance (D): Equation 68
- activity balance (θ): Equation 18
- heat balance (T^*): Equation 17
- initial and boundary conditions: Equations 20a, c, d, 21a, c, 22a, c and $d = 0$; $\tau = 0$; $1 \geqq r \geqq 0$;

  $\partial d/\partial r|_{r=1} = 0$; $\tau = 0$; $r = 0$ (71a, b, c)

  $$-\frac{D_A}{D_D}\mathrm{Sh}\, d_s = \left.\frac{\partial d}{\partial r}\right|_{r=1} \qquad r = 1$$

Series fouling:
- mass balance (A): Equation 15 with $\varphi_d = 0$
- mass balance (B): Equation 67
- mass balance (D): Equation 68
- activity balance (θ): Equation 69
- heat balance (T^*): Equation 17
- initial and boundary conditions: Equations 20a, c, d, 21a, c, 22a, c, 70a, b, c, 71a, b, c

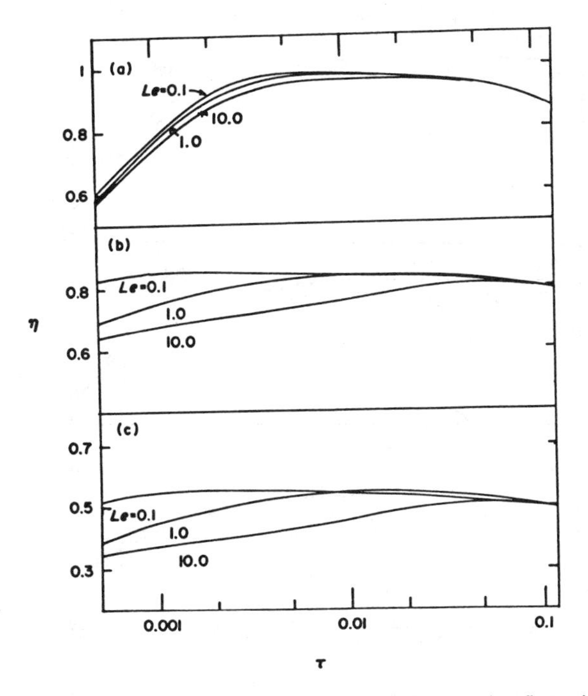

Figure 23. Effectiveness factor for parallel fouling. Nonisothermal transient first-order model. Influence of reciprocal Lewis number. Lc = $\mathcal{L}$ = 1; τ = t_d; σ = 100; Sh = 250; Bi = 25; $\gamma = \gamma_d$ = 20; β = 0.02; (a) φ = 1; (b) φ = 5; (c) φ = 9. (From Kam et al. [11].)

Table 2
Time at Which the Simplified (Psuedo Steady-State) Model Can Be Applied*

	ϕ		
ϕ	**1**	**10**	**100**
1	8×10^{-1}	8×10^{-2}	8×10^{-3}
5	5×10^{-1}	5×10^{-2}	5×10^{-3}
9	7×10^{-1}	7×10^{-2}	7×10^{-3}

* *From Kam et al.* [11]

This model was solved numerically by Kam et al. [11]. In the original paper the mass balance for the component D is omitted in the mathematical formulation, as well as the Thiele modulus for the deactivation reactions (φ_d and φ_f) in the mass balance of reactive A and product B. Therefore the results for parallel and series fouling hold for $\varphi_d = 1$ and $\varphi_f = 1$. Moreover they do not provide values for D_D/D_A and C_{D0}, we suppose that equimolal diffusivities and bulk concentration equal to zero were assumed.

The model proposed can be simplified neglecting the mass balance on the gaseous product B and even withdrawing the consumption for A. The time at which this simplified model

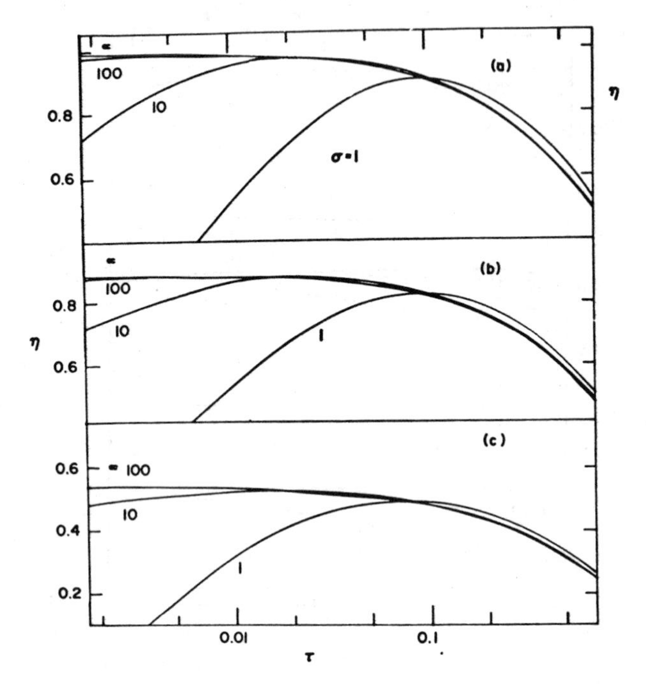

Figure 24. Effectiveness factor for parallel fouling. Nonisothermal transient Langmuir-Hinshelwood model. Influence of parameter σ. $\mathcal{L} = 1$; $\tau = t_d$; Bi = 25; Sh = 250; $\beta = 0.02$; $\gamma = \gamma_d = 20$; $K_A' = K_C' = 10$; $S_A = S_C = 5$; $\varphi_d = 1$; a) $\varphi = 1$; b) $\varphi = 5$; c) $\varphi = 9$. (From Kam et al. [11].)

can be applied is shown in Table 2. The parameter σ is given by Equation 66 for parallel fouling and

$$\sigma = \frac{C_T^*}{\varepsilon_p C_{A0} \varphi y} \quad \text{series fouling}$$

Figures 24 (a), (b), and (c) give the influence of the parameter σ, for parallel fouling, and Figures 25 (a), (b), and (c) for series fouling.

Three values of σ (0.1, 1, and 10) were examined. The results obtained are almost identical and they can be represented by Figures 24 and 25 for parallel and series fouling respectively. There is no noticeable effect of the reciprocal Lewis number on transient fouling in the case of Langmuir–Hinshelwood model.

IMPURITY POISONING

Impurity poisoning is observed when an impurity in the reaction mixture adsorbs on active centers and thereby poisons the catalyst. This effect may be related to the competitive reversible adsorption of the poison precursor, to the irreversible adsorption or reaction of the impurity on or with the surface, or to the physical blockage of the pore structure.

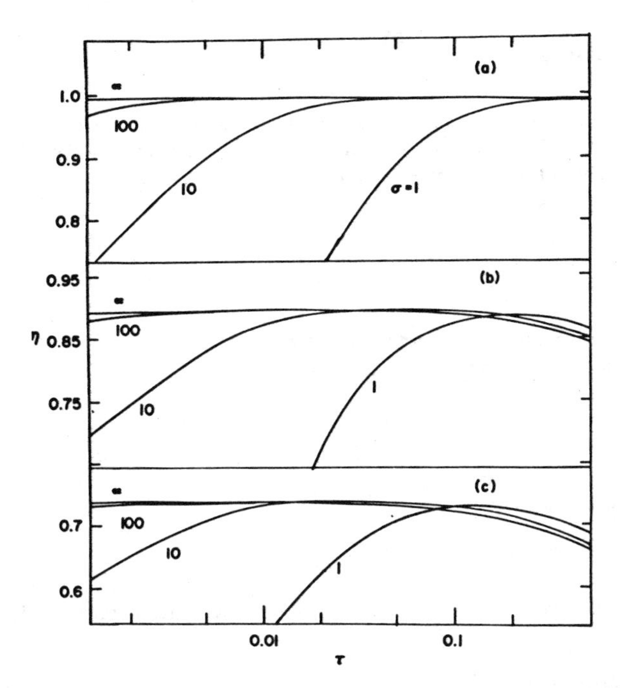

Figure 25. Effectiveness factor for series fouling. Nonisothermal transient Langmuir-Hinshelwood model. Influence of parameter σ. $\mathcal{L}$ = 1; τ = t_d; Bi = 25; Sh = 250; β = 0.02; γ = γf = 20; K_A' = K_C' = 10; S_A = S_C = 5; φ_d = 1; a) φ = 1; b) φ = 5; c) φ = 9. (From Kam et al. [11].)

The General Equations of Irreversible Poisoning

The system is represented by the following stoichiometric equations:

$$\text{Main reaction A} \xrightarrow{k_A} \text{B}$$

$$\text{Poisoning reaction P} \xrightarrow{k_p} \text{W}$$

The problem of predicting the catalyst activity within a pellet where both reaction and internal diffusion are important can be defined by the continuity mass equations for the reactant A, the poison precursor P, and energy balances.

$$D_A \nabla^2 C_A - R_A(C_A, T, \theta) = \varepsilon_p \frac{\partial C_A}{\partial t'} \tag{72}$$

$$D_p \nabla^2 C_p - R_P(C_P, T, \theta) = \varepsilon_p \frac{\partial C_P}{\partial t'} \tag{73}$$

$$k\nabla^2 T + (-\Delta H)\, R_A(C_A, T, \theta) = \varrho C_p \frac{\partial T}{\partial t'} \tag{74}$$

where the diffusivities, D_A and D_p, are assumed to be independent of concentration and extent of poisoning. This assumption excludes pore plugging from our considerations.

The concentration of poison deposited on the surface, C_W, can be related to the unpoisoned active surface by

$$\theta = 1 - \frac{C_W a_W}{a_0} = 1 - \alpha_1 C_W$$

where a_W is the active surface rendered inactive by 1 mole of the poison W.

Hence, the deactivating equation is

$$\frac{\partial C_W}{\partial t'} = R'_P(C_P, T, \theta) \tag{75}$$

Assuming ideal surface [1]:

$$R_i(C_i, T, \theta) = r_i(C_i, T) f(\theta) \tag{76}$$

where $r(C_i, T)$ is the reaction rate at $t = 0$ and $f(\theta)$ is the activity function. The reaction rates are written as

$$r_i(C_i, T) = k_i g_i(C_i, T) \tag{77}$$

Equations 72–75 can be expressed in dimensionless form as

$$\frac{\partial^2 A}{\partial \xi^2} + \frac{\alpha}{\xi}\frac{\partial A}{\partial \xi} - \varphi^2 g_A(A, T^*) \exp\left[\gamma\left(1 - \frac{1}{T^*}\right)\right]\theta^m = \frac{\partial A}{\partial \tau} \tag{78}$$

$$\frac{D_P}{D_A}\left[\frac{\partial^2 P}{\partial \xi^2} + \frac{\alpha}{\xi}\frac{\partial P}{\partial \xi} - \varphi_P^2 g_P(P, T^*) \exp\left[\gamma_P\left(1 - \frac{1}{T^*}\right)\right]\theta^q\right] = \frac{\partial P}{\partial \tau} \tag{79}$$

$$\frac{\partial^2 T^*}{\partial \xi^2} + \frac{\alpha}{\xi}\frac{\partial T^*}{\partial \xi} - \varphi^2 \beta g_A(A, T^*) \exp\left[\gamma\left(1 - \frac{1}{T^*}\right)\right]\theta^n = \mathscr{L}\frac{\partial T^*}{\partial \tau} \tag{80}$$

$$-\frac{\partial \theta}{\partial \tau} = \varepsilon_P \varphi_P^2 g_P(P, T^*) \exp \gamma\left\{\left(1 - \frac{1}{T^*}\right)\right\}\theta^q \tag{81}$$

where α characterizes the geometry of the pellets.

By solving these equations by numerical procedure with the adequate initial and boundary conditions the system behavior can be obtained in terms of a time dependent effectiveness factor, $\eta(\tau)$.

In impurity poisoning, the effectiveness factor is function of two Thiele modulus, φ and φ_P, and of the time.

Since the temperature profile is uniform within the pellet ($T^* = 1$) the Equations 78, 79, and 81 take the form

$$\frac{\partial^2 A}{\partial \xi^2} + \frac{\alpha}{\xi}\frac{\partial A}{\partial \xi} - \varphi^2 g_A(A, 1)\,\theta^m = \frac{\partial A}{\partial \tau} \tag{82}$$

$$\frac{D_A}{D_P}\left\{\frac{\partial^2 P}{\partial \xi^2} + \frac{\alpha}{\xi}\frac{\partial P}{\partial \xi} - \varphi_P^2 g_P(P, 1)\,\theta^q\right\} = \frac{\partial P}{\partial \tau} \quad (83)$$

$$-\frac{\partial \theta}{\partial \tau} = \varepsilon_P \varphi_P^2 g_P(P, 1)\,\theta^q \quad (84)$$

Isothermal, Pseudo Steady-State Analysis—Numerical Solutions

First-order system. The simplest case is that in which reactions are first order in reactant and first order in active sites.

Masamune and Smith [2] analyzed the impurity poisoning in a isothermal spherical pellet with uniform initial activity.

The rate of poisoning is taken to be slow compared with the main reaction. Hence the right term in Equations 82 and 83 can be neglected, considering a quasi steady state. The system is expressed by

$$\frac{\partial^2 A}{\partial r^2} + \frac{2}{r}\frac{\partial A}{\partial r} - \varphi^2 A^n \theta^m = 0 \quad (85)$$

$$\frac{\partial^2 P}{\partial r^2} + \frac{2}{r}\frac{\partial P}{\partial r} - \varphi_P^2 P^p \theta^q = 0 \quad (86)$$

$$\frac{\partial \theta}{\partial t} = P^p \theta^q \quad (87)$$

where $m = n = p = q = 1$

Assuming negligible external resistance, the initial and boundary conditions become

$$t = 0, \quad \theta = 1, \quad A = \frac{\sinh(\varphi r)}{r \sinh \varphi}, \quad P = \frac{\sinh(\varphi_P r)}{r \sinh \varphi_P} \quad 0 \leqq r \leqq 1$$

$$r = 1, \quad A = P = 1$$

$$r = 0, \quad \frac{dA}{dr} = \frac{dP}{dr} = 0 \quad (88a, b, c)$$

Equations 85 through 87 were integrated numerically with the initial and boundary conditions already shown. Results are showed as effectiveness factor versus dimensionless time for values of φ and φ_P from 1 to 25 (Figure 26).

The effect of the diffusion resistance of the poisoning reaction is illustrated on Figure 27. The upper bound here represents no deactivation and is independent of time. The lower limit represents no diffusion resistance ($\varphi_P = 0$) and the η can be obtained analytically as

$$\eta = e^{-t}\left\{\frac{3(\varphi \coth \varphi - 1)}{\varphi^2}\right\}$$

The maximum activity is retained by the catalyst when the diffusion resistance is high for the poisoning reaction and low for the main reaction.

Langmuir–Hinshelwood kinetics. An extension of the Masamune and Smith [2] analysis has been made by Chu [4] using Langmuir–Hinshelwood rate forms.

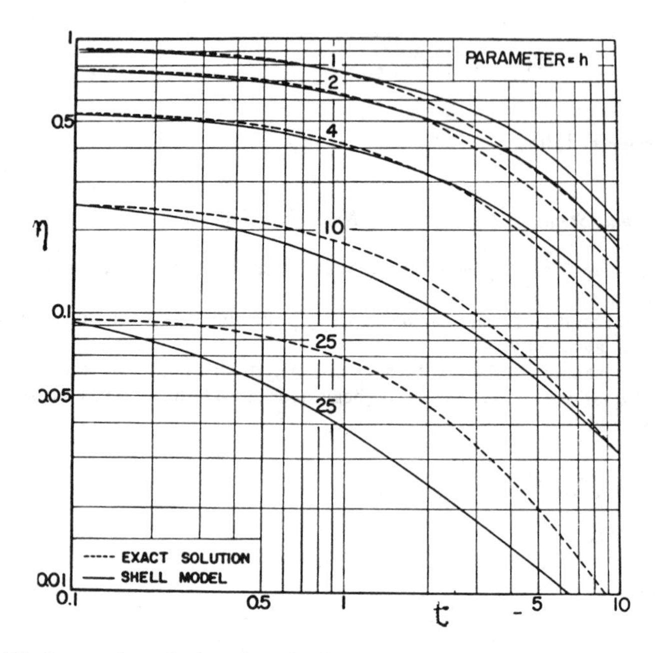

Figure 26. Effectiveness factor for impurity poisoning, $\varphi_p = 10$. Comparison of shell model and numerical solution. $h = \varphi$. (From Masamune and Smith [2].)

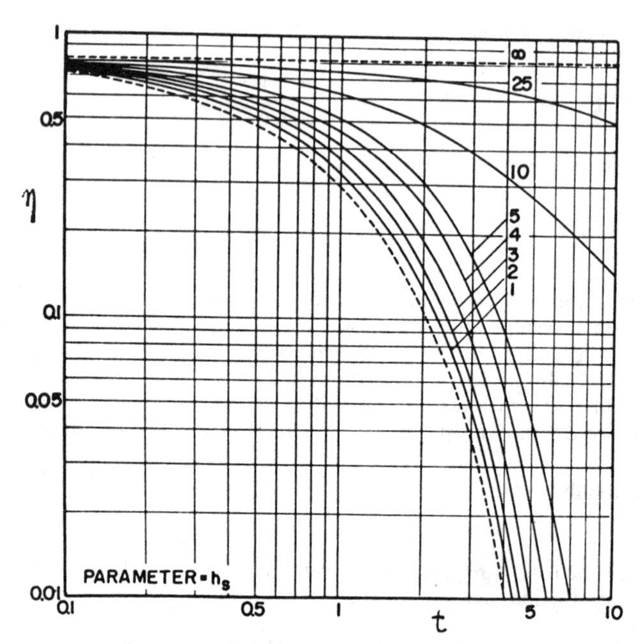

Figure 27. Effect of diffusion resistance for poisoning reaction on effectiveness factor. $h = \varphi = 2$. $h_s = \varphi p$. (From Masamune and Smith [2].)

Main reaction rate is expressed by

$$g_A(A, T^*) = \frac{K'_A A}{(1 + K'_A A + K'_B B + K'_P P)} \tag{89}$$

and poisoning reaction rate by

$$g_P(P, T^*) = \frac{K'_P P}{(1 + K'_A A + K'_B B + K'_P P)} \tag{90}$$

The effectiveness factor, η, can be calculated from Equation 44. It is a function of t, φ, φ_P and K'_i. The η decreases with the time of poisoning. To provide a measure for the extent to which the effectiveness factor is reduced at certain time t, a relative effectiveness factor, ε, is defined as

$$\varepsilon = \frac{\eta}{\eta_0} \tag{91}$$

where η_0 is η at $t = 0$.

Figure 5 shows the exponential relation of ε with the time t. However, there are exceptions like case B.

The parameters values are listed in Figure 5. The rate of deactivation increases when the adsorption equilibrium constant of the poison precursor or its partial pressure at the catalyst surface increases. With the other components of the reaction mixture the effect is reversed.

External mass transfer resistance. The analysis to show the effect of internal and external mass resistance on the poisoning of catalyst pellet has been made by Hegedus [12]. He assumes an irreversible nonlinear reaction for the reactant A and the poison P. The system can be expressed by Equations 85, 86, and 87.

Table 3
Parameters and Calculated Initial Effectiveness Factors*

Curve Designation	h_1	h_2	β_A	β_P	$\epsilon(0)$	λ
1	1	1	1	1	0.719	2
2	10	10	10	10	0.151	2
3	10	10	0.1	0.1	0.00343	2
4	0.1	0.1	10	10	0.999 ...	2
5	1	0.1	10	10	0.912	2
6	1	0.1	0.1	0.1	0.231	2
7	1	10	0.1	0.1	0.231	2
8	1	10	10	10	0.912	2
9	1	1	1	1	0.621	1
10	1	1	1	1	0.446	0
11	10	10	75	75	0.243	2
12	0.1	0.1	0.3	0.3	0.989	0
13	0.1	0.1	75	75	0.999 ...	0
14	1	5	1	1	0.721	2
15	$\sqrt{2}$	5	1	1	0.566	2
16	$\sqrt{10}$	5	1	1	0.212	2
17	$\sqrt{2}$	$\sqrt{2}$	1	1	0.566	2
18	$\sqrt{10}$	$\sqrt{10}$	1	1	0.212	2

* $m = n = p = q = 1$
$h_1 = \phi$, $h_2 = \phi_p$, $\beta_A = sh$, $\beta_p = Sh_p$, $\epsilon(0) = \eta_0$, $\lambda = \alpha$
From Hegedus [12].

Under combined external and internal mass transfer resistances, the boundary conditions at the surface of the catalyst are, in dimensionless form

$$r = 1, \quad \frac{\partial A}{\partial r} = Sh(1 - A_S)$$

$$\frac{\partial P}{\partial r} = Sh_P(1 - P_S) \tag{92}$$

the initial and boundary conditions at the center pellet are given by Equations 88a, c.

The system was solved for all three-dimensional geometries. The effectiveness factor is function of the following parameters

$$\eta = \eta(Sh, Sh_P, \varphi, \varphi_P, t)$$

The results correspond to the parameters range given in Table 3.

The variation of effectiveness factor with time for different combinations of external and internal mass transfer control in spherical pellets is given in Figure 28.

When the poisoning reaction is less than or as equally influenced by internal mass transfer resistance as the main reaction, the η increases from its initial value to unity (curves 1, 4, 5, and 6). The value of unity represents the completely poisoned catalyst pellet with distribution of the reactive species.

If the poisoning process is more strongly influenced by diffusional control than the main reaction (curves 2, 3, 7, 8) the effectiveness factor initially decreases with time, then pass through a minimum, and rises to the unity.

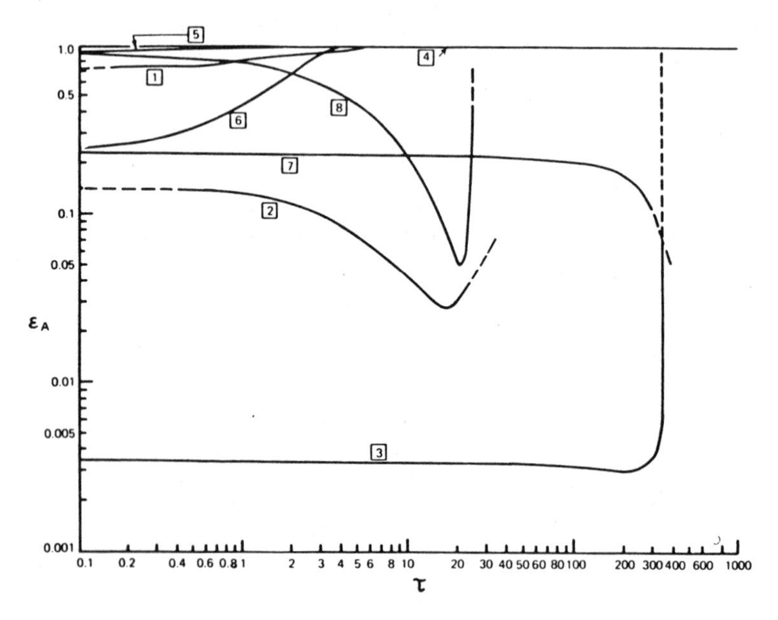

Figure 28. Time-dependent variation of the effectiveness factor of spherical catalyst pellets. Influence of external and internal mass transfer effects (Table 3) during impurity poisoning. $\epsilon_A = \eta$. (From Hegedus [12].)

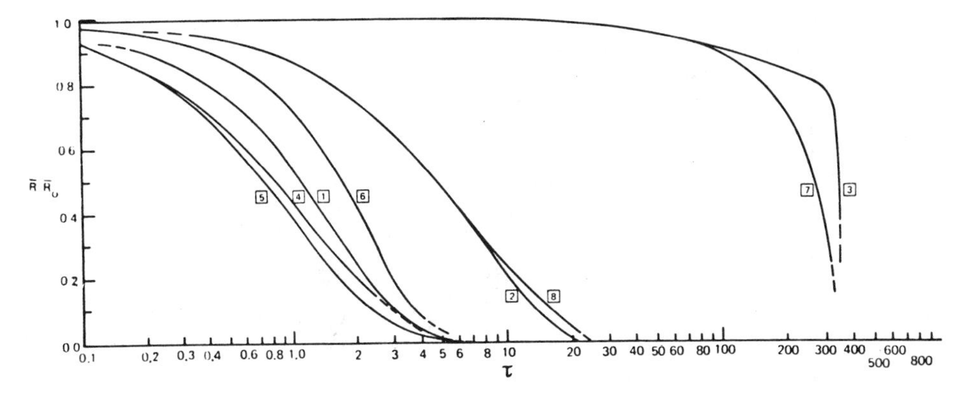

Figure 29. Time-dependent decay of the activity in spherical catalyst pellets. $\bar{R}/\bar{R}_o = \Lambda$. (From Hegedus [12].)

When the mass transfer resistance for the poison precursor increases, it also usually increases the film resistance for the reactant and the effectiveness factor diminishes significantly.

Figure 29 shows the effect of internal and external resistances on the time dependent decay of the overall reaction rate, normalized against its initial value, Λ, for the conditions shown in Table 3. It also appears that catalysts where the poisoning process is hindered by an internal and external mass transfer resistance have a longer lifetime than those with a kinetically controlled poisoning reaction.

The effect of pellet geometry on the lifetime of catalysts is showed in Figure 30. The time dependence of Λ shows that for the same values of φ and Sh, an infinite slab has the longest life.

First-order nonuniform initial activity. Becker and Wei [13] have investigated the effect of nonuniform activity distribution in a spherical pellet where the main reaction is first order. For this analysis one third of the volume of a support sphere is impregnated with catalytic material.

A constitutive equation relates the first-order rate constant for main reaction to the amount of poison on the surface.

$$\varphi_m^2 = \varphi^2 \exp(-\lambda p W) \tag{93}$$

Hence, assumed pseudo steady-state, the Equation 85 is expressed as

$$\frac{\partial^2 A}{\partial r^2} + \frac{2}{r}\frac{\partial A}{\partial r} = \varphi_m^2 A \tag{94}$$

Poison rate is given by a nonselective first-order deposition process without saturation. The mass balance for P is

$$\frac{\partial^2 P}{\partial r^2} + \frac{2}{r}\frac{\partial P}{\partial r} = \varphi_P^2 P \tag{95}$$

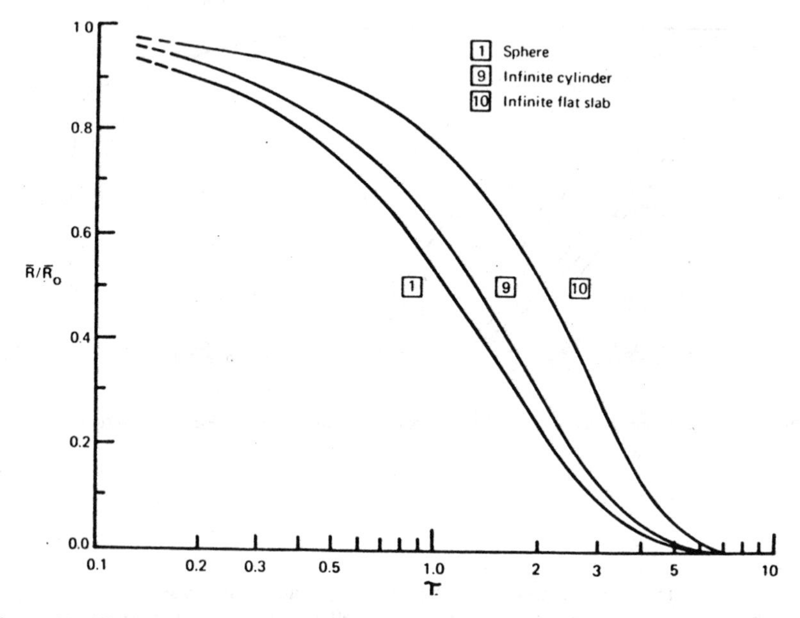

Figure 30. Effect of the geometry of catalyst pellets on their lifetime. Table 3 lists the parameters in detail. Impurity poisoning. $\bar{R}/\bar{R}_o = \Lambda$. (From Hegedus [12].)

Poison accumulation is expressed by

$$\frac{\partial W}{\partial \tau} = \varphi_P^2 P \tag{96}$$

Initial and boundary conditions are

$$\tau = 0, \quad W = 0, \quad 0 \le r \le 1$$

$$r = 1, \quad A = P = 1, \quad \tau > 0$$

$$r = 0, \quad \frac{dA}{dr} = \frac{dP}{dr} = 0 \tag{97a, b, c}$$

The effectiveness factor is a function of the following parameters: φ_m, φ_P, $\lambda_p\tau$ and catalyst distribution. It is written as:

$$\eta(\tau) = \frac{3\left(\frac{dA}{dr}\right)_{r=1,\tau}}{\varphi_m^2} \tag{98}$$

The Thiele modulus for the nonuniformly impregnated support pellets is defined by an average value of the rate constant over the entire support sphere, k_{A_v}

$$\varphi_m^2 = \frac{R^2 k_{A_v}}{D_A} \tag{99}$$

Calculations were realized for the range of parameters

$$0.2 \leq \varphi_P \leq 20 \quad \text{and} \quad 0.1 \leq \varphi_m \leq 12$$

and resulting η-vs.-τ curves were reported by Becker [14].

A catalyst selection chart is shown in Figure 31.

The criterion for catalyst selection was based on the longest useful life, $\tau_{0.4}$, under different experimental conditions. The $\tau_{0.4}$ is the time taken to reduce initial catalyst efficiency to 0.4.

In Figure 31 the boundaries between the region of optimum catalyst distribution are fitted by the broken line curves.

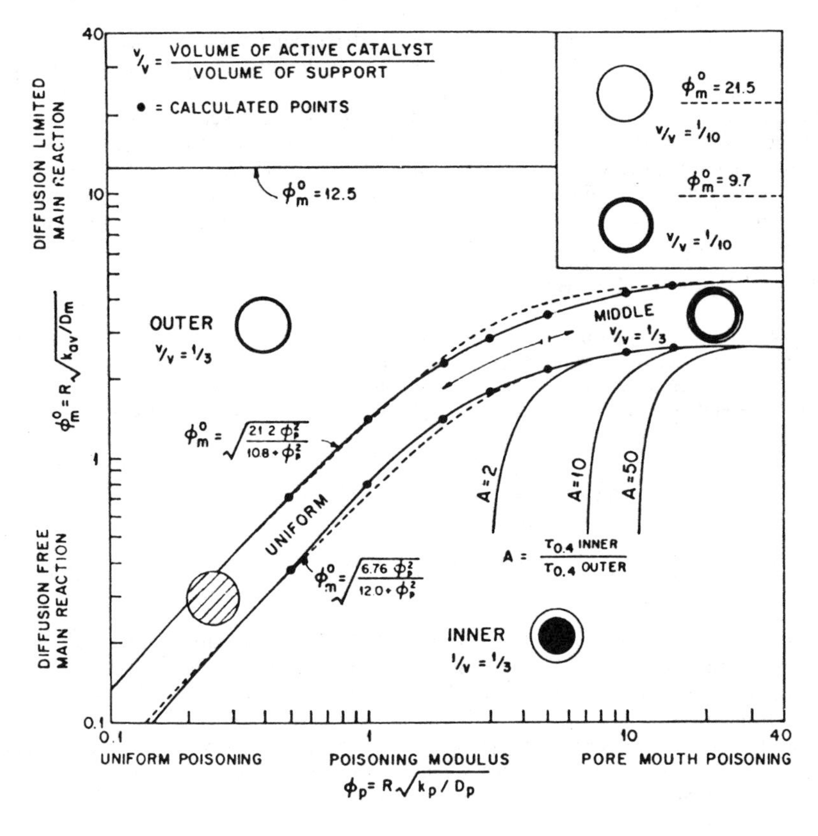

Figure 31. Catalyst selection chart for four catalyst designs based on a criterion of longest useful life. (From Becker and Wei [13].)

In Figure 32 the CO conversion is plotted for three stages during the catalyst's life. The conversion is given by

$$x = 1 - \exp(Vk_m\eta/F)$$

The maximum activity is obtained by different designs depending on the duration of the poisoning experiment; for fresh performance the egg-shell design is better than uniform and middle catalyst. This last configuration was proposed for long term durability.

These results show that the highest effectiveness is obtained when the catalyst is placed close to the sphere center.

Corbett and Luss [8] analyzed the case of spherical pellet having different activity distributions, for the consecutive first-order reaction, $A \rightarrow B \rightarrow C$, with impurity deactivation mechanism.

Now, the rate constant in Equation 77 is expressed as

$$k_i(r) = k_i u(r) \tag{100}$$

where k_i is the volume-averaged rate constant and u(r) is the initial activity distribution.

In this work is analyzed the effect of the pore-mouth poisoning in terms of effectiveness factor and selectivity.

Figure 33 described the behavior of a pellet where the initial diffusional resistance is small. Catalysts which concentrate the active component near the interior of the pellet are more poison resistant.

When the diffusional resistance is appreciable ($\varphi = 10$) the pellet with the active component near to the surface has higher initial activity but deactivates faster (Figure 34).

Here, for shorter process time, catalysts with exterior enrichment are better. However, where it is important to maintain a high level of activity for a long period of operation, it may be desirable to use a catalyst in which the activity is confined to an inner core.

The General Equations of Reversible Poisoning

One of the most common mechanisms for impurity poisoning is through competitive reversible adsorption of poisonous species with reactants. The reversibility is defined by recovery of the activity upon removal of the poison from the reaction mixture.

If the reversible poisoning by the poison precursor is modeled, $k_P = 0$ the Equations 72 and 73 will simplify to

$$D_A\nabla^2C_A - R_A(C_A, C_P, T) = \varepsilon_P\frac{\partial C_A}{\partial t'} \tag{101}$$

$$D_P\nabla^2C_P - f(C_{Pa}, t') = \varepsilon_P\frac{\partial C_P}{\partial t'} \tag{102}$$

where $R_A(C_A, C_P, T)$ corresponds to the separable or nonseparable representation of poisoning kinetics and $f(C_{Pa}, t')$ is a function of the time and adsorbed poison concentration.

In order to describe the poisoning, specifications must be made about the adsorption law of the poison, the kinetics of the main reaction, and thermal conditions of the system.

If the adsorption of the poison is rapid, equilibrium conditions prevail and the concentrations between the gas and adsorbed poison phase concentrations are related by a Langmuir adsorption isotherm:

$$\frac{P_a}{P_{max}} = \frac{K_P'P}{(1 + K_P'P)} \tag{103}$$

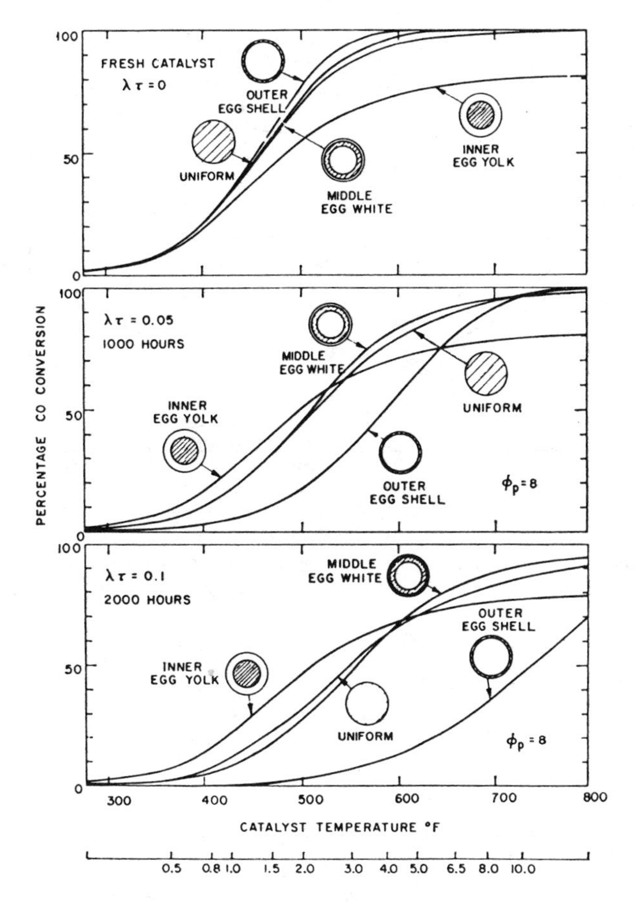

Figure 32. CO conversion over base metal oxide catalyst at different times of operation with low-lead fuel. $\chi = 1 - \exp(V\, k_m\eta/F)$, $\lambda = \lambda_p$. (From Becker and Wei [13].)

Pseudo Steady-State—Numerical Solutions

Isothermal first-order system. Gioia et al. [15] have studied the problem of simultaneous reversible poisoning and first-order main reaction in spherical pellet with uniform initial activity distributions.

The system obeys the Equations 101 and 102 where

$$R_A(C_A, C_P, T) = kC_A(1 + K_P C_P)^{-1} \tag{104}$$

and

$$f(C_{Pa}, t') = \frac{\partial C_{Pa}}{\partial t'} \tag{105}$$

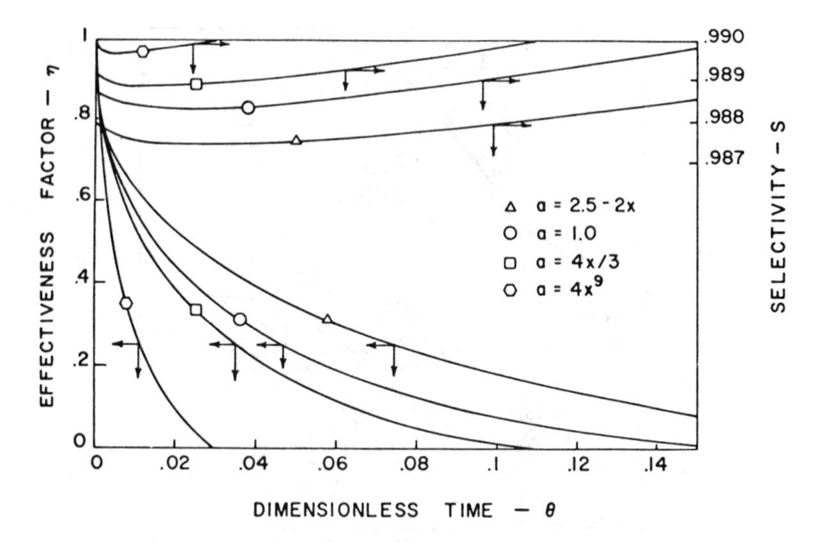

Figure 33. The effect of the nonuniform initial activity distribution on the effectiveness factor and selectivity, spherical pellets in which occur impurity poisoning. $\varphi = 1$; $\varphi_B = 0.1$; $a = u(r)$; $\theta = t$. (From Corbett et al. [8].)

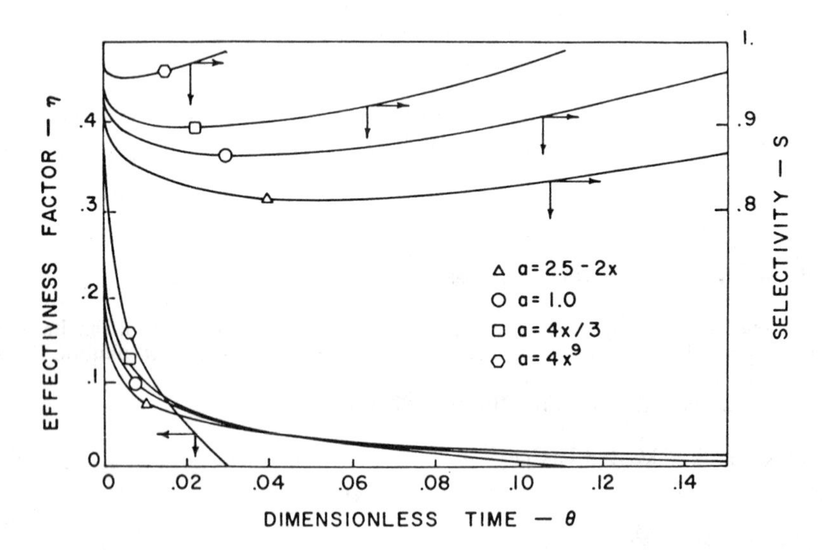

Figure 34. The effect of the nonuniform initial activity distribution on the effectiveness factor and selectivity, spherical catalyst. Impurity poisoning. $\varphi = 10$; $\varphi_B = 1$; $a = u(r)$; $\theta = t$. (From Corbett and Luss [8].)

Equations 101 through 105 can be combined and expressed in dimensionless form as

$$\frac{\partial^2 A}{\partial r^2} + \frac{2}{r}\frac{\partial A}{\partial r} = \varphi^2(1 + K'_P P)^{-1} A \tag{106}$$

$$\frac{\partial^2 P}{\partial r^2} + \frac{2}{r}\frac{\partial P}{\partial r} = \left[\frac{\varepsilon_P}{P_{max}} + \frac{K'_P}{(1 + K'_P P)^2}\right]\frac{\partial P}{\partial \tau} \tag{107}$$

which are solved simultaneously with the initial and boundary conditions

$$\tau = 0, \quad A = 1, \quad P = 0, \quad 0 \le r \le 1$$

$$\tau > 0, \quad A = P = 1, \quad r = 1$$

$$\frac{dA}{dr} = \frac{dP}{dr} = 0, \quad r = 0 \tag{108a, b, c}$$

Considering that the main reaction is pseudo steady-state, time does not appear as an independent variable in Equation 106 but as a parameter through the brackets right term.

The set of equations have been solved numerically for values of $\varphi = 1, 10, 100$, K'_P from 10^{-1} to 10^4 and values of P_{max} between 10 and 10^5.

The results have been given in terms of ratio between overall reaction rate during the poisoning process and initial reaction rate, Λ.

Figures 35, 36, and 37 show the Λ-vs.-τ curves for different values of φ and K'_p. The solid curves are for $P_{max} = 10$, the dotted curve for $P_{max} = 10^5$. In the same figures the curves according to a pore-mouth poisoning model described by Sada and Wen [16] are reported. The results show a slight dependence of Λ and P_{max}, meaning a proportionality between

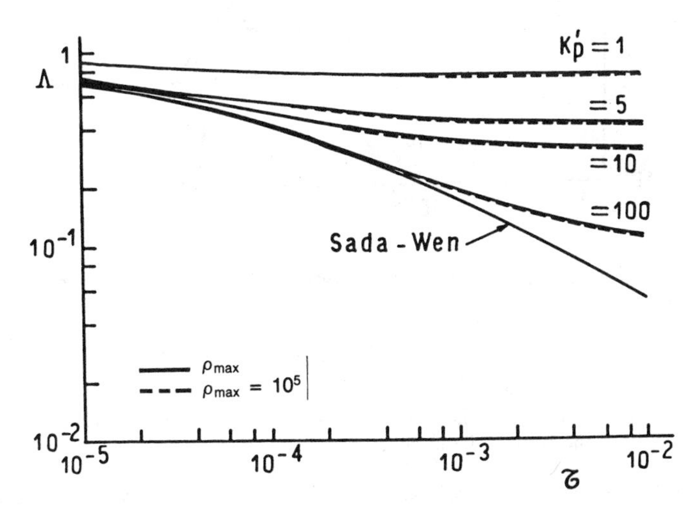

Figure 35. Time-dependent decay of the activity for various values of K'_p at $\varphi = 1$. Comparison with the formulae proposed by Sada and Wen for the shell model. (From Gioia et al. [15].)

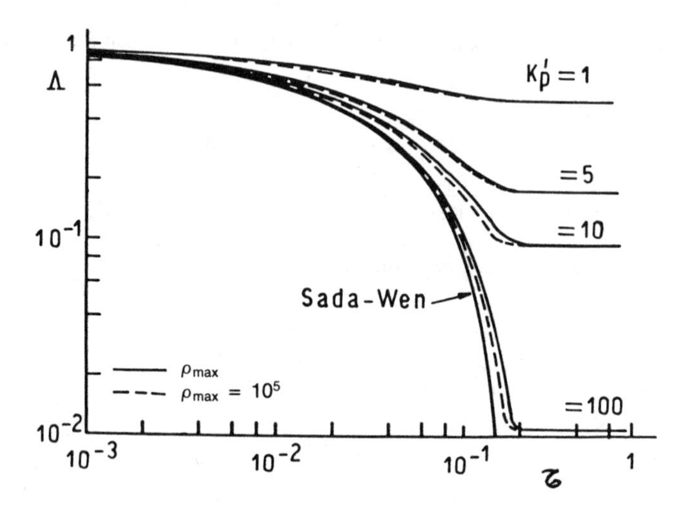

Figure 36. Time-dependent decay activity for various values of K_p' at $\varphi = 10$. Comparison with the shell model. (From Gioia et al. [15].)

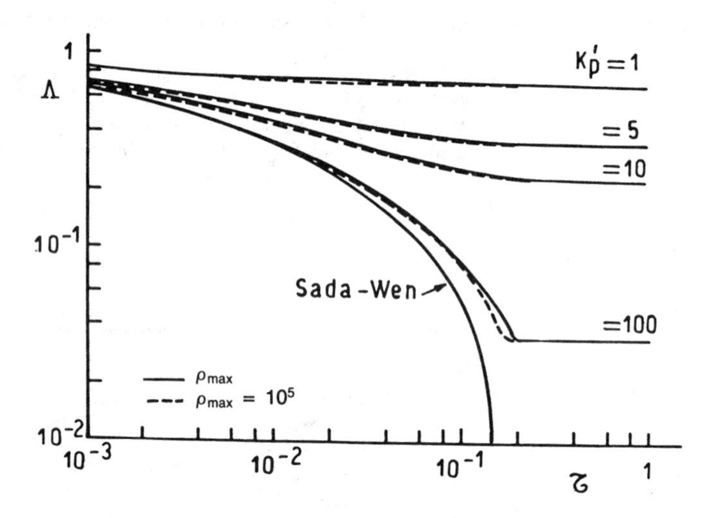

Figure 37. Time-dependent decay activity for different values of K_p' at $\varphi = 100$. Comparison with the shell model. (From Gioia et al. [15].)

a given value of Λ and the time to reach a given relative coverage of surface. The asymptotic value for large K_p' with low φ is reached abruptly. It corresponds to a step poison penetration reaching the center of the pellet.

Isothermal Langmuir–Hinshelwood kinetics. Valdman et al. [17] extended the analysis of Gioia [15] to the situation where the main reaction follows Langmuir–Hinshelwood kinetics, as

$$R_A(C_A, C_P, T) = \frac{kC_A}{(1 + K_A C_A + K_P C_P)} \tag{109}$$

where Equation 109 represents the nonseparable poisoning kinetics.

They assume that the gas-phase accumulation term $\partial C_P/\partial t'$ in Equation 102 is small compared with that in the adsorbed phase and can be neglected.

The dimensionless mass balance for A and P for a flat slab:

$$\frac{\partial^2 A}{\partial r^2} = \varphi^2 \left(\frac{A}{1 + K'_A A + K'_P P} \right) \tag{110}$$

$$\frac{\partial^2 P}{\partial r^2} = \varphi_P^2 \left[\frac{1}{(1 + K'_P P)^2} \right] \frac{\partial P}{\partial \tau} \tag{111}$$

Valdman et al. integrated the Equations 110 and 111 numerically, for different geometries, for values of φ within the range of 0.1 to 20 and K'_i equal 1 and 10. Results are given as effectiveness factor which is function of the following parameters

$$\eta = \eta(\tau, \varphi, K'_A, K'_P)$$

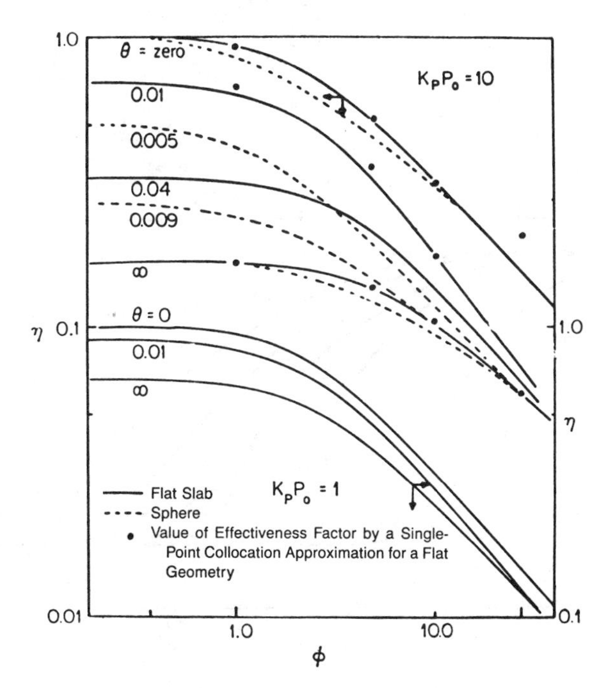

Figure 38. Effectiveness factor versus Thiele modulus at various times of poisoning. $\theta = \tau$. (From Valdman et al. [17].)

and is mathematically expressed for this case as

$$\eta = \frac{1 + K'_A}{\varphi} \frac{dA}{dX}\bigg|_{X=1,\tau} \tag{112}$$

In Figure 38 shows the variation of η vs. φ at different values of dimensionless time, τ, for slab and spherical geometry. For intermediate values of τ, the posioning is faster in a sphere than in a slab.

The relative effectiveness factor, η/η_0, as a function of real time is shown in Figure 39 for five values of φ. These results mean the relative drop in the reaction rate as a function of time.

For low φ the poisoning is completed in lesser time than for the larger φ, and it is accompanied by a larger drop in the activity.

Figure 40 represents the reactivation of the catalyst for two situations. The time necessary for the catalyst regeneration is roughly six fold that for deactivation.

Nonisothermal; external transport resistance; Langmuir–Hinshelwood kinetics. A more complex analysis of reversible poisoning has been given by Valdman and Hughes [18], where they consider the effect of temperature and external mass and heat transfer resistances over the system just described.

In a nonisothermal system, the energy balance is added to the mass balance of A and P, defined by Equations 110 and 111.

The energy balance, in dimensionless form, is

$$\frac{\partial^2 T^*}{\partial x^2} = \varphi^2 \beta g_A(A, P, T) \tag{113}$$

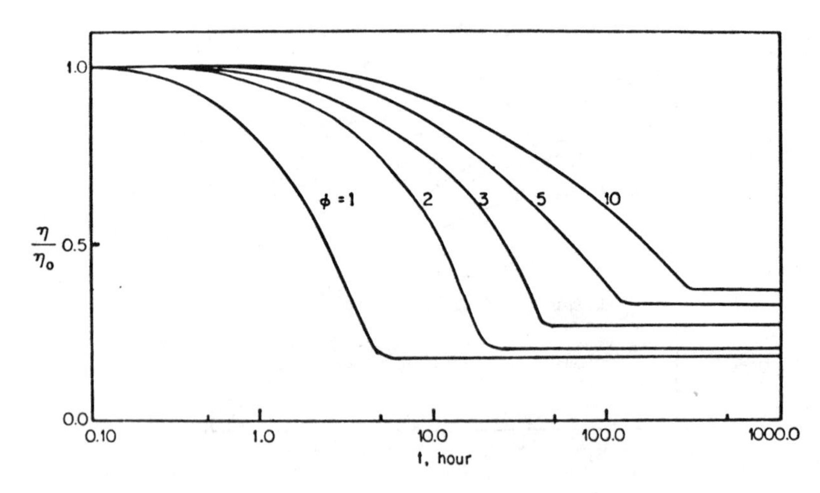

Figure 39. Time-dependent variation of ϵ during poisoning of flat slab catalyst pellets. (From Valdman et al. [17].)

where $g_A(A, P, T)$ takes the form of the Langmuir–Hinshelwood expression

$$g_A(A, P, T) = \frac{A \exp\left[(\gamma - \delta_A)\left(1 - \frac{1}{T^*}\right)\right]}{1 + K'_A \exp\left[\delta_A\left(1 - \frac{1}{T^*}\right)\right] A + K'_P \exp\left[\delta_P\left(1 - \frac{1}{T^*}\right)\right] P} \tag{114}$$

Under external resistance the boundary conditions at the catalyst surface are defined in a dimensionaless form as

$$X = 1 \qquad -\frac{dA}{dX} = Sh(1 - A_S)$$

$$-\frac{dP}{dX} = SH_P(1 - P_S)$$

$$\frac{dT}{dX} = Bi(1 - T_S) \tag{115a, b, c}$$

The initial and boundary conditions at center pellet are described by the Equations 108a, c.

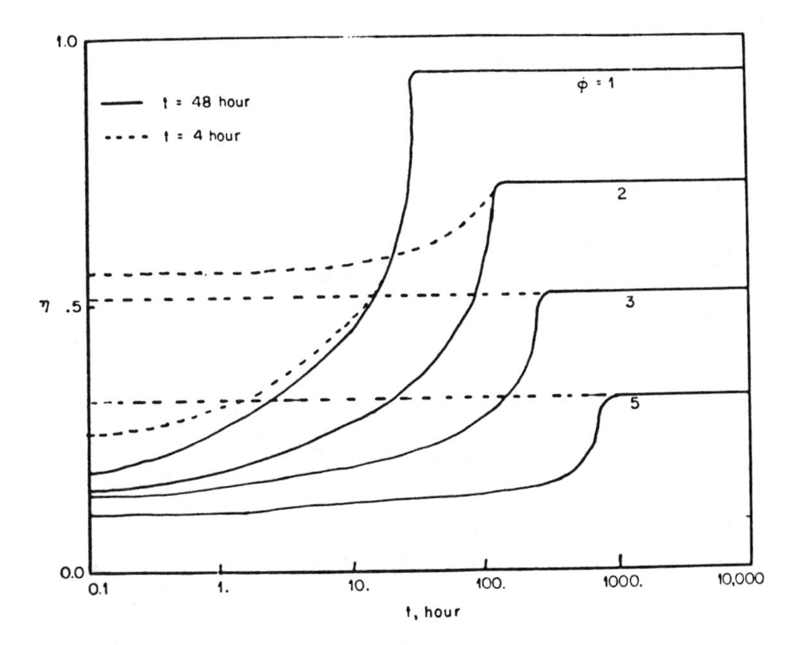

Figure 40. Time-dependent variation of η during poison desorption for various values of Thiele moduli. (From Valdman et al. [17].)

Equation 113 is solved by numerical procedure simultaneously with the pseudo steady-state mass conservation equations to obtain the system behavior in terms of a time-dependent effectiveness factor based on bulk stream condition.

In nonisothermal pellets the effectiveness factor is function of φ, K_i', τ, β, γ, Sh, Bi, δ_i. The number of equations can be reduced by one using the Prater relations Equation 33. Now, the problem consists of solving simultaneously by the Equations 115 and 33 subjected to boundary conditions.

In Figure 41 the effectiveness factor is plotted versus Thiele modulus for three different values of dimensionless time. This figure shows the effect of external mass and heat transfer resistance for the limiting case of nonpoisoned, uniformly and nonuniformly poisoned nonisothermal pellet.

The η is based on fluid conditions for the values of the parameters $\beta = 0.01$, $\gamma = 20$, Sh = 25, Bi = 0.5. The values of K_A' and k_P' were kept constant at 1 and 10, respectively. For small φ the η is approximately equal to unity for $\tau = 0$. For values of Thiele modulus within the range of 0.5 and 3.0 the effectiveness facıor has a value larger than the unity.

The broken lines show the η for unpoisoned pellets for $\beta = 0$ and $\beta = 0.01$. For the last value of β the η takes a large value as a consequence of the accumulation of heat within the particle due to the film resistance.

The analysis made by Valdman and Hughes [18] of the limiting case of initial ($\tau = 0$) and ($\tau = \infty$) steady state shows that when reaction occurs with adsorption of reactant on poison the sensitivity of the system to temperature changes is decreased while the effect of the external resistances has the inverse effect. The system subjected to external heat and mass transfer resistances becomes more sensitive to thermal changes.

Figure 41 shows that the drop in η is very pronounced for low φ, and little effect is observed at very large Thiele modulus.

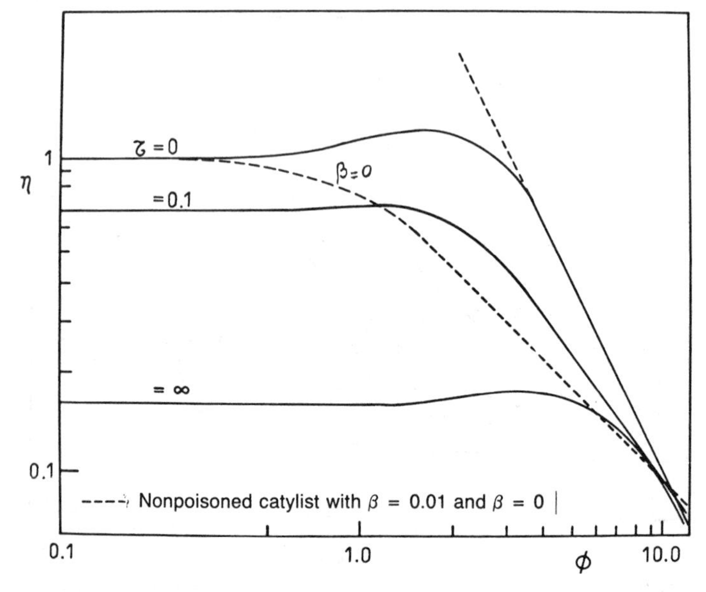

Figure 41. Effect of external heat and mass transfer resistances on a nonisothermal catalyst slab. (From Valdman and Hughes. [18].)

CATALYST DEACTIVATION—ANALYTICAL SOLUTIONS FOR THE EFFECTIVENESS FACTOR

Isothermal—Pseudo Steady-State Analysis

In the first part of this chapter, the results were obtained by direct solution of the mass, activity, and energy balances. Now, approximate solutions for the effectiveness factor are given for various models. In all cases the assumptions are: isothermal pellet and pseudo steady state.

The model of *uniform deactivation* Essumes the precursor concentration (A, B, or P) to be independent upon the pellet coordinate ξ.

First-Order Reaction

The effectiveness factors for parallel, series, and independent mechanisms, are as follows:

$$\eta = \exp(-t_d) \tag{116}$$

$$\eta = \exp(-t_f) \tag{117}$$

$$\eta = \exp(-t)\frac{3\varphi \coth \varphi - 1}{\varphi^2} \tag{118}$$

The effectiveness factor for a spherical pellet calculated by Equations 116 through 118 are given in Figures 1, 2, and 27 [2].

First-Order Reaction, Nonuniform Initial Activity

The effectiveness factor for series fouling, slab geometry, and initial activity distribution, $u(x) = x^z$ is given by

$$\eta = \frac{\sum_{m=1}^{\infty} (m/P)\, a_m}{\varphi^2} \tag{119}$$

and

$$\frac{\partial \omega}{\partial t_f^*} = (1-\omega)\left[\frac{1}{z+1}\sum_{m=0}^{\infty}\frac{a_m}{m/P+z+1}\right] \tag{120}$$

where

$$P = (z+2)^{-1} \quad \text{and}$$

$$a_m = \frac{[P\varphi(1-\omega)^{0.5}]^{2m-P}}{m!\,(m-P)!\, I_{-P}[2P\varphi(1-\omega)^{0.5}]}$$

The results for two values of z are shown in Figures 17 and 18 [9].

Langmuir–Hinshelwood Kinetics, Reversible Poisoning.

For the case of large time of impurity poisoning, an analytical solution is developed for calculating the effectiveness factor for the limiting cases for large and low Thiele moduli [17].

Large value of Thiele moduli. At large times when the poison can be considered uniformly adsorbed over the entire catalyst, the effectiveness factor can be calculated by

$$\eta = \frac{1+K_A'}{\varphi}\left\{\frac{2}{K_A'}\left[1-\frac{\ln(1+K)}{K}\right]\right\}^{1/2} \tag{121}$$

where $K = K_A'/(1+K_P')$. This formula is generally valid for $\varphi > 3$.

Low value of Thiele moduli. In this case, when $\tau \to \infty$, the concentration profile of C_A tends to unity, and the effectiveness factor for uniform poison distribution is given by

$$\eta(\tau \to \infty) = \frac{1+K_A'}{(1+K_A'+K_P'} \tag{122}$$

This equation is valid for any value of K_P' and $\varphi < 0.5$. As K_P' increases this limit also increases.

Shell model. In the *shell model* it is supposed that there is a sharp boundary between completely deactivated and fresh catalyst at a position which changes with time.

For parallel and independent mechanisms the model is a spherical shell (thickness, $R\text{-}r_i$) of completely fouled catalyst ($\theta = 0$) and a sphere of fresh catalyst ($\theta = 1$) of radius r_i. For series mechanism there is a central core, the radius r_i, which is completely fouled ($\theta = 0$), and an outer layer of fresh catalyst ($\theta = 1$).

First-order reaction. The equations needed to calculate the effectiveness factor for spherical pellet are:

Parallel fouling

$$\eta = \frac{3}{\varphi^{12}}\frac{r_i}{1-r_i}(1-A_i) \tag{123}$$

$$t_d = -\ln\left[\frac{\varphi r_i \cosh(\varphi r_i) - \sinh(\varphi r_i)}{\varphi \cosh \varphi - \sinh \varphi}\right] + \varphi^{12}\left[\frac{r_i^3}{3} - \frac{r_i^2}{2} + \frac{1}{6}\right] \tag{124}$$

$$A_i = \frac{1/(1-r_i)}{1/(1-r_i) + {}'r_i \coth(\varphi r_i) - D_A/D_A'} \tag{125}$$

In order to test the shell model, the authors calculated the effectiveness factor for the same conditions as used to illustrate parallel fouling, the results are given in Figure 1, the dotted lines correspond to numerical solutions. When $\eta > 0.1$ the deviation is generally less than 15% [2].

In series fouling Masamune and Smith [2] found that the deviations for the shell model are much larger than for parallel fouling.

For independent poisoning when $\varphi_P \to \infty$ and φ finite, the effectiveness factor values for the main reaction can be obtained by using Equations 123 and 125 and

$$t = \varphi_P^2\left(\frac{r_i^3}{3} - \frac{r_i^2}{2} + \frac{1}{6}\right) \tag{126}$$

The effect of relative diffusion resistance (φ) for the main reaction is shown in Figure 26. The shell model gives results which agree reasonably well with the numerical solution for values of $\varphi < 10$ and $\varphi_P = 10$. For effectiveness factor greater than 0.1 the deviatons are approximately the same as for parallel self fouling.

First order reaction. Adsorption of precursor on catalyst and support. For slab geometry, the effectiveness factor can be estimate with the following equations:

$$\eta = \frac{\tanh(\varphi - \varphi x_i)}{\varphi} \tag{127}$$

$$\frac{dx_i}{dt_P} = (1 - x_i) - \frac{1}{\varphi}\tanh\varphi(1 - x_i) \tag{128}$$

This equation is solved numerically and the solution is used in Equation 127. The results are shown as a dotted line in Figure 16.

Approximate Solutions

First-Order system. parallel, series, and independent mechanisms. Tai and Greenfield [19] have developed approximate solutions for a time-dependent effectiveness factor for parallel, series, and impurity deactivation in a spherical pellet.

The main and the deactivation reactions are both assumed to be first order with respect to the activity and to the reactant and poison concentrations respectively.

The method of solution of the activity and mass balance involves a successive approximation technique combined with linearization of the activity around a properly chosen expansion point (x^+ or θ^+).

The effectiveness factor for parallel and independent deactivation is given by

$$\eta = \frac{3\lambda^{1/3}}{\varphi^{4/3}}\left\{\frac{(A_i'(Z_1)B_i(Z_0) - B_i'(Z_1)\ A_i(Z_0))}{A_i(Z_1)B_i(Z_0) - A_i(Z_0)B_i(Z_1)}\right\} - \frac{3}{\varphi^2} \tag{129}$$

where $A_i(Z)$, $B_i(Z)$, $A_i'(Z)$ and $B_i'(Z)$ are Airy functions and their derivatives respectively.

$$\lambda = \frac{\theta^+ t}{\sinh\varphi}\frac{\sinh(\varphi x^+)}{x^{+2}} - \frac{\varphi\cosh(\varphi x^+)}{x^+}$$

$$\theta^+ = \exp\left[-\frac{t\sinh(\varphi X^+)}{X^+\sinh(\varphi)}\right]$$

$$Z_0 = \left(\frac{\varphi}{\lambda}\right)^{2/3}$$

$$Z_1 = \left(\frac{\varphi}{\lambda}\right)^{2/3} (\lambda + \mu)$$

$$\mu = \theta^+ \left\{ 1 + \frac{td}{\sinh \varphi} \left[\varphi \cosh (\varphi X^+) - \frac{\sinh (\varphi X^+)}{X^+} \right] \right\}$$

For $0 < \varphi < 10$ the best solution is obtained with $X^+ = 1$, and for $\varphi \geq 10$ is much better use $\theta^+ = 0.14$.

For impurity poisoning, the expansion parameter is chosen according to the value of φ_P.

The approximate solutions in Equation 129 was compared with the numerical solutions for parallel and independent deactivation respectively. Agreement with the numerical solution is excellent.

For series deactivation, the effectiveness factor is obtained using the pellet surface ($X^+ = 1$) as the expansion point.

$$\eta = \frac{3\lambda^{1/3}}{\varphi^{4/3}} \left[\frac{f'(Z_1) g(Z_1)}{f'(Z_b) - f(Z_b)} + \frac{Bi'(Z_1)}{Bi(Z_1)} \right] - \frac{3}{\varphi^2} \tag{130}$$

where

$$Z_1 = \left(\frac{\varphi}{\lambda}\right)^{2/3} (\lambda + \mu)$$

$$Z_b = \left(\frac{\varphi}{\lambda}\right)^{2/3} (\lambda x_b + \mu)$$

$$f(Z) = A_i(Z) - \frac{A_i(Z_1)}{Bi(Z_1)} Bi(Z)$$

$$f'(Z) = \frac{df}{dZ} = A_i'(Z) - \frac{A_i(Z_i)}{Bi(Z_i)} Bi'(Z)$$

$$g_i(Z) = v \frac{Bi(Z_b)}{Bi(Z_1)} - v^{1/3} \frac{Bi'(Z_b)}{Bi(Z_1)}$$

$$v = \frac{\varphi^{1/3} \theta^{1/2}}{\tanh (\varphi \theta^{1/2} X_b)}$$

$$\lambda = \frac{tf}{b_0^*} \exp(-tf) \left(\frac{\varphi}{\tanh \varphi} - 1 \right)$$

$$\mu = [\exp(-tf) - \lambda]$$

$$\theta = \exp\left[-tf\left(1 + \frac{1}{b_0^*}\right) + \frac{tf}{b_0^* \sinh \varphi} \right]$$

$$X_b = \frac{\theta - \mu}{\lambda}$$

For series deactivation two functions are necessary to approximate the activity profiles adequately.

In this case, the approximate solution underestimates the η for intermediate values of the Thiele modulus ($3 < \varphi < 10$) at large times. Agreement with the numerical solution is good in the other ranges of Thiele moduli.

For all cases, this approximate solution was found to be superior to the shell model.

First order system. Parallel and Series Mechanisms. The method is based on the development of a differential equation for the cumulative species concentration and the application of a single-point collocation to the resulting equation. For values of $\varphi < 5$ a collocation procedure can be used, while for $\varphi > 5$ a modified method based on the concept of an effective reaction zone has been developed [20].

For *parallel fouling* and low values of φ, the effectiveness factor can be approximated as follows:

$$\eta = (1+\alpha)\left[\frac{W_1 \exp(-\psi_1)}{(\varphi^2/B_{12})\exp(-\psi_1)+1} + W_2 \exp(-td)\right] \tag{131}$$

The Equation 131 coupled with

$$td = \frac{\varphi^2}{B_{12}}[1-\exp(-\psi_1)] + \psi_1 \tag{132}$$

provides an implicit analytical solution for the catalyst effectiveness factor. The constants B_{12}, W_1, and W_2 to be used are given in Table 4.

For $\varphi < 5$ and spherical geometry the results have an excellent agreement with the numerical solution given by Masamune and Smith [2] (the maximum error less than 5%).

For *parallel fouling* with large values of φ;

$$\eta = (1+\alpha)(1-\lambda)\int_0^1 [y(1-\lambda)+\lambda]^\alpha \frac{d\psi}{d\,td} \exp(-td\,y^2)\,dy \tag{133}$$

the derivative $d\psi/dt$ can be evaluated by

$$\frac{d\psi}{d\,td} = y^2 + 2y\,td\,\frac{y-1}{1-\lambda}\frac{d\lambda}{d\,td} \tag{134}$$

where $d\lambda/d\,td$ can be obtained by differentiating the following equation

$$\frac{2\,td}{(1-\lambda)^2} + \frac{2\alpha\,td\,y_c}{(1-\lambda)y_c[(1-\lambda)+\lambda]} - \varphi^2[1-\exp(-td\,y_c^2)] = 0 \tag{135}$$

Table 4
Collocation Constants

Geometry	α	B_{12}	W_1	W_2
Slab	0	2.5	0.833	0.1667
Cylinder	1	6.0	0.375	0.125
Sphere	2	10.5	0.233	0.1

From Villadsen and Stewart [22]

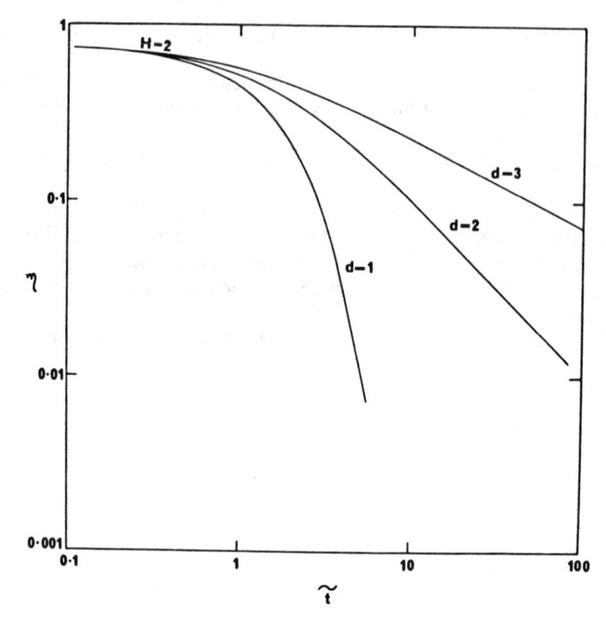

Figure 42. Effectiveness factor for parallel fouling. Isothermal half-order kinetics and various deactivation order. $\tilde{t}$ = td, H = 2. (From Do and Weiland [21].)

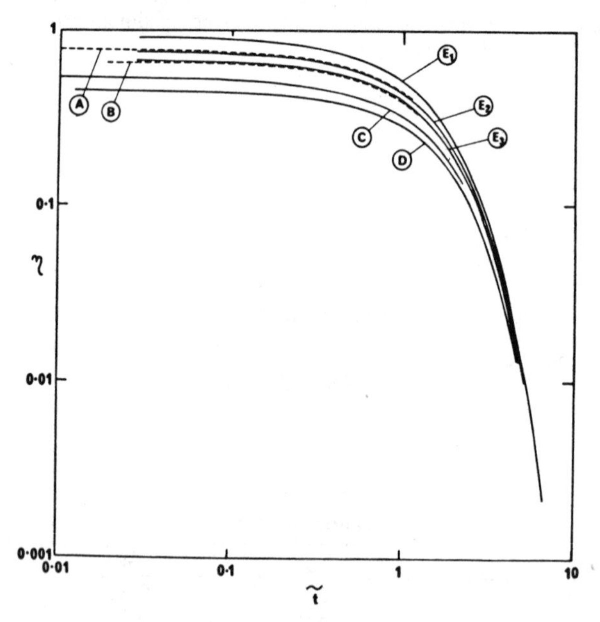

Figure 43. Effectiveness factor for parallel fouling, Isothermal pellet. Effect of order and type of kinetics. $\tilde{t}$ = t_d, H = 2, (A, n = 1/2; B, n = 1; C, n = 2; D, n = 3; E_1, β_M = 0.1; E_2, β_M = 1; E_3, β_M = 10). (From Do and Weiland [21].)

where

$$y_c = \begin{cases} 0.464 & \text{slab geometry} \\ 0.53 & \text{spherical geometry} \end{cases}$$

λ = fractional penetration distance in the pellet

For $\varphi = 10$ and 25 the results agree with the numerical solution of Masamune and Smith [2] with a maximum error of 10%.

For *series fouling* with small values of φ;

$$\eta = (1+\alpha)\frac{W_1 \exp\left[-\left(b_0+\frac{1}{P}\right)tf_A + \psi_1/P\right]}{1+(\varphi^2/B_{12})\exp\left[-\left(b_0+\frac{1}{P}\right)tf_A+\frac{\psi_1}{P}\right]} + W_2\exp(-b_0\, tf_A) \tag{136}$$

and

$$tf_A = \frac{1}{\left(b_0+\frac{1}{P}\right)}\ln\left[\frac{\left[\varphi^2\left(b_0+\frac{1}{P}\right)/B_{12}b_0\right][1-\exp(-b_0\psi_1)]+1}{\exp\left\{-\psi_1\left(b_0+\frac{1}{P}\right)\right\}}\right] \tag{137}$$

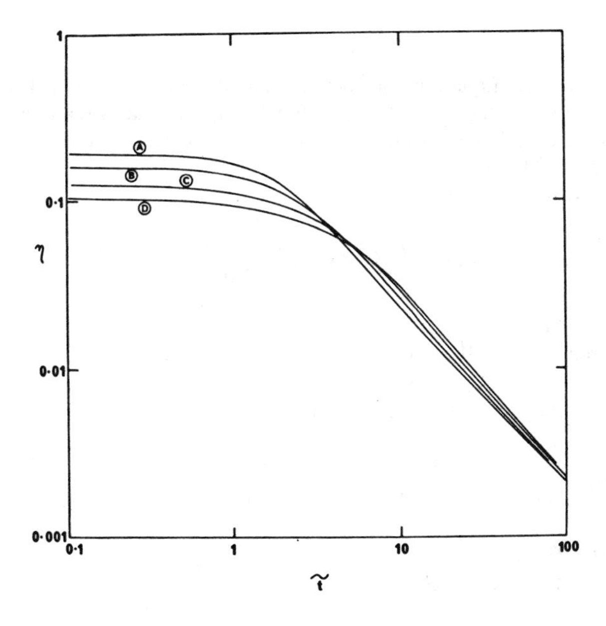

Figure 44. Effectiveness factor for series fouling. Isothermal pellet. Effect of kinetic order. $\tilde{t} = tf$, H = 2, $b_{Ao} = 0$ (A, n = 1/2; B, n = 1; C, n = 2; D, n = 3). (From Do and Weiland [21].)

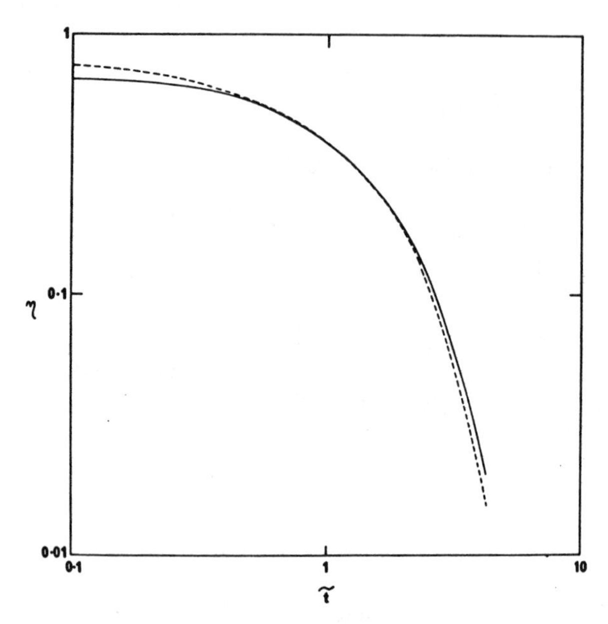

Figure 45. Comparison between approximate solution (—) and that of Masamune and Smith[2] (---), for parallel fouling. $t^{\sim} = td$, $\varphi = 2$, $H = \pi^2/4$. (From Do and Weiland [21].)

The results obtained by Equation 136 were compared with the numerical solutions, for $\varphi = 1, 2$, and 5 and $0.1 \leqq tf_A \leqq 3$. The Equation 136 approximates the numerical results with reasonable accuracy. At large values of tf_A there is some deviation, the maximum error being of the order of 20%.

For *series fouling* with large values of φ;

$$\eta = (1+\alpha)(1+\lambda)\int_0^1 [y(1-\lambda)+\lambda]\alpha\left(y^2 + 2\,tf_A\,y\,\frac{y-1}{1-\lambda}\frac{d\lambda}{d\,tf_A}\right)$$

$$\exp\left[-\left(b_0+\frac{1}{P}\right)tf_A + tf_A\,\frac{y^2}{P}\right]dy \tag{138}$$

where $d\lambda/d\,tf_A$ is given by the following equation

$$\left[\frac{4\,tf_A}{(1-\lambda)^3} + 2\,tf_A\,\alpha\left\{\frac{y_c-1}{(1-\lambda)^2[y_c(1-\lambda)+\lambda]} - \frac{y_c(y_c-1)}{(1-\lambda)\,[y_c(1-\lambda)+\lambda]^2}\right.\right.$$

$$\left.- \frac{y_c[1-2\lambda-2y_c(1-\lambda)]}{[y_c(1-\lambda)^2+\lambda(1-\lambda)]^2}\right\}$$

$$-2\varphi^2\,tf_A\,y_c\,\frac{y_c-1}{1-\lambda}\exp\left[-\left(b_0+\frac{1}{P}\right)tf_A + tf_A\,\frac{y_c^2}{P}\right]\frac{d\lambda}{d\,tf_A}$$

$$= \varphi^2 y_c^2 \exp\left[-(b_0 + 1/P)\, tf_A + tf_A\, y_c^2/P\right] - \frac{2}{(1-\lambda)^2}$$

$$+ \frac{2y_c\alpha}{(1-\lambda)\left[y_c(1-\lambda)+\lambda\right]} \tag{139}$$

The results obtained for the case of $P = 1$ and $b_0 = 1$ were compared with numerical results of Masamune and Smith [2] for $\varphi = 10$. The method appears to predict the effectiveness factor with sufficient accuracy.

Table 5
Catalyst Effectiveness Factor for Various Chemical Kinetics and Orders of Parallel Deactivation

Kinetics	d = 1	d = 2	d = 3
Half-order	$\ln\left(\frac{M}{\eta\sqrt{A^+}}\right) + \frac{1}{2}\left(\frac{1}{A^+} - \frac{1}{M}\right) = td$	$\frac{2M-1}{\frac{\eta}{H}M^{1/2}} - \frac{H(2A^+-1)}{A^+} + \ln\left[\frac{(1-\sqrt{A^+})(1+M^{1/2})}{(1+\sqrt{A^+})(1-M^{1/2})}\right] = Htd$	$\ln\left(\frac{M}{\eta\sqrt{A^+}}\right) + \frac{H}{\eta} - \frac{1}{N} + \frac{H^2}{\eta^2} - \frac{1}{N^2} = 2H^2td$
1st-order	$\ln\left(\frac{M}{\eta}\right) + \frac{1}{A^+} - \frac{1}{M} = td$	$\ln\left(\frac{M}{\eta}\right) + \frac{H}{\eta} - \frac{1}{N} = H\,td$	$\frac{H^2}{2\eta^2} - \frac{1}{2N^2} = H^2td$
2nd-order	$\ln\left(\frac{A^+M}{\eta}\right) + 2\left(\frac{1}{A^+} - \frac{1}{M}\right) = td$	$\ln\left\{\frac{(A^+)^2}{\eta}\right\} + \frac{H}{\eta} - \frac{1}{N} = Htd$	$1 - \frac{\eta}{H} - A^+ + \ln\left[\frac{\eta}{(A^+)^2}\right] + \frac{1}{N} - \frac{H}{\eta} + \frac{H^2}{2\eta^2} - \frac{1}{2N^2} = H^2td$
3rd-order	$\ln\left[\frac{(A^+)^2M}{\eta}\right] + 3\left(\frac{1}{A^+} - \frac{1}{M}\right) = td$	$2(A^+ - M) + \ln\left[\frac{(A^+)^3}{\eta}\right] + \frac{H}{\eta} - \frac{1}{H} = Htd$	$\frac{2}{3}[M^3 - (A^+)^3] + \frac{3}{2}[M^2 - (A^+)^2] + 3(M - A^+) + 2\ln\left[\frac{\eta}{(A^+)^3}\right] + \frac{2}{N} - \frac{2H}{\eta} + \frac{H^2}{2\eta^2} - \frac{1}{2N^2} = H^2td$
Michaelis-Menten	$\frac{1-\beta_M}{\beta_M}\ln\left(\frac{A^+}{M}\right) + \frac{1}{A^+} - \frac{1}{M} + \ln\left[\frac{H(1-A^+)}{\eta}\right] + \frac{1}{\beta_M}\ln\left(\frac{1+\beta_M-(\eta/H)}{A^+ + \beta_M}\right) = t_d$	$\frac{1+\beta_M}{\beta_M}\ln\left(\frac{M}{A^+}\right) + \ln\left[\frac{H(1-A^+)}{\eta}\right] + \left(\frac{H}{\eta} - \frac{1}{1-A^+}\right) + \frac{1}{\beta_M}\ln\left(\frac{A^+ + \beta_M}{M + \beta_M}\right) + \frac{1}{M+\beta_M} - \frac{1}{A^+ + \beta_M} = Htd$	$\frac{1}{1+\beta_M}\ln\left\{\frac{\left(1+\beta_M-\frac{\eta}{H}\right)(1-A^+)H}{\eta(A^+ + \beta_M)}\right\} + \frac{1}{1+\beta_M}\left(\frac{H}{\eta} - \frac{1}{1-A^+}\right) + \frac{1}{2}\left[\frac{H^2}{\eta^2} - \frac{1}{(1-A^+)^2}\right] + \frac{\beta_M}{1+\beta_M}\left(\frac{1}{A^+ + \beta_M} - \frac{1}{M+\beta_M}\right) + \frac{\beta_M}{2}\left[\frac{1}{(A^+ + \beta_M)^2} - \frac{1}{(M+\beta_M)^2}\right] = H^2td$

* $M = 1 - \eta/H$; $N = 1 = A^+$
From Do and Weiland [21]

N-th order Michaelis–Menten reactions, general deactivation order. Do and Weiland [21] using a lumping technique analytically obtained the effectiveness factor for small values of the Thiele modulus ($\varphi < 2$).

The effectiveness factor for parallel fouling ($m = 1$) when the main reaction is n-th order of Michaelis–Menten kinetics, are given in Table 5, for various values of deactivation order d.

Similar equations for series fouling for first-order deactivation ($f = 1$) are given in Table 6.

The parameters needed are

$$H: \left[\frac{\gamma}{\varphi(1+1/Sh)}\right]^2 \quad \text{with} \quad \begin{array}{ll} \gamma = \pi/2 & \text{slab} \\ \gamma = 2.4048 & \text{cylinder} \\ \gamma = \pi & \text{sphere} \end{array}$$

A^+: quasi steady-state reactant concentration given in Table 7.

For n-th-order reaction $g(A, 1) = A$ and for Michaelis–Menten,

$g(A, 1) = A(1+\beta_n)/A+\beta_n$.

For the case of *parallel fouling*, Figure 42 shows the effect of the deactivation order d on the time dependence of the effectiveness factor for half-order kinetics. The effect of reaction order and type of chemical kinetics are shown in Figure 43.

Table 6
Catalyst Effectiveness Factor for Various Chemical Kinetics and First-Order Series Deactivation

Kinetics	η	
n-th-order	$\frac{n}{1+b_0}\ln\left(\frac{M}{A^+}\right) + \frac{1}{b_0}\ln\left(\frac{H\,N}{\eta}\right) + \frac{1-(n-1)b_0}{b_0(1+b_0)}\ln\left(\frac{b_0+\frac{\eta}{H}}{N+b_0}\right) = tf_A$	for $b_0 \neq 0$
	$n\ln\left(\frac{M\,N}{A^+\frac{\eta}{H}}\right) + \frac{H}{\eta} - \frac{1}{N} = tf_A$	for $b_0 = 0$
Michaelis-Menten	$\frac{1}{1+b_0}\ln\left(\frac{M}{A^+}\right) + \frac{1}{b_0}\ln\left(\frac{H\,N}{\eta}\right) + \frac{1}{1+b_0+\beta_M}\ln\left(\frac{A^+\beta_M}{M+\beta_M}\right) + \frac{(b_0+1)^2+\beta_M}{b_0(1+b_0)(1+b_0+\beta_M)}\ln\left(\frac{b_0+\frac{\eta}{H}}{N+b_0}\right) = tf_A$	for $b_0 \neq 0$
	$\ln\left(\frac{M}{A^+}\right) + \frac{\beta_M}{1+\beta_M}\ln\left(\frac{H\,N}{\eta}\right) + \frac{H}{\eta} - \frac{1}{N} + \frac{1}{1+\beta_M}\ln\left(\frac{\beta_M+A^+}{M+\beta_M}\right) = tf_A$	for $b_0 = 0$

* $M = 1 - \eta/H$
$N = 1 - A^+$
From Do and Weiland [21]

Table 7
Pseudo Steady-State Reactant Concentration for Various Chemical Kinetics

Kinetics	A*
n = 1/2	$\left[-\frac{1}{2H}+\left(\frac{1}{2H}\right)^2+1^{1/2}\right]^2$
n = 1	$\frac{H}{1+H}$
n = 2	$-\frac{H}{2}+\left[\left(\frac{H}{2}\right)^2+H\right]^{1/2}$
n = 3	$\left(\frac{H}{2}\right)^{1/3}\left\{\left[\left(1+\frac{4H}{27}\right)^{1/2}+1\right]^{1/3}-\left[\left(1+\frac{4H}{27}\right)^{1/2}-1\right]^{1/3}\right\}$
Michaelis-Menten	$-\frac{1}{2}\left(\frac{1+\beta_M}{H}+\beta_M-1\right)+\left[\frac{1}{4}\left(\frac{1+\beta_M}{H}+\beta_M-1\right)^2+\beta_M\right]^{1/2}$

* *From Do and Weiland [21]*

For first-order chemical kinetics and first-order parallel deactivation the agreement between an approximate solution and a numerical one is quite close for small values of the Thiele modulus, but becomes poorer as Thiele modulus increases.

For *series fouling*, the effect of reaction order on the effectiveness factor is shown in Figure 44. Comparison between analytical and numerical solution are given in Figure 45 for $b_{A0} = 1$.

NOTATION

A	C_A/C_{A0} dimensionless concentration of A
A_S	C_{AS}/C_{A0} dimensionless concentration of A at pellet surface
A^+	C_A^+/C_{A0} dimensionless concentration of A, pseudo steady state, fresh catalyst.
A'	dimensionless concentration of A at $\xi = \xi_i$
a_0	initial active surface
a_w	active area poisoned by 1 mol of P
b	C_B/C_{A0} dimensionless concentration of B
b_0	C_{B0}/C_{A0}
b^*	$C_BD_B/C_{A0}D_A$
b_0^*	$C_{B0}D_B/C_{A0}D_A$
b_S^*	$C_{BS}D_B/C_{A0}D_A$
B	C_B/C_{B0} dimensionless concentration of B
Bi	h R/K, Biot number
C_A	concentration of A
C_B, C_D	concentration of B and C
C_C	concentration of deposited material on catalyst
Cf_C	concentration of active sites deactivated by C_C
$C_T(\xi)$	initial concentration of active sites, $C_T^* \mu(\xi)$
C_T^*	constant defined by Equation 9)
C_{A0}, C_{B0}, C_{D0}	concentration of reactant or products in bulk
C_{AS}, C_{BS}, C_{DS}	concentration of reactant or products at pellet surface
C_p	heat capacity
C_A^+	concentration of A, pseudo steady-state, fresh catalyst
CD	quantity of coke deposited on the catalyst
Ci	concentration of specie i
C_{Pmax}	maximum adsorbed poison concentration

C_P concentration of precursor poison
C_{Pa} adsorbed poison concentration
C_W concentration of deposited poison on catalyst
d deactivation for parallel fouling
d_A C_D/C_{A0} dimensionless concentration o f D
d_{A0} C_{D0}/C_{A0}
D_A, D_B, D_D, D_P effective diffusivity of A, B, D, and P
D'_A, D'_B, D'_D, D'_P effective diffusivity of A, B, D, and P in fouled region of catalyst (shell model)
E activation energy, main reaction
Ed activation energy, parallel reaction
Ef activation energy, series reaction
Ep activation energy, impurity deactivation
f deactivation order for series fouling
F flow rate
$g(A, T^*)$ function used in Equations 15 through 17
$g_d(A, T^*)$ function used in Equations 15 and 18
$g_f(B, T^*)$ function used in Equations 16 and 19
$g_p(P, T^*)$ function used in Equation 81
h heat transfer coefficient
I_n modified Bessel function of the first kind and order n
k effective thermal conductivity
k_A rate constant, main reaction
k_{A0} rate constant, main reaction, $T = T_0$
k_d parallel fouling rate constant (deposition of coke)
k_{d0} parallel fouling or consecutive reaction-rate constant, $T = T_0$
k_f series fouling rate constant (deposition of coke)
k_{f0} series fouling rate constant, $T = T_0$
k_g mass transfer coefficient
k_m $k_A \exp(-\lambda_p C_W)$: first-order rate constant with poisoning
k_p impurity poisoning rate constant
k_{p0} impurity poisoning rate constant, $T = T_0$
K_A, K_D adsorption equilibrium constant of A and D
K_{A0}, K_{D0} adsorption equilibrium constant a $T = T_0$
K_P adsorption equilibrium constant of precursor poison
K'_i K_iC_{i0}, dimensionless adsorption equilibrium constant
K'_D $K_{D0}C_{A0}$
K_M Michaelis constant
$\mathscr{L}$ $D_AC_PP/(\varepsilon_p k)$ reciprocal Lewis number
L thickness of slab/2
m, q activity order, main and fouled reactions
n, p reactant order, main and fouled reactions
p D_B/D_A
P C_p/C_{p0}
P^+ $C_p^+/_{p0}$
P_{max} C_{pmax}/C_{p0}
Q heat of adsorption
r r'/R
r_i radius of central core for shell model
r' radial coordinate
$r(Ci, T)$ rate of reaction, fresh catalyst
R radius of pellet or gas constant
$\mathscr{R}_A(C_A, T, \theta)$ rate of main reaction, catalyst deactivated
$\mathscr{R}_d(C_A, T, \theta)$ rate of parallel reaction, catalyst deactivated
$\mathscr{R}_f(C_B, T, \theta)$ rate of series reaction, catalyst deactivated
$\mathscr{R}(Ci, T, \theta)$ rate of reaction, catalyst deactivated
$\mathscr{R}'_P(C_p, T, \theta)$ rate of poisoning reaction

- s — selectivity k_{d0}/k_{A0} (parallel fouling), k_{f0}/k_{A0} (series fouling)
- Sh — Sherwood number modified, k_gR/D_A (spehere), k_gL/D_A (slab)
- Sh_P — Sherwood number modified, k_gR/D_P
- t' — time
- t — $t(k_Pa_Wa_0^{q-1}C_{P0}^P)$, dimensionless time
- t_d — $t'k_{d0}C_{A0}y/C_T^*$, dimensionless time
- t_f — $t'k_{f0}C_{B0}y/C_T^*$, dimensionless time
- t_{fA} — $t'k_{f0}C_{A0}y/C_T^*$, dimensionless time
- t_f^* — $t'k_{f0}C_{B0}D_Ay/(D_B\ C_T^*)$, dimensionless time
- T — temperature
- T_0 — bulk temperature
- T_S — surface pellet temperature
- T^* — T/T_0
- T_S^* — T_S/T_0
- $u(r)$ — distribution function of activity
- V — volume of catalyst
- x' — distance from center plane to any point in the slab
- x — x'/L
- x_i — dimensionless position of the poison boundary
- y — active sites fouled by 1 mol of coke
- Z — parameter showing the distribution of activity

Greek Symbols

- α — geometric shape parameter (0 slab, 1 cylinder, 2 sphere)
- β — $(-\Delta H)D_AC_{A0}/kT_0$, thermicity factor
- ΔH — heat of reaction
- γ — E/RT_0, dimensionless activation parameter, main reaction
- $\gamma_d, \gamma_f, \gamma_P$ — dimensionless activation parameters, deactivating reaction
- δ_i — Q_i/RT_0, dimensionless adsorption equilibrium parameter
- ε — relative effectiveness factor
- ε_p — catalyst porosity
- η — effectiveness factor
- η_B — effectiveness factor for reactant B
- η_0, η_{Bi} — effectiveness factor of the unpoisoned pellet
- θ — fraction of activity left, fraction of site-actives left
- Λ — ratio between poisoned and unpoisoned reaction rates
- λ — $k/\varrho\ C_p$, thermal diffusivity
- λ_P — $a_W\varepsilon_PC_{PS}/a_0\varrho$, dimensionless constitutive coefficient
- ξ — dimensionless space variable
- ϱ — density of catalyst pellet
- σ — defined in Equation 66
- τ — $t'D_A/R^2\varepsilon_p$ or $t'D_A/L^2\varepsilon_p$, dimensionless time, parallel, series or irreversible poisoning
- τ — $t'D_P/R^2\varepsilon_p$ or $t'D_p/L^2\varepsilon_p$, dimensionless time, reversible poisoning
- φ — $R(k_{A0}/D_a)^{1/2}$ or $L(k_{A0}/D_A)^{1/2}$, Thiele modulus, main reaction
- φ — $R(k_{A0}K_{A0}/D_A)^{1/2}$ or $L(k_{A0}K_{A0}/D_A)^{1/2}$, Thiele modulus
- φ_B — $R(k_{d0}/D_B)^{1/2}$, or $L(k_{d0}/D_B)^{1/2}$, Thiele modulus consecutive reaction
- φ' — $R(k_{A0}/D'_A)^{1/2}$ or $L(k_{A0}/D'_A)^{1/2}$, Thiele modulus, shell model
- φ_d — $R(k_{d0}/D_A)^{1/2}$ or $L(k_{d0}/D_A)^{1/2}$, Thiele modulus, parallel fouling
- φ_f — $R(k_{f0}/D_A)^{1/2}$ or $L(k_{f0}/D_A)^{1/2}$, Thiele, modulus, series fouling

φ_m — $R(k_{Av}/D_A)^{1/2}$, Thiele modulus for nonuniform catalyst

φ_P — $R(k_{P0}/D_P)^{1/2}$ or $L(k_{P0}/D_P)^{1/2}$, Thiele modulus, poisoning reaction

χ — conversion

ψ_1 — cumulative concentration at the interior collocation point

ω — amount of coke per unit volume/maximum amount of coke per unit volume

REFERENCES

1. Butt, J. B., et al., "On the Separatibility of Catalytic Deactivation Kinetics," *Chem. Eng. Sci.,* Vol. 33, No. 10 (1978), pp. 1321–1329.
2. Masamune, S., and Smith, J. M., "Performance of Fouled Catalyst Pellets," *A. I. Ch. E. Journal,* Vol. 12, No. 2 (1966), pp. 384–394.
3. Lamba, H. S., and Duduković, M. P., "A New Technique to Solve Models for the Reaction of Solid Particles and for Parallel Catalyst Deactivation," *Chem. Eng. J.,* Vol. 16 (1978), pp. 117–135.
4. Chu, C., "Effect of Adsorption on the Fouling of Catalyst Pellets," *I & EC Fundamentals,* Vol. 7, No. 3 (1968), pp. 509–514.
5. Kam, E. K. T., et al., "The Effect of Film Resistances on the Fouling of Catalyst Pellets-I," *Chem. Eng. Sci.,* Vol. 32 (1977), pp. 1307–1315.
6. Sagara, M., et al., "Effect of Nonisothermal Operation on Catalyst Fouling," *A. I. Ch. E Journal,* Vol. 13, No. 6 (1967), pp. 1226–1229.
7. Weisz, P. B., and Hicks, J.S., "The Behaviour of Porous Catalyst Particles in View of Internal Mass and Heat Diffusion Effects," *Chem. Eng. Sci.,* Vol. 17 (1962), p. 265.
8. Corbett, W. E., and Luss D., "The Influence of Non-Uniform Catalytic Activity on the Performance of a Single Spherical Pellet," *Chem. Eng. Sci.,* Vol. 29, No. 6 (1974), pp. 1473–1483.
9. Shadman-Yazdi, F., Petersen, E. E., "Changing Catalyst Performance by Varying the Distribution of Active Catalyst Within Porous Supports," *Chem. Eng. Sci.,* Vol. 27 (1972), pp. 227.
10. Murakami, Y., et al. "Effect of Intraparticle Diffusion on Catalyst Fouling," *I & EC Fundamentals,* Vol. 7, No. 4 (1968), pp. 599–605.
11. Kam, E. K. T., et al., "The Effect of Film Resistances on the Fouling of Catalyst Pellets–II," *Chem. Eng. Sci.,* Vol. 32 (1977), pp. 1317–1325.
12. Hegedus, L. L., "On the Poisoning of Porous Catalysts by an Impurity in the Feed," *I & EC Fundamentals,* Vol. 13, No. 3 (1974), pp. 190–196.
13. Becker, E. R., and Wei, J., "Nonuniform Distribution of Catalysts on Supports–II. First Order Reactions with Poisoning," *J. Cat,* Vol. 46, No. 3 (1977), pp. 372–381.
14. Becker, E. R., Ph D. Dissertation, Chem. Eng. Dept, Univ of Delaware, 1975.
15. Gioia, F., et al., "Simultaneous Reversible Poisoning and First Order Reaction in Porous Spherical Pellets," *Chem. Eng. Sci.,* Vol. 27 (1972), pp. 1745–1748.
16. Sada, E., and Wen, C. J., "Effect of Catalyst Poisoning on the Overall Selectivity and Activity," *Chem. Eng. Sci.,* Vol. 22 (1976), pp. 559–571.
17. Valdman, B., et al., "Impurity Poisoning of Catalyst Pellets," *J. Cat.,* Vol. 42, No. 2 (1976), pp. 303–311.
18. Valdman, B., and Hughes, R., "An Analysis of Reversible Poisoning of a Single Catalyst Pellet in the Presence of External Resistance," 6 Simposio Ibero-americano de Catálise, Río de Janeiro, Brasil, 1978, p. 24.
19. Tai, N. M., and Greenfield, P. F., "Approximate Solutions for Catalyst Deactivation within a Particle," *Chem. Eng. J.,* Vol. 16 (1978), pp. 89–100.

20. Kulkarni, B. D., and Ramachandran, P. A., "A Simple Method for Calculation on Effectiveness Factors under Conditions of Catalyst Fouling," *Chem. Eng. J.*, Vol. 19 (1980), pp. 57–66.
21. Do, D. D., and Weiland, R. H., "Interphase and Intraphase Effectiveness Factors for Catalysts with Self-Poisoning," *Chem. Eng. J.*, Vol. 21 (1981), pp. 133–148.
22. Villadsen, J. V., and Stewart, W.E., *Chem. Eng. Sci.*, Vol. 22, (1967), pp. 1483.

SECTION II

DISTILLATION AND EXTRACTION

CONTENTS

CHAPTER 10

GENERALIZED EQUATIONS OF STATE FOR PROCESS DESIGN

Helmut Knapp
Institute of Thermodynamics and Plant Design
Technical University of Berlin
Federal Republic of West Germany

CONTENTS

INTRODUCTION

When the process of a production plant is conceived, designed, revised, analysed, or optimized, accurate and reliable information is required about the properties of all materials handled in various operations. Beginning with the specifications for feed and products, the process engineer must calculate mass and energy balances, changes of state, phase equilibria, etc.

The basis of equipment design is the knowledge of the operating conditions such as temperature, pressure, composition, transfer of mass, heat, and work at all important points in the process. The transfer surfaces required for heat and mass transfer can then be determined.

As an example, consider a heat exchanger where one stream is cooled and partially condensed while a countercurrent stream is heated and partially evaporated. The required surface can be determined if the cooling and heating curves (i.e., the Q–T or H–T diagrams) and the heat transfer coefficients are known.

As another example, consider a distillation column where heat and mass is transferred between the vapor and the liquid phase. The number of theoretical stages can be determined if the condition of the final thermodynamic equilibrium is known.

All these calculations require information about the thermodynamic properties such as density, heat capacity, enthalpy, heat of vaporization, distribution coefficients in phase equilibrium and heat of reaction.

The engineer engaged in process design and in need of the information should consider all possible methods and all available data:

- Experimental data, often correlated by empirical and graphical methods
- Reliable and proven equations describing the thermodynamic properties
- Plausible, mostly simplified, physical, or chemical models as a basis for proposing adequate correlations
- General thermodynamic laws

In this chapter a review is given of generalized methods that can be used to predict the required information on volumetric and caloric properties of a great number of technically important substances and mixtures thereof.

The correlations discussed in this chapter are based on the application of equations of state. The described methods offer several advantages:

- The effect of pressure at any temperature on all thermodynamic properties can be calculated with an equation of state
- The principle of corresponding states allows the use of generalized equations of state. Only a few parameters are needed to characterize a pure substance or a mixture
- Electronic computers can be easily used as the information is available in the form of analytical equations

The application of the method and its advantages are limited to fluid materials consisting of so-called normal fluids such as noble gases, nitrogen, oxygen, carbon monoxide, hydrocarbons, and certain hydrocarbon derivatives. Carbon dioxide, hydrogen sulfide, hydrogen and—with reservations—some polar substances can also be included.

At the end of the chapter references are listed offering more detailed information for each paragraph.

EQUATIONS OF STATE

Equations of states (EQS) are material equations that describe the PVT behavior of pure substances or the PVTX behavior of mixtures in the fluid state. Background reading can be found in References 1 through 3.

There are various analytical forms of EQS for pure substances

Example for a pure substance:

$$p = p(T, v, A_{ik})$$

or $$v = v(T, p, A_{ik})$$

or $$z = z(T, p, A_{ik})$$

or $$z = z(T, v, A_{ik})$$

or $$z = z(T_r, v_r, A_k)$$

where A_{ik} = specific coefficient for substance i, k = 1, 2, ...
A_k = general coefficient, k = 1, 2, ...
n = number of moles in a system
p = pressure
R = universal gas constant, 8.314 J/mol K
T = temperature
T_r = reduced temperature, T/T_c
T_c, v_c = critical properties
V = total volume of a system
v = molar volume, V/n
v_r = reduced volume, v/v_c
x = concentration of a component in a mixture, e.g., mole fraction

PRESENTATION OF PVT PROPERTIES

The information on the PVT properties of a pure substance can be presented in various formats

- In tables: for example, Table 1, a steam table of water.
- In diagrams: for example, Figure 1, a schematic p–v diagram with isotherms; Figure 2, a t–v diagram of H_2O with isobars; Figure 3, a schematic p–T diagram with isochores
- By multivariable mathematical functions, so-called equations of state (EQS) For example, EQS for the ideal gas state:

$$pv = RT \quad \text{or} \quad z = 1 \tag{1}$$

where z = compressibility factor, pv/RT

or EQS for a real gas proposed by van der Waals [1]:

$$\left(p + \frac{a}{v^2}\right)(v - b) = Rt \quad \text{or} \quad z = \frac{v}{v-b} - \frac{a}{RT\,v} \tag{2}$$

where a, b = specific constants of a substance

The corrective term a/v^2 accounts for the effect of the forces acting between the molecules, and b accounts for the "hard" volume occupied by the molecules. In the equation for the compressibility the first term is often called the "repulsive" and the second the "attractive" term. These meaningful corrections give a simple, qualitively correct, but quantitatively less accurate analytical description of the behavior of a real gas.

Example: Virial EQS

$$z = 1 + B(T)\varrho + C(T)\varrho^2 + \ldots \tag{3}$$

where ϱ = molar density, 1/v
B(T) = second virial coefficient
C(T) = third virial coefficient

In the virial EQS proposed by Kamerlingh Onnes [5] corrections are made for the effect of interaction between two, three, and more molecules. The coefficients for pure substances

Table 1
Properties of Water and Steam

T °C	p bar	v^L m³/kg	v^V m³/kg
0	0.0061	0.00100	206.29
50	0.1233	0.00101	12.045
100	1.0132	0.00104	1.673
200	15.550	0.00115	0.127
300	85.92	0.00140	0.021
374.15	221.20	0.0031	0.0031

T = temperature; p = pressure; v^L = specific volume of saturated liquid; v^V = specific volume of saturated vapor.

depend only on temperature and can be calculated by the methods of statistical mechanics based on knowledge about the interaction energy of the molecules. The virial EQS is theoretically important but fails at higher densities.

Example: EQS with "Simple" and "Reference" Fluids

$$z = z^{(0)} + \omega z^{(1)} \tag{4}$$

where $z^{(0)}$ = compressibility of simple fluid
$z^{(r)}$ = compressibility of reference fluid
$z^{(1)}$ = corrective term, $(z^{(r)} - z^{(0)})/\omega^{(r)}$
ω = acentric factor

According to Pitzer's [6] suggestion the compressibility consists of a term for simple spherical molecules and a corrective term that accounts for the effect of nonspherical shape of the molecules. $z^{(1)}$ can be calculated as the difference between the compressibility of a reference and of a simple fluid.

Example: Augmented Rigid Body EQS

$$z = z^{\text{repulsive}} - z^{\text{attractive}} \tag{5}$$

where
$$z^{\text{rep}} = \frac{1 + \varrho^* + \varrho^{*2} - \varrho^{*3}}{(1 - \varrho^*)^3} \tag{5a}$$

$$z^{\text{attr}} = \frac{f_1(\varrho^*)}{T} + \frac{f_2(\varrho^*)}{T^2} \tag{5b}$$

with ϱ^* = reduced density, ϱv^*
v^* = volume of hard sphere molecules
f_1, f_2 = functions of reduced density

As a result of computer calculations for the behavior of hard bodies, simple analytical equations were developed (e.g., Equation 5a by Carnahan and Starling [17]. The attractive term is developed in powers of (1/T) modeling the results of perturbation theory.

Example: Theoretical Nonanalytical EQS

$$z = 1 + \varrho\left(\frac{\partial A^+}{\partial \varrho}\right)T \tag{6}$$

where A^+ = molar residual Helmholtz free energy

Starting with our knowledge of the properties of molecules and the laws governing the motion and interaction of molecules it is possible with the help of statistical mechanics to calculate macroscopic properties such as the free energy. For process calculations however this theoretical approach is too complicated, too time consuming, and too inaccurate.

EQUATION OF STATE FOR MIXTURES

The materials treated in chemical production plants are sometimes pure substances but are usually mixtures of many components. Thermodynamic properties of multicomponent

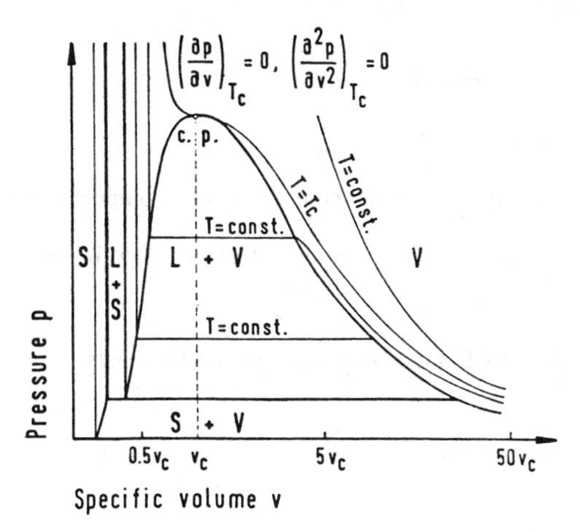

Figure 1. Typical schematic diagram showing pressure vs. specific volume with isotherms and boundaries of one- or two-phase areas.

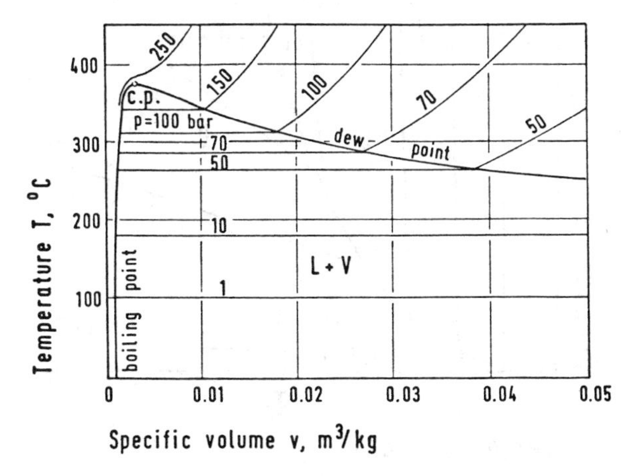

Figure 2. Diagram showing temperature vs. specific volume of water with isobars and saturation curves for vapor and liquid.

systems depend not only on temperature and pressure but also on composition. The PVTX behavior of the mixture can be described by an equation of state for a fluid whose coefficients depend on the composition.

Example: Structure of EQS for Mixtures

$$z = z(T, v, A_{mk}) \tag{7}$$

where A_{mk} = coefficients for a specific mixture of constant composition, k = 1, 2, ...

The coefficients could be found by fitting to experimental PVTX data or—in lack of sufficient experimental data—more conveniently by combining the coefficients of the pure components in the appropriate manner.

Example: Structure of Mixing Rule

$$A_{mk} = A_{mk}(A_{ik}, x_i, k_{ij}) \tag{8}$$

where k_{ij} = binary parameter for the combination of component i and j

The mixing and combination rules are empirical considering, however, theoretical knowledge.

Example: Mixing Rules and Combination Rule for the Constants a and b in Equation 9

$$a_m = \sum\sum x_i x_j a_{ij} \qquad a_{ij} = \sqrt{a_i a_j} \tag{9}$$

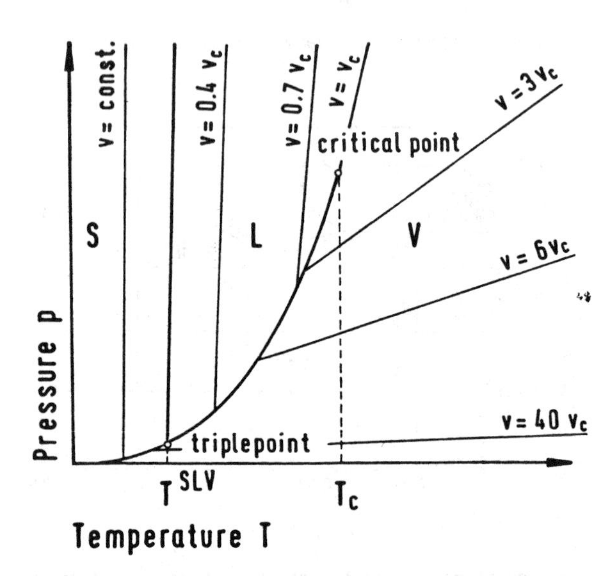

Figure 3. Schematic diagram showing pressure vs. temperature with isochores and coexistence curves.

$$b_m = \sum x_i b_i \tag{10}$$

Liquid mixtures of strongly interacting substances (e.g. by chemical association or solvation) can be better described with models for excess functions. The equation of state fails mainly because simple mixing rules are inadequate. More sophisticated mixing rules, however, often improve the accuracy of the calculations.

MODIFICATIONS OF EQUATIONS OF STATE

During the past century hundreds of modifications to the basic equations and to the mixing rules have been proposed in order to improve the accuracy and to extend the range of validity. The development is characterized by increasingly complicated structures—especially after the advent of fast electronic computers. Most of the equations have been arranged empirically and the numerable coefficients (sometimes more than fifty) are determined by fitting to all available experimental data using the criterion of minimum deviation between experimental and calculated data. The sets of coefficients are not unique. For the description of mixtures, mixing rules must be given for each coefficient.

GENERALIZED EQUATIONS OF STATE (GEQS)

Experience indicates and molecular theory explains that many fluids, pure substances or mixtures, have a similar PVT behavior, i.e., isotherms in p–v diagrams or vapor pressure curves look alike. According to the principle of corresponding states, proposed originally by van der Waals [8], a fluid can be characterized by the coordinates of one outstanding point on the PVT surface, such as the critical point or the triple point. The critical point is preferable as it contains more "information" about the fluid state than the triple point.

In accordance with a few "general" rules the coefficients of an equation of state can be calculated for a specific substance from the coordinates of the "correspondence point." It is also possible to introduce reduced variables, e.g., $T_r = T/T_c$, $p_r = p/p_c$ and $v = v/_c$ into an equation of state which then has a "universal" set of coefficients once and for all fitted to a "reference" fluid. A mixture can be represented by "pseudo critical" coordinates calculated with a few mixing rules.

Example: GEQS with Specific Coefficients for Pure Components or Mixtures

$$z = z(T, v, A_{ik}) \tag{11}$$

where $A_{ik} = A_{ik}(T_{ci}, p_{ci}, v_{ci} \quad \text{or} \quad \omega_i)$ (12)

If the GEQS describes the PVTX behavior of a multicomponent fluid, mixing rules are required for all k coefficients, A_{ik}.

Example: GEQS with Reduced Variables and "Universal" Coefficients

$$z = z(T_r, v_r, \omega, A_k) \tag{13}$$

where A_k = "universal" constants

For multicomponent fluids pseudocritical values are calculated for all correspondence parameters with empirical mixing and combination rules, e.g.,

$$T_{cM} = \sum\sum x_i x_j T_{cij} \quad \text{with } T_{cij} = \sqrt{T_{ci}T_{cj}}\,(1-k_{ij}) \tag{14}$$

$$v_{cM} = \sum\sum x_i x_j v_{cij} \quad \text{with } v_{cij} = (1/8)\,(v_{ci}^{1/3} + v_{cj}^{1/3}) \tag{15}$$

$$\omega_{cM} = \sum x_i \omega_i \tag{16}$$

The reader may refer to References 9 and 10 for additional discussions.

CHOICE OF EQUATION OF STATE

It is important to consider the purpose for which an equation of state is to be used: (see Table 2).

Case 1

Sometimes a pure substance such as water, nitrogen, methane, or ammonia has been thoroughly investigated, i.e., many pvT data, vapor pressures, heat capacities, velocities of sound have been measured. If the properties of the substance should be represented in tables or diagrams within the experimental accuracy, it is practical to interpolate the data with an equation of state containing many adjustable coefficients (often more than 40). A smaller set of coefficients might be sufficient if different sets are used for interpolations in limited areas of the p–T field.

Case 2

Only few experimental points (e.g., a few vapor pressures or densities) are known. Sometimes only the chemical structural formula is known. A simple but "safe" equation of state, mostly of the van der Waals type, allows the estimation of PVT properties.

Table 2
Input and Output of Various Types of Equations of State

Information	Comp.	Result	Type
H–C–C–H (structural formula of ethane, with H above and below each C)	1	Estimate PVT	EQS: VdW
Few Q(T, p) Few $p^{LV}(T)$	1	Calculate PVT	EQS: VdW
> 100 {PVT, caloric} data	1	Calculate PVT, $h-h^0$ Construct table or diagram	Empirical virial EQS with fitted coeff. A_{ik}, k=20–60
T_c, p_c, v_c or ω	1	Calculate PVT, $h-h^0$	GEQS with calculated coeff. $A_{ik}(T_{ci}, p_{ci}, v_{ci}$, or $\omega_i)$
T_c, p_c, v_c or ω VLE → k_{ij}	2, 3...	Calculate PVT, $h-h^0$ VLE, LLE	GEQS: VdW or virial +mix. rules $A_{mk}(A_{ik}, k_{ij}, x)$

Info—Input information that is or should be available. Comp—Number of components in the fluid. Result—Information that is requested and can be calculated with the equation. Type—Equation that is recommended for the specified in- and output.

Case 3

If process calculations for a great variety of multicomponent mixtures must be done, it is most practical to use a generalized equation of state. Only a few specific parameters for each substance and only one or two parameters for each binary combination is required.

The further discussion will be concerned with Case 3.

THERMODYNAMIC FUNCTIONS

EQS give information how the specific volume or the density depend on temperature, pressure, and composition. In addition, however, many more thermodynamic properties can be calculated from EQS by referring to general thermodynamic relations.

Example: Enthalpy Departure $h - h^0$

The enthalpy of a substance in the state of an ideal gas depends only on temperature:

$$h^0(T, p^0) = h^0(T^+, p^0) + \int_{T^+}^{T} c_p^0(T)\, dT \tag{17}$$

where h^0 = standard enthalpy ideal gas state
T^+ = arbitrary reference temperature
p^0 = pressure in standard state, e.g., $p^0 = 0$
$c_p^0(T)$ = molar heat capacity in ideal gas state as function of temperature, listed in standard thermophysical tables

$$c_p^0(T) = \sum C_m T^m \tag{18}$$

The effect of pressure on the enthalpy, the so-called enthalpy departure $h - h^0$, can be calculated if the PVTX behavior of the fluid is known:

$$h - h^0 = \int_{p^0}^{p} \left(\frac{\partial h}{\partial p}\right)_{T,x} dp \tag{19}$$

$$h - h^0 = \int_{p^0}^{p} \left[v - T\left(\frac{\partial v}{\partial T}\right)_{p,x}\right] dp \tag{19a}$$

$$h - h^0 = \int_{\infty}^{v} \left[T\left(\frac{\partial p}{\partial T}\right)_{v,x} - p\right] dv + pv - RT \tag{19b}$$

Useful equations of state are structured such that the differentiation and integration can be carried out analytically.

The knowledge required to calculate HTPX is therefore (see Figure 4):

- $c_p^0(T)$
- $z(T, v, x)$

The enthalpy or the molar heat capacity of an ideal gas mixture can be calculated with ideal mixing rules:

$$h^0(T, p^0, x) = \sum x_i h_i^0(T) \tag{20}$$

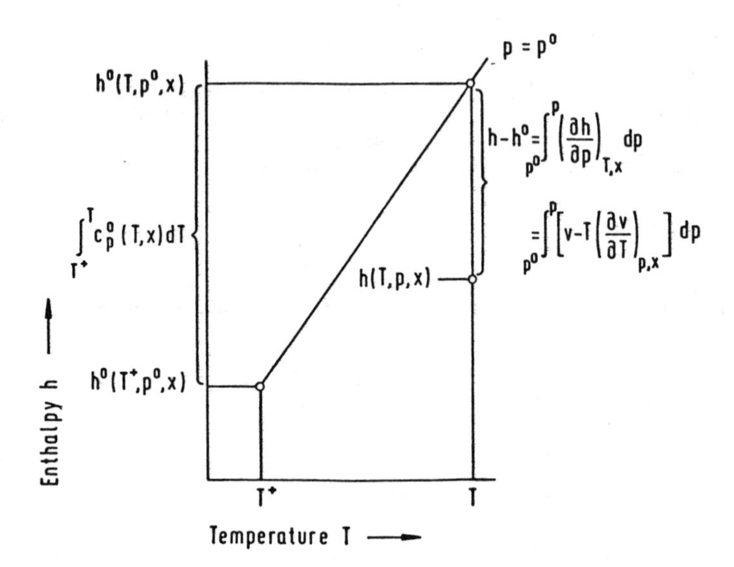

Figure 4. Enthalpy-temperature diagram of fluid with constant composition showing the isobar p = p° for the standard state (ideal gas) and the isothermal pressure effect at one temperature.

$$c_p^0(T, x) = \sum x_i c_{p,\,i}^0(T) \tag{21}$$

Example: Entropy Departure $s - s^0$

The effect of pressure on the entropy can be calculated in accordance with a so-called Maxwell relation by referring to an EQS:

$$\left(\frac{\partial s}{\partial p}\right)_{T,x} = -\left(\frac{\partial v}{\partial T}\right)_{p,x} \tag{22}$$

The entropy in the standard state (ideal gas at $p^0 = 0.1013$ MPa) at the reference temperature $T^+ = 25°$ C = 298. 15 K is usually listed. The entropy of a mixture can then be calculated:

$$s(T, p, x) = \sum x_i s_i^0(T^+, p^0) + R \sum x_i \ln x_i + \int_{T^+}^{T} \sum x_i c_{p,i}^0(T)\, d \ln T - \int_{p0}^{p} \left(\frac{\partial v}{\partial T}\right)_{p,x} dp \tag{23}$$

Example: Fugacity of a Pure Substance or of a Component in a Mixture

Knowledge of the fugacity helps to find important information, e.g., vapor pressures, distribution coefficients, or K-values for components in coexisting phases. Thermodynamic equilibrium in coexisting phases is characterized by equal temperature, pressure, and fugacity of each component in all phases.

The fugacity of a single- or multi-component fluid, $f = \varphi p$, and of a component i for a mixture, $f_i = \varphi_i p_i$, can be calculed provided an EQS is known:

$$RT \ln \varphi = \int_{p^0}^{p} \left(v - \frac{RT}{p} \right) dp \tag{24a}$$

$$= \int_{\infty}^{v} \left(\frac{RT}{v} - p \right) dv + pv - RT - RT \ln z \tag{24b}$$

$$RT \ln \varphi_i = \int_{p^0}^{p} \left[\left(\frac{\partial V}{\partial n_i} \right)_{T,p,n_j \neq i} - \frac{RT}{p} \right] dp \tag{25a}$$

$$= \int_{\infty}^{v} \left[\frac{RT}{v} - n \left(\frac{\partial p}{\partial n_i} \right)_{T,p,n_j \neq i} \right] dv - RT \ln z \tag{25b}$$

where φ = fugacity coefficient, $\varphi = 1$ for $p = 0$
φ_i = fugacity coefficient of component i, $\varphi_i = 1$ for $p = 0$
p_i = partial pressure, $x_i p$

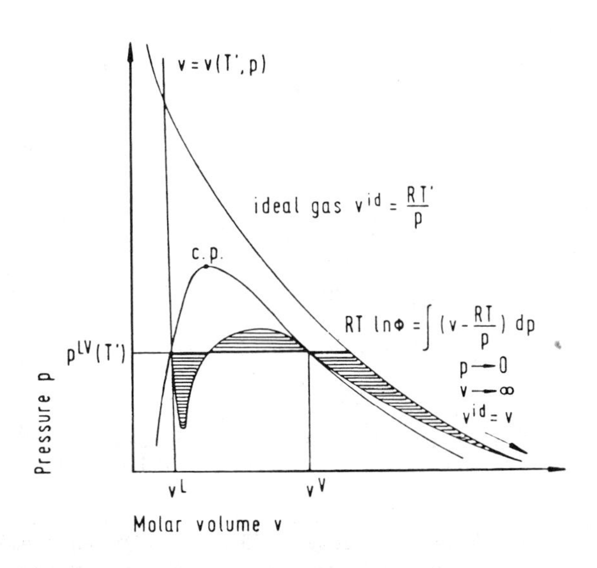

Figure 5. Schematic pressure-volume diagram with subcritical isotherm $T = T' < T_c$ for the ideal gas and for a real fluid. The shaded area represents $\int (v^{real} - v^{ideal})dp$ and is an indication of the "non-ideality" of the fluid.

Example: Vapor Pressure of a Pure Substance

If vapor and liquid coexist, the temperature, pressure and the chemical potential μ or the fugacity ($f = \varphi p$) and hence the fugacity coefficient φ in both phases are equal

$$f^L(T, p) = f^V(T, p) \tag{26a}$$

or

$$\varphi^L(T, p) = \varphi^V(T, p) \tag{26b}$$

For a given temperature $T = T'$ a pressure must be found where the EQS has two solutions v^L and v^V for which the fugacities φ^L and φ^V calculated with Equation 24 are equal. As shown in Figure 5 the equilibrium condition $\varphi^L = \varphi^V$ means that the shaded areas between $v^L < v < v^V$ are equal (Maxwell criterion).

Example: Phase Equilibrium in Multicomponent Systems

When mixtures are cooled or heated or separated it is important to know dew or bubble points or compositions and fractions of all coexisting phases. The problems can be solved if the distribution coefficients or the K-values of all components in a mixture are known.

In the thermodynamic equilibrium the fugacities of each component in all coexisting phases are equal.

Example: Conditions for Vapor-Liquid Equilibrium

$$f_i^L = f_i^V \tag{27a}$$

$$\varphi_i^L x_i p = \varphi_i^V y_i p \tag{27b}$$

and

$$K_i = y_i/x_i = \varphi_i^L/\varphi_i^V \tag{28}$$

where x_i = mole fraction, φ_i^L = fugacity coefficient } of component i in liquid

y_i = mole fraction, φ_i^V = fugacity coefficient } of component i in vapor

As fugacity coefficients of components in a mixture can be calculated with Equation 25 it is possible to calculate K-values. The procedures are all "trial and error," and often there are several stacked iterative loops.

The bubble point pressure of a liquid mixture at a given temperature is calculated by first assuming a pressure. With T, p, x Equation 25 and an EQS can be used to calculate φ_i^L. Assuming $\varphi_i^V = 1$, the vapor mole fractions $y_i = x_i/\varphi_i^L$ are calculated. With T, p, y_i values for φ_i^V and for a new $y_i = x_i\varphi_i^V/\varphi_i^L$ are calculated until there is no more change. The deviation from the criterium $\sum y_i = 1$ is an indication for the next iteration of the pressure.

GENERAL PROCEDURES FOR PROCESS CALCULATIONS

In basic process design, energy- and mass-balances are performed. Therefore information is required about the thermodynamic properties of the materials as a function of temperature, pressure, and composition such as density, vapor pressure, dew point, bubble point, state (vapor, liquid, multiphase), heat capacity, enthalpy, entropy, free energy, fugacity, fugacity coefficient, or equilibrium constant for phase or chemical equilibria.

Nowadays it is popular, most convenient, and efficient to do the process calculations with the aid of electronic computers.

It is possible and opportune to calculate volumetric and caloric properties of a fluid material for the state of an ideal gas first and then correct for the effect of pressure.

The properties in the state of an ideal gas can be calculated with two material equations:

$$pV = n\,RT \tag{29}$$

where $n = m/M$

$$c_p^0(T) = C_0 + C_1 T + C_2 T^2 + \ldots \tag{30}$$

For each specific substance the values of M, C_0, C_1, C_2 and the values of h_{293}^0 and s_{298}^0 must be known and stored. For mixtures ideal mixing rules are valid.

The effect of pressure can be calculated in accordance with Equations 19, 22, 24, and 25. The only material equation required for specific substances or mixtures is an equation of state including mixing rules, viz.

$$z = z(T, p, A_{ik}) \quad \text{for pure substances}$$

or

$$z = z(T, p, A_{mk}) \quad \text{for mixtures}$$

For each specific substance or mixture the coefficients in the equation of state and in the mixing and combination rules must be known and stored. Generalized equations of state GEQS are very practical because only a few specific parameters (e.g., T_c, p_c, v_c, or ω) for each substance and often only one specific parameter (k_{ij}) for each binary combination have to be known and stored.

EXAMPLES OF GEQS

Four frequently used and "popular" GEQS were selected. These equations all require 3 parameters to characterize a pure substance. One binary parameter is required to achieve better accuracy in the description of binary or multicomponent mixtures. A detailed review is given by Oellrich et al. [9] and Knapp et al. [10].

Example: RKS Redlich, Kwong, and Soave

A considerable improvement of the accuracy of the van der Waals EQS could be attained by a modification of Redlich and Kwong [11].

Soave [12] has generalized the RK-EQS by suggesting rules for determining a and b from T_c, p_c, and ω.

$$z = \frac{v}{v-b} - \frac{a}{RT\,(v+b)} \tag{31}$$

where

$$a = 0.42747\alpha R^2T_c^2/p_c \quad (31a)$$

$$\alpha = [1+m(1-\sqrt{T/T_c})]^2 \quad (31b)$$

$$m = 0.48+1.574\omega-0.176\omega^2 \quad (31c)$$

$$b = 0.0866\ RT_c/p_c \quad (31d)$$

For mixtures, a and b are calculated with mixing rules

$$b = \sum x_i b_i \quad (32a)$$

$$a = \sum x_i x_j a_{ij} \quad (32b)$$

and the combination rule

$$a_{ij} = \sqrt{a_i a_j}\,(1-k_{ij}) \quad (32c)$$

where $k_{ij} = 0$ for $i = j$
k_{ij} for $i \neq j$ can be found by fitting to experimental VLE data

The structure of useful GEQS permits the closed differentiation and integration when calculating departure functions. Real gas properties can be calculated with RKS as follows:

$$h-h_0 = \frac{1}{b}\left(T\frac{da}{dT}-a\right)\ln\frac{v+b}{v}+pv-RT \quad (33)$$

$$\ln\varphi = z-1-\ln\frac{p(v-b)}{RT}-\frac{p(v-b)}{b\,RT} - \frac{a}{b\,RT}\left(2\frac{\sum_j x_j a_{ij}}{a}-\frac{b_i}{b}\right)\ln\frac{v+b}{v} \quad (34)$$

Example: PR (Peng and Robinson)

The equation was proposed by Peng and Robinson [13] as a modification of the attractive term of the van der Waals EQS:

$$z = \frac{v}{v-b}-\frac{av}{RT\,(v(v+b)+b(v-b))} \quad (35)$$

Parameter a and b for pure substances can be determined according to general rules

$$a = 0.45724\ R^2T_c^2\alpha/p_c \quad (35a)$$

$$\alpha = [1+\kappa(1-\sqrt{T/T_c})]^2 \quad (35b)$$

$$\kappa = 0.37464+1.54226\omega-0.26992\omega^2 \quad (35c)$$

$$b = 0.07780\, RT_c/p_c \tag{35d}$$

The mixing and combination rules are the same as for RKS. Departure functions are also closed expressions and can be found in literature.

Example: BWRS (Benedict, Webb, Rubin, and Starling)

Benedict, Webb and Rubin [14] had developed an empirical virial equation with [14] coefficients. Sets of coefficients were determined by fitting to volumetric and caloric data of individual substances, mostly hydrocarbons. Starling [15] extended the EQS to 11 coefficients and proposed general rules for determining these coefficients from T_c, P_c, and ω.
The EQS gives the pressure as a function of temperature and density:

$$\begin{aligned} p &= RT\varrho \\ &+ \left(B_0 RT - A_0 - \frac{C_0}{T^2} + \frac{D_0}{T^3} - \frac{E_0}{T^4}\right)\varrho^2 \\ &+ \left(b\,RT - a - \frac{d}{T}\right)\varrho^3 \\ &+ \alpha\left(a + \frac{d}{T}\right)\varrho^6 \\ &+ \frac{c}{T^2}\varrho^3(1+\gamma\varrho^2)e^{-\gamma\varrho^2} \end{aligned} \tag{36}$$

It can be rewritten for the compressibility factor $z = pv/RT$ as a function of temperature and molar volume.

$$z = 1 + \frac{B}{v} + \frac{C}{v^2} + \frac{D}{v^5} + \frac{C'}{v^2}\left(1 + \frac{\gamma}{v^2}\right)e^{-\gamma/v^2} \tag{37}$$

where B, C, D, C′ are abbreviations for the expressions in Equation 36.
The 11 coefficients $B_0 \ldots \gamma$ can be determined by using the following 11 rules from T_c, ϱ_c, and ω for individual substances. Table 3.

$$\begin{aligned} \varrho_c B_0 &= A_1 + B_1\omega \\ \varrho_c A_0/(RT_c) &= A_2 + B_2\omega \\ \varrho_c C_0/(RT_c^3) &= A_3 + B_3\omega \\ \varrho_c^2 \gamma &= A_4 + B_4\omega \\ \varrho_c^2 b &= A_5 + B_5\omega \\ \varrho_c^2 a/(RT_c) &= A_6 + B_6\omega \\ \varrho_c^3 \alpha &= A_7 + B_7\omega \\ \varrho_c^2 C/(RT_c^3) &= A_8 + B_8\omega \\ \varrho_c D_0/(RT_c^4) &= A_9 + B_9\omega \\ \varrho_c^2 d/(RT_c^2) &= A_{10} + B_{10}\omega \\ \varrho_c E_0/(RT_c^5) &= A_{11} + B_{11}\omega \exp(-3.8\omega) \end{aligned}$$

The 2 sets of 11 general constants A_j and B_j were found by simultaneous fitting to volumetric and caloric data of alkanes.

Eleven different mixing rules are given to calculate the eleven coefficients of a mixture.

Example: LKP (Lee, Kesler, and Plöcker)

The volumetric and thermodynamic functions correlated by Pitzer and co-workers [6] in accordance with the 3-parameter-corresponding-states principle originally presented in tables were presented in an analytical form by Lee and Kesler [16]:

$$z = z^{(0)} + \frac{\omega}{\omega^{(r)}}(z^{(r)} - z^{(0)}) \tag{38}$$

The compressibility factors of both a simple and a reference fluid are given as functions of reduced variables $T_r = T/T_c$, $p_r = p/p_c$, and $v_r = p_c v/(RT_c)$.

$$z = pv/RT = p_r v_r / T_r$$

$$= 1 + \frac{B}{v_r} + \frac{C}{v_r^2} + \frac{D}{v_r^5} + \frac{C_4}{T_r^3 v_r^2}\left(\beta + \frac{\gamma}{v_r^2}\right) e^{-\gamma/v_r^2} \tag{39}$$

where $B = b_1 - \frac{b_2}{T_r} - \frac{b_3}{T_r^2} - \frac{b_4}{T_r^3}$

$$C = c_1 - \frac{c_2}{T_r} + \frac{c_3}{T_r^3}$$

$$D = d_1 + \frac{d_2}{T_r}$$

Experimental data for argon, krypton, and methane were used for the optimization of the 12 "general" constants $b_1, b_2, \ldots, \gamma$ for the simple fluid, and experimental data of octane were used for the reference fluid.

Mixtures are characterized by pseudocritical properties in accordance with three mixing rules suggested by Plöcker, Knapp, and Prausnitz [17]:

$$T_{cm} = \left(\frac{1}{v_{cm}}\right)^{\eta} \sum\sum x_i x_j (v_{cij})^{\eta} T_{cij} \tag{40}$$

where $T_{cij} = \sqrt{T_{ci} T_{cj}}\, k_{ij}$ (40a)

$$k_{ii} = k_{ij} = 1$$

k_{ij} for $i \neq j$ can be found by fitting to binary VLE data. It can also be correlated for certain groups of substances.

$$v_{cm} = \sum\sum x_i x_j v_{cij} \tag{40b}$$

where $v_{cij} = \frac{1}{8}(v_{ci}^{1/3} + v_{cj}^{1/3})^3$ (40c)

$$v_{ci} = z_{ci}\, RT_{ci}/p_{ci} \tag{40d}$$
$$z_{ci} = 0.2905 - 0.085\omega_i \tag{40e}$$
$$p_{cm} = RT_{cm} z_{cm}/v_{cm} \tag{40f}$$
$$z_{cm} = 0.2905 - 0.085\omega_m \tag{40g}$$
$$\omega_m = \sum x_i \omega_i \tag{40h}$$

The optimal value of η was emprically found to be 0.25.

The isothermal departure functions showing the effect of pressure on the thermodynamic properties of a fluid can also be calculated in accordance with Equations 19 through 25.

RESULTS OF PROCESS CALCULATIONS

It might be interesting to give a few examples illustrating the application of GEQS for the calculation of thermodynamic properties. The accuracy of results can be presented by showing experimental and calculated data on diagrams.

Example: PVT Properties of Pure Methane

In Figure 6 three different generalized equations of state show good accuracy at low densities. At higher densities only the LK–GEQS represents experimental data well. It is not surprising as the 11 constants for the simple fluid were fitted to methane data.

Example: Vapor Pressure of Pure Ethane

In Figure 7 the original experimental work was done by Loomis and Walters [18]. The results can be represented by the Antoine equation with a standard deviation of 1.5 Torr.

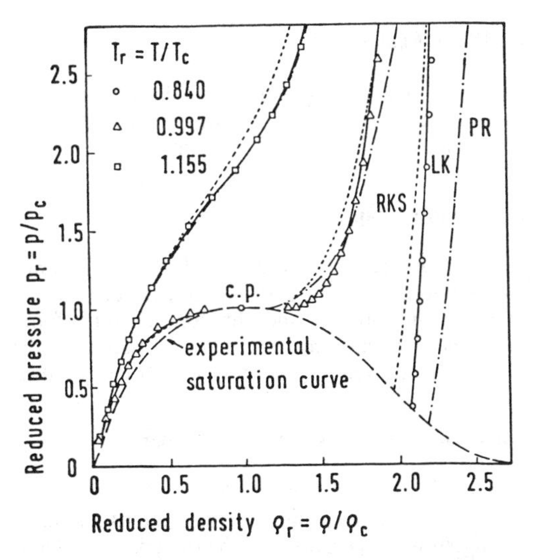

Figure 6. Diagram showing reduced pressure $p_r = p/p_c$ vs. reduced density $\rho_r = \rho/\rho_c$ for pure methane at three reduced temperatures $T_r = T/T_c$. The points are experimental; isotherms were calculated with three generalized equations of state: RKS Equation 31, PR Equation 35 and LK Equation 38.

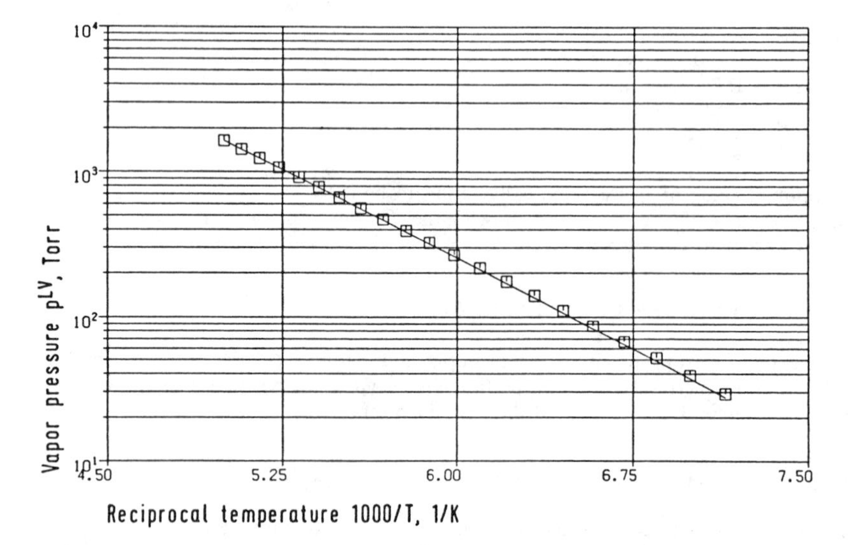

Figure 7. Vapor pressure vs. reciprocal temperature for pure ethane. The curve is based on experimental data interpolated with the Antoine equation: The points are calculated with a bubble-point procedure on the basis of RKS-GEQS.

The coefficients of the equation are listed in a standard handbook [19]. It is obvious that the RKS–GEQS can reproduce the vapor pressure satisfactorily although only 2 critical data and the acentric factor must be known to characterize the properties of the pure fluid.

Example: Vapor-Liquid-Phase Equilibrium

The design of separation processes is based on our knowledge about phase-equilibria, i.e., about the K-values. It is important to judge the accuracy of a correlation. In Figure 8 experimental data and calculated isothermes are compared.

Example: Enthalpy-Temperature Diagram

For the design of heat exchangers, Q – T or H – T diagrams contain the information required by the process engineer (see Figure 9). At temperatures below the dew point it is first necessary to determine the amount and the composition of the vapor and liquid fraction. The enthalpy of the two-phase mixture depends mainly on the liquid/vapor ratio i.e., on the result of the flash calculation preceding the enthalpy calculation.

Example: Coefficient of Performance of Heat-Pump Cycles

For process studies for heat-pump operation the coefficient of performance should be known, i.e., the energy offered in the condenser as heat to the energy required in the compressor. By combining all unit-operations of the cycle in a computer program, characteristic curves can be produced provided the caloric properties of the working fluid are known. When a GEQS is used for the PVT properties of the fluid, the results of the calculations can be compared with results obtained with information on h – p diagrams.

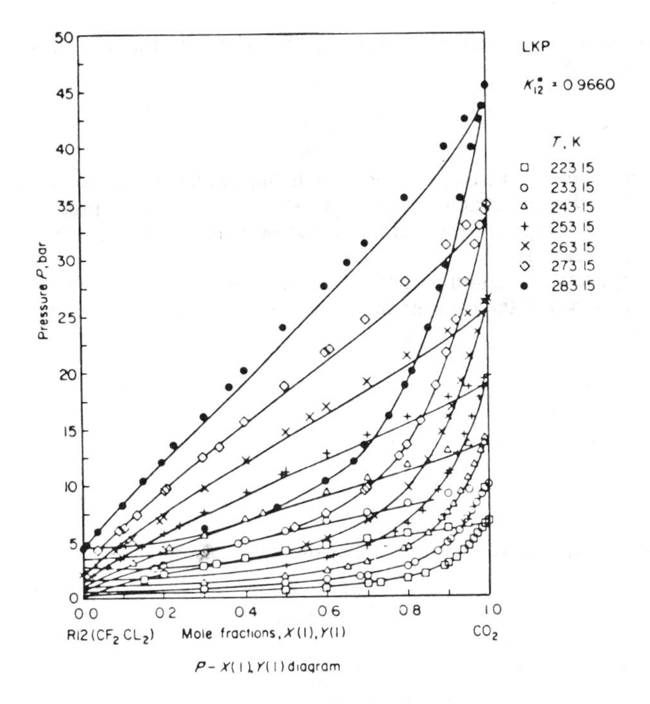

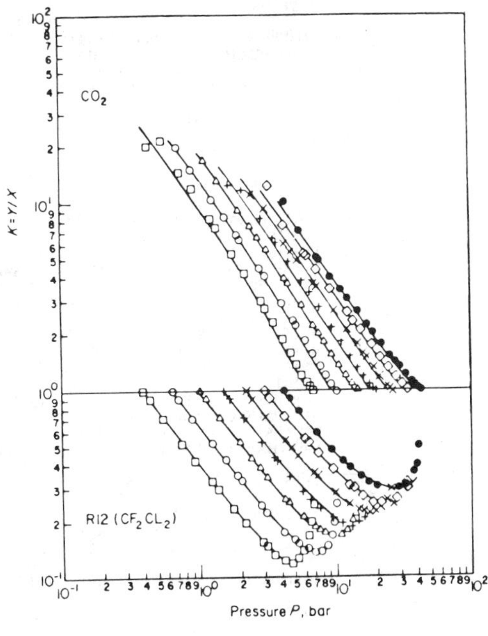

Figure 8. Diagram showing dew- and bubble-point pressure vs. vapor and liquid composition at various temperatures and diagram showing K-values vs. pressure at various temperatures for the binary mixture $CO_2 - CF_2Cl_2$. The points are experimental data taken by Dorau et al [20]. The isotherms were calculated with a flash procedure. The PVTX properties were represented with LKP-EQS.

The deviations in the example shown in Figure 10 were 3%–5%; however, they can also be due to inaccuracies of the diagram.

ADDITIONAL COMMENTS

All the required procedures, most of the thermodynamic relations, many useful correlations for material properties, and many characteristic coefficients are included in process compilers or process calculation packages. Process calculation software is available for small and large electronic computers [22, 23].

The structure of a typical process calculation package can be demonstrated by explaining the contents of a package developed at the Institute of Thermodynamic and Plant Design, TU Berlin/West.

Table 3
List of Modules Contained in a Process Calculation Package

Levels			
5 SYSTEMS	SYST	ENTHALPY DEPARTURE LKP	
4 MULTISTAGE OPERATION	MUOP	ENTHALPY DEPARTURE PR	
3 UNIT OPERATION	UNOP	ENTHALPY DEPARTURE VIR	
2 THERMODYN. PROPERTIES	THEP	ENTHALPY OF MIXING	
1 MATERIAL PARAMETERS	CHAP	ENTHALPY OF VAPORIZ.	DPHS
		ENTROPY DEPARTURE LKP	
Description of Levels		ENTROPY DEPARTURE PR	
		ENTROPY DEPARTURE VIR	
SYSTEMS	SYST	ENTROPY OF MIXING	
		MIXTURE ENTHALPY STAND.	COTX
COMPRESSOR HEATPUMP	HPCC	MIXTURE ENTROPY STAND.	
ABSORBER HEATPUMP	HPAB		
		FUGACITIES IN V AND L	FUGA
MULTISTAGE OPERATIONS	MUOP		
		ACTIVITY COEFF. GE MODELS	
RECTIFICATION MATRIC	MADE	VAN LAAR, MARGULES, WILSON,	
Y/X-NTS PLOT	STYS	NRTL, UNIQUAC, UNIFAC	ACTI
N-STAGE COMPRESSION	MUCO	FUGACITY COEFF. IN V	FCIV
		POYNTING CORRECTION	POYN
UNIT OPERATION	UNOP		
		GEN. EQUATION OF STATE	GEQS
UNIT OPERATION MAN.	MAUO	LEE KESLER PLOECKER EQS	
FLASH F+T+P	FLTP	MIXING RULES FOR PSEUDO	MIXL
FLASH F+Q+P	FLQP	REDUCED VOLUME	VTPL
FLASH F+T+VY	FLTV	FUGACITY COEFF. LKP	FCIL
FLASH F+T+LX	FLTL	PENG ROBINSON EQS	
FLASH F+P+VY	FLPV	PARAM. A+B, MIXING RULES	MIXP
FLASH F+P+LX	FLPL	MOLAR VOLUME	VTPP
BUBBLE POINT X+T	BUBT	FUGACITY COEFF. PR	FCIP
BUBBLE POINT X+P	BUBP		
DEW POINT T+X	DEWT	COMPONENT PROPERTIES	COPA
DEW POINT P+Y	DEWP	VAPOR PRESSURE	VAPP
ISOBARIC HEAT OR COOL	QTPX	LIQUID VOLUME	VOLT
ADIABATIC COMPRESSION	ADCO	SECOND VIRIAL COEFF.	BMTX
ADIABATIC EXPANSION	ADEX		
ISENTHALPIC THROTTLING	IREX	MATERIAL PARAMETERS	CHAP
		MANAGEMENT FOR PARAM:	MAPA
THERMODYN. PROPERTIES	THEP	REQUEST PARAMETERS	PUDA
		RETRIEVE NAMES	NAME
THERMODYN. EQUILIBRIA	THEQ	ARRANGE SUBSTANCES	NORG
VLE CONSTANT K=Y/X	KVLE	M, TLV, TC, PC, VC, OMEGA	PURE
CHEM. REACT. CONSTANT	KCHE	CHEM. FORMULA, NAMES	NAMS
		REQUEST MIXTURE PARAM.	MIPA
CALORIC PROPERTIES	CALO	RETRIEVE KIJ PARAM.	KIJD
		4000 KIJ FOR LKP	KIJL
MIXTURE ENTHALPY MIXTURE ENTROPY	MACA	220 KIJ FOR PR	KIJP

BPBP = Berliner Prozeß Berechnungspaket

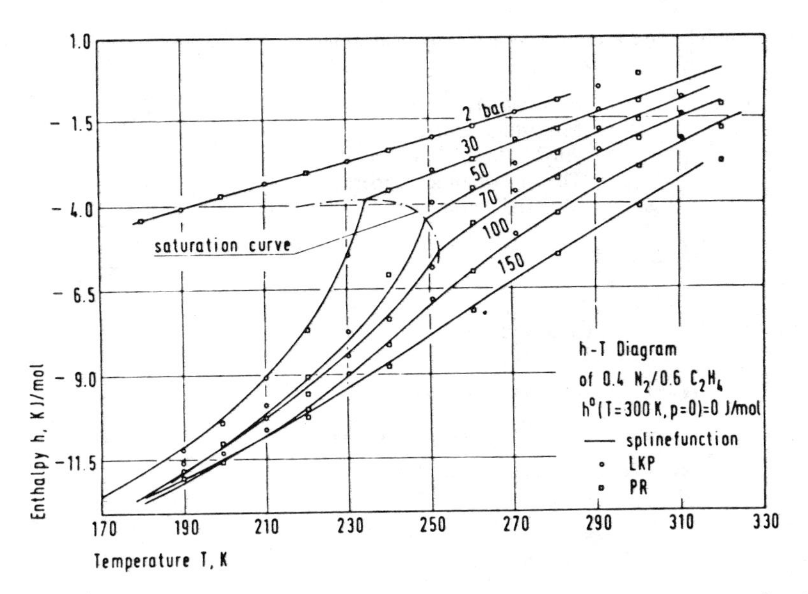

Figure 9. Enthalpy-temperature diagram for a nitrogen-ethylene mixture. The isobars are experimental data taken by Schmid [21] and interpolated by a spline function. The points were calculated with Equation 19 for enthalpy departures on the basis of two GEQS: PR Equation 35 and LKP Equations 38 through 40.

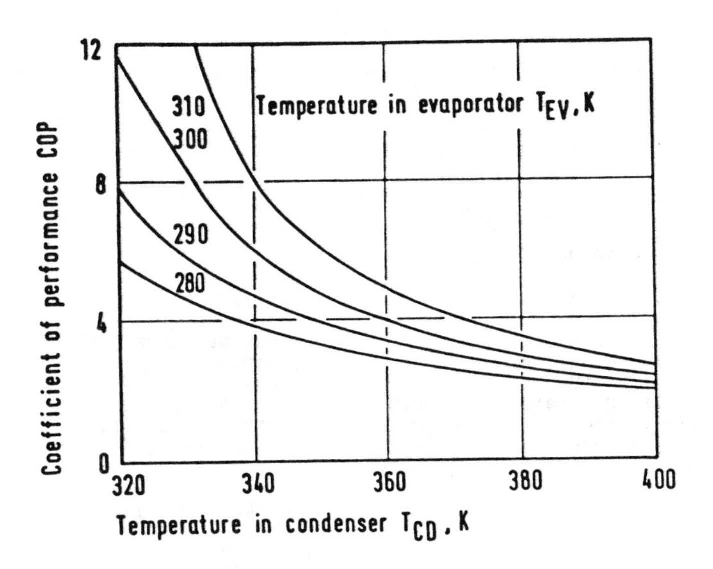

Figure 10. Diagram showing the coefficient of performance of a compression heat pump cycle as a function of condenser and evaporator temperature. The working fluid is R11; the isentropic efficiency of the compressor is 0.75; the PVT properties of R11 were represented with LKP-EQS.

Program modules are arranged in a hierarchical order

Level	*Description*	*Code*
4	Multistage Operations	MUOP
3	Single Stage Unit Operations	UNOP
2	Thermodynamic Properties	THEP
1	Material Parameters	CHAP

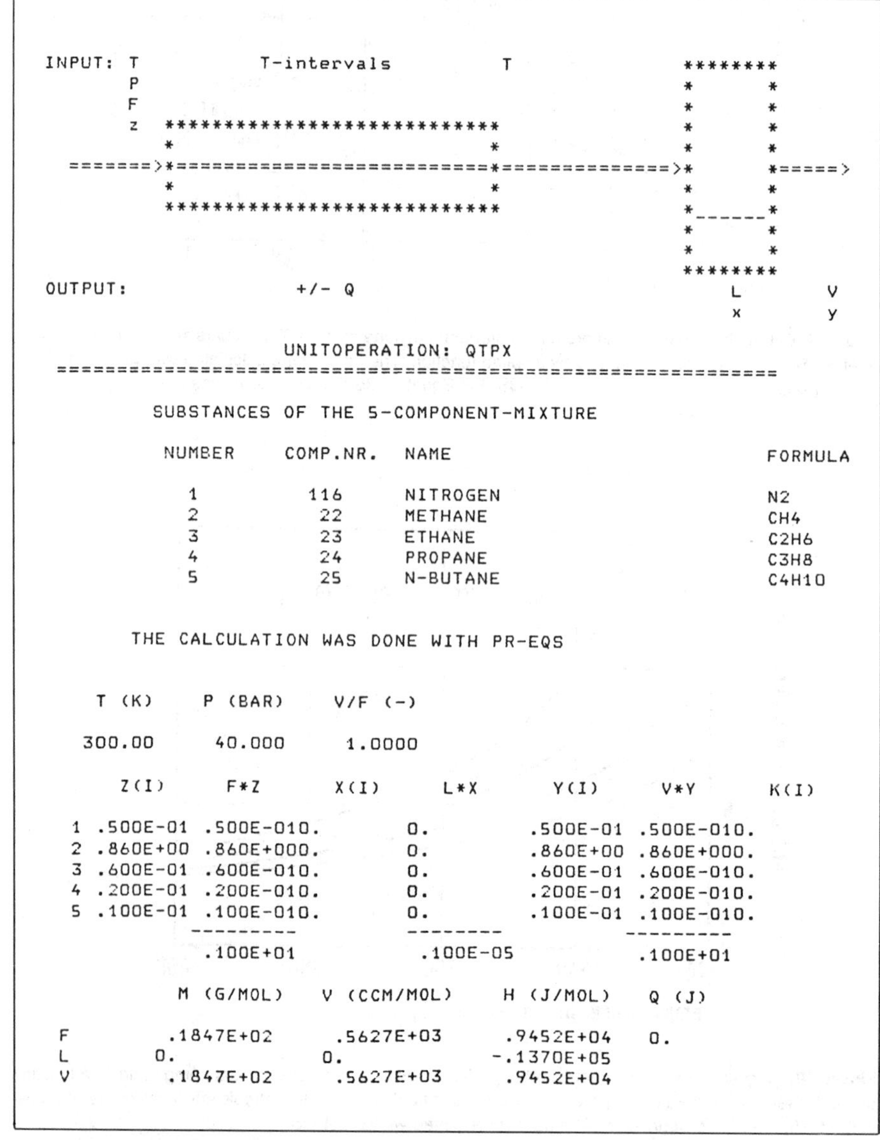

```
INPUT: T            T-intervals          T              *******
       P                                                *     *
       F                                                *     *
       z  **************************                    *     *
          *                        *                    *     *
=======>*========================*==============>*     *=====>
          *                        *                    *     *
          **************************                    *-----*
                                                        *     *
                                                        *     *
                                                        *******
OUTPUT:                  +/- Q                             L        V
                                                           x        y

                    UNITOPERATION: QTPX
==============================================================

      SUBSTANCES OF THE 5-COMPONENT-MIXTURE

       NUMBER     COMP.NR.   NAME                            FORMULA

         1          116      NITROGEN                        N2
         2           22      METHANE                         CH4
         3           23      ETHANE                          C2H6
         4           24      PROPANE                         C3H8
         5           25      N-BUTANE                        C4H10

     THE CALCULATION WAS DONE WITH PR-EQS

   T (K)     P (BAR)     V/F (-)

  300.00     40.000      1.0000

      Z(I)      F*Z        X(I)      L*X       Y(I)      V*Y       K(I)

 1 .500E-01 .500E-010.        0.         .500E-01 .500E-010.
 2 .860E+00 .860E+000.        0.         .860E+00 .860E+000.
 3 .600E-01 .600E-010.        0.         .600E-01 .600E-010.
 4 .200E-01 .200E-010.        0.         .200E-01 .200E-010.
 5 .100E-01 .100E-010.        0.         .100E-01 .100E-010.
           ---------         --------             ---------
            .100E+01          .100E-05             .100E+01

        M (G/MOL)   V (CCM/MOL)    H (J/MOL)    Q (J)

 F       .1847E+02    .5627E+03     .9452E+04    0.
 L      0.           0.            -.1370E+05
 V       .1847E+02    .5627E+03     .9452E+04
```

Figure 11. Printout of results of module QTPX. The calculations can be done at specified temperature intervals. However only 3 points at T = 300, 200, and 130°K are printed as examples.

In Table 3 all modules of the four levels are listed. In this table the significance of GEQS can be easily recognized:

Example: Q–T Diagram of a Process Stream

In this example refer to Figures 11 and 12. Module QTPX in level 3 will provide a management program for the calculation of the enthalpy of a mixture of specified com-

```
      Z(I)       F*Z        X(I)       L*X        Y(I)       V*Y        K(I)

 1 .500E-01 .500E-01 .500E-01 .500E-010.        0.         0.
 2 .860E+00 .860E+00 .860E+00 .860E+000.        0.         0.
 3 .600E-01 .600E-01 .600E-01 .600E-010.        0.         0.
 4 .200E-01 .200E-01 .200E-01 .200E-010.        0.         0.
 5 .100E-01 .100E-01 .100E-01 .200E-010.        0.         0.
           ---------           ---------        ---------
            .100E+01            .100E+01        .100E-05

          M (G/MOL)    V (CCM/MOL)    H (J/MOL)    Q (J)

F          .1847E+02     .3574E+02    -.4918E+04    .1437E+05
L          .1847E+02     .3574E+02    -.4918E+04
V         0.            0.            -.6388E+04

      THE CALCULATION WAS DONE WITH PR-EQS

   T (K)      P (BAR)      V/F (-)

  200.00      40.000        .8129

      Z(I)       F*Z        X(I)       L*X        Y(I)       V*Y        K(I)

 1 .500E-01 .500E-01 .124E-01 .232E-02 .587E-01 .477E-01 .472E+01
 2 .860E+00 .860E+00 .662E+00 .124E+00 .906E+00 .736E+00 .137E+01
 3 .600E-01 .600E-01 .183E+00 .343E-01 .317E-01 .257E-01 .173E+00
 4 .200E-01 .200E-01 .911E-01 .170E-01 .363E-02 .295E-02 .399E-01
 5 .100E-01 .100E-01 .514E-01 .962E-02 .469E-03 .381E-03 .912E-02
           ---------           ---------          ---------
            .100E+01            .187E-00           .813E+00

          M (G/MOL)    V (CCM/MOL)    H (J/MOL)    Q (J)

F          .1847E+02     .2270E+03     .3424E+04    .6028E+04
L          .2348E+02     .5287E+02    -.2561E+04
V          .1731E+02     .2671E+03     .4802E+04

      THE CALCULATION WAS DONE WITH PR-EQS

   T (K)      P (BAR)      V/F (-)

  130.00      40.000        .0000

Example for the calculation of a cooling curve:    QTPX
===============================================

The calculation can be done for any specified temperature-intervals.

         (Heat-exchanger)                            (Phase-seperator)
```

Figure 11. Continued.

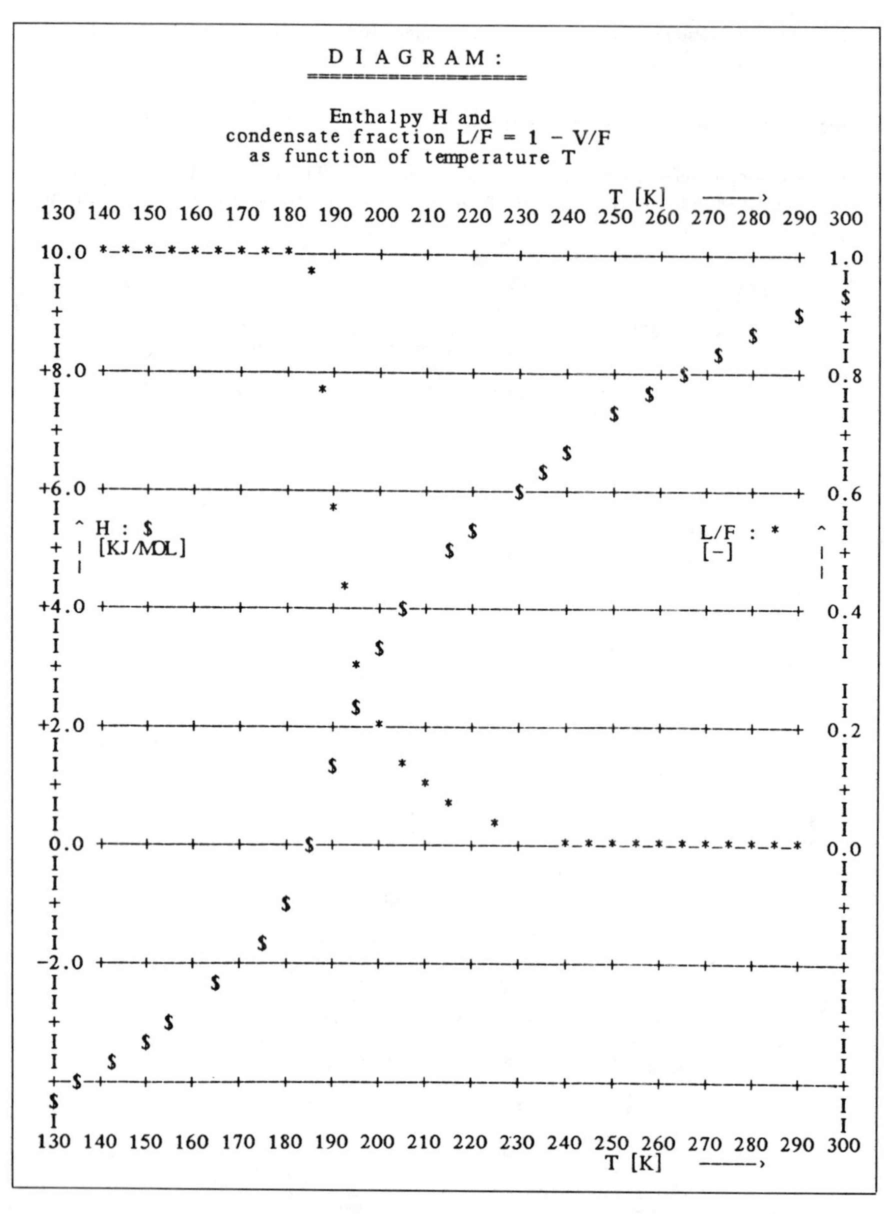

Figure 12. Print-plot showing the enthalpy H and the condensate fraction L/F (liquid/feed) as a function of temperature for natural gas with a composition specified in Figure 11. The print plot illustrates the results of a QTPX calculation.

position as a function of temperature, in specified temperature intervals at a specified pressure. If the option PR–EQS is selected, the enthalpy departure $h-h^0$, i.e., module HDEP in level 2, will be calculated with the chosen GEQS. Modules EQSL, MIXL, VTPL, and FCIL will be called in combination with FLTP to find the state of the mixture at each process point. MACA will call the modules for the standard enthalpy of the mixture and add the enthalpy departure. From level 1 the modules MAPA, PURE, MIPA, and KIJL will supply all parameters that are needed in level 2 as coefficients of the specified component of the mixture.

Figures 11 and 12 show printouts of QTPX for a 5-component mixture being cooled from 300° to 130° K.

References 19, 22, and 23 provide further headings.

Generalized equations of state can be very useful in describing analytically the PVTX behavior of pure substances and of mixtures. GEQS not only give information about the specific or molar volume as a function of temperature, pressure, and composition but also help to calculate departure functions, i.e., the effect of pressure on thermodynamic properties such as enthalpy, entropy, and fugacity.

Beginning with our knowledge about the property in the state of the ideal gas, all important thermodynamic properties at higher pressures (i.e., higher densities) can be calculated as long as the validity of the GEQS can be accepted. In many cases the PVTX information of the GEQS might not be very exact, but it is the only available information. The responsible process engineer has to be aware that at critical process points results of predictive methods must be checked by experimental evidence.

NOTATION

- A_k coefficient
- A_{mk} coefficients for a specific mixture of constant composition
- A^+ Helmholtz free energy
- a, b constants
- B(T) second virial constant
- C(T) third virial constant
- c_p^0 molar heat capacity in ideal gas state
- f_1, f_2 functions of reduced densities
- h enthalpy
- h^0 standard enthalpy for ideal gas
- k_{ij} binary parameter for the combination of component i and j
- n number of moles
- p_i partial pressure
- p^0 pressure at standard state
- p pressure
- R universal gas constant
- s entropy
- T temperature
- T_r reduced temperature
- T^+ reference temperature
- T_c critical temperature
- V volume
- v molar volume
- v_r reduced volume
- v^* volume of spherical molecules
- x concentration
- x_i mole fraction of component i in liquid
- y_i mole fraction of component i in vapor
- z compressibility factor

Greek Symbols

- γ specific heat ratio
- ϱ molar density
- ϱ^* reduced density
- φ fugacity coefficient
- ω acentric factor

REFERENCES

1. Leland, T .W., "Phase Equilibria and Fluid Properties in the Chemical Industry," EFCE Publ. Series No. 11, (1980) Dechema, Frankfurt-Main, FRG, 281–332.
2. Martin, J. J., *Ind. Eng. Chem. Fundam.* Vol. 18, (1979), p. 81.
3. Vidal, J., *Fluid Phase Equilibria,* Vol. 13 (1983), pp. 15–34.
4. van der Waals, J. D., Dissertation, Leiden, (1873).
5. Kamerlingh Onnes, H., *Comm. Phys. Lab. Leiden,* Vol. 71 and 74 (1901).
6. Pitzer, K. S., et al., *J. Am. Chem. Soc.,* Vol. 77 (1955), p. 3433.
7. Carnahan, N.,F., and Starling, K. F., *AIChE j.,* Vol. 18 (1972), p. 1184.
8. van der Waals, J. D., Verband.Kon. Akad. Wetensch., Amsterdam 20 (1880).
9. Oellrich, L., et al., *Intern. Chem. Eng.,* Vol. 21, No. 1 (1981), pp. 1–17.
10. Knapp, H., et al., "Vapor-Liquid Equilibria for Mixtures of low Boiling Substances", Chemistry Data Series VI, Dechema, Frankfurt-Main, FRG., 1982.
11. Redlich, O., and Kwong, J. N. S., *Chem. Reviews,* Vol. 44 (1949), p. 233.
12. Soave, G., *Chem. Eng. Sci.,* Vol. 27 (1972), p. 1197.
13. Peng, D. Y., and Robinson, D. B., *Ind. Eng. Chem. Fundam.,* Vol. 15 (1976), p. 59.
14. Benedict, M., Webb, R. B., Rubin, L. L., *J. Chem. Phys.,* Vol. 8 (1940), p. 314.
15. Starling, K. F., *Hydroc. Proc.,* Vol. 50 (1971), p. 101; and subsequent articles in *Hydroc. Proc.*
16. Lee, B. L., and Kesler, M. G., *AIChE J.,* Vol. 21 (1975), pp. 510, 1040.
17. Plöcker, U., Knapp, H., and Prausnitz, J. M., *Ind. Eng. Chem. Proc. Des. Dev.,* Vol. 17 (1978), p. 324.
18. Loomis, A. G., and Walter, J. E., *J. Am. Chem. Soc.,* Vol. 48 (1926), p. 2051.
19. Reid, J. C., Prausnitz, J. M., Sherwood, T. K., *The Properties of* Gases and Liquids, 3rd ed., Mc Graw Hill Inc., New York (1977).
20. Dorau, W., Al-Wakeel, I. M., and Knapp, H., *Cryogenics,* Vol. 1 (1983), p. 29–36.
21. Schmid, O., Disseration, TU-Berlin, (1980).
22. Eckermann, R., "Phase Equilibria and Fluid Properties in the Chemical Industry," EFCE Publ. Series No. 11, (1980), Dechema, Frankfurt-Main, FRG. 739–758.
23. Neumann, K. K., and Futterer, E., "Phase Equilibria and Fluid Properties in the Chemical Industry," EFCE Publ. Series No. 11, (1980), Dechema, Frankfurt-Main, FRG. 807–820.
24. Curl, R. F., and Pitzer, K. S., *Ind. Eng. Chem.,* Vol. 50 (1958), p. 265.
25. Starling, K. F., *Fluid Thermodynamic Properties for Light Petroleum Systems,* Gulf Publ. Comp., Houston (1973).

CHAPTER 11

MIXTURE BOILING

John R. Thome

Department of Mechanical Engineering
Michigan State University
East Lansing, Michigan, USA

CONTENTS

INTRODUCTION

This chapter on the boiling of liquid mixtures presents a review of the effect of composition on the boiling heat transfer process and is followed by a discussion of the methods for predicting the heat transfer coefficients for nucleate pool boiling. Prior knowledge of single-component boiling is assumed. Only liquids forming miscible mixture systems will be considered here.

It is a well-established fact from numerous experimental investigations that the boiling heat transfer coefficients of mixtures can be considerably lower than those of equivalent pure fluids with the same physical properties as the mixtures. The scope of this chapter, then, is to explain physically why the actual heat transfer coefficients are lower and to describe the available methods for predicting their variation with composition.

The topics to be discussed are:

- Binary mixture phase equilibria
- Mixture nucleate pool boiling
- Prediction of nucleate pool boiling heat transfer coefficients

For a more detailed review, the reader is referred to Thome and Shock [1] and to Volume 1 of this handbook.

The material presented here deals primarily with binary mixture systems but is extendable to multicomponents which are more typical in industrial processes.

PHASE EQUILIBRIA FOR BINARY SYSTEMS

A vapor pressure curve giving the relationship between the saturation temperature and pressure is sufficient to describe the vapor-liquid saturation locus for pure or single-component fluids. For a liquid mixture at equilibrium with its vapor, the mole fraction of a component in the vapor phase, $\tilde{y}$, can be either greater than, less than, or equal to the mole fraction of the same component in the liquid phase, $\tilde{x}$. Hence, the saturation pressure alone does not uniquely define a saturation temperature, T_{sat}, for a mixture system. Phase equilibrium diagrams are therefore useful in depicting visually the relationship between $\tilde{y}$ and $\tilde{x}$ for varying T_{sat} at constant pressure.

Figure 1 is a phase equilibrium diagram for an ideal binary mixture system. Letting $\tilde{x}$ and $\tilde{y}$ refer to the liquid and vapor mole fractions of the more volatile component (the more volatile component is the component with the lower boiling point), the bubble point and

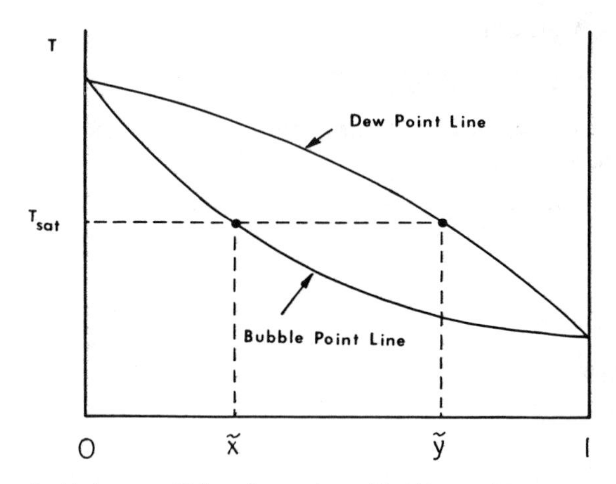

Figure 1. Vapor-liquid-phase equilibrium diagram for an ideal binary mixture system.

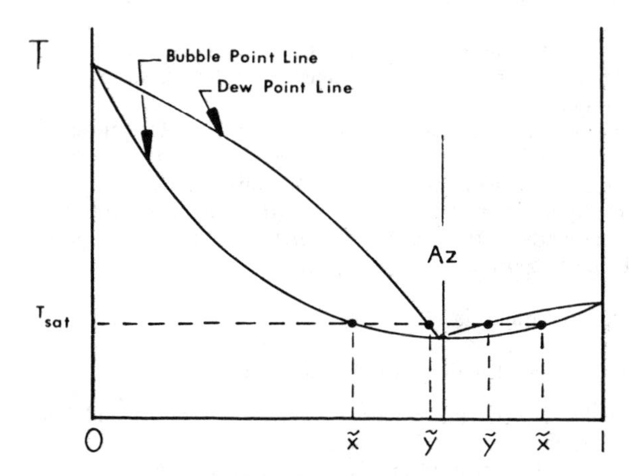

Figure 2. Vapor-liquid-phase equilibrium diagram for an azeotropic mixture system.

dew point lines depict their variations with the saturation temperature at a given pressure as illustrated. Note that $\tilde{y}$ is always greater than $\tilde{x}$ except at the two end points.

Figure 2 shows a phase equilibrium diagram for a binary system forming an azeotrope. The azeotrope is the intermediate point where the dew point and bubble point lines meet. To the left of the azeotrope $\tilde{y} \geqq \tilde{x}$, while to the right $\tilde{y} \leqq \tilde{x}$. A mixture at its azeotrope behaves like a single-component fluid.

MIXTURE NUCLEATE POOL BOILING

Nucleate pool boiling curves at a number of compositions for a hypothetical binary mixture system are shown in Figure 3A. These curves are typical of experimental findings for aqueous, organic, refrigerant, and liquified gas mixture systems. At low heat fluxes there is no mixture effect other than the effect of the variation in the pertinent physical properties on single-phase natural convection. Boiling begins as the heat flux level increases and the boiling curves of the mixtures deviate more significantly from the single-component curves. Figure 3B shows the variation in the wall superheat, ΔT, with composition at a fixed intermediate heat flux. It is seen that a larger wall superheat is required for the mixtures than is expected from a simple linear interpolation between the two single-component values

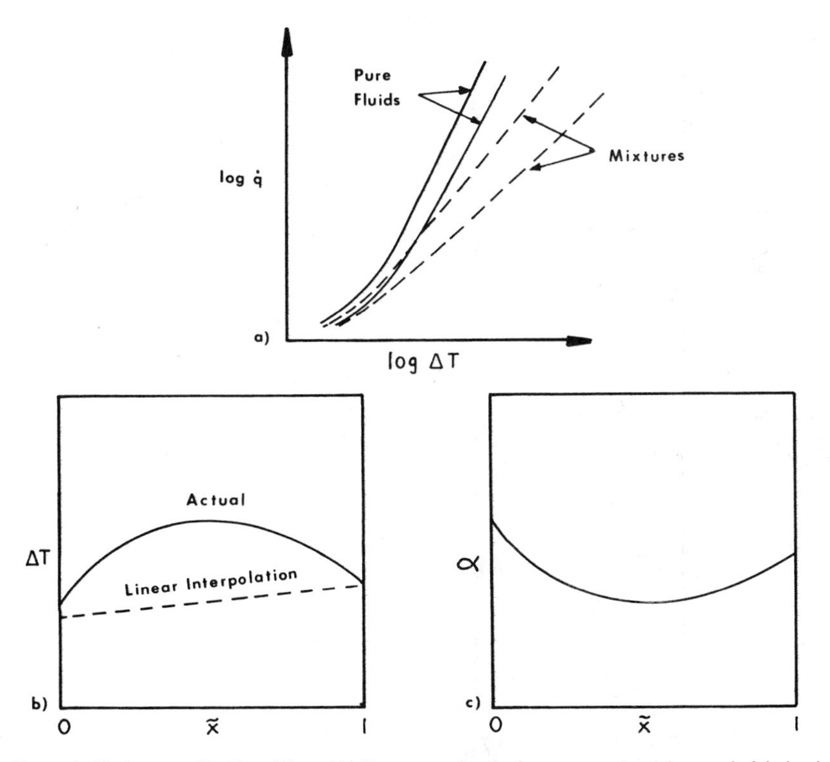

Figure 3. Nucleate pool boiling: (A) pool boiling curves for single components and several of their mixtures; (B) wall superheat as a function of composition; (C) boiling heat transfer coefficient as a function of composition.

in order to transfer the same heat flux. This causes a marked reduction in the boiling heat transfer coefficient as shown in Figure 3C.

Van Wijk, Vos, and Van Stralen [2] gave the first physical explanation for these lower heat transfer coefficients. Assuming phase equilibrium is maintained during the growth of vapor bubbles at a heated wall, they noted that the vapor mole fraction of the more volatile component, $\tilde{y}$, in the bubbles is higher than that in the surrounding bulk liquid, which is $\tilde{x}$. Consequently, more of the more volatile component in the liquid must evaporate to make up the composition difference $(\tilde{y}-\tilde{x})$ as a bubble grows. This in turn causes a reduction in the liquid mole fraction of the more volatile component in the liquid layer adjacent to the bubble. A rise in the local boiling point temperature results as can be deduced from Figure 1. Thus, the heated wall temperature must rise correspondingly in order to transfer heat at the same rate and the heat transfer coefficient (based on the wall temperature minus the bulk saturation temperature) decreases. A similar explanation is valid at either side of the azeotrope for an azeotropic system and also for multicomponent systems.

Sternling and Tichacek [3] in addition attributed the lower mixture heat transfer coefficients to be due to a significant change in the mixture physical properties with composition.

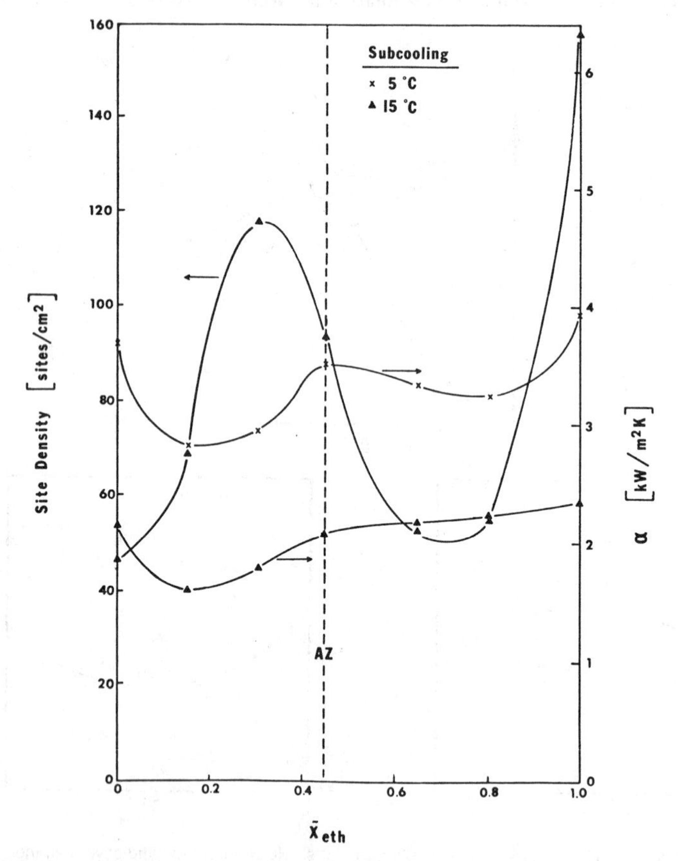

Figure 4. Boiling site density and heat transfer coefficient as a function of composition for ethanol-benzene mixtures at 1.0 bar at 75 kW m^{-2} (from Hui [4]).

They also intuitively concluded them to be the result of a reduction in the activation of new boiling sites on the heated wall caused by accumulation of the less volatile liquid in the vicinity of the boiling sites. However, Hui [4] later observed experimentally that the number of boiling sites per unit area for an azeotropic system could form a maximum to the left of the azeotrope and a minimum to the right at some heat fluxes while the heat transfer coefficient went to a minimum on both sides (see Figure 4). Thome [5] also showed that the two principal heat transport mechanisms active in nucleate pool boiling, bubble evaporation, and cyclic thermal boundary-layer stripping are lower at individual boiling sites in mixtures and are thus partially responsible for the lower mixture heat transfer coefficients.

In summary, it is seen that the reduction in the mixture heat transfer coefficients is a combination of the effect of composition on a number of boiling phenomena.

PREDICTION OF NUCLEATE POOL BOILING HEAT TRANSFER COEFFICIENTS

Nucleate pool boiling in itself is not of practical interest since convective boiling, which is the superposition of boiling heat transfer on flow-induced convection, is the rule rather than the exception in industrial process equipment. However, nucleate pool boiling heat transfer coefficients are utilized to predict the boiling contribution to the overall convective boiling heat transfer coefficient. Consequently, an accurate method for calculating the variation in the heat transfer coefficient as a function of composition in nucleate pool boiling is required. The ultimate method should:

- Incorporate all of the composition-dependent physical mechanisms into the boiling model.
- Include the effect of heat flux on the lowering of the mixture heat transfer coefficient.
- Be sufficiently reliable for all mixture systems over a wide pressure range.
- Be easily implemented with a minimum number of mixture physical properties.

Palen and Small [6] developed perhaps the earliest method for predicting the boiling heat transfer coefficients for use in reboiler design. Note that earlier Kern [7] suggested a constant value of 1.70 kW m^{-2} be assigned to the heat transfer coefficient for hydrocarbon mixtures with wall superheats greater than 4.5 K. Palen and Small give the following correlation for the mixture effect:

$$\frac{\alpha}{\alpha_I} = \exp\left[-0.027(T_{b0} - T_{bi})\right] \tag{1}$$

where α = the predicted heat transfer coefficient
α_I = the ideal heat transfer coefficient calculated using a suitable single-component correlation evaluated with mixture properties
$T_{b0} - T_{bi}$ = the temperature difference between the dew point and bubble point temperatures at the bulk composition (see Figure 5)

Using the excess function approach commonly utilized to predict mixture physical properties, Stephan and Korner [8] give a simple method for correlating binary mixture boiling heat transfer coefficients. They observed that the variation in the absolute values of the vapor-liquid composition difference, $|\tilde{y} - \tilde{x}|$, was similar to the variation in the wall superheat with composition at a constant heat flux as illustrated in Figure 6. Defining the ideal superheat, ΔT_I, using a linear molar mixing law at constant heat flux as

$$\Delta T_I \equiv \tilde{x}_1 \Delta T_1 + \tilde{x}_2 \Delta T_2 \tag{2}$$

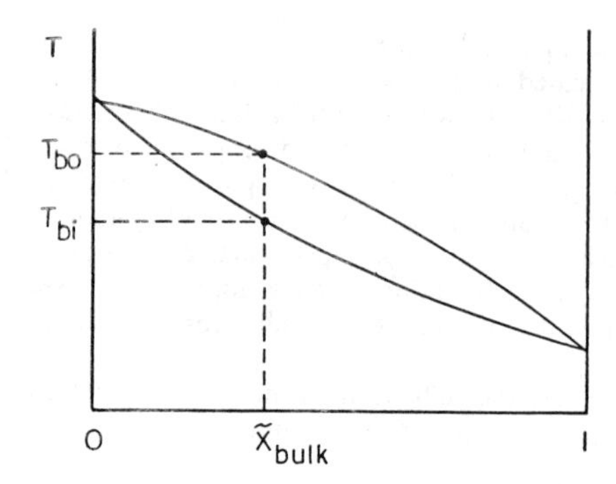

Figure 5. Phase equilibrium diagram showing temperatures to be used in the Palen and Small correlation.

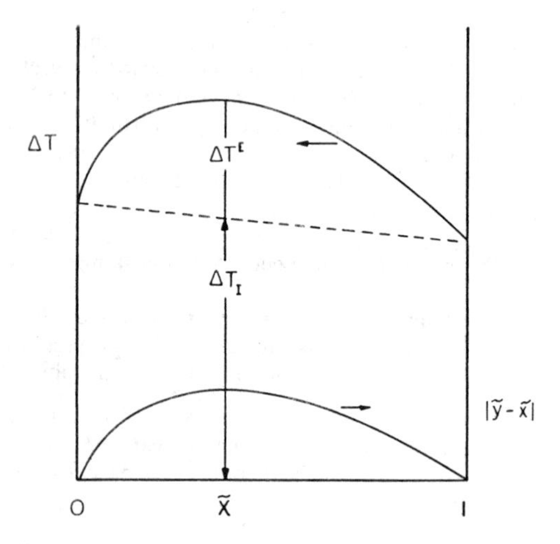

Figure 6. Variation in wall superheat and vapor-liquid composition difference with composition.

Table 1
Numerical Values of A_o*

Mixture	Numerical Values, A_o
Acetone/ethanol	0.75
Acetone/butanol	1.18
Acetone/water	1.40
Ethanol/benzene	0.42
Ethanol/cyclohexane	1.31
Ethanol/water	1.21
Benzene/toluene	1.44
Heptane/methylcyclohexane	1.95
Isopropanol/water	2.04
Methanol/benzene	1.08
Methanol/amyl alcohol	0.80
Methylethyl ketone/toluene	1.32
Methylethyl ketone/water	1.21
Propanol/water	3.29
Water/glycol	1.47
Water/glycerol	1.50
Water/pyridine	3.56

* *From Stephan and Korner [8]*

and correlating the excess superheat, ΔT^E, shown in Figure 6 as

$$\Delta T^E = A \mid \tilde{y} - \tilde{x} \mid \Delta T_I \tag{3}$$

they obtained an expression for the actual wall superheat as

$$\Delta T \equiv T_W - T_{sat} = \Delta T_I + \Delta T^E \tag{4}$$

where the empirical quantity A pertains to the particular mixture system and pressure. The wall superheats ΔT_1 and ΔT_2 correspond to the two single-component values and are predicted using a suitable single-component correlation. Stephen and Korner give the following empirical equation to account for the pressure dependency of A, valid over a range of 1-10 bar:

$$A = A_0(0.88 + 0.12p) \tag{5}$$

with p in bars and A_0 being the value at 1.0 bar. Their listed values of A_0 for 17 binary mixture systems are tabulated in Table 1 (their suggested value for an unlisted mixture system is A_0 equal to 1.53).

Thome [9] derived a completely analytical expression for nucleate pool boiling of binary liquid mixtures based on a thermal model which assumed that the cyclic thermal boundary-layer stripping mechanism is the dominant mode of heat transport. A schematic of this model is shown in Figure 7. The high rate of heat transfer in boiling via this mechanism is due to the cyclic removal of the thermal boundary layer which is carried away by the wake of departing bubbles. Thome arrived at the following equation for the heat transfer coefficient when inertia forces control bubble departure from the heated wall:

$$\frac{\alpha}{\alpha_I} = \frac{\Delta T_I}{\Delta T} = \left[1 - (\tilde{y} - \tilde{x})\left(\frac{\kappa_\ell}{\delta_\ell}\right)^{1/2}\left(\frac{c_{p\ell}}{\Delta h_v}\right)\left(\frac{dT}{d\tilde{x}}\right)\right]^{-7/5} \tag{6}$$

where ΔT_I is calculated using Equation 2. The term to the right of the equal sign is always less than or equal to 1.0 since the product $(\tilde{y} - \tilde{x})$ $(dT/d\tilde{x})$, where $(dT/d\tilde{x})$ is the slope of the

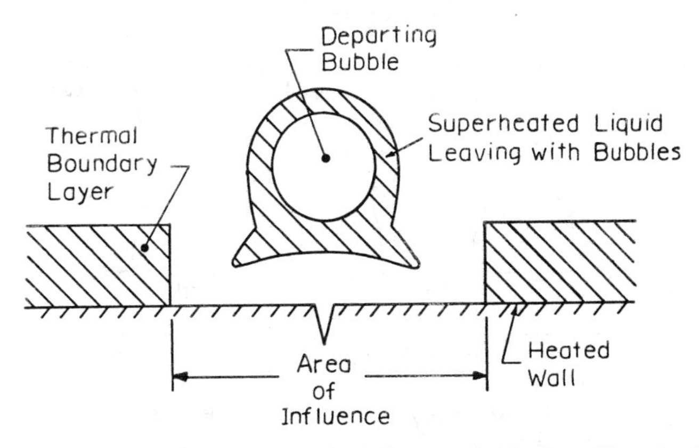

Figure 7. Pool boiling model of Thome [9] illustrating the cyclic removal of the thermal boundary layer by a departing vapor bubble.

bubble point line, is always negative as can be seen from inspection of Figures 1 and 2. The ideal heat transfer coefficient is defined as

$$\alpha_I \equiv \frac{1}{(\tilde{x}_1/\alpha_1)+(\tilde{x}_2/\alpha_2)} \tag{7}$$

which is an expression equivalent to Equation 2. Thome [1] later derived a similar expression for the case where surface tension forces control bubble departure from the heated surface:

$$\alpha\alpha_I = \frac{\Delta T_I}{\Delta T} = \left[1-(\tilde{y}-\tilde{x})\left(\frac{\kappa_\ell}{\delta_\ell}\right)^{1/2}\left(\frac{c_{p_\ell}}{\Delta\tilde{h}_v}\right)\left(\frac{dT}{d\tilde{x}}\right)\right]^{-3/2}\left\{\frac{\sigma \sin\beta}{\sigma_I \sin\beta_I}\right\} \tag{8}$$

Here, the ideal surface tension σ_I and the ideal contact angle β_I are calculated as

$$\sigma_I = \tilde{x}_1\sigma_1 + \tilde{x}_2\sigma_2 \tag{9}$$

and

$$\beta_I = \tilde{x}_1\beta_1 + \tilde{x}_2\beta_2 \tag{10}$$

For a mixture where $\sigma = \sigma_I$ and $\beta = \beta_I$, Equation 8 gives essentially the same prediction as Equation 6.

Thome [10, 11] then postulated a new simple analytic approach to the problem in order to eliminate the difficulty involved in accurately predicting the liquid mass diffusivity, δ_ℓ, appearing in Equations 6 and 8. He noted that the rise in the local boiling point of the mixture adjacent to the heated wall is controlled by the total rate of evaporation at the wall. Thus, as can be seen from Figure 3A, the rise in the local boiling point varies from zero at low heat fluxes (where only single-phase natural convection occurs) to its highest value at the peak nucleate heat flux. It is then assumed that all of the liquid arriving at the heated surface is evaporated at the peak heat flux as in the Zuber hydrodynamic instability

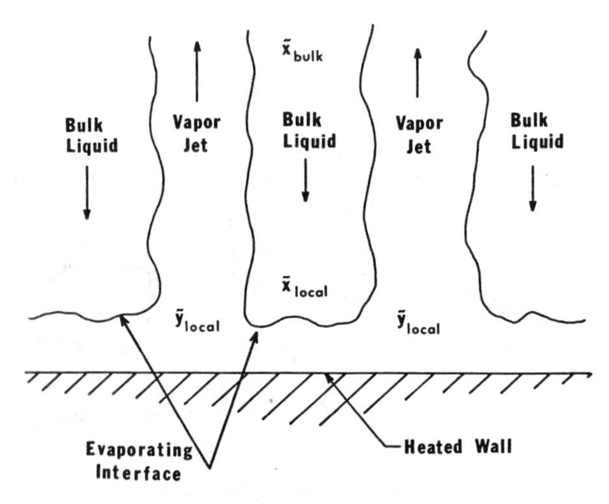

Figure 8. Zuber hydrodynamic instability model.

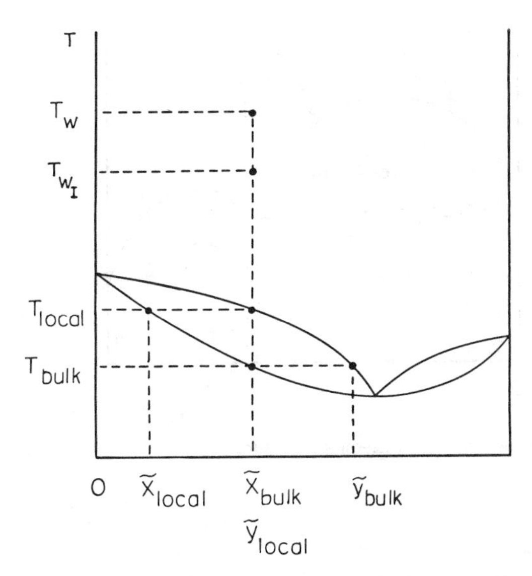

Figure 9. Phase equilibrium diagram showing the maximum rise in the local bubble point, ΔT_{bp}.

model [18], illustrated schematically in Figure 8. For steady-state conditions the composition of the vapor leaving in the vapor jets, $\tilde{y}_{local}$, must be equal to the bulk liquid arriving at the heated wall, $\tilde{x}_{bulk}$, to satisfy the conservation of molecular species. As depicted in Figure 9, the maximum rise in the local bubble point temperature is ΔT_{bp} where

$$\Delta T_{bp} = T_{local} - T_{bulk} = T_{bo} - T_{bi} \tag{11}$$

and is hence the same value used in the Palen and Small correlation, Equation 1. The resulting expression for the heat transfer coefficient is obtained by substituting ΔT_{bp} for ΔT^E in Equation 4 to give

$$\frac{\alpha}{\alpha_I} = \frac{\Delta T_I}{\Delta T} = \frac{\Delta T_I}{\Delta T_I + \Delta T_{bp}} \tag{12}$$

This method will always give a conservative estimate of α for heat fluxes below the peak nucleate heat flux. The method is easily extendable to multicomponent mixtures by using the appropriate dew point and bubble point temperatures and using the following expression in place of Equation 2:

$$\Delta T_I = \sum_{i=1}^{n} \tilde{x}_i \Delta T_i \tag{13}$$

where n = the number of components

ΔT_i = the corresponding single-component values

An approach to modeling the ideal mixture boiling behavior different from the linear mixing law as defined in Equations 2 and 13 is to use an applicable single-component

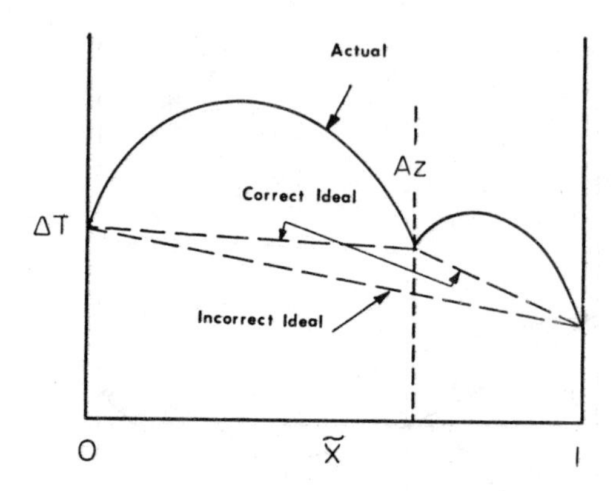

Figure 10. The ideal wall superheat for an azeotropic mixture system.

correlation as suggested in Equation 1 by Palen and Small. The correlation by Stephan and Abdelsalam [19] is recommended.

Using a single-component correlation in principle is more "correct" than utilizing a linear mixing law for predicting the ideal mixture boiling behavior, since the correlation can incorporate nonlinear variation in mixture physical properties. Thus, this approach is recommended when the various mixture physical properties are available. Also, for azeotropic mixture systems the ideal mixing law should be evaluated between the azeotrope point and the appropriate single component in order to have the mixture effect go to zero at $\tilde{x}_{az}$. For instance, to the left of the azeotrope point for a binary mixture system the ideal wall superheat is prorated by Thome [1] as

$$\Delta T_I = \left(\frac{\tilde{x}_1}{\tilde{x}_{az}}\right)\Delta T_{az} + \left(\frac{\tilde{x}_{az} - \tilde{x}_1}{\tilde{x}_{az}}\right)\Delta T_{\tilde{x}_1 = 0} \tag{14}$$

and illustrated in Figure 10.

COMPARISON TO EXPERIMENTAL RESULTS

The mixture nucleate pool boiling heat transfer expressions presented in the last section have been compared to a common set of data by Thome [11] to determine the most accurate method. The mixture system of ethanol-water at 1.0 bar was chosen because aqueous mixtures tend to be the most difficult to predict and because numerous independent data sets are available in the literature. Equation 12 was found to be the most accurate and Figure 11 shows its comparison to experimental data. Note that Equation 14 with $\tilde{x}_{az} = 0.89$ was evaluated using the experimental values.

Mixtures which have wide boiling ranges are also important to accurately predict because the actual heat transfer coefficient can become as small as 1/20 of the ideal heat transfer coefficient since the boiling range ΔT_{bp} can become much larger than ΔT_I. Figure 12 shows the comparison of Equation 12 to the liquid-nitrogen/liquid methane mixture boiling data of Ackermann [20]. This system has a maximum boiling range of 23°K at 5.0 bar but this is low compared to some hydrocarbon mixtures.

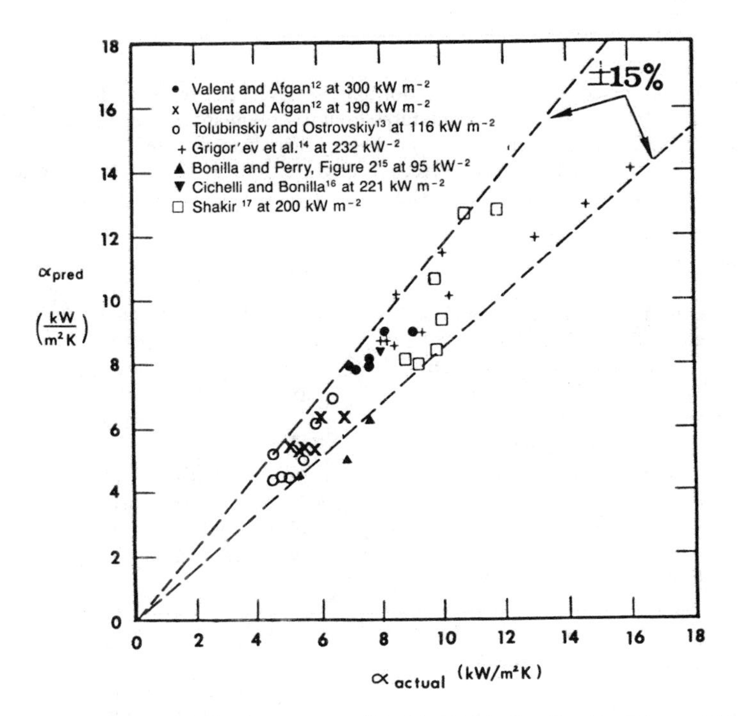

Figure 11. Comparison of ΔT_{bp} theory to experimental data for ethanol-water mixtures at 1.0 bar.

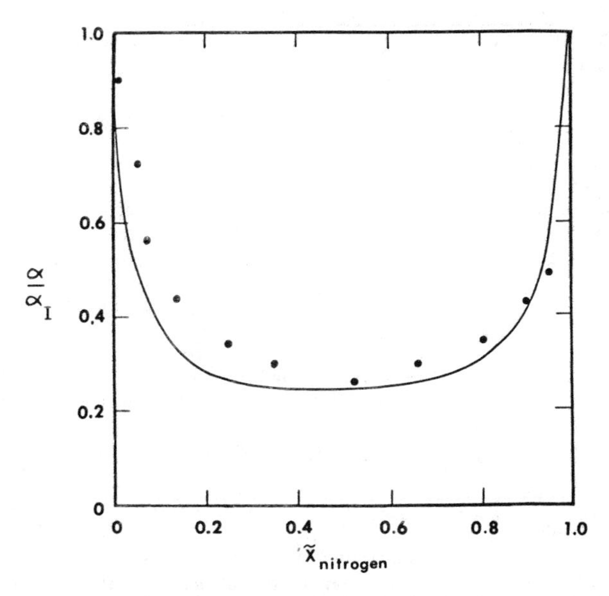

Figure 12. Comparison of ΔT_{bp} theory to experimental data for nitrogen-methane mixtures at 5.0 bar and 210 kW m^{-2}. Solid line is Equation 12.

SUMMARY

The effects of composition on nucleate pool boiling have been reviewed for boiling on "smooth" surfaces (for a discussion of mixtures boiling on enhanced surfaces see Arshad and Thome [21]) and various methods for predicting their heat transfer coefficients have been described. For the prediction of convective mixture boiling the reader is referred to Thome and Shock [1] for a detailed discussion.

NOTATION

A	empirical coefficient
A_0	empirical coefficient at 1.0 bar
$c_{p\ell}$	liquid specific heat (J/(kgK))
Δh_v	latent heat of vaporization (J/kg)
n	number of components
N/A	boiling site density (sites/cm^2)
p	pressure (bar)
$\dot{q}$	heat flux (W/m^2)
T	temperature (K)
T_{bi}	bubble point temperature (K)
T_{bo}	dew point temperature (K)
T_{sat}	saturation temperature (K)
T_W	wall temperature (K)
$dT/d\tilde{x}$	slope of bubble point line (K)
ΔT	$= (T_w - T_{sat})$, wall superheat (K)
ΔT_{bp}	boiling range (K)
ΔT^E	excess superheat (K)
$\tilde{x}$	liquid mole fraction
$\tilde{y}$	vapor mole fraction

Greek symbols

α	heat transfer coefficient (W/(m^{2}K))
α_{actual}	experimentally measured heat transfer coefficient (W/(m^{2}K))
$\alpha_{predicted}$	predicted heat transfer coefficient (W/(m^{2}K))
β	contact angle (rad)
δ_ℓ	liquid mass diffusivity (m^2/s)
κ_ℓ	liquid thermal diffusivity (m^2/s)
σ	surface tension (N/m)

Subscripts

az	azeotrope
bulk	bulk quantity
eth	ethanol
i	component i
I	ideal linear mixing law value
local	local value
N_2	nitrogen
1, 2	components 1, 2

REFERENCES

1. Thome, J. R., and Shock, R. A. W., "Boiling of Multicomponent Liquid Mixtures," In *Advances in Heat Transfer*, Vol. 16, pp. 59–156, T. J. Irvine, Jr. and J. P. Hartnett (Eds.) New York: Academic Press, 1984.
2. Van Wijk, W. R., et al., "Heat Transfer to Boiling Binary Liquid Mixtures," *Chem. Eng. Sci.*, Vol. 5 (1956), pp. 68–80.
3. Sternling, C. V., and Tichacek, L. J., "Heat Transfer Coefficients for Boiling Mixtures," *Chem. Eng. Sci.*, Vol. 16 (1961), pp. 297–337.
4. Hui, T. O., M. S. Thesis, Michigan State University, 1983.
5. Thome, J. R., "Latent and Sensible Heat Transport Rates in the Boiling of Binary Mixtures," *J. Heat Transfer*, Vol. 104 (1982), pp. 474–478.

6. Palen, J. W., and Small, W., "A New Way to Design Kettle and Internal Reboilers," *Hydrocarbon Processing,* Vol. 43, No. 11 (1964), pp. 199–208.
7. Kern, D. W., *Process Heat Transfer,* New York: McGraw-Hill, 1950.
8. Stephan, K., and Korner, M., "Calculation of Heat Transfer in Evaporating Binary Liquid Mixtures," *Chemie-Ingr-Tech.,* Vol. 41, No. 7 (1969), pp. 409–417.
9. Thome, J. R., "Nucleate Pool Boiling of Binary Liquids—An Analytical Equation," *A. I. Ch. E. Symp. Ser.* Vol. 77, No. 208 (1981), pp. 238–250.
10. Thome, J. R., "Prediction of Binary Mixture Boiling Heat Transfer Coefficients Using Only Phase Equilibrium Data," AERE-RS449, 1982.
11. Thome, J. R., "Prediction of Binary Mixture Boiling Heat Transfer Coefficients Using Only Phase Equilibrium Data," *Int. J. Heat Mass Transfer,* Vol. 26 (1983), pp. 965–974.
12. Valent, V., and Afgan, N. H., "Bubble Growth Rate and Boiling Heat Transfer in Pool Boiling of Ethyl Alcohol-Water Mixture," *Wärme- und Stoffubertragung,* Vol. 6 (1973), pp. 235–240.
13. Tolubinskiy, V. I., and Ostrovskiy, Y. N., "Mechanism of Heat Transfer in Boiling of Binary Mixtures," *Heat Transfer-Soviet Res.* Vol. 1, No. 6 (1969), pp. 6–11.
14. Grigor'ev, L. N. et al., "An Experimental Study of Critical Heat Flux in the Boiling of Binary Mixtures," *Int. Chem. Eng,* Vol. 8, No. 1 (1968), pp. 39–42.
15. Bonilla, C. F., and Perry, C. W., "Heat Transfer to Liquids Boiling Binary Mixtures," *Trans. A. I. Ch. E.,* Vol. 41 (1945), pp. 755–787; Vol. 42 (1946), pp. 411–412.
16. Cichelli, M. T., and Bonilla, C. F., "Heat Transfer to Liquids Boiling Under Pressure," *Trans. A. I. Ch. E.,* Vol. 41 (1945), pp. 755–787; Vol. 42 (1946), pp. 411–412.
17. Shakir, S., Ph. D. Thesis, Michigan State University, 1985.
18. Zuber, N., "Hydrodynamic Aspects of Boiling Heat Transfer," AEC Report AECU-4439, Physics and Mathematics, 1959.
19. Stephan, K., and Abdelsalam, M. A., "Heat Transfer Correlations for Natural Convection Boiling," *Int. J. Heat Mass Transfer,* Vol. 23 (1980), pp. 73–87.
20. Ackermann, J. et al., "Heat Transfer in Liquid Nitrogen-Methane Mixtures Under Pressure," *Cryogenics,* Vol. 15 (1975), pp. 657–659.
21. Arshad, J., and Thome, J. R., "Enhanced Boiling Srufaces: Heat Transfer Mechanism and Mixture Boiling," ASME/JSME Thermal Engineering Joint Conference, Honolulu, Vol. 1 (1983), pp. 191–197.

CHAPTER 12

ESTIMATING VAPOR PRESSURE FROM THE NORMAL BOILING POINTS OF HYDROCARBONS

N. V. K. Dutt and D. H. L. Prasad

Regional Research Laboratory
Hyderabad, India

CONTENTS

INTRODUCTION

Determining the vapor pressure of substances is needed:

- For direct use in equipment and process design
- In the calculation of Pitzer's acentric factor [1] and Stiel's polar factor [2], which have been used extensively in the development of estimating techniques for various properties.

In the absence of an experimentally determined vapor pressure-temperature realtionship, the designer has to resort to estimation techniques. The merit of any estimation technique lies in its capability to predict the desired property from other easily determinable properties —obtainable either by experimentation or by a simple calculation procedure. Hydrocarbons and related compounds occur in several industrial systems of importance. Normal boiling point is a property easily determined experimentally as well as by simple calculations. This chapter will therefore be devoted to a state-of-the-art exposition of the methods of estimating vapor pressure from normal boiling point.

EARLY EFFORTS

One of the earliest attempts to relate the saturated vapor pressure of hydrocarbons (with molecular weight greater than 30) to molecular weight, normal boiling point, and critical properties resulted in the following scheme proposed by Cox [3]:

$$\log P = A[1-(B/T)] \tag{1}$$

$$\log (A/A_c) = E(1-T_r)(F-T_r) \tag{2}$$

$$E = 0.0008\,B - 0.04895 \tag{3}$$

$$\log A_c = 0.61076 + 0.0005\ m \tag{4}$$

$$(\log m\ P_c/C) = 0.766484 - 0.0000842\ m - 0.030506 \log m \tag{5}$$

$$\log B = 1.07575 + 0.949128 \log m - 0.101\ (\log m)^2 \tag{6}$$

where T = temperature (°K)
P = pressure (atm)
C = critical temperature (°K)
P_c = critical pressure (atm)
T_r = reduced temperature (= T/C)
m = molecular weight

Assuming F = 0.85 and the values of the other constants (B, E, and A_c) calculated from Equations 2 through 6 in a suitable order, the value of A can be calculated by applying Equation 1 to the normal boiling condition. Tables 1 through 3 give the constants and the results of application of the method to several hydrocarbons.

Perry and Smith [4] suggested the use of:

- Ramsey and Young equation

$$\log T_A = \log T_B + C \tag{7}$$

- Duhring rule

$$T_A = KT_B + C \tag{8}$$

- An equation derived from the Clausius–Clapeyron equation

$$\frac{1}{T_A} = \left(\frac{\Delta H_B}{\Delta H_A}\right)\frac{1}{T_B} + C \tag{9}$$

Table 1
Principal Constants for Normal Paraffin Series for Cox Method*

	n = 3	4	5	6	7	8
Calculated B, °K	230.57	272.51	309.16	341.87	371.54	398.75
Calculated C, °K	373.27	426.34	470.24	507.64	540.20	569.00
Observed C, °K	373.27	426.36	470.36	507.96	540.01	569.36
Calculated P_c, atm	43.77	37.51	33.02	29.60	26.89	24.68
Observed P_c, atm	43.78	37.50	33.04	29.64	26.90	24.70
Calculated log A_c	0.63276	0.63976	0.64676	0.65376	0.66076	0.66776
Observed log A_c	0.63268	—	0.64704	0.65371	0.66105	0.66720
Calculated mP_c/C	0.71264	0.70780	0.70376	0.70023	0.69705	0.69414
Observed mP_c/C	0.71264	—	0.70376	0.70023	0.69705	0.69324
Calculated E	0.13551	0.16907	0.19837	0.22455	0.24828	0.27005

* *From Cox [3].*

Table 2
Calculated Values of A of Cox Method for Normal Paraffin Series

T_r	n = 3	4	5	6	7	8
0.400	4.670	4.846	5.016	5.181	5.343	5.504
0.500	4.534	4.670	4.803	4.932	5.061	5.188
0.600	4.429	4.536	4.641	4.745	4.848	4.952
0.700	4.354	4.440	4.526	4.612	4.698	4.785
0.780	4.314	4.389	3.365	4.542	4.619	4.698
0.925	4.285	4.353	4.422	4.493	4.564	4.637
0.85, 1.0	4.293	4.363	4.434	4.506	4.579	4.653

* *From Cox [3].*

Table 3
Calculations for Miscellaneous Hydrocarbons Using Cox Method*

	Isobutane	Isopentane	Diisopropyl	Diisobutyl	Cyclohexane	Benzene
Observed B, °K	261.08	301.01	331.25	382.37	353.98	353.37
Observed C, °K	406.86	460.96	500.51	549.96	553.11	561.66
Observed P_c, atm	36.54	25018 mm Hg.	23360 mm Hg	18660 mm Hg	30260 mm Hg	36395 mm Hg
Calculated log A_c	0.63578	0.64079	0.64335	0.65913	0.64781	0.65620
Calculated E	0.15990	0.19185	0.21605	0.25695	0.23423	0.23375
Calculated A						
TR = 0.400	—	—	—	5.352	5.141	5.240
0.500	4.611	4.725	4.791	5.060	4.884	4.979
0.600	4.485	4.571	4.623	4.840	4.691	4.782
0.700	4.395	4.461	4.496	4.685	4.554	4.642
0.780	4.347	4.403	4.433	4.604	4.481	4.569
0.925	4.314	4.362	4.387	4.547	4.431	4.517
0.85, 1.0	4.323	4.373	4.399	4.562	4.444	4.531

* *From Cox [3].*

for the calculation of temperatures at equal vapor pressure of two substances, A and B. High values are obtained from interpolation formulae using the minimum amount of experimental data—usually an additional value of vapor pressure, apart from the normal boiling point is required for an estimate.

Thodos [5] proposed

$$\log P = A + (B/T) + (C/T^2) + D((T/T_d) - 1)^n \tag{10}$$

with $$(1/T_d) = 1.3713\ (1/T_b)^{1.0805} \tag{11}$$

and other constants A, B, C, D, and n expressed in terms of normal boiling point by simple graphs. The outlines of the application of the method to n-paraffins is given in Table 4. Extension of the scheme may be attempted for other homologous series of interest.

ADDITIONAL PROPERTY CORRELATIONS

Li and Rossini [6] described a method which employs the number of carbon atoms in the normal alkyl radical (m) as the additional correlating parameter. Antoine equation

$$\log P = A - B/(t + C) \tag{12}$$

Table 4
Physical Constants and Coefficients for Vapor Pressure Equation for Light Normal Saturated Aliphatic Hydrocarbons for Use in Thodos Approach*

	Temperature (°K)			Pressure (mm Hg)		Vapor Pressure Equation Coefficients				
Hydrocarbon	Normal boiling	Divergence	Critical	Divergence	Critical	A	B	C	D	n
Methane	111.49	118.83	191.04	1324	35,008	6.18025	− 296.1	− 8,000	0.2570	1.32
Ethane	184.467	204.74	305.44	2013	36,632	6.73244	− 624.24	− 15,912	0.1842	1.963
Propane	231.105	261.20	369.98	2429	31,928	6.80064	− 785.6	− 27,800	0.2102	2.236
n-Butane	272.666	312.30	425.18	2774	28,477	6.78880	− 902.4	− 44,493	0.4008	2.40
n-Pentane	309.238	357.79	469.84	3103	25,547	6.77767	− 988.6	− 66,936	0.6550	2.46
n-Hexane	341.905	398.79	507.88	3397	22,754	6.75933	−1054.9	− 92,720	0.9692	2.49
n-Heptane	371.591	436.34	540.18	3673	20,520	6.74242	−1108.0	−121,489	1.3414	2.50
n-Octane	398.830	471.00	569.36	3937	18,730	6.72908	−1151.6	−152,835	1.7706	2.50
n-Nonane	423.961	503.14	595.95	4193	17,158	6.72015	−1188.2	−186,342	2.2438	2.50
n-Decane	447.288	533.13	621.12	4445	16,138	6.71506	−1219.3	−221,726	2.7656	2.50
n-Dodecane	489.443	587.61	663.13	4941	14,058	6.71471	−1269.7	−296,980	3.9302	2.50

* *From Thodos [5].*

where P = pressure (mmHg)
t = temperature (°C) is used to represent the vapor pressure of n-haloalkanes. The constant C can be determined from

$$C = 239 - 0.19\, t_b \tag{13}$$

If an additional data point is known, the constants A and B can be determined by applying Equation 12 to the particular substance.

Alternatively,

$$t = t_0 + C_t^I(m-1) - d_t\left(\int_{b/t}^{x} e^{-u} d \ln u - \int_{u}^{x} e^{-u} d \ln u\right) \tag{14}$$

where $u = b_t(m-1)$
$b_t = 0.1505$
$c_t^I = 540$

can be formulated and used for each homologous series of interest. First, one has to evaluate the constants t_0 and d_t from the available normal boiling point data of the particular homologous series. Unavailable boiling points of the compounds of the homologous series can then be predicted from Equation 14. Antoine constants for normal haloalkanes are given in Table 5.

Smialek and Thodos [7] proposed

$$\log P = 5.113 - 3.951\, m \tag{15}$$

$$\text{with } m = \frac{(1/T) - 0.0220 \times 10^{-3}}{(1/T_b) - 0.220 \times 10^{-3}} \tag{16}$$

for the prediction of vapor pressures in the range 5–600 psia. T and T_b are the temperature and normal boiling point in degrees Rankine.

Kudchadker and Zwolinski [8] related the constants of the Antoine equation to normal boiling point according to the following scheme.

$$\log(1078 - T_b) = 3.03191 - 0.049901\, m^{2/3} \tag{17}$$

$$\frac{B}{T_b} = 3.53813 - 9.7736 \times 10^{-5} T_b - 6.66695 \times 10^{-7} T_b^2 \tag{18}$$

$$\frac{C - 273.15}{T_b} = -4.49159 \times 10^{-2} - 2.68408 \times 10^{-4} T_b - 5.18608 \times 10^{-8} T_b^2 \tag{19}$$

The constants for higher hydrocarbons are given in Table 6.

Gomez-Nieto and Thodos [9] proposed

$$\ln P_r = \alpha + (\beta / T_r^m) + \gamma T_r^n \tag{20}$$

The constants are defined in terms of the characterization parameter

$$s = T_b \ln P_c / (T_c - T_b) \tag{21}$$

The other constants are defined by

$$\alpha = -(\beta + \gamma); \quad m = 0.64837 e^{0.10982\, s} - \frac{2725.2}{e^{2.0133\, s}} \tag{22}$$

Table 5
Constants of the Antoine Equation in the Range of 10–1500 mm Hg for n-Haloalkanes*

Compound	Values of Constants		
	A	B	C
1-Fluoroalkanes			
1-Fluoromethane	7.09761	740.218	253.89
1-Fluoroethane	6.97853	854.211	246.16
1-Fluoropropane	6.9533	965.18	239.5
1-Fluorobutane	6.9581	1081.71	232.8
1-Fluoropentane	6.9857	1190.03	227.1
1-Fluorohexane	7.0305	1299.19	221.6
1-Fluoroheptane	7.0835	1405.79	216.6
1-Fluoro-octane	7.1411	1509.34	212.0
1-Fluorononane	7.1977	1608.48	207.6
1-Fluorodecane	7.2542	1704.75	203.6
1-Fluoroundecane	7.308	1797.8	200
1-Fluorododecane	7.357	1885.6	196
1-Fluorotridecane	7.406	1969.1	193
1-Fluorotetradecane	7.449	2048.3	190
1-Fluoropentadecane	7.488	2123.4	187
1-Fluorohexadecane	7.520	2194.8	184
1-Fluoroheptadecane	7.556	2262.5	181
1-Fluorooctadecane	7.586	2327.4	179
1-Fluorononadecane	7.612	2389.7	176
1-Fluoroeicozane	7.636	2450.1	174
1-Chloroalkanes			
1-Chloromethane	6.99445	902.451	243.60
1-Chloroethane	6.94914	1012.771	236.67
1-Chloropropane	6.93111	1121.123	230.20
1-Chlorobutane	6.93790	1227.433	224.10
1-Chloropentane	6.96617	1332.890	218.50
1-Chlorohexane	7.0115	1437.05	213.4
1-Chloroheptane	7.0650	1539.35	208.8
1-Chlorooctane	7.1231	1639.20	204.4
1-Chlorononane	7.1802	1736.11	200.4
1-Chlorodecane	7.2372	1829.68	196.6
1-Chloroundecane	7.2917	1919.62	193.0
1-Chlorododecane	7.3394	2005.72	190.0
1-Chlorotridecane	7.391	2087.9	186.0
1-Chlorotetradecane	7.434	2166.1	184.0
1-Chloropentadecane	7.474	2240.5	180.0
1-Chlorohexadecane	7.506	2311.4	178.0
1-Chloroheptadecane	7.543	2378.8	175.0
1-Chlorooctadecane	7.573	2443.5	173.0
1-Chlorononadecane	7.600	2505.7	170.0
1-Chloroeicozane	7.624	2566.1	168.0

Compound	Values of Constants		
	A	B	C
1-Bromoalkanes			
1-Bromomethane	6.95965	986.590	238.32
1-Bromoethane	6.91995	1090.810	231.71
1-Bromopropane	6.91065	1194.889	225.51
1-Bromobutane	6.92254	1298.608	219.70
1-Bromopentane	6.95580	1401.634	214.38
1-Bromohexane	7.0023	1503.52	209.5
1-Bromoheptane	7.0582	1603.71	205.0
1-Bromooctane	7.1179	1701.61	200.8
1-Bromononane	7.1761	1796.73	196.9
1-Bromodecane	7.2336	1888.67	193.3
1-Bromoundecane	7.2882	1977.14	189.8
1-Bromododecane	7.3390	2061.93	186.6
1-Bromotridecane	7.386	2143.0	184
1-Bromotetradecane	7.430	2220.2	181
1-Bromopentadecane	7.470	2293.8	178
1-Bromohexadecane	7.506	2364.0	175
1-Bromoheptadecane	7.540	2430.9	173
1-Bromooctadecane	7.570	2495.2	170
1-Bromononadecane	7.597	2557.2	168
1-Bromoeicosane	7.621	2617.5	166
1-Iodoalkanes			
1-Iodoethane	6.87991	1093.235	230.94
1-Iodomethane	6.83198	1175.709	225.26
1-Iodopropane	6.81603	1267.062	219.53
1-Iodobutane	6.82262	1358.860	214.20
1-Iodopentane	6.85172	1454.028	209.17
1-Iodohexane	6.8954	1549.17	204.55
1-Iodoheptane	6.9488	1644.29	200.25
1-Iodooctane	7.0070	1738.53	196.23
1-Iodononane	7.0645	1830.37	192.5
1-Iododecane	7.1220	1919.75	188.9
1-Iodoundecane	7.1772	2006.28	185.5
1-Iodododecane	7.2290	2089.47	182.3
1-Iodotridecane	7.277	2169.2	179
1-Iodotetradecane	7.322	2245.4	176
1-Iodopentadecane	7.360	2318.3	174
1-Iodohexadecane	7.401	2388.0	171
1-Iodoheptadecane	7.437	2454.6	168
1-Iodooctadecane	7.469	2518.7	166
1-Iodononadecane	7.498	2580.6	164
1-Iodoeicosane	7.524	2640.9	162

* *From Li and Rossini [6].*

Table 6
Antoine Constants and Normal Boiling Points of Higher n-Alkanes*

Formula	Description	Antoine Constants			Normal Boiling Point, °C
		A	B	C	
$C_{21}H_{44}$	n-Heneicosane	7.0770	2022.5	125.5	356.5
$C_{22}H_{46}$	n-Docosane	7.0842	2054.0	120.1	368.6
$C_{23}H_{48}$	n-Tricosane	7.0911	2083.8	114.8	380.1
$C_{24}H_{50}$	n-Tetracosane	7.0976	2112.0	109.6	391.3
$C_{25}H_{52}$	n-Pentacosane	7.1038	2138.8	104.6	401.9
$C_{26}H_{54}$	n-Hexacosane	7.1096	2164.3	99.6	412.2
$C_{27}H_{56}$	n-Heptacosane	7.1152	2188.5	94.8	422.2
$C_{28}H_{58}$	n-Octacosane	7.1205	2211.6	90.0	431.6
$C_{29}H_{60}$	n-Nonacosane	7.1256	2233.6	85.4	440.8
$C_{30}H_{62}$	n-Triacosane	7.1304	2254.6	80.9	449.6
$C_{31}H_{64}$	n-Hentriacosane	7.1356	2276.9	75.9	459
$C_{32}H_{66}$	n-Dotriacontane	7.1400	2296.1	76.1	468
$C_{33}H_{68}$	n-Tritriacontane	7.1442	2314.4	67.3	476
$C_{34}H_{70}$	n-Tetratriacontane	7.1482	2331.9	63.1	483
$C_{35}H_{72}$	n-Pentatriacontane	7.1521	2348.6	59.1	491
$C_{36}H_{74}$	n-Hexatriacontane	7.1558	2364.6	55.1	498
$C_{37}H_{76}$	n-Heptatriacontane	7.1593	2380.0	51.2	505
$C_{38}H_{78}$	n-Octatriacontane	7.1627	2394.7	47.4	512
$C_{39}H_{80}$	n-Nonatriacontane	7.1660	2408.8	43.7	518
$C_{40}H_{82}$	n-Tetracontane	7.1691	2422.3	40.1	525
$C_{41}H_{84}$	n-Hentetracontane	7.1721	2435.3	36.5	531
$C_{42}H_{86}$	n-Dotetracontane	7.1750	2447.7	33.0	537
$C_{43}H_{88}$	n-Tritetracontane	7.1777	2459.7	29.6	543
$C_{44}H_{90}$	n-Tetratetracontane	7.1804	2471.2	26.3	548
$C_{45}H_{92}$	n-Pentatetracontane	7.1829	2482.3	23.0	554
$C_{46}H_{94}$	n-Hexatetracontane	7.1854	2493.0	19.8	559
$C_{47}H_{96}$	n-Heptatetracontane	7.1878	2503.3	16.7	565
$C_{48}H_{98}$	n-Octatetracontane	7.1900	2513.1	13.7	570
$C_{49}H_{100}$	n-Nonatetracontane	7.1922	2522.7	10.7	574
$C_{50}H_{102}$	n-Pentacontane	7.1944	2531.8	7.8	579
$C_{51}H_{104}$	n-Henpentacontane	7.1964	2540.7	4.9	584
$C_{52}H_{106}$	n-Dopentacontane	7.1984	2549.2	+2.1	588
$C_{53}H_{108}$	n-Tripentacontane	7.2003	2557.5	−0.7	593
$C_{54}H_{110}$	n-Tetrapentacontane	7.2021	2565.5	−3.3	597
$C_{55}H_{112}$	n-Pentapentacontane	7.2039	2573.1	−6.0	601
$C_{56}H_{114}$	n-Hexapentacontane	7.2056	2580.6	−8.5	605
$C_{57}H_{116}$	n-Heptapentacontane	7.2073	2587.7	−11.1	609
$C_{58}H_{118}$	n-Octapentacontane	7.2089	2594.7	−13.5	613
$C_{59}H_{120}$	n-Nonapentacontane	7.2104	2601.4	−16.0	617
$C_{60}H_{122}$	n-Hexacontane	7.2119	2607.9	−18.3	620
$C_{61}H_{124}$	n-Henhexacontane	7.2134	2614.2	−20.6	624
$C_{62}H_{126}$	n-Dohexacontane	7.2148	2620.2	−22.9	628
$C_{63}H_{128}$	n-Trihexacontane	7.2161	2626.1	−25.1	631
$C_{64}H_{130}$	n-Tetrahexacontane	7.2174	2631.8	−27.3	634
$C_{65}H_{132}$	n-Pentahexacontane	7.2187	2637.4	−29.5	637
$C_{66}H_{134}$	n-Hexahexacontane	7.2199	2642.7	−31.6	641
$C_{67}H_{136}$	n-Heptahexacontane	7.2211	2647.9	−33.6	644
$C_{68}H_{138}$	n-Octahexacontane	7.2223	2652.9	−35.6	647
$C_{69}H_{140}$	n-Nonahexacontane	7.2234	2657.8	−37.6	650
$C_{70}H_{142}$	n-Heptacontane	7.2245	2662.5	−39.5	653
$C_{71}H_{144}$	n-Henheptacontane	7.2256	2667.1	−41.4	655
$C_{72}H_{146}$	n-Doheptacontane	7.2266	2671.6	−43.3	658
$C_{73}H_{148}$	n-Triheptacontane	7.2276	2675.9	−45.1	661
$C_{74}H_{150}$	n-Tetraheptacontane	7.2286	2680.1	−46.9	663
$C_{75}H_{152}$	n-Pentaheptacontane	7.2295	2684.2	−48.7	666
$C_{76}H_{154}$	n-Hexaheptacontane	7.2304	2688.2	−50.4	668
$C_{77}H_{156}$	n-Heptaheptacontane	7.2313	2692.0	−52.1	671
$C_{78}H_{158}$	n-Octaheptacontane	7.2322	2695.8	−53.7	673
$C_{79}H_{160}$	n-Nonaheptacontane	7.2330	2699.4	−55.3	676
$C_{80}H_{162}$	n-Octacontane	7.2338	2702.9	−56.9	678
$C_{81}H_{164}$	n-Henoctacontane	7.2346	2706.4	−58.5	680
$C_{82}H_{166}$	n-Dooctacontane	7.2354	2709.7	−60.0	682
$C_{83}H_{168}$	n-Trioctacontane	7.2361	2713.0	−61.5	684
$C_{84}H_{170}$	n-Tetraoctacontane	7.2369	2716.2	−63.0	687
$C_{85}H_{172}$	n-Pentaoctacontane	7.2376	2719.2	−64.4	689
$C_{86}H_{174}$	n-Hexaoctacontane	7.2383	2722.2	−65.8	691
$C_{87}H_{176}$	n-Heptaoctacontane	7.2389	2725.2	−67.2	693
$C_{88}H_{178}$	n-Octaoctacontane	7.2396	2728.0	−68.6	694
$C_{89}H_{180}$	n-Nonaoctacontane	7.2402	2730.8	−69.9	696
$C_{90}H_{182}$	n-Nonacontane	7.2408	2733.5	−71.3	698
$C_{91}H_{184}$	n-Hennonacontane	7.2415	2736.1	−72.5	700
$C_{92}H_{186}$	n-Dononacontane	7.2420	2738.7	−73.8	702
$C_{93}H_{188}$	n-Trinonacontane	7.2426	2741.2	−75.1	704
$C_{94}H_{190}$	n-Tetranonacontane	7.2432	2743.6	−76.3	705
$C_{95}H_{192}$	n-Pentanonacontane	7.2437	2746.0	−77.5	707
$C_{96}H_{194}$	n-Hexanonacontane	7.2443	2748.0	−78.7	709
$C_{97}H_{196}$	n-Heptanonacontane	7.2448	2750.6	−79.8	710
$C_{98}H_{198}$	n-Octanonacontane	7.2453	2752.8	−80.9	712
$C_{99}H_{200}$	n-Nonanonacontane	7.2458	2755.0	−82.1	713
$C_{100}H_{202}$	n-Hectane	7.2463	2757.1	−83.1	715

* *From Kudchadker and Zwolinski [8].*

Table 7
Physical Properties and Correlating Parameters for Gomez-Nieto and Thodos Approach*

	Physical Properties							
Hydrocarbons	M	T_b(°K)	T_c(°K)	P_c(atm)	s	β	γ	m
Methane	16.04	111.67	191.04	46.06	5.3885	−4.51239	0.12050	1.11752
Ethane	30.07	184.53	305.43	48.20	5.9151	−4.37985	0.13234	1.22579
Propane	44.09	231.09	369.97	42.01	6.2199	−4.36761	0.14092	1.27762
Butane	58.12	272.65	425.17	37.47	6.4773	−4.42382	0.14194	1.30784
Pentane	72.15	309.22	469.78	33.31	6.7520	−4.36868	0.17322	1.35913
Hexane	86.17	341.89	507.86	29.94	7.0019	−4.29220	0.19131	1.41574
Heptane	100.20	371.59	540.17	27.00	7.2647	−4.30804	0.20022	1.45354
Octane	114.22	398.81	569.36	24.64	7.4936	−4.41938	0.17823	1.47145
Nonane	128.25	423.94	593.80	22.60	7.7824	−4.35362	0.22096	1.52245
Decane	142.28	447.27	616.10	20.73	8.0316	−4.27085	0.25005	1.57794
Undecane	156.30	469.04	636.00	19.18	8.2983	−4.38840	0.25533	1.58803
Dodecane	170.33	489.44	653.90	17.83	8.5739	−4.31913	0.29734	1.63814
Tridecane	184.36	508.59	670.10	16.64	8.8543	−4.41719	0.31163	1.64998
Tetradecane	198.38	526.72	684.90	15.58	9.1438	−4.21163	0.38483	1.73048
Pentadecane	212.41	543.78	698.20	14.64	9.4507	−4.02817	0.45506	1.81212
Hexadecane	226.44	559.94	710.40	13.79	9.7653	−4.08617	0.49323	1.82630
Heptadecane	240.46	574.97	721.30	13.14	10.1209	−3.85532	0.59128	1.92210
Octadecane	254.49	589.27	731.20	12.31	10.4228	−3.61280	0.67425	2.02991
Nonadecane	268.51	602.85	740.30	11.67	10.7764	−3.43623	0.77022	2.11628
Eicosane	282.54	615.85	748.70	11.09	11.1536	−3.22674	0.87213	2.22677

* *From Gomez-Nieto and Thodos [9].*

$$n = 7.0; \quad \beta = -4.39474 - \frac{9.168 \times 10^{12}}{e^{5.95\,s}} + \frac{3.6529}{e^{5.473/s^{3.5}}} \tag{23}$$

and

$$\gamma = -1.0668 - 0.33056\,s - 1.6363e^{0.1106\,s} \tag{24}$$

The parameters and coefficients for n-alkanes are given in Table 7.
Lee and Kesler [10] proposed

$$\ln P = \ln P_n + \sqrt{(SG/SG_n) - 1}\,[(T_b/T)^3 - 1] \bullet [3.214 - 3.765(T_b/T_b^1)^2] \tag{25}$$

T_b^1, the reference substance (n-Hexane) normal boiling point is equal to 341.9° K. SG and SG_n are both represented by

$$SG_n = 1.0475 - 1.511 \times 10^{-4} T_b + 7.127 \times 10^{-8} T_b^2 - 116.4/T_b \tag{26}$$

Subscript n refers to the reference substance. The method has been tested for n-paraffins, olefins, naphthalenes and aromatics. The agreement of the predicted values with experimental data is better compared to the earlier methods.

METHODS USING NORMAL BOILING POINT ONLY

Neumann and Ostertag [11] expressed vapor pressure (in bars) as a function of temperature (in Kelvins) in the form

$$\ln P = A - T_b \frac{A}{T} + 0.013163 \tag{27}$$

with $A = A_1 + A_2 T_b + A_3 T_b^2$ (28)

Table 8
Coefficients of Vapor Pressure Equation*

Homologous Series	Temperature Range (°C)	A	m	n	m′	n′
n-Paraffins	−181 to 379	6.9281	958.08	2.9068	273.19	−0.2857
n-Monoolefins	−153 to 378	6.9047	970.45	2.7360	241.34	−0.3213
n-Alkylcyclopentanes, C_6–C_{22}	− 40 to 406	6.9498	1025.82	2.7611	252.81	−0.3247
Alkylcyclopentanes, C_5–C_7	− 12 to 129	6.8529	959.76	3.1187	241.33	−0.2114
Alkylcyclopentanes, C_8	−0.3 to 158	6.8607	947.12	3.3084	240.91	−0.1933
Alkylcyclohexanes, C_8	10 to 157	6.8365	997.58	2.8413	253.06	−0.2886
n-Monoolefins, C_2–C_5	−153 to 60	6.8525	947.76	3.4077	238.62	−0.1473
Diolefins, C_3–C_6	− 99 to 69	6.8186	878.40	5.1529	224.27	+0.2557
Alkylbenzenes, C_6–C_9	+ 6 to 205	6.9914	747.20	4.9464	186.85	+0.1683
Alkylbenzenes, C_{10}	50 to 235	7.0036	925.56	3.6261	224.53	−0.1198
Paraffins, C_1–C_5	−181 to 57	6.7753	932.10	3.2231	238.91	−0.1766
Paraffins, C_6	− 41 to 92	6.8281	948.79	3.0846	242.71	−0.2574
Paraffins, C_7	− 18 to 123	6.8481	992.98	2.6683	253.16	−0.3603
Paraffins, C_8	+ 3 to 153	6.8681	1040.02	2.4225	262.63	−0.4091
Paraffins, C_9	12 to 178	6.8681	959.94	3.0990	244.31	−0.2493
n-Alkanes, C_{21}–C_{37}	203 to 504	7.5895	1117.54	3.8520	234.91	−0.1797
n-Alkylnaphthalenes C_{10}–C_{22}	87.5 to 415	7.1222	873.13	3.8272	220.67	−0.1460
Tetrahydronaphthalenes	63 to 254	7.0181	936.83	3.5766	238.28	−0.1871
Styrenes	33.6 to 220.8	7.0132	939.09	3.5342	237.66	−0.2026
n-Alkylcyclopentanes, C_{22}–C_{41}	218 to 537	7.5752	1074.49	3.9419	253.96	−0.2146
Cyclopentanes, C_6–C_7	− 44 to +106.3	6.8537	965.78	3.0474	242.11	−0.2220
n-cyclohexanes, C_6–C_{42}	− 17 to 551	6.9961	978.47	3.3492	239.53	−0.1912
Cyclohexenes, C_6–C_8	− 15 to 137	6.8732	955.79	3.2233	239.26	−0.1913
n-Alkenes, C_{21}–C_{40}	204 to 523	7.5555	1083.60	3.8824	237.56	−0.1851

* *From Dutt [12].*

$$A_1 = 8.4289$$

$$A_2 = 0.008553$$

$$A_3 = 0.6247 \times 10^{-5}$$

The method is applicable to substances which do not possess strong hydrogen bonds (in the temperature range of 77° to 575° K).

Dutt [12] proposed

$$\log P = A - (m + nt_b)/(t + m^1 + n^1 t_b) \tag{29}$$

A, m, n, m^1, and n^1 are constants for each homologous series as given in Table 8. The units of vapor pressure and temperature are mm Hg and °C respectively. The method is capable of predicting the vapor pressure within 1% to 3% for most of the substances studied.

In a recent study, Twu [13] developed a modified form of the Benedict–Webb–Rubin equation.

$$Z = 1 + \varrho^*(B_1 - B_2T^{*-1} - B_3T^{*-3} + B_9T^{*-4} - B_{11}T^{*-5})$$
$$+ \varrho^{*2}(B_5 - B_6T^{*-1} - B_{10}T^{*-2}) - \varrho^{*5}(B_7T^{*-1} + B_{12}T^{*-2})$$
$$+ B_8\varrho^{*2}T^{*-3}(1 + B_4\varrho^{*2}) \exp(-B_4\varrho^{*2}) \tag{30}$$

With $B_i = a_i + \beta(T_b)b_i$ (31)

$$T^* = KT/\varepsilon$$

$$\varrho^* = \varrho\sigma^3 \tag{32}$$

$$\sigma^3 = 0.3189/\varrho_c$$

$$(\varepsilon/K) = T_c/1.2593 \tag{33}$$

The coefficients of Equation 30 are given in Table 9.
Critical temperatures can be calculated from

$$(T_b/T_c) = 0.533272 + 0.191017 \times 10^{-3}T_b + 0.779681 \times 10^{-7}T_b^2 - 0.284376 \times 10^{-10}T_b^3 + 0.959468 \times 10^{28}/T_b^{13} \tag{34}$$

and critical density

$$\rho_c = (1 - 0.419869 - 0.505839\alpha - 1.56436\alpha^3 - 9481.70\alpha^{14})^8 \tag{35}$$

with $\alpha = (1 - T_b/T_c)$ (36)

$$\beta(T_b) = -0.0793627 + 0.526852 \times 10^{-3}T_b + + 0.218337 \times 10^{-9}T_b^3 + 0.53326 \times 10^{-13}T_b^4 - 0.486459 \times 10^{12}/T_b^6 \tag{37}$$

In this scheme, temperatures are expressed in degrees Rankine, pressures are expressed in psia, and density (ϱ) in lb mole ft^{-3}. Vapor pressures as well as several other thermodynamic properties of the normal paraffins (C_1 to C_{100}) are represented accurately using this method.

SUMMARY AND RECOMMENDATIONS

Methods available in the published literature for estimating vapor pressure from normal boiling point and other easily determinable properties have been described in the preceding sections. It is recommended that:

- If the compound under consideration is listed in one or more of the tables presented, the parameters and coefficients listed should be used to determine the vapor pressure-temperature relationship, giving preference to the methods in the reverse chronological order.

Table 9
Generalized Parameters of the Modified Benedict-Webb-Rubin Equation*

Parameter i	$B_i = a_i + \beta(T_b)\, b_i$	
	a_i	b_i
1	1.48899	0.454054
2	5.06436	−2.70728
3	2.19018	11.5231
4	4.89597	0
5	4.47728	2.48542
6	4.91620	10.2970
7	11.4665	11.7112
8	9.51138	20.6286
9	0.0927705	2.73392
10	1.48921	−2.75461
11	0.0139887	0.188888
12	3.05376	3.00936

* *From Twu [13].*

- If the compound under consideration is not listed in any of the tables, the method which has been tested for the particular homologous series (to which the compound belongs), or other chemically similar series should be used.

n-Alkanes are the series of hydrocarbons most extensively investigated. Other homologous series have been investigated to varying degrees. Methods such as the Cox method and the Smialek and Thodos method give estimates within 3% to 5%, while the generalized Benedict–Webb–Rubin equation proposed by the Twu, Kudchadker and Zwolinski method and Dutt's method give estimated values usually within 1% to 3%. The principles of any of the methods can be extended to the other homologous series either directly or through the use of the reference substance data and characteristic parameters such as acentric factor [1], polar factor [2], critical compressibility factor [14], and molar polarization [15].

NOTATION

A, A_c, B	constants
C	critical temperature or constant
C_t	coefficient
H	enthalpy
m	molecular weight
n	constant
P	pressure
P_c	critical pressure
s	characterization parameter, Equation 21
T	absolute temperature
T_r	reduced temperature
t	temperature
u	constant
z	compressibility factor

Greek Symbols

α, β, γ	constants
ϱ	density
σ, ε	potential parameters

REFERENCES

1. Pitzer, K. S., "The Volumetric and Thermodynamic Properties of Fluids. I-Theoretical Basis and Virial Coefficients," *J. Am. Chem. Soc.*, Vol. 77 (1955), pp. 3427–3433.
2. Halm, R. L., and Stiel, L. I., "A Fourth Parameter for the Vapor Pressure and Entropy of Vaporization of Polar Fluids." *AIChE Journal*, Vol. 13 (1967), pp. 351–355.
3. Cox, E. R., "Hydrocarbon Vapor Pressures," *Ind. Eng. Chem.*, Vol. 28 (1936), pp. 613–616.
4. Perry, J. H., and Smith, E. R., "A Method for Interpolating Data, Based on Duhring's Rule," *Ind. Eng. Chem.*, Vol. 25 (1933), pp. 195–199.
5. Thodos, G., "Vapor Pressures of Normal Saturated Hydrocarbons," *Ind. Eng. Chem.*, Vol. 42 (1950), pp. 1514–1526.
6. Li, J. C. M., and Rossini, F. D., "Vapor Pressures and Boiling Points of the 1-Fluoroalkanes, 1-Chloroalkanes, 1-Bromoalkanes, and 1-Iodoalkanes," *J. Chem. Eng. Data*, Vol. 6 (1961), pp. 268–270.
7. Smialek, R J., and Thodos, G., "A Simple Vapor Pressure Relationship for the Normal Paraffins," *J. Chem. Eng. Data*, Vol. 9 (1964), pp. 52–53.
8. Kudchadker, A. P., and Zwolinski, B. J., "Vapor Pressures and Boiling Points of Normal Alkanes, C_{21}–C_{100}" *J. Chem. Eng. Data*, Vol. 11 (1966), pp. 253–255.
9. Gomez–Nieto, M., and Thodos, G., "A New Vapor Pressure Equation and Its Application to Normal Alkanes," *Ind. Eng. Chem. Fundam.*, Vol. 16 (1977), pp. 254–259.

10. Lee, B. I., and Kesler, M. G., "Improve Vapor Pressures Prediction," *Hydrocarbon Processing,* Vol. 59, No. 7 (1980), pp. 163–167.
11. Neumann, K. K., and Ostertag, G., "A New Method for the Approximation of the Vapor Pressure–Temperature Functions Using the Normal Boiling Point Only," *Chem.-Ing.-Tech.,* Vol. 48 (1976), p. 491.
12. Dutt, N. V. K., "Estimation of Vapor Pressure from Normal Boiling Point of Hydrocarbons," *Can. J. Chem. Eng.,* Vol. 60 (1982), pp. 707–709.
13. Twu, C. H., "Prediction of Thermodynamic Properties of Normal Paraffins Using Only Normal Boiling Point," *Fluid Phase Equilibria,* Vol. 11 (1983), pp. 65–81.
14. Meissner, H. P., and Sefarian, R., "PVT Relations of Gases," *Chem. Eng. Progress,* Vol. 47 (1951), pp. 579–584.
15. Viswanath, D. S., and Prasad, D. H. L., "A New Third Parameter for the Correlation of Thermodynamic Properties," In *Proceedings of the Fifth Symposium on Thermophysical Properties,* C. F. Bonilla (Ed.) New York: ASME, 1970, pp. 236–247.

CHAPTER 13

ESTIMATING LIQUID AND VAPOR MOLAR FRACTIONS IN DISTILLATION COLUMNS

Pierucci Sauro, Laura Pellegrini, and Eliseo Ranzi

Department of Chemical Engineering
Politecnico of Milan
Piazza Leonardo da Vinci
Milano, Italy

CONTENTS

INTRODUCTION

In the following pages we will deal with the problem of designing multicomponent distillation columns, both those considered "conventional" and "complex."

In this context there are two main classes of methods for solving the set of equations describing the process, according to the different type of approach. To the first class belong methods that provide a simultaneous solution of all the governing equations; they are the so called "global methods." In the second class are collected all those methods that have been developed as an alternative to the global methods, which generally provide an extremely time-consuming routine. These methods apply a "partitioning" of the whole system of equations to be solved.

The aim of this chapter is to provide a computational basis, from which the user can start to build up a computer routine for solving distillation columns.

It is worth pointing out that no general rule can be given—for instance in the choice of the promoter—because it must be the user who, according to the type of column to be solved, adopts a particular method.

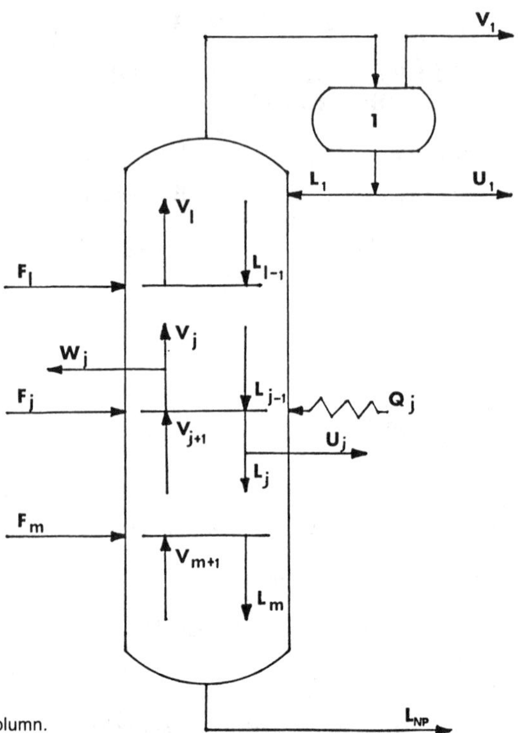

Figure 1. Complex column.

MATHEMATICAL MODEL

Figure 1 is a sketch of a complex column. The trays are numbered from top to bottom with the condenser as the first stage and the reboiler as the NP-th stage.

It is assumed that one feed stream, F_j, one vapor sidestream, W_j, one liquid side-stream, U_j, and one inter heater, Q_j, are provided at each stage. A simpler column model can be obtained setting the undesired quantities to zero.

The mixture to be separated contains NCP components. Each stage is assumed to be an equilibrium stage; that is to say that the vapor leaving the stage is in equilibrium with the liquid stream leaving the same stage. An ideal equilibrium stage is shown in Figure 2.

The following is a system of equations describing the model.

Component material balance around each tray:

$$-L_{j-1}X_{j-1,i}+(L_j+U_j)X_{j,i}+(V_j+W_j)Y_{j,i}-V_{j+i}Y_{j+1,i}-F_jZ_{j,i}=0 \quad (1)$$

Equilibrium equation:

$$Y_{j,i}-K_{j,i}X_{j,i}=0 \quad (2)$$

Total material balance around each tray:

$$-L_{j-1}+(L_j+U_j)+(V_j+W_j)-V_{j+1}-F_j=0 \quad (3)$$

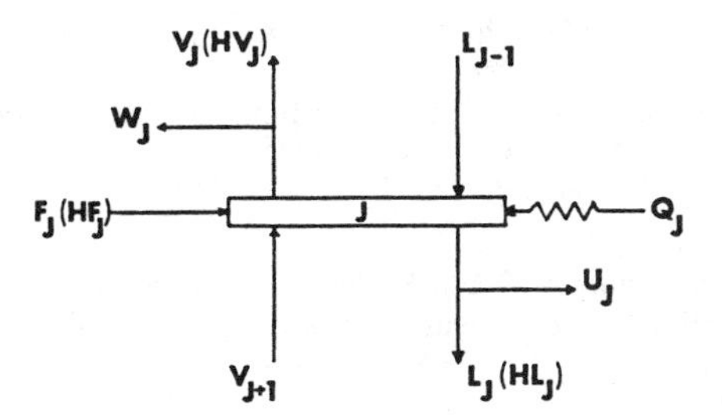

Figure 2. Ideal stage.

Summation equation:

$$\sum Y_{j,i} - X_{j,i} = 0 \tag{4}$$

Enthalpy balance around each tray:

$$-L_{j-1}HL_{j-1} + (L_j + U_j)HL_j + (V_j + W_j)HV_j - V_{j+1}HV_{j+1} - F_jHF_j - Q_j = 0 \tag{5}$$

Specification equations: any consistent set of equations which saturates the degrees of freedom of the model. (6)

A method for determining the variables to be specified in the model consists in counting the number of independent constraining equations and the total number of variables that completely describe the process. Subtracting the former from the latter we have the number of "degrees of freedom," or the number of equations that can be specified arbitrarily.

Let the feed-state variables (F_j, TF_j, PF_j, $Z_{j,i}$) be known, as well as the number of trays and all pressures, P_j.

The number of unknown variables in the model is NP (2 NCP + 6) while the equations number is NP(2NCP + 3). The degrees of freedom number NG is therefore 3 NP, which is clearly related to the unknowns U_j, W_j and Q_j.

The specification equations set can be arbitrarily chosen, but the equations are commonly divided into two classes:

- *Standard specification equations,* if the whole equations set contains linear equations in the unknowns V_j, W_j, L_j, U_j, and Q_j

  $j = 1, NP$

- *Non-standard specification equations* in all the other cases.

The choice of a particular variables set doesn't generally represent a significant problem when you are going to apply a solving algorithm; thus the preceding equations set may be listed by substituting some equations with their algebraic combinations: material or enthalpy balance equations are obtained from a balance of all stages from the condenser to the j-th stage; the summation equation is substituted by bubble point equation, etc.

PARTITIONING ALGORITHM

Component material balances Equation 1 and Equilibrium equations (Equation 2) can be combined to give

$$-L_{j-1}X_{j-1,i}+[(L_j+U_j)+K_{j,i}(V_j+W_j)]X_{j,i}-V_{j+1}K_{j+1,i}X_{j+1,i}-F_jZ_{j,i}=0 \tag{7}$$

The summation equation is substituted by the bubble-point equation

$$\sum K_{j,i}X_{j,i}-1=0 \tag{8}$$

Total material balances Equation 3 and Enthalpy balances Equation 5 are combined and solved to obtain a function of side-stream total flow rates and thermal duties [1]:

$$V_{j+1}=\sum_{k=1}^{NP}(\alpha_{k,j+1}U_k+\beta_{k,j+1}W_k+\gamma_{k,j+1}F_k+\delta_{k,j+1}Q_k) \tag{9}$$

$$L_j=\sum_{k=1}^{NP}(A_{k,j}U_k+B_{k,j}W_k+C_{k,j}F_k+D_{k,j}Q_k) \tag{10}$$

where the influence factors are easily calculated by means of the following analytical formula. By means of an overall enthalpy balance coupled with an overall material balance around the column, the following relation is obtained:

$$V_1(HV_1-HL_{NP})=\sum_{k=1}^{NP}Q_k+\sum_{k=1}^{NP}F_k(HF_k-HL_{NP})-\sum_{k=1}^{NP}W_k(HV_k-HL_{NP})-\sum_{k=1}^{NP}U_k(HL_k-HL_{NP}) \tag{11}$$

The same two balance equations around the section containing the trays 1 to j give:

$$V_{j+1}(HV_{j+1}-HL_j)=V_1(HV_1-HL_j)+\sum_{k=1}^{j}U_k(HL_k-HL_j)+\sum_{k=1}^{j}W_k(HV_k-HL_j)-\sum_{k=1}^{j}F_k(HF_k-HL_j)-\sum_{k=1}^{j}Q_k \tag{12}$$

The two Equations 11 and 12 may be combined to eliminate V_1 from Equation 11, which may be then written in terms of influence factors which may be rearranged as follows:

$$\alpha_{k,j+1}=\begin{cases}-\delta_{k,j+1}(HL_k-HV_1) & 1\leqq k\leqq j\\ -\delta_{k,j+1}(HL_k-HL_{NP}) & j+1\leqq k\leqq NP\end{cases}$$

$$\beta_{k,j+1}=\begin{cases}-\delta_{k,j+1}(HV_k-HV_1) & 1\leqq k\leqq j\\ -\delta_{k,j+1}(HV_k-HL_{NP}) & j+1\leqq k\leqq NP\end{cases}$$

$$\gamma_{k,j+1}=\begin{cases}-\delta_{k,j+1}(HF_k-HV_1) & 1\leqq k\leqq j\\ -\delta_{k,j+1}(HF_k-HL_{NP}) & j+1\leqq k\leqq NP\end{cases}$$

$$\delta_{k,j+1} = \begin{cases} -\dfrac{HL_j - HL_{NP}}{(HV_1 - HL_{NP})(HV_{j+1} - HL_j)} & 1 \leqq k \leqq j \\ -\dfrac{HL_j - HV_1}{(HV_1 - HL_{NP})(HV_{j+1} - HL_j)} & j+1 \leqq k \leqq NP \end{cases}$$

The influence factors of the variable V_1 are:

$$\delta_{k,1} = \frac{1}{HV_1 - HL_{NP}}$$

$$\alpha_{k,1} = -\delta_{k,1}(HL_k - HL_{NP})$$

$$\beta_{k,1} = -\delta_{k,1}(HV_k - HL_{NP})$$

$$\gamma_{k,1} = \delta_{k,1}(HF_k - HL_{NP})$$

At this point it is possible to write the influence factors of the liquid streams as a combination of those calculated for the vapor streams. In fact, the total material component balance around the section containing the tray 1 and j gives:

$$L_J = V_{j+1} + \sum_{k=1}^{j} (F_k - W_k - U_k) - V_1 \qquad 1 \leqq j \leqq NP$$

$$L_{NP} = \sum_{k=1}^{NP} (F_k - W_k - U_k) - V_1$$

These relations may be written as follows:

$$L_j = \sum_{k=1}^{NP} (A_{k,j}U_k + B_{k,j}W_k + C_{k,j}F_k + D_{k,j}Q_k)$$

where for $NP > j \geq 1$

$$A_{k,j} = \alpha_{k,j+1} - \alpha_{k,1} - 1$$

$$B_{k,j} = \beta_{k,j+1} - \beta_{k,1} - 1$$

$$C_{k,j} = \gamma_{k,j+1} - \gamma_{k,1} + 1$$

$$D_{k,j} = \delta_{k,j+1} - \delta_{k,1}$$

and for $j = NP$

$$A_{k,NP} = -\alpha_{k,1} - 1$$

$$B_{k,NP} = -\beta_{k,1} - 1$$

$$C_{k,NP} = -\gamma_{k,1} + 1$$

$$D_{k,NP} = -\delta_{k,1}$$

The system of Equations 7, 8, 9 and 10 is solved according to an iterative procedure which refers to the following steps:

Step 1. If an initial set of L_j, U_j, V_j, W_j, $K_{j,i}$ is assumed the subsystem Equation 7 can be solved for the liquid compositions $X_{j,i}$. The following recurrent formula is used:

$$X_{j,i} = \frac{K_{j+1,i}V_{j+1,i}X_{j+1,i} + E_{j,i}}{L_j + G_{j,i}}$$

for $1 \leqq j < NP$, and

$$X_{NP,i} = \frac{E_{NP,i}}{1 + G_{NP,i}}$$

where $E_{1,i} = F_1 \cdot Z_{1,i}$

$$G_{1,i} = U_1 + K_{1,i}(W_1 + V_1)$$

and for $j = 2, \ldots NP$

$$E_{j,i} = F_j \cdot Z_{j,i} + E_{j-1,i}\frac{L_{j-1}}{L_{j-1} + G_{j-1,i}}$$

$$G_{j,i} = U_{j,i} + K_{j,i}\left(W_j + V_j\frac{G_{j-1,i}}{L_{j-1} + G_{j-1,i}}\right)$$

Step 2. Liquid compositions from Step 1 are normalized and introduced into the bubble point equations (Equation 8). They are solved to give the tray temperature: a new estimate of the variables $K_{j,i}$, HV_j, HL_j is also obtained. Many methods are available in the literature for solving the bubble point equation: Watson [2], Muller [3], Newton–Raphson, etc. Ching-Tsan Lo [4], for instance, proposes the following iterative procedure:

$$T_j^{n+1} = T_j^n\left[1 - \frac{T_j^n}{A}\left(1 - \sum K_{j,i}X_{j,i}\right)\right]$$

A is an adaptive parameter:

$$A = \sum X_{f,i}a_i$$

where $X_{f,i}$ is the feed composition of i-th component and a_i the Clausius–Clapeyron parameter for i-th component

$$\ln P_{v,i} = \frac{a_i}{T} + b_i$$

where $P_{v,i}$ is the vapor pressure of i-th component. For hydrocarbon mixtures it can be assumed

$$T_j^{n+1} = T_j^n[1 - C(1 - \sum K_{j,i}X_{j,i})]$$

where C may vary from .1 and .2.

Value C = .15 is suggested as default one. Ching-Tsan Lo method derives from well known iterative procedures which require two estimates—S^n and S^{n-1}—of $\ln\{\sum K_{j,i}X_{j,i}\}$ at two different temperatures T_j^n and T_j^{n-1}; then the bubble point temperature can be iterated by assuming:

$$\ln \sum K_{j,i}X_{j,i} = A + \frac{B}{T_j^{n+1}} = 0$$

so that:

$$T_j^{n+1} = -\frac{B}{A}$$

The A and B parameters can be easily estimated from the S values at T_j^n and T_j^{n-1}. The resulting n+1 approximation for T_j is the following:

$$T_j^{n+1} = T_j^n - \frac{S^n T_j^n (T_j^n - T_j^{n-1})}{S^n T_j^n - S^{n-1} T_j^{n-1}}$$

Step 3. The NG degrees of freedom of the model are saturated by means of standard specification equations that are linear equations in the unknowns V_j, W_j, L_j, U_j, and Q_j. Let

$$L_k(\bar{V}_j, \bar{L}_j, \bar{U}_j, \bar{W}_j, \bar{Q}_j) = 0 \tag{13}$$

$k = 1, \ldots NG$

the system of specification equations; then it may be reduced by means of Equations 9 and 10 to the following:

$$L'_k(\bar{U}_j, \bar{W}_j, \bar{Q}_j) = 0 \tag{14}$$

where only the NG variables $\bar{U}_j$, $\bar{W}_j$, $\bar{Q}_j$ are present. The linear system Equation 12 is solved to give the values of the NG unknowns whose values, after introducing inside Equations 9 and 10 give the profiles of vapor and liquid streams.

Step 4. An error measure related to the whole procedure is estimated by weighting the discrepancies of only component material balances hence all the remaining model equations are satisfied at the end of Step 3.

$$E.\,M. = \sum_{j=1}^{NP} \sum_{i=1}^{NCP} \frac{|\text{discrepancy Equation 1}|}{WT_i}$$

where the weighting factor may be satisfactorily defined as:

$$WT_i = \sum_{j=1}^{NP} F_j Z_{j,i}$$

Step 5. When E. M. $\leqq \varepsilon$ then convergence is met; otherwise the procedure restarts from Step 1.

Initial Guess Profiles

The iterative procedure has to be initialized by providing $K_{j,i}$, L_j, U_j, V_j, W_j profiles.

Boston and Sullivan [5] suggest an estimating technique that may be of value. Most of the simulation packages [6] use proprietary algorithms derived from short-cut methods. At this moment no general rule can be recommended except the following one which is based on the user's knowledge of the problem to be solved.

Temperature and vapor profiles have to be guessed by the user. These values may be given for some trays, say top, feed, and bottom; in the intermediate ones a linear interpolation can be applied. Values of the withdrawal rates U_j and W_j have to be initially guessed, too.

Total material balances equations are then solved for liquid profile, while K-values profiles may be estimated by thermodynamic models only depending on T and P.

Accelerating Convergence of Partitioning Algorithm

The convergence behavior of partitioning algorithm is related to its prediction capability of liquid compositions. Had they been exactly known, Steps 2 and 3 would then provide the solution of the problem.

The convergence speed, therefore, may be enhanced by correcting the calculated liquid molar fractions computed by Step 1 in order to make them closer to the solution values.

The pioneer idea in this area was given by Lyster et al. [7], and after that it has been largely circulated in the literature as the θ method.

For convenience let us partially redefine the adopted nomenclature:

$$L_{j,i} = X_{j,i}L_j = \text{component liquid flow rate}$$

$$V_{j,i} = Y_{j,i}V_j = K_{j,i}X_{j,i}V_{j,i} = \text{component vapor flow rate}$$

where D = total product distillate
d_i = total distillate product of component i
b_i = total bottom product of component i
$F_i = \sum_{j=1}^{NP} F_j Z_{j,i}$ = total amount of component i fed to the column

The correction applied by θ promoter [8] is:

$$(L_{j,i})_{co} = (L_{j,i})_{ca} \cdot R_i \cdot \psi_j, \quad (j = 1, \ldots NP) \tag{15}$$

$$(V_{j,i})_{co} = (V_{j,i})_{ca} \cdot R_i \cdot \sigma_j, \quad (i = 1, \ldots NCP) \tag{16}$$

where ψ_j and σ_j are multipliers independent of i, and

$$R_j = \frac{(d_i)_{co}}{(d_i)_{ca}}$$

The subscript co stands for the corrected values, and ca for the calculated ones by the solution of component material balances.

The main task of the promoter is then to give a better estimate of the liquid and vapor molar fractions according to the following expressions:

$$(X_{j,i})_{co} = \frac{(L_{j,i})_{ca} \cdot R_i}{\sum_{i=1}^{NCP} (L_{j,i})_{ca} \cdot R_i}, \quad \begin{array}{l} j = 1, NP \\ i = 1, NCP \end{array} \tag{17}$$

$$(Y_{j,i})_{co} = \frac{(V_{j,i})_{ca} \cdot R_i}{\sum_{i=1}^{NCP} (V_{j,i})_{ca} \cdot R_i}, \quad \begin{array}{l} i = 1, NP \\ i = 1, NCP \end{array} \tag{18}$$

This correction substantially accelerates convergence in those problems which would converge without it and brings about convergence in most problems which would not converge otherwise [9].

For conventional distillation columns (one feed, no sidestreams) (see Figure 3):

$$R_i = \frac{(di)_{co}}{(di)_{ca}} = \frac{Fi}{(di)_{ca} + \theta(bi)_{ca}}$$

while for complex ones (multifeeds and sidestreams)

$$\frac{(di)_{co}}{(di)_{ca}} = \frac{Fi}{(di)_{ca} + \theta(bi)_{ca} + \theta_2^U(U_{2,i})_{ca} + \theta_2^W(W_{2,i})_{ca} + \ldots}$$

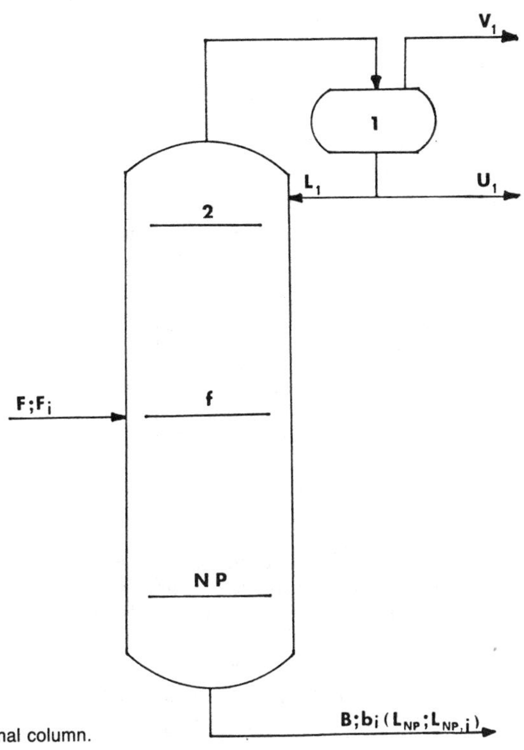

Figure 3. Conventional column.

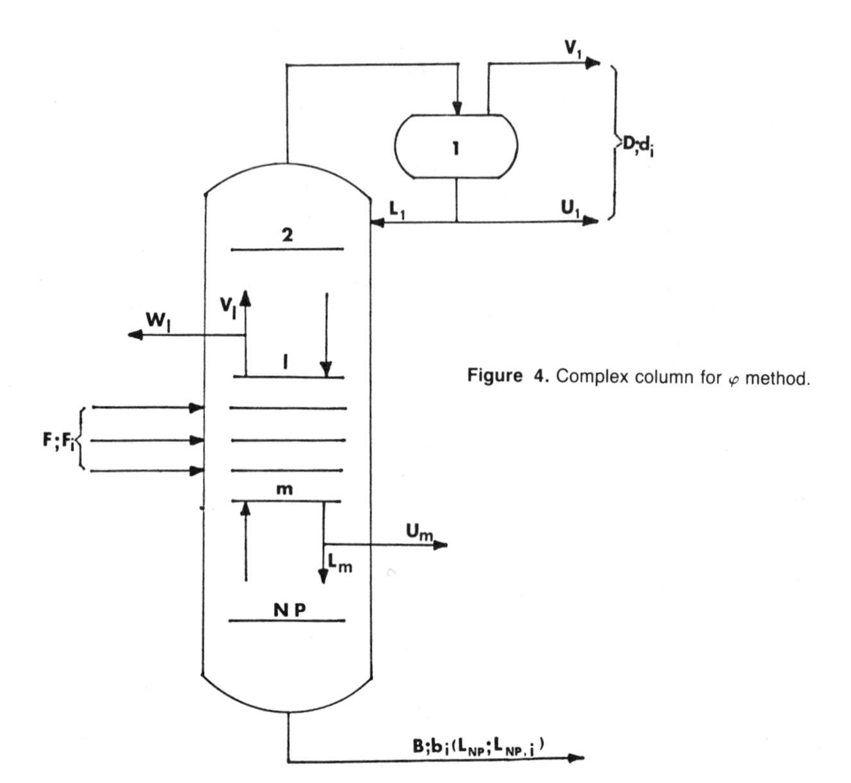

Figure 4. Complex column for φ method.

The theta multipliers θ, θ_2^U, θ_2^W, ... are defined as follows:

$$\left(\frac{b_i}{di}\right)_{co} = \theta\left(\frac{b_i}{di}\right)_{ca}$$

$$\left(\frac{U_{2,i}}{di}\right)_{co} = \theta_2^U\left(\frac{U_{2,i}}{di}\right)_{ca}$$

$$\left(\frac{W_{2,i}}{di}\right)_{co} = \theta_2^W\left(\frac{W_{2,i}}{di}\right)_{ca}$$

According to the theta method, for each side stream a new additional multiplier is introduced in analogy with the definition of the primary one.

For the column of Figure 4, for instance, two new θ's are introduced following the definition:

$$\left(\frac{W_{l,i}}{di}\right)_{co} = \theta_l^W\left(\frac{W_{l,i}}{di}\right)_{ca} \quad \text{and} \quad \left(\frac{U_{m,i}}{di}\right)_{co} = \theta_m^U\left(\frac{U_{m,i}}{di}\right)_{ca}$$

As a consequence, the factors R_i are defined as:

$$R_i = \frac{(di)_{co}}{(di)_{ca}} = \frac{F_i}{(di)_{ca} + \theta(bi)_{ca} + \theta_l^W (W_{l,i})_{ca} + \theta_m^U (U_{m,i})_{ca}}$$

where F_i indicates the total feed-flow rate of component i.

The parameters θ, θ_l^W, θ_m^U are usually calculated by imposing each component in agreement with the assigned values D, W_l, U_m, namely:

$$\sum (di)_{co} - D = 0$$

$$\sum (W_{l,i})_{co} - W_l = 0$$

$$\sum (U_{m,i})_{co} - U_m = 0$$

which according to the θ multipliers definition become:

$$\sum R_i \cdot (di)_{ca} - D = 0$$

$$\sum \theta_l^W \cdot R_i (W_{l,i})_{ca} - W_L = 0$$

$$\sum \theta_m^U \cdot R_i (U_{m,i})_{ca} - U_m = 0 \tag{19}$$

Equation 19 is in the θ unknowns; it can be easily solved by means of successive linearizations according to the Newton–Raphson algorithm. Had the system been solved, the values R_i would then be introduced in Equations 17 and 18 which give the corrected molar fractions.

The θ promoter is reliable when applied to conventional columns, though it may cause instability or the entire procedure to diverge when it is applied to complex columns where multifeeds and multiside-streams are present. Another disadvantage is found in the limited class of standard specification equations which can be treated by the promoter; Equation 19, in fact, requires that the withdrawal and distillate flow rates should be known or have, at least, a value close to the solution.

Standard specifications involving total flow rates inside the column can't be directly utilized due to the unknown values of the multipliers ψ_j and σ_j which appear in (15) and (16). (Pierucci [10] proved that for a conventional column ψ_j and σ_j are 1 in the enriching section, while they both are equal to θ in the stripping one. However these results can't be applied to complex columns, hence the multipliers ψ_j and σ_j may show a component dependence).

For the conventional column shown by Figure 3 Pierucci pointed out that in principle the corrected component flow rates may be calculated in a different way, i.e., by solving tray-by-tray component material balances having properly corrected the separation factors $S_{f,i}$ of the feed tray.

Provided the separation factors are defined as follows:

$$S_{j,i} = \frac{V_{j,i}}{L_{j,i}} = K_{j,i} \frac{V_j}{L_j}$$

then the component material balances stated in terms of liquid flow rates are represented by the following set of equations:

$$L_{1,i}\left(1+\frac{U_1}{L_1}+S_1\right)-S_{2,i}L_{2,i}=0$$

$$-L_{NP-1,i}+L_{NP,i}(1+S_{NP,i})=0$$

$$-L_{f-1,i}+L_{f,i}(1+S^{f}_{f,i})-S_{f+1,i}L_{f+1,i}-F_i=0$$

$$-L_{j-1,i}+L_{j,i}(1+S_{j,i})-S_{j+1,i}L_{j+1,i}=0 \qquad (j \neq f \neq 1 \neq NP)$$

For any given set of values $S_{j,i}(j = 1, NP)$, these equations are readily solved for $L_{j,i}$ by use of well-known recurrence formulas applied to tridiagonal matrices. Let $(S_{j,i})_{ca}$ be the values of the separation factors which give the $(L_{j,i})_{ca}$, and $(S_{j,i})_{co}$ those giving the values $(L_{j,i})_{co}$. The corrected molar liquid flow rates are obtained by solving this system by means of the following set of separation factors:

$$(S_{j,i})_{co} = (S_{j,i})_{ca} \qquad (j \neq f)$$

$$(S_{f,i})_{co} = (S_{f,i})_{ca} \cdot \theta^{-1} \tag{20}$$

As the multiplier θ^{-1} is calculated on the basis of a standard specification equation involving total flow rates, usually this equation is

$$\sum (d_i)_{co} - D = 0$$

From Equation 20 we can correctly assume that the primary θ multiplier, while modifying the feed tray separation factors, allows us to match the specification of the output stream D in terms of component material balances. This formulation is the starting point of the analogue applied to the next distillation configurations.

A multifeeds (conventional) column will be defined as the one which satisfies all the conditions of a conventional one except for the presence of more than one feed stream.

Let's suppose that the feed streams are located between the trays p and q (Figure 5).

We define an enriching section above the tray p, a stripping section under the tray q, and a feed section containing the trays from p to q both included.

Pierucci et al. [11] showed that a unique multiplier θ^* is defined for the separation factors $S_{j,i}$ of each component in the trays of the feed section; its value is searched according to the following definition: "Leaving unchanged all the $S_{j,i}$ factors ($j < p$ and $j > q$) find the unique multiplier θ^* of all the separation factors $S_{j,i}$ of the feed section ($p \leqq j \leqq q$) so that the distillate flow rates, obtained by the solution of component material balances in the whole column, agree with a standard specification involving total flow rates (usually an assigned value to D)."

This formulation may be considered as an obvious extension of the one found for conventional columns, hence they both coincide when $p = q$.

The present procedure may be summarized as follows:

1. Define for all the components $i = 1, NCP$

$$(S_{j,i})_{co} = (S_{j,i})_{ca}\theta^* \qquad \text{for} \quad p \leqq j \leqq q$$

$$(S_{j,i})_{co} = (S_{j,i})_{ca} \qquad \text{for} \quad j < p \text{ and } j > q$$

2. Solve

$$L_{1,i}\left[1+\frac{U_1}{L_1}+(S_{j,i})_{co}\right]-(S_{2,i})_{co}L_{2,i}=0$$

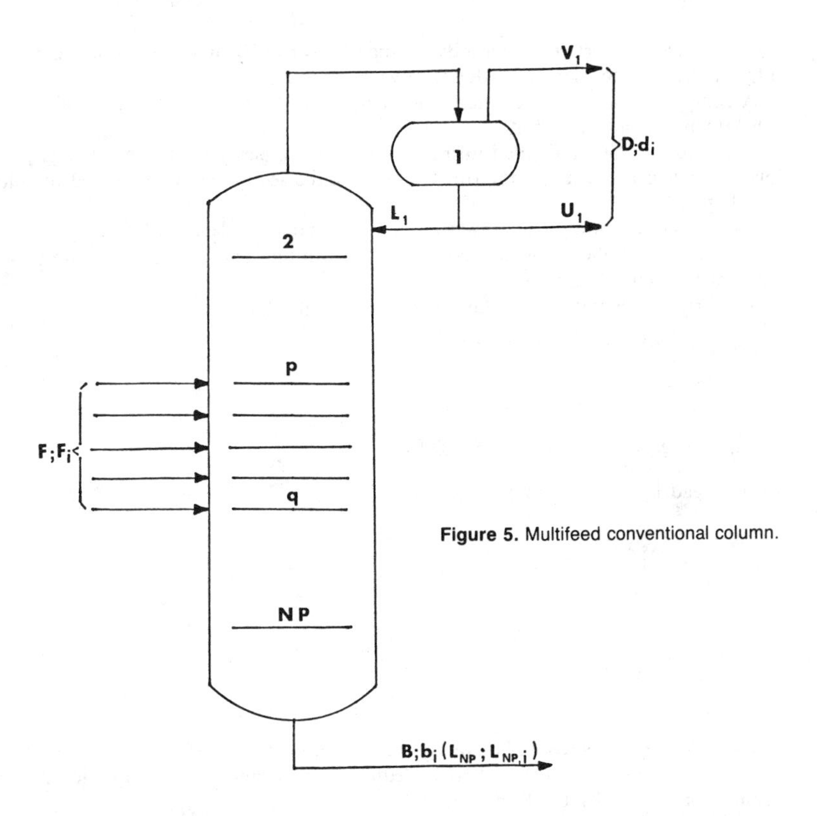

Figure 5. Multifeed conventional column.

$$-L_{NP-1,i} + L_{NP}[1 + (S_{NP,i})_{co}] = 0$$

$$-L_{j-1,i} + L_{j,i}[1 + (S_{j,i})_{co}] - (S_{j+1,i})_{co}L_{j+1,i} = 0 \quad \text{for} \quad j < p \text{ and } j > q$$

$$-L_{j-1,i} + L_{j,i}[1 + (S_{j,i})_{co}] - (S_{j+1,i})_{co}L_{j+1,i} - F_{j,i} = 0 \quad \text{for} \quad p \leqq j \leqq q$$

3. Calculate the discrepancy of the standard specification equation. Restart the sequence with a new θ^* if an unacceptable disagreement is found. When the standard specification equation assigns the D value, then:

$$f(\theta^*) = \sum (di)_{co} - D = \sum (L_{1,i})_{co}\left(\frac{U_1}{L_1} + S_{1,i}\right) - D$$

The value of θ^* may be updated by means of several techniques capable of rooting a single equation in one unknown. A Newton–Raphson technique is convenient, because it proved to be stable and efficient as a consequence of the fact that $f(\theta^*)$ is always (as intuitively assumed) monotonously dependent on θ^* with positive first derivatives.

The values of the derivatives were numerically calculated. As a rule, few iterations are required to root the equation. Therefore, the overall computing time spent is in general an insignificant quantity when compared with the time needed by the convergence of the entire model.

Nevertheless, alternative methods as quasi–Newton, Broyden, or others can be successfully applied if necessary in order to save computing time.

A complex column is one where one or more feed streams are present with at least one side-stream (liquid or vapor).

The column may be divided into three sections, upper ($j < 1$), central ($1 \leqq j \leqq m$), and lower ($j > m$); 1 and m are the trays delimiting the section containing feed or side-streams (see Figure 4).

The corrected component flow rates are calculated as described below:

A unique multiplier θ^* is defined for the separation factors $S_{j,i}$ of each component in the trays of the central section.

The original definitions of the additional θ multipliers

$$(W_{j,i})_{co} = \theta_j^W (W_{j,i})_{ca} \cdot R_i$$

and

$$(U_{j,i})_{co} = \theta_j^U (U_{j,i})_{ca} \cdot R_i \qquad 1 \leqq j \leqq m \tag{21}$$

are replaced by the following:

$$\left(\frac{W_j}{V_j}\right)_{co} = \theta_j^{*W} \left(\frac{W_j}{V_j}\right)_{ca}$$

and

$$\left(\frac{U_j}{L_j}\right)_{co} = \theta_j^{*U} \left(\frac{U_j}{L_j}\right)_{ca} \tag{22}$$

The two definitions would be exactly the same if Equations 15 and 16 were true. Therefore Equation 22 may be considered as an equivalent formulation of Equation 21 within the assumptions made by the θ method.

In fact, according to Equation 16

$$(W_{j,i})_{co} = (V_{j,i})_{co} \cdot \left(\frac{W_j}{V_j}\right)_{co} = (V_{j,i})_{ca} R_i \psi_j \left(\frac{W_j}{V_j}\right)_{co}$$

For θ_j^W definition it results:

$$(W_{j,i})_{co} = \theta_j^W (W_{j,i})_{ca} R_i = \theta_j^W (V_{j,i})_{ca} \left(\frac{W_j}{V_j}\right)_{ca} R_i \frac{\psi_j}{\psi_1}$$

The combination of the two equations gives the proof:

$$\left(\frac{W_j}{V_j}\right)_{co} = \frac{\theta_j^W}{\psi^1} \left(\frac{W_j}{V_j}\right)_{ca} = \theta_j^{*W} \left(\frac{W_j}{V_j}\right)_{ca}$$

The second of Equation 22 may be proved by analogous steps.

The values θ^*, θ_j^{*W}, $\theta_j^{*U}(1 \leqq j \leqq m)$ are searched through the following steps:

1. Define for all components i:

$$(S_{j,i})_{co} = (S_{j,i})_{ca} \theta^*, \qquad 1 \leqq j \leqq m$$

$$(S_{j,i})_{co} = (S_{j,i})_{ca}, \qquad j < 1 \text{ and } j > m$$

2. Define:

$$(W_j/V_j)_{co} = \theta_j^{*W}(W_j/V_j)_{ca}, \quad 1 \leq j \leq m$$

and

$$(U_j/L_j)_{co} = \theta_j^{*U}(U_j/L_j)_{ca}$$

3. Solve tray-by-tray component material balances for the entire column.
4. Compare the discrepancies of the standard specification equations involving total component flow rates. If they assign total product flow rates then:

$$f_1(\theta^*, \theta_j^{*W}, \theta_j^{*U}, \ldots) = \sum (di)_{co} - D$$

$$f_2(\theta^*, \theta_j^{*W}, \theta_j^{*U}, \ldots) = \sum (W_{j,i})_{co} - W_j$$

$$f_3(\theta^*, \theta_j^{*W}, \theta_j^{*U}, \ldots) = \sum (U_{j,i})_{co} - U_j \tag{23}$$

Restart the sequence with new values θ^*, θ_j^{*W}, θ_j^{*U}, if an unacceptable disagreement is found.

A Newton–Raphson linearization technique has been found to be a stable and efficient way to solve this system Equation 23 in the unknown θ multipliers. The derivatives of the functions for the unknowns have been calculated numerically and updated after each iteration.

Generally, few iterations are required to solve the system; therefore, the overall computing time spent to use this methodology normally is a small quantity when compared with the time needed by the convergence of the entire model.

Nevertheless, if necessary, alternative methods for the solutions of the system Equation 23 like quasi–Newton, Broyden, or others may be successfully applied to save computing time.

Comparative Examples

Four examples will be presented to substantiate the improvement given to the θ-convergence promoter. For this purpose, a θ-method program for the simulation of distillation columns has been built up according to these main rules:

- Component material balance equations are solved by the Boston and Sullivan [5] algorithm.
- Bubble point equations are solved for tray temperature by the Kb method stated by Holland [12].
- Enthalpy and total flow balance equations are solved by the "constant composition" method as described by Holland [12]. In this case, partial molar enthalpies have been analytically calculated.

To this structure, either the original θ-convergence promoter (Holland [12]) or the improved one (θ^* promoter) may be applied.

The entire iterative procedure is synthetically shown in Figure 6.

An error measure of the computation procedure has been defined after the solution of the subset Equation 3 according to the following relation:

$$\text{E.M.} = \frac{\sum_{j=1}^{NP} \left\{ \left| \left(\sum_{j=1}^{NCP} L_{j,i} \right)_{ca} - L_j \right| + \left| \left(\sum_{i=1}^{NCP} V_{j,i} \right)_{ca} - V_j \right| + \sum_{i=1}^{NCP} \left| S_{j,i}(L_{j,i})_{ca} - (V_{j,i})_{ca} \right| \right\}}{\sum_{j=1}^{NP} F_j}$$

The calculation scheme, Figure 6 (if the convergence promoter is disregarded), is basically common to all the methods which adopt a partitioning of the whole equations system describing the distillation column. Distinction being the way each subset is solved.

As the θ promoter or the improved θ^* promoter is structurally independent of the algorithm chosen for the solution of each equation subset, a practical possibility exists of implementing both convergence promoters and a B.P. structure. (This type of implementation is not a new once since the resulting calculation procedure (if referred to the θ-promoter) is similar to a pioneer version of the θ-methods (Lyster et al. [7]).)

In this way, we mainly wanted to verify that better convergence characteristics of the θ^* promoter over the θ promoter are kept when these promoters are applied to a method different from the θ one.

Each example reported here has been solved by five different runs according to the following procedure:

1. θ method with a θ-promoter
2. θ method with θ^* promoter
3. B.P. method
4. B.P. method with θ promoter
5. B.P. method with θ^* promoter

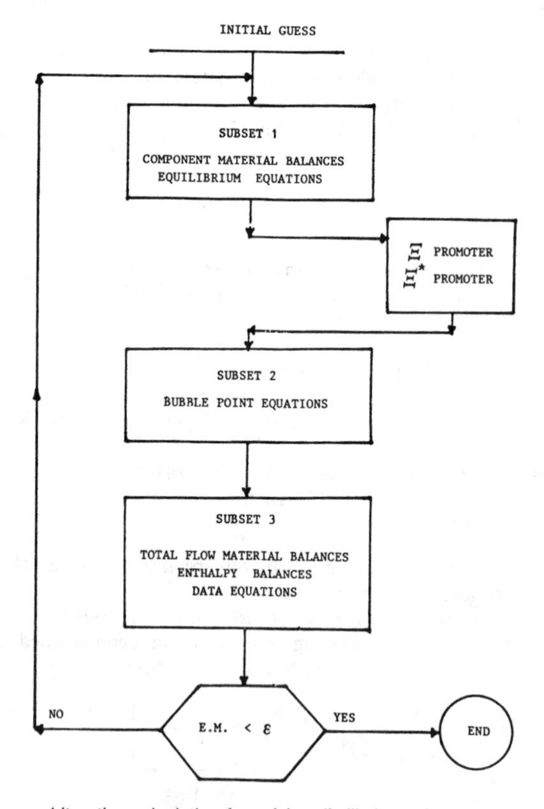

Figure 6. Decomposition and iterative calculation for solving distillation column models.

The set of data for four examples has been reported in Tables 1, 3, 5, and 7, while the corresponding comparative results for ten iterations are shown in Tables 2, 4, 6, and 8. Table 1 is a one-feed two-side-streams column; Table 3 refers to a two-side-streams column taken from Henley [13].

Table 5 (two-feeds column) and Table 7 (one-feed and one-side-stream column) are taken from Fredenslund [14].

Equilibrium constants and molar enthalpies have been calculated by RKS [15] equation of state for Tables 1 and 3. Unifac method for liquid activity coefficient combined with a virial equation of state [16] for vapor fugacity coefficient are used for the calculation of equilibrium constants in the remaining examples. In these cases, molar enthalpies have been estimated by polynomial expressions only depending on temperature.

All the data have been taken from Reid et al. [17]. The data set of each example also includes the initial guess of temperature and total vapor flow rate in some trays; in the intermediate trays a linear interpolation has been assumed. The initial guess of total liquid flow rates has been derived by means of overall material balances, while the equilibrium constants have been initiated by assuming the ideal behavior of the mixture.

Table 1
Data Set for Example 1

Feeds	Flow Rates Tray 5
Ethane	0.05
Propane	0.2
N-Butane	0.4
N-Pentane	0.2
N-Hexane	0.1
N-Decane	0.05
Temperature	378.1
Pressure	20.7

Number of trays = 10; total condenser U_1 = 0.2; liquid reflux L_1 = 10; column pressure = 20.7; Liquid Side-stream U_3 = 0.1 and U_7 = 0.2.

Initial Guess Profiles
T_1 = 324.0; T_5 = 378.0; T_{10} = 378.0; V_1 = 0.0; V_5 = 1.2; V_{10} = 1.2. Units: T(K); P(1.E + 5 Pa); flow rates (mol/time).

Table 2
Comparative Results from Example 1

θ Method	θ-Promoter				θ*-Promoter			
Iter.	θ	θ_3	θ_7	E.M.	θ^*	θ_3^*	θ_7^*	E.M.
1	0.64107	0.50582	0.49047	17.784	1.1739	0.58501	0.76428	11.882
2	58.112	1.7363	9.2047	10.867	0.81425	1.0193	1.0026	2.4115
3	0.46370	0.94422	0.50989	8.8627	1.0272	1.0271	1.0079	0.25617
4	1.4631	1.0162	1.4368	3.1677	0.99765	0.99635	1.0001	0.02826
5	0.86174	0.99865	0.86364	1.5976	1.0002	1.0003	0.99988	0.00457
6	1.0638	1.0003	1.0641	0.63108	0.99996	1.0000	0.99999	0.00247
7	0.97542	1.0001	0.97519	0.26855	0.99996	1.0000	0.99999	0.00013
8	1.0101	0.99996	1.0102	0.10812				
9	0.99599	1.0000	0.99595	0.04402				
10	1.0016	0.99999	1.0016	0.01782				
CPU	3.9301				3.9298			

B.P. Method			θ-Promoter				θ*-Promoter			
Iter.	D_{ca}	E.M.	θ	ϑ_3	θ_7	E.M.	θ^*	θ_3^*	θ_7^*	E.M.
1	0.0562	19.58	0.6410	0.5058	0.4904	13.41	1.173	0.5850	0.7641	12.12
2	0.1505	11.09	2.414	1.264	2.418	5.832	0.8069	1.0213	1.009	4.249
3	0.2331	2.459	0.7637	0.9911	0.7689	3.010	1.042	1.0291	1.005	0.4377
4	0.2174	0.7018	1.071	0.9987	1.070	0.7010	0.9951	0.9946	1.000	0.0587
5	0.2076	0.2862	0.9821	1.000	0.9820	0.1985	1.000	1.000	0.9997	0.0088
6	0.2035	0.1381	1.004	0.9999	1.004	0.0524	0.9999	0.9999	1.000	0.0016
7	0.2017	0.0649	0.9986	0.9999	0.9987	0.0157	1.000	1.000	1.000	0.0007
8	0.2008	0.3121	1.000	0.9999	1.000	0.0034				
9	0.2004	0.0149								
10	0.2001	0.0069								
CPU	5.928		5.868				5.817			

Table 3
Data Set for Example 2

Feeds	Flow Rates Tray 6	Tray 9
Ethane	2.5	0.5
Propane	14.0	6.0
N-Butane	19.0	18.0
N-Pentane	5.0	30.0
N-Hexane	0.5	4.5
Temperature	350.0	383.0
Pressure	20.7	19.0

Number of trays = 16; partial condenser = V_1 = 15.0 and U_1 = 5.0; reflux L_1 = 150; column pressure = 16.5; side-streams U_3 = 3.0 and W_{13} = 37.0; thermal duties Q_3 = −211.0 E + 6.

Initial Guess Profiles
T_1 = 255.0; T_6 = 305.0; T_{16} = 333.0; V_1 = 15.0; V_2 = 170.0; V_{16} = 170.0. Units: T(K); P(1E + 5Pa); flow rates (mol/s), duty (J/s).

* *From Henley [13].*

Table 4
Comparative Results from Example 2

θ Method Iter.	θ	θ Promoter θ_3	θ_{13}	E.M.	θ^*	θ* Promoter θ_3^*	θ_{13}^*	E.M.
1	0.00087	0.00298	0.04539	6919.2	2.3527	1.4704	0.12771	53.195
2	28.280	14.481	1.2862	21.898	0.81998	0.62267	1.0182	10.277
3	0.34293	0.42363	1.0858	21.665	1.0709	1.1392	1.0199	2.6689
4	2.0629	1.9194	1.0376	8.7342	0.97192	0.95859	0.99216	1.0810
5	0.72558	0.74324	0.99156	5.6848	1.010	1.0131	1.0038	0.37079
6	1.1764	1.1669	1.0054	2.4757	0.99591	0.99586	0.99845	0.13567
7	0.93034	0.93294	0.99871	1.2028	1.0015	1.0013	1.0006	0.04787
8	1.0333	1.0324	1.0007	0.53057	0.99946	0.99957	0.99979	0.01682
9	0.98575	0.98606	0.99975	0.23752	1.0002	1.0001	1.0001	0.00584
10	1.0063	1.0062	1.0001	0.10288				
CPU	4.9152				5.1032			

B.P. Method Iter.	D_{ca}	E.M.	θ	θ-Promoter ϑ_3	θ_{13}	E.M.	θ^*	θ*-Promoter θ_3^*	θ_{13}^*	E.M.
1	2.493	18.17	0.0009	0.0030	0.0454	7617	2.353	1.470	0.1277	53.19
2	3.313	8.237	3.377	3.202	0.9541	13.77	0.9901	0.9532	0.8743	2.517
3	9.486	3.863	0.8595	0.8670	1.047	2.403	1.001	1.002	0.9836	0.1411
4	15.07	3.387	1.023	1.023	1.006	0.4136	0.9990	1.001	0.9995	0.0194
5	16.81	2.304	0.9951	0.9951	1.000	0.0912	0.9999	1.000	1.000	0.0041
6	17.65	1.545	1.001	2.000	0.9999	0.0094	1.000	1.000	1.000	0.0017
7	18.15	1.126	0.9999	0.9999	9.9999	0.0028				
8	18.51	0.8632								
9	18.77	0.6829								
10	18.98	0.5497								
CPU	6.511		5.747				5.350			

Table 5
Data Set for Example 3*

Feeds	Flow Rates Tray 8	Tray 24
Ethanol	0.25	
Propanol	0.125	0.125
Water	0.125	0.125
Acetic acid		0.25
Temperature	354.0	368.0
Pressure	1.013	1.013

Number of trays = 30; partial condenser V_1 = 0.4; liquid reflux L_1 = 0.7828; column pressure = 1.013; side-stream U_{21} = 0.2.

Initial Guess Profiles
T_1 = 355.0; T_{21} = 361.0; T_{24} = 367.0; T_{30} = 370.0; V_1 = 0.4; V_2 = 1.2; V_8 = 1.2; V_9 = 0.67; V_{24} = 0.67; V_{25} = 0.7; V_{30} = 0.8. Units: T(K); P(1.E + 5 Pa); flow rates (mol/time).

* *From Fredenslund [14].*

Tables 2, 4, 6, and 8 report the results obtained by executing the examples by five different runs as just described. Exception is made by the fourth example which diverged in the cases adopting the θ method, due, probably, to the use of the Kb method. Better results (Table 8) were obtained, in fact, by substituting it with the bubble equations of B.P. method.

For each procedure adopted, the error measure (E.M.) and the following values of its main characteristic parameters are reported: θ or θ^* when adopting θ or θ^* convergence promoter, and the value of the calculated top distillate flow rate (D_{ca}) when the pure B.P. method is used. The total computing time (in s) is reported on the last row denoted by the item CPU. The runs were executed by an Univac 1100/60 computer.

All the examples show the benefit of adopting the θ^* promoter.

The values quoted by CPU rows show that the computing time increment required by the θ^* promoter is a nonsignificant figure when compared with the time spent to conclude the calculations. In fact, the θ^* promoter demands something more which becomes vanishingly small along with the iterations. (In order to have meaningful comparisons, the θ^* and θ promoters have to be compared in the context of the same method, B.P. or θ.).

Moreover, due to better convergence characteristics, the computing time increment may be compensated either by a minor time amount required by the remaining phases of the calculation procedure, or by a reduced number of iterations. Thus, a smaller global computing time may be obtained.

The convergence of the partitioning method may be enhanced by adopting the T convergence promoter [24] which may be considered a product of a deeper and more detailed analysis of the θ convergence promoter. In particular, attention has been paid in reaching the following two objectives:

- To find a physical meaning of the θ parameters so that a better estimation of the corrected component flow rates may be obtained.
- To substitute the primary θ parameter with a more effective variable.

The T convergence promoter maintains the same meaning and definition of the additional θ_j^{*U} and θ_j^{*W} parameters, while substitutes the primary θ^* with a more comprehensive variable. In fact the correction applied by θ^* results in an uniform translation of the

Table 6
Comparative Results from Example 3

θ Method Iter.	θ Promoter θ	θ_{21}	E.M.	θ^* Promoter θ	θ^*_{21}	E.M.
1	1.9426	2.3207	28.164	0.82752	1.7927	15.250
2	1.5934	1.8429	20.964	0.90949	1.3957	13.130
3	1.2761	1.3691	12.544	0.94386	1.2720	9.6978
4	1.2454	1.3282	11.821	0.95774	1.1853	7.0610
5	1.3045	1.3876	13.296	0.97100	1.1119	4.7530
6	1.0813	1.1013	5.1510	0.99354	1.0324	2.1523
7	1.1732	1.2268	9.5031	0.99281	1.0300	1.7268
8	1.2263	1.2562	10.480	0.99163	1.0266	1.4455
9	1.2726	1.3494	13.081	1.0062	0.98176	1.4991
10	1.2240	1.2687	15.703	1.0012	0.99709	1.1792
CPU	262.5			255.4		

B.P. Method Iter.	D_{ca}	E.M.	θ Promoter θ	ϑ_{21}	E.M.	θ^* Promoter θ	θ^*_{21}	E.M.
1	0.2627	30.73	1.943	2.321	29.30	0.8275	1.793	46.72
2	0.5011	11.81	1.581	1.821	22.17	0.9272	1.385	32.66
3	0.5119	12.88	1.280	1.374	12.92	0.9645	1.174	15.96
4	0.4788	10.01	1.244	1.326	13.12	0.9814	1.089	8.806
5	0.4705	11.20	1.297	1.378	14.69	0.9923	1.037	4.687
6	0.4460	6.305	1.092	1.115	6.312	1.001	1.004	1.925
7	0.4904	18.09	1.175	1.223	11.23	1.004	0.9883	1.697
8	0.4383	12.02	1.215	1.242	11.67	1.003	0.988	1.667
9	0.5523	43.21	1.323	1.417	16.09	1.002	0.9911	1.382
10	0.5175	40.16	1.221	1.262	17.23	1.001	0.9933	0.9070
CPU	273.4		274.1			266.1		

Table 7
Data Set for Example 4*

Feeds	Flow Rates Tray 10	Tray 15
Methylcyclohexane		0.1163
Toluene		0.1163
Phenol	0.7674	
Temperature	454.0	376.0
Pressure	1.013	1.013

Number of trays = 21; total condenser U_1 = 0.1224; liquid reflux L_1 = 0.9914; column pressure = 1.013.

Initial Guess Profiles
T_1 = 373.0; T_{21} = 425.0; V_1 = 0.0; V_2 = 1.114; V_{21} = 1.114. Units: T(K); P(1.E + 5 Pa); flow rates (mol/time).

* *From Fredenslund [14].*

Table 8
Comparative Results from Example 4

	Iter.		B.P. Method					θ Method		
Iter.	D_{ca}	E.M.	θ	E.M.	θ*	E.M.	θ	E.M.	θ*	E.M.
1	0.2037	17.02	3.049	20.05	0.8062	29.54	3.049	15.60	0.8062	31.07
2	0.1384	16.54	0.1321	138.8	1.224	4.130	0.0377	341.9	1.609	0.452
3	0.1146	7.827	9.732	25.28	0.7803	2.512	40.19	24.92	0.5449	2.812
4	0.1335	5.427	0.2120	77.47	0.8933	1.301	0.0190	732.1	1.031	1.178
5	0.1496	7.997	6.809	23.42	0.9482	0.6338	24.84	26.34	0.9849	0.7299
6	0.1459	5.585	0.2470	64.85	0.9703	0.3349	0.0687	191.7	0.9766	0.4475
7	0.1310	2.449	5.523	22.27	0.9846	0.1607	24.50	23.14	1.018	0.2531
8	0.1281	0.9643	0.2683	59.11	0.9951	0.0762	0.0500	274.3	1.013	0.0791
9	0.1300	1.454	4.501	20.95	1.000	0.0513	21.73	24.39	0.9996	0.0337
10	0.1294	1.213	0.3117	48.41	1.000	0.0358	0.0640	211.3	0.9982	0.0283
CPU	29.46		29.73		29.83		11.73		11.79	

Table 9
Design Data for Examples A, B, C

Example	n	f	C	U_1(D)	L_1	Column Press.	Feed Flow Rates		Feed Temp.	Feed Press.	Profiles
A	11	6	5	0.489	1.26	120.	Propane	0.050			T_1=6.40
							Isobutane	0.150			T_6=640.
							n-Butane	0.250	640.	120.	T_{11}=690.
							Isopentane	0.200			V_2=1.75
							n-Pentane	0.350			V_{11}=1.75
B	20	11	4	0.6	1.35	14.7	Cyclohexane	0.10			T_1=619.
							Ethanol	0.40	631.	14.7	T_{11}=631.
							Propanol	0.40			T_{20}=666.
							Toluene	0.10			V_2=1.95
											V_{20}=1.95
C	13	4	11	0.316	0.632	30.	Methane	0.020			
							Ethane	0.100			
							Propylene	0.060			T_1=508.
							Propane	0.125			T_{13}=812.
							Isobutane	0.035	391.	30.	V_2=.948
							n-Butane	0.150			V_{13}=.948
							n-Pentane	0.152			
							n-Hexane	0.113			
							n-Heptane	0.090			
							n-Octane	0.085			
							n-Tetradecane	0.070			

Units: T = °R; P = psia; flow rates (lb-mol/time).

separation factors for all components in the central section, irrespective of their physico-chemical behavior.

Better results can be obtained by adopting the same overall policy provided a variable is schosen, which may account for the differences of chemical components; that is the temperature profile in the trays of the central section (hence T method). (See for instance the Smith–Brinkley [25] short cut method).

Then the actual temperature values $(T_j)_{ca}$ are corrected according to the following equation:

$$(T_j)_{co} = (T_j)_{ca} + \Delta T \quad \text{for} \quad 1 \leqq j \leqq m \tag{24}$$

where ΔT stands for an unknown increment independent of the tray of the section.

The separation factors $S_{j,i}$ depend both on composition and on tray temperature via vapor-liquid equilibrium constants. Nevertheless they are corrected accounting for only temperature dependence, therefore:

$$(S_{j,i})_{co} = S_{j,i}((T_j)_{co}, (X_j)_{ca}, (Y_j)_{ca}) \tag{24a}$$

The equilibrium constants are computed at $(T_j)_{co}$ while the compositions are the "calculated" ones and the ratio V_j/L_j is maintained unchanged. At this point the convergence promoter is based on the correction of component flow rates by means of a set of unknown parameters, namely: a ΔT increment defined by Equation 24 and the multipliers θ_j^{*W} and θ_j^{*U} defined by Equation 22.

T convergence promoter singles out a set of proper values of the unknown parameters ΔT, θ_j^{*U}, θ_j^{*W} so that the resulting component flow rates satisfy both material balances (Step 1 calculation) and standard specification equations involving total flow rates.

The overall procedure to be applied for the estimate of the corrected component flow rates is depicted in the block diagram of Figure 7.

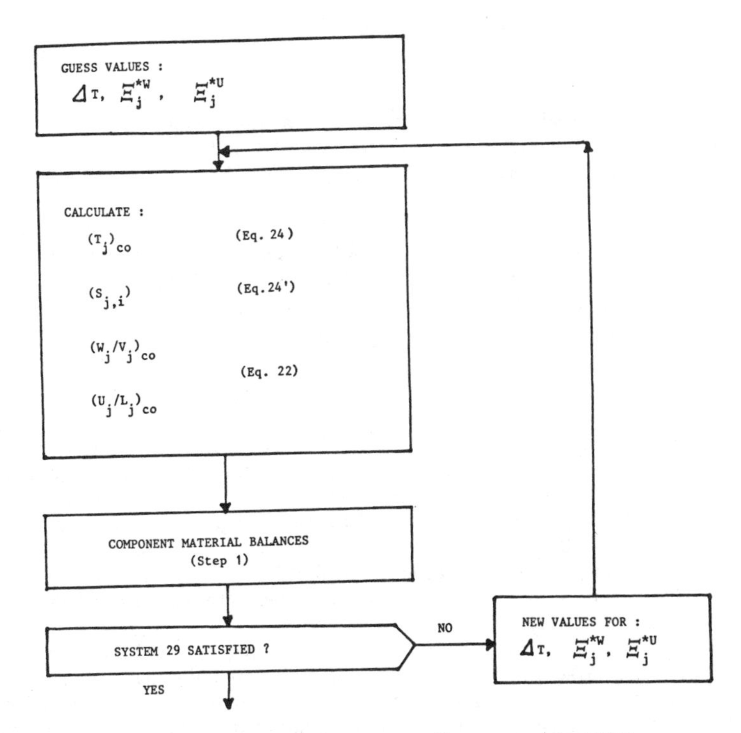

Figure 7. Overall procedure for the estimation of "corrected" component flow rates.

Table 10
Convergence Behavior for Examples A, B, C

Example	Iteration	E.M.	T_1	E.M.	θ	E.M.	D_{ca}
A	1	33.563	879.81	32.436	.1048	8.4735	.357
	2	3.999	627.08	5.4423	1.662	3.9249	.396
	3	.7107	661.91	.9039	.9133	2.4491	.428
	4	.7065−1	655.73	.8947−1	1.007	1.6145	.449
	5	.9866−2	656.34	.6315−1	1.000	1.0436	.463
	6	.2033−2	656.29	.1975−2	1.000	.65678	.473
	7	—	—	—	—	.40151	.479
	8	—	—	—	—	.24033	.483
	9	—	—	—	—	.1419	.485
	10	—	—	—	—	.0832	.487
B	1	1327.75	1110.8	1419.93	2.436−3	61.118	.426
	2	38.301	606.15	44.800	4.854	22.35	.455
	3	12.042	677.08	15.550	.4870	18.59	.481
	4	2.435	651.20	3.528	1.206	15.69	.497
	5	.363	656.32	.694	.961	14.36	.509
	6	.343−1	655.58	.978−1	1.006	13.48	.526
	7	.483−2	655.64	.127−1	.9992	10.77	.551
	8	—	—	—	—	5.493	.578
	9	—	—	—	—	1.472	.594
	10	—	—	—	—	.3790	.599
C	1	1860.	300.19	.655+7	1.795+7	95.071	.639
	2	12.328	582.14	26.667	8.21−3	39.636	.598
	3	7.200	451.	29.647	20.798	8.754	.515
	4	1.256	502.45	5.422	.197	10.675	.494
	5	.2486	487.61	2.7976	2.038	10.925	.491
	6	.5034−1	490.73	1.0703	.731	13.738	.489
	7	.1376−1	490.06	.4683	1.1391	20.958	.484
	8	.3707−2	490.22	.19132	.94686	29.948	.457
	9	.1335−2	490.18	.7835−1	1.022	13.115	.392
	10	—	—	.2954−1	.9913	3.525	.360

Several techniques capable of solving nonlinear equations systems may be applied for finding the unknown values of the parameters, ΔT, θ_j^{*W} and θ_j^{*U}.

The Newton–Raphson linearization approach proved to be stable and efficient in all the examined cases.

The values of the derivatives were numerically calculated. As a rule few iterations are required to solve the system, so that the overall computing time spent in this phase may be neglected when compared with the time needed by the convergence of the whole model.

Nevertheless, faster alternative methods, such as quasi-Newton, Broyden, or other can be successfully applied, if necessary, in order to save computing time.

Examples

Three sample problems examined are described in Table 9. Problem A is derived from Holland [12], while problem B is based on a hydrocarbon-alcohol system described by Fredenslund [15]; lastly, problem C is a gasoline stabilizer solved by Eckert [18].

Equilibrium constants have been calculated by means of the RKS method (case A, C) and Raoult's law (case B).

Molar enthalpies have been estimated either by the Chao–Seader [19] method (case A, C) or by polynomial expressions only depending on temperature (case B). All the data have been derived from Reid [17].

Also included in Table 9 is the initial guess of the temperature and the total vapor flow rates in some trays; in the intermediate ones, a linear interpolation has been assumed. The initial guess of total liquid flow rates has been calculated by means of overall material balances, while the equilibrium constants have been initiated by assuming (where necessary) in ideal behavior of the systems at hand.

Each problem has been solved by applying three different convergence accelerators; namely, T, θ, and B.P. inside the same overall calculation procedure which has been previously summarized when describing the T method.

In other words, with reference to the previously mentioned procedure, Step C is substituted, when using a θ promoter, by the calculation of the θ parameter by which the component flow rates are redefined. The same step is bypassed when dealing with a B.P. promoter. Comparative results for the three cases studied are given in Table 10.

For each method, the error measure according to Step 4 is reported, and the value of its characteristic parameter: T_f and θ for the T and θ promoters while the value of the calculated liquid distillate, for the B.P. promoter, is given.

A few generalizations can be drawn:

- The B.P. method shows a slow, steady convergence, though the initial errors are small.
- The T and θ promoters show a quicker oscillating convergence, though the initial errors may be large.
- As for a distinction between T and θ, Example A shows that for narrow boiling systems having almost constant relative volatilities, the two promoters have the same trend. This is not surprising because the two promoters must coincide when the relative volatilities are strictly constant. Example B, dealing with a mixture showing more variable volatility properties, puts into evidence that the two promoters are no longer coincident and that the T appears more convenient. This benefit, in convergence rate, is enhanced in Example C, where apart from the previous considerations, the nonideality of the mixture plays a role in modifying the volatility of each component.
- The computing time for each iteration is almost the same when applying θ or B.P. promoters, therefore θ is in general preferable to the latter. The T method is more elaborate and tends to become vanishingly small along with the iterations. While neglecting the small amount of time associated with the solution of the tridiagonal system of Step 1, a rough estimate of the time required can be obtained by applying a further boiling or dew point calculation. That is, it is the same as solving for more trays in the column.

Both the modified θ accelerating procedure and the T one introduce some multipliers inside the component material balances for trays belonging to the control section of the column.

In fact the modified θ procedure uses

$$-L_{j-1,i}+L_{j,i}\left\{\left[1+\theta_j^U\left(\frac{U_j}{L_j}\right)\right]+\theta^*S_{j,1}\left[1+\theta_j^W\left(\frac{W_j}{V_j}\right)\right]\right\}$$

$$-\theta^*S_{j+1,i}L_{j+1,i}-F_{j,i}=0$$

while T uses

$$-L_{j-1,i}+L_{j,i}\left\{\left[1+\theta_j^U\left(\frac{U_j}{L_j}\right)\right]+(S_{j,i})_{co}\left[1+\theta_j^W\left(\frac{W_j}{V_j}\right)\right]\right\}$$

$$-(S_{j+1,i})_{co}L_{j+1,i}-F_{j,i}=0$$

Both the equations may be rewritten in terms of molar fractions and total flow rates:

$$-X_{j-1,i}L_{j-1}+X_{j,i}[(L_j+\theta_j^U U_j)+\theta^*K_{j,i}(V_j+\theta_j^W W_j)]$$

$$-\theta^*K_{j+1,i}X_{j+1}V_{j+1}-F_jZ_{j,i}=0 \tag{25}$$

$$-X_{j-1,i}L_{j-1}+X_{j,i}[(L_j+\theta_j^U U_j)+(K_{j,i})_{co}(V_j+\theta_j^W W_j)]$$

$$-(K_{j+1,i})_{co}\cdot X_{j+1,i}V_{j+1}-F_jZ_{j,i}=0 \qquad (26)$$

They have to be compared with the original component balance

$$-X_{j-1,i}L_{j-1}+X_{j,i}[(L_j+U_j)+K_{j,i}(V_j+W_j)]-K_{j+1,i}X_{j+1,i}V_{j+1,i}-F_jZ_{j,i}=0 \qquad (27)$$

so that the way these promoters work is clearly evidenced. The corrections, in fact, are operated by introducing multiplying factors, so that the promoter might be called a "multipliers type" to distinguish it from the "additive type" promoter which is commonly known as the "relaxation method" [20].

The governing component material balances for all the column trays are written as follows:

$$X_{j-1}L_{j-1}+X_{j,i}[(L_j+U_j)+K_{j,i}(V_j+W_j)+\omega]-K_{j+1}X_{j+1,i}V_{j+1}-F_jZ_{j,i}-\omega X^*_{j,i}=0$$

where ω is an adaptive parameter having a fixed value and $X^*_{j,i}$ is the liquid composition profile at the previous iteration.

Equation 28 has been derived by simulating a dynamic behavior of trays, so that

$$HU\frac{dX_{j,i}}{dt}=F_jZ_{j,i}+V_{j+1}K_{j+1}X_{j,i}+L_{j-1}X_{j-1,i}-X_{j,i}[(L_j+U_j)+K_{j,i}(V_j+W_j)] \qquad (28)$$

where HU is the constant liquid hold-up on the tray, t the time. The left side term of the previous equation is written as

$$\frac{HU}{dt}(X_{j,i}-X^*_{j,i})=\omega(X_{j,i}-X^*_{j,i})$$

which enables simple manipulations of Equation 28.

Note that for $\omega=0$ Equation 28 coincides with Equation 27 and for $\omega=\infty$ $X_{j,i}=X^*_{j,i}$, i.e., no modification is operated to composition profiles.

The relaxation promoter is a valuable alternative to the θ and T ones, even if a sort of arbitrarity exists in assigning a value to the ω parameter. Jelinek [21] suggests the rule of assigning $\omega=100$ times the total flow rate fed to the colums for highly nonideal mixtures and 10 times the same figure for low nonideality.

Unpublished tests made by Jelinek and Pierucci [22] aiming to estimate the ω parameter by minimizing standard specification equations gave unsatisfactory results.

SIMULA package [23] uses for hydrocarbon mixtures a default value of $\omega=.25$ times the total flow rate fed to the column.

Use of Nonstandard Specification Equations

A subsequent task of accelerating convergence by θ, θ^*, or T promoter is the possibility of using nonstandard specification equations only depending on component flow rates. Each promoter, in fact, introduces some parameters whose values are obtained by fitting an equal number of specification equations which till now, for convenience, have been considered of standard type. Hence the main task of the promoter is to predict corrected component flow rates, then the possibility of fitting promoter-parameters on specification equations containing component flow rates is easily foreseen. A certain cumbersomeness, however, arises in using nonstandard specification equations; this is due to the constraint imposed by the partitioning method which demands solving together: total material balances, enthalpy balances, and standard specification equations in Step 3 of its iterative procedure. As a consequence, the convergence promoter has to perform two main subtasks:

- Predict corrected component flow rates on the basis of nonstandard specification equations. Let NUP be their number.
- Build up an equivalent number NUP of standard specification equations which will be used in the subsequent Step 3 of the iterative procedure.

These two subtasks are now described in some detail.

Should NUP be the number of unknown parameters introduced by the convergence promoter, then an equal number of specification equations, S_l, only depending on component flow rate variables can be accepted by the promoter, namely:

$$S_l(\bar{L}_{j,i}, \bar{V}_{j,i}, \bar{W}_{j,i}, \bar{U}_{j,i}) = 0 \quad (l = 1, \ldots, NUP) \tag{29}$$

Moreover the flow rates contained in these equations have to satisfy component material balances in each tray of the column, as they result after a full Step 1 calculation.

In the area of distillation columns, from the definitions of NG and NUP, it results:

$$NG \geqq NUP \geqq 1$$

as a consequence of relation

$$NG - NUP = (\text{number of thermal duties-1}) + \begin{cases} 1 & \text{if a liquid distillate exists} \\ 0 & \text{otherwise} \end{cases}$$

(The case NG = 0 is typical of absorbers or liquid-liquid extractors; they both are not covered by the present description).

The model's NG degrees of freedom of the column have to be saturated by an equal number of specification equations.

As mentioned earlier, the partitioning algorithm requires these equations to be a standard type (see Step 3 of the overall calculation procedure of the methods), that is, linear equations in the unknown total flow rates and thermal duties.

$$L_k(\bar{V}_j, \bar{L}_j, \bar{U}_j, \bar{W}_j, \bar{Q}_j) = 0, \quad k = 1, \ldots, NG \tag{30}$$

When nonstandard specification equations are used, then NUP equations of Equation 30 (let us assume for k = 1, ..., NUP) are substituted by Equation 29. The whole set of equations provided to the problem is therefore:

$$\begin{aligned} &S_l(\bar{L}_{j,i}, \bar{V}_{j,i}, \bar{W}_{j,i}, \bar{U}_{j,i}) = 0 \quad l = 1, \ldots, NUP \\ &L_k(\bar{V}_j, \bar{L}_j, \bar{U}_j, \bar{W}_j, \bar{Q}_j) = 0 \quad k = NUP+1, \ldots, NG \end{aligned} \tag{31}$$

Let us suppose that the L_k equations in Equation 31 do not imply the following NUP assignments

$$U_1 + V_1 = \sum_{i=1}^{NCP} (U_{1,i})_{co} + (V_{1,i})_{co}$$

$$U_j = \sum_{i=1}^{NCP} (U_{j,i})_{co} \quad \text{for } j \neq 1$$

$$W_j = \sum_{i=1}^{NCP} (W_{j,i})_{co} \quad \text{for any } j \tag{32}$$

This means that the set of NUP Equation 32 plus the L_k equations in Equation 31 form a consistent set of "standard specification equations."

The convergence promoter then uses the S_l-type equations for correcting component flow rates of each iteration; after that it generates the assignments (Equation 32) which are added to the L_k-type equations in Equation 31.

This set of "standard specification equations" is subsequently used in Step 3 of the iterative procedure.

Examples

In order to evaluate the convergence behavior when dealing with nonstandard specifications, three illustrative examples with complex columns are now examined.

The design data of a two-feeds one-side-stream column are reported in Table 11.

Equilibrium constants and molar enthalpy deviations have been calculated by means of RKS [15] equation of state. Ideal molar enthalpies have been calculated by means of polynomial expressions only depending on temperature.

All the data have been taken from Reid at al.

Table 11 also includes the initial guess of temperature and total vapor flow rate in some trays: in the intermediate ones, as usual, a linear interpolation has been assumed. The initial guess of liquid flow rates has been derived by means of overall material balances, while the equilibrium constants have been initialized by assuming an ideal behavior of the mixture.

The degrees of freedom have been saturated by assigning .08 (Lbmol/h) to the total flow rate of the liquid side-stream; 76.5% to a top recovery of propane, and .785 to the reflux ratio $L_1/(L_1+V_1)$.

Table 12 reports the behavior of convergence for ten iterations. Besides the values ΔT, θ_{16}^{*U}, and the error measure E.M. the total top product flow rate (namely V_1) predicted by the convergence promoter has also been quoted.

For sake of comparison, the same case study has been solved by the same overall calculation procedure but substituting the present convergence promoter with the θ one.

The columns θ and θ_{16}^{U} report the values of the two parameters characteristic of the promoter. It is worth noting that even if the T promoter shows a quicker convergence, the

Table 11
Design Data for Example D

Feeds	Flow Rates Tray 26	Tray 41
Ethane	.015	.005
Propane	.240	.100
N-Butane	.165	.220
N-Pentane	.075	.145
N-Hexane	.005	.030
Temperature	628.	680.
Pressure	263.	263.

Number of trays = 52; distillate in vapor phase; reflux ratio = .785; column pressure = 263.

Initial Guess Profiles

T_1 = 570; T_{26} = 630; T_{52} = 735; V_1 = .28; V_2 = 1.30; V_{52} = 1.30. Units: T = R; P = psia; flow rates = (lb-mole/time).

Table 12
Convergence Behavior for Example D

T Promoter			
ΔT	θ_{16}^{*V}	E.M.	V_1
108.87	.64099	243.03	.27941
−66.87	.97386	28.741	.27924
29.688	1.0718	17.978	.27920
−1.869	1.0622	9.2170	.27919
3.8037	1.0144	2.4750	.27920
.75733	1.0028	.46917	.27920
.21867	1.0002	.16901	.27920
.01823	.99988	.07082	.27920
−.0255	1.0000	.02328	.27920
−.0136	1.0000	.01651	.27920

θ Promoter			
θ	θ_{16}^{V}	E.M.	V_1
.05384	.67094	601.22	.27939
2.5108	.98310	46.848	.27922
.98388	1.0015	11.754	.27921
.95456	1.0139	6.1517	.27920
.93676	1.0054	3.6709	.27920
.96999	1.0007	1.8577	.27920
.98895	.99956	1.0397	.27920
.99901	.99943	.40773	.27920
1.0019	.99968	.15837	.27920
1.0015	.99975	.09111	.27920

Table 13
Design Data for Example E

	Flow Rates	
Feeds	Tray 8	Tray 24
Ethanol	.25	.0
Propanol	.125	.25
Water	.125	.25
Acetic acid	.0	.5
Temperature	628.	680.
Pressure	263.	263.

Number of trays = 30; distillate in vapor phase; liquid reflux (L_1 = .8; liquid side-stream (U_{21}) = .2; column pressure = 263.

Initial Guess Profiles
$T_1 = 640$; $T_{23} = 650$; $T_{24} = 660$; $T_{30} = 666$; $V_1 = .4$; $V_2 = 1.2$; $V_8 = 1.2$; $V_9 = .67$; $V_{24} = .67$; $V_{25} = .7$; $V_{30} = .8$. *Units: T = R; P = psia; flow rates (lb-mol/time).*

difference between the two promoters is not quite evident due to the fact that the mixture has an almost ideal behavior.

In fact we expect that the increasing of the nonideality of the mixtures may lead to a more distinguishable convergence.

The two-feeds, one-side-stream column reported in Table 13, has been solved imposing the following set of data:

- .2 mol/time to the total flow rate of the liquid side-stream
- .8 mol/time to the liquid reflux (L_1)
- .055 mol/time to the top product flow rate of water.

Table 14 reports the convergence behavior for both θ and T promoters.
A mixture with a higher nonideality has been tested by the case study shown by Table 15.
The specification equations set was:

$$V_1 = 0.$$

$$L_1 = .9914$$

Table 14
Convergence Behavior for Example E

T Promoter			
ΔT	θ_{21}^{*V}	E.M.	V_1
5.1167	.81818	13.170	.30681
−.2981	.96033	2.6635	.30632
.03718	1.0000	.48695	.30620
.0	1.0000	.0137	.30621
θ Promoter			
θ	θ_{21}^{V}	E.M.	V_1
.71875	.81357	18.181	.30679
.91043	.87182	7.1851	.30639
.95503	.96108	2.2353	.30621
.97850	.97667	1.669	.30621
.97240	.97064	1.4193	.30621
.97366	.97218	1.3429	.30621
.97513	.97365	1.2891	.30621
.97783	.97658	1.1184	.30631
.98356	.98264	.82509	.30621
.98817	.98752	.59040	.30621

Table 15
Design Data for Example F

	Flow Rates	
Feeds	Tray 10	Tray 15
Methylcyclohexane	.0	.1163
Toluene	.0	.1163
Phenol	.7674	.0
Temperature	818.	677.
Pressure	14.7	14.7

Number of trays = 21; total condenser U_1 = .1224; liquid reflux L_1 = .9914; column pressure = 14.7.

Initial Guess Profiles
T_1 = 673; T_{21} = 766; V_1 = 0; V_2 = 1.114; V_{21} = 1.114. Units: T = R; P = psia; flow rates (lb-mol/time).

Toluene distillate product = 1.111 − 02 mol/time
Table 16 reports the behavior of convergence for ten iterations by using T and θ promoter.

Constant Composition Convergence Promoter

Total flow rates are calculated from total mass and enthalpy balances as previously shown in the preceding pages. Far from convergence the obtained total flow rates do not satisfy component material balances, so that a possible improvement of the convergence behavior of the whole iterative procedure is expected by coupling total mass and enthalpy balances with the component ones.

This way of approaching the total flow rates estimation is commonly known as "constant composition method." In the context of the partitioning method, it may be classified as a further "convergence promoter."

The basic idea of this promoter is to write the total enthalpy content of each stream (liquid or vapor) in terms of component flow rates and partial molar enthalpies. Component flow rates are then imposed to satisfy component material balances. The equations involved are Equations 3, 5, and 6, even if they are rewritten in different form.

The total enthalpy balance around the column is:

$$\sum_{k=1}^{NP} (F_k HF_k + Q_k) = V_1 HV_1 + L_{NP} HL + \sum_{k=1}^{NP} (U_k HL_k + W_k HV_k)$$

Table 16
Convergence Behavior for Example F

ΔT	T Promoter E.M.	D
−27.978	50.038	.12602
.11899	10.307	.11899
.12004	4.7666	.12004
−5.721	2.1044	.12151
−1.541	.66049	.12189
−.4299	.25984	.12217
−.1492	.09563	.12225
.00475	.03844	.12229
.00280	.01385	.12229
.00142	.00892	.12229
θ	**θ Promoter E.M.**	**D**
.38562	48.546	.20374
.80955	6.9386	.20374
1.8147	14.408	.20374
.72687	10.293	.20375
1.4351	9.2657	.20375
.74118	8.6536	.20376
1.3019	6.9952	.20376
.78932	7.4253	.20376
1.2269	5.5372	.20376
.83671	5.4901	.20376

Since

$$L_{NP}HL_{NP} = \sum L_{NP,i}\overline{HL}_{NP,i}$$

where $\overline{HL}_{NP,i}$ are partial molar enthalpies, the component material balances around the column give

$$L_{NP}HL_{NP} = \sum_{i=1}^{NCP} \overline{HL}_{NP,i} \sum_{k=1}^{NP} (F_{k,i} - U_{k,i} - W_{k,i} - V_{1,i}) \tag{33}$$

By substitution of:

$$F_{k,i} = Z_{k,i}F_k$$

$$U_{k,i} = X_{k,i}U_k$$

$$W_{k,i} = Y_{k,i}W_k$$

$$V_{1,i} = Y_{1,i}V_1$$

we obtain:

$$L_{NP}HL_{NP} = \sum_{k=1}^{NP} (F_kHLZ_{k,NP} - U_kHLX_{k,NP} - W_kHLY_{k,NP} - V_1HLY_{1,NP}) \tag{34}$$

where $HLZ_{k,NP} = \sum_{i=1}^{NCP} \overline{HL}_{NP,i}Z_{k,i}$

$$HLX_{k,NP} = \sum_{i=1}^{NCP} \overline{HL}_{NP,i}X_{k,i}$$

$$HLY_{k,NP} = \sum_{i=1}^{NCP} \overline{HL}_{NP,i} Y_{k,i}$$

$$HLY_{1,NP} = \sum_{i=1}^{NCP} \overline{HL}_{NP,i} Y_{1,i}$$

From a general point of view we can define the following variables:

$$HLZ_{k,j} = \sum_{i=1}^{NCP} \overline{HL}_{j,i} \cdot Z_{k,i}$$

$$HLX_{k,j} = \sum_{i=1}^{NCP} \overline{HL}_{j,i} \cdot X_{k,i}$$

$$HLY_{k,j} = \sum_{i=1}^{NCP} \overline{HL}_{j,i} \cdot Y_{k,i}$$

When Equation 34 is substituted in Equation 33, V_1 total flow rate can be rewritten as function of the remaining unknowns

$$V_1 = \sum_{k=1}^{NP} (\alpha_{k,1} U_k + \beta_{k,1} W_k + \gamma_{k,1} F_k + \delta_{k,1} Q_k) \tag{35}$$

where the influence factors are:

$$\alpha_{k,1} = -\delta_{k,1}(HL_k - HLX_{k,NP})$$

$$\beta_{k,1} = -\delta_{k,1}(HV_k - HLY_{k,NP})$$

$$\gamma_{k,1} = \delta_{k,1}(HF_k - HLZ_{k,NP})$$

$$\delta_{k,1} = \frac{1}{HV_1 - HLY_{1,NP}}$$

For any tray j the total enthalpy balance enclosing the top of the column and tray j gives:

$$V_{j+1} HV_{j+1} + \sum_{k=1}^{j} (F_k HF_k + Q_k) = V_1 HV_1 + \sum_{k=1}^{j} (U_k HL_k + W_k HV_k) + L_j HL_j \tag{36}$$

The term $L_j HL_j$ is written through partial enthalpies and component flow rates which satisfy material balances, as indicated for Equation 34.

$$L_j HL_j = \sum_{k=1}^{j} (F_k HLZ_{k,j} - U_k HLX_{k,j} - W_k HLY_{k,j}) - V_1 HLY_{1,j} \tag{37}$$

Equations 35 and 37 are substituted in Equation 36 which may be rearranged in terms of influence factors:

$$V_{j+1} = \sum_{k=1}^{NP} (\alpha_{k,j+1} U_k + \beta_{k,j+1} W_k + \gamma_{k,j+1} F_k + \delta_{k,j+1} Q_k) \tag{38}$$

where

$$\delta_{k,j+1} \rightarrow = \delta'_{k,j+1} + \delta''_{k,j+1}, \quad 1 < k \leqq j$$
$$\searrow = \delta'_{k,j+1}, \quad j < k \leqq NP$$

$$\delta'_{k,j+1} = \frac{(HV_1 - HLY_{1,j})}{(HV_1 - HLY_{1,NP})\,(HV_{j+1} - HLY_{j+1,j})}$$

$$\delta''_{k,j+1} = \frac{-1}{HV_{j+1} - HLY_{j+1,j}}$$

$$\alpha_{k,j+1} = -\delta'_{k,j+1}(HL_k - HLX_{k,NP}) - \delta''_{k,j+1}(HL_k - HLX_{k,j}) \; (1 < k \leqq j)$$

$$= -\delta'_{k,j+1}(HL_k - HLX_{k,NP}) \quad (j < k \leqq NP)$$

$$\beta_{k,j+1} = -\delta'_{k,j+1}(HV_k - HLY_{k,NP}) - \delta''_{k,j+1}(HV_k - HLY_{k,j}) \; (1 < k \leqq j)$$

$$= -\delta'_{k,j+1}(HV_k - HLY_{k,NP}) \quad (j < k \leqq NP)$$

$$\gamma_{k,j+1} = \delta'_{k,j+1}(HF_k - HLZ_{k,NP}) + \delta''_{k,j+1}(HF_k - HLZ_{k,j}) \; (1 < k \leqq j)$$

$$= \delta'_{k,j+1}(HF_k - HLZ_{k,NP}) \quad (j < k \leqq NP)$$

The influence factors of the liquid stream of the following equation!

$$L_j = \sum_{k=1}^{NP} (A_{k,j}U_k + B_{k,j}W_k + C_{k,j}F_k + D_{k,j}Q_k)$$

are then calculated as a combination of those calculated for the vapor streams. The steps to be followed have been previously explained. Once the influence factors have been evaluated the new total flow rates are computed according to the suggestion given in the preceding paragraph.

The constant composition promoter is recommended when partial molar enthalpies are easily and accurately calculated. This happens, of course, for ideal or almost ideal solutions.

When partial molar enthalpies are not easily available due to the complexity of the thermodynamic model adopted, some attempts to estimate them may be done [26] through the use of "virtual values."

The enthalpy HL_j^0 of one mole of the mixture at the standard state pressure and at the temperature of the mixture may be easily estimated as:

$$HL_j^0 = \sum HL_{j,i}^0 \cdot X_{j,i}$$

where $HL_{j,i}^0$ are the standard enthalpies of each component. The virtual value of the partial molar enthalpy is defined as:

$$\bar{H}L_{j,i} \cong \hat{H}L_{j,i} = HL_{j,i}^0 + \Omega \tag{39}$$

where $\hat{H}L_{j,i}$ = virtual value of the partial molar enthalpy

$$\Omega = HL_j - HL_j^0$$

The approximation implied by Equation 39 consists in neglecting the dependence of Ω on the mixture composition at fixed temperature and pressure. Fortunately the neglected terms do not approximate the calculation of the molar enthalpy of the mixture.

In fact the corrected enthalpy of the mixture may be obtained by use of the virtual values,

$$HL_j = \sum X_{j,i}\, HL^0_{j,i} + \Omega$$

which is the right value of the enthalpy of the mixture.

This means that when using the constant composition method, with virtual values of partial enthalpies, approximate estimates of the terms $HLZ_{k,j}$, $HLX_{k,j}$, and $HLY_{k,j}$ are expected, while right values of the molar enthalpies HL_j are conserved.

GLOBAL METHOD

The global approach via the Newton–Raphson method consists in linearizing the system of algebraical nonlinear equations which characterize the distillation column and in solving the obtained system.

The Newton–Raphson global method is a very powerful algorithm; this is due to its speed in reaching the solution and to its ability to allow the solution of problems with a different choice of the specified and unspecified variables.

However the method presents some difficulties. One is the necessity of storing the coefficients of the Jacobs matrix. If the number of coefficients is too large it may be necessary to use auxiliary memory.

The second problem is the computation time required for solving the system of linearized equations, which may be prohibitive for large values of NCP and NP.

And last but not least of the disadvantages of the Newton–Raphson method is its requirement of having good starting data. A change in these data may cause a process to take less or more time before reaching the solution or it may even result in a loss of convergence at all. Therefore it is often used as a final method within a two-stage strategy, where a different algorithm has been previously used in order to reach the proximity of the solution.

Mathematical Model of the Newton–Raphson Method

Let's denote the set of the independent variables $\bar{X}$, $\bar{Y}$, $\bar{L}$, $\bar{U}$, $\bar{W}$, $\bar{T}$, and $\bar{Q}$ as $\bar{Z}$. The system of (2 NCP + 3) NP + NG equations describing the model of the column is expanded in the Taylor series truncated after the first term at the point $\bar{Z}^n$. An increment ΔZ of each variable is obtained by the Newton–Raphson procedure so that new points are taken according to:

$$\bar{Z}^{n+1} = \bar{Z}^n - \Gamma(\bar{Z}^n)^{-1}\bar{f}(\bar{Z}^n) = \bar{Z}^n + \Delta\bar{Z}$$

where $\Gamma(\bar{Z}^n)$ is the Jacobian matrix of $\bar{f}$.

As we'll see later it is a normal and advisable procedure to search for an optimal step along the Newton–Raphson procedure; in this case the new points are given by:

$$\bar{Z}^p = \bar{Z}^0 + p\Delta\bar{Z} \tag{40}$$

where p is generally different from 1.

The superscripts indicate the position on the proposed direction; for instance $\bar{Z}^0$ at $p = 0$ and $\bar{Z}^1$ at $p = 1$.

The system obtained is the following:

Component material balance around each tray:

$$-L_{j-1}\Delta X_{j-1,i}+(L_j+U_j)\Delta X_{j,i}+(V_j+W_j)\Delta Y_{j,i}-V_{j+1}$$

$$\cdot \Delta Y_{j+1,i}-X_{j-1,i}\Delta L_{j-1}+X_{j,i}\Delta L_j+Y_{j,i}\Delta V_j-Y_{j+1,i}$$

$$\cdot \Delta V_{j+1} = b_{j,i} \tag{41}$$

where $j = 1, NP$
$i = 1, NCP$

- Equilibrium equation:

$$-\Delta Y_{j,i}+K_{j,i}\Delta X_{j,i}+X_{j,i}\frac{\partial K_{j,i}}{\partial T_j}\Delta T_j+\sum_{n=1}^{NCP} X_{j,i}\frac{\partial K_{j,i}}{\partial X_{j,n}}$$

$$\cdot \Delta X_{j,n}+\sum_{n=1}^{NCP} X_{j,i}\frac{\partial K_{j,i}}{\partial Y_{j,n}}\Delta Y_{j,n} = e_{j,i} \tag{42}$$

$j = 1, NP$
$i = 1, NCP$

- Summation equation

$$\sum_{n=1}^{NCP}(\Delta Y_{j,n}-\Delta X_{j,n}) = s_j \tag{43}$$

$j = 1, NP$

- Total material balance around each tray:

$$-\Delta L_{j-1}+\Delta L_j+\Delta V_j-\Delta V_{j+1} = m_j \tag{44}$$

or $j = 1, NP$

$$-\Delta V_1+\Delta L_j+\Delta V_{j+1} = m_j' \tag{44a}$$

- Enthalpy balance around each tray:

$$-L_{j-1}\sum_{n=1}^{NCP}\frac{\partial HL_{j-1}}{\partial X_{j-1,n}}\Delta X_{j-1,n}+(L_j+U_j)\sum_{n=1}^{NCP}\frac{\partial HL_j}{\partial X_{j,n}}\Delta X_{j,n}$$

$$+(V_j+W_j)\sum_{n=1}^{NCP}\frac{\partial HV_j}{\partial Y_{j,n}}\Delta Y_{j,n}-V_{j+1}\sum_{n=1}^{NCP}\frac{\partial HV_{j+1}}{\partial Y_{j+1,n}}\Delta Y_{j+1,n}$$

$$-HL_{j-1}\Delta L_{j-1}+HL_j\Delta L_j+HV_j\Delta V_j-HV_{j+1}\Delta V_{j+1}$$

$$-L_{j-1}\frac{\partial HL_{j-1}}{\partial T_{j-1}}\Delta T_{j-1}+\left\{(L_j+U_j)\frac{\partial HL_j}{\partial T_j}+(V_j+W_j)\frac{\partial HV_j}{\partial T_j}\right\}T_j$$

$$-V_{j+1}\frac{\partial HV_{j+1}}{\partial T_{j+1}}\Delta T_{j+1} = h_j \tag{45}$$

$j = 1, NP$

- Linearized specification equations. Corresponding to each new specification equation one of the variables, which in the "standard" problem were fixed, needs to be released. In this case it is necessary to introduce, in each equation of the system, the corrections ΔF_j or ΔU_j or ΔW_j or ΔQ_j of the new unknowns.

  The corrections $\Delta X_{j,i}$, $\Delta Y_{j,i}$, ΔT_j, ΔV_j, ΔL_j (i = 1, NCP; j = 1, NP) are the unknowns variables and $b_{j,i}$, $e_{j,i}$, s_j, m_j, h_j (and d_l for the specification equations with l = 1, NG) are the residuals all equal to zero when the solution is reached.

Some Possible Arrangements of the Linearized System

The equations can be solved following different procedures, according to the specific problem to be considered and to the experience of the users.

In this context we'll propose three possible arrangements; in all the cases the system of the equations is considered as being constituted of two different groups of equations. The variables are also divided in two different groups, ξ and η so that the following system can be obtained:

$$M\xi + N\eta = m \tag{46}$$

$$P\xi + Q\eta = n \tag{47}$$

The solution procedure involves the following steps:

Step 1. By using Equation 46 we obtain the variables ξ as a function of the variables η:

$$\xi = m' + N'\eta \tag{48}$$

Step 2. It is now possible to eliminate the ξ variables from Equation 47 so as to obtain this equation:

$$Q'\eta = n' \tag{49}$$

Step 3. Equation 49 is solved with respect to the η variables.

Step 4. Finally the ξ variables are obtained from Equation 48.

For the three arrangements ξ and η are respectively:

- Arrangement 1 [27]

$$\xi = \Delta Y_{j,i}(i = 1, NCP; j = 1, NP); \Delta V_1; \Delta T_j, \Delta X_{j,i}$$

$$(i = 1, NCP), \Delta V_{j+1}, \Delta L_j(j = 1, NP-1); \Delta T_{NP}$$

$$\eta = \Delta X_{NP,i}(i = 1, NCP), \Delta L_{NP}, \Delta SP_n(n = 1, NG)$$

- Arrangement 2 [28]:

$$\xi = \Delta Y_{j,i}(i = 1, NCP; j = 1, NP); \Delta X_{j,i}(i = 1, NCP),$$

$$\Delta T_j, \Delta V_j, \Delta L_j(j = NP)$$

$$\eta = \Delta SP_n(n = 1, NG)$$

- Arrangement 3 [29]:

$$\xi = \Delta Y_{j,i}(i = 1, NCP;\ j = 1, NP);\ \Delta L_j(j = 1, NP);$$

$$(\Delta X_{j,i}(j = 1, NP)\ (i = 1, NCP))$$

$$\eta = \Delta T_j, \Delta V_j(j = 1, NP);\ \Delta SP_n(n = 1, NG)$$

In the following pages we'll explain in some detail Arrangement 1 (proposed by Buzzi Ferraris) and we'll give some traces for solving the system following the other two arrangements.

Arrangement 1

The proposed method can be employed in the case of a large number of components and stages.

In the following procedure it is assumed that the equilibrium ratios K are not strongly dependent upon both liquid- and vapor-phase composition; therefore the derivatives

$$\frac{\partial K_{j,i}}{\partial X_{j,n}} \quad \text{and} \quad \frac{\partial K_{j,i}}{\partial Y_{j,n}}$$

are negligible when compared to other terms in Equation 42 so that the following equation:

$$-\Delta Y_{j,i} + K_{j,i}\Delta X_{j,i} + X_{j,i}\frac{\partial K_{j,i}}{\partial T_j}\Delta T_j = e_{j,i} \quad \begin{matrix} i = 1, NCP \\ j = 1, NP \end{matrix} \tag{50}$$

can be used instead of Equation 42.

This is well applicable to the cases when either the number of stages, NP, or the number of components, NCP, is large. In fact when the number of stages is very large, the components of the mixture are quite similar and, therefore, the equilibrium ratios are not too far from ideality; on the other hand, if we have a large number of components, it is unrealistic to think that the derivatives could play an important role.

It is worthwhile noting that if in Equation 42 we neglect the partial derivatives of the equilibrium ratios with respect to the composition, it does not mean that the K ratios are independent of the composition. It is necessary only that the terms

$$\Delta Y_{j,i}, K_{j,i}\Delta X_{j,i} \quad \text{and} \quad X_{j,i}\frac{\partial K_{j,i}}{\partial T_j}\Delta T_j$$

are sufficiently more important than the other terms in Equation 42.

The system of equations (Equation 41 through 45) is rearranged as follows:

$EQ_{j,i}$: equilibrium equation $\quad j = 1, NP$

$$\Delta Y_{j,i} = -e_{j,i} + X_{j,i}\frac{\partial K_{j,i}}{\partial T_j}\Delta T_j + K_{j,i}\Delta X_{j,i} \tag{51}$$

$i = 1, NCP$

OB_1: total material balance of Stage 1

$$\Delta V_1 = m_1 - \Delta L_1 + \Delta V_2 \tag{52}$$

ST_j: summation equation $\quad j = 1, NP-1$

$$\Delta T_j = \left[\sum_{n=1}^{NCP} (1 - K_{j,n}) \Delta X_{j,n} + S_j + \sum_{n=1}^{NCP} e_{j,n} \right] \Big/ \sum_{l=1}^{NCP} X_{j,l} \cdot \frac{\partial K_{j,l}}{\partial T_j} \tag{53}$$

$MB_{j+1,i}$: component material balance around Stage $j+1$

$$\Delta X_{j,i} = \Big\{ [(L_{j+1} + U_{j+1}) + (V_{j+1} + W_{j+1}) K_{j+1,i}] \Delta X_{j+1,i}$$

$$- V_{j+2} K_{j+2,i} \Delta X_{j+2,i} + (V_{j+1} + W_{j+1}) X_{j+1,i} \frac{\partial K_{j+1,i}}{\partial T_{j+1}} \Delta T_{j+1}$$

$$- V_{j+2} X_{j+2,i} \frac{\partial K_{j+2,i}}{\partial T_{j+2}} \Delta T_{j+2} - X_{j,i} \Delta L_j + X_{j+1,i} \Delta L_{j+1}$$

$$+ Y_{j+1,i} \Delta V_{j+1} - Y_{j+2,i} \Delta V_{j+2} - b_{j+1,i} - (V_{j+1} + W_{j+1}) e_{j+1,i}$$

$$+ V_{j+2} e_{j+2,i} \Big\} / L_j \tag{54}$$

$i = 1, NCP$

OB_{j+1}: total material balance of the $(j+1)$-th stage

$$\Delta V_{j+1} = m_{j+1} + \Delta L_j - \Delta L_{j+1} + \Delta V_{j+2} \tag{55}$$

EB_{j+1}: enthalpy balance of the $(j+1)$-th stage

$$L_j \frac{\partial HL_j}{\partial T_j} \Delta T_j + L_j \sum_{n=1}^{NCP} \frac{HL_j}{X_{j,n}} \Delta X_{j,n} - HV_{j+1} \Delta V_{j+1} + HL_j \Delta L_j$$

$$= \sum_{n=1}^{NCP} \left[(L_{j+1} + U_{j+1}) \frac{\partial HL_{j+1}}{\partial X_{j+1,n}} + (V_{j+1} + W_{j+1}) \frac{\partial HV_{j+1}}{\partial Y_{j+1,n}} K_{j+1,n} \right] \Delta X_{j+1,n}$$

$$- V_{j+2} \sum_{n=1}^{NCP} \frac{\partial HV_{j+2}}{\partial Y_{j+2,n}} K_{j+2,n} \Delta X_{j+2,n} + HL_{j+1} \Delta L_{j+1} - HV_{j+2} \Delta V_{j+2}$$

$$+ \Bigg[(L_{j+1} + U_{j+1}) \frac{\partial HL_{j+1}}{\partial T_{j+1}} + (V_{j+1} + W_{j+1}) \frac{\partial HV_{j+1}}{\partial T_{j+1}} + (V_{j+1} + W_{j+1})$$

$$\sum_{n=1}^{NCP} \frac{\partial HV_{j+1}}{\partial Y_{j+1,n}} X_{j+1,n} \frac{\partial K_{j+1,n}}{\partial T_{j+1}} \Bigg] \Delta T_{j+1} - \Bigg(V_{j+2} \frac{\partial HV_{j+2}}{\partial T_{j+2}}$$

$$+ V_{j+2} \sum_{n=1}^{NCP} \frac{\partial HV_{j+2}}{\partial Y_{j+2,n}} X_{j+2,n} \frac{\partial K_{j+2,n}}{\partial T_{j+2}} \Bigg) \Delta T_{j+2} - h_{j+1}$$

$$- (U_{j+1} + W_{j+1}) \sum_{n=1}^{NCP} \frac{\partial HV_{j+1}}{\partial Y_{j+1,n}} e_{j+1,n} + V_{j+2} \sum_{n=1}^{NCP} \frac{\partial HV_{j+2}}{\partial Y_{j+2,n}} e_{j+2,n} \tag{56}$$

ST_N: summation equation at the NP-th stage

$$\Delta T_{NP} = \left[\sum_{n=1}^{NCP} (1 - K_{NP,n}) \Delta X_{NP,n} + s_{NP} + \sum_{n=1}^{NCP} e_{NP,n} \right] / \sum_{l=1}^{NCP} X_{NP,l} \cdot \frac{\partial K_{NP,l}}{\partial T_{NP}} \tag{57}$$

$MB_{1,i}$: component material balance on the whole equipment

$$\begin{aligned} & - Y_{1,i}\Delta V_1 - V_1 K_{1,i} \Delta X_{1,i} - V_1 X_{1,i} \frac{\partial K_{1,i}}{\partial T_1} \Delta T_1 - L_{NP} \Delta X_{NP,i} \\ & - X_{NP,i} \Delta L_{NP} - \sum_{j=1}^{NP} (U_j + W_j K_{j,i}) \Delta X_{j,i} - \sum_{j=1}^{NP} W_j X_{j,i} \frac{\partial K_{j,i}}{\partial T_j} \Delta T_j \\ & = V_1 e_{1,i} - \sum_{j=1}^{NP} W_j e_{j,i} + Y_{1,i} V_1 + X_{NP,i} L_{NP} + \sum_{j=1}^{NP} (U_j X_{j,i} \\ & + W_j Y_{j,i} - F_j Z_{j,i}) \end{aligned} \tag{58}$$

$i = 1, NCP$

EB_1: enthalpy balance of the first stage

$$\begin{aligned} & - HV_1 \Delta V_1 - \sum_{n=1}^{NCP} \left[(L_1 + U_1) \frac{\partial HL_1}{\partial X_{1,n}} + (V_1 + W_1) \frac{\partial HV_1}{\partial Y_{1,n}} K_{1,n} \right] \Delta X_{1,n} \\ & - HL_1 \Delta L_1 + HV_2 \Delta V_2 + V_2 \sum_{n=1}^{NCP} \frac{\partial HV_2}{\partial Y_{2,n}} K_{2,n} \Delta X_{2,n} \\ & + \left(V_2 \frac{\partial HV_2}{\partial T_2} + V_2 \sum_{n=1}^{NCP} \frac{\partial HV_2}{\partial Y_{2,n}} X_{2,n} \frac{\partial K_{2,n}}{\partial T_2} \right) \Delta T_2 \\ & - \left[(L_1 + U_1) \frac{\partial HL_1}{\partial T_1} + (V_1 + W_1) \frac{\partial HV_1}{\partial T_1} + (V_1 + W_1) \sum_{n=1}^{NCP} \frac{\partial HV_1}{\partial Y_{1,n}} X_{1,n} \right. \\ & \left. \cdot \frac{\partial K_{1,n}}{\partial T_1} \right] \Delta T_1 = - h_1 - (V_1 + W_1) \sum_{n=1}^{NCP} \frac{\partial HV_1}{\partial Y_{1,n}} e_{1,n} + V_2 \sum_{n=1}^{NCP} \frac{\partial HV_2}{\partial Y_{2,n}} e_{2,n} \end{aligned} \tag{59}$$

Each specification equation is linearized and then added after the last of the previous equations. The sequence in which the unknowns of the linear system are ordered is: $\Delta Y_{j,i}$ ($i = 1, NCP$; $j = 1, NP$); ΔV_1; ΔT_1; $\Delta X_{1,i}$ ($i = 1, NCP$); ΔV_2; ΔL_1; ΔT_2; $\Delta X_{2,i}$ ($i = 1, NCP$); ... ; ΔT_j; $\Delta X_{j,i}$ ($i = 1, NCP$); ΔV_{j+1}; ΔL_j; ... ; ΔL_{NP-1}; ΔT_{NP}; $\Delta X_{NP,i}$ ($i = 1, NCP$); ΔL_{NP}; and then all the "nonstandard" unknowns ΔSP_1; ...; ΔSP_{NG}.

The structure of the resulting coefficient matrix is shown in Figure 8, in which it has been assumed $NCP = 2$, $NP = 5$, the following "nonstandard specification equations":

$$V_1 = 0$$

$$X_{2,1} = 0.05$$

and the "nonstandard unknowns" U_1, Q_2.

Equations 58, 59 and the NG specification equations constitute then the last (NCP+1+NG) equations of the linearized system.
This system can be easily solved by means of the following procedure:

1. The variable ΔV_1 is eliminated from the last (NCP+1+NG) equations by using Equation 52.
2. The variables ΔT_j, $\Delta X_{j,i}$, ΔV_{j+1} are eliminated from Equation 56 by making use of Equations 53, 54 and 55 respectively. Thus, one obtains:

$$\Delta L_j = CEB_j + DEBTJ1_j \Delta T_{j+1} + \sum_{n=1}^{NCP} DEBXJ1_{j,n} \Delta X_{j+1,n}$$
$$+ DEBVJ2_j \Delta V_{j+2} + DEBLJ1_j \Delta L_{j+1}$$
$$+ DEBTJ2_j \Delta T_{j+2} + \sum_{n=1}^{NCP} DEBXJ2_{j,n} \Delta X_{j+2,n}$$
$$+ \sum_{n=1}^{NG} DEBSJ1_{j,n} \Delta SP_n \quad (j = 1, NP-2) \tag{60}$$

or

$$\Delta L_{NP-1} = CEB_{NP-1} + DEBTJ1_{NP-1} \Delta T_{NP}$$
$$+ \sum_{n=1}^{NCP} DEBXJ1_{NP-1,n} \Delta X_{NP,n}$$
$$+ DEBLJ1_{NP-1} \Delta L_{NP} + \sum_{n=1}^{NG} DEBSJ1_{NP-1,n} \Delta SP_n \tag{60a}$$

The terms CEB_j, $DEBTJ1_j$, ... $DEBSJ1_{j,n}$ are memorized because they are necessary in the following backward procedure. The variables ΔT_j, $\Delta X_{j,i}$ (i = 1, NCP), ΔV_{j+1}, ΔL_j, (j = 1, NP − 1) are then eliminated from the last (NCP+1+S) equations by substituting Equations 53 through 55 and Equation 60 or 60a.

3. We obtain:

$$\Delta T_{NP} = CST_{NP} + \sum_{n=1}^{NCP} DSTXJ_n \Delta X_{NP,n} \tag{61}$$

by using Equation 57. The variable ΔT_{NP} is then eliminated from the last (NCP+1+NG) equations.

4. The last (NCP+1+NG) equations are solved with respect to the variables $\Delta X_{NP,i}$ (i = 1, NCP), ΔL_{NP} and the NG "nonstandard" variables ΔSP_1, ... ΔSP_{NG}.

5. It is now possible to obtain the value of all the other variables with the following backward procedure: j = NP; ΔT_{NP} from Equation 61; $\Delta Y_{NP,i}$ (i = 1, NCP) from Equation 51; j = NP − 1,1; ΔL_j from Equation 60a if j = NP − 1; from Equation 60 if j ≠ NP − 1; ΔV_{j+1} from Equation 55

$$\Delta X_{j,i} = \{-b_{j+1,i}+(L_{j+1}+U_{j+1})\Delta X_{j+1,i}+(V_{j+1}+W_{j+1}) \cdot \Delta Y_{j+1,i}-V_{j+2}\Delta Y_{j+2,i}-X_{j,i}\Delta L_j+X_{j+1,i}\Delta L_{j+1} + Y_{j+1,i}\Delta V_{j+1}-Y_{j+2,i}\Delta V_{j+2}\}/L_j \quad (i = 1;\ NCP)$$

ΔT_j from Equation 53; $\Delta Y_{j,i}$ (i = 1, NCP) from Equation 51; j = 1; ΔV_1 from Equation 52.

Figure 9 shows the matrix coefficient structure for the backward procedure.

The previous procedure is not applicable when the derivatives $\partial K_{j,i}/\partial Y_{j,n}$ in Equation 42 cannot be neglected, that is, when the equilibrium ratios, $K_{j,i}$, are strongly dependent on the vapor composition. In this case, provided that it is possible to neglect the derivatives $\partial K_{j,i}/\partial X_{j,n}$ in Equation 42, the procedure has to be modified. It is easy to see that, if we start from the bottom of the separator and we use Equation 42 to obtain $\Delta X_{j,i}$ instead of $\Delta Y_{j,i}$ the structure of the resulting coefficient matrix and then of the applied solving procedure is similar to the previous one.

Arrangement 2 (Naphtali and Sandholm)

Equations 41 through 45 are arranged stage by stage. Eliminating $\Delta Y_{j,i}$ from the linear system by using Equation 50 the coefficient matrix has the structure shown in the Figure 10.

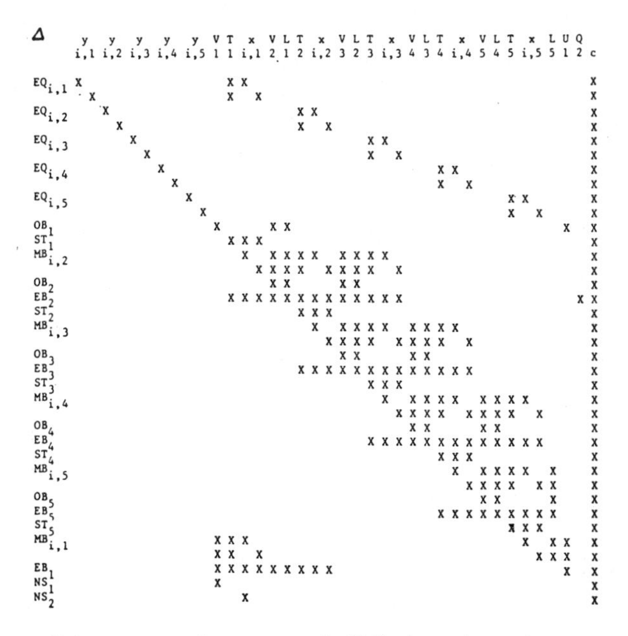

Figure 8. Matrix coefficient structure of arrangement 1 with the forward procedure.

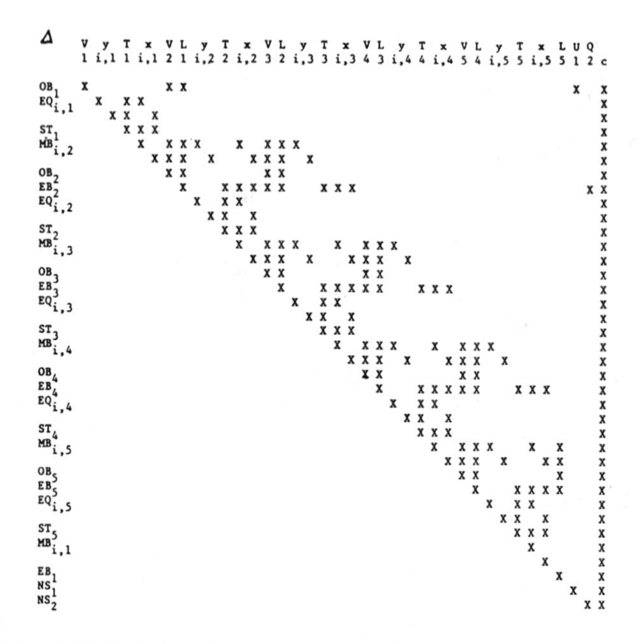

Figure 9. Matrix coefficient structure of arrangement 1 with the backward procedure.

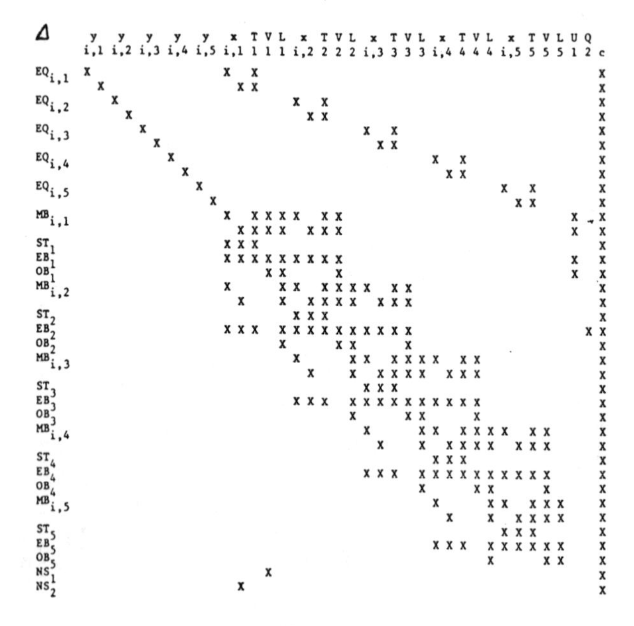

Figure 10. Matrix coefficient structure of arrangement 2.

Arrangement 3 (Donati–Buzzi Ferraris)

In this case the variables $\Delta Y_{j,i}$ and ΔL_j are eliminated from the remaining equations by using Equations 50 and 44a respectively. The system is arranged in this way: material balances of the i-th component and j-th stage Equation 41; stoichiometric equation for all the stages Equation 43; enthalpy balances for all the stages Equation 45; "nonstandard" specification equations. The variables are ordered in this way: $\Delta Y_{j,i}$ ($i = 1, NCP; j = 1, NP$); ΔL_j ($j = 1, NP$); $\Delta X_{j,i}$ ($i = 1, NCP; j = 1, NP$); ΔT_j ($j = 1, NP$), ΔV_j ($j = 1, NP$) "nonstandard variables."

The matrix of the coefficients is shown in Figure 11.

Comparison between the three arrangements

The comparison between the three arrangements has been well developed by Buzzi Ferraris [27].

Figures 12 through 13 show the memory occupation as a function of the components number and of the stages number, respectively. DNR1, DNR2, DNR3 represent the core memory occupation for Arrangements 1, 2, and 3, respectively; DNR1′, DNR2′, DNR3′ represent the core memory that remains necessary when some of the matrices and vectors are transferred on an auxiliary memory.

It's clear that Method 3 is not able to solve a problem in which the number of stages is large, while Method 2 may present some difficulties when the number of stages or the number of components becomes large.

For the three methods Figures 14 and 15 report the computation time as a function of the components number and of the stages number.

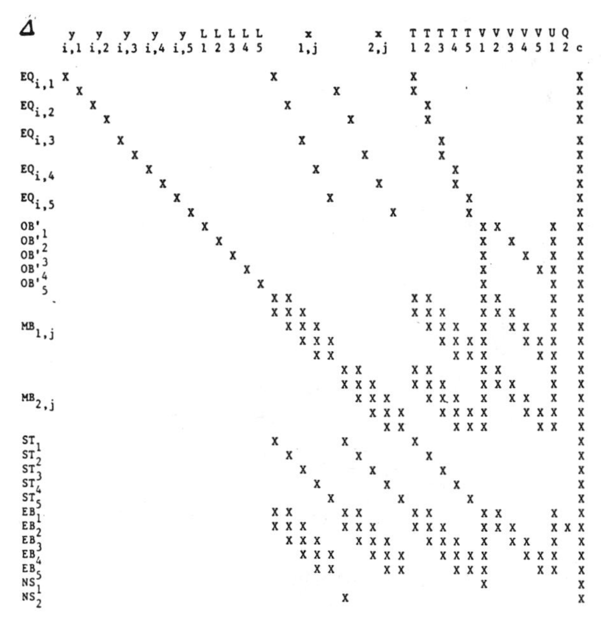

Figure 11. Matrix coefficient structure of arrangement 3.

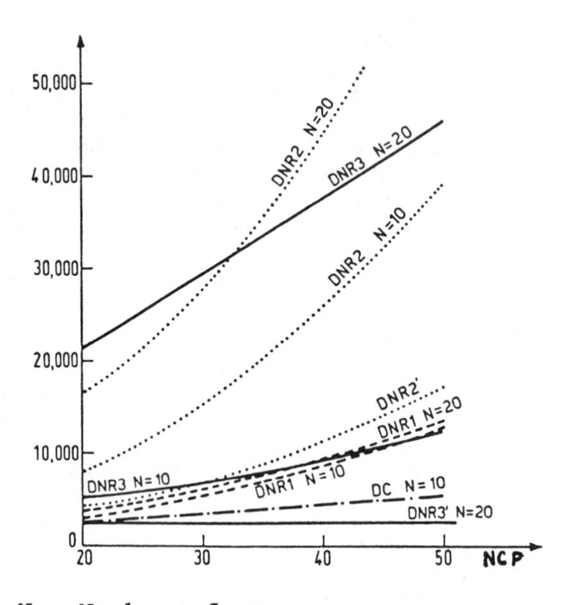

Figure 12. Memory occupation as a function of the component's number.

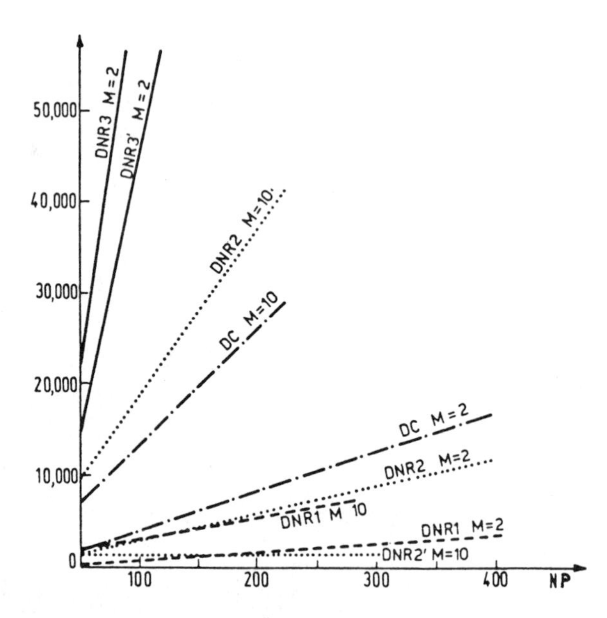

Figure 13. Memory occupation as a function of the stage's number.

Optimal Step Along the Newton–Raphson Direction

Experience shows that Newton–Raphson approach cannot always guarantee convergence when initial values of unknowns are too far removed from final solution. Therefore the Newton–Raphson method is normally used as the final approach when we are close to the solution, and it is used to determine an optimal step along the Newton–Raphson direction obtained from the solution of the linearized system.

To this scope a procedure is followed that foresees the solution of a one-variable minimization problem [30]. The objective function depends on the sum of the squares of each equation residuals, the unknown being the Newton step.

Proof

Let us consider a second-order algebraic function f(x, y) in the unknowns x and y:

$$f(x, y) = a + bx + cy + dxy + ex^2 + gy^2 \tag{62}$$

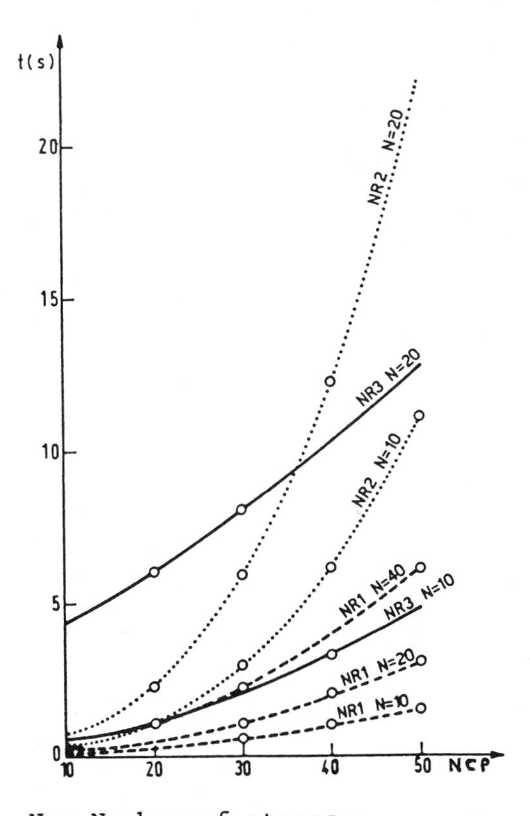

Figure 14. Computation time as a function of the component's number.

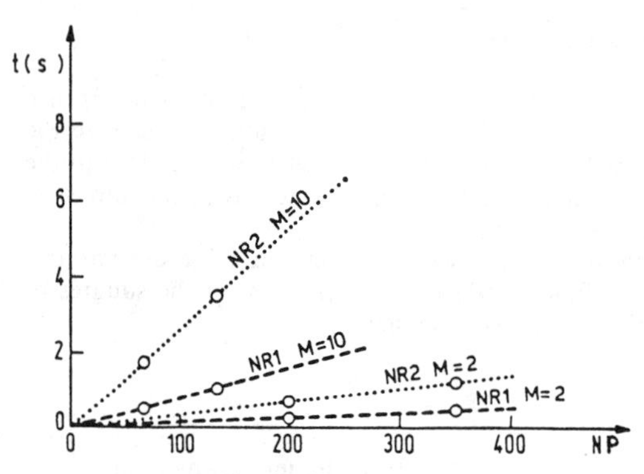

Figure 15. Computation time as a function of the stage's number.

The residual calculated at the point (x_0, y_0) being

$$f_0 = f(x_0, y_0) = a + bx_0 + cy_0 + dx_0y_0 + ex_0^2 + gy_0^2 \tag{63}$$

while the linearized function around the same point reads:

$$f_0 + b\Delta x + c\Delta y + dx_0\Delta y + dy_0\Delta x + 2ex_0\Delta x + 2gy_0\Delta y = 0 \tag{64}$$

The point (x_1, y_1) predicted through Newton–Raphson method is related to (x_0, y_0) by the well-known relations

$$x_1 = x_0 + \Delta x$$

$$y_1 = y_0 + \Delta y$$

defining the "position" of the unknowns x and y along the "Newton direction"

$$x_p = x_0 + p\Delta x$$

$$y_p = y_0 + p\Delta y \tag{65}$$

when $p = 1$.

Along that direction the residual of Equation 62 is proved further on as depending on p as per

$$f_p = f(x_p, y_p) = (1-p)f_0 + p^2 f_1 \tag{66}$$

where f_0, f_1 are the calculated residuals when $p = 0$ and $= 1$, respectively. Proof of Equation 66 can be easily obtained as follows: the point (x_p, y_p) defined by Equation 65 is introduced in Equation 62 so:

$$f_p = f(x_p, y_p) = a + b(x_0 + p\Delta x) + c(y_0 + p\Delta y) + d(x_0 + p\Delta x) \cdot$$

$$(y_0 + p\Delta y) + e(x_0 + p\Delta x)^2 + g(y_0 + p\Delta y)^2$$

is rearranged into:

$$f_p = a + bx_0 + cy_0 + dx_0y_0 + ex_0^2 + gy_0^2$$
$$+ p(b\Delta x + c\Delta y + dx_0\Delta y + dy_0\Delta x + 2ex_0\Delta x + 2gy_0\Delta y)$$
$$+ p^2(d\Delta x\Delta y + e\Delta x^2 + g\Delta y^2) \quad (67)$$

According to Equations 63 and 64, Equation 67 is rewritten so as to give:

$$f_p = f_0 - pf_0 + p^2(d\Delta x\Delta y + e\Delta x^2 + g\Delta y^2) \quad (68)$$

Should $p = 1$, Equation 68 gives:

$$f_1 = d\Delta x\Delta y + e\Delta x^2 + g\Delta y^2 \quad (69)$$

Equations 69 and 68 will then be combined so as to form Equation 66. It is worth noting that Equation 66 analytically relates the function residuals along the Newton direction through a second-order relation on p, f_0 and f_1 being its coefficients. Equation 69 shows an alternative calculation of function residual when $p = 1$. This result may be easily generalized in the following statement: *the residual of a second-order algebraic equation when $p = 1$ may be calculated by substituting each variable for its increment into the equation second-order terms.*

Statement of the Problem

Let's consider the equation:

$$\bar{Z}^p = \bar{Z}^0 + p\Delta\bar{Z}$$

The p value is reached so that the sum of the squares of "properly weighted" residuals of the equations is minimized. The weights render residuals of the same order of magnitude. In this paper we are going to follow the suggestion of Ishii and Otto [31] to normalize all the residuals to unity.

$$b_{j,i} = \frac{b_{j,i}}{(L_i^0 + U_j^0)X_{j,i}^0 + (V_j^0 + W_j^0)Y_{j,i}^0}$$

$$e_{j,i} = \frac{e_{j,i}}{K_{j,i}^0X_{j,i}^0}$$

$$s_j = \frac{s_j}{\sum X_{j,i}^0}$$

$$m_j = \frac{m_j}{(L_j^0 + U_j^0) + (V_j^0 + W_j^0)}$$

$$h_j = \frac{h_j}{(L_j^0 + V_j^0)HL_j^0 + (V_j^0 + W_j^0)HV_j^0}$$

$$d_1 = \frac{d_1}{\text{magnitude of the values of the most dominant variables involved}}$$

The values of the weighted residuals along the direction ΔZ, may be also denoted by the superscript p, so that for instance, $b_{j,i}^{0}$ is calculated at $p = 0$ and $b_{j,i}^{1}$ is the corresponding value for $p = 1$.

The objective function to be minimized may be therefore formulated as:

$$PH(p) = \sum_{j=1}^{NP} \left\{ \sum_{i=1}^{NCP} [(b_{j,i}^{p})^2 + (e_{j,i}^{p})^2] + (s_j^p)^2 + (m_j^p)^2 + (h_j^p)^2 \right\} + \sum_{l=1}^{NG} (d_l^p)^2 \quad (70)$$

Assumptions and Solution Procedure

The equilibrium constants $K_{j,i}$ and the liquid and vapor molar enthalpies HL_j and HV_j are functions of temperature, pressure and phase composition, therefore they vary along the direction $p\Delta Z$ according to particular thermodynamic models.

For the purpose of minimizing the PH function, the following linearization formulas are used:

$$\Delta K_{j,i} = \left.\frac{\partial K_{j,i}}{\partial T_j}\right|_{p=0} \Delta T_j + \sum_{i=1}^{NCP} \left(\left.\frac{\partial K_{j,i}}{\partial X_{j,i}}\right|_{p=0} \Delta X_{j,i} + \left.\frac{\partial K_{j,i}}{\partial Y_{j,i}}\right|_{p=0} \Delta Y_{j,i} \right) \quad (71)$$

$$\Delta HV_j = \left.\frac{\partial HV_j}{\partial T_j}\right|_{p=0} \Delta T_j + \sum_{i=1}^{NCP} \left.\frac{\partial HV_j}{\partial Y_{j,i}}\right|_{p=0} \Delta Y_{j,i} \quad (72)$$

$$\Delta HL_j = \left.\frac{\partial HL_j}{\partial T_j}\right|_{p=0} \Delta T_j + \sum_{i=1}^{NCP} \left.\frac{\partial HL_j}{\partial X_{j,i}}\right|_{p=0} \Delta X_{j,i} \quad (73)$$

The linearized equations include the partial derivatives with respect to all variables which affect the values of the properties. For the value $\bar{Z}^p$ the corresponding values of the properties read as follows:

$$K_{j,i}^{p} = K_{j,i}^{0} + p\Delta K_{j,i} \quad (74)$$

$$HV_j^p = HV_j^0 + p\Delta HV_j \quad (75)$$

$$HL_j^p = HL_j^0 + p\Delta HL_j \quad (76)$$

It should be noted that the system to be solved is of second-order.

Accordingly all residuals along the Newton direction may be expressed as per Equations 71 through 76

$$b_{j,i}^{p} = (1-p)b_{j,i}^{0} + p^2 b_{j,i}^{1} \quad (77)$$

$$e_{j,i}^{p} = (1-p)e_{j,i}^{0} + p^2 e_{j,i}^{1} \quad (78)$$

$$s_j^p = (1-p)s_j^0 \quad (79)$$

$$m_j^p = (1-p)m_j^0 \quad (80)$$

$$h_j^p = (1-p)h_j^0 + p^2 h_j^1 \quad (81)$$

Assuming that the specification equations are either linear expressions in the variables or that they contain nonlinear second-order terms only, their residuals depend on p according to

$$d_l^p = (1-p)d_l^0 + p^2 d_l^1 \tag{82}$$

When p = 1 the residuals may be estimated according to Equation 69

$$b_{j,i}^1 = -\Delta L_{j-1}\Delta X_{j-1,i} + (\Delta L_j + \Delta U_j)\Delta X_{j,i} + (\Delta V_j + \Delta W_j) \cdot Y_{j,i} - \Delta V_{j+1}\Delta Y_{j+1,i} \tag{83}$$

$$e_{j,i}^1 = -\Delta K_{j,i}\Delta X_{j,i} \tag{84}$$

$$h_j^1 = -\Delta L_{j-1}\Delta HL_{j-1} + (\Delta L_j + \Delta U_j)\Delta HL_j + (\Delta V_j + \Delta W_j) \cdot \Delta HV_j - \Delta V_{j+1}\Delta HV_{j+1} \tag{85}$$

By defining f_K^p the residual of a generic model equation

$$f_K^p = (1-p)f_K^0 + p^2 f_K^1$$

in that (according to the following final expression) the PH function depends on p:

$$PH(p) = \sum_K \left\{(1-p)f_K^0 + p^2 f_K^1\right\}^2 \tag{86}$$

This equation is of the fourth-order in p and may be written as:

$$PH(p) = Ap'^4 + Bp'^3 + Cp'^2 + Dp' + E' \tag{87}$$

where $A' = \sum_K (f_K^1)^2$

$$B' = \sum_K (-2f_K^0 f_K^1)$$

$$C' = \sum_K [2f_K^0 f_K^1 + (f_K^0)^2]$$

$$D' = \sum_K -2(f_K^0)^2$$

$$E' = \sum_K (f_K^0)^2$$

It follows that the optimal Newton step is the one which minimizes PH(p) function viz. a. root of equation:

$$\frac{dPH}{dp} = 4Ap'^3 + 3Bp'^2 + 2Cp' + D' = 0$$

In principle, this equation can well be analytically rooted so that three real values are obtained and consequently two points of PH minimum value have to be compared. It happens, all too often, that only one real solution is practically found, and, where two minimum points are existing, one of them corresponds to a nonfeasible solution outside the physical constraints imposed on each variable.

As it will be shown later, Equation 87, written in terms of calculated (weighted) residuals, fits very well the PH function behaviour as shown by Equation 70.

Care has to be taken while evaluating the terms contained in this equation.

Where feasible the f_K^l terms have to be evaluated using proper physical-property values calculated by thermodynamic models rather than by Equations 74 through 76. When $\Delta\bar{Z}$ increments are so "wild" as to violate the physical or practical constraints of the variables, the f_K^l may be approximately estimated as per Equation 83 and other similar equations belonging to the model.

Alternatively, even when less satisfactory results are expected, the f_K^l values may be obtained from Equations 77 through 82 by evaluating the residuals f_K^p at one value of p within the physical domain of the unknown variables.

Application

A mixture containing methylcyclohexane, toluene, and phenol is separated in a two-feed column. The design data are reported in Table 17.

Table 17
Design Data for Application 1 of the Optimal Newton Step Procedure

Feeds	Flow Rates	
	Tray 10	Tray 15
Methylcyclohexane		.1163
Toluene		.1163
Phenol	.7674	
Total	.7674	.2326
Temperature	454.	376.
Pressure	1.013E+5	1.013E+5

Number of trays = 21; total condenser U_l = .1224; liquid reflux L_l = .9914; column pressure = 1.013E+5.

Initial Guess Profiles

*T_l = 373; T_{21} = 425; V_l = 0; V_2 = 1.114; V_{21} = 1.114; Q_l = −.5+8; Q_{21} = .5+8. Units: T = K; P = N/M**2; Q = J/time; flow rates (mol/time).*

Table 18
Finite Increments of Unknown Variables for Application 1

Tray	DT	DV	DL	SDX	SDY
1	−1.8930	.00000	−.15984−03	.67870	.52863
2	−7.1961	−.15981−03	−.43206−02	.76652	.67870
3	−11.530	−.43206−02	−.11605−01	.75343	.75782
4	−14.571	−.11605−01	−.20892−01	.65187	.74728
5	−16.406	−.20892−01	−.28342−01	.49949	.65616
6	−18.197	−.28342−01	−.37243−01	.34271	.51636
7	−19.720	−.37243−01	−.46294−01	.20614	.36820
8	−21.127	−.46294−01	−.63756−01	.97268−01	.23302
9	−20.775	−.63756−01	−.15915	.25651−01	.12072
10	−2.9727	−.15915	−.35108	.17637	.47880−01
11	−7.1531	−.35108	−.35575	.24256	.62903−01
12	−10.960	−.35575	−.35381	.26183	.70161−01
13	−14.991	−.35381	−.34102	.23240	.60473−01
14	−20.477	−.34102	−.30613	.13755	.21616−01
15	−30.361	−.30613	−.36425	.98123−01	.69636−01
16	−42.665	−.36425	−.27587	.39157	.27455
17	−65.093	−.27587	−.21290	.80283	.66811
18	−92.232	−.21290	−.27839	1.0694	1.2162
19	−102.28	−.27839	−.48697	.95252	1.6288
20	−70.381	−.48697	−.66007	.53691	1.5061
21	−1.2661	−.66007	.23842−06	.17868	.84019

Increments on specifications: .42451−07; 84939.0; −88118.0.

First-trial vapor and temperature values for some trays are also presented in the same table, assuming a linear interpolation in the intermediate trays. First trial K-values have been estimated by assuming an ideal behavior of the mixture according to Antoine vapor pressure correlation. The solution of total and component mass balances gives first-trial composition profiles. Guessed values for thermal duties to the condenser and the reboiler are also reported in Table 17.

The simulation is run through standard specifications, that is, assigning V_1, U_1, and L_1, values. Due to the nonideal behavior of the mixture, K values are estimated by the UNIFAC method through successive steps of calculations; liquid and molar enthalpies have been calculated using polynomial expressions depending solely on temperature. All data have been taken from Reid et al. [17].

Table 18 reports for each tray the finite increments resulting from solving the linearized system of equations. For sake of brevity the columns SDX and SDY simply report the values

$$\sum_{i=1}^{NCP} |\Delta X_{j,i}| \quad \text{and} \quad \sum_{i=1}^{NCP} |\Delta Y_{j,i}|$$

The residuals of each equation for p = 1, viz. f_K^1 values, have been calculated by evaluating physical properties at the predicted conditions of temperature and compositions.

The coefficients of Equation 87 are: A = 3.860036; B = −.0439377; C = 164.0858; D = −328.0838; E = 164.0419.

One PH function minimum point exists for p = .95865.

Figure 16 reports plots of PH function for $-1. < p < 1.5$, viz. values obtained by Equations 87 (solid lines) and 70. The agreement of the predictions is fairly good.

A second case study refers to a demethanizer column, a particularly sensitive problem in terms of initial guess profiles.

Table 19 shows the design data.

The simulation is carried out by using the same technique as per the preceding case except for the physical properties, K values, and molar enthalpies, estimated by the RKS equation of state.

Tables 20 and Figure 17 report the foreseen variable increments by solving the linearized system and plots of PH function, respectively.

Table 19
Design Data for a Demethanizer Column

	Flow Rates			
Feeds	Tray 6	Tray 9	Tray 16	Tray 21
CO	.44−01	.882−01	.154	.110
H_2	1.168	4.772	16.83	16.82
Methane	20.37	30.33	45.27	29.39
Acetylene	.66−01	.6182	2.449	2.84
Ethylene	38.03	156.4	404.6	371.2
Ethane	8.29	68.31	278.7	343.2
Propene		.2204	4.517	16.25
Propane		.22−01	.4847	2.16
n-Butane			.8995	14.290
Benzene			.39−01	7.41
Total	67.97	260.8	753.9	803.6
Temperature	146.	176.	210.	239.
Pressure	32.7E+5	33.3E+5	33.7E+5	34.1E+5

Number of trays = 58; V_1 = 111; U_1 = 100.7; L_1 = 153; column pressure = 31.E+5

Initial Guess Profiles
*T_1 = 178; T_4 = 178; T_{21} = 238; T_{58} = 272; V_1 = 111; V_2 = 365; V_6 = 365; V_{58} = 400; Q_1 = −1.+9; Q_{58} = 3.5+9. Units: T = K; P = N/M**2; .Q = J/time; flow rates (mol/time).*

The coefficients of Equation 87 are in this case: A = 32.49896, B = −1.44741, C = 6.311241, D = −9.727675, E = 4.863837. "p" optimal value is .3546252.

Figure 17 describes the values of PH function too, when f_K^l residuals are estimated by means of Equation 83 and similar equations.

In this case the optimal p value is .3239304.

The closer the approach to the solution, the smaller the disagreement among the diverse predictions of optimal p. This trend is clearly expanded in Figure 18 which reports the PH function for the fourth and fifth iteration of the same run.

Table 20
Finite Increments of Unknown Variables for Demethanizer at First Iteration

Tray	DT	DV	DL	SDX	SDY
1	−31.197	−.49989	−.72122	.11436	.33671
2	−1.0662	−1.2205	12.433	.13277	.33548
3	6.1514	11.934	−15.712	.31734	.29167
4	22.042	−16.211	−72.449	.68461	.24330
5	48.821	−72.948	−141.14	1.1790	.28639
6	88.641	−141.64	−115.59	1.1394	.45182
7	111.48	−116.09	−116.77	1.2989	.62193
8	117.99	−117.27	−95.424	1.1188	.71358
9	104.30	−95.923	21.145	.43159	.59758
10	124.27	20.646	36.115	.56273	.80843
11	134.53	35.616	44.805	.65813	.98222
12	137.06	44.306	57.779	.66212	1.1005
13	129.40	57.280	71.396	.56736	1.1199
14	111.35	70.897	79.414	.39743	1.0091
15	86.300	78.915	78.572	.19107	.76003
16	59.254	78.373	267.53	.62263−01	.39970
17	63.391	267.03	303.97	.66843−01	.45806
18	61.782	303.47	314.41	.70265−01	.51540
19	58.422	313.91	319.46	.78539−01	.54644
20	53.140	318.96	319.17	.54661−01	.52250
21	45.968	318.67	426.48	.64920−01	.39630
22	43.296	425.98	416.02	.59893−01	.30419
23	39.948	415.52	401.54	.53449−01	.27193
24	37.321	401.04	392.68	.49044−01	.26501
25	35.243	392.18	388.07	.48270−01	.26686
26	33.450	387.57	385.43	.49250−01	.27182
27	31.786	384.93	383.84	.50198−01	.27774
28	30.182	383.34	382.64	.51122−01	.28394
29	26.607	382.14	381.67	.52024−01	.29010
30	27.048	381.17	380.81	.52906−01	.29617
31	25.500	380.31	380.08	.53769−01	.30210
32	23.958	379.58	379.48	.54622−01	.30793
33	22.422	378.98	378.77	.55463−01	.31370
34	20.882	378.27	378.37	.56287−01	.31927
35	19.349	377.87	377.91	.57116−01	.32485
36	17.818	377.41	377.47	.57944−01	.33038
37	16.286	376.97	377.09	.58774−01	.33586
38	14.752	376.59	376.71	.59609−01	.34132
39	13.216	376.21	376.38	.60457−01	.34672
40	11.677	375.88	376.11	.61330−01	.35209
41	10.135	375.61	375.81	.62229−01	.35747
42	8.5899	375.31	375.55	.63162−01	.36283
43	7.0464	375.05	375.27	.64146−01	.36821
44	5.5048	374.77	375.02	.65197−01	.37362
45	3.9749	374.52	374.73	.66342−01	.37912
46	2.4630	374.24	374.51	.67613−01	.38472
47	.98857	374.01	374.28	.69456−01	.39058
48	−.42578	373.78	374.21	.72942−01	.39680
49	−1.7356	373.71	374.37	.76836−01	.40379
50	−2.8722	373.87	374.83	.81331−01	.41214
51	−3.7432	374.33	376.07	.86665−01	.42265
52	−4.1997	375.57	378.54	.93221−01	.43674
53	−4.0474	378.04	383.00	.10144	.45606
54	−3.0581	382.50	390.42	.11172	.48213
55	−1.0416	389.92	401.99	.12412	.51462
56	2.0078	401.49	418.20	.13737	.54741
57	5.6096	417.70	437.66	.14500	.55716
58	8.7382	437.16	.49957	.12170	.46950

Increments on Specifications: U_l = .60160−03; Q_l = .25363+06; Q_{58} = −.24252+06.

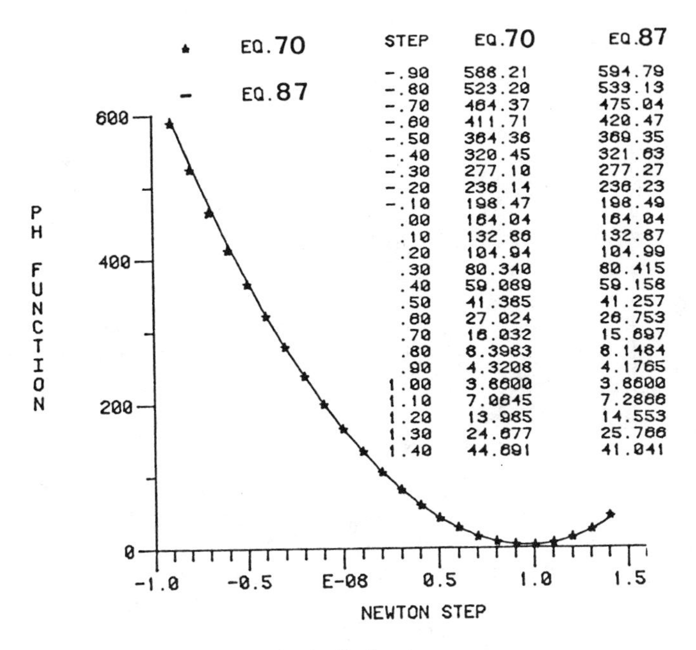

Figure 16. PH function versus Newton step for Application 1.

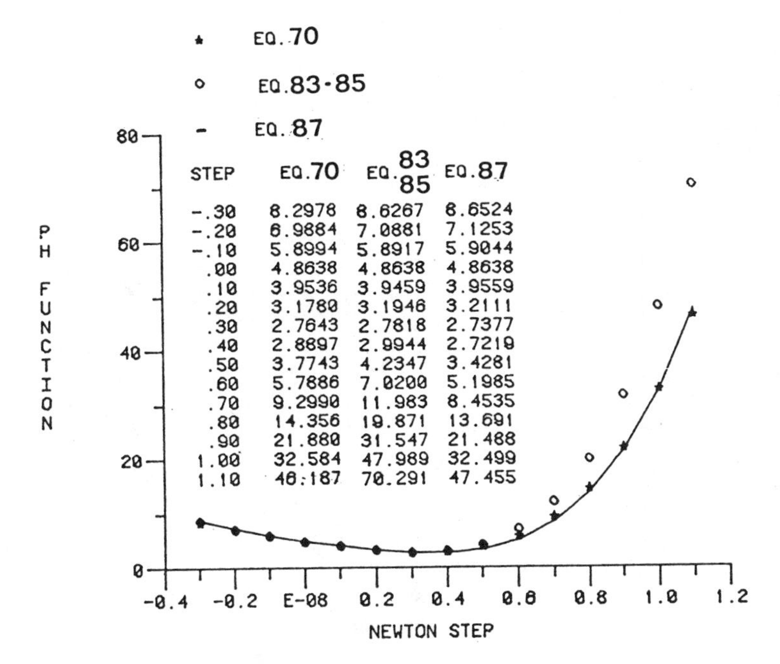

Figure 17. PH function versus Newton step for demethanizer at first iteration.

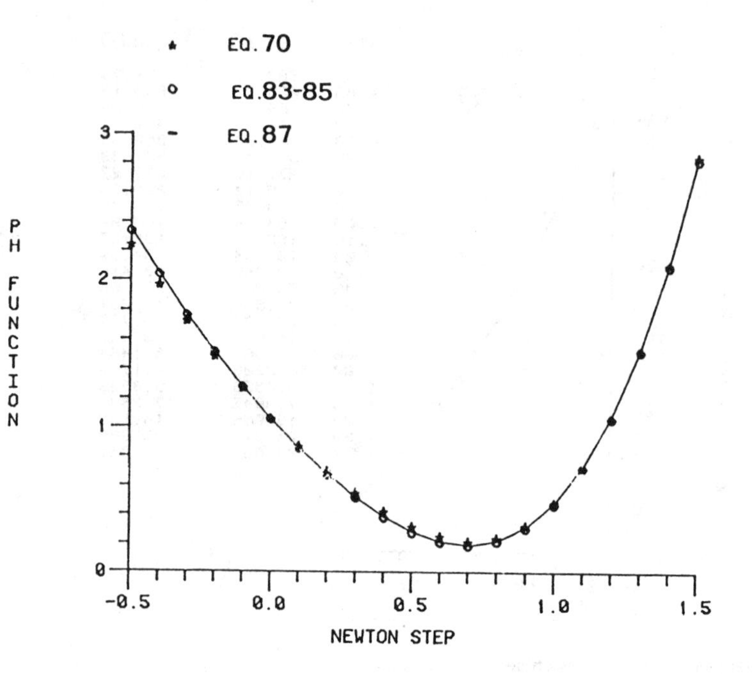

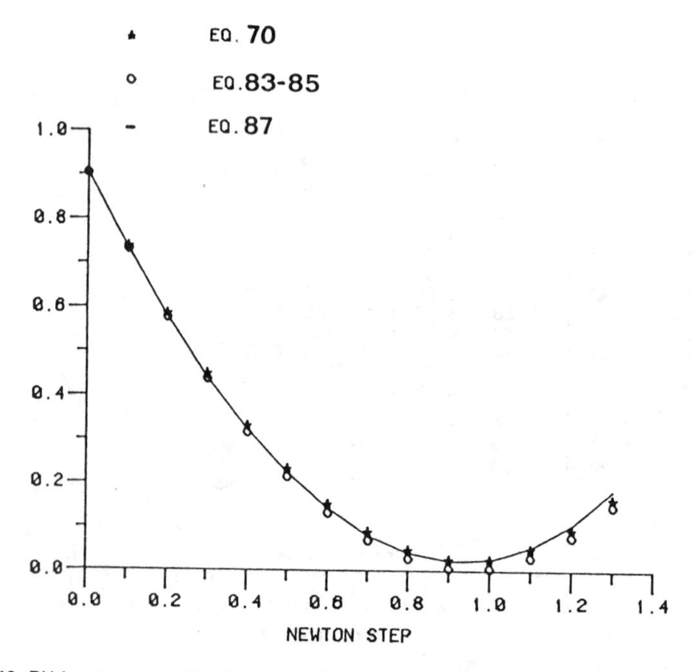

Figure 18. PH function versus Newton step at fourth and fifth iteration for demethanizer.

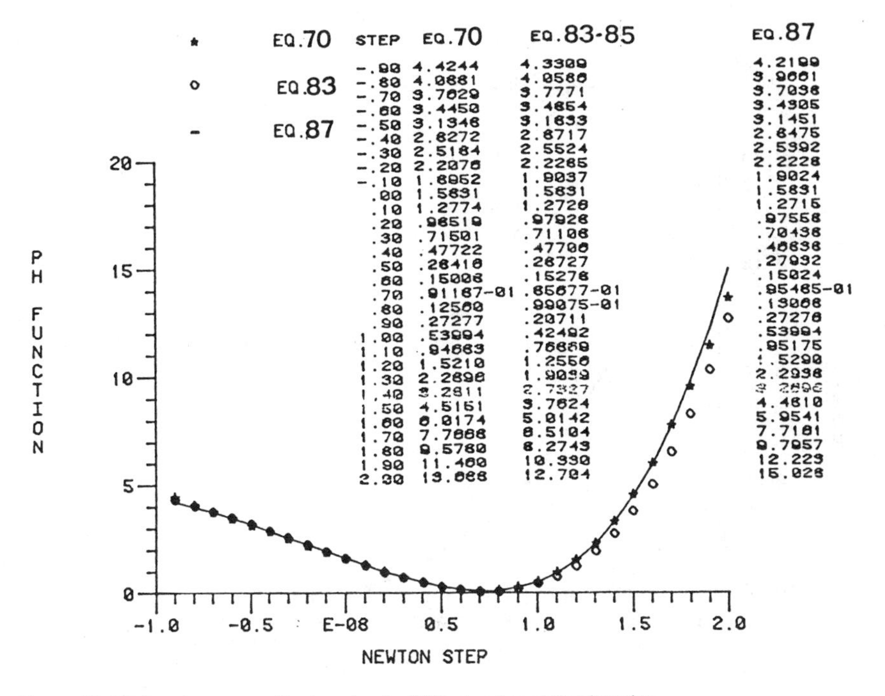

Figure 19. PH function versus Newton step for LPG absorber at first iteration.

Table 21
Design Data for LPG Absorber

Feeds	Flow Rates Tray 1	Flow Rates Tray 6
N_2		.18122−02
CO_2		.57388−02
Methane		.23136
Ethane		.42286−01
Propane		.14196−01
i-Butane		.15102−02
n-Butane		.30202−02
i-Pentane		.60408−03
n-Pentane		.60408−03
n-Hexane	.20931−02	.60408−03
n-Heptane	.16482−02	.30199−03
n-Ottane	.18618−01	
n-Nonane	.95409−02	
n-Decane	.37919−01	
Total	.69820−01	.30204
Temperature	310.	297.
Pressure	33.5E+5	33.5E+5

Number of trays = 6; column pressure = 33.5E+5.

Initial Guess Profiles
$T_1 = 277$; $T_3 = 283$; $T_6 = 277$; $V_1 = .2$; $V_6 = .2$. *Units: T = K; P = N/M**2; flow rates (mol/time).*

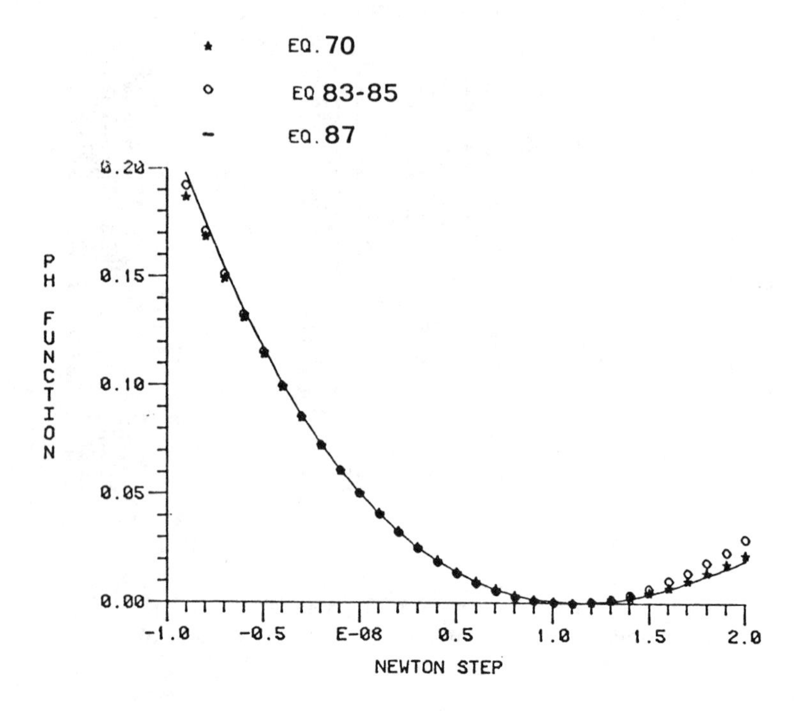

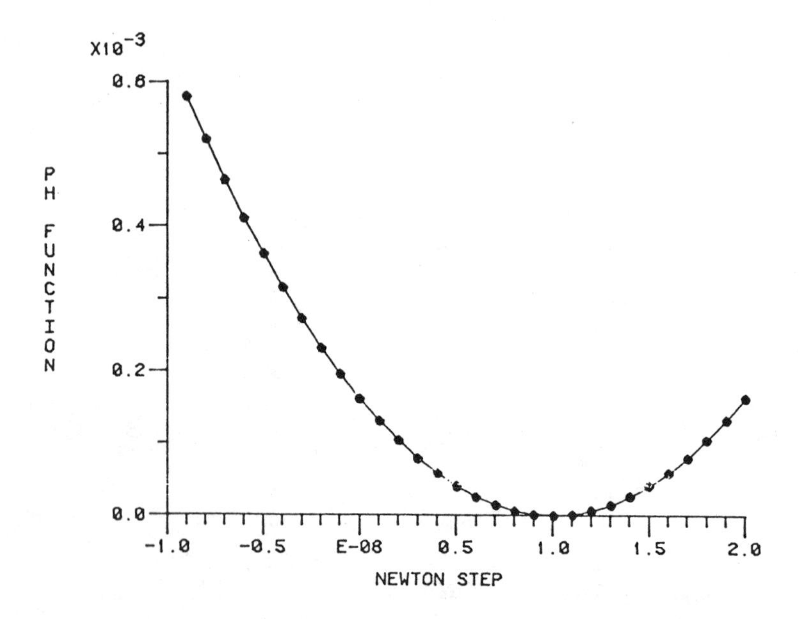

Figure 20. PH function versus Newton step for LPG absorber at successive iterations.

Table 22
Finite Increments of Unknown Variables for LPG Absorber After First Iteration

Tray	DT	DV	DL	SDX	SDY
1	94.881	.65134−01	.21428−01	.29534	.18428
2	103.06	.86562−01	.23492−01	.46617	.17723
3	103.15	.88626−01	.25282−01	.53352	.14516
4	111.72	.90416−01	.26227−01	.60829	.13121
5	112.78	.91361−01	.28631−01	.62591	.11694
6	100.81	.93765−01	.36906−01	.49239	.98996−01

Design data of a final case study have been reported in Table 21, a typical absorber of an LPG plant.

Increments in Table 22 are the ones predicted by linearized system solution and point out a "wild" initial guess.

Figure 19 presents the PH function plots for the points obtained by Equations 70, 83 and 87.

The coefficients of Equation 87 are: A = .4249225; B = 1.080437; C = .502642; D = −3.166157; E = 1.583078. Optimal value p = .7351580.

Figure 20 reports PH function behavior for successive iterations.

NOTATION

- A′, B′, C′, D′, E′ — coefficients of Equation 87
- b — residual of Equation 41
- D — distillate
- d — residual of linearized specification equations
- e — residual of Equation 42
- F — feed molar flow rate
- f — residual of a generic equation of the model
- HF — feed molar enthalpy
- HL — liquid molar enthalpy
- HV — vapor molar enthalpy
- h — residual of Equation 45
- K — equilibrium constants
- L — liquid stream molar flow rate
- m, m′ — residuals of Equation 44 and Equation 44a
- NCP — total number of components
- NG — number of degrees of freedom of column model
- NP — total number of theoretical trays (condenser and reboiler included)
- NUP — number of unknown parameters for the T method
- P — pressure
- p — Newton step
- Q — thermal duty to a tray
- R — ratio between the corrected and the calculated component distillate flow rate
- S — separation factor
- s — residual of Equation 43
- U — liquid side-stream molar flow rate
- W — vapor side-stream molar flow rate
- WT — weighting factor
- X — liquid molar fraction
- Y — vapor molar fraction
- x — generic variable
- y — generic variable
- Z — generic unknown of the model for Newton–Raphson method

Greek Symbols

- α, β, γ, δ — influence factor for a vapor flow rate
- Δ — correction of the variable, estimated by the Newton–Raphson method
- ψ — correcting factor of the component liquid molar flow rate
- φ — generic specification equation

σ	correcting factor of the component vapor molar flow rate	θ	theta multiplier
		Ω	$HL_j - HL_j^0$

Superscripts

0	standard value	*	modified parameter
U	referred to a liquid sidestream	0, 1, p	values of Newton steps
W	referred to a vapor sidestream	n	number of iterations

Subscripts

ca	calculated property	i	component
co	corrected property	l	upper tray of the central section of the column specification equation
j	tray	m	lower tray of the central section of the column
K	generic equation of the model for Newton–Raphson optimal step		
k	tray		

REFERENCES

1. Pierucci, S. J., Ranzi, E. M., and Biardi, G. E., "Possibility of Enlarging Standard Specifications in Distillation Columns," *I & EC Process Design & Development*, Vol. 21: 604 (1982).
2. Hougen, O. A., Watson, K. M., and Ragatz, R. A., *Chemical Process Principles*, 2nd ed., J. Wiley & Sons Inc., New York, London (1964).
3. Muller, D. E., "Math. Table Aids Comp." 10: 208 (1956).
4. Ching-Tsan Lo, "A Method of Temperature Estimation in the Bubble Point Method for Iterative Distillation Calculations," *AIChE J.* 21 (6): 1223 (1975).
5. Boston, J. F., and Sullivan, S. L., Jr., "An Improved Algorithm for Solving Mass Balance Equations in Multistage Separation Processes," *Can. J. of Chem. Eng.*, 50: 664 (1972).
6. PROCESS by Simulation Science Inc., Regent House, Heaton Lane, Stockport SK4 1BS England; PRISMA by Pyrotec N. V., 26 Bredewater, 2700 AB Zoetermeer, P. O. Box 86, The Netherlands; SIMULA by Technimont S.p.A., Via Monte Grappa, 2, 20100 Milano, Italia; GMB by Badger B. V., Prinses Beatrixlaan 9, The Hague, The Netherlands.
7. Lyster W. M., et al. "Figure Distillation This New Way.—Part 1—New Convergence Method Will Handle Many Cases," *Petroleum Refiner*, 38 (6): 221–230 (1959).
8. Holland, C. D., and Pendon, P. G., "Solve More Distillation Problems.—Part 1—Improvents give Exact Answer," *Hydrocarbon Processing*, 53: 7 (1974).
9. Billingsley, D. S., "On the Numerical Solution of Problems in Multicomponent Distillation of the Steady State II," *AIChE J.*, 16: 3 (1970).
10. Pierucci, S. J., et al., "T-method Computes Distillation," *Hydrocarbon Processing*, 179 (1981).
11. Pierucci, S. J., Ranzi, E. M., and Biardi, G. E., "Corrected Flowrates Estimation by Using θ Convergence Promoter for Distillation Columns," *AIChE J.*, 29 (1): 113 (1983).
12. Holland, C. D., *Multicomponent Distillation*, Prentice-Hall Inc., Englewood Cliffs, N. J. (1963).
13. Henley, E. J., "Equilibrium Stage Separation Operations in Chem. Engineering," Wiley, N. Y. (1981).

14. Fredenslund, A., Gurehling, J., and Rasmussen, P., "Vapor-Liquid Equilibria Usıng UNIFAC," Elsevier, Amsterdam (1977).
15. Soave, G., "Equilibrium Constants from a Modified Redlich–Kwong Equation of State," *Chem. Eng. Sci.*, 27: 1197 (1972).
16. Hayden, J. G., and O'Connel, J. P., "A Generalized Method for Predicting Second Virial Coefficients," *Ind. Eng. Chem. Process Des. Dev.*, 14: 3 (1975).
17. Reid, R. C., Prausnitz, J. M., and Sherwood, T. K., *The Properties of Gases and Liquids*, 3rd ed., McGraw Hill (1977).
18. Eckert, E., and Hlavaček, V., "Calculation of Multicomponent Distillation of Non-Ideal Mixtures by a Short-Cut Method," *Chemical Engineering Science*. 33: 77 (1978).
19. Chao, K. D., and Seader, J. D., "A General Correlation of Vapor-Liquid Equilibria in Hydrocarbon Mixtures," *AIChE J.*, 7 (4): 596 (1961).
20. Rose, A., Sweeny, R. F., and Schrodt, V. N., "Continuous Distillation Calculations by Relaxation Method," Industrial and Engineering Chemistry, 50, (5): 737 (1958).
21. Jelinek, J., Hlavaček, V., and Kubiček, M., "Calculation of Multistage Countercurrent Separation Processes. I. Multicomponent Multistage Rectification by Differentiation with Respect to an Actual Parameter," *Chem. Eng. Sci.*, 28: 1555–1563 (1973).
22. Jelinek, J., and Pierucci, S., private conversation.
23. SIMULA by Technimont S.p.A., Via Monte Grappa, 2, 20100 Milano.
24. Pierucci, S. J., and Ranzi, E. M., "T Method for Distillation Columns," *Chem. Eng. Commun.*, 14: 1–22 (1982).
25. Smith, B. D., and Brinkley, W. K., "General Short-Cut Equation for Equilibrium Stage Processes," *AIChE J.*, 6: 446 (1960).
26. Holland, C. D., and Eubank, P. T., "Solve More Distillation Problems. Part 2—Partial Molar Enthalpies Calculated," *Hydrocarbon Processing*, 53: 11 (1974).
27. Buzzi Ferraris, G., "A Powerful Improvement of the Global Newton–Raphson Method for Multistages Multicomponent Sepators," *Comput. Chem. Engng.* 7 (2): 73–85 (1983).
28. Naphtali, L., and Sandholm, D. S., "Multicomponent Separation Calculations by Linearization," *AIChE J.*, 17 (1): 148–153 (1971).
29. Donati, G., and Buzzi Ferraris, G., *Ing. Chem. Ital.*, 10: 157 1974).
30. Pierucci, S., et al., "Optimal Newton Step in Distillation Models Solved by Global Approach," in press—*Comput. Chem. Eng.*
31. Ishii, Y., and Otto, F. D., "A General Algorithm for Multistage Multicomponent Separation Calculations," *Can. J. Chem. Eng.*, 51: 601–606 (1973).

CHAPTER 14

PRINCIPLES OF MULTICOMPONENT DISTILLATION

Yuji Naka

Tokyo Institute of Technology
Yokohama, Japan

Takeshi Ishikawa

Tokyo Metropolitan University
Tokyo, Japan

CONTENTS

PRINCIPLES OF ORDINARY DISTILLATION

Before computers were available for distillation system design, engineers approximated multicomponent systems as binary systems. With the advent of large, high-speed computers in the 1960s, many researchers developed simulation programs of multicomponent distillation systems. Simulation programs generally include functions to precisely estimate physical properties. Unfortunately, computer programs are not the ultimate tool for solving the problems of system synthesis, design, and operation. It is necessary to have imagination on composition profiles, attainable regions of products, etc., under given conditions. In other words, we must fully understand the principles of distillation separation, the effects of vapor-liquid equilibrium and operating conditions on composition profiles.

Innovative design procedures for multicomponent distillation systems have been developed by computers, But, the principles of multicomponent distillation have not changed. In dealing with design and operation problems about multicomponent distillation, we pay particular attention to:

- Specification
- Vapor-liquid equilibrium
- Mass and heat balances
- Computational method

The specification for substances, the feed, products, equipment, etc., may be given in various forms according to the problem types, design, or operation. Specification of products and/or operating conditions is completely related to the degree of freedom in balance equations which represent the distillation system and physical properties such as vapor-liquid equilibria. First of all, prediction methods of vapor-liquid equilibrium relationships for the specified substances must be explained. Mass and heat balance equations for a distillation column are widely used in plate-to-plate-type forms. The structure analysis is performed based on graph theory in order to determine the computational strategies.

On the other hand, critical operations such as total reflux and minimum reflux provide important information about the attainable region of products and the propriety of specifications.

Vapor-Liquid Equilibria

Representation of Vapor-Liquid Equilibria

General Description. For gas and liquid as an idea mixture, the vapor-liquid relationship is given by Raoult's and Dalton's laws.

$$p_i = y_i P = x_i P_i^0(t) \tag{1}$$

where P, p_i, and P_i^0 are the operating pressure, and the partial pressure and the vapor pressure of the pure component i at temperature, t. For representation of the vapor pressure of pure component, Antoine's equation is most popular even though other expressions have been developed by Cox [1], Stull [2], and Prausnitz [3].

$$\log P_i^0 = A_i - B_i/(C_i + t) \tag{2}$$

where A_i, B_i, and C_i are constants for component i, (values are listed in References 4 and 5).

The K-value is defined as the ratio of vapor composition to liquid composition.

$$K_i = y_i/x_i \tag{3}$$

and the relative volatility, α_{ir} is defined by

$$\alpha_{ir} = K_i/K_r = P_i^0(t)/P_r^0(t) \tag{4}$$

Although the relative volatility is a function of temperature, the temperature dependency tends to decrease due to the cancelation of the temperature variations in the numerator and the denominator. In such a case the relative volatility can be assumed to be constant.

For liquid as a nonideal mixture, the activity coefficient of the liquid should be introduced to Equation 1. This assumption is acceptable for an ordinary distillation system because the operating pressure and the molar density are low.

$$y_iP = x_i\gamma_iP_i^0(t) \tag{5}$$

The activity coefficient of liquid is one of the factors that indicates the degree of deviation from an ideal mixture to a nonideal mixture.

The most general form of the vapor-liquid equilibria is

$$y_iP\varphi_i = x_i\gamma_iP_i^0(t) \tag{6}$$

where φ_i is the activity coefficient of vapor. The φ is the factor to indicate the degree of deviation from the ideal gas mixtures.

Activity coefficient of liquid. Many estimation methods for the activity coefficient have been developed: Margules equation [6] and the van Laar equation [7] can be applied only to a mixture with weak nonideality. Also, the Margules equation is hardly extended to a multicomponent system. Wohl [8] proposed an empirical equation including the Margules and van Laar equations. They have been widely accepted in spite of the appearance of the Wilson equation [9].

- *Wilson equation:* The Wilson equation is

$$\ln \gamma_i = 1 - \ln\left(\sum_j x_j\Lambda_{ij}\right) - \sum_k \left(x_k\Lambda_{ki} \Big/ \sum_j x_j\Lambda_{kj}\right) \tag{7}$$

$$\text{where} \quad \Lambda_{ij} = (v_j^L/v_i^L)\exp\left(-(\lambda_{ij}-\lambda_{ii})/RT\right) \tag{8}$$

$$= (v_i^L/v_j^L)\exp\left(-(\lambda_{ji}-\lambda_{jj})/RT\right) \tag{9}$$

The term of $(\lambda_{ij}-\lambda_{ii})$ is related to the interaction energies between the indicated pairs but is empirically determined. The v^L is the molar volume of liquid and T is the absolute temperature. Equation 7 includes only binary constants. In a real case where the operation temperature range is not wide, Λ_{ij} and Λ_{ji} are used as constant parameters. The parameters are collected by Gmehling et al. [10], Hirata et al. [11], Holmes et al. [12], and Nagata [13]. The Wilson equation has the disadvantage of not being applicable to a mixture with phase separation. Orye et al. [14] and Tsuboka et al. [15] modified the Wilson equation, and apply it to a highly nonideal mixture.

- *NRTL Equation:* Renon and Prausnitz [16] proposed the NRTL (non-random, two-liquid) equation developed for phase separation.

$$\ln \gamma_i = \sum_j \tau_{ij}G_{ji}x_j \Big/ \sum_k G_{ki}x_k + \sum_j \left[x_jG_{ij}\left(\tau_{ij} - \sum_k x_k\tau_{kj}G_{kj} \Big/ \sum_\ell G_{\ell j}x_\ell\right) \Big/ \sum_m G_{mk}x_m \right] \tag{10}$$

$$\text{where} \quad \tau_{ij} = (g_{ij}-g_{jj})/RT \quad (g_{ij} = g_{ji}, \tau_{ij} = \tau_{ji}) \tag{11}$$

$$G_{ij} = \exp(-\alpha_{ij}\tau_{ij}) \quad (\alpha_{ij} = \alpha_{ji}) \tag{12}$$

where g denotes a free-energy parameter related to the $i-j$ interaction, which is similar to the Wilson equation.

- *UNIFAC Equation:* Abrams and Prausnitz [17] derived a powerful equation to evaluate activity coefficient. This equation can include the characteristics of all the previously mentioned equations and is called the "universal quasichemical equation (UNIQUAC)." Furthermore, Fredenslund et al. [18] derived UNIQUAC based on the group model (ASOG by Derr et al. [19] and proposed "UNIQUAC functional-group activity coefficient (UNIFAC)." This method treats a molecule as several groups, that is, a molecule is decomposed into several groups. The UNIFAC is

$$\ln \gamma_i = \ln \gamma_i^C + \ln \gamma_i^R \tag{13}$$

The first term in Equation 13 is related to the differences in size and shape of the molecules interaction between groups. That is

$$\ln \gamma_i^C = \ln (\varphi_i/x_i) + (zq_i/2) \ln (\theta_i/\varphi_i) + l_i + (\varphi_i/x_i) \sum_j x_j l_j \tag{14}$$

$$l_i = (z/2)(r_i - q_i) - (r_i - 1) \quad (z = 10) \tag{15}$$

the θ is the volume fraction,

$$\theta_i = q_i x_i \Big/ \sum_i q_j x_j \tag{16}$$

and φ_i is the surface fraction

$$\varphi_i = r_i x_i \Big/ \sum_j r_j x_j \tag{17}$$

The parameters r_i and q_i are specified by component i, and are obtained for van der Waals group volumes and surface areas.

$$r_i = \sum_k \nu_k^{(i)} R \tag{18}$$

$$q_i = \sum_k \nu_k^{(i)} Q \tag{19}$$

The $\nu_k^{(i)}$ is the number of k-type group in molecule i. The Q_k and R_k are the constant parameters obtained from atomic and molecular structure data [20].

The second term in Equation 13 is

$$\ln \gamma_i^R = \sum_k \nu_k^{(i)} (\ln \Gamma_k - \ln \Gamma_k^{(i)}) \tag{20}$$

where Γ_k is the residual activity coefficient of k-type group in a solution:

$$\ln \Gamma_k = Q_k \left[1 - \ln \left(\sum_m \Xi \psi_{mk} \right) - \sum_m \left(\Xi \psi_{km} \Big/ \sum_n \Xi \psi_{nm} \right) \right] \tag{21}$$

and Ξ_m and X_m are the group surface fraction and the group interaction, respectively,

$$\Xi_m = Q_m X_m / \sum_n Q_n X_n \tag{22}$$

$$X_m = \sum_i x_i \nu_m^{(i)} / \sum_n \sum_i x_i \nu_n^{(i)} \tag{23}$$

and the ψ_{mn} is defined by

$$\psi_{mn} = \exp(-a_{mn}/T) \quad (a_{mn} \neq a_{nm}) \tag{24}$$

The a_{mn} is named the group-interaction parameter, which must be evaluated from phase equilibrium data. The $\Gamma_k^{(i)}$ is the residual activity coefficient of k-type group in a reference solution containing only molecules of type i. This value is evaluated by using Equation 21. The UNIFAC is widely accepted, and many researchers collect the group-interaction parameters [10].

If for the UNIFAC the $\Gamma_k^{(i)}$ is neglected and the decomposition of the molecule is not allowed, that is, $r_i = R_i$, and $q_i = Q_i$, the UNIQUAC coincides with the UNIFAC.

Bubble And Dew Point Calculations

Bubble point calculation. When the liquid composition of a mixture and its operating pressure are known, the bubble point calculation is carried out so as to satisfy the following condition:

$$1 = \sum_i y_i \tag{25}$$

The vapor composition y is represented in terms of the relative volatility as follows: Using Equations 1 and 25, the vapor composition can be reformed as

$$y_i = x_i P_i^0(t) / \left(\sum_j p_j\right) \quad \left(P = \sum_j p_j\right) \tag{26}$$

$$= x_i P_i^0(t) / \left[\sum_j x_j P_j^0(t)\right] \tag{27}$$

Dividing Equation 27 by the vapor pressure of the reference component, the vapor composition can be represented by

$$y_i = \alpha_{ir} x_i / \left(\sum_j \alpha_{jr} x_j\right) \tag{28}$$

The bubble point calculation based on Equations 1 and 3 is obtained by putting the vapor composition derived from Equation 1 into Equation 25;

$$1 = \sum_i x_i P_i^0(t)/P \tag{29}$$

Consequently, Equation 29 is a function of temperature only. In the case of introducing the activity coefficient, since it is a function of liquid composition as well as temperature

$$1 = \sum_i x_i \gamma_i P_i^0(t)/P \tag{30}$$

this equation is also a function of temperature only. It can be easily solved by using the one-dimensional convergence method.

Dew point calculation. The dew point calculation is executed under the given condition of vapor composition and operating pressure as follows: As the liquid composition is calculated from Equation 1

$$1 = \sum_i x_i = \sum_i y_i P / P_i^0(t) \tag{31}$$

this equation is a function of temperature only. Furthermore, Equation 31 is reformed by using the relative volatility.

$$x_i = (y_i/\alpha_{ir}) \Big/ \left(\sum_i y_{jr}/\alpha_{jr} \right) \tag{32}$$

If the relative volatility is constant, the vapor composition is directly obtained by Equation 32. But, for a nonideal liquid mixture, the activity coefficient which depends on the composition creates a convergence problem.

$$1 = \sum_i y_i P / \gamma_i(x) P_i^0(t) \tag{33}$$

One convergence method is the "direct substitution method:" Assume that the activity coefficient of each component is equivalent to unity and calculate each corresponding liquid composition. As the sum of the calculated liquid composition may not be equal to unity, each liquid composition should be normalized by

$$x_i = x_i(\text{cal}) \Big/ \sum_i x_j(\text{cal}) \tag{34}$$

Then, the activity coefficients are evaluated based on the normalized values. The computation is carried out continuously until the condition of $\sum_i x_i = 1$ is satisfied.

The direct substitution method is the simplest and its stable convergence may not cover the wide range of vapor composition with high nonideality. The Newton–Raphson method is applicable in general.

Normal Operation Procedure

Figure 1 shows a conventional distillation column, which has a condenser, N-2 plates, and a reboiler. The condenser and a reboiler are numbered as 1 and N, respectively. If the feed is introduced at the f-th plate, the column is divided into an upper part and a lower part: the former is called the "enriching section (or rectifying section)," and the latter is the "stripping section."

Reflux is defined as a part of the condensed stream which returns to the top of the column. The reflux ratio, is defined as

$$R = L/D \tag{35}$$

where L is the reflux and D is the top product flow rate. The minimum of the reflux ratio indicates one of the limitations of realizable distillation with given specifications. This means that the reflux ratio should be greater than the minimum reflux ratio. The other limitation of the reflux ratio is the "total reflux ratio," that is, the value of R is infinite. (The physical significance of minimum reflux and total reflux will be explained later.) The reflux ratio in a real column is determined between the minimum reflux and the total reflux in terms of economy, operability, and so on.

First of all, we consider a composition profile in a column with a normal reflux ratio.

Mass and Heat Balance

Mass Balance.

- *Conventional column:* A conventional distillation as shown in Figure 1 is equiped with a total condenser which completely condenses vapor from the top of the column, or a partial reboiler which vaporizes a part of liquid withdrawn from the column. It is assumed that each plate is an ideal plate, where the liquid and vapor compositions removed from some plates are in the equilibrium state and the heat capacities of liquid and vapor mixtures are constant. Denoting that x and y are the column vectors of liquid and vapor mole fractions, $\underset{\sim}{x} = (x_1, \ldots, x_{C-1})^T$ and $\underset{\sim}{y} = (y_1, \ldots, y_{C-1})^T$, respectively, mass balance equations are obtained around the condenser, each plate, and the reboiler as follows:

  For a total condenser,

$$D = V - L \tag{36}$$

$$D\underset{\sim}{x}_1 = V\underset{\sim}{y}_2 - L\underset{\sim}{x}_1 \tag{37}$$

For the j-th plate in the enriching section, $2 \leqq j \leqq f-1$,

$$D\underset{\sim}{x}_1 = V\underset{\sim}{y}_{j+1} - L\underset{\sim}{x}_j \tag{38}$$

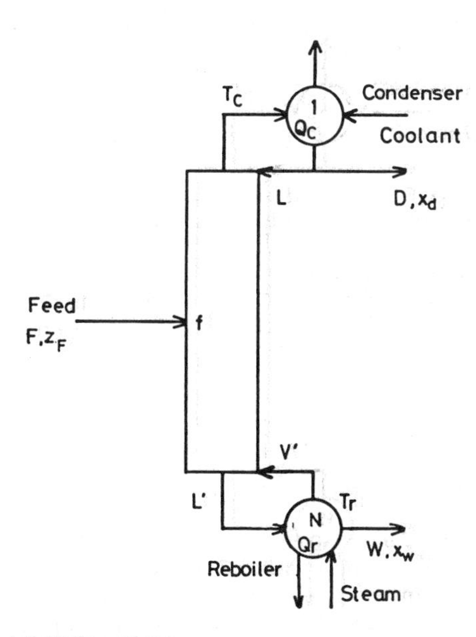

Figure 1. Conventional distillation column.

At the feed plate, denoting liquid and vapor flow rates in the stripping section, as $\bar{L}$ and $\bar{V}$, respectively,

$$L+qF = \bar{L} \tag{39}$$

$$V = \bar{V}+(1-q)F \tag{40}$$

$$D\underset{\sim}{x}_1+\bar{L}\underset{\sim}{x}_f = \bar{V}\underset{\sim}{y}_{f+1}+F\underset{\sim}{z}_F \tag{41}$$

where, $\underset{\sim}{z}_F$ is the mole fraction vector of the feed. The q is defined by

$$q = (H_F-H^L)/(H^V-H^L) \tag{42}$$

Where the H_F is the enthalpy of the feed and, H^V and H^L are the enthalpies of the dew point and the bubble point for the feed composition.

For the stripping section the mass balance equations are given around each plate and the reboiler:

For the k-th plate in the stripping section, $f+1 \leqq k \leqq N-1$

$$W = \bar{L}-\bar{V} \tag{43}$$

$$W\underset{\sim}{x}_N = \bar{L}\underset{\sim}{x}_{k-1}-\bar{V}\underset{\sim}{y}_k \tag{44}$$

For the reboiler

$$Wx_N = \bar{L}x_{N-1}-\bar{V}y_N \tag{45}$$

Pressure in the reboiler of a real column is different from that in a condenser because of the pressure drop at each plate. To simplify discussions, let us neglect the pressure drop. The feasible range of operating pressure, P, is generally determined in terms of the thermodynamic properties of substances, reaction, decomposition, crystallization, and the utility conditions. Finally, the optimal pressure is determined from the standpoint of economy. The pressure effects on the vapor-liquid equilibrium relationships as described earlier is controlled at the condenser.

When the liquid composition $\underset{\sim}{x}_j$ is obtained from a mass balance equation, the vapor composition $\underset{\sim}{y}_j$ is obtained by the bubble point calculation. On the other hand, when $\underset{\sim}{y}_j$ is given, the liquid composition $\underset{\sim}{x}_j$ is calculated by the dew point calculation. Consequently, the composition profile can be calculated successively.

- *Side-cut stream:* A product is sometimes taken out in the middle of a column as a side-cut vapor or liquid stream. In such a case when the liquid flow rate, S, is withdrawn from the j-th plate as shown in Figure 2, the mass balance equations are obtained as

$$L = L'-S \tag{46}$$

$$D\underset{\sim}{x}_1 = V\underset{\sim}{y}_{j+2}-L'\underset{\sim}{x}_{j+1}-S\underset{\sim}{x}_{j+1} \tag{47}$$

The degree of freedom for these equations is discussed in the next subsection.

- *Partial condenser:* A total condenser is generally used rather than a partial condenser as shown in Figure 3. But, for example, if the temperature of the distillate is close to or below the surrounding temperature, a partial condenser is sometimes used in order to reduce the quantity of low temperature coolant. The mass balance Equation 36 is confirmed, but Equation 37 should be substituted by

$$D\underset{\sim}{y}_1 = V\underset{\sim}{y}_2 - L\underset{\sim}{x}_1 \tag{48}$$

The compositions of $\underset{\sim}{x}_1$ and y_1 are in an equilibrium state. Moreover, there are some cases when the noncondensed vapor and a part of the liquid removed from a condenser are withdrawn as the top product.

Heat balance. To delete the assumption of the equimolal transfer between vapor and liquid streams on each plate, the heat balance equation should be added to the mass balance expression

$$Q_c + DH^L(\underset{\sim}{x}_1, T_1) + L_jH^L(\underset{\sim}{x}_j, T_j) = V_{j+1}H^V_{j+1}(\underset{\sim}{y}_{j+1}, T_{j+1}) \tag{49}$$

$$WH^L(\underset{\sim}{x}_N, T_N) + V_kH^V(\underset{\sim}{y}_k, T_k) = L_{k-1}H^L_{k-1}(\underset{\sim}{x}_{k-1}, T_{k-1}) + Q_r \tag{50}$$

Since the enthalpies of vapor and liquid streams depend on their compositions and temperature, solving mass and heat balance equations is much more difficult than solving mass balance equations alone. Where, Q_c and Q_r denote heat loads at a condenser and a reboiler, respectively, they are determined as follows: For a total condenser, when the temperature of a top product, T', is below or equal to the dew point of $\underset{\sim}{y}_2$, the heat duty is obtained from the heat balance equation

$$Q_c = V[H^V(\underset{\sim}{y}_2, T_2) - H^L(\underset{\sim}{x}_1, T')] \tag{51}$$

For a partial condenser, similarly, the heat duty is determined as

$$Q_c = VH^V(\underset{\sim}{y}_2, T_2) - DH^V(y_1, T_1) - LH^L(x_1, T_1) \tag{52}$$

Feed Plate Location

Gilliland [21] proposed a determination method of a feed plate location, which is developed based on the concept of a binary system. The feed plate location is required as the place where the key component ratio in a column is close to that in the feed as follows:

For a liquid feed, the prior-mentioned relationship is represented by

$$(x_{LK}/x_{HK})_f \leqq (x_{LK}/x_{HK})_{z_F} \leqq (x_{LK}/x_{HK})_{f-1} \tag{53}$$

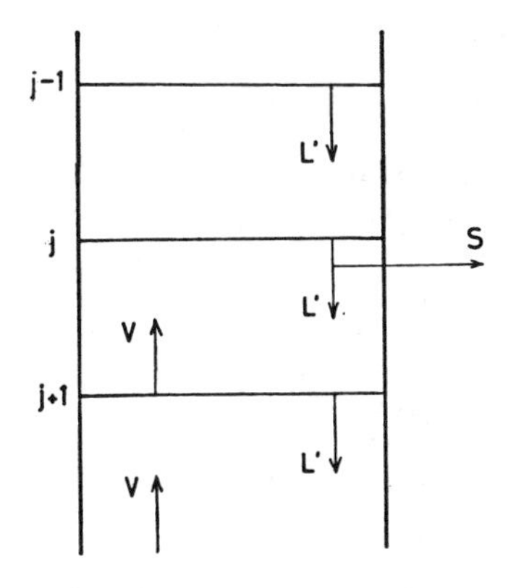

Figure 2. Side-cut stream at the j-th plate.

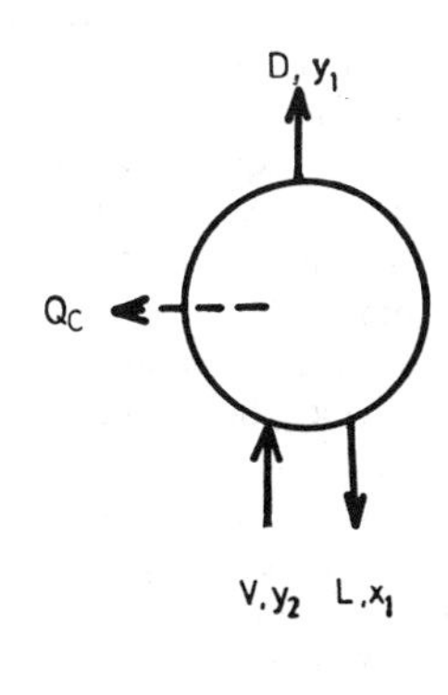

Figure 3. Partial condenser.

For a vapor-liquid feed, using the key component ratio of the feed evaluated for z_F

$$(x_{LK}/x_{HK})_f \leqq (x_{LK}/x_{HK})_{z_F^L} \leqq (x_{LK}/x_{HK})_{f-1} \tag{54}$$

for a vapor feed, using the key component ratio evaluated for z_F^V.

$$(x_{LK}/x_{HK})_{f-1} \leqq (x_{LK}/x_{HK})_{z_F^V} \leqq (x_{LK}/x_{HK})_{f-2} \tag{55}$$

This relationship is useful under the condition that the relative volatility between key components is significantly closer to unity than either the relative volatility between the light key and the next lighter component or the relative volatility between the heavy key and the next heavier component [22]. But, this method is very simple and gives very important information in designing a distillation column in the rough.

Graphical Representation

An example of the relationships of the mass balance equation and vapor-liquid equilibrium helps to scope the feasible composition profiles in the composition space. Consider the vapor-liquid equilibrium relationships in a ternary system in the composition space. As shown in Figure 4, each arrow denotes the bubble point relationship, referred to as the "vapor-liquid tie line." The bottom of an arrow and the arrowhead indicate the liquid composition and the vapor composition, respectively. A reversed arrow denotes the dew point relationship, and is termed the "reversed vapor-liquid tie line." It is apparent that the direction of the arrows follows decreasing temperature gradients. The stream of arrows seems to be the liquid stream. The arrow stream relates to composition profiles in columns. The top products may be located in an area where the arrowheads gather, but the bottom products may be located in an opposite area according to the total mass. The direction can obviously control composition profiles.

The total mass balance is derived from Equations 36 through 45.

$$F = D + W \tag{56}$$

$$F\underset{\sim}{z}_f = D\underset{\sim}{x}_1 + W\underset{\sim}{x}_N \tag{57}$$

The feed, the top, and the bottoms are on a straight line as shown in Figure 5; called the "mass balance line." The feed is located at the point which divides the distance between x_1 and x_N, $[x_1x_N]$, into $[x_1z_F]$ and $[z_Fx_N]$ by the interior division ratio, W/D. In other words, the point z_F is at the gravity of $[x_1x_N]$.

For each plate in the enriching section according to Equation 38, the distance ratio of $[x_1y_{j+1}]$ to $[y_{j+1}x_j]$ is equal to L/D. Also, the flow ratio of L to V equals the ratio of $[x_1y_{j+1}]$ to $[x_1x_j]$. The point of y_{j+1} is at the interior division point of distance $[x_1x_j]$ and the arrows are the vapor-liquid tie line. The liquid composition, x_j, can be obtained by the dew point calculation from y_j. The β is the proportionality coefficient. Similarly, Equation 43 in the stripping section can be mapped on the composition plane. The point x_{k-1} is the gravity of $[x_Ny_k]$ and divides the distance $[x_Ny_k]$ into $[x_Nx_{k-1}]$ and $[x_{k-1}y_k]$ by the ratio, $\bar{V}/W$. The vapor composition y_{k-1} is obtained by the bubble point calculation for x_{k-1}. The ratio of $\bar{L}$ to $\bar{V}$ is equivalent to the distance ratio of $[x_Ny_k]$ to $[x_Nx_{k-1}]$. Mass balance Equations 39 through 41 around the feed plate are illustrated in Figure 5. The intersection of these lines, $[x_1x_{f+1}]$ and $[z_Fy_{f+1}]$, splits the distance, $[x_1x_{f+1}]$ by the ratio of L to D, and the distance, $[z_Fy_{f+1}]$ by the ratio of $\bar{V}$ to F.

When the ratio, L/V, becomes close to unity, the composition profile is swelling and is finally represented as the successive arrows, the number of which corresponds to the number of plates. The final reflux ratio is known as the total reflux ratio.

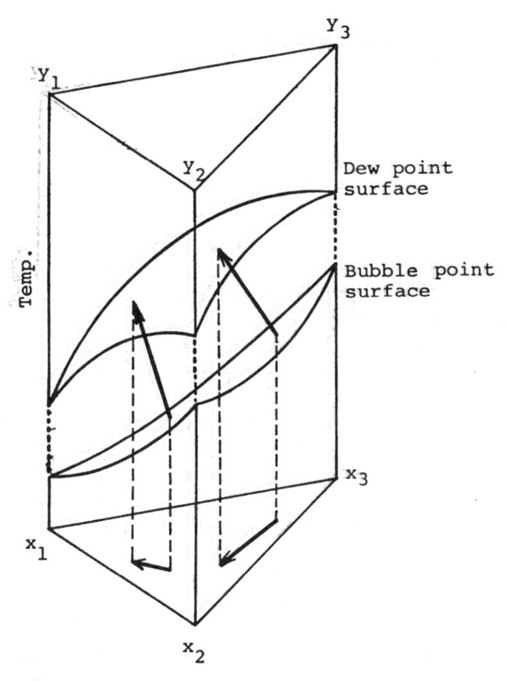

Figure 4. Vapor-liquid tie line.

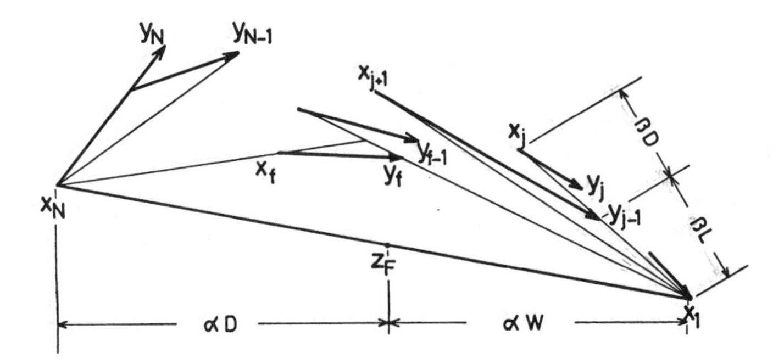

Figure 5. Relationships between composition and streams in a distillation column.

When an arrow exists on the mass balance line defined by Equations 38 or 44, distillation separation does not succeed at the plate. This state indicates the pinch condition, and its reflux ratio is called the "minimum reflux ratio."

Structure Analysis

The structure analysis of a set of mass balance equations related to a conventional column is given by Kwauk [23] and Hanson [24]. They analyzed the structure for a conventional

Table 1
Normalization Matrix for Streams

Equation Number	F	q	D	L	V	L	V	W
36			1	1	*1			
39	1	1		1		*1		
40	1	1		1			*1	
43						1	1	*1

column counting the numbers of equations and independent variables. Recently, graph theory has been used for determining the computational strategy and investigating the over- or under-specification given in process simulations [25, 26, 27]. As we know, the equations used for determination of hard design variables and for determination of the operating conditions are indentical. The difference between both situations is based on constraints related to given problems only. It is significant to differentiate clearly by using the graph theory.

We shall illustrate the use of graph theory by evaluating flow rates. The equations related to the flow rates are Equations 36, 39, 40 and 43. Table 1 shows the normalization matrix, this indicates the relationships between given equations and the variables. If a variable has a unique solution, the corresponding element in the normalization matrix is set as unity. For example, since the value of D can be uniquely determined from Equation 36, the related element is marked with unity. After this investigation is continued for all the equations and variables, one element with unity is chosen from each row and each column, because a variable solved from some equations should be independent of the others. This operation fixes the solving relation between one equation and one variable to be solved. For example, we choose elements identified by an asterisk in the normalization matrix. Figure 6 illustrates the identified relationships between the equations and independent variables. Consequently, if the values of the feed flow rate, F, and vapor-liquid ratio, q, are given, two degrees of freedom are left. If the top product flow rate, D and the reflux flow rate, L, are given, all the other flow rate variables can be calculated as following the arcs in Figure 6. There are several possibilities of choosing a feasible set of two independent variables. For example, W and V can be employed instead of D and L.

Meanwhile, Figures 7A and B show the general representation of the bubble point and dew point calculations, respectively. The black circles indicate the initial fixed values. For the bubble point calculation, following the direction of the arcs shown in Figure 7A, all of the vapor mole fractions are evaluated by using pure component pressure at assumed temperature. The sum of the vapor mole fractions should be equal to unity and this condition is indicated by the double circles. If the condition is not satisfied, the value of temperature

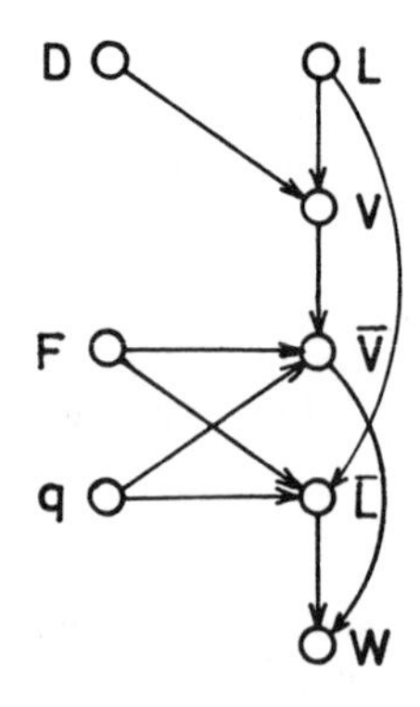

Figure 6. Graph of relationships among flow rates.

should be assumed again and the calculation is repeated. On the other hand, the dew point calculation including the activity coefficients of liquid is complicated as shown in Figure 7B. Assuming the values of the liquid mole fractions and temperature, the vapor mole fractions are calculated. If the calculated values of the vapor mole fractions do not coincide with the values initially assumed, the assumed values should be changed.

Let us consider the graph for all of the mass balance (Equations 36 through 44). Figure 8 gives one example of the graphical representations of solving equations relating to the mass balance and vapor-liquid equilibrium mentioned earlier. The graph corresponds to that of Lewis–Matheson [28]. Here, mole fraction is used instead of mole flow rate for each component. Although operating pressure, the feed plate location, and total number of plates are fixed, the graph for a conventional distillation column alters, subject to the conditions of initial variables as well as the variables to be solved. When the feed conditions, F, q, and $\underline{z}_F$ are given, two degrees of freedom are left. So, we can specify two independent variables. For the example shown in Figure 8, if the feed conditions are known, the two values can be chosen from variables D, L, and $\underline{\underline{x}}_N$.

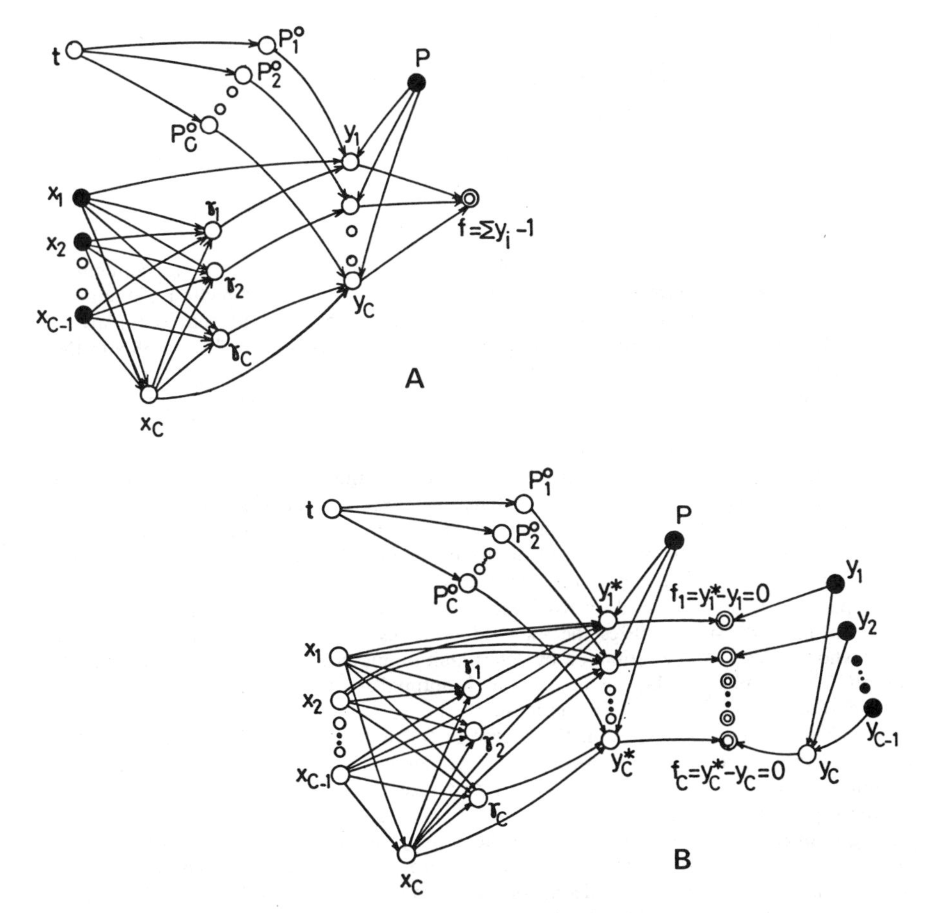

Figure 7. (A) Graph of bubble point calculation; (B) graph of dewpoint calculation.

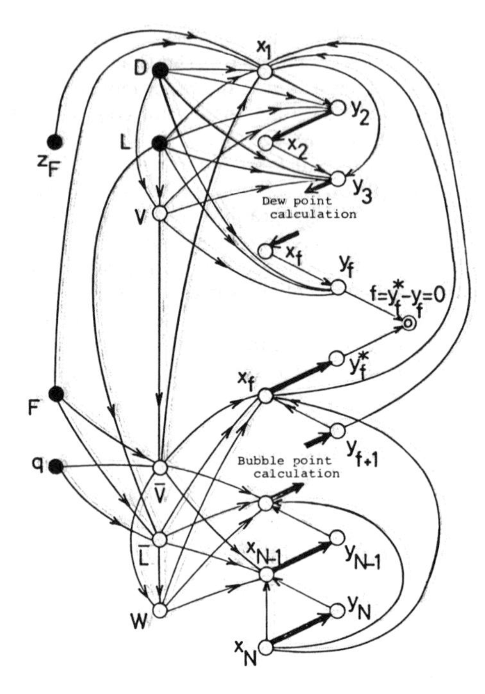

Figure 8. Graph of conventional distillation for simulation.

In the case of simulation problems, two flow rates, for instance, D and L, are sometimes chosen. As shown in Figure 8, the calculation procedure follows the direction of the arcs: First, assuming a liquid composition, $\underset{\sim}{x}_N$, the vapor composition at the feed plate is calculated from the bottom up in the stripping section (using Equations 37 and 38 and the vapor-liquid equilibrium relationship), while $\underset{\sim}{y}_f$ is computed from the top, down ward in the enriching section. Both values of $\underset{\sim}{y}_f$ should be the same. If they are not equal, the assumed value of $\underset{\sim}{x}_N$ is changed. Of course, it is possible to take one of the other plates instead of the feed plate for composition matching.

For design, two independent compositions in the top and/or the bottoms are specified according to two degrees of freedom, for instance, for key component fractions the recovery ratios, Dx_i/Fz_{Fi} of the two independent components in products and so on. Assuming suitable values of the feed plate, f, and the total number of plates, N, all other variables can be calculated by using one of the convergence methods such as the simplex method.

The prior discussion is based on fixed values of f and N. Both values must be given so that a distillation system realizes the specified products. There are many feasible sets of values whose ranges can be estimated by determination methods for the feed plate location and minimum number of plates. One can search optimal solutions in terms of economy by changing the values of f and N in the feasible sets.

When there is a side-cut stream, the degree of freedom increases by one assuming the side-cut plate known. As the composition of the side-cut stream is equal to $\underset{\sim}{x}_j$ according to Figure 2, even if the value of the side-cut flow rate, S, is given, the liquid flowrate, L', and the vapor composition, $\underset{\sim}{y}_{j+2}$, are obtained from Equations 46 and 47. In other words, this degree of freedom can be used as a specification of the side-cut product in a theoretical sense. But, as the attainable region of the side-cut product is hardly estimated, it is difficult to specify a composition of the side-cut product. Consequently, this degree of freedom is used as a specification of a side-cut flow rate in many cases. In addition to the variable,

S, as the side-cut plate, N, is used as a variable, the desired solution can be searched by changing the values of S and N.

Specification of Products

Specification

As there are only two degrees of freedom left for a conventional column in the case of design these degrees of freedom are used, for example, for specifying two compositions. In general, two key components are chosen as the specified variables: "light key component" and "heavy key component." But, in a real case, we fall across more than two specifications on compositions, flow rates, recovery ratio, etc. For example, when the following specifications are noted: for main components in the top or bottoms

$$x_{1i} \geqq \sigma_{1i}, \quad x_{Nk} \geqq \sigma_{Nk} \tag{58}$$

and for impure components

$$x_{1j} \leqq \sigma_{1j}, \quad x_{N\ell} \leqq \sigma_{N\ell}. \tag{59}$$

from the restriction of the degrees of freedom, the equality in all specifications cannot be satisfied simultaneously. It seems that the conditions given by Equations 58 and 59 are over-specified at a glance. But if the restricted composition regions of the top and the bottoms exist in separable regions, both products can be realized. In using the computational method described earlier, two equality specifications are usually put into mass balance equations. In this case, we must distinguish the two dominant specifications from all of the specifications. This aspect is illustrated in Figure 9. For the mass balance line (1), the dominant specifications are x_{1A} for the top and x_{NA} for the bottoms, respectively. On the other hand, for line (2), they are x_{1C} for the top and x_{NC} for the other.

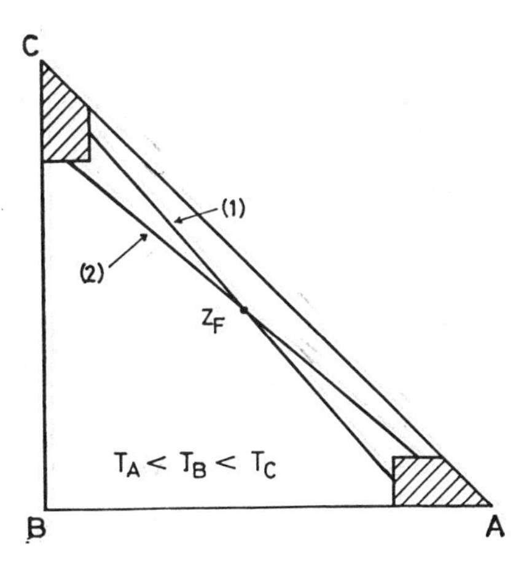

Figure 9. Relationship between specified composition areas of the top and the bottoms.

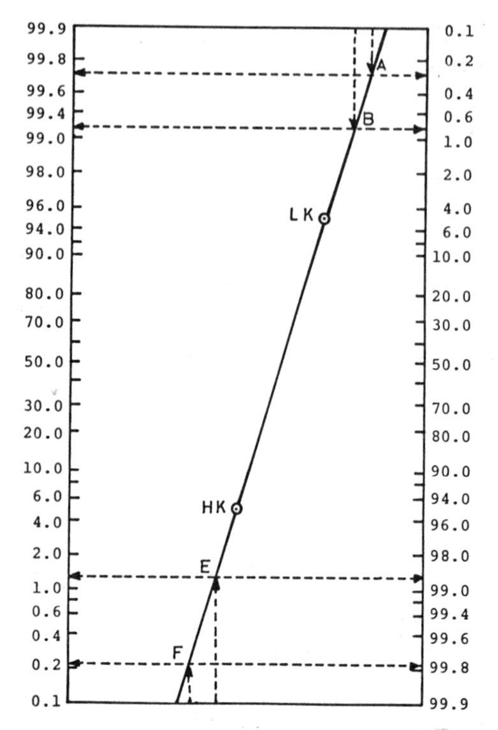

Figure 10. Estimation of recovery of nonkey components (A, B, E, F) (Excerpted by special permission from *Chemical Engineering* (Jan. 29, 1979, pp. 101–104). Copyright (c) (1979), by McGraw-Hill, Inc., New York, N.Y. 10020.)

Propriety of Specification by Short-Cut Methods

The calculation procedure described earlier is not readily applied to a system with a large number of components. Short-cut methods make it easy to investigate the property of given specifications, because of their quick evaluation of the top and the bottom products.

One short-cut approach is the *graphical method.* The distribution of components in products is given by the Hengstebeck–Geddes equation [29, 30]

$$\log (D_i/W_i) = a + b \log \alpha_i \tag{60}$$

where D_i and W_i are the mole flow rates of component i in the top and the bottoms. The a and b are correlation constants. Yaws et al. [31, 32] modified Equation 60 in terms of recovery and proposed a short-cut graphical method.

$$\log (Y_{li}/(100 - Y_{li})) = a + b \log \alpha_i \tag{61}$$

$$\log ((100 - Y_{Ni})/Y_{Ni}) = a + b \log \alpha_i \tag{62}$$

Where, Y_{li} and Y_{Ni} are the percent recovery of component i in the top and in the bottoms. This equation gives a straight line as a plot of $\log [Y_{li}/(100 - Y_{li})]$ vs. $\log \alpha_i$ (shown in

Figure 10). We first plot the relative volatility and the percent desired recovery for the key components and draw a straight line through two points. The recoveries of the other components are read as the values on the straight line corresponding to their relative volatilities.

Another approach is the *Smith–Brinkley method,* Smith and Brinkley [33] presented a short-cut method to simulate distillation, absorption, and extraction processes, based on finite-difference equations. After this, Fleisher et al. [34] extended this method to complex columns with side streams and intermediate heat exchange. When constant vapor and liquid flow rates are fixed and the numbers of plates are given as $f-1$ in the enriching section and $N-f+2$ in the stripping section, the fraction equation of component i recovered in the bottoms is (each plate is numbered from the top down. A condenser and a reboiler have the numbers, 1 and N, respectively.):

$$f_i = (a+bR)/\{a+bR+P_i(b/d)S_{li}^{f-1}c\} \tag{63}$$

where
$$a = (1-S_{li}^{f-1})$$
$$b = (1-S_{li})$$
$$c = (1-S_{Ni}^{N-f+1})$$
$$d = (1-S_{Ni})$$

The stripping factor S is defined by

$$S_{li} = K_iV/L \quad \text{in the enriching section} \tag{64}$$

and

$$S_{Ni} = \bar{K}_i\bar{V}/\bar{L} \quad \text{in the stripping section} \tag{65}$$

The reflux ratio, R is

$$R = R/K_{Ni} \quad \text{for a partial condenser} \tag{66}$$

$$R = R \quad \text{for a total condenser.} \tag{67}$$

The P_i is evaluated by

$$P_i = (K_{fi}/K_i)(L/\bar{L}) \quad \text{for liquid feed} \tag{68}$$

or

$$P_i = L/\bar{L} \quad \text{for vapor feed} \tag{69}$$

The K_i and $\bar{K}_i$ are the K-values in the enriching and the stripping sections, respectively, and functions of temperature only and are evaluated by

$$K_i = K_i[(T_1+T_f)/2] \tag{70}$$

and

$$\bar{K}_i = \bar{K}_i[(T_f+T_N)/2] \tag{71}$$

where T_1 and T_N are the dew point temperature of the top and the bubble point temperature of the bottoms. The T_f is the only one tuning parameter which adjusts the balance of the whole column. The calculation procedure is:

1. Assume temperatures of the top and the bottoms and T_f equal to the feed temperature. Calculate the separation factor f_i.
2. Repeat Step 1 with the slightly higher temperature than the feed temperature.
3. Calculate either

$$x_{Ni} = f_i F z_{Fi}/W \qquad \text{if } (W < D) \tag{72}$$

or

$$x_{1i} = f_i(F z_{Fi} - W x_{Ni})/D \qquad \text{if } (W > D) \tag{73}$$

If the following criterion is satisfied

$$\varepsilon = 1 - \sum_i x_{Ni} \qquad \text{if } W < D \tag{74}$$

or

$$\varepsilon = 1 - \sum_i x_{1i} \qquad \text{if } W > D \tag{75}$$

go to Step 5. If not, the next step is executed.
4. Assume the value of T_f by

$$T_f^{(n+1)} = T_f^{(n)} - \varepsilon^{(n)}(T_f^{(n-1)} - T_f^{(n)})/(\varepsilon^{(n-1)} - \varepsilon^{(n)}) \tag{76}$$

and return to Step 2.
5. Calculate the dew point T_1 and the bubble point T_N based on the compositions $\underset{\sim}{x}$ and $\underset{\sim}{x}_N$. If these new temperatures are equal to the assumed values, the calculation stops. If not, go to Step 2.

The calculation procedure can be extended to a complex column. In such a case, the column is divided into sections by external interacting points, for example, side-cut streams and intermediate heat exchangers. The temperature to be used as the matching parameter is put at every interacting point. Since this procedure is a calculation method based on an operation-type concept, the attainable region of the products can be obtained by changing the values of N, f, V, and L.

This procedure is available for distillation systems with vapor-liquid equilibrium depending upon temperature only. If the vapor-liquid eqilibrium has the dependency of composition, there are few acceptable short-cut methods for calculation of the products.

Critical Operation

There are two critical operations: total reflux and minimum reflux. Total reflux condition determines the minimum number of plates and under the fixed number of plates realizes the maximum distance between x_1 and x_N, the maximum separation. The minimum reflux determines the minimum heat loads removed at a condenser and supplied to a reboiler.

Total Reflux

Representation of reflux. Total reflux is constituted under the conditions of D, F, W, and the infinite value of R. Resulting from the conditions, Equations 36 and 37 are reformed to

$$V = L \tag{77}$$

$$\underset{\sim}{y}_2 = \underset{\sim}{x}_1. \tag{78}$$

Similarly, Equation 38 becomes

$$\underset{\sim}{y}_{j+1} = \underset{\sim}{x}_j \tag{79}$$

As $L/D \rightarrow \infty$ and $L/F \leftarrow 0$, the following equation in the stripping section becomes as the same as Equation 79.

$$\underset{\sim}{y}_k = \underset{\sim}{x}_{k-1} \tag{80}$$

But, the feed composition $\underset{\sim}{z}_F$, the top $\underset{\sim}{x}_1$ and the bottoms $\underset{\sim}{x}_N$ are still on the mass balance line. Assuming the bottom composition $\underset{\sim}{x}_N$, the composition on each plate is calculated from the vapor-liquid equilibrium relationship as shown in Figure 11. The calculated value of $\underset{\sim}{x}_1$ must be consistent with the value of $\underset{\sim}{x}_1$ obtained from the total mass balance equation for convergence.

On the other hand, there is another definition from the situation that after the initial feed is supplied to a column with the fixed number of plates all the input and the output flow rates are cut. In this case, the sum of all the holdup in a condenser, accumuation, plates, and a reboiler must equal the initial feed.

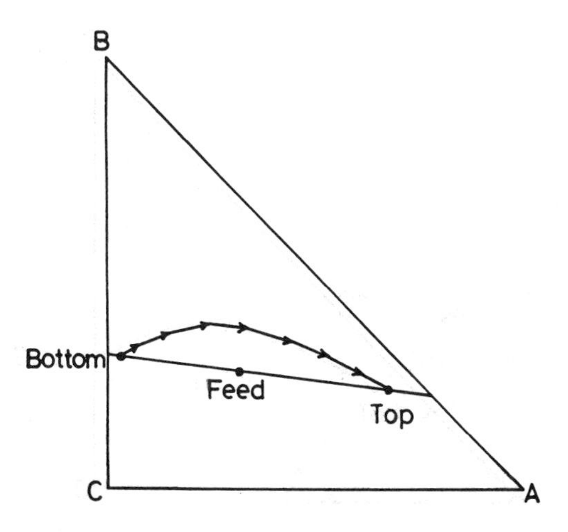

Figure 11. Composition profile under total reflux ($T_A < _B < T_c$).

Separable regions. These regions of the top and the bottom products are obtained by changing the ratio D/F under the total reflux condition. According to total mass balance, as the value of D/F becomes zero, the bottom composition is infinitely approaching the feed composition. The composition profile at the critical state can be obtained from Equation 78, that is, it is calculated from the feed composition by successively combining the vapor-liquid tie lines together. In reverse, as the value of D/F approaches unity, the top composition approaches the feed composition. The composition profile at the critical state can be obtained by successively connecting the reversed vapor-liquid lines. In addition, as the total balance equation can be represented as a straight line, the separable regions of the top and the bottom products for ternary systems are shown in Figure 12.

Minimum number of plates.

- *Fenske equation:* A column with the minimum number of plates which satisfies the specified separation operates at total reflux condition. For a multicomponent mixture in which the Raoult's law is confirmed the Fenske Equation 35 can be used to the minimum number of theoretical plates as follows:

  For the minimum number of plates in the enriching section

$$N_e^{min} = \log [(x_{LK}/x_{HK})_1(x_{HK}/x_{LK})_f]/\log (\alpha_{LK/HK})_e \tag{81}$$

  For the minimum number of plates in the stripping section

$$N_e^{min} = \log [(x_{LK}/x_{HK})_f(x_{HK}/x_{LK})_N]/\log (\alpha_{LK/HK})_s \tag{82}$$

  where, a partial condenser and a reboiler are recognized as one equilibrium stage, respectively. The relative volatility between the light and heavy key components are evaluated as follows:

  The geometric average as common relative volatility

$$(\alpha_{LK/HK})_{av} = (\alpha_{LK/HK})_e = (\alpha_{LK/HK}) \tag{83}$$

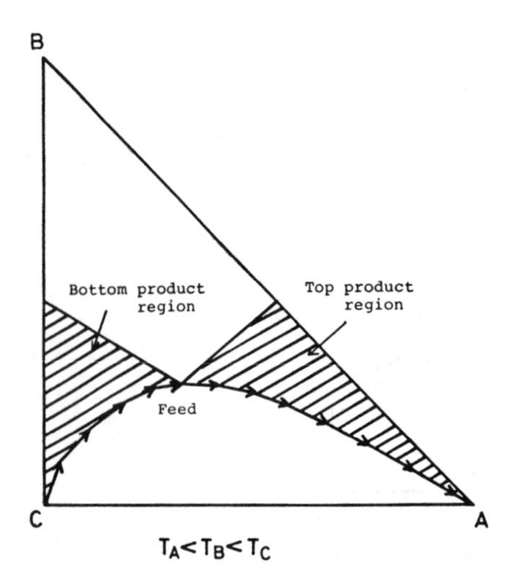

Figure 12. Separable region.

$$= [(\alpha_{LK/HK})_1(\alpha_{LK/HK})_N]^{1/2} \quad (84)$$

or

$$= [(\alpha_{LK/HK})_1(\alpha_{LK/HK})_f(\alpha_{LK/HK})_N]^{1/3} \quad (85)$$

In this case, the minimum number of plates in the whole column is derived from Equations 81 and 82

$$N^{min} = N_e^{min} + N_s^{min}$$

$$= \log [(x_{LK}/x_{HK})_1(x_{HK}/x_{LK})_N]/\log (\alpha_{LK/HK})_{av} \quad (86)$$

- *Winn equation:* Winn [36] proposed a method for calculating the minimum number of plates which introduces the relative volatility difference between the enriching section and the stripping section. This equation is very similar to the Fenske equation.

Minimum Reflux Ratio

Representation of minimum reflux ratio. The minimum reflux ratio is defined as the minimum value of reflux ratio which can realize the specified products theoretically. Obviously, the specification of products must be based on two degrees of freedom. It is very difficult to determine the minimum reflux ratio directly by means of design-type approaches because of the following peculiar characteristics for the composition on profile at the minimum reflux: The two zones of constant composition are located within the enriching section and within the stripping section. An infinite number of plates are consumed in both zones. These zones are referred to as the "pinch point." Moreover, very volatile components and very heavy components may disappear in the zones of constant composition. These components are called nondistributed components. The locations of both zones depend strongly upon the vapor-liquid equilibrium relationships and product specification. The relationships between the location of zones of constant components are shown in Figure 13 [37, 38, 39].

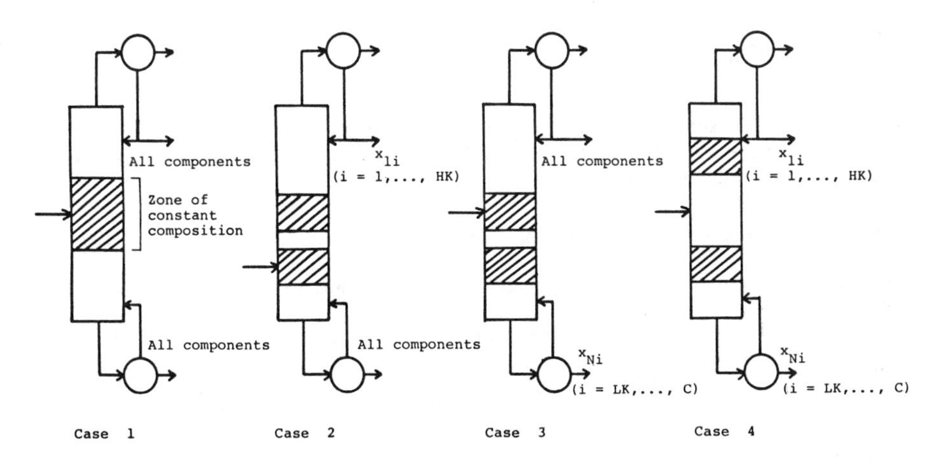

Figure 13. Location of pinch-point zones.

Case 1: The top and the bottoms consist of all components in the feed. Both zones of constant components exist adjacent to the feed plate. In this case course separation is performed.

Case 2: The top contains no component heavier than the heavy key, but the bottoms contains all components. The zone of constant components in the enriching section is apart from the feed plate but the other zone is just below the feed plate.

Case 3. This case is in complete contrast to case 2, that is, the bottoms contains no component more volatile than the light key component and the top consists of all components.

Case 4. The components heavier than the heavy key component and the components more volatile than the light key disappear in the top and in the bottoms, respectively. In this case, both zones of constant component are located in the middle of the enriching section and in the middle of the stripping section.

Calculation. Colburn [40] proposed a method to determine the minimum reflux, assuming a multicomponent mixture as a binary system which is composed of the key components. The value of the minimum reflux obtained was corrected in terms of the influence of the components more volatile than the light key component in the zone of constant composition in the stripping section and of the components heavier than the heavy key component in the zone of constant composition in the enriching section.

- *Underwood equation:* Underwood [41, 42, 43] proposed an equation for the minimum reflux as follows: All components in the feed are numbered in order of decreasing volatility, $\alpha_1 > \alpha_2 > \ldots > \alpha_C$.

$$1 - q = \sum_i \alpha_i z_{Fi}/(\alpha_i - \varphi) \tag{87}$$

When the key components are adjacent, the root exists between the relative volatilities of them, that is, (HK = LK + 1)

$$\alpha_{LK} > \varphi > \alpha_{HK} \tag{88}$$

The minimum reflux ratio can be obtained by putting this root into the following equation.

$$R^{min} + 1 = \sum_i \alpha_i x_{1i}/(\alpha_i - \varphi) \tag{89}$$

Note that there are some distributed components between both key components, and the number of the roots is obtained by (HK − LK). The relationship of relative volatilities and the roots are denoted by

$$\alpha_{LK} > \varphi_1 > \alpha_{LK+1} > \varphi_2 > \ldots > \alpha_{HK-1} > \varphi_{HK-LK} > \alpha_{HK} \tag{90}$$

The mole fractions of distributed components in the top are not unknown, but the relation $\sum_{i=1}^{HK} x_{1i} = 1$ should be satisfied. Let us consider the case when there are two distributed components. The minimum reflux is determined by solving the following simultaneous equations:

$$R^{min} + 1 = \sum_{i \neq s,t} \alpha_i x_{1i}/(\alpha_i - \varphi_1) + \alpha_s x_{1s}/(\alpha_s - \varphi_1) + \alpha_t x_{1t}/(\alpha_t - \varphi_1) \tag{91}$$

$$R^{min}+1 = \sum_{i \neq s,t} \alpha_i x_{1i}/(\alpha_i - \varphi_2) + \alpha_s x_{1s}/(\alpha_s - \varphi_2) + \alpha_t x_{1t}/(\alpha_t - \varphi_2) \quad (92)$$

$$R^{min}+1 = \sum_{i \neq s,t} \alpha_i x_{1i}/(\alpha_i - \varphi_3) + \alpha_s x_{1s}/(\alpha_s - \varphi_3) + \alpha_t x_{1t}/(\alpha_t - \varphi_3) \quad (93)$$

Then, it is assumed that the top product consists of at least the light key component and the more volatile components. The light key component exists in the top according to

$$x_{1LK} / \sum_{j < LK} x_{1j} = z_{FLK} / \sum_{j < LK} z_{Fj} \quad (94)$$

The term related to the heavy key component is omitted from Equations 92 and 93, as it is negligible. If the quantity of the heavy key component in the top is appreciable, an appropriate term should be added to Equations 91 through 93.

The short-cut methods are useful in avoiding tedious calculations. However, if a given mixture displays nonideality, constant relative volatilites are not acceptable and the composition of the top and the bottoms may not be estimated easily. For this reason, minimum reflux problems are solved as an operating problem of a distillation column with an infinite number of plates in the enriching and stripping sections [44, 45]. The minimum reflux is defined at the condition of two specified components on the products withdrawn from a column with an infinite number of plates. This calculation is iteratively executed with changing reflux ratio until the calculated products are recognized in the specified composition regions, respectively. New calculation methods based on the computational procedure for conventional columns are extended to complex columns: Barnes et al. [46] developed a calculation method for distillation columns with multiple feeds. McDonough et al. [47] and Sugie et al. [48] proposed calculation methods of minimum reflux for distillation columns with side-cut streams.

Stage–Reflux Correlation

The critical operations of total reflux and minimum reflux offer two limiting conditions in an economical sense. The minimum reflux conditions indicate the minimum requirement of heat loads and allows the minimum size of a condenser and a reboiler under the infinite number of plates. On the other hand, total reflux determines the minimum number of plates under infinite heat loads. The optimal design of a distillation system is determined from the standpoint of economics and lies between both critical states. The following methods are convenient for quick evaluation of a column.

Gilliland Correlation. Gilliland [49] correlated the Fenske–Underwood equation with the Gilliland minimum reflux equation, empirically. The correlation is represented in terms of $(N - N_{min})/(N+1)$ and $(R - R_{min})/(R+1)$ and is shown in Figure 14. The N and R denote the number of plates and reflux ratio in a real system, and N_{min} and R_{min} are the minimum number of plates and the minimum reflux ratio.

Erbar–Maddox correlation. This correlation [50] is derived from the Underwood method for minimum reflux calculations and the Winn method [51] for the minimum of plates. The correlating lines shown in Figure 15 are calculated by using the Lewis–Menthon method

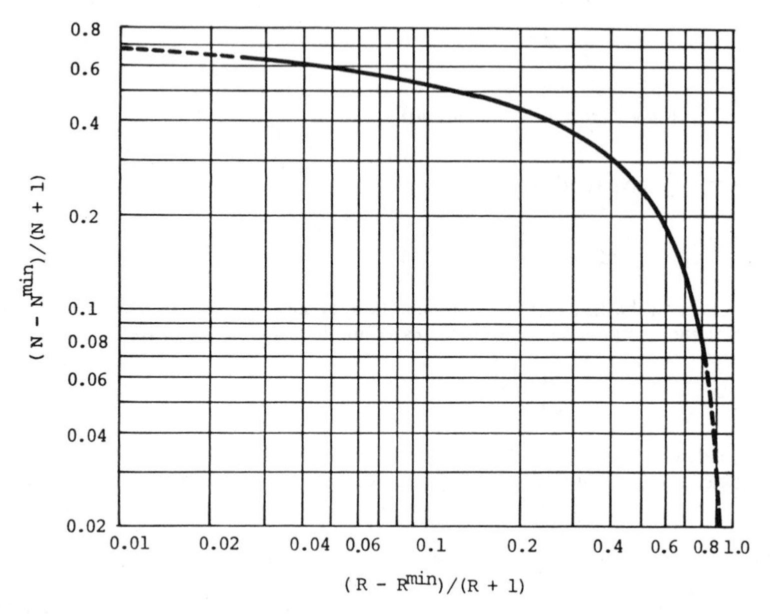

Figure 14. Correlation curve between reflux ratio, R, and the number of plates, N. (Printed with permission from Ref 49. Copyright 1940 American Chemical Society.)

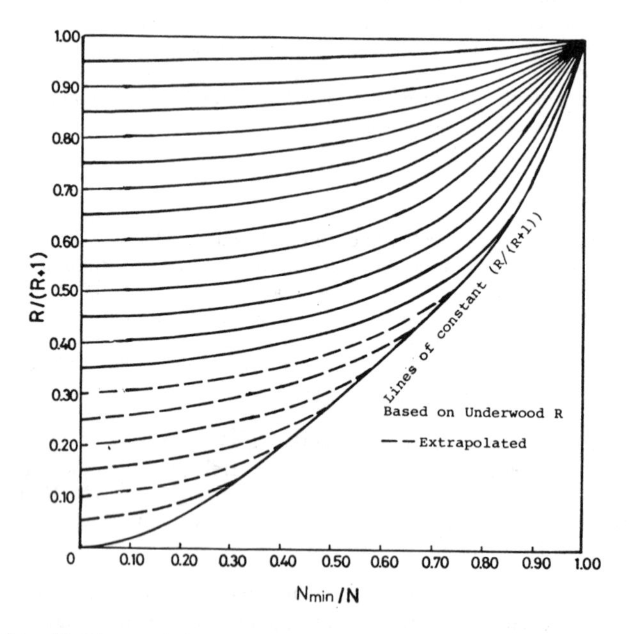

Figure 15. Erbar-Maddox correlation between flux rate and number of plates [50].

[28]. Strangio et al. [51] performed detailed calculations for minimum number of plates and minimum reflux for binary systems and extended the results to multicomponent designs.

Thermodynamic Analysis

Distillation systems are evaluated from the standpoint of thermodynamics, in particular in terms of available energy [52, 53].

The dead state of available energy is defined as the temperature, T_0, pressure, P_0, and composition $x_0 = (x_{01}, \ldots, x_{0C})$ of the surroundings or the dead state. As shown in Figure 16, the available energy loss for a conventional distillation column with a feed flow rate F and composition z_F at T_0 and P_0 and with products at T_0 and P_0 is as follows (where, key components, k and k+1, are separated into nearly pure components):

$$\Delta e_k = \Delta e_{COL} + \Delta e_{EX} + Wp \tag{95}$$

where Wp is the mechanical work required to transport the process stream and to adjust the stream pressure in the column. In general, the mechanical work is very much less than the other available energy losses in conventional distillation systems. The first term is the available energy loss for the column, neglecting heat exchangers at the condenser and reboiler. Denoting the available energy of the feed, the distillate, the bottoms product, and the heat flows supplied and removed as e_F, e_D, e_W, e_r and e_c, respectively, the available energy loss is:

$$\Delta e_{COL} = (e_r - e_c) + (e_F - e_D - e_W) \tag{96}$$

$$= T_0[(Q_c/T_D - Q_r/T_W) - Fc_F \ln (T_F/T_0) + Dc_D \ln (T_D/T_0) + (Wc_W \ln (T_W/T_0)] + \text{const.} \tag{97}$$

where

$$\text{const.} = \tilde{R}T_0 \left\{ D \sum_{i=k}^{k} x_{Di} \ln (x_{Di}/z_{Fi}) + W \sum_{i=k+1}^{C} x_{Wi} \ln (x_{Wi}/z_{Fi}) \right\} \tag{98}$$

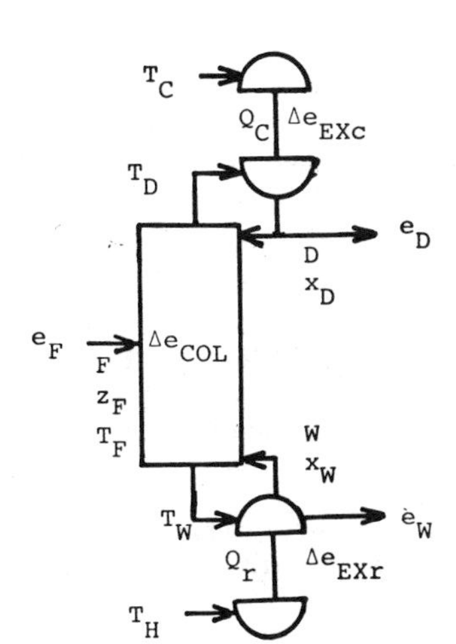

Figure 16. Available energy flow.

The value of const. is the maximum available energy required for the separation of components into two products at T_0 and is determined only by the top and bottoms specified. Accordingly, the avoidable available energy loss of the column, depending on the operating conditions, is as follows:

$$\Delta e_{COL} = \Delta e_{COL} - \text{const.} \tag{99}$$

The second term in Equation 95 is the sum of the available energy losses generated by a heat transfer operation at the condenser and reboiler attached to the column.

$$\Delta e_{EX} = \Delta e_{EXc} + \Delta e_{EXr} \tag{100}$$

where,

$$\Delta e_{EXc} = T_0(1/T_c - 1/T_0)Q_c \tag{101}$$

$$\Delta e_{EXr} = T_0(1/T_W - 1/T_0)Q_r \tag{102}$$

The values of Q_c and Q_r are determined from the conditions of the minimum reflux ratio, R^{min}, and the degree of clearance, θ, to realize the separation apart from the minimum reflux condition. Since the smaller the reflux ratio becomes, the smaller the available energy loss of a column, and it may be expressed that the value of Δe_R becomes the minimum value when $R = \theta R^{min}$. The available energy losses, Δe_{COL}, Δe_{EXc}, and Δe_{EXr} are shown in Figure 17.

For a multicomponent distillation system with N columns, substituting Equations 97, 98, 101 and 102, and the overall heat balance equation of a column into Equation 95, the following is obtained:

$$\Delta E_R = \sum_{j=1}^{N} \Delta e_R(j) \tag{103}$$

$$= \sum_{j=1}^{N} [(1 - T_0/T_H)Q_r - (1 - T_0/T_C)Q_c]|_j - \sum_{j=1}^{N} \{Fc_F[T_F - T_0 - T_0 \ln (T_F/T_D)]$$

$$- Dc_D[T_D - T_0 - T_0 \ln (T_D/T_c)] - Wc_W(T_W - T_0 - T_0 \ln (T_W/T_0))\}|_j$$

$$- F\tilde{R}T_0 \sum_{i=1}^{C} z_{Fi} \ln z_{Fi} \tag{104}$$

The first term is the available energy loss of the heat source and sink streams. When considering isobaric operations, the second term is related to the sensible heat change depending on the state of the system input and output, and has a very small value compared with the first term, in general. In other words, available energy loss in a multicomponent distillation system tends to be determined by the heat source and sink streams.

PRINCIPLES OF AZEOTROPIC DISTILLATION

Generally, this special distillation system is employed as a separation process for nonideal mixtures which have one of the following physical properties:

- The relative volatility between components to be separated in the entire or any part of the composition range is nearly equal to unit; in the later case it is called a tangent azeotrope.
- The mixture forms an azeotrope.

Such mixtures cannot be separated to nearly pure components by using conventional distillation systems. If the relative volatility is changed by adding chemical species to the mixture, it may be separated into specified products. When the additional species is usually selected so as to form azeotropes with more than one components in the feed, it is referred to as a "solvent" or "entrainer." The added solvent should be recovered by distillation, decantation, or another separation methods and is returned to suitable positions in the column. The quantity of lost solvent is continuously fed to the system as make-up solvent or make-up entrainer. The entire system is called "azeotropic distillation." Solvent recycling problems arise concerning the additional quantities and the locations of the recoverd and the make-up solvents. For extractive distillation, additional species are selected as components having bubble points exceeding the feed components.

An azeotropic mixture has some peculiar characteristics in the vapor-liquid equilibria which are denoted as "ridge" and "valley" in ternary systems, and, as characterized surfaces in four-component systems. The most difficult problem is how these characteristics are related to the system structure of azeotropic distillation. We must consider the selection of the solvent and the design by reviewing the entire system.

Since azeotropic distillation is complicated due to the nonideality of the vapor-liquid equilibrium relationship, in reverse, we may find a better separation system from the standpoint of energy savings by mafing use of the nonideality.

Ridge and Valley

Phase Equilibria

Predicting the azeotrope. A solution which is highly nonideal often forms an azeotrope. At the azeotrope, the liquid composition equals the vapor composition, that is

$$\alpha_{ij} = 1.0 \tag{105}$$

and conventional distillation is not possible. To separate such a mixture into desired products, we must first investigate the composition and temperature at the azeotrope.

A binary azeotropic mixture is classified into two types based on its deviation from ideality: a positive deviation and a negative deviation. A positive deviation means that the activity coefficient is greater than 1.0 and forms a minimum boiling point. A negative deviation means that the activity coefficient is less than 1.0 and forms a maximum boiling

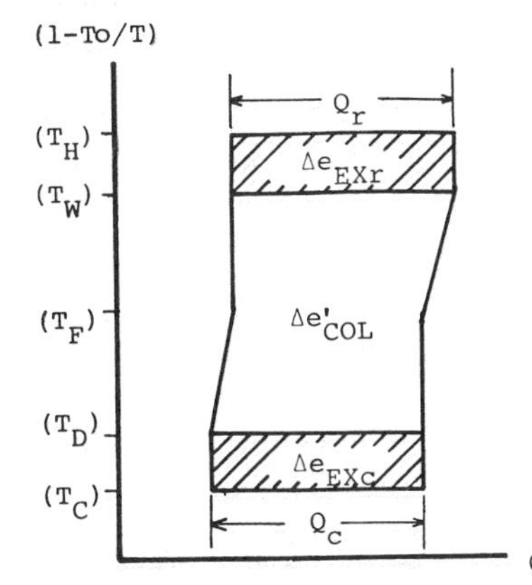

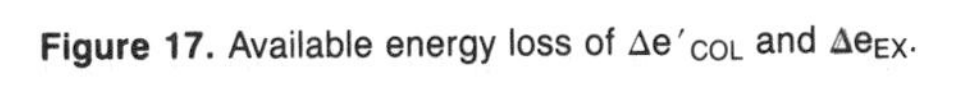
Figure 17. Available energy loss of $\Delta e'_{COL}$ and Δe_{EX}.

point. Figures 18 and 19 show a minimum boiling point azeotrope and a maximum boiling point azeotrope. An azeotropic mixture with a large deviation sometimes produces a partial immiscible composition range. Such a system is called a heterogeneous (binary) azeotrope. An example is illustrated in Figure 20. For ternary systems there are three types of ternary azeotropes: positive azeotrope (minimum boiling point), negative azeotrope (maximum boiling point) and positive-negative azeotrope (saddle-point azeotrope). One example of the saddle point azeotrope is that of acetone/chloroform/methanol system. (This system will be discussed in detail later.)

Azeotropic data are compiled in References 5 and 54 through 57.

The azeotropic conditions are as follows:

$$y_i = x_i, \quad i = 1, \ldots, C \tag{106}$$

Substituting Equation 1 to the vapor-liquid equiliblium relationship, the following equation results

$$y_i/x_i = \gamma_i P_i^0/\varphi_i P = 1 \tag{107}$$

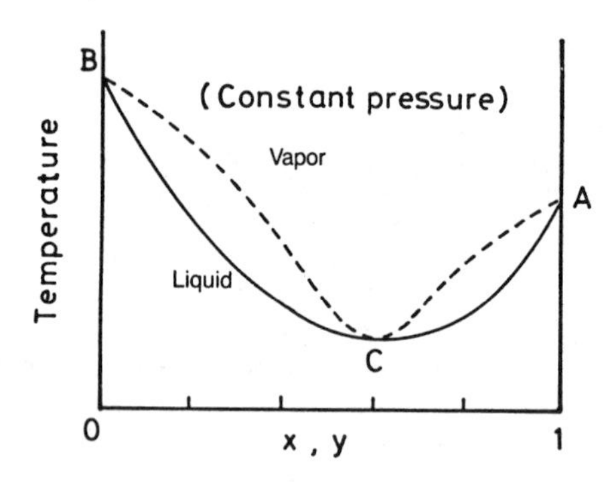

Figure 18. Binary mixture with minimum-boiling azeotrope.

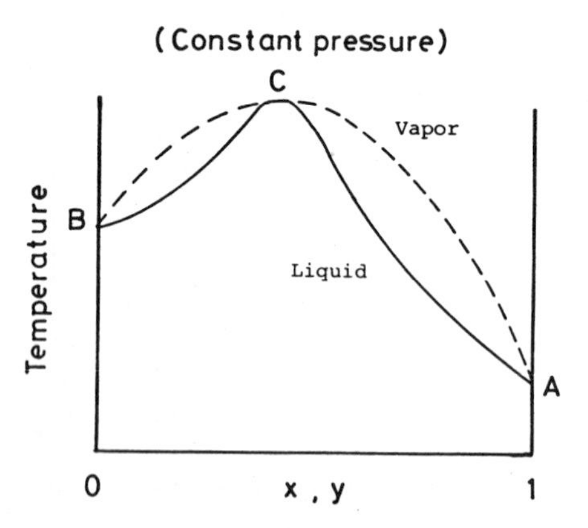

Figure 19. Binary mixture with maximum-boiling azeotrope.

where γ_i and φ_i are the activity coefficients of liquid and vapor phase, respectively. Attempts to predict ternary azeotropes and vapor-liquid equilibrium based on the parameters obtained from binary data have been made [12 and 58 to 60]. Although the vapor-liquid equilibrium relationship based on some activity coefficient can simulate binary data, ternary azeotropes cannot be predicted within adequate accuracy from an engineering standpoint. This is particularly true notes calculating a ridge and a valley. Several short cut prediction methods for azeotropes are reported in references 61 to 63.

Matsuyama et al. [64, 65] proposed a method for investigating the presence and types of ternary azeotropes. This method is based on the thermodynamic and topological properties derived from the residue curves. The residue curve is defined as the behavior of the still-pot composition of the simple batch distillation:

$$dx_i/dx_j = (x_i - y_i)/(x_j - y_j) \tag{108}$$

In other words, residue curves are developed individually from the initial feeds so as to contact vapor-liquid tie lines on the composition plane. Figures 21 to 23 show all patterns of residue curves around pure components and around binary azeotropes. The proposed method can investigate the existence of a ternary azeotrope from the combination of the patterns of pure components and binary azeotropes which constitute the ternary system. The criterion is:

$$2S_1 + 4S_2 + 8S_3 = 2 \tag{109}$$

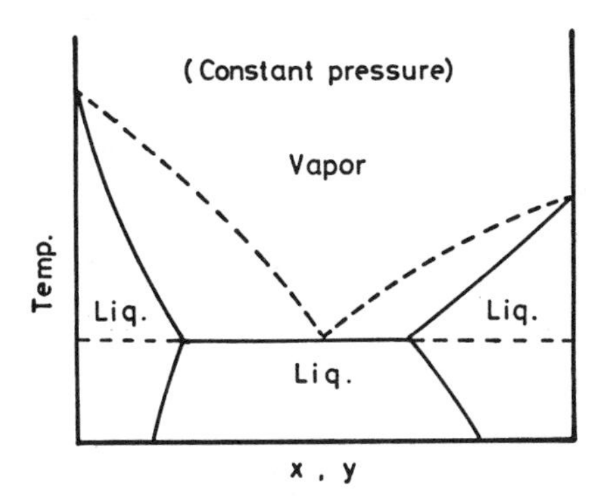

Figure 20. Binary mixture with heterogeneous azeotrope.

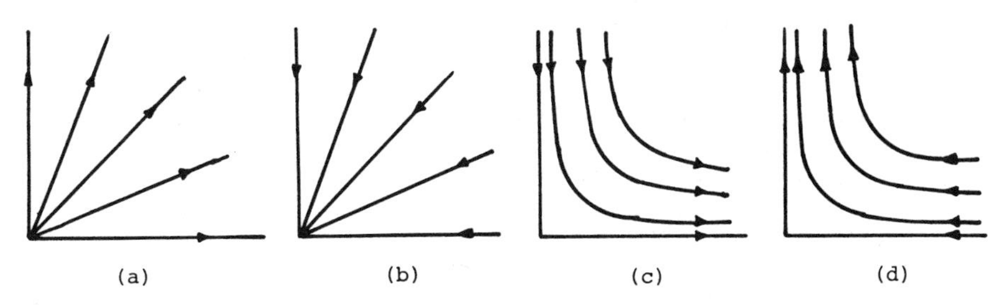

Figure 21. Patterns of residue curves in the neighborhood of pure materials [64].

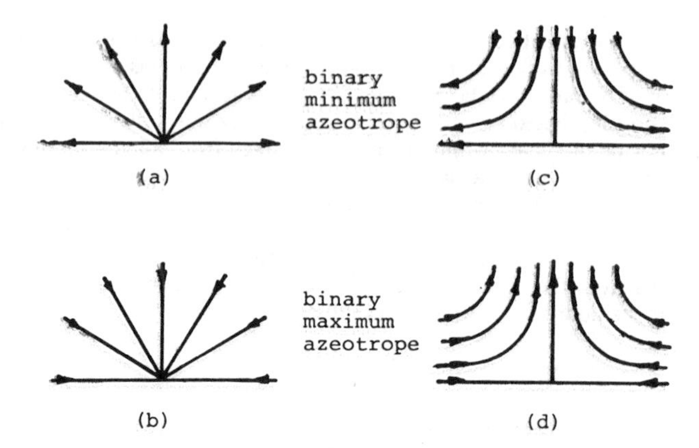

Figure 22. Patterns of residue curves in the neighborhood of binary azeotropes [64].

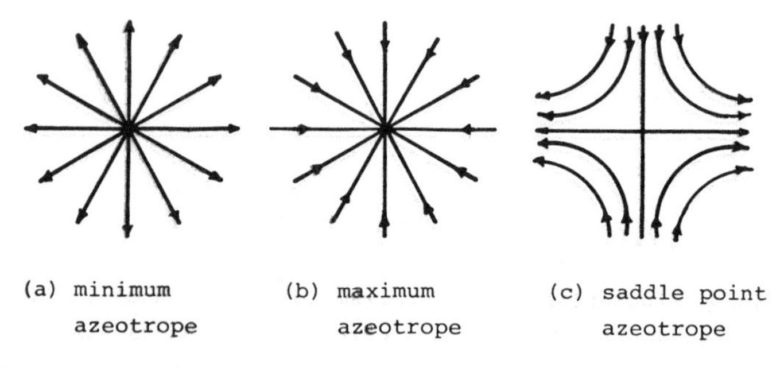

Figure 23. Patterns of residue curves in the neighborhood of ternary azeotropes [64].

where S_1 is the sum of indices of pure materials and, S_2 and, S_3 denote the sums of indices of binary and ternary azeotropes. Indices for S_1, S_2, and S_3 are listed in Table 2. These indices are defined by the residue curves near pure, binary and ternary components, respectively. For example, let us apply to Equation 109 the following system: the ternary system of ethanol (78.3° C), benzene (80.2° C), and water (100.0° C) has three binary azeotropes ethanol/benzene (68.24° C), benzene/water (69.25° C), and ethanol/water (78.174° C), and the ternary azeotrope (64.86° C). From Table 2, as the indices of the pure components are equal to unity, the value of S_1 is 3. Meanwhile, the S_2 must take one of three values, 1, −1, and −3 according to the combination of three binary-systems to make-up ternary systems. As the patterns of residue curves near all of the binary azeotropes correspond to Case c in Figure 22, the value of S_2 is −3. (If a binary azeotrope temperature is the lowest or highest in a system, the index of a binary azeotrope should be set as +1.) The value of S_3 is obtained from Equation 109:

$$S_3 = (2-2S_1-4S_2)/8 = [2-(2)(3)-(4)(-3)]/8 = 1$$

The ternary azeotrope should correspond to Case a in Figure 23 and be a ternary azeotrope with the minimum boiling point. The proposed method cannot estimate the composition and temperature of the azeotrope but can check the temperature consistency of the temperature pattern in a ternary system.

Predicting the liquid-liquid equilibria. Many azeotropic mixtures are partially immiscible at normal temperatures, Partial miscibility appears not only in binary mixtures but also in ternary or more-than-ternary mixtures. Data are reported in References 5 and 66. The characteristic of phase splitting is often applied in azeotropic distillation systems as a solvent recovery system. If liquid-liquid equilibrium data are unavailable, a predictive technique can be used. In multicomponent liquid-liquid equilibria, the key equation for each component i, is

$$(\gamma_i x_i)_u = (\gamma_i x_i)_\ell \tag{110}$$

$$\sum_{i=1}^{C} (x_i)_u = 1 \quad \text{and} \quad \sum_{i=1}^{C} (x_i)_\ell = 1 \tag{111}$$

where the subscripts, u and ℓ, indicate the upper layer and lower layer, respectively. The activity coefficient is expressed as an equation relating the excess molar Gibbs energy of a mixture to its composition $\underset{\sim}{x}$ The following equations are effectively applied to the partially immiscible mixtures: UNIFAC [18, 67], UNIQUAC [17], ASOG, [19] and NRTL [16]. (See the explanation for these equations in the first section.) Substituting one of these equations into the activity coefficients in Equation 110 and using Equation 111, the liquid compositions of both layers can be obtained. Since the liquid-liquid equilibria calculations are more sensitive than the vapor-liquid equilibria to composition changes, attention should be given to selectiving an effective convergence method. Several schemes have been proposed [20].

Definition of Ridge and Valley

The strong nonideality of the vapor-liquid equilibrium appears as the composition-bubble point surface waved up or down (or both). These shapes restrict the composition profiles in azeotropic distillation columns [see References 68 through 71]. To clearly evaluate restrictions, the nonideality has been extracted as "ridge" and "valley." A few definitions of "ridge" and "valley" are used and are classified into the following three types:

Definition 1: The vapor and liquid compositions which are in equilibrium with one another are joined by a series of tie lines (vapor-liquid tie lines). The terms, "valley"

Table 2
Patterns of Residue Curves in the Neighborhood of Pure Materials and Azeotropes and Their Indices*

	Pattern of Residue Curves	Index
Pure material	Figure 21(a)	1
	(b)	1
	(c)	−1
	(d)	−1
Binary azeotrope	Figure 22(a)	1
	(b)	1
	(c)	−1
	(d)	−1
Ternary azeotrope	Figure 23(a)	1
	(b)	1
	(c)	−1

* *Reprinted by special permission from* J. Chem. Eng. Japan, 10*(3): 181–187 (1977).*

and "ridge" are then defined by the curves which divide the patterns of vapor-liquid tie lines. The vapor-liquid tie lines near a valley turn towards the valley, and those near a ridge turn away from the ridge [71].

Definition 2: The valley and ridge are defined by the shape of the composition vs. bubble point surface [68, 72 through 74].

Definition 3: The separatrix to residue curves of a simple batch distillation is called "characteristic lines" [75, 76].

Definition 3 extracts the nonideality of the vapor-liquid equilibria from the standpoint of a differential batch distillation (See "PRINCIPLES OF AZEOTROPIC DISTILLATION.") However, it is proved that the characteristic lines do not agree with the ridge and the valley defined by Definition 1 or by Definition 2 and do not always limit the composition profile of a continuous distillation column.

Using the liquid-composition-vs.-bubble-point diagram of acetone/chloroform/methanol of Figure 24, let us consider the difference between Definition 1 and Definition 2. This mixture has the binary maximum azeotrope of chloroform/acetone at 64.4° C (64.5° C), two binary minimum azeotrope chloroform/acetone at 53.5° C (53.5° C) and methanol/acetone at 55.4° C (54.6° C), and the ternary azeotrope at 57.1° C (57.5° C). These bubble points are calculated using the Antoine and Wilson equations where the values in the parentheses are experimental data [73]. The isothermals are the dot-dash-lines and each vapor-liquid tie line indicated by an arrow shows the relationship between vapor composition and the liquid composition in the equilibrium state. The ridge and the valley defined by Definition 1 are denoted by the solid lines and those defined by Definition 2 are drawn as broken-lines

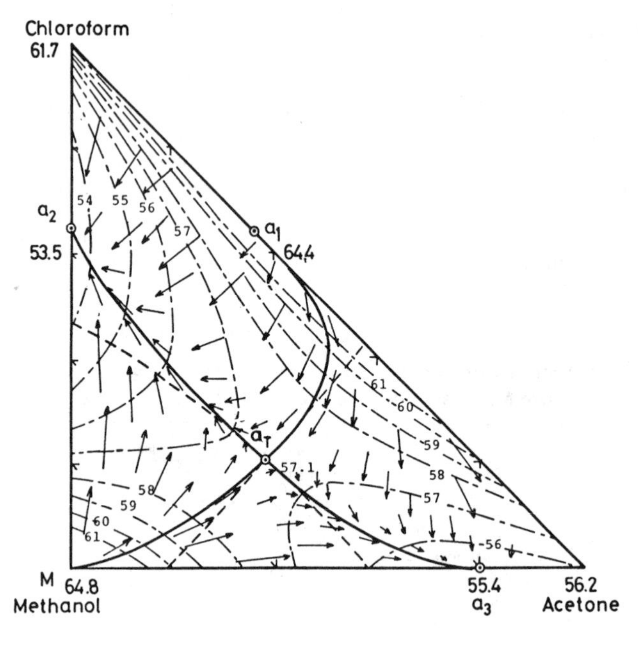

Figure 24. Vapor-liquid equilibrium, valley, and ridge.

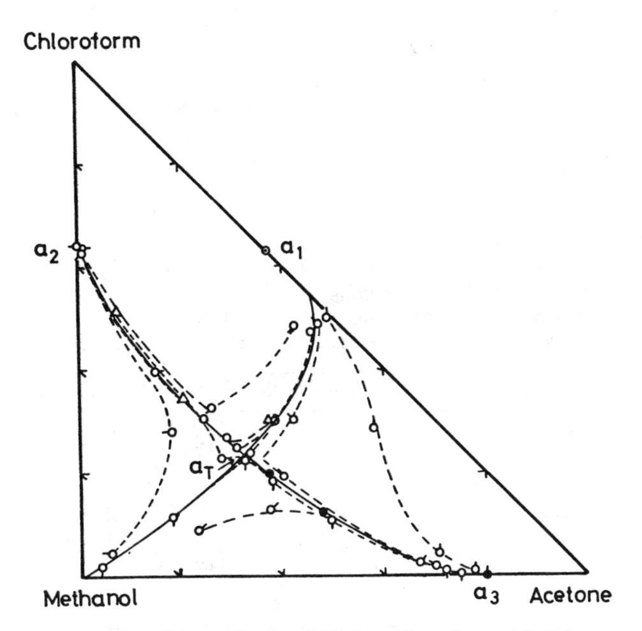

Figure 25. Composition profiles of the total reflux distillation column by experiment.

calculated from the maximum descent method. From this figure, the ridge and the valley defined by Definition 1 are obviously different from those defined by Definition 2. As the ridge and the valley based on Definition 1 are defined only from the directional pattern of the vapor-liquid tie lines, these lines clearly relate to the composition profile of a continuous distillation column. Definition 2 is accepted in many references, but this fact may be due to misunderstandings that the ridge and the valley related to Definition 1 are completely equivalent to the others. (Hence Definition 1 is employed.)

Determination of Ridge and Valley

According to Definition 1, the ridge and the valley split the directions of vapor-liquid tie lines and appear concretely as the composition profile at total reflux. For the acetone/chloroform/methanol system, the composition profiles at total reflux experimentally obtained are shown in Figure 25 [77]. From the results, the ridge and valley are approximately determined as a_1a_TM and $a_2a_Ta_3$. There is however, no method to theoretically determine these quantities. The following procedure evaluates the ridge and valley with sufficient accuracy: Figure 26 indicates the possibility of a ridge and/or valley. It may be possible to easily predict the existence of them using azeotropic data. The ridge is approximately obtained by successive use of the dew point calculation, assuming the initial temperature near the lower temperature of either pure component and azeotrope, or two azeotropes. On the other hand, the valley is obtained approximately from the bubble point calculation using an assumed temperature near the higher temperature of a possible set of temperatures as shown in Figure 26. The results from the computational procedure are little affected by the deviation of the starting point.

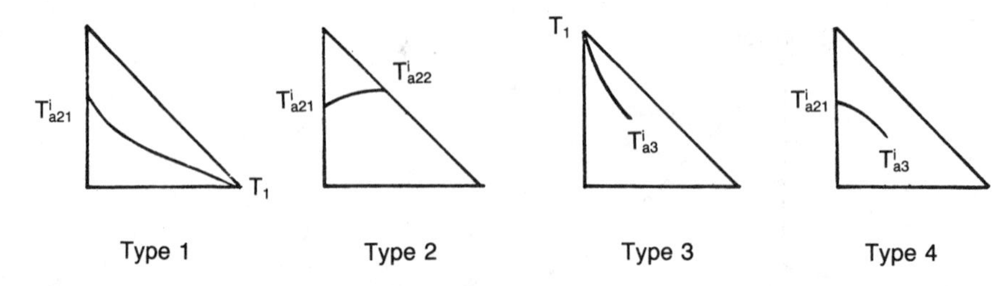

T_1 : Bubble point of pure component
T^i_{a21} or T^i_{a22} : Binary azeotrope (min. temp. i = 1, max. temp. = 2)
T^i_{a3} : Ternary azeotrope (min. temp. i = 1, max. temp. = 2, saddle = 3)

Computational Method	Type	Starting Point of Computation
Bubble point Valley	1	$*T_1 > T^1_{a21}$ or $*T^1_{a21} > T_1$
	2	$*T^1_{a22} > T^1_{a21}$
	3	$*T_1 > T^1_{a3}$
	4	$*T^1_{a21} > T^1_{a3}$
Dew point Ridge	1	$T_1 > *T^2_{a21}$ or $T^2_{a21} > *T$
	2	$T^2_{a22} > *T^2_{a21}$
	3	$T_1 > *T^3_{a3}$ or $T^2_{a3} > *T_1$
	4	$T^2_{a3} > *T^2_{a21}$ or $T^2_{a21} > *T^3_{a3}$

* Assumed value

Figure 26. Relationship between types of valley or ridge and selection of computational methods.

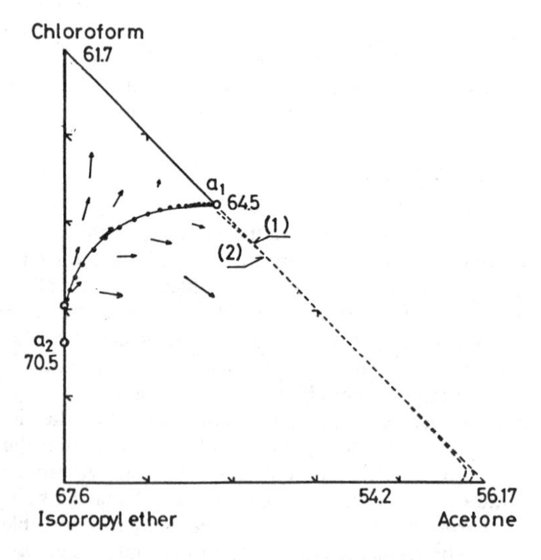

Figure 27. Calculated value of ridge (acetone/chloroform/isopropyl ether).

Example 1: For the acetone/chloroform/iso-propyl ether system

- Calculate the ridge between two binary azeotropes of acetone/chloroform and acetone/iso-propyl ether.
- Investigate whether a valley exists or not, within two azeotropes, acetone/chloroform and acetone/iso-propyl ether.
- If there is a ridge, it belongs to Type 2 according to Figure 26. After the starting point very close to the acetone/chloroform azeotrope is assumed, the dew point calculation is sequentially carried out. The result is shown in Figure 27.
- If there is a valley, it may be classified as Type 2. The bubble point calculation is executed based on the assumed value close to the acetone/chloroform azeotrope. Curves 1 and 2 are calculated based on the starting points (A = 0.365, C = 0.630, IPE = 0.005) and (A = 0.355, C = 0.630, IPE = 0.005), respectively. Curve 1 is located closely along the line of IPE = 0.0. Curve 2 lies in parallel near curve 1 and does not approach it except in an area around the acetone/iso-propyl ether azeotrope. We may conclude that a valley does not exist.

On the other hand, the distillate changes of batch distillation make the valley clear. But, the ridge cannot be obtained by batch distillation.

Restrictions of Ridge and Valley on Composition Profiles

Homogeneous System [71]

For a continuous azeotropic distillation column, the relationships between the valley and ridge, and the composition profiles play an important role in design and control. To gain confidence in the relationships between the ridge and valley, and the composition profiles in a conventional column, we must consider a straight mass balance line and divide the composition plane of the acetone/chloroform/methanol system into eight regions as shown in Figure 28.

Feed Composition in Region I_R. The composition profiles of Runs E1 and E2 whose operating conditions are different are shown in Figure 29. The composition profiles of Run E1 with a large reflux ratio and the feed composition in Region I_R exists in Region IV_0. However, for Run E2 with a smaller reflux ratio than that of Run E1, the composition profile remains in Region I_0. Note that the composition profile never intersects the ridge, a_1a_T. The relationship between the composition profile in Region IV_0 and the feed composition in Region I_R obviously depends on the operating conditions and the number of plates in the column. The region where the composition profiles changes in relation to the reflux condition is shown in this figure. This is due to the curvature of the a_1a_T ridge and it analogously follows the following calculation: When the feed composition is in the region bounded by an arc such as a valley or ridge, and a chord, (for example, I_R, I_V, II_R, and IV_V), the composition profile can be located in a region outside this region by varying the operating conditions and the number of plates in the column, due to the curvature of the valley or ridge.

Feed Composition in Region I_0. The feed and composition profiles of Run E4 are shown in Figure 30. For the feed composition in Region I_0, the composition profile exists in Region I ($= I_R \cup I_0 \cup I_V$) bounded by the a_1a_T ridge and the a_2a_T valley. Similarly, when the feed composition is in a region bounded by the chords only, for example, I_0, II_0, III_0, and IV_0, the composition profile will be in the region which is bounded by the ridges and/or the valleys related to the chords, I, II, III, and IV.

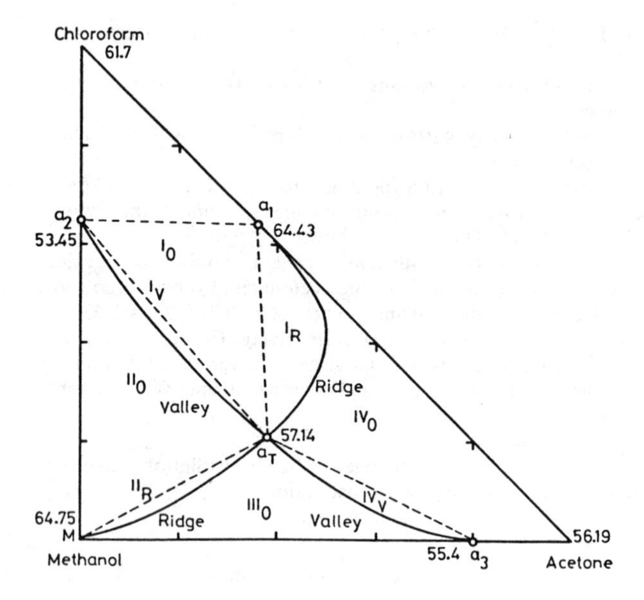

Figure 28. Division of composition plane into subregions (acetone/chloroform/methanol).

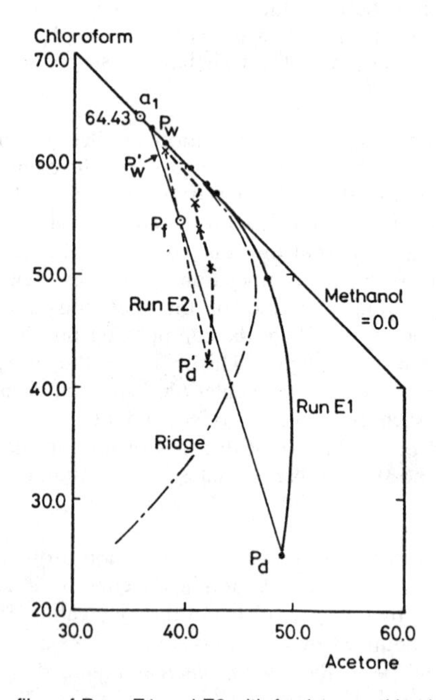

Figure 29. Composition profiles of Runs E1 and E2 with feed composition in Region I_R.

Heterogeneous System [71]

A decanter is often used in a heterogeneous azeotropic distillation system. For primitive azeotropic distillation, the vapor from the top plate is cooled in a condenser and is split to the upper layer (solvent-rich layer) and the lower layer in a condenser according to the liquid-liquid equilibrium relationship. The solution in the upper layer returns to the top of the column. In this case, the reflux composition does not agree with the vapor composition from the top plate. For the benzene/isopropyl alcohol/water system, the valley [71] and the binodal curve at 25° C [78] experimentally obtained are denoted by the broken line and the curved solid line in Figure 31, respectively. The straight lines bounded by the binodal curve correspond to the liquid-liquid tie lines.

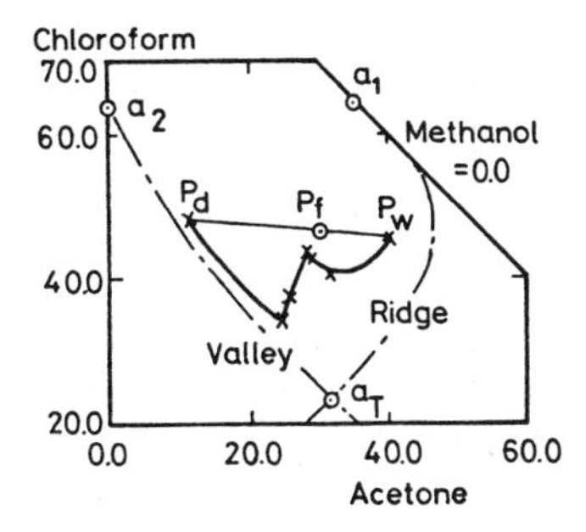

Figure 30. Composition profiles of Run E4 with feed composition in Region I_o.

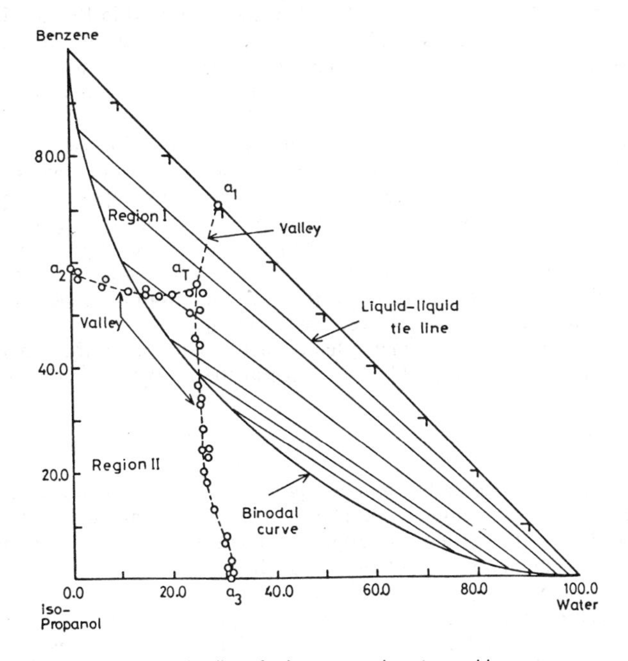

Figure 31. Liquid-liquid equilibrium and valleys for iso-propanol, water and benzene.

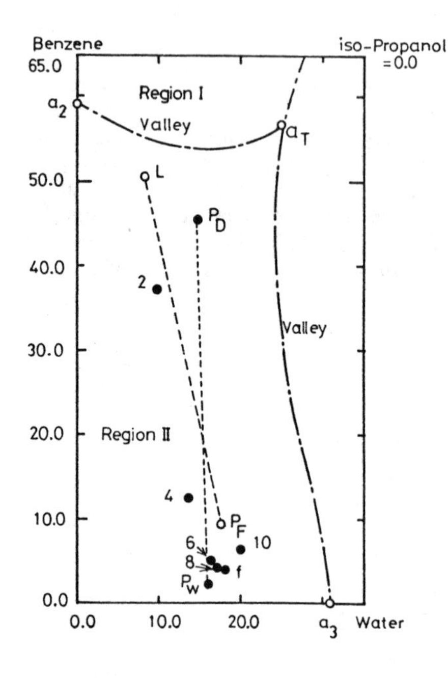

Figure 32. Composition profiles of Run E2-1 with feed and reflux composition in Region II.

Reflux composition outside the region with the feed composition. As shown in Figure 33, the composition profile can intersect the a_2a_T valley and is situated in Regions I and II. When as shown in Figure 32. The result obtained by simulation resembles that of Run E2−1. From the simulation results, the top composition turns towards the binary azeotrope a_2 along the a_2a_T valley as the reflux flow rate decreases, but does not intersect the valley.

Reflux composition outside the region with the feed composition. As shown in Figure 33, the composition profile can intersect the a_2a_T valley and a situated in Regions I and II. When the reflux composition is in the region outside of the feed composition, the composition profile is either in the region with the feed composition or in two regions with the reflux and feed compositions across the valley. The region of the composition profile is located by manipulating the reflux flow rate and/or the boil-up flow rate. By increasing the ratio of the reflux flow rate to the boil-up flow rate, the composition profile invades the region of the reflux from the region of the feed composition.

The boiling point of the reflux may often be higher than that of the vapor from the top. In this case, the temperature profile may bulge in the middle of a column near the top.

Critical Operation

Minimum number of plates. The nonideality of vapor-liquid equilibria makes it difficult to evaluate the minimum number of plates. The Fenske equation and other methods available for ideal mixtures cannot be applied to nonideal mixtures. For ternary azeotropic mixtures, the minimum number of plates can be approximately obtained by using the concept of the composition profile at total reflux. The iteration numbers of the vapor-liquid equilibrium relationship are counted for two following cases: The curve composed of the sequential vapor-liquid tie lines from the bottoms intersect the half-line from the bottom, which is the mass balance extended through the top, at some point beyond the feed.

Meanwhile, the curve of the sequential reverse vapor-liquid tie lines from the top crosses the half-line with the top as the origin beyond the feed. When the curve encloses its corresponding end point, the iteration number denotes the minimum number of plates needed to achieve a specified separation at total reflux.

Minimum reflux ratio. The assumption of the constant relative volatility is not acceptable. To determine the minimum reflux ratio, we cannot apply the Underwood equation to azeotropic mixtures and must resort to one of the mote rigorous calculation methods [79]. For a ternary azeotropic mixture, it is not always necessary to solve complicated equations. Yorizane et al. [80] proposed a simplified method to obtain the minimum reflux ratio. The mass balance equation for the i component around the pinch-point plate, p, and a reboiler, N, is

$$Lx_{pi} = Vy_{pi} + Wx_{Ni}, \quad i = 1, 2, 3 \tag{112}$$

Approximating that the second component of the bottom product is pure

$$x_{N1} = x_{N3} = 0 \tag{113}$$

the following equations are derived from Equation 112:

$$Lx_{p1} = Vy_{p1} \quad \text{and} \quad Lx_{p3} = Vy_{p3} \tag{114}$$

and

$$y_{p1}/x_{p1} = y_{p3}/x_{p3} = L/V \tag{115}$$

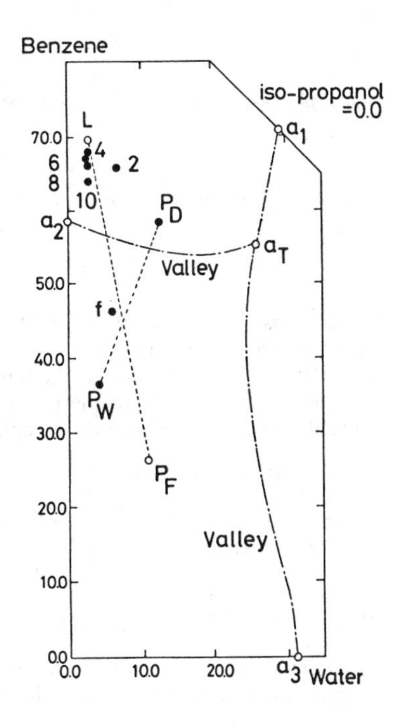

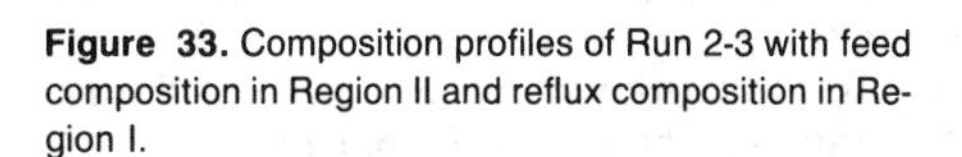
Figure 33. Composition profiles of Run 2-3 with feed composition in Region II and reflux composition in Region I.

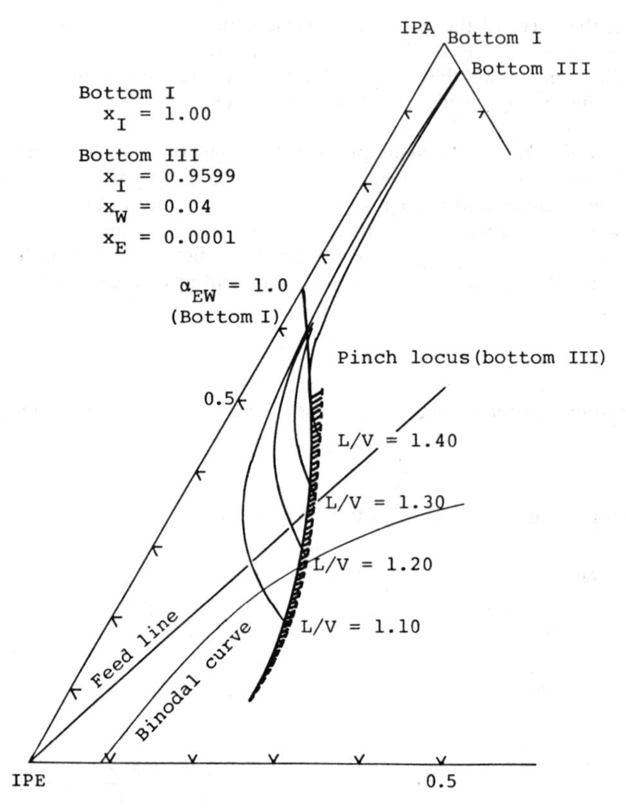

Figure 34. Pinch loci for each reflux ratio [80].

Meanwhile, the relative volatility of the first component to the third component at the pinch point is

$$(\alpha_{13})_p = (y_{p1}/x_{p1})\,(x_{p3}/y_{p3}) = 1 \tag{116}$$

The composition at the pinch point is given as the intersection of the locus related to Equation 116 and the line with the ratio of the first component to the third in the feed. After the vapor composition at the pinch point is calculated using vapor-liquid equilbrium data the minimum value of the internal reflux ratio can be obtained from Equation 115. Then, the minimum reflux ratio is derived from the over-all mass balance as follows:

$$L/V = [R + q(F/D)]/[(R+1) - (1-q)\,(F/D)]. \tag{117}$$

Figure 34 illustrates an example of the iso-propyl alcohol dehydration system which uses iso-propyl ether as a solvent.

Dynamic Behavior

Continuous distillation [81]. In general, when the feed stream with disturbance of the composition is introduced into a distillation column, the top and bottoms are controlled by manipulating the reflux flow rate and the boil-up rate so as to keep the specified values of the products. Accordingly, in designing an azeotropic distillation control, it is important to understand the relationships between the changes of the top and bottoms, along with the manipulative variables. The following results were obtained experimentally with a distillation column composed of 40 Oldershow plates (35 mm diameter), a total condenser, and a reboiler.

Figure 35 shows the changes of the top and the bottoms when the reflux flow rate was varied stepwise. The composition profiles at the initial and final steady states are denoted by the curve of x_1x_N with open circles and by the curve of $x_1'x_N'$ with dark circles, respectively. Time is denoted by t. The top composition moved very rapidly along the valley, but the movement of the bottoms was extremely slow. Both composition profiles and the movements of both ends never intersected the ridge and the valley, and existed in the region with the feed composition.

Meanwhile, when the composition of the feed was changed by step, the dynamic behavior of the top and the bottoms was measured as shown in Figure 36. During the initial steady state, the feed composition z_F and the composition profile, x_1x_N are denoted by the open circles and the black circles, respectively. When the feed composition was changed from z_F to z_F', the top moved rapidly and widely beyond the ridge and, within 60 min, almost finished approaching the final steady state. The response curve is denoted by black triangles. The bottom moved slowly beyond the ridge and settled down at the final steady state. It took about 3 hours to reach the final state. In the case of such disturbance on the feed composition, it is impossible to return one of the final products to the specified product in the initial region by manipulating the reflux flow rate and/or the vapor flow rate. This limitation can be explained from the results mentioned earlier. Accordingly, it is necessary to analyze the feed composition and control the composition in order to keep the feed composition in the initial region.

Batch distillation. Ewell et al. [73] conducted experiments on azeotropic batch distillation of the acetone/chloroform/methanol system. They first divided the feed composition into six regions using "ridge," a_1a_TM, and simple straight lines, a_1a_2, a_2a_T, a_1a_3, and a_Ta_3, which are denoted by the broken lines in Figure 37. The definition of the "ridge" was not clear

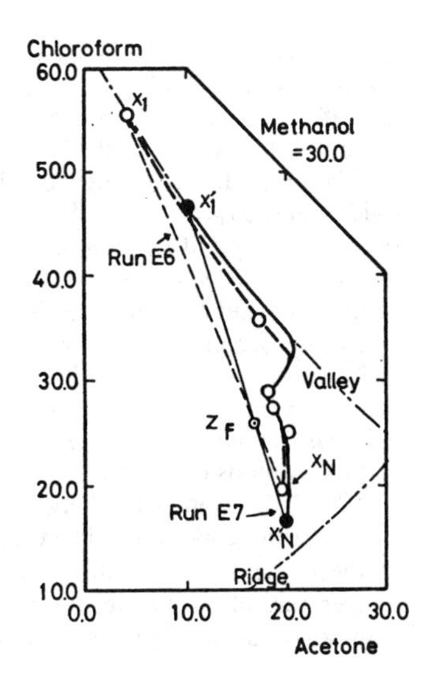

Figure 35. Composition changes of the top and bottom products related to reflux step change (Run E6 → Run E7).

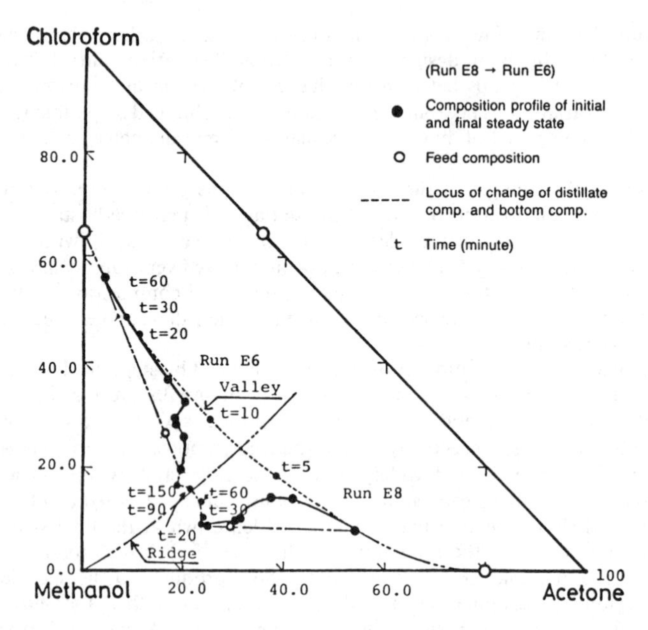

Figure 36. Step change of feed composition (Run E8 → Run E6).

in this study. Changes of the distillate for the still-pot compositions in each region were investigated in terms of temperature and composition. It is clear that there are examples in which the distillate crosses over the ridge. Naka et al. [82] pursued in detail the changes of the distillate composition in each region. The relationships between the distillate composition and the still-pot composition established by experiment and simulation are noted in the following.

- *Still-pot composition in Region II:* Figure 38 shows the changes in the distillate compositions with the still-pot composition in Region II by the solid lines. The broken lines indicate changes in still-pot compositions. The composition changes for all of Runs 2, 2″, and 9, follow nearly the same path from a_2 to a_T, that is, the valley. In general, the distillation paths are changeable due to differences in the number of plates, reflux ratio, holdup and so on. In all cases, the final distillate composition has a tendency to cross the a_1a_T ridge and enter Region V for the larger the number of plates and the greater reflux ratio. As can be seen from the changes in the distillation paths of Run 2 and Run 9, the overhead temperature rises from $T(a_2)$ to $T(a_T)$, but then drops passing over a_T when approaching a_3, and rises up again. This temperature trend agrees with Ewell's results.

- *Still-pot composition in Region V:* Figure 39 shows the distillation paths related to the still-pot composition in Region V. The distillation paths are not similar to those of Region II. The distillation paths between a_3 and a_T move along the valley with little effect of the operating conditions. Then, the distillate compositions, after approaching a_T, return in the direction of a_3 and go to a_T. They never cross the a_1a_T ridge. The overhead temperature of all runs in this region rises between a_3 and a_T and then drops once away from a_T, and then again rises as it approaches a_1.

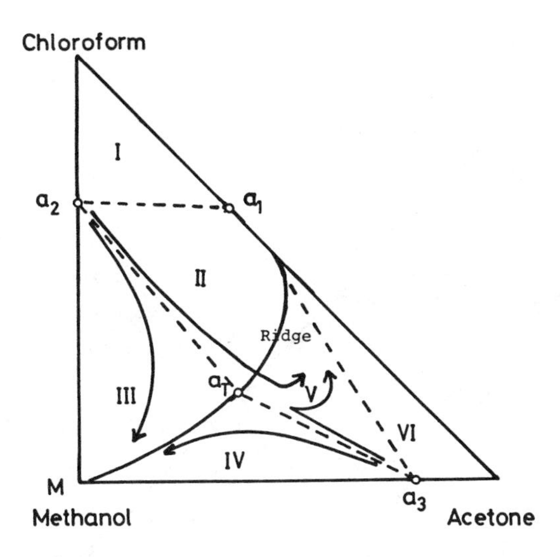

Figure 37. Rectification regions.

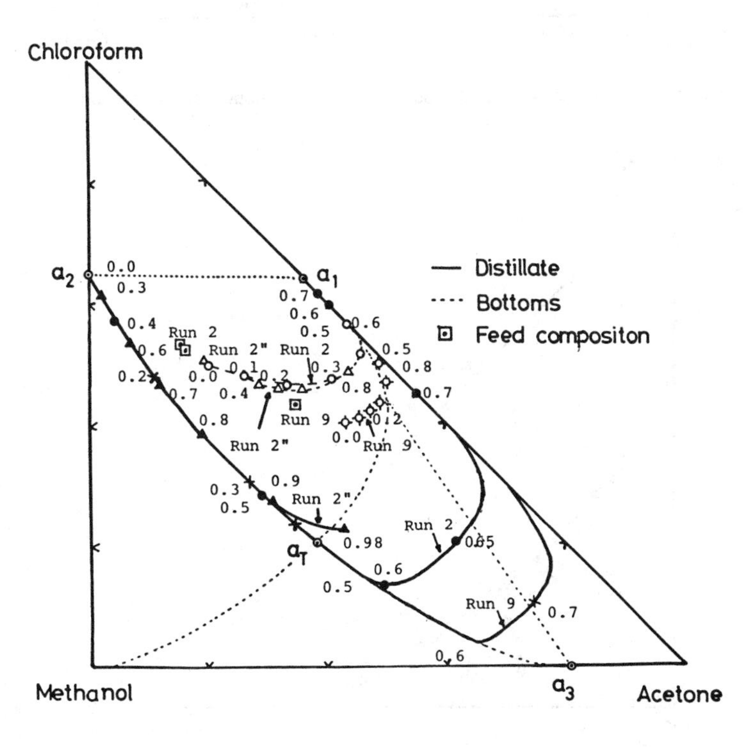

Figure 38. Composition changes of distillates and bottoms by the azeotropic batch distillation.

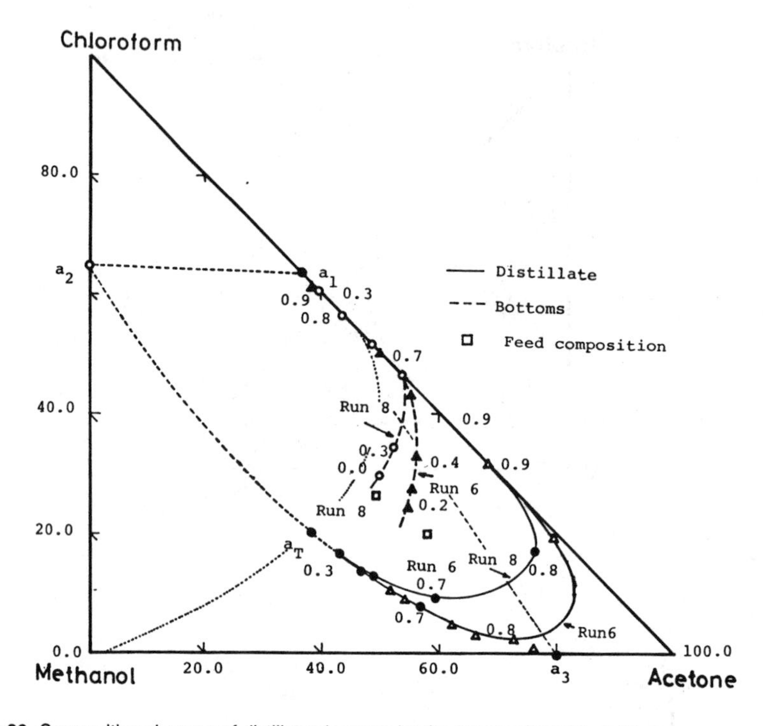

Figure 39. Composition changes of distillates bottoms by the azeotropic batch distillation.

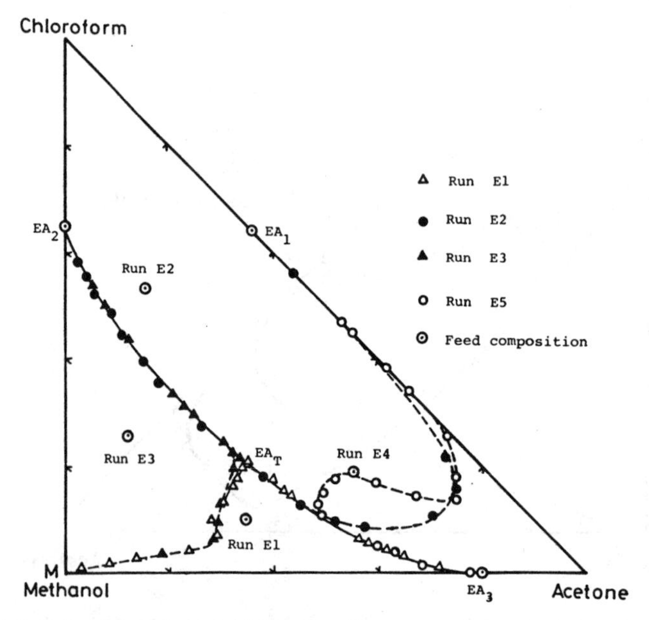

Figure 40. Experimental results of distillate composition changes in azeotropic batch distillation.

- *Batch distillation paths in the other Regions:* The solid line in Figure 40 shows the distillation paths obtained from the previously mentioned experiments. The distillate composition of Run E3 with the still-pot composition in Region III moves along the a_3a_T valley and then approaches M passing over the a_TM ridge. For Run E1 related to the still-pot composition in Region IV, the distillate composition follows the a_3a_T valley and then goes to M but does not cross over the a_TM ridge.

As noted previously, it is evident that the distillation path is strongly related to the region where the still-pot composition exists and is restricted by the valley and the ridge defined by Def. 1. In particular, the valley can be explicitly obtained from batch distillation whereas the ridge cannot. This fact is explained from the relationships between the directions of vapor-liquid tie lines and the distillation path for simple batch distillation.

Effects of Ridge and Valley on System Structure

Azeotropic Distillation

From the prior results, if to clear that a valley or ridge should be accounted for when designing an azeotropic distillation system.

Two examples are shown in Figure 41 which explain the effects of a valley and ridge on the system structure. In comparison with two separation systems of the feed compositions, z_{F1} and z_{F2}, the former separation system produces product A and the intermediate product A′ but product B cannot be distillated from intermediate product A′ by using additional distillation columns because of the distillation barrier of the ridge or valley. The latter separation system separates the feed composition z_{F2} into the product B and the intermediate product B′. Furthermore, the intermediate product B′ can be separated into the desired compositions in Region Aa_1a_2 by using the curvature of a ridge or valley. Hence, product A can be produced.

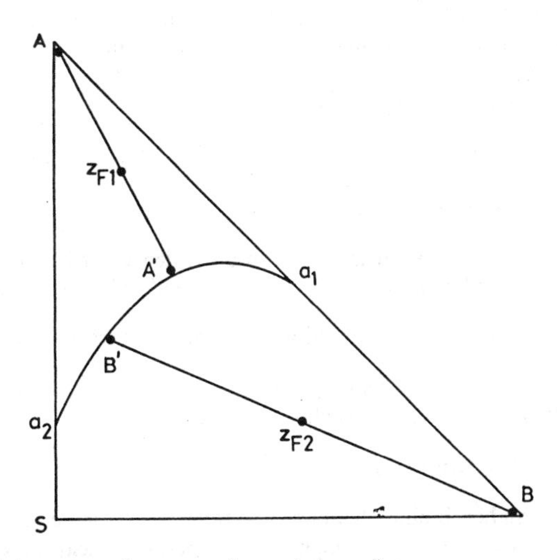

Figure 41. Relationship between valley a_1a_2 and mass balance line.

The following five methods overcome the limitations associated with a ridge or valley:

Method 1: Using a mixing operation.
Method 2: Using the curvature of the ridge and valley in distillation.
Method 3: Using decantation.
Method 4: Using an extractive operation.
Method 5: Using composition change with operating pressure.

Method 1 is used in order to change the region with a feed composition. For example, consider the ethanol/water/benzene system. An azeotropic distillation with a water-rich feed adapts Method 1, where a part of the ethanol product is returned to the feed stream in order to let it shift into the ethanol-rich region. Method 2 is employed by several azeotropic distillation systems. But, if a valley or ridge is regarded as a straight line, Method 2 cannot be used. The possibility of Method 3 is determined from the relationship between the locations of a binodal curve and a valley. As shown by the example in Figure 31, a part of a valley, a_Ta_3 exists in the region of the two-liquid phase and intersects the liquid-liquid tie lines at sufficient angles. Method 3 is ordinarily used for alcohol dehydration systems. Method 4 as well as Method 3 is used in many azeotropic distillation systems, however it requires more than two solvents. The four methods mentioned are based on isobaric conditions. Since some azeotropes change with operating pressure, the barrier of the ridge and valley may also shift. Method 5 may utilize the characteristic for separating an azeotropic mixture into specified products.

Selection of Solvents

Solvent selection is an important and difficult problem in designing an azeotropic (or extractive) distillation system. Much of the azeotropic data are tabulated (see Horsley [54]). Unfortunately as references sometimes accept wrong data, it is neccessary to investigate the consistency of temperature relationship among pure components and azeotropes [65]. Such references may offer good guidelines for screening potential solvents. It is desirable in general, that solvents for azeotropic (and extractive) distillation should:

1. Change the vapor-liquid equilibria between the components to be separated.
2. Be nonreactive with the other components in the feed composition.
3. Be compatible with equipment materials of construction.
4. Have a low latent heat.
5. Be a stable compound to temperature.
6. Be inexpensive.
7. Be nontoxic.

Moreover, for azeotropic distillation,

8. When the distillate with solvent is cooled in a condenser, it is effective for the solvent to separate into two phases: a solvent-rich layer and solvent-poor layer.

The solvent which satisfies the 8th condition must be easily recovered.

9. When the distillate is homogeneous at normal temperatures, a ridge or valley related to the components to be separated should curve.

Conversely, for extractive distillation:

10. Solvent has the higher boiling temperature than the other components in the feed mixture. If its temperature is too high, the temperature heat duty in a reboiler will

increase because of the contribution of the solvent in a column due to its latent heat. This condition should be avoided.

The previous list in general, represents optimum conditions for solvents. In the last two decades, procedures for screening solvents have been developed in terms of vapor-liquid and liquid-liquid equilibria. More reliable equations to estimate the activity coefficients have been developed which greatly facilitates solvent selection. Leaving several candidates through quantitative filters is preferred when calculating the phase equilibrium relationships of a given mixture for a solvent chosen at random. Ewell et al. [61] proposed a quantitative selection method for potential solvents based on the deviations from the nonideality of mixtures. They took into account the hydrogen bonds as the main cause of their deviations and classified all liquids into the following five classes:

Class I: Liquids capable of forming three-dimensional networks of strong hydrogen bonds.
Class II: Liquids containing both active hydrogen atoms and donor atoms (O, N, F) except Class I.
Class III: Liquids composed of molecules containing donor atoms but no active hydrogen atoms.
Class IV: Liquids composed of molecules containing active hydrogen atoms but no donor atoms.
Class V: All other liquids, i.e., liquids having no hydrogen-bond-forming capabilities.

Table 3 gives examples of each class. The deviations of class combinations from Raoult's law are given in Table 4. In addition, Ewell et al. discusses the possibilities of forming maximum boiling azeotropes.

PRINCIPLES OF EXTRACTIVE DISTILLATION

When a mixture has a relative volatility of approximately unity or that of an azeotrope, the specified products are achieved by the use of several columns which may require a large number of plates. Moreover, such a distillation system may not be established in special cases. To overcome such difficulties, distillation systems using solvents and azeotropic and extractive distillation systems have been developed. If a solvent forms no azeotrope with the dominant components in the feed to be separated, and has a relatively low volatility, the separation is referred to as an "extractive distillation." This is a unit operation which effectively utilizes the advantages of liquid-liquid extraction and rectification.

Table 3
Examples According to Ewell's Classification*

Class I	Class II	Class III	Class IV	Class V
Water	Alcohols	Ethers	$CHCl_3$	Hydrocarbons
Glycol	Acids	Ketones	CH_2Cl_2	Carbon disulfide
Glycerol	Phenols	Aldehydes	CH_3CHCl_2	Sulfides
Amino alcohol	Primary amine	Aldehydes	CH_2ClCH_2Cl	Mercaptans
Hydroxylamine	Secondary amines	Esters	$CH_2ClCHClCH_2Cl$	Nonmetallic elements such as (I, P, S)
Hydroxyl acids	Oximes	Tertiary amines	$CH_2ClCHCl_2$	Halo-hydrocarbons not in Class IV
Polyphenols	Nitro compounds and nitriles with α-hydrogen atoms	Nitro compounds and nitriles without α-hydrogen atoms		
Amides	Ammonia			
	Hydrazine			
	Hydrogen fluoride			

* *Excerpted by special permission from R. H. Ewell, J. M. Harrison and L. Berg,* Ind. Chem. Eng., 36(10): 871–875 (1944), Copyright (1944) American Chemical Society

Table 4
Deviations from Raoult's Law*

Classes	Deviations	Hydrogen Bonding
I + V II + V	Always + deviation; I + V, frequently limited solbility	H bonds broken only
III + IV	Always − deviations	H bonds formed only
I + IV II + IV	Always + deviations; I + IV, frequently limited solubility	H bonds both broken and formed, but dissociation Class I or II liquid is more important effect
I + I I + II I + III II + II II + III	Usually + deviations, very complicated groups, some − deviations give some max. azeotropes	H bonds both broken and formed
III + III III + V IV + IV IV + V V + V	Quasi-ideal systems, always + deviations or ideal; azeotropes, if any, will be minima	No H bonds involved

* *Reprinted with permission from R. H. Ewell, J. M. Harrison, and L. Berg,* Ind. and Eng. Chem., *36(10): 871–875 (1944). Copyright (1944) American Chemical Society.*

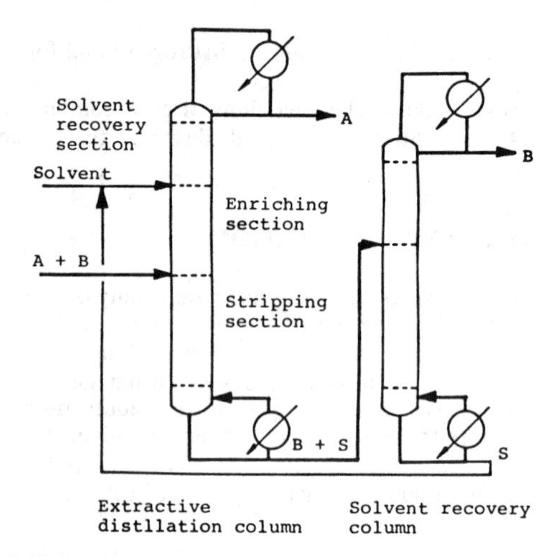

Figure 42. Extractive distillation system.

An extractive distillation is different from azeotropic distillation in terms of the volatility of the solvent; in the case of an azeotropic distillation, having a lower volatility than those of the components in the feed and, in the case of an extractive distillation, a higher volatility than the others.

A typical extractive distillation system is shown in Figure 42. Such a system is composed of at least two columns: an extractive distillation column and a solvent recovery column. The extractive distillation is divided into three sections, the solvent-recovery section, the enriching section, and the stripping section, from the top of the column. The enriching and stripping sections have the same role as those in a conventional column. The solvent-recovery section serves to concentrate the solvent up to a level where the solvent effectively extracts the specific substance in the sections below the solvent feed plate. Accordingly, the solvent is always supplied to some plate over the feed plate. Also, this section serves to prevent the solvent from contaminating the top product. In the case of an extremely low relative volatility of the solvent, the solvent-recovery section may be deleted.

Vapor-Liquid Equilibria

Representation of Vapor-Liquid Equilibria

In distilling a mixture of A and B into two products by the use of a solvent, if the solvent has the highest bubble point of the ternary components and affinity for B, the B component cannot be vaporized readily. Therefore, as the relative volatility between A and B increases, the separation required can be easily achieved. When a solvent has strong intermolecular forces between its own molecules and those of the feed, the vapor-liquid equilibrium relationship for the entire system is much more complex, even if the feed mixture is ideal. In this case, the vapor-liquid equilibrium relationship must be described by one of the methods using the activity coefficient of the liquid phase. As noted earlier, the Wilson equation, NRTL, ASOG, UNIQUAC, and UNIFAC are useful for estimating vapor-liquid equilibria.

Dependence of Vapor-Liquid Equilibria on Solvent Concentration

The solvent concentration in an extractive distillation strongly affects the relative volatility between A and B. Here, A and B components can be denoted as the key components in a multicomponent mixture.

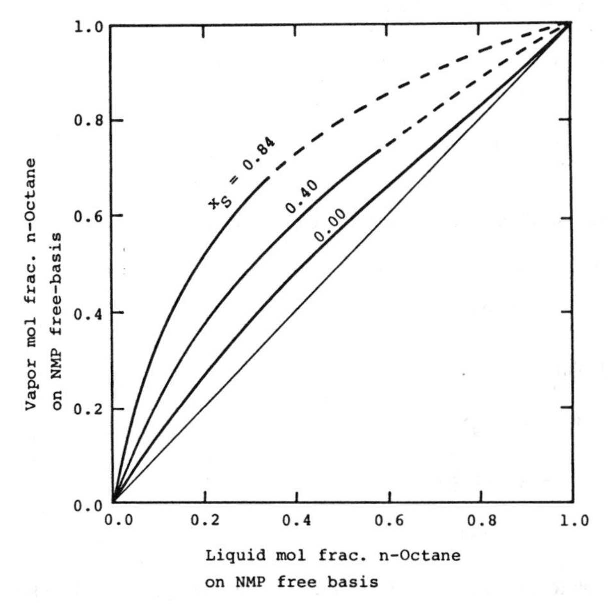

Figure 43. Effect of solvent on vapor-liquid equilibrium relationship: n-octane/ethylbenzene/N-methyl pyrrolidone.

Let's consider the effects of the solvent concentration on the vapor-liquid equilibria. Figure 43 shows the effects of n-octane, ethylbenzene, and N-methylpyrrolidone as a solvent [83] in the x-y diagram. Although the original binary mixture is not nonideal, the vapor-liquid equilibrium line with constant solvent concentration looses the nonideality as the solvent concentration increases. This is a general tendency.

Now let's investigate the dependence of the relative volatility on the solvent concentration. The relative volatility of A and B in the ternary system which includes solvent S, is defined by

$$(\alpha_{AB})_S = (y_A/y_B)\,(x_A/x_B) = (\gamma_A/\gamma_B)\,(P_A^0/P_B^0) \tag{118}$$

A typical relationship between the activity coefficients and the solvent concentration is shown in Figure 44 [84]. As the solvent concentration increases, the activity coefficients become larger but the gradient of the relative volatility becomes flat. That is, the relative volatility in the region of high solvent concentration is independent of the solvent concentration. This characteristic can be applied to the solvent selection problem.

The increase of the relative volatility, $(\alpha_{AB})_S$, makes the separation of the A/B mixture easier. Accordingly, to introduce such a feature in an extractive column, the solvent concentration should be kept in the effective region. In addition, attention should be given to relative volatility, α_{BS}, in the column which recovers the solvent from the B/S mixture.

Normal and Critical Operations

Normal Operation

Mass balance. The basic mass balance equation for an extractive distillation column should correspond to that for a conventional multicomponent distillation. The mass balance equations for the column shown in Figure 45 are:

For the condenser

$$V = L + D \tag{119}$$

$$Vy_{2i} = Lx_{1i} + Dx_{1i} \tag{120}$$

Denoting the solvent-feed plate as K, for the k-th plate in the solvent recovery section, $2 \leqq k \leqq K-1$

$$Vy_{k+1i} + Lx_{k-1i} = Vy_{ki} + Lx_{ki} \tag{121}$$

at the solvent-feed plate, K,

$$L' = L + q_S S \tag{122}$$

$$V' = V - (1 - q_S)S \tag{123}$$

$$V'y_{k+1i} + Lx_{k-1i} + Sz_{Si} = Vy_{ki} + L'x_{ki} \tag{124}$$

Denoting the feed plate as f, for the j-th plate in the enriching section, $K+1 \leqq j \leqq f-1$,

$$V'y_{j+1i} + L'x_{j-1i} = V'y_{ji} + L'x_{ji} \tag{125}$$

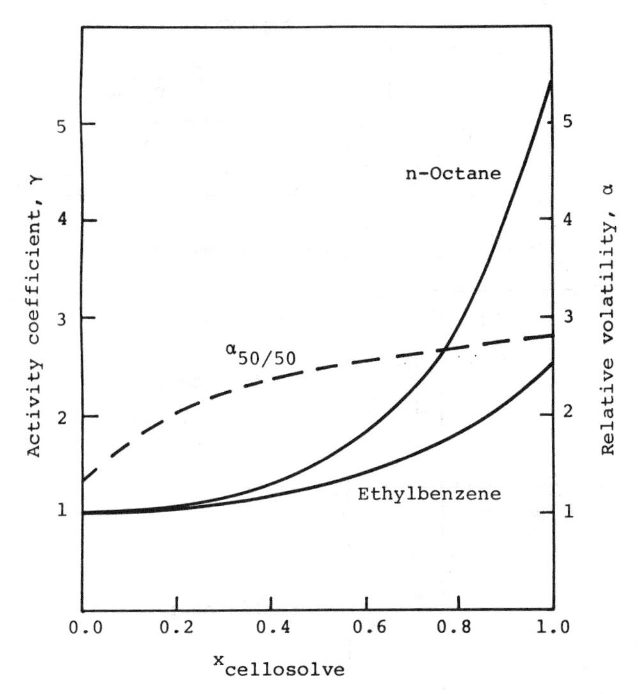

Figure 44. Relationship between the activity coefficients and solvent concentration (calculated from the data in Reference 84).

at the feed plate, f,

$$L'' = L' + q_F F \tag{126}$$

$$V'' = V' - (1 - q_F)F \tag{127}$$

$$V'' y_{f+1i} + L' x_{f-1i} + F z_{Fi} = V' y_{fi} + L'' x_{fi} \tag{128}$$

Denoting the reboiler as N, for the n-th plate in the stripping section, $f+1 \leqq n \leqq N-1$

$$V'' y_{n+1i} + L''_{n-1i} = V'' y_{ni} + L'' x_{ni} \tag{129}$$

at the reboiler

$$V'' = L'' - W \tag{130}$$

$$V'' y_{Ni} = L'' x_{N-1i} + W x_{Ni} \tag{131}$$

In addition to the mass balance equations, the constraints of the vapor and liquid mole fractions at every plate, and the vapor-liquid equilibrium relationship, are necessary for solving the mass balance equations. When the conditions on the feed and the solvent are given, the degree of freedom for all equations is three, equaling that of a conventional column.

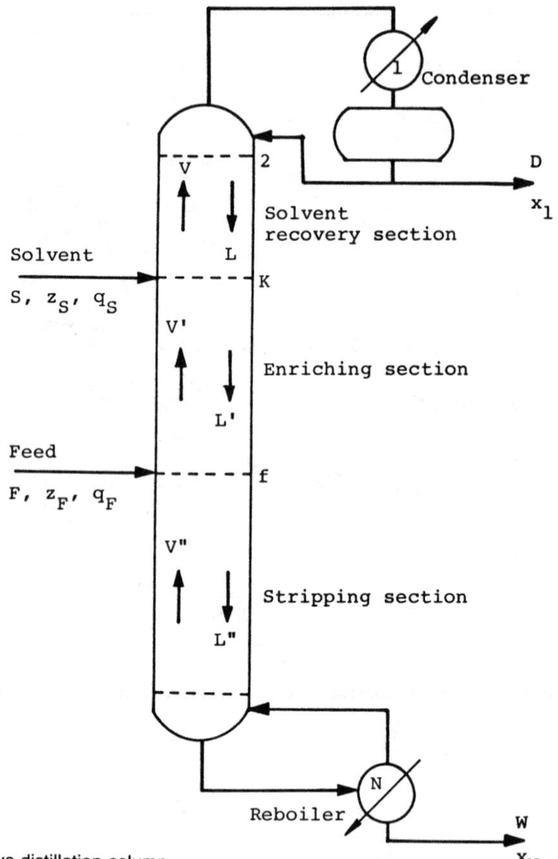

Figure 45. Extractive distillation column.

Solvent quantity. The sum of Equations 119 to 125 related to the condenser, the solvent recovery and enriching sections, derives

$$V' + S = L' + D \tag{132}$$

$$V' y_{j+1i} + S z_{Si} = L' x_{ji} + D x_{1i} \tag{133}$$

Defining the relative volatility of the solvent to the other components, α', as

$$\alpha' = [y_S/(1 - y_S)]/[(1 - x_S)/x_S] \tag{134}$$

The vapor mole fraction of the solvent is obtained from Equation 134:

$$y_S = \alpha' x_S/[1 + (\alpha' - 1)x_S] \tag{135}$$

Substituting Equations 132 and 135 into Equation 133 and assuming $x_j = x_{j+1}$ in the enriching section, the following equation is obtained as

$$Sz_{SS} + (L' + D - S)\alpha' x_{jS}/(1 - x_{jS} + \alpha' x_{jS}) = L' x_{jS} + D x_{1S} \tag{136}$$

If the solvent supplied is pure and the distillate contains almost no solvent, the x_{jS} is

$$x_{jS} = S/[(1 - \alpha')L' - \alpha' D/(1 - x_{jS})] \tag{137}$$

In addition to this assumption, as the solvent volatility is usually very low, the value of α' is close to zero. Consequently, the solvent concentration in the enriching section is

$$x_{jS} = S/L' \tag{138}$$

If the solvent concentration in the enriching section is specified, the quantity of the solvent can be calculated from Equation 139. The experimental relationship between x_{jS} and S/L' is shown in Figure 46 [85]. The value of x_{jS} is empirically fixed in the range of 60 to 80 mol % of the solvent.

Solvent-feed plate. The solvent-feed plate is determined as follows: As shown in Figure 46, the solvent concentration in the enriching section can be roughly estimated from Equation 138. So, once the solvent concentration at every plate, which is calculated from the top by a plate-to-plate calculation method, equaling to the value obtained from Equation 138, the solvent-feed plate can be determined.

Critical Operation

As noted earlier, even if the entire system, including the solvent, is nonideal, the vapor-liquid equilibrium of the key components, A and B, in a multicomponent mixture,

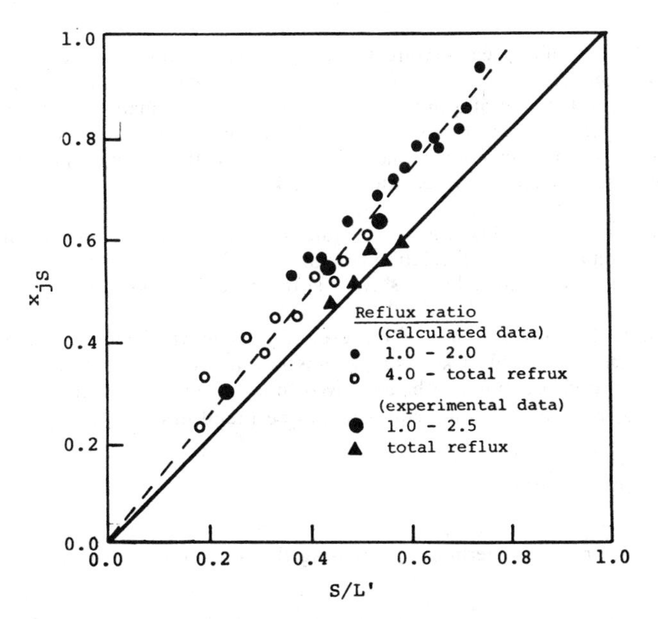

Figure 46. Experimental relationship between x_{js} and S/L'[85].

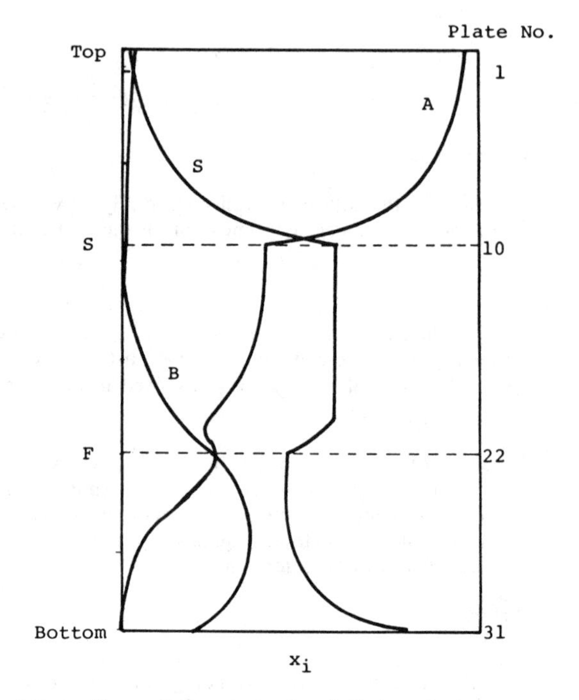

Figure 47. Typical composition profile in an extractive distillation column.

may be regarded as that of an ideal mixture with a high concentration of solvent. Moreover, from the composition profile shown in Figure 47, the solvent composition in an extractive distillation column, in the case of a high solvent concentration, remains fairly constant in each section. Thus, the design methods based on the feature of an ideal mixture is readly available using the assumptions noted earlier. Note, however, that a detailed design should be performed by using a rigorous calculation method.

Minimum number of plates. There is no absolute method for determining the minimum number of plates in an extractive distillation column. The Fenske equation may be applied to this problem if a mixture including a solvent is regarded as an ideal mixture.

Minimum reflux. This involves use of the extended Underwood [86] equation. If the psuedo-binary system composed of key components in a mixture with solvent is regarded as ideal, the Underwood equation can be employed to calculate the minimum reflux ratio. The root obtained from the following equation can be found in $\alpha_{ALK} < \varphi < \alpha_{AHK}$.

$$\sum_{i=1}^{C} z_{Fi}/(\varphi - \alpha_{Ai}) = (1 - q)/\varphi \tag{139}$$

The minimum reflux ratio is determined by putting the root, φ, into

$$R^{min} = \sum_{i=1}^{C} \alpha_{Ai} x_{1i}/(\varphi - \alpha_{Ai}) - \sum_{i=1}^{C} \alpha_{Ai}(S/D) Z_{Si}/(\varphi - \alpha_{Ai}) - S/D \tag{140}$$

Another approach is to use the graphical method [87].

When all components in the feed and solvent appear in the top and bottom products, the pinch points, that is the zones of constant component, exist adjacent to the feed plate. To determine the minimum reflux ratio, two of the independent variables of the product composition as well as the compositions and flow rates of the feed and the solvent, should be given. Under these conditions the minimum reflux ratio can be estimated as follows:

Denoting the mole fractions and flow rates of the vapor and liquid at the pinch point as y^*, x^*, V''^*, and L''^*, respectively

$$y_i^* = y_{fi} = y_{f+1i} \tag{141}$$

$$x_i^* = x_{fi} = x_{f+1i} \tag{142}$$

the component mass balance is

$$L''^* x_i^* = V''^* y_i^* + W x_{Ni} \tag{143}$$

The following equation is derived as

$$V''^*/W = (x_{Ni} - x_i^*)/(x_i^* - y_i^*) \tag{144}$$

Equation 143 indicates that three components, x_i^*, y_i^* and x_{Ni} should be on one straight line. Also, the x^* and y^* are in equilibrium. (Refer to Figure 48.)

1. Draw two straight lines: the feed line, 'SF', which ties z_F and z_S, and the bottom product line specified, 'SO', which ties z_S and x_N
2. Assume x^* on the feed line.
3. Require y^* related to x^* by using the vapor-liquid equilibrium relationship.
4. Obtain the intersection of the bottom product line and the extended line, 'RW', which unites y^* and x^*. The intersection must correspond to x_N specified. If not, return to Step 2.
5. Calculate the minimum reflux ratio from the following equation,

$$R^{min} = \varrho(F + S - D)/D + (1 - q_F)F/D - 1 \tag{145}$$

where $\varrho = \text{'RW'}/\text{'QR'} = V''^*/W$ (146)

and it is assumed that $q_S = 1$.

If it is too difficult to assume x^*, the relationship between R^{min} and S should first be calculated. That is, neglecting the specified condition on the solvent flow rate, S, and changing x^*, the values of R^{min} and S are calculated. There is no way to return to Step 2 in Step 4. Then, in Step 5, the R^{min} and S are evaluated from Equation 144 and the following equation, respectively

$$S = F\,\frac{(z_{FA} - x_{NA})/(x_{1A} - x_{NA}) - (z_{FB} - x_{NB})/(x_{DB} - x_{NB})}{x_{NA}/(x_{1A} - x_{NA}) - x_{NB}/(x_{1B} - x_{NB})} \tag{147}$$

The correlation between R^{min} and S can be obtained by varying the value of x^*. The required minimum reflux can be determined by putting the value of S initially neglected into the correlation.

Selection of Solvents

Representation of Selectivity

The general conditions on a solvent for an extractive distillation were noted in the previous section, however it is difficult to select solvents quantitatively. Solvent selectivity is one of the important indices by which we can evaluate a substance as to its effectiveness.

For the ternary system composed of A, B, and S, the selectivity is defined as

$$S_{AB} = (\alpha_{AB})_S/(\alpha_{AB}) \tag{148}$$

$$= [(\gamma_A/\gamma_B)(P_A^0/P_B^0)]_S/[(\gamma_A/\gamma_B)(P_A^0/P_B^0)] \tag{149}$$

where (α_{AB}) and $(\alpha_{AB})_S$ are the relative volatilities of A to B without and with solvent, respectively. As the vapor pressures of each pure component are almost independent of temperature, the following equation derived

$$S_{AB} \cong (\gamma_A/\gamma_B)_S/(\gamma_A/\gamma_B) \tag{150}$$

If the binary mixture of A and B is regarded as an ideal solution, the selectivity is

$$S_{AB} \cong (\gamma_A/\gamma_B)_S \tag{151}$$

As the ratio of the activity coefficients, $(\gamma_A/\gamma_B)_S$, depends on the solvent concentration, when a large amount of the solvent exists in comparison with those of A and B, the selectivity

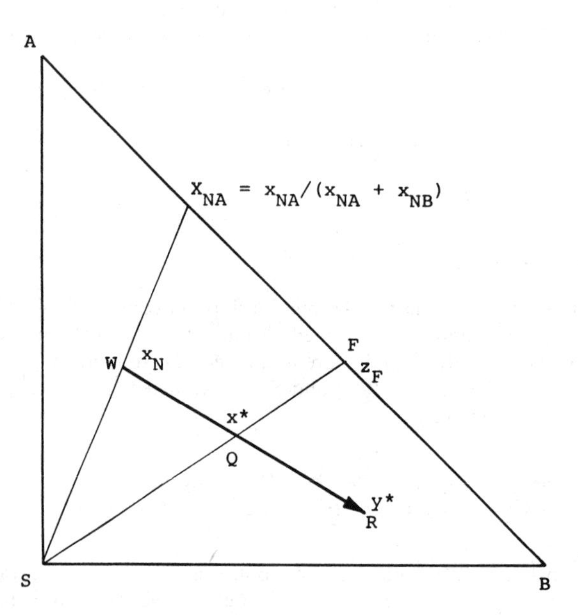

Figure 48. Representation of mass balance equations in the composition plane.

can be defined by using the activity coefficients of A and B in an infinitely diluent solution as

$$S^0_{AB} \cong \gamma^0_A \,|_{x_S \to 1} / \gamma^0_B \,|_{x_S \to 1} \tag{152}$$

Evaluation of Selectivity

Hirata's method. Hirata et al. [88] investigated the effects of the solvent concentration on solvent selectivity. As a result of their experiments, they derived two rules under the assumption that some ternary mixture of A, B, and S is homogeneous in the range of fully high solvent concentration:

Rule 1: The vapor-liquid equilibrium relationship of A and B is regarded as that of an ideal solution. The relative volatility, α_{AB}, is constant and is independent of the A and B concentrations.

Rule 2: The relative volatility, $(\alpha_{AB})_S \,|_{x_S \to 1}$ for a ternary mixture with solvent at infinite dilution is nearly equal to the ratio of $\alpha_{AS} \,|_{x_S \to 1}$ to $\alpha_{BS} \,|_{x_S \to 1}$ for binary mixtures, A/S and B/S, respectively at infinite dilution

$$(\alpha_{AB})_S \,|_{x_S \to 1} = \alpha_{AS} \,|_{x_S \to 1} / \alpha_{BS} \,|_{x_S \to 1} \tag{153}$$

According to Rule 1, the effect of some solvents on a binary mixture can be easily analyzed from the vapor-liquid equilibrium of a mixture which is composed of 90% of S and 10% of A and B. For the acetone/methanol/water system the experimental results are illustrated in Figure 49. On the basis of these two rules, the representative value of selectivity can be evaluated from

$$S_{AB} \,|\, x_S \to 1 = (\alpha_{AB})_S \,|\, x_S \to 1 / \alpha_{AB} \,|\, X_A = 0.5 \tag{154}$$

where $X_A = x_A/(x_A + x_B)$ and the value of 0.5 is representative. The calculated values for S_{AB} are compared with the experimental results for ternary systems in Table 5 [83]. These rules can be applied not only to binary mixtures but also to multicomponent mixtures.

Ochi–Kojima's method. Ochi et al. [89, 90] indicated an example which derived from the Hirata's rules as shown in Figure 50 and evaluated the relative volatility for the infinitely diluent solution of A and B with S

$$(\alpha_{AB})_S \,|_{X_B \to 1} = (P^0_V/P^0_B) \,|_{T_S} / [\gamma^0_A/(\gamma_B)_{BS}] \tag{155}$$

$$(\alpha_{AB})_S \,|_{X_A \to 1} = (P^0_A/P^0_B) \,|_{T_S} / [(\gamma_A)_{AS}/\gamma^0_B] \tag{156}$$

where $X_A = x_A/(x_A + x_B)$. The γ^0_i and $(\gamma^0_i)_{jS}$ are the activity coefficients for the ternary mixture infinite at dilution and for the binary mixture of j on S, respectively. The T_S is the bubble point of the solvent. The pressure of each pure component and the $(\gamma^0_i)_{jS}$ can be easily obtained, but the γ^0_i should be estimated from experiments or by calculations. For example, the value of the activity coefficient at infinite dilution can be obtained by using the Wilson equation as follows: if $x_A = 0$

$$\ln \gamma^0_A = -\ln[x_S\Lambda_{AS} + (1 - x_S)\Lambda_{AB}] + 1$$
$$- x_S\Lambda_{SA}/[x_S + (1 - x_S)\Lambda_{SB}] - (1 - x_S)\Lambda_{BA}/[x_S\Lambda_{BS} + (1 - x_S)] \tag{157}$$

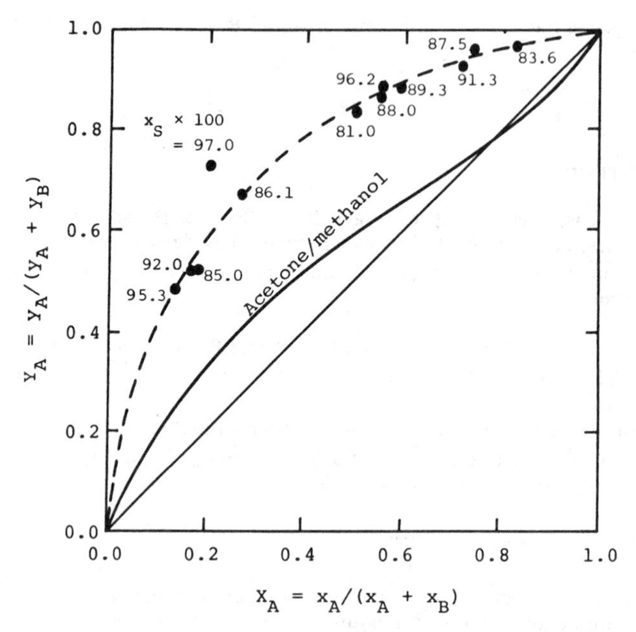

Figure 49. x-y diagram for acetone/methanol system diluted with water [87].

Table 5
Comparison of Solvent Selectivities Between Experimental and Calculated Values by Hirata's Method

Components			$X_A = 0.5$		at $x_S \to 1$			$(\alpha_{AB,S})$	S_{AB}	
A	B	S	α_{AB}	α_{AS}	α_{BS}	α_{AS}/α_{BS}	x_S	$x_S \to 1$	Exp.	Cal.
Carbon tetrachloride	Benzene	Toluene	1.10	2.38	2.36	1.01	0.766 −0.788	1.02	0.92	0.93
Carbon tetrachloride	Benzene	Methanol	1.10	3.42	2.25	1.52	0.913	1.49	1.38	1.35
Carbon tetrachloride	Benzene	Ethanol	1.10	3.68	3.33	1.10	0.904	1.05	1.00	0.96
Carbon tetrachloride	Benzene	n-Propanol	1.10	5.11	4.44	1.15	0.764	1.17	1.05	1.06
Water	Methyl ethyl ketone	Butyl cellosolve	0.73	1.29	3.35	0.39	0.862	0.58	0.54	0.80
Benzene	Cyclohexane	Furfural	1.07	13.3	71	0.19	0.875	0.45	0.18	0.42
Benzene	Cyclohexane	Methyl cellosolve	1.07	7.15	16	0.45	0.892	0.54	0.42	0.51
Methyl cyclohexane	Toluene	Phenol	1.22	22	11.2	1.96	0.899 −0.864	2.32	1.60	1.40
Acetone	Chloroform	Methyl isobutyl ketone	1.27	4.82	2.21	2.28	0.957 −0.819	2.11	1.79	1.56
Acetone	Ethanol	Water	1.27	37.8	11.5	3.28	0.950	2.80	2.58	2.20
Acetone	Methanol	Water	1.37	37.8	7.96	4.75	0.962 −0.810	5.51	3.47	4.02
Acetone	Methyl ethyl ketone	Water	1.84	37.8	44	0.86	0.898	1.31	0.47	0.66
Methanol	Ethanol	Water	1.72	7.96	11.5	0.69	0.956	0.64	0.40	0.37
n-Propanol	Ethanol	Water	0.45	12.7	11.5	1.10	0.952	1.26	2.34	2.80
iso-Propanol	Ethanol	Water	0.86	21.0	11.5	1.83	0.965	1.59	2.13	1.85
n-Butanol	Ethanol	Water	0.26	20.0	11.5	1.74	0.779	1.88	6.70	7.23
iso-Butanol	Ethanol	Water	0.43	29	11.5	2.52	0.956	2.03	5.87	4.72
Methyl ethyl ketone	Ethanol	Water	0.99	44	11.5	3.82	0.920	3.50	3.86	3.54
n-Butanol	n-Propanol	Water	0.47	20	12.7	1.57	0.973	1.12	3.34	2.38
iso-Propanol	n-Propanol	Water	1.8	21	12.7	1.65	0.910	1.39	0.92	0.77
Ethyl acetate	Ethanol	Butyl acetate	1.07	3.79	6.90	0.55	0.816 −0.829	0.50	0.51	0.47

For the calculation of S_{AB}, see text.

Pierotti–Deal–Derr's method [91]. Butler et al. [92] denoted that the activity coefficient, γ_i^0 at infinite dilution for alcohol with water as a solvent, regularly increases in the order of the number of carbons involved by an alcohol molecule. Based on their results, Pierotti et al. proposed a general explanation with five parameters in the case of a solution composed of R_1X_1 and R_2X_2 compounds as a solute and a solvent, respectively, at a fixed temperature.

$$\log \gamma_i^0 |_S = A_{1,2} + B_2(n_1/n_2) + c_1/n_1 + D(n_1 - n_2)^2 + (F_2/n_2) \tag{158}$$

where, the n_1 and n_2 are the number of carbons in the hydrocarbon radicals, R_1 and R_2, respectively. The five parameters, $A_{1,2}$, B_2, C_1, D, and F_2 are determined according to the combination of two of X_1, R_1, X_2, and R_2, of course, except for two conbinations of R_1X_1 and R_2X_2. They tabulated the values of the parameters obtained from experimental data. The selectivity can be evaluated using Equation 152 related to $\gamma_i^0|_{xS \to 1}$. As the value of $\gamma_i^0|_{xS \to 1}$ can be estimated within the accuracy of about 8% error on an average, Equation 158 is useful in the case of selecting solvents.

Weimer–Prausnitz's method [93]. Weimer et al. proposed a method for estimating the activity coefficient at infinite dilution. The excess free energy of mixing, at a fixed pressure is represented precisely based on the excess free energy of mixing for the regular solution [94–96] and the enthropy of mixing developed by Flory–Huggins [97, 98]:

$$\Delta G_P^E = \varphi_1\varphi_2(x_1v_1 + x_2v_2)\,[(\lambda_1 - \lambda_2)^2 + (\tau_1 - \tau_2)^2 - 2\psi_{12}] + RT[x_1 \ln(\varphi_1/x_1) + x_2 \ln(\varphi_2/x_2)] \tag{159}$$

where φ is the volume fraction ($\varphi_1 = x_1v_1/\sum x_iv_i$), the v is the molar volume [cm^2/g-mol]. $\lambda[(cal/cm^3)^{1/2}]$ and $\tau[(cal/cm^3)^{1/2}]$ are the solubility parameters for nonpolar and the inter-

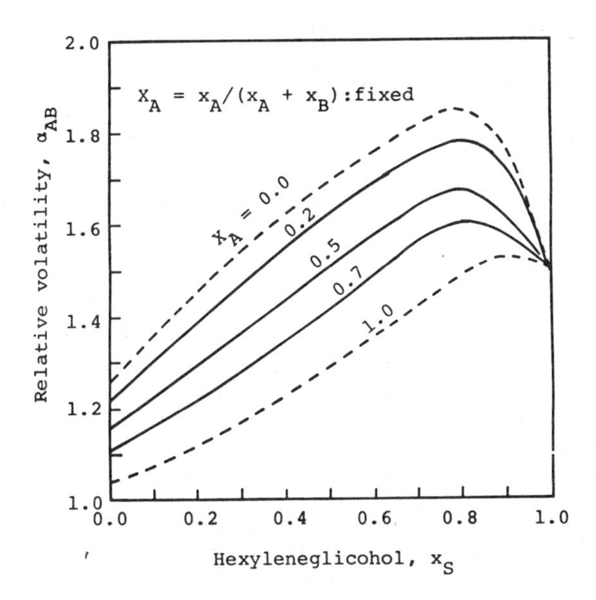

Figure 50. Relative volatility of ethylcyclohexane/ethylbenzene system with hexyleneglicohol [89].

molecular induced-force. Differentiating Equation 159 and setting $x_2 \rightarrow 0$, the activity coefficient at infinite dilution for the i component is obtained by

$$RT \ln \gamma_2^0 = v_2[(\lambda_1 - \lambda_2)^2 + (\tau_1 - \tau_2)^2 - 2\psi_{12}] + RT \ln + [(v_2/v_1) + (1 - v_2/v_1)] \tag{160}$$

The ψ_{12} in Equation 160 is determined by experiment:

For saturated hydrocarbons

$$\psi_{12} = 0.399\ (\tau_1 - \tau_2)^2 \qquad 0°\ C \leqq t \leqq 125°\ C$$

For unsaturated hydrocarbons

$$\psi_{12} = 0.388\ (\tau_1 - \tau_2)^2 \qquad 0°\ C \leqq t \leqq 45°\ C$$

For aromatic hydrocarbons

$$\psi_{12} = 0.477\ (\tau_1 - \tau_2)^2 \qquad 25°\ C \leqq t \leqq 100°\ C$$

The selectivity of A and B in a solvent S is evaluated by using the activity coefficients at infinite dilution for binary systems, A/S and B/S as follows [99]:

$$RT \ln S_{AB}^0 = v_A(\lambda_S - \lambda_A)^2 - v_B(\lambda_S - \lambda_B)^2 + v_A(\tau_S - \tau_A)^2 - v_B(\tau_S - \tau_B)^2$$

$$+ RT\,[\ln (v_A/v_B) + (v_B - v_A)/v_S] + 2(v_B\psi_{SB} - v_A\psi_{SA}) \tag{161}$$

For A and B which are hydrocarbons, as τ_A and τ_B are equal to zero, Equation 161 simplities to [93]:

$$RT \ln S_{AB}^0 = \tau_S^2(v_A - v_B) - [v_A(\lambda_S - \lambda_A)^2 - v_B(\lambda_S - \lambda_B)^2]$$

$$+ 2v_B\psi_{SB}v_A\psi_{SA} + RT\,[\ln (v_A/v_B)$$

$$+ (v_B - v_A)/v_S] \tag{162}$$

Consequently, when the values of the solubility parameters and the molar volumes for A, B, and S components are given, the selectivity can be calculated by Equations 161 or 162.

Solubility parameter: The solubility parameter is evaluated based on the molar latent heat [cal/g-mol] by

$$\delta = (\Delta E^V/v)^{1/2} \tag{163}$$

Assuming the gas to be ideal the value of E^V is theoretically estimated by using the vapor pressure of a pure component such as the Antoine (Equation 2):

$$\Delta E^V = 2.303\ RBT^2/(t + C)^2 - RT \tag{164}$$

where, the B and C are the parameters in the Antoine equation and T is the absolute temperature (= t + 273.15). R is the gas law constant [cal/K,g-mol]. If the pure component vapor pressure is unknown, the empirical equation for the molar latent heat is [100]:

$$H_{298}^V = -2950 + 23.7\ T_b + 0.020\ T_b^2 \tag{165}$$

Table 6
Solubility Parameters*

Solvents/Hydrocarbons	Formulas	(°C)	V(cm³/g-mol)	λ(cal/cm³)^1/2	τ(cal/cm³)^1/2
1. Acetophenone	$C_6H_5COCH_3$	202	117.4	9.44	3.69
2. Tetrahydrofuran	CH_2—CH_2 / O / CH_2—CH_2	66	81.7	8.32	3.71
3. Pyridine	N	115.4	80.9	9.88	3.71
4. Cyclohexanone	O ‖	155.6	104.2	8.84	4.04
5. Chloroethane	C_2H_5Cl	12.3	74.1	7.38	4.32
6. Diethyl ketone	$CH_3CH_2COCH_2CH_3$	101.5	106.4	7.75	4.44
7. Diethyl carbonate	$(C_2H_5O)_2C{=}O$	126	121.9	7.89	4.49
8. Bromoethane	CH_3CH_2Br	38.4	75.1	7.63	4.83
9. Nitrobenzene	—NO_2	210 ~ 211	102.7	9.70	4.89
10. Bis (2-chloroethyl) ether	$ClCH_2CH_2OCH_2CH_2Cl$	178	117.8	8.34	5.22
11. Trimethyl phosphate	$(CH_3O)_3PO$	197	116.2	8.46	5.22
12. Iodoethane	C_2H_5I	71	81.1	7.66	5.24
13. Methyl ethyl ketone	$CH_3COC_2H_5$	79.6	90.1	7.64	5.33
14. Cyclopentanone	O ‖	130.6	89.5	8.70	5.37
15. 2,4-Pentanedione	$CH_3COCH_2COCH_3$	140.5	103.0	8.06	5.69
16. 2,5-Hexanedione	$CH_3COCH_2CH_2COCH_3$	188	117.7	8.45	5.88
17. Diethyl oxalate	$CH_2H_5OOCCOOC_2H_5$	185.7	136.2	8.37	5.94
18. 2-Nitropropane	$CH_3CH(NO_2)CH_3$	120.3	90.7	7.95	6.02
19. Methoxyacetone	$CH_3COCH_2OCH_3$	—	93.2	7.91	6.11
20. Acetone	CH_3COCH_3	56.5	74.0	7.66	6.14
21. Dimethyl carbonate	$(CH_3)_2CO_3$	90 ~ 91	85.0	7.77	6.20
22. Butyronitrile	$CH_3CH_2CH_2CN$	117.5	87.9	7.96	6.28
23. 2,3-Butanedione	O O / CH_2—CH—CH—CH_2	138	87.8	7.73	6.35
24. Aniline	NH_2	184 ~ 186	91.5	9.85	6.37
25. 1-Nitropropane	$CH_3CH_2CH_2NO_2$	131.6	89.5	8.06	6.40
26. N-Methylpyrrolidone	CH_2—CO / NCH_3 / CH_2—CH_2	—	96.6	9.15	6.55
27. Acetic anhydride	$(CH_3CO)_2O$	139	95.0	7.85	7.11
28. Propionitrile	CH_3CH_2CN	97.2	70.9	7.97	7.17
29. Citric acid anhydride	CH_3C—CO / ‖ O / HC—CO	—	89.7	9.42	7.22
30. Methoxy acetonitrile	CH_3OCH_2CN	—	75.2	8.06	7.33
31. Furfural	O—CHO	161.8	73.2	9.04	7.62
32. Nitroethane	CH_3NO_2	101.2	72.1	8.04	7.66
33. Dimethylacetamide	$CH_3CON(CH_3)_2$	163 ~ 165	93.2	8.29	7.69
34. γ-Butyrolactone	O =O	206	77.1	9.50	8.01
35. Dimethylformamide	$HCON(CH_3)_2$	153	77.4	8.29	8.07
36. 3-Chloropropionitrile	$ClCH_2CH_2CN$	176	77.7	8.44	8.73
37. Acetonitrile	CH_3CN	81.6	52.6	8.03	8.98
38. Ethylenediamine	$H_2NCH_2CH_2NH_2$	116 ~ 117	67.3	8.10	9.40
39. Nitromethane	CH_3NO_2	101.2	54.3	8.08	9.44
40. Dimethyl sulfoxide	$(CH_3)_2SO$	189	71.3	8.56	9.47

* *Stecher, P. G., et al.:* The Merck Index *Eighth Edition, Merck & Co., Inc., U.S.A. (1968).*

where, the T_b is the absolute temperature of the standard bubble point [°K]. Meanwhile, for the polar liquid as a solvent, the nonpolar and polar solubility parameters are needed. The following relationship is satisfied:

$$\tau^2 = \delta^2 - \lambda^2 \tag{166}$$

For hydrocarbons, as $\tau = 0,\ \lambda = \delta$ (167)

As the correlation between λ^2 and v for the polar molecules is shown in Reference 93, the τ can be determined from Equations 163 and 166. Values of the solubility parameter for hydrocarbons and solvents are tabulated in Table 6.

Gaschromatograph method [101, 102, 103]. If the extent of interaction between the stationary phase as a solvent and a substance as a solute largely depends on the property of a solute, its retention time through a column in the gaschromatograph is different from that of other solutes. Denoting the retained volumes of solutes, A and B, and the carrier gas as V_R^A, V_R^B, and V_C, the relative volatility of the infinite dilution with solvent S is represented by

$$(\alpha_{AB}^0)_S = (V_R^B - V_C)/(V_R^A - V_C) \tag{168}$$

When the carrier gas flow rate is constant

$$(\alpha_{AB}^0)_S = t_m^B/t_m^A \tag{169}$$

where t_m^i is the retention time from the injection time to the appearance time of the peak of i component. The selectivity is calculated by using the pressures of the pure components, A and B, at the operating temperature of the column

$$S_{AB}^0 = \gamma_A^0/\gamma_B^0 = (\alpha_{AB}^0)_S(P_B^0/P_A^0) \tag{170}$$

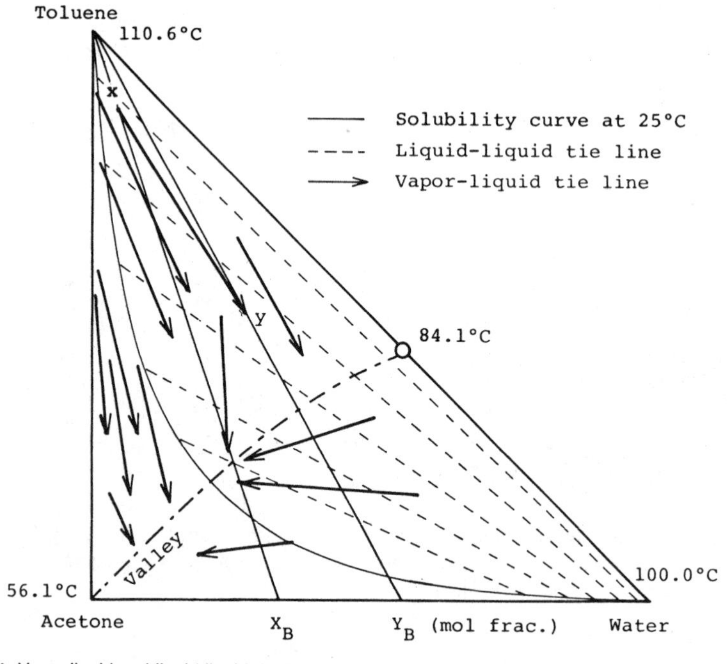

Figure 51. Vapor-liquid and liquid-liquid tie lines for water/acetone/toluene system.

Advanced Extractive Distillation

Thus for extractive distillation systems with homogeneous mixtures have been described. Many recent texts and papers on azeotropic distillation systems give greater attention to system structures which use a minimum azeotrope with the lowest temperature in a mixture. Thorough investigation of the vapor-liquid equilibrium for heterogeneous systems in which the temperature of an azeotrope is not at the lowest temperature of the pure components results in a system which features extraction and extractive distillation. For example, consider the dehydration of acetone by toluene as a solvent [83].

Phase Equilibrium

Figure 51 shows the vapor-liquid tie lines, the binodal curve and the liquid-liquid tie lines. The valley lies between the minimum azeotrope (84.1° C) of the toluene/water mixture and acetone (56.1° C), but there is no ternary azeotrope.

Let us investigate the vapor-liquid equilibrium for the liquid composition, x, near pure toluene. The vapor composition, y, in equilibrium related to the x, contains a significantly larger amount of water than x does, that is, $X_A < Y_A$. The liquid-liquid tie line through

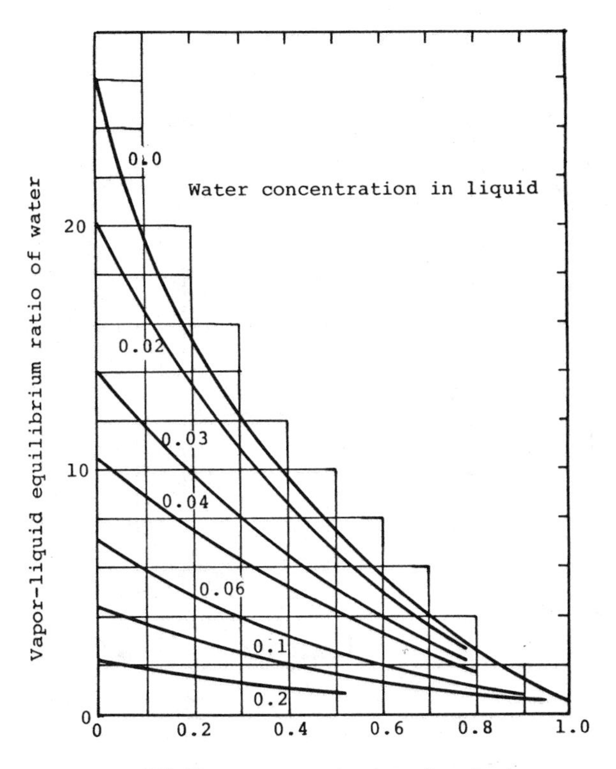

Figure 52. Vapor-liquid equilibrium ratio of water in water/acetone/toluene system.

y shows that the stream condensed with the composition y split into two layers, a water-rich layer and a toluene-rich layer. The liquid-liquid tie lines intersect the valley and transfer the composition in the toluene-rich region to the other. The water in the water-rich layer can be easily removed because, in the water-rich composition range, the vapor-liquid equilibrium relationship for the pseudo-binary system of water and acetone is completely separated from the diagonal line in the x-y diagram.

The toluene-rich layer contains a small amount of water. As shown in Figure 52, fortunately, the vapor-liquid equilibrium ratio of water increases fully as the water concentration in the liquid decreases.

System Structure

First, the original feed of water and acetone is diluted with toluene. The ternary mixture is split into two layers by decantation. This operation corresponds to the extraction of acetone from the acetone/water mixture with toluene. The toluene-rich layer returns to the column and is separated into a toluene/acetone mixture and an intermediate mixture. The separating pattern depends on the flow pattern of the vapor-liquid tie lines as noted in the first section. The water-rich layer is supplied to the second column and is dehydrated. The toluene/acetone mixture is transfered to the third column, and the toluene and acetone are separated. The entire system is illustrated in Figure 53. The first column is the extractive distillation column, the second is the dehydration column, and the third is the solvent

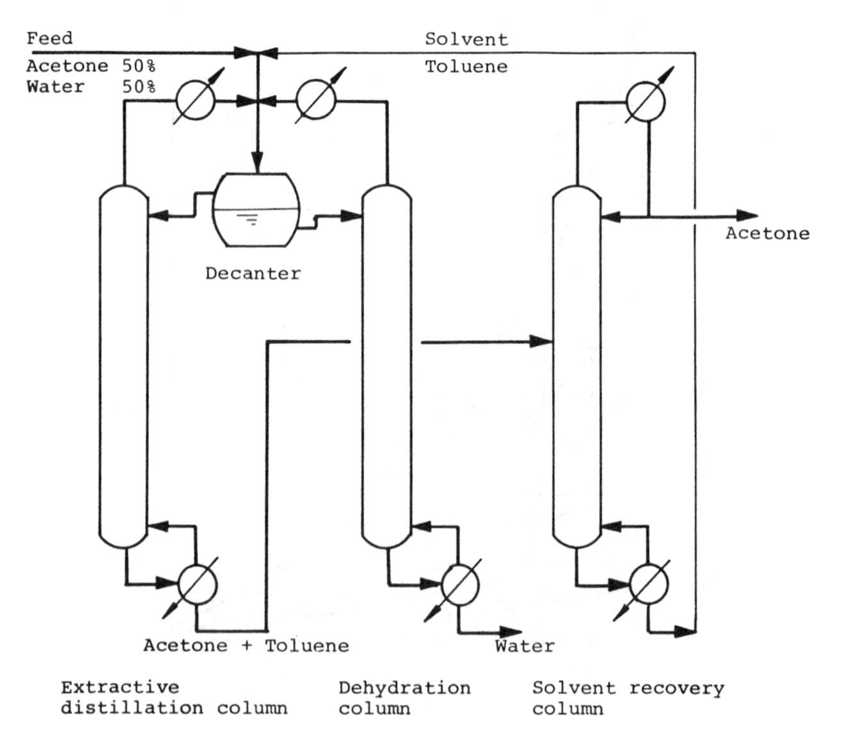

Figure 53. Extractive distillation system with two liquid phases.

recovery column. Figure 54 illustrates the composition profile in the first column. It is clear that toluene exists at a high concentration level in the column. This profile is a feature of the solvent in an extractive distillation.

PRINCIPLES OF BATCH DISTILLATION

Batch distillation is essentially an unsteady-state process; that is, the compositions and flow rates in the column are time variant. The liquid mixture (feed) is separated into its components in a batch-wise mode. The analysis of batch distillation is complex in comparison to continuous distillation. It is, however, a more flexible and versatile process.

Batch distillation is often employed when a relatively small product is to be manufactured and where different kinds of mixtures are to be handled by the same distillation column.

Simple Batch Distillation

The simplest case of batch distillation is commonly called a simple, or a differential, or a single-stage batch distillation. The simple batch distillation process, as shown in Figure 55, consists of a still, a total condenser, and one or more product receivers. The feed mixture to be separated is initially charged to a still (or kettle), and is continuously heated by a heating medium, such as steam. The vapor generated is condensed and then collected in a receiver. The distillate contains a greater portion of the more volatile components (low

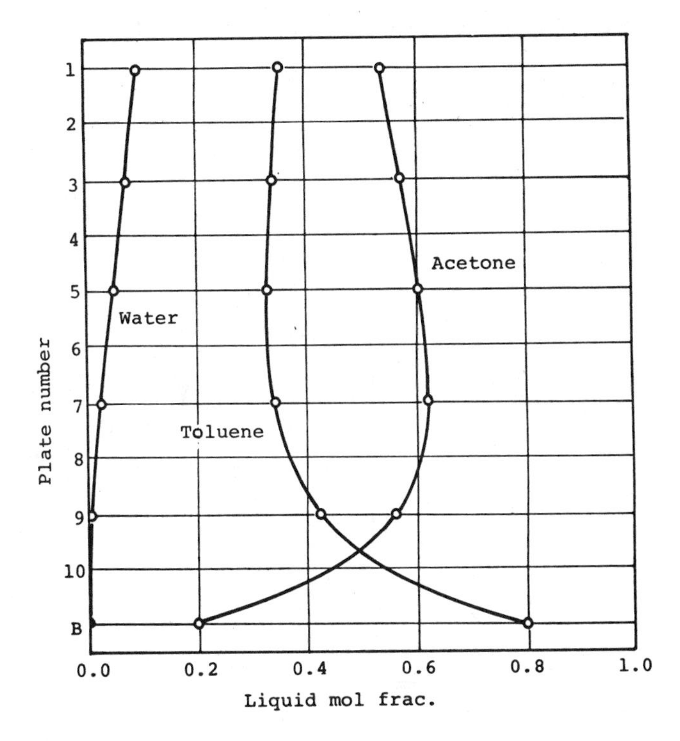

Figure 54. Composition profile in extractive distillation column.

boiling materials) than the residue. Since simple batch distillation provides only one theoretical stage of separation, the refluxing phenomena hardly affects the separation characteristics and consequently, it becomes difficult to achieve a sharp separation of the mixture. Therefore, as a separation method, the simple batch distillation could be generally used for the separation of a mixture with high purity of the product, but low recovery, or a mixture with high recovery, but low purity of the product.

Binary Systems

In a binary system consisting of component A and B at any instant of the simple batch distillation process for vaporizing a differential amount of the still liquid, dV, of composition y_A, two mass balances in differential form can be written as follows:

$$dV = -dW \tag{171}$$

$$\begin{aligned} y_A &= -d\,(Wx_A) \\ &= -W\,dx_A - x_A\,dW \end{aligned} \tag{172}$$

where W and x_A are the moles of liquid in the still and the mole fraction in the liquid, respectively. Substituting Equation 171 into Equation 172:

$$-dW/W = dx_A/(x_A - x_A) \tag{173}$$

Integrating gives the well-known Rayleigh Equation (Reference [104]):

$$\int_{W_F}^{W} (dW/W) = \int_{x_{FA}}^{x_A} dx_A/(y_A - x_A) \tag{174}$$

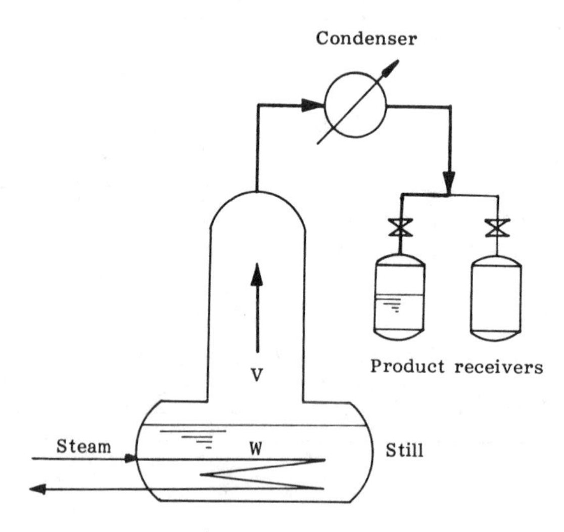

Figure 55. Simple batch distillation.

$$\ln(W/W_F) = \int_{x_{FA}}^{x_A} dx_A/(y_A - x_A) \tag{175}$$

where the subscript F represents initial liquid feed charged into the still. Equation 175 relates the composition of the liquid to the moles of the remaining liquid in the still pot.

In order to use Equation 175, a relationship between y_A and x_A is needed. Assuming that the vapor-liquid equilibrium exists between the liquid and vapor of the still pot, and that both phases are well mixed, the right-hand side of Equation 175 can be obtained by plotting $1/(y_A - x_A)$ vs. x_A as shown Figure 55 and by integrating graphically between the limits x_{FA} and x_A. The area under the curve of Figure 56 is the value of integration between x_{FA} and x_A.

The graphical integration method to obtain explicitly a relationship between (W/W_F) and x_A is rather tedious. There are several approximate methods of integrating Equation 175. First, for the case when the vapor-liquid equilibrium ratio, K_A, can be assumed as constant during operation, Equation 175 can be integrated analytically:

$$\ln(W/W_F) = (dx_A/x_A)/(K_A - 1)$$

$$= \ln(x_A/x_{FA})/(K_A - 1) \tag{176}$$

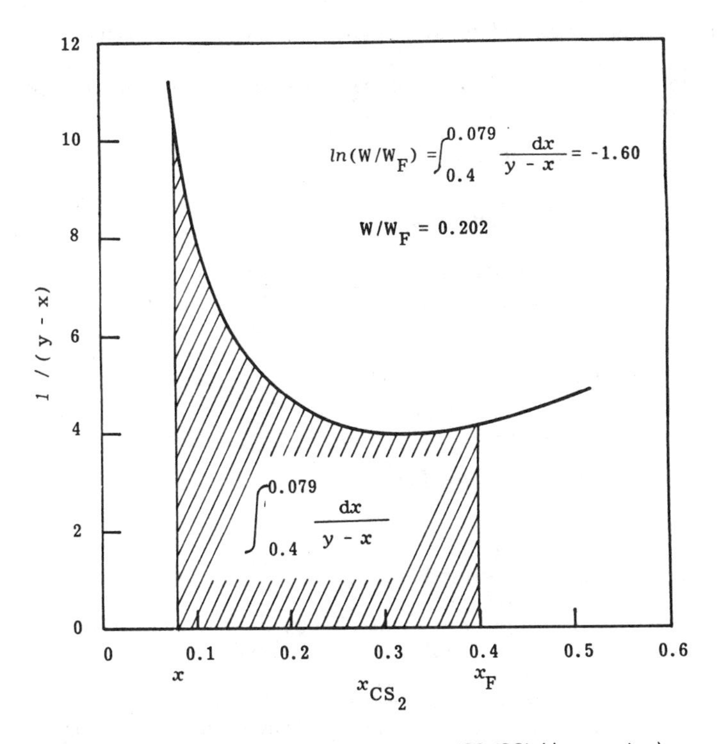

Figure 56. Graphical integration for simple batch distillation (CS_2/CCl_4 binary system).

Also, if the relative volatility, $\alpha_{A/B}$ is constant, the vapor-liquid relation can be defined as follows:

$$y_A = \frac{\alpha_{A/B} x_A}{1 + (\alpha_{A/B} - 1) x_A} \tag{177}$$

Substituting Equation 177 into Equation 175 and integrating analytically gives:

$$\ln (W/W_F) = \frac{1}{\alpha_{A/B} - 1} \ln \frac{x_A(1 - x_{FA})}{x_{FA}(1 - x_A)} + \ln \frac{1 - x_{FA}}{1 - x_A} \tag{178}$$

Using Equations 175, 176, or 178, the residue ratio, W/W_F, can be calculated if the initial feed mole fraction, x_{FA}, and the final still liquid mole fraction, x_A, of component A are known. Conversely, if x_{FA} and W/W_F are specified, x_A could be calculated by a trial-and-error method.

Multicomponent Systems

For a multicomponent system, even for a constant relative volatility, Equation 175 is usually not suitable, because a vapor composition of component A, y_A, may be related not only to the liquid composition of component A, x_A, but also to the compositions of other components, such as x_B, x_C, etc. A more convenient expression descriptive of multicomponent systems can be derived.

Consider a mixture composed of M components, A, B, C, ..., which have constant relative volatilities α_A, α_B, α_C, ..., respectively. We may describe the multicomponent simple batch distillation by defining the component mass balance equations and the vapor-liquid equilibrium relations as follows:

$$y_i \, dV = -d\,(W x_i) \qquad (i = A, B, C, \ldots) \tag{179}$$

$$y_i = (\alpha_i / \textstyle\sum \alpha_j x_j) x_i \qquad (i = A, B, C, \ldots) \tag{180}$$

Dividing this expression by the combination of Equations 171, 179, and 180, for component A and for corresponding equations for other components, the following relationship between component A and component i is obtained:

$$-d\,(W x_A)/-d\,(W x_i) = \alpha_{A/i}(W x_A / W x_i) \qquad (i = B, C, \ldots) \tag{181}$$

This equation can be integrated to

$$\ln (w_A/w_{FA}) = \alpha_{A/i} \ln (w_i/w_{Fi}) \qquad (i = B, C, \ldots) \tag{182}$$

or

$$(w_A/w_{FA}) = (w_i/w_{Fi})^{\alpha_{A/i}} \qquad (i = B, C, \ldots) \tag{183}$$

where w_A and w_i refer to the molar amounts of component A and i in the still liquid, respectively. Equation 183 is a Rayleigh equation for multicomponent systems having constant relative volatilities.

If the relative volatility varies during the batch distillation with respect to the liquid composition, Equation 182 or Equation 183 must be solved for every component simultaneously, when the component residue ratio of any one component is specified.

Distillation Curves

The curve showing the changes in still-liquid composition for a simple batch distillation has been defined as the residue curve [105]. For the multicomponent simple distillation, the mass balance equations shown in Equations 171 and 179 give:

$$W\,dx_i = (x_i - y_i)\,dV \tag{184}$$

At the phase plane, for any two components i and j

$$dx_i/dx_j = (x_i - y_i)/(x_j - y_j) \quad (i = A, B, \ldots) \tag{185}$$

These equations give the liquid composition in the still for a simple batch distillation. The slope of the tie line at a bubble point is equal to the derivative of the liquid compositions. If vapor-liquid equilibrium is well defined, at least for ternary systems, the change in the still liquid composition can be easily estimated by using the tie-line on the phase plane. Naka et al. [82] proposed a calculation method of the residue curve for strongly nonideal ternary systems such as acetone/chloroform/methanol and acetone/chloroform/isopropylether and discussed in the relationship between the residue curve, and the valley or the ridge in detail.

Multistage Batch Distillation

As noted earlier, simple batch distillation has little commercial significance as a separation method for obtaining products which have high purity and also high recovery. In this application it is necessary to use a multistage column with a still and a condenser as shown Figure 57.

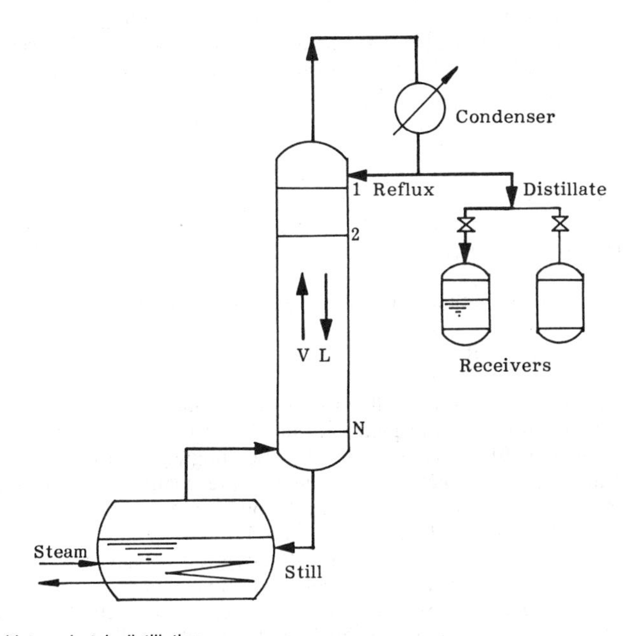

Figure 57. Multistage batch distillation.

In contrast with simple batch distillation, the refluxing phenomena play an important role in the operation. In addition to the liquid in the still, there is liquid holdup, in the column which makes it difficult to solve the problem rigorously.

For the sake of simplicity, the calculation procedures neglecting the column holdup for the multistage batch distillation will be discussed first, and then the effects of holdup on the degree of separation and on capacity will be presented.

Binary Systems

In binary systems, an overall differential mass balance gives:

$$dD = -dW \tag{186}$$

where D and W are the moles of the liquid residue in the still and of the distillate, respectively.

A component mass balance is:

$$x_D\, dD = -d\,(Wx_W)$$

$$dD/W = dx_W/(x_W - x_D) \tag{187}$$

This equation describes the instantaneous change of the liquid composition in the still in terms of the instantaneous withdrawing rate of the distillate and is similar to the Rayleigh equation shown in Equation 173, but differs in that the denominator is $x_W - x_D$ instead of $x_A - y_A$.

In batch distillation the use of this equation to determine the independent variables, such as the recovery ratio of the products to residue or distillate, purity of the products, reflux ratio, or number of theoretical plates reguires certain assumptions and specifications.

First, it is assumed that the vapor and liquid flows are equimolal, and that the liquid holdup in the column is negligible. From mass balances taken for the section within the envelope including the top, the following operating line equations at any point with respect to time are obtained:

$$V = L + D \tag{188}$$

$$Vy_{n+1} = Lx_n + dx_D \tag{189}$$

$$y_{n+1} = (L/V)x_n + (D/V)x_D \tag{190}$$

using the definition of the reflux ratio, R = L/D, Equation 190 gives:

$$y_{n+1} = (R/R+1)x_n + (1/R+1)x_D \tag{191}$$

where V and L are vapor and liquid flows in moles per unit time, respectively; D is moles per unit time of the distillate, and plates are numbered from top to bottom.

Note that Equation 191 is identical with the operating equations for continuous binary distillation, but, as described before, V, L, and D are quantities in moles per unit time instead of flow rate as in continuous distillation. To distinguish them more clearly, differential forms of those variables; dV, dL, and dD have been used (see Reference 105).

The problem of binary multistage batch distillation can be solved by using a McCabe–Thiele graphical method by changing the column compositions.

In order to solve Equations 186, 187, and 191, an operation method or mode must first be selected from among the number of methods for batch distillation.

Normal operations. For the *constant reflux operation,* both compositions of the distillate and still liquid are changing with time during batch distillation, because more volatile component A is removed preferentially from the still.

Combining Equations 186 and 187, and integrating between two limiting values gives:

$$\int_{W_F}^{W} (dW/W) = \int_{x_{WF}}^{x_W} dx_W/(x_D - x_W) \tag{192}$$

or

$$\ln (W/W_F) = \int_{x_{WF}}^{x_W} dx_W/(x_D - x_W) \tag{193}$$

where W_F and W are the feed and final amounts of still liquid respectively; x_{WF} and x_W are feed and final liquid composition in mole fraction, respectively.

Under the conditions of initial feed composition, reflux ratio, and number of plates, Equation 193 can be integrated graphically if one of the items, such as recovery ratio, average distillate composition $(x_D)_{av}$, or required final still composition x_{WF} is specified.

From the overall component balance, the average composition of the distillate is obtained:

$$(x_D)_{av} = (W_F x_{WF} - W x_W)/(W_F - W) \tag{194}$$

where $(x_D)_{av}$ is the average distillate composition. On the other hand, the total vapor requirement, V_{tot} can be found by,

$$V_{tot} = (R+1)D = (R+1)(W_F - W) \tag{195}$$

and alternatively, the time requirement for distillation to yield the specified average composition of the product gives:

$$\theta = (R+1)(W_F - W)/V_{tot} \tag{196}$$

Or, using Equation 193 and the integrated value of the righthand side can be found by the following equation

$$\theta = (R+1)\frac{W_F}{V_{tot}}\frac{e^I - 1}{e^I} \tag{197}$$

where I is the integrated value. These equations are derived by Block [106]. To apply Equations 192 through 197, the following iterative method is needed:

In the case when $(x_D)_{av}$ is specified, we have to determine the boundary values x_{WF} and x_W of the region to be integrated by the graphical method shown in Figure 58. In this figure, the procedure of finding x_W with respect to x_D for an arbitrarily chosen operating line is repeated until the latest value of x_W becomes less than its assumed value. And then the curve for $1/(x_D - x_W)$ vs. x_W is plotted to evaluate the area under the curve and between the limiting values of x_{WF} and x_W (assumed).

Eventually, $(x_D)_{av}$ is calculated by Equation 194 to compare it with its specified value. If they are found agreeable, then V_{tot} and θ can be determined. Otherwise, the initial values of x_W are to be reassumed and the aforementioned calculation procedure repeated.

When x_W (desired) is given, integration of Equation 194 can be carried out in a straightforward manner to determine W/W_F, D, $(x_D)_{av}$, V_{tot} and θ.

For the *variable reflux operation*, the reflux ratio must be continuously adjusted so as to provide a constant distillate composition which satisfies the specification during the distillation. As the more volatile component is preferentially removed the still liquid becomes rich in the less volatile substance. Thus, reflux ratio must be continously increased to hold the distillate composition constant.

Bogart [107] originally developed the following equation under the condition when the holdup on the plate can be neglected.

The overall mass balance is

$$dD = (1/R+1)\,dV \tag{198}$$

and a component mass balance is

$$W = W_F(x_{WF} - x_D)/(x_W - x_D) \tag{199}$$

where x_D is the distillate composition with the specified value. It should be noted that R in Equation 198 is a reflux ratio at any moment.

Substituting Equations 198 and 199 into Equation 187 gives:

$$dV = [W_F(x_{WF} - x_D)/(x_W - x_D)^2]\,(R+1)\,dx_W \tag{200}$$

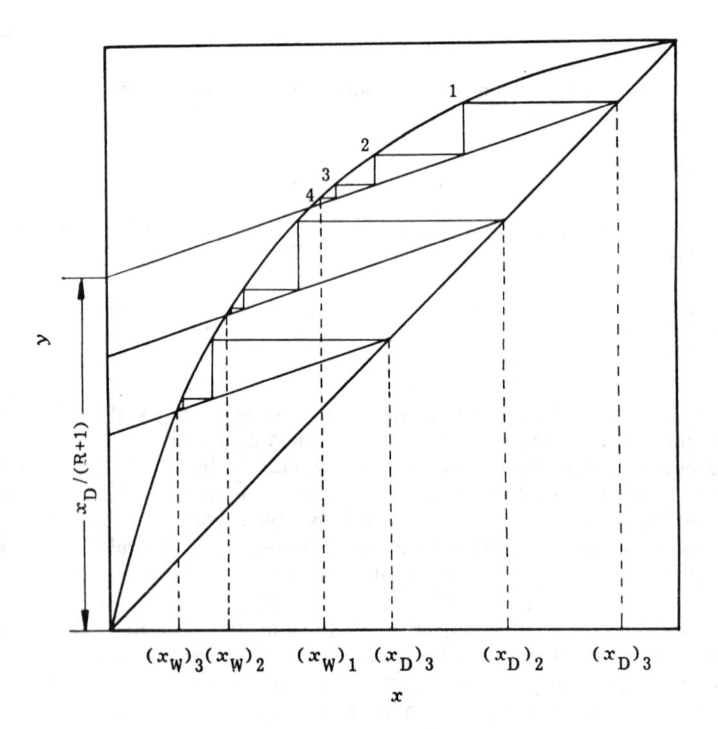

Figure 58. Graphical method for constant reflux operation.

Or, by integrating between the initial feed composition, x_{WF}, and the specified final composition in the still, x_W, the total vapor requirement for the distillation is obtained:

$$V_{tot} = W_F (x_{WF} - x_D) \int_{x_{WF}}^{x_W} \frac{(R + 1)\, dx_W}{(x_W - x_D)^2} \tag{201}$$

The V_{tot} can be calculated using the x-y diagram, Figure 59, as follows.

Plotting $(R+1)/(x_W - x_D)^2$ vs. x_W and then integrating graphically, the integration term on the right-hand side in Equation 201 can be obtained, and finally, V_{tot} can be obtained. If the vapor generating rate is given, the time for batch distillation can be determined directly by dividing V_{tot} with vapor rate.

In addition to the constant reflux and constant composition operation (variable reflux), a cycling procedure can be used [5]. As one of the cycling procedures, a periodic total reflux operation is commonly used to remove trace amount of volatile impurity from the desired components. The unit operates at total reflux until equilibrium is achieved. The distillate is withdrawn without reflux for a short period at time, and then the column is again returned to a total reflux condition. This cycle is repeated until the specification of the residue product is met.

Robinson and Gilliland [108] proposed another operation procedure in order to remove trace amounts of the less volatile impurity from the most volatile product. In this case, the

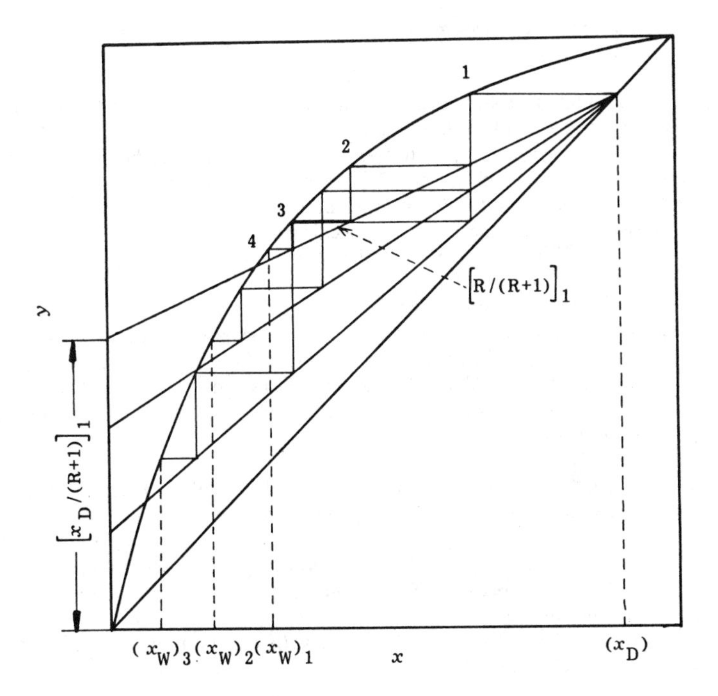

Figure 59. Graphical method for variable reflux operation (constant composition of the distillate).

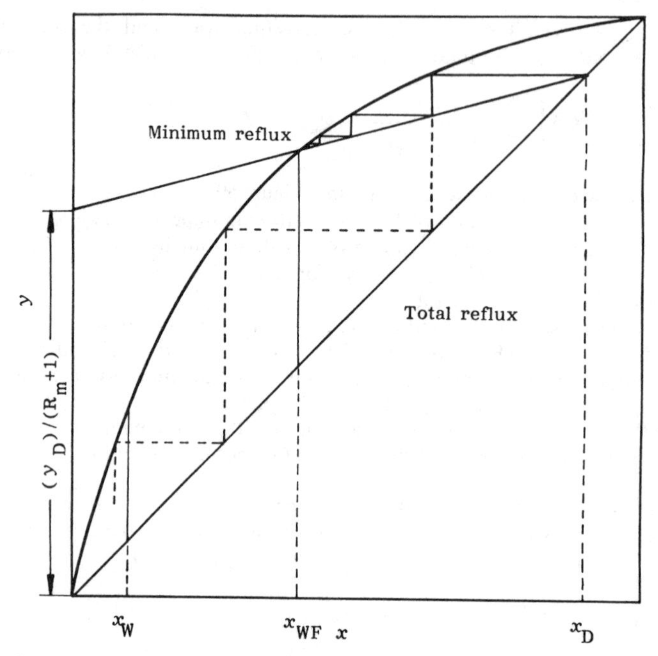

Figure 60. Graphical determination of minimum reflux ratio, R_m, and minimum number of plates, N_m.

feed is charged to an accumulator and liquid is continuously added to the top of the column from the accumulator, and the less volatile component is removed from the bottom.

Kojima [109] also proposed a new batch operation method where the liquid holdup is established in an accumulator which is placed just after the condenser and the operation is carried out under total reflux condition until the equilibrium composition is attained. This holdup may be adjusted to meet the recovery specification of the top product. It can be used to remove trace impurities of the less-volatile product at the bottom, or to purify the most volatile component at the top of the still.

Another possibility of the batch distillation is a optimal reflux operation which will give the desired separation with in a minimum batch-time. For the case of a binary mixture with negligible column liquid holdup, Conward [110] compared the optimal reflux method with the constant reflux and constant product composition operations, and found that a saving of the batch time up to five percent was possible and that the optimal reflux operation gave the lowest vapor requirement.

Limiting operations [5, 108]. / There are two limiting operations in the batch distillation as in the continuous distillation: total reflux and minimum reflux.

- *Minimum number of plates:* The total reflux condition determines the minimum number of theoretical plates required to meet the specification of the desired product-recovery of component A, in the distillate.

  For multistage batch distillation at total reflux with constant relative volatility, the following equation is obtained:

$$\ln (W/W_F) = \frac{1}{\alpha^{N_m+1}+1} \ln \frac{x_W(1-x_{WF})}{x_{WF}(1-x_W)} + \ln \frac{1-x_{WF}}{1-x_W} \quad (202)$$

where N_m is the minimum number of theoretical plates. This equation is identical to Equation 178 except that α^{N_m+1} replaces α. From Equation 202 it is found that when the number of theoretical plates increase the recovery of the top product increases exponentially.

For the case when the relative volatility is not constant, N_m can be determined graphically using the diagonal operating line shown in Figure 60. The minimum reflux operation is another limit for a given separation and recovery at which an infinite number of plates will be required.

- *Minimum reflux:* As shown in Figure 60, at the minimum reflux ratio, the operating line must touch the equilibrium curve. In this case, the minimum reflux ratio can be determined by the value of the intercept, or the slope of the limiting operation line.

  For variable reflux operation with constant relative volatility the minimum vapor requirements corresponding to Rm derived by Robinson and Gilliland are as follows:

$$V_{tot} = \frac{W_F(x_{WF} - x_D)}{2(\alpha - 1)}\left[\frac{2(\alpha-1)+(1+x_D)}{x_D - 1}\ln\frac{(1-x_W)(x_{WF}-x_D)}{(1-x_{WF})(x_W-x_D)} - \ln\frac{x_W^2(x_{WF}-x_D)(1-x_{WF})}{x_{WF}^2(x_W-x_D)(1-x_W)}\right] \quad (203)$$

Stage-reflux correlation. For batch distillation, a correlation between the number of theoretical plates and the reflux ratio is complex in contrast to continuous distillation, because of the effect of column holdup.

Up to now, many analytical and graphical methods have been developed [111–116] for the case of negligible column holdup and constant reflux operations.

Among them, Kojima and Aoyama [112] recommend the empirical correlation, shown in Figure 61, for the number of stages to the reflux ratio when the average distillate composition, (x_W)av and the recovery ratio, β, of the most volatile component are given.

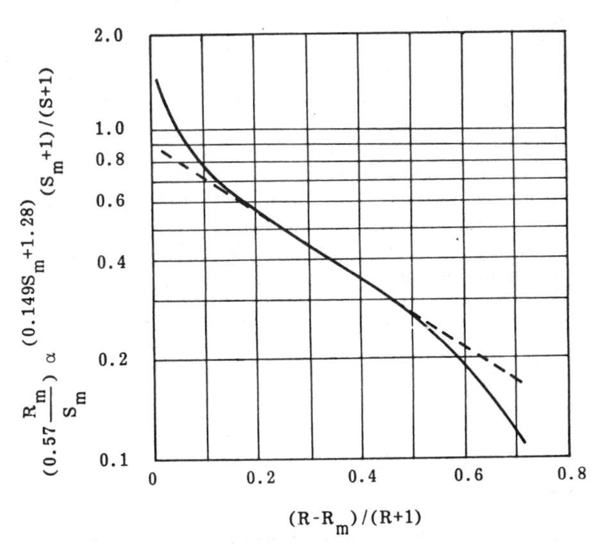

Figure 61. Correlation curve between the reflux ratio, R, and the number of plates, N, for batch distillation [109].

For the limiting range of $(R - R_m)/(R + 1)$ $(0.15 - 0.5)$, the following equation can be used to obtain the number of plates.

$$\log [0.57(R_m/S_m)\alpha^{0.149S_m + 1.28}(S - S_m)/(S + 1)] = -(R - R_m)/(R + 1) - 0.056 \quad (204)$$

where $S = N + 1$
$S_m = N_{m+1}$
R_m = the minimum reflux ratio
N_m = the minimum number of plates

N_m and R_m can be calculated by the following equations

$$N_m + 1 = (1/\log \alpha) \log [\log (1 - \beta)/\log (1 - \delta)]$$

$$R_m = [\log (1 - \beta) - \alpha \log (1 - \delta)]/[(\alpha - 1) \log (1 - \eta)] \quad (205)$$

where $\delta = [x_{WF}/(1 - x_{WF}) (1 - (x_D)_{av})/(x_D)_{av}]$

$$\eta = x_{WF}/(x_D)_{av}$$

It should be noted that the dotted line in Figure 61 is plotted from Equation 204.

Multicomponent Systems

For batch distillation of multicomponent systems, the mass balance equations developed for binary systems can also be applied to both constant reflux and the constant distillate composition situations. Therefore, there is not much difference in the principles of binary and multicomponent batch distillations, except in the determination of the composition change in the column at any moment.

Generally, in order to obtain the relationship between the composition of the top product and the still product in multicomponent batch distillation, stage-by-stage calculations or numerical integration is needed, and sometimes, a trial-and-error procedure is required. However, for the case of total reflux with no column liquid holdup for a mixture of constant relative volatility, the following calculation procedures can be carried out.

Total reflux operation. For the case of a negligible column liquid holdup, Equation 187 of the total reflux operation can be applied for each component in multicomponent batch distillation, but, again it is not easy to evaluate x_D as a function of x_W, in contrast to the binary system, because during the distillation the concentration change in the still cannot be given arbitrarily. After arranging the mass balance, the following equations can be written for a component i of a multicomponent system at total reflux.

For the still:

$$\frac{dx_{Wi}}{d\beta} = \frac{1}{1 - \beta}(x_{Ni} - y_{Wi}) \quad (206)$$

For any plate:

$$y_{n+1,i} = x_{ni} \quad (n = 1, 2, \ldots, N) \quad (207)$$

Where β is a recovery ratio of the distillate and is defined by $(1 - W/W_F)$; N is a bottom plate of the column and n is the plate number.

Using Equations 206 and 207 with the vapor-liquid equilibrium relationship, the liquid composition pattern in the still can be numerically obtained by a stepwise integration with β and then x_{Di} can be calculated by the stage-by-stage method using Equation 207, starting with x_{Wi}. As an example, Figure 62 shows the change of the composition profiles with respect to the amount of the distillate in the column at total reflux for the distillation of a mixture of A, B, and C with relative volatilities, $\alpha_{A/C}$ (5.1) and $\alpha_{B/C}$ (2.2). The column has two theoretical plates and a total condenser.

Further analysis of Equation 206 or 207 gives:

$$\frac{dx_{Wi}}{dx_{Wj}} = \frac{x_{Di} - x_{Wi}}{x_{Dj} - x_{Wj}} \quad (i = A, B, \ldots) \tag{208}$$

These equations give the liquid compositions in the still for a multicomponent-multistage batch distillation and are similar to the residue Equation 185, for the simple multicomponent distillation, but they differ in that the composition differences for component i, are $x_{Wi} - x_{Di}$ instead of $x_i - y_i$. Also, it can be interpreted that slope of the line connecting x_{Di} with x_{Wi} is equal to the derivative of the liquid composition in the still, as shown in Figure 62.

Robinson and Gilliland [108] developed the following general equations for the case that all the relative volatilities are constant.

$$w_A/w_{FA} = (w_i/w_{Fi})^{\alpha_{A/i}^{N+1}} \quad (i = B, C, \ldots) \tag{209}$$

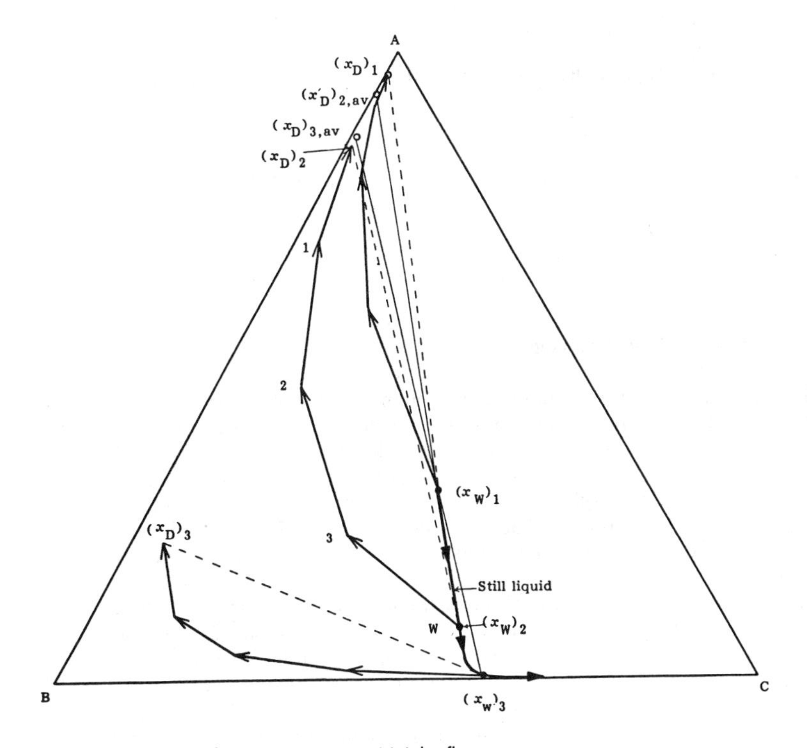

Figure 62. Distillation curves for ternary system at total reflux.

where w_A and w_i are the molar amounts of component A and i in the still, respectively; N is the number of the theoretical plates. By using Equation 209 for a given fraction of A distilled, it is possible determine the recovery fraction, the average composition of total distillate, and the instantaneous composition of the distillate.

Finite reflux operation. Robinson and Gilliland [108] also developed an approximate method for finite reflux operation with an almost analogous method for the total reflux. It is derived as follows:

The operating lines for any component can be written by an expression similar to Equation 191. Dividing the equation for component A, by the corresponding equation for any other component, for example, component B, the following relationship is obtained:

$$\frac{y_{n+1,A}}{y_{n+1,B}} = \frac{1+(x_{DA}/Rx_{nA})}{1+(x_{DB}/Rx_{nB})}(x_{nA}/x_{nB}) \tag{210}$$

And if $\xi_{A/B}$ represents the term of $(1+x_{DA}/Rx_{nA})/(1+x_{DB}/Rx_{nB})$, Equation 210 becomes:

$$(x_{DA}/x_{DB}) = (\alpha_{A/B})_1(\alpha_{A/B}/\xi_{A/B})_2 \cdots$$

$$\cdots(\alpha_{A/B}/\xi_{A/B})_W(x_{WA}/x_{WB}) \tag{211}$$

Using an average value of α/ξ gives:

$$x_{DA}/x_{DB} = (\alpha/\xi)_{av}^{N+1}(x_{WA}/x_{WB}) \tag{212}$$

where for the average value, we can use either

$$(\alpha/\xi)_{av} = [(\alpha)_1 + (\alpha/\xi)_W]/2 \tag{213}$$

or the ratio of (α_{av}/ξ_{av}) which can be respectively calculated by

$$\alpha_{av} = (\alpha_1 + \alpha_W)/2$$

$$\xi_{av} = (\xi 1 + \xi_W)/2$$

Effect of Column Liquid Holdup

The effect of the liquid holdup in the column on the sharpness of separation in batch distillation has been investigated both experimentally and theoretically, by many workers.

In early work, Rose et al. [117] concluded that holdup was detrimental at total reflux operation, while Colburn and Stern [118] indicated that in some cases of finite reflux operation holdup improved the sharpness of separation. In other cases the effect was observed to be the opposite. Robinson and Gilliland [108] studied a number of cases analytically and found that in general holdup in the column is undesirable for batch distillation.

Kojima [119] investigated the effect of the holdup on the degree of separation, defined by the average distillate composition at the 80-percent nominal cut point, that is, 0.8 W_Fx_F, for the mixture of methyl cyclohexane and toluene and found that the holdup effect on the separation is ether detrimental or advantageous depending on the extent of the reflux ratio, as is shown in Figure 63. It can be concluded that increasing the percentage total holdup, which can be defined as the ratio of the total column holdup and the total initial feed, is beneficial at low reflux ratio and is detrimental at high reflux ratio.

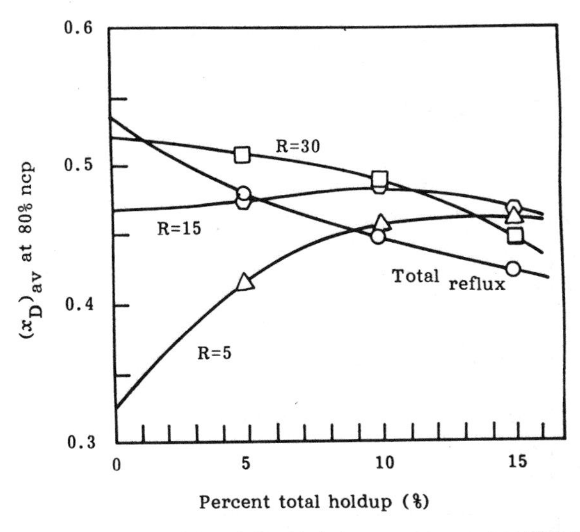

Figure 63. Experimental results of the relationship between average composition of total distillate and percent total holdup [119].

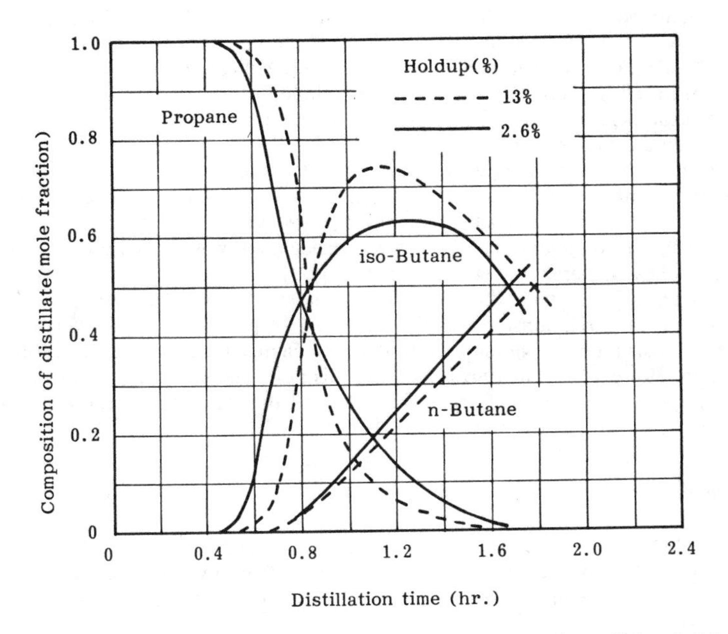

Figure 64. Theoretical distillation curves of the multicomponent system for multistage batch distillation at finite reflux ratio (R = 4, N = 12).

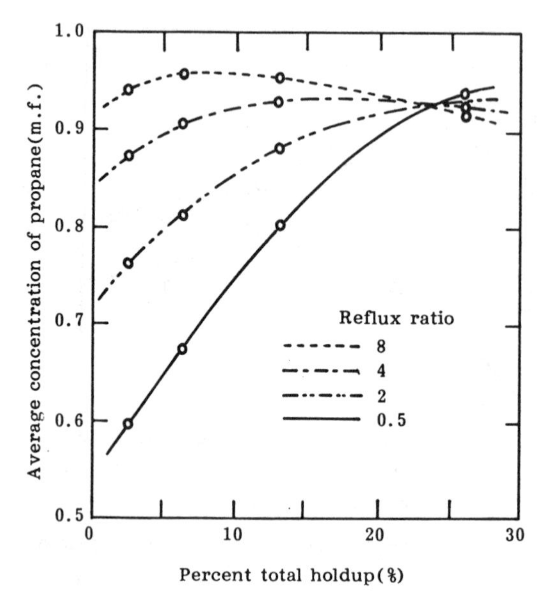

Figure 65. Effect of percent total holdup on average concentration of propane in distillate.

More recently, Steward et al. [120] presented a comprehensive model and experimental data for multicomponent batch distillation to determine the effect of total holdup, in addition to that of the reflux ratio and the effect of the number of theoretical plates on the degree of separation as measured by the average distillate concentration. The study concluded that the effect increasing the percent total holdup at a given number of the plates is either beneficial, negligible, or detrimental depending on the reflux ratio, thus supporting the cases of Colburn and Kojima for binary systems.

As an illustrative example, Figure 64 and 65 show the effect of the percent total holdup for the equimolar mixture of five components; propane, iso-butane, n-butane, iso-pentane, and n-pentane. The column has twelve theoretical plates, a total condenser, and a still and operates at a constant reboiler duty. The rigorous calculation was carried out by using the Liniger–Willoughby numerical integration method [121].

As can be seen in Figure 63 and 64, the effect of the holdup on the degree of separation is clear, however, their relationship is rather complex. In particularly, it is worthwhile to note that there exists a percent holdup which does not change the degree of separation with the reflux ratio. However, for holdups preceding this point the increment in reflux ratio is expected to improve the degree of separation. Beyond this point the phenomena can be opposite.

NOTATION

a_i, B_i, C_i constants
a_{mn} group interaction parameter
C_F specific heat of fluid
D mass rate in condenser
e energy loss
F feed mass rate
G_{ij} thermodynamic parameter defined by Equation 12
g free energy parameter
H^v molar latent heat
K ratio of vapor and liquid compositions

L liquid flow rate
N number of plates
P pressure
P_i partial pressure
P_i^0 partial pressure
Q heat duty
R universal gas constant; radius
R_m minimum reflux ratio
S stripping factor
T absolute temperature
t temperature
V vapor mass rate
v molar volume
W mass rate in stripping section
x_i liquid composition
x_{js} solvent concentration in enriching section
x_m group interaction
x^*, y^* molar fractions of liquid and vapor at pinch point, respectively vapor composition

Greek Symbols

α relative volatility
Γ_k residual activity coefficient
γ activity coefficient
Θ_i volume fraction
Λ_{ij} parameter defined in Equations 8 and 9
λ solubility parameter for non-polar-induced force
Ξ group surface fraction
ξ parameter defined in Equation 210
τ solubility parameter for intermolecular induced-force
τ_{ij} thermodynamic parameter defined by Equation 11
φ activity coefficient
φ_i surface fraction

REFERENCES

1. Cox, E. R., *Ind. Eng. Chem., 15:* 592 (1923).
2. Stull, D., *Ind. Eng. Chem., 39:* 517 (1947).
3. Prausnitz, J. M., et al., *Computer Calculations for Multicomponent Vapor-Liquid Equilibria,* Prentice-Hall, Englewood Cliffs, N. J. (1968).
4. Dreisbach, R. R., *Physical Properties of Chemical Compounds, I, II, and III,* Am. Chem. Soc. Adv. Chem. Ser., Nos. 15 (1955), 22 (1959), and 29 (1961).
5. Perry, R. H., *Chemical Engineers' Handbook,* 5th ed., McGraw-Hill, New York, (1973).
6. Margules, M., *Sitzungsber. Akad. Wiss. Wien. Math. Naturwiss. Kl. II,* 104: 1243 (1895).
7. Van Laar, J. J., *Z. Phys. Chem.,* 185: 35 (1929).
8. Wohl, K., *Trans. Am. Inst. Chem. Eng., 42:* 215 (1946).
9. Wilson, G. M., *J. Am. Chem. Soc., 86:* 127 (1964).
10. Gmehling, J., and Onken, U., *Vapor-Liquid Equilibrium Data Collection,* DECHEMA Chemistry Data Series; *Volume 1–8,* Frankfurt, Starting (1977).
11. Hirata, M., Ohe, S., and Nagahara, K., *Computer-Aided Data Book of Vapor-Liquid Equilibria,* Elsevier, Amsterdam (1975).
12. Holmes, M. J., and Van Winkle, M., *Ind. Eng. Chem., 62:* 21 (1970).
13. Nagata, I., *J. Chem. Eng. Japan, 6:* 18 (1973).
14. Orye, R. V., and Prausnitz, J. M., *Ind. Eng. Chem., 57*(5): 18 (1965).
15. Tsuboka, T., and Katayama, T., *J. Chem. Eng. Japan, 8:* 181 (1975).
16. Renon, H., and Prausnitz, J. M., *Am. Inst. Chem. Eng. J., 14:* 135 (1968).
17. Abrams, D. S., and Prausnitz, J. M., *AIChE J., 21:* 116 (1975).
18. Fredenslund, A., Jones, R. L., and Prausnitz, J. M., *AIChE J., 21:* 1086 (1975).
19. Derr, E. L., and Deal, C. H., *Int. Chem. Eng. Symp. Ser., 32*(3): 40 (1969).

20. Bondi, A., *Physical Properties of Molecular Crystals, Liquids, and Glasses,* Wiley, New York (1968).
21. Gilliland, E. R., *Ind. Eng. Chem., 32:* 918 (1940).
22. Waterman, W. W., Frazier, J. P., and Brown, G. M., *Hydrocarbon Processing, 47*(6): 155 (1968).
23. Kwauk, M., *Am. Inst. Chem. Eng. J., 2:* 240 (1956).
24. Hanson, D. N., Duffin, J. H., and Somerville, G. F., *Computation of Multicomponent Separation Processes,* Reinhold, New York, (1962).
25. Steward, D. V., *J. SIAM Numer. Anal., Ser. B., 2*(2): 345 (1966).
26. Himmelblau, D. M., *Chem. Eng. Sci., 22:* 883 (1967).
27. Takamatsu, T., Hashimoto, I., and Tomita, S., *Kagaku Kogaku Ronbunshu, 8*(4): 500 (1982).
28. Lewis, W. K., and Matheson, G. L., *Ind. Eng. Chem., 24:* 494 (1932).
29. Hengestebeck, R. J., *Trans. AIChE J.,* 309–329 (1946).
30. Geddes, R. L., *AIChE J.,* 389–392 (Dec. 1958).
31. Yaws, C. L., Fang, C. S., and Patel, P. M., *Chem. Eng.,* 101–104 (Jan. 1979).
32. Yaws, C. L., et al., *Hydro. Process.,* 99–100 (Feb. 1979).
33. Smith, B. D., and Brinkly, W. K., *AIChE J., 6:* 446 (1960).
34. Fleisher, M. T., and Prett, D. M., *Chem. Eng. Commun., 10:* 243–260 (1981).
35. Fenske, M. R., *Ind. Eng. Chem., 24:* 482 (1932).
36. Winn, F. W., *Pet. Refiner, 37*(5): 216 (1958).
37. Van Winkle, M., *Distillation,* McGraw-Hill, New York (1971).
38. King, J. C., *Separation Processes,* McGraw-Hill, New York (1971).
39. McCabe, W. L., and Smith, J. C., *Unit Operation of Chemical Engineering,* McGraw-Hill, New York (1975).
40. Colburn, A. P., *Tras. AIChE J., 37:* 805 (1941).
41. Underwood, A. J. V., *Inst. Pet., 31:* 111–118 (1945).
42. Underwood, A. J. V., *Inst. Pet., 32:* 598–613 (1946).
43. Underwood, A. J. V., *Chem. Eng. Prog., 44:* 603–613 (1948).
44. Holland, C. D., *Multicomponent Distillation,* Prentice-Hall, New York (1963).
45. Yamada, I., et al., *J. Chem. Eng. Japan, 10:* 440–445 (1977).
46. Barnes, F. J., Hanson, D. N., and King, C. J., *Ind. Eng. Chem. Process Des. Develop., 11:* 136–140 (1972).
47. McDonough, J. A., and Holland, C. D., *Hydrocarbon Processing & Petro. Ref., 41:* 135–140 (1962).
48. Sugie, H., and Lu, B. C–Y., *Chem. Eng. Sci., 25:* 1837–1846 (1970).
49. Gilliland, E. R., *Ind. Eng. Chem., 32:* 1220 (1940).
50. Erbar, J. H., and Maddox, R. N., *Pet. Refiner, 40*(5): 183 (1961).
51. Strangio, V. A., and Treybal, R. E., *Ind. Eng. Chem., Process Des. Develop., 13:* 279 (1974).
52. Naka, Y., et al. *J. Chem. Eng. Japan, 13*(2): 123 (1980).
53. Naka, Y., Terashita, M., and Takamatsu, T., *AIChE J., 28*(5): 812 (1982).
54. Horsley, L. H., *Azeotropic Data,* Advances in Chemistry Series, No. 6, (1952), No. 35 (1962) and No. 116, (1973), Am. Chem. Soc., Washington.
55. Dean, J. A., *Lange's Handbook of Chemistry,* 11th ed., McGraw-Hill, New York (1973).
56. Weast, R. C., *Handbook of Chemistry and Physics,* 56th ed. Chemical Rubber Company, Cleveland, Ohio (1976).
57. Chu, Ju Chin, *Vapor-Liquid Equilibrium Data,* J. W. Edwards Pub., Michigan (1956).
58. Prausnitz, J., et al., *Computer Calculations for Multicomponent Vapor-Liquid and Liquid-Liquid Equilibrium,* Prentice-Hall, New Yersey (1980).
59. Reid, R. C., Prausnitz, J. M., Sherwood, T. K., *The Properties of Gases and Liquids,* 3rd ed. McGraw-Hill, New York (1977).
60. Severns, W. H., Jr, et al., *AIChE J., 1:* 401 (1955).

61. Ewell, R. H., Harrison, J., and Berg, L., *Ind. Eng. Chem., 36:* 871 (1944).
62. Meissner, H. P., Greenfield, S. H., *Ind. Eng. Chem., 40:* 438 (1948).
63. Horvath, P. J., *Chem. Eng. 68:* 159 (March, 1961).
64. Matsuyama, H., and Nishimura, H., *J. Chem. Eng. Japan, 10:* 181 (1977)
65. Shiozaki, J., and Nishimura, H., *The Memoirs of the Faculty of Eng., Kyushu Univ., 41*(1): 49 (1981)
66. Schafer, K., and Lax, E., (Eds.), *Landolt–Bornstein Zahlenwerten und Funktionen,* Vol. II, Springer-Verlag, Berlin (1964).
67. Fredenslund, A., Gmehling, J., and Rasmussen, P., *Vapor-liquid Equilibria Using UNIFAC,* Elsevier Sci. Pub., Amsterdam (1977).
68. Oliver, E. D., *Diffusional Separation Processes,* John Wiley & Sons, New York (1966).
69. Severns, W., et al., *AIChE J., 1:* 401 (1955).
70. Van Winkle, M., *Distillation,* McGraw-Hill, New York (1970).
71. Naka, Y., *I. Chem. Eng. Japan, 16*(1): 36–42 (1983).
72. Ostwald, W., *Lehrbuch der allgemeinen Chemie,* Engelmann, Leipzig (1896–1902)
73. Ewell, R. H., and Welch, L. W., *Ind. Eng. Chem., 37:* 1224 (1945).
74. Wilson, R. Q., et al., *AIChE J., 1:* 220 (1955).
75. Severns, W. H., and Pigford, R. I., *AIChE J., 1:* 401 (1955)
76. Hoffman, E. J., *Azeotropic and Extractive Distillation,* Interscience Publishers (1969)
77. Naka, Y., Kobayashi, K., and Takamatsu, T., *Kagaku Kogaku Ronbunshu, 1:* 186 (1975).
78. Yorizane, M., Yoshimura, N., and Hase, S., *Kagaku Kogaku, 29:* 229 (1964).
79. Yamada, I., et al., *J. Chem. Eng. Japan, 10*(6): 440 (1977).
80. Yorizane, M., and Yoshimura, S., *Kagaku Kogaku, 32*(4): 382 (1968).
81. Naka, Y., doctoral thesis, Kyoto Univ. Kyoto (1979).
82. Naka, Y., et al., *Kagaku Kogaku, 38:* 501 (1974), (*Int. Chem. Eng., 16:* 272 (1975)).
83. Ishikawa, T., doctoral thesis, Tokyo Metropolitan Univ., Tokyo (1974).
84. Murti, P. S., and Van Winkle, M., *AIChE J., 3*(4): 517 (1957).
85. Yorizane, M., and Sadamoto, S., *The Context on Azeotropic and Extractive Distillation,* SCE Japan, Kansai Division (1976).
86. Takamatsu, T., *Shin Kagakukogaku Koza, 6:* 1, Nikkan Kogyo Shinbunsha (1957).
87. Tanaka, S., and Yamada, J., *Kagaku Kogaku, 28:* 661 (1964).
88. Hirata, M., *Kagaku Kogaku, 28:* 661 (1961).
89. Ochi, K., doctoral thesis, Nipon Univ., Tokyo (1975).
90. Kojima, K., and Ochi, K., *J. Chem. Eng. Japan, 7*(2): 71 (1974).
91. Pierotti, G. J., Deal, C. H., and Derr, E. L., *Ind. Eng. Chem., 51:* 95 (1958).
92. Butler, J. A. V., and Ramchandani, C. N., *J. Chem. Soc., 280:* 952 (1935).
93. Weimer, R. F., and Prausnitz, J. M., *Hydrocarbon Processing, 44:* 237 (1965).
94. Scatcherd, G., *Chem. Revs., 8:* 321 (1931).
95. Hildebrand, J. H., and Scott, R. L., *Solubility of Nonelectrolytes,* 3rd ed., Dover Publications (1946).
96. Hildebrand, J. H., and Scott, R. L., *Regular Solutions,* Prentice Hall (1962).
97. Flory, P. J., *J. Chem. Phys., 9:* 660 (1941).
98. Huggins, M. L., *J. Phys. Chem., 9:* 440 (1941).
99. Helpinstill, J. G., and Van Winkle, M., *Ind. Eng. Chem., 7:* 213 (1968).
100. Shinoda, K., *Solution and Solubility,* Maruzen (1966).
101. Tassios, D. P., *Hydrocarbon Processing, 49:* 114 (1970).
102. Sheets, T. R., and Marchello, J. M., *Hydrocarbon Processing, 42:* 99 (1963).
103. Genkin, A. N., Ogorodnikov, S. K., and Kogan, V. B., *Zh. Prik. Khim., 36:* 142 (1963).
104. Rayleigh, *Phil. Mag., 6th Series, 4,* 521 (1902).
105. Hengstebeck, R. J., *Distillation-Principles and Design Procedures,* Reinhold, New York (1961).
106. Block, B., *Chem. Eng.,* 87 (Feb 6, 1961).
107. Bogart, M. J. P., *Trans. Inst. Chem. Engr., 33,* 139 (1937).

108. Robinson, C. S., and Gilliland, E. R., *Elements of Fractional Distillation*, 4th ed., McGraw-Hill, New York (1950).
109. Kojima, K., *Kagaku Kogaku, 21*, 803 (1957).
110. Coward, I., *Chem. Eng. Sci., 22*, 503 (1967).
111. Rose, A., *Ind. Eng. Chem., 33*, 594 (1941).
112. Kojima, K., and Aoyama, I., *Kagaku Kogaku, 23*, 393 (1959).
113. Ellis, S. R. M., *Ind. Eng. Chem., 46*, 279 (1954).
114. Cichelli, M. T., et al., *Ind. Eng. Chem., 42*, 2502 (1950).
115. Zuiderwerg, F. J., *Chem. Ing. Tech., 25*, 297 (1953).
116. Houtman, J. P. W., and Husdin, A., *Chem. Eng. Sci., 5*, 178 (1956).
117. Rose, A., and Welshans, L. M., *Ind. Eng. Chem., 32*, 668 (1940).
118. Colburn, A. P., and Stearns, R. F., *Trans. A.I.Ch.E., 37*, 291 (1941).
119. Kojima, K., *Kagaku Kogaku, 22*, 492 (1958).
120. Stewart, R. R., et al., *Ind. Eng. Chem. Process Des. Develop., 12*, 130 (1973).
121. Distifano, G. P., *AIChE J., 14*, 190 (1968).

CHAPTER 15

GENERALIZED DESIGN METHODS FOR MULTICOMPONENT DISTILLATION

Ikuho Yamada

Nagoya Institute of Technology
Nagoya, Japan

Takaitsu Iwata

Mitsui Engineering and Shipbuilding Co., Ltd.
Tokyo, Japan

Hideki Mori

Nagoya Institute of Technology
Nagoya, Japan

Yuji Naka

Tokyo Institute of Technology
Yokohama, Japan

CONTENTS

RIGOROUS CALCULATION METHODS

In general, multicomponent distillation problems can be classified into two types, design and operation. Design-related problems establish the total number of plates and the feed plate location and/or the number of plates in each section of the distillation column for the required separation under the given reflux ratio at the top or the bottom of the column (usually the former is given). In contrast, operation type problems establish the composition profile in the column including the distillate and bottom products for specified operating conditions as well as the number of plates in each section of the column.

In the following pages standard calculation methods are described for both types of problems.

Generalized Design Method

In this subsection, a generalized method [1] for solving the design problem of an ordinary distillation column is presented for the doublet separation of a multicomponent system. The term "doublet separation" means that the two key components, a heavy key component (h) and a light key component (l), are beside each other. For example, for the separation of a hydrocarbon system composed of m components, $nC_k(1) \sim nC_{k-1}(2) \sim \ldots nC_{k-j-1}(j) \sim nC_{k-j}(j+1) \sim \ldots nC_{k-m-1}(m)$, $h = j$ and $l = j+1$. (Note the components, $1, 2 \ldots j-1, j \ldots m$, are numbered in order of reciprocal relative volatility hereafter.) Furthermore, the molar rates of the heavy key component in the distillate $Dx_{D,h}$ and the light key component in bottom $Wx_{w,l}$ are usually specified. The principal features of the method are:

- It is applicable not only to nonideal systems but also to nonideal plates.
- It provides complete plate matching for all components by an accurate determination of extremely small amounts of molar rates of nondistributed components existing in the distillate (heavier than the heavy key) and bottoms (lighter than the light key) in the plate-to-plate calculation procedure within allowable significant figures of the computer used [2].
- It employs a condition of minimum total number of plates to determine the number of plates in each section of the column by the same approach as in the Murdoch–Holland's method [3].

General Description

Column model. Consider an ordinary distillation column with a single feed, an over-head product stream, and a bottom product stream with plates in the enriching and stripping sections numbered towards the feed plate from both ends of the column, respectively, (see Figure 1). The flow streams around the feed plate are shown in Figure 2. The following assumptions are applied:

- Adiabatic column.
- Over-head total condenser.
- Saturated vapor and liquid withdrawn from all plates, the total condenser, and the bottom reboiler.
- Uniform composition of vapor entering the plate.

Bubble point and dew point calculation for the general system. For an ideal plate the vapor and liquid leaving the plates are in equilibrium each other, and for a given set of liquid compositions $x_{j,i}$'s, a set of vapor composition $y_{j,i}$'s can be calculated by the bubble point method. The equations used are as follows;

$$y_{j,i} = K_{j,i}x_{j,i} = \gamma_{j,i}K^0_{j,i}x_{j,i} \tag{1}$$

$$\sum_i K_{j,i}x_{j,i} = \sum_i \gamma_{j,i}K^0_{j,i}x_{j,i} = 1 \tag{2}$$

where $K_{j,i}$ denotes the component K-value, $\gamma_{j,i}$ the component activity coefficient in liquid phase, which can be calculated from the given liquid composition (i.e., from Wilson's equation [4]), and $K^0_{j,i}$ is the ratio of component vapor pressure to the operating pressure, which is a function of the temperature of the liquid leaving the plate. Therefore, $y_{j,i}$'s are calculated by Equation 1 with $\gamma_{j,i}$'s calculated, and $K^0_{j,i}$ is obtained by solving Equation 2

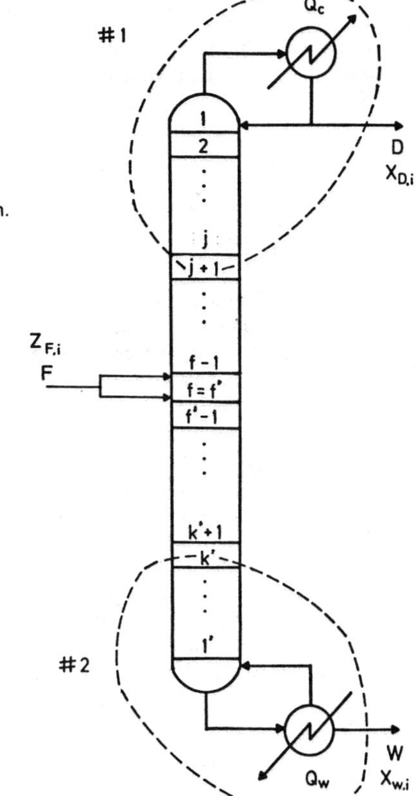

Figure 1. Column model for ordinary distillation column.

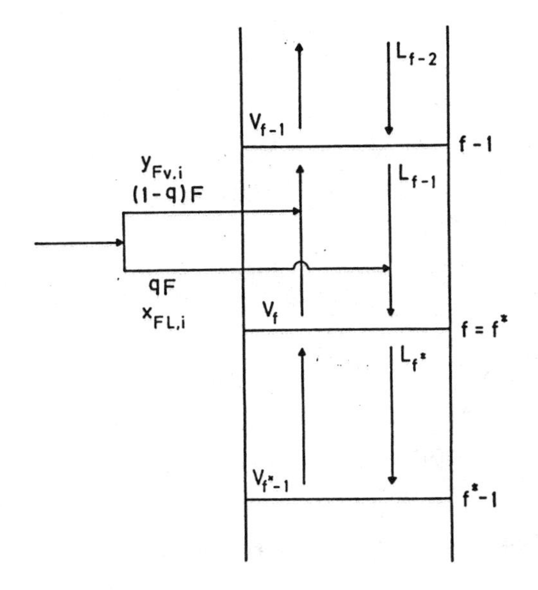

Figure 2. Flow streams around feed plate.

with respect to temperature. If the system is ideal, the calculation procedure to evaluate $\gamma_{j,i}$ is omitted, since $\gamma_{j,i} = 1$ for all components.

On the other hand, for a given set of $y_{j,i}$'s, $x_{j,i}$'s can be calculated by the dew point method. The equations used are:

$$x_{j,i} = \frac{y_{j,i}}{K_{j,i}} = \frac{y_{j,i}}{\gamma_{j,i}K^0_{j,i}} \tag{3}$$

$$\sum_i \frac{y_{j,i}}{K_{j,i}} = \sum_i \frac{y_{j,i}}{\gamma_{j,i}K^0_{j,i}} = 1 \tag{4}$$

For the nonideal system, Equations 3 and 4 are both a function of liquid composition and temperature. However, if the $\gamma_{j,i}$'s are assumed constant for the computed $x_{j,i}$'s obtained in the previous calculation step, these equations are reduced to functions of temperature only, and $x_{j,i}$'s are calculated from Equation 3 with $K^0_{j,i}$ obtained by solving Equation 4 with respect to temperature with $\gamma_{j,i}$'s recalculated. In this case, the values of the $\gamma_{j,i}$'s calculated are not equal to those of the previous step, therefore, the alternative calculation by means of the direct iteration of $\gamma_{j,i}$ is continued until a given convergence criterion with respect to $\gamma_{j,i}$ or temperature has been satisfied; where it is convenient to assume $x_{j,i} = y_{j,i}$ for evaluating the initial values of $\gamma_{j,i}$'s. Of course, the direct iteration loop for $\gamma_{j,i}$'s is omitted for the ideal system.

Plate efficiency. The above calculation procedure fails to determine the relationship between $y_{j,i}$'s and $x_{j,i}$'s when a nonideal plate is considered. In this case, it is convenient to determine the relationship between the vapor and liquid compositions by the Murphree plate efficiency based on the vapor phase $E^{MV}_{j,i}$ [5], defined as follows –

$$E^{MV}_{j,i} = \frac{y_{j,i} - y_{p,i}}{K_{j,i}x_{j,i} - y_{p,i}} \tag{5}$$

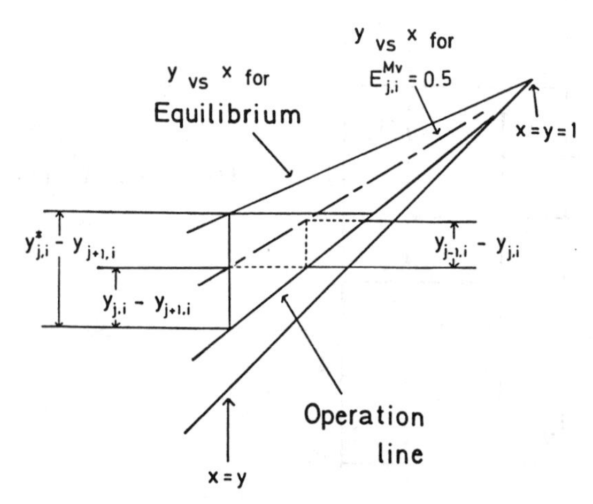

Figure 3a. Vapor composition leaving the plate by considering $E^{MV}_{j,i}$ = 0.5 in enriching section.

where $y_{p,i}$ denotes the component molar fraction of vapor phase entering the plate j, that is, $p \equiv j+1$ for the enriching section and $p \equiv j-1$ for the stripping section with plates numbered as in Figure 1.

We can state that n plates with $E_{j,i}^{MV} = 1/n$ are equivalent to an ideal plate ($E_{j,i}^{MV} = 1$), which is used to determine the actual number of plates from the number of ideal plates required for a specified separation. However, as can be observed in Figures 3A and 3B, it is obvious that two plates with $E_{j,i}^{MV}$ of 0.5 are never equivalent to an ideal plate for either section. Therefore, in order to accurately determine the actual number of plates a calculation procedure involving the plate efficiency is needed.

Many attempts have been made to evaluate $E_{j,i}^{MV}$ for multicomponent systems, but accurate methods are not yet available for correlating experimental results. The reasons for this dilemma arise from the complexity of the heat and mass transfer mechanisms in the froth or bubbles and the behavior of liquid flow patterns on the plate. However, for a more accurate concentration profile, the assumption that $E_{j,i}^{MV}$ is constant throughout the column and is independent of components and equivalent to the overall column efficiency obtained empirically is a more realistic assumption than that of ideal plates.

Plate-to-plate calculation procedure. Lewis and Matheson [6] first suggested a plate-to-plate calculation procedure for the purpose of determinating the number of ideal plates in the enriching and stripping sections for a specified separation of an ideal multicomponent system.

A generalized method for plate-to-plate calculations applicable not only to nonideal systems but to real plates, as well, is described.

We first consider the *enriching section.* For enclosure #1 in Figure 1, the following balance equations are obtained:

Overall material balance:

$$V_{j+1} = L_j + D \tag{6}$$

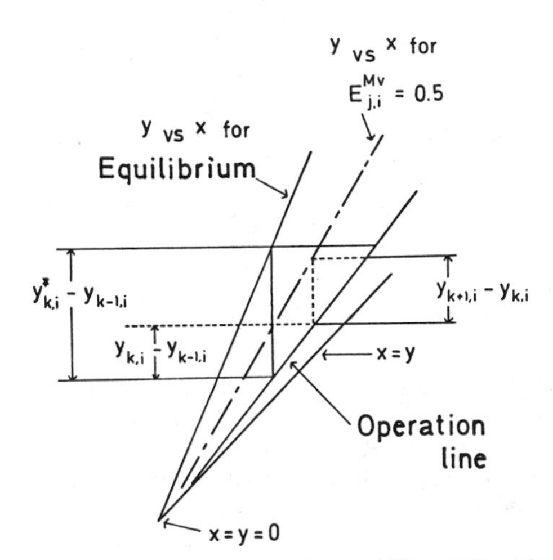

Figure 3b. Vapor composition leaving the plate by considering $E_{j,i}^{MV} = 0.5$ in stripping section.

Component material balance:

$$V_{j+1}y_{j+1,i} = L_j x_{j,i} + D x_{D,i} \tag{7}$$

Enthalpy balance:

$$V_{j+1}H_{j+1} = L_j h_j + D h_D + Q_c \tag{8}$$

where Q_c denotes the heat duty at the overhead total condenser, which is obtained by the enthalpy balance around the total condenser

$$Q_c = (H_1 - h_D)V_1 = (H_1 - h_D)(R_D + 1)D \tag{9}$$

where R_D is the reflux ratio at the top of the column. Moreover, h_j and H_j are computed from the following equations

$$h_j = \sum_i x_{j,i} \int_0^{T^L} C_{pL,i}\, dT^L \tag{10}$$

$$H_j = \sum_i y_{j,i} \int_0^{T^V} C_{pL,i}\, dT^V + \sum_i \lambda_i y_{j,i} \tag{11}$$

where λ_i and $C_{pL,i}$ are the molar heat of vaporization and the molar heat capacity, respectively. Although there are more rigorous functions for evaluating λ_i and $c_{pL,i}$, the assumption of quadratic formulations with respect to temperature is convenient:

$$C_{pL,i} = \bar{a}_{0,1} + \bar{a}_{1,i}T^L + \bar{a}_{2,i}(T^L)^2 \tag{12}$$

$$\lambda_i = \bar{b}_{0,i} + \bar{b}_{1,i}T^V + \bar{b}_{2,i}(T^V)^2 \tag{13}$$

where T^L and T^V are computed from Equations 2 and 4, respectively.

Substitution of V_{j+1} and L_j, (obtained by simultaneously solving Equations 6 and 8) into Equation 7 gives the following operating line.

$$y_{j+1,i} = a_j x_{j,i} + b_j x_{D,i} \tag{14}$$

where a_j and b_j are defined by

$$a_j \equiv \frac{Q_c - D(H_{j+1} - h_j)}{Q_c + D(h_D - h_j)} \tag{15}$$

$$b_j \equiv \frac{D(H_{j+1} - h_j)}{Q_c + D(h_D - h_j)} \tag{16}$$

The operating line given in Equation 14 is not linear, since a_j and b_j are functions of both H_{j+1} and h_j which depend on $y_{j+1,i}$'s and $x_{j,i}$'s. However, assuming $H_j = H_{j+1}$, $h_D = h_j$, and $\lambda_D = H_{j+1} - h_j$ and denoting $R = R_D$, Equation 14 reduces to a linear form, which has frequently been used in the calculation procedure based on the assumption of equimolar rates of vapor and liquid throughout in the enriching section, and Equation 14 can be rewritten as:

$$y_{j+1,i} = a x_{j,i} + b x_{D,i} \tag{14a}$$

where

$$a \equiv \frac{R}{R+1} \tag{15a}$$

$$b \equiv \frac{1}{R+1} \tag{16a}$$

Furthermore, an elimination of $y_{j+1,i}$ with Equations 5 and 14 gives the following equations

$$x_{j,i} = \frac{y_{j,i} - b_j(1 - E_{j,i}^{MV})x_{D,i}}{\gamma_{j,i}K_{j,i}^0 E_{j,i}^{MV} + a_j(1 - E_{j,i}^{MV})} \tag{17}$$

$$\sum_i \frac{y_{j,i} - b_j(1 - E_{j,i}^{MV})x_{D,i}}{\gamma_{j,i}K_{j,i}^0 E_{j,i}^{MV} + a_j(1 - E_{j,i}^{MV})} = 1 \tag{18}$$

where $y_{1,i}$ ($= x_{D,i}$) is assigned by the boundary condition at the total condenser, and $x_{D,i}$, R_D, and $E_{j,i}^{MV}$ are assigned as noted previously for the design type of problems. Moreover, Equations 17 and 18 are reduced to Equations 3 and 4, respectively, when $E_{j,i}^{MV} = 1$ is assumed for all components. For carrying out the generalized plate-to-plate calculation procedure in the enriching section, the important point is how to calculate $x_{j,i}$'s for the given set of values of $y_{j,i}$'s with Equations 17 and 18, since $x_{j,i}$ cannot be obtained directly. Since $\gamma_{j,i}$, a_j, and b_j are the functions of $x_{j,i}$, and H_{j+1} and h_j must be evaluated from unknown $y_{j+1,i}$'s and $x_{j,i}$'s, respectively, as shown by Equations 15 and 16. However, by employing a direct iteration with respect to $\gamma_{j,i}$, $x_{j,i}$, and $y_{j+1,i}$ obtained in the previous iteration step with Equations 14, 17 and 18, new values of $x_{j,i}$'s and $y_{j+1,i}$'s can be obtained simultaneously. Note that it is convenient to use Equations 3, 4 and 14a to evaluate initial values of $x_{j,i}$'s and y_{j+1}'s. The scheme for the plate-to-plate calculation procedure in the enriching section is summarized in the following steps.

1. Calculate the initial values of $x_{j,i}$'s by Equations 3 and 4.
2. Calculate the initial values of $y_{j+1,i}$'s by Equation 14a with $x_{j,i}$'s obtained in Step 1.
3. Calculate h_j with $x_{j,i}$'s and bubble point, and H_{j+1} with $y_{j+1,i}$'s and dew point.
4. Calculate a_j and b_j by Equations 15 and 16, and $\gamma_{j,i}$ from Wilson's equation [4] if the system is nonideal.
5. Calculate the new values of $x_{j,i}$'s by Equation 17 with $K_{j,i}^0$ obtained by solving Equation 18 with respect to temperature.
6. Calculate $y_{j+1,i}$ by Equation 14.
7. Judge the convergence of $x_{j,i}$'s and $y_{j+1,i}$'s. If convergence is achieved, move to the next plate. Otherwise return to Step 3.

These alternative calculations are continued until the composition of the vapor leaving the feed plate has been obtained, (e.g. $j = 1, 2, \ldots f-1.$).

We now consider the *stripping section*. For enclosure #2 in Figure 1, the expressions corresponding to Equations 6 through 8 are as follows:

Overall material balance:

$$L_{m'+1} = V_{m'} + W \tag{19}$$

Component material balance:

$$L_{m'+1}x_{m'+1,i} = V_{m'}y_{m',i} + Wx_{w,i} \tag{20}$$

Enthalpy balance:

$$L_{m'+1}h_{m'+1} = V_{m'}H_{m'} + Wh_w - Q_w \tag{21}$$

where Q_w denotes the heat duty at the bottom reboiler, which is from an enthalpy balance around the reboiler

$$V_w(H_w - h_{1'}) + W(h_w - h_{1'}) = \{R_w(H_w - h_{1'}) - (h_w - h_{1'})\}W \tag{22}$$

where $R_w (\equiv V_w/W)$ denotes the reflux ratio at the bottom of the column.

Substitution of $L_{m'+1}$ and $V_{m'}$, which are obtained by simultaneously solving Equations 19 and 21 into Equation 20 gives the following operating line.

$$x_{m'+1,i} = a'_{m'}y_{m',i} + b'_{m'}x_{w,i} \tag{23}$$

where $a'_{m'}$ and $b'_{m'}$ are defined by;

$$a'_{m'} \equiv \frac{Q_w - W(h_w - h_{m'+1})}{Q_w + W(H_{m'} - h_w)} \tag{24}$$

$$b'_{m'} \equiv \frac{W(H_{m'} - h_{m'+1})}{Q_w + W(H_{m'} - h_w)} \tag{25}$$

The operation line given in Equation 23 is also nonlinear, since $a'_{m'}$ and $b'_{m'}$ are functions of $h_{m'+1}$ which must be evaluated from unknown $x_{m'+1,i}$'s. However, assuming $H_{m'+1} = H_{m'}$, $h_{1'} = h_{m'+1}$, and denoting $R' = R_w$, Equation 23 reduces to a linear relation, as used in the calculation procedure based on the assumption of equimolar rates of vapor and liquid throughout the stripping section, and Equations 23 through 25 are rewritten as follows.

$$x_{m'+1,i} = a'y_{m',i} + b'x_{w,i} \tag{23a}$$

$$a' \equiv \frac{R'}{R'+1} \tag{24a}$$

$$b' \equiv \frac{1}{R'+1} \tag{25a}$$

The plate-to-plate calculation procedure for the stripping section for the previously assigned $x_{w,i}$'s, R_w, and $E^{MV}_{j,i}$ can be carried out by computing $x_{m'+1,i}$'s for given values of $y_{m',i}$'s with a direct iteration of $a'_{m'}$ and $b'_{m'}$. The scheme for the plate-to-plate calculation procedure in the stripping section is as follows:

1. Calculate the initial values of $x_{m'+1,i}$'s by Equation 23a.
2. Calculate $h_{m'+1}$ with $x_{m'+1,i}$'s obtained from Step 1.
3. Calculate $a'_{m'}$ and $b'_{m'}$ from Equations 24 and 25.
4. Calculate new values of $x_{m'+1,i}$'s from Equation 23.
5. Judge $x_{m'+1,i}$'s. If convergence is achieved, move to the next plate after $y_{m'+1,i}$ is calculated from Equation 5. Otherwise return to Step 2.

These additional calculations are continued until the vapor composition on the feed plate is obtained.

Plate-matching and assignment of molar rates of nondistributed component. In solving the design-type problem with doublet separation, the molar rates of small amounts of nondistributed components in the distillate and bottoms cannot be assigned a priori. Usually, for convenience the molar rates of nondistributed components $Dx_{D,i}$ $(i < h)$ and $Wx_{w,i}$ $(i > 1)$ are assumed to be zero. This enables a rough estimate of the number of plates in each section of the column as suggested by the Lewis–Matheson method [6]. However, if these values are once set equal to zero, the values of $x_{j,i}$'s and $x_{m',i}$'s resulting from the plate-to-plate calculation procedure in each section of the column are always equal to zero. This result obviously conflicts with the actual concentration profile in the column, since the liquid composition around the feed plate is greatly affected by the feed composition which also includes the nondistributed components. Therefore, one must consider an extremely small amount of unspecified $Dx_{D,i}$ $(i < h)$ and $Wx_{w,i}$ $(i > 1)$ within allowable significant figures of the computer. These values must be determined by the condition of "plate-matching" which is expressed by the following equations.

For the key component:

$$g(k) \equiv \left[\frac{y_{f,1}}{y_{f,h}}\right]_D - \left[\frac{Y_{f,1}}{Y_{f,h}}\right]_W = 0 \tag{26}$$

For other component:

$$g(i) \equiv [y_{f,i}]_D - [y_{f,i}]_w = 0$$

where $[\ ]_D$ and $[\ ]_w$ denote the values obtained by the plate-to-plate calculation procedure for the enriching and stripping sections, respectively. Moreover, for an ideal solution, the condition of "plate-matching" becomes:

For the key component:

$$g(k) = \left[\frac{x_{f,1}}{x_{f,h}}\right]_D - \left[\frac{x_{f,1}}{x_{f,h}}\right]_w = 0 \tag{26a}$$

For other components:

$$g(i) = [x_{f,i}]_D - [x_{f,i}]_w = 0 \quad (i \neq h, 1) \tag{27a}$$

Determination of the number of plates. Hitherto three concepts for determining the number of plates in the enriching and stripping sections, n and s, have been suggested by Lewis and Matheson [6], Murdoch and Holland [3], and Waterman and Frazier [7].

Lewis and Matheson suggested that the ratio of liquid molar fraction of the light key component to that of the heavy key component on the feed plate, resulting from the plate-to-plate calculation procedure in the enriching and stripping sections, is equal to that in the liquid portion of feed. That is,

For the enriching section:

$$\frac{x_{f,1}}{x_{f,h}} = \frac{x_{FL,1}}{x_{FL,h}} \tag{28}$$

For the stripping section:

$$\frac{x_{f',1}}{x_{f',h}} = \frac{x_{FL,1}}{x_{FL,h}} \tag{29}$$

They determined n and s by the number of plate-to-plate calculation procedures required to satisfy Equations 28 and 29.

Waterman, Frazier, and Brown calculated n and s by using the following equation

$$\frac{d}{dn}\left[\frac{x_{f,l}}{x_{f,h}}\right] = -\frac{d}{ds}\left[\frac{x_{f',l}}{x_{f',h}}\right] \tag{30}$$

where Equation 30 implies that two slopes, $x_{j,l}/x_{j,h}$ vs. n and $x_{m',l}/x_{m',h}$ vs. s are equal at the feed plate (note the plates in the stripping section are numbered from the bottom end of the column). In the following lines, however, the concept by Murdoch and Holland, which is characterized by the superiority to eliminate the excess plates, is used. The required condition to determine n and s is as follows.

$$n+s \rightarrow \min. \tag{31}$$

Calculation steps for determining the number of plates in each section of the column are as follows:

1. Assume n (or s) (the important point in selecting n or s is to carry out steps 2 through 4 and to find the minimum value of n+s easily).

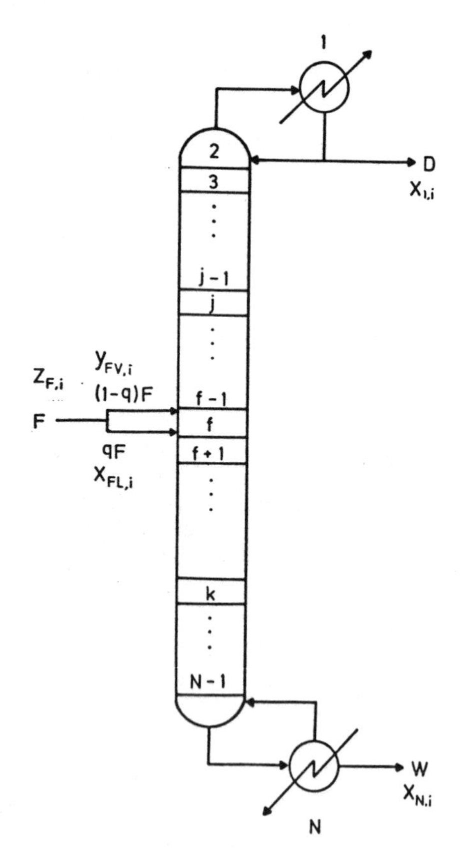

Figure 4. Column model for matrix method.

2. Assume small increments $x_{D,i}$ ($i < h$) and $x_{w,i}$ ($i > l$), and normalize them for all components. (Note that this assumption is not substantially influenced over the normalized $x_{D,i}$ ($i \geqq h$) and $x_{w,i}$ ($i \leqq l$) values, respectively.)
3. Calculate $[y_{f,1}/y_{f,h}]_D$ and $[y_{f,i}]_D$ ($i \neq h,l$) (or $[y_{f',1}/y_{f',h}]_w$ and $[y_{f,i}]_w$) by the plate-to-plate calculation procedure in the enriching (or stripping) section for n (or s) assumed.
4. Determine s (or n) by repeating the plate-to-plate calculation procedure for the stripping (or enriching) section until Equation 26 has been achieved, where the fraction of s (or n) is calculated by extrapolation of the quadratic approximation of $[y_{f',1}/y_{f',h}]_w$ (or $[y_{f,1}/y_{f,w}]_D$) with respect to s (or n).
5. Judge Equation 27. If this equation is satisfied within a given accuracy for all components, the calculation to determine s (or n), and $x_{D,i}$ ($i < h$) and $x_{w,i}$ ($i > l$) have just converged, then go to the next step. Otherwise return to Step 3 after $x_{D,i}$ ($i < h$) and $x_{w,i}$ ($i > l$) are corrected. The regula falsi method or the well-known conjugate gradient method is useful in finding the corrected values.
6. Repeat Steps 1 through 5 for various n (or s).
7. Determine n and s by Equation 31.

Of course this method is applicable when $E_{j,i}^{MV} = 1$. In this case $[y_{f,1}/y_{f,h}]_D$, $[y_{f',1}/y_{f',h}]_w$, $[y_{f,i}]_D$ and $[y_{f',i}]_w$ are replaced by $[x_{f,1}/x_{f,h}]_D$, $[x_{f',1}/x_{f',h}]_w$, $[x_{f,i}]_D$ and $[x_{f',i}]_w$, respectively.

Calculation Methods by Operation Type

As noted earlier, when the operation type of multicomponent distillation problems for the ordinary distillation column are solved, the molar rates of distillate and bottom, D and W, reflux ratio R_D at the top of the column, and the number of plates in enriching and stripping sections, n and s, or the total number of plates and the feed plate location are given beforehand. In the following discussions, the matrix method, θ-method and relaxation method are described.

Matrix Method (Quasilinearlized Method for Component Material Balance)

Bubble point method. For an adiabatic ordinary distillation column, let the plates and stages be numbered from the overhead total condenser, and assign the condenser, top plate ... feed plate ... last plate, and reboiler the symbols 1, 2 ... f ... N − 1, and N, respectively, as schematically shown in Figure 4. The component vaporization plate efficiency $E_{j,i}^{V}$ ($\equiv y_{j,i}/K_{j,i}x_{j,i}$) [8] is used for solving the operation type of multicomponent distillation problems by the matrix method, because the use of $E_{j,i}^{V}$ has remarkable merit which provides the same simple algorithm as the original matrix method proposed by Amundson and Pontinene [9].

By using $E_{j,i}^{V}$, the relationship between $y_{j,i}$ and $x_{j,i}$ is written as follows.

$$y_{j,i} = E_{j,i}^{V}K_{j,i}x_{j,i} = E_{j,i}^{V}\gamma_{j,i}K_{j,i}^{0}x_{j,i} \quad (2 \leqq j \leqq N-1) \tag{32}$$

$$\sum_i y_{j,i} = \sum_i E_{j,i}^{V}K_{j,i}x_{j,i} = \sum_i E_{j,i}^{V}\gamma_{j,i}K_{j,i}^{0}x_{j,i} = 1 \tag{33}$$

Moreover, $E_{j,i}^{V}$ is related to $E_{j,i}^{MV}$ by the following equation,

$$E_{j,i}^{V} = \frac{y_{j+1,i}}{K_{j,i}x_{j,i}} + E_{j,i}^{MV}\left(1 - \frac{y_{j+1,i}}{K_{j,i}x_{j,i}}\right) \tag{34}$$

Of course $E_{j,i}^{V} = E_{j,i}^{MV} = 1$ when the ideal plate is assumed.

The material and enthalpy balances around the plates in the column are as follows:

Component material balance

$$-(D+L_D)x_{1,i}+V_2y_{2,i} = 0 \tag{35}$$

$$L_{j-1}x_{j-1,i}-(V_jy_{j,i}+L_jx_{j,i})$$

$$+V_{j+1}y_{j+1,i} = \xi_{j,i} \quad (2 \leqq j \leqq N-1) \tag{36}$$

$$L_{N-1}x_{N-1,i}-(V_Ny_{N,i}+Wx_{N,i}) = 0 \tag{37}$$

where $\xi_{j,i}$ is defined by

$$\xi_{j,i} = 0 \quad (j \neq f-1, f)$$

$$\xi_{f-1,i} = -(1-q)Fy_{Fv,i} \quad \text{and} \quad \xi_{f,i} = -qFx_{FL,i} \tag{38}$$

Substituting Equation 32 into Equations 35 through 37 gives the following equations

$$-(D+L_D)x_{1,i}+V_2E_{2,i}^{V}\gamma_{2,i}K_{2,i}^{0}x_{2,i} = 0 \tag{39}$$

$$L_{j-1}x_{j-1,i}-(V_jE_{j,i}^{V}\gamma_{j,i}K_{j,i}^{0}+L_j)x_{j,i}$$

$$+V_{j+1}E_{j+1,i}^{V}\gamma_{j+1,i}K_{j+1,i}^{0}x_{j+1,i} = \xi_{j,i} \quad (2 \leqq j \leqq N-1) \tag{40}$$

$$L_{N-1}x_{N-1,i}-(V_N\gamma_{N,i}K_{N,i}^{0}+W)x_{N,i} = 0 \tag{41}$$

where the ideal stage is assumed for the reboiler, and Equations 39 through 40 are composed of the number of components.

Enthalpy balance

$$-(D+L_D)h_1+V_2H_2-Q_c = 0 \tag{42}$$

$$L_{j-1}h_{j-1}-(V_jH_j+L_jh_j)+V_{j+1}H_{j+1} = \zeta_j \tag{43}$$

$$L_{N-1}h_{N-1}-(V_NH_N+Wh_w)+Q_w = 0 \tag{44}$$

where $\zeta_j = 0 \quad (j \neq f-1, f)$

$$\zeta_{f-1} = -(1-q)FH_{Fv} \quad \text{and} \quad \zeta_f = -qFh_{Fl} \tag{45}$$

where Q_c and Q_w are the heat duty at the condenser and reboiler, which are given by Equations 9 and 22, and the same functional forms as Equations 10 and 11 are used for the evaluation of enthalpy of liquid and vapor, h_j and H_j.

As shown in Equations 39 through 44, the operation type of multicomponent distillation problems can be described by a set of complicated nonlinear simultaneous equations. However, if V_j, L_j, $E_{j,i}^{V}$, $\gamma_{j,i}$, and $K_{j,i}^{0}$ are assumed to be constant at the values obtained in the previous calculation step, Equations 39 through 41 are reduced to a set of linear simultaneous equations with respect to $x_{j,i}$'s $(1 \leqq j \leqq N)$, and can be rewritten by the flowing matrix form.

$$M\ X = \xi \tag{46}$$

where

$$x \equiv [x_{1,i} \ldots x_{f-1,i} x_{f,i} \ldots x_{N,i}]^T \tag{47}$$

$$\xi \equiv [0 \ldots -(1-q)Fy_{Fv,i} - qFx_{FL,i} \ldots 0]^T \tag{48}$$

Moreover, M denotes the tridiagonal matrix with the following elements

$$m_{1,1} = -(D+L_D) = -V_1,$$

$$m_{j,j} = -(V_j E^V_{j,i} \gamma_{j,i} K^0_{j,i} + L_j) \quad (2 \leqq j \leqq N-1)$$

$$m_{N,N} = -(V_N K_{N,i} + W),$$

$$m_{j,j-1} = L_j \quad (2 \leqq j \leqq N)$$

$$m_{j,j+1} = V_{j+1} E^V_{j+1,i} \gamma_{j+1,i} K^0_{j+1,i} \quad (1 \leqq j \leqq N-1) \tag{49}$$

$x_{j,i}$'s can be obtained by solving Equation 46 with the well-known Thomas algorithm, and $y_{j,i}$'s are calculated from Equation 32.

However, in general, the obtained sets of $x_{j,i}$'s and $y_{j,i}$'s will not satisfy $\sum_i x_{j,i} = 1$ and $\sum_i y_{j,i} = 1$. Therefore, the values of $x_{j,i}$'s and $y_{j,i}$'s are normalized after Equation 46 has been solved. On the other hand, by assuming that h_j and H_j in Equations 42 through 44 are constants calculated from the normalized values of $x_{j,i}$'s and $y_{j,i}$'s, and by using the overall material balance in Equations 6 and 9 for both sections, new values of V_j's are computed successively by the following equations:

$$V_j = \frac{V_{j-1}(H_{j-1} - h_{j-2}) - D(h_{j-1} - h_{j-2})}{H_j - h_{j-1}} \quad (3 \leqq j \leqq f-1)$$

$$V_f = \frac{V_{f-1}(H_{f-1} - h_{f-2}) - D(h_{f-1} - h_{f-2}) - (1-q)FH_{Fv}}{H_f - h_{f-1}}$$

$$V_{f+1} = \frac{V_f(H_f - h_{f-1}) + Dh_{f-1} - Wh_f - qFh_{FL}}{H_{f+1} - h_f}$$

$$V_j = \frac{V_{j-1}(H_{j-1} - h_{j-2}) + W(h_{j-1} - h_{j-2})}{H_j - h_{j-1}} \quad (f+2 \leqq j \leqq N) \tag{50}$$

where $V_2 (= L_1 + D = (R_D + 1)D)$ is given previously for the operation type of problem. New values of L_j's are computed by Equations 6 and 19. Then Equation 46 is solved again by using new values of matrix elements. Namely, this procedure is the successive iteration of the matrix elements, and the final solution can be obtained by repeating the procedure until a given criterion has been satisfied. This concept was first proposed by Amundson and Pontinene [9], although the ideal plates ($E^V_{j,i} = E^{MV}_{j,i} = 1$ for all components and $T^V_j = T^L_j$ throughout the column) and the ideal system ($\gamma_{j,i} = 1$ for all components) were assumed.

In order to initiate the calculations, values of variables must first be assumed. In most problems, it is sufficient to establish the initial set of V_j's and L_j's based on the assumption of constant molar flow rates of vapor and liquid throughout enriching and stripping sections; and the assumption of a linear variation of temperature with plate location is sufficient to give an initial set of T_j^L, and $E_{j,i}^V = 1$ for all components throughout the column to the initial set of $E_{j,i}^V$'s.

The matrix is named the "E^V-matrix method," since $E_{j,i}^V$'s appear in the matrix elements $m_{j,j}$ $(2 \leqq j \leqq N-1)$ and $m_{j,j+1}$ $(1 \leqq j \leqq N-1)$ as shown in Equation 49, and the successive iteration with respect to $E_{j,i}^V$ is carried out to convergence. In order to obtain a rapid solution, Yamada et al. [10] improved this method in which the value of the convergence criterion gets smaller as the convergence process advances.

Although the assumption of the ideal plate was used, Naka et al. [11] suggested a matrix method which is characterized by an effective method of correcting the set of $x_{j,i}$'s based on the component material balance instead of normalization. Wang and Henke [12] developed the original matrix method applicable to a complex distillation column with an arbitrary number of feed and drawoff streams, side reboilers, and side condensers.

The E^V-matrix method can also be applicable to the operation type for Petlyuk's distillation process [13] which has a merit of energy saving for the separation of ternary liquid mixtures. For the Petlyuk's process in which the plates are numbered as shown in Figure 5, M in Equation 46 almost becomes a tridiagonal matrix which can be solved by the Kubicek algorithm [14]. The matrix elements $m_{j,i}$'s except those in Equation 49 are given as follows.

$$m_{N1,N3} = V_{N3}E_{N3,i}^V\gamma_{N3,i}K_{N3,i}^0, \quad m_{N2,N4} = L'_{N4}$$

$$m_{N3,N1} = L'_{N1}, \quad m_{N4,N2} = V'_{N2}E_{N2,i}^V\gamma_{N2,i}K_{N2,i}^0$$

$$m_{NS,NS} = -(V_{NS}E_{NS,i}^V\gamma_{NS,i} + L_{NS} + S) \tag{51}$$

The methods mentioned so far are referred to as the bubble point method, since in each iteration step a set of liquid temperature which is required to evaluate a new set of matrix elements is computed from the bubble point equation. (Note that the dew point equation is used to evaluate H_j in the E^V-matrix method, since $T_j^V \neq T_j^L$ when the plate efficiency is not equal to unity.)

It is well-known that the bubble point method fails to obtain stable solutions for problems with a relatively wide range of $K_{j,i}$ (so called the wide-boiling mixture) because the liquid temperature calculated is sensitive to $x_{j,i}$'s. Also the enthalpy balance is very sensitive to the liquid temperature compared with V_j's and L_j's.

Sum-rate method. For the problem with a wide-boiling mixture, Friday and Smith [15] proposed a useful method termed the sum-rate method, which employs the successive procedure by Sujata [16]. This sum-rate method was further developed by Burmingham and Otto [17]. However, this method is only applicable to the problem based on the assumption of ideal plates ($T_j^V = T_j^L = T_j$).

In the sum-rate method for which the value of Q_c and Q_W are specified, the way to obtain $x_{j,i}$'s by solving Equation 46 is the same as in the bubble point method, but the values of $x_{j,i}$'s obtained are not normalized in this step, and a new set of L_j's is calculated by the following sum-rate equation instead.

$$\overset{k+1}{L_j} = \overset{k}{L_j}\sum_i x_{j,i} \tag{52}$$

The values of V_j's are calculated by the overall material balance equations for both sections. Next, in order to calculate h_j and H_j given in Equations 10 and 11, $x_{j,i}$'s are normalized, and corresponding values of $y_{j,i}$ are first computed by Equation 1 then normalized. A new set of values for the liquid temperature T_j ($T_j = T_V = T_L$, since the ideal plate is assumed throughout the column) is obtained by solving the set of simultaneous nonlinear equations of enthalpy balance for N stages with the Newton–Raphson method. Equation 46 is again solved with the new set of matrix elements given in Equation 49. These alternative calculations are repeated until a given convergence criterion is satisfied.

Newton–Raphson method. The uses of both bubble point and sum-rates methods are limited to systems involving ideal solutions and/or solutions which slightly deviate from the ideal one, since the employed successive iteration method of component activity coefficients of the liquid phase is valid only for those systems. Generally, highly nonideal solutions are characterized by the fact that the component activity coefficient and partial molar enthalpy in the liquid phase are strongly affected by even a small change in the liquid composition. Therefore, the equations describing the material and enthalpy balances around the plate are highly nonlinear. In order to solve this operation type problem we first consider a case in which the $x_{j,i}$'s are assumed to be independent variables and apply the Newton–Raphson method.

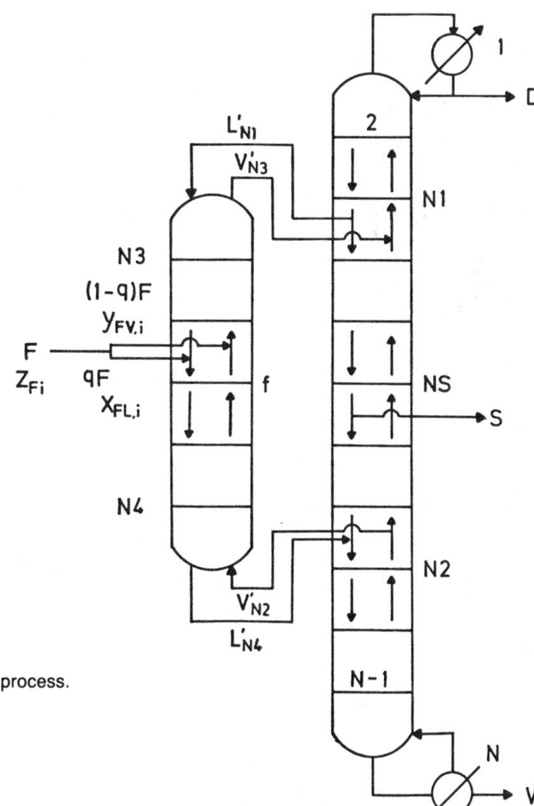

Figure 5. Column model for Petlyuk's process.

Assuming an ideal plate, define the residual functions from Equations 39 through 41

$$f_{1,i} = -(D+L_D)x_{1,i}+V_2y_{2,i} \tag{53}$$

$$f_{j,i} = L_{j-1}x_{j-1,i}-(V_j\gamma_{j,i}K^0_{j,i}+L_j)x_{j,i} + V_{j+1}\gamma_{j+1,i}K^0_{j+1,i}x_{j+1,i}-\xi_{j,i} \tag{54}$$

$$f_{N,i} = L_{N-1}x_{N-1,i}-(V_N\gamma_{N,i}K^0_{N,i}+W)x_{N,i} \tag{55}$$

After applying a Taylor series expansion with respect to $x_{j,i}$ on Equations 53 through 55 and neglecting the second order and higher terms, the Newton–Raphson equations are formulated as follows.

$$|M|' \cdot \Delta x = -f \tag{56}$$

where Δx and ξ consist of subvectors $\Delta\bar{x}_i$ and $\bar{\xi}'_i$

$$\Delta x \equiv [\Delta\bar{x}_1\Delta\bar{x}_2\ldots\Delta\bar{x}_i\ldots\Delta\bar{x}_m]^T \tag{57}$$

$$f \equiv [\bar{f}_1\bar{f}_2\ldots\bar{f}_i\ldots\bar{f}_m]^T \tag{58}$$

and $\Delta\bar{x}_i$ and $\bar{f}_i$ are defined by

$$\Delta\bar{x}_i \equiv [\Delta x_{1,i}\Delta x_{2,i}\ldots\Delta x_{f-1,i}\Delta x_{f,i}\ldots\Delta x_{N,i}]^T \tag{59}$$

$$\bar{f}_i \equiv [f_{1,i}f_{2,i}\ldots f_{f-1,i}f_{f,i}\ldots f_{N,i}]^T \tag{60}$$

M' is a $mN \times mN$ Jacobian matrix and consists of m^2 of $N \times N$ submatrices M'_{ik}'s

$$M' \equiv \begin{bmatrix} \bar{M}'_{11} & & \bar{M}'_{1m} \\ & \bar{M}'_{ik} & \\ \bar{M}'_{m1} & & \bar{M}'_{mm} \end{bmatrix} \tag{61}$$

and $\bar{M}'_{ik}$ is defined by

$$\bar{M}'_{ik} \equiv \frac{\partial\bar{f}'_i}{\partial\bar{x}_k} \tag{62}$$

$$x_k \equiv [x_{1,k}x_{2,k}\ldots x_{j,k}\ldots x_{N,k}]^T \tag{63}$$

The elements of $\bar{M}'_{ik}$ are given as follows.

For $i = k$:

$$\frac{\partial f_{1,i}}{\partial x_{1,i}} = -(D+L_D)$$

$$\frac{\partial f_{1,i}}{\partial x_{2,i}} = V_2\gamma_{2,i}K^0_{2,i}+V_2K^0_{2,i}x_{2,i}\frac{\partial\gamma_{2,i}}{\partial x_{2,i}}$$

$$\frac{\partial f_{j,i}}{\partial x_{j-1,i}} = L_{j-1}$$

$$\frac{\partial f_{j,i}}{\partial x_{j,i}} = \left(V_j \gamma_{j,i} K^0_{j,i} + L_j + V_j K^0_{j,i} x_{j,i} \frac{\partial \gamma_{j,i}}{\partial x_{j,i}} \right)$$

$$\frac{\partial f_{j,i}}{\partial x_{j-1,i}} = V_{j+1} \gamma_{j+1,i} K^0_{j+1,i} + V_{j+1} K^0_{j+1,i} x_{j+1,i} \frac{\partial \gamma_{j+1,i}}{\partial x_{j+1,i}}$$

$$\frac{\partial f_{N,i}}{\partial x_{N-1,i}} = L_{N-1}$$

$$\frac{\partial f_{N,i}}{\partial x_{N,i}} = -\left(W + V_N \gamma_{N,i} K^0_{N,i} + V_N K^0_{N,i} x_{N,1} \frac{\partial \gamma_{N,i}}{\partial x_{N,i}} \right) \tag{64}$$

For $i \neq k$

$$\frac{\partial f_{1,i}}{\partial x_{1,k}} = 0$$

$$\frac{\partial f_{1,i}}{\partial x_{2,k}} = V_2 K^0_{2,i} x_{2,i} \frac{\partial \gamma_{2,i}}{\partial x_{2,k}}$$

$$\frac{\partial f_{j,i}}{\partial x_{j-1,k}} = 0$$

$$\frac{\partial f_{j,i}}{\partial x_{j,k}} = -V_j K^0_{j,i} x_{j,i} \frac{\partial \gamma_{j,i}}{\partial x_{j,k}}$$

$$\frac{\partial f_{j,i}}{\partial x_{j+1,k}} = V_{j+1} K^0_{j+1,i} x_{j+1,i} \frac{\partial \gamma_{j+1,i}}{\partial x_{j+1,k}}$$

$$\frac{\partial f_{N,i}}{\partial x_{N-1,i}} = 0$$

$$\frac{\partial f_{N,i}}{\partial x_{N,k}} = -V_N K^0_{N,i} x_{N,i} \frac{\partial \gamma_{N,i}}{\partial x_{N,i}} \tag{65}$$

The correction $\Delta x_{j,i}$'s of the liquid composition are obtained by solving Equation 56 for the initial values $x^0_{j,i}$'s and the new values of $x_{j,i}$'s for the next trial are calculated by

$$x_{j,i} = x^0_{j,i} + \alpha \Delta x_{j,i} \tag{66}$$

where α is a weighing factor chosen to prevent over-correction. After normalization of $x_{j,i}$'s, the new set of values of $y_{j,i}$ and $T_{j,i}$ are obtained by the bubble point method. The correction of $x_{j,i}$ is continued until the norm of the vector f satisfies a convergence criterion.

When nonequimolar vapor/liquid flow rates are assumed, V_j and L_j are corrected as described before (see Equation 50).

In this discussion, the Newton–Raphson equations are formulated by assuming the compositions $x_{j,i}$'s to be independent variables. More complicated formulations have been proposed for the Newton–Raphson method which uses different independent variables. Some of the choices of the independent variables are listed below.

Independent Variables	References
T_j	Newman [18]
	Tierney and Bruno [19]
T_j, L_j	Boynton [20]
T_j, V_j	Tomich [21]
	Tierney and Yanosik [22]
T_j, L_j, $l_{j,i}$	Gallun and Holland [23]
T_j, V_j, $x_{j,i}$	Ishii and Otto [24]
T_j, l_{ji}, $v_{j,i}$	Naphtali and Sandholm [25]

where $l_{j,i} \equiv L_j x_{j,i}$ and $v_{j,i} \equiv V_j y_{j,i}$

θ-Method

In solving operation-type problems, Lyster and Holland [26] suggested the socalled "θ-method," which is developed by combining the calculation procedures of Lewis and Matheson [6] and Thiele and Geddes [27]. The feature of this method is that the plate-to-plate calculation procedure is used to obtain the component distribution of the column for the assumed liquid composition of distillate and bottom, $x_{D,i,as}$ and $x_{w,i,as}$, and those assumed values, $x_{D,i,as}$'s and $x_{w,i,as}$'s, are corrected by using the overall component material balance and the condition of plate-matching.

Plate-to-plate calculation procedure. The same procedure as in the first section is used.

Correcting method for assumed values of $x_{D,i}$'s and $x_{w,i}$'s. Let the corrected values of $x_{D,i}$ and $x_{D,i,co}$ be $x_{w,i,co}$, $x_{w,i}$, respectively. $x_{D,i,co}$'s and $x_{w,i,co}$'s are calculated by the following equations.

$$x_{D,i,co} = \frac{F z_{F,i}}{D + W \left[\dfrac{x_{w,i}}{x_{D,i}}\right]_{cal} \theta} \tag{67}$$

$$x_{w,i,co} = \frac{F z_{F,i}}{W + D \left[\dfrac{x_{D,i}}{x_{w,i}}\right]_{cal} \theta^{-1}} \tag{68}$$

where $z_{F,i}$ denotes the molar fraction of feed for component i, and $[x_{w,i}/x_{D,i}]_{cal}$ denotes the value which measures the condition of plate matching, and is defined by the following equation

$$\left[\frac{x_{w,i}}{x_{D,i}}\right]_{cal} \equiv \left[\frac{x_{w,i,as}}{x_{D,i,as}}\right] \frac{[y_{f,i}]_D}{[y_{f,i}]_w} \tag{69}$$

where if $[y_{f,i}]_D/[y_{f,i}]_w = 1$ for all components, the condition of plate matching, is satisfied and the assumption of ideal plates used in the plate-to-plate calculation procedure, $[x_{w,i}/x_{D,i}]_{cal}$:

$$\left[\frac{x_{w,i}}{x_{D,i}}\right]_{cal} \equiv \left[\frac{x_{w,i,as}}{x_{D,i,as}}\right] \frac{[x_{f,i}]_D}{[x_{f,i}]_w} \tag{69a}$$

Furthermore, θ is a parameter which provides not only stable convergence but also a rapid solution, and is calculated by

$$\sum_i \frac{Fz_{F,i}}{D + W\left[\frac{x_{w,i}}{x_{D,i}}\right]_{cal} \theta} = 1 \tag{70}$$

The calculations are repeated again by setting the values of $x_{D,i,as}$ and $x_{w,i,as}$ equal to $x_{D,i,co}$ and $x_{w,i,co}$, respectively. These alternative calculations are continued until the value of θ approaches unity within a given tolerance. It is emphasized that Equations 67 and 68 are reduced to the overall component material balance when the values of θ and $[y_{f,i}]_D/[y_{f,i}]_w$ are unity.

Saito, Yamada, and Sugie [28] developed the θ-method of convergence applicable to problems of a highly nonideal system based on finding $x_{D,i,co}$ and $x_{w,i,co}$ with the regula falsi method.

Relaxation Method

Rose et al. [29] proposed the relaxation method for solving multicomponent distillation problems. In this method the steady-state solution is obtained by solving the unsteady-state material balance equations as time approaches infinity. This method is called the original relaxation method and is also known to provide a stable solution for any type of problem involving ideal and nonideal systems as well as for complex columns.

Original relaxation method. The component material balance over a time interval from t to $t+\Delta t$ yields the following integral difference equation:

$$\int_t^{t+\Delta t} [(V_{j+1,i} + L_{j-1}x_{j-1,i}) - (V_j y_{j,i} + L_j x_{j,i})]\, dt = [H_j^L x_{j,i}]_{t+\Delta t} - [H_j^L x_{j,i}]_t \tag{71}$$

where H_j^L denotes the molar liquid holdup on the j-th plate. For the feed plate, the additional term $F \cdot z_{F,i}$ due to the feed stream appears in the first parentheses of the left-hand side in Equation 71.

The mean value theorems of integral and differential calculus are applied on L.H.S. and R.H.S. of Equation 71, respectively, to give the following equations.

$$\text{L.H.S.} = \int_t^{t+\Delta t} [(V_{j+1}y_{j+1,i} + L_{j-1}x_{j-1,i}) - (V_j y_{j,i} + L_j x_{j,i})]\, dt$$

$$= [(V_{j+1}y_{j+1,i} + L_{j-1}x_{j-1,i}) - (V_j y_{j,i} + L_j x_{j,i})]_{av}\, \Delta t \tag{72}$$

$$\text{R.H.S.} = [H_j^L x_{j,i}]_{t+\Delta t} - [H_j^L x_{j,i}]_t$$

$$= \Delta t \left[\frac{dH_j x_{j,i}}{dt}\right]_{t+\eta\Delta t} \qquad (0 \leqq \eta \leqq 1) \tag{73}$$

Substituting Equations 72 and 73 into Equation 71 with the assumption of the constant molar liquid holdup reduces Equation 71 to

$$H_j^L \left[\frac{dx_{j,i}}{dt}\right]_{t+\eta\Delta t} = [(V_{j+1}y_{j+1,i} + L_{j-1}x_{j-1,i}) - (V_j y_{j,i} + L_j x_{j,i})]_{av} \tag{74}$$

Equation 74 is a general basic expression of a relaxation equation. The calculation using Equation 74 with $\eta = 0$ is called the point relaxation method, and with $\eta = 1$ the block relaxation method.

Now, when the value of η is set equal to zero, Equation 74 becomes

$$H_j^L\left[\frac{dx_{j,i}}{dt}\right]_t = [(V_{j+1}y_{j+1,i} + L_{j-1}x_{j-1,i}) - (V_jy_{j,i} + L_jx_{j,i})]_t \quad (75)$$

By applying the approximation of infinite difference on the left-hand side of Equation 75 and labeling the quantities evaluated at t by the superscript k and at $t+\Delta t$ by $k+1$, Equation 75 is rewritten by the following equation.

$$x_{j,i}^{k+1} = x_{j,i}^k + \mu_j[(V_{j+1}y_{j+1}^k + L_{j-1}x_{j-1,i}^k) - (V_jy_{j,i}^k + L_jx_{j,i}^k)] \quad (76)$$

where μ_j is equal to $\Delta t/H_j^L$ and is called a point relaxation factor. The equations for the reboiler and condenser can be obtained by the same derivation as shown previously.

In operation-type problems, the $(k+1)$-th corrected compositions $x_{j,i}^{k+1}$'s are calculated from Equation 76 using the values evaluated in the k-th iteration step, and a new set of $y_{j,i}^{k+1}$'s for the next iteration step is obtained by the boiling point calculation. By repeating this calculation until the values of $x_{j,i}^k$'s have converged, the steady-state solution is obtained. For the case of nonequimolar vapor/liquid flow rates, both flow rates are calculated by using Equation 50 after each iteration step has been completed.

Depending on the choice of the initial values, however, it is possible that the correction term, the second term of the right-hand side in Equation 76, becomes zero, and also the speed of convergence is slow.

Group relaxation method. Ishikawa et al. [30] improved the original relaxation method to facilitate convergence. This is done by first using previously calculated values of $x_{j-1,i}^{k+1}$ instead of $x_{j-1,i}^k$, and secondly by changing the values of η_j in each iteration step, component, and plate. Equation 76 is rewritten as

$$x_{j,i}^{k+1} = x_{j,i}^k + \omega_{j,i}^k[(V_{j+1}y_{j+1,i}^k + L_{j-1}x_{j-1,i}^{k+1}) - (V_jy_{j,i}^k + L_jx_{j,i}^k)] \quad (77)$$

where $\omega_{j,i}^k$ is called the "group relaxation factor" and is given as follows.

The value of $\omega_{j,i}^{k+1}$ for the next iteration step is corrected depending on the value of ϱ_i^k.

$$\omega_{j,i}^{k+1} = \beta_i^k\omega_{j,i}^k \quad (78)$$

$$\beta_i^k = \begin{cases} 1.0, & 0.7 \leqq \varrho_i^k \leqq 1.0 \\ 1.1, & 0.4 \leqq \varrho_i^k < 0.7 \\ 1.2, & 0.0 \leqq \varrho_i^k < 0.4 \end{cases} \quad (79)$$

where ϱ_i^k denotes the logarithmic ratio of ε_i^k to the smallest value ε_b^k

$$\varrho = \left|\frac{\log \varepsilon_i^k}{\log \varepsilon_b^k}\right| \quad (80)$$

and ε_i^k is the error of the component material balance.

$$\varepsilon_i^k = \frac{|F \cdot Z_{F,i} - (D \cdot x_{D,i} + W \cdot x_{w,i})|}{F \cdot Z_{F,i}} \tag{81}$$

This "group relaxation method," is reported to reduce the required number of iterations by 1/4 to 1/6 over the original method of Rose et al.

In another work, Saito [31] discussed the selection of the relaxation factor from the standpoint of stable convergence, and proposed to change the time interval Δt depending on the component material balance.

Method for Solving the Minimum Reflux Problem

Hitherto many attempts have been made to solve the minimum reflux problem for both design and operation types [32–38]. However, except for those described by Holland [38] and Yamada et al. [37], all of these methods are limited to

- The ideal system in which the component relative volatility is assumed to be constant throughout the column.
- Constant molar overflow rates of vapor and liquid in each section of the column.

Although the method [38] described below is limited to solving operationaltype problems, it is applicable to nonideal systems in which the component relative volatility varies throughout the column, as well as the variable molar rates of vapor and liquid in each section of the column.

Real pinch-point location. The heavy and light key components cannot be assigned in the minimum reflux problems for the operation type. However, for an m-component system, the relationships between the existence of the nondistributed component and the pinch point in each section are considered in the following case [34, 39], where the components are numbered in order of reciprocal relative volatility.

Case 1. $h = 1$, $\ell = m$. Nondistributed component exists in neither distillate nor bottoms. The pinch point in each section is across the feed plate. The liquid composition on the feed plate is equal to that of the liquid portion of feed.

Case 2: $h \neq 1$, $\ell = m$. Nondistributed component exists in distillate but no nondistributed component exists in bottoms. The pinch point in the stripping section is adjacent to the feed plate, but the liquid composition on the feed plate differs from that of the liquid portion of feed. The pinch point in the enriching section is split from the feed plate and moves towards the middle region of the enriching section.

Case 3: $h = 1$, $\ell \neq m$. The relationship between existence of the nondistributed component and pinch point in each section is the reverse of Case 2.

Case 4: $h \neq 1$, $\ell \neq m$. The nondistributed component exists in both distillate and bottoms, and the pinch points in both sections are split from the feed plate. Of course the liquid composition on the feed plate differs from that of the liquid portion of feed.

Hypothetical pinch plate. As shown in Figure 6, we consider a hypothetical pinch plate in the section only where the nondistributed component exists. The hypothetical pinch plate is assumed to be located at a finite number of plates from the feed plate, where the number of plates depends on the required accuracy of the solution. Other characteristics of the hypothetical pinch plates are summarized by the following items

- An infinite number of ideal plates is considered between the hypothetical pinch plate and the real pinch point.

- No nondistributed component exists in the real pinch point, but an allowable small amount of nondistributed component exists on the hypothetical pinch plate.
- The liquid composition in the real pinch is very close to that on the hypothetical pinch plate.

Calculation procedure. The calculation procedure for solving the minimum reflux problems for the operation type is summalized by the following items

- For an assumed set of $x_{f,i}$, to calculate $x_{D,i}$ and $x_{w,i}$ by using the characteristics of the hypothetical pinch plate.
- To obtain a corrected set of $x_{f,i}$ which satisfies the overall component material balance.

Determination of existence of the nondistributed component. The existence of the nondistributed component, which was not previously assigned in the operation-type of problems, can be determined from the liquid composition on the feed plate by using the following procedure.

Assuming the real pinch point is adjacent to the feed plate, that is, $x_{f,i} = x_{\infty,i}$, $y_{f,i} = y_{\infty,i}$, $h_f = h_\infty$ and $H_f = H_\infty$ for enriching section, and $x_{f',i} = x_{\infty',i}$, $y_{f',i} = y_{\infty',i}$, $h_{f'} = h_{\infty'}$ and $H_{f'} = H_{\infty'}$ for stripping section, Equations 6 through 8 and Equations 19 through 21 are rewritten as follows.

Enriching section

$$V_\infty = L_\infty + D \tag{6a}$$

$$V_\infty y_{\infty,i} = L_\infty x_{\infty,i} + Dx_{D,i} \tag{7a}$$

$$V_\infty H_\infty = L_\infty h_\infty + Dh_D + Q_c \tag{8a}$$

Stripping section

$$L_{\infty'} = V_\infty + W \tag{19a}$$

$$L_{\infty'} x_{\infty',i} = V_{\infty'} y_{\infty',i} + Wx_{w,i} \tag{20a}$$

$$L_{\infty'} h_{\infty'} = V_{\infty'} H_{\infty'} + Wh_w - Q_w \tag{21a}$$

For a set of $x_{f,i}(= x_{\infty,i}$ or $x_{\infty',i})$, all values of $y_{f,i}(= y_{\infty,i}$ or $y_{\infty',i})$, $h_f(= h_\infty$ of $h_{\infty'})$, and $H_f(= H_\infty$ or $H_{\infty'})$ at the real pinch point can be computed, and Equations 6a through 8a and Equations 19a through 20a are reduced to two sets of nonlinear equations with the variables of $V_{\infty'}$, $L_{\infty'}$ and $x_{D,i}$ for Equations 6a through 8a and the variables of $V_{\infty'}$, $L_{\infty'}$, and $x_{w,i}$ for Equations 19a through 21a. Here h_D and Q_c, and h_w and Q_w are the function of $x_{D,i}$ and $x_{w,i}$, respectively. However, these simultaneous equations can be solved by successive iteration methods with respect to $x_{D,i}$ for Equations 6a through 8a and $x_{w,i}$ for Equations 19a through 21a, using the following calculation steps. (Note the terms in parentheses show those related to the stripping section.)

1. Assume $x_{D,i}$ (or $x_{w,i}$) based on the specification of D (or W) and $Fz_{F,i}$.
2. Calculate h_D and Q_c (or h_w and Q_w).
3. Calculate V_∞ and L_∞ (or $V_{\infty'}$ and $L_{\infty'}$) by solving Equations 6a and 8a (or Equations 19a and 21a), simultaneously.
4. Calculate $x_{D,i}$ (or $x_{w,i}$) with Equation 7a (or Equation 21a).

5. Judge the convergence criterion for $x_{D,i}$ (or $x_{w,i}$). If the criterion has been satisfied, the calculation will terminate. Otherwise return to Step 2 after $x_{D,i}$'s (or $x_{w,i}$'s) are normalized. In this case, if $x_{D,i} < 0$ for the heavy component (or $x_{w,i} < 0$ for the light component) the negative quantity is set equal to zero before the normalization. If $x_{D,i} > 0$ (or $x_{w,i} > 0$) for all components, no nondistributed component exists in distillate (or bottom). Whereas, if $x_{D,i} < 0$ for the heavy component (or $x_{w,i} < 0$ for the light component) the nondistributed component exists in distillate (or bottom), and the real pinch plate is split from the feed plate.

Convergence procedure in each section of column. For a given value of liquid composition on the feed plate, the calculation procedure to converge the component material balance in each section of the column is as follows:

- *Enriching section:* When the liquid composition on the hypothetical pinch plate $x_{H,i}$ is assumed, $x_{D,i}$ can be obtained by solving Equations 6a through 8a in the same manner just described, where $x_{H,i} = x_{\infty,i}$ is used. Using calculated $x_{D,i}$ and assumed $x_{H,i}$, $L_{f-1}x_{f-1,i}$ can be calculated by repeating the plate-to-plate calculation procedure for $E_{j,i}^{MV} = 1$ and an assigned number of finite plates, where the calculated $x_{D,i}$ and $x_{f-1,j}$ must satisfy the following component material balance equation:

$$Vy_{f,i} = L_{f-1}x_{f-1,i} + Dx_{D,i} \tag{7b}$$

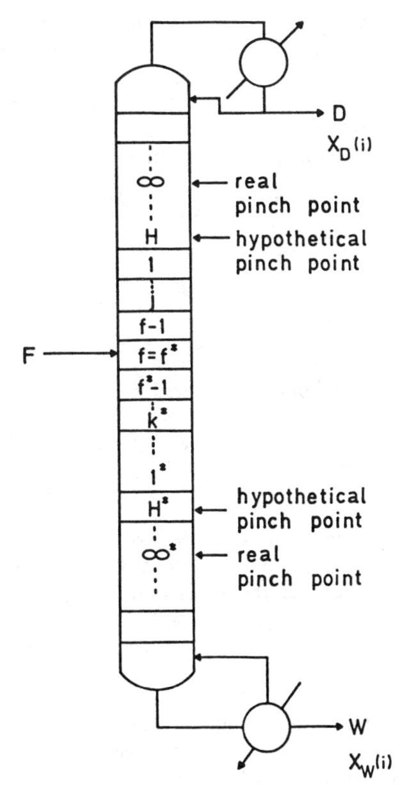

Figure 6. Column model for solving minimum reflux problem for operation type.

where $y_{f,i}$ can be evaluated from $x_{f,i}$ assumed. However, this equation is not satisfied unless the right $x_{H,i}$'s are given. Therefore $x_{H,i}$ is corrected by the normalized θ-method [40] in finding the right $x_{H,i}$'s. The equation is as follows:

$$x_{H,i,co} = \frac{\theta_H x_{H,i,as} V y_{f,i}}{D x_{D,i,ca} + (L_{f-1} x_{f-1,i})_{ca}} \tag{82}$$

where θ_H is calculated by

$$\theta_H = \left[\sum_i \frac{x_{H,i,as} V_f y_{f,i}}{D x_{D,i,ca} + (L_{f-1} x_{f-1,i})_{ca}} \right]^{-1} \tag{83}$$

where $x_{D,i,ca}$ and $(L_{f-1} x_{f-1,i})_{ca}$ are the calculated values based on $X_{H,i,as}$

This calculation process is repeated until convergence is satisfied to a specified tolerance. Also $|x_{H,i} - x_{\infty,i}|$ for the distributed component must be better than the accuracy required. Otherwise one must repeat the calculation again with a new assigned finite number of plate which is larger than the previous one. Furthermore, if a negative $x_{c,i}(l < c < f-1)$ is obtained in the plate-to-plate calculation step, $x_{c-1,i}$ obtained must be used instead of $x_{f-1,i}$, and it will be useful to set the initial value of $x_{H,i}$ equal to $x_{f,i}$. If it is assigned $h = 1$, all components contained in the feed are distributed in distillate, and $x_{D,i}$ can be calculated without any consideration of $x_{H,i}$.

- *Stripping section.* If the liquid composition on the hypothetical pinch plate $x_{H',i}$ is given, $x_{w,i}$ can be calculated by solving Equations 19a through 21a with the same manner as described in the previous paragraph, where $x_{H',i} = x_{\infty',i}$ is used. Using calculated $x_{w,i}$ and assumed $x_{H',i}$, $V_{f'-1} y_{f'-1,i}$ can be calculated by repeating the plate-to-plate calculation procedure. Also the calculated $x_{w,i}$ and $y_{f'-1,i}$ must satisfy the following component material balance equation

$$L_{f'} x_{f',i} = V_{f'-1} y_{f'-1,i} + W x_{w,i} \tag{20b}$$

then, $x_{H',i}$ is corrected by

$$x_{H',i,co} = \frac{\theta_{H'} x_{H,i,as} L_{f'} x_{f',i}}{W x_{w,i,ca} + (V_{f'-1} y_{f'-1,i})_{ca}} \tag{84}$$

where $\theta_{H'}$ is calculated by

$$\theta_{H'} = \left[\sum_i \frac{x_{H,i,as} L_{f'} x_{f',i}}{W x_{w,i,ca} + (V_{f'-1} y_{f'-1,i})_{ca}} \right]^{-1} \tag{85}$$

where $x_{w,i,ca}$ and $(V_{f'-1} y_{f'-1})_{ca}$ are the calculated values based on $x_{H,i,as}$

This calculation process is repeated in the same manner as for the enriching section. Of course, the initial value of $x_{H',i}$, a negative value of $x_{c',i}$ $(1' < c' < f'-1)$, and a finite number of plates assigned are treated in the same manner as in the enriching section. If $\ell = m$, that is no nondistributed component exists in bottoms, $x_{w,i}$ is calculated without any consideration of $x_{H',i}$.

Correction of liquid composition on the feed plate. The calculation procedure outlined in the previous paragraph is based on an assumed $x_{f,i}(= x_{f',i})$. Therefore $x_{f,i}(= x_{f',i})$ is to be corrected to satisfy the overall component material balance. The equation is as follows:

$$x_{f,i,co} = \frac{\theta_f x_{finias} F z_{F,i}}{D x_{D,i,ca} + W x_{w,i,ca}} \tag{86}$$

where θ_f is calculated by

$$\theta_f = \left[\sum_i \frac{x_{finias} F z_{F,i}}{D x_{D,i,ca} + W x_{w,i,ca}} \right]^{-1} \tag{87}$$

This alternative major calculation loop is repeated until a given convergence criterion has been satisfied. Finally, it is useful to set the initial value of $x_{f,i}(= x_{f',i})$ equal to $x_{FL,i}$.

SYNTHESIS OF DISTILLATION SEQUENCES

Multicomponent separation sequences are needed for feed preparation to a reactor, product separation, waste treatment, etc. Distillation systems are used in spite of the large energy consumption as a main unit operation in separation sequences. The objectives of design is as follows: For specified conditions of flow rate, composition, temperature, and pressure of the feed and the utilities, synthesize a process which can separate the feed to the specified products at minimum capital and operating costs. Separation sequence problems have expanded to involve not only distillation but also other separations, e.g. extractive and azeotropic distillation, extraction, adsorption, crystalization.

The aims of process synthesis are the determination of the basic structures as the primary flowsheet in terms of economy. For a distillation sequence, the order of each separation, reflux ratio, and number of plates are approximately determined and then the detailed design of each column is performed. Only the synthesis methods of distillation sequences are considered in the following.

Principles

Difficulties encountered in distillation system synthesis depends on the vapor-liquid equilibrium relationships, product specification, and available operating ranges of separation units. The problems solved are classified into two groups according to whether the multicomponent feed to be separated is regarded as an ideal or nonideal mixture. If the given mixture behaves highly nonideal, additional mass agents may be needed. Accordingly, a solvent recovery system should be included in the separation sequence. Such separation systems are named azeotropic or extractive distillation systems and are discussed later.

Now, let us consider ideal mixtures whose activity coefficients of vapor and liquid phases are equal to unity. The relative volatility between i and j components is represented as a function of temperature by

$$\alpha_{ij} = P_i^0(T)/P_j^0(T) \tag{88}$$

The average value of the relative volatility is evaluated usually by using two or three temperatures of the top, the feed, and the reboiler. These temperatures are obtained from the dew point calculation for the top product and the bubble point for the feed and the bottoms, respectively, for given operating pressures. We can represent a distillation sequence in terms of the relative volatilities between the key components. For example, the relative volatilities are given in a ternary system as follows:

$$\alpha_{AB} > \alpha_{BC}$$

The direct sequence is represented in Figure 7A. But, in exhanging the relative order of α_{AB} and α_{BC} in the direct sequence, the other sequence can be shown in Figure 7B. In other words, when a mixture is regarded as ideal, the system structures can be generated by selecting the order of the key components.

Synthesis Methods

The number of sequences with C components is represented by $\{2(C-1)\}!/C!(C-1)!$ [41]. With increasing components, extreme calculation efforts are needed to determine the optimal sequence. In order to reduce the large number of calculations, many attempts to obtain an optimal or near-optimal sequence have been performed. Synthesis methods developed for distillation sequences are classified into the following three approaches [42]

1. Heuristic
2. Evolutionary
3. Algorithmic.

Heuristic Approach [43–46]:

As rules of thumb have evolved through long and valuable experience, we can narrow the search for an optimal sequence. Many researchers have attempted to select acceptable heuristics and have applied the heuristic rules to the generation of system sequences.

Although there are variations of heuristics employed by each research group, the basic heuristic rules can be generalized as follows:

1. When the relative volatility between key components is in the separable range ($\alpha > 1.05$), the desired component should be isolated in the order of decreasing relative volatility (direct sequence).
2. When the feed contains the predominant component, this component should be separated first.
3. The most difficult separations between the key components should be left for last.

Unfortunately, it is nearly impossible to select the best sequence by using only the heuristic rules because the applicable limitation of heuristics cannot be clarified quantitatively. But, the efforts of accumulating experience and qualifying rules should be continued. Effectiveness of heuristic rules may be increased more and more in the case of restricting the search space before applying the following evolutionary or algorithmic approaches.

Evolutionary Approach [47–50]

The evolutionary approach is composed of two steps: Step 1, generation of an initial sequence as a basic structure, and Step 2, improvement of a basic structure.

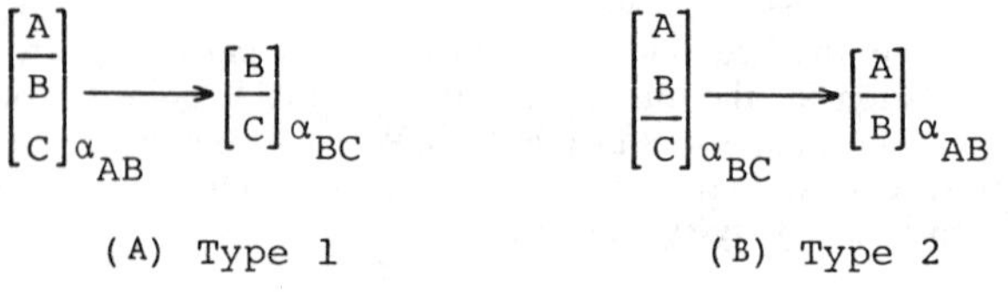

$$\begin{bmatrix} A \\ \hline B \\ C \end{bmatrix}_{\alpha_{AB}} \longrightarrow \begin{bmatrix} B \\ \hline C \end{bmatrix}_{\alpha_{BC}} \qquad \begin{bmatrix} A \\ B \\ \hline C \end{bmatrix}_{\alpha_{BC}} \longrightarrow \begin{bmatrix} A \\ \hline B \end{bmatrix}_{\alpha_{AB}}$$

(A) Type 1 (B) Type 2

Figure 7. Distillation sequences for ternary system.

In the first step, one feasible sequence is provided. In general, heuristic rules are employed not only in with the three rules but also with several rules in order to logically arrive at a feasible sequences [52]:

4. When the mole fraction and relative volatility in the feed do not vary, the direct sequence should be employed.
5. When multicomponent products are required, the sequence which produces the minimum product set should be selected.

Furthermore, for provisional design of distillation columns in the initial sequence, the following rules may be introduced:

6. Operating pressure may be assumed to be close to ambient.
7. The reflux ratio for each column is set to a value of 1.3 times its minimum reflux ratio.

In the second step, the initial sequence is gradually modified with the evolutionary rules based on two concepts [52]:

1. A separation task is substituted to an earlier position in the sequence.
2. The separation method used by a task is changed. With 1, the separating order of separation methods can be altered in the sequence selected. For the previous example of the ternary system, with application of this rule, the direct sequence of a ternary system can be modified to the other.

As for consideration of distillation sequences only, Rule 2 is not necessary. This rule explores the possibilities of introducing other applicable separation methods in addition to distillation. Evolutionary approaches have been developed to improve an existing sequence (or initial sequence) through the preparation of possible separation methods. For each application of the evolutionary rules, total cost of the modified sequence is evaluated by simulation. In addition, engineering judgment must be applied to evaluate whether the modification is successful.

Algorithmic Methods [51–59]:

Several optimization techniques have been applied to synthesis problems; dynamic programing [52], branch and bound [53], etc. Though the former method was often used to determine an optimal sequence, it can be applied to evaluating the best sequence with heat integration. The reason for this is that the synthesis of the heat integrated system is composed of two problems: selecting a distillation sequence, and integrating the heat demand and supply in a sequence. Both problems closely interact with one another. To solve these two problems individually and successively, requires enormous calculation time. On the other hand, the branch and bound method has been developed so as to evaluate an optimal problem for a heat-integrated distillation sequence. First, all possible sequences are represented as a search tree. Each arc controls the developing direction of a sequence and each node corresponds to a distillation subproblem. For example, the search tree for a four-component system is illustrated in Figure 8. The cost of a suitable sequence is evaluated as an upper bound, C_u. Then, for each sequence, the calculation of the lower bound $C_\ell(k)$ is executed. The number k denotes that k columns are combined sequentially and the final number of columns in a sequence is N. The optimal solution lies between:

$$C_\ell(N) \leqq C_{Opt} \leqq C_u \tag{89}$$

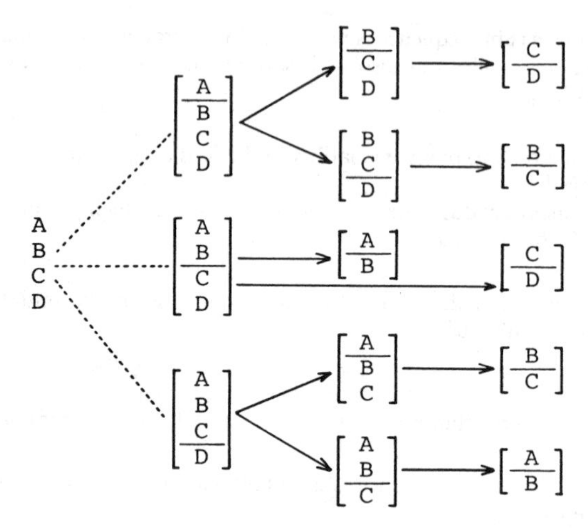

Figure 8. Distillation sequences for four-component system.

If $C_\ell(\mathbf{k})$ for any sequence is greater than C_u, it is not necessary to evaluate the lower bound which includes subsequent distillation columns. The branch developed from the k-th node can be pruned and consequently, the computational period is shortened.

Heat-Integrated Sequence [60]

In this section let us consider synthesizing a multicolumn distillation system with heat integration [41] and [56–60]. The system is to separate an ideal mixture into its pure components. Such a separation is referred to as a "sharp separation."

Synthesis Problem

The problem is defined as follows. Assuming that the upper and lower bounds of the operating pressure and heat and sink source streams are given, synthesize the distillation sequence which minimizes the demand for process utilities. This operating pressure range could be restricted from the physical properties of mixtures and components, e.g., the critical point, the freezing point, and other conditions such as reactivity. The distillation system can be realized theoretically in the desired pressure range by considering the following:

- Heat Source Streams
  H1: Process heat utilities at fixed temperatures.
  H2: Heat source streams available from other processes. (The temperature and heat load of each source stream are given.)
  H3: Heat sources represented by heat removed by condensers in the system. (The temperature and heat load depend on the operating pressure and reflux ratio of each column relating to the condenser.)

- Heat Sink Streams
  C1: Process cold utilities at fixed temperatures.
  C2: Heat sink streams available from other processes. (The temperature and heat load of each heat sink stream are given.)

C3: Heat demands by reboilers. (The temperature and heat load of each column depend on the operating pressure and reflux ratio.)

The synthesis problem is equivalent to the problem of how to arrange the available energy loss of each distillation column closely in order to minimize the available energy loss of required process utilities by maximizing the use of H3 and C3.

$$\Delta E_R(j) \cong \sum_s (1 - T_0/T_{Hs})Q_{rs} - \sum_s (1 - T_0/T_{Cs})Q_{cs} \tag{90}$$

where Q_{rs} and Q_{cs} are the heat loads supplied by H1 and C1, respectively. For H2, H3, and C2, C3, their available energies are already calculated as the removal available energies to be transfered to or from other systems and units. The heat integration for the sensible heat is neglected in the step of the distillation sequences, but it can be readily accounted for in the stage during the detail design.

Synthesis Strategy

The procedure is based on the branch and bound and consists of the following three steps:

1. Investigate the possibility of supplying H2 and C2 to each separation subproblem.
2. Arrange all sequences in the order of a decreasing possibility of heat consumption.
3. Select the optimal operating conditions of the sequence.

Step 1: Possible use of waste heat energy, H2 and C2. Figure 9 shows an example of a five-component separation system. The columns of the matrix correspond to the top products appearing in all sequences and the rows correspond to the bottom products. If a

Top / Bottom	1 A	2 B	3 C	4 D	5 A B	6 B C	7 C D	8 A B C	9 B C D	10 A B C D
1 B	F									
2 C		F			F					
3 D			$H2_1$			$H2_1$		$H2_1$		
4 E				$H2_3$			$H2_3$		$H2_3$	$H2_2$
5 BC	F									
6 CD		$H2_3$			$H2_2$					
7 DE			$H2_2$			$H2_2$		$H2_1$		
8 BCD	F									
9 CDE		$H2_1$			$H2_1$					
10 BCDE	$H2_2$									

$(T_{H2,1} < T_{H2,2} < T_{H2,3})$

Figure 9. Possibility matrix of (H2) heat sources.

separation subproblem is feasible, the mark, F, is indicated at the corresponding element and otherwises nil.

Next, the possibilities of utilization of H2 streams are introduced to the matrix. The H2 streams are numbered in order of temperature as follows: $H2_1, \ldots, H2_m, \ldots, H2_M$, where $T(H2_1) < \ldots < T(H2_M)$.

Assuming that each subproblem can be operated at a pressure between the lower limit, P_l, and upper limit, P_u, i.e.

$$P_\ell \leqq P \leqq P_u \tag{91}$$

If the bottom temperature, T, is higher than $(T(H2_m) - \Delta T_a^{min})$ at the lowest pressure P_ℓ, it is impossible to use the streams $H2_1$ to $H2_m$, and the mark, m, is indicated at the (i, j) element. If a subproblem can utilize any kind of given heat source streams ($H2_M$), F is left. If the result interferes with the utilization of the C1 and C2 streams, other heat sink streams should be provided.

Step 2: Evaluation of the lower bound. In this step, by hypothesizing each sequence with complete heat integration, we can put all of the sequences in the order of the most promising. The hypothetical heat-integrated sequence has N columns with only N reboilers and one condenser, and each column can be operated at any pressure P ($P_l \leqq P \leqq P_u$) independent of the operating conditions of the other columns. According to this assumption, even if the hypothetical j-th sequence is shown in Figure 10, the hypothetical available

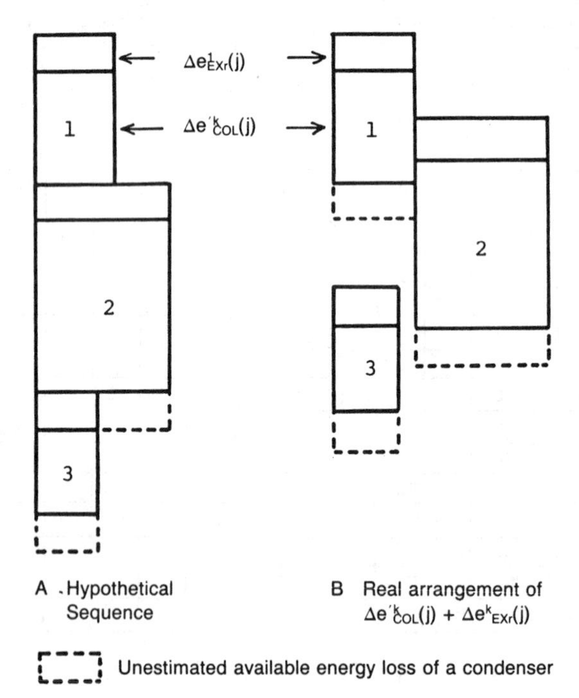

Figure 10. Schematic diagram of a hypothetical sequence in $(1 - T_0/T) - Q$ diagram.

energy loss, $\Delta E_H(j)$ is always evaluated based on the unit arrangement in Figure 10 A, neglecting the temperature levels of each distillation. The minimum available energy loss of the k-th separation subproblem with no condenser in the j-th sequence is defined as follows:

$$\Delta e^{min,k}(j) = \underset{P_k,R_k}{\text{Min}} \; [\Delta e'^{k}_{COL}(j) + \Delta e^{k}_{EXr}(j)] \tag{92}$$

where P_k and R_k are the operating pressure and reflux ratio, respectively. The total minimum avoidable available energy losses for the j-th hypothetical sequence with heat integration is represented by the following equation:

$$\Delta E_H^{min}(j) = \sum_{k=1}^{N} \underset{P_k,R_k}{\text{Min}} \; [\Delta e'^{k}_{COL}(j) + \Delta e^{k}_{EXr}(j)] \tag{93}$$

Though the available energy loss of a condenser involved in the column with the lowest top temperature in the sequence is added to this equation, this value is neglected so as to simplify the evaluation of available energy loss of the sequence. The value of $\Delta e'^{k}_{COL}(j)$ is a function only of operating pressure, if the reflux ratio, $R = \theta R^{min}$ is fixed for each column. If there are some heat source streams of H2, the hypothetical available energy loss should be modified by using the possibility matrix of H2 in Step 1. The hypothetical available energy loss of the j-th sequence which has the possibility of using heat source streams of H2 should be reduced as follows;

$$\Delta E_H(j) = \Delta E_H^{min}(j) - \sum_{m=1}^{M'} (1 - T_0/T_m)Q_m \tag{94}$$

where, T_m and Q_m are the temperature and heat load of the m-th heat source stream in the M' possible source streams checked in Step 1. The operating pressure, P, should be in a feasible region given by Equation 91, and furthermore, the temperatures of the top and bottoms should satisfy the following equations simultaneously,

$$T_W \leqq T_H^{max} - \Delta T_a^{min} \tag{95}$$

$$T_D \leqq T_C^{min} + \Delta T_a^{min} \tag{96}$$

where, T_H^{max} is the highest temperature of heat source streams, H1 and H2, T_C^{min} is the lowest temperature of heat sink streams, C1 and C2. The ΔT_a^{min} is the minimum approach temperature of a heat exchanger.

Here, the other available energy loss is defined as follows: the minimum available energy loss for a real system, $\Delta E_H^{min}(j)$, is the minimum available energy loss for the process utilities supplied to the heat-integrated j-th sequence with feasible operating conditions.

Although the calculation method is presented in detail later, the following equation should be noted:

$$\Delta E_R^{min}(j) > \Delta E_H(j) \tag{97}$$

because $\Delta E_R^{min}(j)$ does not include the available energy loss for any condenser and it may not be feasible to hold all operating pressures of $\Delta E_H^{min}(j)$ when the maximum heat integration is assumed. For the j'-th and j-th sequences, if the following relation holds

$$\Delta E_H^{min}(j') \geqq \Delta E_R^{min}(j) \tag{98}$$

it is not necessary to evaluate $\Delta E_R^{min}(j')$ sequence.

Step 3: Feasible heat integration. This step determines the optimal feasible operating conditions and specifies the heat integrated system, using the $(1-T_0/T)$ vs. Q or the T−Q diagram.

From the definition of $\Delta E_R^{min}(j)$, it is clear that the optimal operating conditions can be obtained by locating the locus of $(\Delta e'^{k}_{COL}(j) + \Delta e^{k}_{EX}(j))$ on the diagram so as to minimize the total available energy loss of the process utilities required in the j-th sequence.

The method to determine the optimal operating pressure is as follows; When there are U process utilities and M heat source streams from other systems, the maximum number of possible utility temperatures which the first column in the j-th sequence can utilize is (U + M). In general, snice the k-th column has the possibility of using removable heat energy from the condensers attached to k − 1 columns, the maximum number of possible utility temperatures is (U + M + k − 1). For example, when there is one process utility H1 and one heat source stream H2, the possible temperatures of the second column are shown in Figure 11. The square with broken lines shows alternative positions at the same temperature of the second column. For a sequence with N columns, the number of the pressure combinations, taking into account the column arrangement, is

$$(U+M)(U+M+1)\ldots(U+M+N-1)(N!)$$

To explore such a large number of pressure combinations would be very time-consuming, so a tree searching method is employed (branch and bound method).

Another assumption is introduced here, namely that the operating pressure P is set equal to the pressure at the bubble point of the bottoms, where T is defined as

$$T_W = T_H(j) - \Delta T_a^{min} \tag{99}$$

where $T_H(j)$ is the temperature of heat source streams. The pressure of the k-th column is calculated by using the bubble point calculation at T_W. From the constraint on the operating pressure

$$\text{if} \quad P \geqq P_u, \quad P = P_u \tag{100}$$

or if

$$P \leqq P_\ell, \quad P = P_\ell \tag{101}$$

Also, the temperature of the top product, T_D, obtained from the dew point calculation should satisfy Equation 96. The previously mentioned assumption is acceptable because reducing the available energy loss of a reboiler and making the temperature of the heat energy removed from a condenser higher results in a greater possibility of reusing heat energy in subsequent columns.

The synthesis tree can thus evolve by adding further columns, taking into account the use of H2 heat sources for heat integration. When using a heat source stream of H2 or H3 at some node, the constraint for the heat load should be checked. If the whole heat load of such a stream is already supplied to reboilers of 1 to k-1 columns, the branches with the node can be pruned. If satisfied, after checking the constraint of the pressure range, the search makes a comparison in the difference between the lower bound (LB) and the upper bound (UB) at every node. If LB > UB, the further branches at some node are pruned away and the search returns to the last node. When all possibilities have been examined, the feasible solution having UB is optimal.

If an initial value of UB is assumed near the optimal value, the computational time can be reduced. For the sequence with the minimum value of $\Delta E_H(j)$ required in Step 2, the

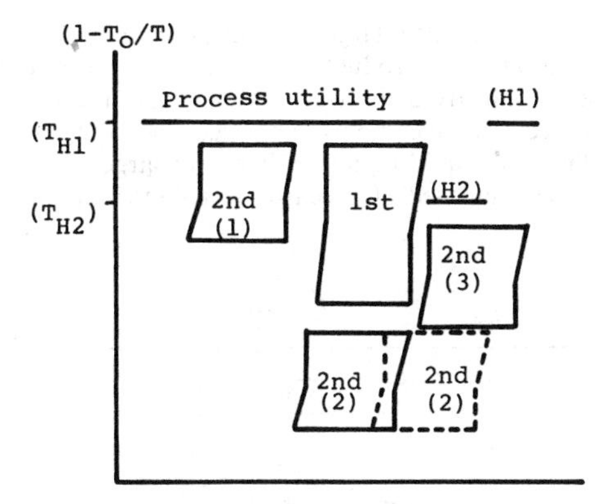

Figure 11. Possible temperature of the second column.

operating pressure for each column is selected which minimizes the available energy of the process utility requirement assuming an additional column. The operating pressure should satisfy Equations 96 and 99–102. Here, this available energy loss is denoted by ΔE_R^{*a} (a = 1, ..., N). The minimum value of ΔE_R^{*a} is then used as the initial value of UB.

$$\Delta E_R^{UB} = \text{Min}\, \Delta E_R^{*a}(j) \tag{102}$$

The synthesis strategy for Step 3 is shown in Figure 12.

Example: As shown in Table 1, two process utilities and one heat source stream such as H2 are available to the separation system, but only the lower temperature utility is employed and the heat source stream is utilized at the reboiler of the 4th column. The optimal sequence using heat integration based on the latent heat is shown in Figure 13. Three heat exchangers, $H_{EX}1$ to $H_{EX}3$, are used for heat integration. The heat integration including sensible heat is shown in Figure 14.

DESIGN OF AZEOTROPIC DISTILLATION SYSTEM

Heterogeneous Distillation

A heterogeneous azeotropic distillation system is composed of more than one column and a decanter to overcome the constraints of the valley and ridge on the composition profile in a distillation column [61]. It is only effective to use a decanter when the angles formed by the intersection of the liquid-liquid tie lines and the valley which obstructs the required separation are sufficiently large. The number of columns required depends on the availability of a decanter and the product specifications.

Classification of System Structures

For existing heterogeneous azeotropic distillation systems, the relationships between valleys and liquid-liquid tie lines are shown in Table 2. Note that for the acetic acid/water/butyl acetate system, there is no binary azeotrope in the acetic acid/water system, but a tangent azeotrope is regarded as a binary azeotrope. This system structure can be classified as Type 1. The allyl alcohol dehydration process is composed of one column and a decanter.

However, as the waste flow contains a large amount of alcohol, this system may not really satisfy the severe specification of the products. Therefore, one more column should be added in order to recover the alcohol from the waste. Meanwhile, the iso-propyl alcohol/water/ iso-propyl ether system needs three columns and a decanter. But, as the angle made by the intersection of the valley and liquid-liquid tie lines are large, the dehydration system of iso-propyl alcohol may be composed of at least two columns and a decanter. In addition,

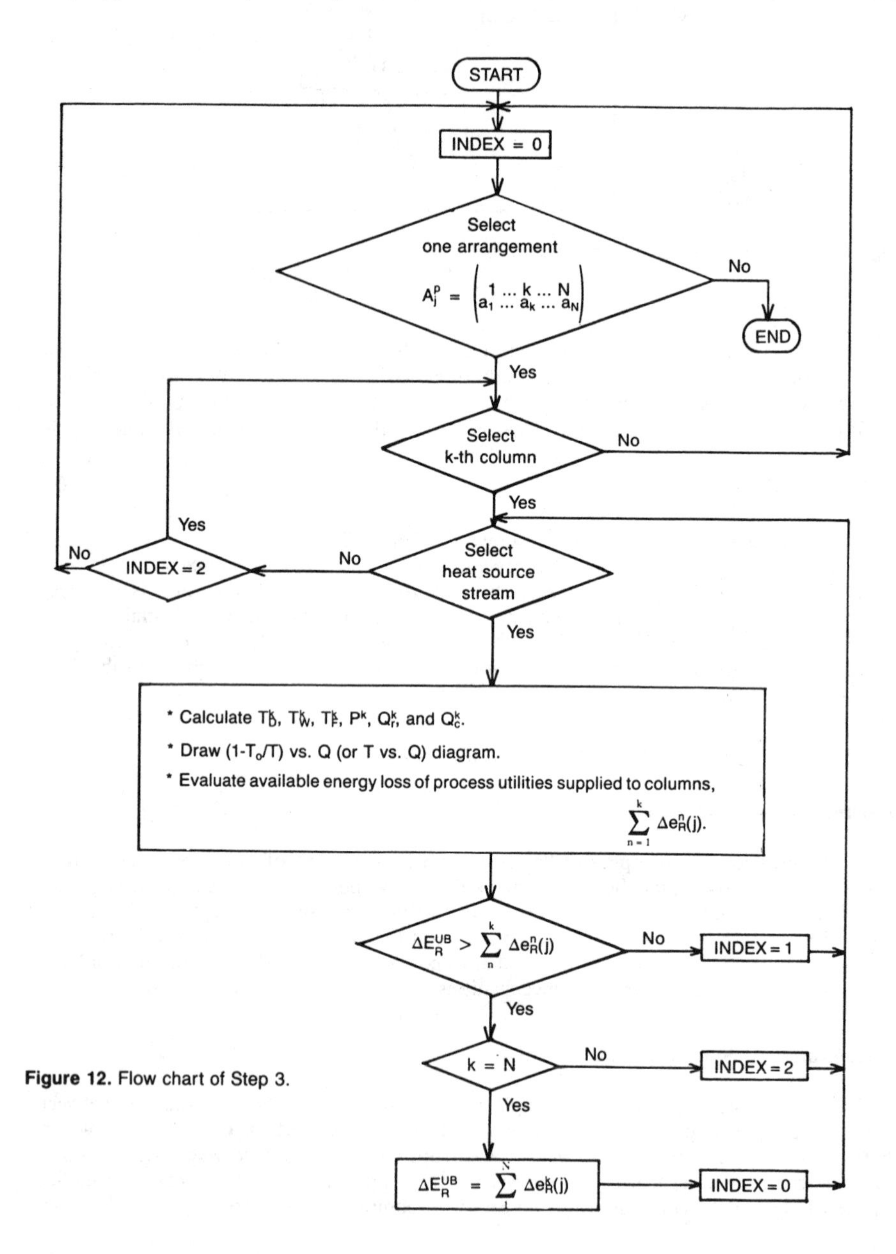

Figure 12. Flow chart of Step 3.

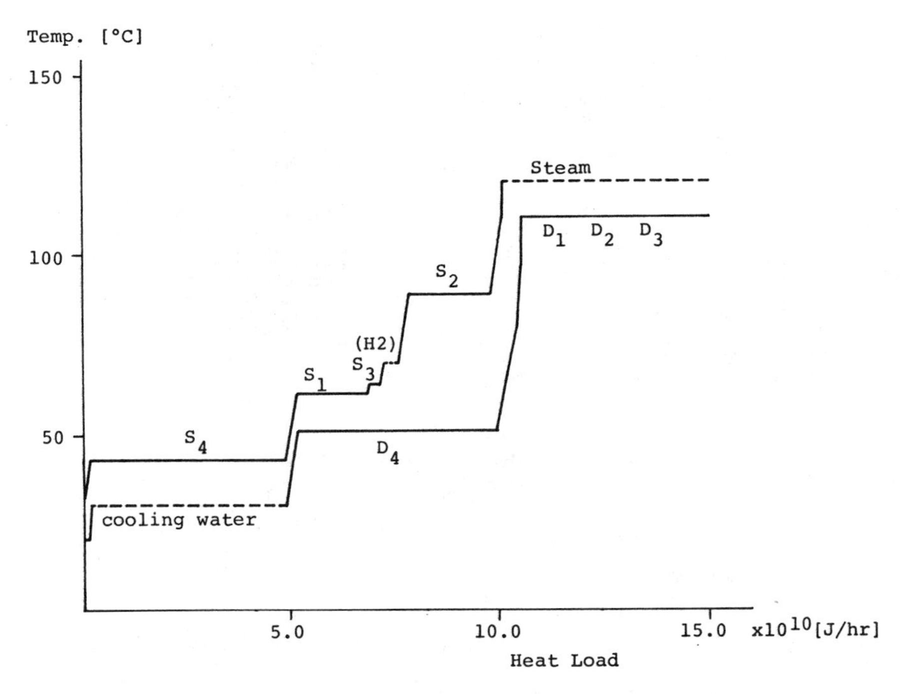

Figure 13. Optimal sequence.

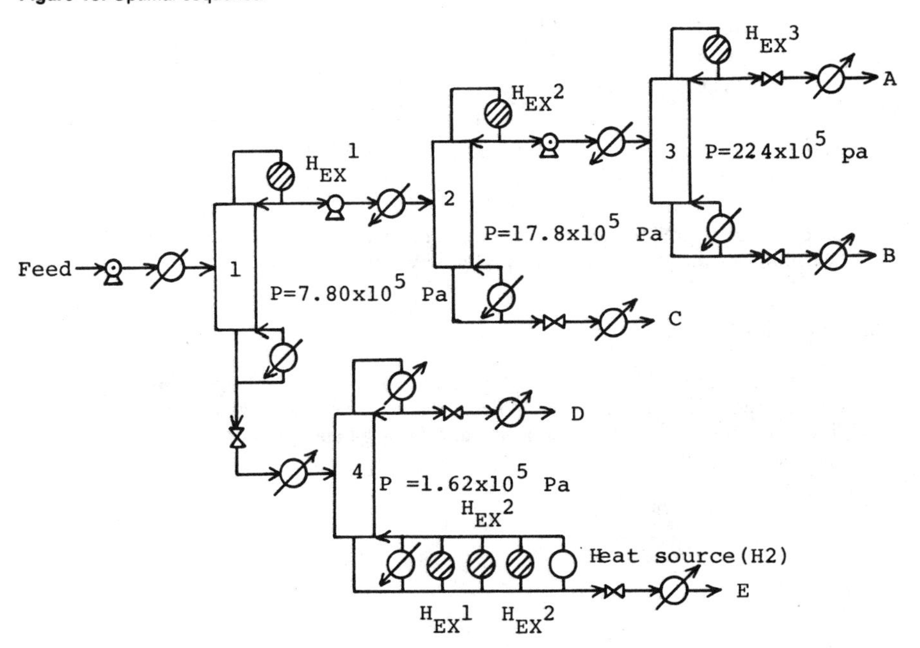

Figure 14. Heat integration of the example.

Table 1
Comparison Between $\Delta E_H(j)$ and $\Delta E_R^{min}(j)$ (10^{10}J/h)

Seq.	$\Delta E_H(j)$	$\Delta E_R^{min}(j)$	Sequence
1	6.86	8.99	(10,1), (2,2), (4,4), (7, 6)
2	7.19	8.65	(5,1), (2,2), (4,4), (7, 8)
3	7.36	12.16	(10,1), (9,2), (7,3), (4, 4)
4	8.15	11.54	(1,1), (7,3), (4,4), (9, 5)
5	8.49	12.92	(1,1), (4,4), (2,5), (7, 8)
6	9.03	---	(10,1), (9,2), (3,3), (4, 7)
7	9.82	---	(1,1), (3,3), (9,5), (4, 7)
8	10,49	---	(10,1), (2,2), (3,6), (4, 9)
9	10.66	---	(10,1), (6,1), (3,3), (4, 9)
10	11.45	---	(8,1), (2,2), (3,6), (4, 10)
11	11.62	---	(8,1), (6,2), (3,3), (4, 10)
12	11.87	---	(5,1), (2,2), (3,8), (4, 10)
13	12.58	---	(1,1), (3,3), (6,5), (4, 10)
14	13.17	---	(1,1), (2,5), (3,8), (4, 10)

(No. of Top product/No. of Bottom product)
The numbers correspond to those in Figure 9.

Table 2
Existing Heterogeneous Azeotropic Distillation

A	B	S	Ref.	Columns	Type
					B (a_1)
Acetic acid	Water	Butyl acetate	62	2	a_2
					S a_3 A
Ethanol	Water	Butul acetate	63, 64 65	2 3	B
i-Propanol	Water	i-Propyl ether	66 67	2 3	a_2 a_1
i-Propanol	Water	Benzene	68	3	a_T S a_3 A
Allyl alcohol	Water	Trichloro-ethylene	69	1	

in a practical sense, preliminary processes must be taken into account from the standpoints of reachability of the specification, controllability, safety, and so on.

Design of Two-Column System

A two-column system [70] shown in Figure 15. To understand the principle of heterogeneous azeotropic distillation systems, let's consider a simple model based on the vapor-liquid and liquid-liquid equilibrium relationships and mass balance equations.

Vapor-liquid equilibrium

$$y_i = \gamma_i(\underset{\sim}{x})x_i P_i^0(t)/P \tag{103}$$

$$\sum_{i=1}^{C} y_i = 1 \tag{104}$$

Where the more volatile component of binary components to be separated and the solvent are denoted as 1 and 3, respectively.

Liquid-liquid equilibrium

$$x_u \gamma_u(\underset{\sim}{x}_u, t_d) = x_\ell \gamma_\ell(\underset{\sim}{x}_\ell, t_d) \tag{105}$$

Subscripts, u and ℓ, are the upper and lower liquid phases, respectively, and the t_d is the decanter temperature. The mole fraction vector is composed of the components, 1 and 3.

Mass balance

The feed is heated up to the boiling point, and the solvent is added to the given binary mixture. Each plate is numbered from the top.

$$F' = F + S \tag{106}$$

$$F'\underset{\sim}{z}'_F = F\underset{\sim}{z}_F + S\underset{\sim}{x}_S \tag{107}$$

The vector of the solvent composition is $x_S = (0, 0)^t$. For the first column:

At the top

$$V^1 + W^1 = L^1 + F' \tag{108}$$

$$V^1\underset{\sim}{y}^1_1 + W^1\underset{\sim}{x}^1_N = L^1\underset{\sim}{x}^1_R + F'\underset{\sim}{z}'_F \tag{109}$$

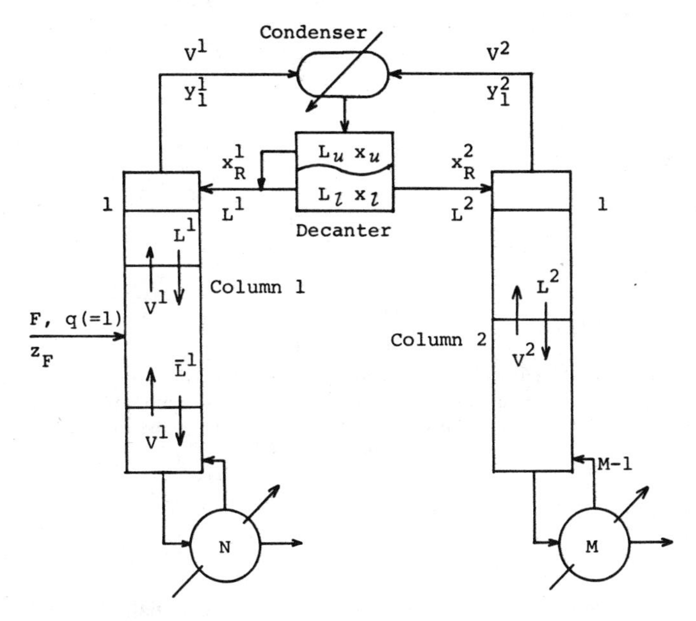

Figure 15. Two-column model.

At the j-th plate above the feed plate

$$V^1\underset{\sim}{y}_j^1 + W^1\underset{\sim}{x}_N^1 = L^1\underset{\sim}{x}_{j-1}^1 + F'\underset{\sim}{z}' \tag{110}$$

At the feed plate

$$V^1\underset{\sim}{y}_f^1 + W^1\underset{\sim}{x}_N^1 = L^1\underset{\sim}{x}_f^1 + F'\underset{\sim}{z}_F' \tag{111}$$

$$V^1 + W^1 = \bar{L}^1 \tag{112}$$

At the j-th plate below the feed plate

$$V^1\underset{\sim}{y}_j^1 + W^1\underset{\sim}{x}_N^1 = \bar{L}^1\underset{\sim}{x}_{j-1}^1 \tag{113}$$

At the reboiler

$$V^1\underset{\sim}{y}_N^1 + W^1\underset{\sim}{x}_N^1 = \bar{L}^1\underset{\sim}{x}_{N-1}^1 \tag{114}$$

For the second column:

At the top

$$V^2 + W^2 = L^2 \tag{115}$$

$$V^2\underset{\sim}{y}_1^2 + W^2\underset{\sim}{x}_M^2 = L^2\underset{\sim}{x}_R^2 \tag{116}$$

At the k-th plate

$$V^2\underset{\sim}{y}_k^2 + W^2\underset{\sim}{x}_M^2 = L^2\underset{\sim}{x}_{k-1}^2 \tag{117}$$

For the condenser:

$$V = V^1 + V^2 \tag{118}$$

$$V\underset{\sim}{x}_c = V^1\underset{\sim}{y}_1^1 + V^2\underset{\sim}{y}_1^2 \tag{119}$$

For the decanter:

$$L_u + L_l = V \tag{120}$$

$$\underset{\sim}{x}_u L_u + \underset{\sim}{x}_l L_l = V\underset{\sim}{x}_c \tag{121}$$

The upper phase is solvent-rich and generally returned to the top of the first column as a reflux flow rate, L^2. If a reflux flow rate is less than the minimum reflux flow rate, a part of the lower flow is added to the reflux flow rate. The reflux flow rate into the first column is

$$L^1 = L_u + \beta L_l \tag{122}$$

$$\underset{\sim}{x}_R^1 = \underset{\sim}{x}_u L_u + \beta\underset{\sim}{x}_l L_l \tag{123}$$

and also the reflux flow rate into the second column, L, must essentially satisfy the following equation.

$$L^2 = (1 - \beta)L_I \tag{124}$$

$$\underset{\sim}{x}_R^2 = \underset{\sim}{x}_I \tag{125}$$

When the feed conditions, F and z_F, the operating pressures of both column and the decanter temperature are given, the degree of freedom of the equations is four, by the structure analysis. Accordingly, for example, when the values of W^1, $x_N^1 = (x_{N1}^1, x_{N1}^1)$, and W^2, in addition to the numbers of plates in both columns and the feed plate are specified, the composition profiles can be obtained. The selection of four independent variables can be altered according to the type of problems to be solved. Moreover, the reflux ratio of the first column, R^1 is defined by

$$R^1 = L^1/W^2 \tag{126}$$

and when its value is fixed, the number of degrees of freedom decreases by one, that is, three. As the minimum value of the reflux ratio is approximately obtained by the Yorizane method [71], the reflux ratio can be determined by taking into account the allowance from the minimum value.

The results of the application of this method for the two-column design for iso-propyl alcohol dehydration is summarized in Table 3. In comparison with the results obtained by Yorizane et al. [68], the two-column system is better with respect to total heat loads to the reboilers and total numbers of plates. Moreover, this system can achieve a more stringent specification.

Consequently, when the angle between the valley and liquid-liquid tie line near the ternary azeotrope is sufficiently large, a two-column system may be possible.

Robinson and Gilliland [72] proposed the computational procedure based on the mass balance with the condition of constant molal overflow and the graphs of the relative volatility data. Note however, that the vapor and liquid flow rates in each column in an azeotropic distillation system often varies significantly because of the nonideality of thermodynamic properties. Heat balances should take this into account along with the material balances. Of course, to increase the accuracy of the vapor-liquid and the liquid-liquid equilibria is essential. Black et al. [73] and Prokopakis et al. [74] solved the mass and heat balance equations, using the Naphtali and Sandholm algorithm [25] and the Ross and Seider algorithm [75], respectively. Block and Hengner [76] developed a simulation model for three-phase distillation as shown in Figure 16. The main feature of this model is that it takes into account the possibility of two-liquid phases at every plate and involes a phase-splitting parameter, η. If $0 \leqq \eta \leqq 1$, the three phases exist, and if $\eta = 1$, the model is equivalent to the ordinary distillation model. Moreover, this model can easily incorporate the Murphree tray efficiencies in the mass balance expressions.

Table 3
Two-Column System of Iso-Propanol Dehydration

Column	Flow Rate (kg-mol/h)		Composition (Mol Frac.) IPA	Water	Benzene	
First column	F	100.0	0.6800	0.3199	0.0001	$q = 1.0$
	W^1	68.04	0.9990	0.00085	0.00015	$R^1 = 10.0$
	V^1	383.52	0.2183	0.2517	0.5300	Number of plates
	L^1	351.56	0.2382	0.1837	0.5781	= 11 + reboiler
Second column	W^2	31.96	0.0009	0.9991	0.1 10	$R^2 = 2.5$
	V^2	21.31	0.1775	0.8185	0.0039	Number of plates
	L^2	53.27	0.0716	0.9268	0.0016	= 2 + reboiler
Total	Vap. = 404.83, Liq. = 404.83					13 + two reboilers

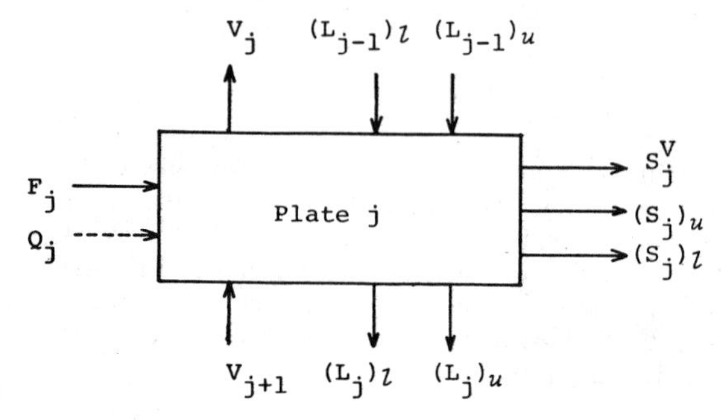

Figure 16. Three-phase plate model.

Phase equilibrium

$$y_{ji} = (K_{ji})_u (x_{ji})_u$$

$$y_{ji} = (K_{ji})_l (x_{ji})_l \quad (1 \leqq i \leqq C)$$

$$(x_{ji}) = K_{ji}(x_{ji})_u \quad (1 \leqq j \leqq N) \tag{127}$$

In many cases of ternary heterogeneous azeotropic systems, in the area where the binodal curve cuts valleys, the arrowheads of the vapor-liquid tie lines whose liquid compositions are in the binodal curve in liquid-liquid equilibrium have the directions towards each other and are very close together.

Stoichiometric equations

$$\sum_{i=1}^{C} y_{ji} = 1$$

$$\sum_{i=1}^{C} (x_{ji})_u = 1 \quad (1 \leqq j \leqq N)$$

$$\sum_{i=1}^{C} (x_{ji})_l = 1 \tag{128}$$

Mass balance in two-liquid phase region

$$\bar{L}_j = (L_j)_u + (S_j)_u + (L_j)_\ell + (S_j)_\ell \quad (1 \leqq j \leqq N) \tag{129}$$

$$(\underset{\sim}{x}_j)_a = \{[(L_j)_u + (S_j)_u] (\underset{\sim}{x}_j)_u + [(L_j)_\ell + (S_j)_\ell] (\underset{\sim}{x}_j)_\ell\}/L_j \tag{130}$$

$$(\underset{\sim}{x}_j)_a = \eta_j(\underset{\sim}{x}_j)_u + (1 - \eta_j) (\underset{\sim}{x}_j)_\ell \quad (1 \leqq j \leqq N) \tag{131}$$

Form Equations 129 through 131, we obtain

$$(L_j)_u + (S_j)_u = \eta_j \bar{L}_j \quad (1 \leqq j \leqq N) \tag{132}$$

$$(L_j)_\ell + (S_j)_\ell = (1 - \eta_j)\bar{L}_j \tag{133}$$

Mass balance for component i

$$V_{j+1}y_{j+1,i}+(L_{j-1})_u\,(x_{j-1,i})_u+(L_{j-1})_\ell\,(x_{j-1,i})_\ell+F_jz_{ji}$$

$$-(V_j+S_j^V)y_{ji}-\bar{L}_j(x_{ji})_a=0 \quad (1 \leqq i \leqq C, \quad 1 \leqq j \leqq N) \tag{134}$$

Where the S^V is a vapor side stream.

Enthalpy balance

$$V_{j+1}h_{j+1}^V+(L_{j-1})_u\,(h_{j-1}^L)_u+(L_{j-1})_\ell\,(h_{j-1}^L)_\ell+F_jh_j^F+Q_j$$

$$-(V_j+S_j^V)h_j^V-\bar{L}_j(\bar{h}_j^L)_a=0 \qquad (1 \leqq j \leqq N) \tag{135}$$

Where, $(\bar{h}_j^L)_a$ is

$$(\bar{h}_j)_a=\eta_j(h_j^L)_u+(1-\eta_j)\,(h_j^L)_l \quad (1 \leqq j \leqq N) \tag{136}$$

and the Q_j is the heat loss from the j-th plate of the column.

Liquid side stream and phase separator

A side-stream composition, x^{SL}, should be equal to the overall composition as

$$\underset{\sim}{x}_j^{SL}=(\underset{\sim}{x}_j)_a \tag{137}$$

$$(S_j)_u=\eta_j(S_j^L)_a \quad \text{and} \quad (S_j)_\ell=(1-\eta_j)\,(S_j^L)_a \tag{138}$$

where the superscript SL is related to a liquid side-stream.
In the case of a single-phase side-stream withdrawn (u or l),

$$\underset{\sim}{x}_j^{SL}=(\underset{\sim}{x}_j)_u \tag{139}$$

$$(S_j)_u=(S_j^L)_a \quad \text{and} \quad (S_j^L)_\ell=0 \tag{140}$$

Homogeneous Distillation [77]

When the appropriate solvent for forming an immiscible composition region with the feed cannot be identified, an attempt to form homogeneous azeotropes should be made.

The synthesis problem for the homogeneous azeotropic distillation of a ternary system is: When the composition and flow rate of the binary feed, A and B, and a solvent are given, an azeotropic distillation system to separate the feed to the specified products should be synthesized.

Process Structures

A single distillation column in an azeotropic system has two inlets for the feed and the solvent in many cases. To simplify the synthesis problem, we may consider an azeotropic distillation system composed of only conventional columns and mixers. After determining the system structure, one can develop the generated system by changing the feed and solvent locations in terms of economics and operability.

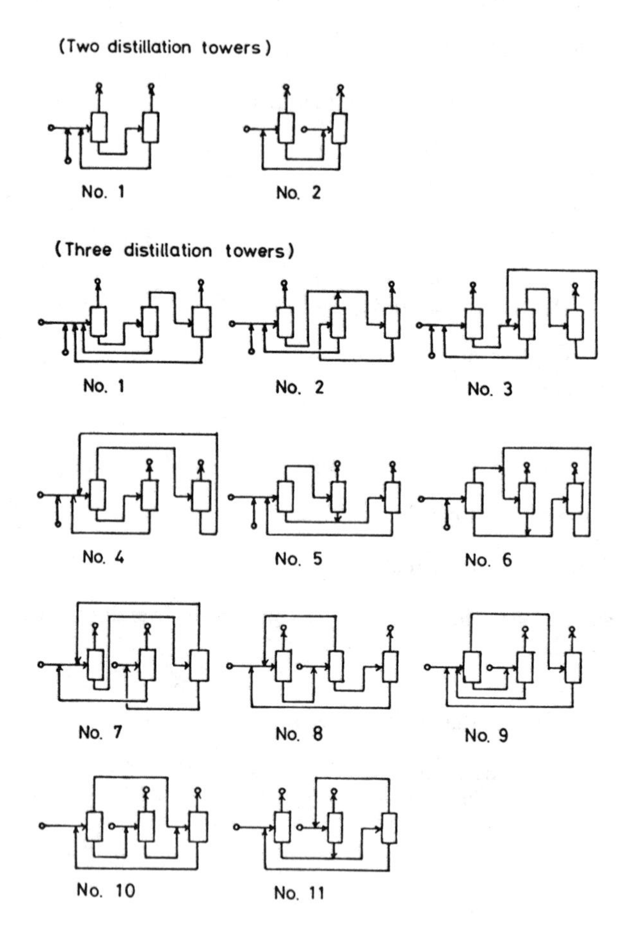

Figure 17. All structures for two-column and three-column system.

Figure 17 shows all system structures with two and three columns, respectively. But, there is no distinction between the feed and solvent for two inlets and between two products for the outlets. These can be generated by using the "incidence matrix" based on the concept of the "block design" [78].

The incidence matrix shown in Figure 18 represents each structure as follows: Each block (column) of the incidence matrix is denoted as a distillation unit or a mixing unit, and each terminal (row) shows the state variables, composition, and flow rate. An element in the incidence matrix represents the connecting relationship between an output and an input. There is a flag "I" in each of the top two terminals, and it represents the input-connecting relationship between a system input and an input of a mixer. The relationships alternate: the feed and the solvent are supplied to one mixer, or to two mixers individually. A flag "Ø" in each of the third and fourth terminals indicates the output connecting relationship between a system output and an output of a distillation. Each of the other terminals has one each of "I" and "Ø" flags. These terminals are named inner terminals. Meanwhile, a column related to a distillation unit has one input and two outputs and each block related to a mixer has two inputs and one output.

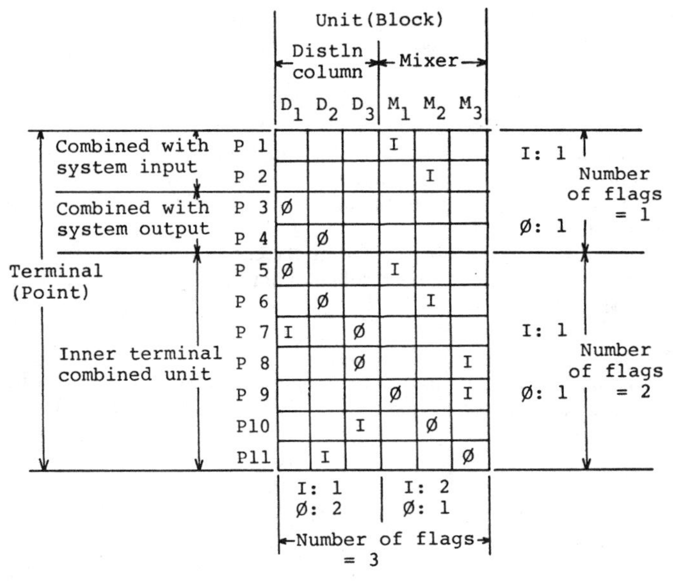

Figure 18. Incidence matrix of Pattern 8.

The numbers of distillation columns and mixers are determined by counting the number of the input and output terminals. As a result, the number of distillation columns, N, is equal to that of the mixers.

Consequently, the generation of system structures can be performed by arranging "I" and "Ø" flags according to the previously mentioned rules.

In addition, as each terminal is represented by composition and flow rate, each block can show mass balances corresponding to a unit, that is,

For the distillation unit (Input = Output)

$$F_1 \begin{bmatrix} x_{F1}^A \\ x_{F1}^B \\ 1 \end{bmatrix} = F_2 \begin{bmatrix} x_{F2}^A \\ x_{F2}^B \\ 1 \end{bmatrix} + F_3 \begin{bmatrix} x_{F3}^A \\ x_{F3}^B \\ 1 \end{bmatrix} \quad (141)$$

For the mixing unit (Output = Input)

$$F_1 \begin{bmatrix} x_{F1}^A \\ x_{F1}^B \\ 1 \end{bmatrix} + F_2 \begin{bmatrix} x_{F2}^A \\ x_{F2}^B \\ 1 \end{bmatrix} = F_3 \begin{bmatrix} x_{F3}^A \\ x_{F3}^B \\ 1 \end{bmatrix} \quad (142)$$

These characteristics are very convenient for filtering feasible structures.

Selection of Possible Process Structures

Rules for the total system. If several heuristics for used units are introduced into the incidence matrix, the existing region of each terminal can be estimated qualitatively.

Fortunately, it is clear that the valley and ridge defined from the vapor-liquid tie lines restrict the separable region of a conventional column [61].

Rule 1: According to the characteristics of distillation and a mixer in a ternary system where the valley and/or ridge is known for each unit, the regions of all inputs and outputs of each unit must be reasonable. For example, the possible locations of a distillation column and a mixer in the composition space are shown in Figure 19. When such information is introduced to the incidence matrix, if there is no feasible set, it is an unrealistic case.

Rule 2: If when the output of the mixer using the curvature of the valley or ridge passes through a distillation unit the distillation unit has the following characteristics, the feasible set is neglected;

- The output is immediately separated
- The output is separated again by the distillation unit which is using the curvature of the valley or ridge.

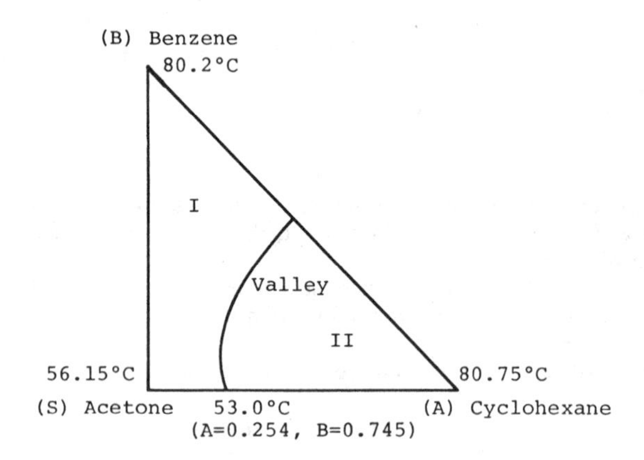

Rule 1

Distillation I—< I, I   II—< II, II   II—< II, II

Mixer I, I >→ I or II*   I, II >→ I or II   II, II >→ II

Rule 2

Following the mixer on the curved area of the valley:

I, I >→ (M) → II*

(1) II* → (D) —< I, I ; II*, I >→ (M) → II → (D) —< I, I ; II*, II* >→ (M) → II → (D) —< I, I

(2) II* → (D) —< II, II

Figure 19. Example of Rules 1 and 2.

By using this rule related to separation efficiency the redundant structures can be pruned. An example is shown in Figure 19.

Rule 3: The mass balance line of a distillation column combines an azeotrope with any point on the specification line of the product. If the mass balance line cuts the valley or ridge with the azeotrope, the product can be produced from the distillation column on the curved region as shown in Figure 20.

These three rules are based on the quantitative information.

Selection from mass balance. The overall mass balance equation is represented as follows:

$$D_1\begin{bmatrix} x(A)_{D1} \\ x(B)_{D1} \\ 1 \end{bmatrix} + D_2\begin{bmatrix} x(A)_{D2} \\ x(B)_{D2} \\ 1 \end{bmatrix} = S\begin{bmatrix} 0 \\ 0 \\ 1 \end{bmatrix} + F\begin{bmatrix} x(A)_F \\ 1-x(A)_F \\ 1 \end{bmatrix} \tag{143}$$

If $x(B)_{D1}$ and $x(A)_{D2}$ are assumed to be suitable values, the values of D_1, D_2, and S can be determined by using Equation 143.

For the degree of freedom, since Equation 143 is derived from the summation of all the 2N unit equations, the number of independent equations for the mass balances is equal to 2N–1. As the number of the inner terminals is equal to 3N–2, the suitable values of N–1 terminals can be assumed to be in the existing region corresponding to the feasible set of the inner terminals; the state vectors of all the other inner terminals are determined from the mass balance equations. It is not difficult to assume suitable values for the N–1 variables by using the information about the ridge and valley. In other words, the existing regions of the inner terminals are obtained by Rule 1, and generally, at least one of the columns in an azeotropic distillation system involves the composition near one of the azeotropes at the top plate (or bottom plate).

Rules for single column. Before the design of each column in a distillation system, the feasibility of each set of the solutions obtained from the mass balance equations should be checked by using the following rule:

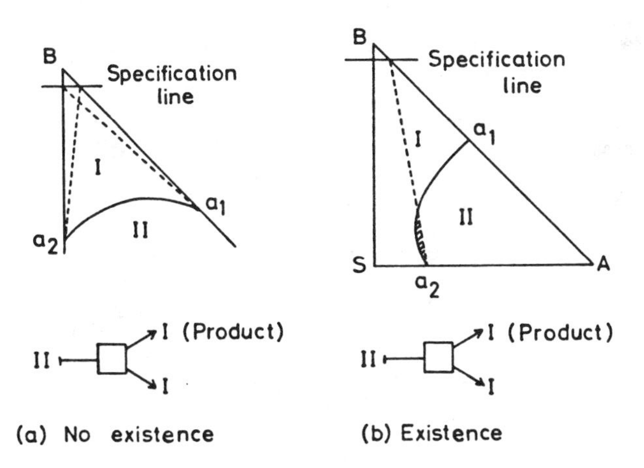

Figure 20. Illustration of Rule 3.

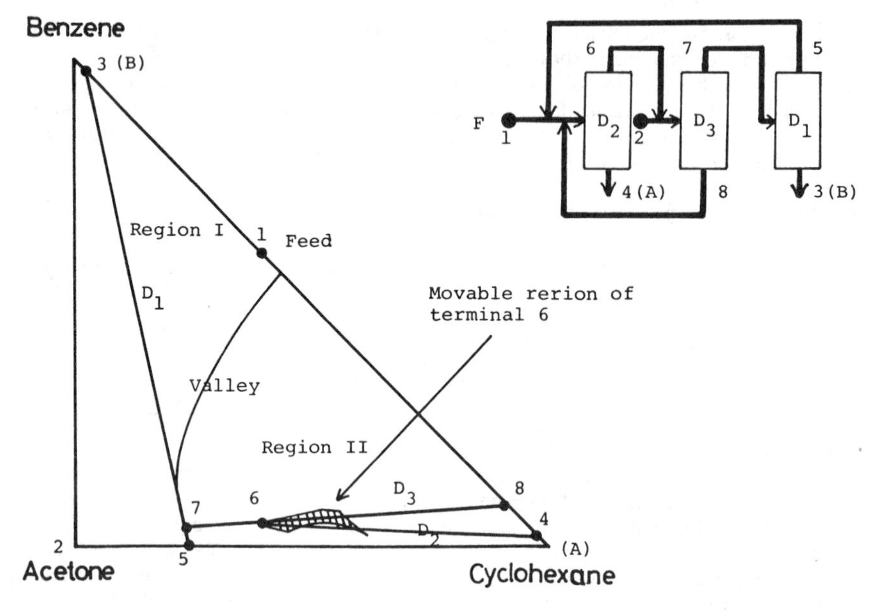

Figure 21. Generated azeotropic distillation sequences.

Rule 4: For a distillation column to separate the given ternary-composition feed to the specified top and bottom products, the following conditions should be satisfied simultaneously;

1. When vapor-liquid tie lines (the bubble point relationship) from the bottoms are obtained successively by using the total reflux condition, $x_j = y_{j+1}$, the curve should intersect the half-line from the feed through the top.
2. When the reverse vapor-liquid tie lines (dew-point relationship) from the top are obtained successively, the curve should intersect the half-line from the feed through the bottoms.

This rule is based on the composition profile at total reflux and requires the top and bottoms to be in the separable region. This can be neadily appied to the investigation of a feasible solution by using the bubble- and dew-point calculations. Furthermore, as the top and bottom temperatures are calculated in applying this rule, the feasibility of the mass balance for a column can be examined by realizable temperature differences between both temperatures. Current technology has permitted the separation of the 3.9° F temperature difference [38].

Several feasible structures may be given by introducing the previous-mentioned rules into all system structures. Note, however, that as the system structures shown in Figure 17 are generated without consideration of how to join the feed and solvent to two inputs and also two products to outputs, the total number of system structures to be investigated is 6 for a two-column system and 32 for a three-column system. For the example of the cyclohexane/benzene system using acetone solvent, two feasible structures are obtained. The locations of the terminals in one of the two structures, Pattern 8, are illustrated in Figure 20. The problem is solved from the condition that the composition of terminal 5 is fixed near the cyclohexane/acetone azeotrope (the shaded area is the movable region of terminal 6). When terminal 6 is in the region, all of the terminals are feasible.

After the compositions and flow rates of all inputs and outputs are obtained, the design of each distillation column can be performed. The design variables for a column are the number of plates, the location of the feed plate, the boil-up flow rate, and q-value. The design calculation can be executed as follows. Set the number of plates and assume the feed plate location as determined by the Gilliland method [79]. The boil-up flow rate and the q-value can be obtained once the specified composition of the top product equals that calculated by plate-to-plate calculations from the bottom up. If the q-value is not acceptable in a real sense, the number of plates and the feed plate location should be changed.

Furthermore, in order to ensure that the system structure is practical, one should consider a multi-feed distillation column in the case where one or more than one mixer is located immediately ahead of the column.

NOTATION

a,b	constants
C_p	specific heat
D	mass rate through distillate
E_{ji}^{MV}	Murphree plate efficiency
E_H^{min}	minimum energy loss term
e	energy loss
F	feed mass rate
$f_{i,j}$	residual functions
g	free energy parameter
H_l^L	molar liquid holdup
H_j, h_j	enthalpy
L	liquid mass rate
n	number of plates in enriching section
Q	heat load
S	number of plates in stripping section
T	temperature
T_H	temperature of heat source streams
T_w	bubble point of bottoms
t	temperature
V	vapor mass rate
$x_{i,j}$	liquid-phase molar fraction
$y_{p,i}$	vapor-phase molar fraction
Z_F	feed composition

Greek Symbols

ε_i^k	error associated with material balance
ζ	parameter defined by Equation 45
Θ	correction factor
μ_j	point relaxation factor

REFERENCES

1. Yamada, I., et al., *Kagaku Kogaku Ronbunshu* (in Japanese), *9:* 389 (1983).
2. Yamada, I., et al., *Kagaku Kogaku Ronbunshu 9:* 205 (1983).
3. Murdoch, P. G., and Holland, C. D., *Chem. Eng. Prog., 44:* 847 (1948).
4. Wilson, G. M., *J. Amer. Chem. Soc., 86:* 127 (1964).
5. Murphree, E. V., *Ind. Eng. Chem., 17:* 747 (1925).
6. Lewis, K., and Matheson, P. L., *Ind. Eng. Chem., 24:* 494 (1932).
7. Waterman, W. W., Frazier, J. P., and Brown, G. M., *Hydrocarbon Processing, 47*(6): 155 (1968).
8. Yamada, I., Matsumoto, N., and Suzuki, M., *J. Chem. Eng. Japan, 6:* 68 (1973).
9. Amundson, N. R., and Pontinene, A. J., *Ind. Eng. Chem., 50:* 730 (1958).
10. Yamada, I., et al., *J. Japan Petrol. Inst., 26:* 467 (1983).

11. Naka, Y., Araki, M., and Takamatsu, T., *Int. Chem. Eng., 19*(1): 137 (1979).
12. Wang, J. C., and Henke, G. E., *Hydrocarbon Processing, 45*(8): 155 (1966).
13. Petlyuk, P. B., *Khim. Prom., 41:* 206 (1965).
14. Kubicek, M., *Commun of ACM, 16:* 760 (1973).
15. Friday, J. R., and Smith, B. D., *AIChE J., 10:* 698 (1964).
16. Sujata, A. D., *Hydrocarbon Processing, 40*(12): 137 (1961).
17. Burmingham, D. W., and Otto, F. D., *Hydrocarbon Processing, 46*(16): 163 (1967).
18. Newman, J. S., *Hydrocarbon Processing, 42*(4): 141 (1963).
19. Tierney, J. W., and Bruno, J. A., *AIChE J., 13:* 556 (1967).
20. Boynton, G. W., *Hydrocarbon Processing, 49*(1): 153 (1970).
21. Tomich, J. F., *AIChE J., 16:* 229 (1970).
22. Tierney, J. W., and Yanosik, H. L., *AIChE J., 15:* 897 (1969).
23. Gallun, S. E., and Holland, C. D., *Hydrocarbon Processing, 55*(1), 137 (1976).
24. Ishii, Y., and Otto, F. D., *Can. J. Chem. Eng., 51:* 601 (1973).
25. Naphtali, L. M., and Sandholm, D. P., *AIChE J., 17:* 148 (1971).
26. Lyster, W. N., et al., *Hydrocarbon Processing, 38*(6), 221 (1959).
27. Thiele, E. W., and Geddes, R. L., *Ind. Eng. Chem., 25:* 289 (1933).
28. Saito, H., Yamada, I., and Sugie, H., *Kagaku Kogaku* (in Japanese), *34:* 1178 (1978).
29. Rose, A., Sweeny, R. F., and Schrodt, V. N., *Ind. Eng. Chem., 50:* 737 (1958).
30. Ishikawa, T., and Hirata, M., *Kagaku Kogaku* (in Japanese), *38:* 865 (1972).
31. Saito, H., Ph. D Dissertation, Tohoku Univ., Sendai, Japan (1975).
32. Underwood, A. J. V., *Ind. Eng. Chem., 41:* 2844 (1949).
33. Acrivos, A., and Amundson, N. R., *Chem. Eng. Sci., 4:* 68 (1955).
34. Yamada, I., et al., *Kagaku Kogaku* (in Japanese), *30:* 513 (1966).
35. Yamada, I., et al., *J. Chem. Eng. Japan, 10:* 440 (1977).
36. Yamada, I., et al., *Kagaku Kogaku Ronbunshu* (in Japanese), *7:* 489 (1981).
37. Yamada, I., et al., *Kagaku Kogaku Ronbunshu* (in Japanese), *9:* 396 (1983).
38. Holland, C. D., *Fundamentals of Multicomponent Distillation,* McGraw-Hill, Inc. New York (1981). pp. 365.
39. Henley, E. J., and Seader, J. D., *Equilibrium-stage Separation Operation in Chemical Engineering,* John Wily & Sons Inc. (1968), pp. 441.
40. Iwata, T., et al. *Kagaku Kogaku Ronbunshu* (in Japanese), *3:* 311 (1977).
41. Rathore, R. N. S., Van Wormer, K. A., and Powers, G. J., *AIChE J., 20:* 491 (1974).
42. Nishida, N., Stephanopoulos, G., and Westerberg, A. W., *AIChE J., 27:* 321 (1981).
43. Lockhart, F. J., *Petrol. Refiner, 26:* 104 (1947).
44. Herbert, V. D., *Petrol. Refiner, 36*(3): 169 (1957)
45. Freshwater, D. C., Henry, B. D., *Chem. Eng.,* 533 (Sep. 1975).
46. Doukas, N., and Luyben, W. L., *Ind. Eng. Chem. Process Design and Development, 17:* 272 (1978).
47. Stephanopoulos, G., and Westerberg, A. W., *Chem. Eng. Sci., 31:* 195 (1976).
48. Thompson, R. W., and King, C. J., *AIChE J., 18:* 941 (1972).
49. Seader, J. D., and Westerberg, A. W., *AIChE J., 23:* 951 (1977).
50. Nath, R., and Motard, R. L., 85th National Meeting of AIChE, Philadelphia (1978).
51. Nishimura, H., and Hiraizumi, Y., *Intrn. Chem. Eng., 11:* 188 (1971).
52. Hendry, J. E., and Hughes, R. R., *Chem. Eng. Prog., 68:* 69 (1972).
53. Westerberg, A. W., and Stephanopoulos, G., *Chem. Eng. Sci., 30:* 963 (1975).
54. Rodrigo, B. F. R., and Seader, J. D., *AIChE J., 21:* 885 (1975).
55. Gomez, M. A., and Seader, J. D., *AIChE J., 22:* 970 (1976).
56. Petlyuk, F. B., Platonov, V. M., and Slavinskii, D. M., *Intern. Chem. Eng., 5:* 555 (1965).
57. Stupin, W. J., and Lockhart, F. J., *Chem. Eng. Prog., 68:* 71 (1972).
58. Rathore, R. N. S., Van Wormer, K. A., and Powers, G. J., *AIChE J., 20:* 940 (1974).
59. Morari, M., and Faith, D. C., III, *AIChE J., 26:* 916 (1980).
60. Naka, Y., Terashita, M., and Takamatsu, T., *AIChE J., 28:* 812 (1982).

61. Naka, Y., et al., *J. Chem. Eng. Japan, 16*(1): 36 (1983).
62. The Soc. Chem. Eng. Japan (Ed.), *Chem. Eng. Handbook,* Maruzen, Tokyo (1978).
63. Hamilton, C. E., *Chem. Eng. Prog., 62*(2): 92 (1966).
64. Harris, R. E., *Chem. Eng. Prog., 68*(10): 56 (1972).
65. Perry, J. H., (Ed.), *Chemical Engineer's Handbook,* 5th ed., McGraw-Hill, New York (1973).
66. Yamada, I., Sugie, H., and Abe, K., *Kagaku Kogaku, 31*(4): 395 (1967).
67. Yorizane, M., Yoshimura, S., and Hase, S., *Kagaku Kogaku, 29*(4): 229 (1965).
68. Yorizane, M., *Handbook of Distillation Engineering,* Hirata, M., and Yorizane, M., (Eds.) Asakura Shoten, Tokyo (1970).
69. Hands, C. H. G., and Norman, W. S., *Trans. Inst. Chem. Engrs, 23:* 76 (1945).
70. Naka, Y., and Takamatsu, T., Private report (1977).
71. Yorizane, M., and Yoshimura, S., *Kagaku Kogaku, 32*(4): 382 (1968).
72. Robinson, C. S., and Gilliland, E. R., *Elements of Fractional Distillation,* McGraw-Hill, New York (1950).
73. Black, C., Golding, R. A., and Dister, D. E., *Extractive and Azeotropic Distillation,* Advances in Chemistry Series 115, ACS (1972).
74. Prokopakis, C. J., Ross, B. A., and Seider, W. D., *Foundations of Compute-aided chemical Process Design,* Mah, R. S. H., and Seider, W. D., (Eds.), AIChE (1981).
75. Ross, B. A., and Seider, W. D., *Comp. and Chem. Eng.,* 5: 1 (1981).
76. Block, U., and Hegner, B., *AIChE J.,* 22: 3 (1976).
77. Takamatsu, T., et al., *AIChE,* submitted (1983).
78. Nagao, H., *Group and Design* (in Japanese) Iwanami, Tokyo (1974).
79. Gilliland, E. R., *Ind. Eng. Chem.,* 32: 918 (1940).

CHAPTER 16

INTERFACIAL FILMS IN INORGANIC SUBSTANCES EXTRACTION

V. V. Tarasov and G. A. Yagodin

Mendeleev Institute of Chemical Technology
Moscow, U.S.S.R.

CONTENTS

INTRODUCTION

Solutions in industrial extraction processes are known to contain suspended particles, micelles, macromolecules of organic substances as well as surface active substances (SAS), which, being accumulated at the interface give rise to the formation of interfacial films. Even in the case of well-purified solutions, such substances and particles may appear due to chemical reactions between the extractant and the extracted compound.

It seems clear that interfacial films formation is quite typical of inorganic substances extraction processes. The film formation results in slower mass transfer rates and worse conditions of emulsion separation and, finally, brings about a dramatic decrease in the process capacity. There is, therefore, a need for extensive studies of extraction conditions so that the negative effects of interfacial film formation be eliminated.

To date much has been done in studying the kinetics and mechanism of chemical reactions accompanying mass transfer across the interface. This may open the way to enhancing extraction rates using controlled variation of interfacial reactions parameters.

Two aspects of the problem of film formation deserve attention. The first is related to the effect of films on reaction kinetics. It is useful to consider whether the above effect is essentially negative and if the phenomenon in question can be used to enhance both the rate of chemical reactions and the selectivity of extraction processes.

The second aspect concerns experimentally obtained kinetic and reaction mechanism data and the possibility of using the latter in describing the kinetics of commercial extraction processes.

This chapter is specifically concerned with the latest advances in the studies of film formation processes during inorganic substances extraction. Consider a simplified model of the interfacial area.

The interface is a region of a sharp variation in physico-chemical properties, which includes the adsorption, δ_1, and interfacial, δ_2, layers [1, 2]. The properties of the layers differ from those of the bulk. When describing processes of the interfacial mass transfer in the stirred systems, the existence of only two other layers is usually considered: a diffusion layer, δ_3, and a hydrodynamic one, δ_4, [3]. The contribution of layers δ_1 and δ_2 resistances to the overall mass transfer rate is generally considered to be negligibly small. In case of a "physical distribution" this assumption is often justified. However this is not so in the case of chemical extraction of inorganic substances or in extraction by surfactants.

In nonstirred systems it is not necessary to consider the layer δ_4. The equilibrium is characterized by the presence of only layers δ_1 and δ_2. The value of δ_1 is comparable with the size of the particles adsorbed, whereas the value of δ_2 may be considerably higher than that of δ_1. Thus, for example, in the amines extraction systems δ_2 depends on the direction and rate of mass transfer and reaches $10^{-4}-10^{-3}$ cm [4]. Such "thick" interfacial layers prove to considerably decrease mass transfer rates [4–9]. These layers may be viewed as a flow-through reactor, its volume depending on the direction and rate of mass transfer. The main reason for the formation of "thick" δ_2 layers is the enrichment of the organic-phase layers adjacent to the interface by the aqueous-phase components during extraction and of the aqueous-phase layers with the organic-phase components during stripping. So, it is unreasonable to consider the layer δ_2 without regarding the processes occurring in the interfacial layers. Under certain conditions irreversible processes take place within δ_2 and δ_3, which often results in the formation of a new microheterogeneous phase, the latter having a drastic effect on mass transfer kinetics. Consequently, unexpected dependencies of extraction rate on reagents concentration, residence time, and stirring intensity are observed [4–9].

METHODS OF STUDYING INTERFACIAL FILM FORMATION PROCESSES

It should be noted that valuable information on the structure and properties of interfacial films at liquid-liquid interfaces may be obtained by measuring interfacial rheological characteristics: viscosity, modulus of elasticity, ultimate strength, etc. [10].

Measuring the Interfacial Viscosity

Rheological characteristics of the interface are important in studying mass transfer kinetics [4, 11].

The interfacial viscosity may be determined [12] by a pendulum-type electromagnetic viscometer. (See Figure 1.) In this device the aqueous and organic phases are poured into the cylindrical glass vessel (1) which contains a thin glass rod (2) (dia. 0.3 mm) secured along its center. The pendulum (3) made of stainless steel in the form of a hollow cylinder fitted with ferrite cores (9) is mounted via the central opening, onto the rod and floats in the binary liquid system crossing the interface so that its upper edge is above the organic-phase level. To avoid liquid penetration into the cylinder, the diameter of the central opening should not exceed 1 mm. The pendulum is operated by means of the additional electromagnet (7) and performs damped torsional oscillations in the field of the main electromagnet (6). The amplitudes are registered by a simple optical system consisting of a light source (4), a mirror (5) fixed on the cylinder coaxially with the central opening, and a scale (8). The temperature

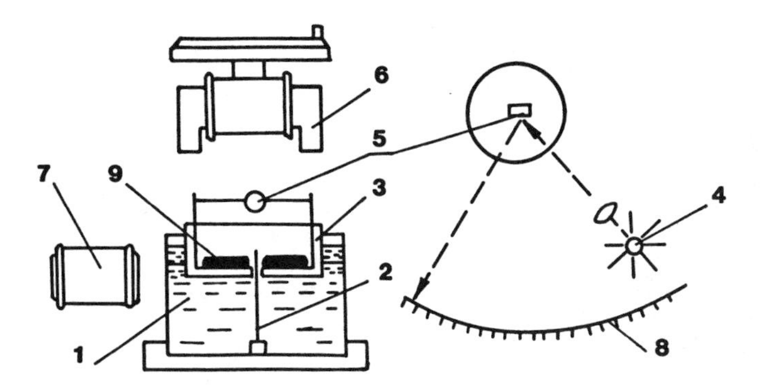

Figure 1. Viscometer [12].

is maintained constant by means of an air thermostat. The calibration of the device has shown that in the case of small deviations of the pendulum from the equilibrium state (< 8°), the oscillations are harmonic and, therefore, the interfacial viscosity could be calculated in the usual way from the logarithmic decrement of the oscillations damping [12, 13]:

$$\eta_i = 2L\pi^{-1}(d_p^{-2} - d_v^{-2})(uT^{-1} - u_f T_f^{-1}) \tag{1}$$

where $u = QT(2L)^{-1}$

$$T = T_0[1 + (u/2\pi)^2]^{0.5}$$

$$T_0 = 2\pi(L/w)^{0.5}$$

The use of the electromagnetic system in lieu of an elastic thread, as is usually employed in similar devices, permitted adjustment of the device over a wide range of operation. The device sensitivity was approximately 10^{-6} kg/s.

It should be noted that pendulum-type devices are rather simple to operate. Still they have a number of drawbacks, the main one being the influence of the bulk phases on the pendulum oscillations. A detailed description of such viscometers has been given in References 14 and 15, and reference has been made as to their relatively low sensitivity.

An attempt has been made [16] to increase the sensitivity of the device by diminishing the moment of inertia of the pendulum. The latter was made of nickel foil in the form of a disc being kept at the interface by the interfacial tension. Although we succeeded in raising the sensitivity of the device, it became less universal; at low interfacial tension values the pendulum would cease floating. Channel-type viscometers are free of the preceding drawbacks [14, 15]. However, they are complicated enough.

Recently a new device has been developed for measuring visco-elastic properties of adsorption layers using light scattering techniques [17]. Originally the device was used for studying monolayers of surfactants on the aqueous surface [17]. Later the method proved to be applicable for liquid-liquid systems as well [18].

Measuring Strength Properties of Interfacial Films

The main component of the device for studying solid-like (rigid) interfacial films is the floating plate. It is made of a thin gold-plated ferromagnetic foil in the form of a needle

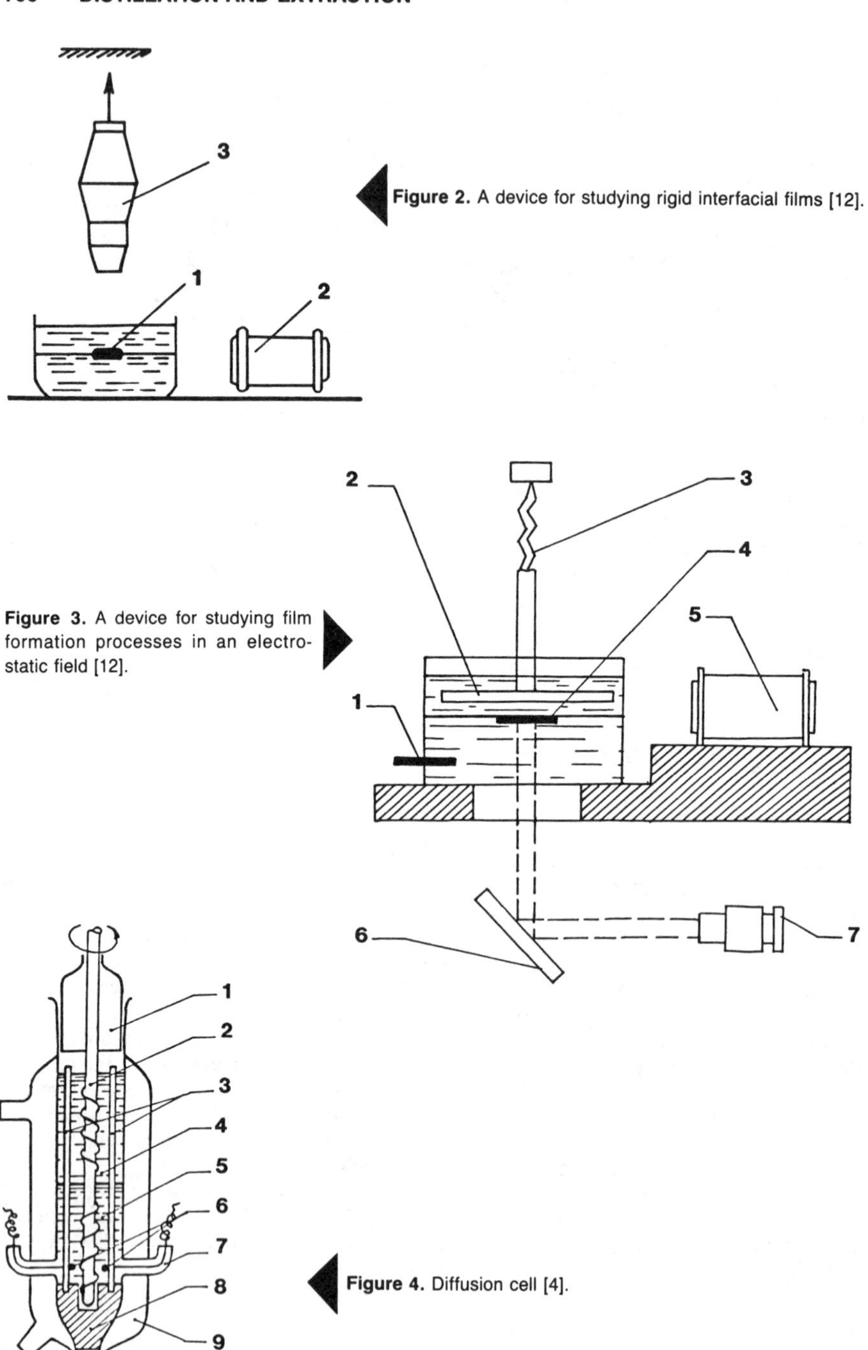

Figure 2. A device for studying rigid interfacial films [12].

Figure 3. A device for studying film formation processes in an electrostatic field [12].

Figure 4. Diffusion cell [4].

(shear deformation) or disc (complex deformation). As shown in Figure 2, the plate (1) is supported at the interface by the interfacial tension. The interfacial film undergoes deformation when the disc or needle is shifted by the electromagnet (2). The deformation Δl is registered by the traveling microscope (3). The device allows various rheological experiments to be performed:

1. Plotting of curves of the relative deformation versus time, t, at a permanent force of the magnet, F.
2. Obtaining strain diagrams, $\Delta l(F)$, and determining the ultimate strength of the film, F_1, from the value of the force resulting in breakage of the interfacial structure, as well as determining the film modulus of elasticity from the slope of the initial straight-line portion of the $\Delta l(F)$ diagram.
3. Studying the kinetics of the interfacial strength, F_1, evolution.
4. Studying the effect on the film formation process of such physico-chemical factors as temperature, electrostatic field, admixtures of surfactants, etc.

Shown in Figure 3 is the device-modification used for studying the film formation process in an electrostatic field. A platinum-wire electrode (1) is inserted into the aqueous phase, whereas the organic phase houses a thin disctype electrode (2) which is lowered to a required depth from the interface by the micrometer screw (3). The floating plate (4) position is registered by means of the mirror (6) and the traveling microscope (7).

It is the rate of mass transfer across the interface that is most of all affected by the presence of interfacial films [4–7]. Therefore the first experimental results indicating the formation of interfacial films during extraction were obtained [19–21] using methods for studying mass transfer kinetics. Among these the stirred diffusional cells and short time contacting method (STCM) have been found to be most effective.

Diffusion Cells Method

The kinetics of extraction via a "flat" stagnant interface between phases positioned one above the other is studied [22–25]. The stirring of the liquids may be performed by stirrers rotating either in the same or in the opposite directions. It is necessary to use baffles to prevent funnel formation and to contribute to the turbulence evolution. A vibrational method of the phase stirring is known too. Figure 4 shows the diffusion cell used in our kinetic experiments [4]. A screw stirrer (2) is inserted through the upper stopper (1). Baffles (3) are adjusted to the Teflon bed (8). The aqueous (5) and organic (4) phase volumes are identical. The electrodes (6) to measure conductivity and to determine the amount of the electrolyte transferred are introduced into the aqueous phase via the tubes (7). The temperature in the cell is maintained constant by means of a water thermostat (9). Solutions are disposed of after the stopper (10) has been removed. The small cell and with screw stirrer diameters of 18 and 8 mm, respectively allows the kinetics to be examined at pms of up to 2,500.

The experiments are carried out under conditions of a relatively stable interface. But the mass transfer kinetics cannot be strictly described, since the distribution law for the rates of local fluid motion is unknown.

The process of quasi-stationary mass transfer under these circumstances is usually described by the simple empirical equation:

$$-dC/dt = Ka(C_{eq} - C) \qquad (2)$$

which, integrated gives:

$$\log(1 - E) = -Kat/2{,}3 \qquad (3)$$

Mass transfer coefficient depending on the stirring intensity can be calculated from the plots of functions $\log(1-E) = f(t)$ in accordance with a well-known [23, 26] equation:

$$Sh = const\, Re^p Sc^q \psi$$

This expression is not the only one available. The data on mass transfer kinetics in diffusion cells have been summarized in References 19 and 23.

In case of a physical distribution in the absence of chemical reactions and interfacial films the value of p is usually 0.7–1.0, and the value of $q = 0.5$.

It seems that a diffusion cell was for the first time used for studying mass transfer with interfacial films formation in Reference 22.

Rigid films have been found to decrease the rate of mass transfer. This phenomenon can be accounted for by the worse conditions of turbulence transfer via the interface. Later work [27] revealed more information on the effect of the gaseous and condensed SAS films on the mass transfer mechanism, the discussion of surface renewal effects having been very beneficial.

A decrease in the rate of physical extraction with SAS forming condensed interfacial films has been also referred to in References 3 and 28–31.

It could be shown that the plot of function $\log(1-E) = f(t)$ will not be a straight line. At the beginning of the film formation process $d^2 \log(1-E)/dt^2 = 0$. Let us assume the film diffusional resistance to be a function of both residence time and the quantity of substance transported through the interface, while the diffusional resistances in the phases remain constant:

$$R_f = \gamma' q \tag{5}$$

$$R = R_f + R_{ph} \tag{6}$$

Then

$$dC/dt = -S[R_{ph} + \gamma' q]^{-1} (C_{eq} - C) \tag{7}$$

Integrating Equation 7 we obtain:

$$\ln(1-E) = -(at + R_f)(R_f^* + R_{ph})^{-1} \tag{8}$$

If $R_f^* \gg R_{ph}$, then

$$\ln(1-E) = -at(R_f^*)^{-1} - E \tag{9}$$

Equation 9 is plotted in Figure 5.

The effect of hydrodynamic conditions on extraction kinetics, as it follows from Equation 9, will be observed only at low values of t, before the diffusion through the film has become rate determining. The slope of the linear part of the plot (Figure 5) will be unaffected by the stirring intensity until the interfacial film has been destroyed.

Short Time Phase Contacting Method (STCM)

The STCM method described in the following has been widely employed by the authors of this chapter when studying interfacial process kinetics, where one of the phases has a considerably higher electric conductivity than the other [4–7, 19, 32–42].

Almost in all cases it is possible to determine conditions when a short-time mass transfer results in a substantial change in the ion concentration of the conductive-phase (aqueous) layers adjacent to the interface. This permits employing a sensitive method for recording the kinetic curves as functions of electric conductivity versus time, which then can be easily transformed to dependencies of the quantity of substance transported through the interface on the diffusion time.

The method is based on the two main principles:

1. The kinetics is studied with nonstirred systems, since in that case it is possible to strictly describe the substance transport in the phases.
2. The kinetics of the interfacial transport is studied beginning with rather small (0.05–0.1 s) values of the phase contacting time to reduce if possible the phase diffusion resistances and to increase the contribution of the film resistance or of the chemical reaction resistance to the overall extraction rate.

Indeed, the one-dimensional rate (r_A) of substance A diffusion across the interface between two quasi-infinite phases is given by

$$r_A = (C^0_{A1} - \alpha C^0_{A2})(D_{A1}/\pi t)^{0.5}[1 + \alpha(D_{A1}/D_{A2})^{0.5}]^{-1} \quad (10)$$

According to Equation 10 the diffusion resistance is:

$$R_{ph} = (D^{0.5}_{A2} + D^{0.5}_{A2})(\pi t)^{0.5}(D_{A1}/D_{A2})^{-0.5} \quad (11)$$

which at $t \to 0$ reduces infinitely.

Thus, the smaller the phase contacting time, the faster the mass transfer.

Integrating Equation 10 we get the expression for the quantity of the substance transferred during the time t over the unit of interfacial area:

$$q_A = 2(C^0_{A1} - \alpha C^0_{A2})(D_{A1}t/\pi)^{0.5}[1 + (D_{A1}/D_{A1})^{0.5}]^{-1} = gt^{0.5} \quad (12)$$

Hence, the plot of the function $q_A = f(t)^{0.5}$ must be a straight line crossing the beginning of the coordinates. Using the value of the slope g one can obtain the diffusivities in the phases. Comparing the latter with the ones known or calculated by other methods one can make a conclusion whether the system is hydrodynamically stable and if convection occurs. In case the experimental diffusivities are greater than the true ones it is necessary to stabilize the system. By contrast, the experimental diffusivities calculated from the initial parts of kinetic curves that are lower than the true ones could be only accounted for either by slow chemical reactions or by the presence of interfacial films.

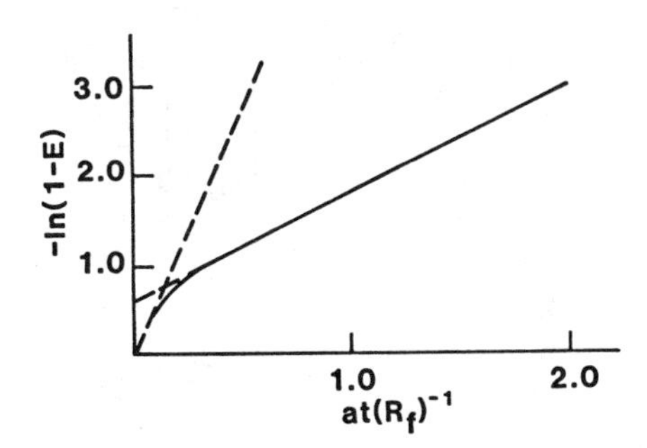

Figure 5. Shape of the kinetic curve in semi-logarithmic coordinates when the interfacial film is being formed [4].

The sensors of the device consist of a tube-like electrode (1) with the inner diameter 1.5–3.0 mm made of a metal (tantalum, platinum, zirconium) (Figure 6). The bottom of the sensor (6) is made of an insulator (Teflon) through which the central (2) electrode is hermetically introduced. A platinum wire with the diameter 0.5 mm is used for the purpose. Thus, a cell (4) with the volume 5–20 mm^3 to measure electric conductivity is formed. The aqueous phase is poured into the cell (4) so that the surface of the liquid is on the level of the tube brim. Two modifications, "a" and "b," of the device have been developed. The first one is used for studying the kinetics of the processes accompanied by a decrease in the aqueous-phase density and an increase of the density of the organic one. The other modification ("b") is recommended for studying the kinetics of the processes accompanied by opposite changes in the density of liquids. The contacting of the phases in the "a"-type sensor takes place after a drop of the organic (5) has been placed on the surface of the aqueous one. The best results are obtained if the drop is placed onto a nonwettable Teflon nozzle (3), the inner diameter of which equals that of the tube-like electrode. In the "b"-type sensor the aqueous phase is kept by capillary forces. The contacting is achieved after the electrode-tube has been shifted enough to touch the Teflon nozzle containing the organic phase (5). At a negligible consumption of reactants (20 mm^3 per experiment) and high output (up to 10 kinetic curves per hour) the method permits us to determine the regime and locale of the reaction; to exactly determine the initial mass transfer rate and, from it, calculate the effective constant of the chemical reaction rate or the film permeability; to estimate the diffusivities of substance in the phases; and to probe rather fine (less than 10^{-3} cm) layers adjacent to the interface.

Let us suppose that interfacial films are absent and diffusion is accompanied by an instantaneous reversible first-order reaction such as: $A \overset{K_{eq}}{\rightleftarrows} B$ (A exists in phase 1 only and B in phase 2 only). In this case the expressions for the extraction rate r_B and the quantity q_B of the substance extracted could be derived from equations (10 through 12), substituting r_B for r_A, q_B for q_A and obtaining K_{eq}, by replacing C^0_{A2} with D_{B2}. The plots of functions

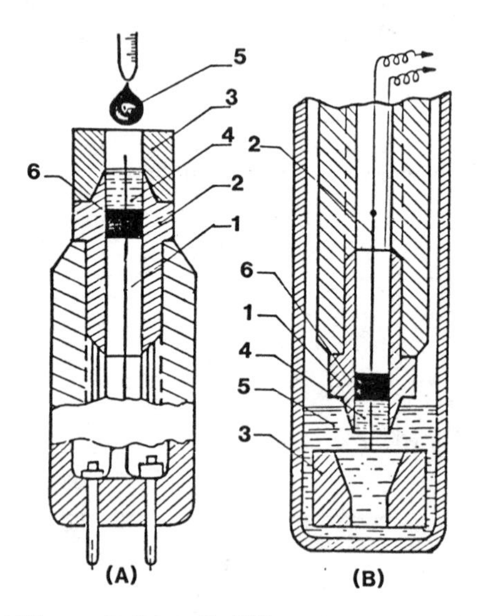

Figure 6. Sensors of short time contacting method [4].

$q_B = f(t)$ and $q_B = f(t^{0.5})$ are given in Figure 7 "a" and "b" (lines 1, the diffusion regime at all t). It can be shown that functions (curves 2, 3) $q_B = f(t)$ and $q_B = f(t^{0.5})$ at low t values should change essentially in the case of slower reactions. The functions of $q_B = f(t^{0.5})$ become linear after a certain period of time, t_{tr}, has elapsed (Figure 7 "b", curves 2, 3). Its existence is due to the fact that the diffusion regime is not achieved immediately, since diffusion resistances increase proportionally to $t^{0.5}$ according to Equation 11.

To estimate t_{tr} value for the irreversible reaction of $A \xrightarrow{k_V} B$ type proceeding in the extracting phase let us suppose that the reaction and diffusion conditions are independent of the diffusion coordinate X, since the interfacial films are absent. Then according to Pitchugin [42]:

$$r_B = C^0_{A1}(k_v D_{A1}/s)^{0.5} \exp(k_v t/s) \{\mathrm{erf}[\alpha(D_{A1}k_v t/D_{A2}s)^{0.5}] - \mathrm{erf}(k_v t/s)^{0.5}\} \quad (13)$$

where $s = (\alpha^2 D_{A1}/D_{A2}) - 1$

$$q_B = C^0_{A1}(k_v D_{A1}/s)^{0.5}\{2(st/\pi k_v)^{0.5} - s \exp(k_v t/s) \cdot (k_v)^{-1} \cdot \mathrm{erf}(k_v t/s)^{0.5} + s \exp(k_v t/s)(k_v)^{-1}\mathrm{erf}(\alpha^2 D_{A1}k_v t/sD_{AZ})^{0.5} - \alpha k_v^{-1}(sD_{A1}/D_{A2})^{0.5}\mathrm{erf}(k_v t)^{0.5}\} \quad (14)$$

when $t \to \infty$, Equation 14 is simplified to

$$q_B = 2C^0_{A1}(D_{A1}/\pi)^{0.5}[t^{0.5} - \alpha(D_{A1}\pi/4D_{A2}k_v)^{0.5}] \quad (15)$$

Then the transient time is:

$$t_{tr} = \pi\alpha^2 D_{A1}(4k_v D_{A2})^{-1} \quad (16)$$

Since $D_{A1} \approx D_{A2}$, $k_v \approx \alpha^2 t_{tr}^{-1}$. So, the less is the solubility of the A reactant in the adjacent phase, the more fast reactions could be studied. The minimum value of t_{tr} that could be

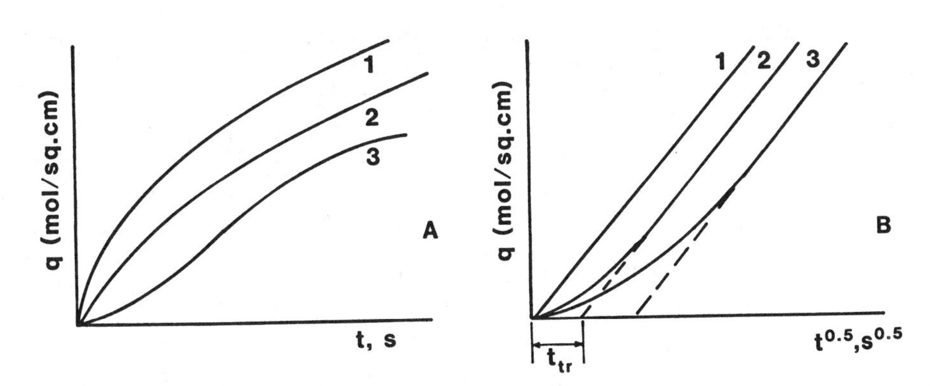

Figure 7. Shapes of kinetic curves obtained by STCM [4].

calculated accurately enough by means of STCM is 0.1 s. Suppose $\alpha = 10^3$, then $k_v = (10^3)^2/0.1 = 10^7\ s^{-1}$. Thus it is possible to calculate the rate constants of very fast reactions [35, 41, 42].

If $D_{A2} \approx 10^{-5}\ cm^2s^{-1}$, then the reaction penetration depth at $k_v = 10^7\ s^{-1}$ will be 10^{-6} cm in an order of magnitude. Considering the fact that the thickness of interfacial films and layers δ_2 may reach 10^{-4}–10^{-3} cm [4, 6, 7, 35], we arrive at the conclusion that the reaction is entirely located within these layers. That is why film formation processes as well as the interfacial layers properties may be studied by measuring the rate constants of fast chemical reactions.

Equation 13 shows the function $r_B = f(t)$ to be running through maximum: at $t = 0$, $r_B = 0$, then increases, and at $t \approx \alpha^2 k_v^{-1}$, the value r_B begins to decrease. In general the kinetic curve $q_B = f(t)$ has the S-like shape (Figure 7 "a," curve 3). However in practice the specific initial portion of this curve is not always observed, as the inflection point of the curve at high k_v values coincides almost with the origin of the coordinates.

Equations10–16 do not account for the fact that diffusion through the absorption layer δ_1 may be accompanied by the overcoming of local energy barriers. Let us assume now that such a barrier exists and the rate of mass transfer through it is proportional to the deviation of the system from the equilibrium, namely [43]

$$-D_{A1}dC_{A1}/dx_{x=0} = k_s[C_{A1}(0, t) - C_{A2}(0, t)] \tag{17}$$

Solving the set of Fick's equations for each of the phases with the boundary condition of Equation 17 as shown in Reference 43 one gets

$$r_A = k_s(C^0_{A1} - C^0_{A2}) \exp(\varphi^2 t) \operatorname{erf} c(\varphi t^{0.5}) \tag{18}$$

$$q_A = 2k_s(C^0_{A1} - \alpha C^0_{A2})\varphi^{-1}(t/\pi)^{0.5} + k_s\varphi^{-2}(C^0_{A1} - \alpha C^0_{A2}) [\exp(\varphi^2 t) \operatorname{erf} c(\varphi t^{0.5}) - 1] \tag{19}$$

where $\varphi = k_s[1 + \alpha(D_{A1}/D_{A2})^{0.5}] \cdot D_{A1}^{-0.5}$

If $t \to 0$, $r_A \to r^0_A$ where

$$r^0_A = k_s(C^0_{A1} - \alpha C^0_{A2}) \tag{20}$$

Equation 20 represents the rate of the process in the kinetic regime, as it does not include diffusivities. If $C^0_{A1} \gg \alpha C^0_{A2}$ tenh $r^0_A \approx k_s C^0_{A1}$. Thus if the reversibility of the process is neglected, the value k_s may be obtained from the simple relationship

$$k_s = r^0_A(C_{A1})^{-1} \tag{21}$$

On the other hand, one can see from Equation 19 that at $t \to \infty$ the function q_A is approximated by the equation:

$$q_A = 2k_s(C^0_{A1} - \alpha C^0_{A2})\varphi^{-1}(t/\pi)^{0.5} - k_s(C^0_{A1} - \alpha C^0_{A2})\varphi^{-2} = g(t^{0.5} - t^{0.5}_{tr}) \tag{22}$$

Therefore κ_s may also be calculated from the plots of functions $q = f(t^{0.5})$ if $t^{0.5}_{tr}$ and g are known:

$$k_s = \pi g[4(C^0_{A1} - \alpha C^0_{A2})\, t^{0.5}_{tr}]^{-1} \tag{23}$$

Thus, in the case of phase reaction the function $q_A = f(t^{0.5})$ at large t is linear (Figure 7 "b", curve 2), t_{tr} being transition time from kinetic to diffusion regime.

It is assumed that the energy barrier occurs due to a reversible reaction ($A \rightleftarrows B$) taking place as a result of the interaction of the interface with A and B substances, the latter being delivered from the bulk (Eley–Rideal mechanism). Then r_B and q_B may be obtained by substituting C^0_{A2} for C^0_{B2}, α for $K_{eq}(K_{eq} = k_s^-/k_s^+)$ in Equations 18, 19, 22.

Equations for the irreversible reaction may be obtained at $K_{eq} = 0$. The reaction in adsorption layer δ_l may proceed according to Langmuir–Hinshelwood mechanism, the adsorbed substances taking part in the interaction. In case of $A_s \xrightarrow{k_s} B_s$ reaction the initial extraction rate may be calculated from the following equation:

$$r^0_B = k_{ss}\theta_A \tag{24}$$

In many cases θ_A and θ_B values can be derived using Langmuir adsorption isotherm:

$$\theta_j = C_j\left[b_j\left(1+\sum_1^n b_k C_k\right)\right]^{-1} \tag{25}$$

Let us consider a theoretical model of molecular diffusion through the layer of a new phase having distinct boundaries. As a rule mechanical properties of this layer differ from those of the bulk. However, mechanical properties alone cannot be expected to solely determine the diffusional permeability of the layer. The most important parameters for the model suggested are as follows: distribution coefficients on both boundaries, the thickness of the layer and diffusivities in it. Therefore, even the layer lacking mechanical properties may substantially vary the rate of substances transfer from one phase into the other. This may be used in the layer detection and investigation.

Suppose species A diffuses from phase 1 into phase 2 without any chemical interactions (Figure 8). When the process is accompanied by an instantaneous formation of the interfacial film with thickness λ the kinetics of transfer from one semi-infinite phase into another at high t values is in good agreement with the equation [4, 5]:

$$q_A = (1 \cdot n^{-1} - m \cdot n^{-2})\,[1 - \exp(n^2 t)\, \mathrm{erf\,c}(n \cdot t^{0.5})] + 2m \cdot n^{-1}(t/\pi)^{0.5} \tag{26}$$

where $1 = C^0_\gamma D_2^{0.5} D_1^{-0.5}(\alpha_1 D_2^{0.5} D_\gamma^{-1} + \alpha_2 D_1^{-0.5})$

$m = 1\alpha_1 D_1^{-0.5}\lambda^{-1}$

$n = (\alpha_1 D_2^{0.5} D_\lambda^{-1} + \alpha_2 D_1^{-0.5})^{-1}\,[\alpha_1\alpha_2 + (D_2/D_1)^{0.5}] \cdot \lambda^{-1}$

It should be noted that $C^0_1 = \alpha_1 C^0_2$, i.e. at $x = \lambda$ the equilibrium is established instantaneously. From Equation 26, at $t \to \infty$ we get the relation:

$$q_A = (1 \cdot n^{-1} - m \cdot n^{-2}) + 2mn^{-1}(t/\pi)^{0.5} = g(t^{0.5} - t^{0.5}_{tr}) \tag{27}$$

Taking into account, that $\alpha_1\alpha_2 = \alpha = C_{1eq}/C_{2eq}$ we obtain the expression for g, which is in very good agreement with that describing the mass transfer in the absence of a film. Thus, at dimensionless time $n^2 t > 1$ the rate values for mass transfer with and without film formation appear to the practically the same. This is due to the fact that phase resistances become dominating in the absence of chemical reactions at high t values.

At $\alpha \gg 1$, $g = 2C_1^0 \cdot \alpha^{-1}(D_1/\pi)^{0.5}$ phase 2 is the limiting one, whereas at $\alpha \ll 1$, $g = 2C_1^0 \cdot (D_2/\pi)^{0.5}$, phase 1 is the limiting one.

It follows from Equation 27 that at large t the plot of function $q = f(t^{0.5})$ is a straight line generally cutting off a segment on the axis to give:

$$t_{tr} = (m - 1 \cdot n)^2 (2mn)^{-2} \geqq 0 \tag{28}$$

Thus, as in the case of a chemical reaction, the formation of a film results in the appearance of transient time t_{tr}. However the value of $t_{tr}^{0.5}$ may be both positive and negative, representing the acceleration of mass transfer (at the initial stages), or its slowing down (Figure 9). At m/n 1 (curve 2) the value of $t_{tr}^{0.5}$ will be positive, which corresponds to a widely accepted view of film retarding effect on kinetics. At $m/n < 1$ the mass transfer rate is initially higher than it is in the absence of a film (curve 3) This has been also referred to in Reference 44.

Figure 8 shows three postulated situations ($m/n > 1$, $m/n = 1$, $m/n < 1$). Suppose, $D_1 = D_\gamma = D_2$. Then, provided the concentration of the diffusing species in layer λ at $t = 0$ is lower than in phase 1 (Figure 8a), the diffusion rate will appear for a period of time to be lower than it is during the contacting without a film. In case the species concentration in the layer equals that in phase I (Figure 8b), the presence of the layer will in no way influence the rate of mass transfer. Acceleration of transfer is apparently due to the layer being temporarily enriched by the diffusing component (Figure 8c). Equation 26 can be simplified at $\alpha_2 \ll (D_2/D_\lambda)^{0.5}$ and $t \to \infty$

$$q_A = 2C_{A1}^0\alpha_1(D_1t/\pi)^{0.5} + C_\lambda^0 \cdot \lambda \cdot (1 - \alpha_1^2 D_1 D_\lambda^{-1}) \tag{29}$$

The result obtained is in good agreement with the observations made in Reference 45, where a special case of the problem under discussion was considered, when $\alpha_1 = 1$, $\alpha_2 = 0$, and $C_{A1}^0 = C_\lambda^0 = 1$. At $t \to 0$

$$q_A = 2C_\lambda^0(D_2t/\pi)^{0.5}[\alpha_2 + (D_2/D_\lambda)^{0.5}]^{-1} \tag{30}$$

Equation 30 shows mass transfer to be solely determined by the properties of phase 2 and layer λ. It follows from Equations 27 and 30 that the curve $q = f(t^{0.5})$ has two asymptotes

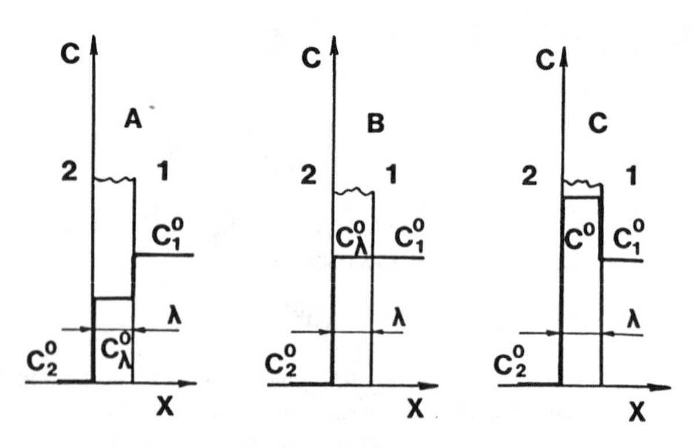

Figure 8. Initial concentration profiles in the bulk phases and in the film [4].

(at $t \rightarrow 0$ and at $t \rightarrow \infty$), which may intersect both under the curve and above it (Figure 9). The data obtained reveal one of the reasons for the appearance of "breaks" on the curves $q = f(t^{0.5})$.

In this model it was assumed that the interfacial film exists before the beginning of the extraction process while its properties remain constant throughout the process. In industrial extraction processes the formation and growth of the film results from the occurrence of chemical reactions and mass transfer. Consider the following situation. Suppose, species A diffuses from phase 1 into phase 2, where it is transformed into two products B and C. Product B turns out to be hardly soluble in phases 1 and 2 and is being accumulated at the interface, thus, forming a film. A schematic representation of the process is given in Figure 10a and b. Cases "a" and "b" differ in the direction of the film growth. The film may grow either in the direction of phase 1 (case "a") or in the direction of phase 2 (case "b"). Let us assume the rate of film growth to be proportional to the diffusion rate of species B into the interface:

$$\text{For case "a" } d\lambda/dt = \mu D_{B2}\, dC_{B2}/dx \text{ at } x = 0 \tag{31}$$

$$\text{For case "b" } d\lambda/dt = \mu D_{B2} dC_{B2}/dx \text{ at } x = \lambda(t) \tag{32}$$

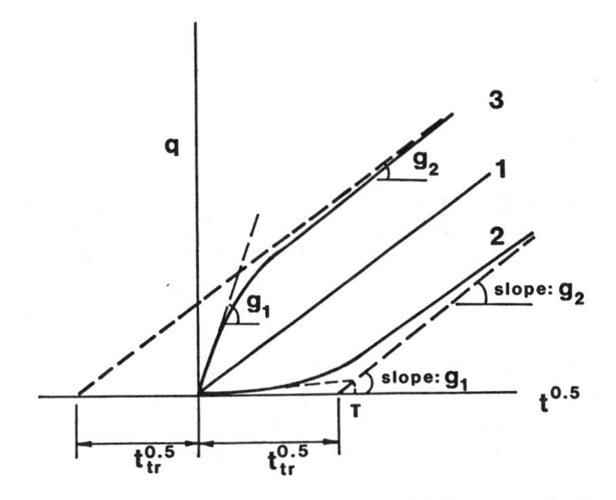

Figure 9. Shapes of kinetic curves in nonstationary molecular diffusion across the interfacial film [4].

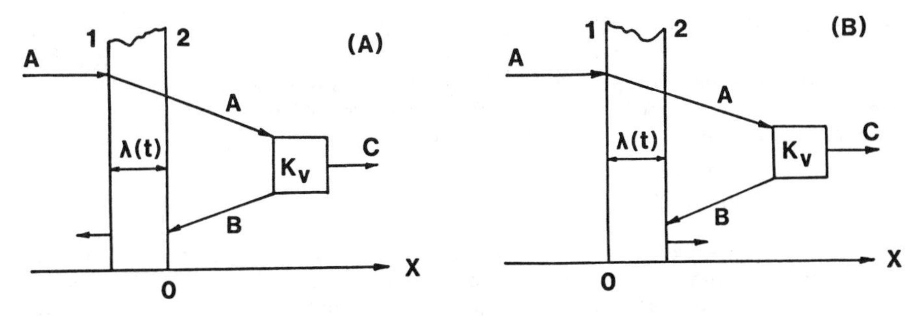

Figure 10. Schemes of diffusion processes with chemical reactions leading to interfacial film formation [42].

The rate of reaction $A \xrightarrow{k_v} B+C$ is:

$$r = k_v C_{A2}(x, t) \tag{33}$$

For case "a" Equation 33 requires no additional explanation, as k_v is not a function of the X coordinate. In case "b" a film bound to alter the reaction conditions is being formed, therefore, $k_v(x \geqq \lambda, t) \neq k_v(x < \lambda, t)$. The type of function $k_v(x < \lambda, t)$ is unknown. To obtain analytical solutions certain approximations of this function are quite necessary. Various simple functions (a linear, a step-wise, etc.) might be useful. Assuming $k_v(x < \lambda, t) = 0$ we consider the reaction within the film to be infinitely slow as compared with the reaction beyond it. The final expressions for $q_c(t)$ at large t values, provided all diffusivities but those of the film are equal, can be written as:

For case "a"

$$q_C = 2C^0_{A1}(Dt/\pi)^{0.5}(1+ZC^0_{A1})^{-1} - \alpha C^0_{A1}(D/k_v)^{0.5}(1+ZC^0_{A1})^{-1} \tag{34}$$

where $Z = \mu \cdot D \cdot D_\lambda^{-1}$

For case "b"

$$q_C = 2C^0_{A1}(Dt/\pi)^{0.5}(\alpha - \alpha ZC^0_{A1})\, 1+\alpha ZC^0_{A1})^{-1}$$

$$-\alpha C^0_{A1}(1+\alpha ZC^0_{A1})^{-1}(D/k_v)^{0.5} \tag{35}$$

It follows then that the type of function q at $t \rightarrow \infty$ holds for all three cases considered (Equations 27, 34, 35). Equations 34 and 35 indicate that coefficient g is function of the extracted compound concentration. Thus, for case "a" the value of g will be:

$$g = 2C^0_{A1}(D/\pi)^{0.5}(1+ZC^0_{A1})^{-1} \tag{36}$$

The plot of function $g = f(C^0_{A1})$ has a "plateau." The limiting value of g at $C^0_{A1} \rightarrow \infty$ is $2D_\lambda/\mu(D\pi)^{0.5}$ On the other hand, the plot of function $C^0_{A1}g^{-1} = f(C^0_{A1})$ allows to obtain the value $1/2(\pi/D)^{0.5}$ in terms of the y-intercept cut off by the straight line. The slope of this line is $\mu(D\pi)^{0.5}/2D_\lambda$. Further the value of μ/D_λ may be calculated.

For the case "b" the value g is:

$$g = 2C^0_{A1}(D/\pi)^{0.5}(\alpha - \alpha ZC^0_{A1})(1 + \alpha ZC^0_{A1})^{-1} \tag{37}$$

The plot of function $g = f(C^0_{A1})$ has a maximum at $C^0_{A1} \approx D_\lambda(\mu D)^{-1}\alpha^{-0.5}$, if $\alpha \gg 1$. At $\alpha ZC^0_{A1} \gg 1$ the plot of function $g/C^0_{A1} = f(C^0_{A1})$ is a straight line, the slope of which is $2(D/\pi)^{0.5}\alpha^{-1}$ and the intercept is $2(D/\pi)^{0.5}$. This allows us to evaluate D and μ/D_λ. Thus, the rate of extraction accompanied by interfacial film formation has been found to decrease even with an increase in the extracted compound concentration.

Conductometric Probing Method

Recently obtained extraction kinetics data are indicative of the existence of rather thick interfacial layers with the properties depending on the rate and direction of diffusion.

To investigate the interfacial layer properties of the organic phase having a marked conductivity we use a recently elaborated method of conductometric probing. Figure 11

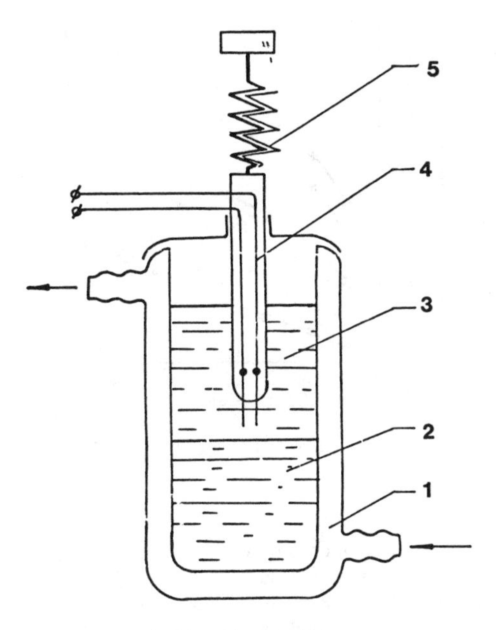

Figure 11. A device for measuring interfacial conductivity [9].

shows a device for measuring interfacial conductivity. The aqueous (2) and the organic (3) phases are poured into the thermostat (1). The value of conductivity is determined by means of a probe (4) approaching the interface with the help of a micrometer screw (5).

The relative increase in conductivity measured at a distance 1 from the interface is given by:

$$R_\psi = (\psi_l - \psi_v)\psi_v^{-1} \tag{38}$$

where $\psi_l = \psi_i + \psi_v$ (39)

The conductometric probe constant varies as the zone approaches the interface. So as to take into account this effect the dependence $\psi_l = f(l)$ is studied after the aqueous phase has been replaced by a solid dielectric.

THE RESULTS OF FILM FORMATION PROCESSES IN LIQUID-LIQUID SYSTEMS

Let us consider the main results obtained during the investigation of two-phase systems using methods just described.

Systems with Neutral Phosphororganic Extractants (NPOE)

It was during the first studies of stripping kinetics of inorganic acids from their solutions in NPOE carried out using STCM that "anomalous" kinetic curves were obtained. Thus, in HNO_3 stripping from the TNP · HNO_3 solutions into the aqueous phase, fixed in a

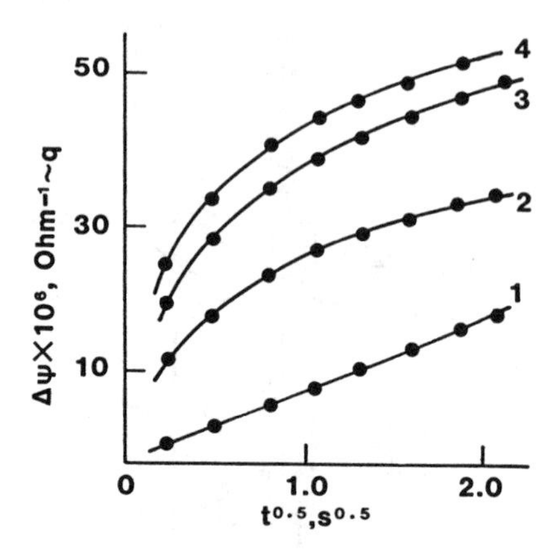

Figure 12. Shapes of kinetic curves obtained by STCM in stripping NNO_3 from TNP.HNO_3 solutions [46].

porous hydrophilic material (strips of chromatographic paper) only those kinetic curves that were obtained during the first contacting had the normal linear shape in the coordinates $\Delta\psi = f(t^{0.5})$ (Figure 12, curve 1, where $\Delta\psi$ is the change in the aqueous phase conductivity proportional to q). During further contacting of the same portion of the organic phase with fresh portions of the aqueous-phase kinetic curves with an upper break similar to those in diffusion through a constant-thickness film (Figure 9, curve 3) were obtained (Figure 12, curves 2, 3, 4).

The values of g for curves 1, 2, 3, and 4 at large t are identical. This fact indicates that the organic phase properties remain unchanged in the bulk. Moreover, the organic phase is the limiting one, since $\alpha \ll 1$ and $K_{eq} \ll 1$ (see Equation 11). Therefore, only a very narrow region adjacent to the interface layer of the organic phase does undergo the changes taking place in the system during the first contacting. These changes get more and more substantial with each contacting.

If the organic phase is mixed thoroughly before each contacting only normal kinetic curves are obtained. This is direct evidence of the existence in the organic phase of a super-saturated layer adjacent to the interface and also of the fact that an important part in the formation of this layer is played by the flux of water leading to the appearance of rather stable microheterogeneity zones (micro-emulsions of "water-in-oil" type). Since $\alpha \ll 1$, HNO_3 is easily stripped into droplets of water, and its interfacial concentration increases.

Taking into account the fact that the value g is stabilized in about 1 s, and assuming the TNP · HNO_3 diffusivity to be 10^{-5} cm^2/s, the thickness of this layer can be evaluated: $\lambda < (Dt)^{0.5} = 3.10^{-2}$ cm. Similar behavior has been also found for $HClO_4$ stripping from TBP · $HClO_4$ and TOPO · $HClO_4$ solutions in benzene. In this case the layer is three or four times narrower than in the system just described. It should also be noted that nonlinear dependencies $g = f(C_{solv})$ were obtained for all these cases, which does not conform to the model of diffusion with an instantaneous irreversible pseudofirst-order reaction.

Stripping kinetics of inorganic salts from their solvates solutions in NPOE into water has been also assumed to be affected by the film formation processes. The coefficient g has been found [4, 47–49] to reach its limiting value, independent of C^0_{solv} over wide solvate concentration ranges $UO_2(NO_3)_2 \cdot 2\,S(C^0_{solv} > 0.1$ M) (Figure 13). The behavior of the plots of function $g = f(C^0_{solv})$ is in poor agreement with the postulated model of a pseudofirst-order irreversible reaction:

$$UO_2(NO_3)_2 \cdot 2\,S \xrightarrow{H_2O} UO^{2+}_{2aq} + 2\,NO^-_{3aq} + 2\,S \tag{40}$$

according to which the rate of stripping and, consequently, the value of g should increase continuously. Assumption that the diffusion of some reaction (Reaction 40) product is a rate-determining step allows us to account for the data plotted in Figure 13. In this case the product concentration in the reaction zone increases as that Reaction 40 can no longer be considered an irreversible one. Calculations [4, 49] show that the values of Reaction 40 equilibrium constants coinciding with those known in the literature may be obtained using the kinetics data, if the "free" extractant concentration in the reaction zone exceeds the solvate concentration in the organic phase. Taking into account the fact that the "free" extractant is almost entirely absent in the bulk phase, we may arrive at the conclusion that the extractant diffusivity is several orders of magnitude less than the true one. This, in turn, implies the existence of a "barrier" hindering the extractant diffusion into the bulk of the organic phase. Supersaturation and microheterogeneity zones, emerging during water transfer into the organic phase by hydrated intermediate complexes of the type $UO_2 \cdot 2\,S \cdot H_2O(NO_3)_2$ or $UO_2 \cdot yH_2O \cdot S(NO_3)_2$, may act as such "barriers." The occurrence of such transfer processes has been recently pointed out in Reference 50. The water supersaturated interfacial layer of the organic phase hinders the removal of the poorly soluble in the extractant from the reacting zone, resulting in supersaturation of the interfacial aqueous phase layers with the extractant, as well as in the formation on this side of the interface of a microheterogeneity zone (emulsion of "oil-in-water" type).

The latter, in its turn, hinders the diffusion of water into the reaction zone, which indicates the occurrence of a reversible second-order reaction:

$$UO_2(NO_3)_2 \cdot 2\,S + H_2O \rightleftarrows UO^{2+}_{2aq} + 2\,NO^-_{3aq} + 2\,S$$

All of this seems to account for the fact that with an increase in the molecular weight of extractants having roughly the same extractability (TPP, TBP, TOP), i.e. with the decrease of their water solubility, the "plateau" in the plots of functions $g = f(C^0_{solv})$ emerges at lower C^0_{solv} values [4].

It is also noteworthy that the extractant basicity has a profound effect on the process involved. Thus, in the case of TOPO forming more stable complexes with $UO_2(NO_3)_2$ the

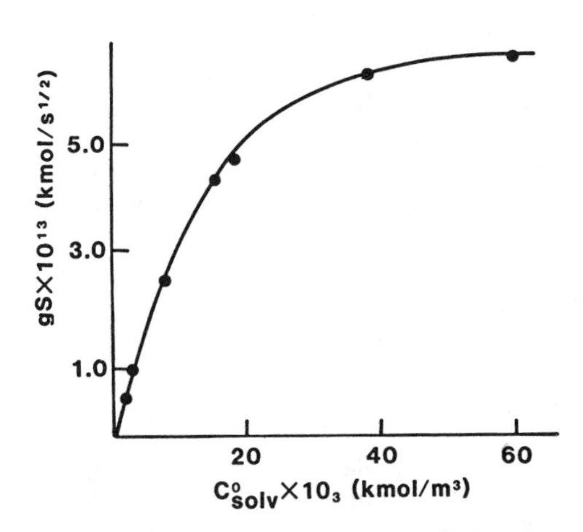

Figure 13. Dependence of the gS slope on $UO_2(NO_3)_2 \cdot$ 2TOPO concentration in theorganic phase [49].

appearance of supersaturated and microheterogeneous layers is observed much earlier [4], in so far as the reverse reaction tends to predominate over considerably lower "free" extractant interfacial concentration.

The preceding mechanism of a "sandwich-like" interfacial structure formation has been first suggested by Tarasov [4]. As will be shown below, the proposed mechanism allows us to account for the data on extraction kinetics, interfacial tension, and rheological properties of the interface.

Amine Systems

The amine extraction of inorganic acids takes place as a result of acid-base interaction giving rise to a salt formation. For example, the following reaction occurs in HX acid extraction with a tertiary amine:

$$R_3N + HX_{aq} \rightleftarrows R_3NHX \tag{41}$$

Stripping corresponds to the reverse reaction. At present it has been conclusively proved that Reaction 41 does not limit the rate of interfacial mass transfer, since all its steps are very fast [4, 33, 35, 36]. Thus, according to STCM results, the plots of the functions $g = f(t^{0.5})$ are usually linear and pass through the origin of the coordinates, whereas the diffusivities of components calculated from g strictly correspond to molecular ones. The extraction rate equation holding for STCM could be obtained if we suppose that extraction is accompanied by an instantaneous reversible second-order reaction [41], which is in a state of equilibrium [4, 33, 36]. The mass transfer coefficients for each component during unsteady-state diffusion have been determined according to Frank–Kamenetsky's method [51]:

$$\beta_j = (D_j/\pi t)^{0.5} \tag{42}$$

The fluxes of components to the reaction zone have been determined according to the equation:

$$r_j = \beta_j(C_j^0 - C_{ij}) \tag{43}$$

In this case the rate of extraction follows the equation:

$$r = (D_{HX}/\pi t)^{0.5} (C^0_{R_3N} C^0_{HX} - K_{eq}^{-1} C^0_{R_3NHX}) [C^0_{R_3N} + K_{eq}^{-1}(D_{HX}/D_{R_3NHX})^{0.5}]^{-1} \tag{44}$$

which coincides in form only with that derived in Reference 52 assuming the occurrence of two slow reactions (ITSCR mechanism) [54]. The diffusion resistance value R has been found experimentally by means of STCM from the slope g of the plots $q = f(t^{0.5})$:

$$R = 2t^{0.5} C^0_{HX} g^{-1} \tag{45}$$

On the other hand, from Equation 44 at $C^0_{R_3NHX} = 0$ it follows that

$$R = (\pi t/D_{HX})^{0.5} [1 + K_{eq}^{-1}(D_{HX}/D_{R_3NHX})^{0.5} (C^0_{R_3N})^{-1}] \tag{46}$$

Thus, the value R at fixed t = t should be a linear function of the inverse amine concentration. This fact is confirmed by Figure 14, which shows the results of the kinetics study of HCl extraction with TOA. The slope of the plot of Equation 46 is K_{eq}^{-1}

$(\pi/D_{R_3NHX})^{0.5}$ and the intercept cut off on the ordinate axis is $(\pi/D_{HX})^{0.5}$ (at $t' = 1$ s). During HCl extraction with TDA ant TLA solutions in toluene the main kinetic regularities are the same. It turns out, however, that two regions of experimental conditions can be singled out for these amines, either of these obeying the equation for the extraction rate of a general type (Equation 44) but with different coefficients. The borderline between these regions proved to be very sharp. It is determined by the value of the acid concentration in the aqueous phase C^*_{HX} which is called critical [4, 33, 36].

Thus, for TDA at $C^0_{HX} < C^*_{HX} = 10^{-3}$M:

$$r = 3.14 \cdot 10^{-5}(C^0_{R_3N}C^0_{HX} - 11.3 \cdot 10^{-2}C^0_{R_3NHX})/(C^0_{R_3N} + 24.0 \cdot 10^{-2})t^{0.5} \quad (47)$$

At $C^0_{HX} > C^*_{HX}$:

$$r = 3.14 \cdot 10^{-5}(C^0_{R_3N}C^0_{HX} - 2.4 \cdot 10^{-2}C^0_{R_3NHX})/(C^0_{R_3N} + 5.1 \cdot 10^{-2})t^{0.5} \quad (48)$$

For TLA at $C^0_{HX} < C^*_{HX} = 10^{-3}$M:

$$r = 3.14 \cdot 10^{-5}(C^0_{R_3N}C^0_{HX} - 16.9 \cdot 10^{-2}C_{R_3NHX})/(C^0_{R_3N} + 38.2 \cdot 10^{-2})t^{0.5} \quad (49)$$

At $C^0_{HX} > C^*_{HX}$:

$$r = 3.14 \cdot 10^{-5}(C^0_{R_3N}C^0_{HX} - 3.0 \cdot 10^{-2}C_{R_3NHX})/(C^0_{R_3N} + 6.8 \cdot 10^{-2})t^{0.5} \quad (50)$$

So, at $C^0_{HX} > C^*_{HX}$ the specific extraction rate, i.e. $gS(C^0_{HX})^{-1}$ exceeds the corresponding values obtained at $C^0_{HX} < C^*_{HX}$ remarkably. These results are indicative of the formation at $C^0_{HX} < C^*_{HX}$ of an interfacial film retarding reaction. Rheological measurements confirm this supposition.

It has been found that the interfacial viscosity η_i of toluene solutions of TOA does not depend on HCl concentration in the aqueous phase (in the range of 0–10^{-2} M of HCl), whereas for 0.1 M TDA-toluene-H_2O-HCl system at HCl concentration of 10^{-3} M an increase of η_i has been observed against η_i values for 0.1 M TDA-toluene-H_2O system. When the aqueous-phase acidity increases to 10^{-2} M, the interfacial viscosities of the previously-mentioned systems do not differ as should be expected in accordance with STCM data. In the case of TLA, system changes in η_i prove to be highly appreciable. Thus, a

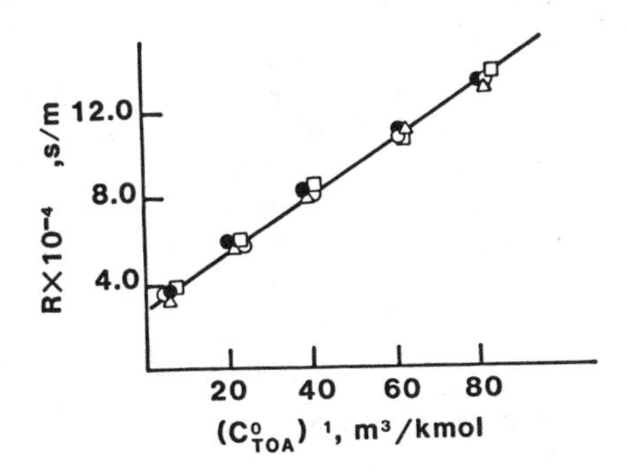

Figure 14. Dependence of phase diffusion resistances on the inverse concentration of TOA during HCl extraction [4].

certain relationship is observed between the tendency tasord interfacial film formation in HCl extraction, the acidity of the aqueous phase, and the molecular weight of the tertiary amine.

Stirring of phases favors a decrease in the film thickness. During the same period of time the interfacial viscosity in nonstirred systems reaches a considerably higher value than in stirred systems. These results are significant, since they point to one of the reasons for the poor agreement of data obtained by different authors who, as a rule, used different methods and worked under different hydrodynamic conditions.

Equations 47 through 50 indicate the occurrence of two stable regimes of the interfacial reactor, both in the presence and in the absence of a film. It should be noted that the formation of "two-dimensional" phases by intermediate interfacial compounds resulting in such kinetic effects as multiplicity of steady-state and stable regimes is a common phenomenon in catalysis.

Judging by the value of the intercept cut off on the ordinate axis the formation of the film does not affect the limiting stage, the latter being the diffusion of the extracted acid in the aqueous phase. It is only the slope of the plots of the $R = f(C^0_{R_3N})^{-1}$ function that is changed; according to the model involved it is equal to $(\pi t/D_{R_3NHX})^{0.5}K_{eq}^{-1}$. Thus, in the presence of a film the rate of interfacial mass transfer decreases because the values D_{R_3NHX} and K_{eq} are diminished either both simultaneously or individually. Let us assume that the effect of the film consists in additional resistance to the diffusion of salt into the bulk of the organic phase only. Then

$$\beta'_{R_3NHX} = \beta_{R_3NHX}\beta_f(\beta_{R_3NHX} + \beta_f)^{-1} \tag{51}$$

In this case the rate of extraction will be described by the same (Equation 47) with β_{R_3NHX} being substituted for β'_{R_3NHX}. However, to observe the linearity of $q = f(t^{0.5})$ plots it is necessary for the β_f value to be a function of the same type as β_{R_3N}, β_{HX} and β_{R_3NHX} values. This requirement means that the thickness of the film, λ, or its resistance, R_f, should increase according to the law ($\lambda \sim R_f \sim t^{0.5}$).

On the other hand, the effect of the film on the unsteady-state process of interfacial mass transfer may essentially consist in changes in K_{eq} value on account of the changes in reaction conditions. In view of interfacial viscosity increase with time one should favor the explanation according to which both β_{R_3NHX} and K_{eq} changes are to be taken into account.

In a range of comparable values of phase resistances the rate of interfacial mass transfer may be appreciably influenced by an organic solvent. Additions of solvating substances (i.e., alcohols) significantly accelerate extraction both in the presence and in the absence of a film. Addition of 5%–7% dodecanol to the organic phase is enough to completely prevent the formation of viscous interfacial films. Then the kinetics are described by the equation of the type of Equation 44, but with different coefficients. As is to be expected, a further increase in alcohol concentration in the organic phase promotes only an earlier achievement of a "plateau" on plots $gS(C^0_{HX})^{-1} = f(C^0_{R_3N})$ without changing the height of the "plateau." According to Equation 46, it is caused by an increase in K_{eq} value and a decrease in diffusion resistance in the organic phase. Unlike HCl extraction, $HClO_4$ extraction with toluene solutions of TLA et all $HClO_4$ concentrations used occurs without the formation of a condensed interfacial film. Thus there is no direct relationship between the association of amine salts and the formation of condensed interfacial films. In $HClO_4$ extraction with dilute TLA solutions kinetic curves $g = f(t^{0.5})$ have a distinct break [4, 49] of the same shape as shown in Figure 9 (curve 2). This feature of interfacial mass transfer is caused by a gradual change in the characteristics of the interfacial layers of the organic phase, which, however, takes place without an appreciable change in their viscosity. It is evidenced by the results of studying the kinetics of $HClO_4$ extraction with concentrated TLA solutions ($C^0_{R_3N} > 0.1$ M), when the organic phase is not rate controlling. As should be expected, in these cases the plots $q = f(t^{0.5})$ do not show breaks, and the specific rate value $gS(C^0_{HX})^{-1}$

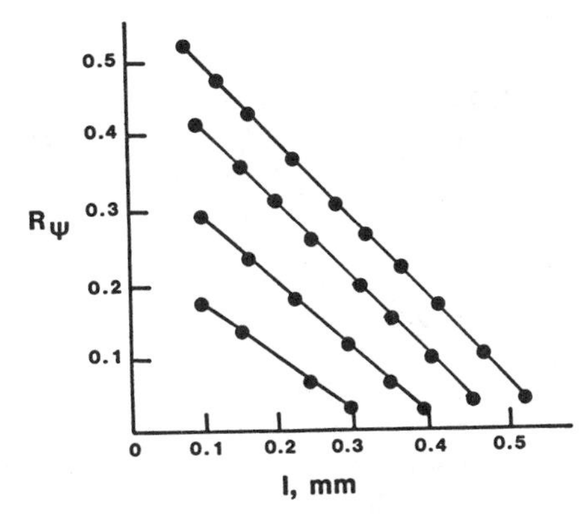

Figure 15. Dependence of relative conductivity increment on a distance from the interface [9].

proves to be the same at all t. It is determined only by the rate of acid diffusion in the aqueous phase.

Addition of high molecular alcohols to the organic phase causes the same consequences as an increase in amine concentration does. Now, the K_{eq} value could be defined. Calculations show that this value characterizes a reaction of amine protonation

$$R_3N + H(H_2O)_x^+ \rightleftarrows R_3NH(H_2O)_y^+ \tag{52}$$

rather than a reaction of neutralization (Reaction 41). Indeed, the value$C_{R_3NHX_i}/C_{R_3N_i} \cdot C_{HX_i}$ is constant, which is not true $C_{R_3NHX_i}/C_{R_3N_i} \cdot C^2_{HX_i}$.

The same consideration accounts for the fact that the extraction rate is independent of Cl^--ions concentration in the aqueous phase. Consequently, the predominant channel of interphase transfer is realized through an instantaneous reversible reaction (Reaction 52), which can be simply explained. Due to the dissociation of R_3NHCl salt into hydrated RNH^+ and Cl^- ions occurring in the interfacial layer of the organic phase the concentration of R_3NHCl at the interface is low as compared with the concentrations of R_3NH^+, H^+-ions, and ion pairs $R_3NH^+(H_2O)_xCl^-$ on this account, which are transfered through the organic phase interfacial layer.

Association of ions into water-separated ion pairs, the formation of contact ion pairs, and R_3NHX molecules proceed at the time of diffusion in the organic phase. These processes are accompanied by dehydration and accumulation of water in the interfacial layer. The existence of the layer of the organic phase enriched in water and ion pairs, as predicted by this model of interfacial processes, is confirmed by a conductometric probing method. It has turned out that even in an equilibrium state there exists a layer, δ_2, possessing a measurable thickness whose conductivity greatly exceeds that in the bulk; Figure 15 shows an increment in the relative value of the changes as the probe approaches the interface (the system: 0.5 M $TOAHClO_4$-toluene-HCl-H_2O). Lines 1, 2, 3, and 4 correspond to probes with different distances between electrodes. In the absence of equilibrium in the system the value of R_ψ becomes different. So, during extraction the higher the value R_ψ is, the greater the mass transfer rate is; during stripping, the opposite situation is observed.

The fact that the interfacial layer enriched in water and in ions has a substantial thickness is indicated by the duration of its formation process (Figure 16). Conditions corresponding to curves 1 and 2 (Figure 16) differ in the extent of deviation from equilibrium. Curve 1 corresponds to nearly equilibrium conditions, while curve 2 corresponds to a more consider-

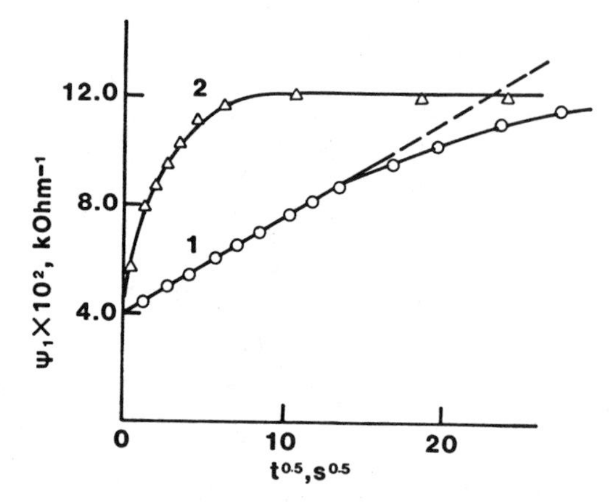

Figure 16. Time dependences of conductivity at a distance 10^{-4} cm from the interface [4].

able deviation from the equilibrium of water. The more dried the organic phase is, the longer time it takes to reach the limit value of conductivity. The layer resembles a through-flow reactor. Near equilibrium the flux of water from this reactor into the bulk of the organic phase becomes negligibly small; therefore the changes take place rapidly.

It can be pointed out that the $HClO_4$ system tends to the same interfacial conductivity value irrespective of the preliminary drying of the organic phase. In all cases the same conductivity value will be obtained in a fixed point from the interface. For instance, it is confirmed by an experimental study of the influence of heating and cooling of a two phase system on the values of ψ_v and ψ_i. As one can see in Figure 17a, irreversible processes occur neither in the bulk of TAO $HClO_4$ solution nor in the interfacial layer, i.e., a true equilibrium exists throughout the system. The interfacial conductivity is connected with amine salt concentration and water activity in the aqueous phase through the following simple relationship:

$$\psi_i = \text{const} \cdot C^{n}_{R_3NHX} \cdot a_w \tag{53}$$

The exponent "n" characterizing the intensity of increment in ψ_i values with increasing amine salt concentration practically does not depend on the nature of a salt cation but rises in the anion series: $ClO_4^- < HSO_4^- < Cl^- < SO_4^{2-}$. The ability of amine salts to extract water is known to increase in this series owing to a rise in anion hydration energy [53]. It is indicative of an equilibrium between the water phase and interfacial layer on the one hand and of the participation of water in the interfacial layers on the other.

Thus at least in reacting liquid-liquid systems, their interface becomes blurred out due to the coexisting phases components being transferred by intermediate complexes, and a rather wide transition region with intermediate properties results. To our mind, inorganic acids extraction with amines is accompanied by the formation of an organic-phase layer enriched in water and capable of dissolving ions in a significant amount, which is responsible for the determining contribution of the protonation reaction to the ionic transfer process. Water mass transfer into the adjacent layers of the organic phase occurs due to the existence of carriers. These are ions that did not undergo dehydration while passing through the interface as well as water-separated ion pairs.

It would be wrong to believe that equilibrium is achieved in the interfacial layers in all cases. The very fact that HCl extraction with tertiary amine solutions under certain conditions gives rise to highly viscous films can be considered to indicate the occurrence of irreversible processes in interfacial layers. The validity of this conclusion is evidenced by the presence of ψ_i hysteresis value (but not ψ_v) upon heating and cooling (Figure 17b). We suppose such components as water and salt to have the limit solubility in the organic-phase layers adjacent to the interface; that is why zones of supersaturation and microheterogeneity may appear during the mass transfer. In accordance with this assumption, the previously mentioned abrupt change in the interfacial transfer rate as well as interfacial viscosity with variation of HCl concentration in the aqueous phase may be connected with a phenomenon of coagulation of dispersed water particles, and, perhaps, those of R_3NHCl under the action of electrolytes. Returning to Figure 16 one can note that the conductometric probing method enabled us to verify the validity of the assumption concerning the "diffusion nature" of cumulative changes in the layers adjacent to the interface, i.e., that $\beta_f \sim t^{-0.5}$

It is not clear whether either thickening the layers adjacent to the surface or changes occurring in them take place. The problem of allowance for interfacial phenomena discovered may prove essential in the case of a description of unsteady-state transfer kinetics with a short-time phase contact. Such conditions are known to be characteristic of mass transfer with a short residence time of reacting phases.

On the other hand the data in Figure 16 show that changes in the layers adjacent to the interface may proceed either very rapidly or rather slowly. Therefore in some cases, a layer formation period may be neglected and interfacial mass transfer may be regarded as diffusion through a layer of constant thickness, while in other cases changes in the layer will affect mass transfer longer. This accounts not only for the presence of a break on $q = f(t^{0.5})$ plots, but also for a significant change of the break point location as the system approaches a state of equilibrium. If changes under discussion taking place in layers adjacent to the interface with time are connected with an increase in their thickness, then it seems possible to offer at least a qualitative explanation of the extraction kinetics.

Further it will be shown that a growing film concept proves to be very promising also with the view of explaining surface active substances (SAS) effects on the kinetics of unsteady-state diffusion transfer. It has already been pointed out that with the same regularities of diffusion penetration and film thickening $q = f(t^{0.5})$ plots are still linear. However g values will be determined not only by phase diffusion of a substance but also by the mass transfer rate in the interfacial film. There may be a case of interest occurring provided film thickening takes place faster than the diffusion penetration. Then extraction

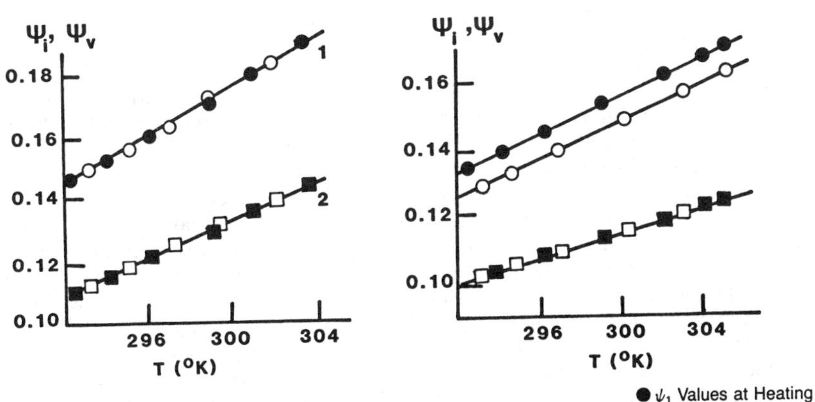

Figure 17. Temperature dependences of ψ_v and ψ_i values [4].
Systems: a.—TOA - $HClO_4$ - H_2O - toluene
b.—TOA - HCl - H_2O - toluene
Heating and cooling rate: 0.1 $K.s^{-1}$

takes place in such a way that it seems as if the whole bulk of the organic phase possessed altered properties. Hence the rate of extraction depends on the dynamic distribution coefficient value between the aqueous phase and interfacial film and not on its equilibrium value. Since the state of the layers adjacent to the interface becomes altered as equilibrium is approached, the transfer conditions will undergo changes too. As one can see from Figure 16, after a certain period of time, the square root dependence of ψ_i value does not hold good any longer: ψ_i changes become slower. In terms of the growing film concept this fact could be viewed as the achievement of an almost invariable film thickness. Naturally, under these conditions the diffusion penetration depth will sooner or later exceed the thickness the film, and under the influence of the latter the kinetics will begin decreasing. Consequently, two linear sections with g_1 and g_2 slopes will appear on the plots of $q = f(t^{0.5})$ function. In the case of amine systems $g_2 > g_1$ i.e., the presence of a film reduces the rate of interfacial mass transfer. HCl-stripping-kinetics results obtained by means of STCM in TOA and TLA systems also suggest the formation of rather thick interfacial layers [4, 49]. The plots of $q = f(t^{0.5})$ functions are straight lines passing through the origin of the coordinates, which indicates an immediate establishment of chemical equilibrium. If during HCl stripping they have breaks in τ point, then $g_2 > g_1$ [4]. However, unlike a case of extraction, these breaks take place during the second and following phase contacts. The situation is similar to the one already described for HNO_3-stripping from TNP solutions. It is apparently one of the most conclusive evidences for change of the properties of the organic-phase layers adjacent to the interface. Moreover the results obtained indicate a relative stability of the occurring changes. Actually after the separation of the aqueous phase from the organic one the layers of the latter adjacent to the interface retain "the past in remembrance" within several minutes.

Since, the abscissa of the break point τ of the plots $q = f(t^{0.5})$ obtained during repeated contacts is equal to $\tau = \pi/4n^2$, and $g_2/g_1 = m/n1$ with $1/m = \lambda/\alpha_1(D_1)^{0.5}$ we can state:

$$\lambda = 2g_1\alpha_1(D\tau)^{0.5}/g_2\pi^{0.5} \tag{54}$$

Assuming $\alpha_1 = \alpha$ during HCl stripping from TOAHCl solutions and knowing that the breaks appear at $\alpha = 0.05/0.25$ we estimated λ values from the experimental values of and g_1/g_2 [4]. The latter ranged from $1.10.^{-3}$ to 5.10^{-3} cm.

Addition of 5%–10% alcohols to the system results in such a considerable thinning of the just mentioned layer that the difference between their conductivity and that of the organic phase completely disappears.

The acids-stripping rate is independent of the intensity of stirring in Lewis diffusion cells [4, 35, 54, 19, 50]. This fact can be regarded as an indication of a limiting role of molecular transfer in layers adjacent to the interface effectively damping the turbulence. The mechanism of turbulence damping is obviously similar to that of capillary waves damping by adsorbing SAS [27–31].

In the cases in question Gibbs' elasticity effects may be expected to be still more pronounced since the local compression and spreading of the layers will be opposed not only by adsorption and desorption processes, but also by processes of the formation and destruction of supersaturation and microheterogeneity zones. This can be illustrated by Figure 18, where the kinetic study results of water dissolving in dried solution of chloride TOA are shown. Plateaus on the curves 2 and 3 are usually considered to indicate that the reaction at the interface is rate controlling. However, as it was shown by means of STCM all the reactions in the system under discussion are instantaneous and the plateaus on the curves are due to the formation of microheterogeneity zones playing the part of a specific diffusion barrier. When the critical stirring intensity ω_{cr} is achieved, these zones are destroyed. A decrease of their diffusion resistances and conductivities is observed at the same time (curve 4). Addition of octanol to the organic phase to prevent supersaturation with water leads to a linear dependence at all stirring intensities (curve 1), whereas ionic surfactants stabilize the microheterogeneous layer and expand the region of the "plateau" existence (curve 3).

It should be mentioned that supersaturation and microheterogeneity zones were observed by other authors as well [55]. On of the most interesting experimental facts obtained using diffusion cells [4, 56] and STCM [4, 38, 49] is an unusual dependence of extraction rate on concentration driving force. The latter increasing the rate of HCl stripping reaches its limit when C_{R_3NHCl} is approximately 0.15 M. The "plateau" at $C_{R_3NHCl} > 0.3$ M is followed by a rate decrease as C_{R_3NHCl} rises. During $HClO_4$ stripping from TOA solutions of $HClO_4$ in toluene the plots of the function show either plateau or maxima [4, 38, 49]. We discussed similar relationships when describing the kinetics peculiarities of $UO_2(NO_3)_2$ stripping from solutions of its TOPO compounds and put forward an explanation with reference to interface "sandwich" structure.

Systems with Acidic and Acidic Chelating Extractants

Perhaps, the concept of blocking the reaction-zone by interfacial films was first successfully applied to explaining the unexpected results obtained in studying the kinetics of zirconium extraction by alkylphosphoric acids [20, 21, 57]. The films were registered then by taking photographs of the interface being deformed by drops of the aqueous phase [20]. Later on, convincing photographs of the films were taken during the extraction of copper by 8-hydroxyquinoline and nickel by sodium salt of HDEHP [58].

The rheological methods employed helped to show that the interfacial films in these cases possessed mechanical strength, in other cases an enchanced viscosity [12, 16]. The film's thickness, as shown by interferometric measurements [16], attains 10^{-4} cm. Particles forming the films are of different nature and origin. However, they all are surface active and capable of forming bonds between each other. In case of extraction by acidic reagents the intermediates and byproducts resulting from the interaction between reagents and various extracted metals species may act as such particles [12]. Interfacial association processes leading to the formation of films are particularly characteristic of the metals apt to hydrolysis and hydrolytic association (Si, Ti, Hf, Ta, Nb, Mo, and others) [4, 12].

The rate of film formation depends to a great extent, on the phase composition and the particular type of organic solvent. An aqueous nitric acid mixture of zirconium-solution of HDEHP in decane will increase with the metal concentration in the aqueous phase and with the extractant concentration in the organic phase. With changing HNO_3 concentration the rate of film formation has the highest values in the range $C_{HNO_3} = 0.7 - 1.5$ kmol/m^3,

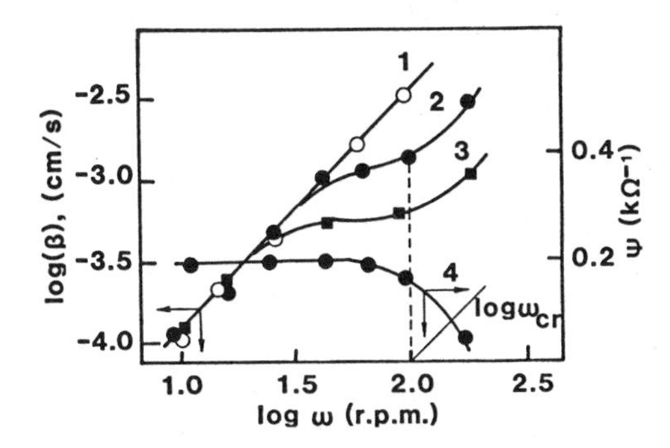

Figure 18. Dependences of mass transfer coefficient and interfacial conductivity on a stirring rate during water extraction by TOAHCl in toluene [4].

whereas it decreases abruptly when the organic solvent decane is replaced by benzene [12]. In the system involved this is due to an increase in the water solubility of the organic phase. The systems: sulphuric acid aqueous solutions of zirconium-solution of HDEHP in nonpolar organic solvent, especially at low concentrations of H_2SO_4, are liable to form viscous interfacial films. The interfacial viscosity increases during the extraction of indium from sulphuric acid solutions by HDEHP [16].

The sulphuric acid aqueous solutions of Ta and Nb at the interface of HDEHP solutions have been also found to produce rigid interfacial films [12, 16]. For systems with Ta, the maximum rate of film formation is observed at $C_{H_2SO_4} = 3\ kmol/m^3$. Figure 19 shows the pictures of the film in the system 0.25 M HDEHP solution in decane—20 g/l $Zr(NO_3)_4$—2 M HNO_3 taken in 45 minutes. A distinct laminated structure indicating the solid-like character of the film is observed. Then the film structure becomes more loose and the droplets are formed within it. These structural changes account for the dependencies obtained in Reference 12.

The interfacial films formed in all cases have had a profound effect on extraction kinetics and the rate of emulsion separation. Mass transfer in extraction apparatus with a relatively low stirring intensity of the phases (diffusion cells, wetted wall columns, horizontal rectangular channels, etc.) are particularly subject to the influence of interfacial films.

The main kinetic particularities of extraction accompanied by film formation are as follows [4, 20, 21]:

1. An abrupt extraction rate decrease long before equilibrium is achieved.
2. The decrease in mass transfer coefficients and even in the absolute extraction rate with the increase in the metal concentration.
3. The existence of conditions when the specific extraction rate is first independent, but then sharply increases with the stirring intensity.

The first of the mass transfer particularities is usually observed at a relatively slow growth of films in the stirred diffusion cells, whereas during the rapid film formation at may be registered by the STCM. The shape of kinetic curves may be accounted for by the change of the phase and film resistances ratio. For example, in 20 minutes after the beginning of Zr extraction by HDEHP its specific rate decreases abruptly and then is independent of the stirring intensity (see the kinetic model of mass transfer with a growing interfacial film). Thus, the film diffusion resistance becomes dominating.

The second particularity is attributed to the fact that the film diffusion resistance may increase faster than the concentration driving force does [4, 38, 56].

The third one is due to the film resistance remaining constant until the hydrodynamic perturbations have destroyed the film. This has already been illustrated by the microheterogeneity zones formation (Figure 18). Diffusion resistance of condensed interfacial films and their resistance to hydrodynamic perturbations in the majority of cases are so profound as to extend the "plateau" region remarkably.

Systems Without Extractants

Interfacial films possessing enhanced viscosity or mechanical strength have been found to form at the interface between the aqueous solutions of salts tending to hydrolysis and hydrolytic association and the organic solvents with low dielectric constant [4, 12, 16]. The film formation in these cases proceeds much slower than in the systems with acid extractants, the limiting values of strength or viscosity being considerably lower. This behavior is most typical of multivalent metals (Si, Ti, Zr, Hf, Ta, Nb, Mo, etc). The existence of interfacial films in the systems with these metals is observed in a broad variety of extraction conditions. However, the less is the metal species association, the less liable to hydrolysis and hydrolytic association, the less is the effect of film formation. For example, solutions of Cu and Al

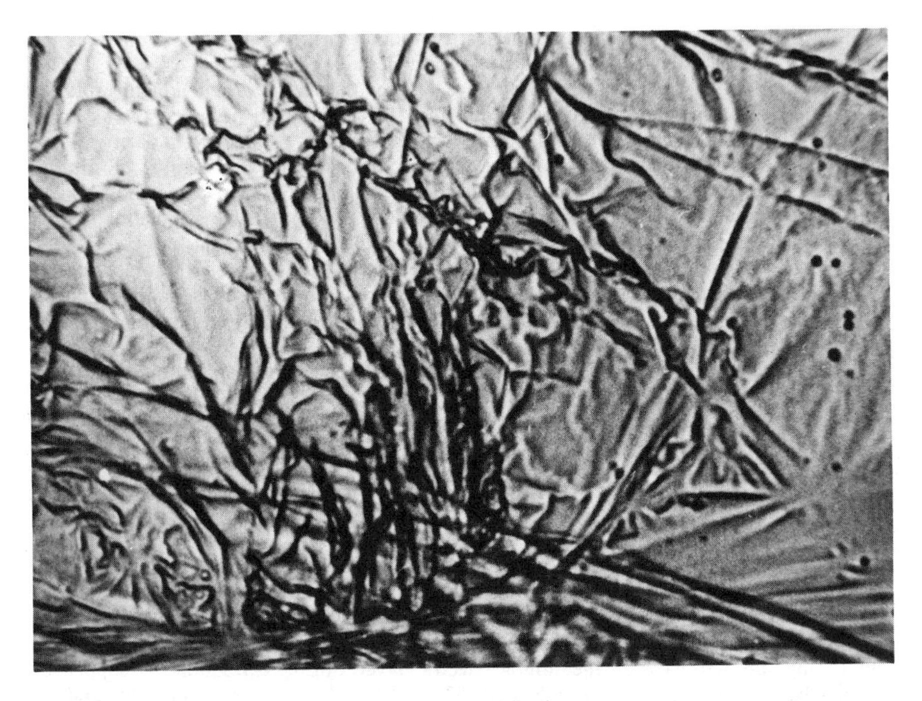

Figure 19. A rigid interfacial film picture at 500-fold multiplication.

sulphates are capable of forming films at the interface between the aqueous phase and alkanes only at pH approaching that of hydroxides precipitation.

Uniqueness of the organic solvent is, undoubtedly, of primary importance. It appears that films are not formed at the interface with polar solvents (e.g., alcohols). The process of film formation is greatly affected by the dryness of a low dielectric constant solvent; the rate of films formation increases sharply in case of a wet solvent.

The phenomena in question are in good agreement with the hydrolytic association mechanism [4] according to which the associates present in the solution and emerging in the water enriched interfacial layer of the organic phase tend to form gel-like interfacial films due to hydrogen bonding. The process is accompanied by a decrease in the interfacial tension. Higher molecular alcohols and esters (e.g., TBP) not only prevent the formation of interfacial films, but also destroy the structures that have already been formed. These reagents are responsible for the increased interfacial polarity, higher solubility of water in the interfacial layers as well as for the destruction of hydrogen bonds between the structurcal elements of the interfacial film.

Besides, the rate of film formation is affected by the following variables:

1. Stirring intensity (as long as the kinetics of the process is dependent on the mass transfer rate of film forming substances).
2. Temperature.
3. The presence of an outer electrostatic field.
4. The presence of surfactants.

From the analysis of the effect of the last two factors on the film formation rate, it is possible to state the sign of the ionic charge of particles adsorbed at the interface [4, 12, 16]. Thus, the rate of film formation at the interface between the sulphuric acid aqueous solution of Ta and decane increases when the electrostatic field is applied provided the electrode immersed into the organic phase is positively charged. The opposite polarity of the electrodes result in the slowing down of the process rate. The admixtures of cationic surfactants (cetyltrimethylammonium bromide), which charge the interface positively, accelerates the formation of films [4, 12]. Anionic surfactants (sodium dodecylsulphate) produce a negative effect. These data indicate that the film is formed by anionic species of Ta.

The opposite picture is observed in the system nitric acid solution of zirconium-decane-H_2O. In this case the film is formed by hydrolized zirconium complexes, sodium dodecyl sulphate accelerating the rate of film formation and cetyltrimethylammonium bromide moderating it.

However, the proposed electrostatic model cannot be considered an ideal one, since SAS may interact with metal containing particles adsorbed. There are added problems when extractant is present in the system.

ACCELERATION AND RETARDATION OF INTERFACIAL CHEMICAL REACTIONS BY THE CHARGED FILMS

Interfacial films formed during inorganic substances extraction are usually charged, apparently owing to the ions present in the systems. There is good evidence that the electrostatic field of interfacial layers has a substantial influence on the mass transfer kinetic. The data on the effect of SAS on the kinetics of chemical reactions occurring in the dense and in the diffused part of the double electric layer during extraction of copper by hydroxyoximes were obtained using STCM [37, 42, 59] and other methods [37, 60] and are in good agreement. Thus SAS have been found to affect not only the hydrodynamics of regions adjacent to the interface layers [27–31], but also the rate and equilibrium of chemical reactions proceeding in layers δ_1 and δ_2 (see Table 1).

Table 1
Catalysis in Copper Extraction and Stripping in the System:
$CuSO_4$-H_2O-toluene-C_8H_{17}

OH NOH

Catalysts	C_{cat} $\frac{kmol}{m^3}$	Extraction			Stripping		
		F_a	P_s	P_v	F_a	P_s	P_v
Sodium dodecylsulphate	10^{-5}	1.1	0.38	0.62	2.1	0.40	0.60
	$5 \cdot 10^{-5}$	1.5	0.45	0.55	3.6	0.43	0.57
	10^{-4}	2.5	0.63	0.37	4.5	0.48	0.42
	$5 \cdot 10^{-4}$	4.3	0.88	0.12	7.2	0.69	0.31
	10^{-3}	4.8	0.91	0.09	8.0	0.74	0.26
Cetyltrimethyl-ammonium bromide	10^{-5}	0.8	0.28	0.72	0.4	0.32	0.68
	$5 \cdot 10^{-5}$	0.63	0.21	0.79	0.3	0.30	0.70
	10^{-4}	0.52	0.15	0.85	0.2	0.25	0.75
	$5 \cdot 10^{-4}$	0.13	0.10	0.90	0.05	0.20	0.80
	10^{-3}	0.11	0.07	0.94	0.04	0.17	0.83

F_a — accelerating factor

The influence of ionic SAS may be regarded as a nonspecific electrostatic catalysis by the field of double electric layer [4, 42, 59] or as a micellar catalysis [61].

CONCLUSIONS

In everything that has been said so far in this chapter, there has been latent a fundamental assumption that interfacial films play a prominent part in extraction processes. On the one hand, the films formed cause an appreciable decrease in extraction rate as well as in the rate of emulsion separation, which in turn results in the lower capacity of extraction apparatus. An adequate treatment of kinetics data is quite impossible without taking into account the effects of film formation. Furthermore, one might arrive at false conclusions about the rate determining step and its mechanism unless the film formation processes are taken into consideration. It is evident that, in a number of cases, the interfacial films formed may "upset" the existing mass balance, since it is the quantity of substance in the organic and aqueous phases alone that is usually estimated, whereas the quantity of substance in the film itself is more often than not neglected. However, it would be a mistake to assume that the influence of interfacial films is essentially negative in character. The phenomenon in question may be used to substantially accelerate extraction processes. Detailed studies requiring a variety of up-to-date analytic techniques become increasingly desirable in order to confirm this practically important conclusion. There is little doubt that such studies capable of throwing more light on the interactions involved in extraction processes and on the rate-controlling mechanisms should be of paramount importance.

NOTATION

a interfacial area per unit volume (m^{-1})
a_W water activity ($kmol/m^3$)
b adsorption constant ($kmol/m^3$)
C concentration ($kmol/m^3$)
D diffusivity (m^2/c)
d_p, d_v diameters of pendulum and vessel (m)
E extent of extraction (dimensionless)
F tangential force (N)
F_a accelerating factor (dimensionless)
F_l ultimate film strength (N/m)
g slope of the $g = f(t^{0.5})$ plot at $t \rightarrow \infty$ ($kmol/m^2/s^{0.5}$)
K mass transfer coefficient (m/s)
k_v rate constant of irreversible first-order volume reaction (s^{-1})
k_s rate constant of first-order irreversible interfacial reaction (m/s^1)
k_{ss}^+, k_{ss}^- forward and back first order interfacial reaction rate constant ($kmol/m^2/s^1$)
K_{eq} equilibrium constant of reversible first-order reaction (dimensionless)
L pendulum moment of inertia ($kg \cdot m^2$)
l, m, and n values in Equation 26
P_v, P_s contributions of chemical reactions in volume and at interface (dimensionless)
q quantity of substance transferred across the unit of interfacial area ($kmol/m^2$)
Q coefficient in Equation 1 (s/m)
r diffusional flux ($kmol/m^2/s$)
R overall diffusion resistance (s/m)
R_f, R_f^* film diffusion resistances at different times (s/m)
Re Reynolds number (dimensionless)
R_{ph} phase diffusion resistances (s/m)
R_ψ relative conductivity increment (dimensionless)
s value in Formulas 13 and 14 (dimensionless)

- S — interfacial area (m^2)
- Sh — Sherwood number (dimensionless)
- Sc — Schmidt number (dimensionless)
- t — time (s)
- t_{tr} — transition time (s)
- T_0, T, T_f — periods of pendulum oscillation in vacuum, at interface without film and at interface with interfacial film (s)
- u — logarithmic decrement of oscillations damping (dimensionless)
- Z — value in Equation 34 and 35 (m^3/kmol)
- w — magnetic field constant (m^{-1})

Greek Symbols

- α — distribution coefficient (dimensionless)
- β — mass transfer coefficient in the phase (m/s)
- β_f — film mass transfer coefficient (m/s)
- γ — coefficient in Equation 7 (m/s/kmol)
- δ_1, δ_2, δ_3, δ_4 — thickness of adsorption, interfacial, diffusion and hydrodynamic layers, respectively (m)
- θ_j — share of interface occupied by adsorbed molecules of j-substance (dimensionless)
- μ — coefficient in Equations 31 and 32 (m^3/kmol)
- λ — interfacial film thickness (m)
- Δl — deformation (m)
- η_i — interfacial viscosity (kg/s)
- ψ_v, ψ_i — volume and interfacial conductivities (ohm^{-1})
- ψ_l — conductivity at a distance l from interface (ohm^{-1})
- τ — time corresponding to breaks in plots $g = f(t^{0.5})$ (s)
- ω — frequency of stirrer rotation (s^{-1})

Subscripts

- A, B, j, k — components
- 1, 2 — phases
- i — at interface
- s — in adsorbed state
- v — in volume
- eq — in equilibrium

Superscripts

- 0 — initial state
- + — for forward reaction
- — for back reaction

Chemical Abbreviations

- HDEHP — di-2-ethylhexylphosphoric acid
- TBP — tri-n-butylphosphate
- TNP — tri-n-nonylphosphate
- TOP — tri-n-octylphosphate
- TPP — tri-n-propylphosphate
- TOPO — tri-n-octylphosphinoxide
- TOA — tri-n-octylamine
- TDA — tri-n-decylamine
- TLA — tri-n-laurylamine

REFERENCES

1. Rusanov, A. I., *Fazovie ravnovesiya i poverhnostnie yavleniya,* Leningrad: Khimiya, 1967, pp. 9–15.
2. Frolov, Yu. G., *Kurs kolloidnoi khimii (proverhnostnie yavleniya i dispersnie sistemi),* Moskva: Khimiya, 1982, p. 10–12.
3. Levich, V. G., *Fisiko-kimicheskaya gidrodinamika,* Moskva: Izd. AN SSSR, 1952, pp. 46–114.
4. Tarasov, V. V., *Mezhfaznye yavleniya i kinetika ekstraktsii neorganitcheskih veshchestv.* Moskva: MKhTI im. D. I. Mendeleeva, 1980, pp. 1–44.
5. Yagodin, G. A., i dr. *Osnovy zhidkostnoj ekstraktsii.* Moskva: Khimiya, 1981, pp. 144–195.
6. Yagodin, G. A., i Tarasov, V. V., "Mezhfaznye yavleniya v sistemah elektrolit-neelektrolit i ih vliyanie na kinetiku ekstraktsii," *V trudah VI Vsesoyuznoj konferentsii po khimii ekstraktsii. Tezisi dokladov.* Tchast I. Kemerovo (1981), pp. 8–9.
7. Tarasov, V. V., i dr. "Mezhfaznye yavleniya pri perenose neorganitcheskih kislot v sistemah s tretichnimi aminami," *V trudah VI Vsesoyuznoj konferentsii po khimii ekstraktsii. Tezisi dokladov,* Tchast I. Kemerovo, 1981, p. 100.
8. Novikov, A. P., Tarasov, V. V., and Yagodin, G. A., "Konduktometritcheskoe issledovanie pripoverhnostnih sloev organitchescoj fasi v protsesse ekstraktsii," *V trudah Vsesoyuznoj konferentsii po ekstraktsii i ekstragirovaniyu. Tezisi dokladov.* Tom 3. Riga (1982), pp. 146–147.
9. Tarasov, V. V., Novikov, A. P., and Yagodin, G. A., "Konductometritcheskoe zondirovanie pripoverhnostnih sloev organitcheskoj fasi v ekstraktionnih sistemah," *Zh. fiz. khimii,* Tom 56, No. 5 (Maj 1982), pp. 1242–1245.
10. Izmajlova, V. N., and Rebinder, P. A., *Strukturoobrazovanie v belkovih sistemah,* Moskva: Nauka (1974), pp. 156–175.
11. Cox, M., and Flett, D. S., "Metal extraction chemistry," *In Handbook of Solvent Extraction,* Teh C. Lo, Malcolm H. I. Baird and Carl Hanson Eds. New York: A Wiley-Interscience Publ., 1982, p. 76.
12. Yagodin, G. A., Tarasov, V. V., and Ivakhno, S. Yu., "Condensed interfacial films in metal extraction systems," *Hydrometallurgy,* Vol. 8 (1982), pp. 293–305.
13. Brown, A. G., Thuman, W. C., and McBain, T. W., "The surface viscosity of detergent solutions as a factor in foam stability," *J. Coll. Int. Sci.,* Vol. 8, No. 5 (Oct. 1953), pp. 491–507.
14. Goodrich, F. G., and Chattergee, A. K., "The theory of absolute surface shear-viscosity," II. The rotating disk problem." *J. Coll. Int. Sci.,* Vol. 34, No. 1 (Sep. 1970), pp. 36–42.
15. Mannheimer, R. J., and Schechter, R. S., "An Improved Apparatus and Analysis for Surface Rheological Measurements," *J. Coll. Int. Sci.,* Vol. 32, No. 2 (Feb. 1970), pp. 195–211.
16. Ivahno, S. Yu., *Adsorbtsiya na granitse razdela faz i ee vliyanie na kinetiku ekstraktsii nekotorih metallov,* Avtoreferat dissertatsii kandidata khimitcheskih nauk, Moskva: MKhTI im. D. I. Mendeleeva, 1979, pp. 1–16.
17. Hard, S., and Nauman, R. D., "Laser Light-Scattering Measurements of Viscoelastic Monomolecular Films," *J. Coll. Int. Sci.,* Vol. 83, No. 2 (Oct. 1981), pp. 315–334.
18. Nauman, R. D., and Gaonkar, A. G., "Laser Light-Scattering Studies of the Interfacial Chemistry in Solvent Extraction Systems," In *Abstracts of Intern. Solv. Extr. Conf.* (ISEC–83), Denver, Colorado, American Inst. of Chem. Engineers, 1983, pp. 16–17.
19. Tarasov, V. V., and Yagodin, G. A., *Kinetika ekstraktsiii. Itogi nauki i tehniki. Neorganitcheskaya khimiya,* Tom 4. Moskva: VINITI, 1974, pp. 91–93.
20. Yagodin, G. A., Tarasov, V. V., and Fomin, A. V., "K voprosu o vliyanii promezutotechnih i pobotchnih productov geterogennoj reaktsii na skorost ekstraktsii,". *Dok. AN SSSR,* Tom 217, No. 6 (Iyun 1976), pp. 1346–1348.

21. Yagodin, G. A., et al. "Structural Mechanical Barriers in Extraction Systems. Effect on Mass Transfer and Coalescence," In *Proceedings of the Intern. Solv. Extr. Conf.* (ISEC–77), Vol. 1 (Can. Inst. Min. Met., Montreal) (1979), pp. 260–264.
22. Lewis, J. B., "The Mechanism of Mass Transfer of Solutes Across Liquid-Liquid Interfaces," *Chem. Eng. Sci.*, Vol. 3 (1954), pp. 248–259 and 260–274.
23. Bulička, J., and Prochazka, J., "Mass Transfer Between Two Turbulent Liquid Phases," *Chem. Eng. Sci.*, Vol. 31, No. 2 (Feb. 1976), pp. 137–146.
24. Bhaduri, M., et al., "A flow-Through Constant Interface Cell to Study the Kinetics of Extraction in the Cu/Lix 64N System," In *Abstracts of Intern. Solv. Extr. Conf.*, Denver, Colorado, American Inst. of Chem. Engineers, 1983, pp. 293–294.
25. Danesi, P. R., et al., "Armolles: An apparatus for Solvent Extraction Kinetic Measurements," *Sep. Sci. Technol.*, Vol. 17, No. 7 (Jul. 1982), pp. 961–968.
26. Pratt, H. R. C., "Interphase mass transfer," In *Handbook of Solvent Extraction*, Teh C. Lo, Malcolm H. I. Baird and Carl Hanson Eds. New York: A Wiley-Interscience Publ., 1982, pp. 113–115.
27. Davies, J. T., and Mayers, G. R. A., "The Effect of Interfacial Films on Mass Transfer Rates in Liquid-Liquid Extraction," *Chem. Eng. Sci.*, Vol. 16, No. 1 and 2 (Dec. 1961), pp. 55–66.
28. Elenkov, D., Boyadziev, Chr., and Boyadziev, L., "Rol poverkhnostno-aktivnih veshests v protsessah massoperedatchi," *Izv. otd. khim. nauk. BAN*, Tom 2, No. 3 (1969), pp. 535–546.
29. Elenkov, D., "Vliyanie dobavok poverkhnostno-aktivnih veschests na massoperedatchu v sistemah gaz-thidkost i zhidkost-zhidkost," *Teor. osn. khim. teknol.*, Tom 1, No. 2 (Fev. 1967), pp. 158–175.
30. Nitsch, W., and Navazio, L., "About the Influence of Soluble Adsorption Layers on the Mass Transfer Between Liquid Phases Controlled by Transport Processes," In *Proceeding of the Intern. Solv. Extr. Conf.*, (ISEC–80), Vol. 1, Rep. 80–220, University of Liege. Belgium, 1980.
31. Scholtens, B. J. R., "Copolymers at a Liquid-Liquid Interface and Their Retarding Effect on Mass Transfer between Both Phases," *Meded. Landbou whogeschool Wageningen.* Nederland, 77–7 (1977), pp. 1–124.
32. Tarasov, V. V., i dr., "Konstruktsiya datchika dlya izutcheniya kinetiki poverhnostnih reaktsij v sistemah zidkost-zidkost." *Izv. Vuzov. Khimiyai Khim. Tekhnol.*, Tom 20, No. 4 (Apr. 1977), pp. 530–532.
33. Yagodin, G. A., i dr., "Nekotorie kinetitcheskie effekti pri ekstraktsii solysnoj kisloti tretitchnimi aminami," *Dokl. AN SSSR*, Tom 249, No. 3 (Mart 1979), pp. 662–665.
34. Tarasov, V. V., i dr., "Issledovanie kinetiki ekstraktsii medi gidrooksioksimami metodom kratkovremennogo kontaktirovaniya faz," *Zh. neorg. khimii*, Tom 25, No. 2 (Fev. 1980), pp. 510–514.
35. Yagodin, G. A., and Tarasov, V. V., "Interfacial Phenomena in Liquid-Liquid Extraction," *Solv. Extr. and Ion Exchange*, Vol. 2, No. 1 (1984).
36. Tarasov, V. V., Nicolaeva, T. D., and Kruchinina, N. E., "Investigation into the Kinetics of Hydrochloric Acid Extraction by Certain Tertiary Amines," In *Proceedings of the Intern Solv. Extr. Conf. (ISEC–80)*, Vol. 3, rep. 80–141, University of Liege, Belgium, 1980.
37. Yagodin, G. D., Ivakhno, S. Yu., and Tarasov, V. V., "Adsorption Phenomena and Their Effects on the Mechanism of Cooper Extraction by Hydroxyoximes," In *Proceedings of the Intern. Solv. Extr. Conf. (ISEC–80)*, Vol. 3, Rep. 80–140, University of Liege, Belgium, 1980.
38. Chekmarev, A. M., et al., "Acid Extraction by Tertiary Amines," In *Proceedings of the Intern. Solv. Extr. Conf.* (ISEC–80), Vol. 2, Rep. 80–158, University of Liege, Belgium, 1980.
39. Tarasov, V. V., i dr. "Kinetika ekstraktsii solyanoj kisloti tri-n-detsilaminom," *Zh. fiz. khimii*, Tom 56, No. 12 (Dek. 1982), pp. 2012–3016.

40. Kizim, N. E., Tarasov, V. V., and Yagodin, G. A., "K razvitiyu metoda kratkovremennogo kontaktirovaniya faz," *Izv. vuzov. Khim. i khim. tekchnol.*, Tom 23, No. 6 (Jyun 1980), pp. 691–695.
41. Yagodin, G. A., Otchkin, A. V., and Tarasov, V. V., "Nekotorie voprosi termodinamiki i kinetiki ekstraktsii aminami," *Khim. prom.*, No. 8 (Avg. 1983), pp. 479–486.
42. Pitchugin, A. A., *Mezfaznie protsessi pri ekstraktsii medi proizvodnimi gidroksioksimov,* Avtoreferat dissertatsii kandidata khimitcheskih nauk. Moskva: MKhTI im. D. I. Mendeleeva (1983), pp. 1–20.
43. Scott, E. J., Tung, L. H., and Drickamer, H. G., "Diffusion Through an Interface," *J. Chem. Phys.*, Vol. 19, No. 9 (Sep. 1951), pp. 1075–1078.
44. Auer, P. L., and Murbach, E. W., "Diffusion across an Interface." *J. Chem. Phys.*, Vol. 22, No. 6 (Jun. 1954), pp. 1054–1059.
45. Duursma, E. K., and Hoede, C., "Theoretical, Experimental and Field Studies Concerning Molecular Diffusion of Radioisotopes in Sediments and Suspended Solid Particles of the Sea. Part A: Theories and Mathematical Calculation," *Netherlands Journal of Sea Research,* Vol. 3, No. 3 (Apr. 1967), pp. 423–457.
46. Yagodin, G. A., Tarasov, V. V., and Kizim, N. F., "The State of Substances in the Organic Phase and Stripping Microkinetics," In *Proceedings of the Intern. Solv. Extr. Conf. (ISEC-74),* Society Chemical Industry, Lyon, France, Vol. 3 (1974), pp. 2541–2548.
47. Fedorov, V. S., and Pushlenkov, M. F., "Izutchenie kinetiki reekstraktsii nitratov nekotorih metallov iz Alkilfosfornih rastvorov v vodu. II. Sistema $VO_2(NO_3)_2 \cdot 2S$ (S—DBEBF, BEDBF, TBFO)," *Radiokhimiya,* Tom 8, No. 2 (Fev. 1966), pp. 132–136.
48. Pushlenkov, M. F., and Fedorov, V. S., "Izutchenie kinetiki reekstraktsii nitrata uranila iz alkilfosfornih rastvorov v vodu. I. Sistema $UO_2(NO_3)_2 \cdot 2TBF$—razbavitel." *Radiokhimiya,* Tom 7, No. 4 (Apr. 1965), pp. 424–430.
49. Krutchinina, N. E., *Issledovanie mezfaznih protsessov v ekstraktsionnih sistemah metodami kratkovremennogo kontaktirovaniya faz.* Avtoreferat kandidata khimitheskih nauk. Moskva: MKhTI im. D. I. Mendeleeva, 1980, pp. 1–16.
50. Tarasov, V. V., Yurtov, E. V., and Yagodin, G. A., "O pritchinah razlitchiya rezimov massoobmena v sistemah zidkost-zidkost pri izmenenii napravleniya perenosa," *Zh. fiz. khimii,* Tom 52, No. 3 (Mart 1978), pp. 596–599.
51. Frank-Kamenetskij, D. A., *Diffusiya i teploperedatcha v khimitcheskoj kinetike.* Moskva: Nauka (1967), pp. 134–135.
52. Danesi, P. R., Chiarizia, R., and Muhammed, M., "Mass Transfer Rate in Liquid-Anion Exchange Processes—I. Kinetics of the Two-Phase Acid-Base Reaction in the System Trilaurylamine-Toluene-HCl-Water," *J. Inorg. Nucl. Chem.,* Vol. 40, No. 8 (Aug. 1978), pp. 1581–1589.
53. Frolov, Yu. G., Ochkin, A. V., and Sergievsky, V. V., "Theoretical Aspects of Amine Extraction," *Atomic Energy Review,* Vol. 7, No. 1 (1969), pp. 71–138.
54. Danesi, P. R., and Chiarizia, R., "The Kinetics of Metal Solvent Extraction," *Critical Reviews in Analytical Chemistry.* Vol. 10, No. 1 (Dec. 1980), pp. 1–105.
55. Augierre F. J., Klinzing, G. E., and Chiang, S. H., "Use of Image Intensification for Simultaneous Mass and Heat Transfer Studies in Liquid-Liquid Systems," In *Abstracts of Intern. Solv. Extr. Conf. (ISEC-83),* Denver, Colorado, American Inst. of Chem. Engineers, 1983, pp. 291–292.
56. Yagodin, G. A., and Tarasov, V. V., Kinetika ekstraktsionnih protsessov. V kn.: *Gidrometallurgiya. Avtoklavnoe vistchesatchivanie, sorbtsiya, ekstraktsiya,"* Moskva: Nauka, 1976, pp. 177–188.
57. Tarasov, V. V., Fomin, A. V., Yagodin, G. A., "Vliyanie mezfaznih plenok di-2-ethylgeksilfosfatov Zr na skorost rasslaivaniya emulsij v ekstrakstionnih sistemah," *Radiokhimiya,* Tom 19, No. 6 (Iuyn 1977), pp. 759–763.
58. Hughes, M. A., "On the Direct Observation of Films Formed at a Liquid-Liquid Interface During the Extraction of Metals," *Hydrometallurgy,* Vol. 3 (1978), pp. 85–90.

59. Tarasov, V. V., et al., "Interfacial Phenomena in Copper Extraction," *In Abstracts of Intern. Solv. Extr. Conf. (ISEC–83),* Denver, Colorado, American Inst. of Chem. Engineers, 1983, pp. 299–300.
60. Miyake, Y., Takenoshita, Y., and Teramoto, M., "Extraction rates of copper with SME 529. Mechanism and effects of surfactants," In *Abstracts of Intern. Solv. Extr. Conf. (ISEC–83),* Denver, Colorado, American Inst. of Chem. Engineers, 1983, pp. 301–302.
61. Osseo–Asare, K., and Keeney, M. E., "Sulfonic acids: catalysts for the liquid-liquid extraction of metals," *Sep. Sci. Technol.,* Vol. 15 No. 4 (1980), pp. 999–1011.

CHAPTER 17

LIQUID-LIQUID EXTRACTION IN SUSPENDED SLUGS

M. H. I. Baird

Chemical Engineering Department
McMaster University
Hamilton, Ontario, Canada

CONTENTS

INTRODUCTION

There is a need for laboratory extraction apparatus in which the rate of extraction can be predicted from basic hydrodynamic and diffusional principles. In most types of extraction equipment involving droplet dispersions, such prediction is difficult because of uncertainties about the droplet behavior and the interfacial area of the droplet dispersion [1]. Even in the case of the gently agitated Lewis cell [2] in which the liquid-liquid interface is flat, the mass transfer coefficients are obtainable only from empirical equations [2, 3]. The ability to predict the extraction rate from first principles is particularly useful when a chemical reaction is involved as in the case of metal ion extraction. In such cases it is important to determine whether the ratecontrolling process is merely diffusional, or dependent on a chemical reaction.

There are relatively few published studies of liquid-liquid mass transfer under hydrodynamically predictable conditions. These have all involved laminar flow in various configurations; jets [4, 5], wetted spheres [6], and wetted wall columns [7].

This short chapter describes a recent addition to these techniques, whereby a cylindrical droplet or "slug" of the lighter liquid phase is maintained stationary in a downflow of the denser liquid phase in a vertical cylindrical tube. Ordinarily the tube can be made of glass, and the denser liquid is an aqueous phase which wets the glass in preference to the light (organic) phase. This apparatus is experimentally very simple and has the added advantage that it can be used to obtain measurements of the effective interfacial tension, as will also be discussed later.

FLOW REGIMES

The rise of a slug of light liquid in a vertical tube, relative to a continuous dense liquid phase is analogous to the rise of a cylindrical gas bubble. Considerable previous work has been done in the flow regimes of cylindrical gas bubbles, and the results [8] are summarized in Figure 1. Under inertially controlled flow conditions, the dimensionless velocity of rise ($u/(gd)^{1/2}$) has the value 0.345; but as the diameter of the containing tube is reduced, the velocity of rise is limited by surface tension rather than by inertia. A bubble will cease to rise at all if the Eotvos number (ratio of buoyancy to capillary forces) is less than about 3.

Recent work with liquid-liquid systems [9, 10] has confirmed that the behavior of slugs in vertical tubes is analogous to that shown for gas bubbles in Figure 1. The relevant dimensionless groups for liquid-liquid systems are:

Froude number, $Fr = u^2(g\Delta\varrho d/\varrho_h)^{-1}$
Eotvos number, $Eo = g\Delta\varrho d^2/\sigma$
Property group, $Y = g\Delta\varrho\, \mu_h^4\, (\varrho_h^2\, \sigma^3)^{-1}$

Thus with the selection of a tube diameter such that the Eotvos number is between 4 and 17, a slug of the lighter liquid will rise at a velocity which is limited primarily by the interfacial tension of the system. Under these conditions, Baird and Ho[9] found that the slug had a smooth and laminar appearance. For larger tube diameters the velocity of the rising liquid slug was so great that the rear surface tended to ripple and become unstable causing partial breakup of the slug into small drops.

Experimentally it is more convenient to maintain a slug in a stationary position, than to permit it to rise through a tube. It was found [10] that the flow velocity of the dense liquid required to hold the slug stationary was expressed by:

$$\sqrt{Fr} = 0.163 \ln(Eo) - 0.222 \tag{1}$$

for $4 < Eo < 17$ and $Y \leq 10^{-8}$

At higher viscosities and Y values, the slope of $\sqrt{Fr}$ versus $\ln(Eo)$ was shallower as expected from Figure 1, but Equation 1 is valid for systems with viscosities in the same order of magnitude as water.

VELOCITY PROFILES

For a slug of large length/diameter ratio, the mass transfer rate depends primarily on the velocity profile in the cylindrical section. Fully developed, steady laminar flow is assumed in both phases, and the Navier–Stokes equation for either phase is written [9]

$$\frac{dP}{dz} - \varrho g = (\mu/r)\frac{d}{dr}\left(r\frac{du}{dr}\right) \tag{2}$$

The boundary conditions are:

- At tube wall, $r = R$ and $u = 0$.
- At the interface, $r = R_i$ and

$$\mu_h\left(\frac{du}{dr}\right)_h = \mu_\ell\left(\frac{du}{dr}\right)$$

- At the centerline of the tube, $r = 0$ and $du/dr = 0$.

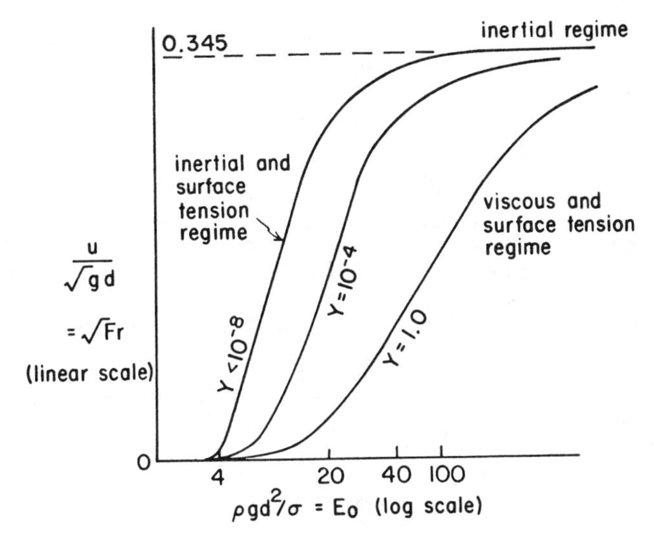

Figure 1. Flow regimes forcylindrical gas bubles in vertical tubes [10] (redrawn from White and Beardmore [8]).

The radius of the slug is not easily measurable, but it can be calculated analytically from Equation 2 with the boundary conditions just shown, given the measured value of the downflow rate of the denser liquid. From the value of R_i and Equation 2, the complete velocity profile in both phases can then be calculated. Details are given by Baird and Ho [9].

MASS TRANSFER

The molecular diffusivities of transferring solutes in liquids are in the order of 10^{-9} m^2 s^{-1}. Under these conditions, diffusional mass transfer at an interface involves a very short range penetration, and the Higbie penetration theory [11] is applicable whenever a "surface contact time" τ can be estimated. Unsteady-state diffusion theory gives the mass transfer coefficient in each phase:

$$k = 2[D/(\pi\tau)]^{1/2} \tag{3}$$

For a slug of length L_S and radius R_i, the surface area can be calculated assuming cylindrical shape; the mass transfer coefficient in each phase is obtained from Equation 3 taking τ to be L_S/u_i with the interfacial velocity u_i obtained from the calculated velocity profile. Hence,

$$k = 2[Du_i/(\pi L_S)]^{1/2} \tag{4}$$

This simplified model was tested for the mass transfer of sparingly soluble organic liquids (e.g. n-butanol) to water, using the simple apparatus shown in Figure 2. The slug consisted of the organic component saturated with water, and pure water was passed downwards. Thus the rate of dissolution of the organic phase in water was controlled by diffusion in the aqueous phase. The rate was measured experimentally by simply monitoring the length of the slug as a function of time.

The penetration mass transfer model predicts that $\sqrt{L_S}$ should decrease linearly with respect to time. Linearity was found in the shrinkage plot over short times, but the slugs were generally found to reach a limiting length L′ below which shrinkage was very slow. This apparently stagnant length was connected with interfacial contamination by trace surfactants; such effects have been observed for gas absorption in jets [12].

The length end-effect could be estimated from the data, and a plot of $(L_S - L')^{1/2}$ versus time was found to be linear over a wide range, as illustrated in Figure 3. The slopes of the plots agreed approximately with the predictions from the penetration mass transfer model.

The slug extraction technique has been extended to ternary systems [13] using the apparatus shown in Figure 4. In this case the organic phase (solvent + transferring solute) is introduced at an organic-wetted disc support and held in place by a downflow of the aqueous phase. The disc support allowed the slug to remain in place while the dense liquid flow was varied.

The principal system studied was iodine(n-heptane)-aqueous sodium thiosulphate. The fast reaction of the iodine in the aqueous phase ensured that the mass transfer was controlled by diffusion in the organic phase. Limitations were found in the usefulness of this technique because of a very strong stagnant end-effect (about 30 mm length), and the tendency of the slugs to become unstable at lengths exceeding 120 mm. However, reasonable agreement with the penetration theory was obtained for slug lengths of 90 and 120 mm.

MEASUREMENT OF INTERFACIAL TENSION

The slug may be held in place by a disc support (Figure 4), by manual adjustment of the downflowing liquid (e.g., by valve A in Figure 1), or by an automatic flow adjustment. The hydrostatic head of the dense and light phases in the vertical tube can itself be used to control the dense-phase flow, using the membrane flow controller illustrated in Figure 5 [10].

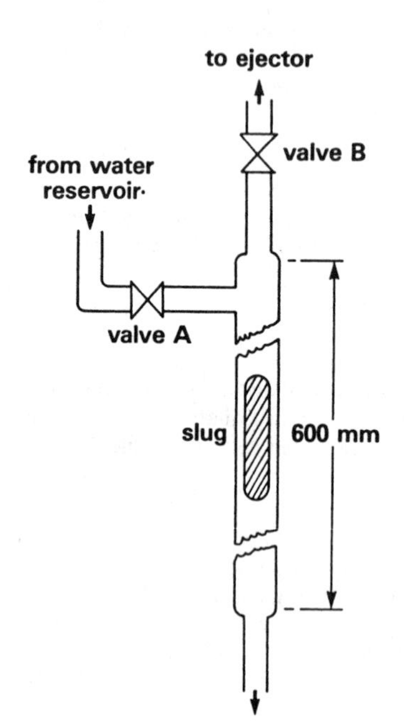

Figure 2. Simple apparatus for stationary slugs [9]. The ejector connection is used to remove trapped air bubbles if necessary.

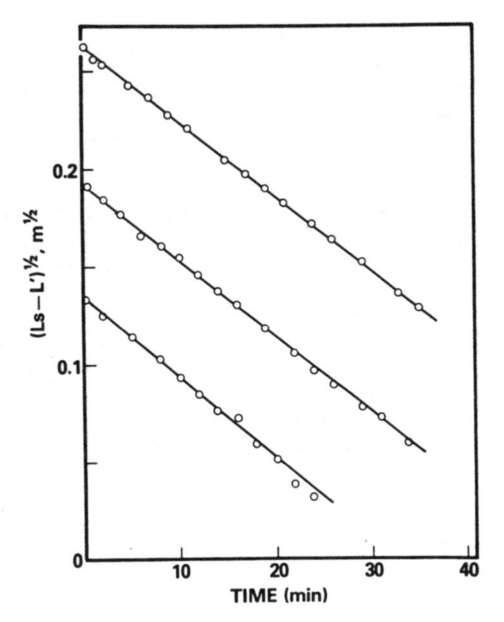

Figure 3. Shrinkage plots for n-pentanol slugs in water in a 4.08 mm diameter tube [9]. The stagnant layer length L'' is 7 mm in this case.

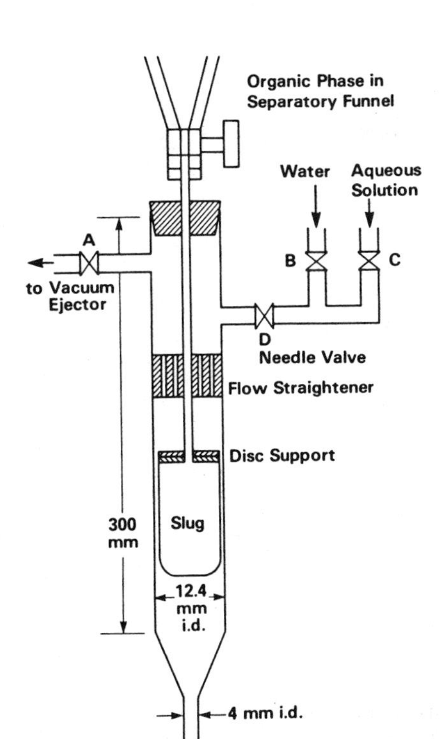

Figure 4. Supported slug apparatus [13]. The disc support consists of a stainles steel upper disc, wet by the aqueous phase, and a Teflon lower disc, wet by the organic phase.

This device has been used to measure the surface tension of gas-liquid systems and the interfacial tension of liquid-liquid systems. Equation 1 relates the measured dense-phase velocity to the interfacial tension of the system, provided $4 < \text{Eö} < 17$ and $Y \leq 10^{-8}$, and it has been confirmed for 22 air-liquid systems and 9 liquid-liquid systems.

In a recent modification of this technique [4], a reservoir of the dense liquid is allowed to drain slowly through a flow restriction into a vertical tube, flared at the bottom, and dipping into the light liquid. As the reservoir level falls and the velocity of the dense liquid in the tube decreases, the point is reached (Equation 1) at which the dense phase begins to rise as a slug. The flow configuration between the reservoir and the tube contains a siphon arrangement, so that flow is terminated soon after the slug begins to rise. It has been found that the reservoir level at which the flow terminates can be calibrated against the interfacial tension of the system used.

It should be emphasized that these slug methods provide *dynamic* interfacial tensions which are lower than the static values when highly surface-active solutes are present. However in the study of liquid-liquid mass transfer systems it is an advantage to know the dynamic interfacial tension, because this is the property that has a strong effect on drop sizes and coalescence phenomena in flowing and agitated systems.

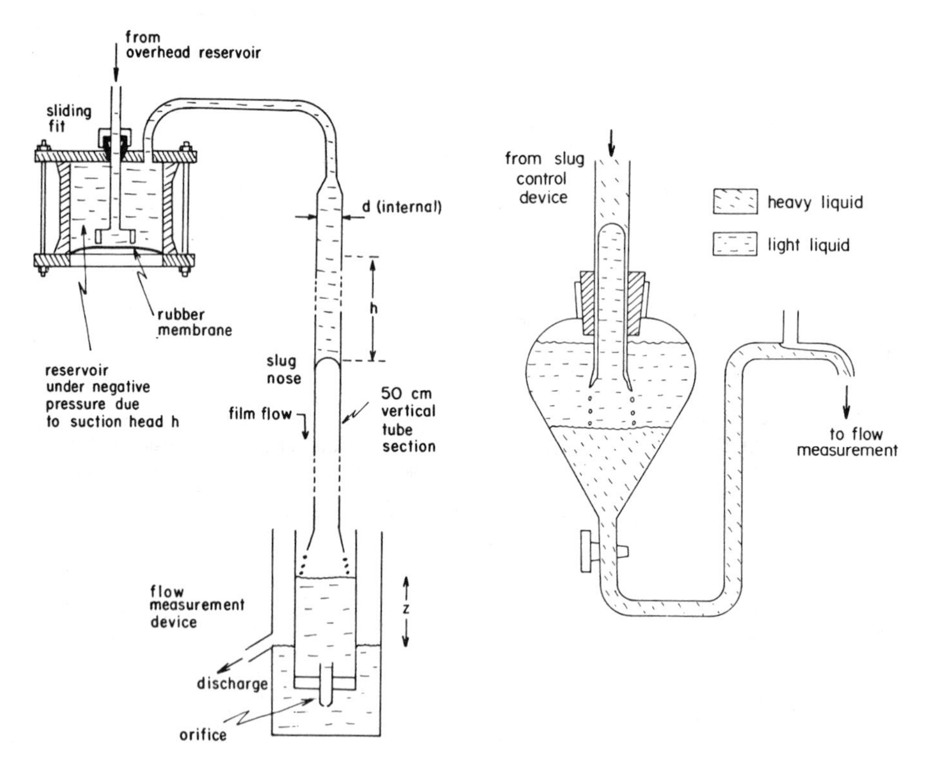

Figure 5. Apparatus to stabilize slugs for continuous mesurement of interfacial tension [10]. The arrangement for gas-liquid systems is shown on the left, with the modification for liquid-liquid systems on the right.

DEVELOPMENT AND APPLICATION

Liquid-liquid slugs are easily formed in inexpensive laboratory glassware and may be used as a tool in evaluating the kinetics and surface behavior of new extraction systems. As noted above, the tube diameter should be selected to give Eotvos numbers in the range 4–17. They accommodate a wide range of systems. It is recommended that an assembly of tubes of internal diameter between about 4 and 15 mm should be available. The greater the interfacial tension, the greater should be the tube diameter. It may be advantageous to make a single tube in sections having different diameters, with the largest diameter at the bottom.

A difficulty in mass transfer work is the analysis of the organic (slug) phase, which can ordinarily be only accomplished after removing the slug from the apparatus. Continuous in situ analysis of the slug may be possible by colorimetric or radiometric means in special cases. Analysis of the flowing dense phase is much easier because it is continuously leaving the apparatus; often an accurate analysis of the dense phase will permit estimation of the slug phase composition by material balance.

A potential new application of liquid slugs is the measurement of the effective interfacial tension during a mass transfer process, i.e., under nonequilibrium conditions.

NOTATION

d	tube diameter	L_S	length of slug
D	molecular diffusivity	L'	length of end-effect
Eo	Eotvos number	P	pressure
Fr	Froude number	r	radial distance
g	acceleration of gravity	u	velocity
k	mass transfer coefficient	z	vertical distance

Greek Symbols

μ	viscosity	σ	interfacial tension
$\Delta\varrho$	density difference $= \varrho_h - \varrho_\ell$	τ	interfacial contact time
ϱ	density		

Subscripts

h	heavy liquid	ℓ	light liquid
i	at interface		

REFERENCES

1. Pratt, H. R. C., "Interfacial Mass Transfer," in *Handbook of Solvent Extraction,* Lo, T. C., Baird, M. H. I., and Hanson, C., (Eds.), New York: Wiley Interscience, 1983, pp. 91–123.
2. Lewis, J. B., "The Mechanism of Mass Transfer of Solutes across Liquid-liquid Interfaces," *Chem. Eng. Sci.,* Vol. 3 (1954), pp. 248–277.
3. Bulicka, J., and Prochazka, J., "Mass Transfer Between Two Turbulent Liquid Phases," *Chem. Eng. Sci.,* Vol. 31 (1976), pp. 137–146.
4. Ward, W. J., and Quinn, J. A., "Diffusion through the Liquid-liquid Interface: Part II. Interfacial Resistance in Three-Component Systems," *A. I. Ch. E. Journal,* Vol. 11 (1965), pp. 1005–1011.

5. Freeman, R. W., and Tavlarides, L. L., "Study of Interfacial Kinetics for Liquid-liquid Systems–I. The Liquid Jet Recycle Reactor," *Chem. Eng. Sci.*, Vol. 35 (1980), pp. 559–566.
6. Ratcliff, G. A., and Reid, K. J., "Mass Transfer into Spherical Liquid Films. Part I–Measurement of Diffusivities of Liquids," *Trans. Instn. Chem. Engrs.* (U. K.), Vol. 39 (1961), pp. 423.
7. Bakker, C. A. P., Van Vlissingen, F. H. F., and Beek, W. J., "The Influence of the Driving Force in Liquid-liquid Extraction—a Study of Mass Transfer With and Without Interfacial Turbulence under Well-defined Conditions," *Chem. Eng. Sci.*, Vol. 22 (1967), pp. 1349–1355.
8. White, E. T., and Beardmore, R. H., "The Velocity of Rise of Single Cylindrical Air Bubbles through Liquids Contained in Vertical Tubes," *Chem. Eng. Sci.*, Vol. 17 (1962), pp. 351–361.
9. Baird, M. H. I., and Ho, M. K., "Liquid-liquid Extraction in Laminar Slug Flow," *Can. J. Chem. Eng.*, Vol. 57 (1979), pp. 467–475.
10. Rao, N. V. R., and Baird, M. H. I., "Continuous Measurement of Surface and Interfacial Tension by Stationary Slug Method," *Can. J. Chem. Eng.*, Vol. 61 (1983), pp. 581–589.
11. Higbie, R., "The Rate of Absorption of a Pure Gas into a Still Liquid during Short Periods of Exposure," *Trans. A. I. Ch. E.*, Vol. 31 (1935), pp. 365–389.
12. Cullen, E. J., and Davidson, J. F., "Absorption of Gases in Liquid Jets," *Trans. Faraday Soc.*, Vol. 53 (1957), pp. 113–120.
13. Baird, M. H. I., and Krovvidi, K. R., "Liquid-liquid Extraction by a Suspended Slug," *Can. J. Chem. Eng.*, Vol. 60 (1982), pp. 569–573.
14. Rao, N. V. R., and Baird, M. H. I., "A Simple Technique for the Measurement of Surface and Interfacial Tension," *J. Phys. E. Sci. Instrum.*, Vol. 16 (1983), pp. 1164–1166.

SECTION III

MULTIPHASE REACTOR SYSTEMS

CONTENTS

CHAPTER 18

REACTION AND MASS TRANSPORT IN TWO-PHASE REACTORS

William J. Hatcher, Jr.

Department of Chemical Engineering
The University of Alabama
University, Alabama, USA

CONTENTS

INTRODUCTION

Reactions involving more than a single phase require that the movement of material from phase to phase be considered in the rate equation. The actual site of reaction may be in one phase; however, if one of the reactants is supplied to the reacting phase from another distinct phase, then mass transfer must be incorporated. Two-phase reacting systems in general may have the reaction occurring in either of the bulk phases. The reaction may also occur at the boundary, or interface, between the phases.

Reactions involving more than one phase are classified as heterogeneous, and they may be catalytic or noncatalytic in nature. Catalytic heterogeneous systems involve at least one phase which is not a reactant but its presence increases the rate of reaction and possibly changes the path of reaction. The catalyst may be in the same phase as one of the reactants; for example, the catalyst and one of the reactants could be in a liquid phase and another reactant could be in the gas phase. In such a case the reacting system would be heterogeneous while the catalysis would be considered homogeneous. Heterogeneous catalytic reactions typically involve a solid catalyst and a fluid phase (gas or liquid) containing the reactants. The reactants migrate to the catalyst surface where the reaction takes place.

Two-phase reactions of the noncatalytic type may involve various combinations of phases:

- Gas-liquid
- Gas-solid
- Liquid-liquid
- Liquid-solid
- Solid-solid

There are a number of industrially important gas-liquid reactions. Typically, gas containing one reactant is contacted with liquid containing a second reactant. The first reactant diffuses to the gas-liquid interface, is absorbed, and then reacts with the second reactant in the liquid phase.

Gas-solid noncatalytic reactions occur at the interface of the phases. In the case of porous solids this may be throughout the solid. In liquid-liquid reactions, there are two liquid phases. A reactant in one liquid phase diffuses to the interface, dissolves in the second phase, and then reacts with another reactant in that phase. The steps involved in liquid-solid reactions are similar to gas-solid reactions. Solid-solid reactions also involve mass transfer from one solid to another followed by chemical reaction.

Because these systems involve two phases the reaction rate will depend upon the degree of contact between the phases as well as the factors which govern homogeneous reaction (temperature, pressure, concentration). Each of the many methods of contacting two phases has a specific overall reaction equation which must be developed for that particular system. The area of contact between the phases is a very important reaction variable. Diffusional steps must be accounted for since one or more reactants must be transported from their original phase to another where the reaction actually occurs. Therefore, factors governing mass transfer become important determinants of the reaction rate.

In this chapter, the effects of transport phenomena upon heterogeneous reaction rates will be described. A general discussion of rate alteration due to diffusion into the external field and then diffusion into the zone of reaction is made. This is followed by a more detailed treatment of some specific two-phase systems.

INTERFACE DIFFUSION AND REACTION

Isothermal

In several cases material transport acts in series with chemical reaction. For example, a flowing fluid containing a reactant is exposed to a fixed nonporous surface which is capable of converting the reactant to products. Under isothermal steady-state conditions, the rate of mass transfer to the surface must equal the rate of reaction. Otherwise, the concentration of reactant at the surface would vary with time, and steady state would be violated. This relationship is:

$$k_g a (C_0 - C_s) = r = kC_s^n \tag{1}$$

when the surface reaction is of intrinsic order n.

Noting that the concentration on the surface C_s cannot be easily measured and considering a first-order reaction, n = 1, we obtain:

$$C_s = \frac{k_g a}{k_g a + k} C_0 = \frac{C_0}{1 + Da} \tag{2}$$

Therefore the global or effective rate r expressed in terms of the bulk fluid concentration is:

$$r = \frac{C_0}{1/k_g a + 1/k} = \frac{kC_0}{1 + Da} \tag{3}$$

This result shows that $1/k_g a$ and $1/k$ are additive resistances and that Da is the ratio of the mass transfer resistance to the intrinsic surface reaction resistance. Thus when diffusion is rapid relative to reaction, the mass transfer resistance is very small compared to the surface reaction resistance and $Da \rightarrow 0$ and $r = kC_0$. On the other hand, when surface reaction is rapid relative to diffusional transport, the surface reaction resistance is very small compared to mass transfer resistance. Da is very large and at the limit $r = k_g a C_0$.

The interphase or external effectiveness is defined by the ratio of the global rate r to the pure chemical reaction control rate r_0, For the first order reaction case:

$$\bar{\eta} = r/r_0 = 1/(1 + Da) \tag{4}$$

For other reaction orders:

$$Da = kC_0^{n-1}/k_g a$$

and Figure 1 illustrates the effects of reaction order on external effectiveness. Note that regardless of the order of the intrinsic surface reaction when Da is very large, the global rate becomes first order.

Often the global rate is measured in terms of the bulk fluid concentration and the experimental rate coefficient ($\bar{k}$) is calculated:

$$r = \bar{\eta} k C_0 = \bar{k} f(C_0) \tag{5}$$

Then Da cannot be calculated because the intrinsic value of k is unknown. However,

$$\frac{\bar{k}}{k_g a} = \bar{\eta} \frac{k}{k_g a} = \bar{\eta}\, Da = \frac{r}{k_g a C_0} \tag{6}$$

Therefore $\bar{\eta}Da$ product can be calculated from experimentally observable parameters. The relationship between external effectiveness and observable parameters is shown in Figure 2.

Nonisothermal

When mass transfer limitations become important in two-phase reactions, heat transfer limitations also can be anticipated. This means, of course, that the temperatures of the two phases can be different. Since the reaction rate coefficient is a function of temperature k(T),

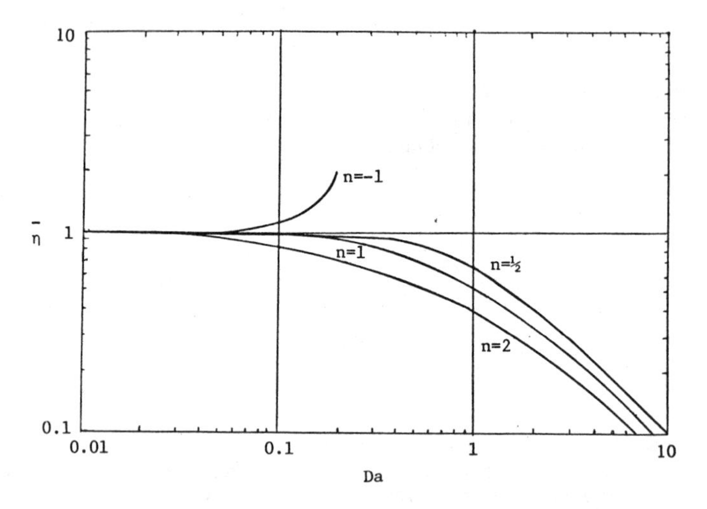

Figure 1. External effectiveness factor. Isothermal case for reaction order n. (G. Cassiere and J. J. Carberry, *Chem. Eng. Educ.*, Winter, (1973) p. 22.)

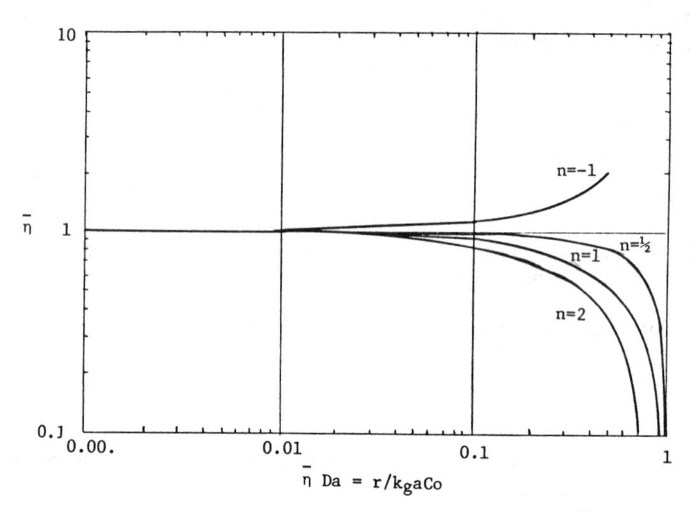

Figure 2. External effectiveness factor versus observables. Isothermal case for reaction order n. (G. Cassiere and J. J. Carberry, *Chem. Eng. Educ.*, Winter, (1973) p. 22.)

then the external effectiveness for the nonisothermal case is a function of the differences of both the concentrations and the temperatures at the surface and in the bulk fluid

$$\bar{\eta} = \frac{k(T_s)C_s^n}{k(T_0)C_0^n} \tag{7}$$

The ratio of the reaction rate coefficients at the surface and bulk temperatures can be written using the Arrhenius relationship as

$$\frac{k(T_s)}{k(T_0)} = \exp\left[-\frac{E}{R}\left(\frac{1}{T_s} - \frac{1}{T_0}\right)\right] \tag{8}$$

The relationship between the surface and bulk temperatures can be derived from a local energy balance. At steady state, the heat generated by surface reaction must be equal to the heat transferred from the surface to the bulk fluid by convection

$$(-\Delta H_R)r = ha(T_s - T_0) \tag{9}$$

Using the definitions of the j factors for mass and heat transfer

$$j_D = \frac{k_g}{u} S_c^{2/3} \tag{10}$$

and

$$j_H = \frac{h}{\varrho u C_p} P_r^{2/3} \tag{11}$$

and making the mass transport—heat transport analogy ($j_D = j_H$), the convection coefficient h may be written

$$h = k_g \varrho C_P (Sc/Pr)^{2/3} \tag{12}$$

substituting Equation 12 in Equation 9 and solving for T_s yields

$$T_s = T_0 + \frac{(-\Delta H)C_0}{\varrho C_P (Sc/Pr)^{2/3}} \frac{r}{k_g a C_0} = T_0 + \alpha \bar{\eta} Da \tag{13}$$

where

$$\alpha = \frac{(-\Delta H)C_0}{\varrho C_P (Sc/Pr)^{2/3}} \tag{14}$$

From Equations 2, 3, and 6

$$\frac{C_s}{C_0} = 1 - \bar{\eta} Da \tag{15}$$

Defining dimensionless groups

$$\beta \equiv \alpha / T_0 \tag{16}$$

and

$$\varepsilon_0 \equiv E/RT_0 \tag{17}$$

then substituting Equations 8, 13, and 15 into Equation 7 yields

$$\bar{\eta} = (1 - \bar{\eta} Da)^{\eta} \exp\left[-\varepsilon_0\left(\frac{1}{1 + \bar{\eta} Da \beta} - 1\right)\right] \tag{18}$$

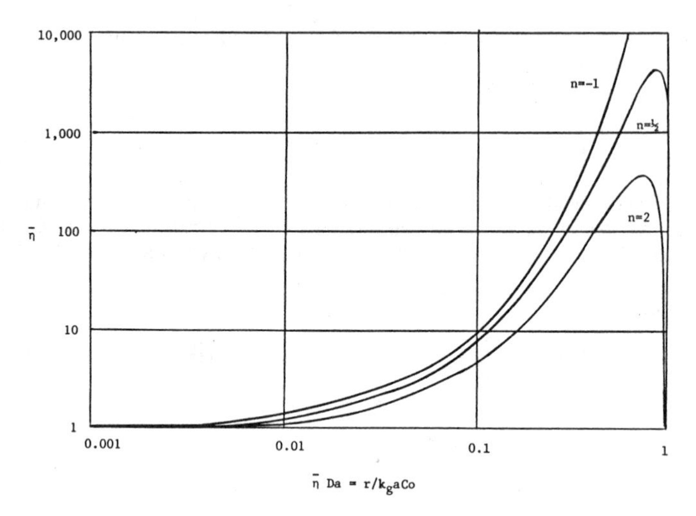

Figure 3. External effectiveness factor versus observables. Nonisothermal case for reaction order n, ϵ_o = 20, β = 1. (J. J. Carberry and A. A. Kulkarni, *J. Catal., 31:* 41 (1973).)

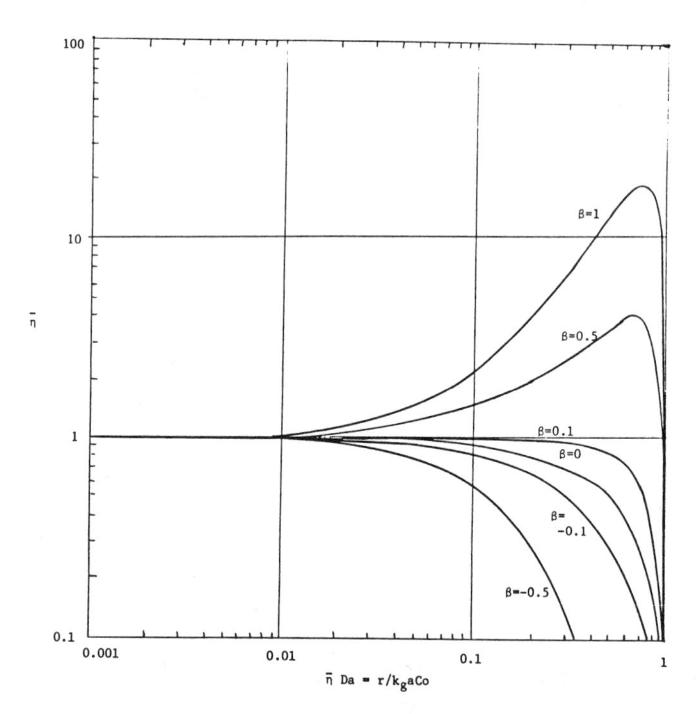

Figure 4. External effectiveness factor versus observables. Nonisothermal case for first order, ϵ_o = 10. (J. J. Carberry and A. A. Kulkarni, *J. Catal.,* 31: 41 (1973).)

This relationship is illustrated for several values of n, ε_0, and β in Figures 3 through 5.

INTRAPHASE DIFFUSION AND REACTION

Isothermal

When mass transport processes occur in the same phase as reaction, the reaction occurs simultaneously with diffusion. Because of the combined series-parallel nature of such processes, mathematical models assume greater complexity than the previously discussed processes in series. An example is diffusion and simultaneous reaction with a porous solid catalyst.

For a flat-plate geometry with half thickness L and first-order reaction, the steady-state component balance is

$$D_e \frac{d^2C}{dZ^2} = kC \tag{19}$$

The boundary conditions are

$$\begin{aligned} &Z = 0, \text{ centerline symmetry} \quad && dC/dZ = 0 \\ &Z = L, \text{ surface concentration} \quad && C = C_0 \end{aligned} \tag{20}$$

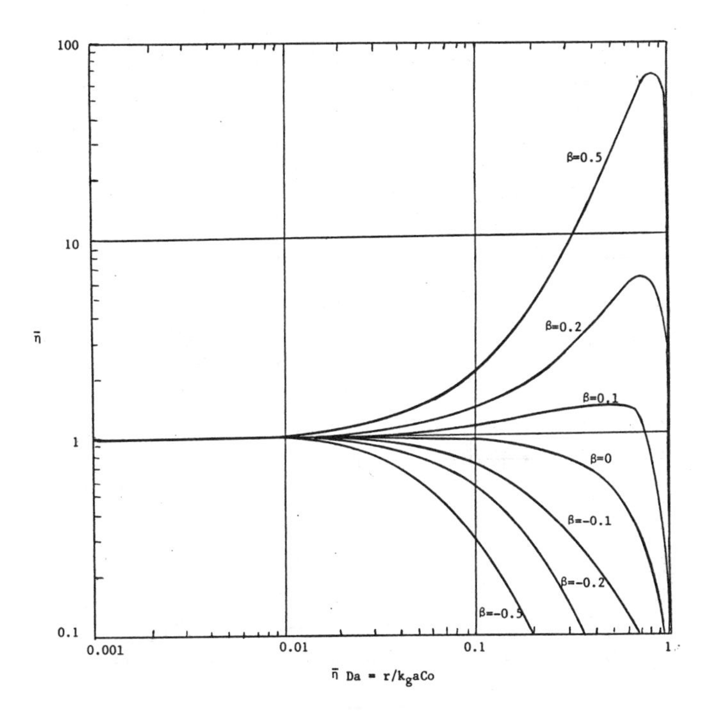

Figure 5. External effectiveness factor versus observables. Nonisothermal case for first order, ϵ_0 = 20. (J. J. Carberry and A. A. Kulkarni, *J. Catal.*, 31: 41 (1973).)

and the solution of Equation 19 is

$$\frac{C}{C_0} = \frac{\cosh \sqrt{(k/D_e Z)}}{\cosh \sqrt{(k/D_e L)}} \tag{21}$$

The intraphase effectiveness is

$$\eta \equiv \frac{\text{reaction rate with pore diffusion}}{\text{reaction rate at surface conditions}}$$

$$= \frac{\dfrac{1}{V_P}\displaystyle\int r\, dV_P}{kC_0} \tag{22}$$

When Equation 21 is substituted into Equation 22 and the integration performed, the result is

$$\eta = \tanh \varphi/\varphi \tag{23}$$

where φ, the Thiele modulus, $= L\sqrt{k/De}$

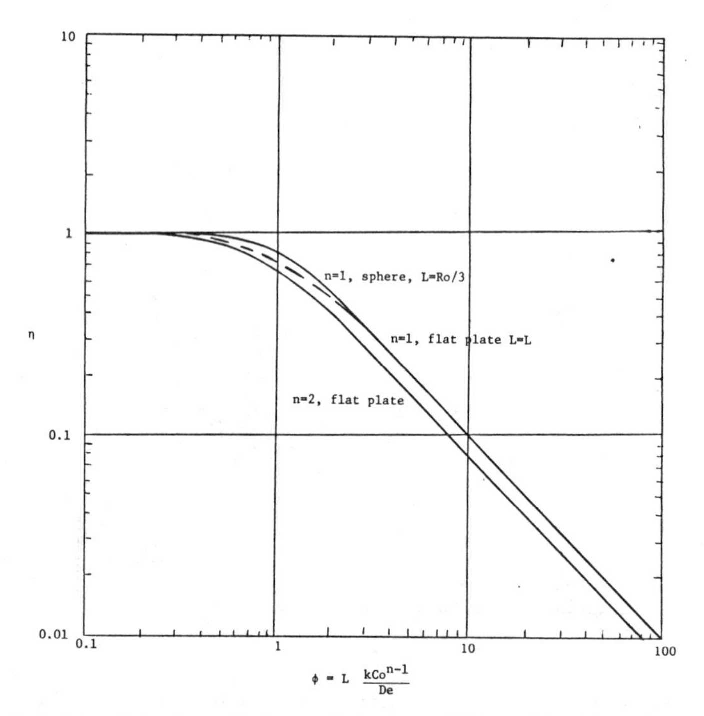

Figure 6. Isothermal intraphase effectiveness factor versus Thiele modulus [1].

For a spherical catalyst pellet of radius R_0, the steady-state component balance is

$$D_e \frac{1}{R^2} \frac{d}{dR}\left(R^2 \frac{dC}{dR}\right) = kC \tag{24}$$

The solution yields

$$\eta = \frac{1}{\varphi} \frac{(3\varphi)\coth(3\varphi) - 1}{3\varphi} \tag{25}$$

where

$$\varphi = \frac{R_0}{3}\sqrt{k/D_e} \tag{26}$$

Figure 6 compares intraphase effectiveness variation with reaction order and geometry.

Nonisothermal

In our example of simultaneous reaction and diffusion in a porous solid catalyst, another complication is the fact that thermal conductivity limitations may cause temperature gradients within the solid. Then the reaction takes place within a field of temperature gradients as well as concentration gradients. To analyze this case, both energy and mass balances must be solved. For a flat-plate of catalyst and first order kinetics the energy balance is

$$-\frac{d}{dZ}\left(\lambda_e \frac{dT}{dZ}\right) = (-\Delta H_R)kC \tag{27}$$

where λ_e is the effective thermal conductivity of the porous solid. The mass balance is Equation 19. Equations 19 and 27 can be combined to eliminate the kC (reaction rate) term. Then two integrations gives the maximum temperature difference in a catalyst particle with complete reaction.

$$\Delta T_{max} = (-\Delta H_R)D_e C_0/\lambda_e \tag{28}$$

This result is true for any catalyst geometry.

Equations 19 and 27 can be solved simultaneously by numerical methods to give the nonisothermal intraparticle effectiveness as a function of three dimensionless parameters, the Thiele modulus at surface conditions φ_0, the Arrhenius number $E = \varepsilon_0/RT_0$, and the dimensionless adiabatic temperature rise $\beta = \Delta T_{max}/T_0$. Results are illustrated in Figure 7.

COMBINED INTERPHASE-INTRAPHASE EFFECTIVENESS

Isothermal

An overall isothermal effectiveness for first-order kinetics within a flat-plate geometry can be derived from a steady-state mass balance. The mass flux to the two-phase interface is set equal to the diffusion rate into the second phase at the interface:

$$k_g(C_0 - C_i) = -D_e \left.\frac{dC}{dZ}\right|_{interface} \tag{29}$$

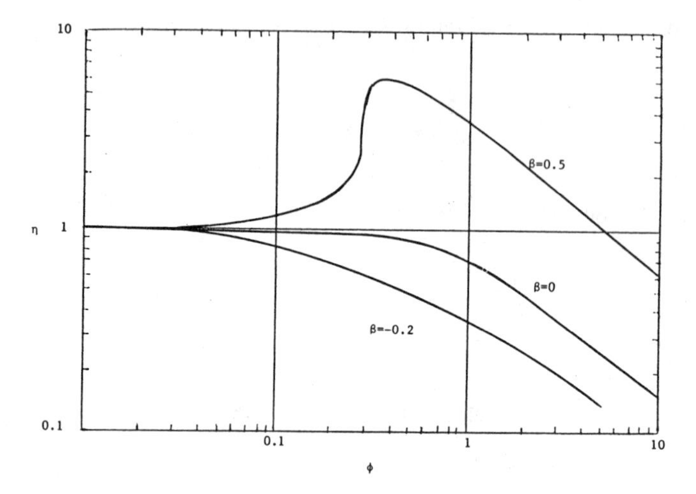

Figure 7. Nonisothermal intraphase effectiveness factor versus Thiele modulus, $\epsilon = 20$ [1].

The overall effectiveness is

$$\eta_0 = \frac{\tanh \varphi}{\varphi[1 + (\varphi \tanh \varphi)/Bi_m]} \tag{30}$$

where Bi_m is the mass Biot number $= k_g L/D_e$. In order to calculate the overall effectiveness using Equation 30 the value of the intrinsic reaction rate coefficient k must be known. The following treatment develops an observable modulus. The observed effective reactions rate is

$$r = \eta_0 k C_0 = \frac{\tanh \varphi \; k C_0}{\varphi[1 + (\varphi \tanh \varphi)/Bi_m]} \tag{31}$$

If the rate is multiplied by L^2/D_eC_0

$$\frac{L^2 r}{D_e C_0} = \frac{\varphi \tanh \varphi}{[1 + (\varphi \tanh \varphi)/Bi_m]} = \eta_0 \varphi^2 \tag{32}$$

The overall effectiveness versus $\eta_0\varphi^2$ modulus is shown in Figure 8 for various Bi_m values.

Three regimes of control of the global reaction rate are defined. If $\varphi \ll 1$, the intrinsic chemical reaction rate is slow compared to the mass transport processes and chemical reaction controls. If $\varphi > 3$ and $\varphi/Bi_m \ll 1$, the intraphase diffusion effect is strong. If $\varphi > 3$ and $\varphi/Bi_m > 1$, interphase diffusion is slow compared to the other processes and it controls the global rate. Table 1 summarizes these regimes.

Nonisothermal

The overall effectiveness for the nonisothermal case can be derived from the steady-state energy and mass balances with, for example, Equation 29 as the appropriate boundary condition for the mass balance at the two-phase boundary. The boundary condition for the energy balance is

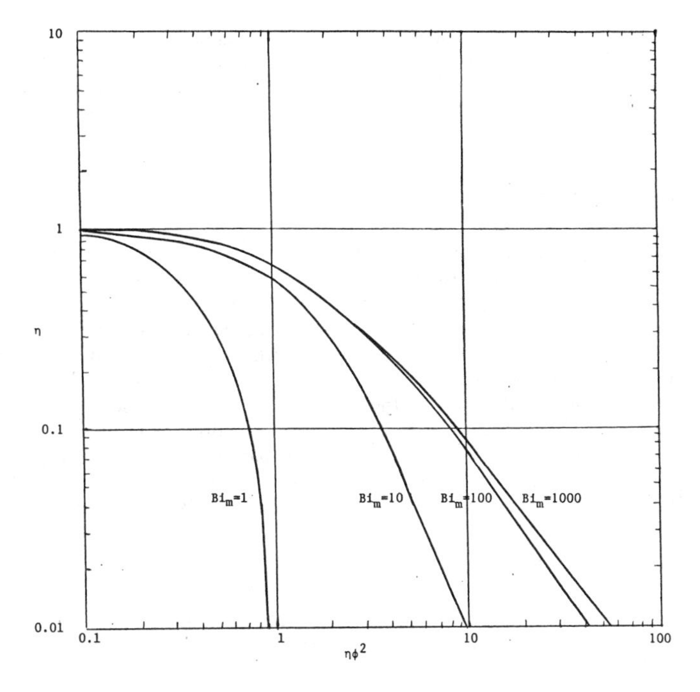

Figure 8. Interphase-intraphase combined effectiveness factor for first order [1].

Table 1
Characteristics of Isothermal Interphase-Intraphase Diffusion and Reaction

Rate-Controlling Step	Experimentally Observed Reaction Order	Experimentally Observed Activation Energy	Effect of Particle Size	Effect of Velocity
Chemical reaction	True order, n	True, E	None	None
Intraphase diffusion-reaction	(n+1)/2	E/2	1/L	None
Interphase diffusion	1	≈ 0	$\left(\frac{1}{L}\right)^x$	u^y

Exponents x and y depend on type of reactor and geometry.

$$h(T_i - T_0) = -\lambda_e \left.\frac{dT}{dZ}\right|_{\text{interface}} \tag{33}$$

The overall effectiveness is dependent on

- The Thiele modulus
- The Arrhenius number
- The dimensionless adiabatic temperature rise
- The mass Biot number
- The thermal Biot number ($Bi_m = hL/\lambda_e$)

Figure 9 illustrates effects of parameters on the overall effectiveness.

For large values of Biot numbers and small values of the Arrhenius number and the adiabatic temperature rise, temperature effects and interphase effects are minor and Figure 6 may be used. For large values of Biot numbers, interphase effects are minor and Figure 7 is appropriate.

GAS-LIQUID REACTIONS

Gas-liquid reactions generally occur by absorption of a gas in solution or by gas absorption followed by reaction with the solvent itself. Such reactions are widely used in the chemical process and petroleum refining industries. The reaction may have as its object the removal of one component from a gas stream. Examples are processes for scrubbing to remove carbon dioxide, carbon monoxide, hydrogen sulfide, sulfur dioxide, chlorine, nitrogen oxides, hydrocarbons, etc.

The object of the reaction also may be the preparation of products such as sulfuric acid, nitric acid, barium carbonate, and barium chloride by liquid-phase processes such as sulfation, nitration, halogenation, sulfonation, hydrogenation, alkylation, and polymerization. Gas-liquid reactions are also important in biological processes such as the production of proteins from hydrocarbons, biological oxidation, and aerobic fermentation.

Gas-Liquid Reaction Models

Several models have been used to describe mass transfer from a gas phase into a liquid phase. The most widely used is the two-film theory. This model pictures a stagnant film

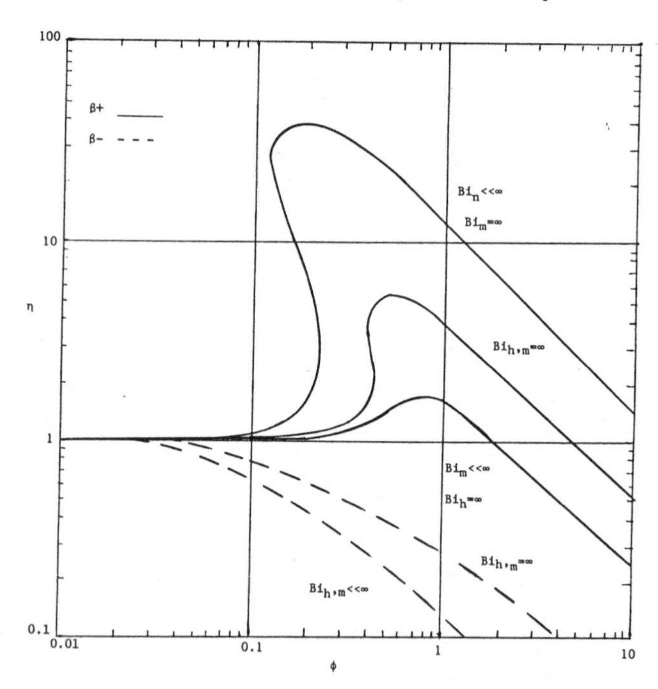

Figure 9. Nonisothermal interphase-intraphase combined effectiveness factor for various mass (Bi_m) and heat (Bi_h) Biot numbers, $\epsilon = 20$ [1].

of fluid existing in both phases along the interface. The model assumes no convection in the films; mass transfer across the films occurs by molecular diffusion alone. At steady state

$$D d^2C/dx^2 = 0 \tag{34}$$

In the absence of reaction, the solution of Equation 34 for the gas and liquid films are respectively

$$N_{AG} = \frac{D_{AG}}{x_G}(P_A - P_{Ai}) \tag{35}$$

and

$$N_{AL} = \frac{D_{AL}}{x_L}(C_{Ai} - C_A) \tag{36}$$

The hydrodynamic properties of the gas-liquid system are accounted for by the gas and liquid film thicknesses, X_G and X_L. These depend on physical properties of the fluids, the degree of agitation, the geometry of the vessel, etc.

The actual existence of these stagnant films is highly unlikely; therefore, their thicknesses are beyond direct measurement. This leads to the introduction of mass transfer coefficients

$$k_G = D_{AG}/X_G \tag{37}$$

and

$$k_L = D_{AL}/X_L \tag{38}$$

Despite its shortcomings the two-film model contains essential features of the real system, that is, gas must enter the liquid by dissolution and molecular diffusion before transfer to the turbulent bulk of the liquid. Also, predictions based on the two-film model are normally very similar to those based on more sophisticated models.

The penetration theory views the mass transfer process in terms of transient diffusion into a stagnant liquid. This situation can be described by

$$D\partial^2C/\partial x^2 = \partial C/\partial t \tag{39}$$

with boundary conditions

$$\begin{aligned} x &= 0; \quad t = t; \quad C = C_{Ai} \\ x &= \infty; \quad t = t; \quad C = 0 \\ x &= x; \quad t = 0; \quad C = 0 \end{aligned} \tag{40}$$

The instantaneous mass flux is

$$N_A = -D(dC/dx)_{x=0} = k_L^1 C_{Ai} \tag{41}$$

where k_L^1 is the instantaneous value of the mass transfer coefficient

$$k_L^1 = \sqrt{D/\Pi t} \tag{42}$$

If Equation 42 is integrated over the range of time from zero to θ, (i.e., the time of exposure) the average value of the mass transfer coefficient is determined

$$k_L = 2\sqrt{D/\Pi\theta} \tag{43}$$

The surface renewal model involves the exposure of liquid elements, or eddies, to the absorbing gas. After a certain time of exposure the eddy is replaced by another eddy coming from the bulk liquid. This replacement could be caused by the flow characteristics of the equipment or by turbulence. If each eddy were exposed for exactly the same time, θ, then the penetration model would apply. At the other extreme, a random replacement or perfectly mixed age-distribution function $\varphi(t)$ could be assumed

$$\varphi(t) = \frac{1}{\tau}\exp(-t/\tau) \tag{44}$$

The average absorption rate is

$$N_A = \int_0^\infty N_A(t)\varphi(t)\,dt \tag{45}$$

Substituting Equations 41, 42, and 44 into Equation 45 yields

$$N_A = \sqrt{D/\tau}\,(C_{Ai}-C_A) \tag{46}$$

Therefore the average mass transfer coefficient based on the surface renewal theory is

$$k_L = \sqrt{D/\tau} = \sqrt{Ds} \tag{47}$$

where s is the rate of surface renewal.

Two-Film Theory

The two-film-theory model is now applied to the situation of mass transfer accompanied by chemical reaction. There can be several distinct regimes of mass transfer-reaction rates. For example, the intrinsic reaction rate can be instantaneous compared to the mass transfer rate. At the other end of the spectrum the reaction could be infinitely slow in comparison to mass transfer.

First consider a reaction infinitely fast.

$$A + bB \rightarrow P$$

In this reaction A is transferred from the gas to the liquid, and the reaction takes place in the liquid phase. Since the reaction is instantaneous, an element of liquid can contain either A or B but not both. Reaction occurs at a plane between the A containing liquid and the B containing liquid as illustrated in Figure 10. In general a balance on component A may be written

$$r_A = D_A d^2C_A/dx^2 \tag{48}$$

Since there is no B in the liquid film from the gas-liquid interface to the reaction plane at x_R, r_A is zero in that zone, and a first integration of Equation 48 yields

$$D_A dC_A/dx = \text{constant} \tag{49}$$

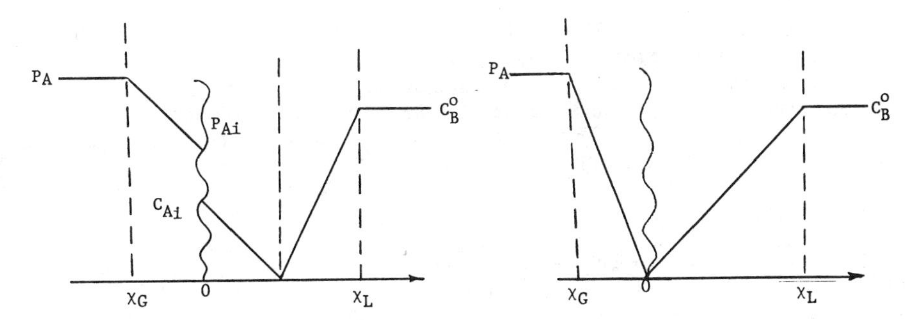

Figure 10. Instantaneous reaction in gas-liquid system. Concentration profiles for A and B; (A) concentration of B low; (B) concentration of B high.

Integration of Equation 49 leads to

$$N_A = D_A C_{Ai}/x_R \tag{50}$$

In a similar manner the flux of B can be derived

$$N_B = D_A C_B^0/(x_R - x_L) \tag{51}$$

By stoichiometry $N_A = -N_B/b$

Eliminating x_R from Equation 50 and 51 and applying Equation 38 we obtain

$$N_A = k_L C_{Ai}\left[1 + \frac{D_B}{D_A}\frac{C_B^0}{bC_{Ai}}\right] = k_L C_{Ai} E \tag{52}$$

Where E is the enhancement factor.

Since $D_A \approx D_B$, the criterion for instantaneous behavior of irreversible reactions of A with B then is

$$\frac{N_A}{k_L C_{Ai}} \gg 1 + \frac{C_B^0}{bC_{Ai}} \tag{53}$$

At the interface the relationship between P_A and C_A is given by the distribution coefficient, Henry's law constant H (m^3 bar/kmol):

$$p_{Ai} = HC_{Ai} \tag{54}$$

When there is appreciable gas-phase resistance, the gas-flux Equations 35 and 54 can be combined with the liquid-flux Equation 52 and the equality of fluxes through the gas and liquid to give

$$N_A = \frac{P_A + \dfrac{D_A}{D_B} H \dfrac{C_B^0}{b}}{\dfrac{1}{k_G} + \dfrac{H}{k_L}} \tag{55}$$

It can be seen from Equation 52 or 55 that increasing the concentration of the liquid-phase reactant C_B^0 will increase the overall rate. This is only true up to a certain point, though. As C_B^0 is increased, the plane of reaction will move toward the interface. At the limit the reaction plane is at the gas-liquid interface and the rate will not be affected by any further increase in the concentration of B. When the reaction plane is at the interface, the gas-phase resistance is controlling and Equation 55 simplifies to

$$N_A = k_G p_A \tag{56}$$

For a reaction first order with respect to the dissolved gas A and zero order with respect to B, the two-film theory's component balance is

$$D_A d^2C_A/dx^2 = kC_A \tag{57}$$

The integral of Equation 57 yields

$$C_A = \frac{C_{Ai} \sinh Ha(1 - x/x_L) + C_A^0 \sinh Ha x/x_L}{\sinh Ha} \tag{58}$$

where Ha, the Hatta number, is

$$Ha = x_L \sqrt{(k/D_A)} = \sqrt{(kD_A}/k_L \tag{59}$$

Using Equation 58 to define the flux at the gas liquid interface ($x = 0$) gives

$$N_A = \frac{Ha}{\tanh Ha}\left[1 - \frac{C_A^0}{C_{Ai}} \frac{1}{\cosh Ha}\right] k_L C_{Ai} \tag{60}$$

when $Ha \gg 1$, C_A^0 tends to zero and

$$N_A = C_{Ai} \sqrt{(D_A k)} \tag{61}$$

Note that the film thickness does not enter this expression since A does not penetrate the whole film and all of the reaction is completed in the film.

The gas-film resistance can be included by using Equations 35 and 54 in Equation 60 to eliminate C_{Ai}:

$$N_A = \frac{P_A - \dfrac{HC_A^0}{\cosh Ha}}{\dfrac{1}{k_G} + \dfrac{H}{k_L} \dfrac{\tanh Ha}{Ha}} \tag{62}$$

The reaction takes place mainly in the bulk liquid when $Ha < 0.3$.

Now consider the case when chemical reaction is very slow with respect to mass transfer. In this situation the amount of reaction occurring in the liquid film is negligible. The gas and liquid films and the reaction in the bulk liquid act as resistances in series, thus

$$N_A = k_G(p_A - p_{Ai}) = k_L(C_{Ai} - C_A^0) = \frac{kC_A^0 C_B^0}{a_i} \tag{63}$$

Regime	Rate Equation	Interface Concentration Profiles
Instantaneous Reaction	$N_A = \dfrac{p_A + \frac{D_A}{D_B} H \frac{C_B^o}{b}}{\frac{1}{k_G} + \frac{H}{k_L}}$ (55)	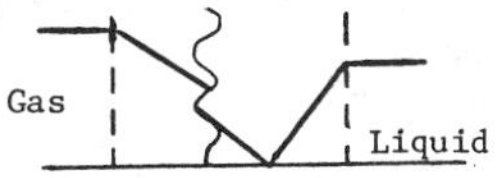 Gas Liquid
Instantaneous at Interface	$N_A = k_G p_A$ (56)	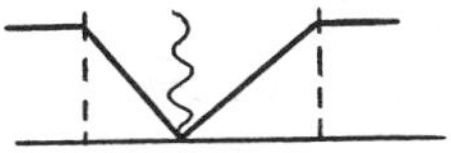
Rapid Reaction	$N_A = \dfrac{p_A}{\frac{1}{k_G} + \frac{H}{E k_L}}$	
Rapid first-order Reaction	$N_A = \dfrac{p_A}{\frac{1}{k_G} + \frac{H}{k_L H_a}}$	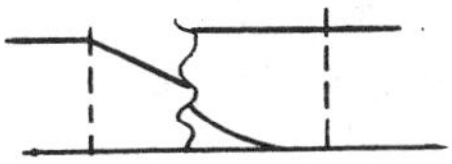
Intermediate Reaction	No general expression	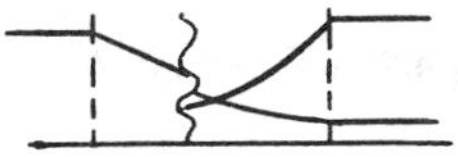
Intermediate Reaction	No general expression	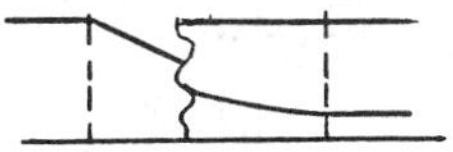
Slow Chemical Reaction	$N_A = \dfrac{p_A}{\frac{1}{k_G} + \frac{H}{k_L} + \frac{H a_i}{k C_B^o}}$ (64)	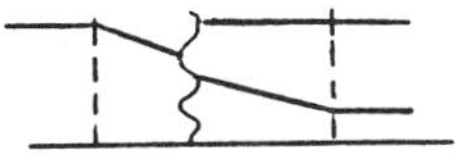
Very Slow Chemical Reaction	$N_A = \dfrac{k}{a_i} \dfrac{p_A}{H} C_B^o$	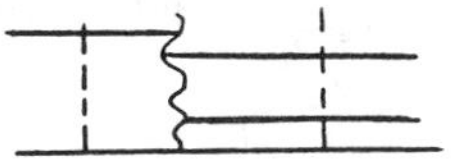

Figure 11. Interface concentration profiles for various kinetic regimes of gas-liquid reactions.

for a reaction first order with respect to each reactant. The term a_i represents the interfacial area per liquid volume. Combining by eliminating intermediate concentrations yields

$$N_A = \frac{P_A}{\frac{1}{k_G} + \frac{H}{k_L} + \frac{Ha_i}{kC_B^0}} \tag{64}$$

Figure 11 illustrates the diverse concentration gradients predicted by the two-film model for various ratios of mass transfer rate to intrinsic reaction rate. As demonstrated by the preceding paragraphs, the overall or global instantaneous rate can be determined analytically for the extreme conditions. The intermediate regime where reaction occurs both within the film and within the main body of liquid cannot be integrated analytically and solutions must be determined by numerical methods.

Penetration Theory

First consider the instantaneous reaction case. In the penetration theory model the reaction plane is not fixed in space, and transients must be taken into account. A component balance on A from the interface to the reaction plane can be written as

$$\partial C_A/\partial t = D_A \partial^2 C_A/\partial x^2 \tag{65}$$

with boundary conditions:

$$t = 0,\ x > 0,\ C_A = 0; \quad t > 0,\ x = 0,\ C_A = C_{Ai}; \quad t > 0,\ x = \infty,\ C_A = 0 \tag{66}$$

In the liquid on the other side of the plane a component balance for B is

$$\partial C_B/\partial t = D_B \partial^2 C_B/\partial x^2 \tag{67}$$

with boundary conditions

$$t = 0,\ x \geqq 0,\ C_B = C_B^0; \quad t > 0,\ x = \infty,\ C_B = C_B^0 \tag{68}$$

The solutions of Equations 65 and 67 may be obtained by the Laplace transform:

$$\frac{C_A}{C_{Ai}} = \frac{\operatorname{erfc}\left(\dfrac{x}{2\sqrt{D_A t}}\right) - \operatorname{erfc}\left(\dfrac{x_R}{2\sqrt{D_A t}}\right)}{\operatorname{erf}\left(\dfrac{x_R}{2\sqrt{D_A t}}\right)} \tag{69}$$

and

$$\frac{C_B}{C_B^0} = \frac{\operatorname{erf}\left(\dfrac{x}{2\sqrt{D_B t}}\right) - \operatorname{erf}\left(\dfrac{x_R}{2\sqrt{D_B t}}\right)}{\operatorname{erfc}\left(\dfrac{x_R}{2\sqrt{D_B t}}\right)} \tag{70}$$

where erf is the error function defined by

$$\operatorname{erf}(x) = \frac{2}{\sqrt{\Pi}} \int_0^x e^{-\beta^2}\, d\beta \tag{71}$$

and $\operatorname{erfc}(x) = 1 - \operatorname{erf}(x)$.

The location of the reaction plane (x_R) moves from the interface towards the interior of the liquid with time and it is proportional to the square root of time. Using the stoichiometric relationship $N_A = -N_B/b$ and writing the fluxes in terms of Fick's law defines the location of the reaction plane as a function of time:

$$\exp(x_R^2/4D_B t)\operatorname{erfc}(x_R/2\sqrt{D_B t}) = \frac{C_B^0}{bC_{Ai}}\sqrt{\frac{D_B}{D_A}}\exp(x_R^2/4D_A t)\operatorname{erf}(x_R/2\sqrt{D_A t}) \tag{72}$$

The flux of A at the interface at any time can be obtained by differentiating Equation 69 to give:

$$N_A(t) = \frac{C_{Ai}}{\mathrm{erf}\left(\frac{m}{2\sqrt{D_A}}\right)}\sqrt{\frac{D_A}{\Pi t}} \tag{73}$$

where m is the proportionality constant

$$x_R = m\sqrt{t} \tag{74}$$

The average rate of absorption at uniform age θ is

$$N_A = \frac{1}{\theta}\int_0^\theta N_{A(t)}\,dt = \frac{k_L C_{Ai}}{\mathrm{erf}\left(\sqrt{\frac{m}{2D_A}}\right)} \tag{75}$$

For a first-order reaction the component balance for A in a unit volume element of liquid with the penetration theory is:

$$\partial C_A/\partial t = D_A \partial^2 C_A/\partial x^2 - kC_A \tag{76}$$

with boundary conditions of

$$\begin{aligned} x &= 0, \quad t > 0, \quad C_A = C_{Ai}; \\ x &= \infty, \quad t > 0, \quad C_A = 0; \\ t &= 0, \quad x > 0, \quad C_A = 0; \end{aligned} \tag{77}$$

Using Laplace transforms the solution of Equation 76 may be obtained as

$$\frac{C_A}{C_{Ai}} = \frac{1}{2}\exp\left(-x\sqrt{\frac{k}{D_A}}\right)\mathrm{erf\,c}\left(\frac{x}{2\sqrt{D_A t}} - \sqrt{kt}\right) + \frac{1}{2}\exp\left(x\sqrt{\frac{k}{D_A}}\right)\mathrm{erf\,c}\left(\frac{x}{2\sqrt{D_A t}} + \sqrt{kt}\right) \tag{78}$$

For large values of kt, Equation 78 reduces to

$$\frac{C_A}{C_{Ai}} \cong \exp\left[-x\sqrt{\frac{k}{D_A}}\right] \tag{79}$$

The instantaneous flux at the interface in an element having surface age t is

$$N_A(t) = \sqrt{kD_A}\left[\mathrm{erf}(\sqrt{kt}) + \frac{\exp(-kt)}{\sqrt{\Pi kt}}\right] \tag{80}$$

the average rate of absorption is given by:

$$N_A = \frac{1}{\theta}\int_0^\theta N_A(t)\,dt = Ha\left[\left(1+\frac{\Pi}{8\,Ha^2}\right)\mathrm{erf}\left(\frac{2}{\sqrt{\Pi}}Ha\right) + \frac{1}{2\,Ha}\exp\left(-\frac{4}{\Pi}Ha^2\right)\right]k_L C_{Ai} \quad (81)$$

Equations for other reaction orders have been derived (see Danckwerts, 1970 [7]).

Surface Renewal Theory

Making a component balance on species A for an instantaneous reaction with the surface renewal model yields the same equation as does the penetration model (Equation 65). The component balance on B is also the same (Equation 67).

The instantaneous flux is the same as the penetration model (Equation 73). The average rate of absorption for the surface-renewal model is found by substituting Equation 73 into Equation 45 and integrating:

$$N_A = \frac{\sqrt{D_A s}C_{Ai}}{\mathrm{erf}\left(\frac{m}{2\sqrt{D_A}}\right)} = \frac{k_L C_{Ai}}{\mathrm{erf}\left(\frac{m}{2\sqrt{D_A}}\right)} \quad (82)$$

Note that the average flux equation for the surface renewal model (Equation 82) is identical to the one for the penetration model (Equation 75).

The equation governing the accumulation, diffusion, and first-order reaction of A in a volume of liquid is the same for the surface-renewal model as it was for the penetration model (Equation 76).

Therefore the solution of the instantaneous flux is the same in Equation 80. The average rate of absorption is determined by substituting Equation 80 in Equation 45 and integrating:

$$N_A = (1+Ha^2)^{0.5} k_L C_{Ai} \quad (83)$$

Comparison of Gas-Liquid Models

It is convenient to compare the flux of A in the diffusion-reaction system to the flux with only physical absorption. The pure mass transfer flux is given by

$$N_A = k_L C_{Ai} \quad (84)$$

whereas the flux with diffusion and reaction may be written as

$$N_A = k_L C_{Ai} E \quad (85)$$

The term E is called the enhancement factor and its functionality depends on the gas-liquid model, the order of reaction, and the ratio of the mass transfer rate to the intrinsic kinetics.

For example the enhancement factor for an instantaneous reaction is listed in Table 2 for each of the models.

When D_A and D_B are equal and the reaction is equimolar Equation 72 yields

$$\frac{1}{\mathrm{erf}(m/2\sqrt{D_A})} = 1 + \frac{C_B^0}{C_{Ai}} \quad (86)$$

Table 2
Comparison of Enhancement Factors for Instantaneous Reaction

Model	Enhancement Factor, E
Two-film	$1+\frac{D_B}{D_A}\frac{C_B^0}{bC_{Ai}}$
Penetration	$[\mathrm{erf}(m/2\sqrt{D_A})]^{-1}$
Surface Renewal	$[\mathrm{erf}(m/2\sqrt{D_A})]^{-1}$

Table 3
Comparison of Enhancement Factors for a First-Order Reaction

Model	Enhancement Factor, E
Two-film	$Ha/\tanh Ha$
Penetration	$Ha\left[\left(1+\frac{\Pi}{8\,Ha^2}\right)\mathrm{erf}\left(\frac{2\,Ha}{\sqrt{\Pi}}\right)+\frac{1}{2\,Ha}\exp\left(\frac{-4\,Ha^2}{\Pi}\right)\right]$
Surface Renewal	$(1+Ha^2)^{0.5}$

Table 4
Comparisons of Enhancement Factor Values for First-Order Reaction

Ha	Enhancement Factor, E		
	Film Model	Penetration Model	Surface Renewal Model
0.01	1.0	1.0	0.94
0.1	1.0	1.0	1.005
1.0	1.31	1.41	1.37
10.0	10.0	10.05	10.39

Thus, for this case the three models predict identical results. Even when D_A and D_B are not equal, the difference in the film and the penetration and surface renewal models is only a few percent.

A comparison of enhancement factors for a first-order reaction is shown in Table 3. Although the enhancement factor expressions in Table 3 appear quite different, they predict values that vary only a few percent. This is illustrated by Table 4.

Summary of Gas-Liquid Reactions

A summary of kinetic regimes, rate equations for the two-film theory, and concentration profiles are shown in Figure 11. Summaries of typical values of mass transfer parameters in various two-phase reactor designs are presented in Table 5.

These reactor types are shown in Figure 12.

When the reaction is very slow compared to mass transfer, it is not necessary to use a great amount of energy to obtain large values of mass transfer parameters. Also, since all or almost all of the reaction takes place in the bulk liquid, the fraction of liquid within the reactor should be high. Therefore, bubble columns, packed columns, and agitated reactors would be suitable.

In situations in which the mass transfer processes and the intrinsic reaction rates are approximately the same, both interfacial area and liquid hold-up should be high. Bubble-cap-plate or sieve-plate columns, ejector-type reactors, or mechanically agitated reactors would be the most desirable choices.

Table 5
Mass Transfer Parameter Ranges in Gas-Liquid Reactors

Type Reactor	Volume Fraction Liquid	k_L cm/s × 10^2	k_G (mol/cm^2-s-bar) × 10^4	ai cm^2/cm^3
a. Hydroclone	0.7-0.9	10-30	-	0.2-0.5
b. Ejector reactor	-	-	-	1-20
c. Venturi	0.05-0.3	5-10	2-10	1-25
d. Submerged jet	0.94-0.99	0.15-0.5	-	0.2-1.2
e. Mechanically agitated	0.2-0.95	0.3-4	-	0.3-8
f. Spray column	0.02-0.2	0.7-1.5	0.5-2	0.1-1
g. Horizontal tube	0.05-0.95	1-10	0.5-4	0.5-7
h. Vertical tube	0.05-0.95	2-5	0.5-8	1-10
i. Bubble column	0.6-0.98	1-4	0.5-2	0.5-6
j. Bubble cap plate	0.1-0.95	1-5	0.5-2	1-4
k. Sieve plate	0.1-0.95	1-20	0.5-6	1-2
l. Counter flow packed column	0.02-0.25	0.4-2	0.03-2	0.1-3.5
m. Cocurrent packed column	0.02-0.95	0.4-6	0.1-3	0.1-17

For cases where the chemical kinetics are much faster than mass transfer, the fraction liquid hold-up is not important. At the same time the larger the interfacial area and mass transfer coefficients are, the smaller the overall reactor size. Venturi reactors, and ejector reactors could be used.

NONCATALYTIC GAS-SOLID REACTIONS

A reaction between a gaseous component and a solid component may occur when the two phases are brought into contact. Examples of such reactions are found in coal gasification, iron ore reduction in the blast furnace, and the roasting of various ores. The regeneration of coked catalysts by means of oxygen-containing gas is another example.

In gas-solid reactions the relative magnitudes of the rate of transport and the rate of reaction determine whether or not there are significant concentration gradients in and around a solid particle. The conditions inside the particle change with time since the solid itself is involved in the reaction. The solid particles may remain unchanged in size during the reaction if they contain large amounts of inert solids which remain in the particle or if the reaction forms a solid product of the same density as the solid reactant. Particles may also shrink in size during reaction if all of the product is gaseous.

Rate equations describing gas-solid reactions are developed in the following paragraphs.

General Model

A gas-solid reaction involving a spherical solid of radius R can be described by a model accounting for interfacial and intraparticle gradients. Assuming that the particle is isothermal, the continuity equation for the gaseous reactant describes the process. This equation contains an accumulation term accounting for the unsteady-state nature of the reaction, a diffusive transport term, and a term for the reaction rate:

$$\frac{\partial}{\partial t}(\varepsilon C_{As}) = \frac{1}{r^2}\frac{\partial}{\partial t}\left(D_e r^2 \frac{\partial C_{As}}{\partial r}\right) - r_A \tag{87}$$

The continuity equation for the solid reactant is

$$\frac{\partial C_B}{\partial t} = -r_B = b(-r_A) \tag{88}$$

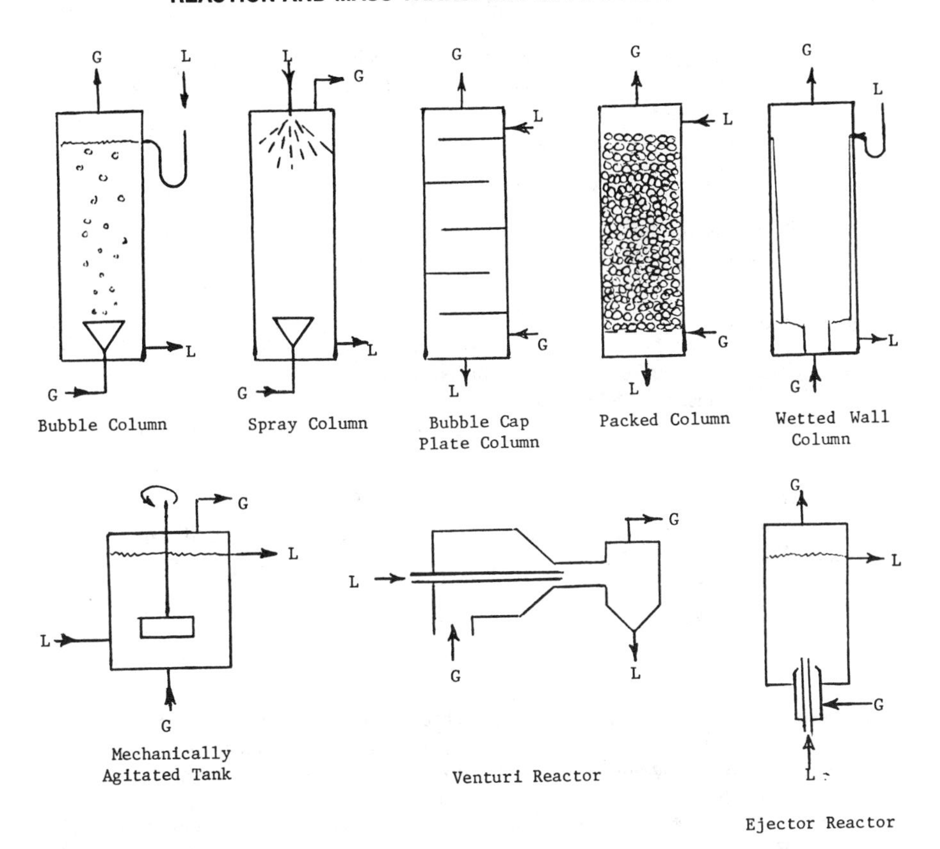

Figure 12. Reactor types for gas-liquid reactions.

The initial and boundary conditions are at

$$t = 0: C_{AS} = (C_{AS})_0; \quad C_B = C_{B0}$$

$$\text{for } r = 0: \partial C_{AS}/\partial r = 0$$

$$\text{for } r = R: D_e(\partial C_{AS}/\partial r)|_{r=R} = k_g(C_{Ag} - C_{As}) \tag{89}$$

The left-hand side of Equation 87 may be neglected in gas-solid reactions by the pseudo-steady-state treatment. Since the gaseous reactant concentration is normally of the order of 10^3 times smaller than solid reactant concentration, the rate at which the reaction layer moves is very small compared to the rate of gas transport.

The effective diffusivity within the solid particle depends on the degree of porosity of the particle. This effect has been determined to be

$$\frac{D_e}{D_{e_0}} = \left(\frac{\varepsilon}{\varepsilon_0}\right)^n \tag{90}$$

where ε_0 is the initial particle porosity and n lies between 2 and 3. The porosity can be related to the degree of solid reactant conversion and the molar volumes of solid reactant and product:

$$\varepsilon = \varepsilon_0 + C_{B0}(v_B - v_P)(1 - \varepsilon_0)\left(1 - \frac{C_B}{C_{B0}}\right) \tag{91}$$

where v_B and v_P are the reactant and product solid molar volumes.

The numerical solution for this general model with reaction $A_g + bB_s \rightarrow$ products, a rate expression of

$$-r_A = kC_A^2 C_B \tag{92}$$

and in the absence of interfacial gradients shows that concentration profiles are influenced by the Thiele modulus, φ_s, where

$$\varphi_s = R\sqrt{\frac{kC_{A0}C_{B0}}{D_{e0}}} \tag{93}$$

This effect is illustrated by Figures 13 and 14 where $\theta = kC_{A0}^2\, t$ is reduced time.

In these figures

$$\frac{C_{B0}(v_B - v_P)(1 - \varepsilon_0)}{\varepsilon_0} = 9$$

$$n = 2$$

X_B is the fractional conversion of the solid reactant in the total particle.

When φ_s is very small, the chemical reaction is rate controlling and there are almost no concentration gradients within the particle. When φ_s is large, the diffusion of gaseous reactant through the solid is rate controlling. In the extreme the overall reaction can be described by a homogeneous model when the chemical reaction is the dominant step. The other extreme is encountered when the diffusion rate is intrinsically very slow relative to the true reaction rate. Then the general model reduces to the heterogeneous model with a shrinking unreacted core.

Figure 13 corresponds to the case when the model approaches a homogeneous one with practically no gradients inside the particle. Figure 14 approaches the case when diffusion becomes rate controlling. The gas and solid concentration profile become vertical in the heterogeneous case since the reaction zone narrows to a plane.

Two stages develop as the reaction proceeds through the solid particle. The first stage lasts until the time at which the solid reactant concentration becomes zero at the outer surface of the particle. During the second stage, only diffusion occurs through the outer zone of the particle where the solid reactant has been completely converted. Then, in the interior of the particle, diffusion and reaction occurs. The generalized model can be simplified by allowing only two diffusivities, De for diffusion through the reactant or partially reacted solid and D_e' for diffusion through completely converted solid.

Then in the first stage

$$\varepsilon\frac{\partial C_{As}}{\partial t} = D_e\left(\frac{\partial C_{As}^2}{\partial r^2} + \frac{2}{r}\frac{\partial C_{As}}{\partial r}\right) - r_A \tag{94}$$

and in the second stage

$$D_e'\left(\frac{\partial^2 C_{As}'}{\partial r^2} + \frac{2}{r}\frac{\partial C_{As}'}{\partial r}\right) = 0 \tag{95}$$

for the completely reacted zone. For a rate expression of

$$r_A = kC_{As} \tag{96}$$

the model can be solved analytically.

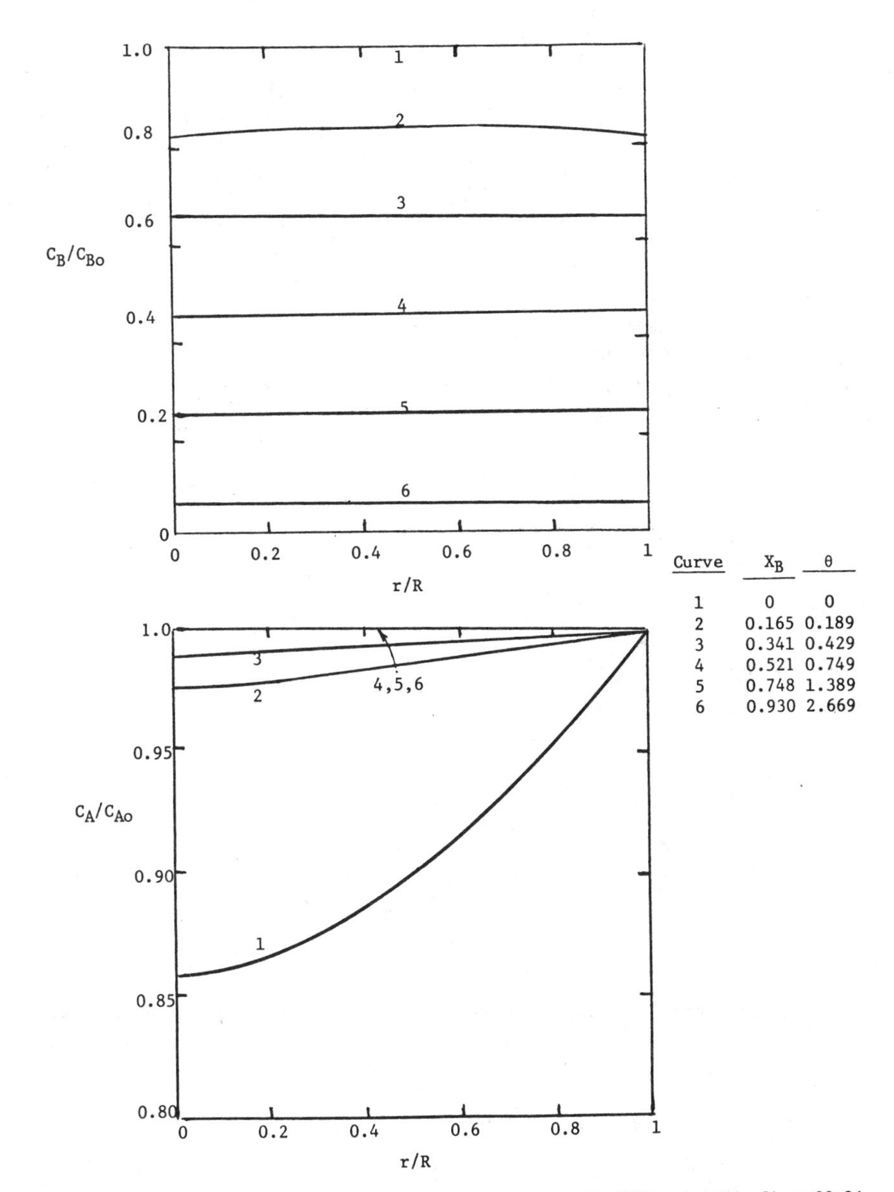

Figure 13. Gas-solid reaction model. Concentration profiles for $\phi_s = 1$. (C. Y. Wen, *Ind. Eng. Chem*, 60: 34 (1968).)

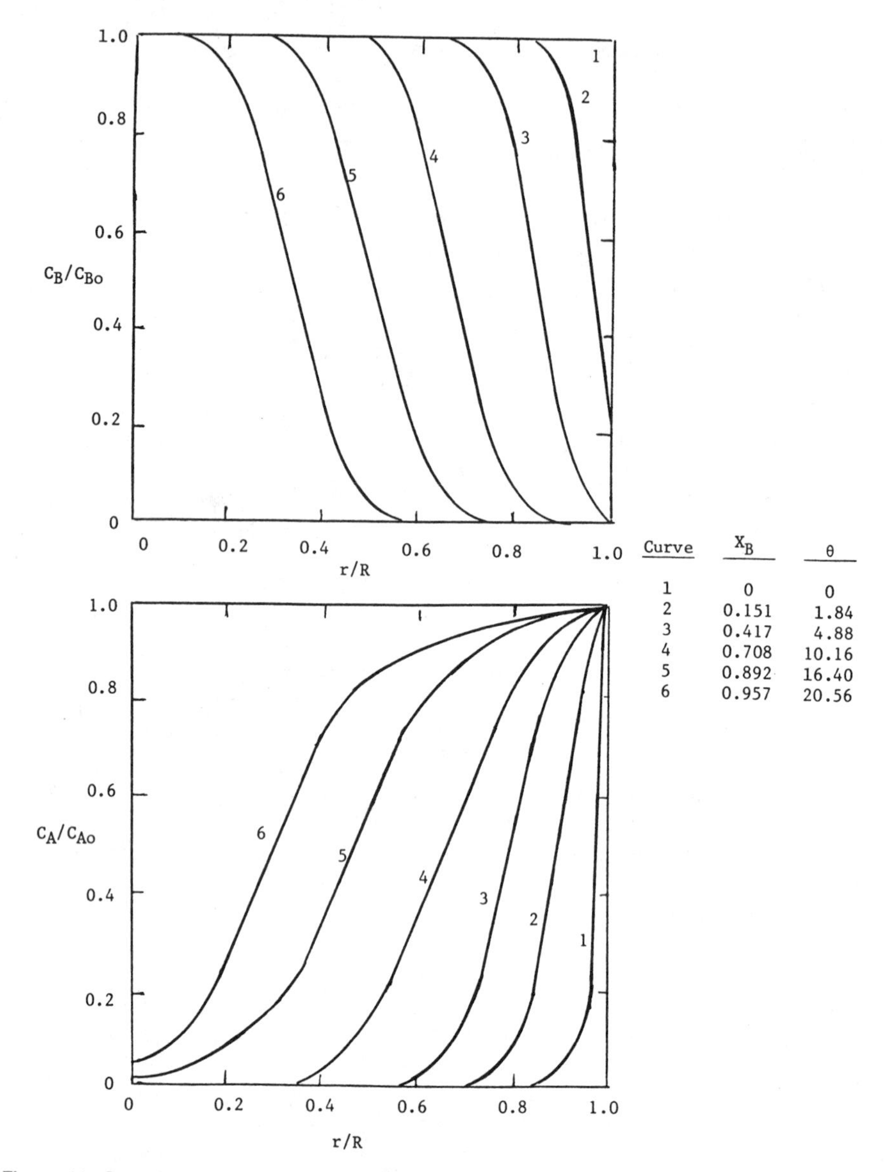

Figure 14. Gas-solid reaction model. Concentration profiles for $\phi_s = 70$. (C. Y. Wen, *Ind. Eng. Chem*, 60: 34 (1968).)

During the first stage the gas concentration profile is

$$\frac{C_A}{C_{A0}} = \frac{1}{\theta_e} \frac{\sinh\left(\varphi \frac{r}{R}\right)}{\frac{r}{R} \sinh \varphi} \tag{97}$$

with

$$\varphi = R\sqrt{\frac{k}{D_e}} \quad \text{and} \quad \theta_e = 1+\left(\frac{D_e}{k_g R}\right)(\varphi \coth \varphi - 1)$$

The solid concentration profile in the first stage is

$$\frac{C_B}{C_{B0}} = 1 - \frac{\sinh\left(\varphi \frac{r}{R}\right)}{\frac{r}{R}\sinh \varphi}\frac{\theta}{\theta_e} \tag{98}$$

with $\theta = kC_A t$. The solid conversion in the total particle can be calculated by integrating the local solid concentration through the total sphere:

$$1 - X_B = \frac{\int_0^R \frac{C_B}{C_{B0}}(4\Pi r^2)\,dr}{\frac{4}{3}\Pi R^3} \tag{99}$$

Then

$$X_B = \frac{3}{\varphi^2}(\varphi \coth \varphi - 1)\frac{\theta}{\theta_e} \tag{100}$$

The time at which the second stage begins is

$$t_e = \frac{\theta_e}{kC_A} = \frac{1}{kC_A}\left[1+\left(\frac{D_e}{k_g R}\right)(\varphi \coth \varphi - 1)\right] \tag{101}$$

Then the solid conversion is given by

$$X_B = 1-\left(\frac{r}{R}\right)_m^3 + \frac{3\left(\frac{r}{R}\right)_m}{\varphi^2}\left\{\left[\varphi\left(\frac{r}{R}\right)_m\right]\coth\left[\varphi\left(\frac{r}{R}\right)_m\right]-1\right\} \tag{102}$$

where the location of the completely reacted solid zone, $\left(\frac{r}{R}\right)_m$, is found from the implicit equation:

$$\theta = 1+\left(1-\frac{D_e}{D_e'}\right)\ln\left[\frac{\left(\frac{r}{R}\right)_m \sinh \varphi}{\sinh\left[\varphi\left(\frac{r}{R}\right)_m\right]}\right] + \frac{\varphi'^2}{6}\left[1-\left(\frac{r}{R}\right)_m\right]^2\left[1+2\left(\frac{r}{R}\right)_m\right]$$

$$+\left\{\left(\frac{D_e}{D_e'}\right)\left[1-\left(\frac{r}{R}\right)_m\right]+\left(\frac{D_e}{k_g R}\right)\left(\frac{r}{R}\right)_m\right\}\left\{\left[\varphi\left(\frac{r}{R}\right)_m\right]\coth\left[\varphi\left(\frac{r}{R}\right)_m\right]-1\right\}$$

$$+\frac{D_e'}{k_g R}\frac{\varphi'^2}{3}\left[1-\left(\frac{r}{R}\right)_m^3\right] \tag{103}$$

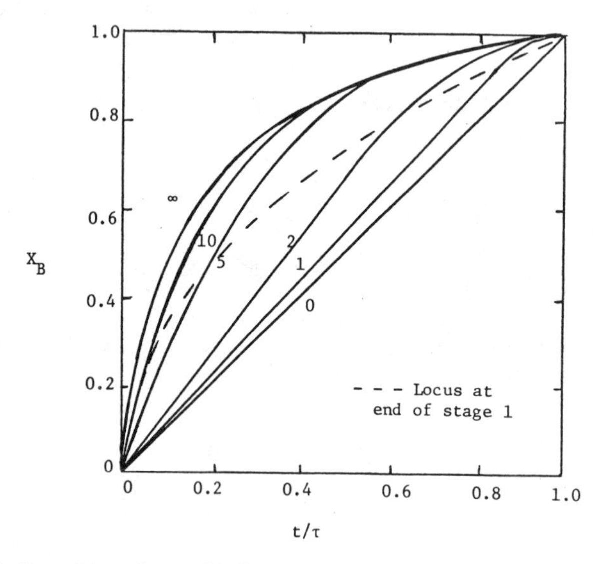

Figure 15. Gas-solid reaction model. Solid reactant conversion versus time for various ϕ_s values. (M. Ishida and C. Y. Wen. *AIChE J., 14:* 311 (1968).)

Figure 15 illustrates how conversion progresses with time and the boundary between the first and second stage.

Homogeneous Model

In a truly homogeneous model, $De = D'_e$, and the continuity equation for the gaseous reactant reduces to

$$r_A = f(C_A, C_B) \tag{104}$$

and Equation 88 is retained for the solid reactant. The boundary conditions are

$$C_{As} = C_{Ag}; \quad \frac{\partial C_{As}}{\partial r} = 0; \quad 0 \leqq r \leqq R \tag{105}$$

Thus Equation 88 can be integrated to obtain a relationship between solid conversion versus time.

Heterogeneous Model

The heterogeneous model with shrinking unreacted core with the pseudo steady-state approximation is

$$\frac{1}{r^2}\frac{\partial}{\partial r}\left(D'_e r^2 \frac{\partial C'_{As}}{\partial r}\right) = 0 \tag{106}$$

with Equation 88 retained for the solid reactant. The reaction is then confined to a front. The reaction rate term does not appear in the right-hand side, but only in the boundary condition at the reaction front

$$r = r_m:\ D_e'\left(\frac{\partial C_{As}}{\partial r}\right) = r_A \tag{107}$$

Solving Equations 106 and 88 yields

$$t = \frac{RC_{B0}}{C_{A0}}\left\{\frac{1}{3}\left(\frac{1}{k_g} - \frac{R}{D_e'}\right)\left[1-\left(\frac{r_m}{R}\right)^3\right] + \frac{R}{2D_e'}\left[1-\left(\frac{r_m}{R}\right)^2\right] + \frac{1}{kC_{B0}}\left(1-\frac{r_m}{R}\right)\right\} \tag{108}$$

The three terms inside the brackets on the right-hand side of Equation 108 represent the three resistances involved in the process. The resistances are external mass transfer, diffusion into the particle, and chemical reaction, respectively. They are purely in series in this process. The equation can be written in terms of the solid reactant conversion, X_B, where

$$X_B = 1 - \frac{\frac{4}{3}\Pi r_m^3}{\frac{4}{3}\Pi R^3} = 1-\left(\frac{r_m}{R}\right)^3 \tag{109}$$

Substituting

$$t = \frac{RC_{B0}}{C_{A0}}\left\{\frac{1}{3}\left(\frac{1}{k_g} - \frac{R}{D_e'}\right)X_B + \frac{R}{2D_e'}[1-(1-X_B)^{2/3}] + \frac{1}{kC_{B0}}[1-(1-X_B)^{1/3}]\right\} \tag{110}$$

The time, τ, for 100 percent conversion is when $X_B = 1$:

$$\tau = \frac{RC_{B0}}{C_{A0}}\left(\frac{1}{3k_g} + \frac{R}{6D_e'} + \frac{1}{kC_{A0}}\right) \tag{111}$$

When external mass transfer is rate controlling, $3\,kg \ll kC_{B0}$ and $kg \ll 2D_e'/R$ and

$$\tau = \frac{RC_{B0}}{3k_gC_{A0}} \tag{112}$$

Then

$$t/\tau = X_B \tag{113}$$

When diffusion through the reacted solid is rate controlling, $2De/R \ll kg$ and $6D_e'/R \ll kC_{B0}$, and

$$\tau = \frac{R^2C_{B0}}{6C_{A0}D_e'} \tag{114}$$

Then

$$t/\tau = 1-3(1-X_B)^{2/3}+2(1-X_B) \tag{115}$$

The third limiting case of chemical reaction rate controlling is not consistent with the shrinking core model with a single diffusivity for both product solid and reactant solid. In this case when the intrinsic chemical reaction was very slow compared to diffusion, the homogeneous model would apply. On the other hand, if the unreacted solid was almost nonporous and the reacted solid was porous, the diffusivity for the unreacted solid could approach zero and the chemical reaction step could be rate controlling as compared to diffusion through the reacted solid. In such a case, $kC_{B0} \ll 6D'_e/R$, $kC_{B0} \ll 3$ kg and

$$\tau = \frac{R}{kC_{A0}} \tag{116}$$

Then

$$t/\tau = (1 - X_B)^{1/3} \tag{117}$$

Grain Model

A model accounting for the structure of the solid is one in which the large particle (pellet) is considered to be a matrix of very small grains. The overall reaction process then includes intrapellet diffusion to the surface of the grains followed by intragrain diffusion and chemical reaction. The continuity equation for the gaseous reactant in the pellet is

$$\varepsilon \frac{\partial C_{As}}{\partial t} = D_{ep} \frac{\partial^2 C_{As}}{\partial r^2} - D_{eg} \left.\frac{\partial C_{Ag}}{\partial y}\right|_Y (1 - \varepsilon) a_g \tag{118}$$

where D_{ep} = effective pellet diffusivity
D_{eg} = effective grain diffusivity
r = pellet radial coordinate
y = grain radial coordinate
Y = radius of the grain
C_{Ag} = concentration of A in the grain
a_g = surface-to-volume ratio of the grain

The factor $(1 - \varepsilon)\, a_g$ is the grain-surface-area-to-pellet-volume ratio.

To obtain the concentration profile with grain, the general model equations (Equations 87 and 88) can be used. Various rate equations can be used in Equations 87 and 88 and numerical integration can be used to solve the system of equations.

An approximate solution is

$$a_g \frac{kC_{A0}}{F_g} t \cong G + \left[\frac{(1-\varepsilon)kC_{B0}F_p}{2D_{ep}a_p^2}\left(\frac{a_g}{F_g}\right)\right] P_p + \frac{kC_{B0}}{2D_{eg}a_g} P_g \tag{119}$$

F_g and F_p are geometric factors for the grain and pellet respectively, and have values of 1 for slabs, 2 for cylinders, and 3 for spheres. The conversion X_B in the grain attained by a shrinking core mechanism is written

$$X_B = 1 - \left(\frac{a_g Y_m}{F_g}\right)^{F_g} \tag{120}$$

The functions of conversion are

$$G = 1 - (1 - X_B)^{1/F_g} \tag{121}$$

$$P_p = P_g = X_B^2 \text{ for } F_p \text{ or } F_g = 1 \tag{122}$$

$$P_p = P_g = X_B + (1 - X_B) \ln (1 - X_B) \text{ for } F_p \text{ or } F_g = 2 \tag{123}$$

$$P_p = P_g = 1 - 3(1 - X_B)^{2/3} + 2(1 - X_B) \text{ for } F_p \text{ or } F_g = 3 \tag{124}$$

CATALYTIC GAS-SOLID REACTIONS

The general concepts of diffusion (of heat and mass) both surrounding and within the locale of a catalytic gas-solid reaction were developed in the earlier sections on interphase and intraphase diffusion and reaction. These phenomena affect the global catalytic reactivity since they cause variations in temperature and composition of reactants at catalytic sites. It is important to note that the concepts apply to a single point in a chemical reactor or to one catalyst pellet because of the variations in fluid phase temperature and composition throughout the reactor.

Key references on the topics covered in this chapter are given below.

NOTATION

a	interfacial area
b	coefficient
Bi	Biot number
C_s	surface concentration
C	concentration
C_p	specific heat
D	diffusion coefficient
D_{eg}	effective grain diffusivity
E	activation energy or enhancement factor
H	enthalpy or Henry's law constant
Ha	Hatta number (see Equation 59)
h	convection coefficient
$j_{D,H}$	mass and heat transfer factors
k	reaction rate coefficient
$k_{L,G}$	liquid and gas mass transfer coefficients
L	thickness or length
m	proportionality constant in Equation 74
N_A	mass flux
n	reaction order
P	pressure
Pr	Prandtl number
r	reaction rate
r, R_0	particle radius
R	universal gas law constant
s	surface renewal rate
Sc	Schmidt number
t	time
T	temperature
V_P	particle volume
x	distance
X	film thickness
z	distance

Greek Symbols

α	thermal diffusivity
β	dimensionless group defined by Equation 16; also dimensionless adiabatic temperature rise
δ_x	film thickness
ε_0	particle porosity, also Arrhenius number Equation 17
$\bar{\eta}$	effectiveness factor
θ	time
λ_e	effective thermal conductivity
ϱ	density
τ	time
φ	Thiele modulus (see Equation 23)

REFERENCES

1. Carberry, J. J., *Chemical and Catalytic Reaction Engineering,* McGraw-Hill, New York, 1976.
2. Froment, G. F., and Bischoff, K. B., *Chemical Reactor Analysis and Design,* J. Wiley and Sons, New York, 1979.

3. Levenspiel, O., *Chemical Reaction Engineering,* 2nd Ed., J. Wiley and Sons, New York, 1972.
4. Smith, J. M., *Chemical Engineering Kinetics,* 2nd Ed., McGraw-Hill, New York, 1970.
5. Astarita, G., *Mass Transfer with Chemical Reaction,* Elsevier, Amsterdam, 1967.
6. Charpentier, J. C., *Advances in Chemical Engineering,* Vol. 11, T.B. Drew, (Ed.) Academic Press, New York, 1981.
7. Danckwerts, P. V., *Gas Liquid Reactions,* McGraw-Hill, New York, 1970.
8. Aris, R., *The Mathematical Theory of Diffusion and Reaction in Permeable Catalysts,* Vol. I and II, Clarendon, Oxford, 1975.
9. Satterfield, C. N., *Mass Transfer in Heterogeneous Catalysis*, MIT Press, Cambridge, Mass., 1970.
10. Szekely, J., Evans, J. W., and Sohn, H. Y., *Gas-Solid Reactions*, Academic Press, New York, 1976.

CHAPTER 19

MASS TRANSFER AND KINETICS IN THREE-PHASE REACTORS

Janez Levec

Department of Chemistry and Chemical Technology
E. Kardelj University
Ljubljana, Yugoslavia

Shigeo Goto

Department of Chemical Engineering
Nagoya University
Nagoya, Japan

CONTENTS

INTRODUCTION

Gas-liquid-solid three-phase reactors are necessary for systems involving a solid catalyst and gaseous and liquid reactants, such as desulfurization of oils and oxidation of organic pollutants. Another application is for removal of pollutants from gas streams such as during the oxidation of sulfur dioxide. There are many extensive reviews of three-phase reactors [1–12].

Gaseous reactants must dissolve into the liquid phase and then transfer to the surface of the solid catalyst before changing to products in the particles. All such three-phase processes involve the following steps: gas-liquid, liquid-solid mass transfer, intraparticle diffusion, and chemical reaction. The relative importance of these individual steps depends on the type of contact in the three phases. Therefore, the choice of reactor is very important for optimum performance.

Several different types of three-phase reactors may be used to obtain contact between the three phases. They may be divided into two main classes, depending on the state of motion of the solid particles.

1. The particles are packed in a fixed bed and the fluid phases (gas and liquid) may be in either cocurrent downflow or upflow.
2. The particles are suspended in the liquid phase by mechanical stirring or bubble movement.

Another interesting type is a catalytic basket reactor in which baskets packed with catalyst particles are rotated as stirrers or kept near baffles.

In this chapter, empirical correlations are first presented in order to estimate the physical parameters of the mass transfer steps. Subsequently, three-phase reactors are analyzed in terms of models which attempt to account for the transport and reaction steps and familiar examples of reaction systems are presented.

It is not difficult to maintain quasi-isothermal conditions in three-phase reactors because the heat capacity of the liquid phase is much higher than that of the gas phase. Hence effects of heat transfer on the performance of three-phase reactors will not be considered in this chapter.

MASS TRANSFER IN THREE-PHASE REACTORS

Gas-Liquid Cocurrent Downflow Reactors (Trickle-Bed Reactor)

Flow Pattern

Figure 1 shows the schematic diagram of a gas-liquid cocurrent downflow reactor in which both gas and liquid flow downward through the fixed packing of catalyst particles. A trickle-bed reactor is a sub-type of this reactor in which there is a low liquid flow rate, that is, the liquid flows in rivulets through a fixed bed.

Conventional operation of a packed bed for gas absorption is in countercurrent flow of gas and liquid phase through the bed, with the liquid descending and the gas moving upward through the bed. When concentrations in both gas and liquid phases are changed, the exit concentration in the liquid phase for the countercurrent operation is higher than that for the cocurrent operation as shown in Figure 2. However, when the concentration in the liquid

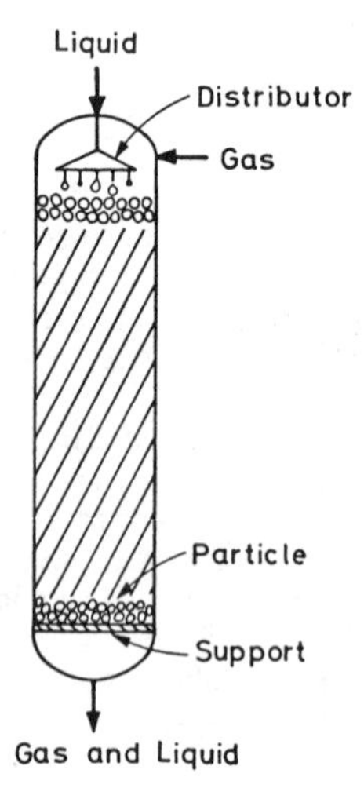

Figure 1. Schematic diagram of gas-liquid downflow reactor (trickle-bed reactor).

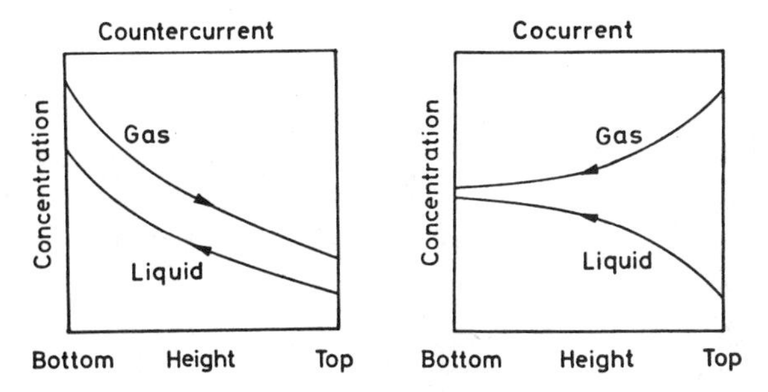

Figure 2. Comparison of profiles between countercurrent and cocurrent flows when concentrations in both gas and liquid phases are varied.

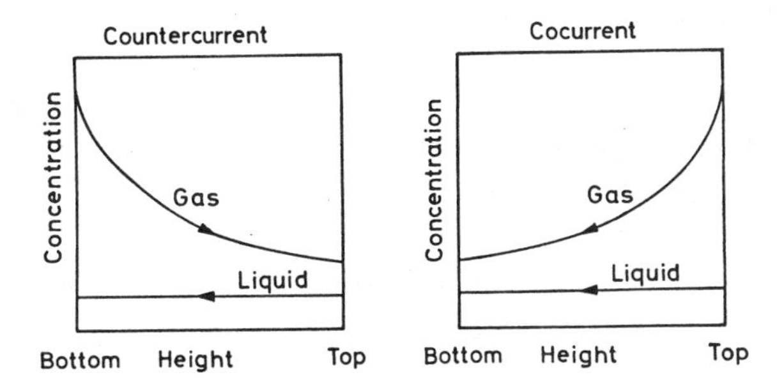

Figure 3. Comparison of profiles between countercurrent and cocurrent flows when the concentration in liquid phase remains constant.

phase remains constant or almost zero due to the chemical reaction, the concentration profile may be independent of the patterns, as shown in Figure 3.

In the countercurrent operation of a packed bed, the flow rate ranges are limited by the flooding point. In contrast, in the cocurrent operation, the column has no such maximum limit of capacity and a high flow of both phases can pass through the packed bed far in excess of the countercurrent flooding point. It is necessary to control the level of the liquid phase at the bottom for the separation of the gas and liquid in the cocurrent operation. However, the gas phase can be separated from the liquid phase outside the bed in the cocurrent operation.

For the sake of these advantages, the cocurrent operation may be mainly used for three-phase reactors. There are two types of cocurrent operation, that is, downflow and upflow. The downflow type will be treated in this section and the upflow type in the subsequent section.

Steps in Mass Transfer

When the gaseous reactant dissolved into the liquid phase, after which the reaction takes place due to the catalytic activity of porous particles, each step of the mass transfer may be illustrated in Figure 4. The outer surface of particles consists of gas-covered and liquid-covered parts.

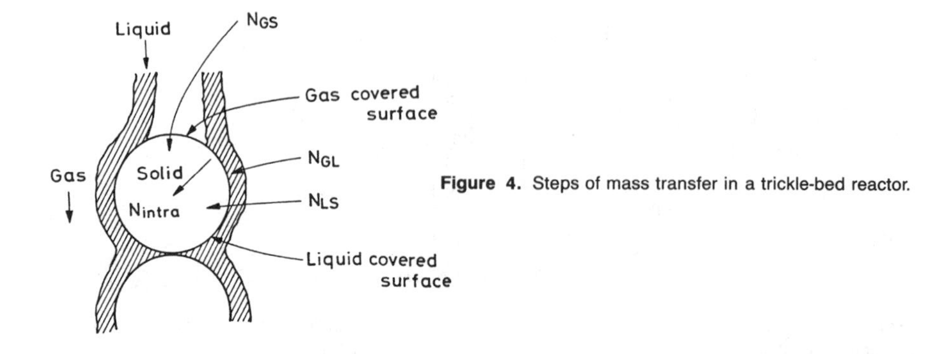

Figure 4. Steps of mass transfer in a trickle-bed reactor.

According to the two-film concept, the rate of mass transfer per unit volume of reactor from gas to liquid, N_{GL}, is given by

$$N_{GL} = (ka)_G (C_G - C_{i,G}) = (ka)_L (C_{i,L} - C_L) \tag{1}$$

It is usually assumed that equilibrium exists at the interface so that Henry's law is applicable:

$$C_{i,G} = H^* C_{i,L} \tag{2}$$

Then the rate N_{GL} can be expressed in terms of overall coefficients $(Ka)_G$ and $(Ka)_L$:

$$N_{GL} = (Ka)_G (C_G - C_G^*) = (Ka)_L (C_L^* - C_L) \tag{3}$$

where $C_G^* = H^* C_L$ and $C_L^* = C_G/H^*$. From Equations 1 to 3,

$$1/(Ka)_G = 1/(ka)_G + H^*/(ka)_L \tag{4}$$

and

$$1/(Ka)_L = 1/[H^*(ka)_G] + 1/(ka)_L \tag{5}$$

Since $(ka)_G$ and $(ka)_L$ are usually within one order of magnitude of each other, the value of Henry's law constant, H^*, determines the relative importance of the transport resistances in the liquid and gas phases. For slightly soluble gases such as O_2, N_2, and H_2, the values of H^* are much larger than unity so that $(Ka)_L$ is approximately equal to $(ka)_L$. On the other hand, for gases very soluble in water, such as NH_3, SO_2, and HCl, the values of H^* are much smaller than unity so that $(Ka)_G \cong (ka)_G$.

The rate of mass transfer from bulk liquid to the liquid-covered surface of the solid particle, N_{LS}, can be expressed in terms of the film theory as

$$N_{LS} = (ka)_{LS} (C_L - C_S) \tag{6}$$

The rate of direct mass transfer from gas to the gas-covered surface of the solid particle, N_{GS}, may be given by

$$N_{GS} = (ka)_{GS} (1 - f_w)(C_G - H^* C_L) \tag{7}$$

where f_w is the fraction of wetting coverage and $(ka)_{GS}$ is the volumetric coefficient of mass transfer from gas to solid, which can be obtained from the measurement of the gas-solid system without the liquid phase.

Since the pores of catalyst particles are assumed to be completely filled with liquid due to capillary force [13], the rate of intraparticle diffusion, N_{intra} may be expressed as

$$N_{intra} = -D_e \, dC/dr \tag{8}$$

and

$$D_e = \varepsilon_p D_m/\tau \tag{9}$$

where the tortuosity factor, τ, is usually in the range of $2 \sim 6$ although there are some exceptions, for example around 1 [15] or around 9 [16].

Gas-Liquid Mass Transfer

Since gaseous reactants in three-phase reactors are usually slightly soluble gases such as O_2 and H_2, the values of Henry's law constants, H^*, are large and the mass transfer resistance in the gas phase is usually negligible. Thus, the estimation of the volumetric coefficient of mass transfer in the liquid phase, $(ka)_L$ is important.

In gas-liquid cocurrent downflow reactors there are various flow regimes such as trickle flow, pulsed flow, spray flow, and bubble flow. The mechanism of mass transfer may be dependent on the flow regime.

Empirical correlations can be divided into two types as follows.

1. Only operating conditions such as gas and liquid flow rates and particle sizes are independent variables in the correlations [17–20].

   For example, Goto and Smith [18] studied absorption and desorption of O_2 in water in a column packed with 0.0541 and 0.219 cm catalyst particles and 0.413 cm glass beads at $u_G = 0.2 \sim 0.8$ cm/s and $u_L = 0.05 \sim 0.5$ cm/s. The flow was in trickle flow regime in these low flow rates. Measured coefficients, $(ka)_L$ were correlated by

$$(ka)_L/D_m = \alpha_L(\varrho_L u_L/\mu_L)^{n_L} Sc^{0.5} \tag{10}$$

   where the constants α_L and n_L are presented in Table 1. The values of $(ka)_L$ were independent of gas flow rates.

2. The pressure drop in the column is involved in the correlations [21–23].

Gianetto et al. [22] investigated the desorption of O_2 from sodium hydroxide solutions at $u_G = 30 \sim 250$ cm/s and $u_L = 0.2 \sim 5$ cm/s in a column packed with four kinds of 6 mm particles. The flow rates are in the pulse and spray flow regimes. Their correlation is

$$\varepsilon_B(ka)_L/(u_L a_L) = 0.0305\{[(\varepsilon_B \delta_{LG})/(a_t \varrho_L u_L{}^2)]^{0.068} - 1\} \tag{11}$$

Table 1

The Values of α_L and n_L

Particles	$d_p \times 10^3$ (m)	α_L (m)$^{n_L - 2}$	n_L
Glass beads	4.13	4,440	0.40
CuO · ZnO	2.91	9,080	0.41
CuO · ZnO	0.541	12,900	0.39

Table 2
The Values of α_s and n_s

Particles	$d_p \times 10^3$ (m)	α_S $(m)^{n_S-2}$	n_S
β-Naphthol	2.41	34,100	0.56
β-Naphthol	0.541	69,900	0.67

where the two-phase, pressure gradient $\delta_{LG} = (\Delta P/\Delta z)_{LG}$ is estimated from Ergun's equation for single-phase flow and the correlation between the two-phase and single-phase pressure gradient.

If a unified correlation for different flow regimes can be obtained in terms of the pressure drop, method 2 may be superior to method 1, although it is necessary to determine the pressure drop.

Liquid-Solid Mass Transfer

Van Krevelen and Krekels [24] studied the dissolution of benzoic acid (0.29 ~ 1.45 cm granular particles) in water, aqueous glycerol, and benzene at $u_G = 0$ (but with a gas phase) and $u_L = 0.012 \sim 0.37$ cm/s.

Their results were well correlated as

$$(ka)_{LS}/(D_m a_t^2) = 1.8\,(Re'_L)^{1/2}\,Sc^{1/3};\quad 0.013 < Re'_L < 12.6 \tag{12}$$

where

$$Re'_L = (\varrho_L u_L)/(\mu_L a_t) \tag{13}$$

Goto and Smith [18] studied dissolution of β-naphthol (0.054 ~ 0.24 cm gramular particles) in the gas-continuous flow regime for $u_G = 0.2 \sim 0.8$ cm/s and $u_L = 0.05 \sim 0.5$ cm/s into water and obtained

$$(ka)_{LS}/D_m = \alpha_S(\varrho_L u_L/\mu_L)^{n_s}\,Sc^{1/3} \tag{14}$$

The constants α_S and n_S were different for each packing material as shown in Table 2.

Hirose et al. [25] measured $(ka)_{LS}$ over a range of flow rates $u_G = 0 \sim 100$ cm/s and $u_L = 0.05 \sim 25$ cm/s covering gas-continuous, pulse and dispersed-bubble flow regimes in a column packed with spheres of three different diameters (0.28 ~ 1.27 cm). The enhancement factor, β, was defined by

$$\beta = (ka)_{LS} \text{ in trickle-bed operation}/\{(ka)_{LS} \text{ in liquid-full operation at the same } u_L\} \tag{15}$$

and increased from 1.2 to 2.0 with the increase in d_p in gas-continuous flow, while β was equal to the reciprocal of the liquid holdup in the pulse and dispersed-bubble flow regimes.

Dharwadkar et al. [26] correlated many published data by the following equation.

$$Sh = 1.367\,Re_L^{0.669}\,Sc^{1/3};\quad 0.2 < Re_L < 2{,}400 \tag{16}$$

Specchia et al. [27] measured the dissolution rate of benzoic acid cylinders and obtained the correlation for $u_G = 0$ (with gas phase).

$$\mathrm{Sh} = 2.79(\mathrm{Re}_L')^{0.70}\,\mathrm{Sc}^{1/3} \tag{17}$$

Yoshikawa et al. [28] used ion-exchange resins to measure $(ka)_{LS}$ for small particles (0.46 ~ 1.3 mm). The values of $(ka)_{LS}$ in gas-liquid cocurrent downflow were somewhat greater than those in liquid-full single-phase flow. The following correlation was proposed by Dwivedi and Upadhyay [29], who reanalyzed published data of liquid-full single flow in fixed and fluidized beds.

$$\varepsilon_B \mathrm{Sh}/\mathrm{Sc}^{1/3} = 0.765\,\mathrm{Re}_L^{0.18} + 0.365\,\mathrm{Re}_L^{0.614}$$

$$10^{-3} < \mathrm{Re}_L < 10^4 \tag{18}$$

The enhancement factor, β, based on Equation 18 could be expressed as

$$\beta = 1 + 0.003\,\mathrm{Re}_L \sqrt{\mathrm{Re}_G} \tag{19}$$

Tan and Smith [30] determined the values of $(ka)_{LS}$ by the dynamic adsorption method and obtained the correlation.

$$\mathrm{Sh} = 4.25\,\mathrm{Re}_L^{0.48}\,\mathrm{Sc}^{1/3} \tag{20}$$

Figure 5 shows the relation between $\varepsilon_B \mathrm{Sh}/\mathrm{Sc}^{1/3}$ and Re_L at $\varepsilon_B = 0.37$. There are some scatters among correlations of various investigators.

Gas-Solid Mass Transfer and Partial Wetting

The volumetric coefficient of mass transfer from gas to solid, $(ka)_{GS}$ can be estimated by Equation 18. This value may be higher by about 1,000 times than that of the liquid-solid system. Therefore, when the limiting reactant is present in the gas phase, the direct mass

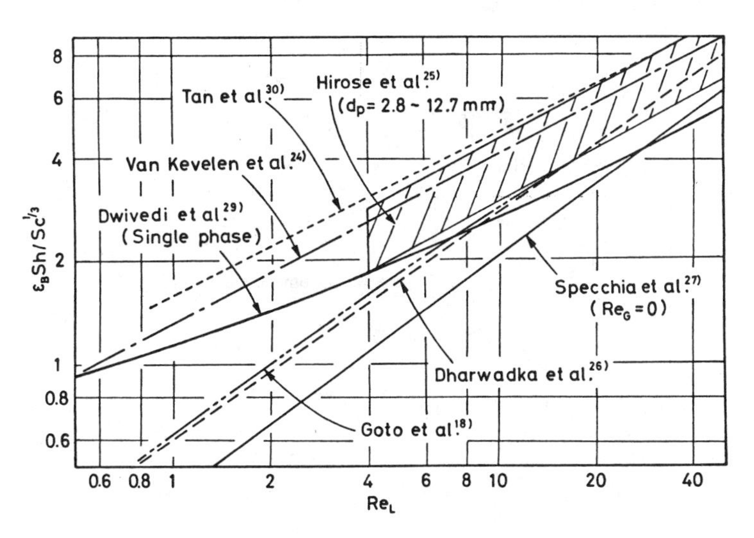

Figure 5. Liquid-solid mass transfer correlations for gas-liquid cocurrent downflow.

transfer from gas to solid through the gas-covered surface may be much higher than the indirect mass transfer from gas to liquid and then from liquid to solid through the liquid-covered surface. Thus, the increase in gas-covered surface causes an increase in the overall reaction rate. Using this effect of a gas-covered surface, Herskowitz et al. [31] and Mata et al. [32] explained their data with minimum reaction rate with respect to the liquid flow rate.

The fraction of partial wetting, f_w, varies from about 0.6 to 1 as the liquid flow rate increases [2, 33]. The catalytic effectiveness factor of the partial wetting catalyst can be approximated by the sum of contributions from the completely liquid-covered and the completely gas-covered surface, using the fraction of partial wetting, f_w, as a weighting factor [34, 35].

Gas-Liquid Cocurrent Upflow Reactors

Flow Pattern

Figure 6 shows the schematic diagram of a gas-liquid cocurrent upflow reactor in which gas and liquid flow upward through the fixed-bed packed with catalyst particles. There are various flow regimes such as bubble flow, spray flow, and slug flow. This type has some advantages over the downflow type in the previous section, such as more uniform distribution of liquid and larger liquid holdup. Disadvantages include a larger pressure drop.

Gas-Liquid Mass Transfer

Empirical correlations can be divided into two types in the same way as the downflow type.

1. Only operating conditions are used [36–38]. Goto et al. [37] compared the values of $(ka)_L$ for the upflow type with those for the downflow type as shown in Figures 7 and 8 for the particle sizes of 0.541 and 2.91 mm, respectively. In the case of the upflow type, the values of $(ka)_L$ were increased as the gas flow rates increased. They were different from the downflow type in which the values of $(ka)_L$ were independent of gas flow rates.

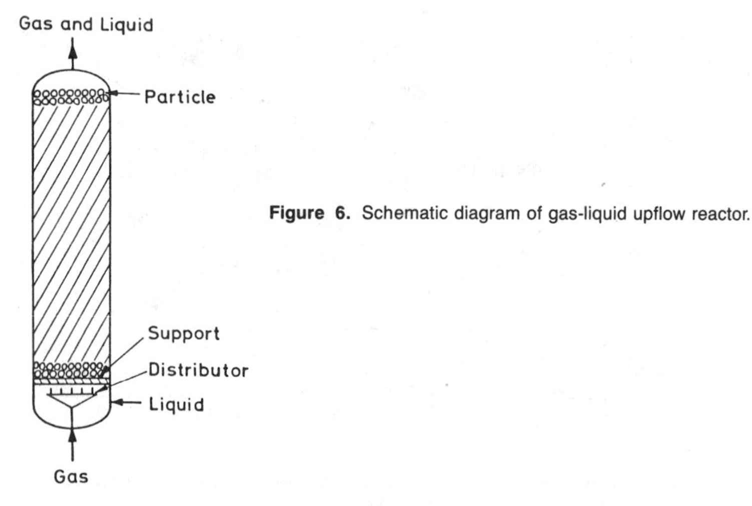

Figure 6. Schematic diagram of gas-liquid upflow reactor.

Although the particle sizes were different by 5.4 times in Figures 7 and 8, the values of $(ka)_L$ were not different. When both gas and liquid flow rates are higher, the upflow type may be superior to the downflow type. Ohshima et al. [38] found that the values of $(ka)_L$ were proportional to the dynamic gas holdups under the conditions studied (less than the minimum fluidization liquid velocity).

2. The pressure drop is involved [39, 40]. Specchia et al. [39] correlated their data as

$$\varepsilon_B(ka)_L/(u_L a_L) = 7.96 \times 10^{-3}\{(\varepsilon_B \delta_{LG})/(a_t \varrho_L u_L^2)\}^{0.275} - 9.41 \times 10^{-3} \quad (21)$$

The values of $(ka)_L$ estimated from Equation 21 for the upflow type were greater than those from Equation 11 for the downflow type at the same pressure drop. However, the pressure drops for the upflow type were greater than those for the downflow under the same operating conditions.

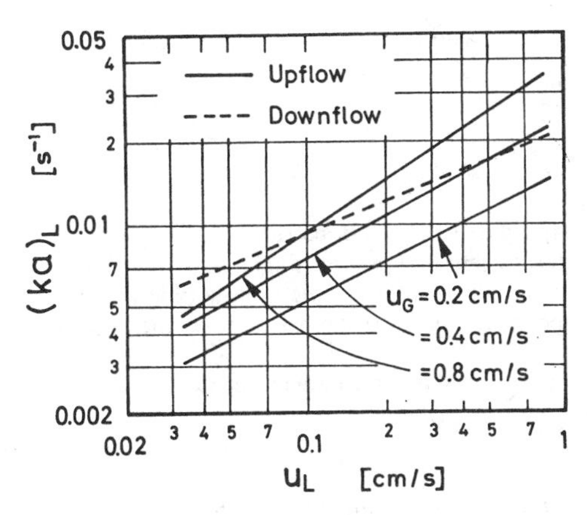

Figure 7. Volumetric mass transfer from gas to liquid at the particle diameter of 0.541 mm.

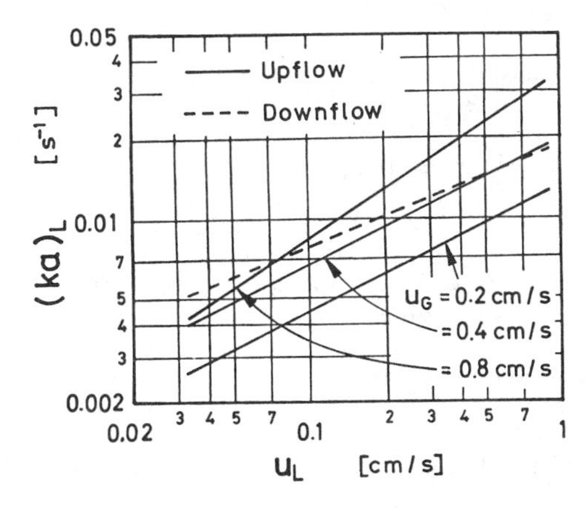

Figure 8. Volumetric mass transfer from gas to liquid at the particle diameter of 2.91 mm.

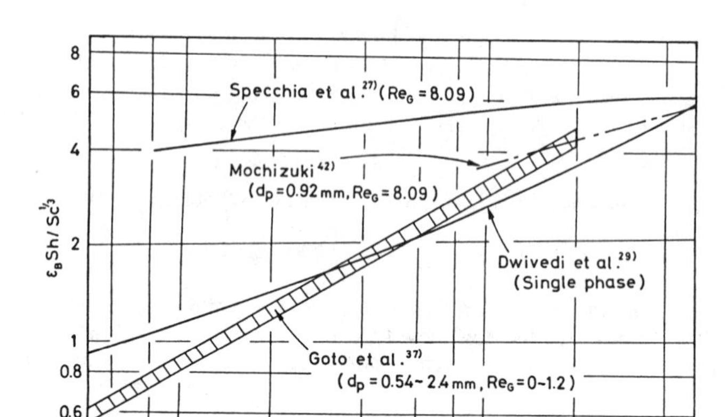

Figure 9. Liquid-solid mass transfer correlations for gas-liquid cocurrent upflow.

Liquid-Solid Mass Transfer

Mochizuki et al. [41–44] divided the relations of Sh vs. Re_L into three regions, that is, gas turbulence dominating, transient and pseudo single-phase regions. Goto et al. [37] found that the mass transfer from liquid to solid for the upflow type was somewhat greater than that for the downflow type, particularly at high gas flow rates and low liquid Reynolds numbers.

Colquhoun-Lee et al. [45] proposed the following correlation for spherical packings,

$$Sh = 0.155(\varepsilon_L d_p^4 \varrho_L^2/\mu_L^3)^{0.28}\, Sc^{1/3} \tag{22}$$

while Delaunay et al. [46] modified Equation 22 by introducing the gas holdup, h_G, as

$$Sh = 0.28\{\varepsilon_L(1-h_G)d_p^4 \varrho_L^2/\mu_L^3\}^{0.218}\, Sc^{1/3} \tag{23}$$

Specchia et al. [27] presented the correlation of $Sh/(Sh)_{liquid\text{-}full}$ as a function of the ratio between the gas and the liquid Reynolds numbers, Re_G and Re_L.

$$\ln [Sh/(Sh)_{liquid\text{-}full}] = 0.480 \ln (Re_G \times 10^2/Re_L) - 0.030[\ln (Re_G \times 10^2/Re_L)]^2 - 0.30 \tag{24}$$

where $(Sh)_{liquid\text{-}full}$ means the Sherwood number for the mass transfer from liquid to solid in the liquid-full reactor and was given by

$$(Sh)_{liquid\text{-}full}/Sc^{1/3} = 2.14(Re_L)^{1/2} + 0.990 \tag{25}$$

The values of $(ka)_{LS}$ for small packings (less than 3 mm) in upflow and downflow operations were almost identical at the same gas and liquid flow rates and correlated by Equations 18 and 19.

Figure 9 shows the relation between $\varepsilon_B Sh/Sc^{1/3}$ and Re_L at $\varepsilon_B = 0.37$.

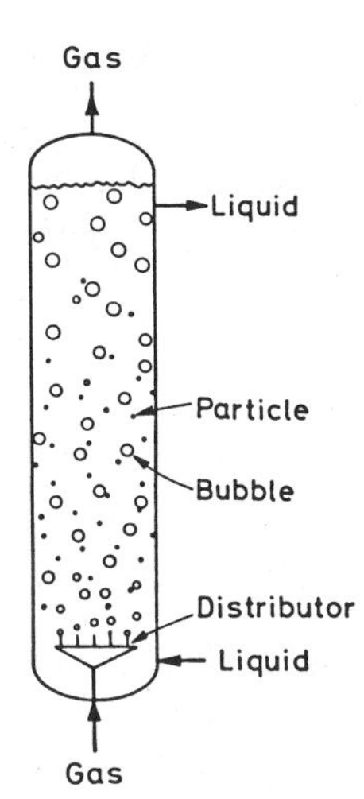

Figure 10. Schematic diagram of bubble-column slurry reactor (BCSR).

Bubble Column Slurry Reactors

Flow Pattern

Figure 10 shows the schematic diagram of a bubble-column slurry reactor in which solid particles are suspended by movement of gas bubbles. This reactor is sometimes referred to as a three-phase fluidized-bed reactor. When particles are absent, the reactor is a bubble-column reactor. The main advantages of the bubble-column reactor are their simple construction, the absence of any moving parts, ease of maintenance, and good mass transfer and heat transfer properties.

There are various flow regimes such as the liquid fluidized, the gas-bubble, and the bubble-wake regimes. In the gas-bubble regime, the liquid phase should be well mixed while the gas phase rises in plug flow.

Gas-Liquid Mass Transfer

For the bubble-column reactor without solid particles, there are many correlations [47–53] on the mass transfer from gas to liquid. Calderbank et al. [47] correlated their data for the bubbles whose diameters were smaller than 2.5 mm as

$$k_L Sc^{2/3} = 0.31[(\varrho_L - \varrho_G)\mu_L g/\varrho_L^2]^{1/3} \tag{26}$$

and for the larger bubbles as

$$k_L Sc^{1/2} = 0.42[(\varrho_L - \varrho_G)\mu_L g/\varrho_L^2]^{1/3} \tag{27}$$

Hikita et al. [53] proposed the following correlation

$$(ka)_L u_G/g = 14.9(u_G \mu_L/\sigma)^{1.76} \{\mu_L^4 g/(\varrho_L \sigma^3)\}^{-0.248} \cdot (\mu_G/\mu_L)^{0.243} Sc^{-0.604} \tag{28}$$

For the bubble-column slurry reactor, Deckwer et al. [54] measured the mass transfer rate in the Fischer-Tropsch slurry reactor. The values of $(ka)_L$ estimated from Equation 26 were in approximate accordance with the data for various amounts of suspended particles (up to 16 wt%). Therefore, the mass transfer rate from gas to liquid in the bubble-column slurry reactor may be idependent of suspended particles and can be estimated from correlations for the bubble-column reactor.

Liquid-Solid Mass Transfer

Since particles may be completely wetted in the bubble-column slurry reactor, the volumetric coefficient, $(ka)_{LS}$ can be expressed as

$$(ka)_{LS} = k_{LS}a_{LS} = k_{LS}(6/d_p)\,(M_s/\varrho_p)/V \tag{29}$$

Equation 29 may also hold for a stirred slurry reactor.

Sano et al. [55] derived the correlation in terms of the energy dissipation rate, ε.

$$Sh = [2 + 0.4(\varepsilon d_p^4/\nu_L^3)^{1/4}\, Sc^{1/3}]\varphi_c \tag{30}$$

where Carman's surface factor, φ_c, was unity for spherical particles. The value of ε in the bubble column slurry reactor can be given by

$$\varepsilon = u_G g \tag{31}$$

Sänger et al. [56] determined the mass transfer coefficient by using ion-exchange resins and proposed the following correlation.

$$Sh = 2 + 0.545(\varepsilon d_p^4/\nu_L^3)^{0.264}\, Sc^{1/3} \tag{32}$$

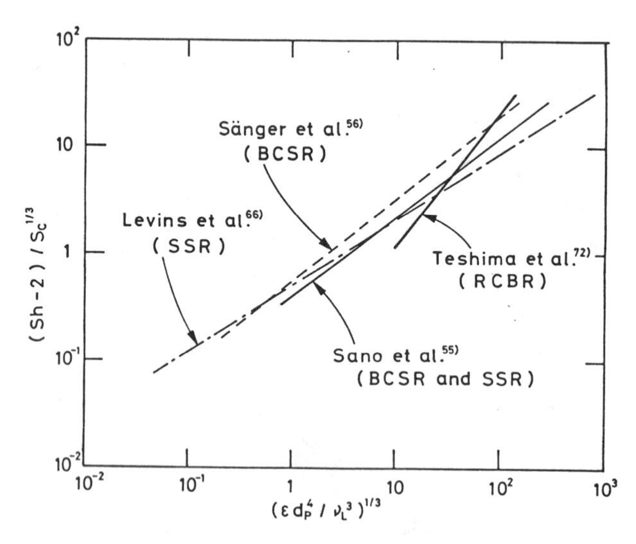

Figure 11. Comparison between correlations of mass transfer coefficients from liquid to solid in various reactors.

Figure 11 shows the relation between $(Sh-2)/Sc^{1/3}$ and $(\varepsilon d_p^4/v_L^3)^{1/3}$ for the bubble-column slurry reactor (abbreviated as BCSR) together with other types of three-phase reactors.

Stirred Slurry Reactors

Flow Pattern

Figure 12 shows the schematic diagram of a stirred slurry reactor in which catalyst particles are suspended by mechanical stirring. There are many types of impellers, such as propellers, paddles, and turbines. The flow pattern may be dependent on the type of impeller. When particles are not present, the reactor is a stirred tank reactor which can be used for gas-liquid reactions. Turbine impellers with radial flow can be used to suspend particles in the stirred slurry reactor. The liquid phase and particles are well mixed in a baffled tank at a sufficiently high rotating speed while the gas phase may be approximated in plug flow.

Gas-Liquid Mass Transfer

For a stirred tank reactor without solid particles, there are many correlations [57–61] on the mass transfer from gas to liquid. Yagi et al. [58] correlated their data for oxygen desorption from Newtonian and non-Newtonian fluids of various concentrations.

$$(ka)_L d_s^2/D_m = 0.060(d_s^2 N \varrho_L/\mu_L)^{1.5}(d_s N^2/g)^{0.19} \cdot Sc^{0.5}(\mu_L u_G/\sigma)^{0.6}(d_s N/g)^{0.32} \tag{33}$$

Equation 33 holds for Newtonian fluids. The values of $(ka)_L$ for non-Newtonian fluids can be estimated by introducing the correlation term. Riet [60] reviewed the measuring method of mass transfer rate from gas to liquid and pointed out the great difference in $(ka)_L$ between non-electrolyte and electrolyte solutions.

For a stirred slurry reactor with suspended particles, Joosten et al. [62] investigated the influence of particles on the mass transfer from gas to liquid. The values of $(ka)_L$ increased somewhat when a small fraction of particles was added. As more particles were added, the values of $(ka)_L$ remained constant at first, and then started to decline at a concentration depending on solid type and particle size. Chapman et al. [63] confirmed these trends. No significant difference in $(ka)_L$ was found between runs without solids and those with 3%

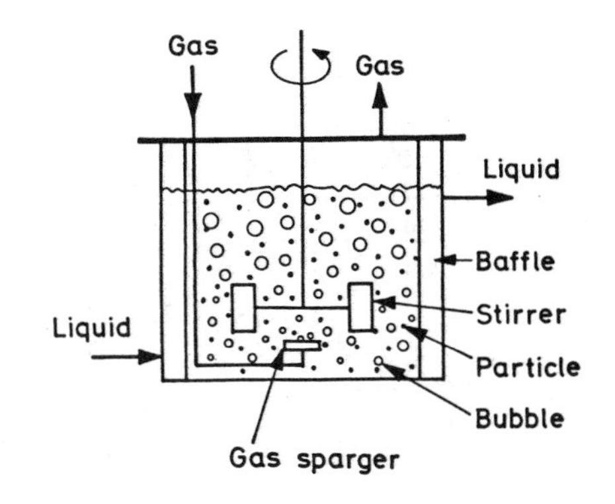

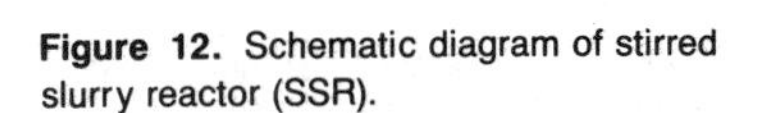
Figure 12. Schematic diagram of stirred slurry reactor (SSR).

solids. Therefore, when the fraction of suspended particles is relatively small, the values of $(ka)_L$ in the stirred slurry reactor can be estimated from correlations for the stirred tank reactor without particles.

Liquid-Solid Mass Transfer

Many investigators [64–68] have correlated the mass transfer coefficient in terms of energy dissipation rate, ε. The value of ε can be related to the torque, T, as

$$\varepsilon = 2\pi NT/(\varrho_L V) \tag{34}$$

where the value of T should be measured by a torque meter.

Otherwise, the value of ε should be estimated from correlations and charts of the power number vs. Reynolds number of rotations. Levins et al. [66] correlated their data as

$$Sh = 2.0 + 0.47(\varepsilon d_p^4/v_L^3)^{0.21}(d_s/d_t)^{0.17} Sc^{0.36} \tag{35}$$

Sano et al. [55] proposed the same Equation 30 as that in the bubble-column slurry reactor. The relation of $(Sh-2)Sc^{1/3}$ and $(\varepsilon d_p^4/v_L^3)^{1/3}$ for the stirred slurry reactor (abbreviated as SSR) is also shown in Figure 11.

On the other hand, Boon-Long et al. [69] used only operating conditions, such as rotating speeds and the dimensions of reactors.

$$Sh = 0.046(2\pi N d_p d_t \varrho_L/\mu_L)^{0.283}(\varrho_L^2 g d_p^3/\mu_L^2)^{0.173} \cdot (M_s/\varrho_p d_p^3)^{-0.011}(d_t/d_p)^{0.019} Sc^{0.461} \tag{36}$$

Catalytic Basket Reactors

Flow Pattern

Figure 13 shows the schematic diagram of a rotating catalytic basket reactor (abbreviated as RCBR) in which baskets packed with catalyst particles are rotated as stirrers. Another type is a stationary catalytic basket reactor (abbreviated as SCBR) in which baskets are kept near baffles. There are various types of catalytic baskets, such as cylindrical and rectangular baskets. The liquid phase may be well mixed while the gas phase rises in plug flow. These reactors have been widely used as laboratory reactors due to higher slip velocities and mass transfer rates between liquid and solid.

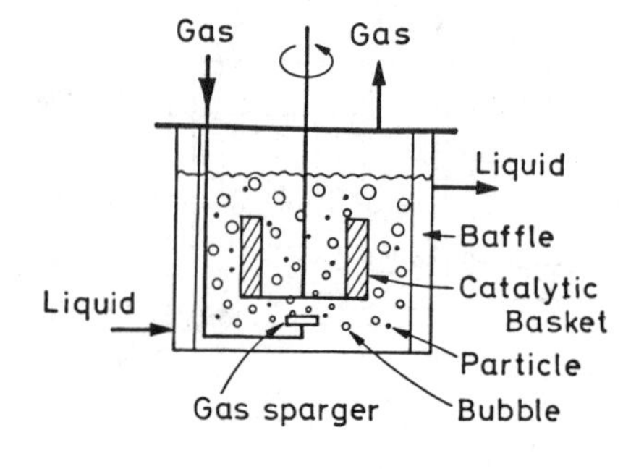

Figure 13. Schematic diagram of rotating catalytic basket reactor (RCBR).

In addition, there is no fear of sedimentation of catalyst particles into the bottom even if large, heavy particles are used [70]. However, powdered catalysts can not be used because it is difficult to keep them in baskets.

Liquid-Solid Mass Transfer

For a rotating catalytic basket reactor without gas bubbles, mass transfer rates from liquid to solid were measured by Suzuki et al. [71] and Teshima et al. [72] The experimental data can be correlated as [72]

$$\mathrm{Sh} = 2+0.012(\varepsilon d_p^4/\nu_L^3)^{0.41}\,\mathrm{Sc}^{0.64} \tag{37}$$

If the values of $\varepsilon d_p^4/\nu_L^3$ are large, the values of Sh in the rotating catalytic basket reactor become greater than those in the stirred slurry reactor as shown in Figure 11. Thus, when the rotating speed is high and the particle size is large, the rotating catalytic basket reactor may be superior to the stirred slurry reactor.

On the other hand, when gas bubbles were introduced, Pavko et al. [73] found that the apparent rate constants in basket type reactors were less than those in the liquid-full reactor and that the inactivity of the catalyst might be due to gas bubbles entrapped inside the catalyst baskets. Goto et al. [74] obtained the effects of gas velocity on the mass transfer coefficient, k_{LS}, and the torque, T, for various rotating speeds in RCBR as shown in Figure 14. For lower rotating speeds, the values of k_{LS} and T were independent of gas velocities. However, for higher rotating speeds, both values of k_{LS} and T were decreased simultaneously as the gas velocities increased. All data almost followed the line of Teshima et al. [72] in Figure 11. Therefore, the values of k_{LS} in RCBR with gas bubbles could be estimated from Equation 37.

If the screen openings are very narrow, it may be difficult for the liquid to go through the baskets and mass transfer coefficients become smaller. The effects of screen openings

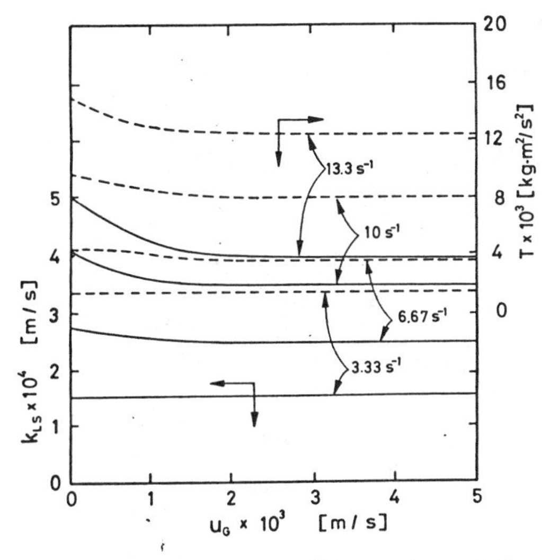

Figure 14. Effects of gas velocities on mass transfer coefficients and torque in the rotating catalytic basket reactor.

on k_{LS} in RCBR are shown in Figure 15. When the openings are narrower than 0.16 mm, the mass transfer rates may retard. The effects of the clearance between the tip of the impeller and the basket on k_{LS} in the stationary catalytic basket reactor are also shown in Figure 15. When the clearance is varied from 15 to 5 mm, the values of k_{LS} are increased, making the clearance narrow as possible. However, the values of k_{LS} in SCBR are about half of those in RCBR. The values of k_{LS} in the stirred slurry reactor are almost the same as those in the stationary catalytic basket reactor as shown in Figure 15.

KINETICS IN THREE-PHASE REACTORS

Overall Effectiveness Factor

A large number of industrially important reaction systems carryed out in three-phase reactors fit the general stoichiometric equation shown below

$$A(g) + \nu B(l) \xrightarrow[\text{surface}]{\text{on catalyst}} \text{products}$$

where the gaseous reactant must first dissolve in the liquid and then both reactants must diffuse or move to the interior surface of the catalyst particle for the reaction to occur. The potential resistances to the reactants are illustrated in Figure 16. Thus, if the reaction rate is expressed in terms of the bulk concentration and temperature of the gas and/or liquid phase, interphase and intraparticle mass transfer resistances both enter the rate equation. Such rates are called global or overall rates, and their use allows us to apply the convention-al conservation equations in the design of a reactor. The concept of an overall effectiveness factor further simplifies the calculation of reaction rate in a three-phase reactor. The global rate of reaction per unit mass of catalyst may be expressed as a product of an overall

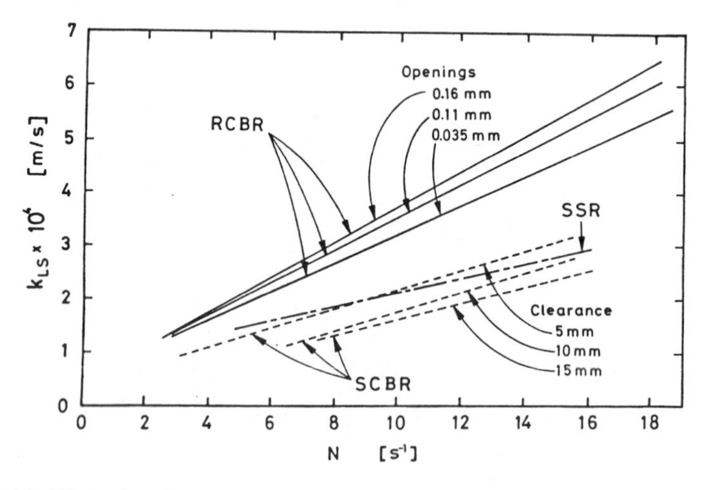

Figure 15. Effects of rotating speeds, screen openings, and locations of baskets on mass transfer coefficients in various reactors.

effectiveness factor, η_0, and the rate evaluated at the bulk conditions. For an isothermal, irreversible, n-th order reaction it is given by

$$R_0 = \eta_0 k_n (C_L^*)^n = \frac{1}{V_p} \int_{V_p} k_n C^n \, dV \tag{38}$$

where C_L^* represents the reactant concentration in the gas phase ($C_L^* = C_G/H^*$) for the case of a gas limiting reactant, or the bulk liquid concentration ($C_L^* = C_L$) when the limiting reactant is present in the liquid phase. From Equation 38 the definition of the overall effectiveness factor can be written as a ratio of the actual (global) rate to the rate based on bulk conditions

$$\eta_0 = \frac{R_0}{k_n (C_L^*)^n} = \frac{1}{V_p} \int_{V_p} C^n \, dV \tag{39}$$

where C is a dimensionless intraparticle concentration ($C = C/C_L^*$) of the limiting reactant within the particle. As indicated in Equation 39 the overall effectiveness factor can be obtained by integrating the rate of reaction over the volume of the whole particle. The concentration profile inside the catalyst particle is the solution of the governing differential equation which results by writing a mass balance for the reactant inside the liquid-filled, porous particle

$$\nabla^2 C - \Phi^2 C^n = 0 \tag{40}$$

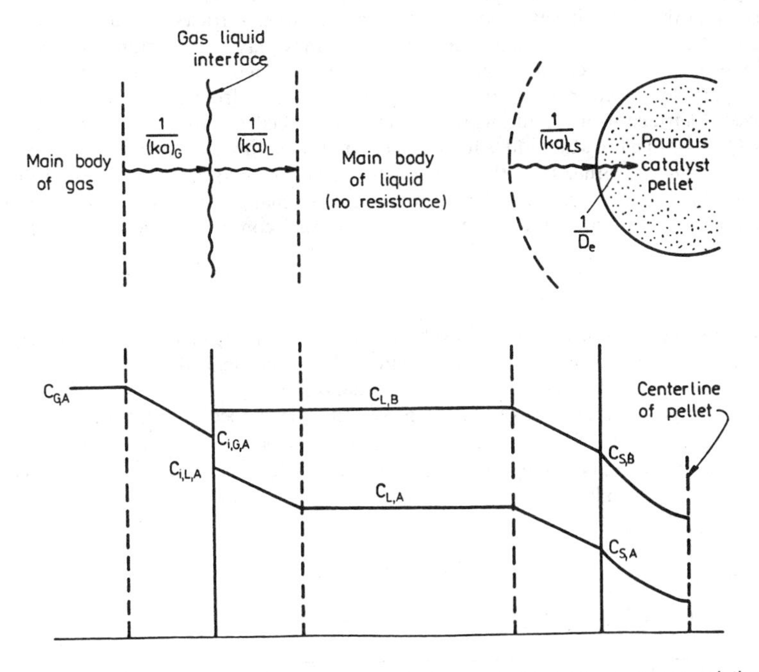

Figure 16. Potential resistances in a three-phase reactor for gas and liquid reactants to reach the interior surface of a solid catalyst.

In this case the Thiele modulus, Φ, is based on bulk conditions. The difficulties associated with the solution of Equation 40 depend strongly on the intrinsic kinetics and boundary conditions. For a first-order reaction and completely wetted particle, for example, it is readily solvable. Although a mechanistically more realistic way of representing the rate of catalytic reaction is in terms of the Langmuir-Hinshelwood model, a power-law model that has the advantage of mathematical simplicity. For the proposed reaction scheme, the intrinsic rate per unit mass of catalyst may be represented by means of the power-law kinetic model as follows

$$R = k_{m+n} C_A^m C_B^n \tag{41}$$

The analysis of the reaction rate in a three-phase system is simplified considerably when one of the reactants is in excess. This is the case when a system with pure liquid B and slightly soluble gas A, or a system with dilute liquid reactant B and highly soluble A under pressure, is used. In the first case the limiting reactant is present in the gas phase while the reaction is liquid-reactant limited in the second. In both cases the variation of unlimiting reactant concentration in the reactor is not significant and is uniform throughout the catalyst particle. Hence the kinetic model can be simplified as

$$R = k_n C^n \tag{42}$$

where k_n is a pseudo n-th order intrinsic rate constant, equal to the product of k_{m+n} and the concentration of unlimiting reactant to the m-th power.

The distinguishing feature of both bubble column and stirred slurry reactors is that small particles (about 100 microns) of catalyst are suspended in a liquid, so they permit operation at catalyst effectiveness factors approaching unity. In gas-liquid cocurrent-flow reactors as well as in both types of catalytic basket reactors large particles are usually used, and consequently, intraparticle mass transfer limitations may play an important role. While mass transfer resistances are encountered in all three-phase reactors, the partial wetting of the catalyst particle, which leads to more direct mass transfer from gas to solid through the gas-covered surface, is a unique feature of the trickle-bed reactor.

Due to its simple use attention is focused here on the methods for prediction of the overall effectiveness factor, rather than the global rate itself. It should be borne in mind that η_0 is defined on the basis of local bulk conditions, so it may change with its position in a reactor. In the analysis only power-law kinetics and isothermal conditions are considered.

Completely Wetted Particles

An effectiveness factor for the first-order reaction in a slurry system, which incorporates interphase and intraparticle mass transfer effects has been defined by Sylvester et al. [75] and Goto [76]. Ramachandran et al. [8, 77, 78] analyzed the slurry reactor also for nonlinear kinetics. Considering plug-flow of the gas phase, they developed an overall rate of reaction and η_0, based on the inlet gas concentration. For the sake of generality let us use the definition of η_0 already expressed by Equation 39. For the slurry reactor it is more convenient to express global rate per unit volume of slurry*, therefore Equation 39 becomes

$$\eta_0 = \frac{R_0}{w k_n (C_L^*)^n} \tag{43}$$

where w is the catalyst loading.

* In a trickle-bed reactor the rate is usually given per unit volume of bed, hence w must be changed to ϱ_B.

Linear kinetics. For linear kinetics with limiting reactant in the gas phase Equation 43 leads to

$$\eta_0 = \eta_s \frac{C_S}{C_L^*} = \eta_s \mathcal{C}_S \tag{44}$$

Catalyst effectiveness factor is given by

$$\eta_s = \frac{1}{\Phi_s}\left(\coth 3\Phi_s - \frac{1}{3\Phi_s}\right) \tag{45}$$

where the Thiele modulus, Φ_s, for the spherical particle is equal to $(d_p/6)\sqrt{k_1\varrho_p/D_e}$. Equating the mass transfer rate through the gas-to-liquid interphase Equation 3 to that from liquid to solid Equation 6

$$(Ka)_L\,(C_L^* - C_L) = (ka)_{LS}\,(C_L - C_S) \tag{46}$$

and the rate from liquid to solid to the diffusion-reaction rate within the catalyst

$$(ka)_{LS}\,(C_L - C_S) = \eta_s w k_1 C_S \tag{47}$$

the unknown concentration at the surface, $\mathcal{C}_S$, can be expressed in terms of the gas concentration

$$\mathcal{C}_S = \frac{C_S}{C_L^*} = \left[1 + \eta_s w k_1\left(\frac{1}{(ka)_{LS}} + \frac{1}{(Ka)_L}\right)\right]^{-1} \tag{48}$$

Equation 48 can also be written in the form

$$\mathcal{C}_S[1 + 6\eta_s\Phi_b^2(1 + Da)/Sh]^{-1} \tag{49}$$

where Da is defined as a ratio of the volumetric liquid-solid transport coefficient to the overall volumetric gas-liquid transport coefficient, $(ka)_{LS}/(Ka)_L$. In Equation 49 Φ_b is a modulus based on bulk conditions and relates to Φ_s as follows

$$\Phi_s = \Phi_b\sqrt{\frac{D_m}{D_e}} \tag{50}$$

From Equation 44 we obtain the expression for η_0

$$\eta_0 = [(1/\eta_s) + 6\Phi_b^2(1 + Da)/Sh]^{-1} \tag{51}$$

which further simplifies when the limiting reactant is present in the liquid phase, or when an equilibrium between the gas and the liquid phase exists ($Da = 0$, $C_L^* = C_L$).

Nonlinear kinetics. The analysis of a system where the reaction with a nonlinear kinetics ($n \neq 1$) takes place is not so straightforward. Difficulty arises because there is no simple relationship between η_s and Φ_s. To avoid the time-consuming solution of a nonlinear differential Equation 40, we can use Bischoff's [79] approximate explicit solution which employs a generalized Thiele modulus

$$\Phi_s = \frac{d_p}{6} R(C_S)\left[2\int_0^{C_S} D_e R(C)\, dC\right]^{-1/2} \tag{52}$$

and catalyst effectiveness factor for the linear kinetics Equation 45. The overall effectiveness factor for n-th-order reaction takes the form

$$\eta_0 = \eta_s C_S^n \tag{53}$$

Since Φ_s from Equation 52 for a n-th-order reaction

$$\Phi_s = \frac{d_p}{6}\left[\left(\frac{n+1}{2}\right)\frac{k_n \varrho_p C_S^{n-1}}{D_e}\right]^{1/2} \tag{54}$$

is also a function of C_S, it is not possible to find an explicit expression for C_S. Therefore, the surface concentration must be obtained simultaneously for η_s and Φ_s. Modifying Equation 47 for an n-th-order reaction, using the relation between C_L and C_S from Equation 46

$$C_L = \frac{1 + \mathrm{Da} C_S}{1 + \mathrm{Da}} \tag{55}$$

and noting that $a_{LS} = 6\,w/d_p\varrho_p$, the closure is obtained by the following expression

$$\frac{\mathrm{Sh}}{1+\mathrm{Da}}(1 - C_S) - \frac{12}{n+1}\eta_s C_S^n \Phi_b^2 = 0 \tag{56}$$

where the modulus, Φ_b, based on bulk conditions, is defined as

$$\Phi_b = \frac{d_p}{6}\left[\left(\frac{n+1}{2}\right)\frac{k_n \varrho_p (C_L^*)^{n-1}}{D_m}\right]^{1/2} \tag{57}$$

Given the values of Sh, Da, C_L^*, and n and the relation between surface, Φ_s, and bulk modulus, Φ_b

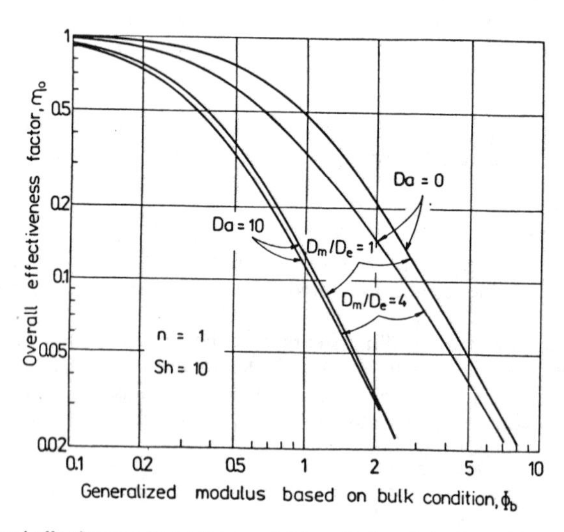

Figure 17. Overal effectiveness factor for completely wetted particle; linear kinetics.

$$\Phi_s = \Phi_b \left[\frac{D_m}{D_e} C_S^{n-1} \right]^{1/2} \tag{58}$$

Equations 45, 56, and 58 can be most easily solved by trial and error (e.g. regula falsi method) for the unknowns: η_s, Φ_s and C_S. Knowing η_s and C_S the overall effectiveness factor can be calculated from Equation 53. In the case of the limiting reactant in the liquid phase (Da = 0), C_L^* in Equation 57 must be replaced with C_L, therefore C_S is equal to C_S/C_L. In Figures 17-19 η_0 are shown for some power-law kinetics as a function of the bulk modulus.

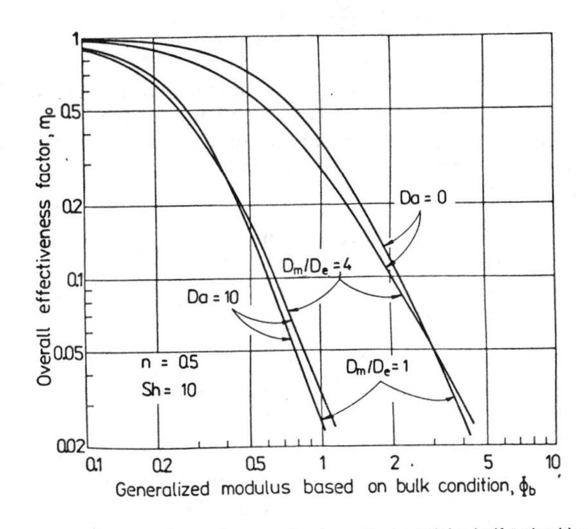

Figure 18. Overall effectiveness factor for completely wetted particle; half-order kinetics.

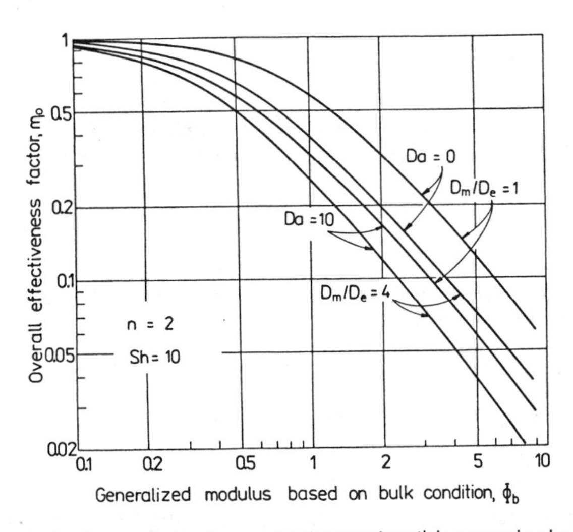

Figure 19. Overall effectiveness factor for completely wetted particle; second-order kinetics.

For nonlinear kinetics, with m = 1 and n = 1, the definition for η_0 Equation 39 yields

$$\eta_0 = \eta_s \frac{C_{S,A}}{C^*_{L,A}} \frac{C_{S,B}}{C_{L,B}} \tag{59}$$

Equation 59 indicates that the surface concentration of both reactants must be calculated simultaneously. Carrying out the integration indicated in Equation 52 gives the Thiele modulus for [1, 1] order reaction as

$$\Phi_s = \frac{d_p}{6}\left[\frac{k_2\varrho_p C_{S,B}}{D_{e,A}}\right]^{1/2}\left[\frac{C_{S,B}/C_{S,A}}{(C_{S,B}/C_{S,A}) - \nu D_{e,A}/3D_{e,B}}\right]^{1/2} \tag{60}$$

This equation is valid when reactant A becomes limiting within the particle. However, the generality of the analysis is not lost since the nomenclature for the components A and B is arbitrary and can be interchanged if necessary. In order to obtain closure we must write two equations similar to Equation 56, one for reactant A

$$\frac{Sh_A}{1 + Da_A}(C^*_{L,A} - C_{S,A}) - \frac{d_p^2}{6}\eta_s \frac{k_2\varrho_p}{D_{m,A}} C_{S,A}C_{S,B} = 0 \tag{61}$$

another for reactant B

$$Sh_B(C_{L,B} - C_{S,B}) - \frac{d_p^2}{6}\eta_s \frac{\nu k_2\varrho_p}{D_{m,B}} C_{S,A}C_{S,B} = 0 \tag{62}$$

and solve them together with Equations 45 and 60 by trial and error for the unknown η_s, Φ_s, $C_{S,A}$ and $C_{S,B}$, respectively. It is obvious that the explicit solution of η_0 for a nonlinear kinetics is not possible except for a few limiting cases [77].

The criterion for significant external mass transfer limitation can be said to correspond to $C_S < 0.95\, C^*_L$. Applying this criterion to

$$\frac{(ka)_{LS}}{1 + Da}(C^*_L - C_S) = R_0 \tag{63}$$

leads to the conclusion that mass transfer from the gas phase to the outer surface of the catalyst pellets will not be significant unless the following inequality holds

$$\frac{10d_p^2}{3} \frac{R_0}{D_m C^*_L} \frac{\varrho_p}{w} > \frac{Sh}{1 + Da} \tag{64}$$

The intraparticle mass transfer resistance may be considered to be negligible when the condition of Equation 65 is satisfied

$$\frac{d_p}{6}\left[\left(\frac{n+1}{2}\right)\frac{k_n\varrho_p(C^*_L)^{n-1}}{D_m}\right]^{1/2} < 0.2 \tag{65}$$

Partially Wetted Particles

A partially wetted catalyst pellet can be modeled as illustrated in Figure 4. The external surface of the pellet is divided into a liquid-covered and gas-covered part, which actually may be covered by a very thin liquid film. It is assumed that the pores of the catalyst pellet are completely filled with liquid due to capillary forces [13, 80]. However, the partial differential Equation 40 together with mixed boundary conditions, which result from the partially wetted catalyst surface, can be solved analytically only for first-order kinetics ($n = 1$). Such calculations have been carried out for various particle shapes: cube, slab, cylinder, and sphere, respectively [31, 81–83]. Herskowitz [83] and Mills et al. [84] showed that the effect of particle shape on the overall effectiveness factor is not significant. The evaluation of η_0 from the previously mentioned models requires extensive computational work. Therefore many attempts have been made to approximate the solution of η_0 for partially wetted particles [34, 35, 82, 85]. The comparison between approximate and exact (numerical) solutions show the discrepancy to be less than 10% for a wide range of mass transfer and kinetics parameters [34, 35, 83]. Usually any shape of the catalyst particles in a trickle-bed reactor is represented by a sphere having an equivalent diameter. Therefore, a simple procedure which leads to an approximate η_0 for the partially wetted spherical catalyst particle is treated in the following.

The overall effectiveness factor, based upon bulk concentration, may be approximated by the sum of contributions from the completely liquid-covered and the completely gas-covered surface using the fraction of partial wetting, f_w, on the external surface of a sphere as a weighting factor. Thus

$$\eta_{0,app} = f_w \eta_{s,w} C_{S,w}^n + (1 - f_w)\eta_{s,d} C_{S,d}^n \tag{66}$$

where $\eta_{s,w}$ represents the conventional catalyst effectiveness factor when the surface of a sphere is entirely covered by liquid, and $\eta_{s,d}$ by gas.

Limiting reactant is gas. For linear kinetics $\eta_{0,app}$ is easily obtained since $\eta_{s,w}$, $\eta_{s,d}$, and $C_{S,d}$ are available as follows

$$\eta_s = \eta_{s,w} = \eta_{s,d} = \frac{1}{\Phi_s}\left(\coth 3\Phi_s - \frac{1}{3\Phi_s}\right) \tag{67}$$

$$C_{S,w} = [1 + 6\eta_s \Phi_b^2 (1 + Da)/Sh]^{-1} \tag{68}$$

$$C_{S,d} = [1 + 6\eta_s \Phi_b^2 / Sh_G]^{-1} \tag{69}$$

The relationship Φ_s vs. Φ_b is given by Equation 50. If there is an equilibrium between the gas and the liquid phase ($Da = 0$) Equation 68 reduces to the form of Equation 69.

For nonlinear kinetics ($n \neq 1$) such a straightforward calculation is not possible in general. Avoiding numerical computation to perform the relationship η_s vs. Φ_s we can employ the same approach as is used in the preceding section. Applying Equation 56 to the partially wetted pellet the following expression

$$\frac{Sh}{1 + Da}(1 - C_{S,w}) - \frac{12}{n+1}\eta_{S,w} C_{S,w}^n \Phi_b^2 = 0 \tag{70}$$

results for a completely liquid-covered pellet, and

$$Sh_G(1 - C_{S,d}) - \frac{12}{n+1}\eta_{s,d} C_{S,d}^n \Phi_b^2 = 0 \tag{71}$$

for a completely gas-covered pellet. Φ_b in the prior equations is defined by Equation 57 and its relation to Φ_s is given by Equation 58. For the particular values of Sh, Da, Φ_b, C_L^*, and n simultaneous solution (by trial and error) of Equations 45, 58, and 70 yields the result for $\eta_{s,w}$, $\Phi_{s,w}$, and $C_{S,w}$, respectively. The same procedure with Equations 45, 58, and 71 leads to $\eta_{s,d}$, $\Phi_{s,d}$, and $C_{S,d}$, respectively, so that Equation 66 can be used to predict $\eta_{0,app}$ if the fraction of partial wetting is known.

Limiting reactant in liquid. When the limiting reactant is present in the liquid phase direct mass transfer from gas to solid does not need to be considered. For this case Dudukovic [86] modified the Thiele modulus as

$$\Phi_s' = \Phi_s/f_w \tag{72}$$

and the approximate overall effectiveness factor takes the form

$$\eta_{0,app}' = \eta_{s,w}' C_{S,w}'^{n} \tag{73}$$

Here $\eta_{s,w}'$ and $C_{S,w}'$ can be calculated from Equations 45, 58, and 70 but replacing Φ_s and Sh with Φ_s' and Sh/f_w, respectively. In Equation 70 Da = 0 and the dimensionless surface concentration is equal to C_S/C_L. The modification of the Thiele modulus and Sherwood number means that the equivalent particle diameter is d_p/f_w, because the reactant can reach the gas-covered part of the pellet only through the liquid-covered surface.

Approximate overall effectiveness factors as a function of the generalized modulus, Φ_b, are shown in Figures 20–22 for the gas and the liquid-limiting reactant. The effect of fractional wetting coverage on $\eta_{0,app}$ is demonstrated in Figure 23.

Kinetics from Laboratory Three-Phase Reactors

The effects of interphase and intraparticle transport processes on the global rate, actually on the overall effectiveness factor, are analyzed in detail in the previous section. Since an intrinsic rate equation (e.g., Equation 41) cannot be predicted, it must be evaluated from laboratory kinetics data. Such data are provided by measurements of the global rate of reaction. The problem is to extract the equation for the intrinsic rate from the global rate

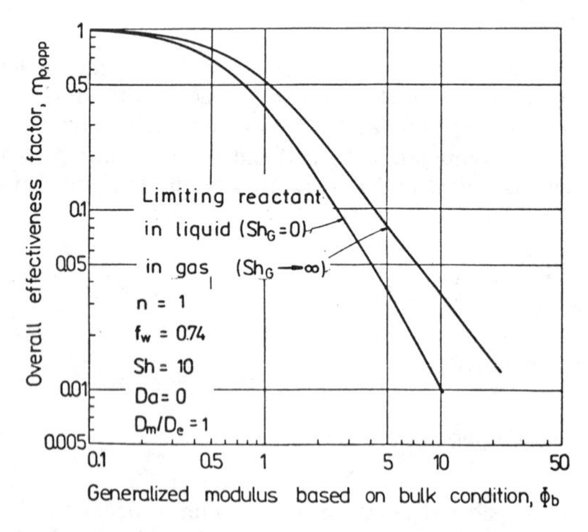

Figure 20. Overall effectiveness factor for partially wetted particle; linear kinetics.

data. Therefore, such construction and operating conditions must be chosen that reduce or eliminate the differences between the global and intrinsic rates.

Choosing the proper type of laboratory reactor to evaluate kinetic parameters for an industrial process is perhaps one of the most crucial decisions that must be made in this area. A wrong choice may lead to erroneous kinetics with disastrous consequences in commercial reactor design. The experimental strategy of studying kinetics in the gas-liquid-solid system, where the chemical reaction takes place on the solid catalyst, usually involves evaluation of the potentially significant external and internal mass transport effects. Thus, the study of a three-phase catalytic process, such as desulfurization of heavy oils or oxidation of organic pollutants in water, requires a laboratory reactor that ensures good contact between phases and reduces or eliminates external mass transfer resistances.

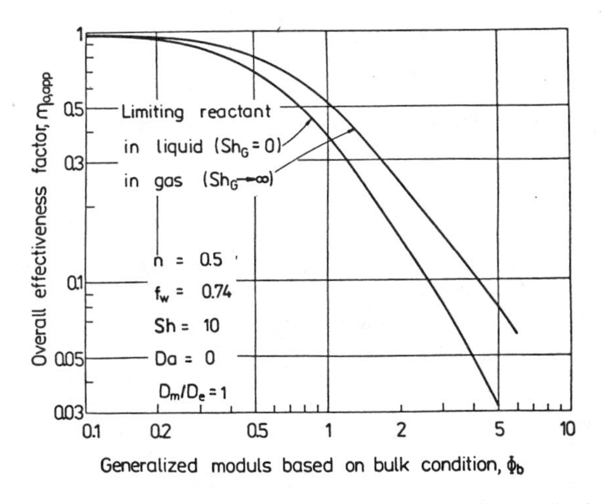

Figure 21. Overall effectiveness factor for partially wetted particle; half-order kinetics.

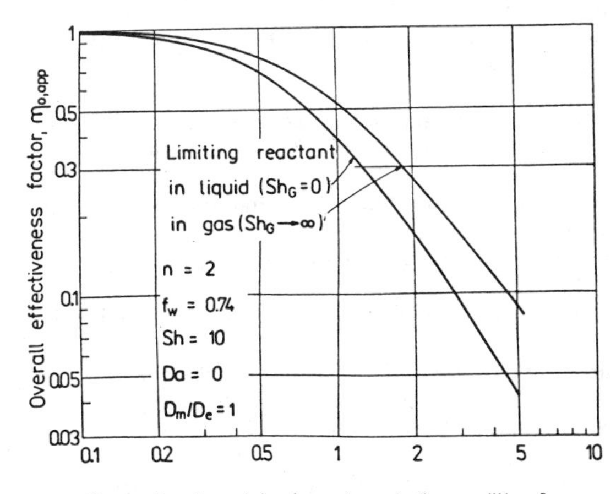

Figure 22. Overall effectiveness factor for partially wetted particle; second-order kinetics.

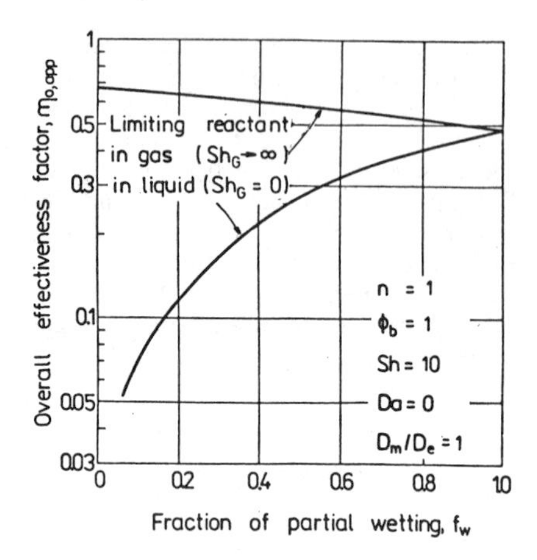

Figure 23. Effect of fractional wetting coverage on overall effectiveness factor.

Weekman [87] suggested that a catalytic basket reactor, as one of the gradientless reactors, might also be used for obtaining kinetics data in a three-phase system. Myers and Robinson [88] employed a rotating catalytic basket reactor (RCBR) to study the desulfurization kinetics. They tested a cross-shaped paddle and annular baskets. However, they concluded that the rotating annular basket is more efficient than the paddle type, and that an RCBR is generally superior to a trickle-bed reactor for the purpose of obtaining kinetics data in a three-phase system. Relatively few studies have been devoted to exploring the usefulness of RCBR [70, 73, 88–91] and stationary catalytic basket reactor (SCBR) [73, 90, 91]. As shown in Figure 15, liquid-to-solid mass transfer resistance can be easily reduced in the RCBR.

Important work which sheds more light on the type of three-phase reactor one should use in studying kinetics is that of Goto and Kojima [90]. They compared a stirred slurry reactor (SSR), RCBR, SCBR, a bubble-column slurry reactor (BCSR), and a bubble-column packed reactor (BCPR) by using oxidation of sulfur dioxide on activated carbon as a test reaction. The best choice can not be recommended directly since it depends on the following operating conditions:

1. The stirred slurry reactor (SSR) is the best choice for fine particles and low gas flow rates.
2. The rotating catalytic basket reactor (RCBR) may be used preferentially for large particles and low gas flow rate.
3. The bubble-column slurry reactor (BCSR) is advantageous when fine particles and a high gas flow rate are used.
4. The bubble-column packed reactor (BCPR) is a good selection for a system with large particles and high flow rates.
5. The stationary catalytic basket reactor (SCBR) should not be used under any conditions.

Figure 24 illustrates the relative contributions of an individual resistance as a function of particle diameter for SSR, RCBR, and SCBR. As indicated, gas-to-liquid mass transfer resistance does not play an important role in these reactors. Pavko and Levec [73] reported negligible external transport effects in RCBR and SCBR, but the kinetic parameters may be erroneous due to incompletely wetted catalyst in the baskets. They found that gas bubbles can be trapped inside the catalyst batch. Measuring the rate of dissolution of the benzoic acid particles with and without the presence of gas, it was found that particles in the rotating

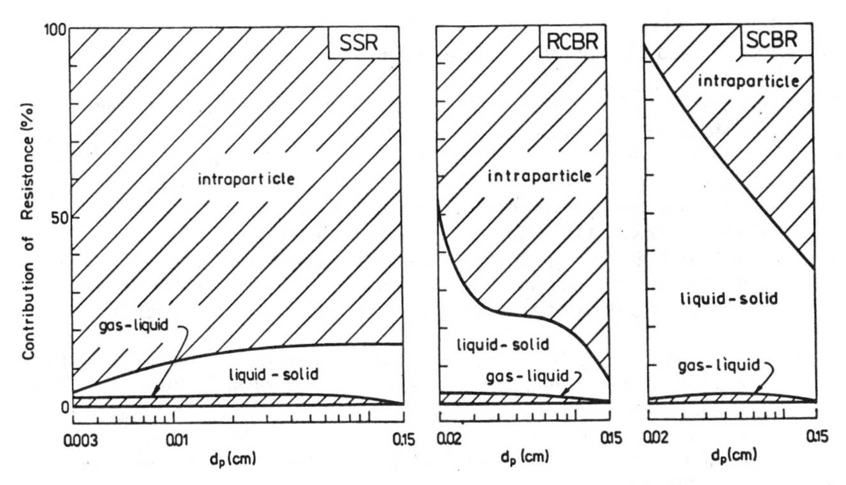

Figure 24. Relative resistances in stirred-type reactors as a function of particle diameter.

baskets are fully wetted as long as the basket thickness to particle diameter ratio is less than 3.0 [91].

Kinetics of Some Hydrogenation and Oxidation Reactions

Most applications of trickle-bed reactors for chemical processing have been for hydrogenation of relatively nonvolatile liquids, notably hydrodesulfurization of petroleum fractions, while a slurry reactor has been used for hydrogenation of unsaturated oils, Fischer-Tropsch synthesis, oxidation of olefins, and polymerization reactions. The growing importance of removing dissolved organic compounds from industrial wastewater and sulfur dioxide from flue gases has also increased the potential use of trickle-bed and slurry reactors. Here we discuss the intrinsic rate equations for a few hydrogenation and oxidation reactions which are usually carried out in three-phase reactors.

Hydrogenation. Petroleum feedstocks include the sulfur-containing compounds such as thiols and thiophenes. Among those taking place in hydrodesulfurization reaction networks are hydrogenolysis of C-S and C-C bonds, and hydrogenation of unsaturated compounds, both consuming hydrogen. As pointed out by Frye and Mosby [92], and Schuit and Gates [93], the rate of hydrodesulfurization of any feed over a cobalt molybdate catalyst is generally consistent with the first-order reaction of each of the sulfur-containing compounds. It is first order with respect to hydrogen partial pressure at low values and zero order at high values. They summarize the intrinsic rate equation as

$$R_{HDS} = \frac{\alpha k_2 C_{SUL} C_{H_2}}{1 + K_{H_2} C_{H_2} + K_{H_2S} C_{H_2S}}$$

where α is the uncreative fraction of sulfur. Inhibition of the reaction by product H_2S is accounted for by the denominator term. The same form of rate equation corresponds to the desulfurization kinetics of model sulfur compounds like benzothiophene [88] and thiophene [94]. It appears to be safe to say that external mass transfer effects do not influence rates of hydrodesulfurization in well-designed pilot-scale and commercial-scale trickle-bed reactors [2, 93]. In contrast, the intraparticle diffusion rate of hydrogen or sulfur-containing oil molecules in liquid-filled pores may be low compared to the intrinsic rate, and therefore rate limiting step.

Hydrogenation of α-methylstyrene over a Pd/Al_2O_3 catalyst is a reaction which is very often used to study the performance of trickle-bed [14, 31, 95, 96] and slurry [97] reactors. While the intrinsic kinetics results are not entirely consistent, over a narrow range of low hydrogen concentration (partial pressure of about 1 atm), and at high styrene concentrations (pure liquid), the reaction appears to be first order in hydrogen [14, 96, 97]. By means of Equation 64 we can conclude that diffusional resistance of the liquid films ($Da = 0$) cannot be expected to be of importance in trickle-bed reactors operating under typical commercial conditions [2], although in a laboratory reactor this might be the case. In a slurry reactor Sherwood and Farkas [97] reported a controlling external mass transfer effect.

Oxidation. A common model reaction for the wet oxidation of organic pollutants is the oxidation of phenol solutions over a copper oxide catalyst [70, 89, 98]. Although the kinetics results disagree slightly, the rate of phenol consumption is first order with respect to phenol concentration (at about 5×10^{-5} mol/cm^3) [98] and one-half order with respect to oxygen (at partial pressures of 4–10 atm) [70, 98]. In a slurry system it may also be dependent on catalyst loading [70].

Other examples are oxidation of formic [15, 99] and acetic acid solutions [100, 101]. In the case of formic acid the rate was found to be proportional to the concentration of acid and to that of oxygen. The data on the rate of acetic acid oxidation showed a half-order dependency on oxygen and a $C/(1+KC)$ functional variation with acetic acid concentration. In both cases the liquid concentrations of oxygen were in the range 3×10^{-7} to 10×10^{-7} mol/cm^3 while the acid was about 50×10^{-7} mol/cm^3. Formic acid was oxidized over a CuO-ZnO catalyst, but acetic acid was very refractory over a catalyst made of copper, manganese, and lanthanum oxides on a zinc aluminate spinel [101]. Oxidation of formic and acetic acid in a laboratory trickle-bed reactor indicated that external mass transfer resistances may retard the reactions [99, 100].

Wet oxidation of SO_2 using active carbon as a catalyst is closely related to the commercial desulfurization of flue gases [32, 102]. Intrinsic rate equations reported in the literature are not at all consistent: the order with respect to oxygen varies from one-half [90, 102] to first order [73, 103], but for SO_2 from zero [73, 103] to first order [103]. The rate can probably be well represented using first order in oxygen (21 mol % in gas) and zero in SO_2 (> 2 mol % in gas), but at low concentration of SO_2, which might be the case in air pollution control, the reaction would depend on the SO_2 concentration. Goto and Kojima [90] found the dependency to be of the power of 0.2.

NOTATION

- a_L specific external surface area of particles for gas-liquid transport (m^{-1})
- a_{LS} outer surface area of particles for liquid-solid transport (m^{-1})
- a_t total geometric external surface area of particles per unit volume reactor (m^{-1})
- C concentration (mol/m^3)
- *C* dimensionless concentration (–)
- Da Damkohler number defined as ratio of volumetric mass transfer coefficients ($= (ka)_{LS}/(Ka)_L$) (–)
- D_e effective intraparticle diffusivity (m^2/s)
- D_m molecular diffusivity in liquid or gas (m^2/s)
- d_p average particle diameter (m)
- d_s stirrer diameter (m)
- d_t tank diameter (m)
- f_w fraction of partial wetting on external surface of particles (–)
- g acceleration of gravity ($= 9.8$ m/s^2)
- H^* Henry's law constant (–)
- h_G gas holdup (–)

$(Ka)_G$ overall volumetric coefficient based on gas phase (s^{-1})
$(Ka)_L$ overall volumetric coefficient based on liquid phase (s^{-1})
k_n pseudo n-th order rate constant ($(m^3/kg \cdot s)(m^3/mol)^{n-1}$)
$(ka)_G$ volumetric coefficient of gas-liquid transport in gas-side (s^{-1})
$(ka)_L$ volumetric coefficient of gas-liquid transport in liquid-side (s^{-1})
$(ka)_{GS}$ volumetric coefficient of gas-solid transport (s^{-1})
$(ka)_{LS}$ volumetric coefficient of liquid-solid transport (s^{-1})
k_{GS} mass transfer coefficient from gas to solid (m/s)
k_L mass transfer coefficient from gas to liquid in liquid side (m/s)
k_{LS} mass transfer coefficient from liquid to solid (m/s)
M_s total mass of catalyst particles in reactor (kg)
N rotating speed (s^{-1})
N_{GL} rate of mass transfer from gas to liquid ($mol/m^3 \cdot s$)
N_{GS} rate of mass transfer from gas to solid ($mol/m^3 \cdot s$)
N_{LS} rate of mass transfer from liquid to solid ($mol/m^3 \cdot s$)
N_{intra} intraparticle diffusion ($mol/m^2 \cdot s$)
m, n order of reaction (–)
n_L constant in Equation 10 (–)
n_s constant in Equation 14 (–)
R intrinsic rate of reaction ($mol/kg \cdot s$)
R_0 overall rate of reaction ($mol/kg \cdot s$)
Re Reynolds number ($= d_p \varrho u/\mu$) (–)
Re′ modified Reynolds number ($= \varrho u/\mu a_t$) (–)
r radial coordinate in particle (m)
Sc Schmidt number ($= \mu_L/\varrho_L D_m$) (–)
Sh Sherwood number ($= k_{LS} d_p/D_m$ for suspended particles, $(ka)_{LS} d_p/a_t D_m$ for partially wetted particles) (–)
Sh_G Sherwood number for gas-covered particle ($= k_{GS} d_p/D_m$) (–)
T torque ($kg \cdot m^2/s^2$)
u_G superficial gas velocity (m/s)
u_L superficial liquid velocity (m/s)
V volume of reactor (m^3)
V_p volume of particle (m^3)
w mass of catalyst per unit volume of slurry (kg/m^3)

Greek Symbols

α_L constant in Equation 10 ($m^{n_L - 2}$)
α_S constant in Equation 14 ($m^{n_S - 2}$)
β enhancement factor defined by Equation 15 (–)
δ_{LG} frictional pressure gradient in liquid-gas two-phase flow ($kg/(m^2 \cdot s^2)$)
ε energy dissipation rate per unit mass of reactor (m^2/s^3)
ε_B void fraction in bed (–)
ε_L energy dissipation rate in liquid per unit volume of reactor ($kg/(m \cdot s^3)$)
ε_p porosity of particles (–)
η_0 overall effectiveness factor defined by Equation 39 (–)
$\eta_{0,app}$, $\eta'_{0,app}$ approximate overall effectiveness factor defined by Equations 66 and 73, respectively (–)
η_s catalyst effectiveness factor (–)
μ fluid viscosity ($kg/(m \cdot s)$)
ν kinematic viscosity ($= \mu/\varrho$) (m^2/s); stoichiometric coefficient in Equation 60 (–)
ϱ density of fluid (kg/m^3)
ϱ_B density of bed (kg/m^3)
ϱ_p apparent density of particle (kg/m^3)
σ surface tension (kg/s^2)
τ tortuosity factor (–)
Φ Thiele modulus based on bulk concentration ($= (d_p/6)\sqrt{k_n (C_L^*)^{n-1}/D_e}$) (–)
Φ_b generalized Thiele modulus based on bulk condition, defined by Equation 57 (–)

Φ_c Carman's surface factor (–)
Φ_s generalized Thiele modulus based on surface concentration, defined by Equation 52 or 54 (–)
Φ_s' modulus defined by Equation 72 (–)

Subscripts

A component A
B component B
d gas-covered surface (dry)
G gas phase
i interphase
L liquid phase
S solid phase
w liquid-covered surface (wetted)

REFERENCES

1. Østergaard, K., "Gas-Liquid-Particle Operations in Chemical Reaction Engineering," *Advances in Chemical Engineering,* Vol. 7 (1968), pp. 71–137.
2. Satterfield, C. N., "Trickle-Bed Reactors," *AIChE Journal,* Vol. 21, No. 2 (1975), pp. 209–228.
3. Charpentier, J. C., "Recent Progress in Two Phase Gas-Liquid Mass Transfer in Packed Beds," *Chem. Eng. J.,* Vol. 11 (1976), pp. 161–81.
4. Goto S., Levec, J., and Smith, J. M., "Trickle-Bed Oxidation Reactors," *Catal. Rev. Sci. Eng.,* Vol. 15, No. 2 (1977), pp. 187–247.
5. Hofmann, H. P., "Multiphase Catalytic Packed-Bed Reactors," *Catal. Rev. Sci. Eng.,* Vol. 17, No. 1 (1978), pp. 71–117.
6. Gianetto, A., et al., "Hydrodynamics and Solid-Liquid Contacing Effectiveness in Trickle-Bed Reactors," *AIChE Journal,* Vol. 24, No. 6 (1978), pp. 1087–1104.
7. Shah, Y. T., *Gas-Liquid-Solid Reactor Design,* New York: McGraw-Hill Book Company, 1979.
8. Chaudhari, R. V., and Ramachandran, P. A., "Three Phase Slurry Reactors," *AIChE Journal,* Vol. 26, No. 2 (1980), pp. 177–201.
9. Van Landeghem, H., "Multiphase Reactors: Mass Transfer and Modeling," *Chem. Eng. Sci.,* Vol. 35, No. 9 (1980), pp. 1912–1949.
10. Rodrigues, A. E., Calo, J. M., and Sweed, N. H. (Eds.), *Multiphase Chemical Reactors Volumes 1 and 2,* Netherlands: Sijthoff and Noordhoff International Publishers, 1981.
11. Goto, S., "Mass Transfer Characteristics in Three-Phase Reactors," *Kagaku Kogaku,* Vol. 46, No. 4 (1982), pp. 228–231.
12. Alper, E. (Ed.), *Mass Transfer with Chemical Reaction in Multiphase Systems Volume 1 and 2,* Hague Netherlands: Martinus Nijhoff Publishers, 1983.
13. Colombo, A. J., Baldi, G., and Sicardi, S., "Solid-Liquid Contacting Effectiveness in Trickle Bed Reactors," *Chem. Eng. Sci.,* Vol. 31, No. 12 (1976), pp. 1101–1108.
14. Satterfield, C. N., Pelossof, A. A., and Sherwood, T. K., "Mass Transfer Limitations in a Trickle-Bed Reactor," *AIChE Journal,* Vol. 15, No. 2 (1969), pp. 226–234.
15. Baldi, G., et al., "Catalytic Oxidation of Formic Acid in Water. Intraparticle Diffusion in Liquid-Filled Pores," IEC Process Des. Develop., Vol. 13, No. 4 (1974), pp. 447–452.
16. Germain, A. H., Lefebvre, A. G., and L'homme, G. A., "Experimental Study of a Catalytic Trickle Bed Reactor," In *Chemical Reaction Engineering-2,* H. M. Hulburt (Ed.) Washington, D. C., 1974, pp. 164–180.
17. Ufford, R. C., and Perona, J. J., "Liquid Phase Mass Transfer with Cocurrent Flow through Packed Towers," *AIChE Journal,* Vol. 19, No. 6 (1973), pp. 1223–1226.
18. Goto, S., and Smith, J. M., "Trickle-Bed Performance Part I. Holdup and Mass Transfer Effects," *AIChE Journal,* Vol. 21, No. 4 (1975), pp. 706–713.

19. Sylvester, N. D., and Pitayagulsarn, P., "Mass Transfer for Two-Phase Cocurrent Downflow in a Packed Bed," *IEC Process Des. Develop.*, Vol. 14, No. 4 (1975), pp. 421–426.
20. Fukushima, S., and Kusaka, K., "Liquid-Phase Volumetric and Mass-Transfer Coefficient and Boundary of Hydrodynamic Flow Region in Packed Column with Cocurrent Downward Flow," *J. Chem. Eng. Japan*, Vol. 10, No. 6 (1977), pp. 468–474.
21. Reiss, L. P., "Cocurrent Gas-Liquid Contacting in Packed Columns," *IEC Process Des. Develop.*, Vol. 6, No. 4 (1976), pp. 486–499.
22. Gianetto, A., Specchia, V., and Baldi, G., "Absorption in Packed Towers with Concurrent Downward High-Velocity Flows-2: Mass Transfer," *AIChE Journal*, Vol. 19, No. 5 (1973), pp. 916–922.
23. Turek, F., and Lange, R., "Mass Transfer in Trickle-Bed Reactors at Low Reynolds Number," *Chem. Eng. Sci.*, Vol. 36, No. 3 (1981), pp. 569–579.
24. Van Krevelen, D. W., and Krekels, J. T. C., "Rate of Dissolution of Solid Substances," *Recueil Trans. Chem.*, Vol. 67 (1948), p. 512.
25. Hirose, T., Mori, Y., and Sato, Y., "Liquid-to-Particle Mass Transfer in Fixed Bed Reactor with Cocurrent Gas-Liquid Downflow," *J. Chem. Eng. Japan*, Vol. 9, No. 3 (1976), pp. 220–225.
26. Dharwadkar, A., and Sylvester, N. D., "Liquid-Solid Mass Transfer in Trickle Beds," *AIChE Journal*, Vol. 23, No. 3 (1977), pp. 376–378.
27. Specchia, V., Baldi, G., and Gianetto, A., "Solid-Liquid Mass Transfer in Concurrent Two-Phase Flow through Packed Beds," *IEC Process Des. Develop.*, Vol. 17, No. 3 (1978), pp. 362–367.
28. Yoshikawa, M., et al., "Liquid-Solid Mass Transfer in Gas-Liquid Cocurrent Flows through Beds of Small Packings," *J. Chem. Eng. Japan*, Vol. 14, No. 6 (1981), pp. 444–450.
29. Dwivedi, P. N., and Upadhyay, S. N., "Particle-Fluids Mass Transfer in Fixed and Fluidized Beds," *IEC Process Des. Develop.*, Vol. 16, No. 2 (1977), pp. 157–165.
30. Tan, C. S., and Smith, J. M., "A Dynamic Method for Liquid-Particle Mass Transfer in Trickle Beds," *AIChE Journal*, Vol. 28, No. 2 (1982), pp. 190–195.
31. Herskowitz, M., Carbonell, R. G., and Smith, J. M., "Effectiveness Factors and Mass Transfer in Trickle-Bed Reactors," *AIChE Journal*, Vol. 25, No. 2 (1979), pp. 272–282.
32. Mata, A. R., and Smith, J. M., "Oxidation of Sulfer Dioxide in a Trickle-Bed Reactor," *Chem. Eng. Journal*, Vol. 22 (1981), pp. 229–235.
33. Mills, P. L., and Dudukovic, M. P., "Evaluation of Liquid-Solid Contacting in Trickle-Bed Reactors by Tracer Methods," *AIChE Journal*, Vol. 27, No. 6 (1981), pp. 893–904.
34. Tan, C. S., and Smith, J. M., "Catalyst Particle Effectiveness with Unsymmetrical Boundary Conditions," *Chem. Eng. Sci.*, Vol. 35 (1980), pp. 1601–1609.
35. Goto, S., Lakota, A., and Levec, J., "Effectiveness Factors of nth Order Kinetics in Trickle-Bed Reactors," *Chem. Eng. Sci.*, Vol. 36 (1981), pp. 157–62.
36. Saada, M. Y., "Assessment of Interfacial Area in Cocurrent Two-Phase Flow in Packed Beds," *Chemie and Industrie Genie Chimique*, Vol. 105, No. 20 (1972), pp. 1415–1422.
37. Goto, S., Levec, J., and Smith, J. M., "Mass Transfer in Packed Beds with Two-Phase Flow," *IEC Process Des. Develop.*, Vol. 14, No. 5 (1975), pp. 473–478.
38. Ohshima, S., et al., "Liquid-Phase Mass Transfer Coefficient and Gas Holdup in a Packed-Bed Cocurrent Up-Flow Column," *J. Chem. Eng. Japan*, Vol. 9, No. 1 (1976), pp. 29–34.
39. Specchia, V., Sicardi, S., and Gianetto, A., "Absorption in Packed Towers with Concurrent Upward Flow," *AIChE Journal*, Vol. 20, No. 4 (1974), pp. 646–653.
40. Alexander, B. F., and Shah, Y. T., "Gas-Liquid Mass Transfer Coefficients for Cocurrent Upflow in Packed Beds—Effect of Packing Shape at Low Flow Rates," *Can. J. Chem. Eng.*, Vol. 54, No. 6 (1976), pp. 556–559.

41. Mochizuki, S., and Mastui, T., "Liquid-Solid Mass Transfer Rate in Liquid-Gas Upward Cocurrent Flow in Packed Beds," *Chem. Eng. Sci.*, Vol. 29, No. 5 (1974), pp. 1328–1330.
42. Mochizuki, S., "Mass Transport Phenomena and Hydrodynamics in Packed Beds with Gas-Liquid Concurrent Upflow," *AIChE Journal*, Vol. 24, No. 6 (1978), pp. 1138–1141.
43. Mochizuki, S., "Particle Mass Transfer and Liquid Holdup in Packed Beds with Upward Concurrent Gas-Liquid Flow," *Chem. Eng. Sci.*, Vol. 36, No. 1 (1981), pp. 213–215.
44. Mochizuki, S., "Empirical Expressions of Liquid-Solid Mass Transfer in Concurrent Gas-Liquid Upflow Fixed Beds," *Chem. Eng. Sci.*, Vol. 37, No. 9 (1982), pp. 1422–1424.
45. Colquhoun-Lee, I., and Stepanek, J. B., "Solid/Liquid Mass Transfer in Two Phase Cocurrent Upward Flow in Packed Beds," *Trans IChemE*, Vol. 56, No. 2 (1978), pp. 136–144.
46. Delaunay, G., et al., "Electrochemical Study of Liquid-Solid Mass Transfer in Packed Beds with Upward Concurrent Gas-Liquid Flow," *IEC Process Des. Develop.*, Vol. 19, No. 4 (1980), pp. 514–521.
47. Calderbank, P. H., and Moo-Young, M. B., "The Continuous Phase Heat and Mass-Transfer Properties of Dispersions," *Chem. Eng. Sci.*, Vol. 16, No. 1 (1961), pp. 39–54.
48. Hughmark, G. A., "Holdup and Mass Transfer in Bubble Columns," *IEC Process Des. Develop.*, Vol. 6, No. 2 (1967), pp. 218–220.
49. Akita, K., and Yoshida, F., "Gas Holdup and Volumetric Mass Transfer Coefficient in Bubble Columns. Effects of Liquid Properties," *IEC Process Des. Develop.*, Vol. 12, No. 1 (1973), pp. 76–80.
50. Akita, K., and Yoshida, F., "Bubble Size Interfacial Area and Liquid-Phase Mass Transfer Coefficient in Bubble Columns," *IEC Process Des. Develop.*, Vol. 13, No. 1 (1974), pp. 84–91.
51. Kowagoe, M., Nakao, K., and Otake, T. "Liquid-Phase Mass Transfer Coefficient and Bubble Size in Gas Sparged Contactors," *J. Chem. Eng. Japan*, Vol. 8, No. 3 (1975), pp. 254–256.
52. Nakanoh, M., and Yohida, F., "Gas Absorption by Newtonian and Non-Newtonian Liquids in a Bubble Column," *IEC Process Des. Develop.*, Vol. 19, No. 1 (1980), pp. 190–195.
53. Hikita, H., et al., "The Volumetric Liquid-Phase Mass Transfer Coefficient in Bubble Columns," *Chem. Eng. Journal*, Vol. 22, No. 1 (1981), pp. 61–69.
54. Deckwer, W. D., et al., "Hydrodynamic Properties of the Fisher–Tropsch Slurry Process," *IEC Process Des. Develop.*, Vol. 19, No. 4 (1980), pp. 699–708.
55. Sano, Y., Yamaguchi, N., and Adachi, T., "Mass Transfer Coefficients for Suspended Particles in Agitated Vessels and Bubble Columns," *J. Chem. Eng. Japan*, Vol. 7, No. 4 (1974), pp. 255–261.
56. Sänger, P., and Deckwer, W. D., "Liquid-Solid Mass Transfer in Aerated Suspensions," *Chem. Eng. Journal*, Vol. 22, No. 3 (1981), pp. 179–186.
57. Prasher, B. D., and Wills, G. B., "Mass Transfer in an Agitated Vessel," *IEC Process Des. Develop.*, Vol. 12, No. 3 (1973), pp. 351–354.
58. Yagi, H., and Yoshida, F., "Gas Absorption by Newtonian and Non-Newtonian Fluids in Sparged Agitated Vessels," *IEC Process Des. Develop.*, Vol. 14, No. 4 (1975), pp. 488–493.
59. Ranade, V. R., and Ulbrecht, J. J., "Influence of Polymer Additives on the Gas-Liquid Mass Transfer in Stirred Tanks," *AIChE Journal*, Vol. 24, No. 3 (1978), pp. 796–803.
60. Riet, K. V., "Review of Measuring Methods and Results in Nonviscous Gas-Liquid Mass Transfer in Stirred Vessels," *IEC Process Des. Develop.*, Vol. 18, No. 3 (1979), pp. 357–364.

61. Nishikawa, M., Nakamura, M., and Hashimoto, K., "Gas Absorption in Aerated Mixing Vessels with Non-Newtonian Liquid," *J. Chem. Eng. Japan,* Vol. 14, No. 3 (1981), pp. 227–232.
62. Joosten, G. E. H., Schilder, J. G. M., and Janssen, J. J., "The Influence of Suspended Solid Material on the Gas-Liquid Mass Transfer in Stirred Gas-Liquid Contactors," *Chem. Eng. Sci,* Vol. 32, No. 5 (1977), pp. 563–566.
63. Chapman, C. M., et al., "Particle-Gas-Liquid Mixing in Stirred Vessels Part 4; Mass Transfer and Final Conclusions," *Chem. Eng. Res. Des.,* Vol. 61 (May, 1983), pp. 182–185.
64. Harriott, P., "Mass Transfer to Particles: Part 1. Suspended in Agitated Tanks," *AIChE Journal,* Vol. 8, No. 1 (1962), pp. 93–102.
65. Brian, P. L. T., Hales, H. B., and Sherwood, T. K., "Transport of Heat and Mass Between Liquids and Spherical Particles in an Agitated Tank," *AIChE Journal,* Vol. 15, No. 5 (1969), pp. 727–733.
66. Levins, D. M., and Glastonbury, J. R., "Particle-Liquid Hydrodynamics and Mass Transfer in a Stirred Vessel Part 2—Mass Transfer," *Trans. Instn. Chem. Engrs.,* Vol. 50, No. 2 (1972), pp. 132–146.
67. Kuboi, R., et al., "Fluid and Particle Motion in Turbulent Dispersion 3 Particle-Liquid Hydrodynamics and Mass-Transfer in Turbulent Dispersion," *Chem. Eng. Sci.,* Vol. 29, No. 3 (1974), pp. 659–668.
68. Conti, R., and Sicardi, S., "Mass Transfer from Freely-Suspended Particles in Stirred Tanks," *Chem. Eng. Commun.,* Vol. 14, No. 7 (1982), pp. 91–98.
69. Boon-Long, S., Laguereie, C., and Couderc, J. P., "Mass Transfer from Suspended Solid to Liquid in Agitated Vessels," *Chem. Eng. Sci.,* Vol. 33, No. 1 (1978), pp. 813–819.
70. Ohta, H., Goto, S., and Teshima, H., "Liquid-Phase Oxidation of Phenol in a Rotating Catalytic Basket Reactor," *IEC Fund.,* Vol. 19, No. 2 (1980), pp. 180–185.
71. Suzuki, M., and Kawazoe, M., "Particle-to-Liquid Mass Transfer in a Stirred Tank with a Basket Impeller," *J. Chem. Eng. Japan,* Vol. 8, No. 1 (1975), pp. 79–81.
72. Teshima, H., and Ohashi, Y., "Particle to Liquid Mass Transfer in a Rotating Catalyst Basket Reactor," *J. Chem. Eng. Japan,* Vol. 10, No. 1 (1977), pp. 70–72.
73. Pavko, A., Misic, D. M., and Levec, J., "Kinetics in Three-Phase Reactors," *Chem. Eng. Journal,* Vol. 21, No. 2 (1981), pp. 149–154.
74. Goto, S., and Saito, T., "Liquid-Solid Mass Transfer in Basket Type Three-Phase Reactors," *J. Chem. Eng. Japan,* Vol. 17, No. 3 (1984), pp. 324–327.
75. Sylvester, N. D., Kulkarni, A. A., and Carberry, J. J., "Slurry and Trickle-Bed Reactor Effectiveness," *Can J. Chem. Eng.,* Vol. 53, No. 3 (1975), pp. 313–316.
76. Goto, S., "Slurry and Trickle-Bed Reactor Effectiveness," *Can. J. Chem. Eng.,* Vol. 54, No. 1/2 (1976), pp. 126–127.
77. Ramachandran, P. A., and Chaudhari, R. V., "Theoretical Analysis of Reaction of Two Gases in a Catalytic Slurry Reactor," *Ind. Eng. Chem. Process Des. Dev.,* Vol. 18, No. 4 (1979), pp. 703–708.
78. Ramachandran, P. A., and Chaudhari, R. V., "Overall Effectiveness Factor of a Slurry Reactor for Non-Linear Kinetics," *Can. J. Chem. Eng.,* Vol. 58, No. 3 (1980), pp. 412–415.
79. Bischoff, K. B., "Effectiveness for General Reaction Rate Forms," *AIChE J.,* Vol. 11, No. 2 (1965), pp. 351–355.
80. Schwartz, J. G., Weger, E., and Dudukovic, M. P., "A New Tracer Method for Determination of Liquid-Solid Contacting Efficiency in Trickle-Bed Reactors," *AIChE J.,* Vol. 22, No. 5 (1976), pp. 894–904.
81. Mills, P. L., and Dudukovic, M. P., "A Dual-Series Solution for the Effectiveness Factor of Partially Wetted Catalyst in Trickle-Bed Reactors," *Ind. Eng. Chem., Fund.,* Vol. 18, No. 2 (1979), pp. 139–149.

82. Mills, P. L., and Dudukovic, M. P., "Analysis of Catalyst Effectiveness in Trickle-Bed Reactors Processing Volatile or Nonvolatile Reactants," *Chem. Eng. Sci.*, Vol. 35, No. 11 (1980), pp. 2267–2279.
83. Herskowitz, Mordechay, "Wetting Efficiency in Trickle-Bed Reactors. The Overall Effectiveness Factor of Partially Wetted Catalyst Particles," *Chem. Eng. Sci.*, Vol. 36, No. 10 (1981), pp. 1665–1671.
84. Mills, P. L., et al., "Some Comments on Models for Evaluation of Catalyst Effectiveness Factors in Trickle-Bed Reactors," *Chem. Eng. Sci.*, Vol. 36, No. 5 (1981), pp. 947–949.
85. Ramachandran, P. A., and Smith, J. M., "Effectiveness Factors in Trickle-Bed Reactors," *AIChE J.*, Vol. 25, No. 3 (1979), pp. 538–542.
86. Dudukovic, M. P., "Catalyst Effectiveness Factor and Contacting Efficiency in Trickle-Bed Reactors," *AIChE J.*, Vol. 23, No. 6 (1977), pp. 940–944.
87. Weekman, Vern W., Jr., "Laboratory Reactors and their Limitations," *AIChE J.*, Vol. 20, No. 5 (1974), pp. 833–840.
88. Myers, E. C., and Robinson, K. K., "Multiphase Kinetic Studies with a Spinning Basket Reactor," Fifth International Sympos. on Chem. Reaction Engng, Houston—ACS Sympos. Series 65, 1978, pp. 447–458.
89. Njiribeako, A. I., Silveston, P. L., and Hudgins, R. R., "A Laboratory Spinning Catalyst Basket Reactor for Multiphase Contacting," *Can. J. Chem. Eng.*, Vol. 56, No. 5 (1978), pp. 643–645.
90. Goto, S., and Kojima, Y., "Oxidation of Sulfur Dioxide in Different Types of Three-Phase Reactors," Proc. of PACHEC'83, Seoul, Korea, Vol. 3, 1983, pp. 172–177.
91. Pavko, A., and Levec, J., "Particle Wetting in a Laboratory Rotating Basket Reactor for Gas-Liquid-Solid Kinetic Measurement," Proc. Third Austrian-Italian-Yugoslav Chemical Engineering Conference, Graz, Austria, Vol. 1 (1982), pp. 527–534.
92. Frye, C. G., and Mosby, J. F., "Kinetics of Hydrodesulfurization," *Chem. Eng. Progr.*, Vol. 63, No. 9 (1967), pp. 66–70.
93. Schuit, G. C. A., and Gates, B. C., "Chemistry and Engineering of Catalytic Hydrodesulfurization," *AIChE J.*, Vol. 19, No. 3, 1973, pp. 417–438.
94. Satterfied, Charles N., Modell, Michael, and Wilkens John A., "Simultaneous Catalytic Hydrodenitrogenation of Pyridine and Hydrodesulfurization of Thiophene," *Ind. Eng. Chem., Process Des. Dev.*, Vol. 19, No. 1 (1980), pp. 154–160.
95. Snider, James W., and Perona, Joseph J., "Mass Transfer in a Fixed-Bed Gas-Liquid Catalytic Reactor with Concurrent Upflow," *AIChE J.*, Vol. 20, No. 6 (1974), pp. 1172–1177.
96. Morita, Shushi, and Smith, J. M., "Mass Transfer and Contacting Efficiency in a Trickle-Bed Reactor," *Ind. Eng. Chem., Fund.*, Vol. 17, No. 2 (1978), pp. 113–120.
97. Sherwood, T. K., and Farkas, E. J., "Studies of the Slurry Reactor," *Chem. Eng. Sci.*, Vol. 21, No. 5 (1966), pp. 573–582.
98. Sadana, Ajit, and Katzer James, R., "Catalytic Oxidation of Phenol in Aqueous Solution over Copper Oxide," *Ind. Eng. Chem., Fund.*, Vol. 13, No. 2 (1974), pp. 127–134.
99. Goto, Shiego, and Smith, J. M., "Trickle-Bed Reactor Performance. Part II. Reaction Studies," *AIChE J.*, Vol. 21, No. 4 (1975), pp. 714–720.
100. Levec, Janez, and Smith, J. M., "Oxidation of Acetic Acid Solutions in a Trickle-Bed Reactor," *AIChE J.*, Vol. 22, No. 1 (1976), pp. 159–168.
101. Levec, Janez, Herskowitz, Mordechay, and Smith, J. M., "An Active Catalyst for the Oxidation of Acetic Acid Solutions," *AIChE J.*, Vol. 22, No. 5 (1976), pp. 919–920.
102. Hartman, Miloslav, and Coughlin, Robert W., "Oxidation of SO_2 in a Trickle-Bed Reactor Packed with Carbon," *Chem. Eng. Sci.*, Vol. 27, No. 5 (1972), pp. 867–880.
103. Komiyama, Hiroshi, and Smith, J. M., "Sulfur Dioxide Oxidation in Slurries of Activated Carbon. Part I. Kinetics," *AIChE J.*, Vol. 21, No. 4 (1975), pp. 664–670.

CHAPTER 20

ESTIMATING LIQUID FILM MASS TRANSFER COEFFICIENTS IN RANDOMLY PACKED COLUMNS

Anthony B. Ponter

Office of the Dean of Engineering
Cleveland State University
Cleveland, Ohio, USA

Patrick H. Au-Yeung

Department of Chemistry and Chemical Engineering
Michigan Technological University
Houghton, Michigan, USA

CONTENTS

INTRODUCTION

For gas absorption or desorption in a packed column, there are resistances to mass transfer in both the gas and liquid phases. In many cases, however, one of the two resistances is negligible compared with the other and the overall rate of mass transfer is controlled by either the gas phase or the liquid phase. Typical examples of liquid-phase-controlled gas absorption in industry include the absorption of carbon dioxide from ammonia synthesis gases (mixture of hydrogen and nitrogen) and the removal of acid gases from sour natural gases using water or physical organic solvents such as propylene carbonate and methanol. For the absorption of gases with a chemical solvent, the mass transfer resistance frequently resides in the liquid phase and the true mass transfer coefficient with chemical reaction (k_L') can be calculated as the product of the liquid film mass transfer coefficient due to physical absorption (k_L) and the enhancement factor due to chemical reaction (E) for the cases in which the latter can be theoretically derived.

The pertinent empirical and theoretical correlations for predicting the volumetric liquid film mass transfer coefficients (k_La) for physical absorption in packed columns are presented in this chapter. The important factors that affect the values of the true mass transfer coefficient (k_L) and the effective interfacial area (a) are discussed with the hope that it will be helpful to the design engineers in selecting the most appropriate correlation for their particular application. Note that S. I. units are used throughout the empirical correlations.

DEFINITION OF MASS TRANSFER COEFFICIENTS

Whitman [1] and Lewis and Whitman [2] first postulated a simple model describing the absorption of a gas into a liquid based on the assumption that the absorption rate is controlled solely by diffusion on each side of the gas-liquid interface and that there is no resistance to mass transfer at the phase boundary, i.e., the gas and liquid phases are in equilibrium with each other. They suggested that the resistances to mass transfer in the two phases may be added to obtain the overall resistance (cf. Ohm's law), namely

$$\frac{1}{K_G} = \frac{1}{k_G} + \frac{H}{k_L} \tag{1}$$

and

$$\frac{1}{K_L} = \frac{1}{k_L} + \frac{1}{Hk_G} \tag{2}$$

when the concentrations at equilibrium can be represented by Henry's law

$$p = Hc \tag{3}$$

In gas absorption the individual mass transfer coefficients are defined in terms of the fluxes N_A and driving forces as follows:

$$k_G = \frac{N_A}{p_A - p_i} \tag{4}$$

$$k_L = \frac{N_A}{c_i - c_A} \tag{5}$$

The overall coefficients are similarly defined in terms of the gas phase and liquid phase driving forces:

$$K_G = \frac{N_A}{p_A - p^*} \tag{6}$$

$$K_L = \frac{N_A}{c^* - c_A} \tag{7}$$

where p^* is the partial pressure of the gas in equilibrium with the bulk liquid concentration c_A and c^* is the liquid concentration in equilibrium with the partial pressure of the gas p_A.

When the gas is only slightly soluble, H is large, making the gas side resistance negligible and the absorption process is said to be liquid-phase controlled. Equation 2 then becomes

$$\frac{1}{K_L} \approx \frac{1}{k_L}$$

or

$$K_L \approx k_L \tag{8}$$

EMPIRICAL CORRELATIONS FOR VOLUMETRIC LIQUID FILM MASS TRANSFER COEFFICIENTS

For Water

Sherwood and Holloway [3] studied the desorption of carbon dioxide, oxygen, and hydrogen from water using a variety of packings in a 0.508-m (20-inch) diameter column packed with different sizes of ceramic Raschig rings and Berl saddles to heights of 0.15 to 1.24 m. It was found that the volumetric mass transfer coefficients $k_L a$ were independent of the gas flow rate up to the loading point. Except at very high water flow rates k_La was found to be a power function of the superficial mass flow rate of water L. Their data could be represented well by the equation:

$$\frac{k_La}{D} = \alpha\left(\frac{L}{\mu}\right)^{1-n}\left(\frac{\mu}{\varrho D}\right)^{0.5} \tag{9}$$

or in terms of the height of a liquid-phase transfer unit H_L:

$$H_L = \frac{1}{\alpha}\left(\frac{L}{\mu}\right)^{n}\left(\frac{\mu}{\varrho D}\right)^{0.5} \tag{10}$$

where $H_L = \dfrac{L}{\varrho k_L a}$ and α and n for the packings studied are given in Table 1.

Table 1
Values of Constants for Use in the Sherwood and Holloway Correlation

	Packings	α	n
Rings	5.08 cm (2.0 in)	341	0.22
	3.81 cm (1.5 in)	383	0.22
	2.54 cm (1.0 in)	426	0.22
	1.27 cm (0.5 in)	1392	0.35
	0.95 cm (3 8 in)	3116	0.46
Saddles	3.81 cm (1.5 in)	732	0.28
	2.54 cm (1.0 in)	778	0.28
	1.27 cm (0.5 in)	686	0.28
Tile	7.62 cm (3.0 in)	503	0.28

From Sherwood and Holloway [3].

Table 2
Values of Constants for Use in the Sherwood and Holloway Correlation

Packing	α	n
2.54 cm rings	523	0.24
2.54 cm saddles	901	0.31
7.62 cm spiral tiles	379	0.21

From Molstad, M. C. et al. [4].

The results of Sherwood and Holloway were confirmed by the experiments of Molstad et al. [4], Vivian and Whitney [5], Whitney and Vivian [6], and Deed et al. [7]. The data of Molstad et al. for 2.54-cm rings, 2.54-cm saddles, and 7.62-cm spiral tiles were correlated by the Sherwood and Holloway correlation, using the constants given in Table 2.

Norman [8] examined the data of Sherwood and Holloway and of Molstad et al., and found that the variation of k_La or H_L for the different packings was remarkably small (Figure 1). He correlated the data for ring and saddle packings by a single equation:

$$\frac{k_La}{D} = 530\left(\frac{1}{\mu}\right)^{0.75}\left(\frac{\mu}{\varrho D}\right)^{0.5} \tag{11}$$

with a maximum deviation of 20%. Norman stated that the deviation was of the same order as the variation among the different sets of data for any particular packing, attributed to differences in liquid distribution or packing arrangements.

Cooper et al. [9] carried out the desorption of carbon dioxide from water in a 0.76 m square tower using very high water flow rates ranging from 18.4 $kg/m^2 \cdot s$ to 76 $kg/m^2 \cdot s$ while the superficial gas velocities were only 2.13 cm/s to 37 cm/s. The height of a transfer unit was found to be two to three times greater than those of Sherwood and Holloway and dependent on the gas rate. The discrepancy was ascribed to the back mixing of the gas within the tower as the gas was carried downwards by the high-velocity water by friction.

Rixon's [10] data on the desorption of carbon dioxide from water in a 0.38-m-diameter column packed with 2.54-cm rings showed that the overall liquid-phase mass transfer coefficient K_La is proportional to the liquid flow rate but independent of the gas flow rate, thus

$$K_La = 0.0017L \tag{12}$$

Hutchings et at. [11] using an acetone-air-water system found that the liquid film coefficients vary with the molar liquid rate to the 0.8 power and independent of packing size and gas rates. They obtained the following correlation:

$$k_La = 0.0155\ L_M^{0.8} \tag{13}$$

Koch et al. [12], studying the absorption of carbon dioxide from air into water in 15.2- and 25.4-cm-diameter towers filled with 0.95-, 1.27-, 1.91-, and 3.18-cm Raschig rings reached the same conclusion that the liquid film coefficient was independent of superficial gas velocity and packing size. The data for all packings were correlated by the equation:

$$k_La = 0.014\ L_M^{0.96} \tag{14}$$

Their results were lower than those obtained by Sherwood and Holloway by 20% to 50% but were in remarkably good agreement with those of Rixon.

The liquid film coefficients obtained by Knoedler and Bonilla [13] who studied vacuum desorption of oxygen from water in a 15.2-cm-diameter tower filled with 0.61 m of Stedman triangular packing, showed no variation with packed height, solute gas concentration, solute gas pressure, or temperature. The results below the loading point were correlated by the equations:

$$H_{0L} = 0.172\, L^{0.3} \tag{15}$$

$$\frac{K_L a}{D} = 3.5 \times 10^5\, L^{0.77} \left(\frac{\mu}{\varrho D}\right)^{0.53} \tag{16}$$

The coefficients were found to be in moderately close agreement with those reported for random packings.

Van Krevelen and Hoftijzer [14] introduced the liquid film thickness into the Sherwood and Holloway correlation to make it dimensionally consistent.

$$\frac{k_L(\mu^2/g\varrho^2)^{1/3}}{D} = c\left(\frac{\Gamma}{\mu}\right)^m \left(\frac{\mu}{\varrho D}\right)^{1/3} \tag{17}$$

Sherwood and Holloway's data were satisfactorily correlated by Equation 17 using the values of 0.025 and 2/3 for the constants c and m respectively.

The data of other investigators for the absorption of oxygen, carbon dioxide, and chlorine in water could be correlated by the same equation, by varying the value of c from 0.0047 to 0.03. Probably due to changes in the liquid distribution as it flows down the packing, the value of c was found to depend on the ratio of the packing diameter d to the packed height Z.

$$c = 0.05\left(\frac{d}{Z}\right)^{1/3} \tag{18}$$

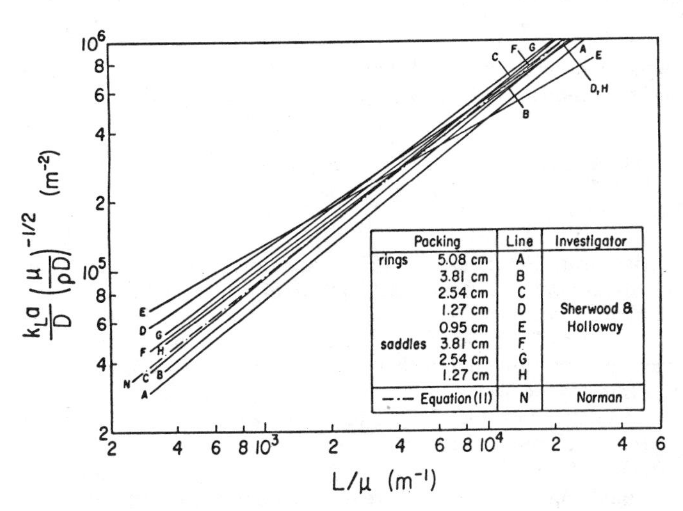

Figure 1. Comparison of empirical correlations for liquid film mass transfer coefficients for different sizes of random packings. (After Norman [8].)

Table 3
Values of c′ for Different Packings for Use in the van Krevelen and van Hooren Correlation, Equation 21

Packing diameter, cm	constant c′
0.6	0.06
0.95	0.25
1.3	0.38-0.66
2.5	0.63-0.84
5.0	0.45-2.38

From Norman [8].

Van Krevelen and Hoftijzer [15] proposed a modified form of Equation 17 using their results for the liquid film resistance in the absence of chemical reaction:

$$\frac{k_L(\mu^2/g\varrho^2)^{1/3}}{D} = 0.015\left(\frac{L}{a_w\mu}\right)^{2/3}\left(\frac{\mu}{\varrho D}\right)^{1/3} \tag{19}$$

The wetted area was estimated from the dry surface of the packing a_t using the empirical relation:

$$\frac{a_w}{a_t} = 1 - e^{-5000\,L/\varrho} \tag{20}$$

While the van Krevelen and Hoftijzer correlation is dimensionally consistent it does not express the correct relationship between the coefficient and the physical properties of the liquid. The coefficient is proportional to $D^{2/3}$ rather than $D^{1/2}$ suggested by Higbie's penetration theory [16] or Danckwert's surface renewal theory [17] and confirmed by Vivian and King [18]. Sherwood and Pigford [19] reported that Equation 19 does not correlate the absorption data of Sherwood and Holloway [3] accurately with deviations as large as 50 percent. The equation indicates a much larger dependency on the size of the packing than is supported by experimental data.

By introducing a characteristic length dimension $1/a_t$ to obtain dimensional consistency, van Krevelen and van Hooren [20] proposed a correlation similar to that of Sherwood and Holloway.

$$\frac{k_L a}{a_t^2 D} = c'\left(\frac{L}{a_t\mu}\right)^{0.8}\left(\frac{\mu}{\varrho D}\right)^{0.4} \tag{21}$$

Hoftijzer [21] has shown that the constant c' in Equation 21 is a function of the size of the packing. The average values for ring packings are presented in Table 3.

Equation 21 correlated satisfactorily the data of Sherwood and Holloway. The large dependency of c' on the packing size in contrast to the findings of Norman [8], however, suggests that the equation does not include all the important variables affecting the liquid-phase mass transfer coefficient.

Shulman et al. [22] determined the effective interfacial areas for Raschig rings and Berl saddles by calculating the ratio k_Ga/k_G using the k_G values obtained by vaporization of naphthalene packings and the k_Ga data of Fellinger [23] for ammonia absorption into water. The values of the liquid film coefficient k_L were then calculated from the ratios k_La/a using the experimentally determined k_La values and the effective area of the packing. The results for several ring and saddle packings were correlated by:

$$\frac{k_L D_p}{D} = 25.1 \left(\frac{D_p L}{\mu}\right)^{0.45} \left(\frac{\mu}{\varrho D}\right)^{0.5} \tag{22}$$

where D_p is the diameter of a sphere having the same surface area as the packing.

Onda et al. [24] developed a general dimensionless correlation for liquid-phase mass transfer coefficients for columns packed with Raschig rings using the two-film theory and the penetration theory. The liquid-phase mass transfer coefficients were obtained by dividing the volumetric coefficients by the wetted areas assuming that the actual interfacial areas were proportional to the wetted areas calculated from the expression given by Fijita et al. [25]:

$$\frac{a_w}{a_t} = 1 - 1.02\, e^{-0.278\,(L/a_t\mu)^{0.4}} \tag{23}$$

Based on the two-film theory, the authors assumed that the characteristic length 1 in the modified mass transfer Nusselt number ($k_L 1/D$) was proportional to the thickness of the flowing liquid film resulting in the following equation:

$$k_L(\varrho/\mu g)^{1/3} = 0.021 \left(\frac{L}{a_t\mu}\right)^{0.49'} \left(\frac{\mu}{\varrho D}\right)^{-1/2} \tag{24}$$

About 90% of the data reported including the authors' agrees within ±20%.

A similar equation with a different constant for the Reynolds number was derived on the basis of the penetration theory:

$$k_L(\varrho/\mu g)^{1/3} = 0.013 \left(\frac{L}{a_t\mu}\right)^{1/2} \left(\frac{\mu}{\varrho D}\right)^{-1/2} \tag{25}$$

Equation 24 can be rearranged in the form:

$$\frac{k_L(\mu^2/\varrho^2 g)^{1/3}}{D} = 0.021 \left(\frac{L}{a_t\mu}\right)^{0.49} \left(\frac{\mu}{\varrho D}\right)^{1/2} \tag{26}$$

and compared to that of van Krevelen and Hoftijzer:

$$\frac{k_L(\mu^2/\varrho^2 g)^{1/3}}{D} = 0.015 \left(\frac{L}{a_w\mu}\right)^{2/3} \left(\frac{\mu}{\varrho D}\right)^{1/3} \tag{19}$$

It can be seen that both equations have the same functional form and differ only in the constant and the exponents of the Reynolds and Schmidt numbers. The higher dependency of k_L on the Reynolds number in the van Krevelen and Hoftijzer equation is probably due to the use of the wetted area which is a function of the liquid rate L. The exponent of 1/3 on the Schmidt number in the van Krevelen and Hoftijzer equation is theoretically inferior to that of 1/2 in Onda's equation as predicted by the penetration theory and surface renewal theory and supported by the majority of investigations. Shulman et al. [26] contended that Onda's correlation is not general enough as its application is limited to Raschig rings and water systems since the correlation must be used with the wetted area data of Fujita [25]. Using the data of Fellinger [23], they further showed that the effective interfacial area is not directly proportional to the wetted area as assumed by Onda et al.

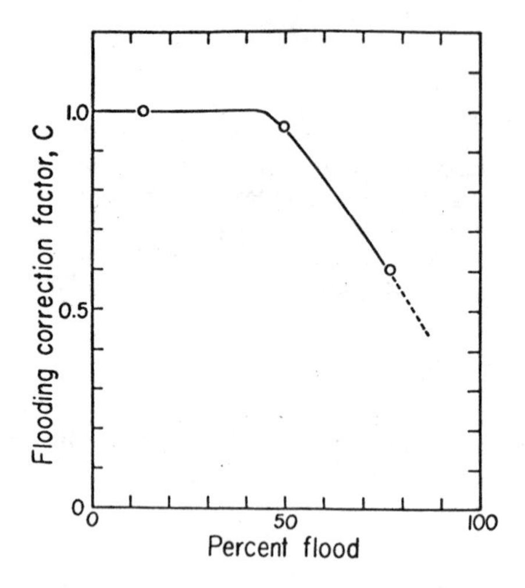

Figure 2. Liquid film correction factor for operation at high percent of flood [27].

Cornell et al. [27] presented a generalized empirical correlation for the height of a liquid-phase transfer unit based on available data obtained from 1938 to 1956 by various investigators:

$$H_L = \varphi\, Sc^{1/2}\, C(Z/3.05)^{0.15} \tag{27}$$

where φ is the packing parameter which is a function of the packing type, size, and superficial mass velocity, C is the correction factor for H_L at high gas rates, and Z is the packed height. The packing parameters φ and the correction factor C as reported by Cornell et al. are given in Figures 2 to 4.

The correlation agreed with the reported data satisfactorily with a maximum average deviation of 12.2%.

Onda et al. [28] developed a dimensionally consistent correlation for estimating liquid film coefficients k_L for both water and organic solvents to within 20 percent:

$$k_L(\varrho/\mu g)^{1/3} = 0.0051\left(\frac{L}{a_w\mu}\right)^{2/3}\left(\frac{\mu}{\varrho D}\right)^{-1/2}(a_t d)^{0.4} \tag{28}$$

The k_L values used in the correlation were obtained by dividing the k_La values in the literature by a_w values calculated from the following equation which takes into account the liquid surface tension and the surface energy of the packing material [29]:

$$\frac{a_W}{a_t} = 1 - \exp\left[-1.45\left(\frac{\sigma_c}{\sigma}\right)^{0.75}\left(\frac{L}{a_t\mu}\right)^{0.1}\left(\frac{L^2 a_t}{\varrho^2 g}\right)^{-0.05}\left(\frac{L^2}{\varrho\sigma a_t}\right)^{0.2}\right] \tag{29}$$

Onda's correlation is convenient for design purpose, when combined with Equation 29 for wetted areas, since it is applicable to different packing types and sizes. Its main weakness is the assumption that the effective interfacial area is the same as the wetted area which has been shown to be incorrect by Shulman et al. [26]. Hence the correlation does not express the correct fundamental dependency of k_L on the system properties and operating conditions and it does not provide reliable values outside the range of operating variables for which it was obtained.

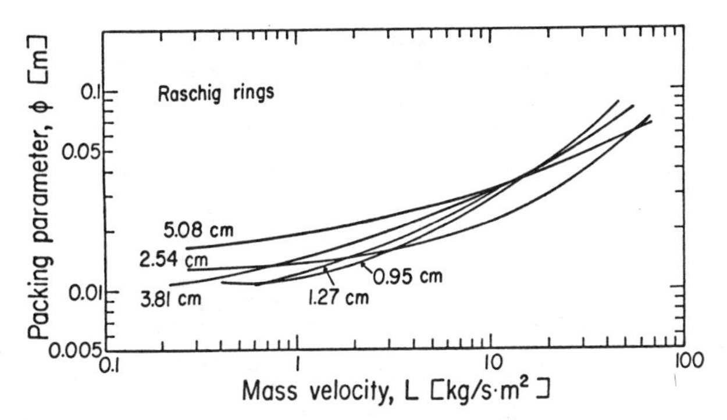

Figure 3. H_L correlation for various-size Raschig rings: 3.05 m packed height and less than 50% flooding [27].

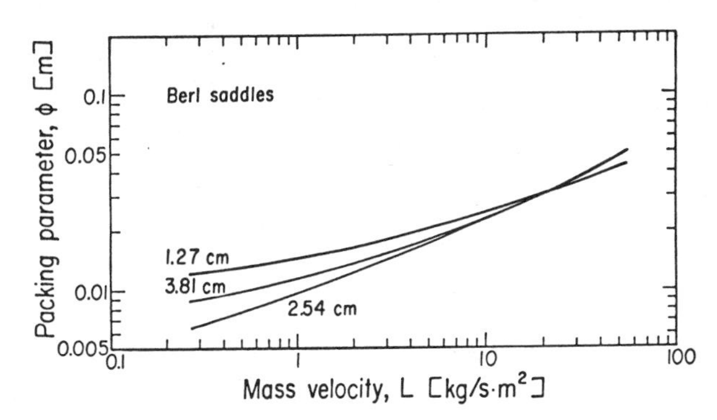

Figure 4. H_L correlation for various size Berl saddles: 3.05 m packed height and less than 50% flooding [27].

Mohunta et al. [30] developed a generalized correlation for large ranges of column diameter and packing size:

$$k_L a \left(\frac{a_t \mu}{\varrho g}\right)^{2/3} \left(\frac{\mu}{\varrho g^2}\right)^{1/9} = 0.0025 \left(\frac{\mu U^3 a_t^3}{\varrho g^2}\right)^{1/4} \left(\frac{\mu}{\varrho D}\right)^{-1/2} \tag{30}$$

The correlation fits the bulk of the available data in the literature and those of the authors on the desorption of oxygen from water and also carbon dioxide from aqueous glycerol solutions to within $\pm 20\%$. The range of variables and physical properties included in the correlation are summarized in Table 4.

Bolles and Fair [31] improved the Monsanto mass transfer model due to Cornell et al. [27] for the prediction of the height of a transfer unit for both the liquid phase and the vapor phase with new correlation constants, i.e. the packing parameter φ and the coefficient for the effect of approach of flood point on liquid phase mass transfer C_{fL}.

$$H_L = \varphi \, Sc^{1/2} \, C_{fL}(Z/3.05)^{0.15} \tag{31}$$

Table 4
Range of Variables and Physical Properties for the Mohunta Correlation, Equation 30

Variables	Range
L	0.1–42 $kg/m^2 \cdot s$
G	0.015–1.22 $kg/m^2 \cdot s$
μ	0.7–1.5 $mPa \cdot s$
$\mu/\varrho D$	142–1,033
D_c	6–50 cm
d	0.6–5.1 cm
D_c/d	5–40

From Mohunta, D. M., Vaidyanathan, A. S., and Laddha, G. S., Indian Chem. Eng., *Vol. 11, No. 3 (1969), pp. 73–79.*

Table 5
Range of Variables for the Improved Mass Transfer Model

Variables	Range
Independent	
Column diameter, m	0.25–1.2
Column diameter/packing diameter	8–64
Liquid density, kg/m^3	480–1,025
Liquid diffusion coefficient, cm^2/s	7.6×10^{-6}–1.5×10^{-4}
Liquid/vapor ratio, kg/kg	0.45–485
Packed height, m	0.15–10.7
Packing size, cm	1.5–7.6
Surface tension, mN/m	5–75
Liquid viscosity, $mPa \cdot s$	0.09–1.5
Dependent	
Flood velocity, m/s	0.073–4.82
Pressure gradient, mm Hg/m	0.12–29.8

From Bolles and Fair [31].

The improved mass transfer model was obtained with a documented databank which comprises 545 observations from 13 different sources. The revised parameters φ and C_{fL} are presented in Figures 5 and 6.

The range of variables included in the data bank for the development of the improved mass transfer model is given in Table 5.

Based on a model reliability study with the data in the data bank, the authors claimed that the safety factor at 95% confidence level for the improved mass transfer model is 1.70 while it is 1.87 for the Monsanto model [27] and 2.23 for the Onda model [28].

From the range of variables given in Table 5 it can be seen that the correlation covers liquid viscosities up to 1.5 $mPa \cdot s$ (cp) only. Hence the model is probably better suited to distillation than absorption or stripping.

For Liquids with High Viscosities

Few investigators have used liquids other than water in their absorption studies since Sherwood and Holloway published their well-known correlation which is based on results obtained with water as the irrigating medium over forty years ago. Yoshida and Koyanagi [32] and Onda et al. [28] have used organic liquids in their absorption studies, however, although these organic liquids have very different surface tensions, most viscosities reported are similar to water, and few of the correlations presented can be used to predict the liquid phase mass transfer coefficient with confidence when the viscosity of the absorbing liquid differs considerably from that of water.

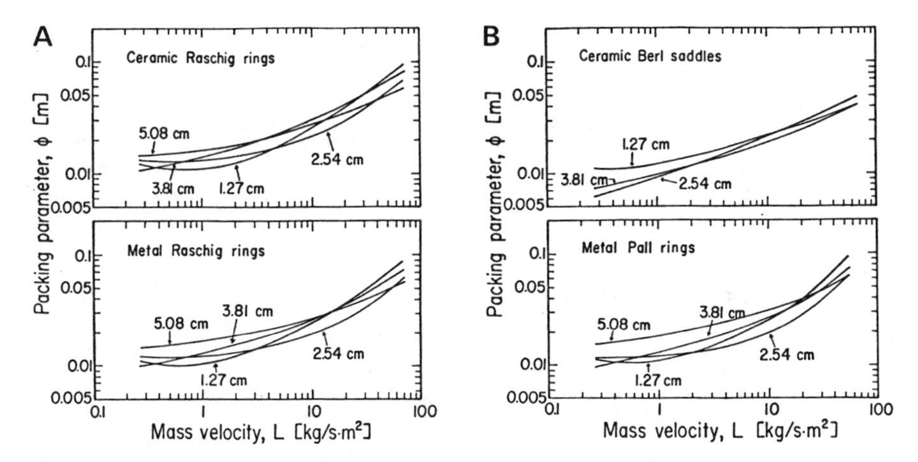

Figure 5. (A) Improved packing parameters for liquid-phase mass transfer; (B) Improved packing parameters for liquid-phase mass transfer [31].

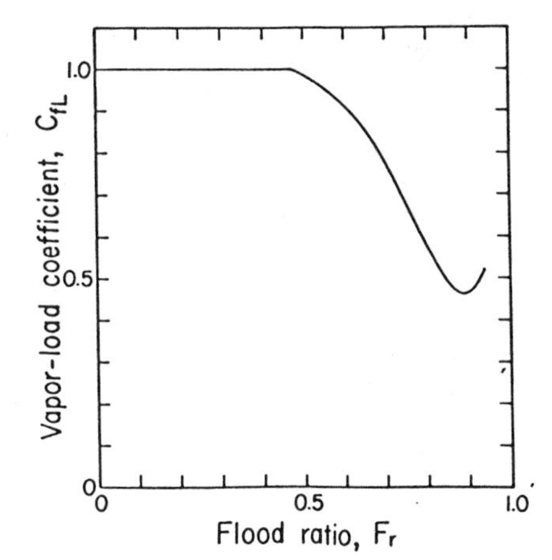

Figure 6. Vapor-load coefficient for liquid-phase mass transfer [31].

Norman and Sammak [33, 34] compared the correlation of van Krevelen and Hoftijzer [15]:

$$\frac{k_L(\mu^2/g\varrho^2)^{1/3}}{D} = 0.015\left(\frac{L}{a_w\mu}\right)^{2/3}\left(\frac{\mu}{\varrho D}\right)^{1/3} \tag{19}$$

with that of Shulman et al. [22]

$$\frac{k_L D_P}{D} = 25.1\left(\frac{D_p L}{\mu}\right)^{0.45}\left(\frac{\mu}{\varrho D}\right)^{0.5} \tag{20}$$

and found that while in Equation 19 the mass transfer coefficient is inversely proportional to the viscosity, Equation 22 suggested that the coefficient is essentially independent of viscosity. They made an important contribution to the understanding of the influence of viscosity on mixing and the mass transfer coefficient by carrying out experiments in a modified disc column using systems with a wide range of liquid viscosities from 0.5 to 20 mPa · s. Mangers and Ponter [35] carried out a comprehensive investigation of the effect of viscosity on the liquid film resistance to mass transfer in a randomly packed column. The liquid film mass transfer coefficients were determined for the countercurrent absorption of carbon dioxide into pure water and aqueous glycerol solutions at 25° C in a 10-cm glass column packed randomly with 1-cm glass Raschig rings to a depth of 50 cm. The systems covered a wide range of viscosity from 0.9 to 26 mPa · s (cp) while the surface tensions are high and relatively constant, 67 to 72 mN/m (dynes/cm).

The authors showed that when $k_La/D\, Sc^{0.5}$ was plotted against L/μ, a sharp change of slope was observed for each aqueous glycerol solution with a sharp increase in pressure drop. The transition points at which the slopes changed were shown to be a function of the minimum wetting rate estimated by Ponter et al. [36].

The change in slopes was attributed to the transition from a partially wetted packing to a fully wetted one as L/μ was increased. The experimental volumetric coefficients obtained for water exhibited good agreement with the Sherwood and Holloway correlation. However, the data demonstrated the limitations of the Sherwood and Holloway equations as the liquid viscosity increases with deviations sometimes as high as 100% (Figures 7 to 10).

The discrepancy could be attributed to the use of volumetric coefficients k_La in the Sherwood and Holloway correlation. The effects of the operating variables on the mass transfer coefficient k_L and the interfacial area a are different; a being dependent on the wetting behavior of the liquid while k_L is not. Since the dependence of k_L on the effect of mixing of the liquid film at the packing junctions (which is a function of viscosity) and the dependence of a on the wetting characteristics of the liquids (which are functions of the viscosity, surface tension, and contact angle) were not taken into account in the Sherwood and Holloway correlation, it is not surprising that the correlation fails to predict the correct

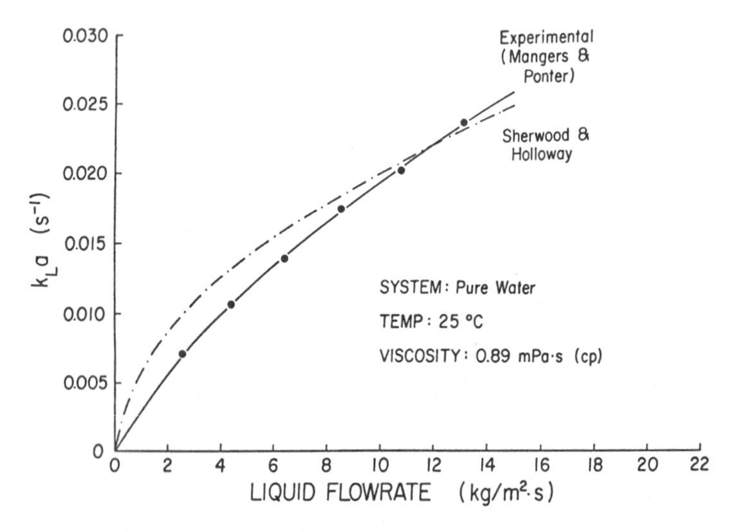

Figure 7. Comparison of experimental data for water with those predicted by the Sherwood and Holloway correlation.

volumetric coefficients when the wetting properties of the liquid used are different from that of water.

The k_La values for the fully wetted packing were correlated by the equation:

$$\frac{k_La}{D} = 26.8\left(\frac{L}{\mu}\right)^{1.44}\left(\frac{\mu}{\varrho D}\right)^{0.50}\left(\frac{\varrho^2 g d^3}{\mu^2}\right)^{-0.183} \tag{32}$$

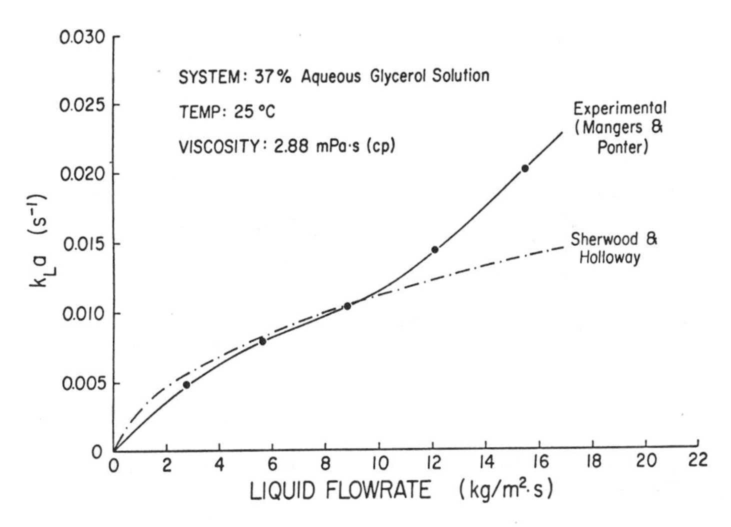

Figure 8. Comparison of experimental data for 37% aqueous glycerol solution with those predicted by the Sherwood and Holloway correlation.

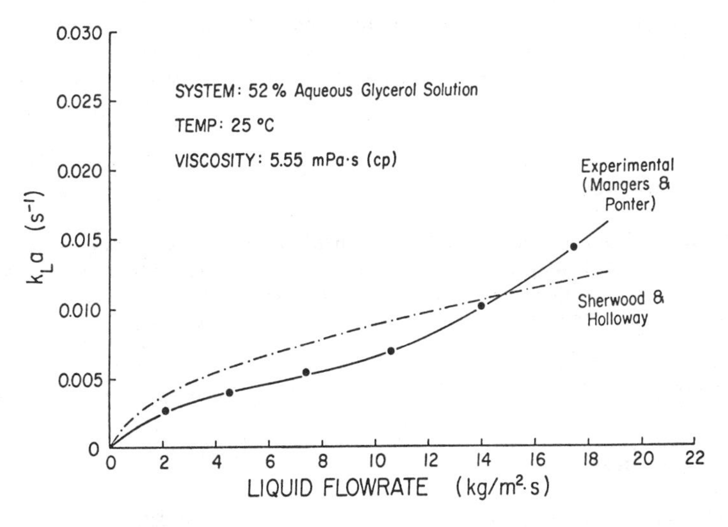

Figure 9. Comparison of experimental data for 52% aqueous glycerol solution with those predicted by the Sherwood and Holloway correlation.

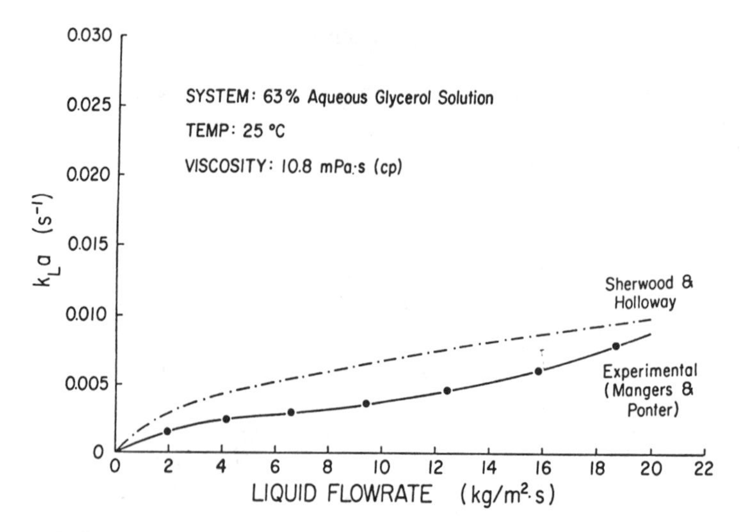

Figure 10. Comparison of experimental data for 63% aqueous glycerol solution with those predicted by the Sherwood and Holloway correlation.

THEORETICAL CORRELATIONS FOR VOLUMETRIC LIQUID FILM MASS TRANSFER COEFFICIENTS

Davidson [37] first provided a formal theoretical basis between the liquid flow over packing pieces and the liquid-phase mass transfer coefficient using Higbie's penetration theory. Three statistical models of a random packing were proposed. In all three models the packing was assumed to be formed by a large number of flat plates. The liquid was assumed to run down the surfaces of the flat plates in laminar flow and to be mixed completely at each packing junction before passing to the next surface. The three models yield three different expressions for the prediction of the height of a liquid phase transfer unit in a packed column.

The Vertical Surface Model

In this model the packing was assumed to be made up of a large number of vertical surfaces of height d which were completely and equally wetted. The wetted perimeter per unit cross-section of the column is equal to a, the surface area per unit volume. This model gives an expression for the height of a liquid film transfer unit H_L:

$$\frac{H_L}{d} = 0.345\,Sc^{1/2}\,Re_l^{2/3}\,Gr^{-1/6} \tag{33}$$

The Random Angle Model

The packing was assumed to consist of a large number of surfaces of length d and inclined at φ to the horizontal with an equal probability of all angles from 0 to $\pi/2$. The equation derived from this model is:

$$\frac{H_L}{d} = 0.244 \frac{Sc^{1/2} Re^{2/3}}{Gr^{1/6}} \tag{34}$$

The Random Length and Random Angle Model

In addition to the assumption of random angles as in the second model, the packing surfaces were assumed to have different lengths with an equal probability of all lengths up to a maximum d which is the physical dimension of the packing element. This model yields an equation similar to that of the random angle model with 0.1833 as the constant instead of 0.244.

$$\frac{H_L}{d} = 0.1833 \frac{Sc^{1/2} Re^{2/3}}{Gr^{1/6}} \tag{35}$$

A comparison of the three equations with experimental data for the volumetric liquid coefficients and the interfacial areas from Shulman et al. [22] showed that the random length and random angle model was in better agreement than the other two models as expected (Figure 11). However, at low Reynolds numbers the model underpredicts the heights of transfer units probably due to lack of complete mixing at the packing junctions. At a high Reynolds number, the theoretical heights of the transfer unit are too high and this was ascribed to the effect of rippling of the liquid film on the packing surfaces.

Bridgewater and Scott [38] reasoned that for a three-dimensional distribution of surfaces it is more appropriate to consider the distribution of a random vector in space. The probability of a flat surface being inclined between φ and $\varphi + d\varphi$ to the horizontal is then $\sin \varphi \, d\varphi$ instead of $d\varphi/(\pi/2)$ as considered by Davidson. This modification of Davidson's random angle model yields the equation:

$$\frac{H_L}{d} = 0.334\, Sc^{1/2}\, Re_l^{2/3}\, Gr^{-1/6} \tag{36}$$

where $Re_l = 4L/a\mu$, the Reynolds number for vertical surfaces

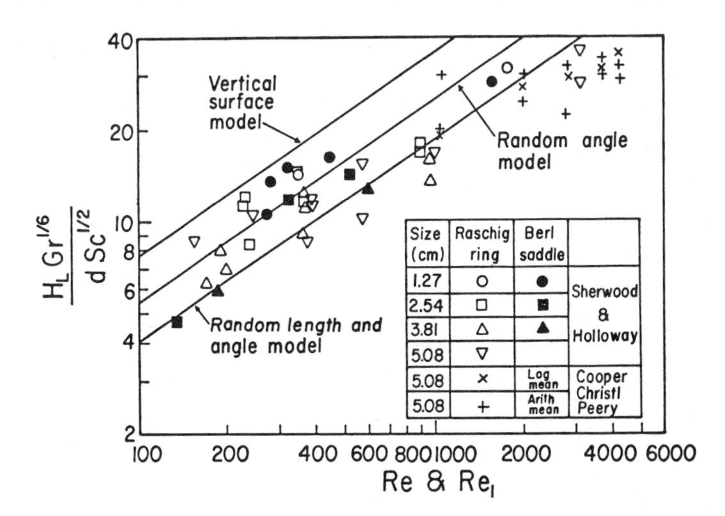

Figure 11. Comparison of the height of a liquid film transfer unit predicted by Davidson's theoretical statistical models with experimental data [37].

Rearranging Equation 36 which uses the Reynolds number for random packings ($2\pi L/a\mu$) and was used by Davidson, gives:

$$\frac{H_L}{d} = 0.247 \frac{Sc^{1/2}\, Re^{2/3}}{Gr^{1/6}} \tag{37}$$

which is similar to Equation 34 derived by Davidson. Incorporating the concept of random length into the random angle model, the authors showed that H_L/d was three-quarters of the value given in Equation 37, an observation consistent with Davidson's.

Bridgewater and Scott also considered the situation where the flow rate per unit width changes from one packing piece to another. For the case where the liquid mass velocity per unit width of plate has a mean value of Γ_m and a maximum value of $2\Gamma_m$ and the flow has a random value from 0 to $2\Gamma_m$, the authors showed that H_L/d for the different models was enhanced by a factor of 1.058. For a different flow distribution:

$$f(\Gamma) = (1/\Gamma_m)\exp(-\Gamma/\Gamma_m) \tag{38}$$

where $0 \leqq \Gamma/\Gamma_m < \infty$

it was shown that H_L/d was increased by a factor of 1.120. Thus the authors have shown the H_L/d is always raised when the flow is different in different parts of the packing or when there is channeling of the liquid in the packed column. However, from the factors calculated for the two liquid flow distributions, the effects of channeling on mass transfer do not seem to be considerable for liquid-film-controlled gas absorption. This is consistent with the finding of Danckwerts and Gillham [39] that there was no loss of effectiveness of absorption up to seven column diameters from the top of the packing.

Zech and Mersmann [40] presented a method to calculate the interfacial area a and the mass transfer coefficient k_L in irrigated packed columns, by visualizing the packing as a system of parallel cylindrical channels which has the same specific area a_t and the same voidage ε as the packing. The irrigating liquid was assumed to flow down the surface of the channels in rivulets and mass transfer was conjectured to take place with perfect mixing at the ends of the packing elements. Combining experimental results obtained in packed absorption and distillation columns with theoretical considerations based on the penetration theory and the fluid dynamics of the rivulets they developed the following equations:

$$k_L = K\left(\frac{6D}{\pi d}\right)^{1/2}\left(\frac{\varrho g d^2}{\sigma}\right)^{-0.15}\left(\frac{Udg}{3}\right)^{1/6} \tag{39}$$

and

$$\frac{a}{a_t} = K\, Re^{1/2}\,(We/Fr)^{0.45}\,(da_t)^{-1/2} \tag{40}$$

where

$$Re = \frac{U}{\nu a_t} \tag{41}$$

$$(We/Fr) = \frac{\varrho g d^2}{\sigma} \tag{42}$$

The values of K for different packings are given in Table 6.

The authors claimed that the method was able to describe the mass transfer process in packed columns accurately. Ponter et al. [41] tested the validity of the Zech and Mersmann model using the data of Mangers and Ponter [35] and found that the model consistently underestimates the volumetric coefficients for the water-glycerol mixtures. The model was found to be no more reliable than the purely empirical correlation of Sherwood and Holloway as neither of them has taken the effect of viscosity on mixing and wetting into account.

Ponter and Au-Yeung [42] developed a semitheoretical model for predicting liquid-film mass transfer coefficients for columns packed with Raschig rings and operating below the minimum liquid wetting rate. The model was based on the random length and random angle statistical model of Davidson [37] and covers a viscosity range of 0.89 to 10.8 mPa · s, where the surface tensions are high and relatively constant (68 to 72 mN/m). By introducing a mixing factor $\alpha_l(k_{Lcm}/k_{Lexpt})$ into Davidson's model, the authors have taken into account the incomplete mixing at the packing junctions at low Reynolds number. The model also enables the maximum effective length d' of the packing elements to be calculated. The model is given by

$$\frac{H_L}{d} = 0.1833\,\alpha_l\left(\frac{d'}{d}\right)^{1/2}\frac{Sc^{1/2}\,Re^{-0.61}}{Gr^{1/6}} \tag{43}$$

Rearranging Equation 43 in a form corresponding to Davidson's model gives

$$\frac{H_L}{d} = 0.1833\,\frac{\alpha_l}{Re^{1.28}}\left(\frac{d'}{d}\right)^{1/2}\frac{Sc^{1/2}\,Re^{2/3}}{Gr^{1/6}} \tag{44}$$

Equation 44 shows the dependency of the mixing factor on the Reynolds number while the exponent of 2/3 ascribed to the Reynolds number represents the contribution to mass transfer by unsteady state diffusion into the liquid film in laminar flow on the packing surface.

Using the data of Mangers and Ponter [35] for the carbon dioxide-aqueous glycerol systems and 1-cm glass Raschig rings, the authors expressed the mixing factor as a function of the minimum wetting rate expression developed by Ponter et al. [36].

$$\alpha_l = 3.46\times10^2[(1-\cos\theta)^{0.6}(\varrho\sigma^3/\mu^4 g)^{0.2}]^{0.728} \tag{45}$$

The maximum effective lengths of the packing elements were shown to be functions of the wetting properties of the liquid and the packing material in addition to the Reynolds number:

$$d' = \beta\,Re^{\psi} \tag{46}$$

Table 6
The Values of K for Use in the Zech and Mersmann Model, Equations 39 and 40

Packing	K
Berl saddles	0.0222
Raschig rings	0.0155
Spheres	0.0085

From Zech and Mersmann [40].

where

$$\psi = 0.732[\sigma(1-\cos\theta)]^{-0.209} \tag{47}$$

or

$$\psi = 3.0[(1-\cos\theta)]^{0.6}(\varrho\sigma^3/\mu^4 g)^{0.2}]^{-0.126} \tag{48}$$

and

$$\beta = 3.13\times10^{-18}[\sigma(1+\cos\theta)]^{-12.3} \tag{49}$$

The average deviation of the model from experimental data is ±8% (Figure 12).

Application of the Correlation to Other Packing Sizes

Considering that the L/μ values at the transition points are similar functions of the minimum wetting rate for all sizes of packings, the irrigating liquid is expected to cover the same fraction of packing surface, d'/d at the same L/μ value. Hence d'/d for different ring sizes can be correlated with the modified Reynolds number (L/μ):

$$d'/d = \omega[0.01(L/\mu)]^{\eta} \tag{50}$$

where ω and η are correlated with the minimum wetting rate and the Kapitsa number respectively giving:

$$\eta = 2.0[(1-\cos\theta)^{0.6}(\varrho\sigma^3/\mu^4 g)^{0.2}]^{-0.126} \tag{51}$$

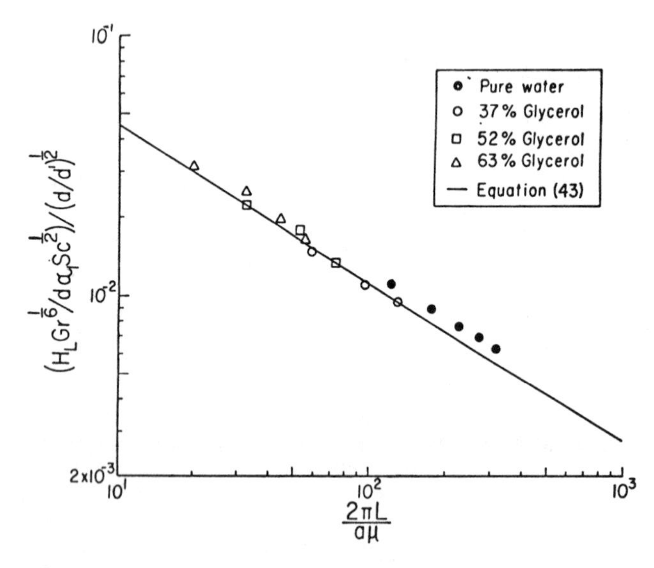

Figure 12. Comparison of the semitheoretical model of Ponter and Au-Yeung with experimental data for 1 cm Raschig rings [42].

and

$$\omega = 0.155(\varrho\sigma^3/\mu^4 g)^{-0.171} \tag{52}$$

There is good agreement between the correlation and the experimental data of Sherwood and Holloway [3] for four different sizes of packings (1.27–5.08 cm) with the average deviation being ±20% (Figure 13).

VOLUMETRIC LIQUID FILM MASS TRANSFER COEFFICIENTS FOR PLASTIC PACKINGS

Recently mass transfer data have been presented for small plastic packings. Merchuk [43] measured volumetric mass transfer coefficients for the absorption of carbon dioxide into water in a 10-cm diameter PVC column packed to a height of 60 cm with different plastic packings. The results were correlated in the form of the Sherwood and Holloway correlation. Excellent agreement with the Sherwood and Holloway equations was found for the 0.95 cm ceramic Raschig rings which were used for control. The constants for the Sherwood and Holloway equation for the plastic packings studied are given in Table 7.

Note that only small volumetric liquid flow rates up to 0.01 $m^3/m^2 \cdot s$ were employed as these experiments were used for the design of packed columns for blood oxygenation.

SYSTEM PROPERTIES AND OPERATIONAL EFFECTS ON THE TRUE LIQUID FILM MASS TRANSFER COEFFICIENT AND EFFECTIVE INTERFACIAL AREA

In view of the complicated nature of the process of gas absorption in a multi-junctioned packed column, many investigators have used small-scale laboratory models such as a string

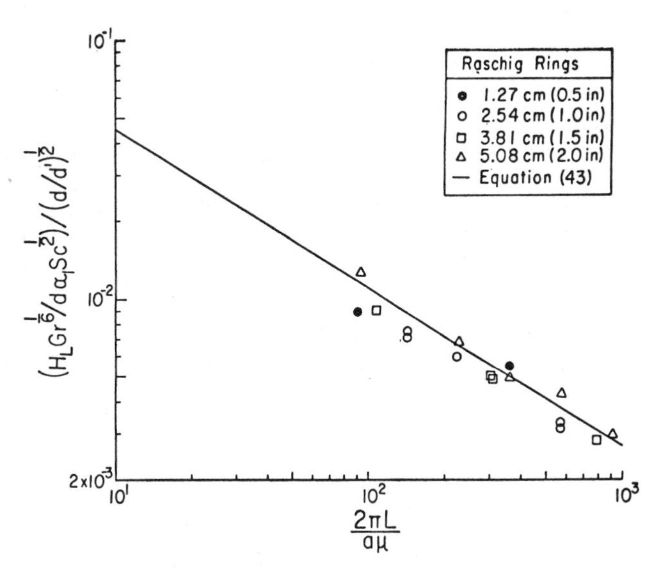

Figure 13. Comparison of the semitheoretical model of Ponter and Au-Yeung with experimental data for different sizes of Raschig rings [42].

Table 7
The Characteristics and Constants for the Plastic Packings

Packing	Void Fraction ε	Apparent Density (kg/m^3)	α	n
8 mm flat PVC rings	0.67	0.37	467	0.309
PVC spirals	0.71	0.32	2,470	0.546
PVC spirals 80% + flat rings 20%	0.70	0.35	644	0.367
9.5 mm PVC Raschig rings	0.68	0.36	1,330	0.412
6 mm PVC Raschig rings	0.58	0.48	803	0.353
4 mm PVC Raschig rings	0.55	0.47	817	0.399
6 mm polypropylene saddles	0.54	0.37	98	0.186
9.5 mm ceramic Raschig ring	0.67	0.62	3,120	0.46

From Merchuk [43].

of discs or a string of spheres to simulate the behavior of packed columns. The laboratory models are constructed to imitate the interruptions that the liquid film encounters at the packing junctions as it flows from one packing piece to the next, and they can provide liquid-film mass transfer coefficient (k_L) and interfacial area (a) data for design purposes if they closely reproduce the conditions actually occurring in a packed column. A few researchers have studied the effect of mixing of the liquid film at packing junctions with these models and the result is important for interpretation and correlation of liquid-phase-controlled mass transfer data in packed columns. It is also important to investigate the effects of system properties and operating conditions on k_L and a and to develop separate correlations for each of them for design since the influence of the hydrodynamic behavior on the two parameters is different. Methods to predict the liquid-film mass transfer coefficient (k_L), the effective interfacial area (a), and the degree of mixing at the packing junctions are now considered.

Determining Liquid-Film Mass Transfer Coefficients From Laboratory Models

Early experiments using the wetted-wall column, a simple laboratory apparatus for the study of mass transfer between a gas and a falling liquid film, have provided valuable data for gas-phase-controlled processes [44, 45]. However, attempts to correlate the results obtained for liquid-phase-controlled processes using wetted-wall columns were unsuccessful. This is due to the inability of the wetted-wall column to reproduce the behavior of the liquid film in a packed column. The performance of the liquid film in absorption processes is determined by the rippling and mixing of the liquid film which occurs at the packing junctions. As there is no junction mixing of the liquid film in a wetted-wall column, the coefficient measured is dependent upon the height of the column. The equation derived on the basis of Higbie's penetration theory for a wetted-wall column contains four dimensionless groups and is given by Norman [8]:

$$\frac{k_L Z}{D} = 0.73\left(\frac{\mu}{\varrho D}\right)^{1/2}\left(\frac{Z^3 g \varrho^2}{\mu^2}\right)^{1/6}\left(\frac{4\Gamma}{\mu}\right)^{1/3} \tag{53}$$

Note that the exponent on the Reynolds number ($4\Gamma/\mu$) is 1/3 which is found to be much lower than that from correlations of experimental data obtained from packed columns. This could be explained by the dependency of mixing at the packing junctions on the Reynolds number.

To simulate the behavior of the liquid film in a packed column, Stephens and Morris [46] developed a disc column in which the liquid film was interrupted at short intervals during its downward flow to the bottom. The disc column consists of 20 to 50 discs about 1.5 cm in diameter, made of metal or unglazed material of low porosity, and threaded on a vertical wire with alternate discs at right angles to each other. Mass transfer coefficients were determined by the absorption of carbon dioxide into water in the disc column at 20° C. The results were correlated by the equation:

$$k_L = 9.5 \times 10^{-4} \Gamma^{0.7} \tag{54}$$

for Γ ranging from 0.025 to 0.16 kg/m · s with Reynolds numbers ($4\Gamma/\mu$) from 100 to 660.

Stephens and Morris concluded that their results were consistent with the Sherwood and Holloway correlation which could be written in the form:

$$\frac{k_L}{D} = \alpha \left(\frac{4\Gamma}{\mu}\right)^n \left(\frac{\mu}{\varrho D}\right)^{0.5} \tag{55}$$

They expressed their data for Reynolds numbers over 100 by the dimensional equation:

$$\frac{k_L}{D} = 6.53 \left(\frac{4\Gamma}{\mu}\right)^{0.7} \left(\frac{\mu}{\varrho D}\right)^{0.5} \tag{56}$$

Later, Morris [47] proposed an equation that is dimensionally consistent:

$$\frac{k_L}{D} \left(\frac{d^3 \mu^2}{\varrho^2 g}\right)^{0.17} = \text{const} \left(\frac{4\Gamma}{\mu}\right)^{0.7} \left(\frac{\mu}{\varrho D}\right)^{0.5} \tag{57}$$

Equation 57 is analogous to Equation 53 for a wetted-wall column but has a different Reynolds number exponent.

Taylor and Roberts [48] repeated the experiments of Stephens and Morris using carbon dioxide and ammonia absorption in the disc column. It was found that the roughened discs gave better wetting characteristics then the smooth discs. The results for carbon dioxide absorption were found to be lower than those of Stephens and Morris by about 30% to 50%. The coefficients were correlated by the equations:

$$\frac{k_L}{D} = 1.90 \times 10^2 \left(\frac{4\Gamma}{\mu}\right)^{0.4} \left(\frac{\mu}{\varrho D}\right)^{0.5} \tag{58}$$

for $\Gamma < 0.0641$ kg/m · s and

$$\frac{k_L}{D} = 6.36 \left(\frac{4\Gamma}{\mu}\right)^{1.0} \left(\frac{\mu}{\varrho D}\right)^{0.5} \tag{59}$$

for $\Gamma > 0.0641$ kg/m · s

The transition at the liquid rate of 0.0641 kg/m · s was attributed to the establishment of a ripple formation on the rim of the discs.

Yoshida and Koyanagi [32], in an effort to study liquid-phase mass transfer rates and effective interfacial areas in packed absorption columns, absorbed carbon dioxide in both water and methanol in bead columns containing spheres made of steel and coated with

Table 8
Values of c″ For the Correlation for Bead Columns, Equation 60

Diameter of Spheres, cm	Number of Spheres	Value of c″
2.54	16	18.2
2.54	8	17.0
1.27	32	16.5
1.27	16	14.7

From Yoshida and Koyanagi [32].

metallic aluminum. The results obtained in the bead columns where the product of the Re and Sc was greater than 40,000 were represented by the equation:

$$H_L \Big/ \left(\frac{\mu^2}{\varrho^2 g}\right)^{1/3} = c''[(Re)\,(Sc)]^{0.5} \tag{60}$$

where the value of c″ was found to depend on the number of spheres and their diameters and is summarized in Table 8.

Norman and Sammak [33] investigated the effect of liquid viscosity on the mass transfer coefficient using a vertical row of modified squared disc packings. The packing element was made of a plate of graphite 2.5 cm long, 2.0 cm wide, 0.4 cm thick, and tapered at the bottom. The sides of the packing element were raised to direct the liquid flow onto the faces of the element, which eliminated the flow around the edge of the packing (this is possible in the disc packings of Stephens and Morris and Taylor and Roberts). Sulfur dioxide was absorbed into water and carbon dioxide was absorbed in water and several organic liquids ranging in viscosity from 0.4 to 20 mPa · s. At a wetting rate Γ of 0.136 kg/m · s, the dependency of the liquid-film mass transfer coefficient on the viscosity could be represented by the equation

$$\frac{k_L}{D^{1/2}} = 0.89\mu^{-0.54} \quad \text{at} \quad \Gamma = 0.136\ \text{kg/m} \cdot \text{s} \tag{61}$$

The experimental data were found to be well-correlated by the equation:

$$\frac{k_L d}{D} = 0.13\left(\frac{\mu}{\varrho D}\right)^{1/2}\left(\frac{d^3 g \varrho^2}{\mu^2}\right)^{1/6}\left(\frac{4\Gamma}{\mu}\right)^{0.61} \tag{62}$$

with a maximum deviation of 38% for values of Reynolds number from 1.2 to 545 and Schmidt numbers from 108 to 43,500. The data of Stephens and Morris for the disc packing were found to be represented by a similar equation but with a constant of 0.062. The lower value of the constant was attributed to less uniform distribution of the liquid on the disc packing. Note that Equation 62 is identical in form to the theoretical equation derived for wetted-wall columns (Equation 53), but with a larger exponent for the Reynolds number which could be explained by the dependency of mixing on the Reynolds number.

The data obtained by Norman and Sammak [34] for the absorption of carbon dioxide in water, amyl alcohol, and two gas-oil fractions using a column of packing elements with vertical spaces of 1 cm between each pair of elements were correlated by the equation:

$$\frac{k_L d}{D} = 0.33\left(\frac{\mu}{\varrho D}\right)^{1/2}\left(\frac{d^3 g \varrho^2}{\mu^2}\right)^{1/6}\left(\frac{4\Gamma}{\mu}\right)^{0.47} \tag{63}$$

Mika [49] noted that none of the laboratory models mentioned previously provides adequate simulation to the flow in packed columns as there is no lateral contact between the elements which is a characteristic feature of packed columns. Hence he carried out absorption of carbon dioxide in water at 20° C on 7 strings of 15-mm spheres each consisting of 12 spheres with the strings arranged in a hexagonal pattern. The touching strings of spheres provided for both vertical and horizontal contact between the packing elements. The experimental results were compared to the data obtained for packed columns in the literature so resulting in the equations

$$\frac{k_L d}{D} = 7.0 \times 10^{-3}\left(\frac{L}{a_w \mu}\right)^{0.83}\left(\frac{\mu}{\varrho D}\right)^{1/2}\left(\frac{d^3 g \varrho^2}{\mu^2}\right)^{1/3} \tag{64}$$

for $Re_w < 30$ and

$$\frac{k_L d}{D} = 1.5 \times 10^{-2}\left(\frac{L}{a_w \mu}\right)^{0.60}\left(\frac{\mu}{\varrho D}\right)^{1/2}\left(\frac{d^3 g \varrho^2}{\mu^2}\right)^{1/3} \tag{65}$$

for $Re_w > 30$.

The wetted areas a_w used in calculating the Reynolds number Re_w were obtained from the work of Shulman et al. [22] and Onda et al. [50]. It was found that for single-string models, the data of Stephens and Morris best fit the results for packed columns. However, the author's seven-string model fit the plots for packed columns better than any other models mentioned before (Figure 14 and 15). Thus it was concluded that a model composed of several strings of packing elements in mutual contact was better than the single-string models in representing the behavior in packed columns.

Alper and Danckwerts [51] discussed the criteria for the simulation of a full-sized packed column by a laboratory model. They stated that the necessary conditions for simulation were that the model and the packed column had the same k_L, k_G, (aZ/U), (h_tZ/U), or (h_t/a) and $(U/\bar{G})$ values. To demonstrate the soundness of these criteria, the authors used a modified string-of-spheres column consisting of spheres 3.72 cm in diameter with a cylindrical pool at the top of each sphere 1 cm in diameter and 1.5 cm deep. A second column with

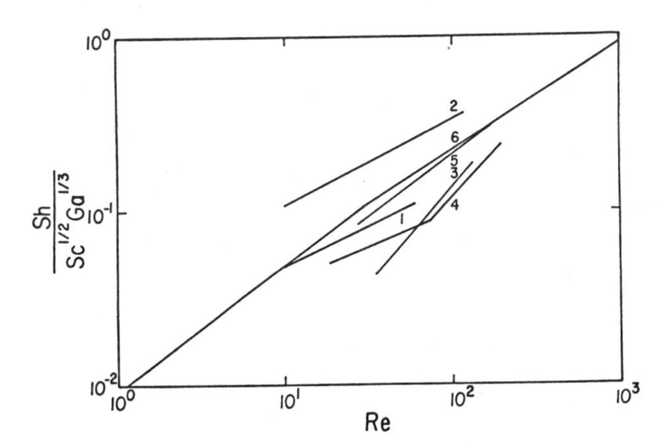

Figure 14. Rate of absorption with liquid-film resistance for models simulating packed columns: (1) wetted tube; (2) string of 16 spheres; (3) string of 27 spheres; (4) string of 61 discs; (5) string of 40 discs; (6) packed column [49].

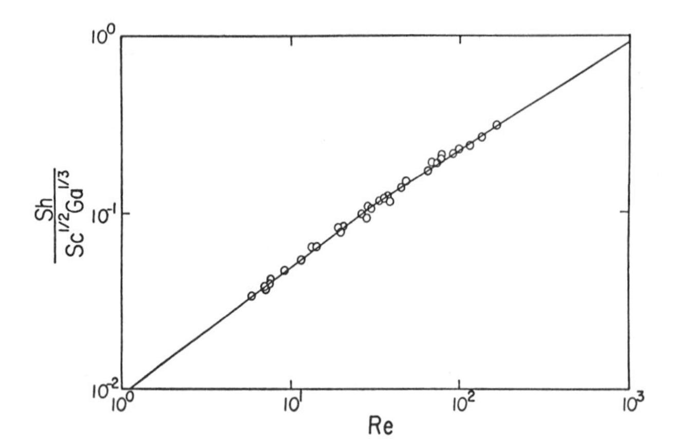

Figure 15. Rate of absorption of CO_2 in water at 20°C using 7 strings of 15 mm spheres, each consisting of 12 spheres [49].

spheres approximately half the size of the first one was also used. The dimensions of the pools in the top of the spheres were chosen such that the value of (h_t/a) was of the same order of magnitude as that of an industrial column. The interfacial area of the sphere column was determined using the sulfite oxidation method and it was found that this area was slightly greater than the geometrical area due to the increase in area by the thickness of the liquid film. Experiments were carried out in the string-of-sphere columns and a packed column 10.2 cm in diameter and packed with 1.27-cm ceramic Raschig rings using five different chemical systems. It was shown that when the liquid and gas flow rates and the number of spheres were properly adjusted to give the same values of k_L, k_G, (aZ/U), $(U/\bar{G})$ and (h_t/a) in the model and the packed column, the predicted absorption rates calculated from the data obtained with the string-of-spheres column always agreed with the measured absorption rates in the packed column within ±7%.

Fukushima and Kusaka [52] investigated the adequacy of using multi-strings of touching spheres models to represent the conditions occurring in a packed column. They measured the interfacial areas and the liquid-phase volumetric mass transfer coefficients in one, two, three, and seven strings of touching spheres and in irrigated columns packed with spheres, Raschig rings, and Berl saddles using the sulfite oxidation method and the physical absorption of carbon dioxide into water, respectively. For the cases where the number of strings was 3 and 7 and the number of spheres per string was greater than 10, it was found that the interfacial area data could be represented by the equation:

$$a_s d = 4.8q^{-1.1} Re_s^{(0.27q^{0.25})} \tag{66}$$

where a_s = the interfacial area per volume of strings of touching spheres
q = the mean number of contact points of touching spheres per sphere
Re_s = the Reynolds number of m strings of touching spheres based on the diameter of spheres, $4\varrho Q/m\pi d\mu$

The authors, using the data of de Waal and Beek [53], Jhaveri and Sharma [54], and Danckwerts and Rizvi [55] showed that Equation 66 was applicable to columns packed with spheres, Raschig rings, and Berl saddles with an accuracy of ±20% if a_s and Re_s were replaced by $a/(1-\varepsilon)$ and Re_p in the previous equation.

The liquid-phase mass transfer coefficients obtained from the volumetric coefficient values using the interfacial areas were found to be correlated well by Equation 67 when the number of touching spheres was greater than 10:

$$\frac{k_{Ls}d}{D} = 7.1\,Re_s^{1/3}\,Sc^{1/3} \tag{67}$$

where k_{Ls} is the liquid-phase mass transfer coefficient for strings of touching spheres.

By replacing k_{Ls} with k_L and 7.1 with $27/\Phi(1-\varepsilon)$ in Equation 67, the authors were able to predict the performance of columns containing packings of surface shape factor Φ. Details of the number of contact points for strings of touching spheres are given in Table 9. The values of the number of contact points and the shape factor for different packings are summarized in Table 10.

Investigation of Mixing at Packing Junctions

Lynn et al. [56] carried out experiments on the absorption of sulfur dioxide into water at 20° C using a string of spheres having a diameter of 2.00 cm. Various numbers of spheres from 3 to 14 were used. It was found that there was little or no mixing of the surface layers with the bulk of the liquid as it flowed from one sphere to the next down the column. It should be noted, however, that a surface-active material (Teepol) had been added to eliminate film rippling which would suppress mixing at the junctions of the spheres.

The results of Yoshida and Koyanagi [32] obtained using bead columns with 2.54-cm and 1.27-cm diameter spheres showed clearly that there was a lack of complete mixing at the packing junctions as the height of a liquid-phase transfer unit increased with an increasing number of spheres in the column (Figure 16).

Davidson et al. [57] studied the effects of mixing and rippling on the liquid-film mass transfer coefficient using a vertical string of spheres (3.78-cm diameter). It was concluded that with a surfactant solution (Lissapol), there was no mixing in the menisci between the spheres and the absorption rate increased above the value predicted by the "no mixing"

Table 9
Values of the Maximum Number of Contact Points q_{max} and the Mean Number of Contact Points q for Strings of Touching Spheres

Number of strings, m	q_{max}	q
1	2	1.8
2	3	2.7
3	4	3.8
7 (opened)	8	5.5
7 (closed)	12	7.7

From Fukushima and K. Kusaka [52].

Table 10
Values of the Mean Number of Contact Points and the Shape Factor for Different Packings

Packing	q	Φ
Spheres	12	3.14
Raschig rings	4.2	6.15
Berl saddles	2.8	4.95

From Fukushima and Kusaka [52].

theory due to the ripples present at higher liquid rates. With pure water, complete mixing occurred within the menisci and the absorption rate increased above the value predicted by the "complete mixing" theory caused by the ripples at high water rates. Their data were compared with those of Yoshida and Koyanagi (Figure 17), which showed clearly that there was incomplete mixing between the spheres at low Reynolds numbers and when small spheres were used. The authors also concluded that the surface-active agent suppressed ripples on the surface of the film and postulated that mixing might also be suppressed in the liquid holdup between the spheres.

Ratcliff and Reid [58] substantiated the conclusions of Davidson et al. [57] by studying the mixing effects in a column of spheres using a dye-trace technique. The laminar flow of liquid through the junction between the spheres is pictured in Figure 18. They showed that there were two mixing effects at the junction:

1. Laminar circulation in the junction holdup between the spheres with nonrippling flow, confined to the lower levels of the liquid and this would not be expected to have any effect on mass transfer into the liquid film with short contact times.
2. Vigorous mixing occurring when ripples pass through the junction.

The degree of mixing is therefore dependent on the frequency and amplitude of the waves, which are functions of the Reynolds number in a rippling system. In a nonrippling system no mixing would be expected at the junctions as far as mass transfer is concerned.

Harrison et al. [59] studied the axial dispersion of liquid in a column of spheres using a dye-trace technique. It was found that the efficiency of mixing at the junctions between the spheres increased with increasing flow rate, and this increase was considerably magnified when rippling occurred in the liquid film.

Atkinson and Taylor [60] investigated the effect of discontinuities on mass transfer to a liquid film using a laboratory model with surfaces arranged in series in a stair-step manner and inclined to the vertical at 45°. It was concluded that the mixing occurring was a result of ripple action at the discontinuities. They recommended that packings used for processes with a high liquid-film resistance have ripple-promoting properties to induce higher degrees of mixing in the liquid film with a corresponding enhancement of mass transfer.

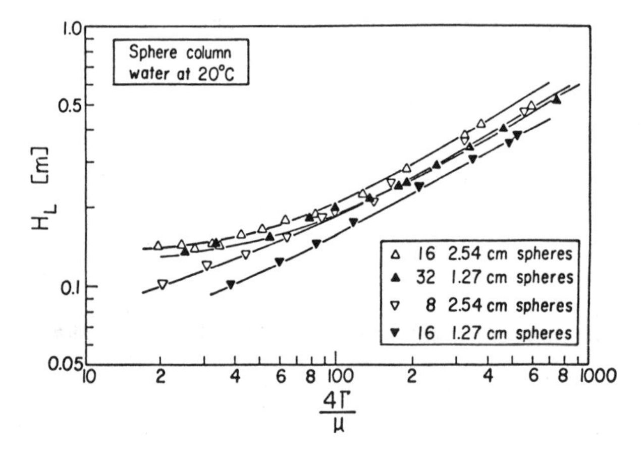

Figure 16. Effects of number and diameter of spheres on H_L for bead columns. (Yoshida, F., and Koyanagi, T., *Ind. Eng. Chem.*, Vol. 50 (1958), pp. 365–374.)

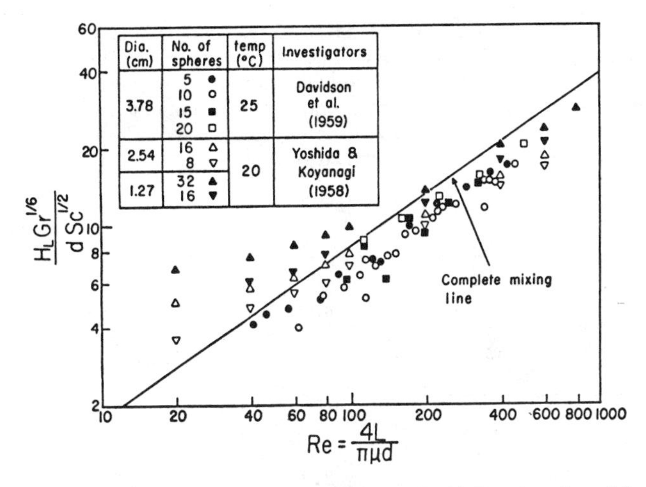

Figure 17. The effect of Reynolds number on the height of a liquid phase transfer unit for sphere columns [57]

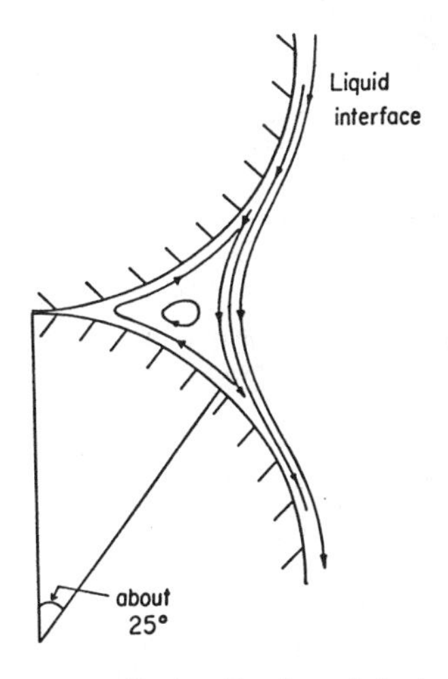

Figure 18. Diagrammatic representation of laminar flow through the junction between two spheres [58].

Norman and Sammak [33, 34] studied the effect of liquid viscosity and mixing on the mass transfer coefficient using a vertical column of square-shaped packings and found that the liquid-film coefficient and hence the degree of mixing was inversely proportional to the liquid viscosity at a constant liquid flow rate. The viscosity of the liquid used ranged from 0.4 to 20 mPa · s. The majority of their absorption data for carbon dioxide in water and organic liquids fell between the "complete mixing" line and the "no mixing" line for values of Reynolds numbers from 1.2 to 545 and Schmidt numbers from 108 to 43,500 (Figure 19). They concluded that the degree of mixing at the packing junctions was a function of the Reynolds number and the spacing between successive packing elements.

Shulman and Mellish [61] studied the liquid flow patterns and velocities existing in packed columns using high-speed motion pictures of fine particles suspended in the flowing water at different points in a 10-cm column packed with 2.54-cm white porcelain Raschig rings. The average point velocity was obtained by averaging the instantaneous velocities of about fifty particles in one small area. The typical points studied are shown in Figure 20. The results showed that laminar flow predominated on the outer wetted surface of the rings

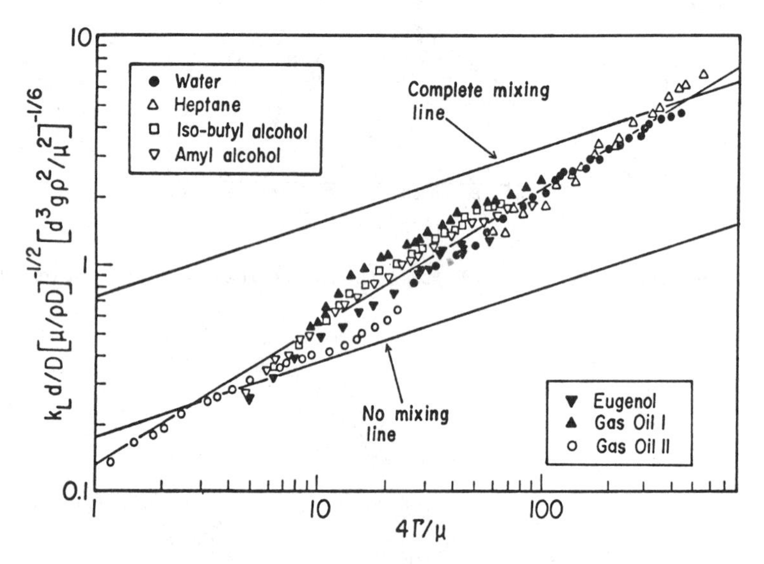

Figure 19. Correlations of data for the absorption of carbon dioxide in water and organic liquids using square-shaped packings. (Norman, W. S., and Sammak, F. Y. Y., *Trans. Inst. Chem. Eng.*, Vol. 41, (1963) pp. 109–116.)

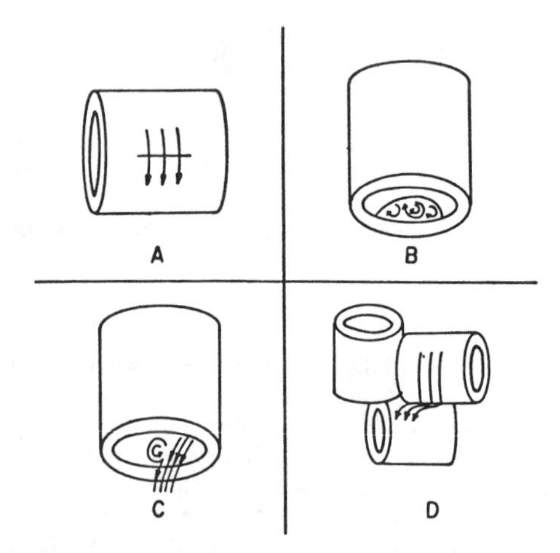

Figure 20. Flow patterns in random packings: (A) film flow; (B) a semistagnant pool; (C) a divided pocket illustrating film flow and a circulating pool; (D) a ring junction [61].

(Figure 20A) with no appreciable mixing. The average velocities of the film flow were found to be high, about 40 to 90 cm/s, which fluctuated rapidly with time. The rapid fluctuations were due to the random feeding of the packing element under observation by the many different sources above the packing piece. Other flow patterns included semistagnant pockets and pockets with film flow and a circulating pool next to each other (Figures 20B and 20C). The liquid at the ring junctions manifested all three flow patterns described, which changed rapidly with time. Visual observation of injected dye solutions indicated thorough mixing at these junctions. The authors suggested the use of a complex model, made up of simple models for different portions of the liquid in the packing to describe the performance of the liquid phase in packed columns.

Copp and Ponter [62] simulated the conditions in a packed column by absorption of carbon dioxide into water and aqueous glycerol solutions in a Macropore similar to the one used by Atkinson and Taylor [60]. It was shown that mixing at the discontinuities decreased with increasing viscosity. The relation between the mass transfer coefficient and the viscosity at a liquid rate of 0.136 km/m · s could be expressed by the equation

$$\frac{k_L}{D^{1/2}} = 0.0535\,\mu^{-0.54} \tag{68}$$

The exponent on the viscosity was found to be identical to that reported by Norman et al. [33] using the squared-shaped packing Equation 61. The corresponding equation for complete mixing, derived from Higbie's penetration theory, is given by

$$\left(\frac{k_L}{D^{1/2}}\right)_{cm} = 24.1\,\mu^{-1/6} \tag{69}$$

where cm = complete mixing

The large discrepancy in the exponential dependency of the mass transfer coefficient on viscosity suggests a viscosity-dependent process affecting the degree of mixing at discontinuities which was not considered in Davidson's derivation of the statistical models (Equations 33 to 35).

Alper [63] described the absorption of carbon dioxide into water in a column of discs made of pyrophylite, 1.5 cm in diameter and 0.5 cm thick. His results showed that k_L decreased with an increasing number of discs initially and then remained constant. As the effect of liquid distribution was eliminated in the disc column, he suggested that the variation of k_L with the number of discs could be explained by surface rejuvenation where mixing occurs up to a certain depth below the surface at the junctions. This being so, the average value of k_L should decrease down the column until the gas diffuses through the stagnant layer into the mixing region after which the coefficient should remain constant. Using an analysis of Mika [64], Alper demonstrated that the thickness of the stagnant layer for a laminar film could be predicted:

$$x = K'N/Q^{2/3} \tag{70}$$

where K' = a constant depending on the disc dimensions, the diffusivity of dissolved gas and the liquid viscosity and density
N = the number of discs over which an appreciable amount of gas has not yet penetrated into the rejuvenation depth
Q = the liquid flow rate

Alper calculated the values of N from the absorption data and found that the thickness of the stagnant layer decreased with an increasing liquid flow rate as expected. He thus concluded that surface rejuvenation rather than surface renewal was more probably taking place at the packing junctions.

Interfacial Areas for Gas Absorption in Packed Columns

Factors that influence the effective interfacial area for gas absorption in packed columns are first discussed.

The effective interfacial area for gas absorption, which includes liquid films on packing surfaces, sprays, rivulets, and drops dripping from packing pieces should not be confused with the wetted area since not all of the wetted area is effective in the mass transfer process. The wetted area can generally be divided into two parts, the fast-moving liquid films on the packing surfaces and the slow-moving liquid layers or stagnant regions. The latter is likely to become saturated with the solute gas in a short period, and since this is renewed very slowly by fresh liquid, it will contribute very little to the gas absorption process. The effective interfacial area provided by the rapid moving liquid film is generally considered dependent on the wettability of the packing surfaces by the irrigating liquid although in some cases rivulets formation especially on poorly wetted surfaces such as teflon (PTFE) and polyethylene has been found to be advantageous [65, 66].

Factors that affect the interfacial area for random packings below loading include:

1. Liquid flow rate.
2. Liquid density.
3. Liquid viscosity.
4. Liquid surface tension.
5. The affinity of the solid for wetting by the liquid which can be characterized by the contact angle or the critical surface tension of the solid.
6. The packing size.

The contact angle θ is defined as the angle made by the edge of a liquid drop with a uniform, flat solid surface on which the drop is resting. At equilibrium the three interfacial tensions must balance along the line of contact (Figure 21):

$$\gamma_{LV} \cos\theta + \gamma_{SL} = \gamma_{SV} \tag{71}$$

Hence

$$\cos\theta = \frac{\gamma_{SV} - \gamma_{SL}}{\gamma_{LV}} \tag{72}$$

The critical surface tension of a solid is the maximum surface tension which allows a liquid to spread over the entire solid surface, i.e., the contact angle is zero or $\cos\theta = 1$. The effective wetted area created in a packing is the result of the interplay between the existing viscous and surface forces. The latter influence is controlled by a combination of the liquid

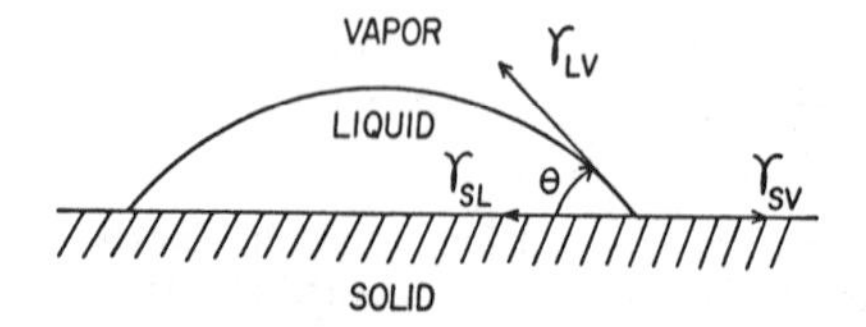

Figure 21. A sessile liquid drop in equilibrium with solid and vapor.

surface tension and the contact angle between the liquid and the underlying solid surface. Ponter et al. [36] have measured and estimated under absorption conditions the minimum wetting rate where a dry patch is formed which reduces the wetted area. It has been demonstrated that σ and θ are surface properties which have to be measured under the appropriate mass transfer condition and values measured at equilibium reported in the literature cannot be used for surface area prediction [67].

Thus the prediction of tower performance behavior using the concept of surface-tension positive and negative characteristics by Zuiderweg and Harmens [68] is only tenable if the surface tension does not go through a minimum with increasing concentration under mass transfer conditions as shown by Ponter et al. [69]. At this time there is no way to quantitatively assess the influence of the surface properties on the effective interfacial area. However, the spreading characteristics have been examined.

Factors that influence liquid film spreading on a vertical plate are now described.

The use of film contacting devices such as packed columns in mass transfer operations is to create a maximum area of contact between the phases in which mass transfer occurs, hence fundamental knowledge of the factors affecting the said objective is important. Peier [70] et al. have developed a model to predict the width of liquid film wetting a vertical plate by incorporating an approximately chosen wetting force term into the Navier-Stokes equations and using the laminar boundary-layer theory. The model is:

$$X(z) = \left(\frac{4D'}{3\varrho}\right)^{1/3} \left(\frac{\mu}{\varrho}\right)^{1/6} \left(\frac{1}{2g}\right)^{1/4} \left(\frac{1}{\sin^2\theta_{st}} - \frac{1}{3} - \frac{\cos\theta_{st}}{\sin^3\theta_{st}}\theta_{st}\right)^{-1/3} z^{-1/12} \tag{73}$$

where $X(z)$ = the half-width of the liquid film
θ_{st} = the stationary contact angle measured at the point where the film has stopped spreading
z = the vertical distance from the liquid feed nozzle at the top of the plate
D' = a constant

The model shows that the half-width of the liquid film depends mainly on the stationary contact angle, the liquid density, and viscosity. The model was shown to agree satisfactorily with experimental results in the laminar flow region obtained using 60% and 70% aqueous glycerol solutions which have viscosities of 10.1 and 20.6 mPa · s (cp), respectively. However, the model does not fit the data obtained with water and mineral oil due to the development of a turbulent boundary layer in the case of water. In a related study, Fabre et al. [71] found a hysteresis effect for the contact angle as the flow rates were successively decreased and increased. They also reported that the film width increased with increasing viscosity and decreased with increasing surface tension in accordance with theory. It was also shown that the spreading of the liquid film immediately after leaving nozzle was influenced by the shape of the nozzle. The best spreading was achieved with a circular nozzle, intermediate spreading with a square-shape, and the poorest spreading with the triangular-shape.

Minimum Liquid Wetting Rate for Packed Columns

The minimum wetting rate is the minimum liquid flow rate per unit periphery (kg/s · m) necessary to wet the packing effectively such that any increase in liquid flow rate would not increase the wetting of the packing. Minimum wetting rates are quoted for design purposes in the literature (e.g. Morris and Jackson [72]), however, these should be used with caution since the minimum wetting rate depends on a variety of factors such as the liquid surface tension, surface energy of the packing material, liquid viscosity, and surface tension gradient that occurs due to the absorption of a surface tension increasing or lowering solute, e.g. absorption of ammonia.

Ponter et al. [36] developed a correlation for predicting the minimum wetting rate for a laminar liquid film on a vertical surface:

$$\frac{\Gamma}{\mu} = 1.12(1-\cos\theta)^{0.6}\left(\frac{\varrho\sigma_e^3}{\mu^4 g}\right)^{0.2} \tag{74}$$

The minimum wetting rate was obtained by reducing the liquid flow until a stable dry patch was formed on the surface. It should be noted that the contact angle and the surface tension used in the correlation were measured under mass transfer conditions, i.e., during absorption of ethanol-air mixtures in the experiments. It has been shown that the predictions by the correlations agreed satisfactorily with the measured minimum wetting rate for all the surfaces, i.e., copper, stainless steel, Perspex, and carbon employed in the experiments. Although the correlation cannot be used directly to predict the minimum rates in packed columns, Mangers and Ponter [35] have shown that the dimensionless groups of physical properties used in the correlation for vertical surfaces can be used to correlate the minimum wetting rates in packed columns.

Correlations for Predicting Interfacial Areas in Packed Columns

The first investigation regarding interfacial areas in packed columns was reported by Mayo et al. [73] who assumed that the entire wetted area was effective for mass transfer. The wetted areas using 1.25-cm and 2.54-cm waxed paper rings were determined by measuring the portions stained by a dye solution irrigating a column packed to a depth of 0.76 m with the paper rings. The percentage of the tower wall wetted was also determined by measuring the area of the dyed surface of the waxed paper liner. The results showed that the wetted area increased with increasing water flow rates up to the flooding point but was independent of the gas mass velocity. The exterior ring surfaces were found to be wetted better than the interior surfaces and the tower wall was wetted considerably more than the ring packing. The percentage of the total surface wetted was found to decrease as the packing size increased. The proportion of water on the wall increased as it flowed down the column while the wetted area of the packing decreased. The authors estimated that about 10% of the total wetted area was inactive due to the stagnant liquid pockets existing at points of contact between the packing elements. The effect of liquid rate on the total surface wetted as given by Mayo et al. is presented in Table 11.

Weisman and Bonilla [74] calculated the wetted areas for spheres and Raschig rings by dividing the k_Ga values obtained from humidification experiments using nonporous packings by the k_G values obtained by the evaporation of water from completely wetted porous packings [75, 76]. The correlations obtained using 1.27-cm spheres and 2.54-cm Raschig rings are given respectively by:

$$\frac{a}{a_t} = 0.018\left(\frac{D_pG}{\mu_G}\right)^{0.31} L^{0.5} \tag{75}$$

for 1.27-cm spheres, and

$$\frac{a}{a_t} = 0{,}54\, G^{0.31}\, L^{0.07} \tag{76}$$

for 2.54-cm Raschig rings.

Note that Equation 76 shows a much smaller dependence of the wetted area on the liquid superficial mass velocity than indicated by the results of Mayo et al. This is probably due to the use of the k_Ga values obtained by McAdams et al. [77] whose data showed a smaller dependency on the liquid rate than other investigators [78].

Table 11
Effect of Liquid Rate on Total Wetted Area

Height of Packing to Tower Diameter Ratio	Liquid Flow Rate, kg/m^2 · s				
	4.23	5.83	6.40	10.8	13.1
	Percentage Wetted Area				
2	44.5	50.9	65.0	68.5	72.0
4	36.1	39.6	51.0	54.5	57.6
6	–	36.2	46.0	48.3	52.8
8	–	34.1	43.1	47.1	49.7
10	28.1	32.6	42.0	45.5	48.0

From Mayo, Hunter, and Nash [73].

Shulman and DeGrouff [78] determined the rates of vaporization of naphthalene Raschig rings using a range of gas rates and the measurements were repeated with water irrigating the packings at different flow rates. The wetted areas of the 2.54-cm naphthalene rings were determined from the ratios of the vaporization rate of the wetted packing and the vaporization rate of the dry packing.

In an extensive research program Shulman et al. [22] obtained the wetted areas for 1.27-cm and 3.81-cm Raschig rings and 1.27-cm and 2.54-cm Berl saddles using the preceding technique. The areas were found to increase with increasing liquid rate but decreased with increasing gas rate below the loading point. Above the loading point, the wetted areas tended to remain constant or increased with increasing gas rate. The fractional wetted areas were correlated by the equations:

$$\frac{a_w}{a_t} = 0.35\left(\frac{L}{G}\right)^{0.2} \tag{77}$$

for Berls saddles, and

$$\frac{a_w}{a_t} = 0.24\left(\frac{L}{G}\right)^{0.25} \tag{78}$$

for Raschig rings.

It is doubtful, however, that these equations could be used to calculate wetted areas for packings made of other materials and systems other than water because of changes in wettability, especially since the naphthalene is so poorly wetted by water.

The k_G values obtained from the vaporization of the naphthalene rings were correlated by the equation:

$$\left(\frac{k_G M_M P_{BM}}{G}\right) = 1.195\left(\frac{\mu_G}{\varrho_G D_G}\right)^{-2/3}\left(\frac{D_p G}{\mu_G(1-\varepsilon)}\right)^{-0.36} \tag{79}$$

The effective interfacial areas for porcelain packings were obtained by dividing the $k_G a$ data of Fellinger [23] for ammonia absorption in water by the k_G values calculated from Equation 79. Fellinger's data were corrected for the liquid film resistance using the Sherwood and Holloway correlation. Below the loading point, the effective interfacial areas were found to increase with increasing liquid rate but were relatively independent of the gas rate. It was found that these effective interfacial areas for gas absorption were much smaller than the corresponding wetted areas. This was explained by considering that the stagnant pockets in the packings became saturated and did not play a further role in the absorption process. Furthermore, the contact angle of water on naphthalene is also very much different from

Table 12
Valuef of m′, n′, and s′ for the Interfacial Area Equation

Packing	Size, mm	L, kg/m^2 · s	m′	n′	s′
Raschig rings	13	0.68-2.0	28.01	0.2323L − 0.30	−1.04
		2.0-6.1	14.69	0.01114L + 0.148	−0.111
	25	0.68-2.0	34.42	0	0.552
		2.0-6.1	68.2	0.0389L − 0.0793	−0.47
	38	0.68-2.0	36.5	0.0498L − 0.1013	0.274
		2.0-6.1	40.11	0.01091L − 0.022	0.140
	50	0.68-2.0	31.52	0	0.481
		2.0-6.1	34.03	0	0.362
Berl saddles	13	0.68-2.0	16.28	0.0529	0.761
		2.0-6.1	25.61	0.0529	0.170
	25	0.68-2.0	52.14	0.0506L − 0.1029	0
		2.0-6.1	73.0	0.0310L − 0.0630	−0.359
	38	0.68-2.0	40.6	−0.0508	0.455
		2.0-6.1	62.4	0.0240L − 0.0996	−0.1355

From Treybal [79].

Table 13
The constants for Calculating the Operating and Static Holdups for 2.54-cm Rings and Saddles

Packing	b′	k′	p′	μ (mPa · s)	b	k	p
Porcelain Raschig rings	0.0486	0.02	0.99	< 12	0.476	0.13	0.1737
				> 12	1.24	0.31	0.1737
Carbon Raschig rings	0.0237	0.02	0.23	< 12	0.476	0.13	0.1737
				> 12	1.24	0.31	0.1737
Porcelain Berl saddles	0.00423	0.04	0.55	< 20	0.525	0.13	0.2817
				> 20	1.24	0.31	0.2817

From Shulman et al. [81].

that on ceramics. The effective areas were presented by Shulman et al. as a series of graphs and correlated by Treybal [79] in the form of an empirical equation:

$$a = m'\left(\frac{808G}{\varrho_G^{0.5}}\right)^{n'} L^{s'} \tag{80}$$

where m', n', and s' as given by Treybal are presented in Table 12 together with the liquid rate range L used in the study.

Shulman et al. [80] explained the discrepancy between the k_Ga values for vaporization and the k_Ga values for gas absorption by the liquid holdup in the packings. He considered that for vaporization, the entire wetted surface is available for mass transfer and hence the interfacial area a is proportional to the total holdup h_t. For gas absorption, there are semistagnant pockets which are proportional to the static holdup and these are not effective for mass transfer. Thus the interfacial area is proportional to the operating holdup. The following relation between the interfacial area and the holdup was derived by Shulman et al. [80].

$$\frac{a_{vap}}{a_{abs}} = 0.85\left(\frac{h_t}{h_0}\right) \tag{81}$$

Total, static, and operating holdup data were obtained by Shulman et al. [80] for 1.27-, 2.54-, and 3.81-cm porcelain Raschig rings; 1.27- and 2.54-cm porcelain Berl saddles; and 2.54-cm carbon Raschig rings.

Shulman et al. [81] have determined the effects of viscosity, surface tension, density, and liquid flow rate on the total, operating, and static holdups. The data for the operating and static holdup for 2.54-cm rings and saddles were correlated in the following forms:

$$h_0 = bL^{0.57}\mu^k\left(\frac{1}{\varrho}\right)^{0.84}\left(\frac{\sigma}{0.073}\right)^{p-0.262\log L} \tag{82}$$

and

$$h_s = b'\mu^{k'}\left(\frac{1}{\varrho}\right)^{0.37}\sigma^{p'} \tag{83}$$

The constants in these equations are given in Table 13.

Shulman et al. [81], noting that the effective interfacial area for gas absorption was proportional to the operating holdup, suggested that Equation 80 could be modified to calculate the effective interfacial areas for other aqueous and nonaqueous systems by using the relation:

$$a = a_{water}\left(\frac{h_0}{h_{0\ water}}\right) \tag{84}$$

Equation 84 is useful as a first approximation for assessing interfacial areas using aqueous systems. It is questionable, however, if the equation could provide accurate values for organic systems with low surface tensions. Examination of Equation 82 reveals that the operating holdup increases with increasing surface tension for small to intermediate liquid flow rates. This implies that a liquid with a higher surface tension provides a larger interfacial area than a liquid with a lower surface tension. This is contrary to the notion that the wetting of the packing should improve as the surface tension is lowered. Buchanan [82] examined the holdup data of several investigators including Shulman and found that the surface tension had no effect on the operating holdup. Hence the assumption that the effective interfacial area is proportional to the operating holdup is disputable for all cases. Situations will exist where a combination of different liquid viscosities and surface tensions will give the same operating holdup but very different interfacial areas. Furthermore, the work of Shulman et al. did not take into account the effect of packing materials on wettability.

Hikita and Kataoka [83] determined the effect of viscosity and surface tension on wetted area in packed columns using the technique of Mayo et al. [73]. Paper Raschig rings 1.5, 2.5 and 3.5 cm in diameter were used in conjunction with a 12.7-cm diameter column. Below the loading point, the gas rate was found to have no measurable effect on the wetted areas. It was also found that the viscosity in the range 0.9 to 3.8 mPa · s had no appreciable influence on the wetted areas. The values of the wetted area were correlated by the equation:

$$\frac{a_w}{a_t} = 0.711L^{1/3}\left(\frac{\sigma}{0.020}\right)^r \tag{85}$$

where r is related to the diameter of the packing by

$$r = -0.056d^{-0.70} \tag{86}$$

The data showed good agreement with those of Fujita and Sakuma [25] and Mayo et al. [73].

Hikita et al. [84] studied the effects of the physical properties of the irrigating liquid on the effective interfacial area for liquid-phase mass transfer in packed columns. The data on the height of a liquid-phase transfer unit H_L were obtained from the absorption of carbon dioxide in various solvents using 7.0-cm and 12.5-cm diameter columns packed with 1.5- and 2.5-cm Raschig rings and 1.3- and 2.5-cm Berl saddles. The effective interfacial areas were deduced from the H_L data by assuming that, at the same Reynolds numbers, the values of k_L for a packed column were equal to those for a single packing piece. The fractional effective areas found were correlated by the following equations:

$$\frac{a}{a_t} = 2.26L^{0.455}\left(\frac{\sigma}{1000}\right)^s \tag{87}$$

$$s = -0.091d^{-0.48} \tag{88}$$

for Raschig rings, and

$$\frac{a}{a_t} = 0.768L^{0.455}\left(\frac{\sigma}{1000}\right)^s \tag{89}$$

$$s = -0.00543d^{-0.98} \tag{90}$$

for Berl saddles.

These equations indicate a strong dependence of the effective interfacial area on the surface tension, but the influence of viscosity was found to be negligible by the authors.

Using the wetted area and effective interfacial area data reported in the literature [22, 25, 32, 83, 84, 85] Mada et al. [86] obtained the following dimensionless correlations for Raschig rings and Berl saddles:

$$\frac{a_w}{a_t} = 9.3 \times 10^{-4}(\mathrm{Fr})^{-1/2}(\mathrm{We})^{2/3}(a_t d)^{3.0} \tag{91}$$

$$\frac{a}{a_t} = 0.071(\mathrm{Fr})^{-1/2}(\mathrm{We})^{2/3} \tag{92}$$

for Raschig rings, and

$$\frac{a}{a_t} = 0.055(\mathrm{Fr})^{-1/2}(\mathrm{We})^{2/3} \tag{93}$$

for Berl saddles.

They also obtained an equation for the wetted areas of Raschig rings, Berl saddles, and spheres:

$$a_w d = 0.61(\mathrm{Fr})^{-1/2}(\mathrm{We})^{2/3} \tag{94}$$

Onda et al. [29] obtained a more general dimensionless correlation for the wetted areas of different packings. Both the effects of surface tension and wettability of the packing material were taken into account. Their study yielded the correlation:

$$\frac{a_w}{a_t} = 1 - \exp\left[-1.45\left(\frac{L}{a_t\mu}\right)^{0.1}\left(\frac{a_tL^2}{g\varrho^2}\right)^{-0.05}\left(\frac{L^2}{\varrho\sigma a_t}\right)^{0.2}\left(\frac{\sigma_c}{\sigma}\right)^{0.75}\right] \quad (95)$$

or

$$\frac{a_w}{a_t} = 1 - \exp\left[-1.45\,Re_t^{0.1}\,Fr^{-0.05}\,We^{0.2}\,(\sigma_c/\sigma)^{-0.75}\right] \quad (96)$$

Equation 96 fit the data obtained from columns packed with a variety of sizes of Raschig rings, Berl saddles, and spheres, and rods made of ceramic, glass, polyvinylchloride and coated with paraffin to within $\pm 20\%$.

The applicable range of this correlation is presented in the following [29]:

$$0.04 < Re_t < 500$$

$$1.2 \times 10^{-8} < We < 0.27$$

$$2.5 \times 10^{-9} < Fr < 1.8 \times 10^{-2}$$

$$0.3 < \frac{\sigma_c}{\sigma} < 2$$

The wettability of the different packing materials was characterized by the critical surface tension σ_c and values are given in Table 14.

Mohunta et al. [87], using the technique of Danckwerts and Gillham [39], i.e., absorption with a rapid second order reaction, obtained values for the interfacial areas using 0.95-, 1.27-, and 1.9-cm Raschig rings. In combination with the published data of Danckwerts and Sharma [88] and Richards et al. [89] they obtained a correlation

$$\frac{a}{a_t} = C'\left(\frac{U\varrho}{a_t\mu}\right)^{1/3} \quad (97)$$

The value of C′ for the area measured by Danckwert's technique was found to be 0.175, whereas the value of C′ from the dye experiments was 0.216. Equation 97 indicates that the effective interfacial area decreases with an increase in viscosity. It should be noted, however, that the range of liquid viscosity investigated was small.

Table 14
Critical Surface Tension of Packing Materials

Packing Material	Critical Surface Tension σ_c, mN/m
Carbon	56
Ceramic	61
Glass	73
Paraffin	20
Polyethylene	33
Polyvinylchloride	40
Steel	75

From Onda, K. Takeuchi, Koyama, [29].

Puranik and Vogelpohl [90] presented a generalized correlation for the values of effective interfacial area during vaporization, absorption with and without chemical reaction, and wetted areas in terms of dimensionless numbers. Extensive data in the literature were analyzed statistically based on the concept of static and dynamic area resulting in the following correlations:

$$\frac{a_p}{a_t} = 1.045(Re_t)^{0.041}(We)^{0.133}(\sigma/\sigma_c)^{-0.182} \tag{98}$$

and

$$\frac{a_{st}}{a_t} = 0.229 - 0.091 \ln(We/Fr) \tag{99}$$

where $a_p = a_v = a_w = a_{AC} = (a_{AP} + a_{st})$

The authors further proposed a general equation for calculating the effective interfacial area irrespective of the type of absorption:

$$\frac{a}{a_t} = \left(\frac{a_p}{a_t}\right) + \left(\frac{Ha}{E} - 1\right)\left(\frac{a_{st}}{a_t}\right) \tag{100}$$

where Ha = the Hatta number, $(D_A k_2 c_{B0})^{1/2}/k_L$
E = the enhancement factor, k_L'/k_L

They claimed that Equations 98 through 100 could be used to predict the interfacial areas for vaporization and absorption with or without chemical reaction within a range of ±20%. The range of variables and physical properties covered by Equations 98 through 100 is given in Table 15.

Many investigators have measured interfacial areas using aqueous systems. Sridharan and Sharma [91], however, have developed new methods for the measurement of effective interfacial area by the chemical method which allow the use of hydrocarbon solvents, polar solvents, and highly viscous systems by considering the reaction of carbon dioxide with selected amines dissolved in these solvents. This important development will allow the effects

Table 15
Range of Variables and Physical Properties for the Correlations of Puranik and Vogelpohl

Variables	Range
a_p/a_t	0.08-0.8
L	0.25-12 $kg/m^2 \cdot s$
μ	0.5-13 $mPa \cdot s$
σ	25-75 mN/m
ϱ	800-1,900 kg/m^3
d	1.0-3.75 cm
We/Fr	0.4-14
We	21×10^{-7}-12×10^{-3}
Re_t	0.5-85
Fr	7.7×10^{-7}-4.7×10^{-3}
σ/σ_c	0.3-1.05

From Puranik and Vogelpohl [90].

of viscosity and surface tension on the effective interfacial area in packed columns to be determined concisely in the future.

Rizzuti et al. [92] noted that in all the work reported the effect of viscosity on the interfacial area has not been studied explicitly. Thus they determined the effective interfacial areas of glass rings (1.0×1.1 cm) packed in a 3.7-cm diameter column by absorbing carbon dioxide in carbonate-bicarbonate-arsenite systems and ozone in aqueous basic phenol solutions. The kinematic viscosity of the absorbing solution was varied from 0.9×10^{-2} to 1.55×10^{-2} cm^2/s by adding different amounts of sugar to the solutions. It was found that the liquid viscosity had a strong influence in the effective interfacial area. Their data were correlated by the the equation:

$$a = 1.1 \times 10^5 v^{0.70} U^{0.326} \tag{101}$$

Bravo and Fair [93] presented the following correlation for predicting the effective area for mass transfer in randomly packed columns

$$a = 0.498 a_t \left(\frac{\sigma^{0.5}}{Z^{0.4}}\right) (Ca_L \, Re_G)^{0.392} \tag{102}$$

The interfacial areas were calculated using a large data bank [94] and the values of k_G and k_L obtained from the correlations developed by Onda et al. [28]. Although absorption data were included in the development of the correlation, it was found that the correlation overpredicted the effective areas for absorption. The authors suggested that the increased area in distillation was due to the simultaneous existence of vapor condensation with liquid boiling, thus providing added effective area within each phase. The lower surface tensions for the liquids used in distillation could also explain the discrepancy.

Influence of Liquid Viscosity on Effective Interfacial Area

It is seen from the last section that many correlations are available in the literature for the prediction of the effective interfacial areas exhibited in packed columns during either absorption without chemical reaction or in the presence of chemical reaction. A careful scrutiny shows considerable discrepancy in the values of the effective interfacial areas and severely conflicting predictions concerning the influence of viscosity on the interfacial areas. For illustrative purposes we will examine two recent investigations by Rizzuti et al. [92] and Mohunta et al. [87]. In the first investigation a correlation predicting effective interfacial area, i.e.

$$a = 1.1 \times 10^5 v^{0.70} U^{0.326} \tag{101}$$

was presented, which indicated that the interfacial area increases as a function of the kinematic viscosities. Using a comparable chemical system and similar packing dimensions Mohunta et al. [87] predicted that the interfacial area in fact decreased as the liquid viscosity increased, i.e.

$$a = 0.175 a_t \left(\frac{L}{a_t \mu}\right)^{1/3} \tag{97}$$

A considerable number of other workers have examined the dependency of interfacial areas on viscosity such as Onda et al. [29], Puranik and Vogelpohl [90], and Shulman et al. [22, 80, 81]. Table 16 highlights the wide range of dependency suggested by these workers. A careful scrutiny of the experimental conditions shows that the correlations were based

Table 16
Dependency of Effective Interfacial Area on Liquid Viscosity

Authors	μ
Rizzuti et al. [92]	+0.70
Shulman et al. [22, 80, 81]	+0.13 (μ < 0.12 g/cm s)
	+0.31 (μ > 0.12 g/cm s)
Onda et al. [29]	$1-[\exp(\mu^{-0.1})]^{-1}$
Puranik and Vogelpohl [90]	−0.041
Mohunta et al. [87]	−0.33

on very narrow viscosity ranges employed during experimentation. Rizzuti varied the viscosity from 0.9 to 1.55×10^{-2} cm^2/s and the correlation of Mohunta et al. was based on viscosities up to only 1.5 mPa · s (cp). With the exception of Onda et al. no importance has been attributed to the wettability of the system which includes the liquid surface tension and the nature of the solid surface characterized by the contact angle. A completely wetted surface will exhibit zero contact angle with nonwetting occurring at 180°. The degree of wettability of the solid by the liquid is therefore indicated by the magnitude of the contact angle. Casual observation shows that better wetting, that is higher wetted areas, are obtained using liquids with low surface tensions. Almost without exception workers have assumed that increased wetted areas will automatically produce an increase in effective interfacial area; however, this general assumption is not valid in all circumstances. This in turn brings into question the desire of engineers to design packings which will allow the largest surface area between the gas and liquid phases. It is our contention that the mass transfer efficiency is controlled primarily, not by flow over the packing surface which contributes to the wetted area value, but by interstitial flow between the packings (the nature of the liquid passage being controlled by the interstitial geometry passage, and not by the shape of the packing surface). The process is complicated by the fact that the liquid does not run uniformly through the packing voids but chooses to distribute itself into a number of liquid channels having high liquid velocities. This qualitative picture was recently reinforced by an excellent paper by Beimer and Zuiderweg [95] who mathematically described this channeling phenomena in their "preferred path" model. One can visualize that an increase in liquid flow rates will result in an increase in wetted areas while the effective interfacial area for physical absorption will be little affected since the additional liquid will be channeled into the already-existing rivulets. The whole process is made more complex by the interaction between the viscous and surface forces as first demonstrated quantitatively by Mangers and Ponter [35] when absorbing carbon dioxide into a series of aqueous glycerol solutions (Figure 22). An examination of these data shows that for a given surface tension, a viscosity increase will result in an increase in mass transfer efficiency to a maximum value followed by a decline. The result demonstrates elegantly that a particular packing will only function at high efficiency for particular combinations of liquid surface tension and viscosity values. This postulate confirms the observations of Norman [8] that the absorption efficiencies using other packings of widely varying shapes tend to be very similar.

As previously mentioned when examining the data of Rizzuti et al. [92] and Mohunta et al. [87] there is considerable difficulty in interpreting from published correlations the effect of viscosity in the absorption process due to the small viscosity ranges used in the experiments. Liquid-film-controlled gas absorption experiments for liquids exhibiting a wide range of viscosities appear to have been conducted by very few workers, e.g. Norman and Sammak [33, 34], Mangers and Ponter [35], and Andrew [96]. Norman and Sammak investigated the effect of liquid viscosity on the mass transfer coefficient using a vertical string of modified squared disc packings made of graphite with the discs touching one another. Carbon dioxide was absorbed into water and several organic liquids covering a viscosity range of 0.4 to 20 mPa · s (cp). Since the discs were either completely wetted or

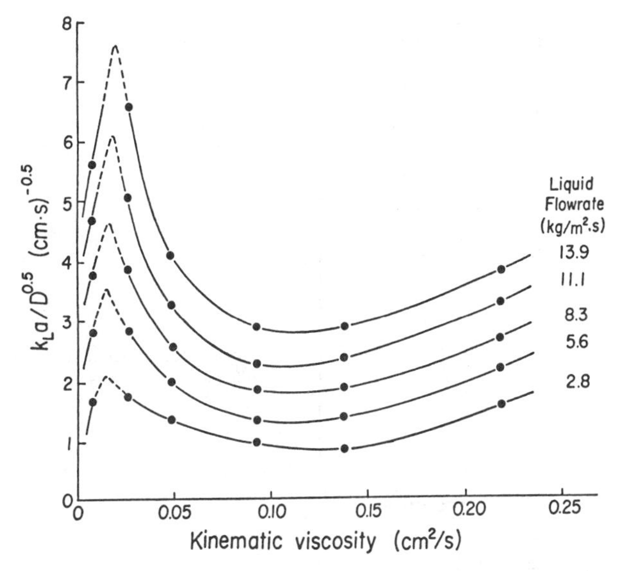

Figure 22. Effect of kinematic viscosity (at constant surface tension) on packed-column mass transfer efficiency for different liquid rates (Mangers, R. J., and Ponter, A. B., *Ind. Eng. Chem. Process Dev.*, Vol. 19 (1980) pp. 530–537.)

their wetted areas determined visually the individual liquid-film mass transfer coefficients were readily calculated using the absorption data and were correlated by the following equation:

$$\frac{k_L d}{D} = 0.13\left(\frac{\mu}{\varrho D}\right)^{1/2}\left(\frac{d^3 g \varrho^2}{\mu^2}\right)^{1/6}\left(\frac{4\Gamma}{\mu}\right)^{0.61} \tag{62}$$

Andrew [96] suggested that a column where the discs touched is a superior model of random packing compared to a sphere column based on mass transfer data obtained in these devices. This justifies the selection of Norman and Sammak's correlation for the determination of effective interfacial areas for a packed column. The effective interfacial area, calculated from:

$$a = \frac{(k_L a) \text{ measured in a packed column}}{k_L \text{ measured in a disc column}}$$

using the k_La data of Mangers and Ponter [35] and the k_L values calculated from Norman and Sammaks' correlation are presented in Figures 23 through 25, which show the dependency of the effective interfacial area on the viscosity at three different liquid flow rates. The values calculated from the correlations of other workers [22, 29, 80, 81, 87, 92] are also given for comparison.

As expected the areas determined during physical absorption and absorption with chemical absorption fall into two distinct groups. When comparing the data for 1-cm glass or ceramic Raschig rings, which have very similar wettabilities, used in the experiments of Shulman et al. and Mangers and Ponter, it is observed that the a/a_t values for water are in good agreement. Similarly using the chemical methods for determining interfacial areas again gave close agreement for the systems using water, reported by Mohunta et al. and

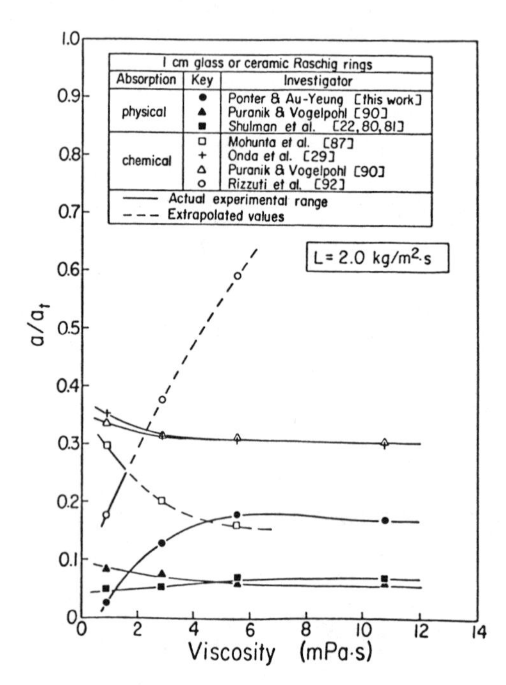

Figure 23. Effect of liquid viscosity on the effective interfacial area for a liquid rate of 2.0 kg/m^2·s.

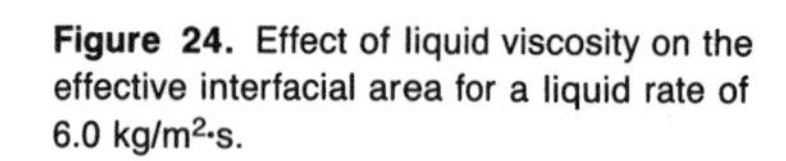

Figure 24. Effect of liquid viscosity on the effective interfacial area for a liquid rate of 6.0 kg/m^2·s.

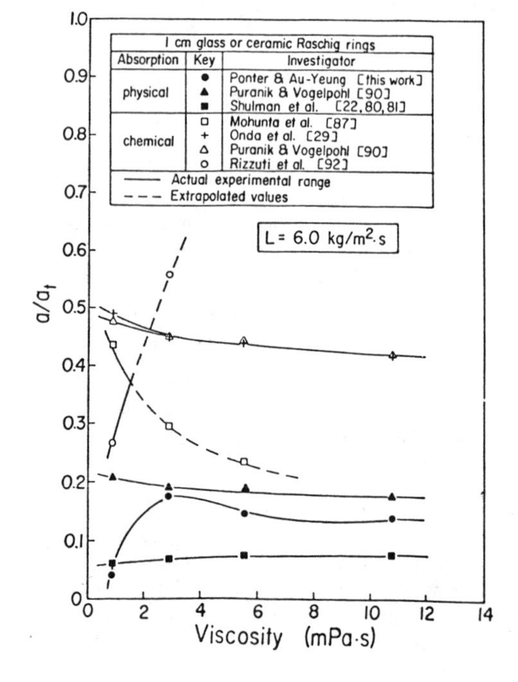

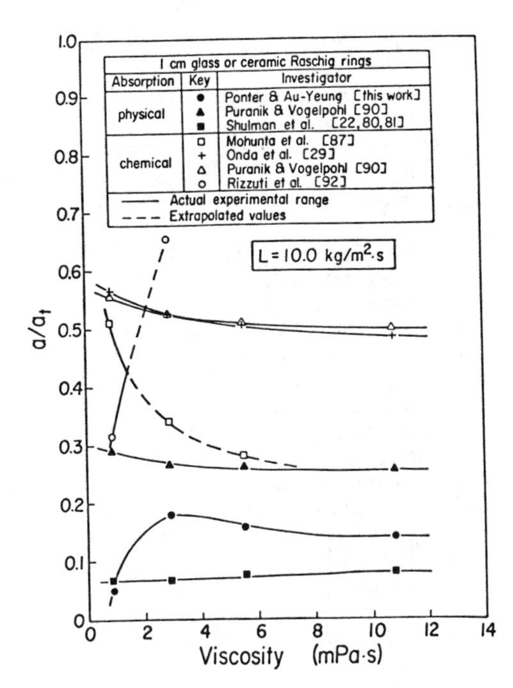

Figure 25. Effect of liquid viscosity on the effective interfacial area for a liquid rate of 10.0 $kg/m^2 \cdot s$.

Puranik and Vogelpohl, although with a/a_t values an order of magnitude higher than those for physical absorption.

The area values calculated from Rizzuti et al.'s correlations lie between those for physical absorption and those for chemical absorption. The correlation indicates a very strong dependency of the effective interfacial areas on the viscosity. Examination of Mangers and Ponter's plot of $k_L a/D^{0.5}$ versus kinematic viscosity in Figure 22 however shows that the range of viscosities studied by Rizzuti et al. falls on the sharply ascending portion of the curves, which explains the high dependency on viscosity in Rizzuti's equation. Hence it is unacceptable to extrapolate Rizzuti's correlation to higher viscosity values for design purposes. Furthermore, when examining the systems with higher viscosities and plotting the experimentally determined values together with the extrapolated values using reported correlations, it is evident that the reported correlations are not valid for higher viscosity liquids. To systematically elucidate the influence of viscosity on the absorption process it is necessary to select liquid systems with fixed wetting properties, i.e., having the same σ $(\cos \theta)$ values where σ is the surface tension and θ is the contact angle, so that the effect of viscosity and the effect of wetting of the packings can be separated and be more readily assessed.

NOTATION

- a interfacial area per unit packed volume (m^2/m^3)
- a_{abs} interfacial area for gas absorption (m^2/m^3)
- a_{AC} interfacial area for absorption with chemical reaction (m^2/m^3)
- a_{AP} interfacial area for physical absorption (m^2/m^3)
- a_p predicted interfacial area per unit volume (m^2/m^3)
- a_s interfacial area per volume of strings of touching spheres (m^2/m^3)

a_{st} static surface area per unit packed volume (m^2/m^3)
a_t total surface area per unit packed volume (m^2/m^3)
a_v interfacial area for vaporization (m^2/m^3)
a_{vap} interfacial area for vaporization (m^2/m^3)
a_w wetted surface area per unit packed volume (m^2/m^3)
b constant in Equation 82
b′ constant in Equation 83
C correction factor for H_L at high gas rate, Equation 27
C′ constant in Equation 97
c concentration of solute in liquid ($kmol/m^3$)
c constant in Equation 17
c^* equilibrium concentration of a gas in a liquid at partial pressure p ($kmol/m^3$)
c′ constant in Equation 21
c″ constant in Equation 60
c_{B0} bulk concentration of reactant B in liquid ($kmol/m^3$)
c_i concentration of a gas in a liquid at the interface ($kmol/m^3$)
c_0 concentration of solute in the bulk of the liquid ($kmol/m^3$)
C_{fL} coefficient for the effect of approach of flood point on liquid phase mass transfer
D liquid diffusivity (m^2/s)
D′ constant in Equation 73
D_c column diameter (m)
D_G gas diffusivity (m^2/s)
D_p diameter of a sphere having the same surface area as the packing (m)
d diameter of packing; size of packing (m)
d′ maximum effective length of packing element (m)
E enhancement factor for liquid phase mass transfer (k_L'/k_L)
G gas mass velocity ($kg/s \cdot m^2$)
$\bar{G}$ mean superficial gas velocity (m/s)
g acceleration of gravity (m/s^2)
H Henry's law constant ($Pa \cdot m^3/kmol$)
H_L height of a liquid phase transfer unit (m)
H_{OL} height of overall liquid phase transfer unit (m)
h_0 operating holdup (m^3/m^3)
h_s static holdup (m^3/m^3)
h_t total holdup (m^3/m^3)
K constant in Equations 39 and 40
K′ constant in Equation 70
K_G overall gas phase mass transfer coefficient ($kmol/s \cdot m^2 \cdot Pa$)
K_L overall liquid phase mass transfer coefficient (m/s)
k constant in Equation 82
k′ constant in Equation 83
k_G gas-film mass transfer coefficient ($kmol/s \cdot m^2 \cdot Pa$)
k_L liquid-film mass transfer coefficient (m/s)
k_{Ls} liquid phase mass transfer coefficient for strings of touching spheres (m/s)
k_2 second-order reaction rate constant ($m^3/kmol \cdot s$)
k_L' liquid-film mass transfer coefficient with chemical reaction (m/s)
L liquid mass velocity ($kg/s \cdot m^2$)
L_M liquid molar velocity ($kmol/s \cdot m^2$)
M_M mean molecular weight of gas (kg/kmol)
m constant in Equation 17
m′ constant in Equation 80
N number of disc over which an appreciable amount of gas has not yet penetrated into the rejuvenation path
N_A molar flux ($kmol/s \cdot m^2$)
n constant in the Sherwood and Holloway equation
n′ constant in Equation 80
P_{BM} mean partial pressure of inert gas in the gas phase (Pa)
p partial pressure of solute gas (Pa)
p constant in Equation 82
p′ constant in Equation 83
p^* partial pressure of gas in equilibrium with liquid of concentration c, (Pa)
p_A partial pressure of solute A, (Pa)
p_i partial pressure of solute in gas at interface, (Pa)
Q liquid volumetric flow rate, (m^3/s)
q mean number of contact points of touching spheres per sphere

q_{max} maximum number of contact points of touching spheres per sphere
r constant in Equation 85
s constant in Equations 87 and 89
s' constant in Equation 80
U superficial liquid velocity (m/s)
x distance from surface of liquid film
X(z) half width of liquid film on a vertical surface as a function of z (m)
Z height of packing; height of wetted-wall column (m)
z vertical distance from the liquid feed nozzle (m)

Subscripts

cm complete mixing value
expt experimental value

Greek Symbols

α constant in the Sherwood and Holloway equation
α_1 mixing factor defined by (k_{Lcm}/k_{Lexpt}) and corresponding to a Reynolds number of 1
β intercept of log-log plot of d' versus Re (m)
Γ liquid mass flow rate per unit perimeter (kg/s · m)
Γ_m mean liquid mass flow rate per unit perimeter (kg/s · m)
γ_{LV} liquid-vapor interfacial tension (N/m)
γ_{SL} solid-liquid interfacial tension (N/m)
γ_{SV} solid-vapor interfacial tension (N/m)
ε void fraction in packed column (m^3/m^3)
η slope of log-log plot of (d'/d) versus L/μ
θ liquid contact angle (degrees)
θ_{st} stationary contact angle (degrees)
μ liquid viscosity (Pa · s)
μ_G gas viscosity (Pa · s)
ν kinematic liquid viscosity (m^2/s)
ϱ liquid density (kg/m^3)
ϱ_G gas density (kg/m^3)
σ liquid surface tension (N/m)
σ_c critical surface tension (N/m)
σ_e effective surface tension during gas absorption (N/m)
Φ surface shape factor
φ packing parameter, Equation 27
ψ slope of log-log plot of d versus Re
ω intercept of log-log plot of d /d versus L/μ

Dimensionless Groups

Ca_L capillary number for liquid ($\mu L/\varrho\sigma$)
Fr Froude number ($L^2 a_t/g\varrho^2$)
Ga Galileo number ($d^3 g\varrho^2/\mu^2$)
Gr modified Grashof number ($d^3 g\varrho^2/\mu^2$)
Ha Hatta number ($(D_A k_2 c_{B0})^{1/2}/k_L$)
Ka Kapitsa number ($\varrho\sigma^3/\mu^4 g$)
Re Reynolds number for random packing ($2\pi L/a\mu$)
Re_l Reynolds number for a vertical surface ($4L/a\mu$)
Re_G Reynolds number for gas phase ($6G/a_t\mu_G$)
Re_p Reynolds number for packed column based on diameter of a sphere having the same surface area as a packing piece ($D_p L/\mu$)
Re_s Reynolds number of m strings of touching spheres based on the diameter of spheres ($4\varrho Q/m\pi d\mu$)
Re_t Reynolds number for packed column based on total packing surface area per unit packed volume ($L/a_t\mu$)
Re_w Reynolds number based on wetted area ($4L/a_w\mu$)
Sc Schmidt number ($\mu/\varrho D$)
We Weber number ($L^2/\varrho\sigma a_t$)

REFERENCES

1. Whitman, W. G., "The Two-Film Theory of Gas Absorption," *Chem. Metall. Eng.*, Vol. 29 (1923), pp. 146–148.
2. Lewis, W. K., and Whitman, W. G., "Principles of Gas Absorption," *Ind. Eng. Chem.*, Vol. 16 (1924), pp. 1215–1220.
3. Sherwood, T. K., and Holloway, F. A. L., "Performance of Packed Towers—Liquid Film Data for Several Packings," *Trans. Am. Inst. Chem. Eng.*, Vol. 36 (1940), pp. 39–70.
4. Molstad, M. C., et al., "Performance of Drip-Point Grid Tower Packings. II. Liquid-Film Mass Transfer Data," *Trans. Am. Inst. Chem. Eng.*, Vol. 38 (1942), pp. 410–434.
5. Vivian, J. E., and Whitney, R. P., "Absorption of Chlorine in Water," *Chem. Eng. Prog.*, Vol. 43 (1947), pp. 691–702.
6. Whitney, R. P., and Vivian, J. E., "Absorption of Sulfur Dioxide in Water," *Chem. Eng. Prog.*, Vol. 45 (1949), pp. 323–337.
7. Deed, D. W., Schutz, P. W., and Drew, T. B., "Comparison of Rectification and Desorption in Packed Columns," *Ind. Eng. Chem.*, Vol. 39 (1947), pp. 766–774.
8. Norman, W. S., *Absorption, Distillation and Cooling Towers,* Wiley, New York, 1961.
9. Cooper, C. M., Christl, R. J., and Peery, L. C., "Packed Tower Performance at High Liquor Rates—The Effect of Gas and Liquor Rates Upon Performance in a Tower Packed with Two-Inch Rings," *Trans. Am. Inst. Chem. Eng.*, Vol. 37 (1941), pp. 979–993.
10. Rixon, F. F., "The Absorption of Carbon Dioxide in and Desorption from Water Using Packed Towers," *Trans. Am. Inst. Chem. Eng.*, Vol. 44 (1948), pp. 119–130.
11. Hutchings, L. E., Stutzman, L. F., and Koch, H. A., "Mass Transfer Coefficients as Functions of Liquid, Gas Rates, Tower Packing Characteristics," *Chem. Eng. Prog.*, Vol. 45 (1949), pp. 253–268.
12. Koch, H. A., et al., "Liquid Transfer Coefficients for the Carbon Dioxide-Air-Water System," *Chem. Eng. Prog.*, Vol. 45 (1949), pp. 677–682.
13. Knoedler, E. L., and Bonilla, C. F., "Vacuum Degasification in a Packed Column. Deoxygenation of Water in Stedman Packing," *Chem. Eng. Prog.*, Vol. 50 (1954), pp. 125–133.
14. Van Krevelen, D. W., and Hoftijzer, P. J., "Studies of Gas Absorption. I. Liquid Film Resistance to Gas Absorption in Scrubbers," *Recl. Trav. Chim. Pays-Bas,* Vol. 66 (1947), pp. 49–66.
15. Van Krevelen, D. W., and Hoftijzer, P. J., "Kinetics of Simultaneous Absorption and Reaction," *Chem. Eng. Prog.*, Vol. 44 (1948), pp. 529–536.
16. Higbie, R., "The Rate of Absorption of a Pure Gas into a Still Liquid During Short Period of Exposure," *Trans. Am. Inst. Chem. Eng.*, Vol. 31 (1935), pp. 365–388.
17. Danckwerts, P. V., "Significance of Liquid-Film Coefficients in Gas Absorption," *Ind. Eng. Chem.*, Vol. 43 (1951), pp. 1460–1467.
18. Vivian, J. E., and King, C. J., "The Mechanism of Liquid-Phase Resistance to Gas Absorption in a Packed Column," *AICHE J.*, Vol. 10 (1964), pp. 221–226.
19. Sherwood, T. K., and Pigford, R. L., Absorption and Extraction, McGraw-Hill, New York, 1952.
20. Van Krevelen, D. W., and Van Hooven, C. J., reported by Norman, W. S., *Absorption, Distillation and Cooling Towers,* Wiley, New York, 1961.
21. Hoftijzer, P. J., "The Performance of Technical Apparatus for Gas Absorption," *Inst. Chem. Eng.*, Joint Symposium on Scaling-up of Chemical Plant and Processes, London (1957), pp. 73–77.
22. Shulman, H. L., et al., "Performance of Packed Columns II. Wetted and Effective Interfacial Areas, Gas- and Liquid-Phase Mass Transfer Rates," *AICHE J.*, Vol. 1 (1955), pp. 253–258.

23. Fellinger, L., "Absorption of Ammonia by Water and Acid in Various Standard Packings," doctoral dissertation, MIT, Cambridge, Mass. (1941).
24. Onda, K., Sada, E., and Murase, Y., "Liquid-Side Mass Transfer Coefficients in Packed Towers," *AICHE J.*, Vol. 5 (1959), pp. 235–239.
25. Fujita, S., and Sakuma, S., "Wetted Area of Raschig Rings in Packed Columns," *Chem. Eng. Japan,* Vol. 18 (1954), pp. 64–67.
26. Shulman, H. L., Press, S., and Whitehouse, W. G., "Liquid-Side Mass Transfer Coefficients in Packed Towers," *AICHE J.*, Vol. 6 (1960), pp. 174–175.
27. Cornell, D., Knapp, W. G., and Fair, J. R., "Mass Transfer Efficiency—Packed Columns—Part 1," *Chem. Eng. Prog.*, Vol. 56 No. 7 (1960), pp. 68–74.
28. Onda, K., Takeuchi, H., and Okumoto, Y., "Mass Transfer Coefficients between Gas and Liquid Phases in Packed Columns," *J. Chem. Eng. Japan,* Vol. 1 (1968), pp. 56–62.
29. Onda, K., Takeuchi, H., and Koyama, Y., "Effect of Packing Materials on the Wetted Surface Area," *Chem. Eng. Japan,* Vol. 31 (1967), pp. 126–134.
30. Mohunta, D. M., Vaidyanathan, A. S., and Laddha, G. S., "Prediction of Liquid-Phase Mass Transfer Coefficients in Columns Packed with Raschig Rings," *Indian Chem. Eng.*, Vol. 11, No. 3 (1969), pp. 73–79.
31. Bolles, W. L., and Fair, J. R., "Improved Mass-Transfer Model Enhances Packed-Column Design," *Chem. Eng.*, Vol. 89, No. 14 (1982), pp. 109–116.
32. Yoshida, F., and Koyanagi, T., "Liquid Phase Mass Transfer Rates and Effective Interfacial Area in Packed Absorption Columns," *Ind. Eng. Chem.*, Vol. 50 (1958), pp. 365–374.
33. Norman, W. S., and Sammak, F. Y. Y., "Gas Absorption in a Packed Column. Part I: The Effect of Liquid Viscosity on the Mass Transfer Coefficient," *Trans. Inst. Chem. Eng.*, Vol. 41 (1963), pp. 109–116.
34. Norman, W. S., and Sammak, F. Y. Y., "Gas Absorption in a Packed Column. Part II: The Effect of Mixing Between Packing Elements on the Liquid Film Mass Transfer Coefficient," *Trans. Inst. Chem. Eng.*, Vol. 41 (1963), pp. 117–119.
35. Mangers, R. J., and Ponter, A. B., "Effect of Viscosity on Liquid Film Resistance to Mass Transfer in a Packed Column," *Ind. Eng. Chem. Process Des. Dev.*, Vol. 19 (1980), pp. 530–537.
36. Ponter, A. B., et al., "The Influence of Mass Transfer on Liquid Film Breakdown," *Int. J. Heat and Mass Transfer,* Vol. 10 (1967), pp. 349–359.
37. Davidson, J. F., "The Hold-up and Liquid Film Coefficient of Packed Towers. Part II. Statistical Models of the Random Packing," *Trans. Inst. Chem. Eng.*, Vol. 37 (1959), pp. 131–136.
38. Bridgewater, J., and Scott, A. M., "Statistical Models of Packing: Application to Gas Absorption and Solids Mixing," *Trans. Inst. Chem. Eng.*, Vol. 52 (1974), pp. 317–324.
39. Danckwerts, P. V., and Gillham, A. J., "The Design of Gas Absorbers—I. Methods for Predicting Rates of Absorption with Chemical Reaction in Packed Columns, and Tests with 1½ in. Raschig Rings," *Trans. Inst. Chem. Eng.*, Vol. 44 (1966), pp. T42–T54.
40. Zech, J. B., and Mersmann, A. B., "Liquid Flow and Liquid Phase Mass Transfer in Irrigated Packed Columns," *Inst. Chem. Eng. Symp. Ser. No. 56* (1979), pp. 2.5/39–47.
41. Ponter, A. B., Pfennigwerth, G. L., and Mangers, R. J., "Liquid Film Mass Transfer Coefficients for Packed Columns Using Organic and Aqueous Systems," *Chem. Ing. Tech.*, Vol. 52 (1980), pp. 656–657.
42. Ponter, A. B., and Au-Yeung, P. H., "Estimation of Liquid Film Mass Transfer Coefficients for Columns Randomly Packed with Partially Wetted Rings," *Can. J. Chem. Eng.*, Vol. 60 (1982), pp. 94–99.
43. Merchuk, J. C., "Mass Transfer Characteristics of a Column with Small Plastic Packings," *Chem. Eng. Sci.*, Vol. 35 (1980), pp. 743–745.
44. Gilliland, E. R., and Sherwood, T. K., "Diffusion of Vapors into Air Streams," *Ind. Eng. Chem.*, Vol. 26 (1934), pp. 516–523.

45. Johnstone, H. F., and Pigford, R. L., "Distillation in a Wetted-Wall Column," *Trans. Am. Inst. Chem. Eng.*, Vol. 38 (1942), pp. 25–50.
46. Stephens, E. J., and Morris, G. A., "Determination of Liquid-Film Absorption Coefficients. A New Type of Column and its Application to Problems of Absorption in Presence of Chemical Reaction," *Chem. Eng. Prog.*, Vol. 47 (1951), pp. 232–242.
47. Morris, G. A., "Application of Absorption Tower Design Methods to Packed Distillation Columns," Proc. Int. Symp. Distill., Brighton, England (1960), pp. 146–152.
48. Taylor, R. F., and Roberts, F., "The Calibration of Disc-Type Laboratory Gas-Absorption Columns," *Chem. Eng. Sci.*, Vol. 5 (1956), pp. 168–177.
49. Mika, V., "Model of Packed Absorption Column. I. Physical Absorption," *Collect. Czech. Chem. Commun.*, Vol. 32 (1967), pp. 2933–2943.
50. Onda, K., et al., "Liquid-side and Gas-side Mass Transfer Coefficients in Towers Packed with Spheres," *Chem. Eng. Japan*, Vol. 27 (1963), pp. 140–145.
51. Alper, E., and Danckwerts, P. V., "Laboratory Scale-Model of a Complete Packed Column Absorber," *Chem. Eng. Sci.*, Vol. 31 (1976), pp. 599–608.
52. Fukushima, S., and Kusaka, K., "Mass Transfer on Multi-Springs of Touching Spheres," *J. Chem. Eng. Japan*, Vol. 11 (1978), pp. 33–39.
53. De Waal, K. J. A., and Beek, W. J., "A Comparison between Chemical Absorption with Rapid First-Order Reactions and Physical Absorption in One Packed Column," *Chem. Eng. Sci.*, Vol. 22 (1967), pp. 585–593.
54. Jhaveri, A. S., and Sharma, M. M., "Effective Interfacial Area in a Packed Column," *Chem. Eng. Sci.*, Vol. 23 (1968), pp. 669–976.
55. Danckwerts, P. V., and Rizvi, S. F., "The Design of Gas Absorbers Part II: Effective Interfacial Areas for Several Types of Packing," *Trans. Inst. Chem. Eng.*, Vol. 49 (1971), pp. 124–127.
56. Lynn, S., Straatemeier, J. R., and Kramers, H., "Absorption Studies in the Light of the Penetration Theory. III. Absorption by Wetted Spheres Singly and in Columns," *Chem. Eng. Sci.*, Vol. 4 (1955), pp. 63–67.
57. Davidson, J. F., et al., "The Hold-up and Liquid Film Coefficients of Packed Towers. Part I. Behavior of a String of Spheres," *Trans. Inst. Chem. Eng.*, Vol. 37 (1959), pp. 122–130.
58. Ratcliff, G. A., and Reid, K. J., "Mass Transfer into Spherical Liquid Films Part 2: Mixing Effects on a Column of Spheres," *Trans. Inst. Chem. Eng.*, Vol. 40 (1962), pp. 69–74.
59. Harrison, D., Lane, M., and Walne, D. J., "Axial Dispersion of Liquid on a Column of Spheres," *Trans. Inst. Chem. Eng.*, Vol. 40 (1962), pp. 214–220.
60. Atkinson, B., and Taylor, B. N., "The Effect of Discontinuities on Mass Transfer to a Liquid Film," *Trans. Inst. Chem. Eng.*, Vol. 41 (1963), pp. 140–145.
61. Shulman, H. L., and Mellish, W. G., "Performance of Packed Columns: Part VIII. Liquid Flow Patterns and Velocities in Packed Beds," *AICHE J.*, Vol. 13 (1967), pp. 1137–1140.
62. Copp, D. A., and Ponter, A. B., "The Effect of Liquid Viscosity on Mixing in Falling Films," *Waerme Stoffuebertrag.*, Vol. 5 (1972), pp. 229–238.
63. Alper, E., "Physical Absorption of a Gas in Laboratory Models of a Packed Column," *AICHE J.*, Vol. 25 (1979), pp. 545–547.
64. Mika, V., "The Application of Higbie Model to Physical Absorption in a Disk Column," *Collect. Czech. Chem. Commun.*, Vol. 24 (1959), pp. 2843–2850.
65. Teller, A. J., "The Rosette: A New Packing for Diffusional Operations Based on High Interstitial Holdup," *Chem. Eng. Prog.*, Vol. 50, No. 2 (1954), pp. 65–71.
66. Ponter, A. B., "Adhesion Effects and Performance of Mass Transfer Equipment Using Teflon Surfaces," *Polym. Eng. Sci.*, Vol. 17, No. 7 (1977), pp. 484–493.
67. Ponter, A. B., et al., "Wetting of Packings in Distillation: The Influence of Contact Angle." *Trans. Inst. Chem. Eng.*, Vol. 45 (1967), pp. T345–T352.
68. Zuiderweg, F. J., and Harmens, A., "The Influence of Surface Phenomena on the Performance of Distillation Columns," *Chem. Eng. Sci.*, Vol. 9 (1958), pp. 89–103.

69. Ponter, A. B., Peier, W., and Fabre, S., "Assessment of Surface Tensions of Binary Liquids Undergoing Distillation at Total Reflux," *Chem. Ing. Tech.*, Vol. 50 (1978), pp. 444–446.
70. Peier, W., Ponter, A. B., and Fabre, S., "Wetting of a Vertical Plate by a Liquid Flowing Through a Vertical Nozzle," *Chem. Eng. Sci.*, Vol. 32 (1977), pp. 1491–1497.
71. Fabre, S., Ponter, A. B., and Peier, W., "Liquid Film Spreading on a Vertical Surface," *Tenside*, Vol. 15 (1978), pp. 16–18.
72. Morris, G. A., and Jackson, J., *Absorption Towers*, Butterworths Scientific Publications, London, 1953.
73. Mayo, F., Hunter, T. G., and Nash, A. W., "Wetted Surface in Ring-Packed Towers," *J. Soc. Chem. Ind.*, Vol. 54 (1935), pp. 375T–385T.
74. Weisman, J., and Bonilla, C. F., "Liquid-Gas Interfacial Area in Packed Columns," *Ind. Eng. Chem.*, Vol. 42 (1950), pp. 1099–1105.
75. Gamson, B. W., Thodos, G., and Hougen, O. A., "Heat, Mass and Momentum Transfer in the Flow of Gases Through Granular Solids," *Trans. Am. Inst. Chem. Eng.*, Vol. 39 (1943), pp. 1–32.
76. Taecker, R. G., and Hougen, O. A., "Heat, Mass Transfer of Gas Film in Flow of Gases Through Commercial Tower Packings," *Chem. Eng. Prog.*, Vol. 45 (1949), pp. 188–193.
77. McAdams, W. H., Pohlenz, J. B., and St. John, R. C., "Transfer of Heat and Mass Between Air and Water in a Packed Tower," *Chem. Eng. Prog.*, Vol. 45 (1949), pp. 241–252.
78. Shulman, H. L., and DeGouff, J. J., "Mass Transfer Coefficients and Interfacial Areas for 1-Inch Raschig Rings," *Ind. Eng. Chem.*, Vol. 44 (1952), pp. 1915–1922.
79. Treybal, R. E., *Mass Transfer Operations*, McGraw-Hill, New York, 1980.
80. Shulman, H. L., Ullrich, C. F., and Wells, N., "Performance of Packed Columns I. Total, Static and Operating Holdups," *AICHE J.*, Vol. 1 (1955), pp. 247–253.
81. Shulman, H. L., et al., "Performance of Packed Columns III. Holdup for Aqueous and Nonaqueous Systems," *AICHE J.*, Vol. 1 (1955), pp. 259–262.
82. Buchanan, J. E., "Holdup in Irrigated Ring-Packed Towers Below the Loading Point," *Ind. Eng. Chem. Fund.*, Vol. 6 (1967), pp. 400–407.
83. Hikita, H., and Kataoka, T., "Wetted Area in a Packed Column," *Chem. Eng. Japan*, Vol. 20 (1956), pp. 528–533.
84. Hikita, H., Kataoka, T., and Nakanishi, K., "Effective Interfacial Area for Liquid Phase Mass Transfer in Packed Columns," *Chem. Eng. Japan*, Vol. 24 (1960), pp. 2–8.
85. Onda, K., Okamoto, T., and Honda, H., "Liquid-Side Mass-Transfer Coefficient in a Tower Packed with Berl Saddles," *Chem. Eng. Japan*, Vol. 24 (1960), pp. 490–493.
86. Mada, J., Shinohara, H., and Tsubahara, M., "Correlation of Previously Reported Data on Wetted and Effective Interfacial Area in Packed Towers," *Kagaku Kogaku* (Abr. Ed. Engl.), Vol. 2 (1964), pp. 111–113.
87. Mohunta, D. M., Vaidyanathan, A. S., and Laddha, G. S., "Effective Interfacial Areas in Packed Columns," *Indian Chem. Eng.*, Vol. 11, No. 2 (1969), pp. 39–42.
88. Danckwerts, P. V., and Sharma, M. M., "Review Series No. 2. The Absorption of Carbon Dioxide into Solutions of Alkalis and Amines (with some Notes on Hydrogen Sulfide and Carbonyl Sulfide)," *Trans. Inst. Chem. Eng.*, Vol. 44 (1966), pp. CE244–280.
89. Richards, G. M., Ratcliff, G. A., and Danckwerts, P. V., "Kinetics of CO_2 Absorption—III. First-Order Reaction in a Packed Column," *Chem. Eng. Sci.*, Vol. 19 (1964), pp. 325–328.
90. Puranik, S. S., and Vogelpohl, A., "Effective Interfacial Area in Irrigated Packed Columns," *Chem. Eng. Sci.*, Vol. 29 (1974), pp. 501–507.
91. Sridharan, K., and Sharma, M. M., "New Systems and Methods for the Measurement of Effective Interfacial Area and Mass Transfer Coefficients in Gas-Liquid Contactors," *Chem. Eng. Sci.*, Vol. 31 (1976), pp. 767–774.
92. Rizzuti, L., Augugliaro, V., and Cascio, G. L., "The Influence of the Liquid Viscosity on the Effective Interfacial Area in Packed Columns," *Chem. Eng. Sci.*, Vol. 36 (1981), pp. 973–978.

93. Bravo, J. L., and Fair, J. R., "Generalized Correlation for Mass Transfer in Packed Distillation Columns," *Ind. Eng. Chem. Process Des. Dev.*, Vol. 21 (1982), pp. 162–170.
94. Bolles, W. L., and Fair, J. R., "Performance and Design of Packed Distillation Columns," *Inst. Chem. Eng. Symp. Ser. No. 56* (1979), pp. 3.3/35–89.
95. Bemer, G. G., and Zuiderweg, F. J., "Radial Liquid Spread and Maldistribution in Packed Columns Under Different Wetting Conditions," *Chem. Eng. Sci.*, Vol. 33 (1978), pp. 1637–1643.
96. Andrew, S. P. S., "Rejuvenation and Renewal: Liquid Film Limited Physical Solution of Gases in Packed Absorbers," *Chem. Eng. Sci.*, Vol. 38 (1983), pp. 9–20.

CHAPTER 21

DESIGNING PACKED-TOWER WET SCRUBBERS: EMPHASIS ON NITROGEN OXIDES

Robert M. Counce and Joseph J. Perona

Department of Chemical Engineering
University of Tennessee
Knoxville, Tennessee, USA

CONTENTS

PACKED-TOWER WET SCRUBBERS

The design of packed-tower wet scrubbers can be divided into two classes: with and without chemical reaction in the liquid phase. While gas-phase reactions may occur in some important systems (e.g., NO_x absorption), these do not affect the interphase mass transfer rate at a point in the tower as the liquid-phase reaction may.

Without Liquid-Phase Reaction

The mass balance for an incremental length of tower is

$$d(Vy_A) = d(Lx_A) = Ra\,S\,dz \tag{1}$$

For a scrubber, mass transfer usually takes place in one direction only (from the gas phase into the liquid phase), and usually in the presence of a non-absorbing gas. The mass balance may then be written in the form

$$L\frac{dx_A}{(1-x_A)} = k_{x0}a(x_{A_i}-x_A)S\,dz \tag{2}$$

$$= K_xa(x_A^*-x_A)S\,dz \tag{3}$$

These equations may be rearranged to provide design equations for the tower height:

$$Z = \int_{x_{A_1}}^{x_{A_2}} \frac{L}{k_{x_0}aS}\cdot\frac{dx_A}{(1-x_A)(x_{A_i}-x_A)} \tag{4}$$

and

$$Z = \int_{x_{A_1}}^{x_{A_2}} \frac{L}{K_xaS}\cdot\frac{dx_A}{(1-x_A)(x_A^*-x_A)} \tag{5}$$

Those factors of Equations 4 and 5 containing concentrations are called "number of transfer units" and defined as

$$H_G = \frac{L}{k_{x0}aS} \tag{6}$$

$$H_{0G} = \frac{L}{K_x aS} \tag{7}$$

Heights of transfer units generally do not change severely from one end of the tower to the other, and often they are taken as constant and placed outside of the integral forms of Equations 4 and 5.

Those factors of Equations 4 and 5 containing concentrations are called "number of transfer units" and defined as

$$N_G = \int_{x_{A1}}^{x_{A2}} \frac{dx_A}{(1-x_A)(x_{A_i}-x_A)} \tag{8}$$

and

$$N_{0G} = \int_{x_A}^{x_{A2}} \frac{dx_A}{(1-x_A)(x_A^*-x_A)} \tag{9}$$

Analogous equations may be written in terms of gas-phase variables. Correlations for prediction of individual-phase mass transfer coefficients k_x and k_y, and interfacial area a are presented in other sections of this handbook.

With Liquid-Phase Reaction

Consider a liquid phase reaction involving component A, transferring from the gas phase to the liquid phase, and a nontransferring liquid-phase component, B:

$$A + \gamma B \rightarrow \text{products}$$

The mass balance equation must now account for disappearance by reaction through an additional term:

$$d(Vy_A) = d(Lx_A) + rh_l S\,dz = RaS\,dz \tag{10}$$

The concentration distribution throughout the tower depends on the relative rates of reaction and interphase mass transfer. The various limiting cases are explained in detail by Astarita [1] and Danckwerts [2].

In the case of a "fast reaction," the concentration of the transferring component is significantly depleted in the liquid film, effectively enhancing the mass transfer coefficient. This concept leads to the introduction of an enhancement factor, E, into the rate equation:

$$R = Ek_{x0}(x_{A_i}-x_A) \tag{11}$$

with

$$E = k_x/k_{x0} \tag{12}$$

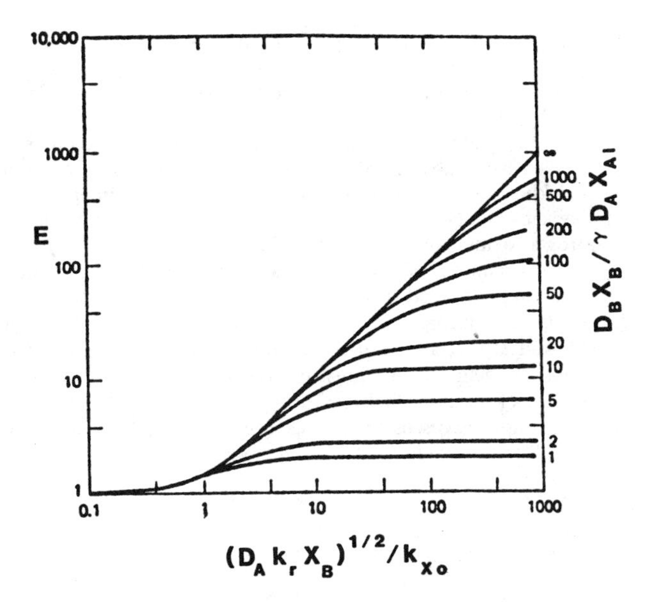

Figure 1. Tower heights are estimated by numerically evaluating design equations such as Equation 13 for fast reactions or more complex expressions involving all the terms in Equation 10. Iterative calculations are required at each point chosen along the tower length to obtain the interface concentrations and rates of transport and reaction.

For the fast reaction case, the bulk liquid concentration may be assumed near its reaction equilibrium value, x_{Ae}, throughout the tower, so that the term $d(Lx_A)$ in Equation 10 may be taken as negligible. Rearrangement yields the design equation for the fast reaction case:

$$Z = \frac{L}{\gamma a S}\int_{x_{B_1}}^{x_{B_2}} \frac{dx_B}{Ek_{x0}(x_{A_i} - x_{Ae})} \tag{13}$$

For a "slow reaction," the concentration of A in the liquid film is not significantly affected by the chemical reaction, so that $E \cong 1$. Two subcases arise:

1. Diffusion control, when the rate of absorption is very slow, so that $x_A \cong x_{Ae}$
2. Kinetic control, when the rate of reaction is very slow, so that $x_A \cong x_{A_i}$

The "instantaneous reaction" regime occurs when the rate of the reaction is diffusion-controlled within the liquid film. The criterion for this condition is

$$\frac{D_B x_B}{\gamma D_A x_{A_i}} \ll \left(\frac{D_A k_r x_B}{k_{x0}^2}\right)^{1/2} \tag{14}$$

The enhancement factor, E, may be estimated for all reaction regimes using the chart in Figure 1.

PACKED TOWERS FOR NITROGEN OXIDE ABSORPTION

The absorption of nitrogen oxides into aqueous media is important in the production of nitric acid as well as in other industrial processes. Interest in this subject is increasing because of the current interest in air pollution abatement and resource recovery. The complexity of nitrogen oxide absorption has produced an enormous amount of literature on this and related kinetic and equilibrium matters; however an integrated understanding of the basic mechanisms involved in this process is available for only the simplest case—the absorption of nitrogen oxides into water and dilute nitric acid [3–7]. The design of nitric acid production towers remains, at least, partly empirical [8, 9]. Part of the complexity of nitrogen oxide scrubbing is related to nitrogen oxide and oxyacid species covering all the positive oxidation states of nitrogen as shown in Table 1. Of the species, only NO, N_2O_3, NO_2 N_2O_4, HNO_2 and HNO_3 are normally important to the modeling of nitrogen oxide scrubbing operations. Modeling efforts must of necessity account for important concentrations and neglect those of insignificance in order to yield a useful mathematical representation. To simplify the following discussion nitrogen oxide (NO_x) "chemical" nitrogen dioxide (NO_2^*) and "chemical" nitric oxide (NO^*) are defined as

$$NO_x = NO_2^* + NO^* \tag{15}$$

$$NO_2^* = NO_2 + 2\,N_2O_4 + N_2O_3 \tag{16}$$

and

$$NO^* = NO + N_2O_3 \tag{17}$$

The chemical reactions of the important nitrogen oxide species is likewise complex; while based on fundamentally correct premises, modeling efforts must be more "overall" in nature at times in order to yield convenient mathematical equations.

Much of the published information about nitrogen oxide scrubbing comes from the nitric acid production industry. At conditions existing in the usual nitric acid production tower [10] the descriptive chemical equations are

$$3\,NO_2(g) + H_2O(\ell) \leftrightarrow 2\,HNO_3(\ell) + NO(g) \tag{18}$$

and

$$2\,NO(g) + O_2(g) \rightarrow 2\,NO_2(g). \tag{19}$$

Table 1
Oxidation States of Nitrogen and Selected Nitrogen Oxides

Compound	Formula	Oxidation State
Nitrous oxide	N_2O	+1
Hyponitrous acid	$H_2N_2O_2$	+1
Nitric oxide	NO	+2
Nitrogen trioxide	N_2O_3	+3
Nitrous acid	HNO_2	+3
Nitrogen dioxide	NO_2	+4
Nitrogen tetroxide	N_2O_4	+4
Nitrogen pentoxide	H_2O_5	+5
Nitric acid	HNO_3	+5

Nitrogen: −3, −2, −1, 0, +1, +2, +3, +4, +5
Nitrogen oxides: +1, +2, +3, +4, +5

From these overall chemical equations, it may be seen that the absorption and reaction of NO_2 and water produces NO, which desorbs to the gas phase. Provided that the gas-phase residence time is adequate, NO may be oxidized in the gas phase and reabsorbed as NO_2. It is helpful to further study Equation 18 to gain insight to this overall reaction. Reaction 18 may be broken down to

$$2\,NO_2(g) \leftrightarrow N_2O_4(g) \tag{20}$$

$$N_2O_4(g) \leftrightarrow N_2O_4(\ell) \tag{21}$$

$$N_2O_4(\ell) + H_2O(\ell) \leftrightarrow H^+(\ell) + NO_3^-(\ell) + HNO_2(\ell) \tag{22}$$

$$3\,HNO_2(\ell) \leftrightarrow H^+(\ell) + NO_3^-(\ell) + H_2O(\ell) + 2\,NO(\ell) \tag{23}$$

$$NO(\ell) \leftrightarrow NO(g) \tag{24}$$

In this scheme of equations, the solubility of NO_2 is neglected because it is much less than N_2O_4 [11] and also because N_2O_4 is much more reactive with water than NO_2 [12, 13]. Counce and Perona [4, 5] used this scheme of reactions to model the absorption of a concentrated gas mixture of NO_2-N_2O_4 in a sieve plate tower. In their work, the concentrations of HNO_3 and HNO_2 acids were sufficiently dilute so that Equation 22 was treated as irreversible. The significance of equation sequence 20 through 24 is illustrated in some data from the work of Makhotkin and Shamsutdinov [14] presented in Figure 2. In this semibatch experiment, 1% NO_2^* in N_2 was bubbled through water. The resulting concentration profiles and ratios show that the ratio of NO^* produced to NO_2^* absorbed reaches 1 : 3 only when a semi-steady-state concentration of HNO_2 is established. This is important in scrubber design and results from the kinetic and mass transfer limitations of the aqueous decomposition of HNO_2 [15] requiring a substantial HNO_2 concentration for a driving force. If this concentration driving force is not established during scrubbing the ratio of

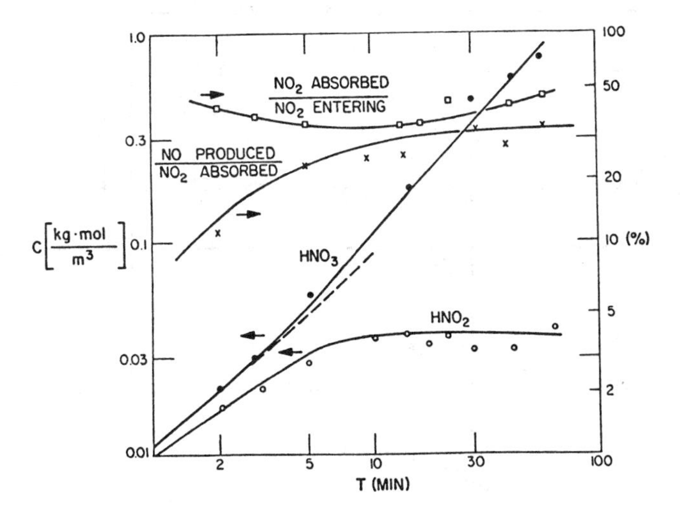

Figure 2. Results of semibatch NO_2^* absorption studies [14].

NO^* produced to NO_2^* absorbed may be very much less than 1 : 3. Although the absorption of N_2O_4 tends to be the dominant absorption route, at least as moderate to high chemical NO_2^* partial pressures, other species such as N_2O_3 are important in nitrogen oxide scrubbing for mixtures of NO^* and NO_2^* containing gases; such a scheme is shown in Figure 3 for scrubbing with water or dilute nitric acid. The gas-phase portion of this scheme shows the oxidation of NO [16] occurring while N_2O_4 and N_2O_3 are in equilibrium with their respective reactant species. [17, 18].

The role of HNO_2 in the scrubbing of gaseous nitrogen oxides is not clear. There is considerable disagreement as to the gas-phase formation and decomposition reaction rate constants [19–21]; experiments with reactors of small ratios of surface area to reactor volume generally yield the smallest rate constants. This suggests the complication of heterogeneous reactions in this determination. Scrubber results of Counce and Perona [6] for NO_x and NO_2^* partial pressures of 0.010 and 0.002 atm, respectively, showed no indication of HNO_2 formation and absorption; this observation is supported by the work of Corriveau [22]. Other work by Andrews and Hanson [7], covering much lower NO_x partial pressures than Counce and Perona [6] and observations by Pigford [23] and Hoftyzer and Kwanten [10] indicate the possibility of HNO_2 formation and absorption at low partial pressures, especially at high NO^*-to-NO_2^* ratios.

In Figure 3 the liquid N_2O_4 and N_2O_3 are shown reacting irreversibly with water in the liquid film (as is usually the case in packed towers when the absorbing solution is water or dilute solutions of aqueous HNO_3), whereas any reaction of NO_2 and water is extremely slow and usually occurs in the bulk phase [13] and may be neglected in many cases. These reactions may be represented by

$$N_2O_4(\ell) + H_2O(\ell) \rightarrow HNO_3(\ell) + HNO(\ell) \tag{25}$$

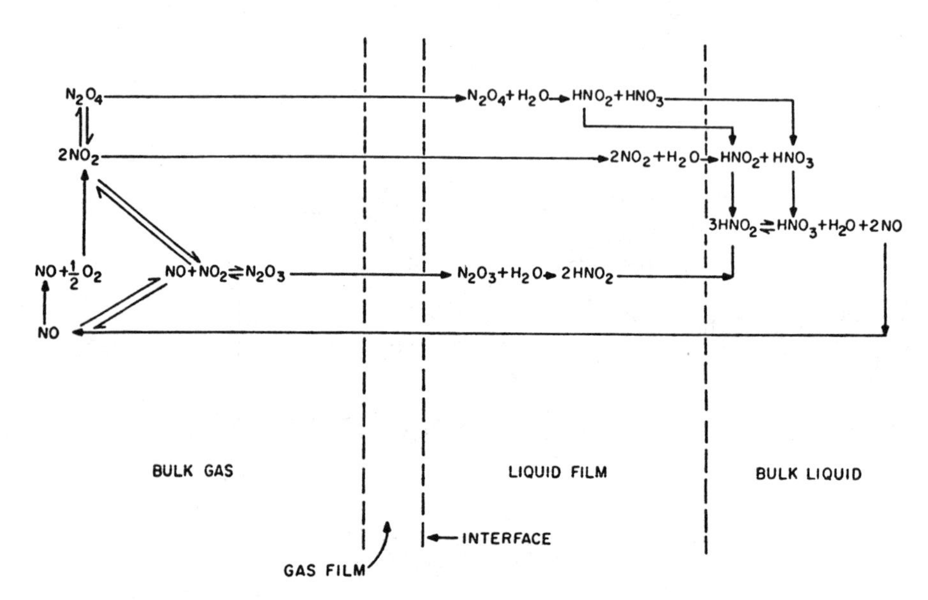

Figure 3. Model for describing mass-transfer and chemical-reaction phenomena.

$$N_2O_3(\ell) + H_2O(\ell) \rightarrow 2\,HNO_2(\ell) \tag{26}$$

$$2\,NO_2(\ell) + H_2O(\ell) \rightarrow HNO_3(\ell) + HNO_2(\ell) \tag{27}$$

The flux equations for the liquid phase are then

$$R_{N_2O_4} = E_{N_2O_4} k_L x_{N_2O_4,i} \tag{28}$$

$$R_{N_2O_3} = E_{N_2O_3} k_L x_{N_2O_3,i} \tag{29}$$

$$R_{NO_2} = E_{NO_2} k_L x_{NO_2,i} \tag{30}$$

Much of the literature data is taken at conditions such that

$$\frac{D_{H_2OH_2O}}{D_{N_2O_4N_2O_4,i}} \gg \frac{\sqrt{D_{N_2O_4} k_{25} x_{H_2O}}}{k_L} \gg 1 \tag{31}$$

$$\frac{D_{H_2O} x_{H_2O}}{D_{N_2O_3} x_{N_2O_3,i}} \gg \frac{\sqrt{D_{N_2O_3} k_{26} x_{H_2O}}}{k_L} \gg 1 \tag{32}$$

$$\frac{D_{NO_2}}{k_L}(k_{27} x^2_{NO_2,i} x_{H_2O}) \ll k_L x_{NO_2,i} \tag{33}$$

Under these conditions the enhancement factors may be calculated as

$$E_{N_2O_4} = \frac{\sqrt{D_{N_2O_4} k_{25} x_{H_2O}}}{k_L} \tag{34}$$

$$E_{N_2O_3} = \frac{\sqrt{D_{N_2O_3} k_{26} x_{H_2O}}}{k_L} \tag{35}$$

and

$$E_{NO_2} = 1 \text{ (no reaction in film)} \tag{36}$$

These enhancement factors will be justified for a wide range of packed tower conditions. The use of Figure 1 for calculation of enhancement factors is very practical as no tests for reaction regimes are required.

The equilibrium decomposition of HNO_2 (Equation 23) [24–27] is shown occurring in the bulk liquid phase with the NO produced in this decomposition transferring back to the gas phase. This phenomena may be represented, assuming no HNO_2 decomposition in the liquid film, by

$$-R_{NO} = k_L(x_{NO} - x_{NO,i}) \tag{37}$$

where

$$x_{NO} = \sqrt{\frac{K_{23} x^3_{HNO_2}}{x_{H+} x_{NO_3^-}}} \tag{38}$$

A very general model for the disappearance of NO^* and NO_2^* in a packed tower is

$$-d(VY_{NO_2^*}) = [(2R_{N_2O_4} + R_{N_2O_3} + R_{NO_2})a - r_{19}\varepsilon]S\,dz \tag{39}$$

$$-d(VY_{NO^*}) = [(R_{N_2O_3} + R_{NO})a - r_{19}\varepsilon]S\,dz \tag{40}$$

$$d(VY_{NO_x}) = d(VY_{NO_2^*}) + d(VY_{NO^*}) \tag{41}$$

A simple model for NO_x absorption efficiency in a packed tower, based on the irreversible absorption of N_2O_4 by Equation 25 is

$$\frac{\sqrt{D_{N_2O_4}k_{25}aP_TSz}}{VH_{N_2O_4}} + 2\left[\ell n\left(\frac{\sqrt{1+2\theta-2\theta X_{NO_2^*}}-1}{\sqrt{1+2\theta}-1}\right) + \frac{\sqrt{1+2\theta-2\theta X_{NO_2^*}}-\sqrt{1+2\theta}}{(\sqrt{1+2\theta-2\theta X_{NO_2^*}}-1)(\sqrt{1+2\theta}-1)}\right] = 0 \tag{42}$$

where $\theta = 4Y_{NO_2,in}P_TK_{P,20}$

This model was originally developed for predicting NO_x removal efficiency for a sieve plate tower [5]; other assumptions implicit in this model are:

1. Ideal gas behavior.
2. Plug flow of the gas phase.
3. No gas-phase resistance.
4. N_2O_4 is the only absorbing component.
5. Little change of the gas results from the absorption rate.
6. NO_2 and N_2O_4 are in an equilibrium state.
7. Isothermal conditions exist.
8. No substantial NO oxidation in the column SZ.
9. No substantial decomposition of HNO_2 occurs in the liquid phase.

This model is a reasonable predictive tool for estimating the NO_x removal efficiency for NO_x gases with high NO_2^* content and $P_{NO_2^*} > 0.01$ atm (the model may also be valid under conceivable conditions at somewhat lower NO_2^* partial pressures) in packed towers with countercurrent flow of gas and liquid and no recycle of the scrubber liquid. In many packed tower designs the heats of reaction are not sufficient to increase the column temperature substantially so the assumption of isothermal conditions is reasonable; this assumption requires some explanation as nitric acid production towers are extensively fitted with heat-removal equipment due to their low liquid throughput per mole of NO_x absorbed. If gas-phase resistance is appreciable for gases of low NO_2^* content or low NO_2^* : NO_x ratios or for recycle of the scrubber liquid, then a more detailed treatment such as presented by Counce and Perona [6] may be necessary. The gas-phase oxidation of NO may also need to be considered if sufficient NO and O_2 concentrations exist in the gas-phase and the residence time is adequate for a nonnegligible extent of reaction. A scrubber model based on an incremental approach is useful in many cases to handle the simultaneous absorption and gas-phase oxidation.

Two modes of scrubber operation, involving recycle and nonrecycle of the scrub solution, are common (see Figure 4). Recycle of the scrub solution is especially important where the

* Efficiency as used here refers to the fraction removal of nitrogen oxides.

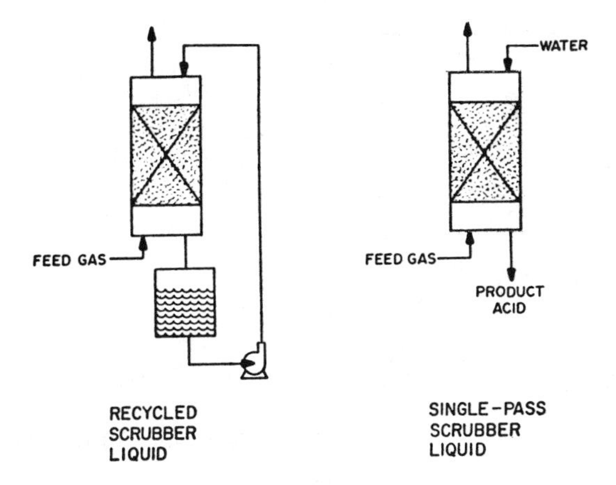

Figure 4. Two modes of scrubber operation involving recycle and nonrecycle of the scrubber liquid.

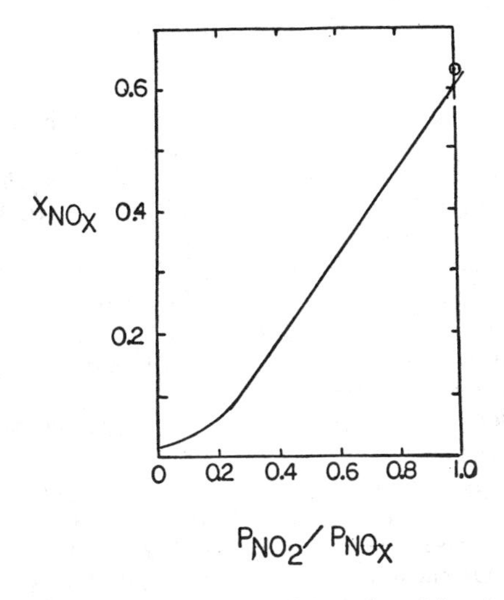

Figure 5. Predicted experimental NO_x scrubber efficiency for a 0.1-m-diameter tower packed with 13-mm Intalox saddles to a height of 0.78 m. Other important information is gas and liquid velocities of 0.29 and 0.0044 m/s, temperature of 298°K. NO_x partial pressure in the feed gas of 0.012 atm and a total pressure of 1.2 atm.

acid product concentration produced from a single pass of the liquid phase through the tower is lower than desired. Recycle of the scrub solution may require some treatment of the scrub solution to destroy any HNO_2 present [28]; if HNO_2 is recycled, then Reactions 23 and 24 [24–27] will occur to a greater extent, releasing gaseous NO and decreasing the scrubber efficiency [4, 5].

If the liquid product from the scrub solution is aqueous HNO_3, then a useful oxidizing agent for HNO_2 is H_2O_2,

$$HNO_2(\ell) + H_2O_2(\ell) \rightarrow HNO_3(\ell) + H_2O(\ell) \tag{43}$$

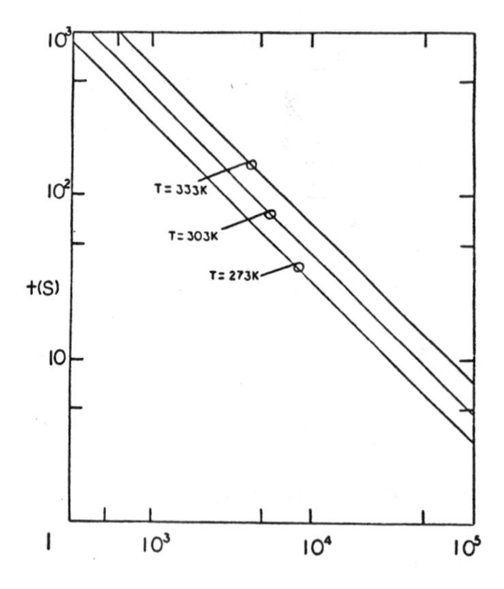

Figure 6. Half-life for NO assuming complete mixing [5].

This reaction is extremely fast and quantitative. Nitrogen oxide scrubber efficiencies* have remained at values near those obtained using a water scrub solution when recycle scrub solutions of up to 3 **M** nitric acid was treated with H_2O_2 before recycle [28, 29]. Similar effects with other reactants for HNO_2 such as NaOH were noted by Koegler [29]; obviously the loss of the acid product results from the use of NaOH.

Chemical NO_2 tends to be more easily absorbed than NO*. This is illustrated in Figure 5 which presents predicted NO_x scrubber efficiencies for a tower packed with 13 mm Intalox saddles to a height of 0.78 m with varying ratios of NO_2/NO_x. This figure illustrates the advantage of pretreatment of the feed gas before aqueous scrubbing. A low NO_2/NO_x ratio in feed gas may be corrected to some extent by oxidation in the scrubber tower; however, the gas phase reaction of NO and O_2 is moderately slow [16]; at low NO concentrations, the absorber size for gases of low NO_2/NO_x ratios may be more influenced by required residence time for the gas-phase oxidation of NO than for the absorption of NO_2.

Oxidation of gaseous NO may be accomplished by the use of several oxidizing species. The particular routes presented here were chosen because they add no new reaction products to the NO_x-HNO_x-H_2O-N_2-O_2 chemical system. The gas-phase oxidation of NO may proceed by

$$2\,NO(g) + O_2(g) \rightarrow 2\,NO_2(g) \tag{19}$$

$$NO(g) + O_3(g) \rightarrow NO_2(g) + O_2(g) \tag{44}$$

$$NO(g) + H_2O_2(g) \rightarrow NO_2(g) + H_2O(g) \tag{45}$$

and

$$NO(g) + 2\,HNO_3(g) \rightarrow H_2O(g) + 3\,NO_2(g) \tag{46}$$

The rate of NO oxidation in air illustrated in Figure 6 for three different temperatures. This reaction is unique in that it is a third-order reaction and the rate of reaction decreases with temperature.

The rate of oxidation is much faster when the reactant is O_3 rather than O_2. Reaction 44 is an extremely fast gas-phase reaction [30]. A simultaneous reaction,

$$NO_2(g) + O_3(g) \rightarrow NO_3(g) + O_2(g), \tag{47}$$

is much slower than Reaction 44 for equivalent NO and NO_2 partial pressures and other conditions [31]. Any NO_3 produced in this reaction will quickly react with NO_2 producing N_2O_5, a very soluble and reactive compound. The scheme of at least a partial oxidation of NO to NO_2 by ozone prior to wet scrubbing has been utilized in some Japanese NO_x removal systems for stack gas treatment [32]. Reaction 45 is fairly slow, however, the reaction of

$$2\,NO_2(g) + H_2O_2(g) \rightarrow 2\,HNO_3(g) \tag{48}$$

is much faster and causes the oxidation of NO to proceed mainly by Reaction 46 [33]. Neither Reaction 45 or 48 is well characterized.

If sufficient HNO_3 exists in the gas phase, such as over-concentrated HNO_3 solutions, the gas-phase oxidation of NO proceeds as Reaction 46, a reaction with autocatalytic routes that are not well understood [34–36]. At lower HNO_3 partial pressures and in the presence of a liquid phase, the reaction shifts to the liquid [24–26, 37].

$$2\,NO(\ell) + HNO_3(\ell) + H_2O(\ell) \rightarrow 3\,HNO_2(\ell) \tag{49}$$

The latter reaction of NO in 25–30 wt% HNO_3 was proposed by Bolme and Horton [38] as a method of removing NO from the tail gas of conventional nitric acid production towers (NO_2 will be removed also in this scheme through absorption and reaction with water). Some equilibrium ratios of $NO:NO_2$ over 70 wt% nitric acid are shown in Table 2. Reaction 46 was gas-phase-limited in studies by Lefers et al. [34] in which gaseous NO was contacted with HNO_3 in a wetted-wall column. The gas-phase reaction of NO and HNO_3 was also demonstrated by Counce [39] to convert the bulk of a 10% NO stream to NO_2 by contact with 70 wt% nitric acid in a conventional tower packed with 13 mm Intalox saddles.

In conclusion, control of gaseous NO_x by aqueous scrubbing is a fairly complex process. However, acceptable NO_x removal for many situations may be achieved by maintaining an acceptable $NO_2:NO_x$ ratio in the feed gas, treatment of the scrubber liquid (if recycle is desired) and proper scrubber design and operation. The described models based on describ-

Table 2
Equilibrium Ratios of Gaseous $NO:NO_2$ Over 70 Wt % Nitric Acid Solutions at 298 K and 1.0 atm*

P_{NO_x} (atm)	P_{NO}/P_{NO_2}
0.001	0.000011
0.01	0.0011
0.11	0.11

* *Calculate using an equilibrium constant, defined by*

$$K = P^2_{HNO_2}\ P_{NO}\ P^{-3}_{NO_2}\ P^{-1}_{H_2O}$$

of 0.013 bar^{-1} at 298 K
From Hoftyzer and Kwanten [10].

ing the nitrogen oxide absorption flux has been shown to be valid where the solution HNO_3 concentrations were < 3 M. Other reviews pertinent to gaseous nitrogen oxide absorption into aqueous media have been done by Counce [40], Sherwood, Pigford, and Wilke [3] and Hoftyzer and Kwanten [10].

NOTATION

- a interfacial area per unit of tower volume
- g gas
- h_ℓ liquid holdup per unit of tower volume
- H_i Henry's law coefficient for component i,
- k_x liquid-side mass transfer coefficient
- k_{x0} liquid-side mass transfer coefficient for physical absorption
- K_x overall mass transfer coefficient based on liquid concentrations
- $K_{p,i}$ chemical equilibrium constant for i-th reaction in units of pressure
- K_i chemical equilibrium constant for i-th reaction
- k_y gas-side mass transfer coefficient
- k_r reaction rate constant
- ℓ liquid
- L molar flow rate of liquid phase
- P_T total pressure
- r rate of chemical reaction per unit of liquid volume
- R rate of absorption per unit of interfacial area
- S tower cross-sectional area
- t time
- T temperature
- V molar flow rate of gas phase
- x_A mole fraction of A in bulk liquid phase
- x_{Ai} mole fraction of A in liquid phase at interface
- x_A^* mole fraction of A in liquid in equilibrium with bulk gas
- X_i conversion or fractional removal of component i
- y_A mole fraction of A in bulk gas phase
- z tower height
- ε void fraction of tower

REFERENCES

1. Astarita, G., *Mass Transfer with Chemical Reaction*, Elsevier, New York, 1967.
2. Danckwerts, P. V., *Gas-Liquid Reactions*, McGraw-Hill, New York, 1970.
3. Sherwood, T. K., Pigford, R. L., and Wilke, C. R., *Mass Transfer*, McGraw-Hill, New York, 1975.
4. Counce, R. M., and Perona, J. J., "Gaseous Nitrogen Oxide Absorption in a Sieve-Plate Column," *Ind. Eng. Chem. Fundam., 18*, 400 (1979).
5. Counce, R. M., and Perona, J. J., "A Mathematical Model for Nitrogen Oxide Absorption in a Sieve-Plate Column," *Ind. Eng. Chem. Process Des. Dev., 19*, 426 (1980).
6. Counce, R. M., and Perona, J. J., "Scrubbing of Gaseous Nitrogen Oxides in Packed Towers," *AIChE. J., 29*, (1983).
7. Andrews, S. P., and Hanson, D., "The Dynamics of Nitrous Gas Absorption," *Chem. Eng. Sci., 14*, 105 (1961).
8. Holma, H., and Sohlo, J., "A Mathematical Model of an Absorber Tower of Nitrogen Oxides in Nitric Acid Production," *Computers & Chem. Engr., 3*, 135 (1979).
9. Kongshaug, G., and Medgdell, G. Th., "Modelling of Absorption Efficiency in Nitric Acid Manufacture," ISMA Technical Conference, Vienna (1981).
10. Hoftyzer, P. J., and Kwanten, F. J. G., "Absorption of Nitrous Gases," *Processes for Air Pollution Control*, CRC Press, Cleveland, Ohio, 165 (1972).
11. Schwartz, S. E., and White, W. H., "Solubility Equilibria of the Nitrogen Oxide and Oxyacids in Dilute Aqueous Solution," *Advances in Envir. Sci. and Engr., 4*, 1 (1981).

12. Kameoka, Y., and Pigford, R. L., *Ind. Eng. Chem. Fundam., 16,* 163 (1977).
13. Lee, Y. N., and Schwartz, S. E., "Reaction Kinetics of Nitrogen Dioxide with Liquid Water at Low Partial Pressure," *J. Phys. Chem., 85,* 840 (1981).
14. Makhotkin, A. F., and Shamsutidinov, A. M., "A Study of the Kinetics of the Absorption of NO_2 and the Effect of Nitrous Acid," *Khim. Khim. Tekhn.* XIX, 1411 (1976).
15. Komiyama, H., and Inoue, H., "Reaction and Transport of Nitrogen Oxides in Nitrous Acid Solutions," *J. Chem. Engr. Japan, 11,* 25 (1978).
16. Bodenstein, M., "Formation and Decomposition of Higher Nitrix Oxides, *Z. Electrochem. 100,* 68 (1922).
17. Verhoek, F. H., and Daniels, F., "The Dissociation Constant of Nitrogen Tetroxide and of Nitrogen Trioxide," *J. Am. Chem. Soc., 53,* 1250 (1931).
18. Beatie, I. R., "Dinitrogen Trioxide," *Prog. Inorg. Chem., 5,* 1 (1963).
19. Kaiser, E. W., and Wu, C. H., "A Kinetic Study of the Gas-Phase Formation and Decomposition Reactions of Nitrous Acid," *J. Phys. Chem., 81,* 1701 (1977).
20. England, C., and Corcoran, W. H., "The Rate and Mechanism of the Air Oxidation of Parts-per-Million Concentrations of Nitric Oxide in the Presence of Water Vapor," *Ind. Eng. Chem. Fundam., 14,* 55 (1975).
21. Wayne, L. G., and Yost, D. M., "Kinetics of the Gas-Phase Reaction Between NO, NO_2, and H_2O," *J. Chem. Phys., 19,* 41 (1951)
22. Corriveau, C. E., Jr., *The Absorption of N_2O_3 into Water,* Master's Thesis in Chemical Engineering, University of California, Berkeley (1971).
23. Pigford, R. L., Personal communication to R. M. Counce (1983).
24. Abel, E., and Schmid, H., "Kinetics of Nitrous Acid I, Introduction and Survey," *Z. Phys. Chem., 132,* 55 (1928a), Translated from German (ORNL-tr-4263).
25. Abel, E., and Schmid, H., "Kinetics of Nitrous Acid II, Orienting Experiments," *Z. Phys. Chem., 132,* 64 (1928b), Translated from German (ORNL-tr-4263).
26. Abel, E., and Schmid, H., "Kinetics of Nitrous Acid III, Kinetics of the Decomposition of Nitrous Acid," *Z. Phys. Chem., 134,* 279 (1928c), Translated from German (ORNL-tr-4265).
27. Abel, E., and Schmid, H., "Kinetics of Nitrous Acid IV, Equilibrium of Nitrous Acid-Nitric Oxide Reaction in Conjunction with its Kinetics," *Z. Phys. Chem., 136,* 430 (1929), Translated from German (ORNL-tr-4265).
28. Counce, R. M., "Nitrogen Oxide Control and Recovery in Nuclear Fuel Reprocessing Operations," paper presented at the American Nuclear Society 1981 Winter Meeting, San Francisco (1981).
29. Koegler, S. C., *Purex NO_x Abatement Pilot Plant,* Rockwell Hanford Operations, RHO-CD-702, Richland, Washington (July 1979).
30. Johnson, H. S., and Crosby, H. J., "Kinetics of the Fast Gas Reaction Between Ozone and Nitric Oxide," *J. Chem. Phys., 22,* 689 (1954).
31. Graham, R. A., and Johnson, H. S., "Kinetics of the Gas-Phase Reaction Between Ozone and Nitrogen Dioxide," *J. Chem. Phys., 60,* 4628 (1974).
32. Rosenberg, H. S., et al., *Control of NO_x Emission by Stack Gas Treatment,* FP-925, Electric Power Research Institute (1978).
33. Grey, D., Lissi, E., and Heicklen, J., "Reaction of Hydrogen Peroxide with Nitrogen Dioxide and Nitric Acid," *J. Phys. Chem., 76,* 1919 (1972).
34. Lefers, J. B., et al., "The Oxidation and Absorption of Nitrogen Oxides in Nitric Acid in Relation to the Tail Gas Problem of Nitric Acid Plants," Sixth International Symposium on Chemical Reaction Engineering (1980).
35. Streit, G. E., et al., "A Tunable Diode Laser Study of the Reactions of Nitric and Nitrous Acids: $HNO_3 + NO$ and $HNO_2 + O_3$," *J. Chem. Phys., 70,* 3439 (1979).
36. Kaiser, E. W., and Wu, C. H., "Measurement of the Rate Constant of the Reaction of Nitrous Acid with Nitric Acid, *J. Phys. Chem., 81,* 187b (1977).

37. Carta, G., and Pigford, R. L., "Absorption of Nitric Oxide in Nitric Acid and Water," *Ind. Eng. Chem. Fundam., 22,* 329 (1983).
38. Bolme, D. W. and Horton, A., "The Humphreys and Glascow/Bolme Nitric Acid Process," *Chem. Eng. Prog., 95* (March 1979).
39. Counce, R. M., personal communication to J. J. Perona (1983).
40. Counce, R. M., *The Scrubbing of Gaseous Nitrogen Oxides in Packed Towers,* Ph. D. Dissertation, The University of Tennessee, Knoxville (1980).

CHAPTER 22

GAS ABSORPTION IN AERATED MIXERS

Masabumi Nishikawa

Department of Nuclear Engineering
Kyushu University
Fukuoka, Japan

CONTENTS

INTRODUCTION

Aerated mixing vessels are often used in chemical and biological industries. When the rate of gas absorption controls the rate of reaction, the gas absorption capacity coefficient k_La is used as the design standard for aerated mixing vessels. A number of correlations have been proposed for the rate of gas absorption in aerated mixing vessels as listed in Table 1, although some correlations seem incompatible with each other. For instance, the exponent of the superficial gas velocity ranged from 0 to 0.76 and that of the impeller speed from 0.7 to 3.0. Nishikawa et al. [1] have proposed the method of estimating k_La in aerated mixing vessels using the two-region model where it is assumed that the region of bubbling-controlling condition and that of agitation-controlling condition coexist in the mixing vessel. This model shows that differences in exponents can be explained by considering the relative intensities of aeration and mechanical agitation. According to this two region model, the overall vessel condition varies from bubbling-controlling condition at relatively high gas rates to agitation-controlling condition at relatively high impeller speeds via intermediate condition. Exponents of the physical properties of the systems for k_La in an aerated mixing vessel also should not be constant in a wide range of operating conditions, because effects of most physical properties on k_La in the aerated tower differ from those in the aerated mixing vessel under the agitation-controlling condition as stated by Nishikawa et al. [2].

Table 1
Comparison of Correlation for k_L a by Various Investigators

Range of Experiment		Experimental Apparatus				Type of Correlation (Exponent of)			Investigators
u_g, cm/s	N, min^{-1}	D, cm	Type	d/D	n_p	u_g	n	P_g/V	
0.165-5.5	60-900	15.3, 44.2	Vaned disk	0.4	16	0.67		0.95	Cooper et al. [4]
0.375-1.77	100-1,250	24.1, 245	Paddle	0.25	2	0.67		0.95	Cooper et al. [4]
0.015-8.5	500-1,680		Propeller			0.4	1.70		Maxon&Johnson [5]
0.03-0.112		305	Turbine	0.2	8	0.40		0.53	Oldshue [6]
0.061-0.61	650-1,780	15.3, 30.6	Turbine	0.33	6	0.76		0.71-0.79	Rushton et al. [7]
-1.69	135-790	13.1	Turbine	0.7	2	0.75	1.67		Johnson et al. [8]
0.08-7.37	300-2,000	15.3	Paddle	0.2, 0.5	4	0	3.0		Friedman et al. [9]
0.68	500-1,000	15.3	Turbine	0.5	6	0.68	2.0		Snyder et al. [10]
-4.07		15.3, 50.3	Paddle	0.5, 0.33	2			0.43-0.95	Karwat [11]
-4.07	500-1,000	15.3, 50.3	Turbine	0.34	8			0.43-0.95	Karwat [11]
-4.07	500-1,000	15.3, 50.3	Propeller	0.3	3			0.43-0.95	Karwat [11]
0.045-7.6	120-650	15, 25, 38.0	Turbine	0.4	12	0.67	2.0		Yoshida et al. [12]
0.049-2.1	645-1,990	10.1	Turbine	0.5	6	0.21	2.04		Hyman&Bogaerde [13]
0.203-2.03	0-3,600	6.0, 19.1	Turbine	0.2-0.7	6	0	1.0		Westerterp et al. [14]
0.203-2.03	0-3,600	6.0, 19.1	Paddle	0.2-0.7	2, 4	0	1.0		Westerterp et al. [14]
0.009-1.9	0-1,500	15.3	Paddle	0.33	2	0.49-0.75	0.70-1.66		Hortacsu [15]
0.2-0.8	300-600	25.0	Turbine	0.4	6	0.28	2.20	0.74	Yagi&Yoshida [16]

It has been also shown by Nishikawa et al. that the two-region model can also be applied for helical ribbon or anchor-type impellers [3].

Mixing vessels are most often used for gas absorption, and various types of mixing techniques for accomplishing intimate contact between liquid and gas bubbles. However, application of the mixing vessel is limited to relatively low gas rates as a decrease in the average density around the impeller causes a lowering of the impeller power which is necessary to break and disperse gas bubbles. In cases where a large k_La is required, the gas-liquid spouted vessel can be applied. In this case power is supplied to water from a pump and gas is introduced at a nozzle attached through the bottom of the vessel. In the gas-liquid spouted vessel, a fine dispersion of gas bubbles is introduced to the vessel through a nozzle. The nozzle creates this dispersion through a choking flow effect resulting in a ten to twenty factor increase in k_La over that for aerated towers.

MEASUREMENT OF THE EFFECTIVE MASS TRANSFER COEFFICIENT

A simple approach to obtaining k_La values is the dynamic method in which the change in dissolved oxygen concentration in the liquid is observed using an oxygen electrode. A schematic of the experimental set-up is shown in Figure 1. The liquid in the vessel is first sparged with nitrogen until an equilibrium value of the dissolved oxygen concentration is obtained. After bubbles of nitrogen have disappeared from the liquid in the vessel, air is sparged, and the change in dissolved oxygen concentration measured using an oxygen analyzer.

A dissolved oxygen balance in liquid is

$$dC/dt = k_La(C_I - C)\psi \qquad (1)$$

where, C, C_I, and ψ are the oxygen concentration in the bulk liquid, the oxygen concentration at the liquid interface which is in equilibrium with the partial pressure of oxygen in bubbles, and a factor to correct for the effect of surface aeration, respectively.

Integration of Equation 1 gives

$$\ln(C_I - C) = -\psi k_La \cdot t + \ln(C_I - C_0) \qquad (2)$$

where C_0 represents the oxygen concentration in the bulk liquid at time zero.

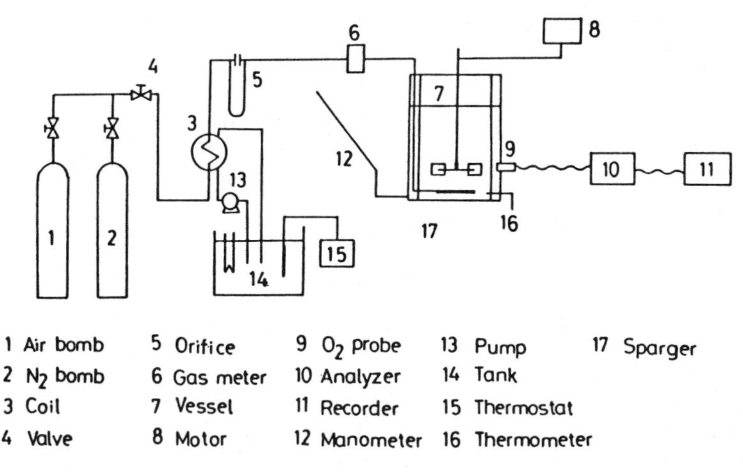

Figure 1. Set-up of the measuring system.

Thus k_La can be obtained from the slope of a plot of $\ln(C_I - C)$ versus time for each measurement when the change of C_I with time is small, after applying the correction factor ψ.

It is possible that aeration from the free liquid surface contributes greatly to gas absorption, expecially when vigorous agitation is applied in a small-scale vessel, as reported by van Dierendonck et al. [17] and Fuchs et al. [18]. However, the surface aeration rate is largely reduced by increasing the aeration rate from a sparger, as reported by Calderbank [19] and Miller [20]. According to Calderbank, the factor ψ, which is introduced to account for the contribution of surface aeration, is

$$\psi^{1/2} = f[(d^2n\varrho/\mu)^{0.7}(dn/u_g)^{0.3}] \tag{3}$$

Usually, the value of ψ will not exceed 1.2.

Since the mixing time is much shorter than the time required for the k_La measurement, it can be assumed that the dissolved oxygen concentration is uniform throughout the vessel.

The error associated with the k_La measurement is due to the time lag in the electrode's response, which is on the order of seconds. As such, when k_La is larger than $0.1\ s^{-1}$, the sodium sulphite oxidation method is considered a more accurate measurement when large k_La values are expected. According to Cooper et al. [4], the reaction rate of oxidation of 0.035–1.0 N sodium sulphite is independent of the sodium sulphite concentration and is controlled by the oxygen dissolution stage to water. Miura [21] has confirmed the applicability of this method to cases where k_La exceeded $6\ s^{-1}$, and Yasunishi [22] also reported that both methods provide comparable values of k_La.

Measured concentration C' is given as follows assuming a first-order time lag in the electrode measurement system.

$$C'(s)/C(s) = 1/(1 + Ts) \tag{4}$$

where C and T are the actual concentration and the first-order time lag, respectively.

The Laplace transform of Equation 1 gives

$$C(s) = k_La \cdot C_I/[s(k_La + s)] \tag{5}$$

where the effect of surface aeration is ignored and C_0 is assumed to be zero.

The inverse Laplace transform of Equation 4 after substitution of Equation 5 produces a step response curve when C is changed from zero to C_I instantaneously.

$$(C_I - C')/C_I = [\exp(-k_La \cdot t) - k_La \cdot T \exp(-t/T)]/(1 - k_La \cdot T) \quad (6)$$

As shown in Figure 2 where Equation 6 is shown for $k_La = 0.1$ s, the effect of the time lag cannot be ignored when it is large because a constant slope is not obtained untill the value of $(c_I - C')/C_I$ becomes so small that the accuracy of the mèasurement decreases.

GAS-PHASE MIXING

It can be assumed that perfect mixing is applied for the liquid phase since its mixing time is much shorter than the time required for the k_La measurement. The degree of mixing in the gas phase is also important since a correct average driving force is required in order to estimate the gas absorption coefficient. It was reported by Westerterp et al. [14] and Hanhart et al. [23] that the residence time distribution of gas passing through an aerated mixing vessel closely resembled that of perfect mixing because of frequent coalescence and redispersion of gas bubbles. This implies that the gas absorption rate should be based on the composition of the gas leaving the vessel and not on some average value between its entrance and outlet composition. Those authors also reported that the k_La value became independent of the superficial gas velocity in the agitation-controlling condition when perfect mixing in the gas phase was assumed.

However, the logarithmic mean value of the entrance and outlet gas compositions assuming plug flow was recommended by Nishikawa et al. [1], because they considered that the coalescence and redispersion of gas bubbles were not frequent enough to provide perfect mixing.

Comparison of gas absorption rates assuming plug flow, k_La, with that of perfect mixing, k_La', calculated from the same experimental data is shown in Figure 3 for the absorption of oxygen into water. As shown values of, k_La and k_La' are comparable when the aeration power per unit mass of liquid in the aerated mixing vessel, P_{av}, is larger than one-tenth of the agitation power per unit mass of liquid in an aerated mixing vessel, P_{gv}. The difference in the types of mixing has no effect on the calculated gas absorption rate. However, in the range where P_{av} is much smaller than P_{gv}, the difference between k_La and k_La' becomes significant. Figure 3 also shows that there is a range where k_La' is almost independent of P_{av}. This range of vigorous agitation coincides exactly with the experimental range examined by Westerterp et al. (shown as the dotted lines in Figure 3). Further decreases in P_{av}, however, result in an increase in k_La' which is not typical. These facts suggest that the calculation of the gas absorption rate using the driving force based on perfect mixing in the gas phase is not valid. Accordingly, it may be more reasonable to use a logarithmic mean value as the driving force, assuming a plug-flow-type mixing in the gas phase for aerated mixing vessels from the viewpoint of the gas absorption rate.

Hanhart et al. concluded from a step response curve that perfect mixing should be assumed in gas-phase mixing. Their result, however, is not conclusive enough assume perfect mixing for the following reasons:

1. The effect of the residence time distribution of gas bubbles on the response curve should be accounted for.
2. The effect of surface aeration should be considered.
3. A time lag in the response curve of one-fourth of the observed average residence time of gas implied that their system was not perfectly mixed.

At present, it is recommended that plug-flow-type mixing be applied to the gas phase in an aerated mixing vessel, as the gas absorption rate is reasonably estimated by using the

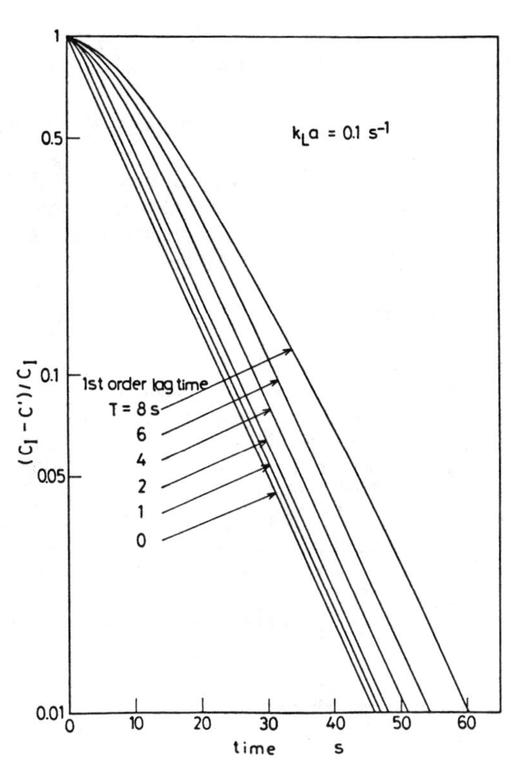

Figure 2. Effect of first-order time lag.

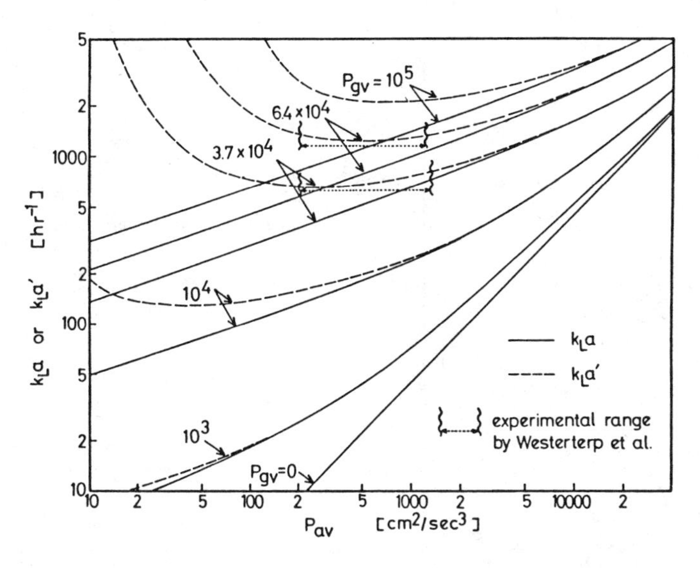

Figure 3. Comparison of k_La with k_La'.

logarithmic mean driving force. The gas absorption rate estimated on the basis of perfect mixing gives unrealistic results when agitation is vigorous and there is little aeration.

CORRELATION OF k_La IN AERATED MIXING VESSELS WITH WATER

Correlations for k_La in aerated mixing vessels by various investigators are listed in Table 1. As can be seen from this table, the exponent of the superficial gas velocity, u_g, and that of impeller speed, n, varies greatly. The large variations in these exponents indicate that the correlation often stated in the form

$$k_La \propto u_g^{\alpha} n^{\beta} \tag{7}$$

is not satisfactory.

Figures 4 and 5. show that constant values cannot be applied to α and β over a wide range of superficial gas velocities and impeller speeds. Since k_La is proportional to the superficial gas velocity and is not affected by the impeller speed when aeration is vigorous, conditions in the vessel resemble those observed in an aerated tower. Mechanical agitation, however, plays the leading role in gas absorption at large impeller speeds, because k_La is greatly affected by impeller speed, and the exponent of the superficial gas velocity shows a constant value of 1/3, as shown in Figure 5. Exponent α varies from 1.0 to 1/3 and β from 0 to 2.4 for the intermediate condition in the case of a turbine-type impeller (refer to Figures 4 and 5). These observations indicate that most of the investigations reported in the literature were performed in a relatively narrow range corresponding to intermediate

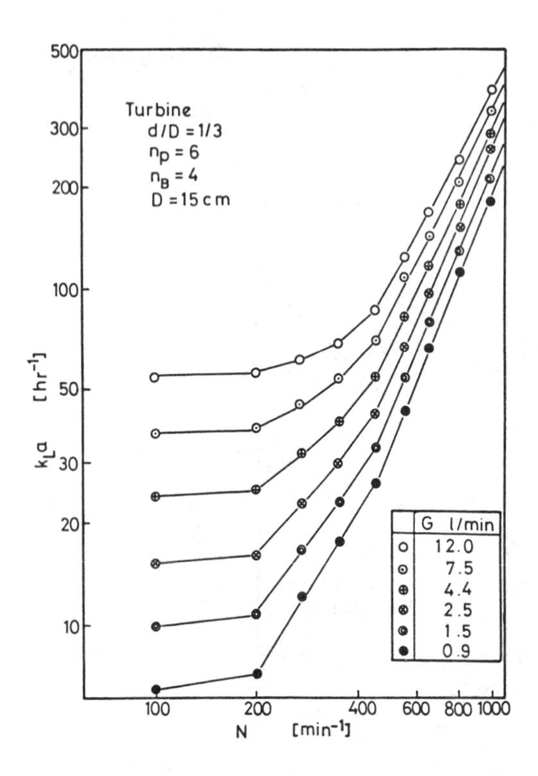

Figure 4. k_La and impeller speed.

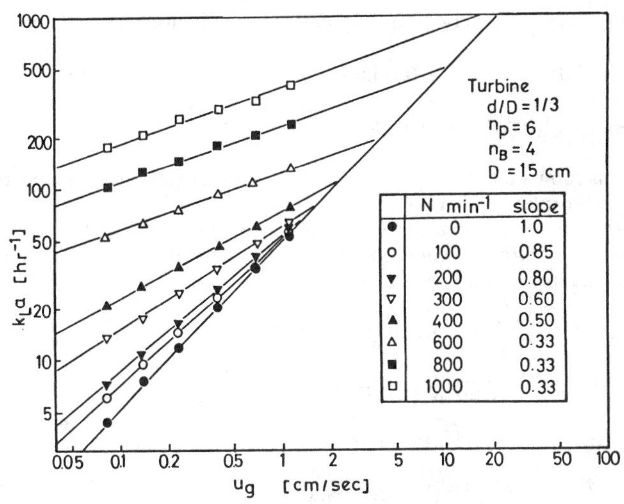

Figure 5. k_La and superficial gas velocity.

Table 2
N_p for Various Impellers

	n_B	d/D	b/d	n_p	θ	N_p (Turbulent)	$N_p \cdot Re$ (Laminar)
Paddle	4		1/6	4	90°	3.08	
Paddle	4		1/5	4	45 (down)	2.62	
Paddle	4		1/5	4	90	3.70	
Paddle	4		1/5	6	90	5.45	70
Turbine	4		1/5	6	90	5.50	70
Ribbon		0.95	1/10	2		0.20	320
Anchor		0.95	1/10			0.29	200

conditions. Accordingly, a correlative equation is required which covers three conditions, i.e., the bubbling-controlling condition, the intermediate condition, and the agitation-controlling condition.

The following equation is obtained for k_La under the bubbling-controlling condition.

$$k_La = 1.25 \times 10^{-5} P_{av}^{1.0} P_{gv}^{0} \tag{8}$$

where P_{av} and P_{gv} are the power consumptions per unit mass of liquid for aeration and agitation, respectively, and are as follows.

$$P_{av} = u_g g \tag{9}$$

$$P_{gv} = N_p n^3 d^5 / V \tag{10}$$

where N_p is the power number. The values of N_p for various impellers used by Nishikawa et al. are shown in Table 2.

In the agitation-controlling condition, the following equations are obtained.

$$k_La = 3.92 \times 10^{-6} P_{av}^{1/3} P_{gv}^{0.8} \quad \text{(for turbine)} \tag{11}$$

$$k_La = 5.69 \times 10^{-6} P_{av}^{1/3} P_{gv}^{0.75} \quad \text{(for paddle)} \tag{12}$$

These equations show that the turbine-type impeller is more effective than the paddle-type impeller for gas absorption into water.

For the intermediate condition, the following equations are obtained by considering a part of the bubbling-controlling condition and part of the agitation-controlling condition to coexist in the aerated mixing vessel.

$$k_La = \{3.92 \times 10^{-6} P_{gv}^{0.8} + 1.25 \times 10^{-5} P_{av}^{2/3} \times [P_{av}/(P_{gv}/N_p + P_{av})]\} P_{av}^{1/3} \quad \text{(for turbine)} \tag{13}$$

$$k_La = \{5.69 \times 10^{-6} P_{gv}^{0.75} + 1.25 \times 10^{-5} P_{av}^{2/3} \times [P_{av}/(P_{gv}/N_p + P_{av})]\} P_{av}^{1/3} \quad \text{(for paddle)} \tag{14}$$

where $P_{av}/(P_{gv}/N_p + P_{av})$ is a correction term for the rapid approach of k_La to the bubbling-controlling condition value at high gas rates and for the agitation-controlling condition at large impeller speeds. It is a measure of the relative liquid volumes of the bubbling-controlling condition part in an aerated mixing vessel.

The k_La values estimated by these equations are compared with the observed values in Figures 6 and 7 for turbine- and paddle-type impellers. As shown the correlations apply over a wide range of superficial gas velocities and impeller speeds extending from the bubbling-controlling regime to the agitation-controlling regime.

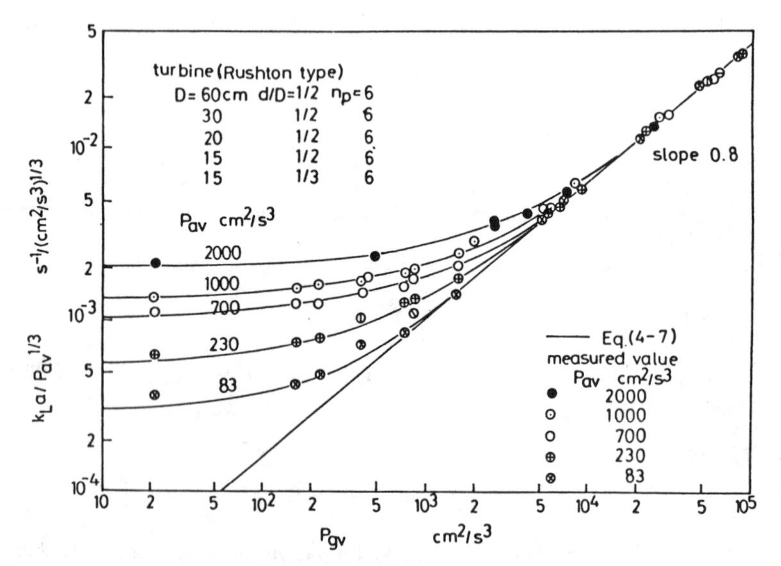

Figure 6. Correlation of k_La for turbine-type impeller.

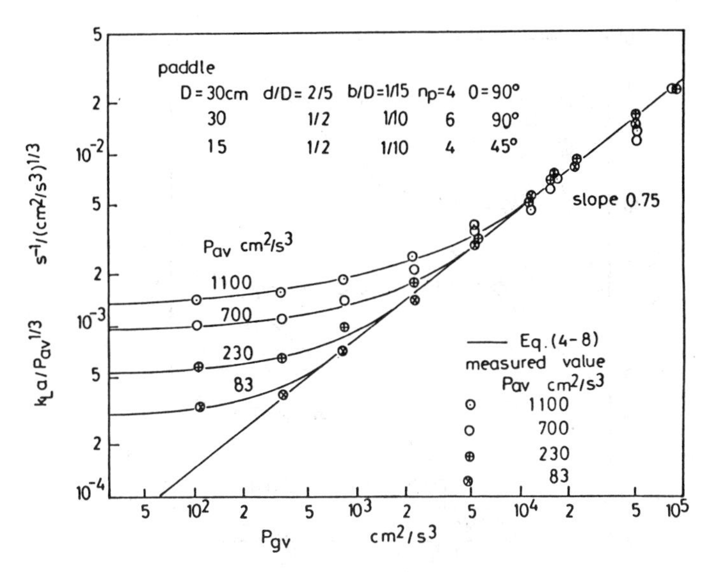

Figure 7. Correlation of k_La for paddle-type impeller.

In general, the value of the superficial gas velocities exponent is smaller at larger impeller speeds and the exponent of the impeller speed is smaller for larger gas velocities (see Table 1).

The critical P_{gv} value, $(P_{gv})_{max}$, above which the agitation-controlling regime is obtained (for a certain P_{av} value) is described by the following equation for turbine-type impellers.

$$(P_{gv})_{max}^{0.8}[1+(P_{gv})_{max}/N_pP_{av}] = 12.7\ P_{av}^{2/3} \tag{15}$$

where it is assumed that k_La from agitation Equation 11 is 80% of the total k_La Equation 13 when P_{gv} is $(P_{gv})_{max}$. In the ordinary aeration rate range, Equation 15 is approximated by

$$(P_{gv})_{max} = 0.19\ P_{av}^{0.90} \tag{16}$$

The critical P_{gv} value, $(P_{gv})_{min}$, below which the bubbling-controlling regime prevails is also approximated by the following equation as in the case of $(P_{gv})_{max}$, though k_La from agitation is 20% of the total k_La at $(P_{gv})_{min}$.

$$(P_{gv})_{min} = 0.514\ P_{av}^{0.84} \tag{17}$$

As shown in Figure 8, there is a wide intermediate condition range and P_{gv} should be more than five times P_{av} to obtain the agitation-controlling condition in the aerated mixing vessel.

The actual agitation power, however, is smaller than P_{gv}, since it decreases with increasing aeration rate as reported by Ohyama and Endoh [24], Michel and Miller [25], and Nagata et al. [26]. Although a reasonable correlation is obtained by using the agitation power for heat transfer from the vessel wall (see Chong et al. [27] and Nagata et al. [28]) agitation power with no aeration provides a much better correlation for the gas absorption rate. Nishikawa et al. [29] recently reported that the heat transfer rate from the vessel wall can

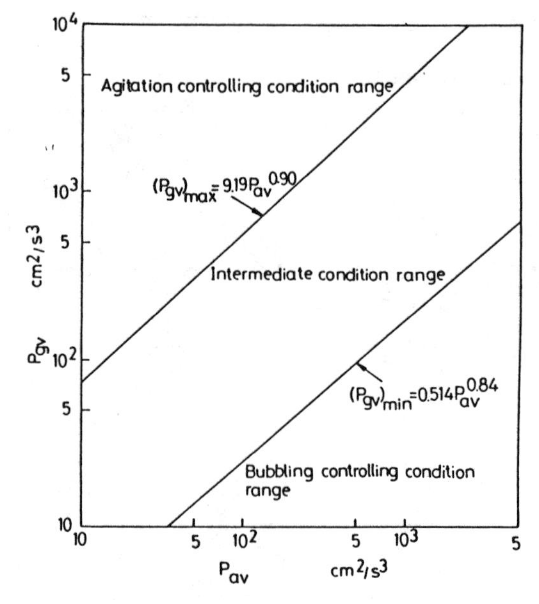

Figure 8. $(P_{gv})_{max}$ and $(P_{gv})_{min}$ versus P_{av}.

be correlated by using the two-region model. Nienow et al. [30] gives extensive discussions on power consumption and flow patterns in aerated mixing vessels. More studies, however, are required.

EFFECT OF PHYSICAL PROPERTIES ON k_La.

For Newtonian liquids in aerated mixing vessels, it is generally accepted that k_La is a function of the liquid viscosity μ, liquid-phase diffusivity D_L, liquid density ϱ, and surface tension σ, in addition to the agitational power and superficial gas velocity.

Sideman et al. [31] proposed the following format for k_La in aerated mixing vessels.

$$(k_Lad^2/D_L) = f(d^2n\varrho/\mu,\ \mu/\varrho D_L,\ \mu u_g/\sigma,\ \mu_g/\mu) \tag{18}$$

where μ_g is the gas viscosity. Unfortunately, they did not study the effect of each dimensionless group independently. This type of correlation cannot be recommended, as the liquid viscosity affects all the dimensionless groups on the right-hand side.

Assuming there is no effect of gas viscosity on k_La, Yagi and Yoshida [16] proposed the following:

$$(k_Lad^2/D_L) = f(d^2n\varrho/\mu,\ \mu/\varrho D_L,\ \mu u_g/\varrho,\ dn^2/g,\ nd/u_g) \tag{19}$$

Their results, however, cannot be used over a wide range of operating conditions as the experiments were conducted only in the intermediate regime which resulted in constant values for α and β. This type of correlation can be applied to correlate k_La under the agitation-controlling regime.

The following correlation can be applied to the bubbling-controlling regime, as Akita [32] and Nakano [33] obtained good results for an aerated tower.

$$(k_L a d^2/D_L) = f(u_g/\sqrt{Dg},\ \mu/\varrho D_L,\ gD^2\varrho/\sigma,\ gD^3\varrho^2/\mu^2) \quad (20)$$

According to the two-region model, $k_L a$ in the aerated mixing vessel is

$$k_L a = (k_L a)_g + [P_{av}/(P_{gv}/N_p + P_{av})]\,(k_L a)_a \quad (21)$$

or in dimensionless form

$$(k_L a D^2/D_L) = f(d^2 n\varrho/\mu,\ \mu/\varrho D_L,\ \mu u_g/\sigma,\ dn^2/g,\ nd/u_g,\ D/d,\ Pg_c/\varrho n^3 d^5) + P_{av}/(P_{gv}/N_p + P_{av})\, f(u_g/\sqrt{gD},\ \mu/\varrho D_L,\ gD^2\varrho/\sigma,\ gD^3\varrho^2/\mu^2) \quad (22)$$

where P_{av} and P_{gv} are the power consumptions per unit mass of liquid for aeration and agitation, respectively, and $P_{av}/(P_{gv}/N_p + P_{av})$ is a term that measures the relative liquid volume of the bubbling-controlling regime and the agitation-controlling regime.

The effect of liquid viscosity and diffusivity on $k_L a$ in the agitation-controlling regime are shown in Figures 9 and 10. The effects of various physical properties on $k_L a$ for turbine-type impellers operating in the agitation-controlling regime are described by the following correlation:

$$(k_L a)_{g,t} = 6.5 \times 10^{-4} \mu^{-1/2} D_L^{1/2} \sigma^{-1/2} \varrho^{1.0} P_{gv}^{0.8} P_{av}^{1/3} \quad (23)$$

or in dimensionless form

$$(k_L a D^2/D_L)_{g,t} = 0.115(d^2 n\varrho/\mu)^{1.5} (\mu/\varrho D_L)^{0.5} (\mu u_g/\sigma)^{0.5} (dn^2/g)^{0.367} (nd/u_g)^{0.167} (d/D)^{0.4} (Pg_c/\varrho n^3 d^5)^{0.8} \quad (24)$$

The effects of liquid viscosity and diffusivity on $k_L a$ for an aerated tower (aerated mixing vessel with no agitation) are shown in Figures 11 and 12, are described by the following relation

$$(k_L a)_a = 8.37 \times 10^{-3} \mu^{-1/3} D_L^{1/2} \sigma^{-0.66} \varrho^{1.0} P_{av} \quad (25)$$

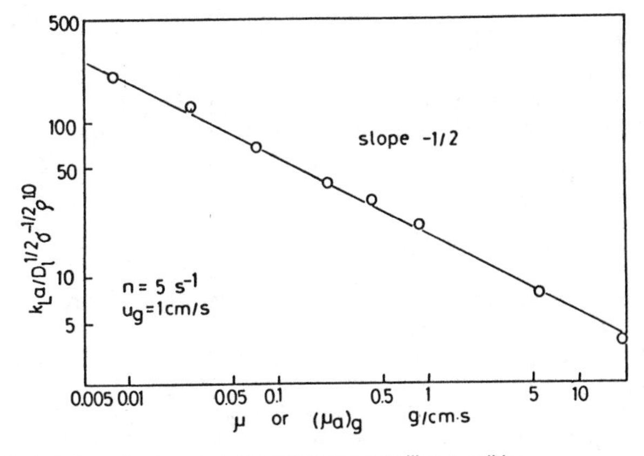

Figure 9. Effect of viscosity on $k_L a$ under agitation-controlling condition.

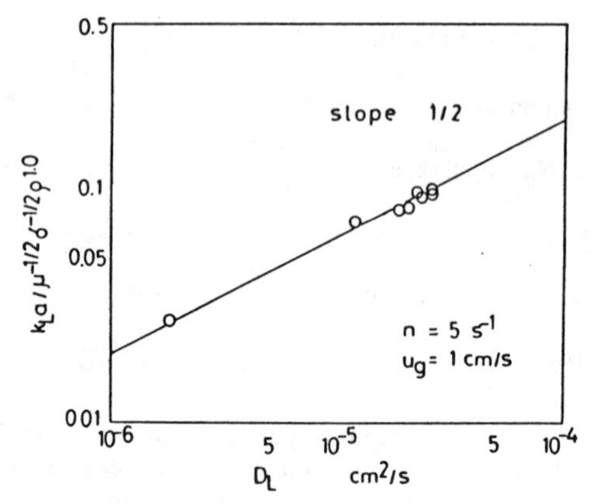

Figure 10. Effect of diffusivity on k_La under agitation-controlling condition.

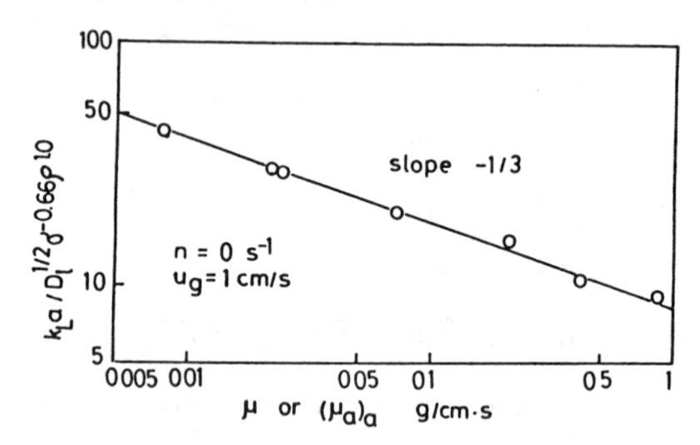

Figure 11. Effect of viscosity on k_La under bubbling-controlling condition.

or in dimensionless form

$$(k_LaD^2/D_L)_a = 0.112(u_g/\sqrt{gD})\,(\mu/\varrho D_L)^{0.5}\,(gD^2\varrho/\sigma)^{0.66}\,\,(gD^3\varrho^2/\mu^2)^{0.42} \tag{26}$$

Accordingly, the overall k_La in an aerated mixing vessel employing turbine-type impellers can be obtained by substitution of Equations 26 and 27 into Equation 22.

$$(k_LaD^2/D_L)_t = (k_LaD^2/D_L)_{g,t} + [P_{av}/(P_{gv}/N_p + P_{av})]\,(k_LaD^2/D_L)_a \tag{27}$$

The k_La values estimated using the prior equation are shown in Figure 13 as solid lines. Good agreement with experimental data is obtained. Consequently, we conclude that the two-region model is applicable to mixing vessels handling Newtonian liquids, over a wide range of operating conditions and physical properties.

Reported exponents for each variable and/or dimensionless group are compared in Tables 3 and 4. The physical properties exponents reported by Nishikawa et al. [1, 2] for the agitation-controlling regime are comparable to those reported by Yagi and Yoshida for

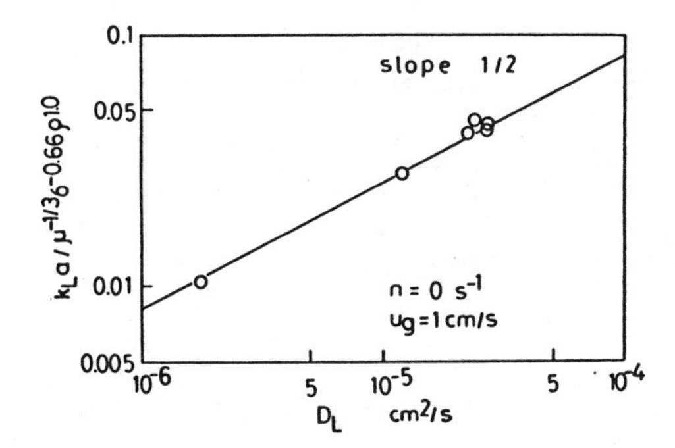

Figure 12. Effect of diffusivity on k_La under bubbling-controlling condition.

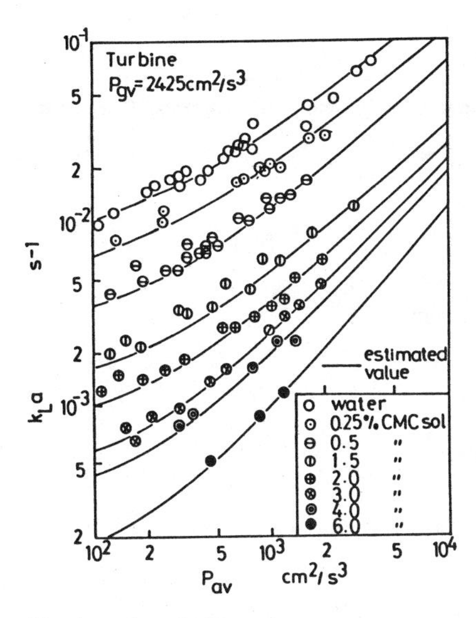

Figure 13. Correlation of k_La for various liquids.

an aerated mixing vessel and those of Akita and Nakano for aerated towers in the bubbling-controlling regime. It can also be seen from these tables that the experiments by Yagi and Yoshida were performed in the intermediate regime. The values observed by Perez and Sandall [34] are different from those of other workers. Some differences in the physical properties exponents can make a large difference in the exponents of the dimensionless groups as is shown. Consequently, the type of correlation proposed by Sideman, where a certain physical property is contained in all dimensionless groups, cannot be recommended.

For paddle-type impellers, the following equation was obtained including the effects of various physical properties:

$$(k_La)_{g,p} = 15{,}4 \times 10^{-4} \mu^{-0.38} D_L^{1/2} \sigma^{-1/2} \varrho^{0.9} P_{gv}^{0.75} P_{av}^{1/3} \tag{28}$$

Table 3
Effect of physical properties on k_La

Type System	μ	D_L	σ	ϱ	P_{gv}	P_{av}
Aerated mixing vessel						
Turbine						
Nishikawa et al.	−0.5	0.5	−0.5	1.0	0.8	1/3
Perez & Sandall	−1.304	0.5	−0.447	0.61	1/3	0.447
Yagi & Yoshida	−0.4	0.5	−0.6	1.0	0.74	0.28
Hattori et al.	−0.488	0.5				
Paddle						
Nishikawa et al.	−0.38	0.5	−0.5	0.9	0.75	1/3
Helical ribbon						
Nishikawa et al.	−0.333	0.5	−0.66	1.0	0.78	1/3
Anchor						
Nishikawa et al.	−0.333	0.5	−0.66	1.0	0.62	1/3
Aerated tower						
Nishikawa et al.	−1/3	0.5	−0.66	1.0	–	1.0
Akita	−0.3	0.5	−0.62	1.0		1.0
Nakano	−0.38	0.5	−0.62	1.0	–	1.0

Table 4
Comparison of Nondimensional Correlation

Type System	$d^2n\varrho/\mu$	$\mu/\varrho D_L$	$\mu u_g/\sigma$	dn^2/g	nd/u_g	d/D	$gD^2\varrho/\sigma$	$u_g/\sqrt{gD}$	$gD^3\varrho^2/\mu^2$
Aerated mixing vessel									
Turbine									
Nishikawa et al.	1.5	0.5	0.5	0.367	0.167	0.4	–	–	–
Yagi & Yoshida	1.5	0.5	0.6	0.19	0.32	–	–	–	–
Paddle									
Nishikawa et al.	1.38	0.5	0.5	0.367	0.167	0.25	–	–	–
Helical ribbon									
Nishikawa et al.	0.053	0.5	–	0.75	–	–	0.66	0.33	–
Anchor									
Nishikawa et al.	0.21	0.5	–	0.51	–	–	0.66	0.33	–
Aerated tower									
Nishikawa et al.	–	0.5	–	–	–	–	0.66	1.0	0.42
Akita	–	0.5	–	–	–	–	0.62	1.0	0.40
Nakano	–	0.5	–	–	–	–	0.62	1.0	0.44

which gives the dimensionless equation for paddle-type impellers under the agitation-controlling regime as follows.

$$(k_LaD^2/D_L)_{g,P} = 0.363(d^2n\varrho/\mu)^{1.38}(\mu/\varrho D_L)^{0.5}(\mu u_g/\sigma)^{0.5}$$
$$(dn^2/g)^{0.367}(nd/u_g)^{0.167}(d/D)^{0.25}N_p^{0.75} \quad (29)$$

Comparison of Equations 25 and 28 or 26 and 29 answers the question of which type of impeller is best for the gas absorption operations in aerated mixing vessels under the agitation-controlling condition.

$$(k_La)_{g,p}/(k_La)_{g,t} = 2.37\mu^{0.12}\varrho^{-0.1}P_{gv}^{-0.05} \quad (30)$$

A comparison between paddle- and turbine-type impellers based on the previous Equation is shown in Figure 14. As shown the turbine-type impeller results in a larger k_La than the paddle-type impeller at a lower viscosity and larger P_{gv}. These observations imply that the disc of a turbine-type impeller produces some disadvantages in the dispersion of gas bubbles for higher-viscosity liquids or low P_{gv}. Visual observations show that coalescence of gas bubbles under the disc of the turbine-type impeller become notable at low impeller speeds. These effects becomes noticeable when the liquid is non-Newtonian.

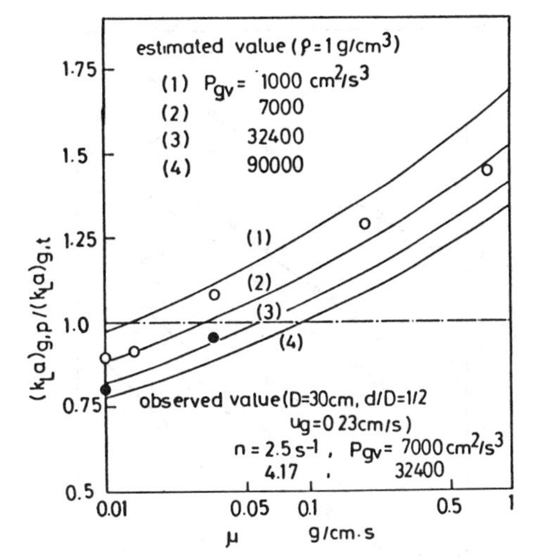

Figure 14. Comparison of turbine- and paddle-type impellers.

Accordingly, it is recommended to use turbine-type impellers for gas absorption under turbulent agitation conditions. Paddle-type impellers may be applicable when milder agitation is required as in the case of some bioreactors. Helical ribbon or anchor-type impellers, however, are more applicable for absorption applications using viscous liquids as shown later.

GAS ABSORPTION IN NON-NEWTONIAN LIQUIDS

Gas absorption in non-Newtonian liquids is often encountered in the chemical and biological industries. There are two major problems that arise when handling non-Newtonians. One is how to estimate a representative viscosity and the other is to evaluate the effect of the elasticity of the liquid. Estimation of the apparent viscosity is first discussed.

As an example, we consider the elasticity of CMC (carboxy methyl cellulose) solutions which diminishes after a long period of aerated mixing. In this manner, the effect of the apparent viscosity on k_La is observed.

Examples of flow curves for CMC solutions are shown in Figure 15. There data were obtained after 12 hours of aerated mixing. As shown, solutions of CMC weaker than 2% display Newtonian behavior, whereas solutions exceeding 2.5% show power-law pseudoplastic behavior. The flow curves of pseudoplastic materials are represented by a power-law model:

$$\tau = k\gamma^m \quad (m < 1) \tag{31}$$

By analogy to Newtonian liquids, an apparent viscosity μ_a is defined.

$$\mu_a = k\gamma^{m-1} \tag{32}$$

The average shear rate in the agitation-controlling regime is related to impeller speed (see Metzner and Otto [35], Calderbank and Moo-Young [36], and Nagata et al. [37]). The value obtained by Nagata et al. for turbine-type impellers is as follows.

$$(\gamma_{av})_g = 11.83n \tag{33}$$

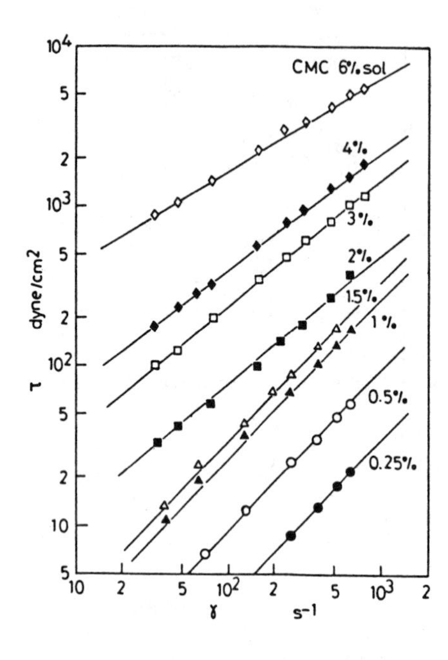

Figure 15. Example of flow curve of CMC solutions.

The average shear rate in the bubbling-controlling regime is related to the superficial gas velocity as shown by Nishikawa et al. [38] in correlating heat transfer coefficients of aerated towers filled with non-Newtonian liquids.

$$(\gamma_{av})_a = 50.0u_g \tag{34}$$

The flow properties of co-polymer solutions like CMC, change with time during aerated mixing. A rapid change is observed during the initial stages followed by a slow change after several hours. Measurement of k_La after 12 hours of aerated mixing of a newly made polymer solution is recommended to exclude errors due to changes in viscosity and elasticity in the course of the measurements and flow curve preparation.

When the two-region model is applied to non-Newtonian liquids, different apparent viscosities should be used for each region. Comparison of apparent viscosities in each region is shown in Figure 16 for a 6% CMC solution; the flow curve of which is shown in Figure 15. The figure shows that the apparent viscosity in the region of agitation-controlling is larger than that in the region of bubbling-controlling. Consequently, estimates of k_La based on the apparent viscosity for the aeration-controlling regime can be misleading.

According to the two-region model, k_La for non-Newtonian liquids can be obtained by using the following apparent viscosity in each region of the aerated mixing vessel.

$$(\mu_a)_g = k(\gamma_{av})_g^{m-1} \quad \text{(agitation-controlling)} \tag{35}$$

$$(\mu_a)_a = k(\gamma_{av})_a^{m-1} \quad \text{(bubbling-controlling)} \tag{36}$$

Estimated values of k_La for non-Newtonian liquids are also compared with observed values in Figure 13. As shown, the agreement is good. Accordingly, the two-region model is also applicable to non-Newtonian liquids.

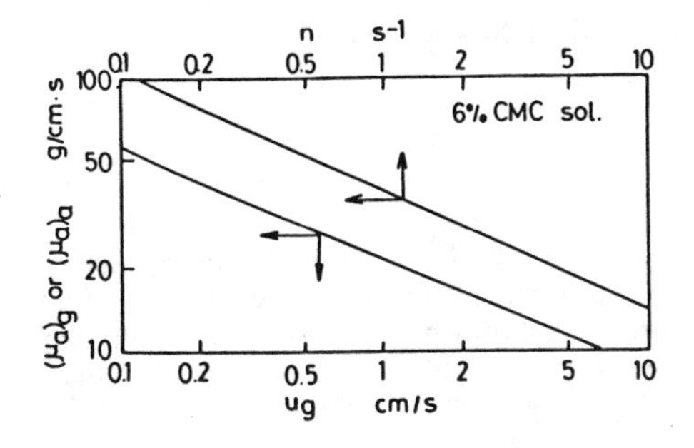

Figure 16. Change of apparent viscosity.

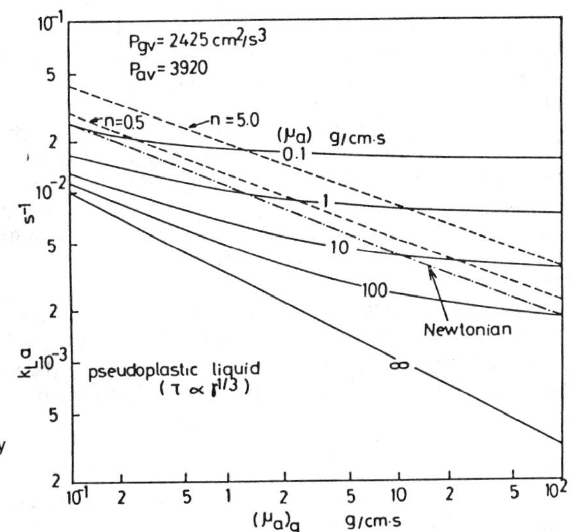

Figure 17. Effect of pseudoplasticity on k_La.

The solid lines in Figure 17 show the effect of $(\mu_a)_g$ and $(\mu_a)_g$ on k_La when P_{gv} and P_{av} are fixed at 2,425 cm^2/s^3 and 3,920 cm^2/s^3, respectively. In the case of Newtonian liquids, both viscosities show the same value, and k_La changes with viscosity as shown by the chain line in Figure 17. As shown, however, k_La for non-Newtonian liquids changes significantly with impeller speed and aeration rate even when P_{gv} and P_{av} are fixed as shown by the broken lines in Figure 17.

Perez and Zandall [34] reported that k_La for non-Newtonian liquids can be correlated using the apparent viscosity calculated by Equation 35, i.e., only the apparent viscosity in the agitation-controlling region was used. The exponent obtained, however, is erroneously larger than the value obtained by other workers, as shown in Table 4. This could be due to the effect of foaming, as noted by Yagi and Yoshida.

Hattori et al. also examined the effect of viscosity on k_La for CMC solutions and reported the following results [39].

$$k_La \propto \mu_a^{-0.35} \tag{37}$$

$$k_La \propto \mu_a^{-1.1} \tag{38}$$

However, the best-fit of their data is given by the following equations

$$k_L a \propto \mu_a^{-0.448} \quad (39)$$

$$k_L a \propto \mu_a^{-1.03} \quad (40)$$

Since this study did not provide details on the method of evaluating the average shear rate used in determining the apparent viscosity, no comments regarding the accuracy of μ_a can be made. It can be said, however, that the result shown by Equation 39 is almost the same as that obtained for the CMC solutions whose apparent viscosity is smaller than several tenths of a centipoise, (note, however; that these show Newtonian behavior for the CMC solutions shown in Figure 15).

For non-Newtonian liquids, a good estimation of $k_L a$ can not be obtained solely by using $(\mu_a)_g$, because $(\mu_a)_a$ has an effect on $k_L a$ (see Figure 17).

The phenomenon of large gas bubble formation (of the size of the disc) immediately below a turbine-type impeller is observed when visco-elastic liquids are agitated under aeration. In such cases impellers have almost no effect on bubble dispersion.

According to Street [40] and Zana and Leal [41], the effect of liquid viscosity on bubble collapse is larger than that of liquid elasticity. In the case of turbine-type impellers, however, this description is not always true, because in the high shear rate region around the impeller tip and in the low shear rate region under the impeller disc, the formation of large gas bubbles is observed. In such liquids with high elasticity another type of impeller (paddle-type may be better than turbine) or reactor (gas-liquid spouted vessel or reactor with static mixer) should be considered.

Yagi and Yoshida reported that the effect of elasticity could not be neglected in calculating $k_L a$ for visco-elastic liquids even when the gas bubbles are well dispersed. They obtained the following correction factor, introducing the Deborah number λ_n, which is the characteristic material time divided by the characteristic process time:

$$\alpha_g = [1 + 2(\lambda_n)^{0.25}]^{-0.67} \quad (41)$$

where the characteristic material time λ_n is the reciprocal of the shear rate at which the reduced complex viscosity, i.e., the ratio of apparent viscosity to zero-shear viscosity, is 0.67 as proposed by Prest et al. [42]. In the experiment where $k_L a$ for CMC solutions was measured after 12 hours of aerated mixing of a newly made CMC solution, however, no effect of the Deborah number could be obtained even for the 6% CMC solution, which should have given a rather large Deborah number from the flow curve. For that 6% CMC solution, the elasticity of liquid could not be observed by the rheogonio-meter even though the flow curve showed non-Newtonian behavior as stated by Nishikawa et al. [2]. Accordingly, the way of determining λ_n from the flow curve may not always be applicable.

For such a strong visco-elastic liquid as PANa (sodium poly-acrylic acid), the correction factor due to elasticity α_g becomes smaller than 0.25 for dense solutions and the effect of elasticity on $k_L a$ is profound. For weak visco-elastic liquids as CMC, however, estimated α_g values are larger than 0.8. Accordingly, such visco-elastic liquids show a weak Weissenberg effect around the impeller shaft, and the correction term can be neglected without large error.

Nakano [33] proposed the following correlation factor for an aerated tower filled with visco-elastic liquids.

$$\alpha_a = [1 + 0.18(\lambda u_b/d_{av})^{0.45}]^{-1} \quad (42)$$

where the average ascending velocity of gas bubbles u_b divided by the Sauter mean bubble diameter d_{av} corresponds to the characteristic process time.

As a result, the following correlation is obtained for k_La in aerated mixing vessels with non-Newtonian liquids for turbine-type impellers.

$$
\begin{aligned}
(k_LaD^2/D_L)_t = {} & 0.115[d^2n\varrho/(\mu_a)_g]^{1.5}\,[(\mu_a)_g/\varrho D_L]^{0.5} \\
& [(\mu_a)_g u_g/\sigma]^{0.5}\,(dn^2/g)^{0.367}\,(nd/u_g)^{0.167} \\
& (d/D)^{0.4} N_p^{0.8} [1+2(\lambda n)^{0.5}]^{-0.67} \\
& + 0.112[P_{av}/(P_{gv}/N_p + P_{av})]\,(u_g/\sqrt{gD}) \\
& ((\mu_a)_a/\varrho D_L)^{0.5}\,(gD^2\varrho/\sigma)^{0.66}\,(gD^3\varrho^2/(\mu_a)_a^2)^{0.42} \\
& [1+0.18(u_b/d_{av})^{0.45}]^{-1}
\end{aligned}
\tag{43}
$$

A similar correlation is obtained for paddle-type impellers using Equations 26, 27, 29, 35, 36, 41, and 42. For the average shear rate in the agitation-controlling condition using paddle-type impellers, the following equation is recommended.

$$(\gamma_{av})_{g,p} = 10.5n \tag{44}$$

GAS ABSORPTION IN VISCOUS LIQUIDS AGITATED BY HELICAL RIBBON OR ANCHOR-TYPE IMPELLERS

For agitation of highly viscous liquids, it is well known that helical ribbon or anchor-type impellers show good mixing performances and are often used under the aerated conditions. Unfortunately, only small gas absorption coefficients have been reported for these impeller-types.

In this section correlation of k_La for aerated mixing vessels agitated by helical ribbon and anchor-type impellers as applied to the two-region model are described.

The effects of impeller speed and superficial gas velocity on k_La for high viscosity liquids are shown in Figures 18 and 19 for a helical-ribbon-type impeller. These figures show that

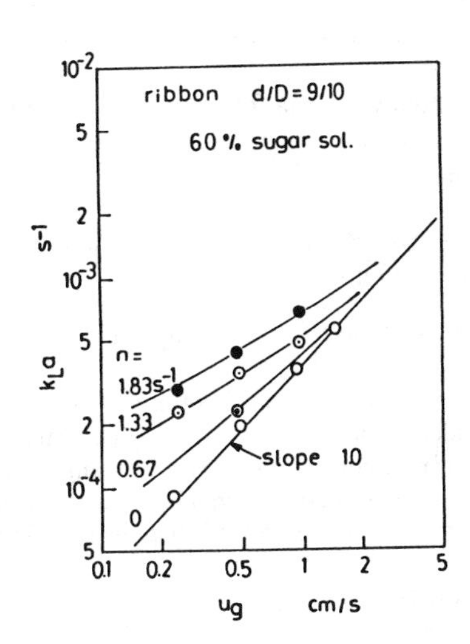

Figure 18. k_La and superficial gas velocity for helical ribbon (60% sugar solution).

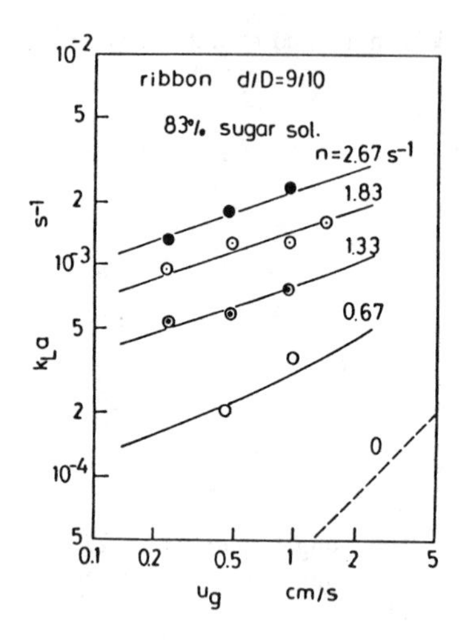

Figure 19. k_La and superficial gas velocity for helical ribbon (83% sugar solution).

the agitation-controlling and bubbling-controlling regions coexist in mixing vessels handling high viscosity liquids under laminar conditions, implying that the two-region model is applicable.

The effects of various physical properties on k_La observed for helical ribbon impellers in the agitation-controlling regime is summarized in the correlation reported by Nishikawa et al. [3].

$$(k_La)_{g,R,L} = 7.54 \times 10^{-3} D_L^{1/2} \mu^{-1/3} \sigma^{-0.66} \varrho P_{av}^{1/3} P_{gv}^{0.78} \quad \text{for} \quad Re \leqq 200 \tag{45}$$

or in dimensionless form

$$(k_LaD^2/D_L)_{g,R,L} = 41.1(\mu/\varrho D_L)^{1/2}(gD^2\varrho/\sigma)^{0.66}(dn^2/g)^{0.75}(d^2n\varrho/\mu)^{0.053}(u_g/\sqrt{gD})^{0.33} \tag{46}$$

where the following relation exists in the laminar region.

$$N_p \cdot Re = \text{const.} \tag{47}$$

Constant values for various configurations of helical impellers are shown as follows by Nagata et al. [43].

$$N_p \cdot Re = 74.3[(D-d)/d]^{-0.5}(n_pd/s)^{0.5} \tag{48}$$

where n_p and s are the number of impeller blades and the impeller pitch length, respectively.

k_La under the agitation-controlling condition is shown in Figure 20 versus P_{gv} for a wide range of mixing conditions. It is interesting to note that k_La in the transition and turbulent mixing coregimes becomes smaller than the value obtained under laminar conditions.

The correction factor introduced by Nishikawa et al. [3] is:

$$A_R = (k_La)_{g,R}/(k_La)_{g,R,L} \tag{49}$$

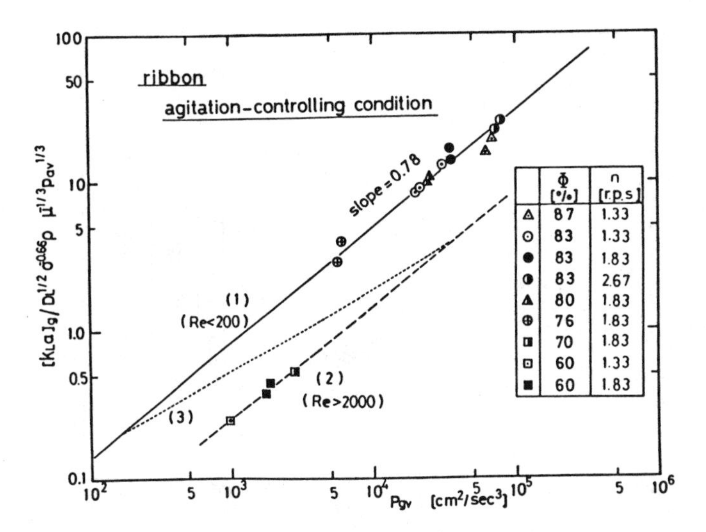

Figure 20. k_La and P_{gv} for helical ribbon under agitation-controlling condition.

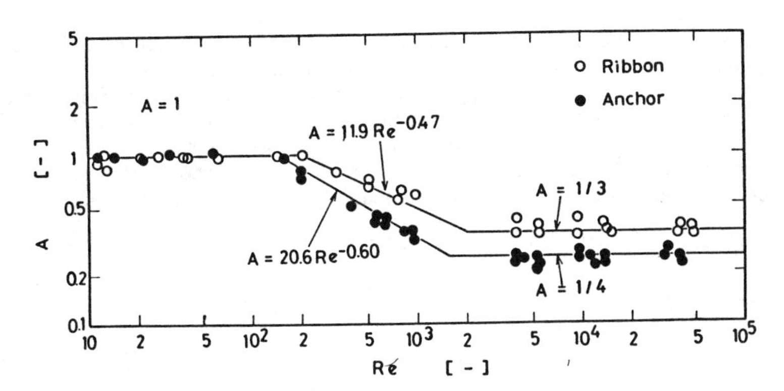

Figure 21. A and Re.

This becomes the following for each mixing condition shown in Figure 21:

$$A_R = 1 \qquad \text{for} \quad Re \leqq 200 \tag{50}$$

$$A_R = 11.9\ Re^{-0.47} \qquad \text{for} \quad 200 \leqq Re \leqq 2{,}000 \tag{51}$$

$$A_R = 1/3 \qquad \text{for} \quad Re \geqq 2{,}000 \tag{52}$$

This tendency of larger k_La values for higher viscosity liquids may be closely related to the gas hold-up in the mixing vessel. The gas hold-up observed for liquid viscosities greater than 2 poise was 13%–20% lower to only a few percent lower when the liquid viscosity was decreased as reported by Nishikawa et al.

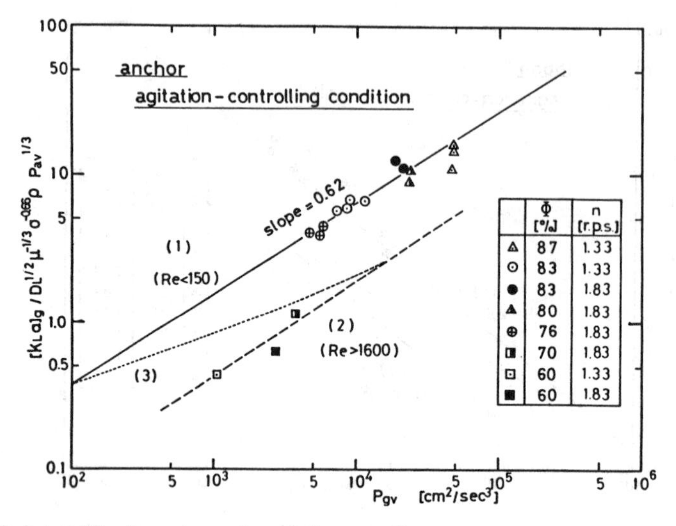

Figure 22. k_La and P_{gv} for anchor under agitation-controlling condition.

When using such a large impeller as a helical ribbon which extends over the entire mixing vessel, the vertical recirculation flow forced by the impeller action through the liquid volume results in longer contact times between gas bubbles and liquid in laminar mixing. The large tangential motion of liquid in turbulent mixing using a helical ribbon impeller may not produce sufficient shear to provide good contact between gas bubbles and the liquid.

The effects of various physical properties on k_La obtained for anchor-type impellers under the agitation-controlling condition for the laminar mixing range is as follows.

$$(k_La)_{g,A,L} = 4.23 \times 10^{-2}\mu^{-1/3}D_L^{1/2}\sigma^{-0.66}\varrho P_{av}^{1/3}P_{gv}^{0.62} \quad \text{for} \quad Re \leq 150 \tag{53}$$

or in dimensionless form

$$(k_LaD^2/D_L)_{g,A,L} = 15.5(\mu/\varrho D_L)^{1/2}(gD^2\varrho/\sigma)^{0.66}(dn^2/g)^{0.51} (d^2n\varrho/\mu)^{0.21}(u_g/\sqrt{gD})^{0.33} \tag{54}$$

As shown in Figure 22, the same tendency as helical ribbon is obtained for the anchor-type impeller.

The correction factor for an anchor is

$$A_A = (k_La)_{g,A}/(k_La)_{g,A,L} \tag{55}$$

This is shown by the following equations, and the values are compared with those for helical ribbons in Figure 21:

$$A_A = 1 \quad \text{for} \quad Re \leq 150 \tag{56}$$

$$A_A = 20.6\,Re^{-0.60} \quad \text{for} \quad 150 \leq Re \leq 1{,}600 \tag{57}$$

$$A_A = 1/4 \qquad \text{for} \quad Re \leq 1{,}600 \tag{58}$$

The dashed lines in Figures 20 and 21 (curve 3 in each figure) show the estimated values of $(k_La)_g$ from laminar to turbulent mixing conditions through the transitional range for a 70% sugar solution.

For the aerated mixing vessel with no aeration, the same equations as Equations 25 and 26 are obtained.

The exponents of each variable or dimensionless group obtained for helical ribbon and anchor-type impellers are given in Tables 3 and 4.

The estimated k_La values applying the two-region model for the intermediate condition using the following equation give good agreement with observed values as shown in Figure 23.

$$(k_LaD^2/D_L) = (k_LaD^2/D_L)_g + [P_{av}/(P_{gv}/N_p + P_{av})]\,(k_LaD^2/D_l)_a \tag{59}$$

The critical P_{gv} value, $(P_{gv})_{max}$, above which the agitation-controlling condition is obtained in the vessel agitated by a helical ribbon impeller is given by the following equation.

$$(P_{gv})^{0.78}_{max,R}[1 + (P_{gv})_{max,R}/N_pP_{av}] = 4.44P^{2/3}_{av} \tag{60}$$

The critical P_{gv} value, $(P_{gv})_{min}$, below which the bubbling-controlling regime prevails in aerated mixing vessels is

$$(P_{gv})^{0.78}_{min,R}[1 + (P_{gv})_{min,R}/N_pP_{av}] = 0.278P^{2/3}_{av} \tag{61}$$

Figure 24 shows the change of agitation-controlling, bubbling-controlling, and intermediate condition regions with the Reynolds number under the laminar agitation condition for a helical ribbon impeller.

This figure implies a simple correlation of the form

$$k_La \propto u_g^{\alpha} n^{\beta} \tag{62}$$

cannot be applied over a wide range of operating conditions.

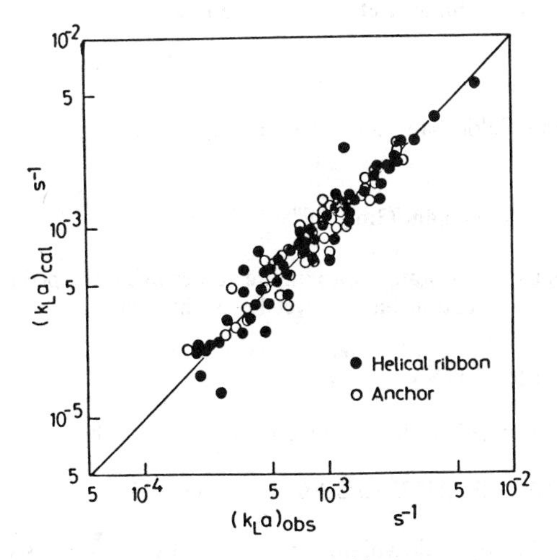

Figure 23. Comparison of observed values with estimated k_La.

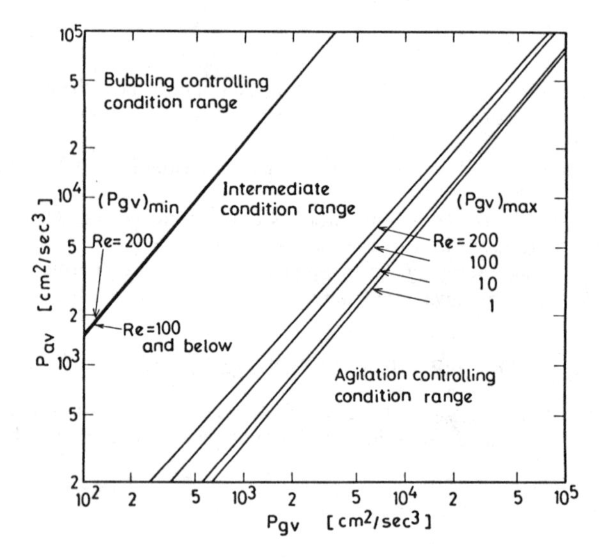

Figure 24. $(P_{gv})_{max}$ and $(P_{gv})_{min}$ and P_{av} for helical ribbon.

A comparison of performances when liquid viscosity varies in gas absorption of turbine, helical ribbon, and anchor-type impellers is shown in Figures 25 and 26.

These comparisons show that turbine-type impellers are good for low viscosity liquids, whereas helical ribbon or anchor-types are good for high viscosity liquids. These plots also show that helical ribbons give larger k_La values than anchor-types when the impeller speed is fixed, although anchors give larger k_La values when compared at constant agitation power per unit mass of liquid under the laminar condition.

The average shear rate in the mixing vessel agitated by a helical ribbon or anchor-type impeller is required as non-Newtonian liquids are often agitated by these impellers.

Nagata et al. [43] and Nishikawa et al. [44] recommend

$$(\gamma_{av})_{g,R} = 30n \tag{63}$$

for helical ribbons, and Calderbank and Moo-Young [36] give the following equation for an anchor:

$$(\gamma_{av})_{g,A} = [9.5 + 9s^2/(s^2 - 1)]n[4m/(3m + 1)]^{m/(1-m)} \tag{64}$$

Equation 43 can be used to obtain approximate values using the prior average shear rate, although experimental verification has not yet been obtained.

SCALE-UP CONSIDERATIONS

The concentration of oxygen transferred from gas to liquid obeys the following equation:

$$\varrho V(k_La)\Delta C = Au_g(P_{in} - P_{out})\,[273/(273 + \theta)]\,(1/22{,}400)\,(M) \tag{65}$$

where V, A, P, θ, and M are liquid volume in the vessel, vessel cross-sectional area, partial pressure of oxygen, temperature of liquid and molecular weight of oxygen, respectively.

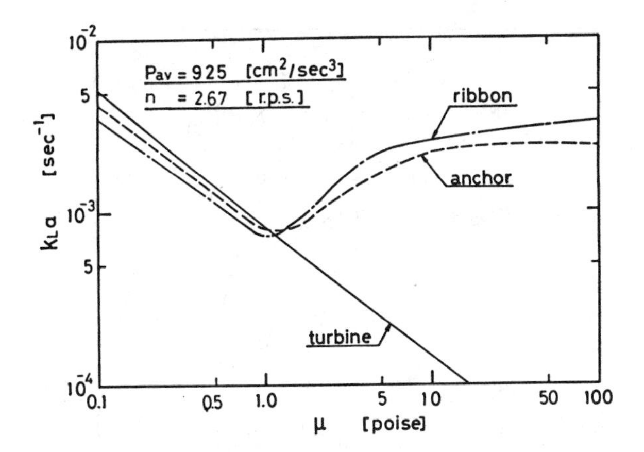

Figure 25. k_La and viscosity for various impellers.

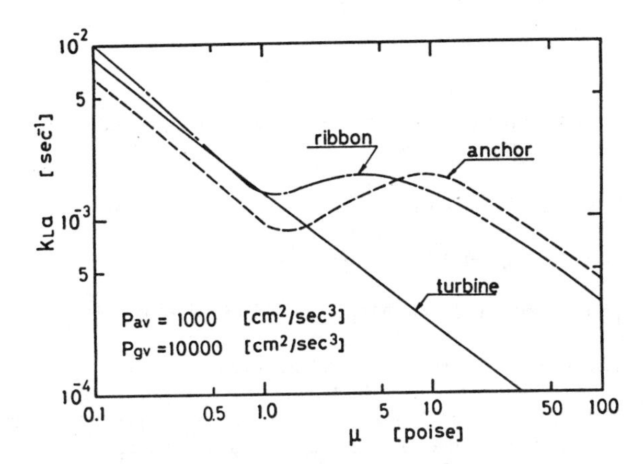

Figure 26. k_La and viscosity for various impellers.

The driving force ΔC is given by the following equation assuming plug-flow-type mixing for the gas phase and perfect mixing for the liquid.

$$\Delta C = (C_{I,in} - C_{I,out})/\ln[(C_{in} - C)/(C_{out} - C)] \quad (66)$$

here C_I is the oxygen concentration at the interfacial liquid film on the gas bubbles.

Assuming Henry's law, ΔC is represented by using the partial pressure of oxygen as

$$\Delta C = (K/H')(P_{in} - P_{out})/\ln[(P_{in} - P)/(P_{out} - P)] \quad (67)$$

where H' and K are the Henry constant and the conversion constant from the volume fraction to weight fraction, respectively.

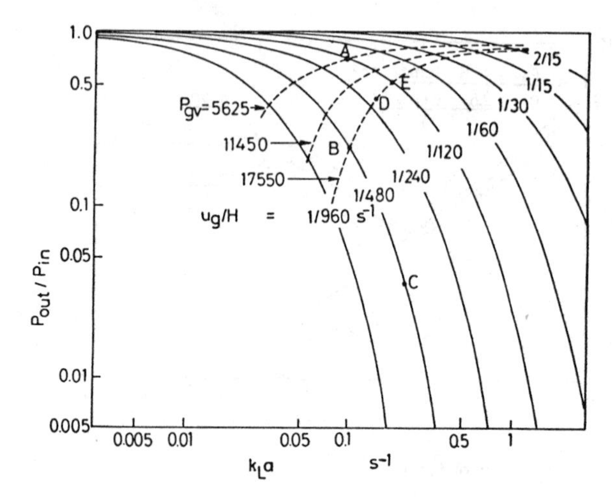

Figure 27. Ratio of gas concentration at outlet to that at entrance.

If rapid consumption of the absorbed gas component in liquid takes place, such as in sulphite oxidation, ΔC can be given by

$$\Delta C = (K/H')(P_{in} - P_{out})/\ln(P_{in}/P_{out}) \tag{68}$$

Substitution of Equation 68 into Equation 65 gives

$$P_{out}/P_{in} = \exp\{-(Hk_La/u_g)(22{,}400\varrho/M)[(273+\theta)/273](K/H')\} \tag{69}$$

where H is the liquid level in the vessel with no aeration.

The ratio of gas concentration at the outlet to that at the entrance can be estimated from Equation 69 and is shown in Figure 27 for the absorption of oxygen into water. The solid lines in Figure 27 show the change of the P_{out}/P_{in} ratio against k_La with u_g/H as a parameter; and the broken lines with P_{gv} as the parameter.

Although the correlations provide large values of k_La for a large P_{gv} at a certain aeration rate, the actual k_La value is limited because the gas concentration at the outlet decreases with increasing k_La as shown in Figure 27. Accordingly, in considering the optimal conditions for gas absorption using aerated mixing vessels, the relative intensity of aeration and agitation should be carefully examined.

Consider the fourfold scale-up of an aerated mixing vessel, keeping geometric similarity. The initial condition is shown by point A in Figure 27.

When both P_{av} and P_{gv} are kept constant in scale-up, i.e., k_La is kept constant; P_{out}/P_{in} in the large-scale vessel is shown by the point B in Figure 27.

The ratio of the gas component absorbed per unit volume of liquid in the large-scale mixing vessel to that in the small-scale vessel is

$$w_2/w_1 = [1-(P_{out}/P_{in})_2](u_{g2}/H_2)/[1-(P_{out}/P_{in})_1] \tag{70}$$

where subscripts 1 and 2 indicate small-scale and large-scale, respectively.

For the effective scale-up, the ratio of w_2/w_1 should be equal to or larger than unity.

$$w_2/w_1 \geqq 1 \tag{71}$$

Consequently, it is insufficient to use only k_La as the scale-up standard for aerated mixing vessels because a constant k_La results in w_2/w_1 values of less than unity (refer to Table 5). In such cases the shortage of the gas component absorbed into liquid should be conpensated by changing the operating conditions to absorb more gas. Three ways to absorb more of the gas component at scale-up are compared in the following:

1. To increase agitation speed to obtain larger k_La and a smaller P_{out}/P_{in} ratio than in the case shown by point B, maintain the ratio of w_2/w_1 as unity (point C in Figure 27).
2. To change the gas velocity to obtain a larger k_La value and a larger P_{out}/P_{in} ratio than in the case shown by point B, keep the ratio of w_2/w_1 as unity (point D in Figure 27).
3. To change the gas velocity keep the VVM ratio (volume of gas per volume of liquid in the vessel per minute) constant. This is often the case in fermentation operations. (Point E in Figure 27.)

A comparison of these three methods is shown in Table 5 for a twofold scale-up of an aerated mixing vessel with a turbine-type impeller. Equations 13 and 70 were used in these estimates. Proper scale-up by maintaining the VVM ratio constant is achieved when mechanical agitation is vigorous, because the ratio of w_2/w_1 is slightly larger than unity as shown in cases A–4 and B–4 in Figure 27. However, too much gas is consumed when the bubbling-controlling condition controls the phenomena in an aerated mixing vessel as in

Table 5
Scale-Up of Aerated Mixing Vessel

Run*	$(k_La)_2/(k_La)_1$	$(P_{av})_2/(P_{av})_1$	$(P_{gv})_2/(P_{gv})_1$	w_2/w_1
A–1	1.0	1.0	1.0	0.742
A–2	–	1.0	impossible	1.0
A–3	1.18	1.38	1.0	1.0
A–4	1.33	2.0	1.0	1.19
B–1	1.0	1.0	1.0	0.847
B–2	1.29	1.0	1.46	1.0
B–3	1.10	1.23	1.0	1.0
B–4	1.44	2.0	1.0	1.29
C–1	1.0	1.0	1.0	0.965
C–2	1.08	1.0	1.19	1.0
C–3	1.04	1.07	1.0	1.0
C–4	1.63	2.0	1.0	1.46
D–1	1.0	1.0	–	0.874
D–2	–	–	–	–
D–3	1.175	1.175	–	1.0
D–4	2.0	2.0	–	1.86

* *From* D = 240 cm, d/D = 0.5, $k_La = 0.2\ s^{-1}$ *to* D = 480 cm.

Initial condition:

	P_{av}	P_{gv}
A	P_{av} = 980 cm^2/s^3,	P_{gv} = 17,550 cm^2/s^3
B	1,960	11,450
C	3,920	5,625
D	8,000	0

Way of scale-up:

1 Keep P_{av} and P_{gv} constant (k_La constant)
2 Keep w_2/w_1 as unity, P_{gv} changed
3 Keep w_2/w_1 as unity, P_{av} changed
4 Keep VVM as constant

cases C–4 and D–4 in Figure 27. It should also be noted that a change in P_{gv} to obtain a larger k_La cannot always be applied because the condition shown by Equation 71 fails when the agitation is vigorous (see case A–2 in Figure 27). Consequently, the method involving a change of impeller speed is applicable only when the condition in an aerated mixing vessel is in the intermediate or bubbling-controlling regimes.

The method to change the superficial gas velocity keeping w_2/w_1 constant may be the most economical way, although the VVM method is more conservative.

The k_La values estimated by using Equation 13 are compared with the experimental values obtained by Fuchs et al. [18] for 10- to 51,000-liter vessels in Table 6 and Figures 28 and 29. It can be concluded from these comparisons that Equation 13 based on the two-region model is applicable to large-scale vessels since the experimental values of Fuchs et al. are properly represented by the estimation curve. Measured values of k_La for a 51,000-liter vessel are much smaller than the values for other vessels or the estimated values. As the measured values of k_La for a 51,000-liter vessel are so close to those obtained for the aerated tower shown in Figures 28 and 29, it is likely that they are erroneously small.

Nishikawa reported that Equation 27 provided good estimates of the gas absorption rates obtained in a 20,000-liter fermentation tank.

GAS-LIQUID SPOUTED VESSELS

The mixing vessel is often used for gas absorption, and various types of mixing techniques to obtain intimate contacting between liquid and gas bubbles. However, its application is limited to relatively low gas rates since the decrease of the average density around the impeller causes a lowering of the impeller power which is necessary to break and disperse gas bubbles throughout the vessel. In the case of the gas-liquid spouted vessel, power is supplied to the liquid via a pump and the gas is dispersed through a nozzle located on the bottom of the vessel. This phenomena is dominated by spouting of high-speed gas-liquid two-phase flow from a nozzle. This results in a choking-flow effect [45, 46, 47].

A schematic of a gas-liquid spouted vessel is shown in Figure 30. Liquid is recirculated from the overflow through a pump, and gas is introduced from a compressor or gas cylinder through the nozzle attached at the vessel bottom. A portion of the power supply is separated from a part of bubble breaking phenomenon, and liquid is concentrated at the nozzle forming a high-speed jet. Liquid energy from the pump is effectively used to break up gas bubbles in the nozzle and disperse them at the nozzle outlet into the turbulent spouted section where vigorous mixing takes place.

The effects of liquid flow rate, gas flow rate, and nozzle diameter on gas holdup are shown in Figures 31 and 32. It is clear from these figures that the gas holdup increases rapidly with increasing liquid flow rate and that it exceeds twice the gas holdup observed in an aerated tower using a single-nozzle sparger with the same gas flow rate. The gas holdup in an aerated

Table 6
Application of Correlative Equation for Large-Scale Vessel (P_{gv} = 1 watt/liter)

u_g	k_La Observed by Fuchs et al. in h^{-1} 10ℓ vessel	200ℓ	550ℓ	3,000ℓ	51,000ℓ	Observed by Nishikawa 20ℓ vessel	k_La Estimated Using Equation 13 P_{gv} = 1 w/ℓ	0 w/ℓ
0 m/h	6	50	8	7	2	–	–	0
14	120	140*	240	210	–	155	163	18
30	190*	200	280	230	60*	210	220	37
60	–	300	370	250	120	302	297	75
100	–	400	440*	300	170	390	393	125
150	–	490	490*	–	200	475	486	188

* *Indicates values obtained from extrapolation*

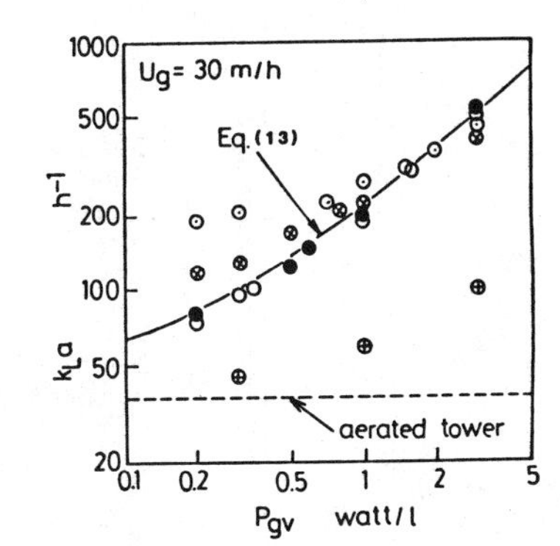

Figure 28. Comparison of k_La estimated using Equation 13 with observed values by Fuchs et al.

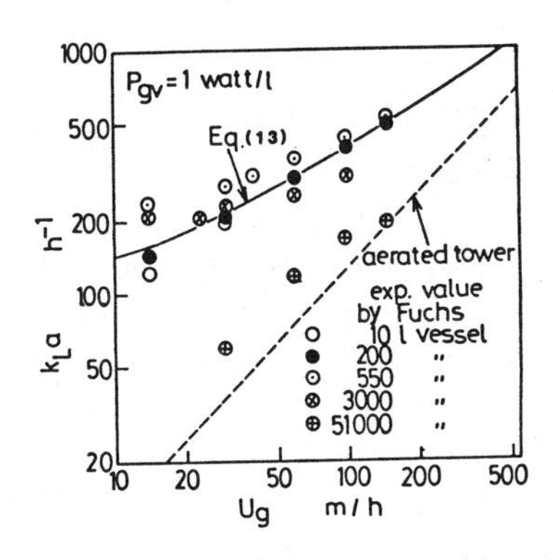

Figure 29. Comparison of k_La estimated using Equation 13 with observed values by Fuchs et al.

tower using a single-nozzle sparger can be derived from these figures because the gas-liquid spouted vessel with zero liquid flow rate corresponds to an aerated tower using a single-nozzle sparger.

Rapid increase of the gas holdup with increasing cocurrent liquid flow rate is one of the characteristics observed in the gas-liquid spouted vessel. On contrast, with a cocurrent aerated tower, gas holdup decreases slightly with increasing liquid flow rate as reported by Kato [48] and Unno [49]. The solid lines in Figure 32 are approximated by the following equations given by Nishikawa et al. [46].

$$H_g = H_{g0} \qquad \text{for} \quad L \leqq L_c \tag{72}$$

$$H_g = H_{g0}(1 + K\log(L/L_c)) \qquad \text{for} \quad L \geqq L_c \tag{73}$$

where L and H_g represent the liquid flow rate and gas holdup in the vessel, respectively, and L_c and K are the critical liquid flow rate for gas spouting and the gas spouting coefficient.

For water in the temperature range of 10° to 60° C, the following equations are obtained.

$$H_{g0} = 1.56 \times 10^{-3} U_g^{0.8} \qquad \text{for} \quad 20 \leqq U_g \leqq 400 \tag{74}$$

$$U_{lc} = 18.1 U_g^{1/5} d_n^0 \tag{75}$$

$$KH_{g0} = 2.35 \times 10^{-3} U_g d_n^{-1} \qquad \text{for} \quad 0 \leqq U_g \leqq 170 \tag{76}$$

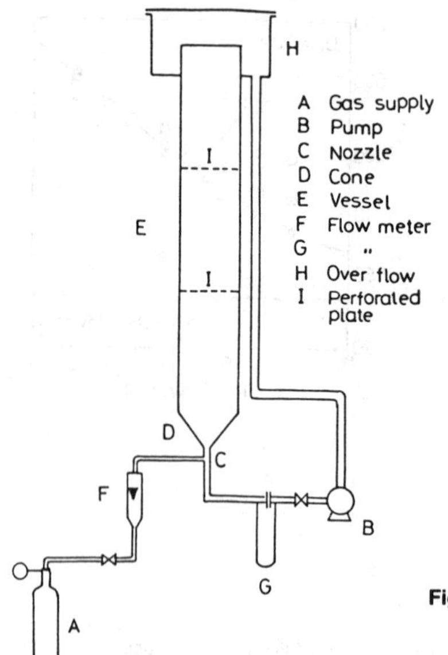

Figure 30. Schematic diagram of gas-liquid spouted vessel.

$$KH_{g0} = 0.4d_n^{-1} \qquad \text{for} \qquad U_g \geqq 170 \tag{77}$$

where U_l, U_g, and d_n are the superficial liquid velocity in m/h, superficial gas velocity, and nozzle diameter, respectively.

According to Govier [50] or Bergles [51], the limit of the bubbly flow regime is at about 0.25 in gas-volume flow-rate ratio and the two-phase flow condition in a pipe changes from bubbly flow to churn turbulent, annular, or slug flow with increasing gas-volume flow-rate ratio. But even when turbulent churn, annular, or slug flow is produced, the initial flow condition soon after the introduction of the gas is bubbly flow. Accordingly, even in the case of a high gas-volume flow-rate ratio, good dispersion of the gas bubbles is obtained in a gas-liquid spouted vessel using choking flow phenomenon.

When NaCl solutions or glycerin solutions are used as liquids, the gas hold-up increases greatly as shown in Figure 33.

Eguchi et al. [52] reported that a droplet population can be treated as one of uniform size by considering mass transfer from droplets in an extraction column if the dimensionless variance $(\sigma/d_{sm})^2$ is less than about 0.03.

The dimensionless variance measured for gas bubbles in the gas-liquid spouted vessel is about 0.01 as reported by Nishikawa et al. [53], and hence the bubble population in the spouted vessel can be treated as one uniform size of around 3 mm. It is interesting to note that the bubble size distribution in gas-liquid spouted vessels is well represented by a normal distribution as shown in Figure 34. The bubble distribution curve in an aerated tower, however, is roughly represented by a logarithmic normal distribution (also shown in Figure 34). The bubble size standard deviation in a gas-liquid spouted vessel is given in the following equation, reported for 6–20-cm diameter columns:

$$\sigma = 0.06(d_n/D)^{0.18}\,[G/(G+L)]^{-2}\,(L/L_c)^{-0.24}[G/(G+L)]^{-2} \tag{78}$$

where G is the gas flow rate in liters per minute.

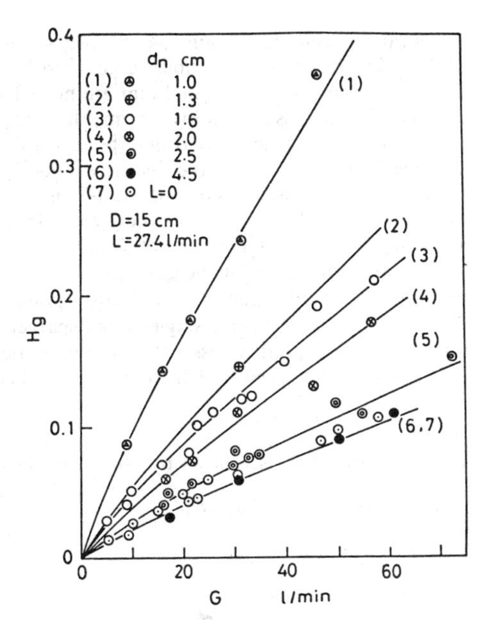

Figure 31. Gas hold-up in a spouted vessel.

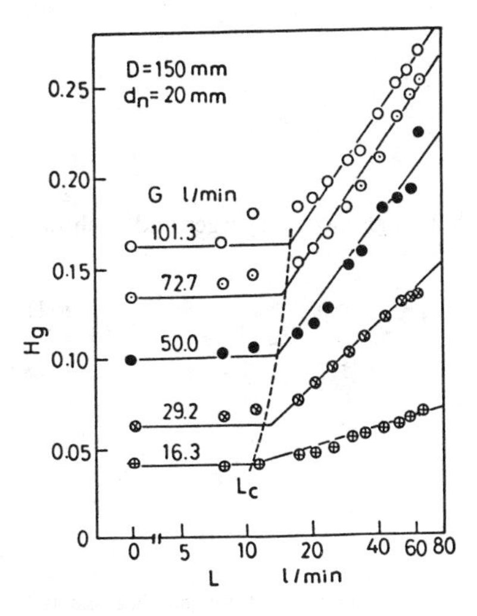

Figure 32. Gas hold-up in a spouted vessel.

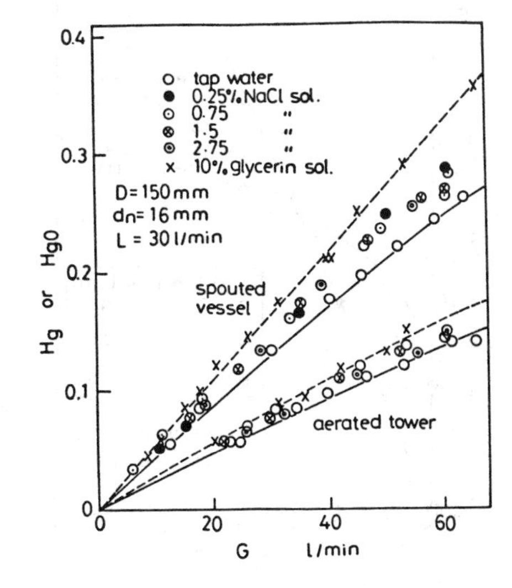

Figure 33. Gas hold-up in a spouted vessel with salt solution.

As can be seen from Figure 34, bubbles smaller than 1 mm account for more than 15% of the voidage, with some bubbles larger than 1 cm in the aerated tower using a single-nozzle sparger. Bubbles smaller than 1 mm are not effective in gas absorption, because they stay dispersed in the vessel for a longer time than needed for complete absorption. Accordingly, these bubbles are inactive and may be called dead bubbles. Bubbles larger than 1 cm are also ineffective because of the fast ascending velocity. Bubbles of 2 to 5 mm in diameter are described by Calderbank and Moo-Young (54) as being highly effective in gas absorption. These bubbles account for more than 95% of the total, and no bubbles smaller than 1 mm nor larger than 1 cm are observed at all in the gas-liquid spouted vessel.

k_La values obtained for the gas-liquid spouted vessel are plotted against the gas flow rate in Figure 35. In this figure, Curve 7, corresponding to no liquid flow rate, gives a value k_La for the aerated tower with a single-nozzle sparger. The data show k_La is 10 to 20 times larger for a gas-liquid spouted vessel than in an aerated tower.

It has been observed that the capacity coefficient per liquid hold-up in the vessel is:

$$k_La' = k_La/(1-H_g) \tag{79}$$

which is correlated as follows by Nishikawa et al. [46]:

$$k_La' = (k_La')_0 - 78.7u_gu_l^2d_n^{-4} \tag{80}$$

$$\text{For} \quad u_g = 0-10 \text{ cm/s}$$

$$u_l = 0-6 \text{ cm/s}$$

$$d_n = 1.0-4.5 \text{ cm}$$

where $(k_La')_0$ is the value for the aerated tower with a single-nozzle sparger and is shown as follows for water:

$$(k_La')_0 = 35.3u_g \tag{81}$$

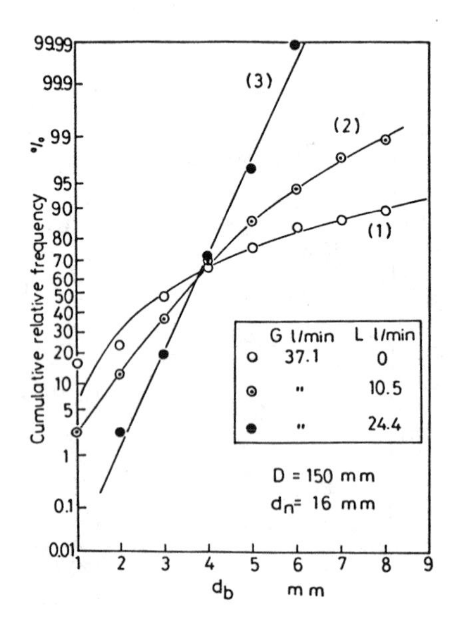

Figure 34. Bubble size distribution in a spouted vessel and aerated tower.

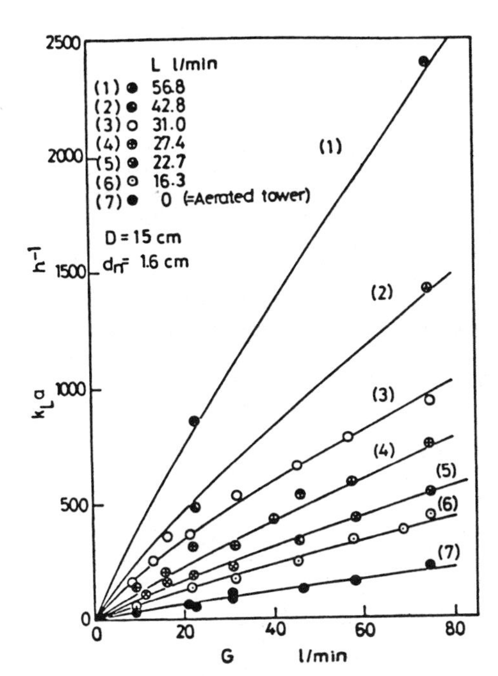

Figure 35. k_La obtained in a gas-liquid spouted vessel.

Change of k_La in the gas-liquid spouted vessel with addition of solid particles is shown in Figure 36. As shown k_La increases significantly the addition of 4.87-mm glass beads, although it decreases with the addition of 1.01- or 2.59-mm particles. These tendencies can be explained by the particle movement through the vessel.

In the case of small particles, the gas-liquid two-phase jet spouts through the particle bed without much of an effect on the movement of the solid particles. That is, the point resulting in the choking-flow phenomena moves to the part in the solid-particle bed and the effective nozzle diameter becomes larger than the actual nozzle size. In addition, the particle bed wall is too soft to keep the friction from spouting. Consequently, not enough pressure is available for the choking flow phenomenon to occur. As a result, k_La for a solid-gas-liquid spouted vessel composed of small particles becomes smaller than that obtained for the gas-liquid spouted vessel. In this case, the movement of solid particles in the spouted bed is not active.

With large particles, the two-phase jet is spouted from the nozzle and the movement of solid particles are active. Moreover, the secondary movement of solid particles appears for a certain fraction of the solids. That is, the part of a three-phase fluidized bed appears over a portion of the three-phase spouted bed in a solid-gas-liquid spouted vessel. The secondary breakage of gas bubbles in that part of a three-phase fluidized bed with large solid particles has the effect of increasing k_La.

The effect of particle size on k_La is shown in Figure 37 for solid-gas-liquid spouted vessels. This shows that particle size should exceed 3 mm in order to obtain large k_La values than those of gas-liquid spouted vessels. In the case of 2- to 3-mm particles, the secondary breakage of gas bubbles is less effective than in the case of larger particles. As the effect of the gas or liquid flow rate on k_La of the solid-gas-liquid spouted vessel is similar to that

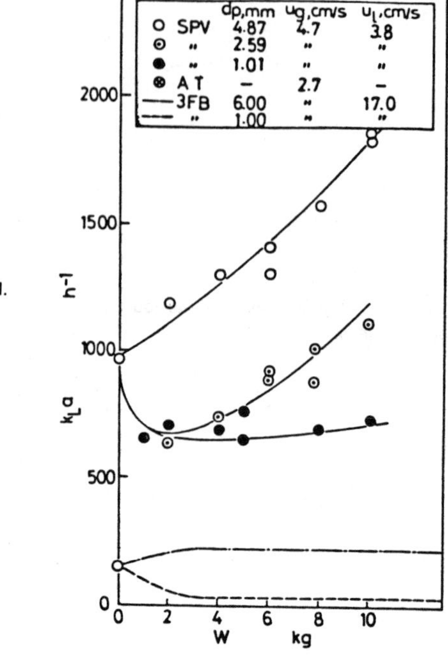

Figure 36. k_La in a solid-gas-liquid spouted vessel.

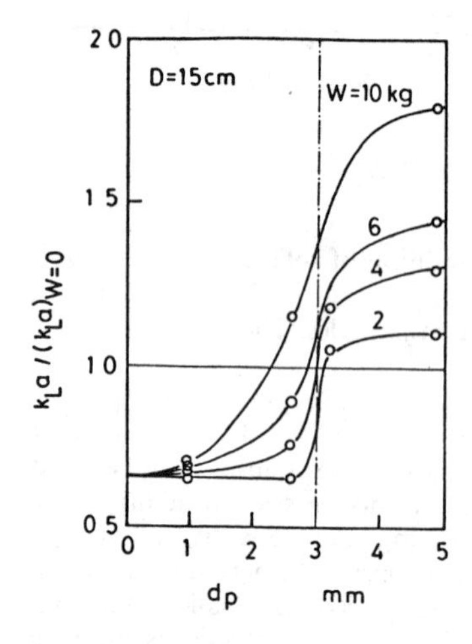

Figure 37. Effect of particle size on k_La.

of the gas-liquid spouted vessel, k_La for the solid-gas-liquid spouted vessel can be estimated by the following equation.

$$(k_La)_{\text{solid-gas-liquid spouted vessel}} = f \times (k_La)_{\text{gas-liquid spouted vessel}} \tag{82}$$

where f is a correction factor and is given in Figure 37.

Ostergaard et al. [55] reported that k_La in the three-phase fluidized bed including 6-mm glass beads was almost twice that observed in the aerated tower, although the liquid superficial velocity was as high as 17 cm/s. It was also reported that k_La in the three-phase fluidized bed decreased to one-fourth or one-fifth of that in the aerated tower when 1-mm

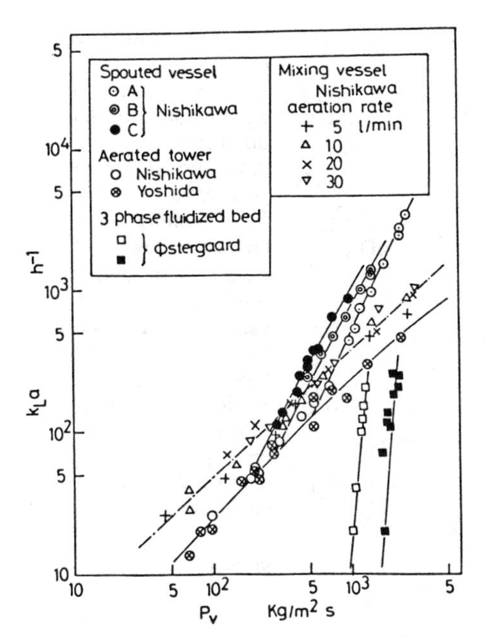

Figure 38. Comparison of k_La for various gas-liquid contactors.

glass beads were used. In the case of the three-phase fluidized bed considerable liquid energy should be used to fluidize solid particles, and the liquid void volume effective for gas absorption decreases due to the large amount of solid particles.

Heuss et al. [56] reports that k_La in vertical or horizontal pipes containing gas-liquid two-phase flow was lower than in an aerated tower despite a high liquid flow rate. In this case, the introduction of a line mixer or static mixer can be effective for gas absorption. Figure 38 shows a plot of k_La versus the power consumption per unit volume of reactor for various gas contactors. The power consumption for a spouted vessel is calculated by multiplying the total flow rate of the two-phase flow and pressure drop at the vessel including friction at the nozzle.

In calculating the power consumption for the aerated tower, only head losses in the tower are considered and friction at the sparging plate is ignored.

In calculating the power for the three-phase fluidized bed, head losses in the bed are assumed to be 1.5 times the loss of the aerated tower considering solid particles, and other friction losses are ignored.

These comparisons show that the gas-liquid spouted vessel is superior to other contactors because of the high capacity coefficient obtained at relatively low power consumption. It is also observed that the efficiency of the aerated mixing vessel is rather good at low gas flow rates, although it cannot be used at high gas flow rates due to loading effects. The three-phase fluidized bed gives the lowest efficiency of power on gas absorption among the four contactors even when 6-mm glass beads are used.

NOTATION

A cross-sectional area (cm)

C oxygen concentration in liquid. (g/cm^3)

C_I C at interfacial liquid film (g/cm^3)

d impeller diameter (cm)

D vessel diameter (cm)

D_L liquid-phase diffusivity (cm^2/s)
dn^2/g Froude number (−)
$d^2n\varrho/\mu$ agitational Reynolds number (−)
g gravitational acceleration (cm/s^2)
$gD^2\varrho/\sigma$ Bond number (−)
$gD^3\varrho^2/\mu^2$ Galileo number (−)
k fluid consistency index (g/s^{2-m})
k_La capacity coefficient based on plug-flow-type mixing (s^{-1})
k_La' capacity coefficient based on perfect mixing (s^{-1})
k_LaD^2/D_L Sherwood number (−)
m flow behavior index (−)
n impeller speed (s^{-1})
n_B number of baffles (−)
n_p number of impeller blades (−)
N impeller speed (min^{-1})
$N_p(= Pg_c/\varrho n^3d^5)$ power number (−)
nd/u_g aeration number (−)
P partial pressure of absorbed gas (atom)
P_{av} aeration power per unit mass of liquid in aerated mixing vessel (cm^2/s^3)
P_{gV} agitation power per unit mass of liquid (cm^2/s^3)
t time (s)
u_g superficial gas velocity (cm/s)
$u_g/\sqrt{gD}$ Froude number (−)
V liquid volume in vessel (cm^3)

Greek symbols

α correction factor for elasticity (−)
γ_{av} average shear rate in vessel (cm/s)
λ_n Deborah number (−)
μ liquid viscosity (g/cm · s)
μ_a apparent viscosity (g/cm · s)
$\mu/\varrho D_L$ Schmidt number (−)
$\mu u_g/\sigma$ capillary number (−)
ϱ specific weight of liquid (g/cm^3)
σ surface tension (g/cm^2)
τ shear stress ($g/cm \cdot s^2$)
φ correction factor for surface aeration (−)

Subscript

a bubbling-controlling condition
g agitation controlling condition
p paddle-type impeller
t turbine-type impeller
r ribbon-type impeller

REFERENCES

1. Nishikawa, M., et al., "Gas Absorption in Aerated Mixing Vessels," *J. Chem. Eng. Japan,* Vol. 14, No. 3 (1981), pp. 219–226.
2. Nishikawa, M., et al., "Gas Absorption in Aerated Mixing Vessels with Non-Newtonian Liquid," *J. Chem. Eng. Japan,* Vol. 14, No. 3 (1981), pp. 227–232.
3. Nishikawa, M., Nishioka, S., and Fujieda, S., "Gas Absorption in Aerated Mixing Vessel with Viscous Liquid Agitated by Helical Ribbon or Anchor," *Kagaku Kogaku Ronbunshu,* Vol. 9, No. 1 (1983), pp. 76–81.

4. Cooper, C. E., Fernstrom, G. A., and Miller, S. A., "Performance of Gas-Liquid Contactors," *Ind. Eng. Chem..*, Vol. 36, No. 6 (1944), pp. 504–509.
5. Maxson, W. D., and Johnson, M. J., "Aeration Studies On Propagation of Baker's Yeast," *Ind. Eng. Chem.*, Vol. 45, No. 11 (1953), pp. 2554–2560.
6. Oldshue, J. Y., "Role of Turbine Impellers in Aeration of Activated Sludge," *Ind. Eng. Chem.*, 48, No. 12 (1956), pp. 2194–2198.
7. Rushton, J. H., Gallagher, J. B., and Oldshue, J. Y., "Gas-Liquid Contacting with Multiple Mixing Turbines," *Chem. Eng. Progr.*, Vol. 52, No. 8 (1956), pp. 319–323.
8. Johnson, D. L., et al., "Effects of Bubbling and Stirring on Mass-Transfer Coefficients in Liquids, "*AIChE Journal*, Vol. 3, No. 3 (1957), pp. 411–417.
9. Friedman, A. M., and Lightfoot, H. N., "Oxygen Absorption in Agitated Tanks, "*Ind. Eng. Chem.*, Vol. 49, No. 8 (1957), pp. 1227–1230.
10. Snyder, J. R., Hagerty, D. F., and Molstad, M. C., "Operation and Performance of Bench-Scale Reactors," *Ind. Eng. Chem.*, Vol. 49, No. 4 (1957), pp. 689–695.
11. Karwat, H., "Distribution of Gasses in Liquids by Means of Stirrers," *Chem. Ing. Tech.*, Vol. 31, No. 10 (1959), pp. 588–597.
12. Yoshida, F., et al., "Oxygen Absorption Rates in Stirred Gas-Liquid Contactors," *Ind. Eng. Chem.*, Vol. 52, No. 5 (1960), pp. 435–438.
13. Hyman, D., and van den Bogaerde, J. M., "Gas-Liquid Contacting Small Bench-Scale Stirred Reactors," *Ind. Eng. Chem.*, Vol. 52, No. 9 (1960), pp. 751–753.
14. Westerterp, K. R., van Dierendonck, L. L., and Vendervos, D., *Chem. React. Eng. Symp.*, Brussels, (1968), pp. 205.
15. Hortacsu, O., M. S. thesis, Oklahoma State Univ., Stillwater, Oklahoma, 1965.
16. Yagi, H., and Yoshida, F., "Gas Absorption by Newtonian and Non-Newtonian Fluids in Sparged Agitated Vessels," *Ind. Eng. Chem., Process Des. Dev.*, Vol. 14, No. 4 (1975), pp. 488–493.
17. van Dierendonck, L. L., Fortuin, J. M. H., and Vandervos, D., "The Specific Contact Area in Gas-Liquid Reactors," *Chem. React. Symp.*, Brussels, (1968), pp. 205–215.
18. Fuchs, R., Ryu, D. D. Y., and Humphrey, A. E., "Effect of Surface Aeration on Scale-up Procedures for Fermentation Process," *Ind. Eng. Chem., Process Des. Dev.*, Vol. 10, No. 2 (1971), pp. 190–195.
19. Calderbank, P. H., "Physical Rate Process in Industrial Fermentation," *Trans. Instn. Chem. Eng.*, Vol. 37, No. 3 (1959), pp. 171–178.
20. Miller, D. N., "Scale-Up of Agitated Vessels Gas-Liquid Mass Transfer," *AIChE Journal*, Vol. 20, No. 3 (1974), pp. 445–453.
21. Miura, Y., Ph. D. thesis, Kyoto Univ., Kyoto, Japan, 1961.
22. Yasunishi, S., Ph. D. thesis, Kyoto Univ., Kyoto, Japan, 1977.
23. Hanhart, J., Westerterp, K. R., and Kramers, H., "The Residence Time Distribution of the Gas in an Agitated Gas-Liquid Contactor," *Chem. Eng. Sci.*, Vol. 18, No. 8 (1963), pp. 503–509.
24. Ohyama, Y., and Endoh, K., "Power Characteristics of Gas-Liquid Contacting Mixers," *Kagaku Kogaku*, Vol. 19, No. 1 (1955), pp. 2–8.
25. Michel, B. J., and Miller, S. A., "Power Requirement of Gas-Liquid Agitated Systems," *AIChE Journal*, Vol. 8, No. 2 (1962), pp. 262–266.
26. Nagata, S., et al., "Power Requirement of Turbine Impellers in Gas-Liquid Mixing Vessels," *Kagaku Kogaku*, Vol. 31, No. 10 (1967), pp. 86–89.
27. Chen. C. C., Lu, W. M., and Chen, P. S., "Heat Transfer in an Agitated Vessel with Aeration," *J. Chinese Inst. Chem. Eng.*, Vol. 3, No. 3 (1972), pp. 7–13.
28. Nagata, S., et al., "Study of Heat Transfer for Aerated Mixing Vessel and Aeration Tower," *Kagaku Kogaku Ronbunshu*, Vol. 1, No. 5 (1975), pp. 460–465.
29. Nishikawa, M., et al., "Heat Transfer to Non-Newtonian Liquid in Aerated Mixing Vessel," *Kagaku Kogaku Ronbunshu*, Vol. 8, No. 4 (1982), pp. 494–499.
30. Nienow, A. W., Widom, D. J., and Middleton, J. C., "The Effect of Scale and Geometry on Flooding, Recirculation, and Power," 2nd European Conf. on Mixing, Cambridge, England, F-1, 1977.

31. Sideman, S., Hortacsu, O., and Fulton, J. W., "Mass Transfer in Gas-Liquid Contacting Systems," *Ind. Eng. Chem.*, Vol. 58, No. 7 (1966), pp. 32–47.
32. Akita, K., Ph. D. thesis, Kyoto Univ., Kyoto, Japan, 1972.
33. Nakano, S., M. Eng. thesis, Kyoto Univ., Kyoto, Japan, 1972.
34. Perez, J. F., and Sandall, O. C., "Gas Absorption by Non-Newtonian Fluids in Agitated Vessels," *AIChE Journal,* Vol. 20, No. 4 (1974), pp. 770–775.
35. Metzner, A. B., and Otto, R. E., "Agitation of Non-Newtonian Fluids," *AIChE Journal,* Vol. 3, No. 1 (1957), pp. 3–10.
36. Calderbank, P. H., and Moo-Young, M. B., "Prediction of Power Consumption in the Agitation of Non-Newtonian Fluid," *Trans. Instn. Chem. Eng.*, Vol. 37, No. 1 (1959), pp. 26–33.
37. Nagata, S., Nishikawa, M., and Tada, H., "Power Consumption of Mixing Impellers in Pseudoplastic Liquids," *J. Chem. Eng. Japan,* Vol. 4, No. 1 (1971), pp. 72–76.
38. Nishikawa, M., Nakamura, M., and Hashimoto, K., "Heat Transfer in Aerated Tower Filled with Non-Newtonian Liquid," *Ind. Eng. Chem. Process. Des. Dev.*, Vol. 16, No. 1 (1977), pp. 133–137.
39. Hattori, K., Yokoo, S., and Imada, O., "Oxygen Transfer in a Highly Viscous Solution," *J. Ferment. Tech.*, Vol. 50, No. 10 (1972), pp. 737–741.
40. Street, J. R., "The Rheology of Phase Growth in Elastic Liquids," *Trans. Soc. Rheol.*, Vol. 12, No. 1 (1968), pp. 103–131.
41. Zona, E., and Leal, G., "Dissolution of a Stationary Gas Bubble in a Quiescent Viscoelastic Liquid," *Ind. Eng. Chem. Fundam.*, Vol. 14, No. 3 (1975), pp. 175–182.
42. Prest, W. M., Prter, R. S., and O'Reilly, J. M., "Non-Newtonian Flow and the Steady-State Shear Compliance," *J. App. Polym. Sci.*, Vol. 14, No. 11 (1970), pp. 2697–2706.
43. Nagata, S., et al., "Power Consumption of Helical Mixer for Mixing of Highly Viscous Liquid," *Kagaku Kogaku,* Vol. 34, No. 10 (1970), pp. 1115–1117.
44. Nishikawa, M., "Heat Transfer of Highly Viscous Non-Newtonian Liquid," *Kagaku Kogaku,* Vol. 42, No. 8 (1978), pp. 420–425.
45. Nishikawa, M., et al., "Studies on Gas Hold-Up in Gas-Liquid Spouted Vessel," *J. Chem. Eng. Japan,* Vol. 9, No. 3 (1976), pp. 214–219.
46. Nishikawa, M., Kosaka, K., and Hashimoto, K., "Gas Absorption in Gas-Liquid or Solid-Gas-Liquid Spouted Vessel," 2nd PACHEC, Denver, Colorado, 1977.
47. Nishikawa, M., et al., "Gas-Absorption in the Multi-Stage Gas-Liquid Spouted Vessel," 3rd PACHEC, Seoul, Korea, 1983.
48. Kato, Y., "Gas Hold-up Average Velocity of Gas Bubbles in Gas Bubbles Column," *Kagaku Kogaku,* Vol. 26, No. 10 (1962), pp. 1068–1075.
49. Inoue, I., and Unno, H., "Concentration Fluctuation in a Cocurrent Bubble Column," *Kagaku Kogaku,* Vol. 36, No. 1 (1972), pp. 65–71.
50. Govier, G. W., Radford, B. A., and Dunn, J. S. C., "The Upward Vertical Flow of Air-Water Mixtures," *Can. J. Chem. Eng.*, Vol. 35, No. 1 (1957), pp. 58–70.
51. Bergles, A. B., and Suo, M., Proc. 1966 Heat Transfer and Fluid Mech. Instn., 1966, pp. 79.
52. Eguchi, W., Shiota, S., and Mori, M., "Effect of the Diameter Distribution in a Continuous Cocurrent Extraction Column," *Kagaku Kogaku,* Vol. 34, No. 8 (1970), pp. 755–762.
53. Nishikawa, M., et al., "Studies of Bubble Size Distribution in Gas-Liquid Spouted Vessel," *J. Chem. Eng. Japan,* Vol. 11, No. 1 (1978), pp. 73–75.
54. Calderbank, P. H., and Moo-Young, M. B., "The Continuous Phase Heat and Mass Transfer Properties of Dispersions," *Chem. Eng. Sci.*, Vol. 16, No. 1 (1961), pp. 39–54.
55. Ostergaard, K., and Suchozebrski, W., 4th European Symp. Chem. React. Eng., Pergamon Press, Oxford, England, 1969.
56. Heuss. J. M., King, C. J., and Wilke, C. R., "Gas-Liquid Mass Transfer in Cocurrent Froth Flow," *AIChE Journal,* Vol. 11, No. 5 (1965), pp. 866–872.

CHAPTER 23

AXIAL DISPERSION AND HEAT TRANSFER IN GAS-LIQUID BUBBLE COLUMNS

B. H. Chen

Department of Chemical Engineering
Technical University of Nova Scotia
Halifax, Nova Scotia, Canada

CONTENTS

INTRODUCTION

The bubble column (Figure 1) is a simple and relatively economical means of achieving excellent contacting between a gas and a liquid. The gas is dispersed as bubbles and rises through a deep pool of liquid, creating a large interfacial area of contact. There is no agitation other than that brought about by the rising bubbles. The liquid is often stagnant (batch operation), but could also move cocurrently or countercurrently with respect to the gas flow (continuous operation).

The bubble column cannot handle the high gas throughputs attainable in packed columns. However, it has found increasing applications as a gas absorber with or without chemical or biological reaction, primarily because of its excellent characteristics:

1. Simple operation due to the absence of moving parts.
2. Low cost.
3. High interfacial area per unit volume.
4. High liquid-phase residence time.
5. Temperature uniformity within the column.

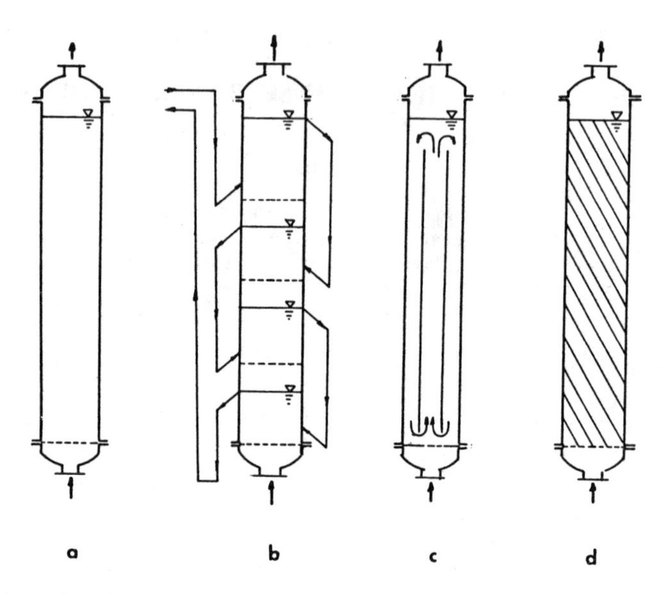

Figure 1. Types of gas-liquid bubble columns: (a) conventional; (b) staged; (c) packed; (d) loop-reactor.

The main disadvantages of bubble columns are:

1. Low gas-phase throughputs.
2. High pressure drop.
3. High degree of backmixing.

Some of these deficiencies can be minimized through design. Where a narrow liquid residence-time distribution is required, the column can be subdivided into stages (Figure 1b) or be packed with solid internals (Figure 1d). If more intense recirculation of the liquid phase is desired, a tube may be incorporated either inside or outside the column to form the so-called loop reactor (Figure 1c). A gas loading of up to 300 $m^3/m^2 \cdot h$ can generally be realized in all of these designs.

Spargers

A sparger is a device for introducing a stream of gas into a continuous liquid in the form of small bubbles. The purpose of sparging is to provide either an excellent interphase contact or simply an agitation of the liquid to various degrees.

Sieve trays or perforated plates are often used for this purpose although recent reports have demonstrated the advantages of using jet devices [1, 2]. Methods are available for sieve trays design [3]. Recently, Ruff et al. [4] suggested the following simple criterion for producing a uniform gas load on all the holes of a perforated plate:

$$We_0 \equiv \frac{u_0^2 d_0 \varrho_G}{\sigma} \geqq 2.0$$

provided that

$$Fr_0' \equiv \frac{u_0^2}{d_0 g}\left[\frac{\rho_G}{\rho_L - \rho_G}\right]^{5/4} \geq 0.37$$

Porous plates are the other means for dispersing gases, particularly in sewage and waste treatment tanks. Bubbles thus generated are generally small, with the actual sizes determined by the pore opening and the pressure drop imposed across the septum. Physical and operating characteristics of porous materials can be found from Perry [5].

Power Consumption

Bernoulli's equation can be used to estimate the agitation power, which is also responsible for the creation of the large interfacial area in bubble columns. Written for the gas between location a, just above the sparger orifice, and location b, at the liquid surface, the energy balance equation is

$$w = \frac{u_a^2}{2} + \frac{P_a}{\rho_{Ga}} \ln \frac{P_a}{P_b} + (z_a - z_b)g$$

where w is the work done by the gas on the vessel contents per unit mass of gas. To estimate the work for the gas compressor, the friction in the piping system leading to the gas distributor, the pressure drop across the orifice, and the compressor inefficiency need to be taken into account.

Lehrer [6] reported that much of the kinetic energy of the gas jet is dissipated as heat; the agitation power is approximately given by

$$w = \frac{c_1 u_a^2}{2} + \frac{P_a}{\rho_{Ga}} \ln \frac{P_a}{P_b}$$

with $c_1 \cong 0.5$ as suggested by Sherwood et al. [7].

Flow Regimes

The flow regime is an important consideration in design as it strongly affects the hydraulics as well as transport phenomena in bubble columns. For example, the column diameter has been reported to have a distinctively different effect on the axial dispersion coefficient depending upon the flow regime as shown below:

$D_L \propto D_t^2$ bubble-flow regime
$D_L \propto D_t^{1.5}$ heterogeneous-flow regime
$D_L \propto D_t$ slug-flow regime

The three flow regimes just mentioned can generally be observed as the gas flow rate increases. Figure 2 depicts the flow phenomenon as observed by Otake et al. [8].

- *Bubbly flow regime (Figure 2a)*. This, usually occurring at low gas flows, is characterized by the appearance of an almost ordered chain bubbling with very little visually observable agitation. Bubbles rise independently with fairly uniform spacing between them. For low-viscosity liquids, this flow pattern generally prevails below $u_G = 0.05$ m/s, although

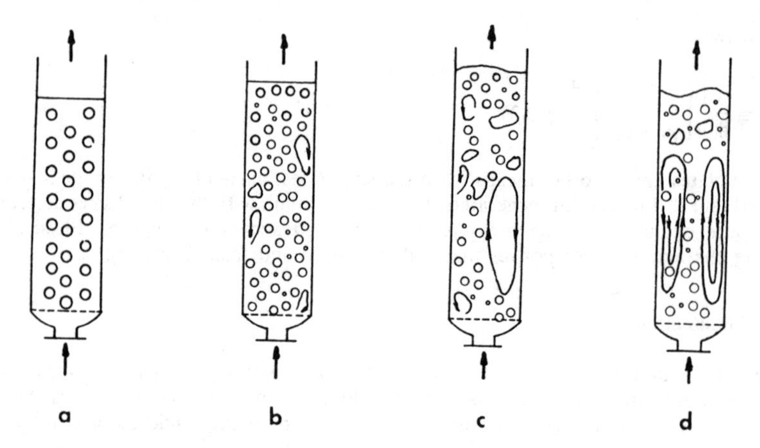

Figure 2. Observed flow patterns in bubble columns [8]: (a) bubble flow; (b,c) transitional flow; (d) churn-turbulent (heterogeneous) flow.

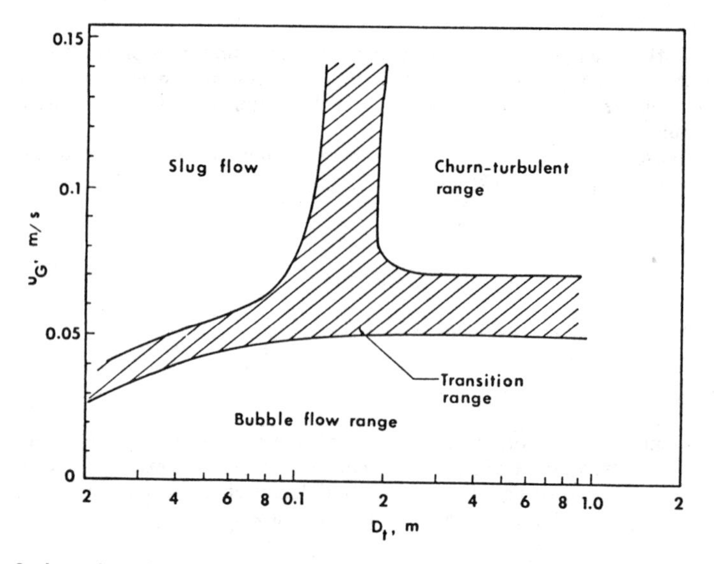

Figure 3. Approximate boundaries for flow regimes in gas-liquid bubble columns [14].

this value may change with the presence of surface-active agents or column geometry. The drift flux model has been shown to describe the flow in this regime [9, 10].

- *Churn turbulent, or heterogeneous regime.* This is characterized by the onset of significant bubble coalescence and local liquid circulation in the column when the gas flow rate is beyond 0.05 m/s. Thus, the bubbles are of sizes far greater than their stable size range between 0.003 and 0.008 m in diameter (Figures 2b and 2c). Close to the upper limit of the regime (Figures 2c and 2d) a symmetrical two-loop liquid circulation, upward in the center and downward near the wall, is well established with large eddies superimposed [8, 11].

- *Slug flow*. In small columns at high gas flow rates, the gas bubbles coalesce to form large slugs with a parabolic outline at the front. Bubble slugs have been observed in columns of diameter up to 0.15 m [12, 13].

The transition from one flow regime to the other is not well defined. An approximate estimate of the boundaries can be obtained from Figure 3 for water and dilute aqueous solutions [14]. Otake et al. [8] have also provided a criterion for the same purpose.

GAS HOLDUP

The gas holdup refers to the volume fraction of a gas-liquid mixture which is occupied by the gas. It is an important design and operating parameter frequently used to describe the hydrodynamics in bubble columns.

Methods of Measurement

Three methods are commonly employed to determine the average gas holdup, ε:

1. *Bed expansion method*. If Z and Z_0 are the aerated liquid height and the initial liquid height (unaerated), respectively, the average gas holdup ε is

$$\varepsilon = (Z - Z_0)/Z$$

This method is satisfactory for many gas-liquid systems where foaming is not a serious problem.

2. *Manometric method*. The average gas holdup is determined by measuring static pressures at several points along the column height, or

$$(1-\varepsilon) = (1/\varrho_L g)(-dP/dx)$$

This method neglects the pressure drop due to fluid friction, acceleration, and deceleration occurring within the bubble column. Cases for which the pressure drop becomes significant have been reported [11].

3. *Transient response with a step function input*. If a step change of tracer concentration from C_0 to 0 is effected at the liquid inlet, the response measured at the column exit will be in the form of an F-curve [15]. The area bound by the curve and the coordinates is related to the average gas holdup as follows:

$$(1-\varepsilon) = (u_L/Z) \int_0^\infty C/C_0 \, dt$$

where u_L = the liquid-phase superficial velocity
Z = the aerated height
t = the time

This method is obviously unsuitable for batch bubble columns and for gas-liquid systems in which a serious adsorption of the tracer takes place.

Table 1
Flow Regime and Flow Index, m_1 in Bubble Columns

Flow Index, n	Flow Regime	u_G, m/s
1.0	Bubble flow	≤ 0.05
0.75	Transitional flow	≤ 0.20
0.55	Churn turbulent	≤ 0.40
0	Blow-off	> 13.5

Experimental Results

General Observations

Three different types of parameters which may influence the gas holdup are

1. Column geometry including diameter and the nature of gas spargers.
2. Operating parameters such as pressure and flow rates.
3. Physical properties of both phases.

Data of Fair et al. [16] and Yoshida and Akita [17] showed a negligible effect of column diameter when the diameter is greater than 0.15 m. Results reported by Ellis and Jones [18] indicated that wall effects increase gas holdup at diameters up to 0.075 m, and then for diameters greater than 0.075 m, gas holdup is independent of diameter, while the data of Botton et al. [19] showed this to occur at $D_t = 0.25$ m. Thus, a critical column diameter should exist. The most accepted value of the critical diameter is 0.15 m.

Bubble behaviors associated with the use of different gas distributors have been investigated by a number of researchers. As a result, the nondependence of gas holdup on the type of distributor is established, provided that the size of the distributor openings is greater than about 0.5 ~ 1 mm [20].

Gas flow rates generally have the effect on gas holdup as shown below

$$\varepsilon \propto u_G^{m_1}$$

The value of m_1 depends on a number of factors. Flow regime, column diameter, and fluid properties may all contribute to the variation of m_1. For air-water systems and column diameter of 0.075 m, the approximate variation of m_1 is given in Table 1.

The literature is much less clear as to the effect of fluid properties such as viscosity andsurface tension on gas holdup. For pure liquids or dilute aqueous solutions of nonelectrolytes, a large majority of the investigators have reported a negligibly small effect of viscosity in the churn-turbulent flow regime, with the exception of Bach and Pilhofer [20] who reported a large effect in the range $0.7 \times 10^{-6} < \nu_L < 124 \times 10^{-6}$ m^2/s.

A small influence of surface tension on gas holdup has been consistently found [20, 21, 22]. In most cases, the gas holdup increases with decreasing surface tension in accordance with the following relation

$$\varepsilon \propto \sigma^{-m_2}$$

where m_2 varies between 0.1 and 0.2.

Few studies have considered the physico-chemical properties of gas. Bhaga et al. [23] and Koetsier et al. [24] concluded from their experimental investigations that the gas holdup increases with increasing gas density while Hikita et al. [22] determined quantitatively the small effect of both gas density and viscosity. Pressure was reported to have an insignificant effect on the gas holdup over the range of 0.1 mPa $< p <$ 1.7 mPa [25, 26].

Empirical Correlations

A large number of correlations are available in the literature. Recent reviews are given by Shah et al. [14], Hikita et al. [22], and Miller [13]. Only some of the generalized correlations, notably those of Hughmark [27], Akita and Yoshida [28], and Kumar [29] are presented here.

Hughmark [27] has measured the gas holdup over a wide range of column dimensions, system properties, and gas flow rates. His data along with those of earlier investigators [16, 17, 18] were found to correlate well with the following equation

$$\varepsilon = 1/[2+(0.35/u_G)(\varrho_L\sigma/72)^{1/3}]$$

Marshelkar [30] reported that this equation, with a slight revision, also correlated the data of Carleton et al. [31] and Reith et al. [32].

The correlation developed by Akita and Yoshida [28] is based on air with 13 different liquids and water with three gases other than air in a 0.1524-m ID column. Dimensional analysis forms the basis of the following correlation:

$$\varepsilon/(1-\varepsilon)^4 = k_1(gD_t\varrho_L/\sigma)^{1/8}(gD_t/\nu_L^2)^{1/12}(u_G/\sqrt{(gD_t)})$$

where $k_1 = 0.2$ for pure liquids and non-electrolytes
$k_1 = 0.25$ for electrolytes

Miller [13] has estimated the multiple correlation coefficient between the linear fit represented by the prior equation and the experimental data to be 0.838. In an extensive study of gas holdup for various gas-liquid systems, Hikita et al. [22] have reported that their data can be correlated with the equation of Akita and Yoshida with a mean deviation of only 11%. Using a semitheoretical approach, Mersmann [33] has obtained a correlation similar to that of Akita and Yoshida.

Another general correlation proposed by Kumar [29] is for air in liquids with surface tension ranging from 0.031 to 0.074 N/m, densities from 800 to 1110 kg/m^3, and viscosities of 0.0009 to 0.012 Pa · s. The correlation is of the form

$$\varepsilon = 0.728\,U - 0.485\,U^2 + 0.0975\,U^3$$

where $U = u_G[\rho_L^2/\sigma(\rho_L-\rho_G)g]^{1/4}$

Although its correlation coefficient is low at 0.724, it is in good agreement with the data of Bach and Pilhofer [20].

Figure 4 shows a comparison of these three empirical correlations along with other more recent data. Table 2 gives details of experimental studies that support the correlations of Hughmark and Akita and Yoshida. It is clear that within the spread of the data, either Hughmark's or Akita and Yoshida's equation gives equally good estimates of gas holdup for less viscous and coalescing liquids. For liquid electrolytes, Akita and Yoshida's correlation is preferred. If more accurate estimates are desired, specific correlations are recommended.

Recent reports by Kelkar et al. [34], Bach and Pilhofer [20], and Schugerl et al. [35] showed that the dilute aqueous solutions of alcohols should be treated separately. The

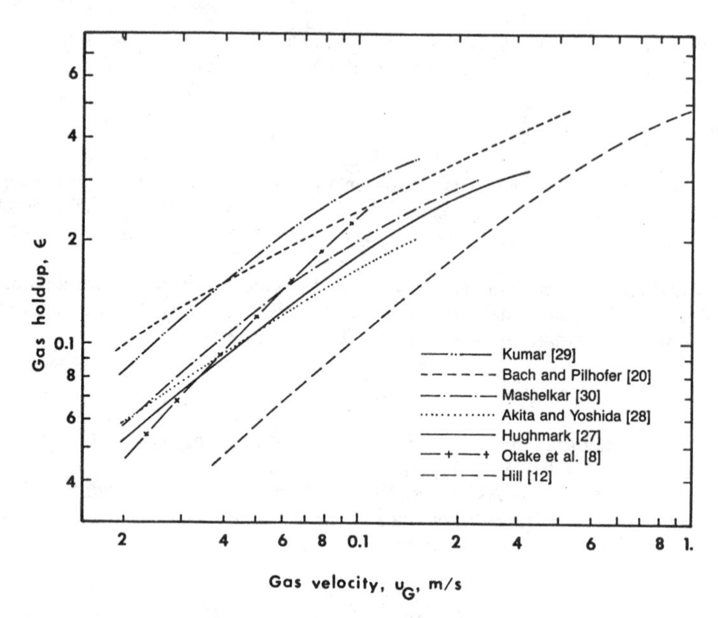

Figure 4. Comparison of empirical correlations of gas holdup in bubble columns.

Table 2
Experimental Details of Gas Holdup Studies Supporting Hughmark's and Akita and Yoshida's Correlations

Reference	System	Physical Properties	D_t (m)	Sparger	u_G (m/s)	u_L (m/s)
Akita and Yoshida [28]	Air-water Air-glycol Air-methanol Air-aq. glycol Air-aq. methanol Water-O_2, H_2, CO_2	$0.79 < \rho_L < 1.59$ g/m^3 $0.58 < \mu_L < 21.1$ CP $22.3 < \sigma < 74.2 \frac{dyne}{cm}$	0.15, 0.60	Single-nozzle $d_0 = .05$ cm	0.0053~ 0.42	0
Hikita and Kikukawa [21]	Air-water Air-aq. methanol Air-aq. sucrose	$0.91 < \rho_L < 1.24$ g/m^3 $0.7 < \mu_L < 13.8$ CP $37.5 < \sigma < 74.8 \frac{dyne}{cm}$	0.1 0.19	Single-nozzle $d_0 = 0.9$ cm $d_0 = 3.62$ cm	0.07 0.338	0
Mersmann [33]	Gas-liquid	Semi-theoretical				
Hughmark [27]	Air-water Air-kerosene Air-light oil Air-aq. glycerol solution Air-aq. $ZnCl_2$ solution	$0.78 < \rho_L < 1.7$ g/m^3 $0.9 < \mu_L < 152$ CP $25 < \sigma < 76 \frac{dyne}{cm}$	> 10 cm	Multi-orifice	0.004~ 0.45	0
Hikita et al. [22]	Air-water Air-organic liquids Air-electrolyte solutions Water-H_2, CO_2, CH_4, C_3H_8, N_2	$0.79 < \rho_L < 1.170$ g/m^3 $0.09 < \mu_L < 17.8$ CP $22.9 < \sigma < 79.6 \frac{dyne}{cm}$ $0.084 \times 10^{-3} < \rho_G < 1.84 \times 10^{-3}$ g/cm^3	0.1	Single-nozzle $d_0 = 1.1$ cm	0.04~ 0.45	0
Otake et al. [11]	Air-water Air-3% ethanol Air-10% ethanol	$0.88 < \rho_L < 1$ CP $44.5 < \sigma < 72 \frac{dyne}{cm}$	0.05	Multi-nozzle, Single-nozzle	7×10^{-3}~ 8.24×10^{-2}	0 ~ 0.15

characteristics of these solutions are fairly representative of those found in bubble-column bioreactors. Kelkar et al. [34] reported the following correlation for low liquid flow rates:

$$\varepsilon = 0.75\, u_G^{0.5} C_N^{0.2}$$

where C_N is the number of carbon atoms in the straight chain of the aliphatic alcohols used.

AXIAL DISPERSION

Dispersion may be defined as the spreading of marked fluid particles as a result of the departure from ideal flow; the fluid particles move forward in the direction of net flow, but at different speeds, thus resulting in the distribution of residence times. Dispersion occurring in a direction opposing the flow, such as that occurring in bubble columns, is commonly known as backmixing.

Axial dispersion always lowers the driving force for mass or heat transfer, and the performance of many contacting devices including bubble columns is adversely affected. A rational design of bubble columns as a chemical reactor or as a simple absorber would therefore require an assessment of the effect of axial dispersion in both the liquid and gas phases. Shah et al. [14] in their review have listed some industrially important gas-liquid reactions which may be carried out in bubble columns.

Liquid-Phase Axial Dispersion

As pointed out previously, the intense mixing in the liquid phase can be clearly observed when a bubble column is in operation Figure 2. Mechanisms responsible for the overall mixing are:

- Turbulent eddies in the main stream as well as those introduced by the stochastic movement of the dispersed phase relative to the continuous phase. This type of mixing has a component in both axial and radial directions.
- A Taylor-type axial dispersion due to the overall circulation of the liquid phase combined with the radial mixing because of mechanism [1].
- Liquid entrainment in the wakes of the bubbles combined with mass exchange between these wakes and the liquid phase.
- Molecular diffusion.

Few studies [36] have dealt with the analysis of the possible contribution from each individual mechanism. Instead, simplified models are often assumed to represent the overall mixing phenomena.

Axial Dispersion Model.

This model draws on the analogy between mixing and molecular diffusion. It assumes that the mixing in bubble columns is of statistical nature and therefore can be represented by a Fick-type diffusion equation. For continuous bubble columns with a uniform intensity of backmixing in the liquid phase, the axial dispersion model is mathematically described by:

$$\partial c/\partial t = D_L \partial^2 c/\partial x^2 - u_L/(1-\varepsilon)\, \partial c/\partial x \tag{1}$$

where c = the concentration of a conserved solute
D_L = the axial dispersion coefficient

u_L = the liquid-phase superficial velocity
ε = the fractional gas holdup

Equation 1 states that the variation of the solute concentration with time is the result of a convective transport with an average velocity $u_L/(1-\varepsilon)$ and a diffusion transport characterized by the axial dispersion coefficient, D_L. It has made two assumptions:

- A negligible radial dispersion.
- Plug flow with a constant velocity.

This model is widely accepted because it is simple, it has successfully correlated experimental data, and it has provided a reliable means to estimate the effect of axial dispersion on the performance of bubble columns and other interfacial contacting devices either as chemical reactors or physical absorbers [37, 38, 39]. Other models are available in the literature [40].

For bubble columns with small or no liquid flows (batch operation), Equation 1 reduces to

$$\partial c/\partial t = D_L\, \partial^2 c/\partial x^2 \tag{2}$$

Determination of axial dispersion coefficients: The determination of D_L requires a solution of either Equations 1 or 2 with appropriate initial and boundary conditions. It involves comparisons of the solution with experimental concentration distributions of a tracer in the liquid phase to infer values of D_L.

The experimental curves are obtained by performing a series of transient-response experiments in which a tracer is injected at the feed inlet and its concentration is monitored as a function of time at one or more downstream positions. Basic requirements for a satisfactory tracer experiment are outlined by Shah [40]. The concentration detection system is obviously dictated by the nature of the tracer used.

The injection of tracer may be in the form of a step, impulse, or sinusoidal function. For bubble columns of finite length, the analytical solution given by Hoffman [41] and Brenner [42] for a step function input of tracer is applicable. Field and Davidson [43] obtained a solution for a pulse input. The solutions relate the dimensionless time T $(= t/\theta)$ and Peclet number, $Pe_L\ \{= (u_L Z)/[(1-\varepsilon)D_L]\}$. c_0 is the concentration if the tracer is evenly dispersed throughout the column. Accurate numerical values were provided by Brenner [42] which are shown graphically in Figure 5. A comparison with the experimental data of Chen and McMillan [44] is indicated.

When a pulse input is used, the method of moments is frequently preferred. Details of this method and others are discussed elsewhere [45].

For batch bubble columns, Equation 2 has been solved and the solution is [46]

$$c/c_0 = 1 + 2 \sum_{n=1}^{\infty} [(\cos n\pi x/Z) \exp(-n^2\pi^2 D_L t/Z^2)] \tag{3}$$

for a pulse input. The boundary conditions used are

$$\partial c/\partial x = 0, \quad x = 0, Z$$

Figure 6 shows a comparison of Equation 3 with the experimental data of Ohki and Inoue [47].

Some investigators [32, 48] have determined D_L from steady-state measurements of axial concentrations. The tracer is introduced from a flat continuous source and its concentrations

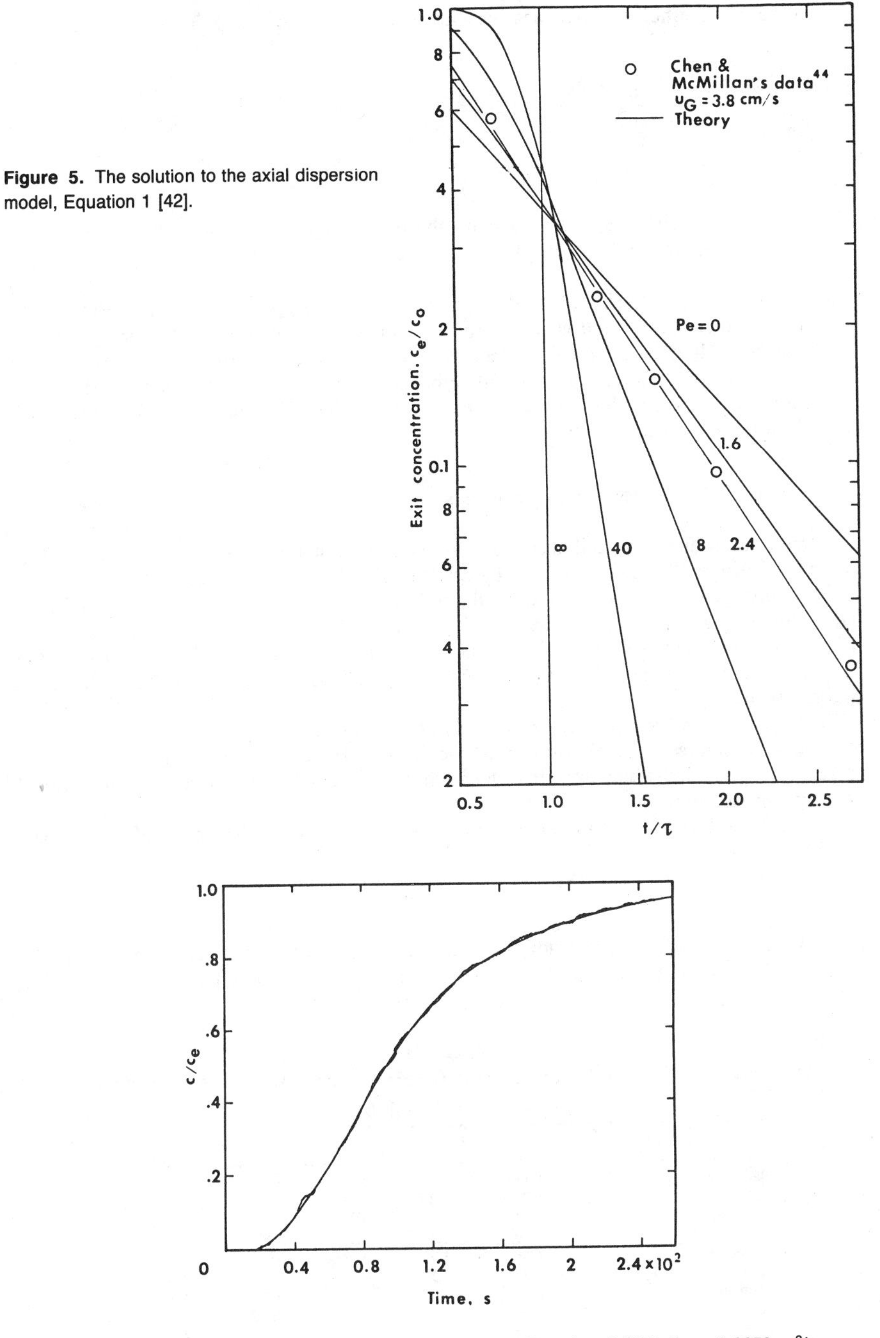

Figure 5. The solution to the axial dispersion model, Equation 1 [42].

Figure 6. A comparison of experimental curve with theory, Equation 3 [47]: D_L = 0.0059 m²/s; u_G = 0.0572 m/s; D_t = 0.04 m.

are measured at different axial locations upstream of the source. A mass balance of the tracer gives

$$u_L/(1-\varepsilon)\,\partial c/\partial x = D_L\,\partial^2 c/\partial x^2$$

Its solution is

$$\ln c/c_e = [-u_L/(1-\varepsilon)D_L]z$$

where z = the distance measured from the tracer source
c_e = the tracer concentration at the exit

A plot of $\ln c/c_e$ vs. z should yield a straight line with slope equal to $u_L/(1-\varepsilon)D_L$.

The unsteady-state method is simple to perform experimentally, but it is unsuitable for batch bubble columns. In addition, it permits only that fraction of the intensity of axial dispersion to be determined which results in backmixing. Nevertheless, this method is quite satisfactory for studies with bubble columns, because in bubble columns the axial dispersion is solely due to the backmixing.

Correlations of Axial Dispersion Coefficients

Numerous studies on axial dispersion in bubble columns have been reported. Data available through 1978 are presented by Shah et al. [49] in their review.

Generally speaking, experimental results indicate that the axial dispersion coefficient is independent of gas-phase properties and liquid-phase flow rate, but depends strongly on column diameter and gas throughput characterized by the superficial gas velocity. It is insensitive to the change in liquid-phase properties such as σ, ϱ, μ, and to the type of gas sparger at high gas flow rates.

There are no general correlations available in the literature which would allow a reasonable estimate of axial dispersion coefficients. However, there are at least three correlations proposed which appear to have limited success in correlating a fair amount of existing experimental data.

A number of investigators have reported their results in the form of power law [21, 50–55].

$$D_L = k_1 D_t^a u_G^b v_L^d \sigma^e \tag{4}$$

Table 3 lists values of the dimensional constant k_1 along with values for the exponents, a, b, d, and e from published studies. As pointed out by Riquarts [56], Equation 4 also

Table 3
Correlations of Axial Dispersion Coefficients in Power-Law Form
$(D_L = k_1 D_t^a u_G^b u^d \sigma^2)$

Reference	k_1	a	b	d	e
Badura et al. [53]	0.688	1.4	0.33	*	*
Baird and Rice [55]	0.35	1.33	0.33	*	*
Deckwer et al. [51]	0.678	1.4	0.3	*	*
Deckwer et al. [52]	2±0.15	1.5	0.5	*	*
Hikita et al. [22]	0.66	1.25	0.38	−0.12	0
Knickle et al. [78]	0.282	1.5	0.37	−0.21	*
Mangartz [54]	0.0155	1.5	0.33	−0.14	*
Riquarts [56]	0.16	1.5	0.375	−0.125	*
Towell and Ackerman [50]	1.22	1.5	0.5	*	*

* *Effects not investigated.*

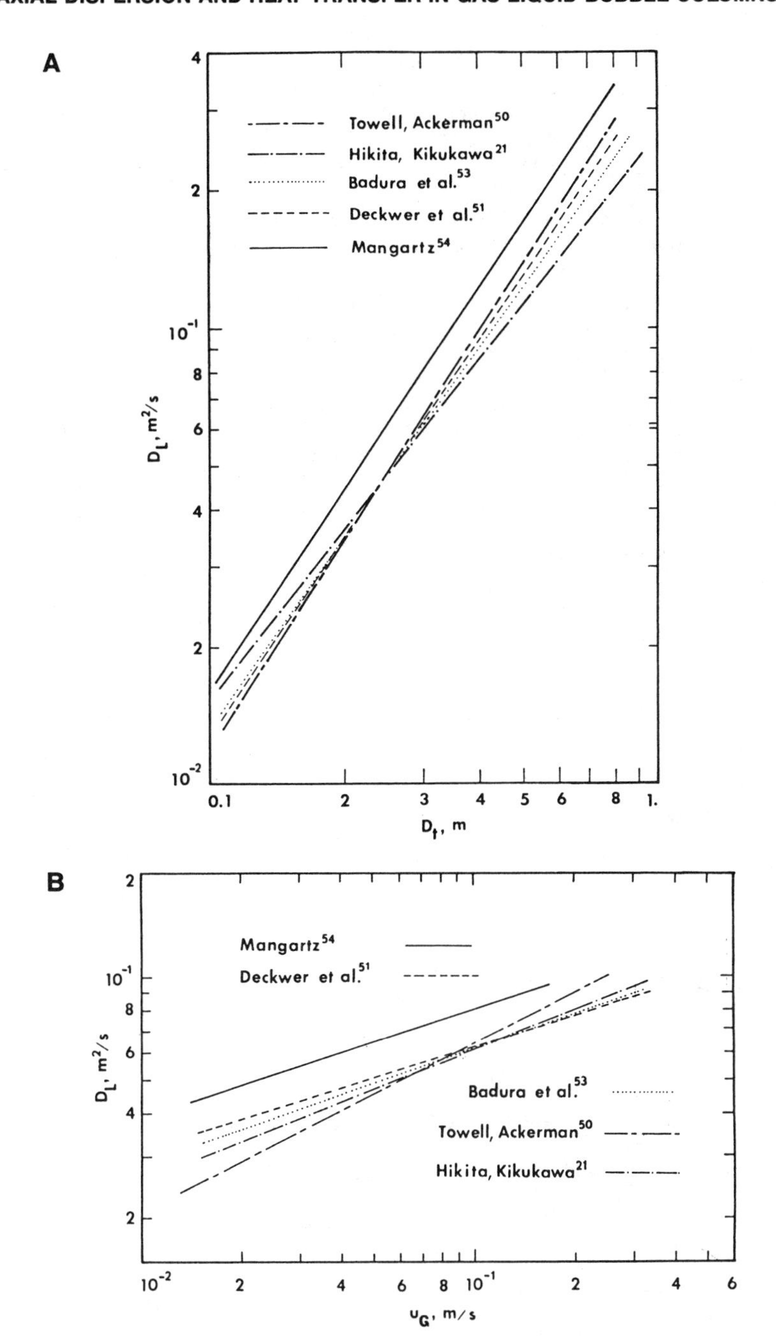

Figure 7. Correlation of liquid-phase axial dispersion coefficients in bubble columns according to Riquarts [56]: (a) effect of gas flow rate, D_t = 0.3 m, ν = 1 × 10^{-6} m²/s; (b) effect of column diameter, u_G = 0.1 m/s, ν = 1 × 10^{-6} m²/s.

correlates results from many other studies which are not given in the power-law form [32, 47, 57, 58]. According to Riquarts [56] this equation permits the calculation of dispersion coefficients with an accuracy of $\pm 30\%$ over the ranges of parameters: $.10 < D_t < 1.0$ m; $0 < u_G < 0.4$ m/s; $1 \times 10^{-3} < \mu_L < .019$ kg/m · s; $0.382 < \sigma < 0.0755$ N/m; and $0.911 \times 10^3 < \varrho_L < 1.233 \times 10^3$ kg/m^3. Figure 7 shows the graphical comparison of these results.

Equation 4 is inconvenient to use because it is a dimensional equation. However, it provides a reasonable estimate of D_L for systems with physical properties substantially different from that of water.

A more rational correlation was proposed by Joshi and Sharma [59] upon considering the circulation velocity within a bubble column. The importance of the effect of liquid circulation on axial dispersion has long been recognized, particularly at high gas flow rates. Freeman and Davidson [60], Rietema and Ottengraf [61], Hills [12], and Whalley and Davidson [62] have developed mathematical models for the purpose of estimating the circulation velocities in bubble columns. Based on a modified version of the energy balance method of Whalley and Davidson [62], Joshi and Sharma [59] have shown that the circulation velocity of the liquid phase in a bubble column can be estimated from the following equation:

$$V_c = 1.31[gD_t(u_G \mp \varepsilon u_L/(1-\varepsilon) - \varepsilon u_{b\infty})]^{1/3} \tag{5}$$

where the plus and minus signs indicate the countercurrent and cocurrent modes of operation, respectively.

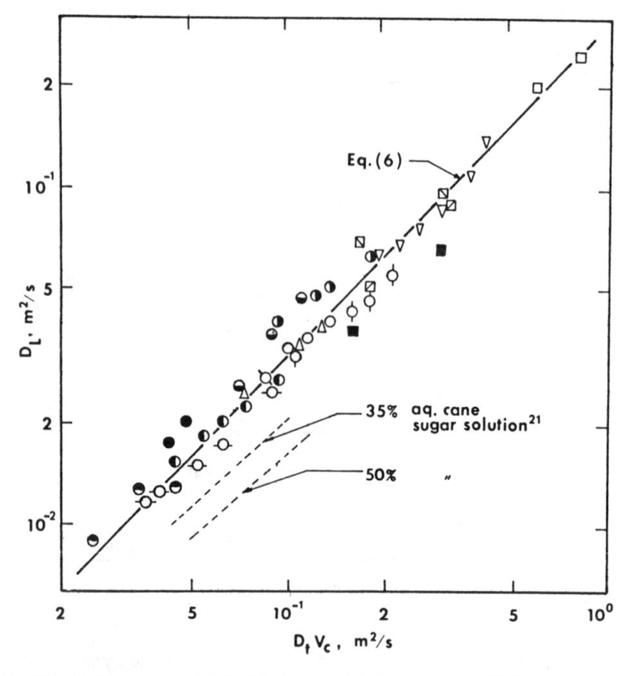

Figure 8. Correlation of experimental liquid-phase axial dispersion coefficients for air-water systems, according to Joshi and Sharma [59].

If this velocity is assumed to be the characteristic velocity of a bubble column, Equation 5 may then be combined with Prandtl's mixing length theory to give

$$D_L = k_2 D_t V_c$$

where k_2 is a proportionality constant, and D_t is the column diameter, which is considered as the characteristic dimension. Joshi [63] and Joshi and Sharma [59] showed that $k_2 = 0.33$. Hence, their final correlation is

$$D_L/D_t V_c = 0.33 \quad (6)$$

which is a dimensionally consistent equation. It has satisfactorily correlated many sets of experimental data for air-water systems. Figure 8 shows the comparison of Equation 6 with some available data [64–66]. The reported standard deviation is 7%. Experimental details of studies included in the comparison are given in Table 4.

Some limitations on Equation 6 should be noted:

1. The calculation of the circulation velocity requires the information on gas holdup which, as reported in previous sections, is difficult to predict with accuracy.
2. Equation 6 applies best when u_G and $\varepsilon u_{b\infty}$ are significantly different, and this occurs only in circulation-dominated bubble columns.
3. It is inadequate for systems other than air and water, as Figure 8 clearly shows.

A consideration based on specific energy consumption has also been suggested as a means to correlate dispersion data in bubble columns. This approach has proved useful in correlating heat and mass transfer data in agitated tanks. For isotropic turbulence, Hinze's theory [67] gives the relation

$$P_m = v'^3/\ell \quad (7a)$$

where P_m = the specific rate of energy dissipation
v' = the eddy fluctuating velocity
ℓ = the scale of turbulence

Table 4
Summary of Liquid-Phase Axial Dispersion Studies Supporting Joshi–Sharma's Correlation System Air-Water

Reference	D_t (m)	Sparger	u_G (m/s)	u_L (m/s)	Remark
Reith et al. [32]	0.14	Sieve plate	0.02~ 0.45	0~ 0.02	•
Aoyema et al. [64]	0.1, 0.2	Porous sieve plates	0.003~ 0.078	0.0018~ 0.0062	•
Vail et al. [65]	0.146	Sieve plate	0.01~ 0.1	0.017~ 0.031	
Kato and Nishiwaki [5]	0.214	Sieve plate	0.04~ 0.20	0.00515	
Towell and Ackerman [50]	0.406 1.067	Single tube, -point spider	0.017~ 0.268	0.003~ 0.015	•
Dekwer et al. [52]	0.2, 0.102	Porous plate	0.05~ 0.068	0.004~ 0.0074	
Deckwer et al. [51]	0.2	56 nozzles (d_0 = 1 mm)	0.05~ 0.12	0.00707	•
Hikita and Kikukawa [21]	0.1, 0.19	Single tube	0.04~ 0.34	0	
Ostergaard [66]	0.214	Wire screen	0.01~ 0.3	0.07 0.12	

• *Also supports Baird and Rice's correlation.*

Equation 7a may be combined with Prandtl's mixing length theory to give

$$D_L = P_m^{1/3}\ell^{4/3}$$

For bubble columns, $P_m = u_G g$, and ℓ is assumed proportional to the column diameter, therefore

$$D_L = k_3(u_G g)^{1/3}D_t^{4/3} \tag{7b}$$

A comparison of Equation 7b with some published data is shown in Figure 9. Baird and Rice [55] recommended $k_3 = 0.35$. The details of the data sources and experimental conditions are summarized in Table 5 [68].

Since some of the power supplied is used to create activities at the gas-liquid interface, Baird and Rice's analysis tends to overestimate the intensity of axial dispersion.

These three correlations cannot be extended without introducing additional uncertainties to gas-liquid systems that contain even a minute amount of surfactants or electrolytes, because such a contamination would seriously alter the bubble behavior in clean solutions. For this reason, specific reports should be consulted for information on dispersion in such contaminated systems.

Recent investigations [34, 48] with air and several dilute alcohol colutions have clearly demonstrated the inadequacy of these correlations. Kelkar et al. [34] reported that their data could only be correlated with the inclusion of gas holdup ε, a dependent variable, as follows:

$$D_L\varepsilon = 1.42D_t^{1.23}[u_G - \varepsilon u_L/(1-\varepsilon)]^{0.73}$$

As reported previously, the gas holdup is dependent upon the number of carbon atoms in the straight chain of an alcohol.

Gas-Phase Axial Dispersion

In bubble columns, the gas is sparged and rises in the form of bubbles. It is the variation in the bubble-rise velocity which causes axial dispersion in the gas phase. The following factors, which determine the bubble rise velocity, should contribute to the intensity of the dispersion in the gas phase:

1. Liquid circulation that causes the bubble to rise with a maximum velocity at the column axis and a minimum near the wall. Some bubbles may even move downward with the liquid.

Table 5
Liquid-Phase Dispersion Studies That Support Baird and Rice's Correlation

Reference	System	D_t (m)	Sparger Type	u_G (m/s)	u_L (m/s)
Argo and Cova [58]	N_2-water NH_3-water	0.10, 0.16 0.448	perforated plate	0.0233, 0.203	0.0038, 0.016
Ohki and Inoue [47]	Air-water	0.16	perforated plate	0.05, 0.25	0
Reith et al. [32]			See Table 4		
Rice et al. [68]	Air-water	0.082	porous plate	0.0038, 0.0135	0.005
Towell and Ackerman [50]			See Table 4		
Deckwer [51]			See Table 4		
Aogama [64]			See Table 4		

2. Frequent coalescence and breakup of bubbles that may alter the bubble-size distribution under turbulent flow conditions. This would become less significant a factor as the gas flow rate decreases.
3. Physical properties of both phases.
4. Gas sparger design, but this is only a factor at the lower limit of the bubbly regime.

The one-dimensional axial dispersion model as described previously for the liquid phase is also found appropriate for the gas phase. The method of determination of the gas-phase axial dispersion coefficient, D_G remains the same, except that in this case a much more sensitive monitoring system is needed, because the transient phenomenon would be over very quickly, perhaps in a few seconds.

Experimental measurements of D_G were reported notably by Kolbel et al. [69], Carleton et al. [31], Towell and Ackerman [50], Menshchikov and Aerov [70], and Field and Davidson [43]. Details of these investigations are summarized in Table 6 [71–73]. All except Mangartz and Pilhofer [73], who investigated the effect of liquid properties, dealt with air-water systems under atmospheric pressure.

Empirical correlations of D_G for specific cases are available in the literature. Recently, Joshi [74], in his review of published in formation, has developed a more rational method of correlation on the assumption that the gas phase moves through the column with a parabolic velocity profile. Following Taylor's approach to the laminar dispersion in circular tubes, Joshi showed that

$$D_G = k_4 u_G^2 D_t^2/\varepsilon \tag{8}$$

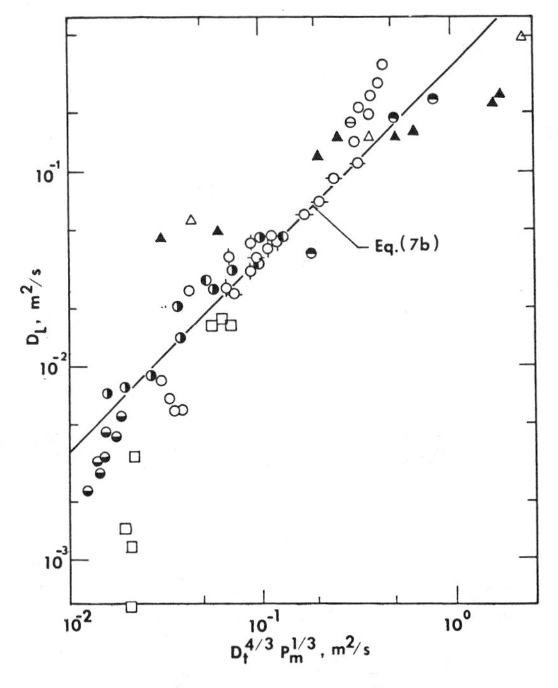

Figure 9. Baird and Rice's correlation of axial dispersion coefficients compared to experimental data [55].

Table 6
Summary of Studies on Gas-Phase Dispersion in Bubble Columns

Reference	D_t (m)	Sparger Type	u_G (m/s)	u_L (m/s)
Kolbel et al. [69]	0.092		0.04-0.07	0
Diboun and Schugerl [71]	0.135	Porous plate	0.01-0.055	0.0012-0.00104
Carleton et al. [31]	0.153, 0.305	Sieve plate	0.03-0.05	
Towell and Ackerman [50]	0.406,	Single-tube	0.0085~ 0.13	0.135
	1.067	α-point spider		0.0072
Menshchiko and Aerov [70]	0.3	Single-tube	0.00762, 0.096	-
Seher and Schumacher [72]	0.45, 1.0	Sieve plate	0.022	0.011
Manquartz and Pilhofer [73]	0.1, 0.14	-	0.01~ 0.13	0~ 0.06
Field and Davidson [43]	3.2	Not given	0.04-0.05	0.03-0.045

with k_4, an empirical constant, equal to 110. The comparison of Equation 8 with experimental data is shown in Figure 10. The standard deviation from the best fit is reported to be 26%. According to Joshi, Equation 8 is valid only when the arbitrarily defined Reynolds number, $R_e' \equiv (\sqrt{\varepsilon}\, D_t u_G \varrho_G/\varepsilon)/\mu_G$ is less than 4,000.

Field and Davidson [43] proposed a correlation of the form

$$D_G = 56.4 D_t^{1.33} (u_G/\varepsilon)^{3.56} \tag{9}$$

which, besides correlating their own data obtained in a column of industrial size, correlates the data of Towell and Ackerman [50], and Carleton et al. [31] with a mean deviation of 46%. The ranges covered are $0.0706 < D_t < 3.2$ m and $0.0085 < u_G$ 0.131 m/s. The liquid-phase flow rate was found to have no significant effect even when u_L and u_G are of comparable magnitude.

In view of the lack of data, Equation 9 is recommended for cases where the gas-phase Reynolds number, as defined by Joshi [66], exceeds 4,000.

HEAT TRANSFER

Bubble columns are often employed to carry out highly exothermic chemical reactions in the liquid phase. A list of such reactions was recently compiled by Shah et al [14]. Therefore, the removal of heat through the wall or an immersed coil may influence the reactor performance. Furthermore, it may be necessary to design a bubble column or its variation for heterogeneous chemical reactions that involves larger heat effects; a knowledge of heat transfer rates within the reactor is then required.

Effective Thermal Conductivity [75–77]

The rate of heat transfer within a batch or continuous-bubble column can be simply characterized by an effective thermal conductivity, k_e [44]. It is defined by

$$q = -k_e dT/dx$$

k_e can be determined experimentally by arbitrarily inducing an axial temperature gradient within the column. Data reported by Chen and McMillan [44] for an air-water bubble column of 0.051-m diameter show extermely large values of k_e, ranging approximately from 10,000 to 40,000 W/m · K over the ranges $0.001 < u_G < 0.04$ m/s, $17 < Z/D < 30$. These values may be compared with $k_e \cong 300$ W/m · K for copper to appreciate realistically the existence of a uniform temperature within the column. From the same study, Chen and

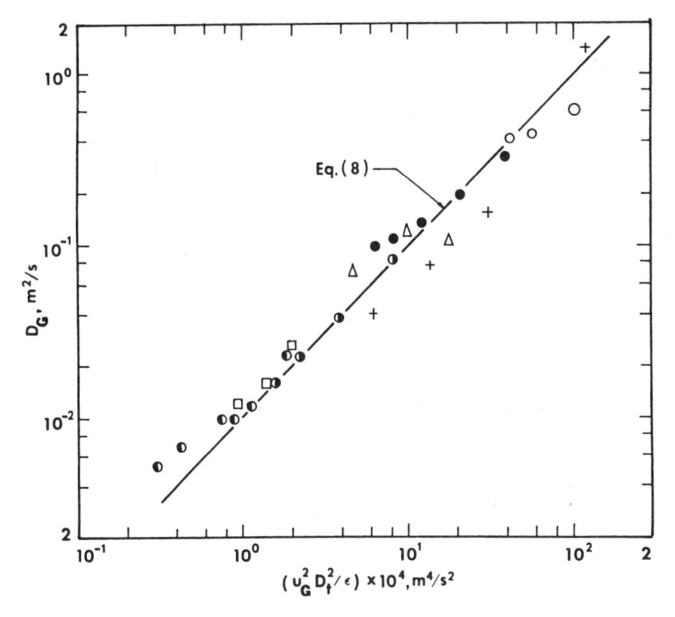

Figure 10. Correlation of gas-phase axial dispersion coefficients [74].

McMillan [44] concluded that the thermal diffusivity α ($= k_e/\varrho_L C_{pL}$) is essentially identical to the liquid-phase axial dispersion coefficient.

Wall Heat Transfer Coefficients

Numerous reports are available in the literature on heat transfer from the wall, inserted rods, or coils to the upward flowing gas-liquid dispersion in bubble columns [78–82]. The experimental results show a surprisingly high degree of consistency: the heat transfer coefficient is affected significantly only by liquid-phase properties and the gas flow rate up to 0.1 m/s, it is unaffected by other factors including gas-phase properties, column diameter, and the type of sparger.

Correlations of experimental data on the wall heat transfer coefficient, h, are invariably presented in a form first suggested by Kast [79] who obtained

$$St = f(ReFrPr^2)$$

where
- St = Stanton number, $h/\varrho_L C_p u_G$
- Re = Reynolds number, $d_B u_G \varrho_L/\mu_L$
- Fr = Froude number, $u_G^2/g d_B$
- Pr = Prandtl number, $(C_p\mu/k)_L$

Table 7 [83–85] shows some of the correlations. This form of correlation was later established theoretically by Deckwer [86] on the basis of the surface renewal theory and Hinze's theory of isotropic turbulence. Deckwer [86] derived the following expression

$$St = 0.1(ReFrPr^2)^{-0.25}$$

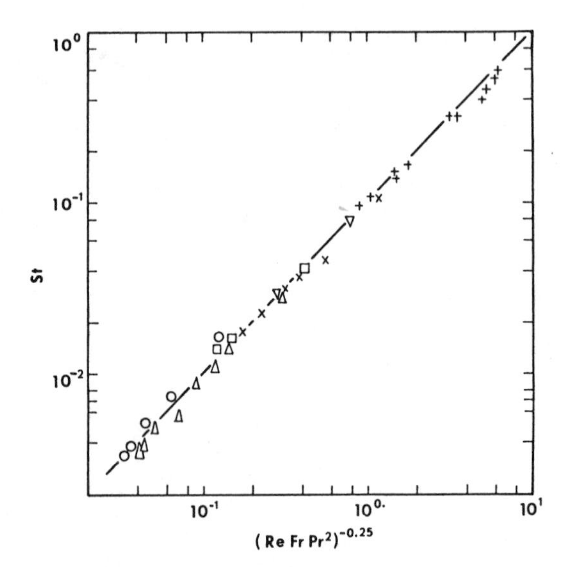

Figure 11. Deckwer's correlation of wall heat transfer coefficients [86].

Table 7
Correlations of Wall Heat Transfer Coefficients in Gas-Liquid Bubble Columns

Reference	Correlation
Kast [79]	$St = 0.1\ (Re\ Fr\ Pr^2)^{-0.22}$
Kolbel and Langemann [82]	$St = 0.124\ (Re\ Fr\ Pr^{2.5})^{-0.22}$
Shaykhutdimov et al. [83]	$St = 0.11\ (Re\ Fr\ Pr^{2.5})^{-0.22}$
Burkel [86]	$St = 0.11\ (Re\ Fr\ Pr^{2.48})^{-0.23}$
Hart [80]	$St = 0.125\ (Re\ Fr\ Pr^{2.4})^{-0.25}$
Steiff and Weinspach [78]	$St = 0.113\ (Re\ Fr\ Pr^2)^{-0.26}$
Louisi [84]	$St = 0.136\ (Re\ Fr\ Pr^{1.94})^{-0.27}$
Deckwer [86]	$St = 0.1\ (Re\ Fr\ Pr^2)^{-0.25}$

The validity of this equation has been checked against the data of Burkel [87], Perner [88], and Kast [79] over the range $6 < Pr < 985$ (Figure 11).

Although most studies have reported a nearly constant heat transfer coefficient for $u_G < 0.1$ m/s, Knickle et al. [85] obtained, in a recent study, data which showed a gradual decrease of h when $u_G < 0.1$ m/s. They also noted a significant effect of the liquid flow, contrary to most other findings.

NOTATION

a, k_1	dimension constants
b, d, e, f	exponential constants
c	tracer concentration (mol/m^3)
c_0	initial concentration (mol/m^3)
c_e	concentration at the exit (mol/m^3)
C_N	number of carbon atoms in the straight chain of an alcohol
$C_{p,L}$	liquid-phase specific heat ($J/kg \cdot K$)
d_0	orifice diameter (m)

d_B bubble diameter (m)
D_G gas-phase axial dispersion coefficient (m^2/s)
D_L liquid-phase axial dispersion coefficient (m^2/s)
D_t column diameter (m)
Fr Froude number (dimensionless)
Fr′ modified Froude number (dimensionless)
g gravitational acceleration (m/s^2)
h wall heat transfer coefficient ($W/m^2 \cdot K$)
k thermal conductivity ($W/m \cdot K$)
k_e effective thermal conductivity ($W/m \cdot K$)
ℓ scale of turbulence (m)
m_1, m_2 exponential constants
p_a pressure at station "a" (N/m^2)
p_b pressure at station "b" (N/m^2)
Pe_L liquid-phase Peclet number (dimensionless)
P_m specific energy dissipation rate (W/kg)
Pr Prandtl number (dimensionless)
q heat flux (W/m^2)
Re Reynolds number (dimensionless)
St Stanton number (dimensionless)
t time (s)
T temperature (°K)
u_0 velocity at the orifice (m/s)
u_a velocity at station "a" (m/s)
u_G gas-phase superficial velocity (m/s)
u_L liquid-phase superficial velocity (m/s)
$u_{b\infty}$ bubble rise velocity (m/s)
v′ eddy fluctuating velocity (m/s)
V_c liquid-circulation velocity (m/s)
We Weber number (dimensionless)
x axial distance from gas inlet (m)
z_a position of station "a" (m)
z_b position of station "b" (m)
Z aerated column height (m)
Z_0 unaerated column height (m)
z distance from tracer inlet (m)

Greek Symbols

α thermal diffusivitiy (m^2/s)
ε gas holdup (fractional)
θ t/τ
μ_L viscosity of liquid ($kg/m \cdot s$)
ν μ/ϱ, kinematic viscosity (m^2/s)
ϱ_G gas-phase density (kg/m^3)
ϱ_L liquid-phase density (kg/m^3)
σ surface tension (N/m)
τ average residence time (s)

REFERENCES

1. Zehner, P., *Chem. Ing. Tech. 47:* 209 (1975).
2. Judat, H., "Gassing of Low Viscosity Liquids," *Int. Chem. Eng., 21:* 188 (1981).
3. Fair, J. R., et al., "Liquid-Gas Systems," in *Chemical Engrs. Handbook,* 5th ed., R. H. Perry and C. H. Chilton (Eds.), McGraw–Hill Book Co., N. Y., 1973, pp. 18–67.
4. Ruff, K., Pilhofer, Th., and Mersmann, A., *Chem. Ing. Tech., 48:* 759 (1976).
5. Fair, J. R., et al., "Liquid-Gas Systems," in *Chem. Engrs. Handbook,* 5th ed., R. H. C. H. Chilton (Eds.), New York: McGraw–Hill Book Co., 1973, pp. 18–72.
6. Lehrer, R. H., "Gas Agitation of Liquids," *Ind. Eng. Chem. Process Des. Dev., 7:* 226 (1968).
7. Sherwood, T. K., Pigford, R. L., and Wilkie, C. R., *Mass Transfer,* New York: McGraw–Hill Co., 1975, p. 655.
8. Otake, T., Tone, S., and Shinohara, K., "Gas Holdup in the Bubble Column with Cocurrent and Countercurrent Gas-Liquid Flow," *J. Chem. Eng.* (Japan), *14:* 338 (1981).
9. Nicklin, D. J., "Two Phase Bubble Flow," *Chem. Eng. Sci., 17:* 693 (1962).
10. Wallis, G. B., *One Dimensional Two Phase Flow,* New York: McGraw–Hill Book Co., 1969.
11. Maruyama, T., Yoshida, S., and Mizushima, T., "The Flow Transition in a Bubble Column," *J. Chem. Eng.* (Japan), *14:* 352 (1981).

12. Hill, J. H., "The Operation of a Bubble Column at High Throughputs I—Gas Holdup Measurements," *Chem. Eng. J., 12:* 89 (1976).
13. Miller, D. N., "Gas Holdup and Pressure Drop in Bubble Column Reactor," *Ind. Eng. Chem. Process Des. Dev. 19:* 371 (1980).
14. Shah, Y. T., Kelkar, B. G., and Deckwer, W. D., "Design Parameters Estimations for Bubble Column Reactors," *AIChE Journal, 28:* 353 (1982).
15. Levenspiel, O., and Bischoff, K. B., "Patterns of Flow in Chemical Process Vessels," in *Advances in Chem. Eng., 4,* T. B. Drew et al. (Eds.), New York: McGraw–Hill Book Co., 1963, pp. 65–192.
16. Fair, J. R., Lambright, A. J., and Anderson, J. W., "Heat Transfer and Gas Holdup in a Sparged Contactor," *Ind. Eng. Chem. Process Des. Dev., 1:* 33 (1962).
17. Yoshida, F., and Akita, K., "Performance of Gas Bubble Columns," *AIChE Journal, 11:* 9 (1965).
18. Ellis, J. E., and Jones, E. L., Two Phase Flow Symposium, Exter, England, June 1965.
19. Botton, R., Cosserat, D., and Charpentier, J. C., "Influence of Column Diameter and High Throughputs on the Operation of a Bubble Column," *Chem. Eng. J., 16:* 107 (1978).
20. Bach, Hans. F., and Pilhofer, Th., "Variations of Gas Holdup in Bubble Columns with Physical Properties of Liquids and Operating Parameters of Columns," *Ger. Chem. Eng., 1:* 270 (1978).
21. Hikita, H., and Kikukawa, H., "Liquid Phase Mixing in Bubble Columns—Effect of Liquid Properties," *Chem. Eng. J., 8:* 191 (1974).
22. Hikita, H., et al., "Gas Holdup in Bubble Columns," *Chem. Eng. J., 20:* 59 (1980).
23. Bhaga, D., Pruden, B. B., and Weber, M. E., "Gas Holdup in a Bubble Column Containing organic Liquid Mixtures," *Can. J. Chem. Eng., 49:* 417 (1971).
24. Koetsier, W. T., Van Swaaij, and Van der Most, M., "Maximum Gas Holdup in Bubble Columns," *J. Chem. Eng.* (Japan), *9:* 332 (1976).
25. Kolbel, H., Borchers, E., and Langemann, H., "Großenverteilung der Gasblasen in Blasensaulen," *Chem. Ing. Tech., 33:* 668 (1961).
26. Deckwer, W. D., et al., "Hydrodynamic Properties of the Fischer–Tropsch Slurry Process," *Ind. Eng. Chem. Process Des. Dev., 19:* 699 (1980).
27. Hughmark, G. A., "Holdup and Mass Transfer in Bubble Columns," *Ind. Eng. Chem. Process Des. Dev., 6:* 218 (1967).
28. Akita, K., and Yoshida, F., "Gas Holdup and Volumetric Mass Transfer Coefficients in Bubble Columns," *Ind. Eng. Chem. Process Des. Dev., 12:* 76 (1973).
29. Kumar, A., et al., "Bubble Swarm Characteristics in Bubble Columns," *Can. J. Chem. Eng., 54:* 503 (1976).
30. Mashelkar, R. A., "Bubble Columns," *Brit. Chem. Eng., 15:* 1297 (1970).
31. Carleton, A. J., et al., "Some Properties of Packed Bubble Column," *Chem. Eng. Sci., 22:* 1839 (1967).
32. Reith, T., Renken, S., and Israel, B. A., "Gas Holdup and Axial Mixing in the Fluid Phase of Bubble Columns," *Chem. Eng. Sci., 23:* 619 (1968).
33. Mersmann, A., "Design and Scale-up of Bubble and Spray Columns," *Ger. Chem. Eng., 1:* 1 (1978).
34. Kelkar, B. G., et al., "Effect of Addition of Alcohols on Gas Holdup and Backmixing in Bubble Columns," *AIChE Journal, 29:* 361 (1983).
35. Schugerl, K., Lucke, J., and Oels, U., "Bubble Column Bioreactors," in *Advances in Biochemical Engineering, 7* T. K. Ghose et al. (Eds). Berlin: Springer-Verlag, 1977, pp. 1–84.
36. Rietema, K., "Science and Technology of Dispersed Two-Phase Systems I and II," *Chem. Eng. Sci., 37:* 1125 (1982).
37. Mhaskar, R. D., "Effect of Backmixing on the Performance of Bubble Column Reactors," *Chem. Eng. Sci., 29:* 897 (1974).
38. Miyauchi, T., and Vermeulen, T., "Longitudinal Dispersion in Two-Phase Continuous-Flow Operations," *Ind. Eng. Chem. Fundam. 2:* 113 (1963).
39. Mecklenburgh, J. C., "Backmixing and Design," *Trans. I. Chem. E., 52:* 180 (1974).

40. Shah, Y. T., *Gas-Liquid-Solid Reactor Design,* New York: McGraw–Hill Book Co., 1979.
41. Hoffman, H., "Der derzeitige Stand bei der Vorausberechnung der Verweilzeit verteilung in Technischen Reaktoren," *Chem. Eng. Sci., 14.* 193 (1961).
42. Brenner, H., "The Diffusion Model of Longitudinal Mixing in Beds of Finite Length," *Chem. Eng. Sci., 17:* 229 (1962).
43. Field, R. W., and Davidson, J. F., "Axial Dispersion in Bubble Columns," *Trans. I. Chem. E., 58:* 228 (1980).
44. Chen, B. H., and McMillan, A. F., "Heat Transfer and Axial Dispersion in Bubble Columns," *Can. J. Chem. Eng., 60:* 436 (1982).
45. Levenspiel, O., *Chemical Reaction Engineering,* New York: Wiley & Sons, 1972.
46. Siemes, W., and Weiss, W., "Flussigkeitsturchischung in engen Blasensaulen," *Chem. Ing. Tech., 29:* 727 (1957).
47. Ohki, Y., and Inoue, H., "Longitudinal Mixing of the Liquid Phase in Bubble Columns," *Chem. Eng. Sci., 25:* 1 (1970).
48. Konig, B., et al., "Longitudinal Mixing of the Liquid Phase in Bubble Columns," *Ger. Chem. Eng., 1:* 199 (1978).
49. Shah, Y. T., Stiegel, G. J., and Sharma, M. M., "Backmixing in Gas-Liquid Reactors," *AIChE Journal, 24:* 369 (1978).
50. Towell, G. D., and Ackerman, G. H., "Axial Mixing of Liquid and Gas in Large Bubble Reactors," Proceedings of the 5th European Chemical Reaction Engineering, B3-1, Amsterdam, 1972.
51. Deckwer, W. D., Burckhart, R., and Zall, G., "Mixing and Mass Transfer in Tall Bubble Columns," *Chem. Eng. Sci., 29:* 2177 (1974).
52. Deckwer, W. D., et al., "Zones of Different Mixing in the Liquid Phase of Bubble Columns," *Chem. Eng. Sci., 28:* 1223 (1973).
53. Badura, R., et al., "Durchmischung in Blasensaulen," *Chem. Eng. Tech., 46:* 399 (1974).
54. Mangartz, K. H., dissertation, Tech U. Munchen, 1977.
55. Baird, M. H. I., and Rice, R. G., "Axial Dispersion in Large Unbaffled Bubble Columns," *Chem. Eng. J., 9:* 17 (1975).
56. Riquarts, H. P., "A Physical Model for Axial Mixing of the Liquid Phase for Heterogeneous Flow Regime in Bubble Columns," *Ger. Chem. Eng., 4:* 18 (1981).
57. Kato, Y., and Nishiwaki, A., "Longitudinal Dispersion Coefficients of a Liquid in a Bubble Column," *Int. Chem. Eng., 12:* 182 (1972).
58. Argo, W. B., and Cova, D. R., "Longitudinal Mixing in Gas-Sparged Tubular Reactors," *Ind. Eng. Chem. Process Des. Dev., 4:* 352 (1965).
59. Joshi, J. B., and Sharma, M. M., "A Circulation Model for Bubble Columns," *Trans. I. Chem. E., 57:* 244 (1979).
60. Freeman, W., and Davidson, J. F., "Holdup and Liquid Circulation in Bubble Columns," *Trans. I. Chem. E., 47:* T251 (1969).
61. Rietema, K., and Ottengraf, S. P. P., "Laminar Liquid Circulation and Bubble Street Formation in a Gas-Liquid System," *Trans. I. Chem. E., 48:* T54 (1970).
62. Whalley, P. B., and Davidson, J. F., Proceedings of the Symposium on Multiphase Flow System, Ser. No. 38, J5, London (1974).
63. Joshi, J. B., "Axial Mixing in Multiphase Contactors—A Unified Correlation," *Trans. I. Chem. E., 58:* 155 (1980).
64. Aoyoma, Y., et al., "Liquid Mixing in Concurrent Bubble Columns," *J. Chem. Eng.,* (Japan) *1:* 158 (1968).
65. Vail, Yu. K., Manakov, N. K., and Manshilin, V. V., "Turbulent Mixing in a Three Phase Fluidized Bed," *Int. Chem. Eng., 8:* 298 (1968).
66. Ostergaard, K., AIChE Symp. Ser. No. 176, *74:* 82 (1978).
67. Hinze, J. O., *Turbulence,* McGraw–Hill, New York, 1958.
68. Rice, R. G., et al., "Reduced Dispersion using Baffles in Column Flotation," *Powder Tech., 10:* 201 (1974).
69. Kolbel, H., Langemann, H., and Platz, J., Dechema Monographien, *41:* 225 (1962).

70. Menshchikov, V. A., and Aerov, M. E., "Longitudinal Mixing of Gas Phase in Bubble-Plate Reactors," *Theor. Found. Chem. Eng., 1:* 739 (1967).
71. Diboun, M., and Schugerl, K., "Eine Blasensaule mit Gleichstron von Wasser und Luft I—Mischungs—vorgange in der Gasphase," *Chem. Eng. Sci., 22:* 147 (1967).
72. Seher, A., and Schumacher, V., "Determination of Residence Times of Liquid and Gas Phases in Large Bubble Columns with the Aid of Radioactive Tracer," *Ger. Chem. Eng., 2:* 117 (1979).
73. Mangartz, K. H., and Pilhofer, Th., *Verfahrenstechnik 14:* 40 (1980).
74. Joshi, J. B., "Gas Phase Dispersion in Bubble Columns," *Chem. Eng. J., 24:* 213 (1982).
75. Yagi, S., Kunii, D., and Wakao, N., "Studies on Axial Effective Thermal Conductivities in Packed Beds," *AIChE Journal, 6:* 543 (1960).
76. Lewis, W. K., Gilliland, E. R., and Gerouard, H., "Heat Transfer and Solids Mixing in Beds of Fluidized Solids," CEP Symp. Ser. No. 38, *58.* 87 (1962).
77. Jain, S. C., and Chen, B. H., "Heat Transfer in a Screen-Packed Fluidized Beds," AIChE Symp. Ser. No. 116, *67.* 97 (1971).
78. Steiff, A., and Weinspach, P. M., "Heat Transfer in Stirred and Non-Stirred Gas-Liquid Reactors," *Ger. Chem. Eng., 1:* 150 (1978).
79. Kast, W., "Untersuchungen zum Warmeubergang in Blasensaulen," *Chem. Ing. Tech. 35:* 785 (1963).
80. Hart, W. F., "Heat Transfer in Bubble Agitated Systems—A General Correlation," *Ind. Eng. Chem. Process Des. Dev., 15:* 109 (1976).
81. Nishikawa, M., Kato, H., and Hashimoto, K., "Heat Transfer in Aerated Tower Filled with Non-Newtonian Liquids," *Ind. Eng. Chem. Process Des. Dev., 16:* 133 (1977).
82. Kolbel, H., Langemann, H., *Erdoel-Zeitschr, 80:* 405 (1964).
83. Shaykhutdinov, A. G., Bakirov, N. U., and Usmanov, A. G., "Determination and Mathematical Correlation of Heat Transfer Coefficients under Conditions of Bubble Flow, Cellular and Turbulent Foam," *Int. Chem. Eng., 11:* 641 (1971).
84. Louisi, Y., Dr. Ing. thesis, Tech. Univ. Berlin, Berlin, 1979.
85. Knickle, H. N., et al., "Backmixing and Heat Transfer Coefficients in Bubble Columns Using Aqueous Glycerol Solutions," AIChE Symp. Ser. No. 225, *79:* 352 (1983).
86. Deckwer, W. D., "On the Mechanism of Heat Transfer in Bubble Columns Reactors," *Chem. Eng. Sci., 35:* 1341 (1980).
87. Burkel, W., "Beitrag zur Dimensionierung von Blasensaulenreaktoren," Dr. Thesis, Tech. Univ. Munich, Munich (1974).
88. Perner, D., "Der Warmeubergang in Druckblasensaulen," Dipl. Arbeit. Tech. Univ. Berlin, Berlin (1960).

CHAPTER 24

OPERATION AND DESIGN OF TRICKLE-BED REACTORS

Jiri Hanika

Department of Organic Technology
Institute of Chemical Technology
Prague, Czechoslovakia

Vladimir Stanek

Institute of Chemical Process Fundamentals
Czechoslovak Academy of Sciences
Prague, Czechoslovakia

CONTENTS

INTRODUCTION

The trickle-bed reactor is a special case of a cocurrent downflow packed column operated under limiting conditions. At present it provides the best means of carrying out a reaction in which gaseous and liquid reactants are to be contacted with small porous particles or inert packing. Three-phase trickle-bed reactors have not been as extensively studied as gas-phase reactors, despite their extensive use in petrochemical processes and organic technologies.

The first industrial application of cocurrent packed columns is reportedly the synthesis of butynediol in Germany [1]. The potential for the petroleum industry stimulated further research of trickle-bed reactors. New papers and patents were published concerning petroleum refining [1–10], various hydrogenations [11–16], selective hydrogenation [17], oxidation of organic compounds in waste waters [18–20], treatment of effluent gases [21–22], production of hydrogen peroxide [23], deuterium exchange reaction [24–25], immobilized enzyme reaction [26] and others. Reviews of the various applications have appeared in Uhlmann's encyclopedia [27] and recent books [28–29].

The general problem of operation and design of trickle-bed reactors is to find a reliable quantitative relationship between the reaction conversion and operating and system variables such as temperature, pressure, gas and liquid feed rates, reactor length and diameter, catalyst particle diameter, and properties of the mobile phase [30]. A realistic model of the trickle-bed reactor should involve the interactions with the fluid dynamics as well as the resistances to mass and heat transfer between individual phases.

Analysis of the fundamental relationships between heat transfer and fluid flow in trickle-bed reactors has been published in the literature [38–42]. In several papers [12, 15, 43–45] it has been shown that evaporation of the liquid phase may occur under certain circumstances with a corresponding marked increase in the reaction rate. The evaporation of the reaction mixture may in turn lead to the formation of hot spots within the bed and the appearance of multiple steady states.

OPERATION OF A TRICKLE-BED REACTOR

Characteristics of the Trickle-Bed Reactor

Three-phase reactions may be carried out either in autoclaves in a batchwise manner, or continuously in tubular reactors. Depending upon the way of feeding the liquid reaction mixture we speak either of bubble reactors (the liquid is fed at the bottom cocurrently with the gas) or trickle-bed reactors where the liquid trickles down the surface of the catalytic bed, usually cocurrently with the reacting gas.

Shah [28] summarized the advantages of trickle-bed reactors and specialized studies [14, 31, 45, 47, 48] examined individual factors affecting the reactor's performance. In the last few years several authors have critically reviewed [23, 30, 46] the literature cocerning the properties of trickle-bed reactors.

The advantages as well as the drawbacks of the trickle-bed versus slurry reactors have also been investigated in a number of review papers [23, 26, 30–34, 46, 49]. The need for filtration of spent powdered catalyst and the higher installation costs of the slurry reactors give and edge to trickle beds over the slurry operation. In addition, the residence time distribution is close to the plug flow, which is indispensable if high degrees of conversion are to be achieved.

On the other hand, the effectiveness factor in trickle beds is usually low due to the large size of the pellets and their mechanical strength having to be greater.

Design and scale up of trickle-bed reactors is generally more difficult [35]. The reactor performance depends in most cases not only on the reaction itself but also, and often strongly, on the fluid dynamics and a number of transport processes of the reaction components between individual phases.

In an example of a hydrogenation reaction we shall examine which of the steps may affect the operation of the reactor or eventually increase its productivity.

In this case the principal elementary steps of mass transfer between the three phases may be identified as follows:

- Transfer of hydrogen from the bulk gas phase toward the external surface of the liquid film flowing down the external catalytic surface.
- Transfer of hydrogen and the hydrogenated substrate present in the flowing liquid film and in liquid menisci, formed at the contact points of neighboring catalyst pellets.
- Diffusion of the reactants and products within the pores of the catalyst pellet.

The mass transfer is also significantly affected by the nonuniform flow of the mobile phases through the catalytic bed.

The evolution of the heat of reaction, typically high for hydrogenations carried out mostly in trickle-bed reactors, markedly affects the thermal regime of the reactor.

Parameters Affecting the Productivity of a Trickle-Bed Reactor

The regime of operation of a trickle bed reactor is affected by a number of factors which may be divided into the following groups:

1. Reactor (its layout is more or less fixed)
   - Tube diameter (or a bunch of tubes).
   - Length of the tubes.
   - Initial liquid distribution.
   - Location of redistributors for liquid.
   - Possibility and efficiency of heat exchange.

2. Catalyst
   - Diameter of pellet.
   - Catalytic activity and its long-term stability.
   - Distribution of the catalytically active component in the volume of the pellet.
   - Mechanical strength of the pellet (resistance to erosion).

3. Controling factors (affecting the productivity directly and variable within relatively broad limits).
   - Inlet temperature of liquid and gas.
   - Operating pressure in the reactor.
   - Feed rate of liquid substrate.
   - Concentration of substrate in the feed.
   - Feed rate of gas.
   - Amount of catalyst in the reactor.
   - Intensity of heat exchange between the reactor and the surroundings.

The instantaneous state and regime of the reactor is further affected by the character of the chemical reaction taking place (reaction rate, reaction order with respect to individual components, heat of reaction, activation energy, etc.) and physico-chemical properties of the reaction mixture (heat capacities, diffusion coefficients, thermal conductivities, viscosities, etc.). These quantities, however, are more or less fixed but knowledge of them, in some cases, permits one to understand the specific behavior of the reactor.

The following analysis regarding the effect of temperature, pressure, and catalyst properties on the productivity may be applied to all types of gas-liquid-solid catalytic systems.

The Effect of Temperature

The rate of a chemical reaction grows, as a rule, exponentially with temperature. In gas-liquid-solid catalytic reactors, however, the situation is more complex as the reaction rate may be affected, on the one hand, by the partial pressure of vapors of the reaction mixture and, on the other hand, by transport phenomena.

Let us now assume that the system operates at a low temperature in the kinetic regime under the constant overall pressure. Upon increasing the temperature in the reactor two situations may occur: In the first case, if the vapor pressure of the liquid phase is negligible compared to the partial pressure of the gaseous reactant, increased temperature causes increased reaction rate and in turn transition to the diffusion regime [50]. While the

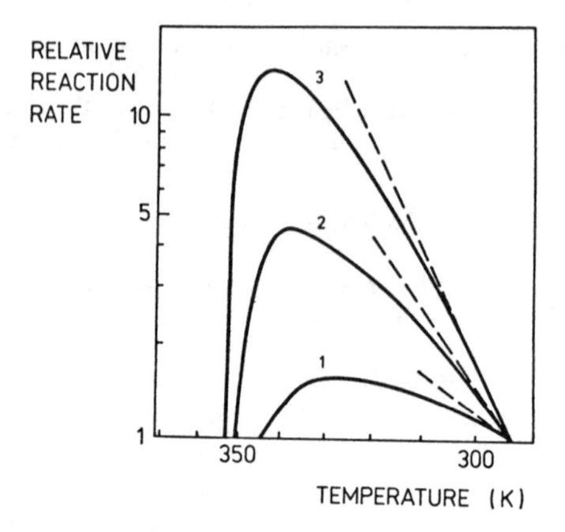

Figure 1. The effect of activation energy on the temperature dependence of the rate of cyclohexene hydrogenation [51] at atmospheric pressure. Curve 1: E = 20 kJ/mol; Curve 2: E = 40 kJ/mol; Curve 3: E = 60 kJ/mol.

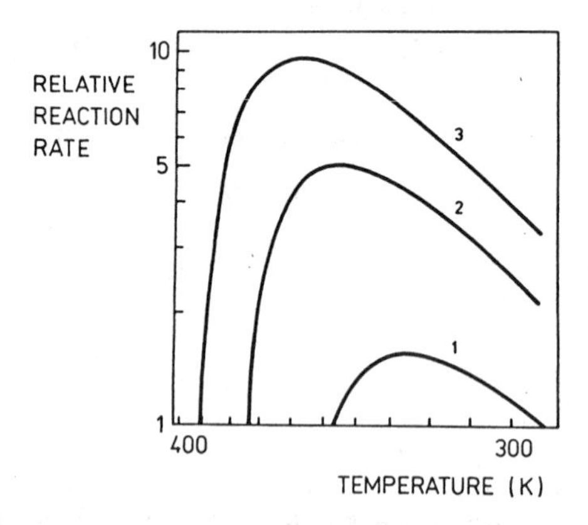

Figure 2. The effect of pressure on the temperature dependence of the rate of cyclohexene hydrogenation [51]; E = 20 kJ/mol. Total pressure: Curve 1—0.1 MPa; Curve 2—0.2 MPa; Curve 3—0.3 MPa.

activation energy of, for instance, hydrogenation reactions amounts to about 40 kJ/mol, the apparent activation energy in the diffusion region is about 5 to 10 times less.

It is thus obvious that further increase of temperature in the diffusion region practically does not provide for an increase of the reaction rate. Equally it would be fruitless to increase catalytic activity or increase the amount of catalyst, for both these measures shift the operating regime of the reactor farther into the diffusion region.

Even more interesting is the other alternative, which may be expected to occur if the vapor pressure of liquid significantly contributes to the overall pressure of the system. Increasing temperature then increases not only the reaction rate but also the partial pressure of the liquid phase. If the reactor is operated under the constant total pressure and the reaction has a positive order with respect to the gaseous reactant (this being a very frequent case, e.g., for hydrogenations), the temperature dependence of the reaction rate exhibits a distinct maximum corresponding to the optimum temperature under the given conditions. A computed example [51] of such a dependence for reaction order equal to unity is shown in Figure 1. This figure shows also the effect of the activation energy of hydrogenation of cyclohexene in benzene solution on the temperature dependence of the reaction rate. It is seen that increasing activation energy shifts the temperature of the maximum reaction rate closer to the boiling point of the liquid phase (353 °K). For illustration the figure shows also by broken lines corresponding dependences free of the impeding effect of the liquid phase.

It is obvious that the total pressure in the reactor has a direct effect on the discussed dependence. Figure 2 plots the relative reaction rate as a function of temperature for three levels of the total pressure in the system (0.1–0.3 MPa) for the hydrogenation of cyclohexene in benzene solution with the activation energy of 20 kJ/mole. It is seen that increasing total pressure causes also the temperature of the maximum reaction rate to increase since the pressure effectively increases the boiling point of the liquid phase.

The Effect of Pressure

The reaction order with respect to the partial pressure of gaseous reactant, i.e., hydrogen in case of hydrogenation and hydrotreating reactions, usually equals unity over a wide interval of conditions [50, 52]. From the standpoint of reactor productivity it is therefore desirable to work at high pressure. Due to the increasing investment costs, however, one has again to choose a certain optimum of the operating pressure.

Properties of the Catalyst

Trickle-bed reactors are packed with pelletized or granulated catalyst. Large sizes of an element of the catalytic bed is mandated by the need for a low hydrodynamic resistance of the reactor. This in turn causes the internal surface of the catalyst pores to be little utilized for the reaction. In view of the low diffusion coefficients in the liquid phase, the factor of internal effectiveness amounts usually to only a few percent [53–55].

The size of the pellets or grains considerably affects also the rate of diffusion of the reaction components and products within the catalyst and thereby the rate of the process as a whole. The effect of internal diffusion, in comparison with the gas-phase reactions, is far more significant as the diffusion in liquid systems is by 3 to 4 orders of magnitude slower. The effect of internal diffusion on the reaction rate in liquid systems can be successfully eliminated only if extremely fine catalyst powder of not excessivelly large catalytic activity is used. With highly active catalysts it is virtually impossible to reach the kinetic region [54] from the standpoint of internal mass transfer and further increase of the content of the active component in the catalyst has naturally no effect.

Typical dependence of the rate constant of hydrogenation of cyclohexene [54] on a medium-size palladium or platinum catalyst using different solvents is depicted in log-log coordinates in Figure 3. The diffusion region in these coordinates has a typical slope of -1. Whether the process is still under the effect of diffusion of the reaction components within the porous structure of the catalyst may be ascertained from an experiment with a catalyst of different grain size. In the negative case the reaction rate is independent of the grain size. Here it is again useless to seek a more active catalyst if the reaction is sufficiently fast as to be under the effect of internal diffusion.

In the paper [54] it was established that for the purpose of describing mass transfer within the catalyst pellets for liquid-phase hydrogenations one can use relationships available for gas-phase reactions with the exception that the rate-controlling step is primarily diffusion of the dissolved hydrogen. Its concentration in the liquid mixture is commonly (under medium pressures) very low.

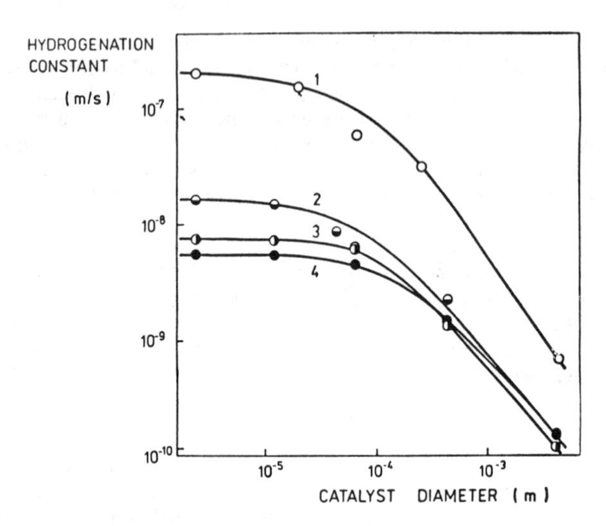

Figure 3. The effect of grain size on the hydrogenation constant of cyclohexene on a palladium catalyst [54]. Curve 1—solvent: carbontetrachloride; Curve 2—methanol; Curve 3—hexane; Curve 4—benzene.

It has been said that the internal surface of the catalyst is little utilized unless a fine powder catalyst (d_p below 0.05 mm) is used. It is thus apparent that in a trickle-bed reactor, with considerably larger catalyst pellets (for a given reaction and catalyst), the principal resistance to mass transfer is concentrated within the catalyst pellet. For this reason it is generally economical to use pellets only superficially treated with the active component and there is no need to increase catalytic activity. The same applies to an eventual increase of the mean temperature in the reactor, which would not contribute significantly to increased productivity either. An exception to this rule would pose the onset of the phenomena such as evaporation of the liquid phase, followed by the reaction on the dry catalyst. The gas-phase reaction is faster because the activation energy of the diffusion process is generally much lower (4–10 kJ/mol) than that of the chemical reaction (liquid-phase hydrogenations, e.g., 30–50 kJ/mol).

The choice of a suitable catalyst for a trickle-bed reactor is further tied to the need to have pellets of sufficient mechanical strength and resistance primarily to the erosion by liquid. The usual carriers for trickle-bed systems are alumina, silica gel, or kiselgur.

The Effect of Solvent

One possibility of affecting the rate of the process is through the use of different solvents. At present, however, there is no unambiguous rule for direct choice of the solvent favorably affecting the given process by its sorption properties. Since "chemical factors" must also be taken into consideration (i.e., solubility of the substrate and reaction products in the reaction mixture), suitable solvents must be found experimentally. A review of this problem in the case of the liquid-phase hydrogenation may be found in the literature [56].

The Flow of Liquid Through a Random Bed of Catalyst Particles

The efficiency and productivity of the reactor strongly depend on the mode in which the mixture is being fed into the column. In column reactors local density of flow (in this case both gas and liquid) often increases in the proximity of the column wall. This wall flow, particularly that of the liquid, strongly decreases the productivity of the reactor for it functions in fact as a bypass of the catalyst bed. The fraction of liquid captured in the wall flow depends mainly on the ratio of the pellet to reactor diameter and further may be affected by the initial distribution of liquid and the depth of the bed. The amount of liquid in the wall region also varies with the physico-chemical properties of the reaction mixture (density, viscosity, surface tension) and the wettability and porosity of the catalyst pellets. Porous materials such as catalyst pellets, and lower surface tension of the liquid reaction mixture contribute to more uniform flow of liquid within the bed.

The magnitude of the wall flow can be directly measured [57, 58, 78] and such data are important for eventual installation of the, so called, flow redistributors. These redistributors are often installed in order to promote effective exchange of the reacting species between the wall flow and the flow down the catalyst surface.

In narrow-tube reactors the wall flow may reach considerable values as may be seen in Figure 4. This figure plots data [57] obtained with water in a bed of nickel catalyst pellets (d_p = 5 mm). It is seen that the plotted dependence displays a sharp upturn in region of low reactor to particle diameter ratio (d_t/d_p).

In practical trickle-bed reactor design the final diameter of the reactor is the result of a compromise. Increased reactor diameter, on the one hand, reduces the amount of liquid in the wall region but, on the other hand, substantially impairs the conditions for the removal of the reaction heat from the reactor. In some cases, when the temperature rise due to the reaction heat would lead to deactivation of the catalyst a narrower tube must be used at the expense of efficiency.

Control of the Regime of a Trickle-Bed Reactor

Inlet Temperature of the Reaction Mixture

The productivity of the reactor may be increased by higher temperature of the feed-reaction mixture. This temperature, however, cannot be increased arbitrarily. In a reactor

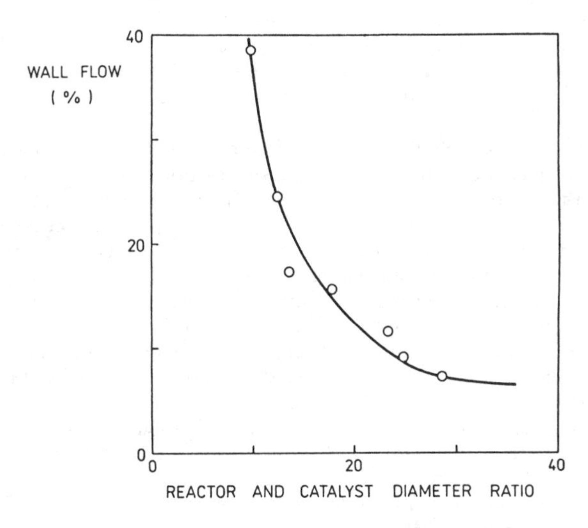

Figure 4. Plot of the wall-flow-versus-reactor-to-pellet-diameter ratio for the water/air/nickel catalyst system [57].

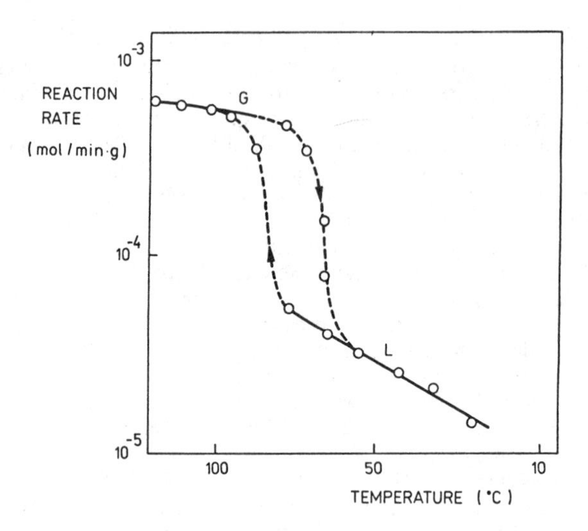

Figure 5. Hysteresis behavior of a trickle-bed reactor for the hydrogenation of cyclohexene in cyclohexane solution at atmospheric pressure following the temperature changes [12]: d_t = 1.5 cm; d_p = 0.4 cm; h = 10 cm; F = 4.5 ml/min; C_0 = 25 % wt.

operated at constant pressure, increasing temperature of the feed increases the vapor pressure of liquid, which in turn reduces the partial pressure of the gaseous component.

In a continuous reactor, where the feed gas is not saturated by liquid vapors higher inlet temperature may produce heating of the reaction mixture up to the boiling point and its total evaporation. The reactor thus becomes a conventional tubular reactor operating in the gas-phase regime. A typical feature accompanying evaporation of the liquid mixture in the trickle-bed reactor is the hysteresis behavior of the whole system. Figure 5 illustrates this interesting phenomenon observed [12] during hydrogenation of cyclohexene on a palladium catalyst (d_p = 4 mm) at atmospheric pressure in the range of inlet temperature between 25° and 125° C. The measurements were carried out in a jacketed reactor (d_t = 15 mm).

Evaporation of the reaction mixture (at a temperature close to the boiling point) is accompanied by a sharp increase of the reaction rate (in the prior case by about a factor of 7) due to the increased degree of utilization of the internal surface of the catalyst in the gas-phase regime. Typical values of the diffusion coefficients of the reaction components in the gas-phase regime are substantially higher than those in the liquid-phase regime.

The dependence of the reaction rate on the temperature of the feed mixture, shown in Figure 5, splits in the region between 60 and 85° C into two branches—one with corresponding low values of the liquid-phase reaction rate, the other with high reaction rates in the gas-phase regime. For the same temperature maintained in the cooling jacket of the reactor, two different values of the reaction rate were observed, depending on whether the reactor was previously operated in the gas- or the liquid-phase regime.

Formation of the hot reaction zones in the reactor, brought about by phase transitions, can be conveniently observed by measuring the axial temperature profiles in the bed. An example of such profiles, detected in a 3-cm diameter externally cooled reactor for hydrogenation of cyclohexene [164] on an 18-cm deep bed of palladium (d_p = 4 mm) catalyst, is depicted in Figure 6. The temperature profiles observed at low inlet temperatures (equalling the temperature in the reactor jacket), i.e. at 40° or 50° C, are flat and point at the isothermal course of the reaction in the liquid-phase regime with correspondingly low conversions. Upon increasing the temperature of the feed to 60° C a hot zone is formed at the inlet end of the catalyst layer where the reaction proceeds in the gas-phase regime practically to 100% conversion. The temperature in this zone considerably exceeds the boiling point of the cyclohexene-cyclohexane mixture, the latter being about 80° C. Past the hot zone the reaction mixture merely cools gradually to the temperature of liquid in the jacket (60° C).

From the practical standpoint the evaporation is a very important and in most instances undesirable phenomenon. Evaporation brings about excessive temperature strain of the catalyst which may lead up to its total poisoning due to various side reactions (isomerization). In addition, a possible shift of chemical equilibria may be also undesirable [44]. Evaporation of the reaction mixture can be effectively supressed by increased total pressure in the reactor, or, by decreased concentration of the hydrogenated substrate (dilution by a suitable solvent), and, of course, by efficient cooling. In contrast, increased flow rate of hydrogen and decreased feed rate of liquid both increase the chance for evaporation of the mixture [12].

This situation is illustrated in Figure 7 in an example of hydrogenation of cyclohexene [164]. The axial temperature profile, measured at low feed rate of the reaction mixture, indicates the appearance of the hot spot at the inlet end of the catalyst bed, while the profiles measured at elevated flow rates of the reaction mixture display an approximately isothermal course typical for the liquid-phase hydrogenation. It is thus obvious that low feed rates of the reaction mixture facilitated evaporation of the reaction mixture through the existence of imperfectly wetted spots within the catalyst bed. In the case of the volatile substrate the reaction can proceed here more rapidly in the gas-phase regime, speeding up liberation of the reaction heat.

Inlet Concentration of Substrate

The reaction rate and hence also the productivity of the reactor are determined by the kinetics of the reaction. For example, liquid-phase hydrogenations are usually zero order

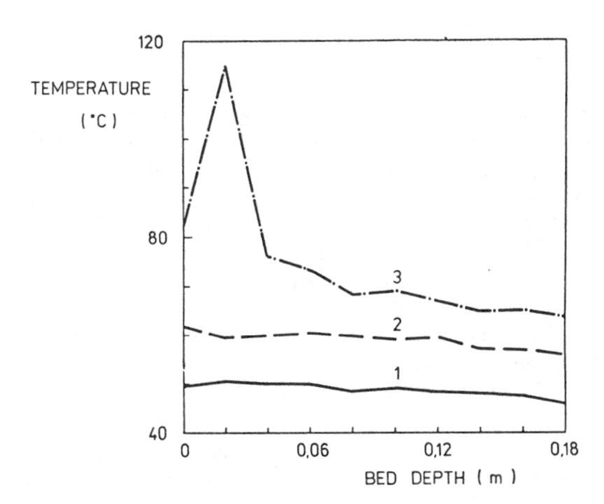

Figure 6. Effect of reactor wall and inlet temperature on the axial temperature profile during cyclohexene hydrogenation at atmospheric pressure in cyclohexane solution [164]: d_t = 3.0 cm; d_p = 0.4 cm; F = 4.5 ml/min; C_o = 42 % wt. Curve 1—T_o = 40°C; Curve 2—T_o = 50°C; Curve 3—T_o = 60°C.

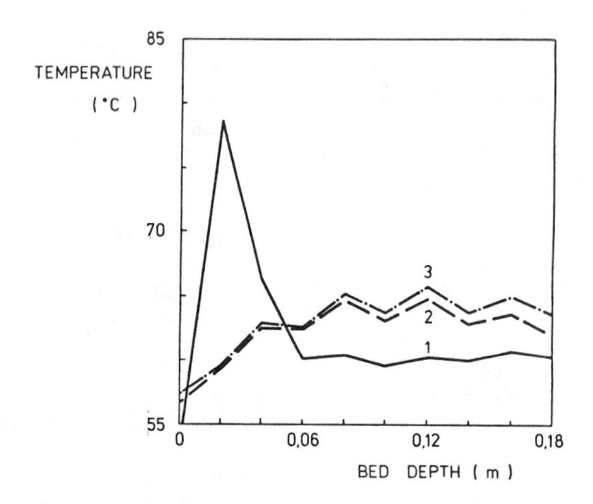

Figure 7. Effect of cyclohexene-cyclohexane mixture (19 % wt) feed rate on the axial temperature profile during cyclohexene hydrogenation at atmospheric pressure [164]. d_t = 3.0 cm; d_p = 0.4 cm; T_o = 60°C. Curve 1—F = 82 ml/h; Curve 2—F = 125 ml/h; Curve 3—F = 238 ml/h.

with respect to the hydrogenated substrate over a broad interval of concentrations. The dependence of the reaction conversion on the kinetic coordinate W_{cat}/F (where F designates the feed rate of the substrate and W_{cat} stands for the weight of the catalyst) at constant liquid flow rate is then a straight line, shown in Figure 8. This figure plots data on the hydrogenation of cyclohexene on a palladium catalyst [62]. The broken straight line corresponds to experiments in the (from the external mass transfer limitations point of view) kinetic regime with the catalyst of equal grain size in a mixed reactor of the type described by Carrbery [53]. In the used reactor [62] the catalyst pellets, protected from abrasion and break down, rotated in a sieve basket which eliminated the effects of external mass transfer. The straight line character of this presumably kinetic dependence appears thus in accord with the found zero reaction order with respect to the hydrogenated substrate.

From this plot one can easily assess the efficiency of the trickle-bed reactor and determine the contribution of the resistance of mass transfer at the external surface of the catalyst. The productivity of the reactor increases significantly by reducing this resistance, i.e. by more uniform distribution of liquid in the bed or eventually increased turbulence in the film trickling down the pellets and by increased gas and/or liquid rates.

Figure 8 further documents that the plotted dependence may change with the depth of the catalytic bed. This observation is no doubt due to the nonuniform distribution of liquid. The reaction mixture was in this case fed into the reactor by a central jet in the axis of the reactor. Accordingly, the efficiency for shorter beds (3 cm) is lower. The depth of the catalytic bed in the reactor, however, cannot be changed arbitrarily as it is one of the principal design parameters of the reactor.

A significant change of productivity of the reactor may be expected when marked temperature gradients due to the imperfect removal of the reaction heat exist. The adiabatic temperature rise and the reactor's productivity increases proportionally with increasing concentration of the substrate in the feed. This growth may become dramatic when it gives rise to evaporation of the reaction mixture, discussed in the previous paragraph. Phase transition, observed during hydrogenation of cyclohexene [12], is illustrated in Figure 9. This figure shows the dependence of the reaction rate on concentration of substrate in the feed. The constant reaction rates, observed experimentally in the liquid-phase regime (line L)

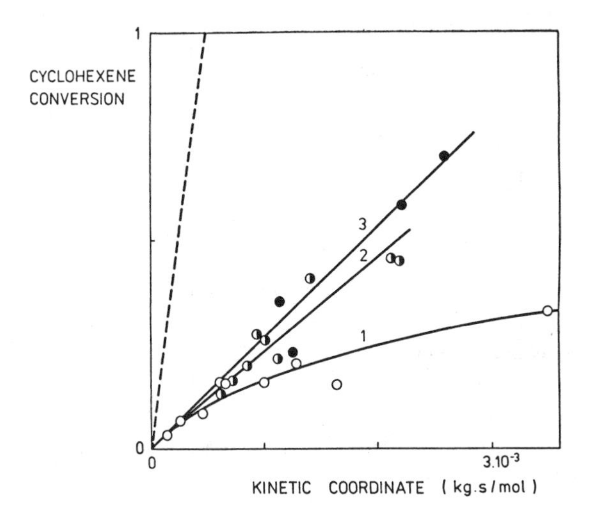

Figure 8. Plot of reaction conversion versus W_{cat}/F coordinate for the hydrogenation of cyclohexene with a different depth of the palladium catalyst bed [62]. d_p = 0.155 cm; T_o = 25°C. Curve 1—h = 3 cm; Curve 2—h = 5 cm; Curve 3—h = 10 cm.

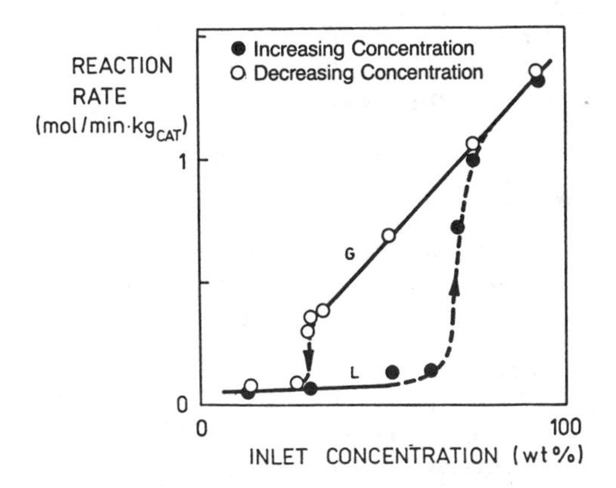

Figure 9. The effect of concentration of substrate on the reaction rate [12]. T_o = 59°C; F = 1 cm^3/min; F_H = 250 cm^3/min; W_{cat} = 6.13 g.

confirm the zero order of the hydrogenation with respect to the hydrogenated substrate. The linear dependence of the reaction rate in the gas phase (line G) then points at the first order with respect to cyclohexene. The experimental results reveal considerable hysteresis of the system and the existence of two stable steady states (liquid and gas/vapor regime) in the region between 30% and 70% by mass of cyclohexene in the feed.

The appearance of these multiple states depends on the previous reaction conditions, i.e., on the reactor history. Further it was observed that evaporation and condensation of the reaction mixture take place always over a narrow concentration range.

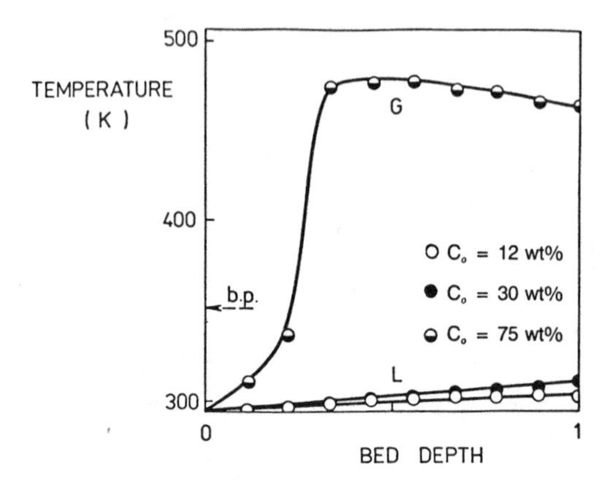

Figure 10. The effect of the initial concentration of cyclohexene on axial temperature profiles in a adiabatic trickle-bed reactor [43]. T_o = 295°K.

The character of the prevailing regime, with respect to the inlet concentration of the substrate, can be monitored also through the axial temperature profiles. Figure 10 shows such profiles measured in an adiabatic trickle-bed reactor [43] for the hydrogenation of cyclohexene. At low concentrations, i.e., 12% and 30% by mass, the reaction proceeded in the liquid phase. At 75% by mass of the inlet concentration a nearly step change of temperature was observed upon evaporation of the reaction mixture. The reaction in this case took place predominantly in the gas-phase regime in the first third of the depth of the catalytic bed.

Pressure and Flow Rate of Hydrogen

Generally it may be stated that increased pressure of hydrogen leads in most cases to increased rate of hydrogenation and hereby to increased productivity. This brings along increased requirements on the removal of the reaction heat and reactor construction. On the other hand, increased total pressure reduces the hazard of catalyst overheating, following the evaporation of the reaction mixture.

Since the increased pressure of hydrogen usually increases both the reaction rate and the rate of diffusion of the reaction components toward the active surface of the catalyst, increased pressure would not help remove the impeding effect of diffusion, which in many instances impairs selectivity of the process.

An alternative way of increasing the reactor productivity is through increased turbulence of the flowing liquid by increased gas flow rate. In Reference 63 it was shown that under such conditions the reactor productivity increases both under the cocurrent and the countercurrent flow arrangement. For illustration Figure 11 plots the conversion of hydrogenation of cyclohexene (in a cyclohexane solution) as a function of the flow rate of hydrogen. The figure was taken over from the same paper [63]. Clearly, the flow rate of hydrogen provides more effective means in case of the countercurrent flow, but the appropriate curve is terminated at the point where flooding of the bed occurs. Under the cocurrent downflow no such limitations were observed (even though here too the flow rate cannot be increased without limits), and this arrangement can be recommended for practical utilization.

Since the increasing flow rate of gas through the reactor also increases the energy costs due to the increased energy losses, an optimum gas flow rate must be found.

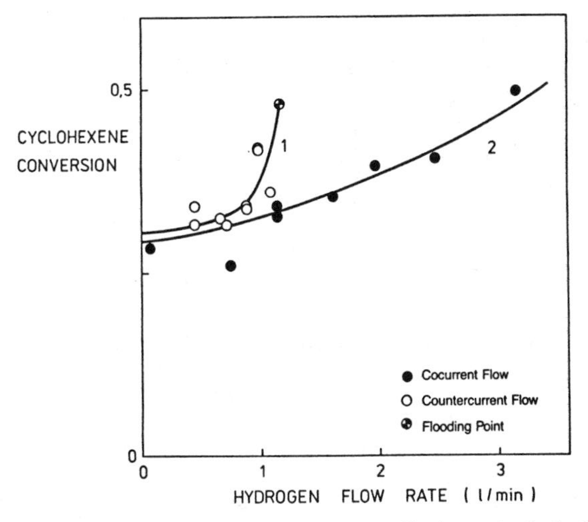

Figure 11. Plot of reaction conversion versus the flow rate of hydrogen for the hydrogenation of cyclohexene in a trickle-bed reactor [63]. $T_o = 25°C$, atmospheric pressure, $h = 5$ cm, $f = 1.5$ ml/min. (1—countercurrent; 2—cocurrent)

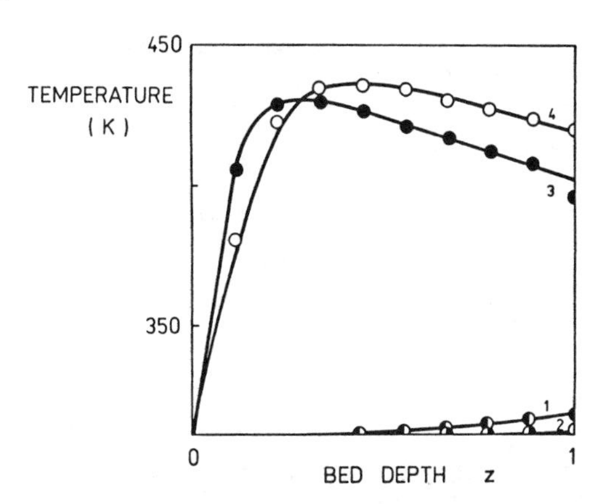

Figure 12. The effect of the hydrogen flow rate on the axial temperature profile during cyclohexene hydrogenation in an adiabatic trickle-bed reactor [43]. $C_o = 35$ wt%; $T_o = 310°K$. Curve 1—$F_H = 1$ l/min; Curve 2—$F_H = 2.7$ l/min; Curve 3—$F_H = 2.8$ l/min; Curve 4—$F_H = 3.7$ l/min.

Increased flow rate of gas may also bring about evaporation of the liquid phase [12]. Similarly as in the case of increased temperature or concentration of substrate one can distinguish three regimes of the reaction: the liquid-phase regime, the gas-phase regime, and the transition regime. In the transition regime the reactor exhibits strong hysteresis.

Information about the instantaneous state of the reactor provides measurements of the temperature profiles along the axis of the reactor. A conspicuous change of the temperature profile in an adiabatic trickle-bed reactor following an increase of the flow rate of hydrogen [43] is illustrated in Figure 12. Corresponding dependence of the reaction conversion on the flow rate of hydrogen is plotted in Figure 13. These results again indicate practically a step change transition from the liquid-phase to the gas-phase regime. Figure 13 shows also the

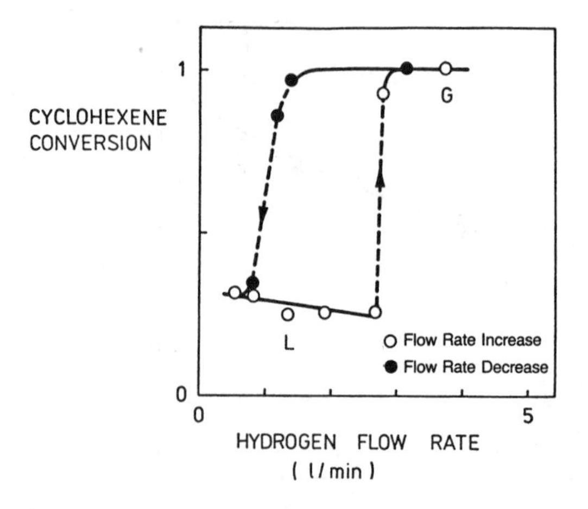

Figure 13. The effect of the hydrogen flow rate on cyclohexene conversion [43] $C_o = 35$ wt%; $T_o = 310°K$.

Table 1
The Efficiency of a Trickle-Bed Reactor from the Standpoint of External and Internal Diffusion [62]

Particle Diameter (cm)	Degree of Utilization of Internal Catalytic Surface η_{int}	External Efficiency of the Catalyst η_{ext}	Overall Efficiency η_{total}
0.052	0.20	0.15	0.03
0.155	0.095	0.16	0.015
0.40	0.022	0.24	0.005

d_t = 1.33 cm; h = 10 cm; F = 3.5 cm^3/min; T_0 = 25 °C; *atmospheric pressure.*

dependence of the conversion on the flow rate of hydrogen at a gradually decreased feed rate of hydrogen with clearly hysteretic features.

Flow Rate of the Liquid Phase

A change of the flow rate of liquid brings along two contradicting results. Increasing flow rate increases the thickness of the liquid film and with it associated resistance to the transport of the gaseous reactant. On the other hand, the liquid is better mixed and thus contributes positively to the reactor productivity. From the published experimental results (e.g., Reference 63) it follows that the reactor productivity depends little on the flow rate of the liquid phase. Only in case of very small catalyst particles was it observed [63] that the reaction rate, or the productivity, increases with increasing flow rate of liquid.

An attempt to increase the productivity of the reactor by increased flow rate of liquid, however, is not very practicable for it increases erosion of the catalyst bed, decreases the degree of conversion and hence increases the requirements on separation of the products and circulation of the reaction mixture.

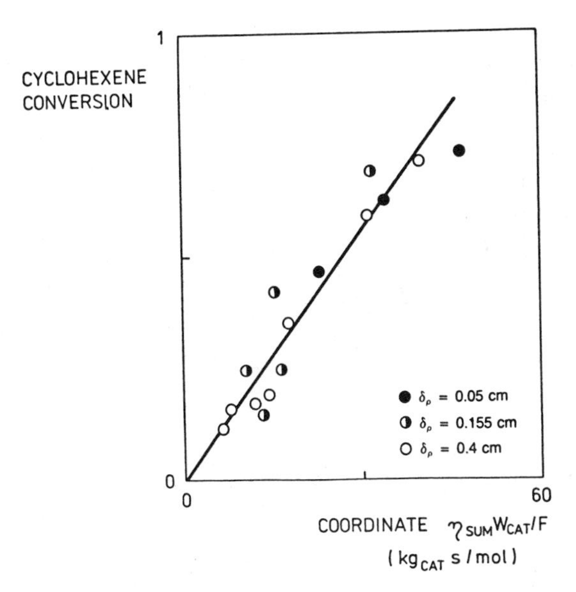

Figure 14. The effect of particle diameter on the dependence of cyclohexene conversion versus η_{sum} W_{cat}/F coordinate [62]. $T_o = 25°C$; atmospheric pressure; h = 10 cm.

Size of Catalyst Pellets

The efficiency of the reaction system from the viewpoint of external mass transfer relates to the diffusion of the reaction components in the liquid film, the turbulence in the menisci between neighboring catalyst particles and eventual nonuniformity of the flow of liquid in the reactor. External efficiency was studied [62] on hydrogenation of cyclohexene on a palladium catalyst and the results are summarized, together with the degree of utilization of the internal catalyst surface, in Table 1. A measure of the overall degree of utilization of the given catalyst is, of course, the product of the two previous quantities, given also in the table.

While the degree of utilization of the internal surface of the catalyst grows with the decreasing size of the catalyst pellet, the efficiency of the reactor from the standpoint of external diffusion diminishes. These results lead to the conclusion that there is an optimum size of the catalyst pellet for which the reaction rate is maximal. In practical situations, however, one must take into consideration also the impact on the pressure losses. The costs of for instance, pumping gas through the reactor strongly depend on the size of catalyst particles.

The effect of particle size on the reaction rate is also apparent from Figure 14 depicting the dependence of the reaction conversion on the coordinate $\eta_{sum}W_{cat}/F$ for three different particle diameters [62]. The figure shows clearly the good correlation between the apparent overall reaction rates measured for different particle sizes. The total effectiveness factor, η_{sum}, was taken from Table 1.

From the data in Table 1 it follows that, as expected, external diffusion of the reaction component markedly affects the overall resistance to mass transfer in the trickle-bed reactor. It is clear, however, that the extent of the effect external resistance varies with the magnitude of the reaction rate constant and will be the more significant the greater the activity of the

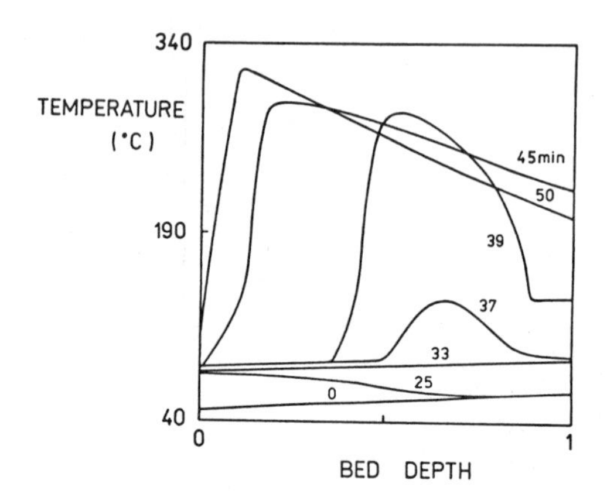

Figure 15. Behavior of the hot spot following the evaporation of the reaction mixture during hydrogenation of 1,5-cyclooctadiene on a palladium catalyst [44] after a feed temperature increase by 32.5°C; $d_t = 30$ mm; $d_p = 4$ mm; $C_o = 50\%$ wt.

catalyst used. It should be noted that the palladium catalyst used in Reference 62 may be rated as one with relatively high activity. Accordingly, the measured total efficiency of the system both from the standpoint of external diffusion and the diffusion within the catalyst pellet is relatively low.

Dynamic Behavior of Trickle-Bed Reactors

Much of the existing literature concentrates on the steady-state operation of trickle-bed reactors. Dynamic properties of laboratory trickle-bed reactors were studied only in a few papers, dealing, for example, with the olefine hydrogenations [43–45, 66].

A very interesting feature of the trickle-bed reactor is its dynamics associated with the transient development of the axial temperature profiles within the bed and the time dependence of concentration of the product in the exit stream [45]. Axial temperature profiles measured in the trickle-bed reactor pointed at the formation of the hot spot as a result of evaporation of the reaction mixture [43–45, 66]. The motion of the hot spot was caused by axial heat transfer and resembled propagation of the flame during combustion (see Figure 15). It was also found [45] that the time required for the reaction system to reach the steady state strongly varied depending on whether the reactor was cooled or operated adiabatically.

The transient axial temperature profiles measured [164] in a laboratory trickle-bed reactor for the hydrogenation of cyclohexene are shown in Figures 16 and 17. The former figure illustrates the situation when the system operated in the liquid regime; the later figure reveals formation of the hot spot in the upper section of the bed. The temperature here exceeded that of the boiling point of the cyclohexene-cyclohexane mixture (82° C). The time required to reach the steady state amounted in this case to nearly 10 minutes.

A special problem in the operation of the trickle beds arises in connection with the evolution of the heat of wetting after the first contact of liquid with the dry layer of the porous catalyst. The heat of wetting reaches considerable values exceeding those typical for the adsorption heats. The heat of wetting is practically proportional to the internal surface of the catalyst and it is therefore of major concern for catalysts of large internal surface such as activated carbon, less so for alumina and silica gel.

The evolution of the heat of wetting gives rise to transient temperature profiles in originally dry catalyst beds [172]. Such profiles were observed [164] after pouring cyclohexane on a palladium-treated activated carbon catalyst and are shown in Figure 18. The peak temperature of the hot spot exceeded the temperature of the boiling point of the hydrogenated mixture.

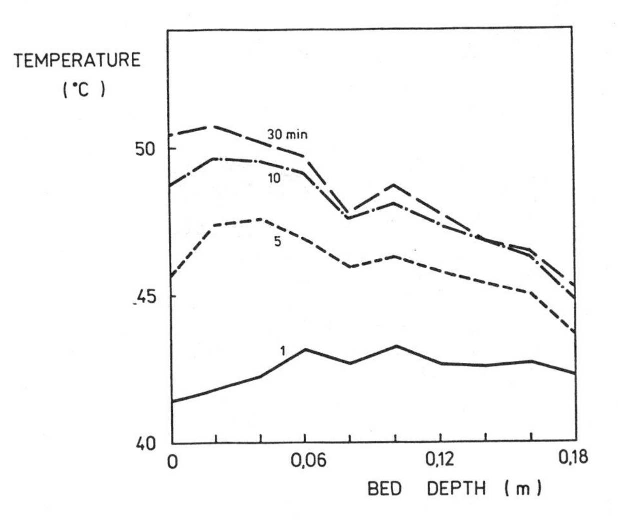

Figure 16. Dynamic axial temperature profiles in a cooled trickle-bed reactor for prior to reaching the steady state in the liquid-phase hydrogenation of cyclohexene on a palladium catalyst [164]; $C_o = 42\%$ wt; $T_o = 40°C$ (temperature of the feed and in the jacket); $d_t = 30$ mm; $d_p = 4$ mm; $F = 0.125$ l/hr.

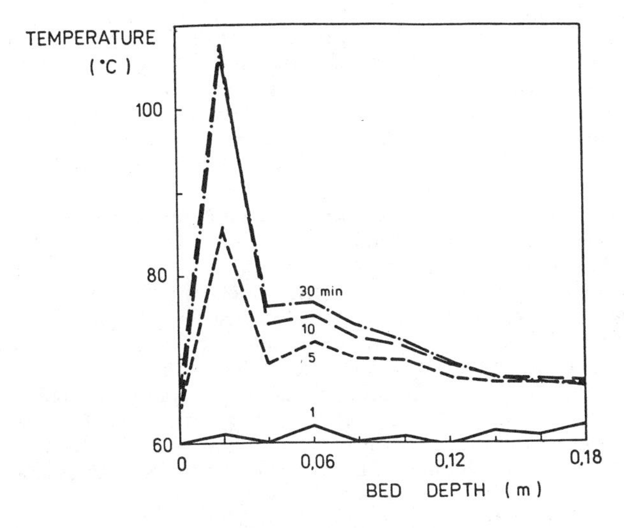

Figure 17. Formation of the hot spot in a cooled trickle-bed reactor prior to reaching the steady state [164] during hydrogenation of cyclohexene on a palladium catalyst; inlet concentration $C_o = 41\%$ wt; $T_o = 60°C$ (temperature of the feed and in the jacket); $d_t = 30$ mm; $d_p = 4$ mm; $F = 0.24$ l/hr.

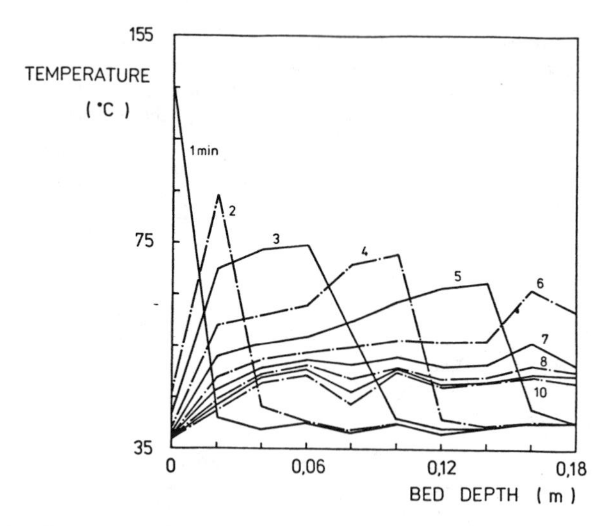

Figure 18. Transient axial temperature profiles following the wetting of the dry palladium catalyst by cyclohexane [164]. $T_o = 40°C$; $d_t = 30$ mm; $d_p = 4$ mm; $F = 0.125$ l/hr.

DESIGN OF TRICKLE-BED REACTORS

Distribution and Mixing of Liquid

Radial Distribution and the Wall Flow of Liquid

An important prerequisite for the uniform distribution of the holdup [83, 73] over the reactor cross section as well as for the uniform distribution of local velocities and catalyst wetting is the knowledge of radial distribution of liquid. Unfortunately even a good function of the liquid distributor, providing a uniform feed of the substrate on the top of the catalytic layer, does not guarantee this uniformity throughout the reactor volume. Particularly the beds where the ratio of the reactor to particle diameter is low tend to transport liquid toward the reactor wall to form there the wall flow. Even though the wall flow is not a total bypass (for there always exists dynamic exchange between the liquid in the bed and that captured in the wall flow) the mere existence of the wall flow may seriously slow down the course of the reaction for the wall is not catalytically active.

Experimental methods of measurement of the distribution of liquid and the wall flow permit one to classify various liquid distributors used to feed the substrate on top of the catalytic layer. Figure 19 shows [174] qualitatively the plot of the fractional amount of the wall flow as a function of the depth of the catalytic section for a central jet (a nozzle located in the axis of the reactor), the uniform distributor, and for the wall source (all liquid on the top is fed onto the reactor wall). It is apparent that after a certain depth of the packed layer (in Figure 19 delimited by the broken line) the wall flow no longer changes, and it is essentially in hydrodynamic equilibrium with the flow within the bed. An example of an experimental profile of the density of irrigation [175] is shown in Figure 20. The curve was computed from the theoretical model presented in the following.

The flow patterns of the reaction mixture in trickle-bed reactors vary also with the physical properties of liquid. Interesting experimental characteristics based on the holdup

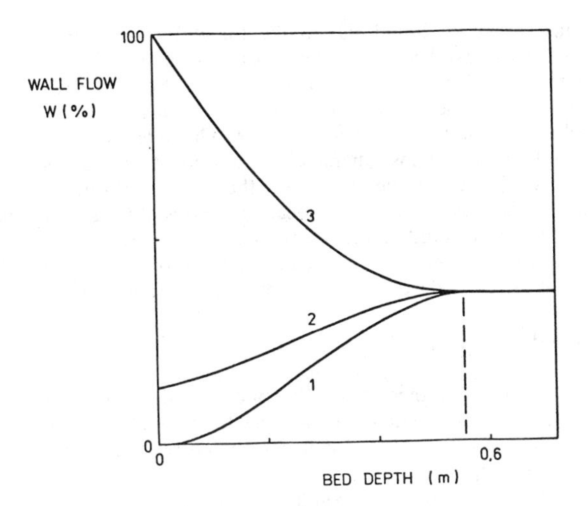

Figure 19. The wall flow as a function of the axial coordinate for various liquid distributors [174]. Curve 1—central jet; Curve 2—uniform distributor; Curve 3—wall distributor.

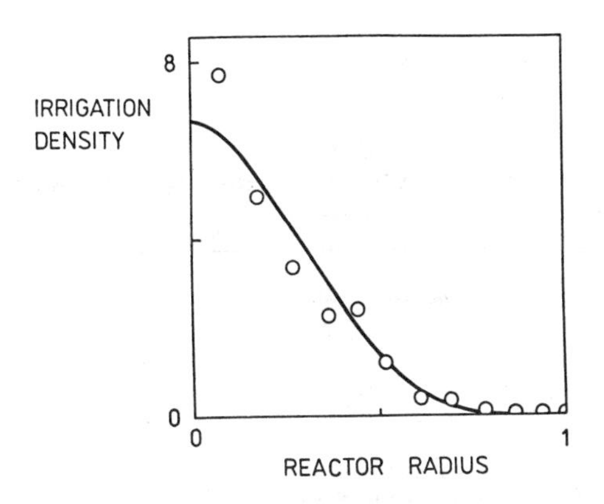

Figure 20. Radial profile of the density of irrigation of hydrocarbon mixture in a trickle-bed reactor packed by cylindrical nickel catalyst pellets irrigated by a central jet [175]. $d_t = 0.25$ m; $d_p = 7$ mm; $h = 0.4$ m; $F = 0.2$ m^3/hr.

measurements for foaming, nonfoaming, and viscous organic liquids were published in Reference 160.

Models to predict the distribution of liquid in a bed of particles have been based so far on the percolation theory [103, 129], or the theory of diffusion. Regarding the radial motion of individual liquid elements in the bed of catalyst as a random process tied to the descending motion through the bed of catalyst, the local density of irrigation, f, expressed as a fraction of the mean superficial velocity of liquid, is governed by the following equation derived by Cihla and Schmidt [155]

$$\left(\frac{Dh}{R^2}\right)\left[\frac{\partial^2 f}{\partial r^2} + \frac{1}{r}\frac{\partial f}{\partial r}\right] = \frac{\partial f}{\partial z} \quad (1)$$

The coefficient of radial spread of liquid, D, unlike the diffusion coefficient, has the dimension of length. Its values have been evaluated mostly on the basis of the solution of Equation 1 for the central jet [80] in a radially unconfined bed. For this reason, however, the results display a seeming dependence on the mean density of irrigation owing to the flooding of the bed below the mouth of the feed nozzle. This effectively transforms the central "point" source into a disc whose diameter is dependent on the feed rate [87]. Correction of data on this effect essentially eliminates the mean feed rate dependence of D, even though certain complexities related to the local conditions of irrigation remain [88].

In consistence with the theory of random processes, D should be proportional to the size of the particle forming the random bed [82] but the results both for nonporous and porous particles are rather inconclusive. Onda et al. [102] proposed the following correlation

$$D = 0.169 d_p^{0.5} \quad (D \text{ and } d_p \text{ in cm}) \tag{2}$$

which yields results close to those of other authors in spite of its dependence on the square root of particle diameter. This agreement is mainly due to the fact that size of catalyst particles is very close to that of the particles for which the correlation was formed. As a rule of thumb one can take $D/d_p = 0.1$.

In the reactor the bed of catalyst particles is confined by the reactors walls. The behavior of the liquid here has been specified by various boundary conditions [79, 81, 96, 131, 155].

Table 2
Theoretical Distributions of the Density of Irrigation for Various Distributors

Distributor	Equation
Central point source	
$f = \begin{cases} \nearrow \infty & r = 0 \\ \searrow 0 & 0 < r \le 1 \end{cases}$	$f = \frac{C}{1+C} + \sum_n \frac{[(q_n^2/B) - 2C]^2 J_0(q_n r) \exp(-q_n^2 zH)}{\{[(q_n^2/B) - 2C]^2 + q_n^2 + 4C\} J_0^2(q_n)}$
$2\int_0^1 rf\,dr = 1; \quad z = 0$	$W = \frac{1}{1+C} - \sum_n \frac{2[(q_n^2/B) - 2C] \exp(-q_n^2 zH)}{\{[(q_n^2/B) - 2C]^2 + q_n^2 + 4C\} J_0(q_n)}$
Disc distributor	
$f = (1/r_1)^2, \quad 0 \le r < r_1$	$f = \frac{C}{1+C} + \frac{1}{r_1} \sum_n \frac{2[(q_n^2/B) - 2C]^2 J_0(q_n r) J_1(q_n r_1) \exp(-q_n^2 zH)}{\{[(q_n^2/B) - 2C]^2 + q_n^2 + 4C\} q_n J_0^2(q_n)}$
$f = 0, \quad r_1 < r \le 1$	$W = \frac{1}{1+C} - \frac{1}{r_1} \sum_n \frac{4[(q_n^2/B) - 2C] J_1(q_n r_1) \exp(-q_n^2 zH)}{\{[(q_n^2/B) - 2C]^2 + q_n^2 + 4C\} q_n J_0(q_n)}$
Circular distributor	
$f = \begin{cases} \nearrow \infty & r = r_1 \\ \rightarrow 0 & 0 \le r < r_1 \\ \searrow 0 & r_1 < r \le 1 \end{cases}$	$f = \frac{C}{1+C} + \sum_n \frac{[(q_n^2/B) - 2C]^2 J_0(q_n r) J_0(q_n r_1) \exp(-q_n^2 zH)}{\{[(q_n^2/B) - 2C]^2 + q_n^2 + 4C\} J_0^2(q_n)}$
$2\int_0^1 rf\,dr = 1; \quad z = 0$	$W = \frac{1}{1+C} - \sum_n \frac{2[(q_n^2/B) - 2C] J_0(q_n r_1) \exp(-q_n^2 zH)}{\{[(q_n^2/B) - 2C]^2 + q_n^2 + 4C\} J_0(q_n)}$
Annular distributor	
$f = \frac{1}{r_2^2 - r_1^2}, \quad r_1 < r < r_2$	$f = \frac{C}{1+C} - \frac{1}{r_2^2 - r_1^2} \sum_n \frac{2[(q_n^2/B) - 2C]^2 J_0(q_n r)[r_2 J_1(q_n r_2) - r_1 J_1(q_n r_1)] \exp(-q_n^2 zH)}{\{[(q_n^2/B) - 2C]^2 + q_n^2 + 4C\} q_n J_0^2(q_n)}$
$f = 0 \begin{cases} \nearrow 0 \le r < r_1 \\ \searrow r_2 < r \le 1 \end{cases}$	$W = \frac{1}{1+C} - \frac{1}{r_2^2 - r_1^2} \sum_n \frac{4[(q_n^2/B) - 2C][r_2 J_1(q_n r_2) - r_1 J_1(q_n r_1)] \exp(-q_n^2 zH)}{\{[(q_n^2/B) - 2C]^2 + q_n^2 + 4C\} q_n J_0(q_n)}$
q_n are roots of	$\left(\frac{2C}{q_n} - \frac{q_n}{B}\right) J_1(q_n) + J_0(q_n) = 0$

The following is the boundary condition that relates the exchange of liquid between the particles in the promixity of the wall and the wall [79, 81]

$$-\frac{\partial f}{\partial r} = B(f - CW) \quad \text{for} \quad r = 1 \tag{3}$$

in terms of the local density of irrigation, f, the wall flow W (as a fraction of the total volume flow rate of liquid) and two dimensionless distribution parameters B and C.

Table 2 summarizes solutions of Equation 1 for various initial conditions, given by the type of the feed distributor used. Solution for an arbitrary type of initial feed of liquid was obtained, too [85].

For deep beds the dimensionless density of irrigation equals uniformly $C/(1+C)$ while the fractional wall flow amounts to $1/(1+C)$. The parameter C of the boundary condition Equation 3 thus controls the magnitude of the wall flow and its values were correlated against the ratio of the reactor to particle diameter ratio by [84]

$$C = k(d_t/d_p) \tag{4}$$

on the basis of measurement of several authors [81, 84, 96, 130, 131] with k equalling 0.365 for spherical particles and 0.181 for Raschig rings. The same source [84] gives B equal to 6.7 and 7 for spheres and Raschig rings, independent of their size.

Special attention should be paid to the design of the liquid distributor, particularly that for low mean densities of irrigation. The requirement of creating a large number of rivulets and hence the large interfacial area calls for additional energy. The kinetic and potential energy of liquid later in the bed is insufficient for this purpose.

Uniform liquid distribution is only the first prerequisite for high contacting efficiency of the catalyst. Apart from the mentioned difficulties encountered at low liquid loads, the high liquid surface tension may cause the liquids to remain largely in the form of rivulets instead of the film reducing thus the coverage of the catalyst surface.

Radial Mixing of Liquid

A significant disadvantage of the trickle-bed reactors is the low degree of radial mixing. This results in a low ability to equalize eventual lateral concentration differences appearing due to the locally different contacting efficiency, catalytic activity, or structural inhomogeneity of the bed. An even more important impact is that on the heat transfer properties of the bed. The bed cannot transport heat rapidly enough to avoid local overheating with subsequent damage to the catalyst.

The major mechanism of radial transport of a species in the reactor is the dispersion with the bulk liquid of the type described in Equation 1. A balance on the species, present in the trickling liquid at a relative concentration c is obtained by writing Equation 1 for the quantity (fc), where f is the relative local density of irrigation. After some arrangement and use of Equation 1 one obtains

$$\left(\frac{Dh}{R^2}\right) f\left[\frac{\partial^2 c}{\partial r^2} + \frac{1}{r}\frac{\partial c}{\partial r}\right] = f\frac{\partial c}{\partial z} - 2\left(\frac{Dh}{R^2}\right)\frac{\partial c}{\partial r}\frac{\partial f}{\partial r} \tag{5}$$

This equation clearly demonstrates the combined radial transport due to the concentration gradients and the gradients of the density of irrigation, through a common parameter D—the coefficient of radial spread. Provided that the irrigation of the catalyst is uniform throughout, the second term on the right-hand side of Equation 5 vanishes and its solution can be obtained analytically [86, 104].

Under nonuniform irrigation and/or the existence of the wall flow one has to resort to solving the complete Equation 5 with the boundary condition obtained by developing Equation 3 into [36]

$$-\frac{\partial(fc)}{\partial r} = B(fc - CWc_w) \quad \text{for} \quad r = 1 \tag{6}$$

or after some arrangement using Equation 3

$$-f\frac{\partial c}{\partial r} = BCW(c - c_w) \tag{7}$$

The symbol c_w designates concentration of species in the wall flow which, under these circumstances, is different from that in the bed.

The characteristic of the radial mixing is thus the coefficient of radial spread, or the ratio d_p/D, which may be taken to be the particle-size-related Peclet (Bodenstein) number. For its values see the paragraph on liquid distribution.

Axial Mixing and the Stagnant Zones in Liquid

The principal mechanisms causing deviations from the plug flow in trickle-bed reactors are the different lengths and tortuosities of paths of individual liquid elements, time delays at particle junctions, and the stagnant zones exchanging mass with the moving liquid.

Initial information about the flow may be obtained from the residence time distribution measurement in the bed [61, 173]. Porous catalyst particles trap part of the liquid, and the RTD curves exhibit then considerable "tailing." The RTD technique has been also used to assess the solid-liquid contacting in trickle beds [161].

Clearly, any deviation from the plug flow, either due to the wall flow, the stagnant zones, or the exchange of liquid with the internal volume of the catalyst changes the productivity and eventually the selectivity of the reaction.

The model that at present appears to best describe all these phenomena is the three parameter cross-flow dispersion model [106, 156]. Liquid holdup is assumed to be split into the dynamic portion, h_d, moving at axially dispersed plug flow and exchanging liquid with the stagnant portion of the holdup, h_s:

$$E_l h_d \frac{\partial^2 c}{\partial z^2} - v_l \frac{\partial c}{\partial z} - q(c - c_s) = h_d \frac{\partial c}{\partial t} \tag{8}$$

The coefficient of exchange between the stagnant and the dynamic region is designated as q and the concentration here in the stagnant zone is c_s. For the latter we may write

$$q(c - c_s) = h_s \frac{\partial c_s}{\partial t} \tag{9}$$

Michelsen and Ostergaard [105] published solution of this model as a response of an infinitely long reactor to an impulse. Bennet and Goodridge [156] obtained the response of a finite reactor to a step change in input. The response to a sinusoidal input has been found by Staněk and Čársky [89, 90].

The cross-flow dispersion model encompasses asymptotically simpler models, such as the axially dispersed model and the cross flow model as well as the plug flow model. It should be borne in mind, however, that the Peclet numbers obtained on the basis of the cross-flow model and the cross-flow dispersion model are fundamentally different and so are their values. This is so because nearly all mechanisms leading to the deviations from the plug flow in trickle-bed reactors can be reasonably accurately described by the diffusional mechanism. The axial dispersion model thus has been found satisfactorily describing the experimental data, but the use of such data for scale-up or for systems with mass transfer and/or chemical reaction is not advisable.

Experimental results yield values of the total holdup that are little sensitive to the choice of the model [111]. However, the stagnant and the dynamic holdups are not generally equal the static holdup (liquid retained after draining the bed) and the operating holdup (sometimes also termed dynamic holdup and equal the difference of the total and the stagnant holdup). Sicardi et al. [112] found the static holdup lower than the stagnant holdup.

The stagnant holdup as a fraction of the total holdup decreases with increasing liquid load [112, 111], even though the total holdup increases.

Numerical values of the fractional stagnant holdup substantially differ for the bed composed of spherical particles or Raschig rings. This points at the strong role of the shape of the packing in affecting hydrodynamic activity of the holdup.

The coefficient of mass exchange between the stagnant and the dynamic liquid is usually correlated [112, 156] as a dimensional group (qd_p/h_s) which strongly increases with the liquid load. The scatter of data from different sources, however, seems to indicate that this group does not properly reflect the effect of particle size and that the effect of the particle shape is also important. Numerical values of this parameter rule out any role of molecular diffusion in the mechanism of the exchange between the stagnant and dynamic regions.

The particle-diameter-related Peclet number reaches values 3 to 10 times smaller for the two-phase downflow compared to the single-phase flow. This points at a much more intensive axial mixing. According to Matsuura et al. [113] the dependence of the Peclet number on the Reynolds number for the two-phase flow has a qualitatively analogous course as in the case of the single-phase flow: Below $Re = 150$ Pe is approximately constant and equal 0.43; above $Re = 400$ Pe equals constantly 1.7. Sicardi et al. [112], however, found lower values of the Peclet number in the lower Re region.

All the previous parameters have been found mostly independent of the gas velocity. Recently, however, Tosun [99] indicated that the relationship between Pe, based on a purely axial dispersion model, and the gas velocity may not be quite simple.

A frequency response analysis of the trickle flow [89, 90, 111] indicates that at low liquid loads the flow exhibits both the dispersion and stagnant zone features. At higher loads the dominant effect is that of the stagnant zones, reducing the previous model to a purely cross-flow model. The time constant for the exchange of liquid in the stagnant zones remains, however, significant compared to the time constant of the contact between the dynamic and the stagnant liquid. The effect of the stagnant zones remains thus important even at extremely large irrigation rates.

Nevertheless, axial dispersion, need not be always of major concern for systems with chemical reaction. No appreciable effect is to be expected if the bed is sufficiently long (large values of the Peclet number based on bed depth), if high conversions are not required and if the reaction does not proceed on only a relatively short section of otherwise deep catalyst bed (large concentration gradients).

Hydrodynamic Behavior

Pressure Drop Under the Two-Phase Downflow

Pressure drop is an important design factor as it affects the energy demand of the process and the dissipated energy also partially contributes to increased intensity of mass transfer and mixing. At low liquid loadings there is practically no interaction between the flow of gas and liquid. Liquid holdup is thus practically independent of gas velocity both at the cocurrent and counter-current flow arrangement. At increased liquid loads the character of liquid flow is modified by interfacial friction and the pressure drop progresively increases. At counter-current operation the upper limit for the liquid loading is the flooding of the bed when a layer of gas/liquid mixture appears on the top of the bed.

Two approaches essentially have been applied to correlate the two-phase pressure drop. One, based on the Lockhart–Martinelli theory [114] of the two-phase flow in pipes extended to packed beds [115, 116, 117, 118, 119, 120, 121] utilizes specific (per unit length) pressure losses for single flow of gas ΔP_g and liquid ΔP_l to calculate the specific two-phase pressure drop, ΔP_{lg}. On defining the Lockhart–Martinelli parameter X as

$$X = (\Delta P_l/\Delta P_g)^{0.5} \tag{10}$$

the two-phase pressure drop, according to Midoux et al. [117] may be computed for a nonfoaming liquid from

$$(\Delta P_{lg}/\Delta P_l)^{0.5} = 1 + \frac{1}{X} + \frac{1.14}{X^{0.54}} \tag{11}$$

For a foaming liquid (index f) the situation is more complicated and the same authors give the following correlation in terms of the Lockhart–Martinelli factor for foaming systems, X′

$$(\Delta P_{flg}/\Delta P_{fl})^{0.5} = 1 + \frac{1}{X'} + \frac{6.55}{X'^{0.43}} \tag{12}$$

where

$$X' = (\Delta P_{fl}/\Delta P_{fg})^{0.5} \tag{13}$$

The definition of the pressure drop for foaming systems, however, was somewhat modified [117].

The second approach relies on the Ergun Equation 124. Other authors modified this equation or recommended slightly different numerical coefficients. These modifications, however, do not seem substantial in view of the fact that the void fraction in this correlation is raised to the third power which in itself causes high sensitivity of results to relatively small variations of the void fraction. Particularly for nonspherical particles both bulk and local voidage vary with the method of packing the particles into the reactor. Also in the proximity of the reactor walls and supporting grids the voidage is altered. This causes local variations of the resistance to the gas flow, which in turn tends to flow preferentially through lower resistance regions. The isobars are then no longer planes perpendicular to the reactor axis. The shape of the isobaric surfaces and the pattern of the streamlines can be computed from the vectorial differential form of the Ergun equation published by Staněk and Szekely [77] in the form

$$-\overrightarrow{\text{grad}\,P} = \frac{150\mu_g(1-\varepsilon)^2}{d_p^2\varepsilon^3}\vec{v}_g + \frac{1.75\varrho_g(1-\varepsilon)}{d_p^2\varepsilon^3}v_g\vec{v}_g = $$
$$= f_1\vec{v}_g + f_2 v_g\vec{v}_g \tag{14}$$

In this form of the Ergun equation $\vec{v}_g$ designates the vector of the local gas superficial velocity and v_g its absolute magnitude. The gradient of pressure on the left hand side is naturally a vector too. The coefficients of the vector quantities may be functions of the spatial coordinated through the locally variable voidage, particle size, etc. Since the flowing gas responds relatively rapidly to the radial variations of the local resistance the radial profile of the axial component of the gas superficial velocity, v_{gz}, can be computed from the following Equation 77:

$$v_{gz} = -(f_1/f_2) + \sqrt{(f_1/f_2)^2 + k/f_2} \tag{15}$$

where the coefficients f_1, f_2 are taken to be functions of spatial coordinates. The constant k is determined so as to make the integral mean of v_{gz} over the whole reactor radius equal mean v_g.

At low gas and liquid loads the two-phase pressure drop can be computed from the Ergun equation with the voidage reduced by the volume occupied by liquid holdup. This is also the approach followed by Hutton and Leung [122] and more recently perfected by Specchia and Baldi [123]

$$\Delta P_{lg} = k_1 \frac{(1-h_l-\varepsilon)^2}{(\varepsilon-h_l)^3} \mu_g v_g + k_2 \frac{(1-h_l-\varepsilon)}{(\varepsilon-h_l)^3} \varrho_g v_g^2 \tag{16}$$

where h_l is the total liquid holdup. Similarly as in the original form of the Ergun equation the coefficients k_1, k_2 depend on particle shape and size but their values are evaluated from pressure drop measurement with the wet packing, where h_l reduces to the static holdup. Resulting values differ significantly from the original Ergun coefficients.

Due to the sensitivity of the two-phase pressure drop to the effective voidage, the former quantity is a sensitive criterion of the instant state of the catalytic bed as a whole or its individual sections if axial pressure profiles, taken through a series of pressure taps along the reactor length, are available. Local increase of the pressure difference is mostly induced by deposits from the liquids or fines carried by the gas phase.

In hydroprocessing operations, under the trickle-flow conditions with small catalyst pellets the following empirical correlation of Clements and Schmidt [158] can be recommended to calculate the two-phase pressure drop

$$\frac{\Delta P_{lg}}{\Delta P_g} = 1{,}507 \mu_l d_p \left(\frac{\varepsilon}{1-\varepsilon}\right)^3 \left(\frac{Re_g We_g}{Re_l}\right)^{-1/3} \tag{17}$$

where d_p is in (m), μ_l in (g/(m · s)), ΔP in (atm/m) and Re_g, Re_l, and We_g designate the Reynolds and the Weber number for gas and liquid.

Liquid Holdup

The literature is relatively rich in data on liquid holdup both for nonporous and porous particles. The interest in the holdup in trickle-bed reactor design stems from its correlation with the efficiency of catalyst wetting, the intraparticle resistance to transport, and two-phase pressure drop.

Liquid holdup is usually defined as a fraction of the volume of the reactor occupied by the liquid or, alternatively, as a fraction of the void volume of the reactor. For porous particles the total holdup is substantially higher and the pore volume accessible to liquid may compose 15 and more percent of the reactor volume. This portion of the total holdup is termed the internal holdup.

The holdup retained in the bed after it was allowed to drain is termed the static holdup. Static holdup was correlated by Mersmann [125] and Charpentier et al. [126] against the Eotvos number and again by Mersmann [127] against the Bond number.

A frequent practice is to present correlations for the dynamic holdup, i.e. total holdup reduced by the static contribution. A typical example is the old, yet widely used correlation of Otake [128]

$$h_d = 1.25\, Re_l^{0.676}\, Ga^{-0.44}\, (a_s d_p) \tag{18}$$

Under the cocurrent flow the gas exercises an accelerating effect on liquid and the gravity factor in the Galileo number must be corrected accordingly.

$$Ga = \frac{d_p^3 \varrho_l(\varrho_l g + \Delta P_{lg})}{\mu_l^2} \tag{19}$$

The credibility of this correlation is further increased by the similarity with the more recent correlation of Specchia and Baldi [123]

$$h_d = 3.86\, Re_l^{0.545}\, Ga_l^{-0.42}\, (a_s d_p/\varepsilon)^{0.65} \tag{20}$$

In the high interaction regime these authors recommend

$$h_d = a'\left[\frac{Re_g^{1.167}}{Re_l^{0.767}\psi^{1.1}}\right]^{b'} (a_s d_p/\varepsilon)^{0.65}$$

$$\psi = \frac{\sigma_{wat}}{\sigma_l}\left[\frac{\mu_L}{\mu_{wat}}\left(\frac{\varrho_{wat}}{\varrho_l}\right)^2\right]^{0.33} \tag{21}$$

with the following values for the constants a', b':

Nonfoaming liquid	$a' = 0.125$	$b' = -0.312$
Foaming liquid	$a' = 0.0616$	$b' = -0.172$

The flow parameter ψ accounts for the difference between surface tension, viscosity, and density of water (subscript wat) and other liquids (subscript 1).

It is stressed that the dynamic holdup, appearing in Equations 20 and 21 is not identical with the dynamic holdup appearing in the formulation of the cross-flow dispersion model. In fact a better term for the holdup in the last two correlations is the operating holdup. For the same reason the static holdup is not the same as the stagnant holdup in the cross-flow dispersion model.

Based on the Lockhart–Martinelli theory and the earlier defined Lockhart–Martinelli parameter $X = (\Delta P_l/\Delta P_g)^{0.5}$, Sato et al. [115] related the total holdup to the specific surface of the bed, a_s, given in this case as

$$a_s = 6(1-\varepsilon)/d^1 \tag{22}$$

where

$$d^1 = d_p\Big/\left(1 + \frac{4d_p}{6d_t(1-\varepsilon)}\right) \tag{23}$$

by

$$h_l = 0.4\, a_s^{1/3}(X)^{0.22} \quad \text{for} \quad 0.1 < X < 20 \tag{24}$$

Midoux et al. [117] extended this approach by using the corrected Lockhart–Martinelli parameter, X', defined by Charpentier et al. [120] to obtain

$$h_l = \frac{0.66(X')^{0.81}}{1+0.66(X')^{0.81}} \tag{25}$$

for nonfoaming liquids and $0.1 < X' < 80$.

For foaming liquids the same authors [117] found

$$h_l = \frac{0.92(X')^{0.3}}{1+0.92(X')^{0.3}} \tag{26}$$

with its validity being confined to $0.05 < X' < 100$.

Using the automodel properties a versatile correlation of the total liquid holdup under the single-phase trickle flow of liquid was formulated by Jiřičný and Staněk [93] in the form

$$h_l/h_{lf} = 0.8216\left(\frac{v_l}{v_{lf}} - 1\right) + 1 \tag{27}$$

where h_{lf} and v_{lf} designate total liquid holdup and superficial velocity of liquid at the flooding point. The flooding point was defined and experimentally determined as a point where first portions of liquid appear on the top of the packed section. The numerical coefficient of the correlation Equation 27 was found versatile (based on experiments with glass spheres and Raschig rings of various sizes) with a standard deviation of the reduced holdup equalling 0.031. The correlation was found to be valid also for two-phase counter-current flow [94].

Practical utilization fo this versatile correlation mandates knowledge of the flooding velocity and holdup. In the absence of such information the correlation can be used in reverse to calculate flooding velocity and flooding holdup from two values of the total holdup.

An equation for the dynamic holdup under cocurrent gas-liquid downflow in packed beds was developed in Reference [162] on the basis of models in which interactions between the liquid holdup, pressure gradient, and liquid flow rate in the gravity-viscosity and the gravity-inertia regimes were considered.

Catalyst Wetting Efficiency

An important factor that may affect the conversion rate in a trickle-bed reactor, associated with its hydrodynamics, is the catalyst wetting [165]. The incomplete wetting gives rise to an apparent increase of the evaluated kinetic rate constant with increasing liquid superficial velocity [2, 5]. Empirical scale-up criteria have been given [2, 149], taking into account contacting efficiency. Mears [151] assumed the reaction rate to be proportional to the effectively wetted fraction of the external geometrical surface of the pellets given by various correlations. Puranik and Vogelpohl [150] recommend for the effective wetted surface area of the packing correlations of the following form, separately for the dynamic contacting efficiency,

$$v_c = 1.05\, Re_l^{0.047}\, We_l^{0.135}\, (\sigma_l/\sigma_c)^{-0.206} \tag{28}$$

and the total contacting efficiency,

$$v_c' = 1.045\, Re_l^{0.041}\, We_l^{0.133}\, (\sigma_l/\sigma_c)^{-0.182} \tag{29}$$

where

$$Re_l = G_l d_p/\mu_l \tag{30}$$

$$We_l = G_l^2 d_p/(\sigma_l \varrho_l) \tag{31}$$

and σ_c designates the critical surface tension.

Under the conditions prevailing in trickle-bed reactors two types of wetting can be defined

1. Internal wetting, or catalyst pore filling, which in most cases is totally due to the capillarity effects

2. External effective wetting, defined as part of the geometrical surface area of the catalyst pellets, effectively contacted by flowing liquid.

External effective wetting may be different from physical external wetting since the pellets are also in contact with the stagnant zones of liquid that contribute little to mass transfer between the internal and the flowing liquid.

Both the internal and the external wetting efficiency of a porous catalyst was considered by Colombo et al. [165] in an analysis of the response curves of the reactor to a step inlet concentration change.

The effect of liquid wetting efficiency on catalyst effectiveness in trickle-bed reactors was considered also by Carpa et al. [166]. These authors concluded that intraparticle concentration profiles are strongly affected by the distribution and size of liquid rivulets trickling down the catalyst bed and that the overall effectiveness factor is not significantly affected by these rivulets.

Theoretical models of the effect of wetting efficiency on the global effectiveness factor are available in recent papers [167–170]. The role of liquid flow rate in affecting the wetting efficiency was assessed in References 13, 64, 165.

Transfer Phenomena

Gas/Liquid Mass Transfer

Gaseous reactants participating in liquid-phase trickle-bed reactions must be transported across the gas-liquid interface to be dissolved in the liquid. The overall rate of interfacial transfer depends generally on the resistance on both sides of the interface. For low-solubility gases, however, as is the case of most gaseous reactants in trickle-bed operations, the major portion of the resistance lies on the liquid side. The interfacial flux can thus be expressed using the liquid-side volumetric mass transfer coefficient $k_l a$, which is a product of the liquid-side mass transfer coefficient and the surface area of the gas-liquid interface per unit reactor volume. Standard mass transfer measurements do not allow separation of the two quantities and the majority of the correlations correlate directly to the product $k_l a$.

In countercurrent operation the volumetric mass transfer coefficient is practically independent of the gas load as may be apparent from the correlation of Onda et al. [133].

$$Sh = 0.0097\, Re_l^{0.67}\, Sc^{0.5}\, Ga^{0.3} \tag{32}$$

where Sh, Re_l, Sc, and Ga designate the Sherwood, Reynolds, Schmidt, and the Galileo numbers.

In cocurrent beds, however, the specific gas-liquid interfacial area is larger than under the countercurrent flow and considerably exceeds the specific external surface of the particles. In addition, it increases with both the gas and the liquid load.

It appears well established that the volumetric liquid-side mass transfer coefficient depends [134, 135] on the rate of energy dissipation $E_L = \Delta P_{lg} v_g$ by the following correlation:

$$k_l a = 0.0011 E_L \frac{D_A}{2.4 \times 10^{-9}} \tag{33}$$

valid for lower rates of E_L between 0.4 and 10 kg force per meter squared per second and by

$$k_l a = 0.0173 E_L^{0.5} \tag{34}$$

for higher rates of dissipation E_L between 5 and 1,000 in the same units.

Separate values of the specific area of gas-liquid interface were obtained by absorption of CO_2 into NaOH solutions and by absorption of oxygen into sodium bisulfite solutions by Shenda and Sharma [136]. The same authors obtained also values of the gas-side volume mass transfer coefficient.

Numerical values of the volume gas-liquid mass transfer coefficients may differ depending on the model used for the evaluation of the experimental data. Moravec and Staněk [92] and Moravec [137] obtained a frequency response for the cross-flow dispersion model for liquid with interfacial mass transfer [137] into the axially dispersed countercurrently flowing gas. They found that a mere change from the plug flow to the axially dispersed flow caused an increase of the $k_l a$ values by about 15%. The partial-volume mass transfer coefficients into the stagnant, $k_{ls}a_{ls}$, and the dynamic part of liquid holdup, $k_{ld}a_{ld}$, exhibited approximately the same trends as the overall coefficient $k_l a$.

A summary of gas-liquid mass transfer coefficient correlations has been presented in a recent paper [163].

Liquid/Solid Mass Transfer

The mass transfer coefficient on the liquid-solid interface, k_s, has been evaluated from data on the rate of dissolution of the particles manufactured from soluble materials, such benzoic acid and naphthalene. Values of k_s have been found dependent on the Reynolds and the Schmidt number in the following way [138]:

$$\frac{k_s}{a_s D_A} = 1.8\left(\frac{\varrho_l v_l}{a_s \mu_l}\right)^{0.5}\left(\frac{\mu_l}{\varrho_l D_A}\right)^{0.333} \tag{35}$$

which is very similar, except for the coefficient, to the single-phase flow case; k_s for trickle flow, however, are greater.

More recently Dharwadkar and Sylvester [139] presented a correlation based on the J-factor for mass transfer, J_D, in the form

$$J_D = (k_s/v_l)\left(\frac{\mu_l}{\varrho_l D_A}\right)^{0.666} = 1.637 Re_l^{-0.331} \tag{36}$$

The effect of the gas rate on the coefficient k_s is practically negligible. In the expressions for the mass flux in a unit volume the coefficient k_s appears again in the product with the specific surface area of the particles, a_s. Especially at low liquid and gas rates the whole particle surface need not be covered by the liquid film and the calculations should be performed actually with the specific effective wetted area surface, a_w. This fact has been respected for instance by Goto et al. [140] in their correlation of k_s based also on the J_D factor. The conditions under the experiments employed so far to obtain k_s, however, are considerably different from those prevailing on catalyst pellets.

The film of liquid on the surface of the catalyst particles in the trickle-bed reactor is usually thin and the liquid-solid mass transfer coefficient may be lumped with the gas-liquid coefficient in the following way

$$\frac{1}{k_{ls}} = \frac{1}{k_l} + \frac{1}{k_s} \tag{37}$$

The resulting overall gas-solid mass transfer coefficient k_{ls} is a measure of the effect of external transport. More details are given by Satterfield [14, 23] who concludes that under the trickle-flow conditions the gas-liquid or liquid-solid mass transfer resistances are less important than the intraparticle resistances, and may be neglected if the Thiele modulus is less than unity [159].

Sato et al. [138] assessed the relative importance of the gas-liquid and gas-solid resistance and concluded that the solid resistance is much greater and very important for 1/2-inch particles. For smaller, 1/8-inch particles, both resistances are approximately equivalent. For even smaller particles the volume liquid-solid mass transfer coefficient would exceed the gas-liquid coefficient, yet the liquid/solid resistance would remain nonnegligible unless the particles were impractically small. The liquid-solid mass transfer coefficient thus plays an important role in catalytic trickle beds. This conclusion [138], however, was made on the basis of data for systems (air-benzoic acid-nonporous particle) not typical for catalytic operations.

Heat Transfer in Trickle Beds

If catalyst deactivation or appearance of side reactions due to the overheating of the reaction mixture by the heat of reaction is of no concern, it is advantageous to design the reactor as one without heat exchange with the surroundings. Adiabatic reactors operate usually with higher productivity compared to the isothermal reactors and as such appear attractive for a number of hydrorefining technologies. The hydrogenated substrate in these cases is usually present in low concentrations. The adiabatic mode of operation is also simpler from the standpoint of the reactor construction.

Reactors with heat exchange are usually constructed as a bunch of cooled tubes. Each tube must have its own liquid distributor and provisions must be made for equal feed rate. Narrow tubes of such reactors are more apt to form the wall flow. The diameter of the tube and the resulting thermal regime must be carefully considered separately in each particular case.

The problem of heat transfer in trickle-bed reactors has been given so far relatively little attention. The research of the thermal behavior has shown that steep temperature gradients

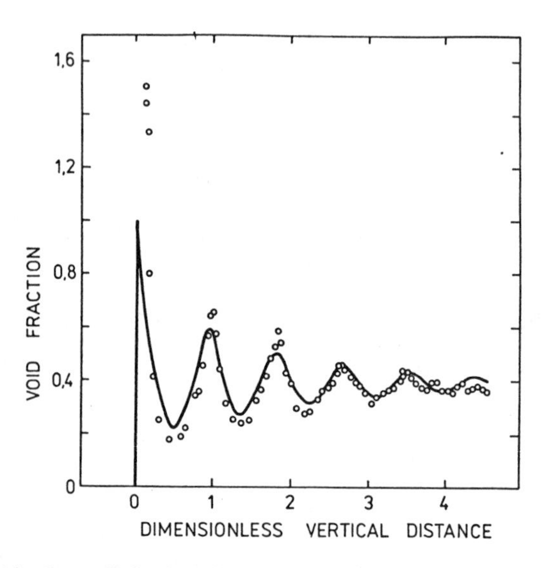

Figure 21. Void fraction profile in a bed of 15-mm spheres dumped into ethanol-filled container [75]. Vertical distance measured in multiples of d_p.

and hot zones with catalyst deactivation, side reactions, and multiple steady states are a reality [12, 27, 43, 65, 66]. Recent studies concentrate on the determination of the principal heat transfer parameters under the cocurrent downflow [39–42].

The mathematical model (the pseudohomogeneous plug flow), describing the heat transfer in a cylindrical reactor contains two parameters: the effective radial thermal conductivity of the bed and the bed-to-wall heat transfer coefficient. Several methods may be found in the literature [67–69] for evaluating these parameters.

The effective radial thermal conductivity, K_r, incorporates the conductivity of the wet bed, K_0 (independent of the hydrodynamic conditions), and the convective contribution of the liquid, K_l, and the gas phase, K_g, due to the radial mixing. It is thus apparent that the effective radial thermal conductivity may vary locally within the reactor, with the variations of local wetting, as well as axially, with eventual change of the flow regime and the progress of the chemical reaction.

According to Weekman et al. [70], the effective radial thermal conductivity may be expressed as a sum of the previous individual contributions, i.e., as

$$K_r = K_0 + K_l + K_g \tag{38}$$

The majority of the so-far-published data were obtained [39–47, 70] for the water-air-glass-beads system or with ceramic packing. From the standpoint of data transfer and trickle-bed reactor design it is safer to obtain the appropriate data on porous pellets wetted by organic liquids with cocurrent downflow of hydrogen. The difference from the former data is considerable due to the higher thermal conductivity of hydrogen, low surface tension of organic liquids, better wettability of the porous pellets, etc.

Structural Properties of the Bed

Trickle-bed reactors are mostly random beds of catalyst particles. Ideally, each volume element of the reactor (small compared to the whole reactor volume yet large compared to a single particle) should exhibit the same void fraction and the same surface area of the catalyst. This is a prerequisite for the uniform flow of gas in each cross section and, under the adiabatic conditions, also for the uniform course of the reaction in each cross section.

Unfortunately, there are several phenomena that cause deviations from the uniformity, or randomness in the previous sense, in real reactors. The presence of the confining walls and supporting grids cause the void fraction displays oscillations in the neighborhood of such containing surfaces. This oscillatory behavior is most conveniently studied on beds of spherical particles where the effect of the method of filling is minimized; this being the third factor causing structural peculiarities of the bed.

Figure 21 shows the area void fraction (void fraction in an infinitesimally thin slice of the bed) obtained experimentally [75] in a bed of 15-mm spheres as a function of the distance from the supporting grid, expressed in multiples of particle diameters. The plot exhibits an oscillatory behavior with the amplitude of the oscillations being damped with increasing distance from the supporting grid.

Figure 22 shows again the area void fraction in the same bed near its top as a function of the distance from the top surface measured by the same coordinate. In this case too, the oscillations appear as a result of leveling and compressing the top surface of the bed. These oscillations, though, are more rapidly damped. In the same work Staněk and Eckert [75] showed that if the bed is prepared by dumping the particles into the reactor by portions, while leveling and compressing each portion, the bed exhibits severe oscillations of the area void fraction throughout its volume with the amplitude peaks reaching down to 0.2 and up to 0.6 and more.

Practically identical observations have been made by Ridgway and Tarbuck [95] who measured void fractions in random beds of spheres in the radial reaction in the proximity of the reactor wall.

Staněk and Eckert [74, 75] published a model describing the course of the void fraction in the proximity of confining surfaces. The lines drawn in Figures 21 and 22 were computed from this model. The idea of the model is that the confining surface forces the particles in its proximity to form a "stratified structure," while the accuracy of the position of each particle in each layer of this structure becomes increasingly randomized with growing distance from the containing surface.

It was found [75], in accord with Ridgway and Tarbuck [95], that the "wave length," or the period of the oscillations of the void fraction for spheres equals about 0.84 of the sphere diameter. The coefficient, characterizing the damping of the oscillations, and the void fraction far away from the confining surface, somewhat depend on the method of filling the bed. Their values, however, indicate that for spherical particles it takes about 4 to 5 diameters for the oscillations to die down. This suggests that in a reactor with the reactor-to-particle diameter ratio equalling 10 or less the bed is seriously nonuniform throughout. Void fraction profiles in such beds do not agree with those for smaller particles, for the maxima and minima cannot be fitted in.

Roblee et al. [143] have shown that for cylindrical particles the oscillations are damped somewhat faster; within about three particle diameters from the wall only. Special types of particles, not typical for catalytic operations (Berl saddles) produce even more uniform beds [143, 144, 147].

In using mixtures of particles, such as may be the case of "diluted" catalyst beds, one should be aware that mixtures of particles may display considerably different void fractions. Binary mixtures of particles of large size disparity tend to decrease the mean void fraction, for the smaller particles fill the interstices between the larger particles. On the contrary, mixtures of close-sized particles tend to form less densely packed beds, for the smaller particles force the larger ones apart.

Also the boundaries between layers of particles of different size (layer of inert particles on top of the catalytic section to improve liquid distribution and saturation by gaseous reactants) may produce void fraction peculiarities. Especially if the particles in the upper layer are smaller an "interfacial region" of lower void fraction appears [76], leading to additional pressure losses [145, 146].

The existence of the void fraction oscillations and generally increased voidage near the containing surfaces such as reactor walls is of particular importance for strongly exothermic reactions. This region becomes one of increased gas velocity at the expense of the flow in the central core of the reactor. Experimental radial velocity profiles, measured at the exit

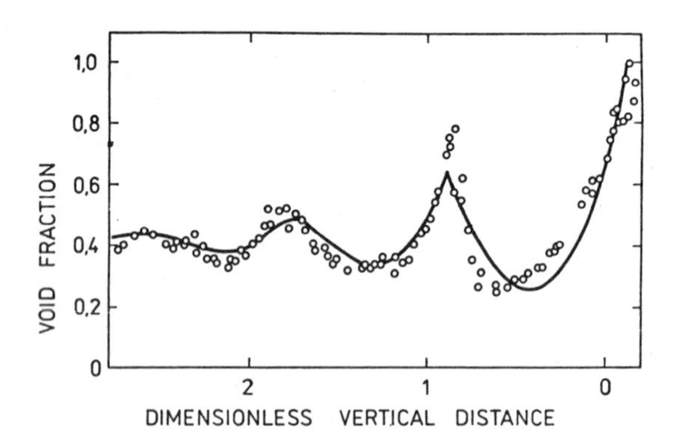

Figure 22. Void fraction profile in a bed of 15-mm spheres near the top compressed by a plane surface [75]. Vertical distance measured in multiples of d_p.

from a packed bed, were published by Drahos and Cermák [98] and a typical result is shown in Figure 23. This velocity profile does not reflect directly the void fraction oscillations, for, as we have mentioned, a differential volume element of the bed must be large compared to a single particle. These velocity profiles thus reflect the real void fraction profile after smoothing (averaging) over several particle diameters.

Drahoš and Cermák [98] report that maximum gas velocity, $v_{g,max}$ has been observed about $0.25d_p$ away from the wall and reached following values expressed as a fraction of the mean velocity, $v_{g,mean}$:

$$\frac{v_{g,max}}{v_{g,mean}} = 1.2 + 0.018(d_t/d_p) \tag{39}$$

while the central core with practically uniform flow begins about $2.25d_p$ away from the reactor wall and the prevailing velocity here is

$$\frac{v_{g,core}}{v_{g,mean}} = 0.83 + 0.0022(d_t/d_p) \tag{40}$$

MODELING OF TRICKLE-BED REACTORS

Pseudohomogeneous Plug Flow Model

A precondition for the validity of all pseudohomogeneous models is that there is no external mass transfer resistance. If this is the case the reaction rate may be conveniently referred to a unit volume of the reactor, a unit mass of catalyst, or a unit volume of catalyst, etc. The reaction rate term is then proportional to the appropriate power of the liquid-phase concentration of the reacting species, c, according to the apparent reaction kinetics.

In hydroprocessing reactors the usual way is

$$\frac{dc}{dz} = -\frac{k_n \eta c^n}{LHSV} \tag{41}$$

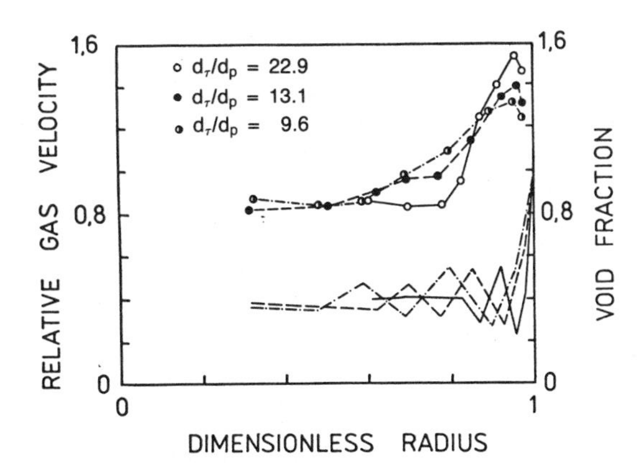

Figure 23. Radial profiles of the relative gas velocity at bed exit [98] and corresponding void fraction profiles.

where z = the dimensionless axial coordinate
k_n = the intrinsic reaction rate constant
n = the reaction order with respect to the liquid substrate
LHSV = the liquid hourly space velocity
η = the effectiveness factor

Additional assumptions involve no axial dispersion effects, no presence of the gaseous reactant in excess, no evaporation of the reaction mixture, no gas phase reaction, etc.

Under the isothermal operation Equation 41 can be easily integrated to obtain, for instance, for the first-order kinetics ($k_n = k_1$) a proportionality between the logarithm of the fractional outlet concentration and the term $k_1\eta/\text{LHSV}$.

Since at low liquid loads the catalyst is not completely wetted by the liquid substrate, the effectiveness varies with the liquid load. We have noted already that the completness of the catalyst wetting is often associated with sufficient liquid holdup. This led Henry and Gilbert [149] to introducing total liquid holdup into the term on the right-hand side of Equation 41. Depending on the type of the correlation used for the holdup, an explicit relation is obtained between the log of the fractional concentration of unconverted species, the geometry of the reactor (pellet diameter, bed depth), liquid load, and physico-chemical properties of liquid. The above authors [149] used Satterfield's correlation [14], which gives satisfactory correlations with LHSV, bed depth, and d_p for the hydrodesulphurization and the hydrocracking data.

The apparent reaction kinetics, however, is more directly related to the degree of wetting rather than liquid holdup. Replacement of the holdup on the right-hand side of Equation 41 by the wetted area, using the correlation of Puranik and Vogelpohl [150], gave Mears [151] a relationship which predicts nearly the same dependence of the log of the fractional concentration of unconverted specie on bed depth, and LHSV, but predicts a different dependence on pellet size and physico-chemical properties of liquid.

Trickle-bed reactions are mostly strongly exothermic and in some instances the reactions follow the adiabatic rather than isothermal course. Local temperature and concentration are then related by the following constraint

$$c + \frac{(c_{pl}\varrho_l + c_{pg}\varrho_g v_g/v_l)}{\Delta H} T = \text{constant} \tag{42}$$

where ΔH is the reaction heat.

The constant on the right side is independent of position within the reactor and may be evaluated, for example, from inlet conditions. This relationship permits one to solve either only the mass or only the heat balance, the remaining quantity being fixed by Equation 42. Nevertheless, the integration becomes complicated due to the temperature dependence of the reaction rate constant.

The underlying differential mass and heat balances were successfully used to describe an experimental laboratory adiabatic trickle-bed reactor for the hydrogenation [157] of cyclohexene.

Axial dispersion effect, neglected in the previous model, need not be always negligible in trickle-bed reactors. The discrepancies between the performance of industrial and pilot-plant reactors are often attributed to axial dispersion. A criterion for the bed depth where axial dispersion is negligible have been given by Mears [4, 151] and Montagna and Shah [5.]

Pseudohomogeneous Plug Flow Model with Phase Transitions

In many hydroprocessing trickle-bed reactors significant evaporation of the liquid-phase may occur [12, 15, 43]. The modeling becomes then very complex. Transition from the liquid- to the gas-phase reaction regime is often accompanied by a marked increase in the reaction rate because the gas surrounding the catalyst pellet offers less mass transfer resistance than the liquid. For the case of an exothermic reaction, this may have an undesirable effect as it gives rise to a narrow reaction zone with steep temperature gradients.

A new approach to the modeling of heat and mass transfer phenomena in a volatile liquid reaction mixture in the trickle-bed reactor at lower wetting efficiency represents the following set of Equation 101.

The hydrogenation reaction of a substrate A to a product B may occur in both the liquid- and the gas-phase following the scheme

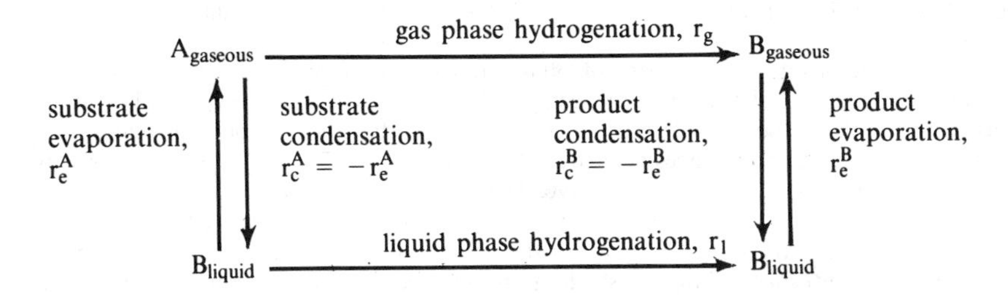

The reaction rates in the gas, r_g, and the liquid, r_l, phase, the rates of evaporation of the substrate, r_e^A, and the product, r_e^B (except for the opposite sign equal the corresponding condensation rates, r_c^A, r_c^B) and the rate of dissolution of hydrogen. r_{dissol}^H, may be introduced into the differential mass balances based on the plug flow to obtain

$$\frac{1}{h}\frac{d(Gy_A)}{dz} = r_e^A - r_g \tag{43}$$

(balance on substrate A in the gas phase)

$$\frac{1}{h}\frac{d(Gy_B)}{dz} = r_e^B + r_g \tag{44}$$

(balance on product B in the gas phase)

$$\frac{1}{h}\frac{d(Gy_H)}{dz} = -r_{dissol}^H - r_g \tag{45}$$

(balance on hydrogen in the gas phase)

$$\frac{1}{h}\frac{d[(1-G)x_A]}{dz} = -r_e^A - r_l \tag{46}$$

(balance on substrate A in the liquid phase)

$$\frac{1}{h}\frac{d[(1-G)x_B]}{dz} = -r_e^B + r_l \tag{47}$$

(balance on product B in the liquid phase)

$$\frac{1}{h}\frac{d[(1-G)x_H]}{dz} = r_{dissol}^H - r_l \tag{48}$$

(balance on hydrogen dissolved in the liquid phase)

The heat balance in the gas/liquid mixture

$$\frac{v_{lg}\varrho_{lg}c_{plg}}{h}\frac{dT}{dz} = \alpha a_s(T_{cat} - T) - r_e^A \Delta H_{evap}^A - r_e^B \Delta H_{evap}^B \tag{49}$$

The heat balance in the catalyst pellets

$$\frac{K_a}{h^2}\frac{\partial^2 T_{cat}}{\partial z^2} = \alpha a_s(T_{cat} - T) - r_l \Delta H_l - r_g \Delta H_g \tag{50}$$

The boundary conditions are fixed by composition of the feed mixture and its thermal state.

A numerical solution of the previous set of differential equations was carried out [101] using data on the hydrogenation of 1, 5-cyclooctadiene to cyclooctane, assuming Langmuir–Hinshelwood-type kinetics. The rates of phase transitions, r_e^A, r_e^B were calculated with the aid of the Raoult–Dalton law and the Clausius–Clapeyron equation for a two-component mixture.

The obtained profiles of the rates of phase transitions along the length of the reactor display a complicated character as is seen in Figure 24. Negative values of the rates of evaporation indicate local condensation at the inlet of the catalytic bed. Increasing temperature of the reaction enhances also the rates of evaporation of the liquid substrate and the evaporation rate becomes again positive. Thanks to the higher value of the reaction rate in the vapor phase, r_g, in comparison with that in the liquid (by approximately an order of magnitude), the product accumulates in the gas phase here to be condensed ultimately. Decreasing concentration of the substrate along the reactor length leads to a diminished gas phase reaction rate and the concentration of the product in the gas phase is decreased. This fact, together with the increasing temperature along the reactor, facilitates evaporation of the product from the liquid phase near the outlet end of the reactor. Mutual competition of phase transition, demonstrated in Figure 24, with the relatively low reaction rate in the liquid phase cause the temperature rise to be mainly due to the intensive gas-phase reaction.

Computed results are naturally very sensitive to the extent of the wetted surface of the catalyst. Clearly, the increased temperature along the reactor length causes the fraction of the wetted catalyst surface to decrease and to become eventually zero upon reaching the temperature of the phase transition.

The previous model was solved using several correlations for the wetted surface [150, 154, 161, 171]. The best agreement between the experimental and predicted profiles was obtained for the correlation of Onda et al. [154], corresponding temperature profiles are shown in Figure 25.

Pseudohomogeneous Radially Dispersed Plug Flow Model with the Wall Flow of Liquid

In order to account for the nonuniform flow of liquid due to the various types of the liquid feed and due to the existence of the wall flow in trickle-bed reactors a pseudohomogeneous radially distributed plug-flow model with the wall flow has been developed by Staněk et al. [36, 37]. In accord with the theory explained in the part on liquid distribution the mass balance on a reacting species undergoing an n-th-order catalytic reaction may be written as

$$f\frac{\partial c}{\partial z} - H\left[f\frac{\partial^2 c}{\partial r^2} + 2\frac{\partial f}{\partial r}\frac{\partial c}{\partial r} + \frac{1}{r}f\frac{\partial c}{\partial r}\right] + Da_n c^n \exp\left(\frac{t}{1 + t/\gamma}\right) = 0 \tag{51}$$

where Da_n is the Damköhler number for an n-th-order reaction, and γ and t are the dimensionless activation energy and dimensionless temperature, defined according to Frank–Kamenetskii as

$$t = \frac{\gamma}{T_0}(T - T_0) = \frac{E}{R_g T_0^2}(T - T_0) \tag{52}$$

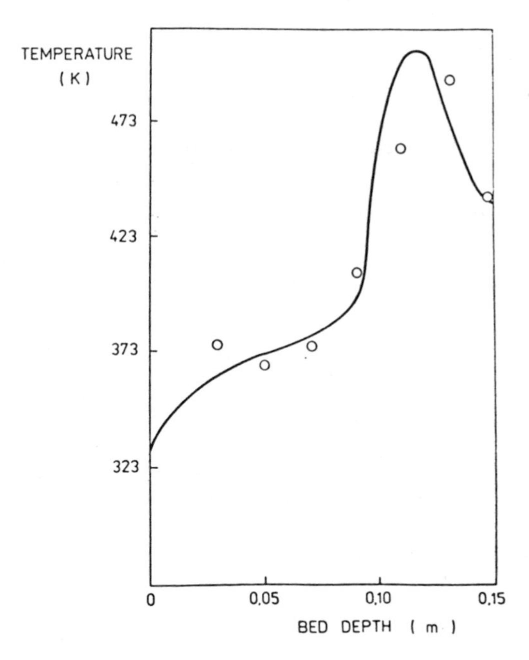

Figure 24. Axial profiles of temperature and the rates of phase transitions of cyclooctadiene and cyclooctane in a trickle-bed reactor [101].

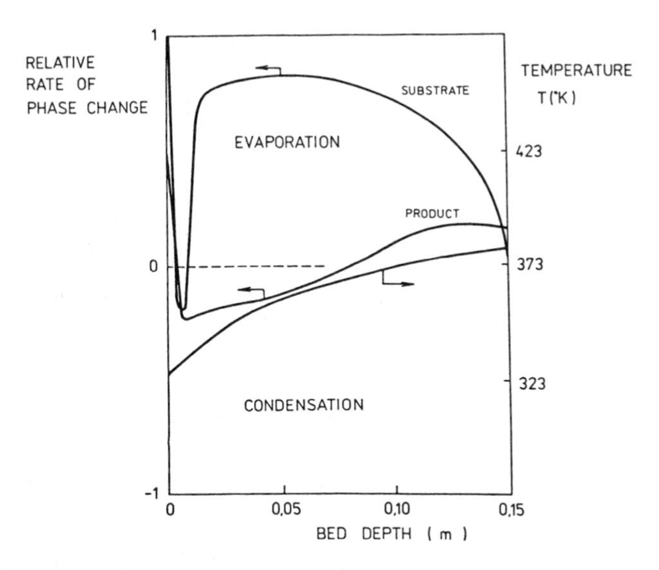

Figure 25. Axial temperature profile in a trickle-bed reactor with phase transitions for the hydrogenation of cyclooctadiene. Comparison of experimental and computed results [101]. The area of wetted surface computed from correlation of Onda [154].

The dimensionless local density of irrigation, f, is given as a solution to Equation 1 for various initial liquid distributors (Table 2).

The model treats the wall flow as a separate entity whose substrate concentration, c_w, differs from that within the bed. Its values, as a function of the axial coordinate, are given by the following equation which represents also the boundary condition for Equation 51.

$$\frac{d(Wc_w)}{dz} = -2Hf\frac{\partial c}{\partial r} = 2HBCW(c-c_w) \quad \text{for} \quad r = 1 \tag{53}$$

H designates the dimensionless bed depth defined by

$$H = \frac{Dh}{R^2} \tag{54}$$

Analogously we may write for the heat balance

$$f\frac{\partial T}{\partial z} - H\left[f\frac{\partial^2 t}{\partial r^2} + 2\frac{\partial f}{\partial r}\frac{\partial t}{\partial r} + \frac{1}{r}f\frac{\partial t}{\partial r}\right] - Da_n c_n U\gamma \exp\left(\frac{t}{1+t/\gamma}\right) = 0 \tag{55}$$

where U is the dimensionless heat of reaction

$$U = \frac{(-\Delta H)c_0}{T_0 \varrho_1 c_{p1}} \tag{56}$$

The subscript zero refers to the conditions at the inlet. The temperature of the wall flow, T_w, is again generally different from that within the bed and the heat balance, under adiabatic operation reads

$$\frac{d(Wt_w)}{dz} = -2Hf\frac{\partial t}{\partial r} = 2HBWC(t-t_w) \quad \text{for} \quad r = 1 \tag{57}$$

The model was built on the assumption that the principal mechanism responsible for the radial transport of species and heat is the convective/diffusional transport jointly with the trickling liquid. This assumption is acceptable for the transport of mass but less so for heat where additional conductive mechanism exists.

The authors [36] have shown that under the adiabatic operation the local temperature and concentration anywhere within the bed or in the wall flow are constrained by an invariant analogous to that written in Equation 42.

Numerical solution of the previous model provides data for the construction of the contours of constant concentration of species and isotherms, as well as the axial profiles of the concentration and temperature of the wall flow. Computed results indicate principally two effects:

1. Strong interaction of the flow pattern, dictated by the type of liquid distributor, with the course of the reaction. Particularly in the entrance region (near the liquid feed), rapid, lower-order reactions, such as hydrogenations, lead to formation of distinctly curved reaction zones. The lateral extent of this zone may be larger than that of the horizontal zone, expected in the plug flow reactor, resulting in a generally slightly higher net reaction rate and higher conversion.

   Computed contours of constant concentration [36], taking place predominantly in the entrance region in a trickle-bed reactor irrigated by a narrow nozzle ($r_1 = 0.1$), are shown in Figure 26. The dimensionless depth ($H = 0.0456$), equivalent to the plotted concentration pattern, sufficed to reduce the inlet concentration of the hydrogenated substrate

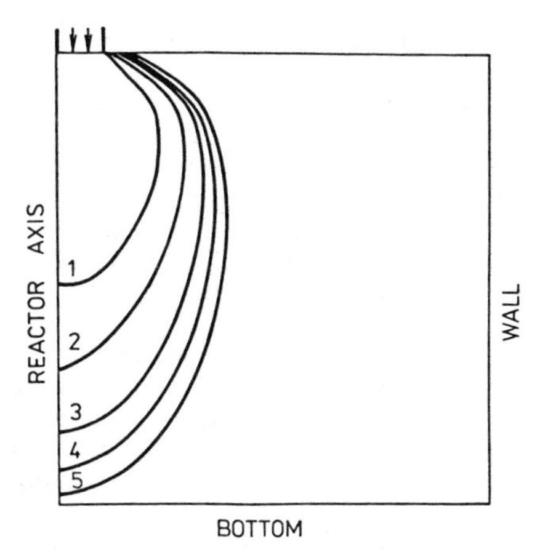

Figure 26. Contours of constant concentration for a bed irrigated by a narrow nozzle and a nearly zero-order reaction [36]. $U = 0.87$; $\gamma = 8.5$; $Da_n = 0.5$; $C = 2$; $n = 0.1$; $H = 0.0456$. Contour 1—90% concentration; Contour 2—70%; Contour 3—50%; Contour 4—30%; Contour 5—10%.

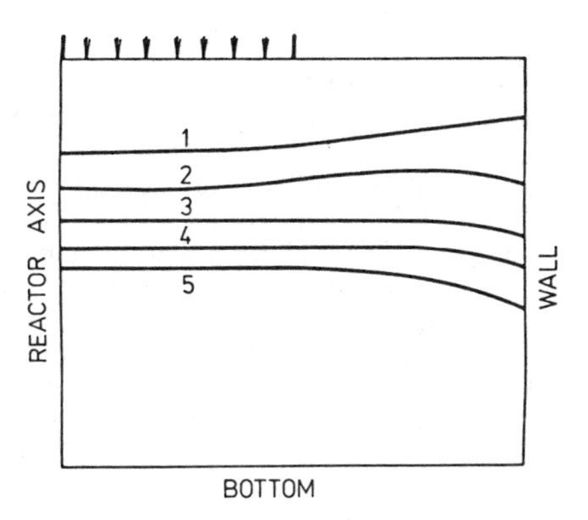

Figure 27. Contours of constant concentration for a bed irrigated by a disc distribution and a first-order reaction [36]. $U = 0.87$; $\gamma = 8.5$; $Da_n = 0.5$; $C = 2$; $H = 0.75$; $r_1 = 0.5$. Contour 1—90% concentration; Contour 2—70%; Contour 3—50%; Contour 4—30%; Contour 5—10%.

down to 1%. Numerical values taken for the computation are typical for the hydrogenation and a reactor with a strong tendency to form the wall flow (low reactor-to-pellet-diameter ratio).

2. The second observed effect is that of the wall flow. Formation of the wall flow removes part of the liquid from the catalyst surface and the reaction there, now at a lower irrigation rate, proceeds faster. If very low reaction conversions are sufficient this effect may lead temporarily to a greater productivity. Farther away from the liquid feed the trickle flow over the catalyst surface is practically in hydrodynamic equilibrium with the wall flow and the reactants partially bypass the catalyst surface via the wall flow. In this major portion of the bed the wall flow always leads to lowered reaction rates, i.e., lowered productivity. The adverse effect of the wall flow is particularly strongly felt if high conversions are required.

Combined effect of the entrance region and the wall flow is illustrated in Figure 27 showing the computed [36] pattern of the contours of constant concentration for a reactor irrigated by a disc distributor ($r_l = 0.5$) and a first-order reaction. The slope of the lines of constant concentration near the wall indicates initially slightly faster progress of the reaction near the wall in the entrance region. This situation, however, is gradually reversed and the reaction in the wall region begins to lag behind the overall rate. The major manifestation of this fact is finally the relatively long section of the reactor depth below the last plotted line of constant concentration. This depth was required to reduce the concentration of the substrate from 10% (for the bottom contour) down to 1% at the reactor exit.

Figure 28 plots the computed [36] cup-mixing mean outlet concentration of the reacting species as a function of bed depth, defined in Equation 54, for a nearly zero-order reaction and the same data as in the last two figures that are typical for liquid-phase hydrogenation. The low value of the parameter C is typical for a reactor with a strong tendency to form the wall flow. Curve 4 in this figure was computed for the uniform feed in a reactor free of the wall flow effects ($C = \infty$). In comparison with Curve 1, computed for a uniform feed in a reactor with wall flow, for a low conversion initially, the reaction may proceed faster. Ultimately, however, the rate of conversion is controlled by the rate of exchange of liquid between the bed and the wall flow and excessively long reactors are necessary to reach high conversion. For narrow central nozzle feed ($r_l = 0.1$, Curve 3) and a disc distributor, reaching over one half of reactor radius ($r_l = 0.5$, Curve 2) the conversion increases more slowly due to higher local irrigation rates but steadily and still essentially no effect of the wall flow is observed.

Similar plots computed [36] for other reaction orders show that for $n = 2$ the peculiarities of the entrance region practically disappear and all distributors provide nearly the same axial profiles of the cup-mixing mean outlet concentration. All these profiles display considerable tailing due to the wall flow, calling for substantially longer reactors if high conversions are needed. For an intermediate reaction order ($n = 0.5$) the results showed the central nozzle and the disc to be more efficient than the uniform feed.

The existence of the tailing on the mean conversion versus bed depth curves was verified experimentally on a laboratory-scale hydrogenation of cyclohexene [37].

Recently [153] the model has been further refined to account for the additional radial heat transfer mechanism due to the conduction via the solid catalyst particles and due to

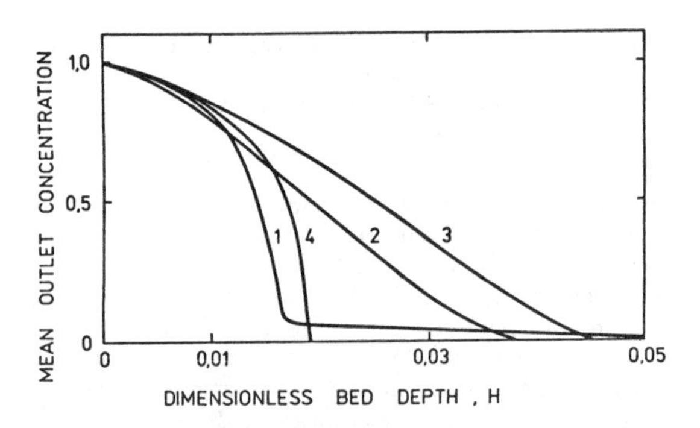

Figure 28. Cup-mixing mean outlet concentration as a function of the dimensionless depth of the catalyst layer in a trickle-bed reactor for nearly zero-order (0.1) reaction [36]. Curve 1—the bed uniformly irrigated on the top; Curve 2—irrigated by a disc ($r_1 = 0.5$) distributor; Curve 3—irrigated by a narrow ($r_1 = 0.1$) nozzle; Curve 4—uniformly irrigated ideal bed free of the wall flow effects.

the convection with the flowing gas. The model was further expanded to permit analysis of the externally cooled reactors. In the presence of additional heat transfer mechanisms, not associated with the dispersion of liquid, the invariants between local concentration and temperature, such as those written in Equation 42 are no longer valid. Numerical results [153] of the expanded model indicate that possible transition of the reactor to the gas-phase regime may also be affected by the type of liquid distributor, by the properties of the pellets (thermal conductivity), and gas flow rate. The possibilities of controlling the regime by external cooling are rather limited and suffice mostly for mere reduction of the extent of hot spots.

Percolation and Mixing Cell Models

In recent papers [100, 103, 129] a new description of fluid flow hydrodynamics in trickle beds was developed emphasizing the random and discontinuous nature of the bed. The bed is represented by an array of randomly connected transport cells. The liquid flow through the cells is described mathematically using the percolation theory [110, 132].

Detailed liquid spreading and maldistribution evolve from the hydrodynamics of the local transport cell. It was shown that highly segregated two-phase flow at low irrigation rates may result either from poor wetting conditions or from strong gas-liquid interactions. Moreover, in the case of highly exothermic reactions, the flow segregation may be brought about by local liquid evaporations [71].

The simultaneous assessment of liquid maldistribution, the concentration field of the reacting components, and the conversion profile in a random bed of catalyst can be made on the basis of simple-cell models. The mutually interconnected mixing cells [141, 142], as a series of stirred tanks, represent perhaps the simplest type of a stagewise model accounting for the backmixing in the reactor. The various aspects of this model have been discussed, among others, by Deans and Lapidus [148]. This method enables also the heat transfer to be examined in case of nonisothermal beds [152].

Heterogeneous models

For catalytic reactions when one of the transport steps becomes the rate limiting step, heterogeneous models must be employed. A heterogeneous axially dispersed plug-flow model, accounting for the mass transfer resistance at the gas-liquid and the liquid-solid interfaces can be formulated as a set of differential mass balances of the following type for one of the reacting species:

$$(\varepsilon - h_l)E_g \frac{d^2c_g}{dz^2} - v_g \frac{dc_g}{dz} - k_l a\left(\frac{c_g}{m} - c_l\right) = 0 \tag{58}$$

$$h_l E_l \frac{d^2c_l}{dz} - v_l \frac{dc_l}{dz} + k_l a\left(\frac{c_g}{m} - c_l\right) - k_s a_s(c_l - c_s) = 0 \tag{59}$$

$$k_s a_s(c_l - c_s) - r_l = 0 \tag{60}$$

Here z designates dimensional axial coordinate, and the reaction rate term is a function of concentrations in the solid catalyst. It is noted that generally three concentrations appear in the statement of the heterogenenous model and, correspondingly, three mass balances in the gas, liquid, and the solid phase must be formulated.

Actually, even more sophisticated and more realistic models could be used, such as the earlier analyzed axially dispersed cross-flow model for the flow of liquid. Four balance equations would then result for what are now generally four concentrations of a single species: in the gas and the solid phase as shown previously and in the stagnant and the dynamic part of the trickling liquid. These models appear, of course, too complex for practical purposes, especially in view of the lack of reliable data, while the relatively simple axial dispersion model performs satisfactorily. Difficulties, however, may arise in scale-up and data transfer.

Upon dropping the terms with the second derivatives in Equations 58 and 59 the simple heterogeneous plug-flow models result. Sato et al. [138] obtained an analytical solution for this case and found that both the liquid- and the gas-phase concentrations decrease exponentially with the depth of the bed, and if the gas is used in excess (in terms of the ratio of liquid and gas velocity and the solubility of the gas), the liquid phase concentration approaches a limiting value.

If the inlet concentration in the liquid is close to this limiting value or if the bed is sufficiently deep, the reactor efficiency may be simply computed from the resistance due to the chemical reaction, divided by the sum of the resistance due to the reaction, the gas-liquid resistance, and the liquid-solid resistance. This simple additive law, however, applies only for the case of fully developed dynamic equilibrium. In their experiments on liquid-phase catalytic oxidation of ethanol Sato et al. [138] found that the depth of the bed typical for bench-scale experiments was not sufficient to eliminate the effect of the entrance region. The efficiency of the reactor can thus be generally improved by presaturating the liquid feed with gas.

CONCLUSIONS

The large number of recently published papers dealing with trickle-bed reactors testifies to their increasing importance in practice. The role of the fluid dynamics and transport phenomena are now much better understood but the lack of pertinent physico-chemical and transport data, necessary for a reliable modeling, design, and scale-up of the system, remains. Moreover, the data available in the literature mostly on the air-water-nonporous-particles system can be often misleading. With sufficient precautions, however, the simple pseudohomogeneous reactor model will serve well for the description of the majority of hydrotreating operations to process petroleum fractions. Such preacutions may involve, for instance, corrections for incomplete wetting of the catalyst.

Special care should be taken if liberations of large reaction heats are to be expected. Under certain conditions this may cause evaporation of the reaction mixture, leading in turn to increased reaction rates and strong overheating of the catalyst with eventual deactivation.

So far not quite adequately acknowledged is the role of the nonuniform flow and the wall flow of liquid affecting the reactor's productivity, yield, selectivity, and the general operating and thermal regime of the reactor, including the appearance of hot spots.

NOTATION

a	interfacial surface area per unit volume
a_s	area of external catalyst surface per unit volume
B, C	dimensionless liquid distribution parameters, Equation 3
c	concentration
d_p	particle diameter
d_t	reactor diameter
D	coefficient of radial spread (dimension of length)
D_A	diffusion coefficient
E_g, E_l	axial dispersion coefficient in gas and liquid phase
Da_n	Damköhler number for n-th order reaction
E	activation energy
E_L	rate of energy dissipation per unit volume
f	dimensionless density of irrigation, defined as a fraction of the mean density
F, F_H	liquid feed rate, hydrogen feed rate

G	mass velocity
h	dimensional bed depth
h_d, h_s, h_l, h_{lf}	dynamic, stagnant, total, and flooding liquid holdup per unit volume of empty reactor
H	dimensionless bed depth, Equation 54
ΔH_g, ΔH_l	reaction heat for gas- and liquid-phase reaction
k_n	intrinsic reaction rate constant
k_l, k_s, k_{ls}	gas/liquid, liquid/solid, and overall gas/solid mass transfer coefficient
K_a, K_r, K_0, K_l, K_g	effective axial, radial thermal conductivity, effective radial conductivity of wet bed, and liquid- and gas-phase contributions to K_r
ΔP_g, ΔP_l, ΔP_{lg}	pressure drop per unit length for gas, liquid, and two-phase flow
ΔP_{fg}, ΔP_{fl}, ΔP_{flg}	pressure drop per unit length for gas, liquid, and two-phase flow for foaming systems
r_g, r_l	gas- an liquid-phase reaction rate per unit volume
r_e^A, r_e^B, r_c^A, r_c^B	rate of evaporation and condensation for substrate and product per unit volume
r_{dissol}^H	rate of hydrogen dissolution per unit volume
r	dimensionless radial coordinate as a fraction of reactor radius
R_g	gas constant
R	reactor radius
t	dimensionless temperature, Equation 52
T	dimensional temperature
U	dimensionless reaction heat, Equation 56
v_g, v_l, v_{lf}	gas, liquid, and liquid flooding superficial velocity
W	dimensionless wall flow as a fraction of overall liquid rate
W_{cat}	weight of catalyst
x	mole fraction in liquid
y	mole fraction in gas
X, X'	Lockhart–Martinelli and modified Lockhart–Martinelli factor
z	dimensionless axial coordinate as a fraction of bed depth

Greek Symbols

α	fluid/solid heat transfer coefficient
γ	dimensionless activation energy, Equation 52
ε	void fraction
μ	viscosity
η	effectiveness factor
ν	contacting efficiency
ϱ	density
σ	surface tension

Subscripts

g	gas
l	liquid
0	initial (inlet) value
w	wall
wat	water

REFERENCES

1. Hofmann, H., "Hydrodynamik, Transportvorgänge und matematische Modelle bei Rieselreaktoren," *Chemie-Ingenieur-Technik,* Vol. 47, No. 20 (1975), pp. 823–868.
2. Bondi, A., "Handling Kinetics from Trickle-Phase Reactors," *Chemical Technology* (US), Vol. 1 (March 1971), pp. 185–188.
3. Hoog, H., Klinkert, H. C., and Schaafsma, A., "New Shell Hydrodesulfurization Process Shows these Features," *Petroleum Refiner,* Vol. 32, No. 5 (May 1953), pp. 137–141.

4. Mears, D. E., "The Role of Axial Dispersion in Trickle-Flow Laboratory Reactors," *Chemical Engineering Science,* Vol. 26, No. 9 (1971), pp. 1361–1366.
5. Montagna, A. A., and Shah, Y. T., "The Role of Liquid Holdup, Effective Catalyst Wetting, and Backmixing of the Performance of a Trickle-Bed Reactor for Residue Hydrodesulfurization," *Industrial Engineering Chemistry, Process Design and Development,* Vol. 14, No. 4 (1975), pp. 479–483.
6. Murphree, E. V., Voorhies, A. Jr., and Mayer, F. X., "Application of Contacting Studies to the Analysis of Reactor Performance," *Industrial Engineering Chemistry, Process Design and Development,* Vol. 3, No. 4 (1964), pp. 381–386.
7. Paraskos, J. A., Frayer, J. A., and Shah, Y. T., "Effect of Holdup, Incomplete Catalyst Wetting and Backmixing during Hydroprocessing in Trickle Bed Reactors," *Industrial Engineering Chemistry, Process Design and Development,* Vol. 14, No. 3 (1975), pp. 315–322.
8. Ross, L. D., "Performance of Trickle-Bed Reactors," *Chemical Engineering Progress,* Vol. 61, No. 10 (Oct. 1965), pp. 77–82.
9. Shah, Y. T., and Paraskos, J. A., "Intraparticle Diffusion Effects in Residue Hydrodesulfurization," *Industrial Engineering Chemistry, Process Design and Development,* Vol. 14, No. 4 (1975), pp. 368–372.
10. Sylvester, N. D., and Pitayagulsarn, P., "Effect of Transport Processes on Conversion in a Trickle-Bed Reactor," *A.I.Ch.E. Journal,* Vol. 19, No. 3 (1973), pp. 640–644.
11. Babcock, B. D., Mejdell, G. T., and Hougen, O. A., "Catalyzed Gas-Liquid Reactions in Trickling-Bed Reactors: Part I. Hydrogenation of α-Methylstyrene Catalyzed by Palladium, Part II. Hydrogenation of α-Methylstyrene Catalyzed by Platinum," *A.I.Ch.E. Journal,* Vol. 3, No. 3 (1957), pp. 366–372.
12. Hanika, J., et al., "Qualitative Observations of Heat and Mass Transfer Effects on the Behavior of a Trickle-Bed Reactors," *Chemical Engineering Communications,* Vol. 2, No. 1 (1975), pp. 19–25.
13. Morita, S., and Smith, J. M., "Mass Transfer and Contacting Efficiency in a Trickle-Bed Reactor," *Industrial Engineering Chemistry, Fundamentals,* Vol. 17, No. 2 (1978), pp. 113–120.
14. Satterfield, C. N., Pelossof, A. A., and Sherwood Thomas K., "Mass Transfer Limitations in a Trickle-Bed Reactor," *A.I.Ch.E. Journal,* Vol. 15, No. 2 (1969), pp. 226–234.
15. Sedriks, W., and Kenney, C. N., "Partial Wetting in Trickle-Bed Reactors—The Reduction of Crotonaldehyde over a Palladium Catalyst," *Chemical Engineering Science,* Vol. 28, No. 2 (1973), pp. 559–569.
16. Turek, F., et al., "Zur Modellierung katalytischer Dreiphasenprozesse," *Chemische Technik,* Vol. 28, No. 3 (1976), pp. 149–152.
17. Krönig, W., "Die Raffination von C_4-Kohlenwasserstoffen durch Kalthydrierung," *Erdöl und Kohle,* Vol. 16, No. 6 (1963), pp. 520–523.
18. Klassen, J., and Kirk, R. S., "Kinetics of the Liquid-Phase Oxidation of Ethanol," *A.I.Ch.E. Journal,* Vol. 1, No. 4 (1955), pp. 488–495.
19. Goto, S., and Smith, J. M., "Trickle-Bed Reactor Performance: Part I. Holdup and Mass Transfer Effects. Part II. Reaction Studies," *A.I.Ch.E. Journal,* Vol. 21, No. 4 (1975), pp. 706–720.
20. Levec, J., and Smith, J. M., "Oxidation of Acetic Acid Solutions in a Trickle-Bed Reactor," *A.I.Ch.E. Journal,* Vol. 22, No. 1 (1976), pp. 159–168.
21. Goto, S., and Smith, J. M., "Performance of Slurry and Trickle-Bed Reactors: Application to Sulfur Dioxide Removal," *A.I.Ch.E. Journal,* Vol. 24, No. 2 (1978), pp. 286–293.
22. Hartman, M., and Coughlin, R. W., "Oxidation of SO_2 in a Trickle-Bed Reactor Packed with Carbon," *Chemical Engineering Science,* Vol. 27, No. 5 (1972), pp. 867–881.
23. Satterfield, C. N., "Trickle-Bed Reactors (Journal Review)," *A.I.Ch.E. Journal,* Vol. 21, No. 2 (1975), pp. 209–228.

24. Enright, J. T., and Chuang, T. T., "Deuterium Exchange Between Hydrogen and Water in a Trickle-Bed Reactor," *Canadian Journal of Chemical Engineering,* Vol. 56 (April 1978), pp. 236–250.
25. Shimizu, M., Kitamoto, A., and Takashima, Y., "New Proposition on Performance Evaluation of Hydrophobic Pt Catalyst Packed in Trickle Bed," *Journal of Nuclear Science and Technology,* Vol. 20, No. 1 (Jan. 1983), pp. 36–47.
26. Goto, S., Levec, J., and Smith, J. M., "Trickle-Bed Oxidation Reactors," *Catalyse Review—Science Engineering,* Vol. 15, No. 2 (1977), pp. 187–247.
27. Joschek, H. I., "Reaktoren für Gas-Flüssig-Fest-Reaktionen," in *Ullmanns Encyklopädie der technischen Chemie,* 4th ed. Weinheim, Verlag Chemie, 1973, Vol. 3, pp. 494–518.
28. Shah, Y. T., *Gas-Liquid-Solid Reactor Design,* New York, McGraw-Hill, 1979.
29. Ramachandran, P. A., and Chaudhari, R. V., *Three-Phase Catalytic Reactors,* London, Gordon and Breach Science Publishers, Ltd., 1982, Chapter 12.
30. Hofmann, H. P., "Multiphase Catalytic Packed-Bed Reactors," *Catalyse Review—Science Engineering,* Vol. 17, No. 1 (1978), pp. 71–117.
31. Gianetto, A., et al., "Hydrodynamics and Solid-Liquid Contacting Effectiveness in Trickle-Bed Reactors," *A.I.Ch.E. Journal,* Vol. 24, No. 6 (1978), pp. 1087–1104.
32. Van Landeghem, H., "Multiphase Reactors: Mass Transfer and Modeling," *Chemical Engineering Science,* Vol. 35, No. 9 (1980), pp. 1912–1949.
33. Ramachandran, P. A., and Chaudhari, R. V., "Predicting Performance of Three-Phase Catalytic Reactors," *Chemical Engineering,* Vol. 87, No. 24 (Dec. 1980), pp. 74–85.
34. Charpentier, J. C., and Favier, M., "Some Liquid Holdup Experimental Data in Trickle-Bed Reactors for Foaming and Nonfoaming Hydrocarbons," *A.I.Ch.E. Journal,* Vol. 21, No. 6 (1975), pp. 1213–1218.
35. Mills, P. L., and Dudukovic, M. P., "Comparison of Current Models for Isothermal Trickle-Bed Reactors with Application to a Model Reaction System," Symposium on Catalytic Reactor Design, Seattle Meeting, March 1983.
36. Staněk, V., et al., "The Effect of Liquid Flow Distribution on the Behavior of a Trickle-Bed Reactor," *Chemical Engineering Science,* Vol. 36, No. 9 (1981), pp. 1045–1067.
37. Staněk, V., and Hanika, J., "The Effect of Liquid Flow Distribution on Catalytic Hydrogenation of Cyclohexene in an Adiabatic Trickle-Bed Reactor," *Chemical Engineering Science,* Vol. 37, No. 9 (1982), pp. 1283–1288.
38. Crine, M., "Heat Transfer Phenomena in Trickle-Bed Reactors," *Chemical Engineering Communications,* Vol. 19, No. 1–3 (1982), pp. 99–114.
39. Specchia, V., and Baldi, G., "Heat Transfer in Trickle-Bed Reactors," *Chemical Engineering Communications,* Vol. 3, No. 6 (1979), pp. 483–499.
40. Matsuura, A., et al., "Radial Effective Thermal Conductivity in Packed Beds with Cocurrent Gas-Liquid Downflow," *Kagaku Kogaku,* Vol. 43, No. 5 (1979), pp. 263–268 (see also *Heat Transfer Japanese Research,* Vol. 8, No. 1 (1979), pp. 44–52).
41. Matsuura, A., et. al., "Apparent Wall Heat Transfer Coefficient in Packed Beds with Cocurrent Gas-Liquid Downflow," *Kagaku Kogaku,* Vol. 43, No. 5 (1979), pp. 269–274 (see also *Heat Transfer Japanese Research,* Vol. 8, No. 1 (1979), pp. 53–60).
42. Hashimoto, K., et al., "Radial Effective Thermal Conductivity in Gas-Liquid Cocurrent Flow Through Packed Beds," *Kagaku Kogaku Ronbunshu,* Vol. 2, No. 1 (1976), pp. 53–59 (see also *International Chemical Engineering,* Vol. 16, No. 4 (1976), pp. 720–727).
43. Hanika, J., et al., "Measurement of Axial Temperature Profiles in an Adiabatic Trickle-Bed Reactor," *The Chemical Engineering Journal,* Vol. 12, No. 1 (1976), pp. 193–197.
44. Hanika, J., et al., "Dynamic Behavior of an Adiabatic Trickle-Bed Reactor," *Chemical Engineering Science,* Vol. 32, No. 5 (1977), pp. 525–528.

45. Hanika, J., Vosecký, V., and Růžička, V., "Dynamic Behavior of the Laboratory Trickle-Bed Reactor," *The Chemical Engineering Journal,* Vol. 21, No. 2 (1981), pp. 108–114.
46. Herskowitz, M., and Smith, J. M., "Trickle-Bed Reactors: A Review," *A.I.Ch.E. Journal,* Vol. 29, No. 1 (Jan. 1983), pp. 1–18.
47. Charpentier, J. C., "Mass Transfer Rates in Gas-Liquid Absorbers and Reactors," *Advances in Chemical Engineering,* Vol. 11, New York: Academic Press, Inc., 1981, pp. 1–133.
48. Tan, C. S., and Smith, J. M., "Mass Transfer in Trickle-Bed Reactors," *Latin America Journal of Chemical Engineering and Applied Chemistry,* Vol. 11, No. 1 (1981), pp. 59–69.
49. Østergaard, K., "Gas-Liquid-Particle Operations in Chemical Reaction Engineering," *Advances in Chemical Engineering,* Vol. 7, New York: Academic Press, Inc., 1968, pp. 71–133.
50. Hanika, J., Pištěk, R., and Růžička, V., "Influence of Reaction Conditions on Toluene Hydrogenation Rate in a Laboratory Autoclave," (Czech), *Chemický průmysl,* Vol. 26, No. 4 (1976), pp. 188–191.
51. Hanika, J., Vostrovský, V., and Růžička, V., "Influence of Reaction Conditions on Hydrogenation Rate in a Stirred Autoclave," (Czech), *Chemický průmysl,* Vol. 22, No. 11 (1972), pp. 549–551.
52. Sporka, K., Hanika, J., and Růžička, V., "Effect of Pressure on Acetophenone Hydrogenation," *Scientific Papers of the Prague Institute of Chemical Technology,* Vol. C27, No. 1 (1981), pp. 83–91.
53. Satterfield, C. N., Ma, Y. H., and Sherwood, T. K., "The Effectiveness Factor in a Liquid-Filled Porous Catalyst," *Institute of Chemical Engineers (London), Symposium Series,* Vol. 28, No. 1, (1968), pp. 26–33.
54. Hanika, J., et al., "Diffusion of Hydrogen in Internal Pores of the Catalyst Grain," *Collection of Czechoslovak Chemical Communications,* Vol. 37, No. 3 (1972), pp. 951–961.
55. Hanika, J., Chlumská, J., and Růžička, V., "Internal Diffusion Effect and Hydrogenation Selectivity of 1,5-Cyclooctadiene," *Collection of Czechoslovak Chemical Communications,* Vol. 45, No. 6 (1980), pp. 1684–1691.
56. Červený, L., and Růžička, V., "Solvent and Structure Effects in Hydrogenation of Unsaturated Substances on Solid Catalysts," *Advances in Catalysis,* New York: Academic Press, Inc., Vol. 30, 1981, pp. 335–377.
57. Prchlík, J., et al., "Liquid Distribution in Reactors with Randomly Packed Porous Beds," *Collection of Czechoslovak Chemical Communications,* Vol. 40, No. 3 (1975), pp. 845–855.
58. Herskowitz, M., and Smith, J. M., "Liquid Distribution in Trickle-Bed Reactors, Part I. Flow Measurements, Part II. Tracer Studies," *A.I.Ch.E. Journal,* Vol. 24, No. 3 (1978), pp. 439–454.
59. Tukač, V., "Catalytic Oxidation of Ethanol by Air in a Trickle-Bed Reactor," Ph. D. Thesis, Prague Institute of Chemical Technology, Prague 1983 (see also Tukač, V., Hanika, J., and Růžička, V., "Nonuniform Liquid Flow in Trickle Bed Reactor and its Efficiency," (Czech), *Chemický průmysl,* Vol. 34 (1984), pp. 566–572.)
60. Kolomazník, K., et al., "Liquid Distribution in Trickle-Bed Reactors II, Experimental Determination of the Spreading Coefficient," *Collection of Czechoslovak Chemical Communications,* Vol. 39, No. 1 (1974), pp. 216–219.
61. Schiesser, W. E., and Lapidus, L., "Further Studies of Fluid Flow and Mass Transfer in Trickle Beds," *A.I.Ch.E. Journal,* Vol. 7, No. 1 (1961), pp. 163–171.
62. Hanika, J., et al., "The Hydrogenation of Cyclohexene in a Laboratory Trickle-Bed Reactor," *Collection of Czechoslovak Chemical Communications,* Vol. 39, No. 1 (1974), pp. 210–215.

63. Hanika, J., Sporka, K., and Růžička, V., "Analyses of Mass Transfer in Trickle-Bed Reactors," (Czech), *Chemický průmysl,* Vol. 22, No. 1 (1972), pp. 1–5.
64. Herskowitz, M., Carbonell, R. G., and Smith, J. M., "Effectiveness Factors and Mass Transfer in Trickle-Bed Reactors," *A.I.Ch.E. Journal,* Vol. 25, No. 2 (1979), pp. 272–282.
65. Weekman, V. W., "Hydroprocessing Reaction Engineering," 4th International Symposium on Chemical Reaction Engineering, Heidelberg 1976, Survey Papers, pp. 615–646.
66. Germain, A. H., Lefebre, A. G., and L'Homme, G. A., "Experimental Study of a Catalytic Trickle-Bed Reactor," *American Chemical Society, Monograph Series,* Vol. 133, 1974, pp. 164–179.
67. Philips, B. D., Leavitt, F. W., and Yoon, C. Y., "Heat Transfer with Molecular Sieve Adsorbent I. Effective Thermal Conductivity II. Heat Transfer to a Packed Bed From Finned Tubes," *Chemical Engineering Progress Symposium Series,* Vol. 56, No. 30 (1960), pp. 219–235.
68. Matsuura, A., Akehata, T., and Shirai, T., "Simplified Calculation Method for Effective Thermal Conductivity and Wall Heat Transfer Coefficient in Packed Beds," *Heat Transfer Japanese Research,* Vol. 4, No. 1 (1975), pp. 79–82.
69. Tsang, T. H., Edgar, T. F., and Hougen, J. O., "Estimation of Het Transfer Parameters in a Packed Bed," *The Chemical Engineering Journal,* Vol. 11, No. 1, (1976), pp. 57–66.
70. Weekman, V. W., and Myers, J. E., "Heat Transfer Characteristics of Cocurrent Gas-Liquid Flow in Packed Beds," *A.I.Ch.E. Journal,* Vol. 11, No. 1 (1965), pp. 13–17.
71. Crine, M., and Marchot, P., "Modelling of Catalytic Trickle-Bed Reactors," in *Chemical Engineering of Gas-Liquid-Solid Catalyst Reactions,* G. A. L'Homme (Ed.), Liege: CEBEDOC, 1979, pp. 134–171.
72. Staněk, V., and Szekely, J., "The Effect of Heat and Gas Flow Interactions on the Performance of Adiabatic Packed-Bed Reactors," *Chemical Engineering Science,* Vol. 25, No. 7 (1970), pp. 1149–1158.
73. Staněk, V., and Kolář, V., "A Model of the Effect of the Distribution of Liquid on Liquid Hold-Up in a Packed Bed and a New Concept of Static Hold-Up," *The Chemical Engineering Journal,* Vol. 5, No. 1 (1960), pp. 51–60.
74. Staněk, V., and Eckert, V., "A Stratified Model of a Random Bed of Equal-Diameter Spheres Confined by a Plane," *Collection of Czechoslovak Chemical Communications,* Vol. 44, No. 3 (1979), pp. 829–840.
75. Staněk, V., and Eckert, V., "A Study of the Area Porosity Profiles in a Bed of Equal-Diameter Spheres Confined by a Plane," *Chemical Engineering Science,* Vol. 34, No. 7 (1979), pp. 933–940.
76. Staněk, V., and Eckert, V., "Course of Porosity in Layered Beds of Spherical Particles," *Archiv für das Eisenhüttenwesen,* Vol. 50, No. 1 (1979), pp. 19–24.
77. Staněk, V., and Szekely, J., "Three-Dimensional Flow of Fluid Through Nonuniform Packed Beds," *A.I.Ch.E. Journal,* Vol. 20, No. 5 (1974), pp. 974–980.
78. Kolář, V., and Staněk, V., "Distribution of Liquid over a Random Packing," *Collection of Czechoslovak Chemical Communications,* Vol. 30, No. 4 (1965), pp. 1054–1059.
79. Staněk, V., and Kolář, V., "Distribution of Liquid Over a Random Packing II. Derivation of Relations for the Distribution of Density of Wetting in a Cylindrical Column," *Collection of Czechoslovak Chemical Communications,* Vol. 32, No. 11 (1967), pp. 4207–4215.
80. Staněk, V., and Kolář, V., "Distribution of Liquid over a Random Packing III. Distribution of Liquid in a Low Bed of Packing Wetted by a Thin Stream of Liquid; Determination of the Coefficient of Radial Spreading of Liquid," *Collection of Czechoslovak Chemical Communications,* Vol. 33, No. 4 (1968), pp. 1049–1061.
81. Staněk, V., and Kolář, V., "Distribution of Liquid Over a Random Packing IV. Verification of the Boundary Condition of Liquid Transfer Between a Packed Bed and

the Wall of a Cylindrical Column and Evaluation of Its Parameters," *Collection of Czechoslovak Chemical Communications,* Vol. 33, No. 4 (1968), pp. 1062–1077.
82. Staněk, V., and Kolář, V., "Distribution of Liquid Over a Random Packing V. Dependence of the Coefficient of Radial Spreading of Liquid D on the Size and Form of the Packing Element and on the Mean Density of Wetting," *Collection of Czechoslovak Chemical Communications,* Vol. 33, No. 8 (1968), pp. 2636–2645.
83. Staněk, V., Kolář, V., and Tichý, J., "Distribution of Liquid Over a Random Packing VI. Effect of Distribution of Liquid on Experimentally Determined Hold-Up of Liquid in a Randomly Packed Bed," *Collection of Czechoslovak Chemical Communications,* Vol. 33, No. 10 (1968), pp. 3235–3243.
84. Staněk, V., and Kolář, V., "Distribution of Liquid Over a Random Packing VII. The Dependence of Distribution Parameters on the Column and Packing Diameter," *Collection of Czechoslovak Chemical Communications,* Vol. 38, No. 4 (1973), pp. 1012–1026.
85. Staněk, V., and Kolář, V., "Distribution of Liquid Over a Random Packing VIII. Distribution of the Density of Wetting in a Packing for an Arbitrary type of Initial Condition," *Collection of Czechoslovak Chemical Communications,* Vol. 38, No. 10 (1973), pp. 2865–2873.
86. Staněk, V., and Kolář, V., "The Radial Spread and the Number of Rivulets in a Trickle Bed," *Collection of Czechoslovak Chemical Communications,* Vol. 39, No. 8 (1974), pp. 2007–2018.
87. Staněk, V., "The Effect of Liquid Spread on the Top of the Bed on Radial Spread within the Packing," *Collection of Czechoslovak Chemical Communications,* Vol. 42, No. 4 (1977), pp. 1129–1140.
88. Staněk, V., and Kolev, N., "A Study of the Dependence of Radial Spread of Liquid in Random Beds on Local Conditions of Irrigation," *Chemical Engineering Science,* Vol. 33, No. 8 (1978), pp. 1049–1053.
89. Staněk, V., and Čarský, M., "Analysis of Frequency Responses of Model Flows of Liquid in Trickle Beds," *Collection of Czechoslovak Chemical Communications,* Vol. 47, No. 11 (1982), pp. 3032–3043.
90. Staněk, V., and Čarský, M., "An Experimental Set-Up for the Identification of Parameters of the Trickle Flow in a Packed Bed," *Collection of Czechoslovak Chemical Communications,* Vol. 48, No. 1 (1983), pp. 258–266.
91. Jiřičný, V., et al., "A Tensometric Method to Determine the Hold-Up of Liquid in Packed Bed Apparatus," *Collection of Czechoslovak Chemical Communications,* Vol. 47, No. 8 (1982), pp. 2190–2199.
92. Moravec, P., and Staněk, V., "Analysis of Observability of Model Parameters for Physical Absorption of Poorly Soluble Gases in Packed Beds," *Collection of Czechoslovak Chemical Communications,* Vol. 47, No. 10 (1982), pp. 2639–2653.
93. Jiřičný, V., and Staněk, V., "A Versatile Correlation of Liquid Hold-Up under the Single-Phase Trickle Flow in a Packed Bed," *Collection of Czechoslovak Chemical Communications,* Vol. 48, No. 12 (1983), pp. 3356–3369.
94. Jiřičný, V., and Staněk, V., "A Versatile Correlation of Liquid Hold-Up in a Two-Phase Counter-Current Trickle-Bed Column," presented for publication in *Chemical Engineering Communications,* 1983.
95. Ridgway, K., and Tarbuck, K. J., "Radial Voidage in Randomly Packed Beds of Spheres of Different Sizes," *Journal of Pharmacy and Pharmacology,* Vol. 18, Suppl. (1966), pp. 168S–175S.
96. Dutkai, E., and Ruckenstein, E., "Liquid Distribution in Packed Columns," *Chemical Engineering Science,* Vol. 23, No. 11 (1968), pp. 1365–1373.
97. Crine, P., Marchot, P., and L'Homme, G. L., "Liquid Flow Maldistributions in Trickle-Bed Reactors," *Chemical Engineering Communications,* Vol. 7, No. 6 (1980), pp. 377–388.
98. Drahoš, J., and Čermák, J., "Statistical Analysis of Local Gas Velocities at the Exit from a Packed Bed," *The Chemical Engineering Journal,* Vol. 24, No. 1 (1982), pp. 71–80.

99. Tosun, G., "Axial Dispersion in Trickle-Bed Reactors. Influence of Gas Flow Rate," *Industrial Engineering Chemistry, Fundamentals,* Vol. 21, No. 2 (1982), pp. 184–186.
100. Crine, M. D., and L'Homme, G. A., "The Percolation Theory: A Powerful Tool for a Novel Approach to the Design of Trickle-Bed Reactors and Columns," *American Chemical Society Symposium Series,* No. 196, Chemical Reaction Engineering—Boston, J. Wei and C. Georgakis (Eds.), 1982, pp. 407–419.
101. Lukianov, V. N., et al. "Modelling of Hydrogenation of Cyclooctadiene in a Trickle-Bed Reactor in the Presence of Phase Transitions," 8th CHISA Congress, Prague, 1984.
102. Onda, K., et al., "Liquid Distribution in a Packed Column," *Chemical Engineering Science,* Vol. 28, No. 9 (1973), pp. 1677–1683.
103. Crine. M., Marchot, P., and L'Homme, G. A., "Mathematical Modelling of the Liquid Trickling Flow through a Packed Bed Using the Percolation Theory," *Publications du Groupe de Chimie Appliquee et de Genie Chimique,* Universite de Liege, Vol. 23, No. 1 (1979), pp. 1–16 (paper presented at the 12th Symposium on Computer Applications in Chemical Engineering, Montreaux, April 8 to 11, 1979).
104. Crank, J., "Mathematics of Diffusion," Oxford University Press, Oxford, 1956.
105. Michelsen, M. L., and Ostergaard, K., "Hold-Up and Fluid Mixing in Gas-Liquid Fluidized Beds," *The Chemical Engineering Journal,* Vol. 2, No. 1 (1970), pp. 37–46.
106. Van Swaaij, W. P. M., Charperntier, J. C., and Villermaux, J., "Residence Time Distribution in the Liquid Phase of Trickle Flow in Packed Columns," *Chemical Engineering Science,* Vol. 24, No. 7 (1969), pp. 1083–1095.
107. Schwartz, J. G., and Roberts, G. W., "An Evalution of Models for Liquid Backmixing in Trickle-Bed Reactors," *Industrial Engineering Chemistry, Process, Design and Development,* Vol. 12, No. 3 (1973), pp. 262–271.
108. Farid, M. M., and Gunn, D. J., "Dispersion in Trickle and Two-Phase Flow in Packed Columns," *Chemical Engineering Science,* Vol. 34, No. 4 (1979), pp. 579–592.
109. Schwartz, J. G., and Dudukovic, M. P., "Liquid Hold-Up and Dispersion in Trickle-Bed Reactors," *A.I.Ch.E. Journal,* Vol. 22, No. 5 (1976), pp. 953–956.
110. Broadbent, S. R., and Hammersley, J. M., "Percolation Processes, I. Crystals and Mazes, II. The Connective Constant," *Proceedings of the Cambridge Philosophical Society,* Vol. 53, No. 3 (July 1957), pp. 629–645.
111. Čársky, M., "Study of the Two-Phase Countercurrent Flow in Packed Column by Harmonic Analysis Method," Ph. D. Thesis, Institute of Chemical Process Fundamentals, Prague, 1980.
112. Sicardi, S., Baldi, G., and Specchia, V., "Hydrodynamic Models for the Interpretation of the Liquid Flow in Trickle-Bed Reactors," *Chemical Engineering Science,* Vol. 35, No. 8 (1980), pp. 1775–1782.
113. Matsuura, A., Akehata, T., and Shirai, T., "Axial Dispersion of Liquid in Cocurrent Gas-Liquid Downflow in Packed Beds," *Journal of Chemical Engineering Japan,* Vol. 9, No. 4 (1976), pp. 294–301.
114. Lockhart, R. W., and Martinelli, R. C., "Proposed Correlation of Data for Isothermal Two-Phase, Two-Component Flow in Pipes," *Chemical Engineering Progress,* Vol. 45, No. 1 (1949), pp. 39–48.
115. Sato, Y., et al., "Pressure Loss and Liquid Hold-Up in Packed Bed Reactor with Cocurrent Gas-Liquid Downflow," *Journal of Chemical Engineering Japan,* Vol. 6, No. 2 (April 1973), pp. 147–152.
116. Larkins, R. P., White, R. R., and Jeffrey, D. W., "Two-Phase Cocurrent Flow in Packed Beds," *A.I.Ch.E. Journal,* Vol. 7, No. 2 (1961), pp. 231–239.
117. Midoux, N., Favier, M., and Charpentier, J. C., "Flow Pattern, Pressure Loss, and Liquid Hold-Up Data in Gas-Liquid Downflow Packed Beds with Foaming and Non-foaming Hydrocarbons," *Journal of Chemical Engineering Japan,* Vol. 9, No. 5 (1976), pp. 350–356.
118. Weekman, V. W., and Myers, J. E., "Fluid Flow Characteristics of Cocurrent Gas-Liquid Flow in Packed Beds," *A.I.Ch.E. Journal,* Vol. 10, No. 6 (1964), pp. 951–957.

119. Turpin, J. L., and Huntigton, R. L., "Prediction of Pressure Drop for Two-Phase, Two-Component Cocurrent Flow in Packed Beds," *A.I.Ch.E. Journal,* Vol. 13, No. 6 (1967), pp. 1196–1202.
120. Charpentier, J. C., Prost, C., and Le Goff, P., "Chute de Pression pour des Ecoulements a Co-Courant dans les Colonnes a Garnisage Arrose: Comparaison avec le Garnissage Noye," *Chemical Engineering Science,* Vol. 24, No. 12 (1969), pp. 1777–1794.
121. Charpentier, J. C., Prost, C., and Le Goff, P., "Ecoulement Ruisselant de Liquide dans Une Colonne a Garnissage. Determination des Vitesses et des Debits Relatifs des Films, Des Filets et des Gouttes," *Chimie et Industrie—Génie Chimique,* Vol. 100, No. 5 (1968), pp. 653–665.
122. Hutton, B. E., and Leung, L. S., "Cocurrent Gas-Liquid Flow in Packed Columns," *Chemical Engineering Science,* Vol. 29, No. 7 (1974), pp. 1681–1685.
123. Specchia, V., and Baldi, G., "Pressure Drop and Liquid Hold-Up for Two-Phase Cocurrent Flow in Packed Bed," *Chemical Engineering Science,* Vol. 32, No. 5 (1977), pp. 515–523.
124. Ergun, S., "Fluid Flow Through Packed Columns," *Chemical Engineering Progress,* Vol. 48, No. 2 (1952), pp. 89–94.
125. Mersmann, A., "Die Trennwirkung von Hohlfüllkörpern," *Chemie-Ingenieur-Technik,* Vol. 37, No. 7 (1965), pp. 672–680.
126. Charpentier, J. C., et al., "Etude de la Rétention de Liquide das une Colonne à Garnissage Arrosé à Contre-Courant et à Co-Courant de Gaz-Liquide," *Chimie et Industrie-Génie Chimique,* Vol. 99 (1968), pp. 803–826.
127. Mersmann, A., "Restflüssigkeit in Schüttungen," *Verfahrenstechnik,* Vol. 6, No. 6 (1972), pp. 203–206.
128. Otake, T., and Okada, K., "Liquid Hold-Up in Packed Towers-Operating Hold-Up Without Gas Flow," *Chemical Engineering Japan,* Vol. 17 (1953), pp. 176–184.
129. Crine, M., Marchot, P., and L'Homme, G. A., "Liquid Maldistribution in Trickle-Bed Reactors," Publications du Groupe de Chimie Apliquee et de Genie Chimique Universite de Liege, Vol. 23, No. 1 (1979), pp. 19–36 (paper presented at the American Institute of Chemical Engineers, 72nd Annual Meeting, San Francisco, November 25–29, 1979).
130. Zarzycki, R., "Liquid Distribution in th Packed Column," Inzy/nieria Chemiczna, Vol. 1, No. 1 (1971), pp. 85–102.
131. Porter, K. E., Barnett, V. D., and Templeman, J. J., "Liquid Flow in Packed Columns, Part I. The Rivulet Model, Part II. The Spread of Liquid over Random Packing, Part III. Wall Flow," *The Transaction of the Institution of Chemical Engineers,* Vol. 46, No. 3 (1968), pp. T69–T94.
132. Kirkpatrick, S., "Percolation and Conduction," *Review of Modern Physics,* Vol. 45, No. 4 (1973), pp. 574–588.
133. Onda, K., Takeuchi, H., and Okumato, Y., "Mass Transfer Coefficients between Gas and Liquid Phases in Packed Columns," *Journal of Chemical Engineering Japan,* Vol. 1, No. 1 (1968), pp. 56–62.
134. Reiss, L. P., "Cocurrent Gas-Liquid Contacting in Packed Columns," *Industrial Engineering Chemistry, Process Design and Development,* Vol. 6, No. 4 (1967), pp. 486–499.
135. Charpentier, J. C., "Recent Progress in Two-Phase Gas-Liquid Mass Transfer in Packed Beds," (Review Paper), *The Chemical Engineering Journal,* Vol. 11, No. 1 (1967), pp. 161–181.
136. Shende, B. W., and Sharma, M. M., "Mass Transfer in Packed Columns: Cocurrent operation," *Chemical Engineering Science,* Vol. 29, No. 8 (1974), pp. 1763–1772.
137. Moravec, P., "Dynamics of the Oxygen Absorption in a Packed Column," Ph. D. Thesis, Institute of Chemical Process Fundamentals, Prague 1983.
138. Sato, Y., et al., "Performance of Fixed Bed Catalytic Reactor with Cocurrent Gas-Liquid Flow," 1st Pacific Chemical Engineering Congress, Kyoto 1972, Paper 8.3., pp. 187–196.

139. Dharwadkar, A., and Sylvester, N. D., "Liquid-Solid Mass Transfer in Trickle-Beds," *A.I.Ch.E. Journal,* Vol. 23, No. 3 (1977), pp. 376–378.
140. Goto, S., Levec, J., and Smith, J. M., "Mass Transfer in Packed Beds with Two-Phase Flow," *Industrial Engineering Chemistry, Process Design and Development,* Vol. 14, No. 4 (1975), pp. 473–478.
141. Hanika, J., Vychodil, P., and Růžička, V., "A Cell Model of the Isothermal Trickle-Bed Reactor," *Collection of the Czechoslovak Chemical Communications,* Vol. 43, No. 8 (1978), pp. 2111–2121.
142. Jameson, G. J., "Model for Liquid Distribution in Packed Columns and Trickle-Bed Reactors," *Transaction of the Institution of Chemical Engineers,* Vol. 4, No. 1 (1966), pp. T198–T206.
143. Roblee, L. H. S., Baird, R. M., and Tierney, J. W., "Radial Porosity Variations in Packed Beds," *A.I.Ch.E. Journal,* Vol. 4, No. 4 (1958), pp. 460–464.
144. Kondelík, P., and Boyarinov, A. I., "Heat and Mass Transfer in Heterogeneous Catalysis XI. Wall Effect in a Tubular Catalytic Reactor I," *Collection of the Czechoslovak Chemical Communications,* Vol. 34, No. 12 (1969), pp. 3852–3861.
145. Szekely, J., and Propster, M., "Resistance of Layer-Charged Blast-Furnace Burdens to Gas-Flow," *Ironmaking & Steelmaking,* Vol. 4, No. 1 (1977), pp. 15–22.
146. Propster, M., and Szekely, J., "The Porosity of Systems Consisting of Layers of Different Particles," *Powder Technology,* Vol. 17 (1977), pp. 123–138.
147. Kondelík, P., Horák, J., and Tesařová, J., "Heat and Mass Transfer in Heterogeneous Catalysis," *Industrial Engineering Chemistry, Process Design and Development,* Vol. 7, No. 2 (1968), pp. 250–252.
148. Deans, H. A., and Lapidus, L., "A Computational Model for Predicting and Correlating the Behavior of Fixed-Bed Reactors," *A.I.Ch.E. Journal,* Vol. 6, No. 4 (1960), pp. 656–668.
149. Henry, H. C., and Gilbert, J. B., "Scale Up of Pilot Plant Data for Catalytic Hydroprocessing," *Industrial Engineering Chemistry, Process Design and Development,* Vol. 12, No. 3 (1973), pp. 328–334.
150. Puranik, S., and Vogelpohl, A., "Effective Interfacial Area in Irrigated Packed Columns," *Chemical Engineering Science,* Vol. 29, No. 2 (1974), pp. 501–507.
151. Mears, D. E., "The Role of Liquid Holdup and Effective Wetting in Trickle-Bed Reactors," 3rd International Symposium on Chemical Reaction Engineering, *Advances in Chemistry,* Series No. 133 (1974), pp. 218–227.
152. Jaffe, S. B., "Hot Spot Simulation in Commercial Hydrogenation Processes," *Industrial Engineering Chemistry, Process Design and Development,* Vol. 15, No. 3 (1976), pp. 410–416.
153. Staněk, V. et al., "On the Influence of Flow Distribution on Phase Transitions in Catalytic Trickle-Bed Reactors," 8th CHISA '84 Congress, Prague 1984.
154. Onda, K., Takeuchi, H., and Kojama, J., "Effect of Packing Materials on the Wetted Surface Area," *Kagaku Kogaku,* Vol. 31, No. 2 (1967), pp. 126–134.
155. Cihla, F., and Schmidt, O., "Studies of the Behavior of Liquids when Freely Trickling Over the Packing of a Cylindrical Tower," *Collection of the Czechoslovak Chemical Communications,* Vol. 23, No. 4 (1958), pp. 569–578.
156. Bennet, A., and Goodbridge, F., "Hydrodynamic and Mass Transfer Studies in Packed Absorption Columns, Part I. Axial Liquid Dispersion, Part II. The Measurement of Total Interfacial Area," *Transactions of the Institution of Chemical Engineers,* Vol. 48, No. 1 (1970), pp. T232–T244.
157. Hanika, J. et al., "Model of the Adiabatic Trickle Bed Reactor," Proccedings of the 5th Symposium Computers in Chemical Engineering, October 5 to 9, 1977, Vysoké Tatry, Czechoslovakia, pp. 265–269.
158. Clements, L. D., and Schmidt, P. C., "Two-Phase Pressure Drop in Cocurrent Downflow in Packed Beds: Air-Silicone Oil Systems," *A.I.Ch.E. Journal,* Vol. 26, No. 2 (1980), pp. 314–317.

159. Specchia, V., Baldi, G., and Gianetto, A., "Solid-Liquid Mass Transfer in Trickle Bed Reactors," in *Proceedings of the 4th International and 6th European Symposium on Chemical Reaction Engineering*, Heidelberg, Federal Republic of Germany, April 6 to 8, 1976, pp. 390–398.
160. Morsi, B. I., Midoux, N., and Charpentier, J. C., "Flow Patterns and Some Holdup Experimental Data in Trickle Bed Reactors for Foaming, Nonfoaming, and Viscous Organic Liquids," *A.I.Ch.E. Journal*, Vol. 24, No. 2 (1978), pp. 357–360.
161. Mills, P. L., and Dudukovic, M. P., "Evaluation of Liquid-Solid Contacting in Trickle-Bed Reactors by Tracer Methods," *A.I.Ch.E. Journal*, Vol. 27, No. 6 (1981), pp. 893–904.
162. Matsuura, A., Akehata, T., and Shirai, T., "Correlation for Dynamic Holdup in Packed Beds with Cocurrent Gas-Liquid Downflow," *Journal of Chemical Engineering of Japan*, Vol. 12, No. 4 (1979), pp. 263–268.
163. Turek, F., and Lange, R., "Mass Transfer in Trickle-Bed Reactors at Low Reynolds Number," *Chemical Engineering Science*, Vol. 36, No. 3 (1981), pp. 569–579.
164. Hanika, J., "Experimental Observation of the Evaporation of the Reaction Mixture during Cyclohexene Hydrogenation in a Bench-Scale Trickle-Bed Reactor," 8th CHISA Congress, Prague 1984.
165. Colombo, A. J., Baldi, G., and Sicardi, S., "Solid-Liquid Contacting Effectiveness in Trickle Bed Reactors," *Chemical Engineering Science*, Vol. 31, No. 9 (1976), pp. 1101–1108.
166. Carpa, V., et al., "Effect of Liquid Wetting on Catalyst Effectiveness in Trickle-Bed Reactors," *The Canadian Journal of Chemical Engineering*, Vol. 60 (April 1982), pp. 282–288.
167. Sakornwimon, W., and Sylvester, N. D., "Effectiveness Factors for Partially Wetted Catalysts in Trickle-Bed Reactors," *Industrial Engineering Chemistry, Process Design and Development*, Vol. 21, No. 1 (1982), pp. 16–25.
168. Dudukovic, M. P., "Catalyst Effectiveness Factor and Contacting Efficiency in Trickle-Bed Reactors," *A.I.Ch.E. Journal*, Vol. 23, No. 6 (1977), pp. 940–944.
169. Mills, P. L., and Dudukovic, M. P., "A Dual-Series Solution for the Effectiveness Factor in Trickle-Beds," *Industrial Engineering Chemistry, Fundamentals*, Vol. 18, No. 2 (1979), pp. 139–149.
170. Ramachandran, P. A., and Smith, J. M., "Effectiveness Factors in Trickle-Bed Reactors," *A.I.Ch.E. Journal*, Vol. 25, No. 3 (1979), pp. 538–542.
171. Kirillov, V. A., and Slinko, M. G., "Some Problems of the Mathematical Modelling of the Heterogeneous Catalytic Liquid Phase Processes, *Chimiceskaja promyslennost*, Vol. 1981, No. 11, pp. 650–656.
172. Hanika, J., and Růžička, V., "Entstehen und Entwicklung der Temperaturwelle in der Schicht des porösen Katalysators bei seiner Benetzung mittels einer Flüssigkeit," *Scientific Papers of the Prague Institute of Chemical Technology*, Vol. C26, No. 1 (1980), pp. 79–86.
173. Hanika, J., et al., "Residence Time Distribution in a Random Bed of the Porous Catalyst Trickled by Liquid," (Czech), *Chemický průmysl*, Vol. 23, No. 11 (1973), pp. 550–554.
174. Hanika, J., "The Intensifications of Chemical Reactors VI; Three-Phase Tubular Reactors with Trickle Bed of Catalyst," (Czech), *Chemický průmysl*, Vol. 26, No. 5 (1976), pp. 256–259.
175. Prchlík, J., et al., "Effect of Physical Properties of the Wetting Liquid on Its Distribution in Trickle-Bed Reactors," *Collection of the Czechoslovak Chemical Communications*, Vol. 43, No. 3 (1978), pp. 862–869.

CHAPTER 25

FOULING IN FIXED-BED REACTORS

R. Hughes

Department of Chemical Engineering
University of Salford
Salford, United Kingdom

CONTENTS

INTRODUCTION

Poisoning, sintering, and fouling are the three types of catalyst deactivation behavior. Sintering occurs as a consequence of inadvertent high temperature operation of the catalytic reactor with consequent temperature excursions of the catalyst. This results in loss of dispersion of the active metal crystallites on a supported metal catalyst or, in extreme cases, loss of support area as well. Poisoning is usually specific, small amounts of material being extremely efficient as deactivating agents. A classic example of poisoning is that of deactivation of nickel catalysts by sulphur compounds, quantities of less than 1 ppm being sufficient to destroy catalyst activity very rapidly. Although in some cases poisoning is reversible and indeed deliberate reversible poisoning has been advocated for some processes (e.g., acetylene hydrogenation) to improve selectivity, in general, poisoning can be regarded as irreversible. In principle, it is possible to purify the feed before this enters the reactor but this may be prohibitively costly. An alternative is to use a guard reactor that removes the poison before the feed enters the reactor proper. However, in general, both sintering and poisoning are generally irreversible and both involve renewal of the catalyst at appropriate time intervals.

Fouling is different in many ways to the other two deactivation processes. Fouling in fixed-bed reactors (and for that matter in other forms of catalytic reactors) may take two forms. The first is due to coke formation on the catalyst, the coking process being a consequence of the main or subsidiary reactions. Because of this intrinsic association with the reactions being processed over the catalyst, it is virtually impossible to eliminate completely fouling by coke formation. All that can usually be achieved is to minimize coke formation by catalyst selection and/or variation of the operating conditions. The second type of fouling is that associated with impurities in the feed. The best known example of this type of behavior is that of metal deposition during the hydrotreatment of petroleum fractions. In this type of operation, the metal deposits arise from metal porphyrins in the oil fractions being processed. Although the amount of these impurities is not large (typically a total of < 150 ppm of vanadium and nickel in Middle East residual oils rising to 250 ppm or higher in Venezuelan residua), continual deposition of these metals on the catalyst can cause severe problems. With the current tendency to hydrotreat increasingly heavier residues the problem is becoming more acute, since the proportion of asphaltenes is increased in the fraction and these may contain up to 2,000 ppm of metals. In many instances coke deposition occurs in parallel with metal deposition thus compounding the problem of deactivation of the catalyst even further.

For both fouling processes the extent of deposition on the catalyst surface can be extensive. For example, coke deposits of up to about 20% by weight can be obtained on reforming catalysts, although more often a level of 1%–4% is taken as the limit at which the catalyst activity is too low for economic operation to continue. Similarly, the metal deposition in hydrodesulphurization catalysts can be extensive resulting in pore blocking and even interparticle bed plugging in some instances. There is, however, one major difference between these two fouling processes. Coke deposits can generally be removed by gas-phase oxidation at temperatures from about 350° to 500° C. This is achieved by passing a gas stream of nitrogen (or a mixture of inert gases) containing about 0.5% to 2% oxygen over the catalyst at these temperatures and sometimes at pressures up to 10 bars. In contrast, metal deposition is irreversible and the catalyst charge in a reactor has to be removed for reprocessing when the activity has dropped to a specified level.

In the following discussion, coke formation, and the effects of metal deposition, as well as the influence of both on reactor design and operation will be considered.

Fouling Due to Coke Deposition

Coke formation on catalysts may take several forms and be caused by different mechanisms. If the temperature is high enough pyrolysis reactions may occur in the gas phase and deposition of gas-phase carbon may occur on the catalyst. At lower temperatures carbon or coke is usually formed directly on the catalyst surface by means of the catalytic processes. Reaction to form coke may proceed in parallel or in series to the main reaction as depicted below:

$$A \rightarrow R \quad \searrow \text{Coke} \qquad \text{Parallel fouling}$$

$$A \rightarrow R \rightarrow \text{Coke} \qquad \text{Series fouling}$$

Here A and R denote reactant and product respectively.

The structure of the deposited coke has stimulated many investigations. It would seem that for oxide catalysts, including cracking catalysts, the coke is in the form of a layered structure consisting of thin filmy aggregates of particle size less than 10 nm [1]. However, recent work using advanced techniques, such as electron microscopy and controlled atmosphere electron microscopy [2] has established that, with metals and supported metal catalysts, coke deposits occur as filaments. This raises some interesting questions concerning the mechanism of carbon formation under these conditions and how this type of growth affects transport of fluid within the catalyst.

In the latter connection there has been considerable discussion in recent years on whether coke deposition causes significant blockage of the pores and therefore limits the transport of reactants to the surface, in addition to covering active sites within the interior pore structure of the catalyst. Hughes and Mann [3] have developed a theory for the blocking of pores based on a "wedge layering" concept and Dautzenberg et al. [4] have presented a similar model for combined coke and metal deposition. It seems, however, that evidence from diffusion and surface area measurements suggests that pore blocking is not a significant problem at moderate coke levels (1%–4%) and only becomes serious when higher coke levels are attained. This conclusion is modified to some extent for zeolite catalysts with smaller pore openings. Results of Butt et al. [5] on the coking of H-Mordenite show an almost 50% decrease in diffusivity for a 4% coke deposit.

The species acting as precursor for the coke has been the subject for much investigation and speculation. Early work by Blue and Engle [6] suggested that olefins were largely responsible for coke deposition at low conversion but aromatics contributed significantly at higher conversions. Eberly et al. [7], however, using infrared techniques showed that aromatic skeletal structures existed on the surface. On the whole, the majority opinion is in favor of aromatic species being the precursor, but there is evidence that olefins can be important in certain circumstances. The chemical composition of the coke is generally accepted as $CH_{0.5}$ to CH, which suggests an aromatic or graphitic basis for the coke deposits.

Metal Deposition on Catalysts

Organic metal compounds exist in many crude oils and in derived liquids from coal liquefaction. For oils the major metallic impurities are vanadium and nickel, while for coal, iron and nickel are predominant. A comparison of the metal compositions of two crudes is given in Table 1. When petroleum residua are treated with hydrogen to remove sulphur, the metal porphyrins react out of the oil to produce solid sulphide deposits. Deposition may occur within the pores of the catalyst pellet (intraparticle pore deposition) or between the pellets constituting the bed (interparticle bed plugging). In general, with metal deposition, it is preferable to use larger pore-sized catalysts to prevent pore blocking, but increased pore diameter leads to a decrease in surface area and to a decrease in pellet strength, so that an optimum pore diameter is required.

The situation is complicated further by the simultaneous deposition of coke which occurs in many instances. This may also occur (as for metal deposition) in the interparticle void space of the bed, resulting in increased pressure drop in the bed. Some results obtained in a trickle-bed hydrodesulphurizer are shown in Figure 1, taken from the data of Kwan and Sato [8]. Coke deposition is greater in the lower part of the catalyst bed (i.e., the exit region),

Table 1
Metals Concentration in 560 °C Residua

	Oils	Resins	Asphaltenes
Khafi crude			
Composition, wt%	26	58	16
Vanadium, ppm	0	139	736
Nickel, ppm	0	36	250
Gach Saran crude			
Composition, wt%	28	62	10
Vanadium, ppm	0	282	1,455
Nickel, ppm	0	100	530

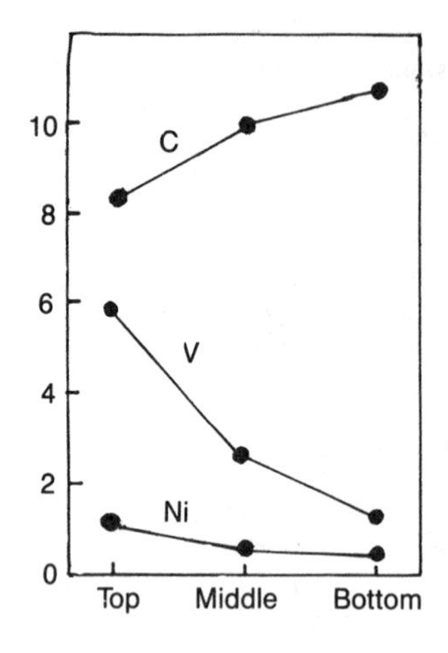

Figure 1. Coke, vanadium, and nickel distribution in catalyst bed (after 1,000 hours) [8].

while metals deposition occurs preferentially in the entrance region and declines monotonically with distance along the bed.

REACTOR FOULING DUE TO COKE DEPOSITION

As a first stage in the discussion of coking in fixed-bed reactors, it is useful to specify what is required in terms of information on coking. Clearly, the level of coke deposition is of basic importance since from experience it should determine the activity of the catalyst bed. However, in addition to the overall amount of coke deposited, it is important to know its distribution. This is important because:

1. The distribution can affect the operation and design of the reactor.
2. Because it is useful in determining the regeneration procedure.

As an example of the first case two types of temperature profiles obtained for different distributions of coke are illustrated in Figures 2A and 2B. These profiles are for an exothermic reaction, and it should be emphasised that temperature profiles are often the only information available to the plant operator, since activity is not easily determinable, sampling to determine concentration profiles is not always readily available, and only the conversion at the reactor exit is measured.

If the reactor operates adiabatically, a steep temperature profile is frequently obtained, if as is usually the case, the catalyst has a high chemical activity. Under these conditions, mass transfer is controlling and the reaction is confined to a small zone in the bed. If coking occurs, it will also be limited to the reaction zone because of the nature of the coking process, and as the zone becomes deactivated the reaction zone will pass down the bed. The temperature profiles indicate the position of the reaction zone and its displacement down the bed with time. The temperature profile therefore serves to estimate the residual activity of the reactor. In Figure 2A the temperature profile "marches" down the bed retaining the same shape with elapsed time. The behavior depicted in Figure 2B is somewhat different in that the temperature profile "reclines" as the catalyst deactivates throughout the bed. Clearly, for the case shown in Figure 2A, provision of sufficient bed length should give satisfactory bed activity since the original catalyst activity should be retained in the bed downstream from the reaction zone. For Figure 2B, however, there will be a loss of activity throughout the bed.

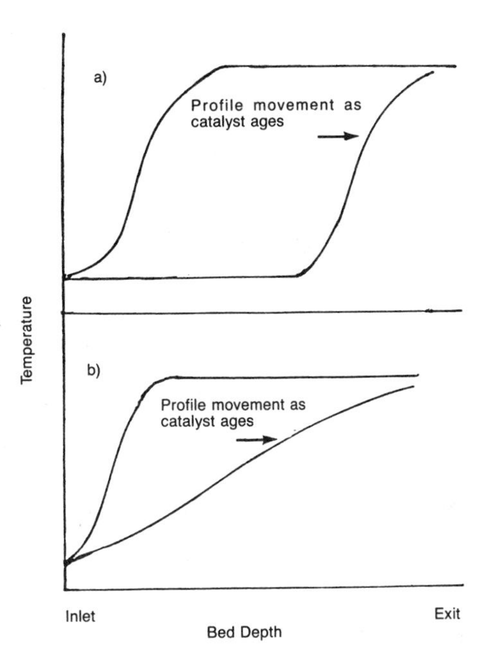

Figure 2. Movement of temperature profile in a catalyst bed as the catalyst ages.

Published information on the coking of industrial reactors is sparse and most laboratory studies have concentrated on thermobalance measurements and differential reactor experiments. Since there is some doubt about the validity of extrapolating thermobalance results to full scale baceuse of the inherent problems of obtaining sufficient mass transport in most thermobalance configurations, these are of limited use. Similarly with differential reactors, these operate at only low conversion, if they are truly differential, so the product concentration is low, unlike that of industrial reactors. Since the product may be the precursor of coke formation, as in the case of series fouling, results from a differential reactor would be unreliable.

Empirical Correlations for Coking

Most other information in the literature refers to modeling aspects and there have been very few experimental studies on integral or pilot-size reactors. A well-known early study which established a general correlation for the time dependence of coke formation was that of Voorhies [9] who used plant data. Some typical results obtained by Voorhies are shown in Figure 3. The correlation obtained was

$$C_c = At^n \tag{1}$$

where C_c = the concentration of coke on the catalyst
t = the process time
A and n constants

A variety of catalysts and operating conditions were used to obtain this correlation, and Voorhies's results indicated that for fixed-bed operation, the coking rate was independent of space velocity, and also had only a slight dependence on temperature. The value of the

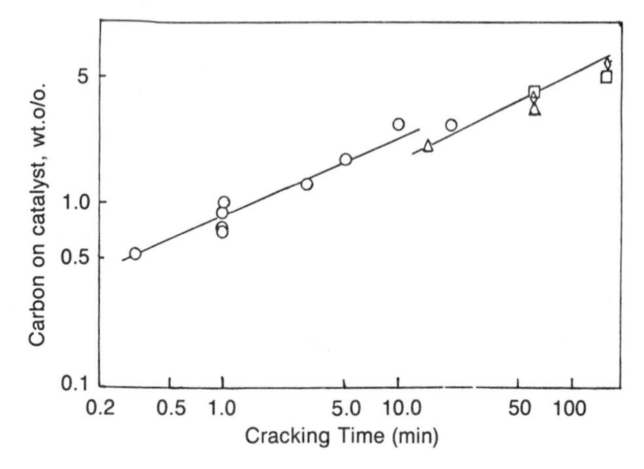

Figure 3. Carbon formation vs. cracking time in fixed-bed catalytic cracking. Symbols represent different space velocities [9].

exponent n varied somewhat but was in the range 0.5–1. This low value of n and the very small temperature dependence led to the suggestion that coking was diffusion controlled in these tests.

The lack of dependence of coke deposition on space velocity has been criticized, and some dependence on space velocity was observed by Eberly et al. [7]. However, the feed was complex in both Eberly's and Voorhies's experiments so this could account for the lack of expected dependence on space velocity. In practice, the Voorhies relation may be used for a large number of reactions where deactivation by coking occurs and can give a first approximation to the amount of coke deposited in a given time.

Another relation obtained by Rudershausen and Watson [10] for the coking rate during the aromatization of cyclohexane was of the form:

$$\frac{dC_c}{dt} = \frac{K}{C_c} \tag{2}$$

This equation can be integrated readily to give the Voorhies relation with n equal to 0.5. However, in contrast to the work of Voorhies a high temperature dependence was observed for coking, and this would rule out a diffusional mechanism. Ruderhausen and Watson were able to explain their data using a Langmuir–Hinshelwood form of chemical kinetics for coke formation.

Another approach to the general problem of catalyst deactivation, but which may be applied to coking processes, is the "time on stream" theory developed by Wojciechowski and coworkers [11]. This assumes that active sites are covered by the deposit and hence are unavailable for reaction; it does not consider pore blocking. Application of the theory leads to deactivation functions of either exponential of hyperbolic form, as described in the next section, and has proved useful in interpreting coking in some oil processing reactions.

Distribution of Coke Deposits

It is well established from the experimental work of Murakami et al. [12] and from modeling studies made by Masamune and Smith [13] that for single-catalyst pellets, coke deposition varies according to whether coking occurs by a parallel or series process. If coke

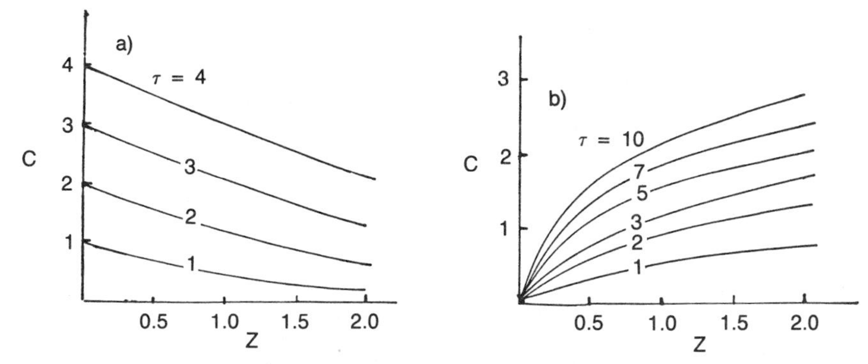

Figure 4. Carbon profiles in a fixed bed reactor: (A) parallel fouling; (B) series fouling [14].

is produced by a parallel reaction to the main reaction, coke will be produced predominantly in the region where the reactant concentration is high, i.e., at the outer part of the catalyst pellet. Conversely, if coke occurs by a reaction in series with the main reaction, then the reaction product is the coke precursor and so deposition will occur preferentially where the product concentration is highest i.e., at the center of the pellet. These conclusions were verified by some convincing experimental evidence produced by Murakami et al. [12] who considered both parallel and series coking reactions.

A similar argument would apply to coke deposits in a fixed-bed reactor and as proposed by Froment and Bischoff [14]. They verified this general conclusion by adoption of an isothermal plug flow homogeneous model. A separable form of kinetics was adopted, as is usual practice, which implies that the overall kinetics can be described by the true (time-independent) intrinsic kinetics multiplied by a deactivation function in which all the deactivation is included. Two forms of deactivation function were used by Froment and Bischoff, an exponential form and a hyperbolic form, defined as follows:

$$\psi = \exp(-\alpha C_c) \tag{3}$$

$$\psi = (1 + KC_c)^{-1} \tag{4}$$

where ψ is the deactivation function. In these relations, the concentration of coke replaces the time dependence, as this is considered to be the true variable.

The appropriate coke profiles obtained by this analysis are shown in Figure 4 for the case of an exponential deactivation function and confirm the general conclusion noted previously. As can be seen from Figure 4A, deposition of coke is greatest at the entrance of the reactor for parallel fouling where the reactant (the coke precursor) has the higher concentration. Conversely for series fouling, deposition is greatest near the reactor exit, where the fouling precursor (in this case the product has the highest concentration Figure 4B). Furthermore, parallel fouling gives a descending coke profile while series fouling gives an ascending profile. Similar results were obtained using the hyperbolic deactivation function. First-order kinetics were assumed for both the main and fouling reactions in this analysis. Experimental confirmation of the descending coke profile has been obtained by Van Zoonen [15]. The reaction investigated was the hydroisomerization of olefins to paraffins over a silica-alumina nickel sulphide catalyst. The coke was formed from the olefin and the coking process therefore occurs by a parallel reaction. The results were in good agreement with theory. Other experimental confirmation of a distribution of coke deposits was obtained by Dumez and Froment [16] for the dehydrogenation of butene to butadiene over a chromia-alumina catalyst and by de Pauw and Froment [17] for the isomerization of n-pentane over a platinum reforming catalyst. In the latter the expected predominant deposition towards the reactor exit was observed confirming that a reaction product was responsible for coking. In the case of butene dehydrogenation, coke was formed from both reactant and the butadiene product and in addition the hydrogen byproduct inhibited coke formation so no clear conclusion would be drawn. However, the experimental results suggest that series fouling was probably the major influence, as might be expected.

Nonisothermal Analysis of Coking

Dumez and Froment [16] also developed a model which predicted temperature profiles for the endothermic dehydrogenation reaction which was considered. Further development which also included a full transient analysis was made by Ervin and Luss [18]. Their main aim was to determine the stability of fixed-bed reactors under conditions where fouling was occurring. A cell model was used for the computations, and account was taken of the particulate nature of the system by using a heterogeneous analysis.

Isothermal treatments of coking behavior have proved useful in determining the general characteristics of coking systems. They are useful when the reaction enthalpy is moderate or small as in many of the isomerization reactions considered. However, certain phenomena, including the dynamics of fixed beds subject to deactivation can only be fully analyzed using a nonisothermal treatment. Inevitably, a nonisothermal analysis complicates the system because of the necessary introduction of additional parameters. The most important of these are the activation energies for the main and deactivation reactions, since the relative magnitude of these will determine whether fouling increases or decreases with changes in temperature within the reactor, but other parameters, such as the heats of adsorption for the various species can become significant if Langmuir–Hinshelwood kinetics are appropriate for the system.

A generalized model has been developed for fixed-bed reactors in recent years by Hughes and co-workers [19, 20, 21, 22]. Initially, Langmuir–Hinshelwood kinetics were considered for the main reaction and coking was considered to be first order. This was modified in the later work to include a more generalized model where the coking reaction was also of Langmuir–Hinshelwood form. The assumptions made in this generalized approach were:

1. Plug flow operation.
2. The fouling reaction occurs much more slowly than the main reaction, so pseudo steady-state conditions apply.
3. The gas stream has a constant average heat capacity and density.
4. The heat capacity of the reactor wall is negligible.
5. Volume changes due to reaction are neglected.
6. External mass and heat transfer coefficients are not influenced by the surface reaction.
7. Particle-particle heat conduction and heat transfer by radiation can be neglected.
8. The particle effective diffusivity and thermal conductivity remain constant.

For a heterogeneous model the heat and mass balance inside the pellets as well as in the gas stream have to be considered. On the other hand, fouling occurs only on and within the catalyst pellets and no deactivation equation is necessary in the external field formulation.

External Field Heat and Mass Balances

With the assumptions stated above, the pseudo steady-state mass and energy balance equations for the adiabatic bed can be written as follows:

$$\frac{\varepsilon' u dC_{A0}}{dz} = -a_v k_c (C_{A0} - C_{As}) \tag{5}$$

$$\varepsilon' u \varrho_g c_{pg} \frac{dT_0}{dz} = a_v h(T_s - T_0) \tag{6}$$

where k_c and h = the film mass and heat transfer coefficients
a_v = the external surface area of the pellets per unit bed volume

By taking a heat and mass balance around the catalyst particles, the concentration and temperature flux terms can be expressed as:

$$-\frac{a_v k_m}{\varepsilon'}(C_{A0} - C_{As}) = (1-\varepsilon')r_A(C_{A0}, T_0)\eta(C_{A0}, T_0, \bar{S}) \tag{7}$$

$$-\frac{a_v h}{\varepsilon'}(T_s - T_0) = (1-\varepsilon')(-\Delta H)r_A(C_{A0}, T_0)\eta(C_{A0}, T_0, \bar{S}) \tag{8}$$

Thus Equations 5 and 6 become:

$$\frac{dC_{A0}}{dz} = \frac{(1-\varepsilon')}{u} r_A(C_{A0}, T_0)\eta(C_{A0}, T_0, \bar{S}) \tag{9}$$

$$\frac{dT_0}{dz} = -\frac{(1-\varepsilon')(-\Delta H)}{u\varrho_g C_{pg}} r_A(C_{A0}, T_0)\eta(C_{A0}, T_0, \bar{S}) \tag{10}$$

In these equations $\bar{S}$ is the mean catalyst activity.
The boundary conditions are:

$$z = 0, \quad C_{A0} = C_{A0}|_{z=0} \tag{11a}$$

$$T_0 = T_0|_{z=0} \tag{11b}$$

These equations can be put into dimensionless form by introducing the following variables:

$$\xi = z/L \tag{12a}$$

$$a_0 = \frac{C_{A0}}{C_{A0}|_{z=0}} \tag{12b}$$

$$\theta_0 = \frac{T_0}{T_0|_{z=0}} \tag{12c}$$

$$\Omega = \frac{k(1-\varepsilon')L}{u}(C_{A0}|_{z=0})^{n-1} \tag{12d}$$

$$\beta' = \frac{-\Delta H}{\varrho_g c_{pg}}\left(\frac{C_{A0}}{T_0}\right)\bigg|_{z=0} \tag{12e}$$

where ξ = the dimensionless coordinate along the packed bed
a_0, θ_0 = the dimensionless bulk concentration and temperature
n = the order of reaction
Ω, β' = the reaction modulus and thermicity factor in the external field, respectively

The reaction modulus Ω is a modified form of the Damköhler number which, as defined, represents the ratio of the reaction rate to the linear flow rate of the reactants.

The dimensionless forms of Equations 9 and 10 become:

$$\frac{da_0}{d\xi} = \Omega f(a_0, \theta_0)\eta(a_0, \theta_0, \bar{S}) \tag{13}$$

$$\frac{d\theta_0}{d\xi} = -\beta'\Omega f(a_0, \theta_0)\eta(a_0, \theta_0, \bar{S}) \tag{14}$$

and the dimensionless boundary conditions are:

$$\xi = 0, \quad a_0 = 1 \tag{15a}$$

$$\theta_0 = 1 \tag{15b}$$

The particle effectiveness factor, $\eta(a_0, \theta_0, \bar{S})$, is obtained by solving the mass and energy continuity equations and may be expressed as:

$$\eta(a_0, \theta_0, \bar{S}) = \frac{\alpha \int_0^1 \delta^{(\alpha-1)} f(a, \theta) S \, d\theta}{f(a_0, \theta_0)} \tag{16}$$

where S and $\bar{S}$ = the point and mean activity of the pellet
α = the geometric shape factor for the pellets

Single-Particle Heat and Mass Balances

For the case of nonisothermal fouling, and with the inclusion of external film resistances, the dimensionless equations are:

$$\frac{d^2a}{d\delta^2} + \frac{\alpha-1}{\delta}\frac{da}{d\delta} = \varphi^2 f(a) S \exp\left[\gamma\left(1-\frac{1}{\theta}\right)\right] \tag{17}$$

$$\frac{d^2\theta}{d\delta^2} + \frac{\alpha-1}{\delta}\frac{d\theta}{d\delta} = -\beta\varphi^2 f(a) S \exp\left[\gamma\left(1-\frac{1}{\theta}\right)\right] \tag{18}$$

$$\tau = 0; \qquad S = 1 \text{ for any } \delta \tag{19}$$

$$\delta = 0; \qquad \frac{da}{d\delta} = \frac{d\theta}{d\delta} = 0 \text{ for any } \tau \tag{20}$$

$$\delta = 1; \qquad \frac{1}{Sh^*}\cdot\frac{da_s}{d\delta} = (a_0 - a_s) \tag{21}$$

$$\frac{1}{Nu^*}\cdot\frac{d\theta_s}{d\delta} = (\theta_0 - \theta_s) \tag{22}$$

where δ is the dimensionless particle radius and γ the Arrhenius number for the main reaction, and where use is made of the following stoichiometric relation between the concentration of reactant A and product B assuming equimolar counterdiffusion:

$$b = b_0 + q(a_0 - a) \tag{23}$$

The temperature may be expressed in terms of concentration via the Prater relation:

$$\theta = \theta_0 + \frac{Sh^*}{Nu^*}\beta(a_0 - a_s) + \beta(a_s - a) \tag{24}$$

The reaction kinetics for the main reaction are taken to be of Langmuir–Hinshelwood form and may be written as:

$$f(a, \theta) = a^n \exp[(\gamma - h_{k_A})(1 - 1/\theta)]/\{1 + K_A^* a \exp[-h_{k_A}(1 - 1/\theta)] + K_B^*(a_0 + b_0 - a)\exp[-h_{k_B}(1 - 1/\theta)]\} \tag{25}$$

The dimensionless rate equation for the deactivation can be assumed to be of the generalized form:

$$\frac{dS}{d\tau} = g_1(a, \theta)S + k_{f_2}/k_{f_1} g_2(b, \theta)S \tag{26}$$

where $g_1(a, \theta)$ and $g_2(a, \theta)$ are functions of concentration and temperature and may be expressed in the form of simple integral kinetics or in complex kinetic form. The term k_{f_1} signifies parallel fouling while k_{f_2} is for series fouling. In the described work two different types of fouling kinetics were studied. In the first case, fouling was assumed to be first order in the concentration of fouling precursor and also first order in active sites concentration:

$$g_1(a, \theta) = a \exp[\gamma_f(1 - 1/\theta)] \tag{27}$$

$$g_2(b, \theta) = b \exp[\gamma_f(1 - 1/\theta)] \tag{28}$$

When Langmuir–Hinshelwood kinetics are assumed for the deactivation reaction, fouling was assumed to be a function of the active sites concentration and the complex kinetics described by g_1 and g_2 above. The functions g_1 and g_2 are now written as:

$$g_1(a, \theta) = \frac{K_A^* \exp[(\gamma_f - h_{k_A})(1 - 1/\theta)]}{1 + K_A^* a \exp[-h_{k_A}(1 - 1/\theta)]} \tag{29}$$

$$g_2(b, \theta) = \frac{K_B^* b \exp[(\gamma_f - h_{k_B})(1 - 1/\theta)]}{1 + K_B^* b \exp[-h_{k_B}(1 - 1/\theta)]} \tag{30}$$

The appropriate equations for external field and pellet were solved numerically.

The prior model has been used to predict temperature, concentration, and activity profiles for an adiabatic reactor operating exothermically and to compare results for fouling due to either a first-order reaction or to Langmuir–Hinshelwood kinetics. Since it was desired to focus attention solely on the deactivating effects of fouling by coke deposition the complicating features of multiple steady states were avoided by appropriate choice of parameters.

Results were obtained for the parameter values of the following:

$Sh^* - 250$

$Nu^* - 1.5$

K_A^*, K_B^* — 10

h_{k_A}, h_{k_B} — −5

γ — 20

γ_f — 40

Ω — 16

The results obtained for an exothermic reaction with fouling by the product are shown in Figure 5. The temperature rise in the reactor, the product concentration, and the activity in the reactor are each plotted as a function of dimensionless reactor length for a value of the dimensionless time τ equal to 1.2. The values of the thermicity factors employed are $\beta = 0.02$ for the pellet and $\beta' = 0.015$ for the packed bed, while the activation energy parameter for fouling, γ_f, is taken as 40 for both first-order and Langmuir–Hinshelwood fouling. The essential feature of Figures 5A and B is that more reaction occurs with first-order fouling compared with Langmuir–Hinshelwood fouling due to the higher effective reaction order in the former case. This is demonstrated by the larger increase in temperature and the increased product concentration. It is also observed that there is no significant difference in product concentration over the first 10% of the reactor length for either type of fouling.

The corresponding activity profiles for the two types of fouling at $\tau = 1.2$ and 0.05 are shown in Figure 5C. For both first-order and Langmuir–Hinshelwood fouling, the activity is highest at the entrance to the reactor and decreases with distance along the reactor. This is in accord with previous findings for a series fouling mechanism, since the fouling precursor is the product, and the concentration of this will be zero at the reactor entrance (for no recycle stream) and will progressively increase along the reactor. However, an important difference between the two types of fouling is evident; Langmuir–Hinshelwood fouling gives a sharp drop in activity close to the reactor entrance, whereas for first-order fouling the activity decreases more gradually from inlet to exit.

One explanation for the initial sharp decrease in catalyst activity for Langmuir–Hinshelwood fouling at the reactor entrance is attributable to the difference in mechanism between this and first-order fouling. At the reactor entrance the main reaction will give only a small amount of product. For Langmuir–Hinshelwood fouling this product can give an immediate coke deposit, but for first-order fouling this product is present in the gas space in the pores of the pellet and may not be immediately available for depositing coke.

It is interesting to note that the initial decrease in activity for Langmuir–Hinshelwood fouling in Figure 5C is very pronounced. Examination of Figure 5A reveals that, although Langmuir–Hinshelwood fouling gives a smaller temperature rise, both types of fouling show small increases in reactor temperature which are relatively close. Because of this relatively small difference in reactor temperature, comparison of Equations 28 and 30 shows that Langmuir–Hinshelwood deactivation is greater than that for first-order by a factor of nearly $K_B^*/(1 + K_B^*)$, provided that the amount of fouling pressursor is the same for both processes. This requirement is fulfilled for the first 10% of reactor length (Figure 5B) and results in greater deactivation by a Langmuir–Hinshelwood mechanism in this region (Figure 5C). Thereafter, the precursor concentration is less than that for first-order fouling, so the activity profile levels off with an increase in reactor length (Figure 5C). In fact, the activity for Langmuir–Hinshelwood fouling under these conditions is always less than that for first-order fouling, as examination of the activity profiles at a reaction time of $\tau = 0.05$ shows (upper part of Figure 5C). For this short process time the precursor (product) concentration and temperature profiles for first-order and Langmuir–Hinshelwood mechanisms are almost identical, but Langmuir–Hinshelwood fouling gives a much more rapid activity decrease, particularly at the front of the reactor.

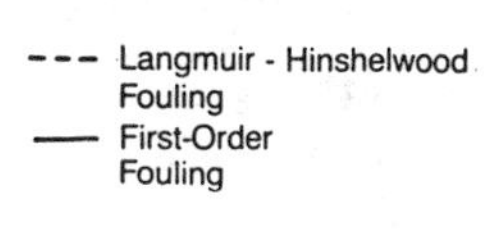

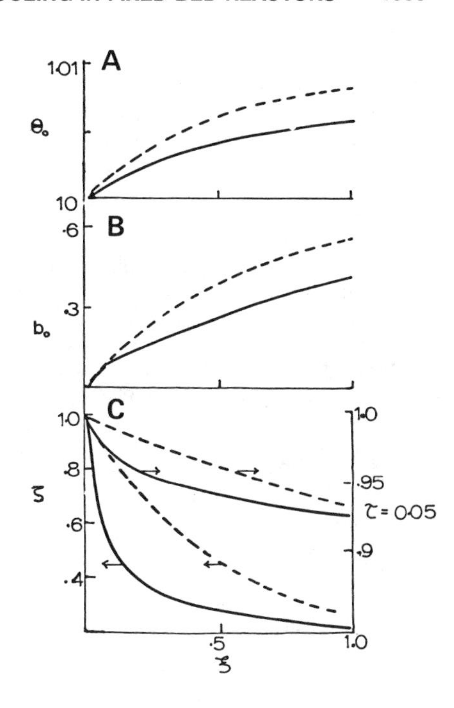

Figure 5. Dimensionless temperature, product concentration, and activity profiles for exothermic reaction with series fouling [23].

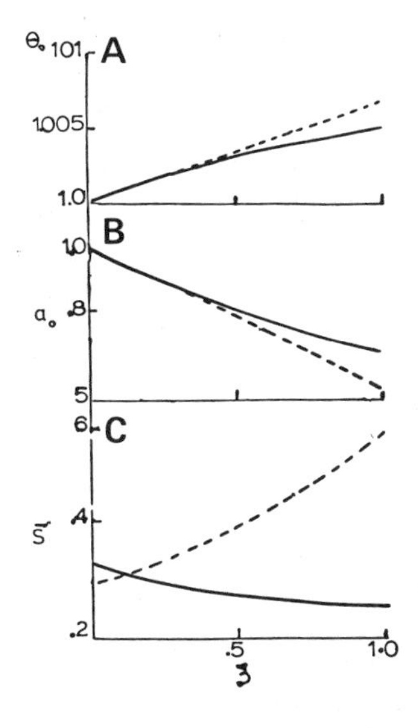

Figure 6. Dimensionless temperature, reactant concentration, and activity profiles for exothermic reaction with parallel fouling [23].

--- Langmuir - Hinshelwood Fouling
— First-Order Fouling

A comparison of the temperature, reactant concentration, and activity profiles for parallel fouling under exothermic conditions is given in Figures 6A, B, and C. The most surprising effect shown by the activity profiles Figure 6C is that Langmuir–Hinshelwood fouling for these parameter values shows behavior completely contradictory to what may be termed "normal" fouling behavior for fouling by a parallel mechanism. "Normal" behavior is shown by the first-order fouling activity profile, in which the activity is least at the reactor entrance and increases with distance along the reactor. This behavior is expected, since the fouling precursor (in this case the reactant) has the greatest concentration at the entrance to the reactor. Therefore, fouling would be expected to be predominant at the entrance and to decrease thereafter as the reactant concentration diminished along the length of the reactor. The Langmuir–Hinshelwood activity profile in Figure 6C, however, shows maximum catalytic activity at the reactor entrance, with a further slight but steady decline with distance along the reactor. Such a profile would in general be recognized as being more typical of series fouling (except that in this case there would be no initial low activity level at the reactor entrance). Furthermore, the overall activity of the bed is much less than that for first-order fouling.

This reversal of the normal parallel fouling activity pattern in a fixed-bed reactor has important consequences if the effect is generally true for Langmuir–Hinshelwood fouling, since a large number of gas-solid reactions obey Langmuir–Hinshelwood kinetics.

The effect of diffusional resistance on activity profiles when both the main and coking reactions obey Langmuir–Hinshelwood kinetics was also determined by Brito–Alayon et al. [23]. The simulated activity profiles were obtained as three-dimensional plots in which the vertical axis represents the mean activity of the pellet at that point in the reactor, while one horizontal axis gives values of the Thiele modulus, φ, plotted on a logarithmic scale. Values of the Thiele modulus ranged from 2.5 to 20 and γ_f was taken as 40. In all figures the fouling time, τ, was equal to 1.2, corresponding to a time on stream of 18 days for the parameters considered.

Figures 7A and 7B show the effect of increasing Thiele modulus on the activity profiles for endothermic reactions with $\beta = -0.015$ for the pellet and $\beta' = -0.020$ for the reactor. For the case of series fouling Figure 7A shows that the reactor activity is always greatest at the inlet of the reactor for the whole range of φ investigated ($\varphi = 2.5$–30). At the reactor outlet the activity was least and almost independent of φ at an activity level of about 0.55.

The decrease in activity from inlet to outlet is greatest for low values of φ and reflects the pronounced effect of Thiele modulus on activity at the reactor inlet. The results show the same pattern as observed previously on single-catalyst pellets obeying first order kinetics for coking [21, 22]. This is to be expected since conditions at the inlet of a plug-flow reactor operating without recycle would be virtually the same as for a single pellet. The effect of an increase in φ is to increase the product concentration inside the pellet. Since this is the precursor of fouling by a series mechanism, the extent of deactivation would be expected to decrease as φ increases, as was observed.

In contrast, when parallel fouling of an endothermic main reaction is considered, the situation is much more complicated, as shown in Figure 7B. At low values of φ (between

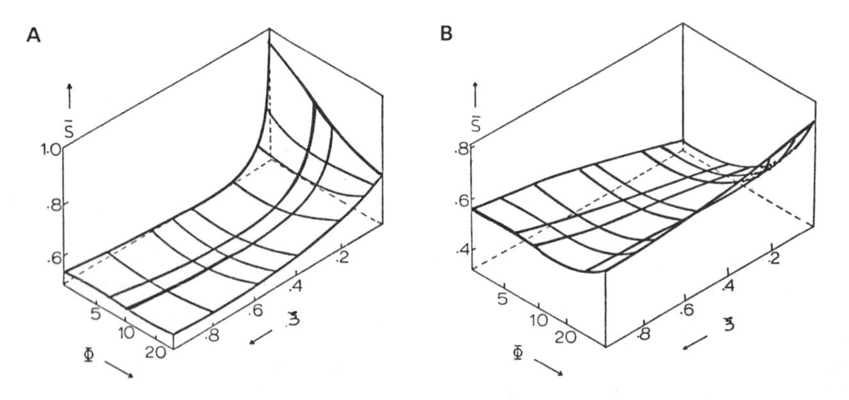

Figure 7. Variation of activity profiles with Thiele modulus for an endothermic reaction: (A) series fouling ($K_A^{\cdot} = K_B^{\cdot} = 10$); (B) parallel fouling ($K_A^{\cdot} = K_B^{\cdot} = 10$) [23].

2.5 and 10) the activity of the reactor is least at $\xi = 0$ and gradually increases throughout the reactor. It should be noted that the activity is very low at the inlet at about 0.4 and is less than 0.6 at the outlet. This behavior is consistent with the usually accepted distribution of activity when parallel fouling occurs, since now the reactant is the fouling precursor and its concentration is greatest at the reactor inlet. At values of the Thiele modulus greater than 10, however, the activity profile is reversed, with the activity now greatest at the reactor inlet and decreasing progressively along the length of the reactor. An explanation for this effect can be given in terms of the distribution of fouling deposit at low and high values of the Thiele modulus. At low values of φ the reactant is distributed uniformly throughout the pellet and therefore deposition of foulant will also occur throughout the pellet. At higher values of the Thiele modulus, reaction is confined to a relatively thin zone near the surface of the pellet, due to the diffusional restriction. Under such conditions it has been shown that the mean activity of the pellet may be greater than for the uniform deactivation obtained with a low value of φ [22]. Another factor contributing to the unexpected activity distribution at $\xi = 0$ is the nonisothermal nature of the reaction. At low values of φ the reaction rate is relatively slow and hence the effect of temperature is not significant. When φ is large, however, the rate of the main reaction is high, and this leads to a greater temperature reduction due to the reaction endothermicity which will favor the main reaction for which γ is 20 while causing a relative decrease in the rate of the deactivation reactions with the larger γ_f value of 40.

At the reactor exit, Figure 7B shows that there is only a small difference in activity for the whole range of φ. This is due to the decrease in temperature caused by the endothermic nature of the reaction reducing the rate of both main and deactivation reactions in this region. It is of interest to note that, at high values of φ for this case of endothermic parallel fouling, the mean activity of the reactor is greater than that at low values of φ. Thus, it may be preferable to operate under diffusion-controlled conditions when γ_f is greater than γ as in this instance. Furthermore, a comparison of Figures 7A and 7B shows that the value of φ selected, affects the deactivation behavior of the reactor more at the inlet than at the exit for both fouling mechanisms.

When the main reaction is exothermic the results obtained for deactivation by a series of fouling mechanisms are shown in Figure 8A. The activity profiles obtained are generally

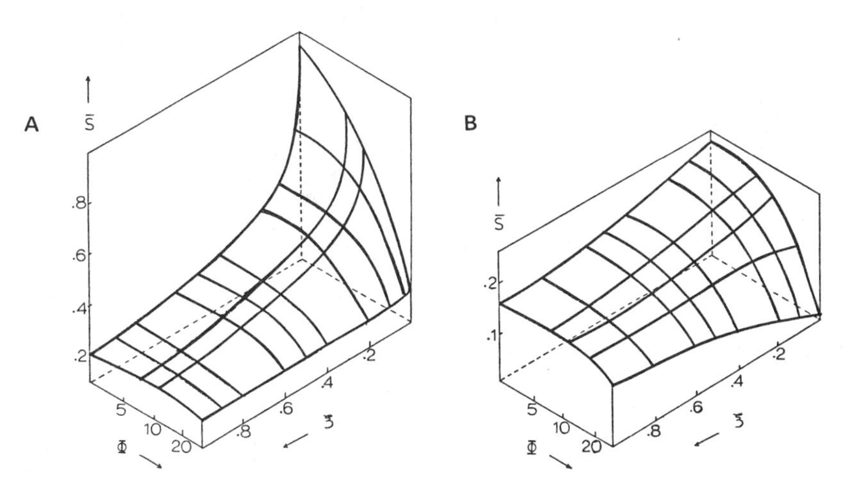

Figure 8. Variation of activity profiles with Thiele modulus for an exothermic reaction: (A) series fouling ($K_A^* = K_B^* = 10$); (B) parallel fouling ($K_A^* = K_B^* = 10$) [23].

Table 2
Particle Surface Temperatures for Endothermic Reactions at $\xi = 0$

Parallel Fouling $\theta_s\|_{\tau=1.0}$	φ	Series Fouling $\theta_s\|_{\tau=1.0}$
0.99946	7	0.99890
0.99876	10.5	0.99783
0.99035	30	0.98878
0.97936	50	

Table 3
Particle Surface Temperatures for Exothermic Reactions at $\xi = 0$

Parallel Fouling $\theta_s\|_{\tau=1.0}$	φ	Series Fouling $\theta_s\|_{\tau=1.0}$
1.00009	2.5	1.00022
1.00067	7.0	1.00150
1.00142	10.5	1.00297
1.00475	30.0	1.00962

similar to those shown in Figure 7A for endothermic series fouling, but now there is a much lower level of activity of about 0.2 at the reactor exit for all values of φ. Also at the inlet of the reactor ($\xi = 0$), a much more increased rate of deactivation occurs as the value of φ is increased compared with the endothermic reaction. As for Figure 7A, an increase in the value of φ will increase the product concentration inside the pellet, and for series fouling this will lead to increased deactivation. However, for endothermic conditions this effect was reduced because of the endothermic temperature decrease which becomes greater as the value of φ increases. In the present exothermic case, however, the temperature increase in the pellets as φ increases will increase the rate of deactivation, since γ_f is greater than γ and will thus give a much more severe decrease in activity at the reactor inlet as φ is increased (Figure 8A). Results for the surface temperature distribution in the first layer of pellets are given in Tables 2 and 3 for endothermic and exothermic reactions respectively.

The increase in temperature along the reactor for the exothermic reaction also accounts for the reduced activity at the exit, since the deactivation reaction is favored by an increase in temperature.

Figure 8B shows the activity profiles for parallel fouling under exothermic conditions. At values of φ between 2.5 and 10 there is a steady decrease of activity from $\xi = 0$ to $\xi = 1$, although the degree of deactivation is severe at all points along the reactor. This behavior is again not typical of the "normal" behavior associated with parallel fouling. When φ is increased above 10, normal-type behavior is observed for parallel fouling, with the activity least at the reactor inlet and increasing along the bed. It was observed previously [23] that, when first-order fouling was used in the simulations at this value of the Thiele modulus, no such anomaly occurred. This effect therefore demonstrates that when Langmuir–Hinshelwood kinetics describe the reactions, the diffusional resistance may have a profound influence on the shape of the activity profile.

All the above modeling results were obtained for values of K_A^* and K_B^* both equal to 10. To investigate whether the anomalous profiles observed previously might be caused by adsorption, simulations where K_A^* and K_B^* were varied from 1 to 40 were made by Brito–Alayon et al. [23].

The effect of variations in K_B^*, the adsorption constant for series fouling, were determined for an exothermic main reaction at values of φ equal to 2.5, where the main reaction would be under chemical control. This value of φ was selected because, for values of φ greater than 10, activity profiles along the reactor were generally much less steep than those for lower φ values, so the effect of parameter variations would be more difficult to determine. K_B^* was varied from 1 to 40 at a constant value of K_A^* equal to 10. It was found that activity profiles showed a continuous decrease from $\xi = 0$ to $\xi = 1$, from a value close to unity at the inlet to the reactor to about 0.3–0.4 at the outlet. With an increase in K_B^*, the decrease of activity in the inlet region of the reactor showed a much steeper decrease than at low values of K_B^*.

The effect of K_A^* on the activity profiles at a value of φ equal to 2.5 is given in Figure 9 for exothermic parallel fouling by a Langmuir–Hinshelwood mechanism. This figure shows that at low values of K_A^* (1 to 5) the "normal" activity characteristics of parallel fouling are observed, i.e., activity is least at the inlet where the reactant concentration is highest. This would be expected because at these low values of K_A^* only a comparatively small amount of reactant will be adsorbed on the active sites of the catalyst, and therefore only small amounts are available for both the main and fouling reactions. This is very similar to the situation for first-order fouling kinetics, and therefore the activity profiles would be excepted to be similar. At $\xi = 0$ there is a fairly steep decrease in activity as K_A^* is increased. It was confirmed that at both short and relatively long fouling times ($\tau = 0.05$ and 1.0 respectively), there is a decrease in surface temperature, when parallel fouling is present, as K_A^* is increased. The decrease in surface temperature is greater, however, for larger values of τ. Therefore, the turnover in activity profiles from a rising profile throughout the reactor to a falling one at K_A^* values greater than 5 must be attributed to increased adsorption of reactant, the fouling precursor, in the first row of catalyst pellets, leading to increased deposition of coke. These results are entirely consistent with the single-pellet simulations made previously [20].

All these results demonstrate that the diffusional resistance and adsorption constants may seriously modify the accepted pattern of activity profiles characterizing series and parallel fouling. The effect is particularly important for parallel fouling where the "normal" activity profile may be reversed. This anomalous behavior occurs only for Langmuir–Hinshelwood rate expressions, but since a large number of important reactions are believed to proceed via such mechanisms, care should be exercised in making predictions of activity profiles for such reactions throughout the whole range of Thiele modulus and adsorption constants.

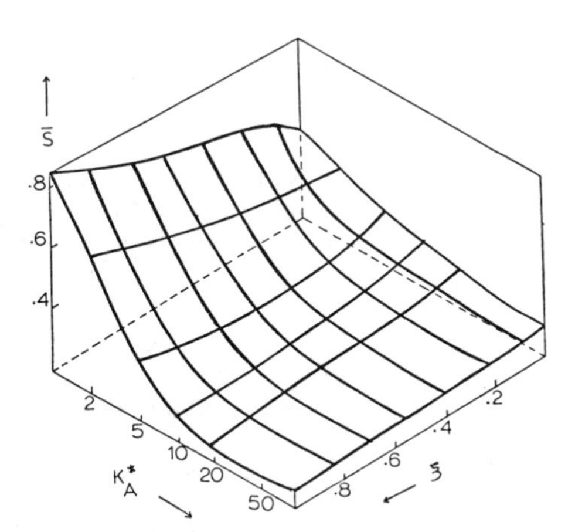

Figure 9. Effect of K_A^* on activity profiles. Exothermic reaction with parallel fouling; $K_B^* = 10$, $\phi = 2.5$ [23].

Coking in Trickle-Bed Reactors

The model described previously can account for coking of a catalytic reactor in terms of active sites deactivation by coke deposition. It does not account for pore blockage which may occur through coke deposition.

Pore plugging caused by metal deposition in trickle-bed reactors used in hydrodesulphurization reactors is considered in the following paragraphs. There is frequently a simultaneous deposition of coke along with the metal. Recent work on the hydrocracking oils derived from coal in trickle-bed reactors has shown the coking can be significant in these systems. The effect of pore size on catalyst performance has been studied by Ahmed and Crynes [24] and Prasher et al. [25] have observed that the effective diffusivities of oils in aged catalysts were severely reduced by coke deposition.

A recent study that has concentrated on coking in trickle-bed hydrodesulphurizer reactors is that of Chang et al. [26]. They demonstrated that the coke deposition was greatest at the reactor inlet, which suggested that deactivation was by a parallel mechanism. Most coke was found to be deposited during the first 40 hours on stream. Some results are shown in Figure 10 and demonstrate that although the maximum amount of coke deposited is probably about 14% by weight, about 30% of this is formed during the first hour of operation. It was found by measurement that coke severely blocked the pore mouths and this was confirmed by Auger spectroscopy. The coke content was also found to be a good indication of catalyst activity for both hydrogenation and hydrodenitrogenation. This is one clearly defined experimental study where pore blocking has been shown to be significant.

FOULING IN FIXED-BED REACTORS DUE TO METAL DEPOSITION

Metal deposition occurs in trickle-bed reactors used for hydrotreating oils and can lead to both catalyst pellet deactivation by pore plugging and to interparticle deposition causing bed plugging.

It is often difficult to determine unequivocally the effect of metal deposition alone as it is frequently accompanied by coke deposition. This is illustrated in Figure 11 [4] where the amount of deposit is plotted against relative time for a typical hydrodesulphurization experiment. It can be seen that coke deposition is initially very rapid but soon reaches a stationary state at about 13% by weight. In contrast, the deposition of the inorganic constituents of the cracked asphaltenes (mainly vanadium and nickel sulphides) continues and gradually blocks the pores in the outer zone of the catalyst particles. This has been confirmed in a number of studies using electron microprobe analysis, one example of which is given in Figure 12 taken from the work of Oxenreiter et al. [27].

It has been suggested by Beuther et al. [28] that the initial high rate of coke deposition is caused by metal components in the catalyst accelerating the dehydrogenation of the oil. This metal dehydrogenation activity is due to incomplete sulphiding of the catalyst.

Metal deposition in hydrotreating units may be divided into two types, namely pore plugging and bed plugging. The latter is manifested by an increase in pressure across the bed. Both effects will be considered in turn.

Pore Plugging in Hydrotreating Units

This concept was first proposed by Himenez [29] on the basis of permeability measurements using fresh and spent catalysts. Because coke deposition occurs simultaneously with metal sulphide deposition, it is important to determine the relative effect of each. Beuther and Schmid [28] compared the surface areas and pore size distribution of fresh and spent catalysts and showed that coke deposition only reduced the pore radius slightly and the pore size distribution maintained the same pattern in spite of a considerable reduction in surface area. The coke content of the catalyst was observed to increase sharply in the first

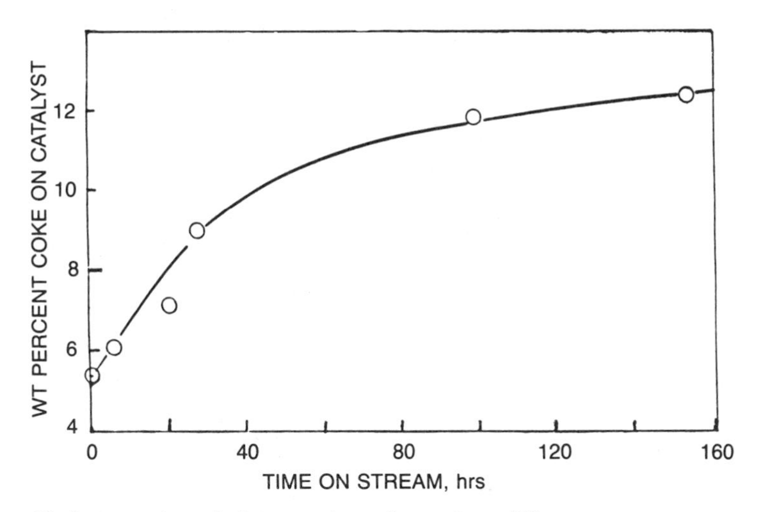

Figure 10. Average coke content over reactor vs. time onstream [26].

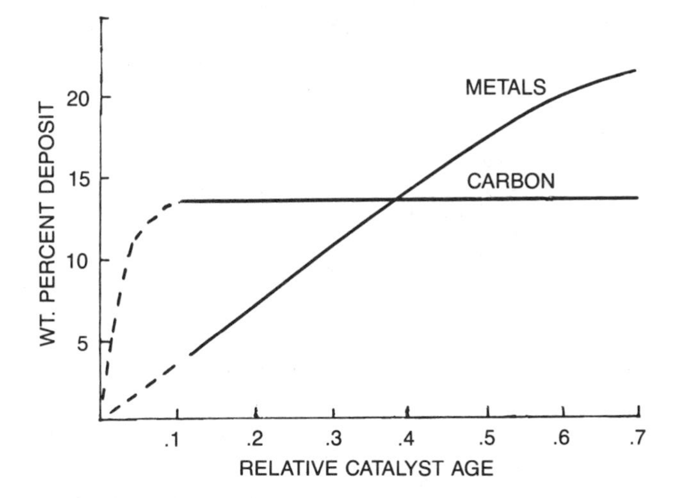

Figure 11. Hydrodesulpherization of a Caribbean long residue [4].

40 hours operation, whereas subsequently it remained almost constant with further process time. This effect has been noted previously. Therefore, the loss in catalyst porosity is made up of three parts:

$$\varepsilon_L = \varepsilon_{C1} + \varepsilon_{C2} + \varepsilon_{MS} \tag{31}$$

Fast coke deposition ε_{C1} makes up about one third of the total porosity, while ε_{C2} (slow coke deposition) and metal sulphide plugging (ε_{MS}) then decrease the remaining catalyst porosity. The following treatment is based on that of Newson [31].

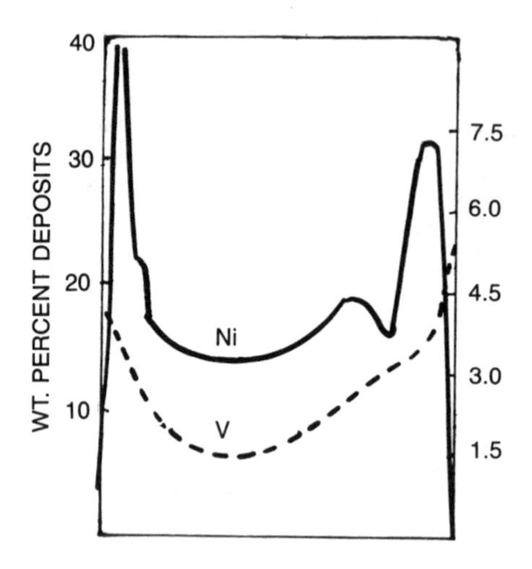

Figure 12. Electronmicroprobe scans of used catalyst pellets [27].

The pore model of Wheeler [30] is used, in which the pellet pore structure is approximated as a composite of N pores each of length L_p where:

$$L_p = \frac{V}{S}2 \tag{32}$$

in which V and S are pellet volume and surface area, respectively.

The nature of the pore plugging process suggests that the number of pores effective for demetallation, N_E, is some fraction of the total given by:

$$N_E = \eta N \tag{33}$$

where η is the time averaged effectiveness factor based on metal profile analysis in spent catalyst.

The pore size distribution may be assumed to Maxwellian:

$$L_M(r) = A_M \frac{r}{r_0} e^{-r/r_0} \tag{34}$$

where $L_M(r)$ is the total length of pores of radius r per gram of catalyst and A_M is the frequency of pore sizes, while r_0 is the most probable pore radius and is linked to the average pore radius, $\bar{r}$, by:

$$\bar{r} = 3r_0 = \frac{2V_g}{S_g} \tag{35}$$

The corresponding pore volume distribution is:

$$\varepsilon_m(r) = \int \pi r^2 L_M(r)\, dr \tag{36}$$

The deactivation rate in a single pore is then considered using the classical treatment of Wheeler [30] giving the differential equation:

$$\pi r^2 D_e \frac{d^2C}{dx^2} = 2\pi r k C(x) \tag{37}$$

When the concentration gradient tends to zero at some distance $x_p \ll L_p$, Equation 37 simplifies to:

$$\pi r^2 D_e \left(\left. \frac{dC}{dx} \right|_{x=0} = 2\pi r k \int_0^{x_p} C(x)\, dx \right) \tag{38}$$

where x_p is the reactant penetration into the pore. Now $\left(\left. \frac{dC}{dx} \right|_{x=0} \right)$ can be approximated by C_s/x and the reaction rate on the pore wall by $2\pi r k(C_s/2)$.

Equation 38 then becomes:

$$\pi r^2 D_e \frac{C_s}{x_p} = 2\pi r x_p k C_s/2 \tag{39}$$

Therefore

$$x_p = \sqrt{\frac{D_{er}}{k}} = K\sqrt{r} \tag{40}$$

Assuming a pseudo steady state (a slow deactivation rate compared with oil residence time), the decrease in pore radius due to metal deposition may be determined. Thus, if the plugging material is spread over all the pores in a pore-mouth type of plugging, the thickness of deposit, $y_{t,\Delta}$, for a finite increment of time Δ is given by:

$$y_{t,\Delta} = \frac{(R_p^*)_t \Delta}{2\pi \sum_m N_{E,m} r_{m,t} x_{m,t}} \tag{41}$$

where R_p^* is the pore plugging rate with:

$$r_{m,t} = r_{m,0} - \sum_0^t y_t \tag{42}$$

and

$$x_{m,t} = K\sqrt{r_{m,t}} \tag{43}$$

for the m-th pore at a time t.

Since the intraparticle metal deposition rate equals the flux of metal containing molecules in the pellet

$$R_D^* = F_t \alpha \sum_m \frac{N_{E,m}(r_{m,t})^2}{x_{m,t}} \tag{44}$$

Where R_D^* is the deposition rate. Also, F_t/F_0 represents the ratio of the flux of metal containing molecules into the pellet at time t compared with that at zero time and is a measure of the decrease in reaction rate due to pore plugging.

The required relation between deactivation and the pore plugging rate is obtained by assuming that deactivation is proportional to the rate of reaction and both are given by power-law expressions of order 2:

$$Y \propto (R_D^*)^2 \propto (R_{HDS}^*)^2 \tag{45}$$

where R_D^* = the deposition rate
R_{HDS}^* = the rate of hydrodesulphurization

Then:

$$\frac{Y_0}{Y} = \left(\frac{R_{D,0}^*}{R_{D,t}^*}\right)^2 = \left(\frac{F_0}{F_t}\right)^2 \tag{46}$$

represents the number of times the initial deactivation rate is increased. F_0 is known from fresh catalyst properties and process conditions; F_t can be calculated for any process time.

The procedure used was to calculate deactivation versus time using Equation 46. The total pore plugging rate is then put into Equation 41 to determine the reduction in pore radii over the whole pore size distribution using Equation 34. The decrease in radii reduces the flux of metal containing models into the pores (Equation 44). This flux is then compared with the initial value to calculate catalyst deactivation versus time.

Newson [31] made predictions of catalyst life under conditions where metal deposition from feedstocks could occur. An example is shown in Figure 13 where the catalyst deactivation rate ratio is plotted against time for various space velocities and differing amounts of sulphur corrosion. The predictions were compared with data from the operation of commercial plants; good order of magnitude agreement was obtained. This method should also be valuable in calculating restricted diffusion due to pore plugging in many other processes.

Bed Plugging in Hydrotreating Units

This effect has been known for some time and occurs because of deposition of metal sulphides (mainly iron, vanadium, and nickel) during residium hydrodesulphurization. The top few feet of the catalyst bed show an appreciable decrease in bed voidage due to interpellet deposition of these sulphides. This is accompanied by an increase in pressure drop across the reactor and when this reaches 50–100 psi the unit has to be shut down [32]. A simple model was developed by Newson to predict this phenomenon. In order to obtain results from the model it was necessary to know the relative rates of deposition of the respective metal sulphides. From a variety of sources it appeared that vanadium deposited faster than nickel by a factor of two while iron deposition was greatest being twice that of vanadium. Since absolute values for the pseudo-rate constants for metal deposition can be deduced from data on Kuwait residium, the rate ratios were taken as 4 : 2 : 1 for iron : vanadium : nickel deposition rates.

The interpellet deposition of sulphides is shown by increasing pressure drop, and the effect of varying the rates of metal deposition on pressure drop is illustrated in Figure 14, in which pressure drop is plotted against time. In commercial practice an increase of the pressure drop ratio from between 5 and 10 would mean the run should be terminated. The figure shows an increase in pressure drop with increased rates of demetallation as expected. Increasing the concentration of metals also decreases the life of the catalyst. The results obtained are comparable to those obtained from commercial units.

An interesting development suggested by Newson [32] is that "graded" beds should be employed to maximize catalyst life. Obviously, increasing the pellet size would reduce bed pressure drop as would increasing bed porosity, and these two together would be expected

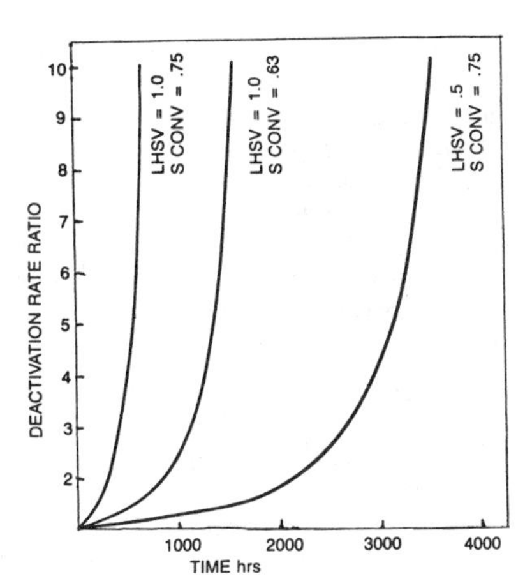

Figure 13. Effect of process conditions on catalyst life [31].

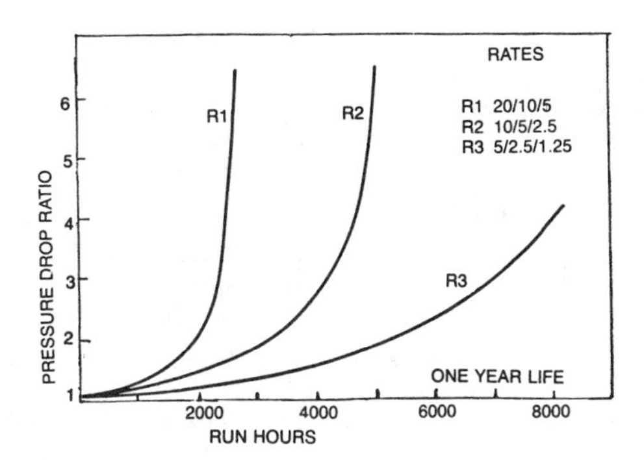

Figure 14. Bed plugging in hydrodesulphurization reactor. Constant concentration Fe/V/Ni = 5/15/5/ ppm [31].

to reduce bed plugging rates. These two effects can be supplemented by a graded hydrogenation effect of the catalyst, wherein the activity is least at the top (inlet) of the reactor and subsequently increases with distance down the bed.

A disadvantage of this approach is that conversion would be lost if the reactor was operated at constant conversion, since the catalyst per unit volume of reactor has been effectively reduced. However, since the conversions in residium hydrodesulphurization are about 75% the effect is not too important and is compensated by less-frequent reactor shut downs. Model predictions using graded beds have been made and have shown that bed plugging is reduced and distributed more evenly over the whole reactor using this approach.

CONCLUSIONS

Fouling in fixed-bed reactors has been considered for both coking and metal deposition types of fouling. For coking it has been shown that the distribution of coke does not always obey the usual predictions that coke laydown is greatest at the inlet of the reactor for parallel fouling and greatest at the outlet for series fouling, since the influence of diffusional resistances and strong Langmuir–Hinshelwood adsorption can reverse this behavior. Nonuniform coke laydown has important implications when catalysts are regenerated by oxidation of the deposited coke, since most predictive models for regeneration assume a uniform deposit of coke.

In many ways, the deposition of metal sulphide is a more serious problem, since the catalyst so deactivated cannot readily be regenerated. Deposition of metal sulphides in hydrotreating units can be considerable giving rise to both pore plugging and bed plugging. A means of avoiding the most severe aspects by grading of the bed is considered. Grading of the bed is achieved by varying the porosity and activity of the catalyst in the bed so as to minimize any pore or bed plugging which may occur.

NOTATION

a	dimensionless reactant concentration
a_v	surface area per unit volume
A	constant preexponential factor
A_m	frequency of pore sizes
b	dimensionless product concentration
C_{pg}	heat capacity of gas mixture
C_{ps}	heat capacity of solid phase
C	concentration
C_A	reactant concentration
C_{A0}	bulk-phase reactant concentration
C_{As}	surface reactant concentration
C_c	coke concentration
h_{KA}	heat of adsorption for species A
h_{KB}	heat of adsorption for species B
h	heat transfer coefficient
ΔH	reaction enthalpy
k	rate constant
k_c	mass transfer coefficient
k_{f_1}, k_{f_2}	rate constants for parallel and series coking, respectively
K	constant
K_A^*, K_B^*	dimensionless Langmuir–Hinshelwood adsorption constants
Lp	pore length
$Lm(r)$	total length of pores/gm of catalyst
n	reaction order
N	number of pores
N_E	fraction of pores effective for demetallation
Nu^*	modified Nusselt number
q	ratio of diffusivities (D_{eA}, D_{eB})
r	radius of catalyst pellet
r_A	rate of reaction
r_0	most probable pore radius
R_D^*	deposition rate of metal
R_P^*	pore plugging rate
S	point activity, surface area
$\bar{S}$	mean activity
Sg	specific surface
Sh^*	modified Sherwood number
t	time
T	temperature
T_0	bulk gas temperature
T_s	surface temperature
u	linear velocity
V	volume of pellet
V_g	specific volume
x	distance
x_p	reactant penetration in pore
y	thickness of metal deposit
Y	ratio between deactivation rate and pore plugging rate
z	axial distance in reactor

Greek Symbols

α	geometric factor ($\equiv$ 2, 1 or 0 for spheres, cylinders or slabs, respectively)
β	thermicity factor $((-\Delta H)D_eC_{A0}/K_eT_0)$
β'	defined by Equation 12e
γ	Arrhenius number
γ_f	Arrhenius number for the fouling reaction
δ	dimensionless pellet radius
ε	voidage of pellet
ε^1	voidage of bed
ε_L,	porosities defined by Equation 31
εc_1, εc_2, ε_{MS}	porosities defined by Equation 31
η	effectiveness factor
θ	dimensionless temperature

ξ dimensionless axial distance
ϱ_g gas density
ϱ_s solids density
τ dimensionless time
φ Thiele modulus
ψ deactivation function
Ω reaction modulus

REFERENCES

1. Haldeman, R. G., and Botty, M. C., *J. Phys. Chem., 63,* 489 (1969).
2. Baker, R. T. K., and Chludzinski, J. J., *J. Catal., 64,* 464 (1980).
3. Hughes, C. C., and Mann, R., ACS Symp. Ser., *65,* 201 (1978).
4. Dautzenberg, F. M., et al., ACS Symp. Ser., *65,* 254 (1978).
5. Butt, J. B., Delgado-Diaz, S., and Muno, W. E., *J. Catal., 37,* 158 (1975).
6. Blue, R. W., and Engle, C. J., *Ind. Eng. Chem., 43,* 494 (1951).
7. Eberly, P. E., Jr., et al., *Ind. Eng. Chem. Proc. Des. Devel, 5,* 193 (1966).
8. Kwan, T., and Sato, M., *Nippon Kagaku Kaishi, 91,* 1103 (1970).
9. Voorhies, A., Jr., *Ind. Eng. Chem., 37,* 318 (1945).
10. Ruderhausen, G. G., and Watson, C. C., *Chem. Eng. Sci., 3,* 110 (1954).
11. Pachovsky, R. D., Best, D. P., and Wojcieckowski, B. W., *Ind. Eng. Chem. Proc. Des. Devel., 12,* 254 (1973).
12. Murakami, Y., et al., *Ind. Eng. Chem. Fundam., 7,* 599 (1968).
13. Masamune, S., and Smith, J. M., *AIChE J., 12,* 384 (1966).
14. Froment, G. F., and Bischoff, K. B., *Chem. Eng. Sci., 16,* 189 (1961).
15. Van Zoonen, D., Proc. 3rd Int. Congr. Catal., Amsterdam, N. Holland Publishing Co., Vol. II, 1319 (1965).
16. Dumez, F. J., and Froment, G. F., *Ind. Eng. Chem. Proc. Des. Devel., 15,* 291 (1965).
17. De Pauw, R. P., and Froment, G. F., *Chem. Eng. Sci., 30,* 785 (1975).
18. Ervin, M. A., and Luss, D., *AIChE J., 16,* 979 (1970).
19. Kam, E. K. T., and Hughes, R., *Chem. Eng. J., 18,* 93 (1979a).
20. Kam, E. K. T., and Hughes, R., *AIChE J., 25,* 359 (1979b).
21. Kam, E. K. T., Ramachandran, P. A., and Hughes, R., *J. Catal., 38,* 283 (1975).
22. Kam, E. K. T., Ramachandran, P. A., and Hughes, R., *Chem. Eng. Sci., 32,* 1317 (1977).
23. Brito-Alayon, A., Hughes, R., and Kam, E. K. T., *Chem. Eng. Sci., 24,* 123 (1981).
24. Ahmed, M. M., and Crynes, B. L., *Prepr. Div. Petr. Chem. Am. Chem. Soc., 23,* 1376 (1978).
25. Prasher, B. D., Gabriel, G. A., and Ma, Y. H., *Ind. Eng. Chem. Des. Devel., 17,* 266 (1978).
26. Chang, H. J., Seapan, M., and Crynes, B. L., ACS Symp. Ser., *196,* 309 (1982).
27. Oxenreiter, M. F., et al., Fuel Oil Desulpherization Symp., Japan Petroleum Institute, Tokyo, November 29th (1972).
28. Beuther, H., and Schmid, B., 6th World Petr. Congr. Frankfurt, Section 3, Paper 20, pp. 297–307 (1963).
29. Himenez, W., Discussion Section 3, Paper 20, 6th World Petr. Congr., Frankfurt, June 21st (1963).
30. Wheeler, A., *Adv. Catal., 3,* 250 (1951).
31. Newson, E. J., *Ind. Eng. Chem. Proc. Des. Devel., 14,* 27 (1975).
32. Newson, E. J., Preprints Vol. 17, No. 2, pp. 49–63, Divn. Fuel Chem., 164th Nat. Mtg. Am. Chem. Soc., New York, Aug. 27–Sept. 1 (1972).

CHAPTER 26

MODELING OF NONCATALYTIC GAS-SOLID REACTIONS

B. D. Kulkarni and L. K. Doraiswamy

National Chemical Laboratory
Poona, India

CONTENTS

INTRODUCTION

The importance of gas-solid noncatalytic reactions becomes apparent when one considers the breadth and diversity of this class of reactions, several examples of which have been presented by Doraiswamy and Sharma [1] and Ramachandran and Doraiswamy [2]. In its general form a gas-solid noncatalytic reaction can be represented by

$$\nu_A A(g) + \nu_B B(s) \rightleftharpoons \nu_c C(g) + \nu_d D(s)$$

where both the reactants and the products contain gas and solid phases and the ν's represent the stoichiometric factors. Typical examples of this general stoichiometric relation come from the class of reactions represented by reduction and roasting of ores. It is possible that for certain schemes the participation or generation of a gaseous or a solid species is not involved. This gives rise to typical situations such as

solid reactants → (fluid + solid) products (decomposition)

solid reactants → fluid products (gasification)

(fluid + solid) reactants → fluid products (oxidation, chlorination of ores)

fluid reactants → (fluid + solid) products

and examples of each of these classes of reactions are encountered in practice. The phenomenological modeling of gas-solid reactions has to recognize this diversity, and appropriate modifications to the general model are therefore necessary.

An important consideration in the modeling of gas-solid noncatalytic reactions, besides the formal distinction between the various types of these reactions, concerns the issues arising out of the participation of the solid species in the reaction. The occurrence of reaction entails changes in the solid phase including changes in physical properties like pore volume, pore dimensions, etc., which in turn influence the effective transport properties such as diffusivities and brings about their dependence on the extent of reaction. Additionally, the system is continuously in a transient state, a situation that is in contrast to that encountered in catalytic systems. The modeling of gas-solid noncatalytic systems therefore involves additional complexities, and simple hypotheses such as quasi-stationarity are often inadequate—and certainly not valid over the entire time course of reaction.

The analogy with gas-solid catalytic systems is, however, helpful in formulating a general methodology for modeling these systems. We can thus conceive of similar basic steps involved, and for a general scheme such as the identification of the following broad steps:

1. Diffusion across the gas film.
2. Adsorption of gaseous reactant on the surface of the solid product.
3. Diffusion through the solid product to reach the interface between the reactants and products.
4. Adsorption at the interface.
5. Diffusion and chemical reaction.

Usually, in a gas-solid reaction one of these steps would be rate limiting. However, due to the transient nature of the operation in a noncatalytic reaction, the controlling regime would be continuously changing from one to another and a specified step (or steps) would control the conversion-time course of the reaction only during a certain time interval. The presence of gaseous product in the reaction system would require consideration of similar additional steps in the reverse direction for the product species. Thermal effects would bring about further complexities due to their effect on the rates of the individual steps.

A rigorous mathematical model for a gas-solid noncatalytic system should take appropriate account of these steps and the simultaneous variations in system properties brought about as a result of reaction. Clearly, a model that considers all these features will be beset with formidable mathematical difficulties and would not be practical. Instead, simpler phenomenological models that account for these complexities in varying degrees are more practical and are available in the literature. In the next section we shall discuss these models in their skeletal form and then consider incorporating additional features in subsequent sections. The present article supplements the earlier reviews on this subject by Ramachandran and Doraiswamy [2] and Doraiswamy and Kulkarni [3].

BASIC MODELS

Sharp Interface Model

The sharp interface model (SIM) is one of the earliest models developed and is well described in the standard text books on chemical reaction engineering. The model is mainly applicable to nonporous solids and assumes that the reaction occurs at a sharp interface that separates the reacted outer shell (or ash layer) and the unreacted inner core of the solid.

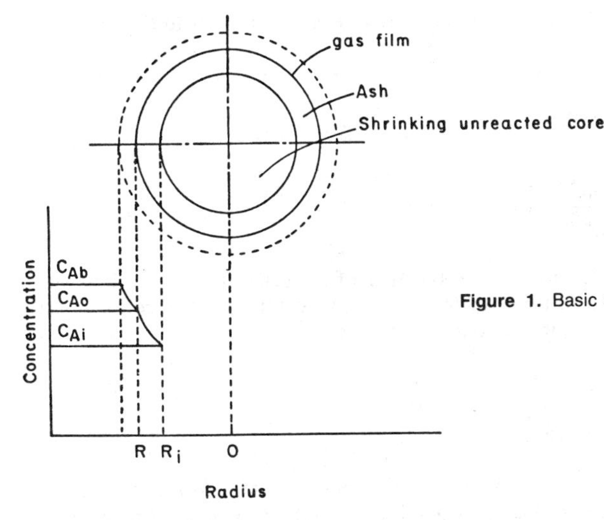

Figure 1. Basic steps in the shrinking-core model.

The steps involved are (Figure 1):

1. Diffusion through the gas film.
2. Diffusion through the ash layer.
3. Reaction at the interface.

The rate of reaction per pellet at any time t accounting for all the three resistances is obtained simply by adding the resistances in series:

$$\frac{1}{R_A} = \frac{1}{R_{Ag}} + \frac{1}{R_{Ad}} + \frac{1}{R_{Ac}} \tag{1}$$

or

$$R_A = \left(\frac{1}{4\Pi R^2 k_g} + \frac{R - R_i}{4\Pi R R_i D_e} + \frac{1}{4\Pi R_i^2 k_s}\right)^{-1} C_{Ab} \tag{2}$$

where k_g = transport resistances in the film
D_e = transport resistances through the solid product
k_s = the resistance at the interface
R_i = the position of the interface in a pellet of radius R

The rate of consumption of A given by Equation 1 can be related to the conversion of B through a stoichiometric balance on solid B as

$$-\frac{d}{dt}\left(\frac{4}{3}\Pi R_i^3 \frac{\varrho_B}{M_B}\right) = \frac{\nu_B}{\nu_A} R_A \tag{3}$$

which on integration subject to the initial condition R = R at t = 0 leads to the following conversion-time relationship:

$$x + \frac{Rk_g}{2D_e}[1 - 3(1-x)^{2/3} + 2(1-x)]$$

$$+ \frac{3k_g}{k_s}[1-(1-x)^{1/3}] = \frac{3k_g C_{Ab} M_B \nu_B}{\nu_A R \varrho_B} t \tag{4}$$

Similar equations can be obtained for other geometries of the pellet. In fact the following general equation can be written in terms of the time required for complete conversion if any one of the three basic steps were to control the rate:

$$t = t_{film} + t_{ash} + t_{reaction}$$

$$= f_0(x)\tau_f + f_1(x)\tau_a + f_2(x)\tau_r \tag{5}$$

where the functions f_0, f_1, and f_2 and the various τ_s' assume different forms for different geometries of the particle and are defined in Table 1 along with other relevant parameters.

The equation in this form helps to identify the relation of conversion to normalized time (t/τ) where τ represents the time required for total conversion. The dependence of conversion on

- The radius of the particle R.
- The concentration in the gas phase C_{Ag}.
- The molal density of solid ϱ_B.
- The stoichiometric coefficient ν.

in addition to k_g or D_e or k_s (depending upon whether film or ash diffusion or reaction controls) is now embodied in a single parameter τ.

It may be noted that the dependence of τ on pellet size is different for different controlling regimes, being first order in R for reaction control, second order for ash diffusion control, and 1.5–2 order for film diffusion control. By varying the size of the particle it is therefore easily possible to delineate the controlling regime.

In view of the differences in the dependence of τ on the size of the pellet, different controlling mechanisms can prevail for different pellet sizes. Normally at the start of the reaction there is no ash layer present and also, unless the reaction is extremely rapid, film diffusion is not rate limiting. The systems therefore begin with the rate controlled by the surface reaction. With progress in time sufficient ash layer builds up and the system passes to ash diffusion control.

The dependence of τ on the gas-phase concentration, while not very important for a single particle, assumes great significance for particles in a reactor.

Gas-solid reactions can also be analyzed, in analogy with catalytic systems, in terms of an effectiveness factor. Such an approach has been used by Ishida and Wen [4], who derived the following expression for the effectiveness factor for a general n-th-order reaction:

$$\eta = \frac{R_A}{4\Pi R^2 k_s C_{Ab}^n} \tag{6}$$

The effectiveness factor will change with time in view of the changing rate R_A, and the following general implicit equation can be derived:

$$\eta = 1 - R_i^2 Da \frac{1}{Sh} + \frac{1-R_i}{R_i} \tag{7}$$

where $Sh = k_g R/D_e$ and $Da = k_s R C_{Ab}^{n-1}/D_e$. The movement of the interface R_i with time appearing in Equation 7 can be obtained by integrating Equation 3 where the R_A term on the rhs can be replaced by R_A given by Equation 6.

The case considered thus far assumes linear kinetics and essentially demonstrates the methodology to obtain the time-conversion relationship. The method could be applied to more complex situations such as involving nonlinear kinetics, represented by power law or Hougen–Watson (H–W) forms. In general, in the presence of such nonlinear kinetics, no analytical expression for the conversion-time relation can be obtained and recourse to numerical integration would be necessary. The general procedures for power law kinetics proposed by Sohn and Szekely [5] and for H–W kinetics proposed by Ramachandran [6] can be followed. Four different H–W forms have been analyzed by Chida and Tadaki [7]. The analysis clearly reveals that for complex H–W forms (such as bimolecular kinetics) the system may possess three quasi-steady states under certain conditions. Also the gas concentration at the reaction interface can change drastically under certain conditions. An important feature of this analysis is that it predicts effectiveness factor values exceeding unity even under isothermal conditions.

The case of the zero-order reaction has been reported by Simonsson [8], while Cannon and Denbigh [9] and Chu and Rahmet [10] report on fractional-order kinetics.

SIM is a phenomenonological model and in the extreme case of reaction control or ash diffusion control requires only one parameter—the relevant time factor—to describe the system. The model in its basic form, while adequate for many practical systems, cannot however account for such features as the leveling off of conversion or the generally observed S-shaped nature of the conversion-time curves. We shall return to this model later with a view to modifying it to take account of some practical considerations.

Volume Reaction Model

The simple SIM discussed in the earlier section is applicable to nonporous or relatively nonporous solids. Many gas-solid reactions in practice, however, involve porous solids where the gas can penetrate within the solid and can react all over within the volume of the solid rather than at the interface. In general, the rate of reaction at the interior points would be lower than at the gross external surface due to diffusional resistance. In a special situation when no diffusion resistance exists, the reaction occurs uniformly all through the

Table 1
Time-Conversion Relationships for SCM for Different Particle Geometries*

Controlling Regime	Flat Plate	Cylinder	Sphere	τ
Film diffusion $f_0(x) =$	x	x	x	$\frac{\varrho_B R}{\nu_B(\beta+1)k_g C_{Ag}}$
Ash diffusion $f_1(x) =$	x^2	$x+(1-x)\ln(1-x)$	$1-(1-x)^{2/3}+2(1-x)$	$\frac{\varrho_B R^2}{2\nu_B(\beta+1)D_e C_{Ag}}$
Reaction $f_2(x) =$	x	$1-(1-x)^{1/2}$	$1-(1-x)^{1/3}$	$\frac{\varrho_B R}{\nu_B k_S C_{Ag}^n}$

* $A(g) + B(s) \rightarrow R(g) + S(s)$

The conversion $x = 1-(r/R)^{\beta+1}$ *where* $\beta = 0, 1,$ *and* 2 *for flat plate, cylinder, and sphere.*

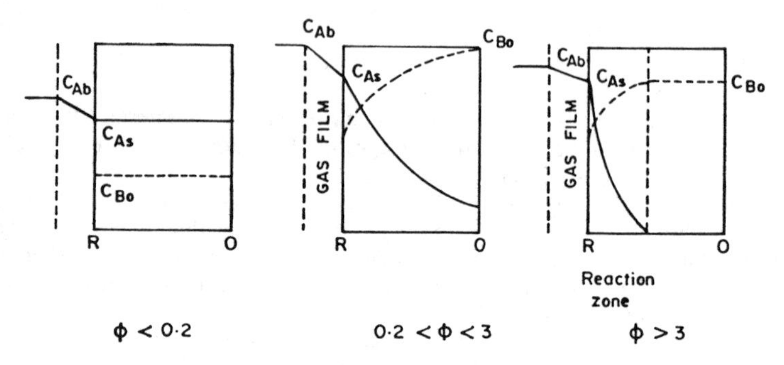

Figure 2. Concentration profiles for the homogeneous model for various values of the Thiele modulus.

pellet leading to the so-called homogeneous model. The rate of reaction per unit volume of the pellet for a general (m, n) order reaction may be represented by

$$R_A = kC_A^m C_B^n \tag{8}$$

where m and n refer to the orders with respect to the gas and solid, respectively. Note that in the volume reaction model the rate of chemical reaction depends not only on the concentration of the gas but also on the activity and concentration of the solid reactant. This means that only for certain values of n ($n < 1$) can the solid reactant be completely reacted within a finite time at any point in a particle and its concentration reduced to zero.

The general conservation equation of heat and mass for the gaseous and solid reactant species for the volume reaction model, illustrated in Figure 2, are given in Table 2. No analytical solution to this set of equations for arbitrary values of m and n seems possible even under isothermal conditions, and recourse to numerical methods in necessary. However, for certain simplified cases, analytical solution can be obtained.

Thus, for the case of low Thiele modulus φ, the concentration profile within the pellet will be nearly uniform; this implies that the conservation equations for the gas and the solid species can be decoupled. Simple integration of the conservation equations for $m = 0$ and 1 (zero and first order, respectively, in gas) would then lead to the following simple conversion-time relations:

$$\begin{aligned} x &= 1 - \exp(-\hat{t}), \quad &\text{for } m = 1 \\ x &= \hat{t} &\text{for } m = 0 \end{aligned} \tag{8}$$

At higher values of $\varphi(\varphi > 0.2)$ the concentration profile within the pellet cannot any more be assumed uniform, and the gas and solid species equations will have to be solved simultaneously. An especially important case from the practical standpoint is when the order with respect to the gaseous component is unity ($m = 1$), while the order n with respect to the solid reactant is arbitrary. In this situation, it is possible to obtain a single equation by defining the Legendre transformation (Del Borghi et al., [11]; Dudukovic and Lamba, [12, 13]:

$$\psi = \int_0^t C_A \, dt \tag{10}$$

where ψ represents a cumulative gas-phase concentration.

The transformed equation for the special case of m = n = 1 can be written in terms of ψ as

$$\nabla^2\psi = \varphi^2[1-\exp(-\psi)] \tag{11}$$

with boundary conditions: $\psi(1) = t$ and $(d\psi/dR)_{R=0} = 0$. This case has been analyzed by Dudukovic and Lamba [12] and approximate analytical solutions to the problem have been proposed by Ramachandran and Kulkarni [14] and by Wen and Wu [15]. The equation of Kulkarni and Ramachandran derived from single-point collocation is given along with other relations in Table 3.

Other analyses, for different values of m and n, have also been presented in the literature. The case of m = 0, n = 1 has been analyzed by Dudukovic and Lamba [12, 13]. The case of m = 1, n = 0 has been analyzed by Ausman and Watson [16] and by Ishida and Wen [14]. In each of these cases, due to the zero-order dependency on either the gas or the solid concentration, the local conversion with respect to gas or solid, in the pellet reaches completion. This leads to the formation of distinct zones in the pellet. Thus, when m = 0 we have a reaction zone between $R_i < R < 1$, while for n = 0 we have a product zone between $R_i < R < 1$ and reaction zone within. The typical situations are sketched in Figure 3.

Ausman and Watson [16] and Ishida and Wen [4] analyzed the case of n = 0 in detail. They divided the total reaction time into two periods, viz. the constant-rate period and falling-rate period, in analogy with the drying process. The time-conversion relations for the two periods are included in Table 3.

The postulation of the zones as noted previously has led to the development of more general zone models [17, 18] in which a reaction zone varying in thickness from zero (corresponding to SIM) to the pellet dimension (corresponding to the homogeneous model) has been considered. The reaction-zone thickness in these models can be related to the intrinsic kinetics and diffusion characteristics of the system and such relations are useful in parameter estimations of the model. The model of Mantri et al. [18] predicts that the

Table 2
Conversion Equations for Heat and Mass for Volume Reaction Model

Model	Governing Equations
Volume reaction model	$N_1 \frac{\partial \hat{C}_A}{\partial \hat{t}} = \frac{1}{\hat{R}^2}\frac{\partial}{\partial \hat{R}}\left(\hat{R}^2\frac{\partial \hat{C}_A}{\partial \hat{R}}\right) - \varphi^2 \hat{C}_A^m \hat{C}_B^n \exp\left[\gamma\left(1-\frac{1}{\hat{T}}\right)\right]$
	$N_2 \frac{\partial \hat{T}}{\partial \hat{t}} = \frac{\partial \hat{T}}{\partial \hat{R}^2} + \frac{2}{\hat{R}}\frac{\partial \hat{T}}{\partial \hat{R}} + \beta m \varphi^2 \hat{C}_A^m \hat{C}_B^n \exp\left[\gamma\left(1-\frac{1}{\hat{T}}\right)\right]$
	$\frac{\partial \hat{C}_B}{\partial \hat{t}} = -\hat{C}_A^m \hat{C}_B^n \exp\left[\gamma\left(1-\frac{1}{\hat{T}}\right)\right]$
	$\hat{R} = 1, \quad \frac{\partial \hat{C}_A}{\partial \hat{R}} = Sh(1-\hat{C}_A), \quad \frac{\partial \hat{T}}{\partial \hat{R}} = Nu(1-\hat{T}), \quad \hat{R} = 0,$
	$\frac{\partial \hat{C}_A}{\partial \hat{R}} = \frac{\partial \hat{T}}{\partial \hat{R}} = 0, \quad \hat{t} = 0, \quad \hat{C}_A = 0, \quad \hat{T} = 1, \quad \hat{C}_B = 1$
	where
	$N_1 = \frac{k_v C_{Ab}^m C_{Bs}^{n-1} f_c R^2 t}{D_{eB}}$
	$N_2 = \frac{\varrho_s C_{ps} k_v C_{Ab}^m C_{Bs}^{n-1} f_0 R^2 t}{k_e'}$
	$\beta_m = \frac{(-\Delta H) D_{eB} C_{Ab}}{k_e' T_b}$

zone thickness is of the order of $(1/\varphi)$ for first-order reactions. The more recent analysis of Do [19] generalizes this result and suggests the zone thickness to be of the order of $(\varphi)^{-2/n+1}$ where n is the order with respect to the gaseous component. The experimental results of Prasannan and Doraiswamy, obtained by the electron probe microanalytical

Table 3
Some Useful Time-Conversion Relationships

Model	System Description	Time-Conversion	Authors
Volume reaction model	Reaction first-order in gas and solid; spherical pellet	$\ln\left[\frac{(1-x)}{0.699} - \frac{\exp(-\hat{t})}{2.33}\right] + \hat{t} = \frac{\varphi^2}{10.5}\left[1 - \frac{1-x}{0.699} + \frac{\exp(-t)}{2.33}\right]$	Kulkarni and Ramachandran [95]
Particle pellet model	First-order reaction	$\hat{t} = 1-(1-x)^{1/3} - \frac{\varphi_1^2}{18}[1-3(1-x)^{2/3}+2(1-x)]$ $+(0.21x-0.31x^2)(1+\varphi_1^2/18)\exp\left[-0.9\left(\ln\frac{\varphi_1^2}{19.44}\right)^2\right]$ where $\hat{t} = \frac{\nu_B M_B A_g k_s(T_0)}{\varrho_B r_{G0}}$ and $\varphi_1 = \left[\frac{3(1-f_{c0})k_s(T_0)R^2}{r_{G0}D_e}\right]$	Evans and Ranade [25]
Particle pellet model	Uniform grains of unchanging size; both pellets and grains can be of arbitrary geometry	$\hat{t} = g_{F_g}(X) + \frac{Da_g}{2F_g} f_{F_g}(X) + \frac{\varphi_p^2}{2F_pF_g} f_{F_p}(X)$ where subscripts g and p refer to grain and pellet, respectively, and F to the shape factor and takes values of 1, 2, or 3 for infinite slab, cylinder, and sphere $g_{F_g}(X) = 1-(1-x)^{1/F_g}$ and $f_{F_i}(x) = \frac{F_i}{F_i-2}\left[1-(1-x)^{2/F_i} - \frac{2}{F_i}x\right]$ where i = g or p	Szekely et al. [26a]
Particle pellet model	Uniform grains of varying sizes; both pellet and grain can have arbitrary geometry	$\hat{t} = gF_g(x) + \frac{Da_g}{2F_g} P_{F_g}(x) + \frac{\varphi_p^2}{2F_pF_g} f_{F_p}(x)$ where g $F_g(x)$ as above and $P_{F_g}(x) = F_g\left\{\frac{1}{F_g-2}[1-(1-x)^{2/F_g}] + \frac{1-[\gamma+(1-\gamma)(1-x)]^{2/F_g}}{(F_g-2)(\gamma-1)}\right\}$	Garza and Dudukovic [62, 63]
Random pore model	Takes account of structural variations through population balance approach; kinetic regime	$x = 1-\left(1-\frac{k_sA_g^m t}{R}\right)^3 \exp\left[-\frac{k_sA_g^m S_0 t}{1-f_{c0}}\left(1+\frac{\psi}{4}\frac{k_sA_g^m S_0 t}{1-f_{c0}}\right)\right]$ where $\psi = 4L_0(1-f_{c0})/S_0^2$; $\hat{t} = \frac{k_sA_g^m S_0 t}{1-f_{c0}}$ This expression can also be expressed as $\frac{dx}{d\hat{t}} = (1-x)[1-\psi\ln(1-x)]^{1/2}$	Bhatia and Perlmutter [33]
	Diffusion control regime	$\frac{dx}{d\hat{t}} = \frac{a(1-x)[1-\psi\ln(1-x)]^{1/2}}{1+\frac{2k_s\varrho_B(1-f_{c0})}{\nu_B M_B D_{eG} S_0}\frac{Z_v}{\psi}[\sqrt{1-\ln(1-\psi)}-1]}$	Bhatia and Perlmutter [13, 14]
SCM	Gasification reaction	$x = \frac{f_{c0}}{1-f_{c0}}\left[\left(1+\frac{k_sA_g^m t}{r_{p0}}\right)\frac{[G'-1-(k_sA_g^m t)/r_{p0}}{G'-1} - 1\right]$ where G' is a solution of $\frac{4}{27}f_{c0}G'^3 - G' + 1 = 0$	Szekely et al. [26a]
Random capillary model	Gasification reaction	$x = 1-\exp[-2\pi(N_0v^2t^2+2N_1vt)]$ where v is the reaction velocity defined as the rate of change of pore radius with time, and N_0 and N_1 are structural parameters $N_0 = \int_{S_{min}}^{S_{max}} p(S_0)\,dS_0$ $N_1 = \int_{S_{min}}^{S_{max}} S_0 p(S_0)\,dS_0$ where $p(S_0)$ is the density distribution based on the initial pore size distribution and S the pore size	Gavalas [84]

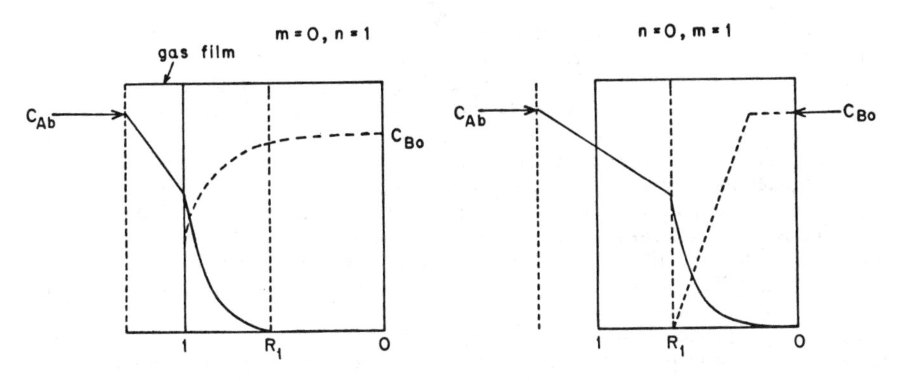

Figure 3. Concentration profiles in the two-zone model.

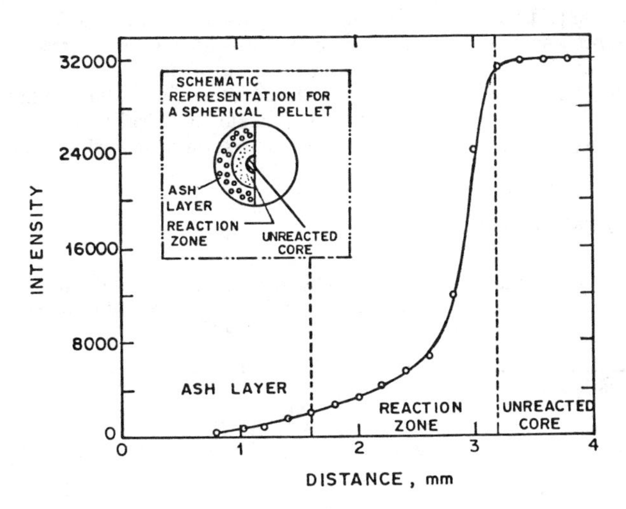

Figure 4. Sulfur profile from electonprobe microanalysis measurements: sintering temperature = 1200°C, X_B = 0.72.

technique (see Figure 4), clearly bring out the validity of the zone model for the oxidation of zinc sulfide.

An interesting case where the reaction rate is independent of the concentration of both the gas and the solid species (m = n = 0) has been analyzed by Ramachandran and Doraiswamy [2]. For no severe diffusional limitation in the pellet, this case leads to the following simple solution:

$$x = \hat{t} \tag{12}$$

When significant resistance to the diffusion of gas A exists, the concentration of A can actually drop to zero within the pellet provided the following criterion is satisfied:

$$\varphi_{cr} < \frac{6}{[1+2/Sh]^{1/2}} \tag{13}$$

For situations of this type a zone of certain thickness forms within the pellet and corresponds to the depth of penetration of the gaseous reactant. The depth of penetration clearly depends on the value of the Thiele modulus φ. For higher values of φ the zone thickness becomes smaller and eventually reduces to a sharp interface as $\varphi \rightarrow \infty$. For finite values of φ where a zone of finite thickness exists, it is necessary to solve the governing diffusion equation within the reaction zone. This reaction zone is characterized by a constant rate all through the zone in view of the zero-order dependency on the reactants. The reaction therefore continues until the whole of the solid reactant in this zone gets exhausted. After this zone is fully exhausted (which happens because both m and n are zero), the gas penetrates further into the pellet and forms another zone where a similar process occurs. Effectively therefore one observes a reaction zone that progressively jumps from one position to the next within the pellet.

The number of zones within the pellet depends on the value of φ and represents a staircase function as shown in Figure 5. The typical poisson-like character of the process is clearly evident from this figure.

A general analysis of the volume-reaction model for the case of arbitrary kinetics has been presented by Do [20]. For large values of φ, using the method of matched asymptotic expansion, it has been shown that the behavior of gas-solid systems corresponds to the shrinking-core model. The method also allows estimation of the time-dependent velocity of the moving interface, which is found to be proportional to the gas flux at that interface for all geometries of solids. A method for an approximate analytical solution to gas-solid systems with arbitrary dependence of rate on the concentration of solids has been proposed by Ramachandran [21].

Particle-Pellet (Grain) Model

In analyzing gas-solid reactions it is possible to conceive of solids as consisting of spherical nonporous equal-sized grains which react according to the sharp interface model.

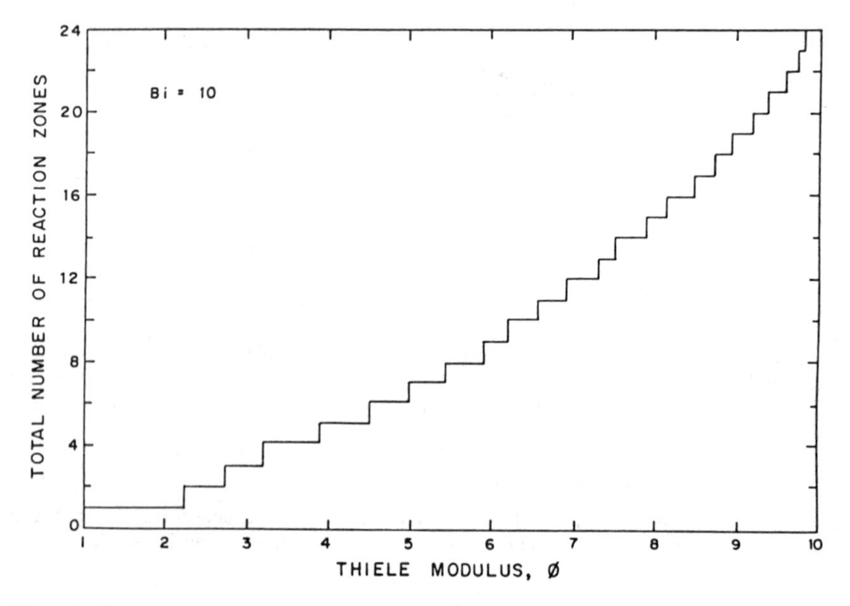

Figure 5. Number of reaction zones that develop in the pellet for various values of ϕ.

The diffusion of gas within the solid occurs in the void space generated between the grains. The basic model is generally referred to as the grain model or particle-pellet model and is shown schematically in Figure 6.

The mathematical formulation of the model for such a physical situation requires consideration of the rate processes within an individual grain, and the overall mass balance for the gaseous reactant in the pellet and its stoichiometric relationship with the extent of solid consumed. The mathematical formulation showing the heat and mass balance equations is presented in Table 4.

A mathematical analysis of the particle-pellet model has been presented by Calvelo and Smith [22] and Szekely and Evans [23, 24] for a simple isothermal first-order reaction. An approximate analytical solution has also been proposed by Evans and Ranade [25]. The constant-size grain model has been generalized to include grain-size distribution by Szekely and Propster [26]. The case of nonlinear kinetics has been examined by Sohn and Szekely [27, 28] for reactions following the H–W rate law.

Let us now analyze the limiting situations permissible within the framework of this model. The basic steps involved concern those occurring within the grain (i.e., reaction at the interface and diffusion through the ash around the grain) and the diffusion of gaseous species through the pellet. In the event the processes within the grain are controlling, the pellet as a whole follows the simple homogeneous model, while the individual grain may be following the reaction control or ash diffusion control process, depending on the concentration surrounding the grain. In view of the controlling processes within the grain, the time-conversion relation will be independent of the pellet dimensions. Increasing the pellet dimension however would result in an increase in the diffusional resistance through the pellet, which would eventually become controlling. At some stage, the processes within the grain would become inconsequential, the pellet as a whole would follow SIM with ash diffusion controlling, and the rate would show the appropriate dependence on the radius of the pellet R.

In the intermediate region where the resistances in the grains as well as the pellet are of the same order of magnitude (note that the resistance within the grain would depend on the concentration surrounding the grain, which would vary with the position of grain

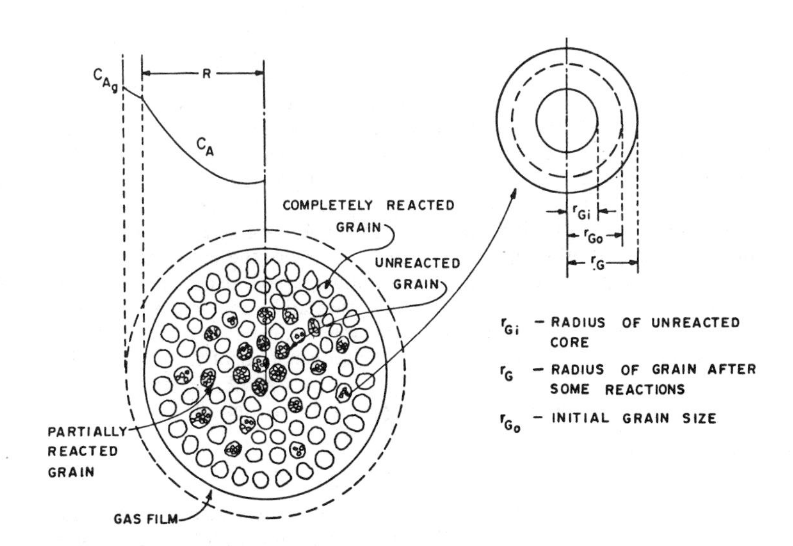

Figure 6. Schematic representation of the grain model.

Table 4
Conversion Equations for Heat and Mass for Particle-Pellet Model

Model	Governing Equations
Particle pellet model	$N_1^{\prime}\frac{\partial \hat{C}_A}{\partial \hat{t}} = \frac{\partial^2 C_A}{R^2} + \frac{2}{R}\frac{\partial \hat{C}_A}{\partial \hat{R}} - \varphi^2\hat{r}_{Gi}^2\hat{C}_A k$
	$N_2^{\prime}\frac{\partial \hat{T}}{\partial \hat{t}} = \frac{\partial^2 \hat{T}}{\partial \hat{R}^2} + \frac{2}{\hat{R}}\frac{\partial \hat{T}}{\partial \hat{R}} + \beta_m\varphi^2\hat{r}_{Gi}^2\hat{C}_A k$
	$\frac{\partial \hat{r}_{Gi}}{\partial \hat{t}} = -\hat{C}_A k$
	where
	$k = \frac{\exp\{\gamma[1-(1/\hat{T})]\}}{1 + Da_G\hat{r}_{Gi}(1-\hat{r}_{Gi})\exp\{\gamma[1-(1/\hat{T})]\}}$
	$N_1^{\prime} = \frac{\nu f_c R^2 M_B C_{Ab} k_s}{D_e \varrho_B r_{G0}}$
	$N_2^{\prime} = \frac{\nu R^2 \varrho_s C_{ps} M_B C_{Ab} k_s}{k_e^{\prime} r_{G0}}$

within the pellet), one could expect the pellet behavior to lie between the limiting cases of the shrinking core model with reaction and ash diffusion controlling.

This simple model is useful in cases where the pellets are formed by compaction of particles of very fine sizes. In most naturally occurring minerals, however, this is not the case and a fictitious grain size will have to be invoked to apply the model. Also the model, in its simple form, does not explain the experimentally observed S-shaped behavior and leveling off of conversion.

Nucleation Models

The models discussed thus far, while adequate to explain the data for many gas-solid reacting systems, do not explain the leveling off of conversion so commonly observed in many gas-solid systems. To remedy this situation one may start on the premise that gas-solid reactions can be represented as topochemical reactions. Further, it can be tacitly assumed that the reaction rate at the start, i.e., at the boundary surface, is far lower than at the interface at some subquent time t. The conjecture is that the initial rate corresponds to the induction period during which nuclei of the solid product phase are formed. This creates additional boundaries between the newly formed solid phase and the solid reactant, which continue to grow until the nuclei touch or overlap each other. This period corresponds to the increasing rate period that is proportional to the solid-solid phase boundary. Once the nuclei start touching and overlapping each other the area of the solid-phase boundary starts falling, and this corresponds to the diminishing rate observed experimentally. These simple postulations explain the S-shape curve for conversion and a maximum in the reaction rate at some intermediate time t in gas-solid reacting systems (as observed in many experimental studies). A typical rate curve observed in the hydrogenation of iron carbide is shown in Figure 7.

The reaction rate for a situation of this type can be written as

$$(R) = (R)_{sp}S \tag{13}$$

where R_{sp} refers to the specific rate per unit solid-solid interface and S to the area of the solid-solid interface that is changing with time. The parameter S can be related to time if the law of nuclei formation is known; this is generally assumed to follow a power law or exponential form of equation. In a nonisotropic solid the specific rate $(R)_{sp}$ in Equation 13 needs to be replaced by an integrated average quantity $(\bar{R})_{sp}$.

An empirical model which relates the conversion to time has been proposed by Avarami [29]:

$$x = 1 - \exp(-Ct^N) \tag{14}$$

C and N in this equation are constants which can be obtained by fitting experimental data. The following three-parameter equation in differential form has been proposed by Erofeev [30]:

$$\frac{dx}{dt} = kx^a(1-x)^b \tag{15}$$

and seems to provide adequate fit to some experimental data [31].

In a more rational approach to nucleation, Ruckenstein and Vavanellos [32] considered the existence of germ nuclei in the form of impurities present and as embryos of the new solid phase. In the process the germ nucleus is transformed into a growth nucleus leading finally to product formation. The growth process is treated as an activated process, and employing the modified Avarami model the following conversion-time relation may be obtained:

$$x = \beta k_{growth}^3 N_{n,0} k_{nf} \int_0^1 \exp(-k_{nf} t_{nf})(t - t_{nf})^3 [1 - x(t_{nf})]\, dt_{nf} \tag{16}$$

where $N_{n,0}$ = the number of germ nuclei per unit solid volume at the beginning
k_{nf}, t_{nf} = the rate constant and time required for nucleus formation, respectively.

Bhatia and Perlmutter [33] have also analyzed this problem using the population balance approach. The probability of both homogeneous nucleation (where no germ nuclei are present) and heterogeneous nucleation around a germ nucleus is considered, and the basic

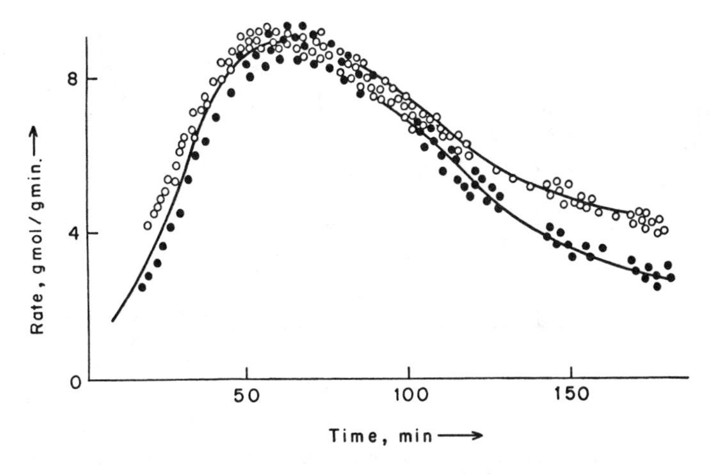

Figure 7. Rate-time curve for hydrogenation of iron carbide [36].

Avarami model has been modified to include a term for the initial volume of a growth nucleus.

Some experimental data on the formation and growth of nuclei in the oxidation of copper iodate and reduction of wustite have been provided by Nueberg [31] and El-Rahaiby and Rao [34]. Some aspects of the theory of nucleation have also been discussed [35].

The models as of now do not account for overlapping of nuclei and changes in the nucleation rate due to consumption of solid reactants. If these effects are taken into account the conversion-time relations soon take a complex form. Some of the models incorporating these features have been discussed by Rozovskii [36].

The rigorous nucleation models should also account for the processes that occur in the prenucleation period which essentially comprise formation of point defects in the form of vacancy or excess atom or ion and movement of these defects to eventually club together and generate clusters. The original solid phase retains these clusters up to the limit of homogeneity. The clusters are then crystallized out as solid-phase nuclei. The typical form of the rate-time profile during the course of nuclei formation is shown in Figure 8. It is possible for a gas-solid reaction to be controlled by prenucleation stage processes, and reduction of copper oxide by carbon monoxide typically follows the rate-time profile shown in Figure 8, which is characteristic of these processes. Deviation from such a physical picture would occur when no solid-phase nuclei are formed or when the processes during nucleation do not control the system behavior. Typical rate-time profiles for these two situations are provided by the reduction of the oxides of vanadium and iron. The experimental data [36] are sketched in Figure 9.

Crackling-Core Model

A phenomenological model that is effectively similar to the nucleation model is the crackling core model of Park and Levenspiel [37]. Similar to the formation of nuclei, this model postulates the formation of a porous pellet by a process of "crackling" from the original nonporous pellet as the first step. The individual grains in the crackled pellet then react according to SIM. The initial period during which crackling occurs corresponds to the induction period. Once the pellet becomes porous, the reaction rate increases due to the enhanced area of contact available but eventually drops as the availability of solid

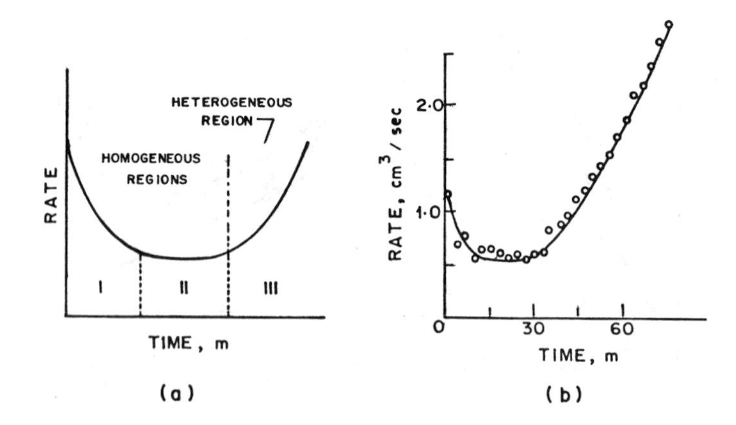

Figure 8. (A) Typical rate-time profile in the prenucleation stages; (B) rate-time curve during reduction of cuprous oxide by carbon monoxide [36].

reactant and area for reaction to occur decreases. This simple model clearly explains the S-shaped curves commonly obtained and has been invoked to explain the data on reduction of oxides such as haematite and manganese oxides [38] and uranium oxide [39].

The model envisages two stages, each of which may control the entire time-conversion behavior for a given system. In the limiting case of fast crackling, the controlling processes correspond to those in the individual grains, where one would observe a simple homogeneous model, and the conversion-time relationship becomes independent of the pellet dimension. The individual grains may follow SIM with ash diffusion or reaction-controlling mechanisms. In the other extreme of slow crackling, the pellet as a whole follows SIM behavior with reaction control. For such situations (see Table 1), the system time constant shows linear dependence on the size of the pellet.

In the intermediate regime where the initial crackling processes and those in the grain are of comparable magnitude, the model yields an algebraic equation for the conversion-time relationship which can be readily used.

An important feature of this model is realized for extremely rapid processes within the grain. The crackling now proceeds faster rendering the pellet porous and takes it to some intermediate level of conversion (x_I). Subsequent conversion to 100% cannot, however, be observed in view of large resistance in the grain. The model thus also explains the leveling off of conversion below 100% as observed in some systems.

A more useful variation of the crackling-core model can be obtained by removing the assumption of no diffusion resistance to gas inside the porous part of the pellet. Also one might consider a volume-reaction model in the porous part of the pellet rather than a shrinking-core model for the individual grain.

A GENERAL MATHEMATICAL FORMULATION

It seems possible to generalize the mathematical models presented thus far. After all, in employing any of these models one is interested in the conversion-time behavior of the system. It seems appropriate therefore to start with a general description, such as the volume-reaction model, and express it in terms of conversion rather than concentration. To illustrate, the general conservation equations of the volume reaction model can be written as

$$\frac{1}{\hat{R}^{\beta}}\frac{\partial}{\partial \hat{R}}\left(\alpha \hat{R}^{\beta}\frac{\partial \hat{C}_A}{\partial \hat{R}}\right) = \text{Rate} = \varphi^2\frac{\partial x}{\partial \hat{t}} \tag{17}$$

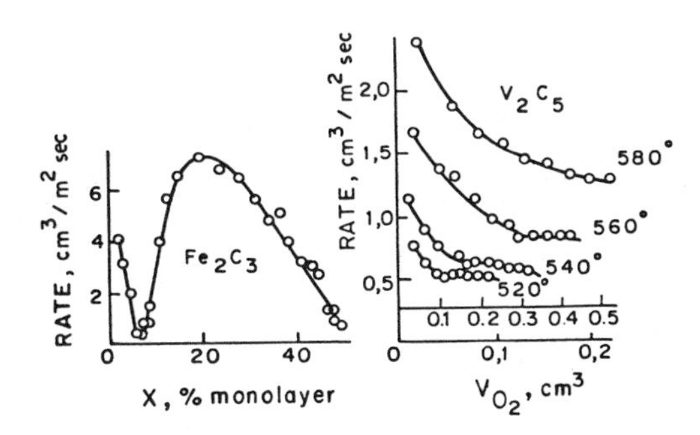

Figure 9. Experimentally measured rates during reduction of oxides [36].

where

$$\frac{dx}{d\hat{t}} = \frac{\hat{C}_A}{f(x)} \tag{18}$$

α in this equation refers to the diffusivity ratio (D_e/D_{e0}). Rearranging and integrating Equation 18 leads to

$$\int_0^{\hat{t}} \hat{C}_A \, d\hat{t} = \int_0^x f(x)\, dx = \psi \tag{19}$$

Equation 19 may be recognized as the Legendre transformation proposed earlier by Del Boughi et al. [11] see Equation 10 and provides the basic relationship between concentration and the extent of conversion. Using Equation 19 in Equation 17 and rearranging, we obtain the following single equation in x:

$$f(x)\frac{d^2x}{d\hat{R}^2} + \beta\frac{f(x)\, dx}{\hat{R}\, d\hat{R}} + f'(x)\left(\frac{dx}{d\hat{R}}\right)^2 - k(x) = 0 \tag{20}$$

where the forms f(x) and k(x) can vary. The general equation (Equation 20) has been proposed using the collocation method for different forms of the functions f(x) and k(x). The general forms of these functions for different models are indicated in Table 5. This generalized mathematical formulation can also account for structural variations through the functions f(x) and k(x).

PRACTICAL MODIFICATIONS OF THE BASIC MODELS

The simple models discussed in the previous section do not consider the complexities that are often present in practical systems. For these models to be of use for such systems, it is necessary to consider the effects of bulk flow, pressure gradient, temperature gradient, and structural changes. In the present section we shall extend the basic models to incorporate these effects.

Bulk Flow Effects

Bulk flow within the solid becomes important for reactions with volume change, and several examples of gas-solid reactions with volume change are encountered in industrial

Table 5
Functional Forms for F(x) for Different Gas-Solid Reaction Models

Functional Form F(x)	Reaction Model
$(1-x)^{-n}$	Volume reaction model
$-1-\frac{(1-x)^{1/3}+(1-x)^{2/3}}{Sh}$	Grain model
$-1-\frac{(1-x)^{1/3}}{Sh}+\frac{(1-x)^{2/3}}{Sh\,[Z_v+(1-Z_v)(1-x)]^{1/3}}$	Grain model with structural variations
$\frac{1}{n}\frac{1}{(1-x)}\ln\left(\frac{1}{1-x}\right)^{(1-n)/n}$	Nucleation model

Table 6
Conservation Equation and Time-Conversion Relation in Presence of Bulk Flow

Condition	Governing Equations
Conservation equation	$\frac{\partial}{\partial \hat{R}}\left(\hat{R}^{\beta}\frac{\partial \hat{C}_A}{\partial R}\frac{1}{1+\Theta\hat{C}_A}\right) = 0$
Boundary conditions	$\hat{R} = 1, \quad \hat{C}_A = 1, \quad \hat{R} = \hat{R}_i,$ $\frac{\partial \hat{C}_A}{\partial \hat{R}} = 2(\beta+1)\varphi^2\hat{C}_A(1+\Theta C_A)$ where $\Theta = \left(\frac{\nu_C}{\nu_A}-1\right)C_{Ab}, \quad \varphi^2 = \frac{kV_P}{2A_PD_e}$
Solid consumption rate	$\frac{d\hat{R}_i}{d\hat{t}} = -\hat{C}_{Ai}, \quad \hat{t} = \frac{kC_tC_{Ab}}{\varrho_B}\frac{A_P}{(\beta+1)V_P}$ $\hat{t}$ is related to x through $x = 1-\left(\frac{\hat{R}_i}{\hat{R}}\right)^{\beta+1}$
Time-conversion relation	$\frac{\ln(1+\Theta)}{\Theta}\frac{\hat{t}}{\varphi^2} = 1 - \frac{(\beta+1)(1-x)^{2/\beta+1}-2(1-x)}{\beta-1} = p(x)$ No external mass transfer $\hat{t} = 1-(1-x)^{1/\beta+1}+\varphi^2\frac{\Theta}{\ln(1+\Theta)}p(x)+\frac{2(1+\Theta)}{Sh}x$ In presence of external mass transfer

practice. General formulations accounting for the effect of bulk flow have been presented by a number of authors [40, 41, 42] for the sharp interface model. The conservation equation for the mass of reactant now includes a term for bulk flow and, in analogy with gas-solid catalytic systems [43] can be written in dimensionless form as shown in Table 6. The time-conversion relations for situations both in the presence and absence of external mass transfer effect are included in the table.

It will be noticed from the time-conversion relation that, in general, the net effect of volume change is to change the time required to attain a given extent of conversion by a factor $\theta/\ln(1+\theta)$ where θ is a volume-change modulus given by

$$\theta = \left(\frac{\nu_C}{\nu_A}-1\right)C_{Ab}$$

In the presence of heat effects significant pressure gradients can develop within the pellet and the simple preceding analysis is no longer valid. In fact, even the conventional Fickian law of diffusion becomes inapplicable and more sophisticated models for diffusion in porous media [44] are necessary. The effect of pressure gradient on the system behavior varies depending on the type of diffusion (Knudsen or bulk) (see, for example, Wong et al. [45], Hite and Jackson), [46].

In the case of gas-solid reactions the presence of structural effects and the inherent transient nature of the systems continually alter the diffusional regime. The effect of pressure gradients cannot therefore be ignored all through the course of conversion, and significant effects on the time-conversion relationship can be expected depending on the regime of operation.

The influence of pressure gradients in gas-solid reactions following SIM has been examined by Turkdogan et al. [47]. Detailed numerical solutions have been obtained to the pressure gradients which arise because of Knudsen flow in the ash layer with small pores and due to the reactant gas having a diffusivity different from that of the product.

Nonisothermal Effects

Many gas-solid reactions are accompanied by evolution of heat leading to significant temperature gradients within the pellet. The existence of higher temperatures within the pellet gives rise to situations where the local reaction rate in the interior of the particle is higher. The higher inside temperature, besides causing difficulties such as sintering, generates nonlinearity in the system. The system behavior now becomes dependent additionally on the initial state, and it could as a whole evolve to different final states for minor variations in parameter values.

The heat effects can be easily incorporated in SIM by writing balance equations which take into account the transfer of heat through the gas film and ash layer and its generation at the reaction interface. The analysis becomes especially simple if the pseudo steady-state assumption is invoked. Such a hypothesis, however, while adequate for mass transfer processes, is inadequate where heat transfer is involved.

The transient heat accumulation term has been accounted for in the analysis of Luss and Amundson [48] who obtained the following equation:

$$\hat{T}_i - 1 = \frac{2}{3} P_1 \sum_{n=1}^{\infty} \frac{\sin(\mu \hat{R}_i)}{\hat{R}_i} f(\mu) \int_{\hat{R}_i}^{1} \sin(\mu Z) \hat{R}_i \exp[-\mu^2 P_2 f(Z)]\, dZ \tag{17}$$

where the μ's are the positive roots of the transcendental equation

$$\mu \cot \mu + Nu - 1 = 0 \tag{18}$$

and the various other parameters are defined as follows:

$$f(Z) = \frac{Z - R_i}{Da \exp[\gamma(1 - 1/\hat{T}_i)]} + \frac{1}{2}(Z^2 - \hat{R}_i^2) + \frac{1}{3}\left(\frac{1}{Sh} - 1\right)(Z^3 - R_i^3) \tag{19}$$

$$P_1 = \frac{3(-\Delta H)\nu \varrho_B}{M_B \varrho_S C_{ps} T}$$

$$P_2 = \frac{\varrho_B k_e'}{M_B \varrho_S C_{ps} D_e C_{Ab}} \tag{20}$$

$$f(\mu) = \frac{(Nu - 1)^2 + \mu^2}{Nu(Nu - 1) + \mu^2} \tag{21}$$

The nonisothermal effectiveness factor can be calculated by relating it to the interfacial movement by

$$\frac{d\hat{R}_i}{dt} = \eta \tag{22}$$

where the interface temperature given by Equation 17 can be used.

SIM has also been used by Shettigar and Hughes [49] and Rehmat and Saxena [50] for the case of coke regeneration and reactions of carbon with steam and oxygen. It has also been employed to describe the decomposition type of gas-solid reactions [51, 52]. As has been pointed out by Levenspiel [53], the controlling processes in these types of reactions are either heat or gas diffusion through the product layer. When heat transfer through the ash layer controls, the reaction interface temperature remains constant and the equation for SIM with ash diffusion control (with D_e replaced by k_e' in the definition of τ in Table 1) represents the conversion-time behavior. For a situation where gas diffusion through the ash layer is controlling, the variation of transport properties with temperature ($D_e \alpha T^{1/2}$ in the Knudsen regime; $D_e \alpha T^{1.5-2}$ in the bulk diffusion regime) should be appropriately accounted for [54].

Similar studies relating to heat effects in the volume-reaction model have been reported by Shettigar and Hughes [49] while Calvelo and Smith [22] have analyzed the particle-pellet model using the pseudo steady-state hypothesis. The governing transient equations for heat and mass for the volume and grain models have already been presented in Tables 2 and 4 along with the appropriate boundary conditions. An analysis of the complete transients for the particle-pellet model indicates a shift in the temperature maximum from the surface to the center with progress of time [55]. Nonisothermal effects in the presence of structural variations have not yet been modeled.

The incorporation of these effects in the basic models can account for the sigmoid nature of the conversion-time profiles observed experimentally. It is also possible in a given situation that the openings of the pores get blocked. Such a phenomenon, known as pore closure, has also been observed experimentally and can be explained theoretically [56–58].

The structural changes arising due to heat effects are usually referred to as sintering, which essentially changes the porosity and tortuosity of the pellet. Here again, empirical relationships accounting for the variation of the transport properties with these changes have been proposed and are convenient for model calculations. The sintering effects using such procedures have been analyzed by Kim and Smith [59], Chen and Smith [60], and Ramachandran and Smith [61], and the empirical relationships proposed are included in Table 7.

Some of the models that include structural variations will now be presented.

The basic grain model has been extended by Garza and Dudukovic [62, 63] to account for variations in the grain size and structure. Incorporating the diffusivity variation with porosity, the model first calculates the time required for complete conversion of the surface layer of the solid. This product layer then moves inside towards the center. Garza and Dudukovic [62, 63] also analyzed systems where pore closure occurs. The model now evaluates the time required for pore closure and obtains the concentration profile of the unreacted solid at this instant. In the subsequent stage the model supposes that the behavior within the pellet is described by SIM which uses the initial solid concentration profile as obtained in the previous stage. Clearly, analytical solutions are no longer possible; however, approximate analytical solutions can be obtained and are presented in Table 3 along with solutions of the other models.

In order to better understand the implications of the structural changes in the modeling of gas-solid reactions, Ramachandran and Smith [64] (see also Chrostowski and Georgakis [65]) focussed attention on a single pore. This model, sketched in Figure 10, assumes a cylindrical pore with a concentric ring of solid B associated with it. Depending upon the difference in the molal volumes of the reactant and product, the pore size increases, decreases or stays constant with progress of reaction. When the molar volume increases there is obviously pore closure. The model yields a simple conversion-time relationship and requires a knowledge of the average structural properties of the solid. The single pore model has been extended to account for Knudsen diffusion [65], and bulk flow and reversibility of chemical reaction [66]. More recently an asymptotic analysis of the single-pore model has been reported by Shankar and Yortsos [67]. The analysis allows closed-form solutions for concentration, pore radius, and conversion profiles in a long, narrow pore of cylindrical

Table 7
Typical Diffusivity and Porosity Variations Due to Structural Changes in Pellet

Relationship	Model	References
$f_c = f_{c0} + \alpha_1(1 - B/B_0)$	Volume reaction model	Fan et al. [103]
$\frac{D_e}{D_{e0}} = \alpha_2 \left\{ \exp\left[1 - \frac{1}{(B/B_0)\alpha_3}\right] + \alpha_2 - 1 \right\}^{-1}$ where α_1, α_2, α_3 are constants and B refers to solid concentration	Volume reaction model	Fan et al. [103]
$f_c = f_{c0} + (1 - f_{c0}) \frac{V_B + V_I}{V_B - v_s V_S}$ $D_e = D_{e0}(f_c/f_{c0})$ where V_B, V_S refer to molar volume of species B and S per mole of solid. V_I refers to molar volume of inert per mol of solid B		Levenspiel [53]
$\frac{(1 - f_c)}{(1 - f_{c0})} = \left(\frac{r_G}{r_{G0}}\right)^3$ where r_G refers to the radius of the grain $\frac{D_e}{D_{e0}} = \left(\frac{f_c}{f_{c0}}\right)$	Particle pellet model	Ramachandran and Smith [64]
Eventually with the progress of time the pore will be completely blocked. The time required for pore closure is obtained as $= \frac{1}{Bi_p} \left\{ \frac{Z}{Z_v - 1} - \frac{1}{(1 - f_{c0})^{2/3}} \left[1 + \frac{Z_v}{1 - Z_v} f_{c0}\right] - \frac{1}{Z - 1} \frac{1}{(1 - f_{c0})^{2/3}} \right\} + 1 - \frac{1}{(1 - f_{c0})^{1/3}} \left(1 + \frac{Z}{1 - Z} f_{c0}\right)$	Particle pellet model	Georgekis et al. [58]
$\frac{D_e}{D_{e0}} = 1 - \left(\frac{1 - f_{c0}}{f_{c0}}\right)(\gamma - 1)(1 - \hat{r}_{Gi}^3)$ where γ refers to the ratio of the volume of solid product formed per unit volume of reactant consumed and $\hat{r}_{Gi}$ refers to the dimensionless position of the shrinking core in the grain. The time required for pore closure is obtained as $\tau = 1 - \bar{\eta} + \frac{Da_g}{2} 1 - \eta^2 + \frac{(\gamma + (1 - \gamma)\bar{\eta}^3)^{2/3}}{1 - \gamma}$ where $\bar{\eta} = \left[1 + \frac{f_{c0}}{(1 - f_{c0})(1 - \gamma)}\right]^{1/3}$	Variable grain model	Garza and Dudukovic [62]

or slab geometry. The analysis provides especially accurate estimates of overall conversion in situations where high pore diffusion resistance (fast reaction) exists. The cases corresponding to incomplete conversion (pore closure) and complete conversion with a moving reaction zone have also been analyzed within the framework of the model.

The single-pore model uses average structural properties and is convenient from a mathematical viewpoint. The simple model cannot therefore take account of factors such as the distribution and variations in the structural properties with reaction. In practice solids have an initial pore size distribution that evolves during the reaction (due to pore intersection and coalescence) and rigorous models should take account of the evolving reaction surface. The net reaction surface is therefore a variable, and some of the models which account for this will now be discussed.

The models of Cristman and Edgar [68, 69] describe the reacting solid as a sphere having a distribution of randomly oriented open pores. The model uses the combination of independent variables of time and location [70] and follows the evolution of pore size distribution using the population balance approach. The results of the population balance are used with the mass balance on the reacting gas to obtain rate and conversion data. The model has been tested against the data of Ulerich et al. [71] on sulfation of limestone and found to fit both the rate-time and conversion-time curves accurately.

The model of Simons and Rawlins [72] is based on the assumption that at the surface of the pellet a distribution of pore radii exists. It is sufficient then to calculate the flux for

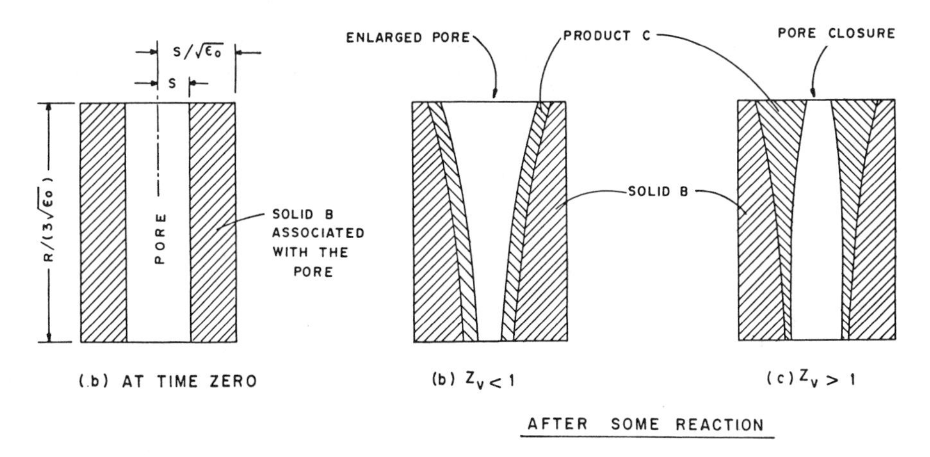

Figure 10. Schematic representation of the single-pore model.

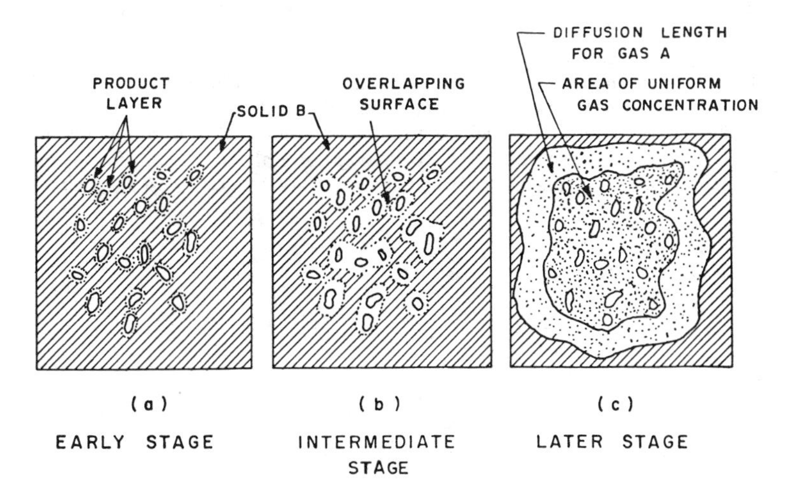

Figure 11. Development of reaction surface according to the random-pore model.

a pore of random r_p and average it over the surface pore size distribution. We can then use it to calculate an average flux of the gaseous reactant at the surface which is then related to the conversion of solid B. The model obviously ignores the variation of reaction surface with time and therefore does not represent the real situation.

In the random pore model of Bhatia and Perlmutter [33, 73] the actual reaction surface of solid B is supposed to arise during the random overlapping of a set of cylindrical surfaces. The development of the reaction surface as envisaged in this model is sketched in Figure 11. The model first calculates the actual reaction surface and uses it subsequently along with intrinsic structural properties of the solid to obtain the conversion-time relationship.

The conversion-time behavior for the cases of kinetic and ash diffusion control using the previous model have been obtained by Bhatia and Perlmutter [33, 73]. The final expressions are given respectively by

$$\frac{dx}{d\hat{t}} = (1-x)\,[1-\psi \ln(1-x)]^{1/2} \tag{23}$$

and

$$\frac{dx}{dt} = \frac{\hat{C}_A(1-x)\,[1-\psi \ln(1-x)]^{1/2}}{1+\dfrac{\beta' Z}{\psi}\{[1-\psi \ln(1-x)]^{1/2}-1\}} \tag{24}$$

where

$$\psi = 4\pi L_0(1-f_0)/S_0^2$$

$$\hat{t} = \frac{k_s C_{Ag}^m S_0 t}{1-f_0} \tag{25}$$

The parameter β' characterizes the diffusional resistance to the flow of the gaseous species in the product layer. The structural parameter ψ generalizes the model in the sense that for $\psi = 0$ one realizes the volume reaction model with $n = 1$, while for $\psi = 1$ behavior typical of the grain model is realized. It is apparent therefore that the concept of reaction order with respect to solids used in the volume reaction models has a relation to the structure of the solid; the random pore model identifies this relation and is graphically illustrated in Figure 12. Bhatia and Perlmutter [74] have analyzed several distributions for the parameter ψ. The analysis indicates that a uniform pore size leads to the lowest reactivity of solid.

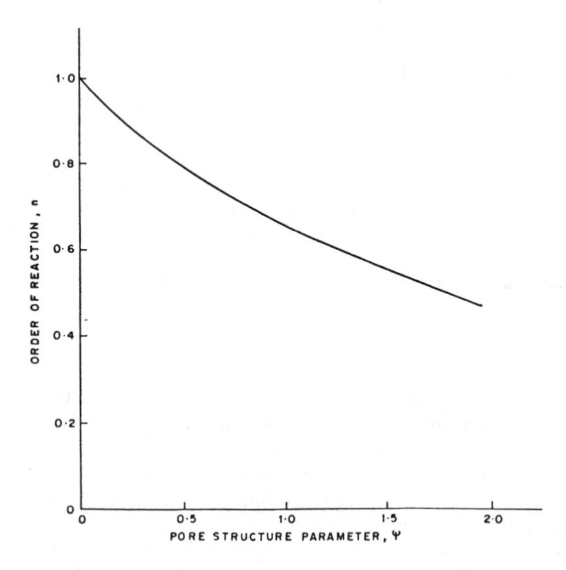

Figure 12. Relationship between the structural parameter of the random pore model and the order of reaction with respect to solids in the volume reaction model [33].

Also, for a given pore size distribution, an optimal structure exists for highest activity of solids.

The experimental data on CO_2-lime reaction has been analyzed by Bhatia and Perlmutter [75] using the random pore model to obtain the kinetic constant as well as the product diffusivity.

The assumption of equal pores and reaction surface areas inherent in the model has been subsequently relaxed by Bhatia and Perlmutter [76]. The rigorous model provides generalizations in terms of a moving pore and reaction surface, and for no intraparticle and boundary-layer diffusional resistances, the following time-conversion relation is reported:

$$\hat{t} = \frac{dx_1}{S^*(x_1)} + \frac{\beta'}{2}\int_0^x \int_0^{x_1} \left[\frac{1}{S^{*2}(x')} + \frac{Z_v - 1}{S_p^{*2}(x')}\right] dx'\, dx_1 \tag{26}$$

For $\beta' \to 0$ and ∞ this equation reduces to the earlier cases of kinetic control and product layer diffusion control. In this equation S^* and S_p^* refer to the dimensionless reaction surface area and pore surface area, respectively, and these quantities can be appropriately evaluated for different models. Thus for the grain model we have

$$S^*(x) = (1-x)^m \tag{27}$$

$$S_p^*(x) = [1+(Z_v-1)x]^m \tag{28}$$

while for the random pore model

$$S^*(x) = (1-x)\sqrt{1-\ln(1-x)} \tag{29}$$

$$S_p^*(x) = [1+(Z_v-1)x]\sqrt{1-\psi \ln[1+(Z_v-1)X]} \tag{30}$$

Substituting these expressions for S^* and S_p^* in Equation 26 and integrating leads to the desired conversion-time relationship. A comparison of predictions obtained using the grain model and the random pore model for a set of parameter values is illustrated in Figure 13.

MODELS FOR GAS-SOLID REACTIONS INVOLVING FLUID PRODUCTS ONLY

The previous sections were concerned with the general description of the various models and their extensions to practical situations for a general gas-solid reaction involving both fluid and solid products. While these models can also be used with appropriate modifications for any other type of gas-solid reaction, elaborate models specifically developed for gasification type of reactions are also available. The present section discusses some of these models. Typical examples of industrial gas-solid reactions where these models can be used include systems such as combustion of carbon, formation of metal carbonyls, chlorination of ores, water gas reaction of carbon, etc.

The modeling of systems involving nonporous solids in this type of reaction is relatively easy—for one need not now bother about the resistance in the ash layer that is normally present when a solid product exists. The system would therefore be controlled either by external mass transfer resistance or by the intrinsic kinetics of the process. It is relatively easy to get rid of mass transfer effects and establish the true kinetics of the system. Doraiswamy et al. [77] used this technique to estimate the kinetics of chlorination of ilmenite by carrying out the reaction in a bed maintained close to the minimum fluidization condition. Under this condition the assumptions of plug flow of gas and complete mixing of solids is reasonably adequate and yields useful results. The method has also been used by Corella et al. [78] subsequently.

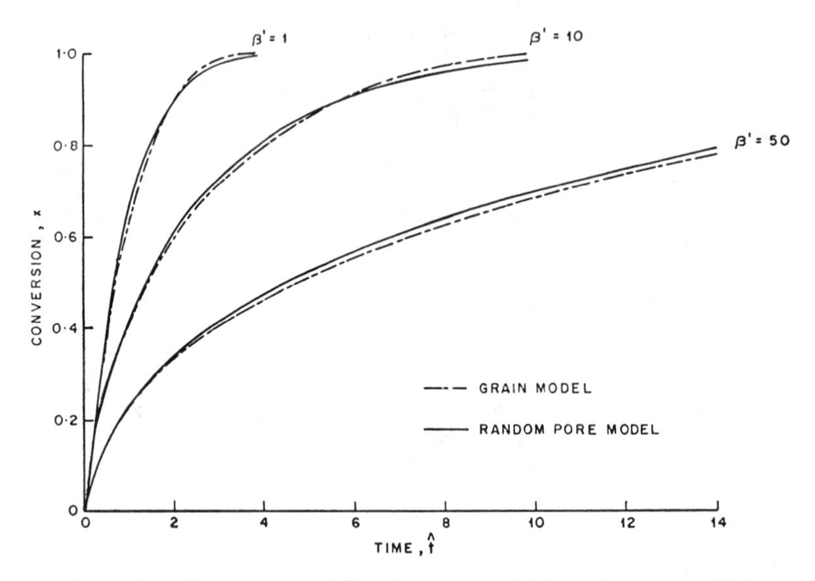

Figure 13. Comparison of conversion predictions for several models for various values of β. Parameters are: $\psi = Z_v = 1$, $\phi = 0$.

Where porous solids are involved the situation becomes more complex due to changes in surface area, pore structure, etc. Petersen [79], on the assumption of random distribution of pores of uniform size and no interparticle diffusion resistance, proposed the following equation for gasification:

$$\frac{\varepsilon_0}{1-\varepsilon_0}\left[\frac{F^2(G-F)}{G-1}-1\right]=0 \tag{31}$$

where

$$F = 1+\left(\frac{k_s C_{Ab}^m t}{r_{p0}}\right) \tag{32}$$

and G is the solution of the cubic equation

$$\frac{4}{27}\varepsilon_0 G^3 - G + 1 = 0 \tag{33}$$

ε_0 and r_{p0} in these equations refer, respectively, to initial porosity and radius of pore.

More rigorous models to account for features such as pore growth, initiation of new pores, coalescence of pores, changes in the density of pore size distribution, changes in the density of pore intersections, length of pore segments, etc., normally present during gasification type of reactions are also available. Some of these features are included in the models of Hoshimoto and Silveston [80, 81], Simons and Pinson [82] and Simons [83]. The model of Hoshimoto and Silveston, which also accounts for interparticle diffusion resistance, predicts a maximum in the reactive surface area vs. solid conversion behavior. The models in general however, require information on a large number of parameters that is usually not known *a priori.*

Gavalas [84] has also proposed a random capillary model that uses a single probability density function related to the pore size distribution. The model gives the conversion in terms of the zero and first moment as

$$x(t) = 1-\exp[-2\pi(\mu_0 v^2 t^2 + 2\mu_1 vt)] \tag{34}$$

where μ_0 and μ_1 are defined as

$$\mu_0 = \int_{\gamma_{Pmin}}^{\gamma_{Pmax}} p(r_{p0})\, dr_{p0} \quad \text{and} \quad \mu_1 = \int_{\gamma_{Pmin}}^{\gamma_{Pmax}} r_{p0}p(rp_0)\, dr_{p0} \tag{35}$$

and v represents the rate of change of pore radius with time. The conversion equation has been tested against the experimental data of Mahajan et al. [85] on char gasification by oxygen by Dutta and Wen [86].

The model of Bhatia and Perlmutter mentioned earlier [33] has also been verified against the data of Hoshimoto et al. [87] on activation of char by steam. A detailed parametric study of char gasification has been presented by Srinivas and Amundson [88, 89]. Effects such as due to the presence of interparticle diffusion and multireactions have also been incorporated.

MODELING OF COMPLEX REACTIONS

The general models discussed thus far have been extended to include more complex reactions. In view of the complexities involved in the general formulation of such models, the studies have mostly been restricted to specific cases. Thus the case of consecutive reactions has been analyzed by Spitzer et al. [90] and Tsay et al. [91]. The case of reactions of two cases with the same solid has been analyzed by Wen and Wei [92]. A single solid reacting with N gaseous components has been modeled by Do [19]. For situations where the rates of all N reactions are large compared to the diffusion rates of the components, it is shown that the solid concentration profile behaves like a traveling wave, and the wave front velocity has linear contributions from all the gaseous reactants. In the instance when only M reactions out of a total of N are faster than the corresponding diffusion rates, the overall pictures is slightly altered. The solid concentration profile still behaves like a traveling wave; however the contribution by the gaseous reactants to the wave front is no longer linear, and a complex relation is operative.

The case of reactions of two components of the solid phase with the same gas has been analyzed by Ramachandran et al. [93] and more recently by Hoshimoto et al. [94]. A typical situation is regeneration of coked catalyst. The case of complex reactions such as dispropor-tion of potassium benzoate to terethalete has been analyzed by Kulkarni and Doraiswamy [95, 96].

CONCLUDING REMARKS

Several mathematical models to simulate gas-solid noncatalytic reactions involving different degrees of complexity have been discussed in this chapter. The chapter begins with a description of the simple sharp interface shrinking-core model, and more elaborate models such as the volume-reaction model and the grain model are then considered. These models, while adequate for explaining certain features, fail to account for a number of other features, and these inadequacies point the way to possible refinements. A new class of models focussing on the single pore were then proposed (see, for example, Szekely and Evans [97] and Chu [98]) and represent perhaps the first attempt to incorporate structural variations as parameters in the model. The single-pore model was subsequently improved upon [61, 99]. The need to simultaneously consider the intersections among the pores and the changes in their structure led to the development of the random pore models [73, 74]. The assumption of average grain or pore size in these models was removed in subsequent models which are generally referred to as distributed pore models and account for variations in pore size distribution with the extent of reaction [69]. Attempts to extend these models to specific situations continue and numerically intensive developments keep appearing in the literature.

The availability of a large number of models capable of explaining a given experimental observation requires stringent tests to discriminate among the models. Also more accurate information on phenomena such as diffusion inside the solid and changes in the structure of the solid (such as pore dimensions, pore volume, pore size distribution) with the extent of reaction is necessary. The recent trend in this area has been to design more rigorous experiments and use sophisticated instrumental techniques to obtain such information. Thus the phenomeon of diffusion in porous solids has been reexamined from a more rigorous point of view recently by Abbasi et al. [44]. Attempts to obtain accurate information regarding the parameters of the models are evident in the studies of Gibson and Harrison [100]. References 101 through 105 are also worth noting.

NOTATION

a	interfacial area for transfer (cm^2/cm^3)
A_p	area of pellet (cm^2)
B	general notation for the solid reactant species
C	dimensionless concentration of gas in the bed
C_A, C_B	concentration of species A, B (mol/cm^3)
$\hat{C}_A$, $\hat{C}_B$	dimensionless concentration of species A, B
C_{Ab}	concentration of gaseous species A in the bulk or in the bubble phase (mol/cm^3)
C_{Ae}	concentration of gaseous species A in emulsion phase (mol/cm^3)
C_{Ai}	concentration of gaseous species A at the interface (mol/cm^3)
C_{As}	concentration of gaseous species A at the surface (mol/cm^3)
C_{B0}	initial concentration of species B (mol/cm^3)
C_p	specific heat of gaseous species (cal/g sec)
$\bar{C}_p$	average specific heat of gaseous stream (cal/g sec)
C_{ps}	molar specific heat of solid (cal/mol sec)
D_e	effective diffusivity of gaseous species A in the solid (cm^2/sec)
D_{eB}	effective diffusivity of gaseous species A in the solid (cm^2/sec)
D_{eG}	effective diffusivity of gaseous species A in the grain (cm^2/sec)
D_{eS}	effective diffusivity of gaseous species A in the solid (cm^2/sec)
D_k	Knudsen diffusivity (cm^2/sec)
D_s	effective axial dispersion coefficient of solid in bed (cm^2/sec)
E	activation energy (cal/mol)
f, f_c	pellet voidage
f_0	initial porosity
$f_{0(x)}$, $f_{1(x)}$, $f_{2(x)}$	functions of conversion defined in Table 1
$f_{(p)}$	fraction of pores removed
F (F_p)	function of fraction of pores removed
h	heat transfer coefficient (cal/sec cm^2 °C)
k	general notation for rate constant $(cm^3/mol)^{n-1}$ (1/sec)
k^*	dimensionless reaction rate constant ($k/k(T_0)$)
k_e'	effective thermal conductivity of solid (cal/sec cm °C)
k_g	phenomenological mass transfer coefficient (cm/sec)
k_p	rate constant for pore removal

k_s surface reaction rate constant (cm/sec)
m general notation for order of reaction with respect to gas species
M_B molecular weight of species B
n general notation for order of reaction with respect to solid species
p partial pressure of species
r_G radial position in the grain (cm)
r_{Gi} radial position of interface in the grain (cm)
r_{G0} initial radius of the grain (cm)
R radius of pellet (cm)
$\hat{R}$ dimensionless radial position
R_A general notation for rate of reaction (mol/sec)
R_{Ac} rate of chemical reaction at the interface (mol/sec)
R_{Ad} rate of diffusion through ash layer (mol/sec)
R_{Ag} rate of diffusion through gas film (mol/sec)
R_g gas constant
R_{GA} rate of reaction per grain (mol/cm^3)
R_i radius of the interface (cm)
$\hat{R}_i$ dimensionless radial position of the interface
R_j radius of the j-th particle
S^* dimensionless reaction surface area
SF shape factor of solid
S_p^* dimensionless pore surface area
t time
$\hat{t}$ dimensionless time
T temperature variable
T_b temperature in the bulk
T_i temperature at the interface
$\hat{T}_i$ dimensionless temperature at the interface
T_s temperature at the solid surface
V_p volume of the pellet (cm^3)
W_1 dimensionless ratio of the pellet size to grain size, R/r_{G0}

Greek Symbols

α diffusivity ratio, D_e/D_{e0}
β shape factor equal to 0, 1, and 2 for slab, cylinder, and sphere
β_m exothermicity factor
γ Arrhenius parameter defined as $E/R_g T_b$; also dimensionless time relative to the process in the pellet defined as $D_e t/R^2$
ΔH heat of reaction (cal/g mol)
η effectiveness factor as defined by Equation 22
ν_A, ν_B stoichiometric coefficients
ϱ_B density of solid species B (gm/cm^3)
φ Thiele modulus
ψ cumulative gas concentration defined by Equation 10 also structural parameter defined by Equation 25

REFERENCES

1. Doraiswamy, L. K., and Sharma, M. M., *Heterogeneous Reactions—Analysis Design and Reactor Design,* Vol. I, John Wiley (1983).
2. Ramachandran, P. A., and Doraiswamy, L. K., *AIChE J.*, 28, 898 (1982).
3. Doraiswamy, L. K., and Kulkarni, B. D., in *Chemical Reaction Engineering Handbook*, J. J. Carberry and A. Varma (Eds.) Marcel Dekker, 1984.
4. Ishida, M., and Wen, C. Y., *AIChE J., 14,* 311 (1968).
5. Sohn, H. Y., and Szekely, J., *Can. J. Chem. Eng., 50,* 674 (1972a).
6. Ramachandran, P. A., *Chem. Eng. J., 23,* 223 (1982).
7. Chida, T., and Tadaki, T., *Int. Chem. Eng., 22,* 503 (1982).
8. Simonsson, D., *Ind. Eng. Chem. Proc. Des. Dev., 18,* 288 (1979).
9. Cannon, K. J., and Denbigh, K. J., *Chem. Eng. Sci., 6,* 145 (1957).

10. Chu, W. F., and Rahmet, A., *Met. Trans. B., 10B,* 401 (1979).
11. Del Borghi, M., Dunn, J. C., and Bishoff, K. B., *Chem. Eng. Sci., 31,* 1065 (1976).
12. Dudukovic, M. P., and Lamba, H. S., *Chem. Eng. Sci., 33,* 303 (1978a).
13. Dudukovic, M. P., and Lamba, H. S., *Chem. Eng. Sci., 33,* 471 (1978b).
14. Ramachandran, P. A., and Kulkarni, B. D., *Ind. Eng. Chem. Proc. Des. Dev., 19.* 717 (1980).
15. Wen, C. Y., and Wu, N. T., *AIChE J., 22,* 1012 (1976).
16. Ausman, J. M., and Watson, C. C., *Chem. Eng. Sci., 17,* 323 (1962).
17. Bowen, J. H., and Cheng, C. K., *Chem. Eng. Sci., 24,* 1829 (1969).
18. Mantri, V. B., Gokarn, A. N., and Doraiswamy, L. K., *Chem. Eng. Sci., 31,* 779 (1976).
19. Do, D. D., *Chem. Eng. Sci., 38,* 1477 (1982).
20. Do, D. D., *Chem. Eng. Sci., 37,* 1477 (1982).
21. Ramachandran, P. A., *Chem. Eng. Sci., 37,* 1385 (1983).
22. Calvelo, A., and Smith, J. M., Chemeca Proceeding, No. 3, 1 (1978).
23. Szekely, J., and Evans, J. W., *Met. Trans., 2,* 1691 (1971a).
24. Szekely, J., and Evans, J. W., *Met. Trans., 2,* 1699 (1971b).
25. Evans, J. W., and Ranade, M. G., *Chem. Eng. Sci., 35,* 1261 (1980).
26. Szekely, J., and Propster, J., *Chem. Eng. Sci., 30,* 1049 (1975).
26a. Szekely, J., Evans, J. W., and Sohn, H. Y., *Gas-Solid Reactions,* Academic Press (1976).
27. Sohn, H. Y., and Szekely, J., *Chem. Eng. Sci., 28,* 1789 (1973a).
28. Sohn, H. Y., and Szekely, J., Chem. Eng. Sci., *28,* 1169 (1973b).
29. Avrami, M., *J. Chem. Phys.,* 8, 212 (1948).
30. Erofeev, B. V., Proc. 7th Int. Symp., Reactivity of Solids, pp. 553. (1961).
31. Neuburg, H. J., *Ind. Eng. Chem. Proc. Des. Dev., 9,* 285 (1970).
32. Ruckenstein, E., and Vavanellos, T., *AIChE J., 21,* 756 (1975).
33. Bhatia, S. K., and Perlmutter D. D., *AIChE J., 26,* 379 (1980).
34. El-Rahaiby, S. K., and Rao, Y. K. *Met. Trans. 10B,* 257 (1979).
35. Rao, Y. K., *Met. Trans., 10B,* 243 (1979).
36. Rozovskii, A. Y., *Kinet. Katal., 22,* 36 (1981).
37. Park, J., and Levenspiel, O., *Chem. Eng. Sci., 32,* 233 (1975).
38. De Bruijn, T. J. W., et al., *Chem. Eng. Sci., 35,* 1591 (1980).
39. Le Page, A. H., and Fane, A. G., *J. Inorg. Nucl. Chem., 36,* 87 (1974).
40. Beveridge, G. S. G., and Goldie, P. J., *Chem. Eng. Sci., 23,* 912 (1968).
41. Gower, R. C., Ph. D. Thesis, Lehigh University (1971).
42. Sohn, H. Y., and Sohn, H. J., *Ind. Eng. Chem. Proc. Des. Dev., 19,* 237 (1980).
43. Weekman, V. W., Jr., and Gorring, R. L., *J. Catal., 4,* 260 (1965).
44. Abbasi, M. H., Evans, J. W., and Abramson, I. S., *AIChE J., 29,* 617 (1983).
45. Wong, R. D., Hubbard, G. L., and Denny, V. E., *Chem. Eng. Sci., 31,* 541 (1976).
46. Hite, R. H., and Jackson, R., *Chem. Eng. Sci., 32,* 703 (1977).
47. Turkdogan, E. T., et al., *SME Trans. AIME, 254,* 9 (1973).
48. Luss, D., and Amundson, N. R., *AIChE J., 15,* 194 (1969).
49. Shettigar, U. R., and Hughes, R., *Chem. Eng. J., 3,* 93 (1972).
50. Rehmat, A., and Saxena, S. C., *Ind. Eng. Chem. Proc. Des. Dev., 19,* 223 (1980).
51. Narisimhan, G., *Chem. Eng. Sci., 16,* 7 (1961).
52. Campbell, R. R., Hills, A. W. D., and Paulin, A., *Chem. Eng. Sci., 25,* 929 (1970).
53. Levenspiel, O., *Chemical Reactor Omni Book,* Oregon State University Press, Oregon (1979).
54. Mu, J., and Perlmutter, D. D., *Chem. Eng. Sci., 35,* 1645 (1980).
55. Sampath, B. S., Ramachandran, P. A., and Hughes, R., *Chem. Eng. Sci., 30,* 125 (1975).
56. Hartman, M., and Coughlin, R. W., *Ind. Eng. Chem. Proc. Des. Dev., 13,* 248 (1974).
57. Hartman, M., and Coughlin, R. W., *AIChE J., 22,* 490 (1976).
58. Georgekis, C., Chang, C. W., and Szekely, J., *Chem. Eng. Sci., 34,* 1072 (1979).

59. Kim, K. K., and Smith, J. M., *AIChE J., 20,* 670 (1974).
60. Chen, S. F., and Smith, J. M., *Indian Chem. Engr., 18,* 42 (1976).
61. Ramachandran, P. A., and Smith, J. M., *Chem. Eng. J., 4,* 137 (1977b).
62. Garza-Garza, O., and Dudukovic, M. P., *Chem. Eng. J., 24,* 35 (1982).
63. Garza-Garza, O., and Dudukovic, M. P., *Chem. Eng. Sci., 6,* 131 (1982b).
64. Ramachandran, P. A., and Smith, J. M., *AIChE J., 23,* 353 (1977a).
65. Chrostowski, J. W., and Georgakis, C., ACS Symp. Ser., *65,* (Chem. React. Eng. Houston) 225–37 (1978).
66. Ulrichson, D. L., and Mahoney, D. J., *Chem. Eng. Sci., 35,* 567 (1980).
67. Shankar, K., and Yortsos, Y., *Chem. Eng. Sci., 31,* 1159 (1983).
68. Christman, P. G., and Edgar, T. F., AIChE Meeting, Chicago (1980).
69. Christman, P. G., and Edgar, T. F., AIChE J., *29,* 388 (1983).
70. Dudukovic, M. P., *AIChE J., 22* (5) (1978).
71. Ulerich, N. H., O'Neill, E. P., and Keairns, D. L., EPRI, FP–426 First Report (1977).
72. Simons, S. A., and Rawlins, W. T., *Ind. Eng. Chem. Proc. Des. Dev., 19,* 565 (1980).
73. Bhatia, S. K., and Perlmutter, D. D., *AIChE J., 27,* 247 (1981a).
74. Bhatia, S. K., and Perlmutter, D. D., *AIChE J., 27,* 226 (1981b).
75. Bhatia, S. K., and Perlmutter, D. D., *AIChE J., 29,* 231 (1983b).
76. Bhatia, S. K., and Perlmutter, D. D., *AIChE J., 29,* 79 (1983a).
77. Doraiswamy, L. K., Bijawat, H. C., and Kunte, M. V., *Chem. Eng. Prog., 55,* 80 (1959).
78. Corella, J., *Chem. Eng. Sci., 35,* 25 (1980).
79. Petersen, E. E., *AIChE J., 3,* 443 (1957).
80. Hoshimoto, K. K., and Silveston, P. L., *AIChE J., 19,* 259 (1973a).
81. Hoshimoto, K. K., and Silveston, P. L., *AIChE J., 19,* 268 (1973b).
82. Simons, G. A., and Pinson, M. L., *Combus. Sci. and Tech., 19,* 217 (1979).
83. Simons, G. A., *Combus. Sci. and Tech., 19,* 227 (1979).
84. Gavalas, G. R., *AIChE J., 26,* 577 (1980).
85. Mahajan, O. P., Yarzal, R., and Walkar, P. L., Jr., *Chem. Eng. Sci., 31,* 779 (1976).
86. Dutta, S., and Wen, C. Y., *Ind. Eng. Chem. Proc. Des. Dev., 16,* 31 (1977).
87. Hoshimoto, K. K., et al., *Ind. Eng. Chem. Proc. Des. Dev., 18,* 73 (1979).
88. Srinivas, B., and Amundson, N. R., *Can. J. Chem. Eng., 58,* 476 (1980a).
89. Srinivas, B., and Amundson, N. R., *AIChE J., 26,* 487 (1980b).
90. Spitzer, R. H., Manning, F. S., and Philbrook, W. O., *Trans. Met. Soc.,* AIME., *242,* 618 (1968).
91. Tsay, Q. T., Ray, W. H., and Szekely, J., *AIChE J., 22,* 1064 (1976).
92. Wen, C. Y., and Wei, J., *AIChE J., 17,* 272 (1971).
93. Ramachandran, P. A., Rashid, M. H., and Hughes, R., *Chem. Eng. Sci., 30,* 1391 (1975).
94. Hoshimoto K. K., et al., *Chem. Eng. J., 27,* 177 (1983).
95. Kulkarni, B. D., and Ramachandran, P. A., *Ind. Eng. Proc. Des. Dev., 19,* 467 (1980).
96. Kulkarni, B. D., and Doraiswamy, L. K., *Chem. Eng. Sci., 33,* 817 (1980).
97. Szekely, J., and Evans, J. W., *Chem. Eng. Sci., 25,* 1091 (1970).
98. Chu, C., *Chem. Eng. Sci., 27,* 367 (1972).
99. Ranade, P. V., and Harrison, B. P., *Chem. Eng. Sci., 34,* 427 (1979).
100. Gibson, J. B., and Harrison, D. P., *Ind. Eng. Chem. Proc. Des. Dev., 19,* 231 (1980).
101. Aris, R., *Ind. Eng. Chem. Fundam., 6,* 315 (1967).
102. Deb Roy, T., and Abraham, K. P., *Met. Trans., 5,* 349 (1974).
103. Fan, L. S., Miyanami, K., and Fan, L. T., *Chem. Eng. J., 13,* 13 (1977).
104. Shen, J., and Smith, J. M., *Ind. Eng. Chem. Fundam., 4,* 9293 (1963).
105. Wenn, C. Y., and Wong, S. C., *Ind. Eng. Chem., 62* (8), 30 (1970).

CHAPTER 27

SELECTIVITY IN COMPLEX REACTIONS IN FLUIDIZED-BED REACTORS

V. S. Patwardhan and B. D. Kulkarni

Chemical Engineering Division
National Chemical Laboratory
Poona, India

CONTENTS

INTRODUCTION

Several models have been presented in the literature to account for the main features of a catalytic fluidized bed. These include the earlier homogeneous models in which the role of mixing is accounted for by a longitudinal dispersion coefficient, two region models, and bubbling-bed models. The last appear to be the most rigorous in that they explicitly recognize the role of the bubble in the analysis of the fluidized bed. Among the bubbling-bed models are those of Davidson and Harrison [1], and Kunii and Levenspiel [2, 3]. These have been modified by several workers over the years, e.g. the Torr-Calderbank [4] modification of the Davidson model [5] and the Fryer–Potter [6, 7] extension of the Kunii–Levenspiel model. Other bubbling-bed models which characterize the fluidized bed include the compartment model of Kato and Wen [8], which incorporates the effect of changing bubble size along the height of the fluid bed, and the Mori–Wen [9] modification of this model. The models presented by Partridge and Rowe [10], Toor and Calderbank [4], Kato and Wen [8], Mori and Muchi [11], and Fryer and Potter [6, 7] consider bubble coalescence in the bed. Other models such as those of Orcutt et al. [12], Kunii and Levenspiel [2], and Kobayashi et al. [13] employ the concept of a single average bubble diameter throughout the bed.

More recently, the models that take account of the end effects in the reactor have been proposed. Thus, Zenz [14] and Behie and Kehoe [15], Grace and De Lasa [16], De Lasa and Grace [17] and Erazu et al. [18] have considered the region in the vicinity of the grid while modeling the fluidized-bed reactor. Also the free board region in the fluidized-bed reactor has been taken into account in the models of Yates and Rowe [19] and de Lasa and Grace [20].

Choice of Model

The salient features of some of the models proposed have been outlined in Table 1. All these models are concerned with simple first-order reactions. The various models simulate the fluid-bed reactor performance at different levels of complexity. Whereas a more complex model serves to present a more realistic picture of the fluidized bed, it has been shown by Mori and Wen [9] that for slow reactions, when the mass exchange from the bubble to the emulsion phase is not controlling, the two-phase concept of a fluid bed could be effectively replaced by a simple homogeneous model incorporating a single phase and even a simple homogeneous model for describing fluid-bed behavior could be termed adequate.

The experimental results of Chavarie and Grace [21] on the ozone reaction (known to conform to the first order reaction scheme $A \rightarrow R$), in which concentration profiles obtained in the bubble, cloud and emulsion phase were compared with those predicted by the various models, clearly showed that the Kunii–Levenspiel (KL) model best characterized the performance of a fluidized-bed reactor. In view of the preceding and the fact that the KL model incorporates the main features of the fluidized bed while retaining an inherent simplicity, the present development uses this model in a variety of complex first-order reaction schemes occurring in a fluid bed and presents a unified development for both conversion and product distribution thereof.

Complex Reactions in Fluidized Beds

The models presented in the literature have mainly been concerned with simple first-order reactions and only a few extensions to complex reaction schemes have been presented. These have been mentioned in the subsequent section. However, simple reactions are the exception rather than the rule in fluidized-bed operation, as can be seen from the examples of industrial fluid-bed reactions, mentioned below:

As an illustration of reversible complex reactions occurring in fluidized-bed operation the isomerization reaction in fluid catalytic reforming is complex and reversible in nature. Fluid coking and fluid catalytic cracking may also feature reversible reactions in addition to being reactions in which a change in number of moles and hence an attendant volume change occur.

The production of phthalic anhydride from naphthalene conforms to a complex reaction scheme, and fluid-bed plants having a convertible feed (naphthalene or o-xylene) are in operation [22]. Other industrial fluid-bed reactions which include the effect of complex reactions or change in number of moles during reaction pertain to the production of acrylonitrile [23], ethylene by ethanol dehydration [24, 25], chloromethanes [26], chlorosilanes [27], high-density polyethylene [28], isophthalonitrile, per- and trichloroethylene, and chlorofluoromethanes.

In a majority of these reactions the intermediate is the desired product and thus selectivity of intermediate formation is an important consideration. Fluidized operation is beset with the problem of lower selectivity of intermediate formation for complex first-order reactions as compared to a fixed-bed reactor, for the same level of conversion. In the next section, we shall consider the selectivity behavior for some of the complex reactions.

Table 1
Salient Features of Some Bubbling-Bed Models Proposed in the Literature

Model	Single Effective/ Variable Bubble Diameter	Mode of Estimation of Bubble Diameter	Bubble/Cloud Phase	Emulsion Phase	Mass Transfer from Bubble to Cloud/ Emulsion	Mass Transfer from Cloud to Emulsion	Remarks
Orcutt et al. [12]	Single effective diameter	Fitted	No cloud; bubble completely mixed	Plug flow or completely mixed	To emulsion	–	Simple model requiring relatively few computations. Reaction occurs in the emulsion phase and both bubble- and emulsion-phase contributions are considered.
Partridge and Rowe [10]	Sectional average size	X-ray data on bubble size distribution	Completely mixed	Plug flow	On the basis of a rigid sphere model	–	X-ray data required for computations. Reaction occurs in cloud and emulsion. Bubble-phase gas contribution is also incorporated.
Toor and Calderbank [4]	Variable bubble diameter	Empirical equation employed for bubble size increase	Bubble/cloud completely mixed	Plug flow; completely mixed or with back mixing coefficient	To emulsion	–	Reaction occurs in the emulsion phase. Contributions from both bubble- and emulsion-phases are considered.
Kunii and Levenspiel [2, 3]	Single effective diameter	Fitted	Bubble in plug flow	Flow reversal in emulsion phase; plug flow	To cloud	To emulsion	Reaction occurs in bubble/cloud and emulsion phases. Representative of back-mixing models. The bubble-phase gas contribution alone is considered.
Kato and Wen [8]	Variable bubble diameter. Bed is divided into compartments	Empirical equation. The size of each compartment corresponds to the bubble diameter at that level	Bubble/cloud phase completely mixed	Completely mixed	From bubble/cloud to emulsion	–	Elaborate computational procedure with no adjustable parameters. Contribution from bubble-phase gas alone is considered.
Fryer and Potter [6]	Variable bubble size	Empirical equation employed	No mixing in bubble and cloud/wake phase	Plug flow	To cloud/wake	To emulsion. Flow reversal in this phase	Reaction occurs in bubble, cloud/wake and emulsion phases. The bubble- and cloud/wake-phase gas contributions are considered.
Behie and Kehoe [15]	Jetting region followed by single effective diameter	Fitted	No cloud; bubble in plug flow	Completely mixed	To emulsion	–	For fast reactions conversion takes place primarily in the jetting region as the mass transfer coefficient in the grid region is higher than for the rest of the bed. Experimental measurements of the mass transfer coefficient in the grid region for various grid designs and gas mixtures are required.
Miyauchi [36]	Single effective diameter	Fitted	Bubble in plug flow	Plug flow	To emulsion	–	Incorporation of a dilute phase above the bubbling region of the bed. Adsorption of gas on the catalyst is also considered.
Mori and Wen [9]	Jetting region followed by variable bubble diameter	Empirical equation presented by Mori and Wen [8]. Compartmental model as per Kato and Wen [7].	Bubble/cloud phase completely mixed	Completely mixed	To emulsion	–	No adjustable parameters. Model on basis of Kato and Wen model but incorporation of jetting region improves prediction of conversion for fast reactions.

COMPLEX REACTIONS IN FLUIDIZED-BED REACTORS

In this section, we consider complex reactions carried out in fluidized-bed reactors. Broadly speaking, reactants are introduced into the fluidized bed through the grid plate, and most of the reactants enter the bubble phase. Most of the chemical reaction takes place in the emulsion phase. The mass transfer resistances between these two phases thus affect the overall selectivities obtained at the reactor exit. The complexity in a chemical reaction scheme can arise due to various factors, i.e., the generation of several products, complex kinetic expressions, change in the number of moles, presence of inerts, etc. Before dealing with some of these cases, it is instructive to consider the relatively simple case of a single irreversible first-order reaction. Reactions near the grid and in the free-board region are neglected. The fluidized bed itself is represented using the Kunii–Levenspiel model as discussed earlier.

Single Irreversible Reaction

Consider the reaction

$$A \xrightarrow{k_1} B \tag{1}$$

with a first-order kinetics which is catalytic and takes place in a fluidized bed of catalyst particles. The simple KL model assumes bubbles of one size which are fast enough to give $U_b \gg U_{mf}$. The main features of the model are shown in Figure 1, which also shows the five resistance steps for the reactant gas to contact and react on the solid surface. The conservation equation for the reactant species A can be written as [2, 3].

$$-U_b \frac{dC_{Ab}}{dl} = \gamma_b k_1 C_{Ab} + K_{bc}(C_{Ab} - C_{Ac}) \tag{2}$$

$$K_{bc}(C_{Ab} - C_{Ac}) = \gamma_c k_1 C_{Ac} + K_{ce}(C_{Ac} - C_{Ae}) \tag{3}$$

$$K_{ce}(C_{Ac} - C_{Ae}) = \gamma_e k_1 C_{Ae} \tag{4}$$

Eliminating C_{Ac} and C_{Ae} by using Equations 3 and 4, one can rewrite Equation 2 as

$$dC_{Ab}/d\tau = -K_1 C_{Ab} \tag{5}$$

where

$$K_1 = k_1 e_1 \tag{6}$$

and

$$e_1 = \left\{\gamma_b + \left[\frac{k_1}{K_{bc}} + \frac{1}{\gamma_c + (k_1/K_{ce} + 1/\gamma_e)^{-1}}\right]^{-1}\right\} \frac{U_0}{(1-\varepsilon_{mf})U_{br}} \tag{7}$$

Equation 5 for a fluidized-bed reactor is similar to that for a fixed-bed reactor with the difference that k_1 gets replaced by K_1. The modified rate constant, K_1, is equal to $k_1 e_1$. The efficiency of contact, e_1, which is given by Equation 7 is a strong function of the ratio of the intrinsic rate constant k_1 and the transport coefficients. The variation of e_1 with k_1 is shown in Figure 2 for specific values of other parameters. The reaction, being catalytic, is

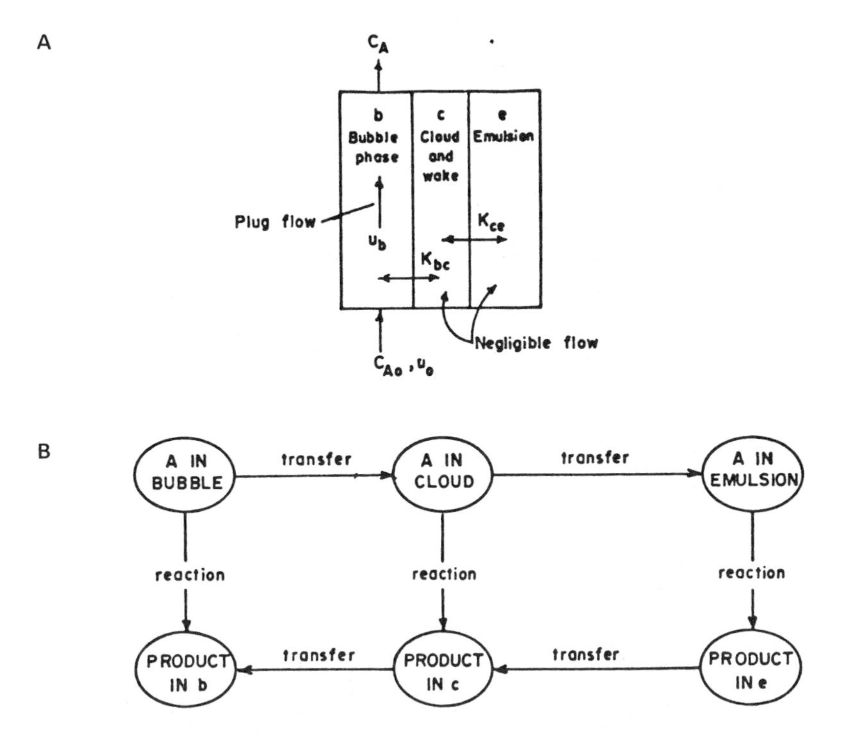

Figure 1. Main features of the simple Kunii-Levenspiel for a bubble bed ($U_b >> U_{mf}$): (A) The model used for chemical conversion calculation; (B) the five mass transfer and reaction steps for the reaction A → R.

mainly confined to the emulsion and the cloud phases. This, in general, reduces the concentration of A in these phases to a value below that in the bubble phase. This reduction is negligible when k_1 is small enough, and a high efficiency of contact is obtained as shown in Figure 2. When k_1 is high, the concentration of A is virtually reduced to zero, and e_1 attains a low value. At intermediate k_1 values, e_1 shows a strong dependence on k_1.

It is seen from the discussion just presented that the hydrodynamics of a fluidized-bed reactor can drastically affect the apparent kinetics, making the modified rate constant less than the intrinsic value. We now consider a complex reaction scheme, where the effect on the kinetics is more complex, and can affect the selectivities in a very significant manner.

Complex Reaction Scheme with Irreversible Steps

Let us consider the reaction scheme

$$\begin{array}{ccccc} A & \xrightarrow{k_1} & R & \xrightarrow{k_3} & S \\ \downarrow k_2 & & \downarrow k_4 & & \\ T & & U & & \end{array} \tag{8}$$

This reaction scheme is commonly known as the Denbigh scheme [29] and leads to several special cases of practical interest when one or more of the rate constants are put equal to

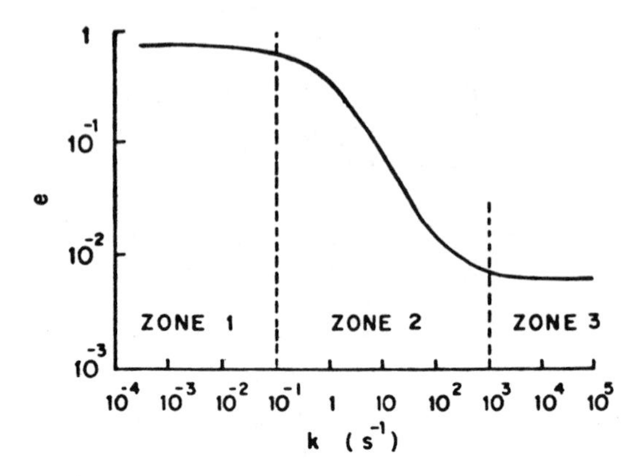

Figure 2. Contacting efficiency as a function of K.

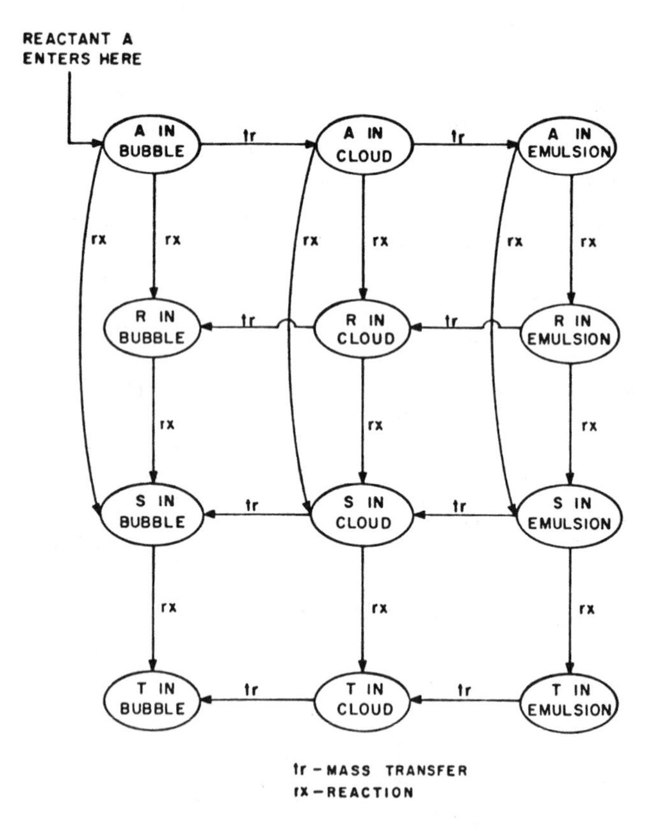

Figure 3. Sketch showing the 20 reaction and mass transfer steps representing the reaction A → R → S → T taking place in a fluidized bed.

zero. Here we are interested not only in the overall disappearance of reactant A, but also in seeing how the concentration of other components changes. The various mass transfer and reaction steps are schematically shown in Figure 3 [30].

For any differential slice of bed of height dl the material balance equation for any specie can be written as

(overall disappearance) = (reaction in bubble)
+ (transfer to cloud-wake)

(transfer to cloud-wake) = (reaction in cloud-wake)
+ (transfer to emulsion)

(transfer to emulsion) = (reaction in emulsion)

Since the diffusion coefficients of different components rarely differ greatly and only influence the K values at most by the one-half power, we often can reasonably take the K_{bc} and K_{ce} values for all components to be identical. However, to keep the derivation general, we here retain the distinction between these different interchange coefficients.

Next, calling $k_1 + k_2 = k_{12}$ and $k_3 + k_4 = k_{34}$ we then have for species A:

- In bubbles

$$-r_{Ab} = \gamma_b k_{12} C_{Ab} + K_{bc}^{A}(C_{Ab} - C_{Ac}) \tag{9}$$

- In cloud

$$K_{bc}^{A}(C_{Ab} - C_{Ac}) = \gamma_c k_{12} C_{Ac} + K_{ce}^{A}(C_{Ac} - C_{Ae}) \tag{10}$$

- In emulsion

$$K_{ce}^{A}(C_{Ac} - C_{Ae}) = \gamma_e k_{12} C_{Ae} \tag{11}$$

For species R we have:

- In bubbles

$$-r_{Rb} = -\gamma_b k_1 C_{Ab} + \gamma_b k_{34} C_{Rb} + K_{bc}^{R}(C_{Rb} - C_{Rc}) \tag{12}$$

- In clouds

$$K_{bc}^{R}(C_{Rb} - C_{Rc}) = \gamma_c k_1 C_{Ac} + \gamma_c k_{34} C_{Rc} + K_{ce}^{R}(C_{Rc} - C_{Re}) \tag{13}$$

- In emulsion

$$K_{ce}^{R}(C_{Rc} - C_{Re}) = -\gamma_e k_1 C_{Ae} + \gamma_e k_{34} C_{Re} \tag{14}$$

Similar expressions can be written for all the other reacting species.

Taking a differential slice of the bed and manipulating and substituting repeatedly to eliminate all cloud and emulsion concentrations leads us eventually to the following expressions, all in terms of bubble concentrations.

$$\frac{dC_{Ab}}{d\tau} = -K_{12} C_{AB} \tag{15}$$

$$\frac{dC_{Rb}}{d\tau} = -K_{34}C_{Rb} + K_{AR}C_{Ab} \tag{16}$$

$$\frac{dC_{Sb}}{d\tau} = \frac{k_3}{k_{34}}(K_{34}C_{Rb} + K_A C_{Ab}) \tag{17}$$

$$\frac{dC_{Tb}}{d\tau} = \frac{k_2}{k_{12}}(K_{12}C_{Ab}) \tag{18}$$

$$\frac{dC_{Ub}}{d\tau} = \frac{k_4}{k_{34}}(K_{34}C_{Rb} + K_A C_{Ab}) \tag{19}$$

where

$$K_{12} = \left[\gamma_b k_{12} + \cfrac{1}{\cfrac{1}{K_{bc}^A} + \cfrac{1}{\gamma_c k_{12} + \cfrac{1}{\cfrac{1}{K_{ce}^A} + \cfrac{1}{\gamma_e k_{12}}}}}\right]\frac{U_0}{(1-\varepsilon_{mf})U_{br}} \tag{20}$$

$$K_{34} = \left[\gamma_b k_{34} + \cfrac{1}{\cfrac{1}{K_{bc}^R} + \cfrac{1}{\gamma_c k_{34} + \cfrac{1}{\cfrac{1}{K_{ce}^R} + \cfrac{1}{\gamma_e k_{34}}}}}\right]\frac{U_0}{(1-\varepsilon_{mf})U_{br}} \tag{21}$$

$$K_A = \frac{\left[\frac{K_{bc}^R K_{ce}^A}{\gamma_c^2} + \left(k_{12} + \frac{K_{ce}^A}{\gamma_c} + \frac{K_{ce}^A}{\gamma_e}\right)\left(k_{34} + \frac{K_{ce}^R}{\gamma_c} + \frac{K_{ce}^R}{\gamma_e}\right)\right]\frac{K_{bc}^A U_0 k_1 k_{34}}{(1-\varepsilon_{mf})U_{br}}}{\left[\left(k_{12} + \frac{K_{bc}^A}{\gamma_c}\right)\left(k_{12} + \frac{K_{ce}^A}{\gamma_e}\right) + \frac{k_{12}K_{ce}^A}{\gamma_c}\right]\left[\left(k_{34} + \frac{K_{bc}^R}{\gamma_c}\right)\left(k_{34} + \frac{K_{ce}^R}{\gamma_e}\right) + \frac{k_{34}K_{ce}^R}{\gamma_c}\right]} \tag{22}$$

$$K_{AR} = \frac{k_1}{k_{12}}K_{12} - K_A \tag{23}$$

By Laplace transformation of Equations 15 to 19 starting with a feed C_{A0}, C_{R0}, C_{S0}, C_{T0}, and C_{U0} we find at the exit of the reactor

$$\frac{C_A}{C_{A0}} = \exp(-K_{12}\tau) \tag{24}$$

$$\frac{C_R}{C_{A0}} = \frac{K_{AR}}{K_{34} - K_{12}}[\exp(-K_{12}\tau) - \exp(-K_{34}\tau)] + \frac{C_{R0}}{C_{A0}}\exp(-K_{34}\tau) \tag{25}$$

$$\frac{C_S}{C_{A0}} = \frac{k_3}{k_{34}}\left\{\frac{K_{34}K_{AR}}{K_{34} - K_{12}}\left[\frac{\exp(-K_{34}\tau)}{K_{34}} - \frac{\exp(-K_{12}\tau)}{K_{12}}\right] + \frac{k_1}{k_{12}} - \frac{K_A}{K_{12}}\exp(-K_{12}\tau) + \frac{C_{R0}}{C_{A0}}[1 - \exp(-K_{34}\tau)]\right\} + \frac{C_{S0}}{C_{A0}} \tag{26}$$

$$\frac{C_T}{C_{A0}} = \frac{k_2}{k_{12}}[1 - \exp(-K_{12}\tau)] + \frac{C_{T0}}{C_{A0}} \tag{27}$$

$$\frac{C_U}{C_{A0}} = \text{same as } \frac{C_S}{C_{A0}} \text{ but with } k_3 \rightarrow k_4 \text{ and } C_{S0} \rightarrow C_{U0} \tag{28}$$

The value of C_{Rmax} and the amount of catalyst needed to achieve this when R is absent in the feed, or $C_{R0} = 0$, is given by

$$\frac{C_{Rmax}}{C_{A0}} = \frac{K_{AR}}{K_{12}}\left(\frac{K_{12}}{K_{34}}\right)^{K_{34}/(K_{34}-K_{12})} \tag{29}$$

and

$$\tau(\text{at } C_{Rmax}) = \frac{W}{v_0} = \frac{\ln(K_{34}/K_{12})}{K_{34} - K_{12}} \tag{30}$$

In the fluidized-bed equations if we put all $K_{bc} = K_{ce} = \infty$ we then find that $K_{12} \rightarrow k_{12}$, $K_{34} \rightarrow k_{34}$, $K_A \rightarrow 0$, $K_{AR} \rightarrow k_1$, and all the fluidized reactor expressions, Equations 24 to 30 reduce to the corresponding fixed bed (plug flow) expressions.

By putting numbers in to these expressions we come up with three important but not unexpected generalizations.

First, with no pore diffusion effects, the fluidized bed always needs more catalyst than the fixed bed to achieve a given conversion or to reach C_{Rmax}.

Second, for reactions in series and with no pore diffusion effects, fluidized beds always give a lower yield of intermediates compared to fixed-bed reactors.

Third, for reactions in parallel, fluidized-bed operations do not affect the product distribution.

In most situations we would not be justified in retaining the distinction between the various interchange coefficients. Thus we would drop the superscripts and use an average K_{bc} and K_{ce} throughout.

Example 1

In the presence of a particular catalyst ($\varrho_s = 2{,}000\ kg/m^3$) reactant A decomposes by first-order reactions as follows

$$A \overset{k_1}{\rightarrow} R \overset{k_3}{\rightarrow} S$$

where $k_1 = 0.8\ s^{-1}$ and $k_3 = 0.05\ s^{-1}$. Using a catalyst bed 2 m in diameter with a feed of A in inert ($U_0 = 0.30$ m/s) we find C_{Rmax} and the amount of catalyst needed to achieve C_{Rmax}

- For fluidized bed operations ($U_{mf} = 0.03$ m/s, $\varepsilon_{mf} = 0.5$) if the estimated effective bubble size is $d_b = 0.32$ m, a rather large value
- For fixed bed operations (assume plug flow of gas)

Additional data: From the literature take $D = 20 \times 10^{-6}\ m^2/s$ for both A and R and $\alpha = 0.33$.

Solution

Let us work everything in SI units.

Fluidized bed. / In essence the KL model says that

$$\begin{matrix} U_{mf} \\ \varepsilon_{mf} + d_b \\ U_0 \end{matrix} \quad \text{gives} \quad \begin{pmatrix} \text{behavior of} \\ \text{the bed} \end{pmatrix}$$

Going through the details of the models we find

$$U_{br} = 0.711(gd_b)^{1/2} = 0.711(9.8 \times 0.32)^{1/2} = 1.26 \text{ m/s}$$

$$U_b = U_0 - U_{mf} + U_{br} = 0.30 - 0.03 + 1.26 = 1.53 \text{ m/s}$$

$$\delta = \frac{U_0 - U_{mf}}{U_b} = \frac{0.30 - 0.03}{1.53} = 0.177$$

$$K_{bc} = 4.50 \frac{U_{mf}}{d_b} + 5.85 \frac{D^{1/2} g^{1/4}}{d_b^{5/4}} = 0.614 \text{ s}^{-1}$$

$$K_{ce} = 6.78 \left(\frac{\varepsilon_{mf} D U_b}{d_b^3} \right)^{1/2} = 0.147 \text{ s}^{-1}$$

$$\gamma_b = 0.001 \sim 0.01$$

$$\gamma_c = (1 - \varepsilon_{mf}) \left(\frac{3U_{mf}/\varepsilon_{mf}}{U_{br} - U_{mf}/\varepsilon_{mf}} + \alpha \right) = 0.24$$

$$\gamma_e = \frac{(1 - \varepsilon_{mf})(1 - \delta)}{\delta} - \gamma_c - \gamma_b = 2.09$$

Next, calculate the needed K values. Since $k_2 = k_4 = 0$, we have $K_{12} \rightarrow K_1$ (put $k_{12} \rightarrow k_1$) and $K_{34} \rightarrow K_3$ (put $k_{34} \rightarrow k_3$). Thus from Equation 20

$$K_{12} = K_1 = \left[0.003(0.8) + \frac{1}{\dfrac{1}{0.614} + \dfrac{1}{0.24(0.8) + \dfrac{1}{\dfrac{1}{0.147} + \dfrac{1}{2.09(0.8)}}}} \right]$$

$$\cdot \frac{(0.3)}{(1. - 0.5)(1.26)} = 0.103 \text{ s}^{-1}$$

Similarly, from Equation 21

$$K_{34} = K_3 = 0.031 \text{ s}^{-1}$$

From Equation 22

$$K_A = 0.026 \text{ s}^{-1}$$

Inserting into Equation 23 gives

$$K_{AR} = K_1 - K_A = 0.0764\ s^{-1}$$

Thus, Equations 29 and 30 give

$$\frac{C_{Rmax}}{C_{A0}} = \frac{K_{AR}}{K_1}\left(\frac{K_1}{K_3}\right)^{K_3/(K_3-K_1)} = 0.44$$

and

$$W = \frac{v_0 \ln (K_3/K_1)}{K_3 - K_1} = 31{,}400\ kg$$

Fixed bed. / Here Equations 29 and 30 reduce to

$$W_{fixed} = \frac{v_0 \ln (k_1/k_3)}{k_1 - k_3} = 7{,}000\ kg$$

$$\frac{C_{Rmax}}{C_{A0}} = \left(\frac{k_1}{k_3}\right)^{k_3/(k_3-k_1)} = 0.83$$

Note: The formation of intermediates is drastically lowered in fluidized-bed operations (44% vs. 83%) even though the bed is much larger (31 tons vs. 7 tons). This all comes from the severe bypassing of the reactant in the large bubbles. Reducing the size of bubbles will greatly improve the performance of the fluidized reactor.

The Denbigh reaction scheme simplifies to the well-known series- and parallel-reaction schemes when appropriate rate constants are put equal to zero.

Complex Reaction Scheme Involving Reversible Steps

Let us consider the following reaction scheme which involves successive reversible reactions

$$A \underset{k_2}{\overset{k_1}{\rightleftharpoons}} R \underset{k_4}{\overset{k_3}{\rightleftharpoons}} S \tag{31}$$

Material balance equations incorporating reaction terms and transport terms for transport between bubble, cloud, and emulsion phases can be written as illustrated earlier for the Denbigh reaction scheme. This has been done by Irani et al. [31] to arrive at the final equation giving concentrations of all species as functions of reactor length. They have also derived equations for the maximum conversion to R and the reaction length required for the same. The equations are lengthy and therefore are not reproduced here.

Example

The following illustrative example gives some idea about the selectivities that can be achieved. The parameters values are: $k_1 = 10.0\ s^{-1}$; $k_2 = 1.0\ s^{-1}$; $k_3 = 1.0\ s^{-1}$; $k_4 = 0.1\ s^{-1}$; $\varrho_s = 2.0\ g/cm^3$; $W/v_0 = 12.0\ g\ \ s/cm^3$; $U_{mf} = 3\ cm/s$; $U_0 = 30\ cm/s$; $\varepsilon_{mf} = 0.4$; $D = 0.2\ cm^2/s$; $v_w/v_b = 0.3$; $d_b = 8$–$30\ cm$. The results of the calculations are presented in Figures 4 through 6.

Figure 4 shows the concentration profiles with varying bubble diameter. As a general observation it can be said that the conversion of A drops with increase in bubble diameter, in analogy with the physical situation. Also, the fractional height of the fluidized bed at which the maximum concentration of R occurs, shifts in the reactor to the right with an increase in bubble diameter. It is noteworthy that in the case of a fluid-bed reactor the extent of the maximum as well as its very presence can be controlled simply by changing the bubble diameter.

Figure 5 shows the variation of $\bar{C}_{Rmax}$ with bubble diameter in the bed. The plug flow reactor [$d_b = 0$] represents the ideal case and corresponds to the highest value of $\bar{C}_{Rmax}$ that is attainable. As can be seen from this figure, $\bar{C}_{Rmax}$ decreases with an increase in bubble diameter. This fact is consistent with the physical situation wherein a larger bubble diameter corresponds to a greater deviation from plug-flow conditions.

The fluid-bed performance is compared with that of a plug-flow reactor (PFR) for $k_1/k_3 = 10$ in Figure 3, which shows the variation of selectivity with conversion for different diameters of bubble. For the same extent of conversion higher selectivities can be realized at lower bubble diameters in the case of a fluid-bed reactor. From the nature of these curves it is apparent that operation at higher conversions in a fluid bed is feasible without serious loss of selectivity for the case when $k_1/k_3 > 1$, except in the vicinity of the maximum conversion possible. In this sense the analysis of the fluid-bed reactor based on the Kunii-Levenspiel model is similar to that of a PFR. It has been found (figure not shown) that for $k_1/k_3 < 1$ operation at lower conversion in a fluid bed is preferable as higher conversions would entail a significant reduction in selectivity, again in analogy with a PFR.

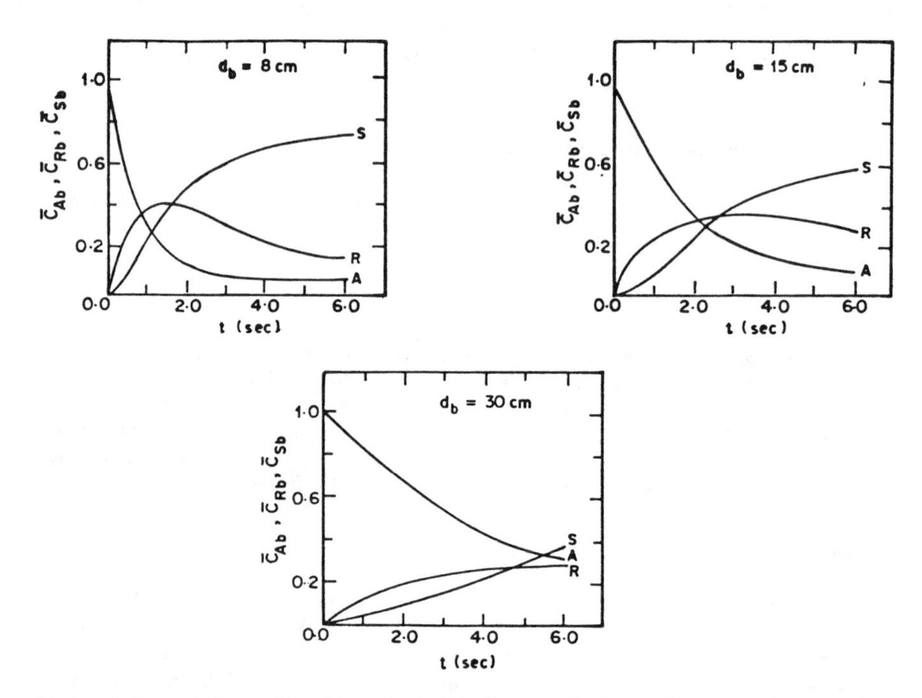

Figure 4. Concentration profiles with varying bubble diameters for the reaction scheme A ⇌ R ⇌ S.

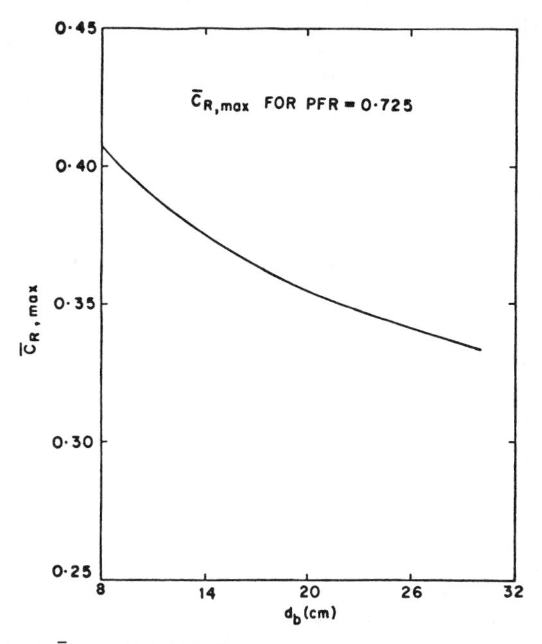

Figure 5. Variation in $\bar{C}_{R,max}$ with bubble diameter for the reaction scheme $A \rightleftharpoons R \rightleftharpoons S$.

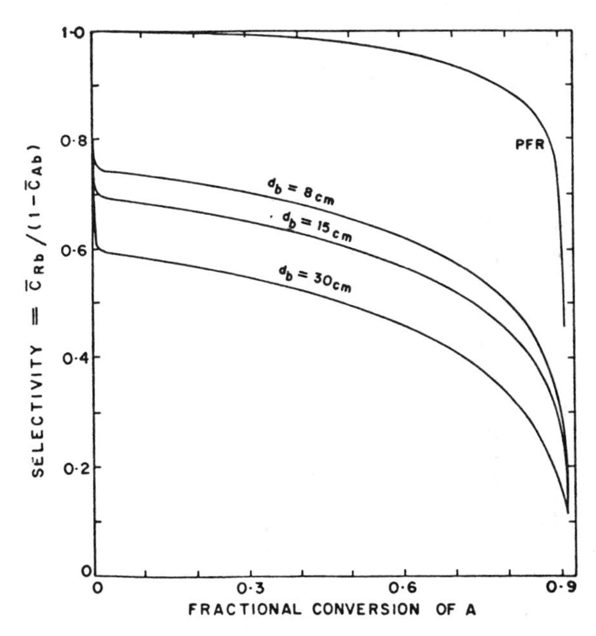

Figure 6. Selectivity-conversion plots with varying bubble diameters for the reaction scheme $A \rightleftharpoons R \rightleftharpoons S$.

Reverse Parallel Reaction Scheme

The reaction scheme incorporating reversible parallel reactions is obtained by rewriting the earlier scheme as

$$\begin{matrix} \rightleftharpoons A \\ \rightleftharpoons S \end{matrix} \tag{32}$$

as indicated by Irani et al. [32]. The performance equations for this scheme can be obtained by minor obvious modifications of the reversible series reaction scheme.

Reaction Scheme Considered by Irani et al.

The following reaction scheme

$$A \xrightarrow{k_1} R \xrightarrow{k_2} S \xrightarrow{k_4} T \quad (A \xrightarrow{k_3} S) \tag{33}$$

has been considered by Irani et al. [32] who have presented the final performance equations for a reactor. They are not presented here, as they are quite lengthy. Using their expressions, selectivities to any of the products can be calculated.

Reactions with Differing Orders

First/Second-Order Parallel Reactions

A simple parallel reaction scheme of the type $A \begin{matrix} \nearrow R \\ \searrow S \end{matrix}$ (34)

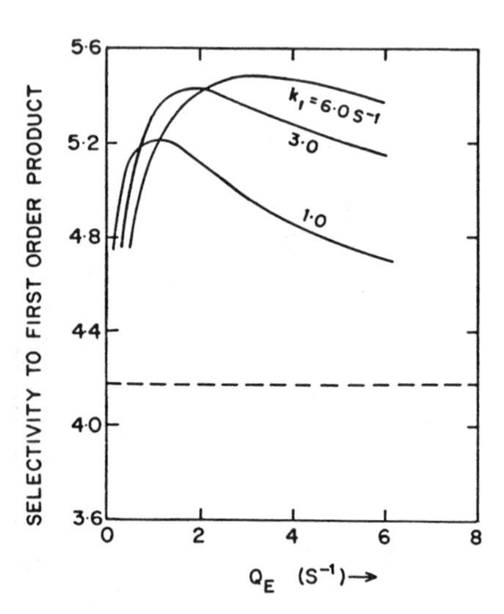

Figure 7. Effect of reaction rate constant on selectivity.

has been considered by Nashaie and Yates [33] where the reaction A → R is first order while the second reaction A → S is second order. In a fluidized-bed reactor where catalytic reactions are conducted, most of the reaction takes place in the emulsion phase. The reactants have to transfer from the bubble phase to the emulsion phase before reaction can take place. Thus, the emulsion phase can be considered as a crossflow reactor where the reactant is added gradually rather than all at once. This results in a lower concentration of A in the emulsion phase thereby improving the selectivity of R. In a study of the isomerization of n-butenal over a fluidizied silica-alumina catalyst [34], which corresponds to this reaction scheme, it was shows that the selectivity to trans-butene—2 is greater than in a fixed-bed reactor under the same conditions.

The mass balance equations incorporating reaction and interphase transport terms cannot be analytically solved in a simple manner in this case, as one of reactions is of second order. Nashaie and Yates [33] therefore resorted to numerical techniques in investigating this reaction scheme theoretically. The performance of the fluidized-bed reactor depends upon factors like the rate constant, bubble velocity, transport coefficients, etc. An interesting feature of such reactions is that the selectivity to R goes through a maximum as the transport coefficient increases. At a very low transport coefficient a given reaction gives very low selectivity depending upon the extent of reaction in the bubble phase. At very high values of transport coefficients the selectivity approaches that for a plug flow reactor. At intermediate values of the transport coefficient, the emulsion phase acts as a reactor with axially distributed feed and gives a higher selectivity. This is illustrated in Figure 7.

Zero/First-Order Consecutive Reactions

The consecutive reaction scheme

$$A \xrightarrow[\text{order}]{\text{zero}} R \xrightarrow[\text{order}]{\text{first}} S \tag{35}$$

has been considered by Irani et al. [35]. Since A reacts with a zero-order kinetics, the concentration of A can fall to zero in a finite length of reaction (unlike the case of a first-order reaction). So the fluidized bed can in general be considered as consisting of two zones. In the first zone both reactions occur simultaneously and the concentration of A drops to zero at the end of this zone. In the second zone only the reaction R → S takes place. Irani et al. [35] presented equations for the reactor performance based on KL model, Miyauchi [36] model and the Fryer–Potter model. The effect of different variables was investigated by the following illustrative example.

Example 2

Consider the following data: $k_1' = 10^{-5}$ g mol/cm^3 sec (zero order); $k_1 = 10.0\ \text{sec}^{-1}$ (first order); $k_2 = 1.0\ \text{sec}^{-1}$; $\varrho_s = 2.0$ g/cm^3; $U_{mf} = 3$ cm/sec; $U_0 = 30$ cm/sec; $\varepsilon_{mf} = 0.4$; $D_c = D_A = 0.2$ cm^2/sec; $W/v_0 = 8.0$ g sec/cm^3; $V_w/V_b = 0.3$; $C_{A0} = 10^{-5}$ g mol/cm^3; $d_b = 8$–30 cm.

The results of the calculations are presented in Figures 8 and 9. Figure 8 shows the concentration profiles based on the KL model for reactant A and product R with varying bubble diameters. As $C_{Sb} = (1.0 - C_{Ab} - C_{Rb})$, the concentration profile for s can be easily obtained. (The corresponding profiles for the Miyauchi and Fryer–Potter models also have been reported by the authors and are in qualitative agreement with Figure 8.)

It is seen from Figure 8 that the concentration profile for A remains unchanged with bubble diameter and hence τ_{cr} (or l_c) is independent of bubble diameter and has a constant value. The value of residence time at which C_{Rmax} occurs corresponds to τ_{cr}, as beyond this

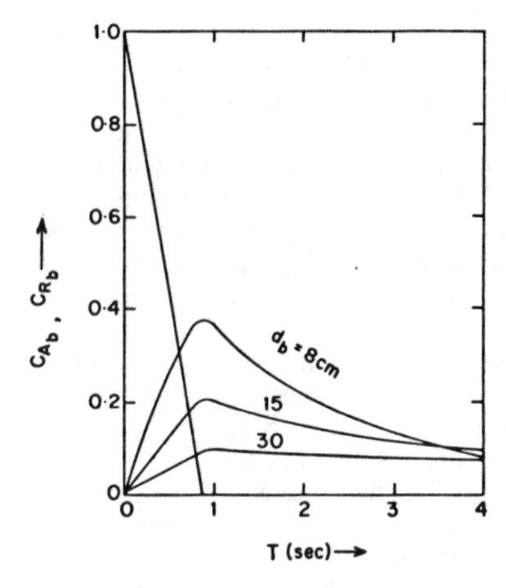

Figure 8. Concentration profiles for the zero/first-order reaction.

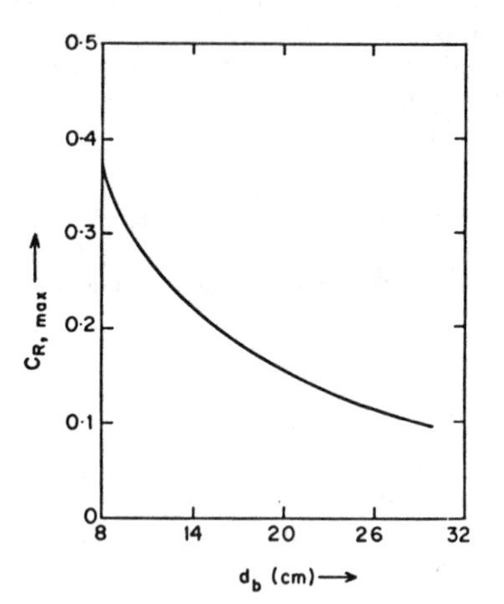

Figure 9. Variation in $C_{R,max}$ with bubble diameter for the zero/first-order reaction.

residence time in the bed only conversion of R to S by a first-order reaction would occur, leading to depletion in the concentration of R.

However, while C_{Rmax} occurs at τ_{cr} the actual value of C_{Rmax} falls with increase in bubble diameter. This is brought out by the plot in Figure 9 and is quite consistent with the physical situation that a larger bubble diameter corresponds to greater deviation from plug flow conditions. As C_{Rmax} occurs at τ_{cr} irrespective of the average bubble diameter employed in order to optimize the production of R the gas residence time in the bed must correspond to τ_{cr} so that C_{Rmax} occurs at the top of the bed. Thus, the magnitude of C_{Rmax} is solely a function of bubble diameter in this case, and bubble diameter has no effect on the height in the bed at which C_{Rmax} occurs (which is dependent only on kinetic considerations such as the rate constant k_1).

Reactions Accompanied by a Volume Change

Depending on the stoichiometry, any reaction, simple or complex, can be accompanied by a change in volume. Here we consider only the simple reaction

$$A \rightarrow \nu B \tag{36}$$

which can lead to a substantial change in the superficial velocity of the gas because the volume of gas varies as conversion varies. This case has been investigated by Irani et al. [31, 32]. They assumed that the volume change affects only the number of bubbles but not the bubble size. Using the KL model they developed an analytical solution for the reactor performance which accounts for the volume change, the presence of inerts in the feed stream and the axial variation of pressure (due to the hydrostatic head) in the reactor. Their main conclusions are exemplified in the illustration presented in the following: The data for fluid-bed operation is: $k = 10\ s^{-1}$; $U_{mf} = 3$ cm/s; $\rho_s = 2.0\ g/cm^3$; $y_{A0} = 0.98, 0.7, 0.5$; $d_b = 8$ cm; $\varepsilon_{mf} \cong \varepsilon_{packed} = 0.5$; $D_e = 0.2\ cm^2/s$; $\alpha_w = 0.33$; $L_f = 100$ cm; $D_t = 100$ cm; $P_t = 1.0$ atm; reaction temperature = 250° C; $\varepsilon_e = 0.5$; $\nu = 0, 1, 2, 3$; $u_0|_{z=0} = 30$ cm/s. The results of the calculations are presented in Figures 10 through 12.

As can be seen from Figure 10 the conversion for a reaction with a change in the number of moles ($\nu > 1$) is always less than that for a reaction with no volume change. The result is not unexpected as the presence of extra moles of product dilutes the reactant species A in the reactor, there by decreasing the conversion. For $\nu < 1$, however, the conversion is always higher than that for a reaction with no volume change. The point is brought out more clearly in Figure 11, which shows the variation of conversion at the reactor exit against the stoichiometric coefficient for different extents of inerts in the feed stream. As is evident from the figure, there can occur a substantial change in the outlet conversion, the maximum decrease being realized when pure reactant A is used in the feed stream. The presence of inerts in the feed stream lowers the extent of decrease in the conversion. The influence of pressure variation along the height of the reactor is shown in Figure 12. The pressure variations do not seem to influence the conversion significantly.

The Effect of Catalyst Dilution

For gas-solid catalytic reactions of first order, the rate constant for the undiluted catalyst may be expressed as k_1' (sec^{-1}). If some quantity of the catalyst is replaced by an equal

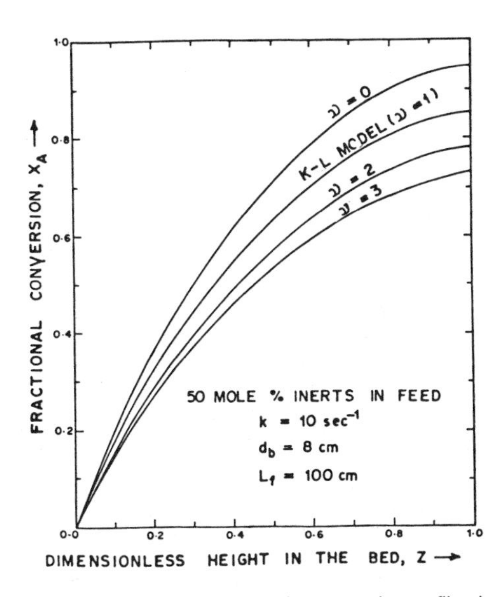

Figure 10. Influence of stoichiometric coefficient on the conversion profile along the bed height.

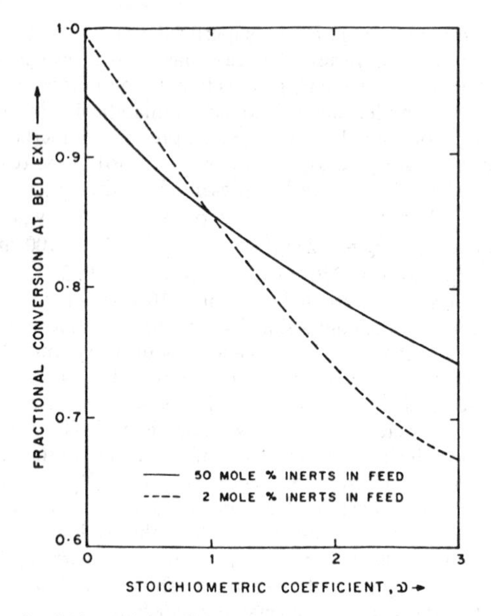

Figure 11. Influence of stoichiometric coefficient and presence of inerts in the feed stream on the exit conversion in the bed.

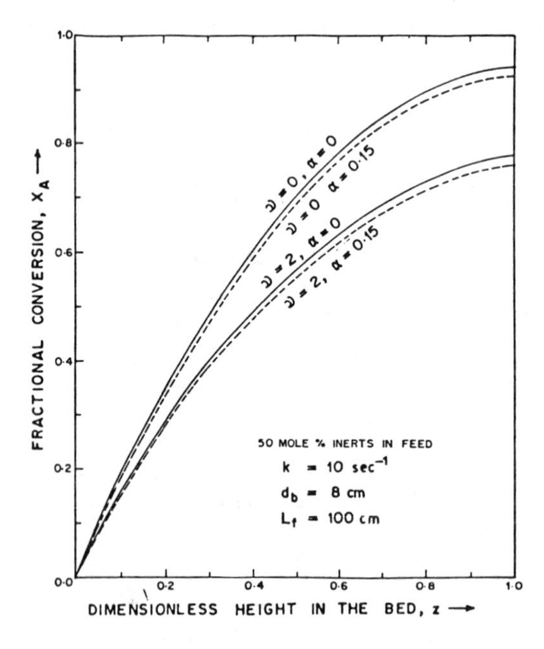

Figure 12. Influence of pressure variation along the bed on the extent of conversion.

quantity of inert solids (of identical size and other properties) and it is assumed to be intimately mixed with the active catalyst remaining, then the overall rate constant for the diluted catalyst can be written as

$$k_1 = k_1'/R' \tag{37}$$

where R' is the catalyst dilution defined as

$$R' = \frac{\text{total weight of solids in the bed}}{\text{weight of active catalyst}}$$

k_1 is also the rate constant that would be realized in a fixed-bed reactor if the same dilution is used. The efficiency of a fluidized-bed reactor, defined by Equation 7 for a simple reaction is dependent on the overall rate constant k_1, which in turn depends upon the dilution. The implications of this as for as selectivity is concerned have been investigated by Irani et al. [37].

For a simple first-order reaction

$$A \xrightarrow{k_1} R$$

the Kunii–Levenspiel model predicts that catalyst dilution will only result in decreasing conversion of A if the same residence time τ is maintained in the bed. However, for complex reactions, the situation is more involved. Consider a consecutive first-order reaction scheme of the type

$$A \xrightarrow{k_1} R \xrightarrow{k_2} S$$

In a plug-flow reactor, the dilution of catalyst does not affect the selectivity or the maximum concentration of intermediate R. This is so because the ratio k_1/k_2 is unaffected by catalyst dilution, as it has the same effect on both the rate constants. Similarly, selectivity or C_{Rmax} will be unaffected by catalyst dilution in a plug-flow reactor for the other reaction schemes mentioned earlier.

However, if we consider the same reaction scheme conducted in a fluid bed, then catalyst dilution does have an effect on the selectivity and C_{Rmax}. Here the ratio K_1/K_2 i.e., k_1e_1/k_2e_2 is important. The ratio k_1/k_2 is unaltered by catalyst dilution, but the corresponding contacting efficiencies e_1 and e_2 are different and hence the ratio of the effective rate constants in the fluid bed is altered with catalyst dilution. This leads to concentration profiles different from that observed in the undiluted catalytic fluid bed.

For establishing the effect of catalyst dilution in a fluid bed in accordance with the theory developed previously, we shall consider the reaction scheme just mentioned and compute the concentration profiles of the various species and also the selectivity and maximum concentration of intermediate R for different conditions. The following data will be used: $k_1 = 10\ s^{-1}$; $k_2 = 1.0\ s^{-1}$; $\varrho_s = 2.0\ g/cm^3$; $W/v_0 = 8.0\ gs/cm^3$; $U_{mf} = 3\ cm/s$; $U_0 = 30\ cm/s$; $\varepsilon_{mf} = 0.4$; $D_e = D_A = 0.2\ cm^2/s$; $V_w/V_b = 0.3$; $d_b = 8,\ 15\ cm$.

Figure 13 shows the variation of the effective rate constant K in a fluid bed for varying values of the rate constant k in a fixed bed for a simple first-order reaction. An interesting conclusion can be drawn from this figure. If the bed is considered to be half diluted with inerts, then the fixed-bed rate constant k is reduced to half its original value, whereas the corresponding effective rate constant K in the fluid bed is reduced to a far lesser extent. In general, then, fluid bed operation is more insensitive (with respect to conversion) to catalyst dilution as compared to fixed bed operations for a simple first-order reaction.

Figure 14 shows the selectivity as a function of conversion for an undiluted fluid bed, diluted fluid bed with different dilution ratios, and the plug-flow reactor, with $k_1/k_2 > 1$.

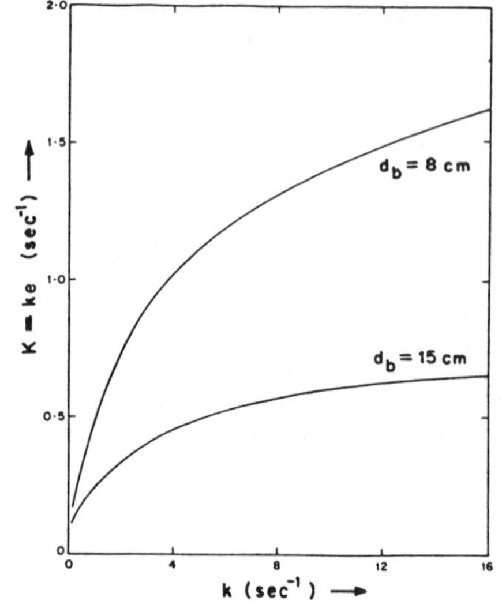

Figure 13. Variation of effective rate constant for a fluid bed with fixed-bed rate constant.

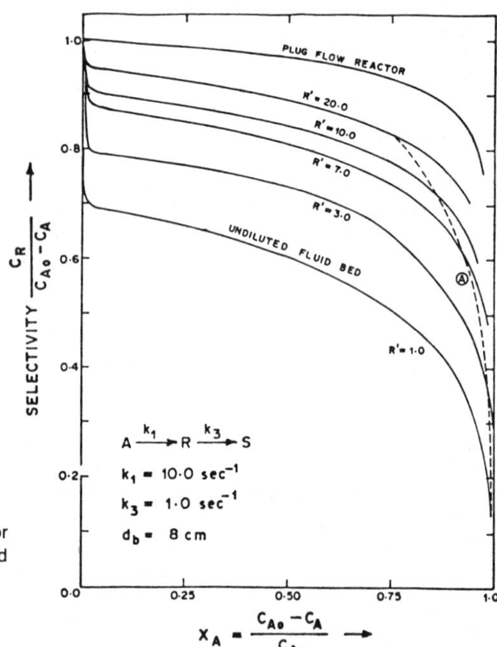

Figure 14. Selectivity-conversion plots for the plug flow; undiluted and diluted fluid-bed reactors.

Obviously, the selectivity is maximum for plug-flow operation and lowest for the undiluted fluid bed. With increasing dilution the selectivity is seen to improve for the same conversion, and for an infinitely diluted bed (corresponding to near-zero conversion) it approaches that for plug flow. Curve A shows the path along which the undiluted fluid-bed operation moves when it is diluted to different extents. For the values of k_1 and k_2 employed, the figure shows that with a dilution ratio of 20 the conversion is reduced from 99% to 75%, whereas the corresponding increase in selectivity of intermediate is from 14% to 83%. Thus, when the intermediate is important, it is preferable to operate at high dilution ratios. Although conversions are slightly decreased, much larger selectivities are realized. It is also preferable to operate such beds at high conversion levels, for at lower conversions the opposite seems to be true; viz. diluting the bed affects the conversion more strongly with relatively little or no improvement in selectivity.

The observations from Figure 14 are particularly useful, for the fixed-bed operation is beset with several operational problems for strongly exothermic reactions. With diluted fluid beds, when $k_1/k_2 > 1$, the reactor can be operated at relatively high conversions (as compared to the undiluted fluid bed) with surprising enhancement in selectivity at any point (determined by economic considerations) along path A straddling the fixed-bed and undiluted fluid-bed operations.

The Effect of Temperature

Let us first consider the simple first-order reaction

$$A \xrightarrow{k_1} R \tag{38}$$

It has been mentioned earlier that the performance equation for a fluidized-bed reactor is the same as that for a fixed-bed reactor except that the intrinsic rate constant is replaced by an effective rate constant as

$$K_1 = k_1 e_1$$

The efficiency of contact, e_1, is given by Equation 7 and depends both upon the k_1 and the transport coefficients. The rate constant depends upon temperature very strongly while the transport coefficients are rather weak functions of temperature. Therefore, the variation of transport coefficients with temperature will not be taken into consideration in the following discussion. Figures 2 and 13 show the dependence of e_1 and K_1 on k_1 (for specific values of other parameters). They also describe, qualitatively, the variation of e_1 and K_1 with temperature, since k_1 increases monotonically with temperature. Figure 2 shows that over a specific temperature range (depending upon the values of other parameters), e_1 changes from one asymptotic value to another. More specifically, e_1, decreases with temperature over this range. Since K_1 is equal to k_1 e_1 it is less sensitive to temperature than k_1 over this temperature range.

Let us consider the first-order consecutive reaction scheme

$$A \xrightarrow{k_1} R \xrightarrow{k_2} S$$

carried out in both a fixed bed and a fluidized bed. The selectivity-conversion plot for a fixed-bed reactor is governed only by the ratio k_2/k_1 which may increase or decrease with temperature depending upon the relative activation energies of the two reactions. However, in the case of a fluidized-bed reactor, the selectivity for a given conversion also depends upon the numerical values of k_1 and k_2 which appear in the expressions for efficiencies. It has been mentioned earlier that for consecutive first-order reactions, the selectivity to R is less than that obtained in a fixed-bed reactor. To illustrate the effect of temperature on selectivities in fixed- and fluidized-bed reactors, we consider an example using the following data: $k_1 = 5.35 \times 10^9 \exp(-9{,}000/T)\ s^{-1}$; $k_2 = 4.61 \times 10^{16} \exp(-15{,}000/T)\ s^{-1}$; $K_{bc} = 3.34\ s^{-1}$; $K_{ce} = 0.6\ s^{-1}$; $U_0/[(1-\varepsilon_{mf})U_{br}] = 0.6$; $\gamma_b = 0.01$; $\gamma_c = 0.42$; $\gamma_e = 0.71$. The equations to be used for the calculation of selectivities are obtained from those given earlier for the Denbigh reaction scheme. The results are shown in Figure 15.

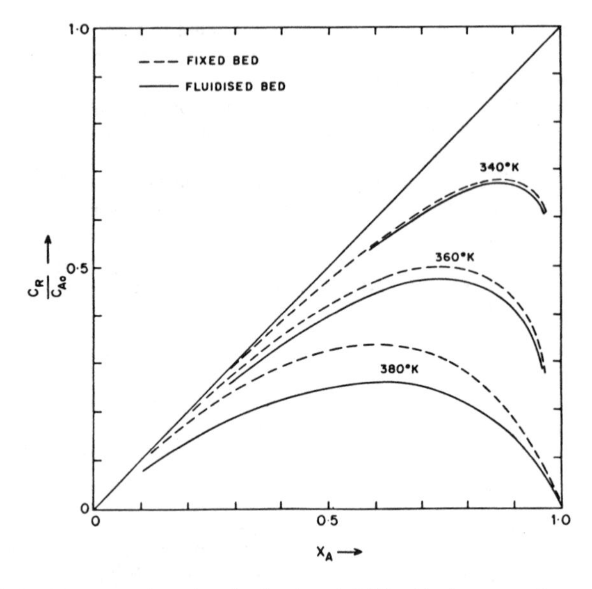

Figure 15. Selectivity-conversion plots for fixed and fluidized-bed reactors for a consecutive reaction scheme.

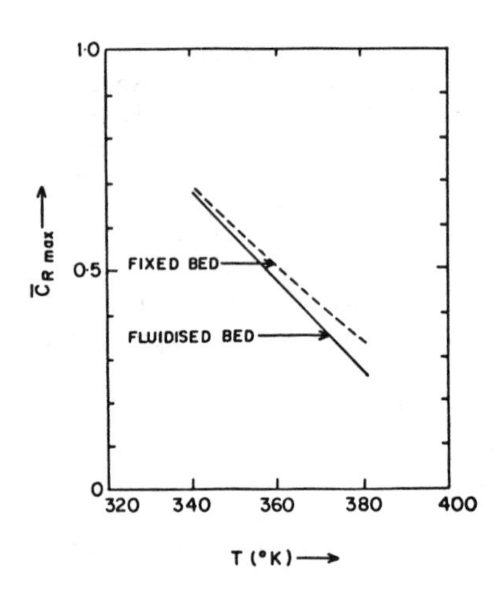

Figure 16. $\bar{C}_{R,max}$ as a function of reactor temperature.

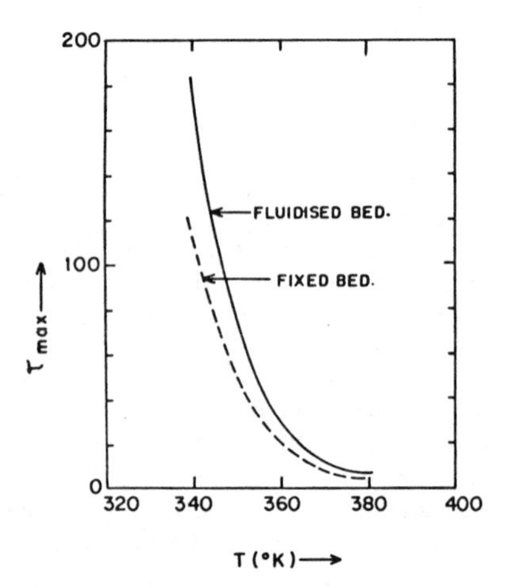

Figure 17. τ_{max} as a function of reactor temperature.

The selectivity-conversion plots for fixed- and fluidized-bed reactors are shown in Figure 15. It is seen that the selectivity is always higher in a fixed-bed reactor, which is indeed expected for this reaction scheme. At low temperatures, the rate constants are low, efficiencies of contact are close to unity, and the difference between fixed- and fluidized-bed reactors is insignificant, as far as selectivity is concerned. At higher temperatures, the contact efficiencies become lower and this difference increases. This trend is shown more explicitly in Figure 16, where the variation of C_{Rmax}/C_{A0} with temperature is shown. Since the second reaction has a higher activation energy in the example, the selectivity to R is seen to decrease with temperature. This trend would be reversed if the opposite was true. Figure 17 shows that for attaining the maximum selectivity to R, a fluidized bed requires much more catalyst than does a fixed-bed reactor.

The Effect of an Optimum Axial Temperature Profile

It is well-known that the selectivity in complex reactions conducted in a fixed-bed reactor can be improved by having the optimum axial temperature profile rather than a uniform temperature. The vigorous mixing of solids which characterizes a fluidized-bed reactor precludes the possibility of an axial temperature profile. However, there is still the choice of using two (or more) fluidized-bed reactors in series, each operating at a different temperature. An upper limit to the improvement that can be achieved can be calculated by considering the case of a large number of fluidized-bed reactors in series. Mathematically, this is equivalent to a single fluidized-bed reactor with an axial temperature profile.

The derivation of equations which describe the optimum temperature profile along a fluidized-bed reactor for first-order consecutive and parallel reactions has been presented by Kulkarni and Patwardhan [38]. Two numerical examples were also presented by them. An interesting feature of the analysis is that a critical temperature exists at which the temperature derivative changes sign. In other words, the optimum temperature profile may be a rising or a falling curve depending upon the inlet temperature. This is in contrast to the case of fixed-bed reactors. These authors investigated wide ranges of kinetic and hydrodynamic parameter values and found that the improvement that can be achieved by having an optimum temperature profile is only marginal. The maximum improvement that they reported was less than one percent in the case of a consecutive reaction scheme.

SELECTION OF PARAMETERS AT THE DESIGN STAGE

In this section we briefly and qualitatively discuss some factors over which the designer of a fluidized-bed reactor has some control. The reaction scheme itself is fixed by the process under consideration. We also assume that the type of catalyst is fixed, which is really a part of the process. In addition there are several physical parameters which can, in principle, be manipulated from the point of view of improving the selectivity to the desired product.

Factors Affecting the Reaction Rate Constant

The reaction temperature affects both conversion and selectivity. The case of simple first-order consecutive reactions has been treated earlier. In more complex reactions, calculations must be conducted to see the effect of temperature. Another alternative is to consider two fluidized beds in series, operating at different temperatures. The additional investment in terms of fabrication, etc. requires that the gains in selectivity should be substantial. This appears unlikely in view of the case of the optimum temperature profile just presented, which fails to show much of an improvement.

Catalyst dilution is another factor which affects the rate constants. The example presented earlier shows that under certain conditions, catalyst dilution can lead to a substantial gain in selectivity without reducing the conversion drastically. Thus, catalyst dilution is a factor which has a potential that needs to be investigated at the design stage.

Factors Affecting Transport Coefficients

The particle size of the catalyst has a strong effect on factors such as the minimum fluidization velocity, bubble velocity, etc., which in turn affect the transport coefficients. Since the relative magnitudes of the rate constant and the transport coefficients determine contact efficiencies, etc., the performance of the fluidized bed can be changed to some extent by changing the particle size.

The bubble size also can have a dramatic effect on the reactor performance. Figures 4 through 6, 8, 9, and 13 demonstrate the effect of bubble size on several other parameters. The distributor must be designed with the desired bubble size in sight. The bubble size also depends upon the internals in the fluidized bed, such as heat exchanger tubes. In tall fluidized beds, it may happen that bubbles grow to a large size by the time they reach the top. In such a case, consideration must be given to the possibility of using an intermediate redistributor (which really breaks the reactor into two fluidized beds in series).

Factors Affecting Reactant Concentrations

Inerts can be used advantageously in some cases, especially if different reactions are of different orders. Feed geometry is another factor which can be exploited in certain cases to advantage. For example, fluidization with one of the reactants in pure form, with the introduction of the other reactants in the form of several side streams, can be beneficial in case of series reactions. The ratio of reactants is another variable. However, both these factors relate to reactions involving several species, and have not been investigated in enough detail so far to evolve general guidelines.

NOTATION

C_A, C_R, C_S, C_T, C_u	concentrations of A, R, S, T, and u
$\bar{C}$	fractional concentration
d_b	bubble diameter
D	gas diffusivity
D_t	reactor diameter
e	contacting efficiency
g	acceleration due to gravity
k_1, k_2, k_3, k_4, k_1'	reaction rate constants
k_{12}	k_1+k_2
k_{34}	k_3+k_4
K_i	modified rate constant for k_i ($= k_ie_i$)
K_A	defined by Equation 22
K_{AR}	defined by Equation 23
K_{bc}, K_{ce}	bubble-cloud and cloud-emulsion gas interchange coefficients, respectively
l	dimensionless height in the fluid bed
l_c	critical value of l
L_f	height of fluidized bed
P_t	total pressure
Q_E	transport parameter of Nashaie and Yates [33]
r	rate of reaction per unit volume
R'	dilution ratio defined by Equation 37
t	time
T	temperature
U_0	superficial gas velocity
U_b	velocity of a bubble
U_{br}	velocity of a bubble with respect to the emulsion phase
U_{mf}	U_0 at minimum fluidization
v_0	volumetric flow rate of gas
v_b	volume of a gas bubble
v_w	volume of bubble wake
W	catalyst weight
X	fractional conversion
y_{A0}	mole fraction of A entering the bed

Greek Symbols

α_w	wake volume/bubble volume	ε_{mf}	bed void fraction at minimum fluidisation
γ_i	volume of solids in region i per unit bubble volume	ν	stoichiometric factor in Equation 36
δ	volume of bubbles/volume of bed	ϱ_s	bulk density of solids bed
		τ	residence time
ε_e	void fraction in emulsion phase	τ_{cr}	τ at which C_A becomes equal to zero

Subscripts

A, R, S, T, u	components A, R, S, T, u, respectively	b	in bubble phase
		c	in cloud phase
		e	in emulsion phase
0	initial value or at entrance	max	maximum value

REFERENCES

1. Davidson, J. F., and Harrison, D., *Fluidized Particles,* Cambridge University Press, NY (1963).
2. Kunii, D., and Levenspiel, O., *Ind. Eng. Chem. Fund., 7,* 338 (1968a).
3. Kunii, D., and Levenspiel, O., *Ind. Eng. Chem. Proc. Des. Dev., 7,* 481 (1968b).
4. Toor, F. D., and Calderbank, P. H., Proc. Int. Symp. on Fluidization, Netherlands University Press, Amsterdam (1967), p. 373.
5. Davidson, J. F., *Trans. Inst. Chem. Engrs., 39,* 230 (1961).
6. Fryer, C., and Potter, O. E., *Ind. Eng. Chem. Fund., 11,* 338 (1972).
7. Fryer, C., and Potter, O. E., Proc. Int. Symp. Fluidization and Its Applications, Toulouse (1973).
8. Kato, K., and Wen, C. Y., *Chem. Eng. Sci., 24,* 1351 (1969).
9. Mori, S., and Wen, C. Y., *Fluidization Technology,* Vol. I, D. L. Keairns (Ed.) Hemisphere, Washington, DC (1976).
10. Partridge, B. A., and Rowe, P. N., *Trans. Inst. Chem. Engrs.* (London), *44,* T35 (1966).
11. Mori, S., and Muchi, I., *J. Chem. Eng., Japan, 5,* 251 (1972).
12. Orcutt, J. C., Davidson, J. F., and Pigford, R. L., *Chem. Eng. Prog. Symp. Ser., 58 (38),* 1 (1962).
13. Kobayashi, H., et al., *Kagaku Kogaku, 33,* 27 (1969).
14. Zenz, F. A., *Inst. Chem. Engrs. Symp. Ser., 30,* 136 (1968).
15. Behie, L. A., and Kehoe, P., *AIChE J., 19,* 1070 (1973).
16. Grace, J. R., and de Lasa, H. I., *AIChE J., 24,* 364 (1978).
17. de Lasa, H. I., and Grace, J. R., paper presented at the 70th AIChE Annual Meeting, New York (1977).
18. Erazu, A. F., de Lasa, H. I., and Sarti, F., *Can. J. Chem. Eng., 57,* 191 (1979).
19. Yates, J. G., and Rowe, P. N., *Trans. Inst. Chem. Engrs., 53,* 137 (1977).
20. de Lasa, H. I., and Grace, J. R., *AIChE J., 25,* 984 (1979).
21. Chavarie, C., and Grace, J. R., *Ind. Eng. Chem. Fund., 14,* 75, 79, 86 (1975).
22. Spitz, P. H., *Hydrocarbon Processing, 47 (11),* 162 (1966).
23. Caporali, G., *Hydrocarbon Processing, 51 (11),* 144 (1972).
24. Winter, O., and Eng, M. T., *Hydrocarbon Processing, 55 (11),* 125 (1976).
25. Bleloch, W., *J. S. Africa Inst. Engrs., 45,* 114 (1946), cf. CA. 41: 4091 g (1947).
26. NCL Annual Report, National Chemical Laboratory, Pune, India (1975).
27. Nagata, S., et al., *Chem. Eng. (Japan), 16,* 301 (1952).

28. Rasmussen, D. M., *Chem. Eng.*, 104 (Sept. 18, 1972).
29. Denbigh, K. G., *Chem. Eng. Sci.*, *8*, 125 (1958).
30. Levenspiel, O., Baden, N., and Kulkarni, B. D., *Ind. Eng. Chem. Proc. Des. Dev.*, *17*, 478 (1978).
31. Irani, R. K., Kulkarni, B. D., and Doraiswamy, L. K., *Ind. Eng. Chem. Proc. Des. Dev.*, 19, *24* (1980).
32. Irani, R. K., Kulkarni, B. D., and Doraiswamy, L. K., *Ind. Eng. Chem. Fund.*, *19*, 424 (1980).
33. Nashaie, S. E., and Yates, J. G., *Chem. Eng. Sci.*, *27*, 1757 (1972).
34. Yates, J. G., *Chem. Eng. (London)*, *303*, 671 (1975).
35. Irani, R. K., et al., *Chem. Eng. Sci.*, *36*, 29 (1981).
36. Miyauchi, T., *J. Chem. Eng., Japan*, *7*, 201 (1974).
37. Irani, R. K., Kulkarni, B. D., and Doraiswamy, L. K., *Ind. Eng. Chem. Fund.*, *18*, 648 (1979).
38. Kulkarni, B. D., and Patwardhan, V. S., *Chem. Eng. J.*, *21*, 195 (1981).

CHAPTER 28

STABILITY ANALYSIS OF FLUID CATALYTIC CRACKERS

Hiroyasu Seko

Research Laboratory of Applied Biochemistry
Tanabe Seiyaku Co., Ltd.
Osaka, Japan

Setsuji Tone

Department of Chemical Engineering
Osaka University
Toyonake, Japan

CONTENTS

MODELING OF CRACKING AND REGENERATION REACTIONS

Introduction

Catalytic cracking converts high-molecular-weight gas oil into such products as gasoline, middle-distillates, and olefins with the aid of a catalyst.

In 1936, the first commercial catalytic cracking process, called the Houdry process, was built at Socony-Vacuum Oil Co., and used fixed beds of catalyst. The catalyst was deactivated by deposition of carbonaceous materials (coke) on its surface, and consequently had to be regenerated periodically by burning off the coke with air.

About 1940, to satisfy a great demand for aviation gasoline, a number of studies were directed toward attemps to develop a continuous catalytic cracking process by taking advantages of the liquidlike flow of catalyst particles in the fluidized bed, and one extension of these attemps led to the fluid catalytic cracking (FCC) unit, which is used in industry today for the production of light oil, gasoline, and cracked gas [1–4]. The latest representative units are the Mobilized UOP FCC, the Exxon Flexicracking unit, the Kellog Ortho-Flow "F", etc.

In the cource of evolution of FCC process, one of the most significant events was the invention of synthetic zeolite catalysts. The invention has made the unit compact and results in high yield of gasoline [5–8]. Further, the application of computer optimization and control to FCC units has produced many economic benefits.

The optimization must be carried out using a detailed model of FCC unit. However, transient behavior of the FCC shows complicating features because of the strong interaction between the reactor and the regenerator. It is not easy to accomplish satisfactory operation and control of the FCC in the presence of disturbances. Therefore, a number of investigators [9–16] have studied the operation and control of FCC. However, the models used in these studies were mostly derived from empirical correlations based on many years' experience. Therefore, in practice, steady-state behavior and stability of the system have not been understood satisfactorily. This may be attributed to both the complexities of chemical reactions in the reactor and regenerator and the lack of an adequate model of the fluidized bed.

One of the essential problems in establishing the rational design and operation procedures of FCC is concerned with the reaction kinetics. For example, in the reactor, the rate constants must be related in terms of coke content on the catalyst because the catalyst activity depends on the coke content [17–20], while in the regenerator the product distribution of CO_2 and CO must also be elucidated because the heat of coke-burning depends on the ratio of CO_2 to CO produced [21]. The other essential problem concerns modeling the fluidized-bed reactor. Recently, a number of two-phase models have been proposed [22–26]. Unfortunately, all of the proposed models of fluidized beds are too difficult to use because of their complexities. A brief model of the fluidized bed that can predict the reactor performance over a wide range of operating conditions is being formulated.

This section provides the modeling of cracking and regeneration reactions as well as of the fluidized bed.

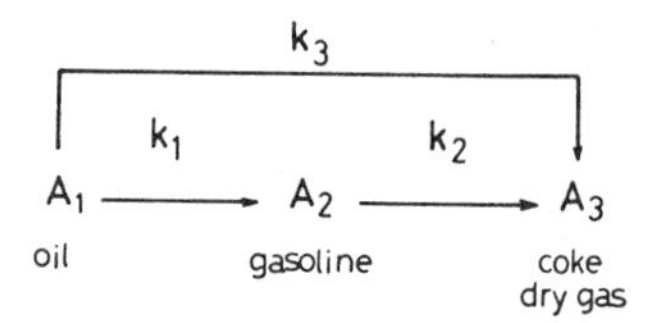

Figure 1. Weekman's cracking model [27].

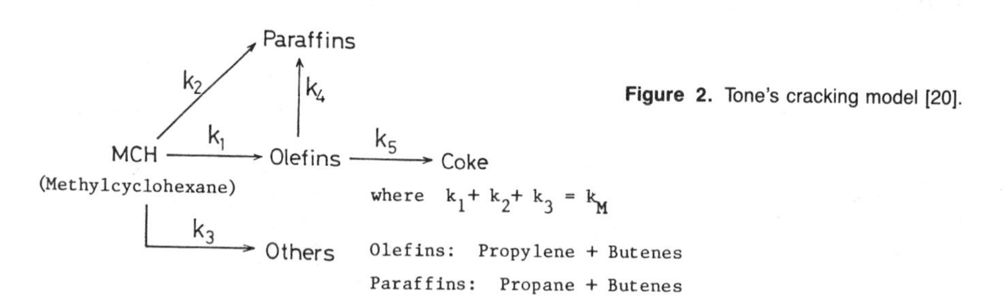

Figure 2. Tone's cracking model [20].

Table 1
Several Expressions of Catalyst Decay

Decay Function	$\varphi = k_{actual}/k_{initial}$
Voorhies [17]	$\varphi = \alpha t_c^{-n}$
Weekman [27]	$\varphi = \exp(-\alpha t_c)$
Lin et al. [18]	$\varphi = \exp(-\alpha n_c)$
Ozawa et al. [19]	$\varphi = 1 - \alpha n_c$
Tone et al. [20]	$\varphi = 1/(1 + \alpha n_c)$

where α = *coefficient*
t_c = *catalyst residence time*
n_c = *coke concentration on catalyst*

Table 2
Reaction Kinetics for Weekman's Model [16, 27]

Reaction Rates	
$r_{oil} = -\delta(k_1 + k_3)A_1^2$	[g/(g-cat · sec)]
$r_{gas} = \delta(k_1 A_1^2 - k_2 A_2)$	[g/(g-cat · sec)]
$r_{coke} = \delta(k_3 A_1^2 + k_2 A_2)$	[g/(g-cat · sec)]
Rate Constants	
$k_1 = k_1^0\varphi = \exp(6.856 - 5{,}096/RT) \cdot \varphi$	$(cm^3/g \cdot sec)$
$k_2 = k_2^0\varphi = \exp(6.646 - 16{,}842/RT) \cdot \varphi$	(1/sec)
$k_3 = k_3^0\varphi = \exp(17.7 - 26{,}127/RT) \cdot \varphi$	$(cm^3/g \cdot sec)$
where $\varphi = \exp(-\alpha t_c)$, $\delta = \varepsilon_c/\varrho_B$	

From Elnashaie and El-Hennawi [16].

Cracking Model

In the catalytic cracking process, carbonaceous material (coke) gradually accumulates on the surface of the catalyst, tending to lower the activity of the catalyst. Since catalyst decay causes a decrease in conversion to the product during the reaction, the factors leading to coke formation are of great importance for the design of a catalytic reactor.

Typical expressions of catalyst decay developed by several investigators [17–20] are listed in Table 1. The decay function φ is expressed by the initial rate constant to yield the actual rate constant. Voorhies [17] showed that the coke on catalyst is a function of the time that the catalyst has been exposed to the feedstock. Weekman [27] advanced a practical correlation of catalyst decay which is a function of catalyst residence time according to the Voorhies' law. On the other hand, Lin et al. [18], Ozawa and Bischoff [19] and Tone et al. [20] expressed the catalyst decay as a function of concentration of coke on catalyst.

The dynamic analysis of FCC must be carried out using a detailed process model which is composed of coke and heat balances around the reactor and regenerator. It is therefore desirable that the reaction products can be predicted as functions of coke content and temperatures in the reactor and regenerator.

Two representative cracking models are shown in Figures 1 and 2, and these kinetic parameters are listed in Tables 2 and 3, respectively. The kinetic model presented by Weekman [27] is formed of three components. A_1 is the gas oil, A_2 is the gasoline, and A_3 is the coke and dry gas. Gas oil and gasoline cracking are assumed to be second and first order, respectively. Decay function φ is a function of catalyst residence time. On the order hand, the kinetic model presented by Tone et al. [20] includes five chemical components. All reactions are first order, and the decay function φ is a function of coke content.

Table 3
Reaction Kinetics for Tone's Model

Reaction Rates	
$r_{MCH} = -(k_1+k_2+k_3)C_{MCH} = -k_M C_{MCH}$	
$r_{Ole} = k_1 C_{MCH} - (k_4+k_5)C_{Ole}$	
$r_{Par} = k_2 C_{MCH} + k_4 C_{Ole}$	
$r_{Other} = k_3 C_{MCH}$	
Rate Constants	
$k_M = k_M^0 \varphi = \exp(25.8 - 38{,}500/RT) \cdot \varphi$	(1/sec)
$k_1 = k_1^0 \varphi = \exp(24.8 - 39{,}100/RT) \cdot \varphi$	(1/sec)
$k_2 = k_2^0 \varphi = \exp(25.2 - 42{,}300/RT) \cdot \varphi$	(1/sec)
$k_4 = k_4^0 = \exp(-5.15 + 7{,}000/RT)$	(1/sec)
$k_5 = k_5^0 = \exp(-5.85 + 3{,}360/RT)$	(1/sec)
$\alpha = \exp(10.5 - 18{,}600/RT)$	(g-cat/mg-coke)
where $\varphi = 1/(1 + \alpha n_c)$	

From Tone et al. [20].

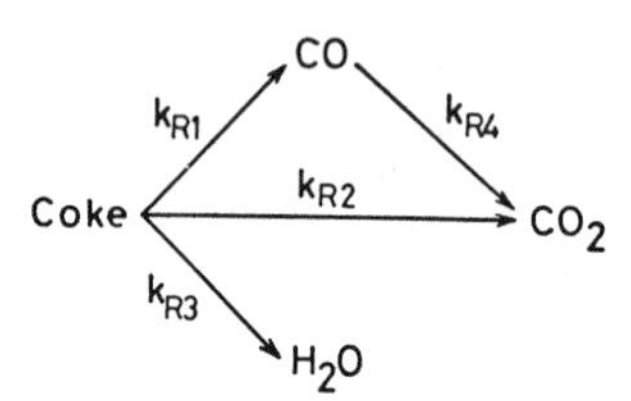

Figure 3. Coke-burning model [21].

Regeneration Model

It is well known that coke consists of carbon and hydrogen and that the coke-burning proceeds as shown in Figure 3 [21, 28].

Let's consider the following assumptions for the development of the regeneration kinetics.

1. In the reactor, a small part of feedstock converts into coke which covers catalyst particles uniformly.
2. In the regenerator, the diffusion of oxygen into the catalyst particles is negligibly small, so the disappearance of coke on the particles is uniform.

With these assumptions, Tone et al. [21] proposed the regeneration kinetics shown in Table 4. On the other hand, Lee et al. [10] used a simplified model of the form of $C \rightarrow CO_2$. Recent operation of the FCC unit has a tendency to convert coke to CO_2 completely by the use of a CO promoter [29, 30]. Therefore, the simplified model has been used for the analysis of the FCC process.

Fluidized-Bed Model

Despite numerous applications of the fluidized bed in industrial process, the design and scale-up of a commercial fluidized bed is still based on many year's experience. Recently,

Table 4
Reaction Rate Equation for the Regeneration

Definitions	
$k_{R1} = 9.923 \times 10^9 \exp(-36{,}540/RT)$	(1/min)
$k_{R2} = 2.578 \times 10^6 \exp(-25{,}640/RT)$	(1/min)
$k_{R3} = 6.557 \times 10^4 \exp(-18{,}210/RT)$	(1/min)
$k_{R4} = 4.003 \exp(-13{,}770/RT)$	(1/g-cat · min)
Reaction Rates	
$r_{CO_2} = k_{R2}C_{O_2}n_c + k_{R4}C_{CO}$	
$r_{CO} = k_{R1}C_{O_2}n_c - k_{R4}C_{CO}$	
$r_{H_2O} = k_{R3}C_{O_2}n_H$	
$r_{O_2} = -r_{CO_2} - 0.5(r_{CO} + r_{H_2O})$	

From Tone, Miura, and Otake [21].

Table 5
Several Two-Phase Models of Fluidized Beds

Investigators	Bubble Phase	Gas Flow BF	Gas Flow EF	Gas Interchange	Estimation of K_j
Partrige and Rowe [32]	B+C	PF	PF	(B+C) ↔ EF	$\frac{K_j d_c}{D_G} = 2 + 0.69\left(\frac{\mu}{\varrho_f D_G}\right)^{1/3} \cdot Re_c^{0.5}$
Kobayashi and Arai [33]	B+CW	PF	PF or BF	(B+C) ↔ EF	$K_j = 11/d_b$
Kunii and Levenspiel [25]	B+CW	PF	PF	B ↔ CE ↔ EF	$K_b = 4.5\left(\frac{u_{mf}}{d_b}\right) + 5.85\left(\frac{D_G^{1/2} g^{1/4}}{d_b^{5/4}}\right)$
Fryer and Potter [24]	B+CW	PF	PF	B ↔ CF ↔ EF (cloud volume = 0)	$K_c = 6.78\left(\frac{\varepsilon_{mf} D_G u_b}{d_b^3}\right)^{0.5}$; $1/K_j = 1/K_b + 1/K_c$
Tone et al. [31]	B+CW	PF	PF	(B+CW) ↔ EF (cloud volume = 0)	$\frac{K_j d_b^*}{u_b^*} = 0.088\left(\frac{D_G}{d_b^* u_b^*}\right)^{0.5}$

B: bubble; C: cloud; CW: cloud-wake; PF: plug flow; BF: backmixing flow; BF: bubble phase; EF: dense (emulsion) phase.

Elnashaie et al. [16] and the authors [14, 15] employed two-phase models of the fluidized bed [25, 31] for analyzing the dynamic behavior of FCC units and obtained valuable evidence that these two-phase models are very useful tools for determining the operation procedure of FCC units.

The two-phase model consisting of bubble and dense phases was first presented by Toomey and Johnstone [22] in which it was assumed that the excess gas flowing above the incipient fluidizing velocity flows through the bed in the form of bubbles. After that, many two-phase models have been proposed. Several examples are shown in Table 5. Partridge and Rowe [32] proposed a model in which a rising bubble is surrounded by a cloud of circulating gas, and some interchange occurs between the cloud and dense phases. Kobayashi and Arai [33] reported a two-phase model describing the gas interchange between the cloud phase involving a cloud-wake overlap and dense phase. Fryer and Potter [24] and Kunii and Levenspiel [25] advanced a model involving an upward flow of solids with the bubbles. Further, Kunii et al. [25] advanced a correlation of successive gas interchange among bubbles, cloud-wake, and dense phase according to the Davidson and Harrison [23] theory and the Higbie-type penetration model [34]. However, these rigorous mathematical models are too complicated for the simulation of such a reaction system as heavy oil cracking with catalyst decay.

For the development of the simplified fluidized-bed model based on bubble behavior, let's consider the following assumptions, according to Figure 4.

1. The fluidized bed consists of the bubble phase with wake solids and the dense phase in which interstitial gas flows at the incipient state.
2. The bubble phase consists of the bubble and wake surrounded by a spherical cloud. The cloud volume is negligibly small compared with the wake volume, and the solids in the cloud are ignored. The cloud volume, therefore, is lumped together with the wake volume. This assumption is supported where u_0/u_{mf} is large [24, 25].
3. Finite interchange of gases occurs between the bubble and dense phases, and the main transfer resistance exists between the cloud boundary and dense phase.
4. The gas composition in the bubble with cloud-wake is uniform.
5. The average size and velocity of the rising bubbles are given for the whole bed.
6. The voidage is uniform over the whole bed.
7. The voidage in the cloud-wake is equal to that of incipient bed.

Further, let's consider the following two assumptions to derive a material balance for each gaseous component in a fluidized bed.

8. The gas interchange coefficient between bubble and dense phases per unit of bubble volume is uniform through the bed.
9. Since the gaseous product traverses the bed so quickly relative to catalyst decay at a definite process time, solids behavior in the fluidized bed can be treated as a pseudo-steady state.

With these assumptions, we may then derive the material balance of gaseous component j for an element of the bed height dZ and of cross-sectional area in both phases as follows [31]:

For dense phase:

$$-u_b(1+\delta_w\varepsilon_{mf})(dC_{j,c}/dZ) = K_j(C_{j,c}-C_{j,e})-\gamma_c r_{j,c}/(1-\varepsilon_c) \quad (1)$$

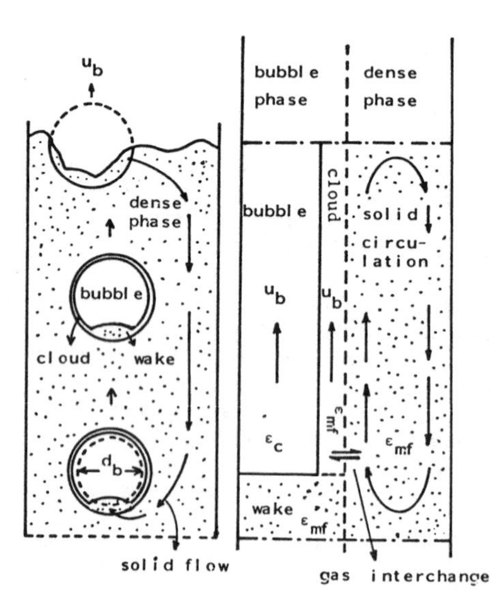

Figure 4. Model to account for gas interchange between bubble and dense phases [31].

For bubble phase:

$$K_j(C_{j,e} - C_{j,c}) - \gamma_e r_{j,e}/(1 - \varepsilon_c) = 0 \tag{2}$$

where suffixes "c" and "e" are used to denote the bubble and dense phases in a fluidized bed, respectively:

$$u_b = (u_0 - u_{mf})/\varepsilon_b \tag{3}$$

$$\gamma_c = (1 - \varepsilon_{mf})\delta_w \tag{4}$$

$$\gamma_e = (1 - \varepsilon_{mf})\,[1 - \varepsilon_b(1 + \delta_w)]/\varepsilon_b \tag{5}$$

Empirical Formulas of Model Parameters in Fluidized Beds

There are unknown parameters such as u_{mf}, ε_{mf}, ε_b, δ_w, and K_j in the two-phase model of the fluidized bed. Especially, it is well known that the value of K_j is the function of bubble diameter d_b which is one of the most important parameters. These values can be experimentaly measured in the majority of cases. A number of studies have been directed toward attempts to correlate these parameters in the fluidized bed, but the ones at high temperature where industrial operations are carried out are little known [40–42]. Table 6 shows some empirical formulas of these parameters. The authors [31] have found that the values of K_j at high temperatures lie in less than about one to ten times those obtained in the room temperature. This may be dependent on the difference in transport properties at low and high temperatures.

MODEL OF FLUID CATALYTIC CRACKING SYSTEM AND ITS TRANSIENT BEHAVIOR

The first section outlined the modeling of the cracking reaction and the regeneration reaction as well as of the fluidized bed. This section is concerned with the fluid catalytic cracking (FCC) system and its transient behavior. FCC units consist of a cracker and a regenerator and are employed in converting heavy hydrocarbon into light oils, gasoline, and cracked gas. But there are a great deal of unknown factors in the transient behavior of the FCC. Therefore, empirical relations have been used for design and operation of commercial FCC.

Table 6
Some Empirical Formulas of Two-Phase Parameters

u_{mf}	Pattipati and Wen [36] (25∼850 °C) $\frac{d_p u_{mf} \varrho_g}{\mu} = \left\{(33.7)^2 + 0.0408 \frac{d_p^3 \varrho_g(\varrho_s - \varrho_g)g}{\mu^2}\right\}^{0.5} - 33.7$	Singh et al. [37] (∼700 °C) $u_{mf} = \propto 1/\mu$
ε_b	Orcutt et al. [38] (∼80 °C) $\varepsilon_b = \frac{u_0 - u_{mf}}{u_0 - u_{mf} + 0.711\sqrt{g d_b}}$	Tone et al. [31] (530∼650 °C) $\varepsilon_b = \frac{L_f - L_{mf}}{L_f} = 0.115\sqrt{\frac{u_0 - u_{mf}}{L_c u_{mf}}}$
δ_w	Woollard and Potter [39] (room temp.) $\delta_w = \frac{654\varrho_f}{\varrho_s(1 - \varepsilon_{mf})} \exp(-66.3 \cdot d_p)$	Otake et al. [40] (400∼650 °C) $\delta_w = 0.33$ {150–170 mesh silica-alumina}
d_b	Mori and Wen [41] (room temp.) $d_b = d_{bm} - (d_{bm} - d_{b0}) \exp(-0.3Z/D_T)$ $d_{bm} = 0.652[A_T(u_0 - u_{mf})]^{2/5}$ $d_{b0} = 0.347G$ (for perforated plate)	Tone et al. [31] (430∼650 °C) $d_b = \frac{0.00142Z + 0.0233}{(T \times 10^{-3})^{1.21}}\left(\frac{u_0}{u_{mf}}\right) + 0.08$ (150–170 mesh silica-alumina)

The recent studies relating to the FCC have tried to explain transient behavior and system stability by mathematical models. Iscol [9] showed that an FCC system consisting of two fluidized beds has multiple steady-state points and may be unstable within the region of practical operations. Lee and Kugelman [10] pointed out that an FCC system having a riser cracker as well as two fluidized beds has a unique steady state point and may be stable. Bromey and Ward [11] proposed a structure analysis control, which is a direct extension of classical multiloop control, and have developed a practical FCC control scheme. However, the models used in these studies were mostly derived from empirical correlations without considering the bubble behavior in a fluidized bed.

The authors [14] and Elnashaie et al. [16] tried to use a two-phase model for analyzing the transient behavior of FCC and have obtained results which successfully simulate the transient behavior of commercial FCC. In this section, attention will be devoted to the analysis of the transient behavior of FCC by using a two-phase model of fluidized bed.

System Description and Developments of FCC

The fluid catalytic cracking (FCC) units are classified into two types as shown in Figure 5. One consists of a fluid cracker of perfectly mixed solids and a fluid regenerator of perfectly mixed solids (B–B type). The other contains a riser that is a plug-flow-type cracker in addition to two fluidized beds (P–B type). In this P–B type, most of catalytic cracking is carried out in the riser.

Heavy hydrocarbon is catalytically decomposed in the cracker or riser. Carbonaceous materials, or coke, (is one of the reaction products) gradually accumulate on the surface of catalyst particles, tending to lower the activity of the catalyst. For recovery of catalyst activity, the spent catalyst is made to flow continuously to the regenerator, where the coke is burnt off with air. The catalyst heated by combustion is sent back to the cracker, where its heat supplies the energy for the cracking reaction.

Some operating conditions in commercial FCC are shown in Table 7. The FCC units have accomplished the major improvements since the middle 1930's. One of the largest improvements was the use of zeolite catalysts. The early silica-alumina catalyst gave 45% gasoline at the 65% conversion level. However, the recent zeolite catalyst containing a CO promoter not only gives 57% gasoline [3], but also makes it possible to operate a riser cracker in a manner similar to a moving bed instead of in a manner similar to a fluidized-bed cracker. In addition, high activity of the zeolite catalyst also makes it possible to eliminate recycle products [13].

The research laboratories of Phillips Petroleum Co. developed a new catalyst containing antimony which reduces the effects of contaminant on FCC catalyst [45]. Shell Co. researched the catalyst attrition rate and found that the attrition rate can be reduced by

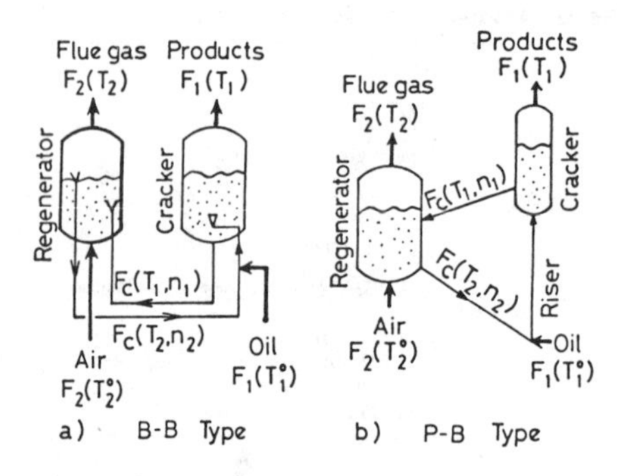

Figure 5. Two types of FCC units.

Table 7
Some Operating Conditions of Commercial FCC

No.	Type	F_1 (ton/hr)	F_2 (ton/hr)	F_c (ton/hr)	T_1 (°C)	T_2 (°C)
1	B–B	40	35	400	450~480	570~600
2	P–B	406	247	3,510	520	640
3	B–B	390	235	6,000	525	632
4	B–B	61	22	240~420	500	704
5	P–B	147	98	1,024	510	689

No.	n_1	n_2	Conv. (wt%)	O_2 (mol%)	CO (mol%)	Ref.
1	9~15	0.5~1.5	70~80	0.5~1.5	9~15	[44]
2	9.9	3.5		0.2		[10]
3	10.4	6.7	55	0.2	9	[43]
4		0.5	85		3	[13]
5	9.5	1.3	46	0.1		[11]

* *mg-coke/g-catalyst*
No. 1: ERE MODEL IV (1960); No. 2: Mobil (1973); No. 3: Standard Oil Model IV (1975); No. 4: ERE MODEL II [High regenerator temperature operation (1979)];
No. 5: Shuldt, S. B., and Proc, F. B., 1971 Join Autom, Control Conf. (1971), 270.

a modification of the regeneration grid, to "two-stage air nozzles" [46]. Considerable decreases in catalyst loss have accomplished. AMCO Oil Co. has developed the Ultra Cat Regeneration unit where the concentration of CO is less than 500 ppm and coke content on regenerated catalyst is about 0.5 mg-coke/g-cat at the regeneration temperature of 660° C. These desirable operating conditions not only have made the process operation safe without afterburning but are also profitable [47]. Pullman Kellog Co. has developed the heavy oil cracking (HOC) process together with Phillips Petroleum Co. [48, 49]. Although the HOC process was the first FCC unit designed especially for charging atomosphere-topped crude, it was applied to the cracking of residual oil and has been employed in the cracking of petroleum including very high-boiling fractions. The Atlantic Richfield Co. has developed the Demet III process where contaminant metals on FCC catalysts are removed by chemical and physical treatments [50].

Reaction Kinetics in the Cracker and the Regenerator

In order to explain the transient behavior of the FCC, let's employ the reaction models of Figure 6 as the reaction schemes in the FCC, where the composition of coke is assumed to be carbon only. In the cracker, the catalyst is deactivated by coke formed by the cracking of methylcyclohexane (MCH) according to the reaction kinetics of Table 3 [20], while in the regenerator the deactivated catalyst is regenerated by burning off the coke with air according to the reaction kinetics of Table 4 [21].

The catalyst particles at the outlet of the cracker and regenerator, respectively, are assumed to have uniform coke content n_1 and n_2 [51, 52], even though there are wide coke distributions owing to extensive catalyst mixing and coke deposition on and removal from the catalyst surface.

B–B Type

For the modeling of the B–B type, let's consider that solids in a fluidized bed are in a perfectly mixed state and that the flow of gas through the bed is a plug flow. Then, according to both the cracking model of Figure 2 and the material balance of gaseous components in the fluidized bed expressed by Equations 1 and 2, the coke-forming rate in the fluidized

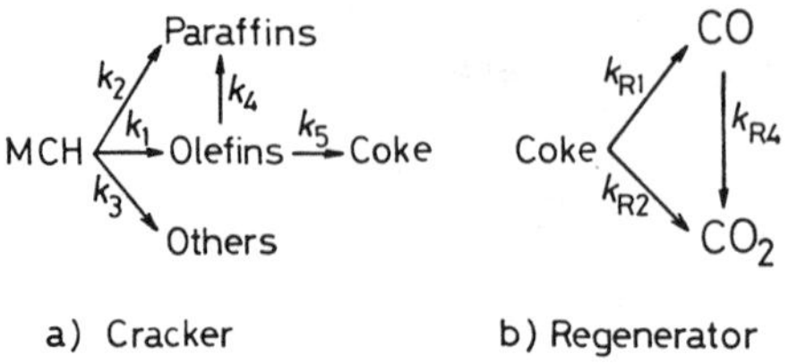

Paraffins
k_2 k_4 k_1 k_5
MCH Olefins Coke
k_3
Others

a) Cracker

CO
k_{R1} k_{R4}
Coke
k_{R2}
CO_2

b) Regenerator

Figure 6. Reaction schemes in the cracker and regenerator [14].

cracker R_l can be shown as the functions of mean concentration of olefins in the bubble and dense phases [14, 31].

$$R_l = k_5(v_1\bar{C}_{Ole,c} + v_2\bar{C}_{Ole,e}) \tag{6}$$

where v_l and v_2 are stoichiometric coefficients for the converting unit and are expressed by

$$v_l = 3.6 \times 10^6 \varrho_f\gamma_c/(\gamma_c + \gamma_e)\varrho_B \tag{7}$$

$$v_2 = 3.6 \times 10^6 \varrho_f\gamma_e/(\gamma_c + \gamma_e)\varrho_B \tag{8}$$

$\bar{C}_{Ole,c}$ and $\bar{C}_{Ole,e}$ denote the mean concentrations of olefins in the axial direction of bubble and dense phases, respectively.

$$\bar{C}_{j,i} = (1/L_f)\int_0^{L_f} C_{j,i}\,dZ \qquad (i = c, e,\ j = \text{gaseous component}) \tag{9}$$

In the same manner, the rate of coke burning R_2 is expressed by the product of coke content in the regenerator and the mean concentration of oxygen.

$$R_2 = -(k_{R1} + k_{R2})\,(v_3\bar{C}_{O_2,c} + v_4\bar{C}_{O_2,e})n_2 \tag{10}$$

where v_3 and v_4 are also stoichiometric coefficients.

$$v_3 = 7.2 \times 10^5 \gamma_c/(\gamma_c + \gamma_e) \tag{11}$$

$$v_4 = 7.2 \times 10^5 \gamma_e/(\gamma_c + \gamma_e) \tag{12}$$

On the other hand, the rate of heat absorption by cracking Q_l is estimated from the conversion of MCH.

$$Q_l = -\Delta H_{MCH}[1 - C_{MCH}(L_f)]F_1 \tag{13}$$

The rate of heat generation by the regeneration reaction Q_2 is likewise estimated from the amount of CO_2 and CO produced in the regenerator.

$$Q_2 = -[\varrho_{CO_2}\Delta H_{CO_2}C_{CO_2}(L_f) + \varrho_{CO}\Delta H_{CO}C_{CO}(L_f)]F_2/\varrho_{Air} \tag{14}$$

Further, the heat resistance between particles and gases is ignored, and the system is assumed to be adiabatic. Consequently, we can derive the following mathematical model of FCC.

Coke balance in the cracker:

$$dn_1/dt = (n_2 - n_1)/\theta_1 + R_1 \tag{15}$$

Coke balance in the regenerator:

$$dn_2/dt = (n_1 - n_2)/\theta_2 + R_2 \tag{16}$$

Heat balance in the cracker:

$$C_{pw}W_1\, dT_1/dt = C_{pw}F_c(T_2 - T_1) + F_1C_{p1}(T_1^0 - T_1) + Q_1 \tag{17}$$

Heat balance in the regenerator:

$$C_{pw}W_2\, dT_2/dt = C_{pw}F_c(T_1 - T_2) + F_2C_{p2}(T_2^0 - T_2) + Q_2 \tag{18}$$

P–B Type

P–B type consists of a riser cracker and a fluidized regenerator. Since the riser is regarded as a plug flow reactor, the coke balance along bed height can be expressed by

$$\partial n_1/\partial t + 3{,}600u_c(\partial n_1/\partial Z) = k_5 v_5 C_{Ole} \tag{19}$$

where v_5 is a stoichiometric coefficient.

$$v_5 = 3.6 \times 10^6 \varrho_f/\varrho_B' \tag{20}$$

Boundary conditions:

$$\text{at} \quad Z = 0, \quad n_1 = n_2 \tag{21}$$

Therefore, the coke content on catalyst at the outlet of the riser can be estimated from the integration of Equation 19. The heat balance along the bed height can be expressed by

$$C_{pw}\varrho_B'(\partial T_1/\partial t)/3{,}600 + (C_{pw}u_c\varrho_B' + C_{p1}u_0\varrho_f)\,(\partial T_1/\partial Z)$$

$$= -\Delta H_{MCH} r_{MCH} \varrho_f \tag{22}$$

Boundary conditions:

$$\text{at} \quad Z = 0, \quad T_1 = (T_2F_cC_{pw} + T_1^0F_1C_{p1})/(F_cC_{pw} + F_1C_{p1}) \tag{23}$$

The catalyst temperature at the outlet of the riser can be also estimated from the integration of Equation 22. On the other hand, the mathematical model for the regenerator is the same as that of the B–B type and may be expressed by Equations 16 and 18.

Steady-State and Transient Behavior

Steady-State Points

At a steady-state point, we can eliminate the time derivative terms in the mathematical model of FCC. For the B–B type, four resulting algebraic equations can be solved for n_1,

n_2, T_1, and T_2 under given values of the operating variables such as catalyst circulation rate F_c, feed rate of feedstock F_1, air flow rate F_2, feed temperature of MCH T_1^0, and feed temperature of air T_2^0. For P–B type, the values of C_j, n_1, and T_1 at the outlet of the riser are successively solved from Equations 19 and 22 using arbitrary values of n_2 and T_2, and the resulting values of n_1 and T_1 are substituted in Equations 16 and 18. This procedure is repeated until the right-hand sides of Equations 16 and 18 are closer to zero. The solutions are the steady-state values corresponding to a given operating condition.

Three selected examples having steady-state points are given in Table 8 together with the values of state variables and scale of units. Case 1 is an example in which the state variables are close to those of Csicsery's commercial FCC [43]. Case 2 is an exceptional example in which the values of $C_{MCH}(L_f)$ and $C_{O_2}(L_f)$ were made as small as possible. Case 3 is an example of the P–B type.

Transient Behavior of the System

To elucidate transient behavior of Cases 1, 2, and 3 near the steady-state points, let's solve the mathematical model of the FCC involving the time-derivate terms for such state variables as n_1, n_2, T_1 and T_2 under some initial conditions.

Figure 7 shows the trajectories of n_1 and T_1 in two-dimensional-phase space for the B–B type. Since the state trajectories shift either toward high values of n_1 and T_1 slowly, or toward low values of T_1 rapidly, it is seen that the steady states of Cases 1 and 2 are unstable. Figure 8 shows the trajectories for the P–B type. It is seen that the state trajectories shift toward the steady-state point. This behavior shows that the steady-state is stable.

To examine whether the state trajectories can be kept near the steady-state points, let's try to calculate the transient behavior of Cases 1, 2, and 3 by changing the values of operating variables stepwise between ±5 and ±12.5%.

Table 8
Three Selected Examples Having Steady-State Points

Case		1	2	3
Type		B–B	B–B	P–B
Operating variables				
F_1 = feed rate of MCH	(ton/hr)	10.0	17.0	30.0
F_2 = air flow rate	(ton/hr)	8.0	8.6	24.0
F_c = cat-circulation rate	(ton/hr)	110	255	600
T_1^0 = feed temperature of MCH	(°C)	200	394	600
T_2^0 = feed temperature of air	(°C)	20	158	640
Scale of unit				
Cracker				
W_1 = weight of catalyst	(ton)	40	200	40
D_1 = diameter	(cm)	900	2,000	500
Regenerator				
W_2 = weight of catalyst	(ton)	200	200	60
D_2 = diameter	(cm)	1,400	1,400	800
State variables				
Cracker				
n_1 = coke content	(mg-coke/g-cat)	6.9	8.3	15.3
T_1 = temperature	(°C)	495	593	587
$C_{MCH}(L_f)$ = MCH conc.	(wt-fraction)	0.12	10^{-5}	10^{-8}
Regenerator				
n_2 = coke content	(mg-coke/g-cat)	2.6	5.5	10.3
T_2 = temperature	(°C)	601	657	646
$C_{O_2}(L_f)$ = oxygen conc.	(mol-fraction)	0.08	0.02	0.02

From Seko, Tone, and Otake [14].

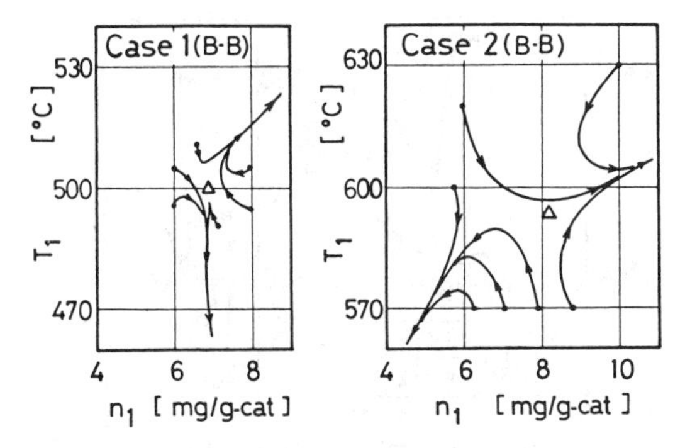

Figure 7. Trajectories of n_1 vs. T_1 for B-B type (Δ: steady state point) [14].

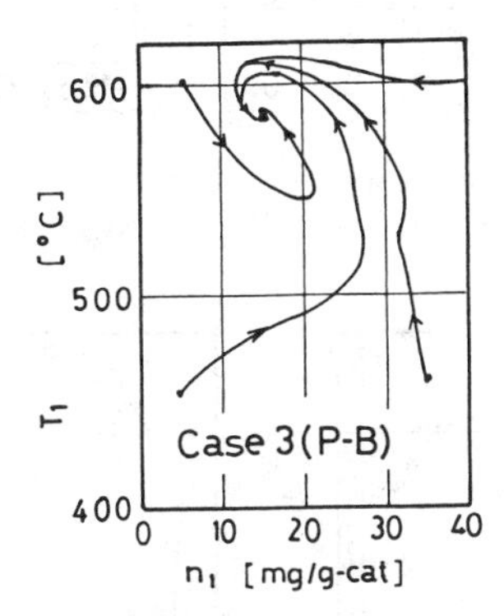

Figure 8. Trajectories of n_1 vs. T_1 for P-B type.

Figures 9, 10, and 11 show transient behavior with respect to a step change in feed temperature of MCH T_1^0 or air flow rate F_2.

In Case 1 of Figure 9, a decrease in T_1^0 leads to drops in T_1 and T_2, while an increase in T_1^0 gives rises in T_1 and T_2. Also when F_2 is decreased, n_1, n_2, T_1 and T_2 increase. Conversely, when F_2 is increased, n_1, n_2, T_1, and T_2 decrease.

In Case 2 of Figure 10, changes in T_1^0 result in the same transient behavior as in Case 1. However, when F_2 is decreased, both n_1 and n_2 increase, and both T_1 and T_2 decrease, contrary to Case 1. Also, when F_2 is increased, both n_1 and n_2 decrease, and both T_1 and T_2 increase. In these cases, although the state trajectories diverge from the steady-state points without control as time increases, it is suggested that the state variables can be easily kept near the steady-state point by using T_1^0 and F_2 as control variables.

Figure 11 shows the transient behavior of the P – B type (Case 3) for various variations of T_1^0 and F_2. It is seen that the trajectories of n_2 and T_2 starting with an arbitrary point A and B converge to new steady-state points in the neighborhood of the original steady-state point. This feature suggests the system is stable.

From such transient behavior, it can be seen that the state trajectories may be arbitrarily moved by appropriate changes in T_1^0 and F_2 and that system stability may not be changed by operating conditions.

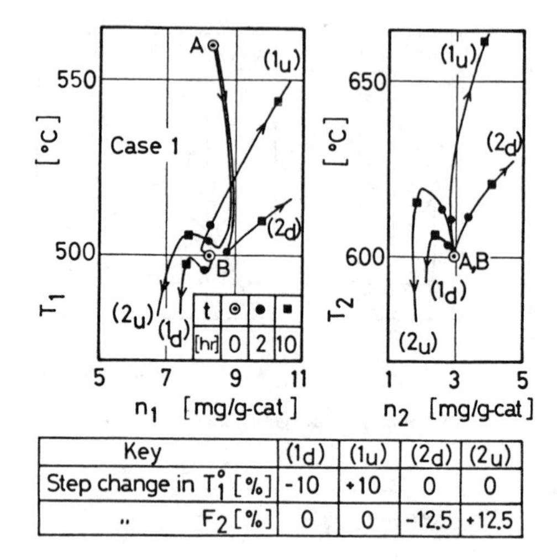

Figure 9. Transient behavior of Case 1 with respect to a step change in T_1° or F_2 (Notation on curves denotes the time from starting point A or B) [14].

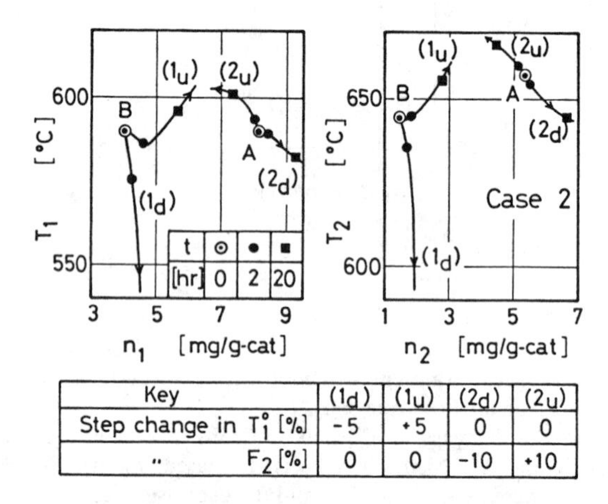

Figure 10. Transient behavior of Case 2 with respect to a step change in T_1° or F_2 (Notation on curves denotes the time from starting point A or B) [14].

Effects of Operating and Reaction Conditions on System Stability

To examine the effects of the operating and reaction conditions on system stability by using an eigenvalue study, the authors [14] tried to linearize the nonlinear terms of Equations 15 through 18 and have obtained the following evidence. That is, in the reaction scheme of Figure 6, as the conversion of MCH increases, the value of $\partial R_1/\partial T_1$ gets smaller. In the same manner, as the concentration of oxygen at the regeneration exit decreases, the value of $-\partial R_2/\partial T_2$ also gets smaller. And when both the values of $\partial R_1/\partial T_1$ and $-\partial R_2/\partial T_2$ are small, the system becomes stable [14].

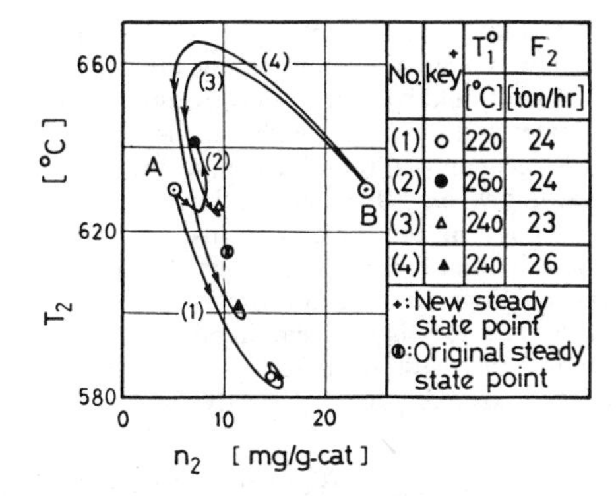

Figure 11. Transient behavior of Case 3 with respect to a step change in T_1^o or F_2.

Further, the authors [14] expanded the stability analysis to a wide range of operating conditions by the use of arbitrary values of $\partial R_1/\partial T_1$ and $-\partial R_2/\partial T_2$, and have obtained the following information.

1. System stability of the FCC is remarkably governed by the values of $\theta_1(\partial R_1/\partial T_1)$ and $-\theta_2(\partial R_2/\partial T_2)$.
2. Operating conditions such as F_1, F_2, F_c, θ_1, and θ_2 also have an effect on system stability.
3. When both the values of $\theta_1(\partial R_1/\partial T_1)$ and $-\theta_2(\partial R_2/\partial T_2)$ are small, the system is stable.

In commercial FCC, the conversion of feedstock is not high because the maximum benefit is in middle conversion, and the oxygen concentration at the regenerator exit is kept at small values to prevent the growth of afterburning.

The relation between the conversion of feedstock and the value of $\partial R_1/\partial T_1$ will be expressed in a variety of shapes because the coke forming rates are influenced by the difference in chemical composition of feedstocks. For instance, in the case of Weekman's model of Figure 1, as the conversion increases, the value of $\partial R_1/\partial T_1$ becomes large. This feature is contrary to Tone's model of Figure 2. In this connection, the authors have found that in Lee's stable FCC [10], $\theta_1(\partial R_1/\partial T_1)$ is nearly zero and $-\theta_2(\partial R_2/\partial T_2)$ is below 0.001, whereas, in Iscol's unstable FCC [9], $\theta_1(\partial R_1/\partial T_1)$ takes a large value because of the peculiarity of a temperature cubic term in the rate of coking [12].

In a fluidized bed it is not easy to obtain very small values of $C_{MCH}(L_f)$ and $C_{O_2}(L_f)$ because of gas by-passing through the bed. In the reaction model of Figure 6, the lower limits of $C_{MCH}(L_f)$ and $C_{O_2}(L_f)$ are about 10^{-5} and 0.02, respectively, and then the values of $\theta_1(\partial R_1/\partial T_1)$ and $-\theta_2(\partial R_2/\partial T_2)$ were 0.013 and 0.008, respectively. This would cause the system to be unstable.

If, as an extreme, we let the gas interchange coefficient between the bubble and dense phases in a fluidized cracker equal infinity, the fluidized cracker can be regarded as a single phase such as P–B type, and the values of $C_{MCH}(L_f)$ can be easily dropped below the order of 10^{-8}. In this case, the value of $\theta_1(\partial R_1/\partial T_1)$ becomes less than zero, which may cause the system to be stable.

STABILITY ANALYSIS

As described in the previous section, the fluid catalytic crackers show complicated feature in the transient behavior with changes in operating conditions. It is therefore important to develop general criterion for stability and multiplicity of steady states. Iscol [9] modeled an FCC by four ordinary differential equations describing the coke and heat balances in reactor and regenerator, and it was found by using a linearization and eigenvalue study that the FCC has multiple steady states and is unstable within the region of practical operations. On the other hand, Lee and Kugelman [10] found a unique and stable steady state by an eigenvalue study in the case of low oxygen concentration at the regenerator exit. The existence of multiple steady states in the FCC was also reported by Elnashaie and El-Hennawi [16], who obtained the number of steady states from the intersections of heat generation and heat removal functions at the steady states.

In the preceeding section, we learned that system stability is influenced by changes in the coke forming and burning rates with temperature. This section provides universal evidence to predict the stability and multiplicity of steady states.

Necessary and Sufficient Conditions for Existence of Steady States

Let's first consider the B–B type. When a steady state exists, the time derivative terms in Equations 15 through 18 vanish. Substituting Equation 15 into Equation 16, we obtain

$$R_1\theta_1 = -R_2\theta_2 \tag{24}$$

By eliminating T_2 from Equations 17 and 18, we obtain

$$\frac{1}{F_c\eta}\left[Q_2+\left(1+\frac{F_2C_{p2}}{F_cC_{pw}}\right)Q_1\right] = \beta_1T_1-\beta_2 \tag{25}$$

where η denotes the heat of coke combustion, which is related to the heat of coke burning, Q_2/F_c, through the following equation.

$$Q_2/F_c = -R_2\theta_2\eta \tag{26}$$

and

$$\beta_1 = \frac{1}{F_c\eta}\left(F_1C_{p1}+F_2C_{p2}+\frac{F_1F_2C_{p1}C_{p2}}{F_cC_{pw}}\right) \tag{27}$$

$$\beta_2 = \frac{1}{F_c\eta}\left(F_1C_{p1}T_1^0+F_2C_{p2}T_2^0+\frac{F_1F_2C_{p1}C_{p2}}{F_cC_{pw}}T_1^0\right) \tag{28}$$

The value of $(1+F_2C_{p2}/F_cC_{pw})Q_1$ in Equation 25 is negligibly smaller than that of Q_2 [1, 14], and then Equation 25 reduces to

$$Q_2/F_c\eta = \beta_1T_1-\beta_2 \tag{29}$$

Further, by eliminating T_2 from Equations 17 and 18 and rearranging it together with Equations 24, 26, and 29, the following relations are obtained:

$$\beta_1T_1-\beta_2 = R_1\theta_1 = -R_2\theta_2 \tag{30}$$

or

$$\frac{\beta_1T_2-\beta_2'}{1+F_1C_{p1}/F_cC_{pw}} = R_1\theta_1 = -R_2\theta_2 \tag{31}$$

where

$$\beta_2' = \frac{1}{F_c\eta}\left(F_1C_{p1}T_1^0 + F_2C_{p2}T_2^0 + \frac{F_1F_2C_{p1}C_{p2}}{F_cC_{pw}}T_2^0\right) \quad (32)$$

Steady states are obtained from the solution of Equation 30 or Equation 31. Therefore, the existence of the solutions becomes the necessary and sufficient conditions for existence of steady states.

Stability Criteron

The FCC model given by Equations 15 through 18 involving nonlinear terms such as R_1, R_2, Q_1, and Q_2 is usually too complicated for deriving the relation of stability directly. When Q_1 in Equation 17 can be set to zero, as previously described, the FCC model can be linearized around a steady state as follows:

$$d\hat{X}/dt = (A+B)\hat{X} \quad (33)$$

where

$$\hat{X}^T = [\hat{n}_1, \hat{n}_2, \hat{T}_1, \hat{T}_2] \quad (34)$$

$$B = \begin{bmatrix} \partial R_1/\partial n_1 & 0 & \partial R_1/\partial T_1 & 0 \\ 0 & \partial R_2/\partial n_2 & 0 & \partial R_2/\partial T_2 \\ 0 & 0 & 0 & 0 \\ 0 & -(\partial R_2/\partial n_2)\eta/C_{pw} & 0 & -(\partial R_2/\partial T_2)\eta/C_{pw} \end{bmatrix} \quad (35)$$

the superscript "^" denotes deviation from steady state, and A is the matrix of the linear terms in Equations 15 through 18. Applying the Hurwitz stability criterion to Equation 33, we can obtain the sufficient conditions for stable steady states such that

$$\theta_1(\partial R_1/\partial T_1) < \beta_1 \quad (36)$$

and

$$-\theta_2(\partial R_2/\partial T_2) < \frac{\beta_1}{1+F_1C_{p1}/F_cC_{pw}} \quad (37)$$

where n_1 and n_2 don't appear. The detailed deviations are given in the "Appendix" section. Both the values of $\partial R_1/\partial n_1$ and $\partial R_2/\partial n_2$ are inherently negative. In the case of considerably low values of $\partial R_1/\partial n_1$ and $\partial R_2/\partial n_2$, at times we can find that the state variables approach the steady state point even though Equations 36 and 37 are not satisfied. However, at any operation conditions, if Equations 36 and 37 are satisfied, steady states are all stable.

It is, on the other hand, noted from Equations 36 and 37 that the sufficient condition for stable steady state is close to the partial derivatives of Equations 30 and 31 with respect to T_1 and T_2, respectively.

Similar behavior is also seen in the graphical steady state analysis of a CSTR described by van Heerden [53]. It is therefore considered that the graphical analysis of Equation 30 or Equation 31 may be useful in detecting the stability and multiplicity of steady state in the FCC.

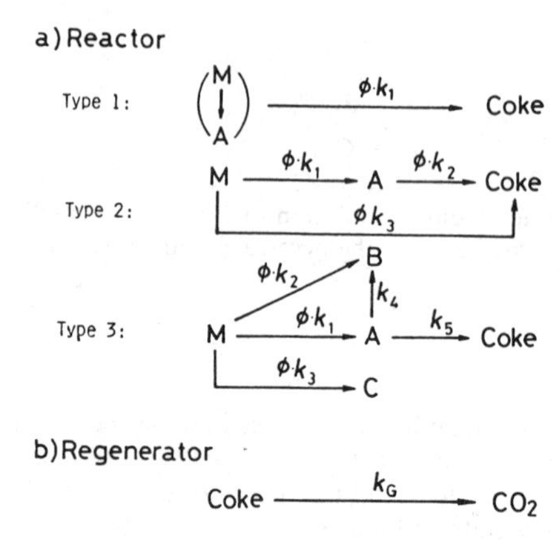

Figure 12. Chemical reaction types in FCC (M: feedstock; A, B, C: products) [15].

Development of Stability and Multiplicity Criteria

Three Types of Chemical Reaction Kinetics

As suggested from Equations 36 and 37, stability may be dependent on the reaction kinetics R_1 and R_2 in the FCC. To examine the effects of reaction kinetics on stability, three types of coke forming rates given in Figure 12 are chosen here. Type 1 is the coking scheme given by Lee and Kugelman [10], where the coke-forming rate is based on catalyst residence time. Type 2 is the cracking of gas oil presented by Weekman [27]. Type 3 is based on methylcyclohexane (MCH) cracking presented by the authors [20]. In the regenerator, on the other hand, the simplified regeneration model, Coke → CO_2, is employed because the kinetics of coke burning rates in the literature [10, 16, 21] were close to each other. For these chemical reaction types in Figure 12, the coke forming and burning rates based on the two-phase model of the fluidized-bed are summarized in Table 9.

Table 10 shows examples of operating conditions in several FCC, along with several dependent variables such as β_1 and $C_{O_2}(L_f)$. In commercial FCC [13, 43, 44] the values of β_1 range between 0.008 and 0.032 mg-coke/g° C, but an exceptionally high value of β_1 has been reported by Elnashaie and El-Hennawi [16].

Graphical Analysis of Equation 30

To obtain the steady state solutions of Equation 30, let's consider the following three functions.

$$y = \beta_1 T_1 - \beta_2 = f(T_1) \tag{38}$$

$$y = R_1\theta_1 = f(n_1, T_1) \tag{39}$$

$$y = -R_2\theta_2 = f(n_2, T_2) \tag{40}$$

If each of the three functions is graphically illustrated as a plane in a y-n_1-T_1 coordinates, the number of steady states will be calculated from the points of intersection of the three

Table 9
Coke Forming and Burning Rates

Coke Forming Rate R_1 (mg-coke/g-cat · hr)	Coke Burning Rate R_2 (mg-coke/g-cat · hr)
Type 1: $R_1 = v_1\theta_1^{n-1}k_1 + \text{const}$	
Type 2: $R_1 = v_2\varphi(k_3\bar{C}_{M,c}^2 + k_2C_{A,c})\gamma_c + v_2\varphi(k_3C_{M,e}^2 + k_2C_{A,e})\gamma_e$	$R_2 = -v_Gk_Gn_2(\gamma_c \cdot \bar{C}_{O_2,c} + \gamma_e \cdot \bar{C}_{O_2,e})$
Type 3: $R_1 = v_3k_5(\bar{C}_{A,c} \cdot \gamma_c + \bar{C}_{A,e} \cdot \gamma_e)$	

where v_1, v_2, v_3, v_G = coefficients, $\bar{C}_{j,i}$ = mean concentration of component j in axial direction of fluidized bed (i = c, e) suffixes: c = bubble phase, e = dense phase

From Seko, Tone, and Otake [15].

Table 10
Operating Conditions in FCC

Source	F_1 (ton/hr)	F_2 (ton/hr)	F_c (ton/hr)	β_1 (mg-coke/g · °C)	C_{O_2} (mol%)
Type 1	406	247	3,510	0.015	0.2-0.3
Type 2	720	405	4,770	0.019	
Type 2*	5,040	608	3,816	0.145	
Type 3	10	8	110	0.012	2.0-4.0
Type 3	40	32	600	0.008	0.2-4.0
Luckenbach [13]	40	22	160-280	0.016-0.032	
Luckenbach [13]	61	22	240-420	0.015-0.027	
Csicsery [43]	390	235	6000	0.008	0.9-3.2
Yamaguchi [44]	40	35	400	0.013	0.5-1.5

* *Exceptional case.*
From Seko, Tone, and Otake [15].

planes, and the stability will be estimated from the slope of the three planes as compares with Equations 36 and 37. A graphical sketch of Equation 38 is illustrated as a plane in Figure 13, where the slope $\partial y/\partial T_1$ corresponds to the value of β_1.

Figure 14 illustrates the sketches of Equation 39 for the three types of Figure 12. Function y of Type 1 is always concave downward for the T_1-axist at any operating condition when $\Delta E_1 = 2{,}400$ cal/mol, whereas when $\Delta E_1 = 3{,}700$ cal/mol, it becomes convex downward. For Type 2, function y is increased with increasing T_1 and is convex downward at any operating condition. Function y of Type 3 is reduced monotonically for the n_1-axis, while for the T_1-axis, it shows a sigmoidal shape over the wide range of operating conditions.

Figure 15 shows a sketch of Equation 40. Though Equation 37 is a function of n_2 and T_2, it can be transformed into a function of n_1 and T_1 by using Equations 41 and 42, which are derived from the partial derivatives of the steady state forms of Equations 15 and 17, respectively.

$$\frac{\partial n_2}{\partial n_1} = 1 - \theta_1 \frac{\partial R_1}{\partial n_1} > 1 \tag{41}$$

$$\frac{\partial T_2}{\partial T_1} = 1 + \frac{F_1C_{p1}}{F_cC_{pw}} - \frac{1}{F_cC_{pw}} \cdot \frac{\partial Q_1}{\partial T_1} > 1 + \frac{F_1C_{p1}}{F_cC_{pw}} \tag{42}$$

As shown in Figure 15, at any operating condition, the function y always increases with increasing n_1, while with increasing T_1, it shows a monotonical increasing shape having three points of inflection. When examining the relation between the values of $-\theta_2(\partial R_2/\partial T_1)$ and C_{O_2} (L_f) for various operating conditions, we can find that when C_{O_2} (L_f) is less than about

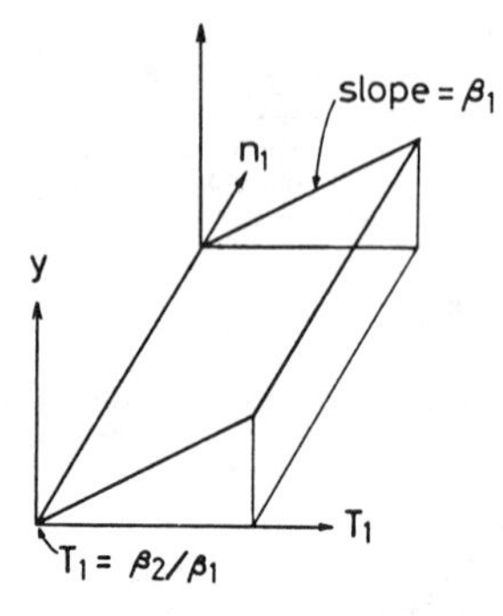

Figure 13. Sketch of Equation 38 [15].

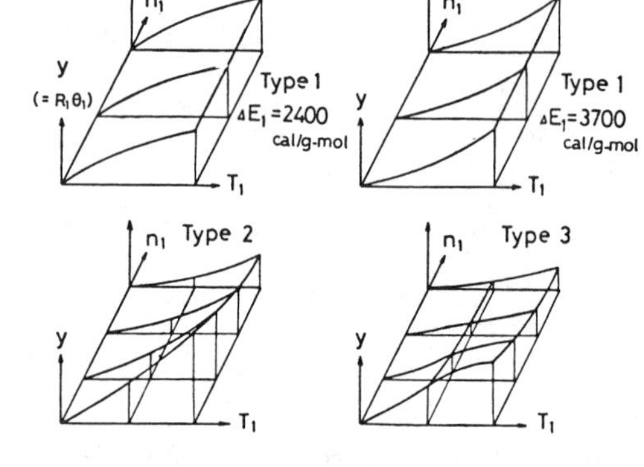

Figure 14. Sketches of Equation 39 for Types 1 to 3 [15].

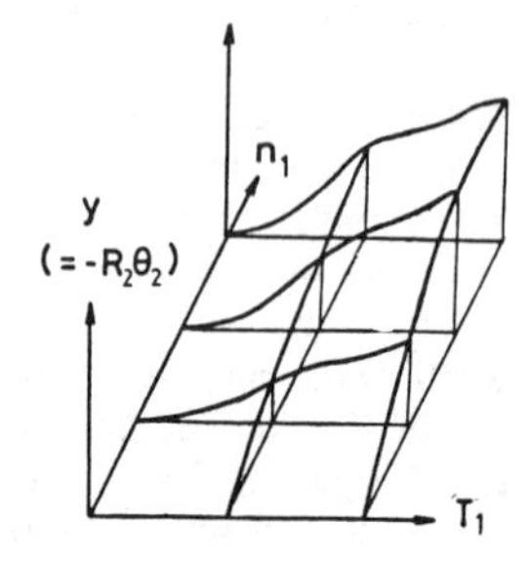

Figure 15. Sketch of Equation 40 [15].

0.3 mol% or greater than about 20 mol%, the value of $-\theta_2(\partial R_2/\partial T_1)$ is smaller than 0.013 mg-coke/g° C which is the average value of β_1 in commercial FCC.

Graphical Analysis

Figure 16 shows the graphical analysis for Type 1, where Steps 1 through 3 show the following procedures.

Step 1: Intersection of the plane of Figure 13 with the plane of Figure 14.
Step 2: Intersection of the plane of Figure 13 with the plane of Figure 15.
Step 3: Overlapping of the two lines obtained by Steps 1 and 2, respectively.

In Step 1, the dotted line AB is the intersection for $\Delta E_1 = 2{,}400$ cal/mol and the solid line AB is for $\Delta E_1 = 3{,}700$ cal/mol. When comparing the slope of Figure 13 with that of Figure 14, we can find that the dotted line AB satisfies Equation 43, but the solid line doesn't satisfy.

$$\theta_1(\partial R_1/\partial T_1) < \beta_1 \tag{43}$$

In Step 2, the curve A′D′E′C′B′ appears only in an exceptionally high value of β_1 [16] because the function $y = -R_2\theta_2$ has three points of inflection for the T_1-axis. In the case of the normal values of β_1 [10, 13, 43, 44], the curve A′C′B′ is obtained as the intersection of Figures 13 and 15. In this step, the dotted lines show the slope condition of $-\theta_2(\partial R_2/\partial T_1) < \beta_1$, which can also be expressed as Equation 44 when Equation 42 is applied.

$$-\theta_2(\partial R_2/\partial T_2) < \frac{\beta_1}{1+F_1C_{p1}/F_cC_{pw}} \tag{44}$$

For the solid lines, the sign of unequality of Equation 44 is inversed.

Step 3 shows the overlapping of the two lines of AB and A′C′B′ obtained by Steps 1 and 2, respectively. A unique steady state exits for each of two different activation energies. Comparison of the conditions of Equations 43 and 44 with the sufficient conditions for stable steady states given by Equations 36 and 37 shows that the point obtained from the two dotted lines is stable and that the other cases are unstable. In the case of $\Delta E_1 = 2{,}400$ cal/mol, if the value of C_{O_2} (L_f) is small, the unique steady state is stable because Equation 44 is satisfied for C_{O_2} $(L_f) < 0.3$ mol% as described previously. This agreed with Lee's result [10].

Figure 17 shows the procedure of Step 3 based on Types 2 and 3. The curve AB is the intersection of the planes given in Figures 13 and 14. On the other hand, the curves A′C′B′ and A′D′E′C′B′ are the intersection of Figures 13 and 15, and are illustrated by using two different values of β_1, one being low and the other high, respectively. In the case of Type 2 having a high value of β_1, two or four possible steady states are recognized. This agreed with the Elnashaie's result [16].

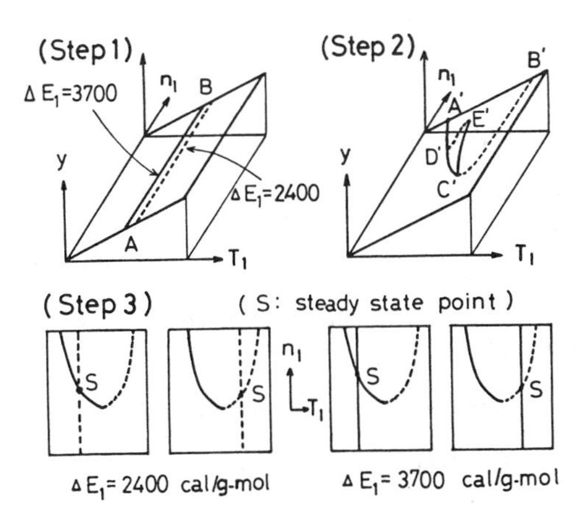

Figure 16. Graphical analysis for type 1 [15].

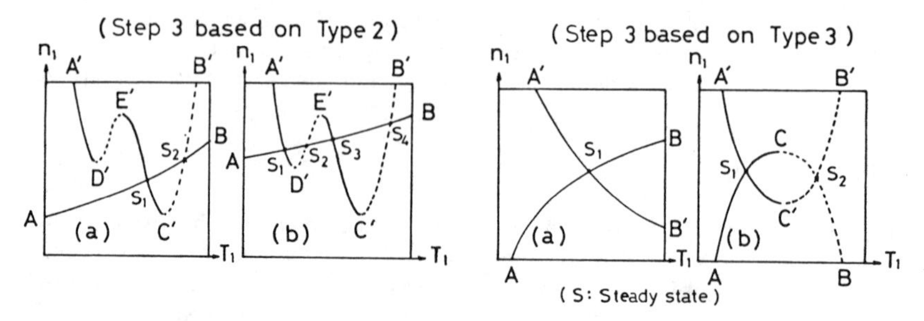

Figure 17. Procedure of Step 3 based on types 2 and 3 [15].

In the preceeding section which was concerned with Type 3, it was found not easy to obtain the slope conditions of Equations 43 and 44 because of large gas by-passing through the fluidized bed. Therefore, the dotted lines are absent as shown in Type 3(a) of Figure 17, and then the steady state obtained is unique and unstable. A decrease in gas by-passing should give drops of the values of $\theta_1(\partial R_1/\partial T_1)$ and $-\theta_2(\partial R_2/\partial T_2)$, and then two possible steady states would be recognized as shown in Type 3(b) of Figure 17.

As, described previously, though the stability and multiplicity are influenced by the chemical reaction types and operating conditions in the FCC, the stability will be predicted from the sufficient conditions for stable steady states given by Equations 36 and 37.

This evidence can also apply to the P-B type because if the gas interchange coefficient K_j is closer to infinity, the fluidized-bed cracker can be regarded as a riser cracker similar to a moving bed.

Examination of Stability Criterion of Steady-State Points

Application of the stability criterion will now be illustrated. For the B-B-type FCC, let us consider the following operation condition such that

$$F_1 = 100 \text{ ton/hr}; \quad \theta_1 = 0.4 \text{ hr}; \quad T_1 = 243.7^\circ \text{ C}$$
$$F_2 = 60 \text{ ton/hr}; \quad \theta_2 = 2.0 \text{ hr}; \quad T_2 = 28.2^\circ \text{ C}$$
$$F_c = 800 \text{ ton/hr}$$

and heat capacities of solids and gas are given:

$$C_{pw} = 0.25 \text{ kcal/kg}^\circ \text{ C};$$
$$C_{p1} = 0.7 \text{ kcal/kg}^\circ \text{ C};$$
$$C_{p2} = 0.27 \text{ kcal/kg}^\circ \text{ C}$$

In addition, let us consider the amounts of coke forming and burning for convenience to develop the system stability.

$$R_1\theta_1 = \omega_1[-10^{-4}(T_1 - 598)^2 + 0.36 - 0.1\omega_3(n_1 - 7)] + 5.3 \tag{45}$$

$$-R_2\theta_2 = \omega_2[-10^{-4}(T_2 - 711)^2 + 0.49 + 0.1\omega_3(n_2 - 1.7)] + 5.3 \tag{46}$$

where ω_1, ω_2, and ω_3 are coefficients.

When the time-derivative terms of Equations 15 through 18 are set to be zero, independent of the values of ω_1, ω_2 and ω_3, we can obtain the steady state solution such that

$$n_1 = 7.0 \text{ mg/g-cat}, \quad T_1 = 538° \text{ C}$$

$$n_2 = 1.7 \text{ mg/g-cat}, \quad T_2 = 641° \text{ C} \tag{47}$$

We learned that system stability of the FCC is governed by the reaction conditions and the sufficient conditions for stable states are given by Equations 36 and 37. Therefore, from Equation 27, the value of β_1 is

$$\beta_1 = 0.0159 \tag{48}$$

and from Equations 45 and 46, partial derivatives of $R_1\theta_1$ and $-R_2\theta_2$ with respect to T_1, T_2, n_1, and n_2 are

$$\theta_1(\partial R_1/\partial T_1) = -2 \times 10^{-4}(T_1 - 598)\omega_1 \tag{49}$$

$$-\theta_2(\partial R_2/\partial T_2) = -2 \times 10^{-4}(T_2 - 711)\omega_2 \tag{50}$$

$$\theta_1(\partial R_1/\partial n_1) = -0.1\omega_1\omega_3 \tag{51}$$

$$\theta_2(\partial R_2/\partial n_2) = -0.1\omega_2\omega_3 \tag{52}$$

We will then select three reaction conditions which can be changed by the use of various values of ω_1, ω_2, and ω_3. These reaction conditions are summarized in Table 11 in which the values of the partial derivatives expressed by Equations 49 through 52 are different with one another. Case A is an example having low values of $\theta_1(\partial R_1/\partial T_1)$ and $-\theta_2(\partial R_2/\partial T_2)$ which satisfies Equations 36 and 37. Case B is an example in which Equation 37 is not satisfied. Case C is the same as Case B, but the absolute values of $\theta_1(\partial R_1/\partial n_1)$ and $\theta_2(\partial R_2/\partial n_2)$ are smaller than those of Case B.

We can find that Case A is stable from the Hurwitz stability criterion and also from Equations 36 and 37. For Case B and Case C, Equations 36 and 37 suggest the possibility of unstable states, but the Hurwitz stability criterion shows that Case B is stable and that Case C is unstable.

In order to confirm these criteria, let us solve Equations 15 through 18 including time-derivative terms under various starting points shown in Table 12 which are near the steady-state point. Figure 18 shows these trajectories of $\hat{n}_1$ and $\hat{T}_1$. In Case A, it is seen that the state trajectories shift toward the steady-state point and that Case A is stable. Further, we can see that the state trajectories approach the steady-state point even though

Table 11
Relation Between Stability and Reaction Conditions

	Case A	Case B	Case C
Coefficients	$\omega_1 = \omega_3 = 1.0$, $\omega_2 = 0.7$	$\omega_1 = \omega_2 = \omega_3 = 1.0$	$\omega_1 = \omega_2 = 1.0$, $\omega_3 = 0.32$
$\theta_1(\partial R_1/\partial T_1)_{ss}$	0.012	0.012	0.012
$-\theta_2\kappa(\partial R_2/\partial T_2)_{ss}$	0.0132	0.0189	0.0189
$\theta_1(\partial R_1/\partial n_1)_{ss}$	−0.1	−0.1	−0.032
$\theta_2(\partial R_2/\partial n_2)_{ss}$	−0.07	−0.1	−0.032
Hurwitz criterion	Stable	Stable	Unstable
Equations 36 and 37	Satisfied	Unsatisfied	Unsatisfied

where $\kappa = 1 + F_1C_{p1}/F_cC_{pw} = 1.35$; $\beta_1 = 0.0159$

Table 12
Starting Points for Solving Equations 15 to 18

Key	Starting Values $\hat{n}_1$	$\hat{n}_2$	$\hat{T}_1$	$\hat{T}_2$
(a)	−2.0	−1.2	10.0	7.0
(b)	2.0	2.8	10.0	7.0
(c)	−2.0	−1.2	−15.0	−18.0
(d)	2.0	2.8	−15.0	−18.0

where $\hat{n}_1 = n_1 - 7.0$; $\hat{n}_2 = n_2 - 1.7$; $\hat{T}_1 = T_1 - 538$, $\hat{T}_2 = T_2 - 641$.

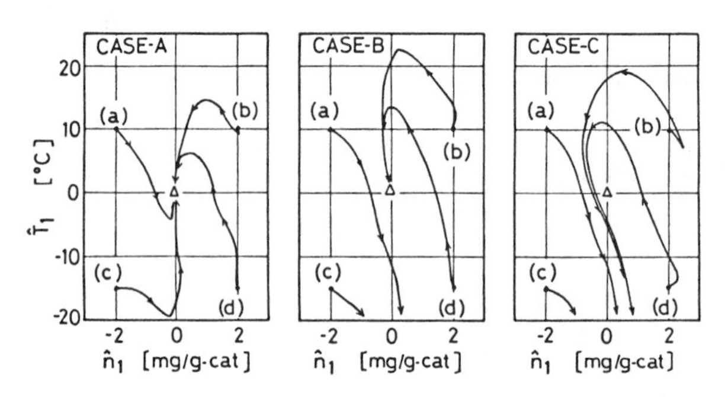

Figure 18. State trajectories for various reaction conditions.

the starting points are considerably for apart from it. In Case B, the trajectories from starting point (b) and (d) approach the steady-state point, but the ones from starting point (a) and (c) diverge from the steady-state point. In such a case, the FCC will not be possible to keep a stable operation without control because there are many disturbances in the FCC. In Case C having low absolute values of $\partial R_1/\partial n_1$ and $\partial R_2/\partial n_2$, we can see that the trajectories from any starting point diverge and that Case C is unstable.

Equations 36 and 37 provide sufficient conditions for stable state, but are not the necessary ones. Though it is not possible to judge rigorously the unstable state in the majority of cases, the useful evidence for design and operation can be obtained from Equations 36 and 37.

APPENDIX

The characteristic equation of Equation 33 is

$$a_0\lambda^4 + a_1\lambda^3 + a_2\lambda^2 + a_3\lambda + a_4 = 0 \quad (1)$$

and the coefficients in Equation 1 are

$$a_0 = 1 \quad (2)$$

$$a_1 = A + C + D \quad (3)$$

$$a_2 = B + CD - \frac{1}{\theta_1\theta_2} + A(C + D) + E\frac{\partial R_2}{\partial T_2} \quad (4)$$

$$a_3 = A\left(CD - \frac{1}{\theta_1\theta_2}\right) + B(C+D) + E\frac{\partial R_2}{\partial T_2}\left(\frac{1+\alpha_1}{\theta_1} + C\right) \tag{5}$$

$$a_4 = B\left(CD - \frac{1}{\theta_1\theta_2}\right) + \frac{E}{\theta_1}\left\{C(1+\alpha_1)\frac{\partial R_2}{\partial T_2} + \frac{1}{\theta_2}\cdot\frac{\partial R_1}{\partial T_1}\right\} \tag{6}$$

where

$$\alpha_1 = \frac{F_1C_{p1}}{F_cC_{pw}} \tag{7}$$

$$\alpha_2 = \frac{F_2C_{p2}}{F_cC_{pw}} \tag{8}$$

$$A = \frac{1+\alpha_1}{\theta_1} + \frac{1+\alpha_2}{\theta_2} + \frac{\eta}{C_{pw}}\cdot\frac{\partial R_2}{\partial T_2} \tag{9}$$

$$B = \frac{\alpha_1\alpha_2+\alpha_1+\alpha_2}{\theta_1\theta_2} + \frac{1+\alpha_1}{\theta_1}\cdot\frac{\eta}{C_{pw}}\cdot\frac{\partial R_2}{\partial T_2} \tag{10}$$

$$C = \frac{1}{\theta_1} - \frac{\partial R_1}{\partial n_1} \tag{11}$$

$$D = \frac{1}{\theta_2} - \frac{\partial R_2}{\partial n_2} \tag{12}$$

$$E = \frac{\eta}{C_{pw}}\cdot\frac{\partial R_2}{\partial n_2} \tag{13}$$

The Hurwitz stability criterion requires following conditions for Equation 1 such that

$$a_0 > 0,\ a_1 > 0,\ a_2 > 0,\ a_3 > 0,\ a_4 > 0 \tag{14}$$

$$a_1a_2 - a_0a_3 > 0 \tag{15}$$

$$a_1a_2a_3 - a_1^2a_4 - a_0a_3^2 > 0 \tag{16}$$

In industrial FCC, as the values of $(\partial R_1/\partial n_1)$, $(\partial R_2/\partial n_2)$, and $(\partial R_2/\partial T_2)$ are negative [10] the use of Equations 14 through 16 yields that

$$\frac{\alpha_1\alpha_2+\alpha_1+\alpha_2}{\theta_1\theta_2} + \frac{1}{\theta_1}\cdot\frac{\eta}{C_{pw}}\cdot\frac{\partial R_1}{\partial T_1} > 0 \tag{17}$$

$$\frac{\alpha_1\alpha_2+\alpha_1+\alpha_2}{\theta_1\theta_2} + \frac{1+\alpha_1}{\theta_1}\cdot\frac{\eta}{C_{pw}}\cdot\frac{\partial R_2}{\partial T_2} > 0 \tag{18}$$

On the other hand, the value of β_1 defined by Equation 27 is written in the form of Equation 19 by using Equations 7 and 8.

$$\beta_1 = \frac{C_{pw}}{\eta}(\alpha_1\alpha_2+\alpha_1+\alpha_2) \tag{19}$$

Substituting Equation 19 into Equations 17 and 18, we obtained Equations 36 and 37, which represent the sufficient conditions for stable steady states.

NOTATION

A_T cross-sectional area of the fluidized bed (cm^2)
$C_{j,c}$, $C_{j,e}$ concentration of gaseous component j in the bubble and in the dense phases, respectively (j = MCH, Ole, Par) (wt-fraction); (j = CO_2, CO, O_2) (mol-fraction)
$C_j(L_f)$ concentration of gaseous component j at the outlet of vessel, respectively (j = MCH, Ole, Par) (wt-fraction); (j = CO_2, CO, O_2) (mol-fraction)
$\bar{C}_j$ mean concentration of gaseous component j at a given level in the bed (j = MCH, Ole, Par) (wt-fraction); (j = CO_2, CO, O_2) (mol-fraction)
C_{pi} specific heat of gas (i = 1, 2) (kcal/kg° C)
C_{pw} specific heat of catalyst (kcal/kg° C)
D_G molecular diffusivity of gas (cm^2/sec)
D_i diameter of vessel i (i = 1, 2, T) (cm)
d_b bubble size (cm)
d_b^* repesentative bubble size (cm)
d_c equivalent cloud diameter (cm)
d_p particle diameter (cm)
F_c catalyst circulation rate (ton/hr)
F_1 feed rate of feedstock (ton/hr)
F_2 air flow rate in the regenerator (ton/hr)
G volumetric gas flow rate through a nozzle (cm^3/sec)
g acceleration of gravity (cm/sec^2)
K_b, K_c gas interchange coefficient between bubble and cloud-wake and between cloud-wake and dense phase, respectively (1/sec)
K_j gas interchange coefficient of component j between bubble and dense phases (1/sec)
k_m rate constant for cracking (m = 1–5) (1/sec)
k_{Rm} rate constant for regeneration (m = 1–3) (1/min)
k_{R4} rate constant for regeneration (1/g-cat min)
L_c, L_f, L_{mf} bed height at static, at fluidization, and at incipient fluidizing conditions, respectively (cm)
n_c, n_H carbon (coke) and hydrogen content on catalyst, respectively (mol/g-cat)
n_i coke content on catalyst in vessel i (i = 1, 2) (mg/g-cat)
Q_1 rate of heat absorption by cracking (kcal ton/kg hr)
Q_2 rate of heat generation by regeneration (kcal ton/kg hr)
R gas constant (1.987) (cal/g-mol° C)
R_1 rate of coking (mg-coke/g-cat hr)
R_2 rate of coke burning (mg-coke/g-cat hr)
Re_p $d_p u_0 \varrho_g/\mu$ particle Reynolds number
Re_c $\varrho_f u_R d_c/\mu$ Reynolds number
r_j reaction rate of component j (j = MCH, Ole, Par) (1/sec); (j = CO_2, CO, O_2) (mol/g-cat min)
$r_{j,c}$, $r_{j,e}$ reaction rate of component j in the bubble and dense phases, respectively (j = MCH, Ole, Par) (1/sec); (j = CO_2, CO, O_2) (mol/g-cat min)
T absolute temperature (°K)
T_i temperature in vessel i (i = 1, 2) (°C)
T_1^0 feed temperature of feedstock (°C)
T_2^0 feed temperature of air (°C)
t process time (hr)
u_b bubble rising velocity (cm/sec)
u_b^* representative bubble rising velocity (cm/sec)
u_c catalyst circulation rate (cm/sec)
u_{mf} incipient fluidization velocity (cm/sec)
u_0 superficial gas velocity based on empty bed (cm/sec)

u_R relative velocity between cloud and interstitial gas (cm/sec)

W_i weight of catalyst in vessel i (i = 1, 2) (ton)

$\underset{\sim}{X}$ vector of dependent variables

Z bed height from the bottom of vessel (cm)

Greek Symbols

α decay constant

β_1 a constant defined by Equation 27 (mg-coke/g° C)

β_2 a constant defined by Equation 28 (mg-coke/g)

γ_c, γ_e ratios of solid in the wake and in the dense regions to volume of bubbles, respectively

ΔE_1 activation energy of coke formation (cal/mol)

ΔH_{MCH} heat of MCH cracking (100) (kcal/kg)

ΔH_{CO_2} heat of formation of CO_2 (kcal/kg)

ΔH_{CO} heat of formation of CO (kcal/kg)

δ_w ratio of wake volume to bubble volume

ε_b bubble volume fraction

ε_c voidage in a packed bed

$\varepsilon_{mf}, \varepsilon_f$ voidage in a bed at incipient fluidizing condition and in a bubble bed as a whole respectively

η heat of coke combustion (kcal g-cat/mg-coke kg)

θ_i catalyst residence time in vessel i (i = 1, 2) (hr)

μ viscosity of gas (g/cm sec)

ν_i coefficient

ϱ_B, ϱ_B bulk density of catalyst at static and at reaction condition, respectively (g/cm^3)

ϱ_f, ϱ_s fluid density and density of solids, respectively (g/cm^3)

ϱ_j fluid density (j = Air, CO_2, CO) (g/cm$_3$)

φ decay function

φ_s sphericity of a particle

Subscripts and Superscripts

1 reactor (cracker)

2 regenerator

c bubble phase

e dense phase

ss steady state

CO carbon monoxide

CO_2 carbon dioxide

MCH methylcyclohexane

Ole Olefins

Par Paraffins

$(\ )^T$ transport of matrix

$(\bar{\ })$ average

$(\hat{\ })$ deviation from steady state ($\hat{A} = A - A_{ss}$)

REFERENCES

1. Nelson, W. L., *Petroleum Refinary Engineering*, McGraw–Hill, New York (1958).
2. "Market & Technical Survey of the International Catalysts and Catalytic Processes Industry," Chemica; Hitech Inc. (1981).
3. Davis, B. H., and Hettinger, W. P., (Eds.) "Heterogeneous Catalysts Selected American Histories," ACS Sym. Ser., Vol. 222 (1983).
4. Kunii, D., and Levenspiel, O., *Fluidization Engineering*, John Wiley and Sons, New York (1969).
5. Meisel, S. L., McCallough, J. P., and Lechtzler, C. H., "Gasoline from Methanol in One Step," *CHEMTECH*, pp. 86, (1976).
6. Planck, C. J., and Rosinski, E. J., U.S. Patent 3, 140, 249.

7. Rollman, L. D., *Inorganic Compounds with Unusual Properties* Vol. II, pp. 387 King, R. B. (Ed.), A. C. S., New York (1979), p. 387.
8. Rollman, L. D., and Walsh, D. E., *Process Catalyst Deactivation* (Figueiredo, J. L. ed.,) Vol. 81, Martinus Nijhoff Publisheres, Hague (1982).
9. Perlmutter, D. D., *Stability of Chemical Reactors,* Prentice-Hall (1972), p. 274.
10. Lee, W., and Kugelman, A. M., "Number of Steady-State Operating Points and Local Stability of Open Loop Fluid Catalytic Cracker," *Ind. Eng. Chem. Process Des. Dev.,* Vol. 12 (1973), p. 197.
11. Bromley, J. A., and Ward, T. J., "Fluidized Catalytic Cracker Control," *Ind. Eng. Chem. Process Des. Dev.,* Vol. 20, (1981) pp. 74.
12. Lee, W., and Weekman, V. W., "Advanced Control Practice in the Chemical Process Industry," *AIChE J.,* Vol. 22 (1976), p. 27.
13. Luckenbach, E. C., "How to Update a Catalytic Cracking Unit," *Chem. Eng. Progr.,* Vol. 75, No. 1 (1979), p. 56.
14. Seko, H., Tone, S., and Otake, T., "Operation and Control of a Fluid Catalytic Cracker," *J. Chem. Eng. Japan,* Vol. 11 (1978), p. 130.
15. Seko, H., Tone, S., and Otake, T., "Criterion of Stability of Steady States and Its Prediction," *J. Chem. Eng. Japan,* Vol. 15 (1982), p. 305.
16. Elnashaie, S. S. E. H., and El-Hennawi, I. M., "Multiplicity of the Steady State in Fluidized Bed Reactor," *Chem. Eng. Sci.,* Vol. 34 (1979), p. 1113.
17. Voorhies, A., Jr. "Carbon Formation in Catalytic Cracking," *Ind. Eng. Chem.,* Vol. 37, No. 4 (1945), p. 318.
18. Lin, C., Parks, S. W., and Hatcher, W. J., Jr., "Zeolite Catalyst Deactivation by Coking," *Ind. Eng. Chem. Process Des. Dev.,* Vol. 22 (1983), p. 609.
19. Ozawa, Y., and Bischoff, K. B., "Coke Formation Kinetics on Silica-Alumina Catalyst" *Ind. Eng. Chem. Process Des. Dev.,* Vol. 7 (1968), p. 72.
20. Tone, S., et al., "Kinetics of Methylcyclohexane Cracking over Silica-Alumina Catalyst," *Bull. Japan Petrol. Instn.* Vol. 13 (1971), p. 39.
21. Tone, S., Miura, S., and Otake, T., "Kinetics of Oxidation of Coke on Silica-Alumina Catalyst," *Bull. Japan Petrol. Instn.,* Vol. 14 (1972), p. 76.
22. Toomey, R. D., and Johnstone, H. F., "Gaseous Fluidization of Solid Particles," *Chem. Eng. Progr.,* Vol. 48 (1952), p. 220.
23. Davidson, J. F., and Harrison, D., "Fluidization," Academic Press, London and New York (1971).
24. Fryer, C., and Potter, O. E., "Countercurrent Backmixing Model for Fluidized Bed Catalytic Reactor: Applicability of Simplified Solutions," *Ind. Eng. Chem. Fundams.,* Vol. 11 (1972), p. 338.
25. Kunii, D., and Levenspiel, O., "Bubble Bed Model," *Ind. Eng. Chem. Fundams.,* Vol. 7 (1968), p. 446.
26. Kato, K., and Wen, C. Y., "Bubble Assemblage Model for Fluidized Bed Catalytic Reactors," *Chem. Eng. Sci.,* Vol. 24 (1969), p. 1351.
27. Weekman, V. W., Jr. "Kinetics and Dynamics of Catalytic Cracking Selectivity in Fixed Bed Reactors," *Ind. Eng. Chem. Process Des. Dev.,* Vol. 8 (1969), p. 385.
28. Massoth, F. E., "Oxidation of Coked Silica-Alumina Catalyst" *Ind. Eng. Chem. Process Des. Dev.,* Vol. 6 (1967), p. 200.
29. Venuto, P. B., and Habib, E. T., "Catalyst-Feedstock-Engineering Interactions in Fluid Catalytic Cracking," *Catal. Rev-Sci. Eng.,* Vol. 18, No. 1 (1978), p. 75.
30. Mauleon, J. L., "Dimersol Scores High in Commercial Operation," *Oil Gas J.,* Vol. 78, No. 9 (1980), p. 63.
31. Tone, S., et al., "Catalytic Cracking of Methylcyclohexane over Silica-Alumina Catalyst in Gas Fluidized Bed," *J. Chem. Eng. Japan,* Vol. 7 (1974), p. 44.
32. Partridge, B. A., and Rowe, P. N., "Chemical Reaction in a Bubbling Gas-Fluidized Bed," *Trans. Inst. Chem. Engrs.,* Vol. 44 (1966), p. 335.

33. Kobayashi, H., and Arai, F., "Effects of Several Factors on Catalytic Reaction in a Fluidized Bed Reactor," *Kagaku Kogaku (Japan)*, Vol. 29, (1965), p. 885.
34. Reid, R. C., and Sherwood, T. K., "Properties of Gases and Liquids," McGraw–Hill, New York (1958), p. 543.
35. Seko, H., et al., "Regeneration of Coked Spent Catalyst in Fluidized Bed," *Kagaku Kogaku Ronbunshu (Japan)*, Vol. 2, (1976), p. 176.
36. Pattipati, R. R., and Wen, C. Y., "Minimum Fluidization Velocity at High Temperatures," *Ind. Eng. Chem. Process Des. Dev.*, Vol. 20 (1981), p. 705.
37. Singh, B., Rigby, G. R., and Callcott, T. G., "Measurement of Minimum Fluidization Velocities at Elevated Temperature," *Trans. Instn. Chem. Engrs.*, Vol. 51 (1973), p. 93.
38. Orcutt, J. C., Davidson, J. F., and Pigford, R. L., "Reaction Time Distribution in Fluidized Catalytic Reactors," *Chem. Eng. Progr. Symp. Ser.*, Vol. 58, No. 38 (1962), p. 1.
39. Woollard, N. M., and Potter, O. E., "Solid Mixing in Fluidized Bed," *AIChE J.*, Vol. 14 (1968), p. 388.
40. Otake, T., et al., "Behavior of Rising Bubbles in a Gas-Fluidized Bed," *J. Chem. Eng. Japan*, Vol. 8 (1975), p. 388.
41. Mori, S., and Wen, C. Y., "Estimation of Bubble Diameter in Gaseous Fluidized Bed," *AIChE J.*, Vol. 21 (1975), p. 109.
42. Geldart, D., and Kapoor, D. S., "Bubble Size in a Fluidized Bed at Elevated Temperatures," *Chem. Eng. Sci.*, Vol. 31 (1976), p. 842.
43. Csicsery, S. M., et al., "Catalyst and Gas Samplers for Fluid Catalytic Cracker Regenerator," *Ind. Eng. Chem. Process Des. Dev.*, Vol. 14 (1975), p. 93.
44. Yamaguchi, T., "Investigation of Fluid Catalytic Cracking: About the Operating Factors," *Sekiyu Gakkaishi (Japan)*, Vol. 3 (1960), p. 629.
45. Tolen, D. F., "Advances in Catalytic Cracking Technology," *Pet. Int.* (Tulsa, Okla), Vol. 39, No. 7 (1981), p. 29.
46. U. S. Patent 3, 974, 091
47. Vasalos, I. A., et al., "New Cracking Process Controls FCC Sulfur Oxide," *Oil Gas J.*, Vol. 75, No. 26 (1977), p. 41.
48. Rush, J. B., "Twenty Years of Heavy Oil Cracking," *Chem. Eng. Progr.*, Vol. 77, No. 12 (1981), p. 29.
49. Finneran, J. A., Murphy, J. R., and Whittington, E. L., "Heavy Oil Cracking Process for Clean Fuel Production," *Oil Gas J.*, Vol. 72, No. 2 (1974), p. 52.
50. Edison, R. R., Siemssen, J. O., and Masologites, G. P., "Crude and Reside Can Be Cat-Cracker Feeds," *Oil Gas J.*, Vol. 74, No. 51 (1976), p. 54.
51. Jacob, S. M., "Carbon Distribution Functions in a Fluid Catalytic Cracker," *Ind. Eng. Chem. Process Des. Dev.*, Vol. 9 (1970), p. 635.
52. Seko, H., Tone, S., and Otake, T., "A Consideration of the Treatment of Coke Distribution in a Fluid Catalytic Cracker," *J. Chem. Eng. Japan*, Vol. 10 (1977), p. 493.
53. Himmelblau, D. M., and Bischoff, K. B., "Process Analysis and Simulation," John Wiley and Sons, New York (1968), p. 148.

CHAPTER 29

HEAT AND MASS TRANSFER IN GAS JET-AGITATED REACTORS

Ana L. Cukierman

Programa de Investigacion y Desarrollo de
Fuentes Alternativas de Materias Primas y
Energia
Ciudad Universitaria
Buenos Aires, Argentina

CONTENTS

INTRODUCTION

Stirred tank reactors have been traditionally used for liquid-phase reactions. However, since the 1960s, there has been renewed interest in using these reactors for carrying out homogeneous and heterogeneous gas-phase reactions.

Advantages of stirred reactors have been extensively discussed elsewhere [1, 2]. Due to the homogeneity of the reaction mixture composition, reaction rates may be obtained directly from a simple mass balance equation. For relatively fast reactions and short residence times, where open reactors are needed, stirred reactors are more adequate than tubular reactors because it is easier to approach perfect mixing than plug flow behavior. Furthermore, stirring makes easier the temperature control, which is not always possible for the other geometries. Consequently, these reactors are appropriate for carrying out kinetic studies.

Herndon [3] published a complete review of homogeneous reactions conducted in gas-phase stirred reactors. From then on, several reactor devices have been presented in the literature. These cover a wide spectrum ranging from simple designs up to sophisticated reactors, particularly for applications to gas-solid reactions [4].

Stirring in gas-phase reactors may be achieved in three ways:

1. Mechanically, with the disadvantages caused by the perturbations of the stirrer, basically in corrosive or unstable media.
2. By molecular or turbulent diffusion; molecular diffusion may be enough for ensuring the medium homogenization at low pressure.
3. By the gas movement created by means of nozzles (turbulent jets), which generate streamlines with internal recycle [1].

The author thanks Dr. Norberto Lemcoff for helpful discussions.

Perhaps, the major advantage of the jet-stirred reactors is that no moving parts are present and therefore, the mechanical complications of the stirrer and its bearing are avoided [5]. Another aspect to be considered is that of power consumption. When the reactor radius is large, power consumption in a jet-stirred reactor is less than in a mechanically stirred one, but the opposite is found when the reactor radius is small. It should be expected that perfect mixing would not be achieved in the mechanically stirred reactor for space times lower than 5 seconds. Conversely, this bad mixing state may be attributed to the low power input per unit volume in the mechanically stirred reactor as compared to the jet-stirred reactor [6].

The design and construction of jet-stirred reactors require some conditions to be fullfilled so that homogenization of the gaseous mixture may be achieved throughout the reactor [5, 7]. Basically these are:

- *Condition of sound limit.* At the nozzle orifice, the velocity of the gas u_N must be lower than the local speed of sound of the gas at the temperature and pressure under consideration:

$$u_N = \frac{4F_{VN}}{\pi d_N^2} < u_s \tag{1}$$

- *Condition of fully developed jet.* This condition has been found experimentally by Liepmann and Laufer [5, 7]:

$$\frac{u_N x \varrho}{\mu} \geqq 7.10^4 \tag{2}$$

- *Condition of recycle limitations.* The reactor content must be recycled by the jets. For a free jet, the recirculation ratio can be evaluated from [8]:

$$\frac{F_V}{F_{VN}} = 0.30\,\frac{x}{d_N} \tag{3}$$

Bush [5] and David and Matras [7] proposed the basis for the design and operation of gas-phase jet-stirred reactors. However, residence-time distribution experiments and direct temperature measurements must be carried out in order to ensure ideal mixing behavior [9].

Gas-phase jet-stirred reactors have been used for both homogeneous and catalytic heterogeneous gas-solid systems. Some applications are listed in Table 1.

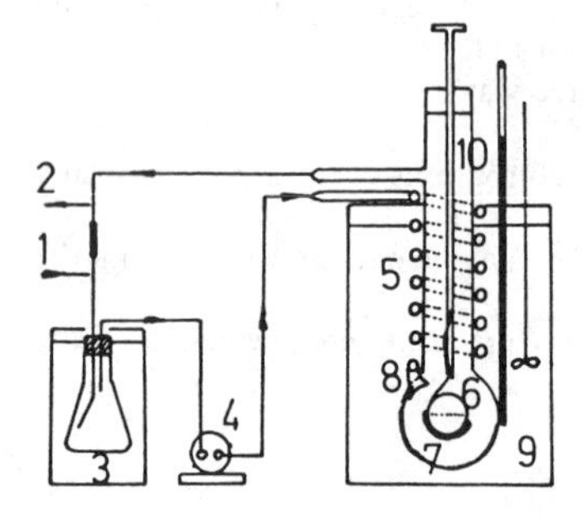

Figure 1. Single-pellet reactor [14].

Table 1
Some Chemical Reactions Carried Out in Gas Agitated Reactors

Reaction	Reaction Characteristics	Operating Conditions	Details of the Reactor		Details of the Injector			Reference
			Geometry	d_R (mm)	Geometry	d_N (mm)	N	
Hydrocarbon combustion	Homogeneous	T > 2,000 °K P = 0.125–1 atm	Spherical	76.2, 152.4	Spherical	1.2, 2.4	68–102	Longwell and Weiss [11]
Reactions of nitrogen atoms	Homogeneous	P = 2.63 × 10^{-3} atm	Spherical	≅ 36	–	5	1	Kistiakowsky and Volpi [12]
Decomposition of di-t-butyl peroxide	Homogeneous	T = 430–550 °K P = 0.013 atm τ = 1–60 sec	Spherical	80	Spherical	0.8	28	Mulcahy and Williams [13]
Catalytic oxidation of H_2	Heterogeneous	T_B = 353–473 °K P = 1 atm	Spherical	44	–	–	2	Maymó and Smith [14]
Radical-reaction: $O + OH \rightarrow O_2 + H$	Homogeneous	T = 228–340 °K	Cylindrical	37	–	1	4	Westenberg et al. [15]
Thermal decomposition of bicycloheptene	Homogeneous	T = 600–685 °K τ = 2 sec	Spherical	140	Cross	1	4	Matras and Villermaux [1]
Dimerization of butadiene	Homogeneous, rapid	T = 773–873 °K τ = 2–4 sec	Spherical	140	Cross	1	4	David and Villermaux [16]
Catalytic oxidation of H_2	Heterogeneous	T_B = 367–453 °K	Spherical	90	–	–	2	Patel and Smith [17]
Reactions of atomic hydrogen:								
• Recombination on the reactor wall	Heterogeneous	P = 0.11–0,98 atm	Spherical	37.4	Cross	–	4	Lede and Villermaux [18]
• On propane	Homogeneous	T = 295 °K P = 0.08–0.83 atm	Spherical	37.4	Cross	–	4	
De hydrogenation by I_2 of isobutane and n-butane	Catalytic homogeneous	T = 723–773 °K P = 1 atm	Spherical	50	Cross	0.3	4	David and Villermaux [19]
Neopentane pyrolisis	Homogeneous	T = 933 °K τ = 0.2 sec	Spherical	36	–	–	8	Azay and Come [9]
Direct thermal decomposition of water*	Catalytic Heterogeneous	T > 2000 °K	Spherical	64	–	0.5	4	Lede et al. [20]

* *Water was injected into a zirconia nozzle located at the focus of a solar furnace simulator. The evolved gases were quenched by rapid turbulent cold jets in a stirred reactor.*

The effect of heat and mass transfer resistances on the overall kinetics of a heterogeneous reaction has been analyzed in a large number of articles [10]. The external heat and mass transfer resistances may produce significant differences in the concentration and temperature values between the bulk fluid phase and the external solid surface. Hence, mass and heat transfer coefficients must be known in order to determine or predict the influence of the diffusional phenomena on the overall kinetics. In this chapter, the field of heat and mass transfer in gas-jet agitated reactors is reviewed. The analysis of two different systems, single-pellet and wall reactors, is carried out.

SINGLE-PELLET JET-STIRRED REACTORS

The single-pellet stirred reactor is useful for measuring activity and selectivity of catalyst or reactive solid pellets, especially those having activity profiles.

Maymó and Smith [14] measured rates of oxidation of hydrogen with oxygen on a single spherical platinum-alumina catalyst pellet. Experiments were carried out in a tank reactor stirred by gaseous jets. The purpose of this work was to carry out a kinetic study and to determine experimental effectiveness factors, by comparing the rates of reaction for catalyst particles and pellets. Heat and mass transfer resistances in the gas phase around the catalyst pellet were also determined. Likewise, intrapellet temperature differences within the pellet were measured and values up to 300° C were reported.

In Figure 1, a scheme of the reactor may be observed. It consisted of a 44 mm-diameter spherical chamber with a 21-mm neck. The catalyst pellet was 18.6 mm in diameter. The reaction mixture was recycled through a pump in order to approach the behavior of an ideal stirred-tank reactor. The recycle and feed gases were preheated before entering the reactor through the nozzles.

Surface concentrations were not measured but the mass transfer resistance was estimated by evaluating the concentration drop between the solid surface and the bulk gas phase according to two equations based on the analogy between heat and mass transfer. Assuming laminar flow, the same transport mechanism and effective gas film thickness for heat and mass transfer, the following equation was first applied to estimate the concentration drop:

$$(C_b - C_s)_{O_2} = \frac{k_e(T_b - T_s)}{D_{O_2}(-\Delta H)} \tag{4}$$

where T_b, T_s are the measured bulk and surface temperatures. From Equation 4, it was found that the concentration drop was always less than 5% of the oxygen surface concentration. Since Equation 4 is only valid for laminar flow, the Chilton–Colburn analogy was also applied. In this case, the calculated concentration drop was somewhat less than 4% of the oxygen surface concentration. Therefore, it was concluded that the gas-film mass transfer resistance could not be large.

On the other hand, it was found that heat transfer resistance was significant. In order to measure surface temperatures, an iron-constantan thermocouple was placed at the pellet surface. For carrying out the measurements, the pellet was placed in the reactor in such a way that the surface thermocouple was on the equator. Hence, surface temperatures could be measured at any position on the equatorial line. Pellet surface temperatures 1° C to 115° C above the bath temperature were reported. Furthermore, it was found that gas-phase temperatures were between bath and surface values. Hence, the results showed that the heat transfer resistance was not negligible.

The surface temperature was found to vary strongly with the position. This fact was attributed to local variations in the heat transfer coefficient between gas and pellet. It was concluded that local variations were due to unsymmetrical mixing and that the radial heat flow model is an approximation.

In order to carry out effectiveness factor calculations, an average surface temperature had to be evaluated. Surface temperatures were measured at different locations of the pellet and local heat transfer coefficients were calculated according to:

$$h_g = \frac{Q}{A(T_s - T_w)} = \frac{(-\Delta H)r_p}{A(T_s - T_w)} \tag{5}$$

An average heat transfer coefficient was used in Equation 5 and then, $\bar{T}_s$ was estimated.

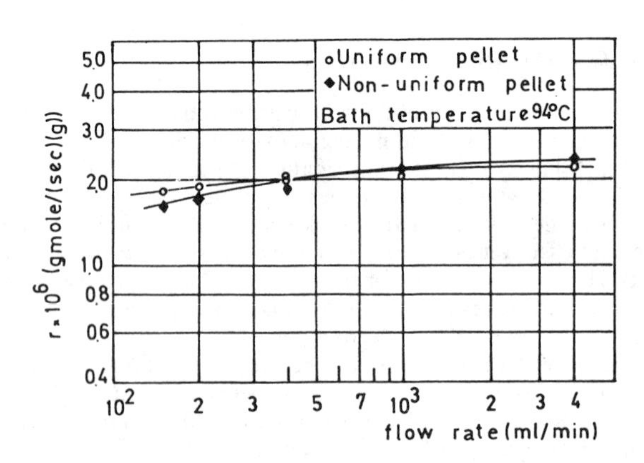

Figure 2. Effect of flow rate on the reaction rate [17].

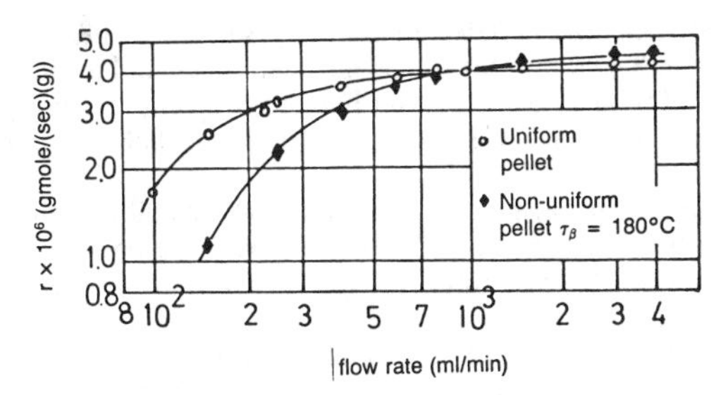

Figure 3. Effect of flow rate on the reaction rate [17].

Patel and Smith [17] studied mass transfer effects in the same kind of reactor as the one used by Maymó and Smith [14]. The catalytic oxidation of hydrogen at atmospheric pressure for uniform and non-uniform catalyst pellets was carried out.

Intrinsic rates were measured in a differential tubular reactor and the following kinetic equation was reported:

$$r_{\infty} = 0.15p_{O_2}^{0.89} \exp(-5{,}300/R_gT) \tag{6}$$

In order to study the effect of mass transfer on the reaction rate, the flow rate was varied from 100 to 4,000 ml/min at two bath temperatures, 94° C and 180° C. In Figures 2 and 3 the reaction rates for uniform and nonuniform pellets are plotted against the flow rate for 94° C and 180° C, respectively. From the uniform pellet curves, it was concluded that mass transfer had a very small effect on the reaction rate for all flow rates at 94° C and for the high flow rates at 180° C. However, mass transfer became more significant for low flow rates at 180° C. This behavior was attributed to an incomplete mixing at low flow rates. At high flow rates, the measured outlet concentration of oxygen might be considered the same as the bulk concentration in the reactor.

In order to evaluate the external mass transfer coefficient the following equations were considered:

$$r = \eta r_{\infty,s} \tag{7}$$

$$r = 2k_g a_m \frac{p}{R_gT}(y_b - y_s)_{O_2} \tag{8}$$

At any flow rate, r, $p_{O_{2,b}}$ and T_s were known. Effectiveness factors were calculated as the ratio between the measured reaction rate and the rate at surface conditions, evaluated from Equation 6 by considering the measured surface temperature and the surface oxygen concentration as the bulk concentration. Solving numerically the mass and energy conservation equations, the relation between the effectiveness factor and the dimensionless groups γ_s, β_s, φ_s was found. Then, from Equation 7, η and $p_{O_{2,s}}$ could be evaluated Hence, the mass transfer coefficient was calculated from Equation 8. For 180° C and at a recirculation rate of 3,000 ml/min, the calculated mass transfer coefficient was 2.5 cm/sec while at a recirculation rate of 4,000 ml/min, k_g was equal to 3.2 cm/sec.

In order to calculate the Reynolds number, the void area in the reactor at the pellet diameter was taken as the area upon which to base the velocity. The pellet diameter was considered as the characteristic length. Hence, flow rates of 3,000 and 4,000 ml/min corresponded to Reynolds numbers of 0.82 and 1.1, respectively.

The mass transfer coefficients evaluated in this way were compared with those arising from the Ranz and Marshall equation for forced convection around a sphere [21]. The mass transfer coefficients were found to be somewhat higher than those resulting from the Ranz and Marshall correlation. This fact was attributed to the probably greater turbulence in the jet-stirred reactor.

Likewise, it was reported that at flow rates below 1,000–2,000 ml/min, gas mixing was incomplete. Hence, the mass transfer coefficient varied with the position and could not be evaluated for low flow rates. It is important to remark that the results discussed up to now are restricted to uniform pellets.

For nonuniform pellets, external mass transfer resistances were very small even at flow rates less than 1,000–2,000 ml/min. At 180° C, mass transfer had no effect on the reaction rate for flow rates greater than 1,000 ml/min (Figure 3). However, at lower Reynolds numbers ($Re < 0.27$) mass transfer resistance became important and caused a greater retardation of the reaction rate for the nonuniform pellets.

In the work mentioned earlier, heat and mass transfer coefficients were not determined directly. Correlations for mass and heat transfer coefficients from a sphere in a reactor stirred by gaseous jets were developed by Cukierman and Lemcoff [22]. The reactor was similar to that described previously. It consisted of a 50 mm-diameter spherical chamber with a 21-mm diameter, 35-mm-long neck and a side-entrance in which either a thermocouple or the nozzle could be placed. Stirring was achieved by means of two turbulent jets emerging from a double orifice bronze nozzle. A schematic diagram of the reactor and the nozzle may be observed in Figure 4. Uniform composition and temperature in the gaseous phase around the pellet were ensured, characterizing the reactor fluid dynamics by a residence time distribution study. Likewise, a flow visualization technique was used to verify the reactor behavior [23, 24]. Both from the tracer experiments and the visualization technique, it was found that at low and intermediate flow rates [35–1,000 ml/min] the reactor closely approached a completely mixed system. At flow rates higher than 2,000 ml/min, channeling of the flow was observed. Consequently, uniformity of the properties around the solid pellet could not be ensured for flow rates higher than 2,000 ml/min.

Mass transfer coefficient measurements were carried out by rigidly mounting, at the center of the reactor chamber, a single naphthalene sphere suspended from a thin steel wire.

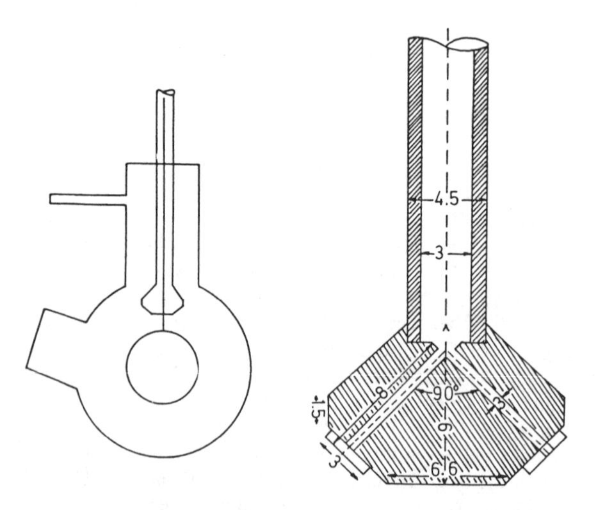

Figure 4. Schematic diagram of reactor and nozzle, dimensions in mm [22].

In order to maintain constant temperature, the reactor and a copper coil preheater were placed in a thermostatic air bath. Air circulated through the preheater before entering the reactor. Experiments were carried out with 1- and 2-cm diameter naphthalene spheres and at flow rates between 50 and 2,000 ml/min.

From a mass balance between the sublimating sphere and the gaseous phase, the mass transfer coefficient could be calculated according to:

$$k_g = \frac{m}{A\Delta t(p_s - p_b)} \cdot \frac{R_g T}{M} \tag{9}$$

where m = the sublimated mass
A = the sphere surface area
Δt = the measured time interval
p_s = the naphthalene vapor pressure
p_b = the naphthalene partial pressure in the bulk fluid

This pressure was calculated from the sublimated mass:

$$p_b = \frac{mR_g T}{\Delta t M F_v} \tag{10}$$

where F_v is the air flow rate. For high flow rates, this value is negligible with respect to p_s.

In order to correlate the experimental results, a Reynolds number was defined in terms of a characteristic velocity. The area upon which to base the velocity was taken as that resulting from dividing the reactor free volume by the most probable path length:

$$S = \frac{d_R^3 - d_p^3}{3(d_R + d_p)} \tag{11}$$

The experimental data for the different pellet diameters were correlated by the following equation:

$$Sh = 2 + 3.851 Re^{0.546} Sc^{1/3} \tag{12}$$

This equation was compared with that obtained by Ranz and Marshall [21] for mass transfer from a sphere to an infinite fluid:

$$Sh = 2 + 0.60 Re^{0.50} Sc^{1/3} \tag{13}$$

The exponent of the Reynolds number is quite similar for both cases; it also agrees with that corresponding to laminar flow around a sphere or to turbulent flow when mass transfer at the front of the sphere is analyzed. The experimental results are plotted in Figure 5 as $(Sh - 2)/Sc^{0.333}$ against the Reynolds number. The Sherwood number values thus calculated were higher than those found in the literature for cylindrical tubes or wind tunnels. This implies a greater turbulence in the spherical reactor stirred by gaseous jets.

For carrying out heat transfer measurements, a 2-cm diameter sphere built in bronze was used. It was heated by an inner electrical resistance, which was connected to a variable voltage source. Experiments were carried out at room temperature and at air flow rates between 35 and 600 ml/min. Experiments with stagnant fluid were also carried out in order to determine the natural convection effect.

Temperature differences between the sphere surface and the bulk gaseous phase were determined with copper-constantan thermocouples connected to a potentiometric recorder.

Heat transfer coefficients were evaluated from:

$$h_g = \frac{Q}{A(T_s - T_b)} \tag{14}$$

where Q is the heat transfer rate and T_s, T_b, the surface and bulk-gas temperatures, respectively.

Heat transfer data at high flow rates involved large errors because temperature differences were very small. Hence, experiments were only carried out up to a flow rate of 600 ml/min.

It was found that at high Reynolds numbers, heat transfer occurred basically by a forced convection mechanism, but when the Reynolds number decreased, the natural convection process became important.

From a regression analysis of the experimental results, the following correlation for natural convection was obtained:

$$Nu^* = 2 + 0.364 Gr^{0.340} Pr^{0.333} \tag{15}$$

Values of $(Nu^* - 2)/Pr^{0.333}$ are plotted against the Grashof numbers (Figure 6). Equation 15 was compared with the equation obtained by Ranz and Marshall [21] for heat transfer by pure natural convection from a sphere to an infinite fluid:

$$Nu^* = 2 + 0.60 Gr^{0.25} Pr^{0.333} \tag{16}$$

The Nusselt number values and the exponent of the Grashof number in Equation 15 are greater than those resulting from Equation 16. This behavior was attributed to the distortion of the streamlines due to the proximity of the reactor walls and the nozzle.

In order to correlate the heat transfer data under a mixed convection mechanism an effective Reynolds number has to be defined. An equivalent Reynolds number, Re^*, was calculated from the natural convection data. It is that value corresponding to the same heat transfer rate as when the prevailing mechanism is forced convection.

The forced-flux direction forms an angle φ with the vertical free-flux direction. Consequently, an effective Reynolds number Re_{eff} for the combined flux was calculated according to the following expression:

$$\begin{aligned} Re_{eff}^2 &= (Re^* + Re\cos\varphi)^2 + (Re\sin\varphi)^2 \\ &= Re^{*2} + 2Re^*\, Re\cos\varphi + Re^2 \end{aligned} \tag{17}$$

Therefore, the Nusselt numbers corresponding to the mixed convection mechanism were calculated from the forced convection equation by replacing the Reynolds number by the effective Reynolds number. Since the heat-mass transfer analogy was valid, the forced convection correlation for heat transfer was:

$$Nu = 2 + 3.851 Re^{0.546} Pr^{0.333} \tag{18}$$

and the equivalent Reynolds number was found from Equations 15 and 18:

$$Re^* = 0.0945 Gr^{0.34} \tag{19}$$

The effective Reynolds number was obtained by adding vectorially the equivalent Reynolds number and the measured one. From the visualization study [23], it was determined that the forced flux around the sphere was, on average, upwards and therefore coincided with the free flux direction. In Figure 7, the experimental values are plotted as

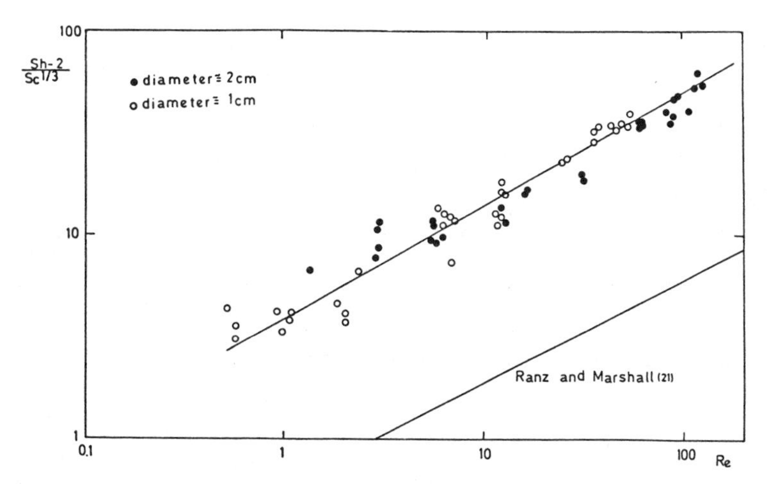

Figure 5. Mass transfer correlation [22].

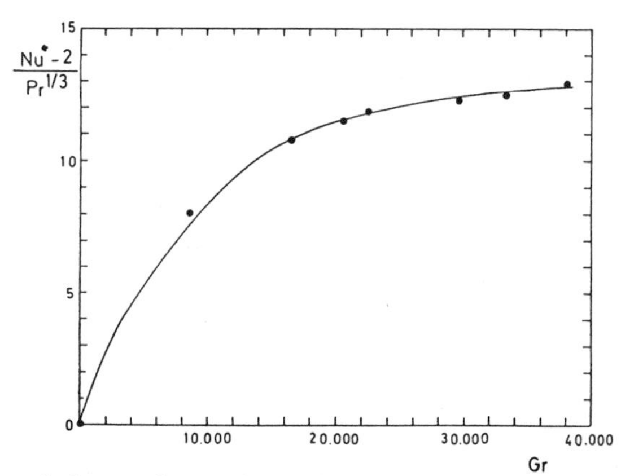

Figure 6. Heat transfer by natural convection mechanism [22].

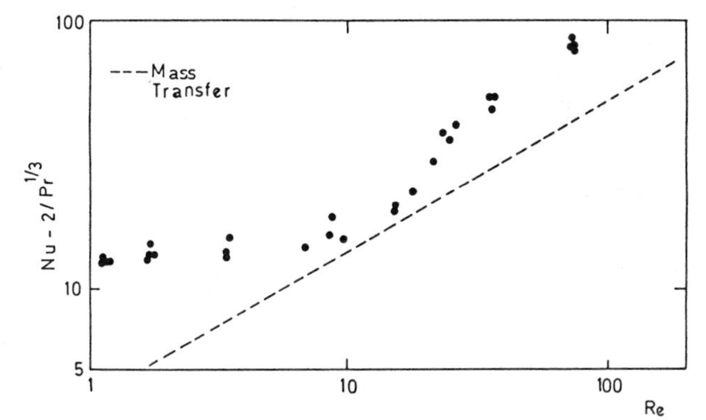

Figure 7. Heat transfer correlation [22].

$(Nu-2)/Pr^{0.333}$ against the effective Reynolds number. It was found that heat transfer results corrected by natural convection were in satisfactory agreement with mass transfer results.

The greater turbulence found in this reactor ensures uniform properties around the spherical solid pellet. As shown before, when a kinetic study is carried out in a single-pellet jet-stirred reactor, the measured parameters are affected by the diffusional resistances. From a knowledge of the heat and mass transfer coefficients, the diffusional resistances can be evaluated and intrinsic reaction parameters determined.

WALL JET-STIRRED REACTORS

Wall jet-stirred reactors with short-residence-times are particularly useful for carrying out highly exothermic homogeneous and heterogeneous reactions (Table 1). Villermaux and co-workers have carried out interesting work in this field.

Houzelot and Villermaux [25] developed an experimental method for determining mass transfer resistances in open reactors with catalytic walls. The method is based upon a chemical destruction at the wall. The internal surface of the reactor under investigation was coated with a deposit of nickel oxide and a flow of ozonized oxygen was fed into the reactor. Hence, a rapid heterogeneous decomposition took place on the nickel oxide.

The overall mass transfer coefficient K_G is given by:

$$\frac{1}{K_G} = \frac{1}{k_g^w} + \frac{1}{k_r} \quad (20)$$

where k_g^w is the mass transfer coefficient in the boundary layer and k_r the heterogeneous reaction rate constant. The latter was determined in a previous study [26].

The method was applied to a spherical reactor stirred by means of four turbulent jets emerging from nozzles. Details of the reactor characteristics may be found elsewhere [1, 16]. The concentration drop across the reactor is given by:

$$\frac{C_S}{C_0} = \frac{1}{1 + 3K_G\tau/R} \quad (21)$$

From Equations 20 and 21 the mass transfer coefficient was evaluated and the results were correlated as Sherwood numbers, $Sh^w = 2k_g^w R/D$, in terms of the nozzle Reynolds number, $Re_N = u\varrho\, d_N/\mu$. The following correlation was reported:

$$Sh^w = 23 + 0.06Re_N \qquad 500 < Re_N < 2{,}000 \quad (22)$$

The same method was applied to other three wall-reactors: annular reactor with baffles, empty tube reactor with partially catalytic wall, and packed-bed reactor with catalytic wall [25].

In order to compare the efficiency of these reactors, controlled by diffusional mass transfer to the wall, three performance criteria were proposed.

The first criterion was based on the comparison of reactors with the same hydraulic radius, r_H, and the same volume and fed with the same flow-rate. By applying this criterion, it was found that the jet-stirred reactor showed the most efficient contacting between the gas and the wall with respect to tubes coated at the wall with inert packing and empty or annular tubes coated at the wall.

The second criterion consisted of comparing reactors with the same hydraulic radius, flow-rate and catalytic wetted surface. In this case, the same conclusions as in the first one were achieved. That is, the jet-stirred reactor was found to behave as the most efficient.

The Le Goff number suggested by Engasser and Horwath [27] was applied as third criterion. This dimensionless group relates the efficiency of mass to momentum transport and is expressed by the ratio of the j-factor to the friction factor:

$$Lf = \frac{\overline{St}(Sc)^{2/3}}{f/2} \tag{23}$$

By applying this criterion, it was reported that annular reactors were more efficient than jet-stirred and inert packed-bed reactors. This behavior was attributed to the fact that mechanical energy is mainly degraded in places where no mass transfer occurs.

David et al. [6] developed a cylindrical reactor stirred by four turbulents jets and studied the heat transfer at the wall. A scheme of the reactor may be observed in Figure 8.

In order to determine the reactor behavior, residence-time distribution measurements were made for both the empty and packed reactor. For the empty reactor, perfect macromixing was found for space times between 0.6–4.5 sec when operating the reactor with air at 20° C under atmospheric pressure. For the reactor packed with glass rods and internal porosity of 0.53, the reactor behaved as perfectly macromixed for space times between 0.6–1.5 sec. For space times above 1.5 sec, a dead space was reported which represented between 0% and 30% of the total free volume.

Heat transfer experiments were carried out by placing the reactor in a water bath at 90° C. In order to measure the temperature, thermocouples were located at the inlet and outlet of the reactor and in the bath. For the empty reactor, measurements were made with five different gasses (N_2, Ar, C_2H_4, i-C_4H_{10}, CO_2) and for space times between 2–4.5 sec.

It was determined that heat transfer resistances in the wall and outside the reactor were negligible, therefore, the overall heat transfer coefficient was considered equal to the wall-gas heat transfer coefficient.

The gas temperature was assumed to be uniform throughout the reactor, and the heat transfer coefficient was calculated according to the following equation:

$$H_G = h_g^w = \frac{C_S F_{vs} c_p}{S_e} \cdot \frac{T_S - T_0}{T_B - T_S} \tag{24}$$

where T_0, T_S, T_B are the measured inlet, outlet, and bath temperatures, respectively. A linear relationship between the Nusselt and the nozzle Reynolds numbers was found:

$$Nu^w = a + b\,Re_N \tag{25}$$

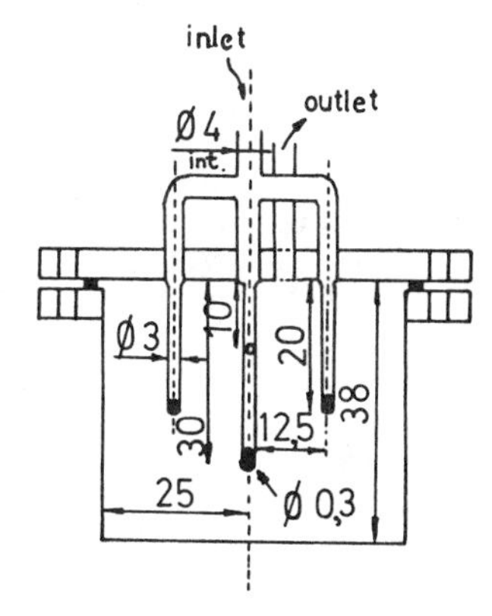

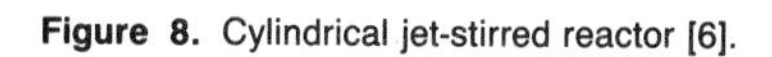
Figure 8. Cylindrical jet-stirred reactor [6].

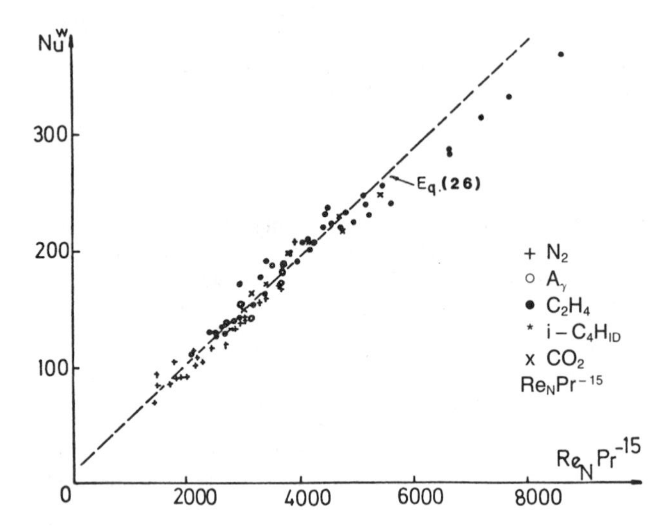

Figure 9. Nu^w vs. $Re_N\ Pr^{-1.5}$ in the empty jet-stirred reactor [6].

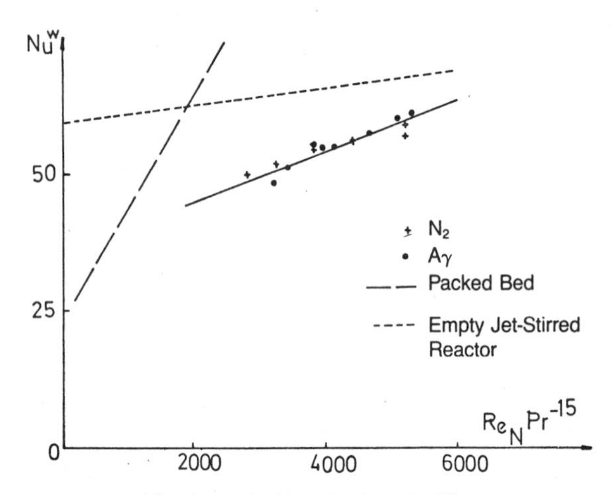

Figure 10. Nu^w vs. $Re_N\ Pr^{-1.5}$ in the packed jet stirred reactor [6].

According to the authors, the parameter a was very uncertain because Equation 25 is only valid for rather high nozzle Reynolds numbers. On the other hand, the parameter b was more significant because it was well determined. Hence, it was correlated with the Prandtl number and was found to vary as $Pr^{-1.5}$. The experimental results for the empty jet-stirred reactor fed with different gases are plotted in Figure 9 as Nusselt numbers against $Re_N Pr^{-1.5}$. The following correlation was obtained:

$$Nu^w = 12 \pm 8 + (0.0460 \pm 0.0025)\, Pr^{-1.5}\, Re_N \tag{26}$$

Likewise, heat transfer results were compared with the mass transfer results obtained by Houzelot and Villermaux [25]. By applying the Chilton–Colburn analogy and considering

$Pr = 0.74$ for oxygen, a very good agreement between the mass and heat transfer results was found.

Heat transfer measurements were also carried out in the cylindrical jet-stirred reactor packed with glass rods. Experiments were made by feeding the reactor with nitrogen and argon. In Figure 10, the results plotted as Nusselt numbers against $Re_N\,Pr^{-1.5}$ may be observed. The following correlation was reported for the cylindrical packed jet-stirred reactor:

$$Nu^w = 35 + 0.005 Re_N\,Pr^{-1.5} \qquad (27)$$

In the same Figure 10, the results obtained for the packed jet-stirred reactor are compared with those corresponding to a packed-bed reactor and the empty jet-stirred reactor. For the comparison with fixed-bed reactors, a correlation relating the wall heat transfer coefficient with the pellet Reynolds number was applied [20]. It was concluded that heat transfer at the wall in the packed jet-stirred reactor is quite similar to the results obtained for fixed beds.

The difference of the Nusselt numbers between the empty and packed jet-stirred reactors was attributed to a large difference in the flow patterns in both reactors. It was also remarked that an increase of the porosity in the jet-stirred reactor favored the heat transfer at the wall and the performance of the empty jet-stirred reactor might be achieved.

In order to compare the heat transfer efficiency of the empty jet-stirred reactor, the three performance criteria reported by Houzelot and Villermaux [25] in the wall mass transfer study were used. From the analogy between heat and mass transfer, the same conclusions for heat transfer were established. By comparing the jet-stirred reactor with an empty tube reactor of the same radius, under the same conditions, it was shown that the wall heat transfer coefficient was approximately 38 times larger in the first case.

Since the jet-stirred reactor behaved as a much better heat exchanger than the empty tube reactor, ten brass cylindrical jet-stirred cells were connected in series in order to approach plug-flow behavior but with better heat and mass transfer properties. Residence-time distribution experiments showed that each cell presented some deviations from perfect mixing.

Furthermore, computer simulations were conducted in order to demonstrate the convenience of carrying out exothermic reactions in a cascade of jet-stirred reactor. A first-order reaction was assumed, and simulations were made for four systems:

- Empty plug flow reactor
- Empty jet-stirred tanks in series
- Packed jet-stirred tanks in series
- Packed plug-flow reactor

The conversion and temperature profiles were calculated for the four systems. The results indicated that in the empty and packed plug-flow reactors, temperature and conversion increased fastly in a short distance generating a hot-spot. This might be avoided in the jet-stirred reactors since it behaves as a better heat exchanger.

In a short communication, Azay and Côme [9] showed that the continuous spherical jet-stirred reactor developed by Villermaux and co-workers may present important temperature gradients, although perfect mixing behavior can be ensured froim residence-time distribution measurements.

Two sets of experiments were carried out: a kinetic study of neopentane pyrolysis and direct temperature measurements. In both sets of experiments, the effect of gas preheating was specially considered. Two reactors of 24 cm and 80 cm were used, and stirring was achieved by means of eight and four nozzles, respectively. Mean space times were about 0.2 sec for the smaller reactor and 2 sec for the larger one. In Figure 11, thermocouple locations may be observed.

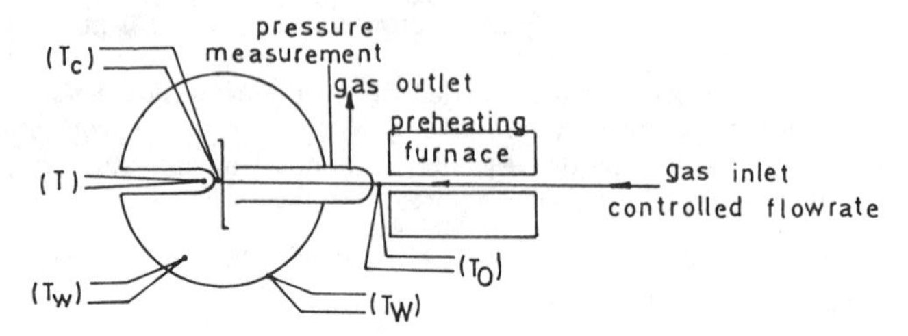

Figure 11. Thermocouple location in the reactor [9].

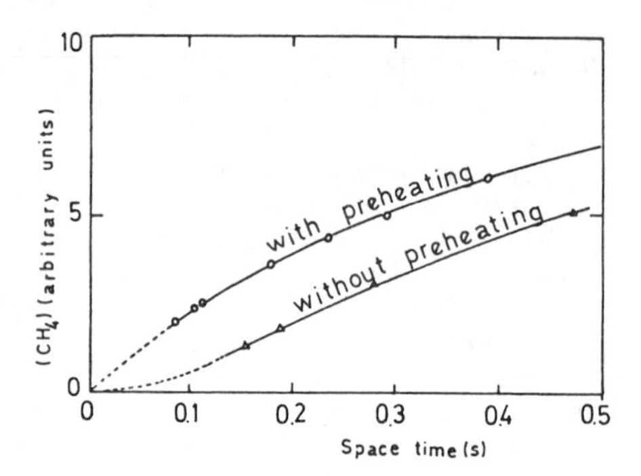

Figure 12. Influence of gas preheating on the methane yield: T_w = 660°C [2]; $T_o \simeq$ 20°C [2]; $T_o \simeq$ 620°C [9].

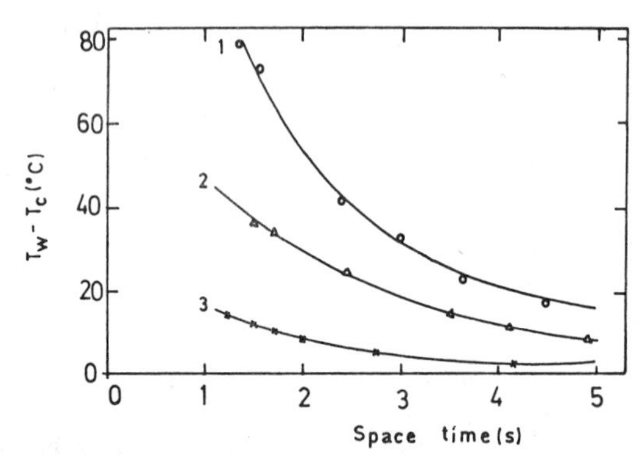

Figure 13. Influence of gas preheating on the thermal gradient: T_W = 500°C; (1) $T_o \simeq$ 20°C; (2) $T_o \simeq$ 210°C; (3) $T_o \simeq$ 420°C [9].

Neopentane pyrolysis was conducted in the smaller reactor at 660° C with and without preheating. The methane yields against the space time for two different inlet temperatures are plotted in Figure 12.

It was found that the methane yield without preheating is smaller than that corresponding to the preheated case. Likewise, it was reported that without gas preheating, methane appeared as a secondary product while with preheating its formation was clearly primary.

These mixed facts were attributed to the existence of a large temperature gradient in the reactor.

In order to check this gradient, direct temperature measurements were conducted in the larger reactor for a constant external wall temperature. In Figure 13 the measured differences between the wall region and the central zone temperatures are plotted against the space time for three different preheating temperatures. It was observed that for a given space time, the gradient decreased as the preheating temperature increased, and that it became negligible if $T_W - T_0$ was less than 40° C. Likewise, it was found, from other experiments, that the thermal gradient decreased with the pressure, when the other operating conditions were kept constant.

It was concluded that residence-time distribution measurements are not sensitive enough for reactors with internal recycles unless good precision at short elapsed times is used. Preheating of the reactants or carrier gas is recommended in order to eliminate temperature gradients and then achieve perfect mixing.

CONCLUSIONS

- For the design and construction of jet-stirred reactors, the conditions given by Equations 1, 2, and 3 must be satisfied. In addition, sensitive methods for detecting deviations from perfect mixing are recommended.
- For single-pellet reactors stirred by gaseous jets, the correlations available for mass and heat transfer around the solid pellet are those developed by Cukierman and Lemcoff [22], given by Equations 12 and 18, respectively.
- For wall-coated jet-stirred reactors, the unique correlations reported are those developed by Houzelot and Villermaux [25] for mass transfer Equation 22 and by David et al. [6] for heat transfer Equations 26 and 27.

Considerably more experimentation and theoretical developments are necessary in order to increase the knowledge in this field.

NOTATION

A	external area of the solid pellet
a	coefficient in heat transfer correlation, Equation 25
a_m	external area of the catalyst pellet per unit mass
b	coefficient in heat transfer correlation, Equation 25
C	concentration
c_p	specific molar heat at constant pressure
D	bulk diffusion
D_e	effective diffusivity
d	diameter
E	activation energy
F_v	volumetric flow rate
f	friction factor
g	aceleration gravity
Gr	Grashof number ($= \varrho^2 \beta g d_p T/\mu$)
ΔH	heat of reaction
H_G	overall heat transfer coefficient
h_g	pellet-gas heat transfer coefficient
h_g^w	wall-gas heat transfer coefficient
K_G	overall mass transfer coefficient
k	thermal conductivity
k_e	effective thermal conductivity of the pellet
k_g	pellet-gas mass transfer coefficient
k_g^w	wall-gas mass transfer coefficient
k_r	heterogeneous reaction rate constant
Lf	Le Goff number ($= [\bar{S}t(Sc)^{2/3}]/(f/2)$)
M	molecular weight
m	sublimated mass
N	number of nozzles
Nu	Nusselt number ($= h_g d_p/k$)
Nu^*	Nusselt number for natural convection
Nu^w	modified Nusselt number ($= 2 h_g^w R/k$)
P	total pressure

p partial pressure
Pr Prandtl number ($= c_p\mu/k$)
Q heat transfer rate
R reactor radius
R_g gas constant
r reaction rate
r_∞ reaction in the absence of external diffusional limitations
Re Reynolds number ($= u\varrho d_p/\mu$)
Re_{eff} effective Reynolds number
Re_N nozzle Reynolds number ($= F_v\varrho/\pi d_N\mu$)
S area defined by Equation 11
S_e exchange surface area, Equation 24
Sc Schmidt number ($= \mu/\varrho D$)
Sh Sherwood number ($= k_g d_p/D$)
Sh^w Sherwood number ($= 2 k_g^w R/D$)
$\overline{St}$ average Stanton number ($= \overline{Nu}/(Re\ Sc)$)
T temperature
T_B bath temperature
T_c central zone temperature
t time
u velocity
u_s local speed of sound
x distance along jet
y mole fraction

Greek Symbols

β thermal coefficient of volumetric expansion
β_s dimensionless group $\left(= \frac{[(-\Delta H)D_e C_{O_2,s}]}{k_e T_s}\right)$
γ_s dimensionless group ($= E/R_g T_s$)
η effectiveness factor of catalyst pellet
μ viscosity
ϱ fluid density
ϱ_c mass of active catalyst particle per volume of pellet
τ mean residence time
φ angle between free and forced flux directions
φ_s dimensionless group $\left(= R_0\left[\frac{(r_s\rho_c)}{2 D_e C_{O_2,s}}\right]\right)$

Subscripts and Supercripts

b bulk phase
N nozzle
0 reaction inlet
p pellet
S reactor outlet
s solid surface
W reactor wall, outer side
w reactor wall, inner side
$\bar{}$ average value

REFERENCES

1. Matras, D., and Villermaux, J., "Un réacteur continu parfaitement agité par jet gazeux pour l'étude cinétique de réactions chimiques rapides," *Chem. Eng. Sci., 28,* 129–137 (1973).
2. Lede, J., and Villermaux, J., "Mesure de constants cinétiques d'espèces tres réactives dans les systemes ouverts—III—Réacteur parfaitement agité et réacteur tubulaire: intérêts pratiques respectifs," *J. Chim. Phys., 74,* 761–766 (1977).
3. Herndon, W. C., "Kinetics in Gas-Phase Reactors," *J. Chem. Ed., 41,* 425–428 (1964).
4. Choudhary, V. R., and Doraiswamy, K. R., "Development of Continuous-Stirred Gas-Solid Reactors for Studies in Kinetics and Catalyst Evaluation," *Ind. Eng. Chem. Process Des. Dev., 11,* 420–427 (1972).
5. Bush, S. F., "The Design and Operation of Single-Phase Jet-Stirred Reactors for Chemical Kinetics Studies," *Trans. Instn. Chem. Engrs., 47,* T59–T72 (1969).

6. David, R., Houzelot, J. L., and Villermaux, J., "A Novel and Simple Jet-Stirred Reactor for Homogeneous and Heterogeneous Reactions with Short Residence Times," *Chem. Eng. Sci., 34,* 867–876 (1979).
7. David, R., and Matras, D., "Règles de construction et d'extrapolation des réacteurs auto-agités par jet gazeux," *Can. J. Chem. Eng., 53,* 297–300 (1975).
8. Hinze, J. O., and van der Hegge Zijnen. B. G., "Transfer of Heat and Matter in a Symmetrical Jet," *Appl. Sci. Res., A1,* 435 (1949).
9. Azay, P., and Côme, G. M., "Temperature Gradients in a Continuous Flow Stirred Tank Reactor," *Ind. Eng. Chem. Process Des. Dev., 18,* 754–756 (1979).
10. Satterfield, C. N., *Mass Transfer in Heterogeneous Catalysis,"* M. I. T. Press, Cambridge, Mass. (1970).
11. Longwell, J. P., and Weiss, M. A., "High Temperature Reaction Rates in Hydrocarbon Combustion," *Ind. Eng. Chem., 47,* 1634–1643 (1955).
12. Kistiakowsky, G. B., and Volpi, G. G., "Reaction of Nitrogen Atoms. I. Oxygen and Oxides of Nitrogen," *J. Chem. Phys., 27,* 1141–1149 (1957).
13. Mulcahy, M. F. R., and Williams, D. J., 'A Stirred-Flow Reactor for Investigating the Kinetics of Gaseous Reactions: Application to the Decomposition of Di-t-butyl Peroxide," *Aust. J. Chem., 14,* 534–544 (1961).
14. Maymó, J. A., and Smith, J. M., "Catalytic Oxidation of Hydrogen—Intrapellet Heat and Mass Transfer," *AIChE J. 12,* 845–854 (1966).
15. Westenberg, A. A., de Haas, N., and Roscoe, J. M., "Radical Reactions in an Electron Spin Resonance Cavity Homogeneous Reactor," *J. Phys. Chem. 74,* 3431–3438 (1970).
16. David, R., and Villermaux, J., "Etude d'une réaction homogène rapide d'ordre deux en phase gazeuse par des méthodes transitoires dans un réacteur autoagité," *Can. J. Chem. Eng., 51,* 630–635 (1973).
17. Patel, K., and Smith, J. M., "Mass Transfer Effects for Uniform and Nonuniform Catalyst Pellets," *J. Cat., 40,* 383–390 (1975).
18. Lede, J., and Villermaux, J., "Mesure de constantes cinétiques d'espèces tres réactives dans les systemes en écoulement. II—Le réacteur autoagité par jets gazeux," *J. Chim. Phys., 74,* 468–474 (1977).
19. David, R., and Villermaux, J., "La déshydrogenation homogène catalytique de l'isobutane et du n-butane en réacteur autoagité continu," *J. Chim. Phys., 75,* 715–722 (1978).
20. Lede, J., et al., "Production of Hydrogen by Direct Thermal Decomposition of Water: Preliminary Investigations," *Int. J. Hydrogen Energy, 7,* 939–950 (1982).
21. Ranz, W. E., and Marshall, W. R., "Evaporation from Drops," *Chem. Eng. Progr., 48,* 141–146; 173–180 (1952).
22. Cukierman, A. L., and Lemcoff, N. O., "Heat and Mass Transfer in a Tank Reactor Stirred by Gaseous Jets," *Chem. Eng. J., 19,* 125–130 (1980).
23. Cukierman, A. L., "Fenómenos de transferencia en un reactor tanque agitado mediante jets gaseosos," doctoral thesis, Universidad de Buenos Aires (1978).
24. Cukierman, A. L., and Lemcoff, N. O., "Visualization of Transport Phenomena in a Simulated Gaseous Phase Stirred Reactors," *Lat. Am. J. Heat Mass Transf., 3,* 63–74 (1979).
25. Houzelot, J. L., and Villermaux, J., "A New Method for the Study and Characterization of Mass Transfer Resistance in Open-Reactors with Catalytic Walls," 4th. Int. Symp. on Chem. React. Eng. Heidelberg, IV–143–151 (1976).
26. Houzelot, J. L., and Villermaux, J., "Etude d'une cinétique de décomposition hétérogène entre un gaz et une paroi catalytique dans un réacteur ouvert parfaitement agité. Application a la décomposition catalytique de l'ozone sur oxyde de nickel," *J. Chim. Phys., 73,* 807–810 (1976).
27. Engasser, J. M., and Horwath, C., "Efficiency of Mass and Momentum Transport in Homogeneous and Two-Phase Flow," *Ind. Eng. Chem. Fundam., 14,* 107–110 (1975).
28. Froment. G. B., "Fixed Bed Catalytic Reactors—Current Design Status," *Ind. Eng. Chem., 59,* 18–27 (1967).

CHAPTER 30

PRINCIPLES AND APPLICATIONS OF CARBON ADSORPTION

Paul N. Cheremisinoff and Emmanuel E. Gonsalves

Department of Civil & Environmental Engineering
New Jersey Institute of Technology
Newark, New Jersey, USA

CONTENTS

INTRODUCTION

Adsorption is becoming an increasingly important and powerful separation tool. It is making significant impact in diversified processing industries. Adsorber systems using activated carbon, activated alumina, silica gel, and molecular sieves are widely accepted in many industries including the natural gas, industrial gas, and nuclear industries. Drying, purification, and bulk separation processes using adsorbents are commercially proven. Commercial units containing an inventory of over 300,000 lb of adsorbent in a single plant are not uncommon. Automatic rapid cycle adsorbers with proven mechanical durability are in broad use. It is estimated that there are several thousand adsorber units in operation at present in the United States alone. The total adsorbent utilization is estimated to be in excess of seventy five million pounds annually.

The use of activated carbon for removal of dissolved organics from water and wastewater as well as from air and gas has long since been demonstrated. In fact, it is one of the most efficient organic removal processes available to the engineer. The increasing need for highly polished effluents from wastewater treatment plants, necessary to accommodate the stringent requirements for both surface water quality and water reuse, has stimulated great interest in carbon treatment systems. Both the great capability for organic removal and the overall flexibility of the carbon adsorption process have encouraged its application in a wide variety of situations. It readily lends itself to integration into larger, more comprehensive treatment systems.

This chapter presents the generalized theory of adsorption, the fundamental aspects to be considered, design principles for adsorption control equipment, and typical applications to which carbon adsorption can be used.

ADSORPTION FUNDAMENTALS

It is already well established that the molecular forces at the surface of a liquid are in a state of unbalance or unsaturation. As a result of this unsaturation, solid and liquid surfaces tend to satisfy their residual forces by attracting onto and retaining on their surfaces gases or dissolved substances with which they come in contact. This phenomenon of concentration of a substance on the surface of a solid (or liquid) is called adsorption. The substance thus attracted to the surface is called the adsorbed phase or adsorbate, while the substance to which it is attached is the adsorbent.

Adsorption is a physical process that deals specifically with the concentration of dispersed materials in a continuous phase, called the carrier stream, on the surface of a highly porous material (see Figure 1). From this figure it can be seen that the rate of adsorption is governed primarily by diffusional resistance [1].

Theory of Adsorption

Theories explaining adsorption are quite complex, simply because it is a function not only of temperature and pressure, but also of concentration and molecular charge orientation. Nearly all of the theories presented are broad based and hold only for specific materials and conditions. As an overall assumption, the quantity of gas adsorbed, under conditions of equilibrium, is a function of the final temperature and pressure [2]:

$$a = f(p, T)$$

Adsorption data is usually presented in the form of an adsorption isotherm, which is a plot of the amount adsorbed versus the pressure (or concentration, if gas) at a constant temperature. Thomas [3] presents five distinct types of physical adsorption isotherms. Langmuir derived the Type I isotherm utilizing the following assumptions:

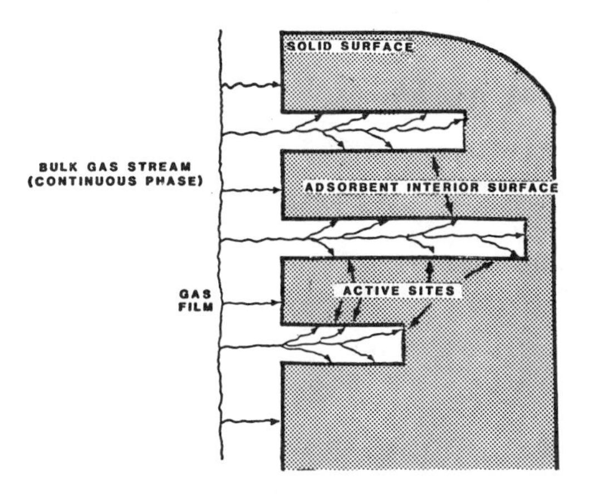

Figure 1. The adsorption process.

1. Adsorption occurs at a definite localized site.
2. Each site can accommodate only one entity.
3. The energy of the adsorbed entity is the same at all sites on the surface and is independent of the presence or absence of other adsorbed entities at neighboring sites.

He presented his findings in the following form

$$X/M = kKP/(1+KP) \quad (1)$$

where X/M = the amount of sorbed vapor per unit weight of adsorbent
k, K = constants for the sorbed vapor
P = partial pressure of the sorbed substance

The assumptions indicate that Langmuir's isotherm is valid for monomolecular adsorption. To describe multilayer physical adsorption, other researchers have presented other isothermal equations in an attempt to fit the various types of behaviors of adsorption.

Adsorption Process

The adsorption process occurs at solid-solid, gas-solid, gas-liquid, liquid-liquid or liquid-solid interfaces. The adsoıption process for each of the aforementioned interaces is very similar in nature.

There are two basic types of adsorption: physisorption and chemisorption. They both are initiated when the molecules of the liquid phase are attracted by the surface of the solid through the attractive forces at the solid surface, thereby overcoming the kinetic energy of the contaminant molecules.

Physisorption occurs when, as a result of energy differences and/or electrical attractive forces (Van der Waals' forces), the adsorbate molecules become physically attached to the absorbent molecules. This type of adsorption is multilayered; that is each molecular layer forms on top of the previous layer with the number of layers being proportional to the contaminant concentration.

On the other hand, when a chemical compound is produced by the adsorbate reacting with the adsorbent, chemisorption has occurred. This process is one layer thick and irreversible because energy is required to initiate the formation of a new compound at the adsorbent surface, and energy will be necessary to reverse the process. Physisorption is reversible because the attractive forces holding the adsorbate to adsorbent is very weak and desorption readily occurs.

The following factors readily affect the adsorption process:

- The physical and chemical characteristics of the adsorbent, i.e., surface area, pore size, chemical composition, etc.
- The physical and chemical characteristics of the adsorbate, i.e., molecular size, molecular polarity, chemical composition, etc.
- The concentration of the adsorbate in the liquid phase.
- The characteristics of the liquid phase, i.e., pH, temperature.
- The residence time of the system.

Nature of Adsorbents

Although it is probable that all solids adsorb gases, or vapors to some extent, adsorption as a rule is not very pronounced unless an adsorbent possesses a large surface area for a given mass. For this reason such adsorbent as silica gel, charcoals, molecular sieves, and aluminium-based compounds are particularly effective as adsorbing agents. These substances possess a very porous structure and with their large exposed surface area can take up appreciable volumes of various gases and liquids. The extent of adsorption can be increased further by "activating" the adsorbents in various ways, that is by distilling out hydrocarbon impurities from the charcoal, thereby leading to the exposure of a larger free surface for possible adsorption.

The following factors directly affect the adsorption capacity of the adsorbents:

- Higher surface areas will give a greater adsorption capacity.
- Larger pore sizes will give a greater adsorption capacity for larger molecules.
- Adsorptivity will increase as the solubility of the solute decreases. Thus for hydrocarbons, adsorption increases with molecular weight.
- For solutes with ionizable groups, maximum adsorption will be achieved at a pH corresponding to the minimum ionization.
- Adsorption capacity will increase with increasing temperature (since adsorption is exothermic).

The aforementioned listing all relate to absorption capacity; the rate of adsorption is also an important consideration. For example, while capacity is increased with the adsorption of higher-molecular-weight hydrocarbons the rate of adsorption is decreased. Similarly, while temperature increases will decrease the capacity, it may—depending on the rate limiting step in the overall process—increase the rate of removal of solute from solution.

Activated Carbon

This applies to any amorphous form of carbon that has been treated to give high adsorption capacities. Typical raw materials include coal, wood, coconut shells, pulp mill residues, petroleum-base residues, and char from sewage sludge pyrolysis. A carefully controlled process of dehydration, carbonization, and oxidation yields a product which, while not pure carbon, is called activated carbon. This material has a high capacity for adsorption, due primarily to the large surface area available for adsorption—500–1,500 m^2/g resulting from a large number of internal pores. Pore size generally ranges from 10–100 Å in radius. Most of the available surface area is nonpolar in nature, but this interaction with oxygen (in production) does produce specific active sites giving the surface a lightly polar nature.

The activation process is achieved by subjecting the raw material to intense heat in the absence of air, followed by steam at extremely high temperatures. Zinc chloride, magnesium chloride, calcium chloride, and phosphoric acid are also used in place of steam as activating agents.

Adsorption on activated carbon occurs when a molecule is brought up to its surface and held there by physical and/or chemical forces. This process is reversible, thus allowing activated carbon to be regenerated (and reused) by the proper application of heat and steam, or solvent.

After initially contacting an activated carbon with a solution, an equilibrium will eventually be reached such that the rates of solute adsorption and desorption are equal. The amount of solute absorbed per unit weight of carbon will increase as the concentration of the solute in the solution is increased.

Some approximate properties of adsorbent carbon include [5].

Bulk density—22–34 lb/ft^3
Heat capacity—0.27–0.36 Btu/16° F
Pore volume—0.56–1.20 cm^3/g
Surface area—500–1,500 m^2/g
Average pore diameter—10–100 Å
Regeneration temperature—100–400° C
Maximum allowable temperature—150° C

Activated carbon has an affinity for organics and its use for organic contaminant removal from waste waters is widespread. The utilization of carbon at a waste water or water treatment facility can be in the form of a powder or granule. As raw water or waste water is passed over granular carbon on a bed, tastes, colors, and odors are removed from potable waters, and dissolved organics, such as phenols, pesticides, organic dyes, sulfactants, etc. are removed from industrial and municipal waste waters. The following is a list of several applications for activated carbon's adsorption abilities [1].

The activated carbon continues to adsorb contaminants until it reaches its adsorption saturation limit. At this point, it is regenerated. The recoverable and waste products are extracted with the regeneration solution, the former being reused and the waste discharged.

Activated Alumina

This substance, also known as hydrated aluminum oxide is produced by special heat treatment of precipitated or native aluminas or bauxite. It is available in either granular or pellet form with the following typical properties [5].

Density in bulk:
  granules—38–42 lb/ft^3
  pellets—54–58 lb/ft^3
Specific heat—0.21–0.25 Btu/lb° F
Pore volume—0.29–0.37 cm^3/g
Surface area—210–360 m^2/g
Average pore diameter—18–48 Å
Regeneration temperature—200–250° C
Stable up to—500° C

Silica Gel

The manufacture of silica gel consists of neutralizing sodium silicate by mixing it with dilute mineral acid and washing the gel formed free from salts produced during the neutralization reaction, followed by the drying, roasting, and grading processes. It is generally used in granular form, although bead forms are available. The material has the following physical properties [5].

Bulk density—44–46 lb/ft^3
Heat capacity—0.22–0.26 Btu/lb° F
Pore volume—0.37 cm^3/g
Surface area—750 m^2/g
Average pore diameter—22 Å
Regeneration temperature—120–250° C
Stable up to—400° C

Silica gel is primarily used in gas drying, although it also finds application in gas desulfurization and purification.

Molecular Sieves

Unlike the amorphous adsorbents, that is, activated carbon, activated alumina, and silica gel, molecular sieves are crystalline, being essentially dehydrated zeolites, i.e., aluminosilicates in which atoms are arranged in a definite pattern. The complex structural units of molecular sieves have cavities at their centers which can be accessed by pores or windows. Due to the crystalline porous structure and precise uniformity of the small pores, adsorption phenomena only takes place with molecules which are of small enough size and of suitable shape to enter the cavities through the pores.

The very strong adsorptive forces in molecular sieves are due primarily to the cations which are exposed in the crystal lattice. These cations act as sites of strong localized positive charge which electrostatically attract the negative end of polar molecules. The greater the dipole moment of the molecule, the more strongly it will be attracted and adsorbed. Polar molecules are generally those which contain O, S, Cl, or N atoms and are symmetrical. For example, molecular sieves adsorb carbon monoxide in preference to argon. Under the influence of the localized, strong positive charge on the cations, molecules can have dipoles induced in them. The polarized molecules are then strongly adsorbed due to the electrostatic attraction of the cations. The more unsaturated the molecule, the more polarizable it is and the more strongly it is adsorbed. Thus, molecular sieves will effectively remove acetylene from olefins and ethylene or propylene from saturated hydrocarbons.

The sieves are manufactured by hydrothermal crystal growth from aluminosilicate gels followed by specific heat treatment to effect dehydration. Table 1 gives the properties of hydrothermal crystal sieves.

Activated Carbon Removal Applications

Acetaldehyde
Acetic acid
Acetone
Activated sludge effluent
Air purification scrubbing solutions
Alcohol
Amines
Ammonia
Amyl acetate and alcohol
Antifreeze
Benzine
Biochemical agents
Bleach solutions
Butyl acetate and alcohol
Calcium hydrochlorite
Can and drum washing
Chemical tank wash water
Chloral
Chloramine
Chlorobenzene
Chlorine
Chlorophenol
Chlorophyl
Cresol
Dairy process wash water
Decayed organic matter

Defoliants
Detergents
Dissolved oil
Dyes
Ethyl Acetate and Alcohol
Gasoline
Glycol
Herbicides
Hydrogen sulfide
Hypochlorous acid
Insecticides
Iodine
Isopropyl acetate and alcohol
Ketones
Lactic acid
Mercaptans
Methyl acetate and alcohol
Methyl-ethyl-ketone
Naphtha
Nitrobenzenes
Nitrotoluene
Odors
Organic Compounds
Phenol
Potassium Permanganate
Sodium hypochlorite
Solvents
Sulfonated oils
Tastes (organic)
Toluene
Trichlorethylene
Trickling filter effluent
Turpentine
Vinegar
Well water
Xylene

Adsorption Equilibria

The relation between the amount of substance adsorbed by an adsorbent and the equilibrium pressure or concentration at constant temperature is called the adsorption isotherm. Five general types of isotherm have been observed in the adsorption process (see Figure 2). In the case of chemisorption only isotherms of Type 1 are encountered, while for physical adsorption all five types occur.

In terms of Type 1, the amount of gas adsorbed per given quantity of adsorbent increased relatively rapidly with pressure and then more slowly as the surface becomes covered with the adsorbent. To explain this phenomenon Langmuir proposed from theoretical consideration an equation to represent Type 1 isotherms.

Langmuir postulated that gases on being adsorbed by a solid surface cannot form a layer more than a single molecule in depth. Further, he visualized the adsorption process as consisting of two opposing actions, a condenstation of molecules from the fluid phase onto the surface and an evaporation of the molecules from the surface back into the fluid. When adsorption first begins every molecule colliding with the surface may condense in it. However, as adsorption proceeds, only those molecules may be expected to be absorbed which strike a part of the surface not already covered by adsorbed molecules. The result is that the initial rate of condensation of molecules on the surface is highest and falls off as the area of surface available for adsorption is decreased. On the other hand, a molecule

Table 1
Properties of Hydrothermal Crystal Sieves

Properties	Anhydrous Sodium Aluminosilicate	Anhydrous Calcium Aluminosilicate	Anhydrous Aluminosilicate
Type	4A	5A	13X
Density ($16/ft^3$)	44	44	38
Specific heat (Btu/lb °F)	0.19	0.19	–
Effective diameter (Å)	4	5	13
Regeneration temperature	200–300 °C	200–300 °C	200–300 °C
Stable up to	600 °C	600 °C	600 °C

From Buonicone and Theodore [5].

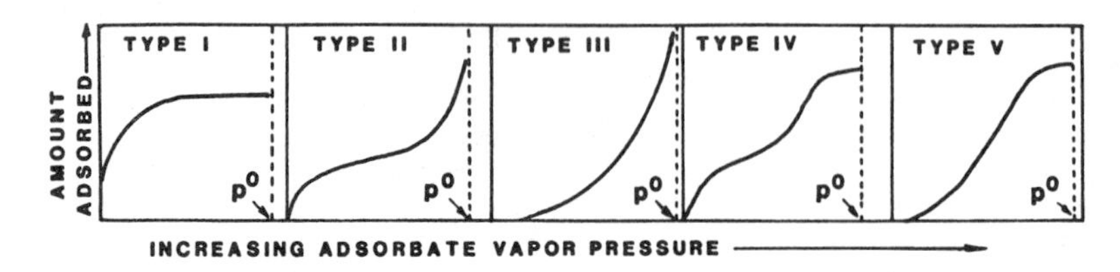

Figure 2. Types of adsorption isotherms. p^0 represents the saturation pressure.

adsorbed on the surface may, by thermal agitation, become detached from the surface and escape into the gas. The rate at which desorption will occur will depend, in turn, on the amount of surface covered by molecules and will increase as the surface becomes more fully saturated. These two rates, condensation (adsorption) and evaporation (desorption) will eventually become equal, and when this happens adsorption equilibrium will be established.

If θ is the fraction of the total surface covered by adsorbed molecules at any instant, then the fraction of surface bare and available for further adsorption is $1-0$. By the kinetic theory, the rate at which molecules strike the unit area of a surface is proportional to the pressure of the gas. The rate of condensation of molecules should be determined both by the pressure and the fraction of bare surface, or $K_1(1-\theta)P$, where K_1 is a proportionality constant. If K_2 is the rate at which molecules evaporate from the unit surface area when the surface is fully covered, then for a fraction θ of surface covered, the rate of evaporation (desorption) will be $K_2\theta$. For the condition of adsorption equilibrium these rates must be equal. Therefore,

$$K_1(1-\theta)P = K_2\theta \tag{2}$$

$$\theta = \frac{K_1 P}{K_2 + K_1 P} = \frac{bP}{1+bP} \tag{3}$$

where $b = K_1/K_2$

Now, the amount of gas adsorbed per unit area or per unit mass of adsorbent, Y, must obviously be proportional to the fraction of surface covered; hence

$$Y = K\theta = \frac{KbP}{1+bP} = \frac{aP}{1+bP} \tag{4}$$

where $a = kb$.

This equation is the Langmuir adsorption isotherm. The constants a and b are characteristics of the system under consideration and are evaluated from experimental data. The magnitudes also depend on temperature. If we divide both sides of the equation by P and take the reciprocal the result is

$$\frac{P}{Y} = \frac{1}{a} + \left(\frac{b}{a}\right)P \tag{5}$$

If (P/Y) is plotted against P, the result is a straight line with slope (b/a) and ordinate intercept (1/a).

For Type II and Type III isotherms, the explanation proposed is that adsorption is multilayered. By assuming that adsorption on solid surfaces takes place with the formation of secondary, tertiary, and finally multilayers upon the primary monomolecular layers postulated by Langmuir, Brunauer, Emmett, and Teller derived a relation for these two types of isotherms [6].

$$\frac{P}{V(P^0 - P)} = \frac{1}{VmC} + \frac{(C-1)}{(VmC)}\frac{P}{P^0} \tag{6}$$

where V = volume, reduced to standard conditions of gas adsorbed at P and T
P = pressure
T = temperature
P^0 = saturated vapor pressure of the adsorbate at temperature T
Vm = volume of gas, reduced to standard conditions, adsorbed when the surface is covered with a unimolecular layer
C = constant at any given temperature.

The constant is approximately given by

$$C = \exp[(E_l - E_L)/RT]$$

where E_l = heat of adsorption of the first adsorbed layer
E_L = heat of liquification of the gas.

The isotherms of Type II follow when $E_l > E_L$, that is when the attractive forces between adsorbed gas and adsorbent are greater than the attractive forces between the molecule of the gas in the liquid state. The isotherms of Type III follow when $E_l < E_L$, i.e., when the forces between the adsorbed and adsorbate are small. By plotting $P/V\ (P^0 - P)$vs P/p^0, a straight line is obtained, with the slope given by $(C-1)/V_mC$ and intercept at $1/VmC$.

To explain Type IV and Type V isotherms it has been suggested that substances exhibiting such behavior undergo not only multilayer adsorption but also condensation of the gas in the pores and capillaries of the adsorbent.

Hysteresis

Adsorption-desorption curves should theoretically coincide. Any point on the equilibrium curve should, therefore, be approachable either by adsorption onto fresh adsorbate or by desorption of a sample with an initially higher adsorbate concentration. However, in practice, different equilibria may result, at least over part of an isotherm, depending on whether the vapor is adsorbed or desorbed. This gives rise to the phenomenon of hysteresis.

DeBoer [7] showed that, in practical cases, the shape of the hysteresis loop may lead to a more or less detailed picture of the shape of the pore present in certain adsorbent. Five different types of hysteresis loops were discovered, three of which have proven to be important in the interpretation of adsorption isotherm. These three are commonly referred to as A, B, and E, respectively (see Figure 3). Type A has two very steep branches and can be related in its simplest form to cylindrical pores open at both ends. A Type B hysteresis loop is characterized by a vertical adsorption branch near saturation and a steep desorption branch at intermediate relative pressures. This type of hysteresis may be caused by either ink bottle structures with very wide bodies and narrow necks, or by slit-shaped pores open at all sides. Type E hysteresis is intermediate between A and B. This type of hysteresis may be attributed to a distribution of pores with narrow necks of rather uniform diameters, but with wide bodies of different diameters.

In every type of histeresis it is clear that the desorption equilibrium pressure is always lower than that obtained by adsorption.

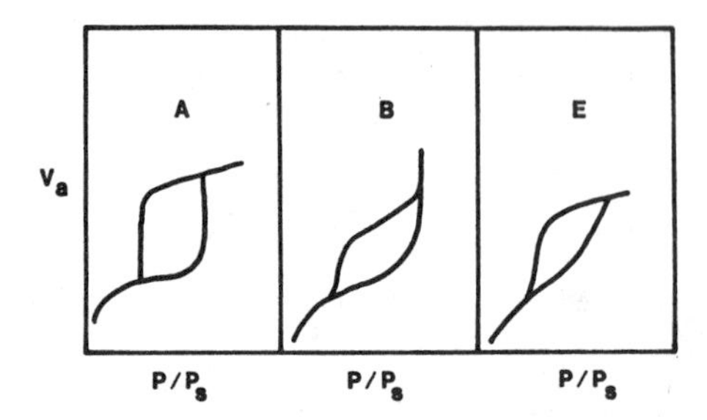

Figure 3. Pore size distribution patterns reflected by N_2-gas isotherms.

Adsorption Phenomena

In any heterogenous system consisting of ions, molecules, or atoms, the interaction between the phases begins with chemical or physical interaction at the phase interface. This phenomenon, molecular or atomic interaction at the surface, can be observed between gas and solid, gas and liquid, liquid and solid, liquid and liquid, and in rare circumstances, between two solid phases. When there is interaction at the interface, whether it is atomic, ionic, or molecular, the particles either bounce back from the surface elastically or the molecule may remain embodied in the surface for a period of time. The type of interaction between the phases of a heterogeneous system depends on the properties and composition of the components of the system. On the standpoint of the distribution of the substances to be sorbed, two types of sorption can occur. Absorption and adsorption. Adsorption takes place when the molecules or atoms sorbed are concentrated only at the interace. On the other hand, absorption takes place when the atoms or molecules are sorbed and are distributed throughout the bulk of the interacting phases. As a general rule, adsorption takes place when one of the phases consists of a solid.

Adsorption of atoms or molecules on a solid from a fluid phase is, therefore, a spontaneous process and as such is accompanied by a decrease in the free energy of the system. Therefore, there is also a loss of entropy:

$$\Delta F = \Delta H - T\Delta S$$

The adsorption process is always exothermic regardless of the type of forces involved.

The adsorption process involves three necessary steps. The fluid must first come in contact with the adsorbent, at which time the adsorbate is preferentially or selectively adsorbed on the adsorbent. Next the unadsorbed fluid must be separated from the adsorbent-adsorbate, and finally, the adsorbent must be regenerated by removing the absorbate or discarding used adsorbent and replacing it with fresh material. Regeneration is performed in a variety of ways, depending on the nature of the adsorbate. Gases or vapors are usually described by either raising the temperature (thermal cycle) or reducing the pressure (pressure cycle) of the adsorbent adsorbate.

The more popular thermal cycle is accomplished by passing hot gas through the adsorption bed in the opposite direction to the flow during the adsorption cycle. This ensures that the gas passing through the unit during the adsorption cycle always meets the most active adsorbent last and that the adsorbate concentration in the adsorbent at the outlet end of the unit is always maintained at a minimum.

In the first step mentioned previously where the molecules of the fluid come in contact with the adsorbent, an equilibrium is established between the adsorbed fluid and that remaining in the fluid phase. Figures 4 through 11 show several experimental equilibrium adsorption isotherms for a number of components adsorbed on various adsorbents, such as silica gel, activated carbon, molecular sieves, and activated alumina.

Heat of Adsorption

The process of a fluid being brought into contact with an evacuated porous solid and part of it being taken away by a solid is always accompanied by a liberation of heat. The extent to which the process is exothermic depends on the type of sorption and the particular system. For physical adsorption, the amount of heat liberated is usually equal to the latent heat of condensation of the adsorbate plus the heat of wetting of the solid by the adsorbate. The heat of wetting is usually only a small fraction of the heat of absorption. On the other hand, in chemisorption the heat involved approximates the heat of reaction.

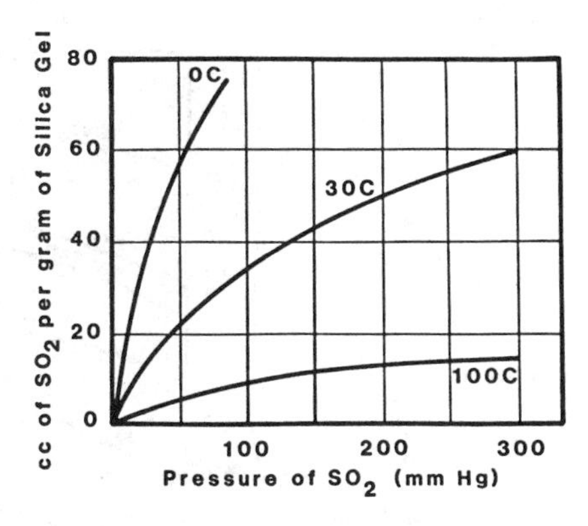

Figure 4. Adsorprion isotherms for SO_2 on silica gel.

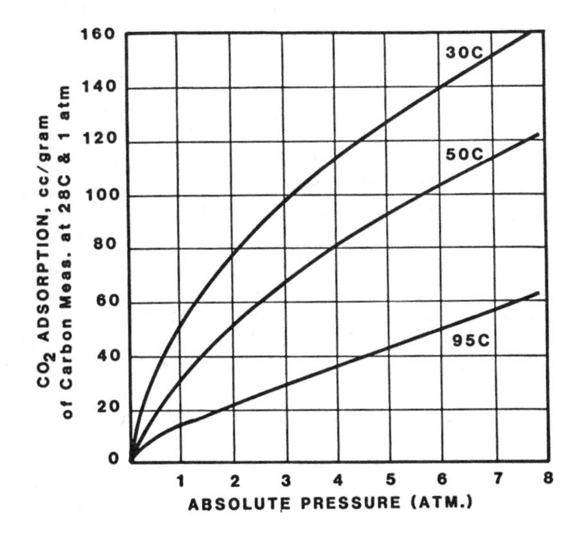

Figure 5. CO_2 adsorption isotherms on activated carbon.

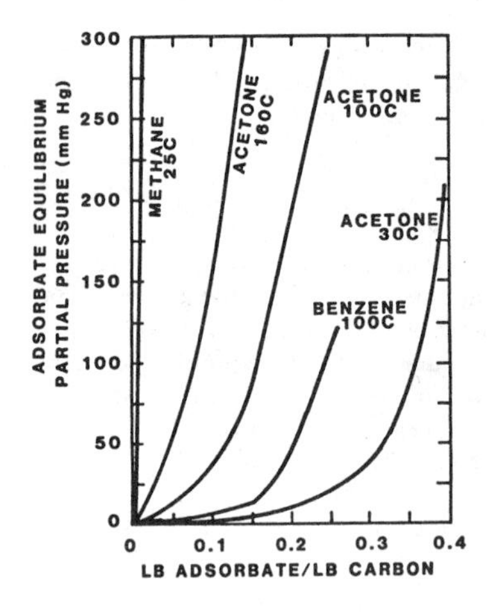

Figure 6. Adsorption isotherms for some hydrocarbons on activated carbon.

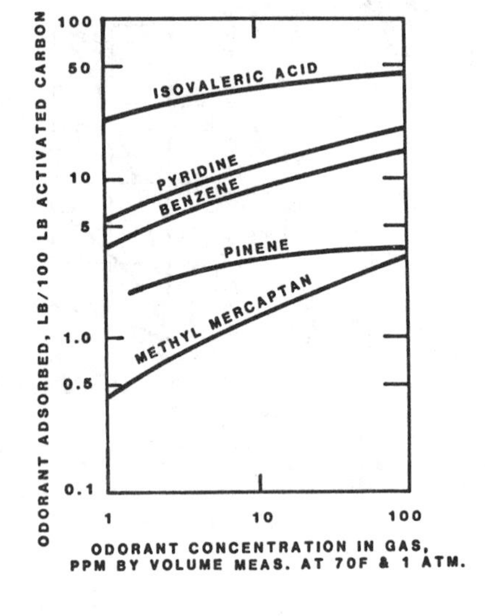

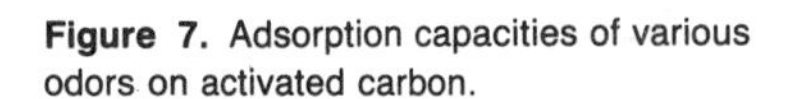

Figure 7. Adsorption capacities of various odors on activated carbon.

Two heats of adsorption may be referred to, namely, the differential and integral. The differential heat of adsorption ($-\bar{H}$) is defined as the heat liberated at constant temperature when the unit quantity of vapor is adsorped upon a large quantity of solid already containing adsorbate. A large quantity of solid is used so that the adsorbate concentration is unchanged. The integral heat of adsorption at any concentration X of adsorbate-adsorbent combination minus the sum of the enthalpies of unit weight of pure solid adsorbent and sufficient pure adsorbed substance (before adsorption) to provide the required concentration X, all at the same temperature. Table 2 lists the integral heats of adsorption for some organic vapors on carbon at various temperatures. The differential and integral heat of adsorption of the system.

Othmer and Sawyer [8] have shown that plots of the type shown in Figure 12 are useful in estimating the heat of adsorption, should it be unavailable. The slope of the isostere is given by

$$\frac{d(\ln p^*)}{d(\ln \bar{p})} = \frac{(-\bar{H})M}{\lambda_R M_R} \tag{7}$$

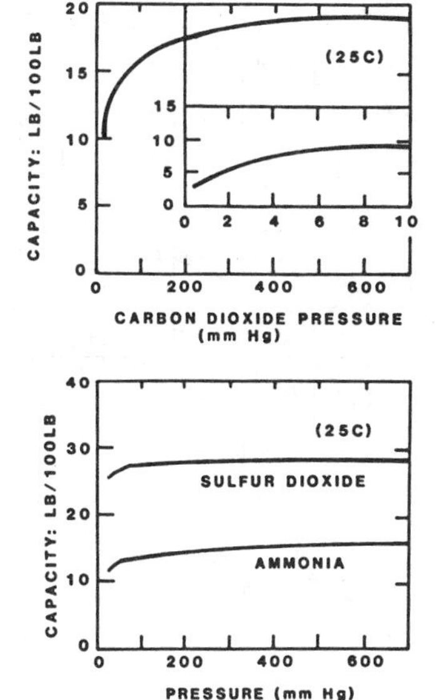

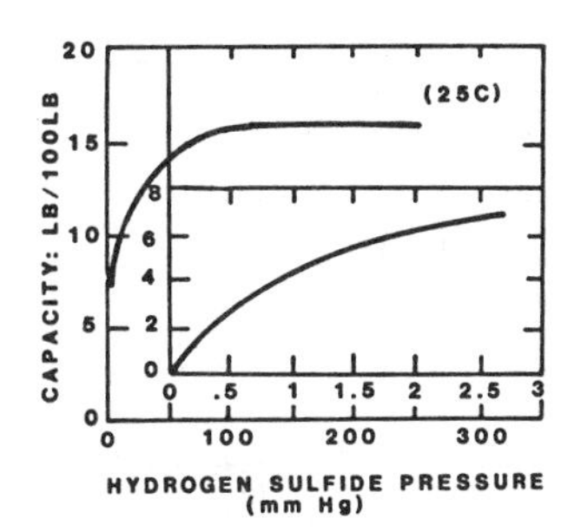

Figure 8. Equilibrium capacity of molecular sieves for some typical gas impurities.

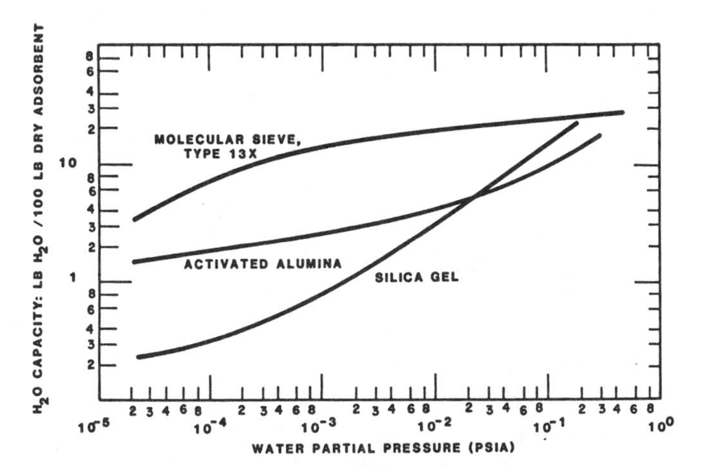

Figure 9. Adsorption isotherms for water at 25°C on various adsorbents.

where $\bar{H}$ Btu/1b of vapor absorbed is referred to the pure vapor and λ_R is the latent heat of vaporization of the reference substance at the same temperature, Btu/1b. M and M_R are the molecular weights of the vapor and reference substance, respectively. If $\bar{H}$ is computed at constant temperature for each isostere, then the integral heat of adsorption at this temperature may be obtained from

$$\Delta H'_A = \int_0^x \bar{H}\, dx \quad (8)$$

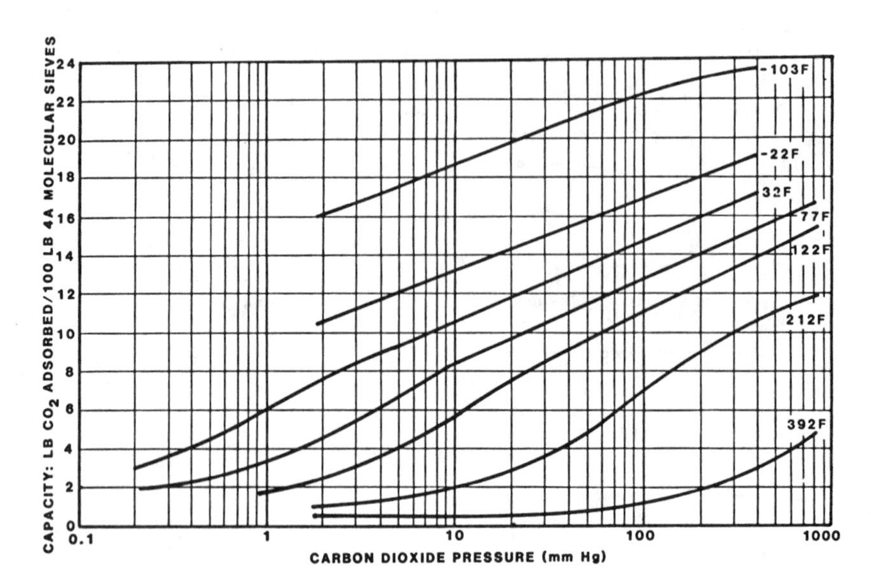

Figure 10. Adsorptive isotherms for CO_2 on Davison 4A molecular sieves.

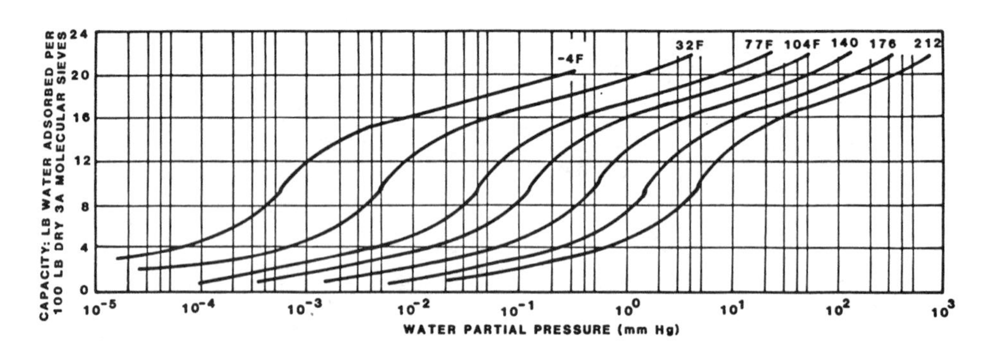

Figure 11. Adsorption isotherms for water on Davison 3A molecular sieves.

H'_A, Btu/1b of absorbate free solid, is referred to the pure vapor and X is the adsorbate concentration, 1b absorbate/1b solid. The integral may be evaluated graphically by determining the area under a curve of $\bar{H}$ vs X. The integral heat of adsorption referred to the solid and the adsorbed substance in the liquid state is given by:

$$\Delta H_A = \Delta H'_A + \lambda X \text{ Btu/1b} \quad \text{of solid} \tag{9}$$

The quantities H, ΔH_A, and $\Delta H'_A$ are negative, since heat is evolved during adsorption.

Mass Transfer Zones (MTZ)

When a fluid is first passed through a bed of adsorbers, most of the adsorbate is initially adsorbed at the inlet part of the bed and the fluid passes on with little further adsorption

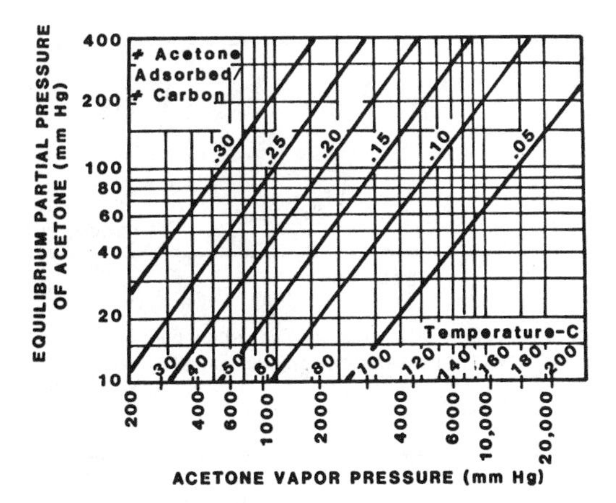

Figure 12. Reference-substance plot of equilibrium adsorption for acetone on activated carbon. (Data of Josefewitz and Othmer, *Ind. Eng. Chem.*, 40, 739 (1948).)

Table 2
Integral Heats of Adsorption of Organic Vapors on Carbon at Various Temperatures

Vapor	Integral Heat of Adsorption (kg-cal/mol*) 0 °C	25 °C
CH_3OH	13.1	13.9
C_2H_5OH	15.0	15.6
CH_3Cl	–	9.2
CH_2Cl_2	–	12.8
$CHCl_3$	14.5	14.5
CCl_4	15.3	15.4
C_2H_5Cl	12.0	15.4
n-C_3H_7Cl	–	15.6
iso-C_3H_2Cl	–	15.8
n-C_4H_9Cl	–	11.6
sec-C_4H_9Cl	–	15.0
tert-C_4H_9Cl	–	12.9
CS_2	12.5	15.4
$(C_2H_5)_2O$	15.5	14.5
C_6H_6	14.7	13.7

* *To convert* kg-cal/g_m to Btu/lb_m, multiply by 1,801.6.

taking place. Later, when the adsorber at the inlet end becomes saturated, adsorption takes place farther along the bed. The situation in the bed may be represented by Figure 13.

As more gas is passed through and adsorption proceeds, the saturation zone moves forward until a breakthrough point is reached, at which the exit concentration begins to rise rapidly above whatever limit has been fixed as a desirable maximum adsorbate level of the fluid. If this continues, saturation sets in. While the concentration when saturated is a function of the material used and the temperature at which it is operated, the dynamic capacity is also dependent on the operating conditions, such as inlet concentration, fluid flow rate, and bed dpeth. The dependence of inlet concentration and fluid flow rate arises from heat effect and mass transfer rates, but the dependence on bed depth, is really

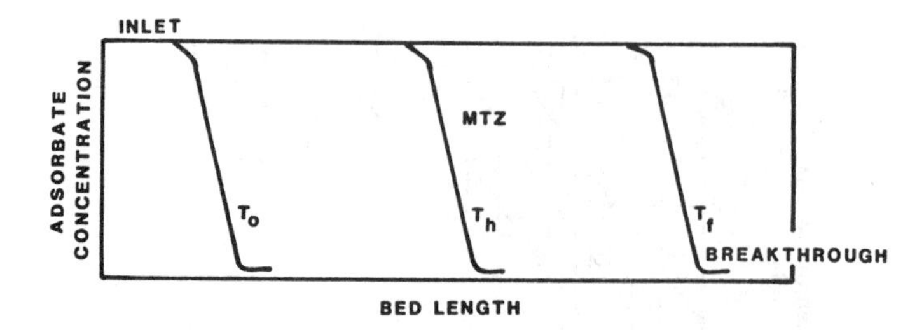

Figure 13. Formation and movement of the MTZ through an adsorbent bed: T_0 = MTZ concentration gradient at the formation of the zone; T_h = MTZ concentration gradient at half-life; and T_f = MTZ concentration gradient at breakthrough.

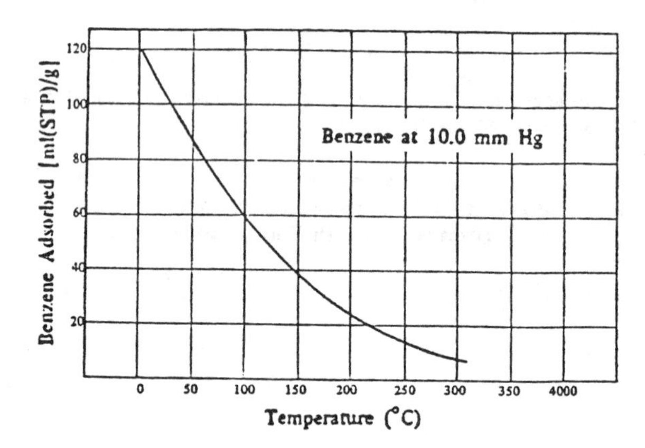

Figure 14. Benzene adsorption isobar on carbon.

dependent on the relative sizes of unsaturated and saturated zones. The zone of the bed where the concentration gradient is present is often called the mass transfer zone (MTZ).

It is extremely important that the adsorber bed should be at least as long as the mass transfer zone length of the key component to be absorbed. Therefore, it is important to know the depth of the mass transfer zone.

The following factors play the most important role in dynamic adsorption and the length and shape of the MTZ:

- The type of adsorbent.
- The particle size of an adsorbent.
- The depth of the adsorbent bed.
- The gas velocity.
- The temperature of the gas stream and adsorbent.
- The concentration of the contaminants to be removed.
- The concentration of the contaminants not to be removed, including moisture.
- The pressure of the system.
- The removal efficiency required.
- The possible decomposition or polymerization of contaminants on the adsorbent.

Effect of Temperature

Adsorption is an exothermic process; hence the concentration of adsorbed fluid decreases with increased temperature at a given equilibrium pressure. Because the equilibrium capacity of adsorbents is lower at higher temperatures, the dynamic or breakthrough capacity will also be lower, and the MTZ (mass transfer zones) is proportionately changed with temperature. Figure 14 shows the variation of the adsorption capacity with temperature.

As already noted, adsorption is an exothermic process. Therefore, as the adsorption front moves through the bed, a temperature front also proceeds in the same direction, and some of the heat is imparted to the gas stream. When the gas leaves the adsorption front, the heat exchange will reverse and the gas will impact heat to the bed. This increase in temperature during the adiabatic operation of the adsorber bed decreases the capacity of the adsorbent [9].

DESIGN PRINCIPLES FOR ADSORPTION CONTROL EQUIPMENT

The engineering design of adsorption equipment must be based on a sound application of the principles of diffusion, equilibrium, and mass transfer. The main requirement in equipment design is to bring the fluid into intimate contact with the adsorbent; that is, to provide a large interfacial area and a high intensity of interface renewal and to minimize resistance and maximize driving force. This contacting of the phases can be achieved in many different types of equipment; the most noteworthy being adsorbers. The final choice of equipment will be determined by the various criteria which must be met.

In most processes involving the adsorption of a fluid from an effluent stream the inlet conditions are known, i.e., flow rate, composition, and temperature. The outlet conditions are specified. The main objectives then in the design of adsorption equipment are the determination of the rate of adsorption and the calculation of the principal dimensions of the equipment.

Selection of Adsorbent

Industrial adsorbents are usually capable of adsorbing both organic and inorganic gases or vapors. However, their preferential adsorption characteristics and other physical properties make them more or less specific for particular applications. Experience has shown that for the adsorption of organics, activated carbon has superior properties, having hydrocarbon selective properties and high adsorption capacities for such materials. Inorganic adsorbents, such as silica gel or activated alumina, can be used to adsorb organic materials, but problems may arise during the regeneration process for the adsorbent. Water vapor with organic contaminants may also be adsorbed by activated alumina, silica gel, and molecular sieves. At times this may be a considerable drawback for organic contaminant removal. Recently, through evaluation of the type of surface oxides present on activated carbon surface, it was found that the preferential adsorption properties of carbon can be partially regulated by the type of surface oxide induced on the carbon.

The normal method of regeneration of adsorbent is by use of steam, hot air, nitrogen, or other gas streams, and this may cause slight decomposition of the organic compound on the adsorbent. As a result, two problems may arise: first, the incomplete recovery of the adsorbate; second, the adsorbent progressively deteriorates in capacity as the number of cycles increases due to the blocking of pores by carbon formed by hydrocarbon decomposition. By using activated carbon, regeneration can be performed by a steaming process. However, this is not feasible, if silica gel or activated alumina are used as the adsorbent, because these substances run the risk of breaking down on contact with liquid water.

In some cases, none of the adsorbents have sufficient retaining capacity for a particular contaminant. In cases like these, the large surface area of the adsorbent is impregnated with inorganic or, in rare cases, with a high molecular weight organic compound which can chemically react with the contaminant in question.

It should be noted that these impregnations serve either a catalytic conversion purpose or are reactions to a nonobjectionable compound or to a more easily adsorbed compound. When impregnation occurs, general adsorption theory no longer applies to the overall effects of the process. Under the particular conditions, an impregnated adsorbent is available for most compounds which are not easily adsorbed by nonimpregnated commerical adsorbents.

Since adsorption takes place at the interface boundary, the surface area of the adsorbent is an important consideration in the adsorption process. As a rule of thumb, the higher the surface area of an adsorbent, the higher the adsorptive capacity. However, the adsorbent's surface area must possess the particular pore size. At low partial pressure, the surface area in the smallest pores in which the adsorbate can enter is the most efficient. At higher pressures the larger pores become more important, while at very high concentrations, capillary condensation will take place within the pores, and the total micropore volume is the limiting factor.

The action of molecular sieves is different from other adsorbents in that their selectivity is determined more by the pore size limitations. In choosing molecular sieves as the adsorbent, care should be taken to ensure that the contaminant to be removed is smaller than the pore size. Hence it is important that the particular adsorbent not only has an affinity for the contaminant in question, but also that it has sufficient surface area and pore size distribution to facilitate the adsorption process.

Design Data

Having chosen the adsorbent, the next step will be to calculate the quantity of adsorbent required and eventually consider other facts such as temperature rise of the fluid stream due to adsorption and what the useful life of the adsorbent might be under operating conditions. The sizing and overall design of the adsorption system depends on the properties and characteristics of both the feed fluid to be treated and the adsorbent. The following information should be known or available for design purposes.

Fluid Stream

1. Adsorbate concentration.
2. Temperature.
3. Temperature rise during adsorption.
4. Pressure.
5. Flow rate.
6. Presence of adsorbent contaminant material.
7. Gas density at operating temperature and pressure.
8. Fluid viscosity at operating temperature and pressure.

Adsorbent

1. Adsorption capacity as used on-stream.
2. Temperature rise during adsorption.
3. Isothermal or adiabatic operation.
4. Adsorbent life cycle and collected contaminant material stability.
5. Possibility of catalytic effects causing adverse chemical reaction in the gas stream or formation of polymers on the adsorbent bed, with consequent deterioration.
6. Bulk density.
7. Particle size (usually reported as a mean equivalent particle diameter). The dimension and shape of particles affect both the pressure drops through the adsorbent bed and

the diffusion rate into the particles. All things being equal, adsorbent beds consisting of smaller particles, although causing a higher pressure drop, will be more efficient.
8. Pore data, which are important because they may permit elimination from consideration of adsorbents whose pore diameters will not admit the desired adsorbate molecules.
9. Hardness, which indicates the care that must be taken in handling the adsorbents to prevent the formation of undesirable fines.
10. Regeneration information.

The Separation Process: Staged vs. Continuous Operation

Adsorption systems for solvent recovery or air purification most frequently employ a fixed adsorbent bed. It is the simplest and most trouble free in its operation. Other systems have been developed in which the adsorbent is moved countercurrent to the fluid stream or the adsorbent is fluidized in the vapor stream. When the adsorbent bed is fixed or moves as a unit, the vapor concentration distribution in the adsorbent bed follows a definite pattern that can be analyzed and then used to determine the design of the adsorption system.

If it is only required to treat a gas stream for a short period, then usually only one adsorption unit is necessary, provided, of course, a sufficient time interval is available between adsorption cycles to permit regeneration. This is usually not the case, in that an uninterrupted flow of treated gas is often required. It is then necessary to use two or more units capable of operating in this fashion. The units designed to handle fluid flows without interruption are characterized by their mode of contact—staged or continuous. By far, the most common type of adsorption system used to remove an objectionable pollutant from a gas stream consists of a number of fixed-bed units operating in such a sequence that the gas flow remains uninterrupted.

Typical of continuous-contact operation for gaseous pollutants adsorption is the fluidized-bed technique. The term fluidization is used to designate the fluid-solid contacting process in which a bed of finely divided solid particles is lifted and agitated by a rising stream of process fluid. Figure 15 gives the various states of passing a gas through a bed of solids:

- *Fixed bed:* Gas velocity is too low to disturb the arrangement of the particles. As the gas passes up through the bed, it exhibits a pressure drop as it flows through the interstills.
- *Fluidized bed (quiescent):* Gas velocity sufficient to lift the adsorbent particles and enable them to move locally. As the gas velocity is increased, further enlargement of the bed results and creates a voidage large enough to allow the particles to move locally.
- *Fluidized bed (boiling bed):* Gas velocity sufficiently large to lift the adsorbent particles and enable them to circulate freely through the bed. The bed appears much like a boiling liquid, with well-defined gas bubbles, giving a distinct interface between the top of the

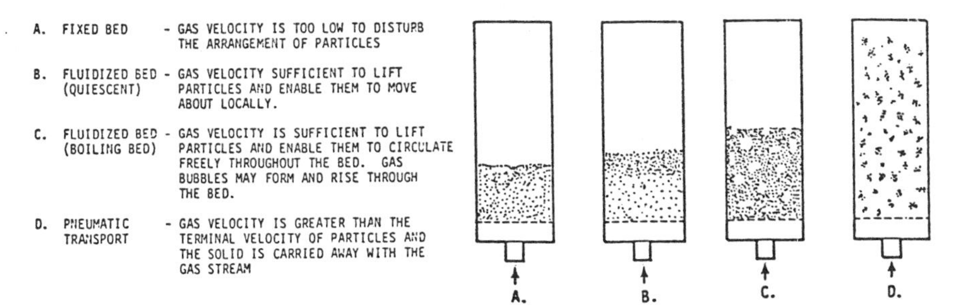

Figure 15. Various sites of passing gas through a bed of solids.

solids and the escaping gas. The gas pressure drop is not much different from that of the quiescentfluidized state. This is the condition most often used for adsorption.

- *Pneumatic pressure:* Gas velocity greater than the terminal velocity of particles and the solid adsorbent is carried away with the gas steam.

During the steady-state staged-contact operation, the gas flows up through a series of successive fluidized bed stages, permitting maximum gas-solid contact on each stage. A typical arrangement is shown in Figure 16 for multistage countercurrent adsorption with regeneration. In the upper part of the column the solids are contacted countercurrently on perforated plates in relatively shallow beds with the gas stream containing the pollutant, the adsorbent solids moving from plate to plate through downspouts. In the lower part of the column, the adsorbent is regenerated by similar contact with hot gas, which desorbs and carries off the pollutant. The regenerated adsorbent is then recirculated by an air lift to the top of the column. A serious disadvantage of such a fluidized-bed technique is the high attrition losses of the adsorbent caused by the fluidization of the beds. Because of this attrition, filtration of the effluent airstream may be required. The influent airstream, however, need not be filtered since no plugging of the fluidized beds can occur.

During continuous-contact operation, the gas and adsorbent are in contact throughout the entire unit, without periodic separation of the phases as is characteristic of the

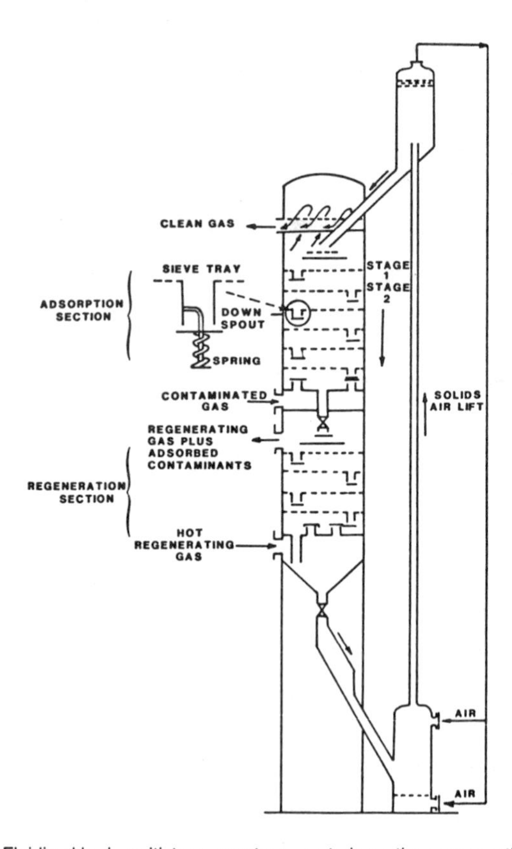

Figure 16. Fluidized-bed, multistage countercurrent absorption regeneration.

steady-state staged-contact operation. The operation may be carried out in strictly continuous, steady-state fashion, characterized by movement of the solid as well as the gas, or owing to the rigidity of the solid adsorbent particles, it may be possible to operate more advantageously in a semicontinuous manner, by a moving as on a stationary solid. This latter method of operation results is unsteady-state conditions, where compositions in the system change with time. Typical of this continuouscontact steady-state mode of adsorption is the *hypersorber* (Figure 17). This unit is specifically used for separating mixtures of low molecular weight hydrocarbons and other gaseous components.

Although various types of operations and various types of adsorber beds have been used to remove gaseous pollutants from process streams, the fixed-bed adsorber has been by far the most common and most successful type of adsorption operation. If intermittent or batch operation is practical, a simple one-bed system will suffice, cycling alternately between the adsorption and regeneration phases. However, since most commercial applications require continuous operation, a two- or three-bed system is usually selected. In these systems, one or two of the beds will serve as by-passes for regeneration while the others are adsorbing. Typical two- and three-bed systems are shown in Figure 18 and Figure 19, respectively.

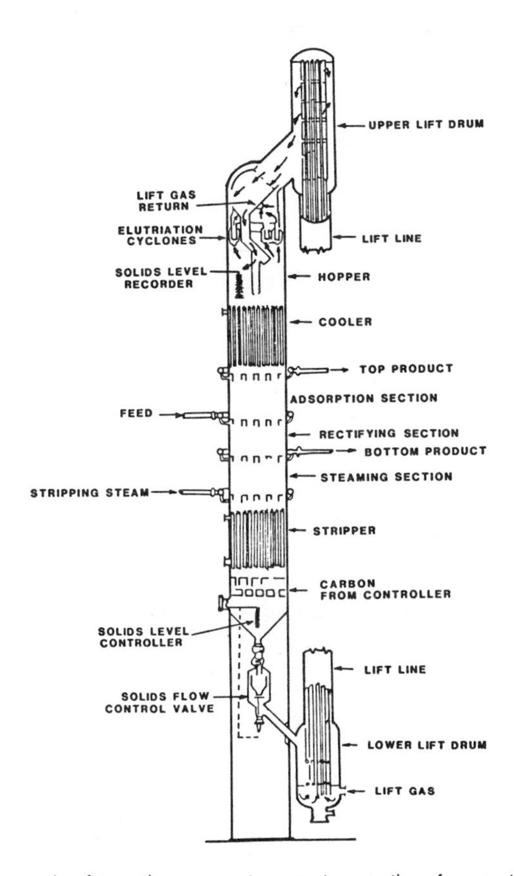

Figure 17. Hypersorber for continuous countercurrent separation of a gas stream into two products.

Stagewise-Contact Operation

Gases are treated continuously for solute removal in a stagewise mode of contact usually using fluidized-bed techniques. A typical system is shown in Figure 16. The adsorption section of this unit may be represented by Figure 20. These type of systems can operate either isothermally or adiabatically, depending on the operating conditions and the materials to be removed.

Isothermal Operation

Similar to the gas adsorption procedure, a material balance for the solute about the N_p stages in Figure 20 is given by

solute in = solute out

$$(G_sY_{n_{p+1}} + 1) + L_sX_0 = G_sY_1 + L_sX_{n_p} \tag{10}$$

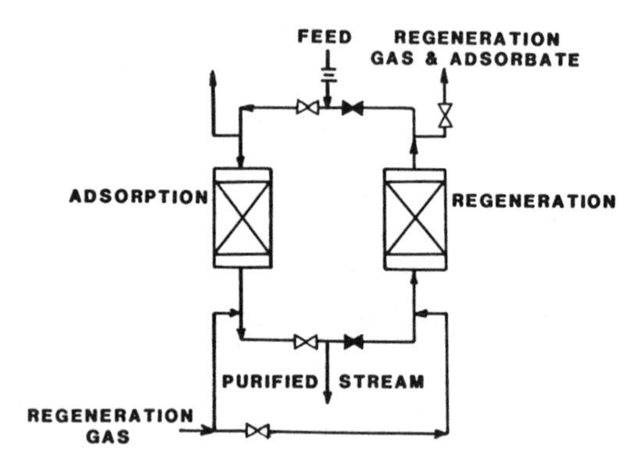

Figure 18. Typical two-bed adsorption system.

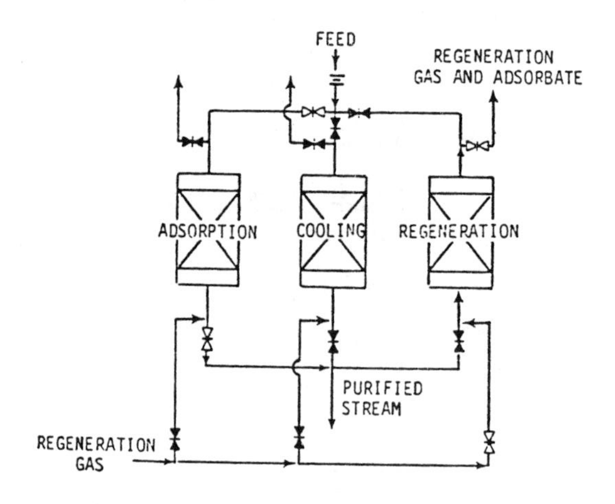

Figure 19. Typical three-bed adsorption system.

or

$$G_s(Y_{n_{p+1}} - Y_1) = L_s(X_{n_p} - X_0) \tag{11}$$

Equation 11 is the equation of the operating line on Figure 21, defined by the coordinates $(X_{n_p}, Y_{n_{p+1}})$ and (X_0, Y_1) and the slope L_s/G_s. The number of theoretical stages necessary to accomplish the desired separation is found, again, by stepping off this number of stages between the equilibrium curve (at a specified temperature and pressure) and operating line as shown in Figure 21. Alternatively, the L_s/G_s (adsorbent/gas) ratio for a predetermined number of stages may be found by trial-and-error location of the operating line. For desorption or adsorbent stripping, the procedure followed is the same only with the operating line falling below the equilibrium curve with the minimum adsorbent/gas ratio, giving the largest number of stages. This operating line touches the equilibrium curve within the specified range of concentrations.

If the equilibrium curve is known, the trial-and-error becomes the redundant. From the given curve, if a form of the Freundlich isotherm equation in concentration is known, i.e.

$$Y^* = mX^n \tag{12}$$

and taking the logarithm of both sides, we have

$$\log Y^* = \log m + n \log X \tag{13}$$

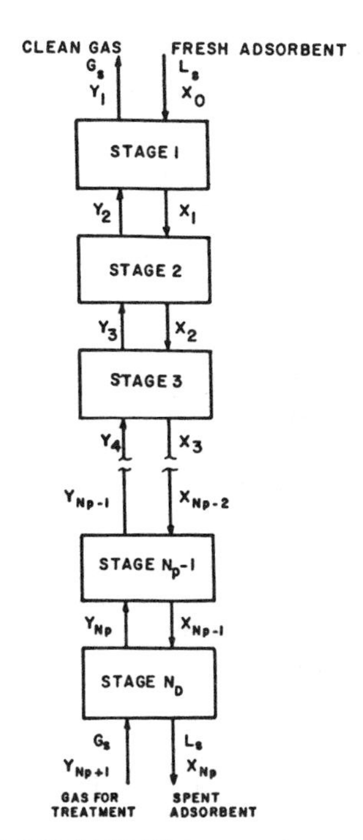

Figure 20. Countercurrent stagewise contact adsorption.

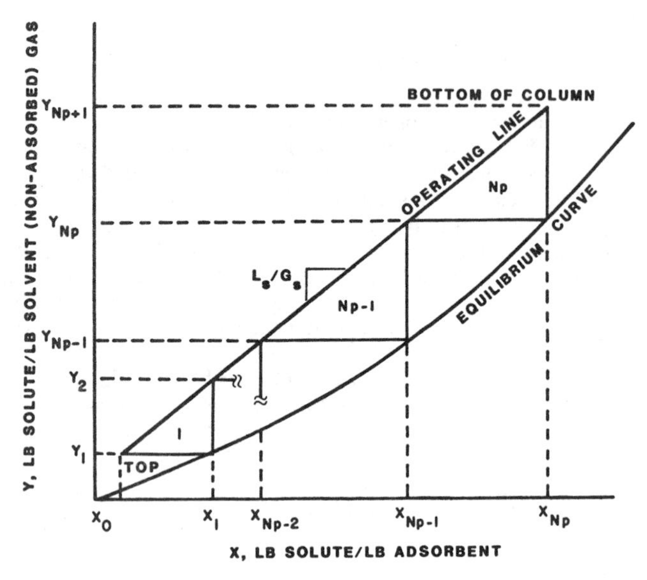

Figure 21. Operating line for countercurrent stagewise contact adsorption.

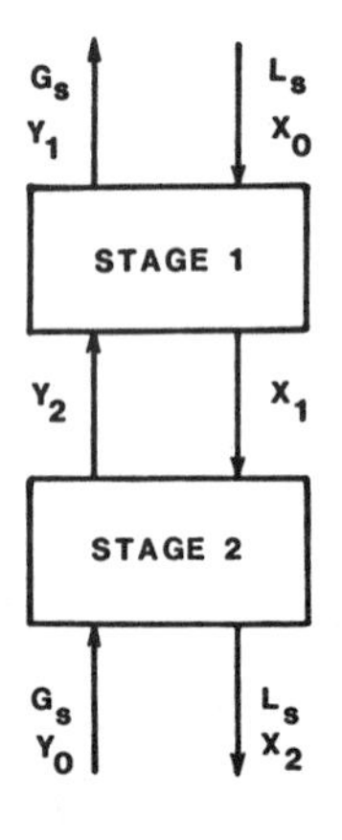

Figure 22. Countercurrent adsorption.

If ($\log Y^*$) is plotted against ($\log X$) a straight line results, with a slope of n and y-intercept of log m. Consider only the two-stage (Figure 22) adsorption section with fresh adsorbent ($X_0 = 0$) entering the top of the column. A solute material balance will yield

$$G(Y_0 - Y_1) = L_s(X_2 - 0) \tag{14}$$

Applying Equation 3 to the effluents from the lower stages, we have

$$X_2 = (Y_2/m)^{1/n} \tag{15}$$

Combining Equations 13 and 14 then

$$L_s/G_s = (Y_0 - Y_1)/(Y_2/m)^{1/n} \tag{16}$$

The operating line of the upper ideal stage is now given by

$$G_s(Y_2 - Y_1) = L_s X_1 = L_s\left(\frac{Y_1}{m}\right)^{1/n} \tag{17}$$

If we eliminate (L_s/G_s) from Equations 15 and 16 we have

$$\left(\frac{Y_0}{Y_1}\right) - 1 = \left(\frac{Y_2}{Y_1}\right)^{1/n}\left(\frac{Y_2}{Y_1} - 1\right) \tag{18}$$

Equation 18 may be solved for the immediate concentration Y_2 for specified terminal concentration Y_0 and Y_1. L_s/G_s is then given by Equation 16. Treybal provides a convenient graphical solution to Equation 17 as shown in Figure 23.

Adiabatic Operation

When adsorption is accompanied by a liberation of large quantities of heat and a subsequent rise in temperature, isothermal operation cannot be assumed. This usually is the case when solutions are not dilute. Under adiabatic conditions stage by stage calculations are required to determine the rate of adsorption. A material balance for the N_p state shown in Figure 20 is

$$G_s(Y_{n_p+1} - Y_1 = L_s(X_{n_p} - X_0) \tag{11}$$

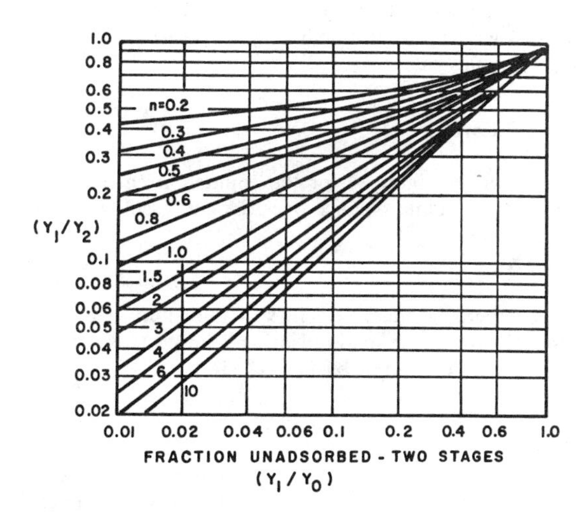

Figure 23. Graphical solution to Equation 17.

and an enthalpy balance for the adiabatic operation is

$$G_s(H_{G,n_p+1} - H_{G,1}) = L_s(H_{L,n_p} - H_{L,0}) \tag{19}$$

where H_G refers to the enthalpy of the adsorbate gas mixture in the Btu/lb solvent gas, and H_L to the enthalpy of the adsorbent solid plus adsorbate in Btu/lb adsorbent. If the enthalpies refer to solid adsorbent, carrier gas (component C) and adsorbate (component A) as a liquid, all at a base temperature, T_0, then.

$$H_G = C_{P_c}(T_G - T_0) + Y(C_{P_A}(T_G - T_0) + \lambda_{A_0}) \tag{20}$$

where C_{P_c} = gas heat capacity (Btu/lb°F)
C_{P_A} = adsorbate (vapor) heat capacity (Btu/lb°F)
λ_{A_0} = latent heat of vaporization of A at T_0 (Btu/lb)

In addition,

$$H_L = C_{P_B}(T_L - T_0) + X c_{P_{AL}}(T_L - T_0) - \Delta H_A \tag{21}$$

where C_{P_B} = heat capacity of the adsorbent (component B) (Btu/lb°F)
$C_{P_{AL}}$ = heat capacity of liquid A (Btu/lb°F)
ΔH_A = integral heat of adsorption at X and T_0 (Btu/lb adsorbent)

To determine the number of ideal or theoretical stages required to facilitate the adsorption process, some trial and error is needed. This usually involves the application of both material and energy balances.

Continuous-Contact Operation

Continuous-contact may be carried out in either a strictly continuous, steady-state or a semicontinuous unsteady-state condition, the latter being the more prevalent for the removal of gaseous pollutants. The following theoretical analysis will be based upon the removal of a single pollutant from a gas stream, this being the usual case. Isothermal operation will be assumed to simplify the analysis.

Steady-State/Moving-Bed Adsorbers

Consider the system shown in Figure 24 for the countercurrent adsorption of a single pollutant from a gas stream. A solute balance over the entire column given

$$G_s(Y_1 - Y_2) = L_s(X_1 - X_2) \tag{22}$$

where G_s and L_s are the superficial mass velocities of the solute free gas (solvent gas or unadsorbed gas) and the adsorbate free solid (adsorbent), respectively. Over the upper part of the column, the solute balance is given by

$$G_s(Y - Y_2) = L_s(X - X_2) \tag{23}$$

These equations establish the operating line in X, Y coordinates, a straight line of slope L_s/G_s between the terminal points (X_1, Y_1) and (X_2, Y_2). The solute concentration X and Y at any level in the column fall upon this line. An equilibrium curve corresponding to the specified temperature and pressure for the system under consideration is also plotted and

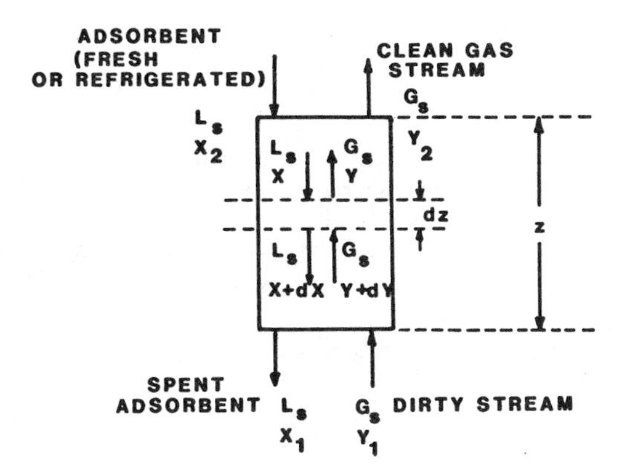

Figure 24. Continuous countercurrent adsorption of a single pollutant.

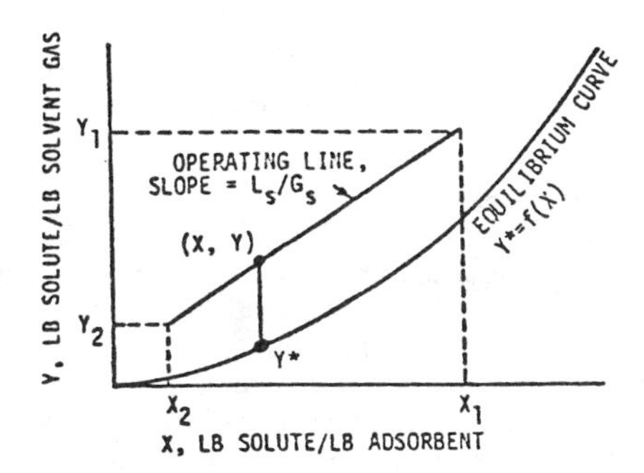

Figure 25. Continuous countercurrent adsorption (isothermal) of a single pollutant.

the minimum solid/gas ratio is determined. It is observed that the equilibrium curve falls below the operating line for adsorption and above it for desorption.

The resistance to mass transfer of adsorbate from the gas to the adsorbed state on the adsorbent will include that residing in the gas surrounding the solid particles, that corresponding to the diffusion of adsorbate through the gas within the pores of the solid, and possibly an additional resistance at the time of adsorption. The latter is usually negligible during physical adsorption. If the remaining resistances may be characterized by an overall gas mass transfer coefficient based on A_p (the outside surface of the solid particles), K_yA_p (the rate of adsorbate transfer over the differential adsorber height dz) may be given as

$$L_s\,dX = G_s\,dY = K_yA_p(Y - Y^*)\,dz \tag{24}$$

where Y^* is the equilibrium composition in the gas corresponding to the adsorbate composition X. The driving force $(Y - Y^*)$ is then represented by the vertical distance between the operating line and the equilibrium curve as shown in Figure 25. Equation 24 can be rearranged, integrated, and the number of transfer unit N_{0g} determined.

$$N_{0g} = \int_{Y_2}^{Y_1} \frac{dy}{Y - y^*} = \frac{K_yA_p\,dz}{G_s} = \frac{Z}{H_{0g}} \dots \tag{25}$$

where

$$H_{0g} = G_s/K_yA_p\ldots \tag{26}$$

Equation 25 is normally evaluated graphically and the active height Z, determined through knowledge of the height of a transfer unit H_{0g} which is characteristic of the system.

The use of an overall coefficient or overall height of a transfer unit implies that the resistance to mass transfer within the pores of the solid particles may be characterized by an individual mass transfer coefficient K_sA_p or height of a transfer unit H_s [11]. Thus,

$$\frac{G_s}{K_yA_p} = \frac{G_s}{K_yA_p} + \frac{mG_sL_s}{L_sK_sA_p} \tag{27}$$

or

$$H_{0g} = H_g + \left(\frac{mG_s}{L_s}\right) H_s \tag{28}$$

where $m = dy^*/dx$, the slope of the equilibrium curve. These equations hold strictly when the mass transfer resistance of the gas surrounding the particles is controlling.

Unsteady-State/Fixed-Bed Adsorbers

Fixed-bed adsorbers are the usual choice for the control of gaseous pollutants when adsorption is the desired method of control. Figure 26 shows the case for a binary solution containing a strongly adsorbed solute (gaseous pollutant) at concentration C_0. The gas stream containing the pollutant is to be passed continuously down through a relatively deep bed of adsorbent, which is initially free of adsorbate. The top layer of adsorbent, in contact with the contaminated gas entering, at first adsorbs the pollutant rapidly and effectively, and what little pollutant is left in the gas is substantially all removed by the layers of adsorbent in the lower part of the bed. At this point in time the effluent from the bottom of the bed is practically pollutant free as at C_1. The top layer of the bed is practically

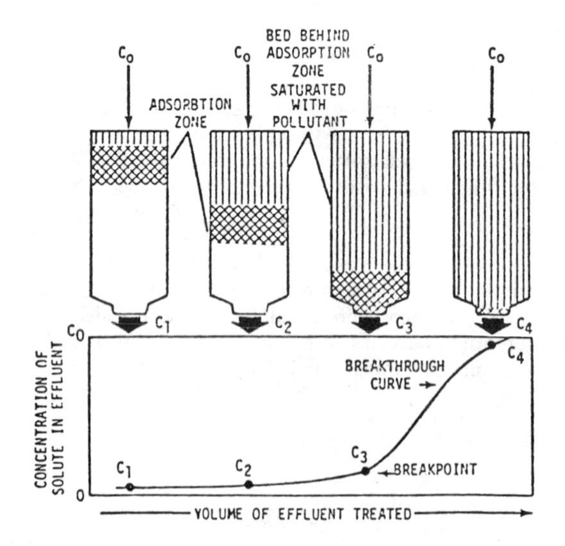

Figure 26. Adsorption wave front.

saturated, and the bulk of the adsorption takes place over a relatively narrow adsorption zone in which there is rapid change in concentration. The saturated bed length is L_{st}. As the gas stream continues to flow, the adsorption zone of length L_z travels downward, similar to a wave, at a rate usually much less than the linear velocity of the gas stream through the bed. At some time later roughly half of the bed is saturated with the pollutant, but the effluent concentration C_2 is still substantially zero. Finally, at C_3, the lower portion of the adsorption zone has reached the bottom of the bed, and the concentration of the pollutant in the effluent has suddenly risen to an appreciable value for the first time. The system is said to have reached its "breakpoint." The pollutant concentration in the gas effluent stream now rises rapidly as the adsorption zone passes through the bottom of the bed and at C_4 has just about reached the initial C_0. At this point the bed is just about fully saturated with pollutant. Between C_3 and C_4 is termed the "breakthrough." If the gas stream continues to flow, little or no additional adsorption takes place since the bed is almost in equilibrium with the gas stream.

The time at which the "breakthrough" curve appears and its shape greatly influence the method of operating a fixed-bed adsorber. The breakpoint time generally decreases with decreased bed height, increased adsorbent particle size, increased flow rate through the bed, and increased initial pollutant concentration in the entering gas stream. It is normal industrial practice to determine experimentally for a particular system the breakpoint and breakthrough curve under conditions as close as possible to those expected in the process.

The service time is one of the prime factors to be considered in sizing adsorption equipment. When deep or long adsorbent beds are used, as in solvent recovery (with activated carbon adsorbent material), L_{st} is large relative to L_z and essentially determines the service time. With usual solvent recovery practices, the solvent air velocity through the bed is maintained at near 100 ft/min. In these cases L_z is on the order of 3 in. and z, the absorbent bed length or depth can range from 16 to 36 in. The capacity of the saturated-bed length, L_{st}, can be calculated with considerable accuracy if an adsorption isotherm has been determined.

The capacity of the adsorption zone, besides varying with C_0 and the temperature, T, also varies with the adsorbent mesh size and flow velocity. The service time, t_b, to penetration concentration, C_b is dependent on the adsorptive capacities of the saturated bed and the adsorption zone as expressed by the equation,

$$t_b = (A/VC_0)(X_{st}L_{st} + X_zL_z) \tag{29}$$

where A = area of adsorbent bed (ft^2 or cm^2)
V = total gas volumetric flow rate (ft3/min or liter/min)
X_{st} = adsorptive capacity per unit adsorbent bed volume in the saturated zone (lb/ft 3 or g/cm3)
X_z = mean adsorptive capacity per unit bed volume in the adsorptive zone (lb/ft 3 or g/cm3)

Also, X_zL_z is a fixed amount wherein the L_z is defined as the bed length in which C decreased from C_0 to essentially zero. L_{st} varies with the C_B/C_0 ration; at $C_B = C_0$, $L_{st} = z$, while at $C_B \approx 0$, $L_{st} = z - L_z$.

In thin beds, 2 in. or less, as used often in air purification, flow velocities near 40 ft/min are usually employed, which shortens L_z down to the 2-in. level. In these applications, the service time is essentially

$$t_B = (A/VC_0)X_LL_z$$

To attain the desired service time, the dimensions of the adsorbent volume, $A(L_{st} + L_z)$ or AL_z are optimized with consideration given to resistance to air flow that can be tolerated and the hardware to contain the adsorbent bed.

The adsorptive capacity and length of the adsorption zone can be calculated if the effluent concentration curve has been determined for the flow system. This is possible only if the adsorption zone has attained a steady state before the vapors penetrate the bed. If the bed is shorter than L_z the vapor distribution curve is still changing and the effluent concentration curve reflects this change.

The design of fixed-bed adsorbers can became quite complex in different situations. The time to the breakpoint and the shape of the breakthrough curve might have to be predicted. Most mathematical models involve the use of numerical integration and simultaneous solution of differential equations derived from material and energy balance. The fundamental equations are obtained by making mass and heat balances over a differential volume element of the packed column for both the gas stream and the adsorbing solid. The equations are linked together by an adsorption rate expression obtained by considering transport at the gas-solid interface and intraparticle diffusion. The model which follows was developed by Collins and Chao [12] to dynamically simulate multicomponent fixed-bed adsorption. It may also be used in the case where only a single component is adsorbed. The following are the chief assumptions made in the development.

1. The state variables (i.e., Y_i', X_i', T_G, T_s and V) are functions of axial-bed length and time.
2. The pressure drop across the bed is negligible compared to absolute pressure.
3. The predominant component of the gas stream (the carrier gas) is inert to the adsorbent.
4. The column is operated adiabatically.
5. Axial dispersion is negligible.

These assumptions are intended to reflect conditions encountered in industrial adsorbers.

Consider a differential section dz of the adsorption bed. The gas stream is made up of an inert carrier and adsorbates (or single adsorbate). The differential mass balance of the inert is given by

$$\frac{\partial G_m}{\partial z} = -\varepsilon \frac{\partial \varrho_m}{\partial t} \tag{30}$$

where G_m = molar flow rate of the inert gas (equal to $\varrho_m V$) ($lb_m/hr\ ft^2$)
Z = height of adsorption bed (ft)
ε = void fraction of the bed
ϱ_m = molar density of the inert gas (lb_m/ft^3)
t = time (hr)

The adsorption or desorption of the adsorbates leads to a change in the gas composition expressed by component mass balances

$$\frac{\partial Y_i'}{\partial z} + \left(\frac{\varepsilon}{V}\right)\frac{\partial Y_i'}{\partial t} = -\frac{R_i}{G_m} \tag{31}$$

where Y_i' = gas concentration expressed in lbm adsorbate i per lbm inert gas.
v = superficial linear velocity of gas stream (ft/hr)
R_i = rate of sorption of i per unit bed volume per unit time (positive value denotes adsorption and a negative value desorption) ($lbm/hr\ ft^3$)

Equations 29 and 30 may be compacted in terms of the substantive derivative defined by

$$\frac{D}{D_t} = \frac{\partial}{\partial t} + \left(\frac{V}{\varepsilon}\right)\frac{\partial}{\partial z} \tag{32}$$

The substantive derivative measures the rate of change for a fixed parcel of gas. It follows that the inert is described by

$$\frac{\partial V}{\partial z} = \left(\frac{\varepsilon}{\varrho_m}\right)\frac{D\varrho_m}{D_t} \tag{33}$$

an the adsorbate by

$$\frac{DY'_i}{D_t} = \frac{R_i}{\varepsilon\varrho_m} \tag{34}$$

For each of the adsorbates the differential mass balance for the adsorbed phase can be expressed as

$$\frac{\partial X_i}{\partial t} = \frac{R_i}{\varrho_s} \tag{35}$$

where X_i = solid loading, lbm of i adsorbed per lbm of solid adsorbent
ϱ_s = bulk density of the adsorbent (lbm/ft^3)

The solid loading X_i is the average value of the concentration of the adsorbate which varies within the solid pellet as a result of diffusion. In order to evaluate interphase transport, the local concentration at the pellet surface (X'_{si}) must be related to X'_i. Glueckauf [13] obtained an approximate analytical solution

$$\frac{\partial X'_i}{\partial t} = \frac{15\bar{D}_i}{\gamma^2}(X'_{si} - X'_i) \tag{36}$$

where $\bar{D}_i$ = intraparticle diffusivity (ft^2/hr)
r = radius of the pellet (ft)

The simplifying assumption is made that initially at time t = 0, the concentration of the adsorbed phase is uniform in pellet, then

$$X'_i(R, t = 0) = \text{constant} \tag{37}$$

Tury [14] verified Equation 37 for $(\bar{D}_i t/r^2) \geqq 0.101$

Yoshida, Ramaswami, and Hongen [15] investigated gas transport to a solid pellet surface and expressed the transfer rate by

$$R_i = K_{Gi}a(P_i - P_{si}) \tag{38}$$

where K_{Gi} = mass transfer coefficient ($lb_m/hr\ ft^2\ atm$)
a = surface area available for mass transfer (ft^2/ft^3)
P_i = partial pressure of component, i, in the gas stream (atm)
P_{si} = partial pressure of component, i, at the pellet surface (atm)

For mass transfer coefficients in gas adsorption, K_G has been correlated in terms of a I_d factor as a function of the Reynold's number. By setting the adsorption rate equal in Equations 34 and 35, an overall rate equation proportional to the partial pressure difference is obtained.

$$R_i = K_{Gi}a(P_i - P_i^*) \tag{39}$$

or a solid concentration gradient

$$R_i = KC_ia(X_i'^* - X_i') \tag{40}$$

where p^* = partial pressure of component, i, in the gas stream in adsorption equilibrium with X_i (atm)
$X_i'^*$ = solid loading in equilibrium with a gas concentration Y_i 1bm i adsorbed/1bm adsorbent

The choice of Equations 39 or 40 depends on calculation convenience for a particular application. The overall mass transfer coefficients K_{Gi} or K_{Ci} are normally evaluated experimentally.

For adiabatic operations, heat balances on the gas stream and solid adsorbent must be considered. Figure 27 shows the control volume with the input and output valves used in the heat balances. For the gas,

$$\frac{DT_G}{Dt} = \frac{U_a(T_s - T_G) - \sum_{j=1}^{j} R_j\bar{c}_{p_j}(T_s - T_G)}{\varepsilon\varrho_m c_p} \tag{41}$$

where U = overall heat transfer coefficient (Btu/hr ft °R)
c_p = molar heat capacity of the gas at T_G (Btu/1bm °R)
$\bar{c}_{p_j}$ = average heat capacity of component, j, between T_G and T_s (Btu/1bm °R)
T_s = average pellet temperature (°R)
j = refers to components that are being desorbed

The summation over j includes all the components for which R_j is negative. The heat balance on the solid is expressed by:

$$\frac{\partial T_s}{\partial t} = \frac{U_a(T_G - T_s) + \sum_{i=1}^{i} R_i\Delta H'_{A_i} + \sum_{K=1}^{K} R_K c_{\varrho_K}(T_G - T_s)}{\varrho_s c_{\varrho_s}} \tag{42}$$

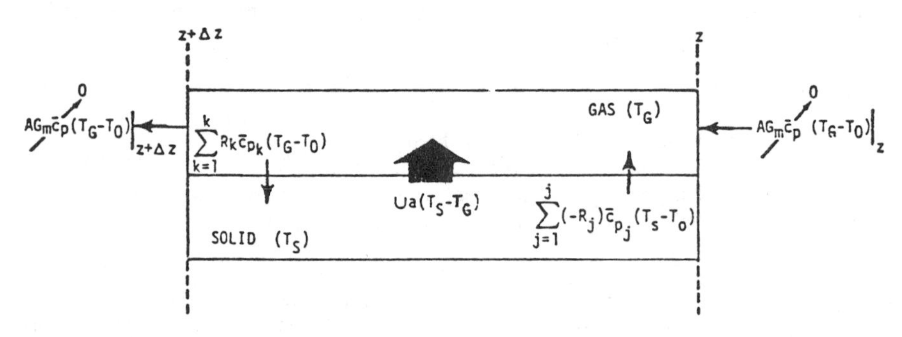

Figure 27. Energy balance on gas stream.

where $\Delta H'_{A_i}$ = heat of adsorption of the component i (Btu/1bm)
c_{ϱ_s} = solid heat capacity (usually considered constant) (Btu/1bm °R)

Finally, the equation of state for the gas can be expressed as

$$\varrho_m = \frac{P_t}{(1 + \sum_{i=1}^{i} Y_i) f R' T_G} \tag{43}$$

where f = compressibility factor (unity for an ideal gas)
P_t = total pressure (atm)
R' = gas constant (atm ft^3/1bm °R)

If f is constant, Equation 43 can be differentiated and substitution into Equation 33 gives

$$\frac{\partial v}{\partial z} = \frac{-fR'T_G}{P_t} \sum_{i=1}^{i} R_i + \left(\frac{\varepsilon}{T_G}\right)\left(\frac{DT_G}{Dt}\right) \tag{44}$$

The foregoing set of equations mathematically describes the dynamics of the adsorber. The use of these requires knowledge of the overall mass transfer coefficient and adsorption equilibria.

The solution to the preceeding equations involve numerical integration using the method of characteristics. Gamson, Thodos, and Hougen [16] have presented solutions to these equations, and their application to adsorption systems have been discussed by Acrivos [17].

A simpler treatment by Michaels [18] and further developed by Treybal [10] is also used. Its scope is limited to isothermal adsorption from dilute feed mixtures and to cases where the equilibrium adsorption isotherm is concave to the solution concentration axis, where the adsorption zone is constant in height as it travels through the adsorption column, and where the height of the adsorbent bed is large relative to the height of the adsorption zone.

The idealized breakthrough curve, as shown in Figure 28, results from the flow of carrier gas through an adsorbent bed at the rate of G_s 1b/hr ft^2, entering with an initial adsorbate concentration Y_0 1b adsorbate/1b carrier gas. The total adsorbate-free effluent after any time is W 1b/ft^2 of bed cross section. The breakthrough curve is steep, and the adsorbate

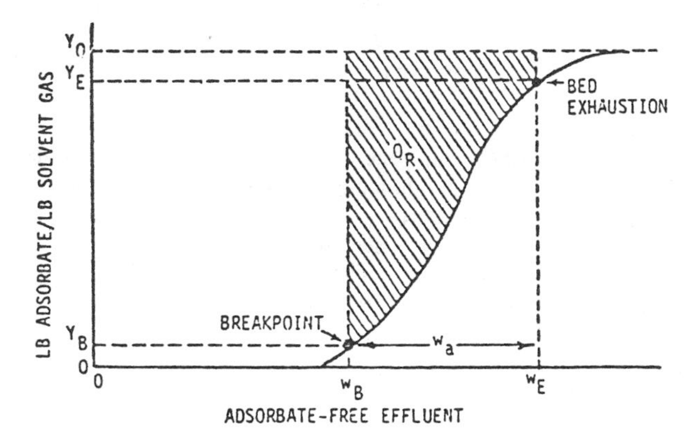

Figure 28. Idealized breakthrough curve.

concentration in the effluent rises rapidly from essentially zero to that in the incoming gas. Some low values of Y_B are arbitrarily chosen as the breakpoint concentration, and the adsorbent is considered as essentially exhausted when the effluent concentration has risen to some arbitrarily chosen value Y_E, close to Y_0. The principal concern is with the quantity of effluent W_B at the breakpoint and the shape of the curve between W_B and W_E. The total effluent accumulated during the appearance of the breakthrough curve is $W_A = W_E = W_B$. The adsorption zone, of constant height L_z ft is that part of the bed in which the concentration change from Y_B to Y_E is occurring at any time.

Letting Θ_a be the time required for the adsorption zone to move its own height down the column, after the zone has been established, then

$$\Theta_a = \frac{W_a}{G_s} \tag{45}$$

Letting Θ_E be the time required for the adsorption bed zone to establish itself and move out of the bed, then

$$\Theta_E = \frac{W_E}{G_s} \tag{46}$$

If the height of the adsorbent bed is z ft, and if Θ_F is the time required for the formation of the adsorption zone

$$L_z = Z\frac{\Theta_a}{\Theta_e - \Theta_F} \tag{47}$$

The quantity of adsorbate removed from the gas in the adsorption zone from the breakpoint to exhaustion is Q_r lb adsorbate/ft^2 of bed cross section. This is given by the shaded area of the figure, then

$$Q_R = \int_{W_B}^{W_E} (Y_0 - Y)\,dw \tag{48}$$

If, however, all the adsorbent in the zone were saturated with adsorbate, it would contain $Y_0\,W_a$ lb adsorbate/ft^2. Consequently, at the breakpoint, when the zone is still within the column, the fractional ability of the adsorbent in the zone still to adsorbate solute is

$$F = \frac{Q_R}{Y_0 W_a} = \left[\int_{W_B}^{W_E} (Y_0 - Y)\,dw\right] / \, Y_0 W_A \tag{49}$$

If $F = 0$ so that the adsorbent in the zone is essentially saturated, the time of formation of the zone at the top of the bed, Θ_F, should be substantially the same as the time required for the zone to travel a distance equal to its own height, Θ_a. On the other hand, if $F = 1.0$ so that the solid in the zone contains essentially no adsorbate, the zone-formation time should be very short, i.e., partically zero. Therefore,

$$\Theta_f = (1 - F)\Theta_a \tag{50}$$

Hence, from Equations 39 and 42,

$$L_z = Z\frac{\Theta_a}{\Theta_E - (1 - F)\Theta_a} = Z\frac{W_a}{W_E - (1 - F)W_a} \tag{51}$$

The adsorption column, z ft tall and of unit-cross-sectional-area, contains $Z\varrho_A$ lb adsorbent, where ϱ_A is the apparent packed density of the solid in the bed. If this were all in equilibrium with the entering gas and therefore completely saturated at an adsorbate concentration X_T lb adsorbate/lb solid, the adsorbate weight would be $Z\varrho_A X_T$ lb. At the breakpoint, the adsorption zone of height L_z ft is still in the column at the bottom, but the rest of the column $Z-L_z$ or L_{st} ft is substantially saturated at the breakpoint, therefore, the adsorbate is $(Z-L_z)_A X_T + L_z\varrho_A(1-F)X_T$ lb. The fractional approach to saturation of the column at the breakpoint is, therefore

$$\text{degree of saturation} = \frac{(Z-L_z)\varrho_A + L_z\varrho_A(1-F)X_T = Z-FL_z}{Z\varrho_A X_T} \tag{52}$$

In the fixed bed of adsorbent, the adsorption zone actually moves downward through the solid, as shown previously. The solids can, however, upward through the column countercurrent at a sufficient velocity so that the adsorption zone remains stationary within the column, as in Figure 29(a). Here the solid leaving the top of the column is shown in equilibrium with the entering gas, and all adsorbate is shown as having been removed from the effluent gas. This, of course, would require an infinitely tall column, but the major concern would be the concentration at the levels corresponding to the extremities of the adsorption zone. The operating line over the entire column is

$$G_s(Y_0-0) = L_s(X_T-0) \tag{53}$$

or

$$\frac{L_s}{G_s} = \frac{Y_0}{X_T} \tag{54}$$

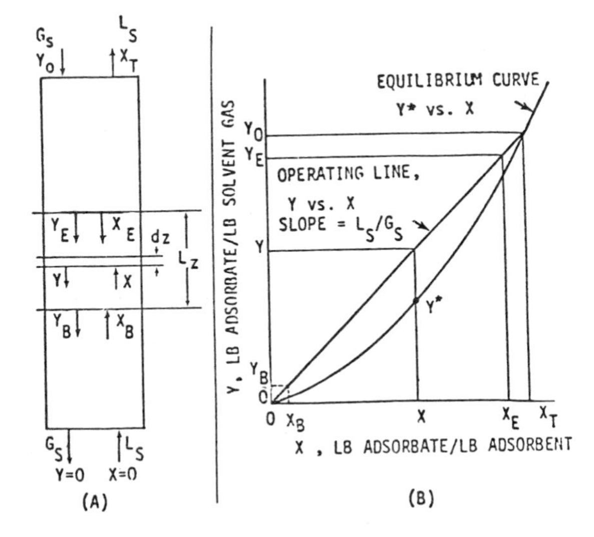

Figure 29. The adsorption zone.

Since the operating line passes through the origin of Figure 29(B), at any level in the column the concentration of adsorbate in the gas Y, and upon the solid X, can then related by

$$G_sY = L_sX \tag{55}$$

Considering the differential height dz, the ratio of adsorption is

$$G_s dY = K_y A_p (Y - Y^*)\, dz \tag{56}$$

For the adsorption zone, therefore

$$N_{0g} = \int_{Y_0}^{Y_E} \frac{dy}{Y - Y^*} = \frac{L_z}{H_{0g}} = \frac{L_z}{G_s/K_y A_p} \tag{57}$$

where N_{0g} is the overall number of gas transfer units in the adsorption zone. For any value of z less than L_z, assuming H_{0g} remains constant, changing concentrations.

$$\frac{L_z \text{ at } Y}{L_z} = \frac{W - W_B}{W_a} = \frac{\int_{Y_B}^{Y} \frac{dy}{Y - Y^*}}{\int_{Y_B}^{Y_E} \frac{dy}{Y - Y^*}} \tag{58}$$

From the Equation 58, the breakthrough curve can be plotted by evaluating the integrals.

In addition to the previously stated limitations of this method, its success largely depends upon the constancy of $K_y A_p$ or H_{0g} for the concentrations within the adsorption zone. This, in turn rests on the relative constancy of the resistances to mass transfer in the gas and within the pores of the solid.

Estimating Pressure Drop Through Fixed Beds

Another vitally important consideration in the design of adsorption control equipment is the ability to estimate the pressure drop across the adsorption bed. Ergun [19] derived a correlation to estimate the pressure drop for the flow of a single fluid through a bed of packed solids when it alone fills the voids in the bed. This relationship is given by

$$\left(\frac{\Delta P}{Z}\right) \frac{g d_p \varepsilon^3}{2 \varrho_f v^2 (1 - \varepsilon)} = \frac{75(1 - \varepsilon)}{N_{Re}} + 0.875 \tag{59}$$

where ΔP = pressure drop of gas ($1b/ft^2$)
Z = depth of packing (ft)
g = conversion constant, $4 \cdot 18 \times 10^8 (ft - 1b/1b_f hr^2)$
d_p = effective particle diameter (the diameter of a sphere of the same surface/volume ratio as the packing in place) (ft)
d_p = $6(1 - \varepsilon)/a_p$
ε = fractional void volume in dry packed bed (ft^3 void/ft^3 packed volume)
ϱ_f = gas density ($1b/ft^3$)
v = superficial velocity of gas through bed (ft/hr)
N_{Re} = Reynolds number
μ_G = gas viscosity (1b/ft hr)

Union Carbide [20] uses a simpler modified form of the Ergun equation to calculate the pressure drop through molecular sieve beds:

$$\frac{\Delta P'}{Z} = \frac{f_t C_t G^2}{\varrho_t d_p} \tag{60}$$

where C_t = presure drop coefficient (ft hr^2/in.^{2})
f_t = friction factor
G = superficial mass velocity (1b/hr ft^2) (= $\varrho_f V$)
$\Delta P'$ = pressure drop (1b$_f$/in.^{2})

The friction factor is determined from Figure 30, as a function of the modified Reynolds number. The pressure drop coefficient, c_t, is also determined from the same figure which has c_t plotted as a function of E. For molecular sieves pellets, the effective particle diameter can be obtained from

$$d_p = \frac{d_c}{(2/3) + (1/3)(d_c/L_c)} \tag{61}$$

where d_c = particle diameter (ft)
L_c = particle length (ft)

The suggested values for ε and d_p for various sizes of molecular sieves are:

	ε	dp (ft)
1/8 in pellets	0.37	0.0122
1/16 in pellets	0.37	0.0061
14 × 30 mesh granules	0.37	0.0033

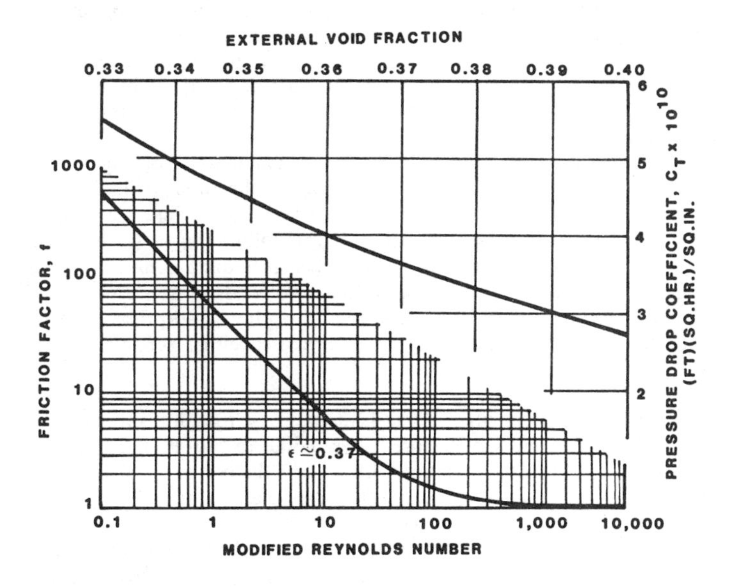

Figure 30. Friction factor as a function of the Reynold's number.

Regeneration

To make granular activated carbon economically feasible for both air- and water-pollution treatment, the spent carbon must be regenerated and reused. When a carbon column or bed has been in operation for some time, the quality of the product air and/or water deteriorates rapidly until it passes some predetermined limit beyond which it is no longer acceptable. The carbon must then be thermally regenerated.

Thermal regeneration of carbon is presently the only feasible procedure for destroying adsorbed organics. Thus maximum effort has been concentrated on optimization of thermal regeneration techniques, using a reducing atmosphere of flue gas and steam. Steam and air are the most common regenerating agents in air and water pollution adsorption systems. Steam has several properties that make it uniquely suited for this purpose. Its almost exclusive use in solvent recovery systems is well known. Its saturated vapor temperature is at a temperature level where many of the solvents of recovery value are poorly adsorbed or become readily desorbable, but the temperature is not too high to cause extensive damage to the solvents. The heat of condensation is high; hence, saturated steam rapidly delivers a large amount of heat to the adsorbent at virtually a constant and moderate temperature.

A great many solvents are insoluble in water; condensation of the steam-vapor mixture and subsequent decantation result in a satisfactory separation and liquid solvent recovery. Steam is fairly inactive with most solvents. Further, steam provides an atmosphere in which combustible vapors can be handled safely at high concentrations.

In air-pollution control applications where the pollutant concentrations are low, and, in addition, the pollutants may have no recovery value, steam regeneration may not be the best regenerating method. When adsorption takes place at low concentrations, the vapors may be held tightly by the adsorbent, and, relative to the amounts of pollutants adsorbed, the amount of steam required to regenerate the carbon may be large. This would cause even the slightly soluble organic compounds to be completely dissolved in the steam condensate. Under these conditions the more economically feasible approach is to regenerate with air or with an inert non-condensable gas such as flue gas. For example, such miscible pollutants as the solvents 4-methyl-2-pentanone and propanone would be completely dissolved in large quantities of steam. Recovering these chemicals by distillation is not economical; therefore, if steam is used, a serious disposal problem will result and thus lead to water pollution. To solve this problem, the adsorbent is regeneratd with an inert noncondensable gas and the vapors released are burned in a small thermal incinerator. If the amount of gas adsorbed is so small so as not to exceed 25% of the lower explosive limit (LEL), then air can be used as the regenerating agent.

If the condensate vapor is insoluble in the steam condensate, or slightly soluble as is trichlorethane, then complete or partial separation of the organic liquid is possible by condensation/decantation. However, when the desorbed organic liquid is soluble and after generation and thermal incineration the vapors can repollute the air (as by release of SO_2, HCl or NO_2), the problem of pollutant disposal becomes quite complex and difficult.

To accomplish regeneration or desorption, the various adsorption processes in practical use, utilize several techniques. Nevertheless, the adsorption-desorption cycles are usually classified into the following four types. These are used either separately or in various combinations.

1. *Thermal swing cycles:* Using either direct heat transfer through a surface, the adsorbent is reactivated by raising the temperature between 300° to 500° F. The bed is flushed with a dry purge gas or reduced in pressures. This returns the bed to its adsorption condition. High design loads on the adsorbent can usually be obtained, but a cooling step is needed to complete the process.
2. Pressure swing cycles: The pressure in the adsorption bed is lowered or a vacuum is created within the bed. This causes desorption. This cycle can be achieved at nearly isothermal conditions with no heating or cooling steps. The advantages include fast

cycling with reduced adsorber dimensions and adsorbent inventory, direct production of a high product, and the ability to utilize gas compression as the main source of energy.
3. Purge gas stripping cycles: A nonadsorbent purge gas is used to desorb the bed by reducing the partial pressure of the adsorbed component. Such stripping is more efficient when the operating temperature is high and the operating pressure is low. Using a condensable purge gas has the additional advantage of reducing the power requirements gained by using a liquid pump instead of a blower and an effluent stream which may be condensed to separate the desorbed material by simple distillation.
4. Displacement cycles: The material already adsorbed on the bed is purged by using another adsorbable material. The stronger the adsorption of the purge, the more completely the bed is desorbed using lesser amounts of purge, but the more difficult it becomes to subsequently remove the adsorbed purge itself from the bed.

Basically, the following sequence is the steps taken for the thermal regeneration of carbon [21].

1. The granular carbon is hydraulically transported in a water slurry to the regeneration station for dewatering.
2. After dewatering, the carbon is fed to a furnace and heated to 1,500° F–1,700° F in a controlled atmosphere which volatizes and oxidizes the adsorbed impurities.
3. The hot regenerated carbon is quenched in water.
4. The cooled regenerated carbon is again hydraulically transported to the adsorption equipment or to storage.

The thermal regeneration process basically requires three steps:

- Drying.
- Baking (pyrolysis of adsorbates).
- Activating (oxidation of the residue from the adsorbate).

The regeneration process itself requires 30 minutes: the first 15 minutes is a drying period during which the water retained in the carbon pores is evaporated—a 5-minute period during which the adsorbed material is pyrolyzed and the volatile portions thereof are driven off and a 10 minute period during which the adsorbed material is oxidized and the granular carbon reactivated.

During regeneration, the overall carbon losses usually vary from 5% to 10% per regeneration cycle. If a 5% loss of carbon per regeneration cycle is assumed, then most of the carbon originally in use will have to be replaced after 20 cycles and the bulk or aggregate properties of the mix will have approached a constant value. This fact, coupled with the relatively high cost of granular activated carbon (14c to 30c per pound) makes it economically necessary to regenerate and reuse the carbon and is, therefore, a very crucial design consideration.

ADSORPTION CONTROL EQUIPMENT

Having considered the fundamental characteristics of the adsorption phenomenon and the design principles used for the design of adsorption control equipment, it is appropriate at this time to discuss the types of control equipment used for control by adsorption. Because the theory has already been discussed independently, the control equipment will now be considered according to its type.

Overall Efficiency

It is possible to determine the overall efficiency of any pollution control device by dividing the amount of pollutants captured by the amount of pollutants available to be captured.

Where it is expressed as a fraction or percent overall efficiency (E_0), it must be expressed in consistent terms (e.g., by mass or volume, area or soiling index, number, etc). It is desirable to state the units from which the efficiency was determined.

It is an established fact that particulate matter in exhaust gas usually has a log-normal size distribution which can be expressed on probability-type coordinates as a straight line [22]. It can also be plotted to give curves such as Figure 31, which is a plot of the cumulative weight percent undersize versus particle diameter for some unspecified dust. Comparing Figure 31A and 31B, it is apparent that overall collection efficiency of particulate matter is a function of both the type of collection device and the type (size and distribution) of dust. This can be expressed mathematically as

$$E_0 = \int_0^\infty F(E)\, d(d) \qquad (62)$$

where E_0 = the determined efficiency for a specified diameter
d = differential operator
(d) = particle diameter (ft)

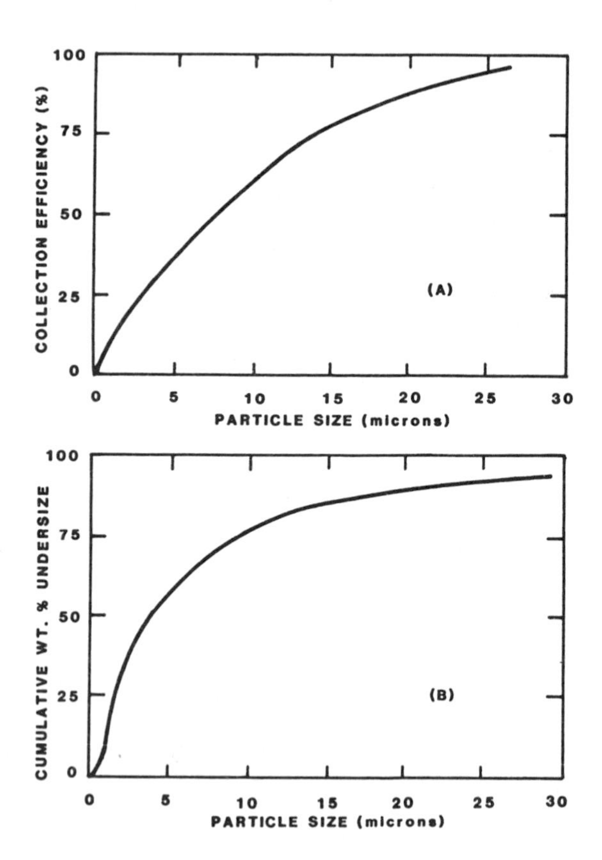

Figure 31. Particle efficiency and distribution curves required to determine overall efficiency: (A) a typical collection efficiency vs. particle diameter curve; (B) particle size distribution for some "unspecified" dust.

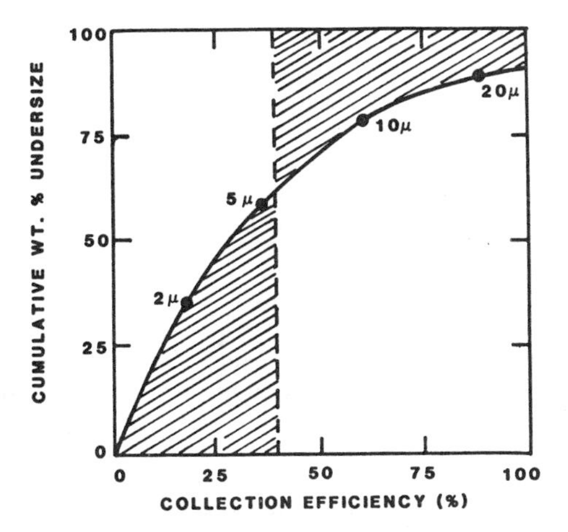

Figure 32. Graphical estimation of overall particulate collection efficiency.

These efficiencies can be obtained graphically by plotting cumulative weight percent undersize versus collection efficiency for various particle diameters. Figure 32 shows this by combining the data from the two previous figures. The overall efficiency for the system can be obtained by balancing the area under the curve on the left with an equal area on the right. For example, the efficiency as shown is 38%. This method is usually accurate enough for design purposes. However, in practice it is best to take measurements and calculate the overall efficiency directly for the given system and operating conditions.

Sometimes it is desirable to place collectors in series to obtain more complete removal of pollutants. The efficiency for such a system can be calculated using physical measurements if the system is operating; it is also possible to estimate the overall efficiency using the overall efficiencies of each individual device. In this case, the equation for a series of collectors becomes

$$E_0 = E_{0,1} + E_{0,2}(1 - E_{0,1}) + E_{0,3}(1 - E_{0,1})(1 - E_{0,2}) + \ldots \quad (63)$$

where $E_{0,1}$ = overall collection efficiency of first collector
$E_{0,2}$ = overall collection efficiency of second collector efficiency, etc.

Therefore, for η units,

$$E_0 = [E_{0,1}] + [(E_{0,2})(1 - E_{0,1})] + \ldots$$

$$[(E_{0,n})(1 - E_{0,1})(1 - E_{0,2}) \ldots (1 - E_{0,n-1})] \quad (64)$$

Collectors in parallel are used to obtain greater through put rates. Their efficiency can be estimated using the weighted individual unit overall efficiencies.

Adsorbers

Commerically available adsorbers composed of thin beds of adsorbent on a support material are able to handle up to 3,000 ft^3/min of gas. These adsorbers may be either stationary of fluidized beds and are either regenerative or nonregenerative systems. Non-regenerative systems find use when concentrations are extremely low (odors at less than 2 ppm) or for small laboratory-type installations. Adsorbers are designed to meet the following criteria:

1. Retention of the gases to permit sufficient contact time between the gaseous and solid phases.
2. The system capacity must exceed the breakthrough point which occurs when excessive pollution leaves the exit gas.
3. The system should give low gas flow resistance.
4. Uniform distribution and packing is necessary to prevent gas channeling.
5. Precleaners are needed to remove particulate matter.
6. A spare adsorber is required in parallel for continuous systems to permit regeneration and/or replacing of spent system.

The cycle time in hours for two different types of commercial adsorbers shows the kinds of operating extremes that can be encountered. In this example, the pressure swing adsorber has a total cycle time of 48 hours and the temperature swing solvent recovery adsorber has a cycle time of 4 hours.

	High Pressure Gas Dryer	*Organic Solvent Recovery Unit*
Hours onstream/offstream	24	2
Purge at 1 atm	2	–
Hot gas drying	10	0.33
Steam stripping	–	0.75
Cold gas drying	5	0.42
Standby	7	0.50
Offstream total	24	2.00

Turk [23] presents an equation for the typical average rekentim time reguired to completely adsurb an organic vapor from air:

$$t = \frac{1.29 \times 10^6 W}{Q Y_i \bar{M}} \tag{65}$$

where t = duration of adsorbent service before saturation breakthrough (hr)
w = weight of adsorbent (lb)
Q = volumetric air flow rate (ft^3/min)
Y_i = vapor inlet concentration in the air (ppm)
$\bar{M}$ = average molecular weight of adsorbed vapors

When there is a mixture of several organic vapors, the value for M can be obtained using the chain rule:

$$\bar{M} = N_A M_A + N_B M_B + \ldots N_x M_x \ldots \tag{66}$$

where N_x = number of moles of component X
M_x = molecular weight of component X

Other Control Equipment

Contactors

The size, number, shape, and configuration of vessels to contain carbon are selected on the basis of physical capacity, hydraulic loading, contact time, feed characteristics, pretreatment, desired product quality, mode of operation, and relative economics.

Economic analysis has shown that there is no apparent advantage of shop-fabricated over field-erected vessels. A maximum diameter of 13 feet is imposed by transportation clearances for shop fabricated vessels. The apparent savings of larger diameter vessels which require less intervessel piping and valving, may be offset by the additional "idle" carbon required to fill a larger spare contacting vessel.

The desired product quality will establish the required contact time and hence the approximate total carbon volume which actively contacts the waste water. The hydraulic loading selected will fix the total cross-sectional area and total carbon bed depth.

The designer has the option of converting the total carbon bed depth into one or more bed depths in series and the overall cross-section of the carbon bed into several beds in parallel.

The height of each vessel must be sufficient to permit expansion of the carbon bed during backwash in a dowflow bed, or to allow proper expansion of an upflow bed during service. The vessel should be designed to permit up to 50% expansion during backwash to assure that accumulated suspended solids can be disengaged from the surface of the carbon particles.

The vessels should be arranged with inlets and outlets oriented to accommodate internal distributors and external piping systems. Although conical bottoms in vessels facilitate removal of carbon by slurry discharge, flat or dish-shaped bottoms provide more efficient distribution during service and backwash operation.

If nozzles are used as the means of collecting the waste water after it passes through the carbon bed, they should be screened so as to retain carbon in the 60- to 80-mesh range.

When flat porous bottoms are used, an arrangement of funnels through the bottom may be employed to remove carbon from the vessel for regeneration. Filter bottom designs used in rapid sand filters should be considered for use in carbon contactors.

Special screens and, on occasion, dual screens have been utilized in lieu of the porous or perforated filter bottom. These screens should be designed to physically support the carbon during service operation, and to allow maximum backwash velocity, thus enhancing the cleansing capability of the system.

Figures 33 through 37 depict the contacting vessels used or designed for the granular carbon waste water treatment plants at Pomona. Lake Tahoe, Colorado Springs and Rock River [21]. Of primary interest is the design of the bed inlet and outlet to assure ease of carbon removal and proper water distribution. These two objectives present conflicting design requirements, for the best design for each cannot be achieved simultaneously.

The Pomona contactor (Figure 33) with a 1.5 bed-depth-to-diameter ratio (L/D) offers good distribution of waste water across the bed during service operation. The Neva–Clog–Screen at the base retains the carbon in the bed and provides a perforated surface through which the waste water passes before entering the underdrain systems. During backwash the Neva–Clog–Screen provides a distribution of the waste water across the bottom of the bed.

This screen system, or a perforated plate, or any other bottom support system must also be designed to distribute the backwash water at the maximum anticipated rate and to withstand the associated uplift force. The maximum backwash velocity at Pomona, 12 gpm/sq ft, although adequate for expanding 16 × 40 mesh carbon, is not sufficient to provide 50% expansion of 8 × 30 mesh carbon. Even if backwash pumps were sized to force an upward flow of 20 gpm/sq ft through the carbon beds, the excessive pressure drop across the screens or plates might force them to warp or to be dislodged. Special consideration is required to assure that these types of bottom support systems can hydraulically handle maximum flows in both directions and are adequately held in place.

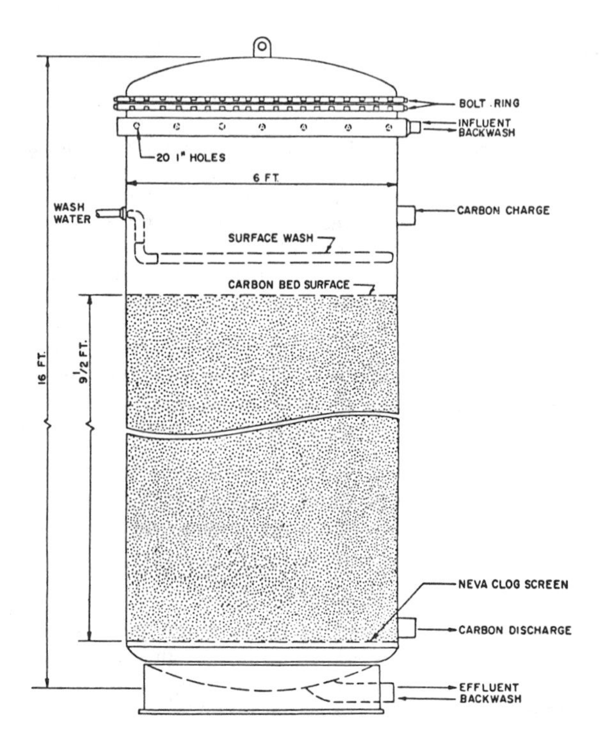

Figure 33. Pressurized contactor I.

Removal of the major part of the carbon from the carbon bed is facilitated by keeping the bed flooded during withdrawal operations. The removal of the last stump or heel of carbon in the farthest corner of the bed from the withdrawal port may be difficult. A supplementary backwash (upflow) on the order of 3 to 6 gpm/sq ft or nozzles in the side of the column just above the underdrain system may aid in flushing this last quantity of carbon from the bed.

The Tahoe design (Figure 34) is applicable to either upflow or downflow operation. The conical bottom offers the greatest east of removing spent carbon from the bed at a possible sacrifice of initial distribution of water across the bed during normal operation.

The angle of repose of granular activated carbon immersed in water is sufficiently steep that conical-bottomed vessels must be designed with a minimum bottom slope of 45 degrees. Although the carbon will flow when completely inundated, the use of shallower angles is not recommended.

The Colorado Springs contractor (Figure 35) and the Rock River contractor (Figure 36) both offer a flat bottom support for the carbon. Funnel-shaped ports through the support are provided for carbon removal. The effectiveness of this type of system in withdrawing carbon has not been proven in actual operation at this time.

The actual bottom support system utilized is different for the two contractors. The Colorado Springs contractor employs a perforated stainless steel plate into which Eimco's Flexkleen nozzles are inserted for backwash control and distribution. In contrast, Rocky River contractor employs a porous tile filter bottom covered with several inches of graded gravel and sand.

Gravity flow contactors are designed similarly to concrete rapid sand filters. A typical gravity filter design is shown in Figure 37. The requirements to be satisfied in carbon contacting are an adequate side wall depth to provide for 50% bed expansion during backwash and a means for drawing off the spent carbon to be regenerated. Carbon can be removed from the contactor through a trough on top of the underdrain system or the installation of funnels similar to those employed at Colorado Springs or Rocky River.

Distribution problems, i.e., channeling, at low flow velocities cannot be determined at this time. The gravity contactors can be designed using existing sand filter technology, with the additional requirement for carbon withdrawal already noted.

A minimum bed-depth-to diameter ratio (L/D) of 1 has been suggested to assure good distribution and adequate protection against backmixing. Deeper beds (L/D greater than 1) are susceptible to higher pressure drops.

Corrosion in carbon treatment systems presents some special problems for the designer. Activated carbon is a highly reactive material, especially in the presence of oxygen and water. Contacting vessels are relatively easy to protect through the use of special coating

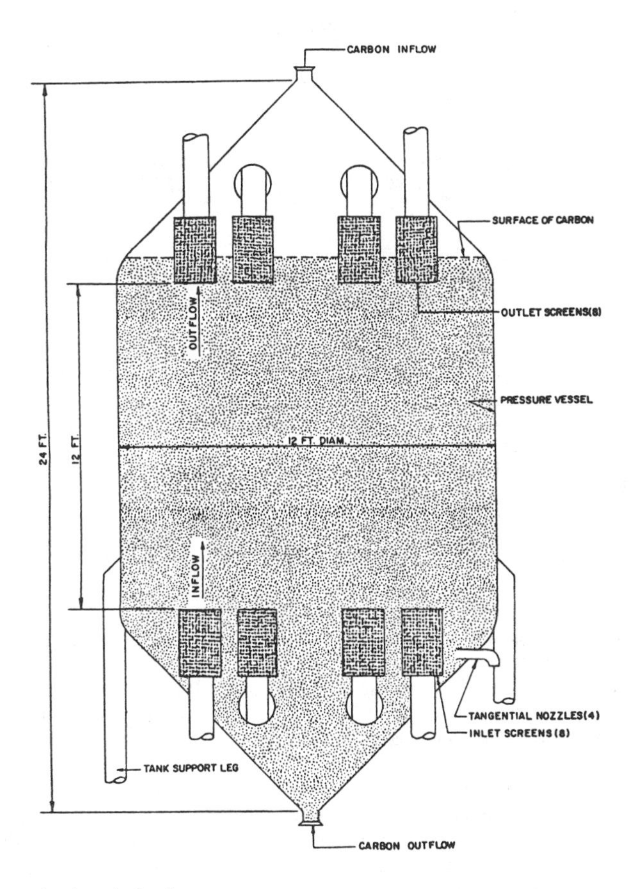

Figure 34. Pressurized contactor II.

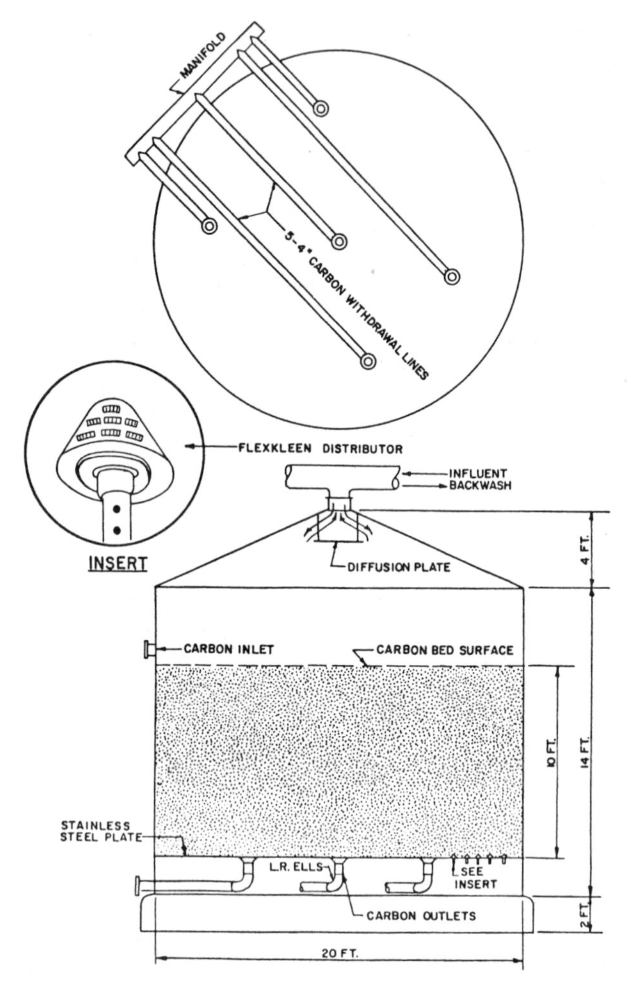

Figure 35. Pressurized contactor III.

materials. Piping, pumps, and valves are more vulnerable, however, and thus deserve special attention from the designer.

Carbon contacting vessels have so far been constructed of materials that resist attack by the waste water to be treated. In the case of municipal wastes, the contractors could be fabricated from carbon steel and coated with a lining to resist corrosion, or be formed of concrete.

The selection and method of application of the corrosion protective lining are both important. The corrosion pits reported at Pomona and Lake Tahoe were more likely caused by imperfect application of the coating material than by failure of the coating itself. Typical coating materials range from a painted coal tar epoxy to laminated rubber linings. Some of the newer polyethylene coatings are being tested for this application.

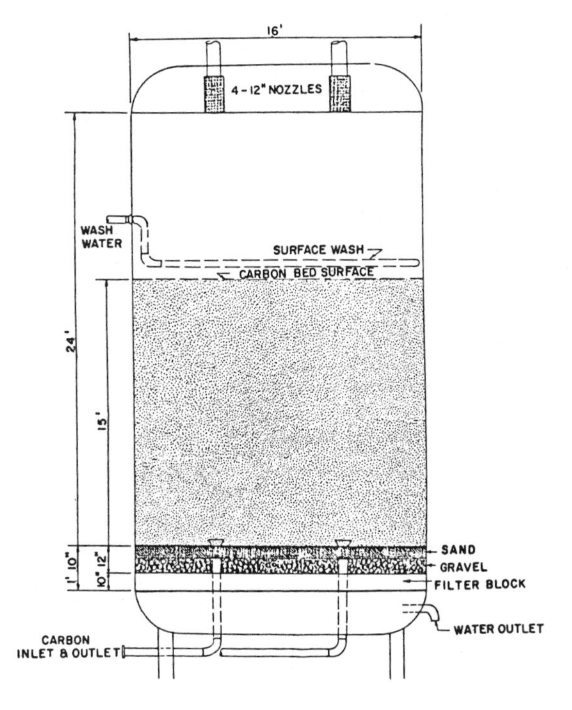

Figure 36. Pressurized contactor IV.

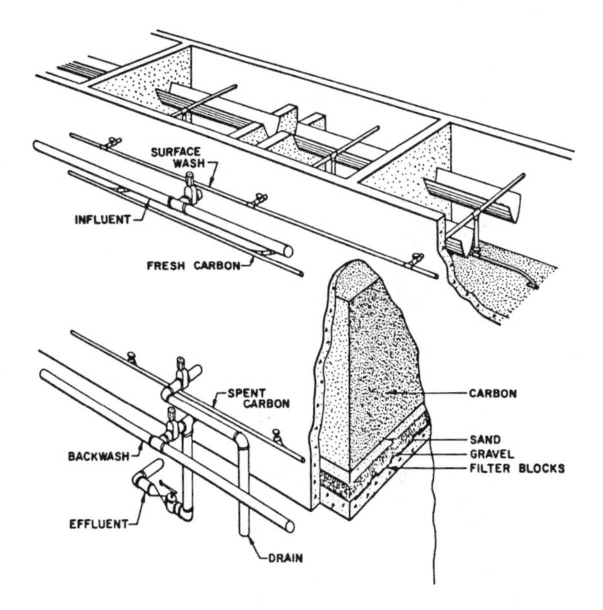

Figure 37. Gravity contactor.

Piping

Carbon is an abrasive material and when hydraulically transported will tend to wear the inside of pipes, particularly in locations where high head loss and excessively turbulent flow are encountered, such as at elbows. Long radius elbows should be used for all bends to reduce wear at these points. Experience with unlined straight pipes, however, has indicated negligible damage to the inside surfaces of the pipes after being in service for several years.

In designing the complete hydraulic system, it is recommended that maximum use be made of manifolding to increase the overall efficiency and flexibility of the installation. The advantages due to a carefully designed manifold system result in a more compact and economical system with lower initial investment for capital equipment, reduction in spare equipment, and in operating and maintenance costs. The reliability of the installation is improved and the downtime is reduced. The flexibility inherent in manifolding arrangements with necessary valves, pumps, controls, etc. should enable an operator to overcome most emergencies by switching to alternate pipe lines. The piping system should be designed with an eye towards easy flushing after each slurry transport operation. This requires the use of sufficient cleanouts, flushing connections, and drains.

Steel pipes have been used satisfactorily in applications where the slurry transport is not continuous. Steel is also advisable when the piping is readily accessible to effect economic repair or replacement. More expensive materials or linings such as rubber, saran, polyvinyl chloride, and stainless steel may be justified only under special conditions, e.g., in the industrial sector where corrosive liquids may be encountered.

Valves used in wastewater disposal systems may be classified in several categories, each of which may be further subdivided according to various design options [21].

- Diaphragm (straight way).
- Globe.
- Rotary (ball, butterfly, cone, plug).
- Slide, (gate valve, shear gate).
- Sphere check.
- Swing check.

In selecting a valve for installation in a slurry transport line, there are four major considerations: the purpose of the valve, its effectiveness in accomplishing functional requirements, the resistance of the valve to the abrasive effects of slurry transport, and the cost.

The globe valve and slide valve, normally used to effect positive shut off of flow in a pipe line are not applicable because they require a positive seating.

The passage of an abrasive slurry through these valve configurations in the open position may wear the seating or perhaps leave carbon granules lodged in the seating grooves, thus preventing positive shutoff.

Preferred valves to assure positive off and on operations are the rotary type such as the ball, cone, and plug valves. These valves should offer no restriction to slurry transport when in the open position. The diaphragm valve, certain variations of which offer limited blockage of the open passage, has a movable element of flexible rubber, leather, or some special composition, which will be worn for a period of use and will require replacement.

Both swing-type and spherical-check valves are suitable for back flow prevention in slurry pumping. Although the seating face against which the closing device rests is susceptible to abrasive wear and the flow is restricted by the configuration of the valve, there is no acceptable substitute that can achieve the same purpose.

Regulation of slurry flow can best be accomplished by either a diaphragm valve or a rotary valve such as the butterfly valve. These valves should have as their only function flow regulation and would not be expected to provide positive flow shutoff. The useful life of the wearing surface or seating face of valves can be extended by the use of rubber lining and stainless steel.

If the carbon slurry piping system is one to two inches in size, the cost of periodically replacing a common valve may be far less than installing a high-priced, specially-lined valve.

Pumps

The motive force to convey slurries through pipelines can be provided by pumps alone or a combination of pumps and eductors. For the range of slurry concentrations used in existing facilities, either centrifugal pumps or by a combination of centrifugal pumps and eductors have proven satisfactory. If carbon is to be transported at higher slurry concentrations, consideration should be given to using diaphragm slurry pumps or double-acting positive displacement pumps of the simplex, duplex, and triplex variety.

For the pumping of a 25% granular carbon slurry, centrifugal pumps should have extra large suction inlets, a nonclogging type of impeller, and an extra large packing box with seal to protect the shaft from wear. Preferred materials of construction include 316-in stainless steel, silicon iron or rubber lining (this is true especially for those components in contact with the abrasive slurry). Field experience indicates, however, that there is a tendency for the rubber linings out to pull out [21].

The eductor serves the two-fold purpose of mixing carbon and water and of accelerating the transport fluid. It must, of course, have a pump associated with it to assure pressure and flow. In such an application, the pump may be selected as though it were intended for pumping clean water.

Control System

Automatic operation of carbon-contacting systems is accomplished by standard equipment which is well developed and reliable. Although automatic control is quite feasible, it is frequently not required because of the extended lengths of time between operations of the equipment; valves serving separate carbon vessels are usually operated only during withdrawal and replacement of carbon for regeneration, except for occasional flow reversals. Since this operation occurs about once every one to two months, it is best to operate the valve manually, with careful observation and attendance. In downflow arrangements in which the carbon beds act as filters in addition to providing adsorption sites, the control and operation of these systems can become somewhat complex and subject to failures resulting in delays and even complete plant shutdown. Basically a surface-type filter, the downflow carbon contactor is vulnerable to all of the problems of this type of filter. Any severe pretreatment upset resulting in a sudden increase in suspended solid may completely blind or clog the surface of the bed, requiring backwashing to restore service.

Typical Plant Operation

So far, we have discussed all the component parts of a carbon treatment plant and several possible flow configurations. However, there are some important aspects of a carbon plant which cannot be adequately described in the preceding format. The arrangement of the carbon contactors in a workable system requires a considerable design effort in the area of internal piping, valves, pumps, and intermediate tanks. These elements are vital in providing the flexibility which characterizes an efficient and economical carbon treatment system. Therefore, in this section we shall illustrate one particular integrated plant design complete with piping, valves, pumps, etc., and attempt to demonstrate how plant flexibility may be obtained.

Figure 38 illustrates a design for a plant with 2-stage contactors. This design provides 3 parallel 2-stage systems in an arrangement intended to provide almost uninterrupted operation by rotation of those 3 systems. Each of the 3 parallel systems can be shifted from the service (treatment) mode to the backwash mode and back again, or to the regeneration mode and back again. At any given time, two 2-stage systems are in service, although one of these can be taken out of service to be backwashed, while one is being regenerated. In Figure 38, contactors A and B are in service, C and D are momentarily out of service so that C can be backwashed, and E and F are in the regeneration mode. E is being emptied of spent carbon and F is being filled with regenerated carbon. E has just been taken out of service to be regenerated, while F will be ready to replace either A or C when one of them is exhausted.

CARBON ADSORPTION APPLICATIONS

Carbon adsorption has long been recognized as a viable method for removing pollutants from waste water and gas streams. Of course, practical problems relating to the scaling up of laboratory prototypes to treatment plants, have hindered its full-scale implementation. Nevertheless, it is clear that carbon could be a widely useful treatment system; one in which the problems of dependability, sensitivity, and cost are no worse and often less than with more conventional treatment technologies, particularly biological.

Liquid-Phase Applications

The principal liquid-phase application of activated carbon includes [24]:

1. Sugar decolorization.
2. Municipal water purification (for water purification systems).
3. Purification of fats, oils, foods, beverages, and pharmaceuticals.
4. Industrial/municipal waste water treatment.
5. Other liquid-phase miscellaneous applications.

There are roughly 100 large-scale systems currently in use for industrial/municipal waste water treatment. The following is a breakdown of the various systems.

Tertiary Treatment Plants

Carbon adsorption when applied to well-treated secondary effluent is capable of reducing COD to less than 10 mg/l and the BOD to under 2 mg/l. Removal efficiencies may be in the range of 30%–90% and vary with flow variations and different bed loadings; for example, recent data from the South Lake Tahoe plant (Site 11, Table 3), show removal efficiencies vary from 30%–60%. Carbon loadings in tertiary treatment plants fall within the range of 0.25–0.87 lb COD removed/lb of carbon, and if the columns are operated downflow, over 90% suspended solids reduction may be achieved.

It should be noted that all the plants described in Table 4 treat large flow rates (1–100 mgd) and require contact time of around 25–35 minutes [25].

Physical-Chemical Treatment (PCT) Plants

Physical-chemical treatment acts on the effluent of a primary treatment to remove BOD, COD suspended solids, and if present, some color. Effluent concentrations of BOD and COD are somewhat higher than for tertiary treatment systems. Carbon adsorption, in conjunction with lime clarification (PCT) has been shown to be capable of achieving over 90° BPD and suspended solids reduction, which is at least equivalent to secondary treatment. Carbon loadings for PCT plants have been measured as high as 1.0 lb of COD removed/lb carbon but will typically be in the range of 0.4 to 0.6 lb COD/lb carbon and 0.15 to 0.3 TPC/lb carbon. It is clear that both tertiary and PCT plants have biological activity taking

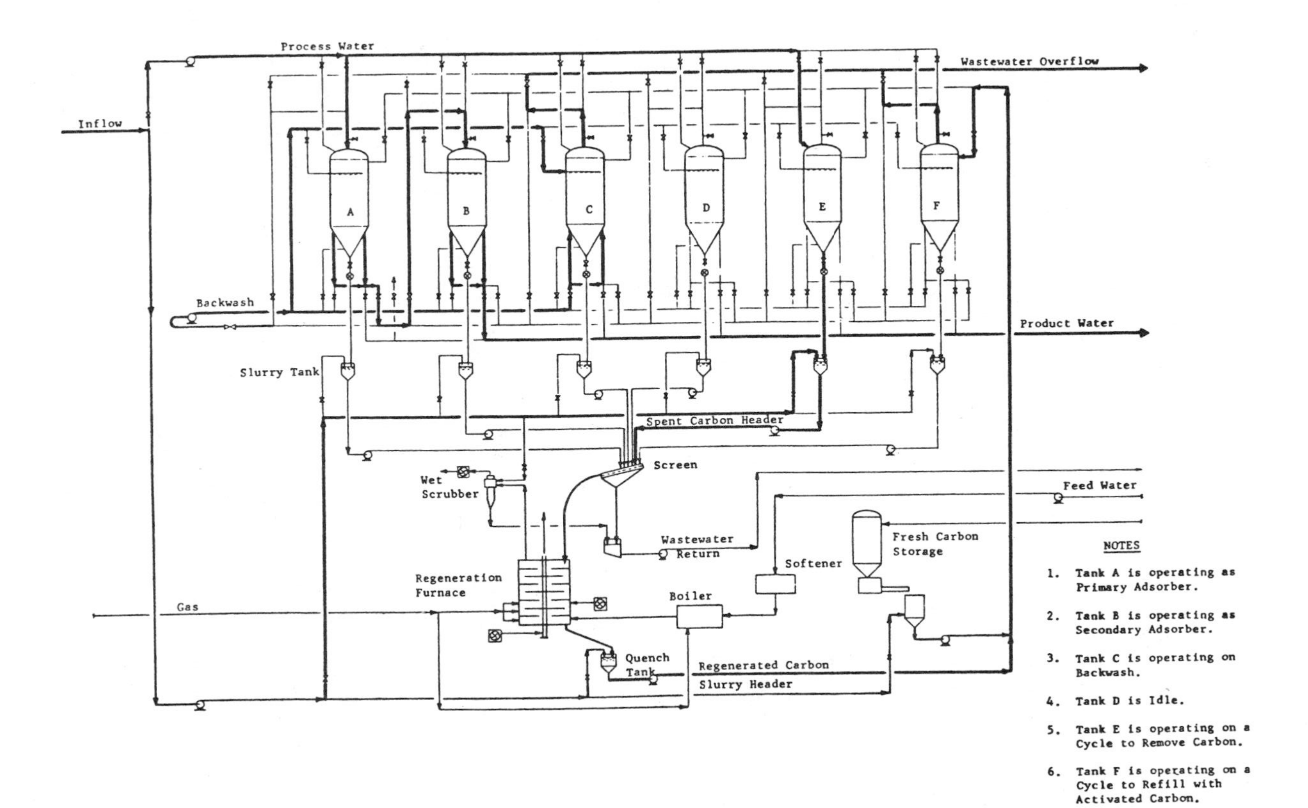

Figure 38. Process flow diagram.

Table 3
Tertiary Treatment Plants

Site	Status 1970	Design Engineer	Average Plant Capacity (mgd)	Contactor Type	No. of Contactors in series	Contact Time[a] (min)	Hydraulic Loading (gpm/ft^3)	Total Carbon Depth (ft)	Carbon Size	Effluent Requirements[b] (Oxygen Demand)
1. Arlington, Virginia	Design	Alexander Potter Assoc.	30	Downflow Gravity	1	38	2.9	15	8 × 30	BOD < 3 mg/l
2. Colorado Springs, Colorado	Operating December 1970 to Present	Arthur B. Chafet & Assoc.	3	Downflow	2	30	5	20	8 × 30	BOD < 2 mg/l
3. Dallas, Texas	Design	URS Forest and Cotton	100	Upflow Packed	1	10	8	10	8 × 30	BOD < 10 mg/l BOD < 5 mg/l (by 1980)
4. Fairfax County, Virginia	Design	Alexander Potter Assoc.	36	Downflow Gravity	1	36	3	15	8 × 30	BOD < 3 mg/l
5. Los Angeles, California	Design	City of Los Angeles	5[c]	Downflow Gravity	2	50	4	26	8 × 30	COD < 12 mg/l
6. Montgomery County, Maryland	Design	CH_2M/Hill	60	Upflow Packed	1	30	6.5	26	8 × 30	BOD < 1 mg/l COD < 10 mg/l
7. Occoquan, Virginia	Design	CH_2M/Hill	18	Upflow Packed	1	30	5.8	24	8 × 30	BOD < 1 mg/l COD < 10 mg/l
8. Orange County, California	Construction	Orange County Water District	15	Upflow Packed	1	30	5.8	24	8 × 30	COD < 30 mg/l
9. Piscataway, Maryland	Operating March 1973 to Present	Roy F. Weston	5	Downflow Pressure	2	37	6.5	32	8 × 30	BOD < 5 mg/l
10. St. Charles, Missouri	Construction	Moran and Cooke	5.5	Downflow Gravity	1	30	3.7	15	8 × 30	
11. South Lake Tahoe, California	Operating March 1968 to Present	CH_2M/Hill	7.5	Upflow Packed	1	17	6.2	14	8 × 30	BOD < 5 mg/l COD < 30 mg/l
12. Windhock, South Africa	Operating October 1968 to Present	National Institute for Water Research Pretoria, S. Africa	1.3	Downflow Pressure	2	30	3.8	15	12 × 40	COD < 10 mg/l

[a] *Empty bed (superficial) contact time for average plant flow.*

[b] *BOD: Biochemical oxygen demand. COD: Chemical oxygen demand.*

[c] *50-mgd ultimate capacity.*

From Reference 21.

Table 4
Physical-Chemical Treatment Plants

Site	Status 1973	Design Engineer	Average Plant Capacity (mgd)	Contactor Type	No. of Contactors in series	Contact Time[a] (min)	Hydraulic Loading (gpm/ft^3)	Total Carbon Depth (ft)	Carbon Size	Effluent Requirements[b] (Oxygen Demand)
1. Cortland, New York	Design	Stearns & Wheler	10	Downflow Pressure	1 or 2	30	4.3	17	8 × 30	TOD < 35 mg/l
2. Cleveland Westerly, Ohio	Design	Engineering-Science	50	Downflow Pressure	1	35	3.7	17	8 × 30	BOD < 15 mg/l
3. Fitchburg, Massachusetts	Construction	Camp Dresser & McKee	15	Downflow Pressure	1	35	3.3	15.5	8 × 30	BOD < 10 mg/l
4. Garland, Texas	Design	URS Forest and Cotton	30[c]	Upflow Downflow	2	30	2.5	10	8 × 30	BOD < 10 mg/l
5. Le Roy, New York	Design	Lozier Engineers	1	Downflow Pressure	2	27	7.3	26.8	12 × 40	BOD < 10 mg/l
6. Niagara Falls, New York	Design	Camp Dresser & McKee	48	Downflow Gravity	1	20	3.3	9	8 × 30	COD < 112 mg/l
7. Owosso, Michigan	Design	Ayres, Lewis, Norris & May	6	Upflow Packed	2	36	6.2	30	12 × 40	BOD < 7 mg/l
8. Rosemount, Minnesota	Construction	Banister, Short, Elliot, Hendrickson, and Associates	0.6	Upflow Downflow Pressure	3 (max.)	66 (max.)	4.2	36 (max.)	12 × 40	BOD < 10 mg/l
9. Rocky River, Ohio	Construction	Willard Schade & Assoc.	10	Downflow Pressure	1	26	4.3	15	8 × 30	BOD < 15 mg/l
10. Vallejo, California	Design	Kaiser Engineers	13	Upflow Expanded	1	26	4.6	16	12 × 40	BOD < 45 mg/l (90% of time)

[a] *Empty bed (superficial) contact time for average plant flow.*

[b] *BOD: Biochemical oxygen demand. COD: Chemical oxygen demand. TOD: Total oxygen demand.*

[c] *90-mgd ultimate capacity.*

From Reference 21.

place in the carbon beds which increases the apparent carbon activity. PCT plants (Table 5) currently are being designed and constructed with similar characteristics to tertiary treatment plants.

Treatment of Industrial Waste Streams

Tables 5 through 8 list the various types of systems current in use to treat industrial/municipal waste water. Frequently, segregated industrial waste streams are treated with activated

Table 5
Full-Scale Activated Carbon Systems Treating Organic Chemical Wastes

Company	Granular Systems	Wastewater Flow (gpd)	Contaminants Removed	Reactivation Scheme	Comments
Allied Chemical Birmingham, Alabama	Hydrodarco 4000	3,000	Phenols	None	
Dow Chemical Midland, Michigan	Witco 718	145,000	Phenols	4% NaOH	Removes phenols from 18% brine solution for use in chloralkali plant. Reactivation solution recycled to phenol plant
East Coast Manufacturer New Jersey	Filtrasorb 4000	58,000	Polyethers	Thermal	Discharges to municipal system
General Electric Pittsfield, Maine	Filtrasorb 300	Batch treatment (varies with use)	Color	Thermal	Used for treating a variety of waste stream in-plant; on-stream, December 1971
General Electric Selkirk, New York	Filtrasorb 300	300,000	Phenolics, toluene, color	Thermal	
Hardwicke Chemicals Elgin, South Carolina	Filtrasorb	20,000	Mixed organics	Thermal	Subscribes to Calgon Adsorption Service
Hercules Chemicals Hattiesburg, Mississippi	Filtrasorb 400	3,200,000	COD, mixed organic, wood sugars	Thermal	Just completing start-up
Monsanto Anniston, Alabama	Filtrasorb 300	65,000	*p*-nitrophenol *p*-chlorobenzene	Chemical	Reactivation chemical with recovered product returned to process
Rhodia, Inc. Portland, Oregon	Filtrasorb 400	150,000	Chlorophenols	Thermal	
Schenectady Chemicals Schenectady, New York	Filtrasorb 300	22,000	Phenols, resorcinol, xylol	Thermal	On-stream, March 1973
Stephan Chemical Co. Fieldsboro, New Jersey	Filtrasorb 300	15,000	Mixed organic associated with liquid detergent intermediates	Thermal	On-stream, June 1972
Stauffer Chemicals Skaneateles Falls, New York	Filtrasorb 300	8,000	COD from washout of batch equipment used to make mixed detergents and wash compounds	Expended carbon saved and shipped to central processing facility where a tank car load is accumulated	Treats segregated waste side-stream. Effluent then joins main waste stream.
Reichhold Chemicals Tuscaloosa, Alabama	Filtrasorb 300	500,000	Phenols, BOD, COD	Thermal	Treats concentrated waste resulting from extensive in-plant water use reduction; on-stream, late 1972
Sherwin-Williams Chicago, Illinois	Nuchar WV-G 12 × 40	65,000	*p*-cresol	10% NaOH	
Vicksburg Chemical Vicksburg, Mississippi	Hydrodarco 4000	10,000	Phenols	NaOH	
Ashland Chemicals Great Meadows, New Jersey	Hydrodarco[a]	140,000	Color, COD	None	Biological treatment with 85 ppm of carbon added to influent
Dow Chemical Midland, Michigan	Nuchar[a]	14,000,000	Mixed organics	None	Biological treatment with carbon added to influent
Koppers Chemical Follunsbee, West Virginia	Nuchar[a]	20,000	Phenolics	None	Used when biological treatment efficiency drops

[a] *Powdered system.*
From Minor [26].

carbons. The flows treated are generally small in comparison with PCT and/or tertiary systems (3,000–20,000 gpd). Some of these systems employ thermal reactivation for larger systems (flow rates above 60,000 gpd). Systems where recovery takes place usually have longer contact times (50–200 minutes).

Where adsorbates are easily recovered, carbon loadings approach 1 lb COD removed/1 b of carbon with 80% reductions, for concentrations of 700 ppm COD and 200 ppm TOC. For lower influent concentrations; example brine waste water containing 150–750 ppm phenol and 1,500–1,800 ppm acetic acid could be reduced to 1 ppm phenol and 100–200 ppm acetic acid with loadings of 0.09–0.16 1b/1b and 0.04–0.06 1b/1b, respectively.

Further Potential Applications

Activated carbon has a wide range of potential applications for waste treatment with the largest number of removal of mixed organic matter from aqueous wastes. Table 9 gives potential areas of applicability.

Table 6
Summary of Industrial Sources Using Granular Activated Carbon Systems

Industry	Location	Principal Product	Contaminant(s) Removed
Velvet Textiles	Blackstone, Virginia	Velvet	Dyes, detergents, organics
BASF Wyandotte Chemical	Washington, New Jersey	Polyethers	Polyethers (MW 1,000-3,000)
ARCO-Watson Refinery	Wilmington, California	Refinery products	COD
Stephen Leedom	Southampton, Pennsylvania	Carpet mill	Dyes
Reichhold Chemicals, Inc.	Tuscaloosa, Alabama	Phenol, formaldehyde, pentacrylthritol, orthophenylphenol, synthetic resins and plastics	COD, phenols
Schenectady Chemicals, Inc.	Rotterdam, New York	Phenolic resins	Phenols
Chipman Div. of Rhodia, Inc.	Portland, Oregon	Herbicides 2,4-D acid, MCPA acid, 2,4-DB acid and esters of these products	COD, phenols
Sherwin Williams Co.	Chicago, Illinois	*p*-Cresol	*p*-Cresol
Mobay Chemical Co.	Houston, Texas		Color
Burlington Army Ammunition Plant	Burlington, Iowa	Explosives	TNT
Stepan Chemical Co.	Bordentown, New Jersey	Intermediate detergents	Color and organics
Georgia Pacific	Conway, North Carolina	Phenolic resins	Phenols
Stauffer Chemical	Skaneateles Falls, New York	Strong alkaline detergents	COD
General Electric Co.	Selkirk, New York	Plastics	Phenols and COD
C. H. Masland & Sons	Wakefield, Rhode Island	Carpet yarn	Color and COD
St. Regis Paper Co.	Pensacola, Florida	Kraft products	Color
Monsanto Industrial Chemicals	Anniston, Alabama	Intermediate organic chemicals (polynitrophenol)	Polynitrophenol
Hercules, Inc.	Hattiesburg, Mississippi	Acid resins, turpines and solvents	Organics
Dow Chemical	Midland, Michigan	Phenols	Phenols and acetic acid
Hardwicke Chemical Co.	Elgin, South Carolina	Intermediate and specialty	COD, color
Crompton and Knowles Corp.	Gibraltar, Pennsylvania	Dyes	Dye, COD

From Reference 27.

Table 7
Industrial Waste Adsorption Treatment Plants Using Carbon

Location	Impurity	Average Flow Rate (gpm)	Reactivation or Regeneration Method
Washington, New Jersey	Polyols	100	Furnace
E. St. Louis, Illinois	Nitrophenol	50	Caustic
Burlington, Iowa	TNT	100	None
Southampton, Pennsylvania	Dye	350	Furnace
Portland, Oregon	Insecticides	100	Furnace
Conway, North Carolina	Phenol	25	Caustic
Wilmington, California	Refinery wastes	2,900	Furnace
Latrobe, Pennsylvania	Cyanide	20	None

From Rizzo [28].

Table 8
Various Industrial Wastewater Treatment Activated Carbon Installations

Industry Location	Installation Date	Design Flow Rate (000 gpd)	Organic Contaminants	Pretreatment	Contact Time (min)	Adsorption System: Adsorber Type	Adsorption System: Carbon Reactivation
1. Carpet Mill, British Columbia	6/73	50	Dyes	Screens		Moving bed	None
2. Textile Mill, Virginia	7/70	60	Dyes	Filtration	57	Moving bed	None
3. Oil Refinery, California	3/71	4,200	COD	Equalization, oil flotation	60	Gravity beds in parallel	Multiple hearth furnace
4. Oil Refinery, Pennsylvania	3/73	2,200	BOD	Equalization, oil flotation, filtration		Moving bed	Multiple hearth furnace
5. Detergent, New Jersey	6/72	15	Xylene, alcohols, TOC	None	540	Downflow beds in series	Multiple hearth furnace
6. Chemicals, Alabama	11/72	500	Phenolics, resin intermediates	Chemical clarification	173	Moving beds	Multiple hearth furnace
7. Resins, New York	3/73	22	Xylene, phenolics, resorcinol	Chemical clarification	30	Downflow beds in series	Rotary kiln
8. Herbicide, Oregon	11/69	150	Chlorophenols, cresol	None	105	Upflow beds in series	Multiple hearth furnace
9. Chemicals, New York	3/69	15	Phenol, COD	Equalization	200	Downflow beds in series	None
10. Chemicals, Texas	11/71	1,500	Nitrated aromatics	Activated sludge filtration	40	Moving beds	Rotary kiln
11. Chemicals, New Jersey		100	Polyols	Equalization, clarification		Moving bed	Multiple hearth furnace
12. Explosives, Switzerland	3/72	5	Nitrated phenols	Equalization	150	Downflow beds in series	Multiple hearth furnaceNone
13. Pharmaceuticals, Switzerland	10/72	25	Phenol	Equalization, pH adjusted settling	90	Downflow beds in series	None
14. Insecticide, England	1962		Chlorophenol	Equalization, clarification		Downflow beds in series	Rotary kiln
15. Wood Chemicals, Mississippi	8/73	3,000	TOC	pH adjustment flotation filtration	50	Moving beds	Multiple hearth furnace
16. Dyestuffs, Pennsylvania	8/73	1,500	Color, TOC	Equalization, clarification, filtration	50	Moving beds	Multiple hearth furnace

From Hager [29].

Table 9
Potential Applications for Waste Treatment with Activated Carbon

Aqueous wastes	With adsorbate concentrations up to about 10,000 ppm; SS ≦ 50 ppm, O & G ≦ 10 ppm
Organics	Prefer chemicals with low solubility, low polarity, low degree of ionization. Problem chemicals include acetone, ethanol, glycol, soaps, etc., and others that are of low molecular weight and/or high solubility.
Regeneration	• *Thermal,* which destroys adsorbates, is economical if carbon usage is above roughly 1,000 lb/day • *Chemical,* may be used if one (or just a few) solute is present, which can dissolve off the carbon. This allows material recovery • *Biological,* if wastes are highly biodegradable. Virgin carbon activity not obtained after reactivation
Disposal	Disposal of the carbon may be required if use is less than ~1,000 lb/day and/or a hazardous component mitigates against any form of regeneration
Inorganics	Cr and CN currently removed in industrial applications. Other possibilities are somewhat limited. Strong electrolytes are not well adsorbed
Organic Wastes	Current applications: None known Potential applications: Removal of color, oil and grease, or other adsorbable impurities from solvents with resaleable potential

Gas/Vapor-Phase Applications

The principal gas/vapor applications of activated carbon consist of the following:

- Odor control.
- Gas masks: protection against hazardous gases in industry.
- Indoor air purification systems.
- Solvent recovery/separation of hydrocarbon gas mixtures.

Table 10
Odor–Threshold Concentrations

Substance	ppm
Carbon tetrachloride	71.8
Ammonia	53.0
Phosgene	5.6
Chlorine	3.5
Acrolein	1.8
Amyl acetate	1.0
Pyridine	0.23
Hydrogen sulfide	0.18
Oil of wintergreen	0.066
Crotonaldehyde	0.062
Benzyl sulfide	0.006
Diphenyl ether	0.0012
Isoamyl mercaptan	0.00043
Ethyl mercaptan	0.00026
Vanilin	0.000079
Butyric acid	0.000065
Artificial musk	0.0000034

Table 11
Retentivity of Vapors by Activated Carbon
(percent retained in a dry airstream at 20° C, 760 m by weight)

Substance	Retentivity (%)	Remarks
Acetaldehyde	7	Reagent
Acetic acid	30	Reagent, sour vinegar
Acetone	15	Solvent
Acetylene	2	Welding and cutting
Acryaldehyde	15	Acrolein, burning fats
Acrylic acid	20	
Ammonia	Negligible	
Amyl acetate	34	Lacquer solvent
Amyl alcohol	35	Fuel oil
Benzene	24	Benzol, paint solvent & remover
Body odors	High	
Bromine	40 (dry)	
Butane	8	Heating gas
Butyl acetate	28	Lacquer solvent
Butyl alcohol	30	Solvent
Butyl chloride	25	Solvent
Butyl ether	20	Solvent
Butylene	8	
Butyne	8	
Butyraldehyde	21	Present in internal combustion exhaust, *i.e.*, diesel
Butyric acid	35	Sweat, body odor
Camphor	20	
Caprylic acid	35	Animal odor
Carbon disulfide	15	
Carbon tetrachloride	45	Solvent, cleaning fluid
Chlorine	15 (dry)	
Chloroform	40	Solvent, anesthetic
Cooking odors	High	
Cresol	30	Wood preservative
Crotonaldehyde	30	Solvent, tear gas

(Continued.)

Substance	Retentivity (%)	Remarks
Decane	25	Ingredient of kerosene
Diethyl ketone	30	Solvent
Essential oils	High	
Ethyl acetate	19	Lacquer solvent
Ethyl alcohol	21	Grain alcohol
Ethyl chloride	12	Refrigerant, anethetic
Ethyl ether	15	Medical ether, reagent
Ethyl mercaptan	23	Garlic, onion, sewer
Ethylene	3	More retentivity by reaction
Eucalyptole	20	
Food (raw) odors	High	
Formaldehyde	Negligible	Disinfectant, plastic ingredient
Formic acid	7	Reagent
Heptane	23	Ingredient of gasoline
Hexane	16	Ingredient of gasoline
Hydrogen bromide	12	
Hydrogen chloride	12	
Hydrogen fluoride	10	
Hydrogen iodine	15	
Hydrogen sulfide	3	Oxidizes to increase retentivity considerably
Indole	25	In excreta
Iodine	40	
Iodoform	30	Antiseptic
Isopropyl acetate	23	Lacquer solvent
Isopropyl alcohol	26	Solvent
Isopropyl chloride	20	
Isopropyl ether	18	Solvent
Menthol	20	
Methyl acetate	16	Solvent
Methyl alcohol	10	Wood alcohol
Methyl chloride	5	Refrigerant
Methyl ether	10	
Methyl ethyl ketone	25	Solvent
Methyl isobutyl ketone	30	Solvent
Methyl mercaptan	20	
Methylene chloride	25	
Naphthalene	30	Reagent, moth balls
Nicotine	25	Tobacco
Nitric acid	20	
Nitro benzene	20	Oil of bitter almonds Oil of mirbane
Nitrogen dioxide	10	Hydrolyzes to increase retentivity
Nonane	25	Ingredient of kerosene
Octane	25	Ingredient of gasoline
Ozone	Decomposes to oxygen	Generated by electrical discharge
Packing-house odors	Good	
Palmitic acid	35	Palm oil
Pentane	12	Light naphtha
Pentylent	12	
Phenol	30	Carbolic acid, plastic ingredient
Propane	5	Heating gas
Propionic acid	30	
Propylene	5	Coal gas
Propyl mercaptan	25	
Propyne	5	
Putrascine	25	Decaying flesh
Pyridine	25	Burning tobacco
Sewer odors	High	
Skatole	25	In excreta
Sulfur dioxide (dry)	10	Oxidizes to sulfur trioxide, compon in city atmospheres
Sulfur trioxide	15	Hydrolyzes to sulfuric acid
Sulfuric acid	30	
Toilet odors	High	
Toluene	29	
Turpentine	32	Solvent
Valeric acid	35	Sweat, body odor, cheese
Water	None	
Xylene	34	Solvent

The first sign of air pollution, in many cases is indicated by the sense of smell. Therefore, it is important in selection or designing air pollution control systems that the odor threshold concentrations of gases/vapors be known. Table 10 summarizes such a consideration. Further, the retentivity of these and other gases/vapors by carbon is again a very important consideration. Table 11 lists the percent retentivity in a dry air stream at 20° C and 760 m by weight.

As a result of the diversity of carbon as an adsorbent, various other applications are possible, i.e.

- Control of volatile toxic vapor in laboratory storage spaces.
- Filters in cigarettes.
- Removal of sulfur from synthesis gases.
- Removal of arsenic from hydrogen chloride.
- Gas chromatography.
- Placed in shipping and storage containers to protect odor-sensitive substances; and to prevent odorous materials from creating a nuisance.
- Kitchen hoods.
- Storage of fruits and flowers.

Outlook for Carbon Adsorption

Carbon adsorption has been demonstrated to be both practical and economical. The major benefits of carbon treatment include:

- High organics removal.
- Applicability to wide variety of organics.
- Capability for inorganics removal (e.g., CN, CR ...).
- Possible material recovery of organics and inorganics in a few cases (e.g., phenols, acetic acid, ethylene diamine, Cr).
- High flexibility (e.g., rapid start-up and shutdown).
- Low sensitivity to feed variations (flow rate or concentration).
- Insensitivity to toxic materials.
- Minimum land areas.
- Destruction of wastes when using thermal regeneration.

Just as there are many benefits, of course, there are some limitations.

- Relatively high capital and operating cost, especially if thermal reactivation is used.
- Limited generally to wastes with less than about 1% organic contaminant, or less than 5% if adsorbates are to be recovered.
- Cost savings associated with thermal reactivation may be realized if carbon usage is above 100 lb/day or if central reactivation facility is available.
- If carbon cannot be regenerated thermally or chemically, then it must be disposed of.
- Low tolerance of suspended solids should be less than 50 ppm to reduce need for frequent backwashing.
- Inability to remove low molecular weight and/or highly soluble organic chemicals (e.g., methanol, ethanol, glycol, soap).
- Problems with thermal reactivation systems.

NOTATION

A area of adsorbent bed

A_p outside surface area of solid particles

a adsorbate/adsorption-quantity, also area for mass transfer

C constant at any given temperature

c_p molar heat capacity

c_{pa} adsorbate (vapor) heat capacity (Btu/lb° F)
c_{pc} gas heat capacity (BTU/lb° F)
c_t pressure drop coefficient
D_i intraparticle diffusivity
d_c particle diameter
E_1 heat of adsorption of first adsorbed layer
E_L heat of liquefaction of gas
E_∂ efficiency for a specified diameter
ft friction factor
G superficial mass velocity
G_m molar flow rate of inert gas
$\bar{H}$ differential heat of adsorption
H_G enthalpy of adsorbate gas mixture
H_L enthalpy of adsorbant solid plus adsorbate
j components being desorbed
k, K constants
K_{Gi} mass transfer coefficient
K_1 proportionality constant
K_2 rate at which molecules evaporate from unit surface area when surface is fully covered
L_c particle length
L_s/G_s adsorbent/gas ratios
L_{st} saturated bed length
L_z adsorption zone length
M molecular weight
MTZ mass transfer zone
N_p number of stages in stagewise adsorption
P partial pressure of sorbed substance
P_i partial pressure
p^0 saturated vapor pressure of adsorbate at temperature T
p pressure
R rate of sorption per unit bed volume per unit time
r radius
T temperature
t time
U overall heat transfer coefficient
V total gas volumetric rate
V volume of gas reduced to standard conditions
V_m volume of gas at standard conditions
X/M amount of sorbed vapor per unit weight of adsorbent
X_{st} adsorptive capacity per unit adsorbent bed volume in the saturated zone
X_z mean adsorptive capacity per unit bed volume in the adsorptive zone
Y gas adsorbed per unit area or unit mass of adsorbent
Y_A gas concentration, lbm adsorbate per lbm inert gas
Y equilibrium composition of gas
Z height of adsorption bed

Greek Symbols

ε void fraction of bed
ΔP pressure drop
θ fraction of total surface covered by adsorbed molecules at any instant
λ_{A0} latent heat of vaporization Btu/lb
λ_R latent heat of vaporization of a reference substance
ϱ_m molar density of inert gas

REFERENCES

1. Cheremisinoff, P. N., and Moressi, A. C., "Carbon Adsorption," *Pollution Engineering*, 6(8): 66–69 (1974).
2. Leatherdale, J. W., "Air Pollution Control by Adsorption," *Carbon Adsorption Handbook*, P. N. Cheremisinoff and F. Ellerbusch (Eds.), Ann Arbor Science, Ann Arbor, MI, 1978.
3. Thomas, J. M., and Thomas, W. J., "Introduction to the Principles of Heterogeneous Catalysis," Academic Press, New York, 1967, p. 46.
4. Langmuir, I., *Journal of American Chemical Society*, 40:1361 (1918).

5. Buonicone, A. J., and Theodore, L., *Industrial Control Equipment for Gaseous Pollutants*, Vol. I, CRC Press, Inc., Ohio, 1975.
6. Brunauer, S., Emmett, P. H., and Teller E., *J. A. Chemical Society*, 60:309 (1938).
7. de Boer, J. H., *The Dynamic Character of Adsorption*, Clarenden Press, Oxford, (1953).
8. Othmer, D. F., and Sawyer, F. G., *Ind. Eng. Chem.*, 35:1269 (1943).
9. Kovack, J. L., "Gas Phase Adsorption and Air Purification," *Carbon Adsorption Handbook*, P. N. Cheremisinoff and F. Ellerbusch (Eds.), Ann Arbor Science Pub., Inc., Ann Arbor, MI, 1978.
10. Treybal, R. E., *Mass Transfer Operations*, lst and 2nd eds., McGraw–Hill Book Co., New York, 1955 and 1959.
11. Eagleton, L. C., and Bliss, H., *Chem. Eng. Progr.*, 49 (1953).
12. Collins, H. W., Jr., and Kwang-Chu Chao, A Dynamic Model for Multicomponent Fixed-Bed Adsorption, paper 82C presented at 74th National American Institute of Chemical Engineers Meeting, New Orleans, LA, March 1973.
13. Glueckauf, E., *Trans. Faraday Society*, 51:1540 (1953).
14. Jury, S. H., *Am. Inst. Chem. Eng. J.*, 13(6): 1124 (1967).
15. Yoshida, F., Ramaswami, D, and Hougen, O. A., *Am. Inst. Chem. Eng. J.*, 8(1): 5 (1962).
16. Gamson, B. W., Thodos, G., and Hougen, O. A., *Trans. Am. Inst. Chem. Eng.*, 39:1 (1943).
17. Acrivos, A., *Ind. Eng. Chem.*, 48(4): 703 (1956).
18. Michaels, A. S., *Ind. Eng. Chem.*, 44: 1922 (1952).
19. Ergun, S., *Chem. Eng. Progr.*, 48: 39 (1952); and *Ind. Eng. Chem.*, 41: 1179 (1949).
20. Union Carbide Corp., Linde Division, Molecular Sieve Dept., New York, Bulletin F-34A, F-34-1, and F-34-2.
21. Process Design Manual for Carbon Adsorption, Technology Transfer, U. S. Environmental Protection Agency (October 1971).
22. Hasketh, Howard E., *Understanding and Controlling Air Pollution*, 2nd ed. Ann Arbor Science Pub. Inc., Ann Arbor, MI (1974).
23. Turk, A., "Source Controlling Gas-Solid Adsorption," *Air Pollution Handbook*, A. C. Stern (Ed.), Academic Press, New York, 1968.
24. Hassler, John W., *Purification with Activated Carbon*, Chemical Publishing Co., Inc., New York, 1974.
25. Lyman, Warren J., "Applicability of Carbon Adsorption to the Treatment of Hazardous Industrial Wastes," *Carbon Adsorption Handbook*, P. N. Cheremisinoff and F. Ellerbusch (Eds.), Ann Arbor Science Pub. Inc., Ann Arbor, MI, 1978.
26. Minor, P. J., "Organic Chemical Industry's Wastewater," *Enviro-Sci. Technol.*, 8(7): 620–625 (1974).
27. "Development Document for Effluent Limitations Guidelines and New Source Performance Standards for Synthetic Resins—Segments of the Plastics and Synthetic Materials Manufacturing Point Source Category," U. S. Environmental Protection Agency, Washington, DC, 1974.
28. Rizzo, J. L., "Removal of Toxic Chemicals by Filtration Adsorption," paper presented at the 3rd Symposium on Hazardous Chemicals Handling and Disposal, Indianapolis, Indiana, April 12, 1972.
29. Hager, D. G., "Industrial Wastewater Treatment by Granular Activated Carbon," *Ind. Water Eng.* (Jan./Feb. 1974).

CHAPTER 31

INTRODUCTION TO FUNDAMENTALS OF ION EXCHANGE TECHNOLOGY

Francis J. Desilva

Water Process Division
Belco Pollution Control Corp.
Parsippany, New Jersey, USA

Paul N. Cheremisinoff

Department of Civil & Environmental Engineering
New Jersey Institute of Technology
Newark, New Jersey, USA

CONTENTS

INTRODUCTION

All natural waters contain in varying concentrations dissolved salts which dissociate in water to form charged particles called ions. These ions are the positively charged cations and negatively charged anions that permit the water or solution to conduct electricity and are therefore called electrolytes. Electrical conductivity is thus a measure of water purity, with low conductivity corresponding to high purity.

The process of ion exchange is uniquely suited to the removal of ionic species from water supplies for several reasons. First, ionic impurities are present in rather low concentrations. Second, modern ion exchange resins have high capacities and can remove unwanted ions preferentially. Third, these modern ion exchange resins are quite stable and readily regenerated, thereby allowing their use for many years.

Other economic advantages that ion exchange equipment offers are:

1. The process and equipment have been tested over many years. Designs are well developed into pre-engineered "off-the-shelf" units that are rugged and reliable.
2. Fully manual to completely automatic units are available.
3. There are many manufactures of ion exchange systems on the market which kepp costs competitive.
4. Temperature effects from 0° C to 35° C are negligible.
5. The process is excellent for both small and large installations, for example, from home water softeners to huge utility applications.

In addition to dissolved ions many waters contain nonionic impurities, such as suspended and colloidal matter, that could be harmful in industrial application. These constituents must also be removed; usually before demineralization to avoid deposition on and subsequent fouling of the ion exchange resins. For this reason, many demineralizer plants normally include pretreatment systems.

IMPORTANCE OF HIGH QUALITY WATER FOR INDUSTRIAL USE

Water problems in cooling, heating, steam generation, and manufacturing are caused in large measure from the kinds and concentrations of dissolved solids, dissolved gasses, and suspended matter in the make-up water supplied. Table 1 lists the major objectionable ionic constituents present in many water supplies that can be removed by demineralization.

Prevention of scale and other deposits in cooling and boiling waters is best accomplished by removal of dissolved solids. Whereas, in municipal water purification such removal is limited to the partial reduction of hardness and the removal of iron and manganese, in industrial water treatment it is often carried much further and may include complete removal of hardness, reduction or removal of alkalinity, removal of silica, or even the complete removal of all dissolved solids.

The two most frequently encountered water problems—scale formation and corrosion—are common to cooling, heating, and steam generating systems. Hardness (calcium and magnesium), alkalinity, sulfate, and silica all form the main source of scaling in heat-exchange equipment, boilers, and pipes. Scales or deposits formed in boilers and other heat-exchange equipment act as insulation, preventing efficient heat transfer and causing boiler tube failures through overheating of the metal. Free mineral acids (sulfates and chlorides) cause rapid corrosion of boilers, heaters, and other metal containers and piping, alkalinity causes embrittlement of boiler steel, and carbon dioxide and oxygen cause corrosion, primarily in steam and condensate lines.

Low quality steam can produce undersirable deposits of salts and alkali on the blades of steam turbines; much more difficult to remove are silica deposits which can form on turbine blades even when steam is satisfactory by ordinary standards. At steam pressures above 600 psi, silica from the boiler water actually dissolves in the gaseous steam and then reprecipitates on the turbine blades at their lower pressure end.

Table 1
Major Objectionable Ionic Constituents

Constituent	Chemical Designation	Problems Caused
Hardness	Calcium and magnesium salts expressed as $CaCO_2$, Ca, Mg	Main source of scaling in heat exchange equipment, boilers, pipe lines, etc. Forms curds with soap, interferes with dyeing
Alkalinity	Bicarbonate (HCO_3), carbonate (CO_3), and hydrate (OH), expressed as $CaCO_3$	Cause of foaming and carryover of solids with steam. Embrittlement of boiler steel. Bicarbonate and carbonate produce CO_2 in steam, a source of corrosion
Free mineral acidity	H_2SO_4, HCl, etc., expressed as $CaCO_3$	Rapid corrosion and deterioration
Chloride	Cl^-	Interferes with silvering processes and increases t.d.s.
Sulfate	$(SO_4)^{--}$	Calcium sulfate scale is formed
Iron and manganese	Fe^{++} (ferrous) Fe^{+++} (ferric) Mn^{++}	Discolors water, deposits in water lines, boilers, etc. Interferes with dyeing, tanning, paper manufacture, and process work
Carbon dioxide	CO_2	Corrodes water lines particularly steam and condensate lines.
Silica	SiO_2	Scale in boilers and cooling water systems. Insoluble scale on turbine blades due to silica vaporization in high pressure boilers (over 600 lbs)

In the operation of every cooling, heating, and steam-generating system, the water changes temperature. Higher temperatures, of course, increase both corrosion rates and scale forming tendency. Evaporation in process steam boilers and in evaporative cooling equipment increases the dissolved solids concentration of the water, compounding the problem.

In addition to the formation of scale or corrosion of metal within boilers, auxiliary equipment is also susceptible to similar damage. Attempts to prevent scale formation within a boiler can lead to make-up line deposits if the treatment chemicals are improperly chosen. Thus, the addition of normal phosphates to an unsoftened feedwater can cause a dangerous condition by clogging the make-up line with precipitated calcium phosphate. Deposits in the form of calcium or magnesium stearate deposits, otherwise known as "bathtub ring" can be readily seen, and are caused by the combination of calcium or magnesium with negative ions of soap stearates.

DEMINERALIZATION THEORY

Ion exchangers are materials that can exchange one ion for another, hold it temporarily, and release it to a regenerant solution. In a typical demineralizer, this is accomplished in the following manner: the influent water is usually passed through a hydrogen cation exchange resin which converts the influent salt (e.g., sodium sulfate) to the corresponding acid (e.g., sulfuric acid) by exchanging an equivalent number of hydrogen (H^+) ions for the metallic cations (Ca^{++}, Mg^{++}, Na^+). These acids are then removed by passing the effuent through an alkali regenerated anion exchange resin which replaces the anions in solution (Cl^-, SO_4^-, NO_3^-) with an equivalent number of hydroxide ions. The hydrogen ions and hydroxide ions neutralize each other to form an equivalent amount of pure water

(CATION EXCHANGE)

$R\text{-}H^+ + C^+ \leftrightarrow R\text{-}C^+ + H^+$

C^+ REPRESENTS COMMON CATIONS:e.g.- Ca^{++}, Mg^{++}, Na^+

(ANION EXCHANGE)

$R^+OH^- + A^- \leftrightarrow R^+A^- + OH^-$

A REPRESENTS COMMON ANIONS:e.g.- $Cl^-, SO_4^=, NO_3^-$

(EFFLUENTS)

$H^+ + OH^- \rightarrow H_2O$

Figure 1. Summary of the principal ion exchange mechanism.

(See Figure 1). During regeneration, the reverse reaction takes place. The cation resin is regenerated with either sulfuric or hydrochloric acid and the anion resin is regenerated with sodium hydroxide.

There are various arrangements or equipment possible, but in all cases, except in mixed-bed demineralization, the water should first pass through a cation exchanger. In mixed demineralization the two exchange materials, that is, the cation exchange resin and the anion exchange resin, are placed in one shell instead of two separate shells. In operation, the two types of exchange materials are thoroughly mixed so that we have, in effect, a number of multiple demineralizers in series. Higher quality water is obtained from a mixed-bed unit than from a two-bed system.

PROPERTIES OF ION EXCHANGE MATERIALS

An ion exchange resin suitable for industrial use must exhibit durable physical and chemical characteristics which are summarized by the following properties:

- *Functional groups* The molecular structure of the resin is such that it must contain a macroreticular tissue with acid or basic radicals. These radicals are the basis of classifying ion exchangers into two general groups:

  1. *Cation exchangers*—in which the molecule contains acid radicals of the HSO_3 or HCO_2 type able to fix mineral or organic cations and exchange them with the hydrogen ion H^+.
  2. *Anion exchangers*—containing basic radicals, for example amine functions of the type NH_2 able to fix mineral or organic anions and exchange them with the hydroxyl ion OH^- coordinate to their dative bonds.

  The presence of these radicals enables a cation exchanger to be assimilated to an acid of form H-R and an anion exchanger to be a base of form OH-R when regenerated.

  These radicals act as immobile ion exchange sites to which are attached the mobile cations or anions. For example, a typical sulfonic acid cation exchanger has immobile ion exchange sites consisting of the anionic radicals SO to which are attached the mobile cations, such as H^+ or Na^+. An anion exchanger similarly has immobile cationic sites to which are attached mobile (exchangeable) hydroxide anions OH. The radicals attached to the molecular nucleus further determine the nature of the acid or base, whether it will be weak or strong.

  Exchangers are divided into four specific classifications depending on the kind of radical, or functional group, attached; strong acid, strong base weak acid, or weak base. Each of these four types of ion exchangers are described in detail later.

- *Solubility* The ion exchange substance must be insoluble under normal conditions of use. All ion exchange resins in current use are high molecular weight polyacids or polybases which are virtually insoluble in most aqueous and nonaqueous media. This is no longer true of some resins once a certain temperature has been reached. For example, some anion exchange resins are limited to a maximum temperature of 105° F.

Liquid ion exchange resins exist also, yet we do not consider their applicability here and they also exhibit very limited solubility in aqueous solutions.

- *Bead size* The resins must be in the form of spherical granules of maximum homogeneity and dimensions so that they do not pack too much, the void volume among their interstices is constant for a given type, and the liquid head loss in percolation remains acceptable. Most ion exchange resins occur as small beads or granules usually between 16 and 50 mesh in size.
- *Resistance to fracture* The ion or ionized complexes that the resins are required to fix are of varied dimensions and weights. The swelling and contraction of the resin bead that this causes must obviously not cause the grains to burst.

  Another important factor is the bead resistance to osmotic shock. This shock will inevitably occur across the bead boundary surface, as there will be a salinity gradient of a different magnitude, during the cycle of the exchange material. The design of ion exchange apparatus also must take into consideration the safe operation of the ion exchange resin and avoid excessive stresses or mechanical abrasion in the bed, which could lead to breakage of the beads.

TYPES OF ION EXCHANGE RESINS

As mentioned previously, ion exchange materials are grouped into four specific classifications depending on the functional group attached: strong acid cation, strong base anion, weak acid cation, or weak base anion. In addition to these we also have inert resins that do not have chemical properties.

Strongly Acid Cation

Strongly acidic cation resins derive their exchange activity from sulfonic functional groups (HSO). the major cations in water are calcium, magnesium, sodium, and potassium and they are exchanged for hydrogen in the strong acid cation exchanger when operated in the hydrogen cycle. The following equation represents the exhaustion phase, is written in the molecular form (as if the salts present were undissociated), and shows the cations in combination with the major anions; the bicarbonate, sulfate, and chloride anions:

$$\begin{matrix} Ca & 2\,HCO_2 & & & Ca & 2\,H_2CO_3 \\ Mg\cdot & SO_4 + & 2\,RSO_3H & \rightleftharpoons 2\,RSO_3 & Mg + & H_2SO_4 \\ Na & 2\,Cl & & & Na & 2\,HCl \end{matrix}$$

Where R represents the complex resin matrix. Because these equilibrium reactions are reversible, when the resin capacity has been exhausted it can be recovered through regeneration with a mineral acid. The strong acid exchangers operate at any pH, split strong or weak salts, require excess strong acid regenerants (typical regeneration efficiency varies from 25% to 45% in cocurrent regeneration), and they permit low leakage. In addition, they have rapid exchange rates, are stable, exhibit swelling less than 7% going from Na^+ to H^+ from, and may last 20 years or more with little loss of capacity.

These resins have found a wide range of application, being used on the sodium cycle for softening and on the hydrogen cycle for softening, dealkalization, and demineralization.

Weakly Acidic Cation

Weakly acidic cation exchange resins have carboxylic groups (COOH) as the exchange sites. When operated on the hydrogen cycle, the weakly acidic resins are capable of removing only those cations equivalent to the amount of alkalinity present in the water, and most efficiently the hardness (calcium and magnesium) associated with alkalinity, according to these reactions:

$$\begin{matrix} Ca & & Ca \\ Mg & (HCO_3) + RCOOH \rightleftharpoons 2\,RCOO & Mg + H_2CO_4 \\ 2\,Na & & 2\,Na \end{matrix}$$

These reactions are also reversible and permit acid regeneration to return the exhausted resin to the hydrogen form.

The resin is highly efficient for it is regenerated with 110% of the stoichiometric amount of acid as comparted to 200% to 300% for strong acid cation exchange resins. It can be regenerated with the waste acid from a strong acid cation exchanger and there is little waste problem during the regeneration cycle. In order to prevent calcium sulfate precipitation when regenerated with H_2SO_4 it is usually regenerated stepwise with initial H_2SO_4 at .5%. The resins are subject to reduced capacity from increasing flow rate (above 2 gpm/ft), low temperatures, and/or hardness-alkalinity ratio especially below 1.0.

Weakly acidic resins are used primarily for softening and dealkalization, frequently in conjunction with a strongly acidic polishing resin. Systems which use both resins profit from the regeneration economy of the weakly acidic resin and produce treated water of quality comparable to that available with a strongly acidic resin.

Strongly Basic Anion

Strongly basic anion exchange resins derive their functionality from the quaternary ammonium exchange sites. All the strongly basic resins used for demineralization purposes belong to two main groups commonly known as Type I and Type II. The principal difference between the two resins, operationally, is that Type I has a greater chemical stability, and Type II has a slightly greater regeneration efficiency and capacity. Physically, the two types differ by the species of quaternary ammonium exchange sites they exhibit. Type I sites have three methyl groups, while in Type II, an ethanol group replaces one of the methyl groups. In the hydroxide form, the strongly basic anion will remove all the commonly encountered inorganic acids according to three reactions:

$$\begin{matrix} H_2SO_4 & & SO_4 \\ 2\,HCl & & 2\,Cl \\ 2\,H_2SiO_3 & + 2\,ZOH \rightleftharpoons 2\,Z & HSiO_3 + 2\,H_2O \\ 2\,H_2CO_3 & & 2\,HCO_3 \end{matrix}$$

Like the cation resin reactions, the anion exchange reactions are also reversible and regeneration with a strong alkali, such as caustic soda, will return the resin to the hydroxide form.

The strong base exchangers operate at any pH, can split strong or weak salts, require excess high-grade NaOH for regeneration (with the typical efficiency varying from 18% to 33%), and are subject to organic fouling from such compounds when present in the raw water, and to resin degradation due to oxidation and chemical breakdown. The strong base anion resins suffer from capacity decrease and silica leakage increase at flow rates above 2 gpm/ft^3 of resin, and cannot operate over 130°–150° F depending on resin type. The normal maximum continuous operating temperature is 120° F, and to minimize silica leakage, warm caustic (up to 120° F) should be used. Type I exchangers are for maximum silica removal. They are more difficult to regenerate and swell more (from Cl to OH form) than Type II. The major case for selecting a Type I resin is where high operating temperatures and/or very high silica levels are present in the influent water or superior resistance to oxidation or organics is required.

Type II exchangers remove silica (but less efficiently than Type I) and other weak anions, regenerate more easily, are less subject to fouling, are freer from the oder of amine, and are cheaper to operate than Type I. Where free mineral acids are the main constituent to be removed and very high silica removal is not required, Type II anion resin should be chosen.

Weakly Basic Anion

Weakly basic anion resins derive their functionality from primary (R-NH), secondary (R-NHR′), tertiary (R-N-R′2), and sometimesquaternary amine groups. The weakly basic resin readily absorbs such free mineral acids as hydrochloric and sulfuric, and the reactions may be represented according to the following:

$$H_2SO_4 + 2\ ZOH = 2\ Z\ SO_4 + 2\ HO$$

$$2\ HCl \qquad\qquad\qquad 2\ Cl$$

Because the preceding reactions are also reversible, the weakly basic resins can be regenerated by applying caustic soda, soda ash or ammonia. The weak base exhanger regenerates with a nearly stoichiometric amount of base (with the regeneration efficiency possibly exceeding 90%) and can utilize waste caustic from following strong base anion exchange resins. Weakly basic resins are used for strongly acid waters (Cl, SO_4, NO_3) and for low alkalinity systems. They do not remove anions satisfactorily at pH's above 6 nor do they remove CO or silica, but they have capacities about twice as great as for strong base exchangers. Weak base resins can be used to precede a strong base anion resin to provide the maximum protection of the latter against organic fouling and to reduce regenerant costs.

Inert Resin

There also exists a type of resin with no functional groups attached. This resin offers no capacity to the system but increases regeneration efficiency in mixed-bed exchangers. These inert resins are of a density between cation and anion resins and when present in mixed-bed vessels help to separate cation and anion resins during backwash.

Inert resins are advantageous because they:

1. Classify cation and anion resins so that little or no mixing of cation or anion resin occurs before regeneration, and a buffering mid-bed collection zone exists.
2. Improve regeneration efficiency thereby reducing resin quantities needed.
3. Protect against osmotic shock since the inert layer effectively prevents the exposure of cation resin to the caustic regenerant solution and the exposure of anion resin to the acid regenerant solution.

RESIN PERFORMANCE

Variances in resin performance and capacities can be expected from normal annual attrition rates of ion exchange resins. Typical attrition losses that can be expected are listed below:

- *Strong cation resin:* 3% per year for 3 years or 1,000,000 gals/cu ft.

- *Strong anion resin:* 25% per year for two years or 1,000,000 gals/cu ft.
- *Weak cation/anion:* 10% per year for two years or 750,000 gals/cu ft.

A steady fall-off of resin exchange capacity is a matter of concern to the operator and is due to several conditions:

- *Improper backwash.* Blowoff of resin from the vessel during the backwash step can occur if too high a backwash flow rate is used. This flow rate is temperature dependent and must be regulated accordingly. Also, adequate time must be allotted for backwashing to insure a clean bed prior to chemical injection.
- *Channeling.* Cleavage and furrowing of the resin bed can be caused by faulty operational procedures or a clogged bed or underdrain. This can mean that the solution being treated follows the path of least resistance, runs through these furrows, and fails to contact active groups in other parts of the bed.
- *Incorrect chemical application.* Resin capacities can suffer when the regenerant is applied in a concentration that is too high or too low. Another important parameter to be considered during chemical application is the location of the regenerant distributor. Excessive dilution of the regenerant chemical can occur in the vessel if the distributor is located too high above the resin bed. A recommended height is 3 inches above the bed level.
- *Mechanical strain.* When broken beads and finess migrate to the top of the resin bed during service, mechanical strain is caused which results in channeling, increased pressure drop, or premature breakthrough. The combination of these resulting conditions leads to a drop in capacity.
- *Resin fouling.* In addition to the physical causes of capacity losses just listed, there are a number of chemically caused problems that merit attention, specifically the several forms of resin fouling that may be found.

1. Organic fouling occurs on anion resins when organics precipitate onto basic exchange sites. Regeneration efficiency is then lowered thereby reducing the exchange capacity of the resin. Causes of organic fouling are fulvic, humic, or tannic acids or degradation products of DVB (divinylbenzene) crosslinkage material of cation resins. DVB is degraded through oxidation and causes irreversible fouling of downstream anion resins.
2. Iron fouling is caused by both forms of iron ions; the insoluble form will coat the resin bead surface and the soluble form can exchange and attach to exchange sites on the resin bead. These exchanged ions can be oxidized by subsequent cycles and precipitate ferric oxide within the bead interior.
3. Silica fouling is the accumulation of insoluble silica on anion resins. It is caused by improper regeneration which allows the silicate (ionic form) to hydrolyze to soluble silicic acid which in turn polymerizes to form colloidal silicic acid with the beads. Silica fouling occurs in weak base anion resins when they are regenerated with silica-laden waste caustic from the strong base anion resin unless intermediate partial dumping is done.
4. Microbiological fouling (MB) becomes a potential problem when microbic growth is supported by organic compounds, ammonia, nitrates, etc. which are concentrated on the resin. Signs of MB fouling are increased pressure drops, plugged distributor laterals, and highly contaminated treated water.
5. Calcium sulfate fouling occurs when sulfuric acid is used to regenerate a cation exchanger after exhaustion by a water high in calcium. The precipitate of calcium sulfate (gypsum) that forms can cause calcium and sulfate leakage during subsequent service runs. Given a sufficient calcium input in the water to treat, calcium sulfate fouling is especially prevalent when the percent solution of regenerant is greater than 5%, or the temperature is greater than 100° F, or when the flow rate is less than

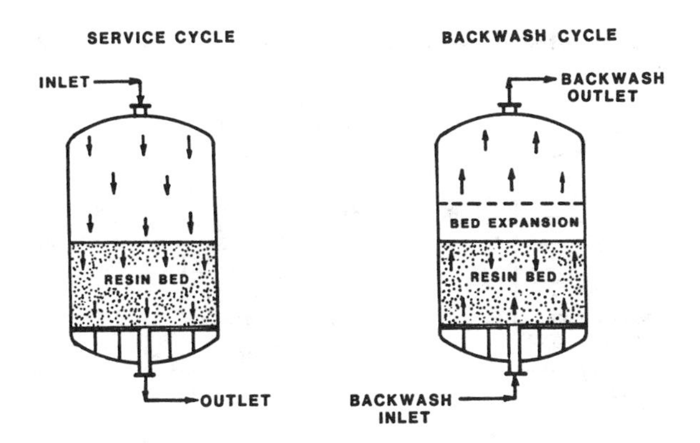

Figure 2. Illustrates the service and backwash cycles.

1 gpm/cu ft. Stepwise injection of sulfuric acid during regeneration can help prevent fouling.

6. Aluminium fouling of resins can appear when aluminum floc from alum or other coagulants in pretreatment are encountered by the resin bead. This floc coats the resin bead and in the ionic form will be exchanged. However, these ions are not efficiently removed during regeneration so the available exchange sites continously decrease in number.
7. Copper fouling is found primarily in condensate polishing applications. Capacity loss is due to copper oxides coating the resin beads.
8. Oil fouling does not cause chemical degradation but gives loss of capacity due to filming on to the resin beads and the reduction of their active surface. Agglomeration of beads also occurs causing increase pressure drop, channeling and premature breakthrough. The oil fouling problem can be alleviated by the use of surfactants.

SEQUENCE OF OPERATION

The mode of operation for ion exchange units can very greatly from one system to the next, depending on the users requirements. Service and regeneration cycles can be fully manual to totally automatic, with the method of regeneration being cocurrent, countercurrent, or external. The most common means of ion exchange found in industry is the cocurrent downflow fixed-bed technique. Figures 2 and 3 illustrate the service run, backwash cycle, regenerant introduction cycle, and slow and fast rinse step of a cocurrent system. These are followed by the service (Figure 4) and regeneration (Figure 5) cycles of a countercurrent unit.

The exhaustion phase is called the service run. This is followed by the regeneration phase which is necessary to bring the bed back to initial conditions to cycle. The regeneration phase includes four steps:

- Backwashing to clean the bed.
- Introduction of the excess regenerant.
- A slow rinse or displacement step to push the regenerant slowly through the bed.
- Finally a fast rinse to remove the excess regenerant from the resin and elute the unwanted ions to waste.

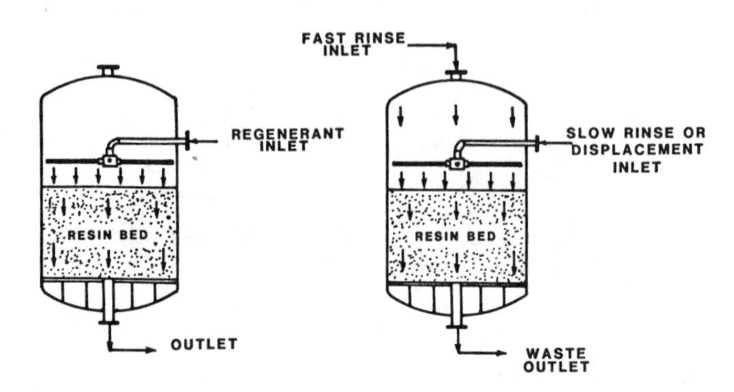

Figure 3. Illustrates the regeneration introduction and rinse/displacement cycles of a cocurrent bed.

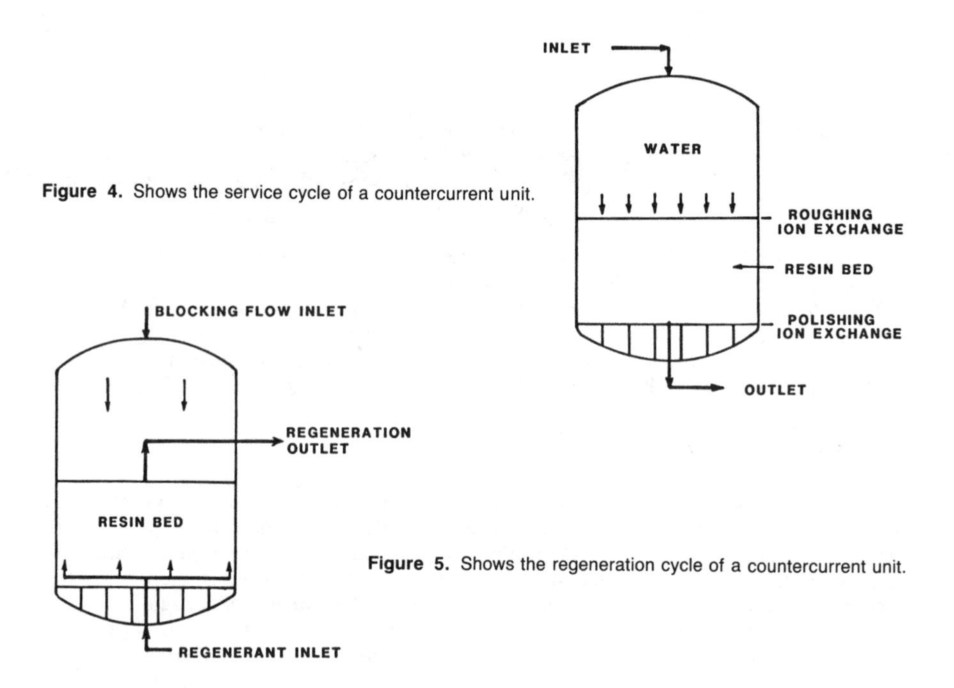

Figure 4. Shows the service cycle of a countercurrent unit.

Figure 5. Shows the regeneration cycle of a countercurrent unit.

Service Cycle

The service cycle is normally terminated by one or a combination of the following criteria:

- High effluent conductivity.
- Total gallons throughput.
- High pressure drop.
- High silica.

- High sodium.
- Variations in pH.

Termination of the service cycle can be manually or automatically initiated.

Backwash Cycle

Normally, the first step in the regeneration sequence is designed to reverse flow from the service cycle using sufficient volume and flow rate to develop proper bed expansion for the purpose of removing suspended material (crud) trapped in the ion exchange bed during service cycle. The backwash waste water is collected by the raw water inlet distributor and diverted to waste via value sequencing. Backwash rate and internal design should avoid potential loss of whole bead resin during the backwash step. (Lower water temperature means more viscous force and more expansion.)

Regenerant Introduction

This introduction of regenerant chemicals can be cocurrent or countercurrent, depending upon effluent requirements, operating cost, etc. Regenerant dosages (pounds per cubic foot), concentrations, flow rate, and contact time are determined for each application. The regenerant distribution and collection system must provide uniform contact throughout the bed and should avoid regenerant hideout. Additional effluent purity is obtained with counter-current systems since the final resin contact in the service will be the most highly regenerated resin in the bed creating a polishing affect.

Displacement/Slow Rinse Cycle

Normally the final steps in the regeneration sequence are generally terminated on acceptable quality. Displacement which precedes the rinse step is generally an extension of the regenerant introduction step. The displacement step is designed to give final contact with the resin removing the bulk of the spent regenerant from the resin bed.

Fast Rinse Cycle

This, the final step, is essentially the service cycle except that the effluent is diverted to waste until quality is proven. This final rinse is always in the same direction as the service flow. Therefore, in countercurrent systems the displacement flow and rinse flows will be in opposite directions.

SEQUENCE OF OPERATION FOR MIXED-BED UNITS

In mixed-bed units both the cation and the anion resins are mixed together thoroughly in the same vessel by compressed air. The cation and the anion resins being next to each other, constitute an infinite number of cation and anion exchangers. The effluent quality obtainable from a well-designed and operated mixed-bed exhanger will readily produce demineralized water of conductivity less than 0.5 mmho and silica less than 10 ppb.

Service Cycle

As far as the mode of operation is concerned, the service cycle of a mixed-bed unit is very similar to a conventional two-bed system in that water flows into the top of the vessel, down through the bed, and the purified effluent comes out the bottom. It is in the regeneration and the preparation for it that the mixed-bed differs from the two-bed equipment. The resins must be separated, regenerated separately, and remixed for the next service cycle.

Backwash Cycle

Prior to regeneration, the cation and the anion resins are separated by backwashing at a flow rate of 3.0 to 3.5 gpm/ft. The separation occurs because of the difference in the density of the two types of resin. The cation resin, being heavier, settles on the bottom, while the anion resin, being lighter, settles on top of the cation resin. After backwashing, the bed is allowed to settle down for 5–10 minutes and two clearly distinct layers are formed. After separation, the two resins are idependently regenerated.

Regenerant Introduction

The anion resin is regenerated with caustic flowing downward from the distributor placed just above the bed, while the cation resin is regenerated with either hydrochloric or sulfuric acid, usually flowing upward. The spent acid and caustic are collected in the interface collector, situated at the interface of the two resins. The regenerant injection can be carried out simultaneously as described or sequentially. In sequential regeneration, the cation resin regeneration should precede the anion resin regeneration to prevent the possibility of calcium carbonate and magnesium hydroxide precipitation, which may occur because of the anion regeneration waste coming in contact with the exhausted cation resin. If this precipitation occurs, it can foul the resins at the interface. This becomes very critical when only the mixed-bed exchanger is installed to demineralize the incoming raw water. In the case of sequential regeneration, during the caustic and acid injection period, blocking flow of the demineralized water is provided in the opposite direction of the regenerant injection. This is required to prevent the caustic from entering the cation resin and acid from entering the anion resin. When regeneration is carried out simultaneously, acid and caustic injection flows act like blocking flow to each other and no additional blocking flow with water is needed. In a few sequential-type regeneration systems, acid is injected to flow downward through the central interface collector which now also acts as an acid distributor.

Rinsing and Air Mix Cycles

After completion of the acid and eaustic injection, both the cation and anion resins are rinsed slowly to remove the majority of the regenerant, without attempting to eliminate it completely. After the use of 7–10 gallons of slow rinse volume per cubic foot of each type of resin, the unit is drained to lower the water to a few inches above the resin bed. The resins are now remixed with an upflow of air. After remixing, the unit is filled completely with water flowing slowly from top, to prevent anion resin separation in the upper layers. The mixed-bed exchanger is then rinsed at fast flow rates. The conductivity of the effluent water may be very high for a few minutes and will then drop suddenly to the value usually observed in the service cycle. This phenomenon is characteristic of mixed beds and is due to the absorption of the remaining acid or caustic in different parts of the bed, by one or the other resin. This, no doubt, results in the loss of resin capacity, but this loss is negligible as compared to the length of the service cycle and the savings in the overall time required for regeneration.

ION EXCHANGE SOFTENING (SODIUM ZEOLITE SOFTENING)

This is one of the ion exchange processes used in water purification. In this process, sodium ions from the solid phase are exchanged with the hardness ions from the aqueous phase. Consider a bed of ion exchange resin having sodium as the exchangeable ion, with water containing calcium and magnesium hardness allowed to percolate through this bed. Let us denote the ion exchange resinous material as RNa, where R stands for resin matrix and Na is its mobile exchange ion. The hard water will exchange Ca and Mg ions rapidly, so that water at the effluent will be almost completely softened. Calcium and magnesium salts will be converted into corresponding sodium salts.

The softening reaction will be as follows:

$$\begin{matrix} Ca \\ \\ Mg \end{matrix} \begin{bmatrix} HCO_3 \\ CO_3 \\ Cl \\ SO_4 \end{bmatrix} + 2\,RNa \quad R_2 \begin{matrix} Ca \\ \\ Mg \end{matrix} + 2\,Na \begin{cases} 2\,HCO_3 \\ CO_3 \\ 2\,Cl \\ SO_4 \end{cases}$$

The reaction will proceed toward the right hand side to its completion until the bed gets completely exhausted or saturated with Ca and Mg ions. In order to reverse the equilibrium so that the reaction proceeds toward the left hand side, the concentration of sodium ions has to be increased. This increase in sodium ions is accomplished by using a brine solution of sufficient strength that the total sodium ions present in the brine are more than the total equivalent of Ca and Mg in the exhausted bed. This reverse reaction is carried out in order to bring the exhausted bed of resin back to its sodium form. This process is known as "regeneration." When the softener with the fresh resin, in the sodium form, is put in service, the sodium ions in the surface layer of the bed are immediately exchanged with calcium and magnesium, thereby producing soft water with very little residual hardness in the effluent. As the process continues, the resin bed keeps exchanging its sodium ions with calcium and magnesium ions until the hardness concentration increases rapidly and the softening run is ended.

This softening process can be extended to a point where the hardness coming in and coming out is the same. When this condition is reached, the bed is completely exhausted and does not have any further capacity to exchange ions. This capacity is called the "total breakthrough capacity." In practice, the softening process is never extended to reach this stage as it is ended at some predetermined effluent hardness, much lower than the influent hardness. This capacity is called "Operating Exchange Capacity." After the resin bed has reached this capacity, the resin bed is regenerated with a brine solution.

The regeneration of the resin bed is never complete. Some traces of calcium and magnesium remain in the bed and are present in the lower bed level. In the service run, sodium ions exchanged from the top layers of the bed form a very dilute regenerant solution which passes through the resin bed to the lower portion of the bed. This solution tends to leach some of the hardness ions not removed by previous regeneration. These hardness ions appear in the effluent water as "leakage." Hardness leakage is also dependent upon the raw water characteristics. If the Na/Ca ratio and calcium hardness is very high in the raw water, leakage of the hardness ions will be higher.

SEQUENCE OF OPERATION FOR SOFTENER UNITS

The following are the basic steps involved in a regeneration of a water softener:

1. *Backwashing.* After exhaustion, the bed is backwashed to effect a 50% minimum bed expansion to release any trapped air from the air pockets, minimize the compactness of the bed, reclassify the resin particles, and purge the bed of any suspended insoluble material. Backwashing is normally carried out at 5–6 gpm/ft. However, the backwash flow rates are directly proportional to the temperature of water.
2. *Brine injection.* After backwashing, a 5%–10% brine solution is injected during a 30-minute period. The maximum exchange capacity of the resin is restored with 10%

strength of brine solution. The brine is injected through a separate distributor placed slightly above the resin bed.

3. *Displacement or slow rinse.* After brining, the salt solution remaining inside the vessel is displaced slowly, at the same rate as the brine injection rate. The slow rinsing should be continued for at least 15 minutes and the slow rinse volume should not be less than 10 gallons/cu ft of the resin. The actual duration of slow rinse should be based on the greater of these two parameters.
4. *Fast rinse.* Rinsing is carried out to remove excessive brine from the resin. The rinsing operation is generally stopped when the effluent chloride concentration is less than 5–10 ppmm in excess of the influent chloride concentration and the hardness is equal to or less than 1 ppm as CaCO.

DEMINERALIZER ARRANGEMENTS

Each arrangement will vary substantially in both operating and installed cost. Important factors for selection:

1. Influent water analysis.
2. Flow rate.
3. Effluent quality.
4. Waste requirements.
5. Operating cost.

Typical arrangements are illustrated in Figures 6 through 11. Figure 6 arrangement is applicable to low alkalinity water with low effluent silica levels required. It generally has

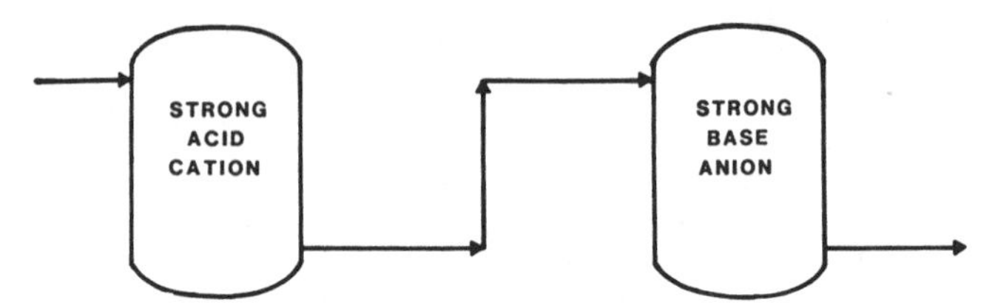

Figure 6. Shows a demineralizer arrangement with low alkalinity water and low effluent silica levels required. Arrangement has high operating costs but produces approximately neutral waste.

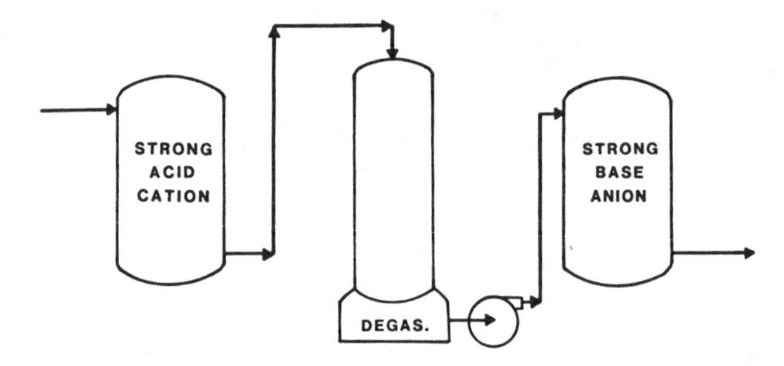

Figure 7. Demineralizer arrangement with application of high alkalinity water with low effluent silica levels required.

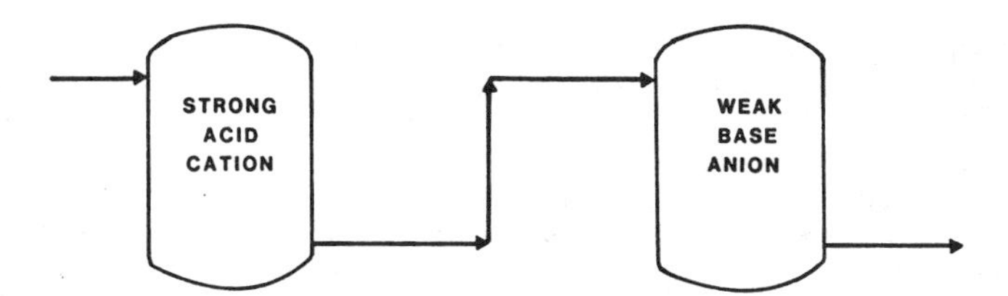

Figure 8. Demineralizer arrangement similar to Figure 6 but with no silica removal.

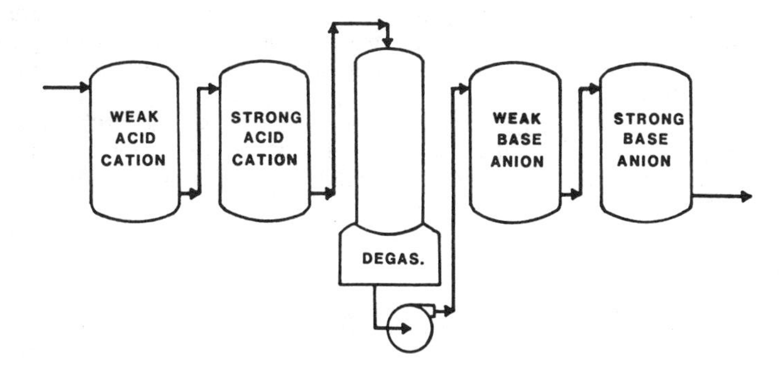

Figure 9. Demineralizer arrangement with application to high alkalinity, hardness, chloride, and sulfate waters with a hardness-to-alkalinity ratio approaching unity.

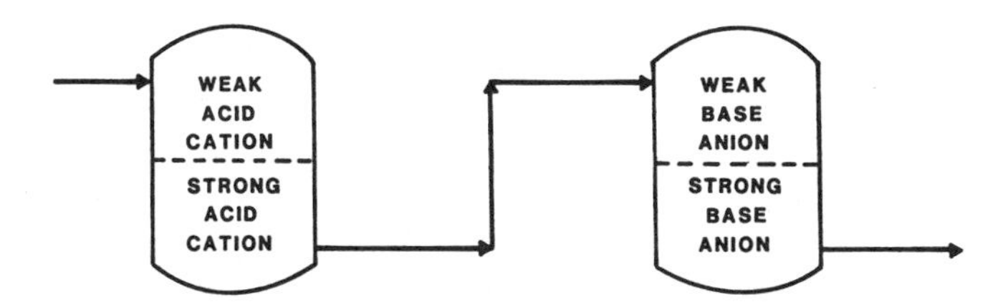

Figure 10. Demineralizer arrangement similar to Figure 9 but with lower alkalinity.

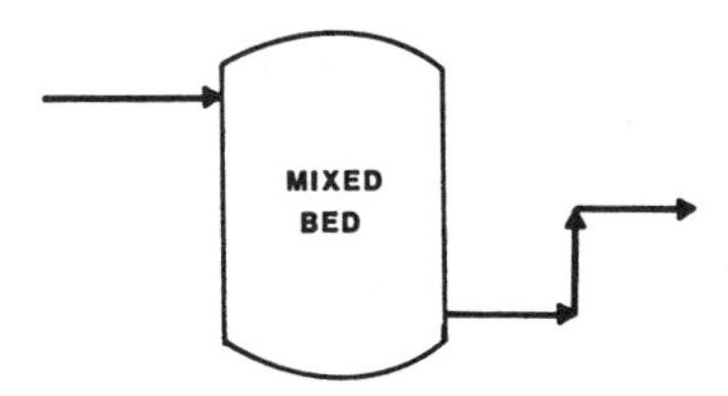

Figure 11. Shows a mixed-bed arrangement that serves as an effluent polisher.

high operating costs but produces almost neutral wastes. The arrangement in Figure 7 is for high alkalinity with low effluent silica levels required. It has a higher initial cost than the arrangement shown in Figure 6 and requires additional chemicals to neutralize wastes. In addition, repumping is required.

The Figure 8 arrangement is similar to that of Figure 6 except that no silica removal is required. It shows excellent removal of strong acids and has low operating costs. It does require additional chemicals to neutralize wastes.

Figure 9 shows an arrangement for high alkalinity, high hardness, and high chloride and sulfide waters. This scheme displays high regeneration efficiences whereby cations and anions are regenerated in series. It produces a low silica effluent, but repumping is required.

Figure 10 is similar to that of Figure 9 but handles lower alkalinity. It is generally a low cost operation with moderate to low initial costs. Resin separation must be maintained in this system. The principal disadvantage is that it is difficult to operate. In particular, it requires accurate backwash rates in order to maintain bed stratification. Also, silica polymerization can occur on the weak base ion resin which often results in resin embrittlement and fracture.

Finally, Figure 11 shows a system used as an effluent polisher to further reduce total dissolved solids. This system produces a high quality water. The cation and anion resins are ultimately mixed during service and are restratified by backwashing prior to regenerant introduction.

CHAPTER 32

PRINCIPLES OF PACKED-BED ELECTROCHEMICAL REACTORS

Yung-Yun Wang

Department of Chemical Engineering
National Tsing Hua University
Hsinchu, Taiwan, Republic of China

CONTENTS

INTRODUCTION

In the recent development of electrochemical reactors, the particulate electrode reactor has attracted much attention. This electrode is one which contains beds of conducting particles. Electrolytes flow through the channels between these discrete particles. There are two types of particulate electrodes: packed-bed and fluidized bed. Packed-bed electrodes have been studied extensively both in theory and in experiments, because they have a large surface area per unit volume, are easier to prepare and to regenerate and have a smaller IR loss.

The basic theories governing the packed-bed electrode, the porous electrode and the fluidized-bed electrode are similar. A systematic and clear treatise on this subject can be found in the Pickett monograph [1]. Newman, in a series of papers [2–16], described the current distribution in one-dimensional porous electrodes from a macroscopic point of view. Alkire [17–22] analyzed the current distribution in flow-through porous electrodes theoretically. He also discussed the two-dimensional current distribution. Fleischmann [23–25] treated the potential distribution in fluidized-bed electrodes. Wroblowa [26–29] studied the behavior of flow-through porous electrodes. Sioda [30–41] investigated the relationship between limiting current and flow rate. Volkman [42–43] analyzed the performance of the packed-bed electrode. Coeuret [44–50] discussed the utilization of flow-through porous electrodes. Kuhn [51] reviewed the design of electrochemical reactors. Wang [52–54] investigated experimentally the potential and the current distribution as well as the performance of packed-bed electrodes.

The author thanks Dr. Chi-chao Wan for helpful discussions throughout the course of this work.

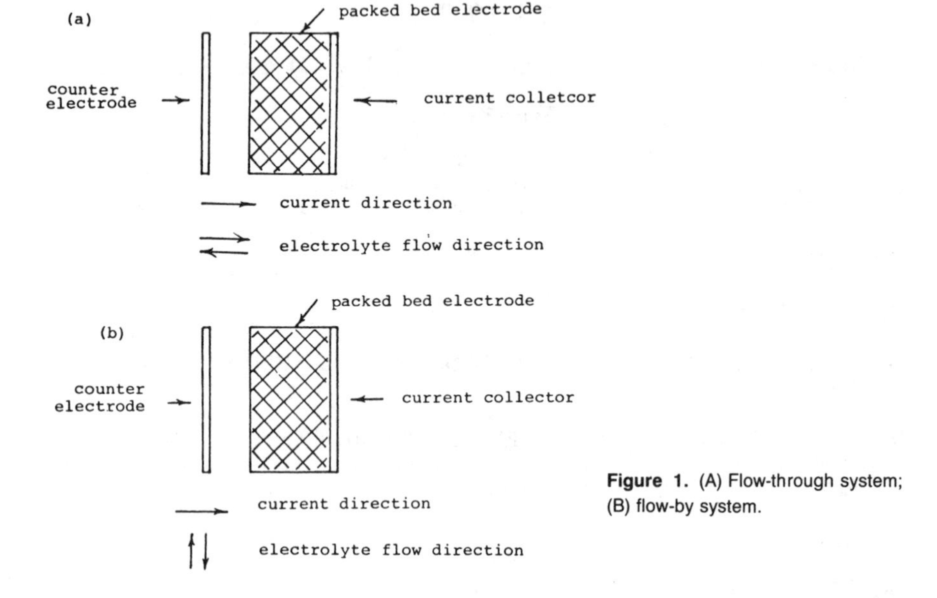

Figure 1. (A) Flow-through system; (B) flow-by system.

Based on the flow directions of current and electrolytes, we can classify packed-bed electrodes into either a flow-through system or a flow-by system (see Figure 1). In the flow-through system, the flow direction of the current is parallel to that of the electrolyte. On the other hand, the flow direction of the current is perpendicular to that of the electrolyte in the flow-by system. Most treatises on packed-bed electrodes emphasize the flow-through system. Newman [15], Alkire [19, 22] and Storck [55–56] etc. have, however, also studied the flow-by system.

There are two major areas of research in packed-bed electrodes: the analysis of the potential and the current distribution, and the studies of mass transfer phenomena.

POTENTIAL AND CURRENT DISTRIBUTION IN PACKED-BED ELECTRODES

The potential and current distribution in packed-bed electrodes can be analyzed under four following conditions:

1. The electrode operates under a lower overpotential and the solid matrix has a finite conductivity.
2. The electrode operates under a lower overpotential and the solid matrix has a very high conductivity.
3. The electrode operates under limiting current and the solid matrix has a very high conductivity.
4. The electrode operates under limiting current and the solid matrix has a finite conductivity.

Case 1

This concerns the condition where the electrode operates under a lower overpotential and the solid matrix has a finite conductivity.

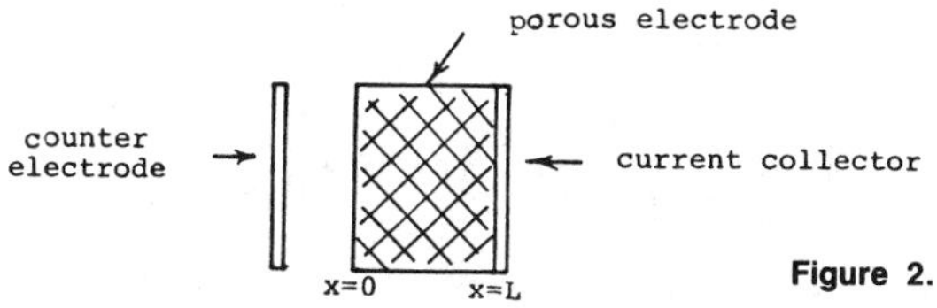

Figure 2. Schematic view of the porous electrode.

From Newman [2], the electrochemical reaction on the porous electrode (Figure 2) is given by

$$\sum_i S_i M_i^{Z_i} \rightarrow ne \tag{1}$$

where S_i = the stoichiometric coefficient of species i
Z_i = the valence of species i

Define φ_m as the potential of the solid matrix, φ_s as the potential of the pore-filling electrolyte, i_m as the current density of the solid matrix phase, and i_s as the current density of the electrolyte.

We now apply the following assumptions:

- The solid matrix obeys Ohm's law, i.e.

$$i_m = -\sigma \nabla \varphi_m \tag{2}$$

where σ = the conductivity of the solid matrix

- N_i, the flux of a mobile ionic species i, is due to migration, diffusion, and convection, hence

$$N_i = -Z_i U_i e C_i \nabla \varphi_s - D_i \nabla C_i + vC_i \tag{3}$$

where U_i = the mobility of species i
C_i = the concentration of species i
D_i = the diffusion coefficient of species i
v = the fluid velocity

- Current density in the pore electrolyte is given by

$$i_s = F \sum_i Z_i N_i \tag{4}$$

- Conservation law for species i implies

$$\frac{\partial C_i}{\partial t} = -\nabla N_i + \frac{S_i}{nF} \nabla i_m \tag{5}$$

- Principle of electroneutrality

$$\sum_i Z_i C_i = 0 \tag{6}$$

- The polarization equation is given by

$$\nabla i_m = af(\varphi_m - \varphi_s, C_i) \tag{7}$$

for an oxidation-reduction equation is given by

$$\nabla i_m = a i_0 \left\{ \frac{C_1}{C_1^0} \exp\left[-\frac{\alpha nF}{RT} (\varphi_m - \varphi_s) \right] - \frac{C_2}{C_2^0} \exp\left[\frac{(1-\alpha)nF}{RT} (\varphi_m - \varphi_s) \right] \right\} \quad (8)$$

where a = the specific interfacial area per unit volume
α = the transfer coefficient in Tafel polarization equation
C^0 = the reference concentration

- Conservation of charge

$$\nabla i_m + \nabla i_s = 0 \quad (9)$$

Furthermore, if the concentration is uniform, then Equation 4 can be written as

$$i_s = -\kappa \nabla \varphi_s \quad (10)$$

where $\kappa = Fe \sum_i Z_i^2 U_i C_i$ is the conductivity of the pore electrolyte.

The boundary conditions are

$$x = 0, \quad i_s = I, \quad i_m = 0, \quad \varphi_s = 0, \quad x = L, \quad i_s = 0$$

where I = the overall current density in one-dimensional porous electrode
L = the thickness of the electrode

From these equations, the current distribution of porous electrodes can be solved. In particular the Tafel polarization range and the linear polarization range.

Tafel Polarization Range

$$\frac{dj}{dX} = \frac{2\theta^2}{\delta} \sec^2 (\theta X - \psi) \quad (11)$$

where $j = \frac{i_m}{I}$

$X = \frac{x}{L}$

$\delta = L\,|I|\,\beta \left(\frac{1}{\kappa} + \frac{1}{\sigma} \right)$

$\beta = (1-\alpha) \frac{nF}{RT}$

$\psi = \frac{\lambda}{2\theta}$

$\lambda = \frac{L\,|I|\,\beta}{\kappa}$

$$\tan\theta = \frac{2\delta\theta}{4\theta^2 - \lambda(\delta - \lambda)}$$

Thus, if δ and λ are very small, i.e., the conductivity of solid matrix and the electrolyte are high, the current distribution of porous electrode is quite uniform. If λ is small but δ is large, i.e., the conductivity of the electrolyte (solid matrix) is high (low, respectively), then the current distribution is not uniform and the current distribution moves towards the current collector. If κ and σ are comparable, then the current tends to be distributed to the ends of the electrode. When the current density increases, or the thickness of the electrode increases, or the values of λ and δ become larger, the nonuniformity of the current distribution becomes more apparent.

Linear Polarization Range

The polarization equation can be simplified to

$$\frac{di_s}{dx} = ai_0 \frac{nF}{RT}(\varphi_m - \varphi_s) \tag{12}$$

Solving for the current distribution of porous electrode, we obtain

$$\frac{dj}{dX} = \frac{\nu\kappa}{(\kappa+\sigma)\sinh\nu}\left[\frac{\sigma}{\kappa}\cosh\nu(1-X) + \cosh\nu X\right] \tag{13}$$

where $\nu = L\sqrt{\frac{ai_0 nF}{RT}\frac{\kappa+\sigma}{\kappa\sigma}}$

When $|I| < ai_0L$, the result of linear analysis is more accurate then that of Tafel analysis.

Case 2

This concerns the case where the electrode operates under lower overpotential and the solid matrix has a very high conductivity.

Considering that a highly conductive fixed bed with spherical grains operates under very low overpotential, Coeuret [44, 48] has shown that (1) the electrical conductivity through the pores of the fixed bed obeys Ohm's low; (2) potential drops occur only in electrolytes when the dispersed phase is equipotential; and (3) the polarization equation is reduced to a linear expression under very low overpotential, and the fixed bed (Figure 3) can be characterized by

$$i = i_0 \frac{nF}{RT}\eta \tag{14}$$

$$\eta(X) = E(X) - E_{eq} \tag{15}$$

where $\eta(X)$ = the overpotential

With the boundary conditions

$$i_s(0) = 0$$

$$i_s(1) = I$$

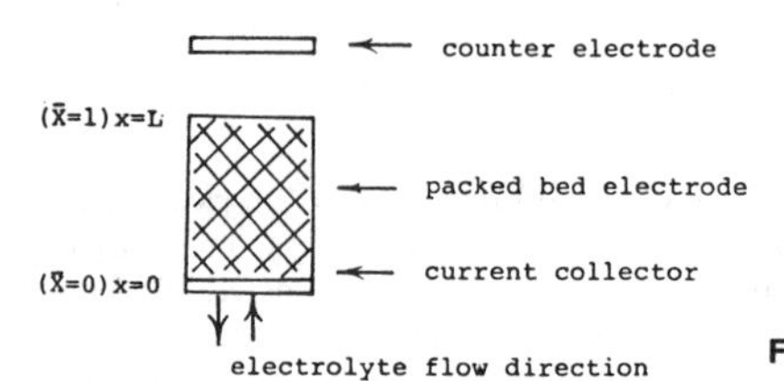

Figure 3. Schematic view of the packed-bed electrode.

we obtain

$$\frac{d\eta(X)}{dX} = \frac{dE(X)}{dX} = -\frac{d\varphi_s(X)}{dX} = \frac{L}{\kappa} i_s(X) \tag{16}$$

Furthermore

$$\frac{di_s(X)}{dX} = i_0 \frac{nF}{RT}(1-\varepsilon)aL\eta(X) \tag{17}$$

where ε = the bed porosity

Combining Equations 16 and 17, the following differential equation is obtained

$$\frac{d^2\eta(X)}{dX^2} = -K^2\eta(X) = 0 \tag{18}$$

where $K^2 = i_0 \frac{nF}{RT}(1-\varepsilon)a\frac{L^2}{\kappa}$

Solving Equation 18, we have

$$\eta(0) = \frac{IL}{\kappa K \sin h\,(K)} \tag{19}$$

$$\frac{\eta(X)}{\eta(0)} = \cos h\,(KX) \tag{20}$$

Thus the potential distribution of the packed-bed electrode is given by

$$\frac{\eta(X)-\eta(0)}{\eta(1)-\eta(0)} = \frac{\cos h\,(KX)-1}{\cos h\,(K)-1} \tag{21}$$

Rewritting the expression of K^2, we have

$$K^2 = \frac{i_0 \frac{nF}{RT}(1-\varepsilon)aL\eta(1)}{\frac{\kappa}{L}\eta(1)}$$

$$K^2 = \frac{\left[\frac{di_s(X)}{dX}\right]_{X=1}}{\frac{\kappa}{L}\eta(1)} \tag{22}$$

In other words, K^2 is simply the ratio of the electrochemical reaction rate at $X = 1$ to the flux of electrical conductance in the electrolyte. Thus the overpotential distribution of the packed-bed electrode is only a function of K: When $K \rightarrow 0$, the overpotential approaches a constant. The electrode is fully utilized in all three dimensions when $K \rightarrow \infty$. The electrode acts like a two-dimensional electrode since the reaction is concentrated on the top of the electrode. When the value of K is intermediate, the depth of reaction decreases as the value of K increases. Therefore, to increase the utilization and the reaction region of the electrode, we should use a thinner electrode, increase the diameter of packing particles, choose an electrolyte of better conductivity or reactions with smaller exchange current density.

Coeuret defined the effectiveness of an electrode as a measure of the utilization of an electrode by

$$\xi = \frac{\text{measured electrolytic current}}{\text{current obtained with an electrode whose overpotential is the same at every point}} = \frac{I}{i_0 \dfrac{nF}{RT}(1-\varepsilon)aL\eta(1)} = \frac{\tan h\,(K)}{K} \tag{23}$$

If the electrode is not operated under very low overpotential, the polarization equation will have to be expressed by a power function instead of

$$i = i_0 \frac{nF}{RT} r^{1-n'} \eta^{n'} \tag{24}$$

where n' = a positive exponent less than one
r = the dimensional coefficient

Then the effectiveness of an electrode becomes

$$\xi_{n'} = \frac{\text{observed electrolytic current}}{\text{current which would be obtained if } \eta(X) = \eta(1) \text{ at every height X}} = \frac{\tan h(K_{n'})}{K_{n'}} \tag{25}$$

where $K_{n'} = \left\{ \dfrac{n'+1}{2} i_0 \dfrac{nF}{RT}(1-\varepsilon)a \dfrac{L^2}{\kappa} \left[\dfrac{r}{\eta(1)} \right]^{1-n'} \right\}^{1/2}$

When $n' = 1$, Equation 25 reduces to Equation 23 and the potential distribution of the electrode becomes

$$\frac{\eta(X)-\eta(0)}{\eta(1)-\eta(0)} = \frac{\cos h(K_{n'}X)-1}{\cos h(K_{n'})-1} \tag{26}$$

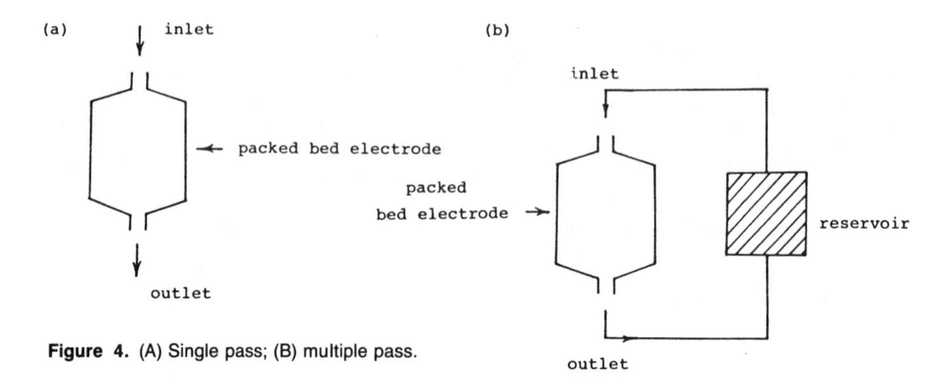

Figure 4. (A) Single pass; (B) multiple pass.

Case 3

This concerns the condition where the electrode operates under limiting current and the solid matrix has a very high conductivity.

Coeuret [49] has shown that when dispersed phase is equipotential, current flowing through the pores of the electrode obeys Ohm's law. Under the condition that the electrode operate at limiting current, Coeuret solves for the potential distribution of the electrode (Figure 3)

$$E(X) = E(0) - \frac{LI}{\kappa}\frac{1-R}{R\ln(1-R)}[(1-R)^{-X} + X\ln(1-R) - 1] \tag{27}$$

where $R = \dfrac{|I|}{nFvC_i}$ = the overall efficiency of the transformation of the reacting ions in the electrode

C_i = the feed concentration

Similar to the ξ defined in Equation 23, R is a measure of the utilization of the electrode. Thus the overall potential drop through the fixed bed under limiting current

$$E(1) - E(0) = -\frac{LI}{\kappa}\left[\frac{1}{R} + \frac{1}{\ln(1-R)} - 1\right] \tag{28}$$

and the potential distribution is given by

$$Y = \frac{E(X) - E(0)}{E(1) - E(0)} = \frac{(1-R)^{-X} + X\ln(1-R)^{-1}}{(1-R)^{-1} + \ln(1-R) - 1} \tag{29}$$

Note that the potential distribution is only a function of R. When $R \to 0$, $Y = X^2$ and the potential distribution varies as the height of the bed changes. When $R \to 1$, $Y \to 0$ except at $X = 1$ and $E(X) = E(0)$, the electrode becomes an ideal three-dimensional electrode which is fully utilized.

Newman [10, 57] has also discussed the operations of electrodes under limiting current theoretically. Under the condition of limiting current, the concentration of reactant on the surface of the electrode approaches zero, migration can be ignored if there exists supporting electrolyte. Let k_m be the local mass transfer coefficient, then

$$v\frac{dC}{dt} = -k_m aC \tag{30}$$

Integrating Equation 30 over the length of the bed, we obtain

$$\frac{C_i}{C_L} = \exp\left(\frac{ak_mL}{v}\right) \tag{31}$$

where C_L = the exit concentration

Using the following three relationships,

$$v\frac{dC}{dx} = \frac{1}{F}\frac{di_s}{dx} \tag{32}$$

Limiting cell current density

$$I_{lim} = vF(C_i - C_L) \tag{33}$$

Ohm's law

$$i_s = -\kappa\frac{d\varphi_s}{dx} \tag{34}$$

The electric potential difference in the solution between the bed inlet and the outlet $\Delta\varphi_s$ can be calculated

$$\Delta\varphi_s = \frac{vFC_i}{\gamma\kappa}[-1 + (1+\gamma L)\exp(-\gamma L)] \tag{35}$$

where $\gamma = \dfrac{ak_m}{v}$

Equation 35 can be used as an electrode design criterion.

Case 4

This concerns the condition where the electrode operates under limiting current and the solid matrix has a finite conductivity.

Graphite granules or carbon grains are often used as packing particles in many packed-bed electrode applications. Constant resistance exists in these systems and the conductivity of the solid matrix is not very high. Similiar effects are resulted in fluidized bed electrodes when they operate under very low expansion [23–25].

Applying Ohm's law to the electrolyte and the solid matrix, Coeuret [50] obtained (Figure 3)

$$E(X) = E(0) + \frac{LI}{\kappa}\left\{\frac{\kappa}{\sigma}X + \left(1 + \frac{\kappa}{\sigma}\right)\frac{1-R}{R\ln(1-R)}[(1-R)^{-X} + X\ln(1-R) - 1]\right\} \tag{36}$$

where $R = \dfrac{|I|}{nFvC_b}$

Thus

$$W(X) = \frac{\kappa}{LI}[E(X) - E(0)] \tag{37}$$

Equation 37 shows that when an electrode operates under limiting current, its potential distribution is only a function of R and κ/σ.

Typically, packed-bed electrodes operate in two methods.

1. *Single-pass*. The electrolyte passes the packed-bed electrode only once (Figure 4A). Thus the electrolyte concentration at the inlet is a constant. Based on Fleischmann [58, 59], Equation 38 can be obtained

$$\frac{C_i}{C_L} = \exp\left(\frac{ak_mL}{v}\right) \tag{38}$$

which is the same as Equation 31.

2. *Multipass*. The electrolyte cycles through a reservoir and the packed-bed electrode (Figure 4B). Thus the electrolyte concentration at the inlet decreases gradually. From the same papers [58, 59],

$$C_i = C_i^0 \exp\left\{-\frac{avt}{V}\left[1-\exp\left(\frac{ak_mL}{v}\right)\right]\right\} \tag{39}$$

where C_i^0 = the initial inlet concentration
V = the volume of the reservoir

Most studies of the packed bed electrodes have so far emphasized the theoretical analysis of the electrodes which were assumed to operate in the state of either activation polarization or limiting current alone. It seems that in order to make the whole electrode reach the range of limiting current, a very thin electrode, a very low concentration of the electrolyte, or a very low flow rate is needed. The experiments of Newman [57], Alkire [20] and Coeuret [49], etc. were carried out under such conditions. However, these systems have probably very limited potentials in the engineering applications.

From a practical point of view, the packed-bed electrode should be applicable when the flow rate and the concentration of the electrolyte are at significant levels and when the electrode is not too thin. Nevertheless, both mass transfer and activation control may exist in such a system. Even in the case of activation control, a part of the electrode may have reached Tafel polarization due to the wide current distribution. Wang [52] showed that such electrodes cannot completely reach mass transfer control. Moreover the current distribution is not uniform and an increase of the flow rate will not only reduce the mass transfer control range but will also make the current distribution less uniform. In contrast, the current distribution actually becomes more uniform with increasing current. This is due to the fact that the front portion of the electrode approaches the limiting current and the high concentration overpotential exerts a leveling effect on the current distribution.

MASS TRANSFER PHENOMENA IN PACKED-BED ELECTRODES

Packed-bed reactors are highly useful in many chemical engineering applications. Chemical reactions and separations processes such as absorption, filtration, ion exchange, chromatography, catalytic reaction, distillation, etc. can benefit from using packed-bed reactors. Consequently, the mass transfer phenomena in packed-bed reactors has been a major research topic. A better understanding of the mass transfer phenomena and the flow of fluid in a packed bed can help us properly design and operate a packed bed more effectively.

A large number of studies of the mass transfer phenomena in packed-bed reactors can be found in the literature. Many workers [60–71] studied the mass transfer phenomena by looking into the sublimation of the solid acid packing particles and the vaporization of liquid from the surface of the porous particles. The mass transfer was also studied [72–80] by analyzing the shape change of the solid acid packing particles when they dissolved in liquid or the acid concentration change of the liquid. Some [80] showed the effect of the different packing methods on the mass transfer rate in packed-bed reactors. Some [68, 83] studied the effect of the porosity of the packed bed on the mass transfer rate. Gupta [83] pointed out that the value of the j-factor for mass transfer (j_d) is proportional to $(\text{porosity})^{-1}$ when the Reynolds number (Re) is high. The heat transfer coefficient and the mass transfer coefficient were compared and verified (62–70).

Several investigators [84–87] discuss the relationship between the mass transfer coefficient, the friction coefficient and the pressure drop when the fluid flows through the packed bed. Kramers [80] proposed that to measure more accurately the mass transfer rate between the fluid and the packing particles, one should place a single active particle in a packed bed made up of inert particles. Rhodes [79] studied the local mass transfer rate at the surface of spherical packing particles by placing a single active particle in an ordered packed bed.

Some [84, 88–95] established theoretical models to discuss the mass transfer on the surface of packing particles. Pfeffer [91] derived, by using the free surface model of Happel [93] and the thin boundary layer solution of Levich [96], the relationship of j_d and Re, when Re is low and the Peclet number (Pe) is high. Sirkar [92] solved the Navier–Stokes equation by using the creeping flow solution of Tam [97]. The result is applicable when porosity is large. Many researchers [77, 93, 95, 98–101] have also studied the effect of axial dispersion, natural convection, etc. of a packed bed on the mass transfer rate under very low Re and high Pe. Mass transfer under low Pe was investigated in References 102–104.

When we consider the efficiency of packed-bed reactors, we should also consider the uniformity of the flow of fluid through the packed bed. In many studies of mass transfer phenomena, researchers assumed that the fluid in the packed bed behaves like plug flow. In reality, the existence of the wall of the packed bed cannot be ignored. Experimental results [105–113] showed that the flow rate near the wall is larger than that found in the center of the bed. This phenomena is particulary acute with large packing particles in a packed bed of smaller diameter [114]. Theoretical investigations of the flow in a packed bed can be found in [115–118]. Weiman [119] suggested that the ratio of the diameter of the packed bed over that of the packing particles should be made greater than 25. Baker [120] pointed out that this ratio affects the flow of fluid, especially when it is less than 8. Porter [117] and Onda [113] have suggested that the effect of the wall on the flow increases as the flow rate decreases.

Hanratty [121–126] developed an electrochemical method to study the flow of fluid in pipe or a packed bed. There are usually two approaches using either a single active particle in a packed bed of inert particles or a packed bed of active particles.

Since

$$\frac{i_{lim}}{nF} \cong k_m C_b \tag{40}$$

where C_b = the bulk concentration

The mass transfer coefficient k_m can be calculated from Equation 40. Thus the relationship between the mass transfer coefficient k_m and Re was obtained.

A Single Active Particle in a Packed Bed of Inert Particles

Hanratty [125] studied the local and the overall mass transfer on the surface of a single spherical active particle which was in a packed bed of inert glass particles. The relationship between Re and the overall mass transfer rate obtained by Hanratty was as follows:

$$\text{Re} > 140 \quad (\text{Sh})(\text{Sc})^{-1/3} = 1.59\text{Re}^{0.56} \tag{41}$$

$$140 > Re > 35 \quad (Sh)(Sc)^{-1/3} = 1.44Re^{0.58} \tag{42}$$

where Sh = the Sherwood number
Sc = the Schmidt number

Note that the mass transfer rate in a packed bed of active particles will be lower than that in a packed bed with a single spherical active particle. The reason is that the effective concentration driving force is smaller when all spheres are active.

Karabelas [126] used a similar method and derived an asymptotic relationship of mass transfer in a packed bed with a single active spherical particle.

When Sc $\cong$ 1,600, they obtained:

- For pure laminar natural convection

$$Sh = 0.46\,(G_rS_c)^{1/4} \tag{43}$$

where Gr = the Grashof number

- For forced convection at low Re (Re < 10) and large Pe

$$Sh = 4.58(Pe)^{1/3} \tag{44}$$

- For forced convection at Re(Re = 90–120) close to the transition to unsteady flow

$$(Sh)(Sc)^{-1/3} = 2.39Re^{0.56} \tag{45}$$

Mandelbaum [98] investigated mass transfer in a packed bed of ceramic Raschig rings, which consisted of a single active ring under low Re, with both free and forced convection. The results were

For aiding flow

$$2 \times 10^{-4} < \frac{Re}{(Gr^{1/2})} < 2.5 \times 10^{-2} \qquad \frac{Sh}{(ScGr)^{1/4}} = 0.645 \tag{46}$$

$$\frac{Re}{(Gr^{1/2})} > 2.5 \times 10^{-2} \qquad \frac{Sh}{(ScGr)^{1/4}} = 1.153\left(\frac{Re}{Gr^{1/2}}\right)^{0.155} \tag{47}$$

For opposing flow

$$2 \times 10^{-4} < \frac{Re}{(Gr^{1/2})} < 5.5 \times 10^{-2} \qquad \frac{Sh}{(ScGr)^{1/4}} = 0.618 \tag{48}$$

$$\frac{Re}{(Gr^{1/2})} > 5.5 \times 10^{-2} \qquad \frac{Sh}{(ScGr)^{1/4}} = 1.334\left(\frac{Re}{Gr^{1/2}}\right)^{0.253} \tag{49}$$

Wang [127] studied the mass transfer in packed beds with a single copper particle by the limiting current method. The experimental results were

For spherical particles

$$10 < Re < 800 \quad \varepsilon j_d = 0.72Re^{-0.51} \tag{50}$$

$$0.3 < \mathrm{Re} < 10 \qquad \varepsilon j_d = 1.2\mathrm{Re}^{-0.67} \tag{51}$$

For cylindrical particles

$$10 < \mathrm{Re} < 250 \qquad \varepsilon j_d = 0.59\mathrm{Re}^{-0.44} \tag{52}$$

$$0.9 < \mathrm{Re} < 10 \qquad \varepsilon j_d = 0.95\mathrm{Re}^{-0.62} \tag{53}$$

where $\mathrm{Sc} \cong 2{,}200$

Where the porosity of the packed bed lies between 0.37 and 0.43.

The mass transfer relationship can also be characterized by

For spherical particle

$$800 < \mathrm{Pe} < 25{,}000 \qquad \mathrm{Sh} = 4.8\mathrm{Pe}^{0.28} \tag{54}$$

$$25{,}000 < \mathrm{Pe} < 2{,}000{,}000 \qquad \mathrm{Sh} = 0.5\mathrm{Pe}^{0.51} \tag{55}$$

For cylindrical particle

$$2{,}000 < \mathrm{Pe} < 25{,}000 \qquad \mathrm{Sh} = 2.1\mathrm{Pe}^{0.36} \tag{56}$$

$$25{,}000 < \mathrm{Pe} < 800{,}000 \qquad \mathrm{Sh} = 0.3\mathrm{Pe}^{0.55} \tag{57}$$

The experiments of Wang showed that a higher mass transfer coefficient can be obtained by using smaller packing particles. Furthermore, the mass transfer coefficient of spherical packing particles is higher than that of cylindrical packing particles when the flow rate is lower and the particles are of the same diameter. The difference diminishes, however, as the flow rate increases.

A Packed Bed of Active Particles

Smith [128] measured the mass transfer coefficient in a packed bed of spherical nickel particles. The following relationship was obtained

$$\mathrm{Re} > 40 \qquad j_d = 0.906\mathrm{Re}^{-0.38} \tag{58}$$

He also discovered that mass transfer rate of a packed bed is higher than that of a fluidized bed under a given Re.

Krishna [129, 130] showed the mass transfer correlation as follows

$$\begin{matrix} 1 < \mathrm{Re} < 2{,}800 \\ 256 < \mathrm{Sc} < 1{,}150 \end{matrix} \qquad j_d = 0.822\mathrm{Re}^{-0.38} \tag{59}$$

Pickett [131] suggested that the major drawback of a packed-bed electrode is the lack of selectivity of reaction because of large electrode potential variation. Consequently, he designed a packed-bed electrode with a single layer of uniform particles and observed its mass transfer data.

$$23 < \mathrm{Re} < 520 \qquad (\mathrm{Sh})(\mathrm{Sc})^{-1/3} = 0.83\mathrm{Re}^{0.56} \tag{60}$$

In many processes (e.g., ion exchange, catalytic, absorption, chromatography) the mass transfer occurs under very low flow rate ($Re << 1$) in the packed bed. Using the limiting current technique, Newman [57] found the mass transfer coefficient under very low flow rate ($Re < 0.1$) in a packed-bed electrode with spherical stainless steel packing particles. The measured data agrees with the results of Wilson [132].

$$\varepsilon j_d = 1.09 Re^{-0.66} \tag{61}$$

when $Pe < 100$, the results are smaller than those of Wilson.

Using a packed bed of spherical conducting grains, Coeuret [45] obtained the mass transfer correlation as follows:

$$0.04 < Re < 30 \quad 1{,}700 < Sc < 11{,}000 \qquad (Sh)(Sc)^{-1/4} = 5.4 Re^{0.33} \tag{62}$$

while Matic obtained [133]

$$(Sh)(Sc)^{-1/3} = 0.5 Re^{0.7} \tag{63}$$

Storck [134] established the following relationship for the mass transfer in a stacked-net electrode

$$4 < Re < 150 \qquad j_d = 6.40 Re^{-0.507} \tag{64}$$

Vögtländer [135] found the mass transfer rate between nets and the liquid is

$$j_d = 0.773 Re^{-0.585} \tag{65}$$

while Sioda [32, 35, 37, 40] obtained

$$j_d = 0.904 Re^{-0.64} \tag{66}$$

Olive [136] analyzed the mass transfer during the recovery of copper in a packed bed with spherical copper particles. The mass transfer correlation is given by

$$0.1 < Re < 3 \qquad (Sh)(Sc)^{-1/4} = 4.3 Re^{0.35} \tag{67}$$

Kinoshita [137] studied the mass transfer correlation of a carbon felt flow-through electrode during the reduction of bromium. The results were

For 0.25 cm thick electrode

$$Sh = 1.29 Re^{0.72} \tag{68}$$

For 0.175 cm thick electrode

$$Sh = 1.01 Re^{0.61} \tag{69}$$

Using the electrochemical technique, Wang [127] showed that the effect of the wall on the mass transfer coefficient of a packed bed increases when the ratio of the diameter of the bed over that of the packing particles decreases. The mass transfer coefficient near the wall is higher than that of the center.

CONCLUSIONS

Packed-bed electrodes are established in applications [138] such as desalting, electroorganic synthesis, and electrowinning. Recent development has focused on the recovery or removal of metallic species from dilute solution, including hydrometallurgical liquor as well as industrial waste water. Some examples of the metallic species are copper, mercury, zinc, cadmium, iron, lead, chromium, molybdenum, silver, gold, platinum, and palladium. Applying packed-bed electrodes to a redox battery and electroanalytical detector is also quite promising.

Many corporations in the United States (e.g. PPG Industries, Inc., Du Pont, Kennecott Corp.) have built pilot plants of packed-bed electrodes. The future of packed-bed electrodes depends on whether one can design more efficient electrodes and produce electricity economically to compete with the nonelectrochemical reactors.

NOTATION

a	specific interfacial area
C	concentration
D_i	diffusion coefficient
E	potential
F	Faraday's constant
Gr	Grashof number
i	current density
I	overall current density
j	current density ratio
j_d	mass transfer factor
$k_{n'}$	effectiveness parameter in Equation 25
k^2	ratio of electrochemical reaction rate to flux of electrical conductance in electrolyte
L	thickness
n	parameter, exponent
n'	exponent
N_i	flux of mobile ionic species
Pe	Peclet number
r	dimensionless coefficient
R	parameter in Equation 36
Re	Reynolds number
S_i	stoichiometric coefficient
Sc	Schmidt number
Sh	Sherwood number
t	time
u_i	mobility
v	velocity
x	distance
X	dimensionless thickness
Y	potential distribution (see Equation 29)
Z_i	valence number

Greek Symbols

α	transfer coefficient
β	parameter in Equation 1
γ	parameter in Equation 35
δ	parameter in Equation 11
ε	voidage
$\eta(x)$	overpotential function
θ	dimensionless group or angle
κ	conductivity
λ	parameter in Equation 11
ν	parameter in Equation 13
ξ'_n, ξ	electrode effectiveness
σ	conductivity
φ_m	potential of solid matrix
φ_s	potential of pore-filling electrolyte
ψ	parameter in Equation 11

REFERENCES

1. Pickett, D. J., *Electrochemical Reactor Design*, Elsevier Scientific Publishing Company, Amsterdam, 1979.
2. Newman, J., and Tobias, C. W., *J. Electrochem. Soc.*, 109, 1183 (1962).
3. Parrish, W. R., and Newman, J., *J. Electrochem. Soc.*, 117, 43 (1970).
4. Dunning, J. S., Bennion, D. N., and Newman, J., *J. Electrochem. Soc.*, 118, 1251 (1971).
5. Dunning, J. S., Bennion, D. N., and Newman, J., *J. Electrochem. Soc.*, 120, 906 (1973).
6. Johnson, A. M., and Newman, J., *J. Electrochem. Soc.*, 118, 510 (1971).

7. Bennion, D. N., and Newman, J., *J. Appl. Electrochem.*, 2, 113 (1972).
8. Tiedemann, W., and Newman, J., *J. Electrochem. Soc.*, 122, 70 (1975).
9. Tiedemann, W., and Newman, J., *J. Electrochem. Soc.*, 122, 1482 (1975).
10. Newman, J., and Tiedemann, W., *AIChE J.*, 21, 25 (1975).
11. Zee, J. V., and Newman, J., *J. Electrochem. Soc.*, 124, 706 (1977).
12. Trainham, J. A., and Newman, J., *J. Electrochem. Soc.*, 124, 1528 (1977).
13. Trainham, J. A., and Newman, J., *J. Appl. Electrochem.*, 7, 287 (1977).
14. Trainham, J. A., and Newman, J., *J. Electrochem. Soc.*, 125, 58 (1978).
15. Trainham, J. A., and Newman, J., *Electrochimica Acta*, 26, 455 (1981).
16. Trainham, J. A., and Newman, J., *J. Electrochem. Soc.*, 129, 991 (1982).
17. Alkire, R., *J. Electrochem. Soc.*, 120, 900 (1973).
18. Alkire, R., and Plichta R., *J. Electrochem. Soc.*, 120, 1060 (1973).
19. Alkire, R., and Ng, P. K., *J. Electrochem. Soc.*, 121, 95 (1974).
20. Alkire, R., and Gracon, B., *J. Electrochem. Soc.*, 122, 1594 (1975).
21. Alkire, R., and Gould, R., *J. Electrochem. Soc.*, 123, 1842 (1976).
22. Alkire, R., and Ng, P. K., *J. Electrochem. Soc.*, 124, 1220 (1977).
23. Fleischmann, M., and Oldfield, J. W., *J. Electroanal. Chem.*, 29, 211 (1971).
24. Fleischmann, M., and Oldfield, J. W., *J. Electroanal. Chem.*, 29, 231 (1971).
25. Fleischmann, M., Oldfield, J. W., and Porter, D. F., *J. Electroanal. Chem.*, 29, 241 (1971).
26. Wroblowa, H. S., *J. Electroanal. Chem.*, 42, 321 (1973).
27. Wroblowa, H. S., and Saunders, A., *J. Electroanal. Chem.*, 42, 329 (1973).
28. Wroblowa, H. S., and Razumney, G., *J. Electroanal. Chem.*, 49, 355 (1974).
29. Razumney, G., Wroblowa, H. S., and Schrenk, G. L., *J. Electroanal. Chem.*, 69, 299 (1976).
30. Sioda, R. E., *Electrochimica Acta*, 13, 375 (1968).
31. Sioda, R. E., *Electrochimica Acta*, 13, 1559 (1968).
32. Sioda, R. E., *Electrochimica Acta*, 15, 783 (1970).
33. Sioda, R. E., *Electrochimica Acta*, 16, 1569 (1971).
34. Sioda, R. E., *J. Electroanal. Chem.*, 34, 399 (1972).
35. Sioda, R. E., *J. Electroanal. Chem.*, 34, 411 (1972).
36. Sioda, R. E., and Kemula, W., *Electrochimica Acta*, 17, 1171 (1972).
37. Sioda, R. E., *Electrochimica Acta*, 17, 1939 (1972).
38. Sioda, R. E., *Electrochimica Acta*, 19, 57 (1974).
39. Sioda, R. E., *J. Appl. Electrochem.*, 5, 221 (1975).
40. Sioda, R. E., *J. Appl. Electrochem.*, 7, 135 (1977).
41. Sioda, R. E., and Piotrowska, H., *Electrochimica Acta*, 25, 331 (1980).
42. Volkman, Y., *J. Appl. Electrochem.*, 8, 347 (1978).
43. Volkman, Y., *Electrochimica Acta*, 24, 1145 (1979).
44. Coeuret, F., Hutin, D., and Gaunand, A., *J. Appl. Electrochem.*, 6, 477 (1976).
45. Coeuret, F., *Electrochimica Acta*, 21, 185 (1976).
46. Coeuret, F., and Goff, P. L., *Electrochimica Acta*, 21, 195 (1976).
47. Coeuret, F., *Electrochimica Acta*, 21, 203 (1976).
48. Paulin, M., Hutin, D., and Coeuret, F., *J. Electrochem. Soc.*, 124, 180 (1977).
49. Gaunand, A., Hutin, D., and Coeuret, F., *Electrochimica Acta*, 22, 93 (1977).
50. Gaunand, A., and Coeuret, F., *Electrochimica Acta*, 23, 1197 (1978).
51. Houghton, R. W., and Kuhn, A. T., *J. Appl. Electrochem.*, 4, 173 (1974).
52. Wang, Y. Y., et al., *J. Electrochem. Soc.*, 129, 347 (1982).
53. Wang, Y. Y., et al., *Hydrometallurgy*, 8, 231 (1982).
54. Wang, Y. Y., et al., *J. Chin. I. Ch. E.*, 13, 133 (1982).
55. Storck, A., Enriquez-Granados, M. A., and Roger, M., *Electrochimica Acta*, 27, 293 (1982).
56. Enriquez-Granados, M. A., Hutin, D., and Storck, A., *Electrochimica Acta*, 27, 303 (1982).

57. Appel, P. W., and Newman, J., *AIChE J.*, 22, 979 (1976).
58. Chu, A. K. P., Fleischmann, M., and Hills, G. J., *J. Appl. Electrochem.*, 4, 323 (1974).
59. Chu, A. K. P., and Hills, G. J., *J. Appl. Electrochem.*, 4, 331 (1974).
60. Gamson, B. W., Thodes, G., and Hougen, O. A., *Trans. AIChE J.*, 39, 1 (1943).
61. Wilke, C. R., and Hougen, O. A., *Trans. AIChE J.*, 41, 445 (1945).
62. Bradshaw, R. D., and Myers, J. E., *AIChE J.*, 9, 590 (1963).
63. Bar-Ilan, M., and Resnick, W., *Ind. Eng. Chem.*, 49, 313 (1957).
64. Bradshaw, R. D., and Bennett, C. O., *AIChE J.*, 7, 48 (1961).
65. Hurt, D. W., *Ind. Eng. Chem.*, 35, 522 (1943).
66. Resnick, W., and White, R. R., *Chem. Eng. Progr.*, 45, 377 (1949).
67. Deacetis, J., and Thodes, G., *Ind. Eng. Chem.*, 52, 1003 (1960).
68. Mcconnachie, J. T. L., and Thodes, G., *AIChE J.*, 9, 60 (1963).
69. Chu, J. C., Kalil, J., and Wetteroth, W. A., *Chem. Eng. Progr.*, 49, 141 (1953).
70. Evnochides, S., and Thodes, G., *AIChE J.*, 7, 78, (1961).
71. Hobson, M., and Thodes, G., *Chem. Eng. Progr.*, 47, 370 (1951).
72. Hobson, M., and Thodes, G., *Chem. Eng. Progr.*, 45, 517 (1949).
73. Williamson, J. E., Bazaire, K. E., and Geankoplis, C. J., *Ind. Eng. Chem. Fundam.*, 2, 126 (1963).
74. Evans, G. C., and Gerald, C. F., *Chem. Eng. Progr.*, 49, 135 (1953).
75. McCune, L. K., and Wilhelm, R. H., *Ind. Eng. Chem.*, 41, 1124 (1949).
76. Gaffncy, B. J., and Drew, T. B., *Ind. Eng. Chem.*, 42, 1126 (1950).
77. Wilson, E. J., and Geankoplis, C. J., *Ind. Eng. Chem. Fundam.*, 5, 9 (1966).
78. Upadhyay, S. N., Agrawal, B. K. D., and Singh, D. R., *J. Chem. Eng. Japan*, 8, 413 (1975).
79. Rhodes, J. M., and Peebles, F. N., *AIChE J.*, 11, 481 (1965).
80. Thoenes, D., and Kramers, H., *Chem. Eng. Sci.*, 8, 271 (1958).
81. Gamson, B. W., *Chem. Eng. Progr.*, 47, 19 (1951).
82. Baumeister, E. B., and Bennett, C. O., *AIChE J.* 4, 69 (1958).
83. Gupta, A. S., and Thodes, G., *Chem. Eng. Progr.*, 58, 58 (1962).
84. Carberry, J. J., *AIChE J.*, 6, 460 (1960).
85. Ergun, S., *Chem. Eng. Progr.*, 48, 227 (1952).
86. Ranz, W. E., *Chem. Eng. Progr.*, 48, 247 (1952).
87. El-Kaissy, M. M., and Homsy, G. M., *Ind. Eng. Chem. Fundam.*, 12, 82 (1973).
88. Payatakes, A. C., Tien, C., and Turian, R. M., *AIChE J.*, 19, 58 (1973).
89. Kusik, C. L., and Happel, J., *Ind. Eng. Chem., Fundam.*, 1, 163 (1962).
90. Neale, G. H., and Nader, W. K., *AIChE J.*, 19, 112 (1973).
91. Pfeffer, R., *Ind. Eng. Chem. Fundam.*, 3, 380 (1964).
92. Sirkar, K. K., *Chem. Eng. Sci.*, 29, 863 (1974).
93. Happel, J., *AIChE J.*, 4, 197 (1958).
94. Happel, J., and Pfeffer, R., *AIChE J.*, 10, 605 (1964).
95. Nelson, P. A., and Galloway, T. R., *Chem. Eng. Sci.*, 30, 1 (1975).
96. Levich, V. G., *Physicochemical Hydrodynamics*, Prentice Hall, New York, 1962.
97. Tam, C. K. W., *J. Fluid Mech.*, 38, 537 (1969).
98. Mandelbaum, J. A., and Böhm, U., *Chem. Eng. Sci.*, 28, 569 (1973)
99. Tardos, G. I., Gutfinger, C., and Abuaf, N., *AIChE J.*, 22, 1147 (1976).
100. Miyauchi, T., and Kikuchi, T., *Chem. Eng. Sci.*, 30, 343 (1975).
101. Wakao, N., and Funazkri, T., *Chem. Eng. Sci.*, 33, 1375 (1978).
102. Miyauchi, T., Matsumoto, K., and Yoshida, T., *J. Chem. Eng. Japan*, 8, 228 (1975).
103. Martin, H., *Chem. Eng. Sci.*, 33, 1043 (1978).
104. Fedkiw, P., and Newman, J., *Chem. Eng. Sci.*, 33, 1563 (1978).
105. Hoftyzer, P. J., *Trans. Instn. Chem. Engrs.*, 42, T109 (1964).
106. Scott, A. H., *Trans Instn. Chem. Engrs.*, 13, T211 (1935).
107. Herskowitz, M., and Smith, J. M., *AIChE J.*, 24, 439 (1978).
108. Kubo, K., et al., *J. Chem. Eng. Japan*, 11, 405 (1978).

109. Kim, B. M., and Harris, T. R., *Chem. Eng. Sci.*, 28, 1653 (1973).
110. Dutkai, E., and Ruckenstein, E., *Chem. Eng. Sci.*, 25, 483 (1970).
111. Dutkai, E., and Ruckenstein, E., *Chem. Eng. Sci.*, 23, 1365 (1968).
112. Farid, M. M., and Gunn, D. J., *Chem. Eng. Sci.*, 33, 1221 (1978).
113. Onda, K., et al., *Chem. Eng. Sci.*, 28, 1677 (1973).
114. Norman, W. S., *Absorption, Distillation and Cooling Towers*, John Wiley & Sons Ins., New York (1961).
115. Gunn, D. J., *Chem. Eng. Sci.*, 33, 1211 (1978).
116. Stanek, V., and Szekely, J., *Canadian J. Chem. Eng.*, 50, 9 (1972).
117. Porter, K. E., *Trans. Instn. Chem. Engrs.*, 46, T69 (1968).
118. Porter, K. E., and Jones, M. C., *Trans. Instn. Chem. Engrs.*, 41, T240 (1963).
119. Weiman, J., and Bonilla, C. F., *Ind. Eng. Chem.*, 42, 1099 (1952).
120. Baker, T., Chilton, T. H., and Vernon, H. C., *Trans. Instn. Chem. Engrs.*, 31, T296 (1953).
121. Hanratty, T. J., and Reiss, L. P., *AIChE J.*, 8, 245 (1962).
122. Shaw, P. V., Reiss, L. P., and Hanratty, T. J., *AIChE J.*, 9, 154 (1963).
123. Reiss, L. P., and Hanratty, T. J., *AIChE J.*, 9, 362 (1963).
124. Shaw, P. V., and Hanratty, T. J., *AIChE J.*, 10, 475 (1964).
125. Jolls, K. R., and Hanratty, T. J., *AIChE J.*, 15, 199 (1969).
126. Karabelas, A. J., Wegner, T. H., and Hanratty, T. J., *Chem. Eng. Sci.*, 26, 1581 (1971).
127. Wang, Y. Y., et al., *Chem. Eng. Sci.*, 37, 939 (1982).
128. Smith, J. W., and King, D. H., *Canadian J. Chem., Eng.*, 53, 41 (1975).
129. Krishna, M. S., and Rao, C. V., *Chem. Age India*, 18, 41 (1967).
130. Krishna, M. S., Jagannadharaju, G. J. V., and Rao, C. V., *Periodica Polytechnica (Chem. Eng.)*, Budapest, 11, 95 (1967).
131. Pickett, D. J., and Stanmore, B. R., *J. Appl. Electrochem.*, 5, 95 (1975).
132. Wilson, E. J., and Geankoplis, C. J., *Ind. Eng. Chem. Fundam.*, 5, 9 (1966).
133. Matic, D., *J. Appl. Electrochem.*, 9, 15 (1979).
134. Storck, A., Robertson, P. M., and Ibl, N., *Electrochimica Acta*, 24, 373 (1979).
135. Vögtländer, P. H., and Bakker, C. A. P., *Chem. Eng. Sci.*, 18, 583 (1963).
136. Olive, H., and Lacoste, G., *Electrochimica Acta*, 24, 1109 (1979).
137. Kinoshito, K., and Leach, S. L., *J. Electrochem. Soc.*, 129, 1993 (1982).
138. Sioda, R. E., *Chem. Eng.*, Feb. 21, 57 (1983).

SECTION IV

SPECIAL APPLICATIONS AND REACTOR TOPICS

CONTENTS

CHAPTER 33

INDUSTRIAL CRYSTALLIZATION

P. A. M. Grootscholten

Unilever Research Laboratory
Vlaardingen, Netherlands

S. J. Jancic

Sulzer Brothers
Winterthur, Switzerland

CONTENTS

INTRODUCTION TO INDUSTRIAL CRYSTALLIZATION

Aims and Problems

Crystallization from solution is widely utilized in the chemical industry, and it is characterized by the formation of a spectrum of differently sized crystalline particles. The control of product crystal size in a crystallizer is a problem of kinetics and not of thermodynamics. Thus, by fixing the energy and material inputs the prescribed weight of a crystalline material is distributed over a few large crystals or over many small ones, and this is determined by the continuous interaction between crystallization kinetics and contraints imposed on the crystallizing system. The result of this interaction is the crystal size distribution which is of great importance for both plant performance and product marketability.

A rational procedure for the design of crystallizers must involve solutions of all relevant conservation equations together with kinetics of all rate processes involved. The knowledge of the flow patterns established for a given crystallizer geometry and mode of operation is a further prerequisite of successful crystallizer design. It is rather unfortunate that the required design information is invariably lacking in the open literature. For example, laboratory experiments which are mainly performed to provide equilibrium and kinetic data are often carried out in regions remote from industrial relevance. On the other hand, comparison of laboratory studies with reliable data from pilot plant and commercial crystallizers is often impossible due to lack of published data. Nevertheless, the fundamental knowledge of the crystallization process still enables some degree of responsibility to be included in the design of crystallizers when evaluating design procedures and when implementing design results in order to meet required product specification.

The objective of this chapter is to elaborate on fundamentals of crystallization and to show how this knowledge can be combined with other technical disciplines to design and operate better crystallizers. Crystallization is a vast subject and no attempt has been made to be exhaustive. The authors have been selective in the work they report and quote, and they have drawn extensively from research that has been undertaken in the crystallization research group in the Laboratory for Process Equipment at Delft University of Technology in the Netherlands.

Crystallization as a Unit Operation

Before crystallization equipment can be rationally designed and operated in the most efficient manner, the principles that control the development of a crystal size distribution in a crystallizing system must be known. In an industrial crystallizer the product crystal size distribution results from particulate processes controlled by both the crystallization driving force and by the manner in which solid-liquid contacting and transport through the crystallizer are being established.

The crystallization driving force determines the rate at which stable nuclei are formed and are subsequently being advanced in size by crystal growth. Nucleation and crystal growth may be referred to as crystallization kinetics and this latter has frequently been the object of theoretical and experimental studies under conditions which are often highly simplified and very often far remote from industrial relevance. This means that most experimental studies have been oriented to the derivation of crystallization kinetics at conditions where the level of driving force was carefully controlled and where particulate processes, other than nucleation at smallest stable size and crystal growth, were suppressed. In an industrial crystallizer, however, solid-liquid contacting and transport of materials through the crystallizer cannot be established without simultaneous occurrence of additional particulate processes. These latter may be referred to as population functions.

The crystallization kinetics and simultaneously occurring population functions, or population events, interact in a most complex and unpredictable manner, and together they determine the development of a population of crystals in a crystallizing vessel. The generation of a crystal size distribution in a crystallizing system may be illustrated by Figure 1 where pure crystallization kinetics with all participating population events are represented over size ranges where they are likely to occur.

The pure crystallization kinetics are as follows:

- *Primary nucleation:* A number flux of particles generated at critical nucleus size, mostly at very high supersaturations, i.e., in the vicinity of cooling surfaces, in the boiling zone, or briefly, in all situations where generation of supersaturation exceeds dissipation of supersaturation.
- *Secondary nucleation:* A number flux of a spectrum of particles generated as a result of the presence of growing parent crystals whereby new particles are being born both at critical nucleus size and into the range of sizes, say up to about 20 μm. This can take place at lowest supersaturations and per definition leaves no visible marks on the surface of parent crystals. It can take many forms, but most likely in industrial crystallization are crystal/crystal and crystal/crystallizer contacts, e.g., agitators, pumps, or any moving part.
- *Crystal growth:* A process of size enlargement which is supersaturation dependent and can also depend on crystal size, similar to the biological aging process of human beings. The process of crystal growth as in humans can exhibit growth dispersion—a situtation where crystals of one and the same size group grow at different rates. Furthermore, crystals can change their shape and appearance (habit modification) through the selective influence of impurities on the growth rate of different crystal faces.
- *Dissolution of crystals:* A process of size reduction and as such is opposite to the process of crystal growth. It leads eventually to the disappearance of crystals and as such is a process opposite to nucleation. It takes place only in the presence of undersaturation and may be expected in all parts of crystallization equipment where this condition is met, e.g., heated surfaces.

With reference to Figure 1 the following population events may be distinguished:

- *Feed population:* A number flux of a spectrum of particles introduced with crystallizer feed.
- *Product population:* A number flux of a spectrum of particles being withdrawn from the crystallizer. The product population can be totally or partially representative of the crystallizer contents, and this may be controlled by adjusting the design features for most crystallizer sizes and design configurations. The classification event indicated in Figure 1 represents preferential withdrawal of large crystals.
- *Attrition:* A number flux of a spectrum of particles generated at all conditions of supersaturation, even in undersaturated solutions. It is a result of mechanical interaction between crystals and particularly between crystals and moving parts of a crystallizer such as agitators and circulating pumps. It leads per definition to visible marks on the surface of parent crystals often resulting in very presentable, rounded-off crystals of free-flowing properties. Crystals resulting from attrition are generally considered to be a little larger then those resulting from secondary nucleation.
- *Breakage:* A number flux of a spectrum of particles resulting from dramatic damage to parent crystals, which often leads to their disappearance with reference to the size group in which the event took place. It occurs in badly designed systems as a result of high energy contacts between crystals and moving crystallizer parts, particularly in clearance spaces between the moving part swept area and static components, e.g., draft tubes, pump housing, etc.
- *Fines dissolving:* A negative number flux of a spectrum of small particles aimed at counteracting an inadvertent number balance caused by excessive positive number fluxes caused by all of the previously mentioned sources. This population event is equivalent to preferential removal of small crystals (fines), and there are several design configurations which make this possible.
- *Agglomeration:* A population event characterized by a negtive number over a range of sizes where crystals agglomerate and a positive number flux into the size range where stable agglomerates are being generated. In physical terms agglomeration takes place when two or more crystals are attracted by any sort of cohesion forces and remain together for a sufficient period of time to grow into stable, self-contained. crystalline entities of highly irregular shape and properties. Dewatering and drying costs of such crystals increase dramatically and storage and free-flowing properties are often imparted.

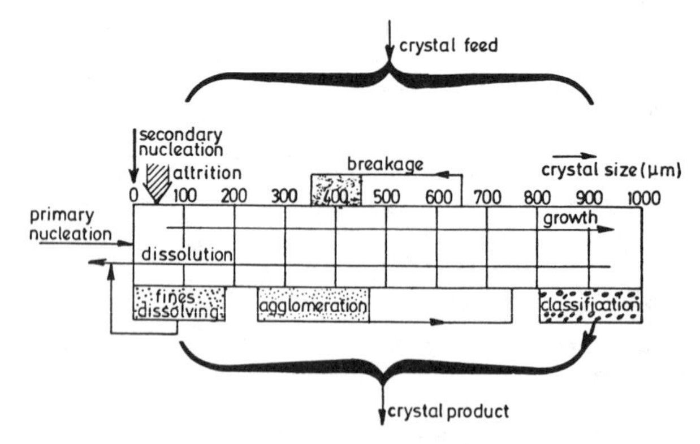

Figure 1. Schematic representation of crystallization kinetics and population events leading to the development of crystal size distribution.

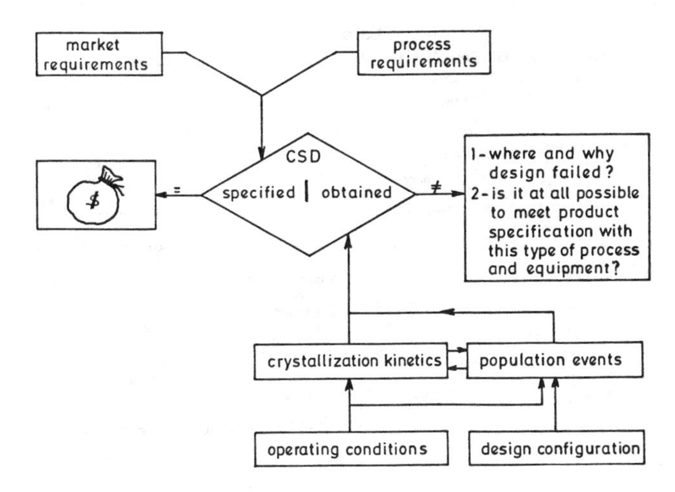

Figure 2. A pictorial representation of aims and problems of industrial crystallization.

A full-scale crystallizer inevitably represents a conglomerate of probably all population events illustrated in Figure 1, and if these can be balanced to meet desired product specification and process requirements, the crystallizer may be said to be well designed. In commercial crystallization, factors like shape, size, strength, and purity of the crystal product are important to the consumer, and the rate of production and costs are important to the producer. As illustrated in Figure 2, a combination of market requirements and process requirements leads to a certain specification of a crystal product and the principal aim of an industrial crystallization is to meet such specifications. The problem of industrial crystallization consists of matching design configurations and operating conditions to the crystallizing system in such a way that the resulting crystallization kinetics and population events favorably interact and eventually lead to design specification of the product. If this latter situation is not met there are at least two questions requiring an answer and these are indicated in Figure 2.

NUCLEATION KINETICS

Mechanisms

Nucleation may be defined as the process in which the smallest stable aggregates of a crystalline phase are formed in a crystallizing system. Depending upon circumstances regarding the crystallization system and its environment, the following nucleation mechanisms may be distinguished:

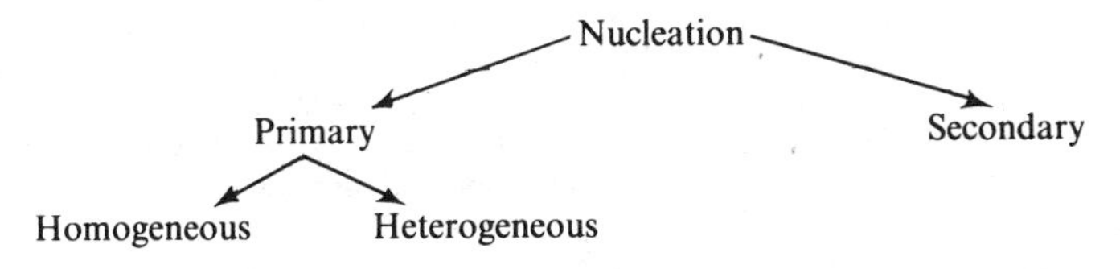

Two types of primary nucleation may be distinguished: homogeneous nucleation, occurring within a homogeneous solution where no foreign surfaces previously existed, and

Table 1
Sources of Nuclei in Industrial Crystallization

Source of Nuclei	Type of Nucleation Process	Prevention or Remedy
Nucleation induced by excessive supersaturation		
Boiling zone	Primary	Reduce specific production rates, increase crystal surface area.
Hot feed inlet	Primary	Enhance heat dissipation, reduce degree of superheating, carefully chose inlet position.
Inlet of direct coolant	Primary	Enhance heat dissipation, reduce temperature of coolant, chose inlet position.
Heat exchangers, chillers, etc.	Primary	Reduce temperature gradients by increasing surface area, increase liquid velocities.
Reaction zone	Primary	Enhance mixing and dissipation of supersaturation, increase crystal surface area.
Cavitating moving parts	Primary	Adjust tip speed, suppress boiling by sufficient static head.
Nucleation induced by crystal interactions		
Crystal/crystallizer contacts		
Collisions with moving parts (impellers, pumps etc.)	Secondary	Adjust tip speed and design configuration, coat impellers with soft materials, reduce if possible magma density and mean crystal size.
Collisions with crystallizer wals	Secondary	
Crystal/crystal contacts		
Crystal grinding in small clearance spaces (impeller/draft tube, pump stator/rotor)	Secondary Attrition Breakage	Carefully specify all clearances and hydrodynamics of two phase flow therein.
Crystal/solution interaction		
Fluid shear, effect of impurities etc.	Primary Secondary	Reduce jetting, get to know the effect of impurities for each particular system, prevent incrustation.

heterogeneous nucleation, occurring in the presence of a foreign surface other than the material to be crystallized (such as dust particles or crystallizer wall surfaces). It is thought that the influence of the foreign surface is catalytic and that the activation energy for nucleation may thereby be reduced.

The presence of a crystal in its supersaturated solution usually induces the formation of further crystals at levels of supersaturation at which primary nucleation would not otherwise occur. This catalyzing effect of parent crystals is now referred to as secondary nucleation.

According to Strickland–Constable and Mason [1] secondary nucleation may occur in a variety of different ways, and these include:

- Initial breeding.
- Needle breeding.
- Collision breeding.

Botsans, Denk, and Chua [2] have proposed a further possible mechanism of secondary nucleation, which they refer to as impurity concentration gradient nucleation. Besides the previously mentioned mechanisms it has been suggested that nuclei can be produced by what has been called "fluid shear," i.e., by flow of the solution relative to the crystal [3].

It is beyond the scope of this chapter to discuss in detail the various nucleation mechanisms. For more information see References 4 and 32.

Nucleation in Industrial Crystallizers

Excessive nucleation rates are common in industrial crystallizers so that it is sometimes difficult to grow large crystals. Unfortunately, the present state of nucleation theory is such that it is not possible to predict nucleation rates. However, the principles of nucleation are useful in estimating the possible sources of excess crystal nuclei. An outline of possible sources of nuclei is presented in Table 1 together with recommendations as to the prevention of excessive nucleation rates. These are given in very general terms since no reliable universal rules can be expected to hold for each crystallizing system and crystallizer configuration.

In industrial crystallization the supersaturation levels are kept below the values at which uncontrolled nucleation occurs. It has also been found that the maximum useful supersaturation is even lower since at even modest supersaturation, crystals of the desired shape and mechanical properties cannot be grown. It is of interest, therefore, to consider the relative importance of various types of nucleation in suspensions of crystals growing at low supersaturations.

Homogeneous nucleation is unlikely to occur in industrial crystallizers since low supersaturations are employed, and the presence of nondissolved impuritites would, probably, induce primary heterogeneous nucleation.

Primary nucleation might occur however in regions where supersaturation is being created. Not being able to dissipate the created supersaturation by distributive mixing with the bulk liquid or by deposition onto the surface of growing crystals, leads to accumulation of supersaturation to a degree where primary nucleation is possible. Regions of such high supersaturation may be expected in boiling zones, external heat exchangers, chillers, etc. (see Table 1). Problems of this kind can generally be overcome by an adequate choice of the specific production rate, by allowing large crystal surfaces to be available for growth in regions where supersaturation is being created, and by allowing for adequate distributive mixing. The tendency of superheated feed solution to flash off while it is being transported to the boiling zone is a common source of excessive primary nuclei. This is especially true for highly viscous media, which in the state of segregation with respect to the bulk of solution penetrate nonsuppressed regions of the crystallizer and consequently give rise to violent local evaporation. Cavitation on crystallizer moving parts leads to similar effects. These problems can be reduced by reducing tip speed, by increasing static suppression, and by improving distributive mixing.

Secondary nucleation appears to be the most likely mechanism of crystal nucleus formation in suspensions of growing crystals. Initial breeding is possible if crystallization has been started by seeding the crystallizer and may therefore be encountered in batch crystallizers or in the start-up period of continuous crystallizers.

Needle breeding is associated with the growth of imperfect crystals, and would not be expected under the conditions in which regular crystals are grown. This type of nucleation is therefore unlikely to be of much significance in industrial crystallizers.

Attrition is of universal occurrence and can take place even under the conditions where very regular crystals are grown, and at supersaturations which are much too low for nucleation. Therefore, attrition would seem to be the most likely source of excess nuclei, even in properly designed and operated crystallizers. With reference to Table 1 various types of nucleation are possible through crystal interactions. Collisions with moving parts such as stirrer and pump impellers invariably enhance secondary nucleation. Also crystal/wall contacts, e.g., collisions with baffles and crystals rolling past the interior of a crystallizer under centrifugal forces are known to increase secondary nucleation rates. These sources of nuclei can be reduced by contolling the frequency and the energy of such contacts.

An interesting and a very important form of crystal/crystal contacts are particulate interactions leading to grinding in small clearance spaces, e.g., between impellers and draft tubes, between moving and stationary pump components, etc. In this use the choice of the clearances must be made according to the specification of product crystals and the hydrodynamics of two-phase flow through the clearance regions should be optimized to reduce such effects.

Scale-up of Secondary Nucleation Kinetics

The overall nucleation rate in a crystallizer is determined by the interaction of the secondary nucleation characteristics of the material being crystallized with the hydrodynamics of the crystal suspension. When crystallizing a given material, crystallizers of different sizes, agitation levels, flow patterns, etc. will produce different nucleation rates. In order to incorporate nucleation kinetics in design models these changes in nucleation rates should be predictable.

Table 2
Scale-up of secondary nucleation kinetics as given by Equations 2 and 3

	Ottens and de Jong (1972)	Evans et al. (1974b)	Jancic (1976)
f(G)	pd_s^5/V_c	$pd_s^{5.5}/V_c$	$p(d_s^5/d_D^2)\,(V_{prop}/V_c)$
b	3	3	4
c	2	2.5	3
i	3	2.5	4
Exponent d in the relation $B \propto \lambda^d$ when crystallizer is scaled up on the basis of:			
Constant ε (power input per unit mass)	0	0.5	0.33
Costant N (stirrer speed)	2	2.5	3
Constant N_{is} (stirrer speed at which suspension of crystals is attained)	−0.55	−0.05	−0.4
Constant v_t (tip speed)	−1	−0.5	−1

The generation of potential nuclei at the surface of the parent crystals and the rate of removal of these nuclei from the surface out into the bulk solution act simultaneously to determine the overall nucleation rate.

Several secondary nucleation mechanisms have been suggested in the literature: collisions of crystals with the crystallizer [5, 6, 7], the collisions of the crystals with one another [5, 6], and fluid shear [8]. Most of the experimental evidence, however, indicates that a dominant mechanism causing secondary nucleation in continuous crystallizers arises from collisions of macro-sized crystals with crystallizer impeller or circulation pumps impeller.

A number of authors have developed mechanistic descriptions of the processes causing secondary nucleation in agitated crystallizers [5–7, 9]. Their approaches have all been based on the assumption that secondary nuclei are produced as a result of collision between crystals and other parts of the crystallizer or other crystals. Hence, only contact nucleation is considered.

The nucleation rate equations resulting from the work of Ottens at al. [5], Evans et al. [6] and Jancic [10] can all be written in the form:

$$B^0 = K_N f(G) N^b M_i \Delta c^n \tag{1}$$

where f(G) is a function of the geometry and size of the crystallizer and agitator [9]. If crystallizers of similar geometry are considered, the size of which is defined by a length scale λ, then Equation 2 can be written:

$$B^0 \propto \lambda^c N^b M_i \Delta c^n \tag{2}$$

Values of f(G), b, c, and i are summarized in Table 2 for the three approaches.

Scale-up rules for secondary nucleation kinetics can be developed on the basis of a number of different criteria, for example, constant power input per unit mass of magma, constant stirrer speed, constant tip speed, or stirrer speed just to maintain all crystals in suspension. The published equations as represented by Equation 2 can be used to predict the resulting relation between the secondary nucleation rate and crystallizer size, these relations being expressed in the form:

$$B^0 \propto \lambda^d \tag{3}$$

Corresponding values of d are also included in Table 2.

The prediction of all three approaches are rather similar. They all show that scale-up on the basis of constant ε should at most produce a modest increase in nucleation rate (d = 0.5 according to Evans et al.). Alternatively scale-up on the basis of either constant tip speed or the particle suspension criteria is in all cases predicted to result in a decrease in nucleation rate with increasing scale.

Work by Grootscholten et al. [11] demonstrates that nucleation rates for sodium-chloride can indeed be correlated by an equation of the form of Equation 1. Since measuring supersaturation Δc in a large crystallizer is nearly impossible and it often varies, Δc was substituted by average solids residence time, this latter being a good measurement of the average level of supersaturation. This gave the following semi-empirical equation:

$$B^0 = 0.023 K_N^{0.6} N^{1.2} \tau_{co}^{-1.6} \bar{M}_s \tag{4}$$

where N = the impeller-speed
τ_{co} = the overall solids residence time
$\bar{M}_s$ = the average crystal concentration in the crystallizer

Nucleation rate constant K_N appeared to be a strong function of crystallizer size and values of K_N, obtained from crystallizers of sizes ranging from 91 liters to 280 m^3 could be successfully correlated by the empirical equation:

$$K_N = 1 \times 10^{16} K_{imp}^{2.25} d_s^2 \tag{5}$$

where d = the agitator diameter
K_{imp} = the dimensionless impeller-number which is defined as:

$$K_{imp} = \frac{2nN\varphi_{vc}^{0.5}}{(gH_m)^{0.75}} \tag{6}$$

where φ_{vc} = the pumping rate
H_m = the total head delivered by the impeller

CRYSTAL GROWTH

After the energy barrier for the formation of a new crystal has been surmounted, the whole system will spontaneously adjust to a state of greater stability. This process is manifested by the regular deposition of the excess solute from the solution on the various faces of a crystal. The result is a decrease in the concentration of the solution, on the one hand, and an increase of crystal size, on the other. The rate of increase in crystal size may be specified by the velocity at which the crystal face moves outward in a direction normal to the face; this velocity is called the linear face growth rate. Different faces of a crystal may grow at different rates, and so the average linear growth rate is often approximated by the increase of the diameter of a sphere having the same volume as the crystal, or by the increase in a characteristic dimension of the crystal.

Processes Involved in the Growth of Crystals

The growth of crystals from solution is at least a three-stage process:

1. The transport of solute from the bulk of the solution to the vicinity of the crystal surface.
2. Some processes at the crystal surface, probably involving adsorption into the surface layer followed by the orientation of the molecules into the crystal lattice (often referred to as the surface integration process).
3. Dissipation of the heat of crystallization liberated at the crystal surface.

It is generally accepted that the first two processes are governed mainly by the concentration driving force which may be interpreted in terms of the concentration profiles in Figure 3.

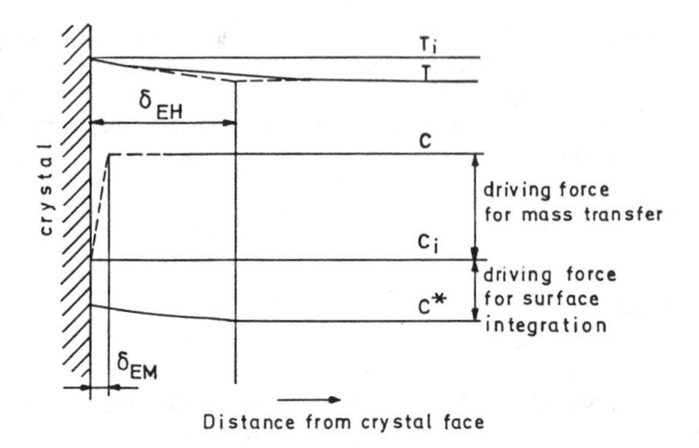

Figure 3. Concentration and temperature profiles in crystallization from solution for a system of exothermic heat of crystallization and normal solubility.

This diagram shows a crystal face growing in the supersaturated solution of concentration c. The transport of solute from the bulk of solution to the crystal surface is effected by diffusion and convection across the effective mass transfer boundary layer, δ_{EM}, surrounding the crystal.

The process of diffusional transfer is followed by an incorporation reaction at the crystal-solution interface. Hence, the corresponding driving forces are $(c - c_i)$, for transport to the interface, and $(c_i - c^*)$, for the surface integration process, as indicated in Figure 3. After the solute molecule has been incorporated into the crystal lattice, the heat of crystallization will be liberated and must be dissipated through the surface film of the bulk of solution. In general, therefore, the surface temperature, T_i, and hence the equilibrium saturation concentration at the crystal surface will be slightly higher than in the bulk of the solution (see Figure 3). This effect will decrease the driving force available for crystallization. However, the rate of heat removal from the crystal surface to the bulk of the solution is usually neglected as an overall crystallization rate controlling factor.

The equation that is conventionally used to describe the mass transfer step is:

$$R_m = k_d(c - c_i) \tag{7}$$

The governing equation for the surface integration process usually takes the empirical form:

$$R_i = k_r(c_i - c^*)^2 \tag{8}$$

The proportionality constant, k_r, depends upon temperature, presence of impurities, and the characteristics of the crystal surface. The values of the exponent, r, have been shown to vary from system to system.

The Overall Crystal Growth Process

Assuming the process of crystal growth to consist of mass transfer through the crystal boundary layer followed by the surface integration process, the overall rate of crystal growth should be a function of the concentration driving force and the corresponding rate coefficients:

$$R_g = f(\Delta c, k_d, k_r) \tag{9}$$

where $\Delta c = (c - c^*)$. Assuming that volume diffusion and the surface integration are two processes occuring in series, under steady-state conditions the rate at which mass is transferred to the crystal surface through the boundary is equal to the rate at which mass is deposited by the process at the crystal surface. Thus, combining Equations 7 and 8, the overall crystal growth rate can be expressed in terms of the overall driving force $(c - c^*)$ and the overall crystal growth rate coefficient. For this purpose it is necessary to eliminate the concentration at the crystal-solution interface, c_i.

In the case of a first-order surface integration process, r = 1, the overall crystal growth rate becomes:

$$R_g = k_g(c - c^*) \tag{10}$$

where

$$\frac{1}{k_g} = \frac{1}{k_d} + \frac{1}{k_r} \tag{11}$$

If the surface integration process is of a higher order, i.e., r > 1, then a simple relationship between the growth rate and the overall driving force cannot be found for an arbitrary value of r. However, many authors have successfully correlated overall crystal growth rates with the overall driving force through a power law:

$$R_g \cong \Delta c^g \tag{12}$$

For example, Mullin and Garside [12] and Garside and Mullin [13] measured growth rates of potash-alum crystals in a fluidized-bed growth cell, and found good agreement between the experimental data and an equation of this form (see Figure 4.) Equation 12 is often used in theoretical studies of crystallizer operation.

Growth rate R_g, which has the dimensions mass per unit area per unit time, can be converted to linear growth rate by the equation:

$$G = \frac{k_a}{3\, k_v \varrho_c} R_g \tag{13}$$

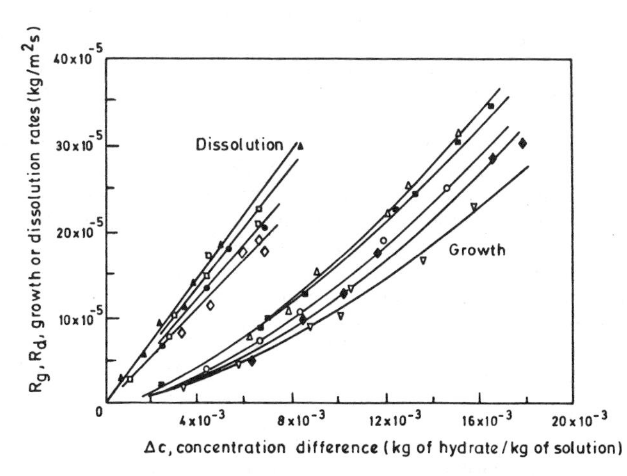

Figure 4. Growth and dissolution rates for potash alum crystals at 32°C (after Garside and Mullin [13]).

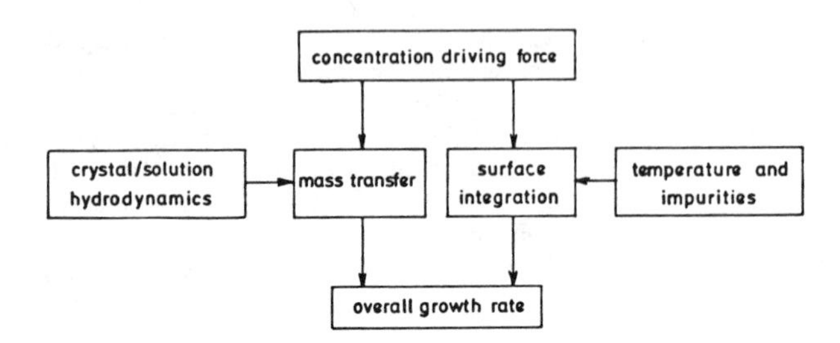

Figure 5. Schematic representation of influences exerted by the growth environment upon processes involved in crystal growth.

where k_a and k_v are shape factors for surface area and volume, respectively.

McCabe [14] postulated that crystals of all sizes grow at the same rate, where the increase in size is measured as an increase in a selected linear dimension. This is often referred to as the "McCabe's ΔL law." For many systems where crystallization from solution occurs McCabe's ΔL law has been found to fit experimental laboratory data well. In a number of systems, however, it has been observed that the growth rate actually increases with increasing crystal size, and it has even been reported that crystals of one size may exhibit variations in the growth rate. The latter effect is referred to as "growth dispersion" [15].

Overall Resistance to Growth

The concept of the crystal growth process occurring in series generates the formulation of total resistance as the sum of resistance to mass transfer and resistance to surface integration (see Equation 11). It is obvious that the mutual contribution of those partial resistances determines the total resistance and hence the kinetics of crystal growth for a given driving force. It is of interest to examine possibilities of selectively influencing resistances to crystal growth in order to be able to attain desired growth rate or to be able to predict it.

It is generally accepted that both the mass transfer and surface integration increase with concentration driving forces. It is also accepted that crystal/solution hydrodynamics selectively influence mass transfer and that factors such as temperature and impurities exert effects solely upon surface integration kinetics. The influence exerted by the growth environment upon processes involved in crystal growth are presented schematically in Figure 5.

Before crystal growth rates can be meaningfully influenced through any of the actions illustrated in the schematic diagram of Figure 5, it is necessary to gain knowledge of the relative contribution of mass transfer and surface integration resistances to crystal growth. This could be done relying upon the use of dissolution data, assuming that the dissolution process is controlled solely by mass transfer. This method appears to be generally accepted [10, 16–18] and consists of measuring growth and dissolution rates under identical hydrodynamic conditions. This enables the surface integration kinetics to be derived from Equations 7 and 8 by eliminating interfacial concentration, c_i:

$$R_i = k_r\left(c - c^* - \frac{R_D}{k_d}\right)^r \tag{14}$$

where R_D is the measured dissolution rate and where it is assumed that the mass transfer coefficient, k_d, measured for dissolution could also be used to estimate the mass transfer resistance present during crystal growth.

With reference to Figure 5 it follows that pure surface integration kinetics can be combined with appropriate mass transfer resistance to yield overall crystal growth rates, and these latter can be predicted for large crystallizers resorting to small-scale laboratory experiments.

THE POPULATION BALANCE CONCEPT

Introduction

Whereas mass and heat balances are often sufficient to adequately characterize one-phase homogeneous systems, additional information is needed to describe dispersed-phase particulate systems. So the concentration of mass in a particular system does not give information on whether this mass is distributed over a large number of small or a small number of large particles.

A breakthrough in the description and prediction of particulate processes has been the introduction of the concept of particle population balance, which was principally developed by Randolph and Larson [19]. In particular the crystallization unit operation has received a tremendous boost forward following this concept. The benefit of the population balance concept is that it has greatly facilitated the mathematical manipulation of crystal size distributions. Moreover, its application has led to the development of experimental techniques, which enable nucleation and growth phenomena to be studied and quantified under conditions where both kinetics proceed simultaneously. Since its introduction at the beginning of the seventies, the descriptive and predictive possibilities of the population balance technique have strongly stimulated research on various aspects of the crystallization process and as such raised the chemical engineering appreciation of crystallization as a unit operation.

In order to be as simple and clear as possible, a detailed description and evaluation of the population balance theory have been deliberately excluded as far as possible. A comprehensive introduction to the topic is given by Randolph and Larson [20].

The General Population Equation

To facilitate formulation of the population balance, the crystal size distribution is represented by what is known as the population density, n, which is defined as the slope of the cumulative number size distribution. The units of n are number per unit length per unit volume. The number of crystals, dN, in size range dL is thus ndL, and the number of crystals present in the distribution between sizes L_1 and L_2 is:

$$N = \int_{L_2}^{L_1} n \, dL \tag{15}$$

When the law of conservation of crystal population is expressed in point population density, $\dot{n}$, (the number of crystals may change from point to point within the crystallizing system), the following relationship is obtained for the configuration shown in Figure 6.

$$\int_V \left[\frac{\partial \dot{n}}{\partial t} + \frac{\partial \left(\dot{n} \frac{dL}{dt} \right)}{\partial L} - B(L) - A(L) - M(L) + \lambda(L) + D(L) + P(L) \right] dV + n \frac{dV}{dt}$$

$$- Q_i n_i + Q_0 n_0 = 0 \tag{16}$$

where $\dot{n}$ = the number of crystals of size L to L + dL found at position (x, y, z)
Q = volume flow rate of suspension

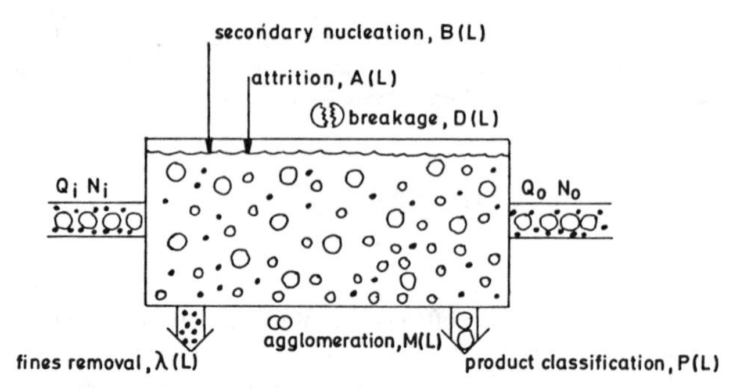

Figure 6. Schematic representation of a flow system containing an arbitrary suspension of particles undergoing various population events.

i = inlet
0 = outlet
B(L) = secondary nucleation operating throughout the volume dV, such that B(L)dL is the number of crystals of size L to L + dL created per unit volume and unit time
A(L) = attrition operating throughout the volume dV, such that A(L)dL is the number of crystals of size L to L + dL created per unit volume and unit time
D(L) = breakage operating throughout the volume dV, such that D(L)dL is the number of crystals of size L to L + dL created through breakage per unit volume and unit time; it could assume negative values for size groups where crystals cease to exist as a consequence of breakage
λ(L) = fines removal operating throughout the volume dV, such that λ(L)dL is the number of crystals of size L to L + dL ceasing to exist due to fines destruction
M(L) = agglomeration operating throughout the volume dV, such that M(L)dL is the number of crystals of size L to L + dL created through agglomeration; it can assume negative values for size groups where crystals cease to exist due to agglomeration
P(L) = preferential product removal operating throughout the the volume dV, such that P(L)dL is the number of crystals of size L to L + dL preferentially withdrawn through the system

Equation 16 represents a general population equation for an arbitrary suspension of particles.

The first term in Equation 16 accounts for the net number rate at which crystals join the specified size group. The second term accounts for changes in the suspension volume, and the last two terms indicate the number rate at which crystals are carried into and out of the system.

In some cases the general population equation can be reduced to a simpler form by imposing additional assumptions which are determined by the specific physical conditions of the system in question. For example, under steady-state conditions all time derivatives are zero, or the inlet stream may not contain any crystal, in which case $n_i = 0$. If the contents of the crystallizer is well mixed, then the point population density, $\dot{n}$, can be replaced by the average density, n, according to:

$$n = \frac{\int_V \dot{n}\, dV}{V} \tag{17}$$

Moments of the Distribution

The population density can be found by solving the general population Equation 16, subject to the appropriate crystallization kinetics and the population events being available.

It is often convenient to use the average population density, for the whole crystallizer, which was defined in Equation 17. The i-th moment of the distribution described by the population density, n, is defined as follows:

$$M_i = \int_0^\infty nL^i\, dL \tag{18}$$

A necessary condition for any realistic population density distribution is that moments of the distribution converge for $i \geq 0$.

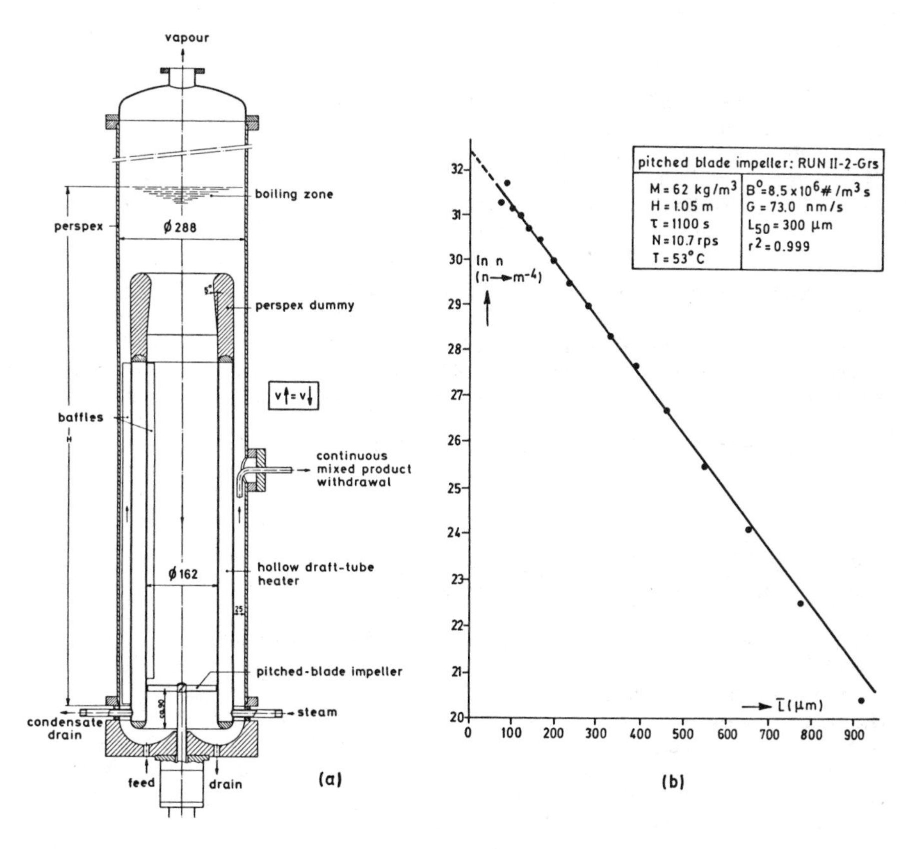

Figure 7. (A) Experimental sodium chloride MSMPR crystallizer: (B) a typical population density plot obtained at steady state.

The physical significance of the several moments of distribution are shown below:

Zero moment (the total number of crystals per unit volume of crystal suspension)

$$M_0 = \int_0^\infty n \, dL \tag{19}$$

First moment (the total length of crystals per unit volume of crystal suspension)

$$M_1 = \int_0^\infty nL \, dL \tag{20}$$

Second moment (multiplied by a surface factor gives the total crystal surface area per unit volume of crystal suspension)

$$M_2 = \int_0^\infty nL^2 \, dL \tag{21}$$

Third moment (multiplied by a volume factor gives the total volume of crystals per unit volume of crystal suspension, i.e. volume fraction of solids)

$$M_3 = \int_0^\infty nL^3 \, dL \tag{22}$$

From a practical point of view the crystal size distribution is frequently characterized by two parameters, the median size, L_{50}, and the coefficient of variation, c.v. The median size is defined by way of the cumulative weight distribution: 50 weight % of the product has a particle size smaller or larger than L_{50}. The coefficient of variation is a measure of the width of a distribution and is in general defined as the ratio of standard deviation to the arithmetic mean size of the distribution.

Mixed-Suspension-Mixed-Product-Removal Crystallizer Concept

The concept of population balance can be conveniently applied to a so-called mixed-suspension mixed-product removal (MSMPR) crystallizer [20]. In this configuration it is assumed that the crystallizer content is perfectly mixed and that the withdrawal stream is representative of its content.

For the MSMPR configuration just described, the general population Equation 16 reduces to:

$$\frac{\partial n}{\partial t} + \frac{\partial\left(n \frac{dL}{dt}\right)}{\partial L} - B(L) - A(L) - M(L) + D(L) - \frac{Q_i}{V} n_i + \frac{n}{\tau} = 0 \tag{23}$$

where the point population density, $\dot{n}$, has been substituted from Equation 17, $n_0 = n$ because of mixed-product removal, dV/dt is zero for constant suspension volume, and Q/V is the reciprocal of the resicence time, τ. Equation 23 can be further simplified by considering clear liquid feed ($n_i = 0$) at steady-state conditions ($\partial n/\partial t = 0$). Furthermore, rate processes such as secondary nucleation, attrition, breakage, agglomeration, and dissolution can be ignored for configurations where these processes can be suppressed or are operative at a very minor level so that their influence on the crystal size distribution can be ignored altogether. In this case Equation 23 reduces at steady state to:

$$\frac{d(nG)}{dL} + \frac{n}{\tau} = 0 \tag{24}$$

where G = the linear growth rate dL/dt.

To obtain a relationship for the number distribution of the crystallizer content, the appropriate growth rate expression must be introduced into Equation 24 and solved for the boundary conditions.

Population Density Distribution for MSMPR Crystallizer—Size-Independent Crystal Growth Rate

For the MSMPR configuration at steady state and for the growth rate independent of crystal size, Equation 24 reduces further to:

$$G\frac{dn}{dL} + \frac{n}{\tau} = 0 \quad (25)$$

which integrates to give:

$$n = n^0 \exp(-L/G\tau) \quad (26)$$

Equation 26 suggests that if the value of ln (n) is plotted against crystal size, L, a straight line results. Indeed many authors have found that the experimental population density plot obtained in laboratory MSMPR crystallizers agrees well with Equation 26. An example is given in Figure 7 where an experimental MSMPR crystallizer for sodium chloride is shown together with a typical product distribution presented in the form suggested by Equation 26.

Suspension Density in an MSMPR Crystallizer

The suspension established in a steady-state MSMPR crystallizer for conditions of size-independent growth may be obtained from the third moment of the product distribution given by Equation 22. This latter combines with the product distribution expressed by Equation 26 to yield:

$$M_s = k_v\varrho_c n^0 \int_0^\infty L^3 \exp\left[-\frac{L}{G\tau}\right] dL \quad (27)$$

and this integrates to give:

$$M_s = 6k_v\varrho_c n^0[G\tau]^4 \quad (28)$$

Median Size and Coefficient of Variation in an MSMPR Crystallizer

The median size in an MSMPR crystallizer at steady state and for conditions of size-independent growth is given by:

$$L_{50} = 3.67G\tau \quad (29)$$

Values of L_{50} can also be determined directly from the cumulative weight distribution. Coefficient of variation of the weight distribution corresponding to the straight line population density distribution according to Equation 26 are always 50% regardless of the kinetics that produce the size distribution.

In view of the relatively large spread in the weight distribution of crystals obtained from a MSMPR-crystallizer it should be noted that such a system should not be used for production purposes since such units often produce a commercially undesired distribution. So a too-wide distribution may cause drying, transport, and storage problems. A MSMPR system should only be used as an experimental tool to derive fundamental kinetic data (for instance the influence of impurities, temperature, or impeller-design on the kinetic rates).

Population Density Distribution for MSMPR Crystallizer—Size-Dependent Crystal Growth Rate

Some crystallizing systems exhibit crystal growth rates which are dependent on size, and in this case different product size distributions are generated at MSMPR conditions. This situation is also implied by Equation 24 since the growth rate, G, may not be exempted from differentiation. In this case knowledge of the dependence of the crystal growth rate on size is essential, and the availability of the appropriate crystal growth rate expression is a prerequisite for successful evaluation of the population density distribution. Size-dependent growth rates are frequently described by the model proposed by Abegg, Stevens, and Larson [21], which reads:

$$G(L) = G_0(1+\gamma L)^b \quad \text{for} \quad \begin{matrix} b < 1 \\ L \geqq 0 \end{matrix} \tag{30}$$

Although the model does not follow directly from known mechanisms of crystal growth, it has been found to represent the overall growth rates rather well. In addition to this, the equation satisfies the following requirements for size-dependent growth models:

1. It is consistent with the population density concept in that all the moments of the population density distribution generated by the model converge.
2. Crystal nuclei grow at a finite rate.

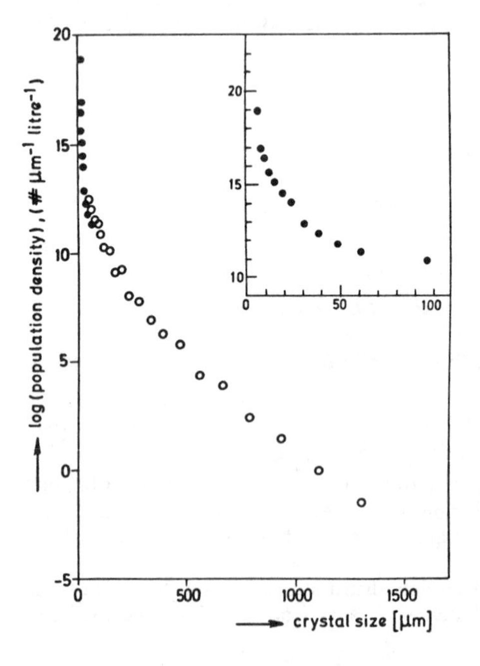

Figure 8. Typical population density plot obtained in the 1.3 liter continuous MSMPR crystallizer for potash alum/water system at 30°C.

3. The growth-rate expression is a continuous function of L and b in a region which includes the point L = 0, b = 0.

The experimental steady-state population density distributions are usually plotted as ln (n) vs. L. When the crystal growth rate is size dependent, however, the population density plot is not a straight line (see Figure 8) and the analytical form may be obtained by substitution of the growth rate model in and subsequent integration of the population balance equation (Equation 24).

The population density distribution may be used to derive the size, area, and weight distributions. In addition to this, the moments of the distributions can yield predictions for characteristic sizes and the coefficients of variation for the distribution involved. Calculation techniques and applications are discussed in detail by Jancic and Grootscholten [4].

POPULATION FUNCTIONS

Deviations from the MSMPR Crystallizer Configuration

The population balance concept has been extensively used in the study of bulk crystallization processes through the MSMPR crystallizer configuration. Both fundamental kinetic data and design-oriented crystallization kinetics have been by now obtained for different systems. Most studies are unfortunately based upon assumption that population functions such as size-dependent secondary nucleation, size-dependent crystal growth, attrition, breakage, agglomeration, and classification are not operative. In this case the analysis of systems undergoing bulk crystallization is considerably simplified by being reduced to the form of Equation 24:

$$\frac{d(nG)}{dL} + \frac{n}{\tau} = 0 \tag{24}$$

The solution of Equation 24 results in a linear population density plot for size-independent crystal growth, and for size-dependent growth a curved population density plot is obtained (see Figures 7 and 8, respectively). In order to simplify the theoretical population density analysis, many experimental configurations have been built with features aimed at meeting the assumptions upon which Equation 24 is based. This meant not only achieving steady-state MSMPR crystallizer conditions but suppressing or ignoring important population functions such as size-dependent nucleation, attrition, breakage, agglomeration, and classification. These latter conditions are usually not met in large-scale commercial crystallizers where at least some deviations from the MSMPR conditions take place and where most population functions are operative at least to some extent. A good example of the population density plot obtained at realistic plant conditions is shown in Figure 9 for the case of sodium chloride crystallization. The crystallizer was a 1.5 m^3, 1 m-diameter evaporator with external circulation and capacity of 10 t/day. The crystallizer featured a tangential inlet and an axial outlet, and its schematic representation together with examples of typical population density plots is shown in Figure 9. The shape of the population density distribution deviates very strongly from the straight line which was obtained for the 55-liter crystallizer (see Figure 7). The features of population density distributions obtained in large sodium chloride crystallizers can be characterized as follows (see Figure 9):

- At sizes below about 100 μm, population density increases sharply as crystal size decreases.
- At sizes between about 100 and 300 μm, a plateau occurs exhibiting a maximum and a minimum.
- At sizes larger than about 300 μm, population density sharply decreases with an increase in crystal size.

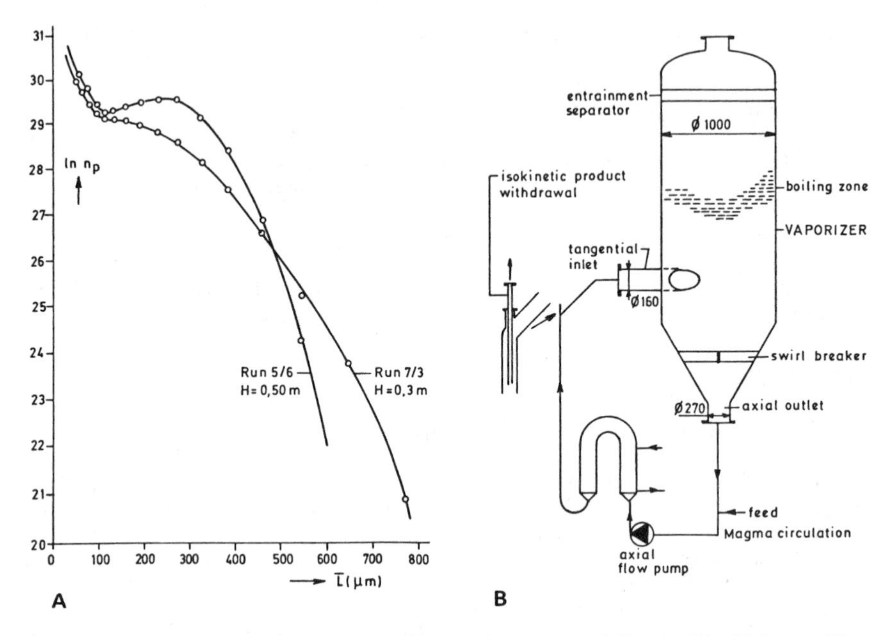

Figure 9. Typical population density plots (A) obtained in the 1,500 liter NaCl crystallizer (B) for two different suspension levels with reference to the center line of tangential inlet.

Depending upon operating conditions the slope at the upper and the lower end of the distribution and the width of the plateau may change, whilst the characteristic shape shown in Figure 9 is retained.

The reasons for the "peculiar" shape of population density plots obtained in large scale NaCl crystallizers are to be sought both in deviation from the MSMPR crystallizer operation and in the enhancement of population functions which were suppressed in experiments involving 55-liter crystallizers. The less-emphasized deviation from the straight line in the case of a lower suspension level in the evaporator body (see Figure 9) suggests that the flow patterns in the crystallizer may be related to population density through some classification effect. Another possibility may be sought in the fact that the external circulation feature may lead to crystals of various sizes being selectively exposed to spatial variations of supersaturation thus giving rise to pseudo size-dependent growth. In what follows classification effects will be examined, since these enhance deviation from MSMPR crystallizer conditions, and population functions likely to occur in industrial units will be discussed, since these complicate further the application of the population balance concept to design, operation, and analysis of large-scale crystallization processes.

Classification

If the density difference between solid and liquid phase is substantial and if large product crystals are produced, it is in generally very difficult to attain a well-mixed suspension with uniform intensive properties. The flow pattern established within a crystallizer often induces crystals to deviate from liquid stream-lines and this leads to classification, which is often size dependent and preferential according to the spectrum of fluid velocities involved. This effect may be called internal classification. If segregation is effected only at the product removal point, it is referred to as external classification. External classification may be selective for small crystals, and in this case we refer to a fines removal system. If external classification is selective for larger crystals, we may refer to it as classified product removal system.

External Classification

There are crystallizers designed to achieve external classification feature devices which exert influence on local hydrodynamic conditions in the region of the off-take point in such a way that a preferential group of crystals is removed. External classification can be selective for small crystals in which case we refer to fines removal. If, however, external classification exerts selectivity for large crystals, we refer to classified product removal.

Fines removal. / These systems are resorted to in cases where excessive creation of nuclei should be counteracted by design features and not by reduction in the driving force since this latter also leads to equivalent reduction in the production rate. However, high production rates are possible at high supersaturations, and this latter gives rise to excessive nucleation rates, often to an extent where production of large crystals is no more possible. In order to be able to run a crystallizer at high supersaturations and to produce large crystals at high production rates the excessive nuclei must not be allowed to compete for supersaturation. Therefore, they should be selectively removed from the crystallizing system, and this should be done while they are very small as to minimize production losses. In this way a high number of crystals is removed, but their mass represents only a small production loss. Fines removal systems feature mostly an annular settling zone (see Figure 10) where small crystals are preferentially situated. The fines are withdrawn from the crystallizer and are dissolved in the heat exchanger loop to be recycled as clear mother liquor. The

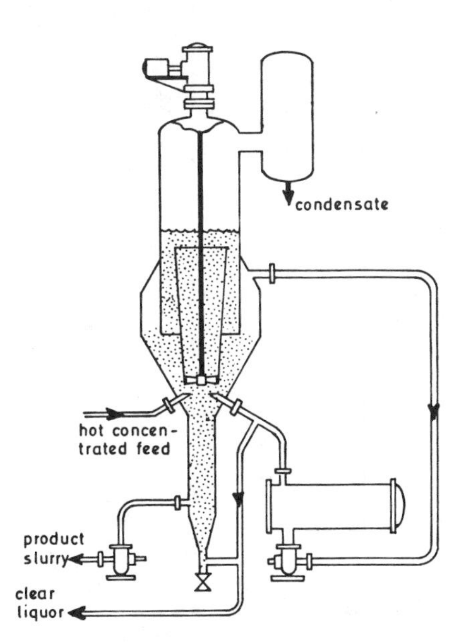

Figure 10. Schematic representation of a crystallizer featuring annular settling zone to effect fines removal and destruction thereof through dissolution (a DTB—Draft tube baffled crystallizer concept).

classification effect is both dependent on crystallizer design features and on operating conditions. In order to study the classification effect in DTB crystallizers, de Leer [22] used a 1 m^3 unit as shown in Figure 11. The annular settling space of a DTB crystallizer is considered as a poorly mixed volume inside the crystallizer body. The classifying action of the annular settling space which depends on the operating conditions, the salt-system used, and the geometry of the DTB-crystallizer can under stationary operating conditions be given as:

$$\lambda_f(L) = \frac{n(L)}{n_A(L)} \tag{31}$$

where $n_A(L)$ = the population density at the outlet of the annular volume
$n(L)$ = the population density in the well-mixed volume of the DTB crystallizer

The effect of fines withdrawal rate on the classification coefficient $\lambda_f(L)$ is shown in Figure 12. The results in Figure 12 show that for the potash alum/water system the

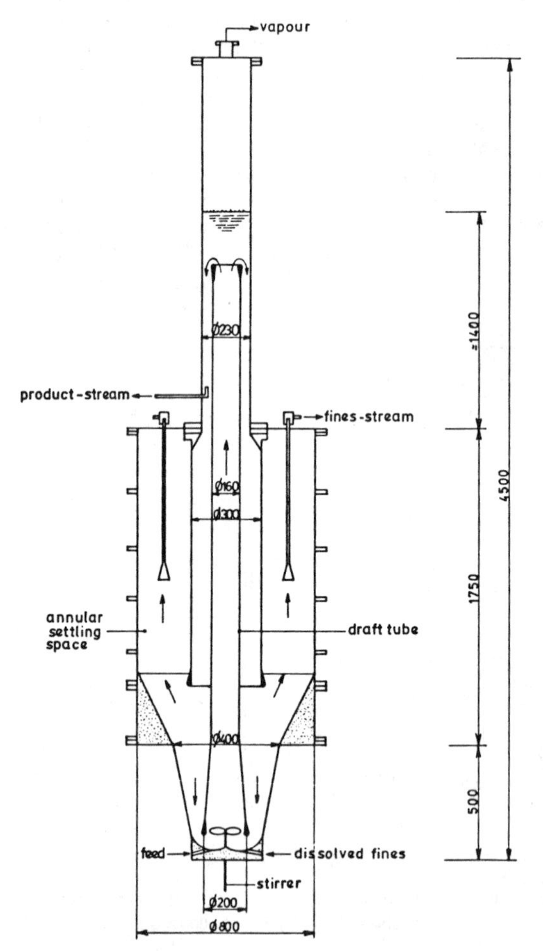

Figure 11. One m³ DTB crystallizer used for the study of fines removal.

classification coefficient is a strong function of crystal size and that it can be significantly influenced by fines withdrawal rate. It can be seen that increased fines withdrawal rate leads to an increase in the maximum size of fines withdrawn from the DTB crystallizer. It is also evident that high withdrawal rates lead to enhanced removal of crystal mass and in this way to increased product losses.

The effect of the fines withdrawal rate on the product crystal size distribution is demonstrated in Figure 13. For conditions where the height of the settling zone was 0.36 m and the impeller speed 800 rpm the product median size increased from 305 to 640 μm by increasing the fines withdrawal rate from 33.3 to 83.3 l/min. This effect is due to a higher number of small crystals as well as a larger portion of the available crystal surface area being removed at the higher withdrawal rates, thus allowing fewer crystals to benefit from the available supersaturation. This latter is forced to a higher level, which underlines the importance of not only removing large numbers of small crystals but also removing sufficient surface area in order to force the supersaturation to a higher level.

Classified product removal system. / Crystallizers featuring product classification facility exert hydrodynamic influences on the way the product is being removed. The classifying action is aimed at withdrawing larger and more uniform product crystals, and this is mostly effected in "salt catchers" and "elutriation legs." The result on product classification achieved by superimposing an elutriation leg on the double radial inlet configuration of a forced circulation evaporative crystallizer for sodium chloride (see Figure 14) has been obtained by Grootscholten (1982) through a joint research venture with Imperial Chemical Industries, United Kingdom. The elutriation leg device was attached to the conical bottom

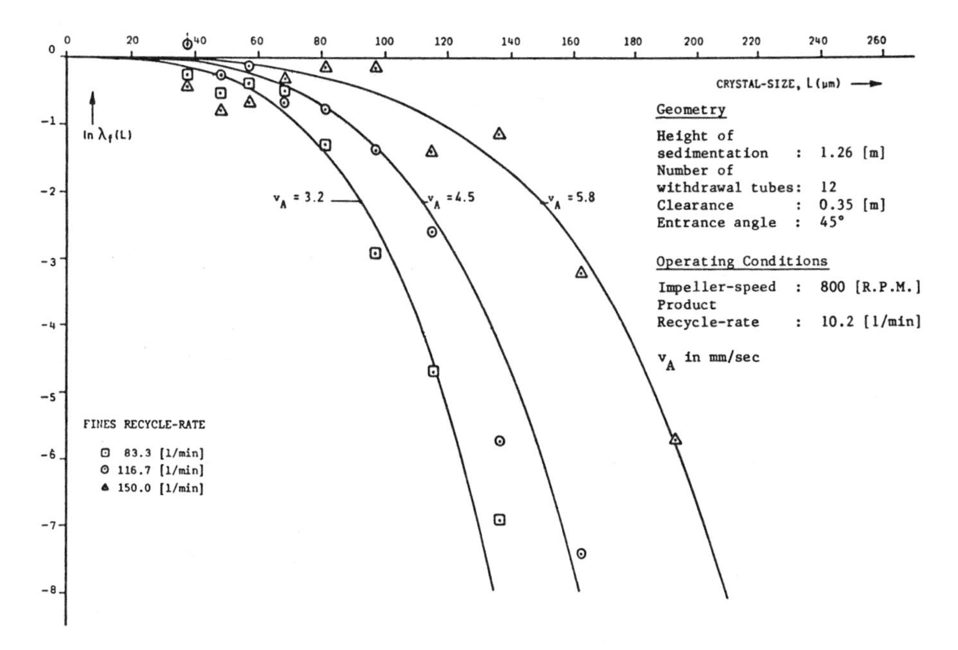

Figure 12. The effect of the fines withdrawal rate on classification coefficient in DTB crystallization.

of the evaporator body and the classifying action has been determined at conditions of continuous crystallization as:

$$\lambda_p(L) = \frac{n_c(L)}{n_p(L)} \tag{32}$$

The elutriation leg device is schematically represented in Figure 15 where the sampling points for population density in the circulation loop of an evaporative crystallizer, $n_c(L)$,

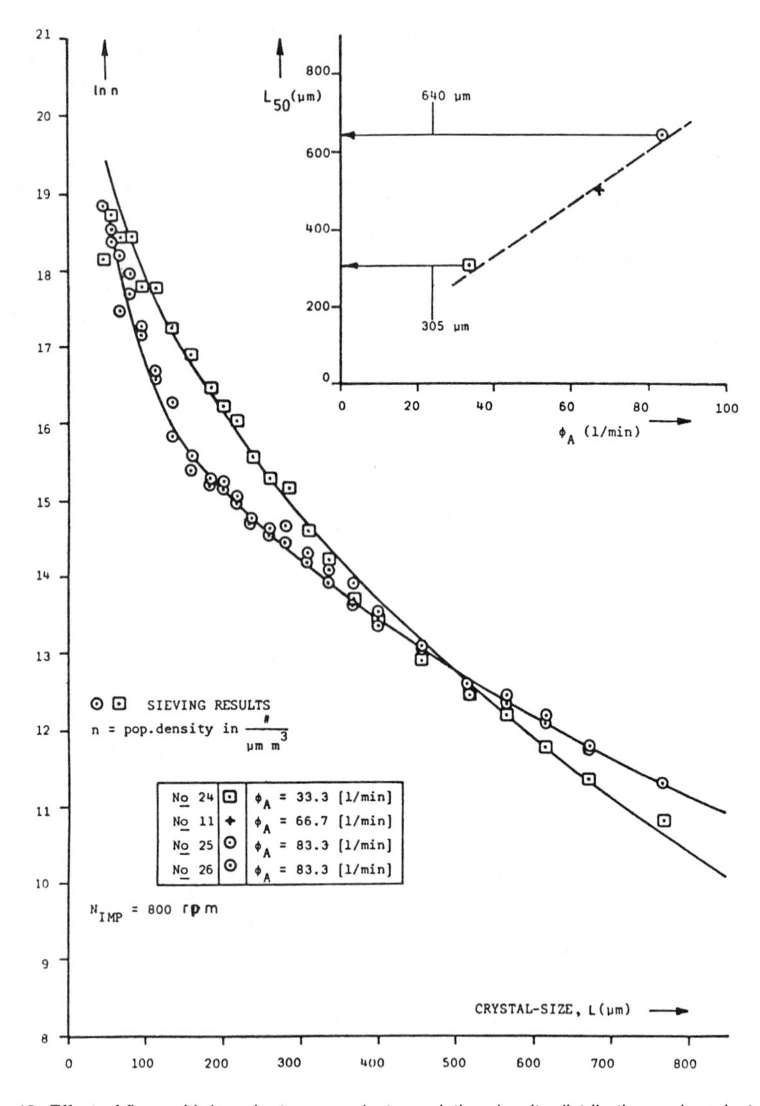

Figure 13. Effect of fines withdrawal-rate on product population density distribution and product median size.

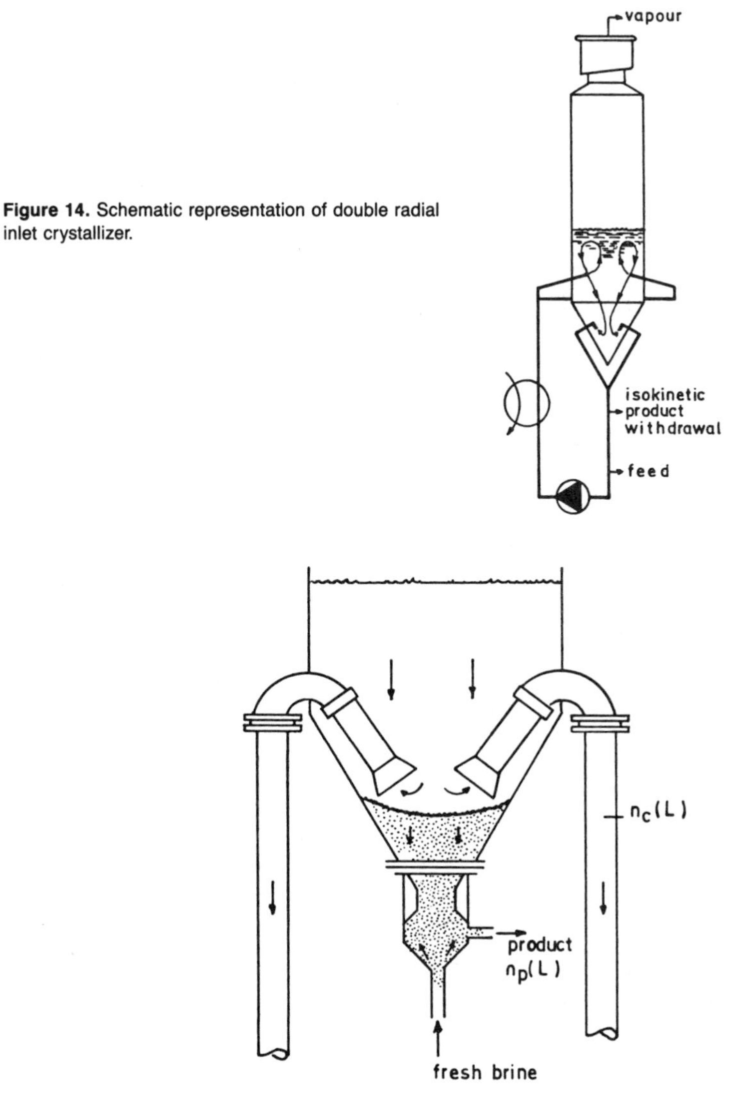

Figure 14. Schematic representation of double radial inlet crystallizer.

Figure 15. Schematic representation of elutriation device, used to classify product in continuous crystallization of sodium chloride.

and for population density of product $n_p(L)$ are shown. For reasons of confidentiality the details given in Figure 15 are not to scale.

The product classification achieved by the use of the device shown in Figure 15 is presented in Figure 16 for different conditions of product removal. For reasons of confidentiality the level of clear liquor advance is indicated as a percentage of the maximum value used in the experiments. The results indicate pronounced selectivity of the classifying device for particles larger than about 300 μm and at the same time underline the importance of

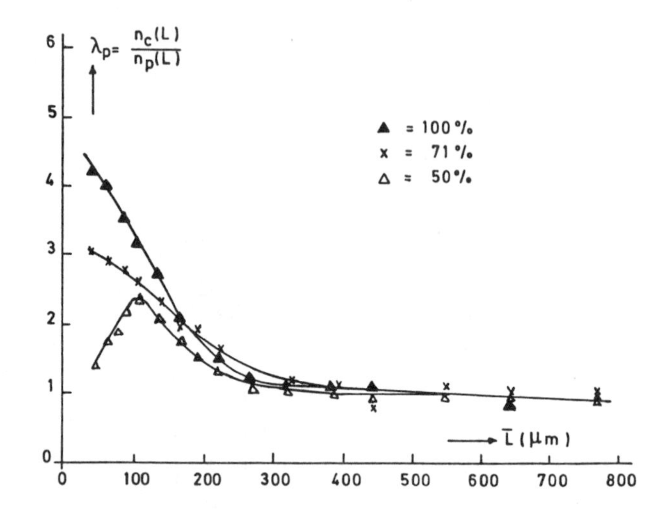

Figure 16. Product classification parameters obtained for elutriation leg device shown in Figure 15 for different levels of clear liquor advance.

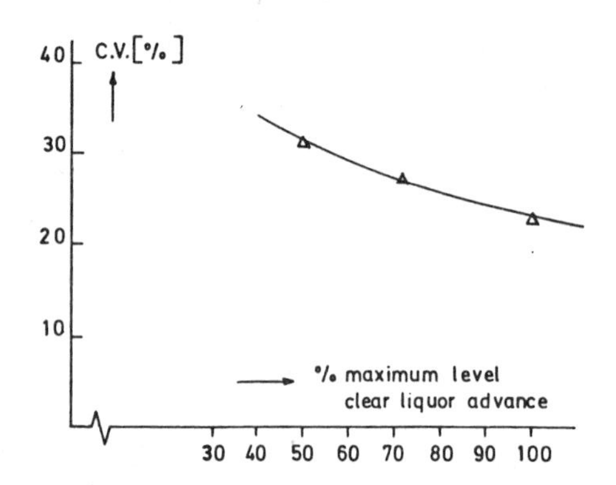

Figure 17. Effect of the level of clear liquor advance to classification device on the coefficient of variation of product withdrawn.

the choice of the level of clear liquor advance on classification parameter. Figure 17 illustrates the reduction in the coefficient of variation of the withdrawn product with an increasing level of clear liquor advance. The median size of the product withdrawn is, however, reduced with an increasing level of clear liquor advance to the classifying device (see Figure 18). This effect may be explained by increasing the selectivity of the product offtake stream for large crystals, which generates an accumulation of small crystals within the crystallizer. This latter situation leads to the reduction in supersaturation caused by the presence of not-withdrawn crystals which compete for their share of supersaturation. This fact points to the importance of combining product classification strategy with fines removal if large and uniform crystals are to be produced and subsequently withdrawn.

Population balance for external classification system. / For systems having both fines removal and product classification the population density distributions in the withdrawal streams differ from those in the crystallizer. At steady state and for conditions of simultaneous fines removal and products classification, the population balance, with reference to Equation 16, may be written as:

$$\frac{d\,(nG)}{dL} = -P(L) - \lambda(L) - \frac{Q_0 n_0}{V} \tag{33}$$

or

$$\frac{d\,(nG)}{dL} = -\frac{Q_p n_p}{V} - \frac{Q_A n_A}{V} \tag{34}$$

where Q_p and Q_A are the product fines withdrawal rates and n_p and n_A are the population densities in the corresponding streams.

A relationship between n, the population density in the crystallizer, and n_p and n_A may be established by defining crystal residence time $\tau_c(L)$ as:

$$\tau_c(L) = \frac{Vn}{Q_p n_p + Q_A n_A} \tag{35}$$

Combining and rearranging Equations 34 and 35 gives the slope of the semi-log population balance in terms of average population density in the crystallizer:

$$\frac{d\ln(n)}{dL} = -\frac{1}{G(L)\tau_c(L)} - \frac{d\ln G(L)}{dL} \tag{36}$$

Population densities in the product and fines removal stream can be found from classification coefficients as defined by Equation 31 and 32, respectively.

If $\lambda_f(L)$ and $\lambda_p(L)$ are known, $\tau_c(L)$ can be easily evaluated from the equation:

$$\tau_c(L) = \frac{V}{\dfrac{Q_p}{\lambda_p(L)} + \dfrac{Q_A}{\lambda_f(L)}} \tag{37}$$

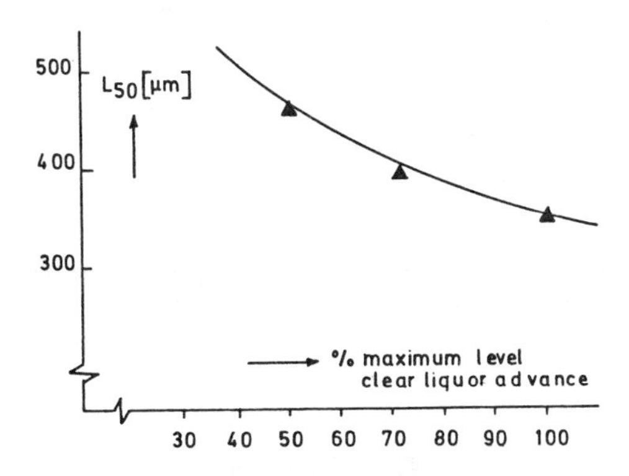

Figure 18. Effect of the level of clear liquor advance on the median size of product withdrawn.

For systems without fines removal ($Q_A = 0$) Equation 37 reduces to:

$$\tau_c(L) = \lambda_p(L) \cdot \tau \tag{38}$$

where V/Q_p is defined as the average liquid residence time τ. In this case it may be derived that the slope of the semi-log product population density distribution is given by:

$$\frac{d \ln (n_p)}{dL} = -\frac{1}{G(L)\tau_c(L)} - \frac{d \ln G(L)}{dL} - \frac{d \ln \tau_c(L)}{dL} \tag{39}$$

To obtain the population density distribution of the product crystals, the appropriate expressions for size-dependent growth rate and crystal residence time distribution must be introduced into Equation 39, and this must be solved for the boundary condition.

Internal Classification of Solids

Crystallizers exhibiting internal solids classification do so as a result of flow patterns established within the crystallizer body for each particular design configuration and way of operation. Internal classification is characterized by preferentially oriented spatial variations of size distribution and solids concentration, which violates assumptions upon which the MSMPR concept is based.

Inherent to local variations in concentration and distribution of the crystals are variations in the crystal surface area. Crystallization is a surface-area-dependent process and thus the driving force necessary for growth and nucleation is inversely proportional to the crystal surface area available for deposition of supersaturation. It is therefore not difficult to visualize that variations in local solids concentration and size distribution may induce local variations in the concentration driving force and as a consequence variations in growth rate. Internal classification can in such cases induce pseudo size-dependent growth rates by selectively exposing the growing crystals to spatial variations in supersaturation.

It is obvious that the overall population balance analysis for internally classified crystallizing systems is no longer straightforward as suggested by Equation 39. In this case a crystallizer may be considered subdivided into a number of packets with volume V_i, each volume having population density n_i, and crystal growth rate G_i. General population balance Equation 16 can now be rewritten to take internal classification into account. This gives:

$$\frac{d}{dL}\left(\sum V_i \dot{n}_i G_i\right) = -Q_p n_p \tag{40}$$

According to definition of crystal residence time (see Equation 35) $\tau_c(L)$ may be written as:

$$\tau_c(L) = \frac{\sum V_i \dot{n}_i}{Q_p n_p} \tag{41}$$

With reference to Equation 17 the average population density in the crystallizer can be found by averaging as follows:

$$n = \frac{\sum V_i \dot{n}_i}{V} \tag{42}$$

Population density, n, is a bucket-average or cupmixing density which can only be evaluated after analysis of the crystallizer content.

The local crystal growth rates G_i may be averaged to give an average crystal growth rate, which may be a function of crystal size. In formula:

$$\bar{G}(L) = \frac{\sum (V_i \dot{n}_i G_i)}{\sum (V_i \dot{n}_i)} \tag{43}$$

Combining and rearranging Equations 40, 41, 42, and 43 gives:

$$\frac{d\,[\bar{G}(L)\tau_c(L)n_p]}{dL} = -n_p \tag{44}$$

From this equation the slope of the population density plot for the product crystals can be derived as:

$$\frac{d \ln (n_p)}{dL} = -\frac{1}{\bar{G}(L)\tau_c(L)} - \frac{d \ln \tau_c(L)}{dL} - \frac{d \ln \bar{G}(L)}{dL} \tag{45}$$

Similarly the slope of the population density plot for the cup-mixing population density within the crystallizer can be written as:

$$\frac{d \ln (n)}{dL} = -\frac{1}{\bar{G}(L)\tau_c(L)} - \frac{d \ln \bar{G}(L)}{dL} \tag{46}$$

It can be shown that the average growth rate as defined by Equation 43 may be written in terms of crystal residence times, $\tau_i(L)$, in individual volume elements:

$$\bar{G}(L) = \sum \left[G_i(L) \frac{\tau_i(L)}{\tau_c(L)} \right] \tag{47}$$

where $\tau_i(L)$ is defined as:

$$\tau_i(L) = \frac{V_i \dot{n}_i}{Q_p n_p} \tag{48}$$

Equation 47 clearly shows that the evaluation of the average growth in a segregated system and hence the solution of the differential population balance are intimately tied up with the question: "During what time and to what driving force is the crystal of size L being exposed?" To answer this question knowledge is required concerning the distribution of concentration driving forces and individual residence times in regions of different driving forces. The successful application of the population balance technique will then depend on the accuracy with which the degree of solids and liquid mixing has to be and can be determined in a realistic system. It is obvious that in a realistic crystallizing system internal and external classification operate simultaneously and this complicates the correlation between crystallizer contents and crystallizer product stream. In such cases both phenomena will be implicit in the crystal residence time distribution $\tau_c(L)$.

Asselbergs [23] and Grootscholten [24] resorted to an extensive experimental program designed to shed light on classification functions in 1,500 liter evaporative crystallizers. A series of experiments was performed in order to derive solids residence time as a function of size. The technique relied upon washing-out transients and these involved experiments with sand/brine systems as well as more realistic salt/brine systems.

The influence of various crystallizer geometries has also been considered, and these are schematically represented in Figure 19. The geometry shown in Figure 19A is a standard design configuration featuring a horizontal tangential inlet of superheated liquor and a bottom axial outlet of crystal suspension. The geometry shown in Figure 19B, is a reversed situation to that shown in Figure 19A, and is achieved by reversing the direction of rotation of the circulation pump. By vertical introduction of the slurry through the axial bottom inlet the crystals were considered to be more effectively carried up to the region of the boiling zone. The last configuration presented in Figure 19C, features a double radial inlet of superheated slurry and double bottom-directed outlets to recycle circulating slurry to calandria.

The results of wash-out experiments clearly indicated that significant classification occurs with all geometries even at small sizes. Furthermore it became obvious that the choice of inlet and outlet configuration plays a decisive role as to which of the residence times is higher. For example, Figure 20 shows results of classification experiments for configurations described in Figure 19A and 19B together with population density distributions developed at identical conditions of classification. Similarly, the results of classification and corresponding crystallization experiments for classical tangential inlet configuration are compared with the double radial inlet configuration (see Figure 19C) in Figure 21.

The results in Figure 20 clearly show the effect of circulation flow reversal. For tangential operation the crystal residence time decreases with size. In the large-size range the curve levels out: crystals larger than a certain size probably all have the same (minimum) circulation time from inlet to outlet of the evaporator body. The same value of the overall solid-phase residence time is almost a factor 2 lower than the liquid residence time and suggests that the magma concentration in the evaporator-body is considerably lower than in the circulation loop and hence in the product stream.

For reversed flow operation large crystals are found to be preferentially kept in the crystallizer. The overall solids residence time is only 13% higher than the liquid residence

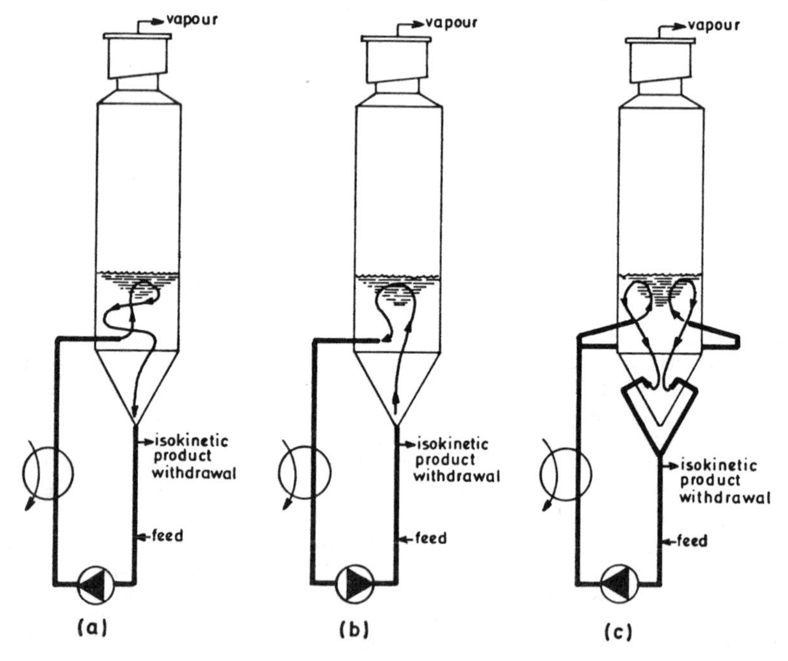

Figure 19. Schematic representation of various geometries concerning inlet and outlet in forced circulation crystallizers.

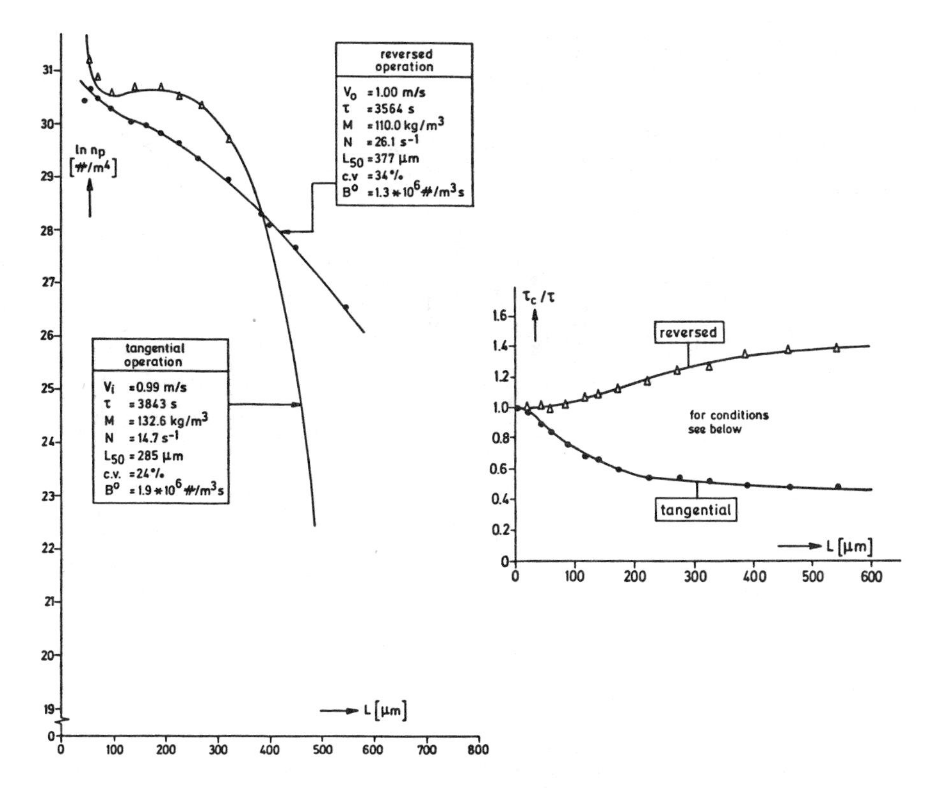

Figure 20. The influence of the inlet and outlet configuration on classification and obtained crystal size distributions at identical conditions for tangential and reversed operation.

time for the preceeding experiment. This observation indicates that reversed operation is a nearer approach to conditions of mixed suspension than tangential inlet operation.

As all measures have been taken to ensure isokinetic and representative product removal from the circulation loop, classification can be attributed to nonuniform suspension of crystal in the evaporator body. This conclusion is evidenced by the results of experiments, which were carried out to determine local solids concentrations in the evaporator body [23]. These measurements demonstrated that for tangential flow, operation solids are accumulated in the wall region. The solids concentration was found to be lower than in the circulation loop. For reversed circulation the hold-up in the evaporator body was much higher and in a majority of the cases even higher than in the external loop. For reversed circulation a spouted bed is created and the larger particles are retained in the evaporator body.

From Figure 21 it can be seen that the difference in solids classification characteristics caused by the tangential and radial inlet configurations are not very dramatic. Although the flow pattern in the double radial inlet system was found to be mainly axial, the average magma concentration in the evaporator body is still lower than in the circulation mains ($\tau_c(L)/\tau = 0.64$).

The results of classification experiments together with corresponding crystallization experiments (see Figures 20 and 21) reveal a relationship between the kind and extent of solids classification influenced by crystallizer geometry at one side and the unusual type of

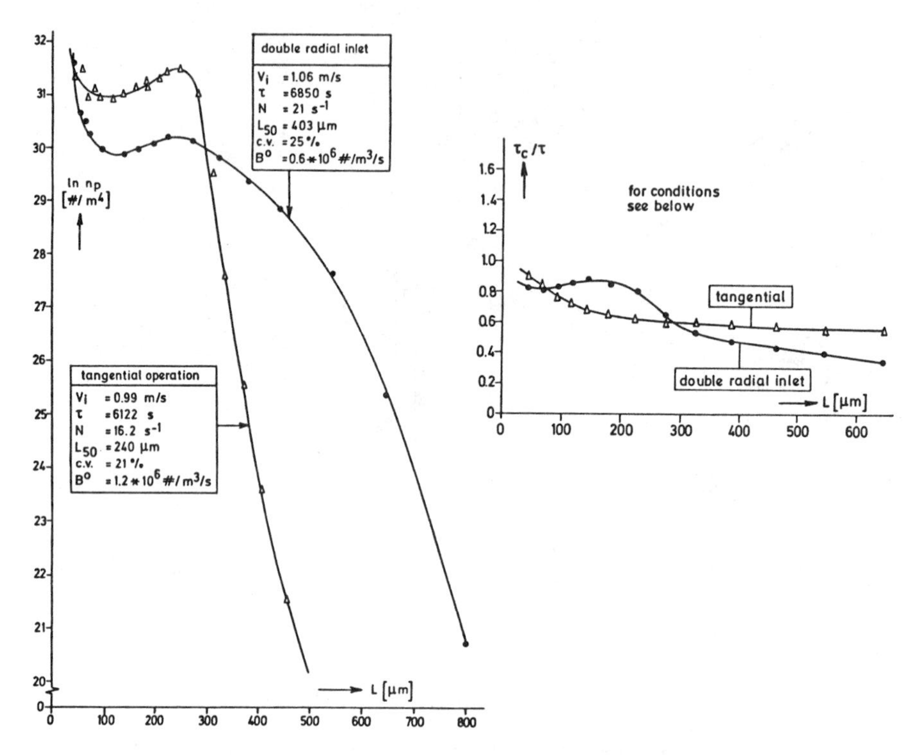

Figure 21. The influence of the inlet and outlet configuration on classification and obtained crystal size distributions at identical conditions for tangential and double radial inlet configurations.

crystal size distribution of the product, at the other. For example, the analysis of Equation 46 indicates that extremes in the population density plot (i.e., $d \ln (n)/dL = 0$) are only possible for negative values of $d \ln (\tau_c(L)/dL$ and this is indeed confirmed by results shown in Figures 20 and 21. For example, the solids residence time distribution for reversed circulation (see Figure 20) implies absence of any extremes in the population density plot and this appears to be supported by crystallization experiments performed for this flow configuration at identical conditions of classification (note the almost straight line of the relevant population density distribution in Figure 20).

Furthermore, the existence of inflection points in experimental classification curves explain, at least qualitatively, the occurrence of more than one extreme in population density plots (see Figure 20 and 21).

Despite this qualitative agreement on the influence of solids classification on the shape of the crystal size distribution (see Figure 20 and 21) it will be clear that the difference in crystallization behavior of the tangential and double radial inlet unit can never be explained on basis of the corresponding classification characteristics of solids alone. In further assessments of the influence of crystallizer-geometry on crystallization behavior the phenomena of liquid maldistribution and the pseüdo size-dependent growth rate has to be taken into consideration.

Internal Maldistribution of Liquid

In the previous section an implication has been made that internal solids classification may induce pseudo size-dependent growth rates by selectively exposing growing crystals to

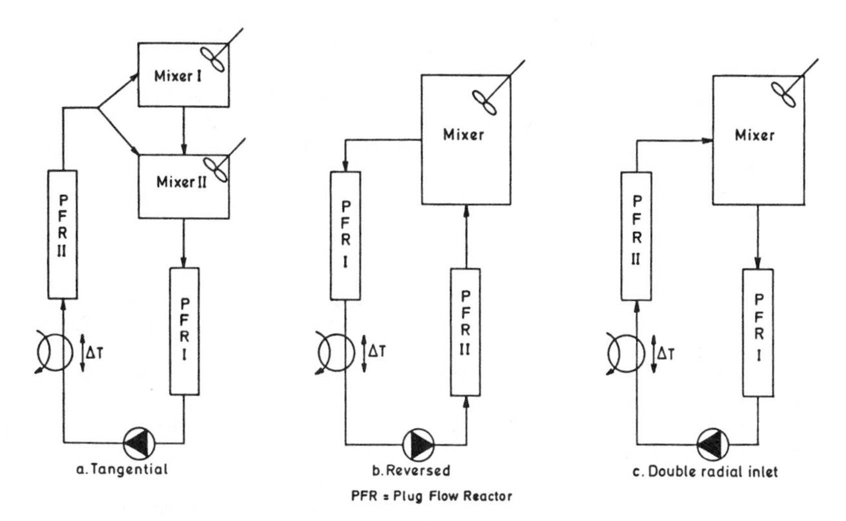

Figure 22. Reactor-in-series concepts proposed to account for liquid flow encountered in inlet/outlet crystallizer configurations presented in Figure 19.

local variations of supersaturation. However, rigorous analysis of the distribution of supersaturation within a crystallizer is very difficult due to both experimental and theoretical limitations. For example, variations in supersaturation are controlled by processes which tend to promote supersaturation decay (i.e., nucleation, crystal growth, liquor superheating in calandria, etc.) and by processes tending to dissipate it (i.e., liquor circulation, axial mixing, etc.). Such situations may be analyzed by resorting to analysis which regards a crystallizer as composed of a number of reactor blocks connected in series to form a closed loop, each reactor featuring idealized but well-defined flow. Application of this chemical reactor engineering approach, combined with supersaturation balances over the individual reactors can lead to an estimation of the driving force encountered in various sections of a crystallizer. Superimposition of such a driving force variation with classification phenomenon leads to the concept of pseudo size-dependent growth rate.

In order to quantitatively describe the extent of liquid maldistribution for the configurations shown in Figure 19, Grootscholten (1982) proposed several reactor-in-series concepts which are based upon experimentally obtained knowledge of velocity and temperature distribution within the crystallizer (see Figure 22).

The tangential inlet crystallizer is represented by the concept proposed in Figure 19A. The loop comprises two ideal mixers and a shortcircuiting bypass over the first mixer. In view of the high liquid velocities (0,5–2,0 m/s) it was assumed that the flow in the external loop is plug-flow (PFR 1 and PFR 2). The fraction of the evaporator inlet stream, which enters the boiling zone and hence the first mixer, follows from heat balance considerations. The volume of the first mixer is taken as the evaporator-body volume above the centerline of the inlet. The second mixer is considered to occupy the region below the centerline of the tangential inlet. The mixer is fed by the outlet flow from the first mixer and the short-circuiting bypass stream from the tangential inlet.

For reversed operation the evaporator body is regarded as an ideal mixer whereas the flow in the external loop resembles plug-flow (see Figure 19B). Short-circuiting experiments have shown that short-circuiting is virtually absent for reversed operation and that the degree of solids mixing is substantially lower than for tangential operation. Considering the high degree of solids mixing and the short-circuiting phenomenon the preceding assumption seems not to be unrealistic.

The evaporator body of the double radial inlet system is also assumed to be a continuous-stirred-tank reactor (see Figure 19C). This concept is further supported by the results of short-circuiting measurements which indicated that hardly any short-circuiting of mass flow occurs. Pseudo size-dependent growth rates have been evaluated from the estimated driving force distribution and sectional residence times (this is discussed in a later section). The results are presented in Figure 23 for different design configurations.

The simulation results clearly indicate that both internal solids classification and pseudo size-dependent crystal growth are contributing to the development of product size distribution in industrial crystallizers. The simulations further show the substantial differences in pseudo size-dependent growth behaviour.

Reversal of the circulation direction diminishes deviations of the individual growth rates from the crystallizer mean value. A comparison between results presented in Figure 23B and D shows that for tangential operation the variations of crystal growth rate are much larger than for the double radial inlet systems. Considering the shape of the τ_c-curve and the nonuniform distribution of the solids concentration in the evaporator body, it is most likely that large particles are not effectively carried up to the boiling zone (first mixer) and preferentially populate the cone section of the evaporator body (second mixer) and the external loop, in which the supersaturation is lower than in the first mixer. It is, moreover, not difficult to visualize that short-circuiting of liquor enhances the driving force in the first mixer, thus promoting the nonuniform distribution of the driving force. For the radial inlet configuration the generated supersaturation is more effectively distributed over a large volume (only one mixer), which reduces local deviations of supersaturation from the crystallizer mean value.

The results further show that the hydrodynamics can be manipulated by crystallizer geometry in order to influence the development of product crystal size distribution (note the difference in height above the tangential inlet in Figures 20 and 21 and the effect thereof on the pseudo size-dependent growth rates of Figure 23). As such, relationships between crystallizer geometry and hydrodynamics should deserve a more prominent place in design strategies.

In order to elaborate on the type and degree of mixing in the evaporator body Grootscholten [24] relied upon tracer injection techniques in order to test the validity of the reactor-in-series concept presented in Figure 22.

Attrition and Breakage

Attrition is common in industrial crystallization and is characterized by visible damage on the surface of parent crystals. It can take place even in undersaturated solutions and often assumes a mechanical character rather than being solely governed by thermodynamic driving forces for crystallization. If properly controlled, attrition can sometimes help produce "rounded off" crystals with desirable dewatering, drying, and storage properties. At the same time, attrition is an additional source of nuclei and as such should be carefully controlled.

Most probable causes of attrition are high energy mechanical contacts with growing crystals, such as impellers, pumps, small clearances between rotating and stationary parts, etc. Grootscholten [24] performed a study to gain more understanding of the fragmentation of crystals undergoing mechanical stresses of the kind likely to be encountered in industrial crystallization. For this purpose a 7.5-liter agitated vessel was used featuring a draft tube and a pitched blade impeller (see Figure 24). Two impeller diameters have been used in order to evaluate the effect of radial clearance between impeller and the draft tube on attrition. In order to eliminate growth and dissolution of parent crystals and of attrition fragments, the experiments have been performed in pure ethanol.

Figure 25 shows the effect of the radial clearance between the impeller and the draft tube on the crystal size distribution resulting from exposing a narrow sieve fraction of seed crystals to attrition. A sieve fraction $419\ \mu m < L < 498\ \mu m$ has been allowed to undergo

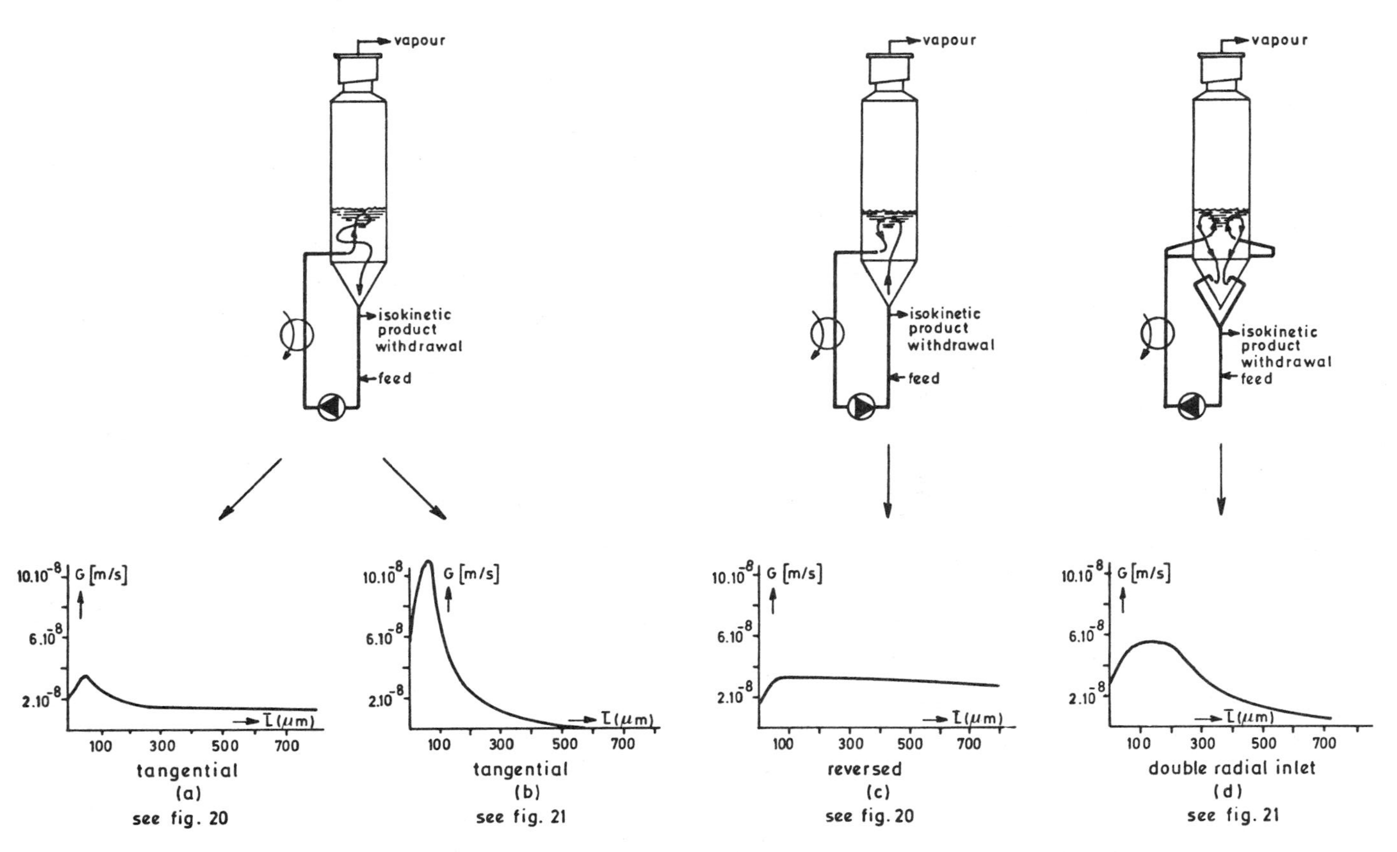

Figure 23. Evaluated pseudo size-dependent growth rates arising from spatial variations in supersaturation for each design configuration.

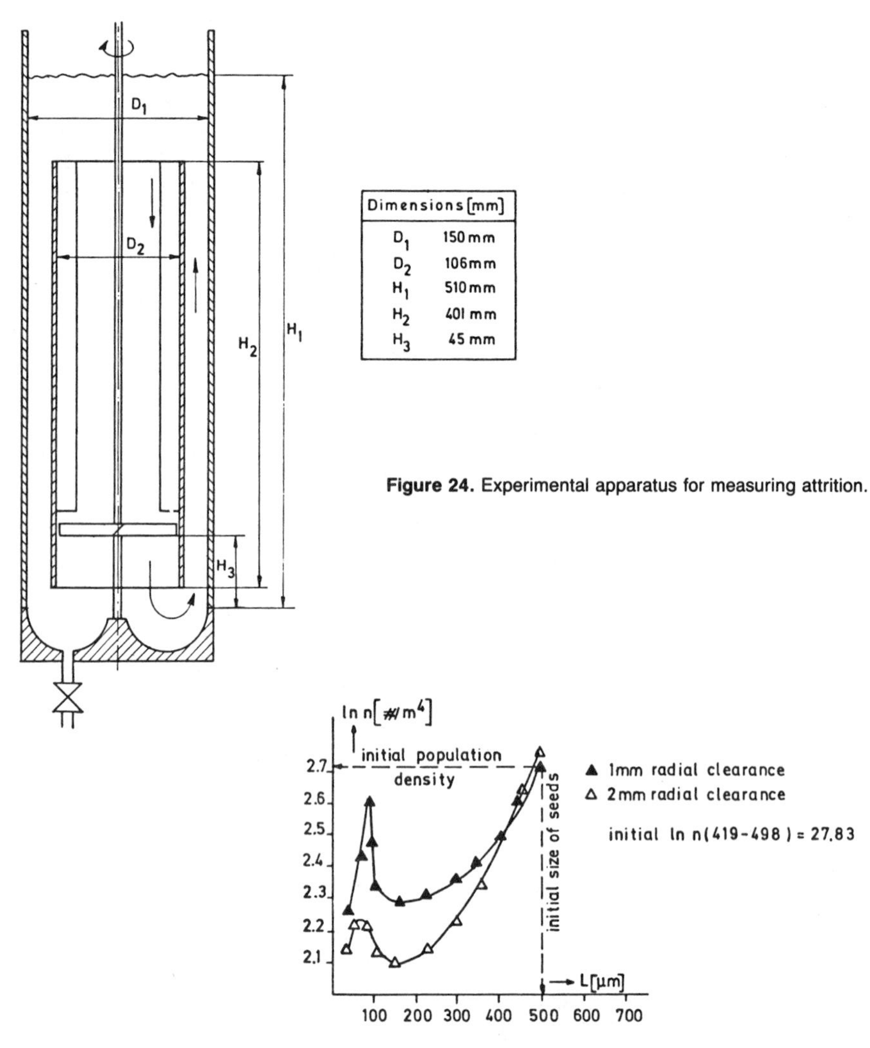

Figure 24. Experimental apparatus for measuring attrition.

Figure 25. Crystal size distribution developed from a narrow fraction of seeds having undergone attrition during 120 seconds at conditions of 540 rpm and with suspension density of 20 kg/m^3.

attrition for a period of 120 seconds at conditions of impeller speed of 540 rpm and with a suspension density of 20 kg/m^3. The results indicate that smaller radial clearances lead to more intensive generation of attrition fragments, these latter being most frequent in the size range between 50 and 100 µm.

In order to assess the influence of seed size, the experiment has been repeated at a constant radial clearance of 1 mm, constant stirrer speed of 540 rpm, and constant suspension density of 20 kg/m^3 but with two different sieve fractions of seed crystals.

The results indicate again that most attrition fragments are generated into the size range between 50 and 100 µm. It further appears that large seed crystals are more effective in generating attrition fragments, at least at small radial clearances.

Figure 26 shows a succession of photomicrographs obtained by electron microscopy for seed crystals and for a number of fractions generated by exposing the seeds to attrition. The photomicrographs indicate the absence of cubic habit in fragmented crystal particles. This implies that attrition occurring within clearances is not capable of generating fragments with the same habit as that of seed crystals.

Population Balance

In the particle phase, space crystal fracture represents discontinuous particle trajectories from the large to the small size ranges. Crystal fracture phenomena are formally represented in the population balance by birth and death functions B(L) and D(L). The population balance Equation 24 extends to:

$$\frac{d\,(nG)}{dL} + D(L) - B(L) = -Q_p n_p \tag{49}$$

The death function D(L), describing the probability of fracture at a given size, will be a complex function of the crystallizer environment, population density, and crystal size. The birth term B(L) comprises a probability function describing the relative occurrence of sizes of the fragments formed by breakage of larger crystals. B(L) is fundamentally related to D(L), since a death event at a given size represents at least two birth events at smaller sizes. It has been shown by several workers [19, 25] that formulation of consistent birth and death functions is a very complicated matter. Two body birth and death functions have been formulated by Randolph [20] assuming that fracture results in birth of two particles of equal size. The solutions of the resulting integro-differential equation showed a narrowing of the crystal size distribution as well as a reduction in median size. Randolph also inferred that the net effect of abrasion (i.e., formation of small fragments with the size of the parent

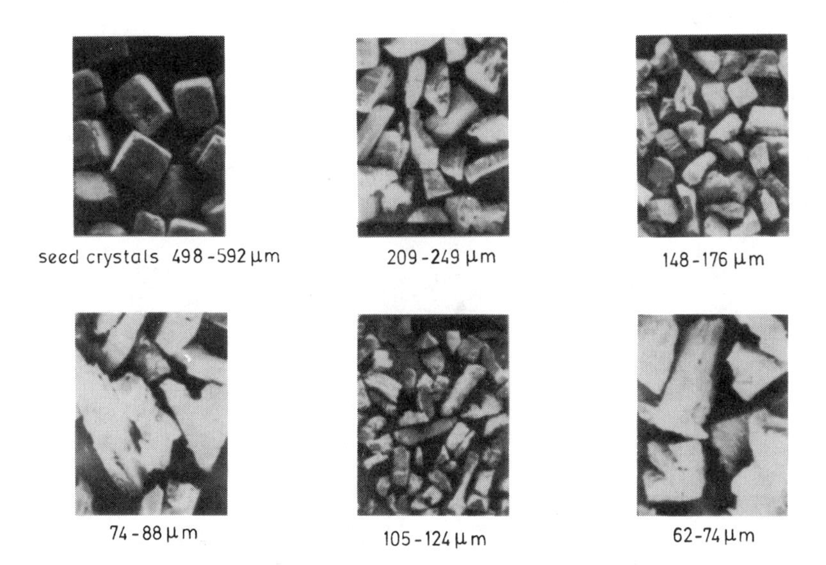

Figure 26. A succession of electron photomicrographs starting from seed crystals undergoing attrition within small radial clearance zone and showing the habit of generated fragments.

crystals remaining constant) would be a higher (apparent) nucleation rate, whereas no major change in the shape of the population density distribution occurred.

Fitzgerald [25] extended this work and modeled breakage on the basis of the more realistic assumption that volume (or mass) rather than size is conserved upon breaking and that all crystals have the same probability of breaking. It was shown that when new particles are formed by breakage alone (i.e., no nucleation), the crystal size distribution may deviate considerably form the semilog population density plots as predicted by Equation 24.

To evaluate the effects of abrasion on the crystal size distribution Juzaszek and Kawecki [26] offered a different approach which does not rely on birth and death functions. An expression for an effective growth rate was proposed, incorporating both the effects of crystal growth and abrasion. It was assumed that the largest fragments produced by abrasion are sufficiently small to be counted as nuclei. Resulting population density distributions were concave downward, exhibiting increased curvature for exacerbated abrasion.

If we consider the spectrum of sizes resulting from the fracture of monosized crystals (see salt mill experiments), it will be clear that the simplification of birth and death functions often results from mathematical reasons rather than through the results of experimentation.

In order to quantify the birth and death function (B(L) respectively D(L)) in sodium chloride crystallization, Grootscholten [24] operated the MSMPR crystallizer, shown in Figure 7, as a batch attrition apparatus, wherein the crystal suspension, obtained under conditions of continuous MSMPR crystallizer, was used as the starting material. By monitoring the population density distribution as a function of time, excess birth and death rates were evaluated from the population balance equation adapted to the batch attrition apparatus:

$$\frac{\delta n}{\delta t} = B_E(L) - D_E(L) \tag{50}$$

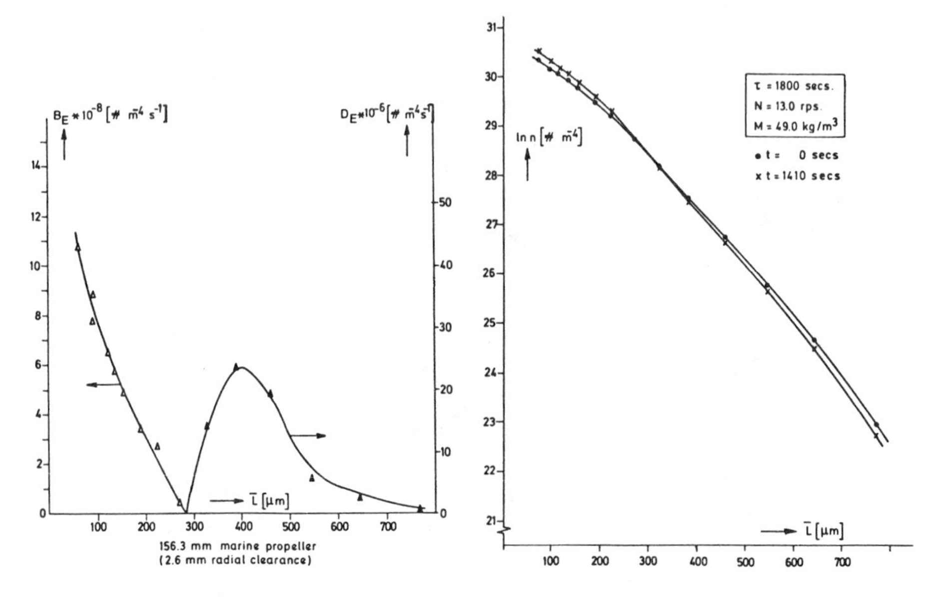

Figure 27. Population density plots and excess birth and death function as function of size, 91 liter MSMPR crystallizer for sodium chloride-water system.

If an increase in population density n was found with time, $\delta n/\delta t$ was denoted as an excess birth function, whereas decreasing population densities were interpreted as excess death. The concept of excess function was used because the possible overlap of both functions at intermediate sizes implies that in that case real birth and death functions cannot be evaluated a priori from the described experiment without assuming one of the two functions.

Typical semilog population density plots and functions derived thereof according to Equation 50 are given in Figure 27.

To preserve internal consistency the mass detached from the large size range should be conserved in the small size range. In mathematical terms:

$$k_v \varrho_c \int_0^{L^*} B_E(L) L^3 \, dL = k_v \varrho_c \int_{L^*}^{\infty} D_E(L) L^3 \, dL \tag{51}$$

where $B_E(L^*) = D_E(L^*) = 0$

Whether birth and death functions can be taken into account in the prediction of crystal size distribution in large crystallizers depends on whether the phenomena of crystal breakage can be scaled-up with a sufficient degree of confidence.

Agglomeration

One of the most ambiguous problems in attempting to relate the morphology of the crystal product to external solution conditions and to crystallization kinetics is the formation of agglomerates. Natural and synthetic particulate systems consist not only of ultimate particles but also of clusters or agglomerates of such particles.

Agglomeration of particles can be quantified either in terms of the degree of agglomeration or by the agglomeration rate. The degree of agglomeration indicates the states of agglomeration in a sample at a given time and is normally expressed, either on a number or weight basis, as the ratio of agglomerated particles to the total particles in the sample. The agglomeration rate indicates the progress of a reaction by which particles are agglomerating as a function of time.

The presence of agglomerated particles in a process could be a nuisance or an advantage depending on the process and the field of technology involved. For example the direct removal of very fine particulate matter from a suspension by means of sedimentation or filtration is impractical on an industrial scale. Consequently, these operations are often preceded by an agglomeration stage where stable particle clusters more amenable to separation are formed. Such processes are commonly found in water and sewage treatment plants where the agglomeration of fines is carried out in large vessels, normally with the aid of a coagulent. Thus in this case, the presence of large agglomerates would be useful. Conversely, in the sugar industry the presence of agglomerates, commonly called "conglomerates" (i.e. objectionable grains, irregularly grown together) delay the exhaustion of the massecuite, and the purging and drying of the resulting sugar. Remelting of conglomerates normally results in increased costs and limitations on the capacity of the refinery. In fact, the suppression of conglomerate formation is a dominating problem in sugar boiling. The presence of conglomerates causes the crystal product to be unnecessarily bulky and results in higher costs in packaging and difficulties in storage and thus must be kept to a minimum.

Very few workers have examined the agglomeration problem in solution crystallization. There are a number of reasons for this. There is a lack of knowledge of the behavior of particulate agglomerates in solution, not only under static conditions, but also under conditions of changing concentration and liquid composition. Further, there are no convenient techniques for measuring the distribution of agglomerate size. From an industrial point of view it has often been easier to overcome agglomeration problems by introducing or removing additives or impurities, by employing ultrasonic irradiation or by altering the operating parameters of the process, rather than finding the basic cause of the agglomeration phenomenon.

Effect of Operating Parameters on Agglomeration Rate

In most previous studies on the crystallization of soluble salts from supersaturated aqueous solutions, the experimental conditions were chosen so as to minimize agglomeration. Consequently little information is available on the precise role played by agglomeration phenomenon in the unit operation. Nevertheless, the following operating parameters may influence agglomeration in crystallizing systems:

- Temperature.
- Liquid turbulence.
- Magma density.
- Supersaturation.
- Particle size distribution.
- Residence time.
- Purity of reagents.
- Vessel and stirrer material.

An increase in temperature for example could have an indirect effect through the liquid viscosity, since an increase in temperature lowers the viscosity of the medium and enhances particle mobility, which in turn may affect agglomeration. Liquid turbulence in a vessel depends on the vessel geometry and the hydrodynamics of the system (stirrer pitch, stirrer speed, presence of baffles, volume of the vessel, etc.). An increase in liquid turbulence would increase the collision frequency between the particles in the suspension and so increase the agglomeration rate. Higher magma or suspension density would also increase the collision frequency between the particles. However the two preceding parameters also have another effect—crystal breakage caused by particle-stirrer and particle-particle collisions. This will tend to break-up agglomerates. A number of cases are known where the supersaturation is the controlling factor in the formation of agglomerates, i.e. the formation was either enhanced or prevented (see for example Table 3).

The particle size distribution could have a large effect on the agglomeration through the size-dependent tendency of the originally well-developed single crystals to agglomerate. The existence of a critical size range where agglomeration preferentially occurs may be a result of two opposing effects, viz., the increasing probability of collision between the growing single crystals and the increasing kinetic energies of the crystals; the latter effect, which has to be taken into consideration above a certain crystal size, favors the tendency of the crystals to separate in spite of the occurrence of collisions.

The residence time of particles could be important in that an increase in residence time increases the total number of collisions between particles, thus offering more opportunity

Table 3
Agglomerate Content in the Fraction 1.4 < L < 1.7 mm for Different Process Conditions in Continuous Sucrose Crystallization

Run	Agglomerate content (%)	Production rate (kg sugar/m^3 h)	Evaporation rate (kg vapor/m^2 h)	Temperature difference: feed—crystallizer (°C)	Remarks
1 (5 runs)	90–100	350–400	200–280	20–24	Adiabatic operation
2 (10 runs)	50–80	150–300	100–180	10–15	Adiabatic operation
3 (15 runs)	90–100	350–700	280–540	7–8	Partial heat input
4 (5 runs)	40–60	150–300	140–280	2–4	Partial heat input

for particles to agglomerate. It is common knowledge that agglomeration problems in many pure or impure systems have been overcome by the introduction or removal of additives or impurities (e.g., addition of $MgSO_4$ and NaCl to KCl brine stops agglomeration). Finally, changing the stirrer material would have an indirect effect through the nucleation rate since it has been shown by many workers that with harder stirrer materials, e.g., stainless steel, the nucleation rate increases with an increase in the total number of particles present in the system. Consequently, an increase in the collision frequency between particles would result, and an increase in agglomeration rate might be expected with harder stirrer material.

Effect of Physical Properties on Agglomeration

Besides operating parameters, which have been mentioned previously, physical properties of the system can also exert influence on agglomeration. For example, one of the main factors that has been suggested as being the cause of conglomerates in sugar boiling is the surface tension of the liquid. It is known that a crystal undergoing mass transfer leads to the establishment of a concentration gradient that can give rise to local differences in surface tension, which may be such as to promote aggregation of colliding particles. Similar to this situation is coalescence induced by the Marangoni effect in various mass transfer separation processes.

The preceding discussion implies that the agglomeration rate can be strongly dependent on many factors. A few quantitative examples will be presented in what follows.

Experimental Quantification of Agglomeration

Published data of Kuijvenhoven [27] represent a quantitative evidence of agglomeration in continuous sugar crystallization. The work has been performed in a 1.5 m^3 continuous crystallizer featuring a draft tube and a marine-type impeller. The operating conditions included variations in the production rate in the way in which the heat of vaporization was supplied. This latter was achieved either adiabatically relying upon highly superheated feed or by partially supplying heat by condensation of steam. The percentage of agglomerated product crystals has been evaluated through photomicrographs by visual counting. The results are presented in Table 3 for lumped groups of experiments. The results obtained at

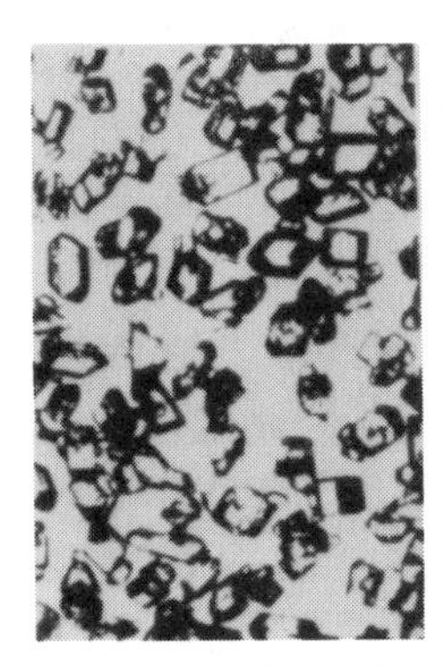

Figure 28. Effect of production rate on appearance of sucrose product crystals (200–250 μm) obtained in continuous crystallization at adiabatic conditions: (A) production rate—430 kg sugar/m^3h; (B) production rate—200 kg sugar/m^3h.

high production rates at adiabatic conditions (see run 1) show that almost all product crystals are agglomerates. This is probably due to high supersaturation levels and to vigorous local flashing of superheated feed liquor. Results obtained in run 2 refer to lower production rates at adiabatic conditions characterized by reduced local flashing of feed. Consequently, the product crystals contain less agglomerates (50%–80%) than product obtained in run 1 (90%–100%). The influence of the level of production on the appearance of product crystals is illustrated in Figure 28.

Results of run 3 show that partial heat input is not a sufficient condition for reduction of agglomerates in product crystals. Although the local flashing of feed solution and, through this, high local supersaturation are suppressed, the high production rate is possible due to high level of supersaturation in bulk solution which appears to have given rise to a high level of agglomerates in the product (90%–100%).

The lowest level of agglomerates obtained in experiments shown in Table 3 is 40% to 60% and refers to a low production rate and a low level of local flashing of superheated feed. This leads to the conclusion that low mean supersaturations and an absence of high local supersaturations leads to less agglomerated product crystals.

DESIGN-ORIENTED CRYSTALLIZATION KINETICS

In one of the previous chapters it has been shown that for size-independent growth $n = n^0 \cdot \exp(-L/G\tau)$ (Equation 26) is the fundamental relation between crystal size and the population density of crystals obtained in an MSMPR crystallizer. If experimental population densities are plotted logarithmically against crystals size, a straight line should be obtained. The crystal growth rate can then be derived from the slope of the straight line: the nucleation rate can be calculated from the intercept, $\ln(n_0)$, and the growth rate, G, according to the following equation:

$$B^0 = n^0G \qquad (52)$$

If the population density distribution is determined from sieve analysis data, the smallest size at which still reliable data points can be obtained is usually between 50 and 100 μm. In that case extrapolation to zero size has to be made through the subsieve size range where generally no data points are available. In most cases linear extrapolation is performed, and this should pose no difficulties if the nature of the phenomenon follows the linear law over the entire region of extrapolation. This, however, cannot be substantiated with any degree of certainty since our understanding of the crystallization process in the micron size range is still limited by both theoretical and experimental limitations. As a consequence the results obtained from such extrapolations are suspect both as far as physical significance and accuracy are concerned.

It has further been pointed out that not all population density plots are straight lines, especially those obtained from large industrial crystallizers. In this latter case various population events can be operative and this results in the deviation from the idealized situation described by the straight line population density plot. It has also been shown in the literature that crystallization kinetics derived from such curved population density plots are very sensitive to the size down to which the size analysis has been performed and this can vary in a most unpredictable manner. This implies that the critical extrapolation step is often performed arbitrarily and according to taste and intuition rather than following a well-defined convention. This fact creates ambiguities for the design engineer when kinetic data from different authors are compared in order to choose the most suitable set of data for a given design problem. There is obviously a need to introduce a unified approach concerning both extrapolation procedures and size ranges through which the extrapolation is to be performed. Extrapolation crystallization kinetics would still lack physical significance and would be based upon trends governed by phenomena occurring outside the extrapolation region. In other words the extrapolation nuclei population densities are a reflection of the extrapolation technique and as such cannot be coupled at some later stage during design with population functions in order to predict the crystal size distribution for a given design configuration. An illustration of this problem is given in Figure 29. It is clear that even if n^0 represented a meaningful physical quantity, it would have been very difficult to predict from it population densities at larger sizes, using Equation 16, since the relevant population functions in the micron size range are not defined due to problems associated with size analysis and our lack of understanding of all phenomena involved. In this small size range the most significant population function is probably that of secondary nucleation, and this has not yet been rigorously quantified.

It is due to both theoretical and experimental limitations that in practice real nucleation rates can only be approximated by fictitious number rates which are frequently termed effective nucleation rates. So, nucleation rates, obtained from linear extrapolation through the subsieve size range, represent fictitious number rates of nuclei, which are born at zero size and which grow at a fictitious constant rate to larger sizes.

In practice a variety of methods can be used to derive effective kinetic data from experimental crystal size distributions. The question of which derivation method should be used will in practice depend on available size analysis techniques as well as the target of the experimental study.

If the size distribution has to be used to derive fundamental nucleation data, then the small size range where nuclei are born will be an important size region. The size analysis technique and the method of deriviation should be adjusted to adequately cover this area.

If the derived kinetics are to be used for chemical engineering purposes (and this is not always in contradiction with the first target), the use of effective kinetic rates will often be satisfactory as long as these kinetics can be used to predict the commercially important part of the crystal size distribution.

A proposal has been put forward by Jancic [28] to standardize procedures involved when deriving crystallization kinetics. Rather than using some fictitious extrapolated value for

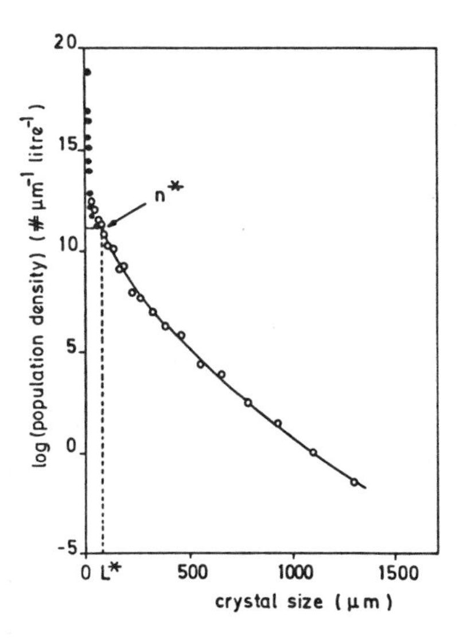

Figure 29. Schematic representation of extrapolation to the fictitious zero size through the size region where no data points are available. The values of L* and n* lie within the range of standard size analysis techniques.

n^0 at $L = 0$ it was suggested to use a finite value for the population density n^* at size L^* (see Figure 29). Size L^* must fall within the range of most routinely used size analysis techniques. If sieves are used to determine size distributions, size L^* could, for instance, be chosen as 100 μm. This procedure allows the uncertain and often arbitrary extrapolation to be substituted with a more reliable interpolation. Effective nucleation rates, obtained in this manner, represent a convective number rate at L^* ($B^* = n^*G^*$) rather than a fictitious nucleation rate at zero size. Kinetics derived in this way are probably less uncertain as far as accuracy and physical significance are concerned than in cases where nonlinear extrapolation to zero size has to be made.

The use of standard crystallization kinetics in crystallizer design has in principle no effect on current design procedures. The effect of this standardization proposal is, in the first instance, to focus attention on that part of the crystal size distribution which is in the range $L^* < L < \infty$. When L^* is chosen at the 100-μm size level the required portion of the experimental crystal size distribution is complete for most purposes encountered in engineering practice. Furthermore the value of L^* lies in this case outside the range that is relevant for prediction of mean sizes and variation coefficients for most distributions developed under moderate mass flux.

Essential for the use of effective kinetic data in predicing size distributions is that these kinetics are used mathematically in the same way as they have been derived. Mathematics involved in derivation and prediction procedures should be uniquely defined according to a well-established convention.

Prediction of Secondary Nucleation Kinetics

The knowledge of secondary nucleation kinetics in cases where this standardization procedure is followed becomes less important as long as reliable values of the standard population density, n^*, are available. The value of n^* is undoubtedly influenced by secondary nucleation kinetics as well as by other difficult-to-measure particulate events occurring in the subsieve micron size range. The net result of all these known and unknown

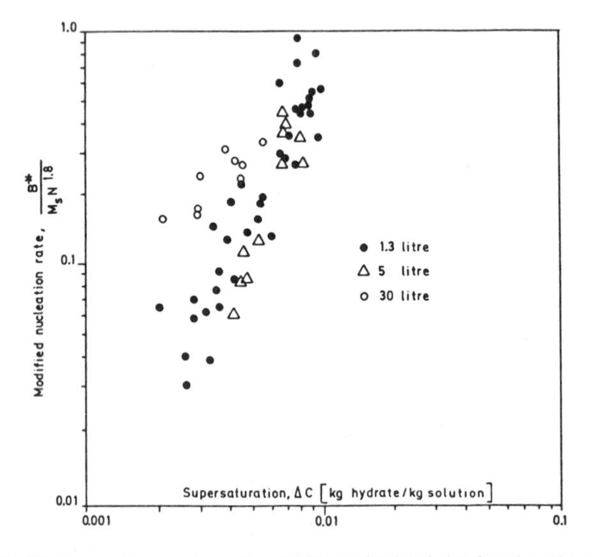

Figure 30. Modified convective number rates at 100 μm size level as a function of supersaturation obtained in three different MSMPR crystallizers at 30°C (after Jancic and Grootscholten [4]).

particulate events is the net convective rate, n^*G^* at which crystals populate the size ranges that determine the commercially important part of the size distribution.

Prediction of Standard Population Densities

Predicting n^* values for commercial crystallizers from small-scale experiments is at least as difficult and unreliable as predicting nuclei population densities. For example, results shown in Figure 30 indicate trends that vary with the scale of operation even at a scale-up ratio as low as 5. This means that either more reliable scale-up procedures or results from commercial units must become available before any reliable design results may be expected. Nevertheless, the introduction of a standard population density concept would be an important contribution to design strategy, since all references to the micron size range where no data points are available are avoided.

Prediction of Various Population Functions

The population functions such as fines removal, internal classification, attrition, etc. are important contributors to the development of product crystal size distribution in large commercial crystallizers. Gaining a means of control over these functions would be an important contribution to design and operation of crystallizers producing specified crystal size distributions. It is therefore important to indentify and quantify these functions for any chosen design configuration and mode of operation. A number of recent studies have been directed towards determining such population functions at commercially relevant conditions. The problems of preferential fines removal in draft-tube baffled crystallizers has been studied by de Leer [22] and the problems of internal classification and attrition in forced circulation crystallizers have been considered by Asselbergs [23] and Grootscholten [24]. The occurrence of population functions in industrial crystallizers has been discussed earlier in this chapter.

THE USE OF FUNDAMENTAL PRINCIPLES IN CRYSTALLIZER DESIGN

Successful crystallizer design prescribes a formulation of crystallizer type and volume with operating conditions suited to match product requirements such as average size and scatter around that size (coefficient of variation) at a prescribed production rate. The main objective of successful crystallizer design is to match the crystallizer type and operating conditions to the crystallizing system in such a way that the resulting crystallization kinetics and population events favorably interact and eventually lead to design specification of the product. Requirements such as the production rate and average crystal size determine crystallizer volume through the mass balance and the population balance, respectively. The required coefficient of variation plays a decisive role in the choice of the type of crystallizer, which should be suited for manipulating population functions, especially fines dissolving and classified product removal. Crystallizer operating conditions are in most cases determined by technical aspects such as the need to reduce incrustation and scaling or energy consumption. The operating conditions are directly related to the manner in which the driving force for crystallization is created and as such affects both mass balance and population balance within a crystallizer.

Interactions Between Crystallization Kinetics and Operating Conditions in Industrial Crystallizers

In a previous section in this chapter the population balance concept leading to the development of the general population equation has been discussed. It has further been shown how pure crystallization kinetics and population events combine to generate the product crystal size distribution. At this stage it is important to note that pure crystallization kinetics and contributing population events cannot be specified and adjusted to this specification independently of each other. Different design configurations will feature different interactions between operating conditions and configuration constraints on one hand, and pure crystallization kinetics and population events on the other. Expressing this interaction for each design configuration in quantitative terms requires detailed knowledge of both crystallizer operation and system constraints together with the kinetics of all ensuing rate processes that define development of a certain population of crystals. The flow diagram presented in Figure 31 illustrates in general and qualitative terms the interrelationship involved in industrial crystallization. The blocks in the diagram represent design specification (hatched area), operating variables (shadow area), and properties referring to driving forces, crystallization kinetics, and population functions (blank area). The blocks can exert influence upon each other through numbers balance (thick lines), through mass balance (thin lines), or through other effects (dotted lines) arising from the kinetics or flow patterns. As indicated on the flow diagram of Figure 31 operating variables such as the production rate of crystallizable solute, hydrodynamics, and crystal residence time can be manipulated to attain design specifications with reference to crystal production rates, size distributions and the purity and habit of product crystals. In order to secure the prescribed crystal production rate, an equivalent production rate of crystallizable solute is effected and this results in a certain supersaturation level and a certain magma (suspension) concentration according to the mass balance. At this point the specified crystal production rate is already secured, but whether the crystal mass is deposited on a few large crystals or onto many small ones, or in other words, whether the properties of product crystal size distribution match specification, is a question of other relationships established within the system. For example the supersaturation level being the driving force influences both nucleation and crystal growth rate and the magma concentration influences attrition, net nucleation rate, and local crystal surface area available for dissipation of supersaturation through deposition of solute. The diagram further implies that the product crystal size distribution is generated through simultaneous interaction between crystallization kinetics, population functions, and crystal residence time according to the relevant numbers balance. The crystal size distribution defines the crystal surface area, and this in turn determines how much of the created supersaturation can be deposited through crystal growth. The supersaturation balance determines the general level of driving forces and through local variations induced by hydrodynamics, these determine further the rate of contributing particulate processes. The hydrodynamics for example play a very important role since they govern both the rate of crystallization and flow patterns within a given design configuration. The flow patterns define spatial variations in the driving force which combine with spatial variation of magma concentration to eventually influence product crystal size distribution both through kinetic effects and through contribution of residence time distribution between crystals and solution.

The functional relationship shown in Figure 31 is probably encountered in most industrial crystallizers. The relative importance of different interactions, however, is dependent on the particular crystallizer type as well as on the material to be crystallized.

DESIGN OF CONTINUOUS-STIRRED TANK CRYSTALLIZERS

Design of Continuous Well-Mixed Crystallizers with Mixed-Product Removal

This class of crystallizers includes the classic MSMPR concept in the absence of any external classification. Again, in the absence of population functions other than nucleation at zero size and the crystal growth rate, the general population equation is reduced to:

$$\frac{d(nG)}{dL} + \frac{n}{\tau} = 0 \tag{24}$$

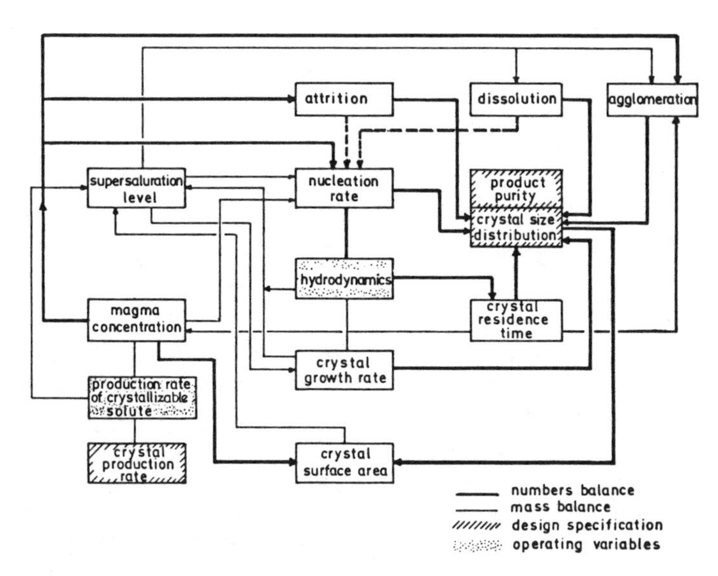

Figure 31. A flow diagram illustrating interrelationships between design specification, operating conditions, and rate processes for an arbitrary design configuration.

which for size-independent growth gives:

$$G\frac{dn}{dL} + \frac{n}{\tau} = 0 \tag{25}$$

with median size:

$$L_{50} = 3.67G\tau \tag{29}$$

and the coefficient of variation amounting to 50%.

The suspension density established in a steady-state MSMPR crystallizer at conditions of size-independent growth may be obtained from the third moment of the product distribution given by Equation 22. This latter combines with the product distribution expressed by Equation 26 to give:

$$M_s = 6k_v\varrho_c n^0[G\tau]^4 \tag{28}$$

The important kinetic consideration in design is the relationship between the growth and nucleation kinetics. As it is difficult to determine beforehand the level of supersaturation required, it is convenient for design purposes to combine equations for nucleation and crystal growth:

Nucleation:

$$B^0 = k_n M_s^j \Delta c^n \tag{53}$$

Growth:

$$G = k_g \Delta c^g \tag{54}$$

which gives:

$$B^0 = k_n M^j G^i \tag{55}$$

The mass balance does not contain B^0 explicitly but:

$$B^0 = n^0 G \tag{52}$$

which combines with Equation 28 to give:

$$M_s = 6k_v\varrho_c \frac{B^0}{G}(G\tau)^4 \tag{56}$$

Equation 56 relates crystallization kinetics, mean retention time, and suspension density (production rate in this case).

It remains only to incorporate a representation of the crystal size distribution to obtain a suitable design equation. For the cases to be discussed in this section it has been shown that the coefficient of variation of the size distribution is always predicted to be 50%. As a consequence the size distribution is completely described by a mean or average size (see Equation 29). For example, substituting $L_{50}/3.67$ for $G\tau$ in Equation 56 gives:

$$M_s = 0.033k_v\varrho_c B^0 \frac{L_{50}^4}{G} \tag{57}$$

This is the basic design equation resulting from the population balance with L_{50} being the representation of the size distribution.

In the classic case the kinetics most often encountered feature a higher order of secondary nucleation kinetics than that of growth. The value of the coefficient i is higher than unity is such cases, and the value of j is in most cases equal to unity. The crystallization kinetics are then:

$$B^0 = k_n M_s G^i \tag{58}$$

which combines with Equation 57 to give:

$$G = \left[\frac{30.3}{k_v k_n \rho_c L_{50}^4}\right]^{1/(i-1)} \tag{59}$$

Specification of L_{50} permits the calculation of the appropriate growth rate, and then from Equation 60 the retention time can be calculated. Thus:

$$\tau = \frac{L_{50}}{3.67\,G} \tag{60}$$

For $i > 1$ the size distribution can be altered to some degree by changing the retention time τ. Clearly there are limits to the degree to which the size distribution can be changed by changing residence time alone. As i approaches unity the range of possible values of L_{50} narrows and approaches one value.

For $j = 1$ the suspension density disappears from the design equation and therefore does not affect the size distribution. It must therefore be specified from other considerations such as the solubility of the solute, and the handling properties of the slurry. Once M_s is determined and the desired production rate is known, the flow rates, crystallizer volume, and energy inputs (or removals in the case of cooling) can be determined from the calculated residence time.

Alternatively, if a growth rate is to be specified because of crystal quality considerations or other reasons, L_{50} can be calculated by rearranging Equation 59 as:

$$L_{50} = \left[\frac{30.3}{k_v k_n \varrho_c G^{i-1}}\right]^{1/4} \quad (61)$$

Alteration in the kinetic constant k_n or order i resulting from changes in temperature or addition of additives can also be evaluated by use of Equations 59 and 61.

Design Example

Design a well-mixed cooling crystallizer with mixed-product removal to produce a crystal product conforming to the following specifications:

- Mean crystal size, L_{50}—9.8×10^{-4} m.
- Production rate, P_c—400 kg/h.
- Suspension density, M_s—200 kg/m^3 suspension.
- Volume shape factor, k_v—0.5.
- Crystal density, ϱ_c—1.75×10^3 kg/m^3.

Relative kinetics are as follows:

$$B^0 = 3 \times 10^{14} M_s G^{1.5} \quad [\text{m}^3 \text{ slurry}]^{-1} \text{s}^{-1}$$

1. Determine required growth rate from Equation 59:

$$G = \left[\frac{30.3}{(0.5)(3 \times 10^{14})(1.75 \times 10^3)(9.8 \times 10^{-4})^4}\right]^{1/(1.5-1)}$$

$$G = 1.6 \times 10^{-8} \text{ m/s}$$

2. Determine retention times:

$$\tau = \frac{L_{50}}{3.67\,G} = \frac{9.8 \times 10^{-4}}{3.67 \times 1.6 \times 10^{-8}} = 16{,}689 \text{ sec} \cong 4.63 \text{ hr.}$$

3. Determine discharge flow and crystallizer volume:

$$Q = \frac{P_c}{M_s} = \frac{400}{200} = 2 \text{ m}^3/\text{h}$$

$$V = Q\tau = (2)(4.63) = 9.3 \text{ m}^3$$

Again the feed rate can be calculated from the mass balance, and with a knowledge of the solubility curve and the heat capacity, the heat exchange surface can be determined and the design completed.

The same equations apply for an evaporative system so long as it is remembered that τ is the residence time of the crystals and the evaporation must be accounted for in the mass balance. For the evaporative case, it will also be necessary to alter the volume of the crystallizer body to allow for vapor disengagement space above the liquor surface. The heat exchange surface will depend on the production rate, the temperature of the feed, and the amount of vapor produced.

It should be mentioned that design configurations utilizing evaporation to produce supersaturation may deviate from the MSMPR concept, particularly for larger units, and this will be further discussed in the following sections.

Material and Energy Balance

In formulating the material balance over a crystallizing system unambiguous definition of various parameters and their dimensions is rather important. For purposes of the present development the definition of symbols is given in Figure 32.

For a binary salt system with a single feed stream containing no crystal particles, a single product stream, and for evaporative operation (parallel feed, parallel product discharge) the general balance equations are:

Overall Mass Balance

$$\Phi_{mf} = \Phi_{mpS} + \Phi_{mpL} + \Phi_{mv} \quad \text{(kg/s)} \tag{62}$$

Solute Balance

$$\varphi_{vf} c_f = \varphi_{vpSL} M_s + \varphi_{vpL} c_p \quad \text{(kg NaCl/s)} \tag{63}$$

The balance equations can be solved to give a relation for the magma concentration which reduces to:

$$M_s = \frac{\varphi_{mv}}{\varphi_{vpSL}} \cdot \frac{c_f}{\varrho_{Lf} - c_f} \quad (\text{kg/m}^3) \tag{64}$$

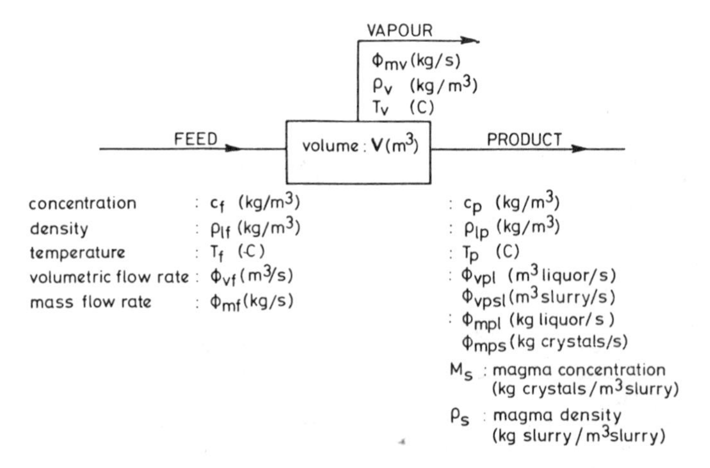

Figure 32. Definition of variables for a crystallizing system undergoing evaporation of solvent.

For composition and liquor density of the feed and product stream that are approximately equal:

$$\frac{\varrho_{Lp}}{\varrho_{Lf}} \cong \frac{c_p}{c_f} \tag{65}$$

Since the rate of salt production, P_c, is given by:

$$P_c = \varphi_{vpSL} \cdot M_s \tag{66}$$

Equation 66 can be directly used to evaluate the rate of evaporation, when P_c and the feed conditions are given:

$$\varphi_{mv} = \frac{\varrho_{Lf} - c_f}{c_f} \cdot P_c \tag{67}$$

The magma concentration is fixed when in addition the rate of product withdrawal has been chosen (or visa versa). The vaporizer diameter can be calculated from the vapor load using entrainment correlations. The maximum allowable vapour velocity is normally expressed in terms of vapor velocity head or the square root of the relative vapor-liquid density difference. The vapor load per unit cross-sectional area of vaporizer may thus be readily evaluated for a given vapor temperature.

Energy Balance

The total heat flow, φ_w, which has to be transferred to the system is described by the energy balance:

$$\varphi_w = \varphi_{mf} c_{pf}(T_f - T_p) + P_c Q_{cr} + \varphi_{mv} r \tag{68}$$

where c_{pf} = the specific heat of the feed brine
Q_{cr} = the heat of crystallization
r = the heat of vaporization

The heat of crystallization is very small for salt and may be neglected. The heat flow necessary for evaporation of the solvent is much larger than the first term on the right-hand side of Equation 68.

DESIGN OF EXTERNAL FORCED-CIRCULATION EVAPORATIVE CRYSTALLIZERS

To complete this chapter a design procedure will be illustrated for an external forced circulation evaporative crystallizer for sodium chloride.

Starting points that will be taken into consideration are:

1. *Production capacity.*
2. *Product specification.*
3. *Raw materials.* The presence of impurities (e.g., carbonates and sulphates) and their possible influence on crystallization behavior will be left out of consideration since the influence of impurities is often very specific to a particular system.
4. *Steam and operating temperature.* As salt is often manufactured in a multiple-effect operation to reduce energy costs, temperatures cannot be arbitrary chosen. Temperatures

should follow from optimization of the temperature distribution over the multiple-effect installation. This aspect will not be discussed here.

Geometry and Principles of Operation

The design strategy chosen and the design rules applied are very much dependent on the specification that a given crystalline material must comply with. In many cases strict specifications are imposed on the size range and the size level of a crystalline product. In order to be able to meet specifications of the consumer, knowledge of design principles, mode of operation, and technical problems is required.

Figure 33 shows an evaporative forced-circulation crystallizer with external calandria as used for the manufacture of vacuum salt. This design features a horizontal tangential inlet and a central bottom outlet.

Alternative designs include a double radial inlet or central bottom inlets. The evaporator body may be fitted with an elutriation leg if washing and classified discharge of crystals is required. The elutriation leg is positioned centrally for evaporators with external circulation, in which case the bottom outlet to the circulation pump is off-center.

In both designs the magma is circulated by means of an axial flow pump, heated up in the calandria and reintroduced into the evaporator body below the liquor level. Water is evaporated in the boiling zone (flashing), causing the slurry to cool down. Supersaturation created in this zone becomes available for deposition onto the surface area of growing crystals.

The liquor level in external circulation machines must be kept appreciably above the inlet in order to suppress boiling and prevent salting within the inlet pipe. This method of operation, however, results in short-circuiting from the inlet to the bottom outlet of liquor

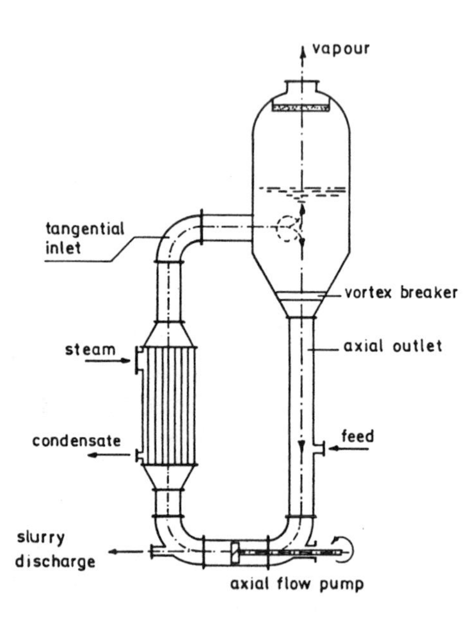

Figure 33. Schematic representation of an evaporative forced-circulation crystallizer with external calandria.

that has not flashed to equilibrium at the pressure in the vaporizer; in other words, part of the hot inlet stream fails to rise to the boiling zone and will instead find its way directly or indirectly to the vaporizer outlet. Short-circuiting results in an increase of the temperature level of the liquor circulating through the heat exchanger, and hence in an undesirable decrease of the logarithmic mean temperature difference available for heat transfer between the heating steam and the circulating magma. For a given rate of circulation this effect can be offset only by increasing the pressure of the heating steam or by increasing the heating surface. Alternatively the capacity of an existing salt plant may be reduced to the extent that short-circuiting takes place.

Although it has been claimed by Newman [29] that the definite pattern of secondary circulation produced by the swirling flow in evaporators with a tangential inlet is capable of preventing short-circuiting, it has been most convincingly demonstrated by Rutten and De Jong [30] that short-circuiting takes place to an appreciable extent in these systems. Comparative studies of short-circuiting [24] have shown that the degree of short-circuiting is much less in radial inlet evaporators.

Solid-Liquid Two-Phase Flow

An aspect of salt evaporator design that has received relatively little attention is the solid-liquid two-phase flow. Conditions of uniform crystal suspension should be ensured in order to effectively utilize the vessel volume and to avoid local regions of high supersaturation.

Also it is of particular importance to elevate sufficient quantities of crystals to the boiling zone for relief of generated supersaturation. In regions where the flow pattern is reasonably well defined (e.g., the circulating line) minimum transport velocities may be predicted from generalized correlations for vertical and horizontal hydraulic transport.

Although uniform distribution of the solids may normally be desirable, it should be kept in mind that a given specified product crystal size distribution may be successfully produced by manipulating the solids residence time for crystals of a given size via (internal) classification (i.e., crystal size distribution and magma density vary throughout the suspension). The removal rate of particles of a given size is also affected by the location and design of the product discharge port: preferential removal may occur when the product slurry is not withdrawn isokinetically and coaxially. The net result will be a difference between the crystal size distribution in the proximity of the point of discharge in the vessel and in the product (external classification).

Design Strategy

Geometry and basic design parameters follow from requirements on heat and mass transfer as well as the desired flow pattern in the evaporator body, which influences the mixing charcteristics of the solid and liquid phase. The established flow pattern depends on the choice of inlet geometry (tangential, radial, or axial-central, for instance).

The main starting points in designing the unit are:

1. The dynamic pressure of the released vapor should not exceed a maximum value in order to prevent excessive entrainment of small brine droplets. This is of particular importance if the released vapor is to be used for heating the circulating slurry in the next effect (in a multiple-effect operation). Brine could cause corrosion problems in the calandria. It is often recommended:

$$\frac{1}{2}\varrho v^2 \leqq 1.75 \quad (\mathrm{N/m^2}) \tag{69}$$

This criterion determines the diameter of the cylindrical part of the evaporator body. It can be easily shown that the preceding criterion is satisfied for:

$$D_v \geqq \sqrt{\left(\frac{0.68\ \varphi_{mv}}{\varrho_v^{0.5}}\right)} \quad \text{(m)} \tag{70}$$

where D_v = the evaporator diameter (m)
φ_{mv} = the vapor load (kg/s)
ϱ_v = the vapor density (kg/m^3)

Equation 70 clearly shows that D_v increases with decreasing vapor density (low pressure boiling).

2. The cylindrical part of the evaporator body, which is for reasons of simplicity assumed to be of constant diameter, is fitted with a bottom cone. For this cone it is often recommended to feature an angle of inclination at the apex of less than 60° to prevent unwanted accumulation of solids on the inclined walls.
3. The height of the vapor space should be about 3/4 D_v to prevent excessive entrainment and possible scale formation in the vapor space due to violent boiling and spouting.
4. The circulating flow rate follows from the temperature rise over the heater (in general $\Delta T \cong 3-8°$ C) and the heat balance. The temperature-rise is limited by the hydrostatic height above the inlet needed to prevent boiling and possibly scale formation in the inlet. Circulating flow rate:

$$\varphi_{vc} = \frac{\varphi_W}{\varrho_{SL} c_{pSL} \Delta T_c} \quad \text{(m}^3\text{/s)} \tag{71}$$

where φ_W = the total heat load (W)
ϱ_{SL} = the slurry density
c_{pSL} = the specific heat of the slurry (J/kg °C)
ΔT_c = the temperature rise over the calandria (°C)

5. The diameter of the inlet is determined by the circulating flow rate and desired velocity. Very often the diameter and hence inlet velocity is chosen to maintain hydrodynamic similarity on the basis of the Froude or Reynolds number, characterizing liquor flow in the body of a studied pilot-plant unit. Reynolds numbers based on inlet velocity and evaporator body diameter have been reported by Newman and Bennett [31] and are in the range 5×10^6 to 10×10^6, where $Re = v_i D_v \rho/\eta$. Froude numbers ($Fr = v_i^2/gD_v$) range from 0.07 to 0.16. In this context it is important to note that similarity is difficult to define without knowing which aspect of crystallizer performance should remain constant. The liquid velocity past the heating tubes is limited only by the pumping power needed or available and by accelerated erosion and corrosion and degradation of the circulating crystals. Tube velocities normally range from a minimum of 1 m/s to about 3 m/s.

Inlet diameter:

$$D_i = \sqrt{\left(\frac{4\varphi_{vc}}{\pi v_i}\right)} \quad \text{(m)} \tag{72}$$

where v_i is the inlet velocity (m/s)

For obvious reasons the diameter of the circulation loop is often adapted to the diameter of the circulation pump.

6. The design of the external heater follows from optimization of costs of pumping power (i.e., pressure drop over the heater) and investment in heater and circulation pump.
7. In order to determine the height of the cylindrical part of the evaporator body, the minimum liquor level leading to suppression of boiling in the tubes must be determined. This minimum level follows from the vapor pressure, the inlet-temperature, and the vapor pressure-temperature curve. The minimum total height becomes:

$$H_{cylinder} = \text{minimum liquor level} + D_i + 3/4D + \text{additional length} \quad (m) \tag{73}$$

The additional length underneath the inlet follows from mechanical considerations, but can be arbitrarily set to 0.5 m.
8. The minimum liquor volume can now be estimated:

$$V = \frac{\pi}{4} D_v^2(h_{min} + D_i + 0.5) + V_{cone} + V_{external\ loop} \quad (m^3) \tag{74}$$

9. From the overall mass balance and estimated volume of the crystallizer a rough estimate can be made of the expected solids concentration. From the total solids hold-up and solids production rate a first-order approximation of the overall solids residence time can be made.

The rate of salt production, P_c, is given by:

$$P_c = \varphi_{vSL} M_s \quad (kg/s) \tag{66}$$

where φ_{vSL} is the volumetric production rate (m^3/s) and M_s the magma concentration (kg/m^3). Under the assumption of mixed-suspension and mixed-product removal (as first approximation) the total solids hold-up is:

$$W = VM_s \quad (kg/s) \tag{75}$$

The overall residence time, τ_c, is given by

$$\tau_{c0} = \frac{W}{P_c} \quad (s) \tag{76}$$

The minimum solids residence time is determined by the maximum allowable level of solution supersaturation, which may result in excessive primary nucleation, uncontrolled growth, impurity incorporation, and incrustation. The maximum residence time will be determined by the size of the crystallizer (large volumes) and too-high solids concentration resulting in partial degradation of the circulating crystals (rounding and breakage of the largest crystals). Rounding of the crystals may have a negative influence on impurity incorporation and the inclusion of solute in the crystals. Moreover the liquid residence time determines the level of impurities in the liquid phase.

If the design following the previous considerations does not satisfy requirements on solids residence time, additional measures should be taken. Extension of the solids residence time by decreasing the volumetric production rate φ_{vSL} is limited by slurry handling and magma concentration restrictions.

It is important to note that the procedure proposed at this stage gives the approximate design of the evaporator without any reference to the proces of crystallization. The question which then remains is whether the designed vessel is able to meet requirements on crystal size distribution (i.e., mean size and uniformity).

Population Balance for External Circulation Crystallizer

In this design example it will be assumed that solids classification influences the crystal size distribution directly, through residence-time variations, and indirectly, through selectively exposing the growing crystals to local variations of supersaturation, thus inducing pseudo size-dependent growth rates. If classification occurs and the growth rate is a direct or indirect function of size, the differential numbers balance can be written in terms of product population density:

$$\frac{d\ln(n_p)}{dL} = -\frac{1}{G\tau_c} - \frac{d\ln(\tau_c)}{dL} - \frac{d\ln(G)}{dL} \tag{39}$$

The solution of this equation characterizes the shape of the distribution of product crystals, which is what we are actually interested in. For the crystallizer average distribution it follows:

$$\frac{d\ln(n)}{dL} = -\frac{1}{G\tau_c} - \frac{d\ln(G)}{dL} \tag{36}$$

Before these equations could be used for design purposes, knowledge is required concerning crystal residence time and growth rates as a function of crystal size. For existing plants information on residence-time distribution can be experimentally obtained from the already-mentioned wash-out experiments. Pilot-scale experiments conducted by Asselbergs [23] and later by Grootscholten [24] revealed that the studied external forced-circulation crystallizers are liable to be very poorly mixed with respect to the solid phase, resulting in strong size dependency of the crystal residence times. From consideration of the various flow regimes in these crystallizers (high liquor velocities in circulation mains and low velocities in the evaporator body) classification was thought to be induced by size-dependent segregation in regions of low liquor velocities. Typical τ_c/τ curves obtained from two differently designed external forced-circulation crystallizers (pilot scale) are given in Figure 34.

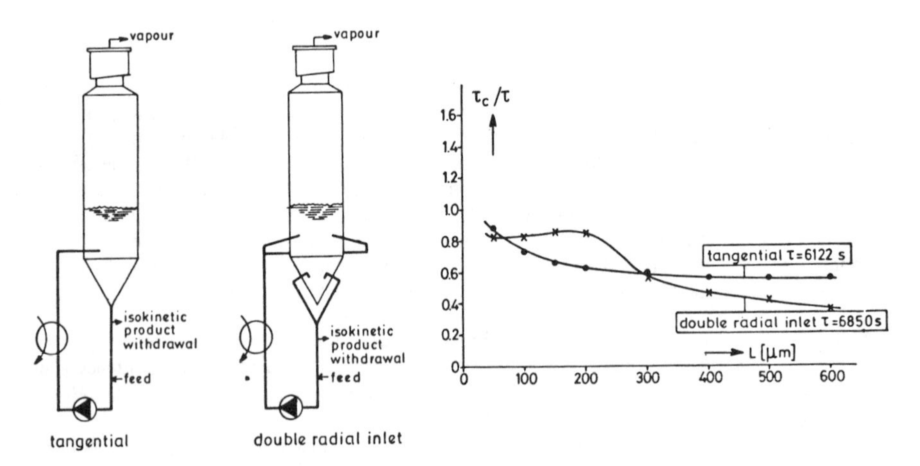

Figure 34. Classification parameters for the tangential and double radial inlet crystallizer configuration (see also Figure 21).

Prediction of pseudo size-dependent growth rates requires knowledge concerning the spatial distribution of the concentration driving force and particle residence times in regions of different driving forces.

Liquid-Phase Flow Model

As has been shown before, the liquid phase in a crystallizer with a double radial inlet may be represented by the network of elements which is shown in Figure 35.

Supersaturation, generated in the boiling zone, becomes available for deposition on the crystals. The driving force, Δc_i, in the inlet flow to the evaporator body is unknown. This problem can been solved by assuming that the entering flow instantaneously drops in temperature due to flashing in the boiling zone. The resulting Δc_i can be calculated from the temperature-drop, which is approximately equal to the temperature-rise over the heat exchanger. Then evaporation of water takes place at a constant temperature. If the evaporator body is considered as a steady-flow stirred tank reactor through which liquid is flowing and in which material is simultaneously being deposited on the surface of the growing crystals, then the material balance in terms of supersaturation becomes:

$$\varphi_{vc}(1-x)\Delta c_i + \varphi_{mv}\frac{c_v}{\varrho_1 - c_v} = \varphi_{v0}(1-x)\Delta c + P_c V_b \tag{77}$$

where x = the solids volume fraction in the inlet and outlet flow
V_b = the volume of the evaporator body
P_c = the supersaturation deposition rate

The deposition rate may be written in terms of the supersaturation as:

$$P_c = \Delta c \int_0^{\infty} k_g(L) k_a n_b L^2 \, dL \tag{78}$$

where $k_g(L)$ is the overall growth rate coefficient, which is a function of the surface integration rate constant and the mass transfer coefficient, the latter being a function of particle size.

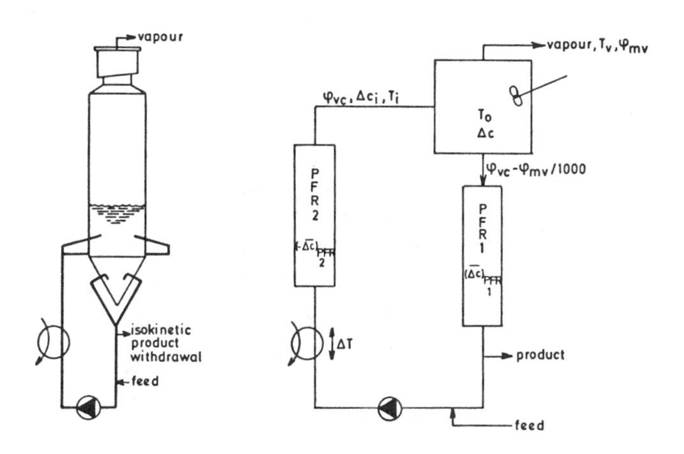

Figure 35. Reactor in series concept proposed to account for liquid flow encountered in double radial inlet crystallizer configuration.

Since the deposition rate depends on the crystal surface area ($\cong \int n_b L^2 \, dL$) available for deposition, the population density distribution n_b in the evaporator body must be known. It will be shown in the next section that this latter can be obtained from the crystal residence time distribution $\tau_c(L)$ and the volumes of the reactor elements. Supersaturation Δc follows directly from Equations 77 and 78.

With reference to Figure 35 the supersaturation is further dissipated onto the surface of growing crystals within the reactor element PFR1. As a result of this an exponential decay of supersaturation is established within this reactor element. The decay rate will be a function of the overall growth rate coefficient k_g and the surface area ($\cong \int n_m L^2 \, dL$) in PFR1.

Integration of the supersaturation profile enables the evaluation of an average supersaturation of the length over PFR1, which is given by:

$$(\Delta \bar{c})_{PFR1} = \frac{(1-x)\varphi_{vc}\Delta c}{k_1 V_{PFR1}} \left[1 - \exp\left(- \frac{k_1 V_{PFR1}}{(1-x)\varphi_{vc}} \right) \right] \tag{79}$$

where k_1 is the product of overall growth rate coefficient and available crystal surface area.

The heat exchanger is thought to produce a step-increase in temperature, which may result in undersaturation. The ensuing dissolution results in an increase in bulk-concentration and an exponential decrease in undersaturation. An average value of the undersaturation in PFR2 follows from:

$$(-\Delta \bar{c})_{PFR2} = \frac{(1-x)\varphi_{vc}(-\Delta c)}{k_2 V_{PFR2}} \left[1 - \exp\left(- \frac{k_2 V_{PFR2}}{(1-x)\varphi_{vc}} \right) \right] \tag{80}$$

where k_2 depends on the mass transfer coefficient for dissolution and the crystal surface area. The value of $(-\Delta c)$ follows from the bulk concentration in the outlet of PFR1, the temperature-rise and the solubility-curve.

To judge whether a particular system is liable to be poorly mixed with respect to the concentration driving force the supersaturation half-life time, which is the time needed to deposit 50% of the available supersaturation on the crystals (in a batch system), could be

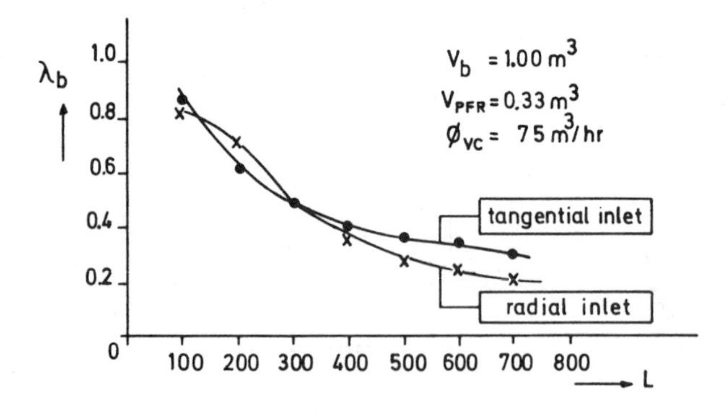

Figure 36. Solids distribution factor for evaporator body.

compared with the slurry turnover time which is defined as the time needed to turn over the crystallizer content once. It can be easily shown that:

$$t_{0.5} = -\frac{\ln 0.5(1-x)}{k_g A_T} \tag{81}$$

where x = the solids volume fraction
k_g = the growth rate coefficient (see Equation 10)
A_T = the available crystal surface area (in forced circulation crystallizers, A_T in general ranges from 500–2,500 m^2/m^3),

Local Residence Times

Evaluation of pseudo size-dependent growth rates requires knowledge concerning particle residence times in regions of different driving forces.

For the present purpose particle residence times in both PFR's are assumed to be equal to their respective liquid residence times, τ_{11} and τ_{12}, and they are assumed to be independent of size. In view of the high liquid velocities in the circulation mains (1–3 m/s) these assumptions seem not to be unrealistic. The particle residence times in the evaporator body, $\tau_b(L)$, can then be readily derived from the overall particle residence times $\tau_c(L)$ using the equation:

$$\tau_b(L) = \tau_c(L) - \tau_{11} - \tau_{12} \tag{82}$$

where $\tau_l = V_{PFR}/\varphi_{vSL}$. For crystallizers featuring product classification facility Equation 82 extends to:

$$\tau_b(L) = \lambda_p(L)^{-1}\tau_c(L) - \tau_{11} - \tau_{12} \tag{83}$$

where $\lambda_p(L)$ relates population density in circulation mains to product population density according to Equation 32.

The population density in the evaporator body can be related to population density in circulation mains by means of an internal classification parameter, $\lambda_b(L)$:

$$n_b = \lambda_b(L) n_c \tag{84}$$

In this equation $\lambda_b(L)$ is defined as:

$$\lambda_b(L) = \frac{\tau_b(L)}{\tau_{lb}} \tag{85}$$

where τ_{lb} is the overall liquid residence time in the evaporator body, and this can be directly calculated from the body slurry volume and the volumetric product withdrawal rate. Figure 36 shows internal classification parameter distributions calculated from the solids residence time distributions in Figure 34.

This distribution profile is thought to be dependent on the turbulence level, density difference between solid and liquid phase, physical properties of the suspension, and the design geometry of the vessel.

In new designs solids classification parameters should follow from considerations of hydrodynamics and particle mechanics. This is, however, a very complicated matter. Alternatively information could be provided by pilot-scale classification experiments. If the parameters $\lambda_b(L)$ and $\lambda_p(L)$ can be deducted from such experiments, evaluation of $\tau_c(L)$ is straightforward (combine Equations 83 and 85), as liquid residence time is one of the preselected process parameters.

Evaluation of Pseudo Size-Dependent Growth Rates

With reference to Equation 47 the average crystal growth rate of a particle of size L follows from:

$$\bar{G}(L) = \frac{1}{\tau_c(L)}\left[G_b(L)\tau_b(L) + \bar{G}_{PFR1}(L)\tau_{11} + (-\bar{G})_{PFR2}\tau_{12}\right] \tag{86}$$

The linear growth and dissolution rates, arising in this equation, can be calculated by making use of the appropriate growth and dissolution rate models. The average growth rate, $\bar{G}(L)$, can now be used together with crystal residence time distribution, $\tau_c(L)$, and appropriate nucleation kinetics to solve the population balance Equations 36 and 39.

Simulation Algorithm

Using an arbitrary start distribution, n_c, in the inlet flow to the evaporator body, the size distribution and hence the crystal surface area ($\cong \int n_b L^2\, dL$) can be estimated using Equation 84. Then the supersaturation balance (Equation 77) can be solved. Estimation of the average driving forces in PFR1 and PFR2 is further straightforward (Equations 79 and 80).

Pseudo size-dependent growth rates can be calculated as function of L from the estimated driving force distribution, growth kinetic data, and sectional residence times. From $\tau_c(L)$ and $\bar{G}(L)$ values of $(d \ln(\tau_c))/(dL)$ and $(d \ln(\bar{G}))/(dL)$ can be computed as functions of L. For the boundary condition $n^0 = B^0/G(0)$ the differential product population density distribution can be numerically integrated with respect to L yielding the population density distribution n_p in the product stream. The cup-mixing magma concentration (required in the nucleation rate equation) follows from the third moment of the cup-mixing population density distribution n, which follows from integration of Equation 22. The overall solids residence time τ_{c0} can be found from the estimated hold-up of solids ($= V\bar{M}_s$) and the solids production rate.

The distribution in the circulation mains ($n_c = \lambda_p n_p$) can be used to reestimate the CSD and surface area in the body. Computed values of $\bar{M}_s$ and τ_{c0} give a new value of B^0. This loop calculation must be iterated until the population density converges at a stationary value at conditions where both material and supersaturation balances are closed. A schematic simulation flow diagram is given in Figure 37.

The simulated crystal size distribution must comply with product requirements as defined in the design targets. If the desired specification (mean crystal size and coefficient of variation) is not obtained, measures should be taken to secure satisfactory product. Such measures may include enhancement or suppression of the population functions (i.e., nucleation, growth, classification of solids, liquid mixing induced growth) which control the development of the crystal size distribution. In general nucleation and growth rates cannot be greatly affected by operating conditions. Impeller speed and pump design are controlled by the desired circulating flow rate. Moreover, the overall solids residence time will in general be prescribed by the maximum allowable supersaturation (fouling, scale-formation, and impurity incorporation are then important aspects) whereas the maximum residence time is limited by the size of the resulting crystallizer volume. Probably the best way to manipulate the crystal size distribution is to affect those population functions which are most intensive in the size ranges which determine the desired properties of the product.

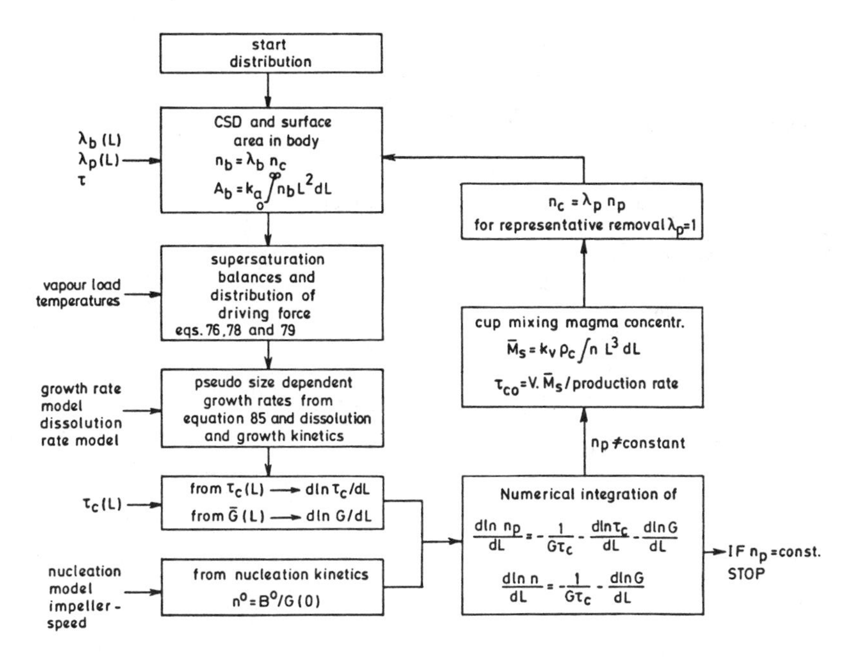

Figure 37. Simulation algorithm.

What measures should be taken to meet product requirements will depend on whether modifications in geometry are able to promote the desired (or hinder the undesired) population events without deteriorating the optimum design which was based on heat transfer, economic, and equipment considerations. In practice the final design will be a trade-off between economics, construction, and requirements on capacity, size distribution, shape, and purity of product crystals.

Design Example

Design a forced-circulation evaporative crystallizer to produce sodium chloride to the following specification:

- *Production capacity,* P_c—20 ton/hr (= 5.5 kg/s).
- *Product specifications:*
  Median size, L_m—400–500 μm.
  Coefficient of variation—20%–25%.
- *Raw materials*—saturated brine at 20° C.
  Feed concentration, c_f—320 kg/m^3 (sat. at 55° C).
  Feed specific heat—3,290 kJ/kg.
  Density of feed, ϱ_{lf}—1,185 kg/m^3.
- *Steam and operating temperature:*
  Evaporator temperature—55° C.
  Boiling point elevation—7° C.
  Additional ΔT_s—2° C.
  Vapor condensation temperature—46° C.

Vapor pressure—0.100 bar.
Vapor density, ϱ_V—0.06869 kg/m^3.
Heat of evaporation—2,400 kJ/kg.
Available steam from former effect—66° C.

- *Product:*

Slurry concentration, M_s—250 kg/m^3.
Representative product removal from circulation loop.
Density of crystals, ϱ_C—2,155 kg/m^3.
Dynamic viscosity of magma—2.3 10^{-3} Ns/m^2.
Specific heat crystals—877 kJ/kg · C.
Specific heat magma, c_{psl}—2,700 J/kg · C.
Slurry density, ϱ_{sl}—1,300 kg/m^3.

- *Kinetics:*

Nucleation rate

$$B_0 = 10^8 K_{imp}^{1.35} d^{1.2} N^{1.2} \tau_{c0}^{-1.6} \bar{M}_s$$

Growth rate

$$R_g = k_g \Delta c$$

$$\frac{1}{k_g} = \frac{1}{k_d} + \frac{1}{k_r}$$

$$k_r = 3.7 \times 10^{-4} \text{ m/s at } 55° \text{ C}$$

$$k_d \text{ from } Sh = 2 + 0.9\, Re^{0.5} Sc^{0.33}.$$

Dissolution rate

$$R_d = k_d \Delta c$$

k_d, see above.

Residence time

Minimum—3,600 secs
Maximum—7,200 secs.

- *Material balances:*

Vapor load

$$\varphi_{mv} = \frac{\varrho_{Lf} - c_f}{c_f} P_c = \frac{1,185 - 320}{320} \cdot 5.5 = 14.8 \text{ kg vapor/s}$$

Overall mass balance

$$\varphi_{mf} = \varphi_{ms} + \varphi_{mc} + \varphi_{mv} = 5.5 + 14.8 = 20.3 \text{ kg feed/s} = 0.01718 \text{ m}^3\text{/s}$$

Volumetric production rate

$$\varphi_{vsl} = \frac{P_c}{M_s} = \frac{5.5}{250} = 0.022 \text{ m}^3\text{/s}$$

- *Energy balance:*

$$\varphi_w = \varphi_{mf} c_{pf}(T_f - T_p) + \varphi_{mv} r = 14.8 \cdot 2.4 \cdot 10^6 = 3.55 \cdot 10^7 \text{ W}$$

For temperature-rise over the heater $\Delta T_c = 4°$ C.

- *Circulating flow rate:*

$$\varphi_{vc} = \frac{\varphi_w}{\varrho_{sl} c_{psl} \Delta T_c} = \frac{3.55 \cdot 10^7}{1{,}300 \cdot 2{,}700 \cdot 4} = 2.5 \text{ m}^3/\text{s}$$

- *Diameter evaporator body:*

$$D \geqq \sqrt{\left(\frac{0.68 \varphi_{mv}}{\varrho_v^{0.5}}\right)} = \sqrt{\left(\frac{0.68 \cdot 14.8}{(0.06869)^{0.5}}\right)} = 6.2 \text{ m}$$

- *Inlet velocity and crystallizer inlet and outlet diameter:* In the present development a double radial inlet is chosen. The slurry is recycled to the heater through a single central bottom outlet. Accoring to Re criterion (assume $Re = 5 \times 10^6$):

$$v_i = \frac{Re \eta}{\varrho D_v} = \frac{5 \cdot 10^6 \cdot 2.3 \cdot 10^{-3}}{1300 \cdot 6.2} = 1.43 \text{ m/s}$$

Total circulating flow rate

$$\varphi_{vc} = 2.5 \text{ m}^3/\text{s}$$

$$2 \frac{\pi}{4} D_i^2 v_i = 2.5 \text{ m}^3/\text{s}$$

$$D_i = \sqrt{\left(\frac{\varphi_{vc} 4}{2 \pi v_i}\right)} = \sqrt{\left(\frac{2.5 \cdot 4}{2 \cdot \pi \cdot 1.43}\right)} = 1.05 \text{ m}$$

To maintain equal velocities in inlets and outlets

$$D_0 = \sqrt{2 D_i^2} = \sqrt{2 \cdot 1.05^2} = 1.50 \text{ m}$$

- *Minimum level above inlets:*

Inlet temperature of evaporator body

$$T_i = T_v + \Delta T_s + \Delta T_c = 55 + 2 + 4 = 61° \text{ C}$$

Corresponding vapor-pressure

$$p_i = 0.2086 \text{ bar}$$

Pressure in evaporator

$$p_v = 0.100 \text{ bar}$$

This gives a difference of 0.1086 bar. At $\varrho_{sl} = 1{,}300\ kg/m^3$ the required hydrostatic head is:

$$h = \frac{(p_i - p_v)}{\varrho g} = \frac{0.1086 \cdot 10^4}{1{,}300 \cdot 9.81} = 0.85\ m$$

- *Minimum total height cylindrical part of body:*

$$H_{cylinder} = h + D_i + \frac{3}{4} D_v + \text{additional height}$$

$$= 0.85 + 1.05 + 0.75 \cdot 6.2 + 0.5 = 7.05\ m$$

- *Minimum liquor volume:*

$$V = \frac{\pi}{4} D_v^2 (h + D_i + 0.5) + V_{cone} + V_{external\ loop}$$

For 60° cone and base diameter 6.2 m

$$V_{cone} = 40\ m^3$$

The volume of the external loop follows from an estimation of the heat-exchanger volume, volume of the circulation pump and ancillary piping. For the present calculation:

$$V_{external\ loop} \cong 30\ m^3$$

$$V = \frac{\pi}{4} \cdot 6.2^2 \cdot (0.85 + 1.05 + 0.5) + 40 + 30 = 142\ m^3$$

- *Overall solids residence time:*

$$\tau_{c0} = \frac{W}{P_c} = \frac{VM_s}{P_c} = \frac{142 \cdot 250}{5.5} = 6{,}455\ \text{sec.}$$

- *Circulation pump:* To circulate $2.5\ m^3/s$ an axial flow pump is chosen with the following specifications:

Impeller diameter—$d = 1.1$ m.
Pump discharge coeff.—$k_p = 0.6$.
Impeller speed—$N = 3.2$ rps.
Dimensionless impeller number—$K_{imp} = 3.1$.

- *The logarithmic mean temperature difference (LMTD):* This follows from:

$$LMTD = \frac{T_i - T_0}{\ln \dfrac{T_c - T_0}{T_c - T_i}} = \frac{61 - 57}{\ln \dfrac{66 - 57}{66 - 61}} = 8.5°\ C$$

For an overall heat transfer coefficient of $U = 2.0\ kW/m^{2}\,°C$ (= 1,718 kcal/m^2° C hour) this results in a surface area for heat transfer:

$$A = \frac{\varphi_w}{U \cdot LMTD} = \frac{3.55 \cdot 10^7}{2 \cdot 10^3 \cdot 8.5} = 2{,}000 \text{ m}^2$$

- *Solids classification:* For the present purpose the internal classification parameter distribution, $\lambda_b(L)$, shown in Figure 36 is assumed to represent the classification process (typical for sodium chloride/brine).

  For $\tau = 6{,}455$ secs Equations 83 and 85 yield the overal particle residence time distribution, $\tau_c(L)$, which is given in Figure 38.
- *Distribution of concentration driving force:* For kg $= 1 \cdot 10^{-4}$ m/s, $A_T = 2000$ m^2/m^3 and $x = 0.11$ (corresponding to $M_s = 250$ kg/m^3) the supersaturation half-life time is:

  $t_{0.5} = 3.2$ secs.

  The turn-over time, TO, of the system is:

  $TO = V/\varphi_{vc} = 142/2.5 = 57$ secs.

  As TO $\gg t_{0.5}$ nonuniform distribution of supersaturation is likely to occur and should therefore be taken into account in evaluating crystal growth rates.

  The liquid-phase flow model, shown in Figure 35, is used to account for nonideal liquid mixing.
- *Sectional residence times:* For $\tau_{11} + \tau_{12} = 30/0.022 = 1{,}364$ secs the crystal residence times in the evaporator body follows directly from the $\tau_c(L)$-curve, given in Figure 38. Resulting values of $\tau_b(L)$ are given in Figure 39.

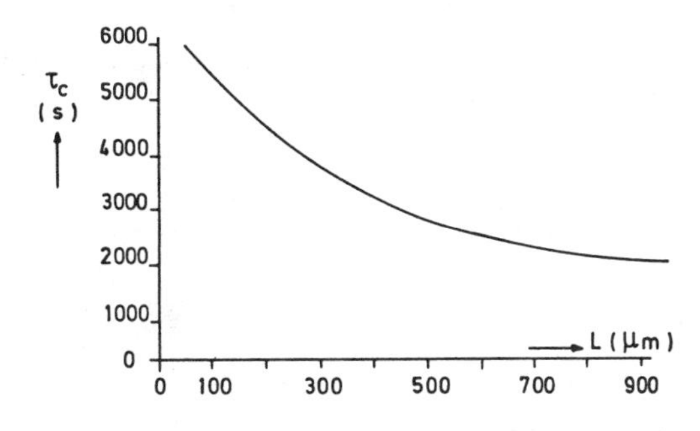

Figure 38. Solids residence time as function of size estimated for proposed crystallizer design.

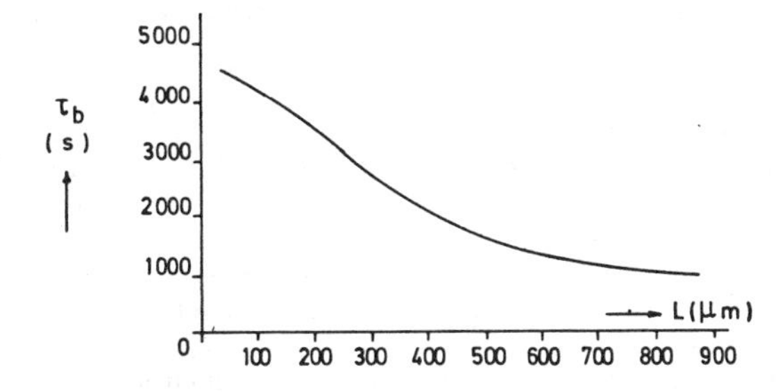

Figure 39. Residence time in evaporator body.

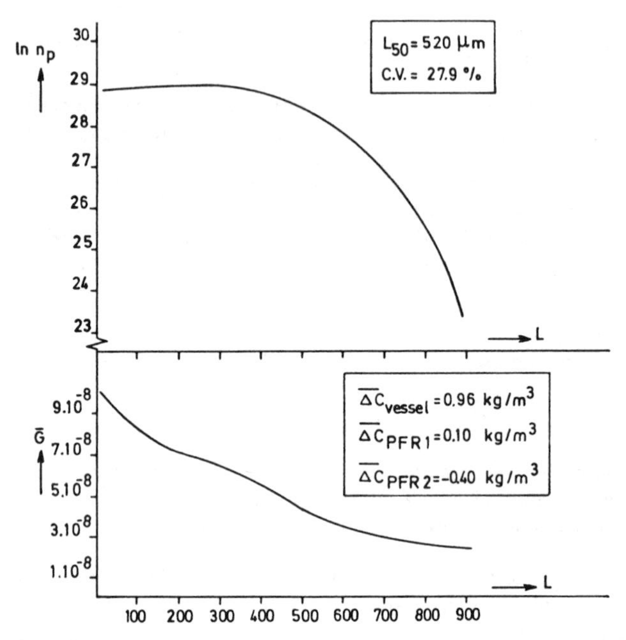

Figure 40. Simulated product population density distribution and computed pseudo size-dependent growth rates (representative product removal).

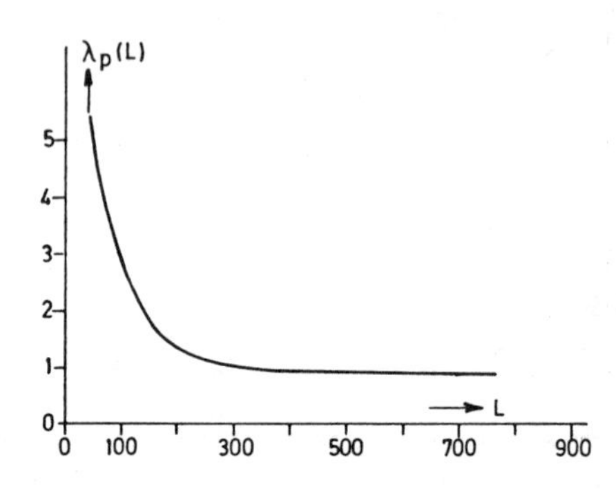

Figure 41. Product classification parameter encountered in an elutriation leg (pilot-scale crystallizer).

- *Simulation results:* The simulation calculations have been performed for the design input data specified in this example for a forced-circulation evaporative crystallizer. The effects of internal classification and liquid mixing have been accounted for by considering the solids residence time distribution of Figure 38 and the reactor-in-series concept to describe liquid maldistribution. The resulting population density distribution is shown in Figure 40 together with corresponding values of pseudo size-dependent growth rates caused by spatial variation in supersaturation.

  The results indicate that the value of median product size complies with the prescribed specification. However the somewhat higher value of the coefficient of variation should

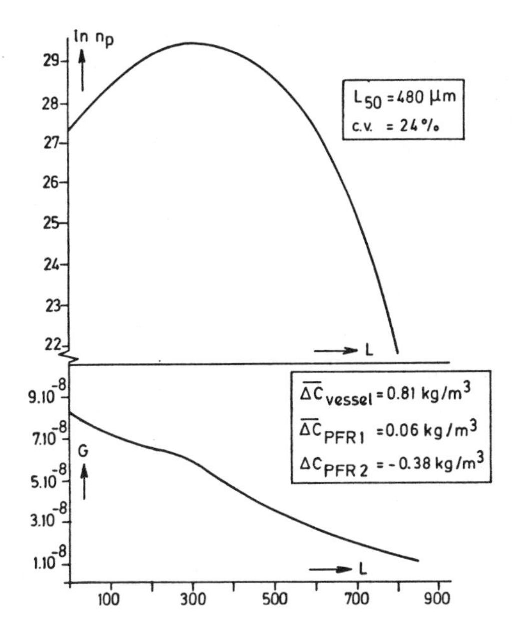

Figure 42. Simulated product population density distribution and computed pseudo size-dependent growth rates under conditions of classified product removal.

be reduced to the specified value of 20%–25%. This could be done by imposing preferential classification on product size by enhancing population function of classified product removal. This could be achieved by an elutration device of the kind presented in Figure 15 and by resorting to corresponding product classification parameters (see Figure 41).

Further simulations have been carried out for the case of classified product removal, all other parameters being kept constant. The results of these simulations are presented in Figure 42. The effect of classified product removal is reflected in a reduction in the coefficient of variation from 27.9% to 24%. As expected this is accompanied by a reduction in the median product size form 520 to 480 μm due to considerations put forward previously (see POPULATION FUNCTIONS).

It may be concluded that the proposed design complemented with appropriate product classification is capable of producing the crystalline product, which complies with the requirements set on both product size and product uniformity.

NOTATION

A	crystal surface area per unit volume (l^2/l^3)
A(L)	number flux caused by attrition ($-/l^4t$)
B^0	nucleation rate ($-/l^3t$)
B(L)	secondary nucleation or birth rate ($-/l^4t$)
c	solution concentration (m/l^3)
c*	solution equilibrium concentration (m/l^3)
c_i	concentration at crystal/solution interface (m/l^3)
Δc	concentration driving force (m/l^3)
c.v.	coefficient of variation
d_D	draft tube diameter (l)
d_s	strirrer diameter (l)

D	diameter (l)
D	diffusion coefficient (l^2/t)
$D(L)$	number flux caused by breakage ($-/l^4t$)
g	overall growth rate order
G	linear growth rate (l/t)
$G(L)$	size-dependent crystal growth rate (l/t)
G_0	growth rate of nuclei (l/t)
k_a	surface shape factor
k_d	mass transfer coefficient (l/t)
k_g	overall growth rate coefficient (SI-units)
k_N	nucleation rate constant (SI-units)
k_r	surface integration rate constant (SI-units)
k_v	volume shape factor
K_{imp}	dimensionless impeller number
L	characteristic crystal size (l)
L_{50}	mass median size (l)
$M(L)$	number flux caused by agglomeration ($-/l^4t$)
M_i	i-th moment of the distribution (SI-units)
M_s	suspension density (m/l^3)
n	population density ($-/l^4$)
$\dot{n}$	point polulation density ($-/l^4$)
n^0	nuclei population density ($-/l^4$)
N	total number of crystals ($-/l^3$)
N	impeller speed ($-/t$)
$N(L)$	cumulative number distribution ($-/l^3$)
$P(L)$	number flux caused by product classification ($-/l^4t$)
P_c	crystal production rate (m/t)
Q	volume flow rate of suspension (l^3/t)
R_D	dissolution rate (m/l^2t)
R_g	growth rate (m/l^2t)
R_i	surface integration rate (m/l^2t)
R_m	mass transfer rate (m/l^2t)
r	order of surface integration process
Re	Reynolds number ($= vL\varrho_L/\eta$)
Sc	Schmidt number ($= \eta/D\varrho_L$)
Sh	Sherwood number ($= k_dL/D$)
t	time (t)
T	temperature (°C)
T_c	steam temperature (°C)
T_i	temperature at crystal/solution interface (°C)
T_i	inlet temperature (°C)
T_0	outlet temperature (°C)
T_v	vapor temperature (°C)
ΔT_s	temperature difference caused by heat short-circuiting (°C)
v	solution/crystal relative velocity (l/t)
v_i	inlet velocity (l/t)
V	volume (l^3)

Greek Symbols

δ_{EH}	effective heat transfer boundary layer (l)
δ_{EM}	effective mass transfer boundary layer (l)
ε	power dissipation per unit mass (l^2/t^3)
$\lambda(L)$	classification parameter –
$\lambda(L)$	number flux caused by fines removal ($-/l^4t$)
η	solution viscosity (m/lt)
ϱ_c	density of crystal (m/l^3)
ϱ_L	density of liquid (m/l^3)
τ	slurry residence time (t)
$\tau_c(L)$	size-dependent residence time of crystals (t)
τ_{c0}	overall solids residence time (t)
φ_{vc}	circulating flow rate (l^3/t)
φ_w	heat flow rate (fl/t)

Note: Letter symbols in dimension columns have the following dimensions:

Dimension/Symbol	l	m	t	f
SI	meters	kilograms	seconds	newton
Others	micrometer	grams	hours	

REFERENCES

1. Strickland-Constable, R. F., and Mason, R. E. A., "Breeding of Nuclei," *Nature, 197* (1963), p. 897.
2. Botsaris, G. D., Denk, E. G., and Chua, J. O., CEP Symp. Ser. No. 118, (1972).
3. Powers, H. E. C., "Nucleation and Early Crystal Growth," *Ind. Chemist., 39* (July 1963), p. 351.
4. Jancic, S. J., and Grootscholten, P. A. M., "Industrial Crystallization," Delft University Press, 1984.
5. Ottens, E. P. K., Janse, A. H., and de Jong, E. J., "Secondary Nucleation in a Stirred Vessel Cooling Crystallizer," *Journal of Crystal Growth, 13/14* (1972), p. 500.
6. Evans, T. W., Margolis, S., and Sarofilm, A. F., "Mechanisms of Secondary Nucleation in Agitated Crystallizers," *AIChE Journal, 20* No. 5 (1974), p. 950.
7. Bennett, R. C., Fiedelman, H., and Randolph, A. D., "Crystallizer Influenced Nucleation," *Chem. Eng. Prog., 69* No. 7 (1973), p. 86.
8. Sung, C. Y., Estrin, J., and Youngquist, G. R., "Secondary Nucleation of Magnesium Sulphate by Fluidshear," *AIChE Journal, 19* No. 5 (1973), p. 957.
9. Garside, J., and Jancic, S. J., "Measurement and Scale-up of Secondary Nucleation Kinetics for the Potasch Alum-Water System," *AIChE Journal, 25* No. 6 (1979), p. 948.
10. Jancic, S. J., "Crystallization Kinetics and Crystal-Size Distribution in Mixed-Suspension Mixed-Product Removal Crystallizers," doctoral thesis, University of London, 1976.
11. Grootscholten, P. A. M., v.d. Brekel, L. D. M., and de Jong, E. J., "Effect of Scale-up on Secondary Nucleation Kinetics for the Sodium Chloride-Water System," *Chem. Eng. Research & Design*, to be published in 1984.
12. Mullin, J. W., and Garside, J., The Crystallization of Aluminium Potassium Sulphate; a Study in the Assesment of Crystallizer Design Data, Part I: Single crystal growth rates," *Trans. Instn. Chem. Engrs., 45* (1967), T285.
13. Mullin, J. W., and Garside, J., "The Crystallization of Aluminium Potassium Sulphate; a Study in the Assesment of Crystallizer Design Data, Part III: Growth and dissolution rates," *Trans. Instn. Chem. Engrs. 46* (1968), T11.
14. McCabe, W. L., "Crystal Growth in Aqueous Solutions; Part I: Theory," *Ind. & Eng. Chem. 21* (1929), p. 30.
15. White, E. T., and Wright, P. G., "Magnitude of Size Dispersion Effects in Crystallization," CEP Symp. Ser. No. 110, *67* (1971), p. 81.
16. Mullin, J. W., and Garside, J., "The Crystallization of Aluminium Potassium Sulphate; a Study in the Assesment of Crystallizer Design Data, Part II: growth in a Fluidized Bed Crystallizer," *Trans. Instn. Chem. Engrs., 45* (1967) T291.
17. Garside, J., Mullin, J. W., and Das, S. N., "Growth and Dissolution Kinetics of Potassium Sulphate Crystals in a Agitated Vessel," *Ind. & Eng. Chem. Fund., 13* No. 4 (1974), p. 299.
18. Phillips, Y. R., and Epstein, N., "Growth of Nickel Sulfate in a Laboratory Scale Fluidized-Bed Crystallizer," *AIChE Journal, 20* No. 4 (1974), p. 678.
19. Randolph, A. D., and Larson, M. A., "Transient and Steady-State Size Distributions in Continuous MS Crystallizers," *AIChE Journal, 8* (1962) p. 639.
20. Randolph, A. D., and Larson, M. A., *Theory of Particulate Processes*, Academic Press, New York, 1971.
21. Abegg, C. F., Stevens, J. D., and Larson, M. A., "Crystal Size Distribution in Continuous Crystallizers when Growth Rate is Size-Dependent," *AIChE Journal 14* No. 1, (1968), p. 118.
22. Leer B. G. M. de, "Draft-Tube-Baffle Crystallizers, a Study of Stationary and Dynamically Behaving Crystal Size Distributions," doctoral thesis, Delft University of Technology, November 1981.

23. Asselbergs, C. J., "Evaporating Crystallizers: A Study of Factors in Operation and Design of Salt Evaporators," doctoral thesis, Delft University of Technology, October 1978.
24. Grootscholten, P. A. M., "Solid-Liquid Contacting in Industrial Crystallizers and Its Influence on Product Size Distribution," doctoral thesis, Delft University of Technology, August 1982.
25. Fitzgerald, T. J., and Yang, T., "Size Distribution for Crystallization with Contributions Growth and Breakage," *Ind. Eng. Chem. Fund. 11* No. 4, (1972) p. 588.
26. Juzaszek, P., and Kawecki, W., 5-th Symp. on Industrial Crystallization, Chisa, Czechoslovakia 1972.
27. Kuijvenhoven, L. J., "Aspects of Continuous Sucrose Crystallization," doctoral thesis, Delft University of Technology, November 1983.
28. Jancic, S. J., in *Industrial Crystallization*, S. J. Jancic and E. J. de Jong (Eds.), North Holland, Amsterdam (1982).
29. Newman, H. H., "Salt Evaporator Design," paper presented at a meeting of the Salt Producers Association, Chicago, 1955.
30. Rutten, J. J., and de Jong, E. J., "Factors Affecting the Rate of Evaporation in a Cylindrical Evaporator-Body with Tangential Inlet," paper presented at the 3-rd CHISA-Congress (Symp. on Industiral Crystallization), Czechoslovakia, 1969.
31. Newman, H. H., and Bennett, R. C., "Circulating Magma Crystallizers," *Chem. Eng. Progress, 55* No. 3 (1959), p. 65–70.
32. Mullin, J. W., *Crystallization*, 2-nd ed., Buttersworth, London (1972).

CHAPTER 34

BATCH COOLING CRYSTALLIZATION

Jaroslav Nyvlt

Institute of Inorganic Chemistry
Czechoslovak Academy of Sciences
Prague, Czechoslovakia

CONTENTS

INTRODUCTION

Before deciding on the mode of crystallization, we must consider two aspects of the problem: i.e., why cooling crystallization and why a batch operation.

Cooling Crystallization

Phase equilibria are essential for the decision on how to perform crystallization most efficiently. In the case where a sufficient amount of crystals is formed as a solution is cooled, then cooling may serve our purpose, though it need not be the only and, sometimes, the most economical approach to the problem. There are several possible ways to perform cooling [1–4]:

1. Indirect cooling
   - External.
   - Internal.

2. Direct cooling
   - By air.
   - Vacuum.
   - Bubbling through.
   - Immiscible liquid.

For substances which do not exhibit a clean-cut tendency to the formation of incrustations, indirect cooling with built-in cooling surfaces (e.g. cooling coils, cooling baskets, tube bundles, cooling draft-tubes, cooling jackets) can be adopted. If, however, there is a danger of incrustations, it is safer to choose indirect cooling with an external cooler, or even better to decide in favor of direct contact cooling. Vacuum cooling, too, may be included in this class: the cooling effect is accomplished by adiabatic evaporation of a part of the solvent, but the amount of the removed solvent is usually less significant and the cooling effect predominates. In order to be effective, vacuum cooling requires that:

1. Steam of the required quality is available in the plant.
2. The solution does not contain large amounts of dissolved and readily evolving gases.
3. The solution boiling temperature is not considerably higher than that of pure solvent.
4. Inlet concentration of the solution is sufficiently high, but not so high that the liberated heat of crystallization could compensate for the heat of evaporation.

Other criteria for evaluating vacuum cooling are the use of condensate waste vapors which should include the examination of the required final temperature and the demands for the vacuum system quality resulting therefrom; temperature difference attainable in the condenser; maximum concentration of the suspension; etc. Temperature as low as 0° C may be reached by vacuum cooling, which, however, is a very demanding task with respect to the vacuum system used; more common are temperatures 20° to 30° C. When lower temperatures are necessary, the following methods may be taken into consideration:

1. Indirect cooling with brine or ammonia.
2. Direct cooling with an immiscible liquid, where either its apparent heat and its heat of vaporization (butane, freons, propane-butane mixture) is utilized or where only its apparent heat is used (e.g., gasoline).

At moderate final temperature requirements it is also possible to consider cooling by introducing a stream of air or other suitable gas over the solution level or by bubbling air or other suitable gas through the solution. In the latter case gentle agitation and evaporation of a part of the solvent are accomplished simultaneously.

Batch Operation

The advantages of a batch crystallizer are:

1. In most cases it is a very simple piece of equipment virtually free of mechanical troubles.
2. Scaling-up of the apparatus does not involve any substantial risks.
3. Incrustations mostly disappear spontaneously when a fresh batch is started up.

4. Automatic process control enables it to produce larger crystals.
5. There are lower requirements for the maintenance and for quality of operators.

On the other hand, there are also drawbacks to batch crystallization:

1. The product quality may vary from charge to charge.
2. There are higher demands for manual work (in particular with more obsolete types of equipment) and for operators time.
3. It requires larger headroom.

It is not possible to determine a generally valid break-even limit between batch and continuous operation. The decision will be largely affected by the crystallized substance proper and by the conditions existing in the user's plant. When considering continuous process, respect the performance criterion and consider 40 to 200 kg/h as the minimum capacity for which a technically feasible apparatus can be reliably employed. If batch crystallization is selected, the performance criterion is not useful; the process can be carried out batchwise in any scale. The essential questions are those of complexity and difficulty of the crystallization process itself and of the retention time needed for the required growth of crystals; e.g., if the crystals are growing very slowly, then a batch process can be controlled more easily. There is one argument against batch crystallization at large capacities, i.e., the demand for a large built-up area.

Several steps [5] are performed in succession in a batch crystallizer. These are:

1. Filling of the crystallizer.
2. Cooling to saturation temperature.
3. Crystallization period.
4. Removal of suspension.
5. Cleaning of the crystallizer.

The time needed to complete a step is determined by the attainable or allowable rate of the rate-controlling mechanism. For instance, the time needed for Step 2 is determined by the attainable heat transfer rate. In Step 3, however, it may be necessary to restrict the heat transfer rate in order to control the nucleation and growth rates or to prevent incrustation on the cooling surfaces. After completion of the crystallization the suspension can be fed to an intermediate storage vessel or directly to a filter. In the latter case the required time for Step 4 is determined by the capacity of the filter. The size of the crystallizer would affect the duration of all the stages of a cycle. The required total cycle time can be calculated if all times required to complete individual periods are known. The available cycle time is equal to the ratio of the production per cycle to the average required production rate. Thus the available cycle time is a function of the crystallizer volume, too. Steps 1, 2, 4, and 5 are given by technical and mechanical properties of the equipment. The duration of Step 3, on the other hand, is determined by physical properties of the system and the requirements on the product quality; we shall thus concentrate on these problems in this chapter.

MATERIALS BALANCE

The overall materials balance of a batch crystallization takes the form

$$m_0 = m_f + m_c + m_g \tag{1}$$

where m_0 = the initial mass of the solution
m_f = the mass of mother liquors
m_c = the mass of crystals
m_g = mass of evaporated solvent

The overall heat balance takes an analogous form

$$m_0c_{p0}T_0+Q = m_fc_{pf}T_f+m_ch_c+m_cc_{pc}T_f+m_gi' \tag{2}$$

where T_0 and T_f = the initial and final temperature, respectively
c_p = the specific heat capacity of the solution
c_{pc} = specific heat capacity of crystals
h_c = heat of crystallization
Q = the heat amount fed into the crystallizer
i' = specific enthalpy of the vapor

Supposing additivity of specific heat capacities we can simplify Equation 2 to

$$m_0c_{p0}(T_0-T_f)+Q+m_ch_c = m_gh_v \tag{3}$$

where $h_v = (i'-c_{psolv}T_f)$ is the evaporation heat of the solvent.

We shall use w = kg of crystallizing substance per kg of free solvent (which is total water after subtraction of equivalent hydrate water); thus, 1 kg of solution contains $w/(w+1)$ kg of solute and $1/(w+1)$ kg of free solvent. The materials balance written for the crystallizing substance thus becomes

$$(m_0w_0)/(w_0+1) = (m_fw_f)/(w_f+1)+m_c \tag{4}$$

and the balance of the solvent

$$m_0/(w_0+1) = m_f/(w_f+1)+m_g \tag{5}$$

so that

$$m_c = m_0(w_0-w_f)/(w_0+1)+m_gw_f \tag{6}$$

Resulting equations for the materials and heat balance are:

- Cooling crystallizer:

$$m_c = m_0(w_0-w_f)/(w_0+1) \tag{7}$$

$$Q = -m_0[(w_0-w_f)h_c/(w_0+1)+c_{p0}(T_0-T_f)] \tag{8}$$

- Vacuum cooling crystallizer:

$$m_c = \frac{h_v(w_0-w_f)/(w_0+1)+c_{p0}(T_0-T_f)w_f}{h_v-h_cw_f} \tag{9}$$

$$m_g = m_0\frac{c_{p0}(T_0-T_f)+(w_0-w_f)h_c/(w_0+1)}{h_v-h_cw_f} \tag{10}$$

Equations 9 and 10 can be used [7] for $h_v \gg h_cw_f$, i.e., either in the case of slightly soluble substances or with systems exhibiting low heat of crystallization (both of these conditions are realized simultaneously in most cases). For $h_v = h_cw_f$ Equations 9 and 10 cannot be used—crystallization is isothermal and adiabatic, $T_0 = T_f$ and $w_0 = w_f$. If $h_v < h_cw_f$, i.e., in extremely concentrated solutions with large heat of crystallization the temperature rises with adiabatic evaporation so that the adiabatic evaporation is not effective enough.

Optimum Temperature in Batch Cooling

The final temperature in batch crystallization is usually chosen so that maximum possible yield is obtained in acceptable batch time or with acceptable consumption of cooling water. The solubility-temperature dependence can be approximated by a parabolic curve in most cases; by cooling at higher temperatures more crystals are gained than by cooling within the same temperature interval at lower temperatures. In the latter case, heat transfer is also lower due to a smaller temperature gradient between the suspension and the coolant. Both of these factors contribute to the fact that efforts for raising the yield of cooling crystallization lead to a higher consumption of cooling water. A lower final temperature gives more product crystals and reduces losses of the substance in mother liquors, but it requires more cooling water and longer batch time and thus higher costs of energy and of attendance. This situation may be characterized by a target function

$$Z = K_c m_c - K_w m_w - \eta K_c m_f - K_E P t_c - K_N t_c$$

$$= K_c(m_c - \eta m_f) - (K_w \dot{m}_w + K_E P + K_N) t_c \quad (11)$$

where Z = overall profit ($)
K_c = price of the product ($/kg)
K_w = cost of cooling water ($/kg)
K_E = cost of electric energy ($/kWh)
K_N = other time-dependent expenses ($/h)
m_c = mass of product crystals (kg)
m_w = mass of cooling water (kg)
$\dot{m}_w$ = flow of cooling water (kg/h)
m_f = mass of the crystallizing substance dissolved in mother liquors (kg)
P = power consumption of the agitator (kW)
t_c = batch time (h)
η = coefficient of economical losses by mother liquors: (this value depends on the treatment of mother liquors: in complete recycling $\eta = 0$, for complete discharge $\eta = 1$, if there are additional costs for adjustment of mother liquors K_z ($/kg substance), then $\eta = 1 + K_z/K_c$).

This target function Equation 11 passes through a maximum, for which

$$Z' = K_c(m_c' - \eta m_f') - (K_w \dot{m}_w + K_E P + K_N) t_c' = 0 \quad (12)$$

where symbols with primes represent the derivative of the particular variable with T_f, i.e.

$-m_c'$ = yield rise due to lowering T_f by 1 K
m_f' = loss in mother liquors reduction due to lowering T_f by 1 K
t_c' = batch time extension due to lowering T_f by 1 K

Having expressed the solubility of the crystallizing substance in the form

$$w_{eq} = a + bT + cT^2 \quad (13)$$

we can write the materials balance equation (Equation 7) in the form

$$m_c = m_0(w_0 - a - bT_f - cT_f^2)/(w_0 + 1) \quad (14)$$

so that

$$m_c' = m_0(b+2cT_f)/(w_0+1) = -m_f' \tag{15}$$

$$m_f = m_0(a+bT_f+cT_f^2)/(w_0+1) \tag{16}$$

Using an empirical equation for the cooling curve

$$\ln[(T_f-T_w)/(T_0-T_w)] = -kt_c \tag{17}$$

where T_0 = the initial temperature of the solution
T_w = the average temperature of cooling water we obtain

$$t_c' = k^{-1}(T_f-T_w)^{-1} \tag{18}$$

After inserting all corresponding expressions into Equation 12 we obtain the value of the optimum temperature of cooling

$$T_f = 1/2(T_w - b/2c) + \left[1/4(T_w + b/2c)^2 + \frac{(K_w\dot{m}_w + K_E P + K_N)(w_0+1)}{2ckK_c(1+\eta)m_0} \right]^{1/2} \tag{19}$$

from which we can find the corresponding batch time using Equation 17.

PROGRAMMED BATCH COOLING

Batch-operated, stirred cooling crystallizers are widely used in the chemical industry, but they usually yield a poor quality nonuniform product. This is mainly due to the use of a high cooling rate in the initial stages of the process, resulting in the formation of large numbers of crystal nuclei which cannot grow to the desired size. In addition, the large temperature drop across the cooling surfaces in the early stages often causes intensive scaling and significantly decreases the cooling capacity of the crystallizer. These disadvantages, normally associated with batch cooling crystallizers, can be overcome by the application of an appropriate temperature control.

Seeded Crystallizer

First attempts to solve the theoretical cooling curve for a simple batch operation were published for the case of a low supersaturation where no additional nucleation proceeds and the growth of seed crystals is sufficient to keep the supersaturation constant [3,8]. We shall consider a batch crystallizer of unit volume V provided with an efficient stirrer maintaining uniformity of temperature and concentration throughout the crystallizer and capable of maintaining the crystals in contact with the entire liquid volume. The crystallizer is filled with solution and seeded with a quantity of crystals sufficient to lead to an acceptable subsequent rate of crystallization and, at the same time, to enable the crystals to grow to the required size:

$$m_{c0} = L_0^3/(L_f^3 - L_0^3) \cdot \Delta m \tag{20}$$

where m_{c0} = the mass of seeds
L_0 = the size of the seeds
L_f = the size of the product crystals
Δm = the crystal mass increase resulting from the materials balance

We now begin to supersaturate the seeded solution at the maximum allowable rate, i.e., so as to enable us to neglect nucleation. This means that the supersaturation has to be kept constant. The balance describing mass transfer between the liquid and solid phases can be written as

$$-(dT/dt) \cdot (dw_{eq}/dT) = k_G A_c \Delta w^g \tag{21}$$

where the left-hand side of this equation describes the supersaturation rate (the first term is the cooling rate, $-\dot{T}$, and the second one is the temperature coefficient of solubility, w_{eq}), and the right-hand-side term describes the desupersaturation rate due to the crystal growth: A_c is the surface area of crystals and Δw is supersaturation. This means that the allowable cooling rate is proportional to the surface area of crystals present at any instant. Supposing that the crystals grow with a constant growth rate, $\dot{L}$, at constant supersaturation, their size will be

$$L = L_0 + \dot{L}t \tag{22}$$

and knowing their number

$$N_c = m_0/(\alpha \varrho_c L_0^3) \tag{23}$$

we can obtain their surface area at time t:

$$A_c = N_c \beta L^2 = m_0 \beta/(\alpha \varrho_c L_0) \cdot (1 + \dot{L}t/L_0)^2 \tag{24}$$

Substitution of this surface area into Equation 21 leads to the expression of the allowable cooling rate

$$-\dot{T} = 3 \frac{m_0}{(dw_{eq}/dT)} (\dot{L}/L_0)(1 + \dot{L}t/L_0)^2 \tag{25}$$

or, after integration of Equation 25, to the equation of the cooling curve

$$(T_0 - T_t)/(T_0 - T_f) = 3(\dot{L}t/L_0) m_0/(dw_{eq}/dT) \cdot [1 + \dot{L}t/L_0 + (\dot{L}t/L_0)^2/3] \tag{26}$$

which can be approximated by the maximum term

$$(T_0 - T_t)/(T_0 - T_f) = (t/t_c)^3 \tag{27}$$

In the preceding equations, α and β are the volume and surface shape factors, respectively, ϱ_c is the crystal density, T_0, T_t, and T_f the starting, instantaneous, and final temperature, respectively, and t_c is the batch time. Passing from Equation 24 to 25 we used the definition of the overall crystal growth rate

$$\dot{L} = k_G \beta \Delta w^g/(3\alpha \varrho_c) = k'_G \Delta w^g \tag{28}$$

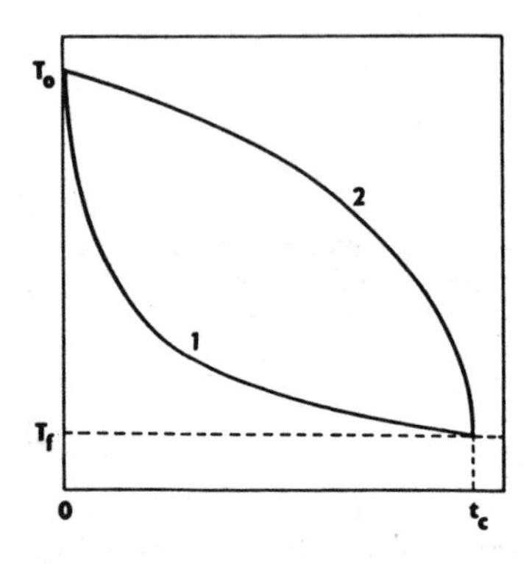

Figure 1. Cooling curves of a batch crystallizer: (1) natural (free) cooling; (2) programmed cooling.

According to Equation 26 or 27, initially cooling should be slow and only after the total crystal surface area has significantly increased should the cooling rate be increased, too. These conclusions show that the free, uncontrolled cooling is unsuitable for batch crystallizers, as the free-cooling curve is of inverse character, the fast rate ocurring initially, and vice versa (Figure 1).

Unseeded Crystallizer

The procedure just described has been later extended for the case of a batch crystallizer without seeding [9, 10]. The supersaturation rate is for constant supersaturation, Δw, just compensated by the rate of crystal growth and nucleation:

$$-\dot{T}(dw_{eq}/dT) = k_G A_c \Delta w^g + k_N \Delta w^n \tag{29}$$

The surface area of crystals, A_c, is not known and is, of course, a function of time

$$A_c = \int_0^t \dot{N}_N \beta L^2 \, dt \tag{30}$$

Number of crystals, $\dot{N}_N$, originating in a unit amount of solution per unit of time, i.e., the numerical nucleation rate, may be expressed by the relation

$$\dot{N}_N = k_N \Delta w^n/(\alpha \varrho_c L_N^3) \tag{31}$$

where L_N is the initial crystal size. This nucleation rate is also constant because of constant supersaturation. After substituting Equations 31 and 22 into Equation 30, a relation for the surface area of crystals is obtained

$$A_c = k_N \Delta w^n \beta/(3\alpha \varrho_c \dot{L}) \, [(L/L_N)^3 - 1] \tag{32}$$

On substituting Equation 32 together with 28 into Equation 29, the following equation is obtained

$$-\dot{T}(dw_{eq}/dT) = k_N \Delta w^n [(\dot{L}t/L_N) + 1]^3 \tag{33}$$

which, after integration for t = 0 till t, gives

$$(T_0 - T_t)\,(dw_{eq}/dT) = k_N \Delta w^n L_N/4\dot{L}[(\dot{L}t/L_N + 1)^4 - 1] \tag{34}$$

An equivalent equation holds for final conditions of crystallization where $t = t_c$ is the batch time and $T = T_f$ is the final temperature. The ratio of both these relations gives

$$(T_0 - T_t)/(T_0 - T_f) = \frac{[(\dot{L}t/L_N + 1)]^4 - 1}{[(\dot{L}t_c/L_N + 1)]^4 - 1} \tag{35}$$

which can be simplified to

$$(T_0 - T_t)/(T_0 - T_f) = (t/t_c)^4 \tag{36}$$

Batch Cooling Control

In fact, the shape of cooling curves described by Equation 27 and Equation 36 is very similar. These equations were derived later by different authors using the population balance technique [11–13] or iterative optimization procedure [14]. Experiments [9, 12, 15, 16] carried out with programmed cooling and free cooling have shown the advantage of the former method (e.g., crystals of sodium nitrate produced [16] in a 200 L agitated batch crystallizer by free cooling with the batch time 10 h attained the mean crystal size of 0.67 mm, whereas crystals obtained under similar conditions by programmed cooling had $L_{50} = 1.70$ mm). Similar conclusions may be drawn from computer simulations, discussed below. All these experiments have shown that, irrespective of the true kinetic parameters of the system, any cooling curve corresponding to Equations 27 or 36 gives better results than a free-cooling curve. This conclusion enables us to design a simple control of batch cooling [17, 18]. Its principle is based on the following:

Starting from Equation 25 we can write for the maximum cooling rate at the end of the batch crystallization

$$-\dot{T}_{max} = 3m_0/(dw_{eq}/dT)\,(\dot{L}/L_0)\,(1 + \dot{L}t_c/L_0)^2 \tag{37}$$

so that

$$(\dot{T}_t/\dot{T}_{max}) = \left(\frac{L_0 + \dot{L}t}{L_0 + \dot{L}t_c}\right)^2 \tag{38}$$

Integration of Equation 25 leads to the expression

$$(T_0 - T_t) = m_0(dw_{eq}/dT)^{-1}\,[(1 + \dot{L}t/L_0)^3 - 1] \tag{39}$$

from which we can extract the time as a function of temperature

$$t = \frac{[(dw_{eq}/dT)\,(T_0 - T_t)/m_0 + 1]^{1/3} - 1}{\dot{L}/L_0} \tag{40}$$

Substituting this value into Equation 38 we obtain

$$\frac{\dot{T}_t}{\dot{T}_{max}} = \left(\frac{1 + (dw_{eq}/dT)\,(T_0 - T_t)/m_0}{1 + (dw_{eq}/dT)\,(T_0 - T_f)/m_0}\right)^{2/3}$$

$$\approx \left(\frac{T_0 - T_t}{T_0 - T_f}\right)^{2/3} \tag{41}$$

The cooling rate, $-\dot{T}$, is proportional to the flow rate of cooling water, $\dot{V}_w$, and to the instantaneous temperature difference $(T_t - T_w)$:

$$-\dot{T}_t = k \cdot \dot{V}_{wt}(T_t - T_w) \tag{42}$$

and

$$-\dot{T}_{max} = k \cdot \dot{V}_{max\,w}(T_f - T_w) \tag{43}$$

Combining Equations 41 to 43 we obtain the final equation

$$\frac{\dot{V}_{wt}}{\dot{V}_{w\,max}} \sim \left(\frac{T_0 - T_t}{T_0 - T_f}\right)^{2/3} \frac{T_f - T_w}{T_t - T_w} \tag{44}$$

describing the optimum throughput of cooling water in every instant of batch cooling.

An example of a simple operation of batch cooling is shown in Figure 2. The resistance thermometer (1) is connected to the electronic switch (2) operating the valves A and B which is adjusted so that, for instance, between 60° and 55° C both valves are closed and only slow natural cooling due to heat losses proceeds; at 55° C it opens the first valve letting through, say, 20% of the maximum total flow of cooling water. Crystallization proceeds under this gentle cooling down to the second adjusted temperature, say 39° C, where the first valve A is closed and the second valve B, giving 80% of the maximum throughput, is opened. Cooling proceeds more rapidly until the third adjusted temperature is reached, for example, 32° C, where the valve A is opened again so that the full amount of cooling water is available for cooling, now. This simple arrangement [18] leads to an improvement of product quality without any substantial additional costs.

Analysis of the Cooling Curve

In many cases there are no specific demands upon the product crystal size, or even that a fine-grained product be required. The capacity of the crystallizer is then limited by the possibility of heat transfer only; we have to consider the surface area of the cooler, acceptable intensity of agitation, consumption of cooling water, scaling etc. Useful information of the existing equipment can be obtained by the analysis of cooling curves [19].

The heat transfer can be described in a usual way,

$$\dot{Q} = KA_f(T - T_w) \tag{45}$$

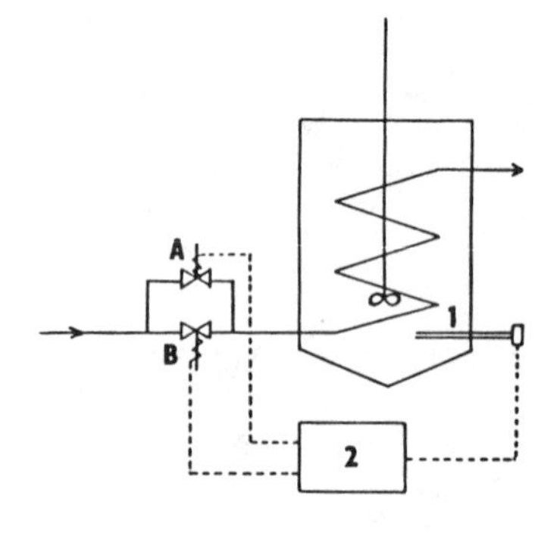

Figure 2. Automatic operation of a batch cooling crystallizer: (1) resistance thermometer; (2) electronic switch; (A, B) valves.

or

$$\dot{Q} = dQ/dT \cdot (-\dot{T}) \tag{46}$$

The total heat to be removed from the suspension can be subdivided into the apparent heat, Q_T

$$dQ_T = m_0 c_{p0} \, dT \tag{47}$$

where m_0 = the mass of initial solution (or that of the suspension)
c_{p0} = specific heat capacity of initial solution and the heat of crystallization Q_c

$$dQ_c = -h_c \, dm_c \tag{48}$$

where m_c is the mass of crystals given by Equation 7. Using Equation 13 for the dependence of solubility on temperature, we have

$$dQ_c/dT = h_c m_0 (b + 2cT)/(w_0 + 1) \tag{49}$$

and thus

$$-\dot{T} = \frac{K(T - T_w)A_f/m_0}{c_{p0} + h_c(b + 2cT)/(1 + w_0)} \tag{50}$$

with formal boundary conditions

$$h_c = h_c \quad (T < T_{0eq})$$

$$h_c = 0 \quad (T \geqq T_{0eq}) \tag{51}$$

Using Equation 50 we can obtain from experimental values $-\dot{T}(T)$ useful information: so far as no crystallization occurs (region of undersaturated solution), the plot of experimental cooling rate, $-\dot{T}$, vs. temperature, T, is linear with the slope

$$\text{slope} = KA_f/m_0 c_{p0} \tag{52}$$

from which the overall heat transfer coefficient, K, or its product KA_f can be obtained. Construction of these linear intercepts is facilitated by the fact that they cross the axis $-\dot{T} = 0$ at values T_w (see Figure 3). If the start of crystallization (near T_{0eq}) causes a break or even a discontinuity of these lines, the slope becomes lesser. From the ratio of both slopes we can assess the value of the heat of crystallization

$$h_c = (\dot{T}_1/\dot{T}_2 - 1)c_{p0}(1 + w_0)/(b + 2cT_{0eq}) \tag{53}$$

SIZE DISTRIBUTION OF PRODUCT CRYSTALS

Theoretical studies [9, 10] as well as results of model and full-scale experiments [20] have shown that under certain circumstances the size distribution of product crystals from a batch- operated crystallizer may be formally described by the same relation as was theoretically derived for the continuous mixed-suspension mixed-product removal (MSMPR) crystallizer. This, of course, greatly simplifies the treatment of data obtained by batch measurements.

For continuous MSMPR crystallizer the following relations have been derived for the crystal population density [21]

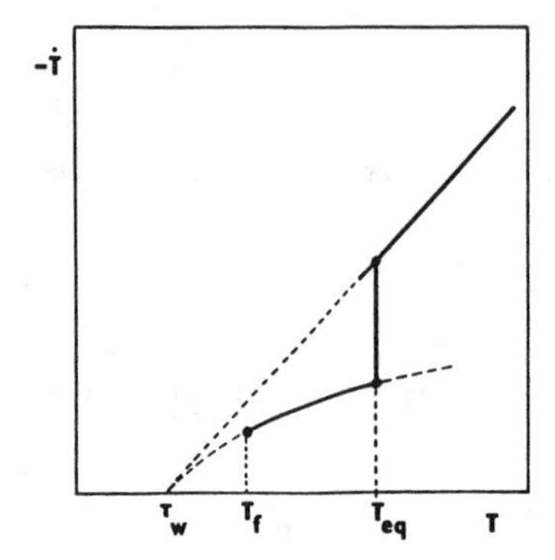

Figure 3. Analysis of the batch cooling curve: plot of the natural cooling rate, $-\dot{T}$, vs. temperature, T.

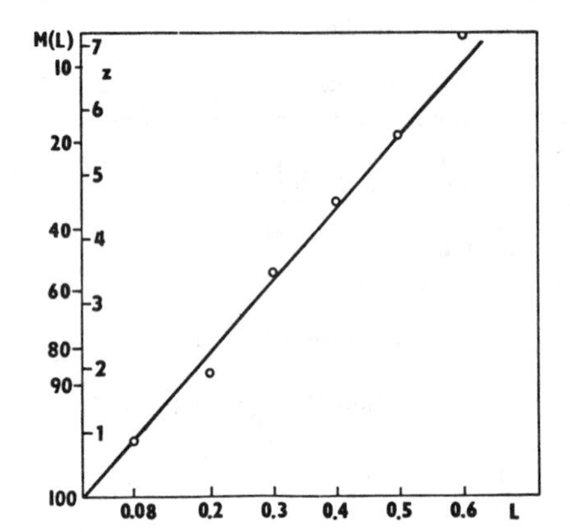

Figure 4. Crystal size distribution in the z-L coordinates system.

$$n(L) = n^0 \exp(-z) \tag{54}$$

and for the cumulative crystal size distribution [21, 22]

$$M(L) = 100(1+z+z^2/2+z^3/6)\exp(-z) \tag{55}$$

where

$$z = (L-L_N)/\dot{L}\bar{t}_1 \tag{56}$$

It follows from Equation 56 that in the z-L coordinate system the size distribution of product crystals from a continuous MSMPR crystallizer will be represented [22] by a straight line (Figure 4). If the size distribution of product crystals from a batch crystallizer fulfills the conditions of Equations 54 through 56, i.e., if it is likewise represented in the z-L coordinate system by a straight line, then further relations derived from Equation 54 for the continuous crystallizer can be expected to be formally valid for the batch process as well. If these conditions don't hold, however, the crystal size distribution cannot be approximated by Equation 55. We shall see later that mathematical simulation of the course of batch crystallization demonstrated that the variation of supersaturation can be represented, as an approximation, by the equation

$$\Delta w = \Delta w_{max}(t/t_{max})^x t^x \exp(-Kt) \tag{57}$$

A general explicit calculation of the crystal size distribution based on Equation 57 is not feasible. We shall therefore consider separately three models involving

1. Constant supersaturation ($x = 0$, $K = 0$).
2. Supersaturation decreasing at a constant rate.
3. Monotonously decreasing supersaturation ($x = 0$).

Batch Crystallization with Constant Supersaturation

For the mass of crystals in unit volume of suspension we can write

$$m_c = \int_0^{L_{max}} \alpha \varrho_c n(L) L^3 \, dL \tag{58}$$

where, at a constant supersaturation,

$$n(L) = \dot{N}_N / \dot{L} = \text{const.} \tag{59}$$

so that

$$m_c = \frac{1}{4} \alpha \varrho_c n(L) L_{max}^4 \tag{60}$$

Similarly, the mass of crystals larger than a certain mesh size L is given as

$$m(L) = \int_L^{L_{max}} \alpha \varrho_c n(L) L^3 \, dL = \frac{1}{4} \alpha \varrho_c n(L) \, (L_{max}^4 - L^4) \tag{61}$$

The oversize is defined as

$$M(L) = 100 \, m(L)/m_c \tag{62}$$

Substituting Equations 61 and 62 into 60, we obtain the same equation as derived previously [10]

$$M(L) = 100[1 - (L/L_{max})^4] \tag{63}$$

which can be rearranged to

$$100 - M(L) = 100(L/L_{max})^4 \tag{64}$$

Taking logarithms,

$$\ln [100 - M(L)] = (4.605 - 4 \ln L_{max}) + 4 \ln L \tag{65}$$

we have a linear equation from which L_{max} characterizing a given product can readily be calculated.

Batch Crystallization with a Constantly Decreasing Rate of Supersaturation

For the supersaturation we can write

$$\Delta w = \Delta w_{max}(1 - t/t_c) \tag{66}$$

Applying the power-law approximation, we can express the nucleation rate as

$$\begin{aligned} \dot{N}_N &= k_N \Delta w^n = k_N \Delta w_{max}^n (1 - t/t_c)^n \\ &= \dot{N}_{Nmax}(1 - t/t_c)^n \end{aligned} \tag{67}$$

In like manner, we obtain for the linear crystal growth rate

$$\dot{L} = k_G\beta/(3\alpha\varrho_c)\Delta w^g = k'_G\Delta w^g = k'_G\Delta w^g_{max} \cdot (1 - t/t_c)^g$$
$$= \dot{L}_{max}(1 - t/t_c)^g \tag{68}$$

By the end of the batch operation the crystals born at time t will have grown to the size

$$L = \int_t^{t_c} \dot{L}\, dt = L_{max}(1 - t/t_c)^{g+1} \tag{69}$$

where

$$L_{max} = \dot{L}_{max}t_c/(g+1) \tag{70}$$

The total mass of crystals is again

$$m_c = \int_0^{t_c} \alpha\varrho_c L^3(t)\dot{N}_N(t)\, dt$$
$$= \alpha\varrho_c L^3_{max}\dot{N}_{Nmax}t_c/(n + 3g + 4) \tag{71}$$

and the mass of crystals born at time t

$$m(t) = \int_t^{t_c} \alpha\rho_c L^3(t)\dot{N}_N(t)\, dt = \alpha\rho_c L^3_{max}\dot{N}_{Nmax}t_c/(n + 3g + 4) \cdot$$
$$[1 - (1 - t/t_c)^{n+3g+4}] \tag{72}$$

Hence, the oversize is

$$M(t) = 100\, m(t)/m_c = 1 - (1 - t/t_c)^{n+3g+4} \tag{73}$$

and after substitution from Equation 68, we obtain the equation

$$M(L) = 100[1 - (L/L_{max})^{(n+3g+4)/(g+1)}] \tag{74}$$

which can be rearranged to the form

$$100 - M(L) = 100\,(L/L_{max})^{(n+3g+4)/(g+1)} \tag{75}$$

analogous to Equation 64.

Batch Crystallization with One-Shot Creation of Supersaturation

For x = 0, Equation 57 yields

$$\Delta w = \Delta w_{max} \exp(-Kt) \tag{76}$$

where Δw_{max} = the maximum (i.e. initial) supersaturation
K = an adjustable parameter

Crystals are born by nucleation throughout the desupersaturation of solution. Using the power-law approximation, we can write

$$\dot{N}_N = k_N \Delta w^n = k_N \Delta w^n_{max} \exp(-nKt)$$

$$= \dot{N}_{Nmax} \exp(-nKt) \tag{77}$$

Similarly,

$$\dot{L} = k'_G \Delta w^g = k'_G \Delta w^g_{max} \exp(-gKt)$$

$$= \dot{L}_{max} \exp(-gKt) \tag{78}$$

By the end of the experiment (i.e. by complete depletion of supersaturation, characterized theoretically by $t \rightarrow \infty$) the crystals born at time t will have grown to the size

$$L = \int_t^\infty \dot{L}\,dt = \dot{L}_{max} \exp(-gKt)/gK \tag{79}$$

The growth will obviously be largest for those crystals which were born first, i.e. at time $t = 0$:

$$L_{max} = \dot{L}_{max}/gK \tag{80}$$

so that we can write

$$L = L_{max} \exp(-gKt) \tag{81}$$

The total crystal mass in unit volume of suspension will be

$$m_c = \int_0^\infty \alpha \varrho_c L^3(t) \dot{N}_N(t)\,dt$$

$$= \dot{N}_{Nmax} \alpha \varrho_c L^3_{max}/(n+3g)K \tag{82}$$

By analogy, we can calculate the mass of crystals older than a certain time t, i.e. the crystals formed within the time interval (0, t);

$$m(t) = \int_0^t \alpha \varrho_c L^3(t) \dot{N}_N(t)\,dt = \dot{N}_{Nmax} \alpha \varrho_c L^3_{max}$$

$$/(n+3g)K \cdot [1 - \exp(-(n+3g)Kt)] \tag{83}$$

Thus, for the oversize fraction of crystals older than time t we can write

$$M(t) = 100\{1 - \exp[-(n+3g)Kt]\} \tag{84}$$

or in terms of the crystal size, L, using Equation 81

$$M(L) = 100[1 - (L/L_{max})^{n/g+3}] \tag{85}$$

Taking logarithms, we obtain

$$\ln[100 - M(L)] = [4.605 - (n/g+3) \ln L_{max}]$$

$$+ (n/g+3) \ln L \tag{86}$$

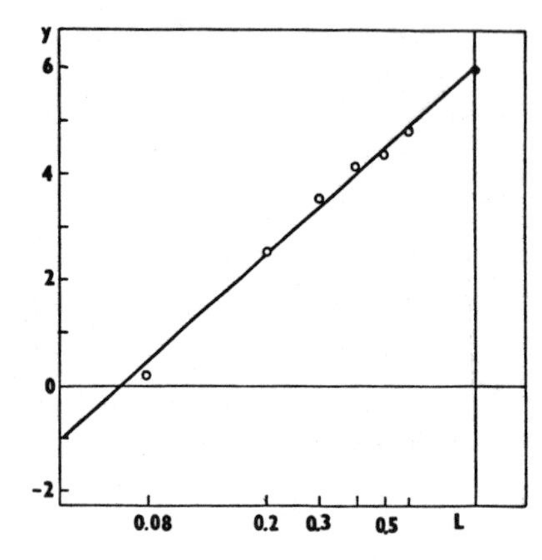

Figure 5. Crystal size distribution in coordinates ln [100–M(L)] vs. L according to the Eq. (87).

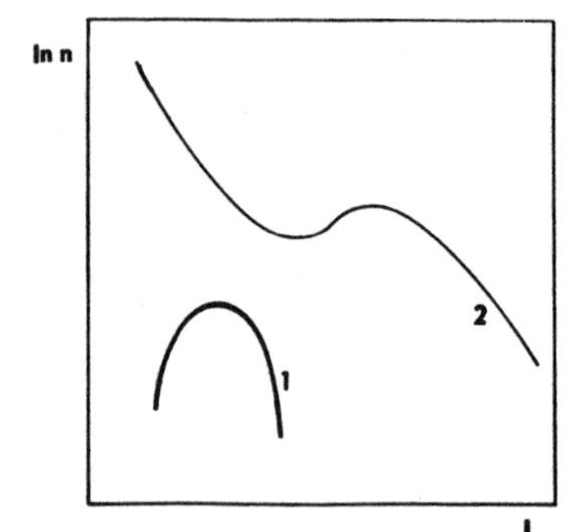

Figure 6. Population density of seeded batch crystals: (1) seeds; (2) product.

Upon comparing Equations 64, 75 and 85 one can see that the crystal size distribution for each of the models considered in this paragraph may be expressed by a relationship of the type

$$\ln [100 - M(L)] = (4.605 - s \ln L_{max}) + s \ln L \tag{87}$$

where s is the only value depending on the choice of the model. This relationship allows the largest crystal size, L_{max}, to be evaluated irrespective of the particular model employed. An alternative data treatment starts from Equations 55 and 56, yielding the mean crystal size, $\bar{L}$, as a characteristic parameter of the product. One cannot guess beforehand which of the two correlations will better fit experimental data for a given product, and therefore a graphical comparison must be made (Figure 4 and Figure 5).

Solution of the problem of size distribution of crystals from batch crystallizers using the population balance technique has been summarized in a recent paper [13]. For the perfectly mixed batch crystallizer with negligible attrition and agglomeration the population balance of a cooling batch crystallization reduces to

$$\frac{\partial n}{\partial t} + \frac{\partial (n\dot{L})}{\partial L} = 0 \tag{88}$$

To simplify solution of this equation a new variable y will be defined such that

$$y = \int_0^t \dot{L}(t)\, dt \tag{89}$$

and so

$$t = \int_0^y \dot{L}^{-1}(y)\, dy \tag{90}$$

The variable y can be thought of as the size of a crystal at any time t, which was originally nucleated at time $t = 0$. With y as a variable, the population balance for a crystal system in which growth rate is independent of size becomes

$$\partial n/\partial y + \partial n/\partial L = 0 \tag{91}$$

The moment equations obtained by moment transformations of Equation 91 are

$$dN_c/dy = n(y, 0) \tag{92a}$$

$$dL_c/dy = N_c \tag{92b}$$

$$dA_c/dy = 2\beta L_c \tag{92c}$$

$$dm_c/dy = 3\alpha\varrho_c A_c/\beta \tag{92d}$$

To determine the batch population density as a function of time and size, the population balance equation must be solved under some known initial conditions. For example, with the assumption of constant n^0, it holds for natural cooling

$$dm_c/dy = 6\alpha\varrho_c(n^0y^3/6 + N_0y^2/2 + L_0y + A_0/2\beta) \tag{93}$$

Using this equation the relationship between y and t can be estimated by solving the supersaturation balance

$$d\Delta w/dy + dw_{eq}/dy + dm_c/dy = 0 \tag{94}$$

numerically. This solution of y with respect to t defines the population density as a function of time. If a constant cooling rate is considered instead of natural cooling, analysis is analogous. The only change is that in the supersaturation balance

$$d\Delta w/dy - \dot{T}dw_{eq}/dT \cdot dt/dy + dm_c/dy = 0 \tag{95}$$

The value of dm_c/dy may be determined by solution of the moment equations and Equation 95 can then be solved numerically to define the relationship between y and t. This in turn may be used to calculate the population density as a function of time.

Effect of Seeding

The problem becomes even more complicated when seeding of the batch is introduced: it has been demonstrated [23] that the population density of seed crystals, well defined at the beginning of batch crystallization, undergoes a strong deformation in the course of the process (Figure 6) and a mathematical model based on usual equations of the population balance as a function of the crystal size and the process time has been found not to be suitable to describe such behavior.

Relatively simple is the calculation of the crystal size distribution in a seeded batch crystallizer under the condition of negligible nucleation [24]. The seeds may be subdivided according to their size into individual size classes with size L_{0i} and mass m_{0i}. Each of the classes passes during the crystallization to a higher size

$$L_i = L_{0i} + \Delta L \tag{96}$$

which is accompanied by a mass increase

$$m_i = m_{0i}(1 + \Delta L/L_i)^3 \tag{97}$$

The increase in size, ΔL, can be calculated by solving the equation for the total mass increase during crystallization, Δm_c:

$$\Delta m_c = \Delta L^3 \sum_i (m_{0i}/L_{0i}^3) + 3\Delta L^2 \sum_i (m_{0i}/L_{0i}^2) + 3\Delta L \sum_i (m_{0i}/L_{0i}) \tag{98}$$

After having inserted this value of ΔL back into Equations 96 and 97 we obtain final crystal sizes, L_i, and corresponding mass of individual fractions, m_i.

As soon as nucleation occurs simultaneously in seeded solution, step-by-step solution of a model of batch crystallizer is almost inevitable [25–27]. First, initial surface area and the number of seeded crystals are computed

$$A_0 = \beta N_0 L_0^2 \tag{99}$$

$$N_0 = m_0/(\alpha \varrho_c L_0^3) \tag{100}$$

The mode of cooling was expressed by Equations

$$T = T_w + (T_0 - T_w) \exp(-Kt) \tag{101a}$$

for noncontrolled cooling and

$$T = T_0 - (T_0 - T_f)(t/t_c)^x \tag{101b}$$

for controlled cooling with $x = 1$ describing constant cooling rate, $x = 3$ and $x = 4$ corresponding to Equations 27 and 36, respectively. By use of the so-determined temperature, the values of kinetic constants of growth, k_G, and of nucleation, k_N, and also the equilibrium solubility were calculated:

$$k_G = k_{G0} \exp(-E_G/RT) \tag{102}$$

$$k_N = k_{N0} \exp(-E_N/RT) \tag{103}$$

$$w_{eq} = w_0 \exp(-E_w/RT) \tag{104}$$

The supersaturation rate is defined as the change in solubility per unit of time

$$s = w_{eq(t-1)} - w_{eq(t)} \tag{105}$$

so that the actual supersaturation at time t is

$$\Delta w_t = \Delta w_{t-1} + \Delta(\Delta w) \tag{106}$$

where

$$\Delta(\Delta w) = s - k_G A_c \Delta w^g - k_N \Delta w^n \tag{107}$$

Then, the number of newly formed crystals is:

$$N_t = k_N \Delta w^n / (\alpha \varrho_c L_N^3) \tag{108}$$

and the linear crystal growth rate corresponding to the given value of supersaturation,

$$\dot{L} = k_G \beta \Delta w^g / (3\alpha\varrho_c) \tag{109}$$

All the crystals present in the suspension will now grow on by the value $\dot{L}$

$$\begin{aligned} L_{0t} &= L_{0,t-1} + \dot{L} \\ L_{it} &= L_{i,t-1} + \dot{L} \\ &\vdots \\ L_j &= L_N \end{aligned} \tag{110}$$

and in the corresponding manner also the mass of individual fractions is rising as well

$$\begin{aligned} m_{0t} &= N_0 \alpha \varrho_c L_{0t}^3 \\ m_{it} &= N_i \alpha \varrho_c L_{it}^3 \\ &\vdots \\ m_j &= N_j \alpha \varrho_c L_N^3 \end{aligned} \tag{111}$$

$$m_t = m_{0t} + \sum_i m_{it} \tag{112}$$

Surface area of crystals is then

$$A_c = \beta m_{0t} / (\alpha \varrho_c L_{0t}) + \sum_i \beta N_i L_{it}^2 \tag{113}$$

As long as the time and temperature have not reached the required final value, next step is calculated starting with Equation 101. After reaching the final values, the product crystals are separated into the chosen number of fractions according to their size, and the crystal size distribution is computed.

The results from these computer simultations can be summarized as follows:

1. Noncontrolled and constant rate cooling had a similar result—considerable nucleation at the beginning of the process, which resulted in fine product crystals.
2. Programmed cooling has shown a considerable dependence on the initial supersaturation—the optimum results were obtained when the supersaturation was almost constant during the run.
3. In nonseeded experiments, the crystal size distribution could be nearly linearized by theoretical Equations 55 or 87. With the increasing amount of seeds there appears a steep

change in the crystal size distribution. The maximum size of product crystals is obtained under cooling which corresponds to theoretically predicted conditions.
4. With reduced batch time, also the maximum size of product crystals is decreasing.
5. The supersaturation-time dependence can be approximated by an empirical equation of the form Equation 57.

For efficient seeding to be carried out it is necessary to know the saturation temperature of the solution and to be close to this temperature in order to introduce a sufficient amount of seeds. In industrial crystallizers where the solution volume is large, the mass of seed crystals according to Equation 20 can be considerably high, particularly if the seeds are large, (e.g., when they are taken as a part of product crystals from the preceding run). In such cases, several hundred kg of seeds may be necessary. Seeding with such a number of crystals is certainly not desirable. An alternative is to reuse a part of the suspension left in the crystallizer from the preceding run [28, 29].

Let us consider a batch crystallizer containing m_0 of a solution saturated at T_0. After completing the cooling down to temperature T_f, the mass of crystals given by Equation 7 is precipitated with the average crystal size equal to $\bar{L}_0$. We shall discharge only a portion $(1-p_0)$ of suspension from the crystallizer so that there remains portion p_0. After filling up by a fresh solution and new cooling in the second run we obtain the same mass of crystals,

$$m_c = m_c p_0 + m_0(1-p_0)(w_0 - w_f)/(w_0 + 1) \tag{104}$$

but, considering negligible nucleation, we get larger crystals: if there is no dissolution of crystals during addition of a fresh solution, it holds

$$\bar{L}_1 = \bar{L}_0(m_c/m_{c0})^{1/3} = \bar{L}_0 p_0^{-1/3} \tag{105}$$

In a similar way, as in preceding paragraphs, we obtain for the maximum allowable cooling rate

$$-\dot{T}_{max} = 3\dot{L}_{max} m_c p_0/(\bar{L}_0 dw_{eq}/dT) \tag{106}$$

Repeating the procedure k times we obtain

$$\bar{L}_k = \bar{L}_0 p_0^{-k/3} \tag{107}$$

$$-\dot{T}_{max,k} = 3(\dot{L}_{max}/L_0)m_c p_0^{(4-k)/3}/(dw_{eq}/dT) \tag{108}$$

or, for two successive runs,

$$\bar{L}_k = \bar{L}_{k-1} p_0^{-1/3} \tag{109}$$

$$-\dot{T}_{max,k} = -\dot{T}_{max,k-1} p_0^{-1/3} \tag{110}$$

and

$$\dot{m}_{c,k} = \dot{m}_{c,k-1} p_0^{1/3} \tag{111}$$

Equation 106 describes the maximum allowable cooling rate; if the actual cooling rate is $-\dot{T}$, then the mass grown on seeds will be

$$m_v = m_c \dot{T}_{max}/\dot{T} \tag{112}$$

and the mass of newly born crystals

$$m_m = m_c(\dot{T} - \dot{T}_{max})/\dot{T} \quad (113)$$

Since the number of crystals, spontaneously formed at the metastability limit, is approximately constant

$$N_m = 9m_m/(2\alpha\rho_c\bar{L}_m^3) = \text{const.} \quad (114)$$

we can estimate the average size of fine crystals

$$\bar{L}_m = \left(\frac{9m_c(\dot{T} - \dot{T}_{max})}{2\alpha\varrho_c N_m \dot{T}}\right)^{1/3} \quad (115)$$

The larger is the mass of coarse fraction, the smaller are the fine crystals, therefore. this condition leads to a limiting crystal size, which can be attained by seeding by a portion of suspension, and this limit can be surpassed only when changing technological parameters. As a special case the alternating of larger and smaller crystals in odd and even runs has been found when the crystallization was very sensitive to seeding [29].

Effect of Agitation

Agitation has two opposite effects in batch crystallization: increased agitation diminishes the thickness of the adjacent layer on the surface of the crystals and thus speeds up the crystal growth rate (this is favorable for obtaining larger crystals); increased agitation also leads to higher nucleation due to crystal contacts and their possible attrition which results in smaller product crystals. The result is a dependence of mean crystal size on an agitation rate exhibiting a maximum. For example, this dependence measured in a pilot-plant crystallizer with sodium nitrate [30] is schematically depicted in Figure 7 for two different positions of the propeller: Curve 1 corresponds to the distance of the propeller from the bottom (270 mm); curve 2 to the distance of 200 mm. The results can be interpreted so that with low agitation speed (up to 200 rpm) the crystals were not suspended completely so that a part of the crystals remained at the bottom and didn't grow. Agitation speed higher than 400 rpm leads obviously to attrition and breaking of crystals so that the average crystal size diminishes again. The optimum agitation speeds corresponded in these cases to 350 and 450 rpm, respectively.

Theoretical treatment of this phenomenon is difficult: the theory of secondary nucleation allows no predicition until now, and the term "secondary nucleation" itself comprises several different mechanisms of the birth of crystals in suspension so that a general description can be hardly found. If we accept that secondary nucleation can be described by an empirical equation [31]

$$\dot{N}_N = k_{NA} m_c^{\sigma c} t_c^{\sigma} \Delta w^{(1-\sigma)n} \quad (116)$$

where
k_{NA} = a formal kinetic constant of nucleation combined with micro-attrition
m_c = suspension density
c = exponent of secondary nucleation
t_0 = batch time
Δw = supersaturation
n = true exponent of nucleation
σ = coefficient of secondary nucleation.

We can try to describe the complex process as follows [32]: the exponent c assumes values [33] c = 0 (primary nucleation and surface layer mechanism of secondary nucleation), c = 1

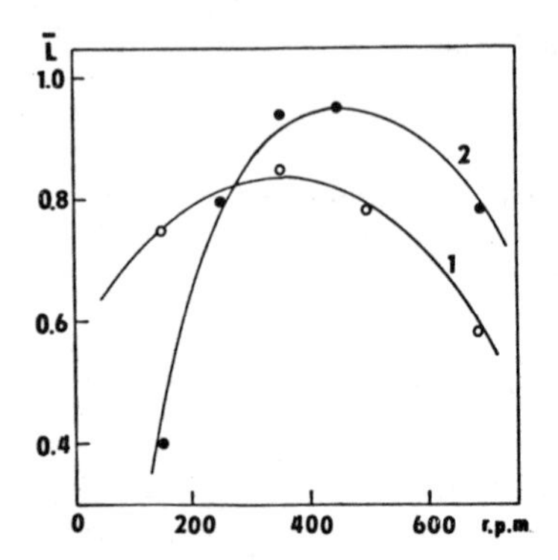

Figure 7. The effect of agitation on the mean crystal size of the product from a batch crystallizer. The Curves 1 and 2 correspond to different insertions of the propeller.

(collisions crystal—impeller or crystallizer) or c = 2 (collisions crystal—crystal); the coefficient σ is between 0 (no attrition) and 1 (only microattrition). From Equation 116 it follows that secondary nucleation depends on supersaturation with the effective power

$$n_A = (1-\sigma)n \tag{117}$$

Due to changes in m_c during a batch run, the coefficient σ depends on mean crystal size, suspension density, and thus on the batch time. This dependence can be formally described [32] by

$$\sigma = t_c/(\theta + t_c) \tag{118}$$

where θ is an adjustable constant. If the batch time is very short, σ approaches 0, and for an extremely long batch time σ is close to 1.

Balance of the number of crystals takes the form

$$N_N t_c = 9m_c/(2\alpha\varrho_c\bar{L}^3) \tag{119}$$

Average crystal size can be defined as

$$\bar{L} = \dot{L}t_c = k'_G \Delta w^g t_c \tag{120}$$

Inserting Equations 120 and 116 into Equation 119 we obtain

$$\bar{L}^{1+3g/n_A} = 3\frac{4.5^{g/n_A}k'_G}{(\alpha\varrho_c k_{NA})^{g/n_A}} m_c^{(1-\sigma_c)g/n_A} \cdot t_c^{1-(1+\sigma)g/n_A} \tag{121}$$

We shall discuss this type of equation later in the paragraph on production rate of batch crystallizers Equation 182. Here, we are interested in the shape of the dependence of $\bar{L}$ on t_c. From the condition

$$d\bar{L}/dt_c = 0 \quad \text{for} \quad L_{max}$$

we obtain as approximation

$$\sigma_{max} = n_A/g - 1 \quad \text{for} \quad L_{max} \tag{122}$$

and, using Equation 118,

$$t_c/\theta = (n_A/g - 1)/(2 - n_A/g) \quad \text{for} \quad L_{max} \tag{123}$$

It can be concluded from Equation 123 that a maximum on the $\bar{L}(t_c)$ curve can be expected for $1 < n_A/g < 2$.

If $n_A/g < 1$, the mean crystal size diminishes with longer batch time, if $n_A/g > 2$ the value of $\bar{L}$ is steadily increasing with t_c.

Effect of Temperature Cycling

The growth and dissolution rates of crystals can be described in many cases using similar equations

$$\dot{m}_c = k_G A_c \Delta w \tag{124}$$

and

$$-\dot{m}_c = k_d A_c \Delta w \tag{125}$$

Nevertheless, these two processes cannot be considered as reciprocal ones: the growth rate constant, k_G, is usually smaller than the dissolution rate constant, k_d, the difference being of an order of magnitude [2, 34]. It is well known from theoretical considerations that very small crystals exhibit a higher solubility than the bigger ones. If the crystallization occurs under such conditions that the temperature periodically varies on both sides of the corresponding saturation temperature, small crystals mostly dissolve and big crystals grow preferentially. The same holds for corners, edges, and faces of crystals, so that rounded crystals are obtained [35]. The dissolution of small crystals results in narrower crystal size distributions [36, 37].

This effect can be also reached by a simple temporary rising of temperature at the end of batch crystallization [38]: consider a unit amount of solution with concentration w_0 at temperature T_0, containing m_{10} of crystals with size L_{10}, and m_{20} of crystals with size L_{20}, where $L_{10} > L_{20}$. Then, we shall raise the temperature by ΔT and slowly cool down back to T_0. The size of the crystals diminishes during the period of dissolution:

$$L_i = L_{i0} - \Delta L \tag{126}$$

and so does their mass

$$m_i/m_{i0} = (1 - \Delta L/L_{i0})^3 \tag{127}$$

Simultaneously, it holds $-\Delta m = \Delta w$. Using the procedure described by Equation 98, for the condition $\Delta L \geqq L_{20}$ we have

$$(m_{10} + m_{20}) > \Delta T(dw_{eq}/dT) \geqq m_{10} \cdot [(L_{20}/L_{10})^3 - 3(L_{20}/L_{10})^2 + 3(L_{20}/L_{10})] + m_{20} \tag{128}$$

from which the necessary temperature rise, ΔT, can be obtained; for calculated or greater ΔT, recrystallization proceeds and crystals smaller than L_{02} will dissolve. This simple procedure leads to narrower crystal size distribution and hence to better separability of product crystals.

KINETICS OF CRYSTALLIZATION FROM BATCH EXPERIMENTS

The measurements of crystallization rate on batch-operated crystallizers are attractive because of the ease of the experiments' performance and the simplicity of the experimental arrangement. However, unlike results from continuous model crystallizers which are easy to evaluate thanks to a simple mathematical description (constant supersaturation, constant suspension concentration, and constant crystal surface area in the steady state), adequate evaluation of results from batch experiments is mathematically complex and mostly very laborious. However, theoretical studies as well as results of model and full-scale experiments have shown that under certain circumstances the size distribution of product crystals from a batch-operated crystallizer may be formally described by the same relation as was theoretically derived for the continuous MSMPR crystallizer. Thus, the relationships holding for a continuous crystallizer may formally be applied to batch crystallization. This, of course, greatly simplifies the treatment of data obtained by batch measurements. In principle, we can distinguish three typical cases:

- The crystallization is conducted in such a way as to keep the supersaturation of the solution constant throughout the experiment.
- The supersaturation is produced at a constant rate throughout the experiment, but as a result of increasing crystal area it reaches a maximum after a time and then gradually decreases.
- An initial supersaturation produced almost instantaneously at the beginning of the run decreases monotonously during the measurement.

Batch Crystallization with Constant Supersaturation

If the crystallization process is conducted in such a way that the supersaturation rate is proportional to the crystal area, the supersaturation can be expected to be more or less constant during the run. This condition is met in the crystallization runs in which cooling is performed at a rate corresponding to the Equation 36. If the nucleation occurs at a constant supersaturation at a steady rate throughout the run, we can define the mean product crystal size (per mass) as

$$\bar{L} = L_N + \dot{L}t_c/4^{1/3} \tag{129}$$

Comparison with the relationship holding for a continuous crystallizer

$$\bar{L} = L_N + 3\dot{L}\bar{t}_l \tag{130}$$

gives the formal equation

$$t_c = 4.76\bar{t}_l \tag{131}$$

The linear crystallization rate, $\dot{L}$, is calculated from Equation 129

$$\dot{L} = 4^{1/3}(\bar{L} - L_N)/t_c \tag{132}$$

From Equations 58 and 132 we can derive for the nucleation rate

$$\dot{N}_N = \frac{m_c}{\alpha \varrho_c t_c (\bar{L} - L_N)^3} \tag{133}$$

Batch Crystallization with a Constant Supersaturation Rate

When the supersaturation is produced at a constant rate during the whole process, it first increases almost linearly until a major portion of the crystals are produced spontaneously. During the rest of the process, i.e., practically for the batch time t_c, the crystals formed grow up, desupersaturating the solution. As the supersaturation decreases, so does the crystal growth rate, and if we replace the latter by the effective average rate $\dot{L}_{av}$, we can write

$$\bar{L} = L_N + \dot{L}_{av} t_c \tag{134}$$

Identifying this average crystal growth rate with the linear crystallization rate Equation 130, we can write the formal equation

$$t_c = 3\bar{t}_l \tag{135}$$

The only kinetic parameter which it seems reasonable to evaluate in this case is the mean linear crystallization rate,

$$\dot{L}_{av} = (\bar{L} - L_N)/t_c \tag{136}$$

while the nucleation rate expressed with respect to the whole batch time is of no real physical significance here. Since, however, the supersaturation-time curves for a series of runs carried out, e.g., at different supersaturation rates, are of similar shapes and can be represented by equations which differ only in values of adjustable parameteres, we can choose one run of the series as a reference base (*) and compare kinetic parameters of the other runs with the reference values. The relative crystal growth rate will be expressed as a ratio of Equation 136 to an analogous equation written for the reference run:

$$\dot{L}_{rel} = (\dot{L}/\dot{L}^*) = (\bar{L} - L_N)t_c^*/(\bar{L}^* - L_N)t_c \tag{137}$$

By analogy to the previous case, the nucleation rate is derived formally

$$\dot{N}_N = \frac{27 m_c}{2\alpha \varrho_c (\bar{L} - L_N)^3 t_c} \tag{138}$$

The relative nucleation rate is then given as

$$\dot{N}_{Nrel} = \dot{N}_N/\dot{N}_N^* \doteq \frac{m_c(\bar{L}^* - L_N)^3 t_c^*}{m_c^*(\bar{L} - L_N)^3 t_c} \tag{139}$$

Batch Crystallization with One-Shot Formation of Supersaturation

Precipitation is a typical example of crystallization with one-shot formation of supersaturation: the supersaturation is built up practically instantaneously on mixing the components. Then crystals are formed after an induction period following which the supersaturation decreases monotonously until an equilibrium concentration is reached. Under these conditions neither the mean crystal growth rate nor the average nucleation rate can be defined since both the processes proceed at different supersaturations and for different and, moreover, unknown times. As in the previous case, however, we can regard the curves for the supersaturation decay as similar to one another and, taking one arbitrary run of the series as a reference base (*), express all the kinetic data with respect to this run.

Again, the values of the mean crystal size $\bar{L}$ (distribution mode) and of $\dot{L}t_c$ (straight line slope) can be found from the graphical representation of product crystal size distribution in the z-L coordinate system. If the distribution is approximately linear, we can assume that the crystal population density is described, at least formally, by Equation 54. The mass of crystals formed is then given as

$$m_c = 2/9 N_c \alpha \varrho_c \bar{L}^3 \tag{140}$$

The nucleation rate is formally expressed as

$$\dot{N}_N = 9m_c/[2\alpha\varrho_c(\bar{L} - L_N)^3 t] \tag{141}$$

and the relative nucleation rate expressed with respect to a reference run (*) is given by

$$\dot{N}_{N\,rel} = \dot{N}_N/\dot{N}_N^* = m_c(\bar{L}^* - L_N)^3/m_c^*(\bar{L} - L_N)^3 \tag{142}$$

For the mass of crystals formed we can write

$$m_c = \int_{L_N}^{\infty} k_G A_c \Delta w^g \, dt \tag{143}$$

where both the crystal area, A_c, and the supersaturation, Δw, are a function of time:

$$A_c = 2/9 \beta N_c (L_N + \dot{L}t)^2 \tag{144}$$

$$d\Delta w/dt = k_G A_c \Delta w^g \tag{145}$$

The linear crystallization rate, $\dot{L}$, in Equation 144 is also a function of supersaturation

$$\dot{L} = k_G \beta \Delta w^g/3\alpha\varrho_c = k_G' \Delta w^g \tag{146}$$

Equation 145 can be rewritten (for $L_N = 0$) into the form

$$d\Delta w/\Delta w^{3g} = 2/9 k_G \beta N_c (k_G/3\alpha\varrho_c)^2 t^2 \, dt$$

$$= K' t^2 \, dt \tag{147}$$

Integrating Equation 147 within the limits $t = 0$, $\Delta w = \Delta w_0$ and $t = t_c$, $\Delta w = 0$, we obtain

$$\Delta w_0^{1-3g} = (g - 1/3) K' t_c^3 \tag{148}$$

Hence, the ratio of the slopes of the linearized particle size distributions for a given and the reference run may be written as

$$(\dot{L}t_c)_{rel} = \dot{L}t_c/(\dot{L}t_c)^* = (\dot{L}/\dot{L}^*)(\Delta w_0/\Delta w_0^*)^{(1-3g)/3} \tag{149}$$

The relative linear crystallization rate is then defined by the equation

$$\dot{L}_{rel} = \dot{L}/\dot{L}^* = (\dot{L}t_c)_{rel} (\Delta w_0)_{rel}^{(3g-1)/3} \tag{150}$$

which gives $\dot{L}_{rel}$ at the same time as a function of the relative supersaturation

$$(\Delta w_0)_{rel} = \Delta w_0/\Delta w_0^* \tag{151}$$

The nucleation rate varies with supersaturation according to the relation

$$\dot{N}_N = k_N \Delta w^n \tag{152}$$

and a similar expression may be written for the relative nucleation rate:

$$\dot{N}_{Nrel} = \dot{N}_N/\dot{N}_N^* = (k_n/k_N^*)\,(\Delta w_0)_{rel}^n \tag{153}$$

Elimination of the relative initial supersaturation, $(\Delta w_0)_{rel}$, from Equations 150 and 153 yields

$$\dot{N}_{Nrel} = \text{const.} \cdot \dot{L}_{rel}^{3n/(3g-1)} \tag{154}$$

In a similar way, relations have been derived for the kinetics of batch crystallization with constant and variable supersaturation, applicable where the crystal size distribution cannot be approximated by the relation holding for continuous MSMPR crystallizers.

Batch Crystallization with Constant Supersaturation

The linear crystal growth rate is given by

$$\dot{L} = L_{max}/t_c \tag{155}$$

and for the nucleation rate one can derive from Equations 59 and 60

$$\dot{N}_N = 4m_c/(\alpha \varrho_c L_{max}^3 t_c) \tag{156}$$

For a series of experiments with different supersaturations, we can use the relationship

$$\dot{N}_N = \text{const.} \cdot \dot{L}^{n/g} \tag{157}$$

allowing the relative kinetic exponent n/g to be estimated. The nuclei population density is given as

$$n^0 = 4m_c/(\alpha \varrho_c L_{max}^4) \tag{158}$$

Batch Crystallization with a Constantly Decreasing Rate of Supersaturation

The linear crystal growth rate is given by

$$\dot{L}_{max} = (g+1)L_{max}/t_c \tag{159}$$

and from Equation 71 it follows that

$$\dot{N}_{Nmax} = (n+3g+4)m_c/(\alpha \varrho_c L_{max}^3 t_c) \tag{160}$$

Hence, we get for the initial nuclei population density

$$n_0^0 = m_c[3+(n+1)/(g+1)]/(\alpha \varrho_c L_{max}^4) \tag{161}$$

It is clear that the absolute values of nucleation and growth rates cannot be evaluated unless we know the values of the nucleation and growth exponents. For the relative quantities with respect to a reference run (*) one can derive from the prior relationships

$$\dot{L}_{rel} = L_{max}t_c^*/L_{max}^*t_c \tag{162}$$

$$\dot{N}_{Nrel} = m_c L_{max}^{*3} t_c^*/(m_c^* L_{max}^3 t_c) \tag{163}$$

$$n_{0rel}^0 = m_c L_{max}^{*4}/m_c^* L_{max}^4 \tag{164}$$

Batch Crystallization with One-Shot Creation of Supersaturation

From Equation 80 the initial linear crystal growth rate can be expressed as

$$\dot{L}_{max} = gKL_{max} \tag{165}$$

and Equation 82 provides an expression for the initial nucleation rate

$$\dot{N}_{Nmax} = K(n+3g)m_c/(\alpha \varrho_c L_{max}^3) \tag{166}$$

Hence, the initial nuclei population density is

$$n_0^0 = \dot{N}_{Nmax}/\dot{L}_{max} = (n/g+3)m_c/(\alpha \varrho_c L_{max}^4) \tag{167}$$

As in the previous case, we must refer the data to a reference run (*), so that

$$\dot{L}_{rel} = KL_{max}/(KL_{max})^* \tag{168}$$

or, assuming $K \sim 1/t_c$,

$$\dot{L}_{rel} = L_{max}t_c^*/(L_{max}^*t_c) \tag{169}$$

Similarly, for the relative nucleation rate we get

$$\dot{N}_{Nrel} = Km_c L_{max}^{*3}/K^* m_c^* L_{max}^3) \tag{170}$$

or

$$\dot{N}_{Nrel} = m_c L_{max}^{*3} t_c^*/(m_c^* L_{max}^3 t_c) \tag{171}$$

The relative initial nuclei population density is then

$$n_{0rel}^0 = m_c L_{max}^{*4}/m_c^* L_{max}^4 \tag{172}$$

The true (absolute) values of the nucleation and crystal growth rates can be determined with confidence only for model experiments with constant supersaturation. For

$$L_{max} = 4^{1/3}\bar{L} \tag{173}$$

the Equations 155 through 158 become identical with Equations 129 through 133, i.e., they are independent of the crystal size distribution function.

For batch experiments with variable supersaturation, the maximum (initial) nucleation and crystal growth rates can be determined providing the values of the nucleation and growth exponents are known. Values of relative growth rates, Equations 162 and 169, and of the relative nucleation rates, Equations 163 and 171, are apparently independent of the choice of the model.

PRODUCTION RATE OF BATCH CRYSTALLIZERS

Starting from the population balance Equation 91 and substituting Equation 89 with constant L for y we obtain

$$dn/dL + n/\dot{L}t_c = 0 \tag{174}$$

with the solution

$$n = n^0 \exp(-L/\dot{L}t_c) \tag{175}$$

The nuclei population density, n^0, can thus be determined from the plot of log n vs. L. The nucleation rate is described by a power-law

$$\dot{N}_N = k_N m_c^c \Delta w^n \tag{176}$$

and the growth rate by

$$\dot{L} = k'_G \Delta w^g \tag{177}$$

Substituting Equation 177 into Equation 176 we obtain

$$\dot{N}_N = n^0 \dot{L} = m_c^c (\dot{L}/k'_G)^{n/g} k_N \tag{178}$$

and thus, nuclei population density is

$$n^0 = k_N k_G'^{-n/g} m_c^c \dot{L}^{n/g-1} \tag{179}$$

The growth rate from Equation 89 is

$$\dot{L} = \bar{L}/t_c \tag{180}$$

For the concentration of suspension holds

$$m_c = \alpha \varrho_c \int_0^\infty L^3 n(L)\, dL = 6\alpha \varrho_c n^0 (\dot{L}t_c)^4 \tag{181}$$

Substituting for $\dot{L}$ from Equation 180 and for n^0 from Equation 179, Equation 181 can be transformed into

$$\bar{L}^{1+3g/n} = 3B_N m_c^{(1-c)g/n} (t_c/3)^{1-g/n} \tag{182}$$

where

$$B_N = k'_G (4.5/\alpha \varrho_c k_N)^{g/n} \tag{183}$$

is a system constant, composed of growth parameters, k'_G and g, nucleation parameters, k_N and n, and the properties of crystals, α and ϱ_c. The value of this system constant can be determined either from individual parameters from Equation 183, or from model experiments using Equation 182. Its advantage consists in the fact that the system constant is not so sensitive to temperature changes or hydrodynamic conditions as are individual nucleation and growth constants, which diminishes the risk in the design of crystallizers [39, 40].

The exponent of secondary nucleation, c, is for batch crystallization in most cases equal to 0 or 1 and can be found by a trial-and-error method. The relative kinetic exponent, g/n, can be determined from at least two experiments carried out with equal initial and final conditions (w_0, T_0, and T_f) but with different batch time, t_c. For these conditions Equation 182 can be transformed into

$$d \log \bar{L}/d \log t_c = (1-g/n)/(1+3g/n) \quad (m_c = \text{const}) \tag{184}$$

which allows us to calculate the value of g/n from two corresponding values of $\bar{L}(t_c)$, or, if g/n is already known, it gives the dependence of $\bar{L}$ on t_c. If secondary contact nucleation predominates, Equation 121 has to be used instead of Equation 182, and the relative kinetic exponent can be found from Equation 123.

When the batch crystallizer is cooled by natural cooling, large temperature differences at the beginning of the run often leads to heavy incrustations on the cooling surface so that the heat transfer is reduced. The cooling then proceeds extremely slowly and the necessary batch times of 20 or more hours are not unusual. Previous studies on incrustation led to the conclusion that if the heat flow intensity is lower than a certain critical value characteristic for the given system, no incrustation arises and cooling proceeds with a high heat transfer coefficient corresponding to a clean surface. Knowledge of this critical value (or corresponding critical temperature difference) enables us to divide the cooling process into two steps and this achieves a surprisingly shorter batch time and higher efficiency of cooling.

It follows from Equation 50 that if h_c is very low or c = 0, the cooling curve simplifies for $T_0 = T_{0\,eq}$ to

$$-\dot{T} = K'(T-T_w) \tag{185}$$

with a simple solution

$$T = T_{wt}+(T_0-T_{w0}) \exp(-K't) \tag{186}$$

Nevertheless, if incrustations form on the cooling surface, K′ depends on time and we can assume, as a first approximation, that [41]

$$K'_t = (K_0'^{-1}+qT)^{-1} \tag{187}$$

where q characterizes the rate of change of the heat resistance of the incrustation with time:

$$q = (K_f'^{-1}-K_0'^{-1})/t_c \tag{188}$$

In this case, Equation 186 modifies to

$$T = T_{wt}+(T_0-T_{w0}) \exp\left(\frac{-K'_0 t}{1+qK'_0 t}\right) \tag{189}$$

The ideal batch time without incrustation is then

$$t_{c0} = K_0'^{-1} \ln [(T_0 - T_{w0})/(T_f - T_{wf})] \tag{190}$$

whereas the batch time with incrustation occurring is longer:

$$t_{cs} = \frac{t_{c0}}{1 - q \ln \dfrac{T_0 - T_{w0}}{T_f - T_{wf}}} \tag{191}$$

For given equipment and a given system, a critical temperature difference (or, with constant cooling water parameters, a critical temperature in the crystallizer, T_m) can be found, below which no incrustation occurs on the cooling surface. From this knowledge, we can postulate the cooling of the solution in two stages [41, 42]. In the first stage, the suspension has to be cooled down from the initial temperature, T_0, to the intermediate temperature, T_m:

$$t_{m1} = \frac{\ln \dfrac{T_0 - T_{w0}}{T_m - T_{wm}}}{K_0' \left[1 - q \ln \dfrac{T_0 - T_{w0}}{T_m - T_{wm}} \right]} \tag{192}$$

Below T_m, no incrustation should occur so it is advantageous to start the second cooling stage with clean cooling surfaces. This can be done either by removing the incrustation that had occurred in the first stage in an appropriate way, e.g., by temperature shocks, or by transferring the suspension to another crystallizer with clean cooling surfaces (the incrustation in the first-stage vessel could be redissolved by the hot fresh solution in the next batch). The second-stage cooling time is

$$t_{m2} = K_0'^{-1} \ln \frac{T_m - T_{wm}}{T_f - T_{wf}} \tag{193}$$

and as

$$t_{m1} + t_{m2} \leqq t_{fs} \tag{194}$$

the total batch time is much shorter. It has been shown [41] that two-stage batch cooling can lead, particularly with substances forming incrustations very easily, to significantly shortened batch times, even by an order of magnitude. If the critical temperature difference $T_m - T_w$ lies between 5 and 15 K, Equation 194 always holds and thus the two-stage batch cooling is advantageous.

NOTATION

- A_c crystal surface area
- A_f heat exchanging surface area
- a, b, c adjustable constants
- B_N system kinetic constant of crystallization
- c secondary nucleation exponent
- c_{p0} specific heat capacity of solution
- g kinetic exponent of crystal growth
- h_c heat of crystallization
- h_v heat of evaporation
- K_i price, costs

K, K' heat transfer coefficient; constant
k constant
k_G, k'_G growth rate constant
k_d dissolution rate constant
k_N, k_{NA} nucleation rate constant
L crystal size
L_0 size of seeds
L_N initial crystal size
L_{max} maximum crystal size
$\bar{L}$ mean crystal size (mode)
$\dot{L}$ linear crystallization rate
$\dot{L}_{av}$ average linear crystallization rate
ΔL increase in crystal size
$M(L)$ cumulative crystal size distribution
m_c crystal mass, suspension concentration
m_0 initial mass of solution
m_{c0} mass of seeds
m_f mass of mother liquor
m_g mass of evaporated solvent
$m(L)$ mass of oversize crystals
$\dot{m}_c$ specific output of the crystallizer
N_c number of crystals
$\dot{N}_N$ nucleation rate
$n(L)$ crystal population density
n^0 nuclei population density
n kinetic exponent of nucleation
n_A apparent kinetic exponent of nucleation
P power consumption
p_0 portion of suspension left in the crystallizer
Q heat fed to the crystallizer
q coefficient defined by Equation 188
s slope; supersaturation rate
T temperature
T_0 initial temperature
T_f final temperature
T_m intermediate temperature
$-\dot{T}$ cooling rate
t time
t_c batch time
$\bar{t}_l$ mean residence time of solution in a continuous crystallizer
$\dot{V}_w$ throughput of cooling water
w concentration (kg/kg of free solvent)
w_{eq} solubility
Δw supersaturation
x empirical exponent
y variable
Z target function
z dimensionless residence time of crystals of size L
z_N dimensionless initial crystal size

Greek Symbols

α volume shape factor
β surface shape factor
ϱ_c crystal density
σ coefficient of secondary nucleation
η coefficient of economical losses
θ constant

Subscripts

0 initial, inlet, seed
f final, product
w cooling water
c crystals
max maximum
t at time t
m small crystals
v large crystals
rel relative

REFERENCES

1. Nývlt, J., *Industrial Crystallization—the State of the Art,* 2nd Ed., Verlag Chemie, Weinheim, 1982.
2. Mullin, J. W., *Crystallization,* 2nd Ed., Butterworths, London, 1972.
3. Nývlt, J., *Industrial Crystallization from Solutions,* Butterworths, London, 1971.
4. Matz, G., *Kristallisation-Grundlagen und Technik,* 2nd Ed., Springer, Berlin, 1969.

5. Janse, A. H., "Nucleation and Crystal Growth in Batch Crystallizers," Ph. D. thesis, University of Technology, Delft, 1977.
6. Nývlt, J., and Kočová, H., "Combined Materials and Heat Balance of a Crystallizer," *Chem. průmysl* (Prague), Vol. 26 (1976), p. 567.
7. Nývlt, J., "Applicability Region of Vacuum Cooling Crystallizers," *Chem. průmysl* (Prague), Vol. 29 (1979), p. 71.
8. Nývlt, J., and Václavů, V., "Cooling Rate in a Batch Crystallizer," *Chem. průmysl* (Prague), Vol. 14 (1964), p. 79.
9. Mullin, J. W., and Nývlt, J., "Programmed Cooling of Batch Crystallizers," *Chem. Eng. Sci.*, Vol. 26 (1971), p. 369.
10. Nývlt, J., Kočová, H., and Černý, M., "Size Distribution of Crystals from a Batch Crystallizer," *Collect. Czechosl. Chem. Commun.*, Vol. 38 (1973), p. 3199.
11. Larson, M. A., and Garside, J., "Crystallizer Design Techniques Using the Population Balance," *The Chem. Engineer*, 1973, No. 274, p. 318.
12. Jones, A. G., "Optimal Operation of a Batch Cooling Crystallizer," *Chem. Eng. Sci.*, Vol. 29 (1974), p. 1075.
13. Tavare, N. S., Garside, J., and Chivate, M. R., "Analysis of Batch Crystallizers," *Ind. Eng. Chem. Process Design Devel.*, Vol. 10 (1980), p. 653.
14. Ajinkya, M. B., and Ray, W. H., "On the Optimal Operation of Crystallization Processes," *Chem. Eng. Commun.*, Vol. 1 (1974), p. 181.
15. Nývlt, J., Křičková, J., and Matuchová, M., "Automatic Measurement of Crystal Growth Rate," *Chem. průmysl* (Prague), Vol. 22 (1972), p. 327.
16. Sůra, J., Míček, F., and Nývlt, J., "The Effect of the Cooling Way in a Batch Crystallizer on the Product Quality," *Chem. průmysl* (Prague), Vol. 19 (1969), p. 302.
17. Nývlt, J., Sůra, J., and Procházka, S., "Automatic Operation of Batch Cooling Crystallizers," *Chem. průmysl* (Prague), Vol. 26 (1976), p. 505.
18. Nývlt, J., Sůra, J., and Procházka, S., "Method of Cooling of a Batch Crystallizer," Czech. patent A0 180 905, 1980.
19. Nývlt, J., "Analysis of the Cooling Curve of a Batch Crystallizer," *Chem. průmysl* (Prague), Vol. 24 (1974), p. 179.
20. Nývlt, J., "Effect of Kinetic Parameters on the Behavior and Product Crystal Size Distribution of Batch Crystallizers," *AIChE Symp. Ser.*, Vol. 72, No. 153 (1976), p. 61.
21. Randolph, A. D., and Larson, M. A., *The Theory of Particulate Processes*, Acad. Press, New York, 1971.
22. Nývlt, J., "Evaluation and Characterization of Crystal Size Distribution of Products from Agitated Crystallizers," *Chem. průmysl* (Prague), Vol. 18 (1968), p. 579; *Intern. Chem. Engng*, Vol. 9 (1969), p. 422.
23. Baumann, K. H., "An Analysis of the Behavior of the Seed Crystals in Batch Crystallization," *Crystal Res. and Technol.*, Vol. 17 (1982), p. 1357.
24. Nývlt, J., "Crystal Size Distribution from Batch Crystallization," *Chem. průmysl* (Prague), Vol. 23 (1973), p. 343.
25. Nývlt, J., "The Effect of the Mode of Cooling on Size Distribution of Crystals Produced in Batch Crystallizer," *Collect. Czechosl. Chem. Commun.*, Vol. 39 (1974), p. 3463.
26. Nývlt, J., "Seeding and Its Effect on Size of Product Crystals in a Batch Crystallizer," *Collect. Czechosl. Chem. Commun.*, Vol. 41 (1976), p. 342.
27. Nývlt, J., "The Effect of Kinetic Parameters on Discontinuous Crystallization Process," *Collect. Czechosl. Chem. Commun.*, Vol. 40 (1975), p. 2592.
28. Veverka, F., and Nývlt, J., "Method of Batch Crystallization," Czech. patent, A0 204 551, 1980.
29. Nývlt, J., and Veverka, F., "Seeding with a Portion of Suspension from the Preceding Run," *Chem. průmysl* (Prague), Vol. 30 (1980), p. 343.
30. Míček, F., Nývlt, J., and Sůra, J., "The Effect of Agitation on the Crystal Size," *Chem. průmysl* (Prague), Vol. 18 (1968), p. 285.

31. Nývlt, J., "Kinetics of Secondary Nucleation and the Product Mean Crystal Size from a Continuous Agitated Crystallizer," *Collect. Czechosl. Chem. Commun.*, Vol. 46 (1981), p. 79.
32. Nývlt, J., "The Effect of Secondary Nucleation on the Product Crystal Size from a Batch Crystallizer," *Chem. průmysl* (Prague), Vol. 31 (1981), p. 334.
33. Ottens, E. P. K., and de Jong, E. J., "Nucleation in Continuous Agitated Crystallizers," *Ind. Eng. Chem. Fundam.*, Vol. 12 (1973), p. 179.
34. Nývlt, J., "The Effect of Periodic Temperature Changes on the Shape of Crystals," *Particle Growth in Suspensions*, Brunel University, SCI Monogr., Vol. 38 (1973), p. 131.
35. Nývlt, J., "Method of Crystallization Leading to Coarse Product Crystals," Czech. patent 105 232, 1962.
36. Skřivánek, J., and Nývlt, J., "Crystal Size Distribution in a Twinned Crystallizer," *Collect. Czechosl. Chem. Commun.*, Vol. 33 (1968), p. 2799.
37. Žáček, S., Hostomský, J., and Skřivánek, J., "Experimental Study of the Recrystallization of Potash Alum Suspension in a Double Crystallizer," *Kristall und Technik*, Vol. 16 (1981), p. 667.
38. Žáček, S., Karel, M., and Nývlt, J., "Reduction of the Number of Small Crystals in the Product From Batch Crystallization by Thermal Treatment of Suspension," *Chem. průmysl* (Prague), Vol. 31 (1981), p. 574.
39. Nývlt, J., "Design of Batch Crystallizers," in *Industrial Crystallization*, Mullin, J. W. (Ed.) Plenum Press, New York, 1976, p. 335.
40. Nývlt, J., and Mullin, J. W., "Design of Batch Agitated Crystallizers," *Kristall und Technik*, Vol. 12 (1977), p. 1243.
41. Nývlt, J., "Two-Stage Batch Cooling Crystallization," *J. Separ. Proc. Technol.*, Vol. 2, No. 2 (1981), p. 19.
42. Veverka, F., and Nývlt, J., "The Method of Batch Cooling," Czech. patent, A0 199 434, 1980.

CHAPTER 35

SEPARATION THEORY IN THERMAL DIFFUSION COLUMNS

Ho-Ming Yeh and Shau-Wei Tsai

Chemical Engineering Department
National Cheng Kung University
Tainan, Taiwan, Republic of China

CONTENTS

INTRODUCTION

If two gases of different concentration initially at the same temperature are allowed to diffuse together, a transient temperature gradient results from the ordinary diffusion. This phenomena was first noted by Dufour [1] and was called the Dufour effect or diffusion thermoeffect. Conversely, if a temperature gradient is applied to a homogeneous solution, a concentration gradient is usually established. The name thermal diffusion or Soret effect is generally applied to the second effect.

The existence of thermal diffusion was recognized in the liquid state by Ludwig [2] as early as in 1856. However, this effect in gases seems to have not been researched, though it was predicted theoretically by Enskog [3] in 1911 from the kinetic theory of gases. Later, Chapman and Dootson [4] confirmed this prediction experimentally in 1917. From then on, thousands of references to the separation of mixtures by thermal diffusion appeared in the literature, and the separable mixtures included biological solutions and suspensions, aqueous solutions, polymer or organic solutions, isotopic solutions, liquid metal solutions, and various gas mixtures [5].

The equipment used in thermal diffusion is generally categorized into two kinds, i.e. static cells (or single-stage convective-free cells) and thermogravitational columns (or Clausius-Dickel columns). In the static cells shown in Figure 1, which were used in the early work of thermal diffusion, attempts have been made to eliminate convective currents. Since the concentration gradient at steady state is such that the flux due to ordinary diffusion just counterbalances that resulting from thermal diffusion, the separation obtained in those cells

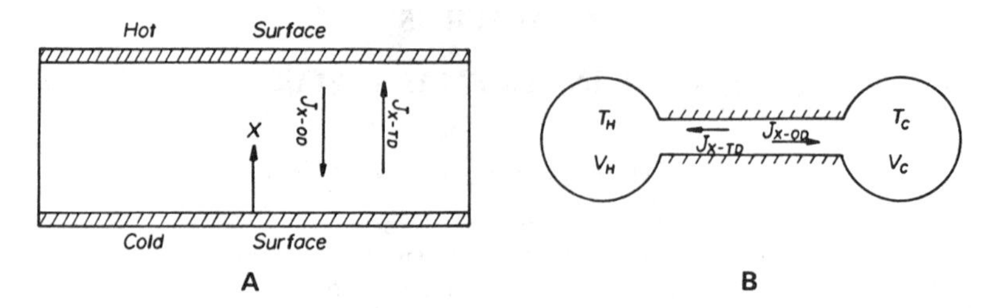

Figure 1. Two types of static cells: (A) flat-plate static cell; (B) two-bulb static cell.

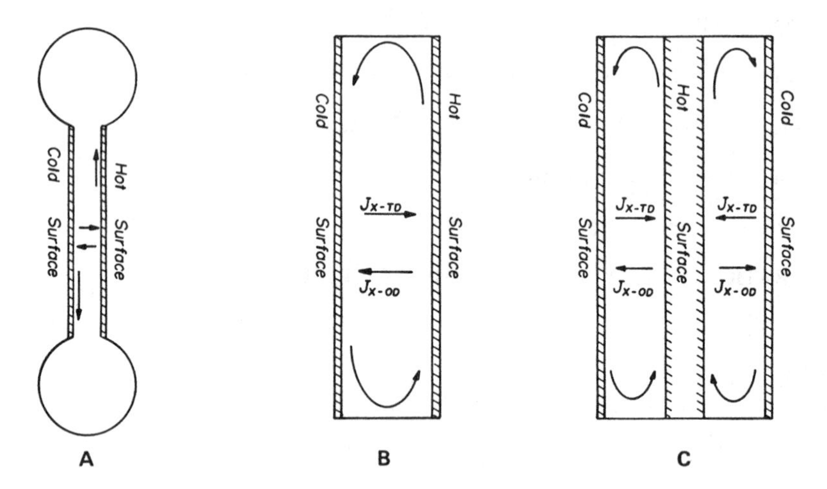

Figure 2. Various types of Clausius-Dickel columns: (A) two-bulb column; (B) flat-plate column; (C) concentric-tube column.

is very small and of no practical use. Thus, measurement made is of theoretical interest only, e.g., to investigate the nature of intermolecular forces based on the kinetic theory of gases. In the thermogravitational columns shown in Figure 2, natural convective currents in the vertical direction are coupled with the thermal diffusion in the horizontal direction to produce a cascading effect analogous to the multistage effects of countercurrent extraction, and thus efficiency is increased by orders of magnitude. Sometimes, thermogravitational column is called Clausius-Dickel column due to its first being devised by Clausius and Dickel [6] in 1938.

A more detailed study of the mechanism of separation in the Clausius-Dickel column finds that the convective currents in the column actually have two conflicting effects: the desirable cascading effect and the undesirable remixing effect. It is believed that proper adjustment of the convective strength might improve the separation. Consequently, based on this idea, some improved columns have been devised [7–25].

Typically, thermal diffusion columns are designed with columns in series-parallel arrangement to obtain larger scale or higher purity product, or even to reduce the time needed to reach steady-state operation. For simplicity, we will consider first the separation theory in binary mixtures.

ESTIMATION OF THERMAL DIFFUSION CONSTANT

The basic rate law for the thermal diffusion in binary gas mixtures is

$$J_{X-TD} = \varrho(\alpha D/T)c(1-c)\,(dT/dx) \tag{1}$$

which was derived theoretically from the kinetic theory of gases [3]. Equation 1 can also be used to define the thermal diffusion constant α in the liquids. Numerous experimental data of α have been published in the literature. However, care must be taken to see whether concentration c is in volume, mole, or weight fraction. Besides, a negative value of α means the component referred will diffuse in the opposite direction.

The estimation of the theoretical value of α for gases requires first an assumption about the forces between two molecular, then the evaluation of the collision integrals for that force law. By substituting these integrals into the expressions as shown by Grew et al. [26], the thermal diffusion constant can be found as the function of temperature, composition, and molecular weights of components. Since for most gas mixtures, it is difficult to describe the collision accurately, the theoretical attempt to predict α has met very limited success.

Several attempts have been made to extend the kinetic theory of gases, or by using the thermodynamics of irreversible processes, to the analysis of the thermal diffusion in liquids [27, 28]. However, due to the complexities involed in the liquids, the prediction of α is at most qualitative in agreement. As a result, it is necessary either to consult the literature or to determine the feasibility experimentally to obtain the value of α. Some experimental data have been listed in Table 1 and Table 2. For an extensive tabulation of the Soret coefficient, α/T, in the liquids, the reader may refer to the report submitted by Von Halle [29].

SEPARATION THEORY IN BINARY MIXTURES

Static Cells

Consider the static cells shown in Figure 1, in which the convective currents are inhibited. The net flux of component 1 in the x-direction due to the thermal and ordinary diffusions is

$$J_x = \varrho D\{(\alpha/T)c(1-c)\,(\partial T/\partial x) - (\partial c/\partial x)\} \tag{2}$$

The solution of Equation 2 depends on the type of equipment and the mode of operation.

Steady-State Operation

At steady state, the net flux of component 1 will be zero, and the solution of Equation 2 is

$$q = c_H(1-c_c)/[c_c(1-c_H)] = (T_H/T_C)^{\alpha} \tag{3}$$

under the assumption of constant α. Sometimes, we would like to find the concentration distribution as a function of position. For the flat-plate cells shown as Figure 1(a), the result is

$$c(x) = [1+\psi\exp(-2A\xi)]^{-1} \tag{4}$$

where

$$A = \alpha\Delta T/(2\bar{T}) \tag{5}$$

$$\xi = x/\ell \tag{6}$$

Table 1
Some Experimental Thermal Diffusion Constants for Binary Gas Mixtures

Mixtures	c (mole%)	T_H (°K)	T_C (°K)	$\bar{T}$ (°K)	α*
H_2-D_2	0.5	373	288		0.173
	0.2**	360	273		0.149
H_2-He	0.5	760	273		0.152
		292	90		0.137
H_2-CH_4	0.5	523	300		0.288
		300	190		0.222
H_2-Ne	0.5	288	128		0.36
		290	90		0.28
		290	20		0.174
H_2-CO	0.5	373	288		0.330
		293	90		0.216
H_2-N_2	0.5	456	288		0.312
		373	288		0.340
		292	90		0.24
H_2-C_2H_4	0.5	523	300		0.277
		200	190		0.241
		373	288		0.32
H_2-O_2	0.5	294	90		0.192
H_2-Ar		456	288		0.28
		286	108		0.22
		292	90		0.191
H_2-C_3H_6	0.5	523	300		0.305
		376	232		0.284
H_2-C_3H_8	0.5	523	300		0.315
		375	231		0.291
H_2-CO_2	0.5	456	288		0.284
		373	288		0.298
		300	400		0.272
H_2-Rn	0.5	373	273		0.31
D_2-N_2	0.5	373	287		0.313
^3He-^4He	0.5	613	273		0.059
He-Ne	0.5	373	288		0.388
		400	300		0.364
		600	200		0.316
		293	90		0.330
		293	20		0.242
H_2-N_2	0.5	373	287		0.36
He-Ar	0.5	373	288		0.372
		400	300		0.42
		273	90		0.31

Mixtures	c (mole%)	T_H (°K)	T_C (°K)	$\bar{T}$ (°K)	α*
He-Kr	0.5	373	288		0.400
He-Xe	0.5	373	288		0.403
He-Rn	0.5	373	273		0.64
^{20}Ne-^{22}Ne	0.5	819	691		0.0346
		638	460		0.0318
		645	302		0.0302
		490	195		0.0254
		296	195		0.0233
		296	90		0.0187
		195	90		0.0162
Ne-Ar	0.5	373	288		0.181
Ne-Kr	0.5	373	288		0.267
Ne-Xe	0.5	373	288		0.253
Ne-Rn	0.5	373	273		0.23
$^{14}N_2$-$^{14\,15}N_2$	0.5	623	195		0.0051
	0.5			293	0.018
				89	<0.001
N_2-Ar	0.5			293	0.071
				89	0.035
N_2-CO_2	0.5	400	288		0.050
		373	288		0.061
				283	0.036
				372	0.051
$^{16}O_2$-$^{16\,18}O_2$	0.5			264	0.0099
				389	0.0128
				443	0.0145
O_2-Ar	0.5			283	0.050
^{36}Ar-^{40}Ar	0.5	835	638		0.0250
		635	455		0.0218
		623	273		0.0182
		495	195		0.0146
		296	195		0.0116
		296	90		0.0071
		195	90		0.0031
Ar-Kr	0.5	373	288		0.055
				185	0.038
Ar-Xe	0.5	373	288		0.077
				185	0.063
Ar-Rn	0.5	373	273		0.024
Kr-Xe	0.5	373	288		0.016

* *α is the mean value in the temperature range of T_H and T_C*
** *For D_2*

Table 2
Some Experimental Thermal Diffusion Constants for Binary Liquid Mixtures

Mixtures	c_f	T (°K)	T (°K)	$\bar{T}$	α	Remark
CCl_4-cyclohexane		302	294	298	$-1.839+0.852c_f-0.607c_f^2+1.496c_f^3-1.814c_f^4$	c_f in weight fraction for CCl_4 [45]
H_2O-KCl	0.2	300.5	285.5	293	0.346	c_f in molarity [46] for KCl
	0.6				0.270	
	1.0				0.296	
H_2O-KBr	0.2	290.5	285.5	293	0.404	c_f in molarity [46] for KBr
	0.6				0.357	
	1.0				0.407	
H_2O-ethanol	5	290.5	285.5	293	−1.225	c_f in weight percent [47] for ethanol. Negative value means the lighter component will diffuse to the colder wall
	10				−1.149	
	20				−0.727	
	30				0.607	
	50				1.312	
	80				0.724	
	90				0.352	
Heptane-octane	10	290.5	285.5	293	0.448	c_f in weight percent [47] for octane
	30				0.492	
	50				0.545	
	70				0.592	
	90				0.642	
Benzene-heptane	50	290.5	285.5	293	−0.331	c_f in weight percent [47]

$$\psi = \{\exp[2A(1-c_0)]-1\}/\{1-\exp[-2Ac_0]\} \tag{7}$$

$$\bar{T} = \{T_H T_C/(T_H-T_C)\}\ln(T_H/T_C) \tag{8}$$

For the two-bulb apparatus shown as Figure 1(b), the solution of c_H in terms of the initial concentration c_0, the volume ratio V_R and the temperature ratio T_R is [30]

$$c_H = \{-a_2+\sqrt{(a_2^2-4a_1a_3)}\}/(2a_1) \tag{9}$$

where

$$a_1 = T_R-1 \tag{10}$$

$$a_2 = -[V_R+T_R+(T_R-1)(V_R+1)c_0] \tag{11}$$

$$a_3 = T_R c_0(V_R+1) \tag{12}$$

$$T_R = (T_H/T_C)^\alpha \tag{13}$$

$$V_R = (V_C/V_H) \tag{14}$$

$$c_c = [c_0(1+V_R)-c_H]/V_R \tag{15}$$

During the derivation of Equation 15, the assumption that the volume of the tube connecting the bulbs is negligible.

Approach to Equilibrium

During the design of static thermal diffusion equipment, it is best to be able to estimate the time required for the cells to approach equilibrium conditions. It is found that the final approach to equilibrium is (for $t \gg t_r$)

$$f = [c(\ell,t)-c(0,t)]/\{c(\ell,\infty)-c(0,\infty)\} = 1-\varphi\exp(-t/t_r) \tag{16}$$

where f = the fractional approach to equilibrium separation
φ = a constant depending on the type of equipment
t_r = the relaxation time needed to attain $(1-\varphi/e)$ of the equilibrium separation

For flat-plate cells, the relaxation time is predicted by

$$t_r = \ell^2/[D(\pi^2 + A^2)] \tag{17}$$

and

$$\varphi = 4\pi^2 c_0(1-c_0)\,[\exp(-2Ac_0)+\exp(-A)]\,[\exp(2A)-1]$$
$$[\exp(A)+\exp(2Ac_0)]/\{A(A^2+\pi^2)\,[1+(\pi^2/A^2)-4c_0(1-c_0)]$$
$$[\exp(2A)-\exp(2Ac_0)]\,[1-\exp(-2Ac_0)]\} \tag{18}$$

For two-bulb apparatus, Jone and Furry [31] have made the assumption that the net flux is, quasi-stationary i.e., J_x is constant along the tube connecting the bulbs. The result is

$$t_r = \{[\exp(2bA)-1]/(2bADA_C)\}\,\{V_H V_C/[V_H \exp(2bA)+V_C]\} \tag{19}$$

$$\varphi = 1.0 \tag{20}$$

During the derivation of Equation 19, the quadratic form of concentration in Equation 2 is linearized as

$$c(1-c) \cong a + bc \tag{21}$$

Clausius-Dickel Columns

Consider the flat-plate Clausius-Dickel column shown in Figure 3. The temperature gradient applied in the horizontal direction has two effects:

1. A net flux of one component of the solution relative to the other is brought about by the diffusions.
2. Convective currents are produced parallel to the plates owing to the density differences. The combination of these two effects is to produce a concentration difference between the two ends of the column, which is generally much greater than that obtained in static cells.

The complete theory of separation in a Clausius-Dickel column was first presented by Furry et al. [31]. They derived the equation of separation by first solving the equations of continuity, equation of motion, and equation of energy simultaneously. The resulting concentration distribution and velocity distribution were then substituted into the transport equation to obtain the net transport of component 1 through the total cross section of the column. Finally, the separation equation was obtained by solving the resulting transport equation. Basically, the form of the separation equation depends on the mode of operation and the type of equipment. The results are presented below. Yet, some important assumptions have been made during the derivation, i.e.

1. Laminar flow conditions existed in the column.
2. End effects in the column were neglected.
3. Heat was transfered by conduction in the horizontal direction only.

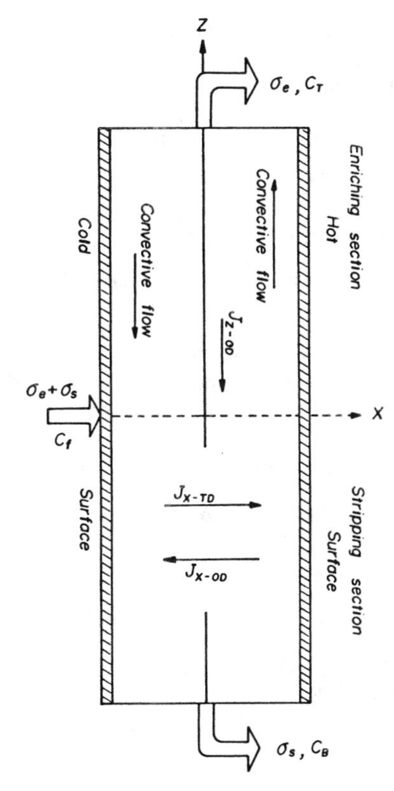

Figure 3. Schematic diagram of the fluxes prevailing in the flat-plate Clausius-Dickel column.

Batch Operation

For the flat-plate column without reservoirs, the separation factor q obtained at steady state is

$$q = \{c_T(1-c_B)/[c_B(1-c_T)]\} = \exp(2A_1) \tag{22}$$

where

$$A_1 = H_0 L/(2K) \tag{23}$$

$$H_0 = BQ^{-3}\int_{T_1}^{T_2} [\varrho D\alpha/(\lambda T)]G(T)\,dT \tag{24}$$

$$K = K_0 + K_1 \tag{25}$$

$$K_0 = BQ^{-7}\int_{T_1}^{T_2} (\varrho D/\lambda)G^2(T)\,dT \tag{26}$$

$$K_1 = BQ^{-1}\int_{T_1}^{T_2} \varrho\lambda D\,dT \tag{27}$$

$$Q = (2\omega)^{-1} \int_{T_1}^{T_2} \lambda \, dT \tag{28}$$

and the function G(T) is the solution of

$$\frac{d}{dT}\left\{\frac{1}{\lambda}\frac{d}{dT}\frac{\mu}{\lambda}\frac{d}{dT}\frac{1}{\varrho\lambda}\frac{d}{dT}\left[\frac{\varrho DG(T)}{\lambda}\right]\right\} = -g\left(\frac{d\varrho}{dT}\right) \tag{29}$$

under the boundary conditions:

$$G(T_1) = G(T_2) = d[G(T_1)]/dT = d[G(T_2)]/dT = 0 \tag{30}$$

Once the dependence of those physical properties on temperature is known, the separation factor can be easily deduced. As an approximation, if all the physical properties are assumed constant, except for the density on the right side of Equation 29, the transport constants H_0 and K can be integrated to give

$$H_0 = (2\omega)^3 \varrho\alpha\beta_T gB(\Delta T)^2/(6!\mu\bar{T}) \tag{31}$$

$$K_0 = (2\omega)^7 g^2 \beta_T^2 \varrho B(\Delta T)^2/(9!\mu^2 D) \tag{32}$$

$$K_1 = 2\omega\varrho DB \tag{33}$$

Of course, all the physical properties in Equations 31 through 33 should be evaluated at the mean temperature, say at the arithmetic mean temperature.

For the concentric-tube column without reservoirs, which is operated at steady state, Equation 22 is still valid. However, some modifications for Equations 23 through 30 should be made

$$H_0 = 2\pi Q_1^{-3} \int_{T_1}^{T_2} [\varrho D\alpha/\lambda T]G(T) \, dT \tag{34}$$

$$K_0 = 2\pi Q_1^{-7} \int_{T_1}^{T_2} [\varrho D/\lambda]G^2(T) \, dT \tag{35}$$

$$K_1 = 2\pi Q_1^{-1} \int_{T_1}^{T_2} r^2\lambda\varrho D \, dT \tag{36}$$

$$Q_1 = [\ln (R_2/R_1)]^{-1} \int_{T_1}^{T_2} \lambda \, dT \tag{37}$$

and

$$\frac{d}{dT}\left\{\frac{1}{\lambda r^2}\frac{d}{dT}\frac{\mu}{\lambda}\frac{d}{dT}\cdot\frac{1}{\varrho\lambda r^2}\frac{d}{dT}\left[\frac{\varrho DG(T)}{\lambda}\right]\right\} = -g\left(\frac{d\varrho}{dT}\right) \tag{38}$$

under the same boundary conditions as Equation 30. Since r is the function of temperature, one should find this function from the equation of energy before solving Equation 38. Again, if all the physical properties are assumed constant except for the density on the right side of Equation 38 one might obtain the approximate solution as

$$H = H_0 F(k) \tag{39}$$

$$H_0 = \alpha\beta_T \rho g(2\pi R_1)(\Delta T)^2 (R_2 - R_1)^3/(6!\,\mu\bar{T}) \tag{40}$$

$$K = K_0 G(k) + K_1(k+1)/2 \tag{41}$$

$$K_0 = \beta_T^2 \varrho g^2 (2\pi R_1)(\Delta T)^2 (R_2 - R_1)^7 / (9!\, \mu^2 D) \tag{42}$$

$$K_1 = 2\pi \varrho D R_1 (R_2 - R_1) \tag{43}$$

where k is the ratio of R_2 to R_1. The modifying factors F(k) and G(k) are presented in Figure 6 with Brinkman number B_r equal to zero [19, 21, 32].

For the flat-plate column with reservoirs at both ends, the solution of the transport equation at steady state is found to be the same as Equations 9 through 15, except that T_R is replaced by exp(2A), c_H by c_T, and c_C by c_B. Again, for the concentric-tube column with reservoirs at both ends, the results are all the same as those in the flat-plate column with reservoirs. However, the transport constants should be modified by Equations 39 through 43.

For the Clausius-Dickel columns operated at transient state, one obtains Equation 44 by taking a material balance for component 1

$$m[\partial c/\partial t] = K[\partial^2 c/\partial z^2] - H_0 \partial[c(1-c)]/\partial z \tag{44}$$

where m is the mass of solution per unit length of the column. One might refer to References 29 and 30 for further information.

Continuous Operation

Usually, the Clausius-Dickel column is designed with the feed introduced from the intermediate of the column when the column is operated continuously. Consequently, the whole column might be divided into two sections: the enriching and stripping sections. In order to obtain the separation equation, one should solve the transport equations for both sections simultaneously. The exact solution, obtained by Furry et al. [31], is implicit and too complicated to analyze. This is due to the nonlinear form of concentration in the transport equations. Numerous investigators have considered this product form of concentration to be constant and obtained a very simple solution in the concentration range $0.3 < c < 0.7$, i.e.

$$c_T = c_f + (4\sigma_e')^{-1}[1 - \exp(-\sigma_e' L_e')] \tag{45}$$

$$c_B = c_f - (4\sigma_s')^{-1}[1 - \exp(-\sigma_s' L_s')] \tag{46}$$

where the dimensionless quantities σ_e', σ_s', L_e', and L_s' are defined as

$$\sigma_e' = \sigma_e/H_0, \qquad \sigma_s' = \sigma_s/H_0 \tag{47}$$

$$L_e' = L_e H_0/K, \qquad L_s' = L_s H_0/K \tag{48}$$

If the flow rates withdrawn from both sections are equal and the feed is introduced from the intermediate of the column, one might obtain the separation equation as

$$\Delta = c_T - c_B \tag{49}$$

$$= (2\sigma')^{-1}[1 - \exp(-\sigma' L'/2)] \tag{50}$$

For the concentration not in the range $0.3 < c < 0.7$, Yeh et al. [33–35] have derived an explicit solution by using the linear approximation, say Equation 21, coupled with the least squares method. The result is

$$\Delta = (-3S)/2 + \sqrt{(9S^2/4 + 12c_f(1-c_f))} \tag{51}$$

where

$$S = 4(L')^{-1}\{\sigma'L' + (1-2c_f)L' \exp[-(1-2c_f)L'/2 - \sigma'L'/2]\}$$
$$\times \{\sigma'L' - (1-2c_f)L' \exp[(1-2c_f)L'/2 - \sigma'L'/2]\}/\{2\sigma'L'$$
$$+ [(1-2c_f)L' - \sigma'L'] \exp[-(1-2c_f)L'/2 - \sigma'L'/2]$$
$$- [(1-2c_f)L' + \sigma'L'] \exp[(1-2c_f)L'/2 - \sigma'L'/2]\} \quad (52)$$

During the derivation of Equation 51, the assumptions of equal withdrawing rates at both ends and the feed introduced from the center of the column have been made. Equation 51 was later confirmed by the experimental data [35].

Multicolumns

Thermal diffusion is generally used to separate the mixtures which are hard to separate by conventional methods, i.e., distillation, extraction, etc. Due to the difficulty of separation, the columns are designed in series-parallel cascades both to increase the separation efficiency and to decrease the time required to begin production at the desired separation. Jone and Furry [31] have given detailed discussion about the design of such columns. Of course, other types of cascading columns have also been proposed in the literature [36]. We will consider those columns shown in Figure 4, as proposed by Jone et al. Investigation of the multicolumns led Jone et al. to the conclusion that, in each stage of a properly designed cascade, the separation should be approximately one half of that obtained in a batch column of the same dimensions under the steady-state conditions. Moreover, they also designed the spacing between the surfaces from the relation $K_c = 10\,K_d$. The resultant equations are

$$N_{j+1} = N_j/\gamma \quad (53)$$

$$N_1 = \sigma_e(C_{Ne} - c_f)(\gamma+1)/[\gamma H_0 c_f(1-c_f)] \quad (54)$$

$$L_e = (2KH_0^{-1}) \ln\{C_{Ne}(1-c_f)/[(1-C_{Ne})c_f]\} \quad (55)$$

$$\Lambda_e = \{(\gamma+1)\ln\gamma/[2(\gamma-1)]\}\Lambda_{ei} \quad (56)$$

$$\Lambda_{ei} = (4\sigma_e K/H_0^2)\{(1-2c_f)(C_{Ne}-c_f)/[c_f(1-c_f)]$$
$$- (1-2C_{Ne})\ln\{C_{Ne}(1-c_f)/[c_f(1-C_{Ne})]\}\} \quad (57)$$

$$L_j = 2KH_0^{-1}\ln\{c_{je}(1-c_{jf})/[c_{jf}(1-c_{je})]\} \quad (58)$$

$$c_{je} = (0.5)\{1 + \sigma_e'(\gamma+1)N_j^{-1} - \sqrt{\{1 + 2(1-2C_{Ne})[(\gamma+1)\sigma_e'N_j^{-1}] + [(\gamma+1)\sigma_e'N_j^{-1}]^2\}}\} \quad (59)$$

where N_j = the number of columns in each sequential stage decreased in a constant ratio γ
L_j = the length of column at stage j
L_e = the height of the enriching section
Λ_e = the total length of the columns in the enriching section
c_{je} = the concentration of component 1 leaving the j-th stage

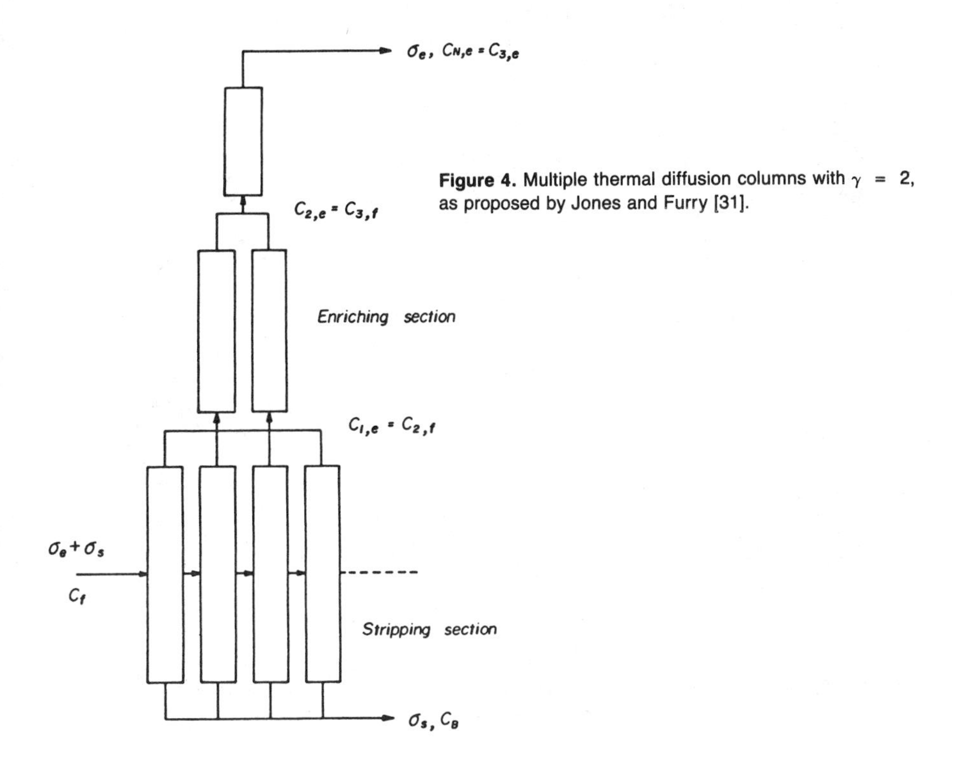

Figure 4. Multiple thermal diffusion columns with $\gamma = 2$, as proposed by Jones and Furry [31].

In order to satisfy the law of the conservation of mass for the whole columns, the designed equations for the stripping section are

$$\sigma_s = \sigma_e[(C_{Ne} - c_f)/(c_f - c_B)] \tag{60}$$

$$\tanh[-L_s H_0 b'/(2K)] = b'(c_B - c_f)/[c_B + c_f + \sigma_s' N_1^{-1}(c_B - c_f) - 2c_f c_B] \tag{61}$$

where

$$b' = \sqrt{\{(1 - \sigma_s' N_1^{-1})^2 + 4\sigma_s' c_B N_1^{-1}\}} \tag{62}$$

The height of stripping section L_s could also be evaluated from simplified Equation 46 if concentration is in the range $0.3 < c < 0.7$. In principle, if all the physical properties, column width, spacing between surfaces, γ, C_{Ne} and c_B are known, we can calculate the number of columns N_j and the height of column L_j in each stage, and the height of stripping sections L_s. Some examples have been given in References 30 and 31.

Improved Columns

As we have mentioned before, the convective currents in the Clausius-Dickel columns have two conflicting effects. The desirable cascading effect, which is necessary to secure high

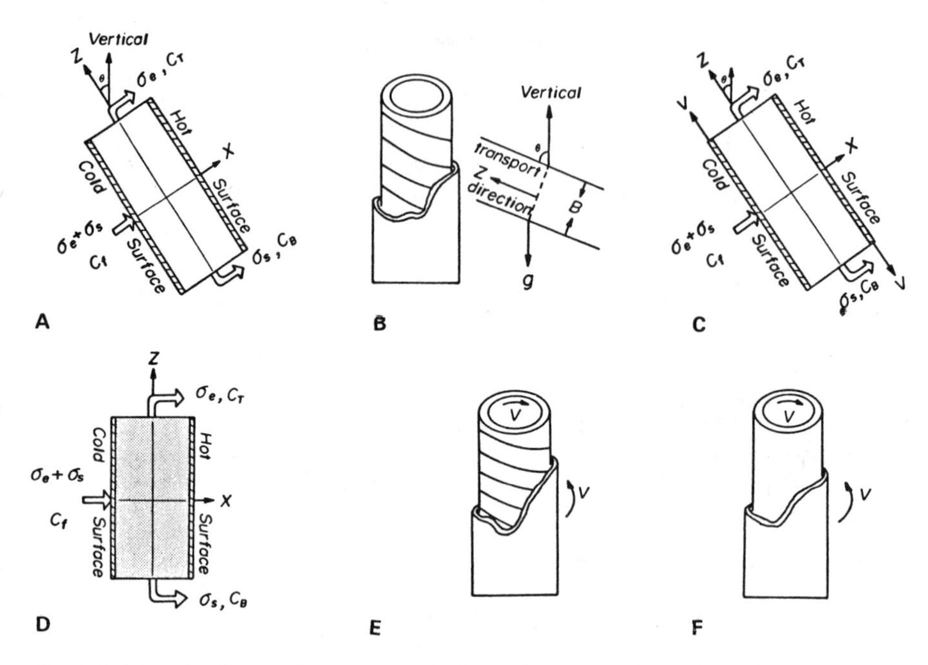

Figure 5. Schematic diagrams of some improved columns: (A) inclined column; (B) wired column; (C) moving wall column; (D) flat-plate packed column; (E) rotated wire column; (F) rotary column.

separation, has a multistage effect. However, the undesirable remixing effect due to the convective currents exists also, which bring down the fluid rich in one component at the top of the column to the bottom rich in the other component. Therefore, it appears that proper adjustment of the convective strengths might effectively suppress this remixing effect while preserving the cascading effect, and thereby lead to improved separation. Based on this concept, numerous improved columns [7–25] have been proposed. Some of them are shown diagrammatically in Figure 5.

The mathematical treatment of those columns is based on the one used by Jone et al. [31]. For simplicity, we consider the cases with all physical properties to be constant except for the density in the derivation of velocity distribution. The resulting separation equations for each improved column with feed introduced from the intermediate of the column and $\sigma_e = \sigma_s$ have the same form as that in the Clausius-Dickel column. Yet, the transport constants have to be modified and are listed in Table 3.

In principle, one might obtain the optimal conditions for maximum separation from the separation equation in each improved column, e.g. the angle of inclination in an inclined or wired column, tube speed of rotation in a rotated wired column, the permeability of packing in a packed column, etc. Since the exact solution of the transport equation is implicit and difficult to analyze, we first consider the simplest cases, i.e., $0.3 < c < 0.7$. The results are presented in Table 3 and illustrated in Figures 6 and 7. Yeh et al. [37] have also derived a simple but precise equation applicable to the whole range of concentration in various types of improved columns. From these equations derived, they concluded that the optimal conditions for the whole range of concentration were exactly the same as those shown in Table 3. However, the separation equation should be modified by

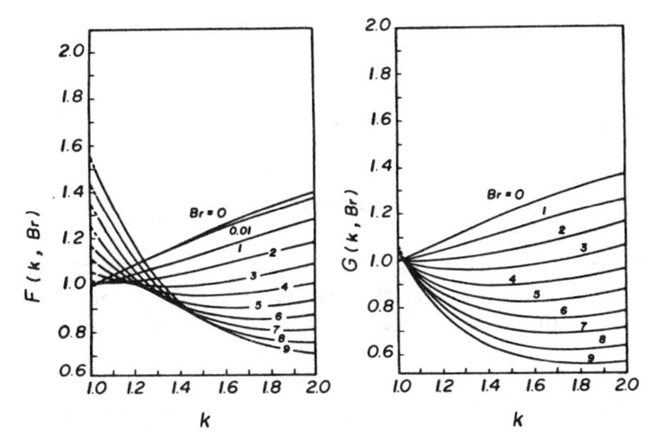

Figure 6. Graphical representation of modifying factors $F(k, B_r)$ and $G(K, B_r)$ in the rotary column with inner wall heated.

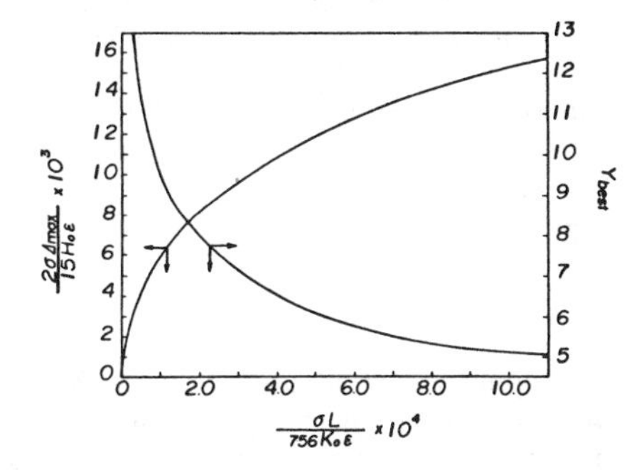

Figure 7. Values of permeability and separation for best performance vs. reduced flow rate.

Table 3
Transport Constants, Optimal Conditions and Maximum Separation of Various Types of Improved Columns

Column Type	H	K	Optimal Conditions	Maximum Separation Δ_{max}
Inclined	$H_0 \cos\theta$	$K_0 \cos^2\theta + K_1$	$\theta_{opt} = \cos^{-1}\sqrt{(\sigma L/2.52K_0)}$ $K_1 \ll K_0, \quad \sigma < 0.52K_0/L$	$0.226\sqrt{(H_0^2L/\sigma K_0)}$
Wired	$H_0 \cos^2\theta$	$K_0 \cos^4\theta + K_1 \cos^2\theta$	$\theta_{opt} = \cos^{-1}(\sigma L/2.52K_0)^{1/4}$ $K_1 \ll K_0, \quad \sigma < 0.52K_0/L$	$0.226\sqrt{(H_0^2L/\sigma K_0)}$
Movingwall	$H_0 \cos\theta - H_1V$	$K_0 \cos^2\theta + K_1 +$ $K_2V\cos\theta + K_3V^2$	$\theta_{opt} = \cos^{-1}[1.096\sqrt{(\sigma L/K_0)}]$ $\sigma < 0.84K_0/L$ $V_{opt} = 5.29\sqrt{[\sigma LD/K_1(2\omega)^2]}$	$0.26\sqrt{(H_0^2L/\sigma K_0)}$
Rotated wired	$H_0 \cos^2\theta - H_1V\sin\theta\cos\theta$	$K_0\cos^4\theta + K_1\cos^2\theta$ $- K_2V\sin\theta\cos^3\theta$ $+ K_3V^2\sin^2\theta\cos^2\theta$	$\theta_{opt} = \cos^{-1}[1.096\sqrt{(\sigma L/K_0)}]$ $\sigma < 0.84K_0/L$ $V_{opt} = 2.35K_0K_3^{-1}[2.192$ $\sqrt{(K_0/\sigma L)} - 2.38]^{-1/2}$	$0.26\sqrt{(H_0^2L/\sigma K_0)}$
Rotary	$H_0F(k, B_r)$, see Figure 6	$K_0G(k, B_r) + K_1(k+1)/2$		
Packed	$15\varepsilon H_0H_2$	$189\varepsilon K_0K_4$	Y_{opt}, see Figure 7	see Figure 7

$H_1 = \alpha\varrho(2\omega)B(\Delta T)/3!\bar{T}$; $H_2 = (Y^{-2})[1 - 3Y^{-2}(Y\tanh Y - 1)]$; $K_2 = \beta_T\varrho g(2\omega)^5B(\Delta T)/1680\mu D$; $K_3 = \varrho(2\omega)^3B/30D$;
$K_4 = (Y^{-4})[1 - 15(4Y)^{-2} - 15Y^{-4} - 15Y^{-1}\coth Y + 15(4Y^3)^{-1}\coth Y + 45(4Y^2)^{-1}\coth^2 Y]$.

For concentric-tube columns, B *and* 2ω *in above transport constants are replaced by* $2\pi R_1$ *and* $(R_2 - R_1)$, *respectively.*

From Yeh and Yeh [37].

$$\dot{\Delta}_{max} = \sqrt{[(1.5/\Delta_{max})^2 + 12c_f(1-c_f)]} - 1.5/\Delta_{max} \quad (63)$$

where Δ_{max} is the maximum separation, listed in Table 3, for each improved column in the concentration range $0.3 < c < 0.7$.

SEPARATION THEORY IN MULTICOMPONENT MIXTURES

The theory of thermal diffusion in multicomponent mixtures is very complicated. Analogous to Equation 2 for the binary mixtures in static cells, we may derive the generalized equation for the net flux of component i in the x-direction as

$$J_{i-x} = \varrho \sum_k D_{ik}\{c_k \sum_j c_j \alpha_{kj} \partial(\ln T)/\partial x - \partial c_k/\partial x\} \quad (64)$$

where α_{kj} = the thermal diffusion constant for components kj pair
D_{ik} = the generalized diffusion coefficient for components ik pair, which can be related to the binary ordinary diffusion coefficients

Besides, these coefficients must satisfy the following conditions:

$$D_{ij} = D_{ij}, \quad \sum_i c_i D_{ij} = \sum_j c_j D_{ji} = 0 \quad (65)$$

$$\alpha_{ij} = -\alpha_{ji}, \qquad \alpha_{ii} = 0 \quad (66)$$

The complexity in the multicomponent mixtures arises from the fact that in a mixture having n components, there are $n(n-1)/2$ independent D_{ij} and $n(n-1)/2$ independent α_{ij}. Hence, only some special cases are considered. For static cells at steady-state, Equation 64 can be reduced to

$$c_k \sum_j c_j \alpha_{kj} \partial(\ln T)/\partial x - \partial c_k/\partial x = 0 \qquad k, j = 1, \ldots, n \quad (67)$$

Since Equation 67 is nonlinear, it is obviously difficult to obtain an exact solution.

For the separation of isotopic gas mixtures in the concentric-tube Clausius-Dickel column, Rutherford [39, 40] has modified the results derived by Jones et al. [31] and obtained the transport equation for component i as

$$\tau_i = c_i \sum_j^{n-1} H_{ij} c_j - (K_c + K_d) \frac{dc_i}{dz} + \sigma c_i \quad (68)$$

under the assumption that all D_{ij} are equal and the concentration is in weight fraction. The transport constants are determined by

$$H_{ij} = 2\pi \int_{T_1}^{T_2} (\alpha_{ij}/T) G(T)\, dT \quad (69)$$

$$K_c = 2\pi Q_1^{-1} \int_{T_1}^{T_2} (\lambda/D\varrho) G^2(T)\, dT \quad (70)$$

$$K_d = 2\pi Q_1^{-1} \int_{T_1}^{T_2} (\lambda D \varrho r^2)\, dT \quad (71)$$

where G(T) is solved from

$$Q_1^3 \frac{d}{dT}\left[\frac{1}{r^2\lambda}\frac{d}{dT}\left(\frac{\mu}{\lambda}\frac{d}{dT}\frac{1}{\varrho\lambda^2 r^2}\frac{dG}{dT}\right)\right] = -g\frac{d\varrho}{dT} \tag{72}$$

The boundary conditions of Equation 72 are the same as in Equation 30. Equation 69 can be further simplified as [38–40]

$$H_{ij} = 2\pi(M_i - M_j)/(M_i + M_j)\int_{T_1}^{T_2}(\alpha_0/T)G(T)\,dT \tag{73}$$

by assuming all isotopic pairs having the same reduced thermal diffusion factor α_0. The separation factor q_{ij}, obtained at total reflux and at steady state, between components i and j was

$$\begin{aligned} q_{ij} &= [(c_i/c_j)_T/(c_i/c_j)_B] \\ &= \exp[H_{ij}L/(K_c + K_d)] \end{aligned} \tag{74}$$

In general, Equation 68 must be solved by numerical method. The reader might refer to Reference 41 for further information. Some experimental values of α_0 have been published in References 42–44.

NOTATION

A	defined by Equation 5 for static cells
A_i	defined by Equation 23 for Clausius-Dickel column
A_c	cross sectional area of the connecting tube in the two-bulb apparatus
a	constant defined by Equation 21
B	column width of the flat-plate Clausius-Dickel column
B_r	Brinkman number, $\mu V^2/\lambda\Delta T$
b	constant defined by Equation 21
c	fraction of component 1 in binary mixtures
c_c, c_H	c in cold and hot bulbs, respectively
c_i	fraction of component i in multicomponent mixtures
c_B, c_f, c_T	c in bottom product, feed stream, and top product, respectively
c_0	c at initial time
c_{je}	fraction of component 1 at the outlet of stage j in the multicolumns apparatus
D	ordinary diffusion coefficient in binary mixtures
D_{ij}	generalized diffusion coefficient
$F(k)$, $F(k, B_r)$	modifying factors defined by Equation 39 and in rotary column, respectively
f	fraction of approach to equilibrium defined by Equation 16
$G(T)$	function solved from Equations 29, 38 or 72
$G(k)$, $G(k, B_r)$	modifying factors defined by Equation 41 and in rotary column, respectively
g	acceleration of gravity
H, H_0	transport constants defined in improved columns and Clausius-Dickel column
H_{ij}	transport constant defined by Equation 69 for multicomponent mixtures
J_X	net flux of component 1 in the x-direction
J_{i-X}	net flux of component i in the x-direction in multicomponent mixtures
J_{X-TD},	flux of component 1 in the

J_{X-OD} x-direction due to the thermal and ordinary diffusions
K, K_0, K_1 transport constants defined by Equations 25, 26 and 27
k R_2/R_1
L total height of Clausius-Dickel columns
L_j height of each column at stage j in the multicolumns apparatus
L'_e, L'_s dimensionless column length defined by Equation 48
ℓ distance between the hot and cold plates in static cells
M_i molecular weight of component i
m mass of mixtures per unit column length
N_j number of columns at stage j in the multicolumns apparatus
Q, Q_1 heat transfer rates per unit area defined by Equations 28 and 37, respectively
q separation factor defined by Equation 22
q_{ij} separation factor for ij pairs, defined by Equation 70
R_1, R_2 outside radius of inner tube and inside radius of outer tube in the concentric-tube Clausius-Dickel column
T_1, T_2 absolute temperature at cold surface and hot surface, respectively
$\bar{T}$ arithmetical mean temperature or defined by equation 8
T_R temperature ratio defined by Equation 13
t time
t_r relaxation time
V tube speed of rotation in improved columns
V_C, V_H volumn of cold and hot bulbs
V_R volumn ratio defined by Equation 14
Y $\omega/\sqrt{k}$, where k is the permeability of the packing in packed columns

Greek Symbols

α thermal diffusion constant in binary mixtures
α_{ij} generalized thermal diffusion constant for ij pairs in multicomponent mixtures
β_T thermal expansion coefficient, $-(\partial\varrho/\partial T)_T$
γ reduced ratio in the multicolumns apparatus
Δ separation efficiency defined by Equations 49 and 51
Δ_{max} maximum separation obtainable in the improved column for the concentration range $0.3 < c < 0.7$
$\dot{\Delta}_{max}$ maximum separation obtainable in the improved column for the whole concentration range
ε void fraction of packing in packed columns
θ angle of inclination in the inclined or wired columns
λ thermal conductivity of the mixtures
μ viscosity of the mixtures
ϱ density of the mixtures
σ mass flow rate in the case $\sigma_e = \sigma_s = \sigma$
σ_e, σ_s mass flow rate in the enriching and stripping sections, respectively
σ'_e, σ'_s dimensionless mass flow rate defined by Equation 47
τ_i net transport of component i in the z direction of a Clausius-Dickel column
φ constant defined by Equation 16
ω one-half of the distance between hot and cold plates in a Clausius-Dickel column

REFERENCES

1. Dufour, L., *Arch. Sci. Phys. et Nat., 26:* 546 (1872).
2. Ludwig, C., *Wien. Akad. Ber., 20:* 539 (1856).
3. Enskog, D., *Phys. Z., 12* (56): 533 (1911).
4. Chapmam, S., and Dootson, *Phil. Mag., 33:* 248 (1917).
5. Grodzka, P. G., and Facemire, B., *Sep. Sci., 12* (2): 103 (1977).
6. Clausius, K., and Dickel, G., *Naturwiss, 26:* 546 (1938).
7. Sullivan, L. J., Ruppel, T. C., and Willingham, C. B., *I.E.C., 49:* 110 (1957).
8. Lorenz, M., and Emery, A. H., Jr., *Chem. Eng. Sci., 11:* 16 (1959).
9. Ramser, J. H., *I.E.C., 49:* 155 (1957).
10. Powers, J. E., and Wilke, C. R., *AIChE J., 3:* 213 (1957).
11. Chueh, P. L., and Yeh, H. M., *AIChE J., 13:* 37 (1967).
12. Emery, A. H., and Lorenz, M., *AIChE J., 9:* 660 (1963).
13. Washall, T. A., and Molpolder, F. W., *I.E.C. Proc. Des. Dev., 1:* 266 (1962).
14. Yeh, H. M., and Ward, H. C., *Chem. Eng. Sci., 26:* 937 (1971).
15. Yeh, H. M., and Tsai, C. S., *Chem. Eng. Sci., 27:* 2065 (1972).
16. Yeh, H. M., and Cheng, S. M., *Chem. Eng. Sci., 28:* 1803 (1973).
17. Yeh, H. M., and Chu, T. Y., *Chem. Eng. Sci., 29:* 1421 (1974).
18. Yeh, H. M., and Ho, F. K., *Chem. Eng. Sci., 30:* 1381 (1975).
19. Yeh, H. M., and Tsai, S. W., *Sep. Sci. and Tech., 17:* 1075 (1982).
20. Yeh, H. M., and Tsai, S. W., *J. of Chem. Eng. Japan, 14:* 90 (1981).
21. Yeh, H. M., and Tsai, S. W., *Sep. Sci. and Tech., 17:* 1075 (1982).
22. Yeh, H. M., *Sep. Sci. and Tech., 18* (6): 585 (1983).
23. Yeh, H. M., and Hsieh, S. J., *Sep. Sci. and Tech., 18* (11): 1065 (1983).
24. Sasaki, K., Yoshitomi, T., and Miura, N., *Bull. of Chem. Soc. Japan, 49:* 363 (1976).
25. Sasaki, K., and Yoshitomi, T., *Bull. of Chem. Soc. Japan, 49:* 367 (1976).
26. Grew, K. E., and Ibbs, T. L., *Thermal Diffusion in Gases,* Cambridge University Press, London (1952).
27. Prigogine, I., de Broukere, L., and Amand, R., *Physica, 16:* 577, 851 (1950).
28. de Groot, S. R., *The Thermodynamics of Irreversible Processes,* Interscience, New York (1950).
29. Von Halle, E., *A New Apparatus for Liquid Phase Thermal Diffusion,* AEC Research and Development Report K-1420, June 24 (1959).
30. Powers, J. E., *Thermal Diffusion in New Chemical Engineering Separation Techniques,* H. M. Schoen (Ed.), Wiley-Interscience (1962).
31. Jones, R. C., and Furry, W. H., *Revs. Mod. Phys., 18:* 151 (1946).
32. Yeh, H. M., *Sep. Sci. and Tech., 11:* 455 (1976).
33. Yeh, H. M., and Chu, T. J., *Chem. Eng. Sci., 30:* 47 (1975).
34. Yeh, H. M., and Chiou, C. F., *Sep. Sci. and Tech., 14.* 645 (1979).
35. Yeh, H. M., and Lu, C. C., *Sep. Sci. and Tech., 13:* 79 (1978).
36. Jones, A. L., *Petrol. Refiner, 36* (7): 153 (1957).
37. Yeh, H. M., and Yeh, Y. T., *The Chem. Eng. J. 25:* 55 (1982).
38. Hirschfelder, J. O., Curtiss, C. F., and Bird, R. B., "Molecular Theory of Gases and Liquids," Wiley, New York, (1954).
39. Rutherford, W. M., *J. Chem. Phys., 58:* 1613 (1973).
40. Rutherford, W. M., *Sep. and Purif. Method, 4* (2): 305 (1975).
41. Rutherford, W. M., *Sep. Sci. and Tech., 16* (10): 1321 (1981).
42. Taylor, W. L., *J. Chem. Phys., 62:* 3837 (1975), *64:* 3334 (1976).
43. Rabinovich, G. D., *Inzh-Fiz. Zh., 15:* 1014 (1968).
44. Rutherford, W. M., *J. Chem. Phys., 59,* 6061 (1973).
45. Horne, F. H., and Bearman, R. J., *J. Chem. Phys., 37* (12): 2857 (1962).
46. Narbekov, A. I., and Usmanov, A. G., *Inzh-Fiz. Zh., 21* (2): 334 (1971).
47. Taylor, W., *J. Chem. Phys., 72* (9): 4973 (1980).

CHAPTER 36

PRINCIPLES OF PARAMETRIC PUMPING

Georges Grevillot

Laboratoire des Sciences du Genie Chimique
CNRS-ENSIC, 1, rue Grandville
Nancy, France

CONTENTS

INTRODUCTION

Parametric pumping is a molecular separation method of fluid mixture discovered in the 1960 s by R. H. Wilhelm and co-workers [1]. It uses the capability of a stationary phase to remove some components of a mobile phase and to restore them under the change of an intensive thermodynamic variable such as temperature. It requires an alternative motion of the mobile phase synchronized with the change of the thermodynamic variable. It is thus a cyclic process in which components are stored during a part of the cycle and released in the other part of the cycle.

Figure 1A presents the basic arrangement of a parametric pump using temperature as the thermodynamic variable. A column packed with a suitable adsorbent is connected to two reservoirs and surrounded by a jacket. The mixture fills the column voids and one reservoir. During the first half-cycle, the column is heated and the fluid phase is forced to flow upward by the motion of the pistons. During the second half-cycle, the column is cooled and the fluid phase flows downward. Suppose the mixture consists of one adsorbable solute in an inert solvent and the solute has a usual behavior of being more adsorbed at

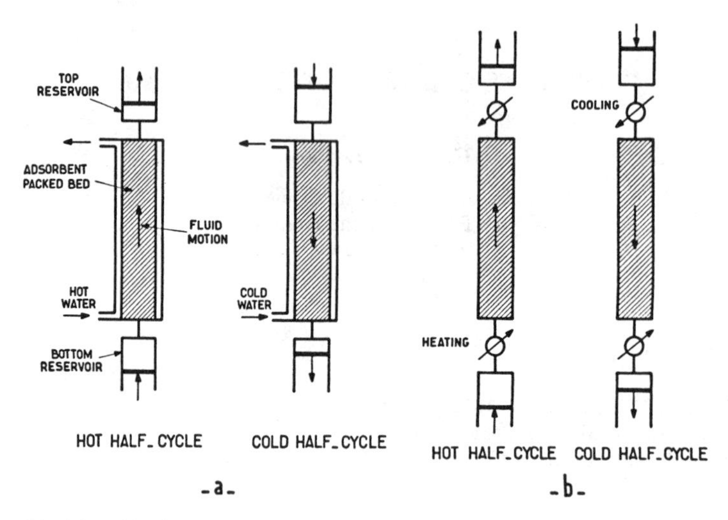

Figure 1. Principles of batch parametric pumping: (A) direct mode; (B) recuperative mode.

the cold temperature than at the hot (Figure 2). Thus during the cold half-cycle, some solute adsorbs and is removed from the fluid while during the hot half-cycle, the solute desorbs and accumulates in the upward flowing fluid. Thus a solute concentration gradient is obtained, the mixture being enriched in solute at the top and depleted at the bottom. During each subsequent cycle, the gradient will increase until a limiting gradient is asymptotically approached. Therefore the separation factor defined as the solute concentration in the top reservoir to that in the bottom reservoir increases as the number of cycles increases. The reason is that a reflux occurs at each end of the bed created by the synchronization of the changes in temperature and in flow direction. The solute is thus "pumped" toward the top reservoir against its concentration gradient in a similar way as in a heat pump heat is pumped against the temperature gradient. The reflux is at the origin of the large separations which can be obtained by parametric pumping even when the difference between the two partition curves at the two operating temperatures is small. Figure 3 shows the now-classical results obtained by Wilhelm et al. [2, 3]. These authors removed toluene from n-heptane using silica gel as the adsorbent and cycling the temperature between 4° C (or 15° C) and 70° C. Clearly the toluene is removed from the bottom reservoir and accumulates in the top one as the number of cycles increases. Also more separation is obtained with longer cycle time which indicates that heat and mass transfer rate limitations have a destructive effect on separation. Parametric pumping is basically an equilibration process for both mass and heat as opposed to rate processes such as gaseous diffusion.

Five extensions of the basic principle just outlined can be envisaged:

- *Mode of operation:* In the process just described, heat is transferred into and out of the bed by means of a jacket surrounding the bed. This mode of heating is called "direct mode." Another way of heating and cooling the bed is by heating or cooling the flowing fluid itself by means of heat exchangers at the two ends of the column before the fluid enters the bed (Figure 1B). This mode is called the "recuperative mode" because it allows a more easy recovery of heat and uses less energy for a given separation. It seems to be more adapted to large-diameter beds where the direct mode would cause radial temperature gradients.
- *Reflux:* The batch parametric pump of Figure 1 may be called a "total reflux" process by analogy with distillation: no feed is introduced, no product withdrawn. By contrast,

the process can be operated at partial reflux, that is, with feed of the mixture at some intermediate point of the column and product withdrawals at the two ends. After a transient period, the pump tends toward a cyclic steady state in which the product concentrations are constant. Just as in distillation, modifying the reflux causes a change in separation. A great number of arrangements are possible depending especially where and when the feed stream occurs. Truly continuous or semicontinuous productions are possible.

- *Multicomponent mixtures:* Parametric pumping is able to remove or to isolate one or several solutes of a mixture or to split the mixture into two groups of components. The type of separation depends especially on the linear or nonlinear character of the equilibrium distribution.
- *Phases used:* Parametric pumping can be worked with a solid or a liquid as the stationary phase and a liquid or a gas as the mobile phase. For liquid-liquid systems, a staged arrangement such as that used in liquid-liquid extraction seems to be more appropriate than a continuous contacting device.
- *Thermodynamic variables:* Any variable which can modify the equilibrium partition of solutes between the two phases can be used (temperature, pH, ionic strength, electric potential and pressure).

Among these five extensions, some combinations have been developed and shown to be capable of separating mixtures. The others make no sense (such as pH with gas-solid systems), are too difficult to devise (such as pH in direct mode), or are inefficient (such as pressure with liquid-solid systems).

Some other processes are related to parametric pumping because they also use the action of an intensive thermodynamic variable together with reflux to perform the separation: the thermofractionation in a moving bed utilizes a cold bed and a hot bed with countercurrent flow of the liquid mixture; the cycling-zone separation processes avoid the alternative motion of fluid; the pressure swing adsorption (PSA) processes appear to be similar to parametric pumping in many respects (pressure being the parameter) in spite of some specificities.

The most recent reviews on parametric pumping are those of Rice [4], Chen [5], and Wankat [6, 7]. The latter include related cyclic separation techniques.

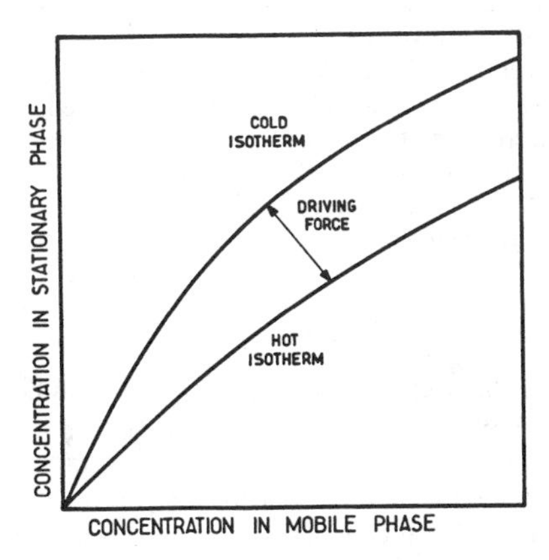

Figure 2. Parametric pumping uses the change of an equilibrium partition induced by a change of an intensive thermodynamic variable.

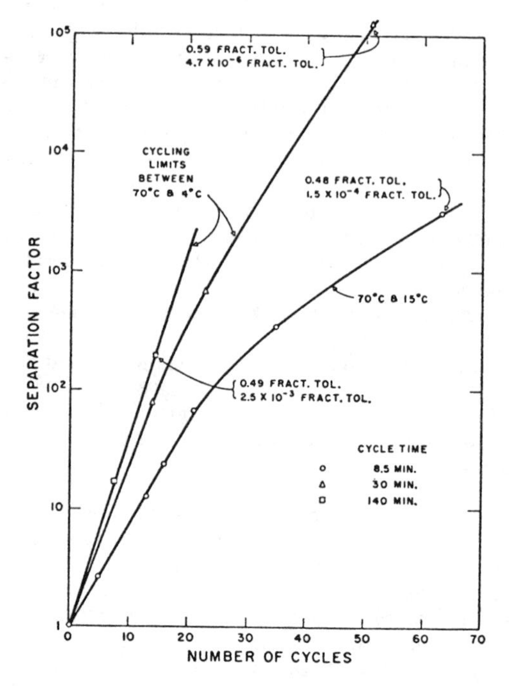

Figure 3. Experimental thermal parametric pumping separation of toluene-heptane mixture with silica gel adsorbent; batch operation [3].

PARAMETRIC PUMPING MODELS

Equilibrium-Staged Models

The objective of the equilibrium-staged model is in part didactic: to bring a better understanding and a visualization of the mechanism of separation in parametric pumping by the use of familiar concepts of chemical engineering. The model is most appropriate for representing real staged systems such as, for example, those used in liquid-liquid extraction parametric pumping. Therefore the two phases called "liquid" and "solid" in the following must be understood in the general sense of "mobile phase" and "stationary phase," respectively. We start with the simplest version of the model: the single-stage total reflux model in direct mode. Some extensions will be made later.

Total Reflux [8, 9]

Consider a pump comprising only one cell (stage) C, containing the solid of volume V_s, a top reservoir TR, and a bottom reservoir BR (Figure 4). The total liquid is divided into two fractions, L_1 and L_2, of equal volumes, V_1. At the start of a cycle, L_1 is in TR, while in the cell, L_2 and the solid are in equilibrium at the temperature T_C. After a transfer down of L_2 into BR and of L_1 into C, the temperature of the cell is changed to T_H and L_2 and solid in the cell are allowed to equilibrate at this temperature. Then, a transfer up brings the liquid fractions back to their initial position (end of cycle). This process may be represented in the plane of the isotherms shown on Figure 5 where X and Y are the

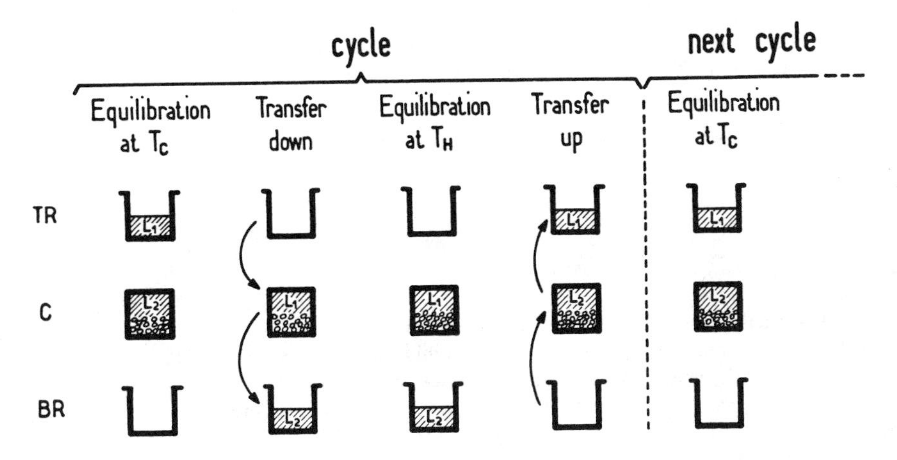

Figure 4. Single-stage parametric pump and its operating cycle.

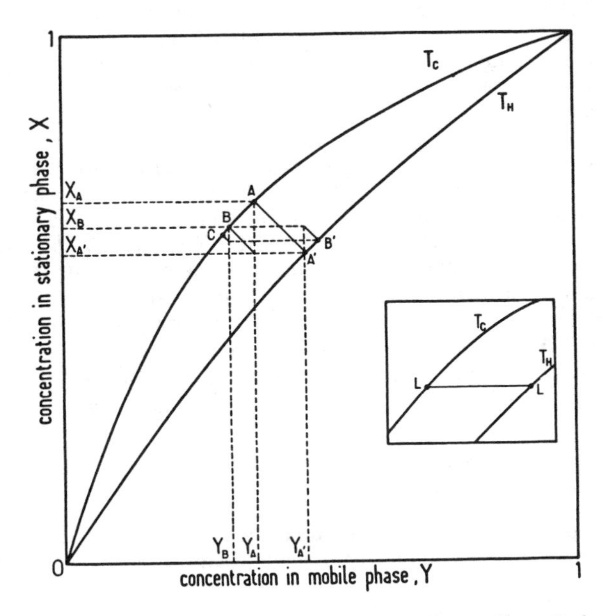

Figure 5. Representation of the first cycles and of the cyclic steady state (framed) of a single-stage parametric pump.

dimensionless concentrations of solute in solid and liquid phases, respectively. The isotherms can be drawn using either experimental data or analytical expressions of the general form:

$$X = g(Y, T) \tag{1}$$

At the start of the first cycle, we suppose that the cell is in equilibrium at T_C represented by point A and that L_1 has the same concentration Y_A as L_2. The first equilibrium at T_H

is represented by point A′ at the intersection of the T_H isotherm and of the operating segment representing the mass balance constraint:

$$\frac{Y_A - Y_{A'}}{X_A - X_{A'}} = -\varrho_C \tag{2}$$

where

$$\varrho_C = \frac{x_r V_l}{y_r V_s} \tag{3}$$

is the cell capacity ratio. The second cycle starts with the mixing of liquid of composition Y_A with solid of composition X'_A. The new equilibrium at T_C is represented by point B, intersection of the T_c isotherm and of the operating segment through (Y_A; $X_{A'}$) and of slope $-\varrho_C$. A certain separation is already observed, as the L_1 and L_2 liquid fractions originally both at Y_A are now $Y_{A'}$ and Y_B, respectively, such that $Y_B < Y_A < Y_{A'}$. The next equilibria are represented by points B′, C, ... showing that the path converges asymptotically toward a simple horizontal segment LL′ as the number of cycles becomes infinite. Then the solid and the liquid fractions have a constant composition at any time of the cycle and each equilibrium is reached without any mass transfer between the two phases. It is clear that the separation obtained with this single-stage pump is limited, rather small, and depends directly on the spacing between the isotherms at the two temperatures.

The N-stage total reflux pump is a generalization of the single-stage pump: it comprises N cells, C_1, C_2 ..., C_N, each containing the same amount of solid, and two reservoirs TR and BR. The total liquid phase is divided here into N + 1 equal fractions. The cycle is similar to that of the previous case, comprising two equilibrations steps of all cells and two transfer steps of all liquid fractions one stage down or up. The transient is more tedious to construct than previously, but at cyclic steady state the construction is simply an N-step staircase between the two isotherms (Figure 6A). Each solid fraction has a constant composition X_i represented by a horizontal segment. Each liquid fraction L_i has also a constant composition Y_i represented by a vertical segment: it is the equilibrium composition in cell C_{i-1} at T_C and in cell C_i at T_H. The limit (or maximum) separation is that of the two extreme fractions, Y_1 and Y_{N+1}. Let us now examine analytically the limit separation when the isotherms are of the mass action law or Langmuir type (nonlinearly sorbing solute):

$$X = \frac{K(T)Y}{1 + [K(T) - 1]Y} \tag{4}$$

where K(T) is the equilibrium constant, a function of temperature only. Based on the known staircase limit pattern, the derivation of the limit separation is straightforward [9]. We obtain:

$$\frac{Y_1}{1 - Y_1} \Bigg/ \frac{Y_{N+1}}{1 - Y_{N+1}} = \left(\frac{K_C}{K_H}\right)^N \tag{5}$$

Thus the cyclic steady-state separation depends on the ratio of the equilibrium constants at the two temperatures and increases exponentially with the number of stages N.

Analogy with Total Reflux Binary Distillation

We first remark that Equation 5 is identical to Fenske's equation for total reflux distillation. It can be used to give the number of stages N required to obtain the desired separation characterized by Y_1 and Y_{N+1}.

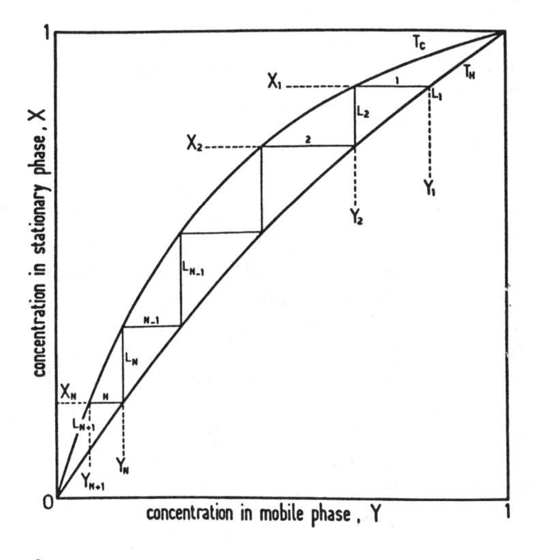

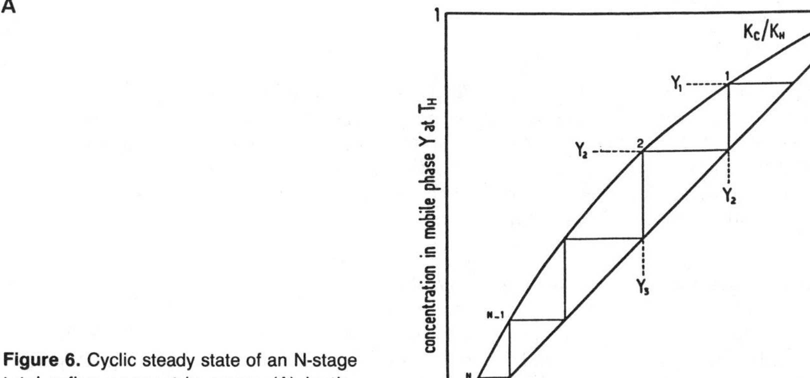

Figure 6. Cyclic steady state of an N-stage total reflux parametric pump: (A) in the isotherm plane; (B) in the (T_H, T_C) plane. Illustration of the analogy with binary total reflux distillation.

From Equation 4 written for each temperature, we can obtain the relation between the two liquid compositions which are in equilibrium with the *same* solid composition:

$$Y(T_H) = \frac{(K_C/K_H)Y(T_C)}{1+[(K_C/K_H)-1]Y(T_C)} \tag{6}$$

Then the staircase construction of Figure 6A between the isotherms leads to another staircase construction in the plane $[Y(T_C), Y(T_H)]$ between the curve represented by Equation 6 and the diagonal, like in total reflux distillation (Figure 6B). The curve is thus analogous to the vapor-liquid equilibrium curve, the mobile phase at T_C and T_H in parametric pumping being analogous to the liquid and vapor phases, respectively, in distillation. The ratio K_C/K_H of the equilibrium constants plays the role of the relative volatility. The reflux in distillation consists of withdrawing material in the vapor phase from the top of the column and reinjecting it as a liquid phase. Here it consists of withdrawing mobile phase at temperature T_H and reinjecting it at temperature T_C. In this way we can define a generalized reflux in the sense that it can use not only a phase change but also any change of a suitable parameter such as temperature.

Partial Reflux

We may carry on the analogy by including the possibility of a partial reflux operation. This can be done by transfering only a part of each liquid fraction down or up and withdrawing the remainder in TR and BR as enriched and depleted products. Compensating fresh feed is added to a stage in the cascade. The partial reflux pump thus comprises an enriching section, a stripping section, and a feed stage. At cyclic steady state, the cascade can be represented on a McCabe–Thiele-type diagram in the $[Y(T_C), Y(T_H)]$ plane as a staircase construction between operating lines and partition lines. The partition lines play the same role as the vapor-liquid equilibrium curve in distillation but they depend here on the reflux ratios. Complete analysis of this pump is available elsewhere [10]. The existence of a minimum reflux and an optimal feed stage location and their calculation have been described. An analytical expression has been derived for the separation factor defined as the ratio of top to bottom product concentration in the case of linear isotherms as:

$$SF \equiv \frac{Y_T}{Y_B} = \frac{d_1 d_2^{N_s} + d_3}{d_4 d_5^{N_e} + d_6} \tag{7}$$

where N_s and N_e are the number of stages in the stripping and enriching sections, respectively, and the coefficients d are known functions of reflux and equilibrium constants. Figure 7 based on Equation 7 illustrates the dependence of the number of stages on reflux for different separations. As in distillation, the optimum reflux would result from an economic balance between the equipment costs, mainly a function of the number of stages, and the operating costs, mainly a function of the reflux.

Extension to Several Transfers per Half-Cycle [11]

An important limitation of the previous model is the single-transfer step per half-cycle, which implies that only one liquid fraction enters or exits the cascade. Therefore the penetration of the mobile phase through the cascade is 1/N and is not independent of the number of stages. For real not staged systems such as a packed bed, this is a too-restrictive limitation. To remove it, we will consider now a batch pump where M liquid fractions are able to penetrate the N-stage cascade (the total number of liquid fractions is thus N + M). By generalization of the cycle described previously (Figure 4), we will have M equilibrations at T_C, each followed by a transfer down (first half-cycle) and M equilibrations at T_H, each followed by a transfer up (second half-cycle). We define a mixed-reservoir pump $[N, M]_m$ when the liquid fractions are well mixed in the top and the bottom reservoir. We may define also a staged-reservoir pump $[N, M]_s$ having M top and M bottom reservoirs: the fractions are thus collected separately during a half-cycle and reinjected in reverse order during the next half-cycle. Such a mode of operation avoids the destruction of concentration gradients coming out of the cascade, or the column, and thus preserves for the reflux successively all the separation already obtained. This idea is due to Thompson and Bowen [12] who applied it to a packed-bed pump; using the local equilibrium model, they showed an increase of the separation relative to a mixed reservoir pump. Wankat [13] discussed later the use of nonmixed reservoir in partial reflux pumps.

The staged-reservoir pump has been studied in detail elsewhere [11]. At cyclic steady state, with linear isotherms, the total reflux pump operation can be represented in the isotherm plane as an assembly of staircases, which clearly show the increase of separation due to the staging of reservoirs and provides an analytical solution for the separation factor:

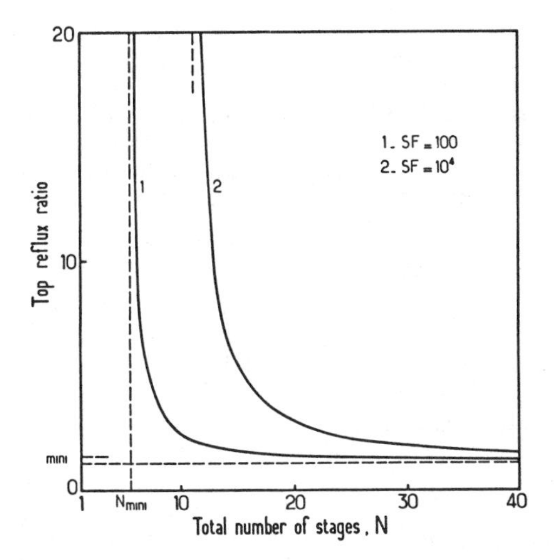

Figure 7. Effect of reflux on number of stages for specified separation with feed stage at the optimum location.

$$SF[N, M]_s = \left(\frac{K_C}{K_H}\right)^N \left(\frac{K_C}{K_H}\frac{\varrho_c + K_H}{\varrho_c + K_C}\right)^{M-1} \tag{8}$$

The separation factor increases exponentially with both the number of stages N and the number of transfers M.

For the mixed-reservoir pump, no simple graphical representation nor analytical expressions for the separation have been developed to date. Cycle-by-cycle computed results show that almost always the limit separation decreases as M increases.

More generally, the staged equilibrium model is well adapted to computer calculations. Separation as a function of the number of cycles can be obtained by computing the concentrations in the cascade, cycle after cycle, starting from any initial state. The cyclic steady-state separation is obtained after convergence of this calculation. A matrix formulation can lead more directly to the solution [14].

Multicomponent Separations

In the case where the solutes adsorb linearly on the solid, they behave independently from each other. Therefore the previous results can be applied to each solute and the multicomponent separation is simply the superposition of separation of each solute. In the case of nonlinearly sorbing solutes, the mass balances are coupled through the equilibrium isotherms and the solution must be calculated on a computer.

Local Equilibrium Models

Linear Equilibria

Direct mode: basic equations. / This model was first developed by Pigford et al. [15] who called it "equilibrium theory." It is a chromatographic model in the sense that it considers propagation of concentration waves through the parametric pumping column and uses it to explain and calculate the separation. It seems to be well adapted to operations using a sorbent in a packed bed, and it has been widely applied not only to parametric pumping but also to cycling zone adsorption.

The basic assumptions of the model are: the solid and fluid phases are everywhere and at any time in equilibrium as a result of a very high rate of transfer of solutes, the bed is isothermal during each half-cycle (for direct mode), the axial dispersion is negligible. Under these conditions, the differential mass balance for one solute in an inert solvent is:

$$u\frac{\partial y}{\partial z}+\frac{\partial y}{\partial t}+\frac{1-\varepsilon}{\varepsilon}\frac{\partial x}{\partial t}=0 \tag{9}$$

where u = the intersticial velocity of the fluid
ε = the void fraction of the packing

If the equilibrium relationship is linear:

$$x = M(T)y \tag{10}$$

where M(T) is the equilibrium constant, a function of temperature only. For convenience, let us define the modified equilibrium constant:

$$m(T)=\frac{1-\varepsilon}{\varepsilon}M(T) \tag{11}$$

which relates the concentrations taken on the same bed volume basis. Eliminating x between Equations 9 and 10 leads at constant temperature to:

$$u\frac{\partial y}{\partial z}+[1+m(T)]\frac{\partial y}{\partial t}=0 \tag{12}$$

This hyperbolic partial differential equation can be solved by the method of characteristics or, more simply, by comparing with the total differential of y:

$$\frac{\partial y}{\partial z}dz+\frac{\partial y}{\partial t}dt=dy \tag{13}$$

We obtain:

$$\frac{dz}{dt}=\frac{u}{1+m(T)} \tag{14}$$

Physically, Equation 14 represents the velocity at which a concentration wave moves through an isothermal bed. This velocity is independent of concentration, and therefore, the whole concentration wave moves at that same velocity (this is not true for nonlinear isotherms); but it varies with temperature through the parameter m(T). In the (z, t) plane, Equation 14 is represented by straight lines, which are the trajectory characteristics of the concentration waves.

When the temperature changes at the beginning of each half-cycle, the local equilibrium changes at a constant amount of solute in the local bed slice. This material balance is:

$$y_C[1+m(T_C)]=y_H[1+m(T_H)] \tag{15}$$

where y_C and y_H are the local concentrations at T_C and T_H, respectively.

Let us now define the important parameter [15] b as:

$$b = \frac{a}{1+m_0}, \quad 0 \leqq b \leqq 1 \tag{16}$$

where

$$m_0 = [m(T_C)+m(T_H)]/2 \tag{17}$$

$$a = [m(T_C)-m(T_H)]/2 \tag{18}$$

and T_C and T_H are the cold and hot temperatures, respectively, considered in the parametric pumping operation. The parameter b is a measure of the relative gap between the two isotherms at the two specified temperatures. The larger b is, the larger the parametric pumping effect will be. Equation 15 then becomes:

$$\frac{y_C}{y_H} = \frac{1-b}{1+b} \tag{19}$$

In the following, we will suppose that upflow (increasing values of z) occurs during the hot half-cycle. Integrating Equation 14 and inserting b leads to the expressions of the penetrations distances first defined by Chen and Hill [16]:

$$L_H = \frac{u_H}{(1+m_0)(1-b)} \frac{\pi}{\omega}, \quad \text{hot half-cycle} \tag{20}$$

$$L_C = \frac{u_C}{(1+m_0)(1+b)} \frac{\pi}{\omega}, \quad \text{cold half-cycle} \tag{21}$$

where π/ω is the half-cycle time.

These quantities are the distances the concentration waves move during each half-cycle of equal duration π/ω. Note that for normal adsorption systems $L_H > L_C$, which gives the "pumping" action. The different intersticial velocities u_H and u_C allow the possibility of unequal volume displacements for the two half-cycles (for partial reflux operation). It would be equivalent for the model to use unequal half-cycle times and a constant velocity.

Chen et Hill [16] defined three regions of operation for the pumps they studied depending on the magnitude of L_H and L_C and the column height (Figure 8). We will use this diagram in the following.

The parametric pumping effect. / At this time we are able to show how the parametric pumping effect is represented by this model. We will consider a total reflux (or batch) pump operating in Region 1. Figure 9A represents the (z, t) plane for the bottom of the pump. During the hot half-cycle of the cycle n+1, the solution in the bottom reservoir of concentration $\langle y_B \rangle_n$ obtained at the end of cycle n flows upward through the column, the wave motion being represented by the characteristic ch1. The solute fluid concentration in the bottom of the bed is thus $y_H = \langle y_B \rangle_n$ and the local equilibrium is represented by point A on Figure 9B. At the end of the half-cycle, the change of temperature leads to a new equilibrium represented by point B, the concentration being reduced to y_C (Equation 19). During the cold half-cycle, the wave motion is represented by characteristic ch2 the slope of which is smaller than ch1 ($L_C < L_H$). Therefore, ch2 does not exit from the bed, and a solution of constant concentration y_C flows into the bottom reservoir. If there is no dead volume, the bottom reservoir concentration at the end of the cycle is $\langle y_B \rangle_{n+1} = y_C$. Thus the cycle results in a decrease in the bottom reservoir concentration such that:

$$\frac{\langle y_B \rangle_{n+1}}{\langle y_B \rangle_n} = \frac{1-b}{1+b} \tag{22}$$

From this first-order equation we can deduce $\langle y_B \rangle_n$ as a function of the cycle number n as:

$$\langle y_B \rangle_n = y_0 \left(\frac{1-b}{1+b} \right)^n \tag{23}$$

where y_0 is the initial uniform concentration of solute at the start of the run. Figure 9B shows the successive equilibria illustrating the way in which the bottom reservoir concentration decreases from cycle to cycle. The two fundamental actions by which the parametric pumping effect occurs are clearly:

1. The equilibrium shift represented by the inclined segments which results in an elementary separation.
2. The reflux represented by the vertical segments that is responsible for the amplification of the separation.

What happens to the top reservoir? As shown in Figure 9A, at cycle n + 2 there is a band of concentration $\langle y_B \rangle_n$ in the column which will never exit at the bottom. This band moves from cycle to cycle by a distance $\Delta z = L_H - L_C$ at each cycle toward the top end of the column, and therefore, solute is transferred from the bottom reservoir and the column into the top reservoir. As pointed out by Pigford et al. [15], the problem of constructing a solution is reduced to the geometrical problem of locating the characteristics in the (z, t) plane. The resulting pattern has to be combined with external (to the bed) equations representing mixing in the reservoir and feed and product withdrawals in order to obtain the complete solution.

The following is a study of the behavior of batch and open systems.

Total reflux. / Pigford et al. [15] used the previous construction in the (z, t) plane to calculate $\langle y_B \rangle_n$ and $\langle y_T \rangle_n$ but for a particular case. Aris [17] generalized their solution.

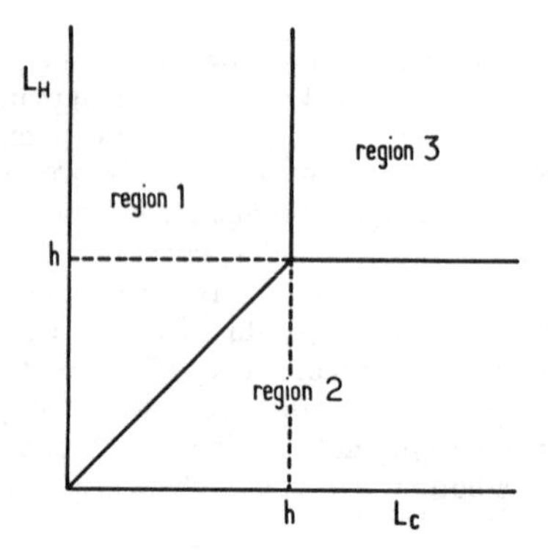

Figure 8. Regions of parametric pump operation.

Further Chen and Hill [16] treated an even more general case which also included the possibility of not completely displaced reservoir volumes. These so-called "dead volumes" represent connecting lines or residual volumes in tanks in a real system. This batch parametric pump is shown on Figure 10 as a particular case of open pumps which will be discussed later. The volume displaced at each half-cycle is $Q\pi/\omega$ where Q is the reservoir displacement rate. At the end of a half-cycle, the dead volume V_T or V_B is assumed to be well mixed with the entering displacement volume $Q\pi/\omega$ before returning to the column (use of nonmixed dead volume has been studied by other authors [12, 18]). The penetration distances are those given by Equations 20 and 21 where $u_H = u_0$, with:

$$u_0 = Q/(A\varepsilon) \tag{24}$$

Therefore for batch pumps we always have $L_H > L_C$ and the operation can occur only in Region 1 or in a portion of Region 3. Concentration transients have been calculated for

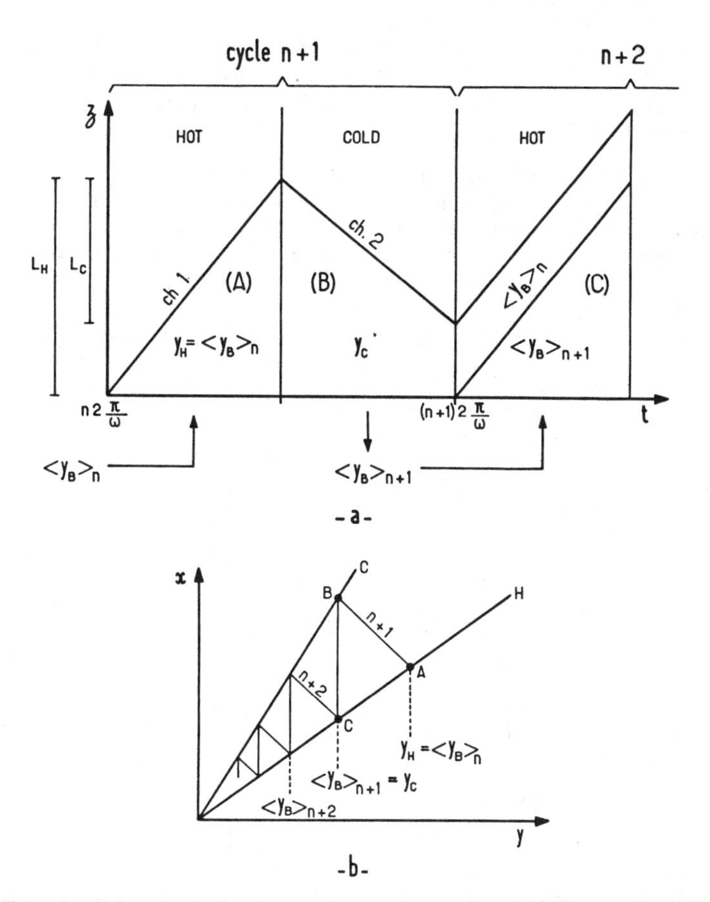

Figure 9. Illustration of the parametric pumping effect by the local linear equilibrium model: (A) in the (z,t) plane; (B) in the isotherm plane.

the two reservoirs and for the two regions [16]. As an example, the bottom reservoir concentration for the more interesting Region 1 is given by:

$$\frac{\langle y_B \rangle_n}{y_0} = \frac{1-b}{1+b}\left[\frac{\dfrac{1-b}{1+b} + \dfrac{V_B}{Q\pi/\omega}}{1 + \dfrac{V_B}{Q\pi/\omega}}\right]^{n-1} \tag{25}$$

which has been obtained in the same way as Equation 23 but with an additional external equation taking into account the bottom dead volume. Figure 11 shows some calculated results illustrating the effect of the volume displaced per half-cycle for top and bottom reservoir concentration transients. When the volume is such that $L_C < h$, the pump operation is in Region 1 and the bottom reservoir concentration decreases exponentially as given by Equation 25. Complete removal of solute from the bottom reservoir is thus possible. When $L_C > h$ (Region 3), the bottom reservoir concentration tends to a finite value at cyclic steady state ($n \to \infty$). Physically this signifies that a wave entering the top of the bed exits from the bottom before the end of the half-cycle and thus solute is transferred directly from the top to the bottom reservoir. At steady state, the top reservoir concentration increases when L_C decreases. The effect of dead volumes has been discussed in detail elsewhere [16]. Dead volume in the bottom reservoir slows down the decrease of the bottom concentration as shown by Equation 25. The top and bottom concentrations at cyclic steady state are reported in Table 1.

Open pumps. / Figure 10 shows two types of open pumps with feed located at the top of the column. Pumps with feed at the bottom have been shown to be less interesting because they cannot remove completely solutes from product streams [16]. The semicontinuous top feed pump operates without feed during upflow and with top feed during downflow. The

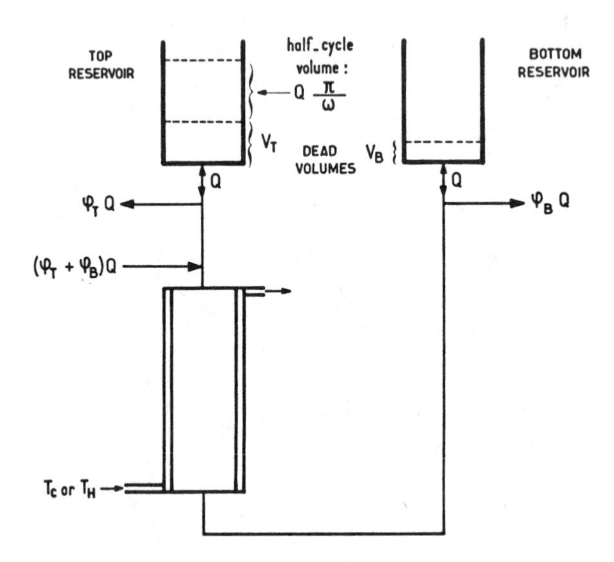

Figure 10. Parametric pump operations: batch ($\phi_T = \phi_B = 0$), semicontinuous top feed (upflow: batch; downflow: $\phi_T \neq 0$, $\phi_B \neq 0$), continuous top feed ($\phi_T \neq 0$, $\phi_B \neq 0$).

Table 1
Steady-State Solutions

Pump Type	Region 1	Region 2
Batch	$\frac{\langle y_B\rangle_\infty}{y_0} = 0$ $\frac{\langle y_T\rangle_\infty}{y_0} = 1 + \frac{2b}{1+b}\frac{1}{1+V_T/(Q\pi/\omega)} \cdot \left(\frac{h-L_c}{L_H-L_c} + \frac{1-b}{2b} + \frac{1+b}{2b}\frac{V_B}{Q\pi/\omega}\right)$	
Continuous top feed	$\frac{\langle y_B\rangle_\infty}{y_0} = 0$ $\frac{\langle y_T\rangle_\infty}{y_0} = 1 + \frac{\varphi_B}{\varphi_T}$	$\frac{\langle y_B\rangle_\infty}{y_0} = \frac{(\varphi_T+\varphi_B)(\varphi_B-b)/\varphi_B}{\varphi_T+\varphi_B-b(1+\varphi_T\varphi_B)}$ $\frac{\langle y_T\rangle_\infty}{y_0} = \frac{(\varphi_T+\varphi_B)(1-b\varphi_B)}{\varphi_T+\varphi_B-b(1+\varphi_T\varphi_B)}$
Semicontinuous, continuous top feed during downflow, batch during upflow	$\frac{\langle y_B\rangle_\infty}{y_0} = 0$ $\frac{\langle y_T\rangle_\infty}{y_0} = 1 + \frac{\varphi_B}{\varphi_T}$	$\frac{\langle y_B\rangle_\infty}{y_0} = \frac{(\varphi_T+\varphi_B)[\varphi_B-b(2+\varphi_B)]/\varphi_B}{\varphi_T+\varphi_B-b(2+\varphi_B-\varphi_T)}$ $\frac{\langle y_T\rangle_\infty}{y_0} = \frac{(\varphi_T+\varphi_B)(1+b)}{\varphi_T+\varphi_B-b(2+\varphi_B-\varphi_T)}$

After Chen and Hill [16].

continuous pump operates with feed all the time. Top and bottom product flow rates are defined as fractions of the reservoir flow rate by means of factors φ_T and φ_B. Material balances show that the flow rates in the column are Q and $(1-\varphi_B)Q$ for the semicontinuous and continuous pumps respectively on upflow, and $(1+\varphi_B)Q$ on downflow. The penetration distances are thus:

$$L_H = \frac{u_0}{(1+m_0)(1-b)}\frac{\pi}{\omega} \quad \text{(semicontinuous)} \tag{26}$$

$$L_H = \frac{u_0(1-\varphi_B)}{(1+m_0)(1-b)}\frac{\pi}{\omega} \quad \text{(continuous)} \tag{27}$$

$$L_C = \frac{u_0(1+\varphi_B)}{(1+m_0)(1+b)}\frac{\pi}{\omega} \tag{28}$$

where u_0 is given by Equation 24.

The pumps can be operated in the three Regions of Figure 8 depending on the magnitude of L_H and L_C. If both L_H and $L_C > h$, the pumps operate in Region 3. If not, the operation is in Region 1 if $\varphi_B \leqq b$ for the continuous pump or $\varphi_B \leqq 2b/(1-b)$ for the semicontinuous pump. In other cases, the pumps operate in Region 2. The concentration transients of the top and the bottom products in the three regions are available elsewhere for both the continuous [19] and the semicontinuous pumps [20]. They are in general similar to those shown on Figure 11 for the batch pump except that overshoot in the top product transient and undershoot in the bottom one can occur. Steady-state solutions ($n \to \infty$) are reported in Table 1. One can see that complete removal of solute from the bottom product is possible for the two pumps when operated in Region 1. This arises because no wave exits from the bottom of the column. Therefore the expression given for the bottom concentration transient of the batch pump Equation 25 still holds true for the two open pumps in Region 1 (for the open pumps, y_0 is also the feed concentration). Note that the value of b does not affect the steady-state concentrations in this region. Specially, high enrichment of the top product can be obtained by decreasing the φ_T value, but the feed flow rate will be decreased in the same time.

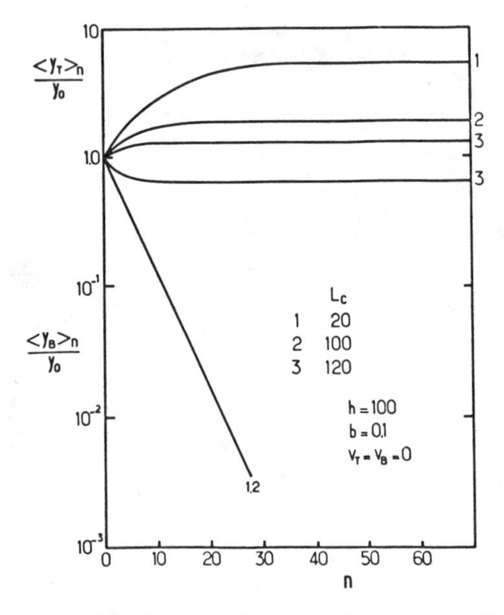

Figure 11. Total reflux separations : if $L_C < h$, infinite separation factor is possible (after Chen and Hill [16]).

The continuous and the semicontinuous pumps are similar in nature, the main difference being the locus of the switching point between Region 1 and Region 2. If the two pumps are operated in Region 1 at their respective switching point, the flow rate of the bottom product is Qb for the continuous pump while it is $2Qb/(1-b)$ for the semicontinuous. This second value must be divided by two to obtain the average flow rate during an entire cycle. Thus the semicontinuous pump has a flow rate of solute-free solution which is $1/(1-b)$ times that of the continuous pump. For large values of b, the advantage of the semicontinuous pump could be significant [20, 21].

Multicomponent separations [21, 22]. / Let us consider a mixture of s adsorbable species in an inert solvent. Suppose that the adsorption isotherm of each species is linear, which signifies that each species adsorbs independently of the others. Thus, the separation of the mixture can be treated as the superposition of the separations of s binary systems, each of them including one species i and the common inert solvent, and being characterized by its own b_i value and associated penetration distances L_{H_i} and L_{C_i}. If we suppose an operating φ_B value such that:

$$b_1 > b_2 \ldots b_k \geqq \varphi_B > b_{k+1} > \ldots > b_s \tag{29}$$

for the continuous pump, or

$$\frac{2b_1}{1-b_1} > \frac{2b_2}{1-b_2} \cdots \frac{2b_k}{1-b_k} \geqq \varphi_B > \frac{2b_{k+1}}{1-b_{k+1}} > \ldots > \frac{2b_s}{1-b_s} \tag{30}$$

for the semicontinuous pump, and

$$L_{C_i} = \frac{u_0(1+\varphi_B)}{(1+m_0)(1+b_i)} \frac{\pi}{\omega} < h, \quad i = 1, \ldots, k \tag{31}$$

then the operation is in Region 1 for only components 1 to k. Accordingly the concentrations of these components will decrease continuously in the bottom product. At steady state, the components 1 to k will appear only in the top product, whereas the remaining components k+1 to s will appear in both top and bottom products. Thus a split can be made which isolates species k+1 to s in the bottom product. On the other hand, it is not possible to isolate species 1 to k in either of the two products. If φ_B is chosen so that $k = s$, complete removal of all solutes from the bottom product is possible.

For a mixture of two solutes, 1 and 2, in a solvent, the pumps must be operated in Region 1 if the goal is to purify the solvent or to concentrate both the two solutes. If the goal is to isolate solute 2, the pumps must be operated in Region 2. It is not possible to obtain solute 1 alone in the solvent. The origin of this problem is the linearity of the isotherms. With nonlinear isotherms, the solutes have to compete for adsorption, and thus if a solute is concentrated in a point, the others must disappear from this point. In case of linear isotherms, the problem can be overcome by using a more sophisticated pump with three temperature levels [23].

Nonlinear Equilibria

Nonlinear adsorption occurs when solutes compete for sites (ion exchange, concentrated nonionic mixtures). We first study the separation of a single independent solute which occurs, for example, with binary ion exchange (only one ion is independent because of the electroneutrality requirement).

The elegant model described here has been proposed by Camero and Sweed [24] and uses the nonlinear equilibrium theory of chromatography [25]. The assumptions are the same as for the linear case. With nonlinear isotherms, two different types of waves occur: dispersive waves when a low concentration displaces a higher concentration and shock waves in the inverse case. For shock waves a finite mass balance must be used in place of the differential mass balance Equation 9 which still hold true for dispersive waves. Dimensionless concentrations are more convenient with nonlinearly sorbing solutes, thus Equation 9 can be rewritten as:

$$u\frac{\partial Y}{\partial z} + \frac{\partial Y}{\partial t} + v\frac{\partial X}{\partial t} = 0 \tag{32}$$

where the adsorption capacity ratio v is given by:

$$v = \frac{1-\varepsilon}{\varepsilon} \cdot \frac{x_r}{y_r} \tag{33}$$

x_r and y_r are reference concentrations, for example, in ion exchange, total exchange capacity of the resin and total concentration in the solution, respectively. Suppose the isotherms have the general form of Equation 1 and are of the favorable type such as given by Equation 4 with $K(T) > 1$. The solution of Equations 32 and 1 at a given temperature is that Y is constant along characteristic curves given by:

$$\left(\frac{dz}{dt}\right)_{Y,T} = \frac{u}{1 + v\left.\dfrac{\partial g}{\partial Y}\right|_{Y,T}} \tag{34}$$

which replaces Equation 14 of the linear model. This equation shows that each concentration has its own velocity and thus a dispersive wave is represented in the (z, t) plane by a fan of characteristic lines. The concentration change, Y_C to Y_H or inversely, associated to a temperature change in the direct mode is related by:

$$Y_C + vg(Y_C, T_C) = Y_H + vg(Y_H, T_H) \tag{35}$$

The solution can be obtained in implicit form as:

$$Y_H = f(Y_C) \tag{36}$$

which replaces Equation 19 of the linear model.

Camero and Sweed [24] showed how the separation builds up from cycle to cycle by constructing the characteristic pattern in the (z, t) plane, but no expressions were discovered for the reservoir concentration transients.

At cyclic steady state, for batch pumps, they showed that there exists four types of wave patterns in the (z, t) plane (Figure 12) corresponding to four types of separation (complete or not in each reservoir). Whatever the type may be, only one wave exists in the bed. It is dispersive on upflow because the bottom reservoir concentration displaces a higher concentration in the column and it is compressive on downflow. In addition, the compression is more effective than the dispersion because the solute is more retained at the cold downflow temperature. Therefore the compressive wave ends in a shock wave before the end of the downflow half-cycle.

Criteria have been developed which unambiguously determine which type of separation will result for any initial, equilibrium and operating conditions. The criteria involves the distance δ of the shock wave from the top of the bed at the end of downflow, and the penetration distance L_H of the concentration Y_1 of the leading edge of the dispersive wave

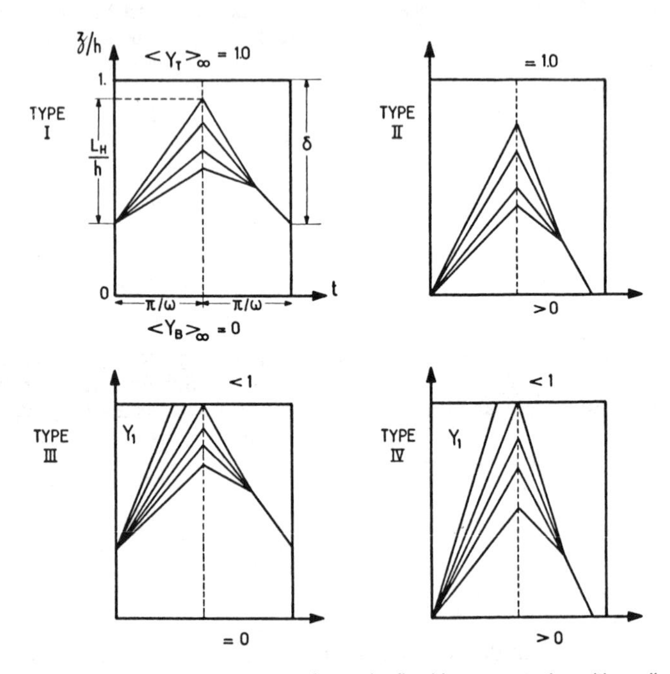

Figure 12. Cyclic steady-state wave patterns for total reflux binary separation with nonlinear equilibria.

($Y_1 = 1$ for Type I and II). The key of the separation type is given by the total amount of solute in the pump, which is known from the initial state. If there is not enough solute to fill the top reservoir with only this solute, then Type III ($L_H/h > \delta$; $\delta < 1$) or Type IV ($L_H/h > 1$; $\delta > 1$) occurs. Type II ($L_H/h < 1 < \delta$) occurs when there is too much solute for filling the top reservoir and the column with the solute, thus the remaining solute must appear in the bottom reservoir. In other cases, the separation is of Type I ($L_H/h < \delta < 1$). Then the separation is perfect, that is, in ion exchange, for example, only the more preferred ion appears in the top reservoir and only the other ion in the bottom reservoir.

A complete fractionation of a multicomponent mixture can be obtained with nonlinear coupled equilibria. Then, at cyclic steady state, the concentration profile in the bed involves $s-1$ waves which are dispersive on upflow and compressive on downflow (Figure 13) and determines s zones, in each of which, only one component appears. The arrangement of the components from the top to the bottom of the bed is in the decreasing order of their affinity, that is, of their selectivity coefficients K, provided that this order does not change in the temperature interval. The other requirements and the detailed criteria can be found in the original paper [24].

Extensions to partial reflux separations have been made using one column with feed at the top or at the bottom. With feed at the top, the column behaves as a stripping column and only separations of Types III and IV are possible. With feed at the bottom, the column is like an enriching column with only Types II and IV possible. A complete fractionation device can be obtained with two columns and feed in a reservoir between them.

Recuperative Mode

The local equilibrium model has been applied to the recuperative mode in batch pumps by Sweed and Rigaudeau [26] and in open pumps by Wankat [13]. Whith respect to the

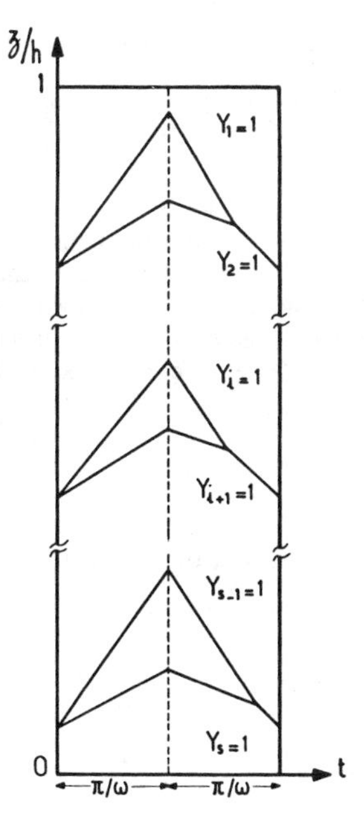

Figure 13. Cyclic steady-state wave pattern for complete separation of an s-component mixture.

direct mode, the assumption that the bed is isothermal must be replaced by the following assumptions:

- The rate of heat transfer is high enough so that solid and fluid are locally in thermal equilibrium throughout the cycle.
- The column is adiabatic.
- The heat of adsorption is negligible.

In addition to the solute mass balance Equation 9 an energy balance must be written as:

$$u\frac{\partial T}{\partial z} + [1+\mu]\frac{\partial T}{\partial t} = 0 \tag{37}$$

where

$$\mu = \frac{1-\varepsilon}{\varepsilon}\frac{\varrho_s}{\varrho_l}\frac{C_s}{C_l} \tag{38}$$

is the heat capacity ratio of the bed. The solution of Equation 37 can be expressed as the velocity of the thermal wave:

$$v_T = \frac{u}{1+\mu} \tag{39}$$

which is analogous to the velocity of the concentration wave Equation 14.

When a thermal wave propagates through the column, which occurs at each half-cycle, a shift in equilibrium occurs at the boundary of the wave. The relation between the concentrations y_C and y_H on both sides of the thermal wave is obtained by writing a finite mass balance expressing that the velocity of the companion concentration wave is equal to the velocity of the thermal wave. In case of linear isotherms, we obtain:

$$\frac{y_C}{y_H} = \frac{m(T_H)-\mu}{m(T_C)-\mu} = \frac{1-b-(1+m_0)(1+\mu)}{1+b-(1+m_0)(1+\mu)} \tag{40}$$

which replaces Equation 19 of the direct mode.

In the (z, t) plane, the characteristics representing the thermal waves have slopes given by Equation 39. Those representing the concentrations waves have slopes given by Equation 14 where m(T) is that of the local temperature. As in the direct mode, the solution could be obtained by a geometrical construction, but this would be tedious in some cases.

As an example, let us consider a batch pump and the common case of liquid-solid phases where in general

$$v_T > v_H > v_C \tag{41}$$

where v_H and v_C are the concentration wave velocity defined by Equation 14, that is:

$$v_H = \frac{u_0}{1+m(T_H)} \tag{42}$$

$$v_C = \frac{u_0}{1+m(T_C)} \tag{43}$$

Let us define the thermal penetration distance L_T as the distance the thermal wave moves during a half-cycle of duration π/ω. From Equation 39, we have:

$$L_T = \frac{u_0}{1+\mu}\frac{\pi}{\omega} \tag{44}$$

If $L_T < h$, the thermal wave does not break through the column end, and in that case, Equation 41 implies that the concentration waves do not intercept the thermal wave. Therefore concentrations never undergo a temperature change and the result is no separation. This holds true for nonlinear isotherms also. Note that the condition $L_T < h$ is equivalent to $\eta < 1+\mu$ where η is the number of column void volume displaced per half-cycle. Thus, if the solid and the fluid have about the same heat capacity, there will be no separation if $\eta < 2$. This explains some experimental results which will be reported further.

If $L_T > h$, thermal breakthrough occurs at each half-cycle. Sweed and Rigandeau [26] give a solution for the bottom reservoir transient with nodead volume in the case where the concentration waves does not travel through the entire bed during a half-cycle (analog of operation in Region 1 of the direct mode). This result can be put in the form:

$$\frac{\langle y_B \rangle_n}{y_0} = \left[\frac{h}{L_T} + \left(1 - \frac{h}{L_T}\right)\frac{m(T_H)-\mu}{m(T_C)-\mu}\right]^n \tag{45}$$

which shows that the bottom reservoir concentration decreases continuously in a way similar to the direct mode.

The higher v_T, the closer the recuperative mode approaches the direct mode. When $v_T \rightarrow \infty$, it is identical to the direct mode. For open pumps having a zero-bottom product concentration at cyclic steady state, the top product concentration is given by a global mass balance, and therefore the separation is the same for the two modes. In this case mixing or unmixing of the reservoirs has no influence on the steady-state results (but it will affect the transient solutions).

An interesting form of Equation 40 is the following:

$$\frac{y_C}{y_H} = \frac{v_H^{-1} - v_T^{-1}}{v_C^{-1} - v_T^{-1}} \tag{46}$$

Wankat [13] used this equation to show that many situations can occur depending on the relative magnitude of v_T with respect to v_H and v_C. In addition to the case above ($v_T > v_H > v_C$), we could have $v_H > v_C > v_T$, for example, for gas-solid systems. From Equation 46 we conclude that $y_C > y_H$. Thus fluid going from cold to hot has its concentration decreased, which is opposite our intuition. If $v_H > v_T > v_C$, the linear theory is no longer valid. The nonlinear theory shows that solute accumulates along the thermal wave. This suggests the possibly promising idea of adjusting the thermal wave velocity by a proper device which seems to be more adapted to cycling zone adsorption.

Wankat [13] has investigated the three prior situations of relative velocities in connection with Equation 46 for a semicontinuous parametric pump with mixed or unmixed reservoirs and linear or nonlinear isotherms.

In conclusion, recuperative mode parametric pumping can achieve large separations in both batch and open operation. It uses also less energy than the direct mode [13].

Other Models

The other models studied to represent parametric pumping separations consider dispersion effects, such as diffusion, axial dispersion, and finite rates for heat and mass transfers.

In fact, all the models which can be developed for conventional adsorption or chromatography (see for example the review of Rodrigues [27] can be adapted to parametric pumping with proper boundary conditions [28]. The simplest models assume thermal equilibrium and neglect axial dispersion. The equations are then a differential mass balance such as Equation 32, an expression for the interphase mass transfer rate which can be written generaly as:

$$\frac{\partial X}{\partial t} = R(X, Y, T) \tag{47}$$

and an equilibrium function (Equation 1). This model was first used by Wilhelm and co-workers [3, 29] to simulate the effects on separation of several parameters (cycle time, fluid displacement, dead volume, phase angle, mass transfer coefficient). An efficient method for solving the set of equations was proposed [29]. The so called STOP–GO algorithm uses two steps: flow without transfer (GO) and transfer without flow (STOP). The time for mass transfer between phases is equal to the time displacement step.

Such dispersive models have also been studied by Rice and co-workers [30–35] using frequency analysis. They investigated mainly the ultimate separation (when $n \rightarrow \infty$) in batch direct mode parapumps. All real batch systems reach an ultimate separation since all systems have dispersive forces acting. Since there is no separation for zero and infinite frequency of the cycles, one would expect the existence of maxima and/or minima in separation factor as frequency increases. Therefore the authors used a frequency analysis by introducing sinusoidal velocity and temperature fields, and obtained analytical expressions for several types of transfer rates and equilibria. Simulation results for gas-solid and liquid-solid systems show that a maximum separation factor occurs as expected. Also, at high frequency, a slight reverse separation occurs. Some coincidences between the results of this approach and of the staged equilibrium model were pointed out by Rice and Foo [36].

PARAMETRIC PUMPING SEPARATIONS

In this section, we will describe typical experimental results and discuss them in connection with the models just presented. Use of temperature, pH, ionic strength, and electric potential as thermodynamic variables will be treated successively. All these results have been obtained at the laboratory scale. The problem of scale-up has been envisaged by Stokes and Chen [37] who proposed a tubular heat exchanger type of device to perform direct-mode thermal parametric pumping. On the example of toluene-heptane separation, they showed that the energy consumption is of the same order as for distillation.

It should be mentioned that the "Sirotherm" ion exchange process [38] in a fixed bed used industrially for water desalination is a form of thermal recuperative mode parametric pumping. This process uses thermally regenerable resins containing both cationic and anionic sites within the same bead-like particles developed by CSIRO in Australia. The two operating temperatures are ambient during loading and about 90° C during regeneration. The latter step is accomplished by refluxing a portion of the desalinated water thus avoiding the need for any chemical regenerant. Such thermally regenerable ion exchange resins have been developed also by other manufacturers [39].

Use of Temperature

Liquid-Solid

Total reflux operation. / In total reflux (or batch) parametric pumping there is neither feed nor product streams as in batch distillation. The volume of liquid and the amount of solutes remain constant throughout a run and equal to their initial values. The separation changes continuously from cycle to cycle and tends toward a constant value (even infinite) at cyclic steady state.

Figure 1A showed a schematic diagram of an experimental apparatus for thermal direct mode batch parametric pumping. The equipment consists of a jacketed column packed with the sorbent particles as in ordinary liquid chromatography. The top and the bottom of the column are connected with a top and a bottom reservoir, respectively. The upflow and downflow percolations use an infusion-withdrawal syringe pump. The syringes are also the reservoirs. Other systems of percolation more adapted to large scale operation will be shown later. The jacket is connected to a cold bath and a hot bath by means of solenoid valves. The system is under command of a timer so that upflow occurs when the jacket is connected to one of the baths and downflow when it is connected to the other bath (usually upflow is preferred during the hot half-cycle to prevent dispersion due to natural convection).

Batch parametric pumping has been used to separate ionic or nonionic species on ion exchange resins or adsorbents. Most of these experiments have been carried out with the direct mode. The first three papers using this mode appeared almost simultaneously and were concerned with separation of aliphatic-aromatic mixtures. Wilhelm et al. [2, 3] use silica gel adsorbent to separate a toluene-n-heptane mixture. Their results were shown in Figure 3. The toluene concentration, the adsorbable component, increases in the top reservoir and decreases in the bottom reservoir, as predicted by the temperature effect on equilibrium. The volume of solution displaced per half-cycle is about the intersticial volume of the bed and therefore a concentration wave entering the bed at the beginning of a half-cycle cannot exit at the end of this half-cycle. This corresponds to Region 1 of the local linear equilibrium model which predicts an exponential decrease of the bottom reservoir concentration. This is almost equivalent to an exponential increase of the separation factor and this is observed on Figure 4. However the results depend on cycle time which is not predicted by the equilibrium model. A good fit of the results was obtained using the STOP–GO method to solve the equations of a model with film control mass transfer rate and nonlinear equilibria [29]. However, the authors had to adjust not only the mass transfer coefficient which depends on relative solid-fluid velocity and thus on cycle time, but also a dead volume parameter which should be constant. Despite this, the nonequilibrium model appears more realistic than the local equilibrium model, principally because it predicts finite separation. The effect of phase angle between changes of column temperature and flow direction has been also investigated experimentally: any deviation from synchronization decreases the final separation factor. The third paper [8] was concerned with the benzene-n-hexane separation on the same adsorbent. The observed separation was considerably less than that obtained by Wilhelm and co-workers because the fluid was displaced very quickly through the bed and therefore there was no time to allow sufficient mass transfer between fluid and solid. The make-up and the propagation of concentration waves are essential to obtaining large separations in packed-bed parametric pumping, which are not obtained here. A good fitting of the experimental results were obtained with the equilibrium-staged model using a few stages. If fast cycles are used, departures from equilibrium can occur not only for mass transfer but also for heat transfer, leading to a phase reversal inside the bed and resulting in a reverse separation [31].

A number of parametric pumping experiments employed ion exchange resins as the solid phase. Sweed and Gregory [40] separated NaCl from water using an ion retardation resin in which both cationic and anionic sites are mixed in the same particle. Thus the resin acts like an adsorbent for NaCl. The isotherms are roughly linear and show that the salt is more retained at cold than at hot temperature. At constant cycle time, the separation decreases as the volume percolated per half-cycle increases. However the fluid velocity increases at once and it is this effect rather than the volume effect which decreases the separation. In fact, the volume displaced is always less than one void volume, and thus the separation should be unaffected if equilibrium is obtained.

A mixture of cations can be separated using cation exchange resins. Such resins contain a limited number of exchanging sites and the ions have to compete for them. Therefore ion exchange equilibria are nonlinear. For example, with the system Ag^+/Cu^{++} with NO_3^- as the anion on a strong acidic resin, the Ag^+ ion is preferred relative to the Cu^{++} ion [41]. This preference is greater at 20° C than at 60° C, and thus a separation can be expected by thermal parametric pumping. Figure 14 shows experimental results where the Ag^+ ion is enriched in the top reservoir and depleted in the bottom. The reverse occurs for the Cu^{++} ion due to electroneutrality requirements. The variation of the total concentration will be discussed further. At constant fluid velocity, it has been shown experimentally that the separation is perfect when the volume displaced per half-cycle is below a critical value. These results agree with the local nonlinear equilibrium theory (Figure 12, Type I). Some dispersion can be taken in account using a graphical approach which combines breakthrough curves obtained experimentally for both the compressive and the dispersive waves [41].

The variation of total concentration shown is Figure 14 was first observed by Butts et al. [42] while separating H^+ and K^+ ions. This was an unexpected result, because ion exchange involves equal fluxes of charges between the resin and the solution. The interpretation was based on the fact that resin swells and absorbs water when temperature increases. But it was shown later that the effect depends on the cations involved and on their relative amount in the pump [43]. On Figure 14, the total concentration increases in the top reservoir; it would increase in the bottom reservoir if the pump contained more copper than silver. These results have been interpreted as a parametric pumping effect on the Donnan partition of electrolytes (cation and anion) between the pores of the resin and the bulk solution. Thus ion exchange parametric pumping results in general in two effects: separation of the two ions and separation of the electrolyte from water.

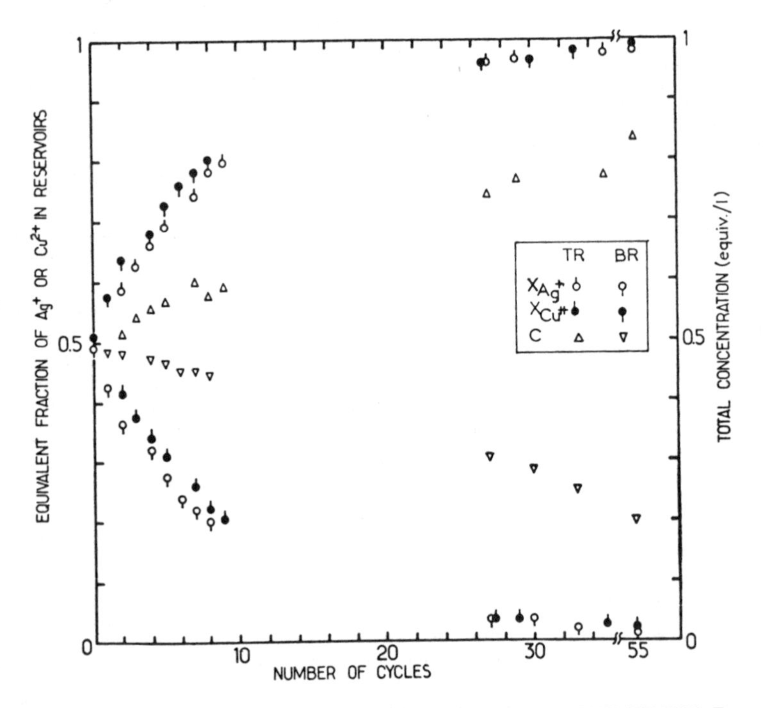

Figure 14. Ag^+ / Cu^{++} total reflux separation on the cation exchange resin DUOLITE C265. T_c = 20°C ; T_H = 60°C ; bed volume = 95 ml ; $Q\pi/\omega$ = 40 ml ; cycle time = 55 min [41].

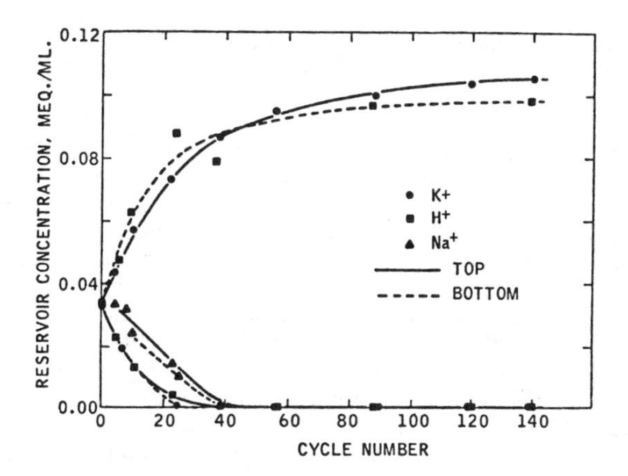

Figure 15. Total reflux separation of a ternary mixture $K^+/Na^+/H^+$ on a cation exchange resin. The ion of intermediate affinity (Na^+) disappears from both reservoirs and thus accumulates in the bed [42].

Multicomponent separations in batch operation have received little attention. Figure 15 shows the separation of a ternary mixture of K^+, Na^+, and H^+ chlorides on a cation exchange resin [42]. We can observe that the K^+ ion builds up in the top reservoir and the H^+ ion in the bottom one. The Na^+ disappears from both reservoirs, and therefore accumulates inside the bed. Thus the ions arrange themselves from the top to the bottom of the pump in the order of the selectivity $K^+ > Na^+ > H^+$, as predicted by the local nonlinear equilibrium model (Figure 13).

Other batch direct mode parametric pumping arrangements have been tried for desalting water. Rice et al. [44] used a dual-column or "back to back" system to study the limit separation at cyclic steady state as a function of cycle time and interpreted their results with a nonequilibrium model taking into account the temperature response of the beds. Goto et al. [45, 46] used a staged system where a thermally regenerable resin is contained in a basket and alternatively immersed in a cold and a hot reservoir.

Partial reflux operation. / A number of open parametric pumping systems with one or two columns have been suggested. They are continuous or semicontinuous depending on the fact that the feed and the product withdrawals occur during the entire cycle or only during part of it. We will examine successively the direct mode separation of one solute in a solvent, the separation of multicomponent mixtures, and recuperative mode operation.

Direct mode: single solute: Chen and co-workers have studied extensively both the continuous [19] and the semicontinuous [20] top feed pumps just shown (Figure 10) using the system toluene-n-heptane/silica gel which exhibits a b value of about 0.22 with temperatures of 4° C and 70° C. The experimental results were in reasonable agreement with the theoretical predictions of the local equilibrium model. In particular they observed the exponential decrease of bottom product concentration in Region 1 (Equation 25). The different behavior of the two types of pumps was also demonstrated by operating them at the same φ_B value such that

$$b < \varphi_B < \frac{2b}{1-b}$$

The bottom product concentration decreases exponentially for the semicontinuous pump which operates in Region 1, whereas the separation is limited and low for the continuous pump (Region 2) [20].

Costa et al. [18] studied the same type of pumps but with a different experimental set-up which avoids the use of syringes and infusion-withdrawal pumps (Figure 14). Feed is added directly into the top reservoir and products are withdrawn from the reservoirs by means of prepositioned suction tubes and pumps. The control of the half-cycle end is made using a photoelectric cell to detect the liquid at the bottom of each reservoir. The problem of column degassing during the hot half-cycle has been solved by means of a capillary which ensures upflow under pressure. Phenol was separated from water using a polymeric adsorbent. This system exhibits a large temperature dependence which leads to a value of the b parameter of about 0.35 for $T_C = 20°$ C and $T_H = 60°$ C. Figure 17 shows experimental results for the continuous version of the pump. When $\varphi_B = 0.72$, the pump is operated in Region 2 and the separation increases toward a flat maximum before decreasing to its cyclic steady-state value. This maximum is predicted by the local equilibrium theory and is due to low values of L_1 and L_2. Then it takes some cycles for solute to move from the top to the bottom and to destroy the transient separation already obtained. When $\varphi_B = 0.145$ the operation is in Region 1 and the bottom product concentration should decrease exponentially. The discrepancy between theory and experiment is explained by the dispersion of the waves, not considered by the theory, which is a determining factor when penetration distances are low (referring to Figure 9, constant concentrations (A) and (B) are not obtained). When L_1 and L_2 are greater, the experimental results agree with the theory as shown on Figure 18 for the semi-continuous version of the pump operated in Region 1.

Similar but not identical pumps with feed at the top of the column were studied by Gregory and Sweed [47]. The cycles are made of four or five steps and thus are symmetric or nonsymmetric. The difference with Chen's arrangements is that some steps are with feed only or with batch reflux only. The theoretical analysis with the local linear equilibrium model leads to results similar to those of Chen. Experimental results were obtained using

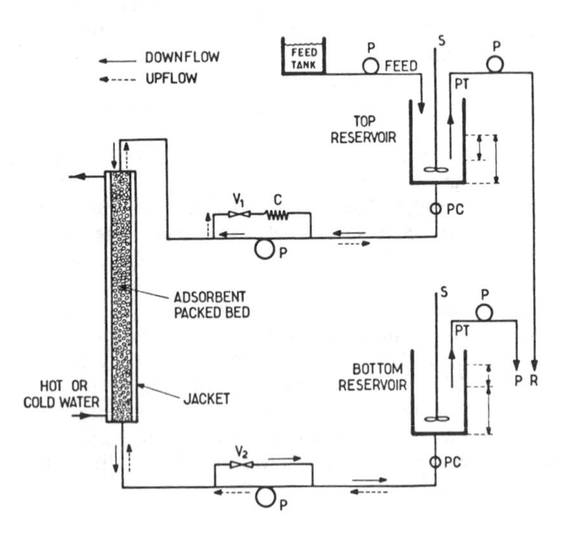

Figure 16. Experimental set-up for thermal direct mode parametric pumping at partial reflux with feed at the top.

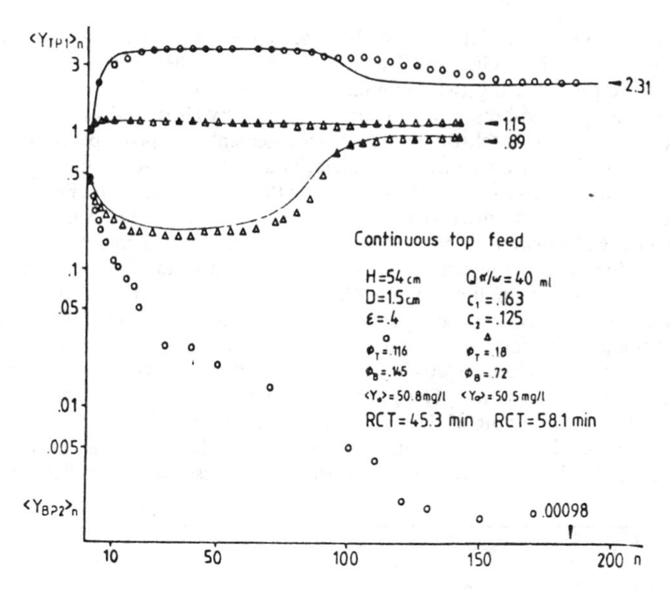

Figure 17. Removal of phenol from water using an adsorbent resin. Experimental results and local linear equilibrium model simulations [18].

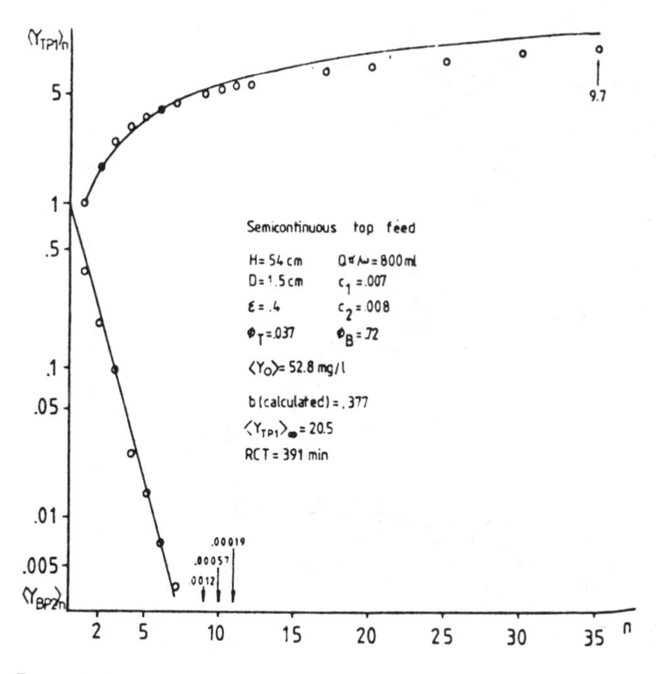

Figure 18. Removal of phenol from water using an adsorbent resin. (semicontinuous top feed; Region 1) [18].

the system NaCl-H_2O-ion retardation resin, but the separations were less than predicted by the theory [48]. The STOP-GO model was used, and then the data fitted very well.

Some parametric pumps using two columns have been proposed. A parametric pump system with feed into a central reservoir was first discussed by Horn and Lin [49]. Rice and Foo [50, 51] extended their batch dual-column arrangement to open operation for continuous desalination of water on a bifunction resin. During a half-cycle, one column is hot while the other is cold and they are inversely so during the second half-cycle. Feed is directed toward the cold column and enriched product is withdrawn from the hot column. The depleted product is withdrawn as a part of the flow between the two columns. At constant enriched product flow rate, the separation decreases linearly with the depleted product flow rate as predicted by a nonequilibrium model. Grévillot et al. [52] used a two-column parapump for separating silver and copper nitrate on a cation exchanger. The separation of two ions is not a multicomponent separation because only one ion behaves independently due to the requirement of electroneutrality. The feed point is between the two columns, which are in phase for temperature. The product enriched in silver (depleted in copper) is withdrawn from the top reservoir while the product enriched in copper (depleted in silver) is withdrawn from the bottom reservoir. Thus the two columns act as enriching and stripping sections. Results show that the separation increases with the reflux ratio. The

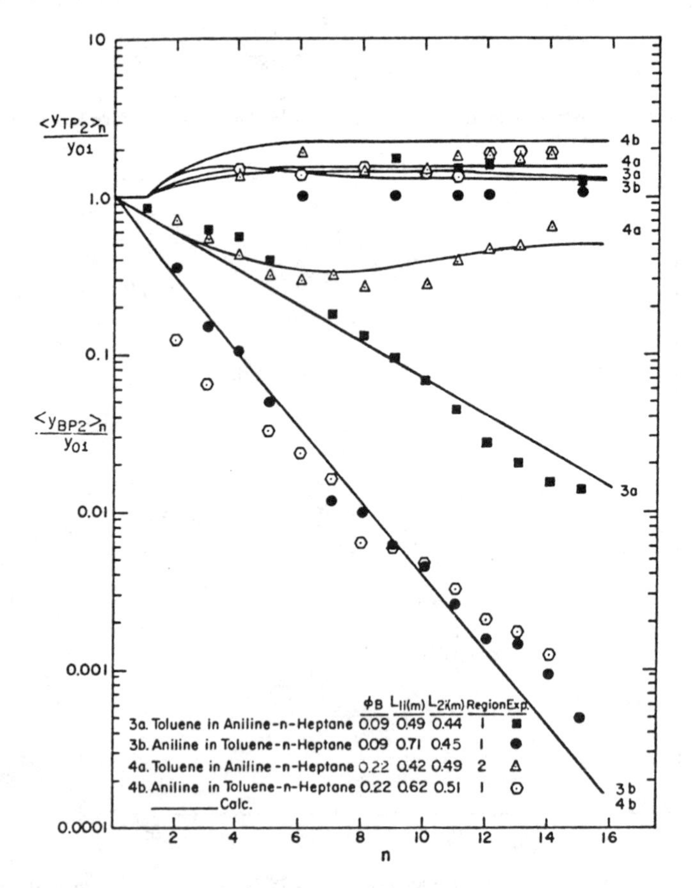

Figure 19. Multicomponent separations with a continuous top feed pump [22].

position of the feed point, that is the relative amount of resin in the two sections, has a little effect on separation in the range studied.

Direct mode: multicomponent mixtures: Such separations have been performed by continuous parametric pumping with top feed in order to test the local linear equilibrium theory [22]. The multicomponent mixture is then considered as pairs of pseudo-binary systems solute-solvent, the superposition of the behavior of each pair giving the complete behavior. Figure 19 shows the results of two runs for the separation of aniline-toluene with n-heptane as the solvent (the adsorbent is silica gel). For run 3, the value of $\varphi_B = 0.09$ is lower than both the b parameters of the two solutes, $b_a = 0.31$ (aniline) and $b_t = 0.15$ (toluene). Therefore the operation is in Region 1 for the two solutes, which can be removed completely from the bottom product as $n \rightarrow \infty$. For run 4, $\varphi_B = 0.22$ is now higher than b_t and therefore the operation switches to Region 2 for toluene which will appear in the bottom product stream even when $n \rightarrow \infty$. In this case the net movement of concentration waves is toward the top for aniline while it is toward the bottom for toluene as illustrated on Figure 20. The separation of glucose from fructose in water on a cation exchanger in calcium form was investigated by Chen and D'Emidio [53]. But because the sorption of glucose is insensitive to temperature (b = 0), the system is rather a one-solute system. When operating the pump in Region 1 for fructose, the bottom product stream contains only glucose in water. Gupta and Sweed [54] have studied theoretically a two-column arrangement with intermediate feed to separate two linearly sorbing solutes. A complete split between the two solutes can be obtained under equilibrium conditions but is only partial when conditions are not at equilibrium. The process needs pure solvent during a part of the cycle and thus is rather a combination of classical adsorption and parametric pumping.

We see that parametric pumps with top feed, which are better than bottom-feed pumps, cannot perform a complete fractionation between even two linearly sorbing solutes: one

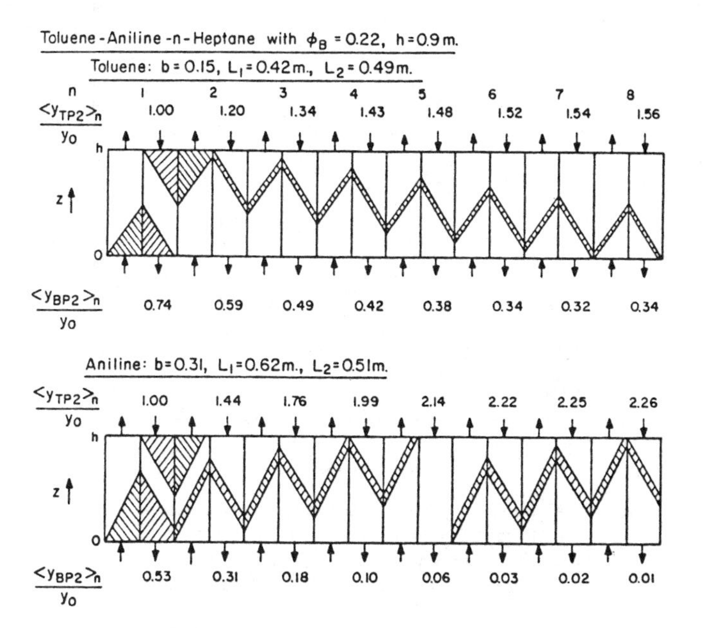

Figure 20. Net movement of concentration waves with aniline and toluene operated in Regions 1 and 2, respectively [22].

solute can be isolated in the bottom product stream but all the solutes always appear in the top product stream. To overcome this problem, a continuous parapump using three temperature levels has been proposed recently by Wu-Xian Zhi and Wankat [23]. The basic cycle comprises feed at the top and downflow percolation at the intermediate temperature T_I. During upflow, the bottom of the column is maintained at T_c in order to keep adsorbed solute A, while the top is maintained at T_H in order to desorb solute B, which is concentrated in the top product stream. Experimental results (Figure 21) show that, after transient, concentrations are above that of the feed, in the top product for pyrene (A), and in the bottom product for acenaphtylene (B) (the common solvent is 2-propanol). Modifying the separation is possible by changing the timing of the cycle as shown after cycle 60. The curves shown are calculated from a staged equilibrium model with 45 stages for pyrene and 27 stages for acenaphtylene. This large difference implies considerably more dispersion for the second solute.

Recuperative mode: Even though the original demonstration of parametric pumping was in the recuperative mode of operation, this mode has received little further attention (when temperature is used as the thermodynamic variable). This is probably due to the low separations obtained in the first experiments and also to the difficulty in preventing heat losses at the column wall at the laboratory scale where small diameters are used. The first experiments reported [1, 3, 28] were for completely open systems, that is, without reflux. Hot feed is introduced at one end of the column during a half-cycle and cold feed is introduced

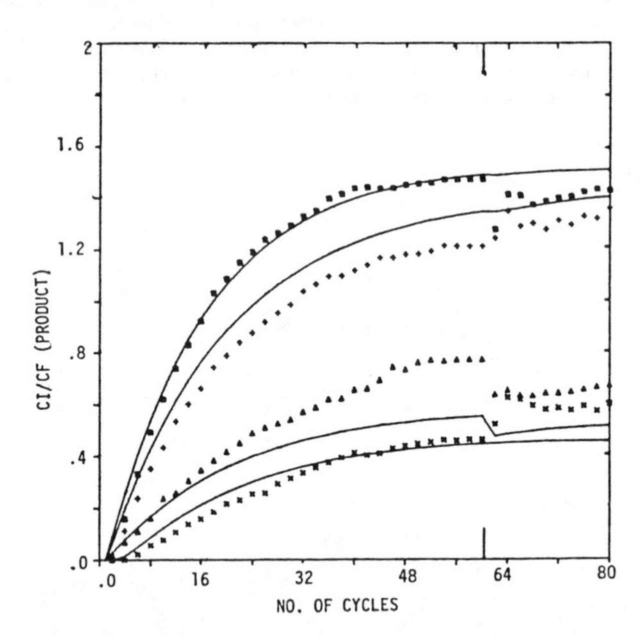

Figure 21. Continuous multicomponent separation using three temperature levels (38°, 100°, and 150°F). Pyrene top (■) and bottom (X) product concentrations. Acenaphtylene top (▲) and bottom (+) product concentrations [23].

at the other end during the second half-cycle. Separation factors below 1.5 were obtained with the mixture NaCl-H_2O on a mixed bed of ion exchange resins. A sophisticated model [28] including heat and mass transfer rates and dispersion was used to simulate successfully the temperature and product concentration profiles during separation runs. It was theoretically shown later [26] (for batch operation) that one of the keys to obtain large separations is to force thermal breakthrough to occur by flowing more than one or two bed void volumes. This was not the case in the first experiments. Also it is evident that the lack of reflux diminishes the relative contribution of cycle-to-cycle accumulation to the overall process. If these two points are taken into consideration, then large separations can be obtained. This is shown on Figure 22 for the system phenol-water-adsorbent resin [55]. The experimental set-up is the same as in Figure 16 except that the cold and hot baths are used with heat exchangers to give the proper temperature to the fluids entering the top and the bottom of the column, respectively, and are not connected to the jacket. The number of bed-void volume percolated is 30 for upflow and 45 for downflow and the reflux ratios are 3 and 2 at the top and the bottom, respectively. A very low phenol concentration is obtained in the bottom product as shown on the figure. Such large separations were predicted by Wankat [13] who showed also that the recuperative mode uses substantially less energy than the direct mode (on the particular example of the separation toluene-n-heptane on silica gel, the direct mode consumes about 15 times more energy). In addition, for large column diameters, the recuperative mode is more efficient to approach thermal equilibrium and therefore to approach equilibrium separations, than the direct mode where radial temperature gradients could be important.

Gas-Solid

Considerably less work has been done on thermal parametric pumping for separations of gases than for liquids. Separations of ethane-propane, argon-propane, and propylene-propane [56, 57] have been studied with activated carbon as adsorbent in a total reflux apparatus consisting of a jacketed column equipped with piston pumps. During the constant volume operation, the gas mixture flows rapidly, with the major portion of the cycle time being used for heating and cooling the column. Although a temperature change of 55° C

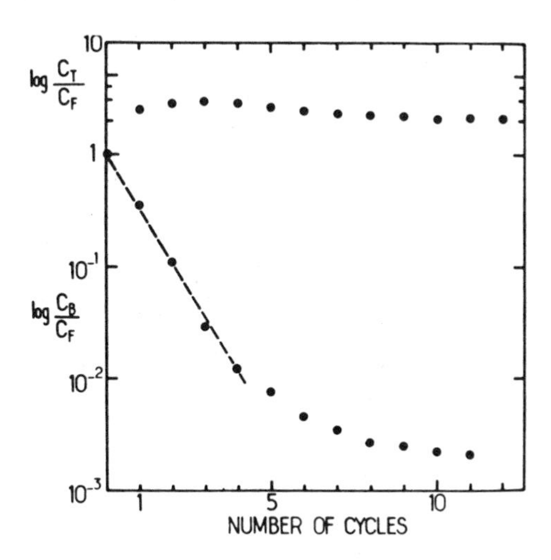

Figure 22. Removal of phenol from water by partial reflux recuperative mode parametric pumping. C_B, C_T and C_F are the phenol concentrations of bottom product, top product and feed, respectively [55].

was imposed on the jacket, its recorded value inside the bed was less than 8° C due to very low gas thermal conductivity. However good separations were obtained for the two first mixtures in which propane is the adsorbed component (temperature has no effect on the adsorption of the propane-propylene mixture). The data were well fitted by a modified STOP-GO model including axial mixing of gases. In such a constant-volume system two contradictory effects seem to occur: an increase in temperature has the effect of desorbing the gas but at the same time causes an increase in pressure, which could have the inverse effect. A constant-pressure batch direct mode apparatus was used by Patrick et al. [58] to separate air from CO_2 on silica gel. High separation factors of the order of 130 were obtained using very large temperature changes. At constant pressure, the change in temperature causes a change in volume, which in turn causes flow of the gas, mixing, and decrease in separation.

A partial reflux process was recently proposed to separate hydrogen isotopes employing vanadium monohydride as adsorbent [59]. The so-called "temperature swing process" (Figure 23) uses two-jacketed columns acting symmetrically in the same arrangement as pressure swing adsorption (PSA) processes. It can be recognized as thermal direct mode parametric pumping in the sense that for each column the temperature change is synchronized with the inversion of gas flow and that reflux is used (at the product end). The gas for the reflux for one column is a part of the product of the other column and inversely, thus avoiding the use of reservoirs which would be difficult to devise with gas at nearly constant pressure (the slight difference $P_H - P_L = 70$ kPa is only to provide for pressure losses across metering and check valves). Effects of cycle time (10 to 50 minutes), purge-to-feed ratio, and temperature difference on the removal of traces of tritium from hydrogen were investigated. Increasing one of the three parameters causes a decrease of the tritium content of the product stream. A local equilibrium model taking into account the variation of the gas velocity during steps 2 and 4 shows the existence of a critical purge-to-feed ratio (and therefore a critical product end reflux ratio) below which complete removal of tritium is not possible.

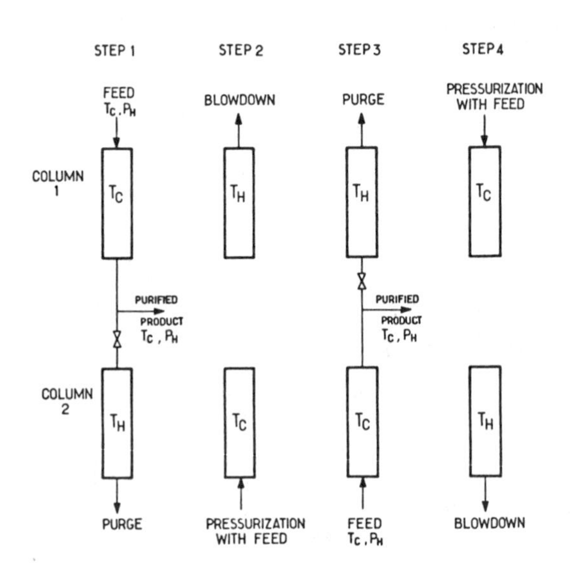

Figure 23. Temperature swing process for hydrogen isotope separation.

For separation of gases, thermal parametric pumping appears less promising than the PSA processes, because it requires comparatively long cycle times, which are due to slow heat transfer.

Liquid-Liquid

Because parametric pumping utilizes solute partition between two phases, it can be used not only in fluid-solid systems but also in fluid-fluid systems. The advantage over conventional extraction and absorption processes is avoiding the solvent recovery section. However, one major difficulty is to hold one liquid phase stationary while the other liquid phase is moved back and forth past the stationary phase.

The first investigations were made by Wankat [60] for the separation of acetic acid from water using diethyl ether as the stationary phase. Two different batch experimental set-ups were used: a horizontal helix with 35 coils contained in a Graham-type condenser where continuous flow of the mobile phase can occur, and a discrete-staged system made of test tubes where the mobile phase is transferred from tube to tube as discrete fractions. Although the separations obtained were very low, they demonstrated the feasibility of liquid-liquid thermal parametric pumping. Two equilibrium-staged models were presented one with discrete transfer and discrete equilibrium steps and the other with continuous flow of the mobile phase through the cascade. Simulations showed that large separations can be achieved. A problem encountered in experimentation was loss of both phases due to evaporation.

Rachez et al. [14] used a batch contacting device adapted from Craig's extractor, which makes the stagewise operation more convenient (Figure 24). It consists of a cascade of Craig's tubes each equipped symmetrically with two transfer reservoirs (classical Craig's tubes involve only one such reservoir and allow transfer in one direction only). The transfer operations are done manually by rotation of the apparatus, which allows transfer of each

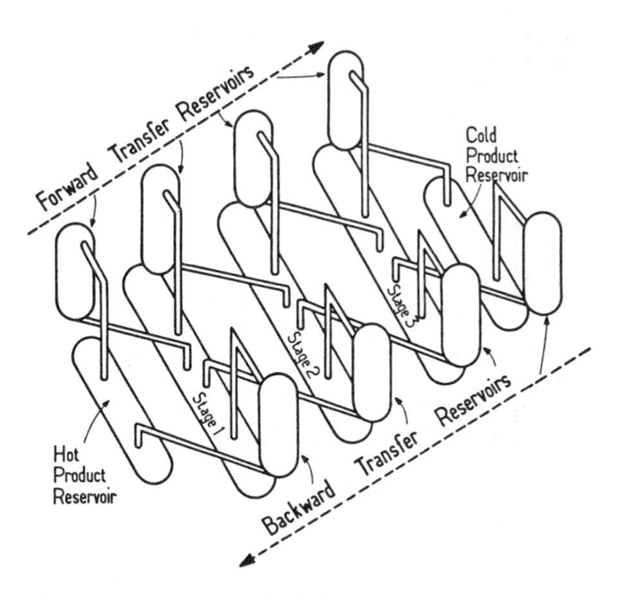

Figure 24. Perspective view of a contacting device for liquid-liquid parametric pumping.

mobile-phase fraction from one stage to a transfer reservoir and finally to the next stage. The whole system is immersed alternately in a hot and a cold bath. The flowsheet is that of Figure 4 extended to five cells (the contactor used in experiments had five stages). The equilibrium-staged model was investigated in detail by introducing a matrix formalism and studying the properties of eigenvalues and eigenvectors. Experimental results for the separation of phenol from toluene using water as stationary phase are reported on Figure 25 with model simulations. A maximum separation factor of 2.4 is obtained at the end of the experiment (n = 21). At that time a net transfer of toluene toward the first stages was observed which was attributed to dilatation of water and a slight dissymmetry of the tubes.

Another way to do liquid-liquid parametric pumping is by impregnating the stationary liquid phase on a porous solid support. The system could then be operated in a column in the same fashion as for liquid-solid pumps. No experiments in this direction have been reported to date.

Use of pH or Ionic Strength

It is known that pH influences the adsorption of some substances on particular adsorbents. This phenomenon is used to separate, for example, proteins on ion exchange resins by first binding the proteins to the resin at a given pH and second eluting them selectively by means of a pH gradient. Therefore pH can be used in parametric pumping as the thermodynamic variable which displaces the adsorption equilibrium. The energy for the separation is thus chemical. Direct mode operation would be difficult to devise because the substrate would be contained in a dialysis tube surrounded by a pH-cycled solution. The recuperative mode is easier to operate and a typical apparatus is shown in Figure 26. Constant pH of the solution entering the column is obtained by using a dialysis cell between each end of the column and the corresponding reservoir.

A number of experiments have been performed with batch and open pumps to separate ions [61], enzymes and proteins. Shaffer and Hamrin [62] remove trypsin from water using a specific adsorbent and show an exponential decrease of the enzyme concentration in one reservoir. This occurs also when the solution contains a second enzyme, α-chymotrypsin. Chen and co-workers have extensively studied the hemoglobin-albumin separation using ion-exchange resins. The two pH levels are choosen so that they frame the isoelectric point of the protein which is to be separated. Thus the protein will bear a negative charge at high pH and positive charge at low pH. It will be taken up by an anion exchanger at high pH

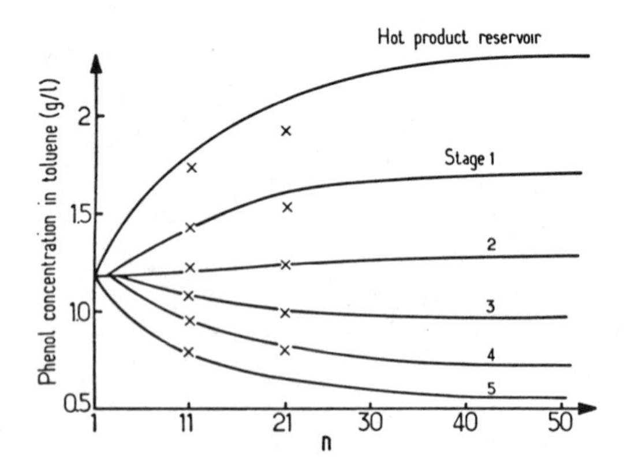

Figure 25. Experimental results (X) and model simulations (–) for batch liquid-liquid parametric pumping (solute: phenol; mobile phase: toluene; stationary phase: water; T_C = 20°C; T_H = 60°C).

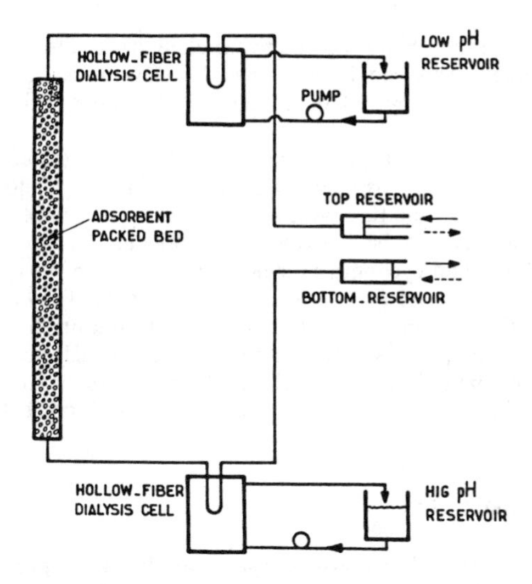

Figure 26. Experimental set-up for pH recuperative mode parametric pumping.

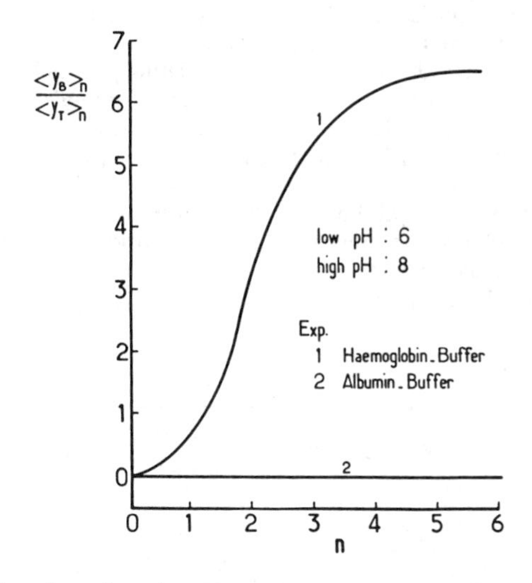

Figure 27. Separation of proteins using pH parametric pumping (after Chen et al. [63]).

and released at low pH (the reverse will occur if a cation exchanger is used). Figure 27 illustrates this effect: in Run 1, the isoelectric point of the solute lies between the two pH levels and the solute accumulates in the bottom reservoir. Whereas, in Run 2, the isoelectric point of albumin is out of the pH interval, and consequently, the parametric pumping has no effect. [63].

Several devices have been developed for protein separations using either one column with feed alternately introduced at the top and the bottom [64] or two columns with central feed [65, 66] or several columns alternately packed with anionic and cationic resins [63, 67]. The equilibrium-staged model adapted to recuperative mode was used in most cases to explain graphically how the separation builds up and to simulate the results. In some cases, a reverse separation was observed. This can be explained by slow mass transfer rates and small displacement volumes per half-cycle [26], though no definitive explanation has been discovered.

Simultaneous changes of pH and ionic strength can be used to separate proteins which have nearby isoelectric points. This effect was used by Chen et al. [68] to purify alkaline phosphatase from a complex mixture with a semicontinuous pump.

The protein separations can be improved by applying an electric field accross the chromatographic column and cycling it in synchronization with the cyclic variation of pH and flow direction [69]. The electric field is applied in the axial direction and acts on the migration velocity of the proteines in the liquid phase and not on the equilibrium partition curve between phases. The process is thus a combination of pH parametric pumping and electromigration rather than electrochemical parametric pumping which is discussed later. Two continuous versions of the process have been tried for the hemoglobin-albumin separation.

Use of Electric Potential

The use of electrical potential as a possible cyclic variable was suggested in the first report on parametric pumping but was only recently demonstrated experimentally by Oren and Soffer [70–72] for water desalting. The mechanism of the electroadsorption seems to be the retention of the ions in the electrical double layer which can be considered as a capacitor. Therefore Faradaic reactions are not necessary and are even undesirable. The electrochemical column comprises two electrodes separated by an isolating porous layer. Each electrode consists of a bed (1 mm thick, 10 mm wide, 96 cm long) of granular graphitized carbon having a specific surface area of 100 m^2/g. The column is connected to two reservoirs as in other parametric pumps. The basic cycle comprises four steps:

1. Charge, during which the potential difference between the two electrodes is increased to a predetermined value (choosen to prevent water decomposition) and induces adsorption of salt at the surface of the carbon.
2. Upflow percolation of the solution.

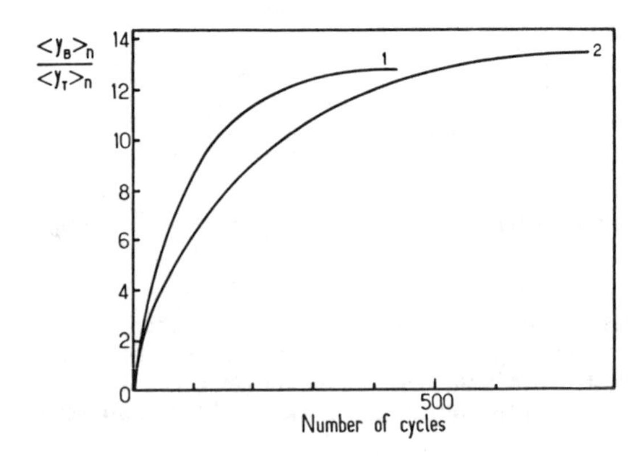

Figure 28. Concentration of NaCl in water by electrochemical parametric pumping (two electrodes ; each electrode is made of 3.1 g of granular graphitized carbon black ; potentials: 0.36 v and 1.28 v ; $<y_T>_n$ = constant = 0.01 N ; cycle time = 8 min ; number of column void volume displaced par half-cycle: 0.5(1) and 0.125(2) (after Oren and Soffer [70]).

3. Discharge—the potential difference is decreased and causes a back transfer of salt into the solution.
4. Downflow percolation.

Figure 28 shows the increase in the bottom reservoir concentration for quasibatch runs (the top reservoir concentration is kept constant). Experimental isopotentiograms, that is salt partition curves between the two phases at constant potential difference between the electrodes, show that the amount of adsorbed salt increases with the potential difference as expected. However it remains small (below 10 μmole/g) in the range of potentials applicable in aqueous solution which explains the considerable number of cycles required to obtain a significant separation. The adsorption capacity can be increased by using activated carbons having a larger surface but the internal porosity of the material increases at once, and this seems to lead to increasing charging and discharging times. Further research on electrochemical parametric pumping would include studies on other ionic and nonionic mixtures and other adsorbents. Also the device could probably be improved by controlling the real driving force which is the potential difference at the solid-liquid interface of each electrode instead of controlling the difference between the two electrodes.

RELATED PROCESSES

Pressure Swing Adsorption

PSA processes are cyclic processes developed during the 1960 s for the separation of gaseous mixtures. As opposed to cryogenic systems and to systems using thermal regeneration, they are also called "heatless fractionation processes" [73]. The more rapid response of a gas-solid system to pressure changes permits shorter cycle times in PSA and hence greater throughput. Additional advantages include low energy requirements and low capital investment costs.

The basic cycle of a PSA process using two columns for continuous production consists of four distinct steps as shown on Figure 29. The feed mixture is separated into two streams:

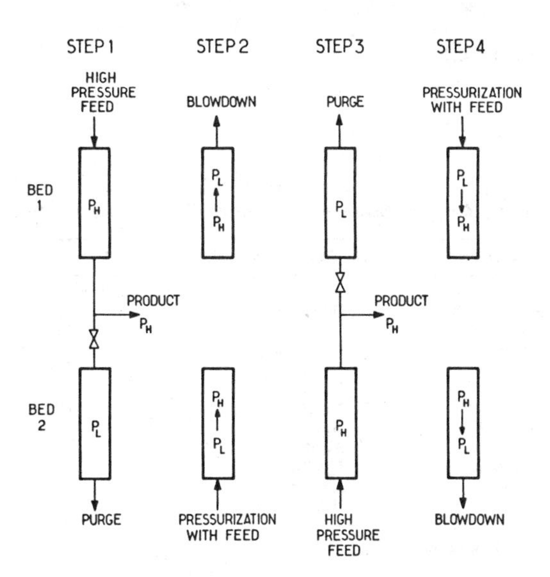

Figure 29. Steps in a pressure swing adsorption cycle.

the purge, containing more of the strongly adsorbed components, and the product, containing the purified less adsorbed component. In the first step, high pressure feed flows through bed 1 while a portion of the purified product flows out at high pressure P_H. The remainder of the product is throttled down to a lower pressure P_L and is passed through bed 2 to purge previously sorbed species. In step 2, bed 2 is repressurized to feed pressure while, in bed 1, the pressure is reduced to P_L (blow-down). Steps 3 and 4 are analogous to steps 1 and 2, respectively, the roles of the two beds being reversed. The similarity with parametric pumping can be shown by considering the operation of one column:

1. The change of a thermodynamic variable, the pressure, is synchronized with the change of the direction of the gas flow.
2. A generalized reflux occurs at the product end of each column as a part of the high pressure product supplied countercurrently at the low pressure.

Although PSA processes have been widely used for many years, few theoretical analyses of their operation have been reported in the literature. An analysis based on a local linear equilibrium model similar to that proposed for parametric pumping has been developed by Shendalman and Mitchell [74]. Isothermal system and ideal compressible gas are assumed. In addition, the changes of pressure (steps 2 and 4) are assumed to be instantaneous. Using the concept of penetration distances, they defined several regions of operation. In one of them, the concentration of the adsorbable species in the product stream decreases exponentially with the number of cycles, as for Region 1 of operation in parametric pumping. Thus complete removal of this species from the carrier gas is possible. The conditions for that have been established as:

$$L_H = \frac{v_H}{1 + k\dfrac{1-\varepsilon}{\varepsilon}} \frac{\pi}{\omega} < h$$

and

$$\gamma > \left(\frac{P_H}{P_L}\right) \varepsilon/[\varepsilon + k(1-\varepsilon)]$$

Table 2
Some Applications of Pressure Swing Adsorption

Adsorbent	Gas Mixture–Operation	Reference
		(73 and 77 are review papers)
Alumine	Oil vapors/air	73
	Air drying	73
Molecular sieves	Air drying	73
	N_2/O_2	73, 77–79
	H_2/CH_4	73
	N_2/CH_4	80
Silica gel	H_e/CO_2	74, 81
	Air drying	73, 82
Activated carbon	H_2/CH_4	73, 78
	Air/CH_4	73
	N_2/ethylene	78
Vanadium hydrides	T_2/H_2	83
Palladium deposited on alumina	D_2/H_2	84

where L_H = the penetration distance of an adsorption wave at the high pressure P_H
k = the partition coefficient
v_H = the intersticial velocity of the gas associated to the volumetric feed flow rate
γ = the ratio of the purge volumetric flow rate to the feed volumetric flow rate.

The results of the equilibrium theory were compared to experimental results for the removal of CO_2 from helium on silica-gel. Although the agreement was poor, the theory does provide a qualitative description of how a PSA process operates. The model was later extended and completed to include the case where the carrier gas is adsorbed [75]. The equilibrium theory was also used to propose two three-column processes for fractionating multicomponent mixtures [76].

PSA processes are widely used industrially in air drying, enrichment of oxygen from air and hydrogen purification. Table 2 presents some examples of applications which have been experimented at least at laboratory scale. The low pressure is usually the atmospheric pressure although some processes use lower pressures, that is purge under vacuum (mainly-one column systems). The high pressure is usually of a few atmospheres but can be up to 50 atmospheres. The processes are usually worked at ambient temperature. A large number of processes have been described (about 150 references exist, mainly in patent form) that differ by the number of columns, the number of steps in the cycle, the phase shift of the columns, the relative direction of flows in the columns in each step, and the type of coupling of the columns. Industrial PSA systems employing up to 20 columns and complex cycles have been described. It seems that there is an increase of interest for pressure swing adsorption in the recent years.

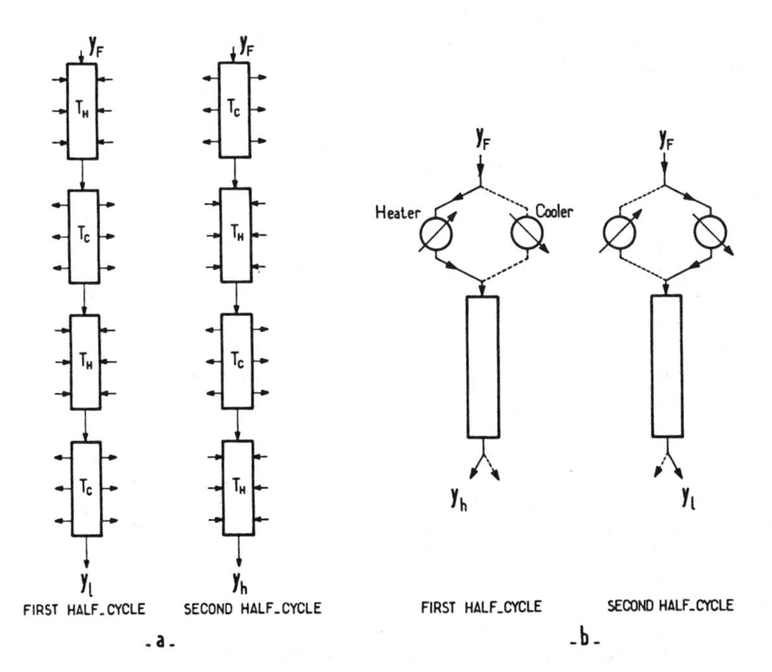

Figure 30. Cycling zone adsorption: (A) direct mode, multiple zones; (B) traveling wave mode, single zone.

Table 3
Cycling Zone Separations

Adsorbent or Stationary Phase	Mixture	Variable	Mode	Reference
Liquid-solid				
Activated carbon	Acetic acid-water	T	DM, TWM	85, 86
Mixed bed of ion exchange resins, or composite resins	NaCl-water	T	DM	95, 96
Polyvinylpyrrolidone resin	Anthracene-pyrene acenaphtylene-2-propanol	T	DM	97
DBAE–celluloze	Glucose-fructose-water	pH, T	DM, TWM	98
		pH	TWM	99
C-18 bonded-phase HPLC adsorbent	Dipeptides-water	IS	TWM	100
Glassy carbon	β-naphtol-water	EP	DM	101, 102
Gas-solid				
Activated carbon	H_e-CH_4	T	DM	85
	H_2-CH_4	T	DM	103
Molecular sieves	N_2-O_2	T	DM	104
	n-pentane-isopentane	T	TWM	91
Liquid-liquid				
Toluene	Diethylamine-water	T	DM	92
2-butanol	Phenol red-bromocresol	IS	TWM	105
	green-water		TWM	94, 106

T = *temperature;* IS = *ionic strength;* EP = *electric potential;* DM = *direct mode;* TWM = *traveling wave mode.*

Cycling Zone Separations

In 1969, Pigford et al. [85, 86] presented a new technique called "cycling zone adsorption" which has been further extended to liquid-liquid systems thus justifying the name "cycling zone separations." As in parametric pumping an intensive thermodynamic variable is cycled to command variations of the distribution of solutes between two phases. But contrarily to parametric pumping, there is no flow reversal; the process needs only one flow direction through a column or a series of columns.

Figure 30 shows the two modes of operation, using temperature as the cyclic variable. In direct mode, the temperature of each bed (or zone) is switched periodically from a cold temperature T_C to a hot temperature T_H. In addition, adjacent columns are out of phase with each other. Considering the first bed, solute is retained on the adsorbent during the cold half-cycle and a depleted solute solution exits from the bed, while during the hot half-cycle desorption occurs and a solution more concentrated than the feed flows out. This elementary separation can be amplified in the second bed, then in the third..., by a proper timing of the temperature changes. It has been shown [86, 87] that three types of separation can occur depending on flow rate, half-cycle duration, and slopes of isotherms:

1. The separation factor is independent of the zone number and is constant at a value of 1.
2. The separation factor increases regularly with the zone number.
3. The separation factor *vs.* zone number exhibits an oscillatory behavior.

For the most interesting case (Case 2), the cyclic steady state ($n \rightarrow \infty$) separation factor increases exponentially with the number of zones in a similar way as the transient separation factor increases with the number of cycles in parametric pumping. The traveling wave mode of operation is analogous to the recuperative mode of parametric pumping: the columns are adiabatic and the fluid entering each column is alternately heated or cooled in a heat exchanger. Depending on the relative velocities of the thermal wave v_T and of the solute concentration waves v_H and v_C, high separation factors can be obtained.

As for parametric pumping, two main types of models have been developed, local equilibrium models [86–91] and stage equilibrium models [92–94], with linear and nonlinear isotherms. Reviews have been made by Wankat [6, 7, 94]. Table 3 shows some example of cycling zone separations with most recent references.

NOTATION

A	cross sectional area of the colum (m^2)
b	dimensionless equilibrium parameter
C	heat capacity (J Kg^{-1}°C^{-1})
D	column diameter (m)
h	bed length (m)
k	partition coefficient, molar volumetric concentration in solid phase/molar volumetric concentration in gas phase
K(T)	equilibrium constant (nonlinear isotherms)
L	penetration distance (m)
m(T)	modified equilibrium constant (dimensionless)
M	number of transfer per half-cycle
M(T)	equilibrium constant (linear isotherms)
n	number of cycles
N	number of stages
P	pressure
Q	reservoir displacement rate (m^3s^{-1})
t	time (s)
T	temperature (°C)
u	intersticial fluid velocity (ms^{-1})
u_0	$Q/(A\varepsilon)$ ($m\ s^{-1}$)
V_B, V_T	dead volume in bottom and top reservoirs, respectively (m^3)
V_f	volume of a mobile-phase fraction (m^3)
V_s	volume of the stationary phase in a cell (m^3)
x	solute concentration in stationary phase (mole m^{-3})
X	solute concentration in stationary phase (dimensionless)
y	solute concentration in mobile phase (mole m^{-3})
Y	solute concentration in mobile phase (dimensionless)
z	axial distance in sorbent bed (z = 0 at the bottom) (m)

Greek Symbols

ε	void fraction in packing
μ	heat capacity ratio
ν	mass capacity ratio in the bed
ϱ	density
ϱ_c	mass capacity ratio in a cell

Subscripts

O	initial or feed condition
B	bottom
C	cold
e	enriching section
f	fluid
H	hot (temperature), or high (pressure)
L	low pressure
r	reference concentration
s	solid, or stripping section
T	top or thermal
$<\ >_n$	value at the end of cycle n

REFERENCES

1. Wilhelm, R. H., Rice, A. W., and Bendelius, A. R., "Parametric Pumping: A Dynamic Principle for Separating Fluid Mixtures," *Ind. Eng. Chem. Fundam.*, Vol. 5, No. 1 (1966), pp. 141–144.
2. Wilhelm, R. H., and Sweed, N. H., "Parametric Pumping: Separation of Mixture of Toluene and n-Heptane," *Science*, Vol. 159 (February 1968), pp. 522–524.
3. Wilhelm, R. H., et al., "Parametric Pumping: A Dynamic Principle for Separating Fluid Mixtures," *Ind. Eng. Chem. Fundam.*, Vol. 7, No. 3 (1968), pp. 337–349.
4. Rice, R. G., "Progress in Parametric Pumping," *Separation and Purification Methods*, Vol. 5, No. 1 (1976), pp. 139–176.
5. Chen, H. T., "Parametric Pumping," in *Handbook of Separation Techniques for Chemical Engineers*, P. A. Schweitzer (Ed.), New York: McGraw-Hill, 1979, Section 1–15, pp. 467–486.

6. Wankat, P. C., "Cyclic Separation Processes," *Separation Science,* Vol. 9, No. 2 (1974), pp. 85–116.
7. Wankat, P. C., "Cyclic Separation Techniques," in *Percolation Processes, Theory and Applications,* A. E. Rodrigues and D. Tondeur (Eds.), Alphen aan den Rijn (The Netherlands): Sijthoff and Noordhoff, 1981, pp. 443–515.
8. Wakao, N., et al., "Adsorption Separation of Liquid by Means of Parametric Pumping," *Kagaku Kogaku,* Vol. 32 (1968), pp. 169–175 (in Japanese).
9. Grévillot, G., and Tondeur, D., "Equilibrium Staged Parametric Pumping. I. Single Transfer Step per Half-Cycle and Total Reflux. The Analogy with Distillation," *AIChE Journal,* Vol. 22, No. 6 (1976), pp. 1055–1063.
10. Grévillot, G., "Equilibrium Staged Parametric Pumping: Part III—Open Systems at Steady-state. McCabe-Thiele Diagram," *AIChE Journal,* Vol. 26, No. 1 (1980), pp. 120–131.
11. Grévillot, G., and Tondeur, D., "Equilibrium Staged Parametric Pumping. II. Multiple Transfer Steps per Half-cycle and Reservoir Staging," *AIChE Journal,* Vol. 23, No. 6 (1977), pp. 840–851.
12. Thompson, D. W., and Bowen, B. D., "Equilibrium Theory of the Parametric Pump. Effect of Boundary Conditions," *Ind. Eng. Chem. Fundam.,* Vol. 11, No. 3 (1972), pp. 415–417.
13. Wankat, P. C., "Continuous Recuperative Mode Parametric Pumping," *Chem. Eng. Sci.,* Vol. 33 (1978), pp. 723–733.
14. Rachez, D., et al., "Stagewise Liquid-Liquid Extraction Parametric Pumping. Equilibrium Analysis and Experiments.," *Sep. Sci. Technol.,* Vol. 17, No. 4 (1982), pp. 589–619.
15. Pigford, R. L., Baker, B., and Blum, D. E., "An Equilibrium Theory of the Parametric Pump," *Ind. Eng. Chem. Fundam.,* Vol. 8, No. 1 (1969), pp. 144–149.
16. Chen, H. T., and Hill, F. B., "Characteristics of Batch, Semicontinuous, and Continuous Equilibrium Parametric Pumps," *Sep. Sci.,* Vol. 6, No. 3 (1971), pp. 411–434.
17. Aris, R., *Ind. Eng. Chem. Fundam.,* Vol. 8 (1969), pp. 603–604.
18. Costa, C. A. V., et al., "Purification of Phenolic Wastewater by Parametric Pumping: Nonmixed Dead Volume Equilibrium Model," *AIChE Journal,* Vol. 28, No. 1 (1982), pp. 73–85.
19. Chen, H. T., et al., "Separations via Continuous Parametric Pumping," *AIChE Journal,* Vol. 18, No. 2 (1972), pp. 356–361.
20. Chen, H. T., et al., "Separations via Semicontinuous Parametric Pumping," *AIChE Journal,* Vol. 19, No. 3 (1973), pp. 589–595.
21. Chen, H. T., and Manganaro, J. A., "Optimal Performance of Equilibrium Parametric Pumps," *AIChE Journal,* Vol. 20, No. 5 (1974), pp. 1020–1022.
22. Chen, H. T., et al., "Separation of Multicomponent Mixtures via Thermal Parametric Pumping," *AIChE Journal,* Vol. 20, No. 2 (1974), pp. 306–310.
23. Wu-Xian-Zhi, and Wankat, P. C., "Continuous Multicomponent Parametric Pumping," *Ind. Eng. Chem. Fundam.,* Vol. 22, No. 2 (1983), pp. 172–176.
24. Camero, A. A., and Sweed, N. H., "Separation of Nonlinearly Sorbing Solutes by Parametric Pumping," *AIChE Journal,* Vol. 22, No. 2 (1976), pp. 369–376.
25. Rhee, H. K., "Equilibrium Theory of Multicomponent Chromatography," in *Percolation Processes, Theory and Applications,* A. E. Rodrigues and D. Tondeur (Eds.), Alphen an den Rijn (NL): Sijthoff and Noordhoff, 1981, pp. 285–328.
26. Sweed, N. H., and Rigaudeau, J., "Equilibrium Theory and Scale-up of Parametric Pumps," *AIChE Symposium Series,* Vol. 71, No. 152 (1975), pp. 1–5.
27. Rodrigues, A. E., "Modeling of Percolation Processes," in *Percolation Processes, Theory and Applications,* A. E. Rodrigues and D. Tondeur (Eds.), Alphen an den Rijn: Sijthoff and Noordhoff, 1981, pp. 31–82.

28. Rolke, R. W., and Wilhelm, R. H., "Recuperative Parametric Pumping," *Ind. Eng. Chem. Fundam.*, Vol. 8, No. 2 (1969), pp. 235–245.
29. Sweed, N. H., and Wilhelm, R. H., "Parametric Pumping: Separations via Direct Thermal Mode," *Ind. Eng. Chem. Fundam.*, Vol. 8, No. 2 (1969), pp. 221–231.
30. Rice, R. G., "Dispersion and Ultimate Separation in the Parametric Pump," *Ind. Eng. Chem. Fundam.*, Vol. 12, No. 4 (1973), pp. 406–412.
31. Rice, R. G., and Foo, S. C., "Thermal Diffusion Effects and Optimum Frequencies in Parametric Pumps," *Ind. Eng. Chem. Fundam.*, Vol. 13, No. 4 (1974), pp. 396–398.
32. Rice, R. G., "Transport Resistances Influencing the Estimation of Optimum Frequencies in Parametric Pumps," *Ind. Eng. Chem. Fundam.*, Vol. 14, No. 3 (1975), pp. 202–208.
33. Rice, R. G., "The Effect of Purely Sinusoidal Potentials on the Performance of Equilibrium Parapumps," *Ind. Eng. Chem. Fundam.*, Vol. 14, No. 4 (1975), pp. 362–365.
34. Foo, S. C., and Rice, R. G., "On the Prediction of Ultimate Separations in Parametric Pumps," *AIChE Journal*, Vol. 21, No. 6 (1975), pp. 1149–1158.
35. Foo, S. C., and Rice, R. G., "Steady State Predictions for Nonequilibrium Parametric Pumps," *AIChE Journal*, Vol. 23, No. 1 (1977), pp. 120–123.
36. Rice, R. G., and Foo, S. C., *AIChE Journal*, Vol. 25, No. 4 (1979), pp. 734–735.
37. Stokes, J. D., and Chen, H. T., "Design and Scale-up of a Continuous Thermal Parametric Pumping System," *Ind. Eng. Chem. Fundam.*, Vol. 18, No. 1 (1979), pp. 147–154.
38. Bolto, B. A., "Sirotherm Desalination," *Chem. Tech.* (May 1975), pp. 303–307.
39. Ackerman, G. R., et al., "Industrial Deionization with Amberlite XD2, a Thermally Regenerable Ion Exchange Resin," *AIChE Symp. Ser.*, Vol. 73, No. 166 (1976), pp. 107–111.
40. Sweed, N. H., and Gregory, R. A., "Parametric Pumping: Modeling Direct Thermal Separations of Sodium-Chloride-Water in Open and Closed Systems," *AIChE Journal*, Vol. 17, No. 1 (1971), pp. 171–176.
41. Grévillot, G., Dodds, J. A., and Marques, S., "Separation of Silver-Copper Mixtures by Ion-Exchange Parametric Pumping. I. Total Reflux Separations," *J. Chromatography*, 201 (1980), pp. 329–342.
42. Butts, T. J., Sweed, N. H., and Camero, A. A., "Batch Fractionation of Ionic Mixtures by Parametric Pumping," *Ind. Eng. Chem. Fundam.*, Vol. 12, No. 4 (1973), pp. 467–472.
43. Grévillot, G., Marques, S., and Tondeur, D., "Donnan Partition Parametric Pumping," *Reactive Polymers*, Vol. 2 (1984), pp. 71–77.
44. Rice, R. G., Foo, S. C., and Gough, G. G., "Limiting Separations in Parametric Pumps," *Ind. Eng. Chem. Fundam.*, Vol. 18, No. 2 (1979), pp. 117–123.
45. Goto, S., Sato, N., and Teshima, H., "Periodic Operation for Desalting Water with Thermally Regenerable Ion Exchange Resin," *Sep. Sci. Technol.*, Vol. 14, No. 3 (1979), pp. 209–217.
46. Matsuda, H., et al., "Periodic Operation for Desalination with Thermally Regenerable Ion Exchange Resin. Dynamic Studies," *Sep. Sci. Technol.*, Vol. 16, No. 1 (1981), pp. 31–41.
47. Gregory, R. A., and Sweed, N. H., "Parametric Pumping: Behavior of Open Systems. Part I: Analytical Solutions," *The Chem. Eng. J.*, Vol. 1 (1970), pp. 207–216.
48. Gregory, R. A., and Sweed, N. H., "Parametric Pumping: Behavior of Open Systems. Part II: Experiment and Computation," *The Chem. Eng. J.*, Vol. 4 (1972), pp. 139–148.
49. Horn, F. J. M., and Lin, C. H., "On Parametric Pumping in Linear Columns Under Conditions of Equilibrium and Nondispersive Flow," *Berichte der Bunsengesellschaft fur Physikalische Chemie*, Vol. 73, No. 6 (1969), pp. 575–580.

50. Rice, R. G., and Foo, S. C., "Continuous Desalination in a Periodically Operated Dual-Column Dual Temperature Process," *Proc. Fifth Australian, Conf. Chem. Eng.*, 1977, pp. 179–183.
51. Rice, R. G., and Foo, S. C., "Continuous Desalination Using Cyclic Mass Transfer on Bifunctional Resins," *Ind. Eng. Chem. Fundam.*, Vol. 20, No. 2 (1981), pp. 150–155.
52. Grévillot, G., Bailly, M., and Tondeur, D., "Thermofractionation," *International Chemical Engineering,* Vol. 22, No. 3 (1982), pp. 440–453.
53. Chen, H. T., and D'Emidio, V. J., "Separation of Isomers via Thermal Parametric Pumping," *AIChE Journal,* Vol. 21, No. 4 (1975), pp. 813–815.
54. Gupta, R., and Sweed, N. H., "Modeling of Nonequilibrium Effects in Parametric Pumping," *Ind. Eng. Chem. Fundam.*, Vol. 12, No. 3 (1973), pp. 335–341.
55. Almeida, F., et al., "Removal of Phenol from Wastewater by Recuperative Mode Parametric Pumping," in *Physicochemical Methods for Water and Wastewater Treatment,* L. Pawlowski (Ed.), Amsterdam: Elsevier, 1982, pp. 169–178.
56. Jenczewski, T. J., and Myers, A. L., "Parametric Pumping Separates Gas Phase Mixtures," *AIChE Journal,* Vol. 14, No. 3 (1968), p. 509.
57. Jenczewski, T. J., and Myers, A. L., "Separation of Gas Mixtures by Pulsed Adsorption," *Ind. Eng. Chem. Fundam.*, Vol. 9, No. 2 (1970), pp. 216–221.
58. Patrick, R. R., Schrodt, J. T., and Kermode, R. I., *Sep. Sci.*, Vol. 7 (1972), p. 331.
59. Hill, F. B., Wong, Y. W., and Chan, Y. N. I., "A Temperature Swing Process for Hydrogen Isotope Separation," *AIChE Journal,* Vol. 28, No. 1 (1982), pp. 1–6.
60. Wankat, P. C., "Liquid-Liquid Extraction Parametric Pumping," *Ind. Eng. Chem. Fundam.*, Vol. 12, No. 3 (1973), pp. 372–381.
61. Sabadell, J. E., and Sweed, N. H., "Parametric Pumping with pH," *Sep. Sci.*, Vol. 5, No. 3 (1970), pp. 171–181.
62. Shaffer, A. G., and Hamrin, C. E., "Enzyme Separation by Parametric Pumping," *AIChE Journal,* Vol. 21, No. 4 (1975), pp. 782–786.
63. Chen, H. T., et al., "Separations of Proteins via pH Parametric Pumping," *Sep. Sci. Technol.*, Vol. 15, No. 6 (1980), pp. 1377–1391.
64. Chen, H. T., Wong, Y. W., and Wu, S., "Continuous Fractionation of Protein Mixtures by pH Parametric Pumping," *AIChE Journal,* Vol. 25, No. 2 (1979), pp. 320–327.
65. Chen, H. T., et al., "Separation of Proteins via Semicontinuous pH Parametric Pumping," *AIChE Journal,* Vol. 23, No. 5 (1977), pp. 695–701.
66. Chen, H. T., et al., "Semicontinuous pH Parametric Pumping: Process Characteristics and Protein Separations," *Sep. Sci. Technol.*, Vol. 16, No. 1 (1981), pp. 43–61.
67. Chen, H. T., et al., "Separation of Proteins via Multicolumn pH Parametric Pumping," *AIChE Journal,* Vol. 26, No. 5 (1980), pp. 839–849.
68. Chen, H. T., Ahmed, Z. M., and Rollen, Victor, "Parametric Pumping with pH and Ionic Strength: Enzyme Purification," *Ind. Eng. Chem. Fundam.*, Vol. 20 (1981), pp. 171–174.
69. Hollein, H. C., et al., "Parametric Pumping with pH and Electric Field: Protein Separations," *Ind. Eng. Chem. Fundam.*, Vol. 21 (1982), pp. 205–214.
70. Oren, Y., and Soffer, A., "Electrochemical Parametric Pumping," *J. Electrochem. Soc.: Electrochemical Science and Technology,* Vol. 125, No. 6 (1978), pp. 869–875.
71. Oren, Y., and Soffer, A., "Water Desalting by Means of Electrochemical Parametric Pumping. I. The Equilibrium Properties of a Batch Unit Cell," *J. of Appl. Electrichem.*, Vol. 13 (1983), pp. 473–487.
72. Oren, Y., and Soffer, A., "Water Desalting by Means of Electrochemical Parametric Pumping. I. The Equilibrium Properties of a Batch Unit Cell," *J. of Appl. Electrochem.*, Vol. 13 (1983), pp. 489–505.
73. Skarstrom, C. W., "Heatless Fractionation of Gases over Solid Adsorbents," in *Recent Developments in Separation Science,* N. N. Li (Ed.), Cleveland: CRC Press, Vol. 2 (1972), pp. 95–106.

74. Shendalman, L. H., and Mitchell, J. E., "A Study of Heatless Adsorption in the Model System CO_2 in Helium: Part I," *Chem. Eng. Sci.,* Vol. 27 (1972), pp. 1449–1458.
75. Chan, Y. N. I., Hill, F. B., and Wong, Y. W., "Equilibrium Theory of a Pressure Swing Adsorption Process," *Chem. Eng. Sci.,* Vol. 36 (1981), pp. 243–251.
76. Nataraj, Shankar, and Wankat, P. C., "Multicomponent Pressure Swing Adsorption," *AIChE Symp. Ser.,* Vol. 78, No. 219 (1982), pp. 29–38.
77. Lee, Hanju, and Stahl, D. E., "Oxygen Rich Gas from Air by Pressure Swing Adsorption Process," *AIChE Symp. Ser.,* Vol. 69, No. 134 (1973), pp. 1–8.
78. Jones, R. L., and Keller, G. E., "Pressure Swing Parametric Pumping. A New Adsorption Process," *J. Sep. Proc. Technol.,* Vol. 2, No. 3 (1981), pp. 17–23.
79. Flores Fernandez, G., and Kenney, C. N., "Modelling of the Pressure Swing Air Separation Process," *Chem. Eng. Sci.,* Vol. 38, No. 6 (1983), pp. 827–834.
80. Turnock, P. H., and Kadlec, R. H., "Separation of Nitrogen and Methane via Periodic Adsorption," *AIChE Journal,* Vol. 17, No. 2 (1971), pp. 335–342.
81. Mitchell, J. E., and Shendalman, L. H., "Study of Heatless Adsorption in the Model System CO_2 in Helium: Part II," *AIChE Symp. Ser.,* Vol. 69, No. 134 (1973), pp. 25–32.
82. Chihara, K., and Suzuki, M., "Air Drying by Pressure Swing Adsorption," *J. Chem. Eng. Jap.,* Vol. 16, No. 4 (1983), pp. 293–299.
83. Wong, Y. W., and Hill, F. B., "Separation of Hydrogen Isotopes via Single Column Pressure Swing Adsorption," *Chem. Eng. Commun.,* Vol. 15 (1982), pp. 343–356.
84. Weaver, K., and Hamrin, C. E., Jr., "Separation of Hydrogen Isotopes by Heatless Adsorption," *Chem. Eng. Sci.,* Vol. 29 (1974), pp. 1873–1882.
85. Pigford, R. L., Baker, B., and Blum, D. E., "Cycling Zone Adsorption, A New Separation Process," *Ind. Eng. Chem. Fundam.,* Vol. 8, No. 4 (1969), pp. 848–851.
86. Baker, B., and Pigford, R. L., "Cycling Zone Adsorption: Quantitative Theory and Experimental Results," *Ind. Eng. Chem. Fundam.,* Vol. 10, No. 2 (1971), pp. 283–292.
87. Gupta, Ramesh, and Sweed, N. H., "Equilibrium Theory of Cycling Zone Adsorption," *Ind. Eng. Chem. Fundam.,* Vol. 10, No. 2 (1971), pp. 280–283.
88. Wankat, P. C., "Multicomponent Cycling Zone Separations," *Ind. Eng. Chem. Fundam.,* Vol. 14, No. 2 (1975), pp. 96–102.
89. Wankat, P. C., "Fractionation by Cycling Zone Adsorption," *Chemical Engineering Science,* Vol. 32 (1977), pp. 1283–1287.
90. Knaebel, K. S., "Multistage Cycling Zone Adsorption for Purification of Binary Mixtures," *AIChE Symp. Ser.,* Vol. 78, No. 219 (1982), pp. 128–135.
91. Jacob, P., and Tondeur, D., "Adsorption Non Isotherme de gaz en lit fixe III—Etude expérimentale des effets de guillotine et de focalisation. Separation n-pentane-isopentane sur tamis 5A," *The Chem. Eng. J.,* Vol. 26 (1983), pp. 143–156.
92. Wankat, P. C., "Cycling Zone Extraction," *Separation Science,* Vol. 8, No. 4 (1973), pp. 473–500.
93. Nelson, C. W., Silarski, D. F., and Wankat, P. C., "Continuous Flow Equilibrium Staged Model for Cycling zone Adsorption," *Ind. Eng. Chem. Fundam.,* Vol. 17, No. 1 (1978), pp. 32–38.
94. Wankat, P. C., Dore, J. C., and Nelson, W. C., "Cycling Zone Separations," *Separation and Purification Methods,* Vol. 4, No. 2 (1975), pp. 215–266.
95. Ginde, V. R., and Chu, C., *Desalination,* Vol. 10 (1972), p. 309.
96. Shih, T. T., and Pigford, R. L., "Removal of Salt from Water by Thermal Cycling of Ion Exchange Resins," in *Recent Developments in Separation Science,* N. N. Li (Ed.) Cleveland: CRC Press Inc., 1977, Vol. 3, Part A, pp. 129–150.
97. Foo, S. C., Bergsman, K. H., and Wankat, P. C., "Multicomponent Fractionation by Direct, Thermal Mode Cycling Zone Adsorption," *Ind. Eng. Chem. Fundam.,* Vol. 19, No. 1 (1980), pp. 86–93.
98. Busbice, M. E., and Wankat, P. C., "pH Cycling Zone Separation of Sugars. A Preparative Separation Technique for Counter-Current Distribution and Chromatography," *J. of Chromatography,* Vol. 114 (1975), pp. 369–381.

99. Dore, J. C., and Wankat, P. C., "Multicomponent Cycling Zone Adsorption," *Chemical Engineering Science*, Vol. 31 (1976), pp. 921–927.
100. Nelson, W. C., and Wankat, P. C., "Application of Cycling Zone Separation to Preparative High-Pressure Liquid Chromatography," *J. of Chromatography*, Vol. 121 (1976), pp. 205–212.
101. Alkire, R. C., and Eisinger, R. S., "Separation by Electrosorption of Organic Compounds in a Flow-Through Porous Electrode. I. Mathematical Model for One-Dimensional Geometry," *J. Electrochem. Soc.*, Vol. 130, No. 1 (1983), pp. 85–93.
102. Eisinger, R. S., and Alkire, R. C., "II. Experimental Validation of Model," *J. Electrochem. Soc.*, Vol. 130, No. 1 (1983), pp. 94–101.
103. Tsai, M. C., Wang, S. S., and Yang, R. T., "Pore Diffusion Model for Cyclic Separation: Temperature Swing Separation of Hydrogen and Methane at Elevated Pressures," *AIChE Journal*, Vol. 29, No. 6 (1983), pp. 966–975.
104. Van Der Vlist, E., "Oxygen and Nitrogen Enrichment in Air by Cycling Zone Adsorption," *Separation Science*, Vol. 6, No. 5 (1971), pp. 727–732.
105. Wankat, P. C., "Thermal Wave Cycling Zone Separation. A Preparative Technique for Counter-Current Distribution and Chromatography," *J. of Chromatography*, Vol. 88 (1974), pp. 211–228.
106. Wankat, P. C., and Ross, J. W., "Partial Fractionation of Dyes by Cycling Zone Separation," *Separation Science*, Vol. 11, No. 3 (1976), pp. 207–213.

CHAPTER 37

HEAT AND MASS TRANSFER IN SOLAR REGENERATORS

P. Gandhidasan

Department of Mechanical Engineering
The University of the West Indies
St. Augustine, Trinidad, West Indies

CONTENTS

INTRODUCTION

Dehumidified air is required for use in many industrial applications with particular reference to chemical, metallurgical, combustion, and air conditioning industries. The dehumidification of air can be accomplished by several methods such as refrigeration, mechanical compression, solid adsorption, and liquid absorption. The liquid absorption system is ideally suitable where moderately low humidities are required and where large quantities of air are to be dehumidified.

A typical commercial liquid absorbent dehumidifier is shown in Figure 1. The strong solution of absorbent is sprayed down over the cooling coils in the absorber chamber. Air

I wish to sincerely thank Prof. M. C. Gupta and Prof. V. Sriramulu of Indian Institute of Technology, Madras, India for their invaluable guidance and encouragement. I also want to express my deep sense of gratitude to Prof. S. Satcunanathan of the University of the West Indies, St. Augustine, Trinidad, for the useful discussions and suggestions, and I am thankful to Dr. K. N. Ramamurthy and Dr. P. Persad for their help in the preparation of the manuscript.

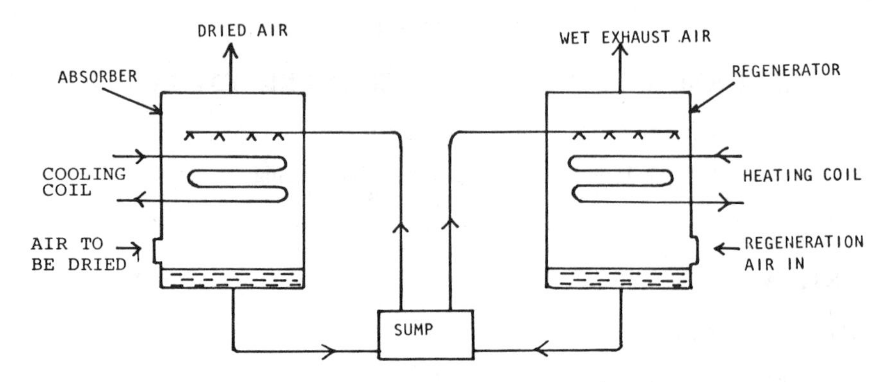

Figure 1. Schematic diagram of a liquid absorbent dehumidifier.

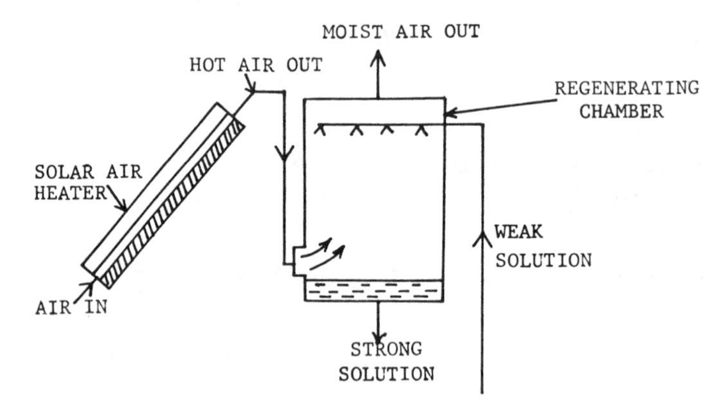

Figure 2. Regeneration using a solar air heater.

to be dried passes over the coils in the chamber, where the solution removes moisture from the air. The coolant in the coil removes the heat that is released when the sorbent takes on moisture. The now-dilute solution is then sprayed over a heating coil in the second chamber called the regenerator. Regeneration air passes over the heating coil in the chamber picking up moisture released by the heated solution and exhausting it. The heat for regeneration has traditionally been drawn from fossil fuels. When calcium chloride, lithium chloride, lithium bromide, glycols, etc. are used as the liquid absorbent, effective regeneration may be obtained at temperatures of the order of 50°–65° C. These temperatures are also effective operating temperatures of simple flat-plate solar collectors. Hence solar energy can be used to advantage for such processes.

Recently attempts have been made to use liquid absorbent in various solar energy applications, such as cooling buildings [1–9], power generation [10], drying [11], food processing [12], etc. In all these applications an important part of the system is the regenerator, in which the weak absorbent solution is concentrated. The heat required for the regeneration process can be supplied either by a solar air heater (Figure 2) or a solar water heater (Figure 3). For example, Löf [1] proposed a solar space cooling system in which the room air is dehumidified by hygroscopic triethylene glycol and the dehumidified air is subsequently evaporatively cooled. The weak glycol is regenerated by solar heated air. In such a system the solar heat is first collected at air (or water) heating collectors and finally transferred to the absorbent. As a result the temperature of regeneration is lower than the collector plate temperature and the rate of regeneration is therefore relatively lower.

On the other hand by using direct solar regenerators where the absorbent solution is itself the heat-collecting fluid, the regeneration process could be made more effective. The absorbent temperature is more or less equal to the collector-plate temperature. The regenerating chamber is also eliminated. Based on the methods of regenerating the absorbent solutions, regenerators can be divided into different categories as follows:

1. Open-type solar regenerator.
2. Closed-type solar regenerator.
3. Wind forced-flow solar regenerator.
4. Forced-flow solar regenerator.

In the subsequent sections, the above four types of regenerators are individually analyzed. Also the performance and operating characteristics of the forced-flow solar regenerator are given, for presently it would seem that this type of regenerator is most effective.

OPEN-TYPE SOLAR REGENERATOR

The open regeneration system is one of the simplest methods of regenerating the weak solution. It is shown schematically in Figure 4. The open regenerator consists of a flat blackened tilted surface over which the weak solution flows as a thin film in contact with the ambient air. The solution is heated by the solar energy absorbed by the surface. If the vapor pressure of the weak solution exceeds the vapor pressure of water in the atmospheric air, mass transfer takes place from the solution to the atmosphere. Energy transfer from the solution to the surroundings occurs by convection, radiation and conduction in addition to evaporation.

Kakabaev et al. [5] described a solar air conditioning system in which the regeneration of lithium chloride solution takes place on an open surface. An analytical procedure for calculating a mass of water evaporated from the weak solution as a function of meteorological quantities and solution initial conditions has been developed by Kakabaev et al. [13]

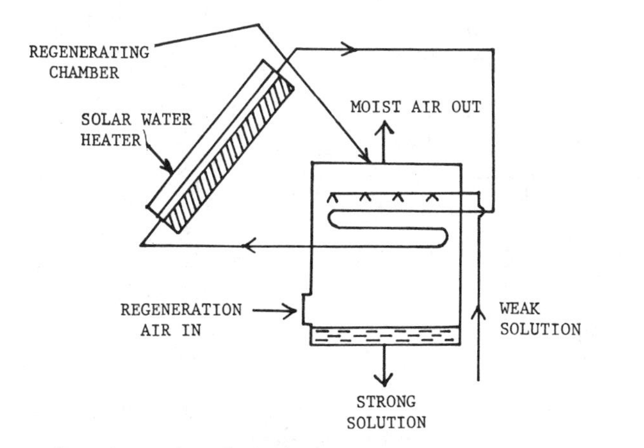

Figure 3. Regeneration using a solar water heater.

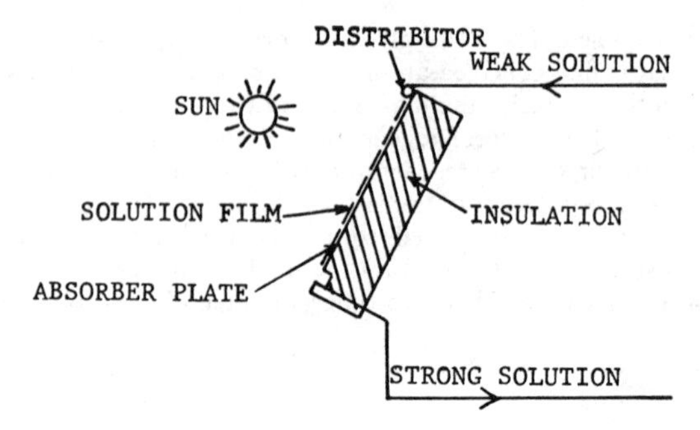

Figure 4. Schematic of an open-type regenerator.

and Collier [14]. The following assumptions have been made to obtain an approximate analytical solution.

1. Steady-state conditions prevail.
2. The temperature gradient in the thin liquid layer is negligible.
3. The absorbent solution flow rate is constant.
4. The solution equilibrium vapor pressure for the absorbent solution is expressed as,

$$P_s = aT_s + \frac{b}{\xi_s} + c \tag{1}$$

   where a, b, and c are determined from available vapor pressure data in the temperature and concentration ranges of interest.
5. The mass transfer coefficient, regenerator overall heat loss coefficient, and the heat of vaporization for water from the absorbent solution are constant.

The amount of water evaporated from the weak solution is expressed as,

$$m = \frac{G_0\left[\frac{A_3}{A_2}k_2 + \beta(P_{s,i} - P_a)\right]}{k_1 - k_2} e^{k_1 x'/G_0} - \frac{G_0\left[\frac{A_3}{A_2}k_1 + \beta(P_{s,i} - P_a)\right]}{k_1 - k_2} e^{k_2 x'/G_0} + \frac{A_3}{A_2} G_0 \tag{2}$$

where $A_1 = \dfrac{\beta b}{\xi_0} + \dfrac{\beta a h'_{fg}}{C_p} + \dfrac{U_L}{C_p}$

$$A_2 = \frac{\beta U_L b}{\xi_0 C_p}$$

$$A_3 = \frac{\beta}{C_p}\left[a(I\alpha' + U_L T_a) + U_L\left(\frac{b}{\xi_0} + c - P_a\right)\right]$$

$$\xi_0 = \frac{m_s}{G_0}$$

$$k_1 = \frac{-A_1 + \sqrt{A_1^2 - 4A_2}}{2}$$

$$k_2 = \frac{-A_1 - \sqrt{A_1^2 - 4A_2}}{2}$$

Taking the derivative of m/x' with respect to G_s/x' and setting the results equal to zero yields an optimum flow rate of absorbent solution per unit collector area for any condition which results in the maximum evaporation of water.

Peng and Howell [15] employed a more exact numerical method for predicting the performance of the open regenerator. The energy and mass balance equations are nondimensionalized using nondimensional variables. They compared their results with results presented by Collier for particular conditions of operation. The numerical model predicts a higher water evaporation rate and lower outlet temperature than that obtained by the analytical model. However, the discrepancy is within 10%. The analytical model may then be considered to be a good approximate method.

Gandhidasan [16] derived a simple expression for the rate of evaporation of water from the weak solution by using the following additional assumptions over the analytical model:

1. The temperature, concentration, and vapor pressure of the weak solution are constant at the arithmetic average of these values at the beginning and end of the regenerator.
2. The relationship between vapor pressure, temperature, and concentration of the absorbent solution is,

$$P_{s,ave} = a + bT_{s,ave} + \frac{c}{\xi_{s,ave}} \tag{3}$$

3. An approximate equation which relates the rate of desorption, solution flow rate, and initial and average value of concentration of absorbent solution is,

$$\frac{1}{\xi_{s,ave}} = \frac{1}{\xi_{s,i}}\left(1 - \frac{m}{2G_s}\right) \tag{4}$$

The mass of water evaporated from the weak solution is expressed as,

$$m = \frac{A(I\alpha' + U_L T_a) + 2G_s C_p T_{s,i} + \frac{1}{b}\left(a + \frac{c}{\xi_{s,i}} - P_a\right)(U_L A + 2G_s C_p)}{\frac{1}{b}\left(\frac{1}{\beta A} + \frac{c}{2\xi_{s,i} G_s}\right)(U_L A + 2G_s C_p) + h'_{fg}} \tag{5}$$

Dividing both sides of the preceding equation by the area, the amount of water evaporated from unit area of the regenerator may be obtained.

Figure 5 shows the diurnal variation of solution temperature and water desorption from calcium chloride solution obtained by using the previous simple approximate method and the analytical model of Collier. The simple approximate method predicts the thermal performance slightly lower than the one predicted by Collier and the variation of the total daily desorption being less than 15% for the particular conditions of the operation. Hence for preliminary design purposes the simple expression derived by Gandhidasan can be used to predict the performance of the regenerator.

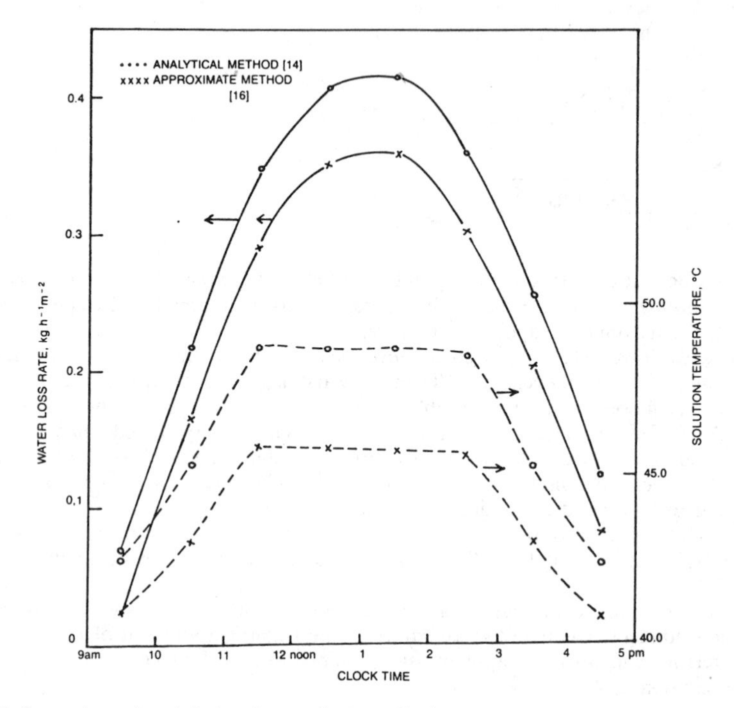

Figure 5. Comparison of analytical and approximate methods.

The open regeneration system is simple to design and cheap to build. It is very suitable for use in dry, dust free, and arid climates. Since the regeneration occurs on an open surface, there is the possibility that the solution may become contaminated with dust. Rain will affect the operation of the system. Further this system is not advisable to use in humid weather conditions.

CLOSED-TYPE SOLAR REGENERATOR

Contamination by dust and influence by rain that affect the open solar regenerator may be eliminated by covering it with a single layer of glass. The regeneration then takes on the appearance of a solar still. Hollands [2] thus investigated the regeneration of lithium chloride solution in a conventional roof-type solar still in which the solution was contained as a shallow pool. The energy balance equations were solved by the numerical method to estimate the mass of water evaporated from the brine. When the still was acting as a desalinator for producing potable water from brackish water, the efficiency was 56% and when it was used for regenerating the lithium chloride solution, the efficiency was in the range of 5%–20% depending upon the insolation and the concentration of the brine. The reduction in efficiency is attributed to the lower partial pressure of lithium chloride solution compared with saline water at a given temperature.

An alternative method of regeneration is possible by covering the open regenerator by a glass and closing both ends of the regenerator. This system is very similar to that of tilted or wick-type solar still [17]. Tilted solar stills or closed-type regenerators perform more efficiently than the basin-type because they respond faster to incident solar radiation and they intercept more energy.

The schematic of the closed type (tilted solar still) regenerator is shown in Figure 6. It consists of a flat, blackened, tilted surface over which the weak desiccant flows as a thin film and is covered by a single glazing with an air gap of about 2.5 cm. The bottom of the regenerator is well insulated. The absorbent solution is heated by solar energy and the water that evaporates from the solution rises to the cover by convection where it is condensed on the underside of the glass cover. The condensate flows along the glass surface by gravity into condensate troughs, and the solution leaving the regenerator becomes strong. Energy transfer from the solution to the glass cover occurs by three different modes, namely convection, radiation, and evaporation-condensation. The glass cover transfers heat to the surrounding atmosphere by convection and radiation. A theoretical study of the tilted solar still as a regenerator for liquid desiccants has been recently carried out by Gandhidasan [18]. To obtain an expression for calculating the amount of water evaporated from the weak solution, the energy balance equations can be solved analytically using the various approximations for open-type solar regenerators. The performance of this type of regenerator depends on the potential for mass transfer, which is the difference in water vapor pressure between the solution film and glass cover. The closed-type regenerator can be used in humid climates since its performance does not depend on the ambient conditions.

WIND FORCED-FLOW SOLAR REGENERATOR

The schematic of the wind forced-flow solar regenerator is shown in Figure 7. This regenerator is somewhat similar to the closed-type regenerator except that both ends of the regenerator are open to atmosphere. It consists of a blackened, inclined absorber plate over which the weak solution flows as a thin film. The absorber plate is heated by solar energy and the water evaporating from the liquid surface is removed by wind gusts. The feasibility

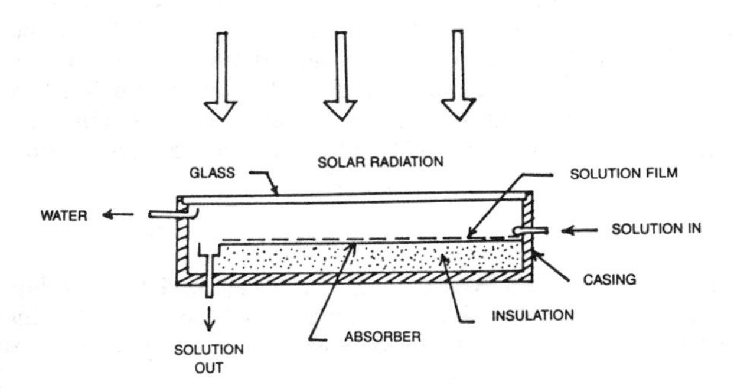

Figure 6. Schematic of the closed-type regenerator.

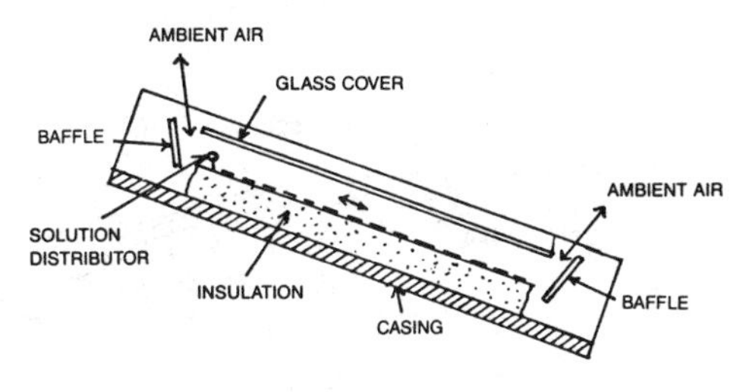

Figure 7. Schematic of a wind-forced-flow regenerator.

of using this type of regenerator for desiccant concentration was demonstrated by Mullick and Gupta [19] who concluded that the removal of water vapor from the regenerator was caused mainly by wind gusts, and thermosiphon action predominated only in the absence of wind [20].

Johannsen and Grossman [21] developed a computer simulation for this type of regenerator in which correlations for heat and mass transfer coefficients and air pressure drop in the regenerator derived from earlier experiments were employed. The influence of various design and operational parameters on the thermal performance of the regenerator with triethylene glycol as the desiccant were carried out using the model. Within the range of 1.2–2.2 m, the effect of regenerator flow length was found to be small.

Though such a system is desirable, it has a drawback because long flow length of the regenerator is limited by the randomness of wind direction. The long flow length of the regenerator would create stagnant pockets of air in the regenerator, which are likely to reduce the effectiveness of the regenerator. On the other hand by maintaining a forced air flow not only is a continuous flow ensured in the unit but also it is made unidirectional. The flow in the unit can be maintained laminar throughout so that the pressure drop in the unit does not become excessive. Depending on the direction of the air flow, it can be classified as a forced-flow cocurrent solar regenerator or a forced-flow countercurrent solar regenerator.

FORCED-FLOW COCURRENT SOLAR REGENERATOR

The schematic of the forced-flow cocurrent solar regenerator is shown in Figure 8. It consists of a flat, blackened surface over which the absorbent solution that is to be regenerated trickles down as a thin film. The bottom of the regenerator is well insulated. In order to reduce the heat losses, the regenerator is covered by a single glazing leaving an air gap of about 5 cm. Water evaporates from the liquid surface as a result of absorption of solar energy by the plate, and it is removed by a laminar forced-air stream, which flows cocurrent to the liquid film and is confined between the absorber plate and the glazing [22]. The operating conditions of the regenerator are such that the surface of the glass is at a higher temperature than the saturation temperature of the vapor air mixture, corresponding to its vapor pressure with the result that the water vapor does not condense on the glass surface. Hence, transmission of solar radiation through the glass cover is not impaired.

Mathematical Model

It has been assumed in the analysis that the air stream and the thin liquid film flow cocurrent to each other and that an intimate contact exists between the solution film and air stream. The thickness of the air stream is large when compared with the thickness of

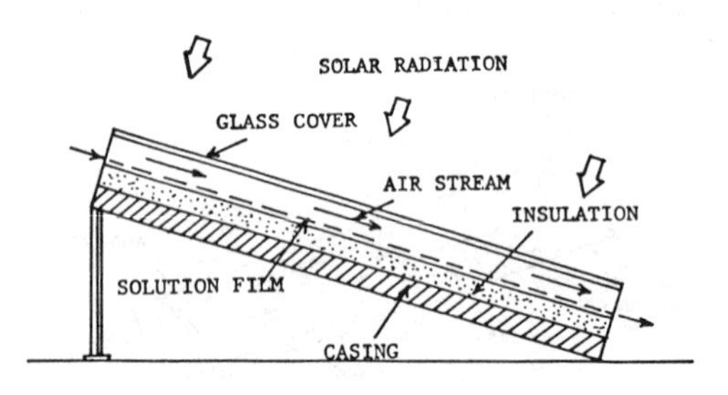

Figure 8. Schematic of a forced-flow cocurrent regenerator.

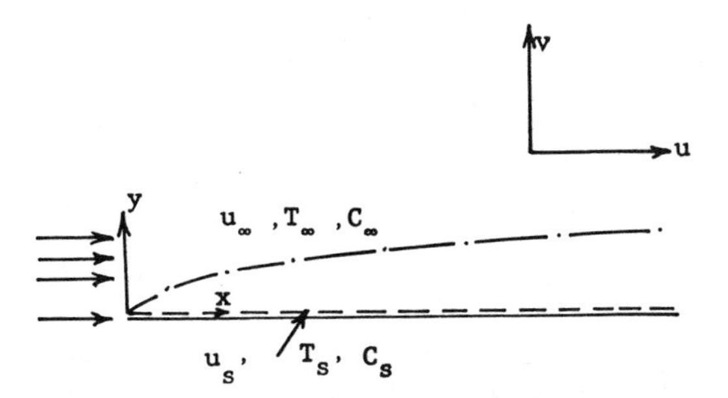

Figure 9. Mathematical model for forced-flow cocurrent regenerator.

the liquid film. The mathematical model considered is shown in Figure 9. Since the glass plate is situated considerably far away from the absorber it is treated as though it were located at infinity. The conditions in a regenerator are such that the boundary layer approximations hold.

It is assumed that there is no slip at the interface. Since the liquid film is of negligible thickness, the resistance to mass transfer stems solely from the air phase, and hence the processes occurring only in the gas phase are dealt with. The solution velocity as used in practice is only a small fraction of the air velocity and, hence, ripples do not appear at all on the solution film. Another assumption implicit in the analysis is that the fluid properties are constant in the range of temperatures generally used in solar regenerators. The flow is considered two dimensional and incompressible.

Steady-State Analysis

In the model just described, there is a moving interface which influences the rates of convective heat and mass transfer between the solution film and air stream. The partial differential equation of momentum for the laminar air stream can be deduced from the Navier–Stokes equations. Equations of continuity, energy, and concentration together constitute a set of appropriate equations for the configuration shown. Assuming that,

- The second derivative of u, namely, $\partial^2 u/\partial x^2$ is negligible compared with $\partial^2 u/\partial y^2$. Similar assumptions have been made in energy and mass equations.
- Viscous friction is negligible.
- The pressure gradient is negligibly small for pure forced convection.

The governing partial differential equations are:

$$\frac{\partial u}{\partial x} + \frac{\partial v}{\partial y} = 0 \tag{6}$$

$$u\frac{\partial u}{\partial x} + v\frac{\partial u}{\partial y} = \nu\frac{\partial^2 u}{\partial y^2} \tag{7}$$

$$u\frac{\partial T}{\partial x} + v\frac{\partial T}{\partial y} = k\frac{\partial^2 T}{\partial y^2} \tag{8}$$

$$u \frac{\partial c}{\partial x} + v \frac{\partial c}{\partial y} = \delta \frac{\partial^2 c}{\partial y^2} \tag{9}$$

The boundary conditions are:

$$\text{At} \quad y = 0, \quad u = u_s, \quad T = T_s, \quad c = c_s$$

$$y \to \infty, \; u = u_\infty, \; T = T_\infty, \; c = c_\infty \tag{10}$$

For this type of problem the Blasius transformation can be successfully employed. The prior set of partial differential (Equations 7 through 9) is reduced to ordinary differential equations by defining a dimensionless variable and a stream function as follows:

$$\eta = y \sqrt{\frac{u_\infty}{\nu x}} \tag{11}$$

$$\Psi = \sqrt{\nu x u_\infty}\, f(\eta) \tag{12}$$

It can be shown that these functions satisfy the continuity equation. Momentum, energy, and diffusion equations are then reduced to the following dimensionless forms:

$$ff'' + 2f''' = 0 \tag{13}$$

$$m_T'' + \frac{Pr}{2} f m_T' = 0 \tag{14}$$

$$m_c'' + \frac{Sc}{2} f m_c' = 0 \tag{15}$$

where

$$m_T = \frac{T - T_\infty}{T_s - T_\infty} \tag{16}$$

$$m_c = \frac{c - c_\infty}{c_s - c_\infty} \tag{17}$$

The boundary conditions are correspondingly reduced to,

$$f(0) = 0, \; f'(0) = \frac{u_s}{u_\infty}, \; f'(\infty) = 1$$

$$m_T(0) = 1, \; m_T(\infty) = 0$$

$$m_c(0) = 1, \; m_c(\infty) = 0 \tag{18}$$

Effect of Normal Convective Diffusive Velocity

The rate of mass transfer during lean hours of sunshine will be low, and hence, the normal convective diffusive velocity, which is contributed by water evaporating from the liquid surface, can be neglected. When the mass transfer rate is high, as in typical mid-day operation, the normal convective diffusive velocity of water evaporating from the liquid surface is always present [23]. This diffusive velocity disappears, if it is countered by a diffusive velocity of air into the solution film. Since this does not occur in a solar regenerator, the normal convective diffusive velocity is taken into consideration. There is a relationship between the normal convective diffusive velocity and the concentration of water vapor at the liquid surface.

For this case, even though the governing (Equations 6 through 9) are valid, the boundary conditions are different. In addition to the boundary conditions as given by Equation 10, one more boundary condition must be specified. The boundary conditions for the situation when a finite normal convective diffusive velocity exists are,

$$\text{At} \quad y = 0,\ u = u_s,\ T = T_s,\ c = c_s$$

$$u = -\delta m_c'(0)\frac{(c_s - c_\infty)}{(m_c - c_s)}\sqrt{\frac{u_\infty}{vx}}$$

$$y \to \infty,\ u = u_\infty,\ T = T_\infty,\ c = c_\infty \tag{19}$$

The boundary conditions (Equation 19) can be transferred to the following dimensionless form:

$$f(0) = \frac{2}{Sc} m_c'(0)\frac{(c_s - c_\infty)}{(m_c - c_s)}$$

$$f'(0) = \frac{u_s}{u_\infty},\ f'(\infty) = 1$$

$$m_c(0) = 1,\ m_c(\infty) = 0 \tag{20}$$

The set of governing equations along with the boundary conditions have been solved on an IBM 370/155 computer by using the Continuous System Modelling Program (CSMP) for the two cases just described. Analysis shows that the ratio of solution velocity to air velocity (u_s/u_∞) exercises a profound influence on the performance. Figure 10 shows typical variations in velocity, temperature, and concentration for $u_s/u_\infty = 0.2$ and Figure 11 their corresponding gradients. The average mass transfer coefficient is evaluated from the following equation, for $Sc = 0.6$.

$$\beta = \delta \frac{[f''(0)]^{Sc}}{\int_0^\infty [f'']^{Sc} d\eta}\sqrt{\frac{u_\infty}{vx}} \tag{21}$$

Figure 12 shows the variation of the mass transfer coefficient with the Graetz number $u_s/u_\infty = 0.4$, which has been computed for a regenerator of length 1 m. Figure 13 depicts the variation of the mass transfer coefficient with the Graetz number for an ethanol-carbon dioxide system. This is chosen for comparison in view of the fact that both experimental and theoretical results have been reported [24, 25] for the case of the liquid stream offering resistance owing to its finite thickness. It is discernible that with the liquid flowing as a thin film, the mass transfer coefficient is higher. The results of the computations that take into account the effect of normal convective diffusive velocity are presented in the same figure, which incidentally also shows the results for a thin film in the absence of the normal convective diffusive velocity. It is obvious that the presence of the convective diffusive

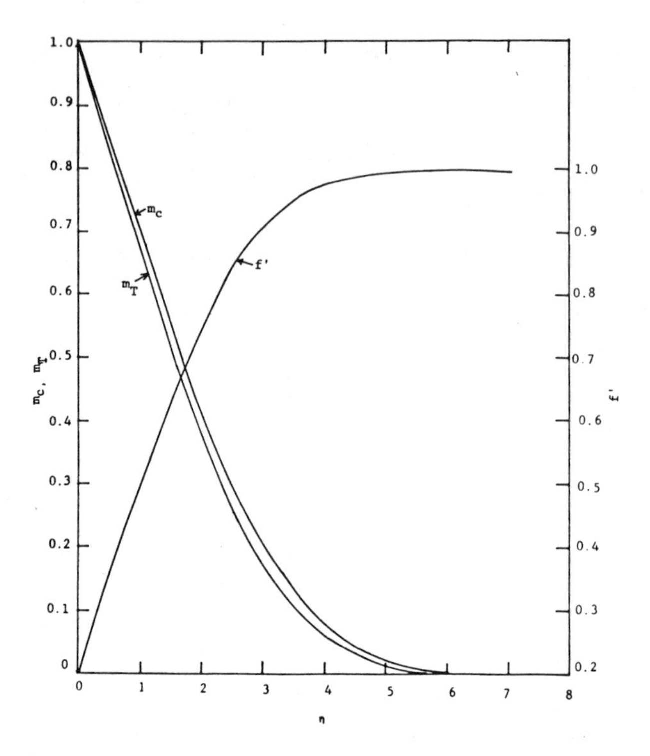

Figure 10. Velocity, temperature, and concentration profiles.

velocity decreases the mass transfer coefficient when compared with that for a thin film, but it is still higher compared with the situation when the solution thickness is finite.

Effect of Buoyancy

In a regenerator, since the absorber plate temperature (or solution temperature) is different from that of the air moving over it, and on account of relatively low velocities used in such devices, buoyancy forces do exist. The performance of a nontracking solar absorber depends to a large extent upon its inclination, and it is common practice to change this inclination seasonally. The inclination that furnishes maximum collection of solar energy in summer is given by rule of thumb, namely that it ought to be less than local latitude by 10° for places located in the northern hemisphere. Accordingly, the optimal collector inclination for stations situated at low latitudes will be negligibly small and the plate may be assumed to be horizontal. The effect of buoyancy is to induce a longitudinal pressure gradient. Buoyancy is not explicitly influenced by the physical properties, and it is assumed that these are constant except for a change of density with temperature which, under the action of the gravitational potential, leads to the buoyancy force. The inclusion of the free convection effect leads to two additional differential Equations 26 and the boundary conditions are as given in Equation 18. Since momentum and energy equations are coupled, the similarity transformation cannot be employed and hence a series solution method has been adopted. The following new variables are now introduced:

$$\Psi = \sqrt{u_\infty \nu x}\,[f_0(\eta) + \zeta f_1(\eta) + \ldots] \tag{22}$$

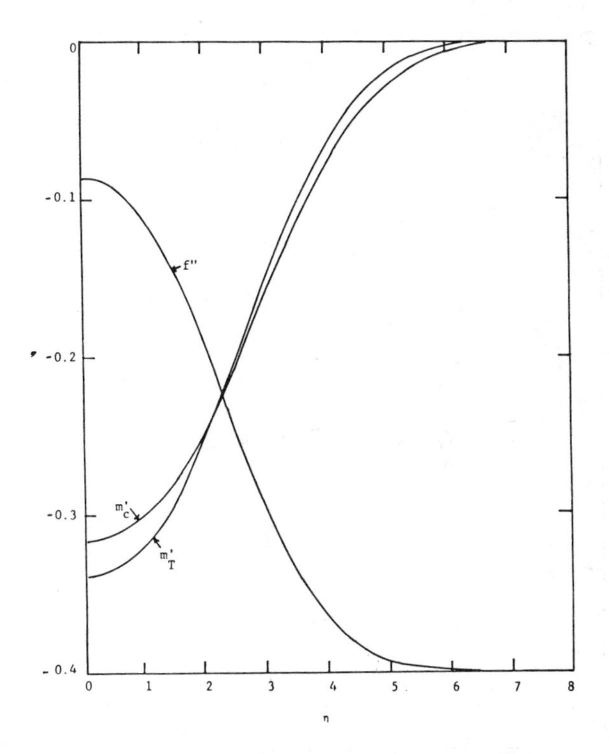

Figure 11. Typical profiles of velocity, temperature and concentration gradients.

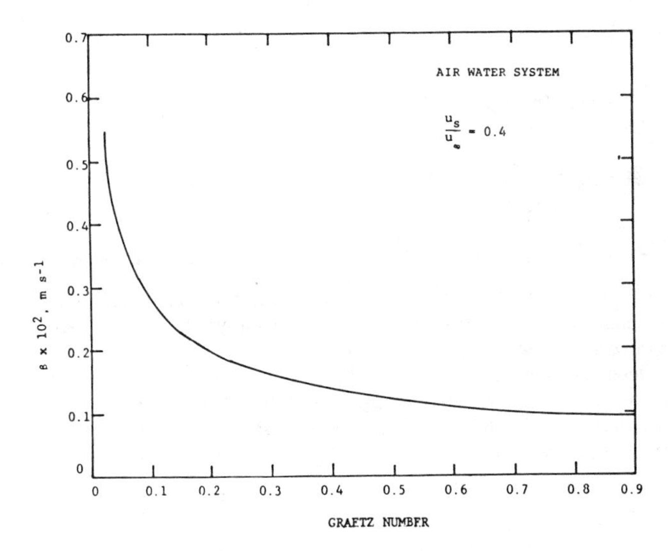

Figure 12. Variation of mass transfer coefficient with Graetz number.

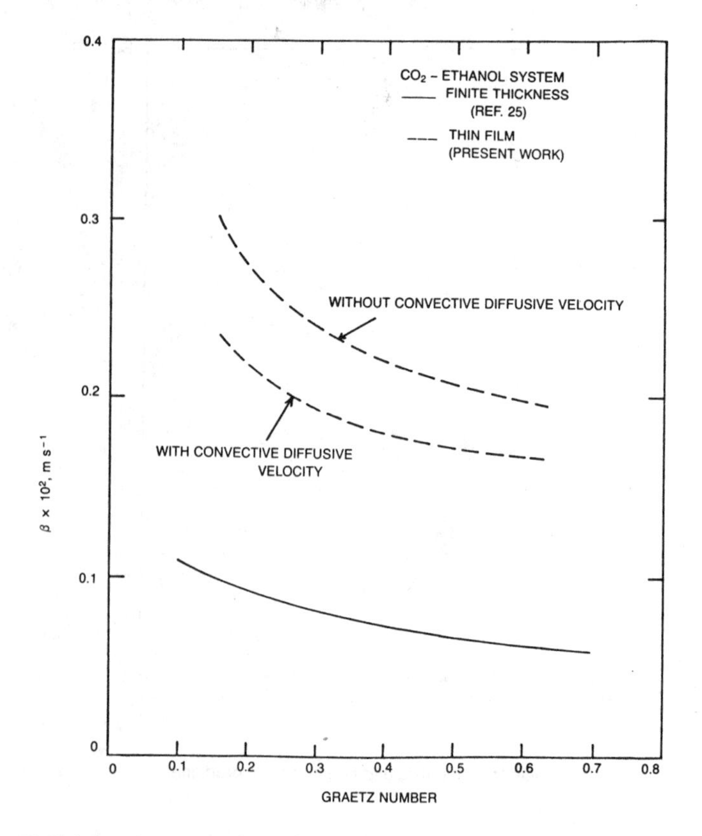

Figure 13. Variation of mass transfer coefficient with Graetz number for CO_2—ethanol system.

$$m_T = \frac{T - T_\infty}{T_s - T_\infty} = m_{T_0}(\eta) + \zeta m_{T_1}(\eta) + \dots \tag{23}$$

$$m_c = \frac{c - c_\infty}{c_s - c_\infty} = m_{c_0}(\eta) + \zeta m_{T_1}(\eta) + \dots \tag{24}$$

where $\zeta = \dfrac{Gr_x}{Re_x^{2.5}}$ is a measure of the strength of the free convection relative to the forced convection. In Equations 22 through 24, the first term in each series represents pure forced convection, whereas the other terms represented the buoyancy-produced deviation from pure forced convective flow.

In order to determine the buoyancy effect, the set of differential equations has been solved for $Pr = 0.7$ and $Sc = 0.6$ for different ratios of solution-to-air velocities. This effect is illustrated for typical values of $Re_x = 1.033 \times 10^3$ and $Gr_x = 2.56 \times 10^6$ which are the typical values encountered in a solar regenerator. The local heat and mass transfer rates are given by

$$\frac{\dot{q}}{\dot{q}_0} = 1 + \zeta \frac{m'_{T_1}(0)}{m'_{T_0}(0)} + \dots \tag{25}$$

$$\frac{\dot{m}}{\dot{m}_0} = 1 + \zeta \frac{m'_{c_1}(0)}{m'_{c_0}(0)} + \ldots \qquad (26)$$

Figures 14 and 15, respectively, show the first-order deviation of local heat and mass transfers for different ratios of u_s/u_∞. It is interesting to note that the effect of buoyancy on local heat and mass transfer is a function of u_s/u_∞, with the buoyancy growing more important as u_s/u_∞ decreases. By keeping the free stream velocity constant and decreasing the solution velocity, the buoyancy effects can be increased. It is also seen that the influence of buoyancy on local heat transfer is greater than local mass transfer.

Transient Response

Attention has been focused so far on processes of heat and mass transfer in a cocurrent flow solar regenerator for steady-state conditions. Since the absorber plate temperature varies with time due to varying insolation, the conditions existing in such a unit may not be steady. It is generally assumed, however, that the actual heat and mass transfer rates at any instant are equal to the rates evaluated by assuming that steady-state conditions exist in the unit. But reflection reveals that since the regenerator is exposed to a continuously varying insolation, there may be a difference between the actual instantaneous values and quasisteady values. Variation in temperature and concentration profiles due to changes in absorber-plate temperature determine the deviation.

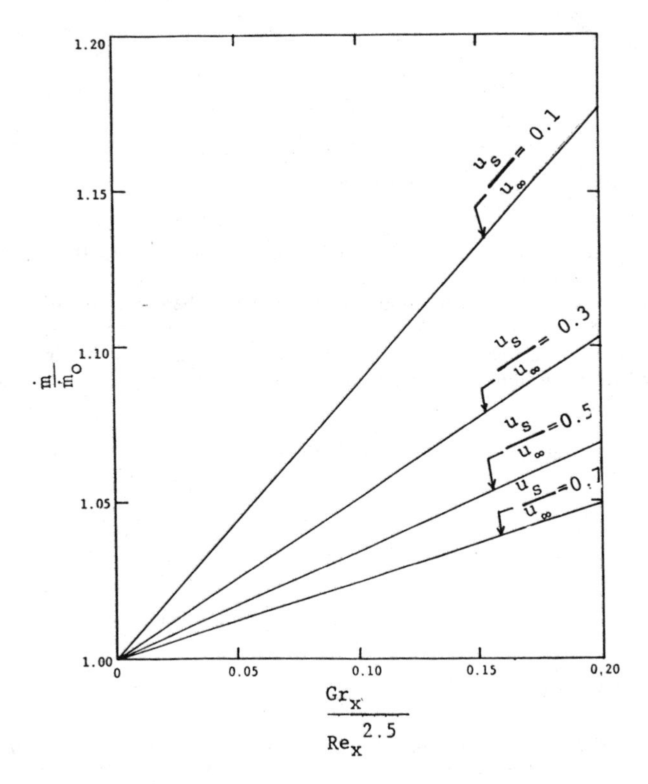

Figure 14. First-order deviation of heat transfer due to buoyancy.

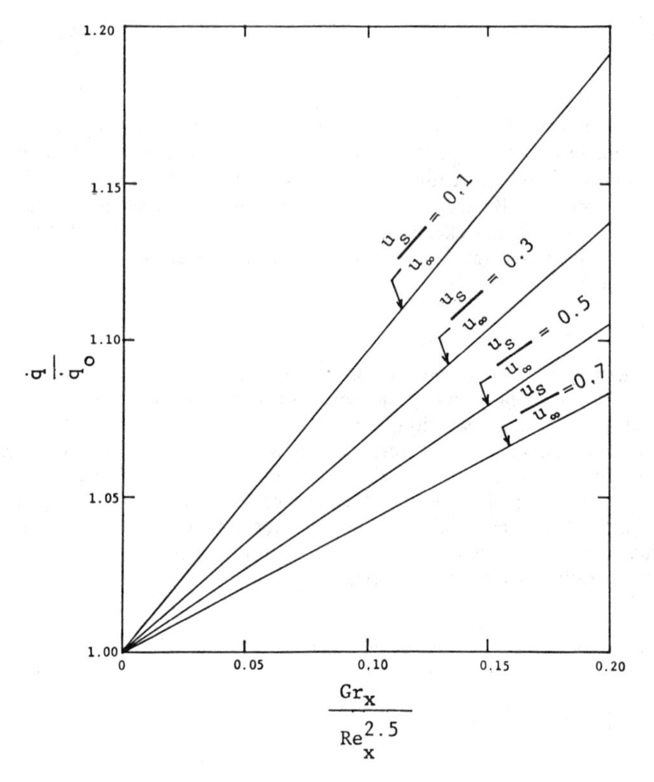

Figure 15. First-order deviation of mass transfer due to buoyancy.

In the ensuing analysis attention has been focused on determining the second-order deviation of instantaneous values from the quasisteady values on a clear day, from morning to noon. It has been assumed that the temperature of the absorber plate is spatially uniform over the surface but may vary in a differential manner with time. The partial differential equations describing convective heat, mass, and momentum transfer for the air stream have been transferred to dimensionless form by introducing the following series expansions:

$$m_T = \frac{T - T_\infty}{T_s - T_\infty} = m_{T_0}(\eta) + a_1 m_{T_1}(\eta) + a_2 m_{T_2}(\eta) + \ldots \tag{27}$$

where

$$a_1 = \frac{\dot{T}_s}{(T_s - T_\infty)}\left(\frac{x}{u_\infty}\right) \tag{28}$$

$$a_2 = \frac{\ddot{T}_s}{(T_s - T_\infty)}\left(\frac{x}{u_\infty}\right)^2 \tag{29}$$

a_1 and a_2 are a measure of the promptness with which the boundary layer responds to impressed changes. Similar analysis has been extended to the mass diffusion equation by defining,

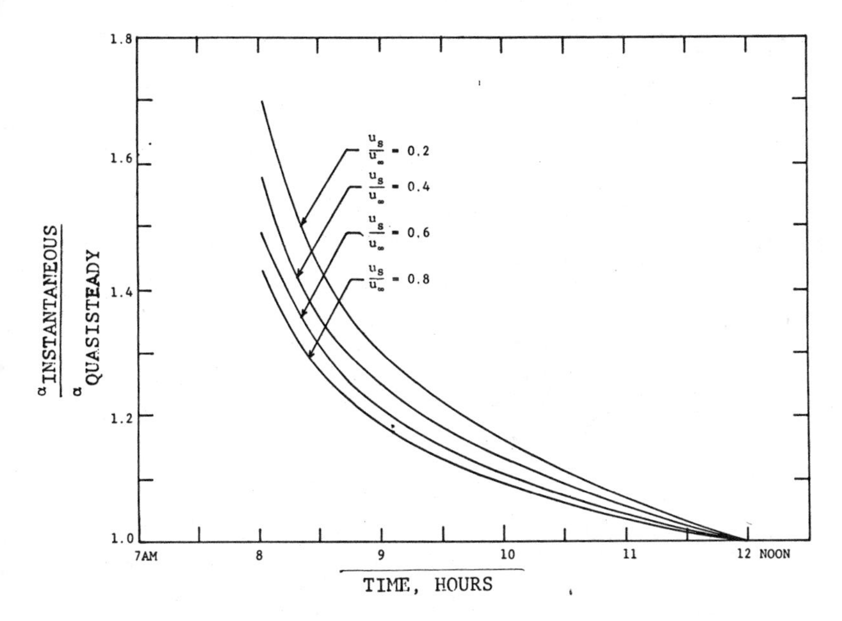

Figure 16. Deviation of instantaneous heat transfer coefficients from a quasisteady value.

$$m_c = \frac{c - c_\infty}{c_s - c_\infty} = m_{c_0}(\eta) + b_1 m_{c_1}(\eta) + b_2 m_{c_2}(\eta) + \ldots \tag{30}$$

where

$$b_1 = \left(\frac{\dot{c}_s}{c_s - c_\infty}\right)\left(\frac{x}{u_\infty}\right) \tag{31}$$

$$b_2 = \left(\frac{\ddot{c}_s}{c_s - c_\infty}\right)\left(\frac{x}{u_\infty}\right)^2 \tag{32}$$

The deviation of instantaneous heat and mass transfer coefficients from quasisteady values has been calculated by the following relations for a tropical climatic data [27]:

$$\frac{\alpha_{inst}}{\alpha_{quasi}} = \left[1 + a_1 \frac{m'_{T_1}(0)}{m'_{T_0}(0)} + a_2 \frac{m'_{T_2}(0)}{m'_{T_0}(0)} + \ldots\right] \tag{33}$$

$$\frac{\beta_{inst}}{\beta_{quasi}} = \left[1 + b_1 \frac{m'_{c_1}(0)}{m'_{c_0}(0)} + b_2 \frac{m'_{c_2}(0)}{m'_{c_0}(0)} + \ldots\right] \tag{34}$$

The deviation of α_{inst} from α_{quasi} and β_{inst} from β_{quasi} are calculated for different u_s/u_∞ (u_∞ = 3 m/s, x = 1 m) by solving the governing equations using appropriate boundary conditions, and the results are shown in Figures 16 and 17, respectively.

$\alpha_{inst}/\alpha_{quasi}$ and $\beta_{inst}/\beta_{quasi}$ vary with u_s/u_∞.

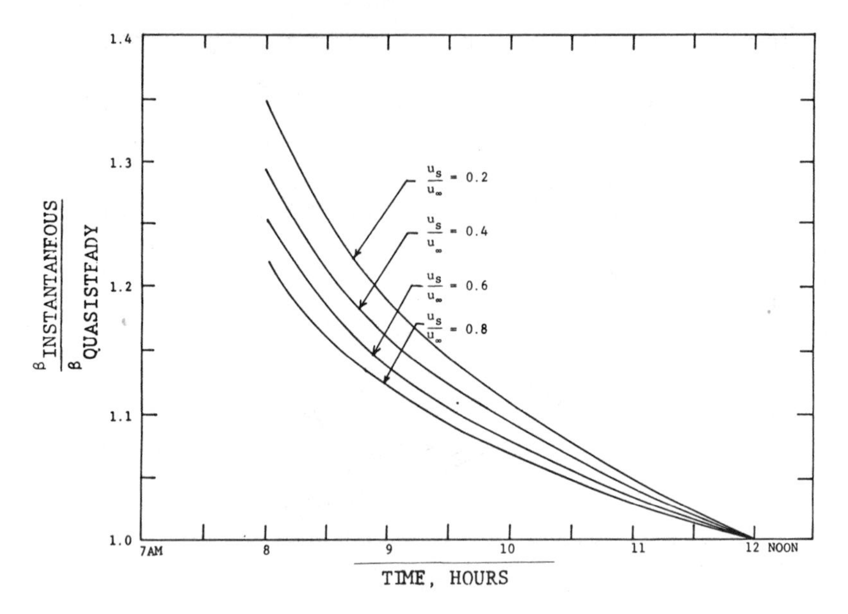

Figure 17. Deviation of instantaneous mass transfer coefficients from a quasisteady value.

Increasing the ratio of solution velocity to air velocity tends to reduce the magnitudes. In other words increasing the velocity ratio decreases the transient effects. The second-order deviation of instantaneous heat and mass transfer coefficients from the quasisteady conditions is higher during the lean hours of sunshine, i.e., between 8 am and 9 am. The performance characteristics of the regenerator are dependent upon the rate of desorption, which is the product of mass transfer coefficient, and the potential for mass transfer, which is given by the difference in partial pressure of water vapor between the solution surface and the free stream. As the plate temperature changes, the mass transfer coefficient and the potential for mass transfer both change, resulting in the maximum rate of desorption at specified flow conditions. The deviation of the instantaneous mass transfer coefficient from quasisteady condition is lower compared to the deviation of the instantaneous heat transfer coefficient from quasisteady conditions for the same operating conditions. The rate of desorption is maximal at noon. As u_s/u_∞ increases the deviation of β_{inst} from β_{quasi} becomes smaller. For operation around noon the deviation is insignificant. Hence the corresponding deviation in the rate of desorption is also negligible.

ANALYSIS OF COUNTERCURRENT SOLAR REGENERATOR

It has been assumed so far that the liquid film and air stream flow cocurrent to each other. However, the other situation, that is, when the solution and air stream are moving in opposite directions is also equally important, in as much as the unit can be operated in countercurrent mode as well. But the processes occurring in countercurrent flow are complex and cannot be analyzed in as ready a fashion as the cocurrent flow.

When the absorbent solution and laminar air stream flow countercurrent to each other there occurs in the air phase a point of flow reversal. In this study, calcium chloride solution is used as the absorbent solution. Since it is more viscous than air, the point of flow reversal usually lies within the air phase, which means that there will be a region in the air phase in the immediate neighborhood of the moving interface where the flow is in the same direction as the liquid film, whilst the bulk of the air stream flows in the opposite direction. The objective of the study has been to predict the rate of mass transfer by determining the point of flow reversal and treating the region between the solution film and the point of flow reversal as a stagnant layer through which mass transfer takes place only by diffusion. This is only an approximate method.

The mathematical procedure for solving the governing equations along with the appropriate boundary conditions is essentially the same as the one used for cocurrent flows. Even for a problem of this nature the Blausius transformation can be successfully employed for limited values of u_s/u_∞. The partial differential equations are reduced to ordinary differential equations by defining a dimensionless variable and a stream function, and the boundary conditions are correspondingly reduced. The details of the computations are given in Reference 28. The set of governing equations is solved for relatively small values of u_s/u_∞ with the primary objective of determining the rate of mass transfer by the numerical method and comparing the results obtained by the approximate method described earlier.

In order to calculate the mass transfer coefficient in countercurrent flow an approximation has been made. As already explained, the region between the interface and the point of flow reversal is assumed as a stagnant layer on account of very low flow velocities prevalent here. The Graetz model is, however, considered for the region outside the reversal zone. The resistance to mass transfer in such cases stems not only from the assumed stagnant layer but also from the bulk phase. Hence, the mass transfer coefficient can be written as,

$$\frac{1}{\beta_{\substack{\text{counter-}\\\text{current}}}} = \frac{1}{\beta_{\substack{\text{cocurrent}\\\text{Graetz model}}}} + \frac{1}{\beta_{\substack{\text{stagnant}\\\text{layer}}}} \tag{35}$$

The mass transfer rate depends mainly on the diffusion coefficient in the stagnant layer and hence it is given by:

$$\frac{1}{\beta_{\substack{\text{stagnant}\\\text{layer}}}} \doteq \frac{r}{\delta} \tag{36}$$

The mass transfer coefficient for the remaining region is calculated from,

$$\frac{1}{\beta_{\substack{\text{cocurrent}\\\text{Graetz model}}}} = \varphi\left[\frac{\delta L}{u_\infty(b'-r)^2}\right] \tag{37}$$

The mass transfer coefficient so determined is compared with that obtained by using the numerical technique, which is similar to the one employed for cocurrent flows. The computations have been carried out up to $u_s/u_\infty = -0.3$.

Figure 18 shows the variation of the mass transfer coefficient with the Graetz number for different ratios of solution velocity to air stream velocity. As observed in cocurrent flow, as the Graetz number increases the mass transfer coefficient decreases for a given value of u_s/u_∞. But, for a particular value of the Graetz number, the mass transfer coefficient decreases as u_s/u_∞ increases, which demonstrates the importance of u_s/u_∞ in the operation of solar regenerators.

It also shows the comparison of cocurrent and counter-current studies. For a given value of u_s/u_∞ the mass transfer coefficient is always higher for cocurrent flow than for counter-current flow irrespective of the method chosen for solution. The rate of mass transfer in countercurrent flow is less compared with that in cocurrent flow because in the former due to the change of flow direction a relatively low-velocity region exists in the immediate vicinity of the liquid surface, as a consequence of which the mass transfer coefficient

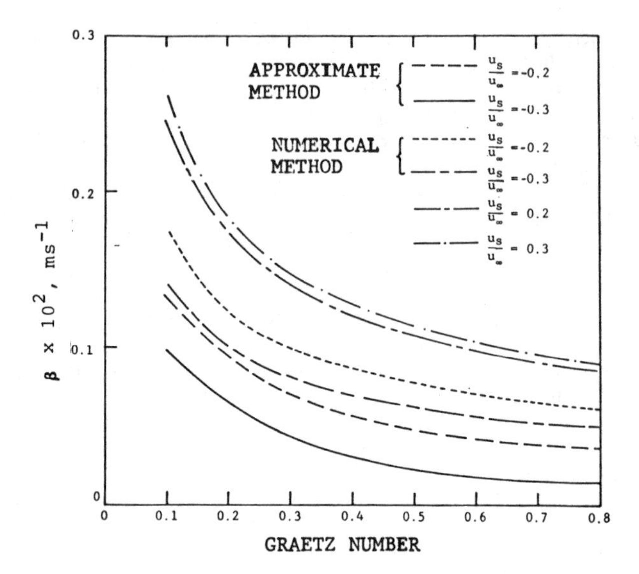

Figure 18. Comparison of cocurrent and countercurrent regenerators.

decreases. In the light of this observation it is clear that for effective operation cocurrent flow in the regenerator is superior to countercurrent flow.

DESIGN CONSIDERATIONS

From the foregoing analysis it is clear that the forced-flow regenerator can be effectively used for regenerating the weak absorbent solutions. Its thermal performance depends on the rate of evaporation of water from the weak solution. For designing such a regenerator a number of factors have to be taken into account. They include climatic variables such as insolation, wind velocity, ambient temperature, and the flow rates and initial parameters of the absorbent solution and air. An analysis has been carried out for predicting the performance characteristics of a forced-flow regenerator by making use of energy balance equations. For the forced flow regeneration system, a general expression was derived [29] to calculate the rate of desorption of water by assuming that the arithmetic mean values of temperature, concentration, and water vapor pressure of solution were constant over the regenerator.

However, in this system the rate of evaporation of water varies along with the flow length of the regenerator. This effect has been taken into account for a cocurrent-flow solar regenerator to examine its thermal performance. By assuming that steady-state conditions prevail in the unit and neglecting the temperature gradient in the thin liquid film, energy balance equations have been written and reduced to a differential equation. The equation along with the accompanying initial conditions was solved [30] in a digital computer for different values of G_s/G_∞. To analyze the thermal performance of the regenerator, hourly weather data for a clear summer day of St. Augustine, Trinidad, were used.

It was realized that the climatic variables can influence the performance of the regenerator. It was found that the windspeed does not have any effect in changing the thermal performance of the regenerator. Further, since the liquid film and air stream are moving relative to each other the ratio of their flow rates would have the effect to change the

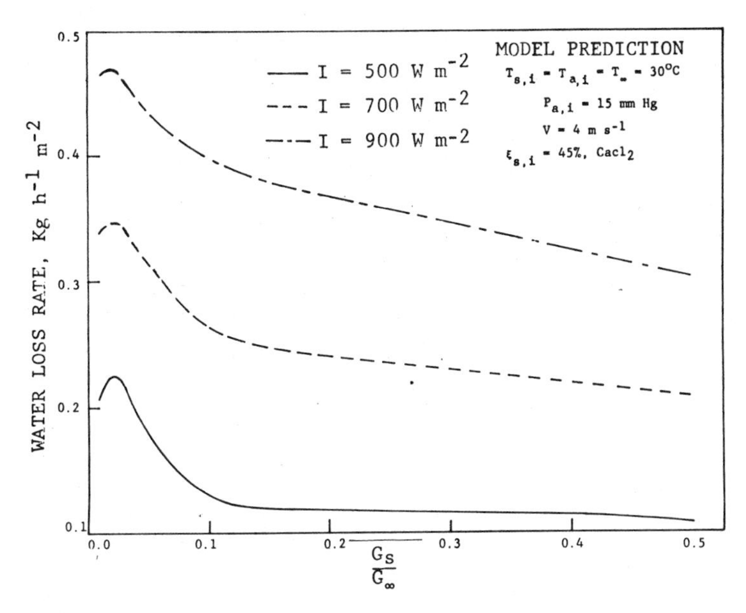

Figure 19. Sensitivity of performance to insolation.

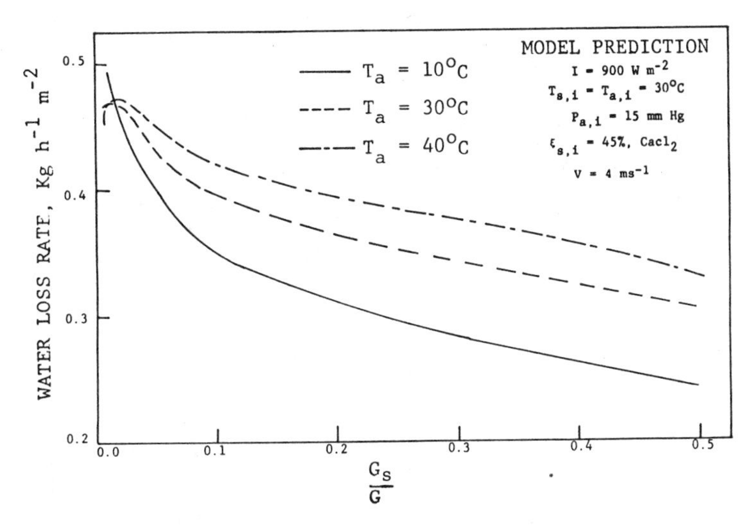

Figure 20. Sensitivity of performance to ambient temperature.

performance of the regenerator. The regenerator performance due to changes in the incident solar flux, as predicted by the model, is shown in Figure 19. As the ratio of G_s/G_∞ increases, the water loss rate decreases. For particular conditions of operation, there will be an optimum ratio of G_s/G_∞, which yields the maximum evaporation of water from the brine. For a parallel flow (cocurrent flow) regenerator, it is found that the optimum ratio would be 0.02. As the insolation increases the water loss rate also increases due to high temperature

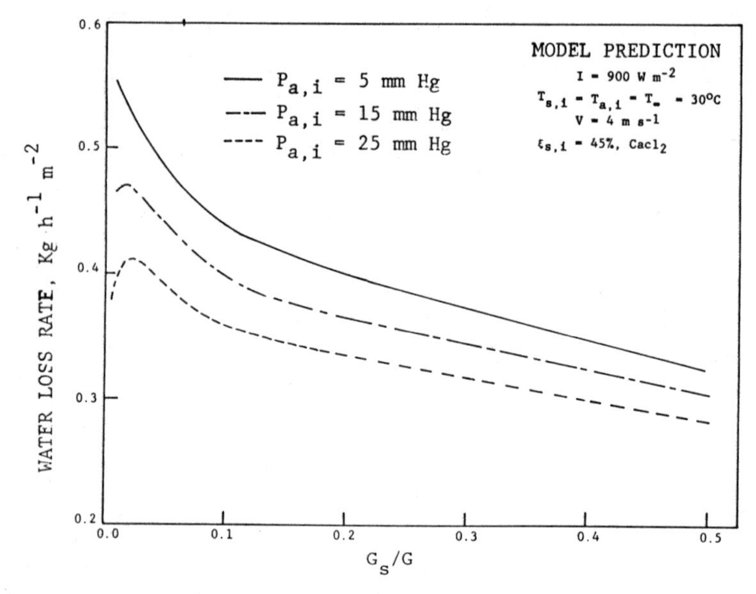

Figure 21. Sensitivity of performance to initial water vapor pressure in air.

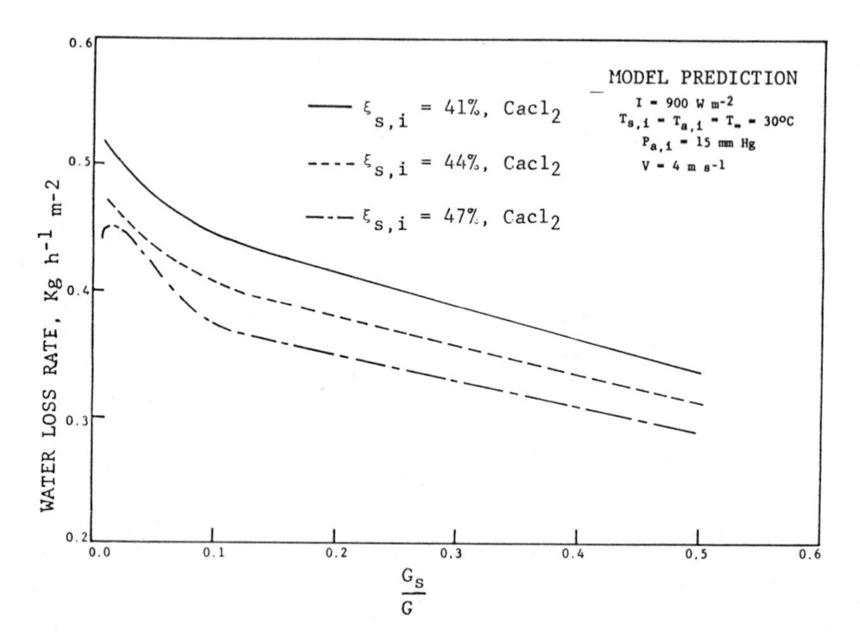

Figure 22. Sensitivity of performance to brine concentration.

of the absorbent solution, which increases the potential for mass transfer. Similar results were obtained with the changes in ambient temperature and are shown in Figure 20. As the ambient temperature increases, the rate of desorption also increases due to high temperature of the absorbent solution.

The regenerator performance due to changes in initial water vapor pressure in air is shown in Figure 21. The evaporation of water from the absorbent solution depends on two factors:

- The convective mass transfer coefficient between the solution film and air stream.
- The difference in water vapor pressure between them.

As the solution flows down the regenerator, its water vapor pressure increases due to an increase in solution temperature. Similarly, as the air stream passes through the regenerator, it picks up moisture from the solution film, and hence, its water vapor pressure increases. The potential for mass transfer is the difference in water vapor pressure between solution film and the air stream. If the initial water vapor pressure in air is less, that is dry air, it tends to absorb more moisture from the solution and hence the water loss rate is more. When the entering air is humid, the absorption rate is less due to high humidity and hence the desorption rate is less.

Figure 22 shows the water loss rate for different brine concentrations of calcium chloride and for different values of G_s/G_∞. It is interesting to note that the less concentrated solution yields more evaporation of water. This is due to the fact for the same operating temperature, the less concentrated solution has higher water vapor pressure which increases the potential for mass transfer.

NOTATION

A	area of the regenerator (m^2)
b'	width of the regenerator (m)
c	concentration ($kg\,m^{-3}$), (moles m^{-3})
C_p	heat capacity of the weak solution ($W\,kg^{-1}\,K^{-1}$)
$\dot{c}$, $\ddot{c}$	first and second derivatives of concentration
G	mass flow rate ($kg\,h^{-1}\,m^{-2}$)
G_0	initial mass flow rate of weak solution per unit collector width ($kg\,h^{-1}\,m^{-1}$)
Gr	Grashof number
h'_{fg}	heat of vaporization for water from the absorbent solution ($W\,kg^{-1}$)
I	insolation ($W\,m^{-2}$)
k	thermal diffusivity ($m^2\,s^{-1}$)
L	length of the regenerator (m)
m	amount of water evaporated from the weak solution per unit time ($kg\,h^{-1}\,m^{-1}$, $kg\,h^{-1}\,m^{-2}$)
m_c	dimensionless concentration $(c-c_\infty)/(c_s-c_\infty)$
m_s	mass rate of flow of absorbent per unit collector width ($kg\,h^{-1}\,m^{-1}$)
m_T	dimensionless temperature $(T-T_\infty)/(T_s-T_\infty)$
$\dot{m}$	mass transfer for combined force and free convection
$\dot{m}_0$	mass transfer for pure forced convection
P	vapor pressure (mm Hg)
Pr	Prandtl number
$\dot{q}$	heat transfer for combined forced and free convection
$\dot{q}_0$	heat transfer for pure forced convection
r	point of flow reversal (m)
Re	Reynolds number
Sc	Schmidt number
T	temperature (deg)
$\dot{T}$, $\ddot{T}$	first and second derivatives of temperature, respectively
U_L	collector loss coefficient ($W\,m^{-2}\,C^{-1}$)

u, v gas-phase velocity components in the x and y direction respectively (ms^{-1})
V wind velocity (ms^{-1})
x distance parallel to the liquid surface (m)
x′ distance traveled by the solution along the solar regenerator (m)
y distance normal to the liquid surface (m)

Greek Symbols

α convective heat transfer coefficient ($W\,h^{-1}\,m^{-2}\,C^{-1}$)
α' absorptance of the liquid layer and the absorber surface .
β convective mass transfer coefficients (ms^{-1}, $W\,h^{-1}\,m^{-2}\,(mm\,Hg)^{-1}$)
δ diffusivity ($m^2\,s^{-1}$)
ξ solution concentration by weight, percentage
ζ strength of free convection relative to forced convection
η dimensionless variable
ν kinematic viscosity ($m^2\,s^{-1}$)
Ψ stream function

Subscripts

a ambient
i initial
s solution
x local value
∞ free air stream

Superscript

′ differentiation with respect to η

REFERENCES

1. Löf, G. O. G., "House Heating and Cooling with Solar Energy," *in Solar Energy Research.* Madison: University of Wisconsin Press, 1955, pp. 33–46.
2. Hollands, K. G. T., "The Regeneration of Lithium Chloride Brine in a Solar Still for Use in Air Conditioning," *Solar Energy,* Vol. 7 (1963), pp. 39–43.
3. Mullick, S. C., and Gupta, M. C., "Solar Air Conditioning Using Absorbents," *Proceedings of the Second Workshop on the Use of Solar Energy for the Cooling of Buildings.* Los Angeles, Calif., Aug. 4–6, 1975, pp. 336–344.
4. Gandhidasan, P., and Gupta, M. C., "A New Method of Sun Powered Air Conditioning," Annex 1976—1, *Bull. I.I.R.* 1976, pp. 665–672.
5. Kakabaev, A., et al., "A Large Scale Solar Air Conditioning Pilot Plate and its Test Results," *International Chemical Engineering,* Vol. 16 (1976), pp. 60–64.
6. Leboeuf, C. M., and Löf, G. O. G., "Analysis of a Licl open-cycle Absorption Air Conditioner Which Utilizes a Packed Bed for Regeneration of the Absorbent Solution Driven by Solar Heated Air," Final Report, for U.S. Department of Energy, Colorado State University, Colo., Contract No. EG-77-5-02-4546, Oct. 1978.
7. Peng, C. S. P., and Howell, J. R., "Optimization of Liquid Desiccant Systems for Solar/Geotechnical Dehumidification and Cooling," *Journal of Energy,* Vol. 5. (1981), pp. 401–408.
8. Factor, H. M., and Grossman, G., "A Packed Bed Dehumidifier/Regenerator for Solar Air Conditioning with Liquid Desiccants," *Solar Energy,* Vol. 24 (1980), pp. 541–550.

9. Robison, H. I., "Open Cycle Chemical Heat Pump and Energy Storage System," Final Report for Solar Energy Research Institute, Coastal Carolina Energy Laboratory of University of South Carolina, Conway, Sub-contract No. ZE-0-9185-1, Jan. 1982.
10. Rao, D. P., et al., "Feasibility Study of a Large Scale Solar Power Generation System Suitable for the Arid and Semiarid Zones," *Solar Energy,* Vol. 27 (1981), pp. 313–322.
11. Merrifield, D. V., and Fletcher, J. W., "Analysis and Development of Regenerated Desiccant Systems for Industrial and Agricultural Drying," Final Report for U.S. Department of Energy, Lockheed—Huntsville Research and Engineering Center, Huntsville, Ala. Contract No. ORNL-Sub-7296-1, Dec. 1977.
12. Schwartzberg, H. G., and Rosenau, J. R., "Use of Solar Concentrated Water Absorbing Brines to Save Energy in Food Processing," *Proceedings of the Second International Conf. on Energy Use Management,* Los Angeles, Calif., Oct. 22–26, 1979, pp. 1922–1931.
13. Kakabaev, A., et al., "Absorption Solar Regeneration Unit with Open Regeneration of Solution," *Applied Solar Energy,* Vol. 5 (1969), pp. 69–72.
14. Collier, R. K., "The Analysis and Simulation of an Open Cycle Absorption Refrigeration System," *Solar Energy,* Vol. 23 (1979), pp. 354–366.
15. Peng, C. S. P., and Howell, J. R., "Analysis of Open Inclined Surface Solar Regenerators for Absorption Cooling Applications—Comparison Between Numerical and Analytical Models," *Solar Energy,* Vol. 28 (1982), pp. 265–268.
16. Gandhidasan, P., "A Simple Analysis of an Open Regeneration System," *Solar Energy,* Vol. 31 (1983), pp. 343–345.
17. Sodha, M. S., et al., "Simple Multiple Wick Solar Still: Analysis and Performance," *Solar Energy,* Vol. 26 (1981), pp. 127–131.
18. Gandhidasan, P., "Theoretical Study of Tilted Solar Still as a Regenerator for Liquid Desiccants," *Energy Conversion and Management,* Vol. 23 (1983), pp. 97–101.
19. Mullick, S. C., and Gupta, M. C., "Solar Desorption of Absorbent Solutions," *Solar Energy,* Vol. 16 (1974), pp. 19–24.
20. Mullick, S. C., and Gupta, M. C., "Wind Forced Air Flow in the Solar Collector—cum—Brine Regenerator Used in Air Conditioning," paper presented at *Sixth Meeting of All India Solar Energy Working Group and Conference,* Allahabad, India, Nov. 1974.
21. Johannsen, A., and Grossman, G., "Performance Simulation of Regenerating Type Solar Collectors," *Solar energy,* Vol. 30 (1983), pp. 87–92.
22. Gandhidasan, P., et al., "The Rate of Mass Transfer in a Solar Regenerator," *Letters in Heat and Mass Transfer,* Vol. 4 (1977), pp. 185–192.
23. Eckert, E. R. G., and Schneider, P. J., "Effect of Diffusion in an Isothermal Boundary Layer," *Journal of the Aeronautical Sciences,* Vol. 23 (1956), pp. 384–387.
24. Byers, C. H., and King, C. J., "Gas-Liquid Mass Transfer with a Tangentially Moving Interface, Part I—Theory," *A.I.Ch.E.* Vol. 13 (1967), pp. 628–636.
25. Byers, C. H., and King, C. J., "Part II—Experimental Studies," *A.I.Ch.E.* Vol. 13 (1967), pp. 637–644.
26. Gandhidasan, P., et al., "Buoyancy Effects in a Solar Regenerator," *Solar Energy,* Vol. 22 (1979), pp. 9–14.
27. Gandhidasan, P., et al., "Transient Response of a Solar Regenerator," *Applied Scientific Research,* Vol. 34 (1978), pp. 259–271.
28. Gandhidasan, P., "Investigations of Heat and Mass Transfer Processes in a Solar Regenerator," Ph.D. thesis, IIT, Madras, India, 1978.
29. Gandhidasan, P., "Simple Analysis of a Forced Flow Solar Regeneration System," *Journal of Energy,* Vol. 6 (1982), pp. 436–437.
30. Gandhidasan, P., "Thermal Performance Predictions and Sensitivity Analysis for a Parallel Flow Solar Regenerator," *Trans. of ASME J. of Solar Energy Engineering,* Vol. 105 (1983), pp. 224–228.

INDEX